CIRCUIT ANALYSIS

CIRCUIT ANALYSIS

Irving L. Kosow, Ph.D.

Professor, Electrical and Computer Engineering Technology
Southern College of Technology

Professor Emeritus, College of Staten Island
City University of New York

JOHN WILEY AND SONS

New York
Chichester
Brisbane
Toronto
Singapore

Production supervised by Frank Grazioli and Cindy Funkhouser
Interior design by Ann Renzi
Copyediting supervised by Joshua Spieler
Cover illustration by Roy Wiemann

Library of Congress Cataloging in Publication Data:

Kosow, Irving L.
Circuit analysis.

Bibliography: p.
1. Electric circuit analysis. I. Title.
TK454.K664 1988 621.319′2 87-8134
ISBN 0-471-03067-8

Printed in the United States of America

10 9 8 7 6 5 4 3 2

To my dear wife, Ruth, who for many, many years encouraged my research, editing of manuscripts, teaching, and writing of books. This dedication is in recognition of the long days throughout many years of quiet loneliness and sacrifices required to see a rough manuscript through to its final published form. She patiently proofread both galley and page proofs several times and also assisted in preparing the index. For one in whom the love of life is exceedingly strong and for whom each moment is very precious, this was a great show of love indeed,

and

to my students, all of them, for their eager desire to learn and for their feedback in the preparation of this book, my gratitude and thanks.

PREFACE

This work is the product of almost 30 years of teaching circuit analysis at both the 2-year and 4-year electrical engineering technology level, along with those courses that require it as a prerequisite: semiconductor electronics, electrical machinery, communications, pulse circuitry, and so forth. It emerged principally from the many handouts developed for student use to supplement the material either missing from or inadequately presented in existing circuit analysis texts. It is intended primarily for use in 2-year and 4-year electrical and electronic engineering technology programs at the technical institute level. It may also be used by engineering schools as either an introductory or a supplementary text in circuit analysis. Because of its many worked examples and the answers to all end-of-chapter problems, it may also be used for self-study since the given answers provide immediate feedback and positive reinforcement to the reader.

There is general agreement among engineering educators that an understanding of and insight into theoretical material are acquired primarily through the solution of problems. Many teachers stress problem solution because they know that the *qualitative* portions of the subject are less difficult and are better understood once the *quantitative* aspects are mastered. Consequently, throughout the text and in the end-of-chapter problems, as well, a conscious attempt has been made to promote and reflect fundamental insights into each topic. In doing this, I have tried to break the "formula habit" wherein students seek to solve problems merely by "number plugging." Instead, both worked examples and problems have been selected and designed in such a way as to develop understanding, confidence, and growth. Often both text examples and end-of-chapter problems are recycled to provide alternative solution methods and ensure both a wider and greater insight into circuit analysis methods. In this way, the student acquires a more fundamental understanding of basic concepts and demonstrates it operationally through use of the various approaches presented. All end-of-chapter problems are cued to text topic sections so that reading assignments by section topics may be coordinated with and referenced to appropriate problems.

It is my deep conviction that pervasive insights, conclusions, and summaries should emerge from a study of the various topics. Most texts assume that the readers make such leaps or integrate such knowledge on their own as they proceed. I have tried to take the student with me on an exciting journey that explores each topic, leaving nothing to chance. As new examples are presented and solved, the insights are summarized in outline form to ensure and enhance student understanding.

Great pains have been taken to ensure that the answers to all end-of-chapter problems are correct because of their importance in providing feedback and reinforcement to the student. To this end, all problems have been solved at least two or three times, using independent methods. A solutions manual for instructor use is available as a supplement.

I have also prepared a set of computer programs written in BASIC for solving many of the text problems. This optional computer supplement contains the programs in printout format with descriptions for their use. A floppy disk for use with IBM-compatible personal computers (PCs) is also available with a "menu" keyed to text sections. This software package is intended to enhance student understanding of circuit theory via the graphical and computational potentials of PCs, as recommended by ABET guidelines.

The mathematical level of the text is comparable to that of existing texts serving 2-year and 4-year engineering technology students. Calculus is used primarily in the appendices for derivation of significant equations, which are both summarized and (in some instances) derived intuitively in the body of the text. Use is made of the concept of the derivative as the rate of change of a particular parameter. The technique of integration appears only in Chapter 14, where it is used to find average and root-mean-square (RMS) values of certain exponential waveforms. I have discovered that students with little or no calculus background can be taught to perform such simple integrations successfully. Since ABET guidelines stress the use of some calculus in accredited engineering technology programs, this relatively minor usage is a long-overdue step in the right direction. Since only a few of the end-of-chapter problems in Chapter 14 require integration, the teacher has the option of either omitting these or using alternative solution methods (trapezoidal or graphical techniques).

A cursory examination shows that this work contains more worked examples and more end-of-chapter problems than are found in most texts. Not all problems need be assigned, but sufficient problems are available so that the teacher need not supplement the text for quiz review and/or final examination review purposes.

Space does not permit a description here of all the ways in which this work differs from others or all the unusual features contained within each chapter. Many of these features (unique to this text) will become evident through personal use. Only a few are summarized below for each chapter.

Chapter 1 defines *all seven* base units of the SI, along with their supplementary and derived units, commonly used in electrical practice. Distinction is made between quantity symbols (in italics) and unit abbreviation symbols (in Roman). *All* the SI unit prefixes are given, including the optimum expression in the most concise form. Emphasis is placed on derivation of conversion factors by the ratio method, as well as rules for using significant figures in calculations.

Chapter 2 provides the bridge between physics and electronics. Metallic, covalent, and ionic bonds are contrasted. Coulomb's and Newton's laws are compared, leading to the theory of Millikan's measurement of electrical charge and explaining why electrostatic forces are so much greater in matter than gravitational forces. Properties of potential difference, current, conductivity, and resistivity are all defined in terms of ion mobility, drift velocity, and free-electron density, providing the essential foundation for later semiconductor work and explaining why germanium and silicon are semiconductors. Distinctions are drawn between current and drift velocity, as well as electrical field intensity and potential difference.

Chapter 3 derives factors affecting ohmic resistance from the E/J ratios presented in Chapter 2, in both SI and English systems, along with an intuitive derivation.

A brief section on American Wire Gage (AWG) tables for copper shows how the entire table may be synthesized from known values for one wire size. Distinctions between positive and negative temperature coefficients are shown in terms of electron mobility and density. Differences are drawn between α_{20} and α_0, including their use and how one temperature coefficient may be found from the other. Use of both resistance measurement and temperature coefficient as two alternative means of identifying unknown materials is shown. The ratio method of problem solving is reinforced.

Chapter 4 draws sharp distinctions between power and energy, quantifying their relation in both worked examples and problems. Tabular summaries of power and energy units and their interconversions through various energy forms are presented, along with practical design applications. Insights into problem-solving procedures and checks for reasonable answers are stressed.

Direct-current circuit analysis is introduced in **Chapter 5**, using only resistors to establish voltage, current, and power relations in simple series and parallel circuits. Using the principle of current-sense direction, both polarity of volt drop and double-subscript notation are introduced and defined, along with concepts of voltage rise and voltage drop and voltage-divider and power-divider rules. An extension of the voltage-divider rule enables derivation of the output of a divider supplied from two different dc sources (useful for later transistor study as well as for unequal sources in parallel). Concepts of practical sources, voltage regulation, open-circuit and short-circuit testing all culminate in the maximum power transfer theorem. Duality is introduced as a way of deriving equations intuitively.

Chapter 6 contains rules for circuit simplification and introduces the concept of equivalent circuits, leading to both analysis and synthesis of series and parallel voltage dividers. Concepts of ground zero and reference voltages are reinforced. Analysis of series–parallel circuits is summarized and extended to both bridge circuits and H-bias networks.

Chapter 7 offers the reader an alternative approach to circuit analysis using network theorems. Bartlett's theorem is included as a means of analyzing symmetrical circuits. Tellegen's theorem is introduced as a power calculation check on the accuracy of solutions of all dc and ac networks, linear and nonlinear, passive and active.

Chapter 8 shows the advantages of mesh and nodal analysis techniques using the format method, which enables equation writing by inspection. A number of shortcuts are shown, using matrix methods to eliminate errors and reduce the number of zeros. A unique set of equations (derived from resistance–conductance duality) is derived for wye–delta and delta–wye conversions that enables use of only one equation set in the same form (reducing the need for memorization). Solution of matrices by the determinant method, using Cramer's rule, is shown as an alternative to solutions involving computer programs. Included are both caveats and useful hints to facilitate algebraic solutions of networks.

Chapter 9, on capacitors, emerges from a review of electronic charge and electrical field intensity presented in Chapter 2. Capacitors as physical components and capacitance as a circuit property are contrasted, along with differences between resistance and capacitance in both circuit action and energy dissipation (or storage). Losses in commercial capacitors are treated from the commercial equivalent series resistance (ESR) approach. The concept of dissipation factor (D) is introduced. A table showing analogs in capacitive and resistive circuits summarizes properties and equations. The apparent inconsistency regarding capacitor energy loss is explained (Section 9-10) and a unique method is presented for determining unknown capacitance by charging a known capacitor. The importance of conservation of charge is stressed in both examples and end-of-chapter problems.

A separate **Chapter 10** on transient *RC* circuit analysis prepares the reader for a course in pulse circuitry. Initial and final steady-state conditions are defined. Rates of change of v_C, i_C, and v_R are presented as decaying exponential functions to eliminate the need for differential equations. Capacitors with initial positive and/or negative charges are also treated in both charge and discharge modes. Quick approximation techniques as well as shortcut equations are given as an aid to plotting transient waveforms and finding specific times for values of interest. A special section on *RC* transient switching analysis (Section 10-9) enables solutions without recourse to differential equations, operational calculus, or Laplace transforms.

Chapter 11 explains magnetic properties and domains as a result of third-shell unbalanced electron spin. Magnetic quantities and their symbols and relations are carefully defined. A table relates quantities in SI to their English and CGS equivalents with appropriate conversion factors. The **B**–**H** curve and the need for it emerges from a discussion of average, incremental, and relative permeability. Magnetic calculations are simplified by classifying magnetic circuit problems into three groups. A unique magnetization curve with primary units in SI but containing scales for CGS and English units enable the reader to convert values of **B** and **H** graphically. Topics essential to transformers and electrical machinery are included.

Chapter 12 uses Faraday's law of electromagnetic induction to develop Lenz's law and self-inductance properties. Mutual inductance and the (important) dot convention are introduced, defined, and illustrated through worked examples, including mutual inductance between three or more coils in parallel. Energy stored in the magnetic field of an inductor and consequences of its release are quantified. A useful summary comparison table shows similarities and differences between resistive, capacitive, and inductive circuits.

Chapter 13 is the counterpart of Chapter 10 covering *RL* transient analysis. All rates of change are found by using exponential relations without recourse to differential equations. Coverage also includes exponential rise and decay of current from initial (steady-state) current values above and below zero. Shortcut equations are given for algebraic solutions of time at specific transition points. Solution of *RL* circuits with multivoltage sources and situations in which inductor current changes direction are included. Power and energy relations during transient current rise and decay are explained both qualitatively and quantitatively, without recourse to calculus. A summary table (Table 13-3) contrasts *L* and *C* at very high frequencies (VHF) and very low frequencies (VLF), setting the stage for ac circuit analysis.

Chapter 14 introduces ac through a study of waveforms in the time domain, including average and RMS values of both sinusoidal and a variety of common nonsinusoidal waveforms, without recourse to calculus. The reader is also shown that elementary integration may be used, as well. Quadrature, apparent, and true power are found to emerge from time-domain relations. A last section shows that maximum instantaneous positive and negative power points may be used to find average and apparent power, respectively.

Chapter 15 on complex algebra prepares the reader for calculations in the frequency domain. It distinguishes between scalars, vectors, and phasors, explaining why the frequency-domain representation is more convenient for solving complex problems but is *not equal* to the domain quantity.

Chapter 16 unifies phasor diagrams of voltage and current with both the impedance triangle and power triangle. Complex power is introduced and Tellegen's theorem is recalled as a check on voltage/current relations in ac series–parallel circuits. The effect of frequency on series–parallel equivalence is shown both graphically and algebraically.

In addition to presenting various ac circuit analysis theorems, **Chapter 17** gives a complete presentation of all eight factors affecting effective resistance, as well as a complete presentation of ac maximum power transfer, including three restricted-load situations that frequently occur in practice. Insights regarding voltage, current, and power ratios expressed in decibels (dB) permit mental computations of these ratios. Power tabulation grids for power factor correction are introduced, as are tabulations for solution of effective resistance and dB measurement. Superposition for sources at different frequencies is introduced for later use in Chapter 22.

Chapter 18 uses mesh and nodal analysis on independent practical voltage and current sources and then extends the analysis to dependent sources in both dc and ac circuits. In addition to a thorough treatment of circuits containing both independent and dependent sources, Tellegen's theorem is used as an alternative method of verification of the analysis. A novel method of analysis for solving ac networks containing both dependent and independent sources is also introduced, using simple conventional series–parallel circuit theory (Section 18-8.1).

Chapter 19 treats series and parallel resonance, while emphasizing both similarities and differences between these dual circuits, both from a change in frequency and from a change in L or C. Particular attention is given to the two-branch parallel-resonant circuit under high-, medium-, and low-Q conditions. A convenient shortcut relation is shown to predict the loading effect for any parallel-connected load, leading to electrical damping to prevent oscillation. Analysis and design of various types of passive wave filters are considered, including resonant series, parallel, and double-resonant filter networks.

Chapter 20 is a relatively complete qualitative and quantitative analysis of both tightly coupled and loosely coupled transformers. Loosely coupled transformers are analyzed by five different methods with discussions of their relative merits. The last method unifies the theory of tightly coupled and loosely coupled transformers. Multicoil loosely coupled transformers are analyzed by using a simplified method of mesh equation writing.

Chapter 21 covers both balanced and unbalanced delta and wye polyphase systems. An equilateral triangle method is used that eliminates the need for more complex approaches. Three-wire unbalanced Y-connected loads are also solved by mesh analysis. The effect of phase sequence on phase voltages of an unbalanced Y-connected load is analyzed and a rule is developed to predict phase sequence depending on the line location of the capacitor of a phase-sequence indicator. Complex power is used to solve three-phase combinational loads, in contrast to methods involving single-line diagrams or delta–wye conversions, both of which are subject to greater error because of calculation complexity. Wattmeter measurement methods for balanced and unbalanced three-phase power are fully treated, including Barlow's method. The two-wattmeter method is explored more extensively because it may be used to determine power factor, phase sequence, total power, system vars, and volt-amperes. The discussion also includes tests to determine whether the lower-reading wattmeter is positive or negative.

Chapter 22 presents a unique method of waveform analysis by identifying three types of symmetry that enable the reader to predict the nature of harmonics present in any periodic waveform, including those containing a dc component and those having no symmetry whatever. Fourier equations of a number of common nonsinusoidal waveforms are presented to verify the method of analysis. Superposition is invoked, once again, to show various circuit responses to nonsinusoidal inputs. Responses of *RLC* series- and parallel-resonant systems show that current is either increased or suppressed, respectively, at the harmonic frequency. The chapter concludes with analysis of series-connected and parallel-connected harmonic sources.

Chapter 23 covers both analog and digital instrumentation, including designs of voltmeters, ammeters, and various types of ohmmeters. Instrumental sensitivity is described via the voltmeter loading effect. Techniques are given to determine proper voltmeter placement in the ammeter–voltmeter method. The Wheatstone bridge and various ac bridges are presented along with equations for computation of unknown resistance, capacitance, or inductance. Various digital instruments are shown and described.

Irving L. Kosow

ACKNOWLEDGMENTS

This book could never have emerged in its present form without the help of many people. First, my appreciation to my colleagues at Southern Tech in the Department of Electrical and Computer Engineering Technologies: to Prof. David E. Summers, Department Head, for his encouragement, inspiring confidence and support in granting leaves to enable the writing; to Prof. Charles L. Bachman, for his notes and problems on polyphase circuits; to Prof. Clifford W. Cowan, for helping to clarify some sticky points; to Prof. Preston A. White III, for editing and revising my handouts; to Prof. John L. Keown, who taught an old dog new computer tricks; and to Profs. Walter E. Burton, Richard L. Castellucis, Caroline Cranfill, Philip T. Moen, Robert W. Robertson, Theresa C. Speake, John E. Tarpley, Yardy T. Williams, Julian A. Wilson, Paul Wojnowiak, and David W. Zimny for collectively providing an atmosphere of dedication to teaching and a spirit of comradeship.

The early reviewers of the initial outline and the preliminary chapters, as well as those who read the complete manuscript (all at that time anonymous), were immeasurably helpful, indeed influential, in shaping the earlier and later versions of this work. The early reviewers were:

Alex Avtgis, Wentworth Institute of Technology

H. J. Bestervelt, Metropolitan State College

Richard D. Morris, Portland Community College

Robert L. Reids, Broome Community College, Binghamton, New York

George B. Rutkowski, Electronic Technology Institute, Cleveland, Ohio

Wayne E. Schlifke, Erie Community College, North Campus

Arthur Seidman, Pratt Institute of Technology

David Summers, Southern College of Technology

The later reviewers of the completed drafts of the manuscript were:

William C. Burns, Greenville Technical College

Gerry Davis, Central Piedmont Community College

Orville Detraz, Indiana University–Purdue University at Fort Wayne

Paul Dingman, Orange County Community College

David Hata, Portland Community College

Larry Jones, Oklahoma State University

Fred Lewallen, University of Houston–Downtown Campus

Carl Morgan, Miami University, Middletown, Ohio

Berthold Scheffield, Trenton State University

James G. Smith, Purdue University, West Lafayette

I gratefully acknowledge the contribution of those in industry who provided both technical data and photographs:

Matthew Simon, Director of Marketing, RTE Aerovox, Inc.

Gary Drinan, Marketing Communications, Allen Bradley Company

Charles A. Soumas, Marketing Manager, Beckman Industrial Corp.

Bob Lewis, Sales Manager, J. W. Miller Div., Bell Industries

Robert E. Roderique, Advertising Manager, Bussman Div., McGraw Edison Co.

Robert E. Voeks, Advertising Manager, Biddle Instruments

Robert Nareau, Sales Manager, Delevan Div., American Precision Industries

Tom McLaren, Manager, Marketing Communications, General Electric Co.
William S. Stewart, Sales Promotion Coordinator, Heath Company
R. E. Delaney, V. P., Marketing, Heinemann Electric Company
Janet Rice, Public Relations Services, Hewlett Packard Company
Albert J. Eisenberg, Marketing Manager, Microtran Company
Robert S. Galvin and Mary Trowbridge, Tektronix, Inc.
David A. Koss, Marketing Communications, Union Carbide Corp.

I sincerely appreciate and acknowledge the efforts of the many people at John Wiley & Sons, Publishers:

Hank Stewart, Editor of Engineering Technology, for his unflagging encouragement, constant reassurance, and helpful support from the initial stages through the final completion of the work; Lilian Brady, Arnie Carolus, Christopher Combest, Elizabeth Doble, Cindy Funkhouser, Roberta Gardner, Frank Grazioli, Lilly Kaufman, Joe Keenan, Maggie Kennedy, Laura McCormick, Sally McCravey, Al Manso, Michael Morgenstern, Kieran Murphy, Joe Morse, Mary Prescott, Ann Renzi, Josh Spieler, Rich Wiley, and Susan Winick.

All of the above plus several others at John Wiley painstakingly brought *Circuit Analysis* from a raw manuscript to the finished product. I sincerely appreciate and acknowledge their diligence, persistence, and patient efforts in helping me with the many details required in producing this book.

Irving L. Kosow
Marietta, Georgia

CONTENTS

Chapter 23

ANALOG AND DIGITAL INSTRUMENTATION 765

APPENDIX

CIRCUIT ANALYSIS

CHAPTER 1

The SI, Scientific Notation, and Unit Conversion

1-1 INTRODUCTION

Why do we begin with the SI and scientific units? The answer lies between the covers of this book. Scientists and engineers know that the terms they use, the quantities they measure with scientific instruments, and the calculations derived from such measurements must all be defined precisely. The need for systems of measurement dates back to Egyptian, Greek, and Roman times. The metric system—the basis of the International System of Units (Système International d'Unités or **SI**)—was adopted at Sèvres, near Paris, in 1875. The **MKSA** (meter–kilogram–second–ampere) **system**, widely used in electrical and electronic work since 1950, was adopted in 1954 by the Tenth General Conference on Weights and Measures in the United States. The MKSA specified the four **base units** from which all other units could be *derived*. Successive conferences in 1960, 1964, and 1971 added *three more* base units (the kelvin, the mole, and the candela), bringing the total number of base units to **seven**. The U.S. Metric Conversion Act of 1975 (Public Law 94-168) established a U.S. Metric Board to "coordinate the voluntary conversion to the metric system" and the SI was accepted as the system of reference "for most scientific and technical work."

On December 17, 1981, the Institute of Electrical and Electronics Engineers (IEEE) Standards Board accepted ANSI/IEEE Standard 268-1982—*American National Standard* (on) *Metric Practice*.[1] This document divides SI units into three classes: **seven base units**, **two supplementary** units, and a sizable number of **derived** units. It also contains rules for using metric prefixes, **conversion factors** between the SI and other systems of units, and a number of useful tables and definitions.

1-2 THE SI

Table 1-1 shows a *modified* ANSI 268 table for the seven SI **base units**. Definitions of the SI base units are given in **Appendix A-1**. Note that base units are defined to be *dimensionally independent* of each other. That is, each base unit is *not* defined in terms of any other base units.

[1] Available in the United States from the American National Standards Institute (ANSI), 1430 Broadway, New York, NY 10018 or the IEEE, 345 East 47 Street, New York, NY 10017. This document supersedes the earlier ANSI Z 210.1 of 1976 and IEEE Standard 268 of 1979.

Table 1-1 SI Base Units

Physical Quantity	Quantity Symbol	Unit of Measure	Unit Abbreviation Symbol
Length	l	meter	m
Mass	m	kilogram	kg
Time	t	second	s
Electric current	I, i	ampere	A
Thermodynamic temperature	T	kelvin	K
Amount of substance	n	mole	mol
Luminous intensity	I	candela	cd

Table 1-2 shows the second class of two *supplementary* units. These two units have been placed in a separate class intentionally because they are *dimensionless derived* units. The radian is the ratio of the *length* of arc to the radius *length* and is dimensionless. The steradian is the ratio between an area and a length squared and is also dimensionless. Such unitless ratios may be used or omitted in expressing units of **physical quantities** without changing the quantity. Definitions of the radian and steradian are given in **Appendix A-2**.

Table 1-2 SI Supplementary Units

Physical Quantity	Quantity Symbol	Unit of Measure	Unit Abbreviation Symbol
Plane angle	θ, ϕ	radian	rad
Solid angle	θ_s, ϕ_s	steradian	sr

Both of the units of Table 1-2 have application in electrical and electronic engineering technology. For example, angular velocity, ω, is expressed in units of either radians per second (rad/s) or s^{-1} (since radians are dimensionless), as shown in Table 1-3. The steradian, widely used in electrical lighting and illumination practice, is outside the scope of this book.

The third and largest class of units in the SI is the *derived* units. Table 1-3 gives some examples of commonly used *derived* electrical units expressed in terms of their base units. With reference to Table 1-3, please note:

1. Some derived units such as magnetization, **M**, and current density, J, are expressed *in terms of base units*. This is also true for such derived units as area in square meters (m^2), volume in cubic meters (m^3), velocity in meters per second (m/s), and acceleration in meters per second squared (m/s^2).
2. Other derived SI units are named for scientists and engineers who have contributed to the body of scientific and mathematical knowledge. When *abbreviated*, these units always appear in *capital* Roman letters, since they represent names or proper nouns. Examples of these are such derived units as the hertz (Hz), newton (N), pascal (Pa), joule (J), watt (W), and degree celsius (°C). There are two such units in the SI base class of Table 1-1: the kelvin (K) and the ampere (A).
3. While **Roman** (upright) uppercase and lowercase type is used for **unit abbreviation** symbols, *italics* are almost always used for *quantity symbols* (see Tables 1-1, 1-2, and 1-3). The only exception to this rule is for **vector** physical quan-

Table 1-3 Examples of Commonly Used Derived Electrical Units Expressed in Terms of Base Units

Physical Quantity	Quantity Symbol	Symbolic Dimensional Analysis	Derived Units	Unit Abbreviation Symbol
Angular velocity	ω	t^{-1}	radian/second or second^{-1}	rad/s or s^{-1}
Capacitance	C	$m^{-1}l^{-2}t^{4}I^{2}$	farad	F
Charge	Q	It	coulomb	C
Current density	J, j	Il^{-2}	ampere/square meter	A/m^2
Electric field strength	$\mathbf{E}$	$mlI^{-1}t^{-3}$	newton/coulomb volt/meter	N/C V/m
Electric flux	$\phi_{\mathbf{E}}$	$ml^{3}I^{-1}t^{-3}$	volt/meter	V/m
Electric potential	V	$ml^{2}I^{-1}t^{-3}$	volt	V
Inductance	L	$ml^{2}I^{-2}t^{-2}$	henry	H
Magnetic field strength	$\mathbf{H}$	Il^{-1}	ampere/meter	A/m
Magnetic flux	$\phi_{\mathbf{M}}$	$ml^{2}I^{-1}t^{-2}$	weber volt-second	Wb V·s
Magnetic induction	$\mathbf{B}$	$mI^{-1}t^{-2}$	tesla	T
Magnetization	$\mathbf{M}$	Il^{-1}	ampere-turns/meter	A/m
Permeability	μ	$mlI^{-2}t^{-2}$	henry/meter	H/m
Permittivity	ε	$t^{3}I^{2}m^{-1}l^{-3}$	farad/meter	F/m
Resistance	R	$ml^{2}I^{-3}t^{-3}$	ohm	Ω
Resistivity	ρ	$ml^{3}I^{-2}t^{-3}$	ohm-meter	Ω · m
Work (energy)	W	$ml^{2}t^{-2}$	joule	J
Power	P	$ml^{2}t^{-3}$	watt	W
Conductivity	σ	$I^{2}t^{3}m^{-1}l^{-3}$	siemens/meter	S/m
Conductance	G	$I^{2}t^{3}m^{-1}l^{-2}$	siemens	S

tities (having both magnitude and direction). These are expressed in Roman **boldface** capitals. Examples are electric field strength, **E**, electric flux, $\phi_{\mathbf{E}}$, magnetization, **M**, or magnetic field strength, **H**, and magnetic induction, **B**.

4. Great care should be taken not to confuse *quantity* symbols (which are usually used in equations and written in *italics* or **boldface**) with unit abbreviation symbols (which **always** appear in Roman lower- or uppercase). Thus the quantity symbol for capacitance is an italic C, while the unit of charge in coulombs is abbreviated with a Roman capital C. A typical problem using both of these "C" symbols might read: "Given $C = 10\ \mu$F, with a charge of $Q = 10$ C, calculate the voltage across the capacitor."
5. Since there are only 52 upper- and lowercase letters in the Roman alphabet and a great many more physical quantities than 52, some duplication is inevitable. As shown in Table 1-3, this is partly avoided by the use of Greek letters. But it has not been avoided completely; for instance, the quantity symbol Q is used for electric charge, as in the above sentence; for quadrature power, as in $Q = 10$ vars; and for the quality factor of a coil, $Q = \omega L/R$. Thus, we may anticipate some duplication in *quantity* symbols. Although unit abbreviation symbols may consist of more than one letter, *no duplication* of *unit abbreviation symbols* is tolerated. Thus, if we read "the circuit Q is 10" and no units are given, we know that the author is implying the quality factor. Alternatively, if we read "a leading power factor synchronous motor produces a Q of 5 kvars,"

we know that the Q implies quadrature power. Consequently, any ambiguity in physical quantity symbols is clarified by use of the unit abbreviation symbols included in the statement.

1-3 SCIENTIFIC NOTATION

Scientists, engineers, technologists, and technicians in all branches of electrical and electronic work must make calculations based on measurements taken in the field or laboratory. As a result of such calculations, base units and derived units are obtained which involve either very large or very small quantities. A capacitance is calculated as 0.000 055 farads. A resistance is calculated as 45 000 000 ohms. To avoid writing numbers with long strings of zeros either before or after the decimal point, **scientific notation** is used.

Scientific notation is merely a form of shorthand notation in which any integral or decimal (fractional) number is expressed as a number between 1 and 10 times an appropriate **power of 10**.

1-3.1 Converting Decimal Numbers to Scientific Notation

The steps for converting *small* numbers to scientific notation are

a. Move the decimal point to the *right* until a number between 1 and 10 is obtained. Record this number.
b. Count the number of places shifted to the right and express this number as a *negative* exponent of 10 to be multiplied by the number recorded in **a**.

EXAMPLE 1-1
Express 0.000 055 farads in scientific notation with SI unit abbreviation.

Solution

0.000 055 farads = $\mathbf{5.5 \times 10^{-5}}$ **F**

1-3.2 Converting Integral Numbers to Scientific Notation

The steps for converting (*large*) integral numbers to scientific notation are

a. Move the decimal point to the *left* until a number between 1 and 10 is obtained. Record this number.
b. Count the number of places shifted to the left and express this number as a *positive* exponent of 10 to be multiplied by the number recorded in **a**.

EXAMPLE 1-2
Express 45 000 000 ohms in scientific notation with SI unit abbreviation.

Solution

45 000 000 ohms = $\mathbf{4.5 \times 10^{+7}\ \Omega = 4.5 \times 10^{7}\ \Omega}$

Note that the solutions to Exs. 1-1 and 1-2 appearing in **boldface** represent a much more concise expression of the originally given values of capacitance and resistance, respectively. But the advantages of scientific notation go much farther, as Ex. 1-3 shows. This example also will illustrate how use of an electronic calculator is made more efficient by using scientific notation.

EXAMPLE 1-3

The time constant, τ, of an electric circuit is defined as the product $\tau = RC$. The circuit resistance is $R = 6500\ \Omega$ and the circuit capacitance is $C = 0.000\,000\,052$ F. Calculate the circuit time constant, τ, using scientific notation. Hint: see Table 1-3.

Solution

Since R and C are expressed in (derived) basic units, their product, τ, will be in seconds in basic units of time. Expressing the numbers in scientific notation yields

$$\tau = RC = (6.5 \times 10^3) \times (5.2 \times 10^{-8}) = \mathbf{3.38 \times 10^{-4}\ s}$$

Key Sequence

Press	Display
6.5 [EE] 3	6.5×10^3
[×] 5.2 [EE] [+/−] 8 [=]	0.000338 or $\mathbf{3.38 \times 10^{-4}}$

Example 1-3 illustrates several important points that may be overlooked by the reader:

1. Each physical quantity in an equation must be expressed in the **basic** units of the SI.
2. The value or result obtained in the *solution* of an equation also emerges in basic units.
3. Every equation may be verified in terms of *dimensional analysis.* In this case the product RC is not expressed in ohm-farad but in seconds. This may be verified by the reader by using the third column of Table 1-3, titled *symbolic dimensional analysis.* Since the product RC produces t as a result, then τ is expressed in seconds.

1-4 SI UNIT PREFIXES

In Exs. 1-1 to 1-3 the units of physical quantities were expressed in terms of their basic units in a type of numerical shorthand called scientific notation. But scientists and engineers have developed an even more efficient means of expression that emerges from (and is based on) scientific notation. The method of SI **unit** (or metric) **prefixes** substitutes a *letter prefix* to the unit abbreviation symbol in lieu of the power of 10. Table 1-4 shows the latest SI prefixes established by ANSI/IEEE Standard 268-1982.

Table 1-4 SI Prefixes

Multiplication Factor	Power of Ten	Prefix	Symbol
1 000 000 000 000 000 000	10^{18}	exa	E
1 000 000 000 000 000	10^{15}	peta	P
1 000 000 000 000	10^{12}	tera	T
1 000 000 000	10^{9}	giga	G
1 000 000	10^{6}	mega	M
1 000	10^{3}	kilo	k
100	10^{2}	hecto[a]	h
10	10^{1}	deka[a]	da
0.1	10^{-1}	deci[a]	d
0.01	10^{-2}	centi[a]	c
0.001	10^{-3}	milli	m
0.000 001	10^{-6}	micro	μ
0.000 000 001	10^{-9}	nano	n
0.000 000 000 001	10^{-12}	pico	p
0.000 000 000 000 001	10^{-15}	femto	f
0.000 000 000 000 000 001	10^{-18}	atto	a

[a] To be avoided where practical except in expressing area and volume, where the prefixes hecto-, deka-, deci-, and centi- may be required. For example, square hectometer and cubic centimeter.

Using the combination of scientific notation and Table 1-4, we can express units of physical quantities in a very concise form, as shown by Ex. 1-4.

EXAMPLE 1-4

Express the following values using SI prefix notation:

a. Quantity and unit	b. Quantity and unit in scientific notation	c. *Solution* using SI prefixes
0.0075 amperes	7.5×10^{-3} A	7.5 mA
75 000 volts	75×10^{3} V	75 kV
0.000 002 farads	2×10^{-6} F	2 μF
99 500 000 hertz	99.5×10^{6} Hz	99.5 MHz
0.000 000 005 seconds	5×10^{-9} s	5 ns
25 000 ohms	25×10^{3} Ω	25 kΩ
0.0028 henry	2.8×10^{-3} H	2.8 mH

Again, the solution of Ex. 1-4 contains a number of distinctions of importance:

1. All **prefix symbols** in Table 1-4 are printed in Roman type (never in italics) with no space between the prefix and the letter symbol(s) for the unit.
2. The distinctions between capital and lowercase letters must be observed. There is a great deal of difference between 2.8 mH and 2.8 MH!
3. Compound prefixes such as micro-micro (μμF) or milli-micro (mμ) should never be used. Only **one** prefix should be applied to a unit abbreviation symbol.
4. When a symbol representing a unit that has a prefix is raised to an exponential power in a given problem, this indicates that the **multiple** or **submultiple** of the unit is raised to the power expressed by the exponent (see Ex. 1-5).
5. It is a common error to assume that because the submultiples are expressed in lowercase, all multiples 10^1 and higher are expressed in uppercase. The prefixes h, k, and da are all in *lowercase* in Table 1-4.
6. The prefix in association with its unit abbreviation represents a shorthand expression for the quantity of the unit. Thus 12 milliamperes is expressed only as 12 mA. It is never expressed as 12 m-amperes or 12 milli-A, etc.
7. Although periods may appear as decimal points in the quantity, the prefix and its unit abbreviation are *never* expressed with periods.
8. There is always a space between the magnitude and the prefixed unit of measure.
9. In selecting the best possible prefix with which to express a given quantity, the guiding rule is the shorthand technique that gave rise to the system in the first place, i.e., to express a physical quantity using the *least* amount of *typesetting* possible and the *fewest* digits and letters. (The reader is invited to attempt alternative solutions to part **c** of Example 1-4 to see if they will yield less typesetting for the seven examples given.)

EXAMPLE 1-5

Convert 9×10^{13} femtofarads to millifarads.

Solution

$$9 \times 10^{13}\ \text{fF} = 9 \times 10^{13} \times 10^{-15}\ \text{F} = 9 \times 10^{-2}\ \text{F}$$
$$= 90 \times 10^{-3}\ \text{F} = \mathbf{90\ mF}$$

EXAMPLE 1-6

Using the equation $P = I^2R$, calculate the power, P, in microwatts (μW) when $I = 5$ μA and $R = 8$ MΩ

Solution

$$P = I^2R = (5\ \mu\text{A})^2 \times 8\ \text{M}\Omega = (5 \times 10^{-6})^2 \times 8 \times 10^6$$
$$= 2 \times 10^{-4}\ \text{W} = 200 \times 10^{-6}\ \text{W} = \mathbf{200\ \mu W}$$

Key Sequence

Press	Display
5 [EE] [+/−] 6 [x^2]	2.5×10^{-11}
[×] 8 [EE] 6 [=]	0.0002 or $\mathbf{2 \times 10^{-4}}$

Several interesting points arise from Exs. 1-5 and 1-6 above. The first is whether the answers are expressed in the simplest possible form with the fewest digits and letters. In the case of 90 mF of Ex. 1-5, we might have written it as 0.09 F (requiring **five** typesettings) or 90 000 μF (requiring **seven** typesettings). The selection of 90 mF requires only **four** typesettings, so it represents the optimum choice and most elegant form of expression of magnitude and unit.

Applying this criterion to Ex. 1-6, we may express 2×10^{-4} W as either 0.2 mW (requiring five typesettings) or 200 μW (also requiring five typesettings). Either form is preferable, in terms of the criterion outlined in step 9 above.

A second point involves use of the electronic calculator in conjunction with equation substitution and solution. We already know that substitutions in equations must be expressed in basic form and that the numerical answer will therefore be in basic units. But I is given in microamperes (μA) and R is given in megaohms

(MΩ) in Ex. 1-6. The solution shows that the values of I and R were converted to basic in Ex. 1-6 before performing the mathematical operations. But the calculator input "sees" $(5\ \mu)^2 \times (8\ \text{M})$. The student "operator" interprets mentally $\mu = 10^{-6}$ and $\text{M} = 10^6$. The key sequence shown in the solution to Ex. 1-6 emerges directly from the given values (of I and R) substituted directly into the equation! This technique is illustrated once more in Ex. 1-7.

EXAMPLE 1-7

Using the equation $W = CV^2/2$, calculate the energy in picojoules (pJ) when $C = 8$ nF and $V = 25$ mV

Solution

$$W = CV^2/2 = \frac{(8\ \text{nF})(25\ \text{mV})^2}{2} = \mathbf{2.5\ pJ}$$

Key Sequence

Press	Display
25 [EE] [+/−] 3 [x^2]	0.000625
[×] 8 [EE] [+/−] 9 [÷] 2 [=]	**2.5 × 10^{-12}**

1-5 CONVERSION OF UNITS

The solutions of Exs. 1-3 through 1-7 were simplified because all values used in all equations were given in the *same system* of units, namely the SI. By definition, a *system* is "a regular order of connected parts." The **English** (or Imperial) **system,** still in use in the United States, is a separate system having its own fundamental and derived units. The **CGS system** (an earlier form of the SI) is another separate system.

Whenever we encounter units expressed in, say, the English system or the CGS system, it is *first* necessary to *convert* these units to corresponding units in the SI before appropriate substitutions in equations can be made. Table 1-5 shows the relation between five units in the SI and corresponding units in the CGS and English systems.

Table 1-5 Relation Between Units of Measure in Various Systems of Units

Physical Quantity	Quantity Symbol	SI Unit (abbreviation)	CGS Unit (abbreviation)	English Unit (abbreviation)
Length	l	meter (m)	centimeter (cm)	foot (ft)
Mass	m	kilogram (kg)	gram (g)	slug (slug)
Time	t	second (s)	second (s)	second (s)
Force	**F**	newton (N)	dyne (dyn)	pound (lb)
Acceleration	a	meter/(second)^2 (m/s^2)	centimeter/(second)^2 (cm/s^2)	foot/(second)^2 (ft/s^2)

Let us consider the first physical quantity in Table 1-5. We know from our unit prefix notation the relation between a meter (in SI) and a centimeter (in CGS), i.e., $1\ \text{m} = 10^2\ \text{cm}$ (see Table 1-4). From this relation we may derive a *conversion factor* between the two systems for units of length:

given 1 meter = 10^2 centimeters and dividing **both** sides by 1 meter

yields: $1 = 10^2\ \text{cm/m}$ and since the reciprocal of unity is unity,

$1 = 10^{-2}\ \text{m/cm}$

The insights behind the above equality (which are true of all conversion factors) are

1. It enables us to multiply lengths in meters to obtain centimeters or multiply lengths in centimeters to obtain meters, using the appropriate ratio.
2. This multiplication does *not* change the length of the given quantity because we are multiplying it by *unity*, i.e., 1 times $x = x$.
3. It is essentially because of this equality between different units (of length, in this case) that we can use conversion factors for *all* physical quantities expressed in different systems. But we must know the equality between them!

EXAMPLE 1-8

The length of a copper wire is 400 cm. Calculate its length in meters, using an appropriate conversion factor, and verify the answer by dimensional analysis.

Solution

$$l = 400\,\cancel{\text{cm}} \times 10^{-2}\,\frac{\text{m}}{\cancel{\text{cm}}} = 400 \times 10^{-2} = \mathbf{4\ m}$$

Note from Ex. 1-8 that multiplying the original length, in cm, by the ratio m/cm resulted in cancellation of cm in both numerator and denominator, leaving only meters in the numerator. Further, since the conversion ratio represents unity, as shown above, the physical quantity of length is *unchanged*.

Let us apply the above principles to obtain conversion factors for units of length among the three systems of Table 1-5. To do this, we must know the relation between 1 m and its length in feet (or 1 ft and its length in meters).

EXAMPLE 1-9

Given that 0.3048 m = 1 ft, calculate the following conversion factors in ratio form: a. ft/m b. ft/cm c. m/ft d. cm/ft

Solution

a. Since 0.3048 m = 1 ft, dividing both sides by 0.3048 m yields

$1 =$ **3.2808 ft/m** (answer rounded to five significant figures)

b. $3.2808\,\frac{\text{ft}}{\cancel{\text{m}}} \times 10^{-2}\,\frac{\cancel{\text{m}}}{\text{cm}} =$ **0.03281 ft/cm** (rounded to four significant figures)

c. **0.3048 m/ft** by inspection (note that this is an *exact* value)

d. $0.3048\,\frac{\cancel{\text{m}}}{\text{ft}} \times 10^{2}\,\frac{\text{cm}}{\cancel{\text{m}}} =$ **30.48 cm/ft** (an exact value)

Note that by obtaining conversion factors in the form of ratios, we can use these ratios to convert lengths directly and easily, as shown by Ex. 1-9. Appendix A-3 contains a list of common conversion factors, for various physical quantities, **in ratio form.**

1-5.1 Rules Regarding Use of Conversion Factors (See Appendix A-3)

a. Conversion factors can be used *only* to change the *units* of a given physical quantity. They *cannot* be used to change the physical quantity itself. Thus, we can change units of a given physical quantity from one system to any other system or within a given system: length to length, area to area, force to force, work to work, power to power, temperature to temperature, etc. Under no circumstances is it possible to use conversion factors to change any physical quantities to other physical quantities.
b. Only one conversion factor (or ratio) is needed for any given physical quantity for conversion *between* systems of units. As shown in Ex. 1-9, if we know the

number of meters per foot, we can easily obtain its reciprocal expressed as the number of feet per meter. This greatly simplifies the number of conversion factors needed.

c. Once we have converted from other (English or CGS) systems to the SI, conversion *within* the SI is greatly simplified by the use of metric prefixes and the fact that only one basic unit is essentially used for each physical quantity.

d. Only one system of units may be used with each equation; i.e., the various factors or elements of a given equation must each be expressed in basic units of the *same* system. This is the only way to obtain identical and correct results, regardless of the system used, for the same physical quantity conditions. This is shown in Ex. 1-10 below which contains *mixed units* from different systems.

EXAMPLE 1-10

A mass of 5 kg is subjected to an acceleration of 30 ft/s^2. Calculate the force required to accelerate the mass in

a. Pounds, using the English system
b. Newtons, using the SI
c. Verify the equality of the force required by converting (**a**) to (**b**)

Solution

a. Since $\mathbf{F} = ma$, we must convert the mass of 5 kg to slugs. From Appendix A-3E we obtain 0.06852 slug/kg, and substituting we obtain $\mathbf{F} = ma = (5\text{ kg} \times 0.06852\text{ slug/kg}) \times 30\text{ ft/s}^2 = \mathbf{10.278\ lb_f}$
b. Since the acceleration is expressed as 30 ft/s^2, we must convert it to m/s^2 or $30\text{ ft/s}^2 \times 0.3048\text{ m/ft} = 9.144\text{ m/s}^2$. Substituting, we obtain $\mathbf{F} = ma = 5\text{ kg} \times 9.144\text{ m/s}^2 =$ **45.72 newtons**
c. From Appendix A-3F we obtain 4.4482 N/lb_f, so using the force calculated in (**a**) we obtain $10.278\text{ lb}_f \times 4.4482\text{ N/lb}_f =$ **45.72 N**

Example 1-10 should be done independently by the reader because it verifies several important points:

1. In the SI the basic unit of mass is the kilogram. Thus, in the substitution of part (**b**) for mass we used 5 kg and *not* 5000 g. Had we used 5000 g in lieu of 5 kg we would have obtained a force of 45 720 N, which is ridiculous. This is a common error frequently made by students and should be avoided.
2. In the English system the basic unit for length is the foot (as shown in Table 1-5) and *not* the yard (a commonly found error). Consequently, the appropriate unit for acceleration must be expressed in ft/s^2 (not yd/s^2) for correct use in the equation $\mathbf{F} = ma$.
3. For identical and correct results using either system, it was necessary to convert **all** quantities to the **basic** units of that system before substituting in the same equation relating mass, force, and acceleration.
4. Example 1-10**c** verifies the above conversion techniques.

1-5.2 Deriving Conversion Factors

Although Appendix A-3 provides many conversion factors, sometimes a particular conversion ratio relation is *not* available. Conversion factors may be derived rather easily from some well-known conversion ratios, as shown by Ex. 1-11.

EXAMPLE 1-11

Using the well-known conversion factor 2.54 cm = 1 in,

a. Derive a conversion factor relating kilometers to miles
b. Express this factor as kilometers per mile and miles per kilometer
c. Using the ratio obtained in (**b**), verify your automobile speed indicator dial to show that a speed of 100 km/h is approximately 62 mi/h (MPH)

Solution

We also require the commonplace ratios 5280 ft/mi and 12 in/ft for solution of the distance in kilometers equivalent to 1 mile. We begin with a length of 1 mi or

a. $l = 1\text{ mi} \times 5280\dfrac{\text{ft}}{\text{mi}} \times 12\dfrac{\text{in}}{\text{ft}} \times 2.54\dfrac{\text{cm}}{\text{in}} \times \dfrac{1\text{ km}}{10^5\text{ cm}} = \mathbf{1.609\ km}$
b. Since 1 mi = 1.609 km, we may write **1.609 km/mi** and its reciprocal **0.6214 mi/km**
c. $v = 100\dfrac{\text{km}}{\text{h}} \times 0.6214\dfrac{\text{mi}}{\text{km}} = \mathbf{62.14\ mi/h}$ or approximately 62 MPH

1-5.3 Significant Figures

No physical measurement is ever absolutely correct. Its numerical value is an approximation. The reliability or accuracy of a measurement is limited by the accuracy of the measuring instrument used. If an ammeter records a current of 25.7 A, this means that the measurement was measured to the *nearest* tenth of an ampere; i.e., its *exact* value lies between 25.65 and 25.75 A. Had the measurement been made to the nearest *hundredth* of an ampere, it would have been recorded as 25.70 A. The value 25.7 A is expressed to three significant figures and the value 25.70 A to four significant figures. By definition, the least significant or right-hand digit is the last one that is known to be reasonably reliable. The following rules apply to the use of significant figures in calculations.

1. *Significance of zeros.* In a current expressed as 0.0580 A, the two zeros before and after the decimal point are *not* significant. The zero after the 8 is a significant figure, however, and the current given is accurate to three significant figures (5, 8, and 0).
2. *Rounding off to a desired number of significant figures.* The least significant digit (to the right) is dropped off and the next significant digit is either increased by one or remains the same, according to the following rules.
   a. When the least significant digit (LSD) is less than 5, the next significant digit remains unchanged.
   b. When the LSD is more than 5, the number 1 is added to the next significant digit.
   c. When the LSD is exactly 5 and is to be dropped by rounding off, the number 1 is added to the next significant digit whenever that next digit is *odd*. If the next significant digit is even, it remains the same. (See Ex. 1-12 below.)
3. *Addition or subtraction of decimal numbers.* When adding numbers containing different numbers of significant digits, the numbers are first added as given (from measurements or data). After adding (or subtracting), the answer is rounded off, retaining digits only as far as the *column* provided by the least significant number. (See Ex. 1-13 below.) Remember that "a chain is only as strong as its weakest link." A calculation is only as accurate as the weakest or *least significant* measurement.
4. *Multiplication and division of numbers.* After performing the required multiplication and/or division of the numbers as originally given, the answer is rounded off to contain only as many significant figures as contained in the least accurate number given. (See Ex. 1-14 below.)

EXAMPLE 1-12

Round off the following numbers to three significant digits: a. 21.75 V b. 31.65 A c. 42.85 W

Solution

a. **21.8 V**; since the next significant digit is odd, 1 is added to 7
b. **31.6 A**; since the next significant digit is even, 0 is added to 6
c. **42.8 W**; since the next significant digit is even, the 8 remains unchanged

---

EXAMPLE 1-13

Add the following columns of voltages and express the answer to the appropriate number of significant figures by rounding off.

1.	2.	3.	4.
50.360	25.0	8.50	876.4
6.432	0.0072	2.7254	7.63
0.575	0.00002	0.028	0.254
57.367	25.00722	11.2534	884.284

Solution

1. **57.367 V** 2. **25.0 V** 3. **11.25 V** 4. **884.3 V**

1. In column 1, since all numbers are expressed to three significant figures beyond the decimal, the answer is accurate to three significant figures beyond the decimal, or a total of five significant figures. This is true even when one number is expressed to only three significant figures.

2. In column 2, since the first number (25.0) is accurate to only one number beyond the decimal point, the answer can be accurate to only one significant figure beyond the decimal. Note that the answer contains three significant figures despite the fact that the last number is accurate to only one significant figure.
3. In column 3, since the first number is accurate to only two significant figures beyond the decimal point, the answer is accurate to only two significant figures beyond the decimal. Note that the answer contains four significant figures even though the smallest number added (0.028) contains only two significant figures.
4. In column 4, since the first number is significant to only one digit beyond the decimal, the answer is expressed to that degree of accuracy. Note that the answer contains four significant figures despite two entries that are accurate to only three significant figures.

EXAMPLE 1-14

Perform the mathematical operations shown below, expressing answers to the appropriate number of significant figures by rounding off.

a. $217.2 \times 5.0 \times 10^{-2}$

b. $\dfrac{365.4 \times 0.00736}{46.25 \times 4.26}$

c. $\dfrac{8.76 \times 10^3}{0.00213 \times 9.4 \times 10^2}$

Solution

a. $10.86 = \mathbf{11}$
b. $0.01365 = \mathbf{0.0136}$
c. $4375.2 = \mathbf{4.4 \times 10^3}$

The insights to be drawn from Ex. 1-14 are:

1. In Ex. 1-14a the answer must be expressed as 11 since one of the numbers is only given to two significant figures.
2. In Ex. 1-14b two of the numbers are expressed to three significant figures, so the answer must be rounded off to three significant figures. The least significant figure, 5, is dropped since the next significant digit is even.
3. In Ex. 1-14c one of the numbers is expressed to two significant figures and the answer is rounded off to two significant figures.

1-6 GLOSSARY OF TERMS USED

Base unit Basic unit of measurement, either fundamental, supplementary, or derived, in which a physical quantity is measured and expressed.

CGS system Units expressed in the centimeter–gram–second system, a version of the metric system adopted in 1875; currently not used by engineers in practical work but generally used in teaching principles of physics.

Conversion factor Multiplying factor that represents an equality (i.e., a dimensionless quantity of unity) used to convert units from one system to another or within the same system without changing the magnitude of the physical quantity. (See Appendix A-3.)

English system Units based on old measures dating from the Roman occupation of Britain, formerly used by many English-speaking peoples but currently being replaced by the SI.

MKSA system Units expressed in the meter–kilogram–second–ampere system, a precursor of the SI, adopted in the United States in 1954.

Multiple Multiplier, expressed as a *positive* power of 10 times the base unit.

Physical quantity Measurable quantity, either fundamental or derived, expressed in terms of a magnitude and a unit.

Power of 10 Positive or negative exponential value of the base 10.

Prefix Letter abbreviation preceding a unit abbreviation and indicating a multiple or submultiple of the base unit, representing a power of 10 times a basic unit.

Prefix symbol Letter symbol (uppercase, lowercase, Roman, or Greek) assigned to a particular power of 10.

Scientific notation Shorthand in which any decimal number is expressed as a number between 1 and 10 times an appropriate power of 10.

SI Système International; system of units derived from the metric system and currently accepted by engineers and scientists; adopted in the United States in 1975.

Significant figure(s) The digits of any decimal number are in order of their significance. The most significant digit (MSD) is at the left and the least significant digit (LSD) is at the right. The LSD is the last value that is known or assumed relevant. The number 1.60 has three significant digits (figures). The number 1.602 has four significant figures. The greater the number of significant figures (or digits) used to express a number, the higher its precision.

Submultiple Multiplier, expressed as a negative power of 10 times the base unit.

1-7 PROBLEMS

Secs. 1-1 to 1-3

1-1 Express the following numbers in scientific notation as a number between 1 and 10 times an appropriate power of 10:

a. 74 500
b. 0.0577
c. 0.000 75
d. 30 000 000
e. 11.5×10^{-1}
f. 0.075×10^{0}
g. 72.5×10^{-3}
h. 0.15×10^{-6}
i. 3575×10^{-3}
j. $0.000\,000\,175 \times 10^{-4}$
k. $0.000\,000\,000\,76 \times 10^{9}$

1-2 For each problem below, perform the following:

1. Convert the numbers to scientific notation
2. Perform the indicated required mathematical operations with a calculator
3. Express the answer in scientific notation to four significant figures

a. $(31 \times 10^2) \times (650 \times 10^3)$
b. $0.000\,078 \times 0.000\,65$
c. $95\,000 \div 0.000\,0375$
d. $(2 \times 10^{-2})^2 \times (4 \times 10^3)^3$
e. $\dfrac{3400 \times 0.0297}{29\,000 \times 0.00\,543}$
f. $(4.5 \times 10^{-5})^2 \div (3.6 \times 10^2)^3$
g. $(8500 \times 10^{-2})^{-2}$
h. $(2975)^{1/3} \times (0.0045)^{1/2}$
i. $(0.075)^2 \times (0.00\,75)^{-0.5}$
j. $(95\,000)^{-3} \div (0.0065)^{1/3}$

1-3 The following problems require the use of an electronic calculator having the y^x and ε^x functions. Perform the indicated operations and express the answer in scientific notation to four significant figures, despite the number of significant figures given in each.

a. $(0.000\,375)^{3/2}$
b. $(75\,000)^{-2.5}$
c. ε^{-3}
d. $\varepsilon^{0.32}$
e. $0.275^{12.2} \times \varepsilon^{-0.2}$

1-4 The following problems require use of the ln and its inverse ε^x keys of an electronic calculator in solving problems involving unknown exponential powers. (See the hint for solution of the first problem below and use the same principle to solve the remaining problems.) This technique will be used later in solving problems involving transient circuits. Calculate x in each of the following, expressing your answer in at least four significant figures.

a. $2^x = 5$
b. $8^x = 35$
c. $\varepsilon^x = 2$
d. $10^x = 6.4$
e. $x = \log_3 175$
f. $x = \log_{3.2}(0.665)$
g. $x = \log_{2.1} 0.75$
h. $\varepsilon^{1x} = 2000$
i. $\varepsilon^{12x} = 2000$
j. $\varepsilon^{-x} = 0.562$
k. $\varepsilon^{-5x} = 0.043$
l. $0.586^x = 0.00075$
m. $(3200)^{-x} = 0.794$
n. $35\varepsilon^{-0.22x} = 0.106$
o. $2.75\varepsilon^{-4.7x} = 0.025$

Hint for solution to Problem 1-4a: if $2^x = 5$, then $x \ln 2 = \ln 5$ and $x = \ln 5/\ln 2$.

Secs. 1-4 to 1-5

1-5 A car travels 200 km in 4 h. Calculate its velocity in

a. kilometers per hour
b. kilometers per second
c. meters per second
d. centimeters per second
e. miles per hour
f. feet per second
g. Which of the above expressions of velocity uses base units of the SI?
h. Which of the above expressions uses base units of the English system?

1-6 Draw the tabulation shown below on a piece of notepaper showing 5 columns, A through E. Label each column as shown in the sample. In column A, enter the following dimensioned quantities vertically: 2500 meters per second, 0.005 siemens, 0.02 webers, 250 000 ohms, 0.0019 henrys, 0.000 000 12 seconds, 0.000 72 siemens per meter, 1525 ampere (turns) per meter, 0.016 volts, 84 000 volts per meter, 0.000 0035 amperes, 0.072 joules, 0.000 000 000 000 24 farads, 6 500 000 watts, 95 000 tesla. Using the information given in Table 1-3 perform the following:

a. In column B of the tabulation, show one possible name of the physical quantity in column A
b. In column C show the quantity symbol
c. In column D show the magnitude and unit abbreviation in powers of 10
d. In column E show the magnitude and unit abbreviation using SI prefix notation

A Dimensioned Physical Quantity	B Physical Quantity	C Quantity Symbol
2500 meters per second	velocity	v

D Magnitude and Unit Abbreviation in Powers of 10	E Magnitude and Unit Abbreviation Using SI Prefix Notation
2.5×10^3 m/s	2.5 km/s

1-7 Current is defined as the rate of charge flow. If a charge of 0.0065 coulombs passes a given point in a conductor in 100 milliseconds, calculate the

a. Current in amperes (Hint: use the relation $I = Q/t$)
b. Current in milliamperes

Express in SI prefix notation, using the least possible typesetting:

c. Charge
d. Given time

1-8 Power is defined as rate of doing work, i.e., $P = W/t$. In doing work, an electrical energy of 7.5×10^4 joules is expended in 5.4×10^2 seconds. Calculate the

a. Energy consumed in watt-seconds. Express answer in powers of 10
b. Average power in watts. Express answer in powers of 10

Express in SI prefix notation, using the least possible typesetting:

c. Energy consumed
d. Power dissipated

Sec. 1-5

1-9 Convert the following quantities to basic units in the SI with unit abbreviation, expressed in scientific notation for possible use in equation substitution:

a. 270 kg
b. 1010 kHz
c. 2800 picowatts
d. 37.5 kilovolt-amperes
e. 2 cubic centimeters (be careful)
f. 8 square centimeters (be careful)
g. 0.025 microseconds
h. 750 microamperes
i. 0.16 attocoulombs (charge on an electron)
j. 0.3 gigameters per second (speed of light)
k. 96.49 kilocoulombs per mole (Faraday's law)
l. 0.00981 km/s^2 (gravity)
m. 1.257 microhenries per meter (μ_0)
n. 0.511 mega-electron volts (electron mass)
o. 9460 terameters (light-year)
p. 0.384 gigameters (earth–moon distance)
q. 29.8 kilometers per second (earth's orbital speed)

1-10 A man weighs 160 lb on earth. Calculate

a. His weight in newtons
b. His mass in kilograms
c. His mass in slugs (use conversion factor from Appendix A-3)

Hint: weight is a force expressed in either pounds, newtons, or dynes, while mass is f/a or f/g expressed in units of kilograms, slugs, or grams (Table 1-5).

d. Verify your answer to (**c**) by computing mass in slugs directly from the weight originally given

1-11 Derive conversion factors expressed as ratios for the following quantities:

a. square inches to square centimeters (Hint: begin with $A = 1\ \text{in}^2$)
b. cubic inches to cubic centimeters (Hint: begin with $V = 1\ \text{in}^3$)
c. grams to pounds (use 2.205 lb/kg; see Appendix A-3E)
d. grams to dynes

1-12 The equation for wavelength of a given frequency is $\lambda = v/f$,

where λ is the length of the wave in meters, feet, etc.
v is the velocity of the wave in meters per second, feet per second, etc.
f is the frequency of the wave in hertz (Hz)

For a radio wave v is the speed of light and the equation becomes

$$\lambda = \frac{3 \times 10^8\ \text{m/s}}{f} = \frac{1.865 \times 10^5\ \text{mi/s}}{f}$$

Using the above relations appropriately, calculate the

a. Wavelength, expressed in centimeters of a TV signal having a frequency of 80 MHz
b. Wavelength in feet of an FM station broadcasting at 102.7 MHz
c. Distance in miles between Tucson and New York if it takes 15 ms for a radio signal to be heard
d. Length in millimeters of red light having a frequency of 4.3×10^{14} Hz
e. Length in mils (milli-inches) of violet light having a frequency of 7.32×10^{14} Hz

1-13 The moment of inertia of a servomotor is 22 slug·ft^2 in English units. Calculate the corresponding inertia values in

a. SI units of kilograms per square meter (Hint: use the ratio 14.6 kg/slug)
b. CGS units of grams per square centimeter

1-14 The servomotor in the preceding problem is a 0.01-hp motor (1 hp = 550 ft·lb/s). Calculate the power developed by the servomotor at full load in

a. SI units of meter-kilogram per second (m·kg/s)
b. CGS units of centimeter-gram per second

1-15 The viscosity of the oil used in the above motor is 1.2 lb·s/ft^2 in the English system. Calculate the equivalent viscosity in

a. SI units
b. CGS units

1-16 The electrical equivalent of heat energy is 1 kWh = 3413 Btu. Using this conversion factor,

a. Derive an equivalent power conversion factor
b. Calculate the Btu/h power equivalent to 2.5 GW of electrical power
c. Convert the power in (b) to joules per second (J/s)
d. Repeat (c) for power in ergs per minute

1-17 The density of concrete is 150 lb/ft^3. Express this density in

a. SI units
b. CGS units

1-8 ANSWERS

1-1 a 7.45×10^4 b 5.77×10^{-2} c 7.5×10^{-4}
d 3×10^7 e 1.15×10^0 f 7.5×10^{-2}
g 7.25×10^{-2} h 1.5×10^{-7} i 3.575×10^0
j 1.75×10^{-11} k 7.6×10^{-1}

1-2 a 2.015×10^9 b 5.07×10^{-8} c 2.53×10^9
d 2.56×10^7 e 6.413×10^{-1} f 4.340×10^{-17}
g 1.384×10^{-4} h 9.648×10^{-1} i 6.495×10^{-2}
j 6.250×10^{-15}

1-3 a 7.262×10^{-6} b 6.492×10^{-13} c 4.979×10^{-2}
d 1.377 e 1.183×10^{-7}

1-4 a 2.322 b 1.710 c 0.6931 d 0.8062 e 4.701
f −0.3507 g −0.3877 h 7.601 i 0.6334
j 0.57625 k 0.6293 l 13.464 m 0.02858
n 26 362 o 1

1-5 a 50 km/h b 1.38×10^{-2} km/s c $13.\overline{8}$ m/s
d $138\overline{8}$ cm/s e 31.07 mi/h f 45.57 ft/s g c h f

1-7 a 0.065 A b 65 mA c 6.5 mC d 0.1 s

1-8 a 7.5×10^4 W-s b $1.3\overline{8} \times 10^2$ W c 75 kJ
d $13\overline{8}$ W

1-9 a 2.7×10^5 g b 1.01×10^6 Hz c 2.8×10^{-9} W
d 3.75×10^5 VA e 2×10^{-6} m^3 f 8×10^{-4} m^2
g 2.5×10^{-8} s h 7.5×10^{-4} A i 1.6×10^{-19} C
j 3×10^8 m/s k 9.649×10^4 C/mol l 9.81 m/s^2
m 1.257×10^{-6} H/m n 5.11×10^5 eV
o 9.46×10^{15} m p 3.84×10^8 m q 2.98×10^4 m/s

1-10 a 711.7 N b 72.62 kg c 49.75 slug

1-11 a 6.452 cm^2/in^2 b 16.387 cm^3/in^3
c 2.205×10^{-3} lb/g d 1.02×10^{-3} g/dyn

1-12 a 3.75×10^2 cm b 9.588 ft c 2798 mi
d 6.98×10^{-4} mm e 1.614×10^{-2} mils

1-13 a 29.84 kg·m^2 b 2.984×10^8 g·cm^2

1-14 a 0.7603 m·kg/s b 7.603×10^4 cm·g/s

1-15 a 5.858 kg·s/m^2 b 0.5858 g·s/cm^2

1-16 a 3413 (Btu/h)/kW b 8.532×10^9 Btu/h
c 2.5×10^9 J/s d 1.5×10^{18} erg/min

1-17 a 2.402×10^3 kg/m^3 b 2.402 g/cm^3

$$D = \frac{\psi}{A} = \epsilon_0 E$$

$$V = \frac{W}{Q} = Ed,\ E = \frac{V}{d}$$

$$I = \frac{Q}{t},\ [(n)(e)(v)(A)]$$

$$V = IR$$

$$R = \frac{\rho l}{A}$$

$$P = VI$$

$$P = \frac{W}{t}$$

$$P = I^2 R = \frac{V^2}{R}$$

$$\alpha_{20} = \frac{R_2 - R_1}{R_1 \Delta T}$$

$$\alpha_0 = \frac{1}{\alpha_{20}} - 20$$

$$R_2 = R_1 \left[\frac{\frac{1}{\alpha_0} + T_2}{\frac{1}{\alpha_0} + T_1}\right]$$

current density

$$J = \frac{I}{A} = [(n)(e)(v)]$$

n = # of charge carriers (m³ at 20°)
e = charge on carrier
v = velocity
A = cross sectional area

$$A = \frac{\pi d^2}{4} \text{ or } \pi r^2$$

$$v = \mu E \text{ or } \mu = \frac{v}{E}$$

$$v = \frac{J}{ne}$$

CHAPTER 2

Current and Voltage

2-1 CHARGE AND MATTER

Matter, as scientists now view it, can be regarded as composed of three *fundamental particles:* the *neutron*, the *proton*, and the *electron*. Table 2-1 shows their symbols, charges, and masses. Note that the neutron and proton are approximately equal in mass but the mass of the electron is approximately 1/1840 the mass of the neutron and proton. While the neutron is *neutral* in charge, the proton is defined as *positively* charged and the electron is oppositely or *negatively* charged, as shown in Table 2-1.

Table 2-1 Relative Properties of the Neutron, Proton, and Electron

Fundamental Particle	Symbol	Polarity of Charge	Mass (kg)	Charge (C)
Neutron	n	0	1.675×10^{-27}	0
Proton	p^{+}	+	1.672×10^{-27}	1.602×10^{-19}
Electron	e^{-}	−	9.108×10^{-31}	-1.602×10^{-19}

The various atoms listed in the periodic table of the elements, from hydrogen (atomic number 1) to lawrencium (atomic number 103), differ only in the number and arrangement of neutrons and protons in their *nucleus* (or center) and in the *shell* structure of electrons *orbiting* the nucleus. Physicists sometimes refer to the surrounding shell structure as an *electron cloud*. The *nucleus* of the lightest atom, hydrogen, is approximately 1×10^{-15} m while the *heaviest nuclei* of atoms are approximately 7×10^{-15} m in diameter. The outer diameter of the electron cloud (which is really the overall diameter of the atom) lies in the range $1\text{–}3 \times 10^{-10}$ m, roughly 10 000 times the diameter of the nucleus.

A complete discussion of atomic structure is outside the scope of this volume.[1] But for our purposes we may summarize the following atomic properties, which have relevance to electricity and electronics:

1. Each *neutral* atom has a number of electrons (in its shell structure) equal to the number of protons in the nucleus.

[1] For more information, see J. Millman and C. C. Halkias, *Integrated electronics*, Ch. 1, McGraw-Hill Book Co., 1972.

2. Electrons comprising the electron cloud are arranged in discrete energy shells called *quanta*, surrounding the nucleus, based on their kinetic energy level.
3. The shells that are *farthest* from the nucleus have the *highest* energy level. The outermost shell, often called the valence shell, may share, lose, or gain electrons most readily.
4. Atoms of different elements have different numbers and arrangements of electrons, protons, and neutrons. This difference accounts for their respective atomic number, atomic weight, and chemical/physical properties.
5. Certain atoms (particularly metals) have a relatively weak attraction between the nucleus and the outermost valence electrons. As a result, at room temperature the additional energy imparted to the valence electrons causes them to escape from their valence shell and exist as *free* electrons in the electron cloud. Copper, for example, contains 8.5×10^{28} free electrons per cubic meter, while silver, an *even better conductor* than copper, contains only 5.8×10^{28} free electrons per cubic meter.[2]
6. The resulting free electrons in metallic conductors account for *metallic bonding* since the atoms that have lost electrons are now positively ionized and repel each other, which causes them to take up equilibrium positions within the electron cloud in the form of an atomic lattice structure. The attraction between the positively charged atoms and the electron cloud is called metallic bonding.
7. When energy is imparted to the atom, say in the form of radiation, electrons receiving such energy are "raised" from a lower (inner) level to a higher (outer) electron shell.
8. Similarly, whenever an electron "drops" from a higher (outer) to a lower (inner) quantum shell, there is an emission of energy in the form of radiation.
9. There are three types of atomic bonds: *metallic* (see item 6 above), *covalent*, and *ionic*.
10. Covalent bonding occurs in atoms such as carbon, silicon, and germanium, which have four valence electrons. Such atoms *share* their valence electrons. Because of this stronger bonding, more energy is required to raise electrons from the valence shell to the free-electron level.
11. Ionic bonding occurs whenever an atom loses electrons to or gains electrons from a different type of atom, as in chemical activity between elements producing chemical compounds. Such bonding is extremely strong and such compounds in their solid state are poor conductors. Conversely, when dissolved in water (if soluble) or in the molten (liquid) state, they are excellent conductors due to ionic dissociation.

2-1.1 Electrostatics

Over 3000 years ago the ancient Greeks were aware of the electrical (*elektros*) properties of amber called elektron (a natural pine resin produced in the same way that coal is produced from carboniferous vegetable matter). They noted that amber *rubbed* with fur becomes electrified and can attract small pieces of straw or bits of "neutral" hair. In our own experience, the shock produced by walking on a rug and touching a doorknob or another person is a common example of "static" electricity. Another example is the effect produced by brushing one's hair when dry and the attraction of the hair to the brush.

---

[2] This apparent discrepancy will be explained in Sec. 2-11. For now, let us say that another factor, electron mobility, accounts for the difference.

These phenomena are no longer a mystery. Simply put, it takes mechanical energy (rubbing) to separate charges from two *dissimilar* bodies. The additional energy imparted to the valence electrons causes them to be liberated from the body that was rubbed. But in accordance with the *law of conservation of charge*, the following phenomena have been verified time and again:

1. When a *glass* rod is rubbed with *silk*, a *positive* charge appears on the glass rod. Measurements show that a *negative* charge of *equal* magnitude appears on the silk. This means that valence electrons have gained sufficient energy to leave the glass rod and be picked up by the silk.
2. When a *rubber* rod is rubbed with cat's fur, the *rubber* rod becomes *negatively* charged and the cat's fur is left *positively* charged. This means that the mechanical energy imparted to the fur has liberated electrons from the fur to the rubber rod.

The positively charged glass rod[3] and the negatively charged rubber rod may now be used to electrostatically charge pith (or Styrofoam) balls or a gold-leaf electroscope to verify the laws of electrostatic repulsion and/or attraction.

These laws (discovered by Charles Du Fay in 1733) are

1. **Like** charges **repel** each other.
2. **Unlike** charges **attract** each other.

The electrical effects, as concluded from studies with charged bodies, are summarized in Table 2-2.

Table 2-2 Electrostatic Interaction between Two Bodies

Charge on Body		Electrostatic
Body A	Body B	Interaction
−	+	Attraction
+	−	Attraction
+	+	Repulsion
−	−	Repulsion
+	0	No effect
−	0	

2-1.2 Coulomb's Law

In 1785 Charles A. de Coulomb first measured, quantified, and deduced[4] the law governing the magnitude of forces of attraction and repulsion between two charged bodies as $\mathbf{F} \propto Q_1Q_2/r^2$. If we replace the proportionality with a constant we obtain

[3] The designations *positive* and *negative* are arbitrary and are attributed to Benjamin Franklin, who stated in 1750 that "electrical matter consists of positive and negative particles." For this reason the charge on an electron is called negative while the charge on a proton is positive. The same applies to a chemical cell, whose anode is called positive and whose cathode is called negative. Had Franklin reversed his terminology, we would have to reverse all polarities of generators, cells, and atomic particles.

[4] Coulomb's law resembles the gravitational law of Newton, which was published more than 100 years before Coulomb's work. Newton's law for the gravitational force between two bodies is $\mathbf{F} = G\dfrac{m_1m_2}{d^2}$. In effect, Coulomb merely replaced Newton's masses with electrostatic charges. But there *are* subtle differences, as noted above.

$$\mathbf{F} = k\frac{Q_1Q_2}{r^2} = \left(\frac{1}{4\pi\varepsilon_0}\right)\frac{Q_1 \cdot Q_2}{r^2} = \frac{8.99 \times 10^9 \cdot Q_1 \cdot Q_2}{r^2} \text{ (newtons)} \quad (2\text{-}1)$$

where k is the $1/4\pi\varepsilon_0 = 1/4\pi \cdot 8.8542 \times 10^{-12} = 8.99 \times 10^9\ \mathrm{N{\cdot}m^2/C^2}$
r is the distance between the charged particles in meters (m)
Q_1 and Q_2 are the respective **charges** on the particles in coulombs (C)

Several important points must be noted with respect to Eq. (2-1), namely

1. Unlike Newton's law, the force **F** in Coulomb's law may be either an *attractive* or a *repulsive* force. (The force of gravitation is always one of attraction.)
2. A *positive* value of **F** in Eq. (2-1) always signifies *repulsion.* (If both Q's are negative, their product is positive. If both Q's are positive, their product is positive.
3. A *negative* value of **F** in Eq. (2-1) always signifies *attraction.*
4. **F** is a vector quantity, having both magnitude and direction.
5. Equation (2-1) is expressed in units of the SI.
6. The value of k is more precisely 8.9875×10^9 as noted and derived above.
7. The value of k appears as 9×10^9 in many textbooks.

From a physical and even a practical point of view, Coulomb's law transcends the simple calculation of forces between two charged bodies. In quantum mechanics it is used to compute the forces of attraction between the nucleus and its orbital electrons. In molecular physics it finds application in determining the forces binding atoms into molecular compounds. If we delve deep enough, we discover that the subject of strength of materials is essentially the study of the *electrical* forces holding the material together! We might go so far as to say that the forces acting on us in our daily lives are predominantly electrical, rather than magnetic or gravitational, as will be shown by the following examples.

EXAMPLE 2-1

Two **electrons** are separated in free space by a given distance. From Table 2-1, the **charge** on each **electron** is $e = -1.602 \times 10^{-19}$ C and the mass of each electron is $m = 9.11 \times 10^{-31}$ kg. If the gravitational constant $G = 6.673 \times 10^{-11}\ \mathrm{N{\cdot}m^2/kg^2}$, calculate the ratio of the *repulsive* electrostatic force, $\mathbf{F}_e$, to the attractive *gravitational* force between them, $\mathbf{F}_g$

Solution

The ratio $\dfrac{\mathbf{F}_e}{\mathbf{F}_g} = \dfrac{kQ_1Q_2/d^2}{Gm_1m_2/d^2} = \dfrac{ke^2}{Gm^2}$

$$= \frac{8.99 \times 10^9\ \mathrm{N{\cdot}m^2/C^2} \times (-1.602 \times 10^{-19}\ \mathrm{C})^2}{6.673 \times 10^{-11}\ \dfrac{\mathrm{N{\cdot}m^2}}{\mathrm{kg^2}} \times (9.11 \times 10^{-31}\ \mathrm{kg})^2}$$

$= \mathbf{4.1\bar{6} \times 10^{42}}$

Note that the solution to Ex. 2-1 shows that the electrostatic forces of *repulsion* between the two electrons are almost infinitely greater than the *attractive* gravitational forces between them.

Now let us assume in Ex. 2-2 that the two electrons are suspended by invisible strings from a given point above the *earth.* How does the repulsive force between them compare with their tendency to fall to earth?

EXAMPLE 2-2

Assume the two electrons of Ex. 2-1 are separated by a distance of 2 cm. Calculate the approximate

a. Repulsive electrostatic force between them, $\mathbf{F}_e$, in newtons
b. Force of gravity acting on them, $\mathbf{F}_g$, in newtons
c. Ratio of (**a**) to (**b**)

Solution

a. $\mathbf{F}_e = \dfrac{ke^2}{d^2} \cong \dfrac{9 \times 10^9(-1.602 \times 10^{-19})^2}{(0.02)^2} = \mathbf{5.774 \times 10^{-25}\ N}$

b. $\mathbf{F}_g = mg = 9.11 \times 10^{-31}\ \text{kg} \times 9.8\ \text{m/s}^2$

$= 8.928 \times 10^{-30}\ \text{kg}\cdot\text{m/s}^2 \times \dfrac{1\ \text{N}}{1\ \text{kg}\cdot\text{m/s}^2}$

$= \mathbf{8.928 \times 10^{-30}\ N}$

c. $\mathbf{F}_e/\mathbf{F}_g = 5.774 \times 10^{-25}\ \text{N}/8.928 \times 10^{-30}\ \text{N} = \mathbf{6.47 \times 10^4}$

Example 2-2 shows that the repulsive force between the two electrons is about 65 000 times greater than their tendency to fall to earth! This clearly explains why relatively light bodies like the charged gold-foil leaves of an electroscope or charged Styrofoam **pithballs** appear to defy gravity indefinitely. (They will slowly lose their charge, of course, to air molecules and ultimately will hang vertically.)

Since Ex. 2-2 shows that the repulsive electrostatic force is much greater than the force of gravity acting on the electrons, the electrons will separate to a distance greater than the 2 cm given in Ex. 2-2. Assuming that the invisible strings supporting the electrons are of infinite length, how far will the electrons separate? Obviously, they cannot go off into space because the force of gravity will take over when the repulsive force is less than the force of gravity. Common sense tells us that the distance at which the two forces are in *equilibrium* and *equal* to each other will be the separating distance, as shown by Ex. 2-3.

EXAMPLE 2-3

Calculate the distance between the two electrons in Ex. 2-2 at the point where their electrostatic and gravitational forces are in equilibrium

Solution

Solving for d in the equation $\mathbf{F}_e = ke^2/d^2$ yields

$$d = \sqrt{ke^2/\mathbf{F}_e}$$

and substituting $\mathbf{F}_g = \mathbf{F}_e = 8.928 \times 10^{-30}$ N from Ex. 2-2 gives

$$d = \sqrt{\frac{9 \times 10^9 \times (-1.609 \times 10^{-19})^2}{8.928 \times 10^{-30}}} = \mathbf{5.11\ m}$$

Example 2-3 shows that the two electrons theoretically will remain suspended in space separated by a distance of approximately 5 m *indefinitely*. (Electrons cannot lose their charge.)

The insights we have just gained from the above considerations of electrostatic charge and force lead quite naturally to a more complete understanding of such concepts as electric field intensity, potential difference, electric flux density, and **electric current**, as we shall see.

2-2 ELECTRIC FIELD INTENSITY, E

The concept of **electric field intensity** emerges from a consideration of how a unit *positive* charge will act when placed in an *electric field*. That is, what is the magnitude of the Coulomb force, $\mathbf{F}_e$, acting on the charge, and in what direction will it move when placed in an electric field?

Figure 2-1a shows the electric field surrounding a unit positive and negative charge. **Figure 2-1b** shows the electrical field (created by a voltage source) between two parallel plates. **Figure 2-1c** shows the *unit positive* (test) charge inserted in the electric field. Let us discuss each in turn.

Note from Fig. 2-1a that the electric (electrostatic) field produced by a unit positive charge is *opposite* to that produced by a unit negative charge. The *positive*

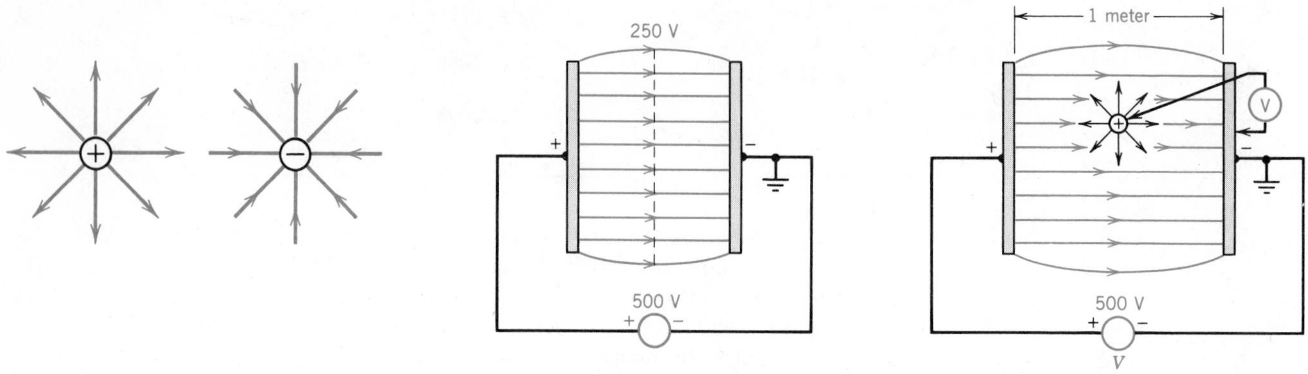

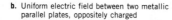

a. Uniform electric fields produced by a unit positive and negative (test) charge

b. Uniform electric field between two metallic parallel plates, oppositely charged

c. Unit positive test charge in uniform electric field

Figure 2-1 Test charges and measuring force exerted on a unit positive charge in a uniform magnetic field

charge is *defined* as producing a field that emerges from the surface of the charged body. Since a **field** is defined as a *force in space*, the direction of force is uniformly emanating from the positive charge. As Fig. 2-1a shows, the reverse occurs with a unit *negative* charge, whose force uniformly is *toward* the charge. It is important to note that the forces shown in Fig. 2-1a are three-dimensional and extend uniformly in all directions from the unit positive charge in much the same way as light emanates from a unit point source, such as a star.

Figure 2-1b shows the electric (electrostatic) field between two metallic parallel plates. The positive plate is 500 V with respect to ground, because the negative plate is grounded as shown.[5] A *voltage gradient* is said to exist uniformly as a line *perpendicular* to the electric field. Halfway between the plates, the voltage with respect to ground is 250 V, as shown in Fig. 2-1b. As we move from the positive plate to the negative plate (ground zero), the voltage decreases uniformly from 500 V to zero.

Figure 2-1c shows a positive *test charge* inserted in the electric field. If the charge is inserted directly at the positive plate, it will be repelled by the positive plate and attracted by the negative plate; that is, it moves from left to right. *Repulsion* is occurring on the *left* side of the test charge and *attraction* is occurring on the *right*. Therefore it moves from left to right.

From the above discussion it is now possible to define *electric field intensity*, **E**, as a measure of the *electric field strength* at any point in an electric field, measured as the force exerted by the field on a unit positive charge placed at that point in the field. **E** is a *vector* quantity, having both *direction* and *magnitude*, measured in units of newtons per coulomb. Quantitatively, **E** is expressed as

$$\mathbf{E} = \frac{\mathbf{F}}{q} \equiv \frac{V}{d} \equiv \frac{kQ}{d^2} \quad \text{(N/C or V/m)} \tag{2-2}$$

where **F** is the Coulomb force defined in Eq. (2-1), in newtons (N), a vector quantity

[5] Had the positive plate been grounded, the negative plate would be −500 V with respect to ground. Either plate could be grounded for obtaining voltages either *above* ground (*positive*) or *below* ground (*negative*).

q is a stationary test charge at any point in the electric field
Q is the total charge in coulombs (C) on a given surface
$k = 1/4\pi\varepsilon_0 \cong 9 \times 10^9\ \text{N}\cdot\text{m}^2/\text{C}^2$ as defined in Eq. (2-1)
V/d is the voltage gradient of the electric field in volts per meter (V/m)

Note the units of **E** and that 1 newton/coulomb ≡ 1 volt/meter

---

EXAMPLE 2-4
Calculate the electric field intensity for the uniform electric field shown in Fig. 2-1c

Solution

$$\mathbf{E} = V/d = 500\ \text{V}/1\ \text{m} = \mathbf{500\ V/m} \equiv \mathbf{500\ N/C}^{6} \qquad (2\text{-}2)$$

EXAMPLE 2-5
Calculate the force exerted by the uniform field on the unit positive test charge (1.6×10^{-19} C) shown in Fig. 2-1c

Solution
Using Eq. (2-2),

$$\mathbf{F} = \mathbf{E}\cdot q = 500\ \text{N/C} \times 1.6 \times 10^{-19}\ \text{C} = \mathbf{8 \times 10^{-17}\ N}$$

Note that the field intensity calculated in Ex. 2-4 has enabled us to calculate the Coulomb force exerted on a unit positive charge, as shown by Ex. 2-5. The net force acting on the charge is in the *same direction* as the electric field intensity, that is, from left to right in Fig. 2-1c, using Eq. (2-2).

But Eq. (2-2) still has another insight in store. Figure 2-1c is, in effect, a parallel plate capacitor. Without knowing the area of the plates, it is also possible to calculate both the charge on the plates and the capacitance of the capacitor, as shown by Ex. 2-6.

EXAMPLE 2-6
Using the data computed in Exs. 2-4 and 2-5, calculate

a. The total charge on the plates
b. The capacitance of the parallel plate capacitor, using Eq. (9-1a)

Solution

a. $$Q = \mathbf{E}d^2/k = \frac{500\ \text{N/C} \times (1\ \text{m})^2}{9 \times 10^9\ \text{N}\cdot\text{m}^2/\text{C}^2} = \mathbf{5.5\bar{5} \times 10^{-8}\ C} \qquad (2\text{-}2)$$

b. $$C = Q/V = 5.5\bar{5} \times 10^{-8}/500 = 1.1\bar{1} \times 10^{-10}\ \text{F} = \mathbf{11\bar{1}\ pF} \qquad (9\text{-}1a)$$

2-2.1 Electric flux density, *D*

A concept frequently confused with **E**, electric field intensity, is the **electric flux density**, *D*. Electric flux density, *D*, is defined as a measure of the electric field strength in terms of electrostatic lines of force, ψ, per unit area. Fortunately, in the SI one line of force is the *same* as one coulomb of charge and so electric flux density is measured in coulombs per square meter. Note that the flux density *D* is related to electric field intensity, **E**, by a factor called the permittivity, ε, or

$$D = \frac{\psi}{A} = \frac{Q}{4\pi r^2} = \varepsilon_0 \cdot \mathbf{E} \qquad \text{(coulombs/square meter) (C/m}^2\text{)} \qquad \mathbf{(2\text{-}3)}$$

where ψ is the electric flux in SI ≡ Q, in coulombs (C)
A is the area in square meters (m²)
E is the electric field intensity (Eq. 2-2) in newtons per coulomb (N/C)

---

[6] Note that 1 volt/meter is the same as 1 newton/coulomb.

r is the distance between a given charge and a smaller unit test charge in meters (m)

ε_0 is the permittivity of the medium between the charges[7] expressed in units of coulombs squared per newton-meters squared ($C^2/N \cdot m^2$) or in units of farads/meter ($1\ F/m \equiv 1\ C^2/N \cdot m^2$)

EXAMPLE 2-7

Two parallel metal plates separated by a distance of 10 cm and measuring 10 cm × 20 cm have a vacuum between them. The plates are oppositely charged and the total charge on each plate is 1.77 picocoulombs (pC). Find the

a. Electric flux density, D, between the plates in coulombs per meter squared
b. Field intensity, $\mathbf{E}$, between the plates in newtons per coulomb and volts per meter
c. Potential difference between the plates in millivolts
d. Force exerted on a test charge of one free electron between the plates in newtons

Solution

a. $D = \psi/A = 1.77 \times 10^{-12}\ C/0.1\ m \times 0.2\ m$
$= \mathbf{8.85 \times 10^{-11}\ C/m^2}$ (2-3)

b. $\mathbf{E} = D/\varepsilon_0 = 8.85 \times 10^{-11}\ C/m^2 / 8.85 \times 10^{-12}\ C^2/N \cdot m^2$
$= \mathbf{10\ N/C \equiv 10\ V/m}$ (2-3)

c. $V = \mathbf{E} \times d = 10\ V/m \times 0.01\ m = 0.1\ V = \mathbf{100\ mV}$ (2-2)

d. $\mathbf{F} = q\mathbf{E} = 1.602 \times 10^{-19}\ C \times 10\ N/C$
$= \mathbf{1.602 \times 10^{-18}\ N}$ (2-2)

Example 2-7 shows how electric flux density is related to electric field intensity **by virtue of the permittivity constant,** ε_0.

We can see that the concept of electric field intensity is extremely useful, and we are now ready to consider a very sophisticated concept: the measurement of the charge on an electron. But since we have been adequately prepared, it should be relatively easy.

2-3 MEASUREMENT OF ELECTRONIC CHARGE

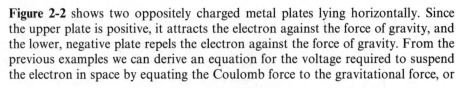

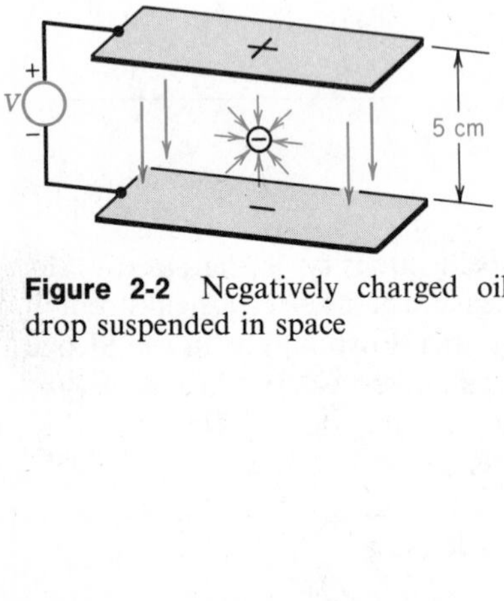

Figure 2-2 Negatively charged oil drop suspended in space

Figure 2-2 shows two oppositely charged metal plates lying horizontally. Since the upper plate is positive, it attracts the electron against the force of gravity, and the lower, negative plate repels the electron against the force of gravity. From the previous examples we can derive an equation for the voltage required to suspend the electron in space by equating the Coulomb force to the gravitational force, or

$$\mathbf{F}_g \text{ (downward)} = \mathbf{F}_e \text{ (upward)}$$

$$mg = \mathbf{E}q = \frac{Vq}{d} \qquad \text{(see Eq. 2-2)}$$

Solving for V in the above relation yields

$$V = \frac{mgd}{q} \quad \text{volts (V)} \tag{2-4}$$

[7] For a perfect vacuum (free space), $\varepsilon_0 = 8.8542 \times 10^{-12}$ farads/meter. ε_0 is an important constant in electrical engineering. It was already used in the quantification of Coulomb's law, Eq. (2-1), and we will later see it used in the definition of capacitance. It is also used in Gauss's law relating an electric field to a magnetic field surrounding a current-carrying wire, and it appears in Maxwell's equations. For most purposes, the permittivity of air is taken as approximately the same as the permittivity of a vacuum. Because of its importance in defining capacitance, permittivity (ε_0) is sometimes called *absolute capacitivity*.

EXAMPLE 2-8
Using Eq. (2-4), calculate the voltage between the plates in Fig. 2-2 if there is a single electron between plates

Solution

$$V = mgd/q = \frac{9.11 \times 10^{-31}\ \text{kg} \times 9.8\ \text{m/s}^2 \times 0.05\ \text{m}}{1.602 \times 10^{-19}\ \text{C}}$$

$$= \mathbf{2.79 \times 10^{-12}\ kg(m/s)^2/C}$$

Note that the units of V are not expressed in volts. But from Appendix A-3H we find that 1 J = 1 kg(m/s)², and we will later learn that 1 V = 1 J/C by definition; therefore,

$$V = 2.79 \times 10^{-12} \frac{\text{kg(m/s)}^2}{\text{C}} \times 1 \frac{\text{J}}{\text{kg(m/s)}^2}$$

$$= 2.79 \times 10^{-12}\ \text{J/C} = \mathbf{2.79\ pV}$$

Example 2-8 shows that an extremely *small* voltage or potential gradient is needed to suspend an electron in space against the force of gravity.[8] Example 2-9 actually shows how Millikan calculated the charge on a single electron.

EXAMPLE 2-9
Using the highest field intensity of 2×10^5 V/m, Millikan calculated the mass of a stationary oil drop as 3.269×10^{-15} kg when balanced in space between two charged plates. Calculate the

a. Field intensity in newtons per coulomb
b. Weight of the oil drop, $\mathbf{F}_g$, in newtons (see Appendix A-3F)
c. Electric force, $\mathbf{F}_e$, maintaining the oil drop stationary
d. Charge on the oil drop in coulombs

Solution

a. $\mathbf{E} = 2.0 \times 10^5\ \text{V/m} \equiv \mathbf{2.0 \times 10^5\ N/C}$ (2-2)
b. $\mathbf{F}_g = mg = 3.269 \times 10^{-15}\ \text{kg} \times 9.8\ \text{m/s}^2$
$= \mathbf{3.204 \times 10^{-14}\ kg{\cdot}m/s^2}$
but from Appendix A-3F, 1 N ≡ 1 kg·m/s²,
so $\mathbf{F}_g = \mathbf{3.204 \times 10^{-14}\ N}$
c. Then to maintain equilibrium, $\mathbf{F}_e = \mathbf{3.204 \times 10^{-14}\ N}$
d. $q_e = \mathbf{F}_e/\mathbf{E} = 3.204 \times 10^{-14}\ \text{N}/2 \times 10^5\ \text{N/C}$
$= \mathbf{1.602 \times 10^{-19}\ C}$ (2-2)

Actually Millikan's calculation was performed somewhat more elegantly than Ex. 2-9. It is possible to find the electronic charge, q_e in one step from the information given in the derivation of Eq. (2-4), that is, $q_e = \mathbf{F}_g/\mathbf{E} = mg/\mathbf{E}$. By using $q_e = mg/\mathbf{E}$, the electronic charge can be found quickly and directly.

2-4 POTENTIAL DIFFERENCE AND VOLTAGE

We have been using the term voltage without really defining it rigorously. From our background with electric fields and charges, we are now ready to attempt a rigorous definition with some degree of insight. (It would have been impossible earlier.) We know from Millikan's experiment (Fig. 2-2) that if the **potential** V is increased, the electron or negative charge will rise vertically. We also know from Exs. 2-8 and 2-9 that the oil drop and the electron have mass. If the voltage V is reduced to zero, then the oil drop rests on the negative plate. The work done in joules is $\mathbf{F}_g \times d$ when the voltage is raised to a value that causes the drop to move from the negative plate, *against gravity*, to touch the positive plate, or

$$W = \mathbf{F}_g d = mgd = \mathbf{F}_e d \qquad \text{(joules) (J)} \qquad (2\text{-}5)$$

[8] This information was used by R. A. Millikan in his classic oil drop experiment. In 1911 Millikan used X-rays to charge atomized oil drops that were squirted between two horizontal metal plates, the lower negative and the upper positive (Fig. 2-2). He adjusted the voltage, V, and thus the field intensity until the oil drop's weight exactly balanced its charge. By measuring the size (volume) of the stationary drop and knowing the density of oil, he was able to calculate the mass (m = density × volume) for a series of stationary oil drops. Since $Q = mg/\mathbf{E}$ (Eq. 2-3), the charge of the drops could also be calculated. He did this to find the smallest possible charge using the highest possible field intensity, **E**. The *lowest* value of charge represented the *charge on one electron*!

But we have seen from Eq. (2-4) that $V = mgd/q$. We can now infer that **potential difference** *is a measure of work per unit charge* since

$$V = \frac{W}{q} = \frac{\mathbf{F}_g d}{q} = \frac{\mathbf{F}_e d}{q} = \mathbf{E}d \qquad \text{(volts) (J/C or V)} \qquad (2\text{-}6)$$

A rigorous definition of potential difference now follows.

Potential V is a measure of the work W (in newton-meters or joules) done to bring a unit positive charge, q, from *infinity* to a given point in an electric field, *working against* the force, $\mathbf{F}_e$, of the field. Note that *work* is only *done* in *overcoming resistance*. In moving the charge q *against* the field force, work is done. The voltage required is the *work done per unit charge*.

Potential difference may also be defined as the work done to carry a given charge, Q, from one point in a circuit (or electric field) to another point in the circuit (or electric field), measured in joules per coulomb or volts. Thus, the potential required to bring a charge, Q, from point b in an electric circuit to point a in that circuit is[9]

$$V_{ba} = V_{b\infty} - V_{a\infty} = V_b - V_a = \frac{W_{b\infty} - W_{a\infty}}{Q} = \frac{W_{ba}}{Q} \qquad \text{(J/C or volts)} \qquad (2\text{-}7)$$

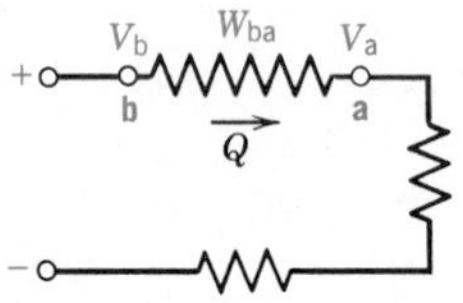

$V_{ba} = V_b - V_a = \frac{W_{ba}}{Q}$

1 volt = 1 joule/coulomb

Figure 2-3 Potential difference in an electric circuit

As shown in **Fig. 2-3**, the potential difference between point b and point a, V_{ba}, in the electric circuit is determined by the energy, W_{ba}, required to move a charge of Q coulombs from point b to point a.

EXAMPLE 2-10

An energy of 20 J is used to move a charge of 2 C from infinity to point **a**. An energy of 30 J is required to bring the same charge to point **b** from infinity. Calculate the

a. Potential of point **a** to infinity
b. Potential of point **b** to infinity
c. Potential difference between points **b** and **a**
d. Which of the two points is at the higher potential and why?

Solution

a. $V_a = W_a/Q = 20\text{ J}/2\text{ C} = 10\text{ J/C} = \mathbf{10\ V}$ (2-6)
b. $V_b = W_b/Q = 30\text{ J}/2\text{ C} = 15\text{ J/C} = \mathbf{15\ V}$
c. $V_{ba} = V_b - V_a = 15\text{ V} - 10\text{ V} = \mathbf{5\ V}$ (2-7)
d. Point b is at the higher potential because more work is required against the force of the field to get to point **b**

EXAMPLE 2-11

Calculate the energy required to move a charge of 60 mC through a potential difference of 5 kV

Solution

$$W = VQ = (5\text{ kV})(60\text{ mC}) = \mathbf{300\ J} \qquad (2\text{-}6)$$

Key Sequence

Press	Display
5 [EE] 3	5×10^3
[×] 60 [EE] [+/−] 3	**300**

Note that in Ex. 2-11 the product of the prefixes kilo and milli yields basic units. The same is true of such products as mega and micro, giga and nano, tera and pico, and so forth. This realization frequently simplifies calculations.

[9] In moving **one** coulomb of charge from point b to point a, whenever **one** joule of energy is expended, then the potential difference between points b and a is **one** volt. This is one definition of the volt in terms of derived units. Expressed in terms of basic units, potential difference is shown in units of kilograms-meters squared/amperes-seconds cubed ($\text{kg}\cdot\text{m}^2/\text{A}\cdot\text{s}^3$).

2-5 CURRENT, *I*

Current is defined as the *rate*[10] of *charge* flow in an electric circuit or in any medium in which *charges are subjected to an electric field.* Since the directions of electric fields are designated from positive to negative, conventional current[11] (by international agreement) is taken as the direction of positive charge flow, that is, from a more positive polarity to a less positive or negative polarity. Current (rather than charge) is taken as a fundamental unit in the SI because it is easier to measure in terms of the forces between two parallel wires (Appendix A-1). Equation (2-8) below shows not only the charge-flow relation for current but also the factors affecting current in various media. (This equation does not include the Ohm's law relation for current.)

$$I = n \cdot e \cdot v \cdot A = \frac{Q}{t} \quad \text{(amperes or coulombs/second)} \qquad (2\text{-}8)$$

where n is the number of free electrons per cubic meter (e^-/m^3)
e is the electronic charge in coulombs per electron (C/e^-)
v is the **drift velocity** in meters per second (m/s)
A is the cross-sectional area in square meters (m^2)
Q is the charge in coulombs (C)
t is the time in seconds (s)

EXAMPLE 2-12

Using dimensional analysis, show that the product of *nevA* in Eq. (2-8) is measured in amperes

Solution

$$\frac{\text{electrons}}{(\text{meter})^3} \times \frac{\text{coulombs}}{\text{electron}} \times \frac{\text{meters}}{\text{second}} \times (\text{meters})^2 = \frac{\text{coulombs}}{\text{second}}$$

$$\equiv \text{amperes}$$

EXAMPLE 2-13

A conductor carries a steady current of 5 mA under the influence of an electric field. Calculate the

a. Total charge passing a given point on the conductor after 2 min
b. Total number of positive ions (or electrons) passing the point in (**a**)

Solution

a. $Q = I \times t = 5$ mA (2 min)(60 s/min) = **0.6 C** (2-8)
b. $p^+ = Q/e = 0.6$ C/1.602 × 10^{-19} C/ion = **3.745 × 10**$^{\mathbf{18}}$ positive ions

---

[10] A *rate* is *any* physical quantity *divided by time.* The rate of water flow may be measured in gallons per minute. It is not the speed of water flow, which is measured in distance per unit time (feet per second, miles per hour, etc.). Current as a rate is not the speed at which charges move in an electric circuit. The velocity of charge flow would be expressed in units of kilometers per second. Other *rates* of importance are power, which is the rate of expenditure of energy or energy per unit time, and angular velocity, measured in radians per second.

[11] Conventional current, or positive charge flow, is in a direction opposite to electron flow in a medium. In a plasma or a gas (neon, argon, sodium or mercury vapor, etc.), positive ions and electrons flow in opposite directions simultaneously. Total current is the sum of the positive and negative flows. Similarly, in a chemical solution both positive ions and negative ions flow when an electric field is applied to the solution. In semiconductors there is both electron and hole flow and frequently the holes are majority carriers, as in the case of *p*-doped materials. We find, therefore, that ions flow in *both* directions in the *gaseous, liquid,* and *solid* states. Only in metals, where the metallic bonding is very strong, do we find the *exception* to the rule, where only electrons are in motion *against* the conventional direction of the electric field.

Note that the solution to part (**b**) of Ex. 2-13 requires the value discovered by Millikan for electronic charge, that is, 1.602×10^{-19} C/ion or electron. If we take the reciprocal of this value, $1/e = 1/1.602 \times 10^{-19}\ \text{C}/e^- = 6.24 \times 10^{18}$ electrons per coulomb. Then if a current of 1 A flows past a given point, there are 6.24×10^{18} electrons per second passing that point!

2-6 Current Density, *J*

Current density, J, is defined as a measure of the quantity of current distributed in a conductor having a cross-sectional area, A, expressed in units of amperes per square meter (A/m²). In terms of the factors in Eq. (2-8) we may write

$$J = nev = \frac{I}{A} \quad \text{(amperes/square meter) (A/m}^2\text{)} \tag{2-9}$$

The concept of current density is sufficiently simple. It is closely related to electric flux density, D, expressed in terms of coulombs per square meter[12] and magnetic flux density, B, expressed in webers per square meter or tesla.[13]

Current density provides a measure of whether a given conductor of given cross-sectional area is capable of carrying a given current safely, as shown by the following example.

EXAMPLE 2-14

A rule of thumb in motor winding and/or transformer design is that J, the maximum current density of the copper windings, should not exceed 100 A/cm². Calculate

a. The maximum current that can be carried by a wire, in a winding 0.05 inches in diameter
b. Repeat (**a**) for a wire of area 2000 circular mils. See Appendix A-3B for conversion factors

Solution

a. Diameter $d = 0.05\ \text{in} \times 2.54\ \text{cm/in} = 0.127\ \text{cm}$

$$A = (\pi/4)d^2 = (\pi/4)(0.127\ \text{cm})^2 = 1.267 \times 10^{-2}\ \text{cm}^2$$

$$I = J \cdot A = (100\ \text{A/cm}^2) \times (1.267 \times 10^{-2}\ \text{cm}^2) = \mathbf{1.267\ A}$$

b. From Appendix A-3B,

$$5.067 \times 10^{-6}\ \frac{\text{cm}^2}{\text{c-mil}} \times 2 \times 10^3\ \text{c-mil} = 0.01\ \text{cm}^2$$

$$I = J \cdot A = (100\ \text{A/cm}^2) \times (0.01\ \text{cm}^2) = \mathbf{1\ A} \tag{2-9}$$

2-7 ION (OR ELECTRON) DENSITY, *n*

Ion or electron **density** is defined as the number of free ions or electrons per unit volume of a given medium or material. Metals are typically good conductors because of their relatively large number of free electrons per unit volume. As noted above, electron density, n, is but one of the factors affecting the magnitude of current, I, and current density, J. We shall see that it is one of the factors affecting **conductivity**, σ, as well.

Table 2-3 lists the electron densities of seven common materials in order of

[12] Not *all* densities in the SI are expressed as some variable physical quantity per unit area. Certainly *mass* density, ρ, is mass/volume. We shall see in Sec. 2-7 that electron density, n, is expressed as the number of free electrons per unit volume, as well.

[13] 1 tesla = 1 weber/square meter

decreasing values of *n*, expressed in terms of free conduction electrons per cubic centimeter and cubic meter.

Table 2-3 Electron Density at 20°C of Common Materials

Material (pure)	Number of Free Conduction Electrons, *n* (per cubic centimeter)	(per cubic meter)
Copper	8.5×10^{22}	8.5×10^{28}
Silver	5.8×10^{22}	5.8×10^{28}
Sodium	2.5×10^{22}	2.5×10^{28}
Potassium	1.3×10^{22}	1.3×10^{28}
Cesium	0.8×10^{22}	0.8×10^{28}
Germanium	2.5×10^{13}	2.5×10^{19}
Silicon	1.4×10^{10}	1.4×10^{16}

Table 2-3 shows one reason (there are others) why germanium and silicon are classified as *semiconductors* in comparison to metals like copper and silver, which are considered good conductors of electricity.

EXAMPLE 2-15

A conductor has a diameter of 3 mm and a length of 30 m. Calculate the number of free electrons in the conductor if it is made of **a.** Copper **b.** Silver

Solution

area, $A = (\pi/4)(d^2) = (\pi/4)(3 \times 10^{-3})^2 = 7.07 \times 10^{-6}\ \text{m}^2$

volume $= A \times l = 7.07 \times 10^{-6} \times 30 = 2.12 \times 10^{-4}\ \text{m}^3$

a. $n\ (\text{copper}) = 8.5 \times 10^{28}\ \text{e}^-/\text{m}^3 \times 2.12 \times 10^{-4}\ \text{m}^3 = \mathbf{1.8 \times 10^{25}\ e^-}$

b. $n\ (\text{silver}) = 5.8 \times 10^{28}\ \text{e}^-/\text{m}^3 \times 2.12 \times 10^{-4}\ \text{m}^3 = \mathbf{1.23 \times 10^{25}\ e^-}$

Example 2-15 poses a question that we still must answer. We know that silver is a better conductor than copper. But we have just seen that silver has a *lower* electron (or ion) density and *fewer* free electrons for the same volume than copper. Will this dilemma be resolved? We must have patience to find out (Sec. 2-11 provides the answer).

2-8 DRIFT VELOCITY, *v*

Drift velocity is defined as the *average* movement of charged particles in any medium when subjected to an electric field. The movement is either opposed to or in the direction of the field, depending on the polarity of charge, and is measured in units of meters per second. Quantitatively, the factors affecting drift velocity are

$$v = \frac{J}{ne} = \frac{I}{Ane} = \mu \mathbf{E} \quad \text{(meters/second) (m/s)} \qquad (2\text{-}10)$$

where μ is the **mobility** of electrons (or ions) in the medium in units of square meters per volt-second ($\text{m}^2/\text{V}\cdot\text{s}$)

and all other terms have been defined above.

Equation (2-10) permits some interesting observations regarding the nature of drift velocity, namely

1. For a given material, n and e are both constant (the latter is always constant). Therefore, drift velocity is a direct function of J, current density; that is, the higher the current density, the higher the drift velocity.
2. For a given electric field intensity or field strength (in volts per meter), the drift velocity is a function of the electron (or hole) mobility of the given material subjected to the electrical field. (See Sec. 2-10.)
3. For a material of given cross-sectional area, drift velocity is directly proportional to current. The higher the current, the greater the drift velocity.[14] Since current is a function of **E**, either I or **E** increases drift velocity.

EXAMPLE 2-16

Copper has an ion mobility of $\mu = 4.26 \times 10^{-3}$ m²/V·s. If a piece of copper is subjected to an electric field intensity of 100 V/m, calculate the drift velocity of the free electrons in meters per second

Solution

$$v = \mu \mathbf{E} = 4.26 \times 10^{-3} \text{ m}^2/\text{V}\cdot\text{s} \times 10^2 \text{ V/m} = \mathbf{0.426\ m/s} \qquad (2\text{-}10)$$

Example 2-16 gives a clue to the nature of drift velocity. It represents the *average* movement of electrons against the direction of the electrical field in a conductor. If we could tag an individual electron in a copper wire, we would find that it moved at about 0.4 m/s in this example.

We can see from Ex. 2-16 that drift velocity is *not* the same as the speed at which free electrons move from atom to atom, that is, the speed of electric current in a wire. If it were, it would take months for a telegraph message or voice communication via telephone to be transmitted over long distances.

2-9 CONDUCTIVITY, σ

Conductivity is defined as the *specific conductance* of a standardized sample of material having a length of 1 m and a cross-sectional area of 1 m² measured at a standard temperature in units of siemens per meter (S/m). As shown by Eq. (2-11), conductivity is directly proportional to the ion or electron mobility, μ, and the electron density, n. Specifically, it may be defined as the ratio of current density to electrical field intensity, that is, $J/\mathbf{E}$, as shown by Eq. (2-11).

$$\sigma = en\mu = \frac{J}{\mathbf{E}} = \frac{1}{\rho} \quad \text{(A/m·V) or (S/m)} \qquad \textbf{(2-11)}$$

where ρ is the resistivity, defined in Sec. 2-10 below, and all other terms have been defined above.

Conductivity, simply put, is a measure of how well the specific sample dimensioned above will pass current easily. Since it is directly proportional to **both**

[14] Occasionally in the literature one finds the assumption that current is the *same* as drift or drift velocity. Equation (2-10) shows that this is not the case.

electron mobility and (free) electron density, the material having the highest conductivity will be the best conductor.

2-10 RESISTIVITY, ρ

As noted from Eq. (2-11), resistivity is the reciprocal of conductivity. Resistivity may be defined as the *specific resistance* of a standardized sample of material having a length of 1 m and an area of 1 m^2 measured at a standard temperature in units of ohm-meters ($\Omega\cdot$m). Stated in equation form,

$$\rho = \frac{\mathbf{E}}{J} = \frac{1}{en\mu} = \frac{1}{\sigma} \quad (\text{m}\cdot\text{V/A or } \Omega\cdot\text{m}) \qquad (2\text{-}12)$$

Resistivity, simply put, is a measure of the degree of opposition to a steady current produced by the specific sample dimensioned above. A material with the *highest* resistivity will be the *best insulator* and the *poorest conductor.*

Appendix Table A-4 lists the resistivity values for 31 common materials expressed in SI units of ohm-meters at various standard temperatures. As shown in the notes to this table, the conductivity of each material may be found by taking the reciprocal of the resistivity (Eq. 2-11).

EXAMPLE 2-17

When subjected to a field strength of $\mathbf{E} = 200$ mV/m, a standard dimensioned sample of material tested at 20°C exhibits a current density of 100 A/cm^2. Calculate the

a. Conductivity
b. Resistivity of the sample
c. Using Appendix Table A-4, determine the material of the sample

Solution

a. $\sigma = J/\mathbf{E} = \dfrac{100 \text{ A/cm}^2 \times 10^4 \text{ cm}^2/\text{m}^2}{0.2 \text{ V/m}}$

$= \mathbf{5 \times 10^6}$ **S/m** or **5 MS/m** (2-11)

b. $\rho = 1/\sigma = 1/5 \times 10^6 \text{ S/m} = 2 \times 10^{-7}\ \Omega\cdot\text{m}$

$= \mathbf{20 \times 10^{-8}\ \Omega\cdot m}$ (2-12)

c. The material of the sample is **steel** (from Appendix Table A-4)

2-11 MOBILITY, μ

Mobility is defined as a measure of the relative drift velocity of a charge carrier for a given electric field intensity, expressed in units of square meters per volt-second (m^2/V·s). Stated in equation form,

$$\mu = \frac{v}{\mathbf{E}} = \frac{\sigma}{ne} \quad (\text{square meters/volt-second}) (\text{m}^2/\text{V}\cdot\text{s}) \qquad (2\text{-}13)$$

Table 2-4 shows the relative free ion mobility in decreasing order for some metals and semiconductors. Note that metallic conductors have only free electrons, whereas the semiconductors (intrinsic germanium and silicon) have free holes as well as free electrons.

We are now in a position to explain why silver is a better conductor than copper despite the fact that copper has a *higher* electron density than silver (see Table 2-3). Silver is a better conductor than copper because it has a higher conductivity,

σ. From Eq. (2-11) we see that conductivity, $\sigma = en\mu$, is directly proportional to the *product* of electron density and electron mobility. This shows that a material having the *highest* density–mobility **product** will be the *best* conductor. But Table 2-4 shows that semiconductors appear to have higher mobilities than metallic conductors. This indicates that a trade-off is taking place between electron density and mobility in determining conductivity of all materials (conductors, semiconductors, and insulators), as shown by Ex. 2-18.

Table 2-4 Relative Mobility of Free Ions in Metals and Semiconductors at 20°C

Pure	Free Ion Mobility ($m^2/V \cdot s$)	
Material	Electrons	Holes
Germanium	0.39	0.19
Silicon	0.14	5×10^{-2}
Silver	6.61×10^{-3}	None
Copper	4.26×10^{-3}	None
Cesium	4.10×10^{-3}	None

EXAMPLE 2-18

Using the data given in Tables 2-3 and 2-4 and Eq. (2-11) for conductivity, calculate the conductivity of a. Copper b. Silver

Solution

a. $\sigma = en\mu = (1.602 \times 10^{-19}\ C/e^-) \times (8.5 \times 10^{28}\ e^-/m^3) \times (4.26 \times 10^{-3}\ m^2/V \cdot s) = \mathbf{5.8 \times 10^7\ S/m}$ (copper) (2-11)

b. $\sigma = en\mu = (1.602 \times 10^{-19}\ C/e^-) \times (5.8 \times 10^{28}\ e^-/m^3) \times (6.61 \times 10^{-3}\ m^2/V \cdot s) = \mathbf{6.14 \times 10^7\ S/m}$ (silver) (2-11)

Example 2-18 shows clearly that silver has the higher density–mobility product and is therefore the better conductor despite its lower electron density compared to copper. It is, in fact, the best metallic conductor known to science at normal ambient temperatures.

EXAMPLE 2-19

Using the calculated values of conductivity in Ex. 2-18, calculate the resistivities of

a. Copper
b. Silver
c. Compare the calculated resistivities with those given in Appendix A-4

Solution

a. $\rho = 1/\sigma = 1/5.8 \times 10^7\ S/m = \mathbf{1.724 \times 10^{-8}\ \Omega \cdot m}$ (2-12)
b. $\rho = 1/\sigma = 1/6.14 \times 10^7\ S/m = \mathbf{1.629 \times 10^{-8}\ \Omega \cdot m}$ (2-12)
c. The resistivities calculated are *identical* to those given in Table A-4

Examples 2-18 and 2-19 verify all of the theory presented in this chapter. Clearly, the resistivities shown in the first column of Appendix A-4 were determined by direct resistance measurement, using very precise electronic bridges and material samples whose dimensions were carefully controlled. Yet because we learned the various factors affecting conductivity and resistivity, we were able to compute these values *independently*. Had our theory been incorrect, our computations would have been in error.

Note also that throughout the given examples we have used the SI basic and derived units consistently. This also verifies the validity and consistency of the SI.

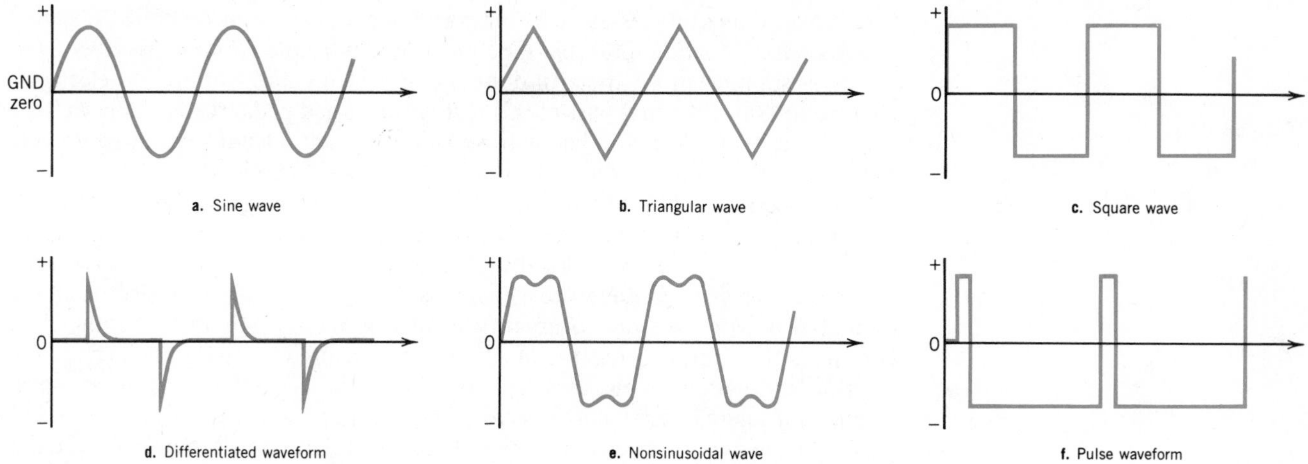

Figure 2-4 Examples of ac waveforms

2-12 SOURCES OF DIRECT VOLTAGE

There are two forms in which electricity is furnished and used commercially: alternating current (ac) and direct current (dc). Examples of ac are shown in **Fig. 2-4**. The one element that all these waves have in common is that the voltage (and current) *varies periodically* by *reversing* positively *above* ground and negatively *below ground, alternately*.[15]

Examples of dc are shown in **Fig. 2-5**. The one element that all these dc voltages have in common is that the voltage (and current) is *unidirectional*. The dc voltage may vary, as shown in Figs. 2-5a, 2-5b, and 2-5d, but it never reverses direction.

The ac voltages may be either stepped up, for purposes of long- or short-distance power transmission, or stepped down, for purposes of energy utilization, relatively easily by using transformers. For this reason, ac is used almost universally in industrial, commercial, and residential applications for power and lighting.

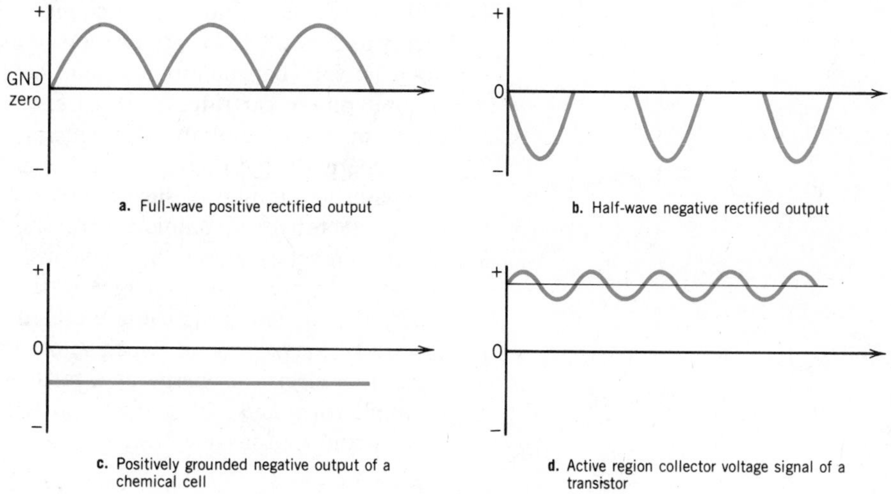

Figure 2-5 Examples of dc voltage waveforms

[15] It will be shown later that some of these waves contain a dc component as well.

In general, dc voltages are used primarily *locally*, as in automobiles, chemical refining, electroplating, portable tape recorders, portable AM/FM receivers, and as power supplies in a variety of domestic, industrial, and commercial electronic equipment. Since the first part of this text is concerned primarily with dc, we shall confine our brief discussion below to some of the methods for generating dc voltages and the ways in which various dc sources are represented.

Most electric circuits contain circuit elements that possess resistance (Sec. 3-2). For this reason, such circuits dissipate or convert electrical energy to heat energy, primarily. As we have seen from Fig. 2-3, when a circuit containing resistance is connected to a dc source, electrical energy is required to displace an electric charge (potential is work per unit charge) from one point in a circuit to another, and current is the average or net drift of charge measured in coulombs per second.

In order to produce electrical energy, it must be *generated* by and converted from other energy forms (*mechanical*, *light*, *heat*, and *chemical*). Alternatively, it may be *converted* from electrical ac sources (rectified) via dc power supplies. It takes energy *originally* to produce electrical energy, since energy can neither be created nor destroyed. Even when the electrical energy is "consumed" in resistance, that energy is converted to other forms (heat, radiation, mechanical, etc.). Such energy conversion may be *irreversible*, as in the case of a burning electric lamp, or *reversible*, as in the case of an automobile battery, which is *discharged* and *recharged*. Let us very briefly consider some of the various sources of dc voltage (work per unit charge).

Sources of dc may be classified in terms of their energy source as

1. Chemical sources, in which chemical energy is converted to electrical energy. Examples of such sources are primary and secondary cells (and batteries) as well as fuel cells.
2. Solar and photovoltaic cells, in which light energy is converted to electrical energy. Examples of such sources are copper oxide and selenium photovoltaic cells as well as *pn*-junction (silicon) solar cells and batteries.
3. Thermoelectric generation, in which heat energy is converted directly to electrical energy. Examples of such sources are thermocouples and semiconductor thermoelectric engines.
4. Piezoelectric generation, in which mechanical energy is converted to electrical energy as a result of the application of mechanical forces to crystals of quartz, Rochelle salt, barium titanate, and so forth. Examples of such sources are ceramic phono cartridges and underwater sound transducers.
5. Electromagnetic generation, in which mechanical energy is converted to electrical energy in the presence of a magnetic field in a rotating machine called a dynamo. Examples of such sources are the various types of dc generators and magnetohydrodynamic generators.
6. Electromagnetic generation, in which electrical energy is converted to electrical energy in the presence of a magnetic field. Examples of such sources are various types of semiconductors using the Hall effect.
7. Electrical conversion, in which alternating electrical energy is converted to direct (dc) electrical energy as a result of rectification or energy conversion. Examples of such sources are dc power supplies, dynamos called rotary converters, and motor–generator sets.

2-13 SYMBOLS USED FOR SOURCES

Regardless of the method of generation outlined in Sec. 2-12, all dc and ac sources are designated by a set of common circuit symbols. There are two major classes of symbols, for *independent* sources and *dependent* sources. Independent sources are designated by a circle. *Dependent* (or controlled) sources are designated by a *diamond.*[16] The following definitions apply:

1. An ideal independent voltage source is one that maintains a specified voltage between its source terminals regardless of the current (load) drawn from it. **Figure 2-6a** shows an ideal independent voltage source.
2. An ideal independent current source is one that maintains a specified current through its terminals regardless of the voltage across the terminals. Figure 2-6c shows an ideal independent current source, designated by an arrow within the enclosed circle.
3. Dependent or controlled sources produce either a voltage or a current as a result of a voltage or current elsewhere in a given circuit. There are four types of dependent sources:
   a. Voltage-dependent voltage sources (VDVS), which produce a voltage as a function of voltages elsewhere in a given circuit. Figure 2-6b shows a typical VDVS.
   b. Current-dependent current sources (CDCS), which produce a current as a function of currents elsewhere in a given circuit. Figure 2-6d shows a typical CDCS.
   c. Current-dependent voltage sources (CDVS), which produce a voltage as a function of a current elsewhere in a given circuit. Figure 2-7b shows a typical CDVS.
   d. Voltage-dependent current sources (VDCS), which produce a current as a function of a voltage elsewhere in a given circuit. Figure 2-7d shows a typical VDCS.

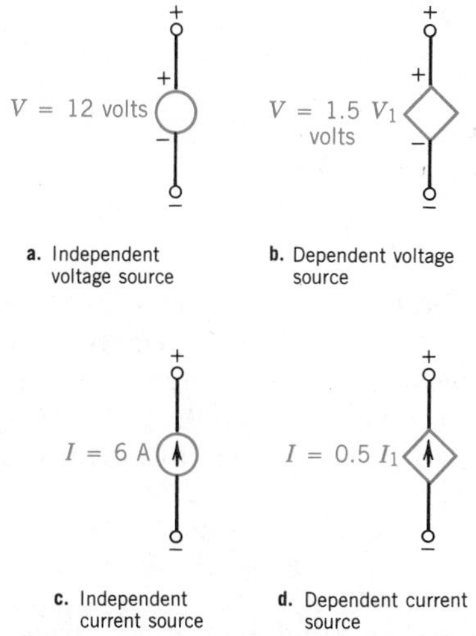

Figure 2-6 Symbols for *ideal* independent and dependent voltage and current sources

Figure 2-7 shows the four types of dependent (controlled) sources for comparison. With respect to both Figs. 2-6 and 2-7, note:

1. Polarities of voltage sources are shown within the symbol and designated by plus and minus signs. These signs indicate the *conventional* direction of the electric field when the source is applied to a load.
2. Polarities of current sources are designated by an arrow shown *within* the symbol. The arrow indicates the *conventional* direction of current when the source is applied to a load.
3. Ideal voltage and current sources contain no associated internal resistance (or impedance). But regardless of the method of generation (Sec. 2-12), no source is ideal; that is, all sources experience a change in terminal voltage whenever a load is connected. The sources shown in Figs. 2-6 and 2-7 are all ideal sources. In most of the work throughout this text, we will assume ideal sources to simplify the development of certain basic laws, theorems, and equations. Practical (nonideal) sources, covered in Chapter 5, represent the behavior of actual (commercial) current and voltage sources in a more realistic way.

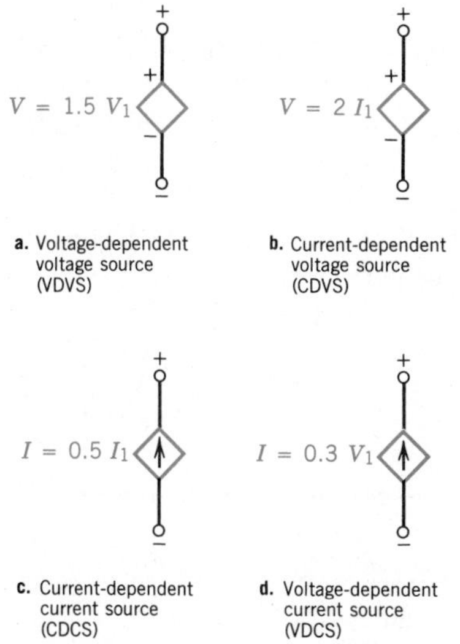

Figure 2-7 Symbols for the four types of dependent (controlled) sources

Figure 2-8a shows one representation of a *practical voltage source.* It consists of an ideal voltage source, V, in series with an internal resistance, r. We shall see later that this simple representation enables us to predict the actual circuit behavior of practical voltage sources in Chapter 5.

Figure 2-8b shows one representation of a *practical current source.* Note that it consists of an ideal current source, I, *paralleled* (shunted) by an internal resistance, r.

---

[16] Since **d**ependent and **d**iamond both begin with the letter "d," the alliteration helps us remember the distinction between dependent and independent source symbols.

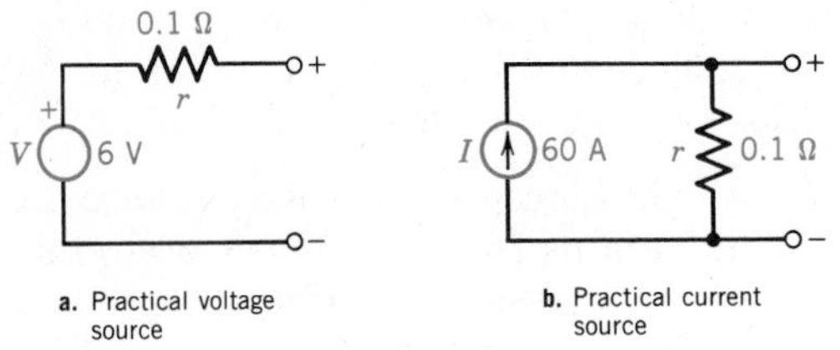

Figure 2-8 Practical dc voltage and current sources

We shall see later that the two circuits in Fig. 2-8 are equivalent and that this modification of the ideal current source more closely represents the behavior of practical current sources. Note in Fig. 2-8 that each graphical symbol displays both the physical quantity (V or I) and its magnitude in appropriate units.

Figure 2-9 shows the distinction between ac and dc source symbols. Symbols for ac sources may be identified in two ways:

1. The physical quantity is associated with the letter (t) in parentheses, implying that the voltage or current is a *time-varying* quantity.
2. The graphical symbol contains within it the **cycle** symbol, indicating a waveform whose voltage or current is varying periodically. The frequency of the periodic variation is sometimes included, as it is in Figs. 2-9a and 2-9b.

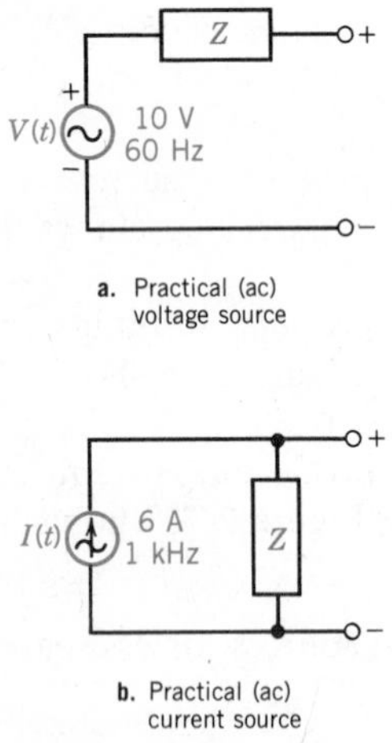

Figure 2-9 Practical ac voltage and current sources

From the foregoing discussion and examination of Figs. 2-6 through 2-9, we may conclude that

1. Regardless of the method of generating electricity (Sec. 2-12), the symbols for such sources remain the same. The 12 V dc source in Fig. 2-6a may have been produced by a battery, a dc generator, a solar battery, or rectification of an ac source.
2. *Independent* sources are represented by *circles*. *Dependent* sources are represented by *diamonds*.
3. A practical voltage source consists of an ideal voltage source and a series-connected element. A practical current source consists of an ideal current source shunted by a parallel-connected element.
4. An ac source may be identified by a cycle symbol and the letter (t), showing that it is a time-varying quantity that changes periodically.

2-14 GLOSSARY OF TERMS USED

Charge, Q Specific quantity of electricity, either positive or negative, measured in coulombs (C).

Conductivity, σ Specific conductance presented to a source by a standardized sample (having a length of 1 m and a cross-sectional area of 1 m^2) of a given material. It is a relative measure of electron density, n, and electron mobility, μ, of free charged particles in the material. Conductivity varies inversely with resistivity, ρ, and is measured in units of siemens per meter (S/m).

Current density, J Measure of the current distributed in a conductor of cross-sectional area, A, expressed in units of amperes per square meter (A/m^2). Current density in any medium is directly proportional to electric field intensity, **E**.

Drift velocity, v Average movement of charged particles in any medium when subjected to an electric field. The movement is either opposed to or in the direction of the field, depending on polarity of charge, and is measured in meters/second (m/s).

Electric current, I Rate of ion or charge flow in any medium subjected to an electric field, in the direction of the field, measured in coulombs per second (C/s) or amperes (A). One coulomb per second ≡ one ampere.

Electric field intensity, E Measure of the field strength at any point in an electric field measured in newtons per coulomb (N/C) or volts per meter (V/m). **E** is a vector quantity having both magnitude and direction.

Electric flux density, D Measure of the field strength in terms of electrostatic lines of force per unit area measured in coulombs per square meter (C/m^2). Note: 1 electrostatic line of force $\equiv$ 1 coulomb of charge.

Electron Smallest unit of matter having a negative charge. The mass of a single electron is 9.1×10^{-31} kg and the charge is -1.602×10^{-19} C.

Electron charge, *e* Charge on a single ion (electron or proton). As measured by Millikan, $e = 1.602 \times 10^{-19}$ C.

Field Force that exists in space between charged bodies, producing either repulsion or attraction, and/or the force of attraction between the bodies by virtue of their mass. Coulomb's law quantifies these forces for an electric field. Newton's law quantifies the forces for a gravitational field.

Ideal current source Current source that maintains a specified current through its terminals regardless of the voltage across its terminals.

Ideal voltage source Voltage source that maintains a specified voltage between the source terminals regardless of the current drawn through it.

Ion density, *n* Number of free ions (electrons, protons, holes) per unit volume (usually per cubic meter). Ion density is a factor in the determination of relative conductivity.

Ionization of matter Separation of matter into positively and negatively charged particles as a result of applied energy (heat, mechanical, electrical, chemical, light). Such charged particles are called ions.

Mobility, μ Measure of the relative drift velocity, v, of a charge carrier for a given electric field intensity, **E**, expressed in units of square meters per volt-second (m^2/V·s).

Permittivity constant, ε_0 Factor used in determining the constant of proportionality in Coulomb's law. $\varepsilon_0 = 8.8542 \times 10^{-12}$ farads per meter for a vacuum. The proportionality constant used in Coulomb's law is $1/4\pi\varepsilon_0 \cong 9 \times 10^9$ N·m^2/C^2.

Pithball Lightweight spherical body made from an excellent insulating material (Styrofoam or pith) capable of storing an electric charge, suspended from a point by a string. One or more pithballs, so suspended, are used to study electric charges.

Potential, *V* Measure of the work required to bring a unit positive charge from infinity to a given point in an electric field, working against the force **F** of the field. V is a scalar quantity, measured in joules per coulomb (J/C) or volts (V). 1 V $\equiv$ 1 J/C.

Potential difference, *V* Measure of the work done to carry a given charge from one point in an electric circuit to another, measured in volts or joules per coulomb (1 V $\equiv$ 1 J/C).

Practical current source Parallel combination of an ideal current source and its internal resistance, r_i.

Practical voltage source Series combination of an ideal voltage source and its internal resistance, r_i.

Proton A small unit of matter having a positive charge. The charge on a proton is equal and opposite to that on an electron, or $+1.602 \times 10^{-19}$ C. The mass of the proton is 1.672×10^{-27} kg or approximately 1837 times that of an electron.

State of matter Physical condition in which matter exists. Depending on temperature, matter may exist in the solid, liquid, or gaseous state. These states, for a given material, determine the nature of the medium (i.e., its permittivity) and its ability to conduct electric charges (i.e., its conductivity) when an electric field is applied across it.

2-15 PROBLEMS

Secs. 2-1 to 2-2

2-1 Two pithballs suspended from a string are simultaneously touched with a negatively charged (rubber) rod. The first picks up a charge of $-100\ \mu$C while the second acquires a charge of $-200\ \mu$C, causing the pithballs to remain suspended in air a distance 25 cm apart. Calculate the electric repulsive force between them in

a. newtons
b. pounds
c. Explain why we may neglect the weight of the pithballs in this problem

2-2 Two positive charges develop a force of repulsion between them of 20 N when they are 10 cm apart. If one charge is three times as large as the other, calculate

a. The smaller charge, Q_1
b. The larger charge, Q_2

2-3 Two parallel metal plates, A and B, having an area of 50 cm^2 are separated by a distance of 10 cm of free space. If 5×10^8 electrons are removed from plate A and transferred to plate B, calculate the

a. Charge on plate A
b. Charge on plate B
c. Total charge and force between the plates
d. Electric field intensity between the plates (newtons per coulomb)
e. Force exerted on a test charge of one free electron between the plates

2-4 A negative test charge of $-25\ \mu$C and a positive charge of $+250\ \mu$C are separated in free space by a distance of 4 m. Calculate the

a. Force between charges
b. Nature of the force (attraction or repulsion)
c. Flux density produced by the greater charge (only) at the point location of the smaller charge

2-5 The plates of a parallel plate capacitor are 2 cm apart and a potential difference of 100 V is applied, charging the plates to opposite polarity. Calculate the

a. Voltage gradient in volts per meter
b. Electric field intensity in newtons per coulomb and kilonewtons per coulomb

c. Force acting on a 1-pC test charge in the free space between the plates expressed in newtons and nanonewtons
d. Flux density of the electric field between the plates in coulombs per square meter and nanocoulombs per square meter
e. Total flux, ψ, in terms of charge, if the plates measure 2×5 cm, expressed in units of coulombs and picocoulombs

Secs. 2-3 to 2-5

2-6 Two protons are separated by a distance of 5 cm in free space. Calculate the

a. Repulsive electrostatic force between them, $\mathbf{F}_{p^+}$
b. Force of gravity acting on them, $\mathbf{F}_g$
c. Ratio $\mathbf{F}_{p^+}/\mathbf{F}_g$
d. Compare this ratio to that of Ex. 2-2 given in the text and give two reasons accounting for differences between them

2-7 Given a point charge of $+400\ \mu$C in space, calculate the

a. Electric field intensity at a distance of 1 m from the charge
b. Force acting on a test charge of $-1\ \mu$C at the given distance of 1 m using Eq. (2-2)
c. Force acting on the test charge, using Coulomb's law (Eq. 2-1)
d. Determine the nature of the force between the charges.

2-8 Two pithballs are suspended independently by separate strings located 3 ft apart. One is given a positive charge of $+100\ \mu$C and the other a negative charge of $-50\ \mu$C. Calculate the

a. Initial force between them in newtons
b. Nature of the force between them
c. Initial force in pounds
d. Explain why the two pithballs develop an increased force between them and immediately touch

2-9 Draw the graphical coordinates X and Y. Locate a positive charge of 500 μC at the origin (intersection) of the coordinates. Calculate the

a. Field intensity at a point where $x = 0$ and $y = +5$ m. Assume free space around the charge
b. Flux density at the point in (**a**)
c. Force exerted on a test charge of $-1\ \mu$C located at the point
d. Direction of the force on the test charge

2-10 A small sphere is located in free space and positively charged to 500 μC. Calculate the force and the direction that its field exerts on a unit test charge of

a. $+2\ \mu$C located 5 m from it
b. $-5\ \mu$C located 2 m from it

2-11 Calculate the electric field intensity and the direction of the field at each location of the test charges given in (**a**) and (**b**) of Prob. 2-10.

2-12 Repeat Prob. 2-11 for the flux densities and the direction of the flux density vectors at each location given in (**a**) and (**b**) of Prob. 2-10. Explain why the flux density is greater for the negative charge.

2-13 A unit positive test charge of 10 μC is repelled by a force of 100 N when located at some point in an electric field created by a large concentrated charge. Calculate the

a. Electric field intensity at the charge location
b. Actual value and nature of the concentrated charge if it is located 5 m from the test charge

2-14 A neutral metallic plate acquires a negative charge of 100 μC. Calculate the

a. Number of electrons added to it
b. Additional mass acquired by the body
c. Number of electrons that would have to be added to raise its mass by 1 gram

2-15 A proton and an electron are 1×10^{-10} m apart. Calculate the

a. Electrostatic force of attraction between them
b. Gravitational force of attraction between them
c. Ratio of (**a**) to (**b**)

2-16 A point charge of -2 mC is suspended in free space. Calculate the

a. Electric field intensity 50 cm from it
b. Flux density at that distance
c. Force exerted on an electron located at the given distance
d. Direction of the electron as a result of the force
e. Field intensity in megavolts per meter

2-17 There are 17×10^{20} free electrons at 20°C in a piece of copper wire. Calculate the

a. Total equivalent charge represented by the free electrons, in coulombs
b. Volume of the copper wire in cubic millimeters

Secs. 2-6 to 2-10

2-18 If 6 C pass a given point in a wire in 2 min, calculate the

a. Current in mA
b. Current density (A/m^2) if the wire has a cross-sectional area of 500 c-mils

2-19 A fuse rated to open at 10 A requires 10 J to change state (solid to vapor) and clear an electric circuit from its supply. If the fuse clears in 500 ms, calculate

a. The charge passing through it in coulombs
b. The potential difference across the fuse

2-20 In charging a lead-acid cell, 220 C of charge is transferred from the negative to the positive electrode in 20 s, requiring 484 J of energy from a dc supply. Calculate the

a. Charging current in amperes
b. Potential difference between the electrodes
c. Voltage gradient (in volts per meter) if the cell electrodes are 1 cm apart

2-21 A current of 10 mA in a transistor releases 500 mJ of heat energy in 10 s. Calculate the voltage across the transistor.

2-22 A power transistor carrying a current of 100 mA converts 60 J of electrical energy into dissipated heat in 1 min. Calculate the

a. Charge transferred to the transistor during the time indicated
b. Voltage drop across the transistor
c. Current density (in A/cm^2) in the transistor collector if its cross section is 2 × 1 cm

2-23 A light-emitting diode (LED) has a voltage drop of 0.7 V and carries a current of 50 μA for 1 ms before producing visible light. Calculate the

a. Light energy in joules needed to produce visible radiation
b. Power rating of the LED in joules per second

2-24 Two wires, one made of silver and one of copper, have identical dimensions: a cross-sectional area of 100 c-mil and a length of 1 ft. If each wire carries a current of 1 μA, calculate the drift velocity, v, in cm/s in

a. The copper wire
b. The silver wire
c. Calculate the time required for the average electron to drift the full length of the copper wire and silver wire, respectively, in hours
d. Explain why it takes longer for the average copper electron than the average silver electron to drift 1 ft

2-25 Using the drift velocities calculated in Prob. 2-24 for silver and copper and the electron mobilities given in Table 2-4, calculate the

a. Electric field intensity for the copper wire in μV/m
b. Voltage across the copper wire
c. Electric field intensity for the silver wire
d. Voltage across the silver wire

Secs. 2-11 to 2-13

2-26 From the original information given in Prob. 2-24 and the answers given to Prob. 2-25, calculate the

a. Current density in each wire
b. Conductivity of copper, using both the $J/\mathbf{E}$ ratio and the value of ρ given in Appendix A-4
c. Conductivity of silver, using both the $J/\mathbf{E}$ ratio and the value of ρ given in Appendix A-4

2-27 A circuit carrying a current of 150 mA dissipates 75 J of electrical energy into heat in 1 min. Calculate the voltage across the circuit.

2-28 It requires 200 mJ to move a charge through a potential of 15 μV against the direction of the field. Calculate the

a. Number of coulombs in motion
b. Current if the charge was moved in 8 min

2-29 The plates of a parallel plate capacitor are 2 cm apart and a potential difference of 100 V is applied, charging the plates oppositely. Calculate the

a. Voltage gradient in volts per meter
b. Electrical field intensity in units of newtons per coulomb
c. Force acting on a 5-μC test charge between the plates

2-30 Given a current of 20 mA, after a duration of 20 s, calculate the number of

a. Coulombs passing through the wire
b. Electrons passing a given point in the wire

2-31 Calculate the currents under the following conditions:

a. 45×10^{12} electrons flowing in 16.5 μs
b. 27.6 mC flowing in 18.24 ks
c. 98 μC in 0.24 ks

2-32 A charge of 25 C is moved from infinity to point p, requiring 100 J of energy. Calculate the

a. Potential difference between infinity and point p
b. Potential difference if an additional 5 J is required to move the same charge to point q from infinity
c. Potential difference between points p and q

2-33 A silver bar measures 0.25 × 0.5 × 4 in. Calculate

a. Its volume in cubic centimeters
b. The number of free electrons it contains at 20°C
c. The charge on the bar if *all* the free electrons are removed, in kilocoulombs
d. The nature of the charge left on the bar

2-34 If the silver bar in Prob. 2-33 carries a current of 100 mA, calculate the

a. Current density in amperes per square meter
b. Drift velocity in nanometers per second (nm/s)
c. Time in hours for the average electron to drift the full length of the bar, assuming the current given is continuously flowing

2-35 Gold is often preferred to silver or copper for electrical contacts because it does not oxidize or tarnish. Its physical constants are: electron mobility, $\mu = 3.79 \times 10^{-3}\ m^2/V \cdot s$, and electron density, $n = 6.75 \times 10^{28}\ e^-/m^3$. When a gold contact measuring 0.5 × 1 × 2 cm carries a maximum current of 2 A, it has a voltage drop of 20 μV across its axial length. Calculate

a. Measured current density, J, in kiloamperes per square meter
b. Measured electrical field intensity, $\mathbf{E}$, in millivolts per meter
c. Conductivity, σ, using the information calculated in **(a)** and **(b)**
d. Conductivity, σ, using the values given in the problem for μ and n
e. Conductivity, σ, given the resistivity of gold in Appendix A-4

2-36 A lamp filament 6 in long carries a current of 0.5 A continuously and has a drop of 120 V across it. Calculate the

a. Charge transferred through the filament in 1 min
b. Current density, in mega-amperes per square meter (MA/m^2), if the area of the filament is 100 c-mils
c. Electric field intensity, in V/m
d. Conductivity of the material of the filament in kilosiemens per meter (kS/m)
e. Resistivity of the material of the filament in micro-ohm-meter ($\mu\Omega \cdot m$)

2-37 A nickel–cadmium cell is charged at its nominal voltage of 1.2 V for 5 h, during which time 9×10^4 C are stored on the electrodes. Calculate the

a. Charging rate of the cell in amperes
b. Work done in charging the cell in kilojoules
c. Equivalent work in newton-meters and foot-pounds
d. Current density in the electrolyte if the electrodes measure 6 × 4 in. Express your answer for J in A/in^2 and A/m^2
e. Explain why the value for current density in English units is so small and the value in SI is so large

2-38 a. Calculate the energy in joules required to move a single electron through a potential of 1 volt.
b. Using the information in (**a**), calculate the number of joules per electron volt and compare this with the value given in Appendix A-3H.

2-39 Points a and b are 5 cm apart. If 10 J of work is done to move +2 C of change from a to b, calculate

a. Which of the two points is at the higher potential (i.e., more positive)
b. Field intensity to move charge from a to b
c. Force required to move charge from a to b

2-16 ANSWERS

2-1 a +2877 N b 646.7 lb_f
2-2 a 2.723 μC b 8.170 μC
2-3 a 80.1 pC b −80.1 pC c −5.768 nN
d 72 N/C e 11.53 aN
2-4 a −3.512 N b attraction c 1.244 $\mu C/m^2$
2-5 a 5 kV/m b 5 kN/C c 5 nN d 44.27 nC/m^2
e 44.27 pC
2-6 a 9.23×10^{-26} N b 1.59×10^{-26} N c 5.805/1
2-7 a 3.6 MN/C b −3.6 N c −3.6 N
d Attraction
2-8 a −53.76 N b Attraction c −12.1 lb_f
2-9 a 179.8 kN/C b 1.592 $\mu C/m^2$ c −0.18 N
2-10 a 0.36 N b −5.62 N toward sphere
2-11 a 180 kN/C b 1.125 MN/C away from sphere
2-12 a 1.594 $\mu C/m^2$ b 9.96 $\mu C/m^2$
2-13 a 10 MN/C b 27.8 mC
2-14 a 6.24×10^{14} b 5.68×10^{-16} kg c 1.099×10^{27}
2-15 a −23.1 nN b 9.875×10^{-48} N c $2.34 \times 10^{39}/1$
2-16 a −71.92 MN/C b −637 $\mu C/m^2$ c 11.52 pN
d Away e −72 MV/m
2-17 a 272.34 C b 20 mm^3
2-18 a 50 mA b 197.4 kA/m^2
2-19 a 5 C b 2 V
2-20 a 11 A b 2.2 V c 220 V/m
2-21 5 V
2-22 a 6 C b 10 V c 0.05 A/cm^2
2-23 a 35 nJ b 35 μJ/s
2-24 a 1.45×10^{-7} cm/s b 2.125×10^{-17} cm/s
c 58.39 kh
2-25 a 0.34 μV/m b 103.7 nV c 0.3215 μV/m
d 98 nV
2-26 a 19.74 A/m^2 b 58 MS/m c 61.4 MS/m
2-27 $8.\bar{3}$ V
2-28 a $13.\bar{3}$ kC b $27.\bar{7}$ A
2-29 a 5 kV/m b 5 kN/C c 25 mN
2-30 a 400 mC b 2.497×10^{18}
2-31 a 437 mA b 1.513 μA c $408.\bar{3}$ nA
2-32 a 4 V b 4.2 V c 0.2 V
2-33 a 8.194 cm^3 b 4.75×10^{23} c 76.13 kC d +
2-34 a 1.24 kA/m^2 b 133.5 nm/s c 211.4 h
2-35 a 40 kA/m^2 b 1 mV/m c 40 MS/m
d 40.55 MS/m e 40.98 MS/m
2-36 a 30 C b 9.868 MA/m^2 c 787.4 V/m
d 12.53 kS/m e 79.8 $\mu\Omega \cdot m$
2-37 a 5 A b 108 kJ c 7.965×10^4 ft · lb, 108 kN · m
d 208.3 mA/in^2, 322.9 A/m^2
2-38 a,b 1.602×10^{-19} J/eV
2-39 a b b 5V c 200 N

CHAPTER 3

Resistance, Conductance, and Resistors

3-1 INTRODUCTION

We now come to the first of the three circuit elements: *resistance*. Resistance may be defined as the *opposition* to a *steady* (dc) current or voltage and the ability, as a circuit property, to dissipate electrical energy.[1] Only resistance (as a circuit property) has the ability to *dissipate* electrical energy. Its unit of measure is the **ohm**.

In Sec. 2-4 we noted that work is done only in overcoming *resistance*. In an electric circuit, we also noted that the work done [in joules per coulomb (J/C)] to move a charge, Q, from one point in a circuit to another point is a measure of potential difference or voltage. We also noted that the amount of charge flow (as a result of such work to overcome resistance) constituted an electric current (Sec. 2-5). Simply put, current (charge flow) occurs in an electric circuit only in response to potential difference. This relationship was first discovered by Georg Simon Ohm in 1827 and is known as Ohm's law.[2]

3-2 OHM'S LAW

Ohm's law quantifies the relationship between potential difference (the cause) and electric current (the effect) in a given electric circuit. Ohm's law states that for a given circuit, a constant of proportionality, R, exists between the voltage applied (*cause*) and the current (*effect*) which results; that is, if the applied voltage is doubled, the current is doubled. If the voltage is tripled, the current is tripled. This opposition or resistance provided by either a circuit or circuit element may be written as

---

[1] The other two circuit elements are *inductance* (Chapter 12) and *capacitance* (Chapter 9), neither of which dissipate electrical energy but rather store it. Inductance is characterized by its ability to oppose a *change* in *current*. Capacitance is characterized by its ability to oppose a *change* in *voltage*.

[2] Actually scientific historians have discovered that the relationship was first written up by Henry Cavendish in 1781 but was never published.

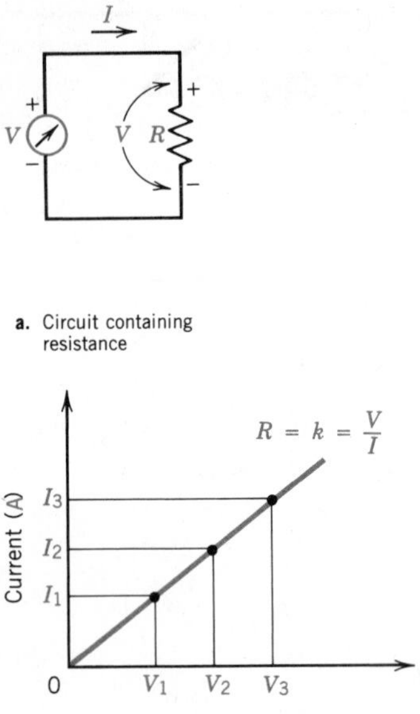

a. Circuit containing resistance

b. Graphical representation

Figure 3-1 Definition of resistance and its graphical representation

$$R = \frac{V}{I} = k = 1/G \qquad \text{ohms } (\Omega) \qquad (3\text{-}1)$$

where V is the potential difference between two points in an electric circuit or across a circuit element, in volts (J/C)

I is the resulting current (in response to V) in amperes (C/s)

G is the conductance in siemens (S)

Several insights may be deduced from Ohm's law which may not be apparent immediately. **Figure 3-1a** shows a circuit containing resistance only. If a voltage, V_1, is applied, a certain current I_1 results, as shown in Fig. 3-lb. If the source voltage is increased to V_2, a current I_2 flows (Fig. 3-1b). A further increase in voltage to V_3 produces current I_3 (Fig. 3-lb). We may conclude the following:

1. The relationship between voltage and current is *linear* (Fig. 3-lb).
2. Since the relationship V/I is linear, resistor R is a *linear resistance.*[3]
3. The independent variable in Fig. 3-lb is the *abscissa,* designated as voltage. The dependent variable or *ordinate* is designated as the current. Since current depends on voltage, current is usually selected as the ordinate.
4. Ohm's law should *not* be interpreted to mean that resistance varies directly with voltage and inversely with current, but rather that the *ratio* of voltage to current, for a *given* resistance, is *constant* (Eq. 3-1).[4]
5. From Eq. (3-1) and Fig. 3-lb we can see that $V_1 = I_1R$; $V_2 = I_2R$; and $V_3 = I_3R$.
6. From Eq. (3-1) and Fig. 3-lb we may also write $I_1 = V_1/R$; $I_2 = V_2/R$; $I_3 = V_3/R$.

EXAMPLE 3-1

A potential of 2.5 V produces a current of 5 mA in a resistor. Calculate its a. Resistance b. Conductance

Solution

a. $R = V/I = 2.5\ \text{V}/5\ \text{mA} = \mathbf{500\ \Omega}$ (3-1)

b. $G = 1/R = 1/500\ \Omega = \mathbf{2\ mS}$ (3-1)

EXAMPLE 3-2

A charge of 800 mC is transferred each second when a potential is applied across the terminals of a lamp filament having a resistance of 150 Ω. Calculate the voltage across the filament.

Solution

$V = I \times R = 0.8\ \text{A} \times 150\ \Omega = \mathbf{120\ V}$ (3-1)

EXAMPLE 3-3

Calculate the current when a potential of 10 V is applied across a 20 kΩ resistor.

Solution

$I = V/R = 10\ \text{V}/20\ \text{k}\Omega = \mathbf{0.5\ mA}$ (3-1)

Figure 3-2 shows a comparison among three different resistors. R_1 has a *low* resistance because it yields a relatively *high* current for a relatively *low* voltage. Conversely, R_3 yields a relatively *low* current for a relatively *high* voltage, while R_2 yields a moderate current for a moderate voltage in comparison to R_1 and R_2.

[3] Not all resistive circuit elements are linear. Commercial resistors are relatively linear within certain bounds of voltage and current. If these are exceeded, resistance increases and the element is nonlinear. Other circuit elements are deliberately nonlinear, such as filaments of incandescent lamps and thyrite resistors. We will learn later that temperature affects the linearity of most resistive materials.

[4] Since $I = V/R = V/k = k'V$, it is possible to say that for a *given* resistance, current is proportional to applied voltage. Also, since $V = I \cdot R = kI$, we may say for a given resistance, voltage across a resistance is proportional to the current in it. Finally, from $I = V/R$ we may say for the *same* applied voltage, as resistance increases, current decreases.

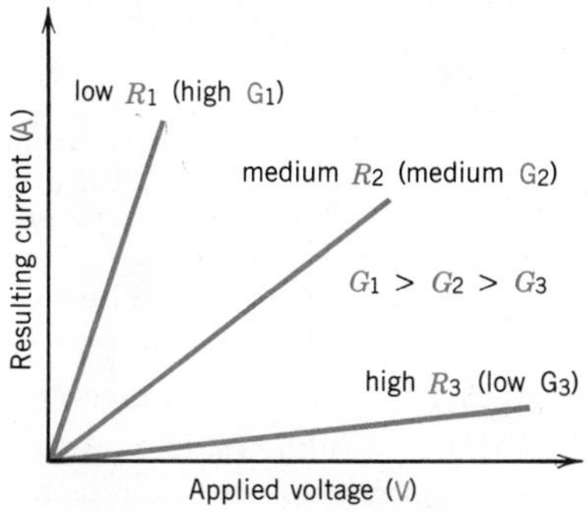

Figure 3-2 Comparisons among linear resistors showing that slope is proportional to conductance (*G*)

In terms of resistance, the family of curves of Fig. 3-2 shows that the slope of a *high* resistance is relatively *small* while the *low* resistance has a *high* slope. For this reason, it is often convenient to express the comparisons in terms of conductances (*G*) rather than resistance. Since $G = 1/R$, note from Fig. 3-2 that the (*low*) resistance, R_1, has a *high* conductance, G_1 and a correspondingly *high* slope. Similarly, high resistance R_3 has a *low* conductance and *low* slope. Consequently, when making comparisons *graphically*, it is more convenient to think in terms of conductance rather than resistance because

1. A *vertical* slope indicates *infinite* conductance and zero resistance.
2. A *high* slope implies a *high* conductance.
3. A *horizontal* slope implies *zero* conductance and infinite resistance.
4. A *low* slope implies a *low* conductance.

One last point emerges from Eq. (3-1) which should not be overlooked—that is, the way it helps us to define a *short circuit* and an *open circuit*.

A *short circuit* is defined as a resistance of *zero* ohms (conductance *infinite*). When a short circuit is applied to a voltage source, however briefly, the current may vary as it rises from zero to some larger value. Since $V = I \cdot R$ and $R = 0$, then the *voltage* across a short circuit *is zero, regardless of the value of current.*

An *open circuit* is defined as an *infinite* resistance (*zero* conductance). Since $I = V/R$ and $R = \infty$, then the current *must always be zero* in an open circuit, *regardless of the voltage* across it.

3-3 VOLT-AMPERE CHARACTERISTICS OF NONLINEAR RESISTORS

Commercial resistors (Sec. 3-7) are deliberately designed to have relatively constant resistance over their working (voltage and current) range. **Figure 3-3a** shows, however, that at extremely high voltages and currents (well beyond the resistor power rating defined in Sec. 3-7.2) the conductance is no longer constant but begins to decrease. This implies that the resistance increases at relatively high voltages, producing high temperatures. This is a desirable nonlinear characteristic and *tends* to prevent damage to the resistor.[5] As shown in Fig. 3-3a, beyond point *k*, the knee of the characteristic curve for commercial resistors, the resistance increases to an almost *horizontal* value (zero conductance, infinite resistance).

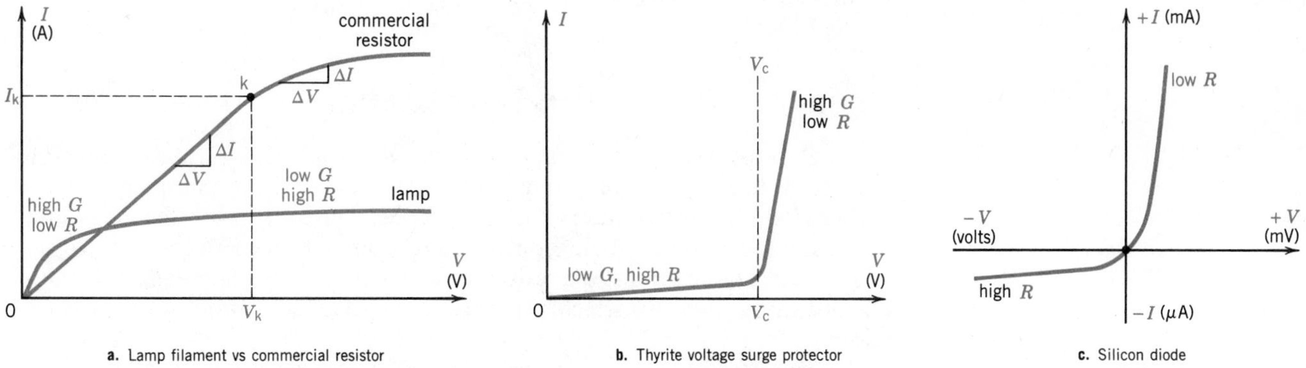

Figure 3-3 Nonlinear volt-ampere characteristics

[5] If the resistance were to decrease with elevated temperature, the decrease in resistance would result in an increase in current (for the same voltage), which would increase the power dissipated and the temperature still more, producing *thermal runaway*.

On the same set of axes, Fig. 3-3a also shows the nonlinear characteristic of an incandescent lamp filament. Note that at low voltages, the filament resistance is low (high conductance). But as voltage (and current) increases and the filament burns more brightly, its temperature increases and its resistance increases. This increase in resistance (again) protects the filament from thermal runaway at excessively higher than rated (normal) working voltages.

Figure 3-3b shows the nonlinear volt-ampere characteristic of a *thyrite* voltage-surge protector. Thyrite is made of carborundum crystals embedded in a clay binding material. Below a critical voltage, V_c, thyrite has a low conductance and very high resistance. Above the critical voltage, thyrite has a high G and very low resistance (i.e., almost a short circuit). When thyrite is connected on the line side of a given circuit (i.e., in parallel with the circuit), it becomes a short circuit in the event of brief excessive voltage surges or transients from the source, which may damage the circuit. In addition to this application, it is also used in lightning arrestors, serving to conduct high-voltage lightning surges from a mast to ground.

The silicon *pn*-diode shown in Fig. 3-3c is a perfect example of a nonlinear resistor. When it is forward-biased in the direction from p^+ to n^-, the conductance is high and the resistance of the diode is very low, normally of the order of 5 to 50 Ω. But when the bias (applied voltage polarity) is reversed, the resistance is extremely high, from 1 to 50 MΩ. It is this nonlinear and nonsymmetrical resistance characteristic that permits the diode to behave as a *rectifier*, that is, its low resistance in one direction and *high* resistance in the *opposite* direction.

Several insights are to be noted regarding nonlinear resistors. The reader may assume, because Eq. (3-1) states the V/I ratio as a constant value, that nonlinear resistors do not obey Ohm's law. This is not so! Although the nonlinear resistor may *not* have a constant resistance over its *entire* characteristic, its *incremental* resistance ($\Delta V/\Delta I$) follows Ohm's law perfectly! This is shown in Fig. 3-3a for the commercial resistor. In the range below the knee, k, the value of $\Delta V/\Delta I$ is the same as that obtained from the ratio V_k/I_k, since the *slope* is the *same* in both cases. Above the knee of the curve, k, the *incremental resistance* is found, as shown, from a higher ΔV divided by a smaller ΔI; this results in a high incremental resistance, as described, and produces the desired commercial resistor characteristic. The incremental resistance is sometimes called the **ac resistance** (Sec. 17-8). The resistance obtained by taking voltages and currents with respect to the *origin* is called the **dc resistance**. Thus in Fig. 3-1b, V_1/I_1, V_2/I_2, and V_3/I_3 all produce the same values of dc resistance because the slope is constant and all voltage/current ratios are measured with respect to the origin.

3-4 FACTORS AFFECTING RESISTANCE

Although resistance can be computed from Ohm's law (Eq. 3-1), using measured values of voltage and current, some difficulties inevitably arise in the process. We will see later that temperature affects resistance. Consequently, if we apply voltage across and cause current to flow through a resistance, heat energy is liberated since work must be done against the resistance to produce charge flow. This heat causes a temperature increase, which, as we will see (Sec. 3-6), produces a *change* in resistance.[6]

There are at least two better ways to determine resistance at a room temperature of 20°C. The first is by Wheatstone bridge measurement (see Chapter 23). The second is by direct calculation, using the resistivity constant, ρ, for the given material. See Eq. (3-2) below.

[6] Ohm's law does not hold if strong electric fields are applied or if the fields vary rapidly.

Resistivity, ρ, and the factors affecting it (Eq. 2-12) were described in Sec. 2-10. Appendix Table A-4 lists the resistivity for 33 common materials in units of ohm-meters. Recall that resistivity is the *specific resistance* of a specified unit length and cross-sectional area of a given material sample. In the SI, the sample dimensions are 1 meter long and 1 square meter in cross-sectional area (i.e., 1 cubic meter). The resistivity in the SI is expressed in ohm-meters. In the English system the sample is 1 ft long and the cross-sectional area is 1 circular mil (c-mil). The resistivity in the English system is expressed in $\Omega \cdot$c-mil/ft.

Clearly, the resistance of a material is directly dependent on the resistivity of the material; that is, for the same length and cross-sectional area, the conductor having the higher resistivity will have the higher resistance.

In addition to the quantification of Eq. (3-1), Ohm also noted that a conductor's resistance varies directly with length and inversely with cross-sectional area (assuming the material remains the same and the temperature is held constant). This "ohmic" relation is summarized at a constant temperature of 20°C as

$$R = \rho l/A \quad |T = k = 20^\circ\text{C}| \quad \text{ohms } (\Omega) \tag{3-2}$$

In units of the SI	In units of the English system
ρ is the resistivity ($\Omega \cdot$m)	ρ is the resistivity ($\Omega \cdot$c-mil/ft)
l is the length in meters (m)	l is the length in feet (ft)
A is the cross-sectional area in square meters (m^2)	A is the cross-sectional area in circular mils (c-mil)

The following insights emerge from a study of Eq. (3-2).

1. Using units expressed in either the SI or the English system, the ratio $\rho l/A$ yields resistance in ohms by dimensional analysis.
2. If the temperature is held constant (say at 20°C), the resistance of a conductor varies *directly* with *resistivity* and *length* and *inversely* in proportion to *cross-sectional area* (Eq. 3-2).[7]
3. In the SI, ρ represents the *specific resistance* of a one cubic meter (1-m^3) sample of a given material. This implies that a conductor of the same material, having an area of one square meter (1 m^2) but a length of 10 meters (10 m), would have 10 times the (specific) resistance, that is, 10(ρ).
4. In the English system, ρ represents the specific resistance of a sample having a length of one foot and a cross-sectional area of one circular mil (1 c-mil). This implies that a conductor of the same material and length but an area of (say) 10 c-mils would have 1/10 the (specific) resistance (i.e., $\rho/10$).
5. The conclusion of items 3 and 4 leads to the insight that once we know the resistance of a *standard* sample of a particular material, we can find the resistance if we are given other dimensions of length and cross-sectional area for that material.
6. We can carry insight 5 a step further by saying that given R, l, and A for a specific conductor (say, copper), we can find R for any other changes in length and area *without* the need for ρ! In effect, the given R, l, and A act as a "known" sample. This is shown in Ex. 3-4, part (**a**).

[7] Equation (3-2) is easily derived as follows. Given $\mathbf{E} = V/l$ (Eq. 2-2) and $J = I/A$ (Eq. 2-9) with resistivity $\rho = \mathbf{E}/J$ (Eq. 2-12), then $\rho = \dfrac{\mathbf{E}}{J} = \dfrac{V/l}{I/A} = \dfrac{RA}{l}$. Solving the relation $\rho = RA/l$ for resistance yields $R = \rho l/A$, which is Eq. (3-2). This relatively simple derivation shows the advantages of and need for the various rigorous definitions established in Chapter 2.

EXAMPLE 3-4

A standardized sample of nichrome wire measuring 1 m^3 has a resistance $R_0 = 0.996\ \mu\Omega$ at 20°C, that is, $l_0 = 1$ m and $A_0 = 1\ m^2$. Calculate the resistance of nichrome wires at 20°C having the following dimensions:

a. $l_1 = 150$ m, $A_1 = 400\ cm^2$
b. $l_2 = 6$ ft, $A_2 = 2500$ c-mil

Solution

a. $$R_1 = \frac{l_1 \times A_0}{l_0 \times A_1} R_0$$

$$= \frac{150\ m \times 1\ m^2 \times 0.996\ \mu\Omega}{1\ m \times (400\ cm^2 \times 10^{-4}\ m^2/cm^2)} = \mathbf{3.735\ m\Omega}$$

Alternative Solution

$$R_1 = \rho l/A = \frac{0.996\ \mu\Omega\cdot m \times 150\ m}{(400\ cm^2 \times 10^{-4}\ m^2/cm^2)} = \mathbf{3.735\ m\Omega} \quad (3\text{-}2)$$

b. $$R_2 = \rho l/A = \frac{0.996\ \mu\Omega \times 6\ ft \times 0.3048\ m/ft}{2500\ c\text{-}mil \times 5.067 \times 10^{-10}\ m^2/c\text{-}mil}$$

$$= \mathbf{1.438\ \Omega} \quad (3\text{-}2)$$

The solutions of Ex. 3-4 illustrate the six points noted above. Example 3-4**a** shows that given original resistance, R_0, original length, l_0, and original cross-sectional area, A_0, we can find the new resistance, R_1, for new values of length (l_1) and area (A_1), by the *RATIO METHOD*. Note that we are merely applying correction factors to the original resistance R_0 when we multiply it by the ratio change in length (the ratio l_1/l_0) and the ratio change in area (A_0/A_1) to find the new resistance R_1. The ratio method used in Ex. 3-4**a** does not recognize the factor of resistivity, ρ, nor require its use, as noted in item 6 above.

The alternative solution of Ex. 3-4a does, in fact, recognize that R_0 is the specific resistance of nichrome in SI units, that is, $\rho = R_0$. Therefore, Eq. (3-2) is used. Note that the numerator and denominator of both solutions to Ex. 3-4a are mathematically equivalent.

The solution to Ex. 3-4b shows point 6 once again. The reasoning is that if we know the specific resistance (ρ) of a standard sample in (say) SI units and we are given new dimensions in English units, we can find the new resistance by unit conversion. Note that Ex. 3-4b was solved without having to find the resistivity of nichrome in English units. Example 3-5 solves Ex. 3-4b directly in English units to verify the insights described above. It also shows that either form of Eq. (3-2) is valid as long as the units are expressed properly.

EXAMPLE 3-5

Solve Ex. 3-4b using the resistivity given in Appendix A-4 for nichrome wire in English units, that is, $\rho = 599.1\ \Omega\cdot$c-mil/ft

Solution

$$R = \rho l/A = \frac{(599.1\ \Omega\cdot c\text{-}mil/ft) \times (6\ ft)}{2500\ c\text{-}mil} = \mathbf{1.438\ \Omega} \quad (3\text{-}2)$$

Examples 3-4 and 3-5 verify the following important points:

1. Whether one uses SI or English units consistently, the resistance of a given length and cross-sectional area will be the same.
2. If one knows the resistance of a given length and cross-sectional area of a conductor, the resistance of any length and cross-sectional area can be found by the ratio method (Ex. 3-4a).[8]
3. Regardless of whether resistivity is expressed in ohm-meters (SI) or $\Omega\cdot$c-mil/ft (English), when units of length and cross-sectional area are properly substituted in Eq. (3-2), the resistance is obtained in ohms.

Example 3-6 illustrates another important use of Eq. (3-2).

[8] In the English system, instead of using square feet for cross-sectional area, the circular mil is used for round wires. The diameter, D, in mils is the diameter in inches × 1000. The area in c-mils, $A_{c\text{-}mil} = (D_{mils})^2$. The rationale for this is that it eliminates the need to express area as $(\pi/4)(D^2)$. See area conversion factors, Appendix A-3**B**.

EXAMPLE 3-6

A meter designer calculates the required resistance value of a manganin shunt as 5 mΩ. The laboratory has a stock of No. 10 AWG manganin wire. Calculate the length in *inches* of wire required to provide the resistance.

Solution

From Appendix Table A-5, the area of No. 10 AWG is 10 381 c-mil, then

$$l = \frac{AR}{\rho} = \frac{10\,381 \text{ c-mil} \times 5 \text{ m}\Omega}{264.6\ \Omega\cdot\text{c-mil/ft}} = 0.1962 \text{ ft} \times 12 \text{ in/ft} = \mathbf{2.354 \text{ in}}$$

With respect to Ex. 3-6, note that information is required from two Appendix tables:

1. Appendix Table A-4 provides the resistivities of materials in SI and English units.
2. Appendix Table A-5 lists AWG number, c-mil area, and resistance for materials in SI and English units.

EXAMPLE 3-7

Appendix Table A-5 provides the following information for No. 20 AWG copper wire:

a. 1000 ft of the wire has a resistance of approximately 10.1 Ω
b. 1000 m of the wire has a resistance of approximately 33.2 Ω

Using Eq. (3-2) and Table A-4, calculate and verify

a. The resistance of 1000 ft of copper wire
b. The resistance of 1000 m of copper wire

Solution

a. $$R = \frac{\rho l}{A} = \frac{10.371\ \Omega\cdot\text{c-mil/ft} \times 1000 \text{ ft}}{1021 \text{ c-mil}} = \mathbf{10.16\ \Omega} \quad (3\text{-}2)$$

b. $$R = \frac{\rho l}{A} = \frac{1.7241 \times 10^{-8}\ \Omega\cdot\text{m} \times 1000 \text{ m}}{5.178 \times 10^{-7} \text{ m}^2} = \mathbf{33.3\ \Omega} \quad (3\text{-}2)$$

Example 3-7 verifies Eq. (3-2) and (simultaneously) verifies the validity of the tables in Appendix A-5.

The previous discussion enables us to classify all solid materials into three large electrical groupings based on resistivity. These are insulators, semiconductors, and conductors.

3-4.1 Insulators

Insulators are defined as materials having resistivities falling between 10^{20} and $10^{8}\ \Omega\cdot$m. Table 3-1 provides a brief list of commonly used insulating materials and their resistivities in ohm-meters.[9]

Table 3-1 Typical Insulating Materials

Name	Resistivity, ρ ($\Omega\cdot$m)
Gases	
Air	∞
Carbon dioxide	∞
Liquids	
Mineral insulating oil	10^{12}
Chlorinated polyphenols	10^{11}
Solids	
Quartz, fused	10^{16}
Mica, muscovite	10^{15} to 10^{11}
Micalex, sheet and rod	10^{12}
Porcelain	10^{11}
Steatite	10^{14}
Cellulose acetate film	10^{13}
Polystyrene	10^{15}
Cloth, varnished cotton	10^{13}

[9] Resistivity ranges are based on data from ITT Staff, *Reference data for radio engineers*, 6th ed., Ch. 4, 1975. H. W. Sams, N. Y.

* 3-4.2 Semiconductors

Semiconductors are defined as materials having resistivities falling between $10^7\ \Omega\cdot\text{m}$ and $10^{-6}\ \Omega\cdot\text{m}$. The "classic" semiconductor materials are **silicon**, with a resistivity of 2300 Ω·m, and **germanium**, with a resistivity of 0.47 Ω·m, in the "pure" state. When these materials are "doped" with controlled amounts of impurities, their resistivity is considerably less.

3-4.3 Conductors

Conductors are defined as materials having resistivities from $10^{-6}\ \Omega\cdot\text{m}$ to $10^{-8}\ \Omega\cdot\text{m}$. Appendix Table A-4 lists a number of common conductors with their resistivities shown in the first column.

Note that the resistivities of a typical insulator (say, $10^{10}\ \Omega\cdot\text{m}$) and a typical conductor (say, $10^{-7}\ \Omega\cdot\text{m}$) are in a ratio of 10^{17}, a difference of 17 orders of magnitude!

3-5 AMERICAN WIRE GAGE TABLES

In the preceding section we referred to Appendix A-5, which is based on the American Wire Gage (AWG) sizes for conductors. At the time these sizes were developed, the English system was used exclusively in the United States and Britain. Since the AWG sizes are standards used by wire manufacturers internationally, countries on the metric or SI system use the equivalent sizes shown in columns 5 and 6 of Appendix A-5. With respect to Appendix A-5, the reader should verify the following:

1. As the AWG numbers *increase*, going down the first column (vertically), the diameter and cross-sectional area (both) *decrease*. Since the length is constant, the decrease in area (reading vertically) produces an *increase* in resistance (columns 4 and 7).
2. The table (more properly) should be read horizontally. For example, No. 10 AWG wire has a cross-sectional area of 10 382 c-mils, and 1000 ft of No. 10 wire has a resistance at 20°C of approximately 1 Ω. Equivalently, No. 10 AWG wire has a cross-sectional area of $5.26 \times 10^{-6}\ \text{m}^2$ and a 1000-m length of this (same) wire has a resistance of 3.27 Ω.[10]
3. Note that the table has been ruled intentionally in groups of three AWG numbers. Going down the table vertically we note that for every three AWG numbers the *area* is approximately *halved* and the resistance is approximately *doubled*. This applies to both English and SI units. For example, No. 3 AWG is approximately half the area of No. 0 and approximately twice the resistance of No. 0.
4. Further, going down the table vertically, note that for every 10 AWG numbers the area is reduced by 1/10 and the resistance multiplied by 10, approximately. This applies to both English and SI units.
5. Knowing Eq. (3-2), therefore, and the rules stated in items 3 and 4, given the data for *one* particular size of wire, it is possible to reproduce the entire Appendix A-5 table, approximately! The wire size usually used is No. 10, which

[10] Since resistance varies in direct proportion to length and the meter is longer than the foot (3.281 ft/m), the resistance per kilometer is 0.999 Ω/kft × 3.281 kft/km = 3.28 Ω, which is close enough.

has a diameter of approximately 100 mils, a c-mil area of approximately 10 380 c-mils, and a resistance of approximately 1 Ω per 1000 ft. These values are not difficult to memorize.

EXAMPLE 3-8
Given the data for No. 10 AWG, calculate resistance per 1000 ft and c-mil area for the following copper AWG wire sizes.
a. No. 1 b. No. 0 c. No. 19 d. No. 20 e. No. 29 f. No. 30

Solution

a. Counting in groups of three, No. 1 AWG requires three multiples of 3 to be reached from No. 10. Since each multiple doubles or halves, the ratio to be used is 2^3 or 8. For No. 1 the $A_{\text{c-mil}} = 10\,382 \times 8 =$ **83 056** and the $R/1000\text{ ft} = 1\ \Omega/8 =$ **0.125 Ω**
b. Going from No. 10 to No. 0 AWG, the ratio is 10. $A_{\text{c-mil}} = 10\,382 \times 10 =$ **103 820** and $R/1000\text{ ft} = 1\ \Omega/10 =$ **0.1 Ω**
c. Going from No. 10 to No. 19 AWG, the ratio is (again) $2^3 = 8$. Therefore $A_{\text{c-mil}} = 10\,382/8 =$ **1298 c-mil** and $R/1000\text{ ft} = 1\ \Omega \times 8 =$ **8 Ω**
d. Going from No. 10 to No. 20, the ratio is (again) 10. Therefore $A_{\text{c-mil}} = 10\,382/10 =$ **1038 c-mil** and $R/1000\text{ ft} = 1\ \Omega \times 10 =$ **10 Ω**
e. Using the above data for No. 20 and going to No. 29, the ratio is again $2^3 = 8$. Then $A = 1038/8 =$ **129.7 c-mil** and $R/1000\text{ ft} = 10\ \Omega \times 8 =$ **80 Ω**
f. Using the above data for No. 20 and going to No. 30, the ratio is 10. So $A = 1038/10 =$ **103.8 c-mil** and $R/1000\text{ ft} = 10\ \Omega \times 10 =$ **100 Ω**

The insights to be gleaned from Ex. 3-8 are

1. Once we have determined the values calculated in Ex. 3-8, it is possible, by moving in steps of three, to find all the remaining values of the table, that is, from 0000 AWG to No. 40 AWG.
2. Once the area in c-mils has been determined for every AWG number, the diameter in mils follows easily from $D_{\text{mils}} = \sqrt{A_{\text{c-mils}}}$.
3. Once the diameter in mils is found, the diameter in millimeters is easily calculated from the ratio 0.0254 mm/mil (Appendix A-3B).
4. The resistance in ohms per kilometer (Ω/km) is calculated from the ratio 3.281 kft/km, as noted in footnote 10.

The above insights enable us to appreciate and understand how Appendix Table A-5 may be derived on the basis of the data for No. 10 AWG copper.

3-6 EFFECT OF TEMPERATURE ON RESISTANCE OF CONDUCTORS

The previous examples and calculations all assumed that the temperature of the various conductors remained constant at 20°C (or 68°F). Even the tables of Appendixes A-4 and A-5 specify a temperature of 20°C. But what if the temperature is higher or lower than 20°C? Is it possible to apply a temperature correction factor to our resistance calculations?

In general, the resistance of most *metallic* conductors tends to increase as the temperature increases.[11] Using the relation $R = \rho l/A$, one's first impression might be that this is due to the linear expansion of length with temperature. But cross-sectional area should also increase as temperature increases. Actually, the increase in cross-sectional area is about the same as the increase in length with temperature; that is, a trade-off exists between these two factors. So we must look elsewhere for our explanation, and the only factor left is the resistivity, ρ.

From Sec. 2-10 we know that resistivity is $1/en\mu$, where e is the electronic charge (a constant, which cannot vary with temperature) of 1.602×10^{-19} C/e^-, n is the electron density or number of free electrons per cubic meter, and μ is the mobility of the free (conduction-band) electrons in the material. When the temperature is increased, more energy is imparted to valence electrons and electrons are "raised" from the valence shell to the conduction "cloud," resulting in an *increase* in electron

[11] There are a few exceptions. For carbon, germanium, and silicon, all semiconductors, in the normal temperature range the resistance tends to decrease with increases in temperature. This is not true near or close to absolute zero (0 K). See Fig. 3-5.

density, n. Further, when the temperature is increased there is an increased random molecular activity coupled with increased electron density. Since electrons tend to repel each other and simultaneously there is increased ionic molecular movement, the *mobility* of the electrons, μ, *drops severely*. This severe *reduction* in μ is *greater than the increase* in n in the relation $\rho = 1/en\mu$, and the *resistivity increases* as the *temperature increases*.[12] These effects hinder charge flow (current) and account for the increase in resistivity or specific resistance with temperature. (We previously learned that resistivity is nothing more than the *resistance* of a specifically dimensioned sample of a given material.)

Most solids exhibit changes in electrical resistance with variations in temperature. Although it is a slight oversimplification, we may generalize the foregoing discussion by classifying *solid materials* in three groups:

1. Conductors whose resistance tends to *increase* with an increase in temperature. Most metals fall in this group. These materials are said to have a *positive* temperature coefficient of resistance, α.[13]
2. Conductors whose resistance tends to *decrease* with an increase in temperature. Semiconductors such as carbon, germanium, and silicon fall in this group. These materials are said to have a *negative* temperature coefficient of resistance $(-\alpha)$.
3. A few special alloys, deliberately produced for this purpose, have a relatively *constant* resistance over a wide range of temperature (constantan, manganin, IaIa, etc.). Such materials are said to have a relatively low (positive) temperature coefficient of resistance $(\alpha \geq 0)$.

Typical of the first group is copper. Since copper is a relatively important material in electrical engineering and technology, its resistance versus temperature characteristic is shown in **Fig. 3-4**.

$$\alpha_{20} = \alpha_{T_1} = \frac{\dfrac{R_2 - R_1}{T_2 - T_1}}{R_1} = \frac{\dfrac{\Delta R}{\Delta T}}{R_1} \text{ in } \frac{\Omega/{}^\circ C}{\Omega}$$

$$\frac{R_2}{R_1} = \frac{K + T_2}{K + T_1}$$

$$K = 1/\alpha_0 = (1/\alpha_{20}) - 20$$

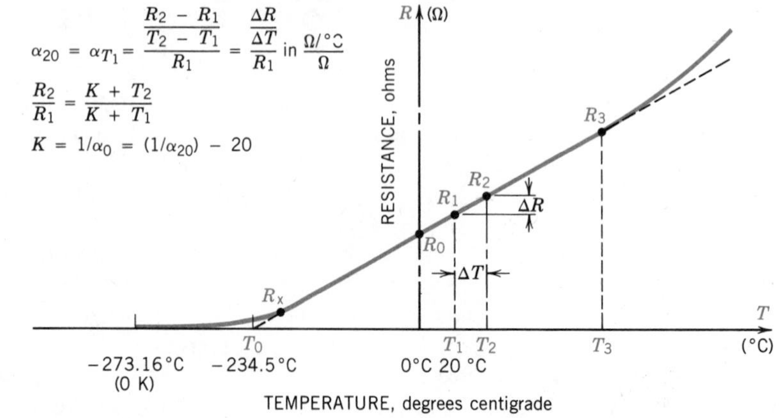

Figure 3-4 Increase in resistance of copper with temperature

---

[12] An equally valid way to account for the increase in resistivity with temperature is presented in D. Halliday and R. Resnick, *Physics, for students of science and engineering*, 2nd ed., John Wiley, 1962, Sec. 31-4. They note that "electrons collide constantly with the ionic cores of the conductor, i.e., they interact with the lattice. . . . In an ideal (pure) metallic crystal at 0°K electron–lattice collisions would not occur, according to the predictions of quantum physics . . . . Collisions take place in actual crystals because (a) the ionic cores at any temperature T are vibrating about their equilibrium positions in a random way, (b) impurities, i.e., foreign atoms, may be present and (c) the crystal may contain lattice imperfections such as rows of missing atoms and displaced atoms. On this view it is not surprising that the resistivity of a metal can be increased by (a) raising its temperature, (b) adding small amounts of impurities and (c) straining it severely, as by drawing it through a die, to increase the number of lattice imperfections."

[13] α is defined in Eq. (3-3).

Figure 3-4 shows that the resistance of copper is relatively *linear* from resistance R_x to R_3. We will develop several equations that will enable us to predict the resistance at various temperatures. But it must be understood that these equations are valid only *within the linear range* of resistance of the material. For copper, that range is from R_x to R_3. Since the range is well below 0°C and well above 100°C, the equations we will derive are sufficiently good for most purposes.

Note that if the *linear* variation is extended (extrapolated) linearly, the resistance of copper becomes zero (superconductive) at a temperature of −234.5°C. In actuality, of course, as shown by Fig. 3-4, at point R_x the resistance decreases as temperature decreases and the resistance of copper *actually* becomes zero at −273.16°C (absolute zero, 0 K). Since we very rarely operate at temperatures so far below 0°C, we may assume that the resistance of *any* piece of copper is zero at −234.5°C, the so-called *inferred zero-resistance temperature* of *copper*, where copper is *superconductive*.

A picture is, indeed, worth a thousand words. From Fig. 3-4, we may define the temperature coefficient of resistance, α_{20}, for copper, taken at 20°C, as follows:

$$\alpha_{20^\circ\mathrm{C}} = \frac{\dfrac{\Delta R}{\Delta T}}{R_1} = \frac{\dfrac{R_2 - R_1}{T_2 - T_1}}{R_1} \quad \text{ohms per degree celsius per ohm } (\Omega/^\circ\mathrm{C}/\Omega) \qquad (3\text{-}3)$$

where R_2 is some resistance value at an elevated temperature T_2

R_1 is the resistance at a temperature $T_1 = 20°C$

From Eq. (3-3) and Fig. 3-4 we may conclude that

1. α_{20} is a coefficient that represents the *ratio* of the linear *slope* of the curve (in Fig. 3-4) to the *resistance* of copper at 20°C, R_1.
2. As a coefficient, α_{20} may be used to find the resistance (along the linear portion of the curve) at temperatures above and below 20°C, once the resistance is **known at 20°C**. This is shown in Eq. (3-4).

Once α_{20} is known for a given conductor or sample of material, and the resistance of that sample has been measured at T_1 or 20°C, we can find its resistance at any other temperature from

$$R_2 = R_1(1 + \alpha_{20}\,\Delta T) = R_1[1 + \alpha_{20}(T_2 - T_1)] \quad \text{ohms } (\Omega) \qquad (3\text{-}4)$$

where all terms have been defined for Eq. (3-3). Values of α_{20} for various common materials are given in the last column of Appendix A-4.

EXAMPLE 3-9

The resistance of a copper wire is 5.2 Ω at 20°C. Calculate its resistance at a. 70°C b. −10°C

Solution

a. $R_2 = R_1[1 + \alpha_{20}(T_2 - T_1)]$
$= 5.2[1 + 0.003\,93(70 - 20)] = \mathbf{6.22\ \Omega}$ (3-4)

b. $R_2 = R_1[1 + \alpha_{20}(T_2 - T_1)]$
$= 5.2[1 + 0.003\,93(-10 - 20)]$
$= 5.2(1 - 0.1179) = \mathbf{4.587\ \Omega}$ (3-4)

Example 3-9 shows that at the higher temperature (70°C) the resistance of the copper wire is definitely higher (6.22 Ω) and at the lower temperature (−10°C) the resistance is definitely lower (4.587 Ω) than at 20°C. Of particular importance, however, are the following insights:

1. When the value of ΔT in Eq. (3-4) is *positive*, the bracketed coefficient factor is *greater* than *unity* and R_2 is greater than R_1.
2. When the value of ΔT in Eq. (3-4) is *negative*, the bracketed coefficient factor is *less than unity* and R_2 is less than R_1.

Although Eq. (3-4) is useful for finding resistance values that depart from 20°C, it has its limitations. In using Eq. (3-4) we must strictly adhere to 20°C as a temperature for making the first resistance measurement, R_1. But frequently room temperature is *not* 20°C (68°F). It is much more convenient to measure resistance and record the room (ambient) temperature, whatever it may be. In short, we require a relation that works for *any* temperature. The clue to finding that relation lies in finding the temperature coefficient of resistance at 0°C rather than at 20°C, that is, finding α_0 from α_{20}.

3-6.1 Determination of α_0

In analyzing Eq. (3-3) we noted that α_{20} represents the **ratio** of the (positive) linear slope of the R versus T curve taken at 20°C with respect to R_1, the resistance at 20°C. Clearly, at 0°C, the positive slope is the *same* (Fig. 3-4) but the value of R_0 is *less*, which results in a different and *higher* value for α_0! Since there is a 20°C difference between α_{20} and α_0, we may write the relation between them as

$$\frac{1}{\alpha_0} = \frac{1}{\alpha_{20}} - 20 = -T_0 \quad \text{degrees celsius (°C)} \qquad (3\text{-}5)$$

where $-T_0$ is the inferred zero-resistance temperature (or inferred absolute temperature).

Since we already know the inferred zero-resistance temperature for copper, let us verify it by using Eq. (3-5).

EXAMPLE 3-10
Given the value of $\alpha_{20} = 0.003\,93$ for copper (Appendix A-4), find the value of **a.** $1/\alpha_0$ **b.** α_0

Solution

a. $1/\alpha_0 = (1/\alpha_{20}) - 20 = (1/0.003\,93) - 20$
$= 234.45 \cong \mathbf{234.5\ \Omega/°C/\Omega}$ (3-5)

b. $\alpha_0 = 1/T_0 = 1/234.45 = \mathbf{0.004\,265\ \Omega/°C/\Omega}$ (3-5)

Several important points emerge from Ex. 3-10:

1. α_0 is greater than α_{20} for reasons anticipated at the beginning of Sec. 3-6.1.
2. Equation (3-5) shows $1/\alpha_0 = -T_0$. Therefore, $T_0 = -234.5$°C, which corresponds to the inferred zero-resistance temperature shown in Fig. 3-4.
3. Since we have a table of values for α_{20}, it is a simple matter to obtain values of α_0 by using Eq. (3-5) and Appendix A-4, as shown by the following examples.

EXAMPLE 3-11
Using Table A-4, calculate the inferred zero resistance temperatures for **a.** Silver **b.** Aluminum **c.** Constantan **d.** Nichrome

Solution

a. $1/\alpha = (1/\alpha_{20}) - 20 = (1/0.0038) - 20 = \mathbf{-243.2°C}$ (3-5)
b. $1/\alpha = (1/\alpha_{20}) - 20 = (1/0.003\,91) - 20 = \mathbf{-235.75°C}$
c. $1/\alpha = (1/\alpha_{20}) - 20 = (1/0.000\,008) - 20 \cong \mathbf{-125\,000°C}$
d. $1/\alpha = (1/\alpha_{20}) - 20 = (1/0.000\,44) - 20 = \mathbf{-2253°C}$ (3-5)

We might have anticipated the answers obtained in parts **(a)** and **(b)** of Ex. 3-11 but the results of **(c)** and **(d)** are somewhat serendipitous. We know for a fact that absolute zero (approximately −273.16°C) is the lowest temperature possible. We also know that *all* materials become *superconductive* at absolute zero. Why should constantan and nichrome require such impossibly low temperatures to reach superconductivity (zero resistance)?

The answer is that they do not, as shown in **Fig. 3-5**. Note that constantan maintains a fairly constant resistance over a very wide temperature range. If we were to *extrapolate* the *linear* portion of the curve to a point where the resistance is zero, we would find that the theoretical temperature is −125 000°C! Similarly, with nichrome, extrapolating the *linear* portion yields a theoretical zero-resistance temperature of −2253°C. Both manganin and nichrome are in that third class of materials whose value of α is positive yet extremely small and whose resistance with temperature is fairly constant (except at extremely low temperatures).

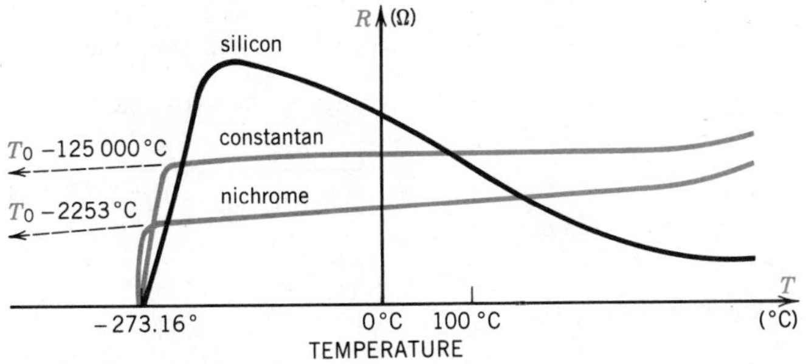

Figure 3-5 Low temperature (superconductive) characteristics

Figure 3-5 also shows silicon, a material of the second group in Sec. 3-6, which has a negative slope and a *negative* temperature coefficient of resistance. It tends to decrease in resistance as temperature increases, or increase in resistance as temperature decreases. But at extremely low temperatures, silicon (like constantan and nichrome), exhibits a resistance drop to zero. Does this mean that we cannot use the values determined in Ex. 3-11? Of course not! It only means that as long as we do not attempt predictions of resistance in either the extremely high temperature or the superconductive range, we can use the factors calculated in Ex. 3-11 (and others, as well).

3-6.2 Using α_0 and T_0 to Determine Resistance at Various Temperatures

If we extrapolate the linear portion of the resistance versus temperature curve to the inferred zero-resistance temperature, we may obtain a simple triangle, as shown in **Fig. 3-6**, for the metal conductor silver. If we examine Fig. 3-6 carefully we find it to consist of a set of **similar** right triangles. If we rotate the diagram 90°

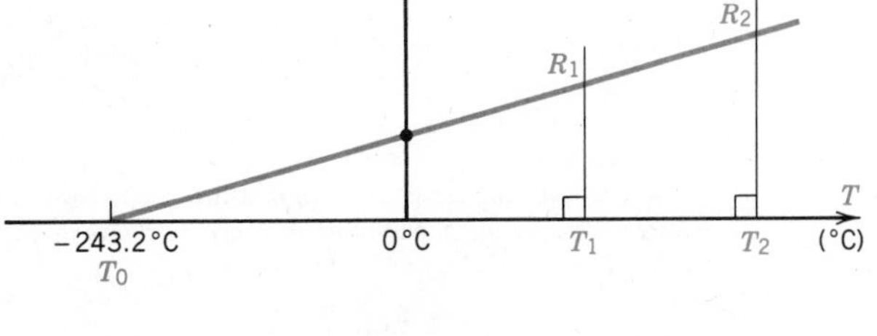

Figure 3-6 Using T_0 to determine any value of resistance for silver at any temperature by the law of similar triangles

clockwise we have a smaller triangle whose base is R_1 and whose altitude is $243.2 + T_1$. We also have a larger similar triangle whose base is R_2 and whose altitude is $243.2 + T_2$. Since we know that the inferred zero resistance for silver (from Ex. 3-11a) is $1/\alpha_0$ (i.e., $-243.2°C = 1/\alpha_0$), from the law of similar triangles we can write for any material

$$\frac{R_2}{R_1} = \frac{|1/\alpha_0| + T_2}{|1/\alpha_0| + T_1} \qquad (3\text{-}6)$$

where R_2 is the resistance at a higher temperature T_2

R_1 is the resistance at a lower temperature T_1

$|1/\alpha_0|$ is the inferred zero resistance temperature of the material as determined from Eq. (3-5), expressed as an absolute (positive) value

The following examples show the extreme versatility of Eq. (3-6).

EXAMPLE 3-12

Verify the solutions to Ex. 3-9, using Eq. (3-6).

Solution

Data obtained from Ex. 3-9 may be tabulated[14]:

R (Ω)	T (°C)	
4.587	−10	
5.20	+20	(given data)
6.22	+70	

Since $1/\alpha_0$ for copper is −234.5 (Fig. 3-4) we may solve for parts (**a**) and (**b**) of Ex. 3-9 as:

a. $$R_2 = R_1 \frac{234.5 + T_2}{234.5 + T_1} = 5.20\ \Omega\,\frac{234.5 + 70}{234.5 + 20}$$
$$= 5.20 \times 1.1965 = \mathbf{6.22\ \Omega} \qquad (3\text{-}6)$$

b. $$R_1 = R_2 \frac{234.5 + T_1}{234.5 + T_2} = 5.20\ \Omega\,\frac{234.5 - 10}{234.5 + 20}$$
$$= 5.20 \times 0.8821 = \mathbf{4.587\ \Omega} \qquad (3\text{-}6)$$

Several important points must be made with regard to Eq. (3-6) and its application in Ex. 3-12:

1. In order to use Eq. (3-6) for any conductor, we must first calculate the value of $1/\alpha_0$, using the values of α_{20} given in Appendix A-4 and the conversion of Eq. (3-5).
2. From Eq. (3-5), we are solving for $-T_0$ and therefore the value of $1/\alpha_0$ is an **absolute** or *positive* value when used in Eq. (3-6).
3. When using Eq. (3-6) with negative values of T_1 or T_2 or both, since these temperatures are below zero, they must be *subtracted* from the *absolute positive value* of $1/\alpha_0$. Note how this is done in Ex. 3-12**b**.
4. Equation (3-6) yields the same values of resistance at corresponding temperatures as Eq. (3-4). But it has the advantage of yielding resistance data at any temperature; that is, we are not restricted to data at 20°C.

EXAMPLE 3-13

The resistance of a bar of gold measures 25 mΩ at 25°C. Calculate its resistance when heated to 75°C.

Solution

$$1/\alpha_0 = (1/\alpha_{20}) - 20 = (1/0.0034) - 20 = 274.1 \qquad (3\text{-}5)$$

$$R_2 = R_1 \frac{(1/\alpha_0) + T_2}{(1/\alpha_0) + T_1} = 25\ \text{m}\Omega\,\frac{274.1 + 75}{274.1 + 25}$$
$$= 25\ \text{m}\Omega \times 1.167 = \mathbf{29.18\ m\Omega} \qquad (3\text{-}6)$$

EXAMPLE 3-14

A nickel wire measures 15 Ω at 25°C and 20 Ω at some higher temperature. Calculate the

a. Temperature of the wire at the higher resistance
b. Temperature increase in degrees celsius
c. Speculate on the use of nickel wire as a means of measuring temperature by measuring its resistance.

[14] Note the technique used in tabulating. The *physical quantity* is at the *top* of the column. The *unit* is directly below it in *parentheses*. **Only** numbers appear in the cells below the physical quantity and the unit.

Solution

a. $1/\alpha_0 = (1/\alpha_{20}) - 20 = (1/0.005\,37) - 20 = 166.2$ (3-5)

$$T_2 = \frac{R_2}{R_1}(166.2 + T_1) - 166.2$$

$$= \frac{20}{15}(166.2 + 25) - 166.2 = \mathbf{88.7\bar{3}°C} \quad (3\text{-}6)$$

b. $\Delta T = T_2 - T_1 = 88.7\bar{3} - 25 = \mathbf{63.7\bar{3}°C}$

c. Nickel has the highest value of α_{20} in Appendix Table A-4. Of all the metals listed, it is the *most sensitive* to temperature change, that is, producing the highest resistance change per degree celsius of temperature change. By measuring the resistance of a nickel wire (using an ohmmeter or a bridge) and using a chart of resistance versus temperature, the temperature of the nickel wire could be determined easily. Alternatively, the ohmmeter could be calibrated directly to read temperature instead of resistance across a given nickel wire located in a temperature probe.

EXAMPLE 3-15

Chemical analysis is expensive and time-consuming. A gold dealer suspects that a bar of gold he has purchased is impure. His daughter, an electronics genius, has an idea. She measures the resistance of the "gold" bar at room temperature (25°C) and finds it to be 30 mΩ. She heats the bar in a temperature bath at 50°C and quickly measures its resistance, finding it to be 35.77 mΩ. Determine

a. The value of α_{20} for the material.
b. The most likely composition of the "gold" bar, using Appendix A-4.

Solution

a. In Eq. (3-6), let $1/\alpha_0 = x$. Then

$$\frac{R_2}{R_1} = \frac{x + T_2}{x + T_1}$$

and substituting given values yields

$$\frac{35.77}{30.00} = \frac{x + 50}{x + 25}$$

Solving for x yields $1.192(x + 25) - x = 50$ and $x = 105.2 = T_0 = 1/\alpha_0$ but

$$\alpha_{20} = \frac{1}{T_0 + 20} = \frac{1}{105.2 + 20} = 0.007\,99 \cong \mathbf{0.008} \quad (3\text{-}5)$$

b. The material is most likely *steel* with a very thin surface plating of gold (see Appendix A-4).

The foregoing examples only partly show the advantages and versatility of Eqs. (3-5) and (3-6). Some of these are the following.

1. We are now no longer restricted to an ambient temperature of 20°C (as with Eq. 3-4) in making resistance–temperature calculations.
2. We can predict high and/or low temperatures with reasonable accuracy from resistance measurements (Ex. 3-14).
3. We can determine the nature of an unknown material by measuring its resistance at two different known temperatures and consulting a table of values of α_{20} (or α_0).[15]
4. Without the use of thermometers (or thermocouples) in contact with the material, we can determine the temperature of a glowing lamp filament (or toaster element) from resistance measurements.
5. We can determine the *internal* temperature of a transformer or a motor winding from resistance measurements. This is an impossible feat unless we intentionally design these devices with built-in heat-sensing devices to detect hot spots.

Before leaving the subject of the effect of temperature on resistance, one last point remains to be made. In Sec. 3-4 we stated Eq. (3-2) for the factors affecting resistance as $R = \rho l/A$, where $T = k = 20°C$. Since the value of ρ is taken at 20°C (Appendix

[15] Actually, we could have used the equation $\rho = AR/l$ to identify an unknown metallic element or alloy. If the cross-sectional area, length, and resistance are carefully measured to 20°C, we could consult a table of resistivities to find the unknown. Both methods could serve as a cross-check in identifying the material tested. If ρ and α_{20} both identify the same material, we can be reasonably sure of its composition without chemical analysis.

A-4), the equation is valid only at 20°C. In light of what we have just learned, however, we may now write the equation for *all four* factors affecting resistance as

$$R = \frac{\rho l}{A}(1 + \alpha_{20}\Delta T) \quad \text{ohms } (\Omega) \tag{3-7}$$

where all terms have been defined.

Note that Eq. (3-7) requires use of α_{20} since resistivity, ρ, is also taken at 20°C.

EXAMPLE 3-16

A tubular wire-wound resistor uses 80 cm of No. 36 nichrome wire. When carrying current, the temperature of the resistor is 45°C. Calculate the resistance of the resistor.

Solution

From Table A-4, at 20°C the resistivity of nichrome is $99.6 \times 10^{-8}\ \Omega\cdot\text{cm}$ and α_{20} is 0.000 44; from Table A-5 the area of No. 36 wire is $1.267 \times 10^{-8}\ \text{m}^2$. Substituting these values yields

$$R = \frac{\rho l}{A}(1 + \alpha_{20}\Delta T)$$

$$= \frac{99.6 \times 10^{-8}}{1.267 \times 10^{-8}} \frac{\Omega\cdot\text{cm} \times 0.8\ \text{m}}{\text{m}^2}(1 + 0.000\,44 \times 25)$$

$$= \mathbf{63.58\ \Omega} \tag{3-7}$$

Key Sequence

Press										Display
	4.4	EE	+/−	4	×	25	+	1	=	1.011
×	99.6	EE	+/−	8	×	0.8	÷			$8.055\,647 \times 10^{-7}$
	1.267	EE	+/−	8	=					63.58

3-7 COMMERCIAL RESISTORS

Commercial resistors may be divided into two large families: *fixed* resistors and *variable* resistors. Fixed resistors may be subdivided into carbon composition types (Sec. 3-7.1), metal film types (Sec. 3-7.3), and wire-wound types (Sec. 3-7.4). Variable resistors may be subdivided into semivariable types (Sec. 3-7.5), variable rheostatic types (Sec. 3-7.6), and variable potentiometer types (Sec. 3-7.7).

3-7.1 Fixed Molded Carbon Composition Resistors

Molded carbon composition resistors are constructed of a resistive element, axial contact leads, and an insulating molded case, as shown in **Fig. 3-7a**. The resistive element contains a mixture of graphite powder and silica plus a binding compound, molded under heat and pressure. Lower-resistance resistors contain a higher proportion of graphite (a semiconductor) to silica (an insulator). Higher resistances are obtained by decreasing the amount of graphite in proportion to the silica and the binder. In this way, resistors varying from 0.1 Ω to 9.1 GΩ are commercially available.

Fixed molded composition resistors, ranging in power rating from 1/8 to 5 W, are generally much too small to include all the information in printed form concerning their rating, which comprises the resistance value, the tolerance range of resistance, and possibly the reliability. Consequently, the Electronic Industries Association (EIA) has developed a five-band color code used in marking molded composition resistors in the following power ratings: 1/8, 1/4, 1/2, 1, 2, 3, 4, and 5 W (Sec. 3-7.2).

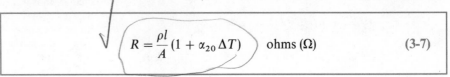

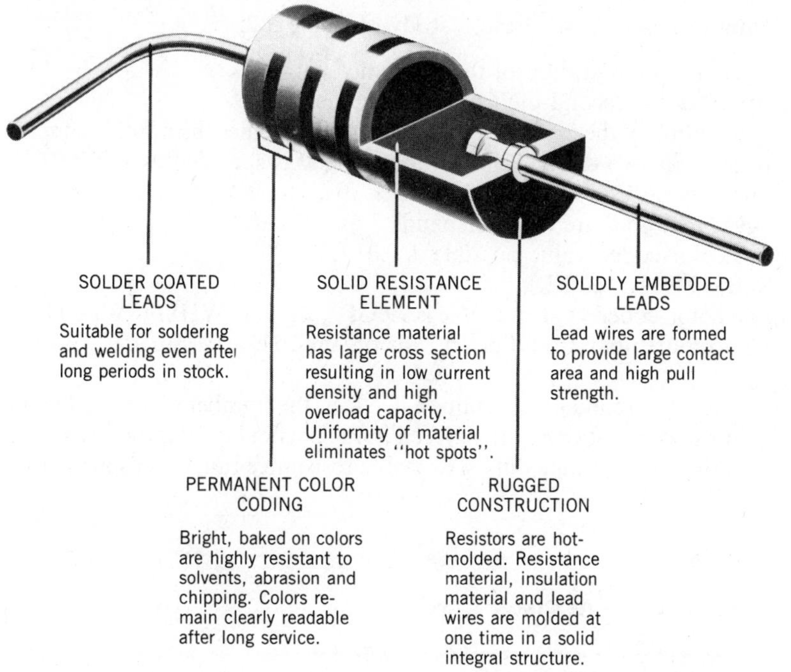

SOLDER COATED LEADS
Suitable for soldering and welding even after long periods in stock.
SOLID RESISTANCE ELEMENT
Resistance material has large cross section resulting in low current density and high overload capacity. Uniformity of material eliminates "hot spots".
SOLIDLY EMBEDDED LEADS
Lead wires are formed to provide large contact area and high pull strength.
PERMANENT COLOR CODING
Bright, baked on colors are highly resistant to solvents, abrasion and chipping. Colors remain clearly readable after long service.
RUGGED CONSTRUCTION
Resistors are hot-molded. Resistance material, insulation material and lead wires are molded at one time in a solid integral structure.

a. Construction (Courtesy Allen-Bradley Co.)

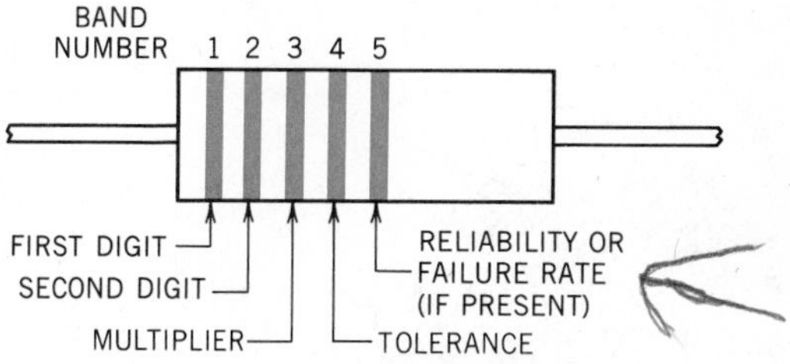

b. EIA resistor color code

Figure 3-7 Fixed molded carbon composition resistors and EIA color code

The five-band color code, read from left to right (**Fig. 3-7b**) on one side of the cylindrical body of the resistor, is designated as follows: the **first** and **second** bands produce a two-digit number; the **third** band designates the multiplier, that is, the power of 10 multiplier or the number of zeros following the two-digit number. The **fourth** band designates the ± percent tolerance range within which the resistor's actual resistance may fall or depart from the *nominal resistance* given by the first three bands. The **fifth** band (*if available*) yields the reliability or percent failure rate per 1000 hours of continuous operation at its rated power.

Note from Fig. 3-7b that the resistance is specified to *two* significant figures, determined from bands 1 and 2. The decimal numbers correspond to the colors given for the first two bands in Table 3-2.

Table 3-2 Color Code for Standard Molded Carbon Composition Resistors

Color	Significant Figures, Bands 1 and 2	Multiplier (power of 10), Band 3	Tolerance (±%), Band 4	Reliability or Failure Rate (%/1000 h), Band 5 (optional)
Black	0	$10^0 = 1$	—	—
Brown	1	$10^1 = 10$	1	1
Red	2	$10^2 = 100$	2	0.1
Orange	3	$10^3 = 1000$	—	0.01
Yellow	4	10^4	—	0.001
Green	5	10^5	0.5	
Blue	6	10^6	0.25	
Violet	7	10^7	0.1	
Gray	8	10^8	0.05	
White	9	10^9	—	
Gold	—	10^{-1}	5	
Silver	—	10^{-2}	10	
None	—	—	20	

The following insights emerge from a study of Table 3-2:

1. Band 1 specifies the first digit of the resistance value.
2. Band 2 specifies the second digit of the resistance value.
3. With the exception of the first color (black), band 3 specifies the number of zeros following the first two digits. For example 2200 Ω is specified as RED-RED-RED. This may be viewed as 22 00 or 22×10^2.
4. Colors silver and gold are used for bands 3 and 4 only.
5. The *lowest* color-coded value possible is 0.01 Ω (BLACK-BROWN-SILVER).[16]
6. The *highest* color-coded value possible is 99 GΩ (WHITE-WHITE-WHITE).[16]
7. If there is no fourth band on a given resistor, the tolerance range is assumed at $\pm 20\%$.
8. The tolerance band (band 4) determines, in effect, the number of **standard commercially available** resistors manufactured in the particular tolerance class. For example, in the $\pm 20\%$ range only six specific resistance range steps are shown in Table 3-3.

Table 3-3 Commercially Available Carbon Composition Resistors

Tolerance (Band 4)		
±5%	±10%	±20%
10	10	10
11		
12	12	
13		
15	15	15
16		
18	18	
20		
22	22	22
24		
27	27	
30		
33	33	33
36		
39	39	
43		
47	47	47
51		
56	56	
62		
68	68	68
75		
82	82	
91		

The following points serve to clarify Table 3-3:

[16] But these are not the lowest and highest values *commercially* available. See Table 3-3 and explanations concerning it. The standard commercial resistance values in Table 3-3 only range from 0.1 Ω to 91 GΩ.

1. The two significant figures in Table 3-3 may be multiplied by any multiplier from 10^{-2} (silver) up to 10^9 (white) in band 3 to find the commercial values of resistors commercially available.
2. The intent of item **1** indicates that the **lowest** *commercially available* standard value is **0.1 Ω** and the **highest** is **91 GΩ**.
3. Resistors made to a higher degree of precision require a wider selection of commercially available values. At ±20% only six values are needed. At ±5%, 24 values are required in Table 3-3 (see following examples).
4. For the three tolerances given in Table 3-3, there is always an overlap between the **upper** and **lower limits** of each commercial value. This is shown in Ex. 3-20. Only in this way can a full range of standard values be obtained at the tolerance given.

EXAMPLE 3-17

A resistor is color-coded RED-RED-ORANGE-SILVER-BROWN. Find:

a. Its nominal value, tolerance, and reliability
b. The upper and lower resistance ranges possible for this resistor by one method of calculation
c. Repeat (**b**) using an alternative (preferred) method of calculation

Solution

a. **22 000 Ω** (22 kΩ); **±10%**; **1% per 1000 h** failure rate
b. Lowest value = 22 kΩ − 0.1 × 22 kΩ = **19.8 kΩ**
   Highest value = 22 kΩ + 0.1 × 22 kΩ = **24.2 kΩ**
c. Lowest value = 22 kΩ(1 − 0.1) = 22 kΩ × 0.9 = **19.8 kΩ**
   Highest value = 22 kΩ(1 + 0.1) = 22 kΩ × 1.1 = **24.2 kΩ**

EXAMPLE 3-18 Specify five color bands for each of the following resistors having a 0.01% failure rate per 1000 h of operation:

a. 0.12 Ω ± 5%
b. 270 kΩ ± 10%
c. 15 MΩ ± 1%
d. 5.5 GΩ ± 2%

Solution

a. BROWN-RED-SILVER-GOLD-ORANGE
b. RED-VIOLET-YELLOW-SILVER-ORANGE
c. BROWN-GREEN-BLUE-BROWN-ORANGE
d. GREEN-GREEN-GRAY-RED-ORANGE

EXAMPLE 3-19

A resistor is color-coded YELLOW-VIOLET-GOLD-GOLD-YELLOW. Find

a. Its nominal value, tolerance, and reliability
b. The extreme upper and lower values possible expressed as a range

Solution

a. 4.7 Ω; ±5%; 0.001% failure rate per 1000 h of operation
b. Lower limit = 4.7 Ω × 0.95 = 4.465 Ω
   Upper limit = 4.7 Ω × 1.05 = 4.935 Ω
   Range = **4.465 Ω** to **4.935 Ω**

EXAMPLE 3-20

The statement was made that for all tolerances given in Table 3-3, there is always an overlap between the upper limit of one resistor and the lower limit of the next higher resistance listed. Show that this is so for a 33-kΩ resistor and its next higher commercial value shown in Table 3-2, at tolerances of a. ±5% b. ±10% c. ±20%

Solution

a. Upper limit of 33 kΩ at ±5% = 1.05 × 33 kΩ = **34.65 kΩ**
   Next higher resistor lower limit = 36 kΩ × 0.95 = **34.2 kΩ**
   This shows an overlap of **450 Ω** between adjacent resistors at 5% for 33 kΩ
b. Upper limit of 33 kΩ at ±10% = 33 kΩ × 1.1 = **36.3 kΩ**
   Next higher resistor lower limit = 39 kΩ × 0.9 = **35.1 kΩ**
   This shows an overlap of **1200 Ω** between adjacent resistors at 10% for 33 kΩ
c. Upper limit of 33 kΩ at ±20% = 33 kΩ × 1.2 = **39.6 kΩ**
   Lower limit of 47 kΩ at ±20% = 47 kΩ × 0.8 = **37.6 kΩ**
   This shows an overlap of **2000 Ω** between adjacent resistors at 20% for 33 kΩ

EXAMPLE 3-21

A designer calculates a collector resistor for a required transistor circuit as 25 kΩ. Calculate the closest standard commercial resistor size the designer should select, if the tolerance of the other resistors in the circuit is a. ±5% b. ±10% c. ±20%. Using Table 3-3 and the overlap technique of Ex. 3-20, the closest value is

a. **24 kΩ** at **±5%**
b. **27 kΩ** at **±10%**
c. **22 kΩ** at **±20%**

(The solution is left as an exercise for the student, using the technique of Ex. 3-20. See Prob. 3-43.)

EXAMPLE 3-22

a. For each of the commercial values selected in Ex. 3-21, calculate the relative percent error between the selected and desired value using the equation

$$r.e. = \frac{\text{nominal commercial value} - \text{specified value}}{\text{specified value}}$$

b. Is the relative error in percent within the tolerance band used in each case? Is this good enough for the designer?

c. Is there a possibility that an extreme upper or lower limit may exceed the allowable tolerance of error? Repeat (**b**), selecting appropriate limit of commercial standard resistors to prove your point.

Solution

a. $$r.e. = \frac{24\ \text{k}\Omega - 25\ \text{k}\Omega}{25\ \text{k}\Omega}$$

$= \mathbf{-4\%}$ (24 kΩ is 4% *below* desired value)[17]

$$r.e. = \frac{27\ \text{k}\Omega - 25\ \text{k}\Omega}{25\ \text{k}\Omega}$$

$= \mathbf{+8\%}$ (27 kΩ is 8% *above* desired value)[17]

$$r.e. = \frac{22\ \text{k}\Omega - 25\ \text{k}\Omega}{25\ \text{k}\Omega}$$

$= \mathbf{-12\%}$ (22 kΩ is 12% *below* desired value)[17]

b. Yes; if the *nominal* commercial value is used, the percent error falls within the allowable tolerance band for that resistor selected. No, it is not (see part **c** below).

c. Lower limit of 24 kΩ at 5% = 24 kΩ × 0.95 = 22.8 kΩ

$$r.e. = \frac{22.8\ \text{k}\Omega - 25\ \text{k}\Omega}{25\ \text{k}\Omega}$$

$= \mathbf{-8.8\%}$ (a greater error than the 5% allowed)

Upper limit of 27 kΩ at 10% = 27 kΩ × 1.1 = 29.7 kΩ

$$r.e. = \frac{29.7 - 25}{25}$$

$= \mathbf{+18.8\%}$ (a greater error than the 10% allowed)

Lower limit of 22 kΩ at 20% = 22 kΩ × 0.8 = 17.6 kΩ

$$r.e. = \frac{17.6 - 25}{25}$$

$= \mathbf{-29.6\%}$ (a greater error than the 20% allowed)

Conclusions and insights to be drawn from Exs. 3-20 through 3-22 are as follows.

1. One cannot select the *nominal* value, as done in Ex. 3-21 and Ex. 3-22**a**, and assume that the selected commercial resistor will actually have that nominal value.
2. It is necessary to use the technique of Ex. 3-22**c** to determine whether the extreme upper or lower limits will fall outside the allowed tolerance band for the standard commercial resistor selected.
3. To maintain a desired tolerance level for all resistors used in a given electronic circuit, it is necessary to specify a tolerance level (in some cases of calculated values) for some commercial resistors that is smaller (better) than the allowable error tolerance of the rest of the resistors, capacitors, and so forth, in the equipment. For example, in Ex. 3-21**a**, **if** a tolerance level of **2%** is specified for the 24-kΩ commercial resistor, there is no chance that the lower limit will exceed the 5% specified for the remaining resistors used. (See solution of Ex. 3-22**c**.)

3-7.2 Commercial Power Rating of Fixed Molded Composition Resistors

Most tubular wire-wound resistors (Sec. 3-7.4) are sufficiently large in diameter and surface area (because of their higher power rating) for the resistance value, tolerance, reliability, and even the power rating to be printed on their surface. But

[17] The advantage of using the relative error equation in the form given is that it specifies whether the error is *above* or *below* the desired value; that is, a plus sign means above and a minus sign means below.

we have seen that commercial fixed molded composition resistors are much too small for such information to be printed and recorded on their surface. If one examines Fig. 3-7b one notes that the power rating is *not* included in the information provided by the five (or four) bands of the EIA resistor color code. Yet power rating (i.e., the ability to dissipate heat) is just as important as resistance value, tolerance, and reliability. Why is it omitted from the color code?

We noted earlier that molded carbon composition resistors are available in the following eight power ratings: 1/8, 1/4, 1/2, 1, 2, 3, 4, and 5 watts. It is unnecessary to record this on the resistor because the *physical size* (i.e., the surface area) *determines the power rating* automatically.

Figure 3-8 shows the relative dimensions for five of the most common sizes in terms of power rating. From this figure we may conclude the following:

1. The larger the diameter and the greater the length of the resistor, the greater the physical surface area and *the better the ability* of the resistor to *dissipate heat.*

Figure 3-8 Physical size (surface area) versus power rating of fixed molded carbon composition resistors from 1/8 to 2 W, respectively

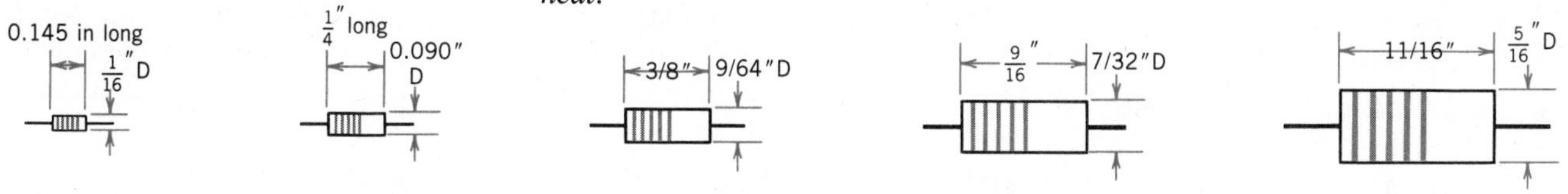

a. $\frac{1}{8}$ watt, 5 kΩ **b.** $\frac{1}{4}$ watt, 5 kΩ **c.** $\frac{1}{2}$ watt, 5 kΩ **d.** 1 watt, 5 kΩ **e.** 2 watt, 5 kΩ

DIAMETER		LENGTH			
SI (mm)	ENG (in)	SI (mm)	ENG (in)	POWER (W)	FIGURE REFERENCE 3-8
1.59	0.0625	3.68	0.145	1/8	a
2.29	0.090	6.35	0.250	1/4	b
3.56	0.140	9.52	0.375	1/2	c
5.72	0.225	14.3	0.5625	1	d
7.94	0.3125	17.5	0.6875	2	e

f. Dimensions, SI and English in decimal equivalents

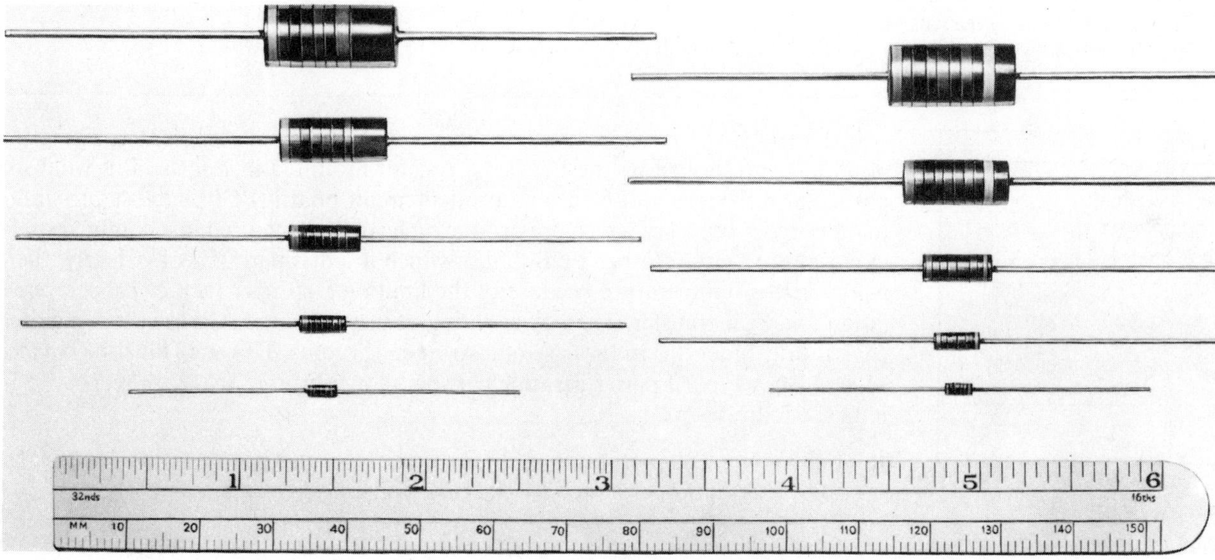

g. Comparisons of physical dimensions of carbon composition resistors (Courtesy Allen-Bradley Co.)

2. The only difference between a 5 kΩ, 1 W and a 5 kΩ, 2 W resistor is that the latter is physically larger and has a greater ability to dissipate its internal heat without excessive temperature rise. (The electrical properties of both resistors are the *same* in an electrical circuit, as we will see.)
3. The *power dissipated* by a resistor is a function of the factors given in Eq. (3-8). The heat energy produced by the resistor (W_r) is a product of the power dissipated and the time ($P_r \times t$). If the resistor surface area is *too small*, the resistor temperature *rises*, causing it to dissipate heat energy at a *higher* rate. The *final* temperature of the resistor is the temperature at which the heat generated ($P_r t$) is equal to the heat dissipated!
4. If this final temperature is too high and the resistor may be damaged as a result, it is necessary to select a resistor having a *higher* power rating. Such a resistor, with its *larger* physical surface area, is capable of dissipating *more* heat energy *without* an accompanying excessive temperature rise.
5. If we consider 20°C as ambient (room) temperature, resistors are permitted to be loaded up to their commercial power rating with a 50°C *rise* in temperature. Consequently, if the ambient surface temperature of the 1-W resistor shown in Fig. 3-8b is 69°C when dissipating 1 W of power in an electric circuit, it cannot be damaged and will not overheat excessively.

Factors affecting the power dissipated by a resistor are[18]

$$P_r = V_r I_r = I_r^2 R = \frac{V_r^2}{R} \quad \text{watts (W)} \qquad (3\text{-}8)$$

EXAMPLE 3-23

Calculate the power dissipated at an ambient temperature of 50°C and select an appropriate power rating for each of the following resistors:

a. A 5-kΩ resistor across a 30-V supply
b. A 2-kΩ resistor carrying a current of 15 mA
c. A resistor having 60 V across it and carrying a current of 20 mA

Solution

a. $P_r = V_r^2/R = 30^2/5000 =$ **180 mW**; use a $\frac{1}{4}$-W resistor
b. $P_r = I_r^2 R = (15\text{ mA})^2 \times 2\text{ k}\Omega =$ **450 mW**; use a $\frac{1}{2}$-W resistor
c. $P_r = V_r I_r = 60\text{ V} \times 15\text{ mA} =$ **900 mW**; use a 1-W resistor

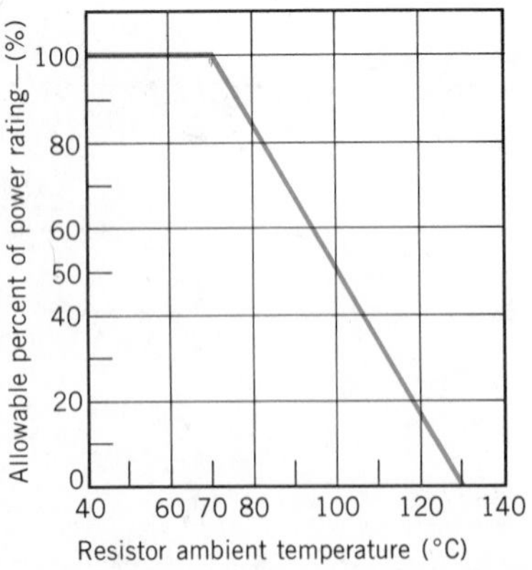

Figure 3-9 Power derating of fixed carbon composition resistors (per MIL-R-39008 and RS 172)

In Ex. 3-23, the resistors were operating within the safe ambient temperature of 50°C as prescribed by the EIA for a resistor mounted in *free air*. But what if a particular resistor is mounted on a printed circuit board (PCB) in close proximity to other resistors, all of which are producing heat, and enclosed in a confined space in close proximity to other PCBs, all of which are dissipating heat? Clearly, there will be rise in temperature because of the limiting factors of lack of proper ventilation and heat transfer. The power rating of resistors mounted in either enclosed chassis or cabinets (as in the case of instruments, radios, TVs, etc.) must be *reduced* or *derated*. A typical **power derating curve** is shown in **Fig. 3-9**. With respect to it, note the following:

1. For ambient temperatures up to 70°C the nominal power rating is the same (100%) as that determined by the physical size of the resistor.

[18] The derivation of Eq. (3-8) stems from the use of Ohm's law, where $I_r = V_r/R$ and $V_r = I_r \times R$. If each of these is substituted separately in the equation $P_r = V_r I_r$, then $P_r = V_r^2/R$ and $P_r = I_r^2 R$. See Eq. (4-3).

2. At ambient temperatures above 70°C the derating curve is absolutely linear. A derating of 50% occurs at 100°C.
3. The derating curve implies that at ambient temperatures above 70°C, the power rating of a given resistor must be decreased. For example, a 1 kΩ, 1 W resistor that dissipates 1 W at 70°C may only dissipate 500 mW at 100°C.
4. This means that the physical size of the above resistor is to be increased if it is to dissipate 1 W at 100°C. Figure 3-9 shows that we would require a 1 kΩ, 2 W resistor, despite the fact that it dissipates only 1 W at 100°C.

EXAMPLE 3-24

For the resistors given in Ex. 3-23, calculate the required *commercial* power rating at an ambient temperature of 98°C

Solution

From Fig. 3-9, the derating factor at 98°C is 0.52 and therefore

a. $P_r = 180\ \text{mW}/0.52 =$ **346 mW**; use a $\frac{1}{2}$**-W** resistor
b. $P_r = 450\ \text{mW}/0.52 =$ **865 mW**; use a **1-W** resistor
c. $P_r = 900\ \text{mW}/0.52 =$ **1.73 W**; use a **2-W** resistor

---

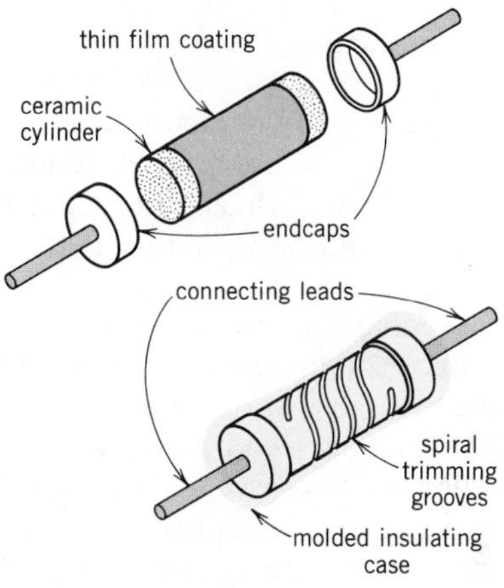

a. Construction

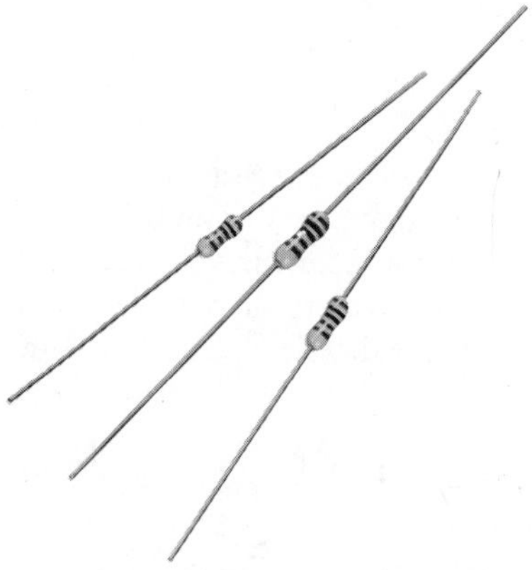

b. Typical metal film resistors (Courtesy Allen-Bradley Co.)

Figure 3-10 Thin-film fixed resistors

Note from Ex. 3-24 that a resistor that dissipates only 180 mW must be commercially rated at 500 mW at an elevated temperature. Similarly, one that dissipates 450 mW must be rated at 1 W at an elevated temperature.

3-7.3 Metal Film Resistors

With the development of integrated circuits, the need arose for resistors having low temperature coefficients of resistance (α_{20}), close precision tolerances (± 0.1 to $\pm 1\%$), relatively high noise immunity, and excellent stability (ability to maintain same value over their entire life). The metal film principle meets these demands. It is used not only for fixed but also for variable resistors (film potentiometers), described in Sec. 3-7.6.

Typical fixed film resistor construction is shown in **Fig. 3-10**. A ceramic cylinder is coated with a thin conductive film of metal or metallic oxide. The end caps containing the connecting leads are then pressed on each end. After a suitable "curing period" to stabilize the film, the resistance is (increased) trimmed to a desired value by means of automatic machine process control. When the desired resistance is obtained, the machine stops cutting its groove. After inspection for proper precision, a silica or glass case is molded around the entire assembly.

Metal film resistors are capable of withstanding higher ambient temperatures without damage (up to 125°C). They are manufactured in power ratings from 50 mW to 2 W. Resistance values from 1 Ω up to 100 MΩ are obtainable, at precision tolerances from 0.1 to 1%.

3-7.4 Wire-Wound Resistors

Wire-wound resistors are manufactured in two types, basically. The first are wire-wound **precision** resistors, which are generally for *low* power rating ($\frac{1}{8}$ to $\frac{1}{2}$ W) and extremely high accuracy (± 0.1 to $\pm 0.01\%$). The second type are wire-wound **ceramic-coated** resistors, generally of *medium* to *high* power rating (2 to 250 W and higher) and ranging in accuracy from 0.1 through 3%.

Figure 3-11a shows the construction of a typical high-power wire-wound resistor, typically 10 to 1000 W and even higher. The resistance wire is wound on a ceramic core. The entire assembly is encapsulated in a combination of ceramic cement and clay, which is fired at a high temperature to provide a ceramic case.

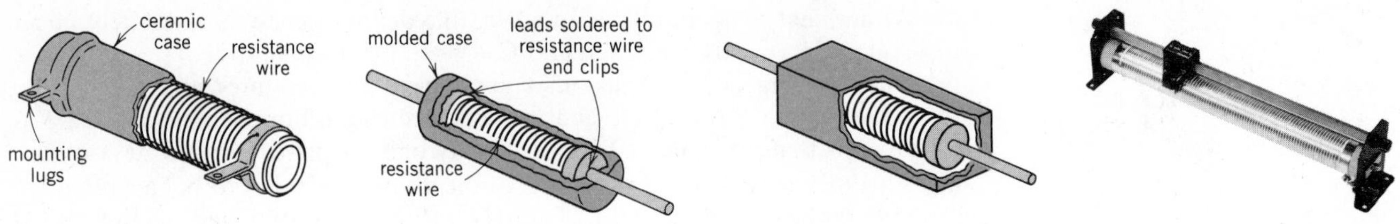

a. Power resistor

b. Medium power precision resistor

c. Encapsulated wire wound precision resistor in ceramic case using refractory cement

d. Wire wound potentiometer (Courtesy Biddle Instruments)

Figure 3-11 Wire-wound resistors

Figure 3-11b shows the typical construction for a medium-power wire-wound resistor of moderate precision. Like the high-power resistors, it is enclosed in a ceramic or plastic molded case to prevent movement and protect and insulate the wire.

Figure 3-11c shows a type of precision wire-wound resistor in which the resistor is first produced by winding resistance wire on an insulated form. After checking the resistor for resistance accuracy (using a Wheatstone bridge, Sec. 23-7), the wire element is then inserted in a ceramic case and held in place by refractory cement. The assembly is then fired at a high temperature to bind and bond the structure.

Figure 3-11d shows another type of wire-wound resistor in which a rectangular ribbon is wound on a ceramic tube. This wire-wound laboratory unit serves as either a *potentiometer* or a *rheostat* (Sec. 3-7.6).

Regardless of type, all wire-wound resistors use (constant temperature) resistance-alloy wire of the kinds listed in Appendix A-4 (advance, constantan, manganin, nichrome, etc.). In almost all cases, the resistance wire is wound on a ceramic insulated core whose form provides contacts at each end for connecting leads. Higher resistances are obtained by using alloys of higher resistivity (nickel base or iron base alloys). For higher-power resistors, the cross-sectional area of the alloy wire is increased. The ceramic case serves not only to prevent movement of wire and/or damage to the resistance element but also to increase the surface area for optimum heat dissipation (as shown in Fig. 3-11c).

3-7.5 Semivariable Resistors

Semivariable resistors are similar in construction to wire-wound resistors with one minor exception. Instead of a fixed value of resistance, one may obtain some resistance selection by choosing appropriate taps. A typical semivariable resistor is shown in **Fig. 3-12a**; it is commonly used in voltage dividers in power supplies to provide the choice of any of several voltages of fixed value (3, 6, 9, and 12 V dc).

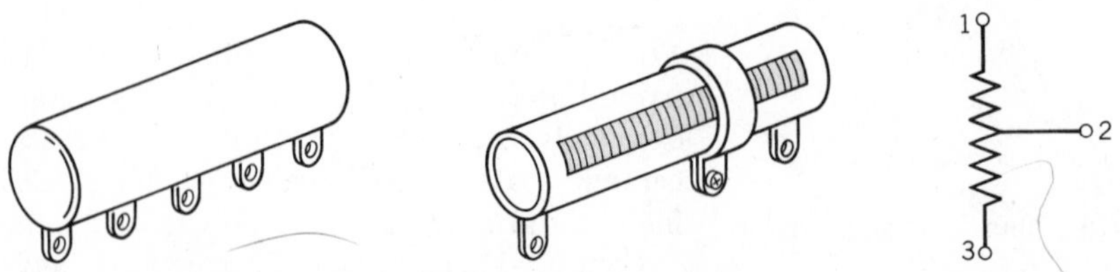

a. Tapped resistor

b. Adjustable resistor

c. Electrical symbol

Figure 3-12 Semivariable resistors

Another type of semivariable resistor is shown in Fig. 3-12b. While preserved in a ceramic case, the resistance element is partially exposed, so that electrical contact can be made with it via an adjustable contact ring. This arrangement permits the choice of any desired voltage with respect to the fixed terminal at either end. Once the desired voltage is obtained, the adjustable contact is firmly tightened and the resistor is wired into the circuit. The electrical symbol for the semivariable resistor is shown in Fig. 3-12c.

3-7.6 Variable Resistors

By exposing the resistive element to a moving contact or wiper arm, which is mechanically rotated (or linearly moved), it is possible to obtain a full range of resistance between one end terminal and the other. Such a device is called either a *potentiometer* (**Figs. 3-13a, b, d, e**) or a *rheostat* (Fig. 3-13c). Potentiometers may have carbon, film, or wire as the resistive element. In the first case they are called carbon potentiometers, in the second film potentiometers, and in the third wire-wound potentiometers. Regardless of construction, the electrical symbol remains as shown in Fig. 3-13b.

The essential difference between a potentiometer (a three-terminal device) and a rheostat (a two-terminal device) is the manner in which they are used in circuitry. Any potentiometer may be used as a rheostat by eliminating one terminal, as shown in Fig. 3-13c.

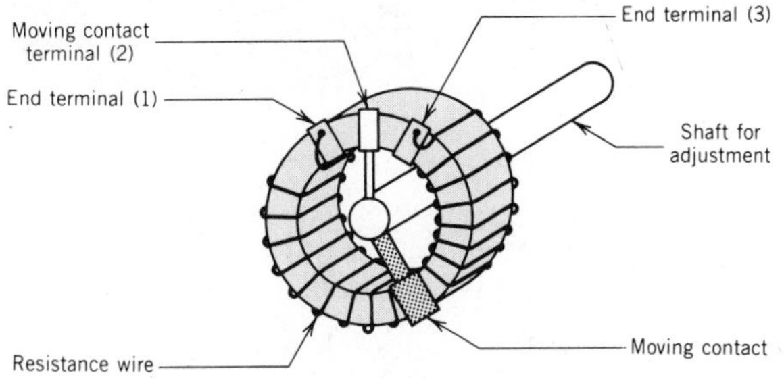

a. Wire wound potentiometer

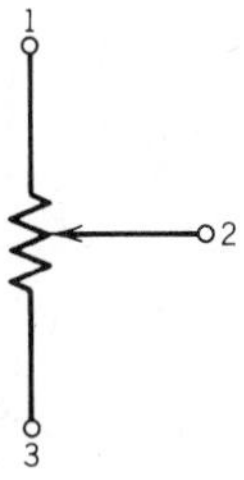

b. Electrical symbol for potentiometer

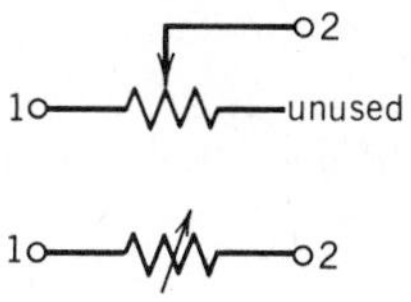

c. Electrical symbols for rheostats

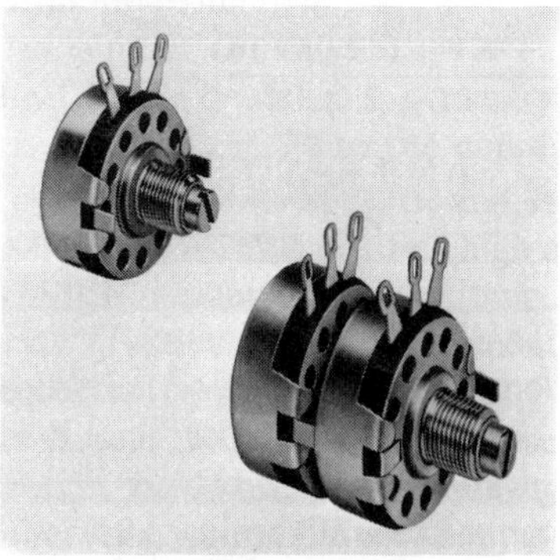

d. Ganged and single panel-mounting potentiometers (Courtesy Allen-Bradley Co.)

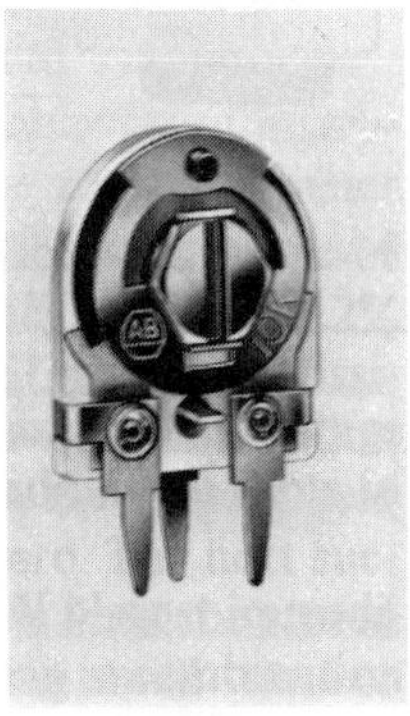

e. Trimming potentiometer, PCB mounting (Courtesy Allen-Bradley Co.)

Figure 3-13 Variable resistors

3-7.7 Adjustable Resistors

The last class of resistors covered in this section comprises *adjustable* resistors. In many laboratory precision measurements, such as in bridge circuits (Secs. 23-7 and 23-14), a wide variety of different accurate values of resistance is needed. For this purpose, *decade resistance* boxes (**Fig. 3-14a**), which have dials that can be rotated to provide any desired value of resistance within limits of the maximum and minimum values obtainable, are used. Decade resistance boxes contain combinations of precision wire-wound resistors of relatively high accuracy ($\pm 0.1\%$). The nine-decade resistance box (Fig. 3-14a) provides resistance values from 1 kΩ to 999 GΩ in 1-kHz steps, with a precision of nine significant figures!

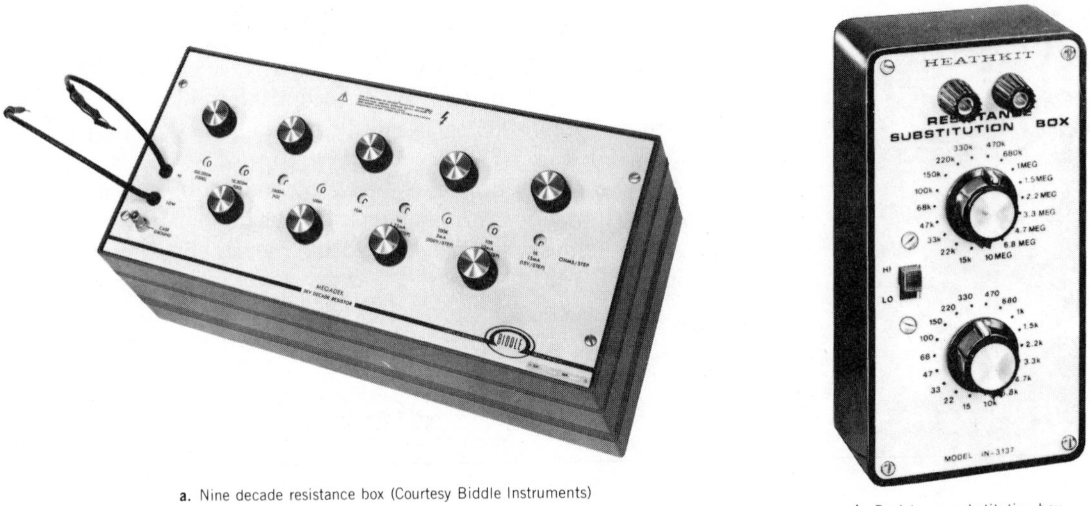

a. Nine decade resistance box (Courtesy Biddle Instruments)

b. Resistance substitution box (Courtesy Heath Co., St. Joseph, Michigan)

Figure 3-14 Adjustable resistors

Resistance *substitution* boxes (Fig. 3-14b) are not the same as decade boxes, nor do they serve the same purpose. They are intended to provide a wide range of commercially standard values of resistance for *substitution* in electronic or electrical circuits. Such substitution boxes provide commercial nominal values of resistance from 15 Ω to 10 MΩ, of approximately 1/2 to 1 W power rating, at a tolerance of $\pm 10\%$. Consequently, they may be used (within limits of their power rating) as electrical loads.

Resistance decade boxes, on the other hand, with their high precision and relatively low power, should *never* be used for electrical loading purposes.

3-8 THERMISTORS AND VARISTORS

Example 3-14 showed that a material having a *high* temperature coefficient of resistance (α_{20}) would produce a *high change* in its resistance with changes in temperature. As suggested in Ex. 3-14, such a material could be used in a temperature-indicating device. It was also shown (Sec. 3-6) that there is a class of *semiconductor* materials whose resistance (and resistivity, ρ) *decreases* with increases in temperature. Such materials are said to have a *negative* temperature coefficient (**NTC**).

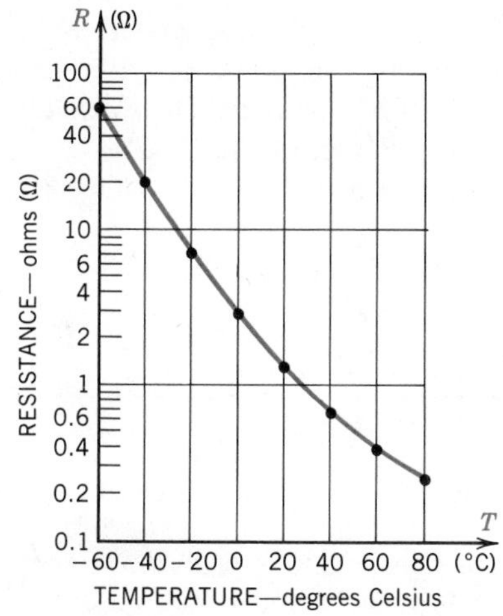

Figure 3-15 Resistance–temperature characteristic of a thermistor

Figure 3-15 shows the nonlinear resistance versus temperature characteristic of a typical NTC *thermistor*, a *therm*ally sensitive res*istor*.[19] The thermistor displays a much higher (negative) temperature coefficient than can be obtained with metals (like nickel) having a positive temperature coefficient. For this reason (and others as well) it is widely used in temperature control devices and electrical indicating thermometers.

Thermistors are made of mixtures of the oxides of manganese, nickel, cobalt, copper, and uranium, with a suitable binder in various proportions, depending on the desired characteristics. The mixture is molded in the shape of a rod, disk, or bead and suitable leads are attached. The mass is then fired to form a hard ceramic material on cooling.

Figure 3-16a shows the electrical symbol for a thermistor. Figure 3-16b shows the thermistor used in a simple circuit. At *low* temperatures, the thermistor resistance is high and therefore the current is *low*. At *high* temperatures, the thermistor resistance is low and the current is *high*. The ammeter, therefore, may be calibrated *directly* in degrees celsius to indicate the temperature.

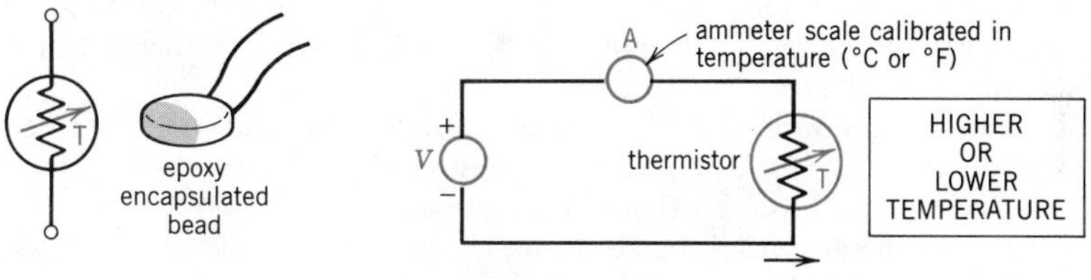

a. Symbol and appearance **b.** Circuit application

Figure 3-16 Symbol and circuitry for a thermistor-controlled electric thermometer

The *thermistor* is a semiconductor having a high sensitivity to **temperature**; that is, its resistance changes markedly with *temperature.*

The *varistor*, on the other hand, is a semiconductor having a high sensitivity to **voltage**; that is, its resistance changes markedly with *voltage.* The term varistor stems from the words *vari*able and res*istor.* Like the thermistor, the varistor is a *nonlinear* semiconductor resistor.

The volt-ampere characteristics of nonlinear resistors were first introduced in Sec. 3-3 to explain that even these resistors obey Ohm's law. Our purpose in briefly including them here is to classify them and show their similarities and differences.

All varistors are *voltage-sensitive* resistors. They can be broken down into two classes: symmetrical varistors and rectifier varistors.

Symmetrical varistors are two-terminal devices made of such materials as thyrite, metrosil, atmite, and silicon carbide. The term "symmetrical" implies that the device will operate in the same way if the two terminals are reversed or if alternating current (ac) is applied. The voltage–current characteristic of a symmetrical varistor was shown in Fig. 3-3b. The electrical symbols for a symmetrical varistor are shown in **Fig. 3-17a.**

Rectifier varistors are equally voltage-sensitive but are *nonsymmetrical*; that is, they tend to have a low resistance in one direction and a high resistance to voltage

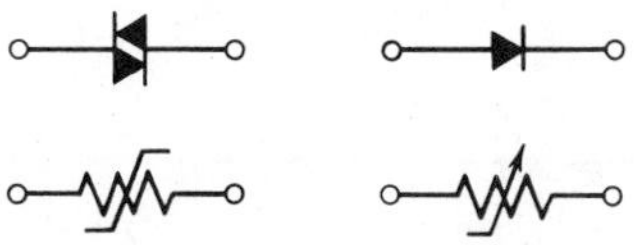

a. Symmetrical varistor **b.** Rectifier varistor

Figure 3-17 Varistor electrical symbol

[19] Note that the use of *semilogarithmic* paper (with the vertical scale logarithmic and the horizontal scale linear) serves two purposes: it makes the curve more linear, and it produces a much wider resistance range on the vertical scale, which shows a resistance change of as much as 600 times.

in the opposite direction. They are made of such semiconductive materials as copper oxide and/or *pn* junctions of selenium, silicon, or germanium. The voltage–current characteristic of a rectifier varistor was shown in Fig. 3-3c. The electrical symbols for a rectifier varistor (or rectifier diode) are shown in Fig. 3-17b.
Applications of these devices were discussed in Sec. 3-3.

3-9 GLOSSARY OF TERMS USED

Area, *A* Cross-sectional area of a material, expressed in square meters (m^2) in the SI or circular mils (c-mils) in the English system.

Conductor Material whose resistivity is between 10^{-6} and 10^{-8} Ω·m.

Circular-mil area, *A* (c-mil) Cross-sectional area expressed as the *diameter* of a round wire, in mils, *squared.*

Error Discrepancy between a measured quantity and the specified (true, standard, nominal, or accepted) value.

Inferred zero-resistance temperature, T_0 Temperature (below 0°C usually) at which the resistance of a given material is zero, obtained by extrapolating the linear portion of the material's resistance–temperature characteristic to a resistance of zero.

Insulator Material whose resistivity is between 10^{20} and 10^8 Ω·m.

Length, *l* Length of a material, expressed in meters (m) in the SI or feet (ft) in the English system.

Linear variation of resistance (with temperature) Portion of the resistance–temperature characteristic that is a straight line.

Negative temperature coefficient of resistance Decrease in resistance as the temperature increases.

Nonlinear resistance Volt-ampere characteristic that does not follow the equation for a straight line.

Nonlinear variation of resistance (with temperature) Range of temperatures over which the resistance–temperature characteristic does not follow the equation for a straight line.

Positive temperature coefficient of resistance Increase in resistance as the temperature increases.

Precision Quality of being exactly (sharply) defined and stated.

Relative error Ratio of the error divided by the specified value, expressed as a percent.

Resistance, *R* Opposition of a material to a steady (unvarying) current (dc) and ability to dissipate (convert) electrical energy to another form, measured in ohms (Ω).

Resistivity, ρ Specific resistance presented to an electrical source by a standardized sample of a material. In the SI that sample has an area of 1 m^2 and a length of 1 m; that is, its volume is 1 m^3.

Resistor color code System of standard colors adopted by the Electronic Industries Association (EIA) for marking and identifying resistance, tolerance, and reliability of resistors.

Semiconductor Material whose resistivity is between 10^7 and 10^{-6} Ω·m.

Temperature, T_{20} Temperature at which all resistivity measurements are made, usually 20°C or 293 K.

Temperature coefficient of resistance, α Measure of the change in resistance with change in temperature for a material, expressed as a ratio to the material's original resistance, measured in units of ohms per degree celsius per ohm.

Thermistor Nonlinear resistor having a high positive or negative temperature coefficient of resistance, that is, a temperature-sensitive resistor.

Tolerance Measure of how closely the color-coded value is approximated or measure of the precision to which the component is manufactured.

Varistor Nonlinear resistor, usually a semiconductor, having a high sensitivity to voltage, producing a nonlinear volt-ampere characteristic that may be symmetrical or nonsymmetrical.

3-10 PROBLEMS

Secs. 3-1 to 3-5

3-1 A resistance sample having a length of 20 cm and a cross-sectional area of 25 cm^2 has a measured resistance of 36 μΩ at 20°C. Calculate its resistivity (ρ) in

a. μΩ·cm
b. Ω·m
c. Ω·c-mil/ft

3-2 A nickel-chromium-cobalt alloy wire (ρ_{20} = 110 μΩ·cm) has a length of 400 mm and a *diameter* of 10 cm (not c-mil area). Calculate its resistance in micro-ohms at 20°C.

3-3 A particular copper wire having a diameter of 2.54 cm has a resistance of 20 Ω. Using the resistivity of this copper as $\rho = 1.75 \times 10^{-6}$ Ω·cm,

a. Calculate its length in kilometers (a theoretical value)
b. Explain why it must be so long

3-4 The resistivity of aluminum ingot is 2.83×10^{-8} Ω·m in SI units and 17.02 Ω·c-mil/ft in English units. Verify the conversion ratio given at the bottom of Appendix A-4 for ρ_{ENG}/ρ_{metric}, using the data given in this problem.

3-5 An aluminum cable has a resistance of 1.4 Ω and a length of 2 km. Calculate its cross-sectional area, *A*, in

a. cm^2
b. Circular mils (c-mil)

(Use resistivities given in Prob. 3-4.)

3-6 In basic SI units, the resistivities of carbon and constantan are 30 and 0.49 μΩ·m, respectively. Convert these resistivity values to

a. Ω·cm
b. Ω·c-mil/ft

3-7 A copper busbar has a rectangular cross section of 2 cm × 4 cm and a length of 1.5 m. Calculate its resistance at 20°C.

3-8 Calculate the resistance of an aluminum conductor having a diameter of 1 mm and a length of 1 mile.

3-9 A strip of copper 3 cm wide and 2 mm thick has a length of 6.562 ft. Calculate its resistance at 20°C.

3-10 Repeat Prob. 3-9 for a strip made of aluminum.

3-11 A particular busbar has a resistance of 20 mΩ at 20°C. Calculate the resistance of a bar of identical material having half the cross-sectional area and three times the length.

3-12 A conductor has a resistivity of 24.6 Ω·c-mil/ft, a length of 25 cm, and a cross-sectional area of 75 cm^2. Calculate its resistance at 20°C.

3-13 An unknown conductor 200 cm long has a diameter of 10 mm and a 20°C resistance of 100 Ω. A second conductor made of the same material has a length of 50 cm and a diameter of 100 mm. Calculate the

a. Resistance of the second conductor
b. Resistivity of the material in ohm-centimeters at 20°C

3-14 A round wire has a diameter of 0.25 in and a resistance of 5 mΩ. The resistivity of the wire is 25 μΩ·cm. Calculate in appropriate SI units

a. The length of the wire
b. The volume of the wire
c. The resistance of the wire if its volume remains the same but the wire is drawn out to half its initial diameter

3-15 The international resistivities established as standard for copper at 20°C are 1.7241×10^{-8} Ω·m in SI units and 10.371 Ω·c-mil/ft in English units. Using your electronic calculator, determine to five significant figures the conversion ratio

a. ρ_{SI}/ρ_{ENG}
b. ρ_{ENG}/ρ_{SI}

3-16 Using the ratios obtained in Prob. 3-15, and given the 20°C resistivity of tungsten wire as 33.2 Ω·c-mil/ft, calculate its resistivity in

a. μΩ·cm
b. Ω·m
c. Ω·cm
d. Check your answer against the value given in Appendix A-4 for tungsten

3-17 The best conductor known is silver. Its 20°C resistivity is 1.63 μΩ·cm. Using any method you choose, calculate its resistivity in English units and compare its value with that given in Appendix A-4.

3-18 An unknown conductor has a diameter of 1 mil (0.001 in), a length of 1 ft, and a 20°C resistance of 1 Ω. Calculate to five significant figures

a. Equivalent area in square meters
b. Equivalent length in meters
c. Resistivity in metric (SI) units
d. Resistivity in English units
e. Ratio of resistivities, ρ_{SI}/ρ_{ENG}
f. Compare the value obtained in (**e**) with that of Prob. 3-15a.

Sec. 3-6

3-19 A wire-wound resistor ($\alpha_{20} = 0.0005$) has a resistance of 100 Ω at 20°C. Calculate the resistance at 600°C.

3-20 The inferred zero resistance of a metal is found to be −550°C. Calculate

a. α_0
b. α_{20}

3-21 The iron grid using in an electric heater element has an inferred zero resistance of −180°C. If its cold (25°C) resistance is 12 Ω, calculate its resistance at 90°C.

3-22 An embedded temperature-sensing element in an electric kiln has a cold (22°C) resistance of 30 Ω and an inferred zero resistance of −258°C. If the kiln temperature rises to 200°C, calculate the expected elevated element resistance, assuming the element is functioning properly.

3-23 A convenient rule of thumb for quickly estimating the *change* in resistance (ΔR) with temperature is to express α_{20} (or α_0) as a percent change in resistance per degree celsius (±%/°C). Using α_{20} from Appendix A-4, calculate this factor for

a. Copper
b. Brass
c. Gold
d. Nickel
e. Which of the above has the greatest resistance change with temperature and which has the least?

3-24 The resistance of a sample of copper is 12.897 Ω at 100°C (in boiling H_2O) and 9.035 Ω at 0°C (in ice H_2O). Calculate from the given data

a. The resistance ratio R_{100}/R_0
b. α_0, the temperature coefficient referenced to 0°C
c. The percent change in resistance per degree celsius
d. α_{20} from the value obtained in (**b**)
e. Explain the discrepancy between your value in (**c**) and that obtained in Prob. 3-23a.

3-25 An aluminum transmission line has a resistance of 0.6 Ω at 55°C. Predict its resistance at

a. 25°C
b. −10°C

3-26 A silver contact has a resistance of 0.05 Ω at −25°C. At what temperature is its resistance 0.1 Ω?

3-27 Using the values of α_{20} given in Appendix A-4, calculate the inferred value of $-T_0$ at which the resistance of the following materials is zero:

a. Constantan
b. Nichrome
c. Gold
d. Which of these temperatures is theoretically impossible?

3-28 The cold resistance of the nichrome element in an electric toaster is 20 Ω measured at 20°C. If the toaster is rated at 120 V and 5 A, calculate the

a. Hot resistance of the heating element
b. Temperature of the element when hot

3-29 Two wire-wound resistors having the same nominal value of 5 kΩ at 20°C are made of constantan and nichrome wire, respectively. Calculate the

a. Resistance of each when the temperature rises to 50°C
b. Tolerance of each resistor over the 30°C range, using 5 kΩ as the standard value

3-30 The copper (stator) armature winding of an alternator is 0.1 Ω at an ambient temperature of 20°C. When the alternator operates for 24 h, the winding resistance increases to 0.12 Ω. Calculate the temperature of the winding after running for 24 h continuously.

3-31 Calculate the precise number of degrees celsius to which the following wires must be heated above an ambient temperature of 20°C in order to increase their resistance by 5% above their resistance at 20°C:

a. Copper c. Constantan
b. Nickel d. Nichrome

3-32 An unknown pure metal sample is to be identified by physical measurement without resorting to chemical analysis. Its resistance at 40°C is 4.27 Ω and its resistance at 20°C is 4 Ω. Using a 5-Ω ordinate scale of resistance,

a. Plot a linear temperature scale as abscissa, and find the temperature $-T_0$ at which the resistance is zero.
b. Verify (**a**) using Eq. (3-3) by solving for $-T_0$, algebraically.
c. Calculate α_{20} for the unknown metal, using the data given originally and Eq. (3-4).
d. What is the possible composition of the metal? Is there more than one possibility?
e. Of all possible metals listed in Appendix A-4, this one has the *most linear* resistance–temperature characteristic at the low-temperature end. Explain.

3-33 The sensing elements used in resistance–temperature detectors (RTDs) are selected from the materials having the most linear variation from 0 to 100°C. Three common materials meeting this criterion, as shown in the table below, are platinum (Pt), nickel (Ni), and nickel–iron alloy (Ni–Fe). Given the information provided in the table, complete *all* missing cells.

Material	α_0 (Ω/°C/Ω)	Resistance Ratio, R_{100}/R_0	Change in Resistance (%)
a. Ni–Fe		1.518	
b. Ni			67.2
c. Pt	0.00385		

3-34 Using the table completed in Prob. 3-33, for each material in the table calculate

a. Percent change in resistance per degree celsius
b. Change in resistance in parts per million (ppm)
c. Which of the three materials is most sensitive to a change in resistance with temperature change?

3-35 A nickel resistance temperature detector (RTD) has a resistance of 512.00 Ω as measured by a Wheatstone bridge at 20°C to five significant figures. Calculate

a. Its resistance at 30°C
b. The temperature when the resistance is 520.00 Ω
c. The smallest temperature rise that can be detected, that is, when the resistance rises to 512.01 Ω
d. Draw conclusions regarding the use of nickel as a resistance–temperature detector.

Secs. 3-7 to 3-8

3-36 Write the resistance value, tolerance, and reliability of the following color-coded resistors:

	Band					Resistance (Ω, kΩ, MΩ)	Tolerance ($\pm$%)	Reliability (%/kh)
	1	2	3	4	5			
a.	brown	violet	yellow	gold	brown			
b.	red	blue	orange	silver	red			
c.	red	violet	blue	no band	no band			
d.	green	blue	green	silver	orange			
e.	brown	black	silver	silver	yellow			
f.	blue	gray	black	red	red			
g.	brown	blue	yellow	gold	yellow			
h.	brown	brown	blue	gold	red			
i.	gray	red	green	silver	no band			
j.	black	red	silver	brown	brown			

3-37 From the resistors coded in Prob. 3-36, select those that are *not* standard commercial values.

3-38 Color code the following resistors to four bands:

a. 3.9 MΩ, $\pm 10\%$
b. 0.51 Ω, $\pm 5\%$
c. 750 kΩ, $\pm 5\%$
d. 270 Ω, $\pm 10\%$
e. 4.3 Ω, $\pm 2\%$
f. 910 Ω, $\pm 1\%$
g. 22 MΩ, $\pm 20\%$
h. 56 kΩ, $\pm 5\%$
i. 160 kΩ, $\pm 5\%$

3-39 A designer calculates the value of a transistor emitter resistor as 3.43 kΩ at $\pm 10\%$. In selecting commercially available resistors, determine

a. Which 3.3-kΩ resistor should be specified
b. Percent error between the lower limit of the resistor specified in **(a)** and the desired value given, as calculated by the designer

3-40 A transistor collector load resistor carries a current of 40 mA continuously with a drop of 50 V across it in a circuit. Calculate the

a. Power dissipated by the resistor
b. Resistor power rating if a derating factor of 1/2 is used to reduce the temperature of the printed circuit board (PCB) on which this resistor is mounted
c. Resistance of the resistor and its commercial power rating
d. If all resistors on the PCB are held to $\pm 10\%$, using a commercial standard resistor give the complete specification: commercial resistance, tolerance, and power rating of the resistor to be used
e. Specify the color code for **(d)** and the power rating

3-41 Four resistors, each 5 kΩ $\pm$ 10%, are selected from a stock of assorted resistors. Resistor A has a power rating of 0.25 W, B is rated at 0.5 W, C is rated at 1 W, and D is rated at 2 W. All of the resistors are to be derated by a factor of 1/3 because of ambient temperature limitations in this particular application ($P_r = 3P_d$). For each resistor above, calculate the

a. Maximum voltage that can be connected across each
b. Maximum current that can flow in each
c. Current and power when 15 V is connected across each separately
d. Is the power rating of any of the resistors exceeded under the conditions of **(c)**? If not, why not?

3-42 In describing Table 3-3, the statement was made that for all tolerances given, there is always an overlap between the upper limit of one resistor and the lower limit of the next higher resistance in the table. Prove that this statement is correct, using the upper limit of a 33-kΩ resistor and its next higher resistance value given in Table 3-3 at tolerances of

a. $\pm 5\%$
b. $\pm 10\%$
c. $\pm 20\%$

Show all calculations.

3-43 A designer calculates a required resistance value as 25 kΩ. Calculate the closest commercial resistance value to be selected if the tolerance of all the resistors used in equipment is to be $\pm$

a. 5%
b. 10%
c. 20%

3-44 For each of the resistors selected in Prob. 3-43,

a. Calculate the relative error between the desired (calculated) value and the nominal commercial value selected.
b. Is the percent error of each resistor in **(a)** within the tolerance band to be used in each case, using the *nominal* standard commercial value?
c. Is there any possibility that an extreme upper or lower limit may exceed the allowable tolerance of error given in Prob. 3-43? Repeat **(b)**, selecting appropriate limits of nominal standard commercial values and calculate relative errors to prove your point.[20]

[20] This problem shows that in selecting nominal commercial values that are closest to a computed (desired) value, the tolerance of the commercial resistor selected may have to be lower (better) than the allowable error of the rest of the equipment to ensure the desired overall tolerance.

3-45 The following resistors are to be used at an ambient temperature of 100°C in an enclosed piece of electronic equipment mounted near a blast furnace. In each case, calculate the power dissipated and the commercial power rating and specify the commercial standard resistor, assuming the tolerance of all components is held to ±5%.

a. A 5-kΩ resistor across a 20-V supply
b. A 2-kΩ resistor carrying a current of 5 mA
c. A resistor carrying 20 mA with 30 V across it

3-46 Three other resistors enclosed in the equipment in Prob. 3-45 have the following nominal values: 22 kΩ, 5 W; 1 MΩ, 2 W; and 5.6 kΩ, 1 W. Calculate the

a. Maximum power each could dissipate in an electric circuit and still not overheat
b. Maximum voltage that could be connected across each to stay within the maximum power specified in (**a**)
c. Maximum current each may carry safely

3-47 Complete all blank cells in the table below. Show all calculations on a separate sheet for the following resistors, whose four bands are coded:

Color Code on a Four-Band Resistor	Nominal Resistance (Ω, kΩ, or MΩ)	Tolerance (±%)	Resistance Range: Minimum Value (Ω, kΩ, MΩ)	Resistance Range: Maximum Value (Ω, kΩ, MΩ)
a. RED-RED-RED-GOLD				
b. BROWN-RED-ORANGE-SILVER				
c. YELLOW-VIOLET-BLUE-NONE				
d. BLUE-GRAY-GREEN-GOLD				
e. BROWN-GRAY-WHITE-SILVER				
f. BROWN-BLACK-SILVER-BROWN				
g. VIOLET-GREEN-GOLD-RED				
h. RED-VIOLET-ORANGE-SILVER				
i. RED-YELLOW-WHITE-BROWN				
j. ORANGE-WHITE-WHITE-RED				

3-48 In column A, show the actual power dissipated, P_d. In column B, show the *commercial* power rating, P_r, using a derating factor of 0.4 for all resistors for each of the following conditions:

a. A 1-kΩ resistor across a 12-V supply
b. A 20-kΩ resistor carrying 3.2 mA
c. A resistor carrying a current of 2 mA when 25 V is across its terminals

	Column A P_d (W)	Column B P_r (W)
a.		
b.		
c.		

3-49 A 2-W resistor is color-coded RED-RED-RED-GOLD. It is used in equipment subjected to a temperature of *over* 100°C and derated by a factor of 1/3. Calculate the

a. Maximum power it should dissipate
b. Maximum voltage that can be connected across it
c. Maximum current it can safely carry without overheating
d. Repeat (**b**) and (**c**), using the *highest* possible resistance value based on the given tolerance
e. Repeat (**b**) and (**c**), using the *lowest* possible resistance value based on the given tolerance
f. Calculate the relative error between the currents in (**c**) and (**e**), using the nominal value current obtained in (**c**) as the standard reference value
g. Explain why the value obtained in (**f**) is not 5%, but a value somewhat lower

3-50 A resistor color-coded YELLOW-VIOLET-BROWN-BROWN is carrying a current of 40 mA in a confined enclosure. Using a derating factor of 0.4, calculate the required *commercial* power rating of the resistor.

3-51 Another resistor in the enclosure in Prob. 3-50 is coded BROWN-GREEN-YELLOW-RED and has a potential of 300 V across it. Calculate the required commercial power rating of the resistor.

3-11 ANSWERS

3-1 a 45 $\mu\Omega\cdot$cm b 450 n$\Omega\cdot$m c 270.7 $\Omega\cdot$c-mil/ft
3-2 56.02 $\mu\Omega$
3-3 579.1 km
3-4 $6.014 \times 10^8 \dfrac{\Omega\cdot\text{c-mil/ft}}{\Omega\cdot\text{m}}$
3-5 a 0.4043 cm^2 b 79 775 c-mils
3-6 a 3000 $\mu\Omega\cdot$cm; 49 $\mu\Omega\cdot$cm b 18.046 $\Omega\cdot$c-mil/ft; 294.75 $\Omega\cdot$c-mil/ft
3-7 32.81 $\mu\Omega$
3-8 57.98 Ω
3-9 0.5747 mΩ
3-10 943.3 $\mu\Omega$
3-11 120 mΩ
3-12 1.363 $\mu\Omega$
3-13 a 0.25 Ω b 0.3927 $\Omega\cdot$cm
3-14 a 63.34 cm b 20.06 cm^3 c 80 mΩ
3-15 a $1.662 \times 10^{-9} \dfrac{\Omega\cdot\text{m}}{\Omega\cdot\text{c-mil/ft}}$ b $6.0153 \times 10^8 \dfrac{\Omega\cdot\text{c-mil/ft}}{\Omega\cdot\text{m}}$
3-16 a 5.518 $\mu\Omega\cdot$cm b 55.18 n$\Omega\cdot$m c 5.518 $\mu\Omega\cdot$cm
d 55.2 n$\Omega\cdot$m
3-17 9.805 $\Omega\cdot$c-mil/ft
3-18 a 5.067×10^{-10} m^2 b 0.3048 m c 1.6624 n$\Omega\cdot$m
d 1 $\Omega\cdot$c-mil/ft e $1.6624 \dfrac{\text{n}\Omega\cdot\text{m}}{\Omega\cdot\text{c-mil/ft}}$ f Same
3-19 129 Ω
3-20 a 0.0018 b 0.001 754
3-21 15.8 Ω
3-22 49.07 Ω
3-23 a 0.393%/°C b 0.21%/°C c 0.34%/°C
d 0.537%/°C e brass; nickel
3-24 a 1.427 45 b 0.004 274 5 Ω/°C/Ω c 0.427 45%/°C
d 0.003 937
3-25 a 0.5381 Ω b 0.4659 Ω
3-26 193.8°C
3-27 a −124 080°C b −2252°C c −274.1°C
d All three
3-28 a 24 Ω b 474.4°C
3-29 a 5001.2 Ω; 5066 Ω b 0.024%; 1.32%
3-30 70.89°C
3-31 a 12.73°C b 9.31°C c 6250°C d 113.64°C
3-32 a −275°C b −276.3°C c 0.003 75
d silver or zinc
3-33 Ni–Fe: 0.005 18; 1.518; 51.8%
Ni: 0.006 72; 1.672; 67.2%
Pt: 0.003 85; 1.385; 38.5%
3-34 a 0.518%; 0.672%; 0.385%
b 5180 ppm; 6720 ppm; 3850 ppm
c nickel
3-35 a 546.41 Ω b 22.325°C c 0.002 91°C
3-36 a 170 kΩ; 5%; 1% e 0.1 Ω; 10%; 0.001%
j 0.02 Ω; 1%; 1%/kh
3-37 a 170 kΩ, 5% b 26 kΩ, 10% c 27 MΩ, 20%
3-38 e yellow-orange-gold-red i brown-blue-yellow-gold
3-39 a 5% b 8.6%
3-40 a 2 W b 4 W c 1250 Ω d 1200 $\Omega \pm$ 5%; 4 W
e brown-red-red-gold, 4 W
3-41 a 20.41, 28.58, 40.82, 57.74 V
b 4.08, 5.72, 8.16, 11.55 mA c 3 mA, 45 mW d No
3-43 a 24 k$\Omega \pm$ 5% b 27 k$\Omega \pm$ 10% c 22 k$\Omega \pm$ 20%
3-44 a 4%, 8%, 12% b Yes c 8.8%, 18.8%, 29.6%
3-45 a 5.1 k$\Omega \pm$ 5%, 0.25 W b 2 k$\Omega \pm$ 5%, 1/8 W
c 1.5 k$\Omega \pm$ 5%, 2 W
3-46 a 2.5 W, 1 W, 0.5 W b 234.5 V, 1000 V, 52.9 V
c 10.66 mA, 1 mA, 9.45 mA
3-48 a 0.5 W b 1 W c 1/8 W
3-49 a 0.66 W b 38.3 V c 17.4 mA
d 39.24 V, 17 mA e 37.33 V, 17.86 mA f 2.64%
3-50 2 W
3-51 2 W

CHAPTER 4

Work, Energy and Power Relations

4-1 INTRODUCTION

From our study of physics we know that work is done only in overcoming an *opposing* force and producing *motion* as a result. Three prerequisites are needed, namely (a) an *applied force* to overcome the *resistance* of the opposing force acting on the body; (b) *displacement* of a body as a result of the applied force, and (c) a *component* of the applied force acting along the direction of the displacement. If the opposing force is measured in newtons and the distance the body moves is measured in meters, then the work done is the product of the force times the distance measured in joules (i.e., 1 joule ≡ 1 newton-meter). This concept provides an insight into electrical circuits in the following way.

Work (and energy) was first introduced in defining potential difference in Chapter 2. We stated (Sec. 2-4) that when a potential is applied to a given circuit, work is done in overcoming the resistance of that circuit. In moving charges, Q, against the field force, work is done. The potential V was defined as a measure of the work per unit charge (in joules per coulomb) expressed in volts (i.e., 1 volt ≡ 1 joule/coulomb).

Work W is done, therefore, in an electric circuit in moving charges Q against the resistance R of the circuit.

In discussing sources of dc (Sec. 2-12) we noted that *electrical* energy may be converted from *other* forms of energy (*mechanical*, *chemical*, *heat*, and *light*) and that energy can be neither created nor destroyed. Such energy may be *potential* energy (by virtue of its condition or position) or *kinetic* (by virtue of its motion or work that is done).

In Chapter 3 we first introduced the term *power* (Sec. 3-7.2) in connection with power dissipated by resistors (Eq. 3-8) when voltage is applied across their terminals and current results. While we used the term power, however, we did not relate it to energy.

The purpose of this chapter is to see the *relation* between *energy* and *power* and to quantify them both in terms of voltage, current, and resistance. Only in this way can we bring together all the concepts we have learned.

4-2 POWER

Power is defined as the time *rate* at which *work* is being done or energy is being expended. Since power is the rate of doing work and a rate is any physical quantity divided by time, we may define the *average* power, P, as

$$P = \frac{W}{t} = \frac{\text{total work done (in joules)}}{\text{total time taken (in seconds)}} \quad \text{in joules per second (J/s) or watts (W)} \qquad (4\text{-}1)$$

With respect to the above, please note that

1. Power is a *scalar* quantity (since time and work are *both* scalar).
2. $1\ \text{W} \equiv 1\ \text{J/s} \equiv 1\ \text{N}\cdot\text{m/s} \equiv 1\ \text{kg}\cdot\text{m}^2/\text{s}^3$ (by definition; they are the *same*).
3. Other power units based on the above definition are ft·lb/s, Btu/s, g·cm/s, erg/s, kg·cal/s, oz·in/s, and so forth (see Appendix A-3I for power conversion factors).

Units of power (electrical, mechanical, and heat) for the SI and the English system are given in Table 4-1. The *basic unit* is shown with its *unit abbreviation* in parentheses. As noted in item 2 above, all units in the SI in Table 4-1 are the *same*, by definition.

Table 4-1 Power Units in the SI and English System

	Electrical Power	Mechanical Power	Heat Power
SI	watt (W)	newton-meter/second (N·m/s)	joule/second (J/s)
English	kilowatt (kW)	horsepower (hp) foot-pound/minute (ft·lb/min)	Btu/minute (Btu/min)

In defining potential difference we learned that $V = W/Q$ (Eqs. 2-6 and 2-7) and in defining current $I = Q/t$ (Eq. 2-8). Substituting $Q = It$ and $W = QV$ in Eq. (4-1) yields

$$P = \frac{W}{t} = \frac{QV}{t} = \frac{ItV}{t} = VI \quad \text{watts (W)} \qquad (4\text{-}2)$$

If we now substitute Ohm's law (Eq. 3-1), yielding $V = IR$ and $I = V/R$, respectively, in Eq. (4-2) we obtain

$$P = \frac{W}{t} = VI = \frac{V^2}{R} = I^2R \quad \text{watts (W)} \qquad (4\text{-}3)$$

where all terms have been defined in the equations cited above.

EXAMPLE 4-1

The hot resistance of a *lamp* filament is 144 Ω and its voltage rating is 120 V as printed on the *bulb*. Calculate the

a. Current drawn by the lamp
b. Power consumed by the lamp, using $P = VI$
c. Power consumed by the lamp, using $P = V^2/R$
d. Power consumed by the lamp, using $P = I^2R$
e. Heat power in Btu/min if all the consumed electrical power is converted to heat (see Appendix A-3I)
f. Heat power in units of kg·cal/min
g. Mechanical equivalent of the heat power in ft·lb/min

Solution

a. $I = V/R = 120\ \text{V}/144\ \Omega = \mathbf{0.8\overline{3}\ A}$
b. $P = VI = 120\ \text{V} \times 0.8\overline{3}\ \text{A} = \mathbf{100\ W}$ (4-2)
c. $P = V^2/R = (120)^2/144 = \mathbf{100\ W}$ (4-3)
d. $P = I^2R = (0.8\overline{3})^2 \times 144 = \mathbf{100\ W}$ (4-3)

e. $P_h = 100 \text{ W} \times \dfrac{0.0569 \text{ Btu/min}}{1 \text{ W}} = \mathbf{5.69 \text{ Btu/min}}$

f. $P_h = 5.69 \dfrac{\text{Btu}}{\text{min}} \times 0.252 \dfrac{\text{kg}\cdot\text{cal}}{\text{Btu}} = \mathbf{1.434 \text{ kg}\cdot\text{cal/min}}$

g. $P_m = 100 \text{ W} \times \dfrac{1 \text{ ft}\cdot\text{lb/min}}{2.260 \times 10^{-2} \text{ W}} = \mathbf{4425 \text{ ft}\cdot\text{lb/min}}$

This example assumed 100% conversion efficiency between electrical power, heat power, and mechanical power. As we shall see, this is impossible, even theoretically.

4-3 EFFICIENCY

According to the law of conservation of energy, the energy (work) expended (in "forcing" charges to flow in an electric circuit) must emerge in some other form(s). In Ex. 4-1, the principal forms are heat and light energy when the filament is connected across the 120-V supply. But the primary purpose of the lamp is essentially to produce light energy and *not* to heat its surrounding area. Quantifying the law of conservation of energy, we may write

$$W_{in} = W_{out} + W_{losses}$$

where W_{in} is the total energy consumed by a system or device in joules
W_{out} is the useful energy produced by the system or device
W_{losses} are the other forms of energy dissipated (or stored) by the system

Substituting $W = P \times t$ from Eq. (4-1) in this relation, we may write $P_{in} \times t = P_{ou} \times t + P_{losses} \times t$, and dividing each term by t, the *same* time, yields

$$P_{in} = P_{out} + P_{losses} \qquad \text{watts (W)} \qquad (4\text{-}4)$$

In the case of the lamp of Ex. 4-1, we would be interested in knowing the ratio of the power equivalent of light produced by the lamp to the power drawn from the electrical source by the lamp, or P_{out}/P_{in}. This ratio would provide us with some measure of the *efficiency* of the lamp as a light conversion *transducer*[1] of electrical energy to light energy. We may write this efficiency, η, as

$$\eta = \frac{\text{useful power output}}{\text{total power input}} = \frac{P_{out}}{P_{in}} = \frac{P_{out}}{P_{out} + P_{losses}} = \frac{P_{in} - P_{losses}}{P_{in}} \qquad (4\text{-}5)$$

Regarding Eq. (4-5) we may conclude that

1. The efficiency ratio, η, is *always less than unity*. Every purposeful energy conversion device dissipates (or stores) energy in other (undesirable) forms, which results in some loss of energy (or power).

---

[1] A transducer may be defined as a device that receives energy from one system or source and transmits that energy, often in a different form, to another system. A microphone, for example, is a transducer that converts air pressure variations (mechanical energy in the form of sound) into variations of electric current (electrical energy).

2. Since power is *usually easier* to measure than energy, the efficiency ratio is usually stated as a power ratio rather than an energy ratio.
3. The greater the losses in any energy conversion device, the lower its conversion efficiency, η.

If we define a machine as any device that does useful *work*, we may rewrite Eq. (4-5) in terms of the ratio of useful output work to input energy:

$$\eta = \frac{\text{useful output work}}{\text{total input energy}} = \frac{W_{out}}{W_{in}} = \frac{W_{out}}{W_{out} + W_{losses}} = \frac{W_{in} - W_{losses}}{W_{in}} \quad (4\text{-}6)$$

Regarding Eq. (4-6) we may conclude that

1. Equation (4-6) is derived from Eq. (4-5) by merely multiplying both numerator and denominator of Eq. (4-5) by the *same* time, *t*.
2. For this reason, the efficiency, η, stated as an energy ratio is always identical to the power ratio.
3. Although the *ratios* of power and energy are the *same*, we cannot convert energy into power and power into energy. Power and energy may be related by Eq. (4-1), *but they are not equal.* (See ratios in Appendixes A-3H and A-3I.)
4. Since $W = P \cdot t$ and $P = W/t$, the *conversion* ratios for units of energy and power, respectively, use the *same* constants. For example, a useful energy conversion factor is 4.186 J/g·cal and a useful power conversion factor is 4.186 W/g·cal/s. The former should be used for energy conversion and the latter for power conversion. They are *not* interchangeable, although they use the *same* constant!
5. Consequently, always be mindful of the distinction between power and energy. Although related by Eq. (4-1), they are two separate and distinct physical quantities.

EXAMPLE 4-2

If 20 N of force is required to overcome the resistance of a mass and move it in the direction of the force a distance of 500 cm, calculate the work done in a. newton-meters b. joules c. foot-pounds d. watt-seconds e. kilowatt-hours.

Solution

a. $W = f \times d = 20\text{ N} \times 500\text{ cm} \times \frac{1\text{ m}}{10^2\text{ cm}} = \mathbf{100\ N \cdot m}$

b. $W = 100\text{ N}\cdot\text{m} \times \frac{1\text{ J}}{1\text{ N}\cdot\text{m}} = \mathbf{100\ J}$

c. $W = 100\text{ J} \times \frac{1\text{ ft}\cdot\text{lb}}{1.356\text{ J}} = \mathbf{73.75\ ft \cdot lb}$

d. $W = 100\text{ J} \times \frac{1\text{ W}\cdot\text{s}}{1\text{ J}} = \mathbf{100\ W \cdot s}$

e. $W = 100\text{ W}\cdot\text{s} \times \frac{1\text{ kW}}{10^3\text{ W}} \times \frac{1\text{ h}}{3600\text{ s}} = \mathbf{2.78 \times 10^{-5}\ kWh}$

EXAMPLE 4-3

If the work done in Ex. 4-2 is accomplished in 1 min, calculate the power required to move the mass in a. newton-meters/second b. joules/second c. watts d. horsepower and e. Btu/min if all the mechanical energy is converted to heat energy, that is, $\eta = 1$.

Solution

a. $P = \frac{W}{t} = \frac{100\text{ N}\cdot\text{m}}{1\text{ min} \times 60\text{ s/min}} = \mathbf{1.\bar{6}\ N \cdot m/s}$ (4-3)

b. $P = 1.\bar{6}\,\frac{\text{N}\cdot\text{m}}{\text{s}} \times \frac{1\text{ J/s}}{1\text{ N}\cdot\text{m/s}} = \mathbf{1.\bar{6}\ J/s}$

c. $P = 1.\bar{6}\,\frac{\text{N}\cdot\text{m}}{\text{s}} \times \frac{1\text{ W}}{1\text{ N}\cdot\text{m/s}} = \mathbf{1.\bar{6}\ W}$

d. $P = 1.\bar{6}\text{ W} \times \frac{1\text{ hp}}{746\text{ W}} = \mathbf{2.234 \times 10^{-3}\ hp}$

e. $P = 1.\bar{6}\text{ W} \times \frac{0.056\,88\text{ Btu/min}}{1\text{ W}} = \mathbf{9.48 \times 10^{-2}\ Btu/min}$

EXAMPLE 4-4

An electric motor connected to a 120-V dc supply draws 10 A and develops an output power of 1.35 hp. Calculate

a. Input power, P_i, in watts
b. Output power, P_o, in watts
c. Motor efficiency, η
d. Motor losses, P_l, in watts
e. Btu/min dissipated by the motor if all the losses are converted into heat power
f. Heat losses in kg·cal/min

Solution

a. $P_i = VI = 120 \text{ V} \times 10 \text{ A} = \mathbf{1200 \text{ W}}$ (4-2)

b. $P_o = 1.35 \text{ hp} \times \dfrac{746 \text{ W}}{1 \text{ hp}} = \mathbf{1007 \text{ W}}$

c. $\eta = \dfrac{P_o}{P_i} = \dfrac{1007 \text{ W}}{1200 \text{ W}} = \mathbf{0.839}$ or $\mathbf{83.9\%}$ (4-5)

d. $P_l = P_i - P_o = (1200 - 1007) \text{ W} = \mathbf{193 \text{ W}}$

e. $P_l = 193 \text{ W} \times \dfrac{0.0569 \text{ Btu/min}}{1 \text{ W}} = \mathbf{10.98 \text{ Btu/min}}$

f. $P_l = 10.98 \text{ Btu/min} \times \dfrac{0.252 \text{ kg·cal/min}}{1 \text{ Btu/min}} = \mathbf{2.767 \text{ kg·cal/min}}$

It is clear that every time energy is fed into a transducer (an energy-transmitting or energy-converting device), it is accompanied by an energy loss in inverse proportion to the transducer efficiency; that is, the higher the transducer efficiency, the smaller the loss. If transducers are *cascaded* (i.e., the output of one is fed as an input to the next), there is a reduction in power each time a *transduction of energy* occurs.

Consider the system shown in **Fig. 4-1**. An aluminum refining plant located 500 km from the generating station requires large quantities of dc for electrolytic processing. The generating plant consists of a fuel-fed boiler that produces steam at high pressures to drive a steam turbine, which in turn drives a three-phase alternator. The alternator output is fed to a three-phase transformer, which steps up voltage for long-distance transmission of electrical energy via transmission lines. At the transformer substation, the high voltage is then transformed to a lower voltage and fed along local distribution lines to a large 3-phase motor which drives a large dc generator to produce the required dc power for chemical refining of aluminum. How many transductions are involved and how can we calculate the overall efficiency of the system?

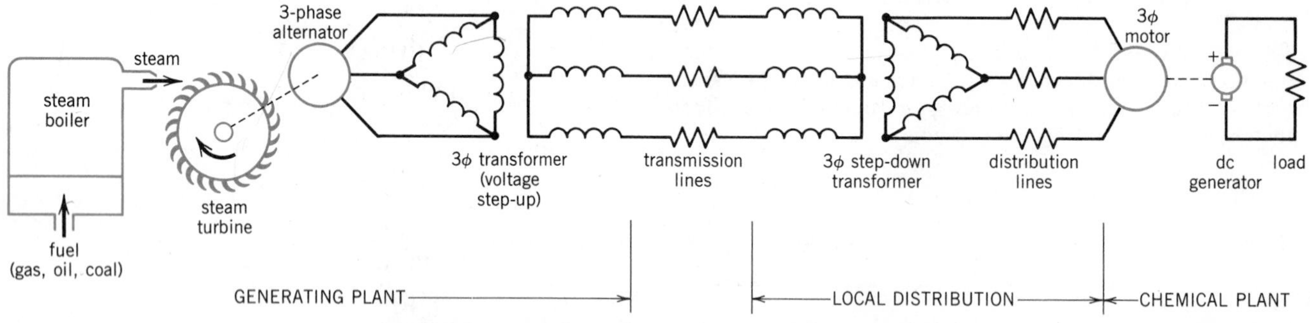

Figure 4-1 Electrical generation, transmission, and distribution system for producing large quantities of dc power

From **Fig. 4-2**, we may draw the following conclusions regarding Fig. 4-1:

1. Nine separate energy transductions are involved.
2. Each transducer operates at its own particular efficiency.
3. The output of one transducer represents the input of the next, that is, $P_{o_1} = P_{i_2}$.
4. The efficiency of each transducer may be represented as $\eta_1 = P_{o_1}/P_{i_1}$, $\eta_2 = P_{o_2}/P_{i_2}$, and so forth.

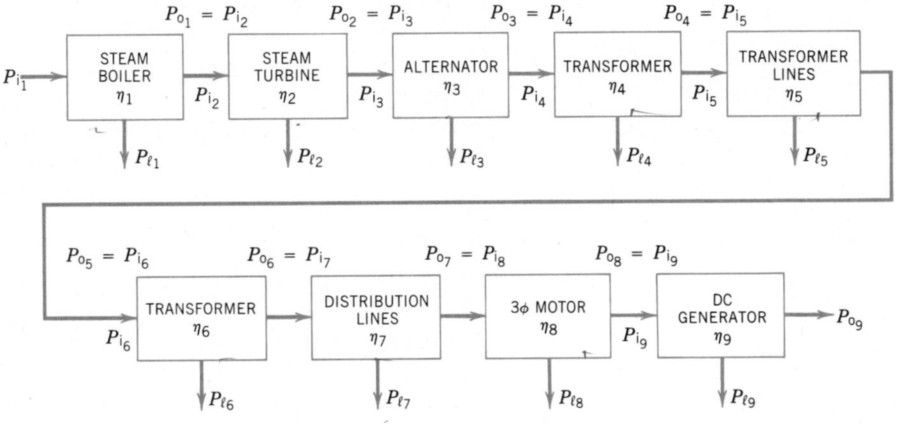

Figure 4-2 Power flow diagram for Fig. 4-1

5. Each transducer, in the process of performing energy conversion, produces P_l, a power loss, which (usually) is not recoverable.
6. From (3) above, we may also write $P_{i_1}\eta_1 = P_{i_2}$, $P_{i_2}\eta_2 = P_{i_3}$, $P_{i_3}\eta_3 = P_{i_4}$, and so forth.
7. Then from (6) we can write by substitution $P_{i_1}\eta_1 \cdot \eta_2 \cdot \eta_3 = P_{i_4}$, and extending this concept to *all* the transducers in Fig. 4-2 we obtain the *overall* system efficiency, η_{so}, for the cascaded transducers as

$$\eta_{so} = \eta_1\eta_2\eta_3\eta_4\eta_5\eta_6\eta_7\eta_8\eta_9 \quad \text{(etc.)} \tag{4-7}$$

A little thought about the nature of Eq. (4-7) reveals the following insights:

1. The *more* transducers cascaded (in series), the *lower* the overall efficiency.
2. If one of the transducer systems has a considerably lower efficiency than the others, it reduces the overall system efficiency tremendously, as shown by the following examples.

EXAMPLE 4-5

For simplicity, assuming that each of the nine transducers in Fig. 4-2 has an efficiency of 0.7, calculate the overall system efficiency.

Solution

$\eta_{so} = (0.7)^9 = \mathbf{0.04} = \mathbf{4\%}$ overall efficiency (4-7)

Example 4-5 proves that the efficiency is considerably lowered because of the large number of cascaded transducers.

EXAMPLE 4-6

Assume that one of the transducers in Ex. 4-5 has an efficiency of 0.25 and each of the others has an efficiency of 0.7. Calculate the overall system efficiency.

Solution

$\eta_{so} = (0.7)^8 \times 0.25 = \mathbf{0.0144} = \mathbf{1.44\%}$ overall efficiency

Key Sequence

Press	Display
.7 [y^x] 8 [×]	5.7648×10^{-2}
.25 [=]	1.44×10^{-2}

Example 4-6 shows how badly *one low* efficiency in the cascaded *chain* can degrade the overall system efficiency. "A chain is only as strong as its weakest link."

4-4 ELECTRICAL ENERGY

Contrary to popular opinion, an electrical utility bills its customers for electrical *energy* consumed each month and *not for electrical power*. There is good sense in this. The power drawn from the utility may vary from zero to extremely large values of instantaneous power. In the latter case, this may occur when an accidental short circuit draws brief but relatively high currents (prior to the operation of a fuse or circuit breaker to clear the circuit from the supply).

Since electrical energy is the product of power, P, and time, t, it is measured in basic units of watt-seconds ($1\ \text{J} \equiv 1\ \text{W}\cdot\text{s}$). In the case of an electrical utility, the cost of electrical energy is billed in terms of kilowatt-hours (kWh) consumed. The consumption of electrical energy (in kWh) is recorded by a **watthour meter (Fig. 4-3a)**, which is installed in every residence (called "occupancy") or in every industrial and commercial establishment, at a point where the utility wires (the service) are brought into the occupancy. The "meter" is basically an *induction motor* that rotates whenever the current drawn by the consumer flows through coils on the motor stator, producing a magnetic field. The speed of motor rotation is directly proportional to the total current drawn by the consumer (the sum of all currents drawn by appliances, lights, etc., switched on in the residence). The motor shaft drives through a gear train, which in turn drives the dials of the watthour meter. Each pointer of each dial is connected by a 10:1 gear to the next dial, and so on. When the dial is read at monthly intervals, the difference between the present and previous reading of the dials represents the energy in kilowatt-hours consumed that month. Also located on the shaft is a visible aluminum disk that moves through an adjustable permanent magnet, for purposes of calibration and to provide visual indication of meter operation. Figure 4-3b shows a typical watthour meter dial. The reading is obtained by (always) taking the *lower* dial number between pointer settings. For example, in Fig. 4-3b the dial reading is 17 234 kWh.

a. Commercial single-phase kilowatt hour meter (Courtesy General Electric Co.)

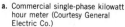

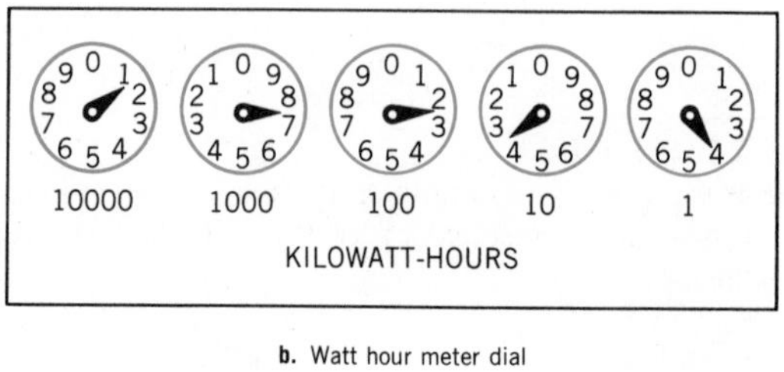

b. Watt hour meter dial

Figure 4-3 Commercial single-phase kilowatt-hour meter and typical dial indications

---

EXAMPLE 4-7

a. Calculate the total energy in kilowatt-hours consumed by running the following household appliances in a 24-h period in a home:
   1. A 2.5-kW air conditioner for 8 h
   2. A 350-W television receiver for 5 h
   3. An electric clock consuming 3 W for 24 h
   4. Six 50-W lamps for 6 h
   5. A 3-kW electric broiler for 30 min
   6. A 600-W clothes washer for 40 min
   7. A 5-kW electric clothes drier for 30 min

b. Calculate the total cost per day for running these appliances if the utility costs are \$0.075/kWh

Solution

a. $W_t = 2.5\ \text{kW} \times 8\ \text{h} + 0.35\ \text{kW} \times 5\ \text{h} + 0.003 \times 24 + 0.3 \times 6 + 3 \times 0.5 + 0.6 \times 0.\bar{6} + 5 \times 0.5 = \mathbf{28.022\ kWh}$

b. Total cost/day = 28.022 kWh × \$0.075/kWh = **\$2.10**

4-5 FUSES AND CIRCUIT BREAKERS

A *fuse* is an overcurrent protective device designed to open (or clear) a circuit whenever excessive current is drawn. The excessive current may be an overload or a short circuit. All fuses contain a *fusible element* that, as a result of excessive current, is heated and melted (severed), thus *clearing* the circuit from its source. Fuses are generally divided into two major classes: low-voltage (600 V or less) and high-voltage (above 600 V).

Several types of low-voltage fuses are shown in **Fig. 4-4**. The *cartridge* type (Fig. 4-4a) and knife-blade type (Fig. 4-4b) are available in voltage ratings of 250 and 600 V. Both have the same internal construction of an insulating powder (talc or other suitable organic insulator) that surrounds the fusible element (a metal of low-temperature melting point). Whenever excessive current occurs, the powder is intended to (1) cool the vaporized metal quickly, (2) absorb the condensed metallic vapor, and (3) quench any arc that may be sustained by the conductive metallic vapor. It is the presence of this powder that gives the fuse its high *interrupting capacity* in the event of sudden short circuits.

Figure 4-4c shows the plug fuse, 600 V or less, available in currents from 0.3 up to 30 A. Such fuses have a medium-screw (or fusestat-screw) base, designed for use in fusible electric lighting panel boxes. Plug fuses are available in two types: *nondelay* (which quickly opens whenever current rating is exceeded) for lighting and other nonmotor circuits and *time delay* (which permits overload starting currents of motors to pass without clearing the circuit). Fuses of the latter type con-

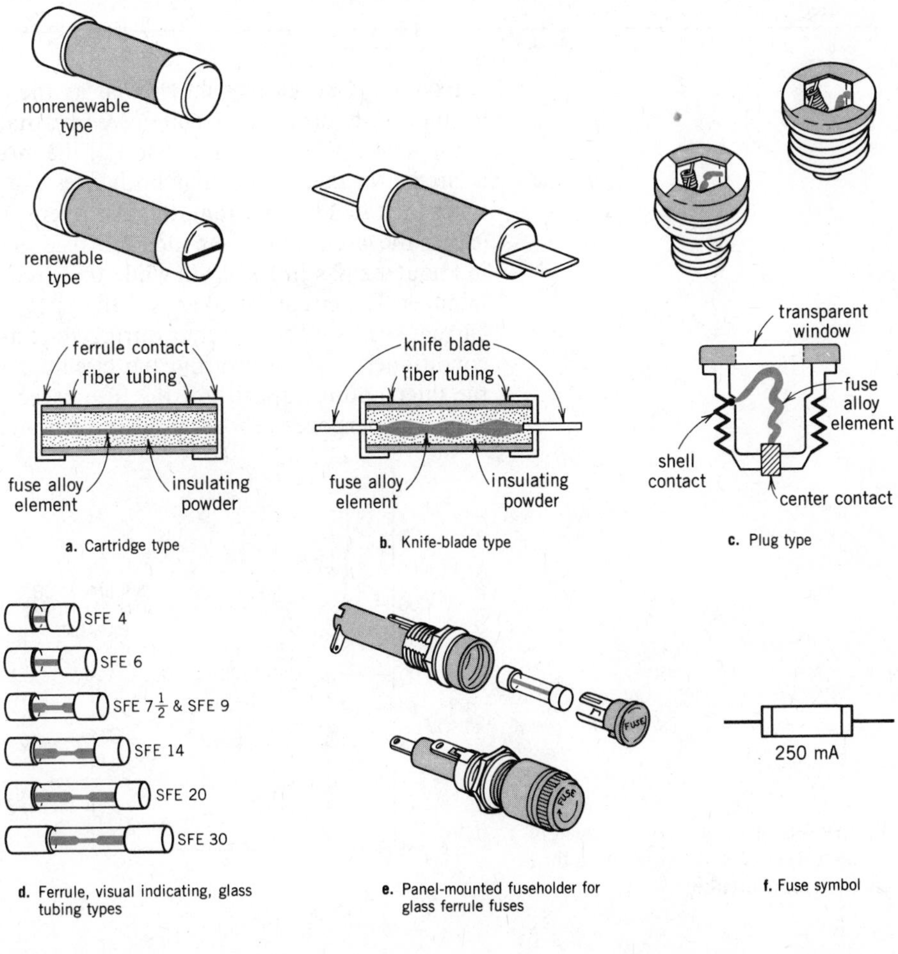

a. Cartridge type **b.** Knife-blade type **c.** Plug type

d. Ferrule, visual indicating, glass tubing types **e.** Panel-mounted fuseholder for glass ferrule fuses **f.** Fuse symbol

Figure 4-4 Types of fuses, their construction, and appearance (Line drawings reprinted with permission Bussman Division, Copper Industries)

tain two elements (one for short circuits and dangerous overloads and the second for sustained overloads) and are called *fusetrons* or *fusestats*.

Figure 4-4d shows small-dimension ferrule (visual indicating) glass tube fuses. These fuses are available in voltage ratings from 32 to 250 V and currents from 1/500 A (2 mA) to as high as 30 A, in both quick-acting and time-lag types. These smaller ferrule fuses are used for protection of electronic as well as automotive equipment. Figure 4-4e shows a typical fuse holder designed for panel mounting, which enables easy fuse removal for examination.

The electrical symbol for a fuse is shown in Fig. 4-4f. In some schematics the current rating appears. When general circuits are discussed, as in this text, the symbol is omitted.

EXAMPLE 4-8

A 20-A fuse is designed to have a resistance element of 25 mΩ to melt in 2 s whenever its energy dissipation exceeds 20 joules. Calculate the

$$P = W/t = I^2/e$$

a. Power required to melt the fuse in exactly 2 s when carrying 20 A
b. Time required for the fuse to clear the circuit when the current is 100 A (i.e., a short circuit) $P = W/t \quad t = W/P = \dfrac{20\,ws}{I^2R}$
c. Time required for the fuse to clear the circuit when the current is 25 A
d. Tabulate current versus time based on the given and calculated data.

Solution

a. $P = W/t = 20 \text{ J}/2 \text{ s} = 20 \text{ W·s}/2 \text{ s} = \textbf{10 W}$ (4-1)
b. $t = W/P = W/I^2R = 20 \text{ W·s}/100^2 \times (25 \times 10^{-3}) \text{ W}$
$= \textbf{80 ms}$ (4-3)
c. $t = W/I^2R = 20 \text{ W·s}/25^2 \times (25 \times 10^{-3}) \text{ W} = \textbf{1.28 s}$ (4-3)
d.

Overload (A)	20	25	100
Clearing time (ms)	2000	1280	80

Example 4-8d clearly shows that as the *current* in the fuse *increases*, the *time* for the fuse to clear the circuit *decreases*; that is, current varies *inversely* with time.

Figure 4-5 shows a comparison of the **inverse-time** characteristics of a fuse and a circuit breaker. Note that both axes are plotted on *logarithmic* scales. At 10 times the rated fuse current, the fuse opens in 10 ms while the circuit breaker trips (clears the circuit) in 10 seconds. Correspondingly, at rated current, the fuse clears in about 1 s (2 s in Ex. 4-8a) while the circuit breaker trips in about 100 s or 1.67 minutes. The circuit breaker is deliberately designed to permit short, momentary current surges (such as occur during starting of the motor of a refrigerator or air conditioner) *without* clearing the circuit. Note that for the same current rating, the interrupting capacity of the fuse exceeds that of the circuit breaker because its major purpose is short-circuit protection.

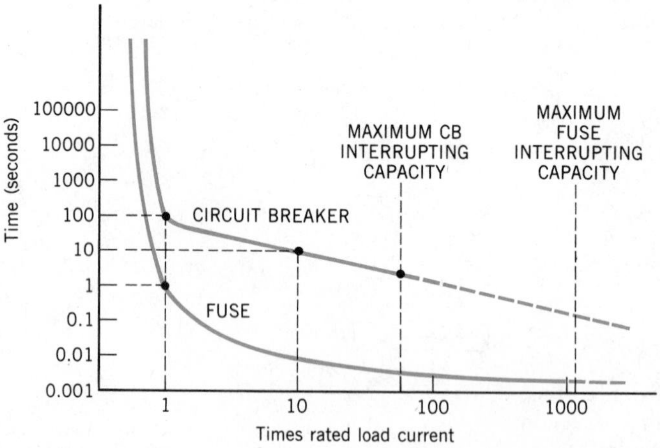

Figure 4-5 Comparison between inverse time curves of a fuse and a time-delay circuit breaker

Unlike a fuse, which clears a circuit as a result of "joule heating" (I^2Rt), a circuit breaker is essentially a mechanical switching device. Circuit breakers contain two elements: (1) a *current-sensing* device and (2) a set of *contacts*, which may be opened or closed mechanically. When the current is excessive, the current-sensing device actuates the mechanical contacts and automatically clears the circuit. Although some circuit breakers are designed to reset themselves automatically, most circuit breakers require *manual* reset (i.e., a person must reset the contacts to bring them to a normally closed position). These two elements are represented in the circuit breaker symbol of **Fig. 4-6a**. Manual reset is advantageous because it tells us that the circuit breaker tripped as a result of some overload condition that previously existed or may persist when the breaker is reset.

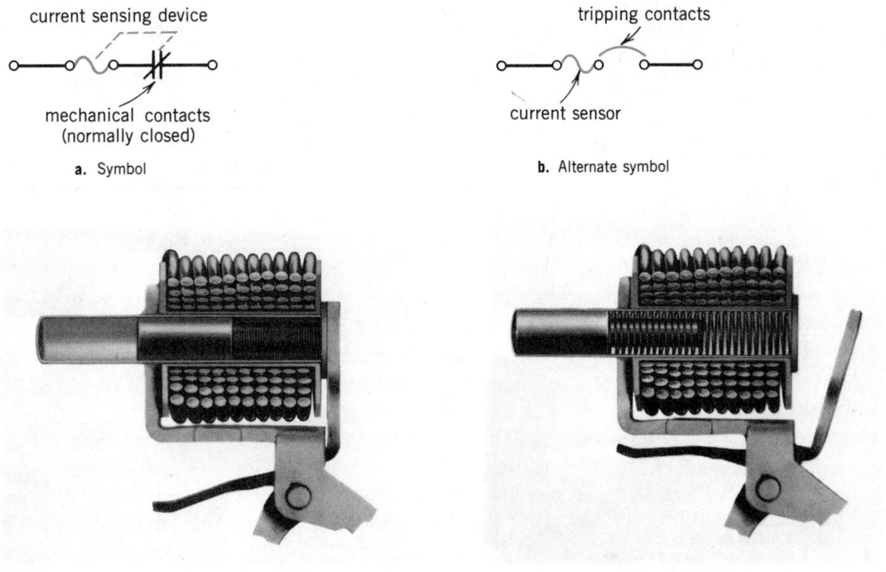

a. Symbol

b. Alternate symbol

c. Magnetic circuit breaker (closed)

d. Magnetic circuit-breaker (tripped or open position)

Figure 4-6 Circuit breaker symbols and internal operation (Courtesy Heinemann Electric Co.)

The current-sensing device may be a bimetallic thermal element, an electromagnetic relay, or a solder-pot thermal melting relay.[2] Some circuit breakers use a combination of thermal and electromagnetic principles to actuate the mechanical contacts.

The cross section of a *magnetic* circuit breaker is shown in Figs. 4-6c and 4-6d. In Fig. 4-6c the current in the coil (wound around a hermetically sealed nonmagnetic tube containing a spring-loaded movable iron core and a silicone liquid fill) is insufficient to produce the magnetic flux needed to move the *iron core* (shown on the left side of the nonmagnetic tube). When the coil current reaches the breaker's trip value, the magnetic flux produced by the coil is sufficient to attract the movable iron core against the coil spring. This increase in magnetic flux also attracts the *armature* (shown at the right in Fig. 4-6d) to the pole piece, thus tripping the breaker contacts (not shown). The silicone liquid in the tube regulates the iron core's speed of travel, creating a controlled time delay that is inversely proportional to the overload current. In instantaneous (nondelay) breakers, the silicone liquid is omitted.

[2] Space does not permit a discussion of various designs. For a complete discussion of fuses and circuit breakers, see I. L. Kosow, *Control of electric machines*, Prentice-Hall, 1973, Ch. 1.

Circuit breakers have an advantage over fuses in that fuses must be replaced after they have cleared the circuit. For this reason, in residences primarily, circuit breakers are installed in new construction, since they are less troublesome.

Fuses, however, have not become obsolete. Because of their *immediate response* to short circuits (Fig. 4-5), they are used for short-circuit protection in larger motors. A motor having a full load current of 100 A is required to have a 125- to 150-A circuit breaker for *overload* protection and a 300- to 500-A fuse for *short-circuit* protection. In this way, using a *combination* of fuse and circuit breaker, the motor (and its control equipment) is protected against either sustained overloads or any short circuits that may occur within the motor itself or the wiring to the motor.

4-6 POWER AND ENERGY SUMMARY

Table 4-2 summarizes the previously presented relationships and distinctions between power and energy.

Table 4-2 Relations between Power and Energy

	Basic Parameter	
	Energy	Power
Definition	Ability to do work	Rate ofdoing work
Quantity symbol	W	P
Defining equation	$W = P \times t$	$P = W/t$
Derived equations	$W = VIt = I^2Rt = (V^2/R)t$	$P = VI = I^2R = V^2/R$
Electrical unit	joule (J)	watt (W)
Equivalent units	$1\ \mathrm{J} \equiv 1\ \mathrm{W \cdot s} \equiv 1\ \mathrm{N \cdot m}$ $\equiv \mathrm{kg(m/s)^2}$	$1\ \mathrm{W} \equiv 1\ \mathrm{J/s} \equiv 1\ \mathrm{N \cdot m/s}$ $\equiv 1\ \mathrm{kg \cdot m^2/s^3}$
Other units (Appendix A-3 H, I)	watt-hour (Wh), erg, Btu, ft·lb, kg·m, hp·h, kg·cal, oz·in	hp, ft·lb/min, Btu/min, g·cm/s, erg/s, kg·cal/s, N·m/s, oz·in/s, kg·m/s

Table 4-3 shows the interrelation between the basic electrical units of potential, current, and resistance and their relation to power and energy.

Table 4-3 Interrelation between Basic Electrical Units, Power, and Energy

Physical Quantity	Defining Equation	Unit Abbreviation Symbol	Definition	Derived Equations
Potential	$V = W/Q$	V	work per unit charge	$V = IR = P/I$ $= \sqrt{PR}$
Current	$I = Q/t$	A	charge per unit time	$I = V/R = P/V$ $= \sqrt{P/R}$
Resistance	$R = V/I$	Ω	ratio of potential to current	$R = V^2/P = P/I^2$ $= W/I^2t = V^2t/W$

4-6.1 Examples Illustrating Relations Given in Tables 4-2 and 4-3

The derived equations and the relations given in Tables 4-2 and 4-3 open a virtual "Pandora's box" of possible application to a variety of design situations.

Only a few are presented below, and the problems at the end of this chapter provide more. The reader should go through these examples carefully and do as many of the end-of-chapter problems as possible. Only through practice is a fundamental understanding of power and energy relations acquired.

EXAMPLE 4-9

Design an electric percolator that will bring 1 liter (approximately 1 qt) of water from a temperature of 20°C to 100°C in 3 min. Assume an efficiency of 90% because of heat transfer losses. The percolator is to operate on 120 V and the resistance element is to be No. 40 AWG nichrome wire. Specify the length of wire needed for winding the element.

The solution involves breaking down the problem analytically to the following *steps*. Calculate the

a. Weight of water in kilograms
b. Heat energy needed to heat the water (including losses) in kilogram-calories
c. Electrical energy in joules
d. Power in watts
e. Hot resistance of the element
f. Length of the element wire in centimeters and inches, assuming nichrome has negligible resistance change due to temperature

Solution

a. One liter of H_2O weighs 1 kg (It takes 1 kg·cal to raise 1 *l* of water by 1°C)

b. $W = 1\ \text{kg} \times \dfrac{1\ \text{cal}}{1°\text{C}} \times \dfrac{80°\text{C}}{0.90} = \mathbf{88.\bar{8}\ kg{\cdot}cal}$

c. $W = 88.\bar{8}\ \text{kg·cal} \times 4186\ \text{J/kg·cal} = \mathbf{3.721 \times 10^5\ J}$
$= \mathbf{3.721 \times 10^5\ W{\cdot}s}$

d. $P = \dfrac{W}{t} = \dfrac{3.721 \times 10^5\ \text{W·s}}{3\ \text{min} \times 60\ \text{s/min}} = \mathbf{2067\ W}$

e. $R = \dfrac{V^2}{P} = \dfrac{120^2}{2067} = \mathbf{6.97\ \Omega}$

f. From Appendix A-4 $\rho = 99.6 \times 10^{-8}\ \Omega{\cdot}\text{m}$, and from Appendix A-5, for No. 40 AWG, $A = 5.014 \times 10^{-9}\ \text{m}^2$

Then $l = \dfrac{AR}{\rho} = \dfrac{5.014 \times 10^{-9}\ \text{m}^2 \times 6.97\ \Omega}{99.6 \times 10^{-8}\ \Omega{\cdot}\text{m}} = \mathbf{3.51 \times 10^{-2}\ m}$

$= 3.51 \times 10^{-2}\ \text{m} \times 10^2\ \text{cm/m} = \mathbf{3.51\ cm}$

$= 3.51\ \text{cm} \times \dfrac{1\ \text{in}}{2.54\ \text{cm}} = \mathbf{1.4\ in}$

The following insights are to be gained from Ex. 4-9:

1. Certain information is given in the problem that does not lead *directly* to the information to be found (i.e., the length of nichrome wire needed).
2. Solving the problem, therefore, is not a matter of merely "plugging" *given* values into a *single* equation.
3. The problem must be broken down into the various steps that lead to the solution.
4. For the given problem, these steps are shown as parts **(a)** through **(f)** in Ex. 4-9.
5. Note, however, that given the individual steps, the solution of each part is a relatively simple process.
6. Relatively complex problems may be solved by breaking them down into a series of relatively simple steps.

EXAMPLE 4-10

An elevator weighing 1 ton (2000 lb) is balanced by a 1-ton counterweight. It is driven through a pulley system by a 220-V motor whose efficiency is 85%. The maximum number of elevator occupants is 10 (at 175 lb per occupant). The elevator must rise at a constant speed of 10 ft/s. Assuming the mechanical losses in the elevator cabling and pulley system are 80%, calculate the motor horsepower required and full-load motor current during operation. The solution involves breaking the problem analytically into the following steps. Calculate the

a. Output power of the motor in foot-pounds per second (ft·lb/s)
b. Output power of motor in horsepower (hp)
c. Input power in kilowatts (kW)
d. Current in amperes drawn from the 220-V supply

Solution

a. $P_{\text{mo}} = \dfrac{W}{t} = 175\ \dfrac{\text{lb}}{\text{occ}} \times 10\ \text{occ} \times \dfrac{10\ \text{ft}}{1\ \text{s}} \times \dfrac{1}{0.8}$

$= \mathbf{2.1875 \times 10^4\ ft{\cdot}lb/s}$ (4-1)

b. $P_{\text{hp}} = 2.1875 \times 10^4\ \dfrac{\text{ft·lb}}{\text{s}} \times \dfrac{1.356\ \text{W}}{1\ \text{ft·lb/s}} \times \dfrac{1\ \text{hp}}{746\ \text{W}}$

$= \mathbf{39.76\ hp}$ (use a **40-hp motor**)

c. $P_{\text{in}} = \dfrac{39.76\ \text{hp} \times 746\ \text{W/hp}}{0.85} = \mathbf{34.895\ kW}$

d. $I = P/V = 34\,895\ \text{W}/220\ \text{V} = \mathbf{158.6\ A}$ (4-2)

Not only does Ex. 4-10 specify the motor required to perform the lift (40 hp), but the solution provides some indication of the motor current drawn from the 220-V mains. This current is useful in determining the motor feeder wiring sizes, the circuit-breaker overload protection, and the motor short-circuit protection needed.

EXAMPLE 4-11

An electric percolator is rated at 120 V, 1000 W and holds 1 liter of water. Calculate the

a. Hot resistance of its heating element
b. Current drawn from the supply
c. Time in minutes required to raise 1 liter of water from a tap temperature of 15°C to boiling, if its efficiency is 92%

Solution

a. $R = V^2/P = 120^2/1000 = \mathbf{14.4\ \Omega}$ (4-3)
b. $I = P/V = 1000/120 = \mathbf{8.\bar{3}\ A}$ (4-2)
c. $P_0 = P_i \times \eta = 1000\text{ W} \times 0.92 = 920\text{ W}$ (4-5)

$$t = \frac{W_0}{P_0} \qquad (4\text{-}3)$$

$$= \frac{(4.186 \times 10^3\text{ J/kg·cal})(1\text{ cal/°C})(1\text{ kg})(100 - 15)\text{°C} \times 1\text{ W·s/J})}{920\text{ W}}$$

$$= 386.75\text{ s}$$

$$t = 386.75\text{ s} \times \frac{1\text{ min}}{60\text{ s}} = \mathbf{6.4\ min}$$

The insights to be gained from a close examination of Ex. 4-11 are

1. Given the simple nameplate data of the device, we can get other useful data by simple calculation with power relation equations. In this case, given the power and voltage rating, we can calculate resistance and current.
2. A kilogram-calorie is the heat energy required to raise 1 kg of water 1°C. We can also write the conversion factor in part (**c**) of the solution as **4.186 kJ/kg·°C** to raise any weight of water (in kilograms) to any temperature (°C) below and up to the boiling point of water. The conversion factor in this form simplifies the dimensional analysis of Ex. 4-11c.
3. In solving any engineering-type problem, we must have some notion, *a priori*, of whether the *answer is reasonable*. We know that if we fill a percolator with a quart of water and plug it into an electric outlet, it will take about 5 min to come to a boil. Consequently, the answer of 6.4 min obtained is reasonably satisfactory. Had it come out either larger or smaller by an order of one magnitude (a factor of 10), it would have been necessary to check our mathematics.[3]

EXAMPLE 4-12

A TV receiver consumes 400 W. It is left on for 24 h with nobody watching it. How many pints of oil were wasted because of this carelessness?[4]

This seemingly insoluble problem is easily solved by breaking it down into a series of simple parts or questions, stated below. Calculate the wasted energy in

a. Joules
b. Foot-pounds (use 0.7375 ft·lb/J)
c. Btu (use 778 ft·lb/Btu)
d. Pounds of fuel oil (use 18 500 Btu/lb)
e. Cubic feet of oil (use density of 50 lb/ft^3)
f. Cubic inches of oil (use 1728 in^3/ft^3)
g. Gallons of oil (use 231 in^3/gal)
h. Pints of oil (use 8 pt/gal)

Solution

a. $400\text{ W} \times 24\text{ h} \times 3600\text{ s/h} = \mathbf{3.456 \times 10^7\ J}$
b. $3.456 \times 10^7\text{ J} \times 0.7375\text{ ft·lb/J} = \mathbf{2.5488 \times 10^7\ ft·lb}$
c. $2.5488 \times 10^7\text{ ft·lb} \times \dfrac{1\text{ Btu}}{778\text{ ft·lb}} = \mathbf{32\,761\ Btu}$
d. $32\,761\text{ Btu} \times \dfrac{1\text{ lb}}{18\,500\text{ Btu}} = \mathbf{1.771\ lb}$
e. $1.771\text{ lb} \times \dfrac{1\text{ ft}^3}{50\text{ lb}} = \mathbf{3.54 \times 10^{-2}\ ft^3}$
f. $3.54 \times 10^{-2}\text{ ft}^3 \times \dfrac{1728\text{ in}^3}{1\text{ ft}^3} = \mathbf{61.2\ in^3}$
g. $61.2\text{ in}^3 \times \dfrac{1\text{ gal}}{231\text{ in}^3} = \mathbf{0.265\ gal}$
h. $0.265\text{ gal} \times \dfrac{8\text{ pt}}{1\text{ gal}} = \mathbf{2.12\ pt}$

The insights to be gained from the solution to Ex. 4-12 are

1. Once again, as in the case of Ex. 4-9, a complex problem becomes relatively simple by *breaking it down* into a *series of simple problems.*
2. The essential clue is in knowing the heat content of fuel oil, part (**d**), expressed as 18 500 Btu/lb. Given the weight of the oil and its density (**e**), we can find the equivalent volume. This leads to other expressions of volume, that is, in gallons or pints.
3. Solving problems, therefore, requires a kind of "road map" in which the problem solver must ask three questions:
   Q. 1—"Where am I?" **A.** I am at a point where I have energy, in joules.
   Q. 2—"Where do I want to go?" **A.** I want to get to pints of fuel oil.
   Q. 3—"How do I get there?" **A.** By finding energy in Btu, using the heat content in Btu/lb, finding the number of pounds of oil, then using the density of oil, finding the volume of oil, and converting this to pints.

[3] Students, on both homework solutions and quizzes, have come up with answers ranging from 84 h to 0.1 s (in solving Ex. 4-11) and were perfectly happy with either result.

[4] The author is indebted to W. H. Hayt and J. E. Kemmerly, *Engineering circuit analysis*, 3rd Ed., McGraw-Hill, 1978 for this problem, as modified above (cf. Ch. 1, Prob. 4, unsolved).

4-7 GLOSSARY OF TERMS USED

Circuit breaker Two-terminal overcurrent protective device containing a current-sensing element and a mechanical switching element to open a set of contacts whenever excessive current passes through it.

Efficiency, η Ratio of useful output to input energy or power.

Energy forms Light, heat, chemical, mechanical, and electrical energy.

Fuse Two-terminal overcurrent protective device with a circuit-opening element that is heated and melted as a result of excessive current passing through it.

Kinetic energy, W Energy possessed by virtue of motion or work done.

Law of conservation of energy Basic physical law which implies that the total energy of the universe is constant. As a result, energy can be neither created nor destroyed; it can only be converted from one form to another.

Potential energy, W Energy possessed by virtue of position or state of energy form.

Power, P Time rate of doing work or expending energy, that is, energy per unit time or work per unit time.

Transducer Device that receives energy from one system or source and transmits that energy, often in a different form, to another system or source.

Watthour meter Instrument for measuring electrical energy consumed by a specific occupancy supplied with electrical energy by a utility.

Work, W Amount of energy converted from one form to another, as a result of motion or conversion of energy from potential to kinetic.

4-8 PROBLEMS

Secs. 4-1 to 4-4

4-1 A resistor dissipates 500 J of energy in 5 min. Calculate the power dissipated by the resistor.

4-2 A resistor dissipates power at a rate of 5 W. How long would it take for the resistor to consume an energy of 400 J?

4-3 A 60-W lamp operates for 6 h/day. Calculate the energy consumed by the lamp in

a. One month, in kilowatt-hours
b. One year

4-4 When a current of 5 A flows in a resistor connected to a 12-V supply for a particular time period, it dissipates 800 joules. How long will it take to dissipate this quantity of energy?

4-5 A motor has an input of 1200 W and an output of 1.5 hp. Calculate its efficiency.

4-6 An electric lamp is rated at 120 V, 60 W. Calculate

a. Its resistance
b. Its current, when connected to a 120-V supply

4-7 A crane lifts a 200-kg weight a distance of 30 m. Calculate the work done in

a. Joules
b. Foot-pounds

4-8 A pump lifts 5000 liters of water a distance of 20 m in 10 min to a reservoir storage tank. Calculate

a. The energy required in kilojoules
b. The power expended in kilowatts

Note: 1 liter of water weighs 1 kg (on earth) = 9.807 N; see Appendix A-3F.

4-9 A transformer winding has an I^2R loss of 900 W. After 6 h of operation, calculate the heat energy loss in

a. Joules
b. Btu
c. Kilogram-calories
d. Kilowatt-hours

4-10 A 220-V motor draws a current of 50 A and delivers 12 hp at its shaft. Calculate

a. The efficiency of the motor
b. The energy consumed by the motor after 8 h of operation, in kilojoules

4-11 A 110-V motor delivers 10 hp and has an efficiency of 85%. What current does it draw from its supply?

4-12 Two transducers are connected in cascade. The first has an efficiency of 90% and the second an efficiency of 60%. If the system provides a power output of 100 mW, calculate the system power input.

4-13 Four transducers in cascade produce an overall efficiency of 40%. If three of the transducers each has an efficiency of 0.8, what is the efficiency of the fourth?

4-14 Two transducers are used in a system, but one transducer is twice as efficient as the other. If the input power to the system is 100 mW and the output power is 40 mW, calculate the efficiency of each transducer. (Hint: $2\eta_1^2$ is the overall efficiency.)

4-15 How long must a 3-kW microwave oven remain on before consuming 5 kJ of energy?

4-16 An electric pump lifts 10 kg of water a height of 30 m each minute. The input to the pump is 100 W. Calculate

a. The overall system efficiency
b. The pump efficiency if the motor efficiency is 85%

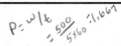

4-17 An electric water heater draws 25 A from a 208-V line, producing 120 kJ of heat energy a minute. Calculate the

a. Efficiency of the water heater
b. Yearly cost of operation at \$0.15/kWh if the heater operates 5 h/day

4-18 If the energy taken from a 220-V supply by a transmitter in 15 min is 1.5 MJ, calculate the

a. Current
b. Charge in coulombs transferred per minute
c. Energy consumed after 100 h of operation

Secs. 4-5 to 4-6

4-19 An electric percolator element has a hot resistance of 42 Ω and draws a current of 6 A. If its overall efficiency is 90%, calculate the

a. Time required to raise the temperature of 2 liters of water from 18°C to its boiling point
b. Power consumed by the device
c. Cost to heat the water at \$0.20/kWh

4-20 An electric furnace is used to raise the temperature of 3 kg of iron from 20 to 750°C in 20 min. If the furnace operates from a 220-V supply and has a 75% efficiency, calculate the

a. Current drawn from the supply
b. Resistance of the heating element
c. Power drawn from the supply
d. Energy transferred from the supply in kilowatt-hours

Hint: The specific heat of iron is 500 J/kg·°C

4-21 A weight of 8000 lb is to be lifted 100 ft by an electric motor. Calculate the

a. Work done in foot-pounds by the motor
b. Equivalent kinetic energy in joules
c. Power required to make the lift in 20 s in joules per second and watts
d. Horsepower of motor required

4-22 A resistor is rated 10 kΩ, 0.5 W. Calculate the

a. Highest voltage that may be applied to it without exceeding its ability to dissipate heat safely
b. Highest current it can safely carry

4-23 In amplifying a circuit voltage, a radio transistor consumes 120 J of energy each minute. Calculate the

a. Power dissipated by the transistor in joules per second and watts
b. Total energy required to operate five such transistors in the radio for 24 h (in joules)
c. Total energy in kilowatt-hours
d. Cost of radio transistor operation only per month, if the radio is operated continuously (24 h/day) and the utility charge is \$0.20/kWh

4-24 A student carries a briefcase of books weighing 5 kg to a classroom located two flights of stairs above ground (7.5 m) in 1 min. Calculate the

a. Energy expended in newton-meters
b. Energy expended in foot-pounds
c. Rate of doing work in joules per second and in horsepower
d. Time required to accomplish the task if the student's rate of doing work is 0.25 hp
e. Efficiency of the student's body if 100 gram-calories is expended in climbing stairs at the rate given in the problem statement
f. Weight of the briefcase in pounds

4-25 It is desired to completely design an electric percolator that heats 2 qt of H_2O at a temperature of 60°F to boiling (212°F) in 5 minutes. Calculate the

a. Weight of H_2O in pounds
b. Total heat energy required in Btu
c. Heat power in Btu/min
d. Consumed electric power in kilowatts if the efficiency of the heating element is 92%
e. Hot resistance of the heating element if used on a 120-V supply
f. Length of resistance element in inches, using No. 36 AWG chromel wire. (See Appendix A-5.) Assume chromel has negligible resistance change due to temperature because of its low α_{20}.

4-26 The resistance of a 120-V percolator heating element, when hot, is 6 Ω. Calculate the

a. Current drawn by the element from the supply
b. Power consumed by the element by three methods ($P = VI = V^2/R = I^2R$)
c. Heat power in Btu/min if the heat transfer power conversion efficiency is 93%
d. Heat power in g·cal/s and kg·cal/s
e. Time required to boil 5 qt of water at a tap water temperature of 15°C

4-27 An elevator motor draws 15 000 J of energy from its power source in order to raise 400 lb of weight a height of 25 ft in 4 s. Calculate the

a. Work done in joules
b. Motor efficiency
c. Output power in kilowatts
d. Motor horsepower

4-28 A deep-well pump is designed to raise 50 gal of H_2O *per minute* from a depth of 30 ft to the surface. Calculate the

a. Weight of 50 gal of H_2O in pounds
b. Power required in foot-pounds per minute
c. Power required in watts
d. Electrical motor power needed in watts, assuming that the overall efficiency of the pump and fluid friction in the pipes is 75%
e. Overall mechanical energy required and expended by the motor after 1 h of operation, in newton-meters

4-29 The overall efficiency of a TV transmitting station is 0.45 and its radiated power is 45 kW, operating 18 h daily. Calculate the

a. Daily energy output in kilowatt-hours
b. Daily energy input in kilowatt-hours
c. Monthly cost of transmission at \$0.10/kWh

4-30 A 30-A fuse is designed to have a resistance element of 8.25 mΩ and clear a circuit whenever its energy dissipation exceeds 30 J. If the fuse melts in exactly 4 s when carrying exactly 30 A, calculate the

a. Power required to melt the fuse
b. Time required for the fuse to "blow" when the current is instantaneously 100 A
c. Time to clear the circuit when the current is 45 A

4-31 A pan containing 1 liter of water at a temperature of 60°F is placed on the heating surface of an electric stove element. It requires 5 min to bring the water up to its boiling point (212°F). Calculate the

a. Energy required in kilogram-calories
b. Energy required in Btu
c. Energy required in joules
d. Input power required by the heating element in watts if the efficiency of heat transfer from element to water in the pan is 82%
e. Repeat (**d**) in Btu/min
f. Repeat (**d**) in kg·cal/min

4-32 A 5-kΩ resistor is rated to dissipate not more than 10 W safely. Calculate the

a. Maximum voltage that may be connected across it
b. Maximum current it can safely carry in milliamperes
c. Heat energy dissipated by the resistor in joules if it operates at maximum voltage and current for 8 h

4-33 A deep-well pump fills a 1000-gal tank located 100 ft above the water level in 20 minutes. If water weighs 8.342 lb/gal, calculate the

a. Work done by the pump in joules
b. Horsepower required to operate the pump, assuming the overall pump efficiency is 70% (including pipe friction and losses)
c. Input power to the electric motor, assuming a motor efficiency of 90.25%
d. Monthly cost of maintaining water in the tank, assuming the pump cycles once every 12 h to fill the tank. Utility energy cost is \$0.10/kWh

4-34 A light-emitting diode (LED) is used as a light source in an emergency panel to locate an emergency switch for a fire alarm. It burns continuously, drawing 30 mA from a 6-V supply. Calculate the

a. Energy in kilowatt-hours to maintain its operation for 1 year
b. Yearly cost at \$0.10 per kilowatt-hour

4-35 An electric motor having an efficiency of 80% draws 20 A from a 120-V supply to operate an air compressor having an overall efficiency of 38.5%. Calculate the

a. Motor horsepower
b. Output horsepower of the compressor

4-36 An electric space heater draws 18 A at 120 V from a supply, cycling 15 min (ON) each hour in order to maintain a room temperature of 25°C. Calculate the

a. Daily cost of operation at \$0.10/kWh
b. Heat power produced in kg·cal/min assuming 95% efficiency
c. Length of time in minutes it takes the heater to produce 1000 Btu
d. Rate at which the room cools down in kg·cal/min
e. Rate at which the room cools down in joules per second (watts)

4-37 An electric hoist is designed to lift 50 tons at a speed of 3 ft/s, using a motor having an efficiency of 85%. If the overall efficiency of the hoist's gears and pulleys is 70%, calculate the

a. Rate of work done in foot-pounds per second
b. Horsepower output of hoist
c. Horsepower output of motor
d. Motor input in kilowatts
e. Current draw by the motor in kiloamperes from a 600-V supply

316.889

4-9 ANSWERS

4-1 $1.\overline{6}$ W
4-2 80 s
4-3 a 10.8 kWh b 131.4 kWh
4-4 $13.\overline{3}$ s
4-5 93.25%
4-6 a 240 Ω b 0.5 A
4-7 a 58.84 kJ b 43 396 ft·lb
4-8 a 980.7 kJ b 1.634 kW
4-9 a 19.44 MJ b 18 430 Btu c 4644.4 kg·cal d 5.4 kWh
4-10 a 81.4% b 316.8 kJ
4-11 67.8 A
4-12 185.2 mW
4-13 78.125%
4-14 0.8944; 0.4472
4-15 $1.\overline{6}$ s
4-16 a 49.035% b 57.7%
4-17 a 38.5% b \$1423.50
4-18 a $1.\overline{6}$ kW b $7.\overline{57}$ A c $454.\overline{54}$ C d $166.\overline{6}$ kWh
4-19 8.4 min
4-20 a 5.53 A b 39.8 Ω c $121\overline{6}$ W d $0.40\overline{5}$ kWh
4-21 a 8×10^5 ft·lb b 1.085 MJ c 54.24 kW d 72.71 hp

4-22 a 70.71 V b 7.071 mA
4-23 a 2 W b 864 kJ c 0.24 kWh d $1.44/mo
4-24 a 368 N·m b 271 ft·lb c 0.00822 hp d 1.97 s
e 87.91% f 11.025 lb_f
4-25 a 4.171 lb b 634 Btu c 126.8 Btu/min
d 2.423 kW e 5.943 Ω f 3.3 in
4-26 a 20 A b 2400 W c 127 Btu/min
d 533.2 g·cal/s, 0.5332 kg·cal/s e 12.57 min
4-27 a 13 650 J b 90.4% c 3.39 kW d 4.544 hp
4-28 a 417.1 lb/min b 10 427 ft·lb/min c 235.7 W
d 314.2 W e 1.131 MN·m
4-29 a 810 kWh/day b 1800 kWh/day c $5400/mo
4-30 a 7.5 W b $0.\overline{36}$ s c 1.79 s
4-31 a $84.\overline{4}$ kg·cal b 335.1 Btu c 353.5 kJ d 1437 W
e 81.73 Btu/min f 20.6 kg·cal/min
4-32 a 223.3 V b 44.72 mA c 288 kJ
4-33 a 1.131 MJ b 1.8 hp c 2 hp d $2.98/mo
4-34 a 1.577 kWh/yr b $0.16
4-35 a 2.574 hp b 1 hp
4-36 a $1.30/day b 29.4 kg·cal/min c 8.57 min
d 9.8 kg·cal/min e 684 J/s
4-37 a 300 000 ft·lb/s b $545.\overline{45}$ hp c 779.22 hp
d 683.9 kW e 1.14 kA

CHAPTER 5

DC Series and Parallel Circuits

5-1 INTRODUCTION

A distinction was made in Sec. 2-12 between *direct* and *alternating* sources of voltage and current. This distinction provides a major division for the treatment of circuit analysis in this text. The fundamental circuit analysis techniques, basic laws, theorems, and relations are first introduced using *direct sources* of voltage and current.

This dc circuit approach and treatment simplifies our circuit analysis considerably and as a result enhances our understanding. Resistance was defined (Sec. 3-1) as the *opposition* produced by a given material (or circuit element) to a *steady* (dc) current or voltage and the ability, as a circuit property, to dissipate electrical energy. Later we will encounter two other (passive) circuit elements (capacitance and inductance), but these only produce opposition to a *change in voltage* and a *change in current*, respectively, in addition to resistance. Furthermore, these two other circuit elements store energy, which complicates the analysis somewhat.

Since resistance as a circuit element is the *only* circuit property to (1) limit a steady dc and (2) dissipate electrical energy in an electric circuit, the next three chapters are confined to dc voltage and current sources (Sec. 2-13) only and resistors as circuit elements only. Unless otherwise indicated, the voltage and current sources used are ideal circuit elements (see definitions in Sec. 2-13), producing constant (unvarying) values of voltage of a given polarity and/or current of a given, unvarying direction. As indicated, current is limited by resistance only, in accordance with Ohm's law. The laws, theorems, relations, and circuit analysis techniques that we develop as a result of this simplification will apply later to more complex ac circuits, with minor modifications.

In this chapter we will investigate the voltage, current, and power relations of two basic types of *circuits*: the *series* circuit and the *parallel* circuit.

5-2 SERIES CIRCUIT

Regardless of the circuit complexity, components are connected in *series* when they carry the **same** current; that is, they are subjected to the *same* rate of charge flow under the influence of an electric field. **Figure 5-1a** shows a voltage source

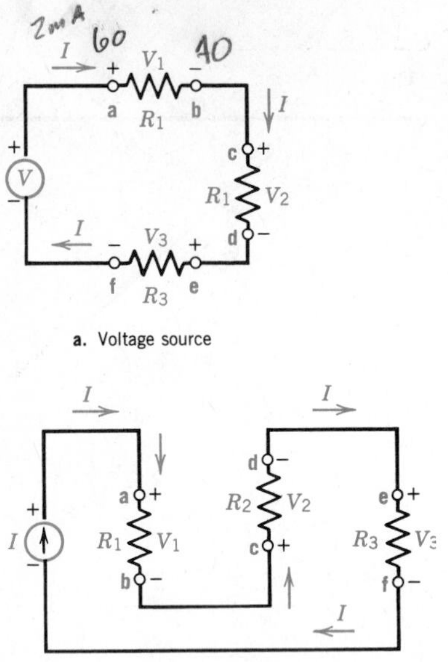

a. Voltage source

b. Current source

Figure 5-1 Voltage and current sources, polarities and current directions in series circuits.

V connected to a *series* circuit consisting of three resistors: R_1, R_2, and R_3. With respect to the *series* circuit, please note that

1. There is only *one* path for the current.
2. The circuit elements are connected in *cascade* (i.e., in a head/tail arrangement), so that the output current of one element becomes the input current for the next.
3. The *same* current I exists in each circuit element (i.e., the rate of charge flow is *identical* in each resistor).

Again, in Fig. 5-1b, a current source is now connected to a series circuit consisting of three resistors: R_1, R_2, and R_3. Note again that (1) there is only one path for the current, (2) the circuit elements are in cascade, and (3) the *same* current I exists in the current source and in each circuit element.[1]

With respect to both circuits shown in Fig. 5-1, we may conclude that the total current, I, furnished by either source is the *same* as the current in each resistor. The definition of the current relations in and of a series circuit is

$$I \equiv I_1 \equiv I_2 \equiv I_3 \equiv \cdots \equiv I_n \quad \text{amperes (A)} \qquad (5\text{-}1)$$

where I_1, I_2, and I_3 are the currents in resistors R_1, R_2, and R_3, respectively
I is the total current drawn from a voltage or current source by the (series) combination of circuit elements

5-3 VOLTAGE DROPS AND POLARITIES

The *direction of current* is our clue to the nature of polarity, voltage drop, and whether a device is either absorbing or delivering power in a dc circuit. Consider resistor R_1 in Fig. 5-1a. The current direction is from left to right. Therefore the *left* terminal is *more positive* than the right (i.e., the potential at **a** is higher than the potential at **b**). If it were not, current would not flow from **a** to **b**, in the conventional direction.

Now consider resistor R_3 in Fig. 5-1a. The current direction is from right to left. Consequently point **e** is more positive than point **f**, otherwise current would not flow from **e** to **f**, in the conventional direction.

Note how these polarity rules apply to Fig. 5-1b. The downward current in R_1 produces an opposite polarity to the upward current in R_2, for the same reasons. Note that in both circuits, the current direction is always the *same* in all elements since there is only one current in a series circuit. Clearly, in Fig. 5-1, each source is supplying energy to the series-connected resistors because the *sources* (*active* devices) are delivering current to the resistors (*passive* devices), which absorb and dissipate electrical energy. (The power absorbed by each resistor is the product of the voltage across it times the current in it.) The voltages V_1, V_2, and V_3 and their respective polarities are the same as those which would be measured by a voltmeter when connected across the respective resistors in Fig. 5-1. This means

[1] *Same* means more than just *equal*. Two nickels may be considered equal to a dime but they are *not the same* as a dime, nor are they even the same as each other. In our later circuit analysis problems, we will have instances where the current in one device is *equal* to the current in another device. But if they are *not* connected in *series*, the current in them is *not the same*.

that we must connect the positive terminal of the voltmeter to point **a** and the negative terminal to point **b** in order for the voltmeter to read *upscale.*

In Fig. 5-1, we may say that V_1 is equal to V_{ab} (i.e., V_1 is the voltage measured from point **a** to point **b**). This voltage will be a *positive* voltage because it is the difference in potential between a *more positive* point **a** and a *less positive* point **b**. But if we were taking the difference in potential from **b** to **a** (i.e., V_{ba}), it would have to be recorded as a *negative voltage* since point **b** is more negative than point **a**, as shown by the following examples.

EXAMPLE 5-1

In Fig. 5-1 the voltage at point **a** is +60 V and point **b** is +40 V. The current is 2 mA. Calculate the value of a. V_{ab} b. R_1 c. V_{ba}

Solution

a. $V_{ab} = V_a - V_b = +60 - (+40) = \mathbf{+20\ V}$
b. $R_1 = V_{ab}/I = +20\ \text{V}/2\ \text{mA} = \mathbf{10\ k\Omega}$
c. $V_{ba} = +40 - (+60) = \mathbf{-20\ V}$

The conclusions to be drawn from Ex. 5-1 are of great importance because they constitute the basis for all later work in circuit analysis. These are

1. The notation V_{ab} has *two* subscripts. It is therefore called **double-subscript** notation. The polarity of V_{ba} is opposite to the polarity of V_{ab} (i.e., $V_{ba} = -V_{ab}$).
2. As noted at the beginning of this section, the direction of current determines the polarity across a component and also the voltage "sense" in terms of double-subscript notation.
3. In determining the value of R_1, we selected the appropriate voltage drop, V_{ab}, which *agreed* with the direction of current through the resistor R_1 (i.e., from **a** to **b**). This produces both a *positive* voltage and a *positive* resistance.
4. Since the voltage drops from point **a** to point **b** (i.e., from +60 to +40 V) the voltage $V_1 = V_{ab} = +20$ V is a voltage *drop*. Such a volt drop is a *positive* voltage.
5. From (4) above, a voltage *drop* occurs when current passes through an element *in the direction* of current; this happens when the voltage drop is from (+) to (−).
6. From (5), therefore, a voltage *rise* occurs when voltage is taken from (−) to (+) *against the direction* of current *in a resistor*.
7. Example 5-1c shows that a *negative voltage drop* is a *voltage rise*, since going from point **b** to point **a** is a rise in voltage.
8. From (7) in Fig. 5-1, therefore, the voltage of the current (or voltage) **source** from point **f** to point **a** must be a *rise* in voltage because it is negative (i.e., V_{fa} must be a negative voltage, which is a *voltage rise*). Note that this voltage rise occurs not against the current but *with the direction of current* in *both* a current source and a voltage source.

We can summarize the reasoning that emerges from steps (1) through (8) above by two simple statements to distinguish between sources (active elements) and resistors (passive elements):

1. A *rise* of potential occurs in an *active* element (source) when going from (−) to (+) through the source, *in the direction of current*. The voltage sense of the potential *rise* is the polarity first encountered (−).
2. A *drop* of potential occurs in a passive element (e.g., a resistor) when going from (+) to (−) through the element *in the direction of current*. The voltage sense of the potential *drop* is the polarity first encountered (+).

5-4 KIRCHHOFF'S VOLTAGE LAW (KVL)

Kirchhoff's voltage law (KVL) states that the *algebraic sum* of *all* voltage *rises* and *drops* taken around any *closed path* (or mesh or *loop*) in any circuit is zero. In symbolic form we may summarize this as

$$\sum_{n=1}^{N} V_n = 0 \quad \text{or} \quad V_1 + V_2 + V_3 + \cdots + V_N = 0 \qquad (5\text{-}2)$$

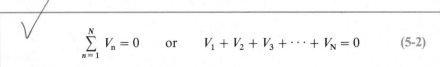

Let us apply KVL to the circuit of Fig. 5-1a. If we do this *in the direction of current* starting at point **a**, we may write

$$V_1 + V_2 + V_3 - V = 0 \quad \text{or} \quad V = V_1 + V_2 + V_3 \quad \text{volts (V)} \quad (5\text{-}3)$$

Note from Eq. (5-3) that the polarity of voltage across each resistor is positive but that the polarity of source voltage is negative, applying rules (1) and (2) from Sec. 5-3 to the statement of KVL in Eq. (5-2).

Equation (5-3) states that the supply voltage is equal to the sum of the voltage drops in a series circuit.[2] Let us now apply KVL to some examples.

EXAMPLE 5-2

Given the series circuit of **Fig. 5-2a** and the values shown,

a. Write the equation for all circuit elements, using KVL
b. Find the value of V_2
c. Find the value of R_2

Solution

a. $V_1 + V_2 + V_3 - V = 0$ (5-2)
b. $V_2 = V - (V_1 + V_3) = 60 - (20 + 30) = \mathbf{10\ V}$
c. $R_2 = V_2/I_2 = 10\ \text{V}/10\ \text{mA} = \mathbf{1\ k\Omega}$

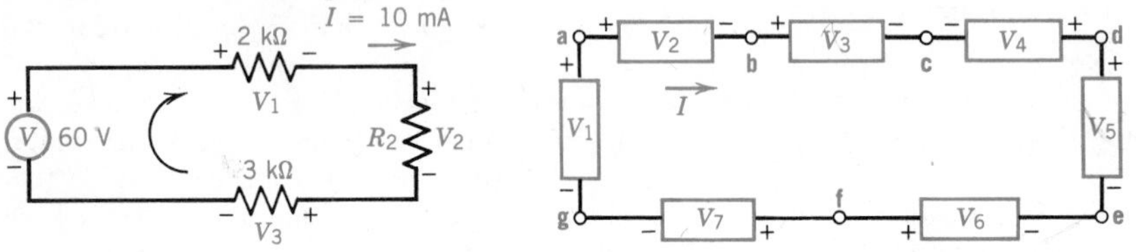

a. Circuit for Ex 5-2

b. Circuit for Ex 5-3

Figure 5-2 Circuits for examples in series relations

EXAMPLE 5-3

Given the series circuit shown in Fig. 5-2b,

a. Write the equation for all circuit elements, using KVL, starting at point **a**
b. Write the elements that are passive for the given current direction
c. Write the elements that are active (sources) for the given current direction

Solution

a. Taking the complete loop abcdefga, we may write

$$V_2 + V_3 - V_4 + V_5 - V_6 + V_7 - V_1 = 0 \quad (5\text{-}2)$$

b. The passive elements are those that are *positive* in (**a**) above, or V_2, V_3, V_5, and V_7
c. The active elements are those that are *negative* in (**a**), or V_1, V_4, and V_6

Examples 5-2 and 5-3 show the unique nature of KVL when properly applied to circuit elements (both active and passive) connected in series.

5-5 SERIES CIRCUIT RELATIONS

We are now ready to summarize current, voltage, resistance, and power relations for series circuits. We already know the current relations that define a series circuit

[2] The right-hand Eq. (5-3) is sometimes shown as a statement of KVL. But this is incorrect. It is, in fact, a relation that *emerges from* KVL, shown in Eq. (5-3).

as given in Eq. (5-1), that is, $I \equiv I_1 \equiv I_2 \equiv I_3$ and so forth. We also used KVL to express the voltage relations, as given by Eqs. (5-2) and (5-3).

From the current and voltage relations, we can derive the resistance relations

$$V = V_1 + V_2 + V_3 + \cdots + V_N \quad \text{volts (V)} \qquad (5\text{-}3)$$

and

$$I \equiv I_1 \equiv I_2 \equiv I_3 \equiv \cdots \equiv I_N \quad \text{amperes (A)} \qquad (5\text{-}1)$$

Substituting the values of I from Eq. (5-1) into Eq. (5-3) by dividing equalities yields $V/I = V_1/I_1 + V_2/I_2 + V_3/I_3 + \cdots + V_N/I_N$, which from Ohm's law yields

$$R_T = R_1 + R_2 + R_3 + \cdots + R_N \quad \text{ohms }(\Omega) \qquad (5\text{-}4)$$

Finally, multiplying each term in Eq. (5-3) by the equalities of Eq. (5-1) yields $VI = V_1I_1 + V_2I_2 + V_3I_3 + \cdots + V_NI_N$ or

$$P_T = P_1 + P_2 + P_3 + \cdots + P_N \quad \text{watts (W)} \qquad (5\text{-}5)$$

EXAMPLE 5-4

Three resistors (R_1, R_2, and R_3) having resistances of 20, 40, and 100 Ω, respectively, are connected in series to an ideal voltage source and draw 100 mA. Calculate the

a. Total resistance of the circuit
b. Volt drop across each resistor
c. Total supply voltage, V_T, by two methods
d. Power dissipated by each resistor by three methods
e. Total power by four methods

Solution

a. $R_T = R_1 + R_2 + R_3 = (20 + 40 + 100)\ \Omega = \mathbf{160\ \Omega}$ (5-4)

b. $V_1 = IR_1 = 0.1\text{ A} \times 20\ \Omega = \mathbf{2\ V}$
$V_2 = IR_2 = 0.1\text{ A} \times 40\ \Omega = \mathbf{4\ V}$
$V_3 = IR_3 = 0.1\text{ A} \times 100\ \Omega = \mathbf{10\ V}$

c. $V_T = V_1 + V_2 + V_3 = 2\text{ V} + 4\text{ V} + 10\text{ V} = \mathbf{16\ V}$ (5-3)
check $V_T = IR_T = 0.1\text{ A} \times 160\ \Omega = \mathbf{16\ V}$ (3-1)

d. $P_1 = V_1I_1 = 2\text{ V} \times 0.1\text{ A} = \mathbf{0.2\ W}$
$V_1^2/R_1 = (2\text{ V})^2/20\ \Omega = \mathbf{0.2\ W}$
$I^2R_1 = (0.1)^2 20 = \mathbf{0.2\ W}$
$P_2 = V_2I_2 = V_2^2/R_2 = I^2R_2$
$= 4 \times 0.1 = 4^2/40 = 0.1^2 \times 40 = \mathbf{0.4\ W}$
$P_3 = V_3I_3 = V_3^2/R_3 = I_3^2R_3$
$= 10 \times 0.1 = 10^2/100 = 0.1^2 \times 100 = \mathbf{1\ W}$

e. $P_T = P_1 + P_2 + P_3 = 0.2 + 0.4 + 1 = \mathbf{1.6\ W}$ (5-5)
$= V_TI = I^2R_T = V_T^2/R_T$
$= 16\text{ V} \times 0.1\text{ A} = 0.1^2 \times 160 = 16^2/160 = \mathbf{1.6\ W}$

Some interesting conclusions may be inferred from Ex. 5-4:

1. R_3 is **five** times the resistance of R_1 and R_2 is **twice** the resistance of R_1.
2. Once we calculated $V_1 = 2$ V and $P_1 = 0.2$ W, we could have found all the other voltages and powers by using the same ratios, that is,

$V_2 = 2 \times V_1 = 2 \times 2\text{ V} = \mathbf{4\ V}$ and
$P_2 = 2P_1 = 2 \times 0.2\text{ W} = \mathbf{0.4\ W}$
$V_3 = 5V_1 = 5 \times 2\text{ V} = \mathbf{10\ V}$ and
$P_3 = 5P_1 = 5 \times 0.2 = \mathbf{1\ W}$

3. Let us extend the above to P_T and V_T by using the ratio $R_T/R_1 = 160/20 = \mathbf{8}$. Then $V_T = 8V_1 = 8 \times 2\text{ V} = \mathbf{16\ V}$ and $P_T = 8P_1 = 8 \times 0.2 = \mathbf{1.6\ W}$

Since the above works so well, let us summarize these ratios in the form of an equation:

$$\frac{R_1}{R_2} = \frac{V_1}{V_2} = \frac{P_1}{P_2} \quad \text{and} \quad \frac{R_1}{R_T} = \frac{V_1}{V_T} = \frac{P_1}{P_T} \text{ yielding two important rules.}$$

5-5.1 Voltage Divider Rule (VDR)

Taking the voltage ratios in proportion to the resistance ratios:

$$V_1 = \frac{R_1}{R_2} V_2 = \frac{R_1}{R_T} V_T \qquad \text{volts (V)} \tag{5-6}$$

5-5.2 Power Divider Rule (PDR)

Taking the power ratios in proportion to resistance and voltage ratios, we obtain

$$P_1 = \frac{R_1}{R_T} P_T = \frac{V_1}{V_T} P_T = \frac{R_1}{R_2} P_2 = \frac{V_1}{V_2} P_2 \qquad \text{watts (W)} \tag{5-7}$$

Note that the *same resistance* ratios are used to find V_1 and P_1 with the VDR and the PDR, respectively. This simplifies problem solving considerably, as shown in Ex. 5-5.

EXAMPLE 5-5

Three resistors $R_1 = 20\ \text{k}\Omega$, $R_2 = 60\ \text{k}\Omega$, and $R_3 = 100\ \text{k}\Omega$ are connected in series across a 90-V supply. *Without* finding the current, calculate the

a. Voltage across each resistor
b. Power dissipated by each resistor
c. Total power, verified by two methods

Solution

a. $V_1 = (R_1/R_T) \times V_T = (20\ \text{k}\Omega/180\ \text{k}\Omega) \times 90\ \text{V}$
$= \mathbf{10\ V}$ (5-6)
$V_2 = (R_2/R_1)V_1 = (60\ \text{k}\Omega/20\ \text{k}\Omega) \times 10\ \text{V} = \mathbf{30\ V}$ (5-6)
$V_3 = (R_3/R_1)V_1 = (100\ \text{k}\Omega/20\ \text{k}\Omega) \times 10\ \text{V} = \mathbf{50\ V}$ (5-6)

b. $P_1 = V_1^2/R_1 = 10^2/20\ \text{k}\Omega = \mathbf{5\ mW}$ (3-8)
$P_2 = (V_2/V_1)P_1 = (30/10) \times 5\ \text{mW} = \mathbf{15\ mW}$ (5-6)
$P_3 = (R_3/R_1)P_1 = (100/20) \times 5\ \text{mW} = \mathbf{25\ mW}$ (5-6)

c. $P_T = P_1 + P_2 + P_3 = (5 + 15 + 25)\ \text{mW}$
$= \mathbf{45\ mW}$ (5-5)
$= V_T^2/R_T = 90^2/180\ \text{k}\Omega = \mathbf{45\ mW}$ (3-8)

The solutions to Ex. 5-5 were obtained without even finding current! The reader may ask, "How is this possible? Isn't current an essential element in a series circuit?"

The answer is found in Eq. (5-1), the basic definition of a series circuit. The current is constant and the SAME in all parts of the circuit. It is this unvarying (constant) property that establishes resistance ratios, voltage ratios, and power ratios *in the same proportion.* Since current is constant, it can be neglected in the solutions as shown in Ex. 5-5.

5-5.3 Extensions of the Voltage Divider Rule (VDR)

Example 5-5 and Eq. (5-6) showed that it is not necessary to solve series circuit relations by finding current. In this subsection we will see that the VDR as expressed by Eq. (5-6) is but a special case of a more general and generic relationship commonly used in circuit analysis. We will do this through a series of developmental examples culminating in Eq. (5-8).

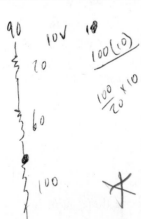

EXAMPLE 5-6

For the circuit given in **Fig. 5-3a**, find the voltage at point x with respect to ground

Solution

The circuit is redrawn in Fig. 5-3b, with the voltage at point x designated as V_x. A little examination of Fig. 5-3b leads to the following conclusions:[3]

1. V_x is closer to +100 V than it is to ground (0 V).
2. V_x is actually 6/8 of +100 V or

$$V_x = V_2 = \frac{R_2}{R_T} V_T = \frac{60\text{ k}\Omega}{80\text{ k}\Omega}(+100\text{ V}) = \mathbf{+75\text{ V}} \quad (5\text{-}6)$$

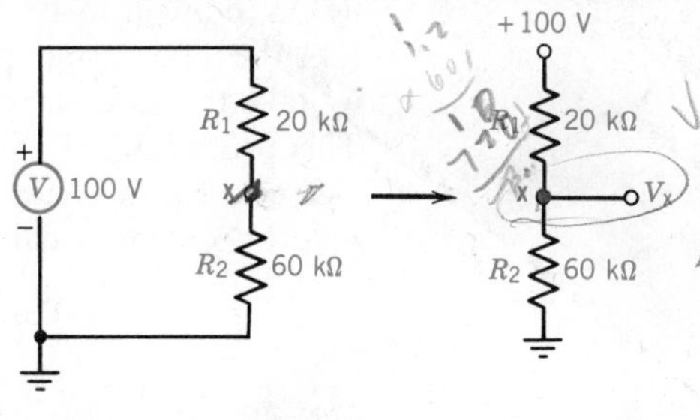

a. Original voltage divider circuit b. Circuit redrawn

Figure 5-3 Circuit for Ex. 5-6

EXAMPLE 5-7

For the circuit shown in **Fig. 5-4a**, find voltage V_x with respect to ground

Solution

The circuit is redrawn in Fig. 5-4b. Note that the positive terminal of the dc supply is grounded, leaving V_x at some negative potential *below* ground. A little examination of Fig. 5-4b leads to the following conclusions:

1. V_x is closer to ground than it is to −100 V
2. V_x, measured with respect to ground, is the voltage across R_1 or

$$V_x = V_1 = \frac{R_1}{R_T} V_T = \frac{20\text{ k}\Omega}{80\text{ k}\Omega}(-100\text{ V}) = \mathbf{-25\text{ V}} \quad (5\text{-}6)$$

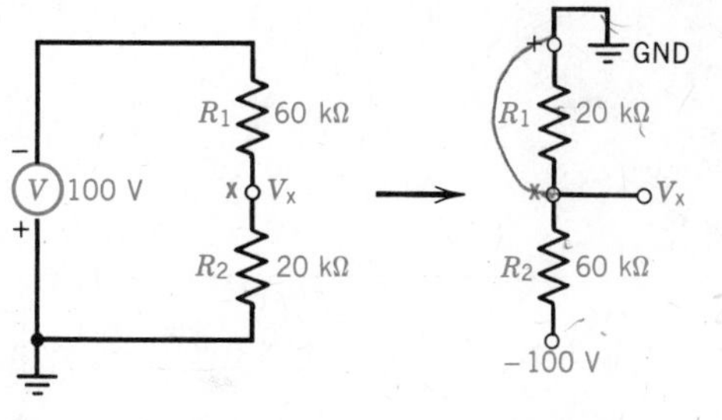

a. Original circuit b. Circuit redrawn

Figure 5-4 Circuit for Ex. 5-7

EXAMPLE 5-8

For the circuit shown in **Fig. 5-5a**, find voltage V_x with respect to ground

Solution

The circuit is redrawn in Fig. 5-5b. With respect to Fig. 5-5b verify the following points:

1. One terminal of R_1 is connected to +100 V and one terminal of R_2 is still connected to −60 V, as in Fig. 5-5a
2. Point **a** is still +100 V with respect to ground, G
3. Point **b** is still −60 V with respect to ground, G
4. No ground terminal is shown in Fig. 5-5b! It is understood that ground, G, lies somewhere between **a** and **b**
5. V_T is now the difference in potential between **a** and **b** or $V_T = V_a - V_b = 100\text{ V} - (-60\text{ V}) = 160\text{ V}$
6. V_x is halfway between +100 V and −60 V because $R_1 = R_2 = R_T/2$. Using the VDR yields

$$V_x = V_2 = \frac{R_2}{R_T}(V_a - V_b) + V_b = \frac{50\text{ k}\Omega}{100\text{ k}\Omega}(160\text{ V}) + (-60\text{ V})$$

and $V_x = 80\text{ V} + (-60\text{ V}) = \mathbf{+20\text{ V}}$

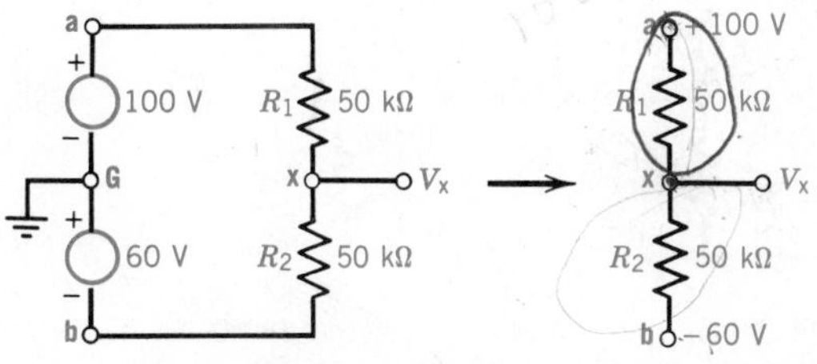

a. Original circuit b. Circuit redrawn

Figure 5-5 Circuit for Ex. 5-8

Some interesting conclusions can be drawn from Ex. 5-8. The solution shows that $V_x = +20$ V, and this raises the question of whether that value is correct. We

[3] There are many advantages to redrawing the circuit in the form shown in Fig. 5-3b. Circuits are shown in this form for single and dual biasing of transistors in communications and industrial electronic circuits. But the major advantage is that it provides some insight into the value of V_x using Eq. (5-8).

noted (point 6 in Ex. 5-8) that V_x is *halfway between* +100 V and −60 V. If we take the difference in potential between V_a and V_x we get +100 − (+20) = 80 V. If we take the difference in potential between V_x and V_b we get +20 − (−60) = 80 V! This proves that V_x must be +20 V in Fig. 5-5b.

We can now write a very useful and important voltage-divider equation for a voltage-divider circuit containing *two* dc supplies as

$$V_o = \frac{R_n}{R_T}(V_p - V_n) + V_n \quad \text{volts (V)} \tag{5-8}$$

where V_o is the output voltage of the divider
R_n is the resistor connected to the *more negative* supply
V_n is the voltage of the *more negative* supply
V_p is the voltage of the *more positive* supply
R_T is the total resistance of the divider

EXAMPLE 5-9

For the circuit shown in **Fig. 5-6**, find output voltage V_o with respect to ground

Solution

V_p, the more positive voltage, is +100 V and V_n is +60 V; substituting these values in Eq. (5-8) yields

$$V_o = \frac{R_n}{R_T}(V_p - V_n) + V_n$$

$$= \frac{20\ \text{k}\Omega}{40\ \text{k}\Omega}(100 - 60)\ \text{V} + 60\ \text{V} = \mathbf{+80\ V} \quad (5\text{-}8)$$

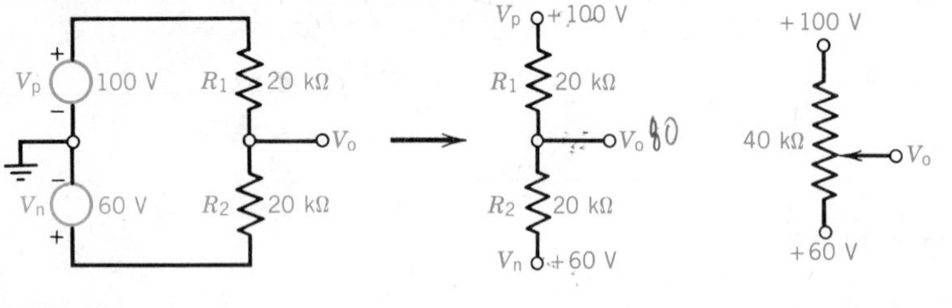

Figure 5-6 Circuit for Ex. 5-9

The validity of Eq. (5-8) is proved by Ex. 5-9. Common sense tells us that V_o should be halfway between +100 V and +60 V since $R_1 = R_2$ and $R_2 = R_T/2$ in Fig. 5-6b.

Another way of looking at the same circuit is shown in Fig. 5-6c. Assume a potentiometer is connected between the two supplies. If the wiper of the potentiometer is connected at the *midpoint* between +100 V and +60 V, then the wiper arm must be at a potential of +80 V.

5-6 PRACTICAL VOLTAGE SOURCES IN SERIES

The foregoing examples all employed *ideal* voltage sources (an ideal voltage source is one that maintains its specified voltage regardless of the current drawn from it), first defined in Sec. 2-13. But every practical dc voltage source known (a battery, a dc generator, a dc power supply, etc.) has some internal resistance associated with it. **Figure 5-7a** shows the symbol for a *practical* dc voltage *source.* Note that a practical voltage source consists of an *ideal voltage source* (of 1.5 V in this case) and an *internal* resistance, r_i, connected *in series* with the ideal source. Let us assume that the practical voltage source of Fig. 5-7a is a chemical cell, shown in Fig. 5-7b. The symbol for a chemical cell is drawn with a *long* line to represent its *anode or positive* terminal and a *short* heavy line to represent its *cathode* or *negative* terminal. The internal resistance of the cell is 0.1 Ω and its *open-circuit voltage* or EMF (electromotive force), E, is 1.5 V, as shown in Fig. 5-7b.[4]

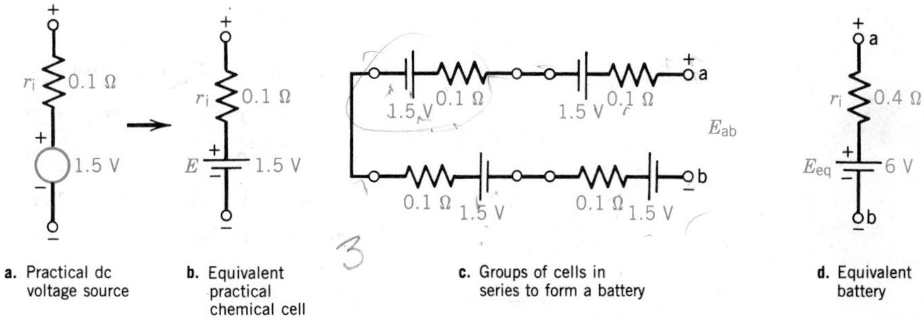

Figure 5-7 Chemical cells connected in series to form a battery

A battery[5] is a group (two or more) of chemical cells connected to produce a desired voltage and power. Such a group of four chemical cells is shown in Fig. 5-7c. In accordance with KVL, the open circuit voltage of the battery, E_{ab}, is the algebraic sum of the individual EMFs of the cells. Since the four cell polarities are aiding, $E_{ab} = 4$ (cells) $\times$ 1.5 V/cell = 6 V. And since the internal resistances are connected in series, from Eq. (5-4) $r_i = 4 \times 0.1\ \Omega = 0.4\ \Omega$. The circuit representing the equivalent battery is shown in Fig. 5-7d.

EXAMPLE 5-10

If one of the cells in Fig. 5-7c is accidentally reversed while making the series connection, calculate

a. The open circuit equivalent voltage
b. The internal resistance of the battery

Solution

a. The algebraic sum is

$$(1.5 + 1.5 - 1.5 + 1.5) \text{ volts} = \mathbf{3\ V} \tag{5-3}$$

b. The internal resistance is unchanged,

$$r_i = 4 \times 0.1 = \mathbf{0.4\ \Omega} \tag{5-4}$$

5-6.1 Terminal Voltage of a Practical Source

Let us connect the battery obtained in Fig. 5-7d to a *load* (a device that receives power from a source). Let us assume that the load is a lamp of resistance R_L as shown in **Fig. 5-8a**. Applying KVL (Eq. 5-2) to the circuit of Fig. 5-8a, starting clockwise at point b, we obtain

[4] In Chapter 2, the boldface letter **E** was used as the symbol for electrical field intensity, measured in newtons per coulomb or volts per meter. In this chapter E is used to designate *open circuit* EMF, measured in volts. The discussion in Sec. 1-2 (item 5) anticipated "some duplication in *quantity* symbols," for reasons indicated. But note that E used for open-circuit EMF is in italics compared to the boldface **E** used for electric field intensity.

[5] Strictly speaking, a battery is defined as a number of *similar* devices arranged as a *group* or *set*. Thus artillerymen speak of a battery of guns and psychologists refer to a battery of tests. In baseball, the pitcher and catcher together are called a battery.

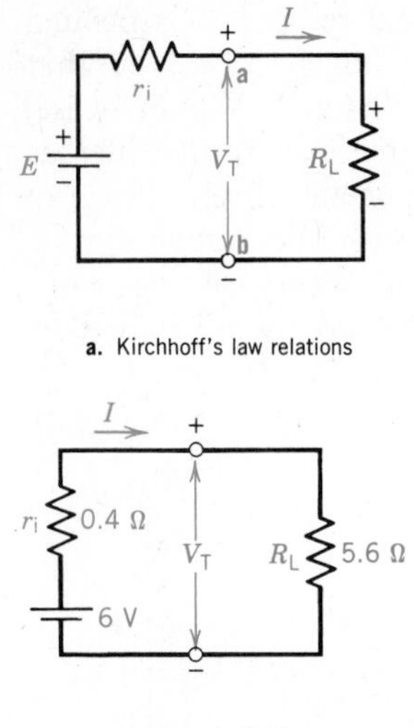

Figure 5-8 Battery under load

$$-E + Ir_i + IR_L = 0 \quad \text{or} \quad E = I(r_i + R_L) \quad \text{volts (V)} \quad (5\text{-}9a)$$

Solving this relation for I, we obtain

$$I = \frac{E}{r_i + R_L} \quad \text{amperes (A)} \quad (5\text{-}9b)$$

Examination of Eq. (5-9b) shows that it is nothing more than an Ohm's law statement for the circuit of Fig. 5-8a. The current in the circuit is the total voltage (open-circuit EMF) divided by the total resistance ($r_i + R_L$).

But going back to Eq. (5-9a), let us define the *terminal voltage* of the battery under load, V_T, as the voltage across the load resistance R_L (or V_{ab} in Fig. 5-8a). We can now rewrite Eq. (5-9a) as

$$E = Ir_i + V_T \quad \text{or} \quad V_T = E - Ir_i \quad \text{volts (V)} \quad (5\text{-}9c)$$

where V_T is the terminal voltage of the practical source (or IR_L)
I is the current drawn from the source
E is the open circuit EMF of the source
r_i is the internal resistance of the source

The insight to be drawn from Eq. (5-9c) is that the terminal voltage, V_T, of a practical dc source *under load* must always be *less* than its open-circuit voltage, E, because of the voltage drop across its internal resistance, Ir_i. The conclusions that emerge from Eq. (5-9) in its various forms are

1. As the load resistance *decreases*, the current drawn from the source *increases* (Eq. 5-9b).
2. As the current drawn from the source increases, the internal volt drop across its internal resistance, Ir_i, increases.
3. As its internal resistance voltage drop increases (with increased load current), its terminal voltage, V_T, decreases (Eq. 5-9c).

EXAMPLE 5-11

A lamp having a resistance of 5.6 Ω is connected across the battery of Fig. 5-7d and the combination is shown in Fig. 5-8b. Calculate

a. The current drawn from the battery
b. The (internal) voltage drop across its internal resistance
c. The terminal voltage of the battery (by two methods)
d. The power dissipated internally within the battery, P_i
e. The power delivered to the load, P_L
f. The total power generated by the battery, P_G, by two methods
g. The efficiency of the battery, η

Solution

a. $I = \dfrac{E}{r_i + R_L} = \dfrac{6\text{ V}}{(0.4 + 5.6)\ \Omega} = \mathbf{1\ A}$ (5-9b)

b. $Ir_i = (1\text{ A})(0.4\ \Omega) = \mathbf{0.4\ V}$

c. $V_T = E - Ir_i = 6\text{ V} - 0.4\text{ V} = \mathbf{5.6\ V}$ (5-9c)
$V_T = IR_L = (1\text{ A})(5.6\ \Omega) = \mathbf{5.6\ V}$

d. $P_i = I^2r_i = (1\text{ A})^2(0.4) = \mathbf{0.4\ W}$

e. $P_L = I^2R_L = (1\text{ A})^2(5.6) = \mathbf{5.6\ W}$

f. $P_G = P_i + P_L = (0.4 + 5.6)\text{ W} = \mathbf{6\ W}$ (5-5)
$P_G = EI = (6\text{ V}) \times (1\text{ A}) = \mathbf{6\ W}$

g. $\eta = P_o/P_{in} = P_L/P_G = 5.6\text{ W}/6\text{ W} = \mathbf{93.\bar{3}\%}$

5-6.2 Voltage Regulation of a Practical Source

Every source of EMF (a dc generator, a battery, or a power supply) is a *practical* source. Consequently, it has internal resistance, and this accounts for the difference between its terminal voltage, V_T, and its no-load or open-circuit voltage, E.

A typical practical voltage source is an automobile 12-V battery, shown in **Fig. 5-9a**. It has an internal resistance of 0.02 Ω. A power rheostat is shown connected as a variable load across the battery terminals to determine its load voltage characteristic or so-called voltage regulation characteristic. A high-resistance voltmeter

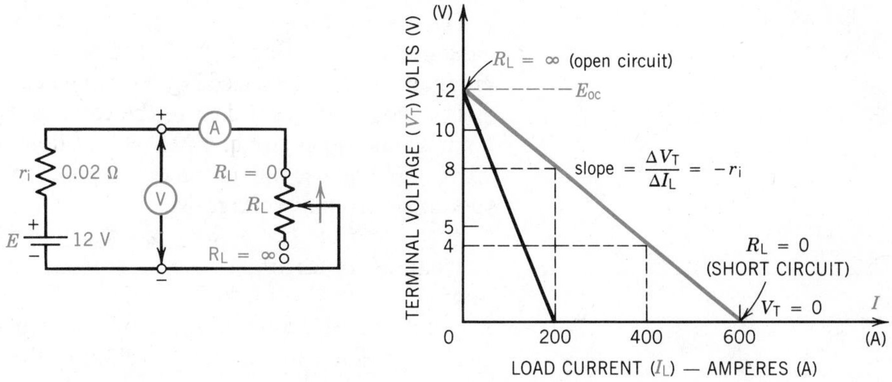

a. Circuit to obtain voltage load characteristic **b.** Voltage regulation characteristic

Figure 5-9 Voltage regulation of a practical source

is connected across its terminals and an ammeter (of negligible resistance) is connected in series with the load, to measure load voltage and load current, V_L and I_L, respectively. The rheostat has an open-circuit position (at the bottom) in Fig. 5-9a and a short-circuit position (at the top), permitting its resistance to be varied from infinity to zero. Common sense and Eq. (5-9c) tell us, in advance of taking data, that

1. When $R_L = \infty$, no current is drawn from the source ($I_L = 0$) and the voltage across the battery is the open-circuit voltage, $E = 12$ V.
2. When $R_L = 0$, maximum current is drawn from the source because the battery is short-circuited. The negligible-resistance ammeter reads $I_{SC} = 12\ \text{V}/0.02\ \Omega = 600$ A. Note that $V_T = 0$ since $R_L = 0$ and $IR_L = 600\ \text{A} \times 0\ \Omega = V_T = 0$ V!
3. Assuming constant internal resistance, the output voltage characteristic may be drawn as shown in Fig. 5-9b between the two coordinates:
   a. Open-circuit condition, $V_T = 12$ V, $I_L = 0$ A
   b. Short-circuit condition, $V_T = 0$ V, $I_L = 600$ A

The linear voltage regulation characteristic shown in Fig. 5-9b permits us to predict the terminal voltage of the storage battery under *any* condition of load. For example, when the load current is 200 A, the terminal voltage is 8 V (Fig. 5-9b), and when the load current is 400 A, the terminal voltage is 4 V (Fig. 5-9b), as verified by Ex. 5-12.

EXAMPLE 5-12

Given an automobile storage battery whose open-circuit EMF is 12 V and whose internal resistance is 0.02 Ω, calculate the terminal voltage for load currents of a. 200 A b. 400 A

Solution

a. $V_T = E - Ir_i = 12 - (200)(0.02) = 8$ V (5-9c)

b. $V_T = E - Ir_i = 12 - (400)(0.02) = 4$ V

5-6.3 Internal Resistance of a Practical Voltage Source

The method shown in Fig. 5-9a to obtain the voltage-load characteristic of a practical source is commonly used for most dc sources. In the case of a lead-acid storage battery, it is permissible to short-circuit the output temporarily since the battery is not damaged by doing so. In other practical voltage sources, such a short circuit would permanently damage the source. This is true, for example, in the case of a dry cell. Consequently, it is necessary to limit the current by using a value of R_L that is sufficiently *high* to stay within the current rating of the practical source, as specified by the manufacturer.

The first method is sometimes called the **open-circuit** and **short-circuit test**, using values of $R_L = \infty$ and $R_L = 0$, respectively.[6] Two separate sets of measurements are made under open-circuit and short-circuit conditions, respectively.

The second method also requires two sets of measurements. The first may be made at the open-circuit EMF (E), although this is not necessarily the case (see Ex. 5-14). The second is a measurement of (safe) load current I_L and its corresponding terminal voltage, V_T.

Regardless of the method of measurement, the same type of calculation for internal resistance applies. Let us consider the open- and short-circuit test method first.

Note that Fig. 5-9b shows that the **(negative)** slope of the voltage regulation characteristic essentially is a function of the internal resistance,[7] or

$$-\Delta r_i = \frac{\Delta V_T}{\Delta I_L} = \frac{E - V_T}{0 - I_L} \quad \text{or} \quad \frac{|V_{T_1} - V_{T_2}|}{|I_{L_1} - I_{L_2}|} \quad \text{ohms } (\Omega) \qquad \textbf{(5-10)}$$

Equation (5-10) shows that whenever I_{L_2} exceeds I_{L_1}, the slope of the voltage regulation characteristic is negative and, consequently, r_i emerges as a negative value. Since the internal resistance of a practical independent voltage (or current) source is a positive value of resistance, we modify Eq. (5-10) to a more useful form, that is,

$$r_i = \frac{V_{T_1} - V_{T_2}}{I_{L_2} - I_{L_1}} \quad \text{ohms } (\Omega) \qquad \textbf{(5-10a)}$$

where V_{T_1} is the *higher* terminal voltage
I_{L_1} is the *lower* load current at V_{T_1}
V_{T_2} is the *lower* terminal voltage
I_{L_2} is the *higher* load current at V_{T_2}

[6] It is essentially this method that is used in testing transformers, alternators, and other types of rotating machinery.

[7] Some practical sources exhibit a *change* in internal resistance with load current. As a consequence, the voltage regulation characteristic is nonlinear (curved). But Eq. (5-10) still holds by taking the slope over *small* incremental ratios of voltage and current. The more vertical the *negative* instantaneous slope, the higher the incremental internal resistance. Finally, the value of r_i in Eq. (5-10) emerges as a negative value because the *slope* is *negative*. The internal resistance, r_i, of any practical voltage source is always a *positive* resistance, as noted in Eq. (5-10a).

EXAMPLE 5-13

Calculate the internal resistance from the load voltage characteristic shown in Fig. 5-9b when the open-circuit voltage is 12 V but the load current at short circuit, I_{L_2}, is **a.** 600 A **b.** 200 A

Solution

a. $$r_i = \frac{V_{T_1} - V_{T_2}}{I_{L_2} - I_{L_1}} = \frac{(12 - 0)\text{ V}}{(600 - 0)\text{ A}} = \mathbf{0.02\ \Omega} \quad (5\text{-}10a)$$

b. $$r_i = \frac{V_{T_1} - V_{T_2}}{I_{L_2} - I_{L_1}} = \frac{(12 - 0)\text{ V}}{(200 - 0)\text{ A}} = \mathbf{0.06\ \Omega} \quad (5\text{-}10a)$$

Both Ex. 5-13 and Fig. 5-9b show that as the internal resistance of a source *increases*, the negative slope also *increases* because the short-circuit current decreases (assuming the open-circuit voltage of the source remains the same).

In many instances, however, it may be damaging to short-circuit a practical voltage source. It is more usual to determine the internal resistance of the source by varying the load current between two (or more) values of load that are *well within* the *rated current* of the source. This method of determining internal resistance is shown in Exs. 5-14 and 5-15.

EXAMPLE 5-14

A load connected across a practical voltage source draws a current of 1.25 A at a terminal voltage of 7.5 V. When another (second) load is substituted, the load current is 2 A and the terminal voltage is 6 V. Calculate the

a. Internal resistance of the practical voltage source
b. Open circuit EMF, E, of the practical voltage source, by two methods
c. Resistance value of the first load
d. Resistance of the second load

Solution

a. $$\mathbf{r}_i = \frac{|V_{T_1} - V_{T_2}|}{|I_{L_2} - I_{L_1}|} = \frac{|7.5 - 6\text{ V}|}{|2 - 1.25|\text{ A}} = 2\ \Omega \quad (5\text{-}10a)$$

b. $E = V_T + Ir_i = 7.5 + (1.25\text{ A})(2\ \Omega) = \mathbf{10\ V}$ (5-9c)

$E = V_T + Ir_i = 6 + (2\text{ A})(2\ \Omega) = \mathbf{10\ V}$ (5-9c)

c. $R_L = V_L/I_L = 7.5\text{ V}/1.25\text{ A} = \mathbf{6\ \Omega}$

d. $R_L = V_L/I_L = 6\text{ V}/2\text{ A} = \mathbf{3\ \Omega}$

EXAMPLE 5-15

A 120 V, 7.2 kW dc generator has a no-load voltage of 130 V. When delivering a current of 12 A, its terminal voltage is 128 V. Calculate the

a. Rated (full load) current of the generator
b. Internal resistance of the generator by two methods
c. Value of current at which the terminal voltage is 124 V

Solution

a. $$I_{L\text{rated}} = \frac{P_o}{V_{l\text{rated}}} = \frac{7200\text{ W}}{120\text{ V}} = \mathbf{60\ A}$$

b. $$r_i = \frac{E_1 - V_{T_2}}{I_{L_2} - I_{L_1}} = \frac{(130 - 128)\text{ V}}{(12 - 0)\text{ A}} = \mathbf{0.1\overline{66}\ \Omega} \quad (5\text{-}10a)$$

$$r_i = \frac{V_{T_1} - V_{T_2}}{I_{L_2} - I_{L_1}} = \frac{(128 - 120)\text{ V}}{(60 - 12)\text{ A}} = \frac{8\text{ V}}{48\text{ A}} = \mathbf{0.16\overline{6}\ \Omega} \quad (5\text{-}10a)$$

c. $$\Delta I_L = \frac{\Delta V_T}{r_i} = \frac{(130 - 124)\text{ V}}{0.1\overline{66}\ \Omega} = \mathbf{36\ A} \quad (5\text{-}10a)$$

The following insights emerge from Exs. 5-14 and 5-15:

1. If it is permissible to short-circuit a practical voltage source, the internal resistance may be obtained from the ratio of its open-circuit voltage to its short-circuit current or $r_i = V_{oc}/I_{sc}$.
2. Alternatively, the methods shown in Exs. 5-14 and 5-15 may be used, without any possible overload and/or consequent damage to the source. The method of Ex. 5-14 uses two safe values of load resistance, R_L, to determine both internal resistance, r_i, and no-load EMF, E.
3. The method of Ex. 5-15b uses the open-circuit voltage, $E = 130$ V, where the load current, I_L, is zero. This method produces essentially the same results.

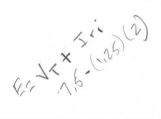

Consequently, if we measure the open-circuit voltage of a practical voltage source and its terminal voltage at one particular load value, we can also find its internal resistance.

5-6.4 Voltage Regulation Calculations

A measure of how well a given practical voltage source maintains its voltage with application of load is important for all practical voltage sources: batteries, power supplies, dc generators, transformers, alternators, and so forth. This measure is called *voltage regulation.*

Voltage regulation, as strictly defined by the IEEE, is the change in output voltage (of a practical source) when the load is reduced from its rated value to zero (load). Expressed as an equation, voltage regulation (**VR**) is

$$\mathbf{VR} = \frac{E_{\text{no load}} - V_{\text{full load}}}{V_{\text{full load}}} = \frac{E - V_{\text{fl}}}{V_{\text{fl}}} \tag{5-11}$$

This ratio (of voltages) may be expressed either as a decimal value or as a percentage. In the latter case, the value is called percent voltage regulation (**PVR**).

EXAMPLE 5-16

Given the data of Ex. 5-15, calculate the percent voltage regulation of the dc generator

Solution

$$\mathbf{PVR} = \frac{E - V_{\text{fl}}}{V_{\text{fl}}} = \frac{130 - 120}{120} = \mathbf{0.077} = \mathbf{7.7\%} \quad (5\text{-}11)$$

Since the numerator of Eq. (5-11) represents the *change in voltage*, the smaller the voltage change, the *lower* the voltage regulation, and the *better* the voltage regulation of the practical source. An ideal source has zero percent regulation; that is, regardless of load it suffers *no change in voltage.*

We know from Eq. (5-9c) that $E = I_{\text{fl}}r_{\text{i}} + V_{\text{fl}}$ and $V_{\text{fl}} = I_{\text{fl}}R_{\text{L}}$. Substituting these in Eq. (5-11) yields

$$\mathbf{VR} = \frac{E - V_{\text{fl}}}{V_{\text{fl}}} = \frac{(I_{\text{fl}}r_{\text{i}} + V_{\text{fl}}) - V_{\text{fl}}}{I_{\text{fl}}R_{\text{L}}} = \frac{I_{\text{fl}}r_{\text{i}}}{I_{\text{fl}}R_{\text{i}}} = \frac{r_{\text{i}}}{R_{\text{L}}} \tag{5-12}$$

Equation (5-12) tells us the distinction between an ideal voltage source and a practical voltage source, first defined in Sec. 2-13. An ideal voltage source has an internal resistance of zero ohms. Such a source has a voltage regulation of **zero**; its terminal voltage remains unchanged regardless of the load current drawn from it.

Equation (5-12) also tells us that if we can keep the internal source resistance, r_{i}, *as small as possible* in proportion to the load resistance, R_{L}, we tend to approach the ideal voltage source.

Finally, it should be noted that some practical voltage sources tend to *increase* their output voltage with application of load (i.e., the full-load voltage is *greater* than the no-load voltage). In this case the voltage regulation and percent voltage regulation are *negative* values.

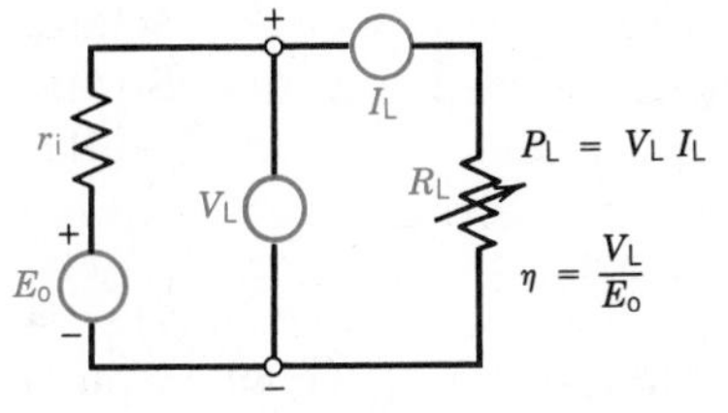

Figure 5-10 Circuit for determining (maximum) power transfer from source to load

5-7 MAXIMUM POWER TRANSFER THEOREM

We are now prepared to consider the conditions under which a practical voltage (or current) source may deliver maximum power to a load, R_L, connected across the terminals of the source. Consider a variable-resistance load, R_L, connected as shown in **Fig. 5-10**, to a practical voltage source whose no-load voltage, E_0, and internal resistance, r_i, are both constant. Our object is to find the value of R_L that will dissipate maximum power in R_L as measured by the product $V_L I_L$. In predicting this, in advance, let us consider two extreme cases and analyze each as shown below.

Parameter	R_L Very High (almost infinite)	R_L Very Low (almost a short circuit)
Load current, I_L	Very small, $I_L \cong 0$	Almost maximum, $I_L \cong E_0/r_i$
Load voltage, V_L	Almost maximum, $V_L \cong E_0$	Very small, $V_L \cong 0$
$P_L = V_L I_L$	Very small product	Very small product

Analysis of the above tabulation shows the following:

1. P_L, the power dissipated in the load, is small when R_L is very high and when R_L is very low, for reasons indicated in the foregoing tabulation.
2. As R_L decreases, I_L increases from approximately zero to its maximum, E_0/r_i.
3. But as R_L decreases, V_L also decreases from its maximum value of E_0.
4. Since P_L is the product of an increasing I_L and decreasing V_L, the point where these two cross should be the maximum value of P_L.

A picture is worth a thousand words. **Figure 5-11** shows the four variables R_L, V_L, I_L, and P_L plotted as a function of load current. To eliminate the need for four vertical ordinate scales, three variables have been normalized to unity. They are expressed as the ratios V_L/E_0, I_L/I_{sc}, and P_L/P_{max}. Note from Fig. 5-11 that when $I_L = 0$, $V_L = E_0$ and the ratio $V_L/E_0 = 1$. Similarly, when the ratio $V_L/E_0 = 0$, then $V_L = 0$ and $I_L = I_{sc} = E_0/r_i$, as shown in Fig. 5-11, and the ratio $I_L/I_{sc} = 1$.

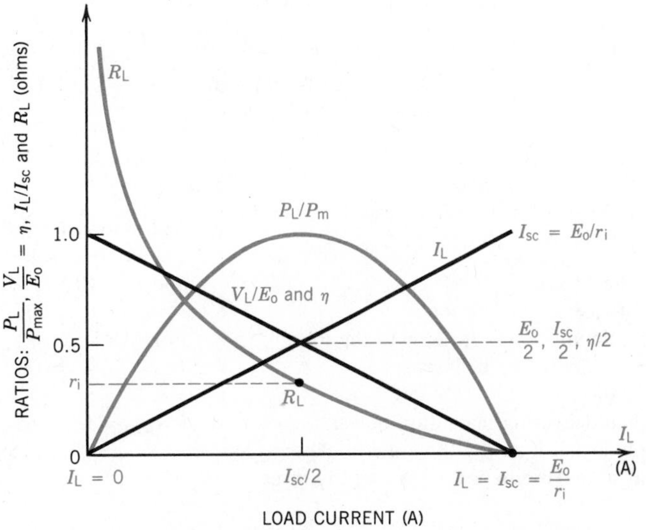

Figure 5-11 Various simultaneous conditions related to maximum power transfer

From a very careful study of Fig. 5-11, it may be concluded that the MAXIMUM POWER TRANSFER (MPT) OCCURS IN R_L WHEN

1. $V_L = E_0/2$ **(5-13)**
2. $I_L = I_{sc}/2 = E_0/2r_i$ **(5-14)**

3. Efficiency, $\eta = P_L/P_T = V_L I_L/E_0 I_L = V_L/E_0 = 50\%$ **(5-15)**
4. $P_L = P_{max} = E_0^2/4r_i = E_0^2/4R_L$ (the product of 1 and 2 above) **(5-16)**
5. $R_L = r_i$ **(5-17)**

That MPT occurs when $R_L = r_i$ is proved in Appendix A-6.1, where the maximum power transfer theorem is derived by techniques of ordinary differential calculus. When the five conditions in Eqs. (5-13) to (5-17) occur simultaneously, the maximum power transfer theorem (MPT) states that

maximum power is transferred from a practical independent dc source to a load when the load resistance is equal to the internal resistance of the practical source.[8]

With regard to the statement of the MPT, the following important practical points should not be overlooked:

1. Maximum power transfer is *not* the same as maximum efficiency.[9] In the case of a variable load connected to a practical voltage source, maximum or 100% efficiency occurs when $R_L = \infty$, that is, the ratio $V_L/E_0 = 1$ (see Eq. 5-15).
2. The intent of maximum power transfer is to maximize the power in R_L and to draw maximum power from the source (Fig. 5-11).
3. In dc circuits MPT always occurs at an efficiency of 50%, and this means that 50% of the power is dissipated internally within the source itself.
4. In Fig. 5-11 the efficiency curve and the curve of V_L are identical for reasons stated in Eq. (5-15).
5. Since the efficiency $\eta = P_L/P_T$ at all times, it is easy to show that the efficiency at any load value may be found more simply from

$$\eta = \frac{P_L}{P_T} = \frac{I_L^2 R_L}{I_L^2(R_L + r_i)} = \frac{R_L}{r_i + R_L} = \frac{V_L}{E_0} \quad \text{(5-15a)}$$

6. Point 5 shows that if we know the value of r_i, for any given value of R_L, we can determine the efficiency. Similarly, if V_L is measured and E_0 is known, the efficiency can be determined from Eq. (5-15a).

EXAMPLE 5-17

A dc power supply has a no-load voltage of 30 V dc and delivers 100 W at a full-load current of 5 A. Calculate the

a. Internal resistance of the source
b. Value of the load resistance, R_L, at full load
c. Load resistance for maximum power transfer to the load
d. Maximum power the dc supply can deliver
e. Full-load voltage regulation
f. Voltage regulation when delivering maximum power
g. Efficiency of the supply at full load (by three methods)
h. Efficiency of the supply at maximum power transfer (by three methods)

Solution

a. At full load $V_L = P_L/I_L = 100\text{ W}/5\text{ A} = 20\text{ V}$ (rated voltage). The internal volt drop at full load $Ir_i = 30\text{ V} - 20\text{ V} = 10\text{ V}$ and $r_i = 10\text{ V}/5\text{ A} = \mathbf{2\ \Omega}$
b. $R_L = V_L/I_L = 20\text{ V}/5\text{ A} = \mathbf{4\ \Omega}$
c. For MPT, $R_L = r_i = \mathbf{2\ \Omega}$
d. $P_m = E_0^2/4r_i = (30)^2/4 \times 2 = \mathbf{112.5\ W}$ (5-16)
e. $VR = (V_{nl} - V_{fl})/V_{fl} = (30 - 20)/20 = 0.5 = \mathbf{50\%}$ (5-11)
f. $VR_{MPT} = (E_0 - E_0/2)/(E_0/2)$
$= (30 - 15)/15 = \mathbf{100\%}$ (5-11)

---

[8] The MPT theorem requires some further refinement and qualification when considering practical **ac** sources. As stated here, the MPT theorem holds for independent practical current and voltage **dc** sources, only.

[9] Maximum *efficiency* occurs in most machines where the fixed and variable losses are equal.

g. $\eta = P_L/P_T = P_L/E_0 I_L$
$= 100\ \text{W}/(30\ \text{V} \times 5\ \text{A}) = \mathbf{66.\bar{6}\%}$ (5-15)
$\eta = V_L/E_0 = 20\ \text{V}/30\ \text{V} = \mathbf{66.\bar{6}\%}$ (5-15a)
$\eta = R_L/(r_i + R_L) = 4\ \Omega/(2 + 4)\ \Omega = \mathbf{66.\bar{6}\%}$ (5-15a)

h. $\eta = V_L/E_0 = 15\ \text{V}/30\ \text{V} = \mathbf{50\%}$ (5-15)
$= P_L/P_T = (15\ \text{V})(7.5\ \text{A})/30 \times 7.5$
$= 112.5\ \text{W}/225\ \text{W} = \mathbf{50\%}$ (5-15)
$= R_L/(r_i + R_L) = 2\ \Omega/(2\ \Omega + 2\ \Omega) = \mathbf{50\%}$ (5-15a)

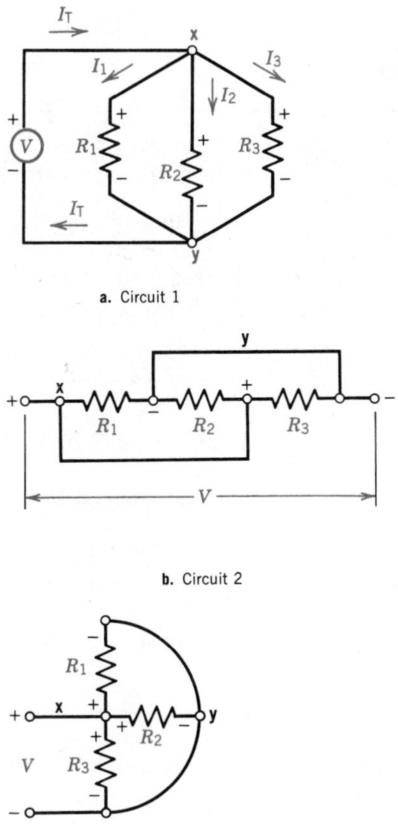

Figure 5-12 Parallel circuits

5-8 PARALLEL CIRCUIT

Regardless of circuit complexity, components are connected in *parallel* when they have the *same* potential applied across their terminals, that is, they are subjected to the *same* electrical field. **Figure 5-12a** shows a voltage source, V, connected across a parallel circuit consisting of three resistors, R_1, R_2, and R_3. With respect to the parallel circuit, please note that

1. There is only one voltage, V, connected across each of the three resistors.
2. There are three paths for division of current, one for each resistor, I_1, I_2, and I_3.
3. The total current, I_T, supplied by the source, V, *divides* among the three resistors and *recombines* in returning to the source.

A second parallel circuit is shown in Fig. 5-12b. At first glance it may not appear as if the three resistors are connected in parallel. But since the conducting wires are assumed to have zero resistance, if we follow each wire to its terminal (or node) and label the polarities of these wires, we discover that *each* resistor has one terminal connected to the positive terminal of the supply and the other terminal to the negative terminal of the supply. Since this is so, circuit 2 is identical to circuit 1.

A third parallel circuit is shown in Fig. 5-12c. Like the two previous circuits, it is also a simple parallel circuit since each resistor is connected directly across the supply (one terminal of each resistor is positive and one terminal of each resistor is negative).

The foregoing discussion provides the basic definition of a parallel circuit. For n devices in parallel across voltage V we may write

$$V \equiv V_1 \equiv V_2 \equiv V_3 \equiv \cdots \equiv V_n \qquad \text{volts (V)} \qquad \textbf{(5-18)}$$

5-8.1 Current Relations of a Parallel Circuit

If we think of current in terms of $I = Q/t$, we may intuitively determine the current relations of a parallel circuit. Returning to Fig. 5-12a, assume that the total current, I_T, shown is 60 A or 60 C/s. Further assume that at junction x, *each second*, there are 30 C entering R_1, 20 C entering R_2, and (you guessed it) 10 C entering R_3. At junction y, these charges recombine to form the total current of 60 C/s = I_T. If we extend this reasoning to n resistors in parallel, we may write that the sum of the branch currents is equal to the total current or

$$I_T = I_1 + I_2 + I_3 + \cdots + I_n = \frac{V_T}{R_T} = V_T G_T \qquad \text{amperes (A)} \qquad \textbf{(5-19)}$$

5-8.2 Resistance Relations of a Parallel Circuit

Equations (5-18) and (5-19) enable us to determine and derive the resistance relations of a parallel circuit by simple use of Ohm's law:

$$I_T = I_1 + I_2 + I_3 + \cdots + I_n \qquad \text{amperes (A)} \qquad (5\text{-}19)$$

$$\frac{V_T}{R_T} = \frac{V_1}{R_1} + \frac{V_2}{R_2} + \frac{V_3}{R_3} + \cdots + \frac{V_n}{R_n}$$

but since *all* voltages are the SAME, dividing by V yields

$$\frac{1}{R_{eq}} = \frac{1}{R_T} = \frac{1}{R_1} + \frac{1}{R_2} + \frac{1}{R_3} + \cdots + \frac{1}{R_n} \qquad \text{siemens (S)} \qquad (5\text{-}20)$$

Equation (5-20) implies that the equivalent (total) parallel resistance is smaller than any of the individual resistances connected in parallel. Another way of writing Eq. (5-20), to prove this, is

$$R_{eq} = \frac{1}{1/R_1 + 1/R_2 + 1/R_3 + \cdots + 1/R_n} \qquad \text{ohms } (\Omega) \qquad (5\text{-}20a)$$

Both Eqs. (5-20) and (5-20a) tell us that the equivalent resistance of n resistors in parallel is the reciprocal of the sum of the reciprocals of the individual resistances in parallel. The reciprocal of resistance is conductance, G, from Eq. (3-1).

But observe that Eq. (5-20) is not expressed in ohms **but in siemens**. This gives an important clue to a simpler relation, which can be written as

$$G_T = G_1 + G_2 + G_3 + \cdots + G_n \qquad \text{siemens (S)} \qquad (5\text{-}21)$$

Equation (5-21) tells us that the total conductance of a parallel circuit is the sum of the individual conductances connected in parallel. Consequently, adding *additional* resistors in parallel *increases* the total conductance, G_T, and *reduces* the total resistance, R_T.

5-8.3 Power Relations of a Parallel Circuit

As in the case of resistance relations, we may derive the power relations from Eqs. (5-18) and (5-19) by simple multiplication or

$$I_T = I_1 + I_2 + I_3 + \cdots + I_n \qquad \text{amperes (A)} \qquad (5\text{-}19)$$

and since the voltages are the *same*, multiplying by a common voltage factor gives $V_T I_T = V_1 I_1 + V_2 I_2 + V_3 I_3 + \cdots + V_n I_n$ or

$$P_T = P_1 + P_2 + P_3 + \cdots + P_n \qquad \text{watts (W)} \qquad (5\text{-}22)$$

Observe that Eq. (5-22) for parallel circuits is exactly the same as Eq. (5-5) for series circuits. This implies that, regardless of whether resistors are connected in series or parallel or any combination of both, the total power dissipated is always the sum of the individually dissipated powers (see Sec. 5-13, Table 5-1).

EXAMPLE 5-18

Six resistors connected in parallel have the following values: 120, 60, 40, 5, 4, and 2 kΩ. Calculate the

a. Equivalent conductance of the parallel combination
b. Equivalent resistance of the parallel combination
c. Circuit voltage if 20 mW is dissipated by the 5-kΩ resistor
d. Current in the 40-kΩ resistor
e. Total current drawn from the supply
f. Total power dissipated by all resistors

Solution

a. $$G_T = \left(\frac{1}{120\ \text{k}\Omega}\right) + \left(\frac{1}{60\ \text{k}\Omega}\right) + \left(\frac{1}{40\ \text{k}\Omega}\right) + \left(\frac{1}{5\ \text{k}\Omega}\right) + \left(\frac{1}{4\ \text{k}\Omega}\right) + \left(\frac{1}{2\ \text{k}\Omega}\right) = 0.001\ \text{S} = \mathbf{1\ mS} \quad (5\text{-}21)$$

b. $R_T = 1/G_T = 1/1\ \text{mS} = \mathbf{1\ k\Omega}$

c. $V = \sqrt{PR} = \sqrt{(20\ \text{mW})(5\ \text{k}\Omega)} = \mathbf{10\ V}$ (4-3)

d. $I = V/R = 10\ \text{V}/40\ \text{k}\Omega = \mathbf{0.25\ mA}$

e. $I_T = VG_T = (10\ \text{V})(1\ \text{mS}) = \mathbf{10\ mA}$

f. $P_T = VI_T = (10\ \text{V})(10\ \text{mA}) = \mathbf{100\ mW}$

EXAMPLE 5-19

Two resistors of 12 and 6 kΩ are connected in parallel. The current in the first resistor is 2 mA. Calculate the

a. Equivalent resistance of the combination
b. Supply voltage
c. Current in the 6-kΩ resistor
d. Total current drawn from the supply
e. Power dissipated by each resistor
f. Total power drawn from the supply, by two methods

Solution

a. $$R_{eq} = \frac{1}{1/12\ \text{k} + 1/6\ \text{k}} = \mathbf{4\ k\Omega} \quad (5\text{-}20a)$$

b. $V = V_1 = I_1R_1 = (2\ \text{mA})(12\ \text{k}\Omega) = \mathbf{24\ V}$

c. $I_2 = V/R_2 = 24\ \text{V}/6\ \text{k}\Omega = \mathbf{4\ mA}$

d. $I_T = I_1 + I_2 = (2 + 4)\ \text{mA} = \mathbf{6\ mA}$

e. $P_1 = V^2/R_1 = 24^2/12\ \text{k} = \mathbf{48\ mW}$ (4-3)
$P_2 = V^2/R_2 = 24^2/6\ \text{k} = \mathbf{96\ mW}$

f. $P_T = P_1 + P_2 = (48 + 96)\ \text{mW} = \mathbf{144\ mW}$ (5-22)
$P_T = VI_T = (24\ \text{V})(6\ \text{mA}) = \mathbf{144\ mW}$

5-9 THE TWO-BRANCH PARALLEL CIRCUIT— SPECIAL CASES AND RELATIONS

Example 5-19 could have been solved much more easily by using certain special relations for the two-branch parallel circuit derived in Appendix A-7. These derived relations are stated below.

Given R_1 and R_2 as two resistors in parallel, then the equivalent resistance, R_{eq}, is

$$R_{eq} = \frac{R_1R_2}{R_1 + R_2} \quad \text{ohms } (\Omega) \qquad \mathbf{(5\text{-}23)}$$

If we know the equivalent resistance and either one of the resistors, the other is

$$R_1 = \frac{R_2R_{eq}}{R_2 - R_{eq}} \quad \text{(5-23a)} \qquad R_2 = \frac{R_1R_{eq}}{R_1 - R_{eq}} \quad \text{(5-23b)}$$

A corollary of Eq. (5-23) (left as a proof for the reader in Prob. 5-61) is the following relation. Given a resistance, R_2, in parallel with another resistance, R_1, of value R_2/n, then the equivalent resistance

$$R_{eq} = \frac{R_2}{1+n} \qquad \text{where } n = \frac{R_2}{R_1} > 1 \qquad \text{(5-23c)}$$

From Appendix A-7C, the branch currents, I_1 and I_2, in any two-branch parallel circuit may be found by using the current divider rule (CDR) from

$$I_2 = I_T \frac{R_1}{R_1 + R_2} \quad \text{(5-24a)} \qquad I_1 = I_T \frac{R_2}{R_1 + R_2} \quad \text{(5-24b)}$$

A corollary of Eqs. (5-24a) and (5-24b) (left as a proof for the reader in Prob. 5-62) is the following relation, known as the resistance ratio rule (RRR). Defining $n = R_2/R_1$, where $R_2 > R_1$, we may write the branch currents as

$$I_2 = \frac{I_T}{n+1} \quad \text{(5-25a)} \qquad I_1 = \frac{n}{n+1} I_T \quad \text{(5-25b)}$$

5-10 SPECIAL CASES OF *N* IDENTICAL RESISTORS IN PARALLEL

Two very simple relationships remain to be recorded here. They are also left as an exercise for the student to derive.

Given a circuit consisting of N identical resistors in parallel, each having the value R_1, then the equivalent resistance, R_{eq}, is

$$R_{eq} = R_T = \frac{R_1}{N} \qquad \text{ohms } (\Omega) \qquad \text{(5-26)}$$

The second relationship stems from the first and has two forms. Given a parallel circuit consisting of N identical resistors, where I_1 is the current in any branch, then the total current, I_T, and total power, P_T, are

$$I_T = N \times I_1 \qquad \text{amperes (A)} \qquad \text{(5-27)}$$

$$P_T = N \times P_1 \qquad \text{watts (W)} \qquad \text{(5-28)}$$

EXAMPLE 5-20

Two resistors, R_1 and R_2, are connected in parallel across a voltage supply, V. The total current drawn from the supply is 50 mA; $R_1 = 20$ kΩ and $R_2 = 5$ kΩ. Calculate the

a. Equivalent resistance of the combination by two methods
b. Supply voltage across the resistors, V
c. Current drawn by R_1 by three different methods

Solution

a. $R_{eq} = \dfrac{R_1 R_2}{(R_1 + R_2)} = \dfrac{20 \times 5}{20 + 5}\ \text{k}\Omega = \mathbf{4\ k\Omega}$ or (5-23)

$R_{eq} = \dfrac{R_1}{1+n} = \dfrac{20\ \text{k}\Omega}{1+4} = \mathbf{4\ k\Omega}$ (5-23c)

b. $V = I_T R_{eq} = 50\ \text{mA} \times 4\ \text{k}\Omega = \mathbf{200\ V}$

c. $I_1 = V/R_1 = \dfrac{200\ \text{V}}{20\ \text{k}\Omega} = \mathbf{10\ mA}$

$I_1 = I_T \dfrac{R_2}{R_1 + R_2} = 50\ \text{mA}\ \dfrac{5\ \text{k}\Omega}{(20+5)\ \text{k}\Omega} = \mathbf{10\ mA}$ (5-24b)

$I_1 = \dfrac{I_T}{n+1} = \dfrac{50\ \text{mA}}{4+1} = \mathbf{10\ mA}$ (5-25a)

EXAMPLE 5-21

A 100-V dc source supplies a current of 5 mA to two resistors in parallel. If $R_1 = 100\ \text{k}\Omega$, calculate the

a. Current in R_2
b. Value of R_2
c. Equivalent circuit resistance by two methods
d. Current in R_2 by three additional methods (excluding **a** above)

Solution

a. $I_1 = V/R_1 = 100\ \text{V}/100\ \text{k}\Omega = 1\ \text{mA}$ and $I_2 = I_T - I_1 = (5 - 1)\ \text{mA} = \mathbf{4\ mA}$

b. $R_2 = V/I_2 = \dfrac{100\ \text{V}}{4\ \text{mA}} = \mathbf{25\ k\Omega}$

c. $R_{eq} = V/I_T = 100\ \text{V}/5\ \text{mA} = \mathbf{20\ k\Omega}$

$R_{eq} = \dfrac{R_1 R_2}{R_1 + R_2} = \dfrac{100 \times 25}{100 + 25}\ \text{k}\Omega = \mathbf{20\ k\Omega}$ (5-23)

d. $I_2 = \dfrac{n}{n+1} I_T = \dfrac{4}{4+1}\ 5\ \text{mA} = \mathbf{4\ mA}$ (5-25b)

$= I_T \dfrac{R_1}{R_1 + R_2} = 5\ \text{mA}\ \dfrac{100\ \text{k}\Omega}{125\ \text{k}\Omega} = \mathbf{4\ mA}$ (5-24a)

$I_2 = V/R_2 = 100\ \text{V}/25\ \text{k}\Omega = \mathbf{4\ mA}$

EXAMPLE 5-22

Fifty identical resistors, each 100 kΩ, are connected in parallel across a 200-V supply. Calculate the

a. Equivalent resistance of the combination
b. Total current drawn by two different methods
c. Current in any single branch by three methods
d. Power dissipated by a single resistor
e. Total power dissipated by four different methods

Solution

a. $R_{eq} = R_1/N = 100\ \text{k}\Omega/50 = \mathbf{2\ k\Omega}$ (5-26)

b. $I_T = V/R_{eq} = 200\ V/2\ \text{k}\Omega = \mathbf{100\ mA}$

$I_T = NI_1 = NV/R_1 = 50 \times 200\ \text{V}/100\ \text{k}\Omega = \mathbf{100\ mA}$ (5-27)

c. $I_1 = V/R_1 = 200\ \text{V}/100\ \text{k}\Omega = \mathbf{2\ mA}$

$I_1 = I_T \dfrac{R_{eq}}{R_1} = 100\ \text{mA}\ \dfrac{2\ \text{k}\Omega}{100\ \text{k}\Omega} = \mathbf{2\ mA}$

$I_1 = I_T/N = 100\ \text{mA}/50 = \mathbf{2\ mA}$ (5-27)

d. $P_1 = VI_1 = 200\ \text{V} \times 2\ \text{mA} = \mathbf{400\ mW}$

e. $P_T = NP_1 = 50 \times 0.4\ \text{W} = \mathbf{20\ W}$ (5-28)

$P_T = V^2/R_{eq} = 200^2/2\ \text{k}\Omega = \mathbf{20\ W}$

$P_T = VI_T = 200\ \text{V} \times 100\ \text{mA} = \mathbf{20\ W}$

$P_T = I_T^2 R_{eq} = (0.1)^2 \times 2\ \text{k}\Omega = \mathbf{20\ W}$

5-11 KIRCHHOFF'S CURRENT LAW (KCL)

A fundamental understanding of Kirchhoff's current law (KCL) requires some preliminary definitions:

1. An electric *network* is the interconnection of two or more circuit elements, either in series or in parallel.
2. An electric *circuit* is an electric network containing at least one *closed path*.
3. A *node* is a point at which two or more elements have a common connection. A simple parallel circuit of the kind we have considered thus far contains only two nodes.
4. A *branch* is a single path in a network. (The network of Ex. 5-22 has 51 branches, including the supply through which total current flows, but only two nodes.)
5. A *path* is a trace taken through an electric network starting at a node and passing through different elements to different nodes.
6. A *loop* is a *closed path* starting at a given node, passing through different elements to different nodes, and returning to the node at which the path began.

7. A *mesh* is a *shortest* possible loop in a network. A network may contain one or more meshes.

We are now ready to consider the second law advanced by Kirchhoff, known as Kirchhoff's current law (KCL). It states that *the algebraic sum of all currents entering and leaving a node is* **zero**. In symbolic form this is summarized as

$$\sum_{n=1}^{N} I_n = 0 \qquad \text{amperes (A)} \qquad \textbf{(5-29)}$$

Kirchhoff defined the currents *entering* a node as *positive* and the currents *leaving* the node as *negative*. In simple terms we may say that "whatever goes into a node must come out of the node" as a statement of KCL.

Kirchhoff's current law is nothing more than the physical law of conservation of charge. Since current is a rate of charge flow, *in a given time*, the total charges *entering* the node must also *leave* the node, since *charge must be conserved*.

Two alternative forms of Eq. (5-29) are

$$\sum I_{\text{entering}} - \sum I_{\text{leaving}} = 0 \qquad \textbf{(5-29a)}$$
$$\sum I_{\text{entering}} = \sum I_{\text{leaving}} \qquad \text{amperes (A)} \qquad \textbf{(5-29b)}$$

EXAMPLE 5-23

In **Fig. 5-13a** the known currents are $I_1 = 4$ A, $I_2 = 3$ A, $I_3 = 2$ A, $I_4 = 5$ A, $I_5 = 1$ A, and $I_6 = 2$ A, each having the current direction shown. Calculate

a. Current in resistor R, I_R
b. Current and direction of I_X

Solution

a. Since there are two unknown currents around node B, we begin with node A. For node A we write $I_1 + I_2 + I_3 - I_R = 0$ and $I_R = (4 + 3 + 2)$ A $=$ **9 A** (5-29a)
b. For node B we can now write: $I_X = I_R + I_6 + I_4 - I_5 = (9 + 2 + 5) - 1 =$ **15 A**; I_X is a current of **15 A** *leaving* node B, since $I_X + I_5 = I_R + I_6 + I_4$

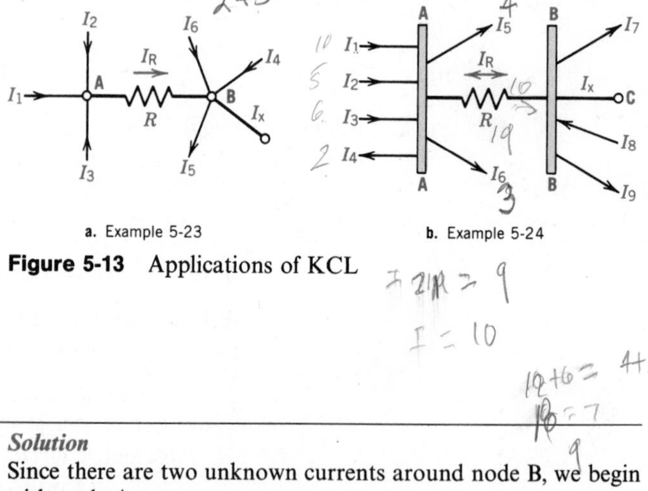

a. Example 5-23 b. Example 5-24

Figure 5-13 Applications of KCL

EXAMPLE 5-24

Figure 5-13b shows two copper busbars, A and B, whose known currents are $I_1 = 10$ A, $I_2 = 5$ A, $I_3 = 6$ A, $I_4 = 2$ A, $I_5 = 4$ A, $I_6 = 3$ A, $I_7 = 4$ A, $I_8 = 6$ A, and $I_9 = 3$ A, each having the current direction shown in Fig. 5-13b. Calculate

a. Current and direction of I_R in resistor R
b. Current and direction of I_X

Solution

Since there are two unknown currents around node B, we begin with node A.

a. $I_R = I_1 + I_2 + I_3 - I_4 - I_5 - I_6$
$= 10 + 5 + 6 - 2 - 4 - 3 =$ **12 A** (5-29a)
$I_R =$ **12 A** *leaving* node A (since this current is -12 A when transposed in the above equation, leaving zero on the right side)
b. $I_X = I_R + I_8 - I_7 - I_9 = 12 + 6 - 4 - 3 =$ **11 A**
$I_X =$ **11 A** *leaving* node B and *entering* node C

5-12 VOLTAGE SOURCES IN PARALLEL

In Sec. 5-6 we considered voltage sources in series. We saw that the purpose of connecting voltage sources in series is to increase the total voltage of the combination. Thus, a battery (consisting of four cells in series, each having an EMF of 1.5 V and an internal resistance of 0.1 Ω) will have an open-circuit EMF of 6 V and an internal resistance of 0.4 Ω. But such a battery, as we saw, is limited by two factors, namely (1) the current it can deliver safely and (2) its internal voltage drop, Ir_i, which reduces its terminal voltage whenever current is drawn.

Voltage sources are connected in parallel to *increase* the *current* and power rating of the battery combination. In connecting such sources in parallel, the internal resistance is also reduced, which improves the voltage regulation of the *battery* simultaneously.

Throughout this section we will use chemical cells and batteries for the purpose of simplification. But the discussion applies equally well to dc generators, dc power supplies, and other dc sources. The discussion that follows is divided into three major considerations of voltage sources *in parallel*: (1) equal EMFs and equal internal resistances, (2) equal EMFs and unequal internal resistances, and (3) unequal EMFs and unequal internal resistances.

5-12.1 Sources in Parallel Having Equal EMFs and Equal Internal Resistances

Figure 5-14a shows a circuit consisting of four batteries, each having an EMF of 6 V and an internal resistance of 0.1 Ω, connected to a load of 2.975 Ω. The batteries are connected in *parallel*. Since each battery has an EMF of 6 V, the combination of the equivalent battery must be 6 V as shown in Fig. 5-14b.[10] But what of the internal resistances?

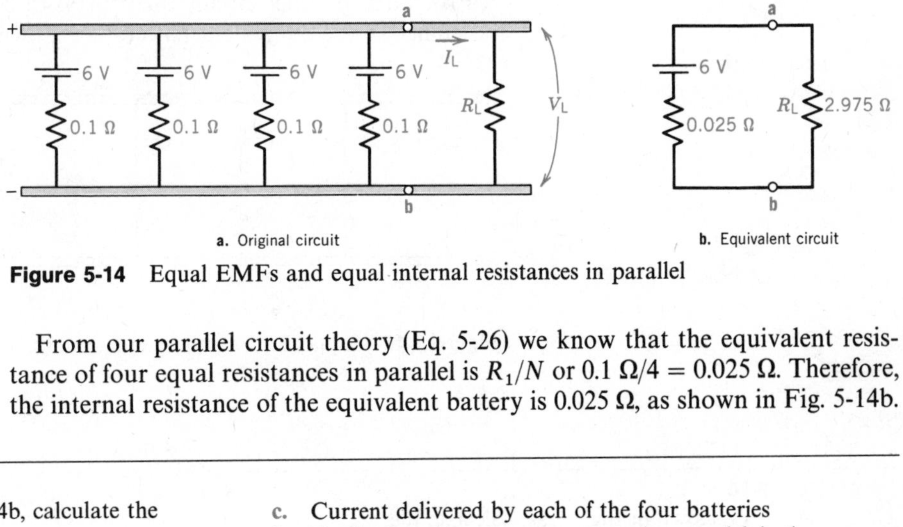

Figure 5-14 Equal EMFs and equal internal resistances in parallel

From our parallel circuit theory (Eq. 5-26) we know that the equivalent resistance of four equal resistances in parallel is R_1/N or $0.1\ \Omega/4 = 0.025\ \Omega$. Therefore, the internal resistance of the equivalent battery is 0.025 Ω, as shown in Fig. 5-14b.

EXAMPLE 5-25

Using the equivalent circuit of Fig. 5-14b, calculate the

a. Load current I_L in resistor R_L
b. Terminal voltage of the battery, $V_{ab} = V_L$
c. Current delivered by each of the four batteries
d. Verify that the voltage across nodes *a* and *b* is the same as the voltage V_L

[10] See Prob. 5-63 for a proof of this statement using Eq. (5-8).

Solution

a. $I_L = \dfrac{E}{R_L + r_i} = \dfrac{6\text{ V}}{(2.975 + 0.025)\ \Omega} = \dfrac{6\text{ V}}{3\ \Omega} = \mathbf{2\ A}$ (5-9b)

b. $V_L = I_L R_L = (2\text{ A})(2.975\ \Omega) = \mathbf{5.95\ V}$

c. Since the batteries are identical,

$$I_1 = I_T/N = 2\text{ A}/4 = \mathbf{0.5\ A} \quad (5\text{-}27)$$

d. Since the current in each battery is 0.5 A, we can calculate the terminal voltage across each battery as

$$V_T = E - Ir_i = 6 - (0.5\text{ A})(0.1\ \Omega) = \mathbf{5.95\ V} \quad (5\text{-}9c)$$

Note that the solution to Ex. 5-25**d** yields the *same* voltage as the voltage across the load, V_L.

There are two nodes in Fig. 5-14a and five branches (see definitions in Sec. 5-11). Each branch is connected in parallel across the two nodes. By definition, branches in parallel MUST have the SAME voltage across them for the simple reason that they are *connected in parallel*. This is the definition of a parallel circuit (Sec. 5-8, Eq. 5-18).

5-12.2 Sources in Parallel Having Equal EMFs and Unequal Internal Resistances

It is well known that as a chemical zinc–carbon cell ages (and its chemical energy is consumed), its internal resistance increases. As long as the cell is not completely consumed, its open-circuit EMF is usually approximately the same as when the cell was brand new.

Figure 5-15a shows four batteries, each having an EMF of 6 V but each having a *different* internal resistance, connected to a load, R_L. As in the previous example, we know that since the open-circuit EMFs of the batteries are 6 V, then the EMF of the equivalent battery must be 6 V. But what of the internal resistances?

From our parallel circuit theory (Eq. 5-20a) we can compute the equivalent resistance of the equivalent battery, as shown by Ex. 5-26.

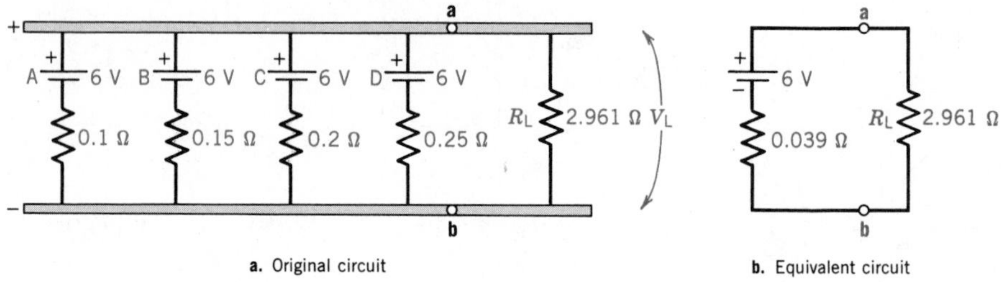

Figure 5-15 Equal EMFs and unequal internal resistances in parallel

---

EXAMPLE 5-26

Given the circuit shown in Fig. 5-15a, calculate the

a. Internal resistance of the equivalent battery
b. Load current, I_L, in resistor R_L
c. Terminal voltage of the battery, $V_{ab} = V_L$
d. Current delivered by each battery, respectively

Solution

a. $$R_{eq} = \frac{1}{1/R_1 + 1/R_2 + 1/R_3 + 1/R_4} = \frac{1}{1/0.1 + 1/0.15 + 1/0.2 + 1/0.25} = \mathbf{0.039\ \Omega} \quad (5\text{-}20a)$$

b. The equivalent battery having an open-circuit EMF of 6 V and an internal resistance of 0.039 Ω connected across R_L is shown in Fig. 5-15b. Then

$$I_L = \frac{E}{R_L + r_i} = \frac{6\text{ V}}{(2.961 + 0.039)\ \Omega} = \mathbf{2\ A} \quad (5\text{-}9b)$$

c. $V_L = I_L R_L = (2\text{ A})(2.961\ \Omega) = \mathbf{5.922\ V}$

d.

Check: I_A +

114 DC SERIES AND PARALLEL CIRCUITS

EXAMPLE 5-27

Given the circuit shown in Fig. 5-16a, ...

a. Internal resistance of the equivalent ba...
b. EMF of the I_L, in ...
c. Load current, ...
d. Terminal voltage ...
e. Current deliv...

Solution

a. Si...

Note from Ex. 5-26**d** that the sum of ... ...ividual batteries is, in fact, the same as the tota... ... the equivalent battery. That these (different) currents pro... ...e as that obtained for the equivalent battery is a verification of our ... regarding treatment of sources in parallel having equal EMFs but unequal internal resistances. The cell having the highest resistance delivers the least current.

5-12.3 Sources in Parallel Having Unequal EMFs and Unequal Internal Resistances

At present, we are equipped to deal with only *two* sources in parallel having unequal EMFs and unequal internal resistances.[11] **Figure 5-16a** shows such a circuit. Now the open-circuit EMF of the *equivalent battery* is an unknown quantity. Let us call that quantity E_x. We can guess that E_x is somewhere between +7.2 and +6.3 V, but in light of the *unequal* internal resistances, we can only guess at its value.

But if we place a ground terminal at node b and remove the load resistance R_L we already have learned a technique for finding E_x, as shown in Fig. 5-16b.

Example 5-27 shows how we can solve problems of this kind, using the general voltage divider Eq. (5-8) developed earlier in this chapter.

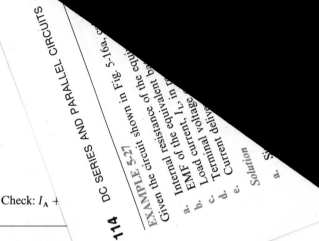

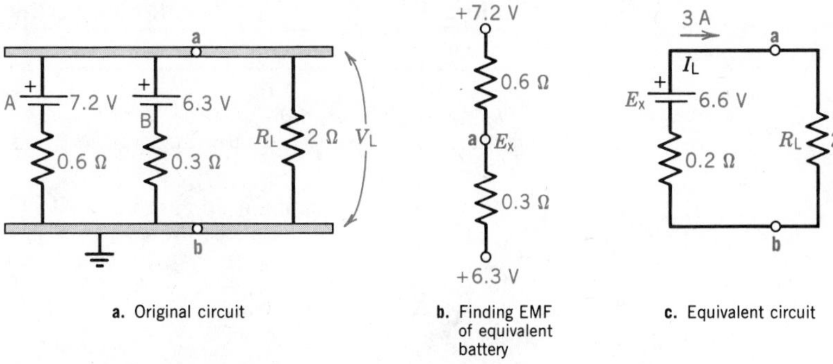

a. Original circuit b. Finding EMF of equivalent battery c. Equivalent circuit

Figure 5-16 Unequal EMFs and unequal internal resistances in parallel

[11] Chapter 7 presents the various techniques, such as Thévenin's and Millman's theorems, and nodal and mesh analysis, that can be used to solve more complex circuits of this kind, including more than two practical sources of unequal EMF and internal resistance connected in parallel to a load.

alculate the

valent battery

ttery, E_x

sistor R_L

of each battery

red by each battery

nce the two batteries are in parallel, with load R_L temporarily removed their equivalent resistance is

$$r_{i(eq)} = \frac{r_1 r_2}{r_1 + r_2} = \frac{(0.6)(0.3)}{0.6 + 0.3} = \mathbf{0.2\ \Omega} \qquad (5\text{-}23)$$

b. Using Fig. 5-16b and Eq. (5-8), we may determine the voltage at E_x, which is the open-circuit voltage at node a of the equivalent battery, as

$$E_x = \frac{R_n}{R_T}(V_p - V_n) + V_n$$

$$= \frac{0.3}{0.9}(7.2 - 6.3) + 6.3 = \mathbf{6.6\ V} \qquad (5\text{-}8)$$

c. We may now replace our equivalent battery having an EMF of $E_x = 6.6$ V and an internal resistance of 0.2 Ω in series with load resistance $R_L = 2\ \Omega$. We calculate the load current exactly as before:

$$I_L = \frac{E}{r_i + R_L} = \frac{6.6\ \text{V}}{(0.2 + 2)\ \Omega} = \mathbf{3\ A} \qquad (5\text{-}9b)$$

d. $V_L = I_L R_L = (3\ \text{A})(2\ \Omega) = \mathbf{6\ V} = V_{ab}$, the terminal voltage of each battery

e. Returning to the original circuit of Fig. 5-16a, we may find the current delivered by each battery as

$$I_A = \frac{E_A - V}{r_{iA}} = \frac{(7.2 - 6.0)\ \text{V}}{0.6\ \Omega} = \mathbf{2\ A}$$

$$I_B = \frac{E_B - V}{r_{iB}} = \frac{6.3 - 6}{0.3} = \mathbf{1\ A} \qquad (5\text{-}9c)$$

Check: $I_L = I_A + I_B = 2 + 1 = \mathbf{3\ A}$ (same as **c**)

Note from Ex. 5-27d that the sum of the individual battery currents is in fact the same as the current supplied to the load and that delivered by the equivalent battery in Fig. 5-16c. This verifies our treatment of sources in parallel having unequal EMFs and unequal internal resistances.

5-13 SUMMARY OF SERIES AND PARALLEL CIRCUIT RELATIONS AND DUALITY BETWEEN THEM

We are now in a position to summarize and compare the relations between series and parallel circuits. Table 5-1 shows such a comparison.

Table 5-1 Summary of Relations between Series and Parallel Circuits

	Series	Parallel
Definition	$I_1 \equiv I_2 \equiv I_3 \equiv \cdots \equiv I_n$	$V_1 \equiv V_2 \equiv V_3 \equiv \cdots \equiv V_n$
Voltage relations	$V_T = V_1 + V_2 + V_3 + \cdots + V_n$	(see definition above)
Current relations	(see definition above)	$I_T = I_1 + I_2 + I_3 + \cdots + I_n$
Resistance relations	$R_T = R_1 + R_2 + R_3 + \cdots + R_n$	$\frac{1}{R_T} = \frac{1}{R_1} + \frac{1}{R_2} + \frac{1}{R_3} + \cdots + \frac{1}{R_n}$
Power relations	$P_T = P_1 + P_2 + P_3 + \cdots + P_n$	

Table 5-1 shows the following significant points:

1. The current relations of a series circuit actually define that circuit.

2. The voltage relations of a parallel circuit actually define that circuit, as well.
3. Comparing voltage and current relations, the remaining equation (apart from the defining equation) is a *summation* of voltages (for series circuits) and currents (for parallel circuits).
4. Resistance is obtained as the sum of individual resistances (for series) and the sum of individual conductances (for parallel).
5. Only *power* relations remain the *same* for *both* series and parallel. Indeed, for this reason, regardless of whether we are dealing with series, parallel, or series–parallel (or even delta or wye) configurations, the *total power is always the sum of the individually dissipated powers.*

5-13.1 Implications of Table 5-1 for Series–Parallel Circuits

In Chapter 6 we encounter series–parallel circuits. Such circuits are nothing more than combinations of series and parallel elements. We will discover that in dealing with the components in series, we apply the series circuit relations of Table 5-1. In dealing with the components in parallel, we apply the parallel circuit relations of Table 5-1. But the power relations, regardless of circuit connection, are the sum of the individually dissipated powers.

5-13.2 Duality between Series and Parallel Circuits

The reader may have observed a certain pattern in Table 5-1. For example, in a *series* circuit the *current* is the *same*, and in a *parallel* circuit, the *voltage* is the *same*. In the series circuit the individual *voltages are added*, and in the parallel circuit the individual *currents are added*. Voltage appears to be taking the place of current and vice versa when comparing series and parallel circuits. Such a pattern is called "duality," and the two circuits (series and parallel) are *duals* of each other.[12]

This implies that any equation that we have obtained in terms of voltage, current, and resistance in a series circuit should have a corresponding *dual counterpart* in terms of current, voltage, and conductance for a parallel circuit. Table 5-2 shows several examples of such duality.

Table 5-2 Examples of Duality in Series and Parallel Circuits

Series	Parallel
Definition $I_1 \equiv I_2 \equiv I_3 \equiv \cdots \equiv I_n$	$V_1 \equiv V_2 \equiv V_3 \equiv \cdots \equiv V_n$
$V_T = V_1 + V_2 + V_3 + \cdots + V_n$	$I_T = I_1 + I_2 + I_3 + \cdots + I_n$
$R_T = R_1 + R_2 + R_3 + \cdots + R_n$	$G_T = G_1 + G_2 + G_3 + \cdots + G_n$
$I \equiv \frac{V_1}{R_1} \equiv \frac{V_2}{R_2} \equiv \frac{V_3}{R_3} \equiv \cdots \equiv \frac{V_n}{R_n}$	$V \equiv \frac{I_1}{G_1} \equiv \frac{I_2}{G_2} = \frac{I_3}{G_3} \equiv \frac{I_n}{G_n}$
(VDR) $V_1 = \frac{R_1}{R_T} V_T$ and $V_2 = \frac{R_2}{R_T} V_T$	(CDR) $I_1 = \frac{G_1}{G_T} I_T$ and $I_2 = \frac{G_2}{G_T} I_T$

[12] Strictly speaking, two networks are duals if the *loop* equations of one (say a series circuit) have the same form as the *node* equations of another (a parallel circuit).

Observe the (perfectly lovely) duality that exists in Table 5-2.[13] Note that in comparing series to parallel circuits, we observe the following *duals*:

Circuit ↓	Series	current	resistance	VDR	voltage	Dual ↑
Dual	Parallel	voltage	conductance	CDR	current	Circuit

Once we know the dual of a particular parameter, we can automatically take the equation for one type of circuit and derive corresponding equations for another type of circuit. For example, the current divider equations for the parallel circuit shown in Table 5-2 were not derived in this chapter, yet they are perfectly valid since they were derived as duals.[14]

5-14 GLOSSARY OF TERMS USED

Battery A number of similar electric sources arranged as a group or set.

Branch Single path in a network consisting of one or more elements in *series*.

Circuit Network containing at least one closed path.

Current-divider rule (CDR) Rule enabling the calculation of current in one resistor connected in parallel with other resistors.

Double-subscript notation Technique for determining the instantaneous voltage across an element and/or the direction of current through the element.

Ground Junction of zero potential obtained by a conducting connection to the earth.

Internal resistance, r_i Resistance contained within any practical source of electrical energy.

Junction Point at which branching occurs (same as a node).

Junction point Same as a junction or node.

Kirchhoff's current law (KCL) The algebraic sum of all currents taken around a given node is zero.

Kirchhoff's voltage law (KVL) The algebraic sum of all voltage rises and drops taken for any closed loop is zero.

Loop Closed path starting at a given node, passing through different elements to different nodes, and returning to the node at which it began.

Maximum power transfer theorem The maximum power is transferred from source to load when the resistance of the load equals the resistance of the voltage or current source.

Mesh Shortest possible loop in a network.

Network Interconnection of two or more circuit elements, either in series or parallel.

Node Point at which two or more elements have a common connection.

Open-circuit voltage Measured potential or EMF of a source when no current flows to an external circuit.

Parallel circuit Circuit containing only two nodes in which all elements are connected so that they have the same voltage across them.

Path Trace taken through an electric network starting at a node and passing through different elements to different nodes.

Power-divider rule Rule that enables calculation of power dissipated by one of a group of resistors connected in series.

Practical voltage source Ideal voltage source containing internal resistance.

Series circuit Circuit containing only one current path in which all elements so connected carry the *same* current.

Terminal voltage Voltage across the terminals of a source.

Voltage-divider rule Rule that enables calculation of voltage across one of a group of resistors connected in series.

Voltage regulation Measure of how well a given practical source maintains its voltage with application of a load.

5-15 PROBLEMS

Secs. 5-1 to 5-5

5-1 Three resistors, R_1, R_2, and R_3, having resistances of 20, 40, and 100 Ω, respectively, are connected in series. Calculate the

a. Total resistance of the series circuit
b. Volt drop across the 20-Ω resistor if the supply voltage is 120 V (use VDR)
c. (Total) circuit current
d. Current in the 20-Ω resistor
e. Voltage across the 40-Ω resistor

[13] The author has not seen a table such as Table 5-2 elsewhere. If the reader can find one, the author would be happy to know of it.

[14] The equations are later derived and appear in Sec. 6-4.

f. Voltage across the 100-Ω resistor
g. Supply voltage as the sum of volt drops in (**b**), (**d**), and (**e**)
h. Power dissipated by the 20-Ω resistor by PDR and one other method
i. Power dissipated by the 40-Ω resistor by PDR
j. Power dissipated by the 100-Ω resistor by PDR
k. Total power as the sum of the individually dissipated powers
l. Total power by using circuit current and supply voltage

5-2 A 40-kΩ and a 20-kΩ resistor are connected in series to a 33-V source. Without finding current, calculate the

a. Voltage drop across the 20-kΩ resistor (use VDR)
b. Power dissipated by the 40-kΩ resistor (use PDR)

5-3 Given $R_1 = 100\ \Omega$, $R_2 = 200\ \Omega$, and $R_3 = 500\ \Omega$ are connected across a 120-V supply. Without finding current, calculate

a. V_1, V_2, and V_3
b. P_1, P_2, and P_3
c. Total power from (**b**)
d. Verify P_T by finding the current and using $V_T I_T$ and $I^2 R_T$

5-4 A battery-operated tape recorder draws 300 mA from a 9-V supply. It is desired to operate the same recorder (at its 9-V rating) from the cigarette lighter 12-V outlet on an automobile dashboard. Calculate the

a. Resistance of the series resistor, R_s, required
b. Minimum commercial power rating of R_s

5-5 Three resistors, R_1, R_2, and R_3, are connected in series to a 220-V supply. The combined voltage drop across R_1 and R_2 is 140 V and that across R_2 and R_3 is 180 V. If the total resistance of the circuit is 11 kΩ, calculate a. R_1 b. R_2 c. R_3

5-6 Three resistors, R_1, R_2, and R_3, are connected in series to a 100-V supply. If $R_1 = 100\ \Omega$, $I_2 = 250$ mA, and $V_3 = 40$ V, calculate a. V_1 b. R_2 c. R_3

5-7 Three resistors, R_1, R_2, and R_3, are connected in series. $R_1 = 5$ kΩ, 2 W; $R_2 = 40$ kΩ, 10 W; and $R_3 = 10$ kΩ, 20 W. Calculate the

a. Maximum supply voltage that can be connected across the series combination without exceeding the power rating of any resistor
b. Maximum power dissipated by each resistor under the conditions of (**a**)

5-8 An unknown resistor R_x is connected in series with a 100-Ω resistor across a 120-V supply. If the known resistor dissipates 25 W, calculate the

a. Value of R_x
b. Power, P_x, dissipated by R_x

5-9 A 120 V, 100 W soldering iron develops too much heat, which tends to oxidize its tip. A resistor, R_s, is connected in series to reduce the power dissipation of the iron to 80 W. Calculate the

a. Resistance of the soldering iron element based on its power rating
b. Voltage across the soldering iron with R_s in series (see hint below)
c. Voltage across the series resistor, R_s
d. Resistance of R_s
e. Power dissipated by R_s
f. Total circuit power consumed, P_T

Hint: the total circuit power is no longer 100 W but somewhat less. Do *not* assume that the series resistor dissipates 20 W and the iron dissipates 80 W. This is a common error made by many students (and some engineers).

5-10 An unknown resistor, R_x, is connected in series with a 100-Ω resistor across a 120-V supply. If the unknown resistor dissipates 25 W, calculate the

a. Two values of R_x that meet this condition (see note below)
b. Power dissipated by the 100-Ω resistor under each condition
c. Verify the validity of each solution using $V_T = \sqrt{P_T R_T}$ for each set of values. (If V_T is not 120 V in each case your solution is in error.)

Hint: two values of R_x emerge from the solution of a quadratic equation. Set up the equality $I = V_T/(R_x + R) = (P_x/R_x)^{1/2}$ and solve for R_x using the quadratic formula. **Note:** The realization that there are **two** values of R_x that produce the same power in each respective value of R_x leads to the maximum power transfer relationship shown in Fig. 5-11. Note in Fig. 5-11 that any number of two values of R_L on either side of P_{max} produce the same power dissipation.

5-11 For the circuit shown in **Fig. 5-17** there are

a. Four sets of series-connected branch circuit resistors. List the letters for each branch and the resistors connected in series in that branch.
b. Only three branch circuit nodes. List the one or more letters comprising each branch circuit node without omitting any letters.
c. Six loops. List the letters and sources comprising each loop.
d. Three meshes. List the letters and/or sources comprising each mesh.
e. Two parallel pairs of branches. List the nodes and the resistors in each parallel branch pair between the nodes.

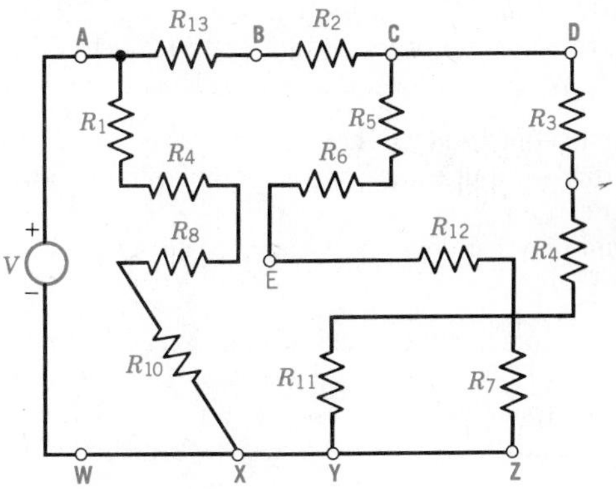

Figure 5-17 Problem 5-11

5-12 A string of eight lamps is connected in series across a 120-V line. Each lamp has a resistance of 12 Ω. Calculate the

a. Total resistance of the string

b. Total current
c. Voltage drop across each lamp
d. Current in each lamp
e. Power dissipated by each lamp
f. When one lamp burns out they all go out. Using a replacement lamp, how do you determine the defective lamp?

5-13 Two voltmeters are connected in series across a 1200 V supply. The internal resistances of the voltmeters are 75 kΩ and 120 kΩ respectively. Calculate the

a. Voltage reading of the lower-resistance voltmeter
b. Voltage reading of the higher-resistance voltmeter
c. Current in each, in milliamperes

5-14 A 300-V (full scale deflection) voltmeter has a resistance of 30 kΩ and is connected in series with a 50-kΩ resistor to an unknown supply voltage. If the voltmeter reading is 220 V, calculate the voltage of the unknown supply.

5-15 A motor armature has a resistance of 0.1 Ω and must operate safely across a 120 V dc supply. Calculate the

a. Current it would draw if instantly connected directly across the supply
b. Value of series resistance, R_s, required to limit the current to 60 A (its full-load current)
c. Power dissipated by R_s during the starting period of the motor

5-16 A series circuit consists of three resistors, $R_1 = 25$ kΩ, $R_2 = 65$ kΩ, $R_3 = 80$ kΩ. If the voltage drop across R_1 is 10 V, calculate the

a. Supply voltage
b. Voltage drop across R_3
c. Power dissipated by R_2

5-17 Three resistors, R_1, R_2, and R_3, are connected in series across a 120-V supply. The combined volt drop across R_1 and R_2 is 70 V and that across R_2 and R_3 is 80 V. If the total resistance of the circuit is 24 kΩ, calculate R_1, R_2, and R_3.

5-18 Three resistors are connected to a 440-V supply; $R_1 =$ 5 kΩ, $I_2 = 10$ mA, and $V_3 = 110$ V. Calculate R_2 and R_3.

5-19 Three resistors are connected in series. $R_1 = 10$ kΩ, 2 W; $R_2 = 30$ kΩ, 2 W; and $R_3 = 40$ kΩ, 5 W. Calculate the

a. Maximum supply voltage that can be connected across the series string without exceeding the power rating of any one resistor
b. Maximum power dissipated by each resistor under the conditions of (**a**)

5-20 Four resistors are connected in series across a dc supply. $R_1 = 1$ kΩ, $V_1 = 15$ V, and $P_1 = 3$ W; $R_2 = 2$ kΩ, $R_3 = 5$ kΩ, and $R_4 = 8$ kΩ. Using voltage division and power division relations *only*, calculate voltage across and power dissipated by

a. R_2 c. R_4
b. R_3 d. The entire circuit

5-21 Calculate the voltage, V_x, with respect to ground for the dual-supply voltage divider shown in

a. Fig. 5-18a
b. Fig. 5-18b
c. Fig. 5-18c

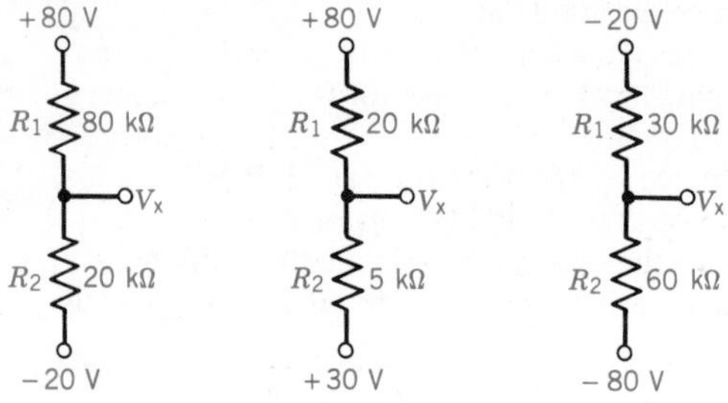

a. Fig. 5-18a b. Fig. 5-18b c. Fig. 5-18c

Figure 5-18 Problem 5-21

Secs. 5-6 and 5-7

5-22 The internal resistance of a single cell is tested by the open-circuit and short-circuit method. The open-circuit voltage is 1.2 V and the short-circuit current is 12 A. Calculate the

a. Internal resistance of the cell
b. Power dissipated internally within the cell during the short-circuit period
c. Explain the disadvantage of this method for measuring internal resistance

5-23 As an alternative to the method in Prob. 5-22, a load resistance R_L is connected in series with an ammeter across an independent voltage source. The voltage across R_L is 1.19 V and the ammeter reads 100 mA. The open-circuit voltage of the cell measured with an electronic voltmeter (EVM) is 1.2 V. Calculate the

a. Resistance of R_L
b. Total circuit resistance ($R_L + r_i$)
c. Internal resistance, r_i
d. Power dissipated within the cell during the test
e. Explain why this test method is preferable to that given in Prob. 5-22 for a dry cell

5-24 A dc generator delivers 40 A to a load when its terminal voltage is 120 V. When the load is increased to 60 A, the terminal voltage drops to 116 V. Calculate the

a. Open-circuit voltage, E_0, of the generator
b. Internal (armature) resistance of the generator
c. Voltage regulation of the generator if its rated voltage is 115 V
d. Current delivered by the generator at rated voltage

5-25 A dc power supply has a no-load voltage of 30 V dc and delivers 100 W at full-load current of 5 A. Calculate the

a. Internal resistance of the source
b. Value of load resistance at full load
c. Maximum power the power supply can deliver
d. Full-load voltage regulation
e. Voltage regulation at maximum power transfer
f. Efficiency of the supply at full load
g. Efficiency of the supply at maximum power transfer

5-26 An electronic power supply has a terminal voltage of 30 V when delivering a current of 5 mA and a terminal voltage of 20 V when delivering a current of 20 mA. Calculate the

a. Internal resistance of the supply, r_i
b. Open-circuit voltage of the supply, E_0
c. Maximum power that the power supply can deliver to a load

Secs. 5-8 to 5-11

5-27 Two resistors, R_1 and R_2, are connected in parallel. If $R_1 = 20$ kΩ and $R_2 = 5$ kΩ and the total current drawn by the parallel combination is 50 mA, calculate the

a. Equivalent (total) circuit resistance
b. Supply voltage across both resistors in parallel
c. Current in R_1 by three different methods

5-28 A 100 V dc source supplies a current of 5 mA to two resistors in parallel. If $R_1 = 100$ kΩ, calculate the

a. Current in R_2
b. Value of R_2
c. Equivalent circuit resistance by two methods
d. Verify the current in R_2 by three different methods

5-29 Five resistors are connected in parallel. Their values are $R_1 = 100\ \Omega$, $R_2 = 25\ \Omega$, $R_3 = 5\ \Omega$, $R_4 = 12\ \Omega$, and $R_5 = 6\ \Omega$. Calculate the equivalent resistance of the combination, using

a. Eq. (5-20)
b. Eq. (5-20a)
c. Eq. (5-21)

5-30 A two-terminal "load box" is constructed of 30 identical 360-Ω resistors, each connected in series to a single-pole, single-throw (SPST) switch to place it across the terminals of the load box. Draw the circuit configuration for three such resistors and their switches, respectively, and calculate the

a. Resistance of the load box when two switches are CLOSED
b. Resistance of the load box when all switches are CLOSED
c. Current drawn from a 120-V supply when only one switch is CLOSED
d. Current drawn from a 120-V supply when ALL switches are CLOSED
e. Power dissipated by each resistor when connected across 120 V
f. Total power capacity of the load box
g. Number of switches that must be closed to produce a load of 3 A when connected to a 120-V supply

5-31 Two resistors must be selected so that the current in one is four times the current in the other. If their equivalent parallel resistance is 5 kΩ, calculate R_1 and R_2.

5-32 A resistance, R_1, of 5 kΩ is shunted by a conductance, G_2, of 50 μS. The combination draws a current of 10 mA from a dc supply. Calculate

a. G_T and R_T
b. I_1 and I_2
c. The voltage across each and the supply voltage

5-33 A current divider having four branches is to be designed to divide a total current of 4 mA as follows:

Branch 1 gets half the current, I_1
Branch 2 gets one-fourth of the current, I_2
Branch 3 gets one-fifth of the current, I_3
Branch 4 gets the rest of the current, I_4, and dissipates 10 mW

For *each* branch, calculate the

a. Current
b. Resistance
c. Conductance
d. Power dissipated
e. Voltage across the divider network

Tabulate for each branch your calculated values from **(a)** to **(d)**.

f. Draw conclusions regarding the current and power dissipation versus conductance
g. Given the conductance, current, and power of one branch, show how it is possible to find the current and power of any other branch, given its conductance

5-34 A 10-mA meter movement having a resistance of 90 Ω is paralleled by a "shunt" whose resistance is 9 Ω. Calculate the

a. Maximum current in the shunt without exceeding the meter deflection
b. Maximum line current to the "shunted" instrument

5-35 Three resistors, $R_1 = 10\ \Omega$, $R_2 = 20\ \Omega$, and $R_3 = 30\ \Omega$, are connected in parallel. Calculate the

a. Total equivalent parallel resistance, R_{eq}
b. Conductances G_1, G_2, and G_3, respectively
c. Total conductance, G_T, by two methods
d. Total current drawn by the parallel combination when connected to a 60-V source using G_T only
e. Verify **(d)** by finding I_1, I_2, I_3, and I_T

5-36 Two conductances, G_1 and G_2, having values of 600 and 900 μS, respectively, are connected in parallel. The combination is then connected in parallel with an 800-Ω resistor. Calculate the

a. Equivalent resistance of the parallel circuit
b. Current in each branch if G_1 carries 2 mA
c. Power dissipated in each branch
d. Voltage across each component
e. Total power dissipated, by two methods

5-37 a. What resistance, R_x, must be connected in parallel with a 25-kΩ resistor to reduce the equivalent resistance to 5 kΩ?

Calculate the

b. Ratio of R_1 to R_x
c. Ratio of currents, I_x to I_1
d. Current in each branch if the total current is 10 mA
e. Total voltage across the parallel combination

5-38 Two resistors, R_1 and R_2, must be selected so that the current in R_1 is three times the current in R_2. If their equivalent resistance is 4.8 kΩ, calculate the values of

a. R_1
b. R_2

5-39 A current of 300 mA flows into a junction of two parallel resistors, $R_1 = 7.2$ kΩ and $R_2 = 2.4$ kΩ, respectively. Calculate the

a. Current, I_1, using the current divider rule (CDR)
b. Current, I_2, using appropriate Eq. (5-25)
c. Equivalent total resistance
d. Total voltage across the parallel combination
e. Sum of the currents calculated in (**a**) and (**b**). Does it yield 300 mA?

5-40 A 6-V battery delivers 10 mA to three resistors in parallel; $R_1 = 1$ kΩ, $R_2 = 2$ kΩ, and $R_3 = R_x$. Calculate the

a. Current in R_x, that is, I_x
b. Power dissipated by R_x and R_1, respectively
c. Resistance ratio R_x/R_1
d. Current ratio I_1/I_x
e. Power ratio P_1/P_x
f. Conductance ratio G_1/G_x
g. Draw conclusions regarding (**c**) through (**f**)

5-41 Fifteen identical 40-kΩ resistors are connected in parallel with a second parallel combination consisting of a 200-kΩ and a 50-kΩ resistor in parallel. Calculate the

a. Equivalent resistance of the second parallel combination
b. Total equivalent resistance
c. Current in each 40-kΩ resistor if the total current is 40 mA
d. Power drawn by one 40-kΩ resistor
e. Total power dissipated by the entire parallel combination

5-42 Two resistors, R_1 and R_2, are connected in series across a 20-V supply; $R_1 = 6$ MΩ and $R_2 = 4$ MΩ. A voltmeter having an internal resistance of 4 MΩ is connected across R_2. Draw the circuit and calculate the

a. *True* voltage across R_2 *before* the voltmeter was connected
b. Equivalent resistance of the parallel combination of R_2 and the voltmeter
c. Actual voltage across R_2 after the voltmeter is connected and voltage recorded by the voltmeter
d. Percent error in voltage. Use:

$$\frac{\text{(Actual–True) voltage}}{\text{True voltage of } R_2} \times 100$$

5-43 A 1-mA meter movement having a resistance of 100 Ω is paralleled by a shunt whose resistance is 0.1 Ω. Calculate the:

a. Maximum current in the shunt (in mA) without exceeding full-scale deflection of 1-mA meter movement
b. Maximum line current to the shunted instrument in amperes
c. Approximate line current measured by the meter when the movement reads 0.5 mA

Secs. 5-12 to 5-13

5-44 Four 1.2-V cells having an internal resistance of 0.1 Ω each are connected (1) in series, (2) in parallel, and (3) in two parallel groups of two in series. For each of the three configurations, calculate (a total of 15 calculations) the

a. Open-circuit voltage of the battery, E_0
b. Equivalent battery internal resistance, r_i
c. Maximum short-circuit current that may be drawn from the combination
d. Current per cell under conditions of (**c**)
e. Power per cell under conditions of (**c**)
f. Draw conclusions regarding the relative power per cell versus the configuration

5-45 A battery consists of eight cells in series, each cell having an internal resistance of 0.1 Ω and an open-circuit EMF of 1.4 V. Calculate the

a. Open-circuit voltage of the equivalent battery if all the cells are series-aiding
b. Equivalent battery internal resistance under conditions of (**a**)
c. Open-circuit voltage of the equivalent battery when two of the cells are opposing the other six
d. Equivalent battery internal resistance under conditions of (**c**)

5-46 A battery consists of four cells in parallel, each cell having an internal resistance of 0.4 Ω and an open-circuit EMF of 2.0 V. Calculate the

a. Open-circuit voltage of the equivalent battery if all cells are connected to the same polarity
b. Equivalent internal battery resistance
c. Load current drawn from the battery by a 0.4-Ω load
d. Current delivered by each cell of the battery

5-47 A battery consists of two cells in parallel, each having an EMF of 1.2 V, but the first cell has an internal resistance of 0.6 Ω and the second cell has an internal resistance of 0.3 Ω. Calculate the

a. Open-circuit voltage of the equivalent battery, E_0, when both cells are connected to the same polarity in parallel
b. Equivalent battery internal resistance
c. Load current delivered to a 0.2-Ω load
d. Current delivered by the first cell
e. Current delivered by the second cell

5-48 The open-circuit voltage of a photovoltaic cell is 0.45 V as measured by an electronic voltmeter (EVM). When an ammeter of negligible resistance is connected across the cell's terminals, the current is 50 mA. Calculate the

a. Internal resistance of the cell
b. Maximum power the cell can deliver to an external load
c. Terminal voltage and load current when a load resistance of 13.5 Ω is connected across the cell's terminals

5-49 Three lead-acid cells each having an open-circuit voltage of 2.1 V and an internal resistance of 0.3 Ω are connected in parallel. Calculate for the battery combination the

a. Equivalent open-circuit voltage

b. Equivalent internal resistance
c. Current delivered to a 2.9-Ω load
d. Terminal voltage across the above load
e. Maximum power the battery is capable of delivering to a load

5-50 An automotive battery has an open-circuit voltage of 12.6 V and a short-circuit current of 300 A. Calculate the

a. Internal resistance of the battery
b. Maximum power it can deliver to a load
c. Power it delivers when its terminal voltage is 12 V
d. Terminal voltage and current delivered to a 1-Ω headlight load

5-51 A storage battery is "recharged" whenever the current through it is reversed, that is, it receives current instead of delivering it. A 3.0-V battery having an internal resistance of 0.05 Ω is charged at a 5.0-A rate from a dc power supply having an open-circuit EMF of 3.6 V and unknown internal resistance.

a. Draw a circuit diagram showing the current direction on this circuit

Calculate the

b. Terminal voltage of the battery during the charging process
c. Terminal voltage of the power supply during the charging process
d. Internal resistance of the power supply
e. Maximum power that can be obtained from the power supply
f. Maximum power that can be delivered by the battery if its open-circuit EMF is 3.2 V

5-52 A battery delivers 8 mA at a terminal voltage of 10 V and 16 mA at a terminal voltage of 9 V. Calculate the

a. Internal resistance of the battery
b. Open-circuit voltage of the battery
c. Maximum power the battery can deliver to a load
d. Battery voltage regulation if 10 mA is the rated battery current

5-53 If the maximum power a battery can deliver to a load is 1984.5 W and its short-circuit current is 630 A, calculate the

a. Open-circuit voltage of the battery (Hint: $V_0/r_i = 630$ A)
b. Internal resistance of the battery
c. Internal power dissipated when delivering short-circuit current
d. Load power dissipated under conditions of (**c**)
e. Internal power dissipated when delivering maximum power to a load
f. Repeat (**e**) when delivering 200 A
g. Repeat (**d**) when delivering 200 A

5-54 A battery whose EMF is 9 V has a terminal voltage of 8 V when a load of 20 Ω is connected across its terminals. Calculate the

a. Circuit current
b. Internal resistance of the battery
c. Voltage regulation, assuming 400 mA is the rated current
d. Efficiency of power transfer to the load
e. Maximum power the battery can deliver to the load
f. Resistance of the load that draws maximum power
g. Current when delivering maximum power
h. Voltage across the load when delivering maximum power

5-55 When a 20-Ω load is connected across a dc power supply it draws 12 A, and when a 10-Ω load is connected it draws 20 A. Calculate the

a. Internal resistance of the supply
b. Open-circuit voltage of the supply
c. Maximum power the supply can deliver to a load
d. Maximum power dissipated internally if the supply is short-circuited
e. Current drawn when delivering maximum power

5-56 An emergency lighting system (used in event of power failure) has 60 cells in series. Each cell has an EMF of 2 V and internal resistance of 1 mΩ. The system is recharged periodically from a 130-V dc supply having negligible internal resistance at a current of 100 A for a period of 5 h. Calculate the

a. Charging ampere-hours per cell and total charging ampere-hours per system
b. Resistance required in series with the battery system to limit the current during charging to 100 A
c. Battery terminal voltage during the charging process
d. Power dissipated by (**b**) during the charging process
e. Power drawn by the battery system during the charging process
f. Energy consumed by the battery during the charging process

5-57 A 22.5-V layer-built battery consists of 15 series-connected cells, each having an EMF of 1.5 V and an internal resistance of 0.05 Ω. Calculate the

a. Maximum current the battery can deliver if momentarily short-circuited
b. Current the battery delivers to a 3.25-Ω load
c. Power delivered to the 3.25-Ω load
d. Terminal voltage across the load and terminal voltage of the battery
e. Maximum power the battery can deliver to an external load
f. Value of external load resistance that will draw maximum power

5-58 Two zinc–carbon cells are connected in parallel to supply a low-voltage alarm system. Since the alarm bell sounds weak when tested, the owner decides to save money by replacing only one cell. The fresh cell has an EMF of 1.5 V and an internal resistance of 0.05 Ω and the remaining (old) cell has an EMF of 1.5 V and an internal resistance of 0.2 Ω. Calculate the

a. Open-circuit voltage of the equivalent battery combination
b. Equivalent battery internal resistance
c. Maximum current the battery can deliver if momentarily shorted
d. Current delivered to a 0.46-Ω load representing the alarm system
e. Current delivered by the fresh (new) cell
f. Current delivered by the remaining (old) cell
g. Explain why it is usually *inadvisable* to replace *only one* cell in a parallel group

5-59 Two batteries are connected in parallel. Battery A has an EMF of 7 V and an internal resistance of 2 Ω. Battery B has an EMF of 5 V and an internal resistance of 2 Ω. The parallel combination is connected to a 5 Ω load. Calculate the

a. EMF of the equivalent battery (Hint: use Eq. 5-8)
b. Internal resistance of the equivalent battery
c. Load current, I_L, in the 5-Ω load and its terminal voltage, V_L
d. Terminal voltage of each battery
e. Current delivered by battery A
f. Current delivered by battery B
g. Explain why battery B neither delivers nor receives current from the nodes.
h. If only the internal resistance of battery B is lower (say 1 Ω), would that improve the situation in enabling battery B to deliver current?

5-60 Two batteries are connected in parallel. Battery A has an EMF of 7 V and an internal resistance of 1 Ω. Battery B has an EMF of 4 V and an internal resistance of 1 Ω. The parallel combination is connected to a 5-Ω load. Calculate the

a. EMF of the equivalent battery (Hint: use Eq. 5-8)
b. Internal resistance of the equivalent battery
c. Load current, I_L, in the 5-Ω load and its terminal voltage, V_L
d. Terminal voltage of each battery
e. Current delivered by battery A
f. Current *received* by battery B
g. Explain why battery B *must* receive current from the nodes

5-61 Given the equation $R_{eq} = \dfrac{R_1 R_2}{R_1 + R_2}$ and $n = R_2/R_1$, prove that $R_{eq} = \dfrac{R_2}{1+n}$ for two resistors in parallel (see Eq. 5-23c).

5-62 Given the relationship $n = I_1 R_1 = I_2 R_2 = I_T R_{eq}$ and the derivation of Prob. 5-61, derive the following:

a. Eq. (5-25a)
b. Eq. (5-25b)

5-63 The statement was made (Sec. 5-12.2) that batteries having *equal* EMFs and *unequal* internal resistances have an equivalent open-circuit EMF that is the same as the individual EMFs, *regardless* of internal resistances. Prove that this is so, using Eq. (5-8).

5-16 ANSWERS

5-1 a 160 Ω b 15 V c 0.75 A d 0.75 A e 30 V f 75 V g 120 V h 11.25 W i 22.5 W j 56.25 W k 90 W l 90 W
5-2 a 11 V b 12.1 mW
5-3 a 15 V, 30 V, 75 V b 2.25 W, 4.5 W, 11.25 W c 18 W d 0.15 A, 18 W
5-4 a 10 Ω b 1 W
5-5 a 2 kΩ b 5 kΩ c 4 kΩ
5-6 a 25 V b 140 Ω c 160 Ω
5-7 a 869.6 V b 1.25 W, 10 W, 2.5 W
5-8 a 140 Ω b 35 W
5-9 a 144 Ω b 107.3 V c 12.67 V d 17 Ω e 9.44 W f 89.44 W
5-10 a 28.8 Ω, 347.2 Ω b 86.8 W, 7.2 W
5-12 a 96 Ω b 1.25 A c 10 V d 1.25 A e 12.5 W
5-13 a 461.5 V b 738.5 V c 6.154 mA
5-14 $586.\bar{6}$ V
5-15 a 1200 A b 1.9 Ω c 6840 W
5-16 a 68 V b 32 V c 10.4 mW
5-17 8 kΩ, 10 kΩ, 6 kΩ
5-18 28 kΩ, 11 kΩ
5-19 a 653.2 V b 0.67 W, 2 W, $2.\bar{6}$ W
5-20 a 30 V, 6 W b 75 V, 15 W c 120 V, 24 W d 240 V, 48 W
5-21 a 0 V b 40 V c −40 V
5-22 a 0.1 Ω b 14.4 W
5-23 a 11.9 Ω b 12 Ω c 0.1 Ω d 1 mW
5-24 a 128 V b 0.2 Ω c 11.3% d 65 A
5-25 a 2 Ω b 4 Ω c 112.5 W d 50% e 100% f 66% g 50%
5-26 a $666.\bar{6}$ Ω b ~~200 V~~ c ~~10 mA~~
5-28 a 4 mA b 25 kΩ c 20 kΩ d 4 mA
5-29 a 2 Ω b 2 Ω c 2 Ω
5-30 a 180 Ω b 12 Ω c $0.\bar{3}$ A d 10 A e 40 W f 1200 W g 9
5-31 6.25 kΩ, 25 kΩ
5-32 a 250 μS, 4 kΩ b 8 mA, 2 mA c 40 V
5-33 a 2 mA, 1 mA, 0.8 mA, 0.2 mA b 2.5 kΩ, 5 kΩ, 6.25 kΩ, 25 kΩ c 400 μS, 200 μS, 160 μS, 40 μS d 10 mW, 5 mW, 4 mW, 1 mW e 5 V
5-34 a 100 mA b 110 mA
5-35 a $5.\overline{45}$ Ω b 100 mS, 50 mS, $33.\bar{3}$ mS c $183.\bar{3}$ mS d 11 A
5-36 a $363.\overline{63}$ Ω b 3 mA, $4.1\bar{6}$ mA c $6.\bar{6}$ mW, 10 mW, $13.\bar{8}$ mW d $3.\bar{3}$ V e $30.\overline{55}$ mW
5-37 a 6.25 kΩ b 4/1 c 4/1 d 2 mA, 8 mA e 50 V
5-38 a 6.4 kΩ b 19.2 kΩ
5-39 a 75 mA b 225 mA c 1.8 kΩ d 540 V e 300 mA
5-40 a 1 mA b 6 mW, 36 mW c, d, e, f: 6/1
5-41 a 40 kΩ b 2.5 kΩ c 2.5 mA d 0.25 W e 4 W
5-42 a 8 V b 2 MΩ c 5 V d −37.5%
5-43 a 1 A b 1.001 A c 500.5 mA
5-44 a 4.8 V, 1.2 V, 2.4 V b 0.4 Ω, 0.25 Ω, 0.1 Ω c 12 A, 48 A, 24 A d 12 A e 14.4 W
5-45 a 11.2 V b 0.8 Ω c 5.6 V d 0.8 Ω
5-46 a 2 V b 0.1 Ω c 4 A d 1 A
5-47 a 1.2 V b 0.2 Ω c 3 A d 1 A e 1 A
5-48 a 9 Ω b 5.625 mW c 20 mA, 0.27 V

5-49 a 2.1 V b 0.1 Ω c 0.7 A d 2.03 V e 11.025 W
5-50 a 42 mΩ b 945 W c 171.43 W
d 12.09 A, 12.09 V
5-51 b, c 3.25 V d 70 mΩ e 46.29 W
f 51.2 W
5-52 a 125 Ω b 11 V c 242 mW d 12.82%
5-53 a 12.6 V b 20 mΩ c 7.938 kW d 0
e 1.9845 kW f 800 W g 1.72 kW
5-54 a 0.4 A b 2.5 Ω c 12.5% d $88.\overline{8}\%$ e 8.1 W
f 2.5 Ω g 1.8 A h 4.5 V
5-55 a 5 Ω b 300 V c 4.5 kW d 18 kW e 30 A
5-56 a 300 Ah b 40 mΩ c 126 V d 400 W
e 600 W f 3 kWh
5-57 a 30 A b 5.625 A c 102.8 W d 18.28 V
e 168.75 W f 0.75 Ω
5-58 a 1.5 V b 0.4 Ω c 37.5 A d 3 A e 2.4 A
f 0.6 A
5-59 a 6 V b 1 Ω c 1 A, 5 V d 5 V e 1 A f 0
5-60 a 5.5 V b 0.5 Ω c 1 A, 5 V d 5 V e 2 A
f −1 A

CHAPTER 6

Series–Parallel Circuits

6-1 INTRODUCTION

In this chapter, we will consider more complex networks containing combinations of both series and parallel elements. Our purpose in analyzing these more complex networks is to reduce them to either a simple series circuit or a simple parallel circuit.

Customary usage in the literature is to assume that any network that is not a simple series circuit or a simple parallel circuit must be a *series–parallel* circuit. In this text, we will make a distinction between a *series–parallel* circuit and a *parallel–series* circuit.

A *series–parallel* circuit is defined as essentially a *series* circuit containing *parallel* elements. Consequently, by a process of circuit reduction, we will reduce this circuit to an equivalent *series* circuit. **Figure 6-1a** shows a typical series–parallel circuit. The circuit has four nodes. The three parallel resistors (R_4, R_2, and R_3) between nodes **b** and **c** may be reduced to a single equivalent resistor, R_p, as shown in Fig. 6-1b. Note that the circuit of Fig. 6-1b is now a simple *series* circuit, enabling use of *series* circuit relations developed in Chapter 5.

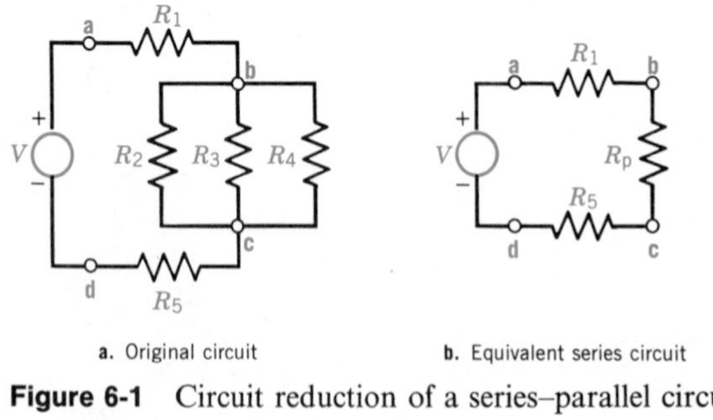

a. Original circuit b. Equivalent series circuit

Figure 6-1 Circuit reduction of a series–parallel circuit

A *parallel–series* circuit is defined as essentially a *parallel* circuit containing *series* elements. Consequently, by a process of circuit reduction, we will reduce this circuit to an equivalent *parallel* circuit. **Figure 6-2a** shows a typical parallel–series circuit. The circuit has two nodes. All series-connected resistances between nodes a and b may be reduced to a single equivalent resistance between the nodes, as shown in Fig. 6-2b. The three series resistances in the first branch have been reduced to R_{S_1}. The two series resistances in the second branch have been reduced to R_{S_2}, and so on. Note that the circuit of Fig. 6-2b is now a simple *parallel* circuit, enabling use of *parallel* circuit relations developed in Chapter 5.

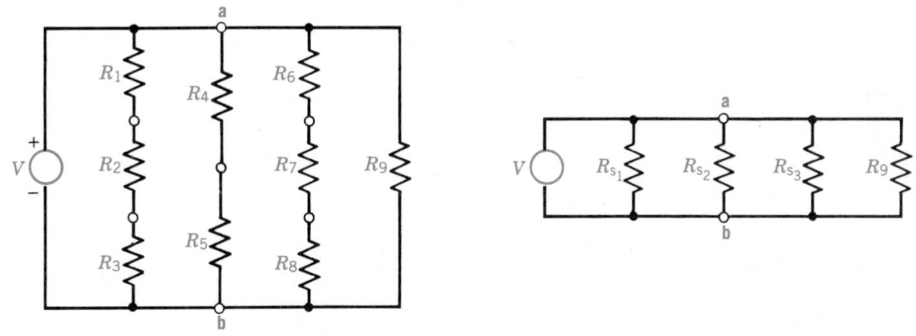

a. Original circuit **b.** Equivalent parallel-series circuit

Figure 6-2 Circuit reduction of a parallel–series circuit

6-2 PRELIMINARY CIRCUIT ANALYSIS RULES FOR CIRCUIT REDUCTION AND SOLUTION

Our goal in this chapter is to emerge with a number of insights regarding circuit reduction and to gain greater familiarity with circuit analysis of series–parallel circuits. We will accomplish this by following a number of simple rules of procedure:

1. Draw the original circuit diagram, labeling all nodes and showing all branches between nodes. Identify currents in each branch by number or letter.
2. Combine the branches containing more than one resistance in series into a single resistance.
3. Combine the resistances in parallel into a single equivalent parallel resistance.
4. The object of the analysis is to reduce the circuit to a simple series or parallel circuit equivalent, as shown in Figs. 6-1 and 6-2.
5. When the desired level of simplification is reached, begin the complete solution by solving for total resistance, current, or voltage, as required.
6. Once voltage, current, and resistance of the equivalent circuit are found, work back to the original circuit, calculating *all* currents, voltages, and powers, as required, to obtain the **complete solution**.

Two more very important rules summarize the above process:

7. *Parallel* circuit laws and relations are used for those portions of the circuit that are connected in *parallel*.
8. *Series* circuit laws and relations are used for those portions of the circuit that are connected in *series*.

6-2.1 Equivalent Circuit Method

The technique of circuit reduction described above is sometimes called the *equivalent circuit* method. It is best illustrated by the following examples.

EXAMPLE 6-1

For the circuit shown in **Fig. 6-3a**, calculate the

a. Single equivalent resistance of the complete circuit
b. Total current and power drawn from the supply
c. Current in, voltage across, and power dissipated by R_1

Solution

The circuit can be reduced to a simple parallel equivalent, as shown in Fig. 6-1b, with four branches across the 60-V supply.

a. $$R_{eq} = \frac{1}{(1/40) + (1/120) + (1/60) + (1/20)}\ \text{k}\Omega \qquad (5\text{-}20a)$$
$= \mathbf{10\ k\Omega}$

b. $I_T = V_T/R_T = 60\ \text{V}/10\ \text{k}\Omega = \mathbf{6\ mA}$
$P_T = V_T I_T = (60\ \text{V})(6\ \text{mA}) = \mathbf{360\ mW}$

c. Current in the fourth branch $I_4 = 60\ \text{V}/20\ \text{k}\Omega = \mathbf{3\ mA}$
$V_1 = I_4 R_1 = (3\ \text{mA})(5\ \text{k}\Omega) = 15\ \text{V}$
$P_1 = V_1^2/R_1 = (15)^2/5\ \text{k}\Omega = \mathbf{45\ mW}$

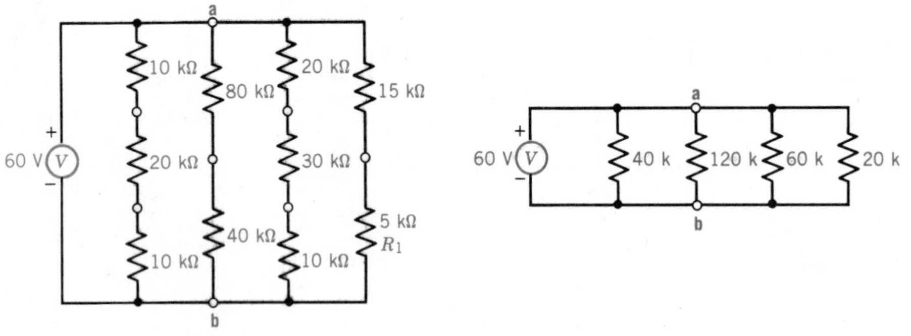

a. Original circuit

b. Simplified equivalent (parallel) circuit

Figure 6-3 Example 6-1

Note that the equivalent circuit obtained in Fig. 6-3b enabled the solution of all parts of Ex. 6-1. We could have reduced Fig. 6-1b to a single resistor of 10 kΩ, but then the calculation of current in the fourth branch would not have been as easily implemented. The point here is that excessive oversimplification is an unnecessary step. We simplify the circuit to the equivalent that is *most useful* for our purposes.

EXAMPLE 6-2

For the circuit shown in **Fig. 6-4a**, calculate the

a. Total circuit current drawn from supply
b. Voltage V_1 across resistor R_1
c. Current in and power dissipated by R_1

Solution

a. We begin the simplification by working from the extreme right to the left in the following steps:
   1. At nodes **a, b** $R_{ab} = (3\ \text{k}\Omega \| 6\ \text{k}\Omega) = 2\ \text{k}\Omega$
   2. At nodes **c, b** $R_{cb} = 10\ \text{k}\Omega + R_{ab} = (10 + 2)\ \text{k}\Omega = 12\ \text{k}\Omega$
   3. At nodes **c, d** $R_{cd} = R_{cd} \| R_{cb} = (6\ \text{k}\Omega \| 12\ \text{k}\Omega) = 4\ \text{k}\Omega$
   4. At nodes **e, f** $R_{ef} = (4\ \text{k}\Omega \| 4\ \text{k}\Omega) = 2\ \text{k}\Omega$. Note that nodes **e, f** are the same as nodes **c, d**
   5. Draw the simple equivalent series circuit shown in Fig. 6-4b to enable calculation of $R_T = (0.3 + 2 + 1.7)\ \text{k}\Omega = 4\ \text{k}\Omega$; then $I_T = V_T/R_T = 80\ \text{V}/4\ \text{k}\Omega = \mathbf{20\ mA}$

b. From Fig. 6-4b, we calculate $V_{cd} = I_T R_{cd} = (20\ \text{mA})(2\ \text{k}\Omega) = 40\ \text{V}$ To find V_1 we return to the original circuit of Fig. 6-4a, using the VDR or $V_1 = V_{ab} = 40\ \text{V}\dfrac{R_{ab}}{R_{cb}} = 40\ \text{V}\dfrac{2\ \text{k}\Omega}{12\ \text{k}\Omega} =$ $\mathbf{6.6\bar{6}\ V}$

c. $I_1 = V_1/R_1 = 6.6\bar{6}\ \text{V}/6\ \text{k}\Omega = \mathbf{1.11\bar{1}\ mA}$
$P_1 = V_1 I_1 = (6.6\bar{6}\ \text{V})(1.11\bar{1}\ \text{mA}) = \mathbf{7.41\ W}$

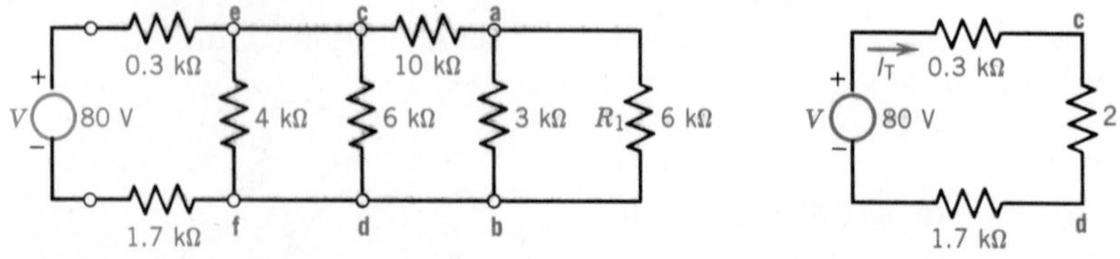

a. Original circuit

b. Simplified equivalent (series) circuit

Figure 6-4 Example 6-2

Note again that the simplification of Fig. 6-4b was just sufficient to enable solution of the given problem.

EXAMPLE 6-3

For the circuit shown in **Fig. 6-5**, calculate a. I_T b. R_x by two methods

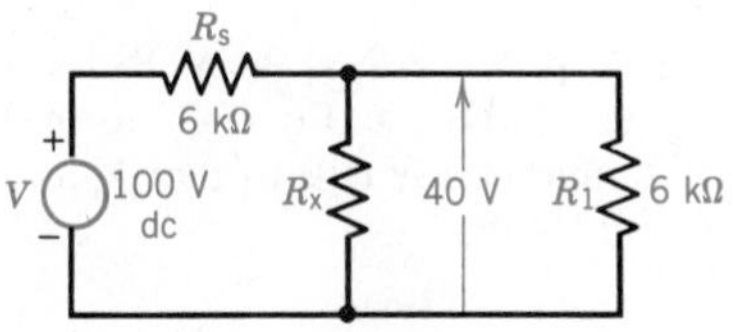

Figure 6-5 Example 6-3

Solution

a. $V_s = V_T - V_x = (100 - 40)\text{ V} = 60\text{ V}$
$I_T = V_s/R_s = 60\text{ V}/6\text{ k}\Omega = \mathbf{10\text{ mA}}$

b. $I_1 = V_1/R_1 = 40\text{ V}/6\text{ k}\Omega = 6.\overline{66}\text{ mA}$
$I_x = I_T - I_1 = (10 - 6.\overline{66})\text{ mA} = 3.\overline{33}\text{ mA}$
$R_x = V_x/I_x = 40\text{ V}/3.\overline{33}\text{ mA} = \mathbf{12\text{ k}\Omega}$

*Alternative Solution for **b***

$$R_T = V_T/I_T = 100\text{ V}/10\text{ mA} = 10\text{ k}\Omega$$

$$R_p = R_T - R_s = (10 - 6)\text{ k}\Omega = 4\text{ k}\Omega$$

$$R_x = R_1R_p/(R_1 - R_p) = \frac{6 \times 4}{6 - 4}\text{ k}\Omega = \mathbf{12\text{ k}\Omega} \qquad (5\text{-}23b)$$

Example 6-3 shows that once the total current, I_T, is determined, there are two different approaches enabling the solution for R_x. The first approach involves current division in the parallel portion of the circuit. The second approach uses resistance relations in series and parallel.

EXAMPLE 6-4

For the circuit shown in **Fig. 6-6a**, calculate

a. Voltage at point **a**
b. Voltage at point **b**
c. V_{ba}
d. V_{ab}
e. How should a voltmeter be connected between points **a** and **b** for the dc voltmeter to read upscale?

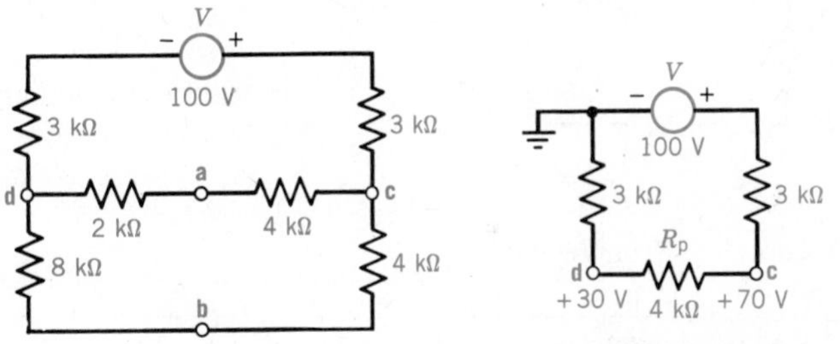

a. Original circuit b. Equivalent series circuit

Figure 6-6 Example 6-4

Solution

Since the circuit is essentially a series circuit having parallel elements, we reduce it to its basic series circuit form by finding the equivalent parallel resistance between **c** and **d** or

$$R_p = \frac{R_{cad} \times R_{cbd}}{R_{cad} + R_{cbd}} = \frac{(6 \times 12)}{(6 + 12)}\text{ k}\Omega = 4\text{ k}\Omega \qquad (5\text{-}23)$$

The equivalent series circuit is shown in Fig. 6-6b. Note that we have "lost" V_a and V_b, but this simplification enables us to find V_c and V_d, which are essential. Using the VDR,

$$V_d = \frac{3\text{ k}\Omega}{10\text{ k}\Omega} \times 100\text{ V} = +30\text{ V}$$

and

$$V_c = \frac{7\text{ k}\Omega}{10\text{ k}\Omega} \times 100\text{ V} = +70\text{ V} \qquad (5\text{-}6)$$

The difference in potential between **c** and **d** is $V_{cd} = V_c - V_d = +70 - (+30) = 40$ V; now that we know that $V_{cd} = 40$ V, we can return to Fig. 6-6a to find V_a and V_b, using the relations of Eq. (5-8) in Fig. 6-6a:[1]

a. $V_a = (V_p - V_n)\dfrac{R_n}{R_T} + V_n$

$$= (70 - 30)\frac{2\text{ k}\Omega}{6\text{ k}\Omega} + 30 = \mathbf{43.\bar{3}\text{ V}} \qquad (5\text{-}8)$$

b. $V_b = (V_p - V_n)\dfrac{R_n}{R_T} + V_n$

$$= (+70 - 30)\frac{8\text{ k}\Omega}{12\text{ k}\Omega} + 30 = \mathbf{56.\bar{6}\text{ V}}$$

---

[1] Again and again, throughout this text, we will encounter the usefulness of the concept involved in this relation. We will see it used in another form in transient circuit analysis, Sec. 10-5, Eq. (10-2a).

c. $V_{ba} = V_b - V_a = (56.\bar{6} - 43.\bar{3})\ \text{V} = \mathbf{13.\bar{3}\ V}$
d. $V_{ab} = V_a - V_b = (43.\bar{3} - 56.\bar{6})\ \text{V} = \mathbf{-13.\bar{3}\ V}$
e. The positive terminal of the voltmeter should be connected to **b** and the negative terminal to **a** for the voltmeter to read upscale.

Several important conclusions may be drawn from Ex. 6-4:

1. The potential at any point in a network is measured with respect to ground zero. In this circuit (Fig. 6-6a), the negative terminal of the supply is zero volts.
2. The potential at point **b** is more positive than point **a** because of the nature of the voltage division. The greater the value of R_n in Eq. (5-8), the more positive the respective voltage between the two resistors of the divider.
3. A voltmeter always measures the *difference in potential* between two points in a circuit. If the potential of **b** was 60 V and the potential at **a** was 60 V, the voltmeter would read zero volts.
4. The entire circuit of Ex. 6-4 was solved for all voltages at all nodes without resorting to current, because the circuit is essentially a series circuit and current is constant in all parts of Fig. 6-6b.
5. Since $V_c = +70$ V and $V_d = +30$ V, we used the VDR for a circuit having *two* dc supplies, derived as Eq. (5-8), to find the potentials V_a and V_b. Common sense tells us that V_a and V_b must be between +30 and +70 V.

EXAMPLE 6-5
For the circuit shown in **Fig. 6-7a**

a. Simplify the circuit, showing only one source and a single equivalent resistance across nodes **a** and **b**
b. Calculate the voltage, V_{ab}, and determine its polarity
c. Calculate the current in the 500-Ω resistor and determine its direction

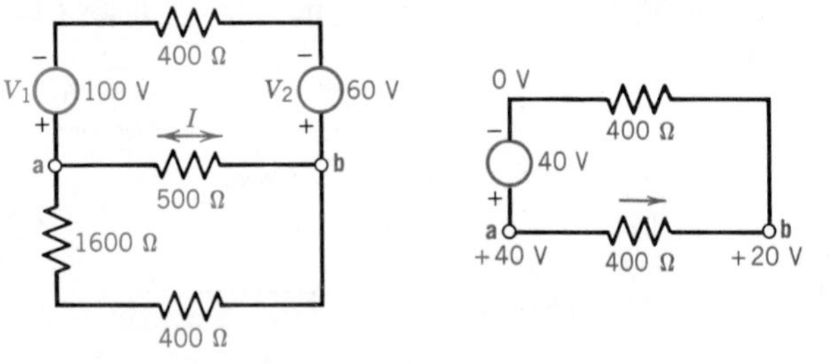

a. Original circuit b. Simplified equivalent series circuit

Figure 6-7 Example 6-5

Solution

a. 1. Since the two sources are connected in series and opposing each other, the net EMF is $V_T = V_1 - V_2 = (100 - 60)\ \text{V} = \mathbf{40\ V}$
2. Equivalent resistance across nodes **a**, **b** is $R_{ab} = (500\ \Omega) \| (2000\ \Omega) = \mathbf{400\ \Omega}$
3. The simplified (series) equivalent circuit is shown in Fig. 6-7b

b. From Fig. 6-7b, $V_{ab} = \dfrac{400\ \Omega}{800\ \Omega} \times 40\ \text{V} = \mathbf{+20\ V}$. Since the negative terminal of the supply is at 0 V, $V_a = +40$ V, $V_b = +20$ V, and $V_{ab} = V_a - V_b = +40 - (+20) = \mathbf{+20\ V}$, as shown in Fig. 6-7b

c. $I_{500} = V_{ab}/500\ \Omega = 20\ \text{V}/500\ \Omega = \mathbf{40\ mA}$ from node **a** to node **b**, on the original circuit in Fig. 6-7a

EXAMPLE 6-6
A load, R_L, draws 20 mA at 25 V from a potentiometer shown in **Fig. 6-8**. The entire potentiometer draws 100 mA from a 100-V supply. Calculate

a. I_2, the current in R_2
b. Resistance values of R_L, R_2, and R_1.
c. Power ratings of R_1 and R_2
d. Explain why the power rating of R_1 must be greater than R_2

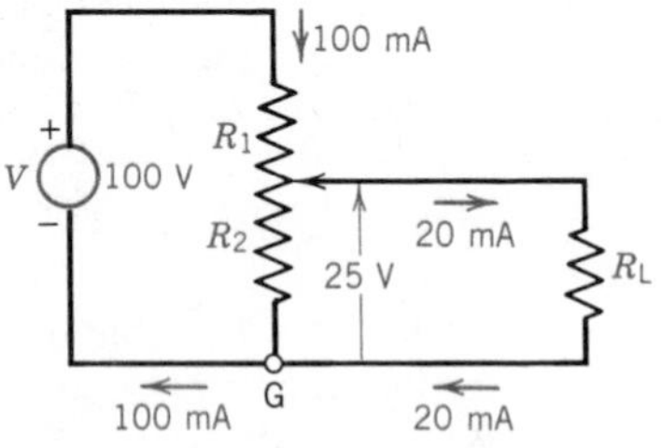

Figure 6-8 Example 6-6

Solution

a. Using KCL, $I_2 = I_1 - I_L = (100 - 20)\ \text{mA} = \mathbf{80\ mA}$ entering node G
b. $R_L = V_L/I_L = 25\ \text{V}/20\ \text{mA} = \mathbf{1.25\ k\Omega}$
$R_2 = V_2/I_2 = 25\ \text{V}/80\ \text{mA} = \mathbf{312.5\ \Omega}$
$R_1 = V_1/I_1 = 75\ \text{V}/100\ \text{mA} = \mathbf{750\ \Omega}$
c. $P_1 = V_1 I_1 = 75\ \text{V} \times 0.1\ \text{A} = \mathbf{7.5\ W}$
$P_2 = V_2 I_2 = 25\ \text{V} \times 0.08\ \text{A} = \mathbf{2\ W}$
d. R_1 has a higher volt drop across it and also carries a larger current than R_2

Note that in solving the above series–parallel circuit, circuit simplification was unnecessary. This is also true of Exs. 6-7 and 6-8, which permit the entire analysis to be performed on the original circuits.

EXAMPLE 6-7

For the circuit shown in **Fig. 6-9**, $I_2 = 2$ mA and all resistances are known and shown. Calculate

a. All remaining currents in all resistors
b. Supply voltage, V_T
c. Voltage V_{ab}

Solution

a. R_3 and R_2 are in parallel (same voltage) and $R_3 = R_2$, therefore $I_3 = I_2 = \mathbf{2\ mA}$; then $I_1 = I_2 + I_3$ (by KCL) $=$ **4 mA** and $V_{cd} = 4\ \text{mA}\cdot[5\ \text{k}\Omega + (20\ \text{k}\Omega)\|20\ \text{k}\Omega)] = (4\ \text{mA})\cdot(15\ \text{k}\Omega) = \mathbf{60\ V}$. But since R_4 is also across V_{cd} (nodes *a* and *d* are the *same* node), $I_4 = V_4/R_4 = 60\ \text{V}/15\ \text{k}\Omega = \mathbf{4\ mA}$. Then the total current $I_T = I_6 = I_5 = (I_4 + I_1) = (I_4 + I_2 + I_3) = (4 + 4)\ \text{mA} = \mathbf{8\ mA}$.
b. Since we know all resistance values and all currents in all resistors, $V_T = V_6 + V_5 + V_{cd} = (8\ \text{mA}\cdot 2.5\ \text{k}\Omega) + (8\ \text{mA}\cdot 1\ \text{k}\Omega) + 60\ \text{V} = (20 + 8 + 60)\ \text{V} = \mathbf{88\ V}$
c. $V_{ab} = V_{db} = V_5 = \mathbf{8\ V}$

Example 6-7 is important in revealing the following:

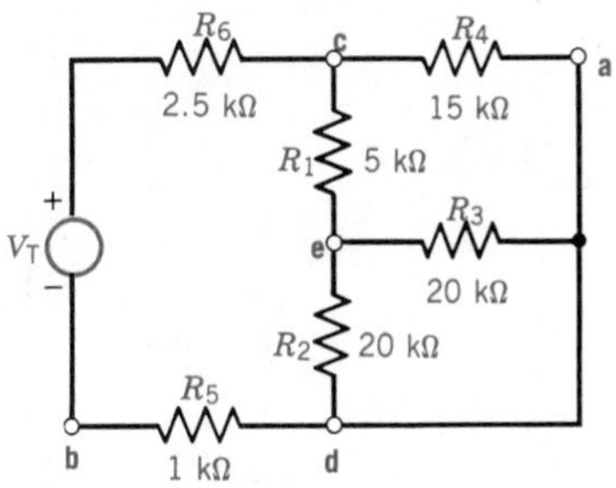

Figure 6-9 Example 6-7

1. The analysis began where *two pieces of information were known about a single resistor*, that is, the resistance of R_2 and current I_2. This led to the voltage V_2 and enabled finding the current I_3, since R_3 and R_2 are in parallel. Knowing the current leaving node e, we found I_1 by using KCL and so forth.
2. Using a combination of KCL to find currents around nodes and KVL to find voltages across nodes, we can synthesize *all currents* and *all voltages*.
3. Since we are adding currents to find and verify total current, no circuit simplification is needed.

EXAMPLE 6-8

For the values given and circuit shown in **Fig. 6-10**, calculate a. R_5 b. R_9 c. V_T

Solution

a. We begin where we know two pieces of information about a resistor: $I_2 = V_2/R_2 = 40\ \text{V}/4\ \text{k}\Omega = 10$ mA, and $I_3 \equiv I_2 = 10$ mA because they are in series. Then by KCL around node c, $I_4 = I_3 - I_5 = (10 - 1)\ \text{mA} = 9$ mA and $V_4 = V_{cG} = (9\ \text{mA})(1\ \text{k}\Omega) = 9\ \text{V} = V_{cd} = V_5$; therefore $R_5 = V_5/R_5 = 9\ \text{V}/1\ \text{mA} = \mathbf{9\ k\Omega}$
b. V_x (measured to ground) $= V_{xG} = V_{xc} + V_{cd} = (50 + 9)\ \text{V} = 59$ V; V_a (measured to ground) $= V_{aG} = V_{ac} + V_{cG} = 80\ \text{V} + 9\ \text{V} = 89$ V. Then $V_{ax} = V_a - V_x = (89 - 59)\ \text{V} = 30\ \text{V} = V_6$ so $I_6 = V_6/R_6 = 30\ \text{V}/8\ \text{k}\Omega = 3.75\ \text{mA} = I_9$ (since they are in series); $I_8 = 3.75\ \text{mA}/(n + 1) = 3.75\ \text{mA}/6 = 0.625$ mA and $V_8 = I_8R_8 = (0.625\ \text{mA})(5\ \text{k}\Omega) = 3.125$ V. So finally, $V_9 = V_x - V_8 = 59 - 3.125 = 55.875$ V and $R_9 = V_9/I_9 = 55.875\ \text{V}/3.75\ \text{mA} = \mathbf{14.9\ k\Omega}$
c. $I_T = I_1 = I_2 + I_6 = 10\ \text{mA} + 3.75\ \text{mA} = 13.75$ mA. $V_1 = I_1R_1 = (13.75\ \text{mA})(2.5\ \text{k}\Omega) = 34.375$ V and $V_T = V_1 + V_{aG} = (34.375 + 89)\ \text{V} = 123.375\ \text{V} = \mathbf{123.4\ V}$

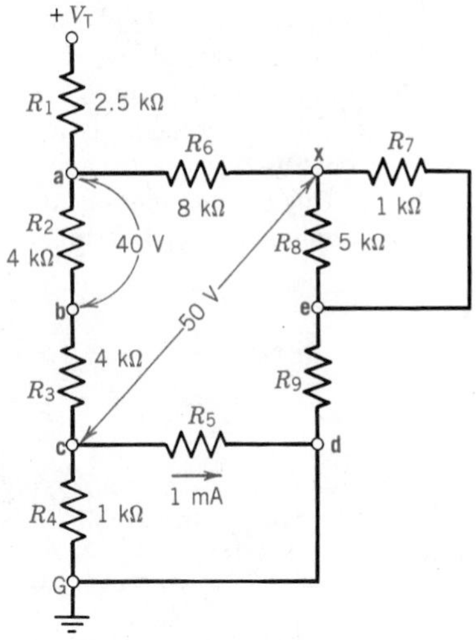

Figure 6-10 Example 6-8

Example 6-8 (like Ex. 6-7) requires no circuit simplification because there are two unknown resistors (R_5 and R_9) and the supply voltage V_T is unknown. Consequently, the analysis must begin wherever two pieces of information are known about a given circuit element. As in the previous case, KVL and KCL are employed to obtain all currents and voltages.

EXAMPLE 6-9

Without using currents, calculate voltages at nodes a, b, c, and d with respect to ground in **Fig. 6-11a**, and find the current in R_1

Solution

Since all supply voltages and resistances are known, we can use the method of circuit reduction. At the same time, we can redraw

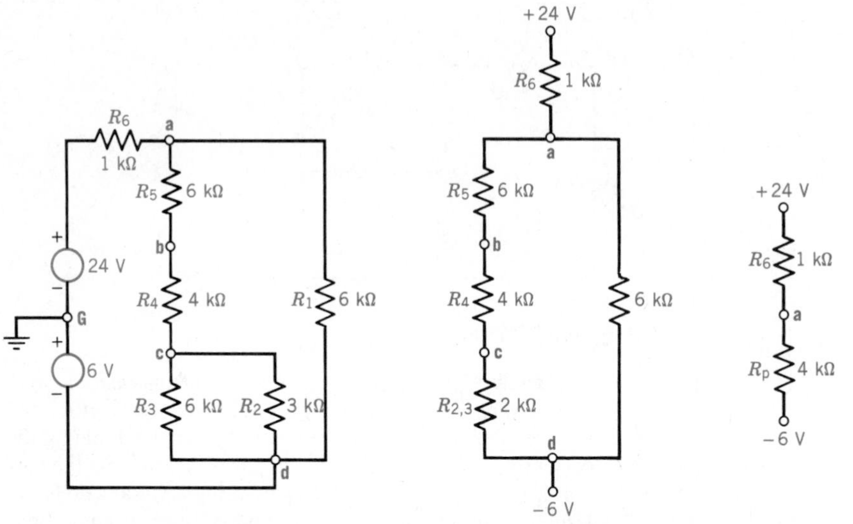

a. Original circuit b. First simplification c. Second simplification

Figure 6-11 Example 6-9

the circuit in a more familiar form, showing a voltage divider with two supplies, as in Fig. 6-11b. Note from Fig. 6-11b that we have also combined R_2 and R_3 into one equivalent resistor with a resistance of 2 kΩ, as shown. Figure 6-11b shows a two-branch parallel circuit having 12 kΩ in one branch and 6 kΩ in the other. The parallel equivalent of the two branches is $R_p = 12\text{ k}\Omega \| 6\text{ k}\Omega = 4\text{ k}\Omega$. This now permits us to draw the second simplification shown in Fig. 6-11c. This second simplification eliminates nodes **b** and **c** but does permit us to find V_a using our (powerful) Eq. (5-8):

$$V_a = (V_p - V_n)\frac{R_n}{R_n + R_6} + V_n = [24 - (-6)]\frac{4\text{ k}\Omega}{(4+1)\text{ k}\Omega} + (-6)$$

$$= \mathbf{+18\ V} = \text{voltage at node } \mathbf{a}$$

We can now return to the circuit of Fig. 6-11b and apply the same Eq. (5-8), but this time the more positive voltage is $+18\text{ V} = V_a$; therefore the voltage at node **b** is

$$V_b = (V_p - V_n)\frac{R_n}{R_n + R_5} + V_n$$

$$= [18 - (-6)]\frac{6\text{ k}\Omega}{(6+6)\text{ k}\Omega} + (-6) = \mathbf{+6\ V} \quad (5\text{-}8)$$

In the same way, using Fig. 6-11b, we can find the voltage at node **c** or

$$V_c = (V_p - V_n)\frac{R_n}{R_n + R_{4,5}} + V_n = 24\text{ V}\frac{2\text{ k}\Omega}{12\text{ k}\Omega} - 6 = \mathbf{-2\ V} \quad (5\text{-}8)$$

By inspection, the voltage at node **d** with respect to **G** is −6 V, that is, $V_d = \mathbf{-6\ V}$. To find the current in R_1 we require voltage V_{ad}. $V_a = +18$ V and $V_d = -6$ V, thus $I_1 = V_{ad}/R_1 = [18 - (-6)]/6\text{ k}\Omega = \mathbf{4\ mA}$

Once again, the extreme power of Eq. (5-8) has been demonstrated. Since we have found all node voltages, observe that it is a simple matter to find all currents.

6-3 VOLTAGE DIVIDER—PRINCIPLES, ANALYSIS, AND SYNTHESIS

In commercial electronic circuits, there is often a need for a variety of dc voltages, both positive (above ground) and negative (below ground), to supply various voltage requirements. We have already seen that two voltage sources may be cascaded (as in Ex. 6-9) to provide a variety of positive voltages and a negative voltage.

Commercial dividers use *only one* dc source to provide a *variety* of *both* positive and negative voltages. The resistors in a commercial divider are selected and designed, deliberately, to provide all the specific dc voltage requirements of a piece of electronic equipment.

Basically, the principle of the commercial voltage divider emerges from the voltage-divider rule (VDR) derived in Sec. 5-5.1 by Eq. (5-6):

$$V_1 = \frac{R_1}{R_2} V_2 = \frac{R_1}{R_T} V_T$$

Let us review the application of the VDR to a simplified voltage divider, analyzed in Ex. 6-10.

EXAMPLE 6-10

For the simplified voltage divider shown in **Fig. 6-12a**, calculate the following voltages with respect to ground: a. V_a b. V_b c. V_c d. V_d

Solution

PRELIMINARY ANALYSIS: Since $V_T = 300$ V, V_a must be less than +300 V with respect to ground because V_a represents the potential across only three resistors, whereas V_T is the potential across *all* four resistors. This is the clue to the analysis of the voltage divider shown in Fig. 6-12a.

a. $V_{aG} = \frac{R_{aG}}{R_T} V_T = \frac{70\ \text{k}\Omega}{100\ \text{k}\Omega} 300\ \text{V} = \mathbf{210\ V}$ (5-6)

b. $V_{bG} = \frac{R_{bG}}{R_T} V_T = \frac{30\ \text{k}\Omega}{100\ \text{k}\Omega} 300\ \text{V} = \mathbf{90\ V}$

c. $V_{cG} = \frac{R_{cG}}{R_T} V_T = \frac{20\ \text{k}\Omega}{100\ \text{k}\Omega} 300\ \text{V} = \mathbf{60\ V}$

d. $V_{dG} = \frac{R_{dG}}{R_T} V_T = \frac{30\ \text{k}\Omega}{100\ \text{k}\Omega} 300\ \text{V} = \mathbf{-90\ V}$

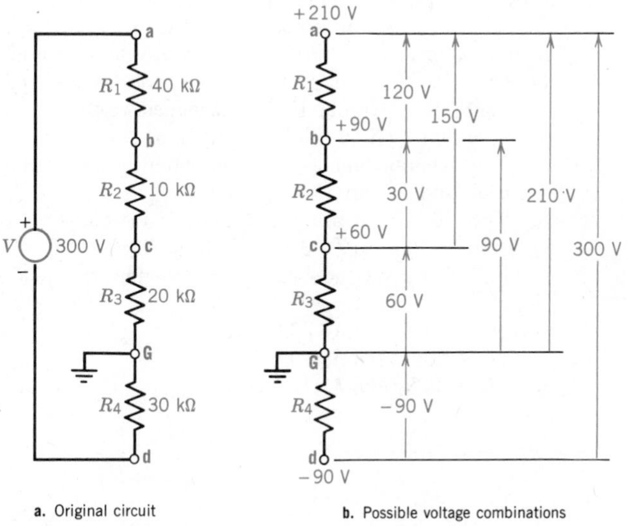

Figure 6-12 Analysis of a simplified voltage divider (Exs. 6-10 and 6-11)

EXAMPLE 6-11

List the possible voltage combinations (without duplication) available from the simplified voltage divider of Ex. 6-10

Solution

Using the voltages obtained in the solution of Ex. 6-10, the diagram of Fig. 6-12b is drawn. Observe that voltage V_{aG} is +210 V (and not +300 V, as we might infer erroneously from Fig. 6-12a) and voltage V_{dG} is −90 V. For the given resistances of Fig. 6-12a, we obtain the following eight possibilities:

$V_{ab} = V_a - V_b = (210 - 90)\ \text{V} = \mathbf{120\ V}$
$V_{bc} = V_b - V_c = (90 - 60)\ \text{V} = \mathbf{30\ V}$
$V_{cG} = \mathbf{60\ V}$
$V_{dG} = \mathbf{-90\ V}$
$V_{ac} = (210 - 60)\ \text{V} = \mathbf{150\ V}$
$V_{bG} = \mathbf{+90\ V}$
$V_{aG} = \mathbf{210\ V}$
$V_{ad} = \mathbf{300\ V}$

These eight possible voltages are shown in Fig. 6-12b.

The voltage divider analyzed in Exs. 6-10 and 6-11 was termed a *simplified* voltage divider because it was assumed that *no loads* were connected across the various nodes with respect to ground. Since the analysis is relatively simple, even with included loads and current requirements, let us now attempt to synthesize (design) a voltage divider based on certain load requirements.

EXAMPLE 6-12

Design a commercial voltage divider that will provide the following load requirements: +50 V at 20 mA for load R_{L_1}, +30 V at 50 mA for load R_{L_2}, −30 V at 10 mA for load R_{L_3}. Specify

a. Supply voltage needed and total current drawn from supply
b. Resistances of the voltage divider required, allowing a 10-mA bleeder current for R_2
c. Commercial resistance and power ratings of R_1, R_2, and R_3

Solution

Since the above voltages are specified with respect to ground, the supply voltage must be the sum of the positive and negative voltage requirements, or +50 V − (−30 V) = +80 V. To determine the supply current:

a. Draw the divider as shown in **Fig. 6-13**, with R_{L_1}, R_{L_2}, and R_{L_3} connected to the respective node voltages and drawing the specified currents. With respect to Fig. 6-13, please note the following:
   1. Currents in R_{L_1} and R_{L_2} are flowing from their positive nodes into ground.
   2. Current in R_{L_3} flows from its more positive ground node (x) into its more negative (−30 V) terminal.
   3. Currents in the resistors R_1, R_2, and R_3 are entered on the diagram only *after* currents in the loads are entered, starting at point x, showing 60 mA entering node G. Given I_2 = 10 mA (approximately 10 to 15% of the total load of 80 mA), we determine I_3 as 70 mA, leaving node G. This yields 80 mA returning to supply and 80 mA leaving the supply of 80 V, as determined above.
   4. Observe that KCL and KVL apply, that is,
      a. The algebraic sum of all currents entering and leaving each node is zero (KCL).
      b. The algebraic sum of all voltage drops and rises is zero $(+20 + 30 + 30 - 80 = 0)$ (KVL).

b. From Fig. 6-13 we may now calculate the values of R_1, R_2, and R_3 as
$R_1 = V_1/I_1 = 20\text{ V}/60\text{ mA} = \mathbf{333.\bar{3}\ \Omega}$
$R_2 = V_2/I_2 = 30\text{ V}/10\text{ mA} = \mathbf{3\ k\Omega}$
$R_3 = V_3/I_3 = 30\text{ V}/70\text{ mA} = \mathbf{428.6\ \Omega}$

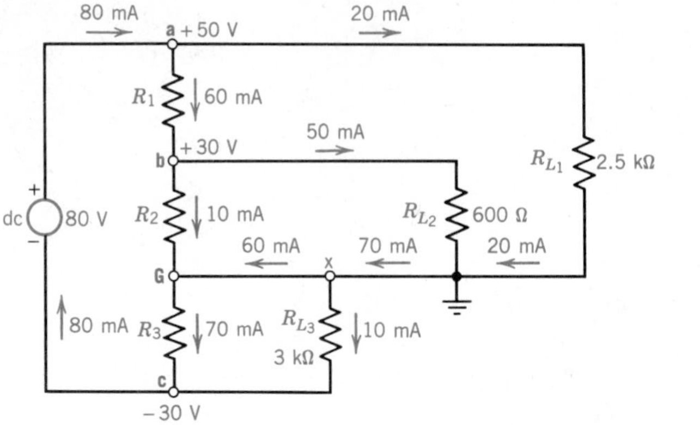

Figure 6-13 Synthesis of commercial voltage divider (Example 6-12)

c. From the voltages and currents given in (**b**) we can determine *commercial* values of all three resistors (from Table 3-3) and their (commercial) power *ratings*:
$P_1 = V_1I_1 = 20\text{ V} \times 60\text{ mA} = 1.2\text{ W}$
(use a **330 Ω, 2 W** resistor)
$P_2 = V_2I_2 = 30\text{ V} \times 10\text{ mA} = 0.3\text{ W}$
(use a **3 kΩ, 0.5 W** resistor)
$P_3 = V_3I_3 = 30\text{ V} \times 70\text{ mA} = 2.1\text{ W}$
(use a **430 Ω, 3 W** resistor)

In Ex. 6-12 we designed a commercial voltage divider to meet certain load requirements at both positive and negative voltages from a single dc voltage supply. But in doing this, certain compromises were made both in selection of bleeder current and in using standard resistance values for commercial resistors. Whenever a circuit is designed, certain trade-offs invariably arise and compromises that must be accepted. After the circuit is synthesized or designed, it is customary to subject it to an *analysis* to see how closely it meets the original specifications. We will do this in Ex. 6-13, using the specified values of Ex. 6-12.

Figure 6-14a shows the voltage divider designed in Ex. 6-12 with the values of load resistances determined from the voltage and current demands of Ex. 6-12.

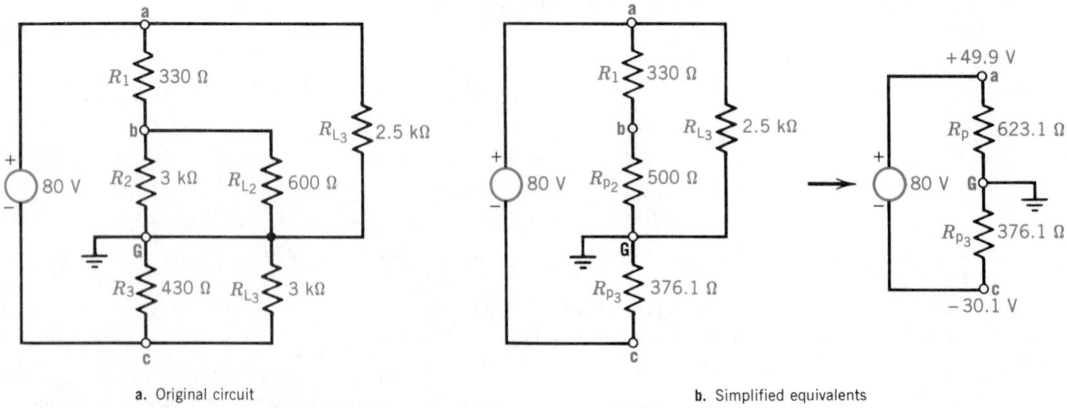

Figure 6-14 Example 6-13

EXAMPLE 6-13

Given the *commercial* voltage divider circuit of Fig. 6-14a, determine the following voltages at nodes with respect to ground: a. V_a b. V_b c. V_c

Solution

The first simplification is shown in Fig. 6-14b. It was found by combining the parallel combinations of R_2 and R_{L_2}, and R_3 and R_{L_3}, into single resistances, respectively, as R_{p_2} and R_{p_3}. The second simplification is also shown at the right in Fig. 6-14b. All of the resistors above ground have been lumped into a single resistance. We may now solve for the node voltages by the VDR as follows:

a. $V_a = \frac{R_p}{R_T} V_T = \frac{623.1}{623.1 + 376.1} 80 \text{ V} = \mathbf{49.9 \text{ V}}$

To find V_b, we return to the first simplification of Fig. 6-14b, where

b. $V_b = \frac{R_{p2}}{R_1 + R_{p2}} V_a = \frac{500}{830} \times 49.9 \text{ V} = \mathbf{30.1 \text{ V}}$

c. $V_c = \frac{R_{p3}}{R_T} V_T = -\frac{376.1}{999.2} \times 80 \text{ V} = \mathbf{-30.1 \text{ V}}$

Example 6-13 proves (conclusively) that the design technique and compromises made did *not* adversely affect the original specifications. Consequently, the commercial values listed in Ex. 6-12c may be used for the commercial voltage divider of Ex. 6-12. Example 6-13 also proves that the method of synthesis shown in Ex. 6-12 is valid.

6-4 CURRENT DIVIDER—PRINCIPLES AND ANALYSIS

The current divider rule (CDR) was first introduced in Eq. (5-24) for current division between two resistors in parallel.

A modification of the CDR was derived by using the principles of duality in Table 5-2, where we note the following relations for n paralleled resistors:

$$G_T = G_1 + G_2 + G_3 + \cdots + G_n \quad \text{siemens (S)} \qquad (5\text{-}21)$$

If the current to conductance G_n is $V_T G_n = I_n$ and $V_T = I_T/G_T$, then

$$I_n = V_T G_n = I_T \frac{G_n}{G_T} \quad \text{amperes (A)} \qquad (6\text{-}1)$$

and from Eq. (6-1), for two *or more* resistors in parallel,

$$I_1 = I_T \frac{G_1}{G_T} \quad (6\text{-}2a) \qquad \text{and} \qquad I_2 = I_T \frac{G_2}{G_T} \quad (6\text{-}2b) \qquad \text{(CDR)}$$

Note that Eqs. (6-2a) and (6-2b) are exactly the same as those derived by the duality between the VDR and the CDR in Table 5-2.

But since we are accustomed to working with resistance values, for n resistors in parallel we may write Eq. (6-1) as[2]

[2] Both Eqs. (6-1a) and (6-2c) are merely Ohm's law statements for individual branch currents of any individual resistor connected in parallel across a common supply, V_T.

$$I_n = I_T \frac{R_T}{R_n} = V_T/R_n \qquad \text{amperes (A)} \tag{6-1a}$$

and Eqs. (6-2) as

$$I_1 = I_T \frac{R_T}{R_1} \quad \text{and} \quad I_2 = I_T \frac{R_T}{R_2} = V_T/R_2 \tag{6-2c}$$

The foregoing equations show that if the total current entering a parallel branch (or any current in a parallel branch) and the individual branch resistances are known, the circuit may be considered a *current divider* and all individual branch currents may be calculated, using the current division principles discussed.

EXAMPLE 6-14

For the practical source shown in **Fig. 6-15**, calculate the

a. Value of R_L
b. Current drawn by R_L
c. Current in r_i
d. Current in r_i when R_L is removed
e. Open-circuit EMF of the practical source

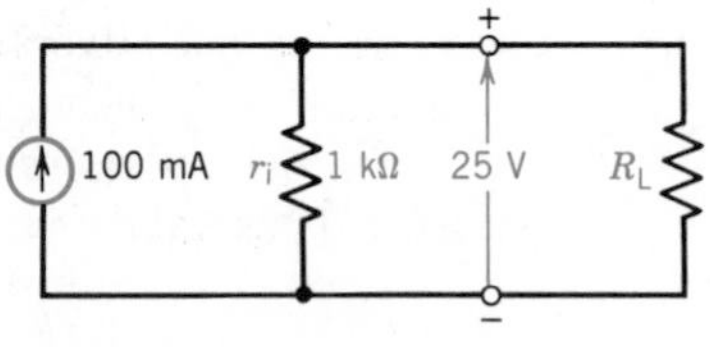

Figure 6-15 Example 6-14

Solution

a. $R_T = V_T/I_T = 25\text{ V}/0.1\text{ A} = \mathbf{250\ \Omega}$

$$R_L = \frac{r_i R_T}{r_i - R_T} = \frac{(1000 \times 250)}{(1000 - 250)} = \mathbf{333.\bar{3}\ \Omega} \tag{5-23b}$$

b. $$I_L = \frac{I_T R_T}{R_L} = \frac{25\text{ V}}{333.\bar{3}\ \Omega} = \mathbf{75\ mA} \tag{6-2c}$$

c. $I_i = V_T/r_i = 25\text{ V}/1\text{ k}\Omega = \mathbf{25\ mA}$ (see note 3 below)
d. $I_i = \mathbf{100\ mA}$ by inspection (where else can the current go?)
e. $E_0 = I_i r_i = (100\text{ mA})(1\text{ k}\Omega) = \mathbf{100\ V}$

Example 6-14 brings out several important points concerning current division and practical current sources:

1. A practical current source behaves exactly like a practical voltage source that has an external load connected across its terminals; that is, the voltage across its terminals decreases with application of load.
2. $R_L = r_i/3$ and drew three times the current of r_i, in accordance with the principles of parallel circuit current division, based on resistance ratios (see Ex. 6-15).
3. When we found I_L in part (**b**) of the solution as 75 mA, we could have subtracted this from the 100 mA of the current source to find I_i in r_i. But we should *always avoid* this! Had we made an error in any step of parts (**a**) and (**b**), then part (**c**) would be wrong as well.
4. In doing current-division problems, therefore, *never* subtract from total current to find remaining currents, unless you are cross-checking your results obtained by other current division methods.

EXAMPLE 6-15

Given the practical voltage source shown in **Fig. 6-16**, calculate the

a. Value of R_L that draws 20% of the source current
b. Total current
c. Current in R_L

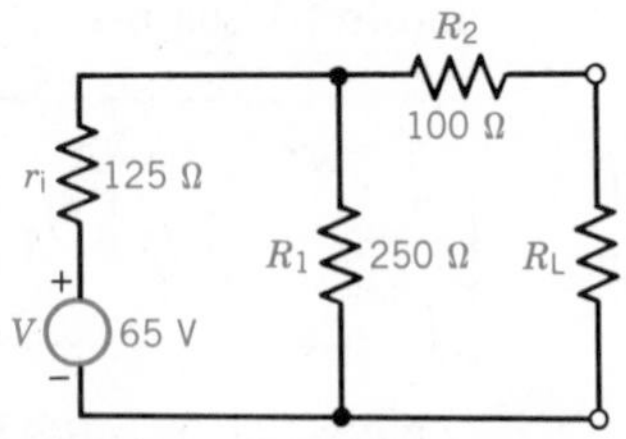

Figure 6-16 Example 6-15

Solution

a. Since current varies *inversely* as resistance, $I_1R_1 = I_2R_2$, in parallel, then $\dfrac{(100 + R_L)\ \Omega}{250\ \Omega} = \dfrac{0.8\not{I}}{0.2\not{I}}$ and solving for R_L yields

$R_L = \mathbf{900\ \Omega}$

b. $R_T = R_p + r_i = \dfrac{1000 \times 250}{1250} + 125 = 200 + 125 = 325\ \Omega$

$I_T = V_T/R_T = 65\text{ V}/325\ \Omega = \mathbf{200\ mA}$

c. $$I_L = I_T\left(\frac{R_1}{R_2 + R_L + R_1}\right) = 200\text{ mA}\,\frac{250\ \Omega}{(100 + 900 + 250)}\ \Omega$$

$= \mathbf{40\ mA}$ (or 20% of the source current) (5-24)

6-5 ANALYSIS OF VARIOUS NETWORKS

EXAMPLE 6-16

For the ladder network shown in **Fig. 6-17**, calculate the voltage V

Solution

We begin where two pieces of information are given, that is, at R_L, which dissipates 40 mW, and $I_L = \sqrt{P_L/R_L} = \sqrt{40 \text{ mW}/40 \text{ k}\Omega} =$ 1 mA. But we also note (quite happily) that the branch containing R_L has the same resistance (50 kΩ) as branch dG in parallel with it. Therefore, we may conclude that (1) $I_{dG} = 1$ mA and (2) $I_{cd} = 2$ mA, and we also note (quite happily) that each vertical 50-kΩ resistor is paralleled by a 50-kΩ equivalent resistance at its right. Therefore we may conclude that (3) $I_{cG} = 2$ mA, (4) $I_{bc} =$ 4 mA, (5) $I_{bG} = 4$ mA, and (6) $I_{ab} = 8$ mA

Applying Kirchhoff's voltage law (KVL) around the first mesh abGa:

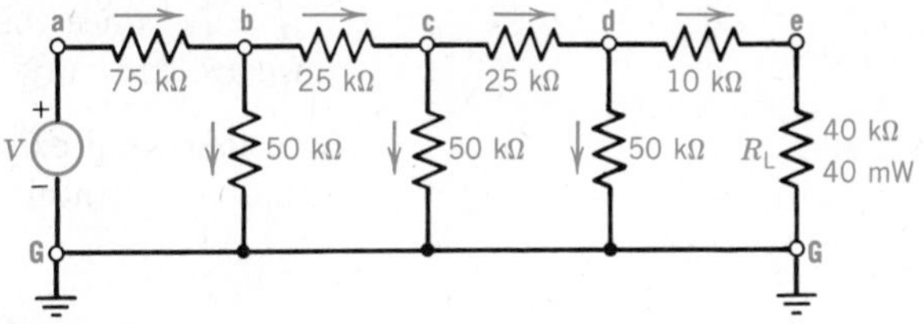

Figure 6-17 Example 6-16; ladder network

$$+(8 \text{ mA} \times 75 \text{ k}\Omega) + (4 \text{ mA} \times 50 \text{ k}\Omega) - V = 0 \quad (5\text{-}3)$$

and $V = 600 + 200 = \mathbf{800\ V}$

EXAMPLE 6-17

Solve Ex. 6-16 *without* using any currents whatever (use only VDR) to calculate voltage V.

Solution

Again we begin where two pieces of information are given at R_L and therefore $V_L = \sqrt{P_L R_L} = \sqrt{40 \text{ mW} \times 40 \text{ k}\Omega} = 40$ V; but R_L is four times as great as its series-connected resistor and therefore $V_{dG} = 50$ V. But we also note (quite happily) that the equivalent resistance $R_{dG} = 25$ kΩ. Consequently, $V_{cd} = V_{dG} = 50$ V, but $V_{cG} = V_{cd} + V_{dG} = (50 + 50)$ V $= 100$ V and $V_{bc} = V_{cG} = 100$ V. This means, then, that $V_{bG} = V_{bc} + V_{cG} = (100 + 100) = 200$ V across $R_{bG} = 25$ kΩ. Then by VDR, since $R_{ab} = 3(R_{bG})$, $V_{ab} =$ 3×200 V $= 600$ V. Applying KVL around the first mesh yields $V_{aG} = V_{ab} + V_{bG} = 600$ V $+ 200$ V $= \mathbf{800\ V}$.

The solutions to Exs. 6-16 and 6-17 show the inherent power of KVL and KCL when applied to complex series–parallel networks. Note that the solution to Ex. 6-16 used KCL exclusively and that of Ex. 6-17 used KVL exclusively. If these two laws are so powerful individually, think of how their capability is magnified *when they are used together*.

EXAMPLE 6-18

Given the ladder network of **Fig. 6-18**, calculate the

a. Equivalent total resistance, R_T
b. Total current, I_T
c. Current in R_4, I_4
d. Voltage across R_3, V_3

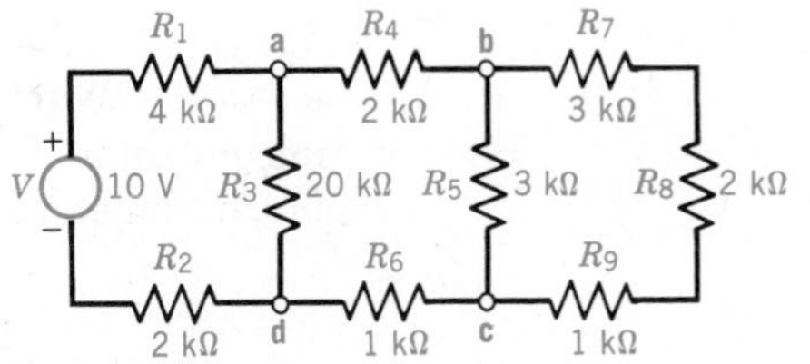

Figure 6-18 Example 6-18

Solution

a. $R_{bc} = \dfrac{(3 + 2 + 1)\text{ k} \times 3\text{ k}}{9\text{ k}} = 2$ kΩ and

$R_{ad} = R_3 \| R_{abd} = (20 \| 5)$ kΩ $= 4$ kΩ

$R_T = R_1 + R_2 + R_{ad} = (4 + 2 + 4)$ kΩ $= \mathbf{10\ k\Omega}$

b. $I_T = V_T/R_T = 10$ V/10 kΩ $= \mathbf{1\ mA}$

c. At node a, $I_3 = I_T/5$ and $I_4 = 4I_T/5 = \mathbf{0.8\ mA}$

d. $I_3 = I_T/5 = 1$ mA/5 $= 0.2$ mA and

$V_3 = I_3 R_3 = (0.2 \text{ mA})(20 \text{ k}\Omega) = \mathbf{4\ V}$

Alternative method using KVL: $V_1 + V_3 + V_2 - V = 0$ and $V_3 = V - (V_1 + V_2) = 10 - (4 + 2)$ V $= \mathbf{4\ V}$ (5-3)

A number of common networks frequently occur in the study of electronic circuits. One of these is the so-called *universal* or *voltage-divider bias* network shown in **Fig. 6-19**. The various resistors are selected and designed to establish desired dc voltage levels at nodes B, C, and E and currents I_C and I_E. Note that in Fig. 6-19

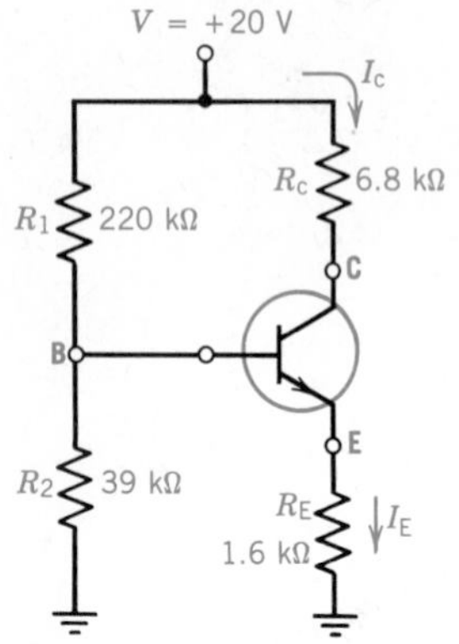

Figure 6-19 Example 6-19; universal transistor bias network

commercial values are used for all resistors. For purposes of analysis we must make two (very valid) assumptions, namely

1. $V_{BE} = 0.6$ V (the voltage drop across the base-emitter junction)
2. $I_C = I_E$ (since the base current is usually small in comparison to I_C and I_E, we are assuming $I_B = 0$)

The following example shows the steps for finding all essential voltages and currents representing the dc analysis of the transistor circuit of Fig. 6-19.

EXAMPLE 6-19

For the transistor network shown in Fig. 6-19, calculate

a. V_{R_1}
b. V_{R_2} and V_B
c. V_{R_E} and V_E
d. I_E and I_C
e. V_{R_C}
f. V_C
g. V_{CE}

Solution

a. $V_{R_1} = \dfrac{R_1}{R_1 + R_2}\,V = \dfrac{220}{220 + 39} \times 20\text{ V} = \mathbf{17\ V}$

b. $V_{R_2} = \dfrac{R_2}{R_1 + R_2}\,V = \dfrac{39}{220 + 39} \times 20\text{ V} = \mathbf{3\ V}$ and
$V_B = V_{R_2} = \mathbf{3\ V}$

c. $V_E = V_B - V_{BE} = 3\text{ V} - 0.6\text{ V} = \mathbf{2.4\ V}$ and
$V_{R_E} = \mathbf{2.4\ V}$

d. $I_E = V_{R_E}/R_E = 2.4\text{ V}/1.6\text{ k}\Omega = \mathbf{1.5\ mA} = I_C$ also

e. $V_{R_C} = I_C R_C = (1.5\text{ mA})(6.8\text{ k}\Omega) = \mathbf{10.2\ V}$

f. $V_C = V - V_{R_C} = 20 - 10.2 = \mathbf{9.8\ V}$

g. $V_{CE} = V_C - V_E = (9.8 - 2.4)\text{ V} = \mathbf{7.4\ V}$

With respect to this solution, please note that

1. The voltage-divider or universal or (H-bias) network is essentially a *parallel–series* circuit (Sec. 6-1).
2. Since there is only a single supply voltage ($V = +20$ V), all voltages are measured easily with respect to ground.
3. The left-hand side of the circuit experiences only one drop in potential while the right-hand branch experiences two drops (at C and E).
4. For the right-hand side, from KVL, we may write $V = V_{R_C} + V_{CE} + V_E = 10.2 + 7.4 + 2.4 = 20$ V
5. In step (**b**), as in the solution to previous examples, we could have obtained V_{R_2} by subtracting V_{R_1} from V. But instead we chose to solve it independently. If we had made an error in calculating V_{R_1} and used subtraction, then V_{R_2}, V_B, and all succeeding calculations would be in error!
6. This type of calculation is very frequently made in semiconductor circuit analysis.

Figure 6-20a shows a typical bridge network. Normally a *null detector* is located in place of the switch between points **X** and **Y** to indicate when they both are at the same potential. For present purposes we may assume that the switch represents an ammeter having negligible resistance. (See Sec. 23-3.)

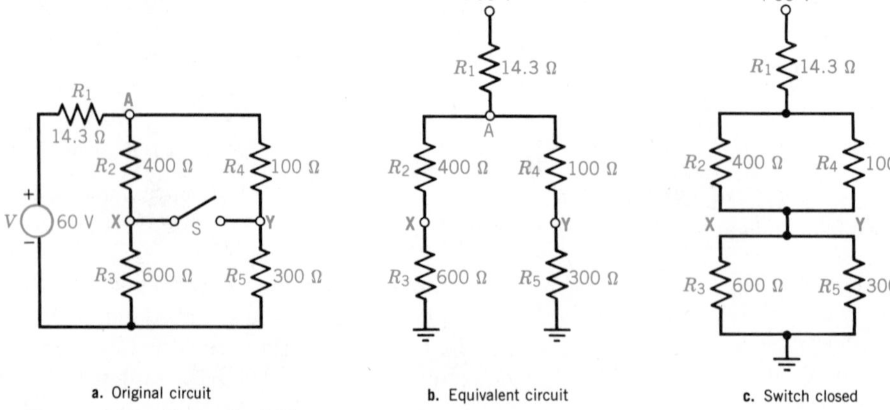

Figure 6-20 Example 6-20

EXAMPLE 6-20

Given the bridge circuit shown in Fig. 6-20a, calculate

a. Voltages at points **X** and **Y** with switch **S** open
b. Potential difference V_{XY} across the open switch
c. V_X and V_Y when the switch is closed

Solution

a. The equivalent circuit (switch open) is shown in Fig. 6-20b. The redrawn diagram clearly shows that the left-hand (LH) branch of 1000 Ω is in parallel with the RH branch of 400 Ω. Then the equivalent parallel resistance,

$$R_p = (1000 \times 400/1400)\ \Omega = 285.7\ \Omega$$

The total circuit resistance is

$R_T = R_1 + R_p = 14.3 + 285.7 = 300\ \Omega$. Consequently,

$$V_A = \frac{R_P}{R_T} V_T = \frac{285.7}{300} 60\text{ V} = 57.14\text{ V}$$

And again using VDR for the LH branch

$$V_X = \frac{R_3}{R_2 + R_3} V_A = \frac{600}{1000} 57.14 = \mathbf{34.28\ V}$$

and for the RH branch

$$V_Y = \frac{R_5}{R_5 + R_4} V_A = \frac{300}{400} 57.14 = \mathbf{42.85\ V}$$

b. $V_{XY} = V_X - V_Y = (34.28 - 42.85)\text{ V} = \mathbf{-8.57\ V}$

c. With switch S closed, the equivalent circuit may be drawn as shown in Fig. 6-20c. This places R_3 and R_5 in parallel but in series with R_2 and R_4.

$$R_T = (R_3 \| R_5) + (R_2 \| R_4) + R_1$$
$$= (600 \| 300) + (400 \| 100) + 14.3 = 200 + 80 + 14.3$$
$$= 294.3\ \Omega$$

$$V_X = V_Y = \frac{(R_3 \| R_5)}{R_T} V_T = \frac{200}{294.3} 60\text{ V} = \mathbf{40.77\ V}$$

With respect to Ex. 6-20 and its solution, please note:

1. Placing an ammeter across terminals X and Y could have some serious effects on the ammeter. V_{YX} is approximately 8.6 V, and if the ammeter has a (practical) resistance of 0.01 Ω, the instantaneous current is approximately 860 A!
2. Once the switch is closed (i.e., terminals X and Y are shorted), they are at the same potential and the current between them is zero. This is shown in the solution to part (**c**).
3. This problem shows that the null detector used for sensing the potentials at X and Y *should have an extremely high resistance* so as not to draw current and to *maintain* the *original voltages* at their respective values.

6-6 GLOSSARY OF TERMS USED

Bleeder resistor Resistor connected across a power source to improve voltage regulation or to protect the remaining load from excessive voltages if part of the load is removed or reduced.

Bridge network Two-port network with an input voltage connected to one port and a null detector connected to the output port. By suitable adjustment of the elements in the bridge, zero voltage output is obtained.

Circuit reduction Process by which any complex circuit is reduced to a single equivalent resistance (or a sufficiently simplified equivalent circuit) to enable the complete solution of the original circuit.

Complete solution For any complex network, all voltages across, currents in, and powers dissipated by every component or device in the circuit network.

Current divider Network consisting of parallel elements to which a voltage is applied and from which one or more desired currents are obtained.

Ladder network Series–parallel configuration of circuit elements resembling the rungs and side bars of a ladder.

Null detector Instrument for measuring and detecting a balance between two opposing electrical quantities.

Parallel–series circuit Essentially a parallel circuit containing series elements.

Port Pair of terminals.

Potentiometer Adjustable voltage divider of the resistance type.

Series–parallel circuit Essentially a series circuit containing parallel elements.

Transistor Active semiconductor device having three or more terminals and two or more junctions.

Voltage divider Network consisting of series elements to which a voltage is applied and from which one or more voltages can be obtained across any portion of the network.

6-7 PROBLEMS

Secs. 6-1 and 6-2

6-1 For the parallel–series circuit shown in **Fig. 6-21**, calculate the

a. Current in, voltage across, and power dissipated by R_1
b. Single equivalent resistance of the complete circuit
c. Total current and power drawn from the supply

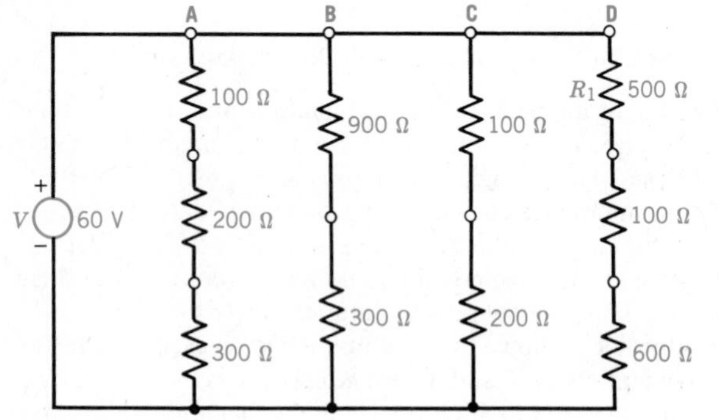

Figure 6-21 Problem 6-1

6-2 Reduce the ladder network of **Fig. 6-22** to a simplified series equivalent and calculate

a. V_{ef} c. V_{ab}
b. V_{cd} d. Power dissipated by R_1

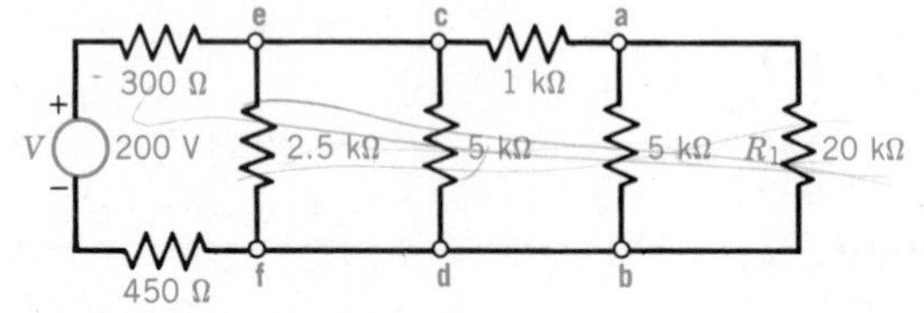

Figure 6-22 Problem 6-2

6-3 Find the complete solution for all voltages across and all currents in the network of **Fig. 6-23**.

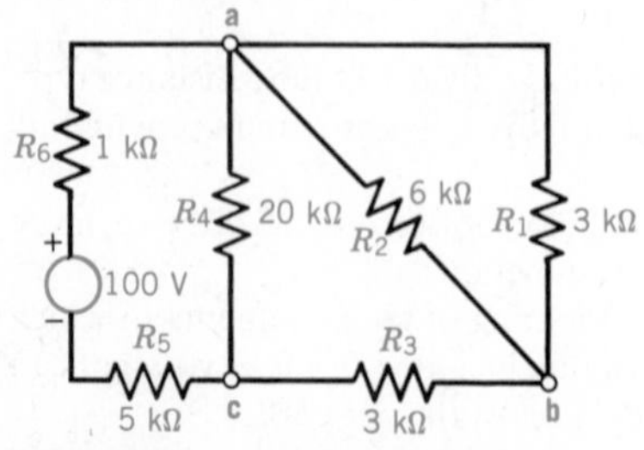

Figure 6-23 Problem 6-3

6-4 For the 9-kΩ potentiometer shown in **Fig. 6-24**, calculate (with respect to ground) the

a. Voltage at point x when switch S is open
b. Voltage at point x when switch S is closed
c. Decrease in voltage at point x due to the loading effect of R_L

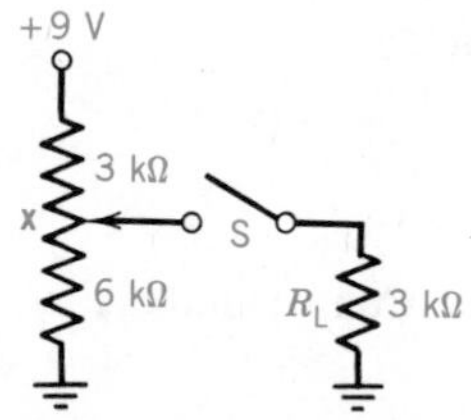

Figure 6-24 Problem 6-4

6-5 Simplify by redrawing the circuit shown in **Fig. 6-25** and calculate the

a. Total current drawn from the 10-V supply
b. Current in the 20-kΩ resistor
c. Power dissipated by the 12-kΩ resistor

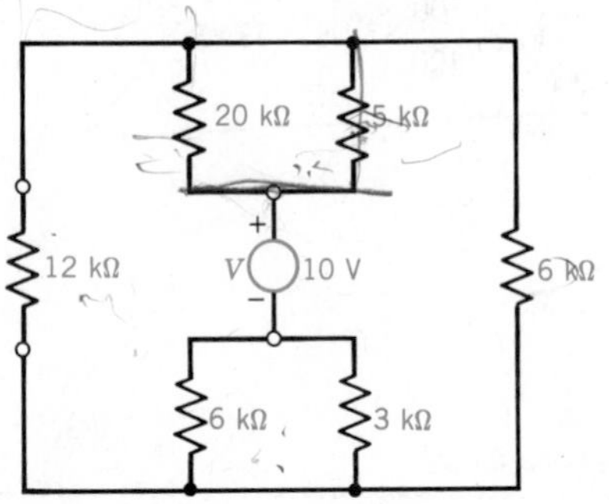

Figure 6-25 Problem 6-5

6-6 For the ladder network shown in **Fig. 6-26**, calculate the

a. Equivalent resistance, R_{ab}, that replaces the network
b. Total power dissipated by the network when a 60-V supply is connected across **a** and **b**
c. Voltages across nodes c–d (V_{cd}) and e–f (V_{ef}), respectively
d. Power dissipated by the 600-Ω load resistor, R_L

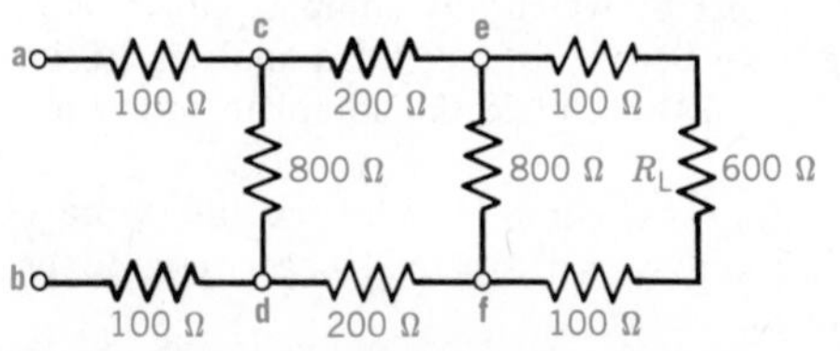

Figure 6-26 Problem 6-6

6-7 For the series–parallel circuit shown in **Fig. 6-27**, calculate the

a. Total current drawn from the supply
b. Voltage across the 16-kΩ resistor
c. Voltage across the 3.3-kΩ resistor

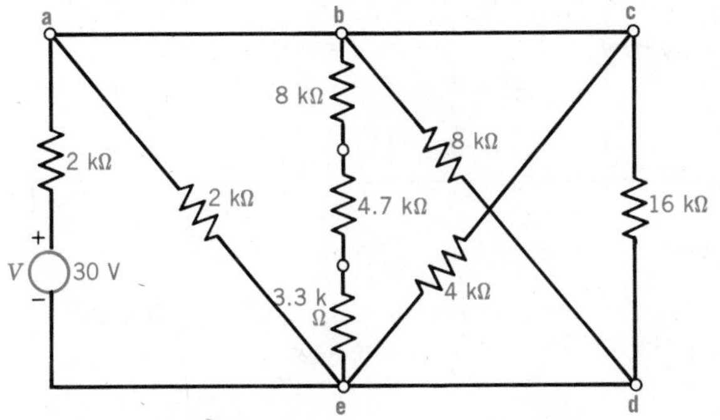

Figure 6-27 Problem 6-7

6-8 Given the circuit shown in **Fig. 6-28**, calculate the

a. Current in the 20-Ω resistor
b. Supply voltage V_T

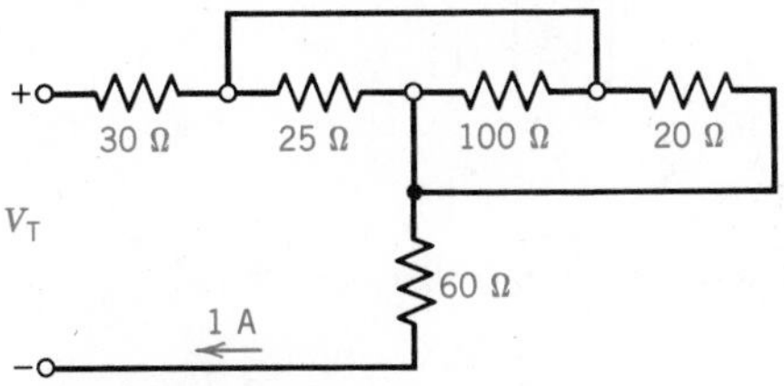

Figure 6-28 Problem 6-8

6-9 For the circuit shown in **Fig. 6-29** calculate the

a. Currents in resistors R_1 through R_8, inclusive
b. Voltage between points x and a, V_{xa}
c. Voltage between points a and b, V_{ab}

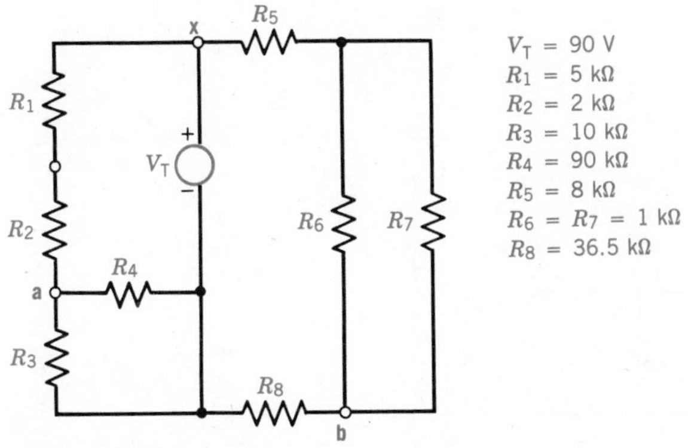

Figure 6-29 Problem 6-9

6-10 For the circuit shown in **Fig. 6-30**, calculate

a. R_B
b. R_A
c. V_{AB}

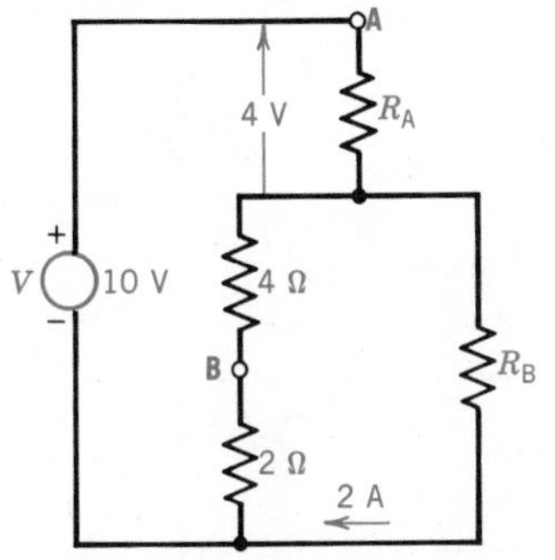

Figure 6-30 Problem 6-10

6-11 For the circuit shown in **Fig. 6-31**, calculate

a. The supply voltage, V_T
b. The current in the 20-Ω resistor, I_{20}

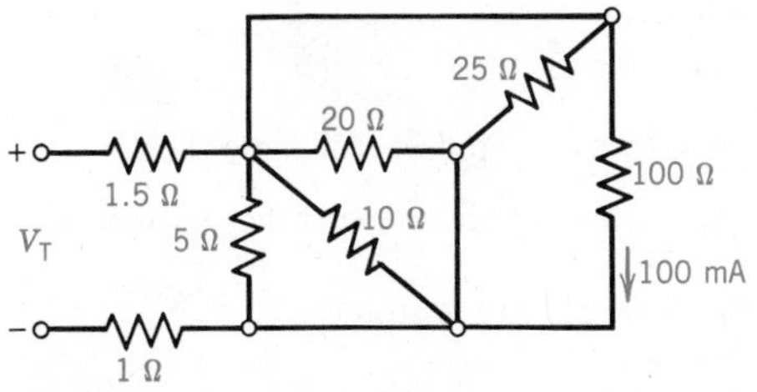

Figure 6-31 Problem 6-11

6-12 All resistors are identical in **Fig. 6-32**. Calculate the supply voltage, V_T, using only the VDR. (Hint: if you cannot solve this algebraically, assume $R = 1\ \Omega$.)

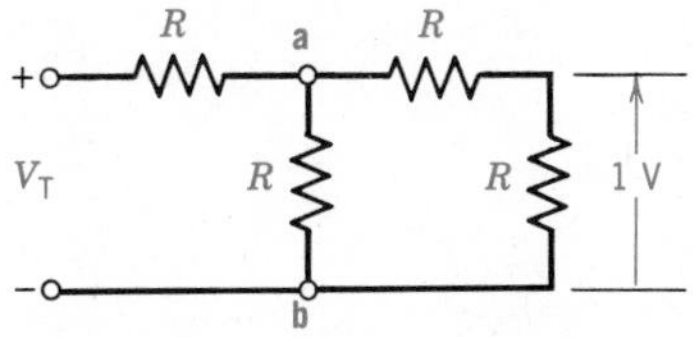

Figure 6-32 Problem 6-12

6-13 All resistors are identical in **Fig. 6-33**. Calculate the supply voltage, V_T, using only the VDR.

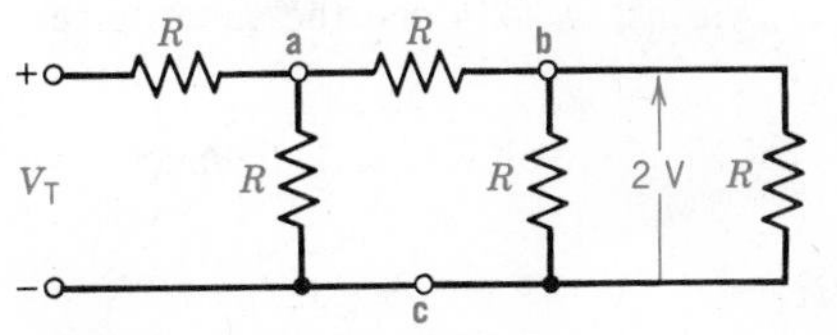

Figure 6-33 Problem 6-13

6-14 Repeat Prob. 6-12, using only the CDR to calculate V_T (assume R = 1 Ω).

6-15 Repeat Prob. 6-13, using only the CDR to calculate V_T (assume R = 2 Ω).

6-16 For the circuit shown in **Fig. 6-34**, calculate

a. V_T
b. V_A
c. V_B
d. V_{AB}
e. V_{BC}
f. Power dissipated by the 6-kΩ resistor

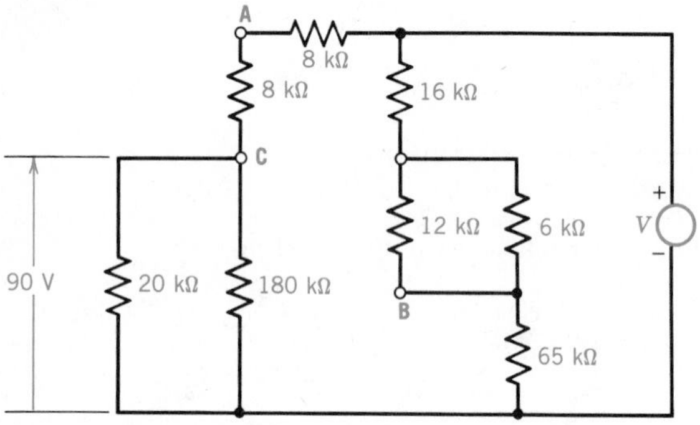

Figure 6-34 Problem 6-16

6-17 For the circuit shown in **Fig. 6-35**, calculate the

a. Equivalent resistance, R_T, of the network across the 60-V source
b. Total current, I_T, drawn from the supply
c. Current in the 900-Ω resistor
d. Voltage, V_{ab}
e. Voltage, V_{bc}

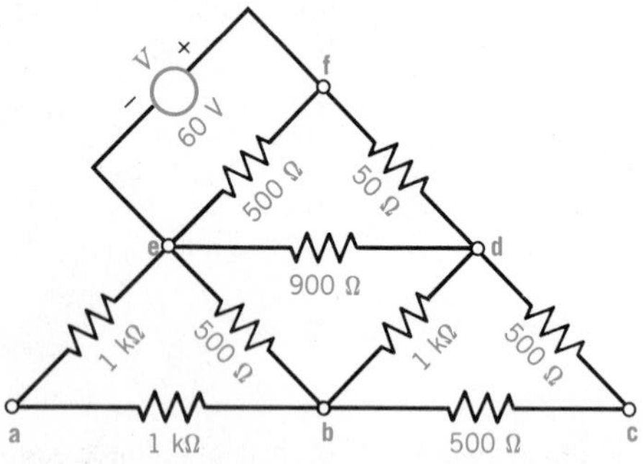

Figure 6-35 Problem 6-17

6-18 For the circuit shown in **Fig. 6-36** and the values given, find the

a. Complete solution, showing all voltages and all currents for all resistors
b. Voltage, V_{cb}
c. Voltage, V_{ab}
d. Total power drawn from the supply
e. Total power dissipated by R_4

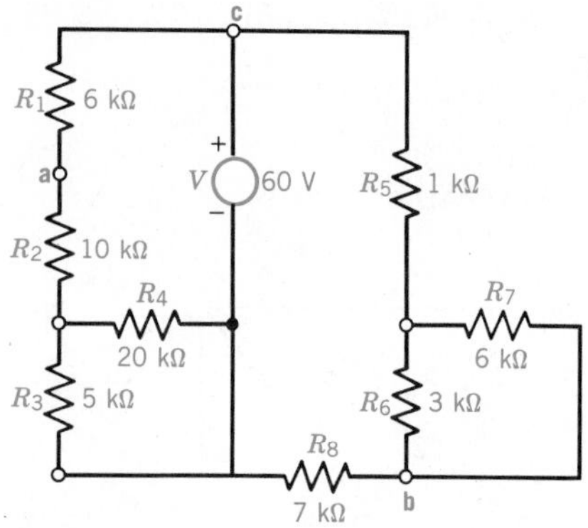

Figure 6-36 Problem 6-18

6-19 For the circuit shown in **Fig. 6-37**, without finding currents and using the VDR exclusively, calculate

a. V_{ab}
b. V_{bc}
c. V_{cd}
d. V_c with respect to ground

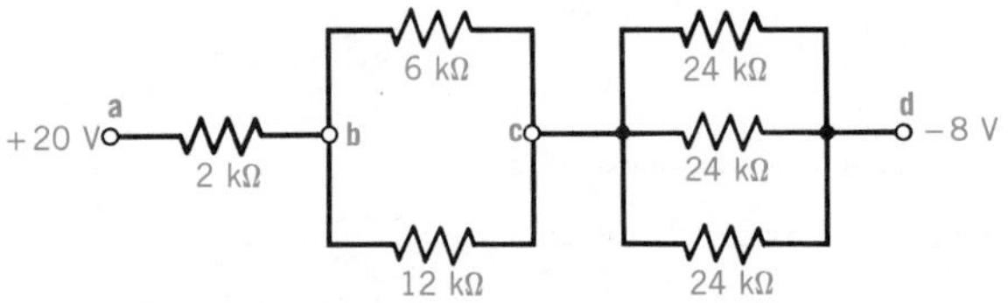

Figure 6-37 Problem 6-19

6-20 For the circuit shown in **Fig. 6-38**, calculate the

a. Total current drawn from the +70 V supply
b. Total current entering the −10 V source
c. Voltage at node x with respect to ground, V_x
d. Voltage at node a with respect to ground, V_a
e. Voltage at node b with respect to ground, V_b
f. Voltage V_{ab}
g. Explain why V_{ab} is a negative voltage

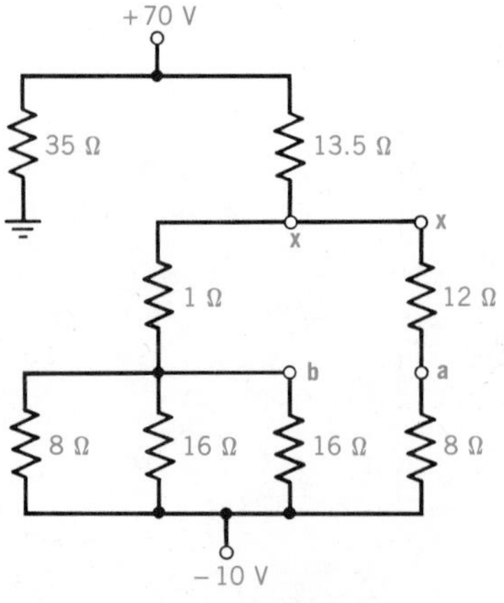

Figure 6-38 Problem 6-20

6-21 For the circuit shown in **Fig. 6-39**, calculate the

a. Total current flowing through the two supplies
b. Voltage at node *a*, V_a
c. Voltage at node *b*, V_b

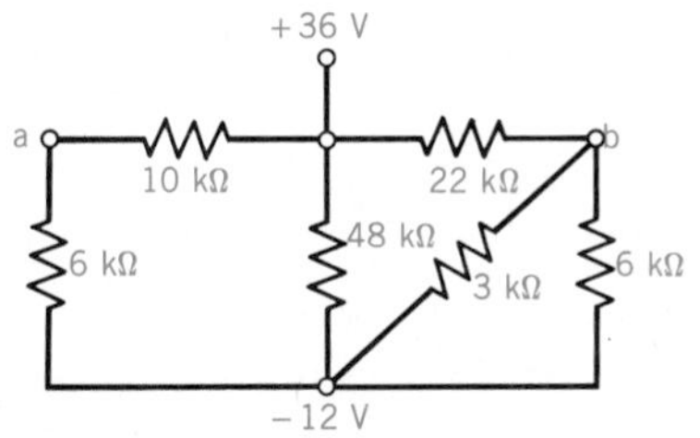

Figure 6-39 Problem 6-21

Sec. 6-3

6-22 A simple voltage divider is used to supply a load drawing 50 mA, at 160 V, from a 220 V dc source. The series dropping resistor is 1 kΩ. Calculate the

a. Current in the bleeder resistor
b. Resistance of the bleeder resistor

6-23 Design a simple voltage divider using a 220-V source delivering 50 mA that supplies a load drawing 144 V at 44 mA. Specify *commercial* resistance values and power ratings of the

a. Series dropping resistor, R_s
b. Bleeder resistor, R_b

6-24 A simple voltage divider consists of two commercial (5%) resistors in series across a 120-V supply to ground: a series resistor ($R_s = 5.6$ kΩ) and a bleeder resistor ($R_b = 4.3$ kΩ). The combination is intended to supply a 50 V, 1 mA load. Calculate the

a. Output voltage if the load is accidentally disconnected
b. Bleeder current when the load is connected and the output voltage is 50 V
c. Current drawn by the load under the conditions of (**b**)
d. Bleeder current when the load is increased and the output voltage is 49 V
e. Current drawn by the load under the conditions of (**d**)
f. Percent increase in load current required to produce the 2% drop in load voltage

6-25 Given the data (and solutions) of Prob. 6-24, calculate the voltage across the load when the load draws exactly 1 mA. Hint: solve for I_s, knowing that $V - I_s R_s = (I_s - 1)R_b$

6-26 Design a commercial voltage divider to supply the following loads from a 40 V dc supply, using a bleeder current of 10 mA: 30 V at 20 mA, 20 V at 22 mA, and 10 V at 10 mA. Specify commercial (5%) resistance values for all four resistors of the divider as well as commercial power ratings for the resistors.

6-27 Using the method of analysis shown in Ex. 6-13, combine load and voltage divider resistances to verify the *actual voltage* across each load of the commercial voltage divider designed in Prob. 6-26. Note: if your actual voltages are not within 5% of the specified voltages, your divider must be redesigned.

6-28 Design a commercial voltage divider to supply the following load requirements: +100 V at 15 mA for load R_{L_1}, +60 V at 30 mA for load R_{L_2}, and −50 V at 10 mA for load R_{L_3}. Allow a bleeder current of 12 mA and use only three resistors in the divider. Specify in your design the

a. Supply voltage needed and total current drawn from the supply
b. Commercial values of R_1, R_2, and R_3 of the divider (use 5% tolerance resistors)
c. Commercial power ratings of R_1, R_2, and R_3, respectively

6-29 Using the method of analysis given in Ex. 6-13, verify the *actual voltage* across each load of the divider designed in Prob. 6-28. (If your actual voltages are not within 5% of the specified voltages, your divider must be redesigned.)

Secs. 6-4 and 6-5

6-30 A simple potentiometer having an overall resistance of 1000 Ω is connected between +100 V and ground. The potentiometer tap is set to feed 50 V at 10 mA to a load, R_L. Calculate the

a. Resistance R_b of the (bleeder) portion of the potentiometer in parallel with R_L
b. Resistance of the series portion of the potentiometer, R_s

Hint: this problem involves solution of a quadratic equation for roots of R_b, using:

Method 1. Equate $I_s = I_b + 10 = (V - V_L)/R_s$ (where $I_b = 50/R_b$ and $R_s = 1000 - R_b$) and solve for R_b.

Method 2. Using VDR, $R_p = R_s$ but $R_s = 1000 - R_b$ and $R_p = 5000\,R_b/(5000 + R_b)$. Substitute and solve for R_b.

6-31 Given the so-called universal (voltage divider or H-parameter) network shown in **Fig. 6-40**, calculate the following voltages:

a. V_{R_1}
b. V_{R_2}
c. V_B
d. V_{R_E}
e. V_E
f. V_C
g. V_{CE}

Note: assume $V_{BE} = 0.6$ V and $I_B = 0$ (i.e., $I_C = I_E$). Check your calculations by $V = I_C R_C + V_{CE} + V_{R_E} = 30$ V

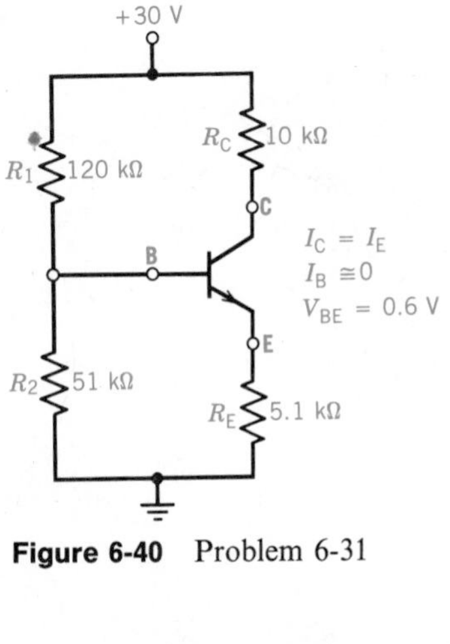

Figure 6-40 Problem 6-31

6-32 A similar type of bias arrangement used for field effect transistors (FETs) is shown in **Fig. 6-41**. Only two differences occur: $V_{GS} = 1.5$ V and the letter subscripts have been changed. Otherwise, the solution and technique are identical to those in Prob. 6-31 and Ex. 6-19. Calculate the following voltages:

a. V_G d. V_D
b. V_S e. V_{DG}
c. $I_S R_S$ f. V_{DS}

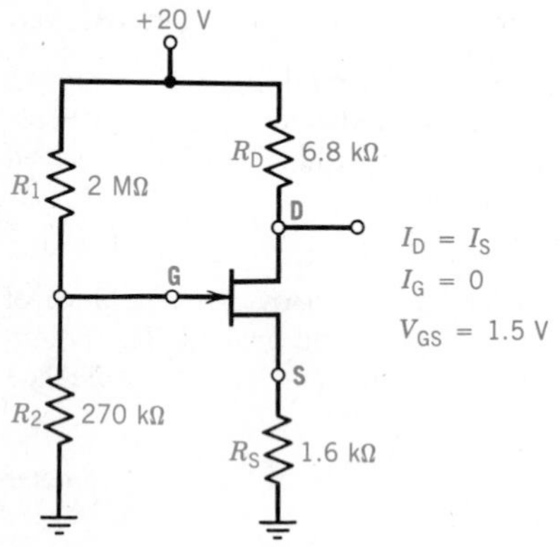

Figure 6-41 Problem 6-32

6-33 Calculate the value of R_L in **Fig. 6-42** that will

a. Reduce the terminal voltage of the current source to 10 V
b. Draw one-third of the source current

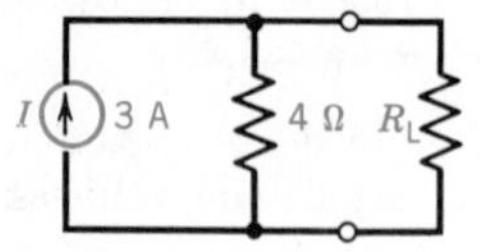

Figure 6-42 Problem 6-33

6-34 Calculate the value of R_L in the *current divider* of **Fig. 6-43** that will

a. Draw 20% of the source current
b. Produce a voltage of 3.5 V across the source terminals

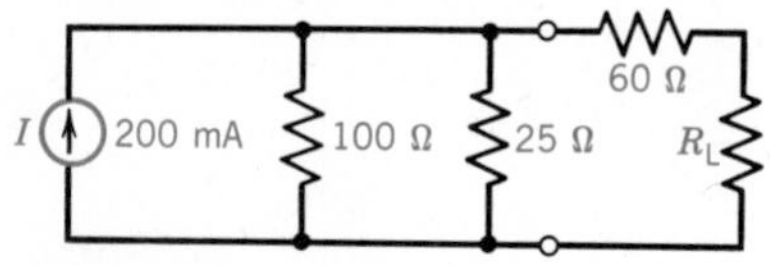

Figure 6-43 Problem 6-34

6-35 For the bridge network connected to two dc supplies shown in **Fig. 6-44**, calculate

a. V_a and V_b with respect to ground when switch S is open
b. Repeat (**a**) when switch S is closed
c. V_{ab} when switch S is opened
d. V_{ab} when S is closed

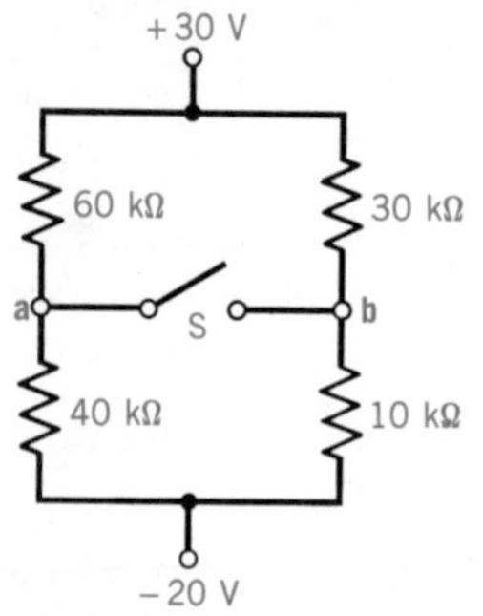

Figure 6-44 Problem 6-35

6-36 Repeat Prob. 6-35 when the more negative supply voltage is changed from −20 V to +10 V, all other values remaining the same.

6-8 ANSWERS

6-1 a 1.25 W b 150 Ω c 0.4 A, 24 W
6-2 a 125 V b 125 V c 100 V d 0.5 W
6-4 a 6 V b 3.6 V c 2.4 V
6-5 a 1 mA b 0.2 mA c $1.\bar{3}$ mW
6-6 a 600 Ω b 6 W c 40, 20 V d 375 mW
6-7 a 10 mA b 10 V c 2.0625 V
6-8 a 0.5 A b 100 V
6-9 b 39.375 V c −22.375 V
6-10 a 3 Ω b $1.\bar{3}$ Ω c 8 V
6-11 a 20 V b 0.5 A
6-12 5 V
6-13 16 V
6-14 5 V
6-15 16 V
6-16 a 170 V b 130 V c 130 V d 0 e 40 V
f $10.\bar{6}$ mW
6-17 a 250 Ω b 240 mA c 60 mA d −12 V
e −15 V
6-18 b 18 V c 0 d 540 mW e 7.2 mW
6-19 a 4 V b 8 V c 16 V d 8 V
6-20 a 6 A b 4 A c 8.286 V d −2.686 V
e 4.6286 V f −7.3146 V
6-21 a 6 mA b 6 V c −8 V
6-22 a 10 mA b 16 kΩ
6-23 a 1.5 kΩ, 4 W b 22 kΩ, 1 W
6-24 a $52.\bar{12}$ V b 11.63 mA c 0.87 mA d 11.4 mA
e 1.28 mA f 47.1%
6-25 49.7 V

6-26 160 Ω, 1 W; 240 Ω, 0.5 W; 510 Ω, 0.25 W; 1 kΩ, 0.125 W all ±5%

6-28 a 150 V, 57 mA b 1 kΩ, 5.1 kΩ, 1.1 kΩ c 2 W, 1 W, 3 W

6-30 a 524.9 Ω b 475.1 Ω

6-31 a 21 V b 8.95 V c 8.95 V d 8.35 V e 8.35 V f 13.6 V g 5.25 V

6-32 a 2.38 V b 0.88 V c 0.55 mA d 16.26 V e 13.88 V f 15.38 V

6-33 a 20 Ω b 8 Ω

6-34 a 20 Ω b 80 Ω

6-35 a 0, −7.5 V b −5.714 V c 7.5 V d 0

6-36 a 18 V, 15 V b 15.714 V c 3 V d 0

CHAPTER 7

Network Theorems

The previous chapter(s) enabled the solution of series, parallel, and series–parallel circuits by means of certain circuit-reduction techniques to produce simpler equivalent circuits. *Series*-connected elements were combined using *series* circuit relations. *Parallel*-connected elements were combined using *parallel* circuit relations.

The goal of these techniques was to produce an equivalent circuit in its simplest form, that is, consisting of a single source and an equivalent load. Whether we were aware of it or not, we were using the concepts underlying certain basic network theorems. Had we introduced them earlier, without some circuitry experience, they would have been meaningless abstractions. Having had some experience in solving simpler series–parallel networks, however, we are now prepared to examine and learn some of the more fundamental network theorems. These theorems find application not only to dc and ac circuit analysis but also to linear and (some) nonlinear electronic circuits, as well. A *theorem*, strictly speaking, is an *insight*.[1] By stating and learning each theorem, we will gain greater insight into the techniques of circuit analysis and also learn to apply each of the theorems.

7-1 THÉVENIN'S THEOREM

Thévenin's theorem states that any combination of linear circuit elements and active (dc) sources, regardless of connection or complexity, connected to a given load, R_L, may be replaced by a simple two-terminal network consisting of a single dc voltage source, V_{TH}, and a single equivalent resistance, R_{TH}, in series with the source and connected across the two terminals of load R_L, where

V_{TH} is the open-circuit voltage measured at the two terminals of interest with load R_L removed.

R_{TH} is the equivalent resistance of the given complex network with all voltage sources shorted (zero resistances) and all current sources opened (infinite resistances) and load R_L removed from the two terminals of interest.[2]

---

[1] From the Greek *theorema*, meaning a view or a sight.

[2] More precisely, the theorem states: the current that will flow through an impedance Z', when connected to any two terminals of a complex network between which there previously existed a voltage E and an impedance Z, is equal to the voltage E divided by the sum of Z and Z'. Thévenin published his theorem in 1883; the concept was originally described by Hermann Helmholtz (1821–1894) in a paper 30 years earlier.

Operationally speaking, to find the Thévenin equivalent between two terminals, T and H, of interest in a complex network, a three-step procedure is required:

1. **Remove load R_L (or any network portion) from across terminals T and H.**
2. **The voltage V_{TH} is the open-circuit voltage measured or calculated across terminals T and H.**
3. **The resistance of the complex network, R_{TH}, is the calculated resistance found by *replacing* all voltage and/or current sources with their equivalent internal resistances. (Voltage sources are shorted and current sources are opened.)**

The following examples illustrate application of Thévenin's theorem.

EXAMPLE 7-1

For the circuit shown in **Fig. 7-1a**, use Thévenin's theorem to determine

a. The current in R_L
b. The voltage across R_L

Solution

The problem solution is accomplished in the following steps:

1. Remove R_L, since we are interested in determining the Thévenin voltage V_{TH} and the Thévenin resistance, R_{TH}, looking back into the original complex network.
2. With R_L removed, the circuit is open and no current flows. Consequently, there is no voltage drop across any of the internal resistances. As shown in Fig. 7-1b, a voltmeter connected across terminals T–H would measure the Thévenin voltage, V_{TH}.
3. The voltage V_{TH} taken counterclockwise from terminal T to H is $+8 +3 +4 -6 -5 = V_{TH} = \mathbf{+4\ V}$ (Fig. 7-1b).
4. The voltage sources shown in Fig. 7-1b are now replaced by their internal resistances as shown in Fig. 7-1c, and an ohmmeter is connected across terminals T and H. The Thévenin resistance from terminal T to H is $(4 + 3 + 2 + 3 + 4)\ \Omega = R_{TH} = \mathbf{16\ \Omega}$
5. The complex network of Fig. 7-1a may now be replaced by its Thévenin equivalent TO THE LEFT of terminals T–H by V_{TH} and R_{TH} as shown in Fig. 7-1d.
6. Replace R_L across terminals T–H and solve for parts (**a**) and (**b**) above, using Fig. 7-1e.

a. $I_L = I = \dfrac{V_{TH}}{R_{TH} + R_L} = \dfrac{4\ \text{V}}{(16 + 4)\ \Omega} = \mathbf{0.2\ A}$

b. $V_L = I_L R_L = (0.2\ \text{A})(4\ \Omega) = \mathbf{0.8\ V}$

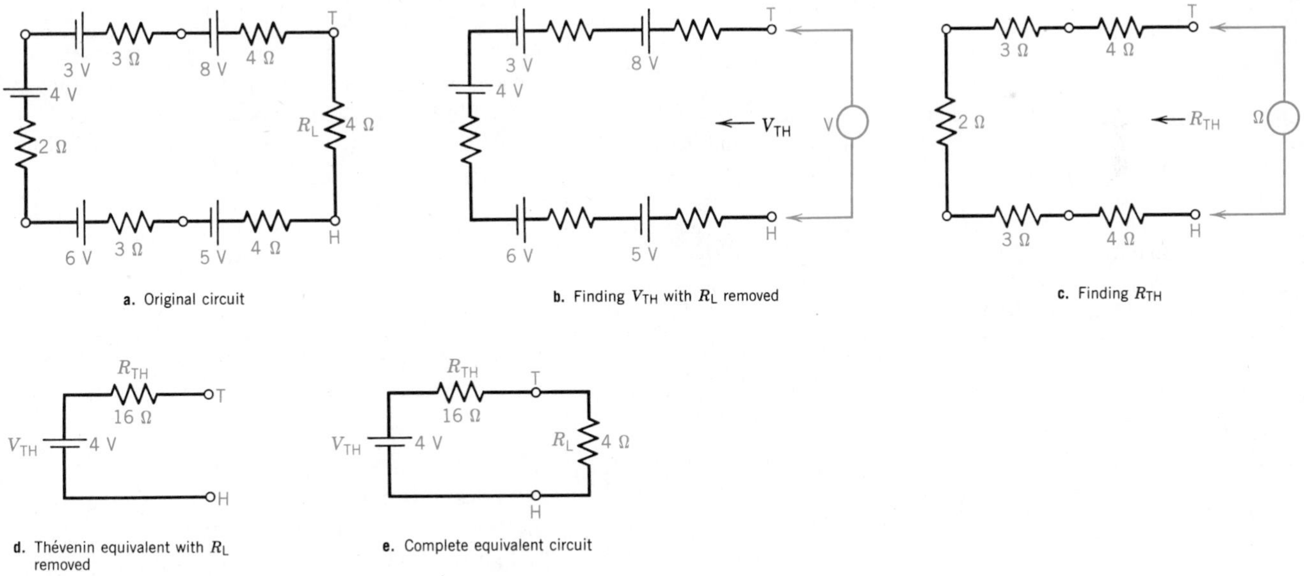

T
3 V
3 Ω
8 V
4 Ω
4 V
R_L
4 Ω
2 Ω
6 V
3 Ω
5 V
4 Ω
H
a. Original circuit
V_{TH}
V
b. Finding V_{TH} with R_L removed
R_{TH}
Ω
c. Finding R_{TH}
16 Ω
d. Thévenin equivalent with R_L removed
e. Complete equivalent circuit

Figure 7-1 Thévenin equivalent of a dc resistive circuit

An examination of the steps shown in Fig. 7-1 reveals a certain previous familiarity with this process. In Sec. 5-6, in considering *practical* voltage *sources*, we used the (same) technique of adding open-circuit voltage sources algebraically to yield an equivalent open-circuit battery voltage and adding the internal resistances to produce an equivalent internal battery resistance as "seen by" the load (cf. Ex. 5-9). In effect, without referring to it by name, we were actually using Thévenin's theorem (as noted in Sec. 7-1).

Let us consider another familiar example, solved in Sec. 5-12.3 by simple series–parallel analysis. **Figure 7-2a** shows two batteries (A and B) of unequal EMF connected in parallel to a 2-Ω load. Let us see how this problem is solved by using Thévenin's theorem.

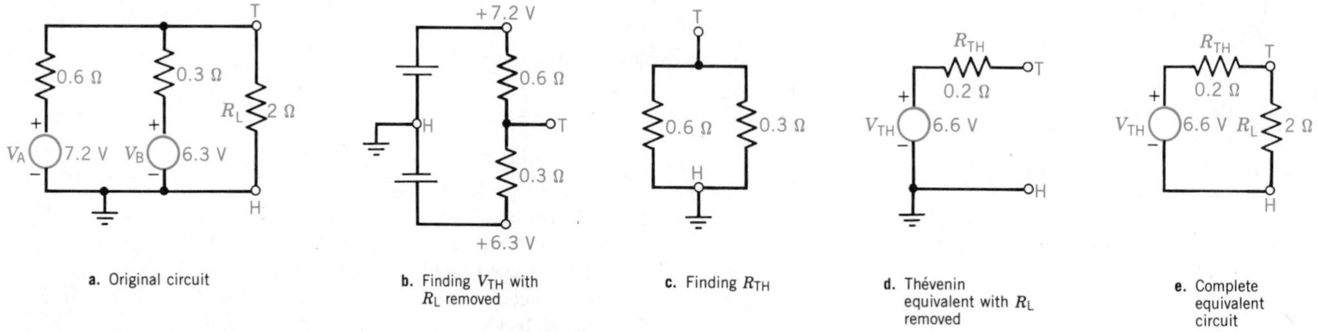

Figure 7-2 Thévenin equivalent of two batteries in parallel

EXAMPLE 7-2

Given the circuit shown in Fig. 7-2a, calculate the

a. Equivalent Thévenin voltage and resistance to the *left* of terminals T–H
b. Load current I_L in resistor R_L
c. Terminal voltage across the three branches in parallel, V_L or V_T
d. Current delivered by each battery

Solution

a. 1. Remove R_L to determine equivalent V_{TH} and R_{TH} to the left of terminals **T–H**. The circuit shown in Fig. 7-2b is redrawn in the familiar form of a voltage divider using dual supplies.

2. Using our "powerful" Eq. (5-8) for finding the voltage at terminal T with respect to ground H, we write

$$V_{TH} = \frac{R_n}{R_T}(V_p - V_n) + V_n$$

$$= \frac{0.3}{0.9}(+7.2 - 6.3) + 6.3 = \mathbf{6.6\ V} \qquad (5\text{-}8)$$

3. The *voltage* sources shown in Fig. 7-2b are now *shorted* as shown in Fig. 7-2c. Clearly, there are two parallel resistors between terminals T–H, and therefore $R_{TH} = \frac{(0.6)(0.3)}{(0.6 + 0.3)}\ \Omega = \mathbf{0.2\ \Omega}$

4. The complex network to the left of terminals T–H in Fig. 7-2a may now be replaced by the Thévenin equivalent shown in Fig. 7-2d.

5. Replacing R_L across terminals T–H in Fig. 7-2e enables a simple solution for the remainder of the problem.

b. $I_L = \dfrac{V_{TH}}{(R_{TH} + R_L)} = \dfrac{6.6\ V}{(0.2 + 2)\ \Omega} = \mathbf{3\ A}$

c. $V_L = V_T = I_L R_L = (3\ A)(2\ \Omega) = \mathbf{6\ V}$

d. $I_A = \dfrac{(E_A - V)}{r_A} = \dfrac{(7.2 - 6)\ V}{0.6\ \Omega} = \mathbf{2\ A}$

$I_B = \dfrac{E_B - V}{r_B} = \dfrac{(6.3 - 6)\ V}{0.3\ \Omega} = \mathbf{1\ A}$

Again, the foregoing problem and its solution have a familiar ring. If we go back to Ex. 5-27, we find that the same problem was solved in the same way by the equivalent battery method. Once again, in effect, we used Thévenin's theorem without being aware of it! Let us now attempt a more difficult example.

EXAMPLE 7-3

Given the original circuit shown in **Fig. 7-3a**, use Thévenin's theorem to find

a. V_{TH}
b. R_{TH}
c. I_L
d. V_L
e. V_{cd}
f. Total current drawn from the source, I_T

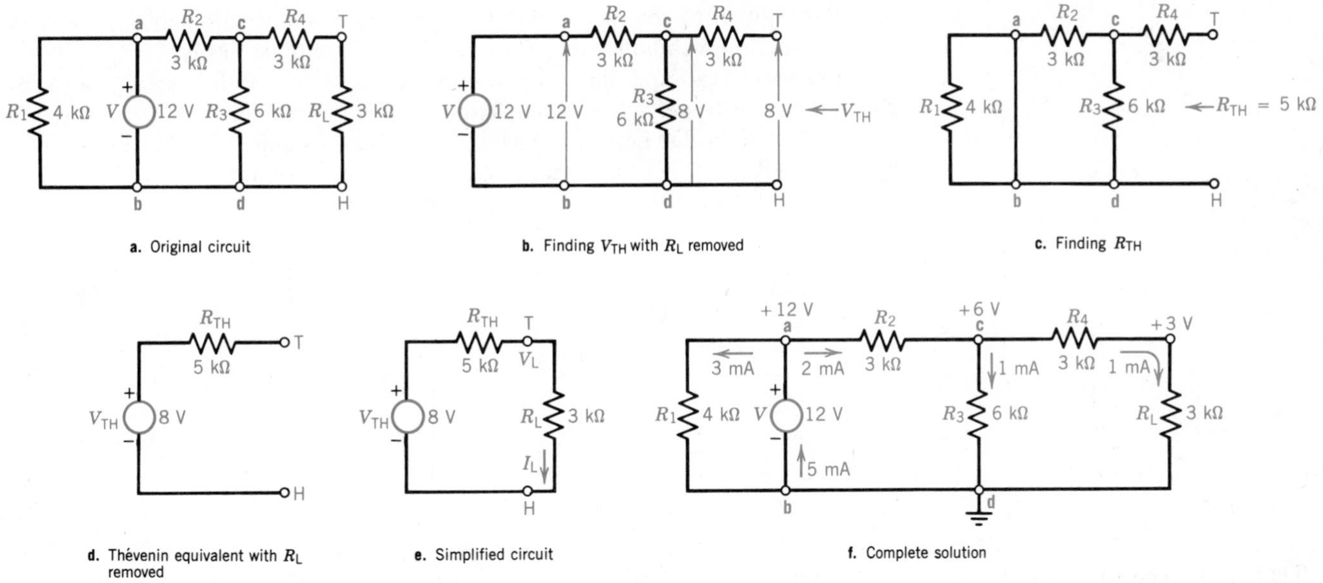

a. Original circuit
b. Finding V_{TH} with R_L removed
c. Finding R_{TH}
d. Thévenin equivalent with R_L removed
e. Simplified circuit
f. Complete solution

Figure 7-3 Example 7-3

Solution

a. 1. Remove R_L to determine equivalent V_{TH} and R_{TH} to the *left* of terminals T–H.

2. In finding V_{TH} we may ignore R_1 since it is in parallel with the source and has 12 V across it. Since it drops no voltage, we ignore it as shown in Fig. 7-3b. Using the VDR for V_{cd}, we find $V_{cd} = \frac{R_3}{R_2 + R_3} \times V_{ab} = \frac{6\text{ k}\Omega}{9\text{ k}\Omega} \times 12\text{ V} = \mathbf{8\ V}$ But since terminals T–H are open, $V_{TH} = V_{cd} = \mathbf{8\ V}$ because there is no drop in voltage across R_4. (R_4 is *dangling* since one terminal is open.) Therefore, $V_{TH} = \mathbf{8\ V}$ as shown in Fig. 7-3b.

3. To find R_{TH} we must short all voltage sources; therefore we connect terminals a and b. But this also shorts out the R_1 resistor, and again we may neglect R_1. The short across terminals a–b in Fig. 7-3c has put R_2 and R_3 in parallel and in series with R_4 across terminals T–H.

b. Therefore, $R_{TH} = 3\text{ k}\Omega + (6\text{ k}\Omega \,\|\, 3\text{ k}\Omega) = \mathbf{5\ k\Omega}$. The Thévenin equivalent may now be drawn in Fig. 7-3d for the network to the left of terminals T–H. Restoring R_L across terminals T–H yields the circuit shown in Fig. 7-3e, which may be used to find I_L and V_L by

c. $I_L = \frac{V_{TH}}{R_{TH} + R_L} = \frac{8\text{ V}}{(5+3)\text{ k}\Omega} = \mathbf{1\ mA}$

d. $V_L = I_L R_L = 1\text{ mA} \times 3\text{ k}\Omega = \mathbf{3\ V}$

e. Once we have found I_L and V_L, we must return to the original circuit (Fig. 7-3a) for the remainder of the problem. Since the current in R_L is 1 mA, the current in R_4 must also be 1 mA (since they are in series across terminals c–d). Therefore $V_{cd} = V_4 + V_L = (3 + 3)\text{ V} = \mathbf{6\ V}$

f. To find the total current drawn from the supply we must find I_2 and I_1, respectively. $I_2 = I_3 + I_4 = \frac{6\text{ V}}{6\text{ k}\Omega} + 1\text{ mA} = 2\text{ mA}$, and $I_1 = V/R_1 = 12\text{ V}/4\text{ k}\Omega = 3\text{ mA}$. Then $I_T = I_2 + I_1 = (2 + 3)\text{ mA} = \mathbf{5\ mA}$. The complete solution, giving *all* currents and voltages, is shown in Fig. 7-3f.

With respect to Fig. 7-3, the following precautions should be observed:

1. The Thévenin voltages found (Fig. 7-3b) are NEVER the true voltages for the complete network solution. Thus, V_{cd} is ACTUALLY **6 V** (Fig. 7-3f) and not 8 V as it appears in Fig. 7-3b.
2. Equivalent voltages and resistances represented in Fig. 7-3d are merely values used as part of the technique involving Thévenin's theorem. Observe that nowhere in the complete solution of Fig. 7-3f is there a voltage of 8 V or a resistance of 5 kΩ such as appears in Fig. 7-3d.
3. The original and final circuits given in Figs. 7-3a and 7-3f with R_L connected represent REALITY. The circuit of Fig. 7-3d, with R_L removed, represents a convenient CONSTRUCT, as does the circuit of Fig. 7-3c with the voltage source shorted.
4. Although the simplified circuit of Fig. 7-3e enables calculation of V_L and I_L precisely, the only realities are the values of I_L, V_L, and R_L to the RIGHT of terminals T–H.
5. In using Thévenin's theorem, therefore, the reader must learn to separate REALITY from constructs.

Up to now, the reader may have examined the circuits of Figs. 7-1, 7-2, and 7-3 with the suspicion that Thévenin's theorem is unnecessary. All three circuits are easily solved by previously developed circuit-reduction techniques. The reader may say, "Why bother to complicate things when I already have fairly straightforward circuit-reduction techniques that I know and can use more easily?"

The following example shows the use of Thévenin's theorem in solving a problem that *cannot* be solved by simple circuit-reduction techniques.

Consider the bridge network shown in **Fig. 7-4a**. Assume that the supply voltage V and all resistance values are given. We are interested in finding the current value, I_L, in load resistor R_L connected across terminals T–H.

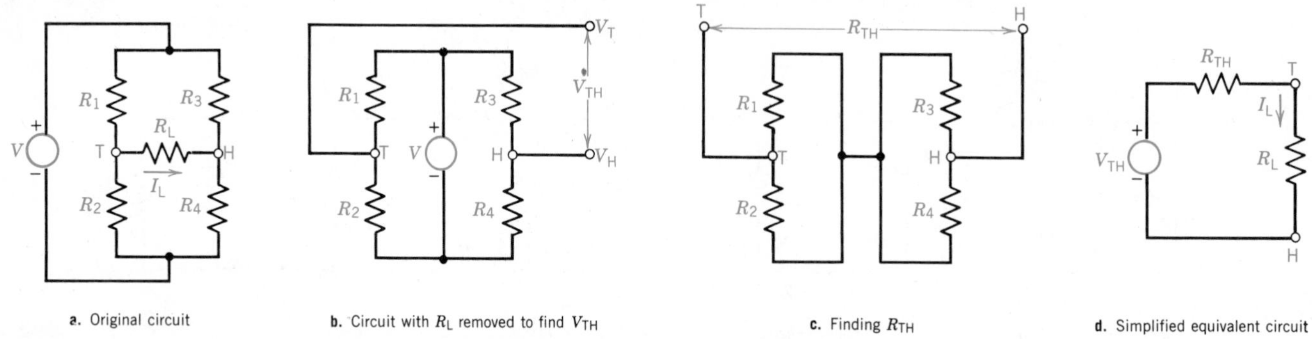

Figure 7-4 Application of Thévenin's theorem to a bridge circuit

The steps in the solution of a bridge network with Thévenin's theorem are

1. Remove resistor R_L from terminals T–H (Fig. 7-4b).
2. Solve for V_{TH} (using Fig. 7-4b).

a. $V_T = \dfrac{R_2}{R_1 + R_2} V$ and $V_H = \dfrac{R_4}{R_3 + R_4} V$ from the voltage divider rule (VDR)

b.

$$V_{TH} = V_T - V_H = V\frac{R_2}{R_1 + R_2} - V\frac{R_4}{R_3 + R_4}$$

$$= V\left(\frac{R_2}{R_1 + R_2} - \frac{R_4}{R_3 + R_4}\right) \quad \text{volts (V)} \qquad (7\text{-}1)$$

3. Solve for R_{TH} by using Fig. 7-4c. Note in Fig. 7-4c that shorting the voltage source produces a series circuit of two parallel branches. Beginning at terminal T and ending at terminal H, we have

$$R_{TH} = (R_1 \| R_2) + (R_3 \| R_4) = \left(\frac{R_1R_2}{R_1 + R_2}\right) + \left(\frac{R_3R_4}{R_3 + R_4}\right) \quad \text{ohms } (\Omega) \qquad (7\text{-}2)$$

4. Replace R_L across terminals T–H as shown in Fig. 7-4d and solve for I_L:

$$I_L = \frac{V_{TH}}{R_{TH} + R_L} = \frac{V\left(\dfrac{R_2}{R_1 + R_2} - \dfrac{R_4}{R_3 + R_4}\right)}{\left(\dfrac{R_1R_2}{R_1 + R_2} + \dfrac{R_3R_4}{R_3 + R_4}\right) + R_L} \quad \text{amperes (A)} \qquad (7\text{-}3)$$

The three equations just derived may be used for *any* Wheatstone bridge circuit, either balanced or unbalanced, as shown by Ex. 7-4.

EXAMPLE 7-4

The resistance of the galvanometer shown in **Fig. 7-5a** is $R_G = 1000\ \Omega$. For the *unbalanced* bridge circuit of Fig. 7-5a, using the given values of voltage and resistance, calculate

a. V_{TH}
b. R_{TH}
c. Current in the galvanometer, I_G
d. Power dissipated by R_G
e. Value of galvanometer resistance that will draw maximum power from the unbalanced bridge circuit of Fig. 7-5a
f. Maximum power that can be drawn from Fig. 7-5a

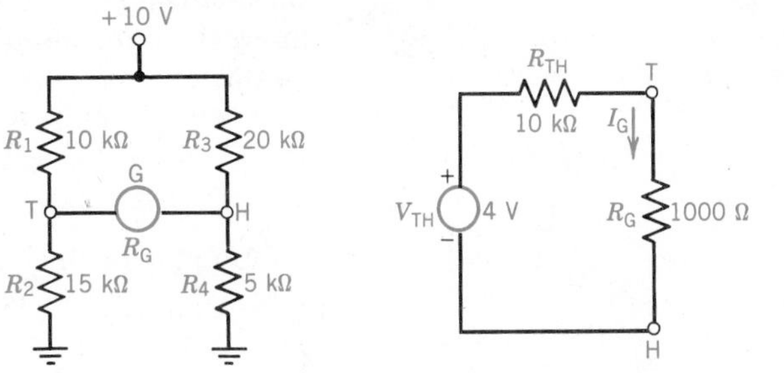

a. Original circuit

b. Thévenin equivalent

Figure 7-5 Example 7-4

Solution

a. $$V_{TH} = V_T - V_H = V\left(\frac{R_2}{R_1 + R_2} - \frac{R_4}{R_3 + R_4}\right)$$
$$= 10\left(\frac{15}{25} - \frac{5}{25}\right) = \mathbf{4\ V} \qquad (7\text{-}1)$$

b. $$R_{TH} = \frac{R_1R_2}{R_1 + R_2} + \frac{R_3R_4}{R_3 + R_4} = \frac{15 \times 10}{25} + \frac{5 \times 20}{25}$$
$$= (6 + 4)\ \text{k}\Omega = \mathbf{10\ k\Omega} \qquad (7\text{-}2)$$

c. $I_L = V_{TH}/(R_{TH} + R_L) = 4\ \text{V}/(10 + 1)\ \text{k}\Omega$
$= \mathbf{0.36\overline{36}\ mA}$ in Fig. 7-5b. (7-3)

d. $P_G = (I_G)^2 R_G = (0.36\ \text{mA})^2 \times 1\ \text{k}\Omega = \mathbf{132.2\ \mu W}$

e. $R_G = R_{TH} = \mathbf{10\ k\Omega}$ by inspection in Fig. 7-5b.

f. $P_{max} = (V_{TH})^2/4R_{TH} = 4^2/4 \times 10\ \text{k} = \mathbf{400\ \mu W}$ (5-16)

The solution to Ex. 7-4 using Eqs. (7-1) through (7-3) developed for the simple Wheatstone bridge circuit shows the extreme power and usefulness of Thévenin's theorem. The circuit of Fig. 7-5a could *not* have been solved by the circuit-reduction techniques presented and developed in Chapters 5 and 6. But Thévenin's theorem is also useful even in solving problems that can also be solved by other circuit-reduction techniques.

7-2 NORTON'S THEOREM

Norton's theorem[3] is the *dual*[4] of Thévenin's theorem. It may be stated as follows.

Any combination of (linear) circuit elements and active (dc) sources, regardless of connection and/or complexity, may be replaced by a two-terminal network consisting of a single current source, I_N, paralleled by an equivalent resistance, R_N.

Because Norton's theorem is the dual of Thévenin's theorem, it broadens our understanding of how Thévenin's theorem may be accomplished experimentally in the laboratory. In using Thévenin's theorem in Sec. 7-1, we deliberately "shorted" voltage sources to determine the Thévenin resistance, R_{TH}. We know that we cannot do this practically because it would damage the practical sources. Furthermore, if we placed an ohmmeter across a circuit containing active elements (dc sources) it would damage the ohmmeter.

---

[3] Credited to E. L. Norton of the Bell Telephone Laboratories.

[4] See Sec. 5-13 on duality. In order to form a *perfect* dual, for dc circuits, the current source, I_N, should be paralleled by an equivalent **conductance**, G_N, where $G_N = 1/R_N$ in Eq. (7-4), as the dual equivalent of a practical voltage source.

In Sec. 5-6.3 we first introduced the concept of the open-circuit and short-circuit test as a generic technique used in obtaining important information about electric networks, devices, and equipment. Let us define these as

Open-circuit test. A determination of the Thévenin voltage (V_{TH}) that exists across any two terminals of an active complex network as measured by an ideal (dc) voltmeter.

Short-circuit test. A determination of the short-circuit (Norton) current (I_N) that would flow in an ideal ammeter connected across the same two terminals of an active complex network.

From the open-circuit and short-circuit test data, the equivalent Thévenin resistance, R_{TH}, and/or Norton resistance, R_N, are

$$R_{TH} \equiv R_N \equiv \frac{1}{G_N} = \frac{V_{TH}}{I_N} = \frac{\text{open-circuit voltage}}{\text{short-circuit current}} \quad \text{ohms } (\Omega) \qquad (7\text{-}4)$$

where V_{TH} is the open-circuit Thévenin voltage measured by the open-circuit test
I_N is the short-circuit current, measured by the short-circuit test
R_{TH} is the equivalent Thévenin resistance between the two terminals of interest across which the test is performed
R_N is the Norton resistance, which is the same as the Thévenin resistance

The significance and value of Eq. (7-4) are best illustrated by its practical application to networks as used in the laboratory. **Figure 7-6a** shows two active (dc) complex networks, **A** and **B**, connected as shown. We wish to replace both networks with a single Thévenin (or Norton) equivalent. No matter how complex each network may be and regardless of the number of active sources in each, the experimental procedure is relatively simple.

As shown in Fig. 7-6b, we connect a (high resistance) voltmeter across the (conveniently selected) terminals T and H and measure the voltage V_{TH} or the open-circuit voltage, V_{OC}, where $V_{TH} = V_{OC}$.

To perform the short-circuit test, as shown in Fig. 7-6c, we place a (low or zero resistance, dc) ammeter across terminals T–H and measure I_N, the Norton current or the short-circuit current, where $I_{SC} = I_N$.

We are now able to draw the Thévenin equivalent of *both* complex active circuits seen to the *left* and *right* of terminals T and H, as shown in Fig. 7-6d. V_{TH} is the

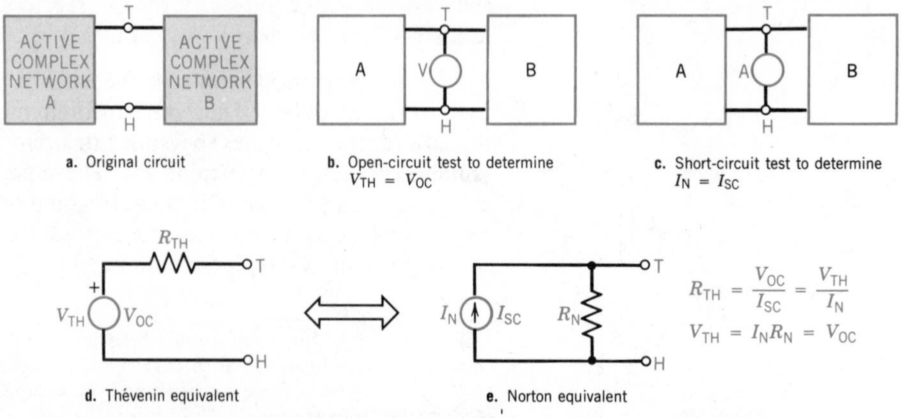

Figure 7-6 Using open-circuit and short-circuit test data to obtain Thévenin and Norton equivalents across two terminals of interest between two active complex networks

open-circuit voltage, V_{OC}, and R_{TH} is the Thévenin resistance calculated from Eq. (7-4).

Given the Thévenin equivalent of Fig. 7-6d, we easily determine the Norton equivalent shown in Fig. 7-6e, where I_N is the short-circuit current, I_{SC}, measured by the ammeter and R_N is the Thévenin resistance, R_{TH}, calculated from Eq. (7-4).

The significance of the procedure shown in Fig. 7-6 is as follows:

1. By means of a simple experimental procedure involving two simple (voltage and current) measurements, we have reduced two or more complex active dc circuits to one simple practical voltage source and/or its equivalent practical current source.
2. Using the equivalent(s) obtained in (1) we can predict the current in any load resistor R_L connected across terminals T–H.
3. We even know that the maximum power which can be drawn from both networks occurs when $R_L = R_{TH} = R_N$! (See Sec. 5-7 on maximum power transfer.)

The procedure just described is best illustrated by the following example.

EXAMPLE 7-5

Given the original circuit shown in **Fig. 7-7a**, use open- and short-circuit test methods to determine the actual current in resistor R_4

Solution

Open-circuit test: remove R_4 and connect a voltmeter across terminals T–H. The voltmeter reads 24 V or $V_{OC} = \dfrac{20\text{ k}}{40\text{ k}} \times 48\text{ V} = 24\text{ V}$ as shown in Fig. 7-7b.

Short-circuit test: with R_4 removed, a short circuit across T–H draws 1.6 mA as shown in Fig. 7-7c. This current may be verified by

$$R_T = R_1 + (R_2 \| R_3) = 20\text{ k} + (20\text{ k} \| 5\text{ k}) = 24\text{ k}\Omega$$

then $$I_T = V_T/R_T = 48\text{ V}/24\text{ k}\Omega = 2\text{ mA}$$

and $$I_3 = I_T\frac{R_2}{(R_3 + R_2)} = 2\text{ mA}\frac{20\text{ k}}{25\text{ k}} = 1.6\text{ mA} = I_N$$

R_{TH} as determined from the above open-circuit and short-circuit test data is

$$R_{TH} = \frac{V_{TH}}{I_N} = \frac{24\text{ V}}{1.6\text{ mA}} = 15\text{ k}\Omega \qquad (7\text{-}4)$$

The Thévenin equivalent circuit is shown in Fig. 7-7d, where $V_{TH} = V_{OC}$ as determined from the open-circuit voltage test and $R_{TH} = 15\text{ k}\Omega$ as determined from Eq. (7-4) with open- and short-circuit test data.

In Fig. 7-7e, R_4 is replaced across terminals T–H enabling us to find the current in R_4, which is, by inspection, **1 mA**.

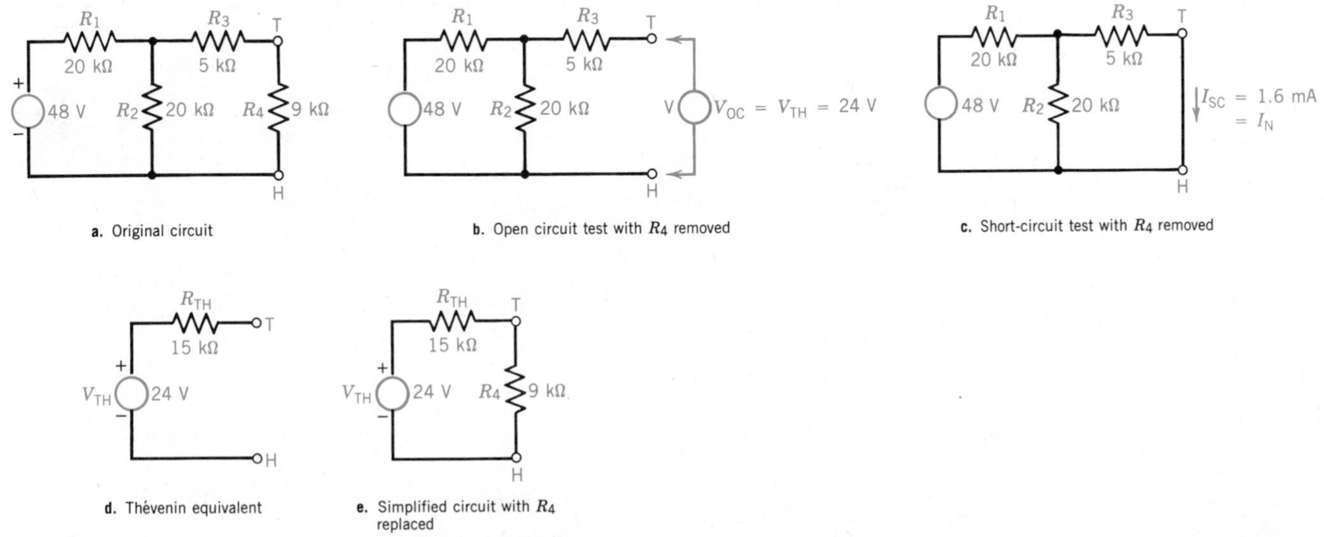

Figure 7-7 Example 7-5; illustration of open-circuit and short-circuit test method for finding Thévenin (and Norton) equivalents

Note that the experimental method used in Ex. 7-5 yields the same values of V_{TH} and R_{TH} as those used in Sec. 7-1 describing Thévenin's theorem. If we short the voltage source in Fig. 7-7b and determine the resistance looking into terminals T–H, we obtain $R_{TH} = 5\ k\Omega + (20\ k\Omega \| 20\ k\Omega) = 15\ k\Omega$. This is exactly the same as that obtained by using open- and short-circuit test data to find R_{TH} in Ex. 7-5.

We have just demonstrated, therefore, that using open-circuit voltage measurements and short-circuit current measurements across any two terminals of a complex network will yield the Thévenin equivalent and/or the Norton equivalent, at the same terminals. Moreover, the two measurements yield R_N or R_{TH} directly from Ohm's law.

7-3 SOURCE-TRANSFORMATION THEOREM

From the relationship given in Eq. (7-4) we may state a corollary, sometimes called the **source-transformation theorem**. It states

Given any complex active network containing one or more practical voltage and/or current sources, any given practical voltage (or current) source may be transformed into an equivalent practical current source (or vice versa), using the Thévenin–Norton conversion.

It was stated earlier that these two circuits are *duals* of each other (review Sec. 5-13) and the source transformation, with appropriate transformation equations, is shown in **Fig. 7-8**. Table 7-1 summarizes this duality along with the transformation equations.

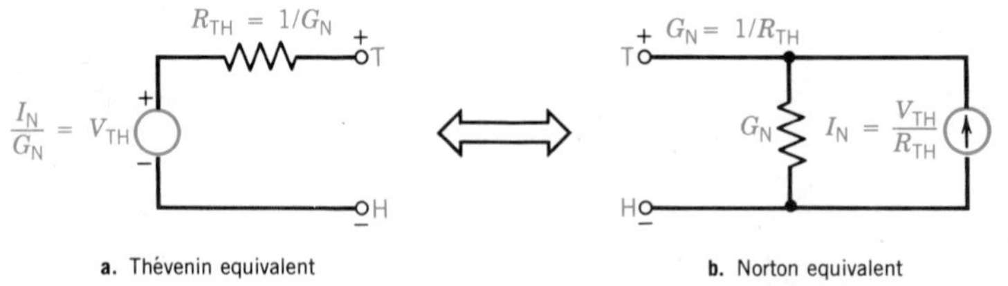

Figure 7-8 Thévenin–Norton source conversion and duality

Table 7-1 Thévenin–Norton Duality

	Thévenin Equivalent	Norton Equivalent
Circuit	Series	Parallel
Source	Voltage, $V_{TH} = V_{OC}$	Current, $I_N = I_{SC}$
Practical opposition	R_{TH}	$1/G_N$
Source transformation	$V_{TH} = I_N/G_N = I_N R_N$	$I_N = V_{TH}/R_{TH}$
Opposition transformation	$R_{TH} = 1/G_N = R_N$	$G_N = 1/R_{TH} = 1/R_N$

The source-transformation theorem, in combination with Thévenin's or Norton's theorem, permits us to simplify networks for relatively easy solution as shown by the following examples.

EXAMPLE 7-6

Convert the network shown in **Fig. 7-9a** to its simplest Thévenin equivalent circuit

Solution

1. Convert the practical current source to a practical voltage source using source transformation (current source to voltage source). $V_{TH} = I_N/G_N = I_N R_N = (5\ mA)(10\ k\Omega) = 50\ V$ and $R_{TH} = R_N = 10\ k\Omega$
2. Find the Thévenin equivalent of the network to the left of nodes **a–b**. $V_{ab} = \frac{9\ k}{12\ k} \times 100\ V = 75\ V$ and $R_{ab} = (3\ k\Omega \| 9\ k\Omega) = 2.25\ k\Omega$

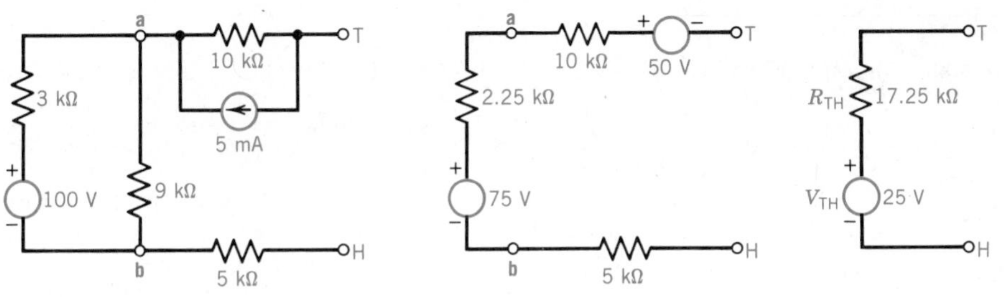

a. Original circuit b. Intermediate circuit c. Final solution

Figure 7-9 Example 7-6

3. The combination of 1 and 2 above produces the series circuit shown in Fig. 7-9b. Note that the 75-V source is being opposed by the 50-V source.
4. Adding series-connected voltages yields $V_{TH} = 75 - 50 =$ **25 V** and $R_{TH} = 10\ k\Omega + 2.25\ k\Omega + 5\ k\Omega =$ **17.25 kΩ**
5. The circuit in its simplest form as a practical voltage source is shown in Fig. 7-9c.

With respect to the original network of Fig. 7-9a, we had the choice of converting one source to either a current or a voltage source. Had we elected to show the sources as current sources, we would have produced two series-connected opposing current sources. This combination could *not* be simplified further into a single current source. Consequently, the conversion to voltage sources, series connected, yields the simplest solution, as shown in Fig. 7-9c. Had the mixed sources been connected in parallel, conversion to current sources would have been indicated, as shown by Ex. 7-7.

EXAMPLE 7-7

Given the circuit shown in **Fig. 7-10a**, find the current in R_L and the voltage **x–G** across all four parallel-connected branches

Solution

1. Since all branches are in parallel, we elect to convert the practical voltage source to a practical current source. $I_N = V_{TH}/R_{TH} = 12\ V/40\ k\Omega = 0.3\ mA$ (note direction of current in Fig. 7-10b). $R_N = R_{TH} = 40\ k\Omega$ in parallel with the current source, as shown in Fig. 7-10b.
2. We now combine current sources and their parallel resistances to produce a single practical current source in parallel with R_L:

$$I_N = 0.8\ mA - 0.3\ mA = 0.5\ mA$$

$$R_N = (40\ k\Omega \,\|\, 60\ k\Omega) = 24\ k\Omega,$$

as shown in Fig. 7-10c

3. We can now find I_L using the CDR, where $I_L = I_T \dfrac{R_N}{R_L + R_N} = 0.5\ mA \dfrac{24\ k}{40\ k} = \mathbf{0.3\ mA}$
4. and $V_L = I_L R_L = 0.3\ mA \times 16\ k\Omega = \mathbf{4.8\ V} = V_{xG}$

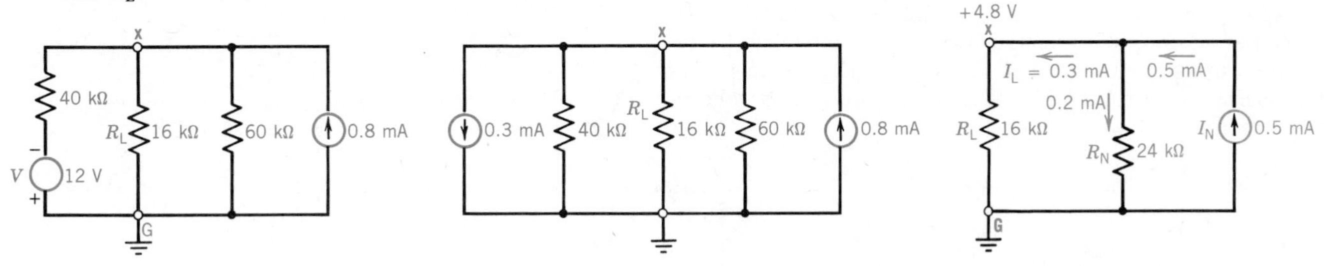

a. Original circuit b. Intermediate circuit c. Complete solution

Figure 7-10 Example 7-7

Note from Fig. 7-10c that the only reality in the final solution consists of R_L, I_L, and V_{xG}. Let us carry this reality over to Ex. 7-8 and see if we can learn something about practical voltage sources in parallel.

EXAMPLE 7-8

Knowing that voltage $V_{xG} = 4.8$ V, convert all sources in Fig. 7-10a to voltage sources and calculate the

a. Generated voltage and internal resistance of the practical voltage source to the right of R_L in Fig. 7-10a
b. Current delivered by the practical voltage source in (**a**) to the node x–G
c. Current received by the practical voltage source to the left of R_L
d. Draw the complete solution for all currents and voltages, using practical voltage sources in parallel with R_L

Solution

a. Using the Norton–Thévenin source conversion, $V_{TH} = I_N R_{TH} = (0.8 \text{ mA})(60 \text{ k}) = \mathbf{48\ V}$ and $R_{TH} = R_N = \mathbf{60\ k\Omega}$ in series with 48 V, as shown in **Fig. 7-11a**.

b. Since the terminal voltage $V = \mathbf{4.8\ V} = V_{xG}$, then $I = \dfrac{E - V}{r_i} = \dfrac{(48 - 4.8)\text{ V}}{60\text{ k}\Omega} = \mathbf{0.72\ mA}$

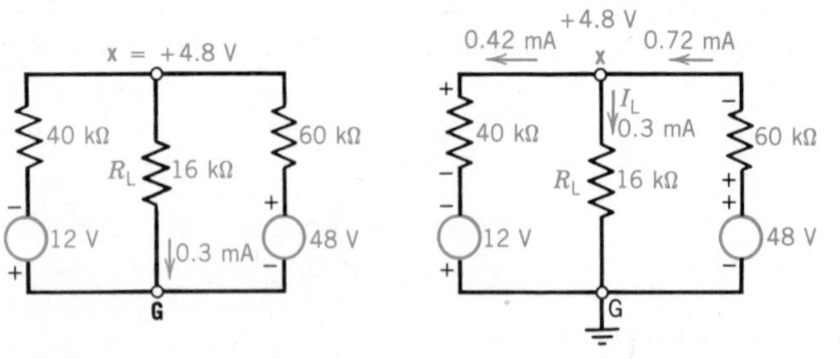

a. Figure 7-10a, converted to voltage sources

b. Complete solution

Figure 7-11 Example 7-8

c. Since node x = +4.8 V, the node polarity *aids* the polarity of the 12-V source. Consequently, $I = \dfrac{E + V}{r_i} = \dfrac{(12 + 4.8)\text{ V}}{40\text{ k}\Omega} = \mathbf{0.42\ mA}$

d. The complete solution is shown in Fig. 7-11b. We already know that $I_L = 0.3$ mA. Note that the *current* around node x, by KCL, shows that the *algebraic* sum is zero.

A number of interesting and important conclusions may be deduced from the network of Fig. 7-11b:

1. The 48-V source is supplying current to the other two branches.
2. The branch containing the practical 12-V source is acting as a load in receiving current from the +4.8-V node.
3. A source whose polarity *opposes* the node voltage delivers a current $I = \dfrac{E - V}{r_i}$
4. A source whose polarity *aids* the node voltage delivers a current $I = \dfrac{E + V}{r_i}$
5. Starting at node x in Fig. 7-11b and tracing a closed loop by KVL in a counterclockwise (CCW) direction, we get $+(0.42\text{ mA} \times 40\text{ k}\Omega) - 12 - (0.3\text{ mA} \times 16\text{ k}\Omega) = 16.8 - 12 - 4.8 = 0$. Note that this verifies KVL for any closed path or loop.
6. Similarly, starting at node x in Fig. 7-11b and tracing by KVL in a clockwise direction, we get $-(0.72\text{ mA} \times 60\text{ k}\Omega) + 48 - 4.8 = -43.2 + 48 - 4.8 = 0$, again verifying KVL.
7. Since KVL and KCL have been verified for Fig. 7-11b, we may conclude that our solution is valid.
8. We have also verified our source-conversion theorem, since the circuit of Fig. 7-11b showing practical *voltage* sources is equivalent to the circuit of Fig. 7-10b, using practical *current* sources. **Both** circuits deliver a current of 0.3 mA to load R_L.

EXAMPLE 7-9

Given the original network shown in **Fig. 7-12a**, find the simplest practical voltage or current source across terminals T–H.

Solution

Inspection shows that we have a practical voltage source (100 Ω in series with a resistance of 9 Ω) in series with a practical current source. Consequently, since current sources in series *cannot* be simplified, we elect to convert the current source to a voltage source in the following steps:

1. $R_N = (8 + 4)\ \Omega \| 6\ \Omega = 12\ \Omega \| 6\ \Omega = 4\ \Omega$ (Fig. 7-12b)
2. Converting the practical current source to practical voltage source: $V_{TH} = I_N R_N = (30\text{ A})(4\ \Omega) = 120$ V and $R_{TH} = R_N = 4\ \Omega$ (Fig. 7-12c)
3. The second simplification of Fig. 7-12c shows two practical voltage sources in series, which when combined yield $V_{TH} = 120\text{ V} - 100\text{ V} = \mathbf{20\ V}$ and $R_{TH} = 9\ \Omega + 4\ \Omega = \mathbf{13\ \Omega}$

The practical voltage source representing the simplest equivalent across terminals T–H is shown in Fig. 7-12d.

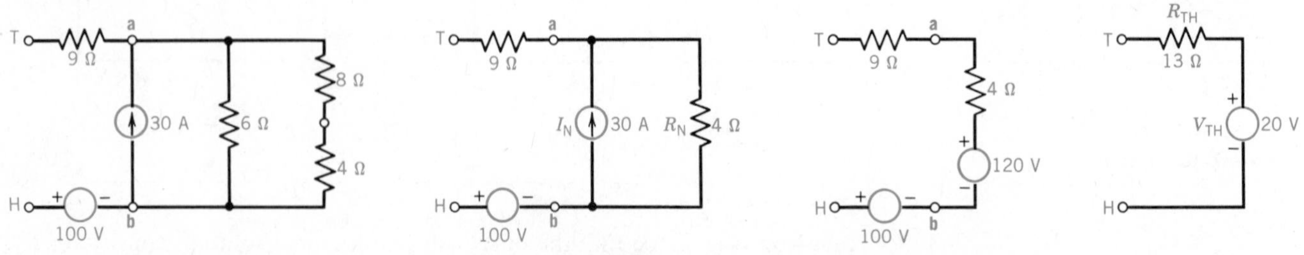

a. Original circuit

b. First simplification

c. Second simplification

d. Thévenin equivalent

Figure 7-12 Example 7-9

7-4 SUPERPOSITION THEOREM

The circuits we have analyzed (and will continue to analyze throughout this text) have been *linear* circuits; that is, the circuit elements (resistances) have a constant voltage–current relation (as noted in Eq. 3-1). The superposition theorem is based on the single assumption that *all* circuit elements are linear and bilateral. Given even one *nonlinear* element in a network, the principle of superposition *cannot* be used. The statement of the superposition theorem is

In any multisource complex network consisting of linear bilateral elements, the voltage across (or current through) any given element of the network is equal to the algebraic sum of the individual voltages (or currents) produced independently across (or in) that element by each source acting independently.[5]

Superposition, as defined above, dictates that the *effect* of *each* voltage and/or current source, *acting independently*, is found on a given element. The *algebraic sum of the individual effects* enables us to find the *total effect* just *as if they were all acting simultaneously*. To perform superposition, therefore, to find the current, I_L, acting on a given load, R_L, we must

1. Select a single voltage or current source and disable all the others. This means *shorting* other *voltage* sources and *opening* other *current* sources. (We did this previously in applying Thévenin's theorem to find the resistance, using an ohmmeter, in Sec. 7-2.)
2. Find the current, I_L, due to the voltage or current source selected.
3. Repeat the above steps for each current and voltage source, recording resultant values of I_L for each, until all the sources have been used.

It should be noted that of all the theorems and methods of analysis that we will learn, superposition is the only one that can be used to find true current in, voltage across, and power dissipated by a given load, when the voltage source(s) contain harmonics (voltages of various frequency components).[6] Let us apply superposition to a number of examples that we have solved by previous methods.

EXAMPLE 7-10

Given the circuit shown in **Fig. 7-13a**, find the current in R_L due to

a. Source V_A acting alone
b. Source V_B acting alone
c. Both sources, and compare answer to Ex. 7-2

Solution

a. We first disable voltage source B by shorting it, leaving only source V_A as shown in Fig. 7-13b. The total resistance of this circuit is $R_T = 0.6\ \Omega + (0.3\ \Omega \| 2\ \Omega) = 0.6 + 0.2609 = 0.8609\ \Omega$.

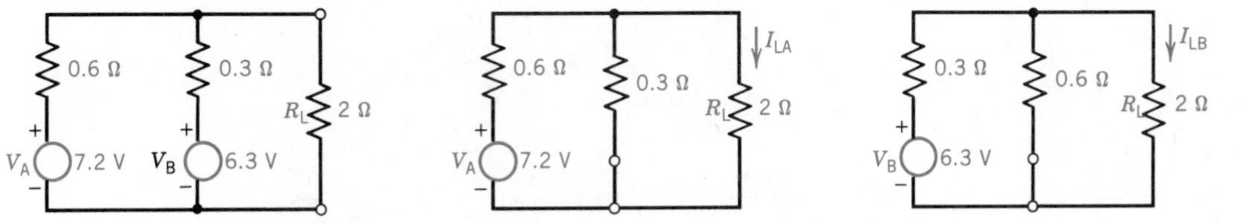

a. Original circuit   b. Source V_A only   b. Source V_B only

Figure 7-13 Example 7-10; superposition solution for current in R_L

[5] The principle of superposition extends to all fields of scientific study including psychology and economics. So extended, it has come to mean that if we know the effect produced by each individual variable acting on a given system, then the net system result of several such variables is the vector sum of the individual effects.

[6] An oft-repeated adage, from instructors to students, "When all else fails, use superposition." See Chapter 22.

The total current from source A, $I_{T_A} = 7.2\ \text{V}/0.8609\ \Omega = 8.\overline{36}$ A. Using the CDR,

$$I_{L_A} = 8.\overline{36}\ \text{A}\,\frac{0.3}{(0.3 + 2)} = 1.\overline{09}\ \text{A}\downarrow$$

b. Now we disable source A in Fig. 7-13a, leaving only V_B as shown in Fig. 7-13c. The total resistance of this circuit is $R_T = 0.3\ \Omega + (0.6\ \Omega \| 2\ \Omega) = 0.7615\ \Omega$. The total current from source B is $I_{T_B} = 6.3\ \text{V}/0.7615\ \Omega = 8.\overline{27}$ A. Using the CDR, $I_{L_B} = 8.\overline{27}\,\frac{0.6}{2.6} = 1.\overline{909}\ \Omega\downarrow$

c. Note that in (**a**) and (**b**) a current direction is shown for the current due to each source; that is, the algebraic sign of the current directions is included. Then the "true" current in R_L is the algebraic sum or $I_L = I_{L_A} + I_{L_B} = 1.\overline{09}\ \text{A}\downarrow + 1.\overline{909}\ \text{A}\downarrow = \mathbf{3\ A}\downarrow$. This is exactly the same answer obtained for both Ex. 7-2 and Ex. 5-24 earlier, using Thévenin's theorem and the equivalent battery method, respectively.

Recall that earlier, when we encountered a network containing both voltage sources and current sources, we were obliged to use source transformation to enable simplification of voltage sources in series and/or current sources in parallel. One of the advantages of superposition is that it uses all sources in their *original* form. This not only saves a conversion step but also eliminates the possibility of error due to source conversion.

EXAMPLE 7-11

Given the circuit shown in **Fig. 7-14a**, find the current in R_L due to

a. The 0.8-mA practical current source
b. The 12-V practical voltage source
c. Both sources, and compare the answer to Exs. 7-7 and 7-8

Solution

a. We first disable the voltage source by *shorting* it as shown in Fig. 7-14b. Since we are interested in R_L, we may combine the other (two) parallel connected resistances by $R_p = (40\ \text{k} \| 60\ \text{k}) = 24\ \text{k}\Omega$. Then by the CDR, the current in R_L due to source A is $I_{L_A} = 0.8\ \text{mA}\,\frac{24\ \text{k}\Omega}{(24 + 16)\ \text{k}\Omega} = 0.48\ \text{mA}\downarrow$

b. We next disable the current source by *opening* its terminals as shown in Fig. 7-14c. Observe that we have a series–parallel circuit whose total resistance is $R_T = 40\ \text{k}\Omega + (16 \| 60)\ \text{k}\Omega = (40 + 12.63)\ \text{k}\Omega = 52.63\ \text{k}\Omega$. Then the current supply by the voltage source is $I_T = 12\ \text{V}/52.63\ \text{k}\Omega = 0.228$ mA. Using the CDR, the current in R_L is $I_{L_B} = 0.228\ \text{mA}\,\frac{60\ \text{k}}{76\ \text{k}} = 0.18\ \text{mA}\uparrow$

c. Note that in (**a**) and (**b**) the current directions are *opposing* due to the respective polarity of the sources. The "true" current in R_L is the algebraic sum of the individual currents or $I_L = I_{L_A} + I_{L_B} = 0.48\ \text{mA}\downarrow + 0.18\ \text{mA}\uparrow = 0.3\ \text{mA}\downarrow$

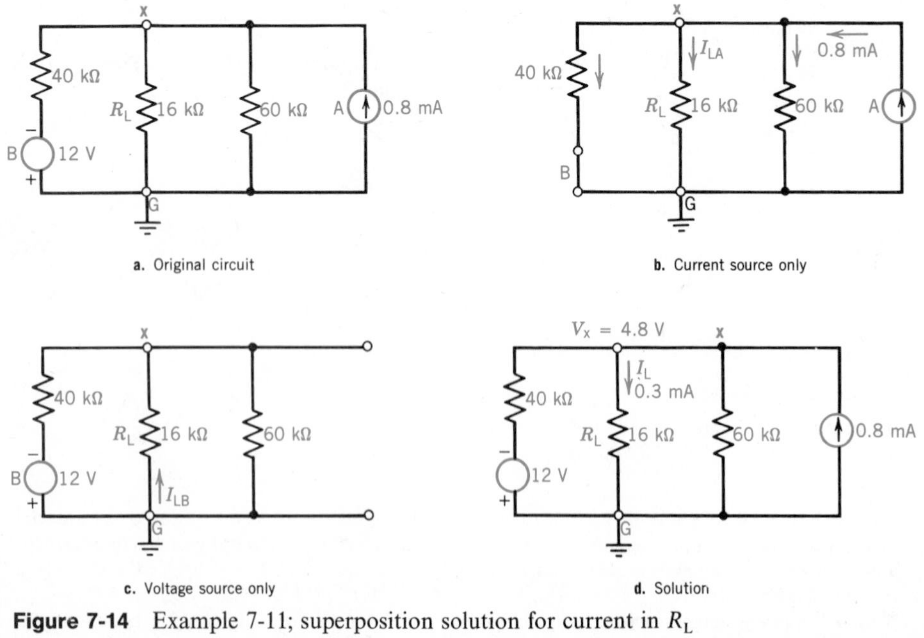

Figure 7-14 Example 7-11; superposition solution for current in R_L

Observe that this is exactly the same answer as that for Exs. 7-7 and 7-8, obtained by source-conversion techniques using Thévenin's theorem. The current direction downward indicates that node x has a positive polarity with respect to ground, as shown in Fig. 7-14d. Once we have found I_L, it is a simple matter to compute V_{xG} and all other currents, using the methods of Ex. 7-8.

In the two previous examples we solved for currents by using superposition techniques. But the statement of the superposition theorem also mentions the algebraic sum of individual *voltages* across nodes of interest due to respective sources. Let us consider a more sophisticated example that will use a variety of techniques learned thus far, in addition to superposition. We should learn a great deal about multisource circuit behavior from Ex. 7-12.

EXAMPLE 7-12

Find the complete solution, showing all currents and voltages, for the network shown in **Fig. 7-15** by calculating

a. R_{TH} with right-hand branch removed from terminals T–H
b. V_{TH}, due to voltage source only
c. V_{TH}, due to current source only
d. Equivalent circuit with right-hand branch restored across terminals T–H
e. All currents and voltages, knowing node voltage V_{TG}

Solution

a. Figure 7-15b shows the right-hand branch removed, the voltage source shorted, and the current source open-circuited. Then $R_{TH} = (2 \| 2)\ \Omega = \mathbf{1\ \Omega}$.
b. Taking only the voltage source and opening the current source, as shown in Fig. 7-15c, to find V_{TH_1}, we obtain (using VDR) $V_{TH_1} = 20\ \text{V}\dfrac{2\ \Omega}{4\ \Omega} = \mathbf{+10\ V}$ as shown.
c. Taking only the current source and shorting the voltage source, as shown in Fig. 7-15d, we observe equal current division using the CDR or $I_{2\ \Omega} = 4\ \text{A}(2\ \Omega/4\ \Omega) = 2\ \text{A}$ in either 2-Ω resistor. Then the voltage across each 2-Ω resistor is $V_{TH_2} = (2\ \text{A})(2\ \Omega) = \mathbf{+4\ V}$, as shown.
d. Using superposition, $V_{TH} = V_{TH_1} + V_{TH_2} = +10\ \text{V} + (+4\ \text{V}) = \mathbf{14\ V}$. We now draw the simplest equivalent showing $V_{TH} = 14$ V and $R_{TH} = 1\ \Omega$ in *series* with the original right-hand branch restored across terminals T–H. The net circuit voltage $V = 14\ \text{V} - 6\ \text{V} = 8$ V. The current in the right-hand branch is $I = \dfrac{V}{R_T} = \dfrac{8\ \text{V}}{2\ \Omega} = \mathbf{4\ A}$ (a true current). The node voltage $V_{TG} = 14\ \text{V} - (4\ \text{A})(1\ \Omega) = \mathbf{+10\ V}$
e. Knowing that the node voltage across all three branches of the original circuit (Fig. 7-15a) is $V_{TG} = +10$ V, we find the current in each branch.
   1. The current delivered to the node by the 20 V source is $I = \dfrac{E - V_T}{r_i} = \dfrac{(20 - 10)\ \text{V}}{2\ \Omega} = \mathbf{5\ A}$ delivered to the node by the LH branch

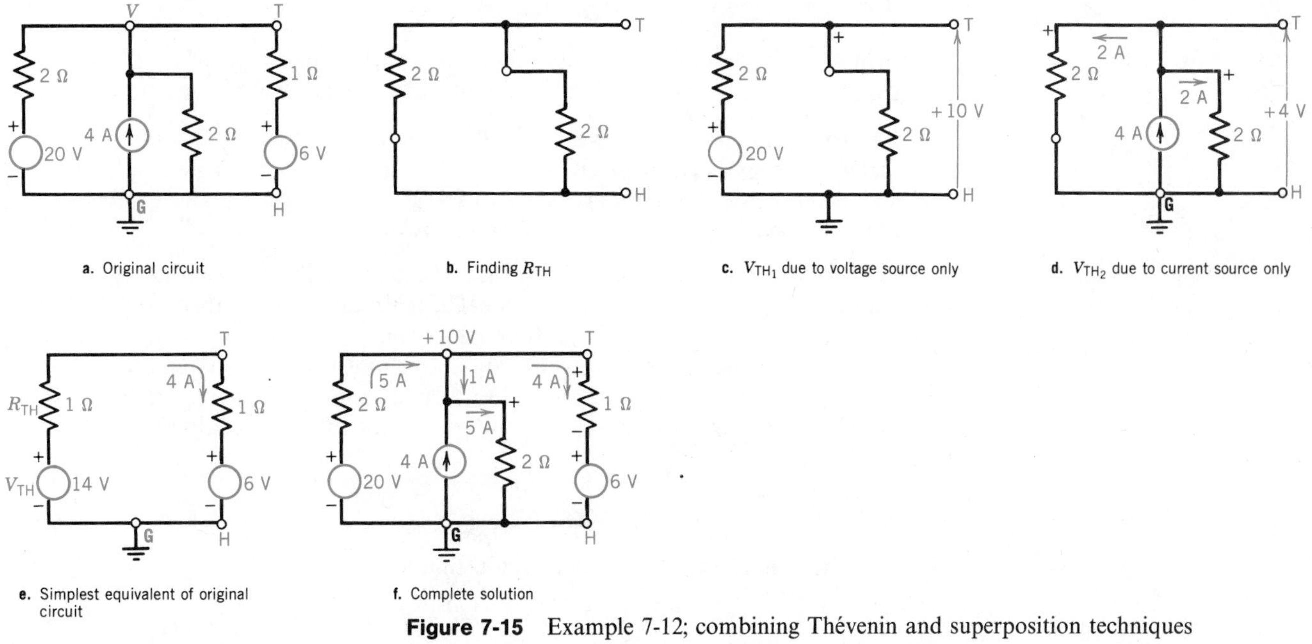

a. Original circuit
b. Finding R_{TH}
c. V_{TH_1} due to voltage source only
d. V_{TH_2} due to current source only
e. Simplest equivalent of original circuit
f. Complete solution

Figure 7-15 Example 7-12; combining Thévenin and superposition techniques

2. Converting the middle branch by the source conversion theorem to a voltage source yields $V_{TH} = 8$ V in series with an internal resistance of 2 Ω. Then $I = \frac{E - V_T}{r_i} = \frac{(8 - 10)\text{ V}}{2\ \Omega} = -1$ A or **1 A** received from the node T.
3. From (**d**), the true current in the right-hand branch is **4 A**

Note that the above solution verifies KCL in that the LH branch supplies 5 A to the node, which, in turn, sends 4 A to the RH branch and 1 A to the middle branch. The complete solution is shown in Fig. 7-15f.

Several important conclusions may be drawn from Ex. 7-12:

1. The combined power of the Thévenin and superposition theorems has turned an extremely difficult (up to now) problem into a simple one.
2. In finding V_{TH} each source was used separately, and V_{TH} was the resultant sum of V_{TH_1} and V_{TH_2}, using superposition, as shown in Ex. 7-12d. This verifies the original statement of the superposition theorem that either voltages or currents may be summed for either a node or component of interest.
3. The Thévenin voltage and resistance to the left of nodes T–H in Fig. 7-15e represent an equivalent construct, but the 4-A current in the RH branch represents REALITY. We may verify this from Fig. 7-15f, where the node voltage, V_{TG}, is (4 A)(1 Ω) + 6 V = 10 V!
4. But the middle branch produced the greatest surprise of all. The 2-Ω resistor shunting the current source receives 1 A from the node and 4 A from the current source, a total of 5 A. This is easily verified from Fig. 7-15f, where the volt drop produces a node voltage of (5 A)(2 Ω) = 10 V across the 2-Ω resistor. Note that the upper terminal of the 2-Ω resistor is *positive*, verifying a *positive* voltage of 10 V at the node T.
5. Also in the middle branch, note that although the current comes *from the node* to the branch, the current in the 4-A current source does *not* reverse. We may conclude that regardless of connection in networks, no matter how complex, the originally given current *directions* of *current sources* and/or the *original polarity of voltage sources will remain the same.*

Example 7-12 is the first instance we have seen of *three* voltage or current sources connected in parallel. If there were four (or more) such unequal voltage and/or current sources, could we solve for the node voltage easily? Perhaps one way of doing so might be indicated by Fig. 7-6a, where we separate the network into groups of two and use our previous methods. But this is time-consuming. If we realize that *all* sources connected in *parallel* MUST (by definition) have a common node voltage, V, this realization leads nicely to the next theorem.

7-5 MILLMAN'S THEOREM

Millman's theorem is essentially a *procedure* rather than a (novel) theorem or insight into circuit analysis. The statement of Millman's theorem is

Any combination of parallel-connected voltage (or current) sources may be represented as a single equivalent source (voltage or current) by applying either Thévenin's or Norton's theorem appropriately to each source individually and combining the results to obtain the voltage across the equivalent combination.

From the above statement, note that Millman's theorem combines the source-transformation theorem with both the Thévenin and Norton theorems. Most important, and not mentioned, is that ALL sources are converted to CURRENT sources as shown by the following equations, that is, Eqs. (7-5a) and (7-5b).

Since Millman's theorem applies to n sources connected in parallel, we may apply parallel circuit rules derived in Sec. 5-8 for the parallel circuit:

$$I_T = I_1 + I_2 + I_3 + \cdots + I_n \quad \text{amperes (A)} \tag{5-19}$$

and

$$G_T = G_1 + G_2 + G_3 + \cdots + G_n \quad \text{siemens (S)} \tag{5-21}$$

plus the definition of a parallel circuit, that is,

$$V \equiv V_1 \equiv V_2 \equiv V_3 \equiv \cdots \equiv V_n \quad \text{volts (V)} \tag{5-18}$$

Suppose *all* our sources in parallel were practical *voltage* sources, then

$$V_{eq} = \frac{V_1G_1 + V_2G_2 + V_3G_3 + \cdots + V_nG_n}{G_1 + G_2 + G_3 + \cdots + G_n} = \frac{I_t}{G_t} \quad \text{volts (V)} \qquad (7\text{-}5a)$$

Now suppose *all* our sources in parallel were practical *current* sources, then

$$V_{eq} = \frac{I_1 + I_2 + I_3 + \cdots + I_n}{G_1 + G_2 + G_3 + \cdots + G_n} = \frac{I_t}{G_t} \quad \text{volts (V)} \qquad (7\text{-}5b)$$

where V_1, V_2, etc. are the voltages of the individual voltage sources
I_1, I_2, etc. are the currents of the current sources
G_1, G_2, etc. are the conductances of the sources

Several warnings and precautions are needed in using Eqs. (7-5a) and (7-5b). Both equations find the node voltage, V_{eq}, by dividing the total current by the total conductance, as shown. Therefore,

1. The numerator of Eq. (7-5a) represents the sum of a series of currents produced by source conversion of each practical voltage source, V_1/R_1, V_2/R_2, etc. The polarity of these sources determines whether the current is positive or negative. If the direction of positive polarity is *toward* the node, the current is *positive.*
2. The same principle applies to the numerator of Eq. (7-5b). If the current from each source is *toward* the (reference) node, the current is *positive.* If it is *away from* the reference node, the current is *negative.*
3. Consequently, the plus sign in both equations should be construed as the *vector sum* of the currents rather than the arithmetic sum.

The foregoing procedures using Millman's theorem are illustrated in the following examples.

EXAMPLE 7-13
Given the original circuit of Fig. 7-15a, find the node voltage, V using Millman's theorem

Solution
Note that in Fig. 7-15a polarities of voltage sources and current sources are *all* positive. We may write a combination of Eqs. (7-5a) and (7-5b) as

$$V = \frac{V_1G_1 + I_2 + V_3G_3}{G_1 + G_2 + G_3} \qquad (7\text{-}5a)$$

$$= \frac{(20\text{ V})(0.5\text{ S}) + 4\text{ A} + (6\text{ V})(1\text{ S})}{(0.5 + 0.5 + 1)\text{ S}}$$

$$= +\frac{(10 + 4 + 6)\text{ A}}{2\text{ S}} = \mathbf{+10\text{ V}}$$

Observe, in effect, that Eqs. (7-5a) and (7-5b) ACTUALLY dictate that voltage sources are converted into current sources for the purpose of adding currents in the numerator! We will use this procedure in the future.

Note that Millman's theorem yielded the node voltage of 10 V almost immediately in Ex. 7-13, whereas Ex. 7-12 required the use of superposition in combination with Thévenin's theorem plus several calculation steps. We may conclude that whenever we encounter a variety of *parallel-connected* sources, Millman's theorem is the most efficient procedure of all.

Note that in the solution of Ex. 7-13, realizing that the numerator represents the sum of currents, we used a combination of Eqs. (7-5a) and (7-5b). Note also that all resistances were *converted into conductances* for purposes of Millman's theorem.

Let us consider sources that may be driving in *opposite* directions and the addition of a load, R_L, as in Ex. 7-14.

EXAMPLE 7-14

Calculate

a. The voltage, V, with respect to ground in the network shown in **Fig. 7-16**
b. The current in the load resistor, R_L
c. The current in each branch, using Norton-to-Thévenin source conversion
d. Draw the complete solution showing all currents and voltages

Solution

a. $$V = \frac{+I_1 - I_2 - I_3 + I_4}{G_1 + G_2 + G_3 + G_4 + G_5} = \frac{(6 - 3 - 2 + 5)\text{ A}}{(0.5 + 0.5 + 0.\bar{3} + 0.\bar{3} + 1)}$$

$$= \frac{6\text{ A}}{2.\bar{6}\text{ S}} = \mathbf{2.25\ V} \qquad (7\text{-}5b)$$

b. $I_L = V/R_L = 2.25\text{ V}/1\ \Omega = \mathbf{2.25\ A}\downarrow$ (leaving node **V**)

c. $$I_1 = \frac{V_1 - V}{r_1} = \frac{12 - 2.25}{2} = \mathbf{4.875\ A}\uparrow \text{ (entering node } \mathbf{V}\text{)}$$

$$I_2 = \frac{V_2 - V}{r_2} = \frac{-6 - 2.25}{2} = \mathbf{-4.125\ A}\downarrow \text{ (leaving node } \mathbf{V}\text{)}$$

$$I_3 = \frac{V_3 - V}{r_3} = \frac{-6 - 2.25}{3} = \mathbf{-2.75\ A}\downarrow \text{ (leaving node } \mathbf{V}\text{)}$$

$$I_4 = \frac{V_4 - V}{r_4} = \frac{15 - 2.25}{3} = \mathbf{4.25\ A}\uparrow \text{ (entering node } \mathbf{V}\text{)}$$

Check, using KCL for all currents entering and leaving node **V**: $\sum I_T = -I_L + I_1 - I_2 - I_3 + I_4$
$= -2.25 + 4.875 - 4.125 - 2.75 + 4.25 = \mathbf{0}$

d. The complete solution is shown in Fig. 7-16d. The reader should verify KCL around points **a**, **b**, and **c** located on node **V**. The reader should also verify that in each of the five branches, the voltage from node **V** to ground is 2.25 V

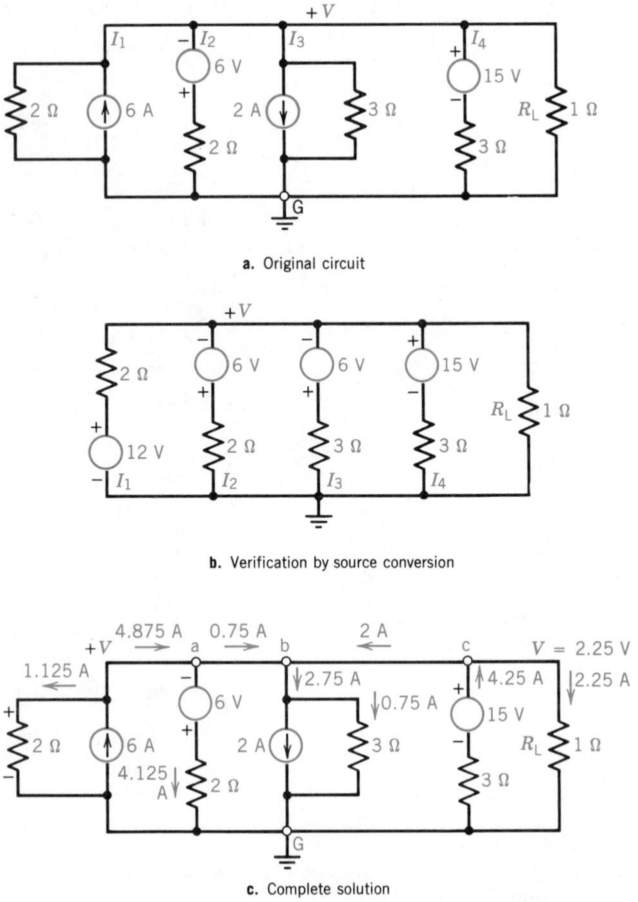

Figure 7-16 Example 7-14; use of Millman's theorem

Example 7-14 shows the extreme power of Millman's theorem in dealing with multisource networks where *all* sources are connected in *parallel*. Since there is only one node voltage, V, with respect to ground in a parallel circuit, we are solving for only one unknown, regardless of the number of parallel-connected practical (voltage or current) sources. We will find this to be a tremendous advantage when considering other methods of analysis in Chapter 8.

Of course, the major disadvantage of Millman's theorem is its limitation to sources in parallel.[7]

[7] This is not strictly true. Millman's theorem could be used with practical current sources in series by first converting them to voltage sources, finding a single equivalent voltage source, and converting it back to a current source. The equation for series-connected practical current sources is $I_{eq} = \dfrac{I_1R_1 + I_2R_2 + I_3R_3 + \cdots + I_nR_n}{R_1 + R_2 + R_3 + R_n}$. The equation, operationally, is the procedure described. Note that it really represents the *dual* of Millman's theorem, Eq. (7-5a).

7-6 RECIPROCITY THEOREM

The reciprocity theorem is occasionally useful but it has its limitations:

1. It can only be used with passive (R, L, and C) elements.
2. The elements must be linear and bilateral.
3. There must only be one active source in the network; that is, it *cannot* be used with multisource networks.

The statement of the reciprocity theorem is

If an EMF, V, at one point in a linear passive network produces a current, I, at any second point in that network, then the *same* voltage, V, acting at that second point will produce the *same* current, I, at the first point (i.e., the original location of V).

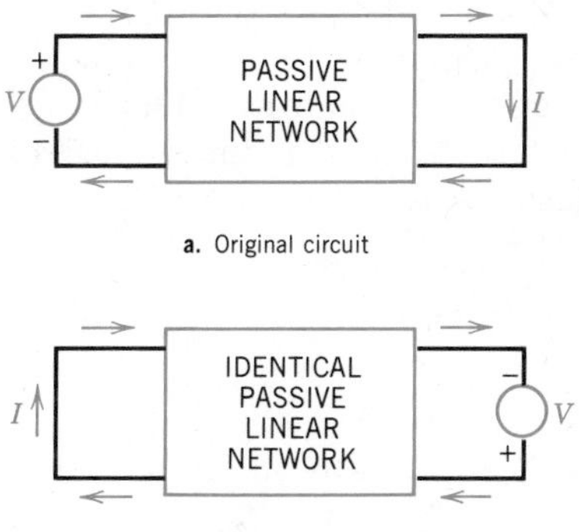

a. Original circuit

b. Reciprocity interchange

Figure 7-17 Reciprocity conditions

The conditions for reciprocity are shown in **Fig. 7-17**. In the original circuit of Fig. 7-17a, the active source V produces a current I at the right. In Fig. 7-17b, the voltage V and current I are *interchanged*. In the process, the network currents and voltages *are also changed*, as we shall prove, *but* the original values of V and I remain the *same*. Most important, in Fig. 7-17, note that the original polarity and current direction remain *unchanged*. We can prove all of the above by the following example.

EXAMPLE 7-15

Given the original circuit shown in **Fig. 7-18a**, show that interchanging V and I by reciprocity yields the same external values of V and I

Solution

Assume we are given $V = 4$ V and all the values of resistance in the network. Then the total resistance seen by the 4-V source is $R_T = (2 \| 2)\ \Omega + 1\ \Omega = 2\ \Omega$. The 2 A drawn from the supply divides at node a equally, yielding 1 A $= I$ as shown in Fig. 7-18a. Now consider the reciprocity interchange shown in Fig. 7-18b. Assume 1 A from the external circuit enters node a. Since the resistance carrying 1 A is 1 Ω, the same volt drop across the 2-Ω resistance yields 1 V/2 Ω = 0.5 A. This sends a total current of 1.5 A to the 4-V supply. But is this correct? What is the total resistance seen by the 4-V supply? $R_T = 2\ \Omega + (2 \| 1)\ \Omega = 2.\bar{6}\ \Omega$, and the total current supplied by the 4 V supply is $I_T = 4\ \text{V}/2.\bar{6}\ \Omega = 1.5$ A!

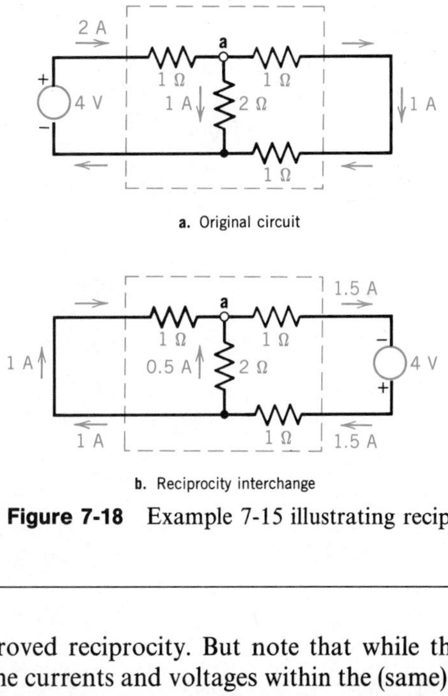

a. Original circuit

b. Reciprocity interchange

Figure 7-18 Example 7-15 illustrating reciprocity

Example 7-15 has proved reciprocity. But note that while the linear network remained unchanged, the currents and voltages within the (same) network actually *did* change. This occurred despite the unvarying reciprocity interchange. Note that the line current directions have been preserved at both input and output. The original polarity of the voltage and the direction of current have both been preserved, at *both ports* of the network.

7-7 COMPENSATION THEOREM

The compensation theorem is sometimes known as the substitution theorem. The statement of the theorem is

Any resistance, R, carrying an initial current, I, in any branch of a linear network may be replaced by an ideal (zero resistance) voltage source V (or any equivalent combination of circuit elements) whose terminal voltage is equal to the voltage developed across R by the current I such that $V = I \times R$. The polarity of V is always the same as that in which I polarizes R by its direction of current.

Figure 7-19a shows branch b–c with a current of 4 A through a 10 Ω resistor producing a voltage across b–c of $V_{bc} = 40$ V. One possible substitution is shown in Fig. 7-19b, and a series of alternative substitutions are shown in Fig. 7-19c. Note that any of these substitutions across branch b–c yield the same voltage of 40 V, with the same polarity as that of the original circuit.

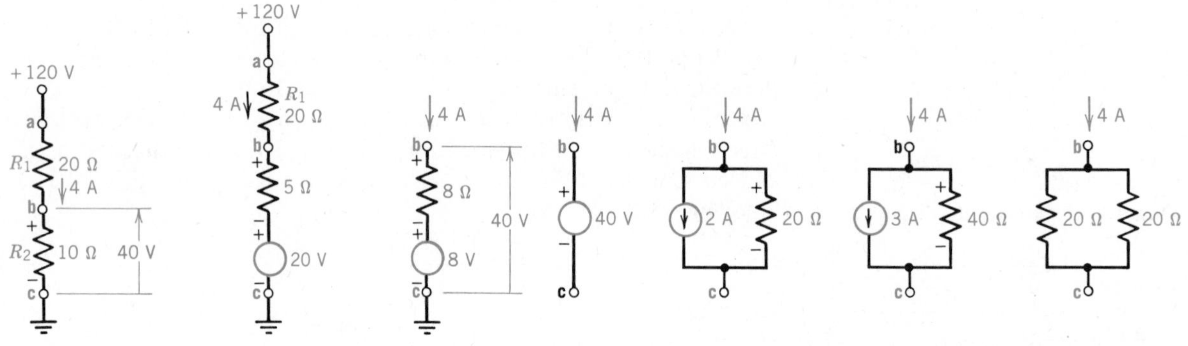

a. Original circuit b. One substitution c. Other possible substitutions across nodes **b** and **c**

Figure 7-19 Various applications of the compensation (substitution) theorem

Consequently, *any of these substitutions* may be made across nodes b–c without producing any change in the original circuit currents or voltages.

EXAMPLE 7-16

For the ladder network shown in the original circuit of **Fig. 7-20a** the current drawn from the supply is 0.2 A. Calculate the current in **a.** R_4 **b.** R_8

Solution

a. $V_{cd} = V - I(R_1 + R_2) = 120\text{ V} - 0.2\text{ A}(200\ \Omega) = 80$ V. Using the compensation theorem, we substitute 80 V across nodes c–d as shown in Fig. 7-20b. I_4 represents the total current drawn from the 80-V supply. Therefore, $R_T = 200 + 200 + (800 \| 800)\ \Omega = 800\ \Omega$ and $I_4 = I_T = 80\text{ V}/800\ \Omega = \mathbf{0.1\ A}$

b. To find the current in R_8 we again use the compensation theorem by substituting a current source of 0.1 A driving into node **e**, as shown in Fig. 7-20c. Since both branches have equal resistances, $I_8 = I_T/2 = 0.1\text{ A}/2 = \mathbf{50\ mA}$

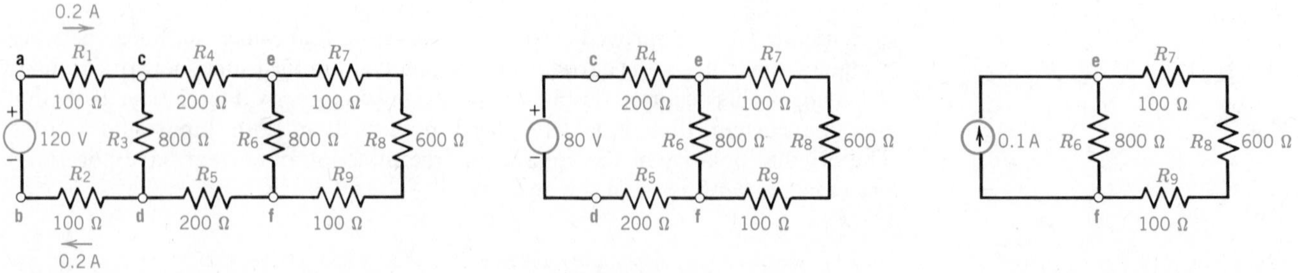

a. Original circuit b. Substitution of 80 V source across nodes c and d c. Substitution of current source of 0.1 A

Figure 7-20 Example 7-16 illustrating compensation theorem

Note that the use of the compensation theorem enables both circuit simplification and clarification of the nature of the problem by concentrating our attention on that portion of interest in any complex network.

Two other types of substitutions frequently occur in dc circuits.[8] The first of these is shown in **Fig. 7-21**, using a bridge circuit (first encountered in Chapter 5), fed by dual power supplies. Using our old friend the voltage-divider theorem, given in Eq. (5-8), for dual supplies, we may verify that the voltage at both nodes T and H is +55 V. We may also verify that the current in the LH branch is 100 V/80 kΩ or 1.25 mA and the current in the RH branch is 100 V/40 kΩ or 2.5 mA. This means that the total current flowing through both supplies is 3.75 mA, as shown in Fig. 7-21a.

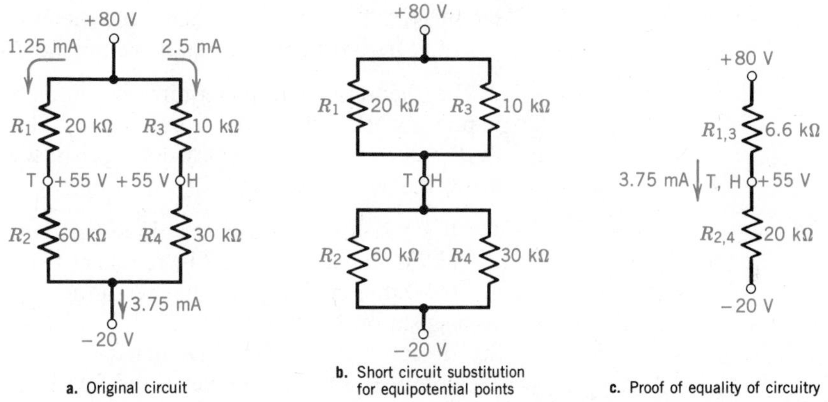

Figure 7-21 Equipotential points replaced by a short circuit

But since points T and H are at the same potential, if we connected a zero-resistance conductor (a short circuit) between them, no current would flow between points T and H. Since it makes no difference, let us connect such a short as shown in Fig. 7-21b. But tying terminals T–H together places R_1 and R_3 in parallel and R_2 and R_4 in parallel, with both groups in series with each other, as shown in Fig. 7-21b. If the short circuit makes a difference, the total current should change, as should the respective voltages for T–H. The parallel equivalent of R_1 and R_3 is $6.\bar{6}$ kΩ and that of R_2 and R_4 is 20 kΩ, as shown in Fig. 7-21c.

Computing the total current in Fig. 7-21c yields 100 V/$26.\bar{6}$ kΩ = 3.75 mA, the same as the total current in Fig. 7-21a.

Computing $V_{T,H} = \dfrac{20\text{ k}}{26.\bar{6}\text{ k}}(100)\text{ V} + (-20) = +55\text{ V}$, yields the same as the original potentials in Fig. 7-21a.

The above insight leads to two corollaries of the compensation theorem:

1. Given any two points in a complex network *at the same* potential, the two points may be short-circuited without producing any changes in the network.
2. Given a single branch in a complex network whose current (and voltage across it) is zero, that branch may be either removed (open-circuited) or short-circuited without producing any changes in the complex network.

Substitution of either an infinite resistance (open circuit) or a zero resistance (short circuit) produces no change in the original circuit conditions, as we observe

[8] Other instances of use of the compensation or substitution theorem occur in ac. At very high frequencies, capacitors are treated as short circuits and inductors as open circuits. In such circuitry, we replace capacitors with shorted conductors and open the inductors. Similarly, at very low frequencies and dc, capacitors are substituted by open circuits and inductors are substituted by short circuits.

from Fig. 7-21. Therefore, if we elected to connect a 1-kΩ, a 500-kΩ, or a 500-MΩ resistor across terminals T–H in Fig. 7-21a, no current would flow in that resistor and the circuit conditions would remain as they were before the connection was made.

7-8 BARTLETT'S THEOREM

Bartlett's theorem applies only to circuits that are *symmetrical*; that is, they contain components connected in such a way that one half of the circuit is the mirror image of the other. The statement of Bartlett's theorem is

Given a symmetrical circuit, it may be separated (cut) into two halves with an infinite resistance between them, and the circuit conditions applying to any branch of either half are the same as those for the original circuit.

The theorem literally requires a symmetrical circuit to be cut in half. What happens when a resistance is cut in half? This depends on its location with respect to the "cutting axis." We may distinguish two conditions and their respective rules using Bartlett's theorem:

1. If a resistor is parallel or along the cutting axis, its resistance in each circuit half is *doubled*; that is, it is treated as two parallel resistances, each having twice the value of the equivalent resistance (a permissible substitution by the compensation theorem).
2. If the resistor is perpendicular to the cutting axis, it is left dangling or open-circuited and may be neglected for the purpose of simplification.

The theorem and the foregoing rules are best illustrated in Exs. 7-17 and 7-18.

EXAMPLE 7-17

Given the symmetrical circuit of **Fig. 7-22a**, calculate

a. The current in the 5-kΩ resistor across nodes **b–e**
b. The potential of point **e** with respect to ground
c. Draw the complete solution showing all currents and voltages at all points

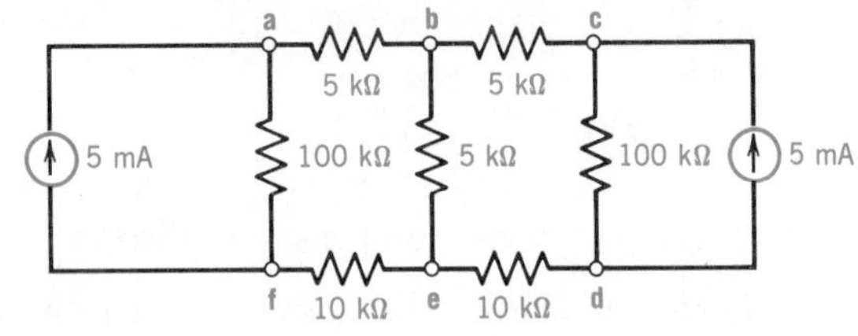

a. Original circuit

Solution

a. The original circuit is "cut" along nodes **b–e**. The 5-kΩ resistor is separated into two paralleled 10-kΩ resistors as shown in Fig. 7-22b. Using the CDR for each half, we note that the current distribution is in the ratio of $n = 4$. Using Eqs. (5-25a) and (5-25b), $I_{100} = 5\text{ mA}/(4 + 1) = 1\text{ mA}$ and $I_{be} = \frac{4}{5}(5\text{ mA}) = 4\text{ mA}$, which is the current in the 10-kΩ resistor. Consequently, if the circuit were rejoined, $I_{be} = 2 \times 4\text{ mA} =$ **8 mA**
b. Since 4 mA flows in the 10-kΩ resistor from **e** to **f**, $V_{ef} =$ (4 mA)(10 kΩ) = **40 V**. This is also the potential of point **e**.
c. The complete solution shown in Fig. 7-22c is left as an exercise for the reader.

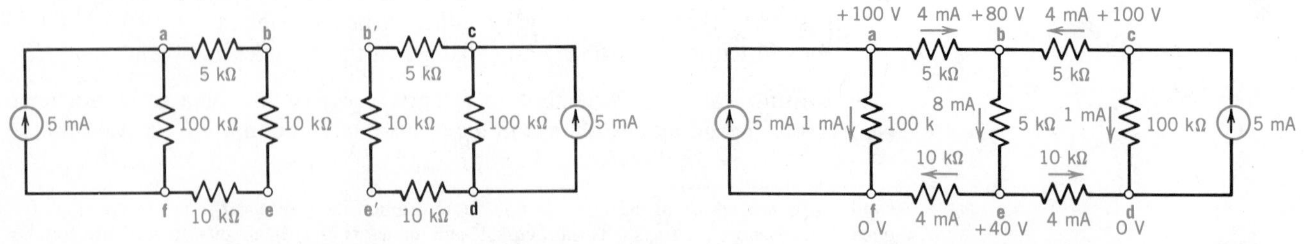

b. Separation by Bartlett's theorem

c. Complete solution

Figure 7-22 Bartlett's theorem for solution of symmetrical circuits (Ex. 7-17)

EXAMPLE 7-18

Given the symmetrical circuit shown in **Fig. 7-23a**, calculate the

a. Voltage at node **c**
b. Voltage at node **d**
c. Current in the 200-Ω resistor across nodes **c–d**
d. Current supplied by each 20-V source

Solution

a. The circuit is cut axially along nodes **c–d** as shown in Fig. 7-23b. Note that the 2000-Ω resistor, which is perpendicular to the cutting axis, is left dangling and draws zero current.[9]

Using the VDR, $V_c = \dfrac{800\ \Omega}{1000\ \Omega} \times 20\ \text{V} = \mathbf{16\ V}$

b. Using the VDR, $V_d = \dfrac{400\ \Omega}{1000\ \Omega} \times 20\ \text{V} = \mathbf{8\ V}$

c. $I_{cd} = (V_c - V_d)/200\ \Omega = 8\ \text{V}/200\ \Omega = \mathbf{40\ mA}$

d. $I_s = 20\ \text{V}/(200 + 400 + 400)\ \Omega = \mathbf{20\ mA}$ or alternatively $\dfrac{I_{cd}}{2} = \dfrac{40\ \text{mA}}{2} = \mathbf{20\ mA}$

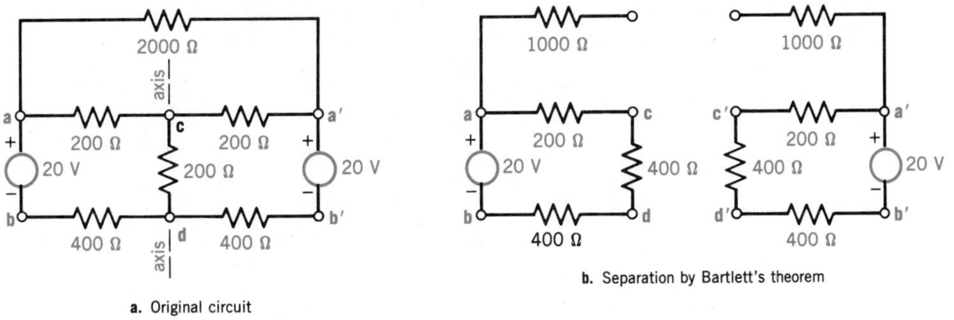

Figure 7-23 Example 7-18 illustrating Bartlett's theorem

7-9 TELLEGEN'S THEOREM

In introducing Kirchhoff's voltage law, we noted that the algebraic sum of all voltage rises and drops taken around any closed path (or mesh or loop) in any circuit is zero, as represented by $\sum_{n=1}^{N} V_n = 0$ in Eq. (5-2).

In presenting Kirchhoff's current law, we also noted that the algebraic sum of all currents entering and leaving a node is zero, as represented by $\sum_{n=1}^{N} I_n = 0$ in Eq. (5-29).

Tellegen's theorem is an extension of KVL and KCL as expressed above. Because it is an insight derived from KVL and KCL, it applies to all networks, linear and nonlinear, as well as passive and active. The statement of Tellegen's theorem is

In any given network, the algebraic sum of the *products* of all branch voltages, V_k and their respective branch currents, I_k, is equal to zero, or

$$\sum_{k=1}^{n} V_k I_k = 0 \qquad \text{or} \qquad V_1 I_1 + V_2 I_2 + V_3 I_3 + \cdots + V_n I_n = 0 \qquad (7\text{-}6)$$

We might infer from Eq. (7-6) that Tellegen's theorem is nothing more than the law of conservation of energy as applied to any network. In effect, it is! It states that the instantaneous and average power flow supplied from the sources in a network *equals* the net instantaneous and average power flow into and absorbed by the remaining network elements at all times.

Recall from Sec. 5-3 that a *negative* voltage *drop* is a voltage *rise* and that a rise of potential occurs in an *active* element (a source) when going from (−) to (+)

[9] Since each end of the 2000-Ω resistor is connected to +20 V, its current is zero in any case. Bartlett's theorem only verifies this for us.

through the source. Consequently, some of the volt-ampere products in Eq. (7-6) (sources) are algebraically negative. If, however, they are transferred to the *right* side of the equation, then the average power flow **to** the circuit elements equals the average power flow **from** the source(s).

Figure 7-24a shows Tellegen's theorem for a source, V_6I_6, supplying power to five circuit elements, V_1I_1 through V_5I_5, respectively. Since power is additive regardless of the method of circuit connection (Table 5-1), we may write Eq. (7-6) as $V_1I_1 + V_2I_2 + V_3I_3 + V_4I_4 + V_5I_5 - V_6I_6 = 0$ for Fig. 7-24a. Note that $k = 6$ for the network of Fig. 7-24a; that is, there are six different devices and sources receiving and supplying energy, represented as *six* separate power *boxes* in Fig. 7-24a.

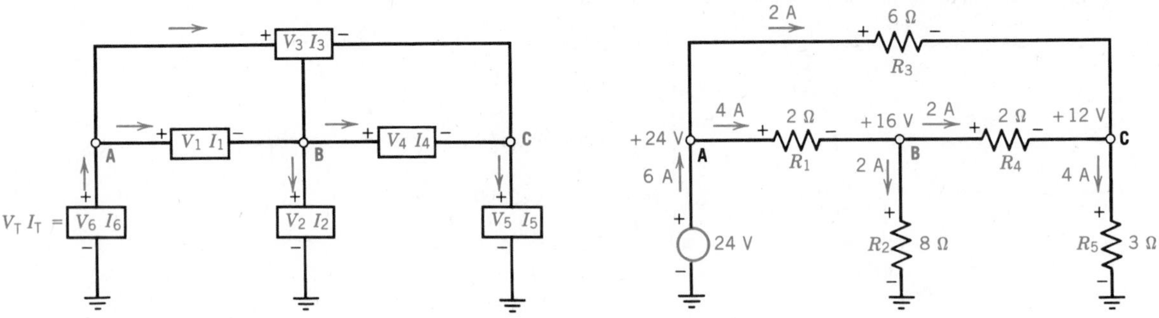

a. Tellegen's theorem

b. Circuit values for Ex. 7-19

Figure 7-24 Tellegen's theorem and Ex. 7-19

EXAMPLE 7-19

Verify Tellegen's theorem, given the complete solution of the network for Fig. 7-24b.

Solution

$$V_1I_1 + V_2I_2 + V_3I_3 + V_4I_4 + V_5I_5 - V_6I_6 = 0 \quad (7\text{-}6)$$

(8 V)(4 A) + (16 V)(2 A) + (12 V)(2 A) + (4 V)(2 A) + (12 V)(4 A) − (24 V)(6 A) = 32 W + 32 W + 24 W + 8 W + 48 W − 144 W = 0

Example 7-19 verifies not only Tellegen's theorem but also the law of conservation of energy, which dictates that in any closed system, the total energy must remain constant. Consequently, when the resistors dissipate power (a total of 144 W), an equivalent amount of power of another form is produced (say heat power exclusively), and therefore the source must *absorb* (**negative**) power from some other energy conversion device in order to maintain the total energy of the system constant (Tellegen's theorem). In Ex. 7-19 the power absorbed (and supplied) must be −144 W. With respect to Fig. 7-24a, the following rules apply to Tellegen's theorem.[10]

The power absorbed by a device (a box) is determined by the voltage polarity encountered as the **current** ENTERS the box. The power is

1. NEGATIVE, when the voltage polarity first encountered is negative.
2. POSITIVE, when the voltage polarity first encountered is positive.

And also,

3. Negative power is power supplied by a box to the system.

[10] The application of Tellegen's theorem to ac circuits and complex power relations is covered in Sec. 16-18.

4. Positive power is power dissipated by the box in the system (e.g., resistors).
5. At all times, the instantaneous and average algebraic power sums must be zero (Tellegen's theorem).
6. Some passive elements (inductors and capacitors) store energy but return that energy to the system. During the energy storage period, the power received is POSITIVE. During the energy discharge period, the power received by the system is NEGATIVE.

Note that these rules also apply to Fig. 4-2, where total power entering and leaving a box is zero.

7-10 GLOSSARY OF TERMS USED

Bartlett's theorem Given a symmetrical circuit, it may be separated into two halves with an infinite resistance between them, and the circuit conditions applying to any branch of either half are the same as those for the original circuit.

Compensation theorem Any resistance, R, carrying an initial current, I, in any branch of a linear network may be replaced by an ideal (zero resistance) voltage source V (or any equivalent combination of circuit elements) whose terminal voltage is equal to the voltage developed across R by the current I, such that $V = IR$. The polarity of V is always the same as that in which I polarizes R by its direction of current.

Galvanometer Instrument for measuring or detecting small electric currents or potential differences.

Millman's theorem Any combination of parallel-connected voltage (or current) sources may be represented as a single equivalent source (voltage or current) by applying the source transformation theorem appropriately to each source individually and combining the results to obtain the voltage across the equivalent combination.

Negative power Power leaving a system. Power delivered by a source to a system is *negative* power with respect to the source.

Norton's theorem Any combination of linear circuit elements and active sources, regardless of connection or complexity, may be replaced by a two-terminal network consisting of a single current source and a single impedance (resistance) connected in parallel with the source.

Open-circuit test Determination of the Thévenin voltage across any two terminals of an active network as measured by an ideal voltmeter, that is, one with infinite impedance (resistance).

Positive power Power *entering* a system. Resistors receive and dissipate *positive* power.

Reciprocity theorem If an EMF, V, at one point in a linear passive network produces a current, I, in a second point in that network, then the same voltage, V, acting at that second point will produce the same current, I, at the first point.

Short-circuit test Determination of the short-circuit current that would flow in an ideal ammeter (i.e., one with zero resistance) connected across the same two terminals of an active network as were used in the open-circuit test.

Source transformation theorem Given any complex active network containing one or more practical voltage and/or current sources, any given practical voltage (or current) source may be transformed into any equivalent practical current source (or vice versa), using the Thévenin–Norton conversion.

Substitution theorem Same as the compensation theorem.

Superposition theorem In any multisource complex network consisting of linear bilateral elements, the voltage across (or current through) any given element of the network is equal to the algebraic sum of the individual voltages (or currents) produced independently across (or in) that element by each source acting independently.

Symmetrical circuit or network One containing components whose values are such that when the network is bisected, one half of it is the mirror image of the other half.

Tellegen's theorem In any given network, the algebraic sum of the products of all branch voltages and their respective branch currents is equal to zero.

Thévenin's theorem Any combination of linear circuit elements and active sources, regardless of connection or complexity, may be replaced by a two-terminal network consisting of a single voltage source and a single impedance (resistance) connected in series with the source.

7-11 PROBLEMS

Secs. 7-1 to 7-3

The first 10 problems are intended as a "rapid-fire" introduction to Thévenin–Norton theorem conversions, to develop insights and skills for the problems that follow.

7-1 For the circuit shown in **Fig. 7-25**, calculate

a. V_{TH}
b. R_{TH}
c. Draw the equivalent practical voltage source across terminals T–H

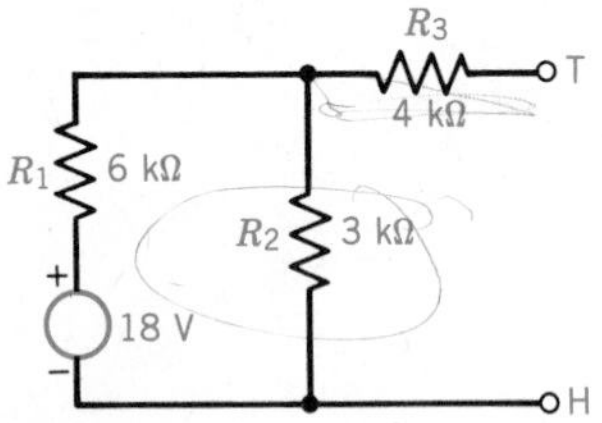

Figure 7-25 Problem 7-1

7-2 For the circuit shown in **Fig. 7-26**, calculate

a. V_{TH}
b. R_{TH}
c. Draw the equivalent practical voltage source across terminals T–H
d. Draw the equivalent practical current source across terminals T–H

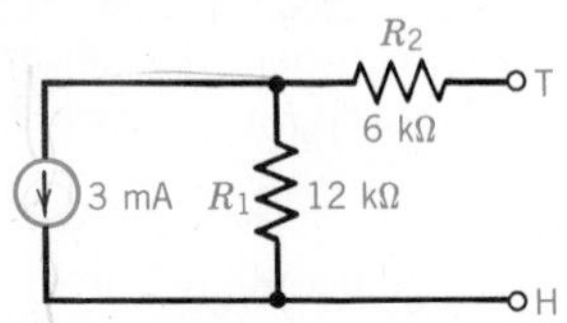

Figure 7-26 Problem 7-2

7-3 For the circuit shown in **Fig. 7-27**, calculate

a. V_{TH}
b. R_{TH}
c. Draw the equivalent practical current source across terminals T–H

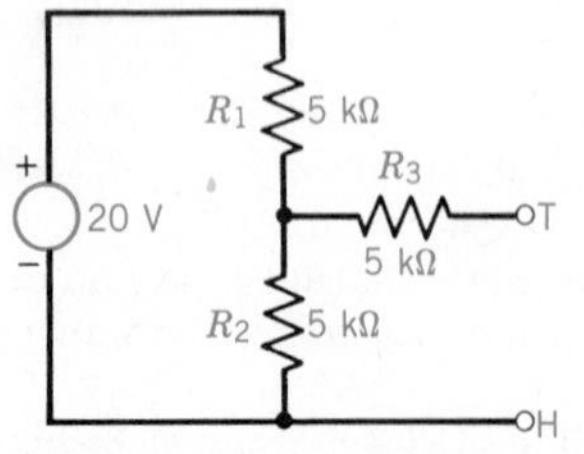

Figure 7-27 Problem 7-3

7-4 For the circuit shown in **Fig. 7-28**, calculate

a. V_{TH}
b. R_{TH}
c. Draw the equivalent practical voltage source across terminals T–H

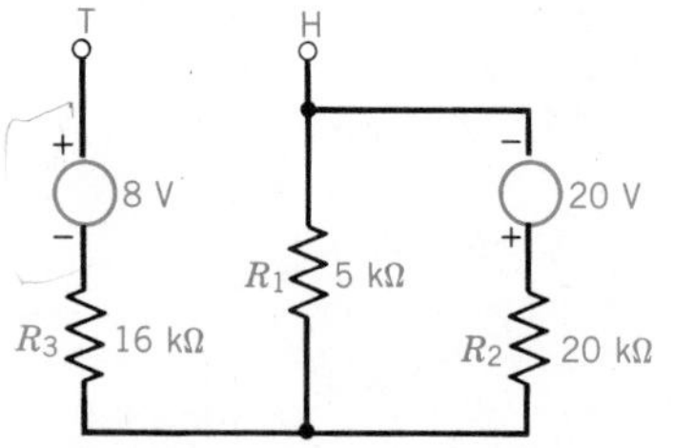

Figure 7-28 Problem 7-4

7-5 For the circuit shown in **Fig. 7-29**, calculate

a. V_{TH}
b. R_{TH}
c. Draw the equivalent practical voltage source across terminals T–H

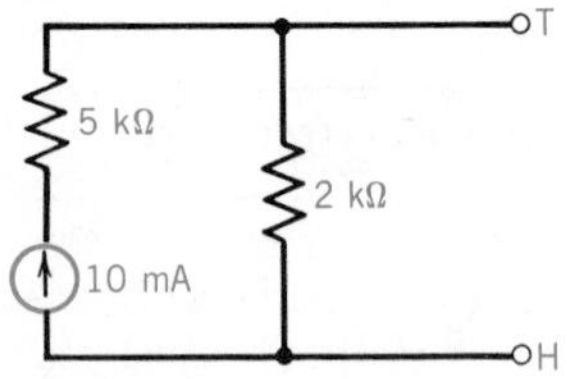

Figure 7-29 Problem 7-5

7-6 For the circuit shown in **Fig. 7-30**, calculate

a. V_{TH}
b. R_{TH}
c. Draw the equivalent practical voltage source across terminals T–H

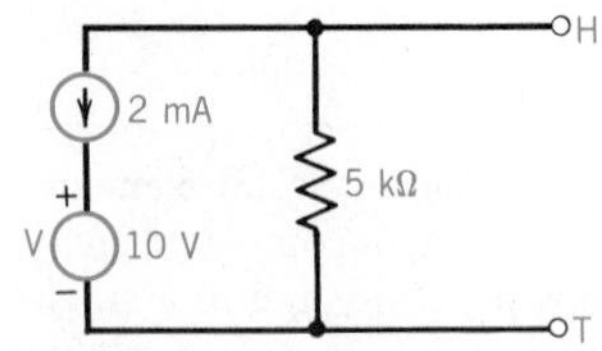

Figure 7-30 Problem 7-6

7-7 For the circuit shown in **Fig. 7-31**, calculate

a. V_{TH}
b. R_{TH}
c. Draw the equivalent practical current source across terminals T–H

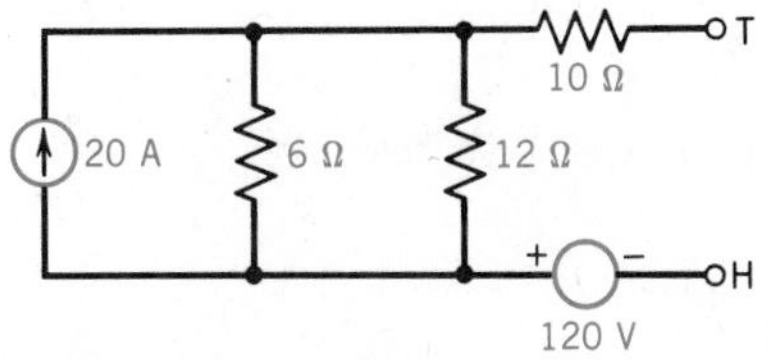

Figure 7-31 Problem 7-7

7-8 For the circuit shown in **Fig. 7-32**, calculate

a. V_{TH}
b. R_{TH} 2.4K

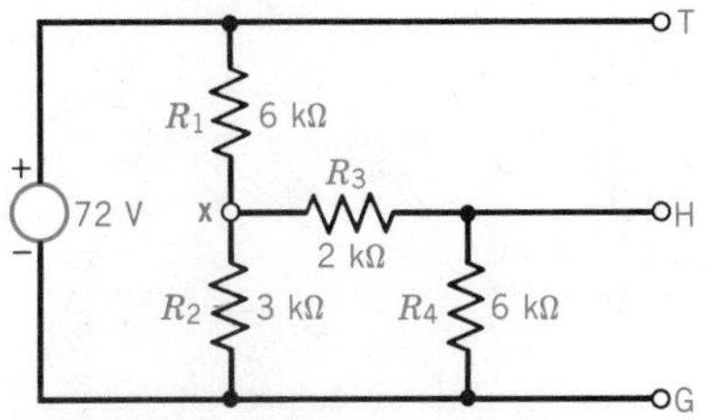

Figure 7-32 Problems 7-8 and 7-9

7-9 For the circuit shown in Fig. 7-32, calculate

a. V_{TG}
b. R_{TG}

7-10 For the circuit shown in **Fig. 7-33**, calculate

a. V_{TH}
b. R_{TH}
c. Draw the equivalent practical current source across terminals T–H

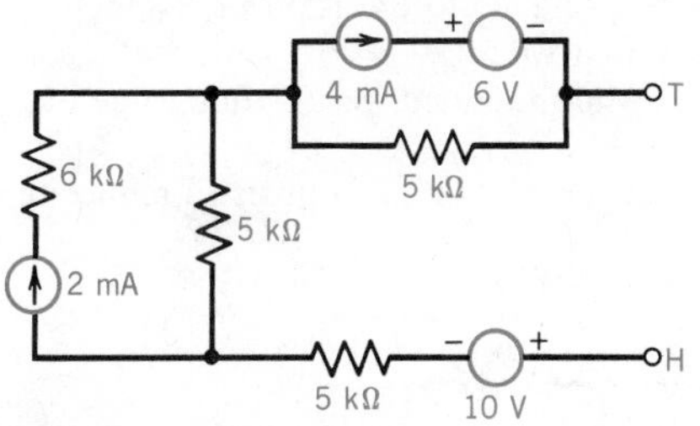

Figure 7-33 Problem 7-10

7-11 The circulating current between the two unequal voltage sources connected in parallel in **Fig. 7-34** is 1 A, as shown. Calculate
a. R_{TH}
b. V_{TH}
c. Draw the equivalent practical voltage source and calculate the
d. Open-circuit voltage as measured by a voltmeter during open-circuit test
e. Short-circuit current as measured by an ammeter shorting terminals T–H
f. R_{TH} from open-circuit voltage and short-circuit current measurements
g. Draw the equivalent practical current source across terminals T–H

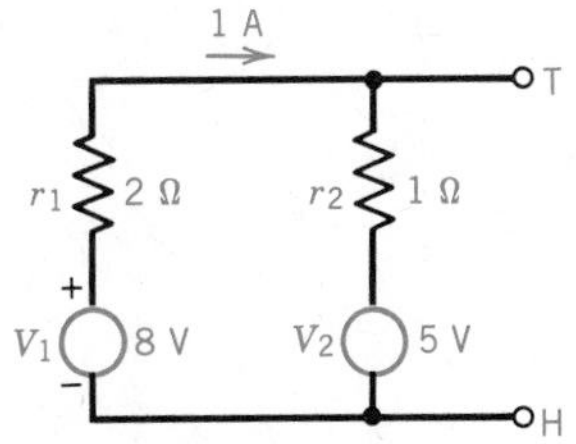

Figure 7-34 Problem 7-11

7-12 Given the circuit shown in Fig. 7-34, convert each voltage source to a practical current source and calculate the

a. Equivalent single current source, I_N
b. Equivalent internal resistance, R_N
c. Open-circuit voltage as measured by an ideal voltmeter connected across T–H
d. Short-circuit current as measured by an ideal ammeter across T–H
e. R_{TH} from open-circuit voltage and short-circuit current measurements

7-13 Given the circuit shown in **Fig. 7-35**, calculate

a. V_L
b. I_L
c. Current delivered by source A
d. Current delivered by source B
e. Power dissipated by R_L
f. Value of R_L that will draw maximum power from sources
g. Maximum possible power that can be drawn from the combination in Fig. 7-35

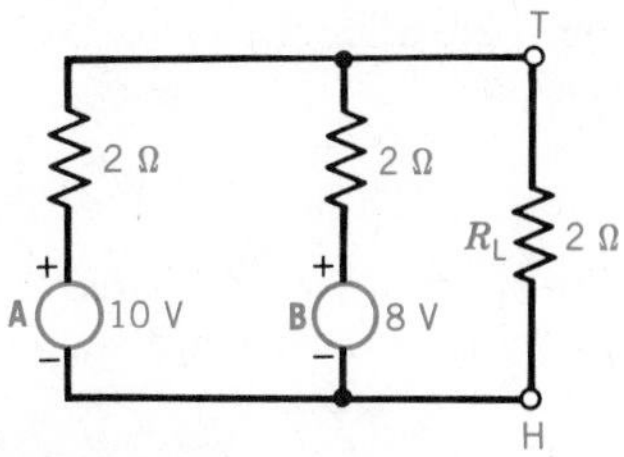

Figure 7-35 Problem 7-13

7-14 Repeat all parts of Prob. 7-13 with the polarity of the 8-V source reversed.

7-15 The galvanometer in the bridge circuit of **Fig. 7-36** has a resistance of 600 Ω. Using Thévenin's theorem, with the galvanometer (temporarily) removed, calculate

a. V_{TH}
b. R_{TH}
c. Current in the galvanometer when connected across terminals T–H
d. V_{TH} with the galvanometer connected

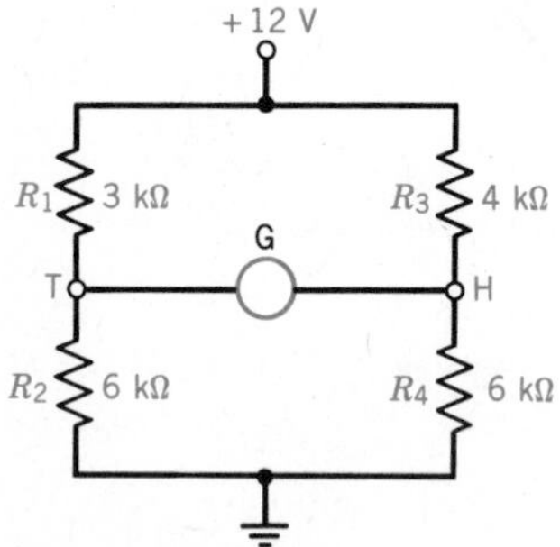

Figure 7-36 Problem 7-15

Secs. 7-4 to 7-6

7-16 Given the circuit shown in **Fig. 7-37**, use the source transformation theorem to convert the practical voltage source to a practical current source and calculate

a. Current in R_3
b. Node voltage V
c. Current delivered to node voltage V by the practical voltage source
d. Current and direction of current in R_1

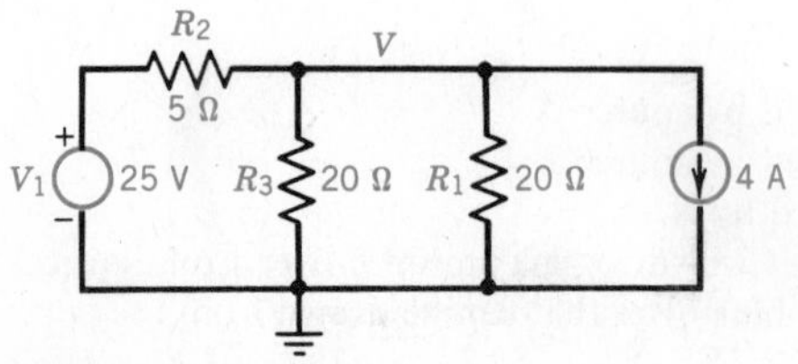

Figure 7-37 Problem 7-16

7-17 For the circuit shown in **Fig. 7-38**, using the source conversion and Millman theorems, calculate

a. Node voltage, V
b. I_L in load R_L
c. Current delivered to node V by practical voltage source V_1
d. Current received from node V by practical voltage source V_2
e. Power dissipated by R_L
f. Value of R_L needed to draw maximum power from node V
g. Using the given diagram, draw the complete solution, showing all currents and voltages

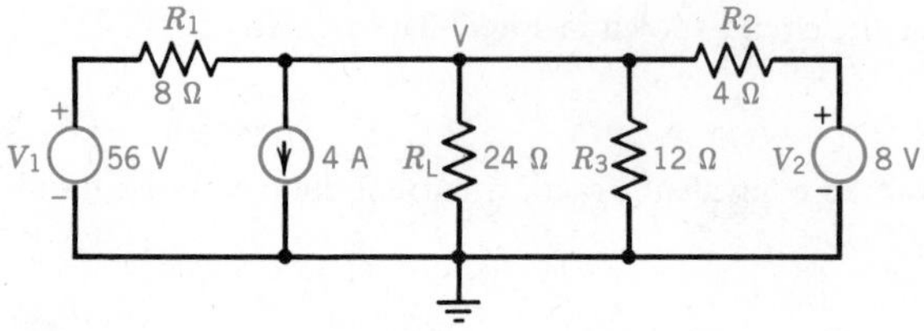

Figure 7-38 Problem 7-17

7-18 For the circuit shown in **Fig. 7-39**, calculate

a. I_4
b. V_4
c. P_4

Hint: remove R_4 and determine V_{TH} and R_{TH}. Replace R_4 and calculate above.

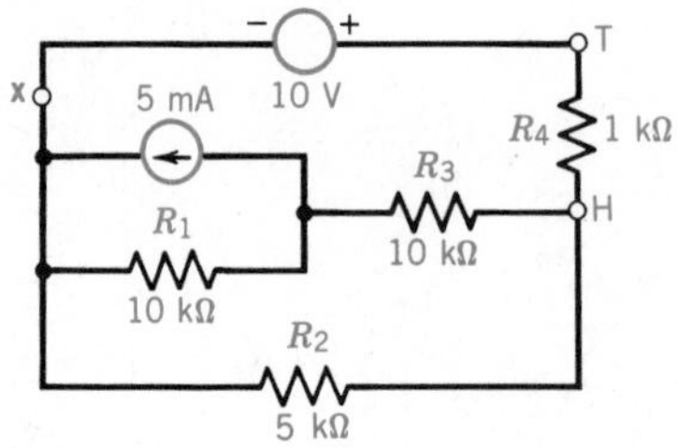

Figure 7-39 Problem 7-18

7-19 For the circuit shown in **Fig. 7-40**, calculate

a. I_3
b. Direction of current in R_3
c. Voltage at node **Y** (with respect to ground)
d. Voltage at node **X**
e. Current delivered to node **X** by V_1
f. Current delivered to node **Y** by V_2
g. I_2 in R_2
h. I_4 in R_4
i. Using the given diagram and the values calculated above, draw the complete solution, showing all currents and voltages. Explain why R_2 receives current from both sources

Hint: remove R_3 and, using Thévenin's theorem, find V_x and R_x to the left of R_3 and V_y and R_y to the right of R_3. Replace R_3 and solve for the required values.

7-20 Using the superposition theorem, verify your value of I_3 obtained in Prob. 7-19.

Hint: use Thévenin equivalent practical sources to reduce the calculations.

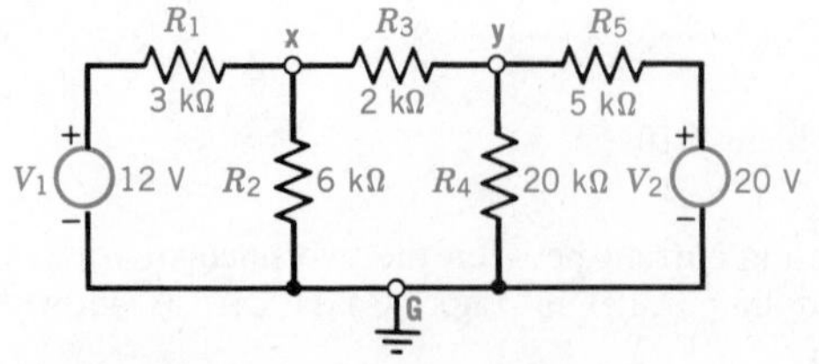

Figure 7-40 Problems 7-19 and 7-20

7-21 For the circuit shown in **Fig. 7-41**, calculate

a. I_4
b. V_4
c. Current delivered by source V_2 to node **T**

Hint: remove R_4 and use Thévenin's theorem to find V_{TG} and R_{TG}.

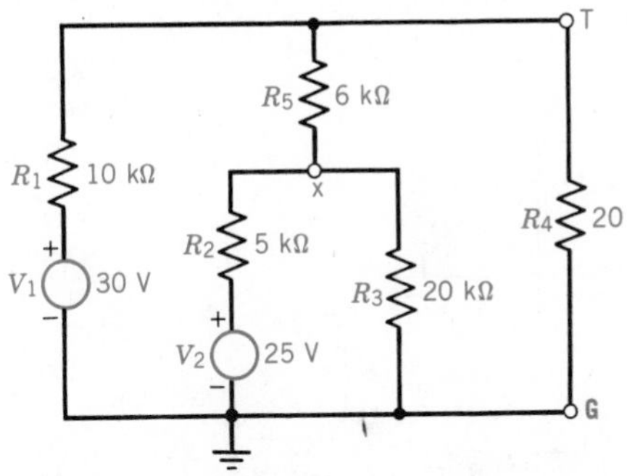

Figure 7-41 Problems 7-21, 7-22 and 7-23

7-22 Repeat Prob. 7-21 by converting all sources to current sources in parallel across nodes **T–G**.

Hint: convert V_2 to a single practical voltage source across nodes **T–G**; then convert both sources to current sources and use Millman's theorem.

7-23 Using superposition, verify the value of V_4 obtained in your solutions to Probs. 7-21 and 7-22. Explain why V_2 neither receives current from nor delivers current to node **T**.

Secs. 7-7 to 7-8

7-24 a. Given the circuit shown in **Fig. 7-42**, solve for the current I using the dual of Eq. (7-5a) for Millman's theorem

b. Convert each practical current source to a practical voltage source and solve for the current I
c. Redraw Fig. 7-42 and show the voltage drop across each source and across R_3
d. Using Fig. 7-42 as given here, show how current divides as it enters each practical current source, calculating the voltage drop across each practical current source and across R_3
e. Compare your results in (**c**) and (**d**) and draw conclusions

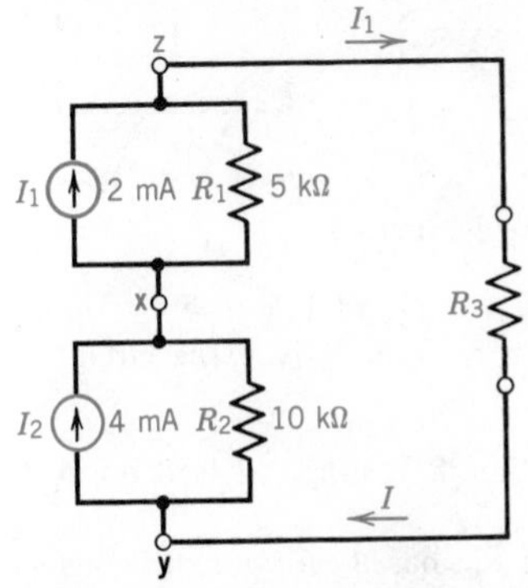

Figure 7-42 Problems 7-24 and 7-26

7-25 Repeat all parts of Prob. 7-24 for the circuit shown in **Fig. 7-43**. Hint: recall from Sec. 5-3 that current through any resistance is from (+) to (−) taken in the direction of current, when doing parts (**c**) and (**d**).

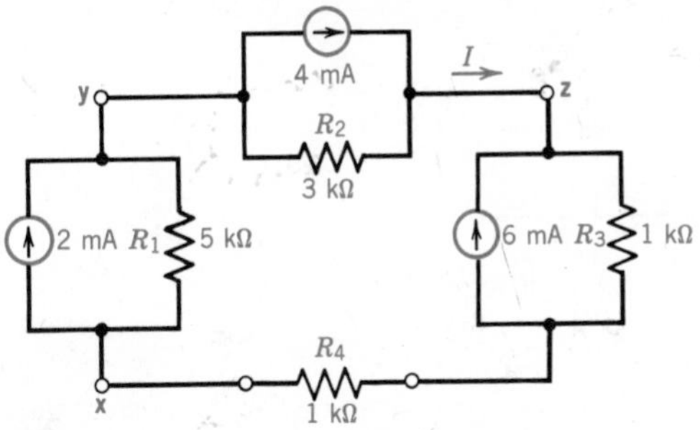

Figure 7-43 Problem 7-25

7-26 Having solved Prob. 7-24, use the compensation theorem to draw *three* equivalent circuits for the voltage and current conditions between nodes **Z** and **X** of Fig. 7-42.

7-27 Given the circuit shown in **Fig. 7-44**,

a. Find the current in R_3
b. Draw three equivalent circuits for current and voltage conditions across nodes **X–Y**, using the compensation theorem

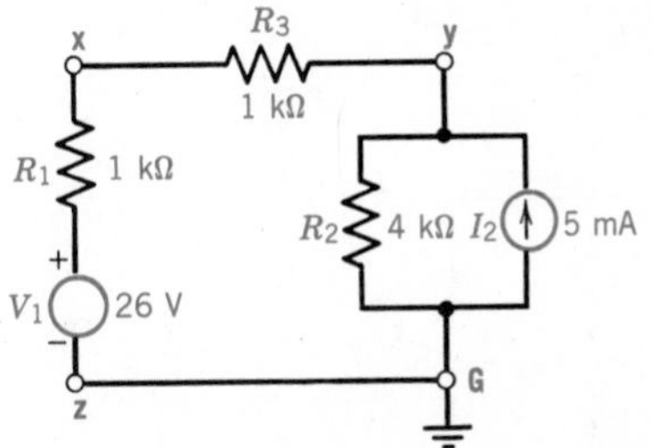

Figure 7-44 Problems 7-27 and 7-28

7-28 Draw three equivalent circuits for current and voltage conditions across nodes **X–Z** in Fig. 7-44.

7-29 Using Millman's theorem and the source transformation theorem for the network of **Fig. 7-45**, calculate the

a. Voltage V_{XG} across R_L
b. Current in R_L
c. Draw three equivalent circuits for current and voltage conditions across nodes **X–G**, using the compensation theorem

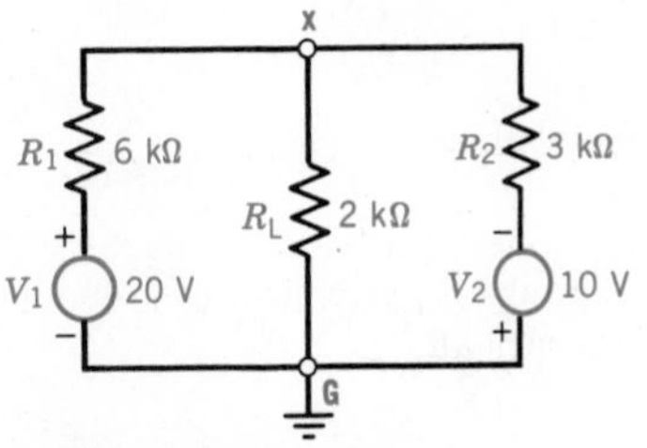

Figure 7-45 Problem 7-29

7-30 Given the circuit shown in **Fig. 7-46**, verify the 1-mA current in R_4 by using the reciprocity theorem. Hint: insert the 48-V source in branch **T–H** and a short in branch **A–G** carrying 1 mA, but be careful of polarities and current directions.

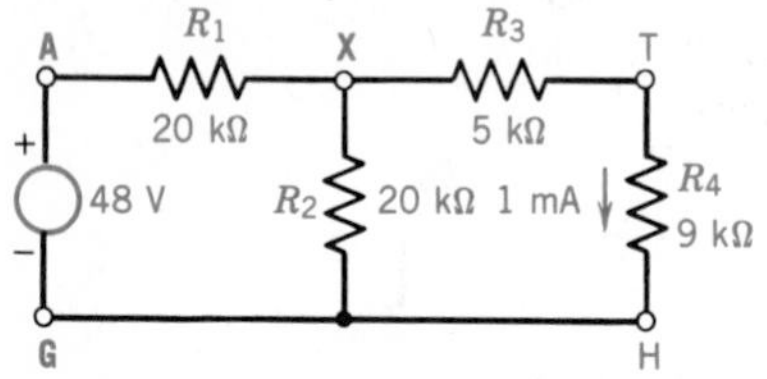

Figure 7-46 Problem 7-30 (circuit of Ex. 7-5) to verify reciprocity theorem

7-31 If we perform the complete solution for the current supplied by the 48-V source in Prob. 7-30, we obtain 1.7 mA, using the given values for R_1 through R_4. **Figure 7-47a** shows the circuit of Prob. 7-30. Assume that R_1 through R_4 are *unknown* (all we know is that $I_4 = 1$ mA and the total current drawn from the 48-V supply is 1.7 mA). Given the circuit shown in Fig. 7-47b, calculate the current I drawn from the 48-V supply when a 28-V source is inserted in series with R_4 as shown. Hint:

1. Use reciprocity to find that component of I produced by the 28-V source acting alone.
2. Use superposition to find I as a result of both sources in Fig. 7-47b.

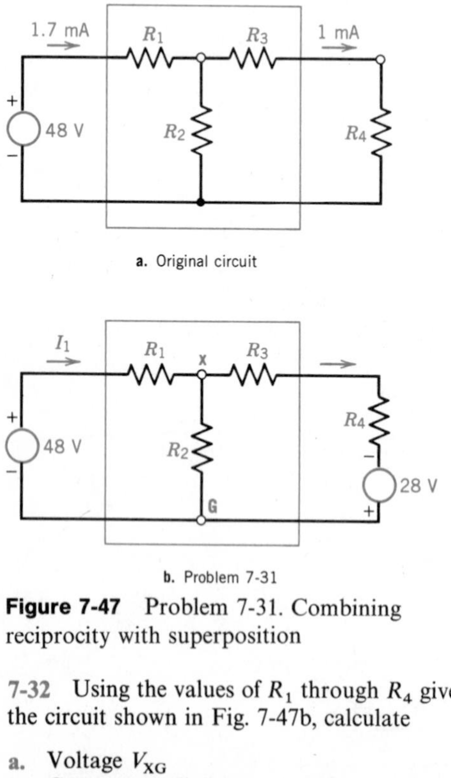

Figure 7-47 Problem 7-31. Combining reciprocity with superposition

7-32 Using the values of R_1 through R_4 given in Fig. 7-46 and the circuit shown in Fig. 7-47b, calculate

a. Voltage V_{XG}
b. Current supplied to node **X** by the 48-V source.

Hint:

1. Use Millman's theorem to find V_{XG}
2. Your answer to part (**b**) should agree with the answer to Prob. 7-31

7-33 The 50-V source shown in **Fig. 7-48a** acting alone produces 5 mA in a load R_L. If a 30-V source is inserted in series with R_L, as shown in Fig. 7-48b, calculate the current, I, drawn from the supply.

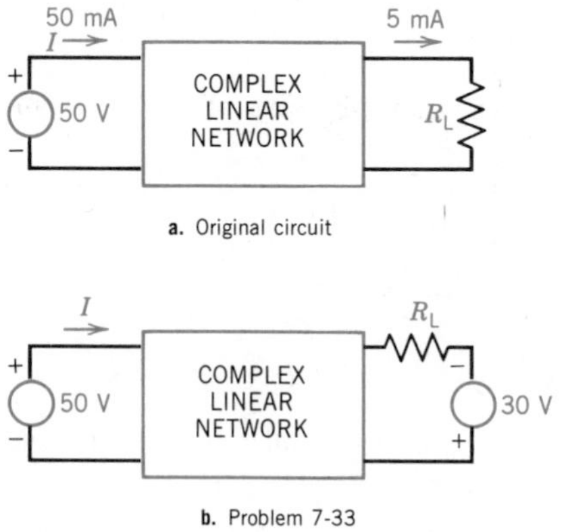

Figure 7-48 Problem 7-33. Combining reciprocity with superposition

7-34 Repeat Prob. 7-33 with the polarity of the 30-V source reversed.

7-35 Given the circuit of **Fig. 7-49a**, find

a. Total current drawn from the 28-V supply
b. Current in R_1, R_2, R_3, and R_4
c. Current I flowing from node **Y** to **X** through a short between **Y** and **X**

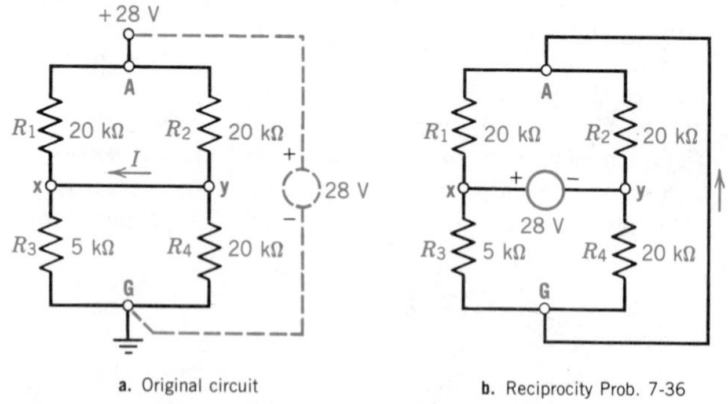

Figure 7-49 Reciprocity in a bridge network

7-36 In Fig. 7-49b, the supply voltage of Fig. 7-49a is interchanged with I calculated in Prob. 7-35c. Given the circuit of Fig. 7-49b, calculate

a. Total current drawn from the 28-V supply, based on your drawing of the circuit
b. Currents in R_1, R_3, R_2, and R_4, based on your drawing
c. Current I flowing from node **G** to **A**

Hint:

1. The short from **G** to **A** parallels R_1, R_3 and also R_2, R_4 in series combination.
2. Your calculated value of I in part (**c**) should agree with that of part (**c**) of Prob. 7-35. If not, recheck your method and calculations.
3. Reciprocity must prevail!

7-37 For the circuit shown in **Fig. 7-50**, calculate using Bartlett's theorem

a. Voltage at node **X**
b. Current in branch **X–G**
c. Draw the complete solution, showing all voltages and currents

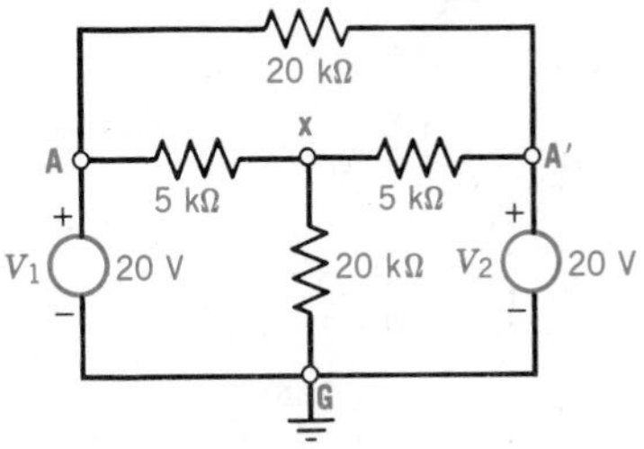

Figure 7-50 Problem 7-37

7-38 Given the circuit shown in **Fig. 7-51**, using Bartlett's theorem calculate

a. V_Y
b. V_X
c. I_{AX}
d. I_{YG}
e. Draw the complete solution, showing all currents and voltages.

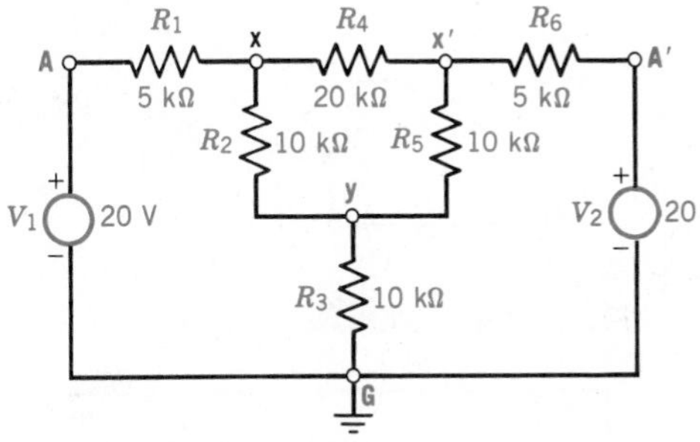

Figure 7-51 Problem 7-38

Sec. 7-9

7-39 For the circuit shown in **Fig. 7-52**:

a. List the positive and negative powers
b. Show that Tellegen's theorem is verified

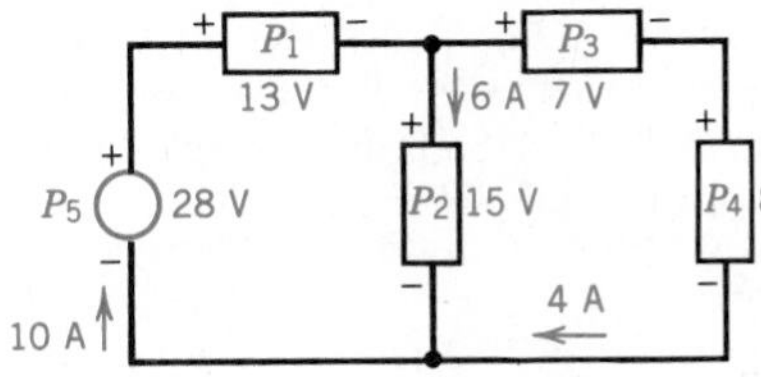

Figure 7-52 Problem 7-39

7-40 For the seven independent voltage and current sources shown in **Fig. 7-53**:

a. List the positive and negative powers
b. List the voltage sources that are being "charged" (i.e., *receiving* energy)
c. List the current sources that are *supplying* energy to the system
d. Show that Tellegen's theorem is verified

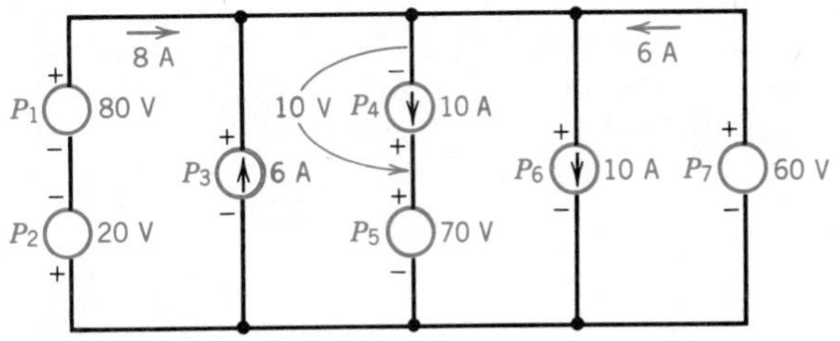

Figure 7-53 Problem 7-40

7-41 For the independent and dependent voltage sources and devices shown in **Fig. 7-54**:

a. List the positive and negative powers
b. Show that Tellegen's theorem is verified
c. Explain the nature of P_2 under the current/voltage conditions shown. Is it necessarily a source?

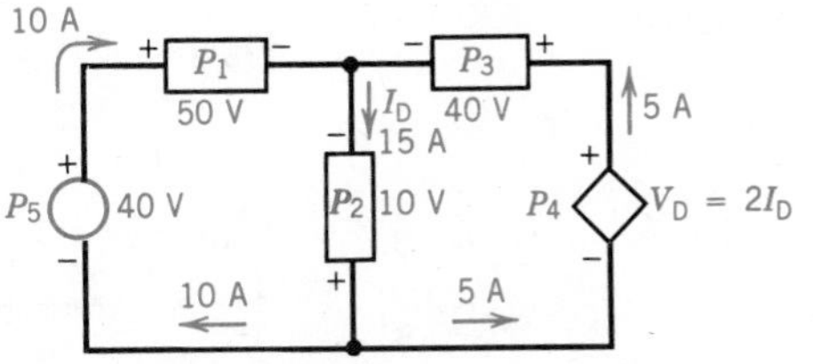

Figure 7-54 Problem 7-41

Problems 7-42 to 7-52

The following problems, shown in **Figs. 7-55** through **7-65**, represent a number of short **review** problems designed to enable the reader to develop facility in redrawing circuits as well as to test understanding and ability to apply the Thévenin, Norton, and Millman theorems appropriately.

For *each* of Probs. 7-42 through 7-52 shown in Figs. 7-55 through 7-65 calculate (**a**) the equivalent Thévenin voltage and (**b**) the equivalent Thévenin resistance, respectively, seen at terminals **T–H**.

The answers to Probs. 7-42 through 7-52 are tabulated below. The reader should test acquired mastery of circuit analysis techniques by using feedback from the answers to correct errors in solution techniques.

Problem No.	Figure No.	V_{TH} (V)	R_{TH} (Ω)
7-42	7-55	12	12
7-43	7-56	−72	28
7-44	7-57	15	20
7-45	7-58	−16	4
7-46	7-59	168	4
7-47	7-60	140	22
7-48	7-61	60	2
7-49	7-62	−10	90
7-50	7-63	207	27
7-51	7-64	92.31	1.846
7-52	7-65	33.6	2.4

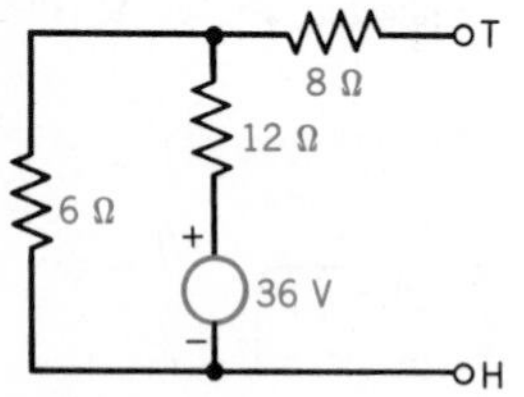

Figure 7-55 Problem 7-42

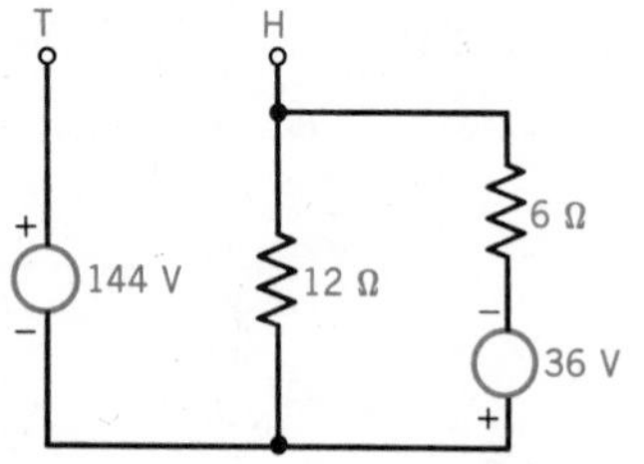

Figure 7-59 Problem 7-46

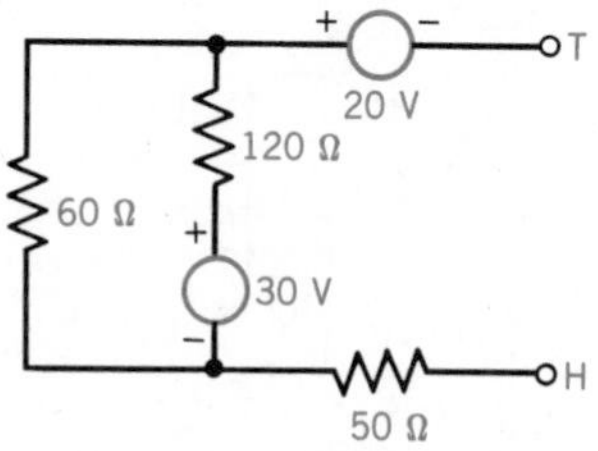

Figure 7-62 Problem 7-49

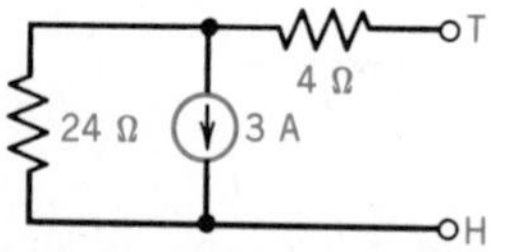

Figure 7-56 Problem 7-43

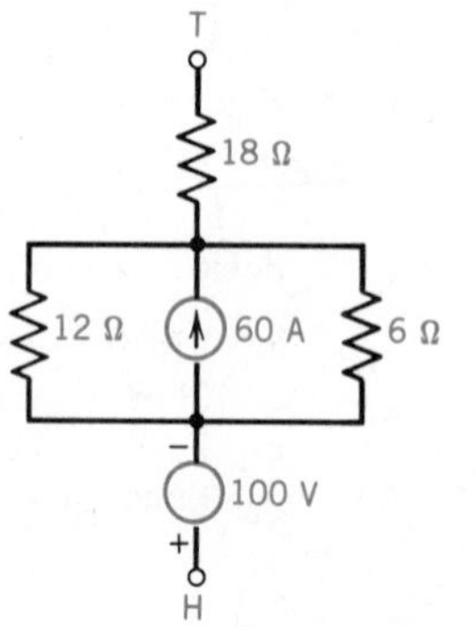

Figure 7-60 Problem 7-47

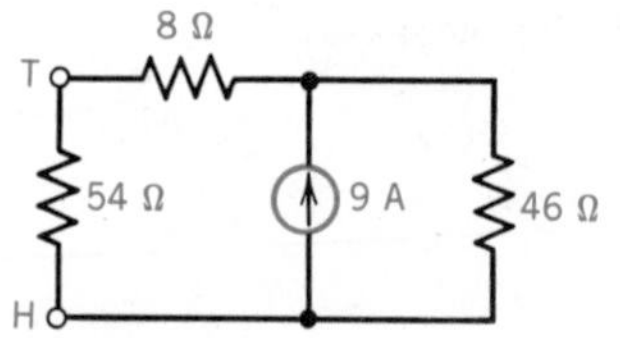

Figure 7-63 Problem 7-50

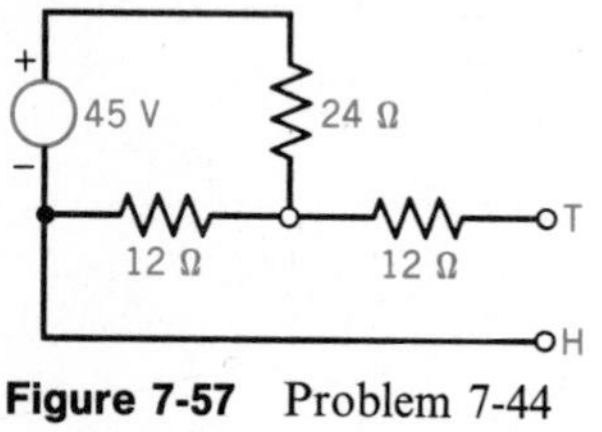

Figure 7-57 Problem 7-44

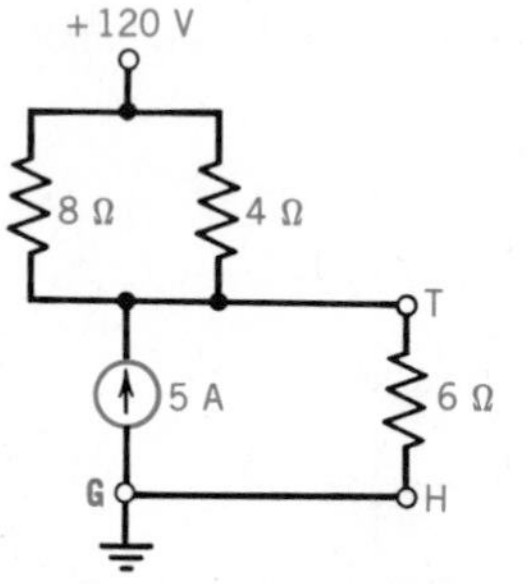

Figure 7-64 Problem 7-51

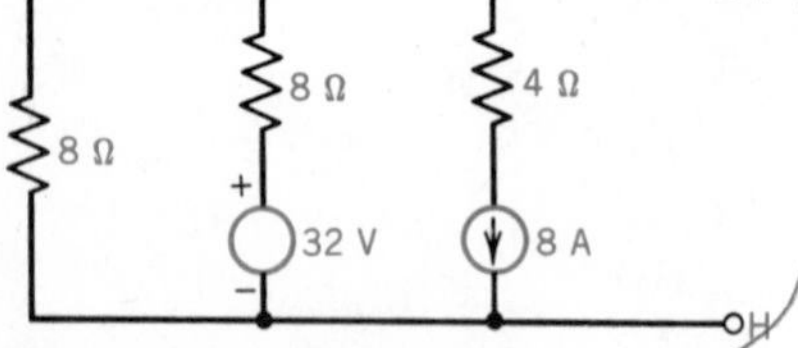

Figure 7-58 Problem 7-45

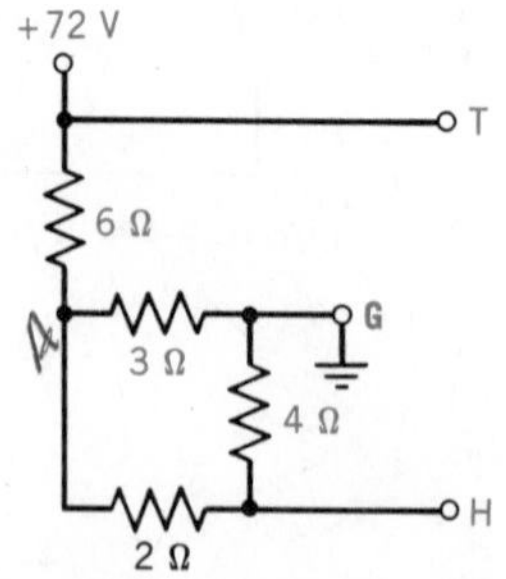

Figure 7-61 Problem 7-48

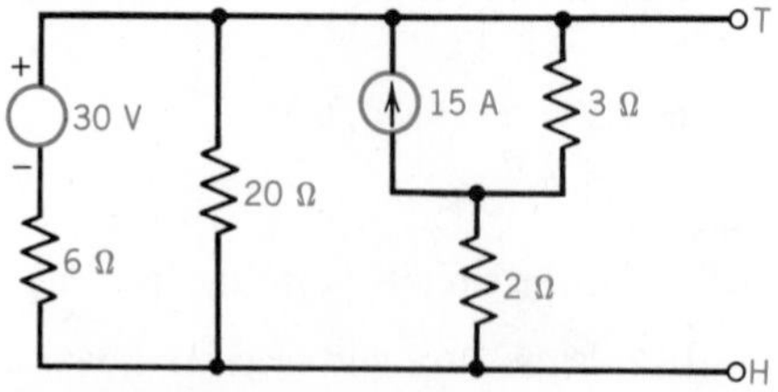

Figure 7-65 Problem 7-52

7-12 ANSWERS

7-1 a 6 V b 6 kΩ
7-2 a −36 V b 18 kΩ
7-3 a 10 V b 7.5 kΩ
7-4 a 12 V b 20 kΩ
7-5 a 20 V b 2 kΩ
7-6 a 10 V b 5 kΩ
7-7 a 200 V b 14 Ω
7-8 a 57.6 V b 3.5 kΩ
7-9 a 72 V b 0 Ω
7-10 a 20 V b 15 kΩ
7-11 a 2/3 Ω b 6 V d 6 V e 9 A f 2/3 Ω
7-12 a 9 A b 2/3 Ω c 6 V d 9 A e 2/3 Ω
7-13 a 6 V b 3 A c 2 A d 1 A e 18 W f 1 Ω g 20.25 W
7-14 a 2/3 V b 1/3 A c $4.\overline{6}$ A d $-4.\overline{3}$ A e $22\overline{2}$ mW f 1 Ω g 250 mW
7-15 a 0.8 V b 4.4 kΩ c 160 μA d 96 mV
7-16 a $0.1\overline{6}$ A b $3.\overline{3}$ V c $4.\overline{3}$ A d $0.1\overline{6}$ A↓
7-17 a 10 V b $0.41\overline{6}$ A c 5.75 A d −0.5 A e $4.1\overline{6}$ W f $2.\overline{1}\overline{8}$ Ω
7-18 a 4 mA b 4 V c 16 mW
7-19 a 1 mA b Y to X c 12 V d 10 V e $0.\overline{6}\overline{6}$ mA f 1.6 mA g $1.\overline{6}$ mA h 0.6 mA
7-21 a 1 mA b 20 V c 0
7-22 a 1 mA b 20 V c 0
7-24 a 2 mA b 2 mA
7-25 a 1.6 mA b 1.6 mA
7-27 a 1 mA
7-29 a 0 b 0
7-32 a $2.\overline{3}$ V b $2.28\overline{3}$ mA
7-33 53 mA
7-34 47 mA
7-35 a 2 mA b 1 mA, 1 mA, 1.6 mA, 0.4 mA c 0.6 mA
7-37 a $17.\overline{7}$ V b $0.\overline{8}$ mA
7-38 a 11.43 V b 17.14 V c 0.5714 mA d 1.143 mA
7-40 a $+P_2, P_5, P_6$ b same as **a**
7-41 a $+P_1, +P_3; -P_2, -P_4, -P_5$ **c** $-P_2$

CHAPTER 8

Circuit Analysis Using Network Equations

8-1 INTRODUCTION

Chapters 5, 6, and 7 stressed a solution of complex networks using circuit-reduction and circuit-simplification methods of analysis. The intent of all the methods discussed was to provide simplified circuit equivalents so that current, voltage, and power relations could be obtained for all branches in a complex network. Methods involving circuit-reduction techniques and various equivalence theorems were presented first because they provided a more solid foundation in circuit behavior.

The methods presented in this chapter are essentially *algebraic* techniques, in which simultaneous network equations are written for the various branches in a given network. The simultaneous equations are solved algebraically for unknown currents or voltages, using either determinants or matrix operation solutions of arrays on personal computers. Once the equations have been solved for unknown currents and voltages, they may be verified by circuit-reduction techniques previously presented.

The advantage of using algebraic methods of circuit analysis is that the *same* general approach may be used regardless of type of circuit or circuit complexity. The basic disadvantage of algebraic methods is that the simultaneous equations obtained are not directly or easily solvable and may require computer assistance, particularly in the case of four or more equations containing four or more unknowns. Prior to the advent of personal computers, this disadvantage was formidable. However, it is now of small consequence since it is anticipated that many of the readers either possess or have ready access to microcomputers or minicomputers.

The first algebraic method that we will consider uses a combination of Kirchhoff's voltage law (KVL) and Kirchhoff's current law (KCL). This is followed by two other algebraic methods, based on KVL and KCL as well, called *mesh analysis* and *nodal analysis.* In all these formal procedures, simultaneous equations are produced that can be represented by matrices, which, when solved, yield either the unknown currents or voltages.

The last algebraic technique we will present is the *delta–wye* conversion. This is a relatively simple algebraic way to produce an equivalent circuit for purposes of circuit reduction and/or simplification.

8-2 SOLVING FOR CURRENTS USING KVL AND KCL

This method may be used to solve any network, no matter how complex. It uses both KCL and KVL to set up sufficient equations representing the unknown currents in a network. (In most instances, values of resistances and practical voltage and/or current sources are known or are given.) The general procedure involves the following steps:

1. Draw a clear circuit diagram showing all the given values for the network.
2. Draw assumed current directions and assign currents in each branch of the network.
3. Insofar as possible, use KCL to reduce the number of unknown currents on your drawn circuit diagram. The fewer the unknown currents, the fewer the equations to be written and the simpler the solution.
4. For the assumed current directions (step 2) indicate voltage polarities across each element in the network drawn.
5. Write equations, using KVL, to record all voltage drops and rises in each (closed) loop. Insofar as possible, use the **sh**ortest loops (**mesh**es) to reduce terms in each equation. The number of equations required is the same as the unknown currents (step 3) in your circuit diagram. Only one equation is acceptable for any one loop.
6. Solve the simultaneous equations written for the unknown currents, using substitution, determinants, or a computer program.
7. If a current emerges as algebraically negative, the assumed direction of current must be reversed. **But** first it must be used as a negative value until **all** unknown currents, both positive and negative, have been found.
8. Correct the current directions for all negative currents on your diagram, if necessary, and enter all current values. Also correct voltage polarities (step 4) if current directions are reversed.
9. Check **each node** to see if KCL prevails. If not, check your work for error.
10. Check **each mesh or loop** to see if KVL prevails. If not, check for error.

These steps are best illustrated by an example.

EXAMPLE 8-1

For the circuit shown in **Fig. 8-1a** find all branch currents and voltages, using Kirchhoff's laws

Solution

The diagram in Fig. 8-1b shows assigned currents and their assumed directions. With respect to Fig. 8-1b please note the following:

1. Only *two* unknown currents are shown rather than three. (See step 3 above.) This reduces the number of equations required to two, rather than three, and simplifies the algebraic solution considerably.
2. Depending on the assumed direction of currents, polarities are assigned for voltage drops across each resistor.
3. Although three possible loops exist, we choose the mes**h**es (inner and **sh**ortest loops) or so-called windows of the network. These have the fewest terms, which also simplifies the equation reduction process. Using KVL we write the two mesh equations for each closed CW loop.

For the LH window the mesh Eq. 1 is:

$$-7.2 + 0.6(I_1 + I_2) + 2I_1 = 0 \qquad \textbf{(1)} \qquad (5\text{-}3)$$

For the RH window the mesh Eq. 2 is:

$$-2I_1 + 0.3I_2 + 6.3 = 0 \qquad \textbf{(2)} \qquad (5\text{-}3)$$

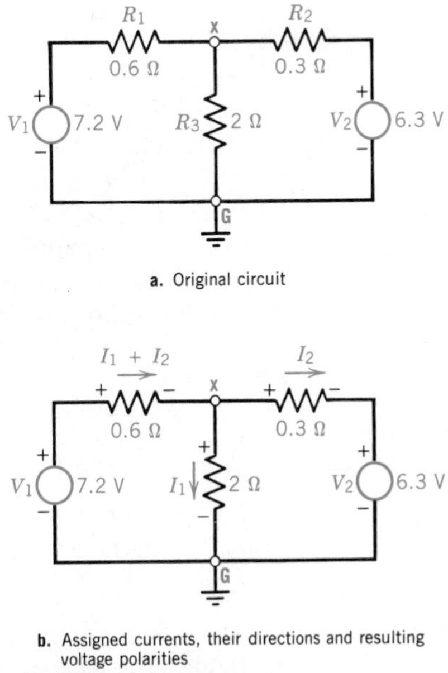

a. Original circuit

b. Assigned currents, their directions and resulting voltage polarities

Figure 8-1 Example 8-1

Collecting terms and transposing, we express each equation in simplest form as

$$2.6I_1 + 0.6I_2 = 7.2 \qquad \textbf{(1)}$$

$$2.0I_1 - 0.3I_2 = 6.3 \qquad \textbf{(2)}$$

This set of simultaneous equations may be solved by one of three methods:

a. The *substitution* method, shown below.
b. The *determinant* method, shown in Appendix B-2.1, and also shown in Ex. 8-3.
c. A computer program for solution of simultaneous equations.

Examination of Eq. (2) shows that if each term in the equation is multiplied by a factor of 2, the equations can be summed and the I_2 term eliminated. This yields

$$2.6I_1 + 0.6I_2 = 7.2 \qquad \textbf{(1)}$$

$$4.0I_1 - 0.6I_2 = 12.6 \qquad \textbf{(2)}$$

$$\text{Adding} \qquad 6.6I_1 = 19.8 \qquad \text{and } I_1 = \mathbf{3.0\ A}$$

We may now *substitute* $I_1 = 3$ A into either Eq. (1) or (2) to find I_2. Since Eq. (1) was untouched, let us substitute in Eq. (1):

$$2.6(3) + 0.6I_2 = 7.2, \text{ from which } I_2 = \frac{7.2 - 2.6(3)}{0.6} = \mathbf{-1\ A}$$

Observe that I_2 is a *negative* value, implying that the *assumed* direction of current was the reverse of what it should have been. NEVERTHELESS, we must use I_2 as a negative value (step 7 above) in finding the current $(I_1 + I_2)$ through the 0.6-Ω resistor in Fig. 8-1b.[1] This current is $I_1 + I_2 = 3 + (-1) = \mathbf{2\ A}$, which is the total current drawn from the supply, V_1.

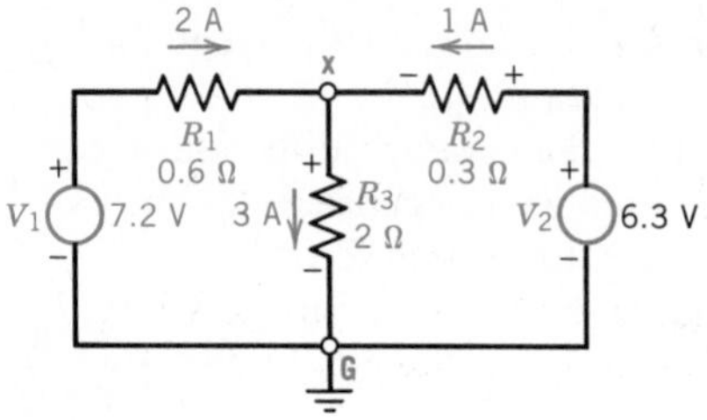

Figure 8-2 Complete solution of Ex. 8-1 for verification

Having found all currents, and their proper (not assumed) directions, we may proceed with preceding steps 8, 9, and 10 to verify our solution. We enter the currents on the original circuit of Fig. 8-1a to show **Fig. 8-2**. Note that at node **X**, Kirchhoff's current law (KCL) prevails. But to ensure that our solution is perfectly valid, KVL must also prevail, as shown in Ex. 8-2.

EXAMPLE 8-2

Verify the solution of Ex. 8-1 shown in Fig. 8-2 by solving for V_{XG} by three KVL methods

Solution

1. $V_{XG} = I_3R_3 = (3\ \text{A})(2\ \Omega) = \mathbf{6\ V}$
2. $V_{XG} = V_1 - I_1R_1 = 7.2 - (2\ \text{A})(0.6\ \Omega) = \mathbf{6\ V}$
3. $V_{XG} = V_2 - I_2R_2 = 6.3 - (1\ \text{A})(0.3\ \Omega) = \mathbf{6\ V}$

Now that we have verified the solution of Ex. 8-1 by using the substitution method to solve the two simultaneous linear equations containing two unknowns, let us turn to a *second* method of solving the same set of equations. The method of solution involves a procedure described in Appendix B-2 for second- and third-order determinants. Readers who are unfamiliar with the procedure should read Appendix B-2 carefully before turning to Ex. 8-3.

EXAMPLE 8-3

Given the following set of simultaneous linear equations, use the determinant method to find unknown currents I_1 and I_2:

$$2.6I_1 + 0.6I_2 = 7.2 \qquad \textbf{(1)}$$

$$2.0I_1 - 0.3I_2 = 6.3 \qquad \textbf{(2)}$$

Solution

Step 1. Write array $\mathbf{A} = \begin{vmatrix} 2.6 & 0.6 & 7.2 \\ 2.0 & -0.3 & 6.3 \end{vmatrix}$ (columns I_1, I_2, V)

[1] This technique of using the currents algebraically, *as found*, is inherent in **all** solutions using KVL and KCL, whenever currents have been combined algebraically to reduce the number of unknown currents.

Step 2. Solve for determinant **D**:

$$\mathbf{D} = \begin{vmatrix} 2.6 & 0.6 \\ 2.0 & -0.3 \end{vmatrix}$$

$$= (2.6 \times -0.3) - (2.0 \times 0.6) = \mathbf{-1.98}$$

Step 3. Solve for I_1, using appropriate terms from array **A**:

$$I_1 = \frac{\begin{vmatrix} 7.2 & 0.6 \\ 6.3 & -0.3 \end{vmatrix}}{\mathbf{D}} = \frac{(7.2 \times -0.3) - (6.3 \times 0.6)}{-1.98}$$

$$= \frac{-5.94}{-1.98} = \mathbf{3\ A}$$

Step 4. Solve for I_2, using appropriate terms from array **A**:

$$I_2 = \frac{\begin{vmatrix} 2.6 & 7.2 \\ 2.0 & 6.3 \end{vmatrix}}{\mathbf{D}} = \frac{(2.6 \times 6.3) - (2.0 \times 7.2)}{-1.98} = \frac{1.98}{-1.98}$$

$$= \mathbf{-1\ A}$$

With respect to the solution of Ex. 8-3, the reader should note the following regarding the solution of simultaneous linear equations by determinant methods:

1. Determinant **D** is the denominator array in the expression for solution of all unknown quantities (I_1 and I_2 in this case).
2. Great care should be exercised, therefore, in solving for determinant **D**. An error in the value of **D** produces an error in all unknown variables. The calculations of step 2 should be checked at least twice before going on (see Sec. 8-9).
3. The values of unknown currents (or voltages) found by this method are the *same* as those found by other methods (substitution, computer programs, etc.). A **negative** current, therefore, implies that the **assumed** direction was **incorrect** and that the (true) current is actually in the **opposite** direction.
4. Nevertheless, if certain branch currents appear on the original circuit diagram in terms of other currents (either their sum or difference), the **negative** current found must be used as a **negative** value (in its original form) to find the actual current and direction of these branch currents. Only when **all** currents have been found algebraically can the directions on circuit or network diagrams be modified, as shown in Exs. 8-1 and 8-2 and described in step 7 in Sec. 8-2.

Let us now turn to an example involving a more complex network to write equations using the KVL/KCL method and solve these equations for unknowns by using determinants.

EXAMPLE 8-4

Given the original circuit shown in **Fig. 8-3a**, find all unknown branch currents and voltages by the KVL and KCL methods

Solution

Since the potential of V_1 is much higher than V_2 or V_3, we assume that source V_1 is charging sources V_2 and V_3. Therefore V_1 is assumed to be delivering current to node **X** while V_2 and V_3 are receiving current from node **X.** (Fig. 8-3b).

Step 1. Only two unknown currents are shown in Fig. 8-3b (rather than three) to reduce the number of unknown currents and simplify the solution. By KCL, $I_1 = I_2 + I_3$

Step 2. For the assumed current directions, polarities of voltage drop across all resistors are shown in Fig. 8-3b.

Step 3. Using KVL, write equations for the two meshes (shortest possible loops) in Fig. 8-3b. For the LH mesh we write

$$-20 + 2(I_2 + I_3) + 2I_2 + 8 = 0 \quad \text{or} \quad 4I_2 + 2I_3 = 12 \ \mathbf{(1)}$$

For the RH mesh, we also write

$$-8 - 2I_2 + 1I_3 + 6 = 0 \quad \text{or} \quad -2I_2 + 1I_3 = 2 \ \mathbf{(2)}$$

Step 4. Write the arrays for simultaneous Eqs. (1) and (2) above, **A** and **D**, respectively:

$$\mathbf{A} = \begin{vmatrix} 4 & 2 & 12 \\ -2 & 1 & 2 \end{vmatrix} \quad (I_2 \;\; I_3 \;\; V) \qquad \mathbf{D} = \begin{vmatrix} 4 & 2 \\ -2 & 1 \end{vmatrix} = 8$$

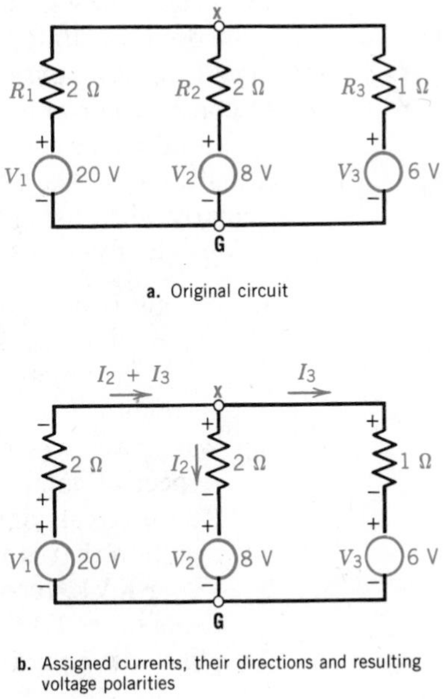

Figure 8-3 Example 8-4

Step 5. Find I_2 and I_3 by the method of determinants:

$$I_2 = \frac{\begin{vmatrix} 12 & 2 \\ 2 & 1 \end{vmatrix}}{\mathbf{D}} = \frac{12 - 4}{8} = \mathbf{1\ A}$$

$$I_3 = \frac{\begin{vmatrix} 4 & 12 \\ -2 & 2 \end{vmatrix}}{\mathbf{D}} = \frac{8 + 24}{8} = \mathbf{4\ A}$$

Step 6. Since both currents above are positive values, the assumed directions of current are correct. Then the current delivered by V_1 to node X is $I_2 + I_3$ or 1 A + 4 A = 5 A

Step 7. VERIFICATION: solve for V_x by three methods:

$$V_x = V_1 - I_1R_1 = 20\ \text{V} - (5\ \text{A})(2\ \Omega) = \mathbf{10\ V}$$

$$V_x = V_2 + I_2R_2 = 8\ \text{V} + (1\ \text{A})(2\ \Omega) = \mathbf{10\ V}$$

$$V_x = V_3 + I_3R_3 = 6\ \text{V} + (4\ \text{A})(1\ \Omega) = \mathbf{10\ V}$$

With regard to the solution of Ex. 8-4, the reader should note:

a. Without step 7, verification, the entire procedure is an exercise in futility. As noted earlier, an error in the determinant **D** yields erroneous values for I_2 and I_3 as well as their sum, $I_2 + I_3$. Only through previously established circuit analysis methods are the currents verified.[2]

b. The determinant method of solving simultaneous linear equations is remarkably quick and efficient BUT the equations must be *properly* written, using KCL and KVL, respectively. Given incorrect equations, the determinant method will solve those equations correctly for their unknowns. Again, the KVL verification procedure of step 7 will show whether any of the previous procedures has been performed incorrectly.

8-3 SOLVING FOR CURRENTS USING MESH ANALYSIS

In Chapter 5 we defined a loop as a closed path in a network. We also defined a *mesh* as the *sh*ortest possible loop in a network.[3] Mesh analysis is the second technique to be used in finding the unknown currents in a network by algebraic methods. We will begin by presenting a generalized mesh analysis technique and verify that it works. Then we will present a so-called **format** method of mesh analysis that enables the mesh equations to be written (almost) *by inspection.*

Before presenting the technique, however, we should note that one of the advantages of mesh analysis over the KVL/KCL method of Sec. 8-2 is that it is unnecessary to combine currents or show a branch as the sum (or difference) of two currents. A second advantage is that all mesh currents are assigned in the *same* direction (usually clockwise, CW). This makes the equation writing somewhat simpler, particularly when using the format method of mesh analysis (Sec. 8-4).

Mesh analysis is accomplished in the following steps:

1. Assign as many CW mesh currents as there are "windows" in the network, one mesh current for each window.
2. Show the voltage polarities across each resistor produced by the (CW) mesh currents.
3. Whenever two (mutual) mesh currents appear in a single resistor, show the respective polarities due to each on each side of the resistor.
4. Resistors carrying mutual currents require two voltage drops in each KVL equation that is written.
5. Using KVL, record one equation representing the volt drops for each mesh, taken in the CW direction. (Using the same direction as the mesh current simplifies the equation writing as well.)

---

[2] This is why Chapter 7 on network theorems precedes Chapter 8 on network equation circuit analysis techniques.

[3] Thus, every mesh is a loop but not all loops are meshes. Also note the letters "sh" in mesh help define it as the *sh*ortest possible loop.

6. As in previous KVL usage, the polarity encountered on *entering* a voltage source is the polarity assigned to the voltage rise or drop.
7. Simplify equations and solve for mesh currents, using simultaneous equation algebraic techniques.
8. Currents through elements common to two windows carrying mutual currents are the difference between the two mesh currents.
9. Currents through elements carrying only one mesh current are the true currents in that element.

These steps are best illustrated by the following example.

EXAMPLE 8-5

For the original circuit shown in **Fig. 8-4a**, find the complete solution showing all currents and voltages by mesh analysis

Solution

The solution to this example follows the steps outlined above.

a. CW mesh currents I_a and I_b are assigned to the two windows of the original network, as shown in Fig. 8-4b.
b. Depending on the "sense" of the current, voltage polarities are assigned to each resistor. Note that only the 2-Ω resistor carrying both I_a and I_b exhibits *two* polarities (step 3 above).
c. KVL for mesh I_a is:

$$-7.2 + 0.3I_a + (2I_a - 2I_b) + 0.3I_a = 0 \qquad \textbf{(1)}$$

KVL for mesh I_b is:

$$+6.3 + 0.1I_b + (2I_b - 2I_a) + 0.2I_b = 0 \qquad \textbf{(2)}$$

d. Simplifying Eqs. (1) and (2) yields

$$2.6I_a - 2I_b = 7.2 \qquad \textbf{(1)}$$

$$-2I_a + 2.3I_b = -6.3 \qquad \textbf{(2)}$$

and the array, **A**, is $\mathbf{A} = \begin{vmatrix} I_a & I_b & V \\ 2.6 & -2 & 7.2 \\ -2 & 2.3 & -6.3 \end{vmatrix}$.

e. From the array **A**, the value of determinant **D** is[4]

$$\mathbf{D} = \begin{vmatrix} 2.6 & -2 \\ -2 & 2.3 \end{vmatrix} = 5.98 - 4 = 1.98$$

$$I_a = \frac{\begin{vmatrix} 7.2 & -2 \\ -6.3 & 2.3 \end{vmatrix}}{\mathbf{D}} = \frac{3.96}{1.98} = \mathbf{2\ A}$$

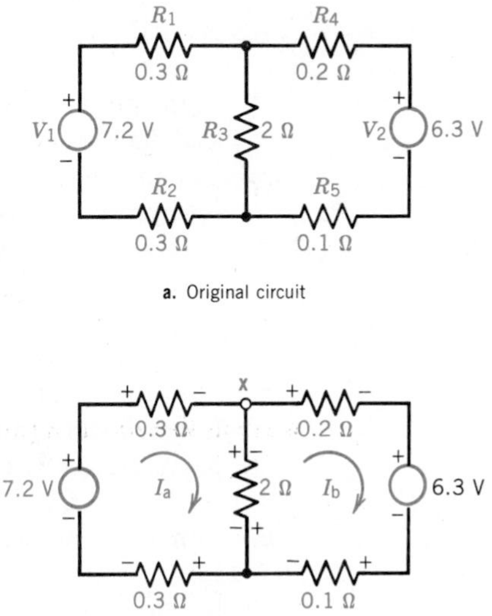

a. Original circuit

b. Assigned mesh currents, their direction and resulting voltage polarities

Figure 8-4 Example 8-5

$$I_b = \frac{\begin{vmatrix} 2.6 & 7.2 \\ -2 & -6.3 \end{vmatrix}}{\mathbf{D}} = \frac{-1.98}{1.98} = \mathbf{-1\ A}$$

f. $I_{R_3} = I_a - I_b = 2\text{ A} - (-1\text{ A}) = \mathbf{3\ A}$ (step 8; see Fig. 8-4b). Since I_a is positive, it is the **true** current in R_1 and R_2 (Fig. 8-4a). Since I_b is negative, the true current in R_4 and R_5 is 1 A in the counterclockwise (CCW) direction.

The complete solution of Ex. 8-5 is shown in Fig. 8-2. (The reader may have noticed that Exs. 8-1 and 8-5 are solving the same network.)

---

[4] In simple **mesh** analysis, the minor terms (minor axis diagonals) of the **determinant** are ALWAYS the SAME. In this case, both the minor terms are (−2). This is one test of the correctness of the **mesh** equations. This is known as *minor axis* symmetry or symmetry about the forcing function axis. This relation does *not* hold for networks containing dependent sources.

8-4 SOLVING FOR NETWORK CURRENTS USING THE FORMAT METHOD OF MESH ANALYSIS

The so-called **format** method enables the reader (literally) to write mesh equations (almost) by inspection. Because it is relatively simpler than the general method of Sec. 8-3, there is *much less* possibility for error. Indeed, the format method of writing mesh equations is probably the most extensively used method of circuit analysis, along with the format method of nodal analysis (Sec. 8-6).

The format method is based on observations regarding the nature of equations resulting from the applications of mesh analysis (Sec. 8-3) plus KVL and KCL to a variety of dc and ac networks. As a consequence, certain "rules" emerged. These rules apply to *each mesh* current for which an equation is written:

1. Sum the total resistance of all resistors encountered by the assigned mesh current in each window. Call this total the *self*-resistance or *common* resistance of the mesh. This represents the total resistance "seen by" or "common to" the assigned mesh current.
2. The product of the given assigned mesh current and the total self-resistance is *always* written as a *positive* volt drop in each mesh equation.
3. Sum the total resistance of all resistors that are *mutually* shared with other mesh currents. Call this total the *mutual* resistance of the mesh.
4. The product of any other assigned mesh current and its total mutual resistance is *always* a *negative* volt drop in each mesh equation.
5. Sum the voltages (forcing functions) in each loop. These will now appear on the right-hand (RH) side of the equation. Consequently, the polarity **LAST** encountered by the mesh current as it LEAVES the source is now the polarity of the voltage.
6. Summarizing, we may write the above procedures in equation form as

$$+I_s(\sum R_s) - I_m(\sum R_m) = \pm\sum V \qquad \text{volts (V)} \qquad \textbf{(8-1)}$$

where I is the particular mesh current assigned

$+\sum R_s$ is the sum of the common resistances encountered by the assigned mesh current in the mesh

$-\sum R_m$ is the sum of the mutual resistances shared with other mesh currents in the mesh

$\pm\sum V$ is the sum of the voltages encountered in the mesh

The format method of mesh analysis is best illustrated by the following examples.

EXAMPLE 8-6

Write the mesh equations for Fig. 8-4b, using the format method of mesh analysis. Compare your results with Ex. 8-5 equations, step (**d**)

Solution

For mesh current, I_a: $R_s = 2.6\ \Omega$, $R_m = 2\ \Omega$ (with I_b) and

$$V = +7.2\ \text{V, then } +2.6I_a - 2I_b = 7.2 \qquad \textbf{(1)}$$

For mesh current, I_b: $R_s = 2.3\ \Omega$, $R_m = 2\ \Omega$ (with I_a) and

$$V = -6.3\ \text{V, then } -2I_a + 2.3I_b = -6.3 \qquad \textbf{(2)}$$

Note that these mesh equations are EXACTLY the same as Eqs. (**1**) and (**2**) in Ex. 8-5d!

With respect to Ex. 8-6, please note:

1. The two mesh equations were written instantly, by inspection.
2. The minor term, R_m, is the same in both equations (-2, in this case), this symmetry of minor axis diagonals represents a *check* when the **determinant** is written. (See footnote 4.)
3. Now that the forcing functions are shown directly on the RH side of the equations, we must reverse our rule regarding source polarity (rule 5, Sec. 8-4). The polarity *last* "seen" by the *mesh current* is the "sign" of the voltage (as opposed to the polarity *first* encountered when using KVL).

EXAMPLE 8-7

Given the circuit shown in **Fig. 8-5a**, using the format method, find the

a. Simplest set of equations to find I_a and I_b
b. Array of coefficients and solve for I_a and I_b, using determinants
c. True branch currents and their true directions (I_1, I_2, and I_3)
d. V_X by three methods to check on the validity of the solutions

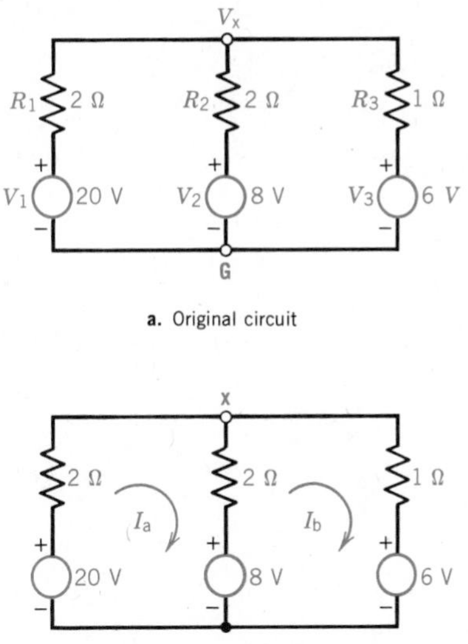

a. Original circuit

b. Assigned mesh currents

Figure 8-5 Example 8-7; format method of mesh analysis

Solution

a. For mesh I_a: $4I_a - 2I_b = 12$ **(1)**
and for mesh I_b: $-2I_a + 3I_b = 2$ **(2)**

b. $$\mathbf{A} = \begin{vmatrix} I_a & I_b & V \\ 4 & -2 & 12 \\ -2 & 3 & 2 \end{vmatrix} \quad \text{and} \quad \mathbf{D} = \begin{vmatrix} 4 & -2 \\ -2 & 3 \end{vmatrix} = 8$$

$$I_a = \frac{\begin{vmatrix} 12 & -2 \\ -2 & +3 \end{vmatrix}}{\mathbf{D}} = \frac{40}{8} = \mathbf{5\ A}, \quad I_b = \frac{\begin{vmatrix} 4 & 12 \\ -2 & 2 \end{vmatrix}}{\mathbf{D}} = \frac{32}{8} = \mathbf{4\ A}$$

Then $I_a - I_b = I_2 = 5 - 4 = \mathbf{1\ A}$

c. $I_1 = I_a = \uparrow 5$ A, $I_2 = \downarrow 1$ A, and $I_3 = I_b = \downarrow 4$ A (to be entered on Fig. 8-5a)

d. $V_X = V_1 - I_1R_1 = 20\text{ V} - (5\text{ A} \times 2\ \Omega) = \mathbf{10\ V}$
$V_X = V_2 + I_2R_2 = 8\text{ V} + (1\text{ A} \times 2\ \Omega) = \mathbf{10\ V}$
$V_X = V_3 + I_3R_3 = 6\text{ V} + (4\text{ A} \times 1\ \Omega) = \mathbf{10\ V}$

With reference to Ex. 8-7, please note:

1. Those branches containing elements carrying **only one** mesh current are "true" currents. I_a and I_b are true currents as noted in the solution of part (**c**) in their assumed (CW) direction in their respective branches.
2. Since I_a is **greater** than I_b in R_2 and I_a is a "downward" current, then $I_2 = I_a - I_b$ must be a **downward** current. In short, the prevailing larger mesh current determines the true current direction for branches containing two opposing mesh currents. BUT if the prevailing larger current is negative, its direction must be reversed!
3. Once more the symmetry of minor axis (upward) diagonals serves as a check on the equation-writing procedure.
4. The use of determinants simplifies solution of the (two) simultaneous equations that were written by inspection.

Up to now, all of the sources used in examples were voltage sources. Since we are solving for unknown (mesh) currents in using **mesh analysis**, the forcing functions (on the RH side of the simultaneous mesh equations) must be **voltages**. How shall we treat current sources whenever they appear in a complex network? The answer is quite simple. We use the source conversion theorem (Sec. 7-3) to convert them to voltage sources, as shown in Ex. 8-8.

EXAMPLE 8-8

Given the circuit originally shown in **Fig. 8-6a**, use mesh analysis to calculate

a. True current in resistor R_L and its direction with respect to nodes **X** and **Y**
b. Voltage at point **X** with respect to ground **G**
c. Voltage at point **Y** with respect to ground **G**
d. Show the complete solution for Fig. 8-6a and verify KCL around nodes **X** and **Y**

Solution

a. Current sources I_1 and I_2 are converted to practical voltage sources as shown in Fig. 8-6b, leaving two windows, in which we assign CW mesh currents I_a and I_b. By inspection, using the format method, we write
for mesh I_a: $+20I_a - 10I_b = -100$
for mesh I_b: $-10I_a + 30I_b = 112$

and the array $$\mathbf{A} = \begin{vmatrix} I_a & I_b & V \\ 20 & -10 & -100 \\ -10 & +30 & 112 \end{vmatrix}$$

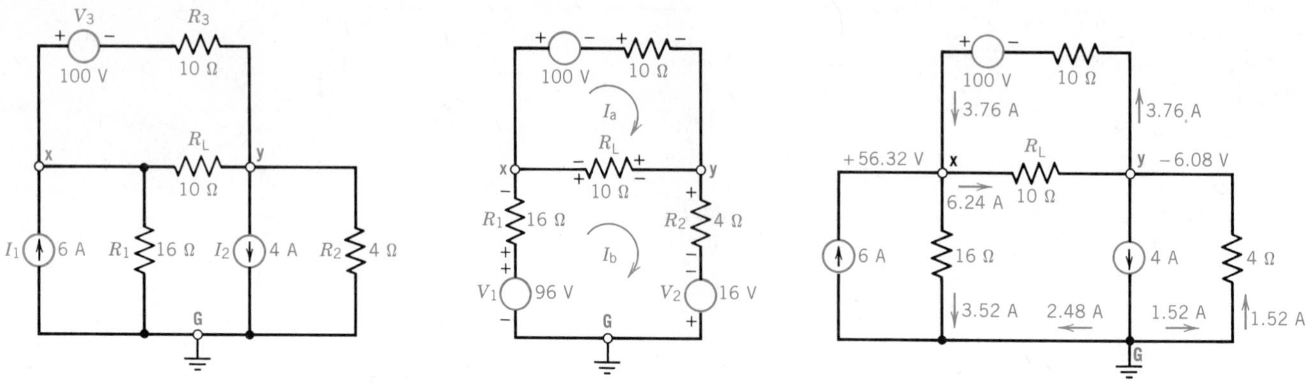

a. Original circuit

b. Source conversions and assigned mesh currents of equivalent circuit

c. Complete solution

Figure 8-6 Example 8-8

$$\text{Then } \mathbf{D} = \begin{vmatrix} 20 & -10 \\ -10 & 30 \end{vmatrix} = 500;$$

$$I_a = \frac{\begin{vmatrix} -100 & -10 \\ 112 & 30 \end{vmatrix}}{\mathbf{D}} = \frac{-1800}{500} = \mathbf{-3.76\ A}$$

$$I_b = \frac{\begin{vmatrix} 20 & -100 \\ -10 & 112 \end{vmatrix}}{\mathbf{D}} = \frac{1240}{500} = \mathbf{2.48\ A};$$

$$I_a - I_b = -3.76 - 2.48 = \overleftarrow{-6.24\ \text{A}} = \overset{\mathbf{x \longrightarrow y}}{\mathbf{6.24\ A}}$$

b. $V_x = V_1 - I_b R_1 = 96 - (2.48 \times 16) = \mathbf{56.32\ V}$
c. $V_y = I_b R_2 - V_2 = (2.48\ \text{A} \times 4\ \Omega) - 16\ \text{V} = \mathbf{-6.08\ V}$
d. The complete solution is shown in Fig. 8-6c. The currents *entering* a node are *positive* while those *leaving* are *negative*, as shown in Eq. (5-29a) for KCL.
Around node **X**: 3.76 A + 6 A − 3.52 A − 6.24 A = 0
Around node **Y**: 6.24 A + 1.52 A − 4 A − 3.76 A = 0

With respect to Ex. 8-8, please note:

1. In finding the current in R_L, since I_a is greater than I_b, I_a predominates in Fig. 8-6b. But $I_a - I_b = -3.76 - 2.48 = -6.24$ A in the direction of I_a in Fig. 8-6b. Therefore the "true" direction of current in R_L is **6.24 A** from node **X** to node **Y**, as shown in Fig. 8-6c.
2. Since I_b emerged positive in the solution for I_b, it represents the "true" current in voltage sources V_1 and V_2 in Fig. 8-6b. This enabled the calculations for V_x and V_y in parts (**b**) and (**c**) of the solution. The calculation of V_x is fairly straightforward since current I_b leaves source V_1 and $V_x = V_1 - I_b R_1$ as shown in part (**b**).
3. The calculation for V_y is more complex and requires some explanation. Since I_b is also a true current in R_2, the voltage drop across R_2 (created by I_b) *opposes* voltage V_2 as shown in Fig. 8-6b. Therefore V_y is $I_b R_2 - V_2$ as shown in part (**c**) of the solution. V_y is NEGATIVE with respect to ground or **−6.08 V**.
4. As further proof of the accuracy and validity of the solution, consider the voltage across R_L, which is $V_x - V_y = 56.32\ \text{V} - (-6.08\ \text{V}) = 62.4$ V. Then by Ohm's law, $I_L = V_L/R_L = (V_x - V_y)/R_L = 62.4\ \text{V}/10\ \Omega = 6.24$ A. This is the same as the value obtained using $I_a - I_b$ in part (**a**) of the solution and that shown on the complete solution of Fig. 8-6c.

We have yet to solve a problem involving *three* simultaneous equations, using determinants to implement the solution for the three unknowns. Example 8-9 poses such a problem.

EXAMPLE 8-9

Using the format method of mesh analysis for **Fig. 8-7a**:

a. Write three mesh equations by inspection
b. Find the current in the 3-Ω resistor R_3
c. Find the current delivered to node **x** by V_1
d. Find the current delivered to node **y** by V_2
e. Calculate V_x and V_y, respectively
f. Verify the solution to (**b**) using V_x and V_y
g. Show the complete solution for Fig. 8-7a

Solution

a. The assigned mesh currents are shown in Fig. 8-7b. By inspection we write
for mesh I_a: $17I_a - 15I_b + 0 = 23$
for mesh I_b: $-15I_a + 19.2I_b - 1.2I_c = 0$
for mesh I_c: $0 - 1.2I_b + 4.2I_c = -12$

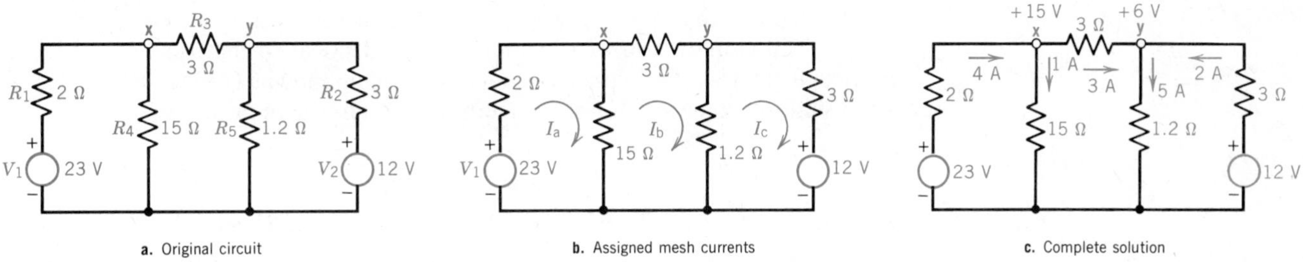

a. Original circuit b. Assigned mesh currents c. Complete solution

Figure 8-7 Example 8-9

$$\text{array } \mathbf{A} = \begin{array}{c} \\ \\ \\ \\ \end{array}\begin{vmatrix} I_a & I_b & I_c & V \\ 17 & -15 & 0 & 23 \\ -15 & 19.2 & -1.2 & 0 \\ 0 & -1.2 & +4.2 & -12 \end{vmatrix}$$

$$\text{Then determinant } \mathbf{D} = \begin{vmatrix} 17 & -15 & 0 \\ -15 & 19.2 & -1.2 \\ 0 & -1.2 & 4.2 \end{vmatrix}$$

$$= 1370.88 - (24.48 + 945) = \mathbf{401.4}$$

b. From array **A**:

$$I_b = \frac{\begin{vmatrix} 17 & 23 & 0 \\ -15 & 0 & -1.2 \\ 0 & -12 & 4.2 \end{vmatrix}}{\mathbf{D}} = \frac{0 - (244.8 - 1449)}{401.4} = \mathbf{3\ A}$$

c. From array **A**:

$$I_a = \frac{\begin{vmatrix} 23 & -15 & 0 \\ 0 & 19.2 & -1.2 \\ -12 & -1.2 & 4.2 \end{vmatrix}}{\mathbf{D}} = \frac{1638.72 - 33.12}{401.4} = \mathbf{4\ A}$$

d. From array **A**:

$$I_c = \frac{\begin{vmatrix} 17 & -15 & 23 \\ -15 & 19.2 & 0 \\ 0 & -1.2 & -12 \end{vmatrix}}{\mathbf{D}} = \frac{(-3916.8 + 414) - (-2700)}{401.4}$$

$$= \frac{-802.8}{401.4} = \mathbf{-2\ A}$$

Then the current *delivered to* node **y** by V_2 is **2 A** (the reversal of I_c)

e. $V_x = V_1 - I_a R_1 = 23 - (4 \times 2) = \mathbf{15\ V}$
$V_y = V_2 - I_2 R_2 = 12 - (2 \times 3) = \mathbf{6\ V}$

f. $I_3 = (V_x - V_y)/R_3 = (15 - 6)\ \text{V}/3\ \Omega = \mathbf{3\ A}$ (same answer as in part **b**, above)

g. The complete solution is shown in Fig. 8-7c

With respect to the solution of Ex. 8-9, please note:

1. The use of determinants to solve the three equations written in part (**a**) of the solution is both expedient and elegant. (See Appendix B-2.)
2. The writing of the three equations in part (**a**) by direct *inspection* (using the format method) is also both expedient and elegant.
3. The mesh written for I_b in part (**a**) contains no voltage source. This leaves a forcing function of zero on the RH side of the second equation.
4. Since I_b is the true current in the 3-Ω resistor between nodes **x–y**, $I_b = I_3 = \mathbf{3\ A}$. Since I_b is positive, this is the *true* direction of current in R_3.
5. Observe the matrix for determinant **D**. Note that each of the three minor upward (negative) diagonals contains two terms that are *identical*, that is, 0, −15, and −1.2. This serves as a check on the equation-writing procedure. In the event that this minor axis symmetry is *not* present, it would be necessary to recheck the equations written by inspection to find the error. (Also see item 7 below.)
6. The number of equations required equaled the number of windows, which in turn equaled the number of mesh currents written.
7. All *mutual* terms in the array are *negative* values (unless zero).
8. One final insight remains. If the vertical *columns* are repeated in the determinant array **D**, an interesting pattern emerges that serves as a *further* check:

17	−15	0	
−15	19.2	−1.2	
0	−1.2	4.2	0
17	−15	0	17
−15	19.2	−1.2	−15
0	−1.2	4.2	0
	−15	0	
		−1.2	

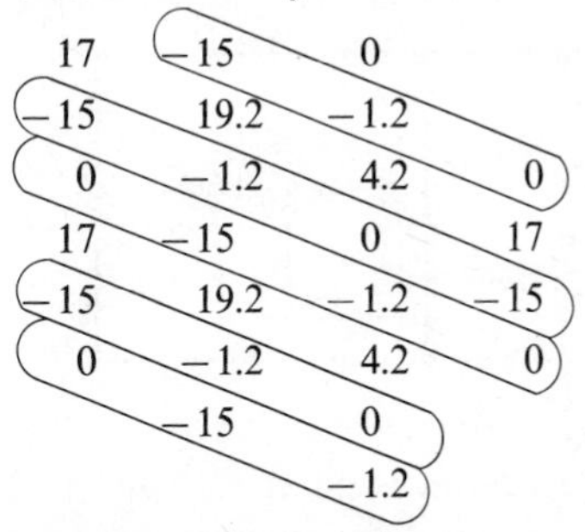

The POSITIVE major axis *diagonals*, beginning with column 1 row 2 and also column 1 row 3, are each IDENTICAL and contain the SAME coefficients (−15, −1.2, 0, −15, −1.2, 0, etc.). Note that the same pattern is repeated again in column 1 rows 5 and 6. This pattern also serves as a final check as to whether the **MESH** equations have been properly written.

9. There is no point in solving for mesh currents unless the array pattern is correct. Therefore, a little time spent in verifying patterns of major downward (positive) diagonals and minor upward (negative) diagonals in the determinant *is well worth the effort*. (See Sec. 8-8, steps 2 and 3.)

We will find that the alternative circuit analysis technique known as *nodal analysis* will employ a format method that is identical to that used in the mesh analysis format method. Since it also solves for unknowns (voltages in this case) using simultaneous equations, the method of determinants will also apply.

8-5 SOLVING FOR VOLTAGES USING NODAL ANALYSIS

In mesh analysis, as we have seen, equations are written using **KVL** to solve for unknown mesh **currents.**

In nodal analysis, equations are written using **KCL** to solve for unknown node **voltages**. In effect, therefore, nodal analysis is the **dual** of mesh analysis.

The general steps for performing nodal analysis are

1. Convert all practical voltage sources to *practical current* sources.
2. Select one node as the reference node: either G (ground) or V_r (reference voltage).
3. Locate all remaining nodes and label them V_1, V_2, V_3, etc.
4. Write KCL equations for the (negative) currents leaving and (positive) currents entering *each* node.
5. Solve the above simultaneous equations for node voltages.

EXAMPLE 8-10

Given the circuit shown in **Fig. 8-8a**, find voltage V by nodal analysis and current in load resistance R_L

Solution

Using the foregoing steps, we convert Fig. 8-8a to 8-8b. Obviously, we have a simple parallel circuit with only two nodes: reference node **G** and node **V**. Since there is only one unknown voltage, only one equation is needed. We encountered this type of circuit previously in discussing Millman's theorem, for which the basic Eq. (7-5) is $V = I_t/G_t$. In nodal form we write for Fig. 8-8b

$$VG_t = V\left(\frac{1}{R_1} + \frac{1}{R_2} + \frac{1}{R_3} + \frac{1}{R_4} + \frac{1}{R_L}\right) = (I_1 + I_4 - I_2 - I_3)$$

Substituting values given in Fig. 8-8a, we write

$$V\left(\frac{1}{2\text{ k}} + \frac{1}{4\text{ k}} + \frac{1}{5\text{ k}} + \frac{1}{20\text{ k}} + \frac{1}{1\text{ k}}\right) = (5 + 1 - 2 - 2)\text{ mA}$$

$$V \times 2\text{ mS} = 2\text{ mA} \qquad \text{and} \qquad V = 2\text{ mA}/2\text{ mS} = \mathbf{1\ V}$$

and $I_L = V/R_L = 1\text{ V}/1\text{ k}\Omega = \mathbf{1\ mA}$

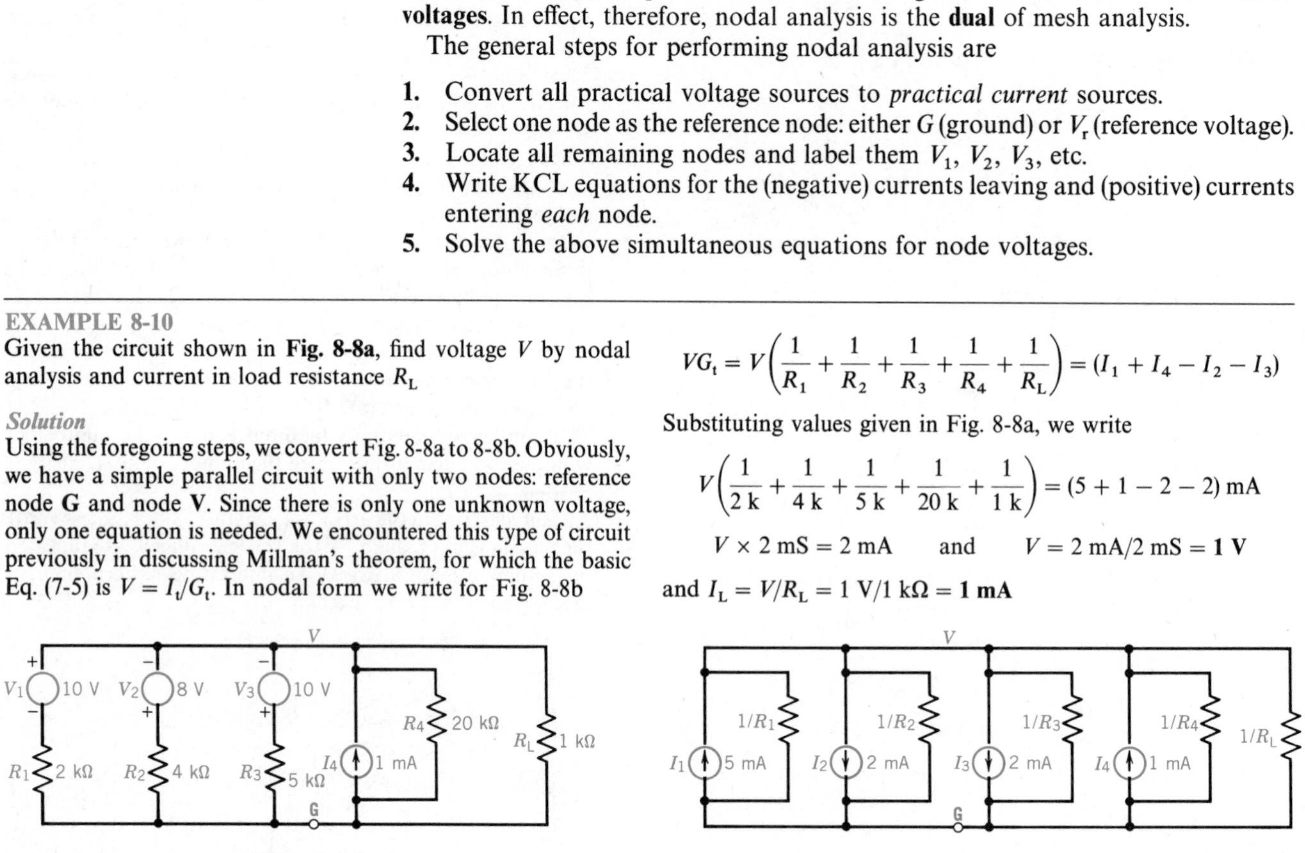

a. Original circuit

b. Nodal analysis equivalent

Figure 8-8 Nodal analysis equivalence; Exs. 8-10 and 8-12

From Ex. 8-10 we might generalize a rule concerning nodal analysis: the product of a given node voltage and the total conductance connected between that node and the reference node equals the total current entering and leaving the given node.

But is that all there is to nodal analysis? Let us consider Ex. 8-11 (previously solved, Ex. 8-9) which has *two* nodes (in addition to the reference node) and attempt to solve it by nodal analysis. To save one step, the practical voltage sources of Ex. 8-9 have been replaced by equivalent practical current sources in **Fig. 8-9a**. Our purpose is to verify Ex. 8-9, which found node voltages V_x and V_y by the format method of mesh analysis.

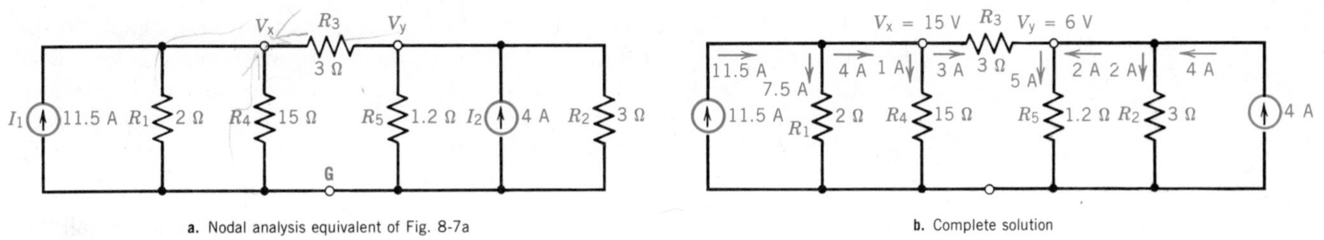

Figure 8-9 Examples 8-11 and 8-13

EXAMPLE 8-11

Using previously developed nodal analysis principles, for the circuit of Fig. 8-9a, calculate

a. V_x
b. V_y
c. Current in R_3
d. Draw the complete solution for Fig. 8-9a

Solution

Unlike the case of Ex. 8-10, there are now two nodes in addition to the reference node. Observe that not all the current from source I_1 returns to ground. Some of the current may flow to R_3 in Fig. 8-9a. This component of current I_1 by Ohm's law is $(V_x - V_y)/R_3$ (see Ex. 8-9f). Let us write the nodal equations for V_x and V_y, therefore, as

for node V_x: $V_x\left(\frac{1}{R_1}+\frac{1}{R_4}+\frac{1}{R_3}\right)-\frac{V_y}{R_3}=I_1$

(current entering node V_x) (8-2)

and for node V_y: $V_y\left(\frac{1}{R_5}+\frac{1}{R_2}+\frac{1}{R_3}\right)-\frac{V_x}{R_3}=I_2$

(current entering node V_y) (8-3)

If the reasoning is correct, substituting values from Fig. 8-9a into the above simultaneous equations should yield the same results as Ex. 8-9:

$$V_x\left(\frac{1}{2}+\frac{1}{15}+\frac{1}{3}\right)-\frac{V_y}{3}=11.5 \quad \text{or} \quad 0.9V_x-0.\bar{3}V_y=11.5 \qquad (1)$$

$$-\frac{V_x}{3}+V_y\left(\frac{1}{1.2}+\frac{1}{3}+\frac{1}{3}\right)=4 \quad \text{or} \quad -0.\bar{3}V_x+1.5V_y=4 \qquad (2)$$

The array for Eqs. (1) and (2) is

$$\mathbf{A}=\begin{vmatrix} V_x & V_y & I \\ 0.9 & -0.\bar{3} & 11.5 \\ -0.\bar{3} & 1.5 & 4 \end{vmatrix} \quad \text{and} \quad \mathbf{D}=\begin{vmatrix} 0.9 & -0.\bar{3} \\ -0.\bar{3} & 1.5 \end{vmatrix}=1.23\bar{8}$$

Using the method of determinants,

a. $$V_x=\frac{\begin{vmatrix} 11.5 & -0.\bar{3} \\ 4 & 1.5 \end{vmatrix}}{\mathbf{D}}=\frac{18.58\bar{3}}{1.23\bar{8}}$$

$= \mathbf{15\ V}$ (same answer as Ex. 8-9e)

b. $$V_y=\frac{\begin{vmatrix} 0.9 & 11.5 \\ -0.\bar{3} & 4 \end{vmatrix}}{\mathbf{D}}=\frac{7.4\bar{3}}{1.23\bar{8}}=\mathbf{6\ V} \text{ (same answer as Ex. 8-9e)}$$

c. Then $I_3=\frac{V_x-V_y}{R_3}=\frac{(15-6)\text{ V}}{3\ \Omega}$

$= \mathbf{3\ A}$ (same answer as Ex. 8-9f)

d. Knowing all node voltages to ground, $V_x = 15$ V and $V_y = 6$ V, we use Ohm's law to obtain all the currents shown in Fig. 8-9b, which represents the complete solution for Fig. 8-9a. Note that all currents obey KCL and all voltages obey KVL. Without this verification, the entire solution is an exercise in futility. In short, unless KCL and KVL are satisfied for all portions of the network, the answers obtained are in error.

8-6 SOLVING FOR NETWORK NODE VOLTAGES USING THE FORMAT METHOD OF NODAL ANALYSIS

Using nodal analysis, we proved that the answers to Ex. 8-11 are the same as those of Ex. 8-9, obtained by the format method of mesh analysis. We are now ready to attempt a *format method* of nodal analysis to enable us to write nodal equations by *inspection*.

We have two clues to the statement of a format method for nodal analysis; the first lies in Eq. (8-1), written for the format method of mesh analysis (Sec. 8-4), and the second lies in Eqs. (8-2) and (8-3), used in the solution of Ex. 8-11. We also know that nodal analysis represents the *dual* of mesh analysis *in every respect.* Armed with these insights, we may write the following rules for the format method of nodal analysis:

1. Sum the total of all conductances connected to a given assigned node. Call this total the *self-conductance* or *common conductance* connected to (or dangling from) the given node.
2. The product of the given (assigned) node voltage and the total self-conductance is ALWAYS written as a positive current in each node equation.
3. Sum the total conductance of all conductances *mutually* shared with *other* node voltages assigned to the network, using a separate shared conductance for each separate other node voltage assigned that is mutual to the given node in rule 1.
4. The product of other assigned node voltages and the mutual conductance is ALWAYS written as a NEGATIVE current in each node equation.
5. Sum the total of all currents entering and leaving the given (assigned) node of interest. These forcing functions will appear on the RH side of the node equation. Entering currents are positive and leaving currents are negative, in accordance with KCL convention.
6. Summarizing, we may write the general equation for the format method of nodal analysis as

$$+V_s(\sum G_s) - V_m(\sum G_m) = \pm\sum I \qquad \text{amperes (A)} \qquad \textbf{(8-4)}$$

where V_s is a particular assigned node voltage

$\sum G_s$ is the sum of common (self) conductances connected to (dangling from) the particular assigned node

V_m is a particular other assigned node voltage whose conductance is mutually shared with the first node voltage of interest, V_s

$\sum G_m$ is the sum of mutual self-conductances shared between nodes

$\pm\sum I$ is the sum of currents entering and/or leaving a given node.

The format method of nodal analysis is best illustrated by the following examples.

EXAMPLE 8-12

For the original circuit given in Fig. 8-8a, write the nodal equation by inspection, using the format method of nodal analysis

Solution

Since there is only one node, V, only one equation is needed. The sum of the conductances dangling from node V is $\left(\frac{1}{R_1}+\frac{1}{R_2}+\frac{1}{R_3}+\frac{1}{R_4}+\frac{1}{R_L}\right)=\left(\frac{1}{2\text{ k}\Omega}+\frac{1}{4\text{ k}\Omega}+\frac{1}{5\text{ k}\Omega}+\frac{1}{20\text{ k}\Omega}+\frac{1}{1\text{ k}\Omega}\right)=2\text{ mS}$

Since there are no other nodes, we may disregard mutual conductances. The nodal equation is $V \cdot 2\text{ mS} = (+5-2-2+1)\text{ mA}$ and $V = 2\text{ mA}/2\text{ mS} = \mathbf{1\ V}$

Note that the solution for node voltage is identical to that of Ex. 8-10 except that the equation was written *directly by inspection.*

EXAMPLE 8-13

Given the circuit of Fig. 8-9a, write the two nodal equations by inspection, using the format method of nodal analysis

Solution

$V_s(\sum G_s)$ self conductances $\quad -V_m(\sum G_m)$ mutual conductances

For node V_x: $\quad V_x\left(\frac{1}{2\ \Omega}+\frac{1}{15\ \Omega}+\frac{1}{3\ \Omega}\right)-\frac{V_y}{3\ \Omega}=11.5\text{ A} \quad$ or $\quad 0.9V_x-0.\bar{3}V_y=11.5 \quad$ (1)

For node V_y: $\quad V_y\left(\frac{1}{1.2\ \Omega}+\frac{1}{3\ \Omega}+\frac{1}{3\ \Omega}\right)-\frac{V_x}{3\ \Omega}=4\text{ A} \quad$ or $\quad -0.\bar{3}V_x+1.5V_y=4 \quad$ (2)

Note that these equations are exactly the same as those in Ex. 8-11.

We are now ready to draw inferences regarding both the format method for mesh analysis and the format method of nodal analysis, its dual.

1. In *mesh* analysis the forcing functions are *voltage* sources, whereas in *nodal* analysis the forcing functions are *current* sources, either positive or negative, depending on polarity.
2. In *mesh* analysis the voltage drops across the *self*-resistances in a closed mesh are *positive*, whereas in *nodal* analysis the current produced by the product of the node voltage and its dangling *self*-conductances is *positive.*
3. In *mesh* analysis voltage drops due to currents in *mutual* **resistances** are *negative.*
4. In *nodal* analysis, currents due to *mutually* shared **conductances** are *negative.*

Table 8-1 summarizes the differences and similarities between the format method of mesh analysis and the format method of nodal analysis. The table reveals the perfect symmetry and duality of the two format methods. In effect, they are the same, insofar as procedure is concerned.

Table 8-1 Format Methods of Analysis of DC Circuits

	Mesh Analysis, Format Method	Nodal Analysis, Format Method
Forcing functions	Voltages	Currents
Required source format	All sources converted to practical voltage sources	All sources converted to practical current sources
Assigned values	Unknown mesh currents	Unknown node voltages
Self terms	Resistances encountered by designated mesh currents in a particular mesh	Conductances connected to designated node voltage in a network.
Polarity of self terms	Positive voltage drops	Positive currents ($+VG_s$)
Mutual terms	Resistances sharing mesh currents	Conductances between two nodes (other than the reference node)
Polarity of mutual terms	Negative voltage drops	Negative currents ($-VG_m$)
Total equations needed	One for each unknown mesh current	One for each unknown node voltage (other than the reference node)
Basic equation	$+I_s(\sum R_s)-I_m(\sum R_m) = \pm\sum V$ (8-1)	$+V_s(\sum G_s)-V_m(\sum G_m) = \pm\sum I$ (8-4)

Let us now attempt a few more examples, which will help us decide the circumstances under which each method is best suited.

EXAMPLE 8-14

Given the original circuit of **Fig. 8-10a**, find

a. Which method of analysis is best suited for the complete solution
b. The network source format required for the method of analysis chosen
c. The unknown voltage(s) and/or currents by the chosen method of analysis
d. The complete solution drawn on the original circuit

Solution

a. Using *mesh* analysis, there are *two* windows, *two* meshes, and *two* unknowns. But using *nodal* analysis, there is only *one* unknown. We choose nodal analysis because it results in *fewer unknowns* and *fewer equations* for this circuit.
b. The sources must be converted to practical current sources. This is shown in Fig. 8-10b. Only one unknown voltage, V_x, exists.
c. By inspection using format method,

$$V_x\left(\frac{1}{60\text{ k}\Omega}+\frac{1}{16\text{ k}\Omega}+\frac{1}{40\text{ k}\Omega}\right)=(0.8-0.3)\text{ mA}$$

$V_x = 0.5\text{ mA}/0.1041\bar{6}\text{ mS} = \mathbf{4.8\ V}$

d. Using the value of V_x from (c), we find from Fig. 8-10a

$$I_1=\frac{(48-4.8)\text{ V}}{60\text{ k}\Omega}=\mathbf{0.72\ mA}\uparrow$$

$$I_2=\frac{[4.8-(-12)]\text{ V}}{40\text{ k}\Omega}=\mathbf{0.42\ mA}\downarrow\text{ (leaving node X)}$$

$$I_L = V_L/R_L = V_x/R_L = 4.8\text{ V}/16\text{ k}\Omega = \mathbf{0.3\ mA}\downarrow$$

and $\quad I_1 - I_2 = (0.72 - 0.42)\text{ mA} = \mathbf{0.3\ mA}$

The complete solution is shown in Fig. 8-10c.

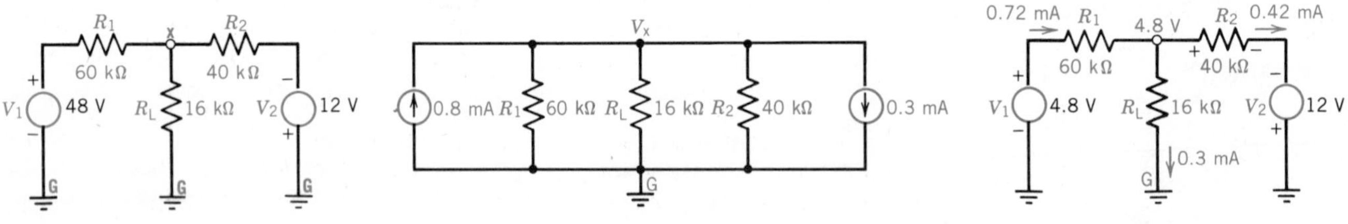

a. Original circuit b. Nodal analysis equivalent c. Complete solution

Figure 8-10 Example 8-14

EXAMPLE 8-15

Repeat all parts of Ex. 8-14 for the circuit shown in **Fig. 8-11a**

Solution

a. Using mesh analysis for each practical voltage source connected to ground, there are three windows, three meshes, and three unknowns. But using nodal analysis, there is only one unknown node, **X**, with respect to ground. Therefore, we choose nodal analysis and solve for V_x.
b. All sources must be converted to practical current sources. This is shown in Fig. 8-11b, along with the one unknown node voltage, V_x.
c. By inspection using nodal format,

$$V_x\left(\frac{1}{1\text{ k}\Omega}+\frac{1}{1\text{ k}\Omega}+\frac{1}{5\text{ k}\Omega}+\frac{1}{5\text{ k}\Omega}\right)=(20+5-1)\text{ mA}$$

$V_x = 24\text{ mA}/2.4\text{ mS} = \mathbf{10\ V}$

d. Using $V_x = 10$ V, from Fig. 8-11a we determine currents to or from the node as

$$I_A=\frac{(20-10)\text{ V}}{1\text{ k}\Omega}=\mathbf{10\ mA}\text{ (toward node)}$$

$$I_B=\frac{(10-5)\text{ V}}{1\text{ k}\Omega}=\mathbf{5\ mA}\text{ (away from node)}$$

$$I_C=\frac{(10)-(-5)\text{ V}}{5\text{ k}\Omega}=\mathbf{3\ mA}\text{ (away from node)}$$

$$I_L = V_x/R_L = 10\text{ V}/5\text{ k}\Omega = \mathbf{2\ mA}\text{ (away)}$$

The complete solution is shown in Fig. 8-11c. Note that KCL is satisfied for all currents entering and leaving node **X**.

a. Original circuit

b. Nodal analysis equivalent

c. Complete solution

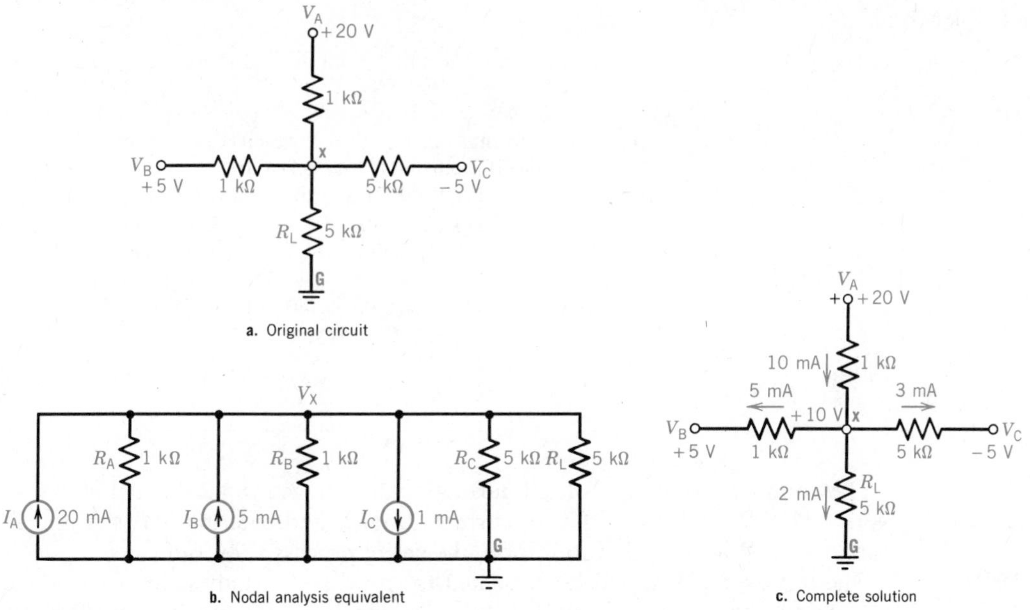

Figure 8-11 Example 8-15

EXAMPLE 8-16

Repeat all parts of Ex. 8-14 for the circuit shown in **Fig. 8-12a**

Solution

If we select mesh analysis, for each practical voltage source connected to ground there are four windows, four mesh currents, and four unknowns. But selecting nodal analysis, there are only two unknown nodes, V_A and V_B, as shown in Fig. 8-12a.

a. We select nodal analysis for solution of the given network.

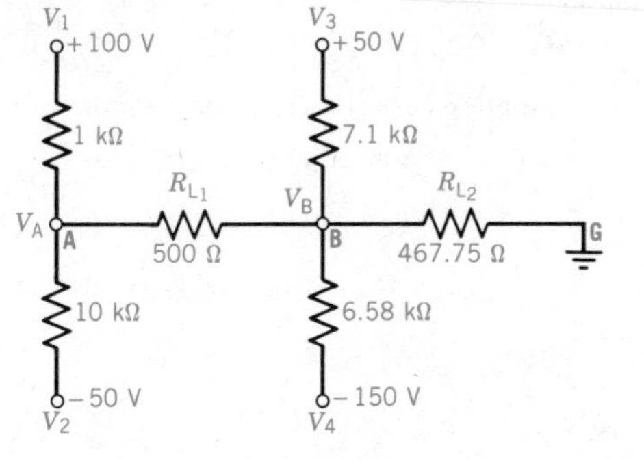

a. Original circuit

b. Nodal analysis equivalent

b. This dictates that *all* practical voltage sources must be *converted* to current sources. The conversion is shown in Fig. 8-12b, using the source conversion theorem. Note that negative sources direct their currents away from the nodes and toward ground.

c. Two nodal equations must be written by inspection, using the nodal format method, for the currents around each node.

Around node **A**:

$$V_A\left(\frac{1}{1\text{ k}\Omega} + \frac{1}{10\text{ k}\Omega} + \frac{1}{0.5\text{ k}\Omega}\right) - \frac{V_B}{0.5\text{ k}\Omega} = (100 - 5)\text{ mA} \quad (1)$$

Around node **B**:

$$\frac{-V_A}{0.5\text{ k}\Omega} + V_B\left(\frac{1}{0.5\text{ k}\Omega} + \frac{1}{7.1\text{ k}\Omega} + \frac{1}{6.58\text{ k}\Omega} + \frac{1}{467.75\text{ }\Omega}\right)$$
$$= (7.042 - 22.8)\text{ mA} \quad (2)$$

c. Complete solution

Figure 8-12 Example 8-16; nodal analysis, format method

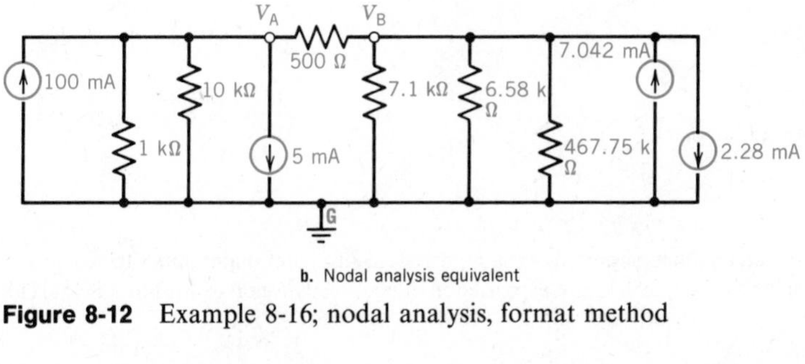

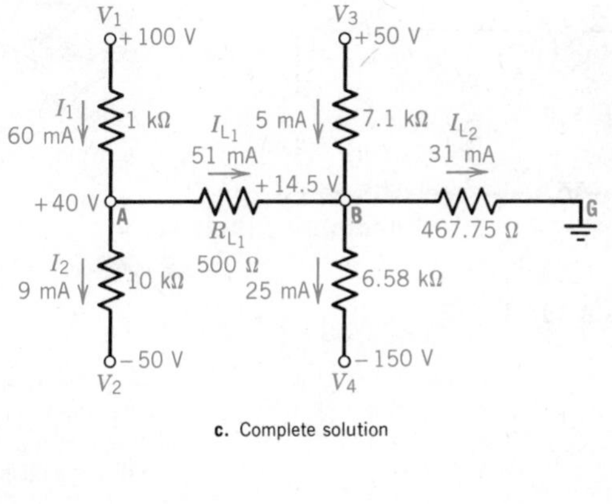

Stating each equation in simplest form,[5]

$$3.1V_A - 2V_B = 95 \qquad (1)$$

$$-2V_A + 4.431V_B = -15.76 \qquad (2)$$

Then the array $\mathbf{A} = \begin{vmatrix} V_A & V_B & I \\ 3.1 & -2 & 95 \\ -2 & 4.431 & -15.76 \end{vmatrix}$

and $\mathbf{D} = \begin{vmatrix} 3.1 & -2 \\ -2 & 4.431 \end{vmatrix} = 9.736$

while $V_A = \dfrac{\begin{vmatrix} 95 & -2 \\ -15.76 & 4.431 \end{vmatrix}}{\mathbf{D}} = \dfrac{389.43}{9.736} = \mathbf{40\ V}$

and $V_B = \dfrac{\begin{vmatrix} 3.1 & 95 \\ -2 & -15.76 \end{vmatrix}}{\mathbf{D}} = \dfrac{141.15}{9.736} = \mathbf{14.5\ V}$

d. The currents delivered (or received) by the respective VOLTAGE sources in Fig. 8-12a are:
$I_1 = (V_1 - V_A)/R_1 = (100 - 40)/1\ \text{k}\Omega = \mathbf{60\ mA}$;
$I_2 = (V_A - V_2)/R_2 = (40 - (-50)/10\ \text{k}\Omega = \mathbf{9\ mA}$;
$I_3 = (V_3 - V_B)/R_3 = (50 - 14.5)/7.1\ \text{k}\Omega = \mathbf{5\ mA}$;
$I_4 = (V_B - V_4)/R_4 = 14.5 - (-150)/6.58\ \text{k}\Omega = \mathbf{25\ mA}$;
while $I_{L_1} = (V_A - V_B)/R_{L_1} = (40 - 14.5)/500\ \Omega = \mathbf{51\ mA}$ and $I_{L_2} = 14.5\ \text{V}/467.75\ \Omega = \mathbf{31\ mA}$

The complete solution with all currents and voltages is shown in Fig. 8-12c. Note that KCL is satisfied for all currents leaving and entering nodes A and B, respectively. Without this verification, the entire process is meaningless.

One last example should be presented to show nodal analysis involving three unknown voltages. The circuit selected is an ideal one for this purpose because of its "internodal" relations. This is the bridged-T network originally solved in Ex. 7-18 with Bartlett's theorem.

EXAMPLE 8-17

For the circuit shown in **Fig. 8-13a**, using nodal analysis (format method), find the current in the 200-Ω resistor between nodes C and D

Solution

Assign node D as the reference node and write three equations for nodes V_A, V_B, and V_C, respectively. The nodal analysis equivalent circuit is shown in Fig. 8-13b. By inspection, using the format method of nodal analysis, we write

for node V_A:

$$+V_A\left(\frac{1}{0.4\ \text{k}\Omega} + \frac{1}{0.2\ \text{k}\Omega} + \frac{1}{2\ \text{k}\Omega}\right) - \frac{V_B}{2\ \text{k}\Omega} - \frac{V_C}{0.2\ \text{k}\Omega} = 50\ \text{mA} \quad (1)$$

for node V_B:

$$-\frac{V_A}{2\ \text{k}\Omega} + V_B\left(\frac{1}{0.4\ \text{k}\Omega} + \frac{1}{0.2\ \text{k}\Omega} + \frac{1}{2\ \text{k}\Omega}\right) - \frac{V_C}{0.2\ \text{k}\Omega} = 50\ \text{mA} \quad (2)$$

for node V_C:

$$-\frac{V_A}{0.2\ \text{k}\Omega} - \frac{V_B}{0.2\ \text{k}\Omega} + V_C\left(\frac{1}{0.2\ \text{k}\Omega} + \frac{1}{0.2\ \text{k}\Omega} + \frac{1}{0.2\ \text{k}\Omega}\right) = 0 \quad (3)$$

These equations are shown in their simplest form below.

$$8V_A - 0.5V_B - 5V_C = 50 \qquad (1)$$

$$-0.5V_A + 8V_B - 5V_C = 50 \qquad (2)$$

$$-5V_A - 5V_B + 15V_C = 0 \qquad (3)$$

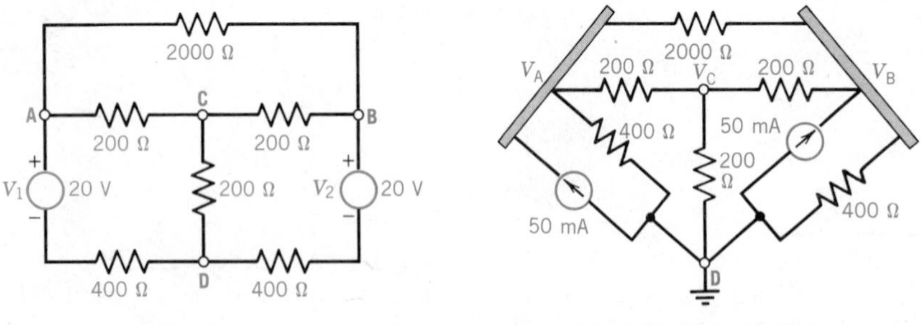

a. Original network

b. Nodal analysis equivalent

Figure 8-13 Example 8-17

[5] By writing the combined conductances in units of millisiemens (mS) and maintaining the current in milliamperes (mA), the units of V_A and V_B are expressed in volts directly (see Exs. 8-14 and 8-15). This simplifies coefficients in determinant form.

The array is

$$\begin{array}{c} \quad V_A \quad\ V_B \quad\ V_C \quad\ I \\ \mathbf{A} = \begin{vmatrix} 8 & -0.5 & -5 & 50 \\ -0.5 & 8 & -5 & 50 \\ -5 & -5 & 15 & 0 \end{vmatrix} \end{array}$$

and determinant

$$\mathbf{D} = \begin{vmatrix} 8 & -0.5 & -5 \\ -0.5 & 8 & -5 \\ -5 & -5 & 15 \end{vmatrix} = 531.25$$

Our only node of interest is

$$V_C = \frac{\begin{vmatrix} 8 & -0.5 & 50 \\ -0.5 & 8 & 50 \\ -5 & -5 & 0 \end{vmatrix}}{\mathbf{D}} = \frac{4250}{531.25} = \mathbf{8\ V}$$

and $I_{CD} = V_C/200\ \Omega = 8\ \text{V}/200\ \Omega = \mathbf{40\ mA}$ (see Ex. 7-18c, where the complete solution for this network is found using Bartlett's theorem).

With respect to the solution of Ex. 8-17, please note:

1. We might have used mesh analysis to solve the network of Fig. 8-13a rather than nodal analysis. But for the *general* case of a bridged-T network, three mesh currents are required since there are three windows. Then for the *specific* case of Fig. 8-13a, since $V_1 = V_2 = 20$ V, the current in the 2000-Ω resistor is zero and we need only write two equations to solve for the two mesh currents in the meshes containing V_1 and V_2.
2. Do not fall into the trap, however, of assuming (similarly) that you can solve Fig. 8-13a with only one node voltage equation because $V_1 = V_2 = 20$ V. The voltages at nodes A and B are unknown voltages and are *not* 20 V. (They are somewhat less because of the volt drop across the 400-Ω resistors in each mesh.)
3. Although Bartlett's theorem was used to solve the network most elegantly in Ex. 7-18, it must be remembered that the bridged-T network of Fig. 8-13a lends itself to such elegant solution because of its *symmetry*. The format method of nodal analysis used in Ex. 8-17 works for *any* case of a bridged T-network having *any* given values of resistance and voltage.
4. The value of $I_{CD} = 40$ mA obtained in Ex. 8-17 is the same as that in Ex. 7-18c, proving the validity of nodal analysis by the format method for three unknown voltages.

8-7 DELTA–WYE AND WYE–DELTA TRANSFORMATIONS

The previous methods presented in this chapter were all *algebraic* methods of solving equations for unknown currents and/or voltages in a network. In the case of mesh analysis, source transformations were required to convert all practical current sources to voltage sources. In the case of nodal analysis, similar source transformations are needed to convert all practical voltage sources to current sources. Once these transformations are accomplished, the equations may be written for the solution of all currents and voltages in any given network.

The delta–wye and wye–delta transformations are special cases of the more general cases of mesh–star and star–mesh transformations. **Figure 8-14a** shows a three-terminal mesh arrangement of three resistors. Note that this circuit (by itself) has only one window and the same mesh current is common to all three resistors. The circuit of Fig. 8-14a is sometimes called a *delta* configuration or alternatively a *pi* or π configuration, as we shall see below.

The mesh arrangement may include any multiple of 3 in commercial polyphase practice. Thus we find the six-terminal or six-phase mesh connection of Fig. 8-14b frequently used.[6] Again, the same mesh (shortest possible loop) current is common to resistors R_1 to R_6 inclusive. The term "mesh" is a general family name for any set of branches forming a closed path in a network.

An alternative arrangement is the *star* configuration. Here one end of *each* component in the network is connected to a *common* terminal. The three-terminal or three-phase star is shown in Fig. 8-14c. The connection is alternatively known as the *wye* (Y) or *tee* (T) connection, as well. The six-terminal star shown in Fig. 8-14d is an alternative configuration to the six-terminal mesh of Fig. 8-14b. The term

[6] Polyphase rectification systems to produce dc frequently convert 3-phase power to 6-phase, 12-phase, 18-phase, etc. using interphase transformers and solid-state rectifiers. This subject is outside the scope of this text.

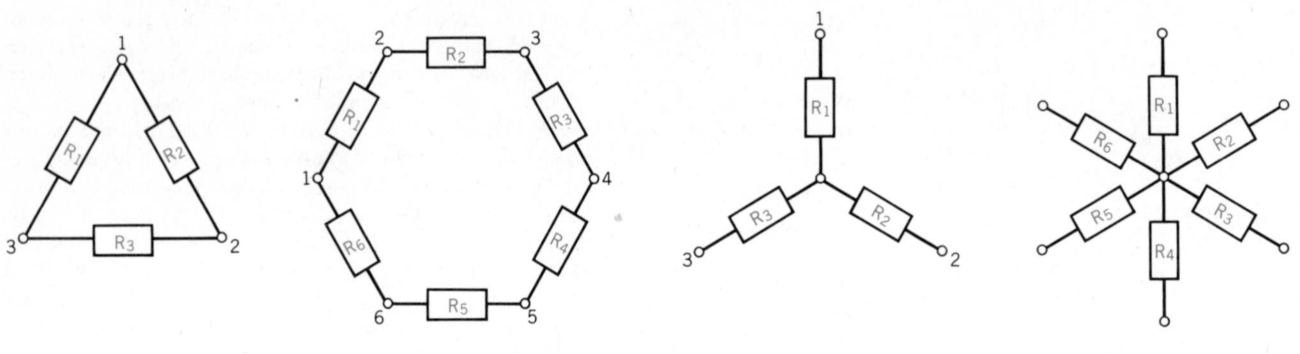

a. 3-terminal mesh (delta or pi) b. 6-terminal mesh c. 3-terminal star (wye or tee) d. 6-terminal star

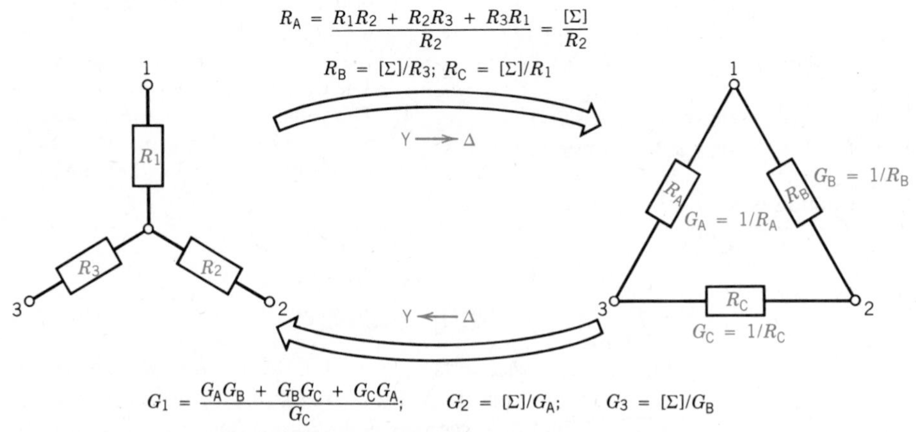

e. Y-Δ and Δ-Y transformations (Appendix B-1) and their dual relationships

Figure 8-14 Comparisons between mesh and star connections and Y–Δ/Δ–Y transformations

"star" is a general family name for any set of branches in which each branch has one end connected to a common terminal.

8-7.1 Wye–Delta Transformation

It is possible to convert or transform each element of a wye-connected star to an equivalent delta-connected mesh, using the wye–delta algebraic transformation shown in Fig. 8-14e. (This transformation is derived in Appendix B-1 for the general case of ac complex impedances and applies, as well, to the specific case of resistances connected in wye as shown in Fig. 8-14e.)

Given wye-connected components R_1, R_2, and R_3, the equivalent delta-connected components, as shown in Fig. 8-14e, are R_A, R_B, and R_C, obtained from

$$R_A = \frac{R_1R_2 + R_2R_3 + R_3R_1}{R_2} = \frac{\sum(R \text{ products})}{R_2} \quad \text{ohms } (\Omega) \qquad (8\text{-}5)$$

$$R_B = \frac{\sum(R \text{ products})}{R_3} \qquad (8\text{-}6)$$

$$R_C = \frac{\sum(R \text{ products})}{R_1} \qquad (8\text{-}7)$$

With respect to these equations, please note:

1. The numerators are always the sum of the products of the *three* known *pairs* of resistances.
2. The *denominator* is always *diagonally opposite* the desired delta equivalent resistance. Diagonally opposite desired R_A (in the delta) is R_2 in the wye. Diagonally opposite desired R_B (in the delta) is the known value of R_3 in the wye. Diagonally opposite desired R_C (in the delta) is the known value of R_1 in the wye.

EXAMPLE 8-18

If each resistor in the wye configuration of Fig. 8-14 is 2 Ω, calculate the resistor of each arm of an equivalent delta

Solution

$R_1 = R_2 = R_3 = 2\ \Omega$ as given. Then the equivalent delta values are

$$R_A = \frac{R_1R_2 + R_2R_3 + R_3R_1}{R_2} = \frac{(2 \times 2) + (2 \times 2) + (2 \times 2)}{2}$$

$$= \frac{12\ (\Omega)^2}{2\ \Omega} = \mathbf{6\ \Omega} \tag{8-5}$$

$$R_B = \sum/R_3 = 12\ \Omega^2/2\ \Omega = \mathbf{6\ \Omega} \tag{8-6}$$

$$R_C = \sum/R_1 = 12\ \Omega^2/2\ \Omega = \mathbf{6\ \Omega} \tag{8-7}$$

As trivial as Ex. 8-18 might appear, it tells us something very important about wye–delta equivalence. If the circuits are symmetrical, that is, **all** components in *either* delta or wye are the SAME, the **delta**-connected equivalents each are **three times as great** as the wye-connected equivalents. Conversely, the wye-connected equivalents are one-third the values of corresponding delta-connected equivalents.

8-7.2 Delta–Wye Transformation

As stated in Appendix B-1, where the delta–wye and wye–delta conversions are derived, the delta and wye configurations are essentially *duals* of each other. From previous experience with duality, we know that the *dual* of resistance (R) is conductance (G). This duality is shown in Fig. 8-14e.

As shown in Fig. 8-14e, given delta-connected resistances R_A, R_B, and R_C, we may convert them to their equivalent conductances, G_A, G_B, and G_C, respectively, by taking their reciprocals. We may then write the delta–wye transformation as

$$G_1 = \frac{G_AG_B + G_BG_C + G_CG_A}{G_C} = \frac{\sum(G\text{ products})}{G_C} \quad \text{siemens (S)} \tag{8-8}$$

$$G_2 = \frac{\sum(G\text{ products})}{G_A} \tag{8-9}$$

$$G_3 = \frac{\sum(G\text{ products})}{G_B} \tag{8-10}$$

Note that Eqs. (8-8) through (8-10) follow the same format as Eqs. (8-5) to (8-7), respectively. This leaves very little to memory except for four essential rules:

1. In converting from wye to delta, use known *resistances* to obtain equivalent unknown resistances.

2. In converting from delta to wye, use known *conductances* to obtain equivalent unknown conductances.
3. The denominator term is always the *known* resistance or conductance *diagonally opposite* the desired resistance or conductance.
4. The numerator is always the sum of products (SOP) of the three *known* pairs of resistances or conductances.

For those who prefer to use resistances *exclusively*, as shown in Appendix B-1, we may *also* write the delta–wye transformation as follows:

$$R_1 = R_A R_B/(R_A + R_B + R_C) = R_A R_B/\sum(Rs) \quad \textbf{(8-11)}$$

$$R_2 = R_B R_C/\sum(Rs) \quad \textbf{(8-12)}$$

$$R_3 = R_C R_A/\sum(Rs) \quad \textbf{(8-13)}$$

With respect to Eqs. (8-11) through (8-13) and Fig. 8-14, note that

1. The denominator is always the sum of the known resistances.
2. The numerator is always the product of the two adjacent known resistance arms on either side of the unknown resistances R_1, R_2, and R_3.

The proof of equivalence for Eqs. (8-8) through (8-13) is shown in Exs. 8-19 and 8-20.

EXAMPLE 8-19
Given the delta configuration of resistances shown in **Fig. 8-15**, calculate the equivalent wye-connected resistances R_1, R_2, and R_3 shown between terminals 1, 2, and 3, respectively, using Eqs. (8-8) through (8-10)

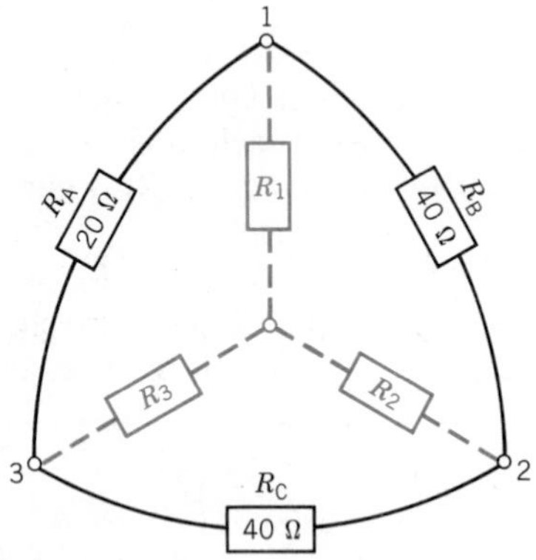

Figure 8-15 Example 8-19, Δ–Y conversion

Solution

$$G_A = 1/R_A = 1/20\ \Omega = 50\ \text{mS};$$

$$G_B = 1/R_B = 1/40\ \Omega = 25\ \text{mS};$$

$$G_C = G_B = 25\ \text{mS}$$

$$\sum G \text{ products} = G_A G_B + G_B G_C + G_C G_A$$

$$= 50 \times 25 + 25 \times 25 + 25 \times 50 = 3125\ (\text{mS})^2$$

$$G_1 = \sum Gs/G_C = 3125\ (\text{mS})^2/25\ \text{mS} = 125\ \text{mS} \quad (8\text{-}8)$$

and $R_1 = 1/G_1 = 1/125\ \text{mS} = \mathbf{8\ \Omega}$

$$G_2 = \sum Gs/G_A = 3125\ (\text{mS})^2/50\ \text{mS} = 62.5\ \text{mS} \quad (8\text{-}9)$$

and $R_2 = 1/G_2 = 1/62.5\ \text{mS} = \mathbf{16\ \Omega}$

$$G_3 = \sum Gs/G_B = 3125\ (\text{mS})^2/25\ \text{mS} = 125\ \text{mS} \quad (8\text{-}10)$$

and $R_3 = 1/G_3 = 1/125\ \text{mS} = \mathbf{8\ \Omega}$

EXAMPLE 8-20
Verify the solutions of Ex. 8-19, using Eqs. (8-11) through (8-13) to find R_1, R_2, and R_3, respectively

Solution

$$\sum Rs \text{ in delta} = R_A + R_B + R_C = (20 + 40 + 40)\ \Omega = 100\ \Omega$$

$$R_1 = R_A R_B/\sum Rs = 20 \times 40/100 = \mathbf{8\ \Omega} \quad (8\text{-}11)$$

$$R_2 = R_B R_C/\sum Rs = 40 \times 40/100 = \mathbf{16\ \Omega} \quad (8\text{-}12)$$

$$R_3 = R_C R_A/\sum Rs = 40 \times 20/100 = \mathbf{8\ \Omega} \quad (8\text{-}13)$$

Examples 8-19 and 8-20 have proved the equivalence of the two sets of equations. Either set is equally valid. The method of Ex. 8-19 has the advantage of using the *same* equation format for delta–wye conversion as used for wye–delta conversion. The method of Ex. 8-20 has the advantage of fewer solution steps to arrive at Y-equivalent converted resistance values. The choice of which set to use is left to the reader.

EXAMPLE 8-21

The bridged-T network of Ex. 8-17 is shown once more in **Fig. 8-16** with the addition of a possible Y-equivalent to replace the existing delta-connected resistors. Using a delta–wye conversion, find

a. The equivalent circuit
b. The current in R_4, using the nodal analysis format method

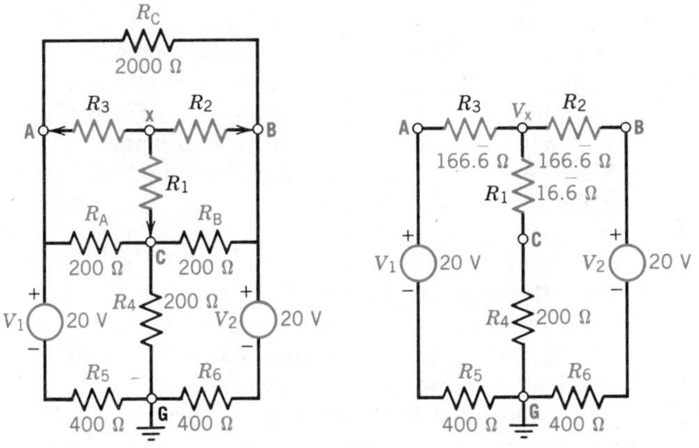

a. Original bridged-T network
b. Equivalent circuit after Δ-Y transformation

Figure 8-16 Example 8-21

Solution

a. $\sum Rs$ in Δ mesh $= R_A + R_B + R_C = (200 + 200 + 2000)\ \Omega$,
$= 2400\ \Omega$

$$R_1 = R_A R_B/\sum Rs = (200)^2/2400 = 16.\bar{6}\ \Omega,$$

$$R_2 = R_B R_C/\sum Rs = 200 \times 2000/2400 = 166.\bar{6}\ \Omega,$$

and $R_3 = R_A R_C/\sum Rs = 200 \times 2000/2400 = 166.\bar{6}\ \Omega.$

The values of R_1, R_2, and R_3 are now used to replace the delta with the equivalent Y in Fig. 8-16b.

b. The circuit of Fig. 8-16b could be solved either by mesh analysis, using two unknown currents for the two windows, or nodal analysis for V_X, using one unknown. We choose nodal analysis, as suggested in (**b**).

Step 1. Convert practical voltage sources to current sources:

$$I_2 = I_1 = V_1/R_{t_1} = 20\ \text{V}/566.\bar{6}\ \Omega = 35.294\ \text{mA}$$

Step 2. Write the nodal equation by inspection:

$$V_X\left(\frac{1}{566.\bar{6}} + \frac{1}{566.\bar{6}} + \frac{1}{216.\bar{6}}\right) = (35.294) \times 2$$

and $V_X = 70.59\ \text{mA}/8.145\ \text{mS} = 8.\bar{6}\ \text{V}$

$$I_4 = I_{XG} = V_{XG}/R_{XG} = 8.\bar{6}\ \text{V}/216.\bar{6}\ \Omega = \mathbf{40\ mA}$$

Note that the value of I_4 is exactly the same as I_{CD} found in Ex. 8-17 by nodal analysis.

EXAMPLE 8-22

For the original circuit shown in **Fig. 8-17a**, using the Δ–Y transformation, calculate in turn

a. V_Y
b. V_Z
c. I_2
d. I_4
e. I_G
f. I_1
g. I_3
h. Show the complete solution for all voltages and currents

Solution

$$\sum Rs = R_1 + R_3 + R_G = (10 + 20 + 1)\ \text{k}\Omega = 31\ \text{k}\Omega;$$

$$R_Z = (20 \times 1)\ \text{k}\Omega^2/31\ \text{k}\Omega = 0.6452\ \text{k}\Omega;$$

$$R_X = R_1 R_3/\sum Rs = (10 \times 20)/31 = 6.452\ \text{k}\Omega;$$

$$R_Y = R_1 R_G/\sum Rs = (10 \times 1)/31 = 0.3226\ \text{k}\Omega$$

The above values of R_X, R_Y, and R_Z are entered in Fig. 8-17b. Note that as a result of the Δ–Y transformation, a relatively simple series–parallel circuit has been produced in which all resistance values are known. Our task now is to find voltages at points **Q**, **Y**, and **Z**.

The equivalent parallel resistance, $R_{QG} = (15.3226\ \text{k}\| 5.6452\ \text{k}) = 4.1253\ \text{k}\Omega$. Using the VDR, $V_Q = \dfrac{R_{QG}}{R_{QG} + R_X}\, 10\ \text{V} = \dfrac{4.1253\ \text{k}\Omega}{(4.1253 + 6.452)\ \text{k}\Omega} \times 10\ \text{V} = 3.9\ \text{V}$

a. $V_Y = \dfrac{R_2}{(R_2 + R_Y)} \times 3.9\ \text{V} = \dfrac{15\ \text{k}\Omega}{15.3226\ \text{k}\Omega} \times 3.9\ \text{V} = \mathbf{3.818\ V}$

b. $V_Z = \dfrac{R_4}{R_4 + R_Z} \times 3.9\ \text{V} = \dfrac{5\ \text{k}\Omega}{5.6452\ \text{k}\Omega} \times 3.9\ \text{V} = \mathbf{3.454\ V}$

c. $I_2 = V_Y/R_2 = 3.818\ \text{V}/15\ \text{k}\Omega = \mathbf{0.2545\ mA}$

d. $I_4 = V_Z/R_4 = 3.454\ \text{V}/5\ \text{k}\Omega = 0.6908\ \text{mA} = \mathbf{0.691\ mA}$

e. $I_G = V_G/R_G = (V_Y - V_Z)/R_G = (3.818 - 3.454)\ \text{V}/1\ \text{k}\Omega$
$= 0.364\ \text{V}/1\ \text{k}\Omega = \mathbf{0.364\ mA}$

f. $I_1 = V_1/R_1 = (V_X - V_Y)/R_1$
$= (10 - 3.818)\ \text{V}/10\ \text{k}\Omega = \mathbf{0.6182\ mA}$
Check: $I_1 = I_2 + I_G = (0.2545 + 0.364) = \mathbf{0.6185\ mA}$

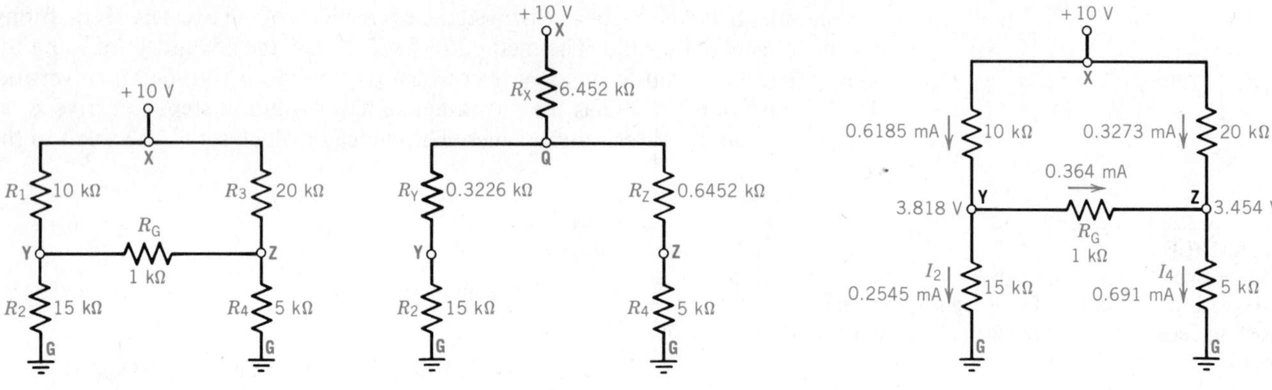

a. Original circuit

b. Δ-Y equivalent

c. Complete solution

Figure 8-17 Example 8-22

g. $I_3 = V_3/R_3 = (V_X - V_Z)/R_3 = (10 - 3.454)/20\ \text{k}\Omega = \mathbf{0.3273\ mA}$
   Check: $I_4 = I_3 + I_G = (0.3273 + 0.364) \cong \mathbf{0.691\ mA}$
h. The complete solution is shown in Fig. 8-17c. Note that currents around nodes **Y** and **Z** may be verified by KCL as proof of the validity of the method and the accuracy of the solutions.

The delta–wye conversion solution of Ex. 8-22 is an important one regarding the solution of unbalanced bridge networks, as noted in the following discussion. Please note that

1. The solution for I_G in Ex. 8-22e compares favorably with the solution of Ex. 7-4 for the same circuit by Thévenin's theorem.
2. The weakness of the Thévenin's theorem method is that it *only* reveals the unbalanced current in the galvanometer. Once this current is found, it is still *impossible* to find the currents in the bridge arms because we do not know the true values of V_T and V_H in Fig. 7-5a.
3. The circuit of Fig. 7-5a and/or Fig. 8-17a may be solved by mesh analysis, but three loops (meshes) are required (at least one encompassing the 10-V source). The circuit does not lend itself easily to nodal analysis because there is no way to convert the 10-V ideal voltage source to a practical current source. Of course, we could use the circuit of Fig. 8-17b for nodal analysis to find V_Y and V_Z after we accomplished the Δ-Y conversion. But the voltage-divider method used above is much simpler, as we have seen.
4. With regard to Fig. 8-17b, the only *reality* in the equivalent circuit shown exists from node **Y** through R_2 and node **Z** through R_4 to ground. Voltage V_Q does *not* exist in the original circuit or in its complete solution. Consequently, once we have found V_Y and V_Z by the methods of (**a**) and (**b**) in the Ex. 8-22 solution, we automatically can find I_2, I_4, and I_G, as shown above. Similarly, knowing V_Y and V_Z, we also know the voltages across R_1 and R_3, leading to the complete solution.
5. The conclusion to be drawn is that if the *complete solution* of the given bridge circuit is desired, the Δ–Y conversion and/or mesh analysis are probably the most efficient methods.

EXAMPLE 8-23

For the three-wire circuit shown in **Fig. 8-18a**, using the Y–Δ conversion, find

a. The current in the ammeter, I_m, assuming the instrument has negligible resistance
b. The current direction in the ammeter
c. Currents supplied by V_1 and V_2

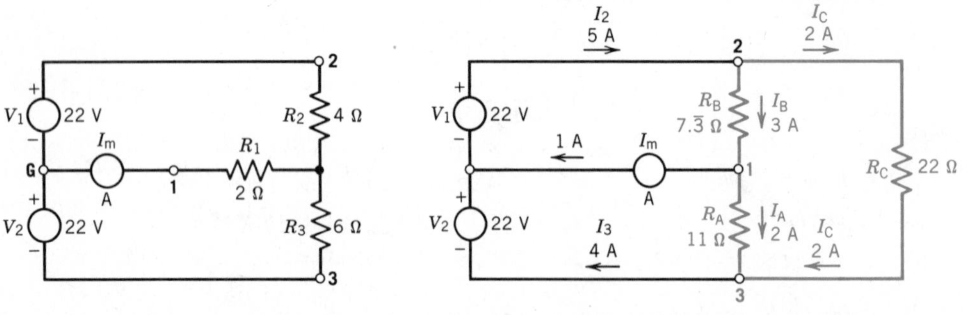

a. Original circuit (Y configuration)

b. Equivalent circuit (Δ configuration)

Figure 8-18 Example 8-23; three-wire system configuration

Solution
Using Eqs. (8-5) through (8-7),

$$\sum(R \text{ products}) = R_1R_2 + R_2R_3 + R_3R_1 = 2 \times 4 + 4 \times 6 + 6 \times 2 = 44$$

$$R_A = \frac{\sum(R \text{ products})}{R_2} = \frac{44}{4} = 11\ \Omega;$$

$$R_B = \frac{\sum(R \text{ products})}{R_3} = \frac{44}{6} = 7.\bar{3}\ \Omega;$$

and

$$R_C = \frac{\sum(R \text{ products})}{R_1} = \frac{44}{2} = 22\ \Omega$$

The equivalent Δ-connected circuit is shown in Fig. 8-18b. Then by Ohm's law, $I_C = 44\ \text{V}/22\ \Omega = 2\ \text{A}$, $I_B = 22\ \text{V}/7.\bar{3}\ \Omega = 3\ \text{A}$, and $I_A = 22\ \text{V}/11\ \Omega = 2\ \text{A}$

a. By KCL around node 1 (Fig. 8-18b), $I_m = I_B - I_A = 3\ \text{A} - 2\ \text{A} = \mathbf{1\ A}$
b. $I_m = \mathbf{1\ A}$ from node 1 to ground (G)
c. $I_1 = I_B + I_C = 3\ \text{A} + 2\ \text{A} = \mathbf{5\ A}$ and $I_2 = I_A + I_C = 2\ \text{A} + 2\ \text{A} = \mathbf{4\ A}$

NOTE: The Y–Δ conversion easily enables the complete solution of any unbalanced wye 3-wire system, using Ohm's law.

Example 8-23 is included to show that we do have occasion to go from Y to Δ and that the Δ configuration is more useful in this case. Note that the ammeter carries the so-called unbalanced current drawn by loads R_B and R_A. Similarly, note that source V_2 carries only 4 A while source V_1 carries 5 A. This is characteristic of unbalanced three-wire systems.

8-8 SOME WORDS OF ADVICE REGARDING SOLUTIONS OF PROBLEMS BY ALGEBRAIC METHODS

Various methods of circuit analysis using network equations have been presented in this chapter, as illustrated by the examples throughout the chapter. But little or nothing has been said regarding ways of ensuring that solution errors are reduced to a minimum. There is nothing so frustrating as to write equations, solve for the variables by determinants or computer techniques, and then discover on attempting the complete solution that the nodal currents do not conform to KCL and the mesh voltages do not conform to KVL. Clearly, an error may have been made in *writing* the equations initially. And whenever the reader solves incorrect equations for unknown variables, the procedure used (and that of a computer program) may *correctly* solve the *incorrect* equations.[7] But the values of the variables obtained do not represent the network and consequently will never stand analysis by KCL and KVL, the true test of any network analysis solution.

Much time can be saved if the reader adheres to the following procedures which seem to work well and save time, in the long run, in finding errors.

1. For mesh analysis, all sources must be converted to voltage sources. For nodal analysis, all sources must be converted to current sources.[8] Therefore, check your source conversions to ensure proper polarity, direction, equivalence, and circuit connection. Draw a neat, large, labeled equivalent circuit to permit equation writing by inspection.
2. Write the equations on a sheet of paper and simplify them algebraically in the form of an array. Then take another sheet of paper and repeat the process. If the arrays on both sheets agree, the probability is that the equations are correctly written.

---

[7] Computer analysts use the acronym GIGO (garbage in, garbage out) for situations in which programs are used to solve equations that do *not* represent the problem.
[8] For an exception to this rule, see Sec. 18-3.1.

3. If determinants are used, solve for **D** on a single sheet of paper. Then take a second sheet and repeat the process. The determinant array should exhibit the minor axis symmetry previously described. An error in the determinant **D** produces errors in the values of *all* unknown variables, since it appears in *all* denominators.
4. When you have found the unknown variables, examine the original circuit for nodes at which you may conveniently check currents for KCL and meshes where you may check volt drops and rises for KVL. If both KVL and KCL are satisfied for at least one test each, the probability is that your variables are correct.
5. But if they are not, you have made an error somewhere. The probability is that you will *not* find it by examining your written work and calculations. The best way is to begin afresh, starting with a clean sheet of paper and going back to step 1 above.

You are now ready to try your wings by "flying through" the various problems given below. Use the answers at the end of the chapter as feedback to verify that the methods used are correct and that the techniques described in this chapter are well understood.

8-9 GLOSSARY OF TERMS USED

Branch Single path in a network containing one or more elements in series.

Bridge network Any network (of relatively few branches) containing only two terminal elements that is not a series–parallel network. Elementary bridges such as the Wheatstone bridge are special cases of a bridge network.

Bridged-T network Tee network consisting of three branches, with a fourth branch across the two series-connected arms of the tee.

Circuit Network containing at least one closed path.

Delta configuration Three mesh-connected elements in the shape of a delta (Δ), also known as a pi (π) network.

Determinant method Algebraic technique for finding the unknown variables in two or more simultaneous linear equations (Appendix B-2.1).

Format method Simplified method of writing either mesh or nodal equations for a network by inspection.

Kirchhoff's current law (KCL) The algebraic sum of all currents taken around a given node is zero.

Kirchhoff's voltage law (KVL) The algebraic sum of all voltage drops and rises taken for any closed loop is zero.

KVL/KCL algebraic method Method of solving for the unknown branch currents of a network by writing simultaneous linear algebraic equations.

Loop Closed path, starting at a given node, passing through different elements to different nodes, and returning to the node at which the path began.

Mesh General classification for any set of branches in a network forming a closed path, representing the shortest possible loop in the network.

Mesh analysis Algebraic technique for writing the simultaneous linear equations in terms of unknown mesh currents in a network.

Network Interconnection of two or more circuit elements in series, parallel, or series–parallel.

Nodal analysis Algebraic technique for writing the simultaneous linear equations in terms of unknown node voltages in a network.

Node Junction of two or more branches of a network.

Path Trace taken through an electric network, starting at a node and passing through different circuit elements to different nodes.

Star General classification for any set of branches in a network in which one end of each branch is connected to a common terminal.

Wye configuration consisting of three star-connected elements in the shape of a wye (Y), also known as a tee (T).

8-10 PROBLEMS

Secs. 8-1 and 8-2

8-1 In the network of **Fig. 8-19**, $V_1 = V_2 = 6$ V, $r_1 = 6\ \Omega$, $r_2 = 3\ \Omega$, and $R_L = 4\ \Omega$; using the KVL/KCL equation method, find

a. I_1
b. I_2
c. I_L
d. V_L

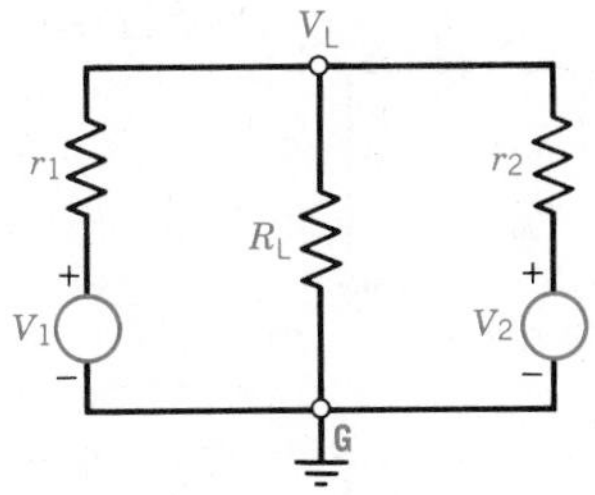

Figure 8-19 Problems 8-1 to 8-7 inclusive

8-2 Repeat all parts of Prob. 8-1, using polarities as shown in Fig. 8-19, when $V_1 = 6$ V, $r_1 = 2\ \Omega = r_2$, $V_2 = 4$ V, and $R_L = 4\ \Omega$

8-3 Repeat all parts of Prob. 8-1, using polarities as shown in Fig. 8-19, when $V_1 = 10$ V, $r_1 = 2\ \Omega$, $V_2 = 2$ V, $r_2 = 1\ \Omega$, and $R_L = 4\ \Omega$

8-4 From your solutions to Probs. 8-1 through 8-3,

a. When does a source deliver current to a node?
b. When does a source receive current from a node?
c. When does a source neither deliver to nor receive current from a node?
d. Is there any change of current in I_2 if r_2 is increased to 100 Ω in Prob. 8-2?
e. Why is the voltage across all three elements in all three problems the same?

8-5 Repeat all parts of Prob. 8-1, using the given values with only the polarity of V_2 reversed in Fig. 8-19. Draw the complete solution showing all currents and voltages, and verify V_L by three methods.

8-6 From your complete solution to Prob. 8-5, explain

a. Why current flows in R_L from ground (G) to node V_L
b. Why current I_1 flows from source V_1 to node V_L
c. Why current I_2 flows from node V_L to source V_2

Secs. 8-3 and 8-4

8-7 Using the format method of mesh analysis, for clockwise mesh currents I_1 and I_2, write two simultaneous linear equations given the circuit and values of Prob. 8-1. Calculate

a. I_1
b. I_2
c. I_L
d. V_L and compare your solution to that of Prob. 8-1

8-8 a. Given the circuit of **Fig. 8-20**, write two simultaneous linear equations and solve for current in R_2, using the format method of mesh analysis.
b. Verify your solution by simple series–parallel circuit theory.

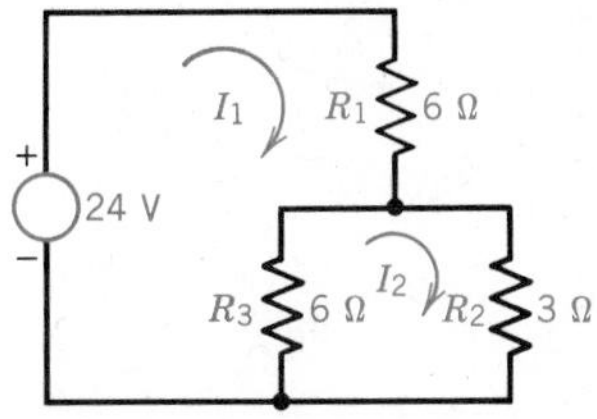

Figure 8-20 Problem 8-8

8-9 Using the format method of mesh analysis, given the circuit of **Fig. 8-21**, calculate

a. Current supplied by source V_1
b. Current supplied by source V_2
c. Voltage V
d. Current in the 10-kΩ resistor

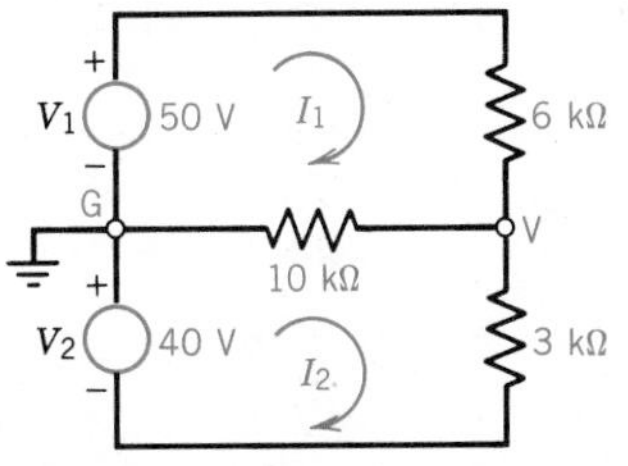

Figure 8-21 Problem 8-9

8-10 Given the circuit of **Fig. 8-22**, use the format method of mesh analysis to find

a. Current I_1 delivered to node V_L by source V_1
b. Current I_2 delivered to source V_2 by node V_L
c. Node voltage V_L
d. Current I_L in R_L
e. Explain why I_L is zero

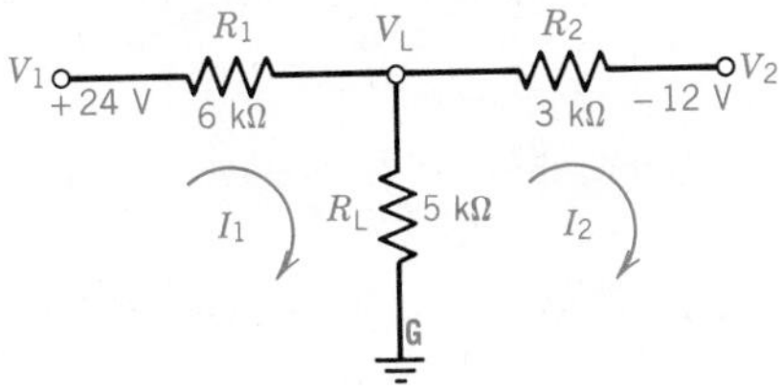

Figure 8-22 Problem 8-10

8-11 Using the format method of mesh analysis on the network of **Fig. 8-23**, find

a. Current delivered to node **X** by the current source, I_2 (Hint: convert source)
b. Current received from node **X** by the voltage source, V_1
c. Current in resistor R_L
d. Current in resistor R_2
e. Verify your solution, using KCL around node **X**
f. Draw the complete solution, showing all voltages and currents on Fig. 8-23

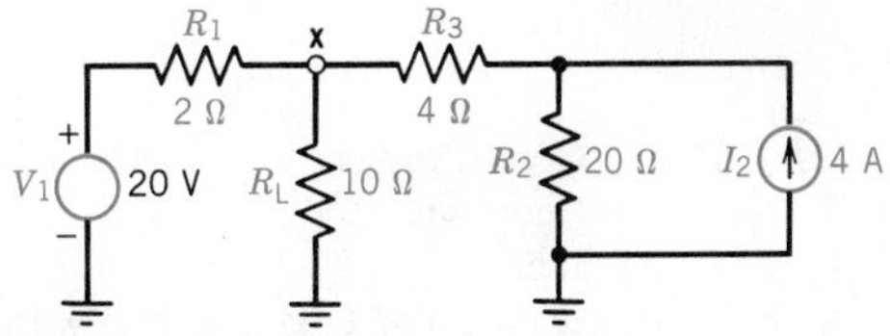

Figure 8-23 Problem 8-11

8-12 Using the format method of mesh analysis on the network of **Fig. 8-24**, find

a. Current in R_5
b. Current in R_4
c. Current in R_3
d. Voltage V_x
e. Verify your solution using KCL around node **X**
f. Verify your solution using KVL to find V_X by three different methods

Hint: In writing mesh equations make sure to add source voltages properly.

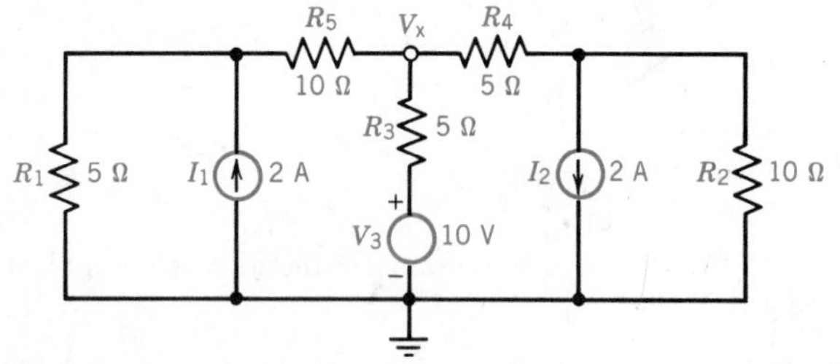

Figure 8-24 Problem 8-12

8-13 Using the format method of mesh analysis on the network of **Fig. 8-25**, calculate

a. Current supplied to the bridge by source V_1
b. Voltage at node **B**, V_B
c. Voltage at node **C**, V_C
d. Current in resistor R_6
e. Draw the complete solution, showing all voltages and currents on Fig. 8-25

Hint: Three mesh currents are required but only one contains the source voltage.

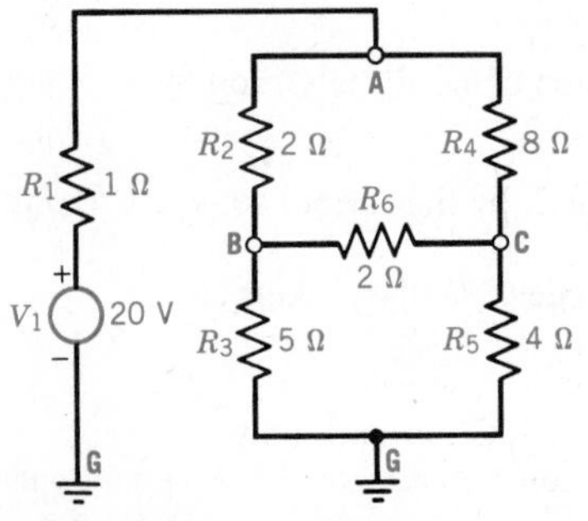

Figure 8-25 Problem 8-13

8-14 Using the format method of mesh analysis on the network of **Fig. 8-26**, find

a. Mesh currents I_1, I_2, and I_3
b. Voltages at nodes **A**, **B**, and **C** with respect to ground, **G**
c. Verify $I_{R_3} = I_2 = 1$ A by two methods
d. Verify $I_{R_4} = 3$ A by two methods
e. Explain why the node voltage at point **C** is negative with respect to ground

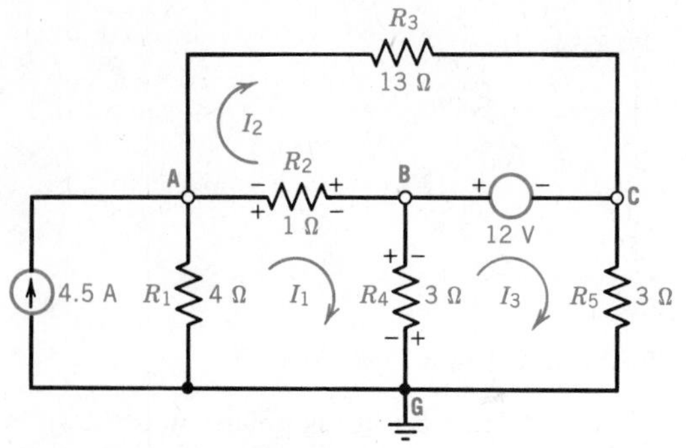

Figure 8-26 Problem 8-14

8-15 Using the format method of mesh analysis on the network of **Fig. 8-27**, find

a. Mesh currents I_1, I_2, and I_3
b. Voltages at nodes **A**, **B**, and **C** with respect to ground
c. Verify $I_{R_2} = 1.\overline{3}$ A by two methods
d. Verify $I_{R_3} = 0.\overline{33}$ A by two methods
e. Verify $I_{R_4} = 1.\overline{6}$ A by two methods

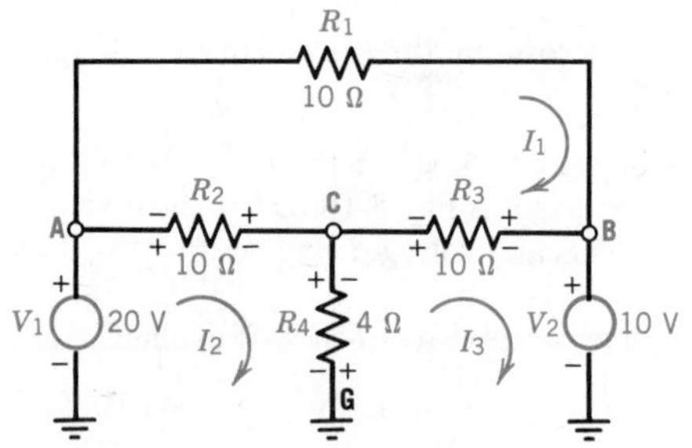

Figure 8-27 Problem 8-15

8-16 Using the format method of mesh analysis on the network of **Fig. 8-28**, find

a. Mesh currents I_1, I_2, and I_3
b. Currents in R_2 and R_6
c. Voltages at points **A**, **B**, **C**, and **D** with respect to ground **G**
d. Potential difference between points **D** and **C**, V_{DC}

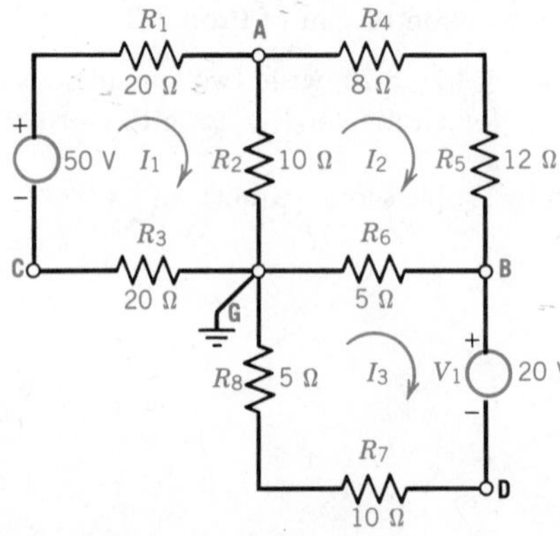

Figure 8-28 Problem 8-16

e. Draw the complete solution showing all voltages and current on Fig. 8-28
f. Use KVL to verify that the sum of all volt drops in mesh I_2 is zero
g. Use KCL to verify that the sum of all currents entering and leaving node **G** is zero

8-17 Using the format method of mesh analysis on the network of **Fig. 8-29**, find

a. Mesh currents I_1, I_2, and I_3
b. Currents in sources V_1, V_2, and V_3 to or from ground **G**
c. Draw the complete solution, showing all currents, their true direction, and voltages at nodes **A**, **B**, and **C**
d. Show that I_{R_1}, I_{R_2}, and I_{R_3} may be found directly without resorting to mesh analysis using Ohm's law

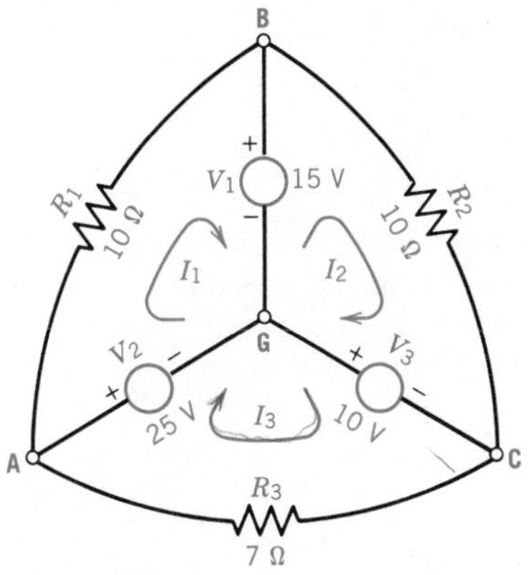

Figure 8-29 Problem 8-17

8-18 Using the format method of mesh analysis on the network of **Fig. 8-30**, find

a. Mesh currents I_1, I_2, and I_3
b. Currents in R_1, R_2, R_3, R_4, and R_5
c. Voltages at nodes **A**, **B**, and **C**

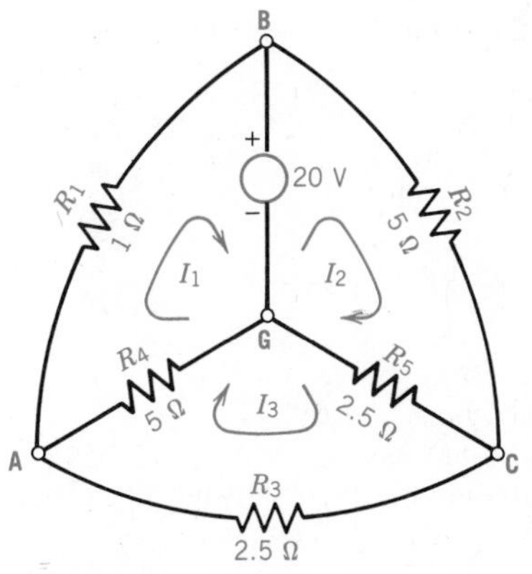

Figure 8-30 Problem 8-18

d. Current supplied by the 20-V source to node **B**
e. Draw the complete solution, showing all currents, their true direction, and all node voltages

Secs. 8-5 and 8-6

8-19 Using the format method of *nodal* analysis on the network of Fig. 8-1a (Ex. 8-1), calculate

a. Voltage V_X
b. Current in R_3
c. Current to node **X** from source V_1
d. Current to node **X** from source V_2

8-20 Using the format method of nodal analysis on the network of Fig. 8-3a (Ex. 8-4), calculate

a. V_X
b. Current to node **X** from source V_1
c. Current from node **X** to source V_2
d. Current to source V_3 from node **X**
e. Verify KCL around node **X**

8-21 Using the format method of nodal analysis on the network of Fig. 8-6a (Ex. 8-8), calculate

a. V_x
b. V_y
c. Current in R_L

8-22 Using the format method of nodal analysis on the network of Fig. 8-7a (Ex. 8-9), calculate

a. V_x
b. V_y
c. All currents leaving and entering nodes **x** and **y**

8-23 Using the format method of nodal analysis on the network of **Fig. 8-31**, calculate

a. V_A
b. V_B
c. Current supplied by V_1 to node **A**
d. Current supplied by V_2 to node **B**
e. Using the original circuit of Fig. 8-31, draw the complete solution showing all currents and voltages

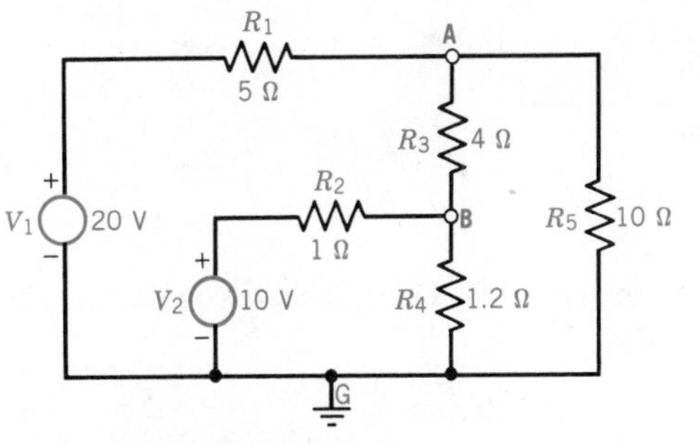

Figure 8-31 Problem 8-23

8-24 Using the format method of nodal analysis on the network of **Fig. 8-32**, calculate

a. Voltage V
b. Current to node **V** from V_1
c. Current to node **V** from V_3
d. Current in R_2 and its direction with respect to ground
e. Using the original circuit of Fig. 8-32, draw the complete solution showing all currents and voltages, ensuring that all currents around node **V** conform to KCL

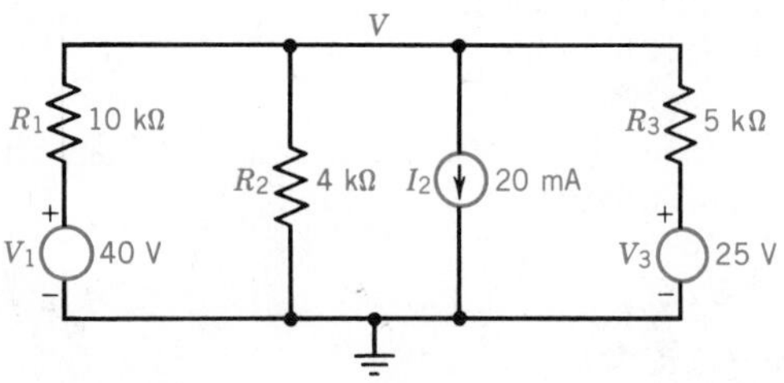

Figure 8-32 Problem 8-24

8-25 Using the format method of nodal analysis on the network of **Fig. 8-33**, calculate

a. Node voltage V
b. Voltage at point x with respect to ground (G)
c. Voltage across ideal current source I_2
d. Voltage across ideal current source I_3

Hint:

1. It is unnecessary to convert V_2 to a practical current source since we already know the current leaving node V in that branch is -10 mA from ideal current source I_2.
2. Verify that $V = 10$ V by superposition, which proves that V_2 has no effect in the third branch because I_2 is open when we desire the effect of V_2.

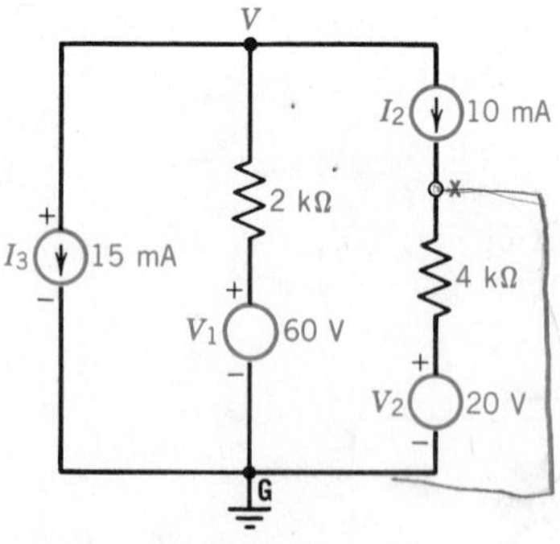

Figure 8-33 Problem 8-25

8-26 Using the format method of nodal analysis on the network of Fig. 8-34, calculate

a. V_1
b. V_2
c. I_L
d. I_3
e. I_4
f. Show the complete solution of all voltages and currents on the original **Fig. 8-34**. Check to see that KCL around nodes V_1 and V_2 and KVL for all three meshes are verified

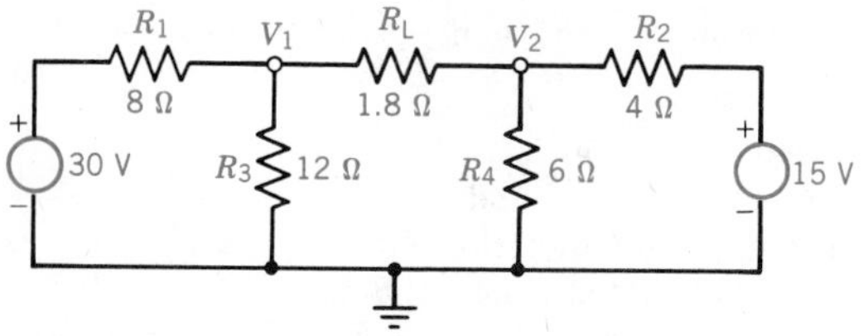

Figure 8-34 Problem 8-26

8-27 Using the format method of nodal analysis on the network of **Fig. 8-35**, calculate

a. V_A
b. V_B
c. I_{R_4}
d. I_{R_2}
e. Current supplied by voltage source V_1 to node V_A
f. Show the complete solution of all voltages and currents on the original Fig. 8-35
g. Explain why the current I_{R_2} is zero by verifying KCL around node A

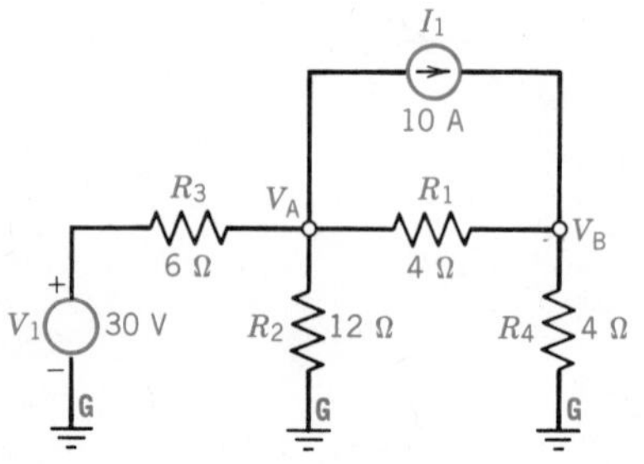

Figure 8-35 Problem 8-27

8-28 Using the format method of nodal analysis on the network of **Fig. 8-36**, calculate

a. V_1
b. V_2
c. V_3
d. I_{R_1}
e. I_{R_2}
f. I_{R_3}
g. Voltage across the ideal 5-mA source
h. Voltage across the ideal 10-mA source
i. Voltage across the ideal 6-mA source
j. Draw the complete solution to Fig. 8-36, showing all currents and voltages
k. Verify KCL around node **G**
l. Verify KVL for the mesh containing the current sources

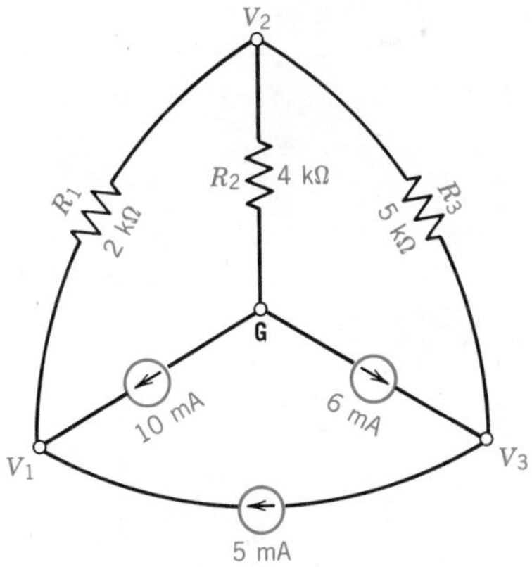

Figure 8-36 Problem 8-28

8-29 Using the format method of nodal analysis on the network of **Fig. 8-37**, calculate

a. V_1
b. V_2
c. V_3
d. I_{R_1}
e. I_{R_2}
f. I_{R_3} and I_{R_4}
g. Voltage across the 4-A source
h. Voltage across the ideal 6-A source
i. Draw the complete solution to Fig. 8-37 showing all voltages and currents
j. Verify KCL around node **G**
k. Verify KVL for the mesh containing the 4-A source

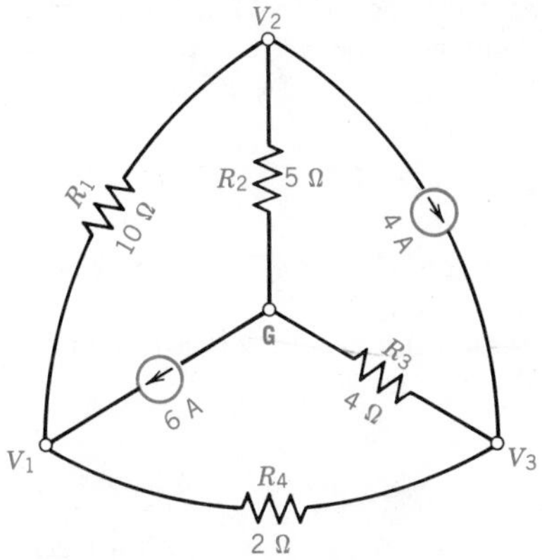

Figure 8-37 Problem 8-29

8-30 Using the format method of nodal analysis on the network of **Fig. 8-38**, calculate

a. V_1
b. V_2
c. V_3
d. I_{R_1}, I_{R_2}, and I_{R_3}
e. Verify KCL for currents entering and leaving node **G**
f. Calculate I_A delivered by source V_A
g. Calculate I_B delivered by source V_B
h. Calculate I_C delivered by source V_C
i. Draw the complete solution to Fig. 8-38 showing all voltages and currents

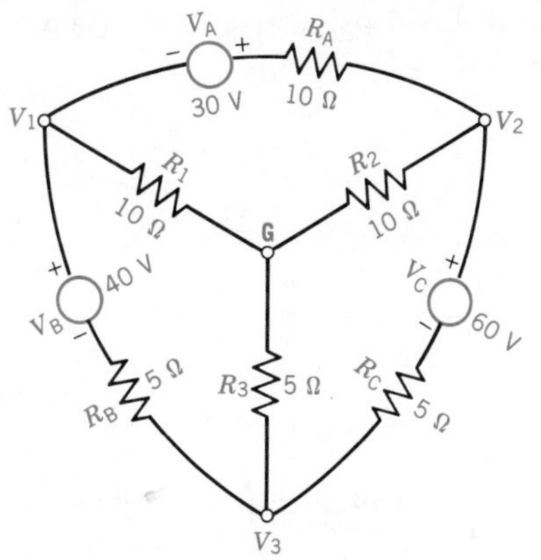

Figure 8-38 Problem 8-30

8-31 Using the format method of nodal analysis on the unbalanced bridge network of Fig. 8-25, previously solved in Prob. 8-13 by mesh analysis, calculate

a. V_A
b. V_B
c. V_C
d. Current in R_6
e. Compare your solution with that of Prob. 8-13

8-32 Using the format method of nodal analysis on the network of **Fig. 8-39**, calculate

a. V_1
b. V_2
c. V_3
d. I_{R_1} and I_{R_3}
e. I_{R_2}
f. I_{R_4}
g. Voltage across the ideal current source I_1
h. Voltage across the ideal current source I_2
i. Show the complete solution of all currents and voltages for Fig. 8-39
j. Using KCL, verify currents entering and leaving node **G**
k. Verify KVL for the mesh containing the ideal current source I_1

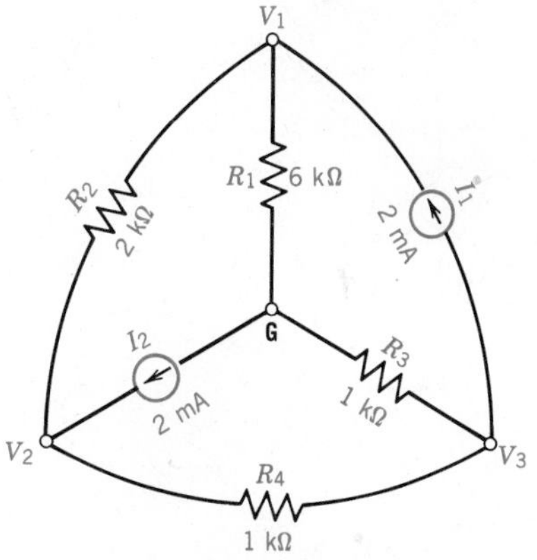

Figure 8-39 Problem 8-32

8-33 Using the format method of nodal analysis on the complex network of **Fig. 8-40**, calculate

a. V_A
b. V_B
c. V_C
d. Current in R_6
e. Current in R_3
f. Current in R_5 delivered by V_2
g. I_2 as the sum of I_6 and I_3
h. Current in R_4
i. Current supplied by V_1 to node **A** (hint: use $V_C + V_1$)
j. Current in R_2
k. Show the complete solution of all currents and voltages for Fig. 8-40
l. Using KCL, verify currents entering and leaving node **A**
m. Repeat (**l**) for node **B**
n. Repeat (**l**) for node **C**
o. Repeat (**l**) for node **G**

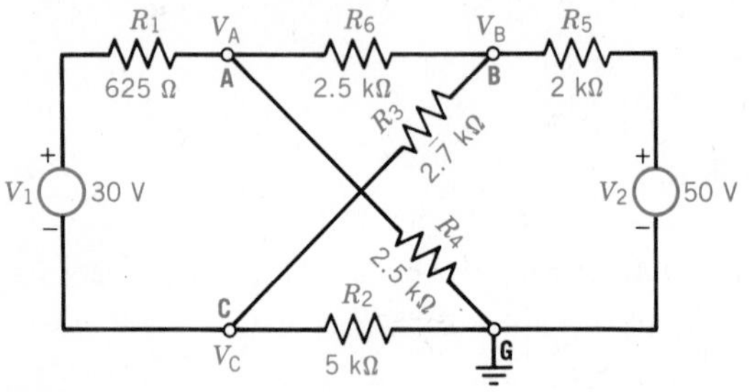

Figure 8-40 Problem 8-33

8-34 Using the format method of nodal analysis on the network of **Fig. 8-41**, calculate

a. V_A
b. V_B
c. V_C
d. Currents in R_1 through R_5
e. Show the complete solution of all currents and voltages for Fig. 8-41
f. Verify your solution, using KCL around nodes **A**, **B**, and **C**
g. Calculate the voltage across the ideal 10-A current source

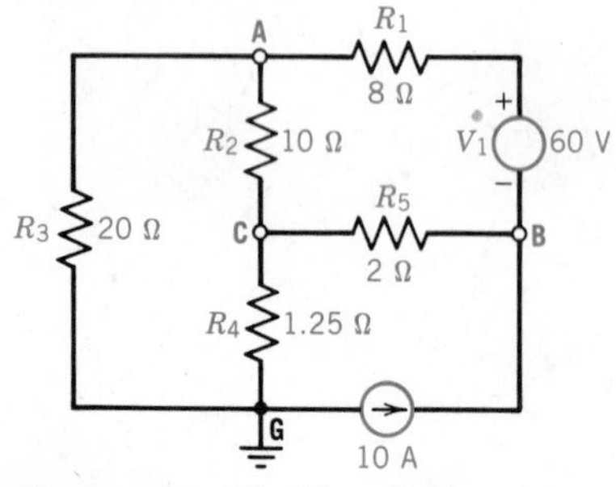

Figure 8-41 Problem 8-34

8-35 Given the network shown in **Fig. 8-42**, using any method you prefer, calculate

a. V_X
b. V_Y
c. V_Z
d. True currents in R_1 through R_7
e. Verify your solution, using KCL around nodes **X**, **Y**, and **Z**
f. Show the complete solution of all currents and voltages for Fig. 8-42

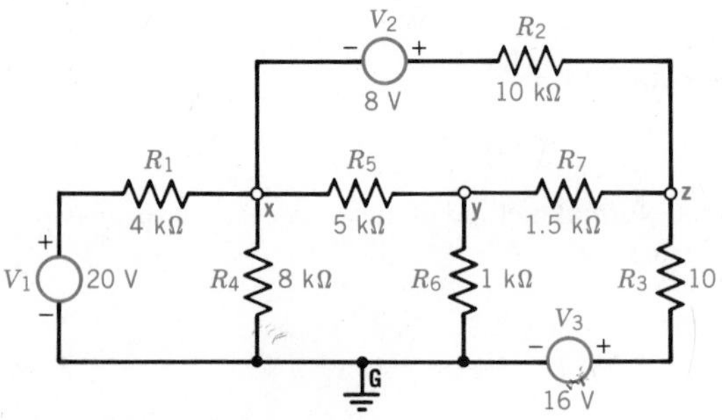

Figure 8-42 Problem 8-35

Secs. 8-7 and 8-8

8-36 Given the unbalanced bridge network of Fig. 8-25, convert the Δ network of **A**, **B**, **C** to an equivalent Y and calculate

a. V_A
b. V_B
c. V_C
d. Current in R_3, R_5, R_2, and R_4
e. Draw the complete solution of all currents and voltages for Fig. 8-25
f. Compare your results with solutions of Prob. 8-13 by mesh analysis and Prob. 8-31 by nodal analysis

8-37 Given the four-mesh, four-node network of **Fig. 8-43**,

a. Convert the network into an equivalent two-mesh network, using the Δ–Y transformation on both mesh ABDA and mesh DCGD
b. Draw the equivalent two-mesh network as a result of the above conversion
c. Combine series-connected elements in (**b**) and, using simple series–parallel theory, calculate voltages at nodes **A**, **B**, **D**, C, and **G**
d. Transfer voltages obtained in (**c**) to original network shown in Fig. 8-43. Find all currents in original components R_1 through R_8
e. Reduce the network of Fig. 8-43 to a single equivalent resistance

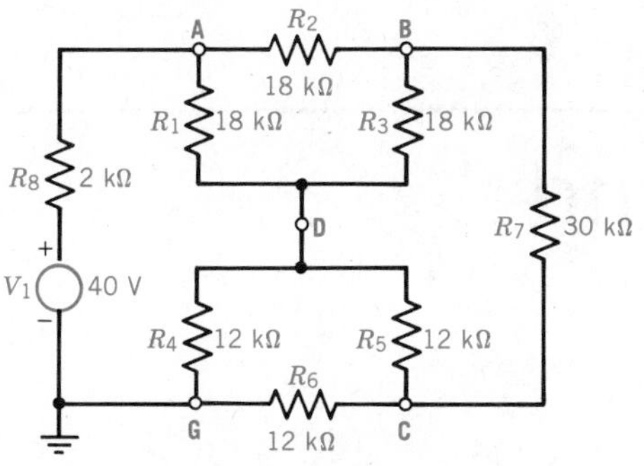

Figure 8-43 Problem 8-37

8-38 Given the asymmetrical bridged-T network of **Fig. 8-44** with three meshes,

a. Convert the delta of R_1, R_2, and R_3 to an equivalent Y
b. Draw the equivalent two-mesh network as a result of the conversion in (**a**)
c. Calculate the current drawn from V_1 and voltages at nodes **A** and **B**
d. Transfer the information calculated in (**c**) to the original network of Fig. 8-44
e. Calculate currents in resistors R_1 through R_5
f. Show the complete solution for Fig. 8-44

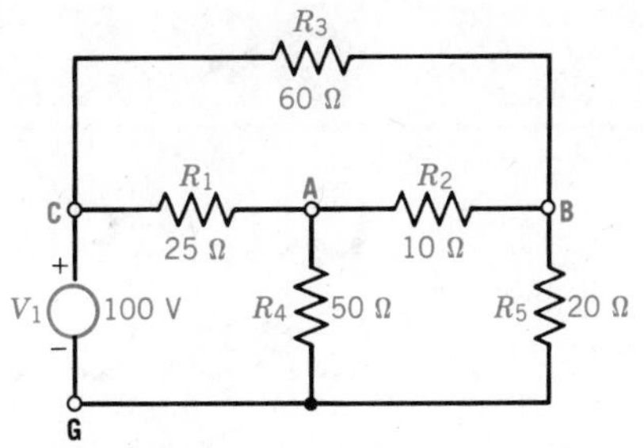

Figure 8-44 Problem 8-38

8-39 Replace V_1 in Fig. 8-44 with an *ideal* current source, $I_1 = 3$ A. Using nodal analysis, find voltages at

a. Node **C**
b. Node **A**
c. Node **B**
d. Compare your answers with those obtained in Prob. 8-38

8-40 Using the format method of mesh analysis on the network of Fig. 8-44, calculate

a. All currents in all resistors, R_1 through R_5
b. Node voltages V_A and V_B

8-11 ANSWERS

8-1 a 1/3 A b 2/3 A c 1 A d 4 V
8-2 a 1 A b 0 c 1 A d 4 V
8-3 a 3 A b 2 A c 1 A d 4 V
8-5 a $1.\overline{2}$ A b $1.\overline{5}$ A c $0.\overline{3}$ A d $-1.\overline{3}$ V
8-7 a 1/3 A b −2/3 A c 1 A d 4 V
8-8 a 2 A
8-9 a $9.7\overline{2}$ mA b $10.\overline{5}$ mA c $8.\overline{3}$ V d $0.8\overline{3}$ mA
8-10 a 4 mA b 4 mA c 0 V d 0 A
8-11 a 2.4675 A b −0.3896 A c 2.078 A d 1.53 A
8-12 a 0.4 A b 1.6 A c 1.2 A d 4 V
8-13 a 4 A b 10 V c 8 V d 1 A
8-14 a 2 A, 1 A, −1 A b 10 V, 9 V, −3 V
8-15 a 1 A, 7/3 A, 5/3 A b 20 V, 10 V, $6.\overline{6}$ V c $1.\overline{3}$ A
8-16 a 1.0315 A, 0.1575 A, −0.9606 A b 0.874 A, 1.118 A
c 8.74, 5.59, −20.63, −14.41 V d 6.22 V
8-17 a 1 A, 2.5 A, −5 A b 1.5 A, 6 A, 7.5 A
c 25 V, 15 V, −10 V
8-18 a −5 A, 2 A, −2 A b 5 A, 2 A, 2 A, 3 A, 4 A
c 15 V, 20 V, 10 V d 7 A
8-19 a 6 V b 3 A c 2 A d 1 A
8-20 a 10 V b 5 A c 1 A d 4 A
8-21 a 56.32 V b −6.08 V c 6.24 A
8-22 a 15 V b 6 V
8-23 a 10 V b 6 V c 2 A d 4 A
8-24 a −20 V b 6 mA c 9 mA d 5 mA to node V
8-25 a 10 V b 60 V c −50 V d 10 V
8-26 a 13.2 V b 11.4 V c 1 A d 1.1 A e 1.9 A
8-27 a 0 b 20 V c 5 A d 0 e 5 A
8-28 a 94 V b 64 V c 69 V d 15 mA e 16 mA
f 1 mA g 25 V h 94 V i 69 V
8-29 a 31.43 V b −2.857 V c 26.286 V d 3.429 A
e −0.571 A f 6.571 A g 29.143 V h 31.43 V
8-30 a 10 V b 30 V c −20 V d 1 A f 1 A
g 2 A h 2 A
8-31 a 16 V b 10 V c 8 V d 1 A
8-32 a 6 V b 4 V c 1 V d, e 1 mA f 3 mA
g 5 V h 4 V
8-33 a 25.92 V b 28.57 V c 1.74 V d 1.059 mA
e 9.657 mA f 10.72 mA g 10.37 mA
h 10.37 mA i 9.315 mA j 0.348 mA
8-34 a 40 V b 20 V c 10 V
8-35 a 8 V b 3 V c 6 V
8-36 a 16 V b 10 V c 8 V d 2 A, 2 A, 3 A, 2 A
8-37 d 1.2 mA, 0.8 mA, 0.4 mA, 1.2 mA, 0.4 mA, 0.8 mA,
0.4 mA, 2 mA e 20 kΩ
8-38 e 2 A, 1 A, 1 A, 1 A, 2 A
8-39 a 100 V b 50 V c 40 V
8-40 a 2 A, 1 A, 1 A, 1 A, 2 A b 50 V, 40 V

CHAPTER 9

Capacitance and Capacitors

9-1 INTRODUCTION

The previous eight chapters used one (and only one) linear, passive, bilateral, and ideal circuit element—resistance. In all of the previously purely linear resistive networks, the circuit responses (in terms of currents and/or voltages) depended only on the magnitude and polarities of the independent voltage (or current) sources (or forcing functions).

In this chapter we introduce a *second* passive, bilateral, linear[1] and ideal circuit element—*capacitance.* Unlike the resistor, which has the property of dissipating electrical energy, the (ideal) capacitor is incapable of dissipating energy. The capacitor is essentially an energy-storing circuit element.

We will discover that because of these unique properties (energy storage and zero power dissipation), circuits containing either capacitance only or a combination of resistance and capacitance exhibit voltage and current responses different from those of purely resistive circuits. The response of capacitive circuits will now depend on both the nature of the sources (forcing functions) and the amount of energy stored in capacitance as a result of its previous history. Analysis of such networks containing linear capacitive elements requires more than simple linear algebraic equations such as those used for ideal resistive networks. Instead, we will use relatively simple equations that are derived from more complex integro-differential equations.

As a trade-off for this added complexity, capacitors yield four very useful circuit properties not found in resistive circuits:

1. They can detect (sense) differences between dc and ac signals as well as differences between ac signals of different frequencies. This frequency-sensitive property lends itself to producing a variety of interesting waveforms (Chapter 10).
2. Capacitors tend to oppose a change in voltage across their terminals whenever the *voltage* is *changing.* This property produces capacitive opposition to alternating current (ac). At very *high* frequencies, capacitors behave as *short*

[1] The subject of *nonlinear* capacitors is outside the scope of this text.

circuits. Conversely, at very *low* frequencies or to direct current, they behave as open circuits.

3. Capacitors are capable of *temporary energy storage* and *energy release*, depending on the polarity and magnitude of voltages across their terminals. This enables them to serve as temporary voltage (and/or current) sources, as in the case of a photographic or stroboscopic flash unit.
4. While doing all of the above, (ideal) capacitors do not dissipate (nor can they generate) electrical energy. An ideal capacitor can only deliver as much energy to a network as was originally stored in its dielectric field. Even practical capacitors are relatively lossless in most network applications.

9-2 ELECTRIC FIELD AND ELECTRIC FIELD INTENSITY—A REVIEW

We began our study of the electric field in Chapter 2, where we introduced the laws of electrostatic replusion and/or attraction (Sec. 2-1a) and Coulomb's law (Sec. 2-1b) quantifying the forces of attraction or repulsion between two charged bodies:

$$\mathbf{F} = k\frac{Q_1Q_2}{r^2} = \left(\frac{1}{4\pi\varepsilon_0}\right) \times \frac{Q_1Q_2}{r^2} \cong \frac{9 \times 10^9\, Q_1Q_2}{r^2} \quad \text{newtons} \qquad (2\text{-}1)$$

In Sec. 2-2 we introduced the concept of electric field intensity, **E**, as a measure of the *electric field strength* at any point in an electric field, measured as the force exerted by the field on a *unit positive charge* placed at that point. We noted that electric field strength, **E**, is a vector quantity, having both magnitude and direction, measured in units of newtons per coulomb (N/C):

$$\mathbf{E} = \frac{\mathbf{F}}{q} \equiv \frac{V}{d} \equiv \frac{kQ}{d^2} \cong \frac{9 \times 10^9\, Q}{d^2} \quad \text{(N/C or V/m)} \qquad (2\text{-}2)$$

where all terms have been defined (Sec. 2-2).

The importance of Eq. (2-2) and its relation to the property of capacitance was shown in Ex. 2-6. Knowing the electric field intensity enabled computation of Q, the charge on the plates of a parallel-plate capacitor, and ultimately the value of C, the capacitance of the capacitor.

In Sec. 2-2.1 we also introduced the concept of electric flux density, **D**, and showed how it differs from electric field intensity, **E**, and how it is related to it. We learned that electric flux density is proportional to electric field intensity, and the constant of proportionality is ε_0, the permittivity of free space:[2]

$$\mathbf{D} = \frac{\psi}{A} = \frac{Q}{4\pi r^2} = \varepsilon_0\mathbf{E} \quad \text{coulombs/square meter (C/m}^2\text{)} \qquad (2\text{-}3)$$

where all terms have been defined.

We learned that one line of electric field flux is the same as 1 C of charge in the SI or that $\psi \equiv Q$ in Eq. (2-3).

Finally, in Chapter 2, we introduced the parallel-plate capacitor (Fig. 2-1) but reserved discussion of the properties of capacitance for this chapter. We are now armed and ready to consider the property of capacitance and the factors affecting it.

[2] See footnote to Eq. (2-3) and note that ε_0 is sometimes called absolute capacitivity.

9-3 CAPACITANCE AS A CIRCUIT PROPERTY

Any time we have two conductors separated by an insulating material (a dielectric) we have the property of capacitance. **Figure 9-1a** shows a typical parallel-plate capacitor with air between the plates as a dielectric. A method of increasing the surface area of the capacitor is shown in Fig. 9-lb, using interleaved plates with air as the dielectric. If these plates are pulled apart, the common area between them is reduced and the capacitance would be reduced. Obviously, if the interleaved plates are *completely* separated, the capacitance would be zero. There are better dielectric materials than air, and these produce higher values of capacitance. Figure 9-1c shows a "sandwich" consisting of two conductors with a dielectric between them. Commercial mica capacitors based on this principle are fairly common.

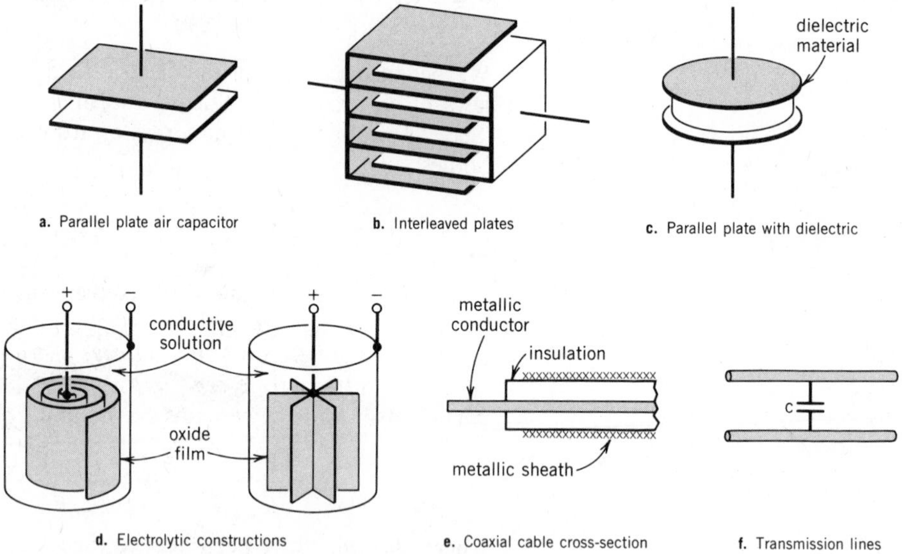

Figure 9-1 Property of capacitance

Aluminum is a fairly good conductor, but when it is oxidized, the aluminum oxide film on the surface is an insulator. Figure 9-1d shows two typical constructions used to increase the surface area, in which one plate is made of aluminum that has been oxidized at its surface and the other is a conductive electrolytic solution, contained in a sealed can. The constructions are designed to maximum the surface area of the aluminum within a given can volume. Figure 9-1e shows that even a coaxial cable can exhibit the property of capacitance since its inner conductor is insulated and the metallic sheath around the outside of the cable serves as a second conductor.

Indeed, two transmission lines running parallel to each other between poles, separated by air, as shown in Fig. 9-1f, have the property of distributed capacitance (i.e., two parallel linear conductors separated by a dielectric).

Figure 9-2 shows the capacitor of Fig. 9-1c connected to an ideal voltage source, V, via a switch, S. When switch S is first closed, conventional current flows from the source to the capacitor. The upper plates of the capacitor are positively charged, as shown in Fig. 9-2, while the lower plates are negatively charged. In effect, electrons have been removed from the upper plate and the same number added to the lower plate. Ultimately, the ammeter, A, reads zero current and the capacitor is fully charged to the source voltage value, as indicated by the ideal voltmeter, which records the same voltage as that of source V. If switch S is opened, the

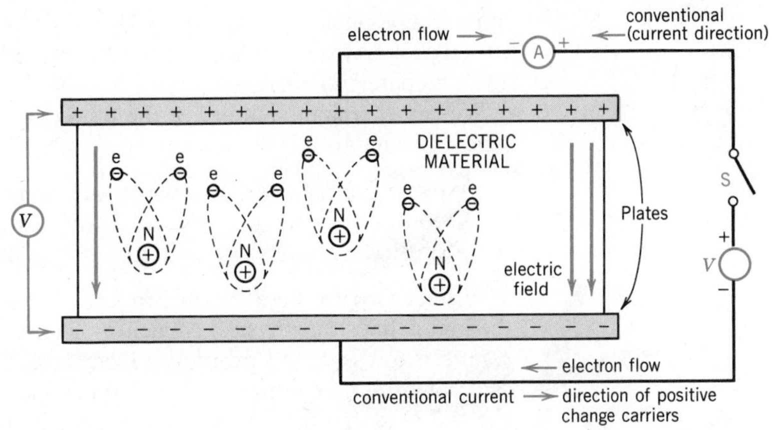

Figure 9-2 Charging current and dielectric stress in a parallel plate capacitor

capacitor remains charged and the ideal (infinite resistance) voltmeter continues to record source voltage, V. Closing the switch a second time produces no current flow from the source since the capacitor voltage is equal and opposite to the source voltage.

But if, somehow, source voltage V could be increased, then when switch S was closed, additional current would flow to increase the number of positive charges on the upper plate and negative charges on the lower plate. Again, when the ammeter current is zero, the voltmeter would record the new value of increased voltage, V, across the capacitor.

Clearly, we can say that the charge on the capacitor, Q, is proportional to the voltage applied to the capacitor. We can write

$$Q \propto V = kV = C \cdot V = Q \qquad \text{coulombs (C)} \tag{9-1}$$

where C is the capacitance of the capacitor, in farads (F)

V is the voltage across the capacitor, in volts (V)

Note that in Eq. (9-1) the value of C is the constant of proportionality between the charge on the plates of the capacitor and the voltage across the capacitor.

Since the capacitor is a circuit element, much like a resistor, we can also write

$$C = \frac{Q}{V} = k \qquad \text{farads (F)} \tag{9-1a}$$

The significance of Eq. (9-1a) is the same as that of Eq. (3-1) expressing Ohm's law for a resistance, $R = V/I = k$. In effect, Eq. (9-1a) states that the ratio of charge on the plates in proportion to the voltage across the plates of a given capacitor *is always constant.*

But Eq. (9-1a) carries a far more important implication. Since the capacitance, C, is a constant for a given capacitor (depending on the geometry of the capacitor and the material of its dielectric), the equation implies that as long as there is a charge on the plates of the capacitor, there will be a voltage, V_C, across it. We observed in Fig. 9-2 that, once charged, whether we open or close switch S, the voltage across the capacitor, V_C, *remains the same.* Indeed, this is an essential property only possessed by capacitors and/or capacitive circuits, namely a tendency to oppose any *change* in *voltage* across their terminals.[3] Since resistance is the only circuit property that can dissipate energy, an ideal capacitor contains no resistance and dissipates no energy.

We can therefore summarize the following circuit properties regarding capacitance. Capacitance is a circuit property that

1. Exists whenever two conductors are separated by a dielectric.
2. Opposes any *change* in *voltage* across its terminals.
3. Stores electrical energy by virtue of the electric field across its dielectric.

With regard to these circuit properties of capacitance, it should be noted that the first defines the physical capacitor. The second, as we will eventually discover, gives rise to the circuit property called capacitive reactance, that is, opposition to ac created by a capacitor. But it also produces a number of interesting effects. In Fig. 9-2 note that the electric field, created by potential difference across the plates of the capacitor, tends to stress the atoms of the dielectric. Electrons in the orbits of each atom of the insulator are repelled from the bottom plate and attracted to the top plate, distorting the electron orbits. (In the absence of an electric field across a dielectric, the nucleus is in the geometric center of all electron orbits.) This phenomenon is known as *dielectric stress* or *polarization.*

If the capacitor is discharged, by temporarily connecting a wire between its positive and negative plates, one would assume that the electric field disappears and the voltage across the plates is zero. But when we remove the short across the plates, unfortunately, some of the atomic orbits remain stressed, producing a small field and a small charge across the plates. This effect, also known as *dielectric absorption,* is very significant, particularly in the case of large capacitors (such as those used in power factor correction). Dangerous electric shocks have been received from capacitors (and even the de-energized windings of large transformers and alternators) that continue to store in their insulation a dielectric charge long after the power has been removed, and such capacitors have been temporarily discharged by shorting their terminal leads. A second effect, known as *dielectric hysteresis,*[4] is also a consequence of this second property (i.e., the tendency to oppose any voltage change across its terminals) which is characteristic only of capacitance.

The third property, that of energy *storage* by virtue of capacitance, is covered in much detail in Sec. 9-10. But if we recall our definition of potential as work per unit charge[5] ($V = W/Q$, Eq. 2-6), we realize that in order to charge a capacitor to some voltage, V_C, across its plates, work must be done and this work is NOT simply the product QV_C. The energy stored in a capacitor as derived in Appendix B-3 is

$$W = \frac{V_C \times Q}{2} = \frac{CV_C^2}{2} = \frac{Q^2}{2C} \quad \text{joules (J)} \tag{9-2}$$

[3] In this sense, capacitance is analogous to resistance, which is the property of an electric circuit to oppose a *steady* current through the circuit as well as to dissipate electrical energy.

[4] Dielectric hysteresis is one of the factors in effective or ac resistance (Sec. 17-8).

[5] During the process of charging the capacitor, half the total energy is dissipated in the connecting leads between the source and the capacitor, and half the total energy ($W_t = QV$) appears in the electric field of the dielectric, by Eq. (9-2). This is but one particular case where the law of conservation of energy appears to be (but is not) violated (see Sec. 9-10).

where V_C is the voltage across the capacitor in volts (V)
Q is the charge on the capacitor plates in coulombs (C)

EXAMPLE 9-1

A 100-V source charges a capacitor fully so that 20 mC of charge is added and removed from each plate, respectively. Calculate

a. The capacitance of the capacitor
b. The energy stored in the magnetic field, using the relation $VQ/2$
c. Repeat (**b**) using the relations $CV^2/2$ and $Q^2/2C$

Solution

a. $C = Q/V = 20\text{ mC}/100\text{ V} = \mathbf{200\ \mu F}$
b. $W = V_C Q/2 = (100\text{ V})(20\text{ mC})/2 = \mathbf{1\ J}$
c. $W = CV^2/2 = (200\ \mu\text{F})(100\text{ V})^2/2 = \mathbf{1\ J}$; $W = Q^2/2C = (20\text{ mC})^2/2 \times 200\ \mu\text{F} = \mathbf{1\ J}$ (9-2)

9-4 FACTORS AFFECTING CAPACITANCE

From our previous study of resistance and factors affecting resistance, recall that in the relation $R = \rho l/A$, two of the factors are geometric (length and cross-sectional area) and the third, resistivity (ρ), is a function of the material comprising the resistor construction. The derivation of the factors affecting capacitance uses the equations from Chapter 2 that appear in Sec. 9-2. We begin with $\mathbf{E} = \dfrac{Q}{(4\pi r^2)\varepsilon_0} = V/d$ from Eqs. (2-3) and (2-2), respectively, from which $V = Qd/\varepsilon_0(4\pi r^2) = Qd/\varepsilon_0 A$. But $C = Q/V$ from Eq. (9-1), and therefore

$$C = \frac{Q}{V} = \frac{Q\varepsilon_0 A}{Qd} = \frac{\varepsilon_0 A}{d} \quad \text{farads (F)} \qquad (9\text{-}3)$$

where ε_0 is the permittivity of vacuum, or 8.8542×10^{-12} F/m (farads/meter)
A is the area of the plates of the capacitor
d is the distance between plates (separated in space by an absolute vacuum)

Note that, as for resistance, the factors affecting capacitance include two geometric factors (area of and distance between plates) and one factor that is a function of the material (or lack of it) between the plates.

We could have derived Eq. (9-3) intuitively, as well. Since $C = Q/V_C$ for a given capacitor of a given plate area, the potential ($V = V_C$) applied to the capacitor will produce a charge, Q, and the same number of electric lines of flux, ψ, since $\psi = Q$. These lines of force repel each other. At a higher voltage there is more repulsion; at a lower voltage there is less charge and less repulsion. But if we now double the area of the plates and apply the same voltage to the capacitor, there is twice as much room for flux lines and charge. For the same applied voltage, the charge is now twice as great to produce the same repulsion on twice the area. Since $C = 2Q/V_C$, the doubling of area has doubled the capacitance.

A similar analysis applies to the distance between the plates, which is determined by the thickness of the dielectric material. Coulomb's law shows that as we decrease the distance between two oppositely charged bodies, the force of attraction between them (and repulsion between electric lines of force) increases. But increased repulsion implies an increase in charge for the same voltage across the capacitor. Consequently, as the distance between plates decreases, the capacitance increases.

9-5 DIELECTRIC CONSTANTS AND DIELECTRIC MATERIALS

Of course, *few* capacitors have a vacuum between their plates. Depending on the dielectric material used in the capacitor, for the same area and distance between plates, there will be a change in capacitance. A correction factor is needed in Eq. (9-3) to enable the prediction of capacitance for dielectrics other than free space. This correction factor is called the dielectric constant, k. It is determined experimentally as shown in **Fig. 9-3**.

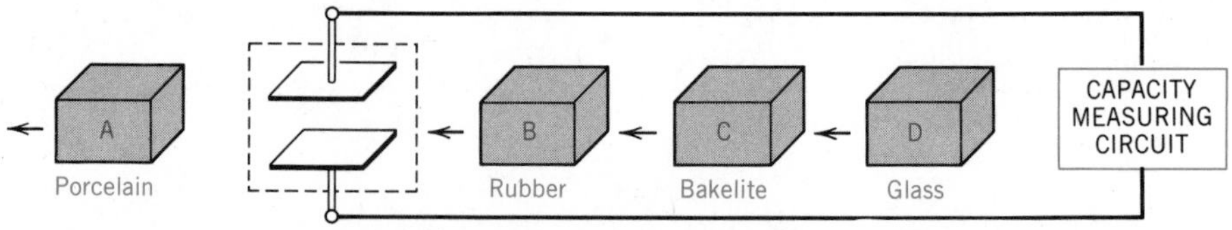

Figure 9-3 Determination of dielectric constant

If the plates of a parallel plate capacitor are rigidly held so that the distance between them is constant and samples of various dielectrics are inserted between them, the change in capacitance is proportional to the dielectric constant of the sample inserted. Let us assume that the air is evacuated from between the plates and that the capacitance so measured is 1 μF. If the air is now pumped back into the measuring chamber, the capacitance increases to 1.0006 μF. If a block of porcelain is now inserted between the plates, the capacitance increases to approximately 6 μF. Bakelite will yield a capacitance of 5 μF and so on, as shown in Fig. 9-3. We can now define the dielectric constant, k, as a unitless ratio[6] where

$$k = \frac{\varepsilon}{\varepsilon_0} = \varepsilon_r \qquad \text{(unitless)} \tag{9-4}$$

where ε is the absolute permittivity of any given material,[7] measured in farads/meter

ε_0 is the absolute permittivity of a vacuum (free space) (F/m)

ε_r is the dielectric constant or relative permittivity of any material compared to that of a vacuum

In effect, the dielectric constant is a measure of how much electric flux or charge is placed on a standard capacitor (Fig. 9-3) with a given material as the dielectric (permittivity ε) compared to the same dimensioned capacitor with vacuum as the dielectric (permittivity ε_0).

We can now write a corrected Eq. (9-3) for any capacitor of any dimensions to include the product $\varepsilon = \varepsilon_0 k = \varepsilon_0 \varepsilon_r$:

$$C = \frac{Q}{V} = \frac{\varepsilon_0 k A}{d} = \frac{\varepsilon_0 \varepsilon_r A}{d} = (8.8542 \times 10^{-12}) \cdot \varepsilon_r \cdot \frac{A}{d} \qquad \text{farads (F)} \tag{9-3a}$$

[6] Note the similarity of ε_r or k to specific gravity, ρ_r. Specific gravity is the *unitless ratio* of two densities, where $\rho_r = D/D_w$, the density of any given material compared to the density of water, D_w, measured respectively in units of kg/m^3. The ratio of the two densities is similar to that of relative permittivity or the dielectric constant.

[7] $\varepsilon = D/E$ from Eq. (2-3) is the ratio of C/m^2 divided by V/m (or C/V·m or F/m).

where ε_r is the relative permittivity (or dielectric constant) for any dielectric (a unitless ratio)

A is the area of the plates in square meters

d is the distance between plates in meters

Equation (9-3a) now permits us to calculate the capacitance of any capacitor of given geometry provided we know the value of relative permittivity, ε_r, of its dielectric. Table 9-1 gives both the relative permittivity and dielectric strength (Sec. 9-6) for various dielectric materials.

Table 9-1 Properties of Various Insulating (Dielectric) Materials[a]

Material	ε_r (−)	E (kV/mm)	Material	ε_r (−)	E (kV/mm)
Vacuum	1	∞	Porcelain	5.7	15
Air	1.0006	3	Pressboard	6.2	7
Asbestos	2	2	Quartz, fused	3.5	13
Bakelite	5	15	Rubber	2.6	18
Cellulose film	5.8	28	Silica, fused	3.6	14
Marble	7	2	Water	70	—
Mica	6	40	Wax, paraffin	2.2	12
Paper (dry)	2.2	5	Barium–strontium titanate (ceramic)	7500	3
Paper (treated)	3.2	15			
Glass	6	6			

[a] Values represent approximate average values depending on relative purity of materials. Adapted from M.G. Say, *Electrical engineers reference book*, 13th Ed., Butterworths, London, 1973, Table 2-7, p. 2-43.

EXAMPLE 9-2

A capacitor consists of 30 sheets of aluminum connected in parallel. Each sheet measures 5 cm × 10 cm and is separated by 29 plates of glass, each plate 2 mm thick. Calculate

a. The capacitance in nanofarads
b. The electric field intensity when 100 V is applied between any two adjacent plates
c. Charge on the capacitor when 100 V is applied
d. Flux density of the electric field
e. Force exerted by the field on the dielectric
f. Permittivity of the dielectric material (glass) used

Solution

a. $C = 8.8542 \times 10^{-12}\, \varepsilon_r A/d$
$= 8.854 \times 10^{-12} \times 6 \times (29 \times 0.05 \times 0.1)/0.002$
$= 3.85 \times 10^{-9}$ F $=$ **3.85 nF** (9-3a)

b. $\mathbf{E} = V/d = 100$ V/0.002 m $=$ **50 kV/m** (2-2)

c. $Q = CV = (3.85$ nF$) \times 100$ V
$=$ **385 nC** (or 385×10^{-9} lines of electric flux ψ) (9-1)

d. $\mathbf{D} = \psi/A = 385 \times 10^{-9}$ lines/$(29 \times 0.05 \times 0.1)$ m^2
$=$ $\mathbf{2.66 \times 10^{-6}}$ **lines/m²** or C/m^2 (2-3)

e. $\mathbf{F} = Q\mathbf{E} = (385$ nC$)(50$ kV/m$) =$ $\mathbf{1.925 \times 10^{-2}}$ **N** (2-2)

f. $\varepsilon = \varepsilon_r \varepsilon_0 = 6.0 \times 8.8542 \times 10^{-12}$
$=$ $\mathbf{5.31 \times 10^{-11}}$ **F/m** (9-4)

Example 9-2 shows how much information can be extracted merely from the geometry of the capacitor and the applied voltage. This is to be expected, however, when we consider the equalities of the derivation of Eq. (9-3a) and realize how all these physical quantities are interrelated.[8]

9-6 DIELECTRIC STRENGTH AND DIELECTRIC LEAKAGE

A second measure of the properties of dielectrics, in addition to relative permittivity (or dielectric constant), is the dielectric strength of the material. Dielectric strengths for various insulating materials are shown in Table 9-1 for comparison

[8] This is precisely why it was possible to calculate the capacitance of the parallel plate capacitor in Ex. 2-6 without knowing the area of the plates or the distance between them.

with dielectric permittivity, ε_r. Note that the symbol for dielectric strength in Table 9-1 is **E**, electric field intensity. Dielectric strength may be defined as that electrical field intensity required to break down a given sample of dielectric material. Consequently, the values shown in the second column of Table 9-1 indicate how much electrical field intensity a given thickness of dielectric can withstand before breaking down. Basically, it is a measure of insulation quality.

With respect to the values given in Table 9-1, note that there is no relation between the dielectric constant and the dielectric strength. A material such as water, which may make a very fine capacitor, unfortunately has no insulating qualities. On the other hand, a vacuum, which is a perfect insulator, does not make a very good capacitor. Extremely high dielectric constants recently have been obtained with barium–strontium titanate ($BaSrTiO_3$), a ceramic material, which can withstand 3000 V/mm of thickness before breaking down, as shown in Table 9-1.

Note that the values of dielectric strength given in Table 9-1 are average values and that variations in material composition produce variations in values of **E**.

EXAMPLE 9-3

A roll of rubber tape used by an electrician to insulate soldered splices has a thickness of 0.5 mm. What maximum voltage can the splices withstand if the electrician wraps them with

a. One layer of tape?
b. Three layers of tape?

Solution

a. From Table 9-1, **E = 18 kV/min;**
$V = \mathbf{E} \times d = (18\ \text{kV/mm})(0.5\ \text{mm}) = \mathbf{9\ kV}$

b. $V = \mathbf{E} \times d = (18\ \text{kV/mm})(3 \times 0.5\ \text{mm}) = \mathbf{27\ kV}$

EXAMPLE 9-4

Calculate the maximum voltage that can be applied to a 0.5-nF capacitor with a plate area of 2 cm × 2 cm and mica as a dielectric.

Solution

$$d = 8.8542 \times 10^{-12}\ \varepsilon_r A/C$$
$$= 8.8542 \times 10^{-12} \times (6) \times (0.02\ \text{m})^2/0.5 \times 10^{-9}$$
$$= 4.25 \times 10^{-5}\ \text{m}$$

$$V = \mathbf{E} \times d = (40\ \text{MV/m})(4.25 \times 10^{-5}\ \text{m}) = \mathbf{1700\ V}$$

With respect to Ex. 9-4, please note that

1. The value of **E** for mica from Table 9-1 is 40 kV/mm, which is the same as 40 MV/m
2. The answer obtained is reasonable since mica capacitors are known as low capacitance, high voltage capacitors
3. The plate area of 4 cm^2 is a reasonable size for a mica capacitor

---

9-6.1 Dielectric Leakage

A certain amount of dc flows in any commercial capacitor whenever dc is applied across its terminals, even when fully charged. This very small current is called dielectric leakage current or just leakage current. Since no dielectric is a perfect insulator, the leakage current varies in inverse proportion to the insulation resistance of the dielectric. Small paper, film, mica, and ceramic capacitors have insulation resistances of the order of 100 000 MΩ. Larger capacitors, usually paper or electrolytic, have lower insulation resistances,[9] usually of the order of 500 to 10 000 MΩ. The larger the capacity of the capacitor, the lower its insulation resis-

[9] This is fairly obvious from $R = \rho l/A$. Larger capacitors have larger areas and smaller dielectric thickness, resulting in lower dielectric insulation resistance.

tance. For this reason manufacturers often specify insulation resistance as an *RC* product in megohm-microfarads (MΩ·μF), since one factor varies inversely with the other. Consequently, the *RC* product of a given type of capacitor is fairly constant at a constant temperature (usually taken at 25°C).

Leakage current is fairly temperature-sensitive, particularly in electrolytic capacitors. The higher the temperature, the greater the leakage due to the lower insulation resistance.

9-6.2 Equivalent Circuit of a Practical Capacitor

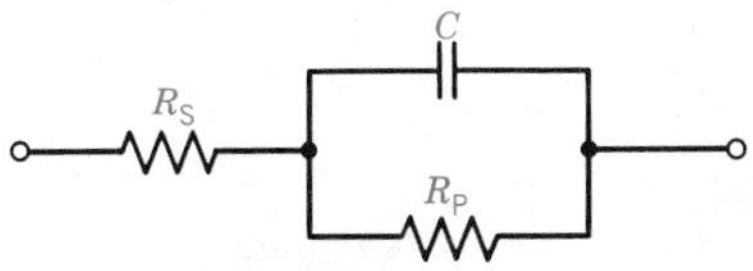

R_S = leads, plates and electrical connection resistance (a low value)
R_P = dielectric leakage, absorption and insulation resistance (a high value of resistance)

Figure 9-4 Equivalent circuit for a practical capacitor

Dielectric leakage is but one resistance component of a commercial capacitor. It is customary to represent this high value of resistance as R_P, a parallel-resistance component in parallel with the capacitance of the capacitor itself, as shown in **Fig. 9-4**.

Another component of loss in a capacitor is sometimes called the heat dissipation loss. This loss is due to the resistance of the capacitor leads, capacitor plates, and electrical contacts between leads, plates, and other interface surfaces. As shown in Fig. 9-4, this relatively *low* value of resistance, R_S, is in *series* with the parallel combination of *C* and R_P.

The distinction between R_S and R_P can be understood more readily when one considers what happens when a capacitor is charged by a dc power supply to (say) 200 V and then placed on a shelf. *Contrary to popular opinion*, the charge slowly leaks away due to continuous (dc) flow through R_P, and *not* because air molecules are continuously removing charge from the capacitor. (A hermetically sealed capacitor also discharges slowly with time.) On the other hand, when a capacitor is continuously connected in an ac circuit where the voltage and current are continuously changing, there is loss due to both R_S and R_P. This loss increases as the frequency increases, hence the need for an equivalent series resistance (ESR) value at a given frequency.[10]

9-7 COMMERCIAL CAPACITORS

Capacitors are commercially available in two large categories: *fixed* and *variable*. Fixed capacitors are available in seven major types: electrolytics, dielectrics, plastic film types, metallized plastic types, glass and ceramics, mica and mica/paper types, and air/vacuum types. Table 9-2 summarizes the major electrical characteristics and ranges of the various major types of fixed capacitors. Since this information appears in the table, it will not be discussed here. Instead, the various applications, advantages, and disadvantages of each appear in the discussion below. **Figure 9-5** shows construction features of various types discussed below.

[10] Capacitor manufacturers frequently combine R_S and R_P into a single equivalent series resistance at a specific frequency of operation. The relation between ESR, the power factor of the capacitor (PF) and the power loss (P_L) at a specific frequency is $\text{ESR} = (PF)/(2\pi f C) = P_L/I^2$.The PF is cos θ, where θ is the phase angle between the actual current drawn by the capacitor and the in-phase or loss component. The complement of θ is the loss angle ϕ. The power dissipation factor (D) is the tan ϕ or the tangent of the loss angle. The lower the D, the smaller the loss in the capacitor.

Table 9-2 Major Electrical Characteristics and Ranges of Fixed Capacitor Types

CAPACITOR TYPE	WORKING VOLTAGE (volts) Min.	Max.	CAPACITY RANGE (farads) Min.	Max.	POWER DISSIPATION FACTOR DF (−)	FREQUENCY (Hz)	CONSTRUCTION (Figure number)
A. ELECTROLYTICS							
1. Aluminum	3	700	0.5 μ	2	3–80	120	
2. Tantalum (foil)	3	450	0.1 μ	8 m	10–20	120	
3. Tantalum (wet)	1.25	630	0.1 μ	8 m	15–50	120	
4. Tantalum (solid)	3	125	1 n	1 m	1–8	120	9-5a and 9-5b
B. DIELECTRIC TYPES							
1. Paper and	50	200 k	100 p	200 μ	0.2–0.6	120	
paper/Mylar					0.2–1.0	1000	
2. Metallized paper	50	4 k	50 p	200 μ	0.8–1.0	1000	9-5c
C. PLASTIC FILM TYPES[a]	30	10 k	20 p	400 μ	0.02–0.8	1 M	9-5d and 9-5e
					0.008–1.0	1 k	
D. METALLIZED PLASTIC	50	1.1 k	1 n	30 μ	0.1–1.0	1 k	9-5e
E. GLASS/CERAMIC	25	30 k	0.1 p	12 μ	0.008–8	1 M	9-5f and 9-5g
					0.008–2	1 k	
F. MICA and	50	100 k	1 p	10 μ	0.05–1	1 M	
MICA/PAPER					0.05–0.7	1 k	
					0.04–0.06	120	
G. AIR AND VACUUM	2	45 k	6 p	0.025 μ	0.008–0.01	1 M	

[a] Plastic film types include polyester, polypropylene, polystyrene, polyethylene, parylene, polycarbonate, teflon and mylar

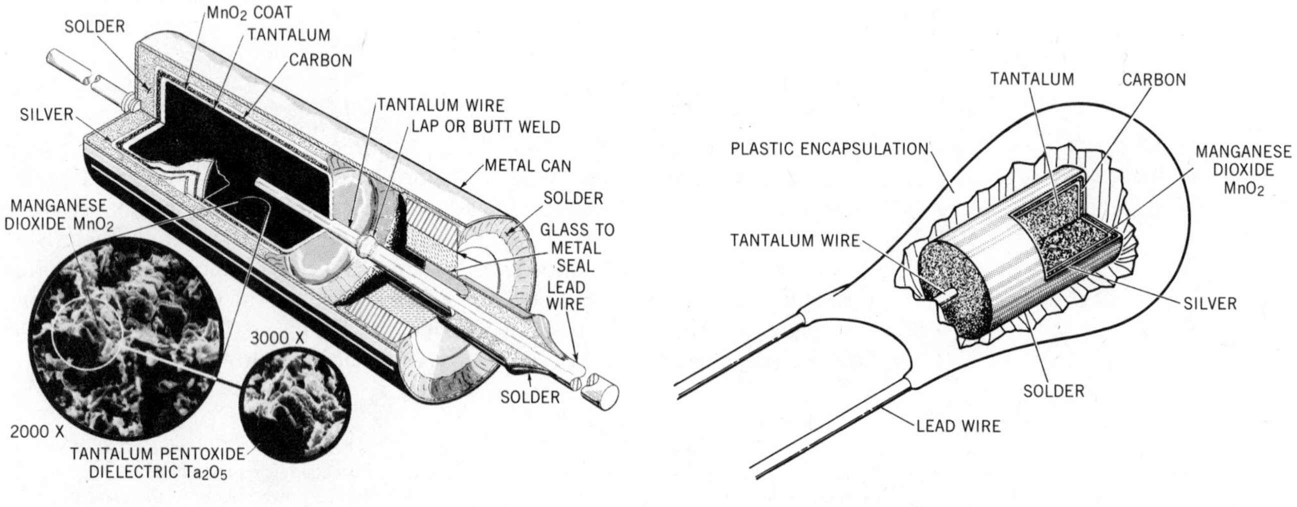

a. Typical axial solid tantalum construction (Courtesy KERMET[R] Capacitors, Electronics Division of Union Carbide Corp.).

b. Typical flat-case solid tantalum construction (Courtesy KERMET[R] Capacitors, Electronics Division of Union Carbide Corp.).

Figure 9-5 Capacitor types and construction features

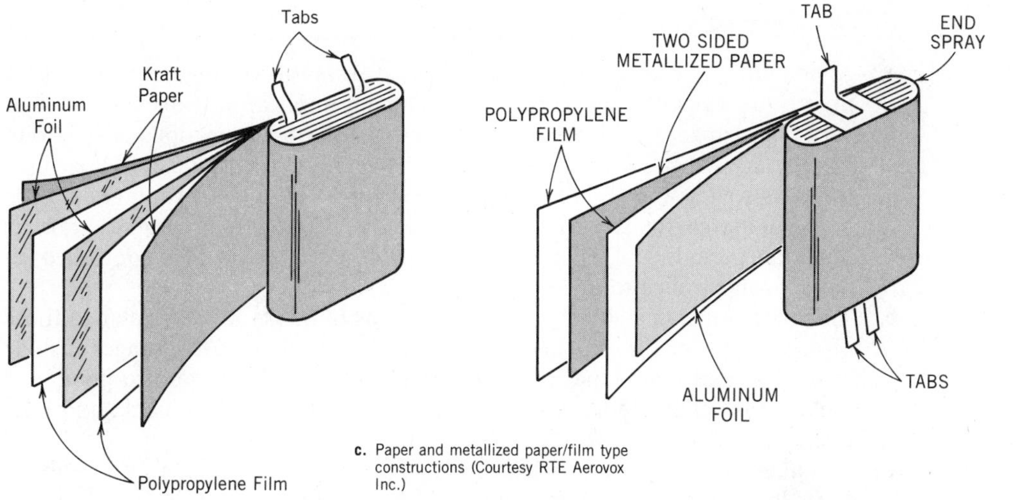

c. Paper and metallized paper/film type constructions (Courtesy RTE Aerovox Inc.)

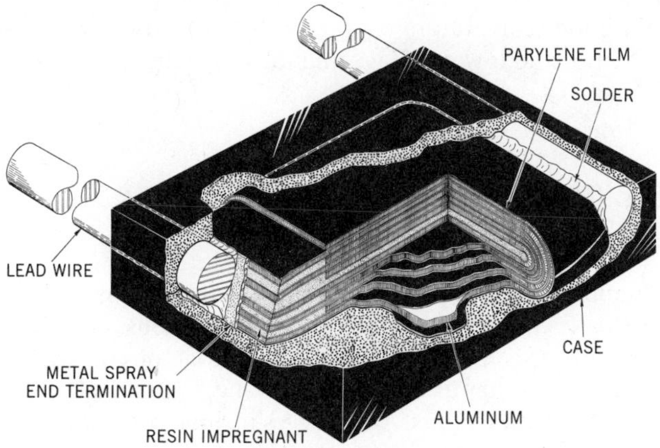

d. Typical flat-case radial molded plastic film construction (Courtesy KERMET[R] Capacitors, Electronics Division of Union Carbide Corp.).

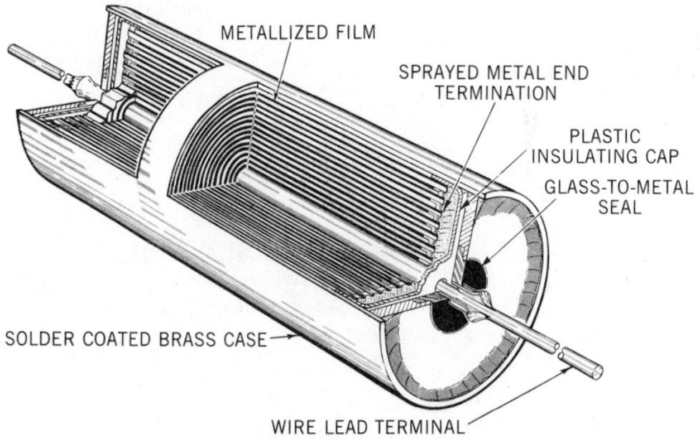

e. Typical axial hermetically sealed metallized plastic construction (Courtesy KERMET[R] Capacitors, Electronics Division of Union Carbide Corp.).

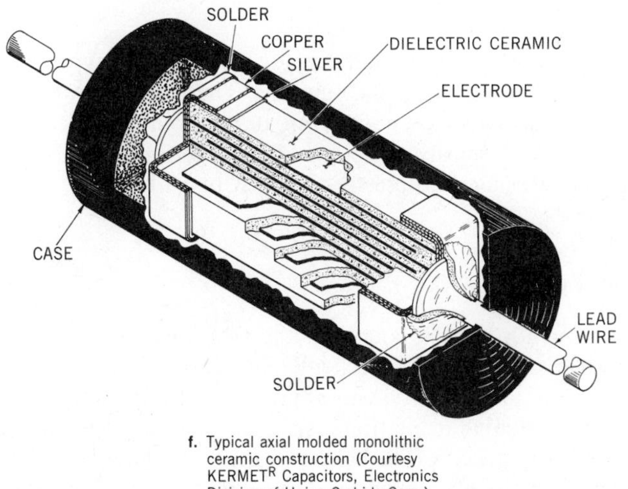

f. Typical axial molded monolithic ceramic construction (Courtesy KERMET[R] Capacitors, Electronics Division of Union Carbide Corp.).

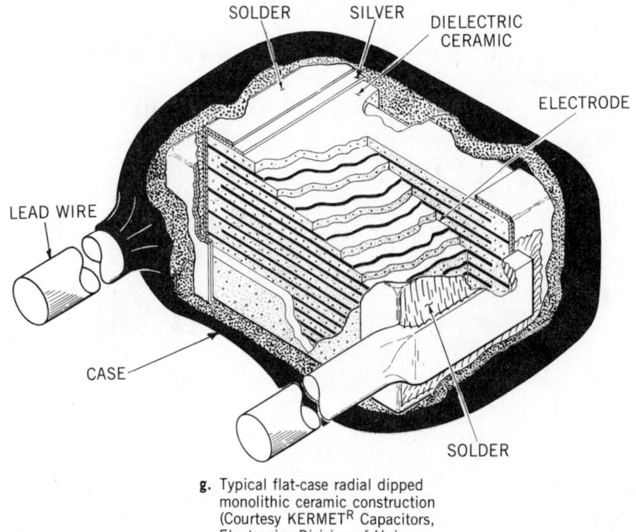

g. Typical flat-case radial dipped monolithic ceramic construction (Courtesy KERMET[R] Capacitors, Electronics Division of Union Carbide Corp.).

Figure 9-5 (Continued) Capacitor types and construction features

9-7.1 Electrolytic Types

a. **Aluminum** This class of capacitors is used primarily in low-frequency filters in various blocking, bypassing, and coupling applications. Because of their very high capacity they are also used in photoflash applications. This group has the advantages of highest voltage and highest capacitance of all electrolytics, as well as the highest capacitance/volume ratio. Disadvantages include high leakage currents, high dissipation factors, need for reforming of dielectric film after periods of long storage, and medium to high cost for controlled reliability of specifications.
b. **Tantalum wet slug and foil** The applications of this type are similar to those of aluminum but limited in maximum capacity (2 mF). Advantages include lower DF and high capacitance/volume ratio. Disadvantages are poor low-temperature characteristics, susceptibility to damage due to shock, and relatively high cost.
c. **Tantalum solid dielectric** This class has applications similar to those above but is limited to a maximum voltage of 125 V. Advantages include high capacitance/volume ratio and good temperature characteristics. Disadvantages include poor RF characteristics and lower voltage and lower maximum capacity than other electrolytics.

9-7.2 Dielectric Types (Paper, Paper and Mylar, and Metallized Paper)

Because of its higher maximum voltage (up to 200 kV) and relatively moderate capacitance (up to 200 μF), this class has many and varied applications, including motor starting and running capacitors, photoflash units, power factor correction, and contact protection. In electronics they are used for blocking, buffering, bypass, coupling, and filtering in low-frequency circuits. Advantages include comparatively low cost in \$/μF, high reliability, moderate stability, and the wide availability of ranges of voltage and capacitance. Disadvantages are the higher dissipation factors at high frequencies.

9-7.3 Plastic Film Types

Applications for these types are essentially the same as for the dielectric types but extended to medium (rather than lower) frequency circuits. The major advantage of this class is high stability coupled with low DFs, even at high frequencies. They are also among the least temperature-sensitive, along with glass ceramic types (Sec. 9-7.4). The major disadvantage is their medium to high cost in \$/μF.

9-7.4 Glass/Ceramic Types

Applications include high-frequency bypass, coupling, and filtering as well as tuning in resonant circuits. Advantages include excellent temperature stability, low losses even at high frequencies, high reliability, and ability to withstand very high temperatures. Because of their very high insulation resistance they have the lowest DFs of all types, even at high frequencies. The ceramics generally have a higher capacitance-to-volume ratio than the glass types and are therefore lower in cost.

9-7.5 Mica and Mica/Paper Types

This group has the advantage of the highest maximum voltage availability (up to 100 kV) coupled with the disadvantage of low capacitance-to-volume ratio. Other advantages include low DFs, good temperature, frequency, and aging characteristics, and low cost. They are used in high-frequency filtering, coupling, and bypassing as well as resonant circuit tuning.

9-7.6 Air/Vacuum Types

This class is used primarily in frequency tuning applications. They have the advantage of very low DFs, which results in high Q. Disadvantages include low capacitance-to-volume ratio, high cost, limitation to low capacitance values, and susceptibility to shock and vibration.

9-8 CAPACITORS IN PARALLEL

Figure 9-6a shows a circuit consisting of three capacitors in parallel: C_1, C_2, and C_3. Let us assume for the moment that they are identical, that is, $C_1 = C_2 = C_3$. If we do this we can intuitively infer the total capacitance, C_T. Since capacitance varies directly with area, by connecting three identical capacitors in parallel, we have increased the total area by a factor of 3, that is, $C_T = 3C_1 = C_1 + C_2 + C_3$. We can infer from this that connecting capacitors in parallel *increases* the total capacitance of the combination.

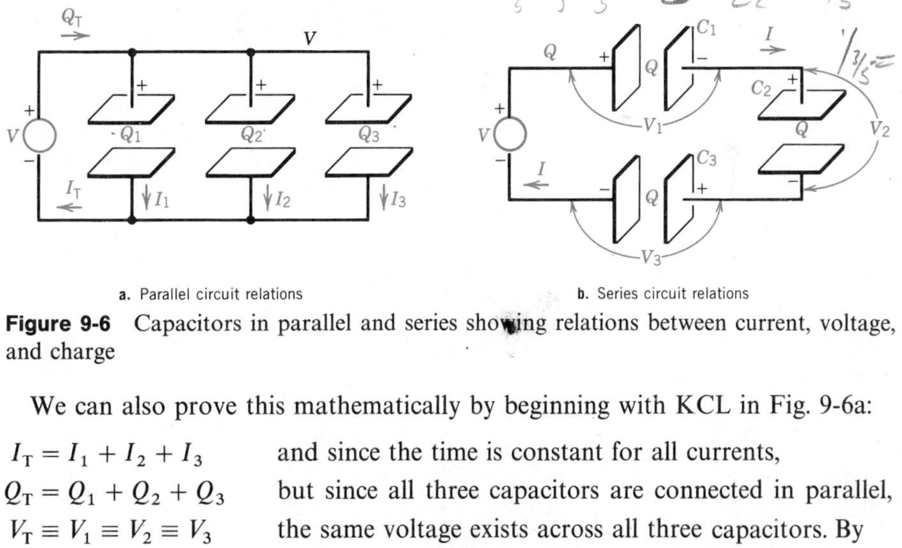

a. Parallel circuit relations   b. Series circuit relations

Figure 9-6 Capacitors in parallel and series showing relations between current, voltage, and charge

We can also prove this mathematically by beginning with KCL in Fig. 9-6a:

$I_T = I_1 + I_2 + I_3$	and since the time is constant for all currents,
$Q_T = Q_1 + Q_2 + Q_3$	but since all three capacitors are connected in parallel,
$V_T \equiv V_1 \equiv V_2 \equiv V_3$	the same voltage exists across all three capacitors. By
$C_T = C_1 + C_2 + C_3$	dividing each Q by the **same** V, we obtain the simple arithmetic sum of the capacitors in parallel since $C = Q/V$

For any number, n, of capacitors in parallel, we therefore can write

$$C_T = C_1 + C_2 + C_3 + \cdots + C_n \qquad \text{farads (F)} \qquad (9\text{-}5)$$

EXAMPLE 9-5

Given $C_1 = 20\ \mu F$, 150 V; $C_2 = 50\ \mu F$, 200 V; and $C_3 = 30\ \mu F$, 100 V, calculate the

a. Total capacitance of the combination when connected in parallel
b. Maximum voltage that can be connected across the combination (use a safety factor of 90%)
c. Total charge on the equivalent capacitor
d. Charge on each capacitor and total charge

Solution

a. $C_T = C_1 + C_2 + C_3 = (20 + 50 + 30)\ \mu F = \mathbf{100\ \mu F}$ (9-5)
b. Since C_3 has the lowest voltage rating, we use $0.9 \times 100\ V = \mathbf{90\ V}$
c. $Q_T = C_T V_T = (100\ \mu F)(90\ V) = \mathbf{9\ mC}$ (9-1)
d. $Q_1 = C_1 V = (20\ \mu F)(90\ V) = \mathbf{1.8\ mC}$
$Q_2 = C_2 V = (50\ \mu F)(90\ V) = \mathbf{4.5\ mC}$
$Q_3 = C_3 V = (30\ \mu F)(90\ V) = \mathbf{2.7\ mC}$
$Q_T = Q_1 + Q_2 + Q_3 = (1.8 + 4.5 + 2.7)\ mC$
$= \mathbf{9\ mC}$ [same as part (**c**)]

9-9 CAPACITORS IN SERIES

Figure 9-6b shows a circuit consisting of three capacitors in series: C_1, C_2, and C_3. During the process of charging the capacitors, we know from our series circuit relations that

$I_T \equiv I_1 \equiv I_2 \equiv I_3$ current is the same in all parts of a series circuit

$Q_T \equiv Q_1 \equiv Q_2 \equiv Q_3$ since time is the same for all currents, charge on each is the same

$V_T = V_1 + V_2 + V_3$ KVL for series-connected components and dividing each voltage by the *same* charge yields $1/C = V/Q$ or

$$\frac{1}{C_T} = \frac{1}{C_1} + \frac{1}{C_2} + \frac{1}{C_3}$$

For any number, n, of capacitors in **series**, we therefore can write

$$\frac{1}{C_T} = \frac{1}{C_1} + \frac{1}{C_2} + \frac{1}{C_3} + \cdots + \frac{1}{C_n} \quad \text{farads}^{-1}\ (\text{F}^{-1}) \qquad \textbf{(9-6)}$$

For two capacitors, C_1 and C_2, in series, we can also write

$$C_T = \frac{C_1 C_2}{C_1 + C_2} \quad \text{farads (F)} \qquad \textbf{(9-7)}$$

Note from Eqs. (9-6) and (9-7) that the equivalent capacitance of series-connected capacitors is always smaller than the capacitance of any of the individual capacitors.

Just as the reciprocal of resistance, R, is conductance, G, we may introduce the reciprocal of capacitance, C as elastance, S, where $S = 1/C$ measured in (farads)$^{-1}$. Equation (9-6) may be rewritten as

$$S_T = S_1 + S_2 + S_3 + \cdots + S_n \quad \text{farads}^{-1}\ (\text{F}^{-1}) \qquad \textbf{(9-6a)}$$

Note that there is no unit for elastance, but it does simplify calculations somewhat and is useful in drawing analogies to electrical and magnetic circuits (see Sec. 9-11).

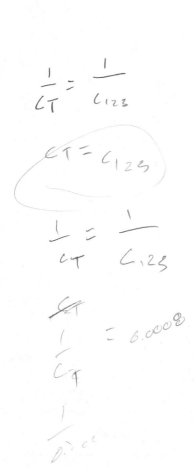

EXAMPLE 9-6

Given the three capacitors of Ex. 9-5 connected in series across a 120 V dc source, calculate

a. Equivalent series capacitance
b. Total charge and charge on each capacitor
c. Voltage across each capacitor in series and total voltage verifying KVL

Solution

a. $$\frac{1}{C_T} = \frac{1}{C_1} + \frac{1}{C_2} + \frac{1}{C_3}$$
$$= \frac{1}{20} + \frac{1}{50} + \frac{1}{30} \quad \text{and} \quad C_T = \mathbf{9.68\ \mu F} \qquad (9\text{-}6)$$

b. $Q_T = C_T V_T = (9.68\ \mu\text{F})(120\ \text{V}) = \mathbf{1.16\ mC} \equiv Q_1 \equiv Q_2 \equiv Q_3$

c. $V_1 = Q_1/C_1 = 1.16\ \text{mC}/20\ \mu\text{F} = \mathbf{58.06\ V}$
$V_2 = Q_2/C_2 = 1.16\ \text{mC}/50\ \mu\text{F} = \mathbf{23.23\ V}$
$V_3 = Q_3/C_3 = 1.16\ \text{mC}/30\ \mu\text{F} = \mathbf{38.71\ V}$
$V_T = V_1 + V_2 + V_3 = (58.06 + 23.23 + 38.71)\ \text{V} = \mathbf{120\ V}$

With respect to Ex. 9-6, please note the following:

1. As indicated in the derivation of Eq. (9-6), the charge on each capacitor is the same as the total charge.
2. The **smallest** capacitor exhibits the **largest** volt drop.
3. Conversely, the **largest** capacitor exhibits the **smallest** volt drop.
4. In comparison to resistors, therefore, capacitors represent an *inverse* relationship with respect to KVL, KCL, and total equivalence, as shown by **Table 9-3**.

Table 9-3 Comparisons between Resistance and Capacitance in Series and Parallel

CIRCUIT RELATION	RESISTANCE	CAPACITANCE
Series circuit	$R_T = R_1 + R_2 + R_3 + \cdots + R_n$	$\frac{1}{C_T} = \frac{1}{C_1} + \frac{1}{C_2} + \frac{1}{C_3} + \cdots + \frac{1}{C_n}$
Parallel circuit	$\frac{1}{R_T} = \frac{1}{R_1} + \frac{1}{R_2} + \frac{1}{R_3} + \cdots + \frac{1}{R_n}$	$C_T = C_1 + C_2 + C_3 + \cdots + C_n$
KVL in series	$I \cdot R_T = I \cdot R_1 + I \cdot R_2 + I \cdot R_3 + \cdots + I \cdot R_n$	$\frac{Q}{C_T} = \frac{Q}{C_1} + \frac{Q}{C_2} + \frac{Q}{C_3} + \cdots + \frac{Q}{C_n}$
KCL in parallel	$\frac{V}{R_T} = \frac{V}{R_1} + \frac{V}{R_2} + \frac{V}{R_3} + \cdots + \frac{V}{R_n}$	$VC_T = VC_1 + VC_2 + VC_3 + \cdots + VC_n$
Series circuit	**Highest** volt drop across **largest** R	**Highest** volt drop across **smallest** C
Parallel circuit	**Highest** current in **smallest** R	**Highest** charge across **largest** C

9-10 ENERGY STORAGE IN CHARGED CAPACITORS

Appendix B-3 derives Eq. (9-2) for the energy stored in the electric field of a capacitor when a potential V is applied, specifically

$$W = Q^2/2C = VQ/2 = CV^2/2 \qquad \text{joules (J)} \qquad (9\text{-}2)$$

But since potential is defined as work per unit charge ($V = W/Q$), it might appear that $W = QV$ represents the work done and energy stored in charging a capacitor (and not $QV/2$). This apparent inconsistency was introduced at the end of Sec. 9-3, where Eq. (9-2) first appears.

Whenever a capacitor is charged, work is done to separate two equal and opposite charges on the plates of the capacitor. The energy resulting from such work is stored in the electric field or the stressed dielectric of the capacitor. This stored energy must be recovered whenever the charges are neutralized or whenever the plates of the capacitor are short-circuited.

Figure 9-7a shows a capacitor, C_1, that has been initially charged by voltage V_0 to a charge Q_0. C_2 in Fig. 9-7a represents a fully discharged capacitor, that is, $Q_2 = 0$ and $V_2 = 0$. What happens when switch S is closed? Capacitor C_1 in effect transfers charge to C_2 and continues to do this as its voltage decreases until the charge and voltage V across C_2 is equal and opposite to the voltage V across C_1.

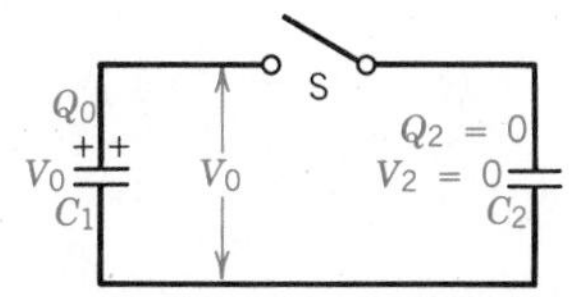

a. C_1 initially charged to Q_0

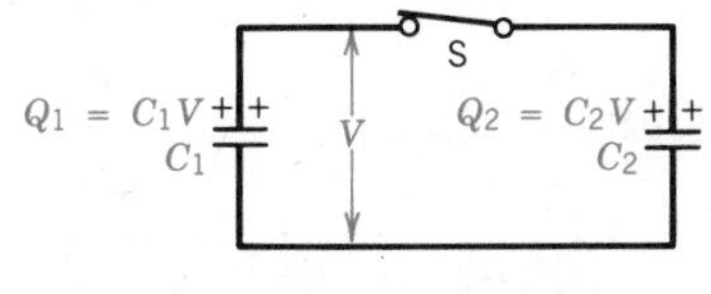

b. Charge distribution after S closes

Figure 9-7 Energy transfer between two capacitors

The charge on C_1 has decreased from Q_0 to Q_1 while the charge on C_2 has increased from zero to Q_2 (Fig. 9-7b). The initial charge, according to the law of conservation of charge, must be preserved, and therefore

$Q_0 = Q_1 + Q_2$ and from Eq. (9-1)

$C_1 V_0 = C_1 V + C_2 V$ solving this relation for V yields by factoring

$$V = V_0 \frac{C_1}{C_1 + C_2} \quad \text{volts (V)} \qquad \textbf{(9-8)}$$

The *initial* energy stored on the plates of C_1 in Fig. 9-7a is $W_0 = C_1 V_0^2/2$, from Eq. (9-2), *before* the switch was closed.

The total energy *after* the switch was closed, from Eq. (9-2), is

$$W_t = \frac{C_1 V^2}{2} + \frac{C_2 V^2}{2} = \frac{(C_1 + C_2)}{2} V^2$$ and substituting for V using Eq. (9-8) yields

$$= \frac{(C_1 + C_2)}{2} \times \left(\frac{V_0 C_1}{C_1 + C_2}\right)^2 = \frac{C_1 \times C_1 V_0^2}{(C_1 + C_2)^2}$$ But $C_1 V_0^2/2 = W_0$ and therefore

$$W_t = \frac{C_1}{(C_1 + C_2)} W_0 \quad \text{joules (J)} \qquad \textbf{(9-9)}$$

Equation (9-9) opens the same "can of worms" introduced at the beginning of this section. How can the **total** energy stored on the **two** capacitors, W_t, become some fraction of the original energy, W_0, on capacitor C_1 before the switch was closed? Energy can neither be created nor destroyed. Where did the extra energy go?

The answer is that the extra energy is dissipated as heat during the time the charges move through the wires when the switch S is first closed and C_2 is charged to voltage V. Clearly, some of the energy transferred from C_1 when S is closed does not get to C_2, as shown by the following examples.

EXAMPLE 9-7

A 6-μF capacitor, C_1, is charged to 90 V and the supply voltage is disconnected. The charged capacitor is then connected across a completely discharged 3-μF capacitor. Calculate the

a. Initial energy stored in the electric field of C_1
b. Equilibrium voltage, V, across the two capacitors after charge transfer
c. Energy stored in the magnetic field of C_1 and C_2 after charge transfer
d. Total energy stored in the two capacitors after charge transfer, by two methods
e. Energy lost in conducting wires during charge transfer

Solution

a. $W_0 = C_1 V_0^2/2 = (6\ \mu\text{F})(90\ \text{V})^2/2 = \mathbf{24.3\ mJ}$ (9-2)

b. $V = \frac{C_1}{C_1 + C_2} V_0 = \frac{6}{9} 90\ \text{V} = \mathbf{60\ V}$ (9-8)

c. $W_1 = C_1 V^2/2 = (6\ \mu\text{F})(60)^2/2 = \mathbf{10.8\ mJ}$ (9-2)
$W_2 = C_2 V^2/2 = (3\ \mu\text{F})(60)^2/2 = \mathbf{5.4\ mJ}$ (9-2)

d. $W_1 + W_2 = (10.8 + 5.4)\ \text{mJ} = \mathbf{16.2\ mJ}$
$W_t = \frac{C_1}{C_1 + C_2} W_0 = \frac{6}{9} 24.3\ \text{mJ} = \mathbf{16.2\ mJ}$ (9-9)

e. $W_L = W_0 - W_t = (24.3 - 16.2)\ \text{mJ} = \mathbf{8.1\ mJ}$

Example 9-7 shows that the law of conservation of energy has not been violated. But what about the law of conservation of charge? Consider Ex. 9-8.

EXAMPLE 9-8

Given the data of Ex. 9-7, calculate the

a. Charge on C_1, initially
b. Charge on C_2, after charge transfer
c. Charge on C_1, after charge transfer
d. Total charge after charge transfer
e. Total capacitance after charge transfer
f. Are the capacitors in parallel or in series? How do you know?

Solution

a. $Q_0 = C_1 V_0 = (6\ \mu\text{F})(90\ \text{V}) = \mathbf{540\ \mu C}$ (9-1)
b. $Q_2 = C_2 V = (3\ \mu\text{F})(60\ \text{V}) = \mathbf{180\ \mu C}$ (9-1)
c. $Q_1 = C_1 V = (6\ \mu\text{F})(60\ \text{V}) = \mathbf{360\ \mu C}$ (9-1)
d. $Q_t = Q_1 + Q_2 = (360 + 180)\ \mu\text{C} = \mathbf{540\ \mu C} = Q_0$
e. $C_T = Q_t/V = 540\ \mu\text{C}/60\ \text{V} = \mathbf{9\ \mu F}$ (9-1)
f. If the capacitors of 6 and 3 μF were in series, $C_T = \frac{6 \times 3}{6+3}\ \mu\text{F} = 2\ \mu\text{F}$; but since $C_T = 9\ \mu\text{F}$, it is obvious that they are in *parallel* even though there is only one path for the charge (Fig. 9-7b)

Example 9-8 shows that *charge must be conserved* and that (unlike energy) no charge is lost when it is transferred from one capacitor to another.

The apparent voltage division that occurs in Eq. (9-8) suggests that it may be possible, experimentally, to determine the capacitance value of *any unknown* capacitor merely by measuring the voltage on a *known* capacitor, V_0, before connecting it across an unknown capacitor and measuring the paralleled voltage, V, again (Fig. 9-7). The voltage-measuring instrument ideally should be an **electrometer** (a very high input-impedance voltmeter) that draws negligible charge from the circuit capacitors.

EXAMPLE 9-9

A 10-μF capacitor is charged by a dc power supply and then removed from the supply. The initial voltage across the 10-μF capacitor is 100 V. It is then connected across an unknown capacitor and the voltage across the combination is 40 V, as measured by an electrometer. Calculate the capacitance of the unknown capacitor.

Solution

From Eq. (9-8), algebraically transposed,

$$C_2 = \frac{V_0}{V} C_1 - C_1 = \frac{100}{40}\ 10\ \mu\text{F} - 10\ \mu\text{F} = \mathbf{15\ \mu F} \quad (9\text{-}8)$$

9-11 ANALOGS BETWEEN ELECTRIC AND CAPACITIVE CIRCUITS

By way of summary, it might be useful to consider analogous[11] relations between capacitive and electric circuits. Table 9-4 shows such a comparison.

Table 9-4 Analogs in Capacitive and Electric Circuits

Dielectric Circuit		Electric Circuit	
Property	Equation	Equation	Property
Capacitance	$C = 1/S$	$G = 1/R$	Conductance
Capacitance	$C = \varepsilon_r \varepsilon_0 A/d$	$G = \sigma A/l$	Conductance
Electric field intensity	$\mathbf{E} = F/Q = V/d$	$\mathbf{E} = V/d$	Electric field intensity
Permittivity	$\varepsilon = \mathbf{D}/\mathbf{E}$	$\sigma = J/A$	Conductivity
Flux density	$\mathbf{D} = \psi/A$	$J = I/A$	Current density
Elastance	$S = d/\varepsilon_r \varepsilon_0 A$	$R = l/\sigma A$	Resistance
Elastance	$S = V/\psi$	$R = V/I$	Resistance
Dielectric flux	$\psi = V/S$	$I = V/R$	Current

[11] An analog is *not* a dual, strictly speaking. But it is useful to draw analogies between the two circuits, even if one cannot be substituted for the other in a network equation (as in the case of duality).

Note how the analogs between capacitive and electric circuits help clarify our thinking. Elastance was introduced briefly in Sec. 9-9 to simplify Eq. (9-6). If resistance is opposition to a steady current, then elastance is opposition to the creation of an electric field. If conductance is the ease with which current flows in a conductor, then *capacitance* is the *ease* with which a *dielectric field* can be set up in a *capacitor*.

Note from Table 9-4 that the analog of current in an electric circuit is dielectric flux, ψ, in a dielectric circuit. Also note that the Ohm's law equivalent of an electric circuit is $\psi = V/S$ in a dielectric circuit, where V is the forcing function and S is the opposition to the creation of an electric field. Readers are urged to study Table 9-4 most carefully to clarify their own concepts regarding the various dielectric quantities introduced in this chapter.

9-12 GLOSSARY OF TERMS USED

Breakdown voltage See dielectric strength.

Capacitance Property of a component or circuit to store a charge in the form of an electric field across a dielectric and oppose any change in voltage across its terminals.

Capacitor Component or device composed of two conductors separated by a dielectric material capable of exhibiting the property of capacitance.

Charge, Q Quantity of charged particles on a body, the number of either positive ions or electrons (negative ions), measured in coulombs (C). Hence 6.2422×10^{18} ions represents 1 C of charge.

Coulomb's law Law quantifying the force of attraction or repulsion between two charged bodies separated by a specific distance (see Sec. 2-1.2).

Current, I Rate of charge flow of ions in an electric medium, measured in amperes (A). One ampere $\equiv$ 1 coulomb per second (1 A $\equiv$ 1 C/s).

Dielectric Material that is a nonconductor of electricity. Application of an electric field to a dielectric only results in displacement of electric charges within the dielectric due to polarization. (See dielectric constant.)

Dielectric absorption Charge retained in a dielectric material after potential has been removed, because of residual stress on the electronic orbits of the dielectric.

Dielectric constant, k or $\varepsilon_r/\varepsilon_0$ Property similar to specific gravity, representing the ratio of the permittivity of a dielectric to the permittivity of free space.

Dielectric flux, ψ Number of electric lines of dielectric force between two oppositely charged bodies. Since dielectric (or electric) flux is proportional to and depends on charge, in the SI, $Q \equiv \psi$; that is, dielectric flux is identical to charge and is measured in coulombs (1 coulomb $\equiv$ 1 line of flux).

Dielectric strength Potential gradient at which electric failure or breakdown occurs in a given thickness of dielectric material subjected to an electric field, measured in volts per meter (V/m).

Electric field intensity, E Measure of the electric field strength at any point in an electric field, measured as the force exerted by the field on a unit positive charge placed at that point in the field, in units of newtons per coulomb (N/C) or volts per meter (V/m). **E** is a vector quantity.

Electric flux density, D Measure of electric field strength in terms of the quantity of flux (coulombs or lines) per unit area. Like **E**, **D** is a vector quantity, having both magnitude and direction.

Electric force, F Force of attraction or repulsion exerted by an electric field on electric charges, measured in newtons (N).

Equipotential line Imaginary line in an electric field connecting the locus of all points having the same potential.

Permittivity, ε Factor relating flux density, **D**, to electric field intensity, **E**; that is, $\varepsilon = \mathbf{D}/\mathbf{E}$. Permittivity of dielectric materials varies with the medium and the state of matter when acted on by an electric field. Permittivity is measured in units of farads per meter (F/m) and also $C^2/N\cdot m^2$. The permittivity of free space, ε_0, is 8.8542×10^{-12} F/m. ε_0 is an important and frequently used constant in electrical engineering.

Polarization Change or distortion of the dielectric atoms as a result of sustaining a steady electric field. It is a measure of the displacement of electron orbits in a dielectric as a result of the force of an electric field.

9-13 PROBLEMS

Secs. 9-1 to 9-6

9-1 A potential difference of 20 V is applied across a 0.05-μF capacitor. Calculate

a. The charge stored on the capacitor
b. The charge stored if the voltage increases to 60 V
c. The electric field intensity in (**a**) if the distance between plates is 6 mm
d. Repeat (**c**) for the conditions of (**b**)

9-2 A potential difference of 100 V is applied to a 1-μF capacitor. Calculate the

a. Charge on the capacitor
b. Energy stored in the dielectric field
c. Electric field intensity if the distance between plates is 5 mm

9-3 The charge on a 100-μF capacitor is increased from 100 to 500 μC in 5 ms. Calculate the

a. Original voltage across the capacitor
b. Final voltage across the capacitor and voltage increase across the capacitor
c. Average current required to increase both the charge and voltage across the capacitor

9-4 A 20-μF capacitor is connected to a 100-V dc constant current source. Calculate the

a. Final charge that appears on the plates
b. Time to reach half charge if current is maintained at a constant rate of 25 mA
c. Time to charge the capacitor completely if the average constant current is decreased to 10 mA
d. Energy stored in the capacitor under the conditions of (**a**)

9-5 A 20-μF capacitor is charged from a constant-voltage dc source. Assume, however, that the current decreases uniformly during the charging period from 50 mA to zero in 0.6 s. Calculate the

a. Average rate of change of current and average current
b. Charge on the capacitor after 0.6 s
c. Voltage of the source
d. Energy stored in the dielectric field

9-6 A 50-pF capacitor having an area of 0.5 cm^2 and distance between plates of 1 mm is fully charged across a 20-V dc supply. Calculate the

a. Charge, in coulombs
b. Flux density, **D**, in appropriate units
c. Electric field intensity, **E**, in appropriate units
d. Absolute permittivity, ε, of the capacitor
e. Dielectric constant, ε_r or k, of the capacitor
f. Total number of electrons stored on the negative plate of the capacitor

9-7 When the impressed voltage across the plates of a capacitor is 600 V, the dielectric flux produced is 3×10^{-3} lines. The distance between plates is 0.5 mm and area of the plates is 2 cm $\times$ 4 cm. Calculate the

a. Flux density, **D**
b. Charge in coulombs
c. Capacity of the capacitor
d. Electric field intensity
e. Absolute permittivity, ε

9-8 The capacitance of two parallel plates held by an insulating clamp and separated by air is measured as 200 pF with a capacity bridge. When the plates are immersed in oil, the capacitance is found to be 600 pF. Calculate the

a. Dielectric constant for oil
b. Capacitance of the oil capacitor if the spacing is reduced to one-third of the original spacing

9-9 A parallel-plate capacitor is constructed of 25 glass plates, each measuring 10 cm $\times$ 12 cm, having a thickness of 5 mm. On either side of each plate is a sheet of aluminum, of which 13 are alternately paralleled as one plate of the capacitor and the other 13 form the second plate. Using a dielectric constant, ε_r, of 7.5 for the particular glass, calculate the

a. Capacity of the capacitor
b. Absolute permittivity of the dielectric
c. Relative permittivity
d. Electric field intensity when a voltage of 600 V is applied to the capacitor
e. Flux density under the conditions of (**d**)
f. Total flux
g. Charge on the capacitor, using the result from (**f**)
h. Charge on the capacitor, using Eq. (9-1)

9-10 It is desired to produce a 1-μF capacitor using the same dimensions of glass and aluminum as in Prob. 9-9. Calculate the

a. Number of glass plates required
b. Number of aluminum sheets required

9-11 An air capacitor consists of two parallel plates of area 0.2 m^2. If the distance between plates is 1 mm, calculate the

a. Capacitance in picofarads
b. The capacitance if a sheet of mica 1 mm thick ($\varepsilon_r = 6$) is introduced between the plates

9-12 An air capacitor measures 20 pF on a capacity bridge. If the distance between the plates is doubled and porcelain ($\varepsilon_r = 5.7$) is inserted between the plates, what is its new capacitance?

9-13 On a printed circuit, two parallel wires of equal length are printed 1 mm apart with a distributed capacitance of 4 pF between them. How close together must the wires be printed to produce a capacitance of 20 pF?

Secs. 9-7 to 9-9

9-14 Three capacitors, of 6, 3, and 2 μF, respectively, are connected

a. In parallel
b. In series
c. With the first two in series but the combination in parallel with the third.
Calculate the equivalent capacitance for each of the above combinations.
d. What combination of these three capacitors yields a total capacitance of 7.2 μF?

9-15 The three capacitors of Prob. 9-14 are *series*-connected to a 100-V supply. Calculate the

a. Charge on each capacitor
b. Voltage across each capacitor
c. Based on your answers in (**b**), verify KVL for capacitors in series

9-16 After being fully charged in series, the three capacitors of Prob. 9-15 are connected in parallel with all positive plates and all negative plates, respectively, tied to two junctions. Calculate the

a. Total charge of the parallel combination
b. Voltage across the parallel combination
c. Which of the three capacitors is charged by the other two?

9-17 Two capacitors of 20 and 10 μF, respectively, are connected in parallel. The combination draws 600 μC from an unknown supply. Calculate the

a. Total capacitance of the combination
b. Voltage across the combination and voltage of the supply
c. Charge on each capacitor as a check against the given total charge

Secs. 9-10 to 9-11

9-18 Thirty capacitors, each 30 μF, rated at 600 V are charged to 500 V when connected in parallel. By means of a complex switch system, they are connected *series-aiding* to produce a high-voltage source for testing the insulation of high-voltage transformers. Calculate the

a. Total capacitance of the series-connected system
b. Potential difference across the series-connected system
c. Charge on each capacitor in series and total charge
d. Energy stored in the series combination
e. Power dissipated, if the energy is released in 5 ms whenever the transformer insulation is defective and breaks down

9-19 Three capacitors, of 5, 20, and 4 μF, respectively, are connected in series across a 40-V supply. Calculate the

a. Charge on each capacitor and total charge
b. Voltage across each capacitor
c. Total energy stored by the combination
d. Total capacitance of the combination

9-20 If the three capacitors of Prob. 9-19 are discharged and connected in parallel across the same supply, repeat all parts of Prob. 9-19.

9-21 The total input capacitance of a cathode ray oscilloscope (CRO) is 6 pF. It is desired to reduce the input capacitance to 2 pF by addition of a capacitor. Find

a. The capacitance that must be added
b. How it must be connected

9-22 A 12-μF capacitor in series with a 6-μF capacitor is paralleled by a 16-μF capacitor. If 5 μF is then connected in series with the combination, calculate the

a. Total capacitance
b. Voltage across each capacitor (in order given) if 100 V is the potential across the combination
c. Charge on each capacitor, respectively
d. Total energy stored by the combination

9-23 A technician discovers that a 12-μF capacitor is needed to replace a defective one in a high-precision, low-voltage electronic circuit. In the parts bin, only two 5-μF and two 10-μF capacitors are available. Using *all* four capacitors, show how the technician solved the problem.

9-24 In **Fig. 9-8**, capacitors C_1 and C_2 are completely discharged. When switch S is thrown to position 1, it remains there until C_1 is completely charged. Switch S is then thrown to position 2, where it remains indefinitely. Calculate

a. Q_0, the initial charge on C_1
b. C_t, the total capacitance in position 2
c. V, the voltage across each capacitor in position 2, by two methods
d. Q_2, the charge on capacitor C_2
e. Q_1, the charge on capacitor C_1
f. Q_t, the total charge on the capacitors, by two methods
g. dV_2, the change in voltage across C_2 for the cycle of switch transition

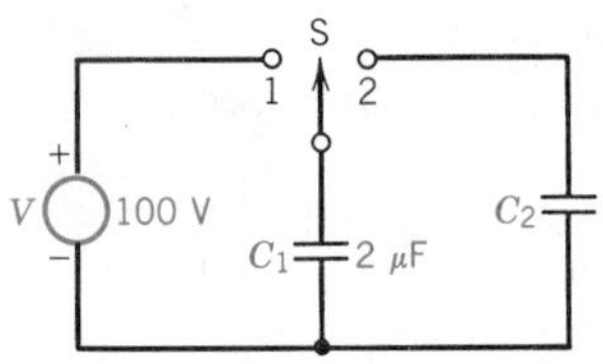

Figure 9-8 Problems 9-24 to 9-26 and 9-28

9-25 **a.** Repeat parts (**c**), (**d**), (**e**), (**f**), and (**g**) of Prob. 9-24 for five additional cycles of switch transition from position 1 to position 2.
b. Tabulate the data for the first six cycles of operation.
c. If it takes six complete cycles to charge C_2 to approximately two-thirds of the applied voltage, will C_2 fully charge in three more cycles? Explain in detail.

9-26 If C_1 in Fig. 9-8 is changed to 10 μF, after five cycles of operation calculate

a. The total charge on the paralleled capacitors, Q_t
b. The voltage across the paralleled capacitors, V
c. Q_2, the charge on capacitor C_2
d. What conclusion can you draw regarding the magnitude of C_1 compared to C_2 versus the number of cycles to fully charge C_2, based on the values computed above?
e. Verify your answer to (**d**) using Eq. (9-8)

9-27 A 20-μF capacitor C_1 is fully charged on a 10-V supply. It is then removed from the supply and connected directly across a 10-μF capacitor, C_2. Calculate the

a. Original charge on C_1 (Q_0)
b. Original energy stored on C_1 (W_0)
c. Charge on C_1 and C_2, respectively, when they are paralleled
d. Voltage across both C_1 and C_2 when paralleled
e. Energy remaining on C_1
f. Energy stored on C_2
g. Total energy stored by both capacitors
h. Compare (**g**) and (**b**) and account for difference between them
i. Calculate energy loss due to charge redistribution

9-28 In Fig. 9-8, C_2 is changed to 1 μF. After five cycles of switch transition, calculate the

a. Total charge on the paralleled capacitors, Q_t
b. Voltage across the paralleled capacitors, V
c. Q_2, charge on capacitor C_2

9-29 A 4-μF capacitor C_1 is charged to 25 V and connected in parallel, with due regard for polarity, to another 8-μF capacitor that has been charged to 50 V. Calculate the

a. Charge on C_1 before connection
b. Charge on C_2 before connection
c. Total charge of the combination in parallel
d. Voltage across the parallel combination
e. Charge on each capacitor in parallel
f. Energy stored in the paralleled system
g. Total energy stored originally by the two capacitors BEFORE the connection is made
h. Compare (**f**) and (**g**) and explain why this is NOT a violation of the law of conservation of energy

9-30 In Prob. 9-29, the voltage of C_2 changed from 50 to $41.\bar{6}$ V but the total charge remained unchanged when the capacitors were paralleled. Since charge IS conserved, the energy loss MUST be due to the voltage change.

a. Derive an equation for the energy loss, W_L, in terms of the drop in voltage, ΔV, the original voltage, V_2, and the original energy, W_{02}
b. Test the validity of the equation derived in (**a**), using the values of Prob. 9-29

Hint: since $W = QV/2$ and Q is constant, $W = kV/2$, where V is expressed as a voltage ratio.

9-31 Given the purely capacitive series–parallel circuit shown in **Fig. 9-9**, calculate the

a. Total capacitance, C_T
b. Total charge, Q_T
c. Charge on capacitors C_2 and C_3
d. Charge on capacitors C_4 and C_5
e. Voltage at point **X**
f. Voltage at point **Y** with respect to ground

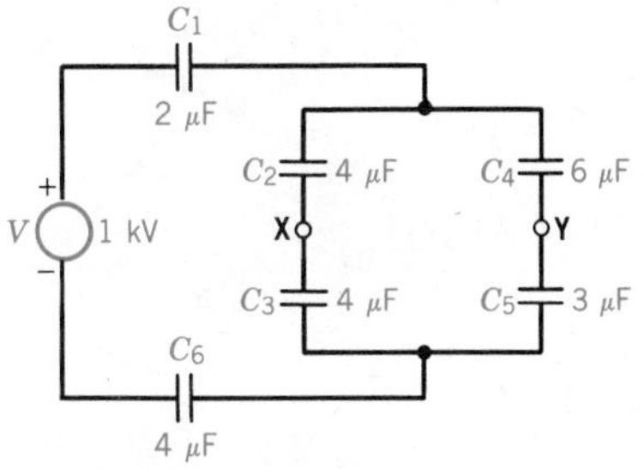

Figure 9-9 Problem 9-31

9-32 Given the purely capacitive series–parallel network shown in **Fig. 9-10**, calculate the total capacitance seen at the input terminals **A**–**B**.

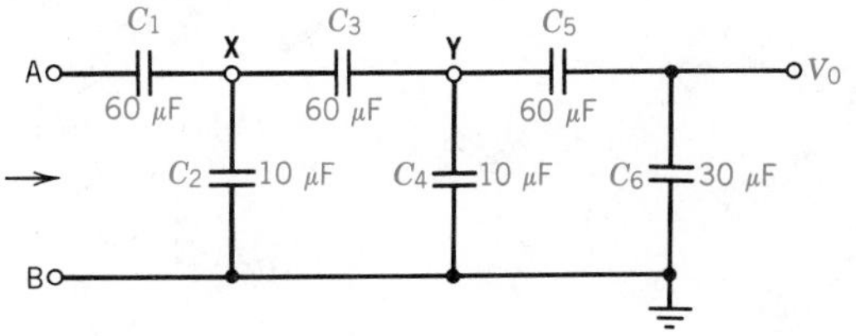

Figure 9-10 Problems 9-32 and 9-33

9-33 If 90 V dc is applied to terminals **A**–**B** in Fig. 9-10, calculate

a. V_X
b. V_Y
c. V_0

9-34 A precision 1-μF capacitor, C_1, is charged to 20 V (exactly) by a dc power supply. It is then removed and quickly connected across an unknown capacitor. The voltage across the parallel combination, as measured by a precision electrometer, is 12.5 V. Calculate the capacitance of the unknown capacitor, C_2 (Hint: use Eq. 9-8, transposed).

9-35 Repeat Prob. 9-34 if all values given are the same except that the electrometer reading is 18.75 V.

9-36 Repeat Prob. 9-34 if the electrometer reading is 1.5 V.

9-14 ANSWERS

9-1 a $1\ \mu$C b $3\ \mu$C c $3.\bar{3}$ kV/m d 10 kV/m
9-2 a $100\ \mu$C b 5 mJ c 20 kV/m
9-3 a 1 V b 5 V c 80 mA
9-4 a 2 mC b 40 ms c 200 ms d 0.1 J
9-5 a $-8\bar{3}$ mA/s, 25 mA b 15 mC c 750 V d 5.625 J
9-6 a 1 nC b $20\ \mu$C/m^2 c 20 kV/m d 1 nF/m e 112.9 f 6.24×10^9
9-7 a 3.75 C/m^2 b 3 mC c $5\ \mu$F d 1.2 MV/m e $3.125\ \mu$F/m
9-8 a 3 b $1800\ \mu$F
9-9 a 3.984 nF b 66.41 pF/m c 7.5 d 120 kV/m e $7.97\ \mu$C/m^2 f, g $2.39\ \mu$C h $2.39\ \mu$C
9-10 a 6275 b 6276
9-11 a 1.77 nF b 10.63 nF
9-12 57 pF
9-13 0.2 mm
9-14 $1\ \mu$F
9-15 a $100\ \mu$C b $16.\bar{6}$ V, $33.\bar{3}$ V, 50 V c 100 V
9-16 a $300\ \mu$C b $27.\overline{27}$ V c $6\ \mu$F
9-17 a $30\ \mu$F b 20 V c 400, 200, 600 μC
9-18 a $1\ \mu$F b 15 kV c 15 mC d 112.5 J e 22.5 kW

9-19 a 80 μC b 16, 4, 20 V c 1.6 mJ

9-20 a 200, 800, 160 μC, 1.16 mC b 40 V c 23.2 mJ
d 29 μF

9-21 a 3 μF b series

9-22 a 4 μF b $6.\overline{6}$ V, $13.\overline{3}$ V, 20 V, 80 V
c 80 μC, 320 μC, 400 μC d 20 mJ

9-24 a 200 μC b 12 μF c $16.\overline{6}$ V d $16\overline{6}$ μC
e $3\overline{3}$ μC f 200 μC g $16.\overline{6}$ V

9-25 $V_2 = 66.51$ V after 6 cycles

9-26 a 1.9375 mC b 96.875 V c 968.75 μC

9-27 a 200 μC b 1 mJ c $13\overline{3}$ μC, $66.\overline{6}$ μC d. $6.\overline{6}$ V
e $44\overline{4}$ μJ f $22\overline{2}$ μJ g $66\overline{6}$ μJ i $33\overline{3}$ μJ

9-28 a 298.8 μC b 99.6 V c $99.\overline{6}$ μC

9-29 a 100 μC b 400 μC c 500 μC d $41.\overline{6}$ V
e $16\overline{6}$ μC, $33\overline{3}$ μC f 10.42 mJ g 11.25 mJ

9-31 a 1 μF b 1 mC c 500 μC d 500 μC
e 375 V f $33\overline{3}$ V

9-32 20 μF

9-33 a 60 V b 40 V c $2\overline{6}$ V

9-34 0.6 μF

9-35 $6\overline{6}$ nF

9-36 $12.\overline{3}$ μF

CHAPTER 10

Transient RC Circuit Analysis

10-1 INTRODUCTION

Two assumptions were made, for purposes of simplicity, in all of the pure (ideal) capacitive circuits covered in Chapter 9, namely

1. that the capacitors were ideal and contained no resistance, and/or no external resistance appeared in any of the circuits; and
2. that the ideal capacitors charged or discharged linearly, that is, at a constant and linear rate.

In practice, we find that *neither* of these assumptions is true. Capacitors do have resistance. Capacitors appear in circuits containing resistance. Nor do capacitors charge and discharge at a linear rate. We will find that the voltage across, current in, power, and energy stored and released all either increase or decrease at an *exponential* rate (Appendix B-4.1) as opposed to a linear rate.

As noted from the title of this chapter, we will be dealing with *transient* (short-lived) changes in voltage, current, power, and energy in R–C circuits. Strictly speaking, a *transient* is the part of the change in a variable that *disappears* in going from one *steady-state* condition to another. The *first* steady-state condition is called the *initial* condition. The *second* steady-state condition will be called the *final* condition. The transient waveform is the *link* between the *initial* and *final* conditions; that is, it represents how we go from one steady state to another steady state.

10-2 TRANSIENT RELATIONS DURING CAPACITOR CHARGE CYCLE

Consider the purely resistive circuit shown in **Fig. 10-1a**. The figure shows that the switch S closes at time $t = 0^+$ s. Is this the initial or final condition? How will we define and designate these?

We know the switch closes exactly at $t = 0$ s from the waveform of Fig. 10-1b. Obviously, then, $t = 0^+$ s represents the instant at which the switch is just closed. Then $t = 0^-$ s must represent the instant when the switch was open and just about ready to close. Let us define these as follows:

$t = 0^-$ s means the instant at which $t \lesseqqgtr 0$ and we still have the *initial* condition.

$t = 0^+$ s means the instant at which $t \geqq 0$ and some source is beginning to *force a change in state* from the initial to the final condition.

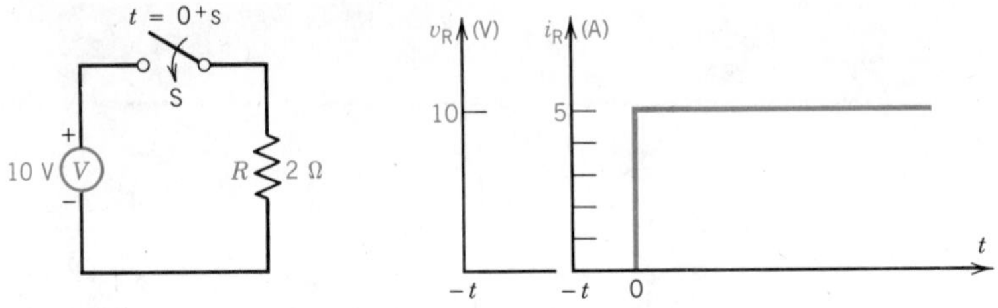

Figure 10-1 Response of purely resistive circuit to a change in voltage

a. Purely resistive circuit

b. Waveform of voltage and current response

Let us see how these definitions apply to Fig. 10-1, a purely resistive circuit. At time $t = 0^-$ s, the switch is opened, $i = 0$, and we have the *initial* condition. At time $t = 0^+$ s, the switch is closed, $i = 0, 1, 2, 3, 4$, and 5 A, **simultaneously**, and we have the *final steady-state* condition, that is, $I_R = 5$ A and $V_R = 10$ V.

In the case of the resistor, the *transient* is the waveform of current i that is rising from 0 to 5 A during the time $t = 0$ to $t = 0^+$.

A word about notation for physical quantities is appropriate here. Transient or *changing* physical quantities are always designated by *lowercase* letters. Quantities that are in their *initial* state or *final* state, commonly called "steady state," are designated by *capital* letters.

Thus, before the switch is closed at $t = 0^-$ s, with respect to R, $V_R = 0$ and $I_R = 0$ are the *initial* conditions. After the switch is closed, at $t = 0^+$ s, $V_R = 10$ V and $I_R = 5$ A are the *final* conditions.

We can see that for a pure resistance, there is *no delay* in time in the rise of current from its original to its final steady-state value in Fig. 10-1b. We defined the property of resistance as opposition to a *steady* current. When the steady current is 5 A in Fig. 10-1b, the resistor has an opposition of 2 Ω. But during the time from $t = 0^-$ to $t = 0^+$ s, there is a *change* of current. The resistance has *no opposition* whatever to a change of current.[1] Therefore the current rises to its steady-state value in zero time, theoretically.

But does this also happen when we apply a voltage to a practical capacitor? Recall that the property of capacitance is defined as *opposition to a change in voltage* across its terminals. Note that this definition does *not* specify opposition to *current*, either changing or steady. Let us now add a capacitor in series with the 2-Ω resistor of Fig. 10-1a, as shown in **Fig. 10-2a**. The resistor will only oppose a *steady* current. The capacitor, as we might guess, will only show opposition to the *change* in voltage; and the greater the *voltage change*, the greater the *opposition*. The capacitor shows *no opposition* to a *change in current*.

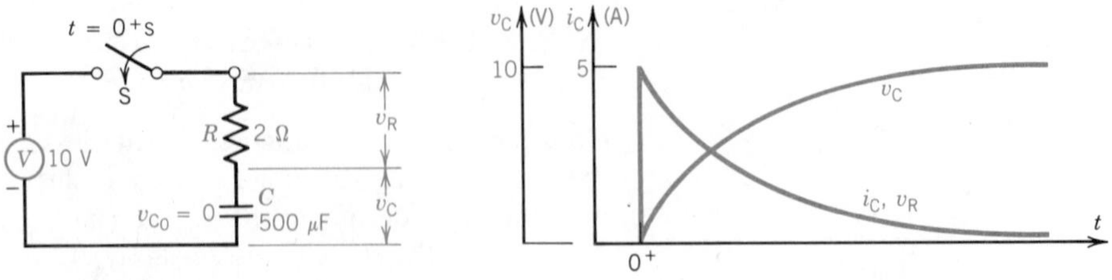

Figure 10-2 Response of a practical capacitive circuit to a change in voltage with capacitor initially discharged

a. Practical capacitive circuit, capacitor initially uncharged

b. Waveform of voltage and current response

[1] There is a circuit property called *inductance* which does have opposition to a *change* in *current*. See Chapter 12.

Figure 10-2b verifies our educated guess. The current rises instantly at $t = 0^+$ to a steady Ohm's law value of 5 A, where it is limited by resistor R. But since the voltage initially tries to change at an infinite rate from 0 to 10 V, the opposition of the capacitor is infinite and the initial voltage across the capacitor, v_C, is zero!

Note from Fig. 10-2b, however, that as time passes beyond $t = 0^+$, the *voltage* v_C *rises* exponentially and the current i_C also *decreases* exponentially. This is *not* a linear relationship and requires some knowledge of exponential functions. Various types of exponential functions are described in Appendix B-4, for readers unacquainted with them, along with a definition of the time constant, τ_c.

Let us summarize the initial and final steady-state conditions for the circuit shown in Fig. 10-2a:

	INITIAL CONDITION		FINAL STEADY-STATE CONDITION
capacitor current, i_C	5 A	decreasing exponentially to	0 A
capacitor voltage, v_C	0 V	increasing exponentially to	10 V
time, t, seconds	0^+		?

In addition to the two steady-state conditions just discussed (initial and final), we may write Kirchhoff's voltage law (KVL) relations that are in effect at all times during the transient that exists between them. This is obvious from Fig. 10-2a, and we may write KVL as

$$V = v_R\downarrow + v_C\uparrow = i_C R\downarrow + v_C\uparrow \qquad \text{volts (V)} \qquad \textbf{(10-1)}$$

Equation (10-1), despite its apparent simplicity, is extremely important. It tells us that

1. *Two transient* conditions are in effect during the time the capacitor is being charged or is undergoing a change in charge.
2. Since V and R are BOTH constant, the two *transient* variables are i_C and v_C (both designated by lowercase italic symbols).
3. Initially, at $t = 0^+$, $i_C = I_0 = V/R$, which is a maximum value for i_C.
4. Initially, $v_C = 0$ in Eq. (10-1).
5. As time goes on during the transient, however, i_C decreases as v_C increases, for reasons explained below.
6. At all times, however, during the transient, the sum $v_R + v_C$ is equal to the supply voltage V and is *constant.*
7. The arrows in Eq. (10-1) show the directions of the magnitudes with time; that is, v_R decreases and v_C increases with time.
8. At all times, V_R is a function of $i_C R$. Consequently, whenever there is a variation of i_C, the *same* variation occurs for v_R. This is such a fundamental relation that it will be repeated often throughout this text.
9. Example 10-3 shows that KVL is verified at all times and especially during the transient period.

But what of i_C, the current flowing instantaneously to the capacitor, and how is it affected by the charge developed on the capacitor? Since $Q = CV$, we can say that the instantaneous increases in charge on the capacitor, dq, are a function of the instantaneous increases in voltage, or $dq/dt = C(dv_C/dt)$. But since current is the rate of charge flow, the instantaneous current is $i_C = C(dv_C/dt)$. We will discover in Eq. (10-6a) that dv_C/dt *decreases* exponentially from an initial value (of V_0/τ) to zero. This explains why the current, i_C, and the resistor voltage, $i_C R = v_R$, are shown to *decrease* in Eq. (10-1) as the voltage (and charge) across the capacitor *increases* and current *decreases*.

The foregoing summary shows that our problem is to find how long it will take for the current to drop to zero and (at the same time) for the capacitor to become fully charged to 10 V. Since v_C and i_C both are a function of time in Fig. 10-2b, we require *two* equations in which these transients vary as a function of time.

Equation (B-5.5) in Appendix B-5 derives the *rising* **transient voltage** across a capacitor during the *charging* period as a *rising* exponential of the form

$$v_C = V(1 - \varepsilon^{-t/RC}) = V(1 - \varepsilon^{-t/\tau_c}) \quad \text{volts (V)} \qquad \textbf{(10-2)}$$

where V is the potential applied to the RC circuit in volts
R is the series resistance in ohms
C is the capacitance in farads
t is the time in seconds
τ_c is the time constant or the product of RC in seconds

The time constant, τ_c, is usually defined as the time for the voltage to rise to 63.2% of its final value. But a more precise definition emerges from Eq. (10-6), where the time constant of an RC circuit is defined as the time for v_C to rise to the same value as applied voltage, V, if it continued to rise linearly and continuously at its *initial* rate of change, dv_{c0}/dt.

Unlike v_C, which undergoes an exponential *rise* during capacitor charge, the current i_C undergoes an exponential *drop* or *decay*. From Eq. (B-4.1) in Appendix B-4, we note that the current is a positive *decaying* exponential function of the form $y = K\varepsilon^{-st} = K\varepsilon^{-t/\tau}$, where $\tau = 1/s = RC$. For the *decaying transient* current in a capacitor, i_C, during the charging period of the capacitor, we may write

$$i_C = C\left(\frac{dv_C}{dt}\right) = I_0\varepsilon^{-t/RC} = I_0\varepsilon^{-t/\tau_c} = \frac{V}{R}\varepsilon^{-t/\tau_c} \quad \text{amperes (A)} \qquad \textbf{(10-3)}$$

where I_0 is the initial current at instant $t = 0^+$ when the switch is first closed and all other terms have been defined above

But as noted in Eq. (10-1), the variation of v_R is always a function of i_C in an RC circuit, since $v_R = i_C R$. Therefore, using Eq. (10-3), we write

$$v_R = i_C R = RC\frac{dv_C}{dt} = \tau_c\frac{dv_C}{dt} = I_0 R\varepsilon^{-t/\tau_c} = V\varepsilon^{-t/\tau_c} \qquad \textbf{(10-4)}$$

where all terms have been defined above

EXAMPLE 10-1

Given the circuit of Fig. 10-2, where $R = 2\ \Omega$, $C = 500\ \mu F$, and $V = 100$ V, calculate the

a. Charging circuit time constant, τ_c
b. Voltage across the capacitor, v_C, 2 ms after switch S closes
c. Capacitor circuit current, i_C, 2 ms after switch S closes
d. Voltage across the resistor, v_R, 2 ms after switch S closes, using Eq. (10-4)
e. Verify Eq. (10-1), using the instantaneous voltages calculated above
f. Verify value of v_R calculated in (**d**), using $i_C R$

Solution

a. $\tau_c = RC = (2\ \Omega)(500\ \mu F) = \mathbf{1\ ms}$
b. $v_C = V(1 - \varepsilon^{-t/\tau_c}) = 10(1 - \varepsilon^{-2/1}) = \mathbf{8.647\ V}$ (10-2)
c. $i_C = I_0\varepsilon^{-t/\tau_c} = \dfrac{V}{R}\varepsilon^{-t/\tau_c} = \dfrac{10}{2}\varepsilon^{-2/1} = 5\varepsilon^{-2} = \mathbf{0.6768\ A}$ (10-3)
d. $v_R = V\varepsilon^{-t/\tau_c} = 10\varepsilon^{-2} = \mathbf{1.353\ V}$ (10-4)
e. $V = v_R + v_C = (1.353 + 8.647)\ V = \mathbf{10\ V}$ (this verifies KVL even at instantaneous values of t)
f. $v_R = i_C R = (0.6768\ A)(2\ \Omega) = \mathbf{1.353\ V}$ (10-4)

Despite its disarming simplicity, Ex. 10-1 is extremely important and the following should be noted:

1. Before time $t = 0$, when the switch is open, $v_C = 0$ and the capacitor is uncharged.
2. After time $t = 0^+$, the voltage v_C rises exponentially, in accordance with Eq. (10-2).
3. At any time, t, during the transient rise of v_C, values of v_C may be computed from the relations given in Eq. (10-2).
4. The circuit (charging) time constant, $\tau_c = RC$, is a factor in determining how long it takes for the capacitor to charge completely. Example 10-1b shows that in $t = 2$ ms, the capacitor charges to approximately 86.5% of its final value of 10 V. Consequently, the ratio t/τ_c is a determining factor. When this ratio is small in Eq. (10-2), v_C is small. When the ratio is zero (whenever $t = 0$), $v_C = 0$. Clearly, if τ_c is large, long values of time, t, are required to charge a capacitor completely.
5. Since the expression ε^{-t/τ_c} is a *decaying* exponential term (see Appendix B-4), both Eqs. (10-3) and (10-4) represent the decay of current, i_C, and the decay of voltage, v_R, with time during the transient period. Note that i_C decays from an initial value of 5 A to 0.6768 A after 2 ms in *solution* 10-1**c**. Similarly, v_R decays from an initial value of 10 V to 1.353 V after 2 ms, as shown in *solution* 10-1**d**.
6. At all times during the transient period, Kirchhoff's voltage law must hold. Note that Eq. (10-1) for the KVL relations of Fig. 10-2 does hold at the instant when $t = 2$ ms, as shown by the solution of Ex. 10-1**e**.
7. It was stated in summarizing Eq. (10-1) in item (8) that at all times $v_R = i_C R$. The solution of Ex. 10-1**f** verifies this at the instantaneous value $t = 2$ ms.
8. It is expected, therefore, that when the transient disappears (and the capacitor is fully charged to 10 V) $i_C = 0$ and, consequently, $v_R = 0$. Again, this is verified by Eq. (10-1), since $v_R = V - v_C = 10 - (10) = 0$, as well as $v_R = i_C R = (0)(R) = 0$.

Frequently, we would like to perform the reverse process shown in Eqs. (10-2) through (10-4). We may ask, *given* the values shown in Fig. 10-2a, how long will it take for the voltage across the capacitor, v_C, to reach 5 V or how long will it take for the current, i_C, to drop from 5 to 3 A? Even more important, we would like to know how long it takes for *any* transient to disappear.

Either Eq. (10-2) or (10-3) may be solved for t algebraically. Appendix Eq. (B-5.8) solves Eq. (10-2) for t, yielding

$$t = \tau_c \ln\left(\frac{V}{V - v_C}\right) = RC \ln\left(\frac{V}{V - v_C}\right) \quad \text{seconds (s)} \qquad \textbf{(10-5)}$$

where $v_C \lneqq V$ and all other terms have been defined above

In effect, Eq. (10-5) furnishes a way of finding t but poses a dilemma. The applied voltage $V = 10$ V, and we are trying to find how long it takes v_C to rise to 10 V. If the denominator of Eq. (10-5) is zero, we cannot evaluate $\ln(\infty)$. We circumvent this by setting v_C to some value *slightly below* 10 V, as shown by Ex.10-2d. Intuition tells us, moreover, that since v_C approaches V as an *asymptotic* value, it cannot ever reach V under any circumstances! Therefore, v_C is actually slightly less than V, even when a capacitor is considered fully charged, verifying the validity of Eq. (10-5).

EXAMPLE 10-2

Given the circuit of Fig. 10-2, calculate the

a. Initial current at the instant switch S closes at $t = 0^+$
b. Initial voltage across the capacitor
c. Charging time constant of the circuit, τ_c
d. Time for the voltage across the capacitor to reach 9.95 V

Solution

a. $I_0 = V/R = 10\ \text{V}/2\ \Omega = \mathbf{5\ A}$
b. From Fig. 10-2b, the initial voltage across the capacitor is zero. Substituting $t = 0$ in Eq. (10-1), $v_C = 10(1 - \varepsilon^0) = \mathbf{0\ V}$, as proof.
c. $\tau_c = RC = (2\ \Omega)(500\ \mu\text{F}) = \mathbf{1\ ms}$
d. $t = \tau_c \ln\left(\dfrac{V}{V - v_C}\right) = (1\ \text{ms}) \ln\left(\dfrac{10}{10 - 9.95}\right)$

$= (1\ \text{ms}) \ln(200) = \mathbf{5.3\ ms}$ (10-5)

Example 10-2d shows that it takes approximately 5.3τ for a capacitor to charge to 99.5% of its fully charged value. Most engineers assume that a capacitor is fully charged when $t = 5\tau_c$. Let us see if we can discover what percentage of full charge this represents.

EXAMPLE 10-3

Calculate

a. The value of v_C when $t = 5\tau_c$ in Ex. 10-1
b. The ratio v_C/V in percent

Solution

a. $v_C = V(1 - \varepsilon^{-t/\tau_c}) = 10(1 - \varepsilon^{-5\tau_c/\tau_c}) = 10(1 - \varepsilon^{-5})$
$= \mathbf{9.933\ V}$ (10-2)
b. $v_C/V = 9.933/10 = \mathbf{99.33\%}$

We can see from Ex. 10-2 that after five time constants, a capacitor is charged to more than 99% of full charge. Let us now turn to some of the uses of Eqs. (10-2) and (10-3), as shown by the following examples.

EXAMPLE 10-4

A 1-MΩ resistor is connected in series with a 10-μF capacitor and applied to a 100-V dc source at time $t = 0^+$ s. Calculate the

a. Circuit time constant, τ_c
b. Capacitor voltage at time $t = 10$ s
c. Charge on the capacitor at $t = 10$ s
d. Capacitor and circuit current at $t = 10$ s, by two methods
e. Voltage drop across the 1-MΩ resistor at $t = 10$ s
f. Verify KVL using **(b)** and **(e)** calculated above

Solution

a. $\tau_c = RC = (1\ \text{M}\Omega)(10\ \mu\text{F}) = \mathbf{10\ s}$
b. $v_C = V(1 - \varepsilon^{-t/\tau_c}) = 100(1 - \varepsilon^{-10/10}) = \mathbf{63.21\ V}$ (10-2)
c. $q_C = C \cdot v_C = (10\ \mu\text{F})(63.21\ \text{V}) = \mathbf{632.1\ \mu C}$
d. $i_C = \dfrac{V}{R}(\varepsilon^{-t/\tau_c}) = \dfrac{100\ \text{V}}{1\ \text{M}\Omega}\varepsilon^{-1} = \mathbf{36.79\ \mu A}$ (10-3)

$i_C = \dfrac{V - v_C}{R} = \dfrac{(100 - 63.21)\ \text{V}}{1\ \text{M}\Omega} = \mathbf{36.79\ \mu A}$ (10-1)

e. $v_R = i_C R = (36.79\ \mu\text{A})(1\ \text{M}\Omega) = \mathbf{36.79\ V}$ (10-4)
f. $V = v_R + v_C = (36.79 + 63.21)\ \text{V}$
$= \mathbf{100\ V}$ by KVL (10-1)

Example 10-4 shows that even during the transient period when the voltage is rising and current is dropping exponentially, KVL applies. Note that in step **(d)** of the solution, current may be found by using Eq. (10-3) but also is found by using KVL and simple series circuit analysis techniques, thereby verifying KVL during the transient period.

10-3 UNIVERSAL TIME CONSTANT GRAPHICAL RELATIONS[2]

A review of Appendix B-4.1 reveals that a rising exponential has the general equation $y = 1 - \varepsilon^{-t/\tau}$ and a decaying exponential has the general equation $y = (1\varepsilon)^{-t/\tau}$ for $K = 1$ and $\tau = 1/s$. The two exponentials, both rising and decaying, are plotted in **Fig. 10-3** as a function of time constant increments (i.e., 1τ through 6τ). For the value $K = 1$, the rising curve shows the *percent* of maximum rise for rising exponentials, while the decaying curve shows the *percent* of decay from the original value when $t = 0$. The abscissa is also normalized in terms of time constant increments rather than time in seconds. These two devices enable the chart to be used to find any values of rising or falling exponentials at any value of time, for all exponential conditions.

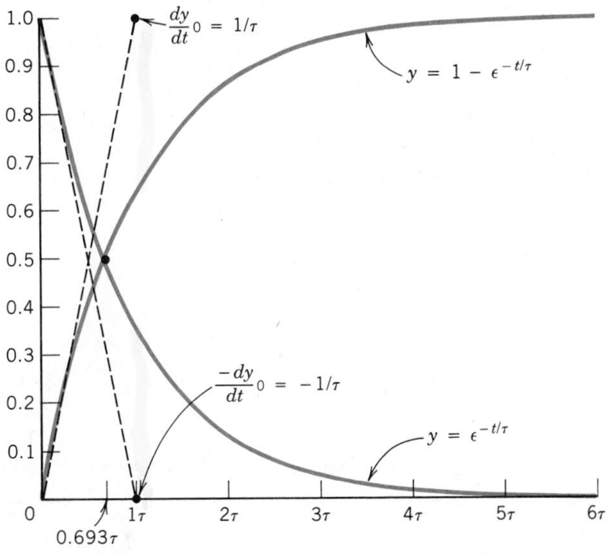

Figure 10-3 Universal time constant chart

Readers may desire to plot their own curves, and the data are given below:

Abscissa, units of τ	0.693τ	1τ	2τ	3τ	4τ	5τ	6τ
$y = 1 - \varepsilon^{-t/\tau}$, *rising*	0.5	0.632	0.865	0.95	0.982	0.9933	0.9975
$y = \varepsilon^{-t/\tau}$, *decaying*	0.5	0.368	0.135	0.05	0.018	0.0067	0.0025

With respect to this data table and the curves drawn in Fig. 10-3, note the following.

1. In 1τ, the rising curve ($y = 1 - \varepsilon^{-t/\tau}$) rises to 63.2% of its maximum *final* value.
2. In 1τ, the decaying curve ($y = \varepsilon^{-t/\tau}$) drops to 36.8% of its original *initial* value.
3. In 0.693τ, $y = 0.5$ of either its final or initial value and the two curves cross each other.
4. The initial rate of change of each curve can easily be determined from tangents (shown dashed in Fig. 10-3) drawn to each curve at its origin. Observe that the slope of each curve has a horizontal value of exactly 1τ and the ordinate of the slope is the normalized value $K = 1$. What does this imply?

[2] The universal time constant chart of Fig. 10-3 may be used equally with capacitive and inductive circuits as well as any other transient relations having the form $y = 1 - \varepsilon^{-x}$ or $y = \varepsilon^{-x}$.

We know that the curve of v_C (in Eq. 10-2) is a rising exponential and is rising to a value of V, the applied voltage. This means that the INITIAL rate of change of voltage, dv_{C_0}/dt, must equal the ordinate V divided by the abscissa, 1τ, or, as shown in **Fig. 10-4a**, it serves to *define* the time constant, τ_c! Initial rate of change of voltage (during capacitor charging) is

$$\frac{dv_{C_0}}{dt} = \frac{V}{\tau_c} = \frac{V}{RC} \quad \text{volts/second (V/s)} \qquad (10\text{-}6)$$

Since the relation described in (3) applies to the decaying curve, as shown in Fig. 10-4b, we may also write a similar equation for the current curve.

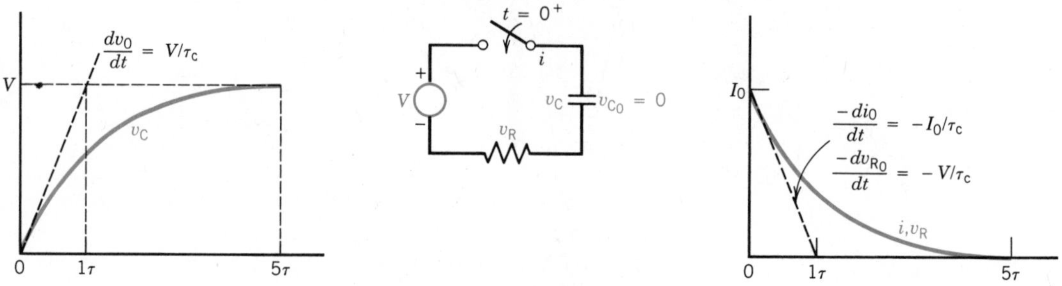

a. Voltage across capacitor, v_C

b. Decay in current, i_C and voltage across resistor, v_R

Figure 10-4 Waveforms for charging conditions in an *RC* circuit whose capacitor is initially uncharged

The initial rate of change of current is

$$\frac{-di_0}{dt} = \frac{-I_0}{RC} = \frac{-I_0}{\tau} = \frac{-V/R}{\tau} \quad \text{amperes/second (A/s)} \qquad (10\text{-}7)$$

5. At every instant, the voltage across the resistor, R, is a function of the current through it, since $v_R = iR$, and R is constant.[3] The decreasing transient voltage v_R was noted in Eq. (10-1). We may also write the initial rate of change of resistor voltage decay, dv_{R_0}/dt, as shown in Fig. 10-4b, as

$$-\frac{dv_{R_0}}{dt} = \frac{-V_0}{\tau} = \frac{-V_0}{RC} = \frac{-I_0R}{\tau} \quad \text{volts/second (V/s)} \qquad (10\text{-}8)$$

6. It should be noted that the *initial* rates of change expressed in Eqs. (10-6) through (10-8) are actually the *maximum* rates of change of both exponentials shown in Figs. 10-3 and 10-4. At approximately 5τ, **all** rates of change in Figs. 10-3, 10-4a, and 10-4b are approximately zero. This is understandable, since the transient has disappeared and we are now in the steady state, where *no*

[3] This relationship holds equally for ac currents as well as currents of a nonsinusoidal nature. Any current in a network may be viewed on a CRO by merely observing the voltage across a resistor carrying the same current, since $v_R = iR$.

change occurs. Tangents drawn to each curve in Fig. 10-3 at $t = 5\tau$ (or more) are zero.

7. Finally, it must be noted that Eqs. (10-6) through (10-8) are valid only for situations in which there is *no charge, initially*, on a capacitor. As we will see in the next section, the initial rates of change are somewhat different for *partially* charged capacitors, as are the equations governing v_C and i_C.

EXAMPLE 10-5

Given the *RC* series circuit of Ex. 10-4, where $R = 1\ \text{M}\Omega$, $C = 10\ \mu\text{F}$, and $V = 100$ V, calculate the

a. Initial rate of change of voltage across the capacitor at $t = 0^+$
b. Initial rate of change of current in the capacitor at $t = 0^+$
c. Initial rate of change of voltage across the 1-MΩ resistor at $t = 0^+$
d. Rate of change of voltage across the capacitor at $t = 60$ s
e. Rate of change of current in the capacitor at $t = 60$ s
f. Rate of change of voltage across the resistor at $t = 60$ s

Solution

a. $$\frac{dv_{C_0}}{dt} = V/\tau_c = V/RC = 100\ \text{V}/10\ \text{s} = \mathbf{10\ V/s} \qquad (10\text{-}6)$$

b. $$\frac{di_{C_0}}{dt} = -I_0/\tau_c = -100\ \mu\text{A}/10\ \text{s} = \mathbf{-10\ \mu A/s} \qquad (10\text{-}7)$$

c. $$\frac{dv_{R_0}}{dt} = -V_0/\tau_c = \mathbf{-10\ V/s} \qquad (10\text{-}8)$$

d, e, f. All rates of change are **0**. The transients disappear after $5\tau = 50$ s

---

Example 10-5 shows that the maximum rates of change occur at $t = 0^+$, that is, when the transient first begins and we go from the initial steady state to the final steady state. The transient ends when all rates of change are zero, the capacitor is fully charged, and the capacitor current is zero.

Example 10-5 raises the question of whether it is possible to determine the rate of change of v_C, i_C, and v_R, respectively, *at any time* during the charging transient. It is possible, and the appropriate equations are derived in Appendix B-5. For the case of dv_C/dt, at any value of t we have from Eq. (B-5.9)

$$\frac{dv_C}{dt} = \frac{V_0}{\tau}\varepsilon^{-t/\tau_c} \quad \text{volts/second (V/s)} \qquad \textbf{(10-6a)}$$

For the case of di_C/dt, at any value of t we have from Eq. (B-5.10)

$$\frac{di_C}{dt} = \left(\frac{-I_0}{\tau}\right)\varepsilon^{-t/\tau_c} \quad \text{amperes/second (A/s)} \qquad \textbf{(10-7a)}$$

And for the case of dv_R/dt, since v_R is iR,

$$\frac{dv_R}{dt} = \left(\frac{-I_0R}{\tau}\right)\varepsilon^{-t/\tau_c} \quad \text{volts/second (V/s)} \qquad \textbf{(10-8a)}$$

The insights to be gained from Eqs. (10-6a) through (10-8a) are that the initial (maximum) rates of change (at time $t = 0^+$) *all decrease exponentially* with time.[4]

---

[4] The advantage of these equations is that we no longer require complex differential equations to find these rates of change. Instead we use simple decaying exponential equations.

Thus as time, t, increases, it forces the rates of change to decrease exponentially, so that at time $t = 5\tau_c$ (or more) the rates are all zero.

Also note that the rate of change of v_C is *positive*, because the slope of tangents to the curve of v_C in Fig. 10-4a is positive. Similarly, note from Eqs. (10-7a) and (10-8a) that the rates of change of i and v_R are both *negative*, since tangents to these curves have *negative slopes.*

EXAMPLE 10-6

Using the data of Exs. 10-4 and 10-5, calculate the rates of change of v_C, i_C, and v_R, respectively, using Eqs. (10-6a) to (10-8a) at the following times:

a. $t = 0^+$
b. $t = 10$ s
c. $t = 25$ s
d. $t = 60$ s

Solution

a. $dv_C/dt = \dfrac{V_0}{\tau}\varepsilon^{-t/\tau_c} = \dfrac{100\text{ V}}{10\text{ s}} \times 1 = \mathbf{10\ V/s}$ (10-6a)

$di_C/dt = (-I_0/\tau_c) \times 1 = -100\ \mu\text{A}/10\text{ s} = \mathbf{-10\ \mu A/s}$ (10-7a)

$dv_R/dt = (-I_0R/\tau_c) \times 1 = (-100\ \mu\text{A})(1\ \text{M}\Omega)/10\text{ s} = \mathbf{-10\ V/s}$ (10-8a)

b. Since $\varepsilon^{-t/\tau_c} = \varepsilon^{-10\text{ s}/10\text{ s}} = \varepsilon^{-1} = 0.3679$ when $t = 10$ s

$dv_C/dt = \dfrac{V_0}{\tau_c}\varepsilon^{-1} = (10\text{ V/s})(0.3679) = \mathbf{3.679\ V/s}$ (10-6a)

$di_C/dt = (-I_0/\tau_c)\varepsilon^{-1} = (-10\ \mu\text{A/s})(0.3679) = \mathbf{-3.679\ \mu A/s}$ (10-7a)

$dv_R/dt = (-I_0R/\tau_c)\varepsilon^{-1} = (-10\text{ V/s})(0.3679) = \mathbf{-3.679\ V/s}$ (10-8a)

c. when $t = 25$ s, $\varepsilon^{-t/\tau_c} = \varepsilon^{-25/10} = 0.0821$

$dv_C/dt = (10\text{ V/s})(0.0821) = \mathbf{0.821\ V/s}$ (10-6a)

$di_C/dt = (-10\ \mu\text{A/s})(0.0821) = \mathbf{-0.821\ \mu A/s}$ (10-7a)

$dv_R/dt = (-10\text{ V/s})(0.0821) = \mathbf{-0.821\ V/s}$ (10-8a)

d. when $t = 60$ s, $\varepsilon^{-t/\tau_c} = \varepsilon^{-60/10} = 0.00248 \cong 0$; consequently, dv_C/dt, di_C/dt, and $dv_R/dt \cong 0$.

Observe the following from Ex. 10-6:

1. As time increases, the rates of change all decrease.
2. Maximum rates of change occur at $t = 0^+$.
3. The rates of change all decrease *exponentially* with time.
4. dv_R/dt is equal and opposite to dv_C/dt at all times.
5. Equations (10-6a) through (10-8a) may be used to find rates of change for any instant of time during the transient period.
6. The rates of change are a function of ε^{-t/τ_c} and consequently may also be determined from the universal time constant graph of Fig. 10-3.

10-4 CAPACITOR CHARGE AND DISCHARGE RELATIONS—CAPACITOR INITIALLY UNCHARGED

We are now ready to observe differences and similarities between charge and discharge of a capacitor in an *RC* circuit. **Figure 10-5a** shows a circuit in which a single-pole double throw (SPDT) switch S may be used either to charge and discharge a capacitor.

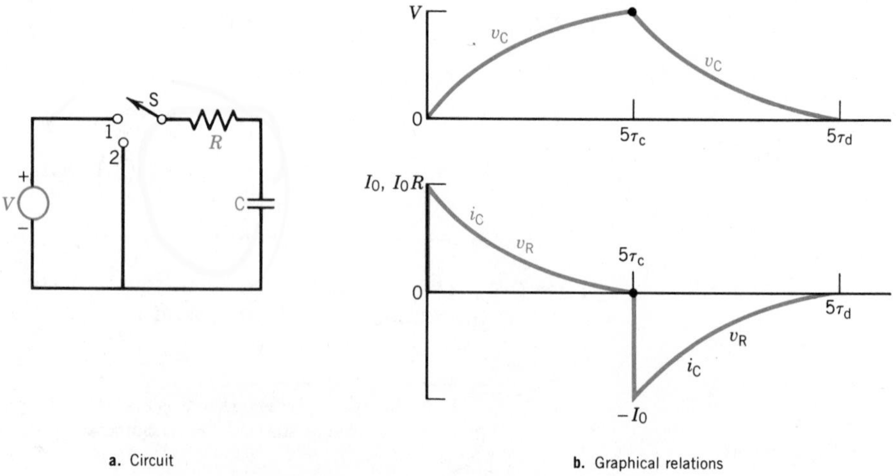

Figure 10-5 Charge and discharge relations and associated waveforms

When the switch is thrown to position 1, the capacitor charges to voltage V in $5\tau_c$. Therefore, as shown in Fig. 5-10b, v_C rises exponentially toward V while i_C decreases from I_0 to 0 and v_R drops from I_0R to 0, exponentially.

At the end of $5\tau_c$, as shown in Fig. 5-10b, *all* transients disappear, $v_C = V$, and both i_C and $v_R = 0$.

The switch is now thrown to position 2. The capacitor, originally charged to V, begins to discharge (exponentially) through resistance R. Since the property of a capacitor is to oppose a change in voltage, the capacitor voltage v_C maintains its polarity as the capacitor discharges from its initial value V to zero.

But during *charge*, the current in R in Fig. 10-5a was from *left to right*. Now, during *discharge*, the current in R is from *right to left*. Clearly, during discharge, the current i_C *reverses* direction. And since $v_R = iR$ at all times, v_R must always be proportional to current in R, as noted earlier. Consequently, during capacitor discharge v_R also reverses direction.

We may summarize the **discharge relations** as follows:

1. v_C, v_R, and i_C all decrease exponentially from their initial values to zero.
2. The curves of i_C and v_R always have a polarity that is opposite to v_C.[5]
3. The charge curves of i_C and v_R are *always* of *opposite* polarity to their discharge curves and vice versa.
4. For the graphical relations shown in Fig. 10-5b, all *rates* of change (during discharge) are of *opposite polarity* to those during charge. Thus, dv_C/dt has a positive rate of change during charge and a negative rate of change during discharge. Similarly, di_C/dt has a negative rate of change during charge and a positive rate of change during discharge.
5. The time constant, $\tau = RC$, is the same (in this circuit) during discharge and charge for Fig. 10-5a.[6]

Table 10-1 summarizes the *charge/discharge* relations for a capacitor (initially uncharged). Particular note should be made of polarity reversals of some of these parameters during the charge cycle and during the discharge cycle.

Table 10-1 Charge and Discharge Relations for a Capacitor (Initially Uncharged)

Parameter	Charging Cycle		Discharging Cycle	
v_C	$V(1 - \varepsilon^{-t/\tau_c})$	(10-2)	$V_0\varepsilon^{-t/\tau_d}$	**(10-9)**
i_C	$I_0\varepsilon^{-t/\tau_c} = \dfrac{V}{R}\varepsilon^{-t/\tau_c}$	(10-3)	$-I_0\varepsilon^{-t/\tau_d} = (-V_0/R)\varepsilon^{-t/\tau_d}$	**(10-10)**
v_R	$I_0R\varepsilon^{-t/\tau_c}$	(10-4)	$-I_0R\varepsilon^{-t/\tau_d}$	**(10-11)**
dv_C/dt	$\dfrac{V_0}{\tau}\varepsilon^{-t/\tau_c}$	(10-6a)	$-\dfrac{V_0}{\tau_d}\varepsilon^{-t/\tau_d}$	**(10-12)**
di_C/dt	$\dfrac{-I_0}{\tau_c}\varepsilon^{-t/\tau_c}$	(10-7a)	$\dfrac{+I_0}{\tau_d}\varepsilon^{-t/\tau_d} = \left(\dfrac{V_0/R}{RC}\right)\varepsilon^{-t/\tau_d} = \dfrac{V_0}{R^2C}\varepsilon^{-t/\tau_d}$	**(10-13)**
dv_R/dt	$\dfrac{-I_0R}{\tau_c}\varepsilon^{-t/\tau_c}$	(10-8a)	$\dfrac{+I_0R}{\tau_d}\varepsilon^{-t/\tau_d}$	**(10-14)**
t	$\tau_c \ln\left(\dfrac{V}{V - v_C}\right)$	(10-5)	$\tau_d \ln\left(\dfrac{V_0}{v_C}\right)$ or $\tau_d \ln\left(\dfrac{I_0}{i_C}\right)$	**(10-15)**

[5] In the event that the capacitor is charged to some *negative* value, during discharge v_C will "decrease" from some *negative* value to zero. But during discharge, i_C and v_R will now be *positive* and opposite to v_C in polarity.

[6] Figure 10-5b is a *special* case. In many instances the time constant during charging is *not* the same as the time constant during discharge, that is, $\tau_c \neq \tau_d$. (See Table 10-1 and Ex. 10-7.)

We are now prepared to deal with more complex networks than those presented up to now, as shown by Ex. 10-6, involving transient RC switching.

EXAMPLE 10-7

In the network shown in **Fig. 10-6a**, the capacitor C is originally discharged, since the switch is in position 2. At time $t = 0^+$ it is thrown to position 1 for 2 ms. It is then returned to position 2, where it remains indefinitely. Calculate the

a. Maximum charging voltage to which capacitor may charge in position 1
b. Charging time constant, τ_1, in position 1
c. Discharging time constant, τ_2, in position 2
d. v_C, i_C, and w_C, energy stored in capacitor at end of 2 ms in position 1, that is, $t_1 = 2$ ms
e. Rate of change of v_C and i_C at the end of 2 ms in position 1
f. v_C and i_C at the instant the switch is returned to position 2, at $t = 2^+$ ms
g. Initial rates of change of v_C and i_C at $t = 2^+$ ms in position 2
h. v_C and i_C after a total of 2 ms has elapsed in position 2
i. Sketch waveforms of v_C and i_C for the first 4 ms of above switching sequence

Solution

a. Recalling Eq. (5-8), we may write

$$V_0 = \frac{R_n}{R_T}(V_p - V_n) + V_n = \frac{60\text{ k}}{90\text{ k}}(400 - 100) + 100$$

$$= \mathbf{300\ V} \qquad (5\text{-}8)$$

b. $\tau_1 = [(R_1 \| R_2) + R_3] \times C$
$= [(30\text{ k} \| 60\text{ k}) + 30\text{ k}] \times 10\text{ nF}$
$= (50\text{ k}\Omega)(10\text{ nF}) = \mathbf{0.5\ ms}$

c. $\tau_2 = R_3C = (30\text{ k}\Omega)(10\text{ nF}) = \mathbf{0.3\ ms}$

d. $v_C = V_0(1 - \varepsilon^{-t/\tau_1}) = 300(1 - \varepsilon^{-2/0.5}) = \mathbf{294.5\ V}$ (10-2)

$$i_C = \frac{V_0}{R}\varepsilon^{-t/\tau_1} = \frac{300\text{ V}}{50\text{ k}\Omega}\varepsilon^{-2/0.5} = (6\text{ mA})\varepsilon^{-4}$$

$$= \mathbf{0.11\ mA} \qquad (10\text{-}3)$$

$w_C = Cv_C{}^2/2 = (50\text{ nF})(294.5)^2/2 = \mathbf{2.17\ mJ}$ (9-2)

e. $$dv_C/dt = \frac{V_0}{\tau_1}\varepsilon^{-t/\tau_1} = \frac{300\text{ V}}{0.5\text{ ms}}\varepsilon^{-4} = 10{,}990\text{ V/s}$$

$$= \mathbf{10.99\ kV/s} \qquad (10\text{-}6a)$$

$$di_C/dt = \frac{-I_0}{\tau_1}\varepsilon^{-t/\tau_1} = \frac{-6\text{ mA}}{0.5\text{ ms}}\varepsilon^{-4} = \mathbf{-0.22\ A/s} \qquad (10\text{-}7a)$$

f. Capacitance always has opposition to a *change* in voltage and $v_{C_1} = \mathbf{294.5\ V}$ at $t = 2$ ms; therefore $v_{C_2} = \mathbf{294.5\ V}$ at $t = 2^+$ ms in position 2. BUT $i_{C_2} = v_{C_2}/R_3 = -294.5\text{ V}/30\text{ k}\Omega = \mathbf{-9.82\ mA} = -I_{02}$ at $t = 2^+$ ms in position 2

g. $dv_C/dt = -V_0/\tau_2 = -294.5\text{ V}/0.3\text{ ms}$
$= \mathbf{-981.\bar{6}\ kV/s}$ (10-12)
$di_C/dt = I_{02}/\tau_2 = +9.82\text{ mA}/0.3\text{ ms} = \mathbf{32.7\bar{3}\ A/s}$ (10-13)

h. $v_{C_2} = V_{02}\varepsilon^{-t/\tau_2} = 294.5\varepsilon^{-2/0.3} = \mathbf{0.375\ V \cong 0}$ (10-9)
$i_{C_2} = -I_{02}\varepsilon^{-t/\tau_2} = -(9.82\text{ mA})\varepsilon^{-2/0.3}$
$= \mathbf{-12.5\ \mu A \cong 0}$ (10-10)

i. The waveforms are sketched in Fig. 10-6b.

Many conclusions emerge from Ex. 10-7, but these are of greatest importance:

1. When the switch is thrown to position 1, we must ask ourselves "What is the equivalent Thévenin voltage and resistance seen by the series combination of R_3C?" The Thévenin voltage is calculated in (a) and the resistance in (b) of the solution above.
2. The charging and discharging time constants are *not* the same. The *discharging* time constant in Ex. 10-6 is *smaller* than the charging time constant. This explains why the capacitor does not charge to its maximum Thévenin voltage of 300 V in 2 ms but does drop practically to zero when discharged.
3. The shorter discharge time constant, τ_2, also accounts for the greater rates of change of voltage and current in (g) than (e) despite the fact that both occur at approximately 2 ms.

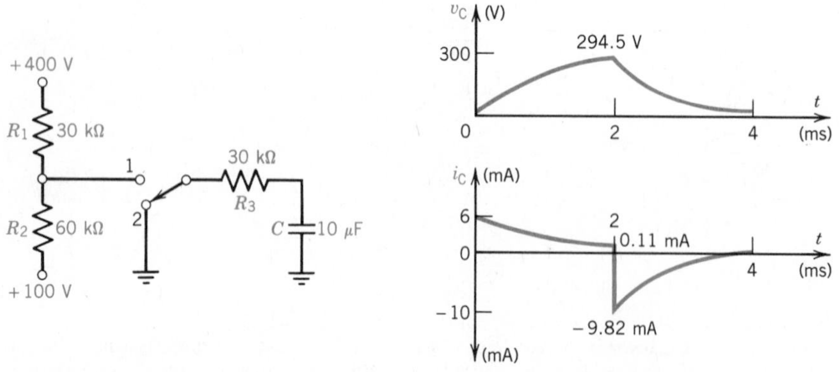

a. Network b. Waveforms

Figure 10-6 Example 10-7

4. Even though $v_C = 294.5$ V at 2 ms and does not change at the instant the switch moves from position 1 to 2, the rate of change of voltage changes considerably in both *magnitude* and *direction* from 10.99 kV/s to −981.6 kV/s.
5. Note that i_C at 2 ms has two distinct values, one positive and one negative, at the *same time*. Also note that the initial negative value of $i_C = -9.82$ mA is greater than the initial positive value of 6 mA. This is due primarily to the reduction in resistance seen by C from a total of 50 kΩ in position 1 to 30 kΩ in position 2.
6. Finally, note once more that in position 1 current is from *left* to *right* in R_3. In position 2, current is from *right* to *left* in R_3. This reversal of current is shown in Fig. 10-6b.
7. The current reversal accounts for the *positive* rate of change of current in (g) above as i_C decreases from −9.82 mA to zero (see Fig. 10-6b).

10-5 CAPACITOR CHARGE AND DISCHARGE RELATIONS WITH AN INITIAL VOLTAGE ACROSS CAPACITOR

The previous discussions and equations all were based on the fact that the capacitor originally was uncharged and that the exponential rise of v_C started from the origin of the v_C versus time curve, that is, $v_C = 0$ at $t = 0^+$.

Let us now consider what happens when a capacitor has a (positive) charge due to some previous history as shown in the circuit of **Fig. 10-7a**. Let us assume that the capacitor voltage due to the initial charge on capacitor C, as represented by voltage V_0 in Fig. 10-7a, is such that V_0 is less than V of the dc supply. When switch S is closed at $t = 0^+$, there is already a voltage $v_C = V_0$ across the capacitor, as shown in Fig. 10-7b.

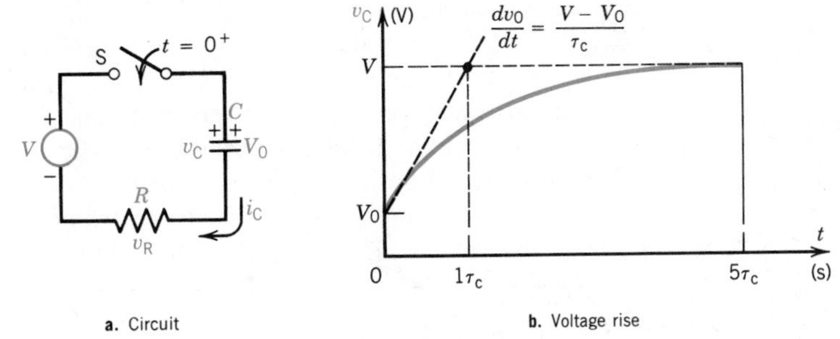

Figure 10-7 Voltage rise across a capacitor with initial (positive) charge

The *initial* rate of rise of v_C is now somewhat less than when the capacitor is uncharged; that is, its slope is not as steep or as high. Since the voltage tries to rise from V_0 to V initially in a time of $1\tau_c$, we may write the initial rate of rise of v_C as shown in Fig. 10-7b as

$$\frac{dv_{C_0}}{dt} = \frac{V - V_0}{\tau_c} = \frac{V - V_0}{RC} \quad \text{volts/second (V/s)} \qquad \textbf{(10-6b)}$$

We may also infer that the instantaneous voltage v_C during its transient rise from V_0 to V (its final or steady-state value) is[7]

[7] One learns from one's students. Many have noticed and told the author (independently) over the years that the form of Eq. (10-2a) used in Ex. (10-8) above is essentially similar to Eq. (5-8). The only difference between the two is that the ratio of resistances in Eq. (5-8) is replaced by the exponential rise factor of $(1 - \varepsilon^{-t/\tau_c})$. The similarity between the two makes it easier to remember the relationships. The reader is advised to write the two equations, one below the other, and verify this personally.

$$v_C = (V - V_0)(1 - \varepsilon^{-t/\tau_c}) + V_0 \qquad \text{volts (V)} \qquad \textbf{(10-2a)}$$

where V_0 is the initial voltage across a capacitor
V is the applied voltage across the RC circuit
τ_c is the time constant of the RC circuit, that is, $R \times C$ in seconds
t is the time during the transient rise in voltage

From our previous knowledge of initial rates of change, we may also write (and derive) the relation for the *initial* rate of change of current as

$$\frac{di_{C_0}}{dt} = \frac{-I_0}{\tau_c} = \frac{-(V - V_0)/R}{RC} = \frac{-(V - V_0)}{R^2C} \qquad \text{amperes/second (A/s)} \qquad \textbf{(10-7b)}$$

where all terms have been defined above and $I_0 = (V - V_0)/R$

Note that the rate of change of current, at all times, is *negative* because current decreases from I_0 to zero as the capacitor voltage increases. Note that the rate of change of current initially is still $-I_0/\tau_c$, but because of the division by R we emerge with R^2C in the denominator of Eq. (10-7b).

But from our earlier analysis of rates of change during those periods **AFTER** the initial transition, we have seen from Eqs. (10-6a) through (10-8a) that the initial rate of change decreases exponentially by the decay factor $\varepsilon^{-t/\tau}$.

For the situation, therefore, of an *initial* voltage across the capacitor due to its *previous history*, we may write the rates of change of Eqs. (10-6a) through (10-8a) at any value of t, measured from t_0, during the capacitor charge cycle *only* as

$$\frac{dv_C}{dt} = \frac{V - V_0}{\tau_c}\varepsilon^{-t/\tau_c} \qquad \text{volts/second (V/s)} \qquad \textbf{(10-6c)}$$

$$\frac{di_C}{dt} = \frac{-I_0}{\tau_c}\varepsilon^{-t/\tau_c} = -\left(\frac{V - V_0}{R^2C}\right)\varepsilon^{-t/\tau_c} \qquad \text{amperes/second (A/s)} \qquad \textbf{(10-7c)}$$

$$\frac{dv_R}{dt} = \left(\frac{-I_0R}{\tau_c}\right)\varepsilon^{-t/\tau_c} = -\left(\frac{V - V_0}{\tau_c}\right)\varepsilon^{-t/\tau_c} \qquad \text{volts/second (V/s)} \qquad \textbf{(10-8c)}$$

With respect to Eqs. (10-6c) through (10-8c), please note:

1. The rates of these equations that provide for some initial charge on a capacitor are only lower for instances where the difference $V - V_0$ yields a lower or smaller value than V. This occurs when V_0 has the *same* polarity as V and tends to *oppose* it.
2. Higher rates of change are produced when V_0 (due to its previous history) has a polarity *opposite* to that of V. In that event, $V - V_0 = V - (-V_0) = V + V_0$, producing a *larger* numerator in Eqs. (10-6c) through (10-8c) and a *higher* rate of change.
3. Extreme care must be taken, therefore, in comparing the nature of the polarity across a capacitor when an *RC* circuit is switched to a *new voltage source*. If

the initial polarity across the capacitor V_0 *opposes* V, the source, we take the *difference* or $V - V_0$. On the other hand, if the polarity across the capacitor *aids* the source polarity, we take the *sum* $V - (-V_0) = V + V_0$ in the numerator of Eqs. (10-6c) through (10-8c).

4. Equations (10-6c) through (10-8c) are general equations that may be used to find the respective rates of change under *any* conditions. For the case of an initially uncharged capacitor, we merely substitute $V_0 = 0$. Similarly, for the *initial* rate of change at $t = 0$, we eliminate the exponential expression ($\varepsilon^{-t/\tau}$) because it is equivalent to unity.
5. Finally, it must be noted that the rates of change in Eqs. (10-6c) through (10-8c) apply *only* during the capacitor *charging* cycle, as summarized in Table 10-2 of Sec. 10-6.

EXAMPLE 10-8

Given the RC circuit of Exs. 10-4 and 10-5, where $R = 1\ \text{M}\Omega$, $C = 10\ \mu\text{F}$, and $V = 100$ V, assume that the capacitor is initially charged to a voltage of 63.2 V with polarity as shown in Fig. 10-7a. When switch S is closed at $t = 0^+$ calculate

a. Initial rate of change of voltage across the capacitor
b. Initial rate of change of current in the capacitor
c. Circuit charging time constant and time to reach $v_C = 100$ V
d. Verify your time in **(c)**, using Eq. (10-2a)
e. Compare your answer in **(a)** to that of Ex. 10-5a and account for differences
f. Compare your answer in **(b)** to that of Ex. 10-5b and account for differences

Solution

a. $$\frac{dv_{C_0}}{dt} = \frac{V - V_0}{\tau_c}\varepsilon^0 = \frac{100 - 63.2}{10\ \text{s}} = \mathbf{3.68\ V/s} \qquad (10\text{-}6c)$$

b. $$-\frac{di_{C_0}}{dt} = \frac{-(V - V_0)}{R^2C}\varepsilon^0 = \frac{(100 - 63.2)}{(1\ \text{M})^2(10\ \mu)} = \mathbf{-3.68\ \mu A/s} \qquad (10\text{-}7c)$$

c. $\tau_c = RC = 10$ s and $5\tau_c = \mathbf{50\ s}$

d. $$v_C = (V - V_0)(1 - \varepsilon^{-t/\tau_c}) + V_0 = (100 - 63.2)(1 - \varepsilon^{-5}) + 63.2 = \mathbf{99.75\ V} \qquad (10\text{-}2a)$$

e. When the capacitor was uncharged in Ex. 10-5 the rate of change of voltage was 10 V/s, compared to 3.68 V/s in **(a)** here. Because the capacitor is partially charged, the (initial) rate of change of voltage is much lower, as are all other rates of change.

f. The same applies to the initial rate of change of current. In Ex. 10-5b we have $-10\ \mu$A/s compared to $-3.68\ \mu$A/s in **(b)** here. Again, because the capacitor is partially charged, the initial rate of change and all other rates of change are lower. But the total time to fully charge the capacitor is the SAME!

With respect to Ex. 10-8, note that despite the lower rates of change of both voltage and current, the final voltage, after $5\tau_c$, across the capacitor is 99.75% of the applied voltage. This verifies the fact that the capacitor fully charges in approximately $5\tau_c$ and that the current decreases to zero in approximately $5\tau_c$, even with an initial charge on the capacitor. In effect, please note that *because* of the *initial* (positive) voltage across the capacitor, the rates of voltage rise and current decrease are both lower to permit the capacitor to still fully charge in $5\tau_c$.

10-5.1 Effect of an Initial Charge on Capacitor

Let us now consider the effect of a capacitor, initially charged, but of *negative* polarity with respect to the source. Such a condition is shown in **Fig. 10-8a**, where the capacitor was initially charged due to a previous history. Note that when switch S is closed, the capacitor polarity is *series-aiding* the source polarity so that current i_C flows from source to capacitor in the direction shown.

The question arises of whether we need a separate set of equations to describe this situation. We can see from Fig. 10-8b that the voltage v_C starts at $-V_0$ (some negative voltage) but rises exponentially toward V when S closes at $t = 0^+$. From Fig. 10-8b, we can see that the initial rate of change of v_C is now steeper than in *both* previous cases shown in Fig. 10-4a (capacitor uncharged) and Fig. 10-7a (capacitor positively charged). From Fig. 10-8b we can see that in a time of 1τ, ΔV, the initial

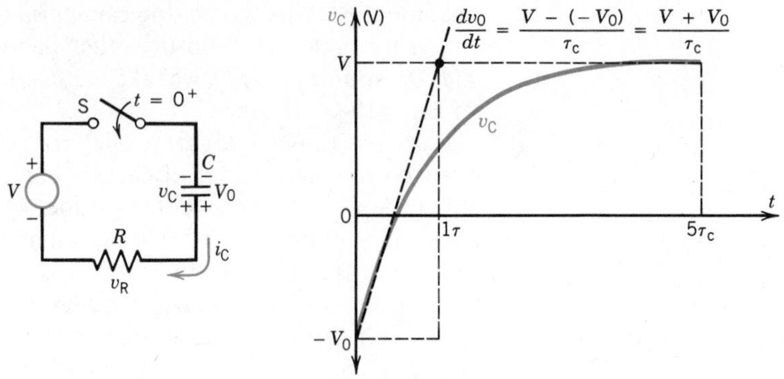

Figure 10-8 Voltage rise across capacitor with initial (negative) charge

a. Circuit b. Voltage rise

change of *voltage*, is $V + V_0$ which is the same as $V - (-V_0)$. This means that we may use Eq. (10-6b) exactly as shown for any case of either positive or negative charge on a capacitor (as well as no charge at all) provided we *preserve the proper polarity* for V_0. If V_0 is a negative voltage with respect to ground it must have a negative sign before it, when substituted in Eq. (10-6c).

The same applies to both Eqs. (10-7c) and (10-8c), as shown by the following example.

EXAMPLE 10-9

Given $R = 1\ \text{M}\Omega$, $C = 10\ \mu\text{F}$, and $V = 100$ V with an initial charge on C of -50 V polarity, as shown in Fig. 10-8a. Calculate the

a. Initial rate of change of voltage across the capacitor
b. Initial rate of change of current in the capacitor
c. Circuit time constant and time to reach $v_C = 100$ V
d. Verify the time in (c) by using Eq. (10-12)
e. Compare your answers to those in Ex. 10-8a and 10-8b and account for differences

Solution

a. $$\frac{dv_{C_0}}{dt} = \frac{V - V_0}{\tau}\varepsilon^0 = \frac{100 - (-50)}{10\text{ s}} = \frac{150\text{ V}}{10\text{ s}} = \mathbf{15\ V/s} \qquad (10\text{-}6c)$$

b. $$-\frac{di_{C_0}}{dt} = \frac{-(V - V_0)}{R^2C} = \frac{-150\text{ V}}{(1\ \text{M}\Omega)^2(10\ \mu\text{F})} = \mathbf{-15\ \mu A/s} \qquad (10\text{-}7c)$$

c. $\tau_c = RC = 10$ s and $5\tau_c = $ **50 s** to reach 100 V

d. $$v_C = (V - V_0)(1 - \varepsilon^{-t/\tau_c}) + V_0 \qquad (10\text{-}2a)$$
$$= [100 - (-50)](1 - \varepsilon^{-5}) + (-50)$$
$$= 150(1 - \varepsilon^{-5}) - 50 = 148.99 - 50 = \mathbf{99\ V}$$

e. With a negative charge on the capacitor (compared) to a positive charge: the initial rates of change of both voltage rise and current decrease are both greater. Nevertheless, it takes only 5τ to charge the capacitor completely and for the current to fall to approximately zero, despite the magnitude of the initial charge.

Example 10-8 proves conclusively that *no special equations* are needed to distinguish positively charged from negatively charged capacitors. Indeed, Eqs. (10-2a), (10-6c), and (10-7c) are general equations and are valid even for the case of no charge on a capacitor, and in this sense *they supersede earlier* equations.

The previous statement implies that we should reexamine the relatively important earlier Eq. (10-5), which enabled us to find any value of t for a given value of v_C. This equation was developed for the special case where $V_0 = 0$ and there was no charge on the capacitor. For this equation we may now write

$$t = \tau_c \ln\left(\frac{V - V_0}{V - v_C}\right) = RC \ln\left(\frac{V - V_0}{V - v_C}\right) \quad \text{seconds (s)} \qquad \mathbf{(10\text{-}5a)}$$

where all terms have been defined above

The use to which Eq. (10-5a) may be put is shown in Ex. 10-10.

EXAMPLE 10-10

Given the conditions of Ex. 10-9, calculate the time when v_C is **a.** -5 V **b.** 0 V **c.** $+50$ V **d.** 99 V

Solution

a. $$t = \tau_c \ln\left(\frac{V - V_0}{V - v_C}\right) = (10\text{ s}) \ln\left[\frac{100 - (-50)}{100 - (-5)}\right]$$
$$= (10\text{ s}) \ln\left(\frac{150}{105}\right) = \mathbf{3.57\ s} \qquad \text{(10-5a)}$$

b. $$t = \tau_c \ln\left(\frac{V - V_0}{V - v_C}\right) = 10\text{ s} \ln\left(\frac{150}{100}\right) = \mathbf{4.055\ s} \qquad \text{(10-5a)}$$

c. $$t = \tau_c \ln\left(\frac{V - V_0}{V - v_C}\right) = 10\text{ s} \ln\left(\frac{150}{100 - 50}\right)$$
$$= \mathbf{10.99\ s} \qquad \text{(10-5a)}$$

d. $$t = \tau_c \ln\left(\frac{V - V_0}{V - v_C}\right) = 10\text{ s} \ln\left(\frac{150}{100 - 99}\right)$$
$$= \mathbf{50.1\ s} \qquad \text{(10-5a)}$$

Example 10-10 shows the extreme power of Eq. (10-5a). From Fig. 10-8b we know that negative values of v_C and the point where $v_C = 0$ all occur when $t < 1\tau_c < 10$ s. Since $\tau_c = 10$ s, the answers to **(a)** and **(b)** are correct. Note also from Ex. 10-9d that when $t = 5\tau_c$, $v_C \cong 99$ V at 50 s. This is verified in Ex. 10-10d as well.

10-5.2 Capacitor Discharge Relations

Due to either a positive or negative initial charge on a capacitor, it was necessary to develop more general equations for the charging condition relations. The question arises of whether it is necessary to correct the equations given in Table 10-1 (Sec. 10-4) for the more general cases of initial charges on capacitors. A little thought may convince us that it is not necessary to do so. All discharge relations, as shown in Table 10-1, drop exponentially from some **initial value** (either positive or negative) **to zero,** *regardless of prior history.* Consequently, it is *unnecessary* to modify the equations for discharge relations shown in Table 10-1 (Eqs. 10-9 through 10-15).

10-6 SUMMARY

We now summarize general equations for charge and discharge relations of a charged capacitor, *under any conditions,* in Table 10-2.

Table 10-2 General Equations for Charge/Discharge Relations of a (Charged) Capacitor

Parameter	Charging Cycle		Discharging Cycle	
v_C	$(V - V_0)(1 - \varepsilon^{-t/\tau_c}) + V_0$	(10-2a)	$V_0\varepsilon^{-t/\tau_d}$	(10-9)
i_C	$\frac{(V - V_0)}{R}\varepsilon^{-t/\tau_c} = I_0\varepsilon^{-t/\tau_c}$	(10-3a)	$-I_0\varepsilon^{-t/\tau_d} = -(V_0/R)\varepsilon^{-t/\tau_d}$	(10-10)
v_R	$(V - V_0)\varepsilon^{-t/\tau_c} = I_0R\varepsilon^{-t/\tau_c} = i_CR$	(10-4a)	$-I_0R\varepsilon^{-t/\tau_d} = -i_CR$	(10-11)
dv_C/dt	$\frac{(V - V_0)}{\tau_c}\varepsilon^{-t/\tau_c}$	(10-6c)	$\frac{-V_0}{\tau_d}\varepsilon^{-t/\tau_d}$	(10-12)
di_C/dt	$\frac{-(V - V_0)}{R^2C}\varepsilon^{-t/\tau_c} = \frac{-I_0}{\tau_c}\varepsilon^{-t/\tau_c}$	(10-7c)	$\frac{I_0}{\tau_d}\varepsilon^{-t/\tau_d} = \frac{V_0/R}{RC}\varepsilon^{-t/\tau_d} = \frac{V_0}{R^2C}\varepsilon^{-t/\tau_d}$	(10-13)
dv_R/dt	$\frac{-(V - V_0)}{\tau_c}\varepsilon^{-t/\tau_c} = \frac{-I_0R}{\tau_c}\varepsilon^{-t/\tau_c}$	(10-8c)	$\frac{I_0R}{\tau_d}\varepsilon^{-t/\tau_d}$	(10-14)
t	$\tau_c \ln\left(\frac{V - V_0}{V - v_C}\right)$	(10-5a)	$\tau_d \ln\left(\frac{I_0}{i_C}\right)$ or $\tau_d \ln\left(\frac{V_0}{v_C}\right)$	(10-15)

Armed with the set of relations in Table 10-2, we are now prepared to engage problems of a more complex nature than we have attempted thus far, as shown by the following examples.

EXAMPLE 10-11

In Fig. 10-9a, capacitor C is completely discharged in the OFF position. Switch S is thrown to position 1, where it remains for 1 ms. It is then thrown to position 2, where it remains until the transient disappears. Calculate

a. τ_1, circuit time constant in position 1
b. Time to reach steady state in position 1
c. Voltage across C at end of 1 ms in position 1
d. τ_2, circuit time constant in position 2
e. Current in capacitor at 1 ms in position 1
f. Initial voltage across and current in capacitor when switch is first thrown to position 2
g. Time for voltage across capacitor to reach 0 V in position 2
h. Voltage across capacitor after 6 ms in position 2
i. Current in capacitor after 6 ms in position 2
j. Plot v_C versus time from 0 to 16 ms for the above switching events
k. Plot i_C versus time from 0 to 16 ms for the above switching events

Solution

a. $\tau_1 = R_1C = (1\ \text{k}\Omega)(1\ \mu\text{F}) = \mathbf{1\ ms}$
b. $5\tau_1 = 5 \times 1\ \text{ms} = \mathbf{5\ ms}$
c. $v_C = V(1 - \varepsilon^{-t/\tau_1}) = -6(1 - \varepsilon^{-1/1}) = \mathbf{-3.793\ V}$ (10-2a)
d. $\tau_2 = (R_1 + R_2)C = (3\ \text{k}\Omega)(1\ \mu\text{F}) = \mathbf{3\ ms}$
e. $i_{C_1} = I_0\varepsilon^{-t/\tau_1} = -6\varepsilon^{-1}\ \text{mA} = \mathbf{-2.21\ mA}$ (10-3a)
f. $V_{02} = V - v_{C_0} = 12 - (-3.793) = \mathbf{15.793\ V}$
$i_{C_{02}} = V_{02}/(R_1 + R_2) = 15.793\ \text{V}/3\ \text{k}\Omega = \mathbf{5.264\ mA}$
g. $t_0 = \tau_2 \ln\left(\dfrac{V - V_0}{V - v_C}\right) = 3\ \text{ms} \ln\left[\dfrac{12 - (-3.793)}{12}\right]$
$= \mathbf{0.824\ ms}$ (10-5a)
h. $v_C = V_0 + (V - V_0)(1 - \varepsilon^{-t/\tau_2})$
$= (15.793)(1 - \varepsilon^{-6/3}) - 3.793 = \mathbf{9.863\ V}$ (10-2a)
i. This is a current decay relation where $I_0 = 5.264$ mA from (f) $i_{C_2} = 5.264\varepsilon^{-6/3} = \mathbf{0.7124\ mA}$

Since $\tau_2 = 3$ ms and $5\tau_2 = 15$ ms, then after 16 ms all transients disappear. The plots of v_C and i_C for the above events are summarized in **Fig. 10-9b**. Note that to each of the times above we must include 1 ms for the time in position 1 to make the total *elapsed* time correct.

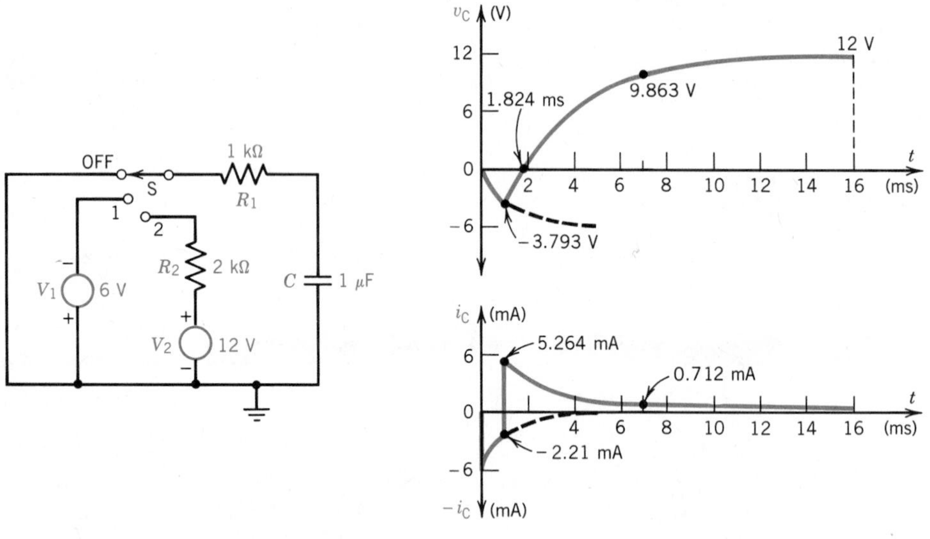

a. Network b. Graphical results

Figure 10-9 Example 10-11

EXAMPLE 10-12

Using the graphs plotted in Fig. 10-9b for v_C and i_C versus time, calculate the *rates of change* at the following instantaneous values:

a. v_{C_0} at $t = 0^+$
b. v_{C_1} at $t = 1$ ms in position 1
c. $v_{C_{02}}$ at $t = 1^+$ ms in position 2
d. v_{C_2} at $t = 7$ ms in position 2 (note t is measured from $t = 0^+$)
e. i_{C_0} at $t = 0^+$
f. i_{C_1} at $t = 1$ ms in position 1
g. $i_{C_{02}}$ at $t = 1^+$ ms in position 2
h. i_{C_2} at $t = 7$ ms in position 2 (note t is measured from $t = 0^+$)

Solution

a. $dv_{C_0}/dt = V/\tau_1 = -6\text{ V}/1\text{ ms} = -6\text{ kV/s}$ (10-6b)
(note $V_0 = 0$ since C is uncharged)

b. $dv_{C_1}/dt = \dfrac{V}{\tau}\varepsilon^{-t/\tau_1} = \dfrac{-6\text{ kV}}{\text{s}}\varepsilon^{-1/1} = \mathbf{-2.21\ kV/s}$ (10-6b)

c. $dv_{C_{02}}/dt = \dfrac{V - V_0}{\tau_2}\varepsilon^{-t/\tau_2} = \dfrac{[12-(-3.793)]}{3\text{ ms}}\varepsilon^{0}$

$= \dfrac{15.793\text{ V}}{3\text{ ms}} = \mathbf{5.26\ kV/s}$ (10-6b)

d. $dv_{C_2}/dt = \dfrac{V - V_0}{\tau_2}\varepsilon^{-t/\tau_2} = \left(\dfrac{5.26\text{ kV}}{\text{s}}\right)\varepsilon^{-6/3}$

$= 0.712\text{ kV/s} = \mathbf{712\ V/s}$ (10-6b)

e. $di_{C_0}/dt = -I_0/\tau_1 = -(-6\text{ mA})/1\text{ ms} = \mathbf{6\ A/s}$ (10-7b)

f. $di_{C_1}/dt = \dfrac{I_0}{\tau_1}\varepsilon^{-t/\tau_1} = \dfrac{6\text{ A}}{\text{s}}\varepsilon^{-1/1} = \mathbf{2.21\ A/s}$ (10-7b)

g. $di_{C_{02}}/dt = \dfrac{-[V_{02}/(R_1 + R_2)]}{\tau_2} = \dfrac{-(15.793\text{ V}/3\text{ k}\Omega)}{3\text{ ms}}$

$= \mathbf{-1.755\ A/s}$ (10-7b)

h. $di_{C_2}/dt = \dfrac{-I_0}{\tau_2}\varepsilon^{-t/\tau_2} = \left(\dfrac{-1.755\text{ A}}{\text{s}}\right)\varepsilon^{-6/3}$

$= \mathbf{-0.2375\ A/s}$ (10-7b)

The conclusions to be drawn from Ex. 10-12 (see Fig. 10-9b) are as follows:

1. At the end of 1 ms in position 1, $v_C = -3.793$ V, when switch S is suddenly connected to position 2 at $t = 1^+$ ms. Despite the fact that the applied circuit voltage *suddenly* changed to +15.793 V, the voltage across the capacitor (by definition) *cannot* change suddenly. Instead, it rose exponentially from its initial condition (−3.793 V) to its final state (+12 V).
2. The same cannot be said for the current, which decreased exponentially from −6 mA to −2.21 mA in position 1 at the end of 1 ms. When the switch was suddenly connected to position 2 at $t = 1^+$ ms, current *instantly jumped* from −2.21 to +5.264 mA in zero time! (Recall that current in a resistive circuit rises instantly as shown in the solution of Ex. 10-10f.)
3. Note that the times shown in both plots of v_C and i_C versus time (in milliseconds) are all measured from time $t = 0^+$, when the switch left the OFF position in Fig. 10-9a. Thus, at the end of 6 ms in position 2, the total ELAPSED time is 7 ms. (This is a source of error for many students and falls under the heading of "bookkeeping errors," as termed by the author.)
4. Note in conclusion (1) above that because the capacitor was initially NEGATIVELY charged, the instantaneous voltage across the circuit in position 2 at $t = 1^+$ ms is +15.793 V. A common error that many students make is to assume that the final voltage to which the capacitor rises is 15.793 V. But an examination of Fig. 10-9a shows that +12 V is the final steady-state voltage, in position 2.

Example 10-12 was concerned with the rates of change of both voltage and current, exclusively, in positions 1 and 2. The conclusions to be drawn from Ex. 10-12 (again, see Fig. 10-9b) are the following.

1. All rates of change of v_C in position 1 are negative. The initial rate of change is always higher and it decreases exponentially. Thus, initially we have −6 kV/s and at the end of 1 ms we have −2.21 kV/s.
2. Despite the fact that the currents are negative in position 1, the rates of change of all currents in position 1 are positive. Again, the greatest rate occurs initially and the rate of change falls off exponentially (from +6 to +2.21 A/s, as shown in the solutions of Ex. 10-12e and f).
3. Similarly, in position 2 (as we can see from Fig. 10-9b) all rates of change of v_C are positive and all rates of change of i_C are negative. Again, the greatest rate of change occurs initially, with the rate falling off exponentially with time.
4. As emphasized earlier, the reader should have some anticipation for "ballpark values" that approximate correct answers. In problems involving transients, "number plugging" into equations can lead to disaster. Drawing the waveforms, as done in Fig. 10-9b, helps in visualizing the transients of interest.
5. Last and most important, from Eq. (10-3), since $i_C = C(dv_C/dt) = dq_C/dt$, from a mathematical point of view it should be possible to plot the curve of i_C once we have the curve of v_C and from Eq. (10-6b) know the rates of changes of v_C.[8] Say that we were given only the curve of v_C versus time in Fig. 10-9b and asked to plot the curve of i_C from it. From Ex. 10-12 we have $dv_{C_0}/dt = \mathbf{-6\ kV/s}$

[8] This is a most important insight. It is easier to determine v_C than i_C. And since dv_C/dt is easily found from Eq. (10-6c) and/or Eq. (10-12), it is then easy to verify i_C by merely multiplying dv_C/dt by C as shown above.

at $t = 0^+$ (position 1), $dv_{C_1}/dt = $ **−2.21 kV/s** at $t = 1$ ms (position 1), $dv_{C_{O2}}/dt =$ **5.26 kV/s** at $t = 1^+$ ms (position 2), $dv_{C_2}/dt =$ **712 V/s** at $t = 7$ ms (position 2).

If we multiply each of these values by $C = 1\ \mu\text{F}$, we obtain corresponding values of i_C automatically. Let us do this, using Eq. (10-3):

$$i_{C_{O1}} = C(dv_{C_O}/dt) = (1\ \mu\text{F})(-6\ \text{kV/s}) = -6\ \text{mA}$$

$$i_{C_1} = C(dv_{C_1}/dt) = (1\ \mu\text{F})(-2.21\ \text{kV/s}) = -2.21\ \text{mA}$$

$$i_{C_{O2}} = C(dv_{C_{O2}}/dt) = (1\ \mu\text{F})(+5.26\ \text{kV/s}) = 5.26\ \text{mA}$$

$$i_{C_2} = C(dv_{C_2}/dt) = (1\ \mu\text{F})(0.712\ \text{kV/s}) = 0.712\ \text{mA}$$

Note that the above values of i_C agree with those shown on the curve for i_C in Fig. 10-9b and also those found by other methods in Ex. 10-11. This insight leads to two interesting and important conclusions:

1. If we **differentiate** the waveform of v_C (versus time), we (should) obtain the waveform of i_C
2. If we **integrate** the waveform of i_C, we (should) obtain the waveform of v_C

This insight, therefore, serves as a complete check on the accuracy of either of the two curves and on our calculations. It also enables us to plot waveforms *easily* and *directly*, as shown by Ex. 10-13.

EXAMPLE 10-13

The voltage, v_C, across a 5-μF capacitor in the network of **Fig. 10-10a** is observed on a CRO to have the waveform of voltage shown in Fig. 10-10b. Find and draw the waveforms of a. dv_C/dt b. i_C

Solution

a. The rate of change of voltage is easily obtained for those portions of the waveform of Fig. 10-10b that have a *constant* rate of change. From

$$0 \text{ to } 10\ \text{ms},\ dv_C/dt = (10 - 0)\ \text{V}/(10 - 0)\ \text{ms} = +1\ \text{kV/s}$$

$$10 \text{ to } 20\ \text{ms},\ dv_C/dt = (10 - 10)\ \text{V}/10\ \text{ms} = 0$$

$$20 \text{ to } 50\ \text{ms},\ dv_C/dt = (0 - 10)\ \text{V}/(50 - 20)\ \text{ms} = -0.\overline{33}\ \text{kV/s}$$

$$50 \text{ to } 60\ \text{ms},\ dv_C/dt = 0 \text{ by inspection}$$

$$60 \text{ to } 70\ \text{ms},\ dv_C/dt = (-10 - 0)\ \text{V}/(70 - 60)\ \text{ms} = -1\ \text{kV/s}$$

$$70 \text{ to } 90\ \text{ms},\ dv_C/dt = [-10 - (-10)]\ \text{V}/(90 - 70)\ \text{ms} = 0$$

$$90 \text{ to } 100\ \text{ms},\ dv_C/dt = 0 - (-10)\ \text{V}/(100 - 90)\ \text{ms} = +1\ \text{kV/s}$$

b. Since $i_C = (C)(dv_C/dt)$ at all times, the waveform of i_C is identical to that of dv_C/dt in which each value of dv_C/dt is multiplied by C. Since $C = 5\ \mu\text{F}$, we multiply each of the values obtained in (**a**) by 5 μF. The waveform of i_C shown in Fig. 10-10c was obtained this way.

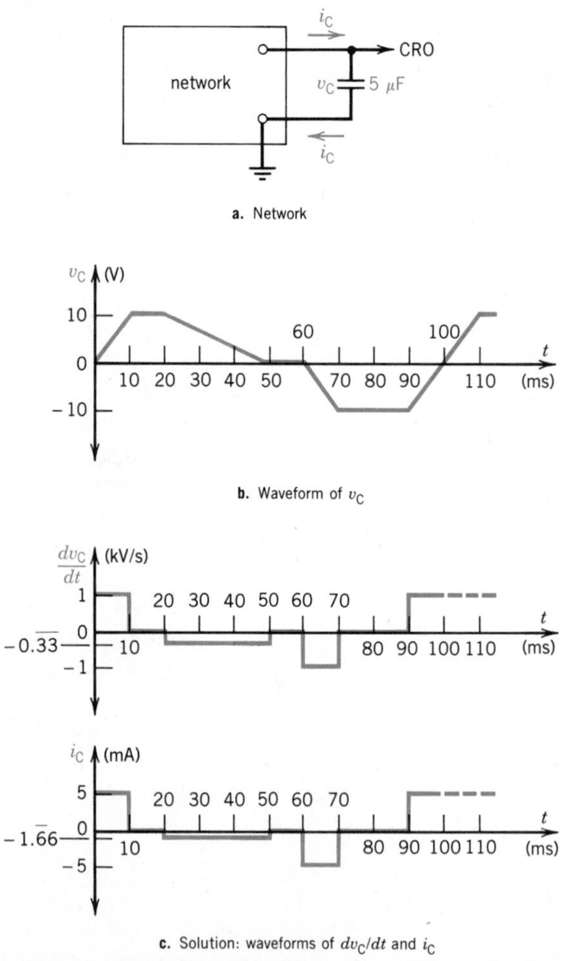

Figure 10-10 Example 10-13

Certain interesting conclusions can be drawn from the waveforms of Fig. 10-10:

1. Whenever the voltage, v_C, is CONSTANT (not changing) the current is zero. Obviously this is because the rate of change of voltage is zero. But we might also say that there is no charge flow to the capacitor once it has been charged to a given voltage. Consequently, when the voltage is a constant 10 V from 10 to 20 ms, the current $i_C = 0$.
2. The current *reverses* (goes *negative*) whenever v_C drops *from a more positive* to a *more negative* value. So when v_C drops from 10 V to 0 or from 0 to -10 V, the capacitor discharges back to the network (i.e., positive charges are removed from the upper plate of the capacitor as the capacitor becomes negatively charged).
3. The current is the same whenever the rate of change of voltage is the same. dv_C/dt is 1 kV/s from 0 to 10 ms and also from 90 to 100 ms. In both cases, the current is a constant $+5$ mA since the slope is constant.
4. While the curve of v_C (Fig. 10-10b) exhibits both positive and negative ramps, the curves of dv_C/dt and i_C exhibit *abrupt changes* or *jumps* in a positive and negative direction. This is because a positive ramp of v_C has a constant rate of change while a negative ramp of v_C has a constant negative rate of change. When v_C is constant its rate of change is zero, hence the abrupt drops to zero.

Point (4) raises the interesting question of what happens to i_C whenever v_C either increases or decreases exponentially. Actually, we can see the answer in Figs. 10-2b, 10-5b, 10-6b, and 10-10b. When v_C increases exponentially in a positive direction, i_C decreases exponentially from some positive value to zero. Conversely, when v_C becomes increasingly negative exponentially, i_C decreases from some negative value to zero exponentially. Consequently, an exponential change in v_C produces an exponential change in i_C at all times. Regardless of whether v_C is an increasing or decaying exponential, i_C **always** decays to zero.[9] And because v_R must follow i_C, resistor voltage v_R always decays to zero when v_C varies exponentially.

10-7 INSTANTANEOUS POWER RELATIONS IN RC CIRCUITS DURING CAPACITOR CHARGE CYCLE

In the previous sections we covered the transient voltages (v_C and v_R) and current (i_C) in an *RC* circuit as well as their rates of change. In this section we will cover the transient *powers* that occur during capacitor charge in an *RC* circuit.

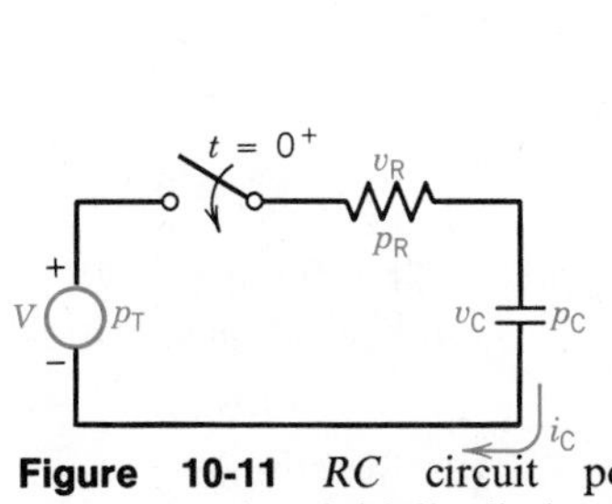

Figure 10-11 *RC* circuit power terms, capacitor initially discharged

Figure 10-11 shows the components *R* and *C* and their respective voltage, current, and power terms, defined. It may come as a surprise that during the cycle of charging the capacitor, that power is expended in charging the capacitor, p_C. Since the capacitor is a *lossless* element, no power should be dissipated by it.[10] But recall that during the charging cycle, energy is *stored* in the dielectric field of the capacitor. Since power is rate at which such energy is stored, power must be furnished by the supply, V, to store that energy, w_C/t.

Moreover, since a current, i_C, flows at the instant the switch is closed at $t = 0^+$ and this current is at a maximum (I_0) at that instant, there is a continuous power loss in the resistor, p_R, during the transient period when a capacitor charges.

[9] This is to be expected since $d(\varepsilon^{-x})/dx = -\varepsilon^{-x}$, and $\pm\varepsilon^{-x}$ is a decaying exponential function.

[10] This is true, and when we study **ac** we will discover that p_C will be *returned to the supply* during the half-cycle when the voltage reverses. This power will later be designated as quadrature reactive power, Q_C.

Equations for each of the losses shown in Fig. 10-11 are derived in Appendix B-3.2. From Eq. (B-3.1) we may write the following equation for the instantaneous power to store energy in the capacitor:

$$p_C = \frac{V^2}{R}(\varepsilon^{-t/\tau_C} - \varepsilon^{-2t/\tau_C}) \quad \text{watts (W)} \qquad \textbf{(10-16)}$$

where all terms have been defined above

A little study of Eq. (10-16) reveals the following:

1. When $t = 0^+$, $p_C = 0$.
2. When $t \geqq 5\tau$, $p_C = 0$.
3. Clearly, then, p_C begins to rise at zero time to some *maximum* during the transient period and returns to zero when the transient disappears.

From Eq. (B-3.4) derived in Appendix B-3.2, we find that the magnitude of $p_{C(max)}$ is

$$p_{C(max)} = \frac{VI_0}{4} = \frac{V^2}{4R} = \frac{I_0^2 R}{4} \quad \text{watts (W)} \qquad \textbf{(10-17)}$$

From the foregoing derivation we discover that $p_{C(max)}$ occurs whenever

1. v_C and i_C cross at $t = \tau \ln 2 \cong 0.693\tau$
2. $v_C = V/2$ and $i_C = I_0/2$ (at $t = \tau \ln 2$).
3. $p_R = p_{C(max)} = p_{R(max)}/4$ as shown by Eq. (10-17).

From Eq. (B-3.5) we may also write the equation for the instantaneous power dissipated in R during the transient rise of v_C and decay of i_C as

$$p_R = i_C^2 R = I_0^2 R(\varepsilon^{-2t/\tau_C}) = (V^2/R)\varepsilon^{-2t/\tau_C} \quad \text{watts (W)} \qquad \textbf{(10-18)}$$

A little study of Eq. (10-18) reveals the following:

1. When $t = 0^+$, $p_R = p_{R(max)} = I_0^2 R = V^2/R$
2. When $t \geqq 5\tau$, $p_R = 0$
3. Since the exponential decay term falls off at the rate of $2t/\tau$, the exponential decay of p_R is *twice as fast* as that of i_C (and also p_T given below), as shown in Fig. 10-12

From Eq. (B-3.8) we also may write the equation for the instantaneous total power drawn by the RC circuit from the supply V during transient capacitor charging cycle as

$$p_T = p_R + p_C = \frac{V^2}{R}\varepsilon^{-t/\tau_C} = I_0^2 R\varepsilon^{-t/\tau_C} \quad \text{watts (W)} \qquad \textbf{(10-19)}$$

Examination of Eq. (10-19) shows that p_T also decays exponentially and that

1. When $t = 0^+$, $p_T = V^2/R = I_0^2 R = p_{R(max)} = p_{T(max)}$

2. When $t \geqq 5\tau$, $p_T = 0$
3. p_T decays at the same rate as i_C (and v_R) during the charging cycle
4. When $t = \tau \ln 2 = 0.693\tau$, $p_T = p_{T(max)}/2 = p_{R(max)}/2 = (V^2/R)/2 = I_0^2R/2$

Figure 10-12 shows the variations of p_C, p_R, and p_T as a function of the time required to charge an RC circuit. (Also plotted in Fig. 10-12 is the instantaneous energy stored in the capacitor, discussed in the next section.)

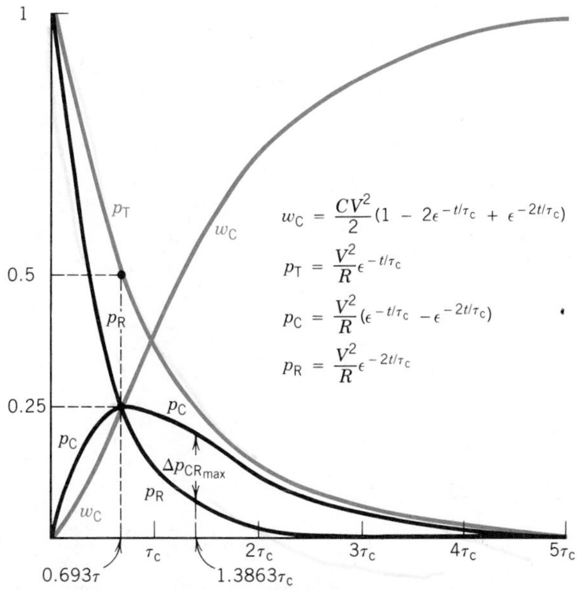

Figure 10-12 Summary of variations of p_R, p_C, and p_T (as well as w_C) as a function of time in charging an RC circuit

With respect to Fig. 10-12, observe that many interesting things happen when $t = \tau_c \ln 2 = 0.693\tau_c$. You may recall that at this time, v_C rises to $V/2$ and i_C decays to $I_0/2$, and that both parameters cross each other (at this time). Note also that at $t = 0.693\tau_c$

1. $p_R = p_C$
2. p_C is at its maximum value where $p_C = p_{R(max)}/4 = p_{T(max)}/4$
3. The energy stored in the capacitor is also one-fourth of its final ($CV^2/2$) value.
4. $p_T = p_{T(max)}/2$ (similarity to i_C)
5. p_R decreases below p_C (prior to this time, $p_R > p_C$) for times greater than $t = 0.693\tau$

Note also that at each instant of time, $p_T = p_R + p_C$. This is easily verified when t is 0.693τ and the sum $p_R + p_C$ is $0.5p_{T(max)}$.

Since p_R and p_C are equal at $t = 0.693\tau$ and also at times $t \geqq 5\tau_c$, there must be some time when the difference between p_C and p_R is a maximum, designated $\Delta p_{CR(max)}$. This is derived in Ex. 3 and Ex. 4 of Appendix B-3.2. These examples show

$$\Delta p_{CR(max)} = p_C - p_{R(max)} = \frac{V^2/R}{8} = \frac{I_0^2R}{8} \quad \text{watts (W)} \qquad \textbf{(10-20)}$$

and that this maximum difference occurs at time (Fig. 10-12)

$$t = \ln 4(\tau) = 1.3863\tau_c \quad \text{seconds (s)} \qquad \textbf{(10-21)}$$

EXAMPLE 10-14

Given $V = 100$ V, $R = 1$ MΩ, $C = 1\ \mu$F in Fig. 10-11 when the switch is first closed at $t = 0^+$ s. Calculate the instantaneous power at $t = 1\tau_c$ for

a. p_C
b. p_R
c. p_T
d. $p_C - p_R$
e. Verify (c) also as the sum of (a) and (b)

Solution

a. $p_C = \frac{V^2}{R}(\varepsilon^{-t/\tau_c} - \varepsilon^{-2t/\tau_c}) = \frac{(100)^2}{1\text{ M}\Omega}(\varepsilon^{-1} - \varepsilon^{-2})$

$= 10\text{ mW}(0.2325) = \mathbf{2.325\ mW}$ (10-16)

b. $p_R = \frac{V^2}{R}\varepsilon^{-2t/\tau_c} = 10\text{ mW}(\varepsilon^{-2}) = \mathbf{1.353\ mW}$ (10-18)

c. $p_T = \frac{V^2}{R}\varepsilon^{-t/\tau_c} = 10\text{ mW}\ \varepsilon^{-1} = \mathbf{3.679\ mW}$ (10-19)

d. $p_C - p_R$ (at $t = 1\tau_c$) = (2.325 − 1.353) mW = **0.972 mW**

e. $p_C + p_R$ (at $t = 1\tau_c$) = (2.325 + 1.353) mW = **3.678 mW**, which compares favorably with the value of p_T calculated in (c)

EXAMPLE 10-15

Repeat all parts of Ex. 10-14 for $t = \tau_c \ln 4 = 1.3863\tau_c$ and (f) also verify Eq. (10-20)

Solution

a. $p_C = \frac{V^2}{R}(\varepsilon^{-t/\tau_c} - \varepsilon^{-2t/\tau_c}) = 10\text{ mW}(\varepsilon^{-1.3863} - \varepsilon^{-2.773})$

$= 10\text{ mW}(0.25 - 0.0625) = \mathbf{1.875\ mW}$ (10-16)

b. $p_R = \frac{V^2}{R}\varepsilon^{-2t/\tau_c} = 10\text{ mW}(\varepsilon^{-2.773}) = \mathbf{0.625\ mW}$ (10-18)

c. $p_T = \frac{V^2}{R}\varepsilon^{-t/\tau_c} = 10\text{ mW}(0.25) = \mathbf{2.5\ mW}$ (10-19)

d. $(p_C - p_R)_{max} = 1.875 - 0.625 = \mathbf{1.25\ mW}$ at $t = \tau \ln 4 = 1.3863\tau$

e. $p_C + p_R = 1.875 + 0.625 = \mathbf{2.5\ mW}$ which verifies value calculated in (c)

f. $\Delta p_{CR(max)} = (V^2/R)/8 = 10\text{ mW}/8 = \mathbf{1.25\ mW}$ (10-20) which verifies value calculated in (d)

Appendix B-3.2 also gives two examples (Exs. 1 and 2) showing how to derive the following relations:[11]

a. When $p_C = 2p_R$, $t = \ln 3\tau_c = 1.0986\tau_c$
b. When $p_R = 5p_C$, $t = \ln 1.2\tau_c = 0.1823\tau_c$

EXAMPLE 10-16

Using the data of Ex. 10-14 for the RC circuit, verify that

a. $p_C = 2p_R$ at $t = 1.0986\tau_c$
b. $p_R = 5p_C$ at $t = 0.1823\tau_c$

Solution

a. $p_C = \frac{V^2}{R}(\varepsilon^{-t/\tau_c} - \varepsilon^{-2t/\tau_c}) = 10\text{ mW}(\varepsilon^{-1.0986} - \varepsilon^{-2.197})$

$= 10\text{ mW}(0.\bar{3} - 0.\bar{1}) = 2.\bar{2}\text{ mW}$ (10-16)

$p_R = \frac{V^2}{R}\varepsilon^{-2t/\tau_c} = 10\text{ mW}\varepsilon^{-2.197} = 1.\bar{1}\text{ mW}$ (10-18)

Note that $p_C = 2p_R$ since $2.\bar{2} = 2(1.\bar{1})$ mW

b. $p_R = \frac{V^2}{R}\varepsilon^{-2t/\tau_c} = 10\text{ mW}(\varepsilon^{-2\times 0.1823})$

$= 6.9\bar{4}\text{ mW}$ (10-18)

$p_C = \frac{V^2}{R}(\varepsilon^{-t/\tau_c} - \varepsilon^{-2t/\tau_c})$

$= 10\text{ mW}(\varepsilon^{-0.1823} - \varepsilon^{-0.3646}) = 1.3\bar{8}\bar{8}\text{ mW}$ (10-16)

Again, note that $p_R = 5p_C$ since $6.9\bar{4}\text{ mW} = 5 \times 1.3\bar{8}\bar{8}\text{ mW}$

Example 10-15 shows that the maximum difference between p_C and p_R occurs when $t = \tau_c \ln 4 = 1.3863\tau_c$. But this example also shows that at this (same) particular time $p_C = 3p_R$. Figure 10-12 also shows that the ratio of p_C/p_R is greatest at $t = \tau_c \ln 4$. Example 10-17 shows how such relations are derived.

[11] See also end-of-chapter Prob. 10-22, which shows $t = \ln(1 + D) \times \tau_c$, where $D = p_C/p_R$, as a means of finding t for any p_C/p_R ratio.

EXAMPLE 10-17

Derive an expression for the time, t, at which $p_C = 3p_R$ during the charge cycle.

Solution

$p_C = 3p_R$ and substituting Eqs. (10-16) and (10-18) yields

$\frac{V^2}{R}(\varepsilon^{-t/\tau_c} - \varepsilon^{-2t/\tau_c}) = \frac{3V^2}{R}\varepsilon^{-2t/\tau_c}$ dividing both sides by V^2/R

$\varepsilon^{-t/\tau_c} - \varepsilon^{-2t/\tau_c} = 3\varepsilon^{-2t/\tau_c}$ and collecting similar terms on each side

$\varepsilon^{-t/\tau_c} = 4\varepsilon^{-2t/\tau_c}$ and taking the natural logarithm of each side yields

$-t/\tau_c = \ln 4 + (-2t/\tau_c)$ and $t/\tau_c = \ln 4$ from which $t = \tau_c \ln 4 =$ $\mathbf{1.3863\tau_c}$

Note that this value of time is the *same* as that given in Eq. (10-21), originally derived in Ex. 3 of Appendix B-3.2.

Figure 10-12 also shows that p_R exceeds p_C by fairly high ratios when t is *far less* than $0.693\tau_c$. Example 10-18 shows the validity of the method used in Ex. 10-17 for one case where the ratio p_R/p_C is fairly high.

EXAMPLE 10-18

Derive an expression for the time, t, at which $p_R = 50p_C$ during the charge cycle.

Solution

$50p_C = p_R$

$50\frac{V^2}{R}(\varepsilon^{-t/\tau_c} - \varepsilon^{-2t/\tau_c}) = \frac{V^2}{R}(\varepsilon^{-2t/\tau_c})$ and dividing both sides by $50V^2/R$ yields

$\varepsilon^{-t/\tau_c} - \varepsilon^{-2t/\tau_c} = 0.02\varepsilon^{-2t/\tau_c}$ and collecting similar terms on each side

$\varepsilon^{-t/\tau_c} = 1.02\varepsilon^{-t/\tau_c}$ then taking the natural log of each side yields

$-t/\tau_c = \ln 1.02 + (-2t/\tau_c)$ or $t/\tau_c = \ln 1.02$ from which $t = \tau \ln 1.02 = 0.0198\tau$

Note from Fig. 10-12 that the value of t at which the ratio $p_R/p_C = 50$ occurs quite close to the *origin* of the p_C curve, where $t \cong \tau \ln 1.0000 = 0$.

10-8 ELECTRICAL ENERGY INSTANTANEOUSLY STORED IN THE CAPACITOR OF AN RC CIRCUIT DURING CAPACITOR CHARGE CYCLE

Appendix B-3.2 derives the instantaneous energy transferred to the capacitor during the charge cycle in Eq. (B-3.7) as

$$w_C = \frac{CV^2}{2}(1 - 2\varepsilon^{-t/\tau_c} + \varepsilon^{-2t/\tau_c}) \quad \text{joules (J)} \qquad (10\text{-}22)$$

A little study of Eq. (10-22) reveals that

1. The equation is an *expanding* (rising) exponential function of the form $y = K(1 - \varepsilon^{-st})$, as described in Appendix B-4. Such functions must begin at zero and expand to some asymptotic limit, K. In this case $K = CV^2/2$.
2. The second term, ε^{-2t/τ_c}, tends to *slow* the rate of rise at *low* values of t as shown in Fig. 10-12, where w_C is also plotted from $t = 0$ to $t = 5\tau_c$.

We have already noted from Fig. 10-12 that when $t = 0.693\tau_c$, both w_C and p_R are one-fourth of their respective maximum values. Let us verify this using Eq. (10-22).

EXAMPLE 10-19

Using Eq. (10-20), show that $w_C = CV^2/8$ when $t = \tau_c \ln 2 = 0.693\tau_c$

Solution

$$w_C = \frac{CV^2}{2}(1 - 2\varepsilon^{-0.693} + \varepsilon^{-2\times 0.693})$$

$$= \frac{CV^2}{2}(1 - 2 \times 0.5 + 0.25) = \frac{CV^2}{8} \quad (10\text{-}22)$$

EXAMPLE 10-20

Verify the value of w_C obtained in Ex. 10-19, using the exponential rise of v_C at $t = \tau_c \ln 2 = 0.693\tau_c$

Solution

We have already seen that when $t = 0.693\tau$, v_C rises exponentially to $V/2$. Therefore, $w_C = Cv_C^2/2 = \dfrac{C(V/2)^2}{2} = CV^2/8$

Note that this is the same as the result obtained in Ex. 10-19. Observe also from Fig. 10-12 that $CV^2/8$ is one-quarter of the final energy of $CV^2/2$ at $t = 0.693\tau_c$.

10-9 TRANSIENT RC SWITCHING ANALYSIS

All the previous sections have presented the necessary theory and equations for this section, in which they are applied. Here, we will consider and solve a variety of problems. Some of these problems show and use mechanical switching for purposes of simplicity. In commercial electronic circuits, such switching may be accomplished by rectangular pulse waveforms, which step from a more negative to a more positive voltage alternately with time. In other instances, such switching may be accomplished with a single electronic semiconductor device: a Shockley diode, a silicon unilateral switch (SUS), a DIAC, an SCR, or a TRIAC. Alternatively, such devices as neon glow lamps and/or active semiconductor devices such as transistors may be used. The latter may be switched from an OFF (cutoff) state to an ON (saturation) state by means of appropriate bias levels. In all instances for our purposes, the semiconductors serve as *switches*. As noted above, mechanical switching simplifies visualization of the connections made in the various states.

It is very important for the reader to develop an ability to visualize such circuits and analyze the initial and final conditions in each switch position. The intent of this section is to show that it is *unnecessary* to use differential equations, operational calculus, or Laplace transforms to obtain a solution for the waveforms that emerge. The solution to each of the examples given below (and the problems at the end of the chapter) all have the same procedure and form:

1. For each switch position, determine the circuit time constants τ_1, τ_2, etc., using Thévenin's theorem, if necessary, to determine equivalent voltage and resistance "seen" by the capacitor.
2. For each switch position, determine the initial and final conditions for the period in which the switch remains in the particular position.
3. Use the previously developed methods of circuit analysis in determining output voltages, v_O, when called for in the problem statement. In some cases, v_O may be a voltage across a resistor. In others, it may be voltage across a capacitor. In still others, it may be the combination of a resistor and a voltage source or capacitor in series.
4. Polarities across resistors are determined by the instantaneous *direction of current* for the respective switch positions (i.e., whenever the capacitor is either charging or discharging).
5. Voltages across resistors are a function of the current in the circuit through them.

6. Use of a tabulation is recommended as shown in Exs. 10-25 through 10-27. The tabulation should allow for times just before and just after new initial conditions are created by new switch positions. This enables calculations of instantaneous current "jumps" created by the *new initial* conditions.

EXAMPLE 10-21

Given the network shown in **Fig. 10-13**, C is initially discharged with switch S open, in position 1. The time duration in position 1, T_1, is 55 ms and the time duration in position 2, T_2, is 30 ms. Switch S closes at $t = 0^+$ and then is opened. It is then closed and opened for three additional successive cycles. Calculate

a. The Thévenin voltage and resistance seen by C in position 2
b. Time constants, τ_1 and τ_2
c. The value of v_C at the beginning and end of each switch throw
d. Sketch v_O, the output voltage waveform versus time (in milliseconds), for the first four cycles of switching
e. Assuming the same time durations, determine maximum and minimum values of v_O
f. Calculate the energy stored in the capacitor at the maximum and minimum values of v_O
g. Calculate the energy dissipated in R_2 each time the switch is thrown to position 1

Solution

a. $V_2 = \dfrac{R_2}{R_1 + R_2} V = \dfrac{500 \text{ k}}{1000 \text{ k}} 100 \text{ V} = \mathbf{50 \text{ V}}$;

$$R_{TH} = (R_1 \| R_2) + R_3 = \left(\frac{500 \text{ k}\Omega}{2}\right) + 50 \text{ k}\Omega = \mathbf{300 \text{ k}\Omega}$$

b. $\tau_1 = (R_2 + R_3)C = (550 \text{ k}\Omega)(50 \text{ nF}) = \mathbf{27.5 \text{ ms}}$;
$\tau_2 = R_{TH}C = (300 \text{ k}\Omega)(50 \text{ nF}) = \mathbf{15 \text{ ms}}$

c. Since capacitor is initially uncharged, v_C is found by:

Time Period		
0 to 30 ms:	$v_{C_2} = V_2(1 - \varepsilon^{-t_2/\tau_2}) = 50(1 - \varepsilon^{-30/15})$ $= \mathbf{43.23 \text{ V}}$	(10-2)
30 to 85	$v_{C_1} = V_{C_2}\varepsilon^{-t_1/\tau_1} = 43.25\varepsilon^{-55/27.5}$ $= \mathbf{5.851 \text{ V}}$	(10-9)
85 to 115	$v_{C_2} = V_{01} + (V_2 - V_{01})(1 - \varepsilon^{-t_2/\tau_2})$ $= 5.851 + (50 - 5.851)(1 - \varepsilon^{-2})$ $= 44.025 \text{ V}$	(10-2a)
115 to 170	$v_{C_1} = V_{C_2}\varepsilon^{-t_1/\tau_1} = 44.025\varepsilon^{-2}$ $= \mathbf{5.958 \text{ V}}$	(10-9)
170 to 200	$v_{C_2} = V_{01} + (V_2 - V_{01})(1 - \varepsilon^{-t_2/\tau_2})$ $= 5.958 + (50 - 5.958)(1 - \varepsilon^{-2})$ $= \mathbf{44.04 \text{ V}}$	(10-2a)
200 to 255	$v_{C_1} = V_{C_2}\varepsilon^{-t_1/\tau_1} = 44.04\varepsilon^{-2}$ $= \mathbf{5.96 \text{ V}}$	(10-9)
255 to 285	$v_{C_2} = V_{01} + (V_2 - V_{01})(1 - \varepsilon^{-t_2/\tau_2})$ $= 5.96 + (50 - 5.96)(1 - \varepsilon^{-2})$ $= \mathbf{44.04 \text{ V}}$ (same as third cycle)	
285 to 340 ms	$v_{C_1} = \mathbf{5.96 \text{ V}}$ (same as third cycle above)	(10-9)

d. The output voltage v_O is the same as v_C calculated above and is shown in Fig. 10-13b.

e. From Fig. 10-13b, $v_O(\text{max}) = \mathbf{44.04 \text{ V}}$ and $v_O(\text{min}) = \mathbf{5.96 \text{ V}}$

f. $w_{C(\text{max})} = Cv_C^2/2 = (50 \text{ nF})(44.04)^2/2 = \mathbf{48.49\ \mu J}$
$w_{C(\text{min})} = Cv_C^2/2 = (50 \text{ nF})(5.96)^2/2 = \mathbf{0.89\ \mu J}$

g. Total energy dissipated $= w_{C(\text{max})} - w_{C(\text{min})}$
$= (48.49 - 0.89)\ \mu\text{J} = \mathbf{47.6\ \mu J}$

$$W_2 = W_t \frac{R_2}{R_1 + R_2} = (47.6\ \mu\text{J})\frac{500 \text{ k}\Omega}{550 \text{ k}\Omega} = \mathbf{43.\overline{27}\ \mu J}$$

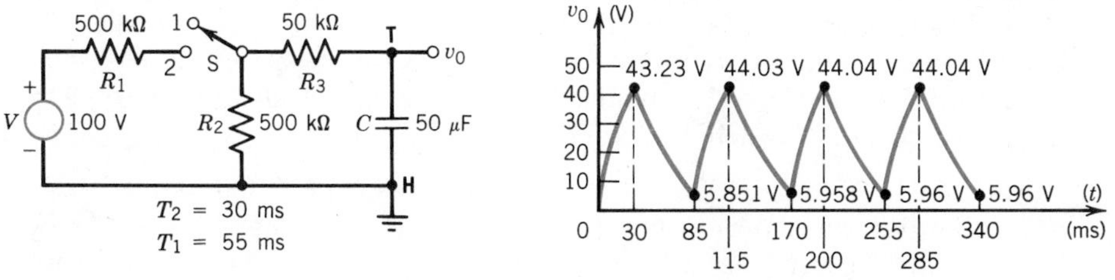

a. Network b. Waveform of v_O vs time

Figure 10-13 Example 10-21; network and solution

The reader may have observed from Fig. 10-13b that after the third cycle of switching, the waveform is repetitive for all remaining switching cycles. This situation is merely a special case resulting from the selection of RC values and duration of switching periods, as we will see from additional examples. It is of some importance to note, however, that whenever the switch is in position 2, energy is supplied to the capacitor. When in position 1, most of that energy is dissipated in R_2 and R_3.

Once we have drawn the waveforms for v_C (or v_O), we are in a position to visualize the rates of change at various time intervals, as shown by Ex. 10-22.

EXAMPLE 10-22

Using the data shown on the waveform of Fig. 10-13b, calculate the

a. Initial rate of change of output voltage, v_O, at $t = 0^+$, in position 2
b. Rate of change of v_O after 30 ms in position 2, at $t = 30$ ms
c. Initial rate of change of v_O after 30 ms in position 1, at $t = 30^+$ ms
d. Rate of change of v_O after 55 ms in position 1 at time, $t = 85$ ms
e. Rate of change of v_O at time $t = 85$ ms in position 2

Solution

a. $\dfrac{dv_{C02}}{dt} = \dfrac{V}{\tau}\varepsilon^0 = 50\text{ V}/15\text{ ms} = \mathbf{3.\bar{3}\ kV/s}$ (10-6b)

b. $\dfrac{dv_{C_2}}{dt} = \dfrac{V}{\tau_2}\varepsilon^{-t/\tau_2} = \dfrac{3.\bar{3}\text{ kV}}{\text{s}}\varepsilon^{-30/15} = \mathbf{451.1\ V/s}$ (10-6b)

c. $\dfrac{dv_{C01}}{dt} = \dfrac{-V}{\tau_1}\varepsilon^0 = \dfrac{-43.23\text{ V}}{27.5\text{ ms}} = \mathbf{-1.572\ kV/s}$ (10-12)

d. $\dfrac{dv_{C_1}}{dt} = \dfrac{-V}{\tau_1}\varepsilon^{-t/\tau_1} = -1572\dfrac{V}{\text{s}}\varepsilon^{-55/27.5}$

$= \mathbf{-212.75\ V/s}$ (10-12)

e. $dv_{C02} = \dfrac{V - V_0}{\tau_2}\varepsilon^0 = \dfrac{(50 - 5.851)\text{ V}}{15\text{ ms}} = \mathbf{2.943\ kV/s}$ (10-6b)

Again note from the solution to Ex. 10-22 that the initial rates of change *at the beginning* of a switch transition are always greater than the rates of change *after* the transition is in effect. Also note the differences between the solutions to (**a**) and (**e**) for both the initial rates of change in position 2; (**e**) is less than (**a**) because the capacitor has an initial charge and positive voltage due to its previous history.

From the solution to Ex. 10-22 please note that

1. All *initial* rates of change are always *greater* at the beginning of the switch transition than the rates after the transition has been in effect for some time.
2. There is a difference between the initial rate of change at $t = 0$, compared to $t = 85$ ms, in position 2. As noted from the solutions to parts (**a**) and (**e**), the latter rate is lower because the capacitor now has an initial charge and a positive voltage as a result of its previous history.
3. Whenever the switch is in position 1, the rates of change of output voltage (and v_{C_1}) are negative because the waveform is decaying exponentially.
4. Drawing the output waveform (and/or waveforms of i_C, v_C, v_R, etc.) versus elapsed time enables us to summarize and visualize how all parameters change as a result of the switching sequence specified.

Examples 10-21 and 10-22 used mechanical switching to produce a triangular sawtooth waveform, shown in Fig. 10-13b. Example 10-23 shows one of the simplest methods of generating such a waveform and a typical commercial application of it, in a neon glow-tube flasher.

EXAMPLE 10-23

A portable self-contained street safety blinker contains a neon glow tube that ignites (ON) at 75 V and extinguishes at 50 V. The simplified relaxation oscillator circuit for the device is shown in **Fig. 10-14a**. When ignited, the resistance of the glow tube is 500 Ω. When in the OFF state, its resistance is assumed infinite.

The blinker battery power supply consists of three 30-V batteries connected in series with a resistance R of 200 kΩ, as shown in Fig. 10-14a, and a 20-μF capacitor.

The waveform of the output voltage across the capacitor and glow tube is shown in Fig. 10-14b. The period t_1 to t_2 represents the time during which the capacitor is being charged from 50 to 75 V. The period t_2 to t_3 is the time it takes for the glow tube to discharge. The period between t_2 and t_4 is the time between successive flashes of the glow tube, as shown in Fig. 10-14b. Calculate

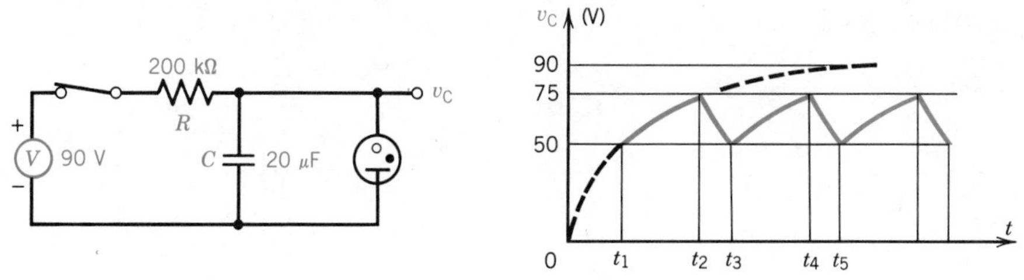

a. Relaxation oscillator circuit

b. Waveform across v_C and glow tube

Figure 10-14 Examples 10-23 and 10-24

a. Time t_1, on Fig. 10-14b
b. Time t_2
c. Time t_3
d. Time for the capacitor to recharge to its firing value of 75 V by two methods
e. Time for the neon glow tube to discharge
f. Time between flashes of the safety blinker

Solution

a. $t_1 = \tau_c \ln\left(\dfrac{V - V_0}{V - v_C}\right) = (200\ \text{k}\Omega)(20\ \mu\text{F}) \ln\left(\dfrac{90 - 0}{90 - 50}\right)$

$= (4\ \text{s})(\ln 2.25) = \mathbf{3.244\ s}$ (10-5a)

b. $t_2 = \tau_c \ln\left(\dfrac{V - V_0}{V - v_C}\right) = (4\ \text{s}) \ln\left(\dfrac{90 - 0}{90 - 75}\right) = \mathbf{7.167\ s}$

c. $t_d = \tau_d \ln\left(\dfrac{V_0}{v_C}\right) = (500\ \Omega)(20\ \mu\text{F}) \ln\left(\dfrac{75}{50}\right)$

$= (10\ \text{ms}) \ln 1.5 = 4.05\ \text{ms} \cong \mathbf{4\ ms}$ (10-15)

$t_3 = t_2 + t_d = 7.167 + 0.004 = \mathbf{7.171\ s}$

d. $t_{(\text{charge C})} = t_2 - t_1 = 7.167 - 3.244 = \mathbf{3.923\ s}$

$t_{ch} = \tau_c \ln\left(\dfrac{V - V_0}{V - v_C}\right) = (4\ \text{s}) \ln\left(\dfrac{90 - 50}{90 - 75}\right)$

$= \mathbf{3.923\ s}$ (10-5a)

e. $t_d = \mathbf{4\ ms}$ (from **c**)

f. $t_{\text{flash}} = t_3 - t_1 = 7.171 - 3.244 = \mathbf{3.93\ s}$

EXAMPLE 10-24

In the previous example, calculate the value of R required to make the time between flashes exactly 2 seconds

Solution

$t = t_{ch} + t_d = 2\ \text{s}$ and $t_{ch} = t - t_d = 2\ \text{s} - 4\ \text{ms} = 1.996\ \text{s}$

$\tau_c = t_{ch} \Big/ \ln\left(\dfrac{90 - 50}{90 - 75}\right) = 1.996\ \text{s}/\ln(2.\bar{6}) = 2.035\ \text{s}$ (10-5a)

$R = \tau_c/C = 2.035\ \text{s}/20\ \mu\text{F} = \mathbf{101.8\ k\Omega}$ (approximately $\mathbf{100\ k\Omega}$)

To test whether the answer is reasonable in Ex. 10-24, observe that when R = 200 kΩ the time between flashes in Ex. 10-23 is approximately 4 s, and when R = 100 kΩ the time between flashes in Ex. 10-24 is approximately 2 s. This suggests also that if R in Fig. 10-14a is a variable rheostat, the time between flashes may be adjusted by varying the time constant of the circuit. (It also may be adjusted by increasing or decreasing V, but this is less practical.)

EXAMPLE 10-25

Given the circuit shown in **Fig. 10-15a**, calculate the

a. Thévenin voltage at switch terminal 1
b. Thévenin resistance at switch terminal 1
c. Circuit time constant τ_1 when switch is in position 1
d. Circuit time constant τ_2 when switch is in position 2

If the time duration of the switch in position 1 is 40 ms and in position 2 is 50 ms, calculate

e. v_C, i_C, and v_0 at the end of 40 ms in position 1
f. v_C, i_C, and v_0 at the end of 50 ms in position 2
g. Tabulate the respective value of v_C, i_C, and v_0 versus time for three successive cycles of switch transition
h. Draw the output waveforms of v_C, i_C, and v_0 versus time for three successive cycles tabulated

Solution

a. $V_{TH_1} = (V_p - V_n)\dfrac{R_2}{R_1 + R_2} + V_n$

$= [-15 - (-30)]\dfrac{5\ \text{k}\Omega}{25\ \text{k}\Omega} + (-30)\ \text{V}$

$= \mathbf{-27\ V}$ (5-8)

b. $R_{TH_1} = (R_1 \| R_2) + R_3 = (20\ \text{k} \| 5\ \text{k}) + 1\ \text{k} = 5\ \text{k}\Omega$

c. $\tau_1 = (R_{TH_1} + R_5)C = (5\ \text{k}\Omega + 3\ \text{k}\Omega)\ 5\ \mu\text{F} = \mathbf{40\ ms}$

d. $\tau_2 = (R_4 + R_5)C = (7\ \text{k}\Omega + 3\ \text{k}\Omega)(5\ \mu\text{F}) = \mathbf{50\ ms}$

e. $v_{C_1} = V(1 - \varepsilon^{-t/\tau_1}) = -27(1 - \varepsilon^{-40/40})$

$= -27(0.6321) = \mathbf{-17.07\ V}$

$$i_{C_{01}} = \frac{V - V_0}{R_{T_1}} = \frac{-27\ \text{V}}{8\ \text{k}\Omega} = -3.375\ \text{mA}$$

$i_{C_1} = (i_{C_0})\varepsilon^{-t/\tau_1} = -3.375\ \text{mA}(\varepsilon^{-1})$

$= \mathbf{-1.242\ mA}$ in position 1 at end of 40 ms

$v_O = (i_{C_1})(R_5) = (-1.242\ \text{mA})\ 3\ \text{k}\Omega$

$= \mathbf{-3.725\ V}$ in position 1 at end of 40 ms

f. $v_{C_2} = V_{C_1}\varepsilon^{-t/\tau_2} = -17.07\varepsilon^{-50/50}$

$= \mathbf{-6.28\ V}$ in position 2 at end of 50 ms

$i_{C_{02}} = v_{C_1}/R_{T_2} = +17.07\ \text{V}/10\ \text{k}\Omega = +1.707\ \text{mA}$

$i_{C_2} = i_{C_{02}}\varepsilon^{-t/\tau_2} = (1.707\ \text{mA})\varepsilon^{-50/50}$

$= \mathbf{+0.628\ mA}$ at end of 50 ms in position 2

$v_O = i_{C_2}R_5 = (0.628\ \text{mA})(3\ \text{k}\Omega)$

$= \mathbf{+1.884\ V}$ at end of 50 ms in position 2

g. The remaining calculations are left as an exercise for the reader. The values for the first three successive cycles are tabulated below.

h. Output waveforms of v_O, v_C, and i_C for the first three successive cycles of switch operation are shown in Fig. 10-15b.

With respect to the above solution, please note the following.

1. The tabulation drawn shows two cells and two time values for each switch position. The reason for this is obvious from Fig. 10-15b, where two values of i_C and v_O are represented each interval of time. The reader is strongly advised to use the tabulation technique shown in part (**g**) of the solution. This technique results in fewest errors and facilitates understanding, simultaneously.
2. Observe that v_C shows the same value at both the end of one switch position and the beginning of the next. This is reasonable when one considers that the property of capacitance is to *oppose a change* in voltage.
3. The waveform of i_C (Fig. 10-15b) is positive whenever v_C has a positive (slope) rate of change and i_C is negative whenever v_C has a negative (slope) rate of change. This is understandable since $i_C = k\,dv_C/dt = (C)\,dv_C/dt$, at all times.
4. The waveform of v_O has exactly the same shape as i_C. This is to be expected since $v_O = i_C R_5$ at all times.
5. Finally, while the values of v_C become slightly more negative with each switch transition, the values of i_C and v_O drift to slightly more positive values with time.

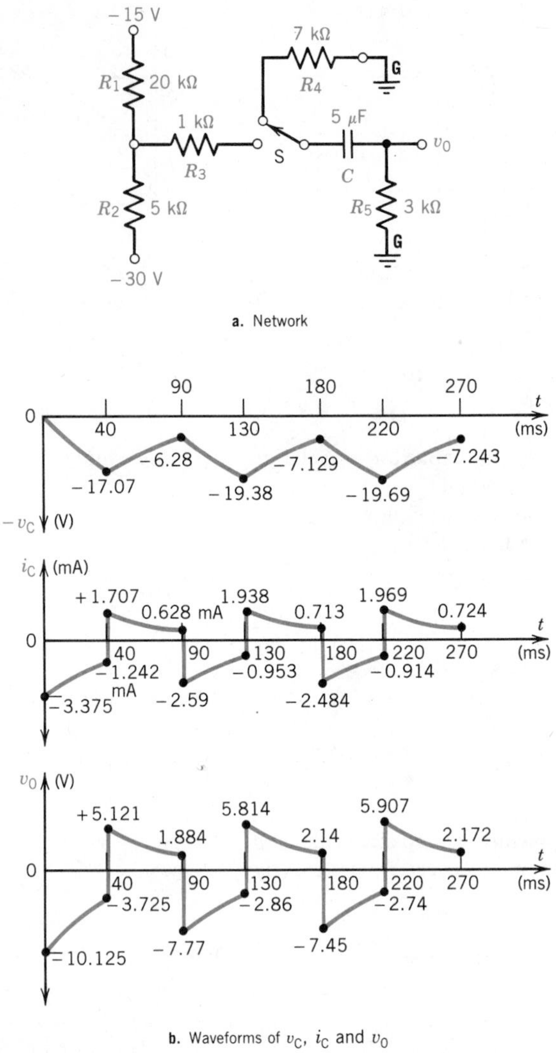

Figure 10-15 Network and solution for Ex. 10-25

Switch position	2	1	1	2	2	1	1	2	2	1	1	2	2
Elapsed time (ms)	0^-	0^+	40	40	90	90	130	130	180	180	220	220	270
v_C (V)	0	0	−17.07	−17.07	−6.28	−6.28	−19.38	−19.38	−7.13	−7.129	−19.69	−19.69	−7.243
i_C (mA)	0	−3.375	−1.242	1.707	0.628	−2.59	−0.953	1.938	0.713	−2.48	−0.914	1.969	0.7243
across R_5, v_O (V)	0	−10.125	−3.73	5.12	1.88	−7.77	−2.86	5.81	2.14	−7.45	−2.74	5.91	2.17

EXAMPLE 10-26

Given the network shown in **Fig. 10-16a**, S_1 closes at $t = 0^+$ ms and S_2 closes at $t = 50$ ms. When $t = 80$ ms, calculate

a. v_C
b. i_C
c. v_O
d. Tabulate all results leading to the calculations in **a**, **b** and **c** above
e. Draw waveforms of v_C, i_C, and v_O versus time from 0 to 80 ms
f. Account for the abrupt change in i_C when S_2 is closed

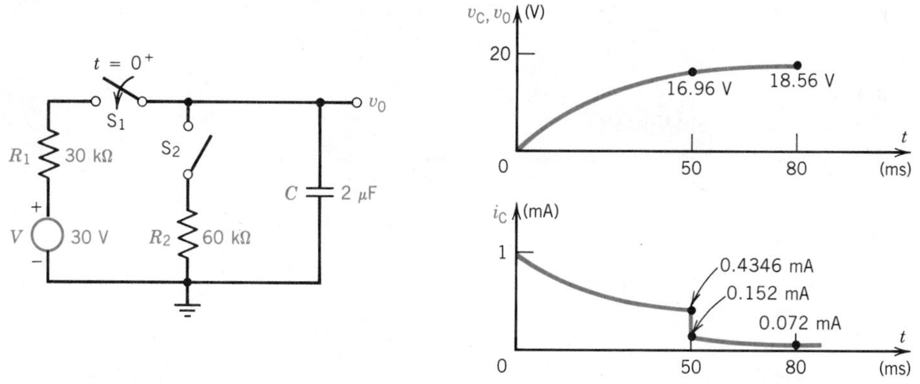

a. Network b. Waveforms

Figure 10-16 Example 10-26

Solution

$$\tau_1 = R_1 C = (30\ \text{k}\Omega)(2\ \mu\text{F}) = 60\ \text{ms};$$

$$\tau_2 = R_2 C = (20\ \text{k}\Omega)(2\ \mu\text{F}) = 40\ \text{ms};$$

$$V = V_1 = 30\ \text{V at } t = 0^+\ \text{ms};$$

$$V = V_2 = 30\ \text{V}\,\frac{60\ \text{k}}{90\ \text{k}} = 20\ \text{V at } t = 50^+\ \text{ms}$$

a. $v_{C_1} = 30\ \text{V}(1 - \varepsilon^{-50/60}) = 16.96\ \text{V}$

b. $i_{C_1} = \dfrac{V}{R_1}\varepsilon^{-t/\tau_1} = \dfrac{30\ \text{V}}{30\ \text{k}}\varepsilon^{-50/60} = 0.4346\ \text{mA}$

c. $v_{C_2} = (20 - 16.96)(1 - \varepsilon^{-30/40}) + 16.96 = \mathbf{18.56\ V}$ (10-2a)

$$i_{C_{O2}} = \frac{V - V_0}{R_2} = \frac{20 - 16.96}{20\ \text{k}} = \mathbf{0.152\ mA}$$

$$i_{C_2} = (i_{C_{O2}})\varepsilon^{-t/\tau_2} = (0.152\ \text{mA})\varepsilon^{-30/40} = \mathbf{72\ \mu A}$$

d. Tabulation:

Switch Closure	S_1			S_2		
t (ms)	0	0^+	50	50^+	80	
v_C (V)	0	0	16.96	16.96	**18.56**	**(a)**
i_C (mA)	0	1	0.4346	0.152	**0.072**	**(b)**
v_O (V)	0	0	16.96	16.96	**18.56**	**(c)**

e. The waveforms of v_C (v_O) and i_C are drawn in Fig. 10-16b.

f. i_C exhibits an immediate response to any (new) initial condition. As shown in Fig. 10-16b, when S_1 is first closed at time $t = 0^+$, i_C rises from 0 to 1 mA in zero time. This is also shown in the tabulation in (**d**). Consequently, when S_2 closes the capacitor has a voltage of 16.96 V in opposition to the Thévenin equivalent voltage of 20 V and a resistance of 20 kΩ. As shown in these calculations, as well as Fig. 10-16b, $i_{C_{O2}}$ responds immediately to the new initial conditions by dropping from 0.4346 to 0.152 mA in zero time.

Example 10-26 teaches us that whenever a switch is thrown, *new* initial conditions are created. Current, i_C, and its counterpart, v_R, respond immediately to the new initial conditions. But the voltage across a capacitor, v_C, by definition does not change immediately. Instead, the response is exponential and is determined by the final conditions set up by the new circuit. This is shown nicely in the relatively simple Ex. 10-27.

EXAMPLE 10-27

Given the network shown in **Fig. 10-17a**, S_1 closes at $t = 0^+$ ms with S_2 open. After 10 ms, S_2 closes. Calculate i_C, v_O, and v_C at times

a. $t = 0^+$
b. $t = 10$ ms with S_2 open
c. $t = 10^+$ ms with S_2 closed
d. $t = 12$ ms
e. $t = 16$ ms
f. Make a complete tabulation similar to that of Ex. 10-26d
g. Plot v_C and v_O versus time from 0 to 16 ms

Solution

$$\tau_1 = (R_1 + R_2)C = (10\ \text{k})(1\ \mu\text{F}) = 10\ \text{ms}$$

a. $I_{01} = \dfrac{V}{R_1 + R_2} = \dfrac{10\ \text{V}}{10\ \text{k}} = \mathbf{1\ mA}$

$$v_O = i_C R_2 = 1\ \text{mA} \times 9\ \text{k}\Omega = \mathbf{9\ V}$$

$$v_C = 0\ \text{V at } t = \mathbf{0^+\ s}$$

b. $v_{C_1} = V(1 - \varepsilon^{-t/\tau_1}) = 10(1 - \varepsilon^{-10/10}) = \mathbf{6.32\ V}$

$$i_{C_1} = I_{01}\varepsilon^{-t/\tau_1} = (1\ \text{mA})\varepsilon^{-1} = \mathbf{0.368\ mA}$$

$$v_O = (i_{C_1})(R_2) = (0.368\ \text{mA})(9\ \text{k}) = \mathbf{3.31\ V}$$

c. At $t = 10^+$ ms as S_2 closes

$v_{C_{02}} = \mathbf{6.32\ V}$ (unchanged)

$$I_{02} = \frac{V - V_0}{R_1} = \frac{(10 - 6.32)\text{ V}}{1\text{ k}\Omega} = \mathbf{3.68\ mA}$$

$$v_{02} = (I_{02})(R_2) = (3.68\text{ mA})(9\text{ k}) = \mathbf{33.11\ V}$$

$$\tau_2 = R_1C = (1\text{ k}\Omega)(1\ \mu\text{F}) = \mathbf{1\ ms}$$

d. $v_{C_2} = (V - V_0)(1 - \varepsilon^{-2/1}) + V_0$

$= (10 - 6.32)(0.8647) + 6.32 = \mathbf{9.5\ V}$

$i_{C_2} = 3.68\varepsilon^{-2/1} = \mathbf{0.498\ mA}$

$v_{02} = i_{C_2}R_2 = (0.498\text{ mA})\ 9\text{ k}\Omega = \mathbf{4.482\ V}$

e. $v_{C_2} = 10$ V since 6 ms exceeds $5\tau_2$

$i_{C_2} = 0$ and $v_{02} = 0$

f. Tabulation:

Switch Closure

t (ms)	0	0^+	10	10^+	12	16
i_C (mA)	0	1	0.368	3.68	0.498	0
v_0 (V)	0	9	3.31	33.11	4.482	0
v_C (V)	0	0	6.32	6.32	9.5	10

g. Waveforms are shown in Fig. 10-17b

With respect to the solution to Ex. 10-27, the tabulation of values in (**f**), and the waveforms shown in Fig. 10-17b drawn from the tabulation, please note the following.

1. Even though the network (Fig. 10-17a) is relatively simple, the waveforms are quite complex. This underscores the *need to show the times* just before and just after the new initial conditions (at 10 ms and 10^+ ms in above tabulation). Observe that the tabulation also shows the times just before and just after the first initial condition (at 0 and 0^+ ms).
2. Note that i_C and its proportional counterpart v_0 both have the same waveform shape. Both rise abruptly at times $t = 0^+$ and $t = 10^+$ ms.

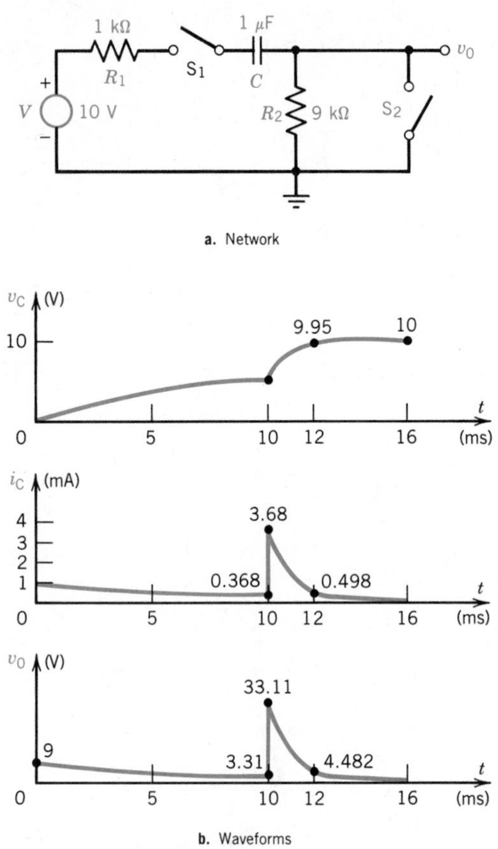

Figure 10-17 Example 10-27

3. When S_2 closes the circuit τ is considerably reduced from 10 to 1 ms. As a result, changes occur more rapidly, and the transient disappears within 15 ms of elapsed time.

10-10 MULTIVOLTAGE SOURCES IN RC TRANSIENT SWITCHING CIRCUITS

The previous circuits used a single source of supply and the switching involved changes in the passive elements of R and C. Circuits in this section will use more than one supply voltage. The changes in the initial conditions, therefore, involve not only changes in τ produced as a result of switching but also changes in both initial and final conditions. We will begin with relatively simple networks and proceed to some more difficult ones. But regardless of complexity, the techniques are essentially the same as those used in Sec. 10-9. A tabulation is used to facilitate the bookkeeping and maintain a record of changes produced by switch transitions.

EXAMPLE 10-28

In the network shown in **Fig. 10-18a**, switch S moves from its OFF position to position 1 at $t = 0^+$ s and remains there for 6 seconds. At the end of 6 s, it is thrown to position 2. Calculate v_C at elapsed times of

a. $t = 0^+$
b. $t = 6$ s
c. $t = 10$ s
d. $t = 15$ s
e. $t = 20$ s
f. $t = 25$ s
g. Calculate the time at which $v_C = 0$ V

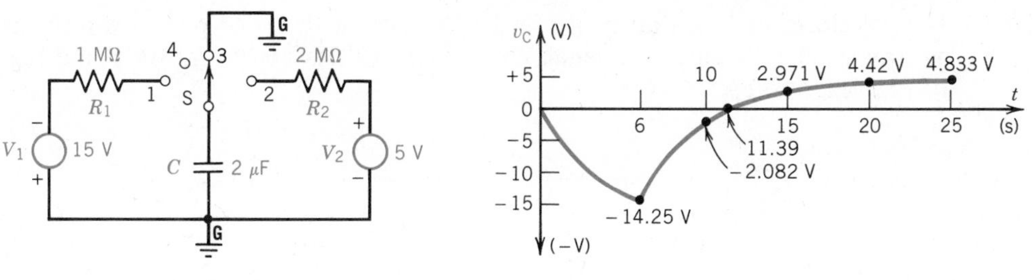

a. Network

b. Voltage across v_C vs time (in seconds)

Figure 10-18 Example 10-28

h. Tabulate results for comparison
i. Draw the resultant waveform for the voltage across the capacitor, v_C, at the given times

Solution

a. $v_{C_0} = 0$ by inspection at $t = 0$

b. $v_C = V_0(1 - \varepsilon^{-t/\tau_1}) = -15(1 - \varepsilon^{-6/2})$
$= \mathbf{-14.25\ V}$ when $t = 6$ s in position 1

c. $v_C = (V - V_0)(1 - \varepsilon^{-t/\tau_2}) + V_0$ (10-2a)
$= (5 - -14.25)(1 - \varepsilon^{-4/4}) + (-14.25) = \mathbf{-2.082\ V}$

d. $v_C = (V - V_0)(1 - \varepsilon^{-t/\tau_2}) + V_0$
$= (19.25)(1 - \varepsilon^{-9/4}) - 14.25 = \mathbf{+2.971\ V}$

e. $v_C = (V - V_0)(1 - \varepsilon^{-t/\tau_2}) + V_0$
$= (19.25)(1 - \varepsilon^{-14/4}) - 14.25 = \mathbf{+4.42\ V}$

f. $v_C = (V - V_0)(1 - \varepsilon^{-t/\tau_2}) + V_0$
$= (19.25)(1 - \varepsilon^{-19/4}) - 14.25 = \mathbf{+4.833\ V}$

g. $t = \tau_2 \ln\left(\dfrac{V - V_0}{V - v_C}\right) = (4\text{ s}) \ln\left(\dfrac{5 - -14.25}{+5 - 0}\right)$
$= (4\text{ s}) \ln\left(\dfrac{19.25}{5}\right) = 5.39$ s after S is in position 2
Therefore $t_0 = 6\text{ s} + 5.39\text{ s} = \mathbf{11.39\ s}$

h. Tabulation

t (s)	0	6	10	11.39	15	20	25
v_C (V)	0	−14.25	−2.082	0	2.971	4.42	4.833

i. These tabulated values are drawn in Fig. 10-18b

Let us analyze the initial and final conditions for each switch position shown in Fig. 10-18. Recall that the final conditions occur when the transient current has decayed to zero and the voltage across the capacitor has reached its asymptotic final value. We may tabulate these for v_C as

	Position 1	Position 2
Final	−15 V	+5 V
Initial	0 V	−14.25 V

Recording these values serves as a check on the curve drawn in Fig. 10-18b. We can see that in position 1, v_C attempts to rise from its initial value of zero to a final value of −15 V, stopping short at −14.25 V. Conversely, in position 2, v_C has an initial value of −14.25 V and a final value of +5 V, stopping short at +4.833 V. From the curve of v_C (Fig. 10-18b) and the knowledge of the initial and final values, we can determine the current waveform, as shown in Ex. 10-29.

EXAMPLE 10-29

Given the waveform of v_C drawn in Fig. 10-18b and the network of Fig. 10-18a, draw the waveform of current in the capacitor, i_C, using values at times

a. 0 s
b. 6 s (two values)
c. 11.39 s
d. 15 s
e. 25 s

Solution

a. $v_C = 0$ at $t = 0^+$ and therefore $I_0 = V_1/R_1 = -15\text{ V}/1\text{ M}\Omega = \mathbf{-15\ \mu A}$

b. After 6 s, $i_C = (-15\ \mu\text{A})\varepsilon^{-6/2} = \mathbf{-0.747\ \mu A}$. BUT when S switches to position 2, $v_C = -14.25$ V, which series *aids* the 5-V supply, AND simultaneously *reverses* capacitor current i_C so that $I_0 = +(5 + 14.25)\text{ V}/2\text{ M}\Omega = \mathbf{+9.625\ \mu A}$ at $t = 6^+$ μA

c. $i_C = I_0\varepsilon^{-t/\tau_2} = (9.625\ \mu\text{A})\varepsilon^{-5.39/4} = \mathbf{2.5\ \mu A}$ when $v_C = 0$ at $t = 11.39$ s

d. $i_C = I_0\varepsilon^{-t/\tau_2} = 9.625\varepsilon^{-9/4} = \mathbf{1.01\ \mu A}$

e. $i_C = I_0\varepsilon^{-t/\tau_2} = 9.625\varepsilon^{-19/4} = \mathbf{0.083\ \mu A}$

The waveform of i_C is drawn in **Fig. 10-19** along with v_C for purposes of comparison in the following comments. With respect to Exs. 10-28 and 10-29 and Fig. 10-19, please note:

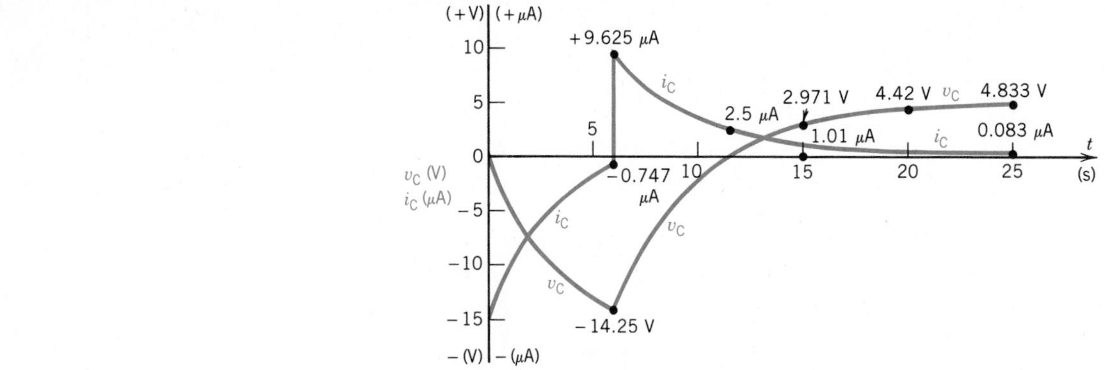

Figure 10-19 Waveforms of v_C and i_C; Example 10-29

1. It is possible to state in general that during *discharge* of a capacitor, i_C is *always* of *opposite* polarity to v_C, and that during *charge* of a capacitor, i_C is of the *same* polarity as v_C. Discharge of the capacitor begins at 6 s, causing abrupt reversal of capacitor current. Charge of the capacitor begins at 11.39 s, where both v_C and i_C are positive.
2. When $v_C = 0$ at 11.39 s, $i_C = 2.5\ \mu A$, in Ex. 10-29c. Is this possible? In position 2, we find v_R as $v_R = i_C R_2 = (2.5\ \mu A)(2\ M\Omega) = 5$ V. But by KVL for any RC circuit, $V_2 = v_C + v_R = 0 + 5 = 5\ V = V_2$. The insight to be gained here is that there is always the possibility of current in a capacitor when the capacitor voltage is zero.[12] Indeed, at time $t = 0^+$ the capacitor current is $-15\ \mu A$ while the capacitor voltage is zero.
3. The best test of whether our values of i_C are valid is to use $i_C = C(dv_C/dt)$ as shown in Eq. (10-3) and Ex. 10-13. Let us apply this test to our values of i_C at $t = 0^+$ and $t = 6$ s. If these values are correct, then our solution is correct.

EXAMPLE 10-30

From the solution data of Exs. 10-28 and 10-29, calculate

a. At $t = 0^+$, dv_C/dt and i_C
b. At $t = 6$ s, dv_C/dt and i_C

Solution

a. $dv_{C_0}/dt = V_0/\tau_1 = -15\ V/2\ s = \mathbf{-7.5\ V/s}$ at $t = 0^+$
$i_C = C\,dv_{C_0}/dt = (-7.5\ V/s)(2\ \mu F) = \mathbf{-15\ \mu A}$ at $t = 0^+$
b. $dv_{C_0}/dt + (V - V_0)/\tau_2 = +(5 - -14.25)/4\ s = \mathbf{4.8125\ V/s}$
$i_C = C\,dv_{C_0}/dt = (2\ \mu F)(4.8125\ V/s) = \mathbf{+9.625\ \mu A}$ at $t = 6$ s

---

The solution of Ex. 10-30 verifies the solution of Ex. 10-29 and the waveforms of i_C and v_C drawn in Fig. 10-19.

We are now prepared to tackle a problem of a more complex nature. We will find that by breaking it up into its constituent elements, the solution is easy. The object of problem is to find the output waveform, v_O. We will do this by finding v_C, i_C, v_R, and v_O, respectively, in each switch position. We will then analyze the data for validity at initial and final conditions.

[12] In **ac** circuits, when an ac source is applied across a "pure" capacitor, the capacitor current is at a maximum when the capacitor voltage is zero.

EXAMPLE 10-31

Given the network shown in **Fig. 10-20a**, the DPDT switch is initially in position 1 for a sufficiently long period to discharge C. It is then alternately switched to positions 2 and 1 for two successive cycles. The duration of the switch in each position is 400 μs. Using increments of 100 μs in each position, calculate for two successive switching cycles

a. v_C
b. i_C
c. v_R
d. v_0
e. Plot the waveform of v_0 versus time in microseconds

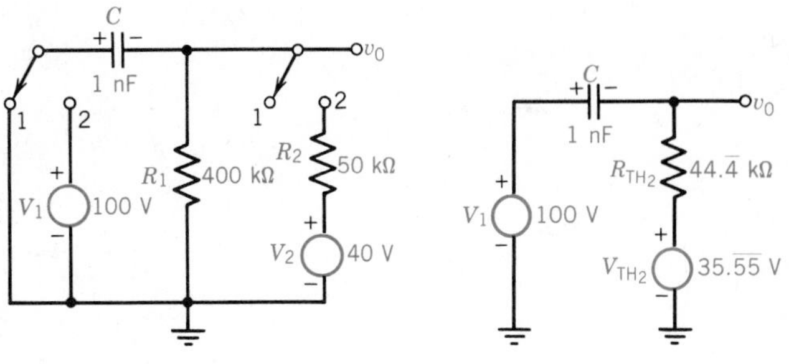

a. Network
b. Equivalent circuit for position 2

Figure 10-20 Example 10-31

Solution

In position 1 the circuit is simple. The capacitor is discharging via a 400-kΩ resistor; $\tau_1 = R_1C = (400\ \text{k}\Omega)(1\ \text{nF}) = 400\ \mu\text{s}$. In position 2, however, we have a very complex dual voltage supply circuit. We must first determine the Thévenin equivalent voltage and resistance seen by C and V_1 to the *right* of the capacitor, C. The equivalent circuit is shown in Fig. 10-20b for position 2, where $V_{TH_2} = \dfrac{400\ \text{k}}{450\ \text{k}} \times 40\ \text{V} = 35.\overline{55}\ \text{V}$ *opposing* V_1 and $R_{TH_2} = (400\ \text{k} \parallel 50\ \text{k}) = 44.\overline{4}\ \text{k}\Omega$. Consequently, $\tau_2 = R_{TH_2}C = (44.\overline{4}\ \text{k})(1\ \text{nF}) = 44.\overline{4}\ \mu\text{s}$. Note that this time constant implies that C charges at a much higher rate than it discharges, as shown by the following tabulation. The individual calculations are left as an exercise for the reader to verify.

Position and τ	2: $\tau_2 = 44.\overline{4}\ \mu$s						1: $\tau_1 = 400\ \mu$s				
t (μs)	0	0^+	100	200	300	400	400^+	500	600	700	800
a. v_C (V)	0	0	57.65	63.73	$64.\overline{44}$	$64.\overline{44}$	$64.\overline{44}$	50.19	39.09	30.44	$23.7\overline{1}$
b. i_C (μA)	0	1450	152.8	16.1	1.7	0.18	−161.1	−125.5	−97.7	−76.1	−59.3
c. v_R (V)	0	$64.\overline{4}$	6.79	$0.71\overline{5}$	$0.07\overline{5}$	0.008	$-64.\overline{44}$	−50.19	−39.09	−30.44	$-23.7\overline{1}$
d. v_0 (V)	0	100	42.35	36.27	35.63	35.56	$-64.4\overline{4}$	−50.19	−39.09	−30.44	$-23.7\overline{1}$

Position and τ	2: $\tau_2 = 44.\overline{4}\ \mu$s					1: $\tau_1 = 400\ \mu$s				
t (μs)	800^+	900	1000	1100	1200	1200^+	1300	1400	1500	1600
v_C (V)	23.71	36.44	40.28	40.69	40.73	40.73	31.72	24.7	19.24	14.98
i_C (μA)	916.5	96.6	10.2	1.07	0.113	−101.8	−79.3	−61.7	−48.1	−37.45
v_R (V)	$40.7\overline{3}$	4.29	0.45	$0.047\overline{5}$	0.005	−40.73	−31.72	−24.7	−19.24	−14.98
v_0 (V)	$76.2\overline{8}$	$39.84\overline{5}$	36	35.6	35.56	−40.73	−31.72	−24.7	−19.24	−14 98

e. The waveform of v_0 is plotted in **Fig. 10-21**

Let us examine the waveform of v_0 shown in Fig. 10-21 and see if it makes good sense at each of the switching transition points. This is reasonable since v_0 was obtained in the tabulation by calculating v_C, i_C, and v_R as steps toward finding v_0. If v_0 is correct, then all the other values are correct.

1. At time $t = 0^+$ the switch is first thrown to position 2 (Fig. 10-20b), but the capacitor, C, is uncharged and v_C (which cannot change) equals zero. The current of 1.45 mA that flows in the $44.\overline{4}$-kΩ resistance produces a v_R drop of 64.4 V, which adds to the 35.5 V equivalent symbolized as V_{TH_2}. This produces an initial voltage of 100 V for v_0. We might also say, alternatively, that since $v_C = 0$, $v_0 = v_C + V_1 = 0 + 100 = 100$ V in Fig. 10-20b.

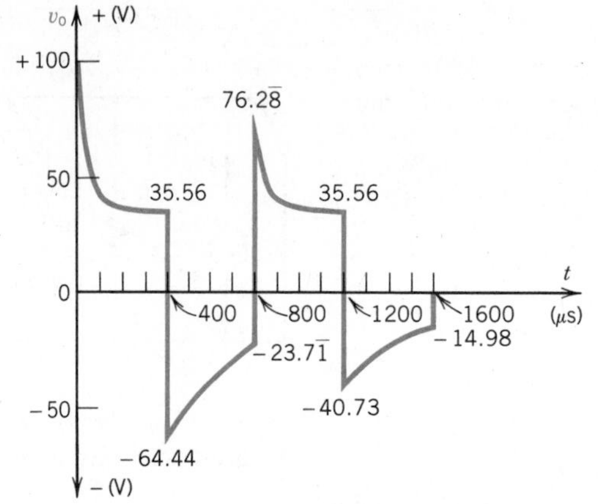

Figure 10-21 Waveform of output voltage, v_O, in Ex. 10-31 for two successive switching cycles

2. In position 2, at time $t = 400$ μs, the capacitor C is completely charged to $(100 - 35.\overline{5})$ V or $64.\overline{4}$ V because of its small τ_2 compared to the period of 400 μs. Since the current is (again) zero, there is no drop across R_{TH_2}, so $v_O = 35.\overline{5}$ V in Fig. 10-20b. Alternatively $V_1 - v_C = 100 - 64.\overline{4} = 35.\overline{5}\text{ V} = v_O$.
3. In position 1, at $t = 400^+$ μs, the capacitor is still charged to 64.4 V as in step 2 above. But the capacitor is charged NEGATIVELY with respect to terminal v_O and the 400-kΩ discharge resistor with respect to ground. Consequently, this makes $v_O = -64.\overline{4}$ V initially and in 400 μs (or $1\tau_1$) the capacitor voltage decays to $-23.7\overline{1}$ V, as shown in Fig. 10-21.
4. At time $t = 800^+$ μs, when the switch is returned to position 2, C is now charged to $23.7\overline{1}$ V in opposition to V_1 in Fig. 10-20b. Therefore $v_O = V_1 - v_C = 100 - 23.7\overline{1} = 76.2\overline{8}$ V.
5. At time $t = 1200$ μs in position 2, because of the relatively short τ_2 compared to its duration period, v_O again decays to $35.\overline{5}$ V, as it did in step 2, because $i_C \cong 0$. BUT THIS TIME, $v_C = 40.73$ V (compared to $64.\overline{4}$ V in step 2).
6. At $t = 1200^+$ μs, in position 1, $v_C = -40.73$ V is applied to R_1 of 400 kΩ. After $1\tau_1$ in position 1 this voltage decays to approximately -15 V.

This analysis shows that our values of v_O are both reasonable and correct. From the data tabulated in the solution of Ex. 10-31 we should also note the following.

1. As noted in the conclusion to Ex. 10-29, the data of Ex. 10-31 show that i_C is of opposite polarity to v_C during discharges (in position 1) and of the same polarity during charge (in position 2).
2. v_O is of EQUAL and opposite polarity to v_C during discharge, in position 1.
3. $v_O = V - v_C$ during charge in position 2. Alternatively, $v_O = V_{TH_2} + i_C R_{TH_2}$ in this position during charge.
4. Once again, we have shown the extreme power of our method of "bookkeeping," generally, by tabulating changes in values at various time intervals, and particularly in showing final and initial conditions at *crucial* time intervals *whenever switching transitions occur.*

10-11 GLOSSARY OF TERMS USED

Exponential function Function of the form $y = K\varepsilon^{st}$ or $y = K(1 - \varepsilon^{-st})$, where K and s are constants and may be real or complex. The graphical plot of an exponential function is always *nonlinear*.

Final conditions Steady-state conditions reached after a transient change in voltage and/or current from the initial conditions.

Initial conditions Values of voltage or current present in a circuit before a change is introduced, either by switching or by changing input voltage or current.

Time constant, τ Value, measured in seconds, for a given transient circuit that determines the time for currents and voltages to reach steady-state conditions in response to any change in input voltage and/or current. The τ for an RC circuit is $\tau = RC$ seconds, i.e., the time to rise to 63.2% of a final value or the time to decay to 36.8% of an initial value. Alternatively, the time constant of an RC circuit may be defined as the time for capacitor voltage to rise to applied voltage, assuming it continues to rise at its initial rate of change and the capacitor is initially uncharged.

Transient That part of the variation in a variable which ultimately disappears during the transition from one steady-state condition to another.

Transient circuit Circuit in which any change in applied voltage (or current) produces transient changes in voltage (and/or current) for a period of time before settling to a steady state.

Transient decay Linear or nonlinear waveform of voltage or current in approaching zero.

Transient period Time required for voltages and currents to reach final steady-state values in response to changes in input voltage and/or current. For RC circuits, the transient period is usually five time constants (5τ).

Transient response Behavior of voltages across and currents in the various components in a transient circuit.

Transient rise Linear or nonlinear waveform of voltage or current during its approach to some finite steady-state value, either positive or negative.

Universal time-constant chart Graph of transient decay for the function $y = 1\varepsilon^{-x}$ and graph of the transient rise for the function $y = 1 - \varepsilon^{-x}$. The universal time constant chart is useful in solving problems involving exponential functions.

10-12 PROBLEMS

Secs. 10-1 to 10-4

10-1 A 1-MΩ resistor is connected in series with an uncharged 10-μF capacitor across a 100-V source at time $t = 0^+$ s. Calculate the

a. Circuit charging time constant, τ_c
b. Instantaneous circuit current, i_C, at $t = 0^+$ s
c. Final value of circuit current when all transients disappear
d. Time required to reach final steady-state values
e. Initial voltage across the capacitor, v_{C_0}, at $t = 0^+$ s
f. Final voltage across the capacitor when all transients disappear
g. Initial rate of change of voltage across the capacitor, dv_{C_0}/dt, at $t = 0^+$ s
h. Initial rate of change of current, di_{C_0}/dt
i. Final rates of change of voltage and current transients at final steady state, after $5\tau_c$

10-2 A completely discharged 40-μF capacitor is connected in series with a 0.4-MΩ resistor to a 400-V supply at $t = 0^+$ s. Repeat all parts of Prob. 10-1.

10-3 For the circuit and conditions given in Prob. 10-1, calculate at time $t = 10$ s

a. Capacitor voltage, v_C
b. Charge on the capacitor, q_C
c. Current in the capacitor and in the circuit, i_C
d. Voltage across the 1-MΩ resistor, v_R
e. Using KVL, verify your answers to parts **(a)** and **(d)**

10-4 Using the circuit and conditions given in Prob. 10-2, calculate all parts of Prob. 10-3 at $t = 20$ seconds.

10-5 Given the network shown in **Fig. 10-22**, at $t = 0^+$ s switch S is thrown to position 1, where it remains for 60 ms, when it is thrown to position 2. If it remains in position 2 for 30 ms before being returned to its OFF position, calculate the

a. Voltage across C at the end of 60 ms
b. Current in C and R at the end of 60 ms
c. Voltage across C at the instant the switch is connected to position 2
d. Current in the capacitor at the instant the switch is connected to position 2
e. Voltage across the capacitor after 30 ms in position 2
f. Current in the capacitor after 30 ms in position 2
g. Voltage across the capacitor when the switch is returned to the OFF position
h. Current in the capacitor when the switch is returned to the OFF position

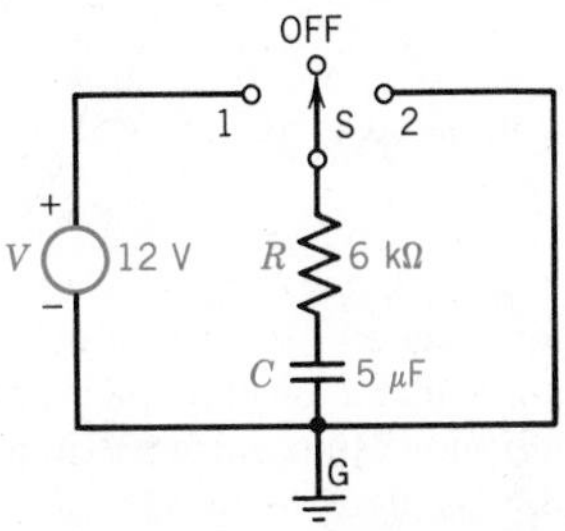

Figure 10-22 Problem 10-5

10-6 For the network shown in **Fig. 10-23**, assume

1. Capacitor C is initially discharged ($v_C = 0$)
2. S is thrown to position 1 at $t = 0^+$ s

Calculate

a. v_C at $t = 35$ ms
b. i_C at $t = 70$ ms
c. v_O at $t = 60$ ms

3. S is now thrown to position 2 after $t = 980$ ms; at the end of 1 s, calculate

d. v_C
e. i_C
f. v_O

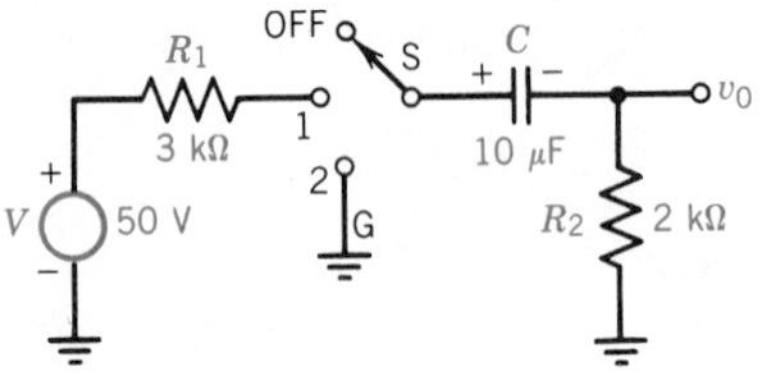

Figure 10-23 Problem 10-6

10-7 The insulation (leakage) resistance of a 60-μF capacitor is 60 MΩ, in parallel with the capacitor (Fig. 9-4). When the capacitor is fully charged from a 300-V supply and placed on a shelf, calculate the

a. Time in hours for the capacitor to fully discharge if removed from the supply at $t = 0^+$ s
b. Time in seconds for v_C to reach 100 V
c. Initial rate of change of voltage in volts/minute
d. Initial charge on the capacitor at $t = 0^+$ s
e. Voltage across the capacitor when $t = 42$ min
f. Charge on the capacitor under conditions in (**e**)
g. Energy stored in the dielectric initially
h. Energy stored under conditions in (**b**)
i. Energy stored under conditions in (**e**)

10-8 A 10-μF capacitor of *unknown* leakage resistance in parallel with the capacitor (Fig. 9-4) is fully charged to 300 V and removed from the supply. It is immediately connected across a 10-MΩ resistor. After 1 min, the capacitor voltage, v_C, is 100 V. Calculate the

a. Discharge time constant, τ_d
b. Leakage resistance of the capacitor

Assume R_s in Fig. 9-4 is negligible compared to the 10-MΩ resistor used in this test.

Secs. 10-5 to 10-9

10-9 A discotheque has a stroboscopic xenon flasher that contains a power supply of 300 V used to charge a 14-μF capacitor. It is desired that the "strobe" fire the xenon lamp every 5 s, in a typical "relaxation oscillator" circuit. Calculate

a. The value of series resistance, R_s, required, assuming the discharge of the capacitor occurs in zero time and the lamp fires at 200 V and extinguishes at 20 V
b. The energy stored in the capacitor when fully charged to 200 V
c. The instantaneous power dissipated in the lamp if the discharge time is 50 μs
d. Draw a curve showing the relaxation oscillator waveform for the first 16 s of operation
e. Ignoring the initial transient (from the origin), calculate the time between successive 5-s flashes for the voltage across the capacitor to reach 150 V

10-10 For the network shown in **Fig. 10-24** and values given therein, calculate the

a. Maximum (final) voltage to which C can possibly charge (use Thévenin's theorem)
b. Circuit time constant, τ_1, in position 1 (use Thévenin's theorem for total equivalent resistance seen by C)
c. Voltage across the capacitor at $t = 20$ ms in position 1
d. Output voltage, v_O, at $t = 20$ ms in position 1
e. Rate of change of v_C at $t = 20$ ms in position 1
f. Circuit current, i_C, at $t = 20$ ms by three independent methods

If the switch moves to position 2 at $t = 80$ ms, calculate the

g. Initial current and voltage across the capacitor in position 2 at $t = 80^+$ ms
h. v_C, v_O, and i_C at $t = 100$ ms
i. Draw output waveforms for v_C, v_O, and i_C from $t = 0^+$ to $t = 100$ ms, making any additional calculations to completely draw all three waveforms

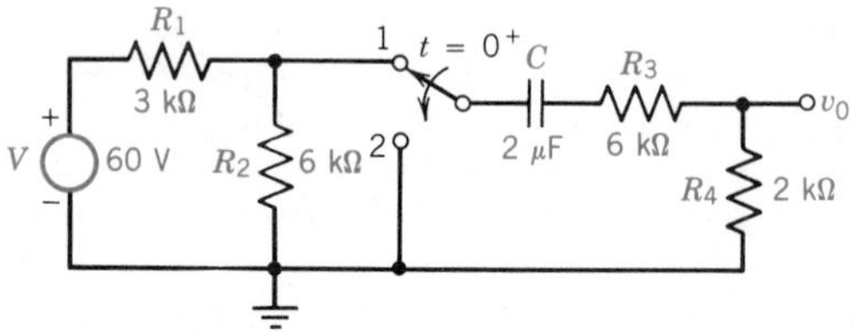

Figure 10-24 Problem 10-10

10-11 Given the network shown in **Fig. 10-25** and the values shown, the action of switch S is as follows. After remaining OFF indefinitely, at $t = 0^+$ it moves to position 1, where it remains for 3 ms. It moves to position 2, where it remains for 3 ms. It then moves back to position 1, where it remains for 3 ms. It then returns to OFF, where it remains indefinitely. Calculate

a. τ_0, τ_1, and τ_2 for each of the three switch positions
b. v_O at the end of 3 ms in position 1
c. v_O at the end of 6 ms in position 2
d. v_O at the end of 9 ms in position 1
e. v_O at the end of 12 ms in the OFF position
f. Tabulate the above data and draw the output waveform, showing v_O versus time for the first 15 ms of the switching sequence

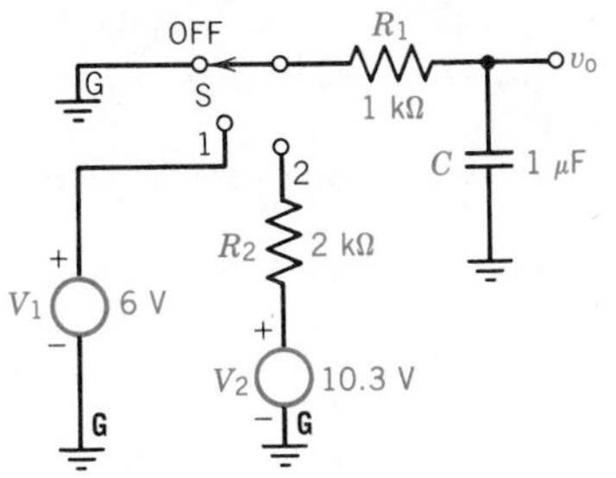

Figure 10-25 Problem 10-11

10-12 For the waveform drawn in Prob. 10-11f, calculate the

a. Time when $v_0 = 0$ when first switched to position 2
b. Time when $v_0 = 0$ when switch returns to position 1 from position 2
c. Minimum time when $v_0 = 0$ after returning from position 1 to the OFF position
d. Include these times in the tabulation you made in Prob. 10-11f
e. Calculate the capacitor current i_C at each of the following critical times in your tabulation: 0^+, 3, 3^+, 4.32, 6, 6^+, 6.55, 9, 9^+, 12, 14, and 15 ms. Use values of v_C and Ohm's law
f. On the same horizontal axis of the output waveform drawn in Prob. 10-11f, also draw the waveform of i_C at each of the recorded times given in (e)
g. Explain why it is necessary to solve for two values of i_C at times $t = 3$, 6, and 9 ms when there is only one value of v_C (and v_0) at these times

10-13 Verify values of i_C calculated in Prob. 10-12e, using the relation $i_C = C\,dv_C/dt$ for each instant of time up to 12 ms. Hint: use Eqs. (10-6b) and (10-12) appropriately to find dv_C/dt in your tabulation of Prob. 10-12d.

10-14 A neon-lamp emergency street flasher is shown in **Fig. 10-26** in its simplest form. The lamp fires at 90 V and is extinguished at 20 V when the switch is ON, and it continues to oscillate between these two values. Assume the resistance of the neon lamp is 50 MΩ when extinguished and 1 kΩ during its firing period. Calculate the

a. Time constant of the circuit during the capacitor charge period, τ_c
b. Time constant of the circuit during firing of the neon lamp, τ_f
c. Charging time, in seconds, during which charge on C builds up to a firing voltage of 90 V from 20 V
d. Firing time, in seconds, during which the neon flasher is discharged
e. Period of oscillation of the relaxation oscillator in seconds
f. Approximate period of oscillation, using the equation $T_0 = \tau_c \ln\left(\dfrac{V - v_d}{V - v_f}\right)$
g. Compare your answer in (**f**) with that in (**e**) and account for differences
h. Approximate value of R to produce an oscillation of one flash per second

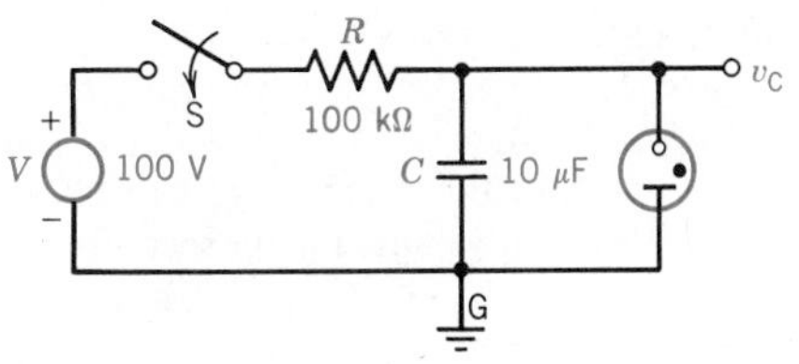

Figure 10-26 Problem 10-14

i. Percent change in the period of oscillation if the battery supply drops to 99 V (i.e., a change of 1% due to battery degradation)
j. Draw the waveform of v_C for three cycles of operation after the switch is first closed, showing values of times at 20 V, 90 V, and decay time during firing of the neon flasher

10-15 From the waveform drawn in Prob. 10-14j, calculate the

a. Rate of rise of voltage across the capacitor at the beginning of the first cycle
b. Rate of rise of voltage for the remaining cycles
c. Account for differences between (**a**) and (**b**)
d. Rate of decay of voltage during all firing periods
e. Explain why the rate of decay of voltage is much higher than the rate of rise of voltage during the charge period

10-16 An electronic photographic high-speed flash unit contains a bank of 10 capacitors in parallel, each of 5 μF, which are charged from a dc power supply of 1500 V through a 100-kΩ resistor. When the capacitors are fully charged they ignite a small neon lamp, indicating that the unit is ready for capacitor discharge through a xenon flash lamp. The camera shutter contacts fire the xenon flash lamp and discharge the capacitors *fully* in 100 μs. Calculate the

a. Time for the neon lamp to light after the flash unit has been discharged
b. Energy in joules stored in the capacitor bank when the neon lamp goes ON
c. Average power dissipated in the lamp during discharge
d. Combined resistance of the shutter contacts and xenon lamp during firing
e. Explain why the average power dissipated in the xenon lamp is so high

10-17 Given the circuit of **Fig. 10-27** with the capacitor C initially discharged, S is thrown to position 1, where it remains for 3 s, and is then thrown to position 2 for 1 s. Calculate

a. V_{TH_1}, equivalent Thévenin voltage at point 1
b. R_{TH_1}, equivalent Thévenin resistance at point 1
c. τ_1, circuit time constant when S is in position 1
d. τ_2, circuit time constant when S is in position 2
e. Using time intervals of 0.5 s, tabulate v_C, i_C, and v_0 for the total elapsed time of 4 s
f. Time in position 1 to reach an output voltage of 5 V
g. Minimum time in position 1 to reach its steady-state output voltage
h. Minimum time in which output voltage is returned to zero in position 2

i. Time for the output voltage to reach 3 V in position 2
j. Value of output voltage 0.7 s after the switch is thrown to position 2 (elapsed time of 3.5 s)
k. Sketch the v_C, i_C, and v_0 values tabulated in (e) as well as those calculated in (f), (i), and (j) versus time in seconds

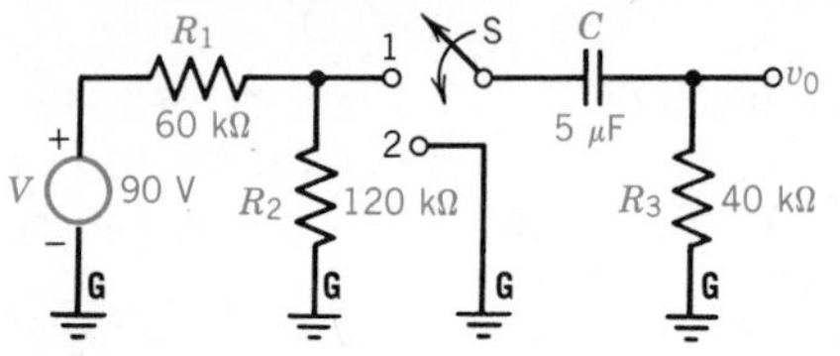

Figure 10-27 Problem 10-17

Sec. 10-10

10-18 In **Fig. 10-28**, after being indefinitely in position 1, switch S moves to position 2 at $t = 0^+$ s and remains there for 1 ms. It then moves to position 3 for 1 ms, after which it returns to position 1, where it remains indefinitely.

a. Using intervals of 0.25 ms, tabulate v_0 and i_C for this switching sequence from $t = 0$ to $t = 3$ ms
b. Sketch the waveforms of v_0 and i_C tabulated in (a)

Based on your tabulation and sketches, calculate the

c. Time, in milliseconds, when $v_0 = 0$ crosses the zero axis in position 3. (Hint: 1 ms $< t <$ 1.25 ms)
d. Rate of change of voltage when the switch is first thrown to position 3 at $t = 1^+$ ms
e. Capacitor current, based on your answer to (d), using $i_C = C\,dv_{C_0}/dt$, at $t = 1^+$ ms

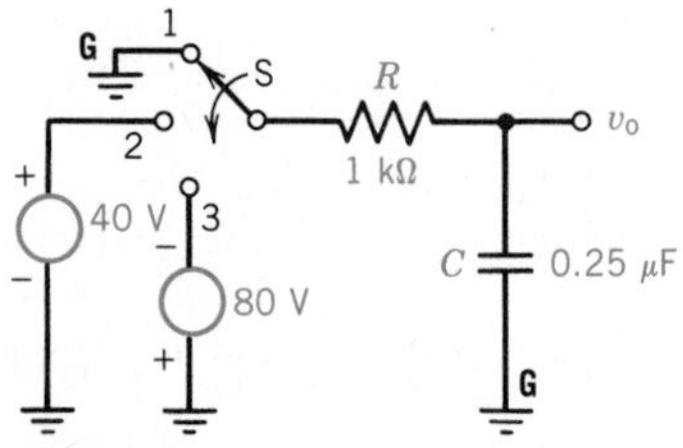

Figure 10-28 Problem 10-18

10-19 Given the circuit shown in **Fig. 10-29**, calculate the

a. Equivalent Thévenin voltage with respect to ground at terminal 1
b. Equivalent Thévenin resistance at terminal 1 before S is closed
c. Complete circuit resistance R_t when S closes at $t = 0^+$ s
d. Circuit time constant when S closes at $t = 0^+$ s
e. Time required for all circuit transients (voltage, current, power, and energy) to disappear
f. Power dissipated in R_t (p_{R_t}) at $t = 0^+$ s; power delivered by the 50- and 10-V sources p_T; and power delivered to C (p_C) at $t = 0^+$ s, in milliwatts (Hint: see Fig. 10-12)
g. Power delivered to the capacitor, p_C; power dissipated in R_t (p_{R_t}); and total power furnished by the source, p_T, when $t = 1.3863$ s after S is closed
h. Energy stored in the capacitor after all transients have disappeared (in millijoules)
i. Energy stored in the capacitor when $t = 1.3863$ s after S is closed, using Eq. (10-22)
j. Repeat (i) by finding v_C at $t = 1.3863$ s and using $w_C = C \cdot (v_C)^2/2$

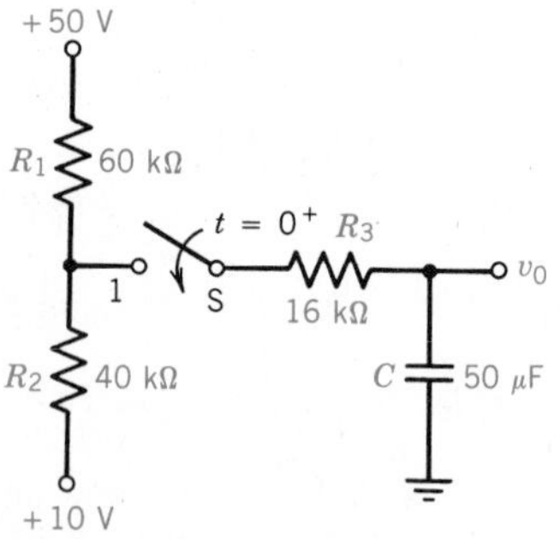

Figure 10-29 Problem 10-19

10-20 Using Fig. 10-12 as a reference and your calculations in solving Prob. 10-19, explain

a. The values and relations between p_{R_t}, p_T, and p_C at $t = 0^+$ (Prob. 10-19f)
b. Repeat (a) at $t = 1.3863$ s (Prob. 10-19g)
c. The relation between v_C and i_C at $t = 1.3863$ s in terms of final and/or initial values and why p_C is a maximum value at this time
d. At what time, in seconds, is the difference between p_C and p_{R_t} a maximum?
e. Using the time found in (d), calculate the values of p_C and p_{R_t}, respectively, and the ratio p_C/p_{R_t}
f. Using the values of p_C and p_{R_t} found in (e), calculate the difference $\Delta p_C - p_{R_t}$
g. Verify the value found in (f), using Eq. (10-20)

10-21 Using the methods shown in Exs. 10-17 and 10-18 as well as those of Appendix B-3.2, derive expressions for the time t in Fig. 10-12 at which

a. $p_C = 1.5p_R$
b. $p_R = 20p_C$
c. $p_R = 10p_C$

10-22 An equation (derived by the author) for instantly finding the time in Fig. 10-12 for any ratio p_C/p_R (either integral or fractional) is $t = \ln(1 + D)\tau_c$ where D is the decimal equivalent of p_C/p_R. Using this equation, repeat all three parts of Prob. 10-21 to verify its validity.

10-23 A capacitor voltage, v_C, rises exponentially from its original value of $V_0 = -150$ V at $t = 0$ to a final value of $V_f = +250$ V. If the circuit time constant is 500 μs, calculate the value of t when the voltage is

a. -50 V
b. 0
c. 100 V
d. 200 V
e. 240 V

10-24 From the data and solutions in Prob. 10-23, calculate the rate of change of voltage when v_C is

a. -150 V
b. -50 V
c. 0
d. 100 V
e. 240 V

10-25 In the circuit shown in **Fig. 10-30**, C is fully discharged in position 1 and $R = 0$. A switching cycle consists of a transition from DPDT position 1 to position 2 and back to 1 again. The total time duration of a switching cycle is 800 μs and the time in each position is 400 μs. If the switching cycle is repeated twice,

a. Calculate τ and final steady-state conditions in *each* switch position
b. Calculate and tabulate v_C, i_C, and v_O in 100-μs intervals in each switch position for a total elapsed time of 1600 μs
c. Plot v_O against time for two switching cycles
d. Explain why v_O decays to zero on positive half-cycles but decays to only about -15 V on negative half-cycles

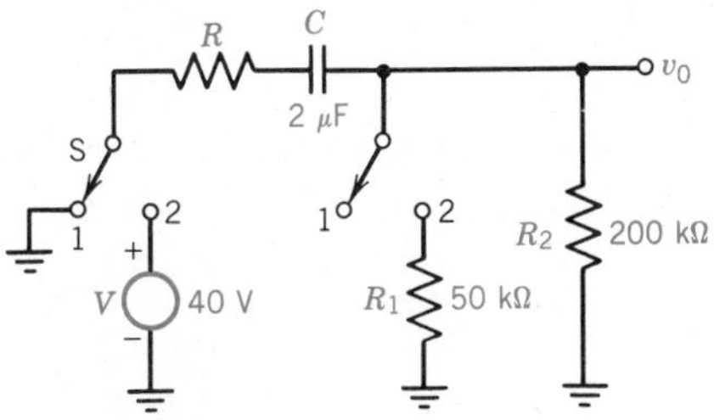

Figure 10-30 Problem 10-25

10-26 Repeat all parts of Prob. 10-25 if $R = 40$ kΩ in Fig. 10-30.

d. Explain why the positive and negative excursions of v_O are now smaller than in Prob. 10-25

10-27 In the circuit shown in **Fig. 10-31**, S has been OFF for a considerable time. At $t = 0$, S goes to position 1, where it stays for 40 ms, and then to position 2, where it stays for 40 ms. It then returns to position 1, where the cycle is repeated a second time. Calculate

a. The equivalent Thévenin resistance seen by capacitor C in positions 1 and 2
b. The time constant of the RC circuit in positions 1 and 2
c. The Thévenin voltage to which the capacitor can fully charge in position 1
d. The Thévenin voltage to which the capacitor can fully charge in position 2
e. Tabulate v_O versus time (in 10-ms intervals) for an elapsed time of 160 ms (or two complete switching cycles) and plot the waveform of v_O versus time
f. Calculate the time for v_O to be zero when switch is first thrown to position 2
g. Repeat (**f**) when the switch is restored to position 1 for a second time
h. If C is reduced in Fig. 10-31 to 50 nF, predict the steady-state values in switch positions 1 and 2. Explain.

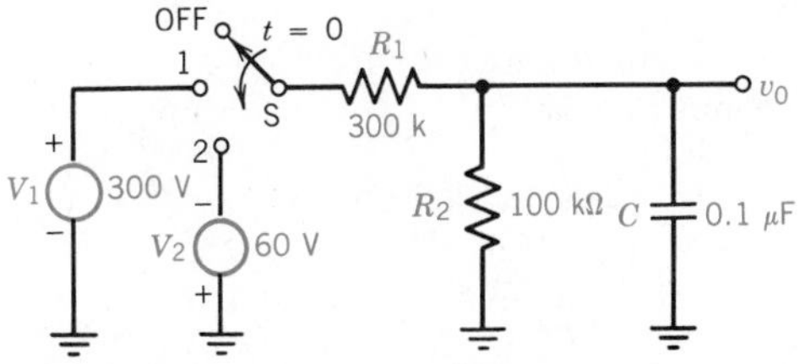

Figure 10-31 Problem 10-27

10-28 In the circuit of **Fig. 10-32**, when S is in the OFF position, the capacitor is completely discharged. The switching cycle alternates S between positions 1 and 2 for a duration of 6 ms in each position. Calculate

a. The RC time constant, τ_1, in position 1
b. The RC time constant, τ_2, in position 2
c. v_C, i_C, and v_O in 2-ms intervals in each switch position for three complete cycles (a total duration of 36 ms) and tabulate the results of such calculations
d. Plot v_C and v_O against time for three complete switching cycles, each on a separate axis for comparison
e. Explain why v_O is not a function of τ_1 in position 1 but is a function of τ_2 and capacitor current in position 2
f. Explain why v_O has a constant magnitude in position 1

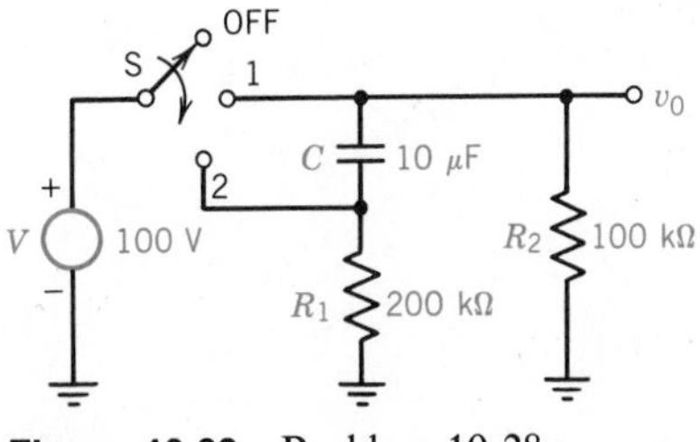

Figure 10-32 Problem 10-28

10-29 From your tabulation of Prob. 10-28c, calculate the times when $v_C = 0$

a. (t_1), after first being switched from position 1 to position 2
b. (t_2), after first being switched from position 2 to position 1
c. (t_3), between 18 and 19 ms
d. (t_4), between 24 and 26 ms

10-13 ANSWERS

10-1 a 10 s b 100 μA c 0 d 50 s e 0 f 100 V g 10 V/s h −10 μA/s i 0
10-2 a 16 s b 1 mA c 0 d 80 ms e 0 f 400 V g 25 V/s h −62.5 μA/s i 0
10-3 a 63.21 V b 632.1 μC c 36.79 μA d 36.79 V e 100 V
10-4 a 285.4 V b 11.42 mC c 286.5 μA d 114.6 V e 400 V
10-5 a 10.38 V b 0.2707 mA c 10.38 V d −1.729 mA e 3.817 V f −0.6361 mA g 3.817 V h 0
10-6 a 25.17 V b 2.466 mA c 6.024 V d 18.39 V e −9.197 mA f −18.39 V
10-7 a 5 h b 1.099 h c −5 V/min d 18 mC e 149 V f 8.94 mC g 2.7 J h 0.3 J i $0.\bar{6}$ J
10-8 a 54.61 s b 12.03 MΩ
10-9 a 346.9 kΩ b 280 mJ c 5.6 kW e 3.03 s
10-10 a 40 V b 20 ms c 25.285 V d 2.943 V e 735.8 V/s f 1.4715 mA g −4.908 mA, 39.27 V h 11.25 V, −1.406 mA
10-11 a 1 ms, 1 ms, 3 ms b −5.7 V c 4.414 V d −5.221 V e −0.26 V
10-12 a 4.32 ms b 6.5514 ms c 14 ms e see 10-13 below
10-13 −6 mA, −0.3 mA; $5.\bar{3}$ mA, 3.435 mA, 1.962 mA; −10.414 mA, −6 mA, −0.5185 mA; 5.221 mA, 0.26 mA
10-14 a 1 s b 10 ms c 2.08 s d 15 ms e 2.095 s f 2.08 s h 48.1 kΩ i 4.43%
10-15 a 100 V/s b 80 V/s d −9 kV/s
10-16 a 25 s b 56.25 J c 562.5 kW d 0.4 Ω
10-17 a 60 V b 40 kΩ c 0.4 s d 0.2 s f 717 ms g 2 s h 1 s i 3.6 s j −1.81 V
10-18 c 1.1 ms d −477.2 kV/s e −119.3 mA
10-19 a 26 V b 24 kΩ c 40 kΩ d 2 s e 10 s f 16.9 mW g 4.225 mW, 8.45 mW h 16.9 mJ i 4.225 mJ j 13 V, 4.225 mJ
10-20 a (16.9 mW) b (4.225 mW) d 2.773 s e 3.169 mW; 1.056 mW; 3/1 f 2.113 mW g 2.113 mW
10-23 a 143.8 μs b 235 μs c 490.4 μs d 1.04 ms e 1.844 ms
10-24 a 800 kV/s b 600 kV/s c 500 kV/s d 300 kV/s e 20 kV/s
10-25 a 400 μs; 80 μs
10-26 a 480 μs; 160 μs
10-27 a 75 kΩ b 7.5 ms c 75 V d −15 V f 53.4 ms g 81.33 ms
10-28 a 2 ms b 1 ms
10-29 a 6.68 ms b 13.38 ms c 18.64 ms d 25.38 ms

CHAPTER 11

Magnetism and Magnetic Circuits

We have learned that electric charges *at rest* produce forces between them which may be quantified by Coulomb's law (Sec. 2-1b). We have also learned that electric charges *in motion* result in an electric current (Sec. 2-5). In this chapter we concern ourselves with magnetism and how it is related to and produced by electricity.

11-1 ELECTRICITY AND MAGNETISM

Electricity has come to mean the study of the properties of electric charges either *at rest* or *in motion*. The term *magnetism* is a study of the properties and results of electric charges that are *exclusively in motion*.

A magnetic field (or *magnetic force in space*) cannot exist without charge motion. Consequently, magnetism is the branch of the study of electricity that deals with forces produced by and between *moving charges only*. Since magnetic fields are due to charges in motion, any time a wire carries current, it is surrounded by a magnetic field (Oersted's law). Devices such as inductors, electromagnetic relays, electromagnets, generators, and motors all operate on the principle of *magnetic force fields* produced by currents in conductors.[1]

Minerals containing iron, nickel, or cobalt were known from ancient times to possess the properties of magnetism and were called *lodestones*. Such *natural magnets* could attract and hold small pieces of iron and steel. Lodestones were also floated on bits of wood and used as magnetic compasses by ancient Chinese mariners. The properties of bits of magnetized iron were known to Thales of Miletus, who noted that like ends of magnetic bars repelled each other and unlike ends attracted each other. In 1600, William Gilbert, in his book *De Magnete*, published a collection of the then-known facts about magnets including his own extensive and careful experiments. He first used the terms north and south poles, noting that *opposite* poles *attract* and *like* poles *repel* each other.

[1] Only three types of fields are known to science. The first is the *gravitational field* between bodies as a result of their relative masses as quantified by Newton (see Ex. 2-1). The second is the *electric field* produced by stationary charged bodies as quantified by Coulomb (Eq. 2-1). The last is the *magnetic field* produced by charges in motion, which is the subject of this chapter. A *field* is a force in space.

But if magnetic fields are only and always produced by charges in motion, several questions arise that we must answer:

1. How can a steel or iron conductor exhibit a magnetic field when no current is flowing in it?
2. Why don't other metals (such as copper or aluminum) exhibit a magnetic field in the absence of current?

Such questions puzzled scientists for many years and were answered only as a result of significant research in nuclear physics in the early part of the 20th century. The answer seems obvious now. Since all matter is composed of charged particles (Sec. 2-1), the magnetic field within an atom is due to the moving electrons spinning about the nucleus. The electrons are believed to have two motions, similar to the motion of the earth about the sun. One motion is the electron orbit around the nucleus. The other motion is the *electron spin about its own axis.*

In most atoms of most materials, all electron spins are *paired* or *compensated.* For each electron spinning in a clockwise (CW) direction, there is a second electron producing a compensating spin in a counterclockwise (CCW) direction. The same paired or compensated situation applies to the orbital motion of the electrons as well. Consequently, the net magnetic field as a result of such charge motion is zero and such materials are said to be *nonmagnetic.*

In atoms of *magnetic* materials, however, there is an *unbalance* of electron spins in the atomic structure. For example, each atom of iron has four uncompensated electrons spinning in the same direction (in the third valence shell). Consequently, whenever a piece of iron is subjected to an external magnetic field, a torque is produced, forcing each uncompensated electron to align its axis of spin parallel to the magnetic field.[2] The magnetic field produced by the magnetized material *increases* and *adds* to the applied external magnetic field. A group of atoms that have been so aligned is called a *magnetic domain*; each such domain is approximately 50 μm in length and just visible to an opaque metallurgical microscope.

Figure 11-1a shows that the domains in an unmagnetized sample of magnetic material are randomly oriented but that *within* the domain all *unpaired* electron spins are *aligned.* When an external magnetic field is applied, as shown in Fig. 11-b, two effects occur: (1) *more and more* domains are *aligned parallel* to the external magnetic field and (2) as this occurs, the individual random domain walls are broken down and the domains *increase in size.* When the external magnetic field is increased further, ultimately almost all the domains are aligned and the magnetic material is said to be *saturated* (i.e., no further increase in magnetism is possible due to magnetic alignment.)

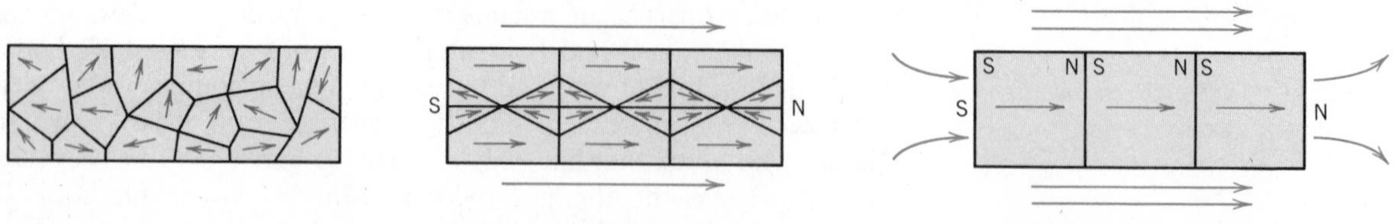

a. Domains in an unmagnetized bar of iron

b. Growth of domains in presence of external magnetic field

c. Saturated material showing domain growth

Figure 11-1 Magnetization of a magnetic material by domain alignment and domain growth

[2] If a conductor is carrying current in a magnetic field, the torque on the conductor is always in such a direction as to align the *field* of the conductor in the same direction as the applied magnetic field (Sec. 11-15.2, Eq. 11-23).

When the applied magnetic field is removed, *some* magnetic materials (like steel) *retain* their domain alignment. Such magnetic materials are said to have a *high retentivity*.[3] Other magnetic materials (like soft iron) return to their original random domain alignment whenever the magnetizing force is removed. Such magnetic materials are said to have a *low retentivity*.[4] Soft iron is used in the cores of large lifting electromagnets so that when the current is switched OFF, the attracted material (iron or steel ingots or bars) is released from the magnet.

11-2 MAGNETIC FIELDS SURROUNDING PERMANENT MAGNETS

The magnetic field surrounding a bar magnet may be traced or patterned by using a small magnetic compass or placing a piece of paper above the magnet and sprinkling iron filings over it. Continuous tapping on the paper will cause the needle-shaped filings to align themselves *parallel* to the magnet's magnetic field. **Figure 11-2a** shows such a magnetic field pattern. In reality the field envelopes the bar magnet in a 3-dimensional shape similar to a football.

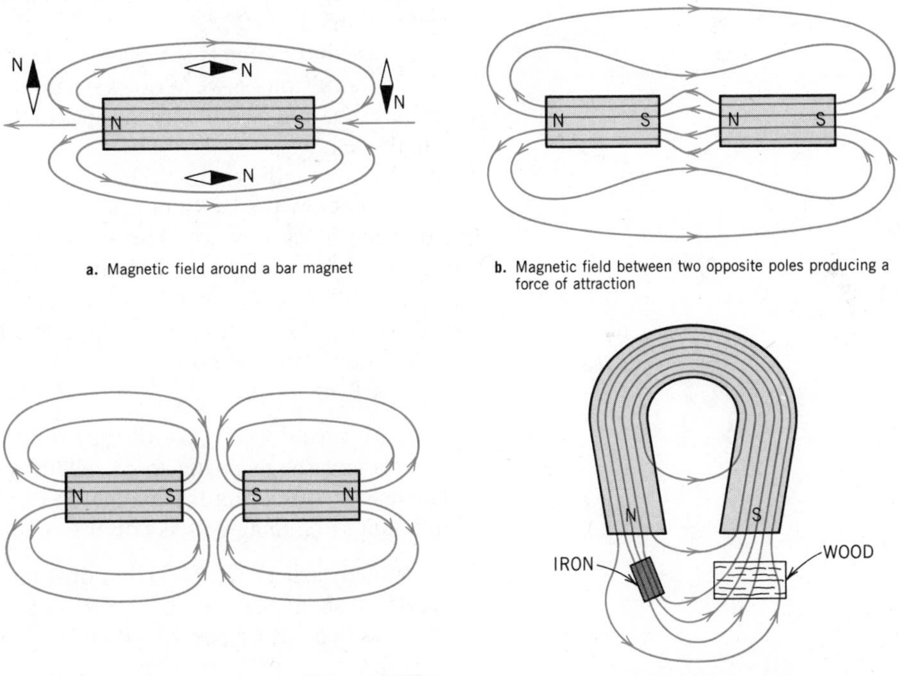

a. Magnetic field around a bar magnet

b. Magnetic field between two opposite poles producing a force of attraction

c. Magnetic field between like poles producing a force of repulsion

d. Magnetic field around a horseshoe magnet

Figure 11-2 Fields surrounding permanent magnets

With respect to Fig. 11-2a, the following points may be noted:

1. The magnetic field is represented by lines of force similar to electric lines of force used for the electric field (Fig. 2-1).

[3] This property is similar to the property of *plasticity*. The material remains in the same shape after the deforming force is removed.

[4] This property is similar to the property of *elasticity*. The material returns to its original state after the deforming force is removed.

2. The magnetic field (unlike an electric field) is continuous, forming a closed loop through the magnetic material. (See point 5 below.)
3. Magnetic lines of force always leave a north pole and enter a south pole.
4. Lines of force in the SAME direction REPEL each other, as noted from the field pattern when the field leaves the north pole. Note that the lines diverge away from each other.
5. Lines of force cannot intersect.
6. Lines of force tend to form the shortest possible magnetic circuit.
7. With respect to the earth's geography, the north pole of the bar magnet, and the compass as well, would face geographic north if the bar were suspended from a string. For this reason, the north pole of a compass is sometimes called a *north-seeking* pole.

Figure 11-2b shows what would happen if a bar magnet were broken in half and the two halves separated. This might be inferred from the domain structure of Fig. 11-1c if the magnet were broken along the lines of the domain boundary, creating separate bar magnets, with opposite poles facing each other. The field pattern shown in Fig. 11-2b attempts to duplicate the field pattern of Fig. 11-2a, but its lines have been "stretched" as a result of separation. If the magnets are free to move, they will come together to form the pattern of Fig. 11-2a with the shortest possible lines.

If one of the bar magnets is turned around so that two *like* poles face each other (Fig. 11-2c) repulsion between poles occurs and the magnets tend to separate. The farther they separate, the shorter the lines of the fields surrounding each magnet. (See point 6 above.)

The field surrounding a horseshoe magnet is shown in Fig. 11-2d. Here, an iron bar is placed below the north pole and a block of wood below the south pole. Observe that the block of wood does *not* disturb the field distribution. The presence of iron, however, tends to concentrate and shorten the field. If the iron bar is free to move, it will be pulled between the poles of the magnet, where it produces the *shortest* possible field path for the magnetic flux.

Two points should be noted with respect to the field patterns shown in Figs. 11-1 and 11-2:

1. Lines of force do not exist (if they did, we would see them). They serve only to represent the *magnetic force field* in space surrounding the bar magnet.
2. The field surrounding the magnet is three-dimensional (as is the bar magnet) and should be imagined as coming out of and also entering the paper plane.

It was noted earlier (Sec. 11-1) that only materials that exhibited uncompensated (third valence shell) electron spins possessed the property of magnetism. All solids may be classified with respect to their magnetic properties as follows.

1. *Ferromagnetic*: unbalanced third shell materials that, in the presence of a magnetic field, enhance or increase the total magnetic flux greatly. These are the materials that may be used to manufacture permanent magnets. They include pure iron, nickel, and cobalt or alloys of these metals with paramagnetic materials. Also included in this class are ferrites (ceramics containing iron oxides).
2. *Paramagnetic*: materials that, in the presence of a magnetic field, increase the total flux only slightly. Platinum and aluminum are paramagnetic.
3. *Diamagnetic*: materials that exhibit a slight reduction in total flux when placed in the magnetic field. Copper and silver exhibit slight diamagnetic properties.
4. *Nonmagnetic*: materials that have no effect on a magnetic field in air or vacuum. When placed in air, they neither increase nor decrease the magnetic

field. Almost all *organic* materials are *nonmagnetic*: plastics, rubber, oil, alcohol, and so forth.

11-3 ELECTROMAGNETISM

Even the best ferromagnetic materials are limited in the strength of the magnetic field they can produce. The discovery of electromagnetism by Oersted in 1820 paved the way for the development of powerful magnetic fields, in comparison to conventional ferromagnetics. It also provided a means for controlling the magnitude of the magnetic field and/or switching it ON or OFF by controlling the current to which the magnetic field is proportional.

Oersted's discovery is represented in **Fig. 11-3a.**[5] He showed that whenever current flows through a conductor, a magnetic field is produced around the conductor and the strength of that magnetic field is proportional to the current. The direction of the magnetic field, shown by the compass bearings in Fig. 11-3a, may be predicted from the current direction. Viewed from the *top*, current is *entering* the conductor in Fig. 11-3a. Since current is *leaving* the observer, we designate this direction by the tail (or barb) of an arrow. Similarly, viewed from the bottom, current is leaving the conductor in Fig. 11-3a but *approaching* the observer's eye. We designate this direction by the point of an arrow (a circle with a dot in it), as shown in Fig. 11-3a.

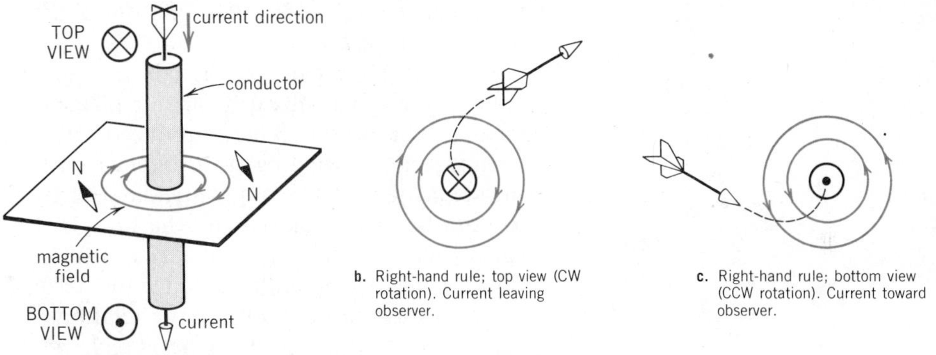

a. Finding field direction using magnetic compasses

b. Right-hand rule; top view (CW rotation). Current leaving observer.

c. Right-hand rule; bottom view (CCW rotation). Current toward observer.

Figure 11-3 Oersted's law for direction of magnetic field around a current-carrying conductor and right-hand rule convention

The relationship between current and its magnetic field is summarized in Figs. 11-3b and 11-3c. Using the *right* hand, let your *thumb* represent the *current direction.* The remaining fingers will point the direction of the magnetic field at right angles to the thumb. In Fig. 11-3b the current direction is into the paper plane and the field around the current has a CW direction. In Fig. 11-3c the current direction is out of the paper plane toward you and the field has a CCW direction.

The significance of Oersted's discovery becomes more obvious if we now wind the straight wire into the shape of a spiral or helix (or a coil), consisting of several loops or *turns*, as shown in **Fig. 11-4a**, wound on a nonmagnetic core. This produces the *same* magnetic effect as that of a bar magnet! How is this possible? Con-

[5] Section 11-15 gives the equation quantifying Oersted's law.

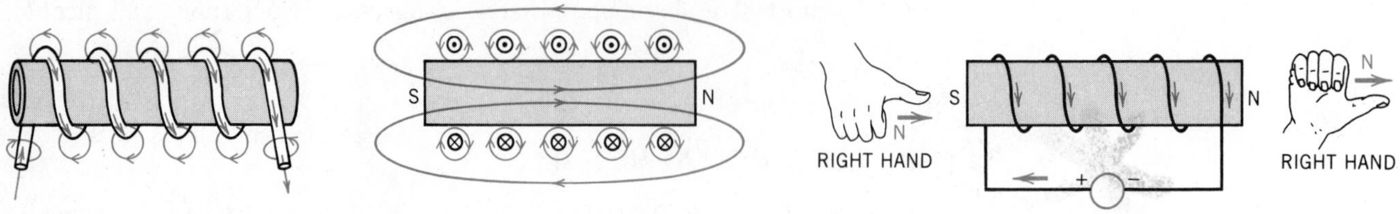

a. Current and magnetic field directions in a helix

b. Cross section of coil showing current and field directions in electromagnet

c. Right-hand rule for electromagnetic polarity

Figure 11-4 Magnetic field produced by a current-carrying coil

sider the top of each loop in Fig. 11-4a and observe that for the given current direction the magnetic field of each loop is CCW. Similarly, the bottom of each loop, because of current direction, exhibits a CW field direction. The helix of Fig. 11-4a is shown in cross section in Fig. 11-4b. This cross-sectional view enables us to see how each turn of the coil, carrying the same current, produces a greater *resultant* flux in the *same* direction. The five conductors in Fig. 11-4b are producing a resultant CCW field at the top of the coil, whereas those below are producing a resultant CW field. The net effect is to produce a field pattern much the same as that of the bar magnet in Fig. 11-1c.

Again, a right-hand rule may be used to predict the direction of the magnetic field produced by a coil, provided we know the direction in which the coil is wound on its core, as shown in Fig. 11-4c. The fingernails of the right hand are now used to designate current direction, and the thumb (this time)[6] points in the direction of the north pole.

It should be noted that the coil of Fig. 11-4a was wound on a nonmagnetic hollow core (made of cardboard or plastic). If a *magnetic* material is inserted in the *core*, the magnetic field produced by the coil is *increased* greatly. We have already seen that winding the straight current-carrying wire into turns also increases the magnetic field. Consequently, the strength of the magnetic field is proportional to both the core material and the turns of the coil. Finally, if the current in the coil is *increased*, the magnetic flux is also *increased*. But by how much? We have no way of knowing until we define our terms and our units of measurement and begin to solve problems involving relations expressed in our equations.

As of this writing, most engineers and scientists in the United States and almost all other parts of the world use the SI system. A table of conversion factors to other magnetic systems of units is given in Table 11-1 and in Appendices A-3Q1 to A-3Q4. The SI will be used throughout this discussion because the use of two or more systems simultaneously (when learning new material) only adds confusion and detracts from a basic understanding of the concepts presented.

11-4 DEFINITIONS OF MAGNETIC TERMS, QUANTITIES, SYMBOLS, AND UNITS OF MEASUREMENT

Magnetic quantities, like those used in electric circuits, are related to each other. Consequently, in defining magnetic quantities, equations will also be given that express these quantities in terms of other quantities (previously or later defined). The more important magnetic terms are now defined below.

[6] Note that this conventional right-hand rule is opposite to that used in Fig. 11-3, where the thumb designates current and the fingers designate the field direction in a straight wire or the cross section of a single conductor (Fig. 11-4b).

11-4.1 Magnetic Circuit

This is the minimal region containing essentially all the magnetic flux or a region at whose surface the magnetic induction is tangential. The magnetic circuit for a closed (transformer) core is shown in **Fig. 11-5a**. Note that the magnetic circuit is a closed path whose average length is indicated by the dashed line. Figure 11-5b shows the four parallel magnetic circuits of a four-pole dynamo. Each magnetic circuit has an average length indicated by the dashed line.

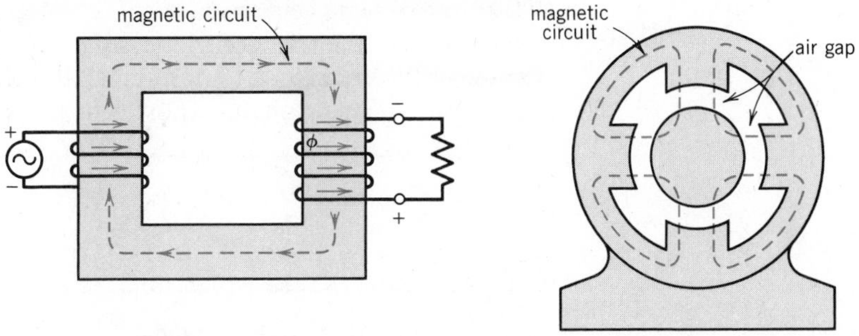

a. Transformer magnetic circuit—single path

b. Dynamo magnetic circuits—4 parallel paths

Figure 11-5 Magnetic circuits

11-4.2 Magnetomotive Force, $\mathcal{F}$

This is the line integral of the magnetizing force (magnetic field intensity) around a closed magnetic circuit, measured in ampere-turns (At):

$$\mathcal{F} \equiv IN = \mathbf{H}l \qquad \text{ampere-turns (At)} \tag{11-1}$$

where I is the current in the coil producing magnetomotive force (MMF)
N is the number of turns of the coil producing MMF

Magnetomotive force (MMF) is the magnetic analog of EMF in an electric circuit.

11-4.3 Reluctance, $\mathcal{R}$

This is the ratio of the magnetomotive force (MMF) to the magnetic flux, ϕ, through any cross section of the magnetic circuit.[7] In the magnetic circuit, reluctance represents the total opposition to MMF in producing a magnetic field. Factors affecting reluctance are

$$\mathcal{R} = \frac{\mathcal{F}}{\phi} = \frac{l}{\mu A} \qquad \text{(ampere-turns per weber) (At/Wb)} \tag{11-2}$$

where l is the average length of the magnetic circuit in meters (m)
A is the cross-sectional area of the magnetic circuit in square meters (m^2)

[7] Reluctance in the magnetic circuit is the analog of resistance in the electric circuit: R is the ratio of electromotive force to current, i.e., $R = V/I = \rho l/A = l/\sigma A$. Observe, therefore, that permeability, μ, is the analog of conductivity, σ.

μ is the absolute magnetic permeability of the material of the magnetic circuit, in henrys per meter (H/m) or teslas per ampere-turn per meter $\left(\frac{\text{T}}{\text{At/m}}\right)$

11-4.4 Magnetic Flux, ϕ

This is a magnetic field in space between two opposite poles (or other regions over which magnetic forces occur) measured in webers. Magnetic flux is the magnetic analog of current in an electric circuit.[8] It is the net effect or result of MMF applied to a given magnetic circuit whose reluctance is $\mathscr{R}$, or

$$\phi = \frac{\mathscr{F}}{\mathscr{R}} = \frac{IN}{\mathscr{R}} = \frac{IN\mu A}{l} \quad \text{(webers) (Wb)} \qquad (11\text{-}3)$$

where all terms have been defined above

11-4.5 Magnetic Flux Density, B

This is the magnetic flux or field passing through a unit area of the magnetic field normal to the direction of the magnetic force measured in units of webers per square meter (Wb/m^2) or

$$\mathbf{B} = \frac{\phi}{A} \quad \text{(webers per square meter) (Wb/m}^2\text{)} \qquad (11\text{-}4)$$

where all terms have been defined above

Magnetic flux density, **B**, is the magnetic analog of the electrical flux density, D (Eq. 2-3) and/or the current density, $J = I/A$ (Eq. 2-9).

11-4.6 Magnetic Field Intensity, H

This is the magnetizing force or magnetic field strength or force that produces flux density at a given point in a magnetic circuit, measured in ampere-turns per meter (At/m).

$$\mathbf{H} = \frac{\mathscr{F}}{l} = \frac{IN}{l} = \frac{\mathbf{B}}{\mu} \quad \text{(ampere-turns/meter)} \qquad (11\text{-}5)$$

Magnetic field intensity, **H**, is the magnetic analog of electrical field intensity, **E**, defined in Sec. 2-2.[9]

---

[8] In the electric circuit, $I = V/R$. In the magnetic circuit $\phi = \mathscr{F}/\mathscr{R}$.

[9] In the electric circuit, $\mathbf{E} = V/d$. In the magnetic circuit, $\mathbf{H} = \mathscr{F}/l$.

11-4.7 Absolute Permeability, μ

This is a measure of the ease of creating flux in a material. It is the ratio of magnetic flux density produced in a given medium to the magnetizing force producing it, measured in units of henrys per meter (H/m).

$$\mu = \frac{\mathbf{B}}{\mathbf{H}} = \mu_r \mu_0 \qquad \text{(henrys per meter)} \qquad (11\text{-}6)$$

where **B** and **H** have been defined above

μ_0 is the permeability of free space $= 4\pi \times 10^{-7}$ H/m

μ_r is the relative permeability, defined in Sec. 11-4.8

11-4.8 Relative Permeability, μ_r

This is the ratio of the magnetic flux density produced in a medium to that produced in a vacuum for the same dimensions by the same magnetizing force. (For air, $\mu_r = 1$.) Mathematically, μ_r is a unitless ratio or[10]

$$\mu_r = \frac{\mu}{\mu_0} \qquad \text{(unitless)} \qquad (11\text{-}7)$$

11-4.9 Permeance, $\mathscr{P}$

This is the ratio of magnetic flux, ϕ, to the magnetomotive force through any cross section of the magnetic circuit. Actually it represents the reciprocal of reluctance or[11]

$$\mathscr{P} = \frac{\phi}{\mathscr{F}} = \frac{\mu A}{l} = \frac{1}{\mathscr{R}} \qquad \text{(webers per ampere turn) (Wb/At)} \qquad (11\text{-}8)$$

Table 11-1 presents a summary of these physical quantities, their units in the SI, and conversion factors to the CGS and English systems of units. The reader must be aware that in dealing with magnetic circuits we must learn a *new vocabulary of magnetic terms.* Tables 11-1 and 11-2, as well as the sections that follow, are designed to help the reader learn this new vocabulary. Our ultimate aim in learning these new terms and physical quantities is to use them to perform magnetic circuit calculations.

[10] In the electric circuit, $\varepsilon_r = \varepsilon/\varepsilon_0$ (Eq. 9-4). In the magnetic circuit $\mu_r = \mu/\mu_0$. Both are unitless ratios that are analogous to specific gravity ρ_r.

[11] In the electric circuit, $G = 1/R$. In the magnetic circuit, $\mathscr{P} = 1/\mathscr{R}$.

Table 11-1 Magnetic Physical Quantities in the SI and Their Conversion Factors

Physical Quantity	Quantity Symbol	Eq. No.	SI Unit	SI Unit Abbreviation	CGS Unit (abbreviation)	CGS Conversion Factor	English Unit (Abbreviation)	Conversion Factor
Magnetomotive force	$\mathscr{F}$	11-1	ampere-turn	At	gilbert (Gb)	$1.257 \frac{\text{Gb}}{\text{At}}$	At	1
Reluctance	$\mathscr{R}$	11-2	ampere-turn per weber	At/Wb	(At/Mx)	$10^{-8} \frac{\text{At/Mx}}{\text{At/Wb}}$	At/line	$10^{-8} \frac{\text{At}/\ell}{\text{At/Wb}}$
Magnetic flux	ϕ	11-3	weber	Wb	maxwell (Mx)	10^8 Mx/Wb	line (ℓ)	10^8 ℓ/Wb
Magnetic flux density	**B**	11-4	tesla or weber per square meter	T or Wb/m^2 (1 T = 1 Wb/m^2)	gauss (G)	$10^4 \frac{\text{G}}{\text{T}}$	ℓ/in^2	6.452×10^4 ℓ/in^2/T
Magnetic field intensity	**H**	11-5	ampere-turns per meter	At/m	oersted (Oe)	$0.01257 \frac{\text{Oe}}{\text{At/m}}$	At/in	$0.0254 \frac{\text{At/in}}{\text{At/m}}$
Absolute permeability	μ	11-6	henry per meter or weber per ampere-meter	H/m (1 H/m = 1 Wb/A·m)	(G/Oe)	μG/Oe	$\frac{\text{lines}}{\text{At·in}}$	3.2 $\mu\ell$/At·in
Permeance	$\mathscr{P}$	11-8	weber per ampere-turn	Wb/At	(Mx/At)	$10^8 \frac{\text{Mx/At}}{\text{Wb/At}}$	$\frac{\text{lines}}{\text{At}}$	$10^8 \frac{\ell/\text{At}}{\text{Wb/At}}$

Table 11-2 Electrical Analogs of Magnetic Physical Quantities

Magnetic Quantity	Quantity Symbol	Unit of Measure	Unit Abbreviation	Defining Equation	Equation Number	Electrical Quantity	Quantity Symbol	Unit of Measure	Unit Abbreviation	Defining Equation
Magnetomotive force	$\mathscr{F}$	ampere-turns	At	$\mathscr{F} = IN$	11-1	Electromotive force	V	volt	V	$V = W/q = IR$
Reluctance	$\mathscr{R}$	ampere-turns per weber	At/Wb	$\mathscr{R} = l/\mu A$	11-2	Resistance	R	ohm	Ω	$R = l/\sigma A$
Magnetic flux	ϕ	weber	Wb	$\phi = IN\mu A/l$	11-3	Current	I	ampere	A	$I = V/R$
Magnetic flux density	**B**	tesla	T	$\mathbf{B} = \phi/A$	11-4	Current density	J	amperes per square meter	A/m^2	$J = I/A$
Magnetic field intensity	**H**	ampere-turns per meter	At/m	$\mathbf{H} = \mathscr{F}/l$	11-5	Electrical field intensity	**E**	volts per meter	V/m	$\mathbf{E} = V/l$
Absolute permeability	μ	henry per meter	H/m	$\mu = \mathbf{B}/\mathbf{H}$	11-6	Specific conductance	σ	siemens per meter	S/m	$\sigma = J/\mathbf{E}$
Permeance	$\mathscr{P}$	weber per ampere-turn	Wb/At	$\mathscr{P} = \mu A/l$	11-8	Conductance	G	siemens	S	$G = \sigma A/l$

11-5 ELECTRICAL ANALOGS OF MAGNETIC PHYSICAL QUANTITIES

Throughout the discussion of Sec. 11-4, where the physical quantities shown in Table 11-1 were introduced, footnotes were used to show their similarity to the electric circuit. Since the previous chapters dealt with electric circuits almost exclusively, they should be familiar ground to the reader by now. If we treat magnetic circuit calculations in the same way as electric circuit calculations, our insights into magnetic circuit relations are developed more quickly. Table 11-2 shows the electrical analogs of those magnetic physical quantities used in magnetic circuits and magnetic circuit calculations.

It appears that, given Table 11-2, we are sufficiently armed to solve all magnetic circuit problems with our experience and insights gained from solving electric circuit problems. Although we are able to solve a number of elementary problems, we are severely limited at this point because of one major problem. In electric circuits, over a normal range of temperature and applied voltage, the conductivity, σ, is fairly constant, and therefore the resistance is fairly constant.

But in magnetic circuits the analogous quantity, permeability, μ, varies drastically with the magnetizing field intensity, **H**.[12] This means it is *nonlinear* and that we must resort to curves (called **B**–**H** curves) to solve most of the magnetic circuit problems using ferromagnetic materials.

11-6 EFFECT OF MAGNETIC FIELD INTENSITY ON PERMEABILITY AND FLUX DENSITY

We can appreciate the foregoing remarks if we consider the **B**–**H** curve of a typical ferromagnetic material, shown in **Fig. 11-6b**. The independent variable is magnetic field intensity, **H**, for both the lower and upper curve. The **B**–**H** curve for a core made of ferromagnetic material is obtained using the method and circuitry shown alongside the **B**–**H** curve. As the potentiometer (Fig. 11-6c) is raised from its zero (or minimum) potential, the current (recorded by the ammeter) is increased. Thus the MMF (in ampere-turns) is easily obtained as the product of the primary turns, N, times the current, I, in the primary. For each potentiometer setting, the magnetic field intensity, **H**, is also computed from IN/l_{av}, where l is the average length of the magnetic core. Opening and closing switch S enables measurement of magnetic flux ϕ in the core by the deflection of fluxmeter, M, which records the magnetic flux in webers. The greater the flux in the core, the greater the deflection of the meter. The flux density, **B**, is then computed at each potentiometer setting from the defining relation $\mathbf{B} = \phi/A$, where A is the cross-sectional area of the ferromagnetic core.

But recall that, by definition, $\mu = \mathbf{B}/\mathbf{H}$ in Eq. (11-6). Consequently, from the **B**–**H** curve of Fig. 11-6b, it is possible to determine permeability, μ, from the *slope* of the **B**–**H** curve, where at any instant $\mu_\Delta = \Delta\mathbf{B}/\Delta\mathbf{H}$.

The initial permeability, μ_1, is shown in Fig. 11-6b by its slope as drawn. Note that as the magnetic field intensity increases, successive slopes (drawn as tangents to the **B**–**H** curve) increase sharply up to a value of x At/m, as in Fig. 11-6a.

Between x At/m and y At/m, the slope of the **B**–**H** curve is constant and the permeability, μ, is constant. This constant permeability is known as μ_{max} in Fig. 11-6a.

[12] In electric circuits, this would be equivalent to a change in resistivity (or conductivity) each time there is a change in voltage. In short, it would mean that resistance would become a voltage-dependent quantity.

a. Permeability curve

b. B H curve

c. Circuit for flux density measurement

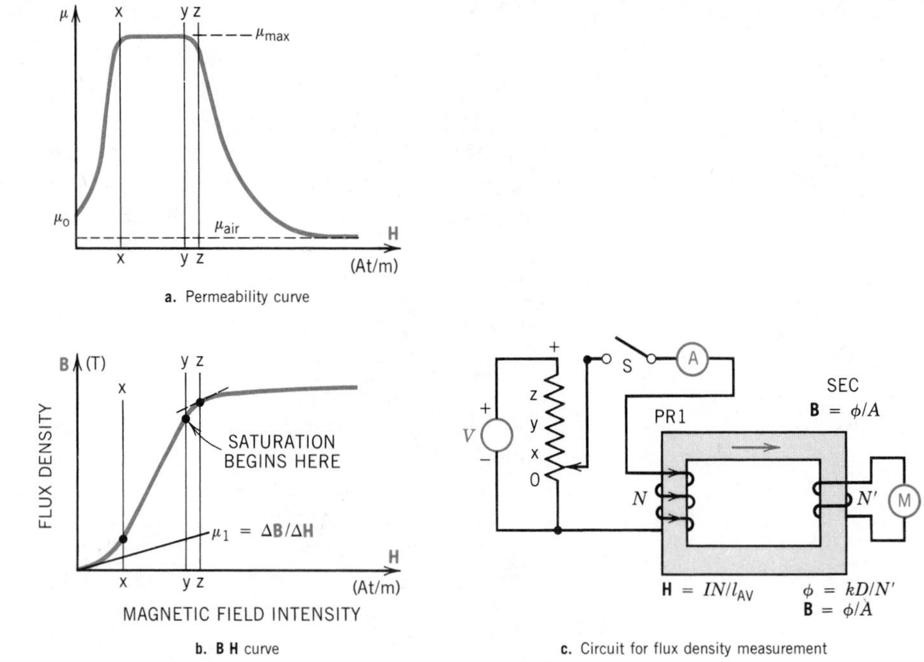

Figure 11-6 Variation in permeability with magnetic field intensity

Where the magnetic field intensity is y At/m, saturation of the magnetic core *begins*. This means that *almost* all the magnetic domains in the ferromagnetic material have been aligned. Consequently, any increase in MMF does NOT produce a proportional increase in flux and flux density. Therefore, μ begins to decrease, as shown by slopes drawn to the **B–H** curve. Observe that beyond z At/m, the permeability μ decreases sharply. At very high magnetic field intensities, *all* the magnetic domains are aligned. Any attempt to set up more magnetic lines of force in the core is repelled by the existing lines and the permeability drops radically to the permeability of air, as shown in Fig. 11-6a.

In summary, then, because of the wide variation in permeability of all ferromagnetics, it is necessary to solve magnetic circuit problems by using a specific **B–H** curve for the specific core material used.

Figure 11-7 shows typical **B–H** curves for three commonly used groups of materials: cast iron, cast steel, and sheet steel. Please note the following with respect to the **B–H** curves of Fig. 11-7.

1. In order of increasing permeability, we find cast iron with the lowest permeability and sheet steel with the highest.
2. Three sets of ordinates are found representing flux density, **B**. The innermost is expressed in SI in units of teslas or webers per square meter ($1\text{ T} = 1\text{ Wb/m}^2$). The next ordinate is expressed in CGS units of kilogauss ($10\text{ kG} = 1\text{ T}$). The outermost ordinate is expressed in English units of kilolines per square inch ($64.52\text{ k-line/in}^2 = 1\text{ T}$). This enables quick comparisons of units of flux density in the three most commonly used systems of measurement.
3. The same sets of systems appear on the abscissa for magnetic field intensity, **H**. The innermost is expressed in the SI in units of ampere-turns per meter (At/m). The centered scale is expressed in oersteds ($1\text{ Oe} = 1\text{ Gb/cm} = 79.58\text{ At/m}$), in the CGS system. The outer scale shows magnetic field intensity in the English system in units of ampere-turns per inch ($1\text{ At/in} = 39.37\text{ At/m}$).

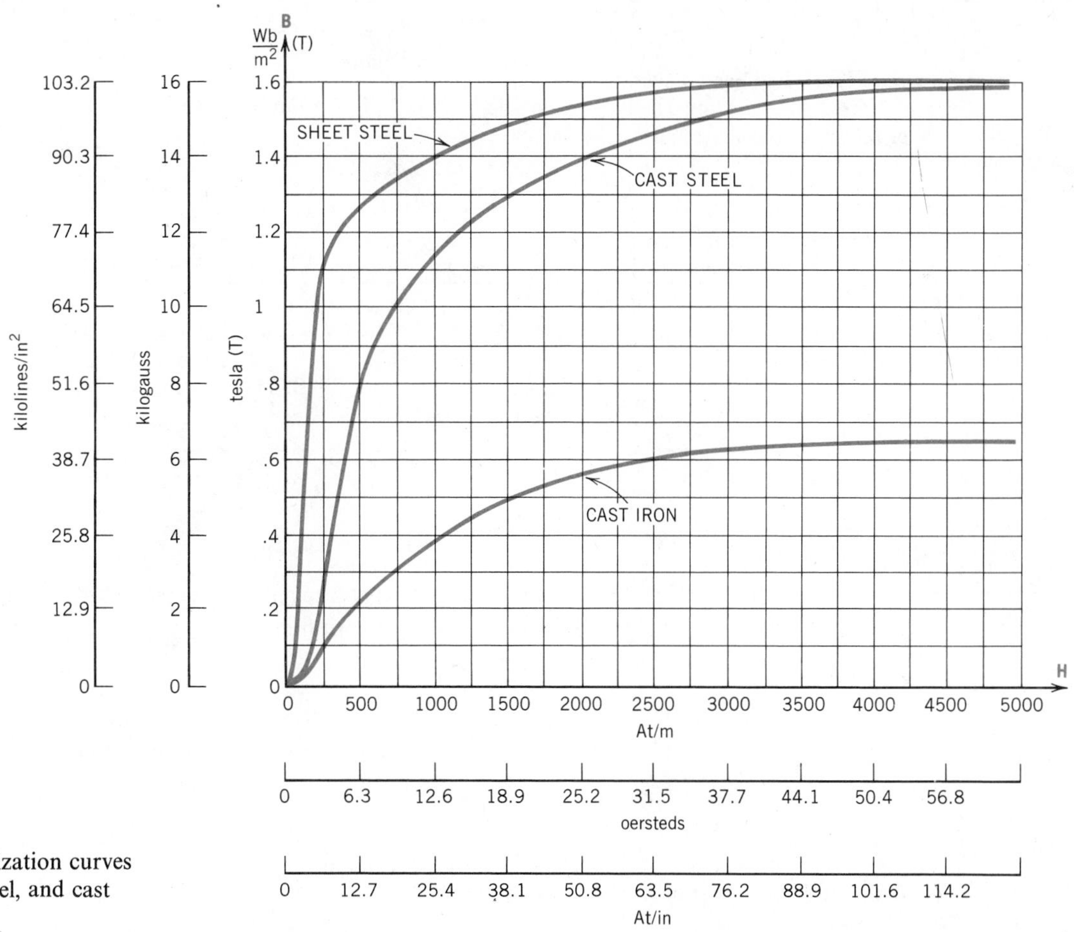

Figure 11-7 Magnetization curves for cast iron, sheet steel, and cast steel

4. The major squares of the graph are spaced at 0.5 cm × 0.5 cm (5 mm × 5 mm). This enables easy interpolation of values of **B** given values of **H** or vice versa, using the SI scales *only*.
5. The scales for the CGS and English systems should *not* be used for interpolation, however, since they are *not* decimally divided. If problems are stated in CGS or English units, it is advisable to convert them to SI *before* obtaining scale readings by interpolation.
6. All three scales show maximum permeability at very low values of **B** and **H**, and μ decreases as **H** increases. At extremely large values of **H** the permeability, μ, is a minimum because the slope of the curves is almost horizontal.
7. A special set of curves would be required for values of **H** below 250 At/m. Such curves are available from the manufacturers of special steels used in commercial application for transformer and dynamo cores.

11-7 USING B–H CURVES

At the simplest possible level, **B–H** curves may be used to find either the required magnetization (in ampere-turns per meter, At/m) to produce a given flux density (in webers per square meter, Wb/m² or tesla, T) or, alternatively, the resultant flux density (in tesla) produced by a given magnetization.

EXAMPLE 11-1

A toroidal core (see Fig. 11-10a) has an average length of 50 cm and no air gap. It is made of cast iron. The desired flux density is 0.6 tesla to be produced by a coil around the core of 500 turns. Calculate the

a. Required magnetic field intensity, **H**
b. Required ampere turns
c. Required current in the coil

Solution

a. From the **B–H** curve (Fig. 11-7) for cast iron,

$$\mathbf{B} = 0.6\ \text{T} \Leftrightarrow \mathbf{2500\ At/m} = \mathbf{H}$$

b. $\mathscr{F} = \mathbf{H} \times l = 2500\,\dfrac{\text{At}}{\text{m}} \times 0.5\ \text{m} = \mathbf{1250\ At}$ (11-1)

c. $I = IN/N = \mathscr{F}/N = 1250\ \text{At}/500\ \text{t} = \mathbf{2.5\ A}$

EXAMPLE 11-2

Given the core and coil of Ex. 11-1, calculate the flux density when the current is reduced to 2 A

Solution

$\mathscr{F} = IN = 2\ \text{A} \times 500\ \text{t} = 1000\ \text{At}$ (11-1)

$\mathbf{H} = \mathscr{F}/l = 1000\ \text{At}/0.5\ \text{m} = 2000\ \text{At/m}$ (11-5)

From the **B–H** curve (Fig. 11-7) for cast iron:

$$\mathbf{H} = 2000\ \text{At/m} \Leftrightarrow \mathbf{0.56\ T} = \mathbf{B}$$

11-7.1 Average Permeability, μ_{av}

The *average* (or static or dc) *permeability* is defined as the ratio of **B** to **H** at any given point on the magnetization curve for a given material, or

$$\mu_{av} = \frac{\mathbf{B}}{\mathbf{H}} \quad \text{henrys per meter (H/m) or webers per ampere-meter (Wb/A·m)} \qquad \text{(11-9)}$$

EXAMPLE 11-3

From the given data and the solution of Ex. 11-1, calculate the average permeability, μ_{av}

Solution

$$\mu_{av} = \frac{\mathbf{B}}{\mathbf{H}} = \frac{0.6\ \text{Wb/m}^2}{2500\ \text{At/m}} = \mathbf{240\ \mu Wb/A\cdot m} = \mathbf{240\ \mu H/m} \qquad \text{(11-9)}$$

EXAMPLE 11-4

From the given data and the solution of Ex. 11-2, calculate μ_{av}.

Solution

$$\mu_{av} = \frac{\mathbf{B}}{\mathbf{H}} = \frac{0.56\ \text{T}}{2000\ \text{At/m}} = \mathbf{280\ \mu Wb/A\cdot m} = \mathbf{280\ \mu H/m} \qquad \text{(11-9)}$$

Examples 11-3 and 11-4 verify point 6 made at the conclusion of Sec. 11-7 (i.e., that the *permeability* is *higher* at lower values of **B** and **H**, respectively, in Fig. 11-7.) Note that the permeability *decreased* from 280 to 240 μH/m as the flux density *increased* from 0.56 T to 0.6 T.

11-7.2 Incremental Permeability, μ_d

The *incremental* permeability, μ_d, is the "slope" permeability during a small change in either flux density or magnetic field intensity. It represents the slope produced by a tangent to the magnetization curve at any given point on the magnetization curve. The incremental permeability is defined as

$$\mu_d = \lim_{\Delta\mathbf{H}\to 0}\left(\frac{\Delta\mathbf{B}}{\Delta\mathbf{H}}\right) = \frac{d\mathbf{B}}{d\mathbf{H}} \quad \text{henrys per meter (H/m)} \qquad \text{(11-10)}$$

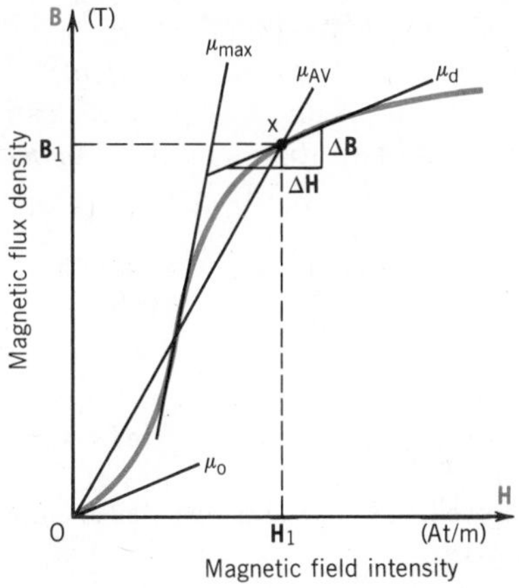

Figure 11-8 Distinctions among μ_{av}, μ_d, μ_O, and μ_{max}

The distinction between *incremental* permeability and *average* permeability is shown in **Fig. 11-8**. Consider point x on the magnetization curve. The average permeability at point x is $\mu_{av} = \mathbf{B}_1/\mathbf{H}_1$. But as shown in Fig. 11-8, μ_{av} is really the *slope* of the line drawn from the *origin*, O, to point x, or $\mu_{av} = OB_1/OH_1 = \mathbf{B}_1/\mathbf{H}_1$.

Now consider the *incremental permeability*, μ_d, at point x, which is found by drawing a tangent to point x on the magnetization curve. As defined in Eq. (11-10), $\mu_d = \Delta\mathbf{B}/\Delta\mathbf{H}$. Observe from Fig. 11-8 that the slope of μ_d is *smaller* than the slope of μ_{av}. Consequently the *average* permeability at point x is *higher* than the incremental permeability at the same point. The *most vertical* slope that can be drawn to the magnetization curve represents the *maximum* permeability, μ_{max}, as shown in Fig. 11-8. The initial permeability represents the *slope* of the tangent to the curve *at the origin*, as shown for μ_1 in Fig. 11-8.

From Fig. 11-8 we can draw the following conclusions:

1. All four values of permeability are found from their respective slopes to the curve as represented by the ratio $\Delta\mathbf{B}/\Delta\mathbf{H}$.
2. The slopes of average permeability (μ_{av}) and the initial permeability (μ_1) *must be drawn through the origin* to determine their values.
3. The slopes of maximum permeability (μ_{max}) and incremental permeability (μ_d) are *not necessarily* drawn through the origin but are found as *tangents* to magnetization curve *at any given point* on the **B–H** curve.
4. Values of μ_1 and μ_{av} are values of the *average permeability* since their slopes *go through the origin.*
5. Values of μ_{max} and μ_d are values of the *incremental permeability* because their slopes do not (necessarily) go through the origin.

EXAMPLE 11-5

Given the magnetization curve for cast steel at the point where the flux density is 1.3 T, calculate the

a. μ_{av}, average permeability
b. μ_d, incremental permeability
c. Given the curve for cast steel, calculate its maximum permeability, μ_{max}
d. List all three permeabilities in order of increasing value

Solution

a. $\mu_{av} = OB/OH = 1.3\text{ T}/1500\text{ At/m} = \mathbf{0.8\bar{6}\ mH/m}$ (11-9)

b. $\mu_d = d\mathbf{B}/d\mathbf{H} = \dfrac{(1.4 - 1.2)\text{ T}}{(2\text{ k} - 1\text{ k})\text{ At/m}}$
$= 0.2\text{ T}/1000\text{ At/m} = \mathbf{0.2\ mH/m}$

c. $\mu_{max} = d\mathbf{B}/d\mathbf{H} = \dfrac{(0.8 - 0.3)\text{ T}}{(500 - 250)\text{ At/m}}$
$= 0.5/250 = \mathbf{2\ mH/m}$ (11-10)

d. In order of increasing value: μ_d, μ_{av}, μ_{max}.

μ_{max} has the *highest* slope and μ_d has the *lowest* slope.

11-7.3 Relative Permeability, μ_r

Consider a coil wound on a nonmagnetic (plastic or cardboard) tube with air as its core. For a given current, area, number of turns, and length of core of the coil, the flux produced in the air may be calculated (Sec. 11-9). If we now insert a ferromagnetic material in the core, the flux increases tremendously, although the current in the coil and its turns is unchanged. Obviously, this increase in flux is due to the (much) greater permeability of ferromagnetics compared to an air core.

A measure of the improvement or increase in flux produced by any ferromagnetic material compared to the same magnetizing force in air is called the *relative permeability*, μ_r, defined in Sec. 11-4.8 and expressed in Eq. (11-7) as $\mu_r = \mu/\mu_O$, a unitless ratio.

μ_O is the permeability of free space, an important physical constant defined as

$\mu_0 = 4\pi \times 10^{-7}$ H/m. For all practical purposes, the permeability of air is the same as that of a vacuum. Therefore *the relative permeability of air*, μ_r, *is unity.*[13] Consequently, if we know the absolute permeability, μ, of any material, we can immediately determine how much better it is than that of air (or vacuum) by using Eq. 11-7.

EXAMPLE 11-6

Given the data of Ex. 11-1 and the solution of Ex. 11-3, calculate the

a. Relative permeability of cast iron compared to air
b. Factor by which the flux density is reduced if the cast iron core is removed from the coil
c. Flux density of the air without a cast iron core in the coil

Solution

a. $\mu_r = \mu/\mu_0 = (240\ \mu\text{H/m})/4\pi \times 10^{-7}$ H/m
$= \mathbf{190.98 \cong 191}$ (11-7)
Note that the permeability of the iron core is 191 times better than that of air.
b. If the iron core is removed, the flux and flux density are reduced by a factor of **1/191**
c. $\mathbf{B} = 0.6\ \text{T}/191 = \mathbf{3.14\ mT}$

From the foregoing we may conclude that relative permeability, μ_r, may be defined as a measure of the *relative ease* with which a magnetic field may be established in a given material *compared to air*, using the same magnetizing force.

11-8 AMPÈRE'S CIRCUITAL LAW OF MAGNETIC CIRCUITS

Table 11-2 shows that magnetomotive force (MMF) in magnetic circuits corresponds to electromotive force (EMF) in electric circuits. Recall that for a series circuit we summarized Kirchhoff's voltage law (KVL) as $\sum_{n=1}^{N} V_n = 0$; that is, the algebraic sum of all voltage rises and drops in any closed path in any electric circuit is zero.

Since MMF in the magnetic circuit is analogous to EMF in the electric circuit, we may infer *Ampère's circuital law of magnetic circuits* or

$$\sum_{n=1}^{N} \mathscr{F}_n = 0 \quad \text{or} \quad I_cN_c - I_1N_1 - I_2N_2 - I_3N_3 - \cdots - I_nN_n = 0 \qquad \mathbf{(11\text{-}11)}$$

where I_cN_c is the MMF produced by a coil of N_c turns carrying I_c amperes
I_1N_1 through I_nN_n are the respective MMF drops required to produce a desired flux[14] in their respective cross-sectional areas and lengths

Ampère's circuital law implies that for each given length of ferromagnetic material of given cross-sectional area, an equivalent MMF may be calculated to produce a desired flux in that material. The sum of the MMFs so computed will represent the total MMF that must be applied to the magnetic circuit to create the desired flux. The algebraic sum of MMFs in a closed magnetic circuit is zero (Eq. 11-11).

The significance of Ampère's (magnetic) circuital law is shown in Ex. 11-7, using a (hypothetical) closed core consisting of three different materials.

---

[13] As noted in the footnote to Eq. (11-7), *relative* permeability is analogous to both specific gravity and relative permittivity. All originate from the same philosophy and all are unitless ratios.

[14] Flux, ϕ, in webers in the magnetic circuit is analogous to *current* in the *electric series* circuit. Since current is the *same* in all parts of the *series electric* circuit, the magnetic flux ϕ is the *same* in all parts of a *series magnetic* circuit.

EXAMPLE 11-7

It is desired to set up a flux of 960 μWb in all parts of the core shown in **Fig. 11-9** having the dimensions given. Find the current necessary in the 300-turn coil to produce the desired flux by performing the following steps. Calculate

a. The flux density in teslas in each core portion
b. $\mathscr{F}$ for the sheet steel portion in ampere-turns
c. $\mathscr{F}$ for the cast steel portion in ampere-turns
d. $\mathscr{F}$ for the cast iron portion in ampere-turns
e. The total MMF required to set up the desired flux
f. The current in the coil

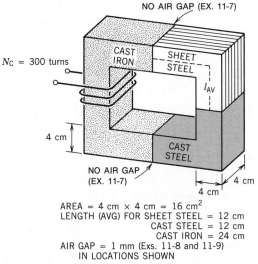

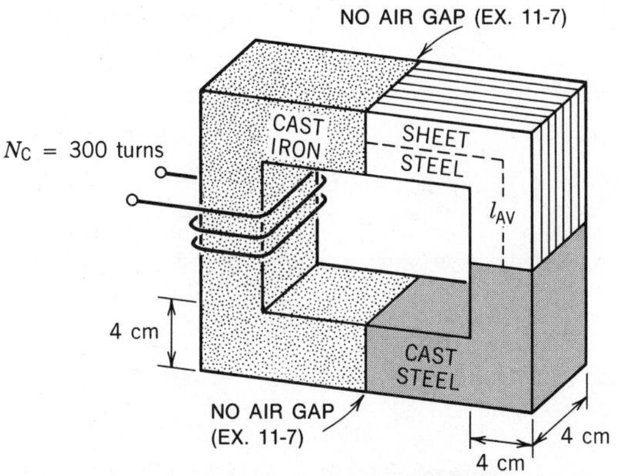

Figure 11-9 Examples 11-7 through 11-9 using hypothetical core

Solution

a. $\mathbf{B} = \phi/A = 960\ \mu\text{Wb}/(16\ \text{cm}^2 \times 10^{-4}\ \text{m}^2/\text{cm}^2)$
$= 0.6\ \text{Wb/m}^2 = \mathbf{0.6\ T}$ (11-4)

b. From the **B–H** curve for sheet steel,
$\mathbf{B} = 0.6\ \text{T} \Leftrightarrow 125\ \text{At/m} = \mathbf{H}$
$\mathscr{F}_1 = \mathbf{H} \times l = (125\ \text{At/m}) \times (0.12\ \text{m}) = \mathbf{15\ At}$ (11-1)

c. From the **B–H** curve for cast steel,
$\mathbf{B} = 0.6\ \text{T} \Leftrightarrow 400\ \text{At/m} = \mathbf{H}$
$\mathscr{F}_2 = \mathbf{H} \times l = (400\ \text{At/m}) \times (0.12\ \text{m}) = \mathbf{48\ At}$ (11-1)

d. From the **B–H** curve for cast iron,
$\mathbf{B} = 0.6\ \text{T} \Leftrightarrow 2500\ \text{At/m} = \mathbf{H}$
$\mathscr{F}_3 = \mathbf{H} \times l = (2500\ \text{At/m})(0.24\ \text{m}) = \mathbf{600\ At}$ (11-1)

e. Using Ampère's circuital law,
$I_c N_c = \mathscr{F}_1 + \mathscr{F}_2 + \mathscr{F}_3 = 15 + 48 + 600$
$= \mathbf{663\ At}$ (11-11)

f. $I_c = I_c N_c / N_c = 663\ \text{At}/300\ \text{t} = \mathbf{2.21\ A}$ (11-1)

11-9 MMFs IN AIR GAPS AND AIR-CORE SOLENOIDS

Up to now we have calculated the magnetomotive force (MMF) for *closed* ferromagnetic cores, such as are used in transformers, shown in Fig. 11-5a. But frequently we encounter air gaps, such as must occur in dynamos (Fig. 11-5b), meter movements (Fig. 11-18), and relays (Fig. 11-20). It is necessary to be able to determine the MMF required to overcome this air gap, using Ampère's circuital law, in order to set up a desired flux or flux density in the gap.

Visualize, if you can, what would happen if the cast iron portion of the magnetic circuit in Fig. 11-9 were separated from the sheet steel and cast steel pieces by an air gap of, say, 1 mm. In effect, there would be two air gaps in series of 1 mm each, or a total gap of 2 mm. Assuming that we wish to maintain the *same* flux in the core, what value of MMF is required to overcome this gap and how much must the current be increased to produce the same flux? (See Ex. 11-8.)

In attacking this problem we must make three basic assumptions:

1. We will assume that the area of the air gap is the same as the area of the ferromagnetic core material, (i.e., that no "fringing" occurs in the gap)
2. We will assume no leakage of magnetic flux from the gap, (i.e., the flux in the air gap is the same as the flux in the ferromagnetic core.)

(Both of these assumptions are reasonable, particularly when the gap is of relatively short length).

3. We will assume that the permeability of air is the same as the permeability of free space, that is, $\mu_{air} = \mu_0 = 4\pi \times 10^{-7}$ H/m

Using Eq. (11-6) then, we may write $\mu_0 = \mathbf{B}/\mathbf{H}$ and therefore

$$\mathbf{H}_{air} = \frac{(\mathbf{B}_{air})}{4\pi \times 10^{-7}} = 7.958 \times 10^5\ \mathbf{B}_{air} \qquad \text{(At/m)} \qquad (11\text{-}12)$$

and

$$\mathscr{F}_{air} = 7.958 \times 10^5\ \mathbf{B}_{air} \times l = 7.958 \times 10^5\ \phi_{air} \times \frac{l}{A} \qquad \text{(At)} \qquad (11\text{-}13)$$

where A is the cross-sectional area of the gap in square meters
l is the length of the gap in meters or average magnetic circuit length in air

---

EXAMPLE 11-8

Calculate the value of current in the 300-turn coil of Ex. 11-7 required to set up a flux of 960 μWb in a total air gap of 2 mm in series with the existing ferromagnetic core (Fig. 11-9)

Solution

$$\mathscr{F}_{air} = 7.958 \times 10^5\ \phi_{air} \times l/A$$
$$= 7.958 \times 10^5 \times 960 \times 10^{-6} \times 0.002/16 \times 10^{-4}\ \text{m}^2$$
$$= \mathbf{955\ At} \qquad (11\text{-}13)$$

$$\mathscr{F}_t = \mathscr{F}_m + \mathscr{F}_{air} = 663\ \text{At} + 955\ \text{At} = \mathbf{1618\ At} \qquad (11\text{-}11)$$

$$I = \mathscr{F}_t/N_c = 1618\ \text{At}/300\ \text{t} = \mathbf{5.39\ A} \qquad (11\text{-}1)$$

Example 11-8 (dramatically) shows for a total air gap of *only* 2 mm in length an additional 955 At are needed. This MMF is greater than the combined MMFs of all three core materials computed in Ex. 11-7, having a total length of 480 mm! Indeed, had the air gap been (say) 2 cm in length (or 20 mm), we might almost ignore the reluctance of the ferromagnetic material entirely! This proves conclusively that it is easier to set up a magnetic field in a ferromagnetic material than in air. This brings us back to the definition of permeability. Permeability is defined as the ease with which a magnetic field can be established in a given material. Fewer ampere-turns and less MMF is required for ferromagnetics than for air because ferromagnetic materials have a lower reluctance due to higher permeability, as noted in the conclusion to Sec. 11-7.3.

EXAMPLE 11-9

From the solutions to Exs. 11-7 and 11-8, calculate the permeability and the reluctance, respectively, for each portion of the core in Fig. 11-9 made of

a. Cast iron (CI)
b. Cast steel (CS)
c. Sheet steel (SS)
d. Air

Use a total gap of 2 mm.

Solution

a. $$\mu_{CI} = \frac{\mathbf{B}}{\mathbf{H}} = \left(0.6\,\frac{\text{Wb}}{\text{m}^2}\right)\Big/(2500\ \text{At/m})$$
$$= \mathbf{0.24\ mWb/A\cdot m} \text{ (from solution 11-7d)} \qquad (11\text{-}6)$$
$$\mathscr{R}_{CI} = \frac{l}{\mu A} = (0.24\ \text{m})/(0.24\ \text{mWb/A}\cdot\text{m})(16 \times 10^{-4}\ \text{m}^2)$$
$$= \mathbf{625\ kAt/Wb} \qquad (11\text{-}2)$$

b. $$\mu_{CS} = \mathbf{B}/\mathbf{H} = \left(0.6\,\frac{\text{Wb}}{\text{m}^2}\right)\Big/(400\ \text{At/m}) = 1.5\ \text{mH/m}$$
$$= \mathbf{1.5\ mWb/A\cdot m} \text{ (from solution 11-7c)}$$
$$\mathscr{R}_{CS} = \frac{l}{\mu A} = 0.12\ \text{m}/(1.5\ \text{mWb/A}\cdot\text{m})(16 \times 10^{-4}\ \text{m}^2)$$
$$= \mathbf{50\ kAt/Wb} \qquad (11\text{-}2)$$

c. $$\mu_{SS} = \mathbf{B}/\mathbf{H} = \left(0.6\,\frac{\text{Wb}}{\text{m}^2}\right)\Big/(125\ \text{At/m})$$
$$= \mathbf{4.8\ mWb/A\cdot m} \text{ (from solution 11-7b)}$$
$$\mathscr{R}_{SS} = l/\mu A = 0.12\ \text{m}/(4.8\ \text{mWb/A}\cdot\text{m})(16 \times 10^{-4}\ \text{m}^2)$$
$$= \mathbf{15.625\ kAt/Wb} \qquad (11\text{-}2)$$

d. $$\mu_{air} = 4\pi \times 10^{-7}\ \text{H/m}$$
$$= \mathbf{4\pi \times 10^{-7}\ Wb/A\cdot m} \text{ by definition a constant} \qquad (11\text{-}10)$$
$$\mathscr{R}_{air} = l/\mu A$$
$$= 2 \times 10^{-3}\ \text{m}/(4\pi \times 10^{-7}\ \text{Wb/A}\cdot\text{m}) \times (16 \times 10^{-4}\ \text{m}^2)$$
$$= \mathbf{994.7\ kAt/Wb} \qquad (11\text{-}2)$$

Example 11-9 shows that the reluctance of the air gap is *greater* than the *combined* reluctances of all the series-connected ferromagnetics in Fig. 11-9. This is remarkable since the air gap has an extremely small length (2 mm) in comparison to the relative lengths of the three ferromagnetics used in the core.

This brings us to the question of calculating the flux that might be set up in a helix or straight solenoid, such as that shown in Fig. 11-4a, wound on a *nonmagnetic* core. Calculating the area of the core is no problem. But what is the *average length of magnetic circuit* to be used in Fig. 11-4b?

The solution to the problem of the straight solenoid may be found by consideration of the ring solenoid wound on a toroidal nonmagnetic core shown in **Fig. 11-10a**. The cross section of the solenoid is circular. Assume that the core is made of a nonmagnetic material, say wood or plastic. Let the cross-sectional area of the core be A in square meters, and the average circumference l_{av}, as shown in Fig. 11-10a. If we now cut the solenoid at x–x and straighten it out, we obtain the same Fig. 11-10b, where the length of the solenoid, l_{av}, is the average circumferential length of the ring. But Fig. 11-10b shows that *leakage* occurs in an *air-core* solenoid! Near the center of the solenoid, the average length of magnetic circuit is l_1. As we move away from the center, the length of magnetic circuit increases to l_2, and so on. A typical map of the field pattern set up in the nonmagnetic core is shown in Fig. 11-10c. Observe that the flux density in the *center* of the solenoid is *greatest* there since all the flux must pass through the center. Obviously, the total MMF of the solenoid is the product of the solenoid coil current, I, and the number of turns, N. Since we know the permeability of the air core, we may write

$$\mu_0 = \frac{\mathbf{B}}{\mathbf{H}} = \frac{\mathbf{B}_m}{IN/l_{av}} = \frac{\mathbf{B}_m \times l_{av}}{IN} = 4\pi \times 10^{-7} \quad \text{henrys per meter (H/m)} \qquad \textbf{(11-14)}$$

where $\mathbf{B}_m$ is the maximum flux density at the center of the solenoid core in teslas
l_{av} is the length in meters of the solenoid or average magnetic circuit length

Note that Eq. (11-14) is the same as Eq. (11-12), developed for MMF in air gaps. Equation 11-14, therefore, may be used for air-core solenoids provided the length of the solenoid is at *least 10* times its diameter.[15]

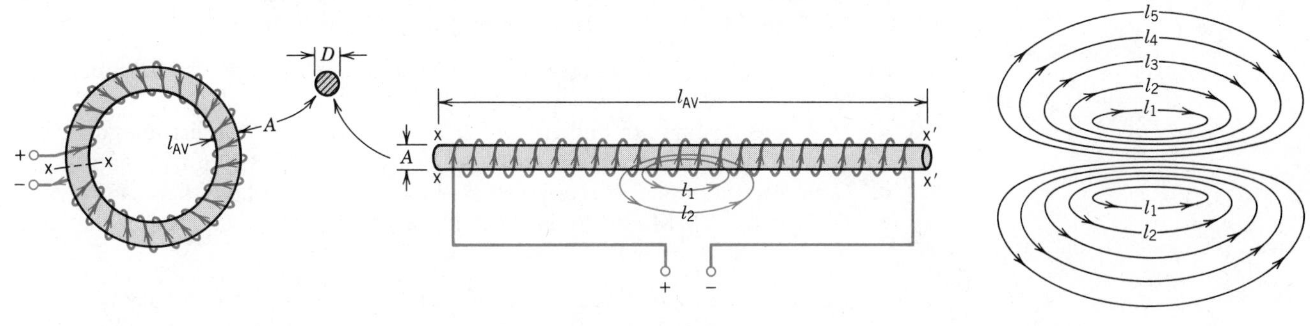

a. Ring solenoid on nonmagnetic toroidal core

b. Straight solenoid on nonmagnetic core

c. Map of field pattern set up in nonmagnetic core

Figure 11-10 Nonmagnetic toroidal core and equivalent straight air-core solenoid

[15] Tables for toroids and solenoids show that when the length of the solenoid is at least 10 times its diameter, the error in Eq. (11-14) is within 0.5%. When the length of solenoid is 20 times its diameter, the error is only 0.1%.

EXAMPLE 11-10

Assume the solenoid in Fig. 11-10b has 300 turns and a length of 10 cm, carries a current of 0.5 A, and is constructed on a nonmagnetic coil form 1 cm in diameter. Calculate the

a. Maximum flux density at the center of the solenoid core, $\mathbf{B}_m$
b. Average flux density at each end of the core, $\mathbf{B}_{av}$

Solution

a. Solving Eq. (11-14) for $\mathbf{B}_m$, we obtain

$$\mathbf{B}_m = 4\pi \times 10^{-7}\, IN/l_{av}$$
$$= 4\pi \times 10^{-7} \times 0.5 \times 300/0.1 = 1.885 \text{ mT}$$

b. $\mathbf{B}_{av} = \mathbf{B}_m/2 = 1.885 \text{ mT}/2 = 0.9425 \text{ mT}$ (see footnote 15)

Example 11-10 shows the following insights regarding air-core solenoids.

1. Maximum flux density occurs at the center of the solenoid since all flux lines must pass through the center (Fig. 11-10c).
2. Because of fringing and leakage, the average flux density at each end of the solenoid is approximately half the flux density at the center.[16]
3. The same Eqs. (11-12) and (11-14) may be used for straight solenoids, for air-core toroids, and for air gaps between ferromagnetic materials. This simplifies and unifies magnetic circuit calculations involving air cores and air gaps because only *one* equation is needed.

11-10 MAGNETIC CIRCUIT CALCULATIONS

We now realize that in working with ferromagnetic core materials we are dealing with *nonlinear* magnetic circuit relations whose saturation and permeability vary with magnetic field intensity, **H**, and/or magnetic flux density, **B**. We have seen that such problems cannot be solved by simple algebraic solution in equations but rather are solved (in part) by resorting to a **B**–**H** curve.

In dealing with air-core solenoids, nonmagnetic toroids, or air gaps between ferromagnetic materials, however, we may use equations exclusively, since the permeability of air is *constant*. Armed with this information, we may subdivide magnetic circuit problems into three major classes:

Class A—Magnetic circuits containing no ferromagnetic materials, air core only.

1. Given either I or IN, find **B** (or ϕ) for a coil of the given dimensions from Eq. (11-13).
2. Given either **B** (or ϕ), find $\mathscr{F}$, N, or I for a coil of the given dimensions from Eq. (11-13) or (11-14).

Class B—Nonlinear ferromagnetic circuits without air gap, magnetic cores only

1. **a.** Given either I or N, compute **H** from the given dimensions and data.
   b. Use **B**–**H** curve to find **B** or ϕ.
2. **a.** Given either ϕ or **B**, find **H** from the **B**–**H** curve.
   b. Using **H**, compute either N or I from the value of $\mathscr{F}$.

Class C—Nonlinear ferromagnetic circuits containing air gaps, using Ampere's law

- Given desired ϕ and A, compute **B** ($\mathbf{B} = \phi/A$).
- Use the **B**–**H** curve to find $\mathbf{H}_m$ and $\mathscr{F}_m$ ($\mathscr{F}_m = \mathbf{H}_m \times l_m$) for ferromagnetic core.
- Given **B** for air gap, compute $\mathscr{F}_a$ ($\mathscr{F}_a = 7.958 \times 10^5\, \mathbf{B}_a l_a$) for air gap(s).
- Compute the total MMF by using Ampère's circuital law, $\mathscr{F}_t = \mathscr{F}_m + \mathscr{F}_a$.
- Use the total MMF to find either the turns or current required in the coil to produce the desired **B**.

In each of the following exercises we will use the same technique. We will (1) read the problem, (2) identify the class into which it falls, (3) solve the problem by appropriate techniques and the steps outlined above, and (4) draw important inferences and insights from the solution.

[16] Proof of this relation for a long solenoid is given in C. L. Dawes, *Electrical engineering*, Vol. I, Fourth Ed., McGraw-Hill Book Co., N.Y., 1952, p. 237.

EXAMPLE 11-11

A straight solenoid has 200 turns wound on a nonmagnetic core whose diameter is 2 cm and whose length is 20 cm. Its coil produces an average flux of 5 μWb at each end. Calculate the

a. Average flux density at each solenoid end
b. Maximum flux density at center of solenoid
c. MMF required to produce the given flux at each end
d. Current in the coil

Solution

This is a class A2 problem.

a. $A = \pi D^2/4 = \pi \times (0.02\text{ m})^2/4 = 3.142 \times 10^{-4}\text{ m}^2$
$\mathbf{B}_{av} = \phi_{av}/A = 5\ \mu\text{Wb}/3.142 \times 10^{-4}\text{ m}^2$
$= 0.0159\text{ Wb/m}^2 = \mathbf{0.0159\ T}$ (11-4)

b. $\mathbf{B}_m = 2\mathbf{B}_{av} = 2 \times 0.0159\text{ T} = \mathbf{0.0318\ T}$

c. $\mathcal{F}_{air} = 7.958 \times 10^5\ \mathbf{B}_m \times l$
$= 7.958 \times 10^5 \times 0.0138 \times 0.2\text{ m} = \mathbf{5066\ At}$ (11-13)

d. $I = \mathcal{F}/N_c = 5066\text{ At}/200\text{ t} = \mathbf{25.33\ A}$ (11-1)

Example 11-11 shows that to set up a relatively low flux in an air-core solenoid requires a relatively high MMF and high current. This is due to the low permeability (and high reluctance) of the air.

EXAMPLE 11-12

A toroidal core, similar to that shown in Fig. 11-10a, has a cross-sectional diameter of 1 cm. The core is made of cast steel. The average circumferential length of the core is 40 cm. A flux of 0.1 mWb is required in the core, which has 1000 turns wound on its periphery. Calculate the

a. Required MMF to set up the desired flux in the core
b. Current I in the coil
c. Reluctance of the magnetic circuit
d. Permeability of the core in henry per meter

Solution

This is a class B2 problem.

a. $$\mathbf{B} = \phi/A = \frac{0.1 \times 10^{-3}\text{ Wb}}{\frac{\pi}{4}(1 \times 10^{-2})^2\text{ m}^2}$$
$= 1.273\text{ Wb/m}^2 = 1.273\text{ T}$ (11-4)

From the **B**–**H** curve for cast steel (Fig. 11-7),

$$\mathbf{B} = 1.273\text{ T} \Leftrightarrow 1375\text{ At/m} = \mathbf{H}$$

and $\mathcal{F}_t = \mathbf{H} \times l = 1375\text{ At/m} \times 0.4\text{ m} = 550\text{ At}$

b. $I = \mathcal{F}_t/N = 550\text{ At}/1000\text{ t} = 0.55\text{ A}$

c. $\mathcal{R} = l/\mu A = \mathcal{F}_t/\phi = 550\text{ At}/0.1 \times 10^{-3}\text{ Wb}$
$= 5.5 \times 10^6\text{ At/Wb}$ (11-2)

d. $\mu = \mathbf{B/H} = (1.273\text{ Wb/m}^2 \times 1\text{ H/Wb/At})/(1375\text{ At/m})$
$= 9.26 \times 10^{-4}\text{ H/m}$ (11-6)

Example 11-12 shows two important points regarding the use of magnetic units.

1. Reluctance is calculated in part (**c**) as the ratio $\mathcal{F}/\phi$, as defined in Eq. (11-2), and *not* in terms of $\mathcal{R} = l/\mu A$. The former is an easier calculation since μ was not determined until part (**d**).
2. In part (**d**) it was desired to express μ in terms of H/m. Consequently, the conversion factor 1 H = 1 Wb/At was used. Alternatively, the unit 1 H/m = 1 Wb/A·m might have been used.

EXAMPLE 11-13

A toroidal core similar to that in Fig. 11-10a has a cross-sectional area of 8 cm² and an average core length of $44.\bar{4}$ cm. The core is made of cast iron. A coil having 400 turns, carrying a current of 0.5 A, is wound on the core. Calculate the

a. MMF
b. Magnetic field intensity, **H**
c. Field flux in core, ϕ
d. Permeability of the core, μ
e. Reluctance of the core, $\mathcal{R}$

Solution

This is a class B1 problem.

a. $\mathcal{F} = IN = 0.5\text{ A} \times 400\text{ t} = \mathbf{200\ At}$ (11-1)

b. $\mathbf{H} = \mathcal{F}/l = 200\text{ At}/(44.4\text{ cm} \times 10^{-2}\text{ m/cm})$
$= \mathbf{450\ At/m}$ (11-5)

c. From the **B**–**H** curve for cast iron (Fig. 11-7):
$\mathbf{H} = 450\text{ At/m} \Leftrightarrow 0.2\text{ T} = \mathbf{B}$

$$\phi = \mathbf{B}A = \left(0.2\frac{\text{Wb}}{\text{m}^2}\right) \times (8\text{ cm}^2 \times 10^{-4}\text{ m}^2/\text{cm}^2)$$
$= 1.6 \times 10^{-4}\text{ Wb} = \mathbf{160\ \mu Wb}$ (11-4)

d. $$\mu = \mathbf{B/H} = \frac{0.2\text{ Wb/m}^2}{450\text{ At/m}} = 4.\bar{4} \times 10^{-4}\text{ Wb/At·m}$$
$= \mathbf{4.\bar{4} \times 10^{-4}\ H/m}$ (11-6)

e. $\mathcal{R} = \mathcal{F}/\phi = 200\text{ At}/1.6 \times 10^{-4}\text{ Wb}$
$= \mathbf{1.25 \times 10^6\ At/Wb}$ (11-2)

Example 11-13 shows that once **H** is computed from given data, we can find **B** and ϕ, which are both used to compute the permeability and reluctance of the closed toroidal core.

EXAMPLE 11-14

The toroidal core of **Ex. 11-12**, having the same dimensions and core material, is modified by sawing a 10-mm-wide air gap in it. It is desired to set up the same flux (0.1 mWb) in the core. Calculate the

a. Required MMF to overcome reluctance of the magnetic core
b. Required MMF to overcome reluctance of the air gap
c. Total MMF needed
d. Required current in the coil to establish the required MMF
e. Reluctance of the magnetic core and the air gap, respectively
f. Explain what would happen to the flux and flux density if the current remained the same in the coil as that found in Ex. 11-12 ($I = 0.55$ A)

Solution

This is a class C problem.

a. From **Ex. 11-12**: $\mathbf{H}_m = 1375$ At/m since the same flux density must be in the core. $\mathcal{F}_m = \mathbf{H}_m l_m = 1375 \text{ At/m} \times 0.39 \text{ m} =$ **536.25 At** (note change in length due to cut) (11-5)

b. $\mathcal{F}_a = 7.958 \times 10^5 \, \mathbf{B}_a l_a$
$= 7.958 \times 10^5 \times 1.257 \text{ T} \times 10^{-2} \text{ m}$
$=$ **10 003 At** (11-13)

c. $\mathcal{F}_t = \mathcal{F}_a + \mathcal{F}_m = (10\,003 + 536.3)$ At
$=$ **10 539 At** (11-11)

d. $I = \mathcal{F}_t / N = 10\,539 \text{ At}/1000 \text{ t} =$ **10.54 A** (11-1)

e. $\mathcal{R}_m = \mathcal{F}_m/\phi = 536.3 \text{ At}/0.1 \text{ mWb}$
$=$ **5.363 × 10⁶ At/Wb** (11-2)
$\mathcal{R}_a = \mathcal{F}_a/\phi = 10\,003 \text{ At}/0.1 \text{ mWb}$
$=$ **1 × 10⁸ At/Wb** (11-2)

f. If the current remains the same in a closed ferromagnetic core and an air gap is cut in the core, the total reluctance increases, causing the flux and the flux density to be correspondingly reduced

Example 11-14 shows that the MMF required to overcome the small air gap (10 mm) is almost 20 times as much as that required for the length of magnetic material that is almost 40 times as long! This means for *many* practical circuits, if the air gap is sufficiently wide, we may ignore the ferromagnetic portion of the circuit and concentrate primarily and exclusively on the air gap. Note from part (**e**) that the *total* reluctance of the circuit (approximately 1.05×10^8 At/Wb) is about 20 times as great as the reluctance of the closed core in Ex. 11-12. It is for this reason that the current must be increased to 10.54 A (from 0.55 A originally) to produce the same flux density in the core.

11-11 PARALLEL MAGNETIC CIRCUIT CALCULATIONS

Up to now our magnetic circuits have been simple *series* circuits. Are parallel magnetic circuits or even series–parallel magnetic circuits possible? The answer is that they are. **Figure 11-11** shows a parallel magnetic circuit consisting of two parallel magnetic paths produced by the *same* MMF. Each path has an average length of $2(l_1 + l_2)$, as shown in Fig. 11-11a. The cross-sectional area seen by any *one* of the two paths is the cross-sectional area, A_0, of the outer leg (*not* the cross-sectional area of the core portion around which the coil is wound). Since the reluctance of each path is the same, the total $\mathcal{R}$ to the total MMF is halved! The total

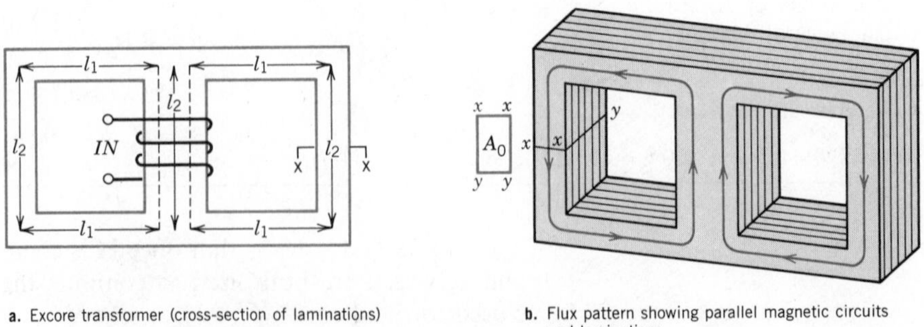

a. Excore transformer (cross-section of laminations)

b. Flux pattern showing parallel magnetic circuits and laminations

Figure 11-11 Parallel magnetic circuit; Example 11-15

flux in the center leg divides so that half the flux appears in each outer leg. Since magnetic flux is the analog of current in an electric circuit, the flux produced by the coil divides into two parallel paths.

EXAMPLE 11-15

An E-core transformer (Fig. 11-11a) is made of sheet steel, with a coil of 500 turns carrying 1 A on its center leg. The cross-sectional area of each outer leg, A_O, is 10 cm^2, which is half the cross-sectional area of the center leg. Lengths are $l_1 = l_2 = 10$ cm. Calculate the

a. Average length of one parallel flux path in meters
b. Total MMF, $\mathscr{F}_t$, produced by the coil
c. Total magnetizing force, $\mathbf{H}_t$, produced by the coil
d. Flux density in center leg and outer core legs
e. Flux in the center leg, ϕ_t
f. Flux in the outer legs, ϕ_O
g. Reluctance of each parallel path, $\mathscr{R}_O$
h. Equivalent reluctance of the entire magnetic circuit, $\mathscr{R}_t$

Solution

This is a class B1 problem.

a. $l_{av} = 2(l_1 + l_2) = 2(10 + 10) = 40 \text{ cm} = \mathbf{0.4 \text{ m}}$
b. $\mathscr{F}_t = IN = 1 \text{ A} \times 500 \text{ t} = \mathbf{500 \text{ At}}$ (11-1)
c. $\mathbf{H}_t = \mathscr{F}_t/l = 500 \text{ At}/0.4 \text{ m} = \mathbf{1250 \text{ At/m}}$ (11-5)
d. From the **B**–**H** curve (Fig. 11-7)

$$\mathbf{H} = 1250 \text{ At/m} \Leftrightarrow \mathbf{1.45 \text{ T}} = \mathbf{B}$$

The flux density in the outer legs must be the same as that in the center leg for the obvious reason that *both* the flux and the cross-sectional area are halved

$$\mathbf{B} = (\phi/2)/(A/2) = \phi/A = \mathbf{1.45 \text{ T}} \quad (11\text{-}4)$$

e. $\phi_t = \mathbf{B}A_t = 1.45 \dfrac{\text{Wb}}{\text{m}^2} \times 20 \text{ cm}^2 \times 10^{-4} \dfrac{\text{m}^2}{\text{cm}^2}$

$$= \mathbf{2.9 \text{ mWb}} \quad (11\text{-}4)$$

f. $\phi_O = \mathbf{B}A_O = 1.45 \dfrac{\text{Wb}}{\text{m}^2} \times 10 \times 10^{-4} \text{ m}^2$

$$= \mathbf{1.45 \text{ mWb}} = \phi_t/2 \quad (11\text{-}4)$$

g. $\mathscr{R}_O = \mathscr{F}_t/\phi_O = 500 \text{ At}/1.45 \text{ mWb}$
$= \mathbf{344.8 \text{ kAt/Wb}}$ (11-2)

h. $\mathscr{R}_t = \mathscr{F}_t/\phi_t = 500 \text{ At}/2.9 \text{ mWb} = \mathbf{172.4 \text{ kAt/Wb}}$ (11-2)

Four important conclusions emerge from Ex. 11-15:

1. The MMF and magnetic field intensity, **H**, produced at the center leg are the SAME as those at the two outer legs. (Recall that the EMF across each branch of a parallel electric circuit is the SAME.) This is why $\mathscr{F}_t$ was used in part (g) of the solution to find the reluctance of each outer leg.
2. The equivalent total reluctance, $\mathscr{R}_t$, of the entire magnetic circuit is EXACTLY HALF of the reluctance of the outer legs. (Recall that the equivalent resistance of two equal parallel resistors is EXACTLY HALF of each individual resistance.)
3. The flux ϕ_t in the center leg is twice that of the flux in the outer legs. Since flux is the magnetic analog of current in the electric circuit, it is clear that the flux produced by the coil divides into two parallel paths.
4. Since **H** is the same in all parts of this circuit and $\mu = \mathbf{B}/\mathbf{H}$, the permeability of the core portions depends on the flux density in each core portion.

11-12 SERIES–PARALLEL MAGNETIC CIRCUIT CALCULATIONS

A series–parallel magnetic circuit is shown in **Fig. 11-12.**[17] Each parallel path is identical. Each path consists of an air gap reluctance in series with the reluctance of the iron core material. Let us begin by assuming that the area of the center leg is twice the area of the outer legs. This means that the same flux density appears in both the center and outer core.

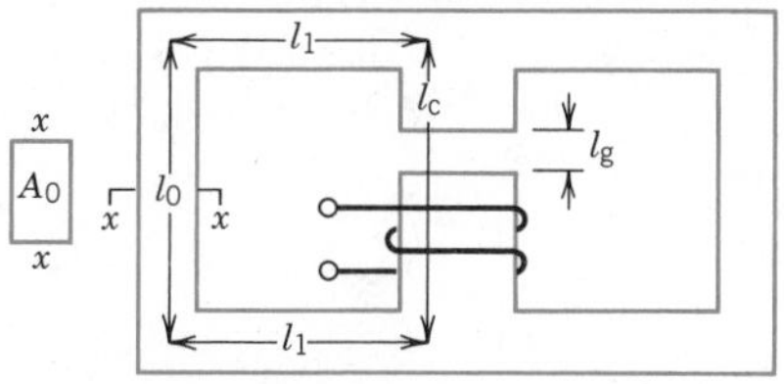

Figure 11-12 E-type, cut-core transformer cross section

[17] Although most texts refer to this as a series–parallel magnetic circuit, to be more precise it is a *parallel–series* magnetic circuit, i.e., essentially a *parallel* circuit with *series* elements.

EXAMPLE 11-16

The core shown in Fig. 11-12 has an air gap 2 mm in length; $l_0 = l_1 = 10$ cm. The material of the core is sheet steel. The outer leg area is 10 cm² and center leg area is 20 cm². It is desired to produce a total flux in the core of 2.9 mWb. The coil on the center core has 500 t. Calculate the

a. Flux density in the gap, **B**
b. MMF required to overcome the reluctance of the air gap, $\mathscr{F}_g$
c. MMF required to overcome the reluctance of the core, $\mathscr{F}_m$
d. Total MMF needed, $\mathscr{F}_t$
e. Current in the coil

Solution

This is a class C problem.

a. $\mathbf{B} = \phi/A = 2.9 \text{ mWb}/20 \text{ cm}^2 \times 10^{-4} \text{ m}^2/\text{cm}^2 = 1.45 \text{ T}$ (11-4)

b. $\mathscr{F}_g = 7.958 \times 10^5 \mathbf{B}_g l_g = (7.958 \times 10^5)(1.45)(0.002 \text{ m}) = 2308 \text{ At}$ (11-13)

c. From the **B–H** curve Fig. 11-7,
$\mathbf{B} = 1.45 \text{ T} \Leftrightarrow 1250 \text{ At/m} = \mathbf{H}$
$l_m = l_0 + 2l_1 + l_c = (10 + 2 \times 10 + 9.8) = 39.8 \text{ cm} = 0.398 \text{ m}$
$\mathscr{F}_m = \mathbf{H} \times l_m = (1250 \text{ At/m}) \times (0.398 \text{ m}) = 497.5 \text{ At}$ (11-5)

d. $\mathscr{F}_t = \mathscr{F}_g + \mathscr{F}_m = 2308 + 497.5 = 2805 \text{ At}$ (11-11)

e. $I = \mathscr{F}_t/N = 2805 \text{ At}/500 \text{ t} = 5.61 \text{ A}$ (11-1)

The following insights are to be drawn from Ex. 11-16 (and Ex. 11-15 preceding it):

1. The two cores are of identical dimensions with the exception of the 2-mm air gap in Ex. 11-16. The two cores carry the same total flux and flux density throughout.
2. In Ex. 11-15, given the coil current and turns, we used total MMF to find total flux and flux density. In Ex. 11-15, given total flux and flux density, we found the total MMF and current needed in the coil.
3. Note that the introduction of a 2-mm air gap has necessitated a current increase from 1 A in Ex. 11-15 to 5.61 A in Ex. 11-16. This is a 561% increase in current for a mere air gap of 2 mm! Once again, this shows that because air has a much higher reluctance than ferromagnetic material (of the same volume and dimensions), it requires a much greater MMF to set up the same flux in it.

Examples 11-15 and 11-16 were simplified because the cross-sectional area of the center leg was exactly twice the cross-sectional area of the outer legs. Since the total flux in the center leg is twice the flux in each of the outer legs, this produces the same flux density throughout the entire core (as noted previously). But if the cross-sectional areas are *not* in a simple 2:1 ratio, the flux density in the outer cores differs from that of the center core. This case of inequality in flux densities is taken up in Ex. 11-17.

EXAMPLE 11-17

Assume a cut core having the same shape as Fig. 11-12 but made of ferrite. Assume the ferrite has the same magnetization curve as cast iron (Fig. 11-7). The air gap, l_g, is 2 mm. The area of the center leg, A_c, is 10 cm² and the area of each outer leg, A_0, is 6 cm². The length of the center leg, l_c, is 14.8 cm; the length of the outer leg, l_0, is 15 cm and l_1 measures 10 cm. The coil of 1000 turns must set up a total flux of 0.2 mWb in the air gap. Calculate

a. $\mathbf{B}_c$, flux density in the center leg of the core
b. $\mathscr{F}_g$, MMF of the air gap
c. $\mathbf{H}_c$ and $\mathscr{F}_c$ for the center leg of the core only
d. $\mathbf{B}_0$, flux density in each outer leg of the core
e. $\mathbf{H}_0$ and $\mathscr{F}_0$ for each outer leg of the core
f. $\mathscr{F}_t$, total MMF required to produce the required flux in the ferrite core and air gap
g. Required coil current, I

Solution

This is a class C problem.

a. $\mathbf{B}_c = \phi_c/A = 0.2 \text{ mWb}/10 \text{ cm}^2 \times 10^{-4} \text{ m}^2/\text{cm}^2 = \mathbf{0.2\ T}$ (11-4)

b. $\mathscr{F}_g = \mathbf{B}_g l_g/4\pi \times 10^{-7} = 0.2 \times 0.002/4\pi \times 10^{-7} = \mathbf{318.3\ At}$ (11-13)

c. From the **B–H** curve (Fig. 11-7) for cast iron,
$\mathbf{B} = 0.2 \text{ T} \Leftrightarrow \mathbf{450\ At/m} = \mathbf{H}$
$\mathscr{F}_c = \mathbf{H} \times l = 450 \text{ At/m} \times 14.8 \text{ cm} \times 10^{-2} \text{ m/cm} = \mathbf{66.6\ At}$ (11-5)

d. The ϕ_0 in each outer leg is half the total flux or 0.2 mWb/2 = **0.1 mWb**
$\mathbf{B}_0 = \phi_0/A_0 = 0.1 \text{ mWb}/(6 \text{ cm}^2 \times 10^{-4} \text{ m}^2/\text{cm}^2) = \mathbf{0.1\bar{6}\ T}$[18] (11-4)

[18] Step **d** of the solution might also be found by saying that the combined cross-sectional area of the two parallel-path outer legs is 2 × 6 cm² or 12 cm² and that $\phi_t = \mathbf{B}_0 A_0 = \mathbf{B}_c A_c$ and $\mathbf{B}_0 = \mathbf{B}_c A_c/A_0 = 0.2 \text{ T} \times 10 \text{ cm}^2/12 \text{ cm}^2 = \mathbf{0.1\bar{6}\ T}$

e. From the **B**–**H** curve (Fig. 11-7) for cast iron,
   $\mathbf{B}_0 = 0.1\bar{6}\ \text{T} \Leftrightarrow 375\ \text{At/m} = \mathbf{H}_0$
   $l_{0T} = 2l_1 + l_0 = (2 \times 10 + 15)\ \text{cm} = 35\ \text{cm} = 0.35\ \text{m}$
   $\mathscr{F}_0 = \mathbf{H}_0 l_{0T} = 375\ \text{At/m} \times 0.35\ \text{m} = \mathbf{131.3\ At}$ (11-5)

f. $\mathscr{F}_t = \mathscr{F}_g + \mathscr{F}_c + \mathscr{F}_{0T} = 318.3 + 66.6 + 131.3 = \mathbf{516\ At}$

g. $I = \mathscr{F}_t/N = 516\ \text{At}/1000\ \text{t} = \mathbf{0.516\ A}$

The insights to be gained from Ex. 11-17 are

1. Regardless of the cross-sectional areas of portions of a magnetic core, the flux will divide equally if each half of the core is symmetrical.
2. For unequal cross-sectional areas, the respective flux densities are found by using $\mathbf{B} = \phi/A$ for each of the ferromagnetic portions of the core.
3. The flux densities found in (2) are used to yield the magnetic field intensity from the magnetization curves of Fig. 11-7 or data from the manufacturer of the magnetic material used.

11-13 MAGNETIC CORE LOSSES

Magnetic cores used in transformers and electrical rotating machinery are made of ferromagnetic alloys, which are also relatively good conductors. In the presence of alternating currents (or even rotation of magnetic armature cores under opposite magnetic poles in dc dynamos), two types of magnetic core losses are produced: *eddy currents* and *hysteresis*. Each of these losses is discussed and quantified in this section.

11-13.1 Eddy Current Loss

Eddy currents are defined as currents that exist as a result of voltages being *induced* in the (magnetic) material because of variations in magnetic flux (Sec. 12-1). The variation of magnetic flux may be produced by either varying the current (and the magnetic field) or by relative movement between the (magnetic) material and a constant magnetic field.

Eddy current **losses** may occur in *all* conductors (magnetic *and* nonmagnetic). Since ferromagnetics are generally good conductors, as a result of changing magnetic field direction, voltages induced in the magnetic cores produce eddy currents. Since all conductors have some resistance, these circulating or "eddy" currents produce an I^2R loss. Not only does this loss result in a reduced efficiency, but the heat produced may be extremely damaging to the insulation of the conductors on the core.

Eddy currents are always set up in planes perpendicular to the direction of the magnetic field.[19] In order to reduce the eddy current loss in ferromagnetics, instead of using solid cores, the ferromagnetic cores are subdivided into thin sheets called *laminations* (perpendicular to the flux path) and the core is called a *laminated core.* The use of laminated cores restricts the magnitude of the circulating currents and so reduces the magnitude of the eddy current loss. (See Fig. 11-11b.) In some instances, insulation is used between the laminations to ensure that currents do not flow across the laminations. If the eddy current loss is still too high, ferromagnetic alloys with higher resistivity are used to restrict the flow of eddy currents. (See Ex. 11-19.) For very high frequency use, the cores are made of sintered or powdered ferromagnetic material. This restricts the eddy currents to individual granules of material and reduces eddy current loss even further.

Experiments have shown that the eddy current power loss, P_e, may be expressed as a loss density in watts per unit volume of lamination (or granule) as

[19] This is dictated by Faraday's law. The induced voltage as a result of changing field flux always occurs at right angles to the field. The eddy currents flow is a result of and in the same direction as the induced voltage.

$$P_e = \frac{\pi}{6\rho}(h \cdot f \cdot \mathbf{B}_m)^2 \quad \text{watts per cubic meter (W/m}^3) \qquad (11\text{-}15)$$

where ρ is the resistivity of the ferromagnetic (or nonmagnetic material) in ohm-meters ($\Omega \cdot$m)

h is the thickness of the lamination in meters (m)

f is the frequency with which the magnetic field is changing in hertz (or cycles per second)

$\mathbf{B}_m$ is the maximum flux density of the varying magnetic field in teslas (Wb/m^2)

The empirical Eq. (11-15) assumes perfect insulation between laminations as well as the *same* flux density throughout all parts of each lamination. To obtain some idea of the magnitude of the eddy current loss, consider Ex. 11-18.

EXAMPLE 11-18

A small isolation transformer operates on 60 Hz and has laminations whose thickness is 0.35 mm. The maximum flux density is 1.5 T and the total volume of the magnetic core is 2000 cm^3. The resistivity of the core lamination material is 50×10^{-8} $\Omega \cdot$m. Calculate the eddy current power loss produced in

a. One cubic meter of lamination material
b. The transformer described above

Solution

a. $P_e = \frac{\pi}{6\rho}(h \cdot f \cdot \mathbf{B}_m)^2$

$= \left(\frac{\pi}{6 \times 50 \times 10^{-8}}\right) \times (3.5 \times 10^{-4} \times 60 \times 1.5)^2$

$= \mathbf{1039\ W/m^3}$ (11-15)

b. $P_t = P_e \times V_t$

$= 1039 \frac{\text{W}}{\text{m}^3} \times (2000\ \text{cm}^3 \times 10^{-6}\ \text{m}^3/\text{cm}^3)$

$= \mathbf{2.08\ W}$

EXAMPLE 11-19

A transformer having laminations which are 0.5 mm thick with a resistivity of 50 $\mu\Omega \cdot$cm has an eddy current loss of 50 W. It is desired to reduce the eddy current loss by using laminations having a resistivity of 85 $\mu\Omega \cdot$cm and a thickness of 0.25 mm. Calculate the anticipated eddy current loss if all other factors are unchanged.

Solution

Since eddy current loss varies *inversely* as ρ and *directly* as the square of lamination thickness, h,

$$P_e = 50\ \text{W}\left(\frac{50\ \mu\Omega \cdot \text{cm}}{85\ \mu\Omega \cdot \text{cm}}\right)\left(\frac{0.25\ \text{mm}}{0.5\ \text{mm}}\right)^2 = \mathbf{7.35\ W}$$

Example 11-18 shows how eddy current loss, P_e, in watts per cubic meter is calculated for various commercial laminations of ferromagnetic material in part (**a**). Once the laminations are "stacked" to form a transformer core and the volume is measured, the power loss is directly computed as in part (**b**) of Ex. 11-18.

Example 11-19 uses the **ratio method** of determining the power loss when a *change* is introduced in one or more of the factors in Eq. (11-15). It is not necessary to substitute in Eq. (11-15), and this technique greatly simplifies computation. Note that the power loss in Ex. 11-19 was reduced to about one-seventh of the original eddy current power loss by selection of a material having half the thickness and a somewhat higher resistivity.

11-13.2 Hysteresis Loss

Hysteresis loss can occur *only* in *ferromagnetic* materials (unlike eddy currents, which occur in *all* conductors having resistance) whenever they are subjected to changing magnetic fields. The term *hysteresis* comes from the Greek word *hysterein*, which means to lag behind. In ferromagnetics, *hysteresis* means a *lag* in the values of *resulting magnetization* due to a changing *magnetizing force*.

We can appreciate this lag if we consider a typical magnetization curve, shown in **Fig. 11-13a**, for a magnetic material that has *not* previously been magnetized or has been demagnetized. This curve is similar to the three magnetization curves shown in Fig. 11-7. The magnetic field intensity **H** is increased to some maximum value, $\mathbf{H}_m$, producing a maximum flux density, $\mathbf{B}_m$, at point **a**. If the magnetic field intensity is now decreased to zero and flux measurements are made, it is found that the material has been magnetized, producing a *residual* magnetic flux density, $\mathbf{B}_r$, at point **b** in Fig. 11-13b. Note that the flux density decreased in Fig. 11-13b but not as rapidly as the decrease in magnetic field intensity, **H** (i.e., **B** lags behind **H**!) In order to reduce $\mathbf{B}_r$ to zero, it is necessary to make **H** negative, at point **c**. At point **c**, the negative value $-\mathbf{H}_c$ is the *coercive* magnetic field intensity required to overcome the residual magnetism of the ferromagnetic material, shown in Fig. 11-13b. We continue to increase **H** in the negative direction and ultimately reach saturation or maximum flux density, $-\mathbf{B}_m$, at point **d** in Fig. 11-13b.

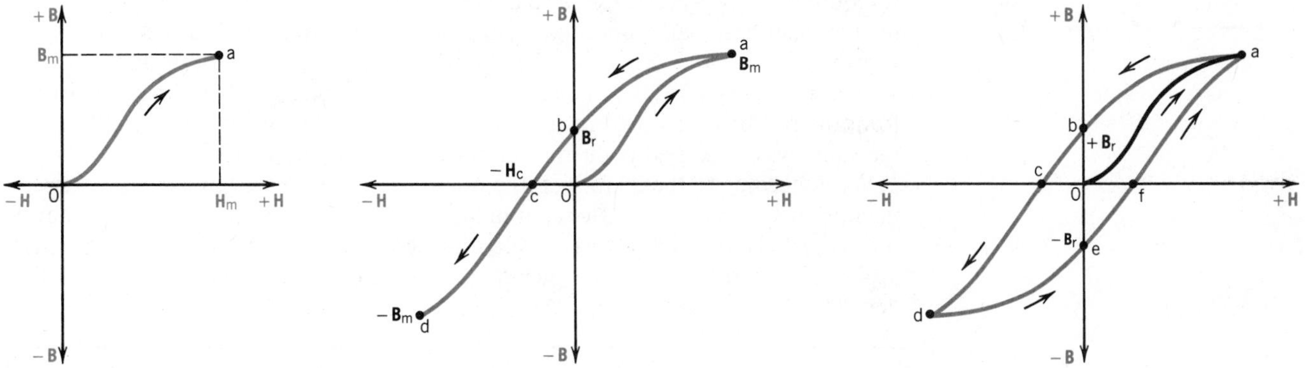

a. Magnetization curve

b. Magnetization in opposite direction

c. Complete loop

Figure 11-13 Development of a hysteresis loop

If we now *reduce* the magnetic field intensity (in the *positive* direction) from its maximum negative value to zero, we discover once more that hysteresis has occurred at point **e**, Fig. 11-13c. Here we have a negative residual magnetic flux density, $-\mathbf{B}_r$, equal and opposite to $+\mathbf{B}_r$. Again, we notice that **B** lags behind **H** because it is necessary to make **H** a positive force at point **f** (Fig. 11-13c) to reduce the residual magnetism to zero. If we continue to increase **H** in the positive direction, we ultimately reach point **a** (Fig. 11-13c), but *not* along the same line (Oa) as we began.

Continued reversals of the direction of **H** will cause continued successive retraces through points **abcdefa** of Fig. 11-13c, producing a *hysteresis loop* as a result of the property of hysteresis exhibited by most ferromagnetic materials.

A circuit for displaying the hysteresis loop of the magnetic core of a transformer is shown in **Fig. 11-14a**. The purpose of the variac is to ensure that maximum flux density (approximately 1 T or more) is reached. The horizontal input to the cathode ray oscilloscope (CRO) represents an analog of the current drawn by the transformer by taking the voltage across R_1. Thus, the horizontal input is proportional to the ampere-turns per unit length or **H**, magnetic field intensity. The output voltage of the low side of the transformer is loaded by R_L and fed through an integrating circuit to the vertical input of CRO. The flux density, **B**, is proportional to this output. The CRO display is shown in Fig. 11-14b. Substitution of various transformers, both of different sizes and containing ferromagnetic cores of different types of sheet steel, produces hysteresis loops of different areas. The *larger the area of the loop*, the *greater the hysteresis power loss*.

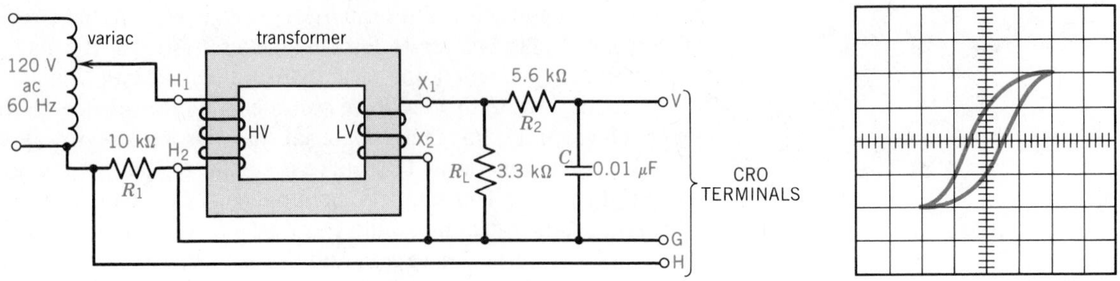

a. Circuit for displaying hysteresis loop (of a small power transformer)

b. CRO display

Figure 11-14 Experimental display of hysteresis loop on a CRO

The question may be asked, "Why does hysteresis produce a power loss in ferromagnetics whenever the magnetic flux density or magnetizing force changes?" The answer can be found by going back to the domain theory shown in Fig. 11-1. To align the magnetic domains of the atoms of a ferromagnetic material, energy is required. (Forces are required to move domains through the distance that results in alignment.) Similarly, once magnetized, energy is required to reverse the magnetization process and break up the residual alignment at points **b** and **e** of Fig. 11-13c and to *realign* them *again* in the *opposite* direction. This energy is drawn from the power source and results in heating the ferromagnetic material in much the same way as the eddy current loss described in Sec. 11-13.1.

We now come to the quantification of hysteresis loss. Based on extensive experimentation at General Electric plant in Schenectady, New York, Steinmetz[20] developed the following hysteresis power loss density equation, which is sufficiently accurate for most engineering purposes over a moderate range of maximum flux densities:

$$P_h = \eta f \mathbf{B}_m^{\ n} \qquad \text{watts per cubic meter (W/m}^3\text{)} \qquad \textbf{(11-16)}$$

where η is the Steinmetz coefficient of hysteresis loss in joules per cubic meter per tesla ($J/m^3/T$)

f is the frequency in hertz (s^{-1}) of the magnetizing force

$\mathbf{B}_m$ is the maximum flux density in tesla (T) or webers per square meter (Wb/m^2)

n is the exponent of $\mathbf{B}_m$ and is usually 1.6 for most flux densities of one tesla. (For higher values of $\mathbf{B}_m$, the value of n tends to increase.)[21]

In Eq. (11-16), values of the Steinmetz coefficient, η, vary from as low as 25 to as high as 18 000 for hard tool steel. Obviously, the material with the lowest coefficient will have the least loss per unit volume, for the same frequency and maximum flux density. An alloy called *Hypernik*, used in instrument transformer cores, has

---

[20] Dr. Charles Steinmetz (1865–1923), who came to the United States from Germany and discovered many theoretical relationships as well as practical principles in electrical engineering. It was Steinmetz who was responsible for long-distance power transmission with alternating voltages.

[21] Although most texts show $P_h = \eta f \mathbf{B}^{1.6}$ as an expression of the equation for hysteresis loss, the value of n may be as low as 1.4 and as high as 2.6, depending on the relative saturation of the magnetic material used.

an η of 25. The best transformer steels have values of about 130 to 150, depending on kVA rating. Cast steel has an η of 2500, and cast iron has an η of approximately 3750. Obviously, from Eq. 11-16, the lower the value of η, the smaller the loss.

Note the similarity between Eq. (11-15) for eddy current loss density and Eq. (11-16) for hysteresis loss density. Both quantify the loss in watts per cubic meter (W/m^3). In both types of losses, the greater the *volume* of material, the greater the *loss*. Thus, larger transformers or dynamos using ferromagnetic materials exhibit correspondingly larger eddy current and hysteresis losses.

EXAMPLE 11-20

The maximum flux density of a transformer is 1.5 T with a total volume of 2000 cm^3. The value of η for the sheet steel laminations is 500 $J/m^3/T$ and the frequency is 60 Hz. Calculate the hysteresis loss produced in

a. One cubic meter of material
b. The transformer described above

Solution

a. $P_h = \eta f \mathbf{B}_m{}^n = 500\ J/m^3/T \times 60\ s^{-1} \times (1.5^{1.6}\ T)$
$= \mathbf{57\ 394\ W/m^3}$ (11-16)

b. $P_h = 57\ 394\ W/m^3 \times (2000\ cm^3 \times 10^{-6}\ m^3/cm^3)$
$= \mathbf{114.8\ W}$

EXAMPLE 11-21

If the frequency of the transformer in Ex. 11-20 is reduced to 50 Hz and the flux density reduced to 1.25 T, calculate the hysteresis loss.

Solution

Using the **ratio method**,

$$P_h = 114.8\ W \times \left(\frac{50}{60}\right) \times \left(\frac{1.25}{1.5}\right)^{1.6} = \mathbf{71.46\ W}$$

Example 11-20 shows that once the hysteresis loss is computed for a unit volume of a given transformer (or dynamo) steel, under specified conditions of flux density or frequency, it can be computed for any size of transformer, depending on the volume of material required.

Example 11-21 shows how the **ratio method** is used whenever change is introduced into one or more of the factors in Eq. (11-16), without the necessity for recomputing the entire equation. This technique is used extensively in the theory of separation of losses to determine how much of the total core loss in a transformer is due to hysteresis and how much is due to eddy current loss.

11-14 MAGNETIC FIELD ENERGY AND INTERACTIONS

We have seen that whenever the magnetic flux changes in a ferromagnetic core, it is accompanied by a certain expenditure of energy due to hysteresis and eddy currents. But this energy expenditure due to core loss is small compared to the energy stored in the magnetic field that is set up in the ferromagnetic core. Consider the solenoid (Fig. 11-10b) wound on a nonmagnetic core. As the current (and magnetic field intensity, **H**) is increased, the flux density, **B**, must increase. **Figure 11-15a** shows the linear nature of the **B–H** curve for air in comparison to that of a ferromagnetic. Note that the **B–H** curve for air is a straight line. The area under this curve is a triangle for any given value of **B** and **H** and it represents the energy density stored in the magnetic field or[22]

[22] Equation (11-17) in units of J/m^3 may be derived intuitively by dimensional analysis if we realize that 1 V = 1 Wb-turn/second from Faraday's law for electromagnetic induction. Since $W = \mathbf{BH}/2$ in units of $\frac{Wb}{m^2} \times \frac{At}{m} = (WbAt/m^3)(1\ V/Wb \cdot t/s) = VAs/m^3 = Ws/m^3 = J/m^3$.

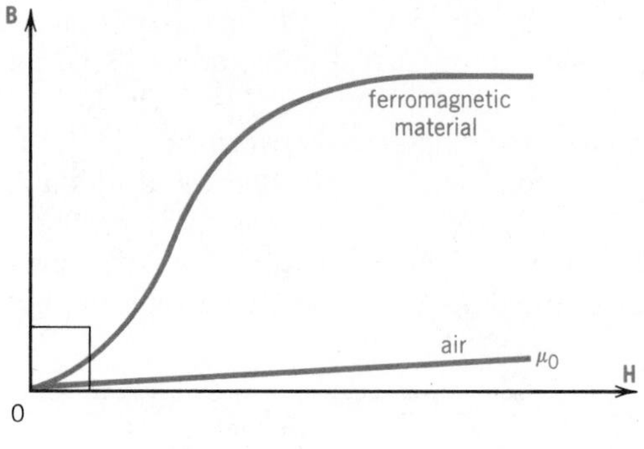

a. Relative permeability of ferromagnetic compared to air

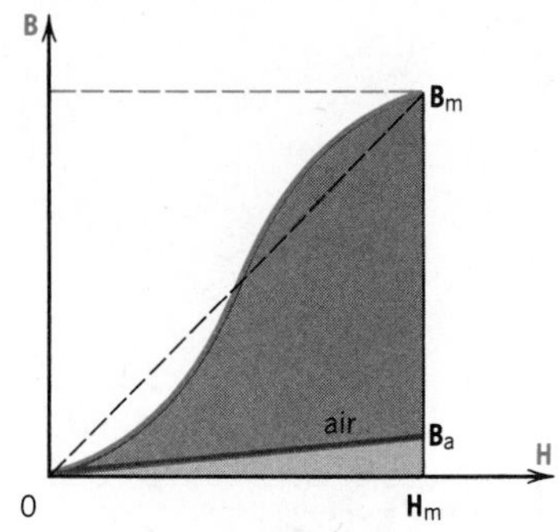

b. Energy stored in magnetic field

Figure 11-15 Comparison between magnetic field energy stored in free space and in ferromagnetic material

$$W = \int_0^{\mathbf{B}_m} \mathbf{H}\,d\mathbf{B} = \frac{\mathbf{BH}}{2} = \frac{\mathbf{B}^2}{2\mu} = \frac{\mathbf{H}^2\mu}{2} \quad \text{joules per cubic meter (J/m}^3\text{)} \qquad (11\text{-}17)$$

But for air $\mathbf{H} = \mathbf{B}/\mu_0$, and substituting for $\mathbf{H}$ in above yields the energy density for air as

$$W = \mathbf{B}^2/2\mu_0 \qquad \text{joules/cubic meter (J/m}^3\text{)} \qquad (11\text{-}18)$$

Since the volume of air is the cross-sectional area, A, times average length of magnetic circuit, l, the energy stored in the air gap is

$$W = \frac{\mathbf{B}^2 l A}{2\mu_0} \qquad \text{joules (J)} \qquad (11\text{-}19)$$

where all terms have been defined above

A comparison between the energy stored in the magnetic field of a solenoid and that in a ferromagnetic core is shown in Fig. 11-15b, where the (shaded) area under the curves represents the stored energy. Note that a straight line drawn from the origin to $\mathbf{B}_m$ on the magnetization curve is also a triangle whose area may be found from Eq. (11-17). Thus, for a given volume of magnetic material and the **B**–**H** curves of Fig. 11-7, it is possible to calculate the energy stored in the magnetic field of a given ferromagnetic material from Eq. (11-17).

EXAMPLE 11-22

A transformer having a core volume of 2000 cm³ has sheet steel laminations described by the magnetization curve in Fig. 11-7. Calculate the energy stored in the magnetic field of the core when the maximum flux density is 1.5 T.

Solution

$$W = \mathbf{BH}/2 = \left(1.5\,\frac{\text{Wb}}{\text{m}^2}\right)\left(\frac{1650\ \text{At/m}}{2}\right) = 1237.5\ \text{J/m}^3 \quad (11\text{-}17)$$

$$W = 1237.5\ \text{J/m}^3 \times 2000\ \text{cm}^3 \times 10^{-6}\,\frac{\text{m}^3}{\text{cm}^3} = \mathbf{2.475\ J}$$

In Fig. 11-15b we compared the relative energy stored in an air-core solenoid with that of a solenoid containing a highly magnetic core. This comparison is somewhat unfair because the flux density in air, $\mathbf{B}_a$, is so much smaller than that in the ferromagnetic material, $\mathbf{B}_m$. Consider a core such as that shown in Fig. 11-12. Here we know from Ex. 11-16 that the flux density in the air gap is the same as that in the center leg of the core. How do the relative energy densities compare? We may be in for a surprise, as shown by Ex. 11-23.

EXAMPLE 11-23

From the given data and the solution of Ex. 11-16, calculate

a. The energy density stored in the magnetic field of the core, W_m
b. The energy density stored in the magnetic field in the air gap, W_g

Solution

a. $W_m = \mathbf{BH}/2 = 1.45 \dfrac{\text{Wb}}{\text{m}^2} \times 1250 \text{ At/m}$

$$= \mathbf{181.25\ J/m^3} \quad (11\text{-}17)$$

b. $W_g = \mathbf{B}^2/2\mu_0 = (1.45)^2/2 \times 4\pi \times 10^{-7}$

$$= \mathbf{8.366 \times 10^5\ J/m^3} \quad (11\text{-}18)$$

Example 11-23 shows that the energy density in the air gap is more than 4600 times the energy density in the iron for the same flux density in both![23] The reason for this, obviously, is that the permeability, μ, in the core is so much greater than it is in the air gap.

Well, if the energy density in the air gap is so much greater, perhaps we can put it to work (since energy is the ability to do work). We know that in terms of mechanical work, that work is expressed as

$$W = \mathbf{F} \times l$$

Consider two opposite poles of permanent magnets separated by an air gap, l, having a cross-sectional area, A, and an energy stored in the gap of $\mathbf{B}^2 lA/2\mu_0$ as expressed by Eq. (11-19). Then the work in pulling the magnets together is

$$W = \frac{\mathbf{B}^2 lA}{2\mu_0} = \mathbf{F}l$$

Solving this relation for **F**, the tractive (or repulsive) force of an electromagnet or permanent magnet, we obtain[24]

$$\mathbf{F} = \frac{\mathbf{B}^2 A}{2\mu_0} = \frac{\mathbf{B}^2 A}{2 \times 4\pi \times 10^{-7}} = 3.979 \times 10^5\, \mathbf{B}^2 A \qquad \text{newtons (N)} \quad (11\text{-}20)$$

where **B** is in tesla, A in meters squared
F is the *force per gap* in newtons

---

[23] In practice, because of fringing and leakage, the flux density in the core is slightly larger than it is in the air gap.

[24] Equation (11-20) also yields the force of repulsion between two identical poles of bar magnets of equal strength, where **B** is the flux density of either bar magnet and A is their cross-sectional area (m^2). It also yields the force of attraction between two opposite poles of equal flux density and cross-sectional area.

It is sometimes more useful to express Eq. (11-20) in terms of the maximum weight, in *kilograms of force* (kg_f), that can be lifted by electromagnets. Since 1 kg_f on earth equals 9.807 newtons, the maximum weight, $\mathbf{F}_w$, that can be lifted is

$$\begin{aligned} \mathbf{F}_w &= 3.979 \times 10^5\, \mathbf{B}^2 A / 9.807\ \text{N/kg}_f \\ &= 4.057 \times 10^4\, \mathbf{B}^2 A \qquad \text{kilograms (kg}_f\text{) per gap} \qquad \textbf{(11-20a)} \end{aligned}$$

where all terms have been defined above

EXAMPLE 11-24

A two-pole horseshoe magnet produces a flux density of 1 tesla (1 T) across each pole face area of 200 cm² when placed close to a block of steel. Calculate the

a. Maximum possible force exerted by the magnet (across both gaps)
b. Maximum weight in kg_f of steel it is capable of lifting

Solution

a. $\mathbf{F} = 3.979 \times 10^5\, \mathbf{B}^2 A$
$= 3.979 \times 10^5 (1)^2 200\ \text{cm}^2 \times 10^{-4}\ \text{m}^2/\text{cm}^2$
$= 7.958\ \text{kN/gap} \times (2\ \text{gaps}) = \mathbf{15.92\ kN}$ (11-20)

b. $\mathbf{F}_w = 15.92\ \text{kN}/9.807\ \text{N/kg} = \mathbf{1623\ kg_f}$
Alternatively, using Eq. (11-20a) directly,
$\mathbf{F}_w = 4.057 \times 10^4\, \mathbf{B}^2 A$
$= 4.057 \times 10^4 (1)^2 \times 200 \times 10^{-4} \times 2$ gaps
$= \mathbf{1623\ kg_f}$ (11-20a)

Example 11-24 shows that we must use Eqs. (11-20) and (11-20a) with some care. The force computed in Eq. (11-20) is the force across *each* gap. If one visualizes a block of steel being lifted with two hands, the force provided by each hand represents only *half* the force needed in accomplishing the lift. Consequently, since Eq. (11-18) provides the *energy density* for each gap, whenever more than one gap occurs in a magnetic circuit, the total force is the sum of the forces developed at each gap, as in the case of a horseshoe magnet. In the case of two bar magnets, however, there is only one gap between them, and the tractive force between opposite poles would be found from Eq. (11-20).

11-15 OERSTED'S LAW AND FORCES ON CURRENT-CARRYING CONDUCTORS

We may now appreciate the quantification of Oersted's law, described in Sec. 11-3 and shown in Fig. 11-3. Oersted noted that the flux density, **B**, of a single (straight) current-carrying conductor may be measured as

$$\mathbf{B} = k \cdot 2I/r = 2 \times 10^{-7}\, I/r \qquad \text{tesla (T)} \qquad \textbf{(11-21)}$$

where k is a constant for the SI of 10^{-7} Wb/A·m (webers per ampere-meter)
I is the current in the single straight conductor, in amperes (A)
r is the perpendicular distance from axis and center of the current-carrying conductor in meters (m)

Equation (11-21) implies that the flux density surrounding a straight single conductor is greatest at the center of the conductor (where the radius distance is almost zero) and varies inversely as the distance from the center of the conductor. It also shows that **B** is directly proportional to I and depends directly on I. If I

is zero, **B** is zero. Conversely, whenever there is current, it is always accompanied by a magnetic field.

EXAMPLE 11-25

A current of 1 A is flowing in a conductor. Calculate the flux density at a perpendicular distance of 1 cm from the center of the conductor.

Solution

$$\mathbf{B} = 2 \times 10^{-7}\, I/r = 2 \times 10^{-7}(1/0.01\text{ m})$$
$$= \mathbf{20\ \mu T = 20\ \mu Wb/m^2} \qquad (11\text{-}21)$$

EXAMPLE 11-26

If the same straight conductor carrying 1 A is wound in the form of a spiral (or helix) on a 2-cm-diameter nonmagnetic core, producing 50 turns and having a length of 20 cm, calculate

a. The maximum flux density at the center of the helix
b. The maximum flux density at each end

Solution

Using Eq. (11-14) for the air-core solenoid yields

a. $\mathbf{B}_m = \mu_0 IN/l_{av} = 4\pi \times 10^{-7} \times 1 \times 50/0.2$
$= \mathbf{314.2\ \mu T} \qquad (11\text{-}14)$
b. $\mathbf{B}_e = \mathbf{B}_m/2 = 314.2\ \mu\text{T}/2 = \mathbf{157.1\ \mu T}$

Example 11-26 shows (once again) that winding a straight current-carrying wire in the form of a helix increases the flux density greatly.

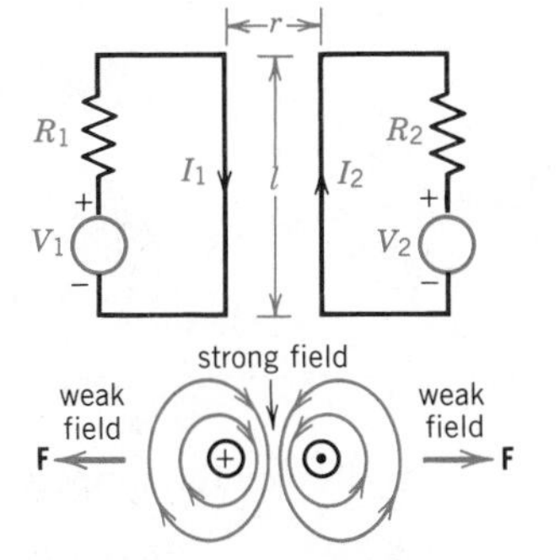

a. Parallel wires. Opposite current directions produce repulsion forces

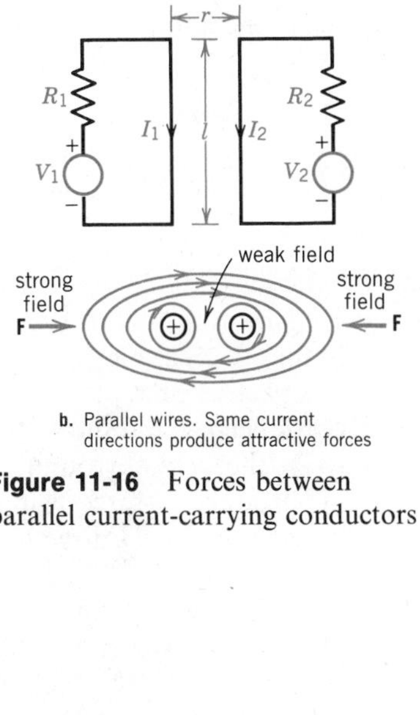

b. Parallel wires. Same current directions produce attractive forces

Figure 11-16 Forces between parallel current-carrying conductors

11-15.1 Force between Parallel Current-Carrying Conductors

In Eq. (11-19) we derived the attractive or repulsive force between opposite or like poles, respectively, of two bar magnets or the ends of two electromagnets. Since straight current-carrying wires have been shown to have a magnetic field around them, what force of attraction or repulsion exists between two parallel straight current-carrying conductors?

Figure 11-16 shows the direction of forces and relative fields about two parallel straight wires carrying current in either the *opposite* or *same* directions. André Ampère's original research on the nature and quantification of force between two parallel current-carrying wires yields

$$\mathbf{F} = 2 \cdot k \cdot I_1 \cdot I_2 \cdot l/r = 2 \times 10^{-7}\, I_1 \cdot I_2 \cdot l/r \qquad \text{newtons (N)} \qquad \mathbf{(11\text{-}22)}$$

where k is 10^{-7} N/A^2 (newtons per ampere squared)
l is the length over which the two conductors are parallel, in meters (m)
I_1 and I_2 are conductor currents, respectively, in amperes (A)
r is the distance between two paralleled conductors, in meters (m)

If the currents are in the SAME direction, **F** is *positive*, indicating a force of *attraction.* If the currents are in the OPPOSITE direction, **F** is *negative*, indicating a force of *repulsion.*[25]

The question arises of whether it is possible to predict the directions of force by using the right-hand rule for fields around a conductor. In Fig. 11-16a we see that the *downward* current, I_1, produces current *into* the paper, resulting in a clockwise

[25] Note that the direction of these forces is *opposite* to that of the forces in Coulomb's law (Sec. 2-1b), where like charges repel and unlike charges attract. In terms of direction of magnetic fields, however, Ampere's relation is similar to Coulomb's law in that like currents produce fields in *opposite directions between conductors*, resulting in *attraction*, as shown in Fig. 11-16b.

(CW) field. Similarly, current I_2 coming *out* of the paper produces a counterclockwise (CCW) field. These two currents (Fig. 11-16a) in opposite directions produce a strong field in the SAME direction *between* the two conductors and a weak *field* around the two conductors. Since lines of force in the SAME direction *repel*, repulsion exists between the two conductors.[26]

Similarly, in Fig. 11-16b, with *both* currents I_1 and I_2 downward into the paper, we observed *both* producing CW fields. But between the conductors the directions of the fields *oppose* each other, resulting in a weak field between conductors and a strong field around both conductors. This produces a force of *attraction* between conductors in Fig. 11-16b because

1. The fields act in such a way as to become uniform and this occurs only when the conductors move closer to each other.
2. Lines of force in *opposite* directions exist and occur *between* the two conductors.

EXAMPLE 11-27

Calculate the force of attraction between two wires carrying currents 10 and 50 A, respectively, whose parallel lengths are 50 cm separated by a distance of 5 cm

Solution

$$\mathbf{F} = 2 \times 10^{-7} I_1 I_2 l/r \tag{11-22}$$

$$= 2 \times 10^{-7}(10)(50) \times 0.5 \text{ m}/0.05 \text{ m}$$

$$= \mathbf{1\ mN}$$

EXAMPLE 11-28

Two parallel conductors, each 1 meter long, are separated in a vacuum by a distance of one meter and produce between them a force of 2×10^{-7} newtons when carrying the same current. Calculate the current in the conductors.

Solution

$$I^2 = \frac{\mathbf{F} \times r}{2 \times 10^{-7} \times l} = \frac{2 \times 10^{-7} \times 1 \text{ m}}{2 \times 10^{-7} \times 1 \text{ m}} = 1 \text{ A}^2$$

and

$$I = \sqrt{1 \text{ A}^2} = \mathbf{1\ A}$$

Example 11-28 provides us with the *empirical* SI definition of the **ampere** as

the current that, if maintained in two straight parallel conductors of infinite length, of negligible cross section, and placed one meter apart in vacuum, would produce between these conductors a force of 2×10^{-7} newton per meter of length.[27]

This definition enables the quantification of a current of one ampere by means of *force* measurement between parallel conductors.

11-15.2 Force Acting on a Current-Carrying Conductor Located in a Uniform Magnetic Field

We come now to one of the more important electromagnetic relations in science. Consider a current-carrying conductor located at *right angles* to a uniform magnetic field, as shown in **Fig. 11-17a**. We have already seen that two parallel current-carrying conductors (Fig. 11-16) develop forces (either attractive or repulsive) at right angles to both the current direction and the fields produced as a result of the current. As shown in Fig. 11-17a, the interaction between the field surrounding the conductor carrying current and the uniform field, ϕ, produces a force **F** in a *downward* direction, at right angles to *both* current direction and *field* direction.

[26] Alternatively, the force fields *tend* to provide a *uniform* field, and this is provided only when the conductors tend to separate from each other.

[27] Note that previously we defined the ampere as a unit of current equivalent to the motion of one coulomb of charge (6.28×10^{18} electrons) passing any cross-section of conductor in one second.

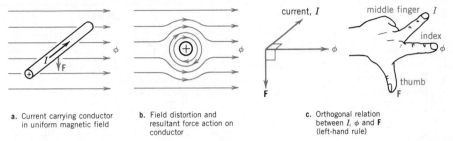

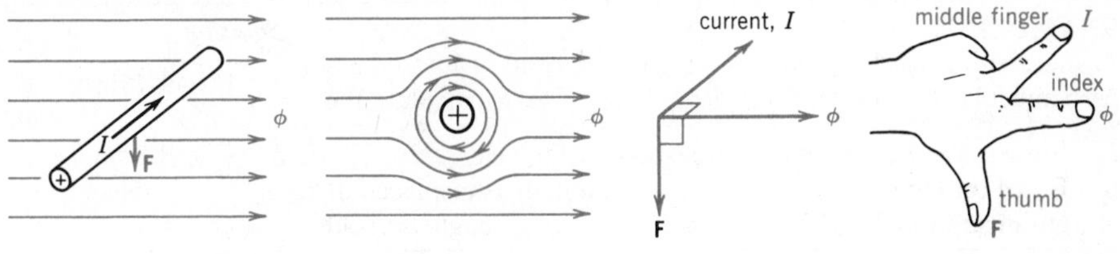

a. Current carrying conductor in uniform magnetic field

b. Field distortion and resultant force action on conductor

c. Orthogonal relation between I, ϕ and **F** (left-hand rule)

Figure 11-17 Force developed by a current-carrying conductor located in a uniform magnetic field and left-hand (LH) rule

How can we predict the direction of this force, and can we express an equation to quantify its magnitude?

The direction of the force is determined from Fig. 11-17b, which shows the CW field produced by the conductor and the resultant distortion of the (previously) uniform field. The field has become *more concentrated* (stronger) *above* the conductor than below. Alternatively, we can also see *repulsion* due to lines in the *same* direction *above* the conductor (Fig. 11-17b) and *attraction* due to lines in the *opposite* direction *below* the conductor. For both these reasons, the force is in a *downward* direction, as shown in Figs. 11-17a and 11-17c. Figure 11-17c also shows the *orthogonal* (mutually perpendicular) relation between the uniform field, ϕ, the current direction, I, and the resultant force on the conductor, **F**. It also shows a left-hand (LH) rule that may be used for this purpose:

- The index finger points the direction of the field flux, ϕ.
- The third (middle) finger points the direction of current, I.
- The thumb points the direction of force produced by interaction of ϕ and I.

The magnitude of force, **F** is obtained from

$$\mathbf{F} = \mathbf{B}\,(I \cdot l)\sin\theta \qquad \text{newtons (N)} \qquad \textbf{(11-23)}$$

where **B** is the flux density of the uniform magnetic field, normally perpendicular to l

l is the length of a conductor acted on by the magnetic field when carrying current

I is a current in amperes in the above conductor

θ is the angle between **B** and $(I \cdot l)$ if not 90° (i.e., if not mutually perpendicular)

Note:

1. Whenever **B** and $(I \cdot l)$ are at right angles (rt/s) to each other, the force developed is *orthogonal*, at rt/s to *both*, and Eq. (11-23) becomes $\mathbf{F} = \mathbf{B}Il$ newtons.
2. The force developed on the conductor is *always* in a direction at rt/s to the reference field, **B**, of uniform flux density, regardless of angle θ. It is the magnitude of the force which decreases as θ decreases.
3. The force is exerted on the conductor in such a way as to align the *field of the conductor* in the same direction as the uniform magnetic field.
4. This relation is fundamental to all moving-coil instruments (Fig. 11-18a) and all motor action (Fig. 11-18b).

EXAMPLE 11-29

A single conductor 10 cm long is carrying a current of 5 A and lies perpendicular to a uniform magnetic field of 1.2 T; calculate the

a. Force on the conductor with respect to the field
b. Force on the conductor when the conductor is displaced at an angle of 60° with respect to the reference magnetic field

Solution

a. $\mathbf{F} = \mathbf{B}Il \sin\theta = (1.2\text{ T})(5\text{ A})\sin 90°$
$= \mathbf{0.6\ N}$ at rt/s to **B** (11-23)

b. $\mathbf{F} = \mathbf{B}Il \sin\theta = (0.6\text{ N})(\sin 60°) = \mathbf{0.52\ N}$ at rt/s to **B**

Figure 11-18 shows but two of the many applications of Eq. (11-23). Figure 11-18a shows the directions of currents in the moving coil of a permanent-magnet-moving-coil (PMMC) meter movement. The conductors under the north pole of the PMMC movement, using the left-hand (LH) rule, produce CW movement; and those under the south pole also produce CW movement whenever current is applied to the meter terminals, as shown in Fig. 11-18a.

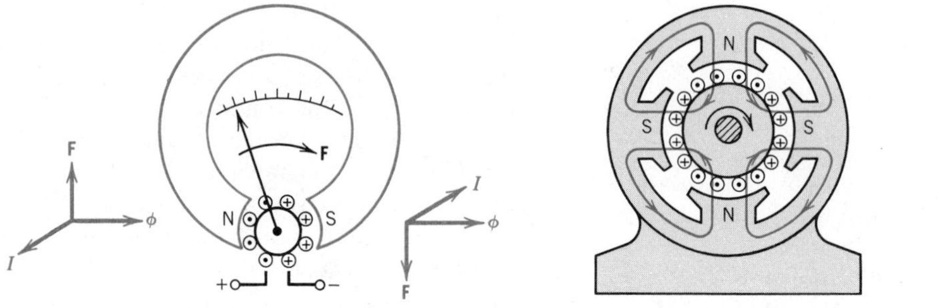

a. Meter movement

b. 4-pole dc motor

Figure 11-18 Forces developed by orthogonal current-carrying conductors in uniform magnetic fields of meters and motors

Figure 11-18b shows the cross section of a four-pole dc motor whose field windings (not shown) produce four parallel magnetic circuit flux paths. *Each* flux path consists of a closed magnetic circuit going through a north pole, an air gap, the armature core, an air gap, a south pole, the armature steel yoke, and back to a north pole. Armature current produces current directions in the conductors as shown in Fig. 11-18b. Each conductor produces an orthogonal CW force tangential to the armature, producing armature rotation in the CW direction.

11-16 GLOSSARY OF TERMS USED

Absolute permeability, μ Ratio of the magnetic flux density produced in a material (**B**) to the magnetizing force (**H**) producing it, measured in SI in units of henrys per meter (H/m) or webers per meter·ampere-turn (Wb/m·At).

Ampère's circuital law The algebraic sum of all MMFs in a closed magnetic circuit is zero. This law is the magnetic circuit analog of Kirchhoff's voltage law for electric circuits.

Average permeability, μ_{av} Ratio of **B** to **H** at a given point on the magnetization curve for a given ferromagnetic material; $\mu_{av} = \mathbf{B}/\mathbf{H}$. (Also static or dc permeability.)

B–H curve Graphical plot of the characteristics of any ferromagnetic material, where flux density, **B**, is the ordinate and magnetic field intensity, **H**, the abscissa. The **B–H** curve is a continuous plot of the permeability of the material since $\mu = \mathbf{B}/\mathbf{H}$.

Diamagnetic materials Materials that, in the presence of a magnetic field, decrease the total flux only slightly. Copper and silver are slightly diamagnetic.

Domain Group of magnetically aligned atoms, so aligned as a result of a magnetizing force.

Eddy currents Electric currents induced in a conductor or a core by a varying magnetic field. Eddy currents are sometimes called Foucault currents.

Eddy current loss Power loss represented by the total I^2R loss produced by the circulating eddy currents in conductors, magnetic cores, or surrounding metallic shielding. Eddy current losses are minimized by using laminations with insulation between them, powdered cores composed of high-resistivity ferrites, and/or dust cores.

Ferromagnetic material Material that, when inserted in an independently established magnetic field, produce a marked increase in magnetic flux. Examples are iron, steel, nickel, cobalt, and alloys of these with other metals such as aluminum and titanium.

Flux density, B Measure of the total magnetic flux passing through a unit area that is perpendicular (normal) to the direction of the magnetic force, measured in SI in units of tesla (T) or webers per square meter (Wb/m^2). Note $1\ T = 1\ Wb/m^2$.

Hysteresis Retardation or lagging of an effect behind its cause.

Hysteresis loss Energy or power loss required to overcome the effect of hysteresis in magnetic materials, that is, work required to realign the domains of ferromagnetic materials having the property of retentivity taken over a complete magnetizing cycle. The power loss due to hysteresis is proportional to the area enclosed by a hysteresis loop.

Incremental permeability, μ_d Permeability measured during a small change in either flux density or magnetic field intensity; $\mu_d = d\mathbf{B}/d\mathbf{H}$.

Linear magnetic circuit Magnetic circuit limited strictly to nonmagnetic materials (wood, air, aluminum, polystyrene, etc.) in which the permeability, μ, is always approximately that of free space ($4\pi \times 10^{-7}$ H/m) and is independent of magnetic field intensity, **H**.

Magnetic circuit Minimal region containing essentially all the magnetic flux or region at whose surface the magnetic induction is tangential.

Magnetic field intensity, H Force that produces flux density at a given point in a magnetic circuit, measured in ampere-turns per meter (At/m); also called *magnetizing force, magnetic field strength or magnetic intensity.*

Magnetic flux, ϕ Magnetic field in space or in a magnetic material, caused by MMF and measured in units of webers (Wb) in the SI.

Magnetomotive force, $\mathscr{F}$ Measured in the SI in units of ampere-turns, the MMF is the line integral of the magnetizing force (magnetic field intensity) around a closed magnetic circuit. The MMF may be viewed as a force capable of producing a magnetic field either in space or in a magnetic material.

Nonlinear magnetic circuit Magnetic circuit containing ferromagnetic material whose permeability, μ, varies with the degree of saturation of the magnetic material.

Nonmagnetic materials Materials that have no effect on a magnetic field in space or air. When placed in an air gap, such materials neither decrease nor increase the magnetic field. Almost all organic materials are nonmagnetic (plastics, rubber, wood, oil, alcohol, etc.).

Oersted's law The flux density around a current-carrying wire is directly proportional to the current and inversely proportional to the distance from the axial center of the wire.

Paramagnetic materials Materials that, in the presence of a magnetic field, increase the total flux only slightly. Metals such as platinum and aluminum are paramagnetic.

Permeability of free space, μ_0 Permeability of a vacuum or the lowest permeability possible. For vacuum, air, and most nonmagnetic materials, μ_0 is a constant of $4\pi \times 10^{-7}$ henrys per meter (H/m) or webers per meter·ampere-turn (Wb/m·At).

Permeance, $\mathscr{P}$ Reciprocal of reluctance, measured in units of webers per ampere-turn (Wb/At) in the SI.

Relative permeability, μ_r Ratio of magnetic flux density produced in a given material or medium to that which would be produced in the same dimensional volume of free space by the same magnetizing force. μ_r is dimensionless since it is a ratio of identical units. Mathematically, $\mu_r = \mu/\mu_0$. For air and most nonmagnetic materials, $\mu_r = 1$.

Reluctance, $\mathscr{R}$ Ratio of the MMF ($\mathscr{F}$) to the magnetic flux (ϕ) through any particular cross section of a magnetic circuit, measured in units of ampere-turns per weber (At/Wb). Reluctance in the magnetic circuit is the analog of resistance in the electric circuit.

Retentivity Property of a ferromagnetic material that is measured by its maximum residual magnetic induction when the magnetizing force is zero.

Saturation Condition of sufficient intense magnetizing force that further increases in magnetic field intensity fail to produce proportional increases in flux density in a ferromagnetic material.

Toroidal core Ferromagnetic core of round cross section, shaped in the form of a ring or a doughnut.

11-17 PROBLEMS

Secs. 11-1 to 11-16

The following Probs. 11-1 through 11-9, based on definitions of magnetic quantities, do not require use of B–H curves.

11-1 A toroidal core (Fig. 11-10a) has a cross-sectional diameter of 1 cm, a spiral coil of 1000 turns carrying a current of 2 A. The average circumferential length of the nonmagnetic core is 40 cm. Calculate the

a. Cross-sectional area of the core in m^2
b. Magnetomotive force (MMF)
c. Magnetic field intensity, **H**
d. Permeability, μ_0, of the nonmagnetic core
e. Flux density, **B**, in the core
f. Total flux within the coil center
g. Reluctance of the magnetic circuit

Note: express all answers in terms of magnitudes and appropriate SI units.

11-2 Assuming that the total flux within the coil center of Prob. 11-1 is 0.5 μWb, calculate the

a. Current in the coil to produce a flux of 2 μWb
b. Coil flux when the current is 4 A

11-3 The permeability of a 20-cm closed core is found to be 0.6 H/m at a flux density of 0.8 T. Calculate the

a. Relative permeability of the material, μ_r
b. Magnetizing force, **H**
c. Magnetomotive force, $\mathscr{F}$
d. Coil current in milliamperes for a coil having 30 turns

11-4 A given closed magnetic core has a permeance of 4×10^{-6} Wb/At, an average magnetic path length of 50 cm, and a uniform cross-sectional area of 50 cm^2. Calculate the

a. Reluctance of the core material
b. Magnetomotive force, $\mathscr{F}$, to produce a flux of 5 μWb in the core
c. Coil current if the coil has 200 turns
d. Permeability of the core, using Eq. (11-2)
e. Permeability of the core, using Eq. (11-6)
f. Relative permeability of the core, μ_r

11-5 A solid aluminum bar 2.5 cm in diameter and 15 cm long has a winding of 100 turns and carries a flux of 10 μWb at the center of its core. Calculate the

a. Reluctance of the core
b. MMF required to produce the desired flux
c. Current in the coil
d. Flux density in the core center
e. Magnetic field intensity producing the specified flux in the core
f. Verify the core permeability, using Eq. (11-6)

11-6 A rectangular steel bar has a cross section measuring 2 cm × 4 cm and a length of 20 cm. Its reluctance is 1×10^6 At/Wb. A coil of 50 turns into which the bar is inserted carries a current of 5 A. Calculate the

a. Permeability of the core
b. Magnetic field intensity, **H**
c. Flux density in the center of the steel core, **B**
d. Flux in the center of the core
e. Relative permeability of the core

11-7 A flux density of 0.5 Wb/m^2 is produced in a ferromagnetic core by a magnetic field intensity of 800 At/m. A coil of 100 turns is wound on the core that has an average magnetic circuit length of 30 cm and a cross-sectional area of 8 cm^2. Calculate in SI units the

a. Magnetomotive force, $\mathscr{F}$
b. Permeability of the core
c. Relative permeability of the core
d. Total flux in the magnetic core
e. Reluctance of the core, using the ratio $\mathscr{F}/\phi$
f. Reluctance of the core, using $\mathscr{R} = l/\mu A$
g. Current in the coil drawn from the dc supply to set up required flux density

11-8 A line drawn from the origin of a magnetization curve to a point on the curve yields $\mathbf{B} = 1.5\text{ T} \Leftrightarrow 2000\text{ At/m} = \mathbf{H}$. Calculate the average permeability and relative permeability in henrys per meter from the given data.[28]

[28] Permeability is the ratio **B/H** in units of tesla per ampere-turn per meter. Since 1 henry = 1 weber per ampere-turn, we may express permeability in units of henry/meter (H/m), where $\mu = \frac{\mathrm{B}}{\mathrm{H}} = \frac{\mathrm{Wb/m^2}}{\mathrm{At/m}} = \frac{\mathrm{Wb}}{\mathrm{At \cdot m}} = \frac{\mathrm{H}}{\mathrm{m}}$ as shown in Eq. (11-6).

11-9 Repeat Prob. 11-8, given the data $\mathbf{B} = 6.2832\text{ mT} \Leftrightarrow 5000\text{ At/m} = \mathbf{H}$. Based on your calculations, describe the nature of the material.

Secs. 11-7 to 11-12

Problems 11-10 through 11-22 require use of the **B–H** curves (Fig. 11-7) for solutions involving nonlinear magnetic circuits. Note that each square measures 5 mm and may be divided into five parts for easy interpolation.

11-10 Given the magnetization curve for sheet steel (Fig. 11-7), determine the

a. Average permeability when the flux density is 1.4 T
b. Incremental permeability when the flux density is 1.4 T
c. Maximum permeability of sheet steel
d. Relative permeabilities for values obtained in **(a)**, **(b)**, and **(c)**

11-11 Given the magnetization curve for cast steel (Fig. 11-7), determine the

a. Average and incremental permeability when the flux density is 0.30 T
b. Repeat **(a)** when the flux density is 1.02 T
c. Repeat **(a)** when the flux density is 1.5 T
d. Using the average permeability values only, determine relative permeabilities for the calculated values in **(a)**, **(b)**, and **(c)**
e. Explain why there is a greater variation in values of μ_d than in μ_{av}, based on your calculations

11-12 The number of turns wound on a closed toroidal core (Fig. 11-10a) is 1000 t. The core is made of cast steel and has a cross-sectional diameter, D, of 1 cm. The average length (average circumference) of the core, l_{av} is 40 cm. A flux of 1 mWb is required in the core. Calculate the

a. Required MMF
b. Current in the coil I_c
c. Reluctance of the magnetic circuit
d. Average (static) permeability of the core **(B/H)** in appropriate units (H/m or Wb/At·m)

11-13 A coil having 200 t and carrying a current of 1 A is wound on a closed toroidal core (Fig. 11-10a). The average core length is 50 cm and the cross-sectional core area is 500 cm^2. If the core is made of cast iron, calculate the

a. MMF
b. Magnetic field intensity, **H**
c. Field flux in the core, ϕ
d. Relative permeability of the core
e. Reluctance of the core

11-14 A closed rectangular core without an air gap has a cross-sectional area of 8 cm^2 and an average magnetic path length of 65 cm. The core is made of laminated sheet steel with a coil of 900 turns carrying 2 A. Calculate

a. MMF
b. **H**
c. **B**
d. ϕ
e. μ_{av}
f. μ_r
g. $\mathscr{R}$ of the core

11-15 A closed toroidal core made of cast steel has an average circumferential length of 30 cm, a cross-sectional area of 5 cm^2, and a coil of 225 turns carrying a current of 1 A. Calculate

a. $\mathscr{F}$
b. **H**
c. **B**
d. ϕ
e. μ_{av}
f. μ_r
g. $\mathscr{R}$ of the core

11-16 A closed rectangular core **(Fig. 11-19)** has a cross-sectional area of 20 cm^2. Lengths A and C of the core are each 50 cm long and made of cast steel. Lengths B and D are each 30 cm long and made of sheet steel. Coil N has 1000 turns and the desired flux in the core is 2 mWb. Calculate

a. **B**, flux density in the core
b. **H**, required for the cast steel portions of the core only
c. **H**, required for the sheet steel portions of the core only
d. $\mathscr{F}_t$, total MMF required to set up the specified flux
e. Current in the coil
f. Average permeability of the sheet steel portions
g. Average permeability of the cast steel portions

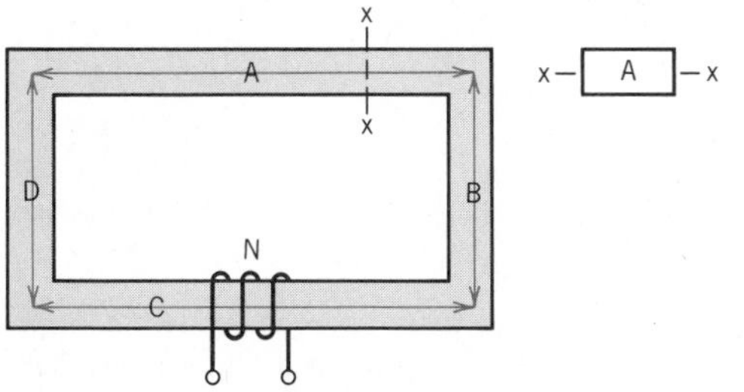

Figure 11-19 Problems 11-16, 11-18, and 11-21

11-17 A closed toroidal core is made of cast iron. Its cross-sectional area is 4 in^2 and its mean magnetic path length is 36 inches. A coil wound around the core has 1000 turns and carries a current of 650 mA. Calculate the

a. $\mathscr{F}$, MMF produced by the coil
b. Magnetic field intensity, **H**, in ampere-turns per inch and ampere-turns per meter
c. Flux density in the core, **B**, in tesla
d. μ_{av} and μ_r in henry per meter

11-18 The closed rectangular core shown in Fig. 11-19 has the same core materials and dimensions as in Prob. 11-16 except that a gap, l_g = 3 mm, has been cut in the A portion of the core. It is desired to establish the same flux and flux density as in Prob. 11-16. Calculate

a. MMF for sheet steel portion
b. MMF for cast steel portion
c. MMF for air gap
d. Total MMF required
e. Required current in the coil
f. Percent increase in current over that of Prob. 11-16 produced by the 3-mm gap

11-19 A cast iron toroidal ring has a cross-sectional area of 2 cm^2, an average length of 40 cm of core material, and an air gap of 1 cm. It is desired to set up a flux of 6 kilogauss (6 kG) in the air gap and the core, using a coil wound on the core carrying a current of 4 A. Calculate the

a. MMF required for the cast iron portion of the core
b. MMF required for the air gap
c. Total MMF required
d. Permeability of the cast iron core, μ_{av}, and relative permeability, μ_r
e. Number of turns on the winding of the coil

11-20 A closed core E-type sheet steel transformer (Fig. 11-11) has a winding on its center leg that carries a current of 0.3 A through 900 turns. The cross-sectional area of its outer leg, A_0, is 3 cm^2 and that of its center leg is 7 cm^2. The dimensions of the core are l_1 = 10 cm and l_2 = 15 cm. Calculate the

a. Magnetic field intensity, **H**, acting on the center leg of the core
b. Flux density and total flux in the center leg of the core
c. Flux density and flux in each outer leg of the core
d. Magnetic field intensity on each outer leg of the core
e. Relative permeability of the sheet steel in center leg
f. Relative permeability of the sheet steel in each outer leg
g. Explain why the relative permeability of the outer-leg portion of the core is necessarily lower than that of the center leg portion

11-21 A rectangular core similar to that shown in Fig. 11-19 has an 800-turn coil and a 5-mm air gap. The total average magnetic circuit length is 30 cm, of which half is made of sheet steel and half of cast steel. The area of the cast steel portion is 8 cm^2, which is also the area of the air gap and cast steel portion. If it is desired to establish a total flux of 0.5 mWb in the air gap and the core, calculate the

a. MMF for the air gap
b. MMF for the sheet steel portion
c. MMF for the cast steel portion
d. Total MMF needed
e. Current in the coil
f. Percent of total MMF needed to overcome air-gap reluctance

11-22 The E-type cut core shown in Fig. 11-12 has a gap in its center core of 5 mm and a flux of 1.2 mWb in the gap. The area of each center leg is 16 cm^2 and that of each outer leg is 6 cm^2. Core dimensions are l_1 = 12 cm, l_0 = 15 cm, and l_c = 14.5 cm. If the core is made of sheet steel and the coil carries a current of 8.5 A, calculate the

a. Flux density in the air gap
b. MMF required to overcome reluctance of the air gap
c. Flux density in the center leg of the core
d. Magnetic field intensity in the center leg of the core
e. MMF required for center leg of the core
f. Flux density in both outer legs of the core
g. Magnetic field intensity in both outer legs of the core
h. MMF required for both outer legs of the core
i. Total MMF required to produce the desired air gap flux
j. Number of turns required on the coil located on the center leg

Secs. 11-13 and 11-14

11-23 A commercial 60-Hz power transformer has a maximum flux density of 1.2 T and a total laminated sheet steel volume of 3500 cm^3. The transformer has laminations 1.0 mm thick with a resistivity of 75 $\mu\Omega\cdot$cm. Calculate the

a. Eddy current loss density, in webers per cubic meter, for the core material used
b. Eddy current power loss for the transformer

11-24 Given a transformer having an eddy current loss of 15 W with the laminations described in Prob. 11-23. It is desired to reduce this loss by using laminations that are 0.5 mm thick, have a resistivity of 100 $\mu\Omega\cdot$cm, and reduce the flux density to 1.0 T. Calculate the anticipated eddy current loss of the transformer, assuming all other factors are unchanged.

11-25 The Steinmetz coefficient, η, for the laminations used in Prob. 11-23 is 330 and n is 1.6 for the given transformer. Calculate the

a. Hysteresis loss density, in webers per cubic meter, for the transformer steel used
b. Hysteresis loss for the given transformer volume

11-26 Reducing the flux density in the transformer to 1 T reduces coefficient n to 1.5 in Prob. 11-25. Calculate the new hysteresis loss in watts, using the ratio method.

11-27 An E-type cut-core transformer (Fig. 11-12) has a volume of 5000 cm^3 of stacked sheet steel laminations. If the maximum flux density in the 5-mm air gap is 1.2 T with the outer leg area (of 20 cm^2) exactly half the center core area of 40 cm^2, calculate the

a. Energy density of the air gap (in J/m^3)
b. Total energy stored in the air gap (in J)
c. Magnetic field intensity of the core
d. Energy density stored in the core material
e. Total energy stored in given volume of core material
f. Ratio of energy density in the air gap to density in the core material

Sec. 11-15

11-28 Each pole of a permanent horseshoe magnet has a pole face area of 25 cm^2 and a flux per pole of 50 μWb. Calculate the

a. Maximum lifting force of the magnet in newtons
b. Maximum weight of iron nails the magnet is capable of lifting in grams

11-29 A bar magnet has a flux of 10 μWb at each pole and a cross section of 1 cm × 5 cm. Calculate the

a. Maximum pull the magnet can exert on ferromagnetic materials
b. Maximum weight the magnet can lift in grams and ounces

11-30 An electromagnetic relay and relay circuit are shown in **Fig. 11-20**, in which a relay current of 0.5 A is sufficient to provide the tractive force to close the armature against the spring to energize a load, R_L, from a 120-V supply. If the flux in the relay gap produced by the relay coil MMF is 50 μWb and the area of the relay pole face is 6 cm^2, calculate the

a. Flux density, **B**, in the air gap, which measures 3 mm
b. Maximum force in newtons that the armature exerts against the spring to close the relay and energize the main 120-V circuit
c. Work done against the retarding spring force
d. MMF of the air gap (and total circuit MMF, assuming that the MMF of the iron and steel portion of the magnetic circuit is negligible compared to air)
e. Number of turns required on the relay coil
f. Power drawn from the 6-V supply by the relay circuit
g. Power controlled by the relay circuit

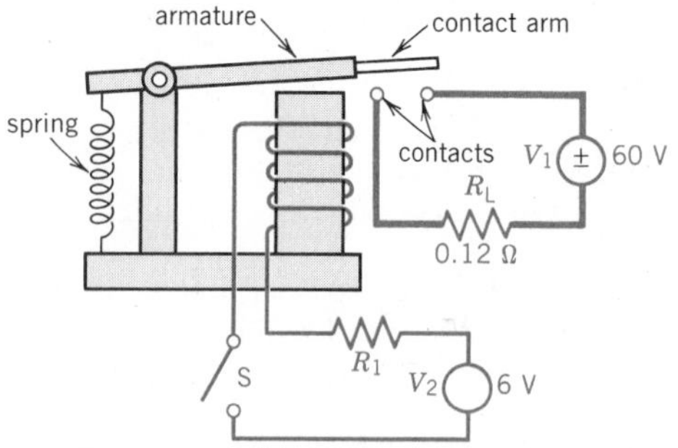

Figure 11-20 Problem 11-30

11-31 A power-line conductor carries a current of 200 A. Calculate the flux density of the field around the conductor at a distance 50 cm away from it.

11-32 Calculate the repulsive force between the two conductors in a 6-ft length of two-wire cord, spaced 5 mm apart, supplying a broiler drawing 15 A in units of

a. Newtons
b. Dynes
c. Pounds (lb_f)

11-33 A busbar 50 cm long on a main power switchboard carries a current of 5 kA. Calculate the

a. Flux density of the field 10 cm from the busbar
b. Force of repulsion (in newtons) between two such bars 10 cm apart each carrying current in opposite directions
c. Force of repulsion in units of lb_f

11-34 A two-wire power transmission line has a length of 800 m between poles. If the lines carry a current of 500 A and the spacing between them is 50 cm, calculate the horizontal force of repulsion between them in

a. Newtons
b. Dynes
c. Pounds (lb_f)

11-35 The vertical component of the earth's magnetic field is 1.5×10^{-4} T in the region of the transmission lines in Prob. 11-34. Calculate the vertical (upward and downward) forces of repulsion between them in

a. Newtons
b. Pounds

11-36 Each conductor of a motor is 30 cm long (active length), carrying a current of 50 A in a uniform magnetic field of 1.6 T. Calculate the force acting on the conductor in units of newtons (N) when it is lying

a. Perpendicular to the field
b. At an angle of 50° from the field
c. Parallel to the magnetic field

11-18 ANSWERS

11-1 a 7.854×10^{-5} m^2 b 2000 At c 5000 At/m d $4\pi \times 10^{-7}$ H/m e 6.283 mT f 493.5 nWb g 4.053 GAt/Wb
11-2 a 8 A b 1 μWb
11-3 a 4.775×10^5 b $1.\overline{3}$ At/m c $0.2\overline{6}$ At/m d $8.\overline{8}$ mA
11-4 a 250 kAt/Wb b 1.25 At c 6.25 mA d,e 4 μH/m f 3.183
11-5 a 243.2 MAt/Wb b 2.432 kAt c 24.32 A d 20.37 mT e 16.21 kAt/m f 1.256 μWb/At·m (or $4\pi \times 10^{-7}$ H/m)
11-6 a 250 μWb/At·m (or 250 μH/m) b 1.25 kAt/m c 312.5 mT d 250 μWb e 199
11-7 a 240 At b 625 μH/m c 497.4 d 400 μWb e,f 600 kAt/Wb g 2.4 A
11-8 750 μWb/At·m (or 750 μH/m), 596.8
11-9 1.257 μWb/At·m (or $4\pi \times 10^{-7}$ H/m); 1.0
11-10 a 1.4 mWb/At b 0.2 mH/m c 60 mH/m d 1114; 159.2; 47 750
11-11 a 1.2 mH/m; 2 mH/m b 1.36 mH/m; 0.4 mH/m c 0.522 mH/m; 0.16 mH/m d 955; 1082; 415.4
11-12 a 550 At b 0.55 A c 55 MAt/Wb d 926 μH/m
11-13 a 200 At b 400 At/m c 9 mWb d 358 e 22.2 kAt/Wb
11-14 a 1800 At b 2769 At/m c 1.49 T d 1.192 mWb e 5.381×10^{-4} H/m f 428.2 g 1.51 MAt/Wb
11-15 a 225 At b 750 At/m c 1.02 T d 510 μWb e 1.36 mH/m f 1082 g 441.2 kAt/Wb
11-16 a 1 T b 725 At/m c 225 At/m d 860 At e 0.86 A f $4.\overline{4}$ mH/m g 1.379 mH/m
11-17 a 650 At b 18 At/in c 0.28 T d 395.1 μH/m; 314.4
11-18 a 135 At b 720 At c 2.387 kAt d 3.242 kAt e 3.242 A f 377%
11-19 a 1 kAt b 4.775 kAt c 5.775 kAt d 191 e 1444 turns
11-20 a 540 At/m b 896 μWb c 448 μWb d 1650 At/m e 1886 f 720
11-21 a 2.487 kAt b 18.75 At c 60 At d 2.566 kAt e 3.21 A f 96.9%
11-22 a 0.75 T b 2984 At c 0.75 T d 150 At/m e 21.75 At f 1 T g 200 At/m h 78 At i 3084 At j 363 t
11-23 a 3.619 kW/m^3 b $12.\overline{6}$ W
11-24 1.953 W
11-25 a 26.51 kW/m^3 b 53 W
11-26 39.6 W
11-27 a 573 kJ/m^3 b 5.73 J c 325 At/m d 195 J/m^3 e 0.975 J f 2938.5
11-28 a 0.796 N b 81.14 g
11-29 a 79.58 mN b 0.286 oz
11-30 a $8\overline{3}$ mT b 1.658 N c 4.974 mJ d 199 At e 398 t f 3 W g 30 kW
11-31 80 μT
11-32 a 16.46 mN b 1.646 kdyn c 3.7×10^{-3} lb$_f$
11-33 a 10 mT b 25 N c 5.62 lb$_f$
11-34 a 80 N b 8 Mdyn c 17.98 lb$_f$
11-35 a 60 N b 15.49 lb$_f$
11-36 a 24 N b 18.4 N c 0

CHAPTER 12

Electromagnetic Induction, Inductance, and Inductors

Resistance, our first circuit property and component, was introduced in Chapter 3. Capacitance, a second circuit property and component, was introduced in Chapter 9. In this chapter we will introduce a third (and last) circuit property, inductance, L. There are only three circuit properties.

Recall that resistance, R, is defined as opposition to a steady (nonvarying) voltage (or current). Capacitance, C, is defined as opposition to a *change in voltage* across the terminals (as well as the ability to store energy in the form of charge or dielectric flux). Inductance, L, as we will see, is the property of a circuit or component to oppose a *change in current* (as well as the ability to store energy in the form of magnetic flux surrounding the *inductor*).

We will also see that a circuit containing inductance has no effect on the current whenever current is constant (nonvarying). But whenever the current undergoes change, even for a short time interval, the property of inductance opposes current by producing an induced electromotive force (EMF) caused by a changing flux linkage with the turns of the inductor. Inductance, defined later, is a measure of the magnitude of induced EMF to the rate of change of current.

12-1 FARADAY'S LAW OF ELECTROMAGNETIC INDUCTION

Whenever current exists in a conductor, it is accompanied by a magnetic field surrounding the conductor. The magnitude of the magnetic field is directly proportional to the magnitude of the current (Oersted's law, Sec. 11-15). Assuming a given magnetic circuit in which all physical factors (length of magnetic circuit, cross-sectional area, turns, and permeability) remain *constant*, we may write

$$\phi = \frac{\mathscr{F}}{\mathscr{R}} = \frac{IN}{\mathscr{R}} = \frac{IN\mu A}{l} = kI \qquad \text{webers (Wb)} \qquad (12\text{-}1)$$

Equation (12-1) is essentially a restatement of Oersted's law (Eq. 11-21) and implies that a magnetic field is one manifestation of current in an electric circuit. Consequently, whenever the current is changing in a circuit, we may write

$$\frac{d\phi}{dt} = k\frac{di}{dt} \qquad \text{webers/second (Wb/s)} \qquad \textbf{(12-2)}$$

Equation (12-2) states that whatever the rate of change of current in a circuit, the rate of change of flux will be proportional to the rate of change of current.

This brings us directly to Faraday's law of electromagnetic (EM) induction. Faraday observed that any change in flux linking a conductor produces an EMF (electromotive force) in that conductor. He noted that this EMF is produced whenever (either)

1. A conductor is moved relative to a stationary magnetic field (Fig. 12-1).
2. The field is varied or moved relative to a stationary conductor (Fig. 12-2).

Faraday's law is quantified for the average EMF, represented as e,

$$e = \frac{\Delta N\phi}{\Delta t} = -N\frac{d\phi}{dt} = -\frac{N\phi}{t} \qquad \text{volts (V)} \qquad \textbf{(12-3)}$$

where N is the number of turns linking ϕ
ϕ is the flux (Wb) that is changing
t is the time during which the flux is changing
$d\phi/dt$ is the time rate of change of flux linkages with turns, N

Observe in Eq. (12-3) the following (not so obvious) points:

1. The minus sign implies that the direction of the average induced EMF, e, is opposed instantaneously and at all times to the applied EMF, which may be responsible for the change in current (and flux) in a coil of N turns (see Lenz's law, Sec. 12-2)
2. Although turns, N, may be constant during current and flux changes, the flux linkages ($N\phi$) *are changing.* This explains why a coil of *fixed* turns *moving* through a *constant, stationary* magnetic field experiences an induced EMF (as in dc generator action).

EXAMPLE 12-1

A bar magnet is located in a coil of 500 turns. The field strength of the bar magnet is 5 mWb. If the bar magnet is totally removed from the coil in 50 ms, calculate the

a. Average EMF induced in the coil
b. Average EMF induced if the time to remove the magnet is 1 s

Solution

a. $e_{av} = N\phi/t = 500(5\text{ mWb})/50\text{ ms}$
$= 50\text{ Wb/s} = \mathbf{50\ V}$ (12-3)

b. $e_{av} = N\phi/t = 500(5\text{ mWb})/1\text{ s}$
$= 2.5\text{ Wb/s} = \mathbf{2.5\ V}$ (12-3)

Example 12-1 shows the following:

1. Whenever there is no voltage to oppose, it is unnecessary to use a minus sign in Eq. (12-3). This does not imply a lack of opposition, however, since work is done in removing the coil *against* the field of the magnet and against the *motion* in removing it (**Lenz's law**). There is opposition, but we have no way of showing it, for now. (See Sec. 12-2.)
2. But the polarity applied to e still holds, for if we connect a center-zero voltmeter to the coil, we obtain voltage of *one* polarity on *removing* the magnet and voltage of the *opposite* polarity whenever we *insert* the magnet in the coil!

Example 12-1 considered the nature of induced EMF when a field is moved relative to a stationary coil. Let us consider the reverse, where the field is stationary and a conductor is moved relative to it as shown in **Fig. 12-1a**. If a force is applied

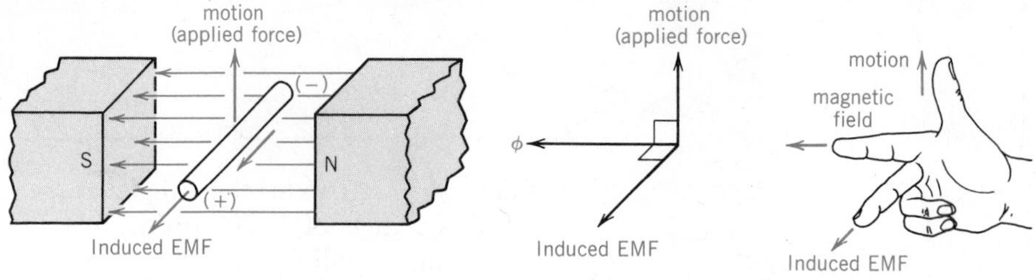

Figure 12-1 Electromagnetic (EM) induction in a moving conductor and RH rule

a. Stationary field, moving conductor

b. Orthogonal relations among directions of magnetic field, motion and induced EMF

c. Fleming's right-hand (RH) rule

in an upward direction to move the conductor (at right angles) through the magnetic field, an orthogonal induced EMF is developed in the direction shown in Fig. 12-1.

The *orthogonal* (mutually perpendicular) relations among the direction of the magnetic field, the motion (direction of force applied to the conductor), and the direction of induced EMF are shown in Fig. 12-1b.

An easy way to remember this relation, known as *Fleming's right-hand* (RH) *rule*, is shown in Fig. 12-1c. The thumb represents the direction of conductor motion (or direction of force applied to the conductor). The index finger points in the direction of the magnetic field. The middle (third) finger shows the direction of EMF induced in the conductor.

The question naturally arises: "Why should an EMF be induced in a conductor moving perpendicular to a magnetic field?" The answer lies (naturally) in the nature of the magnetic field and the nature of the conductor itself. The conductor, by definition, contains relatively many free electrons randomly (Sec. 2-7) and evenly distributed throughout its atomic structure. The magnetic field is a force in space (Sec. 11-1). Now what happens when we *move* a conductor containing many free electrons through a magnetic field? Because of the motion, the force in space is exerted on the moving electrons in such a way as to produce a *separation of charge*. In the case of the conductor shown in Fig. 12-1a, the upward motion of the conductor through the magnetic field causes electrons to be "pushed back" into the paper plane, creating a positive polarity closest to the observer and a negative polarity at the opposite end.

We can easily guess that in the absence of either relative conductor motion or flux linkage change there is no separation of (conductor) charge. Further, we can guess that the faster the conductor moves relative to the stationary field, the more rapidly the charges are separated and the higher the induced voltage. This verifies Eq. (12-3), which is the quantification of Faraday's law.

12-2 LENZ'S LAW FOR DETERMINING POLARITY OF INDUCED EMF

Lenz's law is fundamentally an application of LeChatelier's rule.[1] Lenz's law states:

the polarity and direction of any induced EMF is always such that the current resulting from it produces flux that always tends to oppose any change in the original flux producing the EMF.

[1] LeChatelier, a 17th century French scientist, observed that "nature operates in such a way as to resist a change." Newton's first law of motion and Lenz's law are but two of many examples of LeChatelier's rule.

Stated another way, **Lenz's law** implies that the EMF induced by a changing flux linkage is always of such polarity as to set up a current opposing the change of flux linkage. The law is best understood by considering the two cases shown in **Fig. 12-2**. Figure 12-2a shows that when switch S is closed, a polarity of induced EMF is produced in coil B with (positive) current emanating from the *bottom* terminal. Figure 12-2b shows that when switch S is opened, the polarity of induced EMF reverses in coil B with (positive) current emanating from the *top* terminal.

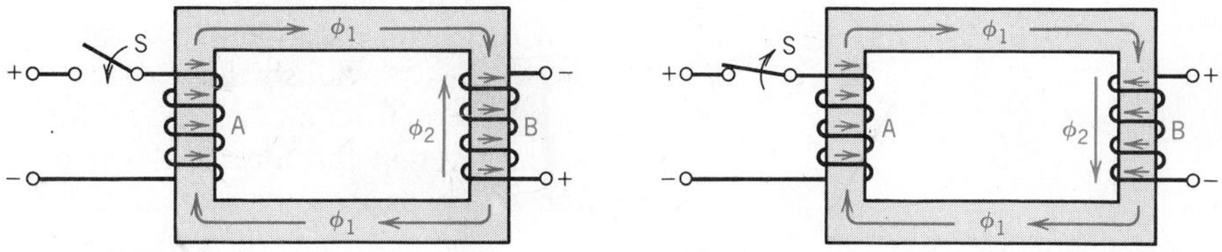

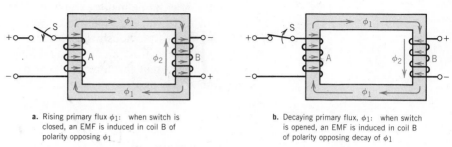

a. Rising primary flux ϕ_1: when switch is closed, an EMF is induced in coil B of polarity opposing ϕ_1

b. Decaying primary flux, ϕ_1: when switch is opened, an EMF is induced in coil B of polarity opposing decay of ϕ_1

Figure 12-2 Lenz's law for determining direction of induced EMF

The proof of both the nature of polarity and polarity reversal stems from an application of Lenz's law. But first note that coils A and B are wound in the *same* direction on the common core. (Prove this by mentally sliding one coil, B, over the other coil, A, in either figure.)

In Fig. 12-2a, when switch S is closed, the following occurs:

1. Current in coil A produces primary flux ϕ_1 having the direction shown by the right-hand magnetic flux rule (Fig. 11-4c).
2. But by Lenz's law, the EMF induced in coil B must oppose this rising flux ϕ_1.
3. Again, using the (magnetic flux) right-hand rule, in order for coil B to oppose ϕ_1 it must produce an opposing flux, ϕ_2, producing the current directions shown in coil B.
4. Mentally superimposing coil B on coil A also shows that the induced EMF in coil B opposes the applied voltage to coil A (since the positive polarity of coil B opposes the positive polarity of the source).[2]

In Fig. 12-2b, when switch S is opened, the following occurs:

1. Current in coil A begins decreasing and primary flux ϕ_1 decreases.
2. But by Lenz's law, the EMF induced in coil B must oppose this decreasing flux, ϕ_1.
3. For coil B to oppose ϕ_1 it must now produce a flux ϕ_2 that *maintains* ϕ_1, that is, is in the *same* direction as ϕ_1.
4. Again, using the right-hand magnetic flux rule, we obtain the current direction in coil B to set up ϕ_2. Note that current now emanates from the top terminal of coil B, making that terminal *positive*.
5. Note that the polarity of coil B when S is *opened* (Fig. 12-2b) is the *opposite* of the polarity of coil B when S is *closed* (Fig. 12-2a). This shows that opening

[2] One cannot refrain from noting that an EMF is also induced in the primary coil A which (like coil B) opposes the applied voltage. It is this primary induced EMF that limits the primary current to a low value for both inductors and open-circuited transformers when ac is applied. This EMF is called a *self-induced* EMF, an EMF of *self-induction* (Sec. 12-3), or a *counter EMF*.

and closing S produces an alternating voltage on coil A that results in an alternating voltage at coil B. (This is transformer action.)

In summary, the foregoing discussion proves Lenz's law; that is, the EMF induced by a changing flux linkage is always in such a direction and polarity as to **oppose** the change of flux linkage.

Lenz's law is also confirmed whenever a moving conductor is displaced through a stationary magnetic field. The moving conductor of Fig. 12-1 is again reproduced in **Fig. 12-3a** with one major exception. This time a resistor R is connected to the conductor to permit a current, I, to flow in response to the generated EMF.

But we learned previously (Sec. 11-15.2) that whenever a conductor carries current in a magnetic field, an orthogonal force is produced as a result (Fig. 11-17a). We also learned that this force is the principle of motor action. Is it possible that generator action and motor action occur *simultaneously* during EM induction?

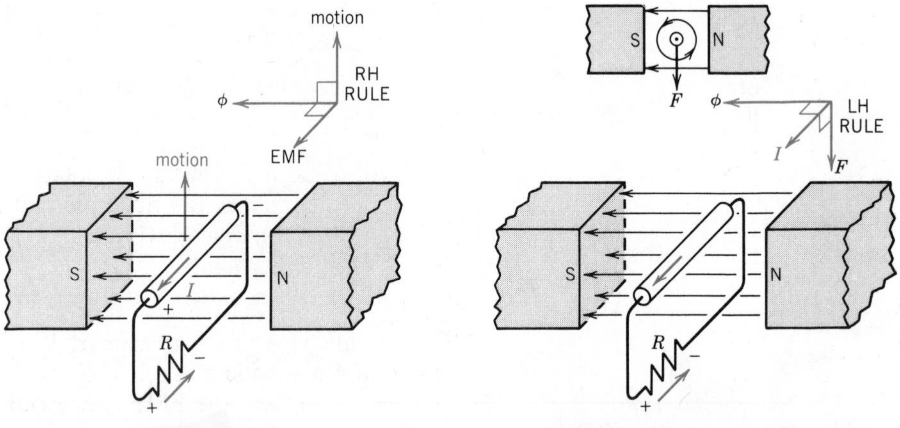

Figure 12-3 Lenz's law during EMF induction in a moving conductor

Lenz's law tells us that indeed this must be so, as shown in Fig. 12-3b. The upward motion imparted to the conductor in Fig. 12-3a produces an EMF that results in current in the conductor. That current, as shown by the LH rule in Fig. 12-3b (originally shown in Fig. 11-17c), produces a force, F, that *opposes* the original motion![3]

Consequently, *all* electromechanical energy conversion devices experience *both* generator and motor action *simultaneously*. An electric motor whose current is drawn from an electric source produces conductor rotation in a magnetic field. But we know that any time a conductor moves in a magnetic field, generation occurs due to Faraday's law. In this instance, using the LH and RH rules simultaneously (with both the thumbs and index fingers pointing in the *same* direction), we find that the induced EMF opposes the current in the motor drawn from the supply. This counter EMF is but another illustration of Lenz's law.

The situation just described is essentially the reason why a dynamo may be operated as *either* a generator or a motor. If mechanical energy is supplied to the rotor of the dynamo, it behaves as a generator (but motor action is also occurring).

[3] The reader can easily demonstrate this by pointing the index fingers of each hand in the *same direction* and rotating the left hand (LH) so that the current is in the same direction as the induced EMF. It will then be observed that the *motion* of the RH (generator action) is opposed by the *force* produced by motor action (LH).

If electrical energy is supplied to a dynamo's armature, it behaves as a motor (but generator action is also occurring).[4]

12-3 SELF-INDUCTION AND SELF-INDUCTANCE

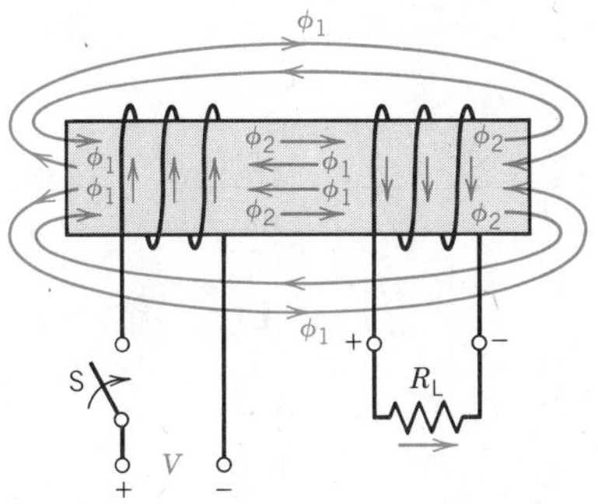

Figure 12-4 Lenz's law for proving direction of self-induced EMF

The previous section noted that while coil A of Fig. 12-2 succeeded in inducing an EMF in coil B whenever the current in coil A was changing, it also produced an opposing or counter EMF against coil A. This may be explained by considering **Fig. 12-4**. Imagine that only a single coil exists connected to a supply V on a core which is shown as the coil on the left. Now imagine that the *same* coil, for purposes of induced EMF, has been slid to the right of the core. When switch S is closed current flows in the (left) coil producing ϕ_1. But by Lenz's law the coil at the right must set up a flux, ϕ_2, which opposes ϕ_1. The induced current and induced EMF in the (right) coil show a polarity and current flow opposing the actual applied voltage and current in the (left) coil. Now if the right coil is slid to the left, we can understand the nature of self-induction. We may define *self-induction* as

the property of an electric circuit or component to oppose any change in current in that circuit or component.

Any circuit or component that possesses the property of self-induction has inductance, L, and is called an inductor.[5] The relation between the EMF of self-induction, e_L, and the property of inductance L may be derived as follows:

$$e_L = k\frac{d\phi}{dt}, \quad \text{induced EMF is proportional to rate of change of flux} \tag{12-3}$$

but

$$d\phi/dt = k\frac{di}{dt}, \quad \text{which in turn depends on rate of change of current} \tag{12-2}$$

therefore $e_L = k\dfrac{di}{dt} = L\dfrac{di}{dt}$, which may be rearranged in terms of L to yield:

$$L = \frac{e_L}{di/dt} = N\frac{d\phi}{di} \quad \text{henry (H)} \tag{12-4}$$

where e_L is the EMF of self-induction in volts (V), produced by di/dt, and

di/dt is the rate of change of current in the circuit in amperes per second (A/s)

Equation (12-4) defines L in terms of given units:

An electric component or circuit has an inductance of 1 henry whenever a current change of 1 A/s induces a voltage in that circuit or component of 1 V

---

[4] It is a general principle of electrical engineering that any device that converts energy (a transducer) from electrical to mechanical will also convert mechanical energy to electrical. This principle of "reversibility" is not only confined to the motor and generator. Other examples are the microphone and the electrodynamic speaker. If sound energy is imparted to a speaker, electrical signals are generated. If electrical signals are imparted to a speaker, sound (mechanical energy) is produced.

[5] Recall that any circuit or component having the property of resistance is called a resistor. Any circuit or component having the property of capacitance is called a capacitor. See Table 12-1 for a comparison of the three properties (Sec. 12-11).

The second term of Eq. (12-4) may be derived from the equality

$$e_L = L\frac{di}{dt} = N\frac{d\phi}{dt} \qquad \text{volts (V)} \tag{12-5}$$

which, when solved for L, yields $L = N\,d\phi/di$.

EXAMPLE 12-2

Assume that current rises linearly from 0 to 5 A in a time of 0.5 s and induces an average EMF of 10 V. Calculate

a. The rate of change of current
b. The inductance of the circuit

Solution

a. $di/dt = (5 - 0)\text{ A}/(0.5 - 0)\text{ s} = \mathbf{10\ A/s}$

b. $L = \dfrac{e_L}{di/dt} = \dfrac{10\text{ V}}{10\text{ A/s}} = \mathbf{1\ H}$ (12-4)

Example 12-2b shows how the magnitude of L is affected by e_L and di/dt. A *large* inductance produces a *high* induced voltage for a *small* rate of change of current. Conversely, a *small* inductance produces a *low* induced voltage for a rapid (*high*) rate of change of current.

EXAMPLE 12-3

An air-core inductor having 500 turns experiences a rise in current from 0 to 0.2 A in 0.1 s. Assume that when the current rises linearly it produces a final flux of 2 μWb. Calculate the

a. Rate of change of current
b. Induced EMF
c. Inductance by two methods

Solution

a. $di/dt = (0.2 - 0)\text{ A}/(0.1 - 0)\text{ s} = \mathbf{2\ A/s}$

b. $e_L = N\,d\phi/dt = 500\text{ t} \times (2 - 0)\ \mu\text{Wb}/(0.1 - 0)\text{ s}$
$= \mathbf{10\ mV}$ (12-5)

c. $L = \dfrac{e_L}{di/dt} = \dfrac{10\text{ mV}}{2\text{ A/s}} = \mathbf{5\ mH}$ (12-4)

$L = N\,d\phi/di = 500\text{ t}\,(2\ \mu\text{Wb})/(0.2\text{ A}) = \mathbf{5\ mH}$ (12-4)

Example 12-3 shows the unity of the two forms of Eq. (12-4) in calculating L. It also shows how inductance varies with such factors as change in flux (and current) and number of turns. But since an inductor is essentially a coil that may or may not have a ferromagnetic core, we would like some indication of the *factors* affecting the magnitude of inductance. This is easily found, beginning with

$$\phi = IN\mu A/l \qquad \text{from the previous chapter} \tag{11-3}$$

Then $d\phi = di(N\mu A/l)$, which may be substituted in Eq. (12-4) yielding

$$L = N\frac{d\phi}{di} = \frac{N^2\mu A}{l} = \frac{N^2}{\mathcal{R}} \qquad \text{henrys (H)} \tag{12-6}$$

where N is the number of turns on the inductor
μ is the *absolute* permeability ($\mu_r\mu_0$) of the core of the coil
A is the cross-sectional area of the coil in square meters (m^2)
l is the average length of the magnetic core or the length of an air-core coil in meters (m)

Equation (12-6) is extremely useful because it permits a computation of inductance in terms of the *physical factors affecting inductance*.[6] It also yields a way of obtaining a desired value of inductance by either adding or removing turns, as shown by the following examples.

EXAMPLE 12-4
A coil is wound with 600 turns on a nonmagnetic core whose relative permeability, μ_r, is unity. It has a length of 20 cm and a cross-sectional area of 4 cm². Calculate the inductance of the coil.

Solution

$$L = \frac{N^2 \mu A}{l} = \frac{(600)^2(1 \times 4\pi \times 10^{-7}) \times 4\ \text{cm}^2 \times 10^{-4}\ \text{m}^2/\text{cm}^2}{20\ \text{cm} \times 10^{-2}\ \text{m/cm}}$$
$$= \mathbf{904.8\ \mu H} \qquad (12\text{-}6)$$

EXAMPLE 12-5
Using the coil dimensions from Ex. 12-4, it is desired to modify the original coil to produce a needed final inductance of 700 μH. Calculate the

a. Number of (final) turns needed on such a coil
b. Number of turns that must be removed from the original coil of Ex. 12-4 to yield the desired 700-μH inductance

Solution
Since we desire turns, N is $\propto (L)^{1/2}$; then using the ratio method

a. $N_f = N_0\left(\dfrac{L_f}{L_0}\right)^{1/2} = 600\,\text{t}\left(\dfrac{700\ \mu\text{H}}{904.8\ \mu\text{H}}\right)^{1/2} = 527.7 = \mathbf{528\ turns}$

b. Turns removed $= N_0 - N_f = 600\ \text{t} - 528\ \text{t} = \mathbf{72\ turns}$

EXAMPLE 12-6
Verify the answers to Ex. 12-5 using Eq. (12-6)

Solution

$$L = N^2\mu A/l = 528^2 \times 4\pi \times 10^{-7} \times 4 \times 10^{-4}/20 \times 10^{-2}$$
$$= \mathbf{700.7\ \mu H} \qquad (12\text{-}6)$$

Examples 12-4 to 12-6 show how precisely inductance may be computed using Eq. (12-6). Example 12-5 shows that it is unnecessary to include in Eq. (12-6) the factors that are held constant. Since only the inductance is changed (reduced), we use the ratio method to find the new (final) turns in the solution to Ex. 12-5**a**. It is important to note that we *cannot* remove a half-turn or a quarter-turn. This is why we rounded the number of final turns from 527.7 to 528 turns in Ex. 12-5**a**. This rounding error produced a slight difference in desired inductance in Ex. 12-6.

The insights to be gained from the foregoing examples and Eqs. (12-6) and (12-5) are as follows:

1. The property of inductance depends on the construction of the inductor.
2. The number of turns is the major property, since doubling the number of turns increases the inductance by a factor of 4.
3. Inductance varies directly as the permeability and the core cross-sectional area and inversely with length of magnetic circuit.
4. Therefore, a low inductance is obtained with coils having few turns, small cross-sectional areas, long magnetic circuits, and cores of nonmagnetic material (low μ). Such a coil having low inductance is shown in **Fig. 12-5a.**
5. A high inductance, conversely, has many turns of fine wire, a large area, an extremely small magnetic circuit path, and a magnetic core of high permeability. Such a "pancake" coil, shown in Fig. 12-5b, has a high inductance. It is also capable of producing extremely high flux, as shown by Eq. (11-3). (The deflection yoke of a TV picture tube is a typical pancake coil.)
6. The induced voltage in a coil (Eq. 12-5) is essentially a function of the rate of change of current. If the rate of change of current is *constant*, the voltage is *constant*. When the rate of change of current is *zero*, the voltage is *zero*. This is shown in Ex. 12-7.

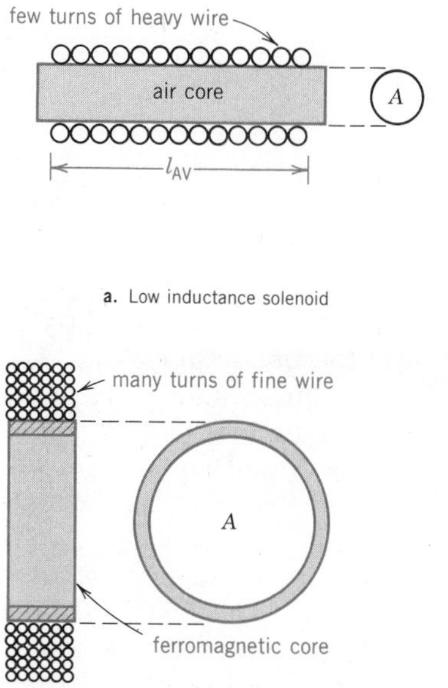

a. Low inductance solenoid

b. High inductance "pancake" coil

Figure 12-5 Comparisons of inductors based on construction and Eq. (12-6)

[6] In this respect it is similar to the equations $R = \rho l/A$ and $C = \varepsilon_0\varepsilon_r A/d$, derived in Chapters 3 and 9.

EXAMPLE 12-7
The rate of change of current over a 10-s period in a 250-mH coil is shown in **Fig. 12-6a**. Calculate and draw the voltage, v_L, across the coil for the 10-s period.

Solution

Interval (s)	0–1	2–4	4–5	5–6	6–7	7–9	9–10
di/dt (A/s)	5	2.5	0	−5	0	−2.5	0
v_L (V)	1.25	0.625	0	−1.25	0	−0.625	0

This table is completed by first calculating di/dt. During the time interval from 0 to 1 s, the current rises from 0 to 5 A. Then $di/dt = 5\ \text{A}/1\ \text{s} = 5\ \text{A/s}$. Using Eq. (12-5), then, $v_L = L\,di/dt = 0.25\ \text{H} \times 5\ \text{A/s} = 1.25\ \text{V}$

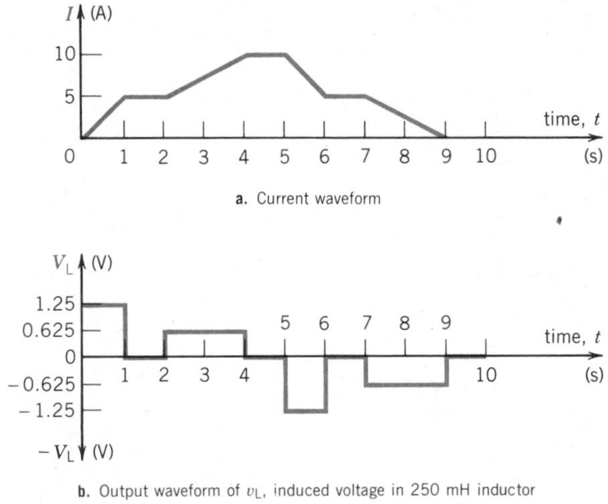

Figure 12-6 Example 12-7

The output v_L waveform, based on the calculations and table in Ex. 12-7, is shown in Fig. 12-6b.

Example 12-7 shows that since the inductor is constant (250 mH), the induced voltage, v_L, across the inductor is only and strictly a function of di/dt (Eq. 12-5). When the *slope*

1. of di/dt is *positive* and *constant*, v_L has a *positive constant* value.
2. of di/dt is *negative* and *constant*, v_L has a *constant negative* value.
3. is zero, v_L is zero.

12-4 COMMERCIAL INDUCTORS AND THEIR APPLICATIONS

An inductor may be defined as a *lumped* circuit element having the property of inductance. An inductor may have either a *ferromagnetic* core or an *air* core. The value of inductance, its ohmic resistance, and the resistance due to core losses are usually measured on modern commercial electronic bridges (Sec. 23-14) at some specified frequency at which the inductor is to be used. These bridges will yield either a series or a parallel equivalent circuit (or both) representing the measured values.

The equivalent circuit of a commercial inductor in the series representation is shown in **Fig. 12-7a**, as measured by an electronic bridge. The symbol for the inductor, L_s, is shown as a coil in series with R_s. R_s represents the combined resistance of the inductor turns of copper (usually), contact resistance at the terminals, and equivalent core losses due to hysteresis and eddy current (Sec. 11-13) at the specific frequency of measurement.[7]

Some types of electrical bridges yield only the parallel circuit representation

[7] Of the three basic circuit components, capacitance, inductance and resistance, only resistance is capable of dissipating electrical energy and producing electrical losses. Capacitors and inductors only store electrical energy.

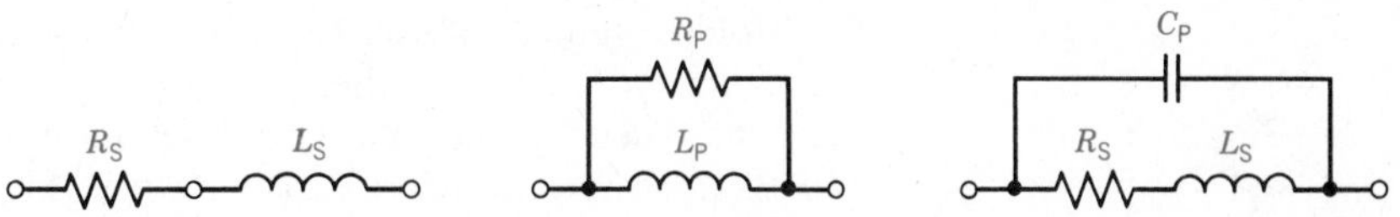

a. Series representation   b. Parallel representation   c. High frequency representation

Figure 12-7 Equivalent circuits of commercial inductors

shown in Fig. 12-7b. The value of L_p thus obtained is somewhat higher than L_s and is related to it by a quality factor, Q. We will discover later (Sec. 16-6) that whether we measure the series or parallel equivalent, we can easily convert from one representation to another.

It should be noted, however, that at high frequencies an inductor also contains *distributed capacitance*. This is due to the insulation between adjacent turns of wire in the coil. (Recall that conductors separated by an insulator have the property of capacitance.) Thus, the equivalent circuit of an inductor is represented as shown in Fig. 12-7c, where C_p represents the total distributed capacitance *between turns* of the coil.[8]

Commercial inductors, like resistors and capacitors, are available in a variety of fixed, adjustable, and variable forms, depending on circuit application. **Figures 12-8a to 12-8f** show the inductor symbols used to represent them in schematic diagrams. Figures 12-8g through 12-8k show the appearances of various commercial types.

a. Fixed (air)

b. Variable (air)

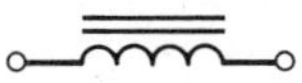

c. Iron core

d. Ferrite core

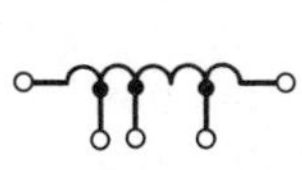

e. Tapped

f. Permeability tuned

g. Toroidal wound ferrite core (Courtesy Bell Industries, J. W. Miller Div.)

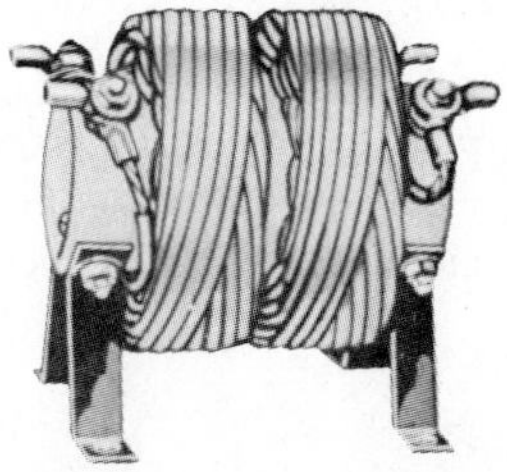

h. High current air choke (Courtesy Bell Industries, J. W. Miller Div.)

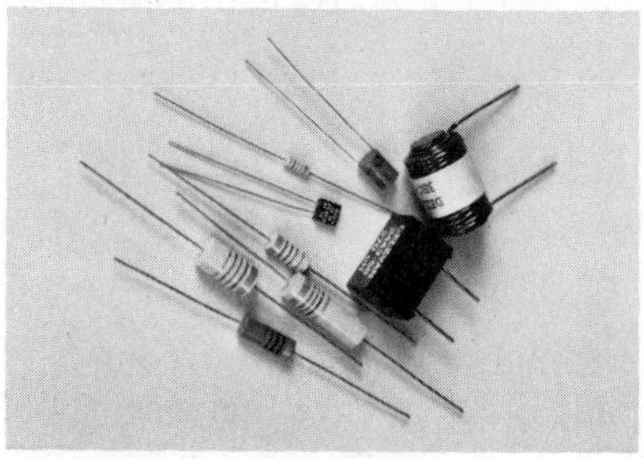

i. Small axial, flatpack and tubular inductors (Courtesy Delevan Div. of American Precision Industries)

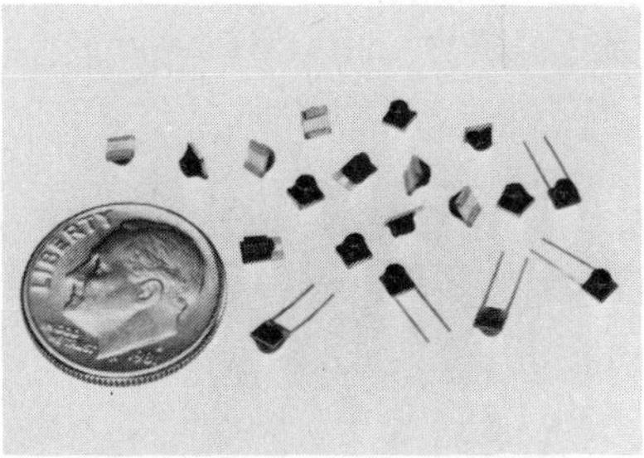

j. PCB surface-mountable inductors (Courtesy Delevan Div. of American Precision Industries)

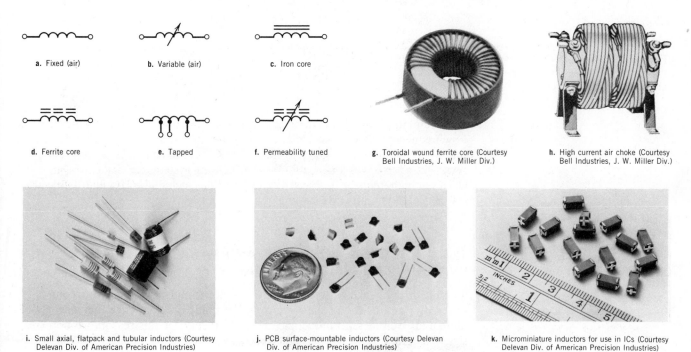

k. Microminiature inductors for use in ICs (Courtesy Delevan Div. of American Precision Industries)

Figure 12-8 Inductor symbols and some commercial types of inductors

[8] Surprising as it may seem, at very high frequencies, *more* current flows in C_p than in L_p; that is, more current flows *between* the wires (through the insulation) than through the coil itself! Figure 12-7c shows how this is possible.

Molded inductors (Fig. 12-8i) are currently available, similar in appearance to resistors for use in a variety of integrated circuits (ICs) and discrete electronic circuits. These smaller inductors vary in inductance from approximately 0.01 μH to 100 mH and are banded, as shown in **Fig. 12-9**, using a color code similar to that for resistors.

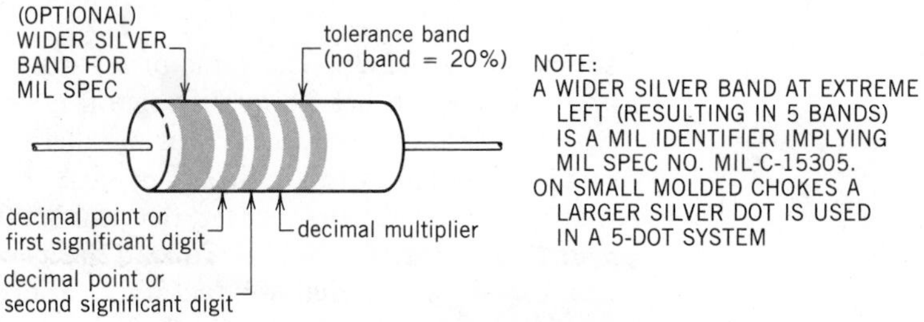

Color	Significant digit	Decimal Multiplier (Put as zeros behind first two digits)		Tolerance (%)
Black	0	1	10^0	
Brown	1	10	10^1	
Red	2	100	10^2	
Orange	3	1000	10^3	
Yellow	4	10 000	10^4	
Green	5	100 000	10^5	
Blue	6	1000 000	10^6	
Violet	7	10 000 000	10^7	
Gray	8	100 000 000	10^8	
White	9	1000 000 000	10^9	
Gold	DECIMAL POINT	—		± 5
Silver	MIL IDENTIFIER	—		± 10
None	—	—		± 20

Figure 12-9 Color coding for small fixed inductors and for flat molded chokes using dot conventions

Inductor applications are many and varied in both power and electronic circuits. Low-frequency applications include current limiting and wave filters for dc power supplies to reduce ripple. Current-limiting inductors are either iron-core "chokes" such as used in low frequency fluorescent lighting systems for ballasts or air-core helixes at high frequencies. Electronic applications include *L-C* oscillators, electromagnetic delays, and waveshaping.

12-5 INDUCTORS IN SERIES AND PARALLEL WITHOUT MAGNETIC COUPLING

One major difference between inductors in series and parallel circuits compared to resistive and/or capacitive circuits is that inductors may have the property of mutual (as well as self-) inductance between them. For this reason, we first consider inductors that have *no magnetic coupling* between them and therefore *no mutual* inductance.

Consider the series circuit shown in **Fig. 12-10a**, where the inductors are physically positioned at right angles to each other to eliminate magnetic coupling. When switch S is closed, since L_1, L_2, and L_3 are in *series*, they each experience the *same* rate of change of current, di/dt. From Kirchhoff's voltage law (KVL) we may write

$$V = v_{L_1} + v_{L_2} + v_{L_3}$$

$$L_{ts}(di/dt) = L_1(di/dt) + L_2(di/dt) + L_3(di/dt), \qquad \text{from which}$$

$$L_{ts} = L_1 + L_2 + L_3 + \cdots + L_n \qquad \text{henry (H)} \qquad \textbf{(12-7)}$$

where L_{ts} is the total inductance of *series-connected* inductors, *assuming* no coupling between them

Observe that Eq. (12-7) is similar to the relation for *resistors* in *series* and capacitors in *parallel* (Eq. 9-5). Clearly, it would be helpful in many instances to treat inductors like resistors.

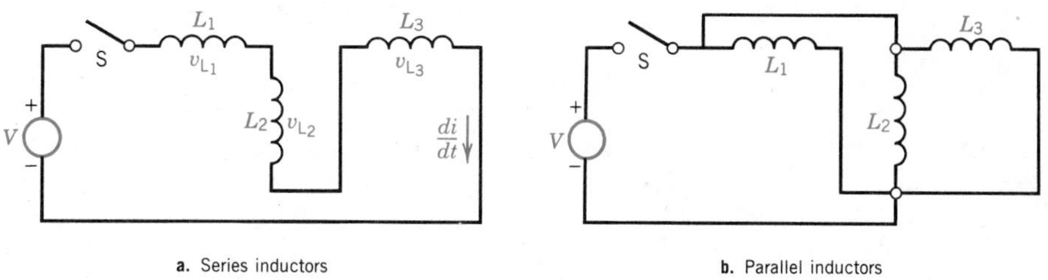

Figure 12-10 Series and parallel inductors with no coupling between them

Now consider the parallel circuit shown in Fig. 12-10b. Again, the inductors have been placed in such a way as to avoid magnetic coupling so that there is no mutual induction between them. Basic parallel-circuit relations from resistors in parallel apply, and therefore when switch S is closed $V = V_{L_t}$ and $i_t = i_1 + i_2 + i_3$ and $di_t = di_1 + di_2 + di_3 = v_{L_t}(dt)/L_t$ from Eq. (12-4), yielding

$$\frac{v_{L_t}(dt)}{L_{tp}} = \frac{v_{L_t}(dt)}{L_1} + \frac{v_{L_t}(dt)}{L_2} + \frac{v_{L_t}(dt)}{L_3} \qquad \text{and therefore}$$

$$\frac{1}{L_{tp}} = \frac{1}{L_1} + \frac{1}{L_2} + \frac{1}{L_3} + \cdots + \frac{1}{L_n} \qquad \text{henry (H)} \qquad \textbf{(12-8)}$$

where L_t is the total inductance of *parallel-connected* inductors, *assuming* no coupling between them

Again, Eq. (12-8) shows that we may treat inductors like resistors insofar as parallel and series circuit relations are concerned, with the understanding that *no magnetic coupling* exists between the connected inductors.

EXAMPLE 12-8

Three coils of 20, 5, and 4 mH are connected, respectively, in series and then in parallel. Assuming no magnetic coupling between coils, calculate the

a. Total series inductance, L_{ts}
b. Total parallel inductance, L_{tp}

Solution

a. $L_{ts} = L_1 + L_2 + L_3 = (20 + 5 + 4) = \mathbf{29\ mH}$ (12-7)

b. $\frac{1}{L_{tp}} = \frac{1}{L_1} + \frac{1}{L_2} + \frac{1}{L_3} = \frac{1}{20} + \frac{1}{5} + \frac{1}{4} = \mathbf{2\ mH}$ (12-8)

12-6 MUTUAL INDUCTANCE

When two (or more) coils are physically located so that part of the flux produced by the turns of any one coil magnetically links the turns of the other(s), and vice versa, such coils are said to be *mutually coupled.*

Mutual inductance between two magnetically coupled coils is shown in **Fig. 12-11**. Assume that switch S has just been closed energizing coil 1. The action that occurs and the definitions that result are:

1. Primary current i_1 builds up in coil 1 producing ϕ_1, which contains two distinct fluxes surrounding coil 1 and due to the turns N_1 of coil 1, described in (2) and (3) below.
2. ϕ_{11} is that component of ϕ_1 produced by coil 1 linking coil 1 *only*, producing self-inductance L_1.
3. ϕ_{12} is that component of ϕ_1 produced by coil 1 linking coil 2 *only* (and coil 1 as well), inducing an EMF in coil 2.
4. But by Lenz's law, the EMF in coil 2 results in a secondary current i_2, whose direction must create flux in opposition to ϕ_{12}. The total flux ϕ_2 produced by i_2 has two components, ϕ_{22} and ϕ_{21}.
5. ϕ_{22} is the flux produced by coil 2 linking coil 2 only, producing self-inductance L_2.
6. ϕ_{21} is the flux produced by coil 2 linking coil 1 only (and coil 2 as well).
7. Observe that ϕ_{21} (per Lenz's law) is in opposition to ϕ_{12}.

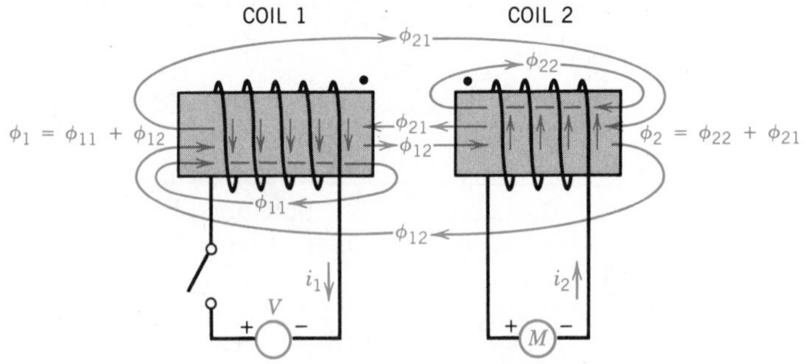

Figure 12-11 Mutual inductance between two magnetically coupled coils

We must now distinguish between self-inductance L and mutual inductance M. *Both are inductances* measured in henrys. But mutual inductance is only that inductance produced by magnetic coupling between two (or more) coils. Therefore we may write from Eq. (12-4) the following expressions for mutual inductance between coils 1 and 2:

$$M_{12} = \frac{N_2\phi_{12}}{i_1} = \frac{e_2}{di_1/dt} \quad \text{henry (H)} \tag{12-9}$$

where M_{12} is the mutual inductance due to ϕ_{12} (from coil 1) on coil 2 (Fig. 12-11)

We may also write

$$M_{21} = \frac{N_1\phi_{21}}{i_2} = \frac{e_1}{di_2/dt} \quad \text{henry (H)} \tag{12-10}$$

where M_{21} is the mutual inductance due to ϕ_{21} (from coil 2) on coil 1 (Fig. 12-11)

And, of course, the *self-inductances* of the two coils, by inspection, are

$$L_1 = \frac{N_1\phi_1}{i_1} \qquad L_2 = \frac{N_2\phi_2}{i_2} \qquad \text{henry (H)} \tag{12-11}$$

Is there a simple relation between mutual and self-inductance in Fig. 12-11? To answer this question, let us first consider the ideal (unity coupling) case where *all* the flux produced by primary current i_1 links coil 2 and *all* the flux produced by secondary current i_2 links coil 1. This means that $\phi_1 = \phi_{12}$ and $\phi_2 = \phi_{21}$. For such an ideal (unity coupling) case since $M_{12} = M_{21} = M$, we may write

$$L_1L_2 = M_{21}M_{12} = M^2 \qquad \text{and solving for } M \text{ yields}$$

$$M = \sqrt{L_1L_2} \qquad \text{henry (H)} \tag{12-12}$$

where all terms have been defined above

Equation (12-12) correctly approximates the situation for commercial iron-core transformers, which are relatively tightly coupled, intentionally and by design. All attempts are made to ensure that the flux from one coil links the flux of another. Such coils have a coefficient of coupling, k, of approximately unity. (See Eq. 12-12a.)

But now consider the possibility of an air-core transformer such as that shown in Fig. 12-11. There is always some leakage flux *not* common to both coils (i.e., ϕ_{11} and ϕ_{22} are leakage fluxes that produce a *reduction* in the mutual inductance from that of the ideal case.) Assuming that the magnetic permeability μ of the mutual flux path remains constant for both coils, we may still assume that $M_{12} = M_{21}$ since they are both produced by the same *mutual* flux. We may now define the coefficient of coupling, k, as the ratio

$$k = \sqrt{\frac{\phi_{12}}{\phi_1} \cdot \frac{\phi_{21}}{\phi_2}}$$

where $\phi_{12} = \dfrac{M_{12}i_1}{N_2}$; $\phi_1 = \dfrac{L_1i_1}{N_1}$; $\phi_{21} = \dfrac{M_2i_2}{N_1}$; and $\phi_2 = \dfrac{L_2i_2}{N_2}$

Substituting these fluxes respectively into the preceding equation yields

$$k = \sqrt{\frac{(M_{12}i_1/N_2)}{(L_1i_1/N_1)} \cdot \frac{(M_{21}i_2/N_1)}{(L_2i_2/N_2)}} = \sqrt{\frac{M_{12}}{L_1} \cdot \frac{M_{21}}{L_2}} = \sqrt{\frac{M^2}{L_1L_2}} \qquad \text{since } M_{12} = M_{21}$$

from which

$$k = \frac{M}{\sqrt{L_1L_2}} \qquad \text{(dimensionless ratio)} \tag{12-12a}$$

EXAMPLE 12-9

Given two coils, $L_1 = 10$ mH and $L_2 = 50$ mH, wound on top of each other on a closed ferromagnetic core, calculate the maximum possible mutual inductance between the two coils

Solution

Assuming unity coupling between the two coils, $k = 1$,
$M = k\sqrt{L_1L_2} = 1\sqrt{10 \times 50} =$ **22.36 mH** (12-12) or (12-12a)

EXAMPLE 12-10
Assuming that the actual coefficient of coupling between the two coils of Ex. 12-9 is 0.9, calculate the mutual inductance M between them

Solution
$M = k\sqrt{L_1 L_2} = 0.9\sqrt{10 \times 50} = \mathbf{20.12\ mH}$ (12-12a)

12-7 THE DOT CONVENTION

A convenient and useful method for predetermining the instantaneous flux direction is to use the *dot convention.* The dots enable us to determine the *direction of flux* produced by a given coil, depending on the direction of current and how the turns are wound on the coil. The *dot* represents the *head* of the flux arrow.

Figure 12-12 shows the case of two *conductively* coupled coils, both wound in the *same* direction on a common core. In Fig. 12-12a the coils are electrically connected in series so that their mutual fluxes, ϕ_{12} and ϕ_{21}, *aid* each other; that is, their mutual fluxes are *additive.* In Fig. 12-12b the coil connections are reversed so that their mutual fluxes, ϕ_{12} and ϕ_{21} *oppose* each other; that is, their mutual fluxes are *subtractive.*

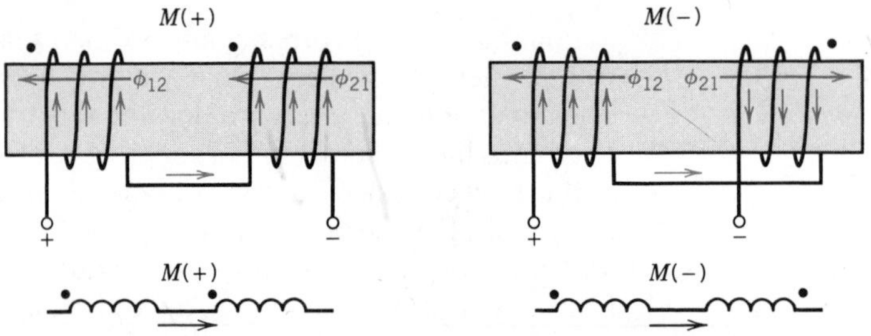

a. Mutual flux aiding additive polarity

b. Mutual flux opposing subtractive polarity

Figure 12-12 Dot conventions for coils wound in same direction on core

If we consider the direction of *currents* as they enter (or leave) each coil, respectively, we might infer from Fig. 12-12 that when current flows into a dot when *entering* each coil, the windings are connected with *additive* polarity. Conversely, when current enters a dot in one coil and leaves a dot in a second coil, the windings are connected with *subtractive* polarity.

But Fig. 12-12 was drawn with both windings wound in the same direction. Does this hold true if the coils are wound oppositely on the same core?

Figure 12-13 shows the case of two conductively coupled coils wound in *opposite* directions on the same core. For the current directions given in Fig. 12-13a, we can see that ϕ_{12} is in a direction *opposite* to ϕ_{21}, yielding *subtractive* polarity. In Fig.

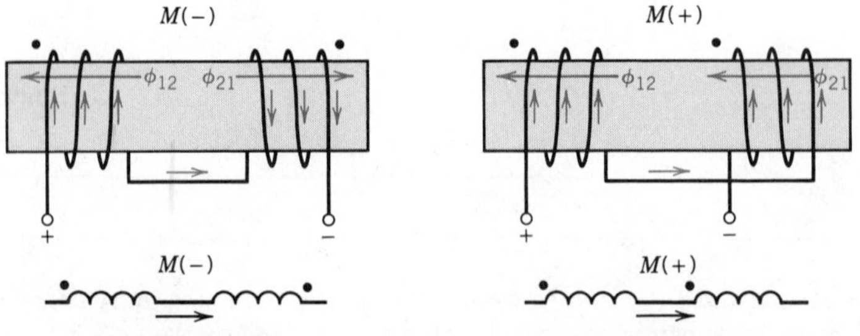

a. Mutual flux opposing subtractive polarity

b. Mutual flux aiding additive polarity

Figure 12-13 Dot conventions for coils wound in opposite directions on core

12-13b the coil connections are reversed, and now the mutual fluxes ϕ_{12} and ϕ_{21} are *aiding* and *additive*.

From Figs. 12-12 and 12-13, respectively, we may write the dot convention as follows:

1. Polarities of conductively coupled coils are *additive* whenever current enters (or leaves) the dotted terminals of two or more coils.
2. Polarities of conductively coupled coils are *subtractive* when current *enters* one dotted terminal of one coil and *leaves* the dotted terminal of a second coil.

12-8 CONDUCTIVELY COUPLED COILS IN SERIES

In Sec. 12-5 we considered series-connected coils (Fig. 12-10a) that had *no* magnetic coupling between them. We are now prepared to reconsider the cases of two or more coils that are *connected in series* but *are also magnetically coupled.*

In **Fig. 12-14a**, two coils are connected in series but there is no coupling between them since their fluxes are at right angles to each other. Thus ϕ_1 induces no voltage in coil 2 and ϕ_2 induces no voltage in coil 1. The total inductance of this combination (as shown in Sec. 12-5) is $L_t = L_1 + L_2$.

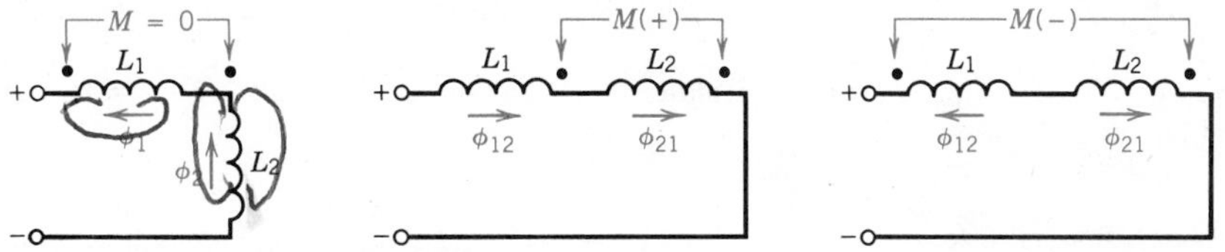

a. No magnetic coupling, $M = 0$
b. Additive mutual coupling, series aiding
c. Subtractive mutual coupling, series opposing

Figure 12-14 Two conductively coupled coils in series with various couplings

In Fig. 12-14b, the current is leaving the dotted terminals of both coils and therefore the polarity is additive (i.e., their mutual fluxes aid each other.) We have shown (Sec. 12-5) that $\phi_{12} = \phi_{21}$ and therefore $M_{12} = M_{21} = M$. The *total inductance* of the combination in Fig. 12-14b, *series aiding*, is

$$L_{ta} = (L_1 + M_{12}) + (L_2 + M_{21}) = L_1 + L_2 + 2M \qquad \text{henry (H)} \qquad \textbf{(12-13)}$$

where all terms have been defined.

In Fig. 12-14c current enters the dot at one coil but leaves at the dot in a second. The polarity is therefore subtractive (i.e., their mutual fluxes *oppose* each other.) The *total inductance* of this combination, *series opposing*, is

$$L_{to} = (L_1 - M_{12}) + (L_2 - M_{21}) = L_1 + L_2 - 2M \qquad \text{henry (H)} \qquad \textbf{(12-14)}$$

In the laboratory (and in the field) the mutual inductance M may be found by connecting the coils in series and measuring the total inductance with an electronic bridge. Then the coil connections are reversed and the total series inductance is again measured with an electronic bridge. Whichever total inductance is higher represents L_{ta}, and the lower total inductance is L_{to}. The value of M is found by Eq. (12-15), derived below.

Since $$L_{ta} = L_1 + L_2 + 2M \qquad (12\text{-}13)$$

and $$L_{to} = L_1 + L_2 - 2M \qquad (12\text{-}14)$$

subtracting the lesser inductance from the greater yields

$L_{ta} - L_{to} = 4M$, from which we may write

$$M = \frac{L_{ta} - L_{to}}{4} \quad \text{henry (H)} \qquad (12\text{-}15)$$

EXAMPLE 12-11

Two magnetically coupled coils are connected in series and their total inductance measures 88 mH. When the coil connections are reversed their total series inductance measures 24 mH. Calculate the

a. Mutual inductance between the coils
b. Value of L_2 if L_1 alone measures 30 mH

Solution

a. $M = (L_{ta} - L_{to})/4 = (88 - 24)\ \text{mH}/4 = \mathbf{16\ mH}$ (12-15)
b. $L_2 = L_{ta} - (L_1 + 2M) = 88 - (30 + 2 \times 16)$
$= \mathbf{26\ mH}$ (12-13)

EXAMPLE 12-12

For the two coils given in Ex. 12-11, calculate the coefficient of coupling between them

Solution

$k = M/\sqrt{L_1 L_2} = 16/\sqrt{30 \times 26} = \mathbf{0.5729}$ (12-12a)

The question now arises, how does one find the total inductance of three (or more) coils that are magnetically coupled and connected in series? The answer lies *intuitively* in the original formats of Eqs. (12-13) and (12-14). For three coils in series we may write the following general equation:

$$L_t = (L_1 \pm M_{12} \pm M_{13}) + (L_2 \pm M_{23} \pm M_{21}) + (L_3 \pm M_{31} \pm M_{32}) \quad \text{(H)} \qquad (12\text{-}16)$$

Note that the mutual terms may be either positive or negative depending on coil polarity, as shown by the dot convention and Ex. 12-13.

EXAMPLE 12-13

Determine the total inductance for the three coils in series having the mutual coupling shown in **Fig. 12-15**. Assume $L_1 = L_2 = L_3 = 20$ mH.

Solution

$$L_t = (L_1 - M_{12} + M_{13}) + (L_2 - M_{21} - M_{23}) + (L_3 + M_{31} - M_{32}) \qquad (12\text{-}16)$$
$$= (20 - 6 + 5) + (20 - 6 - 8) + (20 + 5 - 8)$$
$$= (19 + 6 + 17)\ \text{mH} = \mathbf{42\ mH}$$

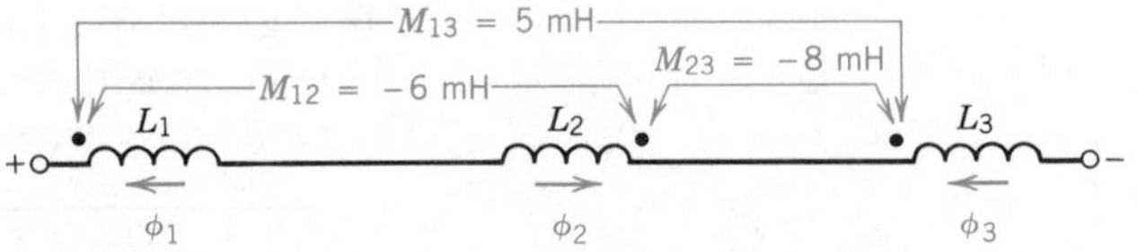

Figure 12-15 Example 12-13

Example 12-13 shows the following important insights:

1. The method just described may be used for *any* number of series-connected inductors with magnetic coupling between them.

2. As long as the permeability of the magnetic paths are unchanged, $M_{12} = M_{21}$, $M_{13} = M_{31}$, and so on. This simplifies calculations so that they can almost be done mentally from the circuit drawing (Fig. 12-15) itself.
3. Note that if there is no coupling whatever between coils in Fig. 12-15 and Ex. 12-13, the total inductance in series is 3×20 mH = 60 mH. The introduction of coupling (in this case) tends to reduce coupling because *four* of the mutual inductances are *negative* and only *two* are *positive* in the solution of Ex. 12-13 using Eq. (12-16).

12-9 PARALLEL-CONNECTED INDUCTORS WITH MUTUAL INDUCTANCE BETWEEN THEM

We have already seen (Sec. 12-5) that if there is no coupling between them, parallel-connected inductors may be treated in the same manner as parallel-connected resistors. For example, in the case of two parallel-connected inductors that are so far apart or are at right angles to each other, we may say $L_t = (L_1 L_2)/(L_1 + L_2) = L_1 \| L_2$.

But what if the inductors have mutual inductance between them? How is the total equivalent inductance found?

The clue lies in applying the same treatment previously used in series application. The polarity marks on the coils tell us whether *each pair* of mutual fluxes is additive or subtractive. If we treat the inductors separately and then take their parallel coupled equivalents, we can find the total equivalent inductance easily. This is shown in Ex. 12-14.

EXAMPLE 12-14

Two inductors, $L_1 = 20$ mH and $L_2 = 5$ mH, are connected in parallel. The mutual inductance $M_{12} = M_{21} = -2$ mH because of the coupling between them and the flux produced by instantaneous currents. Calculate the

a. Total inductance if the coils are far apart and separated so that there is no coupling between them when connected in parallel
b. Total inductance including the effect of coupling when connected in parallel

Solution

a. $L_t = (L_1 L_2)/(L_1 + L_2) = (20 \times 5)/(20 + 5) =$ **4 mH**
b. Including the effect of coupling,
$L_1p = L_1 - M_{12} = (20 - 2)$ mH = 18 mH. Similarly,
$L_2p = L_2 - M_{21} = (5 - 2)$ mH = 3 mH; then
$L_{tp} = L_1p \| L_2p = (18 \| 3) =$ **2.571 mH**

What if three inductors are connected in parallel? How are they treated? Example 12-15 shows this method (which is similar to the previous example) for three parallel-connected inductors having coupling between them. Note the technique used in finding mutual inductance between *pairs* of inductors.

EXAMPLE 12-15

Three inductors, $L_1 = L_2 = L_3 = 20$ mH, are connected in parallel and have instantaneous coil polarities as shown by **Fig. 12-16**. Calculate the equivalent parallel inductance when

a. There is no mutual inductance assumed between them
b. The mutual inductance between them is as shown in Fig. 12-16

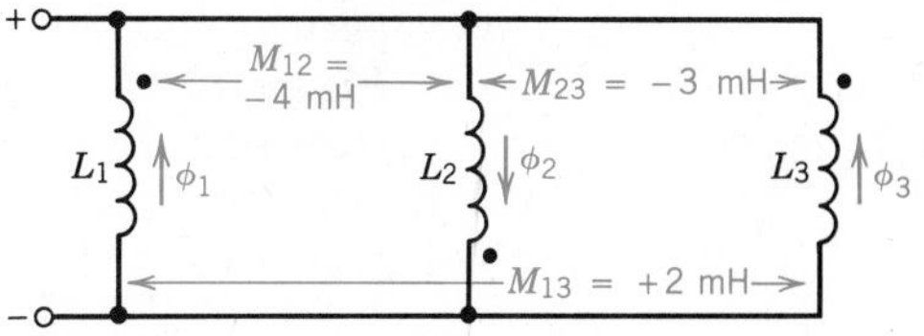

Figure 12-16 Example 12-15 and Prob. 12-24

Solution

a. $L_t = L_1/3 = 20\text{ mH}/3 =$ **6.6̄ mH**
b. $L_1p = L_1 - M_{12} + M_{13} = (20 - 4 + 2) =$ **18 mH**
$L_2p = L_2 - M_{21} - M_{23} = (20 - 4 - 3) =$ **13 mH**
$L_3p = L_3 + M_{31} - M_{32} = (20 + 2 - 3) =$ **19 mH**

Then $$\frac{1}{L_{tp}} = \frac{1}{18} + \frac{1}{13} + \frac{1}{19} = 0.18511$$

and $L_{tp} =$ **5.402 mH** (12-8a)

Examples 12-14 and 12-15 enable us to draw the following conclusions regarding parallel inductors having mutual inductance between them:

1. It is necessary to know whether the mutual fluxes between *each pair* of inductors add or subtract. This is obtained from the dot convention and/or the polarity marks given on the end of each coil.
2. The mutual fluxes either increase or decrease the individual parallel inductance, L_{xp}, depending on the interaction with each of the other coils in parallel.
3. The total equivalent parallel inductance is obtained as the reciprocal of the sum of the reciprocals of each parallel inductor, L_{xp}, for x inductors in parallel.
4. In Ex. 12-14, each individual inductance was *reduced* because of the subtractive mutual inductance between the two coils ($-M = 2$ mH).
5. In Ex. 12-15, two of the three mutual inductances are negative (Fig. 12-13), reducing all three inductors from their original values of 20 mH to some lower values. The total equivalent parallel inductance including the effect of coupling is therefore *reduced* from $6.\bar{6}$ mH to 5.4 mH.

12-10 ENERGY STORED IN THE MAGNETIC FIELD OF AN INDUCTOR

An ideal inductor has negligible or zero resistance. When connected to a constant-current source, I, the current rises from zero to I in a given time, t. Assuming that the current rises exponentially from 0 to I in time t, the energy input to the inductor is [9]

$$W = \frac{LI^2}{2} \quad \text{joules (J)} \qquad (12\text{-}17)$$

where L is the inductance of the inductor in henrys

I is the final current in the inductor in amperes

W is the energy stored in the magnetic field of the inductor in joules

Equation (12-17) is not surprising since it is analogous to the energy stored in the dielectric field (or electric field) of a capacitor, where $W = CV^2/2$ in Eq. (9-2). Neither capacitors nor inductors can dissipate electrical energy. Consequently, the energy stored in the magnetic field of an inductor (like that in the case of a charged capacitor) must either be returned to the supply or dissipated in circuit resistance.

Figure 12-17 shows the energy stored in the inductor, w_L, the instantaneous current rise, the voltage across the inductor, and the instantaneous power drawn from the source during time t. Note the similarity between Fig. 12-17 for the inductor and Fig. 10-12 for the capacitor. Note the following from Fig. 12-17:

1. Power is a maximum where $v_L = V/2$, $i_L = I/2$, and $w_L = W/2$.
2. Instantaneous power is zero whenever the instantaneous voltage or current is zero.

[9] See Appendix B-6.1 for a derivation of Eq. (12-17).

$$v_L = V\epsilon^{-t/\tau}$$

$$w_L = \frac{LI^2}{2}(1 - 2\epsilon^{-t/\tau} + \epsilon^{-2t/\tau})$$

$$p_L = \frac{V^2}{R}(\epsilon^{-t/\tau} - \epsilon^{-2t/\tau})$$

$$i = I(1 - \epsilon^{-t/\tau})$$

$$\tau = L/R$$

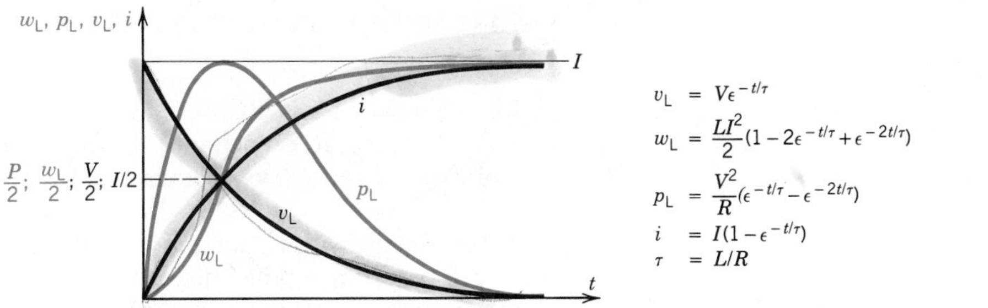

Figure 12-17 Instantaneous energy, current, inductor voltage, and power during energy storage in the magnetic field of an inductor

3. The voltage across the inductor, v_L, decreases exponentially from an initial value, $v_L = L\,di/dt$, to zero.
4. Both the instantaneous current, i, and the stored energy, w_L, increase exponentially from zero to final values, I and W, respectively.
5. The various equations for the instantaneous values of w_L, p_L, i, and v_L shown in Fig. 12-17 are obtained by analogy from Fig. 10-12 for the capacitor.
6. The shaded area under the power curve represents the incremental energy, $w_L = p \times t$, in joules, stored in the magnetic field.

EXAMPLE 12-16

A dc dynamo contains four field poles (Fig. 11-18b), each field coil having an inductance of 15 H, with no coupling between them, connected in series, and carrying a current of 5 A. Calculate the

a. Total field circuit inductance
b. Total energy stored in all four magnetic fields
c. Total energy stored in each magnetic field

Solution

a. $L_t = 4L_1 = 4 \times 15\ \text{H} = \textbf{60 H}$ (12-7)
b. $W = LI^2/2 = (60)(5)/2 = \textbf{150 J}$ (12-17)
c. $W/\text{coil} = 150\ \text{J}/4\ \text{coils} = \textbf{37.5 J/field}$

EXAMPLE 12-17

If the field circuit of the dynamo in Ex. 12-16 is deenergized from its dc supply by means of a single-pole, single-throw (SPST) switch that takes 10 ms to open, calculate the power instantaneously dissipated in the arc across the blades of the switch

Solution

$P = W/t = 150\ \text{W·s}/0.01\ \text{s} = \textbf{15 000 W}$

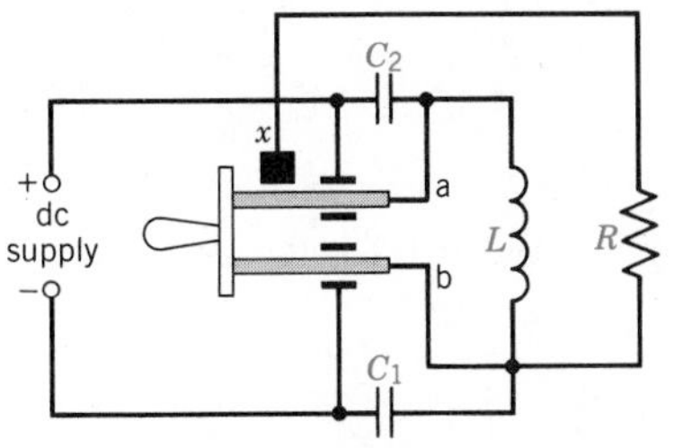

Figure 12-18 Field discharge switch, resistor, and capacitors

Example 12-17 shows that a rapidly collapsing magnetic field not only produces high induced voltages (Eq. 12-3) in the field windings (possibly damaging the insulation) but also must discharge the energy stored in the magnetic field across the contacts of the switch. To prevent damage to the switch blades it is customary to use either *field discharge* resistors or *capacitors* or both, as shown in **Fig. 12-18**. Recall that capacitors have the property of opposing any change in voltage across their terminals. Consequently, when the double-pole, single-throw (DPST) switch of Fig. 12-18 is opened, two actions occur simultaneously:

1. The upper knife blade engages contact arm x, placing the discharge resistor R across the field coil, L, through circuit a-x-R-b. The resistor, R, therefore both retards the rapid inductive decay and absorbs the $LI^2/2$ field energy that would normally appear at the switch blades.
2. Capacitors C_1 and C_2, which bridge the two knife blades and their stationary contacts, absorb the remaining energy, thus preventing severe damage to the blade and contacts.

12-11 SUMMARY COMPARISON OF CIRCUIT ELEMENTS *R*, *C*, AND *L*

In the opening of this chapter it was stated that there are only three circuit elements: resistance, R, capacitance, C, and inductance, L. Now that we have learned the properties and relevant equations for each, we are in a position to compare them.

Table 12-1 provides a summary comparison of the three circuit elements. Space does not permit a detailed discussion of all that can be inferred from Table 12-1. But the reader's attention is directed to some of the following more important insights:

1. The equation for physical realization of the circuit element (row 4) contains two geometric properties (length and area) and one material property for all three elements.

Table 12-1 Comparisons among Properties of Circuit Elements *R*, *C*, and *L*

Row	Property/Relation	Resistance, R	Capacitance, C	Inductance, L
1	Graphical symbol		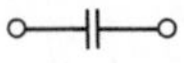	
2	Defining equation	$R \equiv V/I$ in ohms (Ω)	$C \equiv Q/V$ in farads (F)	$L = V\Big/\dfrac{di}{dt}$ in henrys (H)
3	Units of defining equation	ohms = volts/amperes	farads = coulombs/volts	henry = volts/ampere-seconds
4	Equation for physical realization	$R = \rho l/A$	$C = \varepsilon A/d$	$L = \mu N^2 A/l$
5	Material property in above equation	Resistivity, ρ, of conductor	Permittivity, ε, of dielectric	Permeability, μ, of magnetic circuit
6	Energy equation	$W = VIt = I^2Rt = V^2t/R$ (dissipated)	$W = CV^2/2$ (stored)	$W = LI^2/2$ (stored)
7	Electrical effect of element in circuit	Opposes STEADY current	Opposes CHANGE in voltage	Opposes CHANGE in current
8	*Series* circuit of three elements	$R_t = R_1 + R_2 + R_3$	$\dfrac{1}{C_t} = \dfrac{1}{C_1} + \dfrac{1}{C_2} + \dfrac{1}{C_3}$	$L_t = L_1 + L_2 + L_3$ (k and $M = 0$)
9		$I_t \equiv I_1 \equiv I_2 \equiv I_3$	$Q_t \equiv Q_1 \equiv Q_2 \equiv Q_3$	$\dfrac{di}{dt} \equiv \dfrac{di}{dt}_1 \equiv \dfrac{di}{dt}_2 \equiv \dfrac{di}{dt}_3$
10	Kirchhoff's voltage law (KVL)	$V_t = V_1 + V_2 + V_3$	$V_t = V_1 + V_2 + V_3$	$V_t = V_1 + V_2 + V_3$
11	Magnitude in KVL	Highest R has highest V	Lowest C has highest V	Highest L has highest V
12	*Parallel* circuit of three elements	$\dfrac{1}{R_t} = \dfrac{1}{R_1} + \dfrac{1}{R_2} + \dfrac{1}{R_3}$	$C_t = C_1 + C_2 + C_3$	$\dfrac{1}{L_t} = \dfrac{1}{L_1} + \dfrac{1}{L_2} + \dfrac{1}{L_3}$ (k and $M = 0$)
13	Kirchhoff's current law (KCL)	$I_t = I_1 + I_2 + I_3$	$Q_t = Q_1 + Q_2 + Q_3$	$\dfrac{di}{dt} = \dfrac{di}{dt} + \dfrac{di}{dt}_2 + \dfrac{di}{dt}$
14	Defining voltage	$V_t \equiv V_1 \equiv V_2 \equiv V_3$	$V_t \equiv V_1 \equiv V_2 \equiv V_3$	$V_t \equiv V_1 \equiv V_2 \equiv V_3$
15	Magnitude in KCL	Lowest R has highest I	Highest C has highest Q	Lowest L has highest di/dt
16	Circuit time constant	(not applicable)	$\tau = RC$	$\tau = L/R$
17	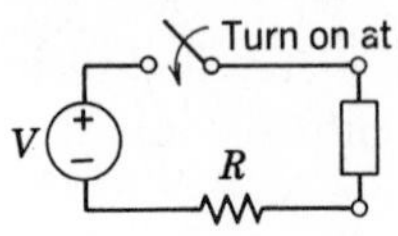	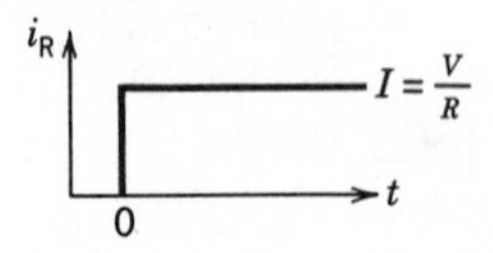	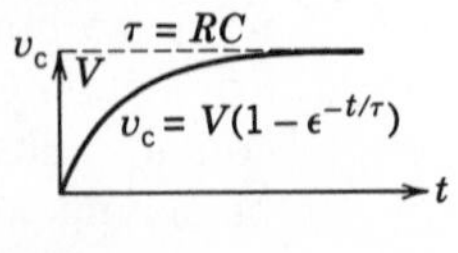	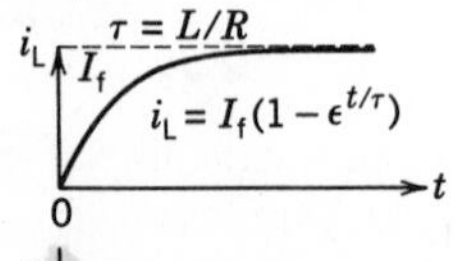
18	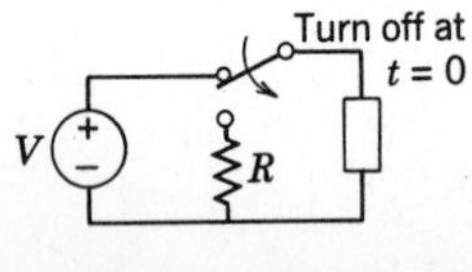	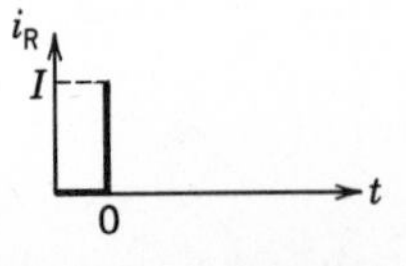	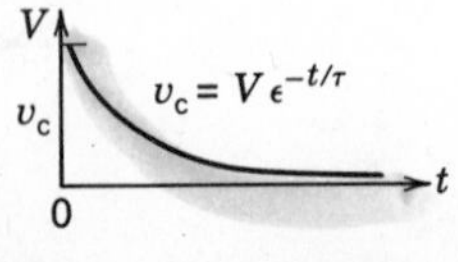	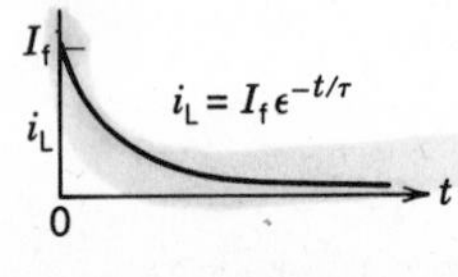

2. Only resistance may dissipate energy (row 6); capacitance and inductance merely store energy.
3. All three components provide opposition (row 7) to some circuit property.
4. The fundamental definition of current being the SAME in all parts of a series circuit (row 9) applies to all three components, since the time, t, is also the same.
5. Kirchhoff's voltage law in row 10 is the same for all three series-connected components.
6. Both R and L behave the same (rows 8, 11, 12, and 15), whereas C is opposite in effect in these summations.
7. The fundamental definition of voltage being the SAME across all parts of a parallel circuit (row 14) applies to all three components.
8. Kirchhoff's current law is essentially the same for all three components in parallel (row 13), since the time and rate of change of time are the same.
9. Since current rises instantly in a resistive circuit (row 16), it has no time constant. Since this is not the case in RC and RL circuits (row 17), we require exponential equations to determine instantaneous values of v_C and i_L, respectively.
10. Row 17 shows the dc transient response of R, C, and L during turn ON.
11. Row 18 shows the dc transient response during turn OFF, when decay occurs.

The reader should study and review Table 12-1 as frequently as necessary to become completely familiar with it. It will be used in the subsequent study of ac circuit properties.

12-12 GLOSSARY OF TERMS USED

Coefficient of coupling, k Ratio of mutual inductance M to the square root of the product L_1L_2, where L_1 is the total inductance of one mesh and L_2 the total inductance of a second mesh mutually coupled to the first.

Electromagnetic (EM) induction Generation of an EMF in a conductor or system of conductors due to a change in flux linkages.

EMF, electromotive force Potential difference produced as a result of EM induction.

Faraday's law of EM induction Any change in flux linking a conductor produces an EMF in that conductor in proportion to the rate of change of flux linkage ($d\phi/dt$).

Fleming's right-hand (RH) rule Convenient finger rule for demonstrating the orthogonal nature and relation between motion (thumb), field direction (index finger), and direction of induced EMF (third finger).

Flux linkages Product of flux (ϕ) and turns (N). Only a *change* in $N\phi$ produces an EMF. Flux linkage changes are produced by physical motion of a conductor relative to the field, or both; or in the case of a transformer by a change in magnitude and direction of the magnetic flux.

Inductor Coil with or without an iron core, having the property of inductance.

Lenz's law The polarity and direction of an induced EMF are always such direction as to oppose the force or change that produced it.

Minimum coupling Locating two or more inductors so that the flux linkage between them is minimized. This is done by placing them at right angles to each other, by magnetic shielding, or by sufficient separation, reducing the coefficient of coupling, k, to approximately zero.

Mutual inductance Common property of two electric circuits wherein an EMF is induced in one circuit by the change of current in another, measured in henrys (H). The symbol for mutual inductance is M.

Self-inductance See self-induction. Self-inductance is measured in henrys (H).

Self-induction Property of an electric circuit or component to oppose any change of current in that circuit or component.

Tight coupling Locating two or more inductors on a common ferromagnetic core so that there is a maximum flux linkage between them. The coefficient of coupling, k, of such circuits is close to unity.

12-13 PROBLEMS

Secs. 12-1 to 12-4

12-1 When the current in an inductor rises linearly from 0 to a final value of 5 A in a time of 1 s, it induces an EMF of self-induction of 2.5 V in the coil. Calculate the

a. Rate of change of current
b. Inductance of the coil

12-2 In Prob. 12-1, when the current reaches its final value (5 A), it produces a flux density of 10 T in a coil having an area of 5 cm^2 and 500 turns. Calculate and verify the

a. Maximum flux produced in the center of the coil
b. Rate of change of flux
c. Induced voltage in the coil, using Faraday's law
d. Inductance of the coil, using $L = N(d\phi/di)$

Note: solutions to (**c**) and (**d**) should verify the data and solution to Prob. 12-1

12-3 The coil used in the two previous problems has a ferromagnetic core whose *relative* permeability (μ_r) is 100 and whose length is π cm. Calculate and verify the

a. Permeability of the core, μ
b. Inductance of the coil, using $L = N^2\mu A/l$
c. Flux density, **B**, using $\mathbf{B} = \mu NI/l$
d. Ratio of L/N and $d\phi/di$, respectively
e. Prove by equation that the two ratios in (**d**) are always equal

Note: Prob. 12-3 verifies the unity of Eqs. (12-1) through (12-6).

12-4 The current in a 50-mH inductor rises to a maximum of 50 mA in 20 ms. Calculate the

a. Rate of rise of current
b. EMF induced in inductor during current rise
c. Energy stored in the magnetic field when the current reaches its final value

12-5 The induced EMF in a 200-turn coil is 200 mV whenever the current is changing at a rate of 100 mA/s. Calculate the

a. Inductance of the coil
b. Ratio L/N in appropriate units
c. Rate of change of flux relative to the rate of change of current in appropriate units

Hint: see Eq. (12-4)

12-6 A 100-t relay coil develops a maximum flux of 10 mWb whenever the current reaches a final value of 50 mA in 2 ms after the relay is energized. Calculate the

a. Inductance of the coil, using only the given parameters I, N, and ϕ
b. EMF of self-induction, using Faraday's law ($e = N\phi/t$)
c. Inductance of the coil, using the value from (**b**) and the given data
d. Ratio of L/N in appropriate units
e. Ratio of ϕ/I in appropriate units

12-7 If the length of magnetic circuit of the relay coil in Prob. 12-6 is 5 cm and the cross-sectional area of the core is 10 cm^2, calculate the

a. Absolute permeability, μ, of the core (in H/m)
b. Relative permeability of the core
c. Flux density in the core (in tesla)
d. Reluctance of the core (in At/Wb)

Secs. 12-5 to 12-10

12-8 Three air-core coils of 600, 300, and 200 mH, respectively, having no magnetic coupling between them, are series connected. Calculate

a. The equivalent inductance of the combination
b. Repeat (**a**) if connected in parallel

12-9 Three inductors are each magnetically shielded so there is no possibility of magnetic interaction between them. Their values are $L_1 = 200$ mH, $L_2 = 50$ mH, and $L_3 = 40$ mH. Calculate the equivalent inductance when they are connected

a. In series
b. In parallel

12-10 A power transformer has a coefficient of coupling of 0.98, a primary inductance of 100 mH, and a secondary inductance of 4 mH. If the primary current changes at the rate of 200 A/s, calculate the

a. Mutual inductance, M, between the windings
b. Secondary induced voltage using the value computed in (**a**)

12-11 Two coils are wound on a common core, having 100 t and 50 t, respectively. The mutual flux between them is 5 mWb when the current in the 100-t (primary) coil rises to a maximum of 500 mA in 0.5 s. Calculate the

a. Primary induced voltage, e_1
b. Secondary induced voltage, e_2
c. Mutual inductance, $M_{12} = M_{21} = M$
d. Rate of change of current in the secondary, di_2/dt
e. Primary inductance, L_1
f. Secondary inductance, L_2
g. Coefficient of coupling, k
h. Primary flux, ϕ_1

12-12 When a current of 150 mA rises in a coil L_1 having 200 t, it creates a flux of 20 μWb. A second coil having 500 t and a coefficient of coupling of 0.75 to the first coil is wound on the same core. Calculate the

a. Inductance of the primary coil, L_1
b. Mutual flux, ϕ_m, common to both coils
c. Mutual inductance, M, between the two coils
d. Self-inductance of the second coil, L_2
e. Total inductance if the two coils are connected series aiding, L_{at}
f. Total energy stored in the magnetic field when the *mutual* flux is 15 μWb

12-13 Two coils having self-inductances of 250 mH and 640 mH, respectively, have a mutual inductance of 150 mH. Calculate the

a. Coefficient of coupling between them
b. Total inductance when connected series aiding
c. Total inductance when connected series opposing

12-14 Two coils are connected series-opposing on a common core. When the current rises from zero to 3 A, it produces a flux of 375 μWb linking the first coil of 250 t and a flux of 900 μWb linking the second coil of 600 t. If the coefficient of coupling between the coils is 0.6, calculate the

a. Self-inductance of the first coil, L_1
b. Self-inductance of the second coil, L_2
c. Mutual inductance between the two coils
d. Equivalent inductance of the combination when connected series opposing
e. Total energy stored in the magnetic field of the entire circuit
f. Equivalent inductance if the coils are now connected series aiding
g. Total energy stored in the magnetic field of the circuit under connection in **(f)** for the same final current of 3 A

12-15 Two coils L_1 and L_2 are carefully measured using an inductance bridge. L_1 measures 20 mH and L_2 measures 50 mH. When placed side by side and connected in series, the combination measures 80 mH. Calculate the

a. Mutual inductance between them
b. Coefficient of coupling of the two coils

12-16 A standard inductor of 50 mH is connected in series with an unknown inductor, L_x, and the equivalent inductance of the combination measures 90 mH. When the connections to L_x are reversed, the combination in series measures 50 mH. Calculate the

a. Value of L_x in mH
b. Mutual inductance between the two coils
c. Coefficient of coupling between the two coils

12-17 Two identical coils when connected series aiding produce a total inductance of 90 mH. If the coefficient of coupling between them is 0.5, calculate the

a. Inductance of each coil
b. Mutual inductance
c. Total inductance, series opposing

12-18 Two coils, 30 mH each, are located on a ferromagnetic common core. When series connected in one configuration, the total inductance as measured by an inductance bridge measures zero! When series connected in a second configuration, the total inductance measures 120 mH. Calculate the

a. Mutual inductance between them
b. Coefficient of coupling between them

12-19 Two coils, one superimposed on the other, are placed on a solid-steel core to produce a coefficient of coupling of unity. The second coil has three times the number of turns of the first, whose inductance is 30 mH. Calculate the

a. Inductance of the second coil, L_2
b. Mutual inductance between coils
c. Total inductance when connected series aiding
d. Total inductance series opposing

12-20 For the series circuit shown in **Fig. 12-19**, $L_1 = 20$ mH, $L_2 = 30$ mH and $L_3 = 25$ mH. The mutual inductances between the coils are $M_{12} = 10$ mH, $M_{13} = 8$ mH and $M_{23} = 10$ mH. Calculate the

a. Total series-connected inductance assuming zero coupling between the coils
b. Total series-connected inductance taking mutual coupling into account

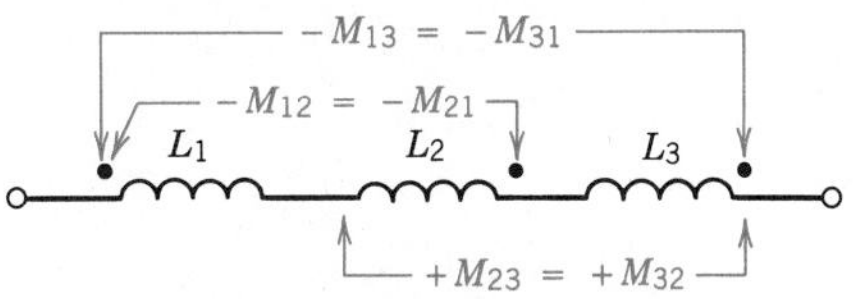

Figure 12-19 Problems 12-20 and 12-21

12-21 In the series-coupled circuit shown in Fig. 12-19, $L_1 = 50\ \mu$H, $L_2 = 40\ \mu$H, and $L_3 = 30\ \mu$H, with mutual couplings of $M_{12} = 10\ \mu$H, $M_{23} = 15\ \mu$H, and $M_{13} = 20\ \mu$H. Repeat both parts of Prob. 12-20.

12-22 In the series circuit shown in **Fig. 12-20** the inductances are $L_1 = 20$ mH, $L_2 = 30$ mH, and $L_3 = 25$ mH. The coupling coefficients are $k_{12} = 0.6$, $k_{23} = 0.7$, and $k_{31} = 0.4$, respectively. Calculate the

a. Mutual inductances M_{12}, M_{23}, and M_{31}, respectively
b. Total series-connected inductance taking **(a)** into account

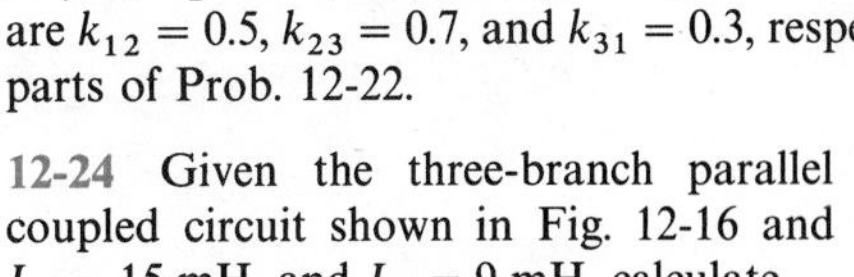

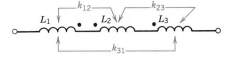

Figure 12-20 Problems 12-22 and 12-23

12-23 In the series-coupled circuit shown in Fig. 12-20, $L_1 = 50\ \mu$H, $L_2 = 40\ \mu$H, and $L_3 = 30\ \mu$H. The coupling coefficients are $k_{12} = 0.5$, $k_{23} = 0.7$, and $k_{31} = 0.3$, respectively. Repeat both parts of Prob. 12-22.

12-24 Given the three-branch parallel inductive mutually coupled circuit shown in Fig. 12-16 and values $L_1 = 10$ mH, $L_2 = 15$ mH, and $L_3 = 9$ mH, calculate

a. Total equivalent parallel inductance assuming no coupling between coils
b. Each individual apparent parallel inductance as a result of the given coupling (L_1p, L_2p, and L_3p, respectively)
c. Total equivalent parallel inductance as a result of the given coupling

12-25 In the parallel-coupled circuit shown in **Fig. 12-21**, $L_1 = 60\ \mu$H, $L_2 = 30\ \mu$H, and $L_3 = 50\ \mu$H. The coupling coefficients are $k_{12} = 0.5$, $k_{23} = 0.7$, and $k_{31} = 0.4$, respectively. Calculate the

a. Equivalent parallel circuit inductance assuming no coupling whatever between coils
b. Values of M_{12}, M_{23}, and M_{31}, respectively for the given coil polarities
c. Values of apparent parallel inductance of each coil taking coupling into account (L_1p, L_2p, and L_3p, respectively)
d. Total equivalent parallel inductance as a result of the coupling given

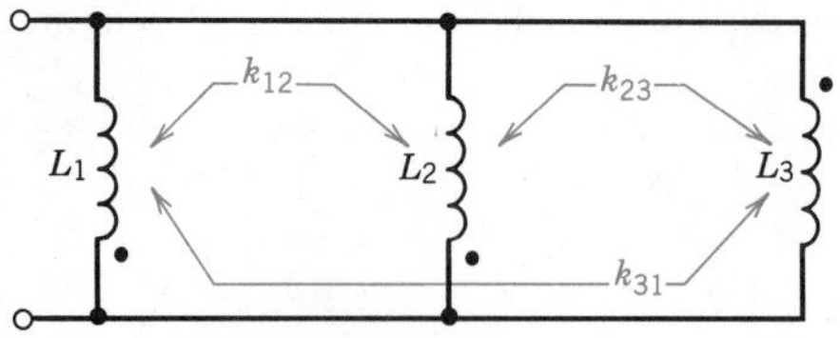

Figure 12-21 Problems 12-25 and 12-26

12-26 In the parallel-coupled circuit shown in Fig. 12-21, $L_1 = L_3 = 120$ mH and $L_2 = 30$ mH. The coupling coefficients are $k_{12} = 0.8$, $k_{23} = 0.2$, and $k_{31} = 0.3$, respectively. Calculate all parts of Prob. 12-25.

12-27 The four field coils of a dc dynamo have a combined resistance of 50 Ω and an inductance of 10 H. Calculate the

a. Field current when connected to a 110-V supply
b. Energy stored in the magnetic field of the dynamo
c. Value of the discharge resistor that must be connected across the field to ensure that the terminal voltage across the coils does not exceed 220 V
d. Energy dissipated in the discharge resistor
e. Energy dissipated within the coil itself

Hint: Since L opposes a change in current, maximum current during discharge *cannot* exceed the value calculated in (**a**).

12-14 ANSWERS

12-1 a 5 A/s b 0.5 H
12-2 a 5 mWb b 5 mWb/s c 2.5 V d 0.5 H
12-3 a 125.7 μH/m b 0.5 H c 10 T d 1 mWb/A
12-4 a 2.5 A/s b 125 mV c 62.5 μJ
12-5 a 2 H b 10 mH/t c 10 mWb/A
12-6 a 20 H b 500 V c 20 H d 0.2 H/t
e 0.2 Wb/A
12-7 a 0.1 H/m b 79 580 c 10 T d 500 At/Wb
12-8 a 1.1 H b 100 mH
12-9 a 290 mH b 20 mH
12-10 a 19.6 mH b 3.92 V
12-11 a 1 V b 0.5 V c 0.5 H d 2 A/s e 1 H
f 0.25 H g 1.0 h 5 mWb
12-12 a $26.\overline{6}$ mH b 15 μWb c 50 mH d $166.\overline{6}$ mH
e $293.\overline{3}$ mH f 3.3 mJ
12-13 a 0.375 b 1.2 H c 0.6 H
12-14 a 31.25 mH b 180 mH c 45 mH
d 121.25 mH e 0.5456 J f 301.25 mH
g 1.356 J
12-15 a 5 mH b 0.1581
12-16 a 20 mH b 10 mH c 0.3162
12-17 a 30 mH b 15 mH c 30 mH
12-18 a 30 mH b 1
12-19 a 270 mH b 90 mH c 480 mH d 120 mH
12-20 a 75 mH b 59 mH
12-21 a 120 μH b 90 μH
12-22 a −14.7 mH; 19.17 mH; −8.944 mH b 66.05 mH
12-23 a −22.36 μH; 24.25 μH; −11.62 μH b 100.5 μH
12-24 a 3.6 mH b 8 mH each c $2.\overline{6}$ mH
12-25 a 14.29 μH b 21.21 μH; −27.11 μH; −21.91 μH
c 59.3 μH; 24.1 μH; 0.98 μH d 0.927 μH
12-26 a 20 mH b 48 mH; -12 mH; -36 mH
c 132 mH; 66 mH; 72 mH d 27.31 mH
12-27 a 2.2 A b 24.2 J c 100 Ω d $1.61\overline{3}$ J e $8.0\overline{6}$ J

CHAPTER 13

Transient *RL* Circuit Analysis

In Chapter 12 **inductance** was introduced as a (third) circuit property that opposes any *change in current* across the terminals. It was assumed that the inductors were pure, that is, that they contained no resistance whatever. We know, however, that inductors must be made of conductive wire and all wires possess some resistance. We also assumed that the rise of current in an inductor was linear and consequently that the rate of change of current (di/dt) was constant. We will discover, however, that *neither* of these assumptions is quite correct as we proceed to study *RL* circuits.

Chapter 10 was devoted to a detailed study of *RC* circuit analysis and the *transient* changes that occurred in going from one *initial* steady state to another *final* steady state. The reader is advised to review Sec. 10-11, the glossary of terms used that apply to *RC* transients, as well as Sec. 13-11, which applies to *RL* transients. Throughout this chapter dealing with *RL* transients comparisons will be made to *RC* circuits so that the reader will have a more comprehensive and unified view of the subject, in terms of similarities and differences. Furthermore, it will be shown that analogs exist between *RL* and *RC* circuits, and this will simplify both the writing of equations and our understanding of the subject.

Unlike most capacitors, which are relatively ideal or pure (i.e., contain little resistance), inductors are *far from ideal.* In drawing and using *RL* circuits throughout this chapter, we will assume that the resistance of the inductor has been "swamped out" by its series-connected resistance.[1] This will enable us to treat the inductor as an "ideal" element at the outset, for purposes of analysis. When we turn later to a study of ac theory, however, we will consider the behavior of a *practical* coil (i.e., one that has resistance).

13-1 FACTORS AFFECTING TIME CONSTANT OF AN *RL* CIRCUIT

Figure 13-1a shows the current rise in a purely resistive circuit when connected to a dc supply, V. Assume that R in Fig. 13-1a is a straight wire made of high-resistivity wire. At the instant the switch is closed, the current rises to its *steady-state* final value, I_f, in practically zero time, as determined by V/R (Ohm's law).

[1] Consider an inductor of 10 mH having a resistance of 20 Ω. If it is connected in series with a resistance of 5 kΩ, then the resistance of the inductor is negligible.

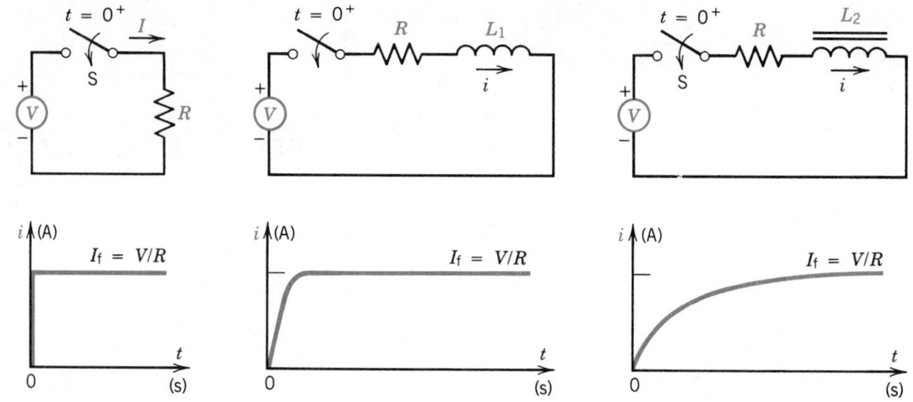

a. Purely resistive circuit ($L = 0$)

b. Same resistance and low inductance L_1

c. Same resistance and higher inductance (L_1)

Figure 13-1 Effect of increasing inductance on time to reach steady state when R and V are constant

Assume now that the straight wire is wound in the form of a *helix* with its resistance unchanged so that it has a small inductance, L_1, as shown in Fig. 13-1b. When switch S is closed, there is now a slight delay in current. Recall that inductance is the property to oppose any change in current. Since the current tries to rise instantly, it induces a counter EMF of self-induction (Sec. 12-3), v_L, where $v_L = L\,di/dt$ (Eq. 12-5). This counter EMF delays the current rise in proportion to the magnitude of inductance, L. Since L_1 is small in Fig. 13-1b, the delay is small.

Assume now that the air-core helix formed in Fig. 13-1b has an iron core inserted in it, forming a higher inductance, L_2, shown in Fig. 13-1c. For the same rate of change of current, the EMF of self-induction of this circuit is greater. This causes the current rise to be delayed still further in proportion to the increased inductance.

We may conclude that the *more inductive* (and less resistive) our *RL* circuit in Fig. 13-1 is, the *longer* the *time* to reach the steady-state Ohm's law value of final current. As in the case of the product *RC* for the capacitive circuit, we may write an expression for the *time constant* of an *RL* circuit as

$$\tau = \frac{L}{R} \quad \text{seconds (s)} \qquad \textbf{(13-1)}$$

where L is the inductance of the circuit in henrys (H)
R is the resistance of the circuit in ohms (Ω)

Equation (13-1) summarizes the conclusions drawn from Fig. 13-1:

1. The *more inductive* the circuit, the *longer* the time to reach steady state.
2. The *less resistive* the circuit, the *longer* the time to reach steady state.

An ideal inductor having zero resistance, therefore, would require an infinite time for its current to rise to some infinite value. Fortunately, such inductors are extremely difficult to construct so we need not be concerned about them.

As in the case of *RC* circuits, the time for all transients to disappear is 5τ. Consequently, for each of the circuits shown in Fig. 13-1, the time to reach steady state is $5\tau = 5L/R$. In Fig. 13-1a, since L is zero, the time to reach steady state is zero. In Fig. 13-1b, since L is very small, the time to reach steady state is a small value. In Fig. 13-1c, since L is greater, the time to reach steady state is longer.

13-2 CURRENT–VOLTAGE RELATIONS DURING TRANSIENT CURRENT RISE

The original *RL* circuit of Fig. 13-1a is shown once again in Fig. 13-2a, along with the two voltage drops in the circuit, v_R and v_L. The voltage drop v_R across the series resistance *R* varies with the current *i*, since $v_R = iR$. The voltage drop across the inductor, v_L, as noted in Sec. 13-1, is a function of the rate of change of current, since $v_L = L\,di/dt$. Let us consider each of these two factors in turn.

Figure 13-2b shows the transient rise in current with time. Directly below it is the transient rise in voltage across the resistor, v_R, with time. Clearly, both waveforms are identical. *i* rises exponentially with time and v_R rises exponentially with time because $v_R = iR$. Consequently, we may conclude that however the current changes (or varies) in an *RL* circuit, the voltage v_R will change in direct proportion to it.

But what of v_L? Figure 13-2c shows the variation of v_L in comparison to *i*, the transient current rise. In effect, v_L *appears* to be a mirror image of *i*, decreasing as rapidly as *i* increases.[2] Since $v_L = L\,di/dt$, it stands to reason that v_L is a maximum when the switch is first closed and the rate of change of current is a maximum. As current *i* increases, successive tangents drawn to the curve of *i* have smaller slopes, indicating a *lesser* value of di/dt and v_L. When the transient disappears, di/dt is zero (slope is horizontal) and v_L is zero.

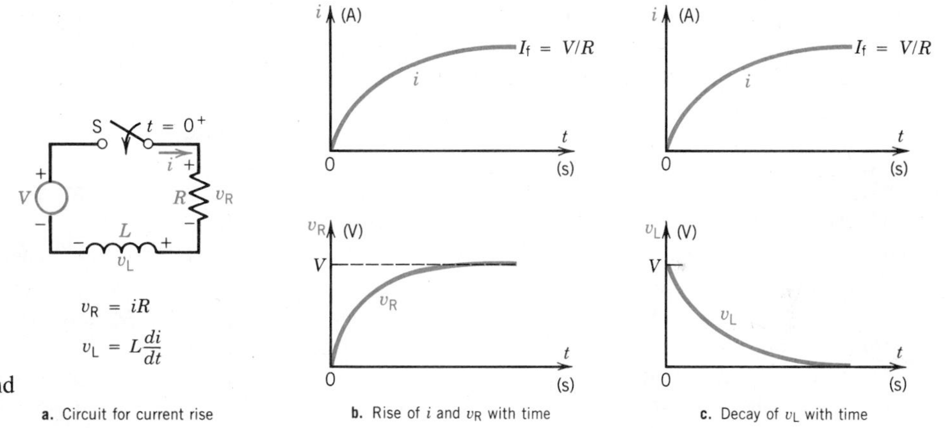

Figure 13-2 Transient rise of *i* and v_R and decay of v_L with time

We may now write Kirchhoff's voltage law (KVL) for the circuit of Fig. 13-2a as

$$V = v_R + v_L = iR + L\frac{di}{dt} = I_f R \qquad \text{volts (V)} \qquad (13\text{-}2)$$

where V is the applied voltage
i is the instantaneous current during the transient period
v_R is the voltage across the series resistor *R*, which varies as *iR*
v_L is the voltage across the inductor, *L*, which varies as $L\,di/dt$
I_f is the final steady-state current after all transients have disappeared

[2] v_L is *not always* a mirror image of *i*, since v_L undergoes a *direction reversal* whenever *i* decays.

If Eq. (13-2) is solved for the instantaneous current i, we obtain[3]

$$i = \frac{V}{R}(1 - \varepsilon^{-t/\tau}) = I_f(1 - \varepsilon^{-Rt/L}) \quad \text{amperes (A)} \qquad (13\text{-}3)$$

where all terms have been defined.[4]

Equations (13-2) and (13-3) are but two manifestations of the *same* equation. The first expresses KVL relations, as noted. The second expresses the transient current i for various values of time, *for the same variables.* The relation between the two is made clear from Ex. 13-1.

From Fig. 13-2b, for the exponential **rise** of v_R we may write

$$v_R = iR = V(1 - \varepsilon^{-t/\tau}) = I_f R(1 - \varepsilon^{-Rt/L}) \quad \text{volts (V)} \qquad (13\text{-}4)$$

where all terms have been defined.

From Fig. 13-2c, for the exponential **decay** of v_L we may write

$$v_L = L\frac{di}{dt} = V\varepsilon^{-t/\tau} = V\varepsilon^{-Rt/L} \quad \text{volts (V)} \qquad (13\text{-}5)$$

where all terms have been defined.

EXAMPLE 13-1

A coil having an inductance of 10 H and a resistance of 8 Ω is connected by means of a switch to a 12-V source. Calculate the

a. Initial current, i_0, at the instant $t = 0^+$ s when the switch is first closed
b. Final steady-state current drawn from the source when all transients have disappeared
c. Time constant of the circuit
d. Time for the current, i, to reach steady state
e. Instantaneous current when $t = 1\tau$ after the switch is first closed
f. v_R at time $t = 1\tau$ by two methods
g. v_L at time $t = 1\tau$ by two methods
h. Rate of change of current at time $t = 1\tau$

Solution

a. We define inductance as a circuit property to oppose a change in current. Since the current $i = 0$ at time $t = 0^-$, the current $i_0 = \mathbf{0}$ at time $t = 0^+$, as well.
b. $I_f = V/R = 12\text{ V}/8\ \Omega = \mathbf{1.5\ A} = I_{ss}$
c. $\tau = L/R = 10\text{ H}/8\ \Omega = \mathbf{1.25\ s}$ (13-1)
d. $t_{ss} = 5\tau = 5 \times 1.25\text{ s} = \mathbf{6.25\ s}$
e. $i = \frac{V}{R}(1 - \varepsilon^{-t/\tau}) = I_f(1 - \varepsilon^{-1/1})$

   $= 1.5\text{ A}(0.632) = \mathbf{0.9482\ A}$ (13-3)
f. $v_R = iR = (0.9482\text{ A})(8\ \Omega) = \mathbf{7.585\ V}$ (13-2)
   $v_R = V(1 - \varepsilon^{-t/\tau}) = 12(1 - \varepsilon^{-1}) = \mathbf{7.585\ V}$ (13-4)
g. $v_L = V - v_R = (12 - 7.585)\text{ V} = \mathbf{4.415\ V}$ (13-2)
   $v_L = V\varepsilon^{-t/\tau} = 12\varepsilon^{-1} = \mathbf{4.415\ V}$ (13-5)
h. $di/dt = v_L/L = 4.415\text{ V}/10\text{ H} = \mathbf{0.4415\ A/s}$ (13-5)

The following conclusions may be drawn from Ex. 13-1:

1. The transients i, v_R, and v_L all occur from the time the switch is closed at $t = 0^+$ until $t = 6.25\text{ s} = 5\tau$.
2. When the final steady state is reached, $i = I_{ss} = V/R = \mathbf{1.5\ A}$, $v_R = V = \mathbf{12\ V}$, and $v_L = \mathbf{0}$.
3. The instantaneous current, i, during the transient period may be found from Eq. (13-3) for any value of time, t.
4. The instantaneous voltage across the resistor R is found as the product $iR = v_R$.
5. The instantaneous voltage across the inductor, v_L, by KVL, is $V - v_R$ at all times.
6. The instantaneous rate of change of current at any time, t, is v_L/L.

[3] A derivation of Eq. (13-3) may be found in Appendix B-6.2

[4] Note the similarity between Eq. (13-3) for instantaneous **current rise** in an *RL* circuit and Eq. (10-2) for instantaneous **voltage rise** in an *RC* circuit. Both are of the form $y = 1 - \varepsilon^{-x}$ (See Eq. B-4.2, Appendix B-4.)

These conclusions raise several interesting questions. Since i, v_L, and v_R are all transients, they undergo a rate of change only during the transient period. The rate of change of *each* is a *maximum* at the instant the switch is closed and decreases to zero when the steady-state is reached after 5τ. How are these rates of change determined, both initially and at any time t during the transient period?

13-2.1 Initial Rates of Change during Transient Current Rise

At the instant switch S is thrown in Fig. 13-2a, the current i is zero. Recall that the property of an inductive circuit is to oppose a change in current. Substituting $i = 0$ in Eq. (13-2) yields $L(di/dt) = v_L = V$ at time $t = 0^+$, as verified by Fig. 13-2c. Since $v_L = L(di/dt)$ at any instant, we may find the **initial rate of change of current** as

$$\frac{di}{dt}_0 = \frac{V}{L} = \frac{I_f}{\tau} \quad \text{amperes per second (A/s)} \tag{13-6}$$

The reader may question how the ratio of V/L (volts per henry) yields units of amperes per second.[5] A simple graphical proof is shown in **Fig. 13-3a**. If a slope is drawn through the origin to the point where it intersects the final or steady-state current, I_f, it is observed that this point occurs in a time of 1τ. Then we can write for the initial slope $\frac{di}{dt}_0 = \frac{I_f}{\tau} = \frac{(V/R)}{(L/R)} = \frac{V}{L}$, which agrees with Eq. (13-6) and verifies our units, as well.

Since the graphical approach works as well as the algebraic, we may apply it to v_R and v_L, as shown in Figs. 13-3b and 13-3c.

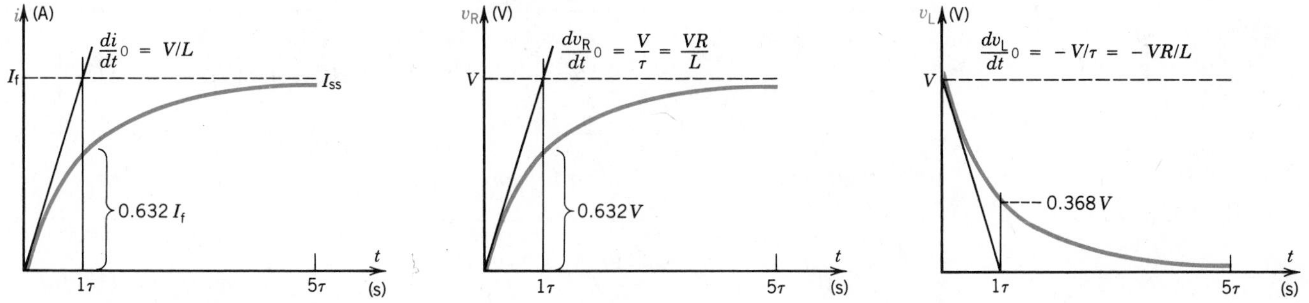

a. Transient i b. Transient v_R c. Transient v_L

Figure 13-3 Initial values di_0/dt, dv_{R_0}/dt, dv_{L_0}/dt during transient current rise

The initial rate of change of voltage across resistor R from Fig. 13-3b is

$$\frac{dv_R}{dt}_0 = \frac{V}{\tau} = \frac{VR}{L} \quad \text{volts/second (V/s)} \tag{13-7}$$

The initial rate of change of voltage across the inductor from Fig. 13-3c is

[5] Recall that L may be expressed in units of Wb/A = (J/A)/A = J/A^2. Then V/L may be written dimensionally as V/(J/A^2) = VA^2/W·s = VA^2/VA·s = A/s.

$$\frac{dv_L}{dt}_0 = \frac{-V}{\tau} = \frac{-VR}{L} \quad \text{volts/second (V/s)} \qquad \textbf{(13-8)}$$

Equation (13-6) actually *defines* the time constant of an RL circuit as the time it takes for current to rise to its final value, assuming it continued to rise at its initial rate of change $\left(\frac{di}{dt}_0\right)$ continuously, that is, $\tau = I_f \frac{di}{dt}_0$.

A close examination of Eqs. (13-7) and (13-8) shows that dv_L/dt is, *at all times* and for any given value of t, merely the **reverse** of dv_R/dt, that is, $dv_L/dt = -dv_R/dt$.

13-2.2 General Equations for Rates of Change during Transient Current Rise

Intuitively, we know that the initial rates of change in Eqs. (13-6), (13-7), and (13-8) are the *maximum* rates of change, respectively, for di/dt, dv_R/dt, and dv_L/dt. We also know from Table 10-1 that these rates of change *decrease* exponentially with time. We may therefore write general equations for the three rates of change for any values of t, including the *initial* rate of change where $t = 0$.

The rate of change of **current** at *any* value of t is

$$\frac{di}{dt} = \frac{V}{L}\varepsilon^{-t/\tau} = \frac{I_f}{\tau}\varepsilon^{-tR/L} \quad \text{amperes/second (A/s)} \qquad \textbf{(13-6a)}$$

The rate of change of voltage across resistor R at *any* value of t is

$$\frac{dv_R}{dt} = \frac{V}{\tau}\varepsilon^{-t/\tau} = \frac{VR}{L}\varepsilon^{-tR/L} \quad \text{volts/second (V/s)} \qquad \textbf{(13-7a)}$$

And since $dv_L/dt = -dv_R/dt$ for *any* value of t, we may write

$$\frac{dv_L}{dt} = \frac{-dv_R}{dt} = \frac{-V}{\tau}\varepsilon^{-t/\tau} = \frac{-VR}{L}\varepsilon^{-tR/L} \quad \text{volts/second (V/s)} \qquad \textbf{(13-8a)}$$

EXAMPLE 13-2

Verify the rate of change of current found in Ex. 13-1g when $t = 1\tau$, using Eq. (13-6a)

Solution

$$\frac{di}{dt} = \frac{V}{L}\varepsilon^{-t/\tau} = \frac{12}{10}\varepsilon^{-1} = \mathbf{0.4415\ A/s} \qquad \text{(13-6a)}$$

EXAMPLE 13-3

Given the data of Ex. 13-1, find the initial rates of change of current, i, voltage across the resistor, v_R, and voltage across the inductor, v_L, using

a. Eqs. (13-6), (13-7), and (13-8), respectively
b. Eqs. (13-6a), (13-7a), and (13-8a), respectively

Solution

a. $$\frac{di}{dt}_0 = \frac{V}{L} = \frac{12\ \text{V}}{10\ \text{H}} = \mathbf{1.2\ A/s} \qquad \text{(13-6)}$$

$$\frac{dv_R}{dt}_0 = \frac{V}{\tau} = \frac{12\ \text{V}}{1.25\ \text{s}} = \mathbf{9.6\ V/s} \qquad \text{(13-7)}$$

$$\frac{dv_L}{dt}0 = -V/\tau = -12\ \text{V}/1.25\ \text{s} = \mathbf{-9.6\ V/s} \quad (13\text{-}8)$$

b. $$\frac{di}{dt}0 = \frac{V}{L}\varepsilon^{-t/\tau} \text{ but when}$$

$$t = 0,\ \varepsilon^{-0} = 1, \text{ and therefore} \quad (13\text{-}6a)$$

$$\frac{di}{dt} = V/L = 12\ \text{V}/10\ \text{H} = \mathbf{1.2\ A/s}$$

$$\frac{dv_R}{dt}0 = \frac{V}{\tau}\varepsilon^{-t/\tau} = \frac{V}{\tau} \text{ when } t = 0 \quad (13\text{-}7a)$$

$$= 12\ \text{V}/1.25\ \text{s} = \mathbf{9.6\ V/s}$$

$$\frac{dv_L}{dt}0 = \frac{-V}{\tau}\varepsilon^{-t/\tau} = -V/\tau \text{ when } t = 0 \quad (13\text{-}8a)$$

$$= -12\ \text{V}/1.25\ \text{s} = \mathbf{-9.6\ V/s}$$

Example 13-3 shows that the *general* equations for rates of change during transient current rise may be used for both initial rates of change and rates of change for any given values of t during the transient. The example also shows that it is possible to find *all* rates of change in RL circuits *without* resorting to differential equations.

The following additional examples are included to show some of the practical applications of the equations presented so far.

EXAMPLE 13-4

The resistance of an air-core solenoid measures 5 Ω on a Wheatstone bridge. If it takes 500 ms for the current in the solenoid to reach its steady-state value (regardless of the voltage applied), calculate the inductance of the solenoid.

Solution

$$\tau = \frac{t}{5} = \frac{500\ \text{ms}}{5} = 0.1\ \text{s}$$

$$L = R\tau = 5 \times 0.1 = \mathbf{0.5\ H} \quad (13\text{-}1)$$

EXAMPLE 13-5

A relay coil having a resistance of 240 Ω and an inductance of 600 mH is designed to operate on 48 V dc and close its contacts when 150 mA is drawn from the supply. Calculate the

a. Time for the relay current to reach steady state
b. Steady-state current in the relay coil
c. Time for the relay to close its contacts

Solution

a. $5\tau = 5L/R = 5 \times 0.6/240 =$ **12.5 ms**
b. $I_f = V/R = 48\ \text{V}/240\ \Omega =$ **200 mA**
c. Solving for the expression $\varepsilon^{-t/\tau}$ in Eq. (13-3) yields $\varepsilon^{-t/\tau} = 1 - \frac{i}{I_f} = 1 - \frac{150\ \text{mA}}{200\ \text{mA}} = 1 - 0.75 = 0.25$, then $\varepsilon^{t/\tau} = 1/0.25 = 4$, and taking the ln of each side yields $t/\tau = \ln 4$, from which $t = \tau \ln 4 = 2.5\ \text{ms} \times \ln 4 =$ **3.466 ms.**

Example 13-4 shows a way of finding the inductance of a coil experimentally by using just an ammeter and a stopwatch. It also shows that the time constant is independent of voltage applied or the final steady-state current.

Example 13-5c shows an important algebraic transposition technique in solving for t in the equation $i = I_f(1 - \varepsilon^{-t/\tau})$. The reader should practice this algebraic technique until it is mastered. (See also Eq. 13-16 in Sec. 13-4.)

13-3 CURRENT–VOLTAGE RELATIONS DURING TRANSIENT CURRENT DECAY

Let us assume that the circuit in Fig. 13-2a has been energized for so long that all transients have disappeared. The current in the RL circuit is $I_{ss} = I_f = V/R$ (Fig. 13-2b). The voltage across the resistance is $v_R = V = I_fR$, (as shown in Fig. 13-2b). Since the rate of change of current, di/dt, is zero, the voltage across the inductor, v_L, is zero (recall $v_L = L\,di/dt$), as shown in Fig. 13-2c.

The circuit of Fig. 13-2a is shown once again in **Fig. 13-4a**. Assume that the switch has remained in position 1 for a very long time, so the conditions of the previous paragraph are the initial conditions. If the switch is rapidly and instantly thrown to position 2 at $t = 0^+$ s, the following *initial* conditions exist:

$I_{ss} = I_0 = V/R$ since the *current cannot change instantly* in an *RL* circuit.
$v_R = I_0R = V_0$ since the voltage across a resistor is a function of current.
$v_L = -V_0 = -I_0R$ since v_L is in parallel with v_R in Fig. 13-4a.

The reader should compare the polarity marks on the inductor in Fig. 13-2a against that of Fig. 13-4a. Note that a *reversal in polarity* has occurred across v_L when the switch is in position 2. The reasons for the polarity reversal are that

1. The stationary field energy ($W = LI^2/2$), which was maintained by the (final) steady-state current, is no longer sustained by the supply voltage. The field therefore collapses (changes suddenly), inducing an EMF in coil L.
2. The EMF of self-induction in coil L is v_L, which opposes the change in current; i.e., it attempts to *maintain* the current in the *same* direction. Recall that the property of L is to oppose a current change.
3. As noted from Fig. 13-4a, the polarity across L is such that it tends (by Lenz's law) to maintain a *clockwise* flow through L and R, maintaining the original current direction.
4. Since v_L is in parallel with v_R when the switch is in position 2, the same polarity obtains in maintaining current i through R and L.

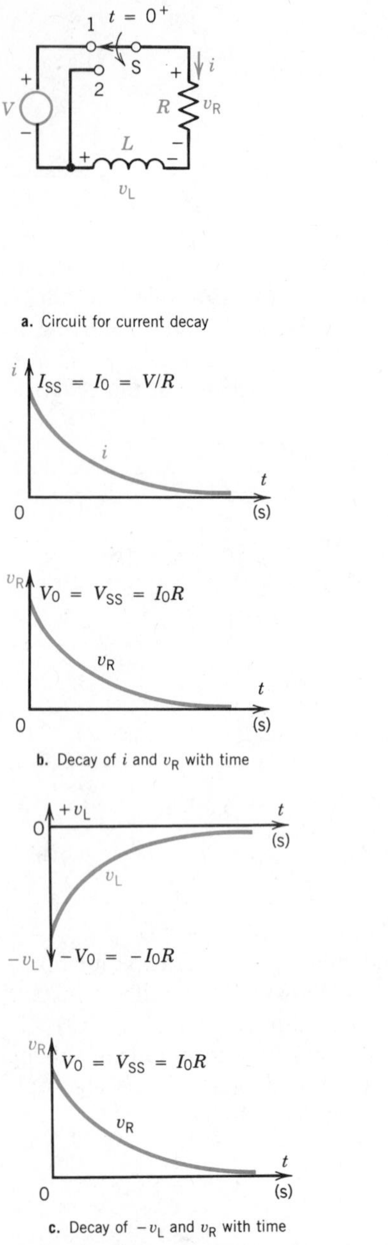

Figure 13-4 Transient decay of i, v_R, and v_L with time

So much for the initial conditions, which are shown for i, v_R, and v_L, respectively in Figs. 13-4b and 13-4c. Note that as the switch remains in position 2, there is a transient decay toward zero in all three parameters. This decay occurs as long as the energy stored in the magnetic field is dissipated in the resistor R. Observe that all three waveforms have the same shape following the equation $y = K\varepsilon^{-x}$. This enables us to write the following equations, which occur during *transient decay*:

$$i = i_L = i_R = I_0\varepsilon^{-t/\tau} = \frac{V}{R}\varepsilon^{-tR/L} \quad \text{amperes (A)} \qquad (13\text{-}9)$$

$$v_R = V_0\varepsilon^{-t/\tau} = I_0R\varepsilon^{-tR/L} \quad \text{volts (V)} \qquad (13\text{-}10)$$

$$(-)v_L = -V_0\varepsilon^{-t/\tau} = -I_0R\varepsilon^{-tR/L} \quad \text{volts (V)} \qquad (13\text{-}11)$$

With respect to Eqs. (13-9), (13-10), and (13-11), note that all three parameters experience a decay toward zero. Both i and v_R decay from a positive (or original) steady-state value to zero. Only v_L experiences a reversal in polarity, moving instantly from zero to some negative value and then decaying with time to zero.

The foregoing summary is perhaps better understood when represented graphically. Assume that the switch, S, in Fig. 13-4a remains in each position (1 and 2) respectively for a duration of exactly 5 time constants. **Figure 13-5** shows the results for two successive complete cycles of switch alternation. With respect to Fig. 13-5, please note that

1. The waveform of v_R follows the waveform of i_L at all times, since $v_R = iR$.
2. **Only** the waveform of v_L alternates *positively* and *negatively*. The waveforms of i_L and v_R, while continuously changing, remain positive.
3. The algebraic sum of v_R and v_L is at all times equal to V *during the current buildup* phase in switch position 1, as noted by Eq. (13-2) (Kirchhoff's voltage law).

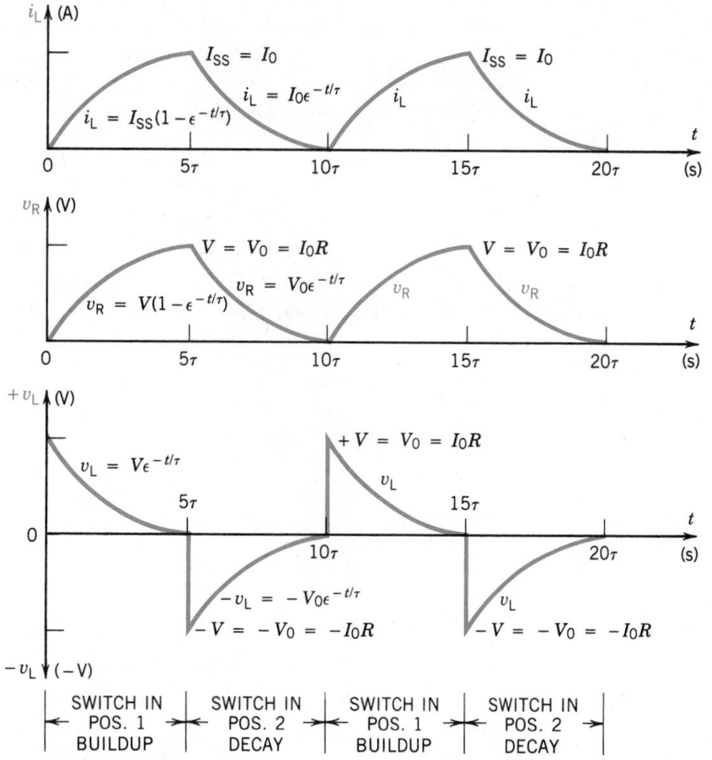

Figure 13-5 Buildup and decay of parameters in an *RL* circuit

4. The algebraic sum of v_R and v_L is at all times **zero** *during the current decay* phase in switch position 2, since they are equal and opposite (Fig. 13-5).
5. All transients, during both buildup and decay, occur in 5τ, after which they remain in their steady state.
6. The final steady-state conditions represent the initial conditions for both i and v_R in going from buildup to decay, and vice versa.
7. The waveform of v_L **always decays** from a maximum positive or maximum negative value to zero.

EXAMPLE 13-6

Given the circuit shown in **Fig. 13-6**, the switch transitions are as follows:

1. At $t = 0^+$ the switch moves from OFF to position 1, where it remains for 15 ms
2. At $t = 15$ ms the switch moves to position 2, where it remains until all transients have disappeared

Calculate:

a. Time constant, τ_1, in position 1 and time to reach steady-state conditions
b. Time constant, τ_2, in position 2 and time to reach steady-state conditions
c. All necessary values to plot i_L versus time
d. All necessary values to plot v_L versus time

Plot i_L and v_L versus time for this switching sequence

Solution

a. $\tau_1 = L/R_1 = 5\text{ H}/2\text{ k}\Omega = \mathbf{2.5\text{ ms}}$ and $5\tau_1 = 5 \times 2.5\text{ ms} = 12.5\text{ ms}$ (13-1)
b. $\tau_2 = L/R_2 = 5\text{ H}/4\text{ k}\Omega = \mathbf{1.25\text{ ms}}$ and $5\tau_2 = 5 \times 1.25\text{ ms} = 6.25\text{ ms}$ (13-1)
c. During current buildup, $I_f = I_{ss} = V/R = 16\text{ V}/2\text{ k}\Omega = \mathbf{8\text{ mA}}$ at $t = 12.5$ ms. During current decay, $I_0 = I_{ss} = 8$ mA which decays to **zero** in **6.25 ms**
d. During buildup, $v_{L_0} = V = \mathbf{16\text{ V}}$, which decays to zero in **12.5 ms.** When S moves to position 2, $v_{L_0} = \mathbf{-16\text{ V}}$ initially, which decays to zero in **6.25 ms**

The waveforms are shown in Fig. 13-6b

With respect to Ex. 13-6 and its solution shown in Fig. 13-6b, note that

1. The switch duration in each position is sufficiently long to reach steady-state conditions for both τ_1 and τ_2.
2. The current rises to its final Ohm's law value of 8 mA in 12.5 ms and decays to zero at time 21.25 ms, after which all transients have disappeared.
3. The voltage across the inductor, v_L, decays in position 1 to zero in 12.5 ms. At exactly 15 ms, when the switch moves to position 2, v_L reverses to -16 V and rapidly decays to zero at time 21.25 ms, after which all transients disappear.

The questions now arise, "What if the time duration in each switch position does *not* permit the transients to reach their steady-state values? What would the waveforms be under such conditions, and how are i_L and v_L computed?" Examples 13-7 and 13-8, using the circuit of Ex. 13-6, answer these questions.

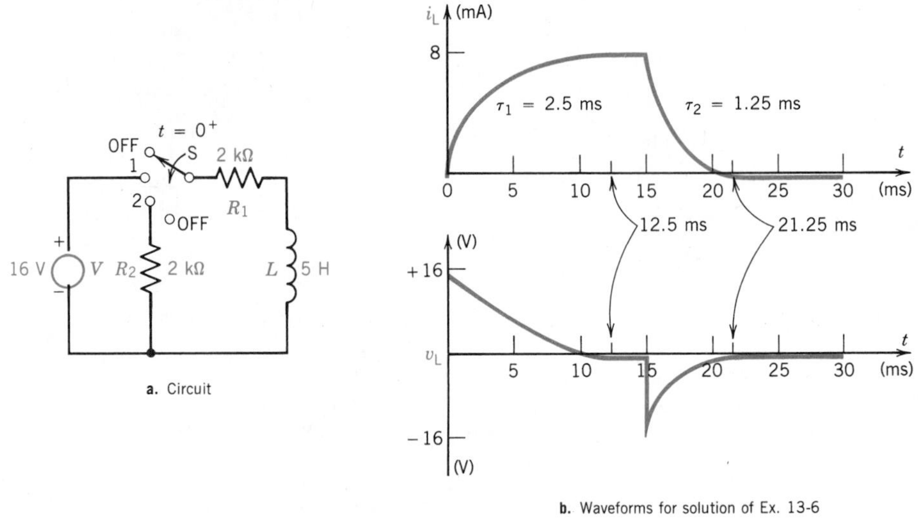

Figure 13-6 Examples 13-6 through 13-8

EXAMPLE 13-7

In the circuit of Fig. 13-6a, switch S moves alternately from position 1 to position 2, remaining for only 1 ms in each position for five successive cycles. Perform all necessary calculations to sketch i_L for the first 5 ms of elapsed time.

Solution

Preliminary calculations given $t_1 = t_2 = 1$ ms. From Ex. 13-6, $\tau_1 = 2.5$ ms and $\tau_2 = 1.25$ ms

Then in position 1, $\varepsilon^{-t/\tau_1} = \varepsilon^{-1/2.5} = \varepsilon^{-0.4}$

While in position 2, $\varepsilon^{-t/\tau_2} = \varepsilon^{-1/1.25} = \varepsilon^{-0.8}$

We first solve for i_L for all switch positions and time durations in each position.[6]

Switch Position	Time (ms)	Calculation for i_L	
OFF		0	
1	1	$i_1 = I_f(1 - \varepsilon^{-0.4}) = 8(0.3297)$	
		$= \mathbf{2.637\ mA}$	(13-3)
2	2	$i_2 = I_0\varepsilon^{-0.8} = 2.637 \times (0.4493)$	
		$= \mathbf{1.185\ mA}$	(13-9)
1	3	$i_3 = (I_f - i_2)(1 - \varepsilon^{-0.4}) + i_2$	
		$= (8 - 1.185)(0.3297) + 1.185$	
		$= \mathbf{3.432\ mA}$	(13-3a)
2	4	$i_4 = i_3\varepsilon^{-0.8} = 3.432 \times (0.4493)$	
		$= \mathbf{1.542\ mA}$	(13-9)
1	5	$i_5 = (I_f - i_4)(1 - \varepsilon^{-0.4}) + i$	
		$= (8 - 1.542)(0.3297) + 1.542$	
		$= \mathbf{3.671\ mA}$	(13-3a)

The solution to Ex. 13-7 used a new equation, (13-3a), which may come as a surprise to the reader. What is the origin of this equation?

The reader may recall Eq. (10-2a), in which an initial voltage exists across a capacitor and the capacitor is attempting to charge to some final value. The

[6] This is an important first step. Since the property of *L* is to oppose any change in current, the *current cannot change* whenever a switch transition occurs. This is not true of v_L, which exhibits *two* values before and after each switch transition. To find v_L we must know i_L. Therefore we first solve for i_L.

analogous situation occurs whenever an inductor is carrying current. The *current cannot change* when the switch transition occurs because the property of an inductor is to *oppose* a change in current. Consequently, the current tries to *rise* to its final value from some *initial* value, I_0, and therefore we may write[7]

$$i = (I_f - I_0)(1 - \varepsilon^{-t/\tau}) + I_0 \qquad \text{amperes (A)} \qquad \textbf{(13-3a)}$$

where I_f is the final value of current determined by Ohm's law as V/R
I_0 is the initial value of current at the instant of switch transition
τ is the inductor time constant, L/R

Having solved Ex. 13-7 for the variation of current with time in the foregoing switching problem, we are now ready for Ex. 13-8. This is somewhat more complex because v_L, as we will discover, has *two distinct values at each switch transition.*

EXAMPLE 13-8

a. Given the circuit of Fig. 13-6a, and the switching sequence specified in Ex. 13-7, calculate values of v_L for the first 5 ms of elapsed time
b. Tabulate values of v_L and i_L (from Ex. 13-7) for the elapsed time of 5 ms
c. Using the tabulation, sketch waveforms of i_L and v_L for the first 5 ms of switching sequence

Solution

We recognize that for *each* switch transition there are *two* values of v_L and must allow for it in our tabulation. (See solution, part **b**, below.)

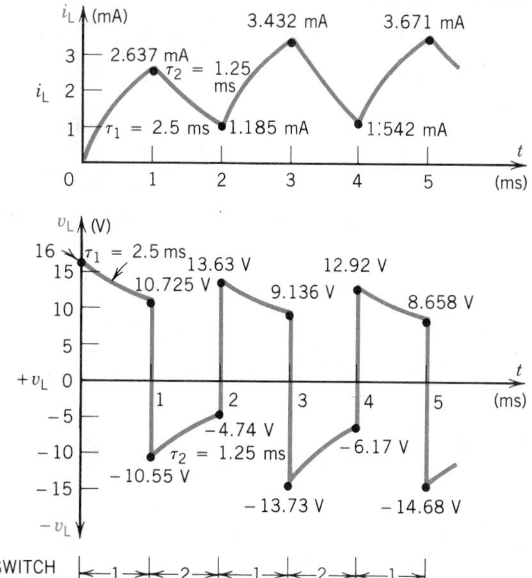

Figure 13-7 Waveforms for solutions of Exs. 13-7 and 13-8

a.

Switch Position	Time (ms)	Calculation for v_L (using values of i_L from Ex. 13-7)
OFF		0 V
1	0^+	16 V (at instant S closes, v_L rises from 0 to **16 V** in 0 s)
1	1	$v_{L_1} = V_0\varepsilon^{-0.4} = 16(0.6703) = \mathbf{10.725\ V}$ (exponential decay)
2	1^+	$-v_{L_1} = -i_{L_1}R_{t_2} = (-2.637\text{ mA})(4\text{ k}\Omega) = \mathbf{-10.55\ V}$ (switch transition voltage)
2	2	$-v_{L_2} = (-v_{L_1})(\varepsilon^{-0.8}) = (-10.55)(0.4493) = \mathbf{-4.739\ V}$ (exponential decay)
1	2^+	$+v_{L_2} = V - i_2R_1 = 16 - (1.185 \times 2) = \mathbf{+13.63\ V}$ (switch transition voltage)
1	3	$+v_{L_3} = (v_{L_2})(\varepsilon^{-0.4}) = 13.63\ (0.6703) = \mathbf{+9.136\ V}$ (exponential decay)
2	3^+	$-v_{L_3} = (-i_{L_3})(R_{t_2}) = (-3.432)(4) = \mathbf{-13.73\ V}$ (switch transition voltage)
2	4	$-v_{L_4} = (-v_{L_3})(\varepsilon^{-0.8}) = (-13.73)(0.4493) = \mathbf{-6.168\ V}$ (exponential decay)
1	4^+	$+v_{L_4} = V - i_4R_1 = 16 - (1.542 \times 2) = \mathbf{12.916\ V}$ (switch transition voltage)
1	5	$+v_{L_5} = (v_{L_4})(\varepsilon^{-0.4}) = 12.916 \times (0.6703) = \mathbf{8.658\ V}$ (exponential decay)
2	5^+	$+v_{L_5} = (-i_{L_5})(R_{t_2}) = (-3.671)(4) = \mathbf{-14.68\ V}$ (switch transition voltage)

[7] The reader should compare Eqs. (5-8), (10-2a), and (13-3a) for similarity in format. This will enhance understanding and insight into their use and application.

b. Tabulation of values of i_L and v_L versus time:

Time (ms)	0	1	1^+	2	2^+	3	3^+	4	4^+	5	5^+
i_L (mA)	0	2.637	2.637	1.185	1.185	3.432	3.432	1.542	1.542	3.671	3.671
v_L (V)	+16	10.725	−10.55	−4.739	13.63	9.136	−13.73	−6.17	12.92	8.658	−14.68

c. Waveforms of i_L and v_L tabulated in **(b)** are shown in **Fig. 13-7.**

Examples 13-7 and 13-8 show that it is possible to calculate values of i_L in switching circuits involving both growth and decay of current with relatively little difficulty. But calculations of v_L require a great deal of care and laborious bookkeeping, as shown in the solution to Ex. 13-8a. The question arises "Is there an *easier* way to obtain the transition voltage of v_L using the transition values of i_L?"

The answer is that there is a simpler way. It only requires knowing the rates of change of current during current rise and current decay. Recall that $v_L = L\,di/dt$, so if we know di/dt and L, we can easily calculate v_L.

13-3.1 General Equations for Transient Rates of Change during Current Decay

Writing general equations for both initial rates of change and rates of change at any time during the transient decay period should not be difficult. All three waveforms (i_L, v_R, and v_L) decay to zero. Consequently, the equations should be similar in form to each other and similar to Eq. (13-8a), derived earlier, for *decay* of v_L during *transient-current rise.* Consequently, for each we may write

$$\frac{di_L}{dt} = -\frac{I_{ss}}{\tau_d}\varepsilon^{-t/\tau_d} = \left(-\frac{I_0}{\tau_d}\right)\varepsilon^{-Rt/L} \quad (-)\text{ amperes/second (A/s)} \qquad (13\text{-}12)$$

$$\frac{dv_R}{dt} = -\frac{V_{ss}}{\tau_d}\varepsilon^{-t/\tau_d} = \frac{-V_0 R}{L}\varepsilon^{-Rt/L} \quad (-)\text{ volts/second (V/s)} \qquad (13\text{-}13)$$

$$\frac{dv_L}{dt} = -\frac{-V_0}{\tau_d}\varepsilon^{-t/\tau_d} = \left(-\frac{-V_{ss}}{\tau_d}\right)\varepsilon^{-Rt/L} \quad (+)\text{ volts/second (V/s)} \qquad (13\text{-}14)$$

With respect to these equations, please note the following *caveats*:

1. The rates of change during current decay of di_L/dt and dv_R/dt are *both negative* at all times. Slopes drawn tangent to the curves of i and v_R in Fig. 13-4b will verify this.
2. But dv_L/dt (Eq. 13-14) always has a positive rate of change since slopes drawn to the (negative) curve of v_L during current decay (Fig. 13-4c) are *always positive.*
3. The initial values or steady-state values are *not always* those obtained after 5τ. Switching may occur *before* the current reaches its final value.
4. The values of τ_d that appear in the denominators may not be the same for decay as during current rise. (See Exs. 13-7 and 13-8.)

Armed with the foregoing relations, we can now return to Ex. 13-8, in which we calculated v_L for a given switching sequence. From Eq. (13-6a) we know that during current rise (from zero) our rate of change of current, $di/dt = \dfrac{I_f}{\tau_r}\varepsilon^{-t/\tau_r}$ at any value of t. But this Eq. (13-6a) does not take into account the situation where the current is rising from some *initial current* to a final value. In such a case the rate of change of current may be written as

$$\frac{di_L}{dt} = \frac{I_f - I_0}{\tau_r}\varepsilon^{-t/\tau_r} = \frac{I_f - I_0}{\tau}\varepsilon^{-Rt/L} \quad (+)\text{ amperes/second (A/s)} \quad (13\text{-}15)$$

where I_0 is the initial current at the beginning of a switch transition
I_f is the final current or Ohm's law value (V/R)
t is the time at which the rate of change is desired
τ_r is the ratio L/R for the given switching situation, during current *rise*

EXAMPLE 13-9

Given only the current waveform shown in Fig. 13-7 for i_L, calculate the

a. Positive rate of change of current (di/dt) and v_L when $t = 1$ ms
b. Negative rate of change of current and v_L when $t = 1^+$ ms
c. Positive rate of change of current and v_L when $t = 2^+$ ms
d. Negative rate of change of current and v_L when $t = 2$ ms

Solution

a. $$di/dt = (I_f/\tau_1)(\varepsilon^{-t/\tau}) = \frac{8\text{ mA}}{2.5\text{ ms}}(\varepsilon^{-1/2.5}) = (3.2\text{ A/s})(\varepsilon^{-0.4})$$
$$= \mathbf{2.145\ A/s} \quad (13\text{-}6a)$$
$$v_L = L\,di/dt = 5\text{ H }(2.145\text{ A/s}) = \mathbf{10.725\ V}\text{ (Fig. 13-7)}$$

b. $$-di/dt = \frac{-I_0}{\tau_2}\varepsilon^{-t/\tau_2} = -I_0/\tau_2 = -2.637\text{ mA}/1.25\text{ ms}$$
$$= \mathbf{-2.11\ A/s} \quad (13\text{-}12)$$
$$-v_L = L\,di/dt = 5\text{ H }(-2.11\text{ A/s})$$
$$= \mathbf{-10.55\ V} \quad \text{(Fig. 13-7)}$$

c. $$di/dt = \frac{I_f - I_0}{\tau_1}\varepsilon^{-t/\tau_1} = \frac{(8 - 1.185)\text{ mA}}{2.5\text{ ms}}\varepsilon^{-0/2.5}$$
$$= \mathbf{2.726\ A/s} \quad (13\text{-}15)$$
$$v_L = L\,di/dt = 5\text{ H }(2.726\text{ A/s}) = \mathbf{13.63\ V}$$

d. $$-di/dt = \frac{-I_0}{\tau_2}\varepsilon^{-t/\tau_2} = \frac{-2.637\text{ mA}}{1.25\text{ ms}}\varepsilon^{-1/1.25}$$
$$= (-2.11\text{ A/s})(\varepsilon^{-0.8})$$
$$= \mathbf{-0.9481\ A/s} \quad (13\text{-}12)$$
$$-v_L = L\,di/dt = 5\text{ H }(-0.9481\text{ A/s})$$
$$= \mathbf{-4.74\ V} \quad \text{(Fig. 13-7)}$$

Each part of Ex. 13-9 warrants explanation. Let us take each step separately and discuss the rationale for the equations used.

Part (**a**): We know that the current is rising positively from 0 to 8 mA. The initial rate of change of current is $I_f/\tau_1 =$ 8 mA/2.5 ms = 3.2 A/s. But this initial rate of change decreases exponentially with time. Consequently, at the end of 1 ms, $di/dt = 2.145$ A/s. This value of di/dt is used to find v_L by using $L\,di/dt$. Note that this value of 10.725 V agrees with the value obtained in Ex. 13-8a when $t = 1$ ms.

Part (**b**): The *current changes direction* at 1^+ ms. The rate of change of current is therefore negative as the current decays exponentially toward zero. The negative slope we are now seeking is the *initial rate of change.* Since the switch is now in position 2, we use the current-decay time constant τ_2 or $-di/dt = -I_0/\tau_2$. The negative rate of change of current yields the $-v_L$ value of -10.55 V. This agrees with the value obtained in Ex. 13-8a when $t = 1^+$ ms.

Part (**c**): The current *again changes direction* at $t = 2^+$ ms. Only now it is trying to *rise* to 8 mA from an initial value of 1.185 mA. Therefore, the initial positive rate of change of current (in position 1) is $di/dt\ 0 = (I_f - I_0)/\tau_1$, since $t = 0$. The positive rate of change of current yields a positive value of $v_L = 13.63$ V. This agrees with the value obtained in Ex. 13-8a when $t = 2^+$ ms.

Part (**d**): From part (**b**) we saw that the initial rate of change of current, $-di/dt = -I_0/\tau_2$, when the current reverses at 2.637 mA in Fig. 13-7. This is the maximum rate of change or -2.11 A/s. Since the current decays exponentially toward zero, we must use Eq. (13-12) or $(-I_0/\tau_2)\,\varepsilon^{-t/\tau_2}$, yielding -0.9481 A/s. The negative rate of change of current yields a negative value of $v_L = -4.74$ V. This agrees favorably with the value obtained in Ex. 13-8a when $t = 2$ ms.

We have learned from Ex. 13-9 the following insights:

1. The waveform of v_L is at all times the waveform of i_L differentiated, since $v_L = L\,di/dt$.
2. The value of v_L in Ex. 13-8 was found by using two different equations. But the value of v_L in Ex. 13-9 was found by using the *same* technique for both positive and negative values.
3. Obtaining the current waveform is relatively simple. From it, we can reconstruct the waveform of v_R, since v_R is at all times the product iR.
4. From the current waveform we can find its rate of change at any time and then obtain v_L from $L\,di/dt$.
5. Using Eqs. (13-6a), (13-12), and (13-15), we have eliminated the need for differential equations in finding v_L.

13-4 ALGEBRAIC SOLUTION OF TIME DURING TRANSIENT RISE AND DECAY OF CURRENT

We can now write two fundamental equations for finding **time** at any point along the curve of current rise and/or current decay, respectively. As in the case of capacitors, we will write these relations for current in the inductor (only) since we have seen that i_L is the fundamental parameter for *RL* circuits. Let us define the time constant during current *rise* as τ_r and the time constant during current *decay* as τ_d, respectively.

Then the time at any point on the curve for exponential *rise* of current is[8]

$$t = \tau_r \times \ln\left(\frac{I_f - I_0}{I_f - I_1}\right) \quad \text{seconds (s)} \tag{13-16}$$

where I_f is the steady-state value of current as determined by V/R_t
I_0 is the initial value of current at which current *rise* begins
I_1 is the current at any point on the curve for which time is desired

Similarly, the time at any point on the curve during exponential *decay* of current is[9]

$$t = \tau_d \times \ln\left(\frac{I_f}{I_1}\right) = \tau_d \times \ln\left(\frac{I_0}{I_1}\right) \quad \text{seconds (s)} \tag{13-17}$$

where I_f is the steady-state or maximum initial value I_0 from which current *decays*
I_1 is current at any point on the *decay curve* for which time is desired

Let us verify Eqs. (13-16) and (13-17) from the curves drawn in Fig. 13-7 and the following examples.

EXAMPLE 13-10

Assume the current rises from 0 to a final value of 8 mA in an *RL* circuit whose time constant is 2.5 ms. How long would it take the current to reach a value of 2.637 mA (as shown in Fig. 13-7)?

Solution

$$t = \tau_r \ln\left(\frac{I_f - I_0}{I_f - I_1}\right) = 2.5 \text{ ms} \ln\left(\frac{8 - 0}{8 - 2.637}\right)$$
$$= 2.5 \text{ ms} \ln(1.4917) = \mathbf{1\ ms} \quad (13\text{-}16)$$

EXAMPLE 13-11

Assume that current rises from an initial value of 1.185 mA to a final value of 8 mA in an *RL* circuit whose time constant is 2.5 ms. How long would it take the current to reach a value of 3.432 mA (Fig. 13-7)?

Solution

$$t = \tau_r \ln\left(\frac{I_f - I_0}{I_f - I_1}\right) = 2.5 \text{ ms} \ln\left(\frac{8 - 1.185}{8 - 3.432}\right) = \mathbf{1\ ms} \quad (13\text{-}16)$$

EXAMPLE 13-12

Assume that current decays from an initial value of 2.637 mA to an instantaneous given value of 1.185 mA in an *RL* circuit whose time constant is 1.25 ms. Calculate the time of decay (see Fig. 13-7).

Solution

$$t = \tau_d \ln(I_0/I_1) = 1.25 \text{ ms} \ln(2.637/1.185) = \mathbf{1\ ms} \quad (13\text{-}17)$$

---

[8] The reader should compare this equation with Eq. (10-5a) for capacitors.
[9] The reader should compare this equation with Eq. (10-15) for capacitors.

EXAMPLE 13-13

How long would it take a waveform of v_L to decay from an initial value of 16 V to a value of 10.725 V in an *RL* circuit whose time constant is 2.5 ms? (See Fig. 13-7.)

Solution

Since Eq. (13-17) is derived for a current *decay* relation, it should apply to a *voltage decay* equally well. Therefore,

$$t = \tau_d \ln(V_0/V_1) = 2.5 \text{ ms} \ln(16/10.725)$$
$$= \mathbf{1\ ms} \qquad \text{(see Fig. 13-7)} \qquad (13\text{-}17)$$

Examples 13-10 to 13-13 enable us to infer the following insights:

1. Equation (13-16) is a generic equation that may be used for *any rising* exponential.
2. Equation (13-17) is a generic equation that may be used for *any decaying* exponential.
3. It is much easier to use Eq. (13-16) than to solve for t in $i = I_f(1 - \varepsilon^{-t/\tau_r})$, during current rise.
4. Similarly, it is much easier to use Eq. (13-17) than to solve for t in $i = I_0\varepsilon^{-t/\tau_d}$, during current decay.

13-5 EQUATION AND WAVEFORM SUMMARIES

We can now present a summary table expressing pertinent equations during current rise and current decay in *RL* circuits. This table should serve as a useful reference for solving problems at the end of this chapter as well as for the examples that follow.

Table 13-1 General Equations for Rise and Decay of Current in RL Circuits

Parameter	Current Rise Cycle		Current Decay Cycle	
i_L	$(I_f - I_0)(1 - \varepsilon^{-t/\tau_r}) + I_0$	(13-3a)	$I_0\varepsilon^{-t/\tau_d} = I_0\varepsilon^{-tR_d/L}$	(13-9)
v_R	$i_L \times R = R[(I_f - I_0)(1 - \varepsilon^{-t/\tau_r}) + I_0]$	(13-4) (13-3a)	$I_0R\varepsilon^{-t/\tau_d} = I_0R_d\varepsilon^{-tR_d/L}$	(13-10)
v_L	$V_0\varepsilon^{-t/\tau_r}$	(13-5)	$-V_0\varepsilon^{-t/\tau_d} = -V_0\varepsilon^{-tR_d/L}$	(13-11)
$\dfrac{di_L}{dt}$	$\dfrac{V}{L}\varepsilon^{-t/\tau_r} = \dfrac{I_f}{\tau_r}\varepsilon^{-tR/L} = \dfrac{I_f - I_0}{\tau_r}\varepsilon^{-Rt/L}$	(13-6a) (13-15)	$-\left(\dfrac{I_{ss}}{\tau_d}\right)\varepsilon^{-t/\tau_d} = \dfrac{-I_0}{\tau_d}\varepsilon^{-Rt/L}$	(13-12)
$\dfrac{dv_R}{dt}$	$\dfrac{V}{\tau_r}\varepsilon^{-t/\tau_r} = \dfrac{VR}{L}\varepsilon^{-tR/L}$	(13-7a)	$-\dfrac{V_{ss}}{\tau_d}\varepsilon^{-t/\tau_d} = \dfrac{-V_0R_d}{L}\varepsilon^{-t/\tau_d}$	(13-13)
$\dfrac{dv_L}{dt}$	$\dfrac{-dv_R}{dt} = \dfrac{-V}{\tau_r}\varepsilon^{-t/\tau_r} = \dfrac{-VR}{L}\varepsilon^{-tR/L}$	(13-8a)	$-\dfrac{-V_0}{\tau_d}\varepsilon^{-t/\tau_d} = \left(\dfrac{V_{ss}R_d}{L}\right)\varepsilon^{-tR_d/L}$	(13-14)
t	$\tau_r \times \ln\left(\dfrac{I_f - I_0}{I_f - I_1}\right)$	(13-16)	$\tau_d \times \ln\left(\dfrac{I_f}{I_1}\right) = \tau_d \times \ln\left(\dfrac{I_0}{I_1}\right)$	(13-17)

We may now make a comparison between the waveforms that occur in *RL* circuits and those in *RC* circuits. Assume that in both circuits shown in **Fig. 13-8** the switch remains in position 1 for 5τ s, after which it moves to position 2, where it remains for 5τ. The left-hand column shows the waveforms for the *RL* circuit. With respect to the *RL* circuit, note that

1. Current, i_L, rises to its final steady-state value in $5\tau_1$ s. When the switch moves to position 2, current decays to zero in $5\tau_2$ s. Since the property of an *RL* circuit is to oppose change in current, the current, i_L, decays to zero with any abrupt change or reversal.
2. The voltage across resistor R, v_R, must always follow current since $v_R = i_LR$. The waveform of v_R is identical to i_L at all times.
3. Only the waveform of v_L undergoes *reversal*, since $v_L = L\,di/dt$. Waveform v_L decays from its maximum positive value of V to zero where the rate of change of current is zero at $5\tau_1$. When the current, i_L, begins to decrease, its rate of change is a maximum negative value. Consequently, v_L abruptly reverses and decays to zero after $5\tau_2$. During both current rise and current decay, v_L is always a decaying waveform.

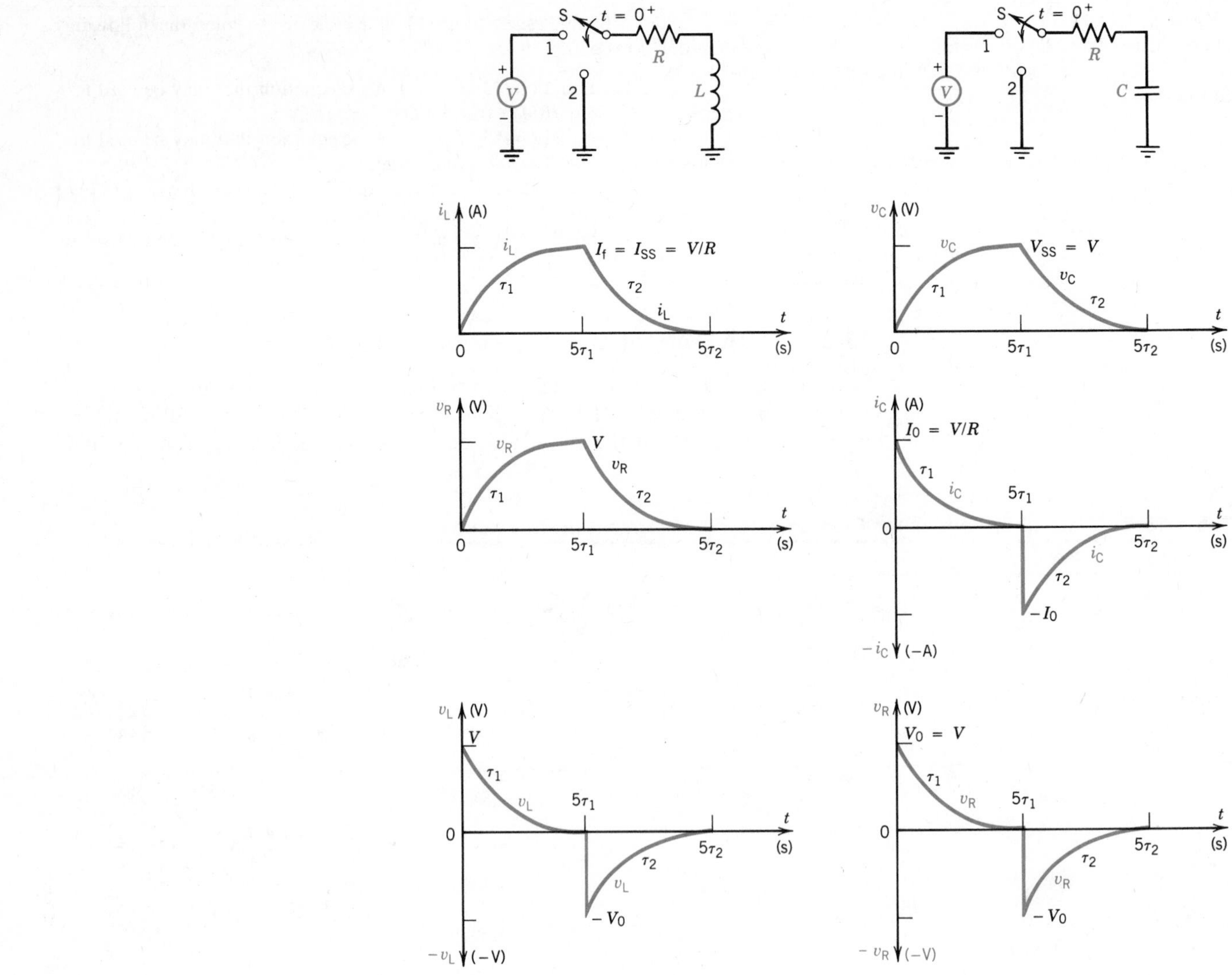

a. *RL* circuit waveforms b. *RC* circuit waveforms

Figure 13-8 Comparisons between waveforms in *RL* and *RC* circuits

In comparison to *RL* circuits, the *RC* waveforms shown in Fig. 13-8 at the right exhibit some similarities and some differences. With respect to the *RC* waveforms, please note:

1. Voltage across the capacitor, v_C, rises to its final steady-state value in $5\tau_1$ and decays to zero in $5\tau_2$. Since the property of an *RC* circuit is to oppose a change in voltage across the capacitor, v_C, the voltage rises and decays without any abrupt change or reversal.
2. But the waveform of current, i_C, in the capacitive circuit decays to zero in position 1 as the capacitor charges. It undergoes an abrupt reversal when switched to position 2 and also decays to zero.
3. The waveform of v_R must always follow current, since $v_R = iR$. The waveform of v_R is identical to the waveform of i_C.

4. Consequently, in *RC* circuits two waveforms exhibit reversal, compared to only one in *RL* circuits. As expected, the **reversal waveforms always decay to zero**, whether they are of positive or negative polarity.

There is one last insight to be gained from an examination of Fig. 13-8. We know that inductance is the property to oppose a change in current, i_L, and capacitance the property to oppose a change in voltage, v_C. Since i_L and v_C are analogs of each other, their waveforms are **identical in shape**. Similarly, v_L, the voltage across the inductor, is the analog of current in the capacitor, i_C. Note also that their waveforms are **identical in shape**. Finally, v_R in either circuit **must always follow current**—which it does.

13-6 *RL* CIRCUIT PROBLEMS INVOLVING MULTIVOLTAGE SOURCES

We are now prepared to consider *RL* circuits of a more complex nature. They should not be difficult because we have already encountered similar problems in Chapter 10 on *RC* circuit transients. Examples 13-14 through 13-17 should be read most carefully and attempted independently by the reader, using the solutions shown as feedback for proper problem-solving techniques.

EXAMPLE 13-14

Given the *RL* circuit shown in **Fig. 13-9a**, calculate the

a. Equivalent Thévenin voltage that exists at terminal 1
b. Equivalent Thévenin resistance at terminal 1
c. Circuit time constant when the switch is in position 1, τ_1
d. Circuit time constant when the switch is in position 2, τ_2

Solution

a. $V_{th} = \frac{R_n}{R_T}(V_p - V_n) + V_n = \frac{10\text{ k}}{50\text{ k}}[36 - (-4)] + (-4)\text{ V}$

$= \mathbf{4\ V}$ (5-8)

b. $R_{th} = R_1 \| R_2 + R_3 = (40 \| 10 + 2)\text{ k}\Omega = \mathbf{10\ k\Omega}$

c. $\tau_1 = L/R_{t_1} = 30\text{ H}/(10 + 5)\text{ k}\Omega = \mathbf{2\ ms}$ (13-1)

d. $\tau_2 = L/R_{t_2} = 30\text{ H}/5\text{ k}\Omega = \mathbf{6\ ms}$ (13-1)

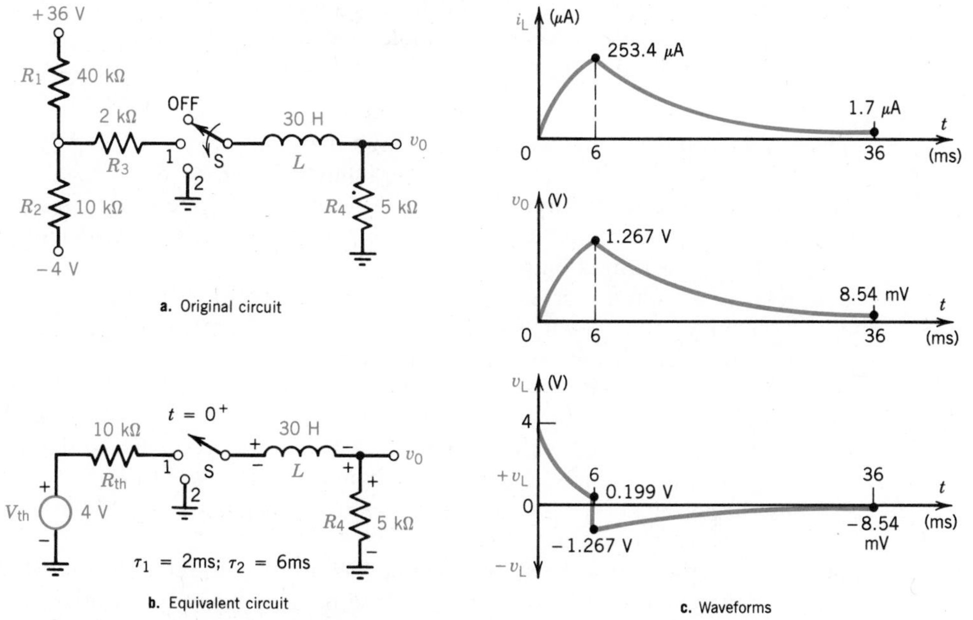

Figure 13-9 Examples 13-14 and 13-15

EXAMPLE 13-15

Given the circuit and data calculated in Ex. 13-14, the switch sequence is that S is thrown to position 1 at $t = 0^+$, where it remains for 6 ms, and S is thrown to position 2 at $t = 6^+$ ms, where it remains for 30 ms. Calculate the

a. Values of i_L at $t = 0^+$, 6, and 36 ms
b. Values of output voltage v_O at $t = 0^+$, 6, and 36 ms. Note R_4 is the output resistor
c. Values of voltage across the inductor, v_L, at $t = 0^+$, 6, 6^+, and 36 ms
d. Draw waveforms showing the variation of i_L, v_O, and v_L with time for 0 to 36 ms

Solution

a. $I_f = I_{ss} = V/R_{t_1} = 4\text{ V}/15\text{ k}\Omega = 266.\bar{6}\ \mu\text{A}$. At $t = 0$, $i_L = 0$. At the end of 6 ms, $i_1 = I_f(1 - \varepsilon^{-t_1/\tau_1}) = 266.\bar{6}(1 - \varepsilon^{-6/2}) = 253.4\ \mu\text{A}$ (13-3)
At the end of 36 ms, $i_2 = I_0\varepsilon^{-t_2/\tau_2} = (253.4\ \mu\text{A})\varepsilon^{-30/6} = 1.71\ \mu\text{A}$ (13-9)

b. Since $v_O = i_L R_O$, at $t = 0$, $v_O = 0$. At the end of 6 ms, $v_{O1} = i_1 R_O = (253.4\ \mu\text{A})(5\text{ k}\Omega) = 1.267\text{ V}$. At the end of 36 ms, $v_{O2} = i_2 R_O = (1.71\ \mu\text{A})(5\text{ k}\Omega) = 8.54\text{ mV}$

c. At $t = 0$, $v_L = V_{th} = V_0 = 4\text{ V}$. At the end of 6 ms, $v_{L_1} = V_0\varepsilon^{-t_1/\tau_1} = 4\varepsilon^{-6/2} = 0.199\text{ V} = \mathbf{199\ mV}$ (13-5)

At 6^+ ms, in position 2, $v_{L_2} = -1.267$ V by inspection since $v_L = -v_O$. The proof of this is as follows:
Initial rate of change of current at $t = 6^+$ ms is

$$\frac{di_{O2}}{dt} = \frac{-I_0}{\tau_2} = \frac{-253.4\ \mu\text{A}}{6\text{ ms}} = -42.2\bar{3}\text{ mA/s} \quad (13\text{-}12)$$

Then $v_{L_{O2}} = L(di/dt) = 30\text{ H}(-42.2\bar{3}\text{ mA/s}) = \mathbf{-1.267\ V}$. At 36 ms, in position 2, $v_{L_2} = -v_{O2} = -8.54$ mV by inspection since $v_L = -v_O$. The proof is as follows:

$$v_{L_2} = v_{L_{O2}}\varepsilon^{-t_2/\tau_2} = (-1.267\text{ V})\varepsilon^{-30/6} = \mathbf{-8.54\ mV} \quad (13\text{-}11)$$

d. The waveforms are drawn in Fig. 13-9c

Several interesting conclusions may be drawn from Exs. 13-14 and 13-15. Let us consider Ex. 13-14 first. The original circuit shown in Fig. 13-9a must first be converted to the equivalent circuit shown in Fig. 13-9b. Once this equivalent is drawn it is a relatively simple matter to find the *L*/*R* time constant in either position 1 or position 2. We may therefore conclude that

In solving multivoltage source problems, it is absolutely essential to calculate and draw the simplest Thévenin voltage and resistance equivalents *before* attempting calculation of i_L, v_R, and v_L.

The solution of Ex. 13-15 actually appears in the waveforms drawn to Fig. 13-9c. Students should attempt to solve parts **(a)**, **(b)**, and **(c)** independently to see if their calculations produce these waveforms. If not, they should turn to the solution portions of Ex. 13-15 to find their sources of error. Several interesting conclusions emerge from Ex. 13-15:

1. The output voltage across resistor R_4 follows the same waveform as i_L. (This should be obvious by now since v_R always follows i_L and is $i_L R$.)
2. When the switch is in position 2, the waveforms of v_O and v_L are reverse images of each other. This occurs because v_L is in series with and also across output resistor R_4, in position 2 only.
3. The polarity of voltage across R_4 is always positive because the current i_L is always positive when S switches from position 1 to position 2.
4. v_L must always act to oppose any change in current.
   a. In position 1 current is *rising* so the polarity of v_L is (+) on the *left* side of *L* in Fig. 13-9b.
   b. In position 2, current in the inductor and in R_4 is decaying. The inductor opposes this decay; (i.e., it tries to maintain current in R_4.) In order to do this, the polarity of v_L is (−) on the left side of *L*, as shown in Fig. 13-9b. This accounts for the reversal of v_L!
5. The polarity of v_L is also found in position 2 as positive on the right side of *L* since inductor *L* is in parallel with R_4, whose polarity is always positive.
6. In the solution for values of v_L, therefore, note that it is possible to write the values of v_L, in position 2 only, *by inspection*, given the values of voltage across R_4 previously calculated. At all times, $v_L = -v_O = -V_{R4}$ in position 2.

EXAMPLE 13-16

For the *RL* circuit shown in **Fig. 13-10a**, switch S contacts position 1 at $t = 0^+$ s and remains there for 1 ms. It then instantly switches to position 2, where it remains until steady-state current is reached. Calculate the

a. Maximum (final) current in position 1
b. Coil current at 1 ms in position 1 (just before switch transition)
c. Maximum (final) current in position 2 and time to reach it
d. Time to reach 0.4 A in position 2
e. v_L at $t = 0^+$ and $t = 1$ ms in position 1
f. v_L at $t = 1^+$ ms in position 2
g. v_L at $t = 1.26$ ms in position 2 when i_L reaches 0.4 A
h. Plot waveforms of i_L and v_L using the values just calculated
i. Explain why v_L never reverses but always has a positive value

Solution

a. $I_{f_1} = V_1/R = 100\text{ V}/100\ \Omega = \mathbf{1\ A}$

b. $\tau_1 = \tau_2 = L/R = 200\text{ mH}/100\ \Omega = 2\text{ ms}$
$i_1 = I_{f_1}(1 - \varepsilon^{-t_1/\tau_1}) = 1\text{ A}(1 - \varepsilon^{-1/2}) = \mathbf{0.3935\ A}$ (13-3)

c. $I_{f_2} = V_2/R = 50\text{ V}/100\ \Omega = \mathbf{0.5\ A}$

d. $t = \tau_2 \times \ln\left(\dfrac{I_f - I_0}{I_f - I_1}\right) = 2\text{ ms} \times \ln\left(\dfrac{0.5 - 0.3935}{0.5 - 0.4}\right)$
$= \mathbf{1.259\ ms} \cong 1.26\text{ ms}$ (13-16)

e. v_L at $t = 0^+$ is 100 V since $i = 0$
$v_L = V\varepsilon^{-t_1/\tau_1} = 100\varepsilon^{-1/2} = \mathbf{60.65\ V}$ (13-5)
Check: $v_L = V - iR = 100 - (0.3935 \times 100) = \mathbf{60.65\ V}$

f. At $t = 1^+$ ms, at the instant the switch touches position 2, the maximum rate of change of current is

$$\frac{di_{02}}{dt} = \frac{I_{f_2} - I_0}{\tau_2} = \frac{(0.5 - 0.3935)\text{ A}}{2\text{ ms}} = 53.25\text{ A/s} \quad (13\text{-}15)$$

Then $v_{L_{02}} = L\dfrac{di_{02}}{dt} = 0.2\text{ H }(53.25\text{ A/s}) = \mathbf{10.65\ V}$

Check: $v_{L_{02}} = V_2 - i_2R = 50\text{ V} - 0.3935 \times 100 = \mathbf{10.65\ V}$

g. $v_{L_2} = V_2 - i_2R = 50\text{ V} - 0.4 \times 100 = \mathbf{10\ V}$

h. Waveforms are plotted in Fig. 13-10b.

i. v_L never reverses because i in either switch position is *always increasing.* Thus, the rate of change of current (di/dt) is *always* positive and therefore v_L is always positive, as shown in Fig. 13-10b

Example 13-16 enables us to draw several important conclusions that verify the theory of *RL* circuits nicely:

1. As shown in Fig. 13-10b, the current attempts to rise to 1 A in position 1 but only reaches 0.3935 A at the end of 1 ms, when the switch is thrown to position 2. From that point on, it tries to rise to a maximum current of 0.5 A gradually since its rate of change is much smaller.
2. v_L always decays toward zero exponentially. In position 1 it decays from 100 to 60.65 V. The rate of change of current here at 1 ms in position 1 is $di_1/dt = v_L/L = 60.65\text{ V}/0.2\text{ H} = \mathbf{303.25\ A/s}$. But when switched to position 2 at 1^+ ms, the rate of change of current drops to **53.25 A/s** (part **f** of the solution). Consequently, v_L drops abruptly at 1 ms because the slope of the current waveform is reduced at 1 ms.
3. As noted in part (i) of the solution, v_L is always positive because the current is always rising, and the rate of change of current is always positive.

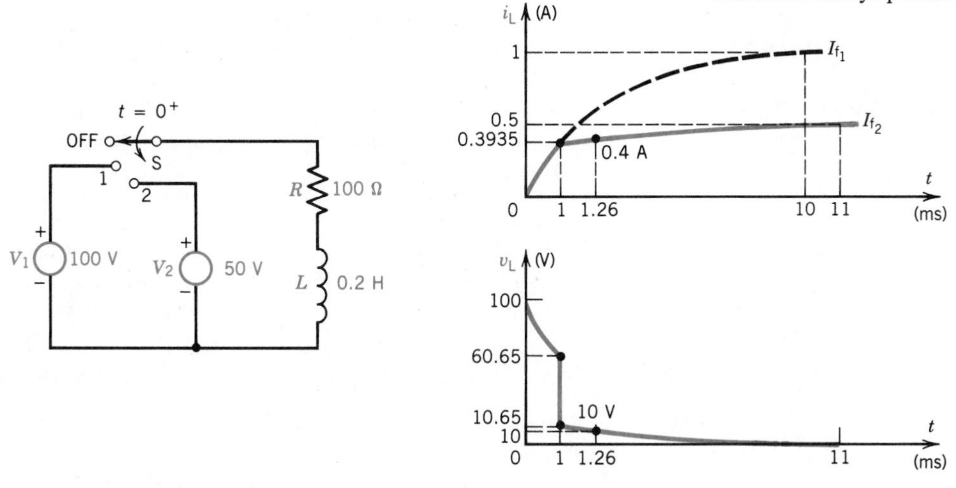

a. Original circuit
b. Waveforms

Figure 13-10 Example 13-16

The circuit of Fig. 13-10a and the waveforms drawn in Fig. 13-10b raise the interesting question, "What would the waveforms of i_L and v_L be if the switching occurred at a later time when the current in position 1 was greater than 0.5 A?" Clearly, the current would then have to decrease to a final value of 0.5 A when switched to position 2. This creates the same kind of situation as occurs when the polarity of the 50-V source is reversed. This problem is treated in Ex. 13-17, which shows that *it is possible* for the *current* to *reverse* under certain circumstances and become *negative*!

EXAMPLE 13-17

Repeat parts **(a)** through **(h)** of Ex. 13-16 with the 50-V source (Fig. 13-10a) reversed. Note: reverse all currents appropriately. Also, in part **(i)** calculate the time when i_L is instantaneously zero.

Solution

a. $I_{f_1} = V_1/R = 100\text{ V}/100\ \Omega = \mathbf{1\ A}$

b. $\tau_1 = \tau_2 = L/R = 200\text{ mH}/100\ \Omega = 2\text{ ms}$ (13-1)
$i_1 = \mathbf{0.3935\ A}$ (from Ex. 13-16b)

c. $I_{f_2} = -V_2/R = -50\text{ V}/100\ \Omega = -0.5\text{ A}$

d. The time for i_L to reach -0.4 A in position 2 is

$$t_2 = \tau_2 \times \ln\left(\frac{I_{f_2} - I_0}{I_{f_2} - I_1}\right) = 2\text{ ms} \times \ln\left[\frac{-0.5 - 0.3935}{-0.5 - (-0.4)}\right]$$

$$= 2\text{ ms} \ln\left(\frac{-0.8935}{-0.1}\right) = 4.38\text{ ms} \qquad (13\text{-}16)$$

Then the time to reach $-0.4\text{ A} = 1\text{ ms} + 4.38\text{ ms} = \mathbf{5.38\ ms}$ (see Fig. 13-11)

e. From Ex. 13-16e, $v_L = \mathbf{100\ V}$ at $t = 0^+$, and **60.65 V** at $t = 1$ ms

f. At $t = 1^+$ ms, at the instant S touches position 2, the *maximum negative* rate of change of current, $-di_{02}/dt$, is

$$\frac{-di_{02}}{dt} = \frac{I_{f_2} - I_0}{\tau_2} = \frac{-0.5 - 0.3935}{2\text{ ms}} = \mathbf{-446.75\ A/s} \qquad (13\text{-}15)$$

Then $$v_{L_{02}} = L\frac{-di_{02}}{dt} = 0.2\text{ H}\,(-446.75\text{ A/s}) = \mathbf{-89.35\ V}$$

Check: $v_{L_{02}} = V_2 - i_{02}R_2 = -50 - 0.3935 \times 100 = \mathbf{-89.35\ V}$

g. $v_{L_2} = V_2 - i_2R_2 = -50 - (-0.4 \times 100) = \mathbf{-10\ V}$
Check:

$$-di_2/dt = (-446.75\text{ A/s})\varepsilon^{-4.38/2} = -50\text{ A/s}$$

$$-v_{L_2} = (-50\text{ A/s})(0.2\text{ H}) = \mathbf{-10\ V}$$

h. Waveforms are plotted in **Fig. 13-11** for v_L and i_L versus time

i. To find the time when $i_L = 0$, that is, when i_L first goes negative,

$$t_0 = \tau_2 \ln\left(\frac{I_{f_2} - I_0}{I_{f_2} - I_1}\right) = 2\text{ ms} \times \ln\left(\frac{-0.5 - 0.3935}{-0.5 - 0}\right)$$

$$= 2\text{ ms} \ln(1.787) = 1.16\text{ ms} \qquad (13\text{-}16)$$

and therefore the time at which t crosses the zero axis is $1\text{ ms} + 1.16\text{ ms} = \mathbf{2.16\ ms}$

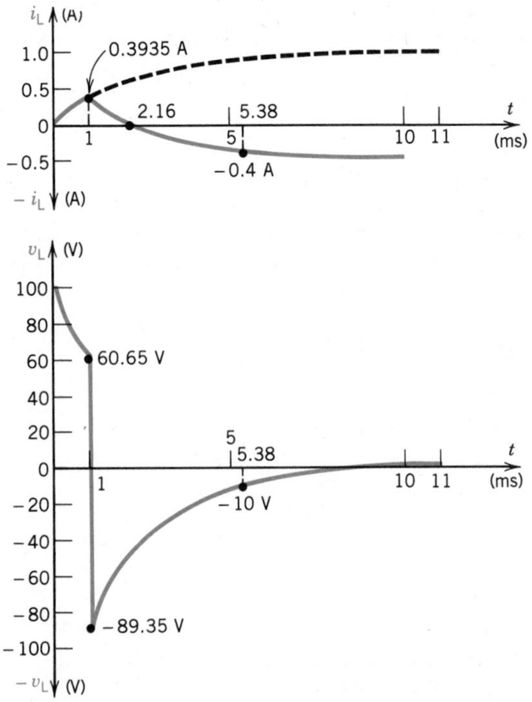

Figure 13-11 Waveforms of i_L and v_L for Ex. 13-17

The following conclusions may be drawn from Ex. 13-17 and Fig. 13-11:

1. It *is possible for i_L to reverse direction* in a multivoltage source network, depending on source polarity and switching circuit arrangement.
2. The current may change its direction but it *never* has *more than one* value at a given time.
3. Conversely, the voltage across the inductor changes direction abruptly and may display more than one value *at the same time.* In Fig. 13-11, v_L shows *two* values at times:

   $t = 0$, $v_L = \mathbf{0\ V}$, and at $t = 0^+$, $v_L = \mathbf{100\ V}$
   $t = 1$ ms, $v_L = \mathbf{+60\ V}$, and at $t = 1^+$ ms, $v_L = \mathbf{-89.35\ V}$

4. Most important, observe from Fig. 13-11 that when in switch position 2 at 1 ms, the current does *not decay* from its positive value of 0.3935 A to its maximum negative value of -0.5 A.[10] Instead, this is a current *rise* in a *negative* direction. This is why Eq. (13-16) is used to compute time in *both* steps of part **(d)** and part **(i)**. (Recall that Eq. 13-16 is used to find time during a current *rise* cycle as noted in Table 13-1.)

---

[10] A decay *always* implies an exponential drop to zero.

13-7 POWER AND ENERGY RELATIONS IN *RL* CIRCUIT DURING CURRENT RISE

We are now prepared to consider power and energy relations in the *RL* circuit. Since we have been exposed to exponential functions in both *RC* and *RL* circuits, this will not be difficult.

Figure 13-12a shows a simple *RL* circuit during transient current rise. From Kirchhoff's voltage law (KVL) we already know that when the switch closes at time $t = 0^+$, KVL must hold at all times. Therefore from Eq. (13-2), $V = v_R + v_L$. We also know, as shown in Fig. 13-12b, that v_L decays exponentially to zero during the transient and v_R rises to its steady-state value of V, the applied voltage, during the transient. We may recall the earlier instantaneous equations for v_R, v_L, and i_L, namely

$$v_R = i_L R = V(1 - \varepsilon^{-t/\tau}) \tag{13-4}$$

$$v_L = V\varepsilon^{-t/\tau} \tag{13-5}$$

$$i_L = \frac{V}{R}(1 - \varepsilon^{-t/\tau}) \tag{13-3}$$

Figure 13-12 Simple *RL* circuit during transient current rise

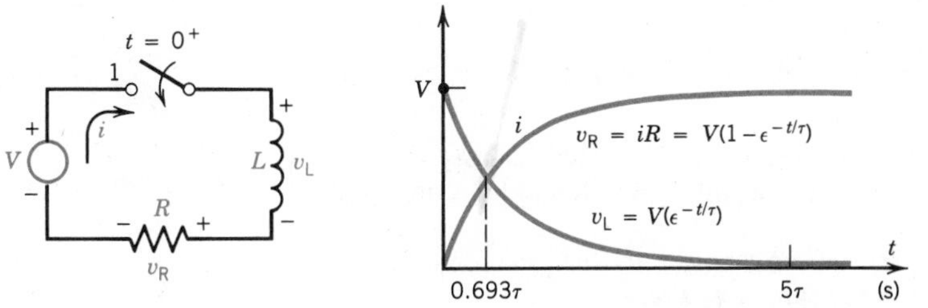

a. *RL* circuit during current rise

b. v_L and v_R during transient current rise

The derivation of instantaneous powers p_R, p_L, and p_T is relatively easy given the above familiar instantaneous equations

$$p_R = v_R \times i_L = V(1 - \varepsilon^{-t/\tau}) \times \frac{V}{R}(1 - \varepsilon^{-t/\tau})$$

$$= \frac{V^2}{R}(1 - 2\varepsilon^{-t/\tau} + \varepsilon^{-2t/\tau}) \quad \text{watts (W)} \tag{13-18}$$

$$p_L = v_L i_L = (V\varepsilon^{-t/\tau})\frac{V}{R}(1 - \varepsilon^{-t/\tau})$$

$$= \frac{V^2}{R}(\varepsilon^{-t/\tau} - \varepsilon^{-2t/\tau}) \quad \text{watts (W)} \tag{13-19}$$

and

$$p_T = p_R + p_L = \frac{V^2}{R}(1 - \varepsilon^{-t/\tau}) \quad \text{watts (W)} \tag{13-20}$$

By substituting for t/τ in each of Eqs. (13-18) through (13-20), we can easily plot the waveforms of p_R, p_L, and p_T during the period of transient current rise in an *RL* circuit. These are shown plotted in **Fig. 13-13**. Note that the ordinate has been "normalized" so that V^2/R represents 1 W. In effect, the exponential portion of each of the three equations is a ratio of V^2/R. When all transients have disappeared in Fig. 13-12 and the current has reached steady state, $p_R = V^2/R = p_T$, the power drawn from the supply. Instantaneous inductor power, p_L, as shown in Fig. 13-13 as well as by Eq. (13-19), is zero since its exponential portion is zero.

EXAMPLE 13-18

In the circuit shown in Fig. 13-12a, the applied voltage V is 10 V, the resistor R is 100 Ω, and the inductance L is 1 H. Calculate the

a. Power dissipated in the resistor, p_R, when the current reaches steady state
b. Steady-state current, I_f
c. Total power drawn from the supply, p_T, at steady state, using VI_f
d. Total power drawn from the supply, using Eqs. (13-20) and (13-18), under steady-state conditions

Solution

a. $p_R = V^2/R = (10)^2/100 = \mathbf{1\ W}$
b. $I_f = V/R = 10\text{ V}/100\ \Omega = \mathbf{100\ mA}$
c. $p_T = VI_f = 10\text{ V} \times 0.1\text{ A} = \mathbf{1\ W}$
d. $$p_T = \frac{V^2}{R}(1 - \varepsilon^{-t/\tau}) = \frac{V^2}{R}(1 - 0) = \frac{V^2}{R} = \frac{10^2}{100} = \mathbf{1\ W} \qquad (13\text{-}20)$$

$$p_T = p_R = \frac{V^2}{R}(1 - 2\varepsilon^{-t/\tau} + \varepsilon^{-2t/\tau}) = \frac{V^2}{R}(1 - 0 + 0) = V^2/R = 10^2/100 = \mathbf{1\ W} \qquad (13\text{-}18)$$

EXAMPLE 13-19

Given the data of Ex. 13-18 and the equations cited in Sec. 13-7, calculate the

a. Time, in terms of time constant τ, at which $v_L = v_R$
b. p_R and p_L, respectively, at the time value in (**a**), using Eqs. (13-18) and (13-19)
c. Total power drawn from supply at the time value in (**a**), using Eq. (13-20) and the sum of p_R and p_L

Solution

a. $v_L = v_R$

$V(\varepsilon^{-t/\tau}) = V(1 - \varepsilon^{-t/\tau})$, canceling V and collecting exponential terms

$2\varepsilon^{-t/\tau} = 1$ from which by taking the reciprocal of each side

$\varepsilon^{+t/\tau} = 2$ and taking the ln of each side yields

$t/\tau = \ln 2$ from which $t = \tau \ln 2 = \mathbf{0.69315\tau}$

b. $$p_R = \frac{V^2}{R}(1 - 2\varepsilon^{-t/\tau} + \varepsilon^{-2t/\tau}) = 1\text{ W}(1 - 2\varepsilon^{-0.69315} + \varepsilon^{-2\times 0.69315}) = 1\text{ W}(1 - 1 + 0.25) = \mathbf{0.25\ W}$$

$$p_L = \frac{V^2}{R}(\varepsilon^{-t/\tau} - \varepsilon^{-2t/\tau}) = 1\text{ W}(\varepsilon^{-0.69315} - \varepsilon^{-2\times 0.69315}) = 1\text{ W}(0.5 - 0.25) = \mathbf{0.25\ W}$$

c. $$p_T = \frac{V^2}{R}(1 - \varepsilon^{-t/\tau}) = 1\text{ W}(1 - \varepsilon^{-0.69315}) = 1\text{ W}(1 - 0.5) = \mathbf{0.5\ W}$$

$$p_T = p_R + p_L = (0.25 + 0.25)\text{ W} = \mathbf{0.5\ W}$$

The following conclusions may be drawn from Exs. 13-18 and 13-19 with reference to Fig. 13-13:

1. When all transients have disappeared, $p_T = p_R = V^2/R$
2. When $v_L = v_R$, then $v_L i = v_R i$ and $p_L = p_R$ at time $t = (\ln 2)\tau = 0.69315\tau$ (Fig. 13-13)
3. At time $t = 0.69315\tau$, $p_L = p_R = 0.25\ V^2/R$ and $p_T = 0.5\ V^2/R$ (Fig. 13-13)
4. Below time $t = (\ln 2)\tau$, p_L is **always** greater than p_R (Fig. 13-13)
5. Above time $t = (\ln 2)\tau$, p_R is **always** greater than p_L (Fig. 13-13)
6. At all times, $p_T = p_R + p_L$

We are now ready to verify Eq. (12-17) for energy stored in the magnetic field of an inductor, which was derived in Appendix B-6. We already know that $W = P \times t$. Then if we integrate the curve of p_L shown in Fig. 13-13 with respect to time we should obtain the total area under the p_L curve, which represents the total energy stored by an inductor. So now we write

$$W_L = \int_0^\infty p_L\,dt = \frac{V^2}{R}\int_0^\infty (\varepsilon^{-t/\tau} - \varepsilon^{-2t/\tau})\,dt$$

$$= \frac{V^2}{R}\left[-\tau\varepsilon^{-t/\tau} + \frac{\tau}{2}\varepsilon^{-2t/\tau}\right]_0^\infty$$

$$= \frac{V^2}{2R}(\tau) = \frac{V^2}{2R}\left(\frac{L}{R}\right) = \frac{L}{2}\left(\frac{V}{R}\right)^2 = \frac{LI^2}{2} \quad \text{joules} \qquad (12\text{-}17)$$

With respect to the derivation of Eq. (12-17), note that

1. It proves that the final energy stored in the magnetic field of an inductor, after all transients have disappeared, is $W = LI^2/2$. (This is proved independently above and in Appendix B-6.)
2. Figure 13-13 shows that the *cumulative* energy stored in the magnetic field during the transient rise of current follows the same waveform and has a similar equation to p_R, the *instantaneous* power dissipated in resistor R. (This is an important insight since *only resistance* can *dissipate* electrical *energy*.)

From the preceding derivation and Fig. 13-13 we can now write the equation for *cumulative* energy stored in the magnetic field due to the transient rise in current as

$$w_L = \frac{LI^2}{2}(1 - 2\varepsilon^{-t/\tau} + \varepsilon^{-2t/\tau}) \quad \text{joules (J)} \qquad (13\text{-}21)$$

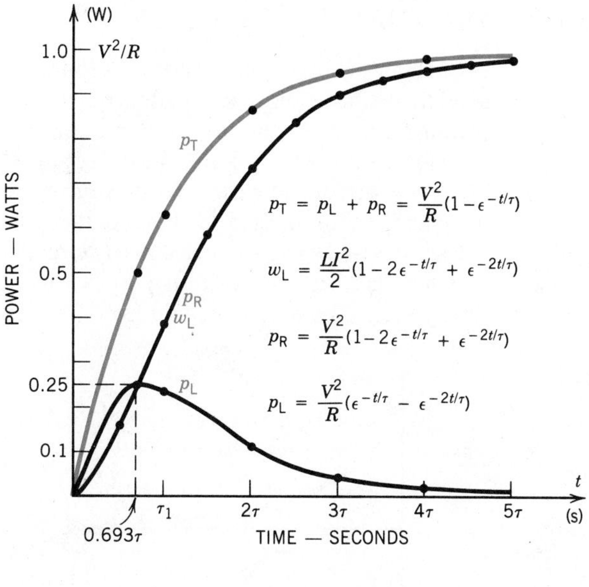

Figure 13-13 Power relations during transient current rise in an *RL* circuit

EXAMPLE 13-20

Given the data of Exs. 13-18 and 13-19, calculate the

a. *Final* energy stored in the magnetic field of the given inductor
b. *Cumulative energy* stored, using Eq. (13-21), when $t = 2\tau$ and $t = 4\tau$
c. *Instantaneous power*, p_L, drawn from the supply to store magnetic field energy when $t = 2\tau$ and $t = 4\tau$
d. *Instantaneous energy* stored in the magnetic field at $t = 2\tau$ and $t = 4\tau$, using $p_L t$
e. *Instantaneous energy* stored in the magnetic field at $t = 2\tau$ and $t = 4\tau$, using the relation $w_{Li} = v_L i_L t$

Solution

a. $W = LI_f^2/2 = (1\ \text{H})(0.1\ \text{A})^2/2 = \mathbf{5\ mJ = 5\ W \cdot s}$

b. When $t = 2\tau$

$$w_L = \frac{LI_f^2}{2}(1 - 2\varepsilon^{-t/\tau} + \varepsilon^{-2t/\tau}) = 5\ \text{mJ}(1 - 2\varepsilon^{-2} + \varepsilon^{-4})$$

$$= 5\ \text{mJ}(0.74765) = \mathbf{3.738\ mJ}$$

When $t = 4\tau$

$$w_L = \frac{LI_f^2}{2}(1 - 2\varepsilon^{-t/\tau} + \varepsilon^{-2t/\tau}) = 5\ \text{mJ}(1 - 2\varepsilon^{-4} + \varepsilon^{-8})$$

$$= 5\ \text{mJ}(0.9637) = \mathbf{4.819\ mJ}$$

c. When $t = 2\tau$

$$p_L = \frac{V^2}{R}(\varepsilon^{-t/\tau} - \varepsilon^{-2t/\tau}) = 1\ \text{W}(\varepsilon^{-2} - \varepsilon^{-4}) = 1\ \text{W}(0.117)$$

$$= \mathbf{0.117\ W} \quad \text{(Fig. 13-13)}$$

When $t = 4\tau$

$$p_L = \frac{V^2}{R}(\varepsilon^{-t/\tau} - \varepsilon^{-2t/\tau}) = 1\ \text{W}(\varepsilon^{-4} - \varepsilon^{-8}) = 1\ \text{W}(0.018)$$

$$= \mathbf{0.018\ W} \quad \text{(Fig. 13-13)}$$

d. $w_{Li} = p_L \times t$, and when $t = 2\tau = 2 \times L/R = 2 \times (1\ \text{H})/100\ \Omega = 20$ ms, the instantaneous energy is: $w_{Li} = 0.117\ \text{W} \times 20\ \text{ms} = 2.34\ \text{mW}\cdot\text{s} = \mathbf{2.34\ mJ}$. When $t = 4\tau = 40$ ms, the instantaneous energy is $w_{Li} = 18\ \text{mW} \times 40\ \text{ms} = \mathbf{0.72\ mJ}$

e. $w_{Li} = v_L i_L t = (V\varepsilon^{-t/\tau}) \times I_f(1 - \varepsilon^{-t/\tau}) \times t$
$= (10\varepsilon^{-2}) \times 0.1(1 - \varepsilon^{-2}) \times 0.02 = \mathbf{2.34\ mJ}$ at $t = 20$ ms
$= (10\varepsilon^{-4}) \times 0.1(1 - \varepsilon^{-4}) \times 0.04 = \mathbf{0.72\ mJ}$ at $t = 40$ ms

Example 13-20 is extremely important in showing the distinction between *instantaneous* energy and *cumulative* energy *stored* in the magnetic field. From Ex. 13-20 we observe that

1. At $t = 20$ ms $= 2\tau$, the instantaneous energy is 2.34 mJ but the cumulative energy stored in the field is 3.738 mJ.
2. At $t = 40$ ms $= 4\tau$, the instantaneous energy is only 0.72 mJ while the cumulative energy stored in the field is 4.819 mJ.
3. Instantaneous energy may be found as the product of $v_L i_L t$ or $p_L t$.
4. Cumulative energy stored in the magnetic field is found from Eq. (13-21).
5. Last, and most important, the equation for cumulative energy stored in the magnetic field is of the same waveform as and is proportional to the power dissipated in the resistor, as shown in Fig. 13-13. (Compare Eqs. 13-18 and 13-21.)

13-8 POWER AND ENERGY RELATIONS DURING TRANSIENT CURRENT DECAY IN *RL* CIRCUIT

The reader at this point may ask, "What happens to the energy stored in the magnetic field since energy can neither be created nor destroyed?" The answer is that as long as the steady-state current remains the same, the magnetic field is suspended in space around the coil and the energy is stored in the field. But in the event that the coil is disconnected from a dc source and connected across a resistive circuit, the energy is discharged in the circuit resistance.

Figure 13-14a shows such an *RL* circuit at its steady-state condition. Assume switch S has been in position 1 for a sufficiently long time that the current is at steady state. We may conclude from Fig. 13-14a that

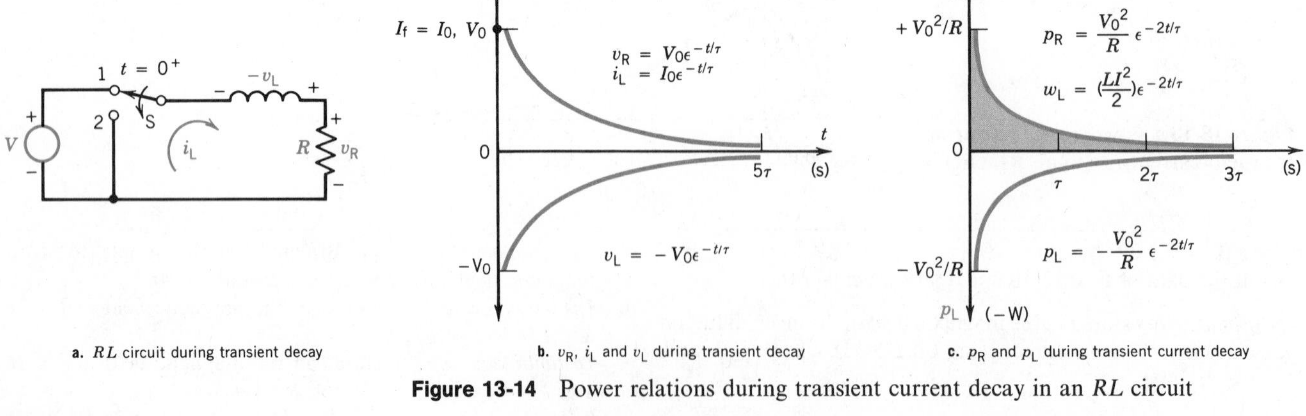

a. *RL* circuit during transient decay

b. v_R, i_L and v_L during transient decay

c. p_R and p_L during transient current decay

Figure 13-14 Power relations during transient current decay in an *RL* circuit

1. At $t = 0$: $I_{ss} = I_f = V/R = I_0$, the initial steady-state current

   voltage across resistor, $v_R = V$
   voltage across inductor, $v_L = 0$, since $di/dt = 0$

2. At $t = 0^+$, S switches instantly from position 1 to position 2. Then

   $v_L = -V_0$ and begins to decay in accordance with $-V_0\varepsilon^{-t/\tau}$
   $i_L = I_0$ and begins to decay in accordance with $I_0\varepsilon^{-t/\tau}$
   $v_R = i_L R = V_0\varepsilon^{-t/\tau}$ as shown in Fig. 13-14b

Of course, this summary represents the same relations as given in Table 13-1 for the current decay cycle of a transient *RL* response. But from the relations given above we can write p_L, stored instantaneous power *returned from* the inductor, as

$$p_L = v_L i_L = (-V_0\varepsilon^{-t/\tau})(I_0\varepsilon^{-t/\tau})$$
$$= -V_0 I_0\varepsilon^{-2t/\tau} = \frac{-V_0^2}{R}\varepsilon^{-2t/\tau} \quad \text{watts (W)} \qquad (13\text{-}22)$$

In the same way, we can write p_R, the total instantaneous power dissipated in the resistor, *R*, as

$$p_T = p_R = v_R i_L = (V_0\varepsilon^{-t/\tau})(I_0\varepsilon^{-t/\tau})$$
$$= +V_0 I_0\varepsilon^{-2t/\tau} = \frac{V_0^2}{R}\varepsilon^{-2t/\tau} \quad \text{watts (W)} \qquad (13\text{-}23)$$

As shown by Eqs. (13-22) and (13-23) as well as Fig. 13-14c, p_R and p_L are equal and opposite at all times. This is understandable since the power *supplied* by the collapsing inductor field is being *dissipated* instantly in the resistor, *R*. Also note that the curves of p_L and p_R in Fig. 13-14c decay faster than the curves of v_R and v_L in Fig. 13-14b. The transients of p_L and p_R are approximately zero at $t = 2.5\tau$, whereas v_R and v_L are approximately zero after 5τ. Finally, note the total power is p_R, as shown in Table 13-2, since only resistance dissipates power.

EXAMPLE 13-21

In the circuit shown in Fig. 13-14a, the applied voltage is 10 V, resistor *R* is 100 Ω, and inductance *L* is 1 H (as in Exs. 13-18 through 13-20). At time $t = 0^+$ when the switch first touches position 2, calculate

a. $i_L = I_0$
b. $v_R = V_0$
c. $v_L = -V_0$
d. p_R, power dissipated in *R*
e. p_L, power returned from the magnetic field
f. Initial energy in the magnetic field, w_{L_0}

Solution

a. $i_L = V/R = 10\text{ V}/100\ \Omega = \mathbf{0.1\ A}$
b. $v_R = V_0 = \mathbf{10\ V}$
c. $v_L = -V_0 = \mathbf{-10\ V}$
d. $p_R = V_0^2/R = (10)^2/100 = \mathbf{1\ W}$
e. $p_L = -V_0^2/R = \mathbf{-1\ W}$ (13-22)
f. $w_{L_0} = LI^2/2 = (1\text{ H})(0.1)^2/2 = \mathbf{5\ mJ}$ (12-17)

EXAMPLE 13-22

Using the data and solutions of Ex. 13-21, calculate the following at time $t = 2.5\tau$ in position 2:

a. t
b. i_L
c. v_L
d. v_R
e. p_R, using $v_R i_L$ and exponential Eq. (13-23)
f. p_L, using $v_L i_L$ and exponential Eq. (13-22)

Solution

a. $t = 2.5\tau = 2.5 \times L/R = 2.5 \times (1\ \text{H})/100\ \Omega = \mathbf{25\ ms}$
b. $i_L = I_0\varepsilon^{-t/\tau} = (0.1\ \text{A})\varepsilon^{-2.5} = \mathbf{8.208\ mA}$
c. $v_L = -V_0\varepsilon^{-t/\tau} = -(10\ \text{V})\varepsilon^{-2.5} = \mathbf{-0.8208\ V}$
d. $v_R = \mathbf{+0.8208\ V}$ by inspection
e. $p_R = v_R i_L = (0.8208\ \text{V})(8.208\ \text{mA}) = \mathbf{6.738\ mW} \cong 0$
$p_R = (V_0^2/R)\varepsilon^{-2t/\tau} = (1\ \text{W})\varepsilon^{-5} = \mathbf{6.738\ mW} \cong 0$ (13-23)
f. $p_L = v_L i_L = (-0.8208\ \text{V})(8.208\ \text{mA}) = \mathbf{-6.738\ mW} \cong 0$
$p_L = -(V_0^2/R)\varepsilon^{-2t/\tau} = -(1\ \text{W})\varepsilon^{-5} = \mathbf{-6.738\ mW} \cong 0$ (13-22)

Example 13-21 shows how the *initial conditions* are determined for all transients (i_L, v_R, v_L, p_R, p_L, and w_L) during *RL* circuit decay of current.

Example 13-22 shows the instantaneous values and their relations during transient decay at a time $t = 2.5\tau$. Note from the solution to Ex. 13-22 that

1. p_R and p_L decay to approximately zero in time $t = 2.5\tau$ because their *rates* of *decay* are *twice as fast* as i_L, v_R, and v_L (Fig. 13-14c versus 13-14b).
2. v_R, p_R, and i_L decay from initial *positive* values to zero.
3. v_L and p_L decay from initial *negative* values to zero.

The reader may now ask, quite properly, "What happens to the energy stored in the magnetic field and at what rate does it decay to zero?" Since an inductor is incapable of dissipating electrical energy, all the energy stored in the magnetic field is dissipated in resistor R when switch S is in position 2 (Fig. 13-14a). If Eq. (13-23) is correct for p_R, then all the area under the curve of p_R in Fig. 13-14c should be equal to $LI^2/2$, since $W_L = \int_0^\infty p_R\,dt$.[11] Let us see if it does. We may now write

$$W_L = \int_0^\infty p_R\,dt = \int_0^\infty \frac{V_0^2}{R}\varepsilon^{-2t/\tau}\,dt = \frac{V_0^2}{R}\left[\frac{\tau}{-2}\varepsilon^{-2t/\tau}\right]_0^\infty$$

$$= \frac{V_0^2\tau}{2R} = \frac{V_0^2}{2R} \times \frac{L}{R} = \frac{LI^2}{2} \qquad (12\text{-}17)$$

Once again we have verified Eq. (12-17) by integrating p_R during current decay. (We did this earlier by integrating a totally different equation for p_L during current rise!) It would appear, therefore, that Eq. (13-23) is correct for p_R and also that the area under the curve of p_R represents the *total* energy stored in the magnetic field, $LI^2/2$.

But earlier, in our conclusions to Ex. 13-20, we observed during transient current rise that the cumulative energy stored in the magnetic field followed the same waveform as the power dissipated in the resistor, R. And in the derivation just given we integrated p_R to find the total area under the curve of p_R, which represents the total energy stored.

From the preceding insight and from Fig. 13-14 we may now write the following equation for energy *remaining* (or left) in the magnetic field as a function of time:

$$w_L = \frac{LI^2}{2}(\varepsilon^{-2t/\tau}) \quad \text{joules (J)} \qquad \mathbf{(13\text{-}24)}$$

Equation (13-24) shows that at time $t = 0^+$ s, the energy remaining in the magnetic field is $LI^2/2$, and that when t is approximately 2.5τ, practically no energy is left in the magnetic field, as shown by Ex. 13-23.

---

[11] Observe that we did this earlier for W_L during transient current rise. In effect, we are verifying Eq. (12-17) three times: once in Appendix B-6, once again during transient current rise, and now during transient current decay.

EXAMPLE 13-23

Using the data and solutions of Exs. 13-21 and 13-22 calculate the

a. Initial energy stored in the magnetic field at $t = 0^+$ s
b. Energy remaining in the magnetic field at $t = \tau = 10$ ms and energy dissipated in R
c. Energy remaining in the magnetic field at $t = 2.5\tau$ and energy dissipated in R

Solution

a. $LI^2/2 = 5$ mJ (from Ex. 13-21f)
b. $w_L = (LI^2/2)\varepsilon^{-2t/\tau} = (5\text{ mJ})\varepsilon^{-2}$
$= \mathbf{0.677\ mJ}$ remaining in the magnetic field
$w_R = W_0 - w_L = 5\text{ mJ} - 0.677\text{ mJ} = \mathbf{4.323\ mJ}$
c. $w_L = (LI^2/2)\varepsilon^{-2t/\tau} = (5\text{ mJ})\varepsilon^{-5} = \mathbf{33.69\ \mu J}$ (13-24)
$w_R = W_0 - w_L = 5\text{ mJ} - 33.69\ \mu\text{J} = \mathbf{4.966\ mJ}$

Example 13-23**c** shows that *practically all* the energy stored in the magnetic field has been dissipated in the resistor, R, after $t = 2.5\tau$. It also explains why the field discharge protection devices shown in Fig. 12-18 are needed.

13-9 SUMMARY OF POWER AND ENERGY RELATIONS IN *RL* CIRCUITS

Table 13-2 presents a summary of power and energy relations in *RL* circuits during both transient current rise and transient current decay.

Table 13-2 Power and Energy Relations in *RL* Circuits

Parameter	Current-Rise Cycle		Current-Decay Cycle	
p_L	$\frac{V_f^2}{R}(\varepsilon^{-t/\tau} - \varepsilon^{-2t/\tau})$	(13-19)	$\frac{-V_0^2}{R}\varepsilon^{-2t/\tau}$	(13-22)
p_R	$\frac{V_f^2}{R}(1 - 2\varepsilon^{-t/\tau} + \varepsilon^{-2t/\tau})$	(13-18)	$\frac{V_0^2}{R}\varepsilon^{-2t/\tau}$	(13-23)
p_T	$\frac{V_f^2}{R}(1 - \varepsilon^{-t/\tau})$	(13-20)	$\frac{V_0^2}{R}\varepsilon^{-2t/\tau}$	(13-23)
w_L	$\left(\frac{LI^2}{2}\right)(1 - 2\varepsilon^{-t/\tau} + \varepsilon^{-2t/\tau})$	(13-21)	$\left(\frac{LI^2}{2}\right)\varepsilon^{-2t/\tau}$	(13-24)

The following points bear repeating with regard to Table 13-2:

1. Both energy *stored* in the magnetic field (during the *current-rise* cycle) and energy *remaining* in the magnetic field (during *current-decay* cycle), w_L, are similar in equation and waveform to the power dissipated in resistance, p_R.
2. Total power, p_T, during the *current-rise* cycle is separate (and distinct) from and is the sum of p_R and p_L.
3. Total power, p_T, during the *current-decay* cycle is the *same* as the power dissipated in resistance, p_R. During decay there can only be one power! The negative power returned from the inductor (but not dissipated in it) must be the same as the (positive) power dissipated in the resistance.
4. The energy remaining in the magnetic field during current decay, $w_L = \frac{LI^2}{2}\varepsilon^{-2t/\tau}$ decays at the same rate as p_L and p_R. Since $w_R = W_0 - w_L$, it is fairly evident that during decay the energy dissipated in resistance is $w_R = \frac{LI^2}{2}(1 - \varepsilon^{-2t/\tau})$. The proof is left as an exercise for the reader (see Prob. 13-17).

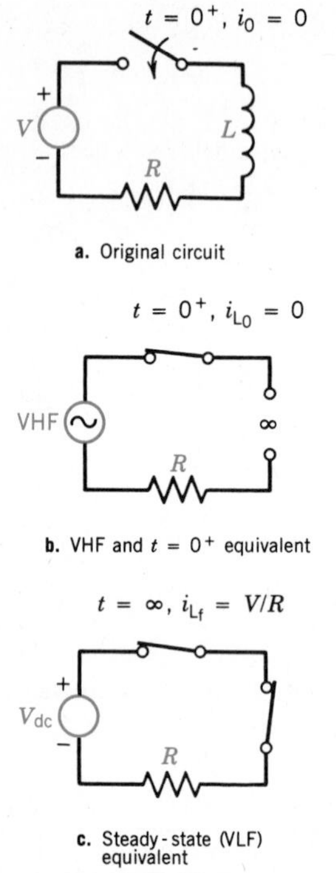

Figure 13-15 Equivalents of *RL* circuit at $t = 0^+$ and $t = \infty$

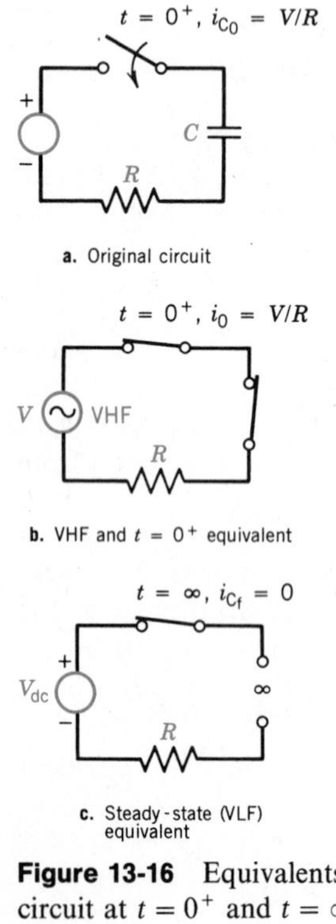

Figure 13-16 Equivalents of *RC* circuit at $t = 0^+$ and $t = \infty$

13-10 SIMPLIFIED *RLC* CIRCUIT ANALYSIS AT $t = 0^+$ AND $t = \infty$

In Chapter 14 we begin our study of ac circuit analysis. We will consider in Chapter 14 the behavior of circuits where applied voltage (and current) is no longer constant but varies periodically at a certain frequency. As the frequency of voltage variation is *decreased*, we will discover that inductors and capacitors approach their respective behaviors under dc (steady-state) conditions. For this reason, dc is generally considered a voltage of zero frequency.

Let us examine inductance and capacitance, respectively, at two specific conditions:

1. At time $t = 0^+$, at the instant the switch is closed.
2. At time $t = \infty$, when the switch has been closed sufficiently long for all transients to disappear.

Figure 13-15a shows an *RL* circuit to which dc is applied at time $t = 0$. We know that at time $t = 0^+$, the initial current, i_{L_0}, is zero. Then an inductor behaves as an ***open circuit*** at time $t = 0^+$.

Assuming that the switch has been closed sufficiently long for all transients to disappear, as shown in Fig. 13-15b, the voltage v_L across the inductor is zero; that is, its volt drop is zero. At steady state, therefore, when $t = \infty$, the inductor behaves as a ***short circuit*** and the supply voltage, V, is across resistor R. The final steady-state current, I_{ss}, is V/R, as shown in Fig. 13-15b.

Let us now turn to *RC* circuits, which were covered in Chapter 10, and apply the same analysis techniques. Assume that switch S in **Fig. 13-16a** closes at time $t = 0^+$. The current initially is $i_{C_0} = V/R$. Thus, at time $t = 0^+$, the **capacitor** behaves as a **short circuit** (in comparison to the inductor which behaves as an open circuit).

Now assume that switch S has been closed for an infinite time so that all transients have disappeared. The voltage across the capacitor is equal to the supply voltage, while the voltage across the resistor is zero. The current is also zero, as shown in Fig. 13-16b. The capacitor, therefore, at time $t = \infty$, behaves as an open circuit of infinite resistance, since it permits no current whatever after all transients have disappeared.

We summarize the behavior of *L* and *C* at times $t = 0^+$ and $t = \infty$ in Table 13-3.

Table 13-3 Behavior of *L* and *C* at $t = 0^+$ and $t = \infty$, VHF and VLF, Respectively

	$t = 0^+$, VHF Behavior	$t = \infty$, VLF Behavior or DC
Inductor, *L*	Acts as an **open circuit**	Acts as a **short circuit**
Capacitor, *C*	Acts as a **short circuit**	Acts as an **open circuit**

We may conclude that at very low frequencies (VLF) or dc, inductors act as short circuits and capacitors as open circuits. We infer this intuitively because

1. Inductors only oppose any *change* of *current*. When steady-state current is reached, the rate of change of current is zero. In the steady state, at $t = \infty$, inductance has no effect and the volt drop across the inductor, $v_L = L(di/dt) = 0$. Therefore the inductor is a **short circuit**.
2. Capacitors only oppose a *change* in *voltage*. In the steady state, when $t = \infty$, the current in the capacitor $i_C = 0$, and the voltage across the capacitor is the

same as the supply voltage, V. Consequently, the capacitor acts as an **open circuit** to dc or very low frequencies (VLF).

What is the effect of inductance and capacitance at time $t = 0^+$ when the switch is first closed to a dc source (or when *very high frequency* voltages are applied)? Table 13-3 concludes that at $t = 0^+$ inductors behave as open circuits and capacitors as short circuits. We infer this intuitively from our previous studies at time $t = 0^+$ for inductors and capacitors because

1. At time $t = 0^+$, $i_L = 0$, since inductors oppose a change in current. Therefore inductors behave as *open circuits* to dc at time $t = 0^+$. Recall also that the current attempts to rise instantly (producing maximum rate of change of current) and so does the applied voltage at time $t = 0^+$. This is analogous to very high frequency (VHF) behavior whenever $t = 0^+$.
2. At time $t = 0^+$, current in a capacitor instantly rises to its Ohm's law value. The capacitor behaves as a short circuit at time $t = 0^+$ or whenever it is subjected to a very high frequency (VHF).

The intuitive inferences of Table 13-3 will be verified when we study the properties of L and C when alternating current (ac) having VLF and/or VHF is applied. The following examples should provide some insights into the behavior of *RLC* circuits at $t = 0^+$ (VHF) and at $t = \infty$ (VLF), when dc is applied at steady state. In each example we will refer to Table 13-3 for the appropriate circuit behavior.

EXAMPLE 13-24

Given the circuit shown in **Fig. 13-17a**,

a. Draw the equivalent circuits at times $t = 0^+$ and $t = \infty$
b. Calculate the current in each branch and total current at time $t = 0^+$
c. Repeat (**b**) for $t = \infty$

Solution

a. The equivalent circuit at $t = 0^+$ is shown in Fig. 13-17b, where the inductor is shown as an open circuit and the capacitor as a short circuit. The equivalent circuit at $t = \infty$, when all transients have disappeared, is shown in Fig. 13-17c. Note that the inductor is shown as a short circuit and the capacitor as an open circuit.
b. When $t = 0^+$, from Fig. 13-17b, $i_t = i_2 = V/(R_s + R_2) = 10\text{ V}/(5 + 1)\ \Omega = \dfrac{10\text{ V}}{6\ \Omega} = \mathbf{1.\bar{6}\ A}$
c. When $t = \infty$, from Fig. 13-17c, $i_t = i_1 = V/(R_s + R_1) = 10\text{ V}/(5 + 5)\ \Omega = \mathbf{1\ A}$

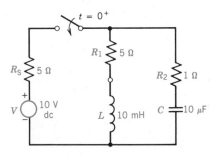

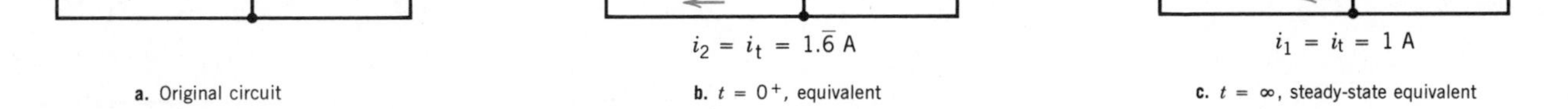

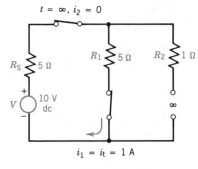

a. Original circuit b. $t = 0^+$, equivalent c. $t = \infty$, steady-state equivalent

Figure 13-17 Example 13-24

EXAMPLE 13-25

Given the circuit shown in **Fig. 13-18a**,

a. Draw the equivalent circuits at times $t = 0^+$ and $t = \infty$
b. Calculate the current in each branch and total current at time $t = 0^+$
c. Repeat (**b**) for $t = \infty$

Solution

a. The equivalent circuit at $t = 0^+$ is shown in Fig. 13-18b, where L_1, acting as an open circuit, disconnects R_3 and R_4. C_1 behaves as a short circuit. The equivalent circuit at $t = \infty$ produces open-circuit capacitors and short-circuit inductors, as shown in Fig. 13-18c.

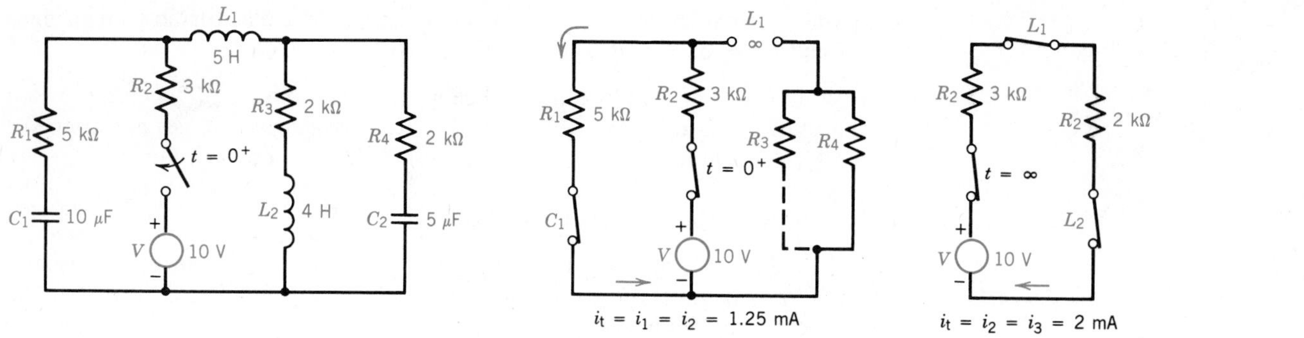

a. Original circuit

b. t = 0+ equivalent

c. t = ∞, steady-state equivalent

Figure 13-18 Example 13-25

b. When $t = 0^+$, from Fig. 13-18b, $i_t = i_1 = i_2 = V/(R_1 + R_2) =$ 10 V/8 kΩ = **1.25 mA**

c. When $t = \infty$, from Fig. 13-18c, $i_t = i_2 = i_3 = V/(R_2 + R_3) =$ 10 V/5 kΩ = **2 mA**

EXAMPLE 13-26

Given the circuit shown in **Fig. 13-19a**,

a. Draw the equivalent circuits at times $t = 0^+$ and $t = \infty$
b. Calculate the current in each branch and the total current at time $t = 0$
c. Repeat (**b**) for $t = \infty$

Solution

a. The equivalent circuit at $t = 0^+$ is shown in Fig. 13-19b. Since L_1 behaves as an open circuit, it disconnects the entire circuit and all currents are zero. The equivalent circuit at $t = \infty$ produces open-circuit capacitors and short-circuit inductors, as shown in Fig. 13-19c.
b. When $t = 0^+$, from Fig. 13-19b, $i_t = 0$ and all currents are zero
c. When $t = \infty$, from Fig. 13-19c, $i_t = i_1 = i_2 = i_4 =$ 10 V/10 kΩ = **1 mA**

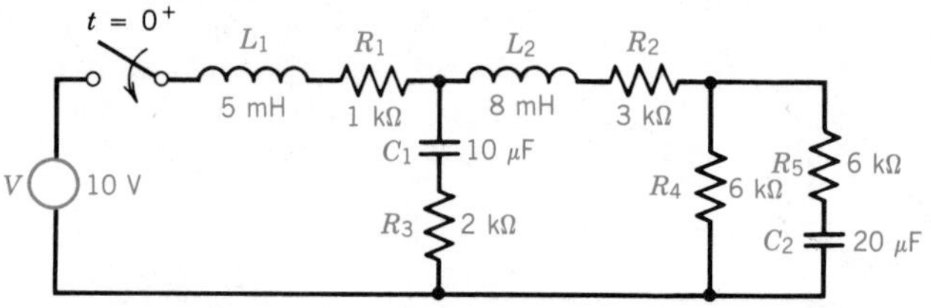

a. Original circuit

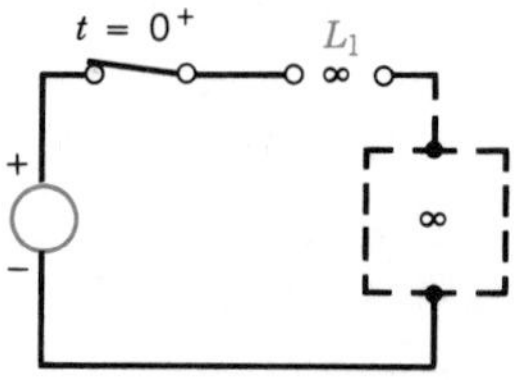

b. $t = 0^+$ equivalent

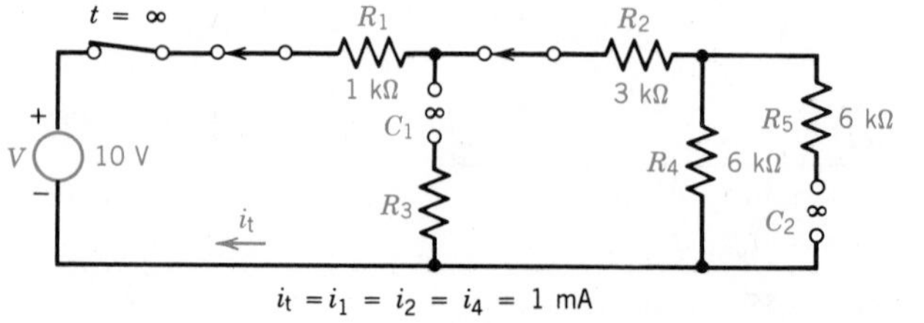

c. $t = \infty$, steady-state equivalent

Figure 13-19 Example 13-26

Examples 13-24 through 13-26 show that for any combination of R, L, and C

1. An equivalent circuit can be drawn at $t = 0^+$ when the switch is first closed and also at $t = \infty$, when all transients have disappeared.

2. At $t = 0^+$, also called the VHF equivalent, we **short** all **capacitors** and **open** all **inductors**.
3. At $t = \infty$, also called the VLF or dc steady-state equivalent, we **open** all **capacitors** and **short** all **inductors**.

13-11 GLOSSARY OF TERMS USED

Exponential function Any one of the forms $y = a\varepsilon^{bx}$, where a and b are constants and may be real or complex (see Appendix B-4). The graphical plot of an exponential function is always nonlinear.

Final conditions Steady-state conditions reached after a transient change in voltage and/or current, from the initial conditions.

Initial conditions Values of voltage or current present in a circuit before a change was introduced, either by switching or changing input voltage or current.

Time constant, τ Time required for an exponential function to rise to 63.21% of its final steady-state value or decay to 36.79% of its original value. For an RL circuit, the time constant is $\tau = L/R$ s. Alternatively, the time constant of an RL circuit may be defined as the time for the current to rise to its final value if it continued to rise at its initial rate of change continuously for the entire time.

Transient The part of the variation in a variable that ultimately disappears during the transition from one steady-state condition to another.

Transient circuit Circuit in which any change in applied voltage (or current) produces (transient) changes in voltage (or current) for a period of time before settling to a steady state.

Transient decay Linear or nonlinear waveform of voltage, current, or power in approaching zero.

Transient period Time required for voltages and currents to reach final steady-state values in response to a change in input voltage or current. The transient period for RL (and RC) circuits is usually at least five time constants (5τ).

Transient response Behavior of voltages across and currents in the various components in a transient circuit.

Transient rise Linear or nonlinear waveform of voltage or current in approaching some finite steady-state value, but never zero.

13-12 PROBLEMS

Secs. 13-1 to 13-5

13-1 A 500-turn inductor has a reluctance of 5×10^4 At/Wb and a resistance of 25 Ω. Calculate the

a. Inductance, L
b. Time constant of the inductor, τ
c. Initial rate of change of current when the inductor is connected to a 100-V supply at $t = 0^+$ s
d. Initial rate of change of voltage across the inductor when connected to the supply
e. Initial current at $t = 0^+$ s
f. Time for *current* to reach steady state and its steady-state value
g. Current at $t = 0.6$ s
h. Time for current to reach 3.2 A

13-2 Using the constants for τ and I_f calculated in Prob. 13-1,

a. Write the expression for the instantaneous value, i, versus time, t
b. Calculate i for t equal to 1, 2, 3, 4, and 5 time constants
c. Repeat **(a)** for the expressions of v_R and v_L
d. Repeat **(b)** for v_R and v_L
e. Sketch waveforms of i, v_R, and v_L on the same axis versus time in seconds
f. Using the values of v_R and v_L calculated in **(d)**, show that $V = v_R + v_L$ at all times to verify Kirchhoff's voltage law

13-3 A 100-mH inductor experiences an initial rate of change of current of 5 kA/s and takes 500 μs to reach its steady-state value. Calculate the

a. Voltage applied to the inductor
b. Resistance of the inductor
c. Final steady-state current
d. i_L, v_R, and v_L when the time is 300 μs

13-4 A coil having a resistance of 100 Ω and an inductance of 150 mH is connected to a 12-V supply via a switch. Calculate the

a. Time constant of the circuit
b. Minimum time for the current to reach steady state when the switch is closed
c. Steady-state current
d. Initial rate of change of current at $t = 0^+$, when switch is first closed
e. v_L and v_R at $t = 0^+$
f. i_L and v_L at $t = 1.04$ ms
g. Time t when i_L is 75.84 mA
h. Time t when v_L is 4.416 V

i. di/dt at $t = 1.04$ ms
j. di/dt at $t = 10$ ms

13-5 In the circuit of **Fig. 13-20**, switch *S* has remained open long enough for all transients to disappear. (Note that V_1 is *opposing* V_2.) With *S* open, calculate the

a. Applied circuit voltage and steady-state current
b. Voltage across coil terminals 1 and 2
c. Voltage v_L across *L*
d. Voltage across R_1

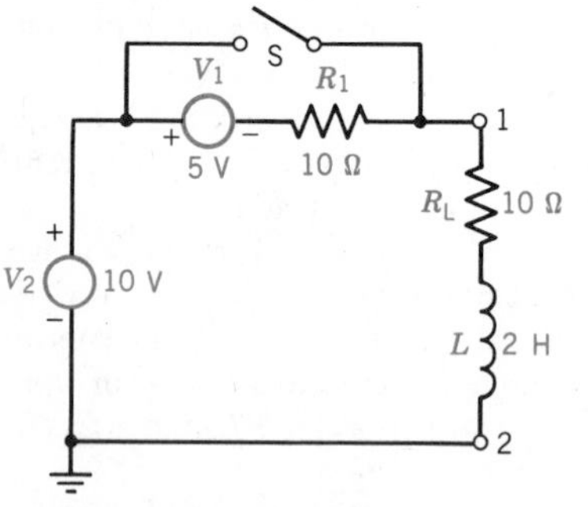

Figure 13-20 Problems 13-5 and 13-6

13-6 Using the initial conditions of Prob. 13-5, assume the switch in Fig. 13-20 closes at time $t = 0^+$ s. Calculate the

a. τ of the circuit
b. Time for steady state to be reached
c. New value of steady-state current
d. Equation for the instantaneous current as a function of time
e. Instantaneous current at $t = 200$ ms after S is closed
f. v_{R_L} at $t = 200$ ms
g. di_0/dt, initial rate of change of current
h. v_L at $t = 200$ ms, using KVL and Eqs. (13-15) and (13-2)

13-7 A practical inductor has a resistance, R_L, of 60 Ω and an inductance, *L*, of 106 mH. It is connected in parallel with a 1-kΩ resistor, R_1, across a 12-V supply for a sufficient time for all transients to disappear. (Hint: draw the circuit to simplify calculations.) With voltage applied *at steady state*, calculate

a. I_{R_1}, current in resistor R_1
b. I_{R_L}, current in the practical inductor
c. v_L, voltage across *L*
d. v_{R_L}, voltage across the coil's resistance and voltage across practical coil terminals

13-8 If the supply voltage is instantly disconnected from the parallel circuit in Prob. 13-7 at time $t = 0$, calculate

a. i_L at $t = 0^+$
b. i_{R_1} and i_{R_L} at $t = 0^+$
c. v_{R_1} at $t = 0^+$
d. The circuit time constant
e. v_{R_1} at $t = 0.3$ ms
f. v_{R_L} at $t = 0.3$ ms
g. i_L at $t = 0.3$ ms
h. The initial rate of change of voltage decay across v_L
i. The initial rate of change of current decay in the inductor
j. The time for all transients to disappear

13-9 In **Fig. 13-21**, switch S_1 is open and SPST switch S_2 is closed at $t = 0$. Calculate

a. V_{th}, equivalent voltage across terminals T–H
b. R_{th}, equivalent resistance seen in series with V_{th} across terminals T–H

If switch S_1 is also closed at $t = 0$, calculate the

c. Current in inductor, i_L, at $t = 0^+$
d. Voltage across inductor, v_L, at $t = 0^+$
e. Final current in inductor, $I_f = I_{ss}$
f. Circuit time constant, τ_r, during current rise cycle
g. Inductor current, i_L, when $t = 500\ \mu s$
h. Voltage across R_L when $t = 500\ \mu s$
i. Time required for i_L to reach 750 μA
j. Minimum time required for all transients to disappear, t_f

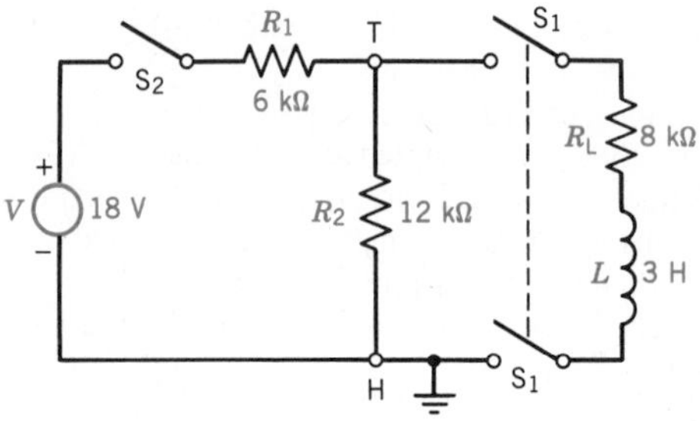

Figure 13-21 Problems 13-9 and 13-10

13-10 Assume S_2 and S_1 are both closed in Fig. 13-21 until all transients have disappeared. If S_2 is opened at $t = 0$, calculate at $t = 0^+$ the

a. Induced voltage in inductor *L*, v_L, with respect to ground
b. Voltage across R_2 with respect to ground, v_{R_2}
c. Voltage across R_L, v_{R_L}
d. Initial rate of change of current, di_0/dt
e. Initial rate of change of voltage across inductor, dv_{L_0}/dt

At time $t = 500\ \mu s$ after S_2 is opened, calculate the

f. Circuit time constant, τ_d, during current decay
g. Current in the inductor, i_L
h. Rate of change of current in the inductor
i. Induced voltage in the inductor, v_L, with respect to ground, using two different methods
j. Voltage across resistor R_2, v_{R_2}, with respect to ground

After S_2 is opened and the current drops to an instantaneous value of 20 μA, calculate the

k. Time *t* after S_2 opens
l. Voltage produced by the inductor, v_L, with respect to ground
m. Voltage across R_2 with respect to ground

13-11 In **Fig. 13-22** switch S is thrown to position 1 at $t = 0$ s and then to position 2 at $t = 10$ ms, where it remains until all transients have disappeared. During the time S is in **position 1**, calculate

a. τ_1, circuit time constant in position 1
b. I_f, final steady-state current
c. v_O, output voltage at steady state
d. W_L, energy stored in the magnetic field at steady state
e. v_L, inductor voltage at $t = 2.5$ ms
f. v_O, output voltage at $t = 2.5$ ms
g. w_L, energy stored in the magnetic field at $t = 2.5$ ms (Table 13-2)

During the time S is in **position 2**, calculate

h. τ_2, circuit time constant in position 2
i. $v_{L_{O2}}$, initial induced voltage at $t = 10^+$ ms
j. $I_{L_{O2}}$, initial current at $t = 10^+$ ms
k. $w_{L_{O2}}$, initial energy remaining in the magnetic field at $t = 10^+$ ms
l. v_O, output voltage at $t = 11$ ms
m. w_{L_2}, energy remaining in the magnetic field at $t = 11$ ms (see Table 13-2)
n. p_{R_1}, power dissipated in R_1 at $t = 11$ ms
o. Using the same time axis from 0 to 12 ms and the data calculated above, roughly plot the following parameters: i_L, v_L, v_O, and w_L

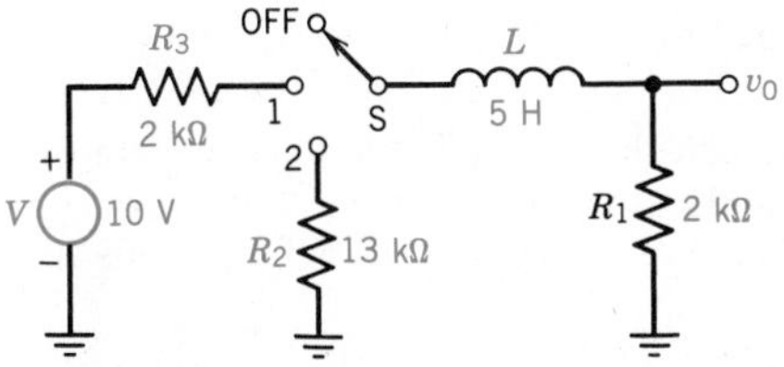

Figure 13-22 Problems 13-11 and 13-12

13-12 In Fig. 13-22, switch S is thrown to position 1 at $t = 0$ ms, where it remains for 1 ms. It is then thrown to position 2, where it remains for 1 ms, after which it is returned to position 1 for 1 ms, and so forth. Perform all necessary calculations to plot curves of i_L, v_O, and v_L for the first two complete alternations (a total of 4 ms) in which the switch duration is 1 ms in each position. (Hint: solve for i_L first, using values of i_L to find v_O and v_L.)

13-13 In the circuit shown in **Fig. 13-23**, switch S closes to position 1 at $t = 0$. Calculate

a. V_{th} and R_{th} in series at position 1 (Hint: draw the simplest equivalent circuit to simplify calculations)
b. τ_1, time constant in switch position 1
c. Steady-state current when switch is in position 1
d. Initial rate of change of current in position 1, di_O/dt
e. Initial voltage across inductor in position 1 at $t = 0^+$ s
f. i_L, v_{R_2}, and v_L when t reaches 300 μs in position 1
g. Time at which v_L reaches 6 V in position 1

If switch S goes to position 2 at 300 μs, calculate

h. τ_2, time constant in switch position 2
i. Initial rate of change of current, $-di_{O2}/dt$
j. Induced voltage across inductor, $-v_{L_{O2}}$, at 300^+ μs
k. i_L, v_{R_2}, and $-v_L$ at $t = 1000$ μs (measured from $t = 0$ μs)
l. Using the same time axis from 0 to 1000 μs and the data calculated above, roughly plot i_L, v_L, and v_O versus time

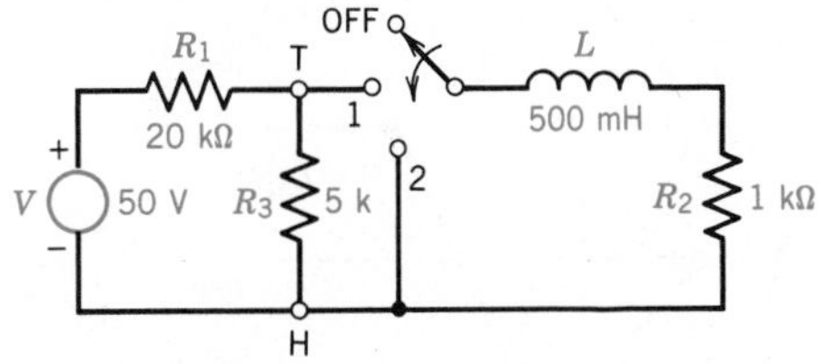

Figure 13-23 Problem 13-13

13-14 Switch S_1 in **Fig. 13-24** closes at time $t = 0$ and switch S_2 closes at $t = 1.0$ s. Calculate the

a. Induced voltage at $t = 0^+$ and 1.0 s after S_1 closes (with S_2 open)
b. Initial rate of change of current when S_2 closes at 1.0 s
c. Induced voltage at $t = 1^+$ s when S_2 closes
d. Induced voltage at $t = 2$ s

(Hint: first calculate τ_1, τ_2, I_{ss_1}, and I_{ss_2} as well as variation of i_L for the times given.)

e. Plot the waveform of v_L versus time for the first 2 s of operation, based on the above calculations
f. Explain why v_L is never negative in this problem

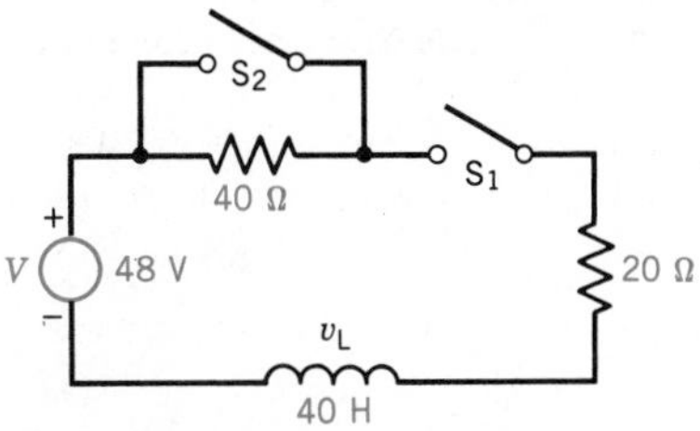

Figure 13-24 Problem 13-14

13-15 The waveform of current in a 1-H inductor is shown in **Fig. 13-25**. Redraw the waveform on a sheet of cross-section paper appropriately scaled. Calculate di_L/dt and v_L at the following intervals of time:

a. From 0 to 1 ms
b. From 1 to 3 ms
c. From 3 to 5 ms
d. From 5 to 6 ms
e. From 6 to 8 ms
f. From 8 to 9 ms
g. From 9 to 10 ms
h. Directly below the waveform of i_L, sketch the waveform of di_L/dt

i. Directly below the waveform of di_L/dt, sketch the waveform of v_L
j. Draw conclusions regarding shapes of waveforms of v_L and di_L/dt versus the waveform of i_L

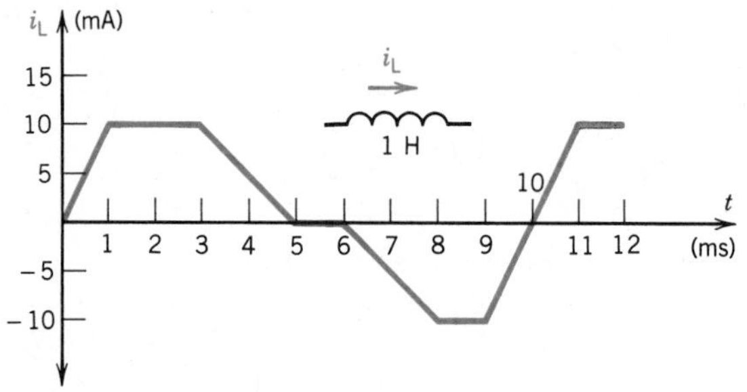

Figure 13-25 Problem 13-15

Secs. 13-6 to 13-9

13-16 In the circuit shown in **Fig. 13-26**, switch S closes to position 1 at $t = 0$ and remains there for 2 ms, after which it moves to position 2 for 3 ms. Calculate

a. V_{th} and R_{th} in series at position 1. (Draw the simplest equivalent circuit to simplify calculations)
b. τ_1, time constant in position 1
c. i_L and v_O at $t = \tau_1$ in position 1
d. i_L and v_O at $t = 2$ ms in position 1
e. τ_2, time constant in position 2
f. i_L and v_O at $t = 2.5$ ms, total elapsed time
g. i_L and v_O at $t = 5$ ms, total elapsed time
h. The energy stored in the magnetic field at times $t = \tau_1$ and $t = 2$ ms in position 1, using values of the current calculated in (**c**) and (**d**)
i. Repeat (**h**) using Eq. (13-21)
j. Draw waveforms of i_L and v_O for the first 5 ms of operation

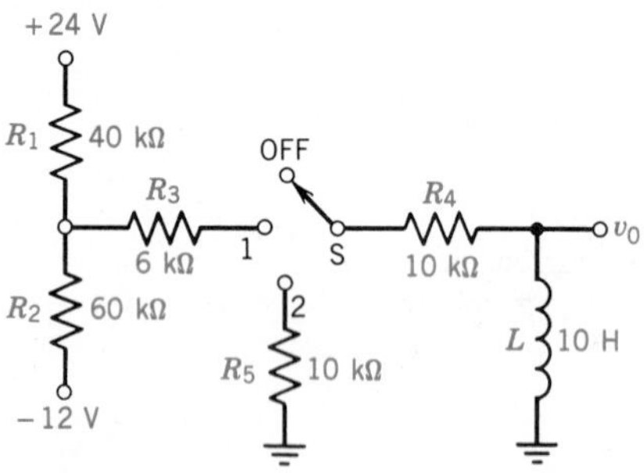

Figure 13-26 Problem 13-16

13-17 In the circuit shown in **Fig. 13-27**, switch S has been closed sufficiently long for all transients to disappear. Calculate the

a. Initial current in inductor L at time $t = 0^-$, before S opens (Hint: treat L as a short circuit; see Sec. 13-10)
b. Output voltage, v_O, at time $t = 0^-$ before S opens
c. Time constant of circuit, τ, after S opens at $t = 0^+$ s
d. Equation for decay of i_L with time after $t = 0^+$ s. (Note: t should be the only unknown in this equation)
e. Output voltage v_O at time $t = 0^+$ and polarity of v_O with respect to ground
f. Energy stored in the magnetic field of L at $t = 0^-$ s, W_0
g. Current in the inductor when $t = 1$ s
h. Energy left in the inductor when $t = 1$ s, w_L
i. Energy dissipated in R_1, using $w_R = W_0 - w_L$
j. Energy dissipated in the resistor, using $w_R = (LI^2/2)(1 - \varepsilon^{-2t/\tau})$
k. Draw the waveform of v_O versus time from $t = 0^-$ s to $t = 3$ s
l. Using the equation $w_R = W_0 - w_L$, derive the equation for w_R given in (**j**)

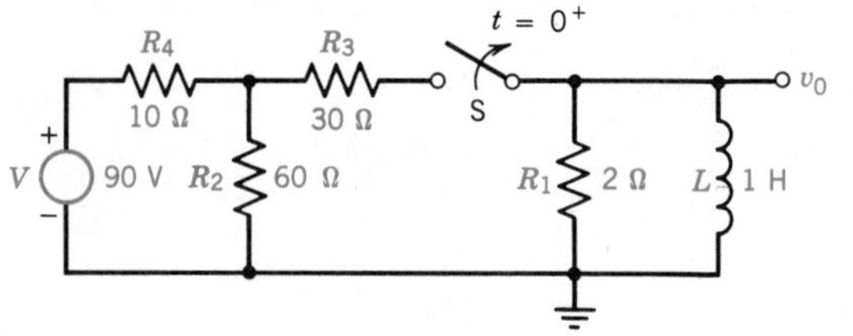

Figure 13-27 Problem 13-17

13-18 Given the *RL* circuit shown in **Fig. 13-28**, switch S moves to position 1 at $t = 0$ s and stays there for 200 ms. It then switches instantly to position 2, where it remains until all transients disappear. Calculate

a. i_L at elapsed times of 0, 100, and 200 ms in position 1
b. i_L at elapsed times of 200^+, 400, and 1200 ms in position 2
c. v_O at the times given in (**a**) and (**b**). (Hint: calculate two values for v_L at each transition time)
d. On the same time axis from 0 to 1.20 s, plot i_L and v_O

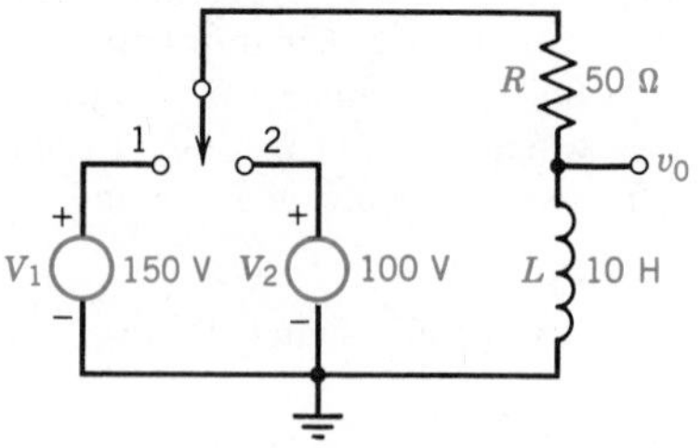

Figure 13-28 Problems 13-18 and 13-19

13-19 Repeat parts (**a**) through (**d**) of Prob. 13-18 with the polarity of source V_2 reversed.

e. Calculate the elapsed time (in milliseconds) when i_L is instantaneously zero in its transient *rise* from a positive to a negative maximum steady-state value.
f. Explain why the value of v_O (and v_L) must be −100 V at the time calculated in (**e**).
g. Explain why v_L (and v_O) is always a positive value in Prob. 13-18.

h. Explain why v_L (and v_0) reverses to a negative value at 200^+ ms in the waveform drawn in part (**d**).

13-20 In the *RL* circuit shown in **Fig. 13-29**, switch S_1 closes at $t = 0$ while switch S_2 opens at $t = 2$ s. Calculate the

a. Final current and time constant at time $t = 0^+$ when S_1 closes
b. Final current and time constant at time $t = 2^+$ s when S_2 opens
c. Current in the inductor, i_L, at 1-s intervals from 0 to 4 s elapsed time
d. Output voltage, v_0, at 1-s intervals from 0 to 4 s, elapsed time
e. On the same time axis, from 0 to 4 s, plot i_L and v_0

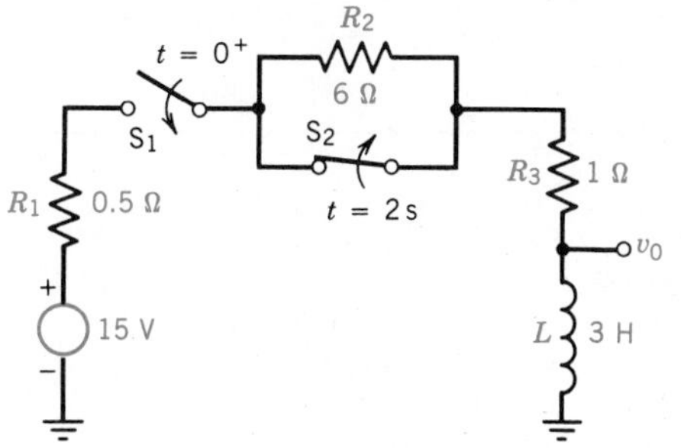

Figure 13-29 Problem 13-20

13-21 In the *RL* circuit shown in **Fig. 13-30**, S closes at $t = 0$ and remains closed until all transients have disappeared. Calculate the

a. Currents i_1, i_2, and i_t at time $t = 0^+$, as well as v_{L_1} and v_{L_2}
b. Currents I_1, I_2, and I_t at time $t = \infty$, as well as v_{L_1} and v_{L_2}

Hint: draw separate equivalent circuits for each time before solving for currents and voltages.

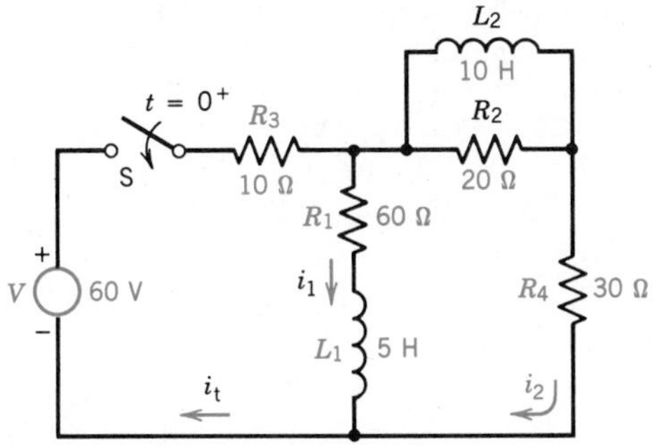

Figure 13-30 Problem 13-21

Sec. 13-10

13-22 In the *RLC* circuit shown in **Fig. 13-31**, S closes at $t = 0$ and remains closed until all transients have disappeared. Calculate the

a. Currents i_1, i_2, i_3, and i_t at time $t = 0^+$ and voltages across L_1 and C_1
b. Currents i_1, i_2, i_3, and i_t at time $t = \infty$ and voltages across L_1 and C_1

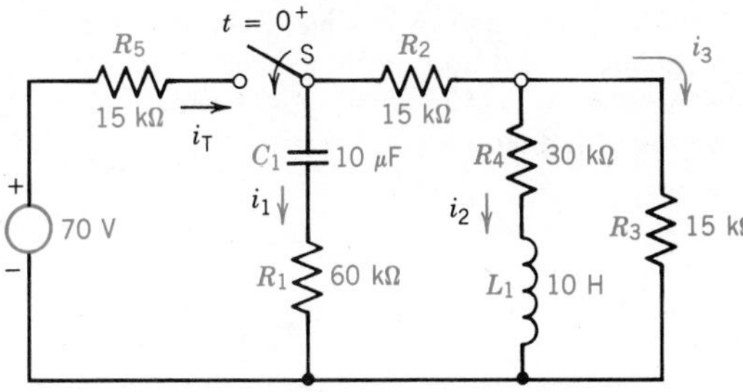

Figure 13-31 Problem 13-22

13-23 In the *RLC* circuit shown in **Fig. 13-32**, switch S closes at $t = 0$ and remains closed until all transients disappear.

a. For time $t = 0^+$, complete the table below, showing *all* currents and voltages in *all* circuit components.
b. For time $t = \infty$, repeat (**a**) above.

	R_1	R_2	L_1	L_2	C_1	C_2
volts						
mA						

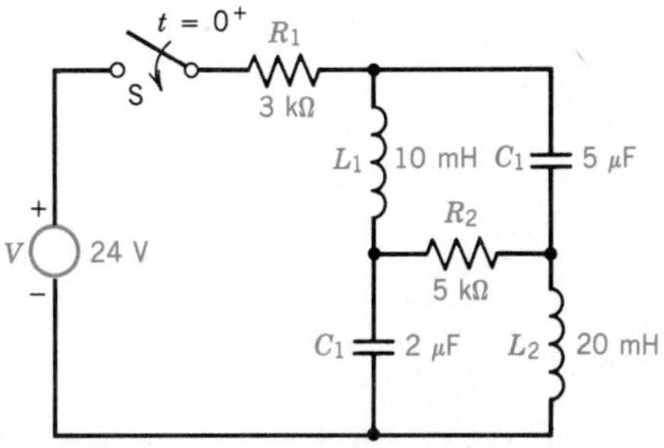

Figure 13-32 Problem 13-23

13-24 In the *RLC* circuit shown in **Fig. 13-33**, switch S closes at $t = 0$ and remains closed until all transients disappear. Construct a table similar to that in Prob. 13-23 and calculate all component voltages and currents at

a. Time $t = 0$
b. Time $t = \infty$

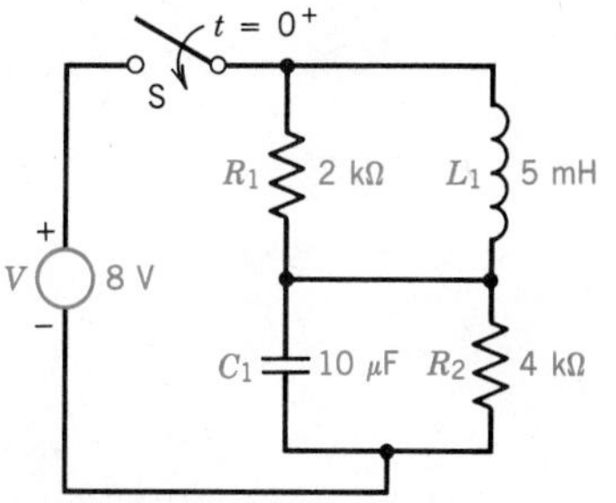

Figure 13-33 Problem 13-24

13-25 In the *RLC* circuit shown in **Fig. 13-34**, switch S closes at $t = 0$ and remains closed until all transients disappear. Construct a table similar to that in Prob. 13-23 and calculate all component voltages and currents at time

a. $t = 0$
b. $t = \infty$

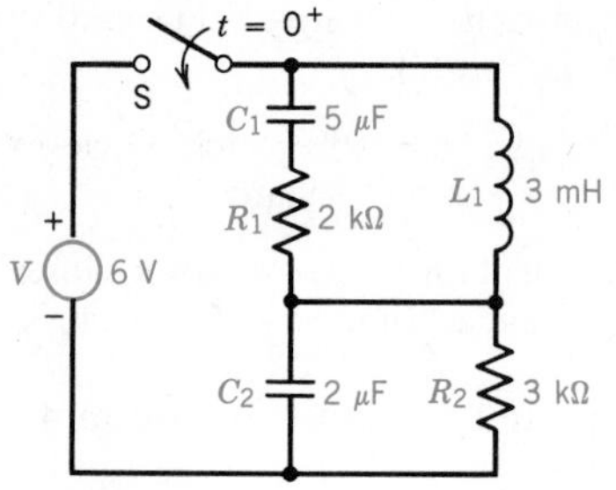

Figure 13-34 Problem 13-25

13-13 ANSWERS

13-1 a 5 H b 0.2 s c 20 A/s d 500 V/s e 0 f 4 A g 3.8 A h 0.3219 s

13-2 b 2.53 A; 3.46 A; 3.8 A; 3.93 A; 3.97 A

13-3 a 500 V b 1 kΩ c 0.5 A d 0.475 A; 475.1 V; 24.9 V

13-4 a 1.5 ms b 7.5 ms c 120 mA d 80 A/s e 12 V; 0 V f 60 mA, 6 V g, h 1.5 ms i 40 A/s j 0.1 A/s

13-5 a 5 V b 250 mA, 2.5 V c 0 d 2.5 V

13-6 a 0.2 s b 1 s c 1 A e 724.1 mA f 7.24 V g 3.75 A/s h 2.76 V

13-7 a 12 mA b 200 mA c 0 d 12 V

13-8 a, b 200 mA c −200 V d 0.1 ms e −9.957 V f 0.5974 V g 9.957 mA h −2.12 MV/s i 2 kA/s j 0.5 ms

13-9 a 12 V b 4 kΩ c 0 d 12 V e 1 mA f 0.25 ms g 0.8647 mA h 6.917 V i 346.6 μs j 1.25 ms

13-10 a −20 V b −12 V c 8 V d $-6.\bar{6}$ A/s e $13\bar{3}$ kV/s f 150 μs g 35.67 μA h −0.2378 A/s i −0.7134 V j −0.428 V k 586.8 μs l −0.4 V m −0.24 V

13-11 a 1.25 ms b 2.5 mA c 5 V d 15.625 μJ e 1.353 V f 4.323 V g 11.68 μJ h $0.\bar{3}$ ms i −37.5 V j 2.5 mA k 15.625 μJ l 0.249 V m 0 n 31 μW

13-13 a 10 V, 4 kΩ b 0.1 ms c 2 mA d 20 A/s e 10 V f 1.9 mA, 1.9 V, 0.498 V g 51.1 μs h 500 μs i −3.8 A/s j −1.9 V k 0.4685 mA, 0.4685 V, −0.4685 V

13-14 a 40 V b 0.88925 A/s c 35.57 V d 21.57 V

13-15 a 10 A/s, 10 V b 0, 0 c −5 A/s, −5 V d 0, 0 e −5 A/s, −5 V f 0, 0 g 10 A/s, 10 V

13-16 a 9.6 V, 30 kΩ b 0.25 ms c 151.7 μA, 3.532 V d 0 e 0.5 ms f 88.3 μA, −1.766 V g 0, 0 h 288 nJ i 288 nJ

13-17 a 2 A b 0 c 0.5 s e −4 V f 2 J g 0.2707 A h 36.63 mJ i, j 1.963 J

13-18 a 0, 1.18 A, 1.896 A b 1.896 A, 1.962 A, 2 A c 150 V, 90.98 V, 55.18 V; 5.2 V

13-19 a 0, 1.18 A, 1.896 A b 1.896 A, −0.5667 A, −1.974 A e 333.4 ms

13-20 a 10 A, 2 s b 2 A, 0.4 s c 0; 3.935; 6.321; 2.355; 2.03 A d 0; 15 V; 9.098 V; 5.518 V; −32.41 V; −2.66 V; −0.218 V

13-21 a 0, 1, 1 A; 50 V, 20 V b $0.\bar{6}$, $1.\bar{3}$, 2 A; 0, 0 V

13-22 a $0.\bar{6}$, 0, $1.\bar{3}$, 2 mA; 20 V, 0 b $0.58\bar{3}$, $1.1\bar{6}$, 1.75 mA; 0, 43.75 V

CHAPTER 14

Waveforms in the Time Domain

14-1 INTRODUCTION

The responses of *RL*, *RC* and *RLC* circuits to a sudden transient change of *input signal* was introduced in Sec. 13-10, along with time responses of these circuits to a steady dc signal. This type of signal, known as a unit-step input signal, is but one of a number of *standard input waveforms* used in circuit analysis.[1] In this chapter we begin the consideration of a second type of input known as the *sinusoidal* waveform. The sinusoidal input is used to determine the *frequency response* of a network, either at a given frequency or over a range of frequencies. The sinusoidal input is of great importance not only for this reason but also because almost all the electrical energy generated throughout the world is produced in the form of a sinusoidal voltage.

The independent voltage (and current) sources that appeared in all previous chapters were dc sources (i.e., either constant-voltage or constant-current sources.) In this chapter we begin our study of alternating sources, that is, sources whose voltages (and/or currents) *vary with time*. These voltage sources, as we will see, change their polarity and magnitude continuously and *periodically*, resulting in currents whose *direction* and *magnitudes* vary in time.

We will see that the network theorems introduced in Chapters 3 to 8 apply equally well to ac circuits, with some minor modifications, namely the effects of alternating current on the three basic circuit elements: resistance, inductance, and capacitance. (This is why we began our study of circuit analysis using simpler dc signal inputs.)

Finally, as a result of the study of sinusoidal waveforms, eventually we will come to a study of nonsinusoidal waveforms (Chapter 22). We will discover that *all* nonsinusoidal *periodic* waveforms are composed of sine waves of different frequencies whose sum yields the original nonsinusoidal periodic waveform. Consequently, we must undertake a detailed study of the properties of a sine wave, as well as its effect on the three basic circuit elements—resistance, inductance, and capacitance.

---

[1] Other input waveforms include the unit ramp, the increasing exponential input, the decreasing exponential input, and the sinusoidal input.

14-2 WAVEFORMS AND APPLICABLE DEFINITIONS

All waveforms undergo variations in their instantaneous values. That is, their instantaneous values *change* with time. Such a change in voltage with time is represented as $v(t)$. Similarly, a change in current with time is represented as $i(t)$.

If the change in voltage or current with time is *random* and exhibits *no regularity* in variation, the waveform is called a *nonperiodic* or *aperiodic* waveform. Waveforms of noise or radiation from outer space are two examples of aperiodic waveforms. An *aperiodic* waveform is shown in **Fig. 14-1a**. The detailed study of aperiodic waveforms is beyond the scope of this text.

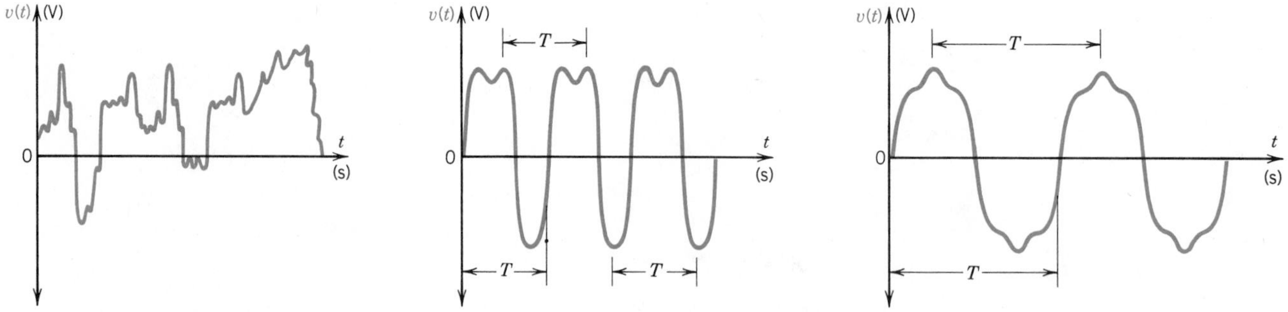

Figure 14-1 Continuous waveforms

Periodic waveforms are those that exhibit variations that recur at *regular intervals*; (i.e., they repeat themselves regularly every T seconds, where T is called the period of the wave.) Figures 14-1b and 14-1c show two periodic nonsinusoidal waveforms. The test of whether a waveform is aperiodic or periodic is given by Eq. (14-1), which defines *periodicity* as

$$v(t) = v(t + T) \qquad \text{volts (V)} \tag{14-1}$$

where $v(t)$ is some specific instantaneous voltage value at time t and the *same* value of voltage at time $(t + T)$ in its cycle of periodic variation

T is the period of the wave in seconds

This definition contains within it the definition of the term *cycle.* A cycle is that interval of time (or space) during which a complete set of nonrepeating events or waveform variations occur. Consequently, one cycle represents the portion of the waveform contained within one period, as shown in Figs. 14-1b and 14-1c as well as Figs. 14-2a, b, and c.

Observe that the title of Fig. 14-1 refers to all waveforms shown as *continuous.* A *continuous* waveform is one whose instantaneous slope changes *gradually* in going from one to the next instantaneous value. No portion of a continuous waveform is ever linear! A sine wave (or sinusoid) is an example of a continuous waveform.

14-2.1 Discrete versus Continuous Waveforms

Figure 14-2 shows three different *discrete* waveforms. Discrete waveforms have two distinct characteristics not possessed by continuous waveforms:

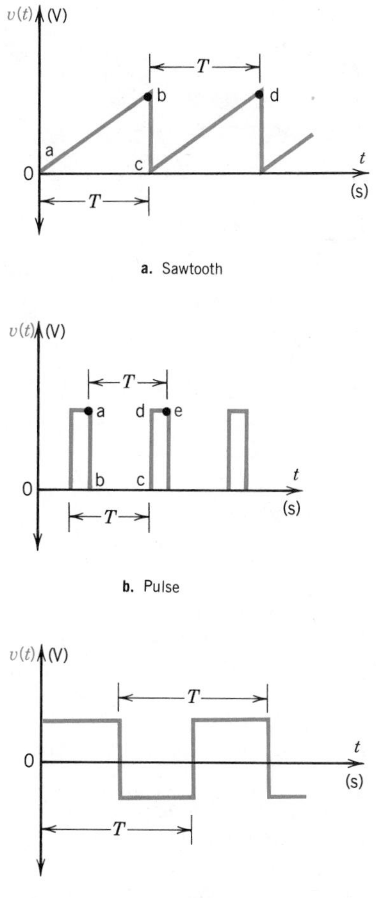

a. Sawtooth

b. Pulse

c. Square wave

Figure 14-2 Discrete periodic waveforms

1. They exhibit *abrupt* changes in slope and/or direction in going from one instantaneous value to another.
2. They usually contain one or more linear portions (where the slope is constant) for some definite portion of the period.

The sawtooth waveform shown in Fig. 14-2a has a *positive* linear slope from **a** to **b** and a *negative* (almost infinite) slope from **b** to **c**. The pulse waveform of Fig. 14-2b exhibits a slope of $-\infty$ from **a** to **b**, constant slopes of zero from **b** to **c** and **d** to **e**, and a slope of $+\infty$ from **c** to **d**.

14-2.2 Alternating versus Pulsating Waveforms

A waveform is considered *alternating* if it has *both* positive and negative values (i.e., it *changes its direction* with respect to some *zero reference*). *All* the waveforms in Fig. 14-1 are alternating. *Only* the waveform of Fig. 14-2c is alternating.

A waveform is considered **pulsating** whenever its voltage (or current) variations exhibit *no change in direction* with respect to the zero reference. A pulsating waveform may have *either* negative-going *or* positive-going pulsations, but *not* both.

The waveforms of Figs. 14-2a and 14-2b are both *positive-going* pulsating waveforms. The waveform of Fig. 14-30 is a *negative-going* pulsating waveform.

14-3 RELATIONSHIP BETWEEN PERIOD AND FREQUENCY

As defined in Eq. (14-1), the **period**, T, is the time (in seconds) to complete one full cycle of variation of a periodic waveform, (i.e., it is the *time per cycle.*) The frequency, f, is the number of complete cycles of any periodic waveform occurring in 1 s, that is, the number of *cycles per unit time*. Obviously, the frequency is the reciprocal of the period, or

$$f = \frac{1}{T} \quad \text{hertz (or cycles/second) or seconds}^{-1} \text{ (Hz, c/s, s}^{-1}\text{)} \qquad (14\text{-}2)$$

Note that frequency is measured in hertz (Hz) or cycles per second (cps or c/s) or simply 1/seconds (s^{-1}).

EXAMPLE 14-1

The waveform of Fig. 14-1c has a positive maximum value of 10 V at 20 ms. The next positive maximum value of 10 V occurs at 60 ms. For the waveform shown, calculate the

a. Period, T
b. Frequency, f

Solution

a. $T = (t + T) - t = 60\text{ ms} - 20\text{ ms} = 40\text{ ms}$ (14-1)
b. $f = 1/T = 1/40\text{ ms} = 25\text{ Hz}$ (14-2)

14-4 GENERATION OF A SINE WAVE

One means of sine wave generation uses the principle of electromagnetic (EM) induction, introduced in Sec. 12-1 and shown in Fig. 12-1. An elementary (hypothetical) single-coil generator (alternator) consisting of one turn is shown in **Fig. 14-3a**. Each end of the coil is connected to a separate slip ring in contact with a

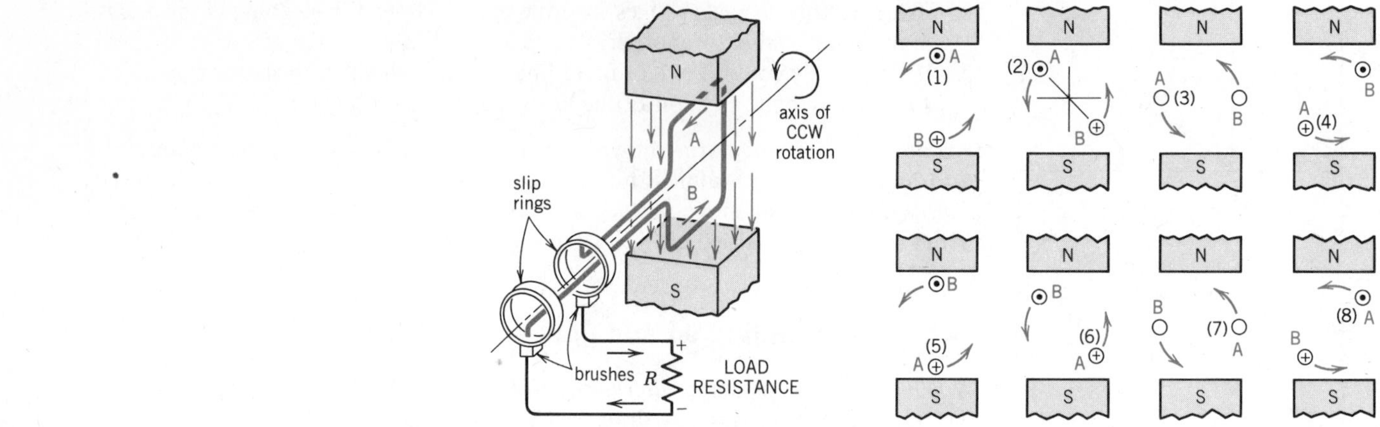

a. Single turn coil rotating CCW in magnetic field

b. Directions of generated voltages for 8 coil positions

Figure 14-3 Generation of sinusoidal voltage

stationary brush. Assume in Fig. 14-3a that the coil is rotating in a counterclockwise (CCW) direction. Applying Fleming's right-hand (RH) rule for EM induction shown in Fig. 12-1c, at the instant of CCW rotation shown in Fig. 14-3a we may conclude the following:

1. Voltage induced in conductor A is *toward* the observer and voltage induced in lower coil side B is away from the observer.
2. Applying the coil side directions of *induced* voltage across load resistor R in the *external circuit*, the brush to which coil side A is connected is positive (+) (and the brush to which B is connected is negative).
3. Since conductor A is directly under the (north) pole center and is moving perpendicular to it, maximum voltage is induced in coil side A in a positive direction, in what we will call position 1, as shown in Fig. 14-3b and Fig. 14-4.

Figure 14-3b shows the effect of continued rotation, in a CCW direction, from positions 1 through 8. (The ninth position is obviously position 1 again.) Using the RH rule and applying it to conductor A *only*, the following may be concluded:

1. In position 2, voltage induced in coil side A is in the same (positive) direction but *is reduced in magnitude* since its direction of motion is no longer perpendicular to the stationary magnetic field.
2. The coil is displaced from its original vertical (90°) position by a displacement angle ϕ, shown in coil position 2.
3. When the coil has rotated 90° CCW, the displacement angle in position 3 is 90° and the voltage inducted in coil side A is zero. The motion of conductor A is now parallel to the magnetic field and no voltage is induced in it.
4. When the coil is rotated to position 4, as shown in Fig. 14-3b, its polarity reverses because its direction of induced EMF reverses. The reversal in polarity and direction of induced EMF occurs because coil side A is moving in the *opposite* direction through a field of the *same* (original) direction. Coil side A now has a negative (−) voltage.
5. Coil side A reaches its maximum negative value in position 5 and continues to have a negative (lesser) voltage in position 6. In position 7 coil side A again shows zero voltage.

6. In position 8, the polarity of coil side A reverses (once again) and becomes positive, since it is moving in its original direction through the stationary (and constant) magnetic field.

The waveform generated by following the successive positions of coil side A from 1 through 8 (to 1 again) is shown in **Fig. 14-4**. The general mathematical expression for this waveform is[2]

$$v(t) = V_m \cos \alpha = V_m \sin(\phi + 90^\circ) = V_m \sin\left(\phi + \frac{\pi}{2}\right) \quad \text{volts (V)} \qquad (14\text{-}3)$$

where V_m is the maximum voltage generated (in coil position 1)
ϕ is the angle of displacement (from coil position 1)
90° is the angular displacement (between positions 7 and 1 in Fig. 14-4) in *degrees*
$\pi/2$ is the angular displacement in radians between the cosine and the sine wave[3]

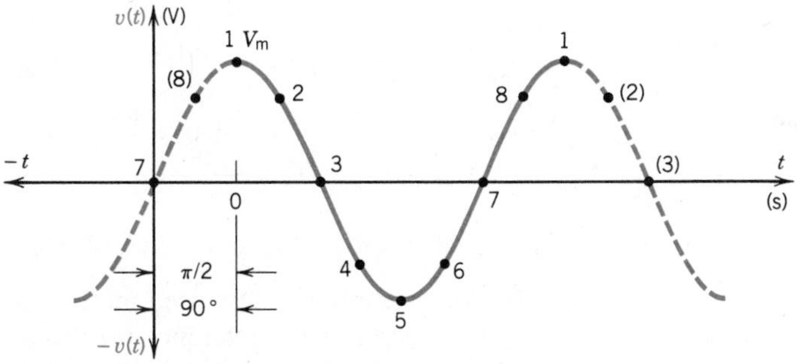

Figure 14-4 Sinusoidal voltage for various coil positions of Fig. 14-3

EXAMPLE 14-2

Assuming $V_m = 10$ V and using both the cosine and sine functions of Eq. (14-3), calculate the instantaneous voltages generated for displacement angles ϕ equal to

a. 45° (coil position 2)
b. 135° (coil position 4)
c. π radians (coil position 5)

Solution

a. $v(t) = V_m \cos \phi = 10 \text{ V} \cos 45^\circ = \mathbf{7.071 \text{ V}}$
$v(t) = V_m \sin(\phi + 90^\circ) = 10 \text{ V} \sin(135^\circ) = \mathbf{7.071 \text{ V}}$
b. $v(t) = V_m \cos \phi = 10 \text{ V} \cos 135^\circ = \mathbf{-7.071 \text{ V}}$
$v(t) = V_m \sin(\phi + 90^\circ) = 10 \text{ V} \sin(135^\circ + 90^\circ) = \mathbf{-7.071 \text{ V}}$
c. $v(t) = V_m \cos \phi = 10 \text{ V} \cos(\pi/2 \text{ radians}) = \mathbf{0 \text{ V}}$
$v(t) = V_m \sin(\phi + \pi/2) = 10 \text{ V} \sin(\pi \text{ radians}) = \mathbf{0 \text{ V}}$

From Ex. 14-2 we may infer the following insights:

1. All three forms of Eq. (14-3) yield the same results.
2. The cosine waveform (Fig. 14-4) produced by CCW rotation *leads* the sine waveform by 90° (or $\pi/2$ radians) as shown by Eq. (14-3). This verifies the trigonometric identity, $\cos \theta = \sin(90^\circ + \theta)$.
3. When the sine or cosine function is expressed in radians, it may be evaluated directly on an electronic calculator in radians, using the RAD key (or switch). Alternatively, the angle in radians may be converted to degrees, using the ratio (π rad/180 deg) appropriately, as shown in the examples below.

[2] The cosine waveform is selected as the reference for the generation of a sinusoid because it is basic to the phasor method introduced in Sec. 14-6.
[3] From elementary trigonometry, we have the identity $\sin(90^\circ + \theta) = \cos \theta$.

EXAMPLE 14-3

Using the conversion ratio (180 deg/π rad), convert the following angles in radians to degrees:

a. $\pi/4$ rad d. π rad
b. $\pi/3$ rad e. 2π rad
c. $\pi/6$ rad f. $3\pi/2$ rad

Solution

a. $\frac{\pi}{4}\text{ rad} \times \frac{180\text{ deg}}{\pi\text{ rad}} = \frac{180}{4}\text{ deg} = \mathbf{45°}$

b. $\frac{\pi}{3}\text{ rad} \times \frac{180\text{ deg}}{\pi\text{ rad}} = \frac{180°}{3} = \mathbf{60°}$

c. $\frac{\pi}{6}\text{ rad} \times \frac{180\text{ deg}}{\pi\text{ rad}} = \frac{180°}{6} = \mathbf{30°}$

d. $\pi\text{ rad} \times \frac{180°}{\pi\text{ rad}} = \mathbf{180°}$

e. $2\pi\text{ rad} \times \frac{180°}{\pi\text{ rad}} = 2 \times 180° = \mathbf{360°}$

f. $\frac{3\pi}{2}\text{ rad} \times \frac{180°}{\pi\text{ rad}} = \frac{3}{2} \times 180° = \mathbf{270°}$

EXAMPLE 14-4

Using the conversion ratio (π rad)/180°, convert the following angles in degrees to radians (leaving the answer as a function of π):

a. 20° d. 150°
b. 90° e. 210°
c. 120° f. 330°

Solution

a. $20° \times \frac{\pi\text{ rad}}{180°} = \frac{\pi}{9}\text{ rad}$

b. $90° \times \frac{\pi\text{ rad}}{180°} = \frac{\pi}{2}\text{ rad}$

c. $120° \times \frac{\pi\text{ rad}}{180°} = \frac{2\pi}{3}\text{ rad}$

d. $150° \times \frac{\pi\text{ rad}}{180°} = \frac{5\pi}{6}\text{ rad}$

e. $210° \times \frac{\pi\text{ rad}}{180°} = \frac{7\pi}{6}\text{ rad}$

f. $330° \times \frac{\pi\text{ rad}}{180°} = \frac{11\pi}{6}\text{ rad}$

Figure 14-5 shows the sine wave of Fig. 14-4 with three minor modifications:

1. The sine function is now used instead of the cosine function; that is, coil position 7 is used as the reference position (see Figs. 14-3 and 14-4).

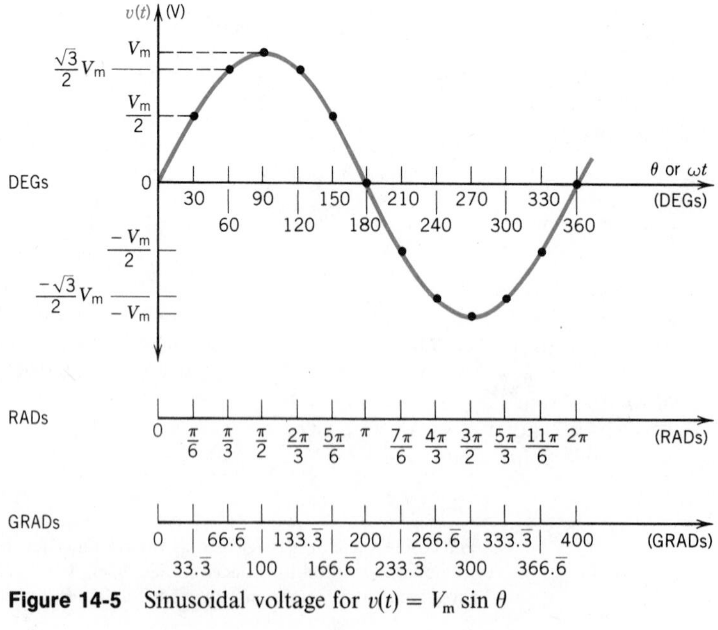

Figure 14-5 Sinusoidal voltage for $v(t) = V_m \sin\theta$

2. Angular displacement is used as the horizontal axis (rather than time) in 30° intervals.
3. Units of angular displacement are shown in degrees, radians, and gradients (grads).[4]

Consequently, the equation for the waveform shown in Fig. 14-5 is

$$v(t) = V_m \sin\theta = V_m \sin\omega t = V_m \sin(2\pi f t) \qquad \text{volts (V)} \qquad (14\text{-}4)$$

where ω is the angular velocity (or angular frequency) in units of degrees/second (°/s), radians/second (s^{-1}), and/or gradients/second (^{g}/s)
f is the frequency in hertz

The last term of Eq. (14-4) shows that a substitution of $2\pi f$ has been made for ω, that is, $\omega = 2\pi f$. This is almost evident from Fig. 14-5, where we can measure the angular frequency as the total number of degrees of CCW rotation per unit time as either $360°/T$ for $2\pi/T$ or $400^g/T$, where T is the period (or the time for one cycle) in seconds. But since $f = 1/T$ in Eq. (14-2), we can write

$$\omega = \frac{\theta}{t} = \frac{2\pi}{T} = 2\pi f \qquad \text{radians/second} \qquad (14\text{-}5a)$$

$$= \frac{360}{T} = 360f \qquad \text{degrees/second} \qquad (14\text{-}5b)$$

$$= \frac{400}{T} = 400f \qquad \text{gradients/second} \qquad (14\text{-}5c)$$

Note that Fig. 14-5 shows the horizontal axis expressed in terms of the (CCW) angular displacement, θ, or the product ωt. From Eq. (14-5a) we can also write

$$\theta = \omega t \qquad \text{degrees, radians, or gradients} \qquad (14\text{-}5d)$$

EXAMPLE 14-5

A sine wave has a maximum amplitude of 10 V and an angular frequency of 200 rad/s. Calculate the

a. Instantaneous voltage, $v(t)$, when $t = 5$ ms
b. Rotation angle θ in radians, degrees, and gradients at the given time
c. Frequency of the sine wave
d. Angular velocity in degrees/second
e. Angular velocity in gradients/second
f. Instantaneous voltage $v(t)$, given the calculated rotation angles in (**b**)

Solution

a. $v(t) = V_m \sin\omega t = 10\text{ V}\sin(200\text{ rad/s}) \times 5\text{ ms}$
$= 10\text{ V}\sin(\mathbf{1\ rad}) = \mathbf{8.415\ V}$ (14-4)

---

[4] Gradients are included because they appear on most electronic calculators. The conversion ratio is 10 grads/9 degrees. The advantage of gradient measure over both radian and degree measures is that it provides a *decimal* division of circular arc. 200 grads = 180 degrees = π radians. It is conventional to abbreviate gradients with a *superscript* "g" and to *omit the unit for radians completely.* Thus, in abbreviation form, $180° = \pi = 200^g$ (see Fig. 14-5).

b. $\theta = \omega t = \mathbf{1\ rad} \times 180°/\pi = \mathbf{57.3°} \times 10^g/9°$
$= \mathbf{63.66^g}$ (14-5d)

c. $f = \omega/2\pi = (200\text{ rad/s})/2\pi\text{ rad/cycle} = \mathbf{31.83\ c/s}$
$= \mathbf{31.83\ Hz}$ (14-5a)

d. $\omega = (200\text{ rad/s})\dfrac{180°}{\pi\text{ rad}} = \mathbf{565.5\ deg/s}$

e. $\omega = (565.5\text{ deg/s}) \times 10^g/9° = \mathbf{628.32\ gradients/second}$

f. $v(t) = V_m \sin\theta = 10\text{ V}\sin(1\text{ rad}) = \mathbf{8.415\ V}$ (14-4)
$v(t) = V_m \sin\theta = 10\text{ V}\sin(57.3°) = \mathbf{8.415\ V}$ (14-4)
$v(t) = V_m \sin\theta = 10\text{ V}\sin(63.66^g)$
$= \mathbf{8.415\ V}$ (14-4)

Example 14-5 should be performed independently by the reader, using the solutions as feedback. Understanding and mastery of the various forms of Eqs. (14-4) and (14-5) are essential to that which follows. Example 14-5 demonstrates that

1. For a given maximum value, angular frequency, and instantaneous time, it is possible to calculate the instantaneous voltage of any sinusoidal voltage waveform.
2. It is also possible to calculate the instantaneous rotation angle in radians, degrees, and gradients.
3. Using $v(t) = V_m \sin\theta$, regardless of the units in which θ is expressed, the *same* instantaneous values of voltage are obtained.

14-5 DEFINITIONS OF AND RELATIONS BETWEEN α, ϕ, AND θ IN SINUSOIDAL WAVEFORMS

Assume that we are given the waveform $v(t) = 120\sin(\omega t + 50°)$ volts. From what we have previously learned, given the value of ω, the angular frequency (or velocity), we should be able to substitute various values of t and plot the instantaneous values of $v(t)$. Is there an easier way? There is!

We already know $\theta = \omega t$, so we can write $v(t) = 120\sin(\theta + 50°)$ volts, where θ, as previously defined, is the *rotation angle*. When $\theta = 0°$, $v(t) = 120(\sin 50°)\text{ V} = 91.93\text{ V}$.

What does this mean? It means that the waveform of $v(t)$ starts at 91.93 V when $t = 0$ and when the rotation angle, $\theta = 0°$. It also means when rotation angle, $\theta = 40°$, $v(t) = 120\sin(40 + 50)° = 120\text{ V}$, which is the maximum positive value of the waveform. The plot of the waveform $v(t) = 120\sin(\omega t + 50°)$ is shown in **Fig. 14-6** for one complete cycle of rotation of angle θ.

From Fig. 14-6 we can see that $v(t) = 0$ when $\theta = 130°$ and 310°. This is verified in Ex. 14-6.

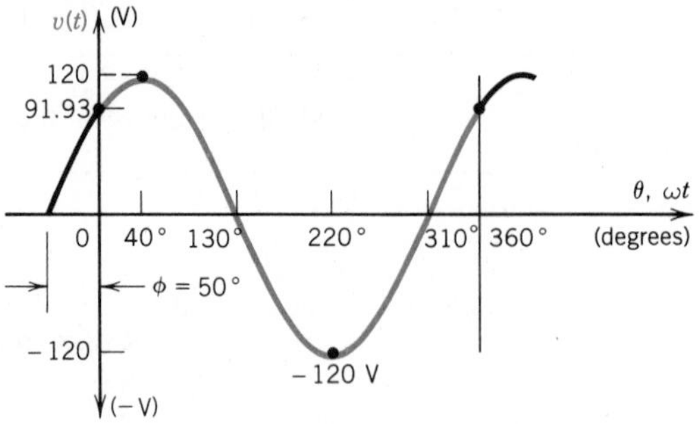

Figure 14-6 Waveform of $v(t) = 120\sin(\omega t + 50°)$ V; Exs. 14-6 and 14-7

EXAMPLE 14-6
Given $v(t) = 120\sin(\omega t + 50°)$ volts, calculate values of $v(t)$ at rotation angles of 130°, 310°, and 360°

Solution

$v(t) = 120\sin(\theta + 50°) = 120\sin(130° + 50°) = \mathbf{0\ V}$
$v(t) = 120\sin(\theta + 50°) = 120\sin(310 + 50)° = \mathbf{0\ V}$
$v(t) = 120\sin(\theta + 50°) = 120\sin(360° + 50°) = \mathbf{91.93\ V}$

EXAMPLE 14-7
Given the waveform $v(t) = 120\sin(\omega t + 50°)$ volts, calculate the rotation angles at which $v(t)$ is instantaneously

a. -120 V
b. $+91.93$ V

Solution

a. $v(t) = 120 \sin(\theta + 50°)$ and substituting $v(t) = -120$ yields $-120 = 120 \sin \alpha$, where $\alpha = \theta + 50°$, from which $\alpha = \sin^{-1}(-120/120) = \sin^{-1}(-1) = -90°$. Then $\theta = \alpha - 50° = -90 - 50 = -140° = \mathbf{+220°}$ (see Fig. 14-6)

b. $v(t) = 120 \sin(\theta + 50°)$
$91.93 = 120 \sin \alpha$ and $\alpha = \sin^{-1}(91.93/120) = 50°$ so
$\theta = \alpha - 50° = 50 - 50 = \mathbf{0°}$ (See Fig. 14-6)

Example 14-6 shows how it is possible to plot a waveform when given its equation containing *both* a **phase displacement** angle, ϕ, and a **rotation** angle, θ. Example 14-7 shows that given some instantaneous value of the waveform, it is possible to find the (positive) rotation angle(s) that produces this instantaneous value. This was done by defining a new angle, α, which represents the angular sum of θ and ϕ. Let us now define each angle in Eq. (14-6), which relates them as

$$\alpha = \theta + \phi = \sin^{-1}\left(\frac{v(t)}{V_m}\right) \qquad (14\text{-}6)$$

where $v(t)$ is any instantaneous value of interest, in volts

V_m is the maximum value of the waveform, in volts

θ is the *rotation angle*, defined as the angular displacement from its zero reference position at time $t = 0$, [i.e., $\theta = \omega t$ from Eq. (14-5d)]

ϕ is the *phase displacement angle*, defined as the *offset angle* by which the waveform has been shifted from the reference zero position. ϕ may either be a positive angle (Fig. 14-6) or a negative angle (Fig. 14-7)

α is the angular algebraic sum of θ and ϕ. Since ϕ may be either positive or negative and $\alpha = \theta + \phi$, as shown in Eq. (14-6), care must be observed when ϕ is negative (see Ex. 14-9)

EXAMPLE 14-8

Given the equation $v(t) = 10 \sin(120\pi t - 45°)$ mV, evaluate at $t = 5$ ms:

a. ω in rad/s
b. θ in degrees
c. ϕ in degrees
d. Instantaneous voltage $v(t)$
e. Plot the waveform as a function of θ for one complete cycle of rotation

Solution

a. $\omega = \mathbf{120\pi}$ **rad/s** by inspection

b. $\theta = \omega t = 120\pi \dfrac{\text{rad}}{\text{s}} \times \dfrac{180°}{\pi \text{ rad}} \times 5 \times 10^{-3}\ \text{s} = \mathbf{108°}$

c. $\phi = \mathbf{-45°}$ by inspection

d. $v(t) = 10 \sin \alpha = 10 \sin(\theta + \phi) = 10 \sin(108° - 45°)$
$= 10 \sin 63° = \mathbf{8.91\ mV}$

e. The waveform is plotted in **Fig. 14-7**

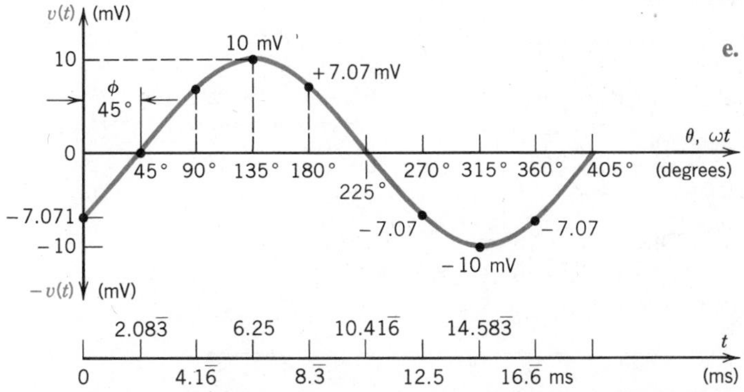

Figure 14-7 Waveform of $v(t) = 10 \sin(120\pi t - 45°)$ mV; Ex. 14-8

An examination of Fig. 14-7 shows the waveform of $v(t) = 10 \sin(\omega t - 45°)$ mV with its negative offset or phase displacement angle of $-45°$. When the rotation angle θ is zero, the instantaneous value of $v(t) = -7.071$ mV, as shown. A comparison of Figs. 14-6 and 14-7 shows that

1. When ϕ is a *positive* phase angle, the waveform begins (at $t = 0$) at some instantaneous *positive* value.
2. When ϕ is a *negative* phase angle, the waveform begins (at $t = 0$) at some instantaneous *negative* value.

Given the *equation* for the waveform (expressed as a positive sinusoidal function), it is fairly easy to calculate instantaneous values and plot the waveform, as was done in Exs. 14-7 and 14-8.

It is quite another matter, however, to determine values of θ, the rotation angle (or values of time, since $t = \theta/\omega$), that will produce a *given instantaneous* value. For example, there is *only one positive maximum* and *one negative maximum* value in *each* cycle of sine wave. But on *either side* of the maximum value, there are *two positive* (and *two negative*) *equal instantaneous* values per cycle. Furthermore, if we assume that the rotation angle, θ, continues to rotate CCW beyond 360° to 720°, 1080°, and so forth, we realize that there are *many* values of θ that will produce the *same instantaneous* value of $v(t)$.

A careful study of the sinusoidal waveform (expressed as a positive sinusoidal function) reveals the following recurring pattern, which is obtained by adding 180° to (+/−) α, modifying Eq. (14-6) as follows:

$$\alpha = \theta_1 + \phi \qquad \textbf{(14-6)}$$

$$180° - \alpha = \theta_2 + \phi \qquad \textbf{(14-6a)}$$

$$360° + \alpha = \theta_3 + \phi \qquad \textbf{(14-6b)}$$

$$540° - \alpha = \theta_4 + \phi \qquad \textbf{(14-6c)}$$

$$720° + \alpha = \theta_5 + \phi \qquad \textbf{(14-6d)}$$

$$900° - \alpha = \theta_6 + \phi \qquad \textbf{(14-6e)}$$

where θ_1 through θ_6 are successive (positive) rotation angles producing the *same* instantaneous value of $v(t)$

ϕ may be either a positive or a negative phase angle

α is $\sin^{-1}(v(t)/V_m)$ or $\sin^{-1}(i(t)/I_m)$

EXAMPLE 14-9

Given the equation $v(t) = 10 \sin(\omega t - 45°)$ mV, plotted in Fig. 14-7, calculate the first three positive rotation angles when $v(t) = 0$. Verify θ_3 by direct substitution.

Solution

$$\alpha = \sin^{-1}(v(t)/V_m) = \sin^{-1}(0/10) = 0° \qquad (14\text{-}6)$$

Then $\theta_1 = \alpha - \phi = 0 - (-45°)$
$= \mathbf{45°}$, first positive rotation angle (14-6)

$\theta_2 = 180° - \alpha - \phi = 180° - 0 - (-45°)$
$= \mathbf{225°}$, second positive rotation angle (14-6a)

$\theta_3 = 360° + \alpha - \phi = 360° + 0 - (-45°)$
$= \mathbf{405°}$, third positive rotation angle (14-6b)

Check: $v(t) = 10 \sin(\omega t - 45°) = 10 \sin(405 - 45)°$
$= 10 \sin 360° = \mathbf{0\ V}$

EXAMPLE 14-10

Given the waveform equation of Ex. 14-9 shown in Fig. 14-7, calculate the first two positive rotation angles when $v(t) = V_m = 10$ mV. Verify θ_2 by direct substitution in the given equation for the waveform.

Solution

$\alpha = \sin^{-1}(10/10) = \sin^{-1}(1) = 90°$

$\theta_1 = \alpha - \phi = 90° - (-45°)$
$= \mathbf{135°}$, first positive rotation angle (14-6)

$\theta_2 = 180° - \alpha - \phi = 180° - 90° - (-45°)$
$= \mathbf{135°}$ (same as θ_1) (14-6a)

$\theta_3 = 360° + \alpha - \phi = 360° - (-45°)$
$= \mathbf{495°}$, second positive rotation angle (14-6b)

Check: $v(t) = 10 \sin(\omega t - 45°) = 10 \sin(495° - 45°)$ mV
$= 10 \sin(450°)$ mV $= \mathbf{10\ mV}$

Examples 14-9 and 14-10 show that the series of Eqs. (14-6) through (14-6e) enable us to find specific successive positive rotation angles yielding the same instantaneous values in any given sinusoidal equation.

Example 14-10 also verifies the fact that there can be only one *positive maximum* value in each cycle. The foregoing equations for θ_1 and θ_2 (the two rotation angles in the first cycle) do indeed yield *the same value* of 135°. It was necessary, therefore, to find θ_3, the rotation angle in the next (second) cycle which yields the maximum positive value.[5]

Frequently, in the laboratory, the display of the sinusoidal waveform is shown on a cathode ray oscilloscope (CRO) as a function of time rather than rotation angle. In such cases we are interested in knowing the specific *times* when $v(t)$ reaches a specific value. It was noted earlier that since $t = \theta/\omega$, if we calculate the rotation angles, we can also calculate the time corresponding to those angles, given the angular frequency, ω.

EXAMPLE 14-11

Given the rotation angles shown in Fig. 14-7 of 0, 45°, 90°, 135°, and so forth in 45° increments, calculate the corresponding values of time for the waveform $v(t) = 10 \sin(120\pi t - 45°)$ mV

Solution

From the given waveform equation, by inspection, $\omega = 120\pi$ rad/s, therefore

1. We first convert this angular frequency to degrees per second from $\omega = 120\pi \dfrac{\text{rad}}{\text{s}} \times \dfrac{180°}{\pi \text{ rad}} = 21\,600$ deg/s
2. Now using the relation $t = \theta/\omega = \theta/(21\,600 \text{ deg/s})$, we can convert all rotation angles into time as follows:

when $\theta = 0$, $t = \mathbf{0}$
$\theta = 45°$, $t = 45°/(21\,600°/\text{s}) = \mathbf{2.08\bar{3}\ ms}$
$\theta = 90°$, $t = 90°/(21\,600°/\text{s}) = \mathbf{4.1\bar{6}\ ms}$
$= (2.08\bar{3} \times 2)$ ms
$\theta = 135°$, $t = 2.08\bar{3}$ ms $\times 3 = \mathbf{6.25\ ms}$, etc.
when $\theta = 360°$, $t = 2.08\bar{3}$ ms $\times 8$
$= \mathbf{16.6\ ms}$, which is the time for 1 cycle

Check: from Eq. (14-5a) $T = 2\pi/\omega = 2\pi/120\pi = \mathbf{16.\bar{6}\ ms}$

The times, in increments of $2.08\bar{3}$ ms, are shown plotted below their corresponding rotation angles in Fig. 14-7.

14-6 READING AND WRITING PHASOR QUANTITIES IN THE TIME DOMAIN

Any alternating quantity may be represented by projecting a rotating line on a fixed axis. This rotating line is called a *phasor* and is assumed to be rotating at a constant angular velocity, ω, in a *counterclockwise* direction. It is conventional to project the rotating phasor on to the horizontal or *real* axis, designated *Re* in

[5] Since there is only one maximum positive value each cycle, solving for θ_4 from Eq. (14-6c) also yields the same value of 495°. This is left as an exercise for the reader.

Fig. 14-8.[6] The imaginary or vertical axis is designated I_m. At time $t = 0$, the phasor projects its length, at point **a**, as the distance Oa. When the phasor rotates to position **b**, its horizontal projection on the real axis is Ob cos θ. When the phasor reaches position **c**, its projection on the real axis is zero because cos 90° is zero. Note that the projections on the real axis in Fig. 14-8 vary with the rotation angle, θ, or ωt since $\theta = \omega t$. If the length Oa represented a voltage, V_m, then the equation for the waveform would be $V_m \cos\theta$ or $V_m \cos\omega t$. Similarly, if length Oa represented a current, I_m, the equation for the waveform would be $I_m \cos\omega t$ or $I_m \cos\theta$.

Let us consider a voltage, V_m, leading a current, I_m, by a phase angle, ϕ, at an instant in time, as shown in **Fig. 14-9a**. Both phasors are rotating at the *same* angular frequency, ωt. Both generate sinusoidal waveforms whose amplitude varies in proportion to V_m and I_m. Observe that the current phasor, I_m, is at rotation angle θ (for the given instant in time) with respect to the reference zero position, where $t = 0$. Similarly, the voltage phasor, V_m, is at rotation angle $\theta + \phi$ or $\omega t + \phi$ (since $\theta = \omega t$). The equation for the current waveform is $i(t) = I_m \cos\omega t$. The equation for the voltage waveform is $v(t) = V_m \cos(\omega t + \phi)$.

Since the cosine wave leads a sine wave by 90°, should we desire to convert these equations to sine functions we merely add 90° to each phase angle, as shown in Fig. 14-9a.

Figure 14-9b shows the two phasors at the same rotation angles originally shown in Fig. 14-9a. We know that in the time domain they are rotating continuously at the angular frequency ωt and projecting a sinusoidal waveform, respectively, on the real axis. But Fig. 14-9b shows all the essential information in Fig. 14-9a, given the relative magnitudes of V_m, I_m, and ϕ.

It is also customary to set one of the phasors as a reference at time $t = 0$.[7] If we select current as a reference, as shown in Fig. 14-9c, voltage is *leading* current by angle ϕ. We could have selected voltage as a reference in which case current would *lag* voltage by angle ϕ.

One last point: it does not really matter whether we express voltages and currents in the time domain exclusively as *sine* functions or exclusively as *cosine* functions.

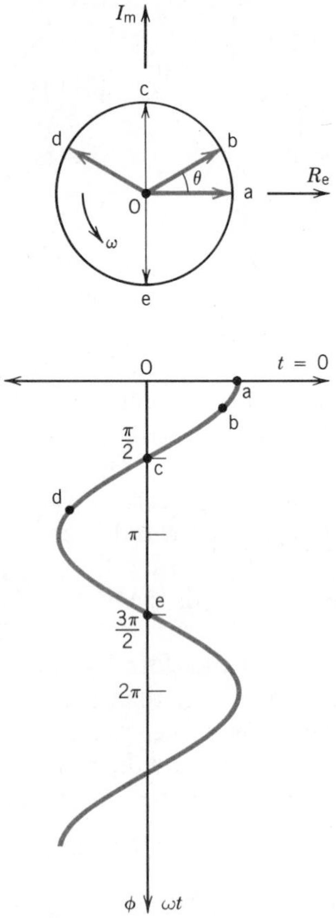

Figure 14-8 Sinusoidal waveform developed by a rotating phasor

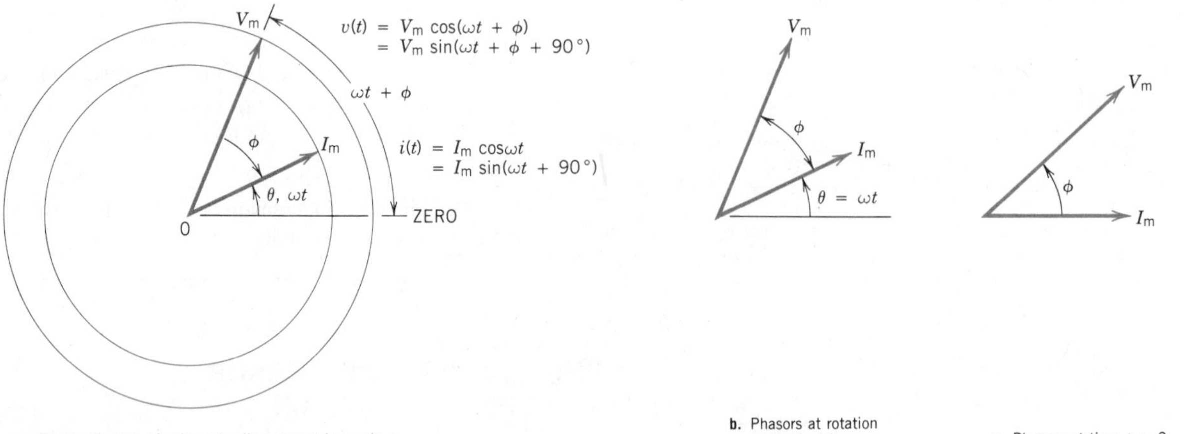

a. Phasor diagram of voltage leading current by angle ϕ

b. Phasors at rotation angles in (**a**)

c. Phasors at time $t = 0$

Figure 14-9 Phasor relations between voltage and current

[6] Actually it is more than a matter of convention. When the horizontal projection is directed on the real axis of the complex plane, phasors are converted *directly* into complex numbers by Euler's identity, $\varepsilon^{j\theta} = \cos\theta + j\sin\theta$, which converts the *exponential* form of a phasor to its *rectangular* form.

[7] We know that ultimately in their counterclockwise rotation, I_m will reach zero and V_m will lead it by angle ϕ. In a sense this is like a photograph of their relation at that instant. Actually, the *same phase relation* holds at *any* instant, since ω is constant.

What is of importance is that we use the *same basis of comparison* so that we can identify the relation between voltages and currents from their respective equations and draw their phasor relationship.

Before we go further, however, we must review some elementary trigonometry. **Figure 14-10a** shows the reference sine axis conventionally at 0°. The cosine axis is the conventional imaginary vertical axis. Thus $\cos \omega t$ is shown in the $+90°$ position. From Fig. 14-10a we may conclude (by way of review) that

1. Positive angles rotate in a CCW direction
2. Negative angles rotate in a CW direction
3. $+\sin \omega t$ is the reference zero axis, and $-\sin \omega t$ is displaced from it by 180°
4. Then $+\cos \omega t = \sin(\omega t + 90°)$; $-\cos \omega t$ is displaced from $+\cos \omega t$ by 180°
5. $-\cos \omega t$ can be written as $\sin(\omega t - 90°)$ or $-\sin(\omega t + 90°)$
6. $-\sin \omega t$ can be written as $\cos(\omega t + 90°)$ or $-\cos(\omega t - 90°)$

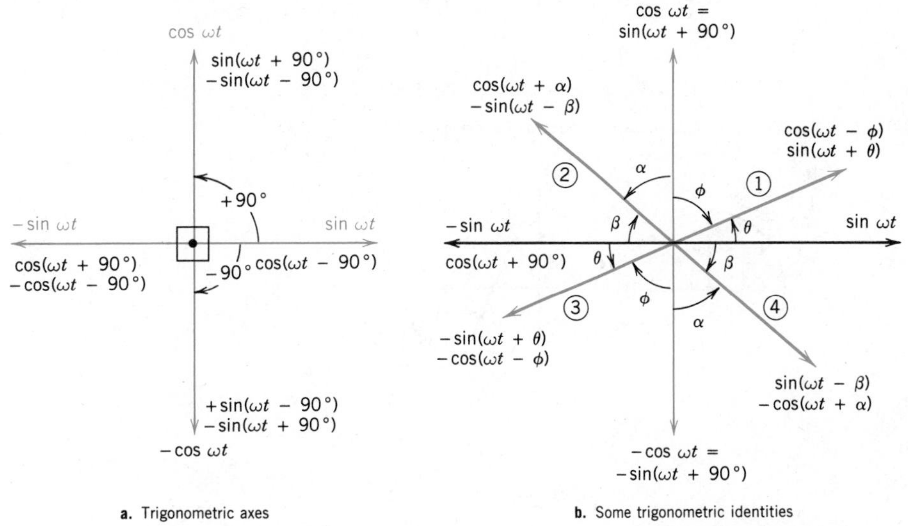

Figure 14-10 Trigonometric relationships

Figure 14-10b shows how equations may be written for phasors that are *not* on the primary reference sine or cosine axes. Note that there are four sets of vertical angles in Fig. 14-10b (α, β, ϕ, θ). Some of the angles are rotated CCW as positive angles and others are rotated CW as negative angles. Using the principles established in Fig. 14-10a, we may write for Fig. 14-10b for phasors 1 through 4, respectively,

1. Phasor 1 $= \sin(\omega t + \theta) = \cos(\omega t - \phi)$
2. Phasor 2 $= \cos(\omega t + \alpha) = -\sin(\omega t - \beta)$
3. Phasor 3 $= -\sin(\omega t + \theta) = -\cos(\omega t - \phi)$
4. Phasor 4 $= \sin(\omega t - \beta) = -\cos(\omega t + \alpha)$

The reader should review Fig. 14-10 to ensure mastery over the techniques involved in writing these trigonometric identities. The techniques are basic and fundamental to the subsequent sections.

14-6.1 Relations of Sinusoidal Waveforms with Respect to a Reference Zero

Frequently in the laboratory a waveform is displayed that does *not begin* its variation at time $t = 0$. The six waveforms of **Fig. 14-11** represent some possibilities the reader may encounter. The question naturally arises, "How can you tell whether a waveform is leading or lagging some reference zero and by how much?" The six waveforms of Fig. 14-11 provide the answer. Let us examine some of them:

Figure 14-11a: The displayed waveform, $v_1(t)$, is *leading* the reference zero waveform, $v_R(t)$, because

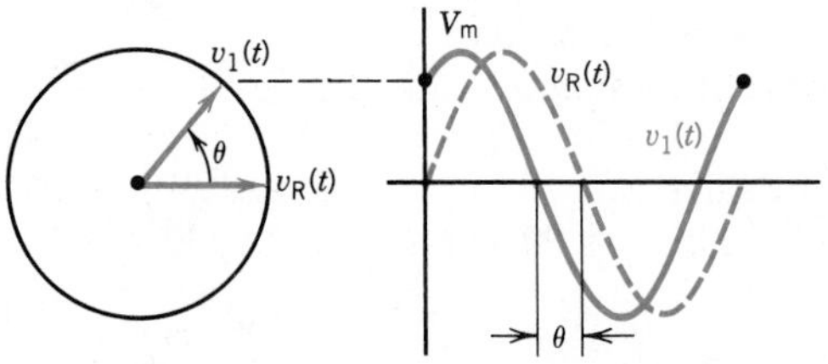

$v_1(t) = V_m \sin(\omega t + \theta)$

$v_R = V_m \sin(\omega t + 0°) = \text{REFERENCE}$

a. $v_1(t)$ leading reference by approximately 50°

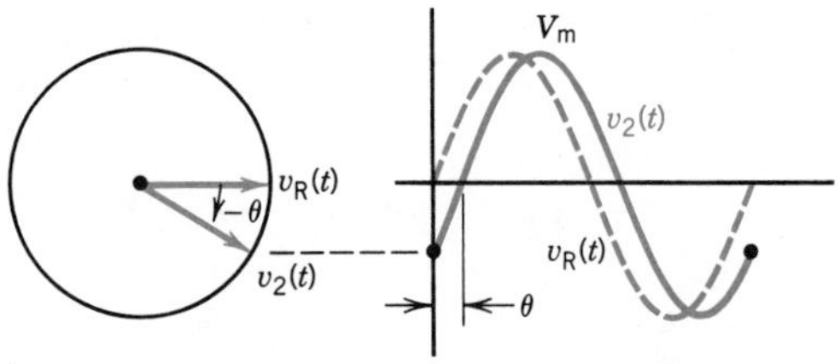

$v_2(t) = V_m \sin(\omega t - \theta)$

b. $v_2(t)$ lagging reference by approximately 30°

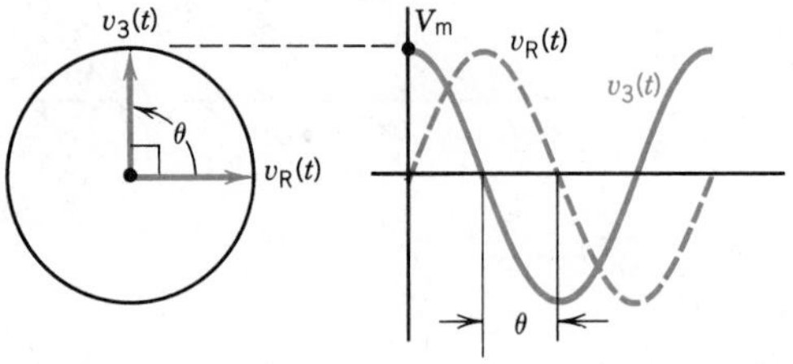

$v_3(t) = V_m \cos \omega t$
$= V_m \sin(\omega t + 90°)$

c. $v_3(t)$ leading reference by 90°

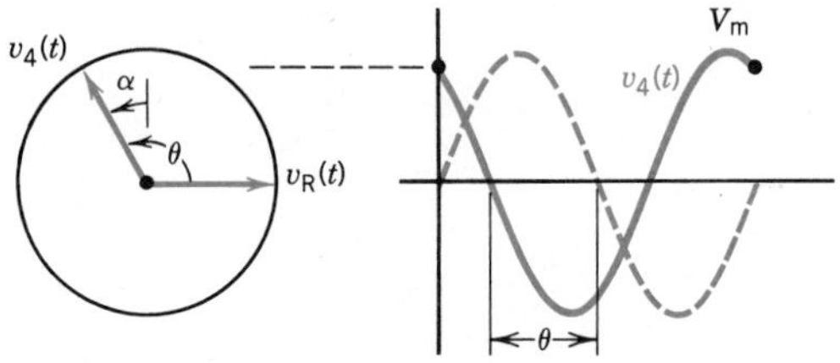

$v_4(t) = V_m \cos(\omega t + \alpha)$
$= V_m \sin(\omega t + 90° + \alpha)$
$= V_m \sin(\omega t + \theta)$

d. $v_4(t)$ leading reference by approximately 120°

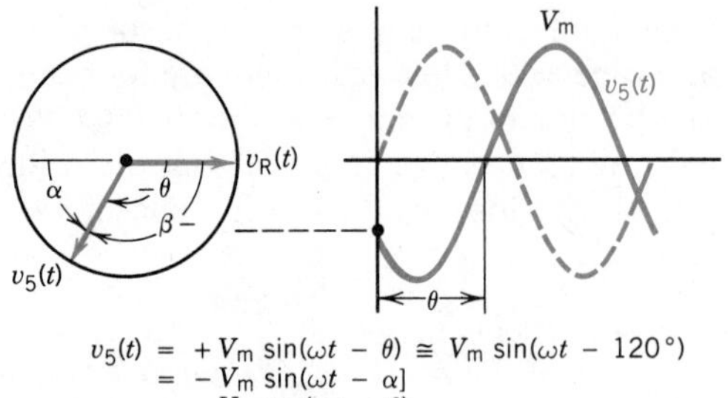

$v_5(t) = +V_m \sin(\omega t - \theta) \cong V_m \sin(\omega t - 120°)$
$= -V_m \sin(\omega t - \alpha]$
$= -V_m \cos(\omega t - \beta)$

e. $v_5(t)$ lagging reference by approximately 120°

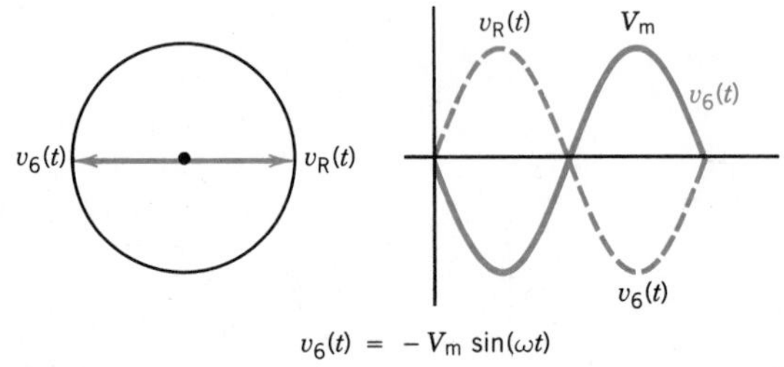

$v_6(t) = -V_m \sin(\omega t)$

f. $v_6(t)$ leading/lagging reference by 180°

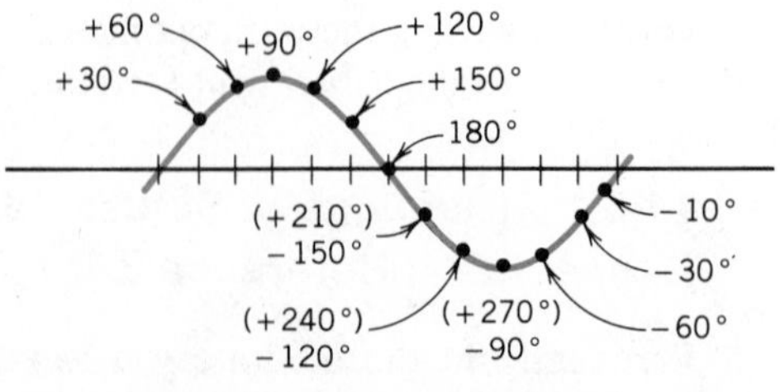

g. Wave point summary to approximate angle of lead or lag

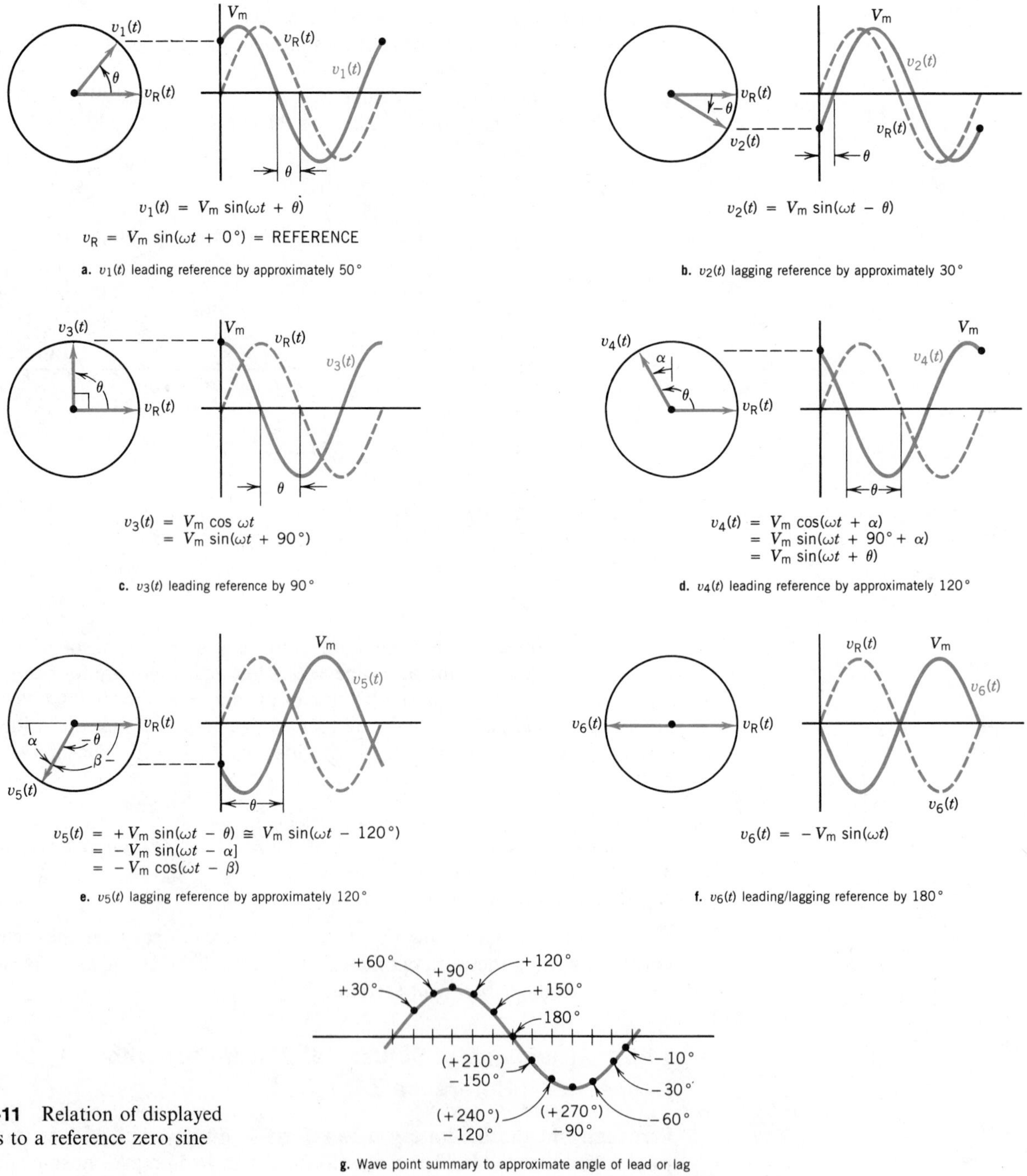

Figure 14-11 Relation of displayed waveforms to a reference zero sine wave

1. It reaches a maximum *before* the reference zero waveform reaches maximum.
2. It crosses the time axis going negative *before* the reference zero does the same.
3. The waveform equation is $v_1(t) = V_m \sin(\omega t + \theta)$ since it is a *leading* waveform.
4. The waveform begins (at $t = 0$) approximately at its $+50°$ wave point (see Fig. 14-11g). Consequently, we assume that it leads the reference waveform by approximately 50°.

Figure 14-11b: The displayed waveform is *lagging* the reference waveform since it comes off its maximum after the reference and crosses the zero axis *after* the reference. At time $t = 0$, the displayed waveform begins at its $-30°$ wave point (Fig. 14-11g). The waveform equation is $v_2(t) = V_m \sin(\omega t - \theta)$, where θ is approximately 30°.
Figure 14-11d: The displayed waveform *leads* the reference waveform because it crosses the zero axis earlier than the reference. The angle of lead, θ, is *greater* than 90°. At time $t = 0$, the displayed waveform begins at approximately its 120° wave point (Fig. 14-11g). The waveform equation may be written in various ways, depending on how it is referenced:[8]

1. If referenced to a sine wave zero, $v_4(t) = V_m \sin(\omega t + \theta) \cong V_m \sin(\omega t + 120°)$
2. If referenced to a cosine zero, $v_4(t) = V_m \cos(\omega t + \alpha) = V_m \cos(\omega t + 30°)$

Figure 14-11e: This is probably the most difficult of all the waveforms to evaluate. But observe that the displayed waveform rises from the zero axis *after* the reference waveform; therefore it is *lagging* the reference by θ, as shown in Fig. 14-11e. Because the waveform starts at $t = 0$ beyond its $-90°$ wave point (Fig. 14-11g), it is lagging by more than 90°. For this waveform we may write $v_5(t) = V_m \sin(\omega t - \theta) \cong V_m \sin(\omega t - 120°)$. Other possibilities are shown in Fig. 14-11, left as an exercise for the reader.

Figure 14-11g represents a summary of wave points to assist the reader in approximating an angle of lead or lag, depending on where the waveform of interest begins at time $t = 0$.

14-6.2 Expressing Equations as Positive Sinusoidal Functions in the Time Domain

We will encounter situations in which a voltage reversal occurs or a voltage in the time domain is preceded by a negative sign. Moreover, some voltages (or currents) may be expressed as cosine functions, whereas others are expressed as sine functions. In order to *compare* their phase relationships properly, it is customary to *convert* them *all* to *positive sine functions*.

There are two methods which the reader may use, depending on personal preference. One is an *algebraic* method, the other is *graphical*. Both yield the same results, as shown by the following examples.

EXAMPLE 14-12
Given the sinusoid $v(t) = -125 \cos(200\pi t - 260°)$ V, express the time domain equation as a positive sinusoidal function, using:

a. An algebraic conversion
b. A graphical conversion

---

[8] Many advanced engineering texts define and reference phasors from the cosine function, as indicated earlier. In this text we will reference phasors to the *sine* function.

Solution

a. (From Fig. 14-10a): $-\cos \omega t = \sin(\omega t - 90°)$
Substituting the above identity into the actual given function,

$$v(t) = -125 \cos(\omega t - 260°) = +125 \sin(\omega t - 260° - 90°)$$
$$= 125 \sin(\omega t - 350°) = 125 \sin(\omega t + 10°)$$
$$= \mathbf{125 \sin(200\pi t + 10°)\ V}$$

b. The graphical conversion method is shown in Fig. 14-12a. It consists of representing the equation in phasor form. The steps are:

1. Represent the waveform $-125 \cos \omega t$ on the cosine scale.
2. Since the phase angle is $-260°$, rotate the phasor $-260°$ in a CW direction.
3. Lay off the same distance $+125 \sin \omega t$ on the reference zero axis.
4. To superimpose the positive sine function on the negative cosine function, rotate the function CCW by 10°, yielding the same results as **(a)** above.

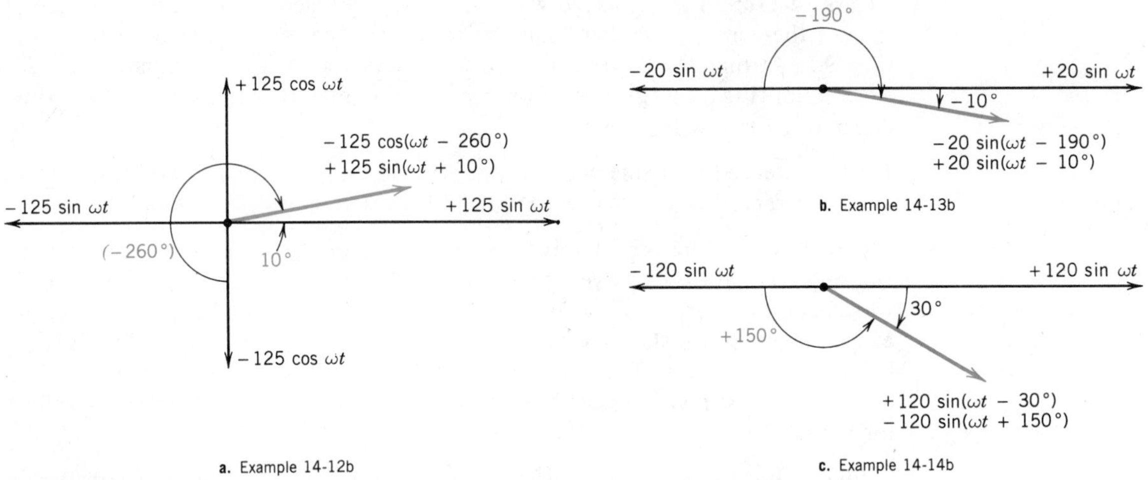

Figure 14-12 Graphical conversion of equations to positive sinusoidal functions

EXAMPLE 14-13

Given the sinusoid $i(t) = -20 \sin(200\pi t - 190°)$ A, express the time domain equation as a positive sinusoidal function using:

a. Algebraic
b. Graphical conversion

Solution

a. $-\sin \omega t = +\sin(\omega t + 180°)$
Substituting the above identity into the actual given function yields

$$i(t) = -20 \sin(\omega t - 190°) = +20 \sin(\omega t - 190° + 180°)$$
$$= 20 \sin(\omega t - 10°) = \mathbf{20 \sin(200\pi t - 10°)\ A}$$

b. The graphical conversion is shown in Fig. 14-12b, where
1. $-20 \sin \omega t$ is represented and rotated CW by $-190°$ to yield $-20 \sin(\omega t - 190°)$ A
2. lay off the same distance $+20 \sin \omega t$ on the reference zero axis, and
3. superimpose the positive sine function on the negative sine function by rotating CW, $-10°$, to yield $+20 \sin(\omega t - 10°)$ A, yielding the same results as **(a)**

EXAMPLE 14-14

Given $v(t) = -120 \sin(1200t + 150°)$ V, express the time domain equation as a positive sinusoidal function using:

a. Algebraic
b. Graphical conversion

Solution

a. $-\sin(\omega t) = \sin(\omega t - 180°)$
Substituting the above identity into the actual given function yields

$$-120 \sin(1200t + 150°) = +120 \sin(1200t + 150° - 180°)$$
$$= \mathbf{120 \sin(1200t - 30°)\ V}$$

b. The graphical conversion is shown in Fig. 14-12c, where
1. $-120 \sin \omega t$ is represented and rotated CCW in a positive direction for $+150°$ to produce $-120 \sin(\omega t + 150°)$
2. lay off $+120 \sin \omega t$ on the reference zero axis, and
3. superimpose the positive sine function on the negative sine function by rotating $+120 \sin(\omega t)$ in the negative CW direction by $-30°$, yielding $120 \sin(\omega t - 30°)$ or **120 sin(1200t − 30°) V**

14-6.3 Sketching Waveforms in the Time Domain

In Sec. 14-6.2 we learned how to express equations as positive sine functions in the time domain. In doing this, we were simplifying the equations for purposes of comparison, for example, as to their phase relation. The solution of Ex. 14-12 yielded the voltage $v(t) = 125 \sin(200\pi t + 10°)$ V. The solution of Ex. 14-13 yielded the current $i(t) = 20 \sin(200\pi t - 10°)$ A. In *reading* the equations we can see that the *phase difference* between the voltage and current is 20°. The maximum value of the voltage is 125 V, while the maximum value of the current is 20 A. The angular frequency of both waveforms is 200π rad/s. Consequently, they continue to maintain the same phase relationship at all times (i.e., **current lags voltage by 20°** or **voltage leads current by 20°**).

In sketching the waveforms, we begin by setting time $t = 0$. For the waveform $v(t) = 125 \sin(200\pi t + 10°)$ V, this means that when $t = 0$, $v(t) = 125 \sin(10°)$ V = **21.7 V**; but it also means that $v(t)$ begins at its $+10°$ value (see Fig. 14-11g).

Similarly, for the current waveform, $i(t) = 20 \sin(200\pi t - 10°)$ A, and when $t = 0$, $i(t) = 20 \sin(-10°)$ A = $\mathbf{-3.473}$ **A**; but it also means that $i(t)$ begins at its $-10°$ value (as shown in Fig. 14-11g).

In *sketching* the waveforms, we are in a sense *writing* the equation in sinusoidal form, as shown by Ex. 14-15.

Certain critical rotation angles (or times) are of interest in sketching the waveform:

1. The *instantaneous value* when the *rotation angle* (*or time*) is *zero*.
2. The *rotation angle* (*or time*) when the waveform is a *positive maximum*.
3. The *rotation angle* (*or time*) when the waveform crosses the *zero* axis and its *instantaneous value* is *zero*.
4. The *rotation angle* (*or time*) when the waveform is a *negative maximum*.
5. The *rotation angle* (*or time*) when the waveform *again* crosses the zero axis and its *instantaneous value* is *zero*.

Finding these rotation angles is not at all complicated because a shortcut can be used, as shown in Ex. 14-15**a**.

EXAMPLE 14-15

Sketch the waveform of $v(t) = 125 \sin(200\pi t + 10°)$ V using as the abscissa:

a. Rotation angle, θ, in degrees
b. Rotation angle, θ, in radians
c. Time, in milliseconds

Solution

$v(t) = V_m \sin(\theta + \phi)$ and when $\theta = 0$,
$v(t) = V_m \sin \phi = 125 \sin 10° = \mathbf{21.7\ V}$

a. In addition to the value when $\theta = 0$ (and $t = 0$), there are three other values of significance needed in sketching the waveform, i.e., when $v(t) = V_m$, when $v(t) = 0$, and when $v(t) = -V_m$. Since the waveform has been shifted to the *left* by 10°, *each* of these values *should appear* 10° *earlier*, as shown in Fig. 14-13a.[9]

b. Corresponding rotation angles in radians are found using the conversion factor π rad/180°:

$$\text{when } v(t) = V_m = 125\text{ V}, \theta = 80° \times \frac{\pi \text{ rad}}{180°} = \mathbf{1.4\ rad}$$

$$v(t) = 0, \theta = 170° \times \frac{\pi \text{ rad}}{180°} = \mathbf{3\ rad}$$

$$v(t) = -V_m = -125\text{ V}, \theta = 260° \times \frac{\pi \text{ rad}}{180°} = \mathbf{4.54\ rad}$$

$$v(t) = 0, \theta = 350° \times \frac{\pi \text{ rad}}{180°} = \mathbf{6.1\ rad}$$

and, of course, 360° corresponds to $\mathbf{2\pi}$ **rads**

[9] Failing this reasoning, the reader may calculate θ by the method shown in Ex. 14-9: $+V_m$ occurs when $\alpha = \sin^{-1}(1) = 90°$, and $\theta = \alpha - \phi = (90 - 10)° = \mathbf{80°}$ from Eq. (14-6). See Fig. 14-13a.

c. Time intervals are found using the relation $t = \theta/\omega$ from Eq. (14-5a), where $\omega = (200\pi \text{ rad/s}) \times \dfrac{180°}{\pi \text{ rad}} = 36\,000$ deg/s:

when $v(t) = V_m$, $\theta = 80°$ and $t = \theta/\omega$
$= 80°/36\,000°/\text{s} = \mathbf{2.\bar{2}\ ms}$

$v(t) = 0$, $\theta = 170°$ and $t = \theta/\omega$
$= 170°/36\,000°/\text{s} = \mathbf{4.7\bar{2}\ ms}$

$v(t) = -V_m$, $\theta = 260°$ and $t = \theta/\omega$
$= 260°/36\,000°/\text{s} = \mathbf{7.\bar{2}\ ms}$

$v(t) = 0$, $\theta = 350°$, and $t = \theta/\omega$
$= 350°/36\,000°/\text{s} = \mathbf{9.7\bar{2}\ ms}$

and when $\theta = 360°$, $t = \theta/\omega = 360°/36\,000°/\text{s} = \mathbf{10\ ms}$, which is the period of the waveform

The three waveforms are sketched in **Fig. 14-13**. Example 14-15 contains some important insights:

1. The angle by which the waveform has been shifted is the angle ϕ in the equation $v(t) = V_m \sin(\omega t + \phi)$.
2. Since ϕ is a *positive* angle in the equation for the given waveform, the shift of the waveform is to the *left*. **Leading positive phase angles shift the waveform to the left.**
3. Whenever ϕ is a *negative* angle, it must *lag* the reference zero, and it shifts the waveform to the *right*. **Lagging negative phase angles shift the waveform to the right.**
4. At time $t = 0$, $\omega t = 0$, and the initial value of $v(t)$ is $V_m \sin \phi$, as shown in Fig. 14-13a. This is always true, regardless of whether ϕ is positive or negative.
5. In sketching the waveforms, it is usually easier to sketch them first as a function of rotation angle, θ, **in degrees**. The degrees can then be converted to either radians or time, as shown in Ex. 14-15.

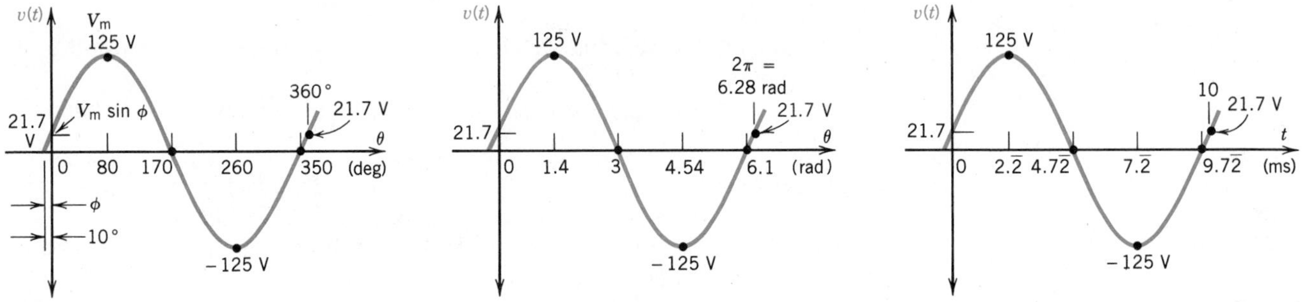

a. $v(t)$ vs rotation angle θ in degrees

b. $v(t)$ vs rotation angle θ in radians

c. $v(t)$ vs time in milliseconds (ms)

Figure 14-13 Example 14-15

14-7 PHASE RELATIONS BETWEEN WAVEFORMS

Two waves are considered "*in phase*" if at any time during their periodic variation they are *both positive-going* and cross the zero axis at the same instant in time. This definition applies even if the two waves are *not of the same frequency*, as shown in **Fig. 14-14a**. Observe that at time $t = 0$, 2, and 4 s, the two sinusoids are in phase. Consequently, $v_2(t)$ is considered in phase with the lower-frequency reference wave, $v_1(t)$.

Two waves are considered 180° out of phase whenever the reference wave crosses the zero axis in a *positive-going* direction and the second waveform crosses the zero axis at the same point in a *negative-going* direction. This definition applies even if the two waves are not of the same frequency, as shown in Fig. 14-14b. Note that $v_2(t)$ is *never* in phase with $v_1(t)$ and is only 180° out of phase at times $t = 0$, 1, 2, 3, etc. seconds.

At angles other than 0° and 180°, the term "phase" applies only to the portion of the period of some reference waveform by which a second waveform of interest advances (*leads*) by some *phase-lead* angle ϕ (or time t) or *lags* by some *phase-lag* angle $-\phi$, (or time t).

When two waves are of the *same frequency*, there is a *constant phase relation* between them. Either of the two waves may be chosen as the reference waveform, as shown by the following examples.

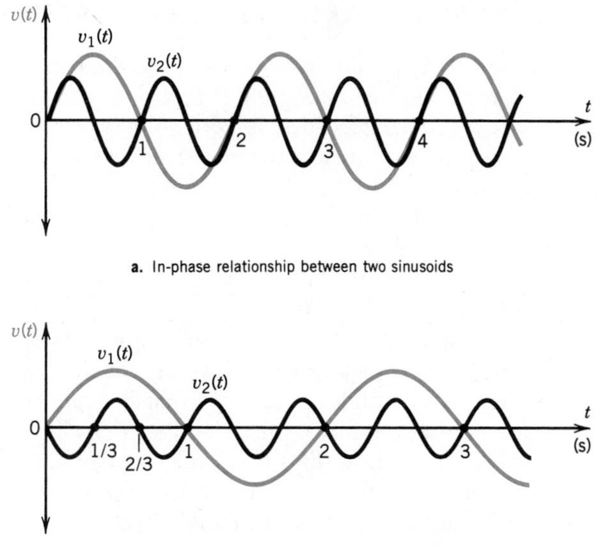

a. In-phase relationship between two sinusoids

b. 180° out-of-phase relation between two sinusoids

Figure 14-14 In-phase and 180° out-of-phase relations between two sinusoids and different frequency

---

EXAMPLE 14-16

The voltage applied to a circuit is $v(t) = -20 \sin(5000t + 120°)$ V and the current drawn from the source is $i(t) = -12 \cos(5000t + 80°)$ A. Find the:

a. Voltage waveform expressed as a positive sine wave function
b. Current waveform expressed as a positive sine wave function
c. Phase relation between the two waveforms using *voltage* as a *reference*
d. Phase relation between the two waveforms using *current* as a *reference*
e. Sketch the two waveforms using as a reference the arbitrary sine wave zero

Solution

Using either the algebraic or the graphical methods shown in Exs. 14-12 through 14-14 we may write[10]

a. $v(t) = 20 \sin(5000t - 60°)$ V
b. $i(t) = 12 \sin(5000t - 10°)$ A
c. By inspection of **(a)** and **(b)**, current is *leading* voltage by 50°
d. By inspection of **(a)** and **(b)**, voltage is *lagging* current by 50°
e. The two waveforms are shown in **Fig. 14-15**

With respect to the solution shown in Fig. 14-15:

1. *Neither* waveform is in phase with the reference zero used to express the positive sine wave functions in **(a)** and **(b)**.
2. The current waveform $i(t)$ lags the reference zero by $\phi = -10°$. Consequently, at time $t = 0$, the instantaneous current is $i(t) = 12 \sin(-10°) = -2.084$ A.
3. The voltage waveform $v(t)$ lags the reference zero by $\phi = -60°$. Consequently, at time $t = 0$, the instantaneous current is $v(t) = 20 \sin(-60°) = -17.32$ V.
4. Current is leading voltage at all times by 50° whenever both waves cross the abscissa.
5. The period of the waveform, $T = 2\pi/\omega = (2\pi \text{ rad/cycle})/(5000 \text{ rad/s}) = 1.257$ ms/cycle.
6. The period of 1.257 ms is equivalent to a rotation angle of 2π radians.

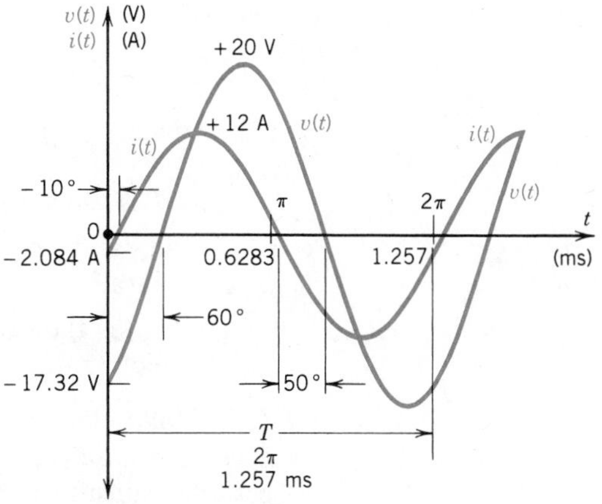

Figure 14-15 Solution to Ex. 14-16e

---

[10] The detailed solutions of parts **(a)** and **(b)** are left as an exercise for the reader. Only the correct answers are shown.

14-8 AVERAGE VALUES OF PERIODIC WAVEFORMS

Up to now, the only waveform considered has been the pure sinusoidal wave. If we attempt to measure a sinusoidal voltage with a dc voltmeter, we discover that the voltage recorded is zero. Similarly, if we attempt to measure a sinusoidal current with a dc ammeter, again we discover that the current is zero. To gain a better understanding of the meaning of the average value, we will have to consider either nonsinusoidal waveshapes or just parts of sine waves. But first some definitions.

We will learn later that dc instruments do record the average or dc value contained within a waveform. Since the average value is sometimes referred to as the dc value, let us incorporate it in our notation. Given any periodic waveform, our object is to determine its average or dc **amplitude**, A_{dc}. From the previous paragraph we might guess that, since a *pure sinusoid* contains no dc, its average amplitude $A_{dc} = 0$. But how does this come about?

In answering this question let us define the **average amplitude**, A_{dc}, as the algebraic sum of all the areas under the curve of the waveform taken for one cycle divided by the period of the waveform.[11] The area *above* the time axis will be designated as *positive* and the area *below* the axis as *negative*, or

$$A_{dc} = \frac{\text{algebraic sum of all areas taken for one full cycle}}{\text{period of the waveform}}$$

$$A_{dc} = A_{av} = \frac{A_{v_1} \cdot t_1 + A_{v_2} \cdot t_2 + A_{v_3} \cdot t_3 + \cdots + A_{v_n} \cdot t_n}{T} \qquad \textbf{(14-7)}$$

where T is the period of the waveform

A_{v_1} is the equivalent average value of any portion of the waveform having duration t_1

A_{v_2} is the equivalent average value of a second portion of the waveform having duration t_2, etc.

The following examples should help us to understand how the average value may be computed without the need for higher mathematics. We will begin with periodic nonsinusoidal waveforms whose average value is greater than (or less than) zero. These waveforms are essentially rectangular in shape and were selected for two reasons:

1. Calculations are simplified when portions of nonlinear (curved) waveforms are reduced to simple rectangles.
2. Rectangular waveshapes have the same average value as their maximum value.

---

[11] In more rigorous terms, if x represents a series of instantaneous values of current or voltage, then the average of the instantaneous values over one full period has a **constant** amplitude, A_{dc}, expressed by

$$A_{dc} = \frac{1}{T}\int_{t_0}^{t_0+T} x\,dt$$

where t_0 is any value of time on the waveform and T is the period.

EXAMPLE 14-17

Given the rectangular nonsinusoidal waveform shown in **Fig. 14-16a,**

a. Calculate the average value of voltage over one period of alternation
b. Draw the simplest rectangular waveform having the same average value

Solution

a. Before we may begin we must determine the time for one cycle, i.e., the period. A careful examination of Fig. 14-16a shows that the period is 20 ms because beyond 20 ms, the waveform repeats the same alternation pattern that occurs after time $t = 0$. Then

$$A_{dc} = \frac{(-3\text{ V}) \times 8\text{ ms} + (3\text{ V}) \times 3\text{ ms} + (-6\text{ V}) \times 4\text{ ms}}{20\text{ ms}}$$

$$= \frac{-39\text{ V}\cdot\text{ms}}{20\text{ ms}} = \mathbf{-1.95\text{ V}} \qquad (14\text{-}7)$$

b. The equivalent average value is shown in Fig. 14-16b as a constant value of **−1.95 V** versus time, i.e., the duration of the original waveform. If a dc voltmeter is connected across a source generating the waveform of Fig. 14-16a, it would record a *constant* dc value of −1.95 V.

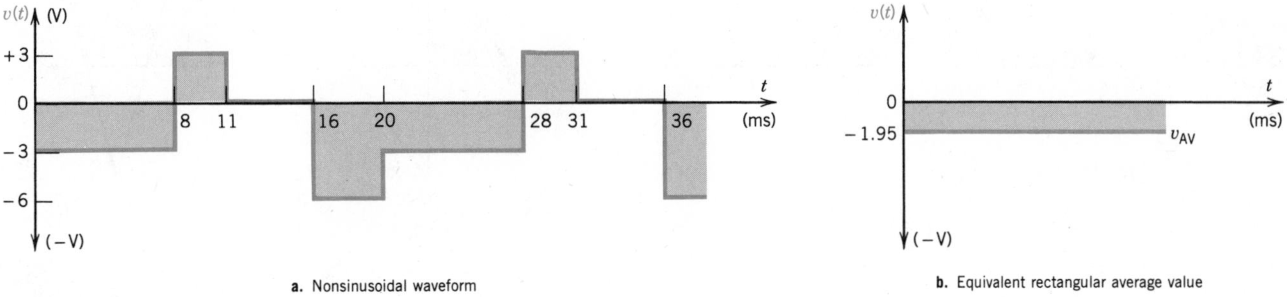

Figure 14-16 Example 14-17

EXAMPLE 14-18

Given the rectangular nonsinusoidal waveform shown in **Fig. 14-17a,**

a. Calculate the average value of current over one period of alternation
b. Draw the simplest rectangular waveform having the same average value

Solution

a. An examination of Fig. 14-17a shows that the period of the waveform is 30 μs. Beyond 30 μs, the waveform repeats the original alternation pattern begun at time $t = 0$.

$$A_{dc} = \frac{(10\text{ mA}) \times 5\ \mu\text{s} + (20\text{ mA}) \times 5\ \mu\text{s} + (30\text{ mA}) \times 5\ \mu\text{s} - (30\text{ mA}) \times 5\ \mu\text{s} - (10\text{ mA}) \times 5\ \mu\text{s}}{30\ \mu\text{s}}$$

$$= \frac{100\text{ mA}\cdot\mu\text{s}}{30\ \mu\text{s}} = \mathbf{3.\bar{3}\text{ mA}} \qquad (14\text{-}7)$$

b. The equivalent average value is shown in Fig. 14-17b.

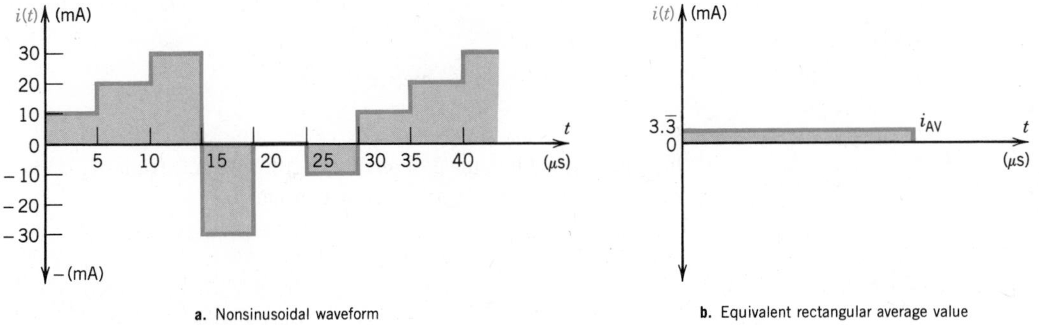

Figure 14-17 Example 14-18

Examples 14-17 and 14-18 obtained the areas under the curves of the rectangular waveforms shown by using the geometric relation $A = b \times h$, where b is the length of the base and h is the length of the height. In effect, this meant that the height (amplitude) of the rectangular waveform becomes its average value, A_{dc}, for its specific duration in time. Is it possible to obtain average values of other geometric figures in terms of their maximum amplitudes? Let us examine other geometric waveforms to determine their areas and average values, respectively, in terms of their maximum amplitudes, A_m. **Figure 14-18a** shows a series of five triangular waveforms. The area of a triangle is $bh/2$, where b is the base (or the duration) and h is the height of the maximum amplitude.

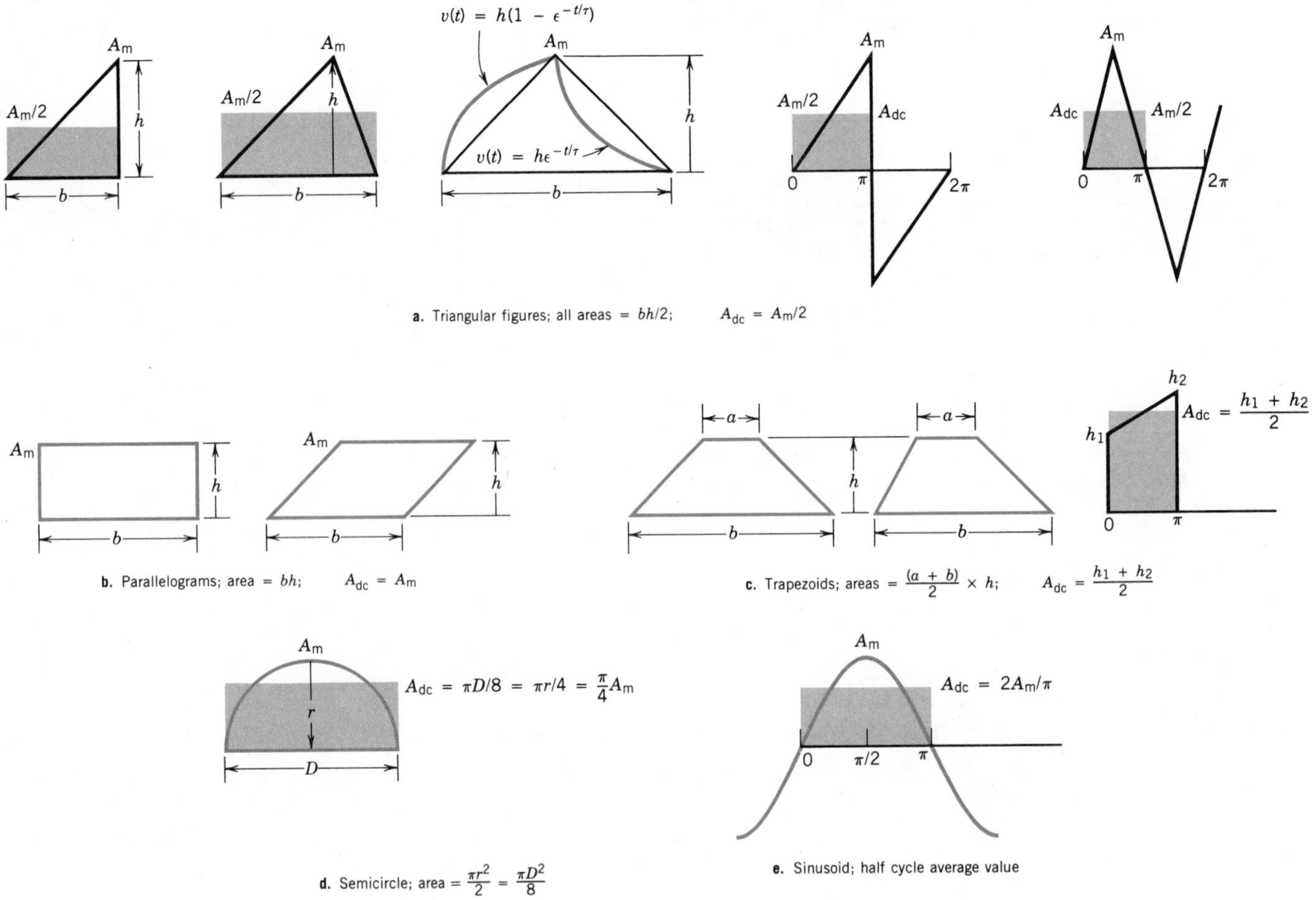

a. Triangular figures; all areas = $bh/2$; $A_{dc} = A_m/2$

b. Parallelograms; area = bh; $A_{dc} = A_m$

c. Trapezoids; areas = $\frac{(a + b)}{2} \times h$; $A_{dc} = \frac{h_1 + h_2}{2}$

d. Semicircle; area = $\frac{\pi r^2}{2} = \frac{\pi D^2}{8}$

e. Sinusoid; half cycle average value

Figure 14-18 Average values of common geometric shapes

It is evident that the area under the curve of *any* triangular waveform is $A_m \times b/2$, from which the average value for *any* triangular waveform is

$$A_{dc} = \frac{A_m}{2} \tag{14-8}$$

Equation (14-8) even applies to the exponential waveform of rise and fall shown in Fig. 14-18a. This waveform is *essentially triangular* because the area *added* to the *left* slope of the dashed triangle is reduced by the area removed from the right slope of the dashed triangle. Consequently, the equivalent area under the curve is the dashed triangle whose average amplitude is a rectangular waveform $A_m/2$. This is an important insight, because the exponential waveform is a common output waveform in *RC* and *RL* circuits. (See Appendix B-8.5 for the average value of the rising exponential portion only.)

We already know that the rectangular waveform (Fig. 14-18b) has an average value $A_{dc} = A_m$. But the rectangle is a special case of the *parallelogram* whose average value also is $A_{dc} = A_m$.

All trapezoidal areas (Fig. 14-18c) are computed as $(a + b)h/2$. The most commonly encountered trapezoidal figure, shown at the far right, has the area $(h_1 + h_2) \times \pi/2$, from which we obtain the (rectangular) average or dc value of a *trapezoid* as

$$A_{dc} = \frac{(h_1 + h_2)}{2} \qquad \textbf{(14-9)}$$

In effect, Eq. (14-9) states that the *average height of the trapezoid* is its *average rectangular amplitude*, A_{dc}. The trapezoid is frequently used in approximating areas under the curves of irregular waveforms or waveshapes whose equations are unknown.

Figure 14-18d shows a semicircle. Computations of average area reveal $A_{dc} = \pi A_m/4$; (i.e., the average amplitude is about three-fourths of the maximum amplitude, A_m). Although the semicircle is not generated electrically, it gives some insight into the average value of a *half-cycle* of sine wave (Fig. 14-18e), whose average amplitude is approximately two-thirds of the maximum amplitude, A_m.

Appendix B-7 shows the derivation of the average amplitude value of **one half-cycle** of sine wave in Eq. (B-7.1) as

$$A_{Hav} = A_{dc} = \frac{2A_m}{\pi} = 0.63662A_m \cong 0.637A_m \qquad \textbf{(14-10)}$$

It should be noted from Fig. 14-18e that the average value for one quarter-cycle, from 0 to $\pi/2$ radians, is also $2A_m/\pi$. (This is proved in Appendix B-7.4.)

Before we consider examples involving the average values of some of the above described waveforms, several very important points must be kept in mind:

1. The average value of a *full cycle* of sine wave is zero. This is obvious because the area below the zero reference cancels the area above the zero reference.
2. The average value of *one full cycle* of a **half-wave rectifier** (HWR) is A_m/π, (i.e., **half** the average value of the half-cycle average). This relation is derived in Appendix B-7.3 and also proved by Ex. 14-21. (Also see Table 14-1.)
3. In obtaining the average value of waveforms shown below, *we will first convert* them to *equivalent rectangular pulses* and then solve for the average value by summation of areas algebraically, using Eq. (14-7), for a period of one cycle. This eliminates the need for integrating the areas under the curves and reduces more complex problems to the simple Eq. (14-7).

EXAMPLE 14-19

Given the waveform shown in **Fig. 14-19a**,

a. Reduce each portion to its equivalent rectangular average value and draw the equivalent waveform
b. Calculate the average value of one cycle of the waveform

Solution

From $t = 0$ to 3 μs,

$V_{av} = V_m/2 = 10\text{ V}/2 = \mathbf{5\text{ V}}$ (14-8)

From $t = 3$ to 5 μs,

$V_{av} = (V_1 + V_2)/2 = [-15 + (-10)]/2 = \mathbf{-12.5\text{ V}}$ (14-9)

From $t = 5$ to 13 μs,

$V_{av} = V_m/2 = 10\text{ V}/2 = \mathbf{5\text{ V}}$ (14-8)

By inspection, $T = \mathbf{13\ \mu s}$

a. The equivalent rectangular values are shown in Fig. 14-19b
b. $V_{dc} = V_{av}$

$$= \frac{(5\text{ V}) \times 3\ \mu\text{s} + (-12.5\text{ V}) \times 2\ \mu\text{s} + (5\text{ V}) \times 8\ \mu\text{s}}{13\ \mu\text{s}}$$

$$= \frac{+30\text{ V}\cdot\mu\text{s}}{13\ \mu\text{s}} = \mathbf{2.308\text{ V}} \quad (14\text{-}7)$$

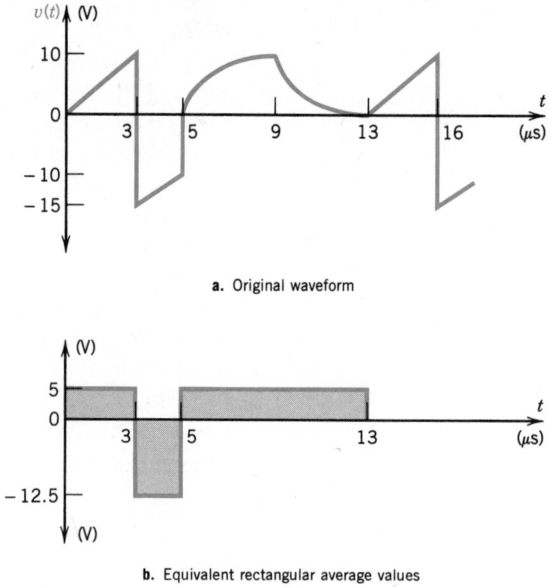

Figure 14-19 Example 14-19

EXAMPLE 14-20

Given the waveform shown in **Fig. 14-20a**,

a. Reduce each portion to its equivalent rectangular average value and draw the equivalent average waveform for calculation of the average value using Eq. (14-7)
b. Calculate the average value of one cycle of the waveform

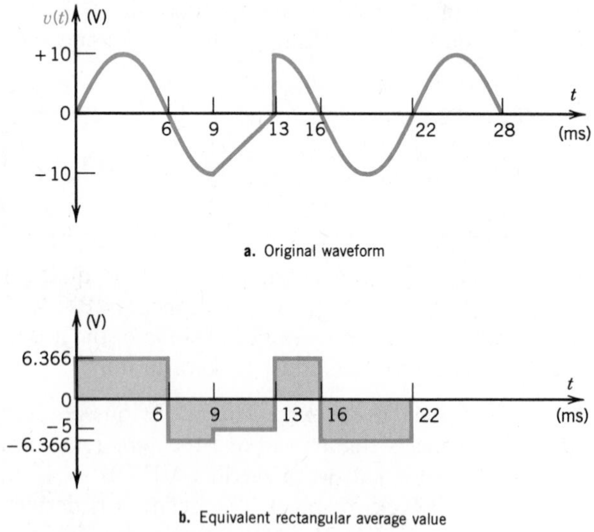

Figure 14-20 Example 14-20

Solution

From $t = 0$ to 6 ms,

$V_{av} = 2V_m/\pi = 2 \times 10\text{ V}/\pi = \mathbf{6.366\text{ V}}$ (14-10)

From $t = 6$ ms to 9 ms,

$V_{av} = -2V_m/\pi = -2 \times 10\text{ V}/\pi = \mathbf{-6.366\text{ V}}$ (14-10)

From $t = 9$ ms to 13 ms,

$V_{av} = -V_m/2 = -10\text{ V}/2 = \mathbf{-5\text{ V}}$ (14-8)

From $t = 13$ ms to 16 ms,

$V_{av} = 2V_m/\pi = 2 \times 10\text{ V}/\pi = \mathbf{6.366\text{ V}}$ (14-10)

From $t = 16$ ms to 22 ms,

$V_{av} = -2V_m/\pi = -2 \times 10\text{ V}/\pi = \mathbf{-6.366\text{ V}}$ (14-10)

By inspection, $T = 22$ ms

a. The equivalent rectangular values calculated above are shown in Fig. 14-20b.
b. $V_{dc} = V_{av}$

$$= \frac{(6.366\text{ V}) \times 6\text{ ms} + (-6.366\text{ V}) \times 3\text{ ms} + (-5\text{ V}) \times 4\text{ ms} + (6.366\text{ V}) \times 3\text{ ms} + (-6.366\text{ V}) \times 6\text{ ms}}{22\text{ ms}}$$

$$= \frac{-20\text{ V}\cdot\text{ms}}{22\text{ ms}} = \mathbf{-0.\overline{909}\text{ V}} \quad (14.7)$$

EXAMPLE 14-21

The current output waveform of a half-wave rectifier (HWR) shown in **Fig. 14-21a** has a maximum value of $I_m = 100$ mA. Calculate the

a. Equivalent rectangular average value for each half cycle and draw the equivalent waveform
b. Equivalent average value of *one cycle* of the waveform
c. Show that the average value of one cycle of HWR output is A_m/π, as derived in Eq. (B-7.2)

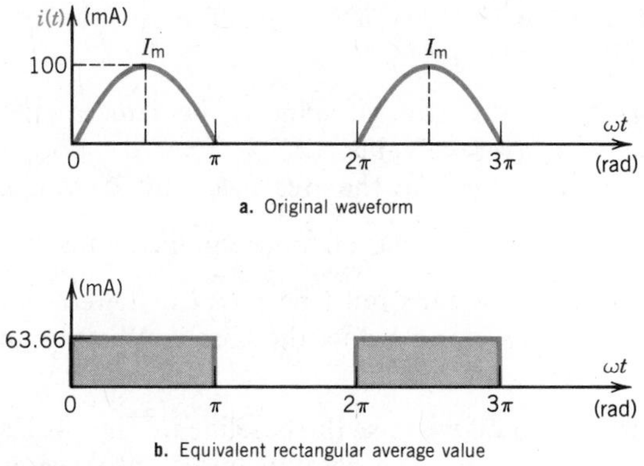

Figure 14-21 Example 14-21

Solution

a. From 0 to π radians,

$$I_{av} = 2I_m/\pi = 2 \times 100 \text{ mA}/\pi = \mathbf{63.66 \text{ mA}} \qquad (14\text{-}10)$$

From π to 2π radians, $I_{av} = 0$, by inspection. The equivalent average waveform is drawn in Fig. 14-21b

b. $$I_{dc} = I_{av} = \frac{(63.66 \text{ mA}) \times \pi + 0 \times \pi}{2\pi} = \mathbf{31.83 \text{ mA}} \qquad (14\text{-}7)$$

c. $$I_{dc} = I_m/\pi = 100 \text{ mA}/\pi = \mathbf{31.83 \text{ mA}} \qquad (\text{B-}7.2)$$

Examples 14-19 through 14-21 show the simplicity and usefulness of converting various waveshapes to equivalent average rectangular values in determining the dc or average value of a nonsinusoidal (or sinusoidal) waveform.

Example 14-21 is particularly interesting because it shows that even if we do not know that $A_{dc} = A_m/\pi$ for an HWR, we can compute it by the method shown in parts (**a**) and (**b**) of the solution.

14-8.1 Symmetrical Nonsinusoidal and Sinusoidal Waveforms Containing a DC Component

Quite frequently, in the laboratory and in industry, waveforms are observed on a cathode ray oscilloscope (CRO) that obviously have an average value either greater or less than zero. Consider the waveform shown in **Fig. 14-22a**. Clearly, the area *above* the zero baseline is *greater than* the area *below* the line. The same is also true of the sawtooth wave shown in Fig. 14-22b and the sinusoidal waveform shown in Fig. 14-22c. All three have positive average values since there is more area above the baseline than below in each case. Let us compute the average value for Fig. 14-22a by the rectangular area method, as shown in Ex. 14-22.

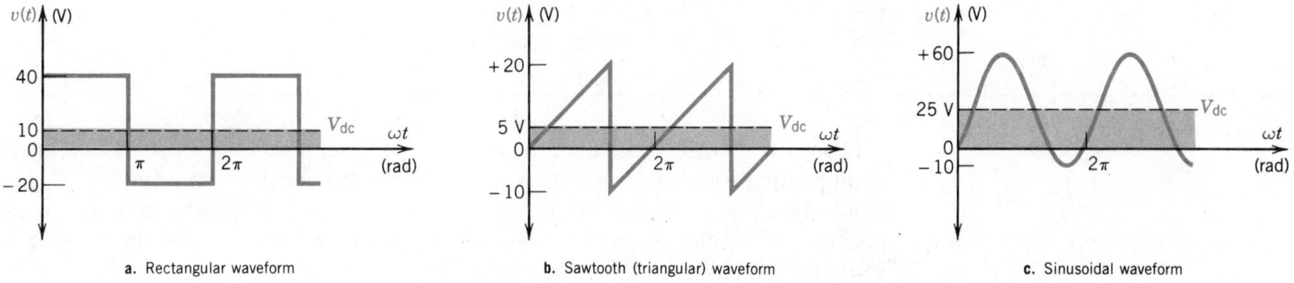

Figure 14-22 Nonsinusoidal and sinusoidal symmetrical waveforms containing an average or dc component

EXAMPLE 14-22

Calculate the average value of the nonsinusoidal rectangular waveform shown in Fig. 14-22a

Solution

$$V_{dc} = V_{av} = \frac{(40 \text{ V} \times \pi) + (-20 \text{ V} \times \pi)}{2\pi} = \mathbf{+10 \text{ V}} \qquad (14\text{-}7)$$

If we draw a dashed line to indicate the 10-V average value, as shown in Fig. 14-22a, several rather interesting insights emerge:

1. The maximun amplitude *above* the dc or average value is the same as the maximum amplitude *below* the dc or average value.
2. The area *above* the dc value is exactly equal to the area *below* the dc value.

A little thought regarding (1) and (2) leads us to an even more significant insight:

3. The dc value is the level by which the baseline must be *raised* or *lowered* to produce a (*symmetrical*) waveform (above and below the baseline) **having an average value of zero**!

The significance of (3) is better understood if we **raise** the baseline in Fig. 14-22a by a factor of +10 V. This leaves +30 V above the baseline and −30 V below the baseline. Clearly, the average value of such a rectangular wave is zero since equal areas are now above and below the baseline.

Some further consideration of this example leads us intuitively to a simple equation for finding the average value or dc component of *any symmetrical waveform*:

$$V_{av} = V_{dc} = \frac{A_m(+) + A_m(-)}{2} \qquad (14\text{-}11)$$

where $A_m(+)$ is the maximum positive value
$A_m(-)$ is the maximum negative value

It would be extremely difficult to use the equivalent average rectangular method with Figs. 14-22b and 14-22c because we do not know the specific angles where the waveform crosses the zero baseline. But when using Eq. (14-11) the average value computations are relatively simple, as shown by Ex. 14-23.

EXAMPLE 14-23

Using Eq. (14-11), calculate the dc component (and average value) of the waveforms shown in Figs. 14-22a, b and c.

Solution

a. $V_{dc} = (V_m^+ + V_m^-)/2 = [+40 + (-20)]/2 = \mathbf{+10\ V}$ (see Ex. 14-22)

b. $V_{dc} = (V_m^+ + V_m^-)/2 = [+20 + (-10)]/2 = \mathbf{+5\ V}$ (14-11)

c. $V_{dc} = (V_m^+ + V_m^-)/2 = [+60 + (-10)]/2 = \mathbf{+25\ V}$ (14-11)

From Fig. 14-22b we can see that setting the baseline at +5 V produces a maximum positive amplitude of +15 V and a maximum negative amplitude of −15 V. Similarly, in Fig. 14-22c, setting the baseline at +25 V produces a maximum positive amplitude of +35 V and a maximum negative amplitude of −35 V, verifying Eq. (14-11).

The reader may now ask whether Eq. (14-11) applies to *any* waveform, such as that shown in Fig. 14-21a. If we attempt to use Eq. (14-11) we obtain an average amplitude of 50 mA, **which is incorrect**. We know from Ex. 14-21 that the average value is 31.83 mA.[12] But the waveform above this average value is **not symmetrical** to the waveform below it. Therefore, Eq. (14-11) may *only* be used with *symmetrical* waveforms of the kind shown in Fig. 14-22.

[12] If this average value is superimposed on Fig. 14-21a, then the area above that value equals the area below it.

It must be noted, however, that the average values of *nonsymmetrical* waveforms (or any periodic waveform of any shape) may be measured experimentally by using the dc (direct-coupled) and ac (alternating-coupled) switches of a CRO,[13] or even by connecting a dc voltmeter across the voltage waveform of interest.

14-8.2 Average Value of the Half-Cycle Output of the Half-Wave (and Full-Wave) SCR

The silicon-controlled rectifier (SCR) or thyristor is extensively used in such applications as motor speed control, heater output controls, battery chargers, inverters, choppers, power supplies, and relay controls. It performs all these functions by controlling the shape of its waveform and consequently its average (and RMS) value (see Sec. 14-9 and Sec. 14-10).

The equation for the half-cycle average, A_{Hav}, or dc value as a ratio of its maximum amplitude for any given value of firing angle, ϕ, or conduction angle, θ, is derived in Appendix B-7 as Eq. (B-7.4). (It is also shown in Table 14-2.) The half-cycle average value is

$$A_{\text{dc}} = A_{\text{Hav}} = \frac{A_{\text{m}}}{\pi}(1 + \cos\phi) = \frac{A_{\text{m}}}{\pi}(1 - \cos\theta) \qquad (14\text{-}12)$$

where A_{m} is the maximum amplitude of the output waveform
ϕ is the firing angle $(180° - \theta)$ producing conduction, in degrees
θ is the conduction angle, (i.e., that portion of the half-cycle during which conduction occurs, in degrees)

EXAMPLE 14-24

A single half-wave SCR is triggered at a firing angle of 45° and produces the output waveform across an electric heater shown in **Fig. 14-23a**. Calculate the

a. Equivalent half-cycle rectangular average values and draw the equivalent waveform
b. Average value of one cycle of the original waveform (Fig. 14-23a)

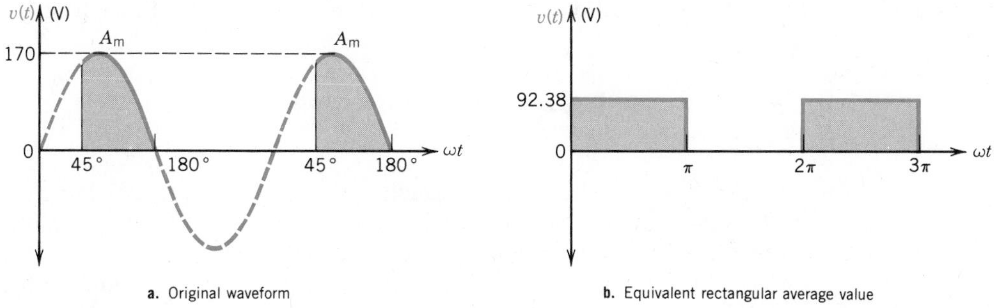

Figure 14-23 Example 14-24

[13] Using direct coupling, both the ac and dc components of the waveform are displayed. Using alternating coupling, the average value or dc component is filtered out and only the ac components are displayed. With ac coupling, the average area above the baseline equals the average area below the baseline. The average or dc component value is a measure of how much the waveform either rises or drops when switching from ac to dc coupling. A positive average value produces a rise or jump in the positive direction. A negative average value is a drop toward the negative direction *when switching from* **ac** *to* **dc** *coupling.* A properly calibrated CRO, by measuring this rise or drop, may record the average value of *any* periodic waveform, quite accurately. See Sec. 23-15, Ex. 23-17.

Solution

$$A_{\text{Hav}} = \frac{A_m}{\pi}(1 + \cos 45°) = \frac{170\text{ V}}{\pi}(1 + 0.7071)$$

$$= \mathbf{92.38\text{ V}} \qquad (14\text{-}12)$$

$$\text{Check: } A_{\text{Hav}} = \frac{A_m}{\pi}(1 - \cos 135°) = \frac{170\text{ V}}{\pi}(1 + 0.7071)$$

$$= \mathbf{92.38\text{ V}} \qquad (14\text{-}12)$$

a. The equivalent half-cycle rectangular average is shown in Fig. 14-23b

b. $$V_{\text{dc}} = \frac{(92.38\text{ V} \times \pi) + 0\text{ V} \times \pi}{2\pi}$$

$$= \mathbf{46.19\text{ V}}\text{, average voltage across heater for one full cycle.} \qquad (14\text{-}7)$$

EXAMPLE 14-25

Two SCRs are used in **full-wave** combination. One is triggered at 40° and the other at 50° to produce the output waveform shown in **Fig. 14-24a**. Calculate

a. Each equivalent half-cycle rectangular average value and draw the equivalent rectangular waveforms
b. The average value of one cycle of waveform

Solution

$$A_{\text{Hav}_1} = \frac{A_m}{\pi}(1 + \cos \phi_1) = \frac{170}{\pi}(1 + \cos 40°)$$

$$= \mathbf{95.565\text{ V}} \qquad (14\text{-}12)$$

$$A_{\text{Hav}_2} = \frac{A_m}{\pi}(1 + \cos \phi_2) = \frac{170}{\pi}(1 + \cos 50°)$$

$$= \mathbf{88.90\text{ V}} \qquad (14\text{-}12)$$

a. The equivalent half-cycle rectangular averages are shown in Fig. 14-24b

b. $$V_{\text{dc}} = \frac{(95.565\text{ V} \times \pi) + (88.90\text{ V} \times \pi)}{2\pi}$$

$$= \mathbf{92.23\text{ V}}\text{, average value of one cycle} \qquad (14\text{-}7)$$

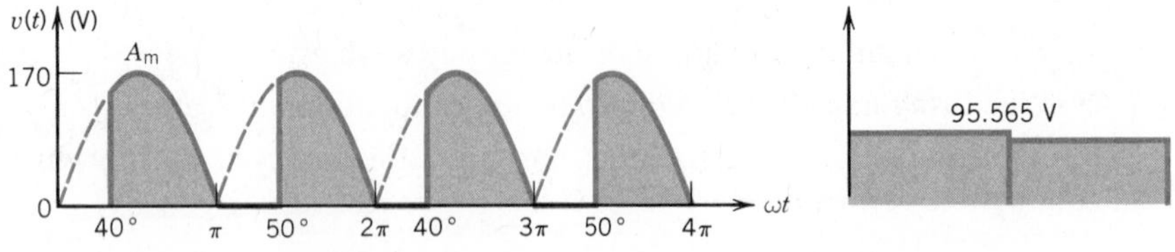

a. Original waveform

Figure 14-24 Example 14-25

Examples 14-24 and 14-25 show that *only one* equation is necessary (for either half- or full-wave combinations) to determine the average values of one cycle.

14-8.3 Average Values Given the Equation of the Waveform

We know that the average value of any periodic waveform is found by dividing the area under the curve (of one cycle) of the waveform by its period.[14] Occasionally, however, we encounter waveforms that do not fall into the pattern of those covered thus far in Sec. 14-8. But if we know (or can approximate) the equation for the waveform, we can find its average value, as shown in Ex. 14-26.

The equation of the waveform also assists us in plotting the waveform so that we can obtain some idea of its appearance. This is useful in testing whether the average value obtained is reasonable.

---

[14] A more precise definition of the average amplitude, $A_{\text{av}} = \int_0^t \frac{a(t)}{T}\,dt$, involves the integral sum of all instantaneous values of $a(t)$.

EXAMPLE 14-26

A periodic waveform has a frequency of 20 Hz and its equation is $v(t) = 50(t - 0.03)^2$ kV

a. Calculate the period of the waveform in milliseconds (ms)
b. Using 5-ms intervals, plot one full cycle of the waveform
c. Using the equation of the waveform and elementary integration, find the average value of the waveform[15]
d. Using the triangular approximation, find the approximate average value of the waveform to determine whether the answer in (c) is reasonable

Solution

a. $T = 1/f = 1/20\text{ Hz} = 50\text{ ms}$

b.

t (ms)	0	5	10	15	20	25	30	35 to 50
$v(t)$ (V)	45	31.25	20	11.25	5	1.25	0	0 0

The waveform is plotted in **Fig. 14-25**

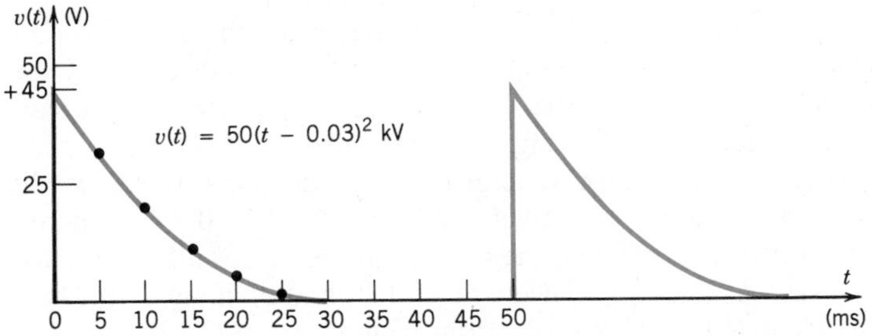

Figure 14-25 Waveform for Ex. 14-26 using data of Ex. 14-26**b**

c. $V_{av} = \int_{t_0}^{t} \frac{v(t)\,dt}{T} = \frac{1}{T}\int_0^{30\text{ ms}} [50(t - 0.03)^2\text{ kV}]\,dt$

$= \frac{50 \times 10^3}{50\text{ ms}} \int_0^{30\text{ ms}} (t - 0.03)^2\,dt$

$= \frac{1 \times 10^6}{3}[(t - 0.03)^3]_0^{30\text{ ms}} = 0.\bar{3} \times 10^6\,[0 - (-0.03)^3]$

$= (0.\bar{3} \times 10^6)(2.7 \times 10^{-5}) =$ **9 V** for 50 ms duration

d. Using the triangular approximation, $V_\Delta = A_{m/2} = 45\text{ V}/2 = 22.5\text{ V}$ for approximately 25 ms. Then

$$V_{dc} \cong \frac{(22.5\text{ V} \times 25\text{ ms}) + 0\text{ V} \times 25\text{ ms}}{50\text{ ms}}$$

$\cong$ **11.25 V** for 50 ms duration

This shows that the average value of 9 V is a reasonable answer since the decay of the waveform is exponential (rather than linear), which *reduces* the area under the curve.

14-9 RMS VALUES OF SINUSOIDAL CURRENT AND VOLTAGE WAVEFORMS

In a dc circuit, the voltage, current, and power consumed by a given resistance are relatively constant, in accordance with Ohm's law, $R = k = V/I = V^2/P = P/I^2$. But in an ac circuit, as we have seen, both voltage and current vary sinusoidally and continuously from zero to maximum positive and negative values. We know that the average values of voltage and current, taken for a full cycle, are both zero. But when an alternating (sinusoidal) source is applied to a (resistive) lamp, the lamp gets hot and produces light. Since the heating effect in a lamp is a function of the *square* of the current times its resistance, the direction of current flow is of little consequence.[16] But since the current (and voltage and power) are *all continuously* changing, we require some means of specifying ac current, voltage, and power. It seems only natural to specify that 1 ampere of ac will produce the *same heating effect* in the same resistance as 1 ampere of dc. By equating the **average ac** power to the (constant) **dc** power consumed by a resistor, we should be able to find the equivalent sinusoidal current and its relation to its maximum sinusoidal value.

[15] The author has found that many students, even those who have *not* had integral calculus, can learn to perform this simple integration.

[16] A negative current squared is positive and a positive current squared is positive.

Figure 14-26 shows an experimental setup for equating the heating effects produced by a dc source and an ac source. A lamp having a hot resistance $R = 100\ \Omega$ is connected to a 100-V dc source through a wattmeter and an ammeter. As expected, the wattmeter, P_1, in the dc circuit records 100 W, and the lamp glows to a certain brightness due to its heating effect.

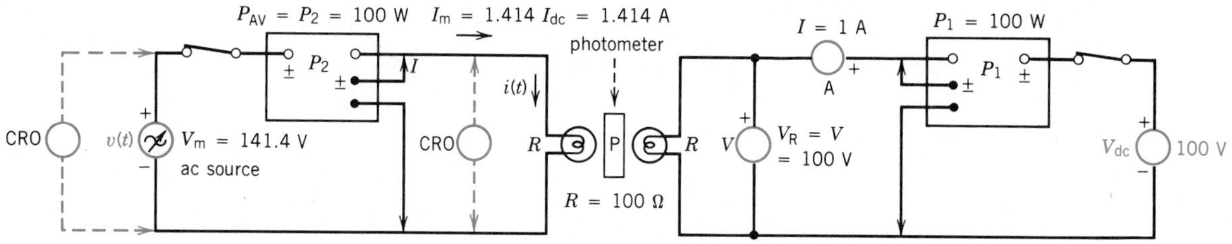

Figure 14-26 Experimental circuits to determine equivalent dc heating effect produced by sinusoidal voltage and current

If we now connect an ac sinusoidal source, we may vary the voltage of the source until the wattmeter, P_2, reads the same power as P_1 in the dc circuit. Using a photometer, we also check to ensure that both identical lamps (having the same resistance, power rating, and voltage rating) of 100 Ω burn with equal brightness. If we use a CRO to display the voltage waveform of the sinusoidal source, we discover that the waveform has a maximum value of $V_m = 141.4$ V. If we divide this voltage by the lamp resistance of 100 Ω, the current in the ac circuit has a maximum value of $I_m = 1.414$ A. Equating powers P_1 and P_2 and dividing by the same resistance, R, yields $P_2 = I_{ac}^2 R = I_{dc}^2 R = P_1$, from which

$$I_{ac} = I_{dc} = 1\text{ A} = \frac{I_m}{1.414} = \frac{I_m}{\sqrt{2}} = 0.7071 I_m \qquad (14\text{-}13)$$

The value I_{ac} that produces the same *heating effect* as I_{dc} is sometimes called the **effective value** of current. It is also called the root-mean-square (RMS) value because it describes the process by which the effective value is derived by calculus, that is, taking the (square) **root** of the **mean** (average) of the sum of instantaneous values of the function **squared**, or

$$I_{ac} = I = I_{rms} = \sqrt{\frac{1}{T}\int_0^t [i(t)]^2\,dt} = \frac{I_m}{\sqrt{2}} = 0.7071 I_m \qquad (14\text{-}14)$$

Equation (14-14), showing current relations, is a special case of the equation derived in Appendix B-7.6 to show that the foregoing relation between the maximum amplitude, A_m, and the RMS amplitude, A_{rms}, of *any sinusoidal* waveform holds as well for **both voltages and currents**. The relation $A_{rms} = A_m/\sqrt{2}$ also holds, as shown in Appendix B-7.6, for

1. The half-cycle of a sinusoid taken from 0 to π radians
2. One quarter-cycle of a sinusoid taken from 0 to $\pi/2$ radians
3. One full cycle of full-wave rectifier output taken from 0 to 2π radians

A further proof of this relation emerges from a consideration of the power produced in the left-hand ac circuit of Fig. 14-26. The instantaneous sinusoidal voltage, $v(t)$, applied across the lamp resistance, R, produces a sinusoidal current, $i(t)$, in phase with the voltage, as shown in **Fig. 14-27a**. But the instantaneous power, $p(t)$, is the product $v \cdot i(t)$ at each instant, shown in Fig. 14-27a. Note that between 0 to π radians, since $v(t)$ and $i(t)$ are *both positive*, the product $p(t)$ is positive.

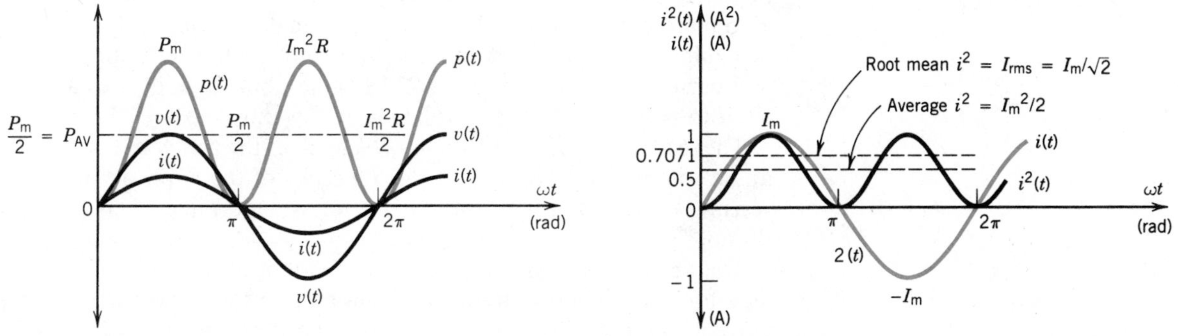

a. Voltage, current and power relations in a resistive circuit

b. Current $i(t)$ and current squared (i^2) when I_m is normalized at $I_m = 1$ A

Figure 14-27 Instantaneous voltage, current, and power; average power and current squared in a pure resistance

Between π and 2π radians, $v(t)$ and $i(t)$ are *both negative*, but their product is *positive*, as shown in Fig. 14-27a. Examination of Fig. 14-27a reveals that

1. The average values of $v(t)$ and $i(t)$ are *both zero* from 0 to 2π radians.
2. The power curve has an average value of $P_m/2$ or $I_m^2R/2$, as shown by the dashed line, since the *area* of the power curve *above* this line *equals* the *area below* the line.
3. The instantaneous power curve $p(t)$ goes through *two* complete cycles from 0 to 2π radians of voltage and current. Its radian frequency is 2ω, compared to ω for $v(t)$ and $i(t)$.
4. The instantaneous power curve begins at its $-90°$ point when $t = 0$ and has the time domain equation $p(t) = P_{av}[1 + \sin(2\omega t - 90°)] = P_{av}(1 - \cos 2\omega t)$, which expands to

$$p(t) = P_{av}[1 + \sin(2\omega t - 90°)] = \frac{I_m^2R}{2}(1 - \cos 2\omega t)$$

$$= \frac{I_m^2R}{2} - \frac{I_m^2R}{2}(\cos 2\omega t) \quad \text{watts (W)} \qquad \textbf{(14-15)}$$

Equation (14-15) is a most important relation because it shows that the power curve has two components:

1. A fixed average (dc) value of $P_{av} = P_m/2 = I_m^2R/2$
2. An ac component, $p_{ac} = -P_{av}\cos 2\omega t$

In deriving Eq. (14-13) we showed that $P_2 = P_1$ or $P_{ac} = P_{dc}$, so we can write

$$P_{av} = P_{dc}$$

$I_m^2R/2 = I_{dc}^2R$; canceling R and solving for I_{dc} yields $I_{dc} = I_m/\sqrt{2}$

But since 1 RMS ampere produces the same heating effect as 1 dc ampere, we again obtain Eq. (14-13), or

$$I = I_{\text{rms}} = I_{\text{m}}/\sqrt{2} \qquad \text{amperes (A)} \tag{14-13}$$

Usage dictates that the subscripts (eff) and (rms) are usually dropped when dealing with RMS currents or voltages. Since 1 RMS ampere equals 1 ampere of dc, we can use the *same* italic I to represent RMS ac current and the **same** italic V to represent RMS ac voltage as we used in dc circuits. It should be noted that all ac equipment (motors, alternators, transformers, toasters, lamps, etc.) has voltage and current ratings in terms of RMS values, but these devices also must be insulated to withstand the peak voltages or currents.

The reader may ask at this point, "Why does the power curve of Fig. 14-27a have an ac component that is *twice the frequency* of the applied voltage and resultant resistive current and why does it have an average value when voltage and current do not?"

The answer to both these questions is found in Fig. 14-27b, where the current waveform, $i(t)$, is shown from 0 to 2π along with the waveform of the instantaneous current squared, $i^2(t)$. Observe that the waveform of $i^2(t)$ in Fig. 14-27b is exactly the same as the waveform of $p(t)$ in Fig. 14-27a. Since $p(t)$ equals i^2R and R is a constant, this is fairly obvious.[17] But squaring the instantaneous current has produced

1. An *inversion* of the current waveform from π to 2π so that all values of $i^2(t)$ are positive.
2. A doubling of the frequency since the waveform of $i^2(t)$ from π to 2π is a repetition of its waveform from 0 to π radians.
3. A dc component since the waveform of $i^2(t)$ has all its area above the baseline and must have an average value greater than zero. As shown in Fig. 14-27b, this average value is $I_{\text{m}}^2/2$, using Eq. (14-11) for a symmetrical waveform.
4. A waveform whose equation may be written as $i^2(t) = (I_{\text{m}}^2/2)(1 - \cos 2\omega t)$, showing both the dc component and ac component of the waveform.

From the foregoing observations we can write an important equation:

$$P_{\text{av}} = P = \text{average } (i^2 \cdot R) = (\text{average } i^2)\, R = \left(\frac{I_{\text{m}}}{\sqrt{2}}\right)^2 R = I^2R \tag{14-16}$$

Equation (14-16) verifies once again the meaning of the RMS or effective value where $I^2 = (\text{average } i^2)$ and $I = \sqrt{(\text{average } i^2)} = I_{\text{m}}/\sqrt{2}$, as shown in the upper dashed line of Fig. 14-27b. Note from Eq. (14-16) that power, P, is **always** *an average* (not an RMS) value, so the subscript (av) is often dropped, since *all power is average power.*

[17] Since $p(t) = v^2/R$, exactly the same waveform of $v^2(t)$ would have resulted had we squared $v(t)$ in Fig. 14-27a and reproduced it as $v^2(t)$ in Fig. 14-27b. Thus we see that the waveform of $p(t)$ results from the exponential (nonlinear) effect of squaring either the voltage or current [or multiplying $v(t)$ by $i(t)$].

14-10 RMS VALUES OF NONSINUSOIDAL VOLTAGE AND CURRENT WAVEFORMS

Equation (14-14) expressed the general equation for the RMS value of current[18] of any period waveform, sinusoidal or nonsinusoidal, as

$$I = \sqrt{\frac{1}{T}\int_0^T [i(t)]^2\,dt}$$

But this would require the use of complex integration, in some cases, for a number of nonsinusoidal waveforms. Instead, as in the case of average values, we will convert the nonsinusoidal waveform to its **equivalent** RMS value in **rectangular form** and sum the resultant rectangular waveforms (using an equation similar to Eq. 14-7 for average value). Since we may be dealing with either voltages or currents, let us use the value R to represent the RMS value in *rectangular* form of each portion of a given waveform. Then the resultant or overall RMS value is an amplitude, A, of

$$A = \sqrt{\frac{R_1^2 t_1 + R_2^2 t_2 + R_3^2 t_3 + \cdots + R_n^2 t_n}{T}} \qquad \text{(14-17)}$$

where R_1, R_2, and R_3 are the **RMS** values of equivalent rectangular waveforms
t_1, t_2, and t_3 are the durations of those waveforms
T is the period or time for one cycle of the waveform

14-10.1 RMS Value of a Rectangular Waveform

The reader may ask, "Why rectangular waveforms in Eq. (14-17); why not some other waveshape?" The answer, by happy coincidence, is that the maximum, average, and RMS values of a rectangular wave are all the *same*. Consequently, if we know the maximum, we also know the average and RMS values. We may write for the rectangular or square wave that the RMS amplitude, A, is the same as the maximum amplitude, A_m, or

$$A \equiv A_m \qquad \text{(14-18)}$$

EXAMPLE 14-27

For the nonsinusoidal rectangular waveform shown in Fig. 14-16, calculate the RMS value

Solution

$$V = \sqrt{\frac{R_1^2 t_1 + R_2^2 t_2 + R_3^2 t_3 + \cdots}{T}}$$

$$= \sqrt{\frac{(-3\text{ V})^2 \times 8\text{ ms} + (3\text{ V})^2 \times 3\text{ ms} + (-6\text{ V})^2 \times 4\text{ ms}}{20\text{ ms}}}$$

$$= \sqrt{12.15\text{ V}^2} = \mathbf{3.486\text{ V}}$$

[18] The general equation for RMS voltage is $V = \sqrt{\frac{1}{T}\int_0^T [v(t)]^2\,dt}$

Note that the average value of this waveform, from Ex. 14-17, was −1.95 V but its RMS value is 3.486 V. Also note that RMS values are **never** *negative* and can *never* be *less* than the *average* value of the waveform. RMS values are usually higher or may be the same as the average value (but never less).[19]

14-10.2 RMS Values of Triangular Waveforms

It was indicated earlier that, because all instantaneous negative values are squared, the RMS value is never negative.

Figure 14-28a shows a triangular waveform whose average value (for one full cycle from 0 to 2π) is zero. The RMS value of the triangular waveform is[20]

$$A_{R\Delta} = \frac{A_m}{\sqrt{3}} \tag{14-19}$$

where A_m is the peak amplitude of the triangular waveform

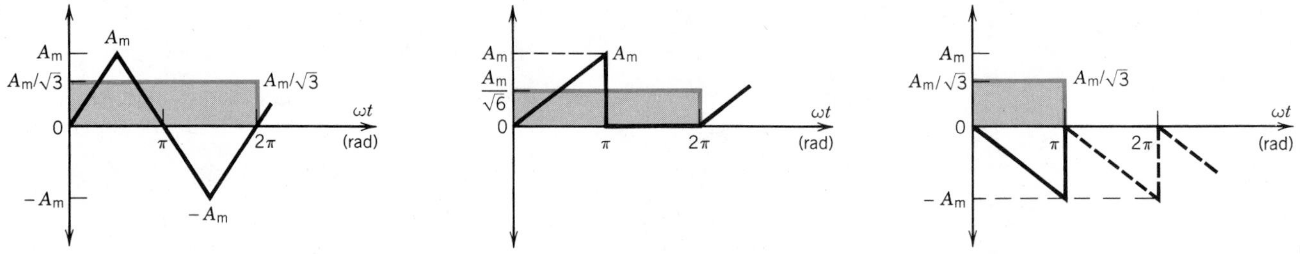

a. RMS value of one full cycle

b. RMS value of one cycle of HWR output (ac and dc components)

c. Half cycle RMS value (ac and dc components) of negative-going waveform

Figure 14-28 RMS values of triangular waveforms

Figure 14-28b shows the *half-wave rectifier* (HWR) output of triangular (sawtooth) waveform. We are interested in determining its RMS value for one full cycle.[21] We can see from Fig. 14-28a and Eq. (14-19) that the RMS value of one half-cycle of a triangular wave, taken for one half-cycle (from 0 to π), is $A_m/\sqrt{3}$. Then, using Eq. (14-17), we may determine the full-cycle RMS value from

$$A_{\Delta HWR} = \sqrt{\frac{(A_m/\sqrt{3})^2 \times \pi + 0 \times \pi}{2\pi}} = \sqrt{\frac{A_m^2}{(2 \times 3)}} = \frac{A_m}{\sqrt{6}} \tag{14-20}$$

This value, obtained by the equivalent rectangular RMS value method, agrees with the calculus derivation of Appendix B-8.4. It also shows the power of the method using Eq. (14-17) as a means of finding RMS values of nonsinusoidal (and sinusoidal) waveforms. (See Table 14-1 for a summary of triangular waveforms.)

Figure 14-28c shows a negative-going sawtooth waveform. Obviously, from Eq. (14-8) the average value of this waveform, taken for one half-cycle from 0 to π, is $-A_m/2$. But the half-cycle RMS value is *always positive* and equal to

[19] As noted, only the rectangular waveform has the same RMS and average values.
[20] See Appendix B-8.3 for a calculus derivation of the RMS value of the triangular waveform.
[21] See Appendix B-8.4 for a calculus derivation of the RMS value of one full cycle of HWR output.

$$A_{\Delta\text{Hrms}} = \frac{A_m}{\sqrt{3}} \qquad \text{(14-19a)}$$

This agrees with the derivation given in Appendix B-8.3 from calculus and with our observation from Fig. 14-28a that the half-cycle and full-cycle RMS values are the same. (See waveform summary in Table 14-1.)

14-10.3 RMS Values of Symmetrical Waveforms Containing a DC Component

The reader may have noticed that Figs. 14-28b and 14-28c were qualified by the parenthetic statement "(ac and dc components)." Unlike the waveform of Fig. 14-28a, whose average value is zero, these waveforms also contain a dc component since the area below the baseline is not equal to the area above it. We will return to these waveforms in Exs. 14-29 and 14-30. But first let us consider the waveform shown in **Fig. 14-29a**. Clearly, the waveform is a sawtooth (triangular) wave containing a dc component. But since we can account for it with Eq. (14-11), we can also find the ac component with Eq. (14-19), to find a rectangular equivalent. The "true" RMS value of the dc and ac component, A, is

$$A = \sqrt{A_{dc}^2 + A_{ac}^2} \qquad \text{(14-21)}$$

where A_{dc} is the amplitude of the dc component that causes A_{ac} to have a zero average value

A_{ac} is the equivalent rectangular ac amplitude of the **symmetrical** ac waveform whose average value is zero

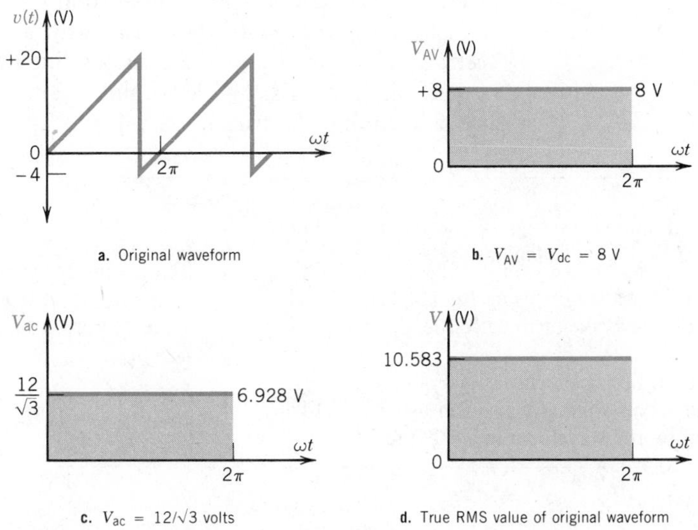

Figure 14-29 RMS value of symmetrical waveform containing a dc component; Example 14-28

EXAMPLE 14-28

Given the waveform and values shown in Fig. 14-29a applied across a 100 Ω resistor, R, calculate the

a. dc component of the waveform
b. ac component of the waveform
c. True RMS value of the waveform
d. Power dissipated in R

Solution

a. The dc voltage, $V_{dc} = \dfrac{+20 + (-4)}{2} = \mathbf{8\ V}$ (14-11)

b. If we **remove** the dc component of 8 V and shift the waveform *down* by 8 V we obtain a pure sawtooth wave whose maximum amplitude is ± 12 V, *having an average value of zero.* Then the ac component is

$$V_{ac} = V_m/\sqrt{3} = 12/\sqrt{3} = \mathbf{6.928\ V} \quad (14\text{-}19)$$

c. The true RMS voltage is

$$V = \sqrt{(8)^2 + (6.928)^2} = \mathbf{10.583\ V} \quad (14\text{-}21)$$

d. $P = V^2/R = (10.583)^2/100 = \mathbf{1.12\ W}$

Table 14-1 summarizes the average values for a full cycle and a half-cycle and the full-cycle RMS values of several common waveforms. It is not an exhaustive list, but it should go a long way toward simplifying many of the problems encountered at the end of this chapter as well as in the laboratory and in industry.

EXAMPLE 14-29

The waveform in Fig. 14-28b has a maximum amplitude of 10 V. Calculate the

a. Average value of the waveform
b. RMS value of the waveform
c. True RMS value of the waveform
d. Power dissipated by the waveform connected across a 10 Ω resistor, R

Solution

a. We cannot use Eq. (14-11) because the waveform is *not* symmetrical above and below the baseline. But we know from previous examples and from Table 14-1 that the full-cycle average of a HWR triangular waveform is

$$V_{dc} = V_{av} = V_m/4 = 10\ \text{V}/4 = \mathbf{2.5\ V}$$

b. $V_{ac} = V_m/\sqrt{6} = 10\ \text{V}/\sqrt{6} = \mathbf{4.0825\ V}$ (14-20)

c. The true RMS value of the waveform (both dc and ac components) is also **4.0825 V** (since we did *not* remove the dc component when using Eq. 14-20! We shall prove this in Ex. 14-30).

d. $P = V^2/R = (4.0825\ \text{V})^2/10\ \Omega = \mathbf{1.\bar{6}\ W}$ (since only the RMS value can be used)

Example 14-29 contains an important insight regarding the application and use of Eq. (14-21). *It should only be used* with waveforms whose ac component is *symmetrical above and below* the zero baseline *whenever the dc component is removed.*

The solution of Ex. 14-29**c** is validated in Ex. 14-30, where two *different* methods are used to find the true RMS value.

EXAMPLE 14-30

The original waveform shown in Fig. 14-28c has a maximum negative amplitude of -20 mA. Calculate the

a. dc component of the waveform using Eq. (14-11)
b. ac component of the waveform with the dc component removed
c. True RMS value of the waveform using Eq. (14-21)
d. True RMS value of the original waveform using Eq. (14-19)
e. Power dissipated by the waveform in a 1 kΩ resistor

Solution

a. $I_{av} = I_{dc} = [0 + (-20\ \text{mA})]/2 = \mathbf{-10\ mA}$ (14-11)

b. Removing the dc component of -10 mA, leaves a triangular wave of ± 10 mA. Then $I_{ac} = 10\ \text{mA}/\sqrt{3} = \mathbf{5.7735\ mA}$, RMS value of pure triangular waveform (no dc).

c. $I = \sqrt{(I_{dc}^2 + I_{ac}^2)} = \sqrt{(-10)^2 + (5.7735)^2} = \mathbf{11.55\ mA}$

d. $I = I_m/\sqrt{3} = 20\ \text{mA}/\sqrt{3} = \mathbf{11.55\ mA}$ (RMS value found *directly*)

e. $P = I^2R = (11.55\ \text{mA})^2 \times (1\ \text{k}\Omega) = \mathbf{133.\bar{3}\ mW}$

Example 14-30 proves conclusively that it is unnecessary to remove the dc component to find the true RMS value, given an equation that includes it. For example, in Table 14-1 the HWR and FWR waveforms shown in columns 2 and 3 all contain dc components, whose average values are shown in columns 4 and 5. But the true

Table 14-1 RMS and Average Values of Various Common Waveforms

	Waveform (Pure)	HWR Waveform	FWR Waveform	V_{av} (full-cycle average)	V_{Hav} (half-cycle average)	V (true RMS)	F_f (V/V_{av})	F_c (V_m/V)
Sine Wave	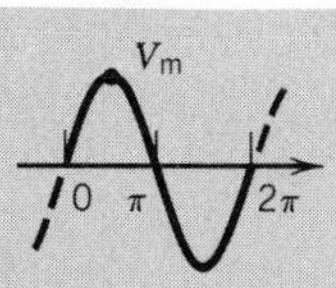			0	$\frac{2}{\pi} V_m$	$\frac{V_m}{\sqrt{2}}$	$\frac{\pi}{2\sqrt{2}}$	$\sqrt{2}$
		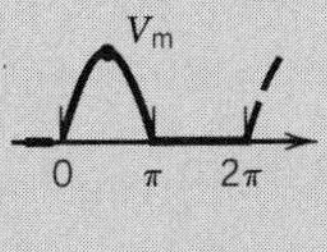		$\frac{1}{\pi} V_m$	$\frac{2}{\pi} V_m$	$\frac{V_m}{2}$	$\frac{\pi}{2}$	2
			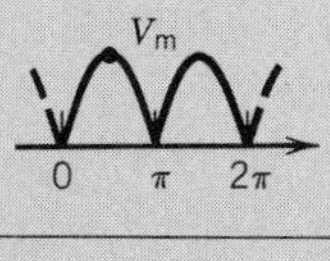	$\frac{2}{\pi} V_m$	$\frac{2}{\pi} V_m$	$\frac{V_m}{\sqrt{2}}$	$\frac{\pi}{2\sqrt{2}}$	$\sqrt{2}$
Square Wave	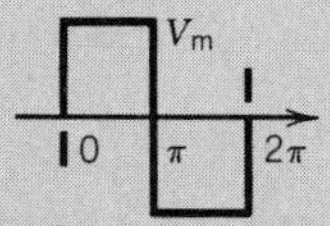			0	V_m	V_m	1	1
		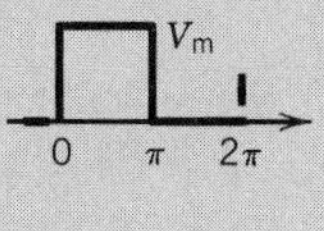		$\frac{V_m}{2}$	V_m	$\frac{V_m}{\sqrt{2}}$	$\sqrt{2}$	$\sqrt{2}$
			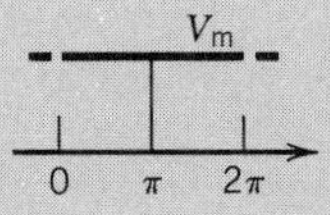	V_m	V_m	V_m	1	1
Triangular Wave	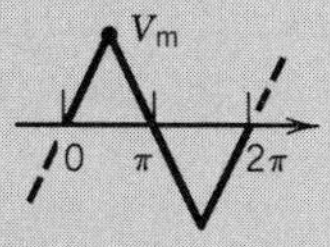			0	$\frac{V_m}{2}$	$\frac{V_m}{\sqrt{3}}$	$\frac{2}{\sqrt{3}}$	$\sqrt{3}$
		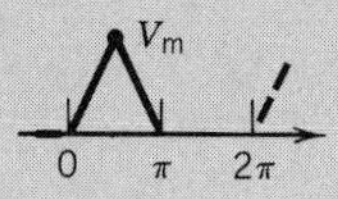		$\frac{V_m}{4}$	$\frac{V_m}{2}$	$\frac{V_m}{\sqrt{6}}$	$\frac{4}{\sqrt{6}}$	$\sqrt{6}$
			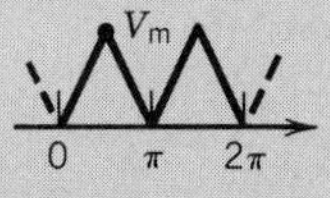	$\frac{V_m}{2}$	$\frac{V_m}{2}$	$\frac{V_m}{\sqrt{3}}$	$\frac{2}{\sqrt{3}}$	$\sqrt{3}$
DC + AC	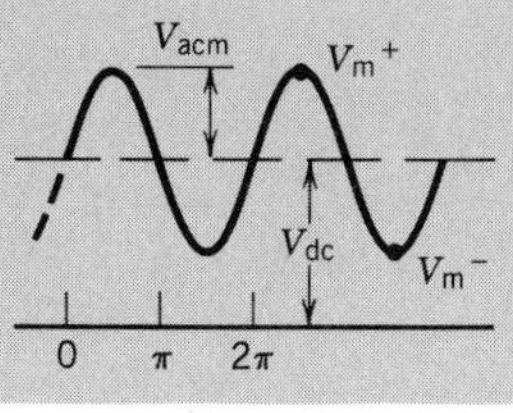	$V_{dc} = \frac{V_m^{(+)} + V_m^{(-)}}{2}$ $V_{ac} = \frac{V_{acm}}{\sqrt{2}}$		$\frac{V_m^+ + V_m^-}{2}$	—	$\sqrt{V_{dc}^2 + \frac{V_{acm}^2}{2}}$	$\frac{V}{V_{dc}}$	$\frac{V_m^+}{V}$

RMS values of these waveforms, as shown in column 6, already *include* the dc component value and may be used directly, as shown in Ex. 14-30**d**. Measurement of these waveforms by using a true RMS voltmeter in combination with a CRO (to record V_m) will verify the values in column 6.[22]

14-10.4 RMS Values of One Half-Cycle and Full Cycle of the HWR Output of an SCR

In Sec. 14-8.2 we determined the half-cycle average values, A_{hav}, of the SCR. We discovered that these values could be used to find average values for either HWR or FWR combinations of SCR output.

Appendix B-7.9 derives the equation for the RMS value of one half-cycle (of the HWR output) of an SCR as

$$A_{Hrms} = \frac{A_m}{\sqrt{2}}\sqrt{1 - \frac{\phi}{180^\circ} + \frac{\sin(2\phi)}{2\pi}} \tag{14-22}$$

Table 14-2 is similar to Table 14-1 for the SCR waveforms in either the HWR or FWR combinations. A very careful examination of Table 14-2 shows the following insights:

1. The equations of V_{Hav} are identical for both the HWR and FWR.
2. The equations of V_{Hrms} are identical for both the HWR and FWR.

Table 14-2 Equations for Determining DC and AC Components of Half-Wave and Full-Wave SCR Output Waveforms

SCR Waveforms	$V_{av} = V_{dc}$ (full-cycle average)	V_{Hav} (half-cycle average)	V (full-cycle RMS)	V_{HRMS} (half-cycle RMS)
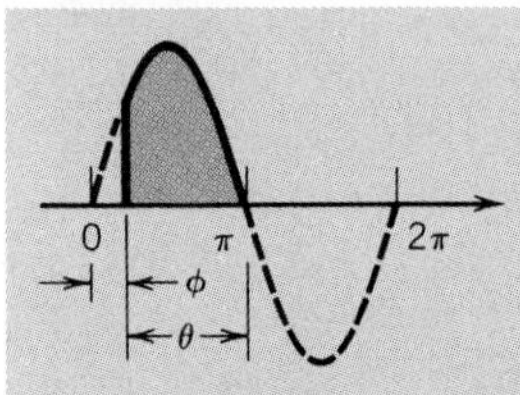 a. HWR, $\theta = 180 - \phi$	$\frac{V_m}{2\pi}(\cos\phi + 1)$	$\frac{V_m}{\pi}(\cos\phi + 1)$	$\frac{V_m}{2}\sqrt{1 - \frac{\phi}{180^\circ} + \frac{\sin 2\phi}{2\pi}}$	$\frac{V_m}{\sqrt{2}}\sqrt{1 - \frac{\phi}{180^\circ} + \frac{\sin 2\phi}{2\pi}}$
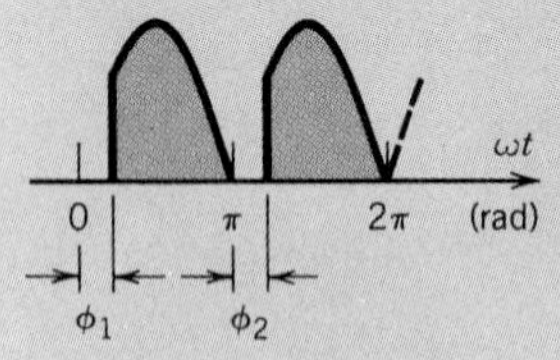 b. FWR, $\phi_1 = \phi_2 = \phi$	$\frac{V_m}{\pi}(\cos\phi + 1)$	$\frac{V_m}{\pi}(\cos\phi + 1)$	$\frac{V_m}{\sqrt{2}}\sqrt{1 - \frac{\phi}{180^\circ} + \frac{\sin 2\phi}{2\pi}}$	$\frac{V_m}{\sqrt{2}}\sqrt{1 - \frac{\phi}{180^\circ} + \frac{\sin 2\phi}{2\pi}}$

[22] Only a true-reading RMS voltmeter may accurately and *directly* measure the "true" RMS value of a nonsinusoidal waveform containing dc and ac (nonsinusoidal) components. Conventional ac electronic voltmeters (called average-responding voltmeters), which rectify the input waveform and use dc voltmeters calibrated for a sinusoidal input, will produce serious measurement errors when attempting to measure nonsinusoidal waveforms. True-reading RMS instruments may contain thermocouples or other devices to respond to the heating effect of the waveform. See Ex. 14-34 for an indirect method of RMS measurement with conventional average-value instruments.

3. Consequently, only *one* equation is neeeded for average values and RMS values, respectively (although several are shown for convenience).
4. True RMS values (for the waveforms shown in Table 14-2) are given because the dc components have *not* been removed.
5. In the event that ϕ_1 is *not* equal to ϕ_2 (see Exs. 14-25 and 14-31), then we must use V_{Hav} and V_{Hrms} to find each average and each RMS value separately and sum them separately.

EXAMPLE 14-31

Given the waveform of Ex. 14-25 (Fig. 14-24a) for two SCRs in FWR combination with one triggered at 40° and the other at 50°, calculate the

a. Equivalent rectangular ac component for the first half-cycle due to ϕ_1
b. Equivalent rectangular ac component for the second half-cycle due to ϕ_2
c. Equivalent value of V taken for one complete cycle from 0 to 2π
d. Power dissipated in a 100 Ω resistor connected across the FWR

Solution

a. $$V_{\text{Hrms}_1} = \frac{V_m}{\sqrt{2}}\sqrt{1 - \frac{\phi_1}{180^\circ} + \frac{\sin 2\phi_1}{2\pi}} = \frac{170\text{ V}}{\sqrt{2}}\sqrt{1 - \frac{40^\circ}{180^\circ} + \frac{\sin 2 \times 40^\circ}{2\pi}} = \frac{170}{\sqrt{2}}\sqrt{0.9345} = \mathbf{116.2\ V} \qquad (14\text{-}22)$$

b. $$V_{\text{Hrms}_2} = \frac{V_m}{\sqrt{2}}\sqrt{1 - \frac{\phi_2}{180^\circ} + \frac{\sin 2\phi_2}{2\pi}} = \frac{170\text{ V}}{\sqrt{2}}\sqrt{1 - \frac{50^\circ}{180^\circ} + \frac{\sin 2 \times 50^\circ}{2\pi}} = \frac{170\text{ V}}{\sqrt{2}}\sqrt{0.87896} = \mathbf{112.7\ V} \qquad (14\text{-}22)$$

c. Then using Eq. 14-17 to sum each half-cycle rectangular equivalent

$$V = \sqrt{\frac{(V_{\text{Hrms}_1})^2 \times \pi + (V_{\text{Hrms}_2})^2 \times \pi}{2\pi}} = \sqrt{\frac{(116.2)^2\pi + (112.7)^2\pi}{2\pi}} = \mathbf{114.46\ V} \qquad (14\text{-}17)$$

d. $P = V^2/R = (114.46)^2/100\ \Omega = \mathbf{131\ W}$

14-10.5 RMS Values Given the Equation of the Waveform

In Sec. 14-8.3 we showed that, *given the equation* for a *nonsinusoidal* waveform, it is possible to

1. Plot the waveform by appropriate substitutions of time.
2. Determine its average values by simple integration techniques.

Let us take the waveform given in Ex. 14-26 which was plotted in Fig. 14-25, and use simple integration to show that it is also possible to determine the RMS value.

EXAMPLE 14-32

Given the waveform of Ex. 14-26 whose equation is $v(t) = 50(t - 0.03)^2$ kV, as plotted in Fig. 14-25, determine the

a. RMS value of the waveform using simple integration
b. Approximate RMS value of the waveform using the triangular approximation

Solution

From Appendix B-7.6, the general equation for the RMS value of any waveform[23] is $A_{\text{rms}} = \sqrt{\frac{1}{T}\int_0^t [a(t)]^2\,dt}$ and squaring each side and expressing the general equation in terms of voltage yields

a. $V_{\text{rms}}^2 = \frac{1}{T}\int_0^t [v(t)]^2\,dt$ where $v(t) = 50(t - 0.03)^2$ kV, $t =$ 30 ms, and $T = 50$ ms, as given in Ex. 14-26. Substituting yields

$$V_{\text{rms}}^2 = \frac{V_k^2}{T}\int_0^t [(t - 0.03)^2]^2\,dt$$

[23] See also Eq. (14-14) for the general equation of RMS current.

$$V_{rms}^2 = \frac{(50\text{ kV})^2}{50\text{ ms}} \int_0^{30\text{ ms}} (t-0.03)^4\,dt$$

$$= \frac{5\times 10^{10}}{5}\left[(t-0.03)^5\right]_0^{30\text{ ms}}$$

$$V_{rms}^2 = (1\times 10^{10})[0-(-0.03)^5]$$

$$= (1\times 10^{10})(2.43\times 10^{-8}) = 243(\text{volts})^2$$

Taking the square root of each side yields

$$V_{rms} = V = \mathbf{15.59\ V}$$

b. Figure 14-25 shows that the waveform is approximately triangular from 0 to 25 ms. Using the triangular approximation $V_\Delta = V_m/\sqrt{3} = 45\text{ V}/\sqrt{3} = 25.98\text{ V}$, from 0 to 25 ms. We may now use the rectangular equivalent to find the approximate RMS voltage from 0 to 50 ms, from Eq. (14-17) as

$$V_{rms} = V = \sqrt{\frac{(25.98)^2\times 25\text{ ms} + 0^2\times 25\text{ ms}}{50\text{ ms}}}$$

$$= \mathbf{18.37\ V} \qquad (14\text{-}17)$$

We would expect the triangular assumption to be slightly higher (as was the case for the average value in Ex. 14-26) because the actual waveform is exponential rather than linear. Thus the actual area under the exponential curve is reduced, reducing the approximate RMS from 18.37 V to the actual RMS value of 15.59 V. This shows that the true RMS value obtained by integration is correct and reasonable.

Table 14-3 shows how average and RMS values of exponentially rising and decaying waveforms are found, given the equation for the waveform. Note that for average values, we begin with the general equation $A_{av} = \int_0^t \frac{a(t)}{T}\,dt$. For RMS values, the general equations shown were derived in Appendix B-8 from $A_{RMS} = \sqrt{\frac{1}{T}\int_0^t [a(t)]^2\,dt}$.

Table 14-3 Equations for Determining Average and RMS Values of Exponentially Rising and Exponentially Falling Waveforms

EXPONENTIAL WAVEFORMS	AVERAGE VALUE	RMS VALUE
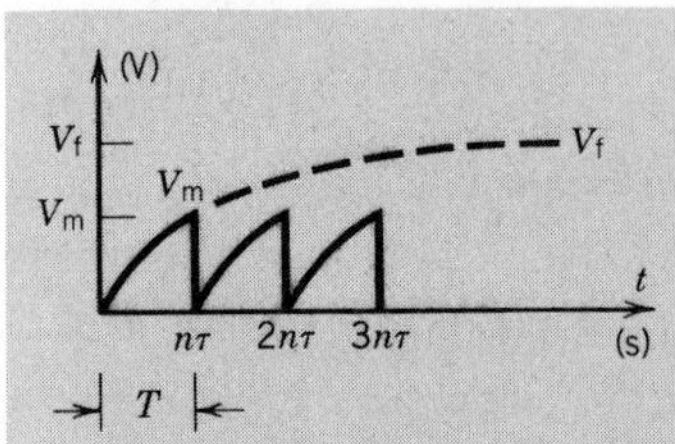	$A_{av} = \int_0^t a(t)\,dt$ $V_{av} = \frac{V_f}{(n\tau)}\left[t + \tau\varepsilon^{-t/\tau}\right]_0^{n\tau}$	$A_{rms} = \sqrt{\frac{1}{T}\int_0^t [a(t)]^2\,dt}$ $V_{rms}^2 = \frac{V_f^2}{n\tau}\left[t + 2\tau\varepsilon^{-t/\tau} - \frac{\tau}{2}\varepsilon^{-2t/\tau}\right]_0^{n\tau}$
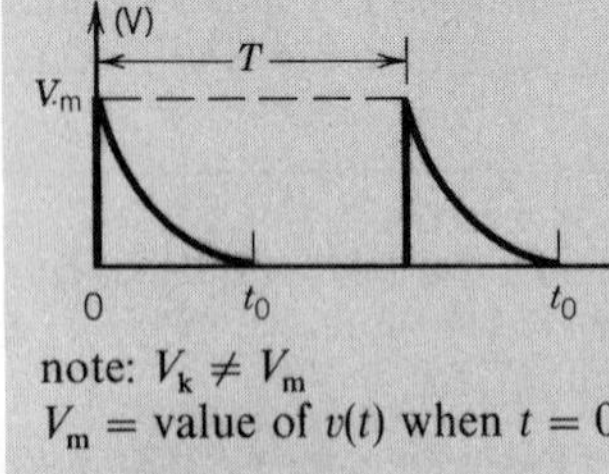	$V_{av} = \frac{V_k}{T}\int_0^{t_0} (t-t_0)^x\,dt$ $V_{av} = \frac{V_k}{xT}\left[(t-t_0)^{x+1}\right]_0^{t_0}$ volts	$V_{rms}^2 = \frac{V_k^2}{T}\int_0^{t_0} [(t-t_0)^x]^2\,dt$ $V_{rms}^2 = \frac{V_k^2}{2x+1}\left[(t-t_0)^{2x+1}\right]_0^{t_0}$ $(\text{volts})^2$
note: $V_k \neq V_m$ V_m = value of $v(t)$ when $t = 0$ in general equation $v(t) = V_k(t-t_0)^x$ volts		
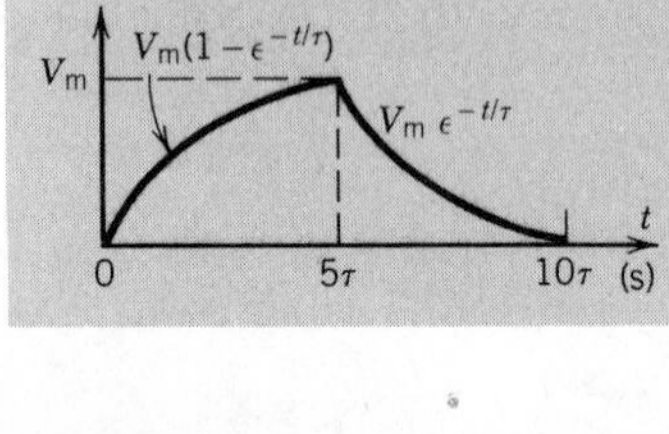	$V_{av} = \frac{V_m}{2}$	$V_{rms} = \frac{V_m}{\sqrt{2}}$

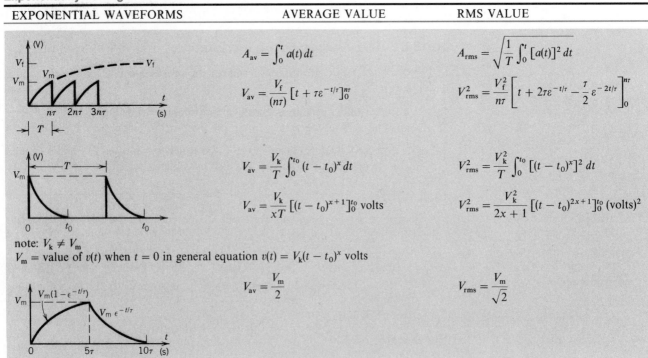

The general equations derived in Table 14-3 *eliminate the need for calculus* in the integration of exponential functions. The reader will discover that with a little practice in the solution of appropriate problems at the end of this chapter, finding the average and RMS values is not difficult. In all cases, however, triangular approximations should be used (as shown in Exs. 14-26 and 14-32) to determine whether the solutions obtained are *reasonable.* Given the equation for the waveform, moreover, the reader should always make a plot of the waveform as an aid in the solution.

14-11 PROPERTIES OF WAVEFORMS

In using instruments to measure sinusoidal and nonsinusoidal waveforms, a number of factors enable us to determine whether the instruments will satisfactorily and accurately measure the waveforms. These are the form factor, F_f, and the crest factor, F_c.

14-11.1 Form Factor, F_f

The form factor of any alternating periodic waveform is defined as the ratio of its RMS value to its half-cycle average value (or average absolute value) taken over a full period of the function. Stated as an equation, the form factor is

$$F_f = \frac{A_{RMS}}{A_{av}} \quad \text{(dimensionless ratio)} \qquad (14\text{-}23)$$

where A_{RMS} is the RMS amplitude of the alternating periodic waveform taken for one full cycle of variation

A_{av} is the half-cycle average amplitude (or average absolute value taken for one full cycle of variation)

Note: The average value of a pure sinusoid (or any symmetrical waveform not containing a dc component) taken for a full cycle is zero.

Column 7 of Table 14-1 shows the values of the form factors for the 10 waveforms illustrated. Each of these values was found by using Eq. (14-23) and the columns 5 and 6 of Table 14-1.

EXAMPLE 14-33

A full-wave rectifier-type ac/dc voltmeter uses a dc voltmeter to record the ac input voltage, applied to the FWR. When a pure sine wave having a maximum value of 170 V is applied to the ac voltmeter, calculate the

a. RMS value
b. Half-cycle average value
c. Form factor, F_f, using (**a**) and (**b**)
d. Form factor using the ratio shown in Table 14-1
e. Ratio of the ac scale to the dc scale of the voltmeter
f. Explain the limitations of the above instrument in measuring square and triangular nonsinusoidal waveforms

Solution

a. $V = V_m/\sqrt{2} = 170\text{ V}/\sqrt{2} = \mathbf{120.2\ V}$
b. $V_{Hav} = 2V_m/\pi = 2 \times 170/\pi = \mathbf{108.225\ V}$
c. $F_f = V/V_{Hav} = 120.2\text{ V}/108.225\text{ V} = \mathbf{1.111}$
d. $F_f = V/V_{av} = \dfrac{\pi}{2\sqrt{2}} = \mathbf{1.111}$
e. Each point on the ac scale is **1.111** times the average values shown on the dc scale
f. The instrument can only measure sinusoidal ac waveforms shown in the first and third **rows** of Table 14-1 only. Depending on the waveform, all others shown in Table 14-1 (col. 7) are *erroneously measured*, either higher or lower, on the **ac** scales.

EXAMPLE 14-34

John Smith, a bright student, decides to overcome the limitations of his average-reading dc/ac voltmeter. He uses the **ac** input terminals to record nonsinusoidal periodic waveforms but reads the rectified (FW) average voltage values on the **dc** scale! He then multiplies each of the average values by the form factor, F_f, of the waveform of interest he observes on the CRO. If the FW (rectified) **dc** value of the ac input (in all cases) is 100 V, calculate the

a. RMS value of a pure triangular sawtooth wave
b. RMS value of a pure square wave
c. RMS value of the output of a half-wave rectifier (row 2, Table 14-1)

Solution

a. $V_{RMS} = F_f \times V_{av} = \frac{2}{\sqrt{3}} \times 100 \text{ V}$
$= \mathbf{115.5\ V}$ (pure triangular wave) (14-23)

b. $V = F_f \times V_{av} = 1 \times 100 \text{ V}$
$= \mathbf{100\ V}$ (pure square wave) (14-23)

c. $V = F_f \times V_{av} = \frac{\pi}{2} \times 100 \text{ V}$
$= \mathbf{157.1\ V}$ (HWR output) (14-23)

Example 14-34 shows that it *is* possible to measure nonsinusoidal waveforms on conventional dc/ac voltmeters without having to use "true-reading" RMS instruments, **provided** (1) the form factor of the waveform measured is known and (2) the measurement is made with the ac input terminals BUT the **reading** (of the FWR average value) is taken from the **dc scale** of the instrument. (But even this method has its limitations, as shown by Prob. 14-50.)

14-11.2 Crest Factor, F_c

The *crest* or amplitude factor of any alternating periodic waveform is defined as the ratio of the highest peak maximum value (either positive or negative) to the RMS value of one cycle of an alternating periodic waveform. Stated in equation form, the crest factor is

$$F_c = \frac{A_m}{A_{RMS}} \quad \text{(dimensionless ratio)} \qquad (14\text{-}24)$$

where A_m is the maximum positive or negative amplitude with respect to the zero reference

A_{RMS} is the RMS amplitude of the periodic waveform taken over one full cycle of variation

The crest factor serves two purposes:

1. It provides a measure of the approximate waveshape of the waveform.
2. It provides an indication of the RMS voltages that a true RMS voltmeter will accept without overloading to the full-scale value of the range being used for measurement.

The last column of Table 14-1 shows the crest factor, F_c, of a number of common waveforms. Note that the triangular HWR waveform (row 8) has a crest factor of $\sqrt{6}$ or 2.45, which is higher than all the others. Short-duration, high-spike pulses may have crest factors of 10 or more. In measuring the true RMS of such pulses, care should be taken not to overload the instrument input, which causes measurement errors.

It should be noted that **both** the form factor, F_f, and the crest factor, F_c, may be *negative* for *negative-going* waveforms, as shown in Ex. 14-35.

EXAMPLE 14-35

An exponential periodic waveform has the equation $v(t) = 25(t - 2)^3$ volts over the duration $0 \lesssim t \lesssim 2$ s and $v(t) = 0$ over the duration $2 \lesssim t \lesssim 4$ s. Thereafter the waveform is periodic, as expressed above

a. Plot the waveform by substituting values of t, and calculate for the waveform
b. V_{av}
c. V_{RMS}
d. Form factor, F_f
e. Crest factor, F_c

Solution

a.

t (s)	0	0.5	1	1.5	2	3	4	
$v(t)$ (V)	−200	−84.4	−25	−3.13	0	0	0/−200	etc.

The waveform is shown plotted in **Fig. 14-30**. Note that $V_m = -200$ V and that the waveform is *negative-going*.

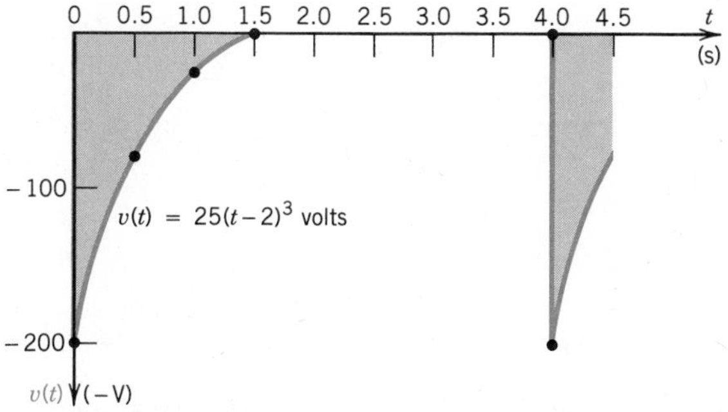

Figure 14-30 Plot of waveform in Ex. 14-35**a**

b. $$V_{av} = \frac{1}{T}\int_0^t v(t)\,dt = \frac{25\text{ V}}{4\text{ s}}\int_0^2 (t-2)^3\,dt$$
$$= \frac{6.25\text{ V}}{4}\left[\frac{1}{4}(t-2)^4\right]_0^2 = 1.5625\text{ V}[0^4 - (-2)^4]$$
$$= 1.5625\text{ V} \times (-16) = \mathbf{-25\text{ V}}$$

c. $$V_{rms}^2 = V^2 = \frac{1}{T}\int_0^t [v(t)]^2\,dt = \frac{(25\text{ V})^2}{4\text{ s}}\int_0^2 [(t-2)^3]^2\,dt$$
$$= \frac{156.25}{7}[(t-2)^7]_0^2 = 22.32[0^7 - (-2)^7]$$
$$= 22.32 \times (+128) = +2857.1\text{ (volts)}^2$$
$$V = \sqrt{2857.1\text{ V}^2} = \mathbf{53.45\text{ V}}$$

d. $$F_f = V/V_{av} = \frac{53.45\text{ V}}{-25\text{ V}} = \mathbf{-2.138}$$

e. $$F_c = V_m/V = \frac{-200\text{ V}}{53.45\text{ V}} = \mathbf{-3.742}$$

The solution of Ex. 14-35 leads to the following insights:

1. When the waveform has a *negative average* value, **both** the crest factor and form factor are **negative**.
2. If the exponential power is an **odd** number in the given equation for $v(t)$, the waveform is *negative-going* and its average value (form and crest factors) must be negative.
3. If the exponential power is an **even** number in the equation for $v(t)$ (as shown in Exs. 14-26 and 14-32), the average value is **positive** and the waveform is a *positive-going* pulse. The crest and form factors are both positive.
4. Although not proved here, the higher the exponent in the equation for $v(t)$, the smaller the average and RMS values and the higher the crest and form factors, all other equation factors remaining the same.

14-12 DERIVATIVES AND INTEGRALS OF SINUSOIDS

Let us now return to our fundamental sine wave, whose equation is

$$v(t) = V_m \sin\theta \qquad \text{volts (V)} \tag{14-4}$$

Suppose we are interested in finding the instantaneous *rate of change* of voltage at any specific angle of rotation, θ. An examination of Fig. 14-5, reproduced once again in **Fig. 14-31a**, leads to the following conclusions:

1. The rate of change of voltage, dv/dt, is zero when θ is $\pi/2$ and $3\pi/2$ radians.
2. dv/dt is a *maximum positive* rate of change at 0 radians and a *maximum negative* rate of change at π radians.
3. For any other values of θ, a tangent slope drawn to the instantaneous value of $v(t)$ should yield positive or negative values greater than zero but less than the maxima in (2).

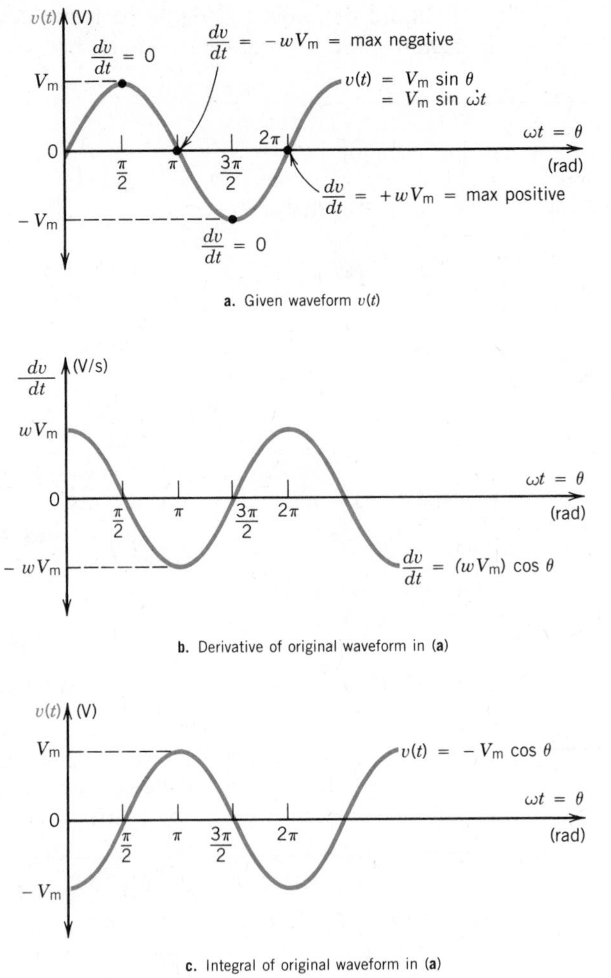

Figure 14-31 Derivative and integral of a sine wave

Since the rate of change of voltage with respect to time is dv/dt, we should be able to determine these rates of change by differentiating our original Eq. (14-4). If we can obtain the derivative of $V_m \sin\theta$, then by substituting values of θ, we should be able to find any rate of change at any given instant.

The derivation in Appendix B-9.1 shows the (interesting and simple) result that

The derivative of the sine of an angle is equal to the cosine of that angle times the derivative of the angle.

Applying Eq. (B-9.1) to Eq. (14-4) yields

$$\frac{dv}{dt} = \frac{d(V_m \sin\theta)}{dt} = \frac{V_m\, d\sin\theta}{dt} = V_m \cos\theta \frac{d\theta}{dt} = \omega V_m \cos\theta \qquad \left(\frac{\text{V}}{\text{s}}\right) \qquad (14\text{-}25)$$

where $\omega = d\theta/dt$ is the angular velocity in radians/second (or s^{-1})

V_m is the original maximum value of the waveform

θ is any rotation angle or the *sum* of the phase and rotation angle (Eq. 14-6) in radians or degrees

The waveform of Eq. (14-25) is shown as the derivative of the original waveform in Fig. 14-31b. Note that the first two conclusions above are verified:

1. dv/dt is zero when θ is $\pi/2$ and $3\pi/2$.
2. dv/dt is maximum positive at 0 radians and maximum negative at π radians.

EXAMPLE 14-36

A sinusoidal waveform has a maximum value of 100 V and a frequency of 1000 Hz. Calculate the rates of change of voltage at rotation angles of

a. 0° d. 180°
b. 45° e. 260°
c. 90° f. 310°

We begin by calculating ω in rad/s or s^{-1}. We do *not* convert ω to deg/s.[24]

$$\omega = 2\pi f = 2\pi \times 1000 \text{ Hz} = 2000\pi \text{ rad/s} = 2000\pi \text{ s}^{-1}$$

a. $$\frac{dv}{dt} = \omega V_m \cos\theta = (2000\pi \text{ s}^{-1})(100 \text{ V})\cos 0° = 200\,000\pi \text{ V/s}$$
$$= \mathbf{6.283 \times 10^5 \text{ V/s}} \text{ when } \theta = 0°$$

b. $$\frac{dv}{dt} = \omega V_m \cos\theta = (200\,000\pi \text{ V/s})(\cos 45°)$$
$$= \mathbf{4.443 \times 10^5 \text{ V/s}}, \text{ when } \theta = 45°$$

c. $$\frac{dv}{dt} = \omega V_m \cos\theta = (200\,000\pi)(\cos 90°)$$
$$= \mathbf{0 \text{ V/s}}, \text{ when } \theta = 90°$$

d. $$\frac{dv}{dt} = \omega V_m \cos\theta = (200\,000\pi)(\cos 180°)$$
$$= \mathbf{-6.283 \times 10^5 \text{ V/s}}, \text{ when } \theta = 180°$$

e. $$\frac{dv}{dt} = \omega V_m \cos\theta = (200\,000\pi)(\cos 260°)$$
$$= \mathbf{-1.091 \times 10^5 \text{ V/s}}, \text{ when } \theta = 260°$$

f. $$\frac{dv}{dt} = \omega V_m \cos\theta = (200\,000\pi)(\cos 310°)$$
$$= \mathbf{4.039 \times 10^5 \text{ V/s}}, \text{ when } \theta = 310°$$

Example 14-36 shows that the *maximum positive* rate of change does occur at a rotation angle of 0° and the *maximum negative* rate of change occurs at 180°. A negative rate of change occurs when $\theta = 260°$ that is somewhat less than the negative maximum at 180° but not quite zero. Finally, note that at 310° the rate of change is positive but not quite the value that would occur at 360° (or 0°).

A question now arises concerning sinusoids that do *not* begin at 0°. How do we find their rates of change and at what angles would their maxima or minima occur?

EXAMPLE 14-37

The current in a 20-Ω resistor is $i(t) = 12 \sin(300\pi t - 40°)$ A. Calculate or identify the

a. Peak current, I_m
b. Angular velocity, ω, in rad/s
c. Phase angle of current with respect to reference zero sine wave, ϕ
d. Frequency, f, and period of waveform, T
e. Sketch the waveform versus rotation angle, θ, and time, t, in ms
f. Maximum rate of change of current and $i(t)$ at di/dt_{max}
g. Rotation angle, θ, and time at which (f) occurs
h. Rates of change of current at $t = 2.407$ ms and 2.6 ms
i. Average value of one half-cycle of alternation, I_{Hav}
j. RMS value of one full cycle of alternation, I
k. Power dissipated in 20-Ω resistor

Solution

a. By inspection, $I_m = \mathbf{12 \text{ A}}$
b. By inspection, $\omega = \mathbf{300\pi \text{ rad/s}}$
c. By inspection, $\phi = \mathbf{-40°}$ (note $\alpha = \theta + \phi$, where $\theta = 300\pi t$ and $\phi = -40°$)
d. $f = \omega/2\pi = 300\pi/2\pi = \mathbf{150 \text{ Hz}}$ and $T = 1/f$
$$= (1/150)\text{ s}^{-1} = \mathbf{6.\bar{6} \text{ ms}} \qquad (14\text{-}5)$$
e. See **Fig. 14-32**
f. $$\frac{di_{max}}{dt} = \omega I_m = (300\pi \text{ s}^{-1}) \times 12 \text{ A} = 3.6\pi \text{ kA/s} = \mathbf{11.31 \text{ kA/s}}$$

$i(t)$ is always **0** whenever di/dt is a maximum value for a pure sinusoid

[24] This is an error made by many students, particularly when rotation angles are given in degrees. They fail to realize that the radian, a *unitless* measure of degrees of arc subtended divided by the radius of the circle, is a ratio of two lengths. The value of ωV_m in Eq. (14-25) must be expressed in volts per second (*not* volt-degrees per second).

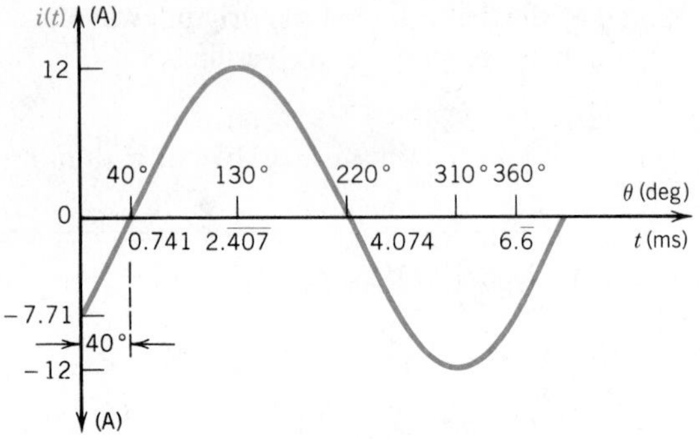

Figure 14-32 Solution to Ex. 14-37e

g. $\theta = 40°$ when di/dt is first a positive maximum (see Fig. 14-32) and $i(t) = 0$. Since $\theta = 300\pi t = 40°$, solving for $t =$

$$40° \bigg/ \left(300\pi \times \frac{180°}{\pi}\right) = \mathbf{0.74\overline{074}\ ms \cong 0.741\ ms}$$

h. $(di/dt) = \omega I_m \cos(300\pi t - 40°)$

$$= (3.6\pi\ \text{kA/s}) \cos\left(300\pi \times \frac{180°}{\pi} \times 2.\overline{407}\ \text{ms} - 40°\right)$$

$$= \mathbf{0\ A/s} \qquad (14\text{-}25)$$

$$di/dt = \omega I_m \cos(300\pi t - 40°)$$
$$= (3.6\pi\ \text{kA/s}) \cos(300 \times 180 \times 2.6\ \text{ms} - 40°)$$
$$= (3.6\pi\ \text{kA/s}) \cos 100.4° = \mathbf{-2042\ A/s}$$

i. $I_{Hav} = I_m \times 2/\pi = 12\ \text{A} \times 2/\pi = \mathbf{7.639\ A}$ (14-10)

j. $I_{rms} = I = I_m/\sqrt{2} = 12\ \text{A}/\sqrt{2} = \mathbf{8.485\ A}$ (14-14)

k. $P = I^2R = (8.485)^2 \times 20 = \mathbf{1440\ W}$ (14-16)

In a sense, Ex. 14-37 is a summary of many of the important relations presented in this chapter. Readers should attempt the solution independently, using the solution shown above as feedback to correct their errors and improve their understanding of the basic theory. Some important conclusions emerge from Ex. 14-37:

1. At all times, as shown by Eq. (14-25), the maximum rate of change of voltage, $dv_{max}/dt = \omega V_m$. Similarly, the maximum rate of change of current, $di_{max}/dt = \omega I_m$, as shown in part (**f**) of solution.
2. The above maximum rate of change relation holds true even for waveforms displaced by positive or negative phase angles ($\pm\phi$).
3. For sinusoids that are offset by some phase angle ($\pm\phi$), for θ in Eq. (14-25) we substitute the angle α (where $\alpha = \theta + \phi$). This concept was introduced in Sec. 14-5 and in Eq. (14-6).
4. In finding the time at which the maximum rate of change of current occurs, part (**g**), we used the relation $di/dt = \omega I_m \cos(300\pi t - 40°)$. But the angle α must equal 0° at the maximum rate of change of current. This automatically gives $300\pi t = 40°$, enabling the solution of t. It also tells us that the rotation angle, θ, must be 40°, as shown in Fig. 14-32.
5. Note from solution part (**h**) that the zero rate of change occurs at $t = 2.\overline{407}$ ms, but di/dt is a *negative* rate of change (−2042 kA/s) at $t = 2.6$ ms, slightly later. This is not as great as the negative maximum rate of change (−11.31 kA/s), which from part (**f**) occurs at $4.\overline{074}$ ms, as shown in Fig. 14-32.
6. As in the past, the average power dissipated in a resistor is always found from the RMS current (I^2R) or RMS voltage (V^2/R), where V is the voltage across the resistor.

14-13 RESPONSE OF PURE RESISTOR TO SINUSOIDAL VOLTAGE AND CURRENT

In Sec. 14-9 we first encountered the response of a purely resistive circuit to a sinusoidal voltage. From Fig. 14-27a, given a pure resistor, R, we may summarize:

1. Sinusoidal current, $i_R(t)$, *is always in phase* with applied sinusoidal voltage, $v_R(t)$.
2. The magnitude of sinusoidal current may be expressed by

$$i_R(t) = \frac{v_R(t)}{R} = \frac{V_m \sin \omega t}{R} = I_m \sin \omega t \qquad \text{amperes (A)} \qquad (14\text{-}26)$$

from which
$$I_m = \frac{V_m}{R} \qquad \text{amperes (A)} \qquad (14\text{-}27)$$

and
$$I_R = \frac{V_R}{R} \qquad \text{amperes (A)} \qquad (14\text{-}28)$$

where I_R and V_R are expressed in RMS values

3. The instantaneous power $p(t)$ is a *positive pulsating* sinusoidal waveform of twice the frequency of $v(t)$ and $i(t)$ and may be expressed by

$$p(t) = P_{av}[1 + \sin(2\omega t - 90°)] = \frac{I_m^2 R}{2}(1 - \cos 2\omega t) \qquad \text{watts (W)} \qquad (14\text{-}15)$$

from which, extracting the active or true or average power, P_{av}, we get

$$P = P_{av} = \frac{I_m^2 R}{2} = \frac{(\sqrt{2} I_R)^2}{2} = (I_R^2)R = I_R^2 \cdot R \qquad \text{watts (W)} \qquad (14\text{-}29)$$

and substituting for $I_R = V_R/R$ given in Eq. (14-28) yields

$$P = P_{av} = I_R^2 \cdot R = \left(\frac{V_R}{R}\right)^2 \cdot R = \frac{V_R^2}{R} \qquad \text{watts (W)} \qquad (14\text{-}30)$$

4. Observe that Eqs. (14-29) and (14-30) are identical to those for dc power relations expressed in Eq. (4-3) with the exception that V_R and I_R are RMS (rather than dc) values, equated by definition of the RMS ampere. Since resistance is the *only* circuit property capable of *dissipating* electrical energy, we would expect Eqs. (14-29) and (14-30) to be equivalent to Eq. (4-3).

14-14 RESPONSE OF PURE INDUCTOR TO SINUSOIDAL VOLTAGE AND CURRENT

We are now interested in determining the current, voltage, and power relations for a *pure* (ideal) inductor, having *zero* resistance. Intuition tells us that such a circuit has a high (long) time constant ($\tau = L/R$), dictating that when voltage is applied, current takes a long time to rise; i.e, current should lag voltage with respect to time.

Figure 14-33a shows the current, $i_L(t) = I_m \sin \omega t$, applied to an inductor, L, that is ideal.[25] Since the current is sinusoidal and continuously changing, it produces a counter EMF in proportion to the rate of change of current [i.e., $v_L = L(di/dt)$, first shown in Eq. (12-5)]. Since the inductor is pure, it is the only

[25] Current is selected as the reference since the property of an inductor is to oppose a change in current, and we are interested in finding the opposition created by the inductor, X_L.

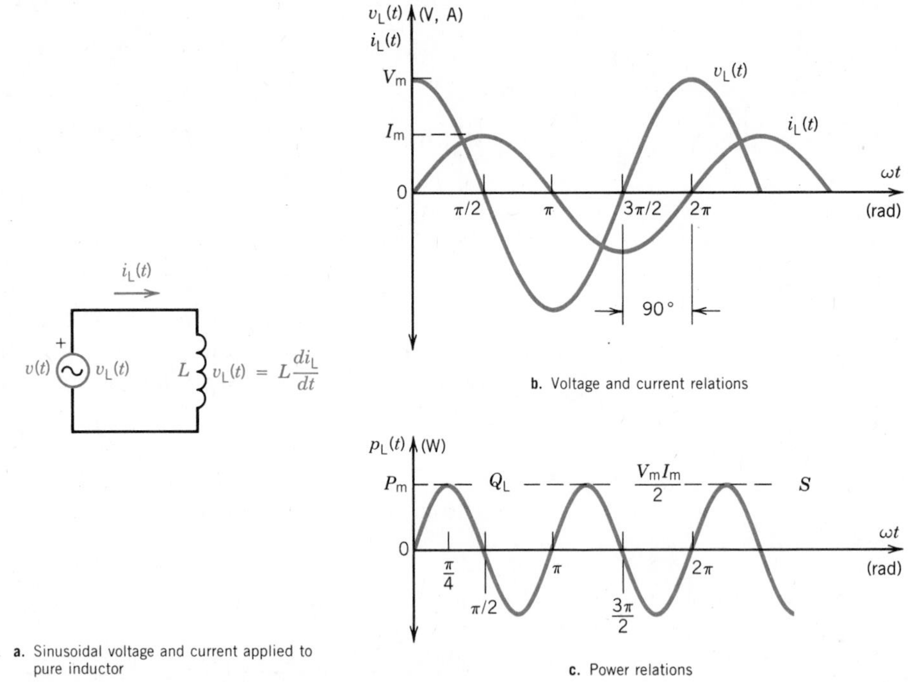

Figure 14-33 Voltage, current, and power sinusoidal relations in a pure inductor

source of opposition to the applied voltage, $v(t)$, and consequently, as shown in Fig. 14-33a, $v(t) = v_L(t)$. We learned earlier, as shown in Figs. 14-31a and 14-31b, that the *derivative* of a waveform *leads* the waveform by 90°. Since $v_L(t)$ is proportional to the *derivative* of the current, it *leads* the current by 90° as shown in Fig. 14-33b. We may also say that the current, $i_L(t)$, **lags** the voltage, $v_L(t)$, by 90°, also shown in Fig. 14-33b.

Applying this analysis to the waveforms shown in Fig. 14-33b yields the following equations:[26]

$$v_L(t) = L\frac{di}{dt} = L\frac{d(I_m \sin \omega t)}{dt} = \omega I_m L \cos \omega t = V_m \cos \omega t \quad \text{volts (V)} \tag{14-31}$$

But from Eq. (14-27) we observe that the opposition to ac created by a pure resistor is $R = V_m/I_m$. Then taking the ratio of V_m/I_m in Eq. (14-31) yields the *opposition to ac* produced by a *pure inductor*, called **inductive reactance**, X_L, where

$$X_L = \omega L = 2\pi f L = \frac{V_m}{I_m} = \frac{V_L}{I_L} \quad \text{ohms } (\Omega) \tag{14-32}$$

where f is the frequency of the sinusoidal voltage across and current in the inductor

[26] See Appendix B-12.1 for a more rigorous derivation.

L is the inductance in henrys

V_L is the RMS voltage across the inductor

I_L is the RMS current in the inductor

Note that Eq. (14-32) shows that the *higher* the frequency, the *greater* the opposition to ac created by a given inductor. We can anticipate this intuitively from two viewpoints:

1. Since the counter EMF of an inductor is proportional to the rate of change of current ($v_L = L\,di/dt$), the greater the rate of change of current, the higher the counter EMF and the greater the opposition to ac.
2. In Table 13-3 (Sec. 13-10) we showed that the VHF (very high frequency) behavior of an inductor is that it acts as an *open* circuit, implying a *very high reactance* at high frequencies.

EXAMPLE 14-38

The current in a 2-mH inductor is $i(t) = 10 \sin 500\pi t$ amperes. Calculate:

a. Inductive reactance, X_L, of the inductor
b. Instantaneous maximum voltage across the inductor, V_m
c. Two equations for the waveform of applied voltage across the inductor
d. Frequency of the voltage across the inductor

Solution

a. $X_L = \omega L = (500\pi)(2\text{ mH}) = \mathbf{3.1416\ \Omega}$ (14-32)
b. $V_m = I_m X_L = (10\text{ A})(3.1416\ \Omega) = \mathbf{31.42\ V}$ (14-32)
c. $v(t) = V_m \cos \omega t = 31.42 \cos 500\pi t$ V, or in terms of the reference current, $i(t)$
$v(t) = V_m \sin(\omega t + 90°)$
$= 31.42 \sin(500\pi t + 90°)$ volts (14-31)
d. $f = \omega/2\pi = 500\pi/2\pi = \mathbf{250\ Hz}$ (14-5)

EXAMPLE 14-39

Calculate the inductive reactance of the 2-mH coil in Ex. 14-38 at frequencies of

a. 500 Hz
b. 1000 Hz
c. 5000 Hz

Solution

Since $X_L = 2\pi f L$, it is directly proportional to frequency, for a given value of L. Consequently we may use the ratio method or

a. $X_{L_1} = \frac{f_1}{f_0} X_{L_0} = \frac{500\text{ Hz}}{250\text{ Hz}} \times 3.1416\ \Omega = \mathbf{6.283\ \Omega}$ at 500 Hz

b. $X_{L_1} = \frac{f_1}{f_0} X_{L_0} = \frac{1000\text{ Hz}}{250\text{ Hz}} \times 3.1416\ \Omega = \mathbf{12.57\ \Omega}$ at 1000 Hz

c. $X_{L_1} = \frac{f_1}{f_0} X_{L_0} = \frac{5000\text{ Hz}}{1000\text{ Hz}} \times 12.57\ \Omega = \mathbf{62.83\ \Omega}$ at 5000 Hz

Example 14-39 shows that when the frequency is doubled, X_L is doubled. When the frequency is increased by a factor of 10, the inductance is also increased by a factor of 10. Consequently, if we know the inductive reactance, X_L, at any frequency, we can find X_L immediately at any higher or lower frequency by the *ratio method* shown in Ex. 14-39.

As in the case of a pure resistor, we are also interested in the power stored in the magnetic field of an inductor and returned to the supply. Multiplying the instantaneous current, $i_L(t)$, by the instantaneous voltage, $v_L(t)$, should yield the instantaneous power, $p_L(t)$. Then by substitution

$$p_L(t) = v_L(t) \times i_L(t) = (V_m \cos \omega t)(I_m \sin \omega t)$$
$$= V_m I_m \cos\theta \sin\theta = (V_m I_m \sin 2\theta)/2$$

$$p_L(t) = \frac{V_m I_m}{2} \sin 2\omega t \qquad \text{watts (W)} \qquad (14\text{-}33)$$

Equation (14-33) is shown plotted in Fig. 14-33c. It shows that

1. The instantaneous power curve has a frequency that is *twice the frequency* of the voltage, v_L, across the inductor and the current, i_L, in the inductor.
2. The average value of the curve of $p_L(t)$ is zero; consequently, $P_{av} = 0$ and the inductor *dissipates no power whatever*.
3. The power curve of Eq. (14-33) is *in phase with the current* since *both* are positive-going at $\theta = 0$ and $\theta = 2\pi$ radians. (See Sec. 14-7.) $p_L(t) = (P_m/2) \sin 2\omega t$, as shown in Fig. 14-33 and Eq. (14-33).
4. The product $+P_m$ is reached when $\theta = \pi/4$ radians $= 45°$. The instantaneous values of

   $i(t) = I_m \sin 45° = 0.7071 I_m = I_m/\sqrt{2} = I_L$, the RMS value of current

   $v(t) = V_m \cos 45° = 0.7071 V_m = V_m/\sqrt{2} = V_L$, the RMS value of voltage

From the foregoing analysis we know that the true power, P_{av}, is zero. Consequently, we speak of the (quadrature) **reactive power**, Q_L, as the RMS product of $V_L I_L$, yielding the following equalities:

$$Q_L = \frac{V_m I_m}{2} = V_L I_L = I_L^2 X_L = \frac{V_L^2}{X_L} = P_m \quad \text{volt-ampere-reactive (var)} \qquad (14\text{-}34)$$

With respect to Eq. (14-34), note that

1. Q_L represents *quadrature* power *stored* in the inductor, not power dissipated or true power. Graphically, it represents the *maximum of the power pulse* (Fig. 14-33).
2. For this reason, Q_L is measured in vars as opposed to watts (true power dissipated in a resistor).
3. Q_L is also found as $I^2 X_L$ or V_L^2/X_L. This is similar to Eq. (14-30) for the resistor.
4. The maximum instantaneous power $P_m = Q_L = V_m I_m/2$. Note that P_m is *not* equal to $V_m I_m$, an error frequently made by students.

The unity of the various elements of Eq. (14-34) is shown in Ex. 14-40.

EXAMPLE 14-40

For the inductor given in Ex. 14-38, calculate

a. RMS voltage
b. RMS current
c. Q_L, quadrature reactive power by four different methods:
   1. $V_m I_m/2$
   2. $V_L I_L$
   3. $I_L^2 X_L$
   4. V_L^2/X_L
d. P_m by two different methods

Solution

a. $V_L = V_m/\sqrt{2} = 31.42\ \text{V}/\sqrt{2} = \mathbf{22.214\ V}$
b. $I_L = I_m/\sqrt{2} = 10\ \text{A}/\sqrt{2} = \mathbf{7.0711\ A}$
c. 1. $V_m I_m/2 = (31.42\ \text{V}) \times (10\ \text{A}/2) = \mathbf{157.1\ vars}$ (14-34)
   2. $V_L I_L = (22.214\ \text{V})(7.0711\ \text{A}) = \mathbf{157.1\ vars}$ (14-34)
   3. $V_L^2/X_L = (22.214)^2/3.1416\ \Omega = \mathbf{157.1\ vars}$ (14-34)
   4. $I_L^2 X_L = (7.0711)^2 \times 3.1416\ \Omega = \mathbf{157.1\ vars}$ (14-34)
d. $P_m = V_m I_m/2 = \mathbf{157.1\ W}$, from c.1 above (14-34)
   $P_m = V_L I_L = \mathbf{157.1\ W}$, from c.2 above (14-34)

Example 14-40 shows that the maximum instantaneous power is indeed $V_m I_m/2$, which is the same as the reactive power, Q_L, obtained by four different methods with Eq. (14-34).

14-15 RESPONSE OF PURE CAPACITOR TO SINUSOIDAL VOLTAGE AND CURRENT

Figure 14-34a shows a sinusoidal voltage applied to a pure capacitor, C. The voltage across the source is now taken as the reference since the property of capacitance is to oppose a change in voltage. The voltage applied to the capacitor, $v_C(t) = V_m \sin \omega t$, is sinusoidal and continuously changing as shown in Fig. 14-34b. But we also know that the instantaneous current to the capacitor, $i_C(t) = C \dfrac{dv_C}{dt}$, from Sec. 10-2. Consequently, for the sinusoidal current waveform we may write[27]

$$i_C(t) = C\frac{dv_C}{dt} = C\frac{d(V_m \sin \omega t)}{dt} = CV_m \frac{d(\sin \omega t)}{dt}$$
$$= \omega C V_m \cos \omega t = I_m \cos \omega t \qquad \text{(14-35)}$$

Note once again, in Eq. (14-35), that the derivative of a sine wave is a cosine wave, as derived in Appendix B-9. Consequently, as shown in Fig. 14-34b, the waveform of $i_C(t)$ **leads** the applied capacitor voltage $v_C(t)$ by 90° and has the form $i_C(t) = I_m \sin(\omega t + 90°)$. Consequently, for $\omega C V_m$ in Eq. (14-35) we may substitute $I_m = \omega C V_m$.

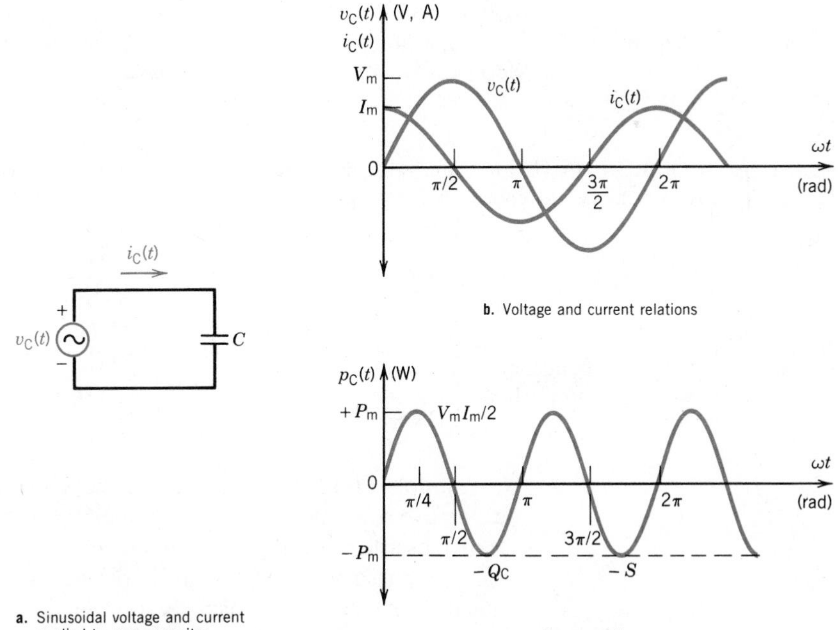

a. Sinusoidal voltage and current applied to pure capacitor

b. Voltage and current relations

c. Power relations

Figure 14-34 Sinusoidal voltage, current, and power relations in a pure capacitor

[27] See Appendix B-12.2 for a more rigorous derivation.

But we are interested in the opposition to the voltage created by the property of capacitance, that is, the ratio V_m/I_m, and therefore we may write for X_C

$$X_C = V_m/I_m = V_m/\omega C V_m = 1/\omega C = 1/2\pi f C = \frac{V_C}{I_C} \quad \text{ohms } (\Omega) \quad (14\text{-}36)$$

where f is the frequency of the sinusoidal voltage across and current in the capacitor
C is the capacitance in farads (F)
V_C is the RMS voltage across the capacitor
I_C is the RMS current in the capacitor

Equations (14-35) and (14-36) lead to the following insights:

1. In a purely capacitive circuit, current leads voltage by 90°.
2. Capacitive reactance varies inversely with both frequency and capacitance. At extremely high frequencies, the capacitive reactance is small. Note from Table 13-3 that a capacitor acts as a short circuit at very high frequencies (VHF).

EXAMPLE 14-41

The voltage across a 5-μF capacitor is $v(t) = 100 \sin 120\pi t$ volts. Calculate:

a. The capacitive reactance of the capacitor
b. Instantaneous maximum current to the capacitor, I_m
c. Two equations expressing the waveform of current, $i_C(t)$
d. The frequency of the voltage across the capacitor

Solution

a. $X_C = 1/\omega C = 1/(120\pi)(5 \times 10^{-6}) = \mathbf{530.5\ \Omega}$ (14-36)
b. $I_m = V_m/X_C = 100\text{ V}/530.5\ \Omega = \mathbf{188.5\ mA}$ (14-36)
c. $i_C(t) = I_m \cos \omega t = \mathbf{188.5(\cos 120\pi t)\ mA}$ (14-35)
$i_C(t) = I_m \sin(\omega t + 90°) = \mathbf{188.5 \sin(120\pi t + 90°)\ mA}$
d. $f = \omega/2\pi = 120\pi/2\pi = \mathbf{60\ Hz}$ (14-5)

EXAMPLE 14-42

For the capacitor given in Ex. 14-41, using the ratio method, calculate the capacitive reactance at frequencies of

a. 300 Hz
b. 600 Hz
c. 1000 Hz

Solution

Since X_C varies inversely with frequency, at higher frequencies the capacitance is reduced proportionately. Consequently

$X_{C_1} = \dfrac{f_0}{f_1} X_{C_0}$ or

a. $X_C = \dfrac{60\text{ Hz}}{300\text{ Hz}} \times 530.5\ \Omega = (1/5)(530.5\ \Omega) = \mathbf{106.1\ \Omega}$

b. $X_C = \dfrac{60\text{ Hz}}{600\text{ Hz}} \times 530.5\ \Omega = (1/10)(530.5\ \Omega) = \mathbf{53.05\ \Omega}$

c. $X_C = \dfrac{60\text{ Hz}}{1000\text{ Hz}} \times 530.5\ \Omega = (0.06)(530.5\ \Omega) = \mathbf{31.83\ \Omega}$

Example 14-42 shows that when the frequency is increased by a factor of 5, the capacitive reactance is reduced to one-fifth of its original value. When the frequency is increased by 10, the capacitive reactance is reduced to one-tenth of its original value. Consequently, if we know the capacitive reactance, X_C, at **any** given frequency, we can find X_C easily and immediately at any higher or lower frequency, using the **ratio method** shown in Ex. 14-42.

As in the case of the pure resistor and pure inductor, we are also interested in the power stored in the electric field of the capacitor and returned to the supply. Again, if we multiply the instantaneous current, $i_C(t)$, by the instantaneous voltage, $v_C(t)$, it should yield the instantaneous capacitive power, $p_C(t)$, by

$$p_C(t) = v_C(t) \times i_C(t) = V_m \sin \omega t \times I_m \cos \omega t$$
$$= V_m I_m \sin \omega t \cos \omega t = \frac{V_m I_m \sin 2\omega t}{2} \quad \text{watts (W)} \qquad (14\text{-}37)$$

Again, as in the case of the pure inductor, Eq. (14-37) is a double-frequency sinusoid whose peak value, P_m, is $V_m I_m/2$. The waveform of $p_C(t)$ is plotted in Fig. 14-34c. From both Eq. (14-37) and the waveform of Fig. 14-34c we conclude that

1. $p_C(t)$ has a frequency twice that of $v_C(t)$ and $i_C(t)$.
2. The average value of $p_C(t)$ is zero; consequently, $P_{av} = 0$ and the capacitor dissipates no power.
3. Whenever the power pulse is *positive*, power is *stored* in the form of an electric field between the capacitor plates. When the power pulse is *negative*, the energy of the field is being *returned to the source.*
4. The power curve of $p_C(t)$ is *in phase* with the voltage since both are positive-going at $\theta = 0°$ and $\theta = 2\pi$ radians (see Sec. 14-7).
5. Since the product, $+P_m$, is reached whenever $\theta = \pi/4$ radians $= 45°$, this is the point where both $v_C(t)$ and $i_C(t)$ cross each other and have instantaneous values that are the same as their RMS values, that is, $P_m = V_C I_C = (V_m/\sqrt{2})(I_m/\sqrt{2}) = V_m I_m/2$.

As in the case of the inductor, we call the (quadrature) capacitive reactive power, Q_C, the product of $V_C I_C$ expressed in reactive volt-amperes (vars). For Q_C we may write the following equalities:

$$Q_C = \frac{V_m I_m}{2} = V_C I_C = I_C^2 X_C = \frac{V_C^2}{X_C}$$
$$= I_m^2 \frac{X_C}{2} = P_m \quad \text{volt-amperes-reactive (vars)} \qquad (14\text{-}38)$$

With respect to Eq. (14-38) and Fig. 14-34c, note that

1. Q_C represents quadrature power stored in the capacitor, not power dissipated or true power consumed. For reasons that we will learn later, graphically it represents the (**negative**) maximum of the power pulse (Fig. 14-34c).
2. Q_C is also found as $I_C^2 X_C$ or V_C^2/X_C. This is similar to Eqs. (14-30) and (14-34) for a resistor and a pure inductor, respectively.
3. The maximum instantaneous power $P_m = Q_C = V_m I_m/2$ (and not $V_m I_m$), as explained above.

EXAMPLE 14-43

For the capacitor given in Ex. 14-41, calculate:

a. RMS voltage
b. RMS current
c. Q_C by four different methods
d. P_m by two methods

Solution

a. $V_C = V_m/\sqrt{2} = 100 \text{ V}/\sqrt{2} = \mathbf{70.71\ V}$

b. $I_C = I_m/\sqrt{2} = 188.5 \text{ mA}/\sqrt{2} = \mathbf{133.3\ mA}$

c. 1. $Q_C = V_m I_m/2 = (100 \text{ V})(188.5 \text{ mA})/2$
$= \mathbf{9.43\ vars}$ (14-38)

2. $Q_C = V_C I_C = 70.71\text{ V} \times 133.3\text{ mA}$ $= \mathbf{9.43\ vars}$ (14-38)

3. $Q_C = I_C^2 X_C = (133.3\text{ mA})^2 \times 530.5\ \Omega$ $= \mathbf{9.43\ vars}$ (14-38)

4. $Q_C = V_C^2/X_C = (70.71)^2/530.5 = \mathbf{9.43\ vars}$ (14-38)

d. $P_m = V_m I_m/2 = (100\text{ V})(188.5\text{ mA})/2 = \mathbf{9.43\ W}$ (14-38)

$P_m = V_C I_C = (70.71\text{ V})(133.3\text{ mA}) = \mathbf{9.43\ W}$ (14-38)

Example 14-43 shows that the maximum instantaneous power is, indeed, $V_m I_m/2$, which is the same as the reactive power, Q_C, obtained by five different methods with Eq. (14-38).

14-16 VARIATION OF INDUCTIVE AND CAPACITIVE REACTANCE WITH FREQUENCY

Based on our study of transient response of L and C, we constructed Table 13-3, which predicted the very low frequency (VLF) and very high frequency (VHF) behavior of L and C. We concluded from Table 13-3 that

1. An inductor behaves as a short circuit at VLF and an open circuit at VHF.
2. A capacitor behaves as an open circuit at VLF and a short circuit at VHF.

Our sinusoidal analysis enables us to verify Table 13-3 by using the previously derived and respective reactance equations:

$$X_L = \omega L = 2\pi f L \quad (14\text{-}32) \qquad X_C = 1/\omega C = 1/2\pi f C \quad (14\text{-}36)$$

Figure 14-35 shows X_L and X_C plotted as a function of frequency. With respect to the plot of X_L, please note the following:

1. X_L increases **linearly** with frequency. Since 2, π, and L are all constants, Eq. (14-32) may be written $X_L = mf$, where m is the slope of the line. Consequently, the greater the frequency, the higher the value of X_L.
2. With regard to the slope of the X_L line, $m = 2\pi L$. This implies that a larger inductance, L, would produce a high (more vertical) slope and a smaller value of L would produce a more horizontal slope.
3. At a frequency of zero, $X_L = 0$. At an infinite frequency, X_L is infinite. This verifies the conclusions in Table 13-3 that the inductor acts as a short circuit at VLF and an open circuit at VHF.

We now come to the plot of X_C versus f shown in Fig. 14-35. Observe that the curve is *asymptotic* to *both* the vertical and horizontal axes. Clearly, Eq. (14-36) is *not* a linear relationship but represents a *hyperbola*, whose equation is $xy = k$.

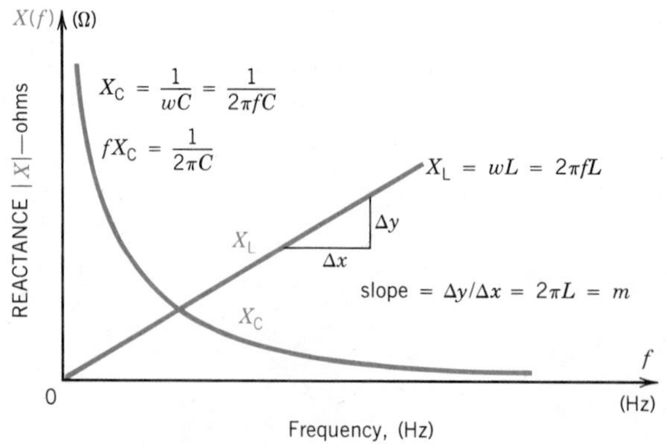

Figure 14-35 Variation of inductive and capacitive reactance vs frequency

Equation (14-36) may be written $(f)(X_C) = 1/2\pi C$, since 2, π, and C are all constants. With respect to the plot of X_C, note that

1. X_C varies *inversely* and *nonlinearly* with frequency. The higher the frequency, the lower the value of X_C.
2. A high value of C will bring the hyperbola closer to the origin, O. A smaller value of C extends the hyperbola farther from the origin.
3. At a frequency of zero, X_C is infinite. At an infinite frequency, X_C is zero. This verifies the conclusions from Table 13-3 that the capacitor acts as an open circuit at VLF and as a short circuit at VHF.

14-17 ACTIVE AND APPARENT POWER, AND POWER FACTOR

We have noticed that in a purely resistive circuit, current is *in phase* with the voltage. In a purely inductive circuit, current lags the voltage by 90°. In a purely capacitive circuit, current leads the voltage by 90°. Neither the ideal (pure) inductor nor the capacitor dissipated power, despite the fact that current was drawn from the supply. Since the phase angle between voltage and current in a resistor is zero, compared to the 90° phase angle in inductors and capacitors, phase angle is clearly related to power.

This relation is derived in Appendix B-10, where we find average or active power for any combination of R-L-C:

$$P = \frac{V_m I_m}{2} \cos\theta = VI \cos\theta = S \cos\theta \qquad \text{(14-39)}$$

where θ is the phase angle between voltage and current waveforms (or phasors)
S is the apparent power or the product of RMS voltage and current
V_m and I_m are the maximum values of time domain sinusoidal waveforms
V and I are RMS values of voltage and current, respectively

The factor $\cos\theta$ in Eq. (14-39) is called the power factor, F_p. The power factor as defined in Appendix B-10.2 is

$$F_p = \cos\theta = P/S = P/VI = 2P/V_m I_m \qquad \text{(14-39a)}$$

where all terms have been defined.

Note that the power factor, F_p, represents the ratio of active (or true) power (in watts), P, to apparent power, S (in volt-amperes).

Equations (14-39) and (14-39a) are general equations that can be used for any combinations of circuits containing either all or some of the basic R, L, and C circuit elements, including the ideal elements themselves.

14-17.1 Power and Power Factor in a Purely Resistive Circuit

Since the current, $i(t)$, *lags* the voltage, $v(t)$, in a pure inductor by 90° at all times, $\theta = 0$ and the power factor $\cos\theta = \cos 0° = 1$ or unity.

Then we may modify our original Eq. (14-30) for a purely resistive circuit to

$$P_{av} = P_R = V_m I_m/2 = V_R I_R = I_R^2 R = V_R^2/R \qquad \text{watts (W)} \qquad (14\text{-}30a)$$

14-17.2 Power and Power Factor in a Purely Inductive Circuit

Since the current, $i(t)$, *lags* the voltage, $v(t)$, in a pure inductor by 90° at all times, as shown in Fig. 14-33b, the phase angle θ is 90° and $\cos\theta = \cos 90° = 0$. As noted in Fig. 14-33c, the average power is zero. We may now modify our original Eq. (14-33) to

$$P_{av} = P_L = \frac{V_m I_m}{2}\cos 90° = VI\cos 90° = S\cos 90° = 0 \quad \text{watts (W)} \quad (14\text{-}33a)$$

14-17.3 Power and Power Factor in a Purely Capacitive Circuit

Since the current $i(t)$ *leads* the voltage $v(t)$ in a pure capacitor by 90° at all times, as shown in Fig. 14-34b, the phase angle θ is 90° and $\cos\theta = \cos 90° = 0$. As noted in Fig. 14-34c, the average power is zero. We may now modify our original Eq. (14-37) to

$$P_{av} = P_C = \frac{V_m I_m}{2}\cos 90° = VI\cos 90° = S\cos 90° = 0 \quad \text{watts (W)} \quad (14\text{-}37a)$$

14-17.4 Power, Power Factor, and Reactive Factor in Practical *RL* and *RC* Circuits

We have defined F_p, the power factor, as the factor by which the apparent power S, must be multiplied to obtain true or active power. Although the pure inductor and pure capacitor dissipate no power, there is a quadrature component of power drawn from the supply and stored in the ideal element (either L or C). This value is obtained from the reactive factor, F_r, defined as

$$F_r = \sin\theta = Q/S \qquad \text{(unitless ratio)} \qquad (14\text{-}40)$$

Equation (14-40) enables us to find Q as well as S from

$$Q = S \sin\theta = \frac{V_m I_m}{2} \sin\theta = VI \sin\theta \qquad \text{(14-40a)}$$

where all terms have been defined.

Equations (14-39) and (14-40) enable us now to examine waveforms whose phase relations may range from more than 0° to less than 90°. These two sets of equations open a host of new insights into the possible information that can be obtained from waveform equations in the time domain. Example 14-44 opens this "Pandora's box."

EXAMPLE 14-44

Given the waveforms of Ex. 14-16 plotted in Fig. 14-15, calculate

a. F_p, circuit power factor, and F_r, circuit reactive factor
b. P, active power of the circuit
c. Q, reactive (quadrature) power of the circuit
d. R, resistance of the circuit
e. X_C, capacitive reactance of the circuit
f. C, capacitance of the circuit

Solution

The equations of $v(t)$ and $i(t)$, respectively, expressed as positive sinusoidal functions are from parts **(a)** and **(b)** of Ex. 14-16: $v(t) = 20\sin(5000t - 60°)$ V and $i(t) = 12\sin(5000t - 10°)$ A, from which $\theta = 50°$

a. $F_p = \cos\theta = \cos 50° = \mathbf{0.6428}$, $F_r = \sin\theta = \sin 50° = \mathbf{0.766}$

b. $P = P_{av} = \left(\frac{V_m I_m}{2}\right)\cos\theta = \left(\frac{20\text{ V} \times 12\text{ A}}{2}\right)\cos 50°$
$= (120\text{ VA})\cos 50° = \mathbf{77.13\ W}$ (14-39)

c. $Q = (V_m I_m/2)(\sin\theta) = (120\text{ VA})\sin 50°$
$= \mathbf{91.925\ vars}$ (14-40a)

d. $R = P/I^2 = P/(I_m/\sqrt{2})^2 = 2P/I_m^2$
$= 2 \times 77.13\text{ W}/(12)^2 = \mathbf{1.071\ \Omega}$ (14-30)

e. $X_C = Q/I^2 = 2Q/I_m^2 = 2 \times 91.925\text{ vars}/(12)^2$
$= \mathbf{1.277\ \Omega}$ (14-38)

f. $C = 1/\omega X_C = 1/(5000)(1.277) = \mathbf{156.6\ \mu F}$ (14-36)

Example 14-44 shows that, given the time domain waveforms of $v(t)$ and $i(t)$, it is possible to calculate **all** the power parameters (S, P, and Q) as well as the equivalent (series) circuit components.

14-17.5 Maximum Positive and Negative Power Points

One final set of insights remain to be demonstrated. These are that the above values of instantaneous active (real), apparent, and quadrature power may be verified by finding the maximum positive and negative excursions of the power waveforms!

For single-phase ac circuits having sinusoidal waveforms of voltage and current, it can be shown that the maximum instantaneous positive value of the power waveform is

$$P_M^{(+)} = \frac{V_m I_m}{2}(\cos\theta + 1) = S(\cos\theta + 1) \qquad \text{watts (W)} \qquad \text{(14-41)}$$

Similarly, it can be shown that the maximum instantaneous *negative* value of the power waveform is

$$P_M^{(-)} = \frac{V_m I_m}{2}(\cos\theta - 1) = S(\cos\theta - 1) \qquad \text{watts (W)} \qquad \text{(14-42)}$$

These maximum positive and negative values of instantaneous power can also be used to find both active (true, real) power, P, and apparent power, S, from

$$P = P_{av} = \frac{P_M^{(+)} + P_M^{(-)}}{2} \quad \text{watts (W)} \qquad (14\text{-}43)$$

$$S = \frac{P_M^{(+)} - P_M^{(-)}}{2} = P_M^{(+)} - P_{av} \quad \text{volt-amperes (VA)} \qquad (14\text{-}44)$$

EXAMPLE 14-45

For the waveforms given for a pure resistor in Fig. 14-27a, algebraically show that

a. $P_M^{(+)} = V_m I_m$
b. $P_M^{(-)} = 0$
c. $P_{av} = P_M^{(+)}/2$
d. $S = P$ (resistive circuit only)

Solution

a. Since $\theta = 0$ for a resistive circuit
$P_M^{(+)} = (V_m I_m/2)(\cos 0° + 1) = \mathbf{V_m I_m}$ (14-41)
b. $P_M^{(-)} = (V_m I_m/2)(\cos 0° - 1) = \mathbf{0}$ (14-42)
c. $P = P_{av} = \dfrac{P_M^+ + P_M^-}{2} = (V_m I_m + 0)/2 = \mathbf{P_M^+/2}$ (14-43)
d. $S = P_M^+ - P_{av} = (P_M^+ - P_M^-)/2 = P_M^+/2 = \mathbf{P_{av}}$ (14-44)

Using Fig. 14-27a, the reader should verify these relations, which apply only to the power relations for a *purely resistive* circuit whose voltage and current waveforms are in phase ($\theta = 0$).

EXAMPLE 14-46

For the waveforms given for a pure inductor in Fig. 14-33b and c, algebraically show that

a. $P_M^{(+)} = V_m I_m/2$
b. $P_M^{(-)} = -V_m I_m/2$
c. $P_{av} = 0$
d. $S = V_m I_m/2$ (inductive circuit only)

Solution

a. Since $0 = 90°$ for a pure inductor
$P_M^{(+)} = (V_m I_m/2)(\cos 90° + 1) = \mathbf{V_m I_m/2}$ (14-41)
b. $P_M^{(-)} = (V_m I_m/2)(\cos 90° - 1) = \mathbf{-V_m I_m/2}$ (14-42)
c. $P = P_{av} = (P_M^+ + P_M^-)/2 = 0$ (14-43)
d. $S = P_M^+ - P_{av} = (V_m I_m/2) - 0 = V_m I_m/2$ (14-44)

Using Figs. 14-33b and 14-33c, the reader should verify these relations for a purely inductive circuit whose current lags the voltage by 90°.

We now come to the final example, which summarizes all the information that can be extracted from equations in the time domain thus far. Let us use the voltage and current equations given in Ex. 14-16, which are plotted in Fig. 14-15.

EXAMPLE 14-47

Given the waveforms $v(t) = 20 \sin(5000t - 60°)$ V and $i(t) = 12 \sin(5000t - 10°)$ A, calculate:

a. Circuit power factor, F_p
b. Circuit reactive factor, F_r
c. Apparent power, S
d. Active (real, average) power using Eq. (14-39)
e. Reactive power, Q, using Eq. (14-40a)
f. Reactance (indicate whether capacitive or inductive), X, and resistance, R, of the circuit
g. Maximum positive excursion of the power waveform, P_M^+
h. Maximum negative excursion of the power waveform, P_M^-
i. Active power, using Eq. (14-43)
j. Apparent power, using Eq. (14-44)
k. Either L or C and draw the practical series equivalent and frequency-independent circuit
l. Using current as the reference waveform,
   1. Redraw current and voltage waveforms originally shown in Fig. 14-15 versus time in ms on one set of axes
   2. Using the values of P_M^+ and P_M^- computed above, draw the power waveform directly below the voltage/current waveform. Show values of S and P on this waveform
   3. Using the value of Q calculated in (e) above, draw the waveform of reactive or quadrature power

Solution

a. $F_p = \cos\theta = \cos 50° = \mathbf{0.6428}$ (14-39a)
b. $F_r = \sin\theta = \sin 50° = \mathbf{0.7660\bar{4}}$ (14-40)
c. $S = V_m I_m/2 = (20\text{ V} \times 12\text{ A})/2 = \mathbf{120\ VA}$ (14-44)
d. $P = S\cos\theta = (120\text{ VA})\cos 50° = \mathbf{77.13\underline{5}\ W}$ (14-39)
e. $Q = S\sin\theta = (120\text{ VA})\sin 50° = \mathbf{91.925\bar{3}\ vars}$ (14-40a)
f. $X = Q/I^2 = 2Q/I_m^2$
$= 2 \times 91.9253/(12)^2 = \mathbf{1.277\ \Omega}$ (14-38)
reactance is **capacitive** because current is leading voltage

$$R = P/I^2 = 2P/I_m^2 = 2 \times 77.135/(12)^2$$
$$= \mathbf{1.0713\ \Omega} \quad (14\text{-}30)$$

g. $P_M^+ = S(\cos\theta + 1) = 120\text{ VA}(\cos 50° + 1)$
$= \mathbf{197.13}$ watts (14-41)
h. $P_M^- = S(\cos\theta - 1) = 120\text{ VA}(\cos 50° - 1)$
$= \mathbf{-42.865}$ watts (14-42)
i. $P = P_{av} = [(P_M^{(+)} + P_M^{(-)})]/2 = (197.13 - 42.865)/2$
$= \mathbf{77.13}$ W (14-43)
see part (**d**) above
j. $S = P_M^{(+)} - P_{av} = 197.13 - 77.13 = \mathbf{120\ VA}$ (14-44)
see part (**c**) above
k. $C = X_C/\omega = 1.277\ \Omega/5000\text{ rad/s} = \mathbf{255.4\ \mu F}$ (Fig. 14-36d)
l. The waveforms are shown using current as a reference in **Figs. 14-36a, b, c**. The equivalent circuit shows the series RC combination representing a 255.4 μF practical capacitor having a resistance of 1.0713 Ω. This model is independent of frequency (for all practical purposes). See Fig. 14-36d.

With respect to the waveforms drawn in 14-36, note that

1. The voltage waveform begins 50° after the current waveform and always lags by 50°.
2. The power waveform only is negative when *either* voltage *or* current is negative (but not both). $p(t)$ is zero when either

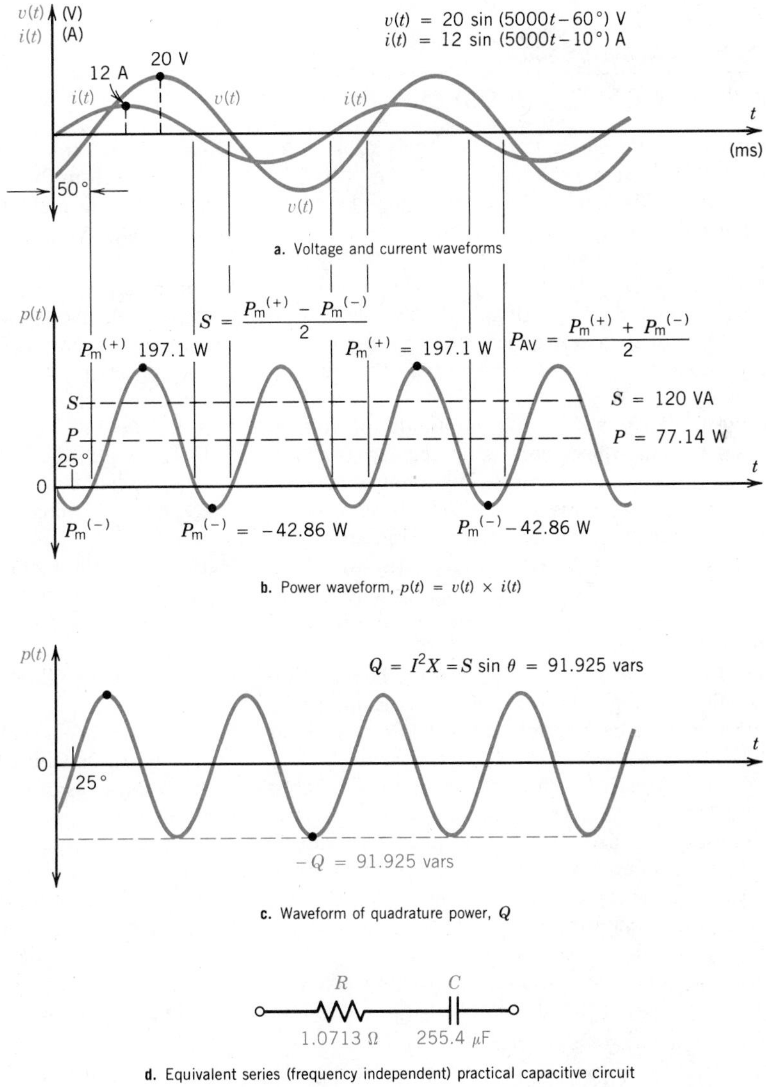

Figure 14-36 Waveforms (Ex. 14-47**l**) and equivalent practical series circuit (Ex. 14-47**k**).

$v(t)$ or $i(t)$ is zero. Since we know the value of $P_M^{(-)}$, this value (of -42.86 W) first occurs when $i(t)$ rises to its 25° value, as shown in Fig. 14-36b.

3. The power waveform is positive when *only* voltage and current are *both positive*. The maximum positive value (of 197.1 W) occurs midway between the points where $v(t)$ is zero and $i(t)$ is zero, that is, the zero power points.
4. The foregoing information enables us to draw the waveform of $p(t)$. Note that for each cycle of voltage (or current), there are always two cycles of power.
5. The average or real power is the average value of the power curve! Since the power curve is symmetrical, its average value is determined from Eq. (14-11) or Eq. (14-43). Note that Eq. (14-43) is a special case of Eq. (14-11)!
6. Since $S = P_M^+ - P_{av}$ and we have both these levels drawn in Fig. 14-36b, we may graphically draw this difference. As a check, we know that S must always be equal to or greater than P.
7. The quadrature power waveform, Q, must have an average value of zero and a maximum positive (and/or negative) value of Q, where $Q = S \sin \theta = I^2X$. This represents the energy stored in the dielectric field of the capacitor and returned to the supply. Note that this waveform leads the power waveform by 50° (at twice the frequency) and lags the current waveform by 25°.
8. The waveforms of Figs. 14-36b and 14-36c were drawn by using the above insights.

14-18 GLOSSARY OF TERMS USED

Alternating waveform Waveform that alternately crosses the zero reference axis, having both positive and negative instantaneous values. Alternating waveforms may be either periodic or aperiodic.

Angular frequency, ω Same as angular velocity, found by multiplying the frequency, f, by 2π radians per cycle, or $\omega = 2\pi f$.

Angular velocity, ω Rate of change of angular displacement, θ, of a rotating phasor or waveform generator, expressed in units of degrees per second, radians per second (rad/s), or revolutions per second (rps).

Aperiodic waveform Any time-varying amplitude (of voltage, current, power, etc.) in which there is no regularity of amplitude variation. Aperiodic waveforms are nonrepetitive.

Average value, A_{av} Level of a waveform, either above or below zero, such that the area enclosed by the waveform above this level exactly equals the area below this level. Any waveform that has either a positive or a negative average value for one full cycle contains a dc component and the level of that component is its average value.

Continuous waveform Waveform whose instantaneous slope changes gradually in going from one to the next instantaneous value. No portion of a continuous waveform is ever linear (see Fig. 14-1).

Crest factor, F_c For a periodic function, the ratio of its crest (peak, maximum) value to its RMS value.

Cycle Interval of time or space contained in one period during which a complete set of events or waveform excursions occur. Consequently, one cycle represents that portion of the waveform contained within one period.

DC component, A_{dc} Same as average value.

Derivative Instantaneous rate of change of any given function (voltage, current, power, etc.) with respect to time, angle, or any other variable.

Discrete waveform Waveform that may change in slope and/or direction abruptly in going from one instantaneous value to another. A discrete waveform usually has one or more linear portions (see Fig. 14-2).

Effective value Same as RMS value.

Form factor, F_f For a periodic function, the ratio of its RMS value to its average value.

Frequency, f Number of complete cycles of any periodic waveform occurring (usually) in 1 s. Frequency is measured in hertz and its equation is the reciprocal of the period ($f = 1/T$), that is, $1\text{ Hz} = 1\text{ s}^{-1} = 1$ cycle per second.

Full-wave-rectified (FWR) waveform Waveform in which the negative half-cycle has been inverted so that both half-cycles are positive-going waveforms for each cycle of alternating input.

Half-wave-rectified (HWR) waveform Waveform resulting from passage of only one half-cycle of each incoming waveform for each input cycle of alternation. A single rectifier provides a HWR.

Instantaneous amplitude, a Value of amplitude of a waveform of voltage, current, power, and so forth at any instant in time.

Peak (instantaneous) value, A_m Maximum (positive or negative) instantaneous value. If a waveform contains a dc component, the amplitudes of its peak values (either positive or negative) may not be equal.

Peak-to-peak value, A_{pp} Difference in amplitude (voltage, current, power, etc.) between the positive and negative peak values.

Period, T Time required to complete one full cycle of variation in the waveform, usually expressed in seconds. The period is the reciprocal of the frequency, that is, $T = 1/f$.

Periodic waveform Waveform that repeats itself in regular cycles, having a defined time interval for the duration of each cycle.

Power, active, P For a single-phase load drawing current from an ac source, the time average of the values of instantaneous power taken over one period of alternation. Active or average power is measured in watts (W).

Power, apparent, S Scalar quantity equal to the product of the RMS values of voltage and current in a two-wire single-phase ac circuit, measured in volt-amperes (VA).

Power factor, F_p Ratio of total active power, P, to total apparent power, S.

Power, reactive, Q Quadrature component of complex power measured in volt-amperes reactive (vars). It is the product of voltage and the quadrature component of current. In a passive network, Q represents the alternating exchange of stored energy (inductive or capacitive).

Pulsating waveform Waveform that does not cross the vertical zero (voltage) axis and whose voltage (current, power, amplitude, etc.) is unidirectional. Also called *direct* waveform or *unidirectional* waveform. Such a waveform may be varying or pulsating, containing both a dc and an ac component.

Radian frequency, ω Same as angular velocity or angular frequency, expressed in units of radians per second (or s^{-1}).

Reactive factor, F_r Ratio of total reactive power, Q, to total apparent power, S.

RMS value, A_{rms} Root-mean-square or effective value of ac voltage (or current) producing a heating effect equivalent to that produced by the *same value* of dc voltage (or current). Quantitatively, the RMS value is equal to the square root of the mean value of the squares of all instantaneous values, taken for one full cycle of alternation of a given waveform. For a half-cycle of sine wave, $A_{rms} = A_m/\sqrt{2}$.

Sinusoidal waveform Continuous, periodic, alternating waveform having an average value of zero, expressed by the general equation $a(t) = A \sin(x + \phi)$, where A is the peak amplitude value, x the independent variable, and ϕ a phase angle that may exist at time $t = 0$.

Slope Rate of change of a waveform measured at some instantaneous value on the waveform, tangent to the value. The slope is zero when horizontal and infinite when vertical. A waveform's slope is constant over the linear region of its shape. A positive slope is represented by an upward positive tangent and a negative slope by a downward negative tangent, $-dy/dx$.

14-19 PROBLEMS

Secs. 14-1 to 14-5

14-1 Given the nonsinusoidal waveform of Fig. 14-16, find the

a. Period, in ms
b. Frequency
c. Maximum amplitude of the waveform

14-2 Repeat Prob. 14-1 for the waveform of Fig. 14-17
14-3 Repeat Prob. 14-1 for the waveform of Fig. 14-19
14-4 Repeat Prob. 14-1 for the waveform of Fig. 14-20
14-5 A periodic waveform has an angular velocity of 200 rad/s Calculate its

a. Frequency
b. Period
c. Value of ωt at the end of the first cycle
d. Value of θ in degrees at the end of 1.5 cycles

14-6 Given a 60-Hz supply, calculate the time required for

a. One cycle
b. Five cycles
c. Ten cycles of alternation

14-7 Calculate the frequency of the waveform that completes 10 cycles in 5 nanoseconds in

a. Cycles per second
b. GHz

14-8 Calculate the period of the waveforms whose frequencies are

a. 10 MHz d. 2 THz
b. 5 GHz e. 60 Hz
c. 4 kHz

14-9 Given the following angles in degrees, convert them to (1) radians and (2) gradients:

a. 55° d. 230°
b. 75° e. 330°
c. 160° f. 410°

14-10 Convert the following angles in radians to (1) degrees and (2) gradients:

a. $\pi/2$ f. 0.6π
b. $\pi/3$ g. 3π
c. $\pi/6$ h. 2.5
d. $3\pi/4$ i. 6
e. $5\pi/6$

14-11 Calculate the angular velocity, ω, in (1) rad/s and (2) deg/s for waveforms having periods of

a. 5 ms d. 0.2 s
b. 4 μs e. 12 ms
c. 8 s

14-12 Calculate the angular velocity, ω, in (1) deg/s and (2) rad/s for waveforms having frequencies of

a. 10 MHz d. 2 THz
b. 5 GHz e. 60 Hz
c. 4 kHz

14-13 Given the following angular velocities, calculate (1) the period and (2) the frequency of each waveform:

a. 500 rad/s d. 1000π rad/s
b. 800 grads/s e. 3×10^2 grad/s
c. 100°/s f. 5×10^3 deg/s

14-14 For each of the following waveforms find (**1**) maximum amplitude, (**2**) frequency, and (**3**) phase at $t = 0$ with respect to a zero-reference sine wave:

a. $i(t) = 10 \cos 377t$ mA
b. $v(t) = 5 \times 10^3 \sin 10^3 t$ mV
c. $v(t) = -2.5 \sin 754t$ V
d. $v(t) = 120 \sin(377t + 30°)$ V
e. $v(t) = 10 \sin(942t - 20°)$ μV
f. $i(t) = 6 \cos(10^5 t - 30°)$ μA

Secs. 14-6 and 14-7

14-15 Express the waveforms of Probs. 14-14**a**, **c**, and **f** as positive sinusoidal functions. (This is a necessary step in order to solve Prob. 14-16 below.)

14-16 For the waveforms shown in Prob. 14-14, **respectively**, calculate the first two *positive* rotation angles in degrees when

a. $i(t) = 5$ mA d. $v(t) = -100$ V
b. $v(t) = -4$ V e. $v(t) = -2$ μV
c. $v(t) = 2$ V f. $i(t) = -6$ μA

Hint: In each solution, verify your angles by substitution in the *original* equation given in Prob. 14-14. It is necessary, however, to use equations as positive sinusoidal functions when applying Eqs. (14-6a) through (14-6e).

14-17 Given the equation $v(t) = -12.5 \cos(500\pi t + 115°)$ V,

a. Express the waveform as a positive sinusoidal function
b. Sketch the waveform for two cycles of rotation, showing the
 1. Values of θ at which positive and negative maximum values occur (a total of four are needed)
 2. Values of θ at which $v(t) = 0$ (a total of four are needed)
c. Calculate the times corresponding to the above eight values of θ and record them on your sketch drawn in part (**b**). Hint: solve for t in Eq. (14-5b).

14-18 For the waveform drawn in Prob. 14-17, calculate the instantaneous voltage $v(t)$ for

a. Rotation angles of (1) 30°, (2) π radians, (3) 270°, (4) $5\pi/6$ radians
b. Times of (1) $t = 0$, (2) $t = 1$ ms, (3) $t = 2$ ms, (4) $t = 4.5$ ms
c. Verify your answers to (**a**) and (**b**) by adding these values to the sketched waveform drawn in the previous problem

14-19 For the waveform sketched and used in Probs. 14-17 and 14-18, calculate the first two positive rotation angles that yield

a. $v(t) = 0$
b. $v(t) = +12.5$ V
c. $v(t) = -5$ V
d. $v(t) = 1$ V

Hint: use the equation found in Prob. 14-17a and Eqs. (14-6) appropriately. Verify your answers, using the waveform and values obtained in Probs. 14-17 and 14-18.

14-20 For the waveform sketched and used in Probs. 14-17 and 14-18, calculate the first two times ($t \geq 0$) when $v(t)$ instantaneously equals

a. 0 c. 4 V
b. 12 V d. -6.25 V

Hint: calculate rotation angles corresponding to given values and then convert them to time, using the method of Prob. 14-17c. Verify your answers by using the waveform sketched.

14-21 Given the waveform $i(t) = 10 \cos 500\pi t$ mA, sketch $1\frac{1}{2}$ cycles of its variation versus the following horizontal (abscissa) scales:

a. Rotation angle, θ, in degrees
b. Rotation angle, θ, in radians
c. Time, t, in appropriate units

14-22 Repeat all three parts of Prob. 14-21, given the equation $v(t) = -5 \sin 1000\pi t$ V

14-23 Repeat all three parts of Prob. 14-21, given the equation $v(t) = 10 \sin(1250\pi t - 60°)$ V

14-24 Repeat all three parts of Prob. 14-21, given the equation $i(t) = -10 \cos(625\pi t - 200°)$ μA (hint: first convert equation to a positive sine function)

14-25 In all cases, using *current* as a reference, give the phase relation between the following pairs of waveforms:

a. $v(t) = -5 \sin(\omega t + 30°)$ V
 $i(t) = 2 \sin(\omega t - 80°)$ mA
b. $v(t) = -5 \cos(\omega t + 90°)$ mV
 $i(t) = 3 \sin(\omega t + 10°)$ μA
c. $v(t) = 5 \cos(\omega t - 20°)$ kV
 $i(t) = 2 \sin(\omega t + 50°)$ A
d. $v(t) = -300 \cos(\omega t - 30°)$ μV
 $i(t) = -12 \sin(\omega t + 50)$ μA

14-26 In all cases, using *voltage* as a reference, give the phase relation between the following pairs of waveforms:

a. $v(t) = -20 \sin(50t + 120°)$ mV
 $i(t) = -12 \cos(50t + 80°)$ mA
b. $v(t) = 85 \sin(1200\pi t + 60°)$ μV
 $i(t) = 16 \cos(1200\pi t - 120°)$ μA
c. $v(t) = 46 \sin(300\pi t - 10°)$ V
 $i(t) = -15 \cos(300\pi t + 95°)$ A
d. $v(t) = -150 \cos(700\pi t - 200°)$ mV
 $i(t) = -100 \sin(700\pi t - 100°)$ mA

14-27 For the waveform shown in **Fig. 14-37a**, write its equation in the time domain with phase angle in

a. Degrees
b. Radians

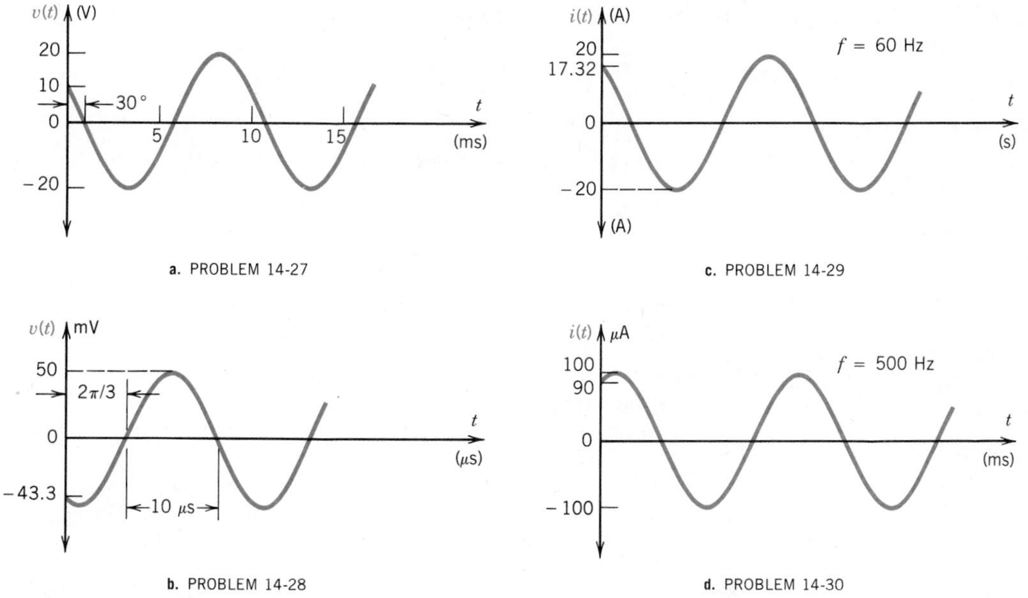

Figure 14-37 Problems 14-27 to 14-30

14-28 For the waveform shown in Fig. 14-37b, repeat Prob. 14-27.

14-29 For the waveform shown in Fig. 14-37c, repeat Prob. 14-27.

14-30 For the waveform shown in Fig. 14-37d, repeat Prob. 14-27.

Sec. 14-8

14-31 For the waveform shown in **Fig. 14-38a**, calculate

a. The average value
b. The frequency in appropriate units

14-32 Repeat Prob. 14-31 for the waveform shown in Fig. 14-38b.

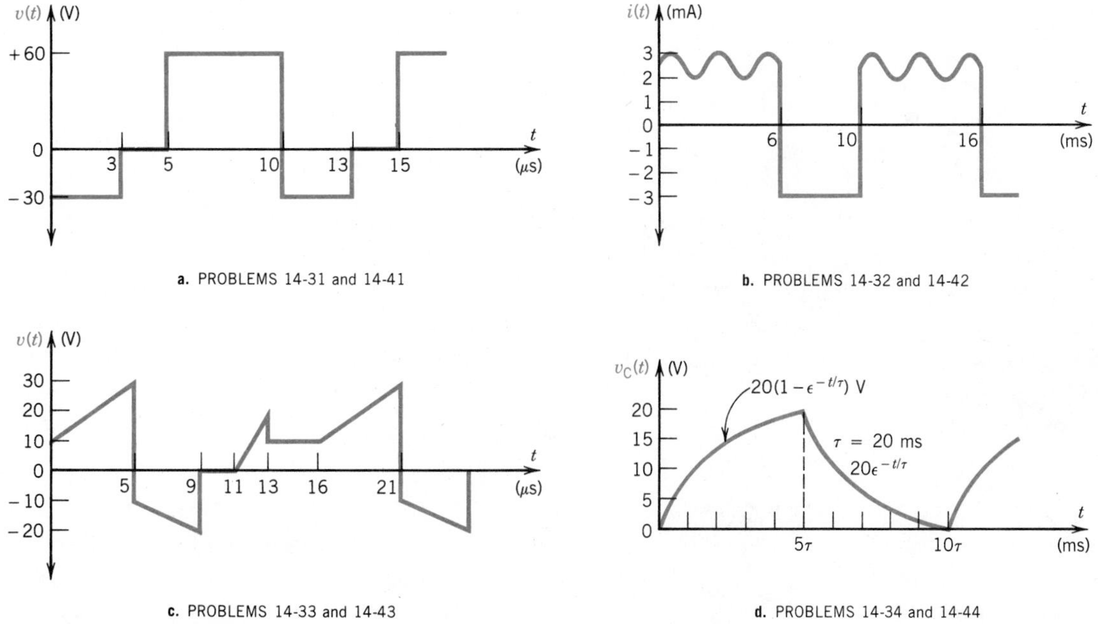

Figure 14-38 Problems 14-31 to 14-34 and Probs. 14-41 to 14-44

14-33 Repeat Prob. 14-31 for the waveform shown in Fig. 14-38c.

14-34 Repeat Prob. 14-31 for the waveform shown in Fig. 14-38d.

14-35 Repeat Prob. 14-31 for the waveform shown in **Fig. 14-39a**.

14-36 Repeat Prob. 14-31 for the waveform shown in Fig. 14-39b, for one cycle of input.

14-37 Repeat Prob. 14-31 for the waveform shown in Fig. 14-39c.

14-38 Repeat Prob. 14-31 for the waveform shown in Fig. 14-39d.

14-39 A periodic voltage waveform has the following equation set: $v(t) = 200(t - 0.05)^2$ volts for $0 \leq t \leq 50$ ms and $v(t) = 0$ for 50 ms $\leq t \leq 80$ ms after which the sequence is periodic.

a. Substitute values sufficient to sketch the waveform (in mV vs ms)
b. Calculate the average voltage by simple integration
c. Check (**b**), using the triangular assumption from 0 to 35 ms for the sketched waveform

14-40 A periodic waveform having a 5 ms period has the equation: $v(t) = 8000t^2$ volts for $0 \leq t \leq 5$ ms

a. Substitute sufficient values to sketch the waveform (in mV vs ms)
b. Calculate the average voltage by simple integration
c. Check (**b**), using the triangular assumption from 1.5 to 5 ms on the sketched waveform

Secs. 14-9 to 14-11

14-41 For the waveform of Prob. 14-31 calculate the

a. RMS value
b. Form factor
c. Crest factor

14-42 For the waveform of Prob. 14-32 calculate the

a. RMS value
b. Form factor
c. Crest factor

14-43 For the waveform of Prob. 14-33 calculate the

a. RMS value
b. Form factor
c. Crest factor

14-44 For the waveform of Prob. 14-34 calculate the

a. RMS value
b. Form factor
c. Crest factor
d. What other waveform has the same form factor equal to its crest factor?
e. Explain why the triangular assumption works for the average value of the waveform in Prob. 14-34 but *not* for the RMS value in (**a**). (See Fig. 14-18a and Table 14.3.)

14-45 For the periodic waveform given in Prob. 14-35, calculate the

a. RMS value
b. Form factor
c. Crest factor

14-46 For the periodic waveform given in Prob. 14-36, calculate the

a. RMS value
b. Form factor
c. Crest factor

14-47 For the periodic waveform given in Prob. 14-38, calculate the

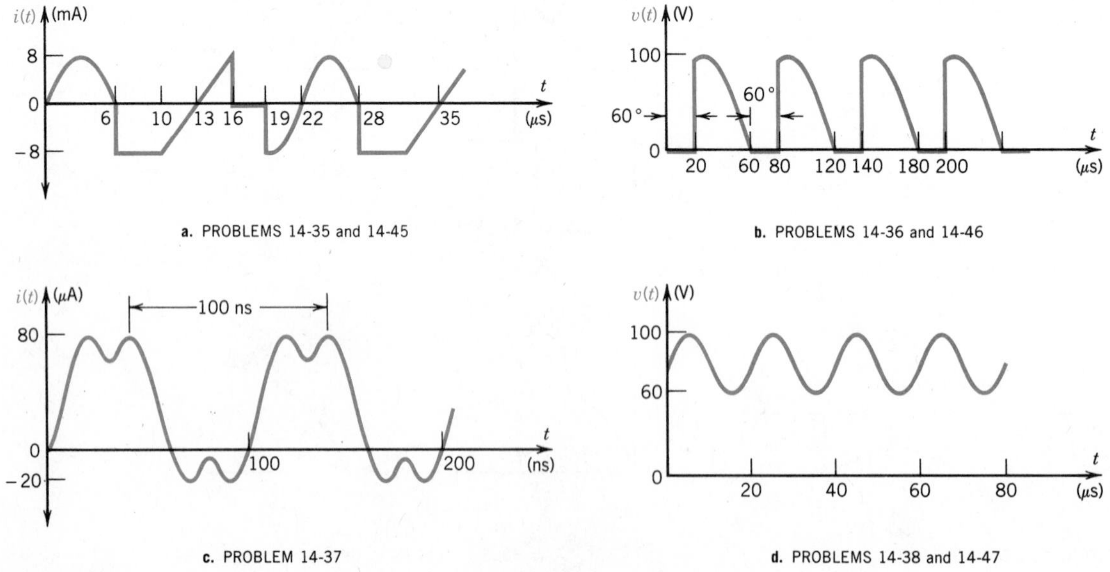

Figure 14-39 Problems 14-35 to 14-38 and Probs. 14-45 to 14-47

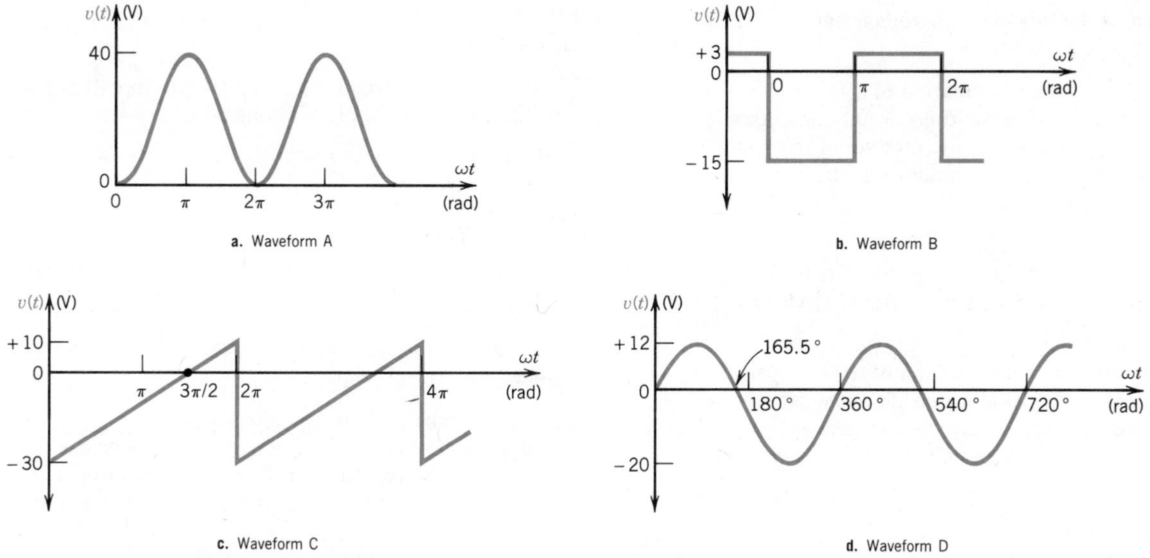

Figure 14-40 Problem 14-50

a. RMS value
b. Form factor
c. Crest factor

14-48 For the periodic waveform whose equation is given in Prob. 14-39, calculate the

a. RMS value
b. Form factor
c. Crest factor

Hint: $V^2 = (1/T)\int_0^t [v(t)]^2\, dt$. Use simple integration to find V.

14-49 For the periodic waveform whose equation is given in Prob. 14-40, calculate the

a. RMS value
b. Form factor
c. Crest factor

14-50 Each of the four waveforms shown in **Fig. 14-40** is measured on the dc and ac voltage scales of an *average-reading* volt-ohmmeter (VOM). For each of the waveforms shown, calculate and tabulate the

a. Average voltage
b. "True" RMS voltage
c. Form factor
d. FWR VOM reading of V_{RMS} on its ac scale (caution: this is not always the average of the input waveform shown, but it **is** 1.1107 times the FWR output *at all times*).
e. Percent error between the VOM reading and the true RMS voltage as measured by a true-reading RMS voltmeter.
f. Can John Smith's method of Ex. 14-34 be used in all cases? Explain when it can and cannot be used.

Sec. 14-12

14-51 A voltage waveform across a load has the equation $v(t) = -150\cos(700\pi t - 200°)$ mV. Express the waveform as a positive sinusoidal function and calculate the

a. Angular velocity of the waveform in radians per second
b. Maximum positive rate of change of voltage
c. RMS value of the waveform
d. Time, t, in milliseconds at which the first maximum positive rate of change of voltage occurs
e. Rate of change of voltage at $t = 1$ ms
f. Instantaneous value of voltage at $t = 1$ ms

14-52 The current waveform produced by the voltage in Prob. 14-51 has the equation $i(t) = -100\sin(700\pi t - 100°)$ mA. Express the waveform as a positive sinusoidal function and calculate the

a. Maximum positive rate of change of current
b. Instantaneous value of current at $t = 1$ ms
c. Rate of change of current at $t = 1$ ms
d. Time, t, in milliseconds at which the first maximum positive rate of change of current occurs
e. Phase relation of current with respect to voltage in Prob. 14-51, using voltage as a reference
f. Instantaneous power at time $t = 1$ ms

14-53 A voltage has the waveform $v(t) = 20 + 10\sin 200\pi t$ mV. Calculate the

a. Average voltage
b. RMS value of the ac component
c. RMS value of the waveform
d. Form factor
e. Crest factor

14-54 For each of the time-domain voltages given in **a** and **b** below applied to a 20-Ω resistor

1. Calculate the time-domain equation of $i(t)$
2. Draw the waveforms of $v(t)$ and $i(t)$ on the same axes
3. Plot the power curve, $p(t)$, as the product of $v(t)$ and $i(t)$ on the same axes, showing the maximum, minimum, and average power values
   a. $v(t) = 100 \sin(\omega t - 60°)$ V
   b. $v(t) = -50 \cos(\omega t - 240°)$ V

14-55 For each of the time-domain currents given in **a** and **b** below flowing in a 50-Ω resistor

1. Calculate the time-domain equation of $v(t)$
2. Draw waveforms of $v(t)$ and $i(t)$ on the same axes
3. Plot the power curve, $p(t)$ on the same axes as the product $v(t) \times i(t)$, showing maximum, minimum, and average power values
   a. $i(t) = -2 \sin(\omega t + 160°)$ A
   b. $i(t) = -3 \cos(\omega t + 180°)$ A

Secs. 14-14 to 14-16

14-56 Calculate the inductive reactance, X_L, of a 5-mH inductor at frequencies of

a. 60 Hz
b. 600 Hz
c. 3 kHz
d. 0 Hz (dc)

14-57 Calculate the inductances, L, that produce the given inductive reactances at the given frequencies:

a. 5 Ω at 20 Hz
b. 100 Ω at 1 kHz
c. 1 kΩ at 1 MHz
d. 10 kΩ at 5 GHz

14-58 For each of the time-domain currents given in **a** and **b** below flowing in a 50-mH inductor

1. Calculate the time-domain equation of $v(t)$
2. Draw waveforms of $v(t)$ and $i(t)$ on the same axes
3. Plot the power curve, $p(t)$, on the same axes showing maximum, minimum, reactive (quadrature), and average power values
   a. $2 \cos(500t - 120°)$ A
   b. $-500 \cos(400t + 120°)$ mA

14-59 Repeat Prob. 14-58, given the following time-domain voltages in **a** and **b** for a 200-mH inductor but calculating $i(t)$ instead of $v(t)$:

a. $-50 \sin(500t - 100°)$ V
b. $10 \cos(1000t - 20°)$ V

14-60 Calculate the capacitive reactance, X_C, of a 5-μF capacitor at frequencies of

a. 60 Hz
b. 600 Hz
c. 3 kHz
d. 0 Hz (dc)

14-61 Calculate the capacitances, C, that produce the given capacitive reactances at the given frequencies:

a. 5 Ω at 20 Hz
b. 100 Ω at 1 kHz
c. 1 kΩ at 1 MHz
d. 10 kΩ at 5 GHz

Sec. 14-17

14-62 For each of the time-domain voltages given in **a** and **b** below applied to a 0.5-μF capacitor:

1. Calculate the time-domain equation of $i_C(t)$
2. Draw the waveforms of $v_C(t)$ and $i_C(t)$ on the same axes
3. Plot the power curve, $p(t)$, on the same axes showing maximum, minimum, reactive (quadrature), and average power values
   a. $-50 \cos(5000t + 40°)$ V
   b. $100 \sin(500t - 20°)$ V

14-63 Repeat Prob. 14-62 but calculating $v_C(t)$ instead of $i_C(t)$, given a 0.1-μF capacitor and the following time-domain currents:

a. $-0.5 \cos(377t - 190°)$ μA
b. $-0.8 \sin(2000t - 100°)$ mA

14-64 Given the following pair of waveforms for voltage across and current in an ac circuit: $v(t) = 50 \cos(628.3t - 30°)$ V and $i(t) = 0.25 \sin(628.3t + 20°)$ A, calculate the

a. Circuit power factor, F_p
b. Circuit reactive factor, F_r
c. Apparent power, S
d. Active (real, average) power, P
e. Reactive power, Q
f. Reactance (indicate whether capacitive or inductive), X, of the circuit
g. Resistance of the circuit, R
h. Maximum positive excursion of the power waveform, $P_m^{(+)}$
i. Maximum negative excursion of the power waveform, $P_m^{(-)}$
j. Active power, using maxima computed in **(h)** and **(i)**
k. Apparent power, using maxima computed in **(h)** and **(i)**
l. Either L or C and draw the *practical series* equivalent (frequency-*independent*) configuration

14-65 Repeat all parts of Prob. 14-64 for the following pair of waveforms for voltage across and current in an ac circuit: $v(t) = -40 \cos(2513.3t + 50°)$ V and $i(t) = -200 \sin(2513.3t + 200°)$ mA

14-66 Repeat all parts of Prob. 14-64 for the following pair of waveforms for voltage across and current in an ac circuit: $v(t) = -120 \sin(1200t - 25°)$ V and $i(t) = -50 \cos(1200t - 100°)$ mA

14-67 Using the current waveform of $i(t)$ drawn as a reference for the given data and calculations of Prob. 14-64, draw waveforms of the following on one set of axes:

a. Voltage, $v(t)$
b. Instantaneous power, $p(t)$

[Hint: use calculated P_m^+ and P_m^- halfway between points where $p(t)$ is zero.]

Directly below, on a separate set of axes, draw to the same abscissa scale

c. Power dissipated in the resistor, p_R (Hint: draw $P_{mR}^+ = I_m^2 R$ in phase with I_m)
d. Quadrature reactive power, p_L, stored in the inductor and returned to the supply [hint: $p_L(t)$ leads the $p_R(t)$ waveform by 90°]
e. Verify active (average) power from the curves drawn in **(b)** and **(c)**
f. Designate S on the curve drawn in **(b)** and Q on the curve drawn in **(d)**

14-68 Using the current waveform of $i(t)$ drawn as a reference for the given data and calculations of Prob. 14-65, draw waveforms of the following on one set of axes:

a. Voltage, $v(t)$
b. Instantaneous power, $p(t)$

Hint: use calculated P_m^+ and P_m^- halfway between points where $p(t)$ is zero. Directly below, on a separate set of axes, draw to the same abscissa scale

c. Power dissipated in the resistor, p_R (Hint: draw $I_m^2 R = P_{mR}$ in phase with I_m)
d. Quadrature reactive power, p_C, stored in the capacitor and returned to the supply [Hint: $p_C(t)$ lags $p_R(t)$ by 90°]
e. Verify active (average) power from the curves drawn in **(b)** and **(c)**
f. Designate S on the curve drawn in **(b)** and Q on the curve drawn in **(d)**

14-20 ANSWERS

14-1 a 20 ms b 50 Hz c −6 V
14-2 a 30 μs b $3\overline{3}$ kHz c 30 mA
14-3 a 13 μs b 76.92 kHz c −15 V
14-4 a 22 ms b $45.\overline{45}$ Hz c 10 V
14-5 a 31.83 Hz b 31.42 ms c 6.283 rad d 540°
14-6 a $16.\overline{6}$ ms b $8\overline{3}$ ms c $16\overline{6}$ ms
14-7 a 2×10^9 c/s b 2 GHz
14-8 a 10 ns b 0.2 ns c 250 μs d 0.5 ps e $1\overline{6}$ ms
14-9 a 0.96 rad, 61.1^g
14-10 a 90°, 100^g
14-11 a 1257 rad/s, 7.2×10^4 deg/s
14-12 a 3.6×10^9 deg/s, 6.283×10^7 rad/s
14-13 a 12.57 ms, 79.58 Hz
14-14 a 10 mA, 60 Hz, +90°
14-15 a $10 \sin 377t$
14-16 a 60°, 300° b 233.13°, 306.87°
c 233.13°, 306.87° d 206.44°, 273.56°
e 8.463°, 211.54° f 210°, 570°
14-17 b1 65°, 425° and 245°, 605°
b2 155°, 335°, 515°, 695° c $0.7\overline{2}$, $1.7\overline{2}$, $2.7\overline{2}$ ms, etc.
14-18 a4 1.09 V, b4 11.75 V
14-19 a 155°, 335° b 65°, 425° c 178.6°, 311.42°
d 150.4°, 339.6°
14-20 a $1.7\overline{2}$, $3.7\overline{2}$ ms b 0.5416, 0.9029 ms
c 1.515, 3.93 ms d $2.0\overline{5}$, $3.3\overline{8}$ ms

14-25 a $v(t)$ lags by 70° b $v(t)$ lags by 10°
c $v(t)$ leads by 20° d $v(t)$ leads by 10°
14-26 a +50° b −90° c +15° d +10°
14-31 a 21 V b 100 kHz
14-32 a 0.3 mA b 100 Hz
14-33 a 5.625 V b 62.5 kHz
14-34 a 10 V b 5 Hz
14-35 a −0.76 mA b $\overline{45}$ Hz
14-36 a 47.75 V b $8.\overline{3}$ kHz
14-37 a 30 μA, 1 MHz
14-38 a 80 V b 50 kHz
14-39 b 104.2 mV c 109.4 mV
14-40 a $66.\overline{6}$ mV b 70 mV
14-41 a 45.5 V b $2.1\overline{6}$ c 1.32
14-42 a 2.725 mA b 9.083 c 1.101
14-43 a 9.0275 V b 1.605 c 3.323
14-44 a 14.14 V b, c 1.414
14-45 a 6.928 mA b −9.12 c ±1.155
14-46 a 63.42 V b 1.33 c 1.577
14-47 a 81.24 V b 1.016 c 1.231
14-48 a 176.8 mV b 1.7 c 2.83
14-49 a 89.44 mV b 1.342 c 2.236
14-50 B −6 V b $10.81\overline{6}$ V c −1.803
d 9.996 V e −7.58%

14-50 D only: a −4 V b 12 V c −3 d 11.55 V
e −3.76%

14-51 a 700π rad/s b 329.9 V/s c 106.1 mV
d 2.3 ms e −317.1 V/s f −41.35 mV

14-52 a 219.9 A/s b −43.84 mA c −197.7 A/s
d $2.\bar{2}$ ms e +10° f +1.813 mW

14-53 a 20 mV b 7.071 mV c 21.21 mV d 1.061
e 1.414

14-56 a 1.885 Ω b 18.85 Ω c 94.25 Ω d 0

14-57 a 39.8 mH b 15.9 mH c 159 μH d 1.59 μH

14-58 a ±50 W b ±2.5 W

14-59 a ±12.5 W b ±0.25 W

14-60 a 530.5 Ω b 53.05 Ω c 10.61 Ω d ∞

14-61 a 1.59 mF b 1.59 μF c 159 pF d 3.183 fF

14-62 a ±3.125 W b ±1.25 W

14-63 a ±3.316 nW b ±1.6 mW

14-64 a 0.766 b 0.643 c 6.25 VA d $4.78\bar{7}$ W
e 4.017 vars f 128.6 Ω g 153.2 Ω h 11.04 W
i −1.46 W j 4.787 W k 6.25 VA

14-65 a 0.5 b 0.866 c 4 VA d 2 W e 3.464 vars
f 173.2 Ω g 100 Ω h 6 W i −2 W j 2 W
k 4 VA l 2.3 μF

14-66 a 0.81915 b 0.5736 c 3 VA d 2.4575 W
e 1.7207 vars f 1377 Ω g 1966 Ω h 5.4575 W
i −0.5425 W j 2.4575 W k 3 VA l 0.6054 μF

14-67 e $4.78\bar{7}$ W f 6.25 VA

14-68 d 3.464 vars e 2 W f 4 VA

CHAPTER 15

Complex Algebra and Phasor Quantities

15-1 INTRODUCTION

The next logical step in our study is the analysis of series and parallel circuits containing R, L, and C elements connected to a sinusoidal voltage or current source. For such analysis, use of waveforms in the time domain is very difficult, unwieldy, and has limitations, particularly with more difficult circuits such as polyphase circuits and those involving dependent sources.

To overcome these objections, we will *convert* our *time-domain* equations to the *frequency* (or *phasor*) domain. This enables us to use complex algebra to solve our problems with relative ease. We will discover that an ac of the form $i(t) = I_m \sin(\omega t + \phi)$ may be easily *transformed* to its *polar* (or phasor) form $\mathbf{I}\underline{/\phi}$. From there, it is a simple matter to convert it to its corresponding *rectangular* form and/or its *exponential* form (by Euler's theorem).

We will learn that operations such as addition or subtraction of our original time-domain equations are best performed in *rectangular* or complex form. Multiplication and division, powers, and roots are more easily performed in either *polar* or *exponential* form.

For these reasons, a separate chapter devoted to the techniques of complex algebra is a prerequisite to the further study of ac.[1] In this chapter we will learn many of the rules governing the mathematical operations described above. Our task, in recent years, has been greatly simplified by the widespread availability of inexpensive electronic calculators, which very easily perform rectangular to polar (R → P) and polar to rectangular (P → R) conversions.[2]

15-2 PHASOR QUANTITIES IN THE TIME AND FREQUENCY DOMAINS

Phasor quantities were introduced in Sec. 14-6. In Fig. 14-9 we showed a current phasor, $\mathbf{I}_m$, lagging a voltage phasor, $\mathbf{V}_m$ by some phase angle, ϕ. Both phasors, rotating counterclockwise (CCW), could generate their respective time-domain

---

[1] In 1893, C. P. Steinmetz first suggested that many advantages would result from the application of complex algebra to ac circuit theory.

[2] Because of the wide varieties of techniques used in such calculators to perform the required conversions, readers should consult their own calculator instruction manual and practice these conversions until they have mastered them.

waveforms. We also noted that since both phasors are rotating (CCW) at the *same* angular frequency, we could "stop the clock" and show their relation in phasor ("vector") form. This relation is again shown in **Fig. 15-1a**.

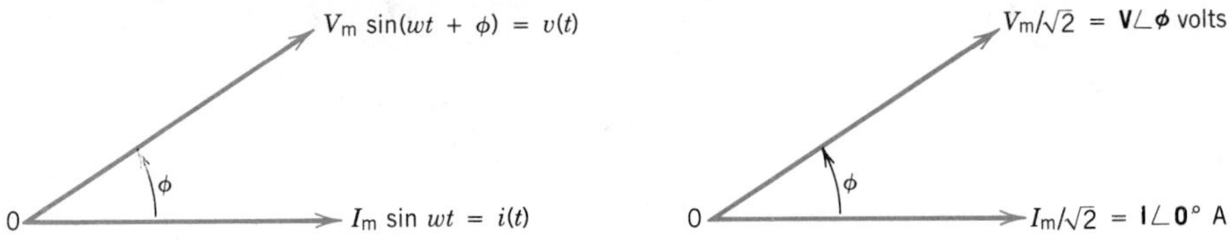

a. Voltage and current phasors b. Frequency-domain representations

Figure 15-1 Frequency-domain representation of rotating voltage, current phasors

When time-varying quantities are *expressed* in the frequency (or phasor) domain, two major changes occur:

1. The quantities are expressed as RMS values (in boldface type) at some phase angle (ϕ or θ).
2. The frequency is *no longer shown* since it is *constant*.

Figure 15-1b shows the frequency-domain representations of $v(t)$ and $i(t)$, the time-domain voltage and current, respectively. Note that since the reference waveform is $i(t)$, its frequency-domain representation is $\mathbf{I}\angle 0°$ A, where **I** is its RMS value. In the same way, $v(t) = V_m \sin(\omega t + \phi)$ may be expressed as $\mathbf{V}\angle\phi$, where ϕ is the constant phase angle by which **V** leads **I**.

But it must be noted that the phasor representation of $\mathbf{V}\angle\phi$ shown in Fig. 15-1b and again in Fig. 15-2 is a *complex number*. If we lay off the phasor $\mathbf{V}_m$ at an angle ϕ to some scale, the point of the arrow may be located with reference to the horizontal (real) axis and the vertical (imaginary or j) axis. Thus, the phasor $\mathbf{V}\angle\phi$ with respect to the real (horizontal) axis is represented by the distance $\mathbf{V}\cos\phi$. Similarly, the phasor $\mathbf{V}\angle\phi$ with respect to the vertical (imaginary) axis, is represented by the distance $\mathbf{V}\sin\phi$. In order to distinguish between the real (horizontal) and imaginary (vertical) axes we place the letter "j" before all *imaginary* quantities. We may then write the relation between the polar and rectangular forms of any complex number, $R\angle\pm\theta$ by[3]

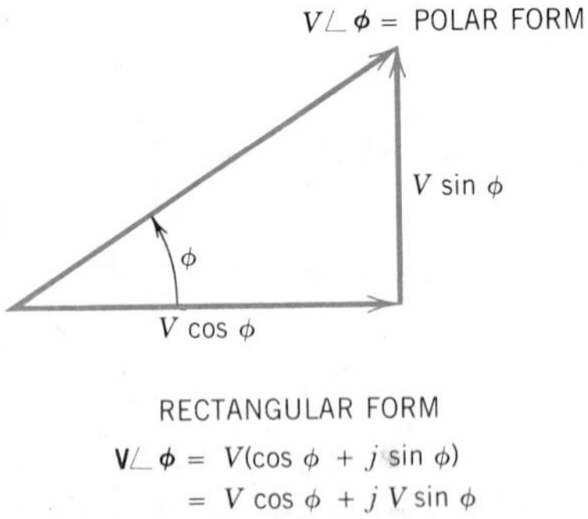

Figure 15-2 Polar and rectangular coordinates of phasor, $\mathbf{V}\angle\phi$

$$R\angle\pm\theta = R(\cos\theta \pm j\sin\theta) = R\cos\theta \pm jR\sin\theta \qquad (15\text{-}1)$$

where j is the operator rotating the complex number by 90° into the imaginary plane $= \sqrt{-1}$

$R\cos\theta$ is the real or horizontal component of the complex number

$jR\sin\theta$ is the imaginary or vertical component of the complex number

Figure 15-2 shows the phasor $\mathbf{V}\angle\phi$ in polar form and its representation in rectangular form as $V\cos\phi + jV\sin\phi$.

[3] The letters ϕ or θ may be used interchangeably to indicate phase angle because we may be dealing with more than one phasor, each of which has its particular phase angle with respect to the reference phasor on the real plane.

EXAMPLE 15-1

Given the waveform $v(t) = 141.42 \sin(\omega t + 30°)$ V, express it in
a. Polar form b. Rectangular form

Solution

a. $\mathbf{V}\underline{/\theta} = (V_m/\sqrt{2})\underline{/\theta} = (141.42/\sqrt{2})\underline{/30°} = \mathbf{100\underline{/30°}\ V}$
b. Convert from polar to rectangular form (P → R) using a calculator: $100\underline{/30°}$ V = **(86.6 + *j*50) volts**

What is the reason or necessity for the P → R conversion? It is used primarily for a number of mathematical operations such as phasor addition. Let us assume that we have two ac voltages, each at its own phase angle, connected in series. By Kirchhoff's voltage law (KVL), the total voltage is the graphical time-domain waveform sum of each. Adding waveforms graphically is a laborious job. But if we add them in the **frequency** domain, it is relatively simple, as shown by Ex. 15-2.

EXAMPLE 15-2

Two waveforms, $v_1(t)$ and $v_2(t)$, are connected in series; $v_1(t) = 141.42 \sin(5000t + 20°)$ V while $v_2(t) = 56.57 \sin(5000t + 70°)$ V

a. Express each waveform as a polar quantity in the frequency domain
b. Convert each waveform to rectangular form
c. Add the two waveforms in rectangular form, leaving answer in rectangular form

Solution

a. $\mathbf{V}_1 = \dfrac{141.42}{\sqrt{2}}\underline{/20°} = \mathbf{100\underline{/20°}\ V}$ and

$\mathbf{V}_2 = \dfrac{56.57}{\sqrt{2}}\underline{/70°}\ \text{V} = \mathbf{40\underline{/70°}\ V}$

b. $\mathbf{V}_1 = 100\underline{/20°}\ \text{V} = \mathbf{93.97 + \mathit{j}34.20}$ volts;
$\mathbf{V}_2 = 40\underline{/70°}\ \text{V} = \mathbf{13.68 + \mathit{j}37.59}$ volts
c. As with all vectors, we sum the horizontal and vertical components separately: $\mathbf{V}_1 + \mathbf{V}_2 = (93.97 + j34.2) + (13.68 + j37.59) = \mathbf{107.65 + \mathit{j}71.79}$ volts

We can see that the answer obtained in Ex. 15-2 as the phasor sum of the two voltages, expressed in rectangular form, does not provide any useful information as to magnitude and phase of the vector sum. Clearly, we must convert the rectangular form back to polar form in the frequency domain and then write its analog as a time-varying function in the time domain. This is shown in Ex. 15-3.

EXAMPLE 15-3

Given the phasor sum in Ex. 15-2c,

a. Convert it to polar form (R → P) using the electronic calculator
b. Convert it from polar form to a time-domain equation, $v_T(t)$

Solution

a. $\mathbf{V}_T = 107.65 + j71.79 = \mathbf{129.4\underline{/33.7°}\ V}$
b. $v_T(t) = \sqrt{2} \times 129.4 \sin(\omega t + 33.7°)$ V
$= \mathbf{183 \sin(5000\mathit{t} + 33.7°)\ V}$

Examples 15-2 and 15-3 are just two cases of many that show the relative advantages of solving ac problems by complex algebra techniques. If we had to add the two waveforms graphically and then find the resultant graphically, it would be both time-consuming and difficult to obtain the answer of Ex. 15-3b with any comparable accuracy. Yet, using complex algebra, the answer is both quickly and accurately obtained.

The following sections are devoted to a series of complex algebra techniques intended to develop the reader's proficiency and understanding, as well as confidence in the use of the electronic calculator. But before we do this, we must develop some definitions and clarify some terms.

15-3 DISTINGUISHING BETWEEN SCALARS, VECTORS, AND PHASORS

A *scalar* (quantity) is a numeric quantity having *magnitude* only. Examples of scalars are length, mass, time, energy, charge, temperature, density, and work.

A *vector* (quantity) is a numeric quantity having *both magnitude* and *direction*. Examples of vector quantities are velocity, acceleration, force, moment of a force, electric field strength, and dielectric polarization. All vectors have coordinates in space and many are expressed three-dimensionally in x, y, and z axes.

A vector having *negative* magnitude is the same as a vector having the same absolute value but pointing in the *opposite* direction. *Vector* quantities are *not* time-varying. This is one important difference between vectors and *phasors*.

A *phasor* (quantity) is a numeric quantity having both magnitude and direction which represents a *time-varying* quantity. Examples of phasors are voltage, current, and power. Phasors may represent periodic sinusoidal and nonsinusoidal waveforms, as long as they are time-varying. Since phasors are expressed in the two-dimensional complex plane, they are *not* space coordinates like vectors. But when a phasor diagram is drawn we must assume that

1. All phasors drawn on the same diagram have the same angular frequency, ω.
2. The magnitude of the phasor is its RMS value.[4]
3. The angle, ϕ or θ, specifying the phasor direction from the origin of the intersection of the real and imaginary axes is its angular rotation from the reference-zero sine wave. (Some texts use a reference-zero *cosine* wave instead.)

Phasors may be *added* graphically by the head–tail method as in vector addition. We will also accomplish this by successive separate addition of horizontal and vertical components of each phasor and converting the rectangular form to polar form for the final sum. (See Exs. 15-5 and 15-6 and also Sec. 15-10.)

Phasors may be *subtracted* graphically by reversing the arrow of the negative phasor and adding it to the other phasors. A negative phasor has its original magnitude but its phase angle is increased (or decreased) by 180°. (See Ex. 15-6.)

The last and perhaps most important point concerns phasor equivalence. We have indicated that a phasor quantity may be used to *express* a time-varying quantity. For example, in the time domain $v(t) = 100 \sin(\omega t + 40°)$ volts. This is expressed by a phasor in the frequency domain as $\mathbf{V} = 70.71\underline{/40°}$ volts. But we can *never* write (as is sometimes shown) an equality sign between these two expressions.[5] We must conclude, therefore, that

1. Any time-varying function in the time domain may be *represented* by a complex number in the frequency domain as a phasor quantity.
2. The phasor quantity *representation* is *not equal* to the time-domain quantity.

Nevertheless, since the phasor quantity yields the RMS value of magnitude and its phase with respect to some reference quantity in the time domain (at the same frequency), it is a much simpler quantity to substitute algebraically in ac circuit equations. We will discover that the use of phasor quantities enables us to treat ac circuits similarly to dc circuits, using the same theorems and laws covered in earlier chapters. Consequently, because the use of complex algebra associated with

---

[4] Some texts use both maximum (peak) and RMS values on phasor diagrams. We will use only RMS values, exclusively.

[5] If the two expressions were equal, we could use the equality to solve for unknown quantities, as we do with any equation.

phasors is such a powerful technique, we turn to the algebra of complex numbers, which is so essential to ac circuit analysis.

15-4 THE *j* OPERATOR

The j operator was first introduced in Eq. (15-1) above as the operator rotating the complex number by 90° into the imaginary plane. Mathematically, $1j = \sqrt{-1}$, as noted in Eq. (15-1). The scales of the axes, both real and imaginary, are shown in **Fig. 15-3**. Note that in both cases, the negative axis is 180° in the opposite direction from the positive axis and increases in the opposite direction. Also note that values on the imaginary j axis are always *preceded* by the letter j. Consequently, we write $-j2$ and not $-2j$.

Since $j = \sqrt{-1}$, it follows algebraically that $j^2 = -1$ and also that $j^3 = -\sqrt{-1} = -j$ and therefore we find $j^4 = +1$ and finally that $j^5 = j^3 \times j^2 = (-j) \times -1 = j$

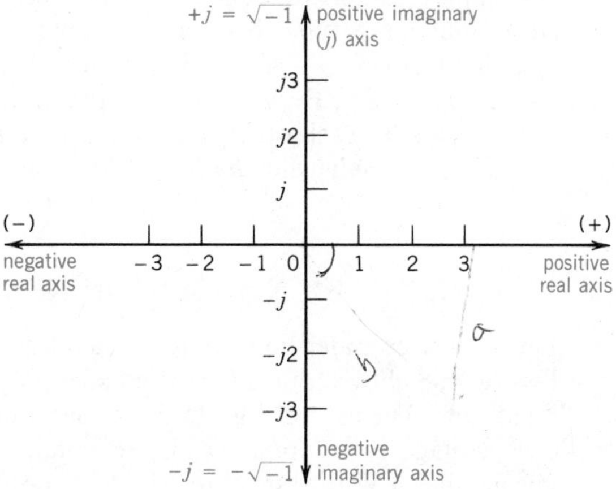

Figure 15-3 Axes of complex numbers

EXAMPLE 15-4

Perform the following mathematical operations on the given expressions containing imaginary quantities:

a. $j5 + j3$
b. $j5 - j3$
c. $j3 - j5$
d. $(j5) \times (j3)$
e. $j5(-j3)$
f. $j5(-j3)j5(j3)$
g. $j6/j3$
h. $j20/(-j50)$

Solution

a. $j5 + j3 = \mathbf{j8}$
b. $j5 - j3 = \mathbf{j2}$
c. $j3 - j5 = \mathbf{-j2}$
d. $(j5)(j3) = j^2 15 = \mathbf{-15}$
e. $j5(-j3) = -j^2 15 = \mathbf{15}$
f. $j5(-j3)j5(j3) = -j^4 225 = \mathbf{-225}$
g. $j6/j3 = \mathbf{2}$
h. $j20/-j50 = \mathbf{-0.4}$

The reader should review each of the solutions shown in Ex. 15-4 and attempt to solve each part separately and independently, using the solutions shown as feedback to correct errors.

Although not a rigorous derivation, **Fig. 15-4** shows a possible empirical proof for the designation of the axes of Fig. 15-3. Consider a radius vector, R, in the real plane at $R\underline{/0°}$. When rotated CCW by 90°, this is equivalent to multiplying R

by j, yielding jR, as shown in Fig. 15-4. Then a rotation of 180° is equal to j^2R or $-R$. Further, rotation by 270° produces $j^3R = -jR$, as shown in Fig. 15-4. Finally, rotation by 360° returns the radius vector R to its zero reference position by $j^4R = 1R = R/\underline{0°}$.

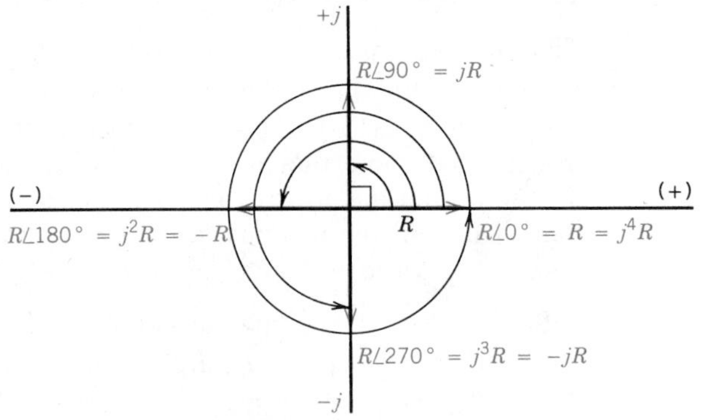

Figure 15-4 Proof of real and imaginary axes by 90° steps of rotation

Complex numbers are called *complex* because they contain both real and imaginary numbers. A typical complex number in rectangular form and its equivalent in polar form is $3 + j4 = 5/\underline{53.13°}$. Another typical complex number might be $4 + j3 = 5/\underline{36.87°}$. Even though the two numbers have the same magnitude, they are *not equal* to each other, since their *directions* are *unequal.* Two complex numbers are equal *only* when *both* their horizontal and vertical components *are equal.*

15-5 COMPLEX NUMBERS IN RECTANGULAR FORM

Let us now consider how complex quantities are written in the rectangular form. **Figure 15-5** shows four different voltages, $\mathbf{V}_1$ through $\mathbf{V}_4$, one in each of the four quadrants. The axes of Fig. 15-5 are the same as those of Fig. 15-3. The values of each voltage are represented in rectangular form. (The reader should verify the given values.) With respect to Fig. 15-5, note that

1. In the first quadrant, *both* horizontal and vertical values are positive.
2. In the second quadrant, the *real* value is *negative* and the *imaginary* is *positive.*
3. In the third quadrant, *both* horizontal and vertical values are *negative.*
4. In the fourth quadrant, the *real* value is *positive* and the *imaginary* value is *negative.*

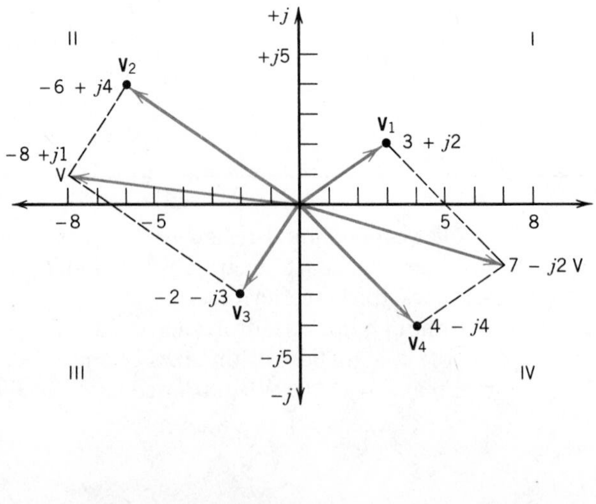

Figure 15-5 Complex numbers in rectangular form in each of four quadrants

5. The real value is always given first, followed by the imaginary value, in expressing the rectangular form. Thus, $\mathbf{V}_1 = 3 + j2$ volts, while $\mathbf{V}_3 = -2 - j3$ volts.

15-5.1 Addition and Subtraction of Complex Numbers in Rectangular Form

Let us now add $\mathbf{V}_1$ and $\mathbf{V}_4$, as shown in Fig. 15-5, and $\mathbf{V}_2$ to $\mathbf{V}_3$.

EXAMPLE 15-5

Add the following complex numbers:

a. $(3 + j2) + (4 - j4)$
b. $(-6 + j4) + (-2 - j3)$

Solution

a. As mentioned in Sec. 15-3, we add the real parts and the imaginary parts separately. Therefore, $\mathbf{V}_1 + \mathbf{V}_4 = (3 + j2) + (4 - j4) = \mathbf{7 - j2}$ **volts**
b. Similarly, $(-6 + j4) + (-2 - j3) = \mathbf{-8 + j1} = \mathbf{V}_2 + \mathbf{V}_3$

The reader should note that this addition may be verified in Fig. 15-5 by the vector parallelogram method or (better still) the head–tail method, for both parts (**a**) and (**b**).

Now suppose we wish to *subtract* $\mathbf{V}_1$ from $\mathbf{V}_4$ (i.e., find $\mathbf{V}_4 - \mathbf{V}_1$). Recalling that a *reversal* of a vector is the same vector taken 180° out of phase, we may write $\mathbf{V}_4 - \mathbf{V}_1 = (4 - j4) - (3 + j2) = (4 - j4) + (-3 - j2) = 1 - j6$ volts.

Observe that the process of *subtraction* results in *addition* of *negative* numbers in both the real and imaginary planes.[6]

EXAMPLE 15-6

For the voltages given in Fig. 15-5, find the following phasor sums:

a. $\mathbf{V}_1 - \mathbf{V}_4$
b. $\mathbf{V}_2 - \mathbf{V}_3$
c. $\mathbf{V}_3 - \mathbf{V}_4$
d. $\mathbf{V}_2 - \mathbf{V}_4$
e. $\mathbf{V}_4 - \mathbf{V}_2$

Solution

a. $\mathbf{V}_1 - \mathbf{V}_4 = 3 + j2 - (4 - j4) = \mathbf{-1 + j6}$
b. $\mathbf{V}_2 - \mathbf{V}_3 = -6 + j4 - (-2 - j3) = \mathbf{-4 + j7}$
c. $\mathbf{V}_3 - \mathbf{V}_4 = -2 - j3 - (4 - j4) = \mathbf{-6 + j1}$
d. $\mathbf{V}_2 - \mathbf{V}_4 = -6 + j4 - (4 - j4) = \mathbf{-10 + j8}$
e. $\mathbf{V}_4 - \mathbf{V}_2 = 4 - j4 - (-6 + j4) = \mathbf{10 - j8}$

Note from Ex. 15-6**d** and **e**, that $\mathbf{V}_2 - \mathbf{V}_4$ is the *reversal* of $\mathbf{V}_4 - \mathbf{V}_2$; i.e., it is a phasor that is exactly 180° out of phase. Furthermore, from all parts observe that subtraction of phasors essentially consists of *adding the reversal* of one phasor to another phasor. The phasor to be *subtracted* is the phasor that is *reversed.*

15-5.2 Multiplication of Complex Numbers in Rectangular Form

We will discover later that multiplication (and division) of complex numbers may be performed more easily in polar form. Nevertheless, in order to understand the process, we will use the rectangular form for multiplication.[7] Inherent to the process are the following realizations:

[6] From a mathematical point of view, subtraction is a special form of addition.

[7] There are currently several inexpensive electronic hand calculators that perform addition, subtraction, multiplication, and division of complex numbers in *rectangular* form. One of these is the Sharp Scientific Calculator, EL506P.

1. Multiplication of two real numbers gives a real number.
2. Multiplication of a real number by an imaginary number gives an imaginary number.
3. The product of two imaginary numbers is a real number.
4. The product of two complex numbers is always a complex number.

EXAMPLE 15-7
Multiply the following complex numbers, leaving the answer in rectangular form:

a. $(4 + j3)(5 - j12)$
b. $(3 - j4)(8 - j6)$

Solution

a. $(4 + j3)(5 - j12) = 20 + j15 - j48 - j^2 36$
$= 20 - j33 + 36 = \mathbf{56 - j33}$
b. $(3 - j4)(8 - j6) = 24 - j32 - j18 + j^2 24 = \mathbf{0 - j50}$

Example 15-7**b** shows that even when the solution is $0 - j50$, we emerge with a complex number whose real value is zero. Consequently, an imaginary number $(-j50)$ is a special case of a complex number $(0 - j50)$.

15-5.3 Conjugate of a Complex Number in Rectangular Form

Unlike the *reversal* of a complex number, which multiplies both the real and imaginary parts by -1, the *conjugate* of a complex number multiplies only the imaginary part by -1, as shown by Ex. 15-8.

EXAMPLE 15-8
Show the (1) conjugate and (2) reversal of the following complex numbers:

a. $4 + j3$
b. $5 - j12$
c. $3 - j4$
d. $0 - j50$

Solution

a. conjugate = $\mathbf{4 - j3}$; reversal = $\mathbf{-4 - j3}$
b. conjugate = $\mathbf{5 + j12}$; reversal = $\mathbf{-5 + j12}$
c. conjugate = $\mathbf{3 + j4}$; reversal = $\mathbf{-3 + j4}$
d. conjugate = $\mathbf{0 + j50}$; reversal = $\mathbf{0 + j50}$ (same value in both cases)

15-5.4 Division of Two Complex Numbers in Rectangular Form

Division consists of *rationalizing the denominator* term of the ratio of two complex numbers. This is done by multiplying *both* the numerator and the denominator by the *conjugate* of the *denominator*. This process, shown in Ex. 15-9, causes the denominator to become a real number that is actually the product sum of the squares of the real and imaginary parts.

EXAMPLE 15-9
Perform the indicated division on the following pairs of complex numbers:

a. $\dfrac{5 - j12}{4 + j3}$

b. $\dfrac{0 - j50}{3 - j4}$

c. $\dfrac{(4 + j3)(5 - j12)}{(3 - j4)(8 - j6)}$

Solution

a. $\dfrac{(5 - j12)(4 - j3)}{(4 + j3) \times (4 - j3)} = \dfrac{-16 - j63}{25} = \mathbf{-0.64 - j2.52}$

b. $\dfrac{0 - j50}{3 - j4} \times \dfrac{(3 + j4)}{(3 + j4)} = \dfrac{200 - j150}{25} = \mathbf{8 - j6}$

c. $$\frac{(4+j3)(5-j12)}{(3-j4)(8-j6)} = \frac{56-j33}{0-j50} \times \frac{j50}{j50}$$
$$= \frac{+1650+j2800}{2500} = \mathbf{0.66 + j1.12}$$

Please note the following from Ex. 15-9 regarding complex number division in rectangular form:

1. Multiplying *both* the numerator and the denominator by the same conjugate does *not* change the value of the original ratio of two complex numbers.
2. The process of using conjugate multiplication (of numerator and denominator) always results in a *real* number in the *denominator*.
3. The real number obtained in the denominator always is the sum of the squares of the real and imaginary parts of the original complex number in the denominator.
4. When either or both the numerator and denominator are products of two (or more) complex numbers, as in Ex. 15-9c, we must first multiply them out to obtain a single complex number in both numerator and denominator. After this, we rationalize the denominator in the usual way.

15-6 COMPLEX NUMBERS IN POLAR FORM

As noted earlier, multiplication and division are more efficiently accomplished in polar form. Other processes, such as obtaining powers, roots, and natural logarithms of complex numbers, are also done more easily using the polar form. The polar form was first shown in Eq. (15-1) and illustrated in Ex. 15-1. The following examples should serve as practice in making P → R (polar to rectangular) conversions and R → P (rectangular to polar) conversions. For reasons that we will discover later, in making R → P conversions, we express the angle in radians as well as degrees.[8]

15-6.1 Rectangular-to-Polar Conversion, Angle in Degrees and/or Radians

Let us first attempt some R → P conversions in which the angle is expressed in degrees. **Figure 15-6** shows four points in the complex plane, one in each quadrant. We will discover that the (absolute) magnitude of all four points in polar form is the *same*. Yet because each is in a different quadrant, the direction (and therefore the angle) must be different. Example 15-10 shows some interesting results regarding the determination of the angle, θ for each of the complex numbers.

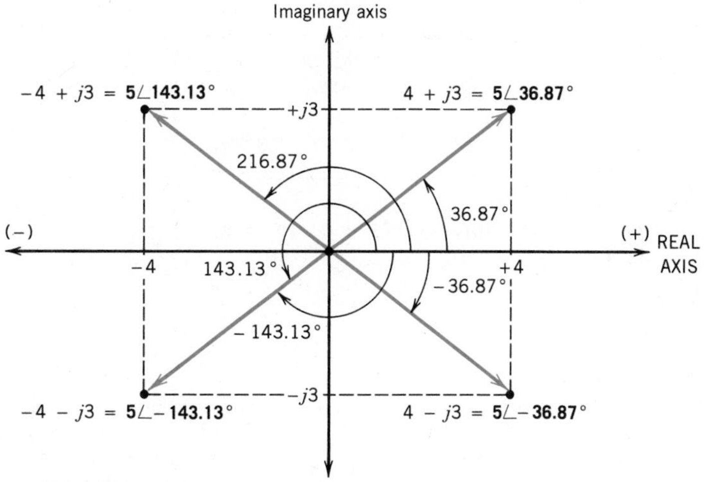

Figure 15-6 Rectangular to polar conversion in all four quadrants. Example 15-10

[8] Almost all modern electronic calculators offer this option, obtained by depressing a radian key or a switch.

EXAMPLE 15-10

Given the following complex numbers in rectangular form, convert them to polar form, angle in **degrees**:

a. $4 + j3$ c. $-4 - j3$
b. $-4 + j3$ d. $4 - j3$

Solution

Using an electronic calculator having R → P conversion capability,

a. $4 + j3 = \mathbf{5\angle 36.87^\circ}$
b. $-4 + j3 = \mathbf{5\angle 143.13^\circ}$
c. $-4 - j3 = \mathbf{5\angle -143.13^\circ}$ or $\mathbf{5\angle 216.87^\circ}$ (depending on calculator)
d. $4 - j3 = \mathbf{5\angle -36.87^\circ}$ or $\mathbf{5\angle 323.13^\circ}$ (depending on calculator)

Example 15-10 yields some interesting conclusions:

1. As noted earlier, the magnitudes are all the same but the directions and therefore the respective angles are different. Consequently, no two complex numbers in Ex. 5-10 are equal!
2. Some calculators always express the angle as a positive value (CCW rotation), even when the angle is greater than 180°. See solution parts (**c**) and (**d**).
3. Other calculators are programmed to convert this angle so that it is always equal to or less than 180°. Consequently *positive* rotation angles *greater than* 180° are expressed as *negative* rotation angles *less than* 180°. See solution parts (**c**) and (**d**).[9] Either answer is acceptable.

EXAMPLE 15-11

Given the following complex numbers in rectangular form, convert them to polar form, angle in **radians**:

a. $4 + j3$
b. $-4 + j3$
c. $-4 - j3$
d. $4 - j3$

Solution

Using an electronic calculator with R → P conversion capability and either a DRG or RAD key or a comparable switch:

a. $4 + j3 = \mathbf{5\angle 0.6435}$
b. $-4 + j3 = \mathbf{5\angle 2.4981}$
c. $-4 - j3 = \mathbf{5\angle -2.4981}$ or $\mathbf{5\angle 3.7851}$
d. $4 - j3 = \mathbf{5\angle 5.6397}$ or $\mathbf{5\angle -0.6435}$

Example 15-11 yields a number of interesting conclusions:

1. Again, the magnitudes are all the same but the angular directions are different.
2. Some calculators express all angles as positive (CCW rotation) values, even when greater than π radians.
3. Other calculators are programmed to convert the angle in radians so that it is always less than π radians. Consequently, positive rotation angles greater than π radians are expressed as negative rotation angles less than π. Either answer is acceptable.
4. Note from the solution that *no units are shown for radian measure* since the radian is unitless.

15-6.2 Polar-to-Rectangular Conversion, Angle in Degrees

The electronic calculator should also yield the conversion of complex numbers, expressed originally in polar form, to rectangular form, by the appropriate P → R conversion technique. Let us begin with complex numbers expressed in polar form, with angle in degrees.

EXAMPLE 15-12

Given the following complex numbers expressed in polar form, convert them to rectangular form using appropriate P → R conversion techniques:

a. $13\angle 67.38^\circ$ d. $41\angle 257.32^\circ$
b. $25\angle 163.74^\circ$ e. $61\angle -280.39^\circ$
c. $17\angle -61.93^\circ$

[9] Still other older calculators are incapable of yielding angles greater than 90° in making these conversions. But they do yield the angle from the negative real axis to the polar radius vector in the given quadrant. Subtracting the obtained angle from either ±180° should yield the above angle. But in all seriousness, because of the many and frequent P → R conversions and R → P conversions that must be performed in ac circuit analysis, the reader would be best advised to trade in such a calculator and obtain one with a little more versatility and usefulness.

Solution

a. $13\underline{/67.38^\circ} = \mathbf{5 + j12}$
b. $25\underline{/163.74^\circ} = \mathbf{-24 + j7}$
c. $17\underline{/-61.93^\circ} = \mathbf{8 - j15}$
d. $41\underline{/257.32^\circ} = \mathbf{-9 - j40}$
e. $61\underline{/-280.39^\circ} = \mathbf{11 + j60}$

Readers should perform the above solutions (again and again) until they have obtained the answers given. This technique is used so frequently in ac circuit analysis that proficiency in it is a basic necessity for solving problems. With respect to Ex. 15-12, there are certain important insights that will help readers determine whether they have obtained a correct or a ridiculous answer. These are:

1. The sum of the rectangular values obtained is always greater than the given polar value by a relatively small factor, ranging from more than 1 to always less than 1.42! For example, in Ex. 15-12**a**, the sum of 5 + 12 exceeds 13 by a factor of 17/13 or 1.3, which shows that the answer is reasonable, since the range of answer acceptability is fairly narrow.
2. The given angle of the polar quantity places the complex number in a particular quadrant. This quadrant determines whether the real and imaginary values are positive or negative, respectively. For example, Ex. 15-12**b**, having an angle of 163.74°, must be in the *second* quadrant, where the real value is negative but the imaginary value is positive, as shown in the solution.
3. Care should be taken with *negative* angles, particularly *large* ones. Example 15-12**e** with an angle of −280.39° turns out to be in the *first* quadrant, where *both* real and imaginary values are *positive*, as shown in the solution.

15-6.3 Polar-to-Rectangular Conversion, Angle in Radians

The electronic calculator should also yield the conversion of complex numbers expressed in polar form, with angle given in radians, to rectangular form, using appropriate P → R techniques. Consider Ex. 15-13, in which the polar values have their angles shown in radians (note the absence of degree signs). Using an electronic calculator with P → R capability and either a [DRG] or [RAD] key or a comparable switch, perform Ex. 15-13 as shown below.

EXAMPLE 15-13

Given the following complex numbers in polar form, angle in radians, convert them to rectangular form using appropriate P → R conversion techniques:

a. $29\underline{/0.761}$
b. $125\underline{/4.3527}$
c. $65\underline{/-4.9611}$
d. $37\underline{/-1.2405}$
e. $65\underline{/-7.4592}$

Solution

a. $29\underline{/0.761} = \mathbf{21 + j20}$
b. $125\underline{/4.3527} = \mathbf{-44 - j117}$
c. $65\underline{/-4.9611} = \mathbf{16 + j63}$
d. $37\underline{/-1.2405} = \mathbf{12 - j35}$
e. $65\underline{/-7.4592} = \mathbf{25 - j60}$

Again, the same insights prevail when the complex number is expressed in polar form angle in radians:

1. The sum of the rectangular values obtained is greater than the polar magnitude by a factor that is more than 1 but always less than 1.42 (a fairly limited and narrow range). All of the answers given in the foregoing solution fall in this range, which leads us to assume that the answers are probably correct.[10]
2. Again, the given angle in polar form, expressed in radians, places the complex number in a particular quadrant: **a** is in the first quadrant, **b** in the third quadrant, **c** in the first quadrant, **d** in the fourth quadrant, and **e** in the fourth quadrant. This serves as a check on whether the real and/or imaginary values are positive or negative, respectively.
3. Care should be taken with angles that are particularly large (or small). In **e** the rotation is negative (CW) and greater than 2π (or 360°), which places it in the fourth quadrant.

[10] The reader may have noted that all the answers to Exs. 15-12 and 15-13 are **integers**. This is not an accident. They are all Pythagorean triples! The most commonly known Pythagorean triple is the 3–4–5 triangle, where $3^2 + 4^2 = 5^2$. But there are many Pythagorean triples and the examples cited have used about 10 of them. Of course, one could use multiples of the 3–4–5 triangle, such as 6–8–10 or 9–12–15, but that would be cheating. The Pythagorean triples used here are all different for the simple reason that the angles are different!

15-7 ARITHMETIC OPERATIONS IN POLAR FORM

15-7.1 Multiplication

We indicated earlier that operations such as addition and subtraction are (best) done in rectangular form and multiplication and division are (best) done in polar form. Let us return to some earlier examples (15-7 and 15-9) in which these processes were performed in rectangular form and compare the polar process for ease and efficiency. Multiplication is accomplished in polar form by multiplying magnitudes and adding angles, algebraically.

EXAMPLE 15-14

Multiply the following complex numbers by

1. Converting them first to polar form
2. Multiplying them in polar form (multiply magnitudes and add angles)
3. Converting the answer to rectangular form as a check for accuracy

a. $(4 + j3)(5 - j12)$ **b.** $(3 - j4)(8 - j6)$

Solution

Multiplying magnitudes and adding angles in polar form:

a. $(4 + j3)(5 - j12) = (5\angle 36.87^\circ)(13\angle -67.38^\circ)$
$= 65\angle -30.51^\circ = \mathbf{56 - j33}$

b. $(3 - j4)(8 - j6) = (5\angle -53.13^\circ)(10\angle -36.87^\circ)$
$= 50\angle -90^\circ = \mathbf{0 - j50}$

With respect to the solution above, please note that

1. Using R → P conversion, each complex number is first converted to polar form.
2. In polar form, multiplication is accomplished by multiplying magnitudes and adding angles algebraically.
3. The answer in polar form is converted to rectangular form by using P → R conversion. These answers are exactly the same as those in Ex. 15-7.

15-7.2 Division

Division is accomplished in polar form by *dividing magnitudes* and *subtracting angles* algebraically. Let us perform each of the rectangular divisions of Ex. 15-9, previously done by conjugate rationalization of the denominator.

EXAMPLE 15-15

Perform the indicated division on the following pairs of complex numbers, using division in polar form:

a. $\dfrac{5 - j12}{4 + j3}$

b. $\dfrac{0 - j50}{3 - j4}$

c. $\dfrac{(4 + j3)(5 - j12)}{(3 - j4)(8 - j6)}$

Solution

a. $\dfrac{5 - j12}{4 + j3} = \dfrac{13\angle -67.38^\circ}{5\angle 36.87^\circ} = 2.6\angle -104.25^\circ = \mathbf{-0.64 - j2.52}$

b. $\dfrac{0 - j50}{3 - j4} = \dfrac{50\angle -90^\circ}{5\angle -53.13^\circ} = 10\angle -36.87^\circ = \mathbf{8 - j6}$

c. $\dfrac{(4 + j3)(5 - j12)}{(3 - j4)(8 - j6)} = \dfrac{(5\angle 36.87^\circ)(13\angle -67.38^\circ)}{(5\angle -53.13^\circ)(10\angle -36.87^\circ)}$
$= 1.3/59.49^\circ = \mathbf{0.66 + j1.12}$

With respect to the solutions shown in Ex. 15-15, please note that

1. Division in polar form is accomplished by dividing magnitudes and subtracting angles algebraically.
2. The answers obtained by the process of polar division are exactly the same as those of Ex. 15-9, obtained by rectangular conjugate rationalization of the denominator.
3. The method uses both R → P and P → R conversions done on an electronic calculator. The calculator simplifies an otherwise cumbersome process.

Readers should perform Exs. 15-14 and 15-15 independently on separate sheets to see whether they can master these techniques. As noted earlier, these processes and operations are fundamental to later ac circuit analysis.

15-7.3 Powers of Complex Numbers

The rules of ordinary algebra apply in taking the powers of complex numbers. Given a complex number in rectangular form, it is first converted to polar form, $r\underline{/\theta°}$. Then we can write the special case of DeMoivre's theorem as

$$(r\underline{/\theta°})^n = r^n\underline{/n\theta°} \tag{15-2}$$

where n is the power to which the complex number is raised
r is the polar magnitude
$\theta°$ is the angle in degrees in polar form[11]

EXAMPLE 15-16
Evaluate the following expressing answer in rectangular form:

a. $(21\underline{/15°})^3$
b. $(21 + j20)^4$
c. $(8 + j15)^{2.2}$

Solution

a. $(21\underline{/15°})^3 = (21^3)\underline{/3 \times 15°} = 9261\underline{/45°} = \mathbf{6548.5 + j6548.5}$
b. $(21 + j20)^4 = (29\underline{/43.6°})^4 = (29)^4\underline{/4 \times 43.6°}$
$= 7.07 \times 10^5\underline{/174.4°} = \mathbf{(-7.04 + j0.69) \times 10^5}$
c. $(8 + j15)^{2.2} = (17\underline{/61.93°})^{2.2} = (17)^{2.2}\underline{/2.2 \times 61.93°}$
$= 509.3\underline{/136.25°} = \mathbf{-367.9 + j352.2}$

Example 15-16 shows that

1. The power, n, to which the complex number is raised may be an integral or some decimal number. Since almost all electronic calculators have a y^x key, this is not a problem in raising the polar magnitude to any decimal number.
2. When the resultant complex number in rectangular form is very large, it may be expressed as shown in the answer to **(b)**.
3. In raising a complex number to a power greater than unity (as in these examples), *there is only one possible answer*. But if the exponent n in Eq. (15-2) is less than unity, there may be several possible answers, as shown in Sec. 15-7.4.

Readers should perform Ex. 15-16 independently to ensure that they have mastered these techniques by emerging with the same answers.

15-7.4 Roots of Complex Numbers

In reality, Eq. (15-2), a special case of DeMoivre's theorem, holds equally well when n is a fractional number (i.e., less than unity). But in this event there may be more than one possible root that satisfies the original complex number. For example, we will see that if $n = 1/2$, there are two possible roots. If $n = 1/3$, there are three possible roots. If $n = 1/4$, there are four possible roots, and so on. The first root obtained is usually called the *principal* root, and the remaining roots are termed *secondary* roots. All roots, however, have the *same magnitude* (or modulus) and differ only by their polar angle (or argument). This difference, $\Delta°$, is found from

$$\Delta° = 360°/x \qquad \text{degrees (°)} \tag{15-3}$$

where x is the reciprocal of n
n is the fractional power in Eq. (15-2) to which the complex number is raised, $n = 1/x$

[11] The angle θ may also be expressed in radians.

EXAMPLE 15-17

Find the principal root and all secondary roots of the following numbers, expressing answers in polar form:

a. $(7 - j24)^{1/2}$
b. $(44 + j117)^{1/3}$
c. $(625\angle 40^\circ)^{1/4}$

a. $(7 - j24)^{1/2} = (25\angle -73.74^\circ)^{1/2} = (25)^{1/2}\angle -73.74 \times \frac{1}{2} =$ $\mathbf{5\angle -36.87^\circ}$ (principal root); $\Delta^\circ = 360^\circ/2 = \mathbf{180^\circ}$. Secondary root $= 5\angle 180^\circ - 36.87^\circ = \mathbf{5\angle 143.13^\circ}$ (second root)

b. $(44 + j117)^{1/3} = (125\angle 69.39^\circ)^{1/3} = \mathbf{5\angle 23.13^\circ}$ (principal root); $\Delta^\circ = 360^\circ/3 = 120^\circ$. Secondary roots $= 5\angle 23.13^\circ + 120^\circ =$ $5\angle 143.13^\circ$ and $\mathbf{5\angle 143.13^\circ + 120^\circ} = \mathbf{5\angle 263.13^\circ}$

c. $(625\angle 40^\circ)^{1/4} = \mathbf{5\angle 10^\circ}$ = principal root; $\Delta^\circ = 360^\circ/4 = \mathbf{90^\circ}$. Three remaining secondary roots $= \mathbf{5\angle 100^\circ}$, $\mathbf{5\angle 190^\circ}$ and $\mathbf{5\angle 280^\circ}$

EXAMPLE 15-18

Show that in Ex. 15-17

a. Part **(a)** has only two possible roots
b. Part **(b)** has only three possible roots
c. Part **(c)** has only four possible roots

Solution

a. Since $\Delta^\circ = 180^\circ$, adding 180° to last (second) root $=$ $5\angle 143.13^\circ + 180^\circ = 5\angle 323.13^\circ$ but $5\angle 323.13^\circ = 5\angle -36.87^\circ$, which is the *same* as the principal root

b. Since $\Delta^\circ = 120^\circ$, adding 120° to last (third) root $=$ $5\angle 263.13^\circ + 120^\circ = 5\angle 383.13^\circ$ but $5\angle 383.13^\circ = 5\angle 23.13^\circ$, which is the *same* as the principal root

c. Since $\Delta^\circ = 90^\circ$, adding 90° to last (fourth) root $=$ $5\angle 280^\circ + 90^\circ = 5\angle 370^\circ$ but $5\angle 370^\circ = 5\angle 10^\circ$, which is the *same* as the principal root

EXAMPLE 15-19

Show that any root yields the original complex number by taking the root to the reciprocal of the given power in Ex. 15-17. In each case, select the final root, as demonstration of validity of the above process.

Solution

a. $(5\angle 143.13^\circ)^2 = 25\angle 286.26^\circ = 25\angle -73.74^\circ$
$= \mathbf{7 - j24}$ (original complex number)

b. $(5\angle 263.13^\circ)^3 = 125\angle 789.39^\circ - 720^\circ = 125\angle 69.39^\circ$
$= \mathbf{44 + j117}$ (original complex number)

c. $(5\angle 280^\circ)^4 = 625\angle 1120^\circ - 1080^\circ$
$= \mathbf{625\angle 40^\circ}$ (original complex number)

The validity of Eqs. (15-2) and (15-3) as well as the insights regarding them are demonstrated in Exs. 15-17 through 15-19. Again, in order to demonstrate mastery of the above techniques, readers must attempt the examples independently, using the answers shown as feedback to correct their errors.

A question frequently asked by students now arises: "Suppose $n = 1/9$, is there some quick way to find the principal root and the last (ninth) root without having to determine the intermediate roots?" The answer is "Yes," and we may write for any complex number in polar form, $(r\angle\theta)^n$:

$$\text{Principal root} = r^n\angle n\theta = r^{1/x}\angle \theta/x \qquad (15\text{-}2a)$$

$$\text{Second root} = r^{1/x}\angle \Delta^\circ + \theta/x$$

$$\text{Third root} = r^{1/x}\angle 2\Delta^\circ + \theta/x$$

$$n\text{th root} = r^n\angle n\theta = r^{1/x}\angle (x - 1)\Delta^\circ + \theta/x \qquad (15\text{-}2b)$$

where x is the number of roots; $x = 1/n$ is an integer
n is the fractional power to which the complex number is raised, $n = 1/x$
Δ° is $360^\circ/x$ or $n \cdot 360^\circ$ (15-3)

EXAMPLE 15-20

a. Find the principal root and last (ninth) root of $(512\underline{/-45^\circ})^{1/9}$
b. Show that the tenth root is the same as the principal root

Solution

a. $(512\underline{/-45^\circ})^{1/9} = 512^{1/9}\underline{/-45^\circ/9} = \mathbf{2\underline{/-5^\circ}}$
$= \text{principal root}$ (15-2a)

$\Delta^\circ = \dfrac{360^\circ}{9} = 40^\circ$ and the ninth root

$= 512^{1/9}\underline{/(9-1)40^\circ + (-45^\circ/9)}$
$= \mathbf{2\underline{/315^\circ}} = \text{ninth root}$ (15-2b)

b. Tenth root $= 2\underline{/315^\circ + 40^\circ} = 2\underline{/355^\circ}$
$= \mathbf{2\underline{/-5^\circ}} = \text{principal root}$

Example 15-20 shows the validity of Eq. (15-2b), which works for all given (n) fractional powers, provided x is an integer and $n = 1/x$.[12]

15-8 EXPONENTIAL FORM OF A COMPLEX NUMBER—EULER'S TRANSFORMATION

We have seen thus far that a complex number may be expressed in either polar or rectangular form. Yet another important form of a complex number is the *exponential* form. We may write the equivalence of all three forms, derived in Appendix B-11, by Euler's theorem:

$$r\underline{/\theta} = r(\cos\theta + j\sin\theta) = r\varepsilon^{j\theta} \qquad (15\text{-}4)$$

where θ is expressed in radians in all terms

ε is the base of natural logarithms, a constant irrational number, approximately $2.7\overline{1828} = \varepsilon^1$

This *transformation* is a significant milestone, not only in mathematics but in engineering as well. Using it, we are able to evaluate the natural logarithm (ln) of complex numbers as well as complex numbers raised to complex powers! We will do some examples of these, but first some practice in the *exponential* forms of complex numbers, $r\varepsilon^{j\theta}$.

EXAMPLE 15-21

Express the complex number $5\underline{/53.13^\circ}$ in rectangular and exponential form

Solution

This problem may be solved by two methods:

Method 1. Using P → R conversion and R → P conversion of the electronic calculator we emerge with the complex number in polar form, angle in **radians**, which enables the writing in exponential form: $5\underline{/53.13^\circ} = 3 + j4 = 5\underline{/0.9273} = \mathbf{5\varepsilon^{j0.9273}}$

Method 2. Using the RAD/DEG conversion factor of π rad/180 deg we write: $53.13^\circ \times \dfrac{\pi \text{ rad}}{180^\circ} = 0.9273$ rad so $5\underline{/53.13^\circ} = 5\underline{/0.9273} = \mathbf{5\varepsilon^{j0.9273}}$

With respect to both methods of Ex. 15-21, note that

1. The polar form, with angle in radians, carries *no units.* This is the way we distinguish between angular measure in degrees and angular measure in radians. It also simplifies writing exponential forms and making calculations with them.
2. The letter "j" in the exponent *cannot* be omitted. Although it does *not* appear in the polar form, it always appears in the exponential form. We will see why this is important when we take the natural logarithm of complex numbers (Sec. 15-9).

[12] Roots of complex numbers raised to noninteger powers such as $n = 1/2.5$ where $x = 2.5$, require special equations (which are outside the scope of this text) to determine the number of roots and the final root. For example, $(100\underline{/25^\circ})^{0.4} = (100\underline{/25^\circ})^{1/2.5} = 6.3096\underline{/10^\circ}$ = principal root. This complex number has a total of *five* roots and the last root is $6.3096\underline{/226^\circ}$, where $\Delta^\circ = 144^\circ$.

EXAMPLE 15-22

Express the complex number $2\varepsilon^{j4.8}$ in rectangular and polar form, angle in degrees

Solution

By inspection, $2\varepsilon^{j4.8} = 2\underline{/4.8} = 0.175 - j1.992 = \mathbf{2\underline{/-84.98^\circ}} = \mathbf{2\underline{/275.02^\circ}}$. The solution to Ex. 15-22 is not as simple as it appears. Note that

1. Writing the polar form, angle in radians, is done by inspection.
2. The rectangular form is done by P → R conversion, using the appropriate RAD key or switch of the electronic calculator.
3. Reconversion of polar form, angle in degrees, is by R → P conversion, using the appropriate DEG key or switch of the electronic calculator.
4. Depending on whether the calculator yields positive angles regardless of size or angles that are less than 180°, one of two answers emerges from step (3). Both answers are acceptable.
5. A good test of whether the procedure is correct is to *reverse the process*, going from *right to left* in the solution of Ex. 15-22.

15-9 MORE ADVANCED CALCULATIONS USING THE EXPONENTIAL FORM OF COMPLEX NUMBERS

We are now prepared to use the exponential form of complex numbers in a variety of ways. The first involves finding the **natural logarithm** of a complex number. To do this, we first *convert* the number from rectangular form or polar form (angle in degrees) to the *exponential* form (angle in radians). Once we find $r\varepsilon^{j\theta}$, we can find its natural logarithm from

$$\ln(r\varepsilon^{j\theta}) = \ln(r) + \ln(\varepsilon^{j\theta}) = \ln(r) + j\theta \qquad (15\text{-}5)$$

where all terms have been defined.

A careful examination of Eq. (15-5) shows the following important insights:

1. We must begin with the natural logarithm of the exponential form, found from the polar form with angle in radians.
2. We emerge, however, with an answer in *rectangular* form, since the first term is real and the second is imaginary.
3. We may convert this answer, if needed, to polar form with angle in degrees, polar form with angle in radians, or even exponential form. We will see the advantages of this, shortly.
4. Eq. (15-5) contains the natural logarithm of a *product* of two terms. We know that $\ln(2 \times 3)$ is $\ln 2 + \ln 3$. Similarly, $\ln(r \cdot \varepsilon^{j\theta}) = \ln r + \ln(\varepsilon^{j\theta})$.
5. Since $\ln(\varepsilon^{x}) = x$, then $\ln(\varepsilon^{j\theta}) = j\theta$.

EXAMPLE 15-23

Evaluate $\ln(3 + j4)$, leaving the answer in rectangular form

Solution

We first find the exponential form before using Eq. (15-5):

$\ln(3 + j4) = \ln(5\underline{/0.9273}) = \ln(5\varepsilon^{j0.9273}) = \ln 5 + \ln(\varepsilon^{j0.9273})$
$= \mathbf{1.609 + j0.9273}$

EXAMPLE 15-24

Evaluate $\ln(-80)$, leaving the answer in rectangular form

Solution

$\ln(-80) = \ln(80\underline{/180^\circ}) = \ln(80\underline{/\pi}) = \ln(80\varepsilon^{j\pi}) = \ln 80 + \ln \varepsilon^{j\pi}$
$= \mathbf{4.382 + j3.142}$

15-9.1 Finding the Antilogarithm of a Complex Number or $\ln^{-1}(N)$

This is the inverse of the process we performed in finding the natural logarithm of a complex number. To perform the **inverse** operation we *must* have our number,

N in *rectangular* form. Let us make the necessary **inversions**, using the answers to Exs. 15-23 and 15-24.

EXAMPLE 15-25

Given $\ln N = 1.609 + j0.9273$, find the complex number **N** in rectangular form

Solution

$$\mathbf{N} = \ln^{-1}(\ln \mathbf{N}) = \ln^{-1}(1.609 + j0.9273) = \varepsilon^{1.609} \times \varepsilon^{j0.9273}$$

$$= 5\varepsilon^{j0.9273} = 5\underline{/0.9273} = \mathbf{3 + j4}$$

With respect to the solution, please note:

1. Given ln **N**, we must have ln **N** expressed in *rectangular* form.
2. The key step involves putting the real and imaginary terms to respective powers of ε and expressing them as a *product*.
3. The next step recovers the exponential form, from which we can easily write the rectangular form and/or polar form, angle in degrees.

Figure 15-7 shows a *road map* for finding ln **N**, given **N**, and also the inverse road map for finding **N**, given ln **N**. The road map also shows what to do when given a complex number in other forms, different from the form needed to find either ln **N** or $\ln^{-1}$(ln **N**) = **N**. A careful examination of Fig. 15-7 shows that

1. To find ln **N** of a complex number, **N** must be expressed in the exponential form.
2. Given ln **N** and asked to find **N**, the value of ln **N** must be expressed in rectangular form.
3. In finding ln **N** we go from **N** in the exponential form to ln **N** in rectangular form.
4. In finding the inverse of ln **N** = $\ln^{-1}$(ln **N**) = **N**, we go from ln **N** in rectangular form to **N** in exponential form.

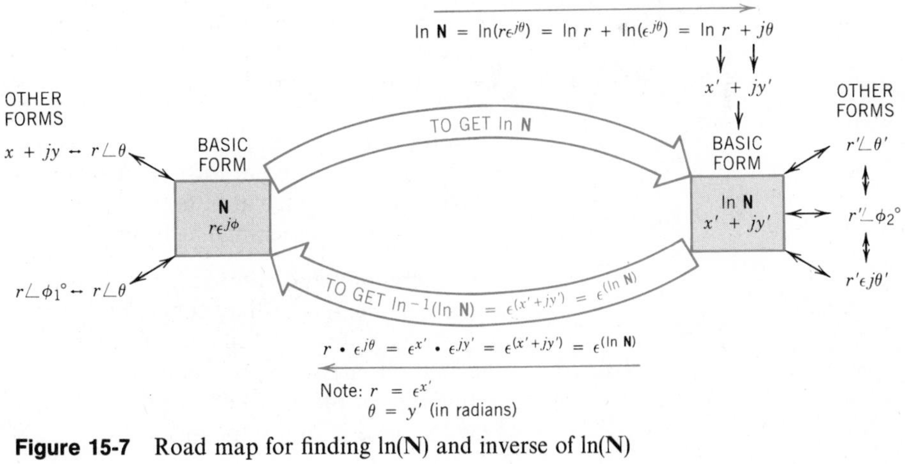

Figure 15-7 Road map for finding ln(**N**) and inverse of ln(**N**)

EXAMPLE 15-26

Given ln **N** = $4.382 + j\pi$, find the complex number **N** in rectangular form.

Solution

$$\mathbf{N} = \ln^{-1}(4.382 + j\pi) = (\varepsilon^{4.382}) \times (\varepsilon^{j\pi})$$

$$= 80\varepsilon^{j\pi} = 80\underline{/\pi} = \mathbf{-80 + j0 = -80}$$

With respect to the solution of Ex. 15-26, note that this is precisely the same value given for **N** in Ex. 15-24.

Although the road map of Fig. 15-7 should go a long way toward helping the reader, there is no substitute for practice, because the technique of finding natural logarithms and their antilogarithms reviews much of what has been presented. Example 15-27 not only provides additional practice but verifies the procedure and value of natural logarithms for the work of the next section.

EXAMPLE 15-27

Evaluate the given data of Ex. 15-16**a**, **b**, and **c**, using natural logarithms and expressing your answer in rectangular form

Solution

a. $\mathbf{N} = (21\underline{/15^\circ})^3 = (21\underline{/0.2618})^3 = (21\varepsilon^{j0.2618})^3$

$\ln \mathbf{N} = 3 \ln(21\varepsilon^{j0.2618}) = 3[\ln 21 + \ln(\varepsilon^{j0.2618})]$
$= 3(3.0445 + j0.2618)$

$\ln \mathbf{N} = 9.1336 + j0.7854$

Then $\mathbf{N} = \ln^{-1}(\ln \mathbf{N}) = \ln^{-1}(9.1336 + j0.7854)$
$= (\varepsilon^{9.1336}) \times (\varepsilon^{j0.7854}) = 9261.0\underline{/0.7854}$
$= \mathbf{6548.5 + \mathit{j}6548.5}$ (see answer Ex. 15-16**a**)

b. $\mathbf{N} = (21 + j20)^4 = (29\underline{/0.761})^4 = (29\varepsilon^{j0.761})^4$

$\ln \mathbf{N} = 4 \ln(29\varepsilon^{j0.761}) = 4[\ln 29 + \ln(\varepsilon^{j0.761})]$
$= 4(3.3673 + j0.761)$

$\ln \mathbf{N} = 13.469 + j3.044$

Then $\mathbf{N} = \ln^{-1}(\ln \mathbf{N}) = \ln^{-1}(13.469 + j3.044)$
$= (\varepsilon^{13.469}) \times (\varepsilon^{j3.044}) = 707281\underline{/3.044}$
$= \mathbf{(-7.04 + \mathit{j}0.69) \times 10^5}$ (see answer Ex. 15-16**b**)

c. $\mathbf{N} = (8 + j15)^{2.2} = (17\underline{/1.08084})^{2.2} = (17\varepsilon^{j1.08084})^{2.2}$

$\ln \mathbf{N} = 2.2 \ln(17\varepsilon^{j1.08084}) = 2.2(\ln 17 + \ln \varepsilon^{j1.08084})$
$= 2.2(2.833 + j1.08084)$

$\ln \mathbf{N} = 6.233 + j2.378$

Then $\mathbf{N} = \ln^{-1}(\ln \mathbf{N}) = \ln^{-1}(6.233 + j2.378)$
$= \varepsilon^{6.233}\varepsilon^{j2.378} = 509.28\underline{/2.378}$
$= \mathbf{-367.9 + \mathit{j}352.2}$ (see answer Ex. 15-16**c**)

With respect to the solutions shown in Ex. 15-26, please note that

1. In order to take ln **N** of any complex number, **N**, it must be expressed in exponential form, which is the basic form (Fig. 15-7).
2. In evaluating ln **N** we emerge with the rectangular form of ln **N**. But this is not the answer, since we are seeking the value of **N**!
3. Therefore, we must take the inverse of ln **N**, which must be expressed in its basic rectangular form (see Fig. 15-7).
4. But the inverse of ln **N** requires expressing the rectangular form as a power of ε, where $\varepsilon^{x'+jy'} = \varepsilon^{x'} \cdot \varepsilon^{jy'}$ as shown in Fig. 15-7, yielding both the exponential form of the complex number and its polar form, angle in radians.
5. Once given the polar form with angle in radians, we use P → R conversion on an electronic calculator, taking care to use the RAD key or switch, in obtaining the answer in rectangular form.

In examining and comparing the solutions to the same problems in Exs. 15-16 and 15-27, the reader may conclude that the latter method is much too laborious and a waste of time. If one has to raise a complex number to an *integral* power, the method of Ex. 15-16 is much easier.

But what if one is asked to raise a complex number to a complex power? Our only recourse is to use the method of Ex. 15-27, as shown in the next subsection.

15-9.2 Raising a Complex Number to a Complex Power

One necessity for the exponential form of a complex number is for use in evaluating a complex number raised to a complex power. This is shown in Ex. 15-28.

EXAMPLE 15-28

Given $\mathbf{N} = (7 - j24)^{(-15+j20)}$, find the value of N, leaving your answer in rectangular form

Solution

	Steps in Solution
$\mathbf{N} = (7 - j24)^{(-15+j20)} = 25\angle{-1.287}^{(25\angle 2.214)}$	Convert to polar form, angle in radians
$\ln \mathbf{N} = (25\angle 2.214) \ln 25\varepsilon^{-j1.287}$	Express ln **N** using exponential form
$= (25\angle 2.214)[\ln 25 + \ln \varepsilon^{-j1.287}]$	Take ln of exponential term
$\ln \mathbf{N} = (25\angle 2.214)(3.219 - j1.287)$	This yields a product in which last term is always in rectangular form
$\ln \mathbf{N} = (25\angle 2.214)(3.467\angle{-0.3803})$	R → P conversion of second term, angle in rads
$\ln \mathbf{N} = 86.67\angle 1.834$	Perform multiplication in polar form
$\ln \mathbf{N} = -22.55 + j83.685$	Conversion to rectangular form to enable $\ln^{-1}(\ln \mathbf{N})$
$\mathbf{N} = \varepsilon^{-22.55+j83.65} = (\varepsilon^{-22.55})(\varepsilon^{j83.685})$	Inverse expression of ln **N**
$\mathbf{N} = (1.61\angle 83.685) \times 10^{-10}$	Evaluate **N** in polar form, angle in rads
$\mathbf{N} = \mathbf{(-6.75 + j14.62) \times 10^{-11}}$	**N** evaluated in rectangular form

The various steps in the solution of Ex. 15-28 are explained, so there is no need to belabor them here. But the reader is in for a surprise when examining the final magnitude of **N**. In effect, we are raising a polar magnitude of 25 to another polar magnitude of 25. We would expect the answer to be a very large number. Instead it is very small! The apparent discrepancy is due to the direction of each of the polar magnitudes. In this case, the direction is *inward toward the origin*, resulting in a very small magnitude! As in the case of the previous examples, the reader is encouraged to attempt the solution independently, using the solution shown as feedback.

15-9.3 Hyperbolic Sine of a Complex Number, A, sinh A

Appendix B-11 derives the hyperbolic sine of a complex number, **A**, as

$$\sinh \mathbf{A} = \frac{\varepsilon^{\mathbf{A}} - \varepsilon^{-\mathbf{A}}}{2} \qquad (15\text{-}6)$$

where **A** is expressed in the rectangular form of the complex number.

EXAMPLE 15-29

Calculate the hyperbolic sine of each of the following complex numbers leaving answer in polar form, angle in radians:

a. $-5.34 + j2.75$
b. $25\angle 73.74^\circ$
c. $5\varepsilon^{j0.5}$

Solution

a. $$\sinh \mathbf{A} = \frac{\varepsilon^{\mathbf{A}} - \varepsilon^{-\mathbf{A}}}{2} = \frac{\varepsilon^{-5.34+j2.75} - \varepsilon^{5.34-j2.75}}{2}$$

$$= \frac{(4.8\varepsilon^{j2.75}) \times 10^{-3} - 208.51\varepsilon^{-j2.75}}{2}$$

$$= -104.3\angle{-2.75} = \mathbf{104.3\angle 0.392}$$

b. $\mathbf{A} = 25\angle 73.74^\circ = 7 + j24$;

$$\sinh \mathbf{A} = \frac{\varepsilon^{\mathbf{A}} - \varepsilon^{-\mathbf{A}}}{2} = \frac{\varepsilon^{7+j24} - \varepsilon^{-7-j24}}{2}$$

$$= \frac{1097\angle 24 - (9.1 \times 10^{-4})\angle{-24}}{2}$$

$$= \mathbf{548.3\angle 24 = 548.3\angle 5.15}$$

c. $\mathbf{A} = 5\varepsilon^{j0.5} = 5\underline{/0.5} = 4.388 + j2.397;$

$$\sinh \mathbf{A} = \frac{\varepsilon^{\mathbf{A}} - \varepsilon^{-\mathbf{A}}}{2} = \frac{\varepsilon^{4.388+j2.397} - \varepsilon^{-4.388-j2.397}}{2}$$

$$= \frac{80.47\underline{/2.397} - 0.0124\underline{/-2.397}}{2}$$

$$= \mathbf{40.2\underline{/2.4}}$$

With respect to the solutions shown in Ex. 15-29, please note:

1. In part **a** two terms emerge in the numerator, of which the first is negligible. Consequently, only the second (negative) term is evaluated as $-104.3\underline{/-2.75}$. Since the negative sign preceding it implies a reversal of the number, adding π radians to the angle yields a positive value of the hyperbolic sine.
2. In part **b** the second term in the numerator is negligible and may be ignored, yielding $548.3\underline{/24}$. But 24 radians is a very large angle. By subtracting 2π multiples (three times) we arrive at the answer shown.
3. In part **c** the second term is less than three orders of magnitude compared to the first. Consequently, it may be ignored, and the answer emerges as shown.
4. Occasionally a second term may *not* be ignored, as shown by Ex. 15-30**a**.

15-9.4 Hyperbolic Cosine of a Complex Number, A, cosh A

Appendix B-11 derives the hyperbolic cosine of a complex number, **A**, as

$$\cosh \mathbf{A} = \frac{\varepsilon^{\mathbf{A}} + \varepsilon^{-\mathbf{A}}}{2} \tag{15-7}$$

where **A** is expressed in the rectangular form of the number.

EXAMPLE 15-30
Calculate the hyperbolic cosine of the following complex numbers leaving the answer in polar form, angle in radians:

a. $(-0.12 - j0.72)$
b. $\ln(-14000)$
c.

Solution

a. $$\cosh \mathbf{A} = \frac{\varepsilon^{-0.12-j0.72} + \varepsilon^{0.12+j0.72}}{2}$$

$$= \frac{0.887\underline{/-0.72} + 1.1275\underline{/0.72}}{2}$$

$$= \frac{(0.667 - j0.585) + (0.848 + j0.7435)}{2}$$

$$= \frac{1.515 + j0.1585}{2} = \frac{1.523\underline{/0.104}}{2}$$

$$= \mathbf{0.762\underline{/0.104}}$$

b. $\mathbf{A} = \ln(-14\,000) = \ln(14\,000\underline{/\pi}) = \ln(14\,000\varepsilon^{j\pi})$
$= 9.55 + j\pi = 9.55 + j3.142;$

$$\cosh \mathbf{A} = \frac{\varepsilon^{9.55+j3.142} + \varepsilon^{-9.55-j3.142}}{2}$$

$$\cong \frac{14\,000\underline{/3.142} + 7.14 \times 10^{-5}\underline{/-3.142}}{2}$$

$$\cong \mathbf{7000\underline{/3.142}} \cong \mathbf{7000\underline{/\pi}}$$

c. $\mathbf{A} = 283\varepsilon^{-j2.13} = 283\underline{/-2.13} = -150.1 - j240;$

$$\cosh \mathbf{A} = \frac{\varepsilon^{-150.1-j240} + \varepsilon^{150.1+j240}}{2}$$

$$= \frac{1.54 \times 10^{65}\underline{/240}}{2} = 7.7 \times 10^{64}\underline{/240}$$

$$= \mathbf{7.7 \times 10^{64}\underline{/1.24}}$$

With respect to the solutions shown in Ex. 15-30, please note that

1. In part **a** the smaller term *cannot* be ignored. Consequently, it is necessary to convert each term to rectangular form, add real and imaginary terms separately, and convert the rectangular sum back to polar before dividing by 2.
2. In part **b**, $\ln(-14\,000)$ must first be evaluated as a complex number in rectangular form. Once **A** is found in rectangular form it is substituted in Eq. (15-7) to yield the final answer of $7000\underline{/\pi}$.
3. In part **c** the answer emerges as a very large number at a very large angle in radians. Dividing 240 by 2π yields **38.2**, so the angle in radians is $240 - 38(2\pi) = $ **1.24** rads.

15-10 PHASORS REVISITED

Sections 15-1 through 15-3 introduced phasor quantities to acquaint the reader with the need for complex algebra and its advantages in performing calculations in the frequency domain rather than the time domain. At this point, however, we may consider ourselves sufficiently familiar with the techniques of complex algebra.[13] We may return, therefore, to the subject of phasors, armed with a powerful arsenal of weapons with which to attack ac circuit analysis. But before we do, let us not forget the *caveats* of Sec. 15-2:

1. Any time-varying function in the *time domain* may be *represented* by a phasor that is a complex quantity in the frequency domain.[14]
2. The phasor quantity is a representation associated with the time-varying function. It is *not equal* to the time-domain quantity.

The following worked examples are intended to carry forward our understanding of how we add (or subtract) sinusoidal functions in the time domain through the use of phasors in the frequency domain. In general, our *modus operandi* will be to

1. Convert the time-varying functions to phasors in complex form.
2. Use complex algebra to add or subtract two (or more) phasor quantities.
3. Convert the phasor in complex form back to a (resultant) time-domain function.

EXAMPLE 15-31

A sinusoidal function consists of the following sine terms: $v(t) = 100\sin(\omega t - 40°) + 80\cos(\omega t - 50°) - 75\cos(\omega t + 20°) - 50\sin(\omega t - 120°)$ volts. Calculate the time domain equivalent of the sinusoidal function.

Solution

We first represent each of the four terms as positive sinusoidal functions with respect to the reference-zero sinusoid, labeling them $v_1(t)$ through $v_4(t)$, respectively:

$$v_1(t) = 100\sin(\omega t - 40°)\text{ V};$$
$$v_2(t) = 80\cos(\omega t - 50°) = 80\sin(\omega t + 40°)\text{ V};$$
$$v_3(t) = -75\cos(\omega t + 20°) = 75\sin(\omega t - 70°)\text{ V};$$
$$v_4(t) = -50\sin(\omega t - 120°) = 50\sin(\omega t + 60°)\text{ V}$$

Next we write the phasors in the frequency domain:

$$\mathbf{V}_1 = (100/\sqrt{2})\underline{/-40°}\text{ V} = 70.71\underline{/-40°}\text{ V};$$
$$\mathbf{V}_2 = (80/\sqrt{2})\underline{/40°} = 56.57\underline{/40°}\text{ V};$$
$$\mathbf{V}_3 = (75/\sqrt{2})\underline{/-70°}\text{ V} = 53.03\underline{/-70°}\text{ V};$$
$$\mathbf{V}_4 = (50/\sqrt{2})\underline{/60°} = 35.36\underline{/60°}\text{ V}$$

Now we may add the above (real and imaginary) terms using complex algebra:

$$\mathbf{V}_1 = 70.71\underline{/-40°} = 54.17 - j45.45;$$
$$\mathbf{V}_2 = 56.57\underline{/40°} = 43.34 + j36.36;$$
$$\mathbf{V}_3 = 53.03\underline{/-70°} = 18.14 - j49.83;$$
$$\mathbf{V}_4 = 35.36\underline{/60°} = 17.68 + j30.62;$$

and $\mathbf{V}_T = \Sigma\mathbf{V}_s = 133.33 - j28.30 = 136.3\underline{/-11.98°}\text{ V}$

Lastly, we convert the phasor in the frequency domain to its time-domain function:

$$v_t(t) = \sqrt{2} \times 136.3\sin(\omega t - 11.98°)\text{ V}$$
$$= \mathbf{192.76\sin(\omega t - 11.98°)\ V}$$

With respect to the solution of Ex. 15-31, please note that

1. Each sinusoidal term was converted to a positive sine function with respect to the zero reference sinusoid.
2. Each sinusoidal term was converted to a phasor in the frequency domain.
3. Each phasor in *polar* form was converted to *rectangular* form.
4. Real and imaginary components of each phasor were added separately.
5. The complex sum in rectangular form was converted to a single phasor in polar form.
6. The resultant phasor in the frequency domain was converted to the time domain as a sinusoid whose peak value is $\sqrt{2}$ times its RMS value at the same phase angle as well as the same frequency, ω, of the given waveforms.
7. Had this problem been attempted by graphical analysis, the plotting would have been long and arduous and the final result hardly accurate.

---

[13] The reader who can solve all the complex algebra problems at the end of this chapter should have little difficulty in applying complex algebra to the solution of ac problems in this and the remaining chapters.

[14] In this text, boldface symbols are used for all complex quantities, even when they are not phasor quantities. Voltage and current are phasor quantities. But impedance is not. Still, because impedance is complex, it is shown in boldface.

EXAMPLE 15-32

An ac parallel circuit consists of two branches. The current in the first branch is $i_1(t) = 5\cos(1000t - 30°)$ mA. The current in the second branch is $i_2(t) = -8\cos(1000t + 50°)$ mA. Calculate the total current expressed as a time-domain function.

Solution

$i_1(t) = 5\cos(1000t - 30°)$ mA $= 5\sin(1000t + 60°)$ mA and
$i_2(t) = -8\cos(1000t + 50°)$ mA $= 8\sin(1000t - 40°)$ mA

$$\mathbf{I}_1 = (5/\sqrt{2})\angle 60° \text{ mA} = 3.5355\angle 60° = 1.768 + j3.062$$
$$\mathbf{I}_2 = (8/\sqrt{2})\angle -40° \text{ mA} = 5.657\angle -40° = 4.334 - j3.636$$
$$\mathbf{I}_T = \mathbf{I}_1 + \mathbf{I}_2 = 6.102 - j0.5742$$
$$= 6.13\angle -5.38° \text{ mA}$$

Then

$$i_T(t) = \sqrt{2}(6.13)\sin(1000t - 5.38°) \text{ mA}$$
$$= \mathbf{8.67\sin(1000t - 5.38°)\ mA}$$

EXAMPLE 15-33

For the circuit shown in **Fig. 15-8**, find the sinusoidal expression for the voltage $v_3(t)$ when $v_T(t) = 100\sin(5000t)$ V; $v_1(t) = 40\cos(5000t - 60°)$ V; $v_2(t) = -80\cos(5000t + 20°)$ V

Solution

$\mathbf{V}_T = (100/\sqrt{2})\angle 0°$ V $= 70.71\angle 0°$ V $= 70.71 + j0$;
$\mathbf{V}_1 = (40/\sqrt{2})\angle 30°$ V $= 28.28\angle 30°$ V $= 24.495 + j14.14$;
$\mathbf{V}_2 = (80/\sqrt{2})\angle -70°$ V $= 56.57\angle -70°$ V $= 19.35 - j53.16$.
From KVL,

$$\mathbf{V}_3 = \mathbf{V}_T - (\mathbf{V}_1 + \mathbf{V}_2)$$
$$= 70.71 + j0 - (24.495 + j14.14) - (19.35 - j53.16)$$

and

$$\mathbf{V}_3 = 26.865 + j39.02 = 47.37\angle 55.45° \text{ V}$$

Then

$$v_3(t) = \sqrt{2} \times 47.37\sin(5000t + 55.45°) \text{ V}$$
$$= \mathbf{67.0\sin(5000t + 55.45°)\ V}$$

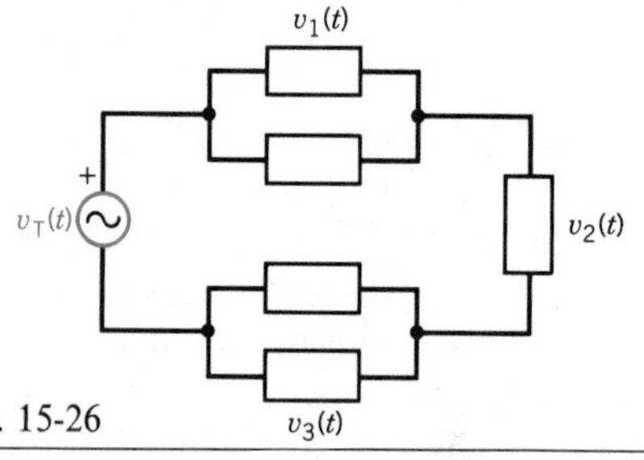

Figure 15-8 Example 15-33 and Prob. 15-26

Examples 15-32 and 15-33 show how easily and elegantly time-domain problems may be solved by using phasors in the frequency domain. The technique involves

1. Converting all time-domain waveforms to phasors in the complex-plane frequency domain.
2. Performing all necessary computations using the phasors in the complex-plane frequency domain to find a solution.
3. Converting the solution back to the time domain.[15]

One final example is given to illustrate the extreme power of this method. Example 15-34 draws on the time-domain reactance relationships of Chapter 14 as well as the frequency-domain principles developed in this chapter.

[15] This method is analogous to the method of Laplace transforms for solving simultaneous linear integrodifferential equations with constant coefficients. It also involves three steps:

1. The equations, a function of time, are converted into their respective Laplace transforms, in the *s* plane.
2. Next comes the solution of the equations to find the Laplace transform of the unknown function, in the *s* plane.
3. Last is the inverse Laplace transform to find the function itself in the time domain.

This subject is beyond the scope of this text.

EXAMPLE 15-34

Given the constant-current source $i(t) = 5\sin(1000t + 50°)$ mA shown in **Fig. 15-9**, calculate

a. $v_R(t)$ in the time domain and $\mathbf{V}_R$ in the frequency domain
b. $v_C(t)$ in the time domain and $\mathbf{V}_C$ in the frequency domain
c. $v_L(t)$ in the time domain and $\mathbf{V}_L$ in the frequency domain
d. $\mathbf{V}_T$ across constant current source in the frequency domain
e. $v_T(t)$ in the time domain
f. Circuit phase angle, θ, and power factor of the circuit
g. Circuit power dissipated using $(V_m I_m/2)\cos\theta$ (see Eq. 14-39)
h. Circuit power dissipated using I^2R (given values)

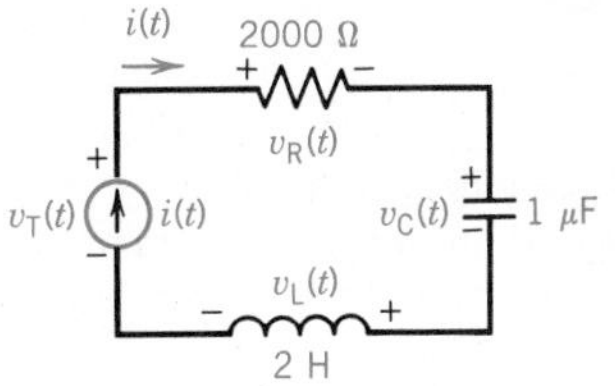

Figure 15-9 Example 15-34 and Prob. 15-27

Solution

a. The resistor voltage is in phase with the current. Therefore,

$$v_R(t) = i(t) \times R = [5\sin(1000t + 50°)\text{ mA}] \times 2000\ \Omega$$
$$= \mathbf{10\sin(1000t + 50°)\ V}$$

and $\quad \mathbf{V}_R = (10/\sqrt{2})\angle 50° = \mathbf{7.071\angle 50°\ V}$

b. The capacitor's voltage requires finding X_C and the knowledge that the capacitor voltage *lags* the current in it by 90°:

$$X_C = 1/\omega C = 1/1000 \times (1\ \mu\text{F}) = 1000\ \Omega;$$
$$v_C(t) = i(t) \times X_C = (5\text{ mA})(1\text{ k}\Omega)[\sin(1000t + 50° - 90°)]\text{ V}$$
$$= \mathbf{5\sin(1000t - 40°)\ V};$$

and $\quad \mathbf{V}_C = (5/\sqrt{2})\angle -40° = \mathbf{3.536\angle -40°\ V}$

c. The inductor's voltage requires finding X_L and the knowledge that voltage across the inductor *leads* the current by 90°:

$$X_L = \omega L = 1000 \times 2\text{ H} = 2\text{ k}\Omega;$$
$$v_L(t) = i(t)X_C = (5\text{ mA})(2\text{ k}\Omega)[\sin(1000t + 50° + 90°)]\text{ V}$$
$$= \mathbf{10\sin(1000t + 140°)\ V};$$

and $\quad \mathbf{V}_L = (10/\sqrt{2})\angle 140° = \mathbf{7.071\angle 140°\ V}$

d. Finding $\mathbf{V}_T$ merely involves adding above phasors in the complex plane or

$$\mathbf{V}_R = 7.071\angle 50° \quad = \quad 4.545 + j5.417$$
$$\mathbf{V}_C = 3.536\angle -40° \quad = \quad 2.709 - j2.273$$
$$\mathbf{V}_L = 7.071\angle 140° \quad = -5.417 + j4.545$$
$$\mathbf{V}_T = \mathbf{V}_R + \mathbf{V}_C + \mathbf{V}_L = \mathbf{1.837 + j7.689} = \mathbf{7.905\angle 76.56°\ V}$$

e. Given $\mathbf{V}_T$ in the complex-plane frequency domain, it is relatively simple to convert it back to the time domain:

$$v_T(t) = \sqrt{2}(7.905)\sin(1000t + 76.56°)$$
$$= \mathbf{11.18\sin(1000t + 76.56°)\ V}$$

f. Comparing $v_T(t)$ against $i(t)$ we find $v_T(t)$ leading $i(t)$ by **26.56°** $= \theta$. Then $F_p = \cos\theta = \cos 26.56° = \mathbf{0.8945}$

g. $$P = P_{av} = \left(\frac{V_m I_m}{2}\right)\cos\theta = \frac{(11.18\text{ V} \times 5\text{ mA})}{2}\cos 26.56°$$
$$= \mathbf{25\ mW}$$

h. Since only resistance is capable of dissipating power, we could have computed it at the outset using the data of Fig. 15-9 or $P = I^2R = (5/\sqrt{2}\text{ mA})^2 \times 2\text{ k}\Omega = \mathbf{25\ mW}$

Many important conclusions and insights emerge from Ex. 15-34. Perhaps the most important are the following:

1. The phasors in the complex plane (or frequency domain) simplify our calculations. For this reason, throughout the rest of this text, they will be used almost exclusively. But they do *not* represent *reality*.
2. The sinusoidal computed voltages and given current expressed in the time domain *do* represent *reality*. They can be observed as waveforms with a cathode ray oscilloscope (CRO) in the laboratory and in the field. The phasors in the frequency domain are merely convenient transformed quantities to simplify calculations.
3. The accuracy and the validity of the technique is proved by step **g** of the solution to Ex. 15-34. The time-domain voltage, $V_m(t)$, was derived from the phasor sum of the three component voltages in the frequency domain. It proves that the procedure is valid and that the time-domain voltages are all correct.
4. It would be a simple matter to plot the four time-domain voltages, $v_R(t)$, $v_C(t)$, $v_L(t)$, and $v_T(t)$, against the given current, $i(t)$, from the solution of Ex. 15-34. These should correspond to the CRO display of these waveforms (or those taken from a printer–plotter in which the given data are programmed into a plotting routine).
5. The circuit of Fig. 15-9 is a series circuit. Current is the same in all parts of a series circuit and there can be only one current, which was given.
6. Voltage relations are:
   a. Voltage across a resistor is always in phase with the current.
   b. Voltage across a pure capacitor lags the current by 90°.
   c. Voltage across a pure inductor leads the current by 90°.
   d. The total voltage may be found as
      1. The phasor sum of all component voltages in the complex plane of the frequency domain.
      2. The *graphical* sum of all component sinusoidal voltages in the time domain.

15-11 GLOSSARY OF TERMS USED

Complex number Number that can be expressed as $x \pm jy$, where x is a real number and $\pm jy$ is an imaginary number. In this text symbols for complex numbers are shown in boldface type.

Complex plane A rectangular coordinate system in which complex numbers are expressed in the form $x + jy$. The horizontal axis represents real numbers (x axis) and the vertical axis imaginary numbers ($\pm jy$ axis). The complex plane is sometimes called the Argand plane.

Conjugate Complex number in which *only* the sign of the *imaginary* term has been *reversed*. Also called complex conjugate. The conjugate of $x + jy$ is $x - jy$. The conjugate of $\mathbf{A}\angle\theta$ is $\mathbf{A}\angle-\theta$. If $\mathbf{Z}$ is the symbol for a complex number, its conjugate is represented as $\mathbf{Z}^*$, where the asterisk denotes use of the conjugate form.

Euler's theorem Identity which states that $\varepsilon^{j\theta} = \cos\theta + j\sin\theta$. Also called Euler's formula. Using this identity, any complex number of the form $A\angle\theta$ may be expressed as $A\varepsilon^{j\theta} = A(\cos\theta + j\sin\theta)$. Euler's theorem is derived in Appendix B-11.

Exponential form Using Euler's theorem, expression of a complex number $A\angle\theta$ in the form $A\varepsilon^{j\theta}$.

Frequency domain Transformation of the time-domain sinusoidal voltages and currents into phasor quantities, in which inductances and capacitances are appropriately converted to reactances. In the frequency domain all calculations are performed with complex algebra.

***j* operator** An imaginary number, $\sqrt{-1}$, representing the ordinate of the complex plane.

Phasor diagram Instantaneous phase relation between two or more phasor RMS quantities at a given instant in time. All phasors on a given circuit diagram are assumed to have the same angular frequency or velocity.

Phasor quantity Complex number used to represent alternating electrical quantities having both magnitude and direction. Voltage and current are phasor quantities. All phasors have constant angular velocity in a given circuit or system. Impedance is *not* a phasor quantity but for convenience may be expressed as a complex number. Symbols for phasor quantities are shown in boldface in this text.

Polar form of a complex number, $A\angle\theta$ Form in which A is the magnitude or absolute value of the complex number and θ is its phase angle with respect to some zero reference. The polar form $A\angle\theta$ may be converted to rectangular form by $A(\cos\theta + j\sin\theta)$ or written as $A\varepsilon^{j\theta}$ in exponential form.

Pure imaginary number Number expressed as $0 \pm jb$. Such a number on the complex plane would be directly on the vertical (imaginary) axis at coordinates $0, \pm jb$.

Reciprocal For a phasor $A\angle\theta$, the reciprocal is $1/A\angle\theta$, or $(1/A)\angle-\theta$. Similarly, for a complex number $\mathbf{Z}$, the reciprocal is $1/\mathbf{Z}$.

Rectangular form Expression of a complex number in the form $a \pm jb$ or $x \pm jy$. The rectangular form is sometimes called the *coordinate* form or *Cartesian* form.

Reversal Change in the direction of a phasor or a complex number by 180°. The reversal of $\mathbf{Z}$ is $-\mathbf{Z}$. The reversal of $2\angle 30^\circ$ is $-2\angle 30^\circ$ or $2\angle -150^\circ$ or $2\angle 210^\circ$, obtained by addition or subtraction of 180°.

Time domain The original ac circuit containing resistances, inductances, and capacitances whose currents and voltages are sinusoidally varying as a function of time.

15-12 PROBLEMS

Secs. 15-1 to 15-7

15-1 Show the following complex numbers as a coordinate plot on an Argand diagram:

a. $2 + j3$	e. $j4$
b. $2 - j3$	f. $-j3$
c. $-3 + j2$	g. 4
d. $-3 - j2$	h. -4

15-2 Using your electronic calculator's P → R conversion capability, convert the following to rectangular form, showing all answers to four significant digits:

a. $52\angle 85^\circ$	f. $165\angle -228^\circ$
b. $4.7\angle -227^\circ$	g. $68.2\angle 198.6^\circ$
c. $87\angle 1.87^{g}$	h. $4870\angle -1.97^\circ$
d. $476\angle -1.07$	i. $650\angle -91.27^\circ$
e. $3.4\angle \pi$	

15-3 Using your electronic calculator's R → P capability, convert the following to polar form (angle in degrees):

a. $-31 - j24$	f. $-5240 + j2050$
b. $72.6 + j78.3$	g. $(0.26 + j4.8) \times 10^{-4}$
c. $1.9 + j152$	h. $(-65 - j489)$
d. $1300 - j14.6$	i. $358 - j38.6$
e. $-247 - j162$	

15-4 Fill in all the missing blanks, giving blank angles in degrees:

a. $155\angle\_\_\_\_ = (\quad - j28)$
b. $215\angle\_\_\_\_ = (-185 + j\quad)$
c. $\quad\angle -334^\circ = (\quad + j3.4)$
d. $\quad\angle -138.6^\circ = (-0.78 - j\quad)$
e. $\quad\angle 187.8^\circ = (-462 - j\quad)$
f. $952\angle\_\_\_\_ = (-238 - j\quad)$

15-5 Write the conjugate of each of the following in the same form as given:

a. $28\angle{-140^\circ}$ c. $(-94 - j26)$
b. $375\angle{184^\circ}$ d. $(-53 + j88)$

15-6 Calculate the reciprocal of each of the following in the same form as given, but answers cannot be in fractional form:

a. $0.4\angle{-85^\circ}$ c. $-16 - j12$
b. $0.16 \times 10^{-5}\angle{-158.4^\circ}$ d. $-70 + j10$

15-7 Evaluate the following sums, leaving answers in rectangular form:

a. $(-20 + j5) - (-3 - j8) - (12 - j14)$
b. $-(7 - j12) - (-3 + j16) + j15$

15-8 Evaluate the following products, leaving answers in polar form, angle in degrees:

a. $(48\angle{-121^\circ})(0.047\angle{106.3^\circ})(3.63\angle{208^\circ})$
b. $(-4 - j5)(3 + j4)(6 - j3)$

15-9 Evaluate the following divisions, leaving answers in polar form, angle in degrees:

a. $$\frac{(18.6\angle{-57.8^\circ})(2.79\angle{48^\circ})}{(46.9\angle{147^\circ})(0.776\angle{-34.24^\circ})(5.8\angle{12.6^\circ})}$$

b. $$\frac{(9 - j40)(11 + j60)(12 - j35)}{(5 - j12)(7 + j24)(-25 - j60)}$$

15-10 Multiply the following in rectangular form, leaving answer in rectangular form:

a. $(-8 - j2)(8 - j6)$
b. $(-3 - j4)(-7 + j8)(5 - j6)$

15-11 Divide in rectangular form, leaving answer in rectangular form (answer cannot be in fractional form):

a. $$\frac{(-5 + j11)(6 - j4)}{(-7 + j1)}$$

b. $$\frac{(5 - j12)(7 + j24)}{(-15 - j20)}$$

15-12 Add the following, leaving answer in rectangular form:

a. $52\angle{85^\circ} + 4.7\angle{-227^\circ} - 68.2\angle{198.6^\circ}$
b. $10\angle{20^\circ} - 20\angle{-30^\circ} + 30\angle{40^\circ} - 50\angle{60^\circ}$

15-13 Raise the following complex numbers to the powers indicated, leaving answers in polar form, angle in degrees:

a. $(-16)^4$ d. $(-0.27 - j1.8)^6$
b. $(-27 + j9)^2$ e. $(\varepsilon^{j3})^3$
c. $(j81)^5$ f. $(-j12)^4$

15-14 Find *all* possible roots of the following complex numbers, leaving answers in polar form, angle in degrees:

a. $(-16)^{1/4}$ d. $(-0.27 - j1.8)^{1/6}$
b. $(-27 + j9)^{1/2}$ e. $(\varepsilon^{j3})^{1/3}$
c. $(j81)^{1/5}$ f. $(-j12)^{1/4}$

15-15 Evaluate the following determinants, leaving answers in rectangular form:

a. $\begin{vmatrix} 9 + j40 & 3 - j4 \\ 3 - j4 & 8 + j15 \end{vmatrix}$ c. $\begin{vmatrix} 11 + j60 & 12 + j35 \\ 25 - j60 & 16 - j63 \end{vmatrix}$

b. $\begin{vmatrix} 3 + j4 & -j5 \\ -j5 & 5 - j12 \end{vmatrix}$

Secs. 15-8 and 15-9

15-16 Calculate the natural logarithm of each of the following, expressing the answer in polar form, angle in radians:

a. $-8 + j3$ d. $1.86\varepsilon^{-j3.2}$
b. $(16 - j7.6)$ e. $16.9\varepsilon^{j54.8^\circ}$
c. $88\angle{-165^\circ}$ f. $682\angle{2.89}$

15-17 Each of the following is the natural logarithm of a complex number **N**. Calculate the value of **N** and express it in rectangular form.

a. $(-28 - j9)$ d. $10.6\angle{138^\circ}$
b. $(-1.74 + j0.649)$ e. $4.83\angle{-108.6^g}$
c. $2.8\varepsilon^{-j1.8}$ f. $19.7\angle{3.08}$

15-18 Calculate the hyperbolic sine of the following complex numbers, leaving answer in polar form, angle in radians:

a. $3.27\angle{28.7^\circ}$ d. $3.86\angle{-88.4^g}$
b. $4.28\varepsilon^{j0.134}$ e. $(-2.1 + j0.48)$
c. $(-4.68 + j3.49)$ f. $8.7\angle{-186^\circ}$

15-19 Calculate the hyperbolic cosine of the following complex numbers, leaving answer in polar form, angle in radians:

a. $7 - j10$ d. $(-0.24 - j0.58)$
b. $2.46 + j5.43$ e. -6
c. $250\varepsilon^{-j1.18}$ f. $\ln(-18\,000)$

15-20 Evaluate the following complex numbers raised to complex powers, leaving answers in polar form, angle in radians:

a. $(8 - j11)^{(-4+j1)}$ d. $[\ln(-2 + j8)]^{(4-j2)}$
b. $(18\varepsilon^{-j2})^{(4.1\angle{-67^\circ})}$ e. $(-22)^{-0.3\angle{-110^\circ}}$
c. $(-j14)^{-j4}$ f. $(3.8\angle{7.2})^{2.5\varepsilon^{-j2.1}}$

Sec. 15-10

15-21 Transform each of the following waveforms in the time domain to the frequency domain:

a. $\sqrt{2}(208)\cos(200\pi t - 50^\circ)$ V
b. $-\sqrt{2}(5)\sin(200\pi t + 140^\circ)$ mA
c. $-75\cos(400t - 20^\circ)$ V
d. $-25\sin(400t + 20^\circ)$ mA
e. $120(\sqrt{2})\cos(377t - 50^\circ)$ V
f. $-50\cos(4000t + 150^\circ)$ A

15-22 Transform each of the following phasors to a waveform in the time domain, assuming $\omega = 1000$ rad/s:

a. $\mathbf{I} = 5\angle{40^\circ}\ \mu$A d. $\mathbf{V} = 120\angle{-350^\circ}$ V
b. $\mathbf{V} = 24\angle{-120^\circ}$ mV e. $\mathbf{I} = 10\angle{400^\circ}$ A
c. $\mathbf{I} = 6\angle{-300^\circ}$ mA f. $\mathbf{V} = 220\angle{\pi}$ V

15-23 Write a single sine term in the time domain for each of the following waveforms:

a. $v(t) = 5\cos\omega t + 12\sin\omega t$ V
b. $i(t) = -7\sin\omega t - 24\cos\omega t$ μA
c. $v(t) = 8\sin(\omega t + 50^\circ) + 15\sin(\omega t - 40^\circ)$ μV
d. $i(t) = 9\cos(\omega t + 70^\circ) + 40\sin(\omega t + 70^\circ)$ mA
e. $v(t) = 5\cos(377t + 50^\circ) - 6\sin(377t - 40^\circ) + \sin(377t + 50^\circ)$ V
f. $i(t) = 8\cos(100t) - 4\sin(100t + 30^\circ) - 6\cos(100t - 60^\circ)$ A

15-24 A series circuit consists of three components whose RMS voltages are $\mathbf{V}_1 = 40\angle{-40^\circ}$ V, $\mathbf{V}_2 = 30\angle{30^\circ}$ V, and $\mathbf{V}_3 = 100\angle{10^\circ}$ V. Calculate the

a. Supply voltage in phasor form
b. Supply voltage in the time domain if the frequency is 60 Hz

15-25 A parallel circuit consists of two branches. The current in the first branch is $\mathbf{I}_1 = 15\angle{-75^\circ}$ mA and the total current is $\mathbf{I}_T = 5\angle{10^\circ}$ mA.

a. Calculate the current in the second branch in phasor form
b. Write the three time-domain currents if the frequency is 400 Hz
c. Explain how it is possible for the total current in an ac circuit to be less than each of its branch currents. (Hint: draw a phasor diagram.)

15-26 Given the circuit shown in Fig. 15-8, find the sinusoidal expression for voltage $v_2(t)$ when

$$v_T(t) = -208\sqrt{2}\sin(\omega t + 150^\circ)\text{ V},$$
$$v_1(t) = -70.71\cos(\omega t + 20^\circ)\text{ V},$$

and

$$v_3(t) = 84.853\cos(\omega t - 150^\circ)\text{ V}.$$

15-27 In the circuit shown in Fig. 15-9, $i(t) = 9\sqrt{2}\sin(2000t)$ mA. Calculate

a. $v_R(t)$ in the time domain and $\mathbf{V}_R$ in the frequency domain
b. $v_C(t)$ in the time domain and $\mathbf{V}_C$ in the frequency domain
c. $v_L(t)$ in the time domain and $\mathbf{V}_L$ in the frequency domain
d. $\mathbf{V}_T$ in the frequency domain and $v_T(t)$ in the time domain
e. Circuit phase angle, θ
f. Circuit power factor, F_p
g. Circuit power dissipated, using time-domain relations
h. Circuit power dissipated, using RMS value of the given current

15-28 From the values computed in Prob. 15-27 draw a phasor diagram (using the given current as a reference) showing phasors $\mathbf{I}$, $\mathbf{V}_R$, $\mathbf{V}_C$, and $\mathbf{V}_L$ to an approximate scale.

15-13 ANSWERS

15-2 a $4.532 + j51.80$ b $-3.205 + j3.437$ c $86.96 + j2.555$ d $228.5 - j417.5$ e $-3.4 + j0$ f $-110.4 + j122.6$ g $-64.64 - j21.75$ h $4867 - j167.4$ i $-14.41 - j649.8$

15-3 a $39.2\angle{-142.3^\circ}$ b $106.8\angle{47.16^\circ}$ c $152\angle{89.28^\circ}$ d $1300\angle{-0.6434^\circ}$ e $295.4\angle{-146.7^\circ}$ f $5627\angle{158.6^\circ}$ g $4.807 \times 10^{-4}\angle{86.9^\circ}$ h $493.3\angle{-97.57^\circ}$ i $360.1\angle{-6.154^\circ}$

15-4 a $\angle{-10.41^\circ}$, 152.4 b $\angle{149.4^\circ}$, $j109.5$ c 7.556, 6.971 d 1.040, $-j0.6877$ e 466.3, $-j63.29$ f $\angle{-104.5^\circ}$, $-j921.8$

15-5 a $28\angle{140^\circ}$ b $375\angle{-184^\circ}$ c $-94 + j26$ d $-53 - j88$

15-6 a $2.5\angle{85^\circ}$ b $62.5 \times 10^5\angle{158.4^\circ}$ c $-0.04 + j0.03$ d $-0.014 - j0.002$

15-7 a $-29 + j27$ b $-4 + j11$

15-8 a $8.189\angle{193.3^\circ}$ b $214.8\angle{-102.1^\circ}$

15-9 a $0.2458\angle{-135.2^\circ}$ b $4.38\angle{37.48^\circ}$

15-10 a $-76 + j32$ b $289 - j298$

15-11 a $-0.24 - j12.32$ b $-8.904 + j9.472$

15-12 a $65.97 + j76.99$ b $-9.943 - j10.6$

15-13 a $65\,536\angle{0^\circ}$ b $810\angle{323.13^\circ}$ c $3.487 \times 10^9\angle{90^\circ}$ d $36.34\angle{128.82^\circ}$ e $1\angle{155.66^\circ}$ f $20\,736\angle{0^\circ}$

15-14 a $2\angle{45^\circ}$, $\angle{135^\circ}$, $\angle{225^\circ}$, $\angle{315^\circ}$ b $5.335\angle{80.78^\circ}$, $\angle{260.8^\circ}$ c $2.408\angle{18^\circ}$, $\angle{90^\circ}$, $\angle{162^\circ}$, $\angle{234^\circ}$, $\angle{306^\circ}$ d $1.105\angle{-16.42^\circ}$, $\angle{43.6^\circ}$, $\angle{103.6^\circ}$, $\angle{163.6^\circ}$, $\angle{223.6^\circ}$, $\angle{283.6^\circ}$ e $1\angle{57.3^\circ}$, $\angle{177.3^\circ}$, $\angle{297.3^\circ}$ f $1.861\angle{-22.5^\circ}$, $\angle{67.5^\circ}$, $\angle{157.5^\circ}$, $\angle{247.5^\circ}$

15-15 a $-521 + j479$ b $88 - j16$ c $1556 + j112$

15-16 a $3.514\angle{0.9141}$ b $2.908\angle{-0.1531}$ c $5.323\angle{-0.5716}$ d $3.260\angle{-1.379}$ e $2.985\angle{0.3262}$ f $7.136\angle{0.4169}$

15-17 a $(-6.3 - j2.85) \times 10^{-13}$ b $0.1398 + j0.1061$ c $-0.4845 - j0.2132$ d $(2.616 + j2.747) \times 10^{-4}$ e $(3.837 + j52.04) \times 10^{-2}$ f $(1.012 + j2.704) \times 10^{-9}$

15-18 a $8.832\angle{1.57}$ b $34.76\angle{0.5718}$ c $53.88\angle{-0.3485}$ d $0.972\angle{2.238}$ e $4.048\angle{2.649}$ f $2861\angle{2.232}$

15-19 a $548.3\angle{2.566}$ b $5.85\angle{-0.85}$ c $1.141 \times 10^{41}\angle{1.28}$ d $0.8709\angle{0.1531}$ e $201.7\angle{0}$ f $9000\angle{\pm\pi}$

15-20 a $7.47 \times 10^{-5}\underline{/6.374}$ b $5.4 \times 10^{-2}\underline{/-14.11}$
c $1.867 \times 10^{-3}\underline{/-10.56}$ d $248.6\underline{/0.796}$
e $0.5665\underline{/1.194}$ f $1.039 \times 10^{6}\underline{/-11.97}$

15-21 a $208\underline{/40^\circ}$ V b $5\underline{/-40^\circ}$ mA c $53.03\underline{/-110^\circ}$ V
d $17.68\underline{/-160^\circ}$ mA e $120\underline{/40^\circ}$ V f $35.36\underline{/60^\circ}$ A

15-22 a $i(t) = 5\sqrt{2}\sin(1000t + 40^\circ)\ \mu$A
f $v(t) = 220\sqrt{2}\sin(1000t + 180^\circ)$ V

15-23 a $v(t) = 13\sin(\omega t + 22.62^\circ)$ V
f $i(t) = 12.93\sin(100t + 120^\circ)$ A

15-24 a $155.24\underline{/2.46^\circ}$ V b $v(t) = 219.5\sin(377t + 2.46^\circ)$ V

15-25 a $15.39\underline{/86.12^\circ}$ mA b $i_t(t) = 7.071(2513t + 10^\circ)$ mA

15-26 $188.3\sin(\omega t - 2.18^\circ)$ V

15-27 e 60.26° f 0.4961 g 162 mW h 162 mW

CHAPTER 16

Series, Parallel, and Series–Parallel Circuit Relations

16-1 INTRODUCTION

The previous chapter showed the advantages of performing circuit analysis and solving problems in the complex plane or frequency domain. The various examples (of Sec. 15-10) and end-of-chapter problems showed that the basic relations derived for dc circuits from Kirchhoff's voltage law (KVL) and Kirchhoff's current law (KCL) hold equally well for ac circuits. Chapter 15 also introduced the technique of transforming the original sinusoidal circuit parameters from the time domain to the frequency domain. Sinusoidal voltage and current waveforms in the time domain were transformed to voltage and current phasors in the frequency domain.

Using frequency-domain techniques, we will discover that reactances may be combined with each other or with resistances easily in much the same way and following the same relations used in our previous solutions of dc circuits. The complex algebra required for such combination and simplification was presented in Chapter 15.

16-2 TIME- AND FREQUENCY-DOMAIN RESPONSES OF BASIC CIRCUIT ELEMENTS *R*, *L*, AND *C*

The responses of a pure resistor (Sec. 14-13), a pure inductor (Sec. 14-14), and a pure capacitor (Sec. 14-15) to a sinusoidal voltage were previously presented. Table 16-1 summarizes the *R*, *L*, and *C* responses (originally presented in Chapter 14 and again in Sec. 15-10), using either sinusoidal current or voltage as a reference (second column).

Once more we may conclude from summary Table 16-1[1] that

1. In a purely resistive circuit, current is in phase with voltage.
2. In a purely inductive circuit, current lags voltage by 90° (or $\mathbf{V}_L$ leads $\mathbf{I}_L$ by 90°).
3. In a purely capacitive circuit, current leads voltage by 90° (or $\mathbf{V}_C$ lags $\mathbf{I}_C$ by 90°).

Table 16-1 shows that when the forcing function is the current ($\mathbf{I} = I\angle 0^\circ$), the responses are $\mathbf{V}_R$, $\mathbf{V}_L$, and $\mathbf{V}_C$, respectively. In the same way, when the forcing function is the voltage ($\mathbf{V} = V\angle 0^\circ$), the responses are $\mathbf{I}_R$, $\mathbf{I}_L$, and $\mathbf{I}_C$, respectively.

[1] Appendix B-12 presents a rigorous derivation of voltage and current relations in pure inductive and capacitive circuits.

Table 16-1 Time- and Frequency-Domain Responses of Basic Circuit Elements

Circuit Element	Time-Domain Response Given $i(t) = I_m \sin \omega t$ A	Response Transformed to Frequency Domain, $\mathbf{I} = \mathbf{I}\angle 0°$
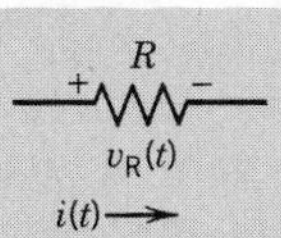	$v_R(t) = R(I_m \sin \omega t)$ volts	$\mathbf{V}_R = \mathbf{I}R\angle 0°$ volts
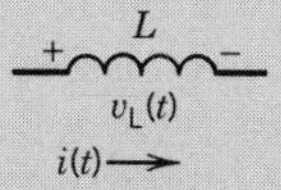	$v_L(t) = (\omega L)\, I_m \sin(\omega t + 90°)$ volts	$\mathbf{V}_L = \mathbf{I}X_L\angle 90° = jX_L \times \mathbf{I}$ volts
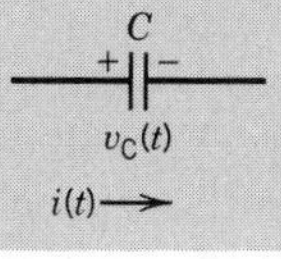	$v_C(t) = (1/\omega C) I_m \sin(\omega t - 90°)$ volts	$\mathbf{V}_C = \mathbf{I}X_C\angle -90° = -jX_C \times \mathbf{I}$ volts
	Time-Domain Response Given $v(t) = V_m \sin \omega t$ V	Response Transformed to Frequency Domain, $\mathbf{V} = \mathbf{V}\angle 0°$
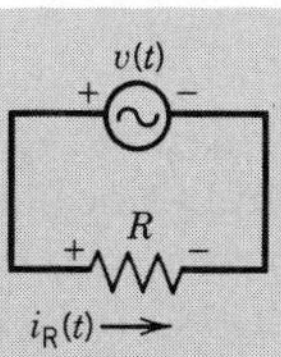	$i_R(t) = \dfrac{V_m}{R}(\sin \omega t)$ amperes	$\mathbf{I}_R = (\mathbf{V}/R)\angle 0°$ amperes
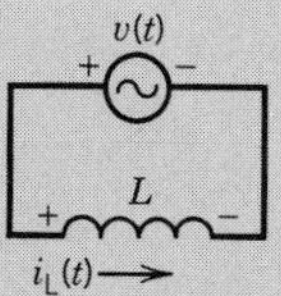	$i_L(t) = \dfrac{V_m}{\omega L} \sin(\omega t - 90°)$ amperes	$\mathbf{I}_L = \dfrac{\mathbf{V}}{X_L\angle 90°} = \dfrac{\mathbf{V}}{jX_L} = \dfrac{\mathbf{V}}{X_L}\angle -90°$ A
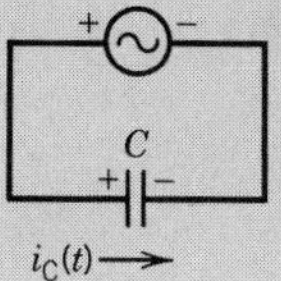	$i_C(t) = \omega C V_m \sin(\omega t + 90°)$ amperes	$\mathbf{I}_C = \dfrac{\mathbf{V}}{X_C\angle -90°} = \dfrac{\mathbf{V}}{-jX_C} = \dfrac{\mathbf{V}}{X_C}\angle 90°$ A

In the (3) cases of the right-hand column of Table 16-1, the circuit element's *opposition* is the ratio of the voltage forcing function divided by its current response (or cause divided by its effect). From the right-hand column of Table 16-1 we can show that

$$\mathbf{R} = \frac{\mathbf{V}_R}{\mathbf{I}_R} = \frac{\mathbf{V}\angle 0°}{\mathbf{I}\angle 0°} = R\angle 0° \quad \text{ohms } (\Omega) \tag{16-1}$$

$$\mathbf{X}_L = \frac{\mathbf{V}_L}{\mathbf{I}_L} = \frac{\mathbf{V}\angle 0°}{\mathbf{I}\angle -90°} = X_L\angle 90° = jX_L = j\omega L \quad \text{ohms } (\Omega) \tag{16-2}$$

$$\mathbf{X}_C = \frac{\mathbf{V}_C}{\mathbf{I}_C} = \frac{\mathbf{V}\angle 0°}{\mathbf{I}\angle 90°} = X_C\angle -90° = -jX_C = -j/\omega C \quad \text{ohms } (\Omega) \tag{16-3}$$

Equations (16-1), (16-2), and (16-3) imply the following:

1. The ratio **V/I** (phasor voltage to phasor current) is defined as the current-limiting opposition or *impedance*, **Z**. **Z** is a complex quantity but it is *not* a phasor quantity.[2] In Eq. (16-1), for a pure resistor, $\mathbf{Z} \equiv R\angle 0°$ ohms.
2. In Eq. (16-2), the impedance, $\mathbf{Z} \equiv jX_L \equiv j\omega L$ ohms, is a complex number that yields the ratio of the voltage and current phasors across a pure inductor. The reactance of the pure inductor, $X_L\angle 90° = j\omega L$, is a complex number having a quadrature relation to resistance. Consequently, in the complex plane, it is measured and defined on the imaginary axis. Inductive reactance, $\mathbf{X}_L$, is a complex quantity but is not a phasor quantity.
3. In Eq. (16-3), the impedance, $\mathbf{Z} \equiv -jX_C \equiv -j/\omega C$ is a complex number that yields the ratio of voltage and current phasors across a pure capacitor. The reactance of the pure capacitor, $X_C\angle -90° = -j/\omega C$, is a complex number having a quadrature relation to resistance. $-jX_C$ is opposite to jX_L (180° out of phase) on the imaginary axis of the complex plane. Consequently, capacitive reactance, $\mathbf{X}_C$, is a complex quantity but is *not* a phasor quantity.

Figure 16-1 shows how the complex impedances **R**, $\mathbf{X}_L$, and $\mathbf{X}_C$ are derived from their time- and frequency-domain responses, respectively. With respect to Fig. 16-1c, please note that

1. Resistance, **R**, is always represented in the complex plane as a *positive* value along the *real* axis. (Negative resistance is represented as a negative value along the real axis; it may occur if the circuit contains *dependent* sources.)
2. Inductive reactance, $\mathbf{X}_L$, is represented in the complex plane as a positive value along the imaginary axis.
3. Capacitive reactance, $\mathbf{X}_C$, is represented in the complex plane as a negative value along the imaginary axis.

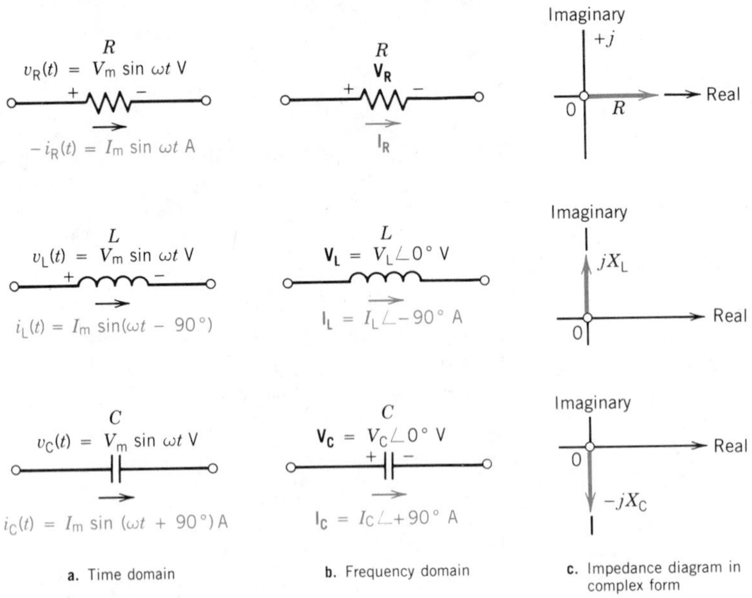

Figure 16-1 Time- and frequency-domain responses and impedances derived for ideal R, L, and C circuit elements

[2] When dividing **V** by **I**, the factor $\varepsilon^{j\omega t}$, which is common to both terms, cancels out. In effect, this cancels out the counterclockwise (CCW) phasor rotation, leaving only the **magnitude** ratio V_m/I_m! But because impedance is a complex quantity it will be shown in boldface throughout the text, even if it is *not* a phasor quantity.

4. **R**, $\mathbf{X}_L$, and $\mathbf{X}_C$, respectively, represent opposition components of impedance, **Z**, and are all measured in ohms.
5. Since **Z** is *not* a phasor quantity, a diagram of impedance factors containing **R**, $\mathbf{X}_L$ and $\mathbf{X}_C$, is called either an *impedance diagram* or an *impedance triangle.* (It *cannot* be called a *phasor* diagram.) The relation between the impedance triangle and the phasor diagram are covered in Sec. 16-3.

16-3 EQUIVALENT (OR TOTAL) IMPEDANCE AND ITS RELATION TO THE PHASOR DIAGRAM

We defined impedance earlier as the ratio of phasor voltage, **V**, to phasor current, **I**. In empirical terms we may write

$$\mathbf{Z} = \frac{\mathbf{V}}{\mathbf{I}} = \frac{V\angle\psi}{I\angle\phi} = \frac{V\angle\pm\theta}{I\angle 0^\circ} = R \pm jX = Z\angle\pm\theta \qquad \text{ohms } (\Omega) \qquad (16\text{-}4)$$

where $V\angle\psi$ is the RMS value of phasor voltage **V** at some angle ψ with respect to reference zero

$I\angle\phi$ is the RMS value of phasor current **I** at some angle ϕ with respect to reference zero

θ is the phase angle between **V** and **I** or the difference $\psi - \phi$

and all other terms have been defined above.

Figure 16-2a shows an unknown impedance that draws a phasor current, **I**, from the input source, **V**. The phasor diagram showing the $V\angle\psi$ and $I\angle\phi$ defined above is represented in two forms in Fig. 16-2b. The phasor diagram on the left shows the voltage and current phasors as they might appear with respect to some zero reference in a circuit. The phasor diagram on the right shows both phasors rotated so that *current* is represented as the *reference* phasor, $I\angle 0^\circ$. Note that voltage, $V\angle\psi$, is now represented as $V\angle\theta$, where voltage still leads current by phase angle θ in Eq. (16-4).

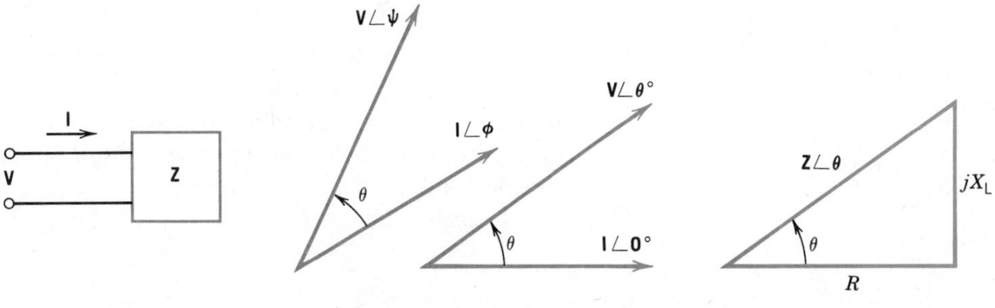

a. Input impedance, Z b. Phasor diagram using current as a reference c. Impedance triangle

Figure 16-2 Input impedance, phasor diagrams, and the impedance triangle emerging from them

Since voltage leads current (or current lags voltage), the unknown impedance contains a combination of resistance and inductive reactance. Consequently, by inference we may draw the impedance triangle $\mathbf{Z} = R + jX_L$ shown in Fig. 16-2c.

The magnitude of **Z**, by definition and from Eq. (16-4), is the ratio **V/I**. The value $Z\angle\theta$, yields the real and imaginary components of **Z**, using θ, the phase angle between **V** and **I**, as shown by Ex. 16-1.

EXAMPLE 16-1

The current drawn by an impedance is $5\angle 20^\circ$ A and the voltage across the impedance is $100\angle 60^\circ$ V. Calculate:

a. The equivalent impedance, **Z**, in complex form
b. The circuit phase angle and power factor, F_p
c. Its component resistance and reactance values, respectively
d. Phasor voltage across the internal resistance component only
e. Phasor voltage across the internal reactance component only
f. Total voltage as the sum of phasor voltages (**d**) and (**e**) above

Solution

a. $\mathbf{Z} = \mathbf{V}/\mathbf{I} = \dfrac{100\angle 60^\circ \text{ V}}{5\angle 20^\circ \text{ A}} = \mathbf{20\angle 40^\circ\ \Omega}$ (16-4)

b. $\theta = \mathbf{40^\circ}$ by inspection; $F_p = \cos\theta = \cos 40^\circ = \mathbf{0.766}$

c. $\mathbf{Z} = R + jX_L = 20\angle 40^\circ = (15.32 + j12.86)\ \Omega$ (16-4)
$R = 15.32\ \Omega;\ jX_L = j12.86\ \Omega$

d. $\mathbf{V_R} = \mathbf{IR} = (5\angle 20^\circ \text{ A})(15.32\angle 0^\circ\ \Omega) = \mathbf{76.6\angle 20^\circ\ V}$

e. $\mathbf{V_L} = \mathbf{IX_L} = (5\angle 20^\circ \text{ A})(12.86\angle 90^\circ) = \mathbf{64.28\angle 110^\circ\ V}$

f. $\mathbf{V_T} = \mathbf{V_R} + \mathbf{V_L} = 76.6\angle 20^\circ + 64.28\angle 110^\circ$
$= (71.98 + j26.20) + (-21.985 + j60.4)$
$= 50 + j86.6 = \mathbf{100\angle 60^\circ\ V}$

Example 16-1 not only verifies Eq. (16-4) but also verifies KVL for an ac circuit (part **f**). With respect to Ex. 16-1, please note that

1. Complex impedance, **Z**, as calculated by the ratio **V/I**, is converted to its real and imaginary components using Eq. (16-4).
2. Ohm's law, $\mathbf{V_R} = \mathbf{IR}$, for an ac circuit requires that we multiply in polar form $5\angle 20^\circ \text{ A} \times 15.32\angle 0^\circ\ \Omega$, yielding $76.6\angle 20^\circ$ V, as shown in part (**d**) of the solution. Similarly, $\mathbf{V_L} = \mathbf{IX_L}$, yielding $64.28\angle 110^\circ$ V, as shown in part (**e**) of the solution.
3. Kirchhoff's voltage law (KVL) is verified since the total voltage is the phasor sum of the voltages across the resistance and the inductive reactance, respectively, in part (**f**).
4. Since we are assuming in Fig. 16-2b that the impedance **Z** is the series equivalent of resistance, *R*, and inductive reactance, jX_L, the respective voltage drops must equal the total supply voltage given.
5. Using current as a reference in a series circuit seems a logical choice since the current is the *same* in all portions of the circuit.
6. The KVL relation used in part (**f**) tells us that in reality $\mathbf{V_T} = V_R + jV_L = 50 + j86.6$ volts. But this insight emerged from the sum $\mathbf{V_T} = \mathbf{V_R} + \mathbf{V_L} = 76.6\angle 20^\circ + 64.28\angle 110^\circ$ V. Let us carry this insight one step further.

Figure 16-3a shows the phasor diagram of Fig. 16-2b with current as a reference. From Euler's theorem we may write for phasor $\mathbf{V}\angle\boldsymbol{\theta}$

$$\mathbf{V}\angle\boldsymbol{\theta} = V(\cos\theta + j\sin\theta)$$
$$= V\cos\theta + jV\sin\theta \quad \text{or substituting KVL} \qquad (16\text{-}5)$$
$$\mathbf{V} = V_R + jV_L \quad \text{but by Ohm's law} \qquad (16\text{-}5a)$$
$$\mathbf{IZ} = IR + jIX_L \quad \text{and dividing by the same value } \mathbf{I}$$
$$\mathbf{Z} = R + jX_L \qquad (16\text{-}5b)$$

The various forms of Eq. (16-5) are shown in Fig. 16-3. Figure 16-3 shows how the impedance triangle may be derived from the phasor diagram of Fig. 16-3a. Since the phase angle θ is **common** to **both** the *phasor diagram* **and** the *impedance triangle*, both are similar "triangles." Consequently $\mathbf{V_T}$, $\mathbf{V_R}$, and $\mathbf{V_L}$ are in the *same ratio* to each other as **Z**, *R*, and jX_L, as shown by Eqs. (16-5), (16-5a), and (16-5b).

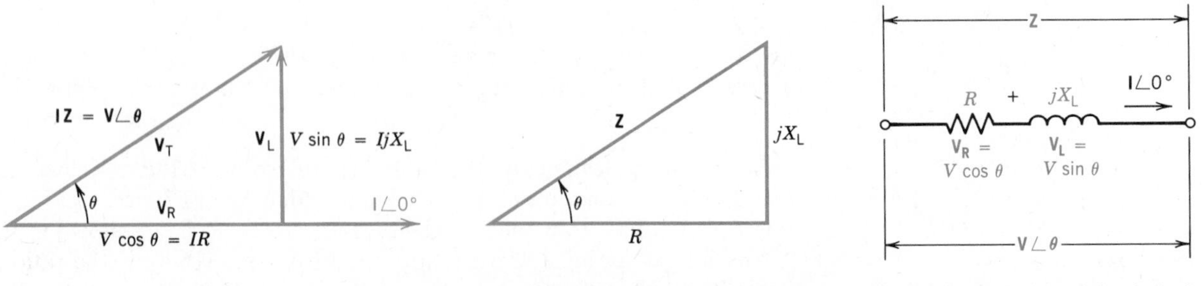

a. Phasor diagram b. Impedance triangle c. Impedance and voltage relations in circuit

Figure 16-3 Derivation of impedance triangle from phasor diagram (Eq. 16-5), for an *RL* circuit

Figure 16-3c shows the circuit impedance and voltage relations, summarized in Eqs. (16-5) through (16-5b). It proves that $\mathbf{Z} = R + jX_L$ from Eqs. (16-4) and (16-5b) and $V\angle\theta = V_R + jV_L = V\cos\theta + jV\sin\theta$ from Eqs. (16-5) and (16-5a). But most important of all, it shows that the impedance triangle *is actually derived* from the phasor diagram *by current division*, i.e., going from Eq. (16-5a) to (16-5b) and from Fig. 16-3a to Fig. 16-3b, merely by dividing each side of the phasor diagram by the common (same) current.

The same analysis that was used for the *RL* circuit of Fig. 16-3 could also apply to an *RC* circuit shown in **Fig. 16-4**. The phasor diagram is shown in Fig. 16-4a. Note from the phasor diagram of Fig. 16-4a for an *RC* circuit that

1. Current leads the capacitor voltage, $\mathbf{V}_C$, by 90°
2. Current is in phase with the voltage across the resistor, $\mathbf{V}_R$
3. The phasor sum of $\mathbf{V}_R$ and $\mathbf{V}_C$ is the supply voltage, $\mathbf{V}_T = \mathbf{V}\angle\boldsymbol{\theta}$
4. Phase angle θ is a negative angle

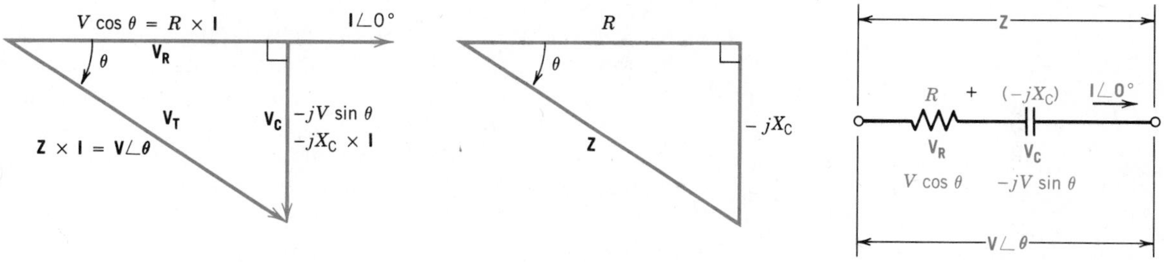

a. Phasor diagram

b. Impedance triangle

c. Circuit impedance and voltage relations

Figure 16-4 Derivation of impedance triangle from phasor diagram of *RC* circuit

By dividing each arm of the phasor diagram of Fig. 16-4a by the common current, **I**, we arrive at the impedance triangle of Fig. 16-4b, where $\mathbf{Z} = R - jX_C$. We might also have inferred this from Fig. 16-1c. The circuit impedance and voltage relations are summarized in Fig. 16-4c.

From Figs. 16-3b and 16-4b, we realize that impedance triangles are useful devices for obtaining

1. The total equivalent impedance from component resistances and reactances
2. The circuit phase angle, θ, and power factor, $\cos\theta$
3. Information as to the nature of the total equivalent impedance. If the total impedance is in the first quadrant, it is inductive. If in the fourth quadrant, it is capacitive.

EXAMPLE 16-2

Given the circuit shown in **Fig. 16-5a**:

a. Draw the impedance triangle and determine total impedance
b. Calculate total current and voltage across each component
c. Draw the phasor diagram using current as a reference

Solution

We first determine the inductive reactance of the coil, $jX_L = j\omega L = j(2000\text{ rad/s})(12\text{ mH}) = \mathbf{j24\ \Omega}$

a. The impedance triangle is drawn in the first quadrant since the circuit is inductive, as shown in Fig. 16-5b. From it we compute the total equivalent impedance: $\mathbf{Z} = 7 + j24 = \mathbf{25\angle 73.74^\circ\ \Omega}$ using R → P conversion.
b. $\mathbf{I} = \mathbf{V}/\mathbf{Z} = 50\text{ V}/25\angle 73.74^\circ\ \Omega = \mathbf{2\angle -73.74^\circ}$ **A**. Note that current lags supply voltage, as expected in an *RL* circuit, by 73.74°. But with current as a reference, the supply voltage *leads* the current by 73.74°. We draw the phasor diagram with the supply voltage $\mathbf{V} = 50\angle 73.74^\circ$ and $\mathbf{I} = 2\angle 0^\circ$ A. Then $\mathbf{V}_R = \mathbf{IR} = (2\angle 0^\circ\text{ A}) \times (7\ \Omega) = \mathbf{14\angle 0^\circ\ V}$ and $\mathbf{V}_L = \mathbf{IX}_L = (2\angle 0^\circ\text{ A}) \times (24\angle 90^\circ\ \Omega) = \mathbf{48\angle 90^\circ}$ **V**.
c. The phasor diagram, using the computed values of **(b)**, is shown in Fig. 16-5c.

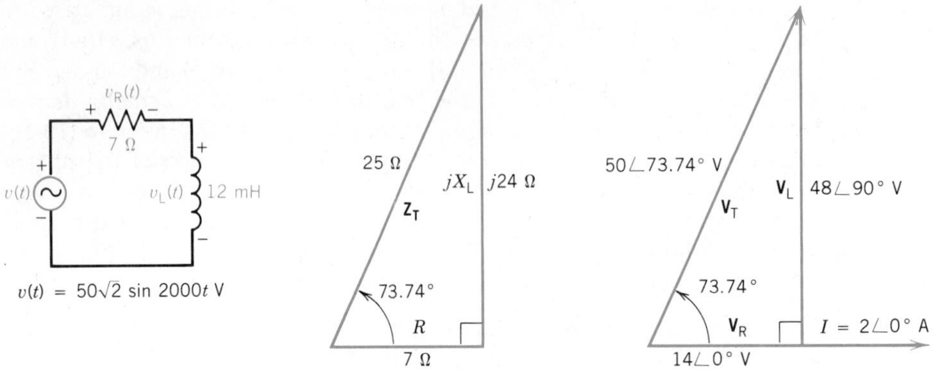

a. Example 16-2 b. Impedance triangle c. Phasor diagram

Figure 16-5 Use of impedance triangle to develop phase relations

The reader should note the similarity between the impedance triangle and the phasor diagram in Fig. 16-5. Both are similar "triangles" and each impedance in the impedance triangle has been multiplied by current to produce the phasor diagram. Most important, note that when *current* is used as a reference, the impedance phase angle is the *same* as the phase angle in the phasor diagram. These two insights are most important for the examples and problems that follow.

16-4 SERIES CIRCUIT RELATIONS

The concept of complex impedance, use of the impedance triangle, and the transformation from the time to the phasor domain enable us to treat ac circuits in much the same way as dc circuits, following the same basic laws and equations.

16-4.1 Current Relations

A series circuit is defined as one in which current is the same in all portions of the circuit. Consequently, for the series ac circuit shown in **Fig. 16-6a**:

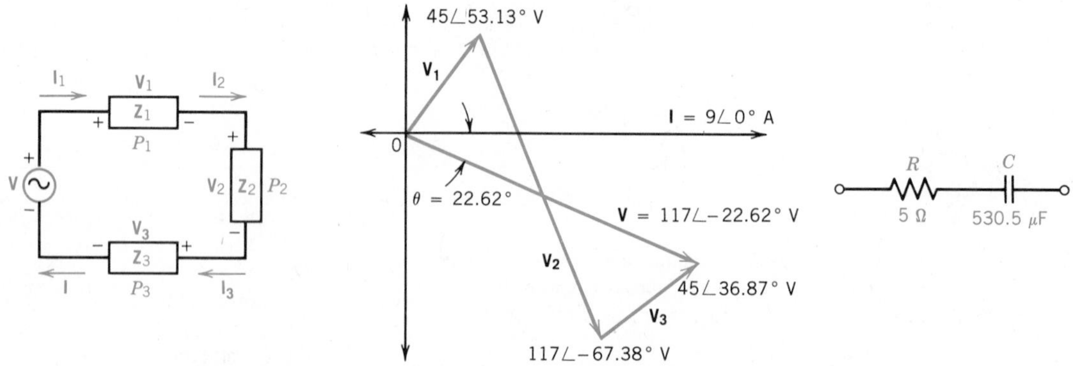

a. Series circuit b. Phasor diagram c. Equivalent series model

Figure 16-6 Series circuit and phasor diagram of all voltages and current (Ex. 16-3) and equivalent series model (Ex. 16-4)

$$\mathbf{I} \equiv \mathbf{I}_1 \equiv \mathbf{I}_2 \equiv \mathbf{I}_3 \equiv \cdots \equiv \mathbf{I}_n \qquad \text{amperes (A)} \qquad (16\text{-}6)$$

where all currents are expressed as RMS values.

16-4.2 Impedance Relations

As in the case of dc circuits, the total impedance is the (complex) sum of the individual impedances added in rectangular form:

$$\mathbf{Z} = \mathbf{Z}_1 + \mathbf{Z}_2 + \mathbf{Z}_3 + \cdots + \mathbf{Z}_n \qquad \text{ohms } (\Omega) \qquad (16\text{-}7)$$

16-4.3 Voltage Relations

By Ohm's law, voltage is the product of current times impedance. And by Kirchhoff's voltage law (KVL), the (complex) sum of the individual voltages across each component is the total (supply) voltage or

$$\mathbf{IZ} = \mathbf{IZ}_1 + \mathbf{IZ}_2 + \mathbf{IZ}_3 + \cdots + \mathbf{IZ}_n$$
$$\mathbf{V} = \mathbf{V}_1 + \mathbf{V}_2 + \mathbf{V}_3 + \cdots + \mathbf{V}_n \qquad \text{volts (V)} \qquad (16\text{-}8)$$

16-4.4 Power Relations

Recall that resistance is the only circuit property capable of dissipating power. In Chapter 14 we found that the (average) power dissipated by a resistor, R, is

$$P = P_{av} = \mathbf{I}_R^2 R = \frac{\mathbf{V}_R^2}{R} = \mathbf{V}_R \mathbf{I}_R \qquad \text{watts (W)} \qquad (14\text{-}30a)$$

Again from Chapter 14, for any complex circuit containing R–L–C in combination,

$$P = VI \cos\theta = S \cos\theta \qquad \text{watts (W)} \qquad (14\text{-}39)$$

where θ is the phase angle between phasors **V** and **I**.

Then multiplying Eqs. (16-6) and (16-8) we get

$$P = V_1 I \cos\theta_1 + V_2 I \cos\theta_2 + V_3 I \cos\theta_3 + \cdots + V_n I \cos\theta_n$$

$$P = S_1 \cos\theta_1 + S_2 \cos\theta_2 + S_3 \cos\theta_3 + \cdots + S_n \cos\theta_n \quad \text{watts (W)} \quad (16\text{-}9)$$
$$P = I^2 R_1 + I^2 R_2 + I^2 R_3 + \cdots + I^2 R_n = P_1 + P_2 + P_3 + \cdots + P_n \quad (16\text{-}10)$$

EXAMPLE 16-3

The complex impedances in the series circuit of Fig. 16-6a are $\mathbf{Z}_1 = 3 + j4$, $\mathbf{Z}_2 = 5 - j12$, and $\mathbf{Z}_3 = 4 + j3$ ohms, respectively. The supply voltage is 117 V at 60 Hz. Calculate:

a. Total equivalent circuit impedance, **Z**, and phase angle, θ
b. Circuit current in each component, **I**
c. Voltage across each component, $\mathbf{V}_1$, $\mathbf{V}_2$, and $\mathbf{V}_3$, and total voltage, **V**, using current as a reference
d. Power dissipated in each component, P_1, P_2, and P_3 and total power (by two methods)
e. Draw the complete phasor diagram showing all voltages and current, using current as a reference

Solution

a. $\mathbf{Z} = \mathbf{Z}_1 + \mathbf{Z}_2 + \mathbf{Z}_3 = (3 + j4) + (5 - j12) + (4 + j3)$
$= \mathbf{12 - j5 = 13\underline{/-22.62^\circ}}\ \Omega$;
$\theta = \mathbf{-22.62^\circ}$

b. $\mathbf{I} = \mathbf{V}/\mathbf{Z} = 117\ \text{V}/13\underline{/-22.62^\circ}\ \Omega = \mathbf{9\underline{/22.62^\circ}\ A}$. Since current is leading voltage (by 22.62°) the circuit is essentially *capacitive*, despite the fact that there is only one capacitive element in it. When we set current as a reference, the total voltage is $\mathbf{117\underline{/-22.62^\circ}\ V}$.

c. $\mathbf{V}_1 = \mathbf{IZ}_1 = 9 \times (3 + j4) = (9\underline{/0^\circ}\ \text{A})(5\underline{/53.13^\circ}\ \Omega)$
$= \mathbf{45\underline{/53.13^\circ}\ V}$
$\mathbf{V}_2 = \mathbf{IZ}_2 = 9 \times (5 - j12) = (9\underline{/0^\circ}\ \text{A})(13\underline{/-67.38^\circ}\ \Omega)$
$= \mathbf{117\underline{/-67.38^\circ}\ V}$
$\mathbf{V}_3 = \mathbf{IZ}_3 = 9 \times (4 + j3) = (9\underline{/0^\circ}\ \text{A})(5\underline{/36.87^\circ}\ \Omega)$
$= \mathbf{45\underline{/36.87^\circ}\ V}$

KVL check:
$\mathbf{V} = \mathbf{V}_1 + \mathbf{V}_2 + \mathbf{V}_3$
$= 45\underline{/53.13^\circ} + 117\underline{/-67.38^\circ} + 45\underline{/36.87^\circ}$ (16-8)
$= (27 + j36) + (45 - j108) + (36 + j27)$
$= \mathbf{108 - j45 = 117\underline{/-22.62^\circ}\ V}$ (see **b** above)

d. $P_1 = I^2R_1 = (9)^2 3 = \mathbf{243\ W}$;
$P_2 = I^2R_2 = (9)^2 5 = \mathbf{405\ W}$;
$P_3 = I^2R_3 = (9)^2 4 = \mathbf{324\ W}$.
$P = P_1 + P_2 + P_3 = 243 + 405 + 324$
$= \mathbf{972\ W}$ and (16-10)
$P = VI\cos\theta = 117 \times 9 \times \cos 22.62^\circ = \mathbf{972\ W}$ (14-39)

e. The complete phasor diagram is shown in Fig. 16-6b

With respect to the solution of Ex. 16-3, please note that

1. Since current leads the (supply) voltage, the circuit is capacitive (see Ex. 16-4).
2. The phasor sum of the individual voltage drops across each complex impedance is the supply voltage (see part **c** of the solution and Fig. 16-6b).
3. Total power dissipated is the sum of the individually dissipated powers. In part **d** of the solution, only the resistive portions of the given complex impedances were used.
4. The head–tail vector method was used in drawing the phasor diagram of Fig. 16-6b. This was done to show that the phasor sum of the individual voltages across each complex impedance is the supply voltage.[3] This is proved in the KVL check of part **c** of the solution, as well.
5. The phasor diagram of Fig. 16-6b contains a phasor crossover (i.e., $\mathbf{V}_2$ crosses **V**). This occurs because we added the phasors in the order $\mathbf{V}_1$, $\mathbf{V}_2$, and $\mathbf{V}_3$. Had we added them in the order $\mathbf{V}_2$, $\mathbf{V}_1$, and $\mathbf{V}_3$ (with $\mathbf{V}_2$ beginning at the origin), no crossover would have occurred. (The reader should prove this by redrawing Fig. 16-6b.) Regardless of the order of addition, the same resultant is always obtained.

EXAMPLE 16-4

From the solution and given information of Ex. 16-3:

a. Calculate the simplest series combination of basic circuit elements to replace the given impedances at the given frequency of 60 Hz
b. Draw the equivalent circuit component model (which only holds for the given frequency)[4]

Solution

a. Since $\mathbf{Z} = 12 - j5$ (at $f = 60$ Hz), the equivalent series model consists of a resistance of 12 Ω and a capacitive reactance of $-j5\ \Omega$. Then $C = 1/\omega C = 1/2\pi f X_C = 1/2\pi 60 \times 5 = 1/377 \times 5 = \mathbf{530.5\ \mu F}$
b. The equivalent series model is shown in Fig. 16-6**c**

[3] As long as the original *magnitude* and *direction* of a phasor are preserved, as in the case of vectors, we may move it in space freely as we choose. All phasors might have been drawn from the origin, but we would not have had the advantage of seeing, graphically, the verification of KVL applied to an ac series circuit.

[4] Had our original series circuit in Ex. 16-3 contained *only* combinations of practical inductors *or only* practical capacitors, we could have drawn a single equivalent series model that is *independent* of frequency. But the impedances of Ex. 16-3 contain a "mixed bag" of *R–L* and *R–C*. Consequently, they *cannot* be replaced by only *two* basic circuit elements.

16-4.5 Voltage Divider Rule (VDR) Applied to AC Circuits

Example 16-3 served to show that the same laws and rules governing dc circuits may be extended to ac circuits, using impedances (in complex form) instead of resistors and using voltage and current phasors in the frequency domain. From our earlier derivation of the VDR in Sec. 5-5.1, we may modify Eq. (5-6) to show

$$\mathbf{V}_1 = \frac{\mathbf{Z}_1}{\mathbf{Z}_2}\mathbf{V}_2 = \frac{\mathbf{Z}_1}{\mathbf{Z}_T}\mathbf{V}_T \quad \text{volts (V)} \qquad (16\text{-}11)$$

EXAMPLE 16-5

Using the given data of Ex. 16-3 and $\mathbf{Z}_T = 12 - j5$, calculate

a. $\mathbf{V}_1$
b. $\mathbf{V}_2$
c. $\mathbf{V}_3$ using the VDR
d. Compare the values obtained by VDR with those of Ex. 16-3**c**

Solution

a. $$\mathbf{V}_1 = \frac{\mathbf{Z}_1}{\mathbf{Z}_T}\mathbf{V}_T = \frac{(3+j4)}{12-j5}(117\angle -22.62^\circ)$$
$$= \frac{5\angle 53.13^\circ\ \Omega}{13\angle -22.62^\circ\ \Omega}(117\angle -22.62^\circ\ \text{V})$$
$$= \mathbf{45\angle 53.13^\circ\ V} \qquad (16\text{-}11)$$

b. $$\mathbf{V}_2 = \frac{\mathbf{Z}_2}{\mathbf{Z}_T}\mathbf{V}_T = \frac{5-j12}{12-j5}117\angle -22.62^\circ$$
$$= \frac{13\angle -67.38^\circ}{13\angle -22.62^\circ}(117\angle -22.62^\circ\ \text{V})$$
$$= \mathbf{117\angle -67.38^\circ\ V}$$

c. $$\mathbf{V}_3 = \frac{\mathbf{Z}_3}{\mathbf{Z}_T}\mathbf{V}_T = \frac{4+j3}{12-j5}117\angle -22.62^\circ$$
$$= \frac{5\angle 36.87^\circ}{13\angle -22.62^\circ}(117\angle -22.62^\circ\ \text{V})$$
$$= \mathbf{45\angle 36.87^\circ\ V}$$

d. These values are exactly the same as those obtained in Ex. 16-3**c**

Example 16-5 points up two major insights (one of which we may have forgotten by now):

1. In a series circuit, it is possible to find all voltages and powers without ever having to solve for current. This is because the current is the same and constant in all parts of the circuit.
2. Unlike the dc circuit, it is possible for the voltage across one component in an ac series circuit to exceed the supply voltage in magnitude! (We will encounter this frequently in series resonance.) Note that $\mathbf{V}_2$ is the same magnitude as $\mathbf{V}_T$ in Exs. 16-3 and 16-5.

16-4.6 Power Divider Rule (PDR) Applied to AC Circuits

Of the three basic circuit properties (R, L, and C), only resistance is capable of power dissipation. A little reflection causes us to realize that we cannot always (and blindly) substitute impedance for resistance, as we did in developing the voltage divider rule (VDR) in Eq. (16-11).

From our earlier PDR relation derived in Sec. 5-5.2 for dc circuits, we may modify these to show the power relations in ac circuits as

$$P_1 = \frac{R_1}{R_T}P_T = \frac{V_{R_1}}{V_{R_T}}P_T = \frac{R_1}{R_2}P_2 = \frac{V_{R_1}}{V_{R_2}}P_2 \quad \text{watts (W)} \qquad (16\text{-}12)$$

where R_1, R_2, etc. are the real, resistive portions of the complex impedances, $\mathbf{Z}_1$, $\mathbf{Z}_2$, etc., respectively

V_{R_1}, V_{R_2}, etc. are the ac voltages across the real, resistive portions *only* of the complex impedances, respectively

Although the various forms of Eq. (16-12) are valid, use of V_{R_1}, V_{R_2}, and V_{R_T} may be somewhat difficult for the reader to remember and apply. The easiest and most viable method is to use the dc power divider rule where $P_1 = (R_1/R_T)P_T$ or $(R_1/R_2)P_2$, as shown by Ex. 16-6.

EXAMPLE 16-6

Using the given data of Ex. 16-3 and $\mathbf{Z}_T = 12 - j5$, calculate

a. P_1
b. P_2
c. P_3 using the PDR
d. Compare the values obtained by PDR with those of Ex. 16-3**d**

Solution

a. Since $\mathbf{I} = 9$ A and $\mathbf{Z}_T = 12 - j5$, $P_T = I^2R_T = 9^2 \times 12 =$ 972 W and $P_1 = \dfrac{R_1}{R_T}P_T = (3/12)\,972\text{ W} = \mathbf{243\ W}$
b. $P_2 = (R_2/R_1)P_1 = (5/3)(243\text{ W}) = \mathbf{405\ W}$
c. $P_3 = (R_3/R_1)P_1 = (4/3)(243\text{ W}) = \mathbf{324\ W}$
d. These values are exactly the same as those obtained in Ex. 16-3**d**

The following insights are to be derived from Ex. 16-6:

1. Use of the PDR enables another set of cross checks against the methods of calculating ac power using V_R^2/R, I_R^2R, or $VI\cos\theta$.
2. The PDR is valid only for series circuits where all series-connected impedances carry the *same* current.

16-4.7 Power Factor, F_p

In Sec. 16-3 we noted that both the impedance triangle and the phasor diagram were similar "triangles" because the angle θ was common to both. It was also stated that V_T, V_R, and V_L are in the same ratio to each other (on the phasor diagram) as Z, R, and jX (on the impedance triangle). By definition, therefore, the power factor or $\cos\theta$ is

$$F_p = \cos\theta = \frac{R_T}{Z_T} = \frac{V_{R_T}}{V_T} = \frac{P}{VI} = \frac{P}{S} \quad \text{(unitless ratio)} \qquad \text{(16-13)}$$

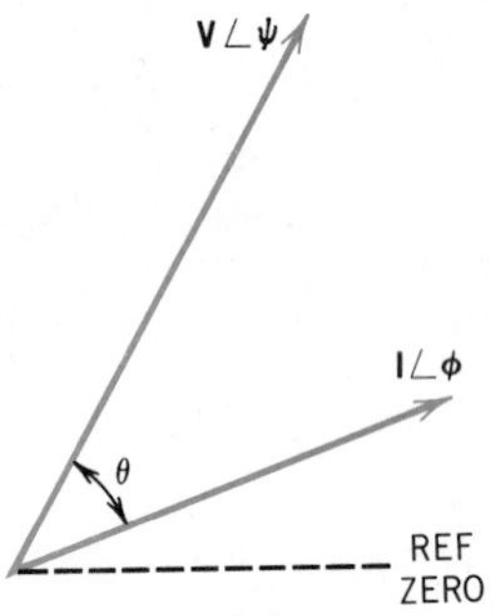

a. Phasor diagram

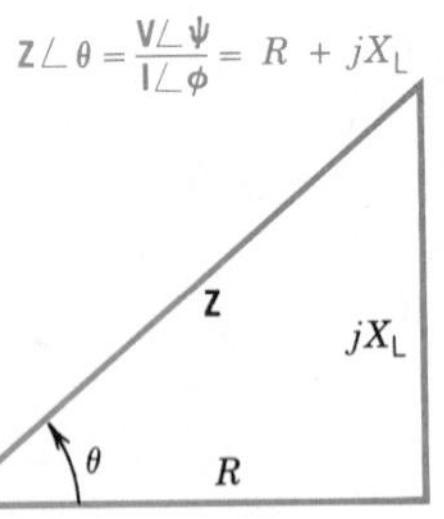

b. Impedance triangle

16-4.8 The Power Triangle and Complex Power

The power triangle emerges quite easily from the phasor diagram and the impedance triangle, as shown in **Fig. 16-7**. Let us treat each of the four diagrams shown separately.

Figure 16-7a Shows the voltage and current phasors with respect to some zero reference. In this case, current $\mathbf{I}\angle\phi$ lags voltage $\mathbf{V}\angle\psi$ by phase angle θ, where $\theta = \psi - \phi$, and the circuit is inductive.

Figure 16-7b Impedance, $\mathbf{Z}\angle\theta$, is computed as the ratio $\mathbf{V}\angle\psi/\mathbf{I}\angle\phi$, as shown in Eq. (16-4). When this complex impedance, $\mathbf{Z}\angle\theta$, is converted from polar to rectangular form, we obtain $\mathbf{Z}\angle\theta = R + jX_L$, enabling us to draw the impedance triangle of Fig. 16-7b.

Figure 16-7c When each of the arms of the impedance triangle is multiplied by the absolute value of current, $|I|$, we obtain the phasor diagram (once more) but now current appears on the reference zero axis! Since current is multiplied by impedance, three voltages emerge:

$\mathbf{V}_R = |\mathbf{I}|\mathbf{R}$, or the voltage drop across the resistance of the complex $\mathbf{Z}$.

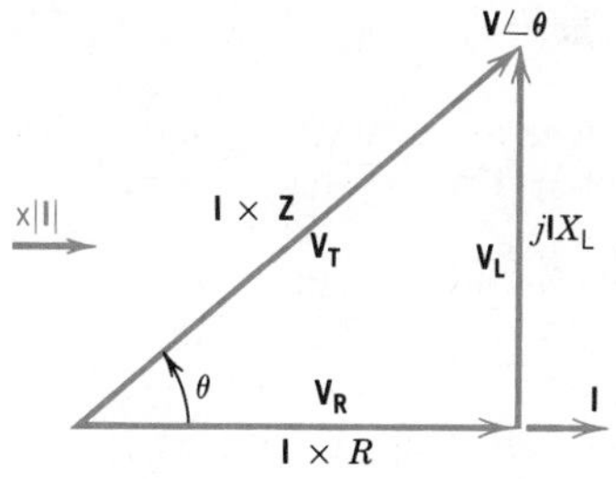

c. Phasor diagram with current as reference

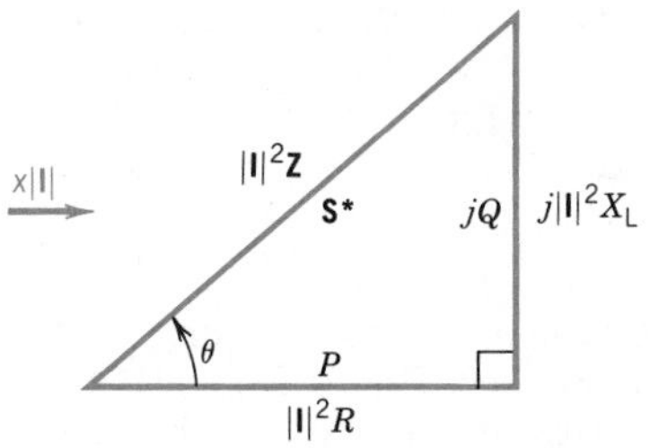

d. Power triangle

Figure 16-7 Evolution of power triangle from phasor diagram and impedance triangle in a series ac circuit

$j\mathbf{V}_L = |\mathbf{I}|j\mathbf{X}_L$, or the voltage drop across the reactance of the complex **Z**.

$\mathbf{V}_T = |\mathbf{I}|\mathbf{Z}$, or the total phasor voltage $V\underline{/\theta}$ across the entire complex **Z**.

Figure 16-7d When each of the arms of the phasor diagram is multiplied by the absolute value of $|I|$ (or the impedance triangle is multiplied by $|I^2|$), we obtain the power triangle. Three power components emerge:

$P = |\mathbf{I}^2|R$, or the real power component, measured in watts.

$jQ = |\mathbf{I}^2|jX_L$, or the quadrature reactive component, measured in vars.

$\mathbf{S}^* = |\mathbf{I}^2|Z$, or the apparent (complex) power, measured in volt-amperes (VA).

Since apparent power $\mathbf{S}^*$ is derived from the phasor diagram, as shown in Fig. 16-7, and is related to complex impedance **Z**, it is sometimes called *complex power*. The value of complex power, $\mathbf{S}^*$ may be computed as

$$\mathbf{S}^* = \mathbf{VI}^* = \mathbf{S}^*\underline{/\theta} = P \pm jQ \qquad \text{volt-amperes (VA)} \qquad (16\text{-}14)$$

where $\mathbf{I}^*$ is the conjugate of the current phasor, that is, $\mathbf{I}^* = (I\underline{/\phi})^* = I\underline{/-\phi}$

V is the phasor $V\underline{/\psi}$

$\mathbf{S}^*\underline{/\theta} = \mathbf{VI}^* = (\mathbf{V}\underline{/\psi})(\mathbf{I}\underline{/-\phi}) = \mathbf{VI}\underline{/\psi - \phi} = \mathbf{VI}\underline{/\theta}$

But since complex power $\mathbf{S}^*$ is obtained as $|I^2|$ multiplied by the arms of the impedance triangle, we may also write complex power as

$$\mathbf{S}^* = \mathbf{VI}^* = (\mathbf{IZ})\mathbf{I}^* = \mathbf{Z}(\mathbf{I} \times \mathbf{I}^*) = \mathbf{Z}(I^2\underline{/0^\circ})$$
$$= |I^2|\mathbf{Z} \qquad \text{volt-amperes (VA)} \qquad (16\text{-}15)$$

Equations (16-14) and (16-15) prove the validity of the process shown in Fig. 16-7.

16-4.9 Typical Worked Examples in Series Circuit Relations

EXAMPLE 16-7

Three impedances are connected in series: $\mathbf{Z}_1 = 25\underline{/-53.13^\circ}\ \Omega$, $\mathbf{Z}_2 = 25\underline{/73.74^\circ}\ \Omega$, and $\mathbf{Z}_3 = 26\underline{/22.62^\circ}\ \Omega$. If the voltage across $\mathbf{Z}_1$ is $125\underline{/-33.13^\circ}$ V, calculate WITHOUT USING current:

a. The applied voltage, **V**
b. Voltages $\mathbf{V}_2$ and $\mathbf{V}_3$
c. Total voltage as the sum of the three series-connected voltages, using KVL

Solution

a. $\mathbf{Z} = \mathbf{Z}_T = \mathbf{Z}_1 + \mathbf{Z}_2 + \mathbf{Z}_3$

$= 25\underline{/-53.13^\circ} + 25\underline{/73.74^\circ} + 26\underline{/22.62^\circ}$ (using P → R)

$= (15 - j20) + (7 + j24) + (24 + j10)$

$= \mathbf{46 + j14 = 48.08\underline{/16.93^\circ}\ \Omega}$

Then $\mathbf{V} = \mathbf{V}_T = \dfrac{\mathbf{Z}}{\mathbf{Z}_1}\mathbf{V}_1 = \dfrac{48.08\underline{/16.93^\circ}}{25\underline{/-53.13^\circ}}\, 125\underline{/-33.13^\circ}$ (16-11)

$\mathbf{V} = \mathbf{240.4\underline{/36.93^\circ}}$ **V**

b. $\mathbf{V}_2 = (\mathbf{Z}_2/\mathbf{Z}_1)(\mathbf{V}_1)$

$= (25\underline{/73.74^\circ}/25\underline{/-53.13^\circ})(125\underline{/-33.13^\circ}\ \text{V})$

$= \mathbf{125\underline{/93.74^\circ}}$ **V**

$\mathbf{V}_3 = (\mathbf{Z}_3/\mathbf{Z}_1)(\mathbf{V}_1)$

$= (26\underline{/22.62^\circ}/25\underline{/-53.13^\circ})(125\underline{/-33.13^\circ}\ \text{V})$

$= \mathbf{130\underline{/42.62^\circ}}$ **V**

c. $\mathbf{V} = \mathbf{V}_1 + \mathbf{V}_2 + \mathbf{V}_3$
$= 125\angle{-33.13^\circ} + 125\angle{93.74^\circ} + 130\angle{42.62^\circ}$
$= (104.68 - j68.32) + (-8.154 + j124.73) + (95.66 + j88.03)$
$= 192.19 + j144.44 = \mathbf{240.4\angle{36.93^\circ}\ V}$

With respect to the solution of Ex. 16-7, please note that

1. Use of the VDR eliminates the need to find current in a series circuit.
2. The solution of part (**b**) uses the original given data rather than the calculated value of **V**. This approach involves less possibility of error.
3. Answers to parts (**a**) and (**c**) prove the validity of Eq. (16-11).

EXAMPLE 16-8

From the given and calculated data of Ex. 16-7, calculate the

a. Circuit current, **I**
b. Power dissipated by each impedance and total power as sum of powers individually dissipated
c. Phase angle θ and $\cos\theta$ by two methods
d. Total power using the relation $VI\cos\theta$
e. Individually dissipated powers using PDR of Eq. (16-12)

Solution

a. $\mathbf{I} = \mathbf{V}/\mathbf{Z} = 240.4\angle{36.93^\circ}\ \text{V}/48.08\angle{16.93^\circ}\ \Omega = \mathbf{5\angle{20^\circ}\ A}$
b. $P_1 = I^2R_1 = 5^2 \times 15\ \Omega = \mathbf{375\ W}$,
$P_2 = I^2R_2 = 5^2 \times 7\ \Omega = \mathbf{175\ W}$,
$P_3 = I^2R_3 = 5^2 \times 24\ \Omega = \mathbf{600\ W}$, and
$P_T = P_1 + P_2 + P_3 = (375 + 175 + 600)\ \text{W} = \mathbf{1150\ W}$
c. From the solution to Ex. 16-7a, the phase angle of the total impedance:
   1. $Z\angle\theta = 48.08\angle{16.93^\circ}\ \Omega$ is $\theta = \mathbf{16.93^\circ}$ and $\cos 16.93^\circ = \mathbf{0.9567}$
   2. From the phasor values of circuit voltage and circuit current, the angular *difference* between voltage and current phasors is $36.93^\circ - 20^\circ = \theta = \mathbf{16.93^\circ}$ and $\cos\theta = \cos 16.93^\circ = \mathbf{0.9567}$
d. $P = VI\cos\theta = (240.4\angle{36.93^\circ}\ \text{V})(5\angle{20^\circ}\ \text{A})(\cos 16.93^\circ)$
$= \mathbf{1150\ W}$
e. $P_1 = (R_1/R_T)(P_T) = (15\ \Omega/46\ \Omega)(1150\ \text{W}) = \mathbf{375\ W}$ (16-12)
$P_2 = (R_2/R_T)(P_T) = (7\ \Omega/46\ \Omega)(1150\ \text{W}) = \mathbf{175\ W}$ (16-12)
$P_3 = (R_3/R_T)(P_T) = (24\ \Omega/46\ \Omega)(1150\ \text{W}) = \mathbf{600\ W}$ (16-12)

With respect to the solution of Ex. 16-8, please note that

1. The phase angle θ may be found from impedance triangle relations or phasor relations. Since $\mathbf{Z} = 48.08\angle{16.93^\circ}$, it follows that $\theta = 16.93^\circ$. Similarly, the angle θ, as defined in Eq. (14-39), is the angle between total voltage and total current phasors; as shown in part **c** of the solution, $\theta = 16.93^\circ$.
2. The powers computed by I^2R in part **b** only use the real, resistive portions of each impedance. This is also true of part **e**, using the PDR of Eq. (16-12).
3. There are many ways of calculating individually dissipated and total power. The reader should use as many as possible because they serve as a cross check on calculations.

EXAMPLE 16-9

Given the phasor current $\mathbf{I} = 5\angle{20^\circ}$ A, $\mathbf{Z} = 48.08\angle{16.93^\circ}\ \Omega$, and total phasor voltage $\mathbf{V} = 240.4\angle{36.93^\circ}$ V from Exs. 16-7 and 16-8, above, calculate the

a. Complex power, $\mathbf{S}^*\angle\boldsymbol{\theta}$
b. Real and quadrature power using (**a**)
c. Complex power using I^2Z
d. Real power using I^2R
e. Quadrature power, using I^2jX_L

Solution

a. $\mathbf{S}^* = \mathbf{VI}^* = (240.4\angle{36.93^\circ}\ \text{V})(5\angle{-20^\circ}\ \text{A})$
$= \mathbf{1202\angle{16.93^\circ}\ VA}$ (16-14)
b. $\mathbf{S}^* = \mathbf{1202\angle{16.93^\circ}\ VA} = \mathbf{1150 + j350\ VA}$ (16-14)
c. $I^2Z = (5)^2(48.08\angle{16.93^\circ}) = \mathbf{1202\angle{16.93^\circ}\ VA}$ (16-15)
d. $I^2R = (5)^2(46\ \Omega) = \mathbf{1150\ W}$
e. $I^2(jX_L) = (5)^2(j14) = \mathbf{j350\ vars}$

With respect to the solution of Ex. 16-9, please note that

1. Defining complex power as $\mathbf{S}^* = \mathbf{VI}^*$ enables us to determine automatically the true phase angle θ as a result of complex multiplication as shown in part (**a**).
2. Complex power also permits us to draw the power triangle $S^*\angle\theta = P \pm jQ$ (Fig. 16-7d).
3. The method of multiplying the arms of the impedance triangle by $|I^2|$ also yields the *same* power triangle as shown in parts **c**, **d**, and **e**.
4. Complex power enables us to determine total real power most easily and quickly. The results of Ex. 16-9 are the same as those of Ex. 16-8. Once total power is obtained, all other powers may be calculated by using the PDR.

EXAMPLE 16-10

A noninductive resistor is connected in series with a practical (RL) coil and an ac ammeter across a 110 V, 60 Hz ac supply. Voltage measurements taken across the resistor and the coil, respectively, are each 75 V and the ammeter reads 0.25 A. Calculate the

a. Absolute value of circuit impedance, $|Z_t|$
b. Absolute value of the impedance of the coil, $|Z_L|$
c. Resistance of the noninductive series resistor, R_s
d. Circuit phase angle, θ_t, using an impedance diagram showing the relation of **a**, **b**, and **c**. (Hint: use cosine law.)
e. Resistance of the coil
f. Phase angle of the coil θ_L
g. Reactance of the coil, jX_L

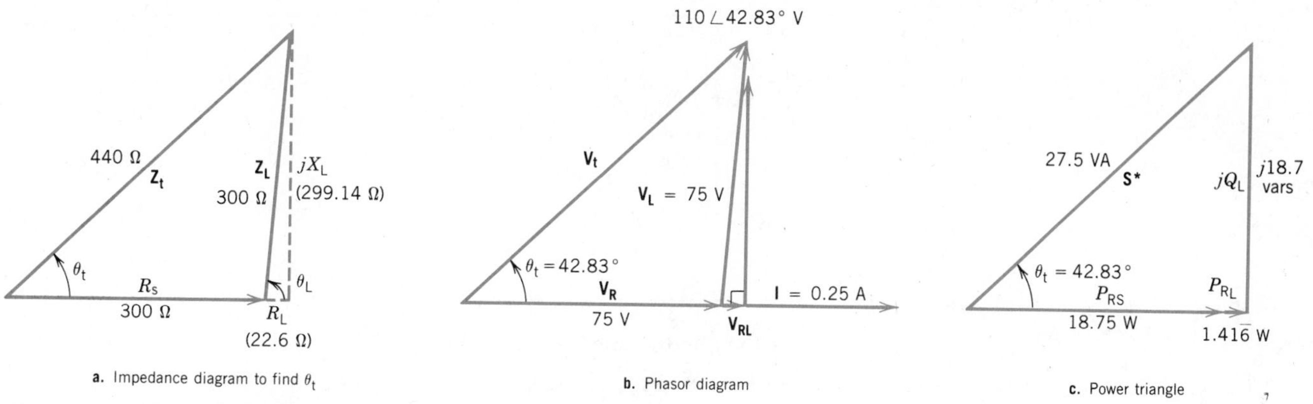

Figure 16-8 Example 16-10

h. Power drawn by series resistor R_s
i. Power dissipated by the coil
j. Complex power of the circuit
k. Quadrature reactive power of the coil

Draw completely labeled diagrams of:

l. Phasor relations, showing $\mathbf{V}_t$, $\mathbf{I}_t$, $\mathbf{V}_L$, and $\mathbf{V}_R$, respectively, labeling each
m. Impedance triangle relations
n. Power triangle showing all relations **h** through **k** above

Solution

a. $|Z_t| = V/I = 110\text{ V}/0.25\text{ A} = \mathbf{440\ \Omega}$
b. $|Z_L| = V_L/I = 75\text{ V}/0.25\text{ A} = \mathbf{300\ \Omega}$
c. $R = V_R/I = 75\text{ V}/0.25\text{ A} = \mathbf{300\ \Omega}$
d. Using impedance diagram drawn in Fig. 16-8a and the cosine law, we have $c^2 = a^2 + b^2 - 2ab(\cos C)$ or $300^2 = 440^2 + 300^2 - 2 \times 440 \times 300 \cos \theta_t$; then

$$\theta_t = \cos^{-1}\left(\frac{440^2}{2 \times 440 \times 300}\right) = \cos^{-1}(0.7\bar{3}) = \mathbf{42.833°}$$

e. From Fig. 16-8, total circuit resistance $R_t = Z_t \cos \theta_t = 440\ \Omega \cos 42.833° = \mathbf{322.\bar{6}\ \Omega}$. Consequently, $R_L = R_t - R_s = (322.6 - 300)\ \Omega = \mathbf{22.\bar{6}\ \Omega}$
f. $\theta_L = \cos^{-1}(R_L/Z_L) = \cos^{-1}(22.\bar{6}/300) = \mathbf{85.\bar{6}°}$
g. $jX_L = Z_L \sin \theta_L = 300 \sin 85.\bar{6}° = \mathbf{j299.14\ \Omega}$
h. $P_{R_s} = I^2R_s = (0.25)^2 300 = \mathbf{18.75\ W}$
i. $P_{R_L} = I^2R_L = (0.25)^2 \times 22.\bar{6} = \mathbf{1.41\bar{6}\ W}$
j. $\mathbf{S}^* = \mathbf{VI}^* = 110\angle 42.833° \times 0.25\angle 0° = \mathbf{27.5\angle 42.833°\ VA}$
$= \mathbf{20.16 + j18.70\ VA}$
k. From (j) above, $jQ_L = \mathbf{j18.7\ vars}$. Alternate check: $jQ_L = I^2jX_L = (0.25)^2 j299.14\ \Omega = \mathbf{j18.7\ vars}$
l. The phasor diagram is drawn in Fig. **16-8b**
m. The impedance triangle is shown in Fig. **16-8a**
n. The power triangle is shown in Fig. **16-8c**

16-5 PARALLEL CIRCUIT RELATIONS

In dc circuit analysis we discovered that the parallel circuit may be considered a *dual* of the series circuit. In Sec. 5-13 we noted that any series circuit equation or relation obtained in terms of voltage, current, and resistance has a dual parallel circuit counterpart in terms of current, voltage, and conductance, respectively. We concluded (Sec. 5-13) that once we know duality exists, we can use the equations for one type of circuit (series) and derive corresponding relations and equations for its dual circuit (parallel).

In dc circuits, conductance (G) in the parallel circuit is the dual of resistance (R) in the series circuit, where $G = 1/R$, as shown in Table 5-2.

In ac circuits we define admittance (**Y**) in the parallel circuit as the *dual* of impedance (**Z**) in the series circuit, where $\mathbf{Y} = 1/\mathbf{Z}$. But (as noted by the **boldface** type), impedance and admittance are **complex** quantities. (The reciprocal of a complex quantity is *always* a complex quantity.) Therefore, given impedance $\mathbf{Z}\angle\boldsymbol{\theta} = R + jX_L$, we may define complex admittance as

$$\mathbf{Y}\underline{/-\theta} = \frac{1}{R + jX_L} = \frac{1}{\mathbf{Z}\underline{/\theta}} = G - jB_L \quad (R\text{–}L \text{ combination}) \quad \text{siemens (S)} \tag{16-16a}$$

$$\mathbf{Y}\underline{/\theta} = \frac{1}{R - jX_C} = \frac{1}{\mathbf{Z}\underline{/-\theta}} = G + jB_C \quad (R\text{–}C \text{ combination}) \quad \text{siemens (S)} \tag{16-16b}$$

where G is the conductance (*not* equal to $1/R$)
$-jB_L$ is the inductive susceptance (*not* equal to $1/jX_L$)
$+jB_C$ is the capacitive susceptance (*not* equal to $1/-jX_C$)

and all other terms have been defined.

EXAMPLE 16-11

A series ac circuit has a resistance of 3 Ω and an inductive reactance of $j4$ Ω. Calculate

a. Its complex impedance in polar form
b. Its complex admittance in polar form
c. Its complex admittance in rectangular form
d. Draw the equivalent parallel circuit

Solution

a. $\mathbf{Z} = R + jX_L = 3 + j4 = \mathbf{5\underline{/53.13^\circ}\ \Omega}$
b. $\mathbf{Y} = 1/\mathbf{Z} = 1/5\underline{/53.13^\circ}\ \Omega = \mathbf{0.2\underline{/-53.13^\circ}\ S}$
c. $\mathbf{Y} = 0.2\underline{/-53.13^\circ} = \mathbf{(0.12 - j0.16)\ S} = G - jB_L$
d. The equivalent parallel circuit is shown in Fig. 16-9b

Example 16-11 and **Fig. 16-9** show that we must use the reciprocal of the polar form of *series* impedance (and *not* the reciprocals of its component resistance and reactance) to find *parallel* admittance of a given series circuit combination.[5]

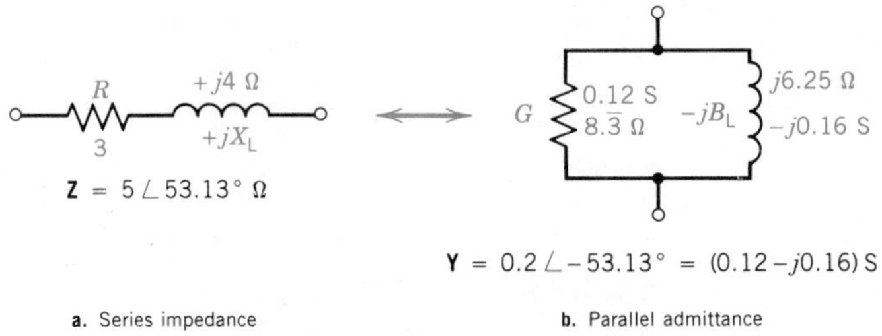

Figure 16-9 Series–parallel equivalence between complex impedance and complex admittance (Ex. 16-11)

Table 16-2 shows a summary of the duality between series-connected ac components and parallel-connected ac components to clarify their relations and terms.

With respect to Table 16-2, please note that

1. In the imaginary plane of a series *RLC* circuit, the net reactance, $\pm jX$, is the difference between $+jX_L$ and $-jX_C$. In the impedance case shown in the second series-connected column, $jX_L > -jX_C$, leaving a net *inductive* reactance of $+jX$, as shown in the impedance triangle.
2. This is shown in the phasor diagram of the *RLC* series circuit as well, since $jV_L > -jV_C$, leaving a net reactance voltage $+jV_X$. The phasor sum of $V_R + jV_X$ is the supply voltage, $\mathbf{V}\underline{/\theta}$, as shown in the phasor diagram.

[5] Our later consideration of series–parallel equivalence explains why the parallel equivalent shows a higher resistance and reactance than the series equivalent in Fig. 16-9b. See Sec. 16-16.

Table 16-2 Duality between *Series* Circuit Impedance and *Parallel* Circuit Admittance

Characteristic	Series-Connected Components	Parallel-Connected Components
Symbol	$\mathbf{Z}\angle\theta$, complex *impedance*	$\mathbf{Y}\angle-\theta$, complex *admittance*
Definition	$\mathbf{Z}\angle\theta = V\angle\psi / I\angle\phi$	$\mathbf{Y}\angle-\theta = I\angle\phi / V\angle\psi$
Complex form	$\mathbf{Z}\angle\theta = R \pm jX$	$\mathbf{Y}\angle\pm\theta = G \pm jB$
Real terms	R = *resistance*	G = *conductance*
Imaginary terms	$\pm jX$ = *reactance*	$\pm jB$ = *susceptance*
Imaginary plane equations	$jX = jX_L - jX_C$	$-jB = -(jB_L - jB_C)$
Treatment of inductance	$+jX_L = j\omega L$	$-jB_L = 1/j\omega L$
Treatment of capacitance	$-jX_C = 1/j\omega C$	$+jB_C = j\omega C$
Impedance triangle *RLC* series circuit / Admittance triangle *RLC* parallel circuit	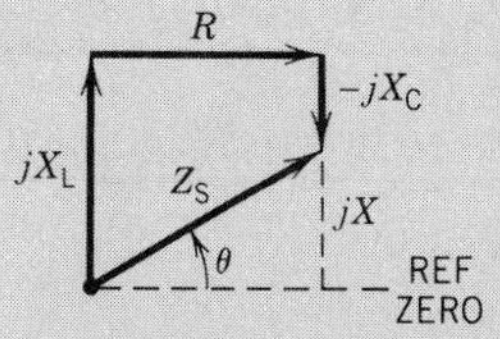	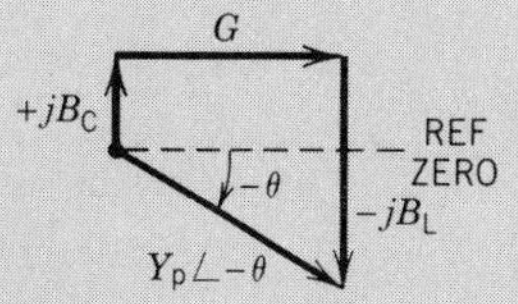
Phasor diagram of currents and voltages	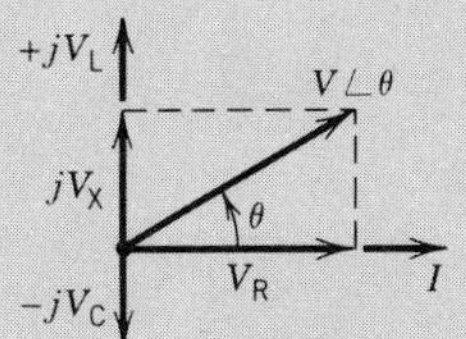	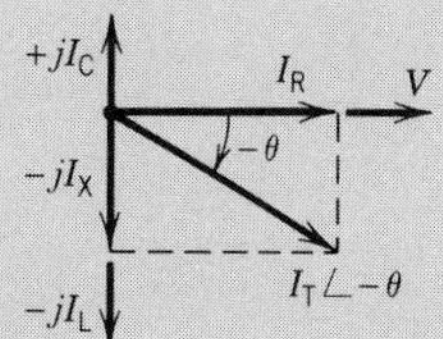
Defining equation	$\mathbf{I} \equiv \mathbf{I}_R \equiv \mathbf{I}_L \equiv \mathbf{I}_C$	$\mathbf{V} \equiv \mathbf{V}_R \equiv \mathbf{V}_C \equiv \mathbf{V}_L$
Phasor sums	$\mathbf{V} = \mathbf{V}_R + \mathbf{V}_L + \mathbf{V}_C$	$\mathbf{I} = \mathbf{I}_R + \mathbf{I}_C + \mathbf{I}_L$
Circuit phase angle	$\theta = \cos^{-1}(R/Z_S)$	$\theta = \cos^{-1}(G/Y_P)$
	$+\theta$ when $X_L > X_C$ and voltage leads current	$+\theta$ when $B_C > B_L$ and current leads voltage
	$-\theta$ when $X_C > X_L$ and current leads voltage	$-\theta$ when $B_L > B_C$ and voltage leads current

3. Observe from the phasor diagram that current lags voltage in an *RLC* series circuit that is predominantly inductive. This makes the phase angle θ a *positive* angle (using *current* as a reference) for an inductive circuit and a negative angle for a capacitive circuit, as shown at the bottom of the left-hand column.
4. The third (parallel-connected) column shows the dual parallel RLC circuit, whose complex admittance is the reciprocal of its dual series *RLC* impedance, $\mathbf{Z}\angle\theta$. Note that the admittance is $\mathbf{Y}\angle-\theta$ because the impedance is $Z\angle\theta$ (for the particular case shown), since $-jB_L > jB_C$ in this *admittance* triangle.
5. The *admittance* triangle of column 3 shows $\mathbf{Y}_P\angle-\theta$, where $-jB_L > jB_C$.
6. The phasor diagram uses voltage as the reference (see defining equation). Here current again lags voltage, indicating (once again) that current predominates. Observe that current, **I**, is the phasor sum of I_R, I_L, and I_C.

7. The circuit phase angle in both the phasor diagram and the admittance triangle is the same, as noted earlier. The phase angle θ is found from the admittance triangle as $\cos^{-1}(G/Y_P)$. In the parallel equivalent, θ is negative because $B_L > B_C$, as shown at the bottom of column 3 for parallel circuits.
8. The *signs* of the imaginary terms are *opposite* as a result of series–parallel duality, as indicated by the arrows crossing each other in the table; that is, in the *imaginary plane*,

   Inductive reactance is positive but inductive susceptance is negative.

   Capacitive reactance is negative but capacitive susceptance is positive.
9. Finally, *since inductance predominates* in *both* columns 2 and 3 of Table 16-2, current lags voltage.

The impedance–admittance relations, comparing two-element series and parallel circuits, are summarized in Table 16-3. These relations, similar to those of Table 16-2, should be studied most carefully by the reader

Table 16-3 Series–Parallel Impedance–Admittance Relations

Circuit Element(s)	Series Impedance Form (Ω)	Parallel Admittance Form (S)	Phasor Diagram
	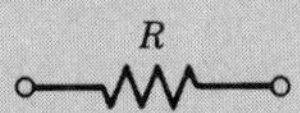	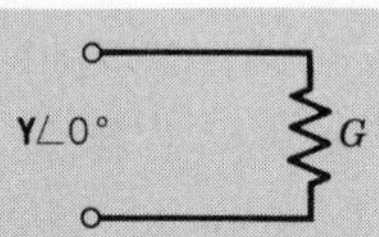	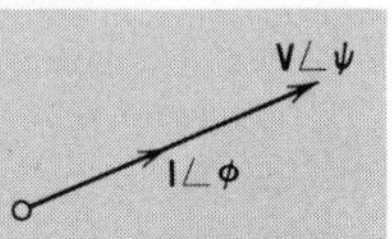
R only	$\mathbf{Z}\angle 0° = R\angle 0°$	$\mathbf{Y}\angle 0° = G\angle 0°$	$\theta = \psi - \phi = 0°$ Current in phase with voltage
	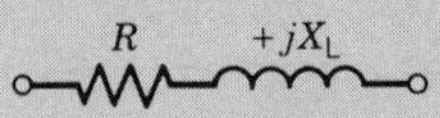	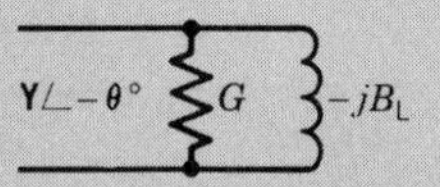	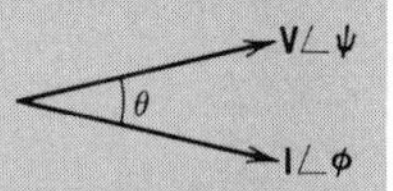
R in combination with L	$\mathbf{Z}\angle\theta = R + jX_L$	$\mathbf{Y}\angle -\theta = G - jB_L$	Current lags voltage by θ
	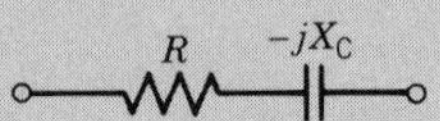	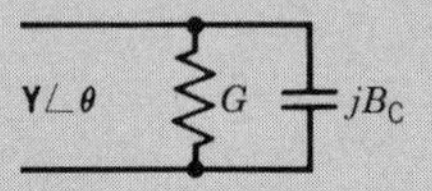	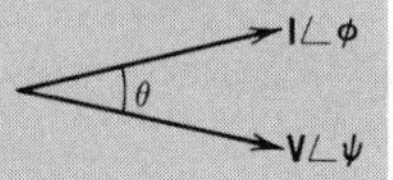
R in combination with C	$\mathbf{Z}\angle -\theta = R - jX_C$	$\mathbf{Y}\angle\theta = G + jB_C$	Current leads voltage by θ

16-5.1 Impedance–Admittance Relations

Admittance is the reciprocal of impedance. A high value of impedance implies a small current and a low value of admittance. A low value of impedance implies a high current and a high value of admittance. In a parallel circuit consisting of n admittances *in parallel*, we may write the *dual* of impedances in series as

$$\mathbf{Y}_T = \mathbf{Y}_1 + \mathbf{Y}_2 + \mathbf{Y}_3 + \cdots + \mathbf{Y}_n \quad \text{siemens (S)} \qquad \textbf{(16-17)}$$

This equation stated in impedance form becomes

$$\mathbf{Y}_T = \frac{1}{\mathbf{Z}_T} = \frac{1}{\mathbf{Z}_1} + \frac{1}{\mathbf{Z}_2} + \frac{1}{\mathbf{Z}_3} + \cdots + \frac{1}{\mathbf{Z}_n} \quad \text{siemens (S)} \qquad \textbf{(16-17a)}$$

For three impedances in parallel, solving for $\mathbf{Z}_T$ in Eq. (16-17a) yields

$$\mathbf{Z}_T = \frac{\mathbf{Z}_1 \cdot \mathbf{Z}_2 \cdot \mathbf{Z}_3}{\mathbf{Z}_1\mathbf{Z}_2 + \mathbf{Z}_2\mathbf{Z}_3 + \mathbf{Z}_3\mathbf{Z}_1} \quad \text{ohms } (\Omega) \qquad (16\text{-}17b)$$

and for two impedances in parallel, solving for $\mathbf{Z}_T$ in Eq. (16-17a) yields

$$\mathbf{Z}_T = \frac{\mathbf{Z}_1 \cdot \mathbf{Z}_2}{\mathbf{Z}_1 + \mathbf{Z}_2} \quad \text{ohms } (\Omega) \qquad (16\text{-}17c)$$

where all complex impedances are added and multiplied in complex form.

The reader should note the following similarities between dc and ac circuits:

1. Eq. (16-17) for ac circuits is similar to Eq. (5-21) for dc circuits.
2. Eq. (16-17a) for ac circuits is similar to Eq. (5-20) for dc circuits.
3. Eq. (16-17c) for ac circuits is similar to Eq. (5-23) for dc circuits.

Aware of these insights into ac circuit parallel impedance relations, we may now add the following equations derived from Secs. 5-9 and 5-10 for dc resistive circuits.

If we know the equivalent impedance of two impedances in parallel, $\mathbf{Z}_T$, and either one of the other impedances, $\mathbf{Z}_2$, then the remaining impedance, $\mathbf{Z}_1$ is

$$\mathbf{Z}_1 = \frac{\mathbf{Z}_2\mathbf{Z}_T}{\mathbf{Z}_2 - \mathbf{Z}_T} \quad \text{ohms } (\Omega) \qquad (16\text{-}17d)$$

where all complex impedances are multiplied and subtracted in complex form.

Given a parallel circuit consisting of N identical impedances in parallel, each having value $\mathbf{Z}_1$, the equivalent impedance, $\mathbf{Z}_{eq}$, is

$$\mathbf{Z}_{eq} = \mathbf{Z}_T = \frac{\mathbf{Z}_1}{N} \qquad (16\text{-}18)$$

16-5.2 Current Divider Rule (CDR)

From Eqs. (5-24a) and (5-24b) derived for two resistors in parallel, we may write the CDR for two impedances in parallel, $\mathbf{Z}_1 \| \mathbf{Z}_2$, as

$$\mathbf{I}_1 = \mathbf{I}_T \frac{\mathbf{Z}_2}{\mathbf{Z}_1 + \mathbf{Z}_2} \qquad (16\text{-}19a)$$

$$\mathbf{I}_2 = \mathbf{I}_T \frac{\mathbf{Z}_1}{\mathbf{Z}_1 + \mathbf{Z}_2} \qquad (16\text{-}19b)$$

where all complex impedances are added in complex (rectangular) form.

EXAMPLE 16-12

A parallel circuit contains two branches. The impedance of the first branch is $\mathbf{Z}_1 = (15 + j20)\ \Omega$ and the impedance of the second branch is $\mathbf{Z}_2 = (24 - j7)\ \Omega$. Calculate

a. Equivalent parallel impedance, using the product-over-the-sum relation of Eq. (16-17c) and equivalent parallel admittance of the circuit.
b. Admittance of each branch and total equivalent circuit admittance
c. Draw the admittance diagram for the circuit

Solution

a. $\mathbf{Z}_1 = 15 + j20 = 25\underline{/53.13^\circ}\ \Omega$ and $\mathbf{Z}_2 = 24 - j7 = 25\underline{/-16.26^\circ}\ \Omega$, while

$$\mathbf{Z}_1 + \mathbf{Z}_2 = 15 + j20 + 24 - j7$$
$$= 39 + j13 = 41.11\underline{/18.435^\circ}\ \Omega.$$

Then $$\mathbf{Z}_T = \frac{\mathbf{Z}_1\mathbf{Z}_2}{\mathbf{Z}_1 + \mathbf{Z}_2} = \frac{25\underline{/53.13^\circ} \times 25\underline{/-16.26^\circ}\ \Omega^2}{41.11\underline{/18.435^\circ}\ \Omega}$$
$$= \mathbf{15.20\underline{/18.435^\circ}\ \Omega} \qquad (16\text{-}17c)$$

and $$\mathbf{Y}_T = 1/\mathbf{Z}_T = 1/(15.20\underline{/18.435^\circ}\ \Omega)$$
$$= \mathbf{65.78\underline{/-18.435^\circ}\ mS} \qquad (16\text{-}16a)$$

b. $$\mathbf{Y}_1 = 1/\mathbf{Z}_1 = 1/25\underline{/53.13^\circ}\ \Omega = 40\underline{/-53.13^\circ}\ \text{mS}$$
$$= 24 - j32\ \text{mS}$$
$$\mathbf{Y}_2 = 1/\mathbf{Z}_2 = 1/25\underline{/-16.26^\circ}\ \Omega = 40\underline{/16.26^\circ}\ \text{mS}$$
$$= 38.4 + j11.20\ \text{mS}$$
$$\mathbf{Y}_T = \mathbf{Y}_1 + \mathbf{Y}_2 = (24 - j32) + (38.4 + j11.2)$$
$$= 62.4 - j20.8$$
$$= \mathbf{65.78\underline{/-18.435^\circ}\ mS} \qquad (16\text{-}17)$$

c. The admittance diagram is shown in Fig. 16-10. Observe that it can be drawn completely from the calculations of part (**b**) of the solution.

Example 16-12 shows the unity of solving parallel circuit problems using either an impedance method or an admittance method. With respect to the solutions of Ex. 16-12, please note that

1. The positive phase angle for $\mathbf{Z}_T$ (and negative phase angle for $\mathbf{Y}_T$) show that the equivalent circuit is inductive (see note at the bottom of Table 16-2); that is, if a voltage is applied to the parallel circuit then the current drawn would lag that voltage by 18.4°.
2. Counting and classifying the number of operations for the impedance method compared to the admittance method shows that neither has any strong advantage over the other. But the admittance method yields the admittance diagram directly, as noted earlier, and (as will be shown in Ex. 16-13) provides a direct clue to the phasor diagram.

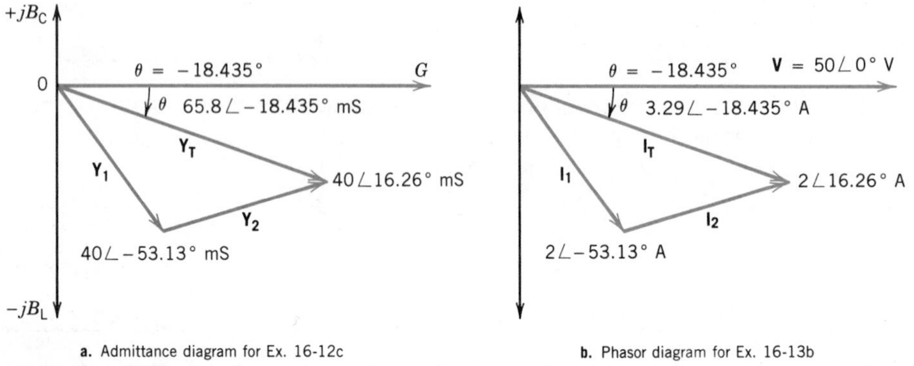

Figure 16-10 Solutions to Exs. 16-12 and 16-13. Relations between admittance diagram and phasor diagram of a parallel circuit using voltage as a reference

EXAMPLE 16-13

Assume a nominal voltage of $50\underline{/0^\circ}$ V across the parallel circuit of Ex. 16-12. Calculate:

a. Circuit in each branch and total current
b. Draw a phasor diagram showing all currents and supply voltage.
c. Total impedance from total voltage and total current
d. Total admittance from total voltage and total current

Solution

a. $$\mathbf{I}_1 = \mathbf{V}/\mathbf{Z}_1 = 50\underline{/0^\circ}\ \text{V}/25\underline{/53.13^\circ}\ \Omega = \mathbf{2\underline{/-53.13^\circ}\ A}$$
$$= 1.2 - j1.6\ \text{A};$$
$$\mathbf{I}_2 = \mathbf{V}/\mathbf{Z}_2 = 50\underline{/0^\circ}\ \text{V}/25\underline{/-16.26^\circ}\ \Omega = \mathbf{2\underline{/16.26^\circ}\ A}$$
$$= 1.92 + j0.56\ \text{A};$$
$$\mathbf{I}_T = \mathbf{I}_1 + \mathbf{I}_2 = (1.2 - j1.6) + (1.92 + j0.56)\ \text{A}$$
$$= 3.12 - j1.04 = \mathbf{3.289\underline{/-18.435^\circ}\ A}$$

b. The phasor diagram is shown in Fig. 16-10b

c. $\mathbf{Z}_T = \mathbf{V}_T/\mathbf{I}_T = 50\underline{/0^\circ}\ \text{V}/3.289\underline{/-18.435^\circ}\ \text{A} = \mathbf{15.2\underline{/18.435^\circ}\ \Omega}$

d. $$\mathbf{Y}_T = \mathbf{I}_T/\mathbf{V}_T = 3.289\underline{/-18.435^\circ}\ \text{A}/50\underline{/0^\circ}\ \text{V}$$
$$= \mathbf{65.78\underline{/-18.435^\circ}\ mS}$$

Example 16-13 leads to some interesting insights regarding ac parallel circuits:

1. By assuming a voltage across a parallel circuit, we may obtain its equivalent impedance and/or admittance. This constitutes a third method of solution in addition to the

impedance method and admittance method, shown in Ex. 16-12.

2. The phasor diagram in Fig. 16-10b shows that the phasor currents are in the same relationship as the admittance diagram of Fig. 16-10a; i.e., the "triangles" are similar and all angles, including phase angle θ, are *equal*.[6]
3. Current is lagging voltage in Fig. 16-10b, showing that the circuit is primarily inductive. This is also verified by the admittance diagram of Fig. 16-10a, which shows that the total admittance, $\mathbf{Y}_T$, has both an imaginary component along the $-jB_L$ axis and a real component along the conductance, G, axis.
4. As shown in Table 16-2 and Fig. 16-10b, when inductance predominates in a parallel circuit, the phase angle is $-\theta$ and voltage leads current. Note that $\mathbf{Y}_T$ is the phasor sum of $\mathbf{Y}_1 + \mathbf{Y}_2$ in Fig. 16-10a and that $\mathbf{I}_T$ is the phasor sum of $\mathbf{I}_1 + \mathbf{I}_2$. Thus, the admittance diagram serves as a cross check on the phasor diagram and vice versa. (We will discover later that the power triangle is similarly related to the admittance diagram and the phasor diagram, as might be expected.)

Above all, we must always bear in mind that the phasor diagram, such as that drawn in Fig. 16-10b, represents the key to the actual waveforms of currents and voltage in the time domain. The RMS values in the phasor or frequency domain at their respective angles are easily transformed to maximum values, leading or lagging their respective reference, expressed in the time domain. As noted in Chapter 15, the *time-domain waveforms represent reality*, in that they are displayed on a CRO by appropriate measurements in the laboratory or in the field. But as we have seen, use of the frequency domain greatly simplifies calculations.

16-5.3 Complex Power in Parallel Circuits

In the series ac circuit, we noted a serial relationship between the impedance diagram, the phasor diagram, and the power triangle, as shown in Fig. 16-7. Since the admittance diagram and the phasor diagram of a parallel circuit appear to be related, the question arises of whether the power triangle is similarly related. But before we answer this question, we must consider how individual complex powers are related to total power in a parallel circuit. Let us return to Ex. 16-13 to find the complex power of each branch and see how they are related to the total complex power of the entire circuit.

EXAMPLE 16-14

Using the individual branch currents and total current, respectively, found in Ex. 16-13 and the given supply voltage, calculate the

a. Complex power for branch 1 in polar and rectangular form
b. Complex power for branch 2 in polar and rectangular form
c. Total complex power as the sum of **a** and **b**
d. Total complex power in polar and rectangular form using total voltage and total current

Solution

a. $\mathbf{S}_1^* = \mathbf{V}\mathbf{I}_1^* = (50\underline{/0^\circ}\text{ V})(2\underline{/-53.13^\circ}\text{ A})^* = \mathbf{100\underline{/53.13^\circ}\text{ VA}}$
$= \mathbf{(60 + j80)\text{ VA}}$

b. $\mathbf{S}_2^* = \mathbf{V}\mathbf{I}_2^* = (50\underline{/0^\circ}\text{ V})(2\underline{/16.26^\circ}\text{ A})^* = \mathbf{100\underline{/-16.26^\circ}\text{ VA}}$
$= \mathbf{(96 - j28)\text{ VA}}$

c. $\mathbf{S}_T^* = \mathbf{S}_1^* + \mathbf{S}_2^* = (60 + j80) + (96 - j28) = \mathbf{156 + j52}$
$= \mathbf{164.44\underline{/18.435^\circ}\text{ VA}}$

d. $\mathbf{S}_T^* = \mathbf{V}_T\mathbf{I}_T^* = (50\underline{/0^\circ}\text{ V})(3.289\underline{/-18.435^\circ}\text{ A})^*$
$= \mathbf{164.44\underline{/18.435^\circ}\text{ VA}} = \mathbf{(156 + j52)\text{ VA}}$

Example 16-14 permits us to generalize from the dc to the ac case, regarding both complex power and true power. In Table 5-1 (Sec. 5-13) we observe that, regardless of whether we are dealing with a series, a parallel, or a series–parallel circuit, the total power is the sum of the individually dissipated powers. The same relation applies to the ac case, regardless of the method of connection (series, parallel, or series–parallel):

[6] This conclusion should come as no surprise. In a series ac circuit, the impedance triangle and the phasor diagram bear a similar relation. Since the parallel circuit is the dual of the series circuit, and **admittance** is the dual of impedance, we might guess that the **phasor** and **admittance** diagrams would be similarly related in the parallel circuit.

$$\mathbf{S}_T^* = \mathbf{S}_1^* + \mathbf{S}_2^* + \mathbf{S}_3^* + \cdots + \mathbf{S}_n^* \quad \text{volt-amperes (VA)} \tag{16-20a}$$

$$\mathbf{V}_T\mathbf{I}_T^* = \mathbf{V}_1\mathbf{I}_1^* + \mathbf{V}_2\mathbf{I}_2^* + \mathbf{V}_3\mathbf{I}_3^* + \cdots + \mathbf{V}_n\mathbf{I}_n^* \quad \text{volt-amperes (VA)} \tag{16-20b}$$

$$P_T \pm jQ_T = (P_1 \pm jQ_1) + (P_2 \pm jQ_2) + (P_3 \pm jQ_3) + \cdots + (P_n \pm jQ_n) \text{ VA} \tag{16-20c}$$

Extracting the real and imaginary components, yields:

$$P_T = P_1 + P_2 + P_3 + \cdots + P_n \quad \text{watts (W)} \tag{16-21}$$

$$\pm jQ_T = \pm jQ_1 \pm jQ_2 \pm jQ_3 \pm \cdots \pm jQ_n \quad \text{vars (vars)} \tag{16-22}$$

Equations (16-20) through (16-22) carry several important precautions that must be brought to the reader's attention:

1. Equation (16-20a) states that the total *complex power* is the sum of all the individual *complex* powers in any complete circuit. Apparent powers that are obtained as $\mathbf{I}_1^2\mathbf{Z}_1$ or $\mathbf{V}_1\mathbf{I}_1^*$ may *not* be added directly since they are not in phase.
2. Complex powers may be added because they represent the polar form of a complex number, which may be converted into its rectangular horizontal and vertical components. As shown in Eq. (16-20c), these real and quadrature components may be added separately (or in rectangular form) to yield the total real and total quadrature component of power. (See Ex. 16-14c also.)
3. Equation (16-21) states that the total **true** power, the **real** component of total complex power, is the (arithmetic) sum of the individually dissipated (true) powers.
4. Similarly, quadrature components of complex power, called reactive power, may be added with due regard for their quadrature direction, as shown in Eq. (16-22).

We are now prepared to consider the power triangle for a parallel circuit. Indeed, it would appear to emerge from the three forms of Eq. (16-20). But we should also realize that the power triangle is related to the phasor diagram, *if we set current as a reference* instead of voltage.[7] **Figure 16-11a** shows the phasor diagram of Fig. 16-10b rotated CCW by +18.435°, so that the total current I_T serves as the reference. This requires adding 18.435° to each phasor quantity in Fig. 16-10b to produce the phasor diagram of Fig. 16-11a.

The impedance triangle of Fig. 16-11b and the power triangle of Fig. 16-11c may be drawn from data computed in Exs. 16-13 and 16-14. But they may be found by alternative methods, as well, as shown in Ex. 16-15.

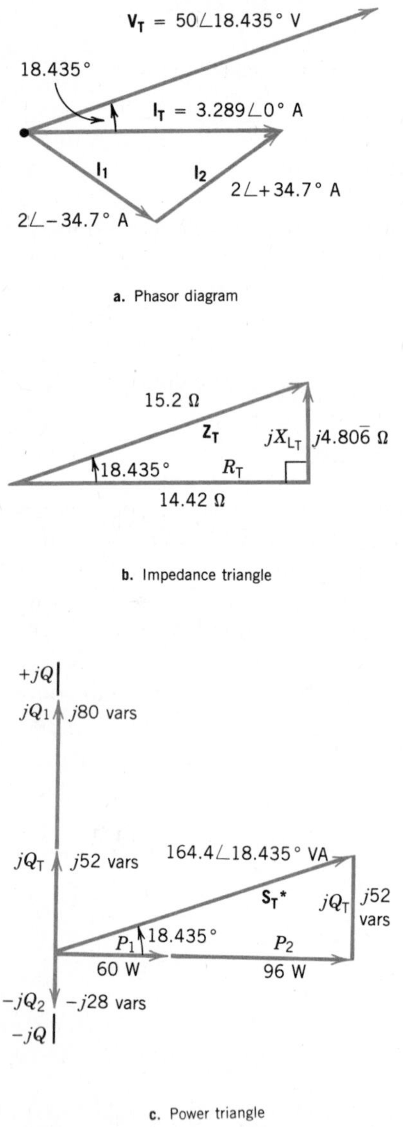

Figure 16-11 Relations among phasor diagram, total impedance triangle, and complex power triangle in an ac parallel circuit

[7] Current is used as the reference in deriving the impedance triangle of a series circuit. Since the power relation equations apply equally to series, parallel, and series–parallel circuits (i.e., they are *not* affected by duality relations), we may use current as the reference for the power triangle and the impedance triangle related to it, as shown in Figs. 16-11b and 16-11c, for a parallel circuit and also a series–parallel circuit. Regardless of the circuit, therefore, current is used as the reference in relating phasor diagrams to impedance triangles and power triangles.

EXAMPLE 16-15

Given the phasor diagram data shown in Fig. 16-11a, calculate

a. Total circuit impedance in polar and rectangular form
b. Total complex impedance, $\mathbf{S}_T^*$, using the method of multiplying arms of the impedance triangle by a factor of $|I^2|$
c. Total true power and reactive (quadrature) power, using the method of (b)
d. Compare the results obtained from the phasor diagram with those of Ex. 16-14, using $\mathbf{S}_T^* = \mathbf{V}_T\mathbf{I}_T^*$, the complex power relations

Solution

a. $\mathbf{Z}_T = \mathbf{V}_T/\mathbf{I}_T = 50\angle 18.435^\circ\ \text{V}/3.289\angle 0^\circ\ \text{A} = \mathbf{15.2\angle 18.435^\circ\ \Omega}$
$= \mathbf{14.42 + j4.807\ \Omega}$

b. $\mathbf{S}_T^* = |I^2|\mathbf{Z}_T = |(3.289)^2| \times 15.2\angle 18.435^\circ\ \Omega$
$= \mathbf{164.4\angle 18.435^\circ\ VA} = \mathbf{(156 + j52)\ VA}$

c. $P_T = |I^2|R = |(3.289)^2| \times 14.42\ \Omega = \mathbf{156\ W}$
$jQ_T = |(3.289)^2| \times j4.80\bar{6}\ \Omega = \mathbf{j52\ vars}$

d. The results obtained by multiplying each of the arms of the equivalent (input) impedance triangle by the absolute value of the current squared are the same as those obtained using complex power $\mathbf{S}_T^* = \mathbf{V}_T\mathbf{I}_T^*$.

The following conclusions are to be drawn from Ex. 16-15:

1. In both the series and parallel circuits (Figs. 16-7 and 16-11, respectively), the phasor diagram is the key to calculations of the impedance triangle and the power triangle, respectively.
2. But the phasor diagram to be used in (1) above should always designate total circuit current as the reference, if it is to be used to find the impedance diagram.
3. The admittance diagram is useful in developing the phasor diagram of a parallel circuit, using voltage as a reference (see Fig. 16-10).
4. The phasor diagram (with current as the reference) yields complex impedance, from which complex power may readily be calculated.
5. Alternatively, complex power may be calculated directly from the phasor diagram as $\mathbf{S}^* = \mathbf{VI}^*$, as well.

16-6 SERIES–PARALLEL EQUIVALENCE

Modern electronic ac impedance bridges (Sec. 23-14) are available that measure the circuit properties of a practical coil and/or a practical capacitor as either a *series model* or a *parallel model* or both. Since an inductor must be wound with wire, it inevitably has resistance. **Figure 16-12a** shows the two possible models for a practical inductor.

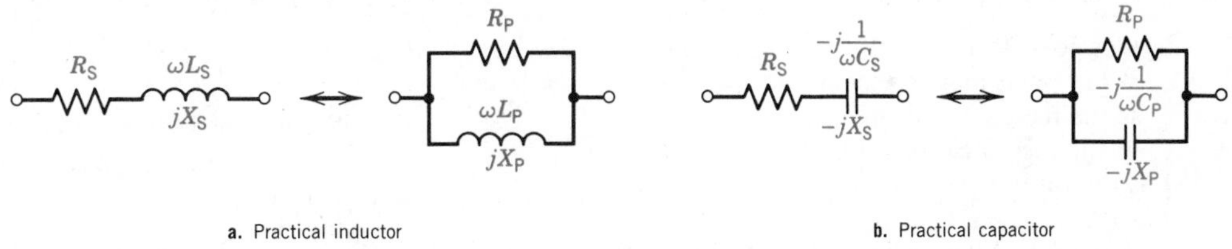

Figure 16-12 Series–parallel equivalence

In the same way, capacitors also have some resistance for reasons outlined in the various subsections of Sec. 9-6. Figure 16-12b shows the two possible circuit models for a practical capacitor, as well.

When equivalent total impedance (or admittance) is measured in the laboratory, no one can say with any degree of certainty whether the ideal inductor or capacitor actually is represented by the *series model* or the *parallel model*.[8] But we can say with complete confidence that *both* models at any given frequency must have the *SAME*:

1. Total input impedance, $\mathbf{Z}_T$, or admittance, $\mathbf{Y}_T$.
2. Current when connected to the same (voltage or current) source.
3. Q, quality factor, and/or D, dissipation factor.
4. Circuit elements (i.e., an R–L series model can only have an equivalent R–L parallel model; similarly, an R–C series model can only have an equivalent R–C parallel model.)

[8] The fact is that in most cases it is a series–parallel combination that may be reduced to the simpler models shown in Fig. 16-12.

Series–parallel equivalence is also useful in reducing more complex series–parallel circuits. We shall see how such reductions are possible when we consider series–parallel ac circuits in Sec. 16-7.

Appendix B-13 derives six basic equations for series–parallel equivalence, which are summarized in a more condensed form as three equations below. The equivalent parallel resistance, R_p, and equivalent parallel reactance, $\pm jX_p$, are

$$R_p = \frac{\mathbf{Z}_s}{\cos\theta} = R_s(1 + Q^2) \quad \text{ohms } (\Omega) \qquad (16\text{-}23)$$

$$\pm jX_p = \frac{\mathbf{Z}_s}{\sin\theta} = \pm jX_s\left(\frac{1 + Q^2}{Q^2}\right) \quad \text{ohms } (\Omega) \qquad (16\text{-}24)$$

where $\mathbf{Z}_s$ is the impedance of the *series* (two-element) circuit (Fig. 6-12)
θ is the phase angle of impedance Z_s in the impedance triangle such that $Z_s\underline{/\pm\theta} = R_s \pm jX_s$
Q is the ratio defined in Eq. (16-25)[9]

$$Q = \frac{X_s}{R_s} = \frac{R_p}{X_p} = \tan\theta = \frac{B_p}{G_p} \quad \text{(unitless ratio)} \qquad (16\text{-}25)$$

EXAMPLE 16-16

A commercial impedance bridge measures a practical inductor's parameters as $R_s = 30\ \Omega$ and $X_s = j60\ \Omega$. Calculate

a. $\mathbf{Z}_s\underline{/\theta}$ in polar form
b. $\mathbf{Y}_p\underline{/-\theta}$ in polar and rectangular form $G_p - jB_p$
c. R_p as the reciprocal of G_p
d. X_p as the reciprocal of $-jB_p$
e. Q as the ratio X_s/R_s
f. Q as the ratio R_p/X_p
g. R_p using $\mathbf{Z}_s/\cos\theta$ and $R_s(1 + Q^2)$ from Eq. (16-23)
h. X_p using $\mathbf{Z}_s/\sin\theta$ and $X_s\left(\frac{1 + Q^2}{Q^2}\right)$ from Eq. (16-24)
i. Draw conclusions about the relative ease of the methods shown in converting a series equivalent to its parallel equivalent circuit

Solution

a. $\mathbf{Z}_s = 30 + j60 = \mathbf{67.082\underline{/63.435^\circ}\ \Omega}$
b. $\mathbf{Y}_p = 1/\mathbf{Z}_s = 1/(67.082\underline{/63.435^\circ}\ \Omega)$
$= \mathbf{14.907\underline{/-63.435^\circ}\ mS} = \mathbf{6.\bar{6} - j13.\bar{3}\ mS}$
c. $R_p = 1/G_p = 1/6.\bar{6}\ \text{mS} = \mathbf{150\ \Omega}$
d. $X_p = 1/-jB_p = 1/-j13.\bar{3}\ \text{mS} = \mathbf{+j75\ \Omega}$
e. $Q = X_s/R_s = 60\ \Omega/30\ \Omega = \mathbf{2}$
f. $Q = R_p/X_p = 150\ \Omega/75\ \Omega = \mathbf{2}$
g. $R_p = \mathbf{Z}_s/\cos\theta = 67.082/\cos 63.435^\circ = \mathbf{150\ \Omega}$
$R_p = R_s(1 + Q^2) = 30\ \Omega(1 + 2^2) = \mathbf{150\ \Omega}$
h. $X_p = \mathbf{Z}_s/\sin\theta = 67.082/\sin 63.435^\circ = \mathbf{+j75\ \Omega}$
$X_p = X_s\left(\frac{1 + Q^2}{Q^2}\right) = +j60\ \Omega\left(\frac{1 + 2^2}{2^2}\right) = \mathbf{+j75\ \Omega}$
i. Although the reciprocal of the individual admittances enables solution of R_p and jX_p, the method using the cosine and sine functions is the simplest. The Q method is also relatively simple and preferable to the admittance method.

Example 16-16 provides a number of interesting other insights that cannot be overlooked:

1. *Both* the *parallel* resistance and reactance are *always higher* than the respective equivalent *series* resistance and reactance. While this is obvious from Eqs. (16-23) and (16-24), logic also tells us that this is necessary in order that both circuits have the same total equivalent impedance and draw the same current from the same (voltage or current) source. (See graphical construction in Appendix B-13.)
2. The circuit Q method also enables us to find the *series* equivalent, given the parallel equivalent, since $R_s = \frac{R_p}{(1 + Q^2)}$ from Eq. (16-23) and $\pm jX_s = \pm jX_p\frac{Q^2}{(1 + Q^2)}$ from Eq. (16-24).

[9] The symbol Q in this case refers to the *quality* ratio of reactance (either capacitive or inductive) to resistance for the series model. The same symbol is used for charge (measured in coulombs) and quadrature reactive power measured in vars.

3. The Q of the series circuit is X_s/R_s, whereas the Q of the equivalent parallel circuit is R_p/X_p. (This is an *important distinction* and will be reiterated when we study parallel resonant circuits.) But for series–parallel *equivalence* to occur, *both Q values must be the same*! This is shown in Eq. (16-25).

4. Finally, it bears repeating that a practical *inductor* in its *series* model can only be equivalent to a practical *inductor* in its *parallel* model, as shown in Fig. 16-12a. The same holds for a practical capacitor, as shown in Fig. 16-12b.

16-6.1 Two-Element Models

It was noted earlier (Ex. 16-4) that, given series or parallel combinations of the three basic circuit elements (R, L, and C), it is impossible to draw a simplified two-element model that is independent of frequency. This was stated even though at some particular frequency the circuit may be reduced to some resultant resistance and reactance. But in dealing with the circuits of Fig. 16-12, we have only two basic circuit elements, one of which is resistance and the other *either* inductance *or* capacitance. In measuring a practical inductor via impedance bridge methods, it would be much more useful to represent it as *either* a series *or* parallel R–L combination, where L is expressed in henrys. Such a model would be useful at a particular frequency but it is *not* independent of frequency, as we will see in Sec. 16-6.2.

Let us assume that resistance of a practical coil is independent of frequency.[10] We already know that

$$X_p = \omega L_p = \omega L_s\left(\frac{1 + Q^2}{Q^2}\right) = jX_s\left(\frac{1 + Q^2}{Q^2}\right) \qquad (16\text{-}24)$$

Dividing both sides by ω:

$$L_p = L_s\left(\frac{1 + Q^2}{Q^2}\right) \quad \text{henrys (H)} \qquad \textbf{(16-24a)}$$

A similar equation may be derived for a practical capacitor where $X_s = X_p\left(\frac{Q^2}{Q^2 + 1}\right)$ or $\frac{1}{\omega C_s} = \frac{Q^2}{\omega C_p(Q^2 + 1)}$; dividing both sides by ω and cross-multiplying,

$$C_p = C_s\left(\frac{Q^2}{Q^2 + 1}\right) = \frac{C_s}{(1 + D^2)} \quad \text{farads (F)} \qquad \textbf{(16-25a)}$$

where D, the dissipation factor, is $1/Q$ (used with capacitors only).[11]

16-6.2 Effect of Frequency on Series–Parallel Equivalence

The significance of Eqs. (16-24a) and (16-25a) is that once a series equivalent model is found *at a given frequency*, for example, it is possible to calculate the equivalent parallel model and vice versa, *but only at that given frequency*. The reason for this

[10] We will discover later that resistance tends to increase at very high frequencies (VHF) because of increased losses. (See Sec. 17-8.)

[11] Capacitor specifications typically include R_p, C_p, and D (rather than Q). If D is less than 0.1, this implies a Q of more than 10. Such a capacitor has a *low* dissipation factor (low D) and is considered a *high*-quality (high-Q) component. See Secs. 9-6.1 and 9-6.2 for additional information on losses in capacitors.

should be obvious, since $Q = X_s/R_s = R_p/X_p$ is a *frequency-dependent* factor that changes with reactance, where reactance is a function of frequency. Most important, however, is the conclusion from Eqs. (16-24a) and (16-25a) that the R–L values of the *series* model are *always different* from the R–L values of the parallel model at the same frequency. The same is true of the R–C *series* model compared to the R–C *parallel* model. Since we have used R–L circuits in the previous examples, let us now switch to an R–C circuit to prove the changes in series–parallel equivalence with frequency. For this comparison we will maintain the parallel R–C model constant and only vary the frequency.

EXAMPLE 16-17

Given the RC parallel model where $C_p = 53$ nF and $R_P = 600\ \Omega$. At frequencies of 0.5 kHz, 1 kHz, 2.5 kHz, 5 kHz, 8 kHz, and 10 kHz, respectively calculate the following equivalent series model parameters:

a. $-jX_{C_p}$
b. Q
c. Q^2
d. R_s
e. $-jX_s$
f. C_s
g. Tabulate the above parameters for ready reference and comparison versus frequency

Solution

(Individual solutions are left as an exercise for the reader. The tabulation of parameters, however, is shown below, along with equations used.)

1. Both R_s and C_s appear to *decrease* as the frequency *increases*. At VHF, both the series and the parallel impedance should be zero (see 7 below).
2. But $-jX_s$ (unlike $-jX_{C_p}$) increases to a maximum (where it equals $R_p/2$) and then decreases!
3. $-jX_s$ is a maximum when the Q is unity ($R_p = -jX_{C_p}$ and $R_s = -jX_s$).
4. $-jX_s$ is a maximum when $R_s = R_p/2 = -jX_s$.
5. $-jX_s$ is a maximum when the radian frequency $\omega = 1/R_pC_p$.
6. When $\omega = 0$ (at dc), $-jX_s = 0$, and $R_s = R_p = 600\ \Omega$. $Z_s = 600\ \Omega$ at $\omega = 0$.
7. At $\omega = \infty$, $-jX_s = 0$, and $R_s = 0$; therefore, $Z_s = 0$ at $\omega = \infty$.
8. At each given frequency, the Q for both the series and parallel models is the *same*. This constitutes a proof of the validity of the calculations in the table as well as Eq. (16-25).

f (kHz)	R_p (Ω)	C_p (nF)	$-jX_{C_p}$ (Ω)	Q (–)	Q^2 (–)	$-jX_s$ (Ω)	R_s (Ω)	C_s (nF)	Equations Used	Equation
0.5	600	53	6006	0.1	0.01	59.46	594	5353	$-jX_{C_p} = 1/2\pi fC$	(14-36)
1.0	↓	↓	3003	0.2	0.04	115.5	576.9	1378	$Q_p = R_p/X_p = Q_s = X_s/R_s$	(16-25)
2.5			1201	0.5	0.25	240.2	480	265	$R_s = R_p/(Q^2+1)$	(16-23)
5.0			600.6	1.0	1.0	300.3	300.3	106	$-jX_s = -jX_p \times Q^2/(Q^2+1)$	(16-24)
8.0			375.4	1.6	2.56	270	168.5	73.7	$C_s = C_p(Q^2+1)/Q^2$	(16-25a)
10.0			300.3	2.0	4.0	240.24	120	66.3		

There are many and varied conclusions to be drawn from the above tabulation. We will concentrate on those that are not so obvious but are most important:

9. Even though the parallel R_p–C_p model is held *constant*, the series R_s–C_s model *changes* and *continuously decreases* as the frequency increases (see 1 above).
10. As the frequency increases, Q increases. When $\omega = \infty$, $C_s = C_p$, since the Q **ratio** in Eq. (16-25a) is almost unity. This shows that Q is frequency-dependent.

The variation of R_s and $-jX_s$ versus frequency, tabulated above, is shown in **Fig. 16-13**, summarizing almost all the foregoing conclusions. Appendix B-13 contains the mathematical proof that explains why the parameters vary as shown in Fig. 16-13.

To summarize, if a fixed 600-Ω resistor is connected in parallel with a fixed 53-nF capacitor and the series impedance parameters R_s, C_s, and Q are measured on an impedance bridge at the frequencies given, we obtain the values shown for R_s, C_s, and Q, summarized in the table in Ex. 16-17. This implies that

1. If any factor in the $R_p \| jX_p$ relation is changed (including frequency), then *both* $R_s + jX_s$ must change correspondingly because the Q changes.
2. The converse is also true. If any factor in the series model ($R_s + jX_s$) is changed, both R_p and jX_p must change because the Q is changed.

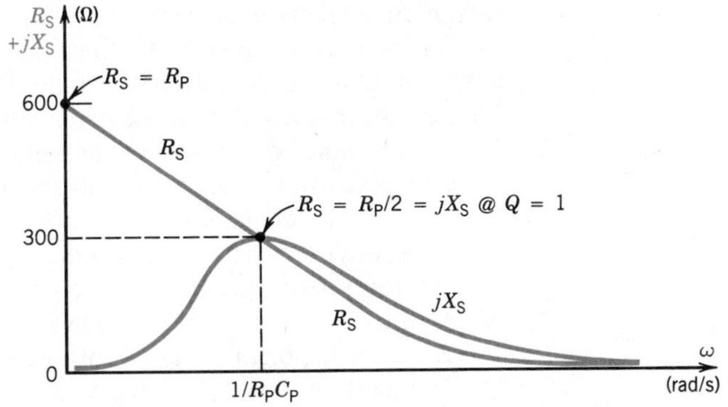

Figure 16-13 Variation in equivalent series resistance and reactance with radian frequency, ω

3. At any given frequency, the series model is just as valid as the parallel model because it has the same equivalent impedance. This is shown in Ex. 16-18.

EXAMPLE 16-18

Using the frequency of 5 kHz and the data tabulated above, show that the parallel circuit has the same impedance as the series circuit by calculating a. $\mathbf{Z}_p$ b. $\mathbf{Z}_s$

Solution

a. $jB_p = 1/-jX_{C_p} = 1/-j600.6 = +j1.\bar{6}$ mS and
$G_p = 1/R_p = 1/600\ \Omega = 1.\bar{6}$ mS;
$\mathbf{Y}_p = G_p + jB_p = (1.\bar{6} + j1.\bar{6})$ mS $= 2.357\angle 45^\circ$ mS;
$\mathbf{Z}_p = 1/\mathbf{Y}_p = 1/2.357\angle 45^\circ$ mS $= \mathbf{424.3\angle -45^\circ\ \Omega}$
$= \mathbf{300 - j300\ \Omega}$

b. $\mathbf{Z}_s = \mathbf{300.3 - j300.3 = 424.7\angle -45^\circ\ \Omega}$

16-7 NETWORK SIMPLIFICATION OF PARALLEL–SERIES AND SERIES–PARALLEL CIRCUITS

It was indicated earlier that *as long as the frequency is fixed*, it is possible to reduce any complex network to a single resistive and single reactive element in either the series or the parallel model. This combination should draw the *same* total current from a voltage or current source as the original complex network. The CDR and VDR (current-divider rule and voltage divider rule) may be used, appropriately, to find all individual component voltages and currents. Knowing all voltages and currents in phasor form, it is a simple matter to draw a phasor diagram and an impedance triangle and power triangle, as well. If necessary, the phasor values in the frequency domain are easily converted into time-domain sinusoidal expressions, from which waveforms may be drawn.

The circuits we will consider are called *series–parallel* circuits. As in the dc cases of such circuits, we will apply series circuit rules to series portions and parallel circuit rules to parallel portions. In effecting network simplification, the series–parallel equivalence methods of Sec. 16-6 should be used whenever possible, since the frequency in these problems is constant and only one source (of voltage or current) is used. The following rules should be helpful:

1. Those portions of the circuit connected in *parallel* involve adding *admittances* (in rectangular form) to produce an equivalent admittance, which may be converted reciprocally to an equivalent impedance.
2. Those portions of the circuit connected in *series* may be combined by adding equivalent *impedances* (in rectangular form) to produce a single equivalent impedance.
3. Steps (1) and (2) may be repeated until a sufficiently simple equivalent is found to enable the complete solution of the network.

4. In some problems it is helpful to *begin* the solution where any two of three factors (current, voltage, and either impedance or admittance) are known. In most cases, however, analysis should begin at the portion that is *farthest* from the source, combining elements and working back toward the source.
5. Check your solution by any of the following methods:
   a. Compute total power as the sum of the individual powers and total power using total voltage and total current.
   b. An alternative check that is sometimes useful is the product-over-the-sum relation for complex parallel impedances to yield a series equivalent.
   c. A third solution check is to use impedance exclusively and then repeat the same problem using admittances exclusively.
   d. Two-element series–parallel conversion techniques (Sec. 16-6) may be used alternatively as a check on network simplification, where applicable.
   e. The final test is that the *simplified* equivalent draws the *same* current from the source as the *original* network. In fact, all of the solution checks first outlined should yield the *same* answers, consistently.
   f. If two different approaches yield two different sets of answers
      1. Use a third approach, or
      2. Rework the problem on a fresh sheet of paper independently.

EXAMPLE 16-19

A parallel-series R–L–C circuit[12] has the values shown in **Fig. 16-14a**. Calculate

a. Component impedance values to reduce it to a three-branch parallel circuit
b. Total admittance of the three-branch parallel circuit in rectangular and polar form
c. Equivalent impedance of the entire network in polar and rectangular form
d. Component impedance values to reduce the three-branch parallel circuit (calculated in **a** above) to the simplest two-branch parallel equivalent circuit
e. Total impedance by parallel-to-series conversion of two-branch circuit found in **d**

Solution

a. Converting the series impedance $Z_s = (4 + j8)\ \Omega$; $Q = X_L/R = 8\ \Omega/4\ \Omega = \mathbf{2}$

$$jX_p = \frac{Q^2+1}{Q^2} \times jX_s = \frac{5}{4}(j8\ \Omega) = \mathbf{j10\ \Omega} \qquad (16\text{-}24)$$

$$R_p = R_s(Q^2+1) = 4\ \Omega\,(2^2+1) = \mathbf{20\ \Omega}\ \text{(See Fig. 16-14b)} \qquad (16\text{-}23)$$

b. $$\mathbf{Y}_p = \frac{1}{20\angle 0^\circ} + \frac{1}{10\angle 90^\circ} + \frac{1}{5\angle -90^\circ} = (50 - j100 + j200)\ \text{mS}$$
$$= \mathbf{50 + j100\ mS} = \mathbf{111.8\angle 63.435^\circ\ mS} \qquad (16\text{-}17a)$$

c. $$\mathbf{Z}_s = 1/\mathbf{Y}_p = 1/111.8\angle 63.435^\circ\ \text{mS} = 8.9443\angle -63.435^\circ\ \Omega = \mathbf{(4 - j8)\ \Omega}$$

d. Converting $j10 \| -j5$ combination in Fig. 16-14b by the product-over-the-sum method:[13]

$$\mathbf{Z}_{eq} = \frac{j10 \times (-j5)}{-j5 + j10} = \frac{10\angle 90^\circ \times 5\angle -90^\circ}{5\angle 90^\circ}$$
$$= \mathbf{10\angle -90^\circ\ \Omega}\ \text{(See Fig. 16-14c)} \qquad (16\text{-}17c)$$

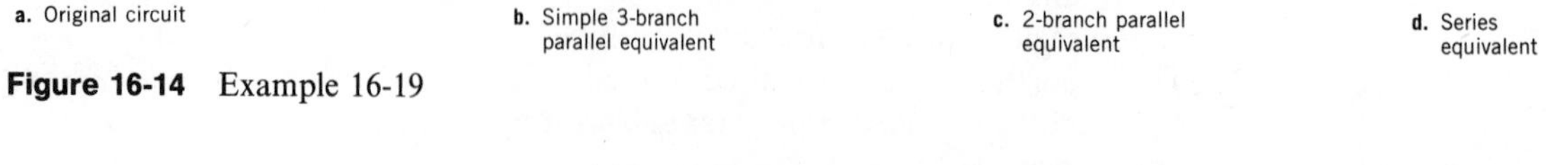

Figure 16-14 Example 16-19

[12] Recall from Sec. 6-1 that a parallel-series circuit is essentially a parallel circuit having series elements.
[13] The notation $j10 \| -j5$ should be read as "$j10$ in parallel with $-j5$."

e. Using parallel-to-series conversion of the circuit shown in Fig. 16-14c,

$$Q = R_p/X_p = 20\ \Omega/10\ \Omega = \mathbf{2} \qquad (16\text{-}25)$$

$$R_s = R_p/(Q^2+1) = 20\ \Omega/(4+1) = \mathbf{4\ \Omega},$$

$$-jX_s = -jX_p\frac{Q^2}{Q^2+1} = -j10\frac{4}{5} = \mathbf{-j8\ \Omega},$$

and $\mathbf{Z}_s = 4 - j8 = 8.9443\underline{/-63.435^\circ}\ \Omega$

With respect to the solution of Ex. 16-19, the following points are worth noting:

1. The original circuit shows a practical coil in parallel with a pure capacitor, commonly used in both oscillator and tuning circuits. Conversion of this circuit to a three-branch parallel circuit is important for the later study of resonant circuits. (See Sec. 19-6.)
2. The solutions of **c** and **e** yield exactly the same answers by independent methods.
3. With $j10 \| -j5$ in Fig. 16-14b, the smaller impedance ($-j5$) predominates. In a parallel circuit, the lower reactance has a higher susceptance, as shown in the solution to part **b**, resulting in an admittance with a positive phase angle. This produces an impedance with a negative phase angle, indicating a net capacitive reactance as shown in part **c**.
4. Note that the solution uses both series-to-parallel and parallel-to-series conversion techniques by the Q method of Sec. 16-6, as well as the product-over-the-sum technique of Eq. (16-17c).
5. Alternatively, by assuming a convenient supply voltage of, say, $20\underline{/0^\circ}$ V, using Fig. 16-14b, the current in each branch and total current yield the total impedance, $\mathbf{Z}_t$.

EXAMPLE 16-20

Given the circuit shown in **Fig. 16-15a**, calculate the

a. Impedance of each branch, $\mathbf{Z}_1$ and $\mathbf{Z}_2$, and their sum, $\mathbf{Z}_1 + \mathbf{Z}_2$ in polar form
b. Current in each branch, $\mathbf{I}_1$ and $\mathbf{I}_2$, using the current divider rule (CDR)
c. Supply voltage, **V**, across the current source, by two methods
d. Draw the phasor diagram, using total current as a reference, showing all currents and voltage
e. Complex power (**VI***) in polar and rectangular form. Draw a power triangle
f. Total power using currents obtained in **b** and using $P = VI\cos\theta$

Solution

a. $\mathbf{Z}_1 = 7 + j24 = \mathbf{25\underline{/73.74^\circ}\ \Omega}$;
$\mathbf{Z}_2 = 15 - j20 = \mathbf{25\underline{/-53.13^\circ}\ \Omega}$;
$\mathbf{Z}_1 + \mathbf{Z}_2 = 22 + j4 = \mathbf{22.36\underline{/10.3^\circ}\ \Omega}$

b. $$\mathbf{I}_1 = \mathbf{I}_T\frac{\mathbf{Z}_2}{\mathbf{Z}_1+\mathbf{Z}_2} = (1.7\bar{8}\underline{/-10.3^\circ}\ \text{A})\frac{25\underline{/-53.13^\circ}\ \Omega}{22.36\underline{/10.3^\circ}\ \Omega} = \mathbf{2\underline{/-73.73^\circ}\ A} \qquad (16\text{-}19a)$$

$$\mathbf{I}_2 = \mathbf{I}_T\frac{\mathbf{Z}_1}{\mathbf{Z}_1+\mathbf{Z}_2} = (1.78\underline{/-10.3^\circ}\ \text{A})\frac{25\underline{/73.74^\circ}\ \Omega}{22.36\underline{/10.3^\circ}\ \Omega} = \mathbf{2\underline{/53.14^\circ}\ A} \qquad (16\text{-}19b)$$

c. $\mathbf{V} = \mathbf{I}_1\mathbf{Z}_1 = (2\underline{/-73.73^\circ}\ \text{A})(25\underline{/73.74^\circ}\ \Omega) = \mathbf{50\underline{/0^\circ}\ V}$
$\mathbf{V} = \mathbf{I}_2\mathbf{Z}_2 = (2\underline{/53.14^\circ}\ \text{A})(25\underline{/-53.14^\circ}\ \Omega) = \mathbf{50\underline{/0^\circ}\ V}$

d. Since current is now being used as a reference, this means **+10.3°** must be added to each phasor quantity. The phasor diagram is shown in Fig. 16-15b.

e. $\mathbf{S}^* = \mathbf{VI}^* = (50\underline{/0^\circ}\ \text{V})(1.7\bar{8}\underline{/-10.3^\circ})^* = 89.\bar{4}\underline{/10.3^\circ}\ \text{VA}$
$= \mathbf{(88 + j16)}\ \text{VA}$

f. $P_1 = I_1^2R_1 = 2^2 \times 7 = \mathbf{28\ W}$
$P_2 = I_2^2R_2 = 2^2 \times 15 = \mathbf{60\ W}$
$P_T = P_1 + P_2 = 28 + 60 = \mathbf{88\ W}$ (see **e** above)
$P = VI\cos\theta = 50\ \text{V} \times 1.7\bar{8}\ \text{A}\cos 10.3^\circ$
$= \mathbf{88\ W}$ (see **e** above)

With respect to this solution please note that

1. As noted in Sec. 16-4.7, drawing the phasor diagram with current as the reference provides a check on the power triangle. Since $\mathbf{V} = \mathbf{IZ}$, then $\mathbf{S}^* = \mathbf{VI}^* = \mathbf{I(IZ)} = \mathbf{I(V)}$, and substituting values $\mathbf{S}^* = (1.78\underline{/0^\circ}\ \text{A})(50\underline{/10.3^\circ}\ \text{V}) = \mathbf{89.\bar{4}\underline{/10.3^\circ}\ VA}$
2. The phasor diagram contains all the information to find the total impedance, $\mathbf{Z}_T = \mathbf{V}/\mathbf{I}$, as well as $\mathbf{Z}_1 = \mathbf{V}/\mathbf{I}_1$ and $\mathbf{Z}_2 = \mathbf{V}/\mathbf{I}_2$. Not only the power triangle, but also the impedance triangle can be drawn from the phasor diagram, as noted earlier.
3. The usefulness of the CDR is also demonstrated in part **b** of the solution, enabling calculation of branch currents from individual branch impedances and their sum.

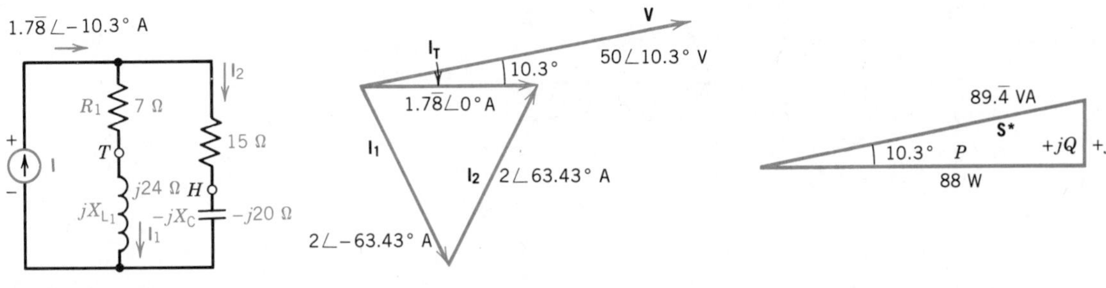

a. Original circuit b. Phasor diagram c. Power triangle

Figure 16-15 Example 16-20

4. In part **e**, the complex impedance **VI***, in rectangular form, yields the true power and quadrature reactive power directly, enabling drawing of the power triangle.
5. Finally, the phasor diagram of Fig. 16-15b contains all the information to write sinusoidal time-domain equations for the voltage and the three currents, given the frequency of the supply.

EXAMPLE 16-21

Given the voltage $V = 50\underline{/0^\circ}$ V in Fig. 16-15a,

a. Find the equivalent Thévenin voltage, $\mathbf{V}_{\mathrm{TH}}$, between terminals **T** and **H** (Hint: use the VDR)
b. Check your answers using the currents found in Ex. 16-20**b**

Solution

a. $$\mathbf{V}_\mathrm{T} = \mathbf{V}_\mathrm{L} = \frac{j24}{7 + j24} \times 50\underline{/0^\circ}\text{ V} = \frac{24\underline{/90^\circ}}{25\underline{/73.74^\circ}} \times 50\underline{/0^\circ}\text{ V}$$
$$= \mathbf{48\underline{/16.26^\circ}} = \mathbf{(46.08 + \mathit{j}13.44)}\text{ volts,}$$
$$\mathbf{V}_\mathrm{H} = \mathbf{V}_\mathrm{C} = \frac{-j20}{15 - j20} \times 50\underline{/0^\circ}\text{ V}$$
$$= \frac{20\underline{/-90^\circ}}{25\underline{/-53.13^\circ}} \times 50\underline{/0^\circ}\text{ V}$$
$$= \mathbf{40\underline{/-36.87^\circ}} = \mathbf{(32 - \mathit{j}24)}\text{ volts, and}$$
$$\mathbf{V}_{\mathrm{TH}} = \mathbf{V}_\mathrm{T} - \mathbf{V}_\mathrm{H} = (46.08 + j13.44) - (32 - j24)$$
$$= \mathbf{14.08 + \mathit{j}37.44 = 40\underline{/69.39^\circ}\text{ V}}$$

b. $$\mathbf{V}_\mathrm{T} = \mathbf{I}_1(jX_\mathrm{L}) = 2\underline{/-73.73^\circ}\text{ A} \times 24\underline{/90^\circ}\ \Omega$$
$$= 48\underline{/16.27^\circ}\text{ V} = \mathbf{(46.08 + \mathit{j}13.44)},$$
$$\mathbf{V}_\mathrm{H} = \mathbf{I}_2(-jX_\mathrm{C}) = 2\underline{/53.14^\circ}\text{ A} \times 20\underline{/-90^\circ}\ \Omega$$
$$= 40\underline{/-36.86^\circ}\text{ V} = \mathbf{(32 - \mathit{j}24)},$$
and, as in **a**, $\mathbf{V}_{\mathrm{TH}} = \mathbf{V}_\mathrm{T} - \mathbf{V}_\mathrm{H} = \mathbf{14.08 + \mathit{j}37.44 = 40\underline{/69.39^\circ}\text{ V}}$

Example 16-21 illustrates the usefulness of the VDR in finding voltages without resorting to currents.

EXAMPLE 16-22

Four complex impedances are connected in a series–parallel arrangement. $\mathbf{Z}_1$ and $\mathbf{Z}_2$ are in parallel and connected in series with the series combination of $\mathbf{Z}_3$ and $\mathbf{Z}_4$ across an unknown supply voltage. Current in $\mathbf{Z}_3$ is $1\underline{/-4.25^\circ}$ A. The impedance values are: $\mathbf{Z}_1 = 12 + j9$, $\mathbf{Z}_2 = 12 - j5$, $\mathbf{Z}_3 = 16 + j12$, and $\mathbf{Z}_4 -$ $8 - j15$ ohms.

a. Draw a circuit diagram showing all values given above. Calculate
b. Current $\mathbf{I}_1$ in impedance $\mathbf{Z}_1$
c. Current $\mathbf{I}_2$ in impedance $\mathbf{Z}_2$
d. Voltage $\mathbf{V}_1$ across $\mathbf{Z}_1$
e. Voltage $\mathbf{V}_2$ across $\mathbf{Z}_2$
f. Voltage $\mathbf{V}_3$ across $\mathbf{Z}_3$
g. Voltage $\mathbf{V}_4$ across $\mathbf{Z}_4$
h. Power dissipated in each of the impedances, P_1 through P_4
i. Total circuit voltage, $\mathbf{V}_\mathrm{T}$
j. Total power using $P_\mathrm{T} = \mathbf{V}_\mathrm{T}\mathbf{I}_\mathrm{T}\cos\theta$ and the sum of P_1 through P_4, inclusive
k. Total power using complex power $\mathbf{S}^* = \mathbf{VI}^*$
l. Draw the power triangle for the entire circuit

Solution

a. See **Fig. 16-16a**, from which the following preliminary calculations are:
$$\mathbf{Z}_1 + \mathbf{Z}_2 = (12 + j9) + (12 - j5) = 24 + j4$$
$$= 24.33\underline{/9.462^\circ}\ \Omega,$$
$$\mathbf{Z}_1 = 12 + j9 = 15\underline{/36.87^\circ}\ \Omega,$$
$$\mathbf{Z}_2 = 12 - j5 = 13\underline{/-22.62^\circ}\ \Omega,\text{ and}$$
$$\mathbf{Z}_\mathrm{p} = \frac{\mathbf{Z}_1\mathbf{Z}_2}{\mathbf{Z}_1 + \mathbf{Z}_2} = \frac{15\underline{/36.87^\circ} \times 13\underline{/-22.62^\circ}}{24.33\underline{/9.462^\circ}} = \mathbf{8.015\underline{/4.788^\circ}}$$
$$= \mathbf{Z}_\mathrm{s} = \mathbf{7.987 + \mathit{j}0.669}$$

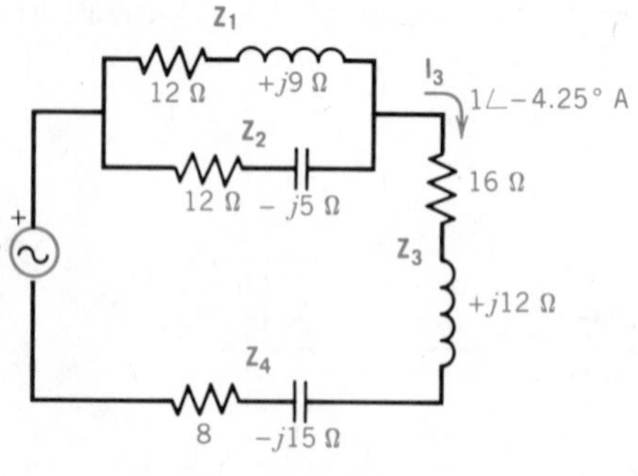

a. Solution to Ex 16-16a

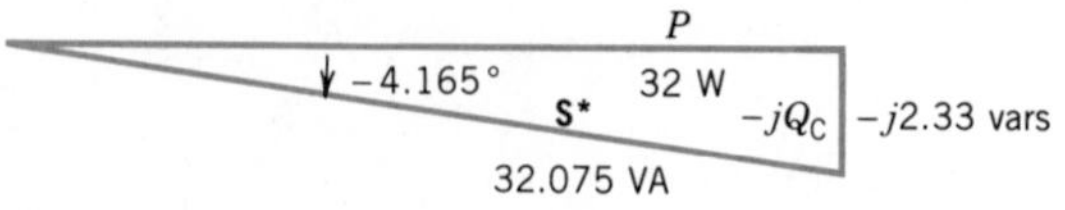

b. Power triangle

Figure 16-16 Example 16-22

b. Using CDR,

$$\mathbf{I}_1 = \frac{\mathbf{Z}_2}{\mathbf{Z}_1 + \mathbf{Z}_2}\mathbf{I}_T = \frac{13\angle -22.62^\circ \times 1\angle -4.25^\circ \text{ A}}{24.33\angle 9.462^\circ}$$

$$= \mathbf{0.5343\angle -36.33^\circ \text{ A}}$$

c. Using CDR,

$$\mathbf{I}_2 = \frac{\mathbf{Z}_1}{\mathbf{Z}_1\mathbf{Z}_2}\mathbf{I}_T = \frac{15\angle 36.87^\circ \times 1\angle -4.25^\circ \text{ A}}{24.33\angle 9.462^\circ}$$

$$= \mathbf{0.6165\angle 23.16^\circ \text{ A}}$$

d. $\mathbf{V}_1 = \mathbf{I}_1\mathbf{Z}_1 = (0.5343\angle -36.33^\circ \text{ A})(15\angle 36.87^\circ\ \Omega)$
$= \mathbf{8.015\angle 0.54^\circ \text{ V}}$

e. $\mathbf{V}_2 = \mathbf{I}_2\mathbf{Z}_2 = (0.6165\angle 23.16^\circ \text{ A})(13\angle -22.62^\circ\ \Omega)$
$= \mathbf{8.015\angle 0.54^\circ \text{ V}}$

f. $\mathbf{V}_3 = \mathbf{I}_3\mathbf{Z}_3 = (1\angle -4.25^\circ \text{ A})(16 + j12)$
$= (1\angle -4.25^\circ \text{ A})(20\angle 36.87^\circ\ \Omega) = \mathbf{20\angle 32.62^\circ \text{ V}}$

g. $\mathbf{V}_4 = \mathbf{I}_4\mathbf{Z}_4 = (1\angle -4.25^\circ \text{ A})(8 - j15)$
$= (1\angle -4.25^\circ \text{ A})(17\angle -61.93^\circ\ \Omega) = \mathbf{17\angle -66.18^\circ \text{ V}}$

h. $P_1 = I_1^2R_1 = (0.5343)^2 12 = \mathbf{3.426 \text{ W}}$,
$P_2 = I_2^2R_2 = (0.6165)^2 12 = \mathbf{4.561 \text{ W}}$,
$P_3 = I_3^2R_3 = 1^2 16 = \mathbf{16 \text{ W}}$, and
$P_4 = I_4^2R_4 = 1^2 8 = \mathbf{8 \text{ W}}$

i. $\mathbf{V}_T = \mathbf{V}_{1,2} + \mathbf{V}_3 + \mathbf{V}_4$
$= 8.015\angle 0.54^\circ + 20\angle 32.62^\circ + 17\angle -66.18^\circ \text{ V}$
$= (8.015 + j0.076) + (16.845 + j10.78) + (6.866 - j15.55)$
$= \mathbf{31.73 - j4.694 = 32.075\angle -8.415^\circ \text{ V}}$

j. $P_T = V_TI_T \cos\theta$
$= (32.075\angle -8.415^\circ)(1\angle -4.25^\circ)\cos(-4.165^\circ) = \mathbf{32 \text{ W}}$
$P_T = P_1 + P_2 + P_3 + P_4$
$= (3.426 + 4.561 + 16 + 8) \text{ W} = \mathbf{32 \text{ W}}$

k. $\mathbf{S}^* = \mathbf{VI}^* = (32.075\angle -8.415^\circ)(1\angle -4.25^\circ)^*$
$= \mathbf{32.075\angle -4.165^\circ} = (\mathbf{32 - j2.33}) \text{ VA} = P - jQ_C$

Example 16-22 stresses the following points:

1. For the two-branch parallel portion of the series–parallel circuit, the CDR is used to find the current, since we know both current and the impedances.
2. The total voltage, $\mathbf{V}_T$, in **i** is found by KVL as the phasor sum of the individual series-connected voltages.
3. Total power is found by three methods, each yielding the same answer. This proves the validity of the solution methods.
4. Since $\mathbf{I}_T$ *leads* $\mathbf{V}_T$, the circuit is primarily *capacitive*, resulting in *negative* reactive power, as shown in the power triangle of Fig. 16-16b. (See solution, part **k**.)
5. Angle θ is the difference angle between the voltage and current phasors in **j**. This difference also emerges when the conjugate is taken in **k**. Hence $\theta = -4.165^\circ$ is the *circuit phase angle* or *power factor angle*.

EXAMPLE 16-23

Calculate the total input impedance of the ladder network shown in Fig. 16-17

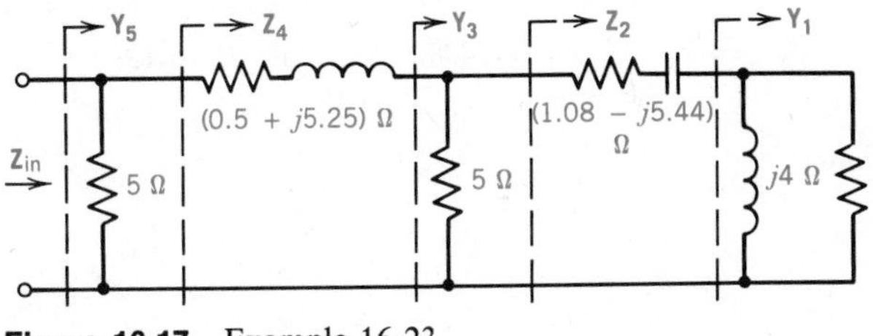

Figure 16-17 Example 16-23

Solution

General approach: Impedances in *parallel* are added as their respective *admittances in rectangular form*. Impedances in *series* are added as *impedances* in *rectangular* form. Using *admittance sums* for *parallel* branches and *impedance sums* for *series* branches, the solution to Fig. 16-17 consists of alternate conversions from admittance to impedance and vice versa. As noted earlier, we begin with the impedances farthest from the supply source, finding in order: $\mathbf{Y}_1$, $\mathbf{Z}_2$, $\mathbf{Y}_3$, $\mathbf{Z}_4$, $\mathbf{Y}_5$, and finally $\mathbf{Z}_{in}$.[14] Let us take each step in turn.

1. $\mathbf{Y}_1 = \dfrac{1}{3\angle 0^\circ} + \dfrac{1}{4\angle 90^\circ} = 0.\bar{3} - j0.25 = 0.41\bar{6}\angle -36.87^\circ \text{ S}$

2. $\mathbf{Z}_1 = 1/\mathbf{Y}_1 = 1/0.41\bar{6}\angle -36.87^\circ \text{ S} = 2.4\angle 36.87^\circ\ \Omega$
$= (1.92 + j1.44)\ \Omega$
$\mathbf{Z}_2 = \mathbf{Z}_1 + (1.08 - j5.44) = (1.92 + j1.44) + (1.08 - j5.44)$
$= 3 - j4 = 5\angle -53.13^\circ\ \Omega$

3. $\mathbf{Y}_3 = \dfrac{1}{5\angle 0^\circ} + \dfrac{1}{\mathbf{Z}_2} = 0.2\angle 0^\circ + 0.2\angle 53.13^\circ$
$= 0.2 + (0.12 + j0.16) = 0.32 + j0.16$
$= 0.35\bar{7}\angle 26.565^\circ \text{ S}$

4. $\mathbf{Z}_3 = 1/\mathbf{Y}_3 = 1/0.357\angle 26.565^\circ \text{ S} = 2.795\angle -26.565^\circ\ \Omega$
$= 2.5 - j1.25\ \Omega$
$\mathbf{Z}_4 = \mathbf{Z}_3 + (0.5 + j5.25) = (2.5 - j1.25) + (0.5 + j5.25)$
$= 3 + j4 = 5\angle 53.13^\circ\ \Omega$

5. $\mathbf{Y}_5 = \dfrac{1}{5\angle 0^\circ} + \dfrac{1}{\mathbf{Z}_4} = 0.2\angle 0^\circ + 0.2\angle -53.13^\circ$
$= 0.2 + (0.12 - j0.16) = 0.32 - j0.16$
$= 0.35\bar{7}\angle -26.565^\circ \text{ S}$

6. $\mathbf{Z}_{in} = 1/\mathbf{Y}_5 = 1/0.357\angle -26.565^\circ \text{ S} = \mathbf{2.795\angle 26.565^\circ\ \Omega}$
$= \mathbf{(2.5 + j1.25)\ \Omega}$

[14] We will discover later that a problem of this kind may also be solved by mesh analysis, involving four unknown currents, requiring a computer program to solve four simultaneous equations in complex form.

Example 16-23 shows the advantages of using admittance sums for parallel branches and impedance sums for series branches. We will use this principle in some of the remaining examples.

EXAMPLE 16-24

Calculate the current drawn from the supply, $\mathbf{V}_T$, given the circuit shown in Fig. 16-18a. Circuit parameters are $\mathbf{Z}_1 = (7 + j24)\ \Omega$, $\mathbf{Z}_2 = (15 - j20)\ \Omega$, $\mathbf{Z}_3 = (12 + j35)\ \Omega$, $\mathbf{Z}_4 = (5 - j12)\ \Omega$, $\mathbf{Z}_5 = (9 + j40)\ \Omega$, $\mathbf{Z}_6 = (4 - j3)\ \Omega$, and $\mathbf{V}_T = 100\underline{/0^\circ}$ V

Solution

1. Converting $\mathbf{Z}_1$, $\mathbf{Z}_2$, and $\mathbf{Z}_3$ to polar form: $\mathbf{Z}_1 = 7 + j24 = 25\underline{/73.74^\circ}\ \Omega$, $\mathbf{Z}_2 = 15 - j20 = 25\underline{/-53.13^\circ}\ \Omega$, $\mathbf{Z}_3 = 12 + j35 = 37\underline{/71.075^\circ}\ \Omega$, and taking the admittance sum yields

$$\mathbf{Y}_{1,2,3} = \frac{1}{25\underline{/73.74^\circ}} + \frac{1}{25\underline{/-53.13^\circ}} + \frac{1}{37\underline{/71.075^\circ}}\ \Omega$$
$$= 40\underline{/-73.74^\circ} + 40\underline{/53.13^\circ} + 27.03\underline{/-71.075^\circ}\ \text{mS}$$
$$= (11.20 - j38.4) + (24 + j32) + (8.76 - j25.57)\ \text{mS}$$
$$= 43.97 - j31.97 = 54.36\underline{/-36.0^\circ}\ \text{mS}$$

and $\mathbf{Z}_{1,2,3} = 1/\mathbf{Y}_{1,2,3} = 1/54.36\underline{/-36.0^\circ}\ \text{mS} = 18.39\underline{/36^\circ}\ \Omega$
$$= (14.88 + j10.81)\ \Omega$$

2. With $\mathbf{Z}_{1-3}$ in rectangular form, we add the series impedances (Fig. 16-18b) to determine the total impedance of the left-hand branch, $\mathbf{Z}_{1-4}$:

$$\mathbf{Z}_{1-4} = \mathbf{Z}_{1-3} + \mathbf{Z}_4 = (14.88 + j10.81) + (5 - j12)$$
$$= 19.88 - j1.19 = 19.92\underline{/-3.43^\circ}\ \Omega$$

and $\mathbf{Z}_5 = 9 + j40 = 41\underline{/77.32^\circ}\ \Omega$

3. Combining $\mathbf{Z}_{1-4}$ with $\mathbf{Z}_5$ using the product-over-the-sum technique (Fig. 16-18c):

$$\mathbf{Z}_{1-5} = \frac{\mathbf{Z}_{1-4}\mathbf{Z}_5}{(\mathbf{Z}_{1-4}) + \mathbf{Z}_5} = \frac{19.92\underline{/-3.43^\circ} \times 41\underline{/77.32^\circ}}{(19.88 - j1.19) + (9 + j40)}$$
$$= \frac{816.72\underline{/73.89^\circ}}{48.38\underline{/53.35^\circ}}$$
$$= 16.88\underline{/20.54^\circ}\ \Omega$$
$$= (15.81 + j5.922)\ \Omega$$

4. As a last step, we may combine $\mathbf{Z}_{1-5}$ with $\mathbf{Z}_6$ (Fig. 16-18d) to yield the total input impedance, $\mathbf{Z}_T$: $\mathbf{Z}_T = \mathbf{Z}_6 + \mathbf{Z}_{1-5} = (4 - j3) + (15.81 + j5.922) = 19.81 + j2.922 = 20\underline{/8.4^\circ}\ \Omega$
5. $\mathbf{I}_T = \mathbf{V}_T/\mathbf{Z}_T = 100\underline{/0^\circ}\ \text{V}/20\underline{/8.4^\circ}\ \Omega = \mathbf{5\underline{/-8.4^\circ}\ A}$

With respect to the above solution, please note that

1. The solution begins with the impedances *farthest* from the supply source.
2. Parallel circuit rules are used for the admittances in parallel and series circuit rules are used for the impedances in series. This is the same as the procedure used for series–parallel dc circuits with the exception that all the parameters are in complex form.
3. The various simplification stages shown in Fig. 16-18 should be studied as a method of attacking and solving series–parallel problems by network reduction techniques.
4. Readers should attempt the solution independently on a separate sheet and use the solutions shown as feedback to correct their errors.

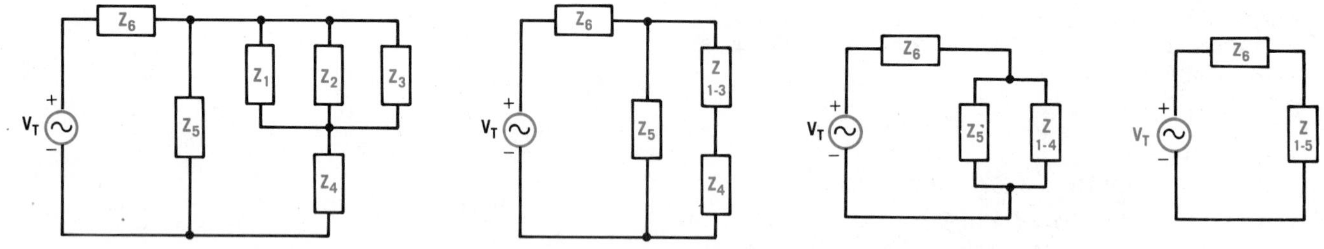

a. Original circuit b. First simplification c. Second simplification d. Third simplification

Figure 16-18 Example 16-24

At this point, one may question whether the foregoing solution, or any solution using network-reduction techniques, yields the correct answers. As indicated earlier, *power calculations* serve as an *excellent cross check* on the accuracy of the solution methods used, as shown in Ex. 16-25.

EXAMPLE 16-25

Using the various given and solution data of Ex. 16-24, calculate the

a. Currents, respectively, in $\mathbf{Z}_1$ through $\mathbf{Z}_6$
b. Powers dissipated individually in $\mathbf{Z}_1$ through $\mathbf{Z}_6$
c. Total power as the sum of the individual powers in **b**, above
d. Total power, using $\mathbf{V}_T\mathbf{I}_T \cos\theta$

Solution

The detailed solution of currents and/or voltages as required is left as an exercise for the reader. Only the answers to each part are shown below to provide feedback as to the correct answer.

a. $\mathbf{I}_1 = 3.117\angle -22.2^\circ$ A, $\mathbf{I}_2 = 3.117\angle 105^\circ$ A, $\mathbf{I}_3 = 2.106\angle -19.51^\circ$ A, $\mathbf{I}_4 = 4.237\angle 15.57^\circ$ A, $\mathbf{I}_5 = 2.06\angle -65.2^\circ$ A, $\mathbf{I}_6 = 5\angle -8.4^\circ$ A
b. $P_1 = \mathbf{68\ W}$, $P_2 = \mathbf{145.73\ W}$, $P_3 = \mathbf{53.22\ W}$, $P_4 = \mathbf{89.76\ W}$, $P_5 = \mathbf{38.19\ W}$, $P_6 = \mathbf{100\ W}$
c. $P_T = P_1 + P_2 + P_3 + P_4 + P_5 + P_6 = \mathbf{495\ W}$
d. $P_T = V_T I_T \cos\theta = (100\ \text{V}) \times (5\ \text{A}) \cos(-8.4^\circ) = \mathbf{495\ W}$

Example 16-25 shows how the power calculations may be used as a cross check to verify the accuracy of the circuit reduction techniques by impedance/admittance methods. In short, if the power is *not* checked, there is no guarantee that the method and/or the solutions are valid. Consequently, even if power is *not* called for, the reader should compute it, as an alternative solution method, to serve as a verification of the accuracy of the solution.[15] This returns us to a consideration of Tellegen's theorem (Sec. 7-9) and its application to ac circuits.

16-8 TELLEGEN'S THEOREM REVISITED FOR APPLICATION TO AC CIRCUITS

In Sec. 7-9 we introduced Tellegen's theorem as applied to dc circuits. We observed that Tellegen's theorem is an extension of KVL and KCL, respectively, where the algebraic sum of the respective products (of all branch voltages and currents) is zero. We may now modify Eq. (7-6) to apply to ac circuits by writing

$$\sum_{k=1}^{y} \mathbf{V}_k\mathbf{I}_k^* = 0 \quad \text{or} \quad \mathbf{V}_1\mathbf{I}_1^* + \mathbf{V}_2\mathbf{I}_2^* + \mathbf{V}_3\mathbf{I}_3^* + \cdots + \mathbf{V}_n\mathbf{I}_n^* = 0$$

$$\mathbf{S}_1^* + \mathbf{S}_2^* + \mathbf{S}_3^* + \cdots + \mathbf{S}_n^* = 0 \qquad (16\text{-}26)$$

The same rules (summarized in Sec. 7-9) apply to complex power in any ac system. Whether the complex power is positive or negative is (again) determined by the voltage polarity encountered as the current ENTERS the box. Complex power is

1. NEGATIVE, when the voltage polarity first encountered by current is NEGATIVE.
2. POSITIVE, when the voltage polarity first encountered by current is POSITIVE.

And also

3. Negative complex power is power *supplied* by a box *to the system.*
4. Positive complex power is power *dissipated or stored* (capacitively or inductively) by a box *within the system.*

[15] Alternatively, Ex. 16-24 may be checked by nodal analysis and/or mesh analysis to verify the voltages and/or currents, respectively. (See Chapter 18.)

5. At all times, the *sums* of *all* complex powers (real and reactive) *must be zero* (Tellegen's theorem).

As in the case of Ex. 16-25, Tellegen's theorem provides an additional technique for verifying calculation methods leading to a complete solution of all voltages and currents in a series–parallel circuit. The elegance of Tellegen's theorem is that it verifies the accuracy of both real and quadrature components of all calculated voltages and currents. If the phasor sum of all complex powers is *not* zero, clearly there is an error and it is the reader's responsibility to find it, by either checking calculations or using an alternative solution technique. Example 16-26 shows how Tellegen's theorem may be used, given the complete solution of all voltages and currents in a series–parallel circuit.

An important distinction must be drawn between $\mathbf{S}_1$, the product of $\mathbf{V}_1$ and $\mathbf{I}_1$, and $\mathbf{S}_1^*$, the product of $\mathbf{V}_1$ and $\mathbf{I}_1^*$. $\mathbf{S}_1$ represents the simple product of the voltage across times the current in a particular circuit impedance. $\mathbf{S}_1^*$, on the other hand, represents the *complex power* in that impedance, defined by $\mathbf{V}_1\mathbf{I}_1^*$. The reader must always be aware of this distinction in reading the literature. In recent years, complex power $\mathbf{S}^*$ appears to be gaining favor and wider use over simple apparent power, $\mathbf{S}$. The reason for use of $\mathbf{S}^*$, as indicated earlier, is that it *relates* the impedance triangle, the phasor diagram (using current as a reference), and the power triangle *so conveniently* in all circuits.

EXAMPLE 16-26

For the four complex power boxes shown in **Fig. 16-19** the voltages are $\mathbf{V}_1 = 500 + j0$ V, $\mathbf{V}_2 = 300 - j200$ V, $\mathbf{V}_3 = 200 + j200$ V, and $\mathbf{V}_4 = 200 + j200$ V. The currents are $\mathbf{I}_4 = 50 - j50$ A, $\mathbf{I}_3 = 50 + j100$ A, and $\mathbf{I}_1 = 100 + j50$ A. Calculate the

a. Complex powers $\mathbf{S}_1^*$ through $\mathbf{S}_4^*$ in rectangular form
b. Total complex power as the phasor sum of $\mathbf{S}_1^*$ through $\mathbf{S}_4^*$
c. Identify the nature of the loads in $\mathbf{S}_2^*$ through $\mathbf{S}_4^*$ as well as in $\mathbf{S}_1^*$
d. Verify Tellegen's theorem by adding real and reactive powers separately

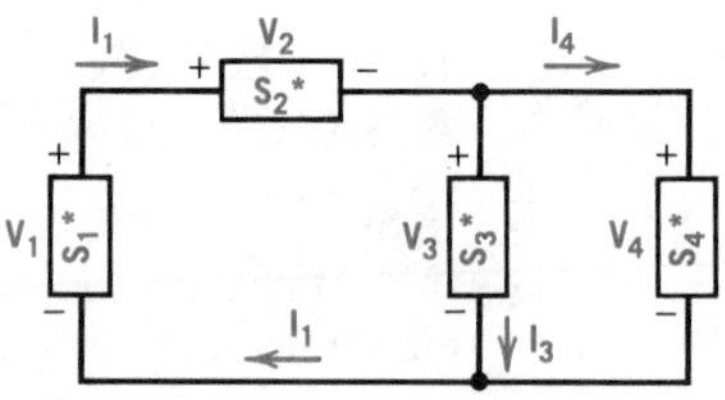

Figure 16-19 Example 16-26

Solution

Complex powers $\mathbf{S}_2$ through $\mathbf{S}_4$ are positive. Complex power $\mathbf{S}_1$ is *negative*.

a. $\mathbf{S}_1^* = -\mathbf{V}_1\mathbf{I}_1^*$
$= -[(500 + j0)(100 - j50)] = \mathbf{(-50\,000 + j25\,000)\ VA}$
$\mathbf{S}_2^* = \mathbf{V}_2\mathbf{I}_1^*$
$= +(300 - j200)(100 - j50) = \mathbf{(20\,000 - j35\,000)\ VA}$
$\mathbf{S}_3^* = \mathbf{V}_3\mathbf{I}_3^*$
$= +(200 + j200)(50 - j100) = \mathbf{(30\,000 - j10\,000)\ VA}$
$\mathbf{S}_4^* = \mathbf{V}_4\mathbf{I}_4^*$
$= +(200 + j200)(50 + j50) = \mathbf{(0 + j20\,000)\ VA}$

b. $\sum\mathbf{S} = \mathbf{S}_1^* + \mathbf{S}_2^* + \mathbf{S}_3^* + \mathbf{S}_4^* = \mathbf{0 + j0}$

c. $\mathbf{S}_2^* = (20\,000 - j35\,000)$ VA means **20 kW** is dissipated and **35 kvars** of reactive capacitive power is stored and returned to the supply, $\mathbf{S}_1^*$, from the load
$\mathbf{S}_3^* = (30\,000 - j10\,000)$ VA means **30 kW** is dissipated and **10 kvars** of reactive capacitive power is stored and returned to the supply $\mathbf{S}_1^*$, from the load
$\mathbf{S}_4^* = (0 + j20\,000)$ VA means load $\mathbf{S}_4$ is a purely inductive load which dissipates no power whatever; **20 kvars** are stored as inductive reactive power and returned to the supply $\mathbf{S}_1$
$\mathbf{S}_1^* = (-50\,000 + j25\,000)$ VA must be interpreted as a source in the light of the negative sign preceding complex conjugate power $\mathbf{S}_1^*$: **−50 kW** is generated power which must be supplied by $\mathbf{S}_1$ to the remaining loads; **+j25 kvars** is returned *capacitive* reactive power stored by the loads in their reactive component elements

d. $\sum\mathbf{P} = \mathbf{P}_1 + \mathbf{P}_2 + \mathbf{P}_3 + \mathbf{P}_4 = (-50 + 20 + 30)\ \text{kW} = \mathbf{0\ kW}$ and $\sum\mathbf{Q} = \mathbf{Q}_1 + \mathbf{Q}_2 + \mathbf{Q}_3 + \mathbf{Q}_4 = (25 - 35 - 10 + 20)$ kvars = **0 kvars**

The following insights may be drawn from Ex. 16-26:

1. Only *sources* (voltage or current) of energy experience a *voltage rise* in a network. Consequently, sources must be preceded by a minus sign before the complex conjugate power product, as shown for $\mathbf{S}_1^*$.
2. The minus sign also reverses the nature of the complex real and quadrature powers. For a source, negative real power is power that must be generated or supplied to the system. Similarly, $+jQ$ implies capacitive reactive power returned by loads to the source and $-jQ$ implies inductive reactive power returned by loads to the source.
3. A check on the accuracy of all network voltage and current calculations is that the complex power phasor sum *must be zero.* This is one of the major contributions of Tellegen's theorem.

16-9 GLOSSARY OF TERMS USED

Admittance, Y Complex quantity expressed as the ratio of steady-state sinusoidal current to voltage across a given component, branch or circuit, measured in siemens (S). Admittance is the reciprocal of impedance. The real part of admittance is conductance, G, and the imaginary part is susceptance, $\pm jB$.

Admittance diagram Triangle, similar to an impedance triangle, showing the relation between conductance, susceptance, and admittance for a given circuit, where all quantities are measured in siemens.

Branch Conductive path where current division occurs in an ac circuit, consisting of one or more two-terminal elements, in series, comprising a (branch) circuit between two nodes or junctions.

Complex number Number that is expressed in the form $a + jb$, where a is a real number and jb an imaginary number.

Complex power Phasor product of the voltage across and the *conjugate* of the current in a given or equivalent impedance in an ac circuit. Complex power is used so that the impedance triangle and the power triangle are similar triangles related by the ratio $|I^2|$.

Conductance, *G* Real part of admittance (**Y**) and reciprocal of resistance, expressed in units of siemens (S), defined as the ability of pure resistance to pass (or conduct) current.

Current divider rule (CDR) As applied to ac circuits, the current in any single branch (in parallel with a second branch) is the ratio of the impedance of the second branch to the sum of the impedances of both branches, as a fraction of the total current entering the junction of the two branches.

Impedance diagram Diagram showing the relation between resistance, reactance, and impedance for any given branch or complete circuit; also called an impedance triangle.

Impedance, equivalent circuit Total opposition to ac created by the phasor sum of circuit combinations of resistance, inductive or capacitive reactance, in a given branch or series, parallel, or series–parallel circuit. Also called input impedance or driving-point impedance.

Junction Point at which branching occurs in a circuit (same as node).

Kirchhoff's voltage law (KVL) As applied to ac circuits, the total applied voltage is proportionally divided among the various impedances connected in series. The phasor sum of the voltages across these impedances in series is the resultant total voltage.

Node Junction of two or more branches.

Phasor Complex number having both magnitude and direction in the complex plane.

Phasor addition Process of adding two or more phasor quantities to produce a final resultant phasor that represents their sum.

Phasor diagram Diagram showing the relation between two or more phasor quantities at a given instant in time, expressed in the frequency domain.

Phasor subtraction Process of reversing one or more phasor quantities in performing phasor addition. Phasor subtraction is a special form of phasor addition.

Power diagram Triangle or construction showing the relation between power, quadrature reactive power, and apparent power for a given branch or circuit.

Power divider rule (PDR) As applied to both dc and ac circuits, the power dissipated by a given resistance in a series branch or circuit is the ratio of that resistance to the total resistance, as a fraction of the total dissipated power.

Susceptance, *B* Imaginary part of admittance (**Y**) and reciprocal of either inductive or capacitive reactance, expressed in units of siemens (S), defined as the ability of pure inductance (or capacitance) to pass ac.

Tellegen's theorem For a given circuit satisfying both Kirchhoff's voltage law and Kirchhoff's current law, the algebraic sum of all branch voltage and branch current **products**, respectively, is zero.

16-10 PROBLEMS

Secs. 16-1 to 16-4

16-1 Given the following voltages across and currents entering a black box, find the simplest two-element series model, using the frequencies given. (Assume a two-terminal passive network within each box.)

a. $\mathbf{V} = 120\underline{/0^\circ}$ V; $\mathbf{I} = 50\underline{/36.9^\circ}$ mA; $f = 60$ Hz
b. $\mathbf{V} = 100\underline{/180^\circ}$ V; $\mathbf{I} = 20\underline{/-180^\circ}$ A; $f = 400$ Hz
c. $\mathbf{V} = 24\underline{/60^\circ}$ V; $\mathbf{I} = 6\underline{/35^\circ}$ mA; $f = 1$ kHz

16-2 The RMS value of collector voltage, $\mathbf{V}_C$, of the common-emitter transistor shown in **Fig. 16-20** is 200 mV. If the voltage across the coupling capacitor is 100 mV, calculate the

a. RMS voltage across the output of the transistor stage, $\mathbf{V}_0$
b. Reactance of the coupling capacitor if the current, $\mathbf{I}_C$, is 5 mA
c. Capacitance of the coupling capacitor if the output waveform has a frequency of 1 kHz

Circuit phase angle, θ, using current as a reference
Resistance of load, R_L

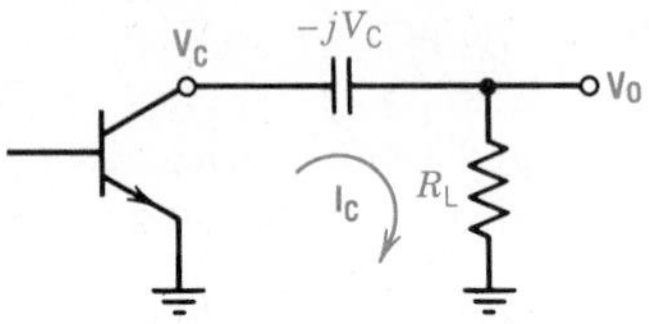

Figure 16-20 Problem 16-2

16-3 The coil of a hi-fi speaker carries an RMS current of $200\angle -40^\circ$ mA when the output stage of an amplifier develops a voltage of $2\angle 0^\circ$ V at a frequency of 5 kHz. Calculate the speaker's

a. Impedance
b. Resistance
c. Reactance
d. Dissipated power
e. Apparent (complex) power
f. Quadrature reactive power
g. Draw the speaker's impedance triangle
h. Draw the speaker's phasor diagram (using current as the reference)
i. Draw the speaker's power triangle showing true power, quadrature reactive power, and apparent power in appropriate units

16-4 A commercial coil is represented by a 50-Ω resistor in series with a 100-mH inductor. Calculate the total input impedance in polar form of the coil at frequencies of

a. 60 Hz
b. 400 Hz
c. 5 kHz
d. 1 MHz
e. 0 Hz (dc)

16-5 The current drawn by a load from a 60-Hz supply is $5\angle 60^\circ$ mA and the voltage across it is $30\angle 20^\circ$ V. Calculate the simplest two-element series model for the load.

16-6 For the model calculated in Prob. 16-5, calculate the frequency in hertz at which current and voltage are out of phase by

a. 20°
b. 60°
c. 80°

16-7 Three impedances, $\mathbf{Z}_1 = 8 + j0\ \Omega$, $\mathbf{Z}_2 = 0 - j25\ \Omega$, and $\mathbf{Z}_3 = 0 + j10\ \Omega$, are connected in series across a 68-V ac supply. Calculate the

a. Total equivalent impedance of the series combination in polar form
b. Circuit current using voltage as a reference
c. Voltage across each impedance using current as a reference zero
d. Circuit phase angle using current as a reference
e. Total circuit power by two methods
f. Phasor diagram showing all voltages and current
g. Impedance triangle
h. Power triangle for the circuit (Hint: use I^2 ratio as multiplying factor)

16-8 For the voltages computed in Prob. 16-7c, using the VDR, calculate

a. $\mathbf{V}_1$
b. $\mathbf{V}_2$
c. $\mathbf{V}_3$

16-9 Three impedances, $\mathbf{Z}_1 = 4 + j3\ \Omega$, $\mathbf{Z}_2 = 12 - j5\ \Omega$, and $\mathbf{Z}_3 = 8 + j9\ \Omega$, are connected in series across a 125-V ac supply. Calculate all parts of Prob. 16-7.

16-10 Find the equivalent impedance of a 0.1-μF capacitor connected in series with a practical coil whose impedance is $3\angle 50^\circ$ kΩ at a frequency of 1 kHz in

a. Complex (rectangular) form
b. Polar form

16-11 Three impedances are connected in series across a 120 V, 60 Hz supply. They are $\mathbf{Z}_1 = 25\angle 30^\circ\ \Omega$, $\mathbf{Z}_2 = 20\angle 60^\circ\ \Omega$, and $\mathbf{Z}_3 = 50\angle -50^\circ\ \Omega$. Calculate all parts of Prob. 16-7.

16-12 What value of capacitance (in μF) must be connected in series with a coil having an impedance of $150\angle 50^\circ\ \Omega$ to provide an impedance of $150\angle -50^\circ\ \Omega$ at a frequency of 1 kHz?

16-13 A commercial relay requires a current of 10 mA (ac or dc) to close its contacts. A dc voltage of 12 V or an ac voltage of 150 V RMS causes the relay to operate. It is desired, however, to operate the relay on 120 V ac, 60 Hz. Calculate the series-connected pure capacitance value to enable such operation.

16-14 A photographic manufacturer requires an adapter to permit a 120 V, 600 W projection lamp (purely resistive) to be used on 220 V, 60 Hz ac. Calculate the

a. Required series resistance and power dissipated by the resistor, using a resistor in series with the projection lamp
b. Series reactance and power dissipated, using an inductor whose resistance is 12 Ω
c. Series capacitance and capacitor voltage rating, using a pure capacitor and power dissipated by the capacitor
d. Which of the three techniques has the most advantages for the application and explain why

16-15 A pure capacitor and a practical coil (R–L) are connected in series across a 12 V, 1 kHz supply and draw a current of 200 mA. Voltage measurements made across each component are 100 V across the capacitor and 90 V across the practical coil. Calculate the

a. Resistance of the coil
b. Inductive reactance and inductance of the coil
c. Capacitive reactance and capacitance of the capacitor
d. Circuit phase angle
e. Power dissipated by the entire circuit

(Hint: see Ex. 16-10 and use the cosine law on drawn impedance diagram.)

16-16 A practical inductor is connected in series with a non-inductive resistor and an ac ammeter across a 120 V, 400 Hz supply. Voltmeter measurements taken across the series resistor and inductor, respectively, are 60 and 80 V. If the ac ammeter reads 0.5 A, calculate the

a. Total circuit impedance, $\mathbf{Z}_T$

b. Resistance of the series resistor, R_s
c. Impedance of the coil, $\mathbf{Z}_L$
d. Circuit phase angle, θ_T (use cosine law)
e. Phase angle of the coil, θ_L (use sine law)
f. Reactance of the coil, X_L
g. Inductance of the coil, L
h. Resistance of the coil, R_L
i. Total power drawn from the supply, P_T
j. Power drawn by the coil only, P_L
k. Phasor diagram for the entire circuit using current as the reference phasor
l. Impedance diagram for the entire circuit showing all angles and values calculated above
m. Power triangle and total complex power for the entire circuit

(Hint: use I^2 ratio times impedance triangle.)

16-17 In Fig. 16-20, the ac collector voltage $\mathbf{V}_C = 10\underline{/0^\circ}$ V, $R_L = 5$ kΩ, and $C = 0.05$ μF. At a frequency of 5 kHz, calculate the

a. RMS collector current, $\mathbf{I}_C$, in polar form
b. Output voltage, $\mathbf{V}_0$, in polar form
c. Power dissipated in R_L
d. Circuit phase angle, θ_T

16-18 What values of resistance (in ohms) and capacitance (in microfarads) must be connected in series with a practical coil having an impedance of $300\underline{/40^\circ}$ Ω to produce a total impedance of $300\underline{/-30^\circ}$ Ω at a frequency of 2 kHz?

16-19 A practical inductor draws a current of 0.8 A from a 208 V, 400 Hz supply. If a precision (noninductive) 100-Ω resistor in series with the inductor causes the current to drop to 0.7 A, calculate the

a. Original circuit impedance
b. New circuit impedance with resistor added
c. Effective resistance of final circuit
d. Inductive reactance of the final circuit
e. Inductance of the coil
f. Resistance of the coil
g. Draw the complete impedance diagram for the final circuit showing all angles and impedance values

Sec. 16-5

16-20 In the three-branch parallel circuit shown in **Fig. 16-21**, the impedances of each branch are $\mathbf{Z}_1 = 30\underline{/0^\circ}$ Ω, $\mathbf{Z}_2 = 40\underline{/90^\circ}$ Ω, and $\mathbf{Z}_3 = 50\underline{/-90^\circ}$ Ω. Calculate in both rectangular and polar forms the equivalent input

a. Conductance
b. Susceptance
c. Admittance
d. Impedance
e. Draw the admittance triangle

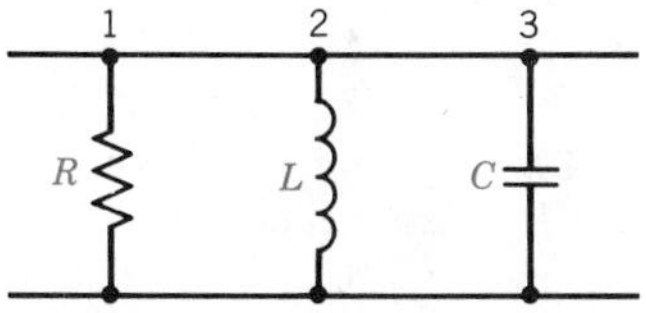

Figure 16-21 Problems 16-20, 16-21

16-21 Repeat all parts of Prob. 16-20, given a frequency of 400 Hz, $R = 400$ Ω, $L = 40$ mH, and $C = 1$ μF in Fig. 16-21.

16-22 A practical inductor consists of a 20-μH inductance in parallel with a resistance of 40 kΩ. When it is connected across a 10 V, 350 MHz supply, calculate the

a. Current in each branch
b. Total current drawn by the device from the supply
c. Total admittance, using (**b**)
d. Total admittance from the sum of the admittances of each branch
e. Draw the admittance triangle for the practical inductor

16-23 A practical capacitor consists of a 25-Ω resistance in parallel with a capacitive reactance of $-j20$ Ω. Calculate the

a. Total admittance of the parallel combination in polar form
b. Impedance of the circuit, using (**a**) in polar form
c. Impedance of the circuit, using the product-over-sum relation, in polar form
d. Draw the admittance triangle for the practical capacitor

16-24 The total current drawn by the practical capacitor of Prob. 16-23 is $2\underline{/51.34^\circ}$ A from a constant-voltage ac source. Calculate the

a. Current in the resistive branch, using the CDR
b. Current in the capacitive branch, using the CDR
c. Source voltage (by two methods), using branch currents
d. Admittance in polar form from (**c**)
e. Impedance of the practical capacitor (compare with Prob. 16-23)
f. Using voltage as a reference, draw a phasor diagram showing all currents and voltages
g. Compare the phasor diagram with the admittance triangle drawn in Prob. 16-23d. What simple relation links the two diagrams?

16-25 A practical capacitor (R–C in parallel) draws 30 mA from a $12\underline{/0^\circ}$ V ac supply. If the current in the resistive branch is 20 mA, draw a current phasor diagram and find the

a. Current in the capacitive branch (magnitude and phase angle)
b. Resistance of the parallel resistor, R_p
c. Reactance of the parallel capacitor, $-jX_p$
d. Admittance of the practical capacitor
e. Power dissipated in the practical capacitor by three different methods (V_R^2/R, $V_R I_R$, and **VI** cos θ)
f. Power dissipated, using complex power $\mathbf{S}^* = \mathbf{VI}^*$ in polar and rectangular form
g. Equivalent impedance in polar form from (**d**)
h. Equivalent series resistance and reactance from (**g**)
i. Repeat (**h**), using the Q method relations of Eqs. (16-23) and (16-24)
j. Verify (**g**), using (**b**) and (**c**) with the product-over-sum method

16-26 An unknown admittance draws a current of $10\underline{/-50^\circ}$ A from a $120\underline{/0^\circ}$ V, 60 Hz supply. If a 150-μF ideal capacitor is connected in parallel with the unknown admittance, calculate the

a. Original admittance in polar and complex form
b. Susceptance of the added capacitor

c. New admittance in polar form
d. New total current drawn from the supply
e. Original complex power drawn from the supply in polar and rectangular form
f. Final complex power drawn from the supply in polar and rectangular form
g. Explain why there is no change in true power in (**e**) versus (**f**)
h. In the light of (**g**), what is the effect of the capacitor? (Explain in terms of reactive vars, circuit phase angle, and total current drawn from the supply.)

16-27 A three-branch parallel circuit consists of the following complex impedances in each branch: $\mathbf{Z}_1 = 100\underline{/20^\circ}\ \Omega$, $\mathbf{Z}_2 = 200\underline{/-40^\circ}\ \Omega$, and $\mathbf{Z}_3 = 400\underline{/50^\circ}\ \Omega$. Assume, for convenience, a supply voltage of $400\underline{/0^\circ}$ V and calculate the

a. Current in each branch
b. Total current
c. Equivalent impedance, in polar form

16-28 a. Verify your solution to Prob. 16-27**c**, using Eq. (16-17**b**).
b. Contrast the methods of Probs. 16-27 and 16-28 in terms of number of operations and possibility of error, in finding the input impedance of a three-branch parallel circuit.

Sec. 16-6

16-29 A practical capacitor has an equivalent admittance of $130\underline{/67.38^\circ}$ mS at 1 kHz. Calculate at the given frequency

a. R_p and C_p
b. Equivalent series impedance in rectangular and polar form
c. R_s and C_s
d. Q
e. Verify C_s, using Eq. (16-25a)

16-30 Given the R–C parallel model where $C_p = 0.1\ \mu$F and $R_p = 100\ \Omega$. At frequencies of 1000, 5000, 15 000, 15 915.5, and 50 000 Hz, respectively, prepare a table similar to that shown in Ex. 16-17. Calculate (without using the table) the

a. Frequency at which $-jX_s$ is a maximum and equal to $R_p/2$ and R_s
b. Value of $-jX_s$ at the frequency in (**a**)
c. Value of R_s at the frequency in (**a**)
d. Value of Q and C_s at the frequency in (**a**)
e. Compare the calculated values with those in your table. They should agree. (If they do not, either the table is improperly prepared or your calculations are in error.)

16-31 The emitter circuit of a common-emitter transistor has a resistance of 250 Ω shunted by a 0.25-μF capacitor. At a nominal frequency of 5 kHz, calculate the

a. Equivalent admittance in rectangular and polar form
b. Impedance in polar and rectangular form ($R_s - jX_s$)
c. Frequency at which jX_s is a maximum
d. Value of R_s and jX_s at the frequency found in (**c**)
e. RMS voltage across the R–C combination when the emitter current is 5 mA RMS at the frequency found in (**c**)

16-32 Calculate for a two-element (R–C) model

a. The capacitor to be added in parallel with a 100-Ω resistor to produce a total admittance of 25 mS at a frequency of 500 Hz
b. The frequency at which $R_s = 50\ \Omega = R_p/2$
c. The equivalent impedance at the frequency found in (**b**)
d. The equivalent series model at the frequency found in (**b**)
e. The equivalent series model at the given frequency in (**a**)

Sec. 16-7

16-33 A two-branch parallel circuit contains an impedance $\mathbf{Z}_1 = 5 + j12$ in the first branch and $\mathbf{Z}_2 = 15 - j8\ \Omega$ in the second branch. Calculate the

a. Admittance $\mathbf{Y}_1$
b. Admittance $\mathbf{Y}_2$
c. Total admittance $\mathbf{Y}_T$
d. Total current if the supply voltage is $100\underline{/0^\circ}$ V
e. Equivalent impedance from (**c**)
f. Equivalent impedance from (**d**)
g. Current in each branch, using the CDR
h. Current in each branch, using the supply voltage
i. Power dissipated in each branch
j. Total power dissipated by two methods
k. Complex power of each branch in polar and rectangular form
l. Total complex power sum by two methods

16-34 Calculate the simplest single input impedance in polar form to replace the network shown in **Fig. 16-22**, using parallel-to-series equivalence techniques.

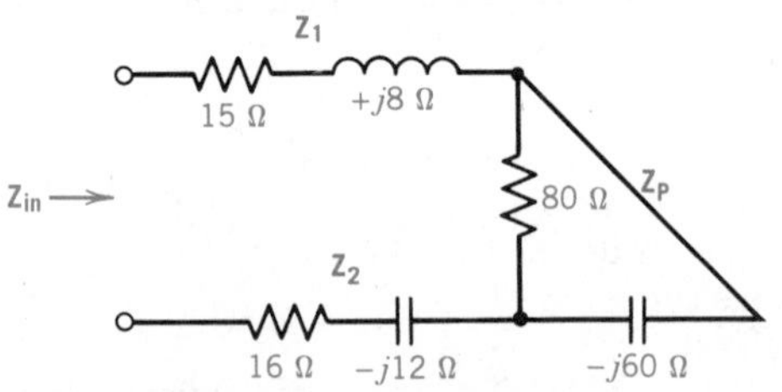

Figure 16-22 Problem 16-34

16-35 Assuming the supply voltage across the network of Fig. 16-22 is $73.31\underline{/-35.34^\circ}$ V, find the

a. Total current
b. Voltage across $\mathbf{Z}_1$, $\mathbf{Z}_2$, and $\mathbf{Z}_p$
c. Current in each branch of $\mathbf{Z}_p$
d. Power dissipated in each of the three impedances
e. Total power as the sum of the individually dissipated powers
f. Total power, using $\mathbf{V}_T\mathbf{I}_T \cos\theta$
g. Complex power of each impedance in polar and rectangular form
h. Total complex power sum by two methods

16-36 Given the network shown in **Fig. 16-23**, calculate the

a. Total current drawn from the $108\underline{/0^\circ}$ V supply, $\mathbf{V}_T$
b. $\mathbf{V}_{ab}$, voltage across two-branch R–L parallel circuit
c. $\mathbf{V}_{cd}$, voltage across two-branch R–C parallel circuit
d. Power dissipated by the 6-Ω resistor
e. Power dissipated by the 8-Ω resistor

f. Total power from $V_T I_T \cos\theta$
g. Total power from total complex power in rectangular form

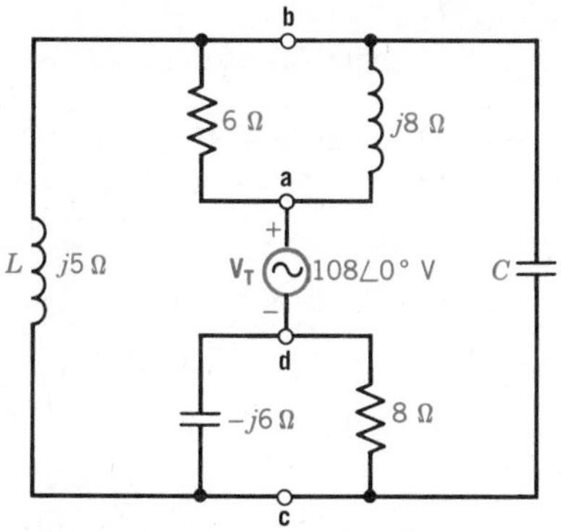

Figure 16-23 Problem 16-36

16-37 A load R_L draws $20\angle 0°$ mA from the capacitive voltage divider shown in **Fig. 16-24**. The divider is connected to a $230\angle -13.1°$ V supply and must produce a voltage of $120\angle 0°$ V across R_L. Calculate the

a. Current in C_2
b. Value of R_L
c. Current drawn from supply, I_{C_1}
d. Voltage across C_1
e. Reactance of C_1, X_{C_1}
f. Values of C_1 and C_2 including dc voltage rating of each

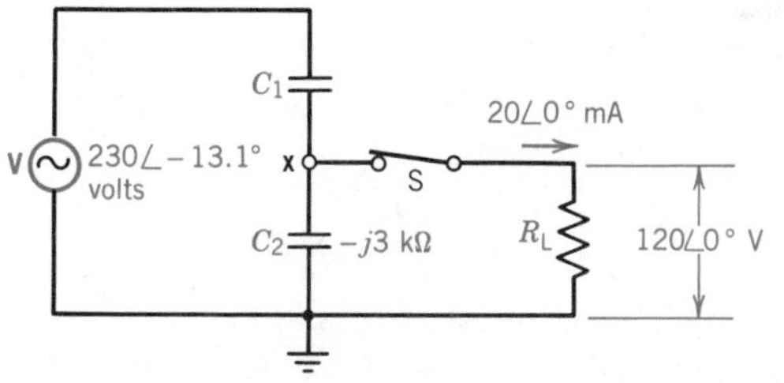

Figure 16-24 Problem 16-37

16-38 The capacitive voltage divider of Fig. 16-24 uses a $9\angle 0°$ V ac supply. X_{C_1} is $-j3$ kΩ and X_{C_2} is $-j6$ kΩ. Switch S is open and the load R_L is $3\angle 0°$ kΩ. Calculate with respect to ground the

a. Voltage at point **X**, $\mathbf{V}_{X_o}$, when S is open (Hint: use VDR)
b. Voltage at point **X**, V_{X_c}, when S is closed
c. From your previous study of resistive voltage dividers, compare the effect of load on voltage $\mathbf{V}_X$, using capacitive dividers
d. Repeat **(c)**, comparing the power loss of capacitive versus resistive voltage dividers

16-39 A full-wave (FW) power supply filter is shown in **Fig. 16-25**. If the FWR has an output frequency of 120 Hz, calculate

a. The input impedance of the filter when $\mathbf{Z}_L = \infty$
b. Repeat **(a)** when $\mathbf{Z}_L$ is $100\angle 0°$ Ω
c. Draw conclusions as to loading effect of the filter on FWR when the filter has no load and when it has a load of 100 Ω

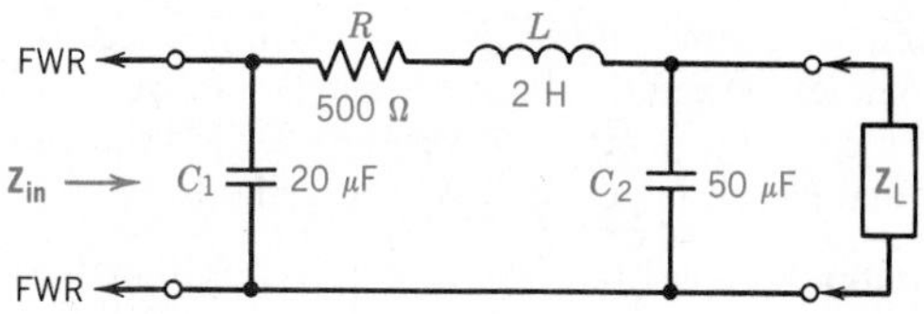

Figure 16-25 Problem 16-39

Sec. 16-8

16-40 Given the series–parallel circuit shown in **Fig. 16-26**, redraw the circuit (if necessary) and calculate the

a. Current in R_2
b. Current in R_1
c. Current in C
d. Current in L
e. Total current drawn from the supply
f. Powers dissipated in R_2 and R_1
g. Total power as the sum of the individual powers
h. Total power, using $\mathbf{V}_T \mathbf{I}_T \cos\theta$
i. Total complex power, $\mathbf{S}_T^* = \mathbf{V}_T \mathbf{I}_T^*$

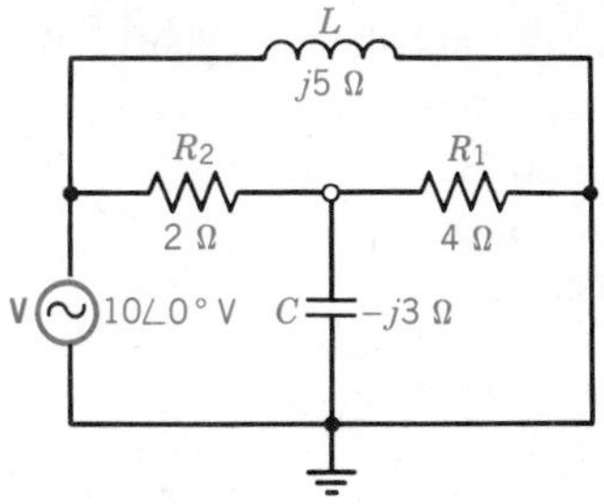

Figure 16-26 Problem 16-40

16-41 In the series–parallel system shown in **Fig. 16-27**, $\mathbf{S}_1 = 600 - j300$ VA, $\mathbf{S}_2 = 500 + j200$ VA, $\mathbf{S}_3 = 600 + j0$ VA, $\mathbf{S}_4 = 0 - j800$ VA, and $\mathbf{S}_5 = 200 + j400$ VA. Calculate

a. $\mathbf{S}_T$ in rectangular and polar form
b. Total reactive quadrature vars, Q_T
c. P_T, total power
d. System power factor, $\cos\theta_T$
e. Total current drawn from supply, $\mathbf{I}_T$
f. Total complex power, $\mathbf{S}_T^* = \mathbf{V}_T \mathbf{I}_T^*$ (not the same as **a**)
g. Draw the overall system power triangle
h. Draw the overall system impedance triangle

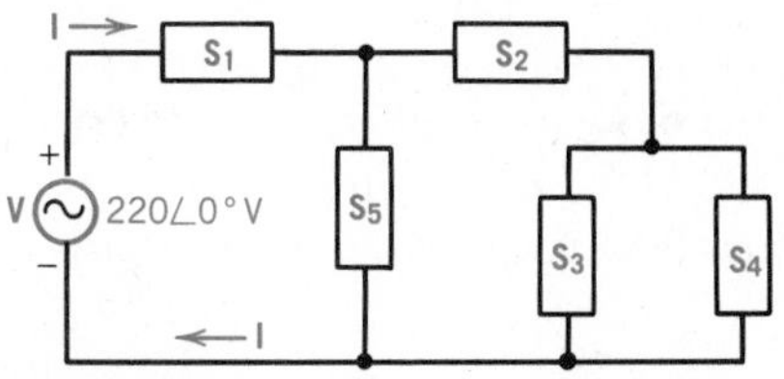

Figure 16-27 Problem 16-41

16-42 For each of the apparent powers, $\mathbf{S}_1$ through $\mathbf{S}_5$, given in Prob. 16-41,

a. Convert each to a complex power, $\mathbf{S}_1^*$ through $\mathbf{S}_5^*$, respectively
b. Add the complex powers to the total complex power (with due regard for voltage polarity) to verify Tellegen's theorem
c. Calculate total circuit impedance by two methods

16-43 For the four complex power boxes shown in Fig. 16-19, the voltages and currents, respectively, in rectangular form are $\mathbf{V}_1 = 250 + j0$ V, $\mathbf{V}_2 = 150 - j100$ V, $\mathbf{V}_3 = \mathbf{V}_4 = 100 + j100$ V, $\mathbf{I}_1 = 10 + j5$ A, $\mathbf{I}_3 = 5 + j10$ A, and $\mathbf{I}_4 = 5 - j5$ A.

a. Calculate complex powers $\mathbf{S}_1^*$ through $\mathbf{S}_4^*$ in rectangular form (with due regard for voltage polarity)
b. Identify the nature of the loads $\mathbf{S}_1^*$ through $\mathbf{S}_4^*$
c. Verify Tellegen's theorem by adding the complex sum of complex powers
d. Draw the system power triangle
e. Draw the system impedance triangle, calculating each term by two methods:
   1. The square of the total current method, $|I^2|$
   2. The ratio of total voltage to total current (input impedance method)

16-11 ANSWERS

16-1 a $R = 1919\ \Omega$, $C = 1.84\ \mu$F b $R = 5\ \Omega$
c $R = 3.625$ kΩ, $L = 0.27$ H
16-2 a 173.2 mV b $20\angle -90^\circ\ \Omega$ c $7.96\ \mu$F d -30°
e $34.64\ \Omega$
16-3 a $10\angle 40^\circ\ \Omega$ b $7.66\ \Omega$ c $j6.428\ \Omega$ d 306.4 mW
e $400\angle 40^\circ$ mVA f $j257.1$ mvars
16-4 a $62.62\angle 37^\circ\ \Omega$ b $256.23\angle 78.74^\circ\ \Omega$ c $5\angle 89.4^\circ$ kΩ
d $628.25\angle 90^\circ$ kΩ e $50\angle 0^\circ\ \Omega$
16-5 a $R = 4.6$ kΩ, $C = 0.688\ \mu$F
16-6 a 138.3 Hz b 29.1 Hz c 8.9 Hz
16-7 a $17\angle -61.93^\circ\ \Omega$ b $4\angle 61.93^\circ$ A
c $32\angle 0^\circ$ V, $100\angle -90^\circ$ V, $40\angle 90^\circ$ V d -61.93°
e 128 W
16-8 a $32\angle 0^\circ$ V b $100\angle -90^\circ$ V c $40\angle 90^\circ$ V
16-9 a $(24 + j7)\ \Omega = 25\angle 16.26^\circ\ \Omega$ b $5\angle -16.26^\circ$ A
c $5\angle 0^\circ$ A: $25\angle 36.87^\circ$ V, $65\angle -22.62^\circ$ V, $60.21\angle 48.37^\circ$ V
d $+16.26^\circ$ e 600 W
16-10 a $(1.9284 + j0.7066)$ kΩ b $2.054\angle 20.12^\circ$ kΩ
16-11 a $64.35\angle -7.57^\circ\ \Omega$ b $1.865\angle 7.57^\circ$ A
c $1.865\angle 0^\circ$ A: $46.62\angle 30^\circ$ V, $37.3\angle 60^\circ$ V, $93.24\angle -50^\circ$ V
d -7.57° e 221.85 W
16-12 $0.6925\ \mu$F
16-13 $0.88\ \mu$F
16-14 a $20\ \Omega$, 500 W b $j25.3\ \Omega$, 300 W
c $70\ \mu$F, 275 V dc, 0 W d capacitor
16-15 a $31.445\ \Omega$ b $448.9\ \Omega$; 71.44 mH c $0.318\ \mu$F
d 58.4° e 1.26 W
16-16 a $240\ \Omega$ b $120\ \Omega$ c $160\ \Omega$ d 36.34°
e 62.72° f $j142.2\ \Omega$ g 56.58 mH h $73.33\ \Omega$
i 48.33 W j 18.33 W m $60\angle 36.34^\circ$ VA
16-17 a $1.984\angle 7.26^\circ$ mA b $9.92\angle 7.26^\circ$ V c 19.68 mW
d 7.26°
16-18 $30\ \Omega$, $0.232\ \mu$F
16-19 a $260\ \Omega$ b $297.14\ \Omega$ c 58.9° d $j254.4\ \Omega$
e 101.2 mH f $53.5\ \Omega$
16-20 a $3\bar{3}\angle 0^\circ$ mS b $5\angle -90^\circ$ mS c $(3\bar{3} - j5)$ mS
d $29.67\angle 8.53^\circ\ \Omega$
16-21 a $2.5\angle 0^\circ$ mS b $4.347\angle -90^\circ$ mS
c $(2.5 - j4.347)$ mS d $19.94\angle 60.1^\circ\ \Omega$
16-22 b $(250 - j227.4)\ \mu$A c $33.8\angle -42.29^\circ\ \mu$S
16-23 a $64.03\angle 51.34^\circ$ mS b $15.62\angle -51.34^\circ\ \Omega$
16-24 a $1.249\angle 0^\circ$ A b $1.562\angle 90^\circ$ A c $31.23\angle 0^\circ$ V
d $(40 + j50)$ mS e $15.62\angle -51.34^\circ\ \Omega$
16-25 a $22.36\angle 90^\circ$ mA b $600\angle 0^\circ\ \Omega$ c $-j53\bar{6}\ \Omega$
d $2.5\angle 48.2^\circ$ mS e 240 mW f $(240 - j268.3)$ mVA
g $400\angle -48.19^\circ\ \Omega$ h $(26\bar{6} - j298.1)\ \Omega$
16-26 a $8\bar{3}\angle -50^\circ$ mS b $j56.55$ mS c $54.06\angle -7.75^\circ$ mS
d $6.487\angle -7.75^\circ$ A e $1.2\angle 50^\circ$ kVA
f $778.45\angle 7.75^\circ$ VA
16-27 a $4\angle -20^\circ$ A, $2\angle 40^\circ$ A, $1\angle -50^\circ$ A b $6\angle -8.13^\circ$ A
c $6\bar{6}\angle 8.13^\circ\ \Omega$
16-28 a $66\angle 8.13^\circ\ \Omega$
16-29 $20\ \Omega$, $19.1\ \mu$F b $7.692\angle -67.38^\circ\ \Omega$
c $2.96\ \Omega$, $22.4\ \mu$F d 2.4 e $22.4\ \mu$F
16-30 a 15.9155 kHz b $-j50\ \Omega$ c $50\ \Omega$ d 1; $0.2\ \mu$F
16-31 a $8.814\angle 63^\circ$ mS b $113.5\angle -63^\circ\ \Omega$ c 2.546 kHz
d $(125 - j125)\ \Omega$ e 884 mV
16-32 a $7.29\ \mu$F b 218.3 Hz c $70.71\angle -45^\circ\ \Omega$
d $50\ \Omega$, $14.6\ \mu$F e $16\ \Omega$, $8.68\ \mu$F
16-33 a $76.92\angle -67.38^\circ$ mS b $58.82\angle 28.07^\circ$ mS
c $92.29\angle -28^\circ$ mS d $9.229\angle -28^\circ$ A
e $10.84\angle 28^\circ\ \Omega$ f $10.84\angle 28^\circ\ \Omega$
g, h $7.692\angle -67.38^\circ$ A; $5.882\angle 28.07^\circ$ A
i 295.8 W; 519 W j 814.8 W
k $769.2\angle 67.38^\circ$ VA; $588.2\angle -28.07^\circ$ VA
l $922.9\angle 28^\circ$ VA
16-34 $73.31\angle -35.34^\circ\ \Omega$
16-35 a $1\angle 0^\circ$ A
b $17\angle 28.07^\circ$ V; $20\angle -36.87^\circ$ V, $48\angle -53.13^\circ$ V
c $0.6\angle -53.13^\circ$ A; $0.8\angle 36.87^\circ$ A
d 15 W; 16 W; 28.8 W e, f 59.8 W

g $17\angle 28.07^\circ$ VA; $20\angle -36.87^\circ$ VA; $48\angle -53.13^\circ$ VA
h $(59.8 - j42.4)$ VA

16-36 a $10\angle 51.54^\circ$ A b $48\angle 88.41^\circ$ V c $48\angle -1.59^\circ$ V
d 384 W e 288 W f, g 672 W

16-37 a $40\angle 90^\circ$ mA b 6 kΩ c $44.72\angle 63.435^\circ$ mA
d $116.3\angle -26.62^\circ$ V e $2.602\angle -90^\circ$ kΩ
f 1 μF, 175 V; 0.9 μF, 175 V

16-38 a $6\angle 0^\circ$ V b $5\angle 33.69^\circ$ V

16-39 a $69.08\angle -89.2^\circ$ Ω b $69.07\angle -89.2^\circ$ Ω

16-40 a $2.538\angle 29.17^\circ$ A b $1.523\angle -23.96^\circ$ A
c $2.03\angle 66.04^\circ$ A d $2\angle -90^\circ$ A e $2.344\angle -19^\circ$ A
f 12.88 W; 9.278 W g, h 22.16 W i $23.44\angle 19^\circ$ VA

16-41 a $1964.7\angle -14.74^\circ$ VA b $-j500$ vars c 1.9 kW
d 0.9671 e $8.93\angle -14.74^\circ$ A f $1964.7\angle 14.74^\circ$ VA

16-42 c $24.63\angle 14.74^\circ$ Ω

16-43 a $(-2500 + j1250)$; $(1000 - j1750)$; $(1500 - j500)$; $(0 + j1000)$ VA e $(20 - j10)$ Ω

CHAPTER 17

Network Theorems Applied to the Frequency Domain and Related Power Relations

The major network theorems introduced in Chapter 7 were used to analyze and simplify dc circuits. This earlier introduction enabled the reader to learn each theorem and its application in the simplest way, since dc circuits are easier to understand and analyze than ac circuits.

The theorems introduced in Chapter 7 all apply equally to ac networks with some minor differences. It is assumed that the reader is already familiar with the statement of each theorem from study of Chapter 7. Consequently, the approach used in this chapter concentrates on applying each theorem to ac circuits in the *frequency* domain.

Not all the theorems described in Chapter 7 will be covered here, although all apply equally to ac networks. The following differences occur, however, in applying these theorems to ac networks:

1. Complex impedances are used in lieu of simple resistances.
2. Mathematical *operations* such as addition, subtraction, multiplication, and division are performed *in complex form.*

Where other important differences arise, they will be noted, as well.[1]

17-1 THÉVENIN'S THEOREM

The statement of Thévenin's theorem (Sec. 7-1) applied to ac circuits is

Any combination of (linear circuit elements) and active ac sources, regardless of connection or complexity, may be replaced by a two-terminal network consisting of a single ac voltage source, V_{TH}, and a single impedance, Z_{TH}, connected in series with the source, where

V_{TH} is the open-circuit RMS voltage measured at the two terminals of interest in the given complex network and

[1] See discussions of the maximum power transfer theorem (Sec. 17-6) and the superposition theorem (Sec. 17-4) applied to ac networks.

Z_{TH} is the equivalent impedance of the given complex network with all voltage sources shorted (zero impedance) and all current sources open (infinite impedance).

The procedure for performing a Thévenin analysis to determine the equivalent Thévenin voltage and impedance is the same as that described in Sec. 7-1. The following examples should serve as both a refresher on Thévenin's theorem and an introduction to its application to ac circuits.

EXAMPLE 17-1

The voltages shown in **Fig. 17-1** are $\mathbf{V}_1 = 12\angle 0^\circ$ V and $\mathbf{V}_2 = 12\angle 45^\circ$ V. The impedances are $\mathbf{Z}_1 = 3 + j4\ \Omega$ and $\mathbf{Z}_2 = 4 - j3\ \Omega$, respectively. Calculate

a. $\mathbf{V}_{TH}$, using the voltage divider rule applied to two sources, Eq. (5-8), modified for ac use
b. $\mathbf{Z}_{TH}$, by shorting all voltage sources

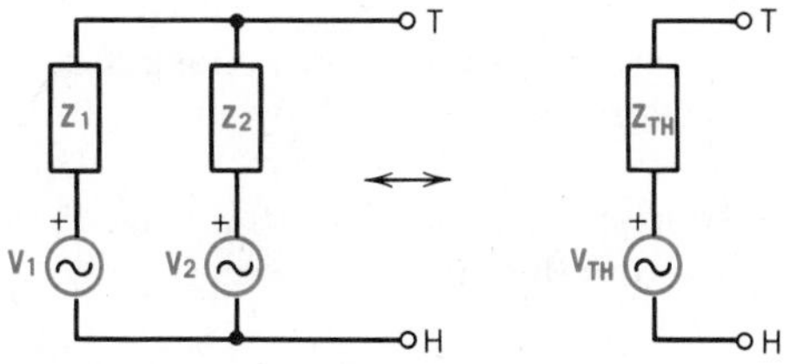

Figure 17-1 Example 17-1

Solution

a. Using Eq. (5-8) modified, applying the VDR, for two sources:

$$\mathbf{V}_{TH} = (\mathbf{V}_1 - \mathbf{V}_2)\frac{\mathbf{Z}_2}{\mathbf{Z}_1 + \mathbf{Z}_2} + \mathbf{V}_2$$

$$= \frac{(12\angle 0^\circ - 12\angle 45^\circ)(4 - j3)}{(4 - j3 + 3 + j4)} + 12\angle 45^\circ \text{ V}$$

$$= (3.515 - j8.485)\frac{(4 - j3)}{(7 + j1)} + (8.485 + j8.485)$$

$$= \mathbf{6.495\angle 22.5^\circ\ V}$$

b. Shorting both voltage sources yields $\mathbf{Z}_1 \| \mathbf{Z}_2$ or

$$\mathbf{Z}_{TH} = \frac{\mathbf{Z}_1\mathbf{Z}_2}{\mathbf{Z}_1 + \mathbf{Z}_2} = \frac{(3 + j4)(4 - j3)}{(7 + j1)} = 3.5 + j0.5$$

$$= \mathbf{3.5355\angle 8.13^\circ\ \Omega}$$

Example 17-1 shows that the Thévenin theorem procedure is exactly the same as that used for dc circuit analysis with the exceptions noted above (i.e., complex impedances substituted for simple resistances and mathematical operations performed in complex form).

EXAMPLE 17-2

In the circuit shown in **Fig. 17-2**, $\mathbf{Z}_1 = 8 - j15\ \Omega$, $\mathbf{Z}_2 = 12 + j5\ \Omega$, and $\mathbf{Z}_L = 6 + j8\ \Omega$. The supply voltage $\mathbf{V} = 20\angle 0^\circ$ V. Calculate:

a. $\mathbf{V}_{TH}$ with $\mathbf{Z}_L$ (temporarily) removed
b. $\mathbf{Z}_{TH}$ with $\mathbf{Z}_L$ (temporarily) removed
c. Current drawn by $\mathbf{Z}_L$ from the source
d. Voltage across $\mathbf{Z}_L$
e. Power dissipated in load $\mathbf{Z}_L$, by two methods

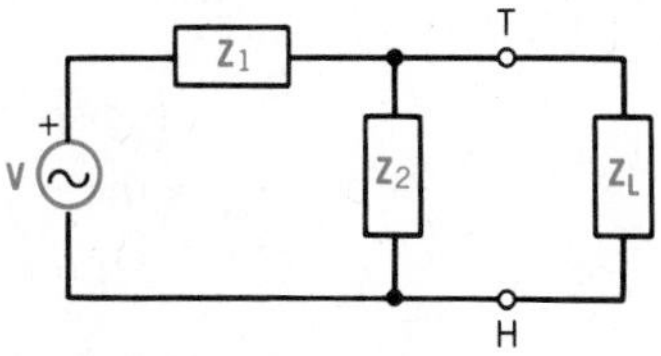

Figure 17-2 Example 17-2

Solution

a. Using the VDR,

$$\mathbf{V}_{TH} = \frac{(12 + j5)(20 + j0)\text{ V}}{(8 - j15 + 12 + j5)} = 7.6 + j8.8 = \mathbf{11.63\angle +49.2^\circ\ V}$$

b. $$\mathbf{Z}_{TH} = \mathbf{Z}_1 \| \mathbf{Z}_2 = \frac{\mathbf{Z}_1\mathbf{Z}_2}{\mathbf{Z}_1 + \mathbf{Z}_2} = \frac{(8 - j15)(12 + j5)}{(20 - j10)}$$

$$= 9.64 - j2.18$$

$$= \mathbf{9.883\angle -12.74^\circ\ \Omega}$$

c. $$\mathbf{I}_L = \frac{\mathbf{V}_{TH}}{\mathbf{Z}_{TH} + \mathbf{Z}_L} = \frac{(20 + j0)\text{ V}}{(9.64 - j2.18) + (6 + j8)} = 1.123 - j0.418$$

$$= \mathbf{1.198\angle -20.42^\circ\ A}$$

d. $$\mathbf{V}_L = \mathbf{I}_L\mathbf{Z}_L = (1.198\angle -20.42^\circ\text{ A})(10\angle 53.13^\circ\ \Omega)$$

$$= \mathbf{11.98\angle 32.71^\circ\ V}$$

e. $$P_L = I_L^2 R_L = 1.198^2 \times 6\ \Omega = \mathbf{8.61\ W};$$

$$P_L = \mathbf{V}_L\mathbf{I}\cos\theta_L = 11.98 \times 1.198\cos[32.71^\circ - (-20.42^\circ)]$$

$$= \mathbf{8.61\ W}$$

Example 17-2 shows how a series–parallel network may be divided, temporarily, to find the Thévenin equivalent voltage and impedance across two terminals of interest. When this is done, the remainder of the circuit is temporarily removed and, after the Thévenin equivalent is found, is restored for purposes of finding voltage across and current in the remainder of the circuit. With respect to the solution of Ex. 17-2, note that

1. The Thévenin equivalent voltage found in (**a**) is *not* the same as V_L found in (**d**).
2. It is entirely possible, in ac circuits, for the voltage across some portion of the circuit to *exceed* the supply voltage or some open-circuit voltage. In this case, with Z_L removed the open-circuit voltage is 11.63 V. Yet when Z_L is restored the (loaded) terminal voltage *rises* to 11.98 V. Although this can never happen with dc circuits, it is a property of ac circuits.[2]

17-2 NORTON'S THEOREM

As indicated in Sec. 7-2, Norton's theorem is the *dual* of Thévenin's theorem. The statement of Norton's theorem applied to ac circuits is: any combination of linear circuit elements and active ac sources, regardless of connection or complexity, may be replaced by an equivalent two-terminal network consisting of a single ac current source, $\mathbf{I}_N$, paralleled by an equivalent impedance, $\mathbf{Z}_N$.

The procedure for performing a Norton analysis to determine the equivalent Norton impedance, $\mathbf{Z}_N$, and Norton current, $\mathbf{I}_N$, is the same as that described in Sec. 7-2, using the open-circuit and short-circuit tests:

Open-circuit test. A determination of the open-circuit Thévenin voltage across any two terminals of an active complex network as measured by an ideal ac voltmeter.

Short-circuit test. A determination of the short-circuit Norton current, $\mathbf{I}_N$, that is measured by an ideal ammeter connected across the same two terminals of an active complex network.

From these test data we may write

$$\mathbf{Z}_N = \mathbf{Z}_{TH} = \frac{1}{\mathbf{Y}_N} = \frac{1}{\mathbf{Y}_{TH}} = \frac{\mathbf{V}_{TH}}{\mathbf{I}_N} = \frac{\text{open-circuit ac voltage}}{\text{short-circuit ac current}} \quad \text{ohms } (\Omega) \tag{17-1}$$

where $\mathbf{V}_{TH}$ is the open-circuit Thévenin voltage measured by the open-circuit test

$\mathbf{I}_N$ is the short-circuit Norton current measured by the short-circuit test

$\mathbf{Z}_N$ is the Norton impedance, which is equivalent to the Thévenin impedance, $\mathbf{Z}_{TH}$, between the two terminals across which the test is performed

As noted in Sec. 7-2, Norton's theorem is important because it provides

1. An experimental procedure enabling us to reduce multisource complex networks to either a single voltage source in series with a single equivalent impedance or a single current source in parallel with a single equivalent impedance.
2. It enables *source transformation* (Eq. 17-1) between practical voltage and current sources, using the Thévenin–Norton conversion.

EXAMPLE 17-3

For the circuit and given data of Ex. 17-2, assume the open-circuit voltage measured across terminals T–H is $11.628\underline{/49.185^\circ}$ V. Calculate the

a. Short-circuit current measured by an ideal ac ammeter across terminals T–H, $\mathbf{I}_N$
b. Equivalent Norton impedance, $\mathbf{Z}_N$, from (**a**) and the given data
c. Compare the value of $\mathbf{Z}_N$ with that found in Ex. 17-2b.

[2] We will discover the reason for this when we study resonant circuits in Chapter 19.

Solution

a. Placing a short-circuit across terminals T–H shorts all impedances across terminals T–H, and the short-circuit (Norton) current is (Fig. 17-2)
$\mathbf{I}_N = \mathbf{V}/\mathbf{Z}_1 = 20\angle 0^\circ \text{ V}/(8 - j15)\ \Omega = \mathbf{1.1765\angle 61.93^\circ\ A}$

b. $\mathbf{Z}_{TH} = \mathbf{Z}_N = \mathbf{V}_{oc}/\mathbf{I}_{sc} = \mathbf{V}_{TH}/\mathbf{I}_N$
$= 11.628\angle 49.185^\circ \text{ V}/1.1765\angle 61.93^\circ \text{ A}$
$= \mathbf{9.883\angle -12.745^\circ\ \Omega}$

c. Above exactly the *same* as that found in the solution to Ex. 17-2**b**

17-3 SOURCE TRANSFORMATION THEOREM

We may now write the Thévenin–Norton duality in terms of the source transformation theorem, which states that

Given any complex active network containing one or more practical voltage and/or current sources, any given practical voltage (or current) source may be transformed into an equivalent practical current (or voltage) source, or vice versa, using the Thévenin–Norton conversion (stated in Eq. 17-1).

Table 17-1 shows the Thévenin–Norton duality between practical ac voltage and current sources, along with the ac source transformation equations.

Table 17-1 Thévenin–Norton Duality in AC Sources

	Thévenin Equivalent	Norton Equivalent
Circuit	Series	Parallel
AC source	Voltage, $\mathbf{V}_{TH} = \mathbf{V}_{OC}$	Current, $\mathbf{I}_N = \mathbf{I}_{SC}$
Practical opposition	$\mathbf{Z}_{TH}$	$1/\mathbf{Y}_N$
Source transformation	$\mathbf{V}_{TH} = \mathbf{I}_N/\mathbf{Y}_N = \mathbf{I}_N\mathbf{Z}_N$	$\mathbf{I}_N = \mathbf{V}_{TH}/\mathbf{Z}_{TH} = \mathbf{V}_{TH}\mathbf{Y}_N$
Opposition transformation	$\mathbf{Z}_{TH} = 1/\mathbf{Y}_N = \mathbf{Z}_N$	$\mathbf{Y}_N = 1/\mathbf{Z}_{TH} = 1/\mathbf{Z}_N$

As in the case of dc circuits, the source transformation theorem enables a simplification of ac networks containing both ac current and ac voltage sources, as shown by Ex. 17-4.

EXAMPLE 17-4
Given the network shown in **Fig. 17-3a** and the values $\mathbf{Z}_1 = 6\angle 30^\circ$ kΩ, $\mathbf{Z}_2 = 9\angle -30^\circ$ kΩ, $\mathbf{Z}_3 = 6\angle -30^\circ$ kΩ, $\mathbf{Z}_4 = 5\angle 40^\circ$ kΩ, $\mathbf{V} = 120/0^\circ$ V, and $\mathbf{I} = 5\angle 0^\circ$ mA, calculate the simplest Thévenin equivalent (consisting of a single practical voltage source) across terminals T–H

Solution
The solution steps are shown in Fig. 17-3. The solution steps involve the following source transformations and/or Thévenin equivalents:

1. Across terminals x, y find the equivalent Thévenin voltage, $\mathbf{V}_{xy}$, and Thévenin impedance, $\mathbf{Z}_{xy}$:

$$\mathbf{V}_{xy} = \frac{\mathbf{Z}_2}{\mathbf{Z}_1 + \mathbf{Z}_2}\mathbf{V} = \frac{(9\angle -30^\circ \text{ k}\Omega)\ 120\angle 0^\circ \text{ V}}{(6\angle 30^\circ + 9\angle -30^\circ)\text{ k}\Omega}$$
$$= \mathbf{82.59\angle -6.59^\circ\ V}$$

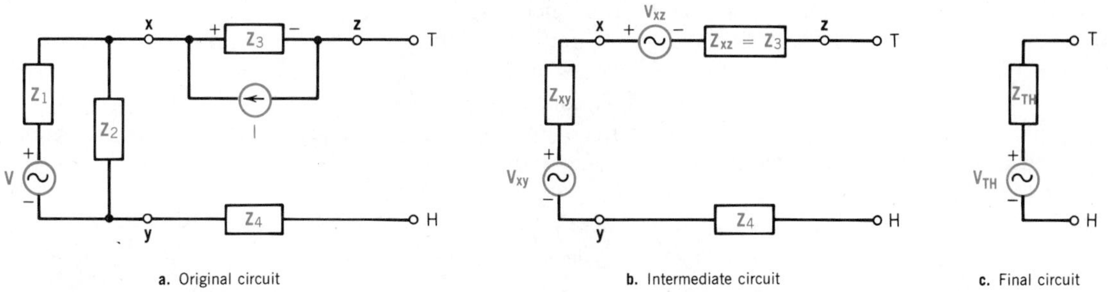

Figure 17-3 Example 17-4 showing source transformation in conjunction with Thévenin and Norton theorems

and $\quad \mathbf{Z}_{xy} = \mathbf{Z}_1 \| \mathbf{Z}_2 = \dfrac{\mathbf{Z}_1\mathbf{Z}_2}{\mathbf{Z}_1 + \mathbf{Z}_2} = \dfrac{9\angle{-30^\circ} \times 6\angle{30^\circ}}{13.08\angle{-6.59^\circ}}$

$$= \mathbf{4.128\angle{6.59^\circ}\ k\Omega}$$

2. Using the source transformation theorem, convert the practical current source across terminals x, z to an appropriate practical voltage source, $\mathbf{V}_{xz}$ and $\mathbf{Z}_{xz}$:

$$\mathbf{V}_{xz} = \mathbf{I}\mathbf{Z}_3 = (5\angle{0^\circ}\ \text{mA})(6\angle{-30^\circ}\ \text{k}\Omega) = \mathbf{30\angle{-30^\circ}\ V}$$

and $\quad \mathbf{Z}_{xz} = \mathbf{Z}_N = \mathbf{6\angle{-30^\circ}\ k\Omega}$

3. The above information is entered into Fig. 17-3b, a simple series circuit, from which:

$$\mathbf{V}_{TH} = \mathbf{V}_{xy} - \mathbf{V}_{xz} = (82.59\angle{-6.59^\circ} - 30\angle{-30^\circ})\ \text{V}$$
$$= \mathbf{56.33\angle{5.6^\circ}\ V}$$

and $\quad \mathbf{Z}_{TH} = \mathbf{Z}_{xy} + \mathbf{Z}_3 + \mathbf{Z}_4$

$$= (4.128\angle{6.59^\circ} + 6\angle{-30^\circ} + 5\angle{40^\circ})\ \text{k}\Omega$$
$$= \mathbf{13.145\angle{3.0^\circ}\ k\Omega}$$

These values, entered into Fig. 17-3c, represent the simplest practical voltage source replacing the entire network of Fig. 17-3a.

Several important insights emerge from the above solution:

1. In using the source transformation theorem, care must be taken to *preserve* the *original* source polarity. From Figs. 17-3a and 17-3b we observe that the practical current source produces a *positive* polarity at point **x**. Consequently, the practical voltage source due to source conversion must have a positive polarity at point **x**.
2. In step 3, $\mathbf{V}_{TH}$ is the *difference* between the source voltages since they are opposing each other. In this case, $\mathbf{V}_{TH}$ is still a positive value since $\mathbf{V}_{xy} > \mathbf{V}_{xz}$, and terminal T is positive. But if the reverse were the case and $\mathbf{V}_{TH}$ emerged as a *negative* voltage, it would mean that terminal **T** is negative and terminal **H** is positive.
3. $\mathbf{Z}_{TH}$ is found in the usual way by shorting voltage sources (and opening current sources) in Fig. 17-3b. But it could have been found directly from Fig. 17-3a as $\mathbf{Z}_{TH} = \mathbf{Z}_4 + \mathbf{Z}_3 + (\mathbf{Z}_1 \| \mathbf{Z}_2)$, which yields the same answer as that shown in the solution.

17-4 SUPERPOSITION THEOREM

The foregoing theorems and those that follow hold only for sources that are *all at the same frequency*, when solved in the frequency domain. The superposition theorem, however, may be used for multisource networks, where the sources may be *either* at the same frequency *or different* frequencies. In reality, this *is the one unique property of the superposition theorem.*[3]

The statement of the superposition theorem is:

In any multisource network, consisting of linear elements and sources at the same or different frequencies, the voltage across or current through any given element of the network is equal to the *complex* RMS sum of the individual voltages across or currents in that element by each source acting independently.

The term complex RMS sum requires explanation. Let us assume that a given load, $\mathbf{Z}_L$, experiences three different RMS currents, $\mathbf{I}_1$, $\mathbf{I}_2$, and $\mathbf{I}_3$, due to three different sources at three *different* frequencies. Then the actual equivalent RMS current, $\mathbf{I}_L$, in $\mathbf{Z}_L$ is

$$\mathbf{I}_L = \sqrt{\mathbf{I}_1^2 + \mathbf{I}_2^2 + \mathbf{I}_3^2} \qquad \text{amperes (A)} \qquad (17\text{-}2)$$

where all terms have been defined above.

Note that this addition is quite different from the dc situation (Sec. 7-4) or even the situation of three sources at the *same* frequency, where currents are either algebraically or vectorially added or subtracted based on current direction through $\mathbf{Z}_L$. Since the RMS currents are squared, the instantaneous *directions* of currents due to each different source frequency *are of no consequence.* To clarify this

[3] If the principle of superposition did *not* have this property, it might only be of minor interest. Other methods of analyzing networks, including mesh and nodal analysis are much easier, as are Thévenin's and Norton's theorems.

important distinction, two examples are shown. Example 17-5 shows two sources at the *same* frequency producing *different* current directions in a given load. Example 17-6 shows the use of Eq. (17-2) when there are two sources at *two different* frequencies.

EXAMPLE 17-5

For the network shown in **Fig. 17-4a**, $\mathbf{V}_1 = 8\angle 20^\circ$ V, $\mathbf{V}_2 = 10\angle 0^\circ$ V, $\mathbf{Z}_1 = 1 + j2\ \Omega$, $\mathbf{Z}_2 = 4 - j8\ \Omega$, and $\mathbf{Z}_3 = 0 + j6\ \Omega$. Çalculate by superposition the

a. Current $\mathbf{I}_3'$ in $\mathbf{Z}_3$ due to $\mathbf{V}_1$ acting alone (Fig. 17-4b)
b. Current $\mathbf{I}_3''$ in $\mathbf{Z}_3$ due to $\mathbf{V}_2$ acting alone (Fig. 17-4c)
c. Actual current in $\mathbf{Z}_3$ as the phasor sum of $\mathbf{I}_3'$ and $\mathbf{I}_3''$ with due regard for current directions

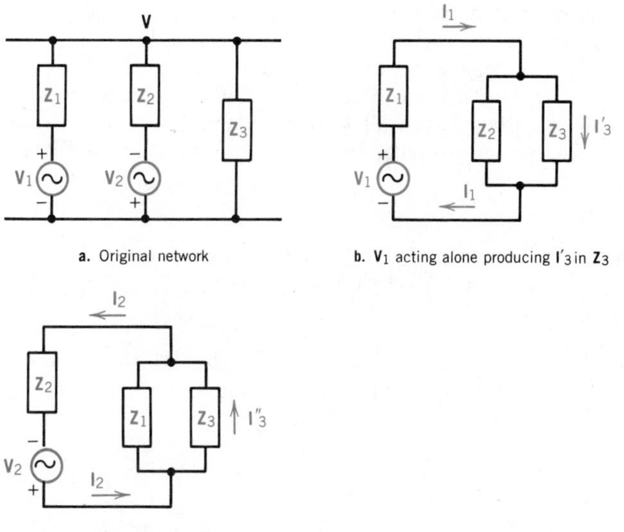

a. Original network

b. V1 acting alone producing I'3 in Z3

c. V2 acting alone producing I''3 in Z3

Figure 17-4 Example 17-5 showing superposition solution

Solution

From Fig. 17-4b,

$$\mathbf{Z}_3 \| \mathbf{Z}_2 = \frac{\mathbf{Z}_3\mathbf{Z}_2}{\mathbf{Z}_3 + \mathbf{Z}_2} = \frac{(4 - j8)(0 + j6)}{(4 - j2)} = 7.2 + j9.6$$

a. $\mathbf{Z}_{T_1} = \mathbf{Z}_1 + \mathbf{Z}_p = (1 + j2) + (7.2 + j9.6) = 8.2 + j11.6 = 14.206\angle 54.74^\circ\ \Omega.$

Then $\mathbf{I}_1 = \mathbf{V}_1/\mathbf{Z}_{T_1} = 8\angle 20^\circ\ \text{V}/14.206\angle 54.74^\circ\ \Omega = 0.5632\angle -34.74^\circ$ A,

and using the CDR, Eq. (16-19):

$$\mathbf{I}_3' = \mathbf{I}_1 \frac{\mathbf{Z}_2}{(\mathbf{Z}_3 + \mathbf{Z}_2)}$$

$$= (0.5632\angle -34.74^\circ\ \text{A}) \times \frac{8.944\angle -63.435^\circ\ \Omega}{4.472\angle -26.565^\circ\ \Omega}$$

$$= \mathbf{1.126\angle -71.6^\circ\ A}$$

b. From Fig. 17-4c,

$$\mathbf{Z}_1 \| \mathbf{Z}_3 = \frac{\mathbf{Z}_1\mathbf{Z}_3}{\mathbf{Z}_1 + \mathbf{Z}_3} = \frac{(1 + j2)(0 + j6)}{(1 + j8)} = 0.55385 + j1.5692,$$

and $\mathbf{Z}_{T_2} = \mathbf{Z}_2 + \mathbf{Z}_p = (4 - j8) + (0.55385 + j1.5692) = 4.55385 - j6.431 = 7.88\angle -54.7^\circ\ \Omega.$

Then $\mathbf{I}_2 = \mathbf{V}/\mathbf{Z}_{T_2} = 10\angle 0^\circ\ \text{V}/7.88\angle -54.7^\circ\ \Omega = 1.269\angle 54.7^\circ$ A.

And using the CDR, Eq. (16-19):

$$\mathbf{I}_3'' = 1.269\angle 54.7^\circ\ \text{A} \frac{(1 + j2)}{(1 + j8)} = \mathbf{0.352\angle 35.26^\circ\ A}$$

c. But since $\mathbf{I}_3'$ is in the *opposite* direction from $\mathbf{I}_3''$, then the true current, $\mathbf{I}_3$ is the *difference between them* (Fig. 17-4b versus 17-4c) or

$$\mathbf{I}_3 = \mathbf{I}_3' - \mathbf{I}_3'' = 1.126\angle -71.6^\circ\ \text{A} - 0.352\angle 35.26^\circ\ \text{A} = \mathbf{1.273\angle -86.95^\circ\ A}.$$

(Note: This example is solved by Thévenin's theorem in Prob. 17-3 as an exercise for the reader and also in Ex. 17-7b to serve as an independent verification of the answer here.)

Example 17-5 shows that the superposition theorem applies equally to both dc circuits and ac circuits where the sources are at the *same* frequency. But what if the sources are *not* at the same frequency? Is it possible to find the current in a given impedance under these conditions? Figure 17-5 and Ex. 17-6 show the importance and value of superposition in solving problems of this kind.[4] It bears repeating that superposition is the *only* method that may be applied to sources of *different frequencies* in network analysis.

[4] It is precisely for this reason that *only* superposition may be employed in dealing with nonsinusoidal waves containing harmonics. See Chapter 22.

EXAMPLE 17-6

For the network shown in **Fig. 17-5a**, calculate by superposition the

a. Current in resistor R due to the 60-Hz source acting independently, I'_R
b. Power dissipated in resistor R due to the 60-Hz source
c. Current in resistor R due to the 120-Hz source acting independently, I''_R
d. Power dissipated in resistor R due to the 120 Hz
e. Total power dissipated in resistor R due to both sources
f. RMS current in resistor R, using I'_R and I''_R
g. Total power dissipated in resistor R, using the actual RMS current in R

Solution

Taking the 60-Hz source independently, the circuit and impedances are shown in Fig. 17-5b. The total impedance, $\mathbf{Z}_{T_1}$, opposing the $40\angle 0^\circ$ V source is

a. $\mathbf{Z}_{T_1} = [(500\angle 0^\circ \,\|\, -j265.25)] + (j377)\ \Omega = 109.81 + j170$
$= 202.38\angle 57.14^\circ\ \Omega$

$\mathbf{I}_{T_1} = \mathbf{V}_1/\mathbf{Z}_{T_1} = 40\angle 0^\circ\ \text{V}/202.38\angle 57.14^\circ\ \Omega$
$= 0.19765\angle -57.14^\circ$ A,

and using the CDR to find $\mathbf{I}'_R$:

$$\mathbf{I}'_R = \mathbf{I}_{T_1}\frac{-jX_C}{(R - jX_C)} = 0.19765\angle -57.14^\circ\ \text{A}\ \frac{265.25\angle -90^\circ}{566\angle -27.946^\circ}$$
$$= \mathbf{92.625\angle -119.19^\circ\ mA}$$

b. $P'_R = (I'^2_R R) = 0.092625^2 \times 500 = \mathbf{4.29\ W}$

c. Taking the 120-Hz source independently, the circuit and impedances are shown in Fig. 17-5c. The total impedance, $\mathbf{Z}_{T_2}$, opposing the $20\angle 0^\circ$ V source is

$\mathbf{Z}_{T_2} = [(500\angle 0^\circ \,\|\, j754)] + (-j132.625)\ \Omega = 347.28 + j97.67$
$= 360.75\angle 15.71^\circ\ \Omega$,

and $\mathbf{I}_{T_2} = \mathbf{V}_2/\mathbf{Z}_{T_2} = 20\angle 0^\circ\ \text{V}/360.75\angle 15.71^\circ\ \Omega$
$= 55.44\angle -15.71^\circ$ mA.

Using CDR, Eq. (16-19):

$$\mathbf{I}''_R = 55.44\angle -15.71^\circ\ \text{mA}\ \frac{754\angle 90^\circ\ \Omega}{904.72\angle 56.45^\circ\ \Omega} = \mathbf{46.2\angle 17.84^\circ\ mA}$$

d. $P''_R = (I''^2_R R) = 0.0462^2 \times 500 = \mathbf{1.067\ W}$

e. $P_R = P'_R + P''_R = 4.29 + 1.067 = \mathbf{5.357\ W}$

f. $\mathbf{I}_R = \sqrt{I'^2_R + I''^2_R} = \sqrt{(92.625)^2 + (46.2)^2}$
$= \mathbf{103.5\ mA}$ (true RMS current in R) (17-2)

g. $P_R = I_R^2 R = (0.1035)^2 \times 500 = \mathbf{5.357\ W}$

Several important insights emerge from Ex. 17-6:

1. Total power is the sum of the powers dissipated in R by *each* source acting *independently*.
2. In this case, when using Eq. (17-2) to find the true RMS current in R, the current direction is unimportant. (Observe from Figs. 17-5b and 17-5c that I''_R has a direction *opposed* to that of I'_R). Since all currents are squared in Eq. (17-2), their directions are eliminated and their squares are added numerically (see **f**).
3. Total power, using the true RMS current in R, is the same as the sum of the individually dissipated powers. This serves as a test for the accuracy of the solution method.
4. All reactances change whenever the frequency changes, as noted in Figs. 17-5b and 17-5c. If the frequency is doubled, inductive reactances are doubled and capacitive reactances are halved.

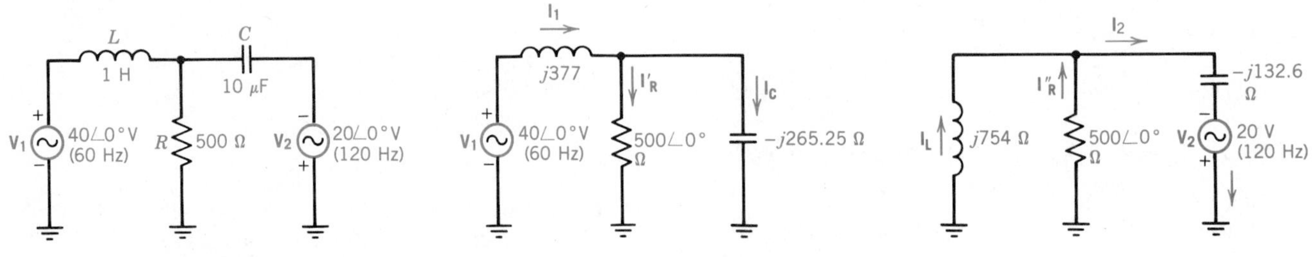

a. Original network b. V_1 acting alone producing I'_R c. V_2 acting alone producing I''_R

Figure 17-5 Example 17-6 showing superposition applied to two sources at different frequencies

17-5 MILLMAN'S THEOREM

Millman's theorem (as noted in Sec. 7-5) is essentially a procedure that combines the source transformation theorem with Thévenin's and Norton's theorems. Applying the procedure to ac circuits, we may modify Eq. (7-4) for the ac voltage $\mathbf{V}$ across n sources in parallel as

$$\mathbf{V} = \frac{\mathbf{V}_1\mathbf{Y}_1 \pm \mathbf{V}_2\mathbf{Y}_2 \pm \mathbf{V}_3\mathbf{Y}_3 \pm \cdots \pm \mathbf{V}_n\mathbf{Y}_n}{\mathbf{Y}_1 + \mathbf{Y}_2 + \mathbf{Y}_3 + \cdots + \mathbf{Y}_n} = \frac{\mathbf{I}_T}{\mathbf{Y}_T} \quad \text{volts (V)} \quad (17\text{-}3)$$

Correspondingly, if all sources in parallel are current sources

$$\mathbf{V} = \frac{\mathbf{I}_1 \pm \mathbf{I}_2 \pm \mathbf{I}_3 \pm \cdots \pm \mathbf{I}_n}{\mathbf{Y}_1 + \mathbf{Y}_2 + \mathbf{Y}_3 + \cdots + \mathbf{Y}_n} = \frac{\mathbf{I}_T}{\mathbf{Y}_T} \quad \text{volts (V)} \qquad (17\text{-}4)$$

The same precautions apply to Eqs. (17-3) and (17-4) as in the case of dc circuits:

1. *Positive* current is assumed to be directed *toward* the node **V** and *negative* currents are assumed to be directed *away from* the node.
2. In the case of voltage sources, the polarity of the sources, $\mathbf{V}_1$, $\mathbf{V}_2$, and so forth, in Eq. (17-3) with respect to **V** determines the total current in the numerator.
3. Consequently, the plus/minus (±) sign in both equations should be construed as the phasor (or vector) sum of the numerator currents rather than the simple arithmetic sum.

EXAMPLE 17-7

Verify the solution of Ex. 17-5 (given Fig. 17-4a) using Millman's theorem by calculating the

a. Node voltage **V**
b. Current $\mathbf{I}_3$ in $\mathbf{Z}_3$

Solution

Preliminary calculations for substitution in Eq. (17-3):

$$\mathbf{Y}_1 = 1/\mathbf{Z}_1 = 1/(1 + j2) = 1/2.236\angle 63.435^\circ = 0.4472\angle -63.435^\circ = 0.2 - j0.4 \text{ S},$$

$$\mathbf{Y}_2 = 1/\mathbf{Z}_2 = 1/(4 - j8) = 1/8.9443\angle -63.435^\circ = 0.1118\angle 63.435^\circ = 0.05 + j0.1 \text{ S},$$

$$\mathbf{Y}_3 = 1/\mathbf{Z}_3 = 1/6\angle 90^\circ = 0.1\bar{6}\angle -90^\circ = 0 - j0.1\bar{6} \text{ S},$$

and $\mathbf{Y}_1 + \mathbf{Y}_2 + \mathbf{Y}_3 = 0.25 - j0.4\bar{6} = 0.5294\angle -61.82^\circ$ S

a. $$\mathbf{V} = \frac{\mathbf{V}_1\mathbf{Y}_1 - \mathbf{V}_2\mathbf{Y}_2}{\mathbf{Y}_1 + \mathbf{Y}_2 + \mathbf{Y}_3} = \frac{(8\angle 20^\circ \text{ V})/2.236\angle 63.435^\circ - (10\angle 0^\circ \text{ V})/8.9443\angle -63.435^\circ \text{ A}}{0.5294\angle -61.82^\circ \text{ S}} \qquad (17\text{-}3)$$

$$= \frac{3.578\angle -43.435^\circ - 1.118\angle 63.435^\circ \text{ A}}{0.5294\angle -61.82^\circ \text{ S}} = \mathbf{7.643\angle 3.05^\circ\ V} \qquad (17\text{-}3)$$

b. $\mathbf{I}_3 = \mathbf{V}_3/\mathbf{Z}_3 = 7.643\angle 3.05^\circ \text{ V}/6\angle 90^\circ\ \Omega = \mathbf{1.274\angle -86.95^\circ\ A}$

The following insights emerge from this solution:

1. Since the source voltage, $\mathbf{V}_2$, directs current *away from* the node, **V**, its current $\mathbf{V}_2\mathbf{Y}_2$ is *subtracted* from $\mathbf{V}_1\mathbf{Y}_1$ in (**a**).
2. Given impedances $\mathbf{Z}_1$, $\mathbf{Z}_2$, and $\mathbf{Z}_3$, great care must be taken in properly finding and summing their respective admittances $\mathbf{Y}_1$, $\mathbf{Y}_2$, and $\mathbf{Y}_3$, as shown in the preliminary calculations prior to (**a**) of the solution. (There are no shortcuts to this procedure.)
3. The value obtained for $\mathbf{I}_3$ with Millman's theorem verifies the superposition technique of Ex. 17-5 and vice versa. (Problem 17-3 at the end of this chapter also serves as an independent verification of the value of $\mathbf{I}_3$.)

17-6 MAXIMUM POWER TRANSFER THEOREM APPLIED TO AC CIRCUITS

Of all the theorems, the maximum power transfer theorem requires special treatment because of the major differences in applying this theorem to ac networks. Recall from Sec. 5-7 that the object of the theorem was to *adjust the load* so that maximum power is transferred from a practical source to a load. Recall that the dc case was fairly straightforward. Given a variable resistive load, R_L, maximum power transfer (MPT) occurs in R_L when each of the five following conditions occurs:

1. The load resistance $R_L = r_i$, the internal source resistance, Eq. (5-17).
2. Load power is maximized and expressed by Eq. (5-16) as $E_0^2/4r_i = E_0^2/4R_L$.

3. Efficiency of power transfer from source to load is 50%, Eq. (5-15).
4. The voltage across the load, V_L, is half the source open-circuit voltage, or $V_L = E_0/2$, Eq. (5-13).
5. The current in the load, I_L, is half the source short-circuit current, or $I_L = I_{sc}/2 = E_0/2r_i$, Eq. (5-14).

Unfortunately, no such neat, simple, cut-and-dried situation applies to the ac case. Consider an ac source, **E**, whose internal impedance is $R_i + jX_i$ as shown in **Fig. 17-6**. A load may be connected across the source terminals, 1–2, and may be represented as $R_L + jX_L$, where

R_i and R_L are the resistances of the source and load, respectively.
jX_i and jX_L are the reactances of the source and load, respectively.[5]

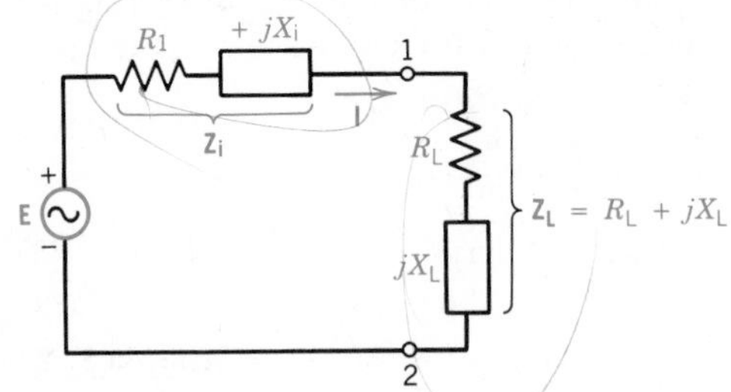

Figure 17-6 General circuit for MPT

Appendix A-6.2 gives the calculus derivation, which shows that maximum power is transferred from ac source **E** to load $\mathbf{Z}_L = R_L + jX_L$ when load impedance, $\mathbf{Z}_L$, is the conjugate of the source impedance, $\mathbf{Z}_i$, or

$$\mathbf{Z}_L = \mathbf{Z}_i^* = \mathbf{R}_i - jX_i \quad \text{ohms } (\Omega) \qquad (17\text{-}5)$$

where $\mathbf{Z}_i = R_i + jX_i$ and all other terms have been defined.

EXAMPLE 17-8

Given the practical voltage source shown in **Fig. 17-7**, $\mathbf{Z}_L$ is an *unrestricted* load and may be adjusted to *any* desired value of resistance and reactance. Calculate the

a. Impedance of the load to draw maximum power from the source
b. Current drawn from the source under the conditions of **a**
c. Power dissipated in the load
d. Voltage across the terminals of the source and across the load

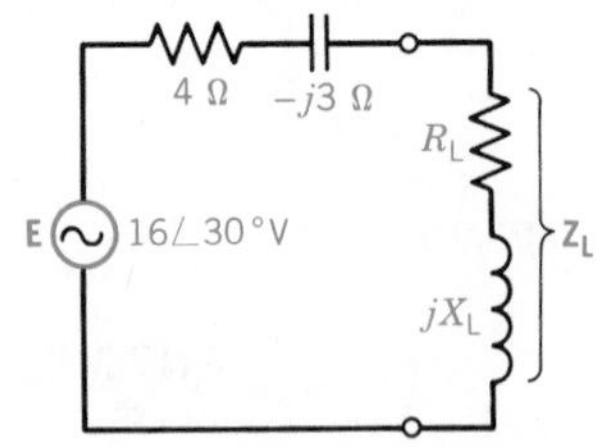

Figure 17-7 Examples 17-8 to 17-11, inclusive

Solution

a. For MPT, $\mathbf{Z}_L = \mathbf{Z}_i^* = (4 - j3)^* = \mathbf{(4 + j3)\ \Omega}$ (17-5)

b. $\mathbf{I} = \dfrac{\mathbf{E}}{(R_i - jX_C) + (R_L + jX_L)} = \dfrac{16\angle 30^\circ\ \text{V}}{(4 - j3) + (4 + j3)\ \Omega} = \dfrac{16\angle 30^\circ\ \text{V}}{8\angle 0^\circ\ \Omega} = \mathbf{2\angle 30^\circ\ A}$

c. $P_L = |I|^2 R_L = (2)^2 4 = \mathbf{16\ W}$

d. $\mathbf{V}_L = \mathbf{V} = \mathbf{I}\mathbf{Z}_L = (2\angle 30^\circ\ \text{A})(5\angle 36.87^\circ\ \Omega) = \mathbf{10\angle 66.87^\circ\ V}$

Several important insights regarding MPT may be drawn from Ex. 17-8, when the load is *unrestricted* and may be adjusted to *any desired value* of resistance and reactance:

[5] We must not make the error of assuming that X_L is always an inductive reactance. In this discussion X_L means *load* reactance, unless otherwise indicated. X_L may indicate either inductive reactance or capacitive reactance in Sec. 17-6.

1. The ac current as drawn from the source is $\mathbf{E}/2R_i$ (as in the dc case) but *not* $I_{sc}/2$.
2. Terminal source voltage is the same as voltage across the load impedance, $\mathbf{Z}_L$. It is found as $\mathbf{I} \cdot \mathbf{Z}_L$. (It is *not* half the source voltage as in the dc case.)
3. Power transfer (from source to load) efficiency is 50% since half the power is dissipated in the load and half in the source itself. (But this only holds for the case of an *unrestricted* load, as will be shown.)
4. MPT is always calculated as $|I|^2 R_L$, even under load restricted conditions, as will be shown.
5. The power factor (PF) of the circuit (*unrestricted load* only) is *unity*, as shown in solution (**b**), since current is in phase with voltage. Consequently, the *highest power* is transferred from source to load when the load can be adjusted to the conjugate of the source impedance. Such an *impedance match* is called a *conjugate match.*
6. The total impedance is a minimum and resistive (unity PF) under conditions of MPT, since the load Z_L is adjusted to the conjugate of the source impedance. Minimum impedance at unity PF ensures MPT to load R_L.

17-6.1 Load Restricted Case A—Fixed X_L but Variable R_L

In many applications, the load reactance is fixed (preset) by the nature of the device.[6] But we have the option of connecting a variable resistance in series with the fixed load reactance to maximize the power transfer. Since the load reactance is fixed, it is rarely the conjugate of the source reactance. Consequently, it is (almost) *impossible to attain a conjugate match.* Under such conditions we settle for the value of R_L that maximizes the power dissipated in R_L. While the power transferred from source to load is not as high as the conjugate match, it represents the greatest power dissipation in R_L, given a **fixed load reactance**, X_L.

For MPT to R_L, given fixed X_L, it can be shown that the *absolute magnitude* of R_L may be found from

$$|R_L| = |R_i + j(X_i + X_L)| \quad \text{ohms } (\Omega) \qquad (17\text{-}6)$$

where all terms have been defined.

Example 17-9 gives this case and shows that the value of R_L obtained in Eq. (17-6) yields MPT under the circumstances of the fixed value of jX_L, given, along with the internal impedance $\mathbf{Z}_i = R_i + jX_i$.

EXAMPLE 17-9

Given the practical source shown in Fig. 17-7 and a fixed value of $jX_L = j10\ \Omega$, calculate the

a. Value of R_L that draws maximum power from the source to the load
b. Power dissipated in the load, $\mathbf{Z}_L$
c. Power dissipated in the load $\mathbf{Z}_L$, using a value of $R_L = 5\ \Omega$
d. Power dissipated in the load $\mathbf{Z}_L$, using a value of $R_L = 10\ \Omega$

Solution

a. $|R_L| = |R_i + j(X_i + X_L)| = |4 + (-j3 + j10)|$
$= |4 + j7| = \mathbf{8.062\ \Omega}$ (17-6)

b. $\mathbf{I} = \dfrac{\mathbf{E}}{(R_i + R_L) + j(X_i + X_L)} = \dfrac{16\angle 30^\circ\text{ V}}{(4 + 8.062) + (j10 - j3)}$
$= \dfrac{16\angle 30^\circ\text{ V}}{12.062 + j7} = \mathbf{1.147\angle -0.13^\circ\ A}$

and $P_L = |I|^2 R_L = (1.147)^2 \times 8.062 = \mathbf{10.61\ W}$

c. Using a lower value of $R_L = 5\ \Omega$,

$$I = \frac{16\angle 30^\circ\text{ V}}{(4 + 5) + j(10 - 3)} = \frac{16\angle 30^\circ\text{ V}}{9 + j7}$$
$$= \frac{16\angle 30^\circ\text{ V}}{11.4\angle 37.87^\circ} = 1.4\angle -7.87^\circ\text{ A}$$

and $P_L = |I|^2 R_L = (1.4)^2 \times 5 = \mathbf{9.85\ W}$ (despite higher current, a lower value of P_L)

[6] A typical application is the case of a highly inductive loudspeaker coil connected to an audio amplifier, where MPT is measured at some central (median) frequency. Another application is a servomotor connected to a servoamplifier at its rated frequency.

d. Using a higher value of $R_L = 10\ \Omega$,

$$I = \frac{16\angle 30^\circ\ \text{V}}{[(4+10)+j7]\ \Omega} = \frac{16\angle 30^\circ\ \text{V}}{(14+j7)\ \Omega}$$

$$= \frac{16\angle 30^\circ\ \text{V}}{15.65\angle 26.565^\circ} = 1.0\bar{2}\angle 3.43^\circ\ \text{A}$$

and $P_L = |I|^2 R_L = (1.0\bar{2})^2 \times 10 = \mathbf{10.45\ W}$ (despite higher R_L, a lower value of P_L)

The following insights emerge from a close inspection of Ex. 17-9:

1. The value of R_L obtained with Eq. (17-6) is the only value that maximizes the power in R_L under the conditions of a **fixed** value of jX_L.
2. The **lower** value of R_L results in a **higher** current but *not as high* a value of P_L as the MPT value of 10.61 W shown in (**b**).
3. The **higher** value of R_L results in a **lower** current and a **lower** value of P_L.
4. For case A (fixed X_L but variable R_L) we *no longer* have at MPT:
   a. Unity PF
   b. 50% power transfer
   c. $\mathbf{I} = \mathbf{E}/2R_i$
   d. Minimum total impedance
5. The MPT in Ex. 17-9 is not nearly as high as the conjugate match as Ex. 17-8c, but since the load is **restricted** to a **fixed** value of reactance, it is the best we can hope for under the circumstances.

17-6.2 Load Restricted Case B—Fixed R_L but Variable X_L (Reactance-Conjugate Match)

This case is the converse of the previous case A. Here R_L is fixed at some predetermined value and we have the option of variable X_L in series with it to maximize the power in R_L.[7] Since R_L is fixed, a conjugate match is rarely possible. But there *must* be *some value* of X_L that maximizes the power dissipated in R_L, given a fixed value of R_L.

It can be shown that when X_L is the conjugate of X_i, MPT to R_L is achieved. For this reason, case B is sometimes referred to as the *reactance-conjugate* match for MPT. Stated in equation form, MPT in R_L occurs when

$$jX_L = (jX_i)^* \quad \text{ohms}\ (\Omega) \qquad (17\text{-}7)$$

EXAMPLE 17-10

Given the practical source shown in Fig. 17-7 and $R_L = 8\ \Omega$ (a fixed value) calculate the

a. Value of jX_L that produces MPT to load R_L
b. Power dissipated in R_L
c. Power dissipated in R_L using a value of $jX_L = j2\ \Omega$
d. Power dissipated in R_L using a value of $jX_L = j4\ \Omega$

Solution

a. $jX_L = (jX_i)^* = (-j3)^* = \mathbf{j3\ \Omega}$ (17-7)

b. $$\mathbf{I} = \frac{\mathbf{E}}{(R_i + R_L) + j(X_i + X_L)} = \frac{16\angle 30^\circ\ \text{V}}{(4+8)+j(3-3)}$$

$$= \frac{16\angle 30^\circ\ \text{V}}{12\angle 0^\circ\ \Omega} = 1.\bar{3}\angle 30^\circ\ \text{A (unity PF condition)}$$

and $P_L = |I|^2 R_L = (1.\bar{3})^2 \times 8 = \mathbf{14.\bar{2}\ W}$

c. $$I = \frac{16\angle 30^\circ\ \text{V}}{12 + j(2-3)} = \frac{16\angle 30^\circ\ \text{V}}{12 - j1}$$

$$= \frac{16\angle 30^\circ\ \text{V}}{12.04\angle -4.76^\circ\ \Omega} = \mathbf{1.329\angle 4.76^\circ\ A}$$

and $P_L = |I|^2 R_L = (1.329)^2 \times 8 = \mathbf{14.12\ W}$ (a lower value of current and power, P_L)

d. $$I = \frac{16\angle 30^\circ\ \text{V}}{12 + j(4-3)\ \Omega} = \frac{16\angle 30^\circ}{12 + j1}$$

$$= \frac{16\angle 30^\circ\ \text{V}}{12.04\angle +4.76^\circ\ \Omega} = \mathbf{1.329\angle -4.76^\circ\ A} \text{ and}$$

$P_L = |I|^2 R_L = (1.329)^2 \times 8 = \mathbf{14.12\ W}$

[7] A typical application of case B would be a bank of lamps in parallel that are to be dimmed (using series reactance) and increased to full brightness. In this case, we are determining the value of jX_L that results in full brightness from a fixed ac supply whose voltage and internal impedance are known.

Example 17-10 shows that given a fixed value of R_L, if we select a lower or a higher value compared to the conjugate of the internal reactance, we do not dissipate as much power in R_L. This proves the validity of Eq. (17-7). Although the MPT achieved by a reactance-conjugate match is hardly as high as the MPT with an impedance-conjugate match, it is the best we can achieve, given a fixed value of R_L, which we are unable to adjust. The unity PF condition is also the MPT case.

17-6.3 Load Restricted Case C—Variable R_L Only ($X_L = 0$) (Magnitude Match)

The last restricted case possible is one where the *load reactance* is *zero*. We have the option of varying the load resistance, R_L, to the value that produces MPT in R_L from the source. It turns out that when the voltage across R_L is of the same *magnitude* as the internal volt drop across the source impedance, then MPT in R_L is achieved. In other words, when the absolute magnitude of R_L equals the magnitude of the source impedance, we achieve MPT. Such a match is called a *magnitude match* and the value of R_L for MPT is

$$R_L = |Z_i| = |R_i + jX_i| \quad \text{ohms } (\Omega) \qquad (17\text{-}8)$$

EXAMPLE 17-11

Given the practical source shown in Fig. 17-7 and $jX_L = 0$, find the

a. Value of R_L that produces MPT in R_L
b. Power dissipated in R_L
c. Power dissipated in R_L using a value of $Z_L = (4 + j0)\ \Omega$
d. Power dissipated in R_L using a value of $Z_L = (6 + j0)\ \Omega$

Solution

a. $R_L = |R_i + jX_i| = |4 - j3| = \mathbf{5\ \Omega}$ (17-8)

b. $\mathbf{I} = \dfrac{\mathbf{E}}{(R_i + R_L) + jX_i} = \dfrac{16\angle 30^\circ \text{ V}}{(4+5) - j3}$

$= \dfrac{16\angle 30^\circ \text{ V}}{(9 - j3)\ \Omega} = \dfrac{16\angle 30^\circ \text{ V}}{9.487\angle -18.435^\circ} = 1.6865 \text{ A}$

and $P_L = |I|^2 R_L = (1.6865)^2 \times 5\ \Omega = \mathbf{14.2\ W}$

c. $\mathbf{I} = \dfrac{16\angle 30^\circ \text{ V}}{[(4+4) - j3]\ \Omega} = \dfrac{16\angle 30^\circ \text{ V}}{(8 - j3)\ \Omega}$

$= \dfrac{16\angle 30^\circ \text{ V}}{8.544\angle -20.56^\circ} = 1.873 \text{ A}$

and $P = |I|^2 R_L = (1.873)^2 \times 4 = \mathbf{14.03\ W}$ (lower P_L despite higher current)

d. $\mathbf{I} = \dfrac{16\angle 30^\circ \text{ V}}{[(6+4) - j3]\ \Omega} = \dfrac{16\angle 30^\circ \text{ V}}{(10 - j3)\ \Omega}$

$= \dfrac{16\angle 30^\circ \text{ V}}{10.44\angle -16.7^\circ\ \Omega} = 1.533 \text{ A}$

and $P = |I|^2 R_L = (1.533)^2 \times 6\ \Omega = \mathbf{14.09\ W}$ (lower P_L despite higher value of R_L)

We are now in a position to compare the three load restricted cases with the *unrestricted* case (*conjugate match*) insofar as the power transferred to R_L is concerned. We may conclude that

1. The *conjugate match* produces the *highest* power transfer to R_L (see Ex. 17-8) and yields a *unity* PF *condition* with the highest current in R_L.
2. The *reactance conjugate match* (Sec. 17-6.2) produces the next highest power transfer to R_L and also yields a *unity* PF *condition* but not a 50% power transfer.
3. The *magnitude match* (Sec. 17-6.3) produces the third highest power transfer to R_L. Consequently, if a conjugate match or a reactance conjugate match is not attainable, use a magnitude match.
4. The lowest MPT may occur with case A (Sec. 17-6.1) except when the fixed value of X_L is the conjugate of X_i or is close to the conjugate of X_i.

17-7 RELATIVE POWER, VOLTAGE, AND CURRENT MEASUREMENT IN DECIBELS

The decibel (dB) originated from the need to measure and compare powers required to produce sound. It was discovered that the human ear detects and responds to sound intensity changes *logarithmically* rather than linearly; that is, a change from 50 to 100 mW or a change from 0.5 to 1 W (a ratio of 2 to 1 in each case) is perceived by the human ear as the *same* change in *intensity*, regardless of sound intensity. Later, as decibel measurement began to be used, other advantages accrued. Since the decibel is a *logarithmic* unit, multiplication and division are reduced to simple addition and subtraction. Further, use of logarithmic ratios compresses a very wide range of numbers into a *much smaller* scale, which is very convenient. Finally, in many practical engineering instances, it is inconvenient to plot power, voltage, and current on a *linear* scale, which restricts the upper and lower limits of the range. For example, a power ratio of 10 million (1×10^7) is (only) represented as 70 dB!

17-7.1 Derivation of the Bel, Power Ratio, and dB_m

The *bel* is a *nondimensional* unit, actually a ratio, used for expressing the relative ratio of two powers, P_2 and P_1. In the United States it is defined as

$$N = \log_{10}\left(\frac{P_2}{P_1}\right) \qquad \text{in bels}$$

(in other countries the *neper* is used) but 1 bel/10 = 1 decibel, abbreviated as 1 dB, or 10 dB = 1 bel

Consequently, the ratio of two powers expressed in decibels is

$$N = 10 \log_{10}\left(\frac{P_2}{P_1}\right) \qquad \text{decibels (dB)} \qquad (17\text{-}9)$$

where P_1 is the reference power against which P_2 is compared.

With respect to Eq. (17-9), please note that

1. If $P_2 = P_1$ the ratio is unity, and since $\log_{10}(1) = 0$, a unity ratio is 0 dB.
2. If $P_2 > P_1$ the ratio is $+n$ dB (this positive value in dB represents a power *gain*).
3. If $P_2 < P_1$ the ratio is $-n$ dB (this negative value in dB represents a power *loss*).

EXAMPLE 17-12

A power amplifier has an output of 1000 W for an input of 1 W. Calculate the power gain (ratio) in dB.

Solution

$N = 10 \log_{10}(1000\text{ W}/1\text{ W}) = \mathbf{+30\ dB}$ (17-9)

EXAMPLE 17-13

A filter has an output of 2 W for an input of 10 W. Calculate the power loss of the filter in dB.

Solution

$N = 10 \log_{10}(2\text{ W}/10\text{ W}) = -6.99\text{ dB} \cong \mathbf{-7\ dB}$ (17-9)

Examples 17-12 and 17-13 show that we generally select P_2 as the *output* and P_1 as the *input*, unless otherwise indicated, in Eq. (17-9). The examples also show that the sign preceding the dB value tells us whether there is power gain or power loss in the process.

In some communications measurements, $P_1 = 1$ mW, a selected reference level across a 600-Ω load. In such instances, Eq. (17-9) is modified to

$$\mathbf{dB_m} = 10 \log_{10}\left(\frac{P_2}{1\text{ mW}}\right)\Bigg|_{\Omega}^{600} \quad \text{in milliwatt decibels } (\text{dB}_\text{m}) \qquad (17\text{-}9a)$$

where P_2 is expressed in milliwatts.

17-7.2 Derivation of Voltage Gain in dB

Like power gain, it is also possible to express voltage gain (or loss) in dB. Consider a voltage $\mathbf{V}_1$ applied as an *input* across impedance $\mathbf{Z}_1$ whose resistance is R_1. This produces an *output* voltage $\mathbf{V}_2$ across output impedance $\mathbf{Z}_2$ whose resistance is R_2.

The relation between the two voltages is often expressed as *relative* voltage gain in dB, derived as

$$\mathbf{dB_{vr}} = N_{\text{rel}} = 10 \log_{10}\left(\frac{P_2}{P_1}\right) = 10 \log_{10}\left(\frac{\mathbf{V}_2^2/R_2}{\mathbf{V}_1^2/R_1}\right)$$
$$= 10 \log_{10}\left(\frac{\mathbf{V}_2}{\mathbf{V}_1}\right)^2 = 20 \log_{10}\left(\frac{\mathbf{V}_2}{\mathbf{V}_1}\right) \qquad (17\text{-}10)$$

where R_2 is the output resistance $= \mathbf{Z}_2 \cos\theta_2$
R_1 is the input resistance $= \mathbf{Z}_1 \cos\theta_1$

This derivation assumes two incorrect and often overlooked factors:

1. It assumes that both impedances, $\mathbf{Z}_1$ and $\mathbf{Z}_2$, are at the same power factors, (i.e., $\cos\theta_2 = \cos\theta_1$).
2. It assumes that both the impedances and their resistances, R_2 and R_1, are equal.

It is possible to write a *correct* expression for the *absolute* voltage gain as

$$\mathbf{dB_{va}} = N_{\text{abs}} = 20 \log_{10}\left(\frac{\mathbf{V}_2}{\mathbf{V}_1}\right) + 10 \log_{10}\left(\frac{R_2}{R_1}\right) + 10 \log_{10}\left(\frac{\cos\theta_1}{\cos\theta_2}\right) \qquad (17\text{-}11)$$

Unfortunately, Eq. (17-11) is such an unwieldy and complicated expression that the last two terms are discarded, yielding the final expression in Eq. (17-10). Consequently, even when the input and output resistances, as well as the power factors, are *unequal*, we use the following **relative voltage gain** expression of Eq. (17-10a):

$$\mathbf{dB_v} = N_{\text{rel}} = 20 \log_{10}\left(\frac{\mathbf{V}_2}{\mathbf{V}_1}\right) = 20 \log_{10}\left(\frac{V_{\text{out}}}{V_{\text{in}}}\right) \quad (\text{dB}) \qquad (17\text{-}10a)$$

EXAMPLE 17-14

An amplifier has an input of 12 mV and an output of 10 V. Calculate the relative voltage gain in dB.

Solution

$$\mathbf{dB_v} = 20\log_{10}(V_{out}/V_{in})$$
$$= 20\log_{10}(10\text{ V}/0.012\text{ V}) = \mathbf{58.42\ dB} \quad (17\text{-}10a)$$

EXAMPLE 17-15

What voltage ratio produces a 20-dB voltage gain?

Solution

$$20\log_{10}(V_{out}/V_{in}) = 20\text{ dB} \quad (17\text{-}10a)$$
$$V_{out}/V_{in} = \log_{10}^{-1}(20/20) = 10^1 = \mathbf{10}$$

17-7.3 Absolute and Relative Current Gain in dB

The equation for **absolute current gain** (or loss) expressed in dB is

$$\mathbf{dB}_{ia} = N_{abs} = 20\log_{10}\left(\frac{I_2}{I_1}\right) + 10\log_{10}\left(\frac{R_2}{R_1}\right) + 10\log_{10}\left(\frac{\cos\theta_2}{\cos\theta_1}\right)$$

But again, as in the case of absolute voltage gain, the last two terms are customarily discarded and we use **relative current gain** (or loss) expressed as

$$N_{rel} = \mathbf{dB_I} = 20\log_{10}\left(\frac{I_2}{I_1}\right) \quad \text{dB} \quad \mathbf{(17\text{-}12)}$$

EXAMPLE 17-16

A transistor has a forward current gain of 40. Express this gain in dB.

Solution

$$\text{dB}_I = 20\log_{10}(40/1) = \mathbf{32\ dB} \quad (17\text{-}12)$$

17-7.4 Relations between Power Ratios and Voltage or Current Ratios, and Their Reciprocals, in Terms of dB

Technicians and engineers who frequently use dB measurement in their work are aware of certain shortcuts and insights that are not immediately obvious. The first of these is that when a voltage ratio is determined from its dB value, the *power ratio is always the square of that voltage ratio.*

The converse is also true. When a **power ratio** is determined from its dB value, *the voltage ratio is always the square root of the power ratio.*[8]

Expressing these in equation form, we may write

$$\frac{P_2}{P_1} = \left(\frac{V_2}{V_1}\right)^2 \quad \text{(dimensionless)} \quad \mathbf{(17\text{-}13)}$$

and

$$\frac{V_2}{V_1} = \left(\frac{P_2}{P_1}\right)^{1/2} \quad \text{(dimensionless)} \quad \mathbf{(17\text{-}13a)}$$

[8] This should be fairly obvious to the reader since power varies as the square of the voltage, and consequently voltage varies as the square root of power.

EXAMPLE 17-17
A power amplifier has a power gain of 6 dB. Calculate its ratio of

a. Output to input power, P_2/P_1
b. Output to input voltage, V_2/V_1

Solution

a. $P_2/P_1 = \log_{10}^{-1}(6/10) = 3.981 \cong \mathbf{4}$ (17-9)
b. $V_2/V_1 = \sqrt{P_2/P_1} = \sqrt{4} = \mathbf{2}$ (17-13a)

A second important insight is that the +3 dB level implies that the *power* has *doubled* and the −3 dB power point is called the **half-power point** (i.e., the power has *halved*.) Similarly (from Eq. 17-13a), in terms of voltage, +3 dB implies a voltage ratio of $\sqrt{2}$ or 1.414 and −3 dB implies a voltage ratio of $1/\sqrt{2} = 0.7071$. (We will discover later that these *half-power points* are important in the definition of *bandwidth* in *resonant* circuits, as well as *break-points* in *filter circuits*).

A third important insight is contained within the previous paragraph: the *same positive* and *negative* dB values represent the *reciprocals* of the current ratios, the voltage ratios and the power ratios, respectively.

EXAMPLE 17-18
Given power gains of +20 dB and −20 dB, respectively, calculate the

a. Respective power ratios, P_2/P_1
b. Respective voltage ratios, V_2/V_1

Solution

a. $P_2/P_1 = \log_{10}^{-1}(20/10) = \mathbf{100}$ for +20 dB power gain (17-13)

$P_2/P_1 = \log_{10}^{-1}(-20/10) = \mathbf{0.01}$ or $\mathbf{1/100}$ for −20 dB power loss (17-13)

b. $V_2/V_1 = \sqrt{P_2/P_1} = \sqrt{100} = \mathbf{10}$ for +20 dB power gain (17-13a)

$V_2/V_1 = \sqrt{P_2/P_1} = \sqrt{0.01} = \mathbf{0.1}$ or $\mathbf{1/10}$ for −20 dB power loss (17-13a)

A fourth and final insight also emerges from the second insight above:

For each **3 dB** increase, *power is doubled*, and for each decrease of **−3 dB**, *power is halved*.

For each **6 dB** increase, *voltage* is *doubled*, and for each decrease of **−6 dB**, *voltage is halved*.

EXAMPLE 17-19
Given +6 dB, +3 dB, 0 dB, −3 dB, and −6 dB, calculate and tabulate the

a. Corresponding voltage ratios
b. Corresponding power ratios

Solution
(Calculations left as an exercise for the reader)

	Gain or loss, in dB	+6	+3	0	−3	−6
a.	Voltage ratio, V_2/V_1	2	1.414	1	0.7071	0.5
b.	Power ratio, P_2/P_1	4	2	1	0.5	0.25

Using the tabulation of Ex. 17-19, the reader should verify each of the four major insights presented above, namely

1. The power ratio *is always the square* of the voltage ratio (and its converse).
2. The −3 dB level is the *half-power* ratio and the +3 dB level is a doubling of the power.
3. The same positive and negative dB values represent the reciprocals of the voltage ratios and power ratios, respectively.
4. a. For each 3 dB *increase*, power is *doubled*. For each −3 dB decrease, power is *halved*.
   b. For each 6 dB *increase*, voltage is *doubled*. For each −6 dB *decrease*, voltage is *halved*.

17-8 EFFECTIVE RESISTANCE

Up to now we have *assumed* that of the three circuit properties (resistance, inductance, and capacitance), the only one that is *unaffected by frequency* is resistance. We did this, in part, in order to concentrate our attention on the circuit properties of inductive and capacitive reactance in various network combinations. Unfortunately, this assumption is *not* a valid one.

Recall that resistance is defined as the (only) circuit property that is capable of dissipating electrical energy when subjected to an electric current, either dc or ac. Consequently, any conditions that lead to *losses* of energy in an electrical circuit will be recorded by a wattmeter (P) for a given effective current (I), enabling the calculation of effective or ac resistance as

$$R_{ac} = R_{eff} = \frac{P}{I^2} \quad \text{ohms } (\Omega) \qquad (17\text{-}14)$$

where P is the power in watts recorded by a wattmeter for a circuit
I is the circuit ac current drawn from a supply and measured by the wattmeter

Up to now, as far as we were concerned, the only resistance factors we knew involved *ohmic* or dc resistance, that is, those covered in Sec. 3-4, where $R = \rho l/A$. Much research enabled physicists and engineers to isolate the various factors that cause a marked *increase* in *effective resistance* (as measured by Eq. 17-14) with *frequency* increases. The following eight factors comprise the effective resistance of an electric circuit, depending on the frequency of the current, the surrounding circuitry, and the adjacent materials. Some of these have been introduced earlier. They are

1. Ohmic (dc) resistance (Sec. 3-4).
2. Eddy current loss (Sec. 11-13.1).
3. Radiation loss (Sec. 17-8.3).
4. Dielectric leakage loss (Sec. 9-6.1).
5. Dielectric hysteresis loss (Sec. 9-3).
6. Magnetic hysteresis loss (Sec. 11-13.2).
7. Skin effect (Sec. 17-8.7).
8. Temperature coefficient effects (Sec. 3-6).

For conductors and materials having positive temperature coefficients of resistance, *each of these factors tends to increase losses* and consequently increase effective resistance as the frequency increases. Let us briefly consider each of these in turn.

17-8.1 Ohmic Resistance

As noted above, ohmic resistance depends on $R_{dc} = \rho l/A$. If the cross-sectional area of a conductor is decreased or its length and resistivity increased due to temperature increases (produced by losses), the ohmic resistance increases. The major factor tending to decrease cross-sectional area is the *skin effect* (Sec. 17-8.7).

17-8.2 Eddy Current Loss

Eddy (or Foucault) currents were described in detail in Sec. 11-13.1. In brief, eddy currents result from voltages induced in the conductors or in adjacent conductors surrounding the electrical circuit due to variations in magnetic flux produced by ac magnetic fields. The I^2R losses that result represent an eddy current power loss, which must be supplied by the source and is recorded by the wattmeter in Eq. (17-14).

17-8.3 Radiation Loss

This loss explains why the effective (radiation) resistance of an antenna may be as high as several hundred ohms while its ohmic (dc) resistance may be as low as 0.1 Ω. The difference is due to the electrical energy supplied to the antenna that is converted into electromagnetic (infrared) light energy. Electromagnetic waves radiated into space are *not* returned to the supply. This loss supplied by the source is recorded by the wattmeter in Eq. (17-14).

17-8.4 Dielectric Leakage Loss

Dielectric leakage was introduced in Sec. 9-6.1. Since no dielectric is a perfect insulator, small leakage currents in the dielectric materials comprising the insulation of a conductor or adjacent to it produce I^2R losses with application of ac electrical fields across them. These losses must be supplied by the source and are recorded by the wattmeter in Eq. (17-14).

17-8.5 Dielectric Hysteresis Loss

Dielectric hysteresis was introduced toward the end of Sec. 9-3. Briefly, it constitutes the energy required to stress the dielectric atoms that are subjected to rapidly reversing alternating electric fields. The time rate at which electric energy is transformed into heat in a dielectric material subjected to a changing electric field is a measure of the dielectric hysteresis loss. Like the previous losses, these too must be supplied by the source and are recorded by the wattmeter in Eq. (17-14).

17-8.6 Magnetic Hysteresis Loss

Magnetic hysteresis was discussed in detail in Sec. 11-13.2. It represents the energy expended in carrying any ferromagnetic material through its complete cycle of magnetization. Consequently, any ferromagnetic material in the vicinity of the electrical circuit will be subjected to hysteresis and heated as a result. This heat loss must be supplied by the source and is recorded by the wattmeter in Eq. (17-14).

17-8.7 Skin Effect

Eddy currents in current-carrying conductors, at right angles to the *axial* current flow, result in and produce a nonuniform current distribution in the conductor. Since the eddy current density is *greatest* at the *center* of the conductor, the *axial* ac currents are forced away from the center and are confined to the outer surface of the conductor. This reduction in cross section results in an increase in ohmic resistance (see Sec. 17-8.1) and consequently in effective resistance.

17-8.8 Temperature Coefficient Effects

Almost all of the foregoing factors, in producing losses, convert electrical energy to heat energy. The resultant heat energy produces an increase in temperature. In Sec. 3-6 we saw how an increase in temperature results in an increase in resistance for all conductors having a *positive* temperature coefficient of resistance.

One final note summarizes all of this discussion. Apart from research studies, it is customary never to attempt to *separate* these losses. Instead, we use Eq.

(17-14), lumping all the losses recorded by the wattmeter–ammeter method of measurement as effective resistance, R_{ac} or R_{eff}. Throughout our discussion, therefore, whenever R is written, it implies ac or effective resistance and *not* dc or ohmic resistance, unless otherwise stated! Consequently, the resistance R that appears in Sec. 14-7.1 and elsewhere, wherever power is recorded by a wattmeter, *represents ac* or *effective resistance.*

17-9 THE POWER TRIANGLE AND TELLEGEN'S THEOREM REVISITED FOR COMPLEX POWER

The power triangle and complex power were first introduced in Sec. 16-4.7, where complex power, $\mathbf{S}^*$, was defined as

$$\mathbf{S}^* = \mathbf{VI}^* = \mathbf{S}\underline{/\theta} = (\mathbf{V}\underline{/\psi})(\mathbf{I}\underline{/-\phi}) = \mathbf{VI}\underline{/\psi - \phi} = \mathbf{VI}\underline{/\theta} = P \pm jQ \quad \text{(VA)} \tag{17-15}$$

where $\mathbf{I}^*$ is the conjugate of the current phasor $I\underline{/\phi}$
$\mathbf{V}$ is the voltage phasor $V\underline{/\psi}$
$\theta = \psi - \phi$

Equation (17-15) contains a number of implications that may not be obvious to the reader:

1. Reactive power Q is *positive* when $\psi > \phi$ (i.e., when phase angle θ is *positive.*) This occurs when current is lagging voltage, and the circuit is INDUCTIVE.
2. Reactive power Q is *negative* whenever $\phi > \psi$ (i.e., when phase angle θ is *negative.*) This occurs when current is leading voltage, and the circuit is CAPACITIVE.
3. This agrees with the selection of $+jQ$ for quadrature reactive power of an *inductive* circuit and $-jQ$ for quadrature reactive power of a *capacitive* circuit.
4. But Eq. (17-15) also suggests both a graphical and empirical method of obtaining the total P, Q, and S^* of several loads in parallel, since $\cos\theta_T = P_T/S_T^*$, where P_T is the sum of the individual average powers of the loads, Q_T is the phasor sum of inductive loads ($+jQ$'s) and capacitive loads ($-jQ$'s), and S_T^* is $P_T \pm jQ_T$.

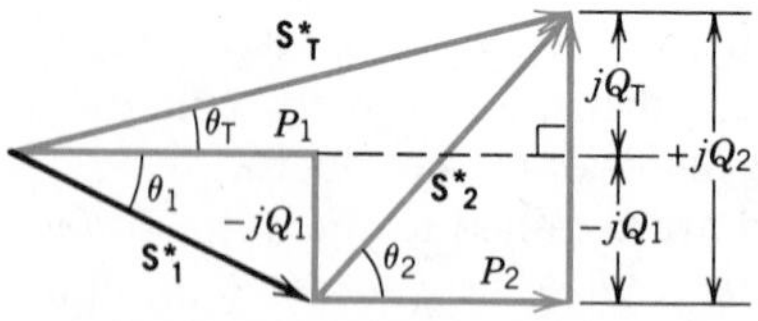

Figure 17-8 Power triangle for two loads in parallel

Figure 17-8 shows the case of two ac loads in parallel. S_1^* is a capacitive load and S_2^* an inductive load. Regardless of the number of loads, we may conclude that

1. Total power is the sum of the individual powers (shown along the *real* axis):

$$P_T = P_1 + P_2 + P_3 + \cdots + P_n \qquad \text{watts (W)} \tag{17-16}$$

2. Total quadrature reactive power ($\pm jQ_T$) is the phasor sum of the individual quadrature reactive powers, where $+jQ$ is inductive and $-jQ$ is capacitive (shown along the imaginary axis), or

$$\pm jQ_T = j(Q_1 \pm Q_2 \pm Q_3 \pm \cdots \pm Q_n) \qquad \text{reactive volt-amperes (vars)} \tag{17-17}$$

3. Complex powers, $\mathbf{S}_1^*$, $\mathbf{S}_2^*$, and so forth, are *always added in rectangular* form. As shown in Fig. 17-8, the only way $\mathbf{S}_1^*$ may be added to $\mathbf{S}_2^*$ is by complex (rectangular) form addition:

$$\mathbf{S}_T^* = \mathbf{S}_1^* + \mathbf{S}_2^* + \mathbf{S}_3^* + \cdots + S_n^* \qquad \text{volt-amperes (VA)}$$

$$(P_T \pm jQ_T) = (P_1 \pm jQ_1) + (P_2 \pm jQ_2) + (P_3 \pm jQ_3) + \cdots + (P_n \pm jQ_n) \qquad \text{(VA)} \qquad (17\text{-}18)$$

And if *all* complex powers in a closed electrical system are summed algebraically, we obtain Tellegen's theorem (originally stated) in Eq. (16-26), which may be modified here as

$$\mathbf{S}_T^* + \mathbf{S}_1^* + \mathbf{S}_2^* + \mathbf{S}_3^* + \cdots + \mathbf{S}_n^* = 0 \qquad \text{volt-amperes (VA)} \qquad (17\text{-}19)$$

where $\mathbf{S}_T^*$ is the only source and $\mathbf{S}_1^*$ through $\mathbf{S}_n^*$ are complex impedance loads

Table 17-2 summarizes the "rules" established in Sec. 16-8 for using Tellegen's theorem with ac sources and loads.

Table 17-2 Summary of Complex Power Polarities for Use with Tellegen's Theorem

Circuit	Complex Power, $\mathbf{S}^* = \mathbf{VI}^*$
 Complex load, motor action	Complex power is positive (current first meets + polarity) P is +, load absorbs real power If Q is +, load absorbs inductive reactive power (**I** lags **V**) If Q is (−), load absorbs capacitive power (**I** leads **V**)
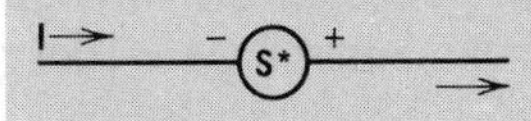 Complex source, generator action	Complex power is negative (current first meets (−) polarity) P is −, real power is supplied by source to the system Q is −, source supplies *inductive* reactive power (**I** lags **V**) Q is +, source supplies *capacitive* reactive power (**I** leads **V**)

EXAMPLE 17-20

Two ideal voltage sources are connected as shown in **Fig. 17-9**; $\mathbf{V}_1 = 50\angle 0^\circ$ V, $\mathbf{V}_2 = 50\angle 30^\circ$ V, and $\mathbf{Z}_3 = 2\angle 90^\circ\ \Omega$. Calculate

a. The current **I**
b. $\mathbf{S}_1^*$
c. $\mathbf{S}_2^*$
d. $\mathbf{S}_3^*$ in rectangular form
Explain which blocks are sources and which are loads, as well as the nature of load in each
e. Verify the accuracy of your results using Tellegen's theorem

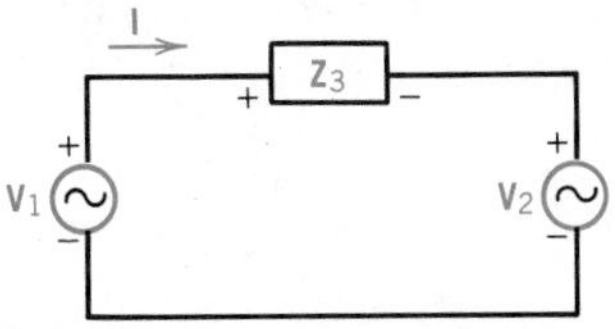

Figure 17-9 Example 17-20

Solution

a. For the given (assumed) direction of current shown in Fig. 17-9,

$$\mathbf{I} = \frac{\mathbf{V}_1 - \mathbf{V}_2}{\mathbf{Z}_3} = \frac{(50 + j0) - (43.3 + j25)}{0 + j2}$$

$$= \frac{6.7 - j25}{0 + j2} = -12.5 - j3.35 = \mathbf{12.94\angle{-165^\circ}\ A}$$

b. $-\mathbf{S}_1^* = \mathbf{V}_1\mathbf{I}^* = 50\angle 0^\circ(12.94\angle -165^\circ)^* = -647\angle 165^\circ$
$= \mathbf{(625 - j167.5)\ VA}$ ($-\mathbf{S}_1^*$ is *not* a source!)

c. $\mathbf{S}_2^* = \mathbf{V}_2\mathbf{I}^* = 50\angle 30^\circ(12.94\angle -165^\circ)^* = 647\angle 195^\circ$
$= \mathbf{(-625 - j167.5)\ VA}$ ($\mathbf{S}_2^*$ is a source!)

d. $\mathbf{S}_3^* = |I|^2 Z_3 = (12.94)^2 \times 2\angle 90^\circ = \mathbf{+j335\ vars}$ (inductive reactive load). $\mathbf{S}_1^*$ and $\mathbf{S}_3^*$ are loads and only $\mathbf{S}_2^*$ is a source. Source $\mathbf{S}_2^*$ supplies real (negative) power to the system and 167.5 vars of quadrature (negative) inductive reactive power to the system. Load $\mathbf{S}_1^*$ absorbs 625 W of real power and 167.5 vars of capacitive reactive power. Load $\mathbf{S}_3^*$ ($\mathbf{Z}_3$) absorbs no real power but stores 335 vars of inductive reactive power.

e. $\sum\mathbf{S}^* = (625 - j167.5) + (-625 - j167.5) + (0 + j335)$
$= \mathbf{0 + j0}$

The extreme importance of Tellegen's theorem is verified since it tells us that our mathematics is correct, as well as the conclusions drawn in (**d**) above. Let us review the "rules" for applying Tellegen's theorem to Ex. 17-20:

1. $\mathbf{S}_1^*$ must be written as a source $-\mathbf{S}_1^*$ since the assumed direction of current first encounters the negative polarity of source $\mathbf{V}_1$, as shown in (**b**).
2. $\mathbf{S}_2^*$ and $\mathbf{S}_3^*$ are treated as loads because positive polarity is first encountered.
3. But $\mathbf{S}_1^*$ turns out to be a load because P_1 is positive power! Similarly, $\mathbf{S}_2^*$ turns out to be a source because P_2 is negative power supplied to the system.
4. No power is absorbed by $\mathbf{S}_3^*$, which only stores inductive reactive power.
5. If the sum of all sources and loads is not zero, we have made a mathematical error. The work should be recalculated on a separate sheet of paper to find the error.

17-10 POWER FACTOR CORRECTION

Loads having moderate to low lagging or leading power factors (below 0.65) result in a severe loss of electrical power to the utility supplying power to a given industrial or commercial occupancy. Lower power factors require the utility to increase their capacity or apparent power (**S**, in volt-amperes) in order to supply a *higher current* for the *lower* power factor loads. This added capacity (and higher current) is needed all along the line, from the generating station, through the transformers and the transmission lines, to the load. The cost of this additional added capacity is kept to a minimum by means of *power factor correction.*

Other advantages also emerge from power factor correction:

1. Since the capacity and line current are both lower, the power losses (I^2R) in the lines are reduced.
2. Similarly, the line volt drop across the line impedance is reduced, making the voltage regulation task easier in maintaining rated voltage to occupancies supplied by the utility.
3. Transmission efficiency from source to load is increased.
4. Utility costs are decreased, reflecting a savings (theoretically) to the consumer.

Almost all commercial, industrial, and residential loads tend to have *lagging* power factors (i.e., current lags voltage) due to inductive reactive loads (motors, fluorescent lights, etc.). Consequently, power factor correction consists of adding capacitive loads in parallel with existing inductive loads to raise the power factor. As we shall see, this has the effect of *reducing* all of the following:

1. The total quadrature inductive reactive power, $+jQ_T$.
2. The total apparent complex power, $\mathbf{S}_T^*$.
3. The total current supplied to the entire system, $\mathbf{I}_T$.

17-10.1 Power Factor (Correction) Improvement Methods

Power factor correction (improvement) consists of ways of *raising* the PF. All PF improvement methods consist of connecting devices across the lines (in parallel with existing inductive loads, $+jQ$) to draw leading currents from the lines and negative quadrature reactive power ($-jQ$). Three types of devices are used commercially:

a. Correction capacitors These are large high-voltage, high-capacity commercial capacitors, connected across the lines of single-phase and three-phase systems. Correction capacitors are rated in *both* kVA (or kvars) and voltage (kV).

b. Synchronous capacitors These are *overexcited* synchronous motors (designed without shaft extensions so that they cannot be coupled to mechanical loads) intended to "float" on three-phase (or single-phase) lines to draw only leading current from the supply for power factor correction. They are rated in both kVA and kV, similar to correction capacitors.

c. Synchronous motors These are constant-speed motors that, when *overexcited*, are capable of driving mechanical loads and simultaneously drawing leading currents from the supply. If not coupled to a load, synchronous motors may be used as synchronous capacitors. In the latter application, almost all their kVA capacity is represented as $-jQ$ correction kvars.

17-10.2 Practical (Economic) Limit to Power Factor Improvement

Ideally, a unity PF load draws the least current from the supply for a given quantity of active power consumed (drawn) from the supply and represents the lowest possible load, **S**, in volt-amperes.

In commercial practice, however, low (lagging) PF loads are rarely, if ever, corrected to unity PF. Instead, they are corrected to approximately 0.85 PF, lagging. We will see that it is uneconomical to correct lagging loads above 0.85 PF because of the relatively high kvars of correction needed in proportion to the actual kW gained (for which the consumer pays) for a system having a constant kVA capacity or load.

17-10.3 Solving PF Correction Problems

At least three possible approaches can be used in solving PF correction problems. These are (1) the admittance (or impedance) approach using the *impedance* triangle, (2) the quadrature-current approach using *current* triangles, and (3) the *complex power* approach using the *power* triangle.

In order to avoid confusion and because the power triangle method is simplest, we will present only the last of the three methods. The complex power triangle approach to PF correction has the following advantages:

1. Since it deals only with power (watts), reactive power (vars), and apparent power (in VA), it may be used *equally* with single-phase or three-phase circuits *without any special changes.*
2. The quadrature var and in-phase watt computations are made independently of line (or phase) voltage.
3. It lends itself to a *power tabulation grid*, as shown in Fig. 17-10b, which provides a *systematic solution-summary* method.

EXAMPLE 17-21

A (polyphase) alternator has a rating of 1000 kVA and supplies its entire capacity to a load at a PF of 0.6 lagging. Calculate the

a. Active (true) power originally dissipated by the load, P_0
b. Inductive reactive quadrature power drawn from and returned to the supply, $+jQ_0$
c. Draw a power triangle showing the *original* load conditions of the system

Solution

a. $P_0 = S^* \cos\theta = 1000\text{ kVA} \times 0.6 =$ **600 kW**
b. $+jQ_0 = S^* \sin\theta = 1000\text{ kVA} \times 0.8 =$ **+*j*800 kvars**
c. The original power triangle is shown in **Fig. 17-10a**

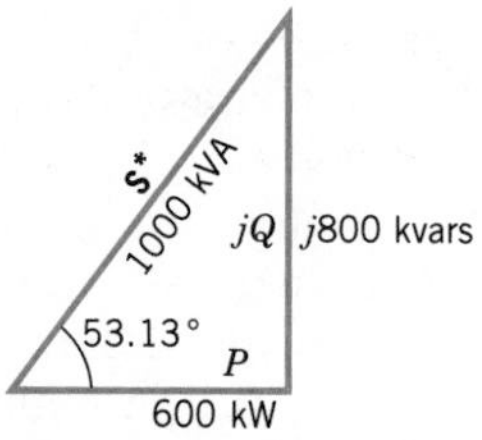

a. Original power triangle

	P (kW)	$\pm jQ$ (kvar)	S* (kVA)	lagging $\cos\theta$
ORIGINAL	600	$+j800$	1000	0.6
ADDED	c 200	d $-j200$	—	—
FINAL	a 800	b $+j600$	a 1000	a 0.8

b. Power tabulation grid

Figure 17-10 Examples 17-21 and 17-22

EXAMPLE 17-22

It is desired to correct the power factor of the system in Ex. 17-21 to 0.8 lagging, in order to supply more active power to consumers, without overloading the alternator (maintaining the same rating of 1000 kVA). Calculate the

a. Final active power supplied by the alternator, P_f
b. Reactive power stored and returned to the supply, $+jQ_f$
c. Additional active power that may be supplied to new consumers, P_a
d. Correction kvars required to raise PF from 0.6 to 0.8 lagging, $-jQ_a$
e. Rating of correction capacitors needed to accomplish above correction (in kVA)
f. Draw a power tabulation grid, summarizing the above calculations

Solution

a. $P_f = S^* \cos\theta_f = 1000\text{ kVA} \times 0.8 =$ **800 kW**
b. $+jQ_f = S^* \sin\theta_f = 1000\text{ kVA} \times 0.6 =$ **+*j*600 kvars**
c. $P_a = P_f - P_0 = 800\text{ kW} - 600\text{ kW} =$ **200 kW**
d. $jQ_a = jQ_f - jQ_0 = j600 - j800 =$ **−*j*200 kvars**
e. $\mathbf{S}_c^* = 0 - j200\text{ kVA} =$ **200 kVA**
f. The power tabulation grid is shown in **Fig. 17-10b**

The following insights emerge from Exs. 17-21 and 17-22 and the grid of Fig. 17-10b:

1. Active power, original and added, is (vertically) summed *arithmetically* to produce final power.
2. Quadrature reactive power is summed *vectorially* with regard to whether it is inductive (positive) or capacitive (negative).
3. In using the power grid, all summations involve the *P*s and *Q*s only; the remaining columns are used only for work entries as part of the solution.
4. Observe that the letters in the corners of the grid spaces correspond to the solution steps in Ex. 17-22. This is done so that the reader can determine the order of the solution steps for the various entries.

EXAMPLE 17-23

Repeat all parts of Ex. 17-22 but instead correct the PF from 0.6 lagging to unity PF

Solution

a. $P_f = S\cos\theta_f = 1000\text{ kVA} \times 1.0 =$ **1000 kW**
b. $jQ_f = S\sin\theta_f = 1000\text{ kVA} \times 0 =$ **0 kvars**
c. $P_a = P_f - P_0 = 1000\text{ kW} - 600\text{ kW} =$ **400 kW**
d. $jQ_a = jQ_f - jQ_0 = 0 - j800 =$ **−*j*800 kvars**
e. $\mathbf{S}_c^* = 0 - j800 =$ **800 kVA**
f.

	P (kW)	Q (kvar)	S^* (kVA)	$\cos\theta$ —
ORIG	600	$+j800$	1000	0.6
ADDED	400	$-j800$	—	—
FINAL	1000	0	1000	1.0

Example 17-23 proves the point raised in Sec. 17-10.2 regarding the economic limit to PF correction. In raising the PF from 0.6 to 0.8, as shown in Fig. 17-10b, we required 200 kvars of capacitive correction to gain 200 kW of additional power. But in going from 0.8 PF to unity, although we gain an *additional* 200 kW of power, it requires $-j800-(-j200)$ kvars or **600 kvars** of capacitive correction. This is three times as much capacitive correction as going from 0.6 to 0.8 PF. For the amount of true power gained, the full correction to unity PF is very uneconomical.

Up to now, the examples used only correction capacitors, which are relatively lossless in comparison to their added correction kvars. Let us consider an example involving a synchronous capacitor, which produces some losses in addition to correction.

EXAMPLE 17-24

A factory draws a 0.6 PF lagging load of 2000 kW from a utility. A synchronous capacitor is purchased to raise the final PF to 0.85 lagging. The synchronous capacitor has losses of 275 kW. Calculate the

a. Original kVA load drawn from the utility, S_0^*
b. Original lagging kvars, jQ_0
c. Final system active power consumed from the utility, P_f
d. Final kVA load drawn from the utility, S_f^*
e. Correction kvars produced by the synchronous capacitor, $-jQ_a$
f. kVA rating of the synchronous capacitor, S_a^*
g. Show the complete power tabulation grid with ALL cells filled

Solution

a. $S_0^* = P_0/\cos\theta_0 = 2000\text{ kW}/0.6 = \mathbf{3333.\bar{3}\ kVA}$
b. $Q_0 = S_0^*\sin\theta_0 = 3333.\bar{3}\times 0.8 = \mathbf{j2667\ kvars}$
c. $P_f = P_0 + P_a = 2000 + 275 = \mathbf{2275\ kW}$
d. $S_f^* = P_f/\cos\theta_f = 2275/0.85 = \mathbf{2676.5\angle 31.8^\circ\ kVA}$
e. $jQ_f = S_f^*\sin\theta_f = 2676.5\times 0.5268 = j1410$ kvars; $-jQ_a = jQ_f - jQ_0 = j1410 - j2667 = \mathbf{-j1257\ kvars}$
f. $S_a^* = P - jQ = 275 - j1257 = \mathbf{1287\angle -77.66^\circ\ kVA}$ ($\cos -77.6^\circ = 0.214$ leading)

g.

	P (kW)	Q (kvar)	S^* (kVA)	$\cos\theta$ —
ORIG	2000	$j2667$[b]	$3333.\bar{3}$[a]	0.6 lag
ADDED	275	$-j1257$[e]	1287[f]	0.214 lead[f]
FINAL	2275[c]	$j1410$[e]	2676.5[d]	0.85 lag[d]

Several important points emerge from Ex. 17-24 and its associated power tabulation grid:

1. The original factory kVA load has been reduced from $333\bar{3}$ to 2676.5 kVA as a consequence of the PF improvement from 0.6 to 0.85 lagging.
2. The kVA reduction permits the utility to supply other consumers without having to add generating capacity, in kVA. (See Ex. 17-25.)
3. The kVA rating and the power factor of the synchronous capacitor, found in step (**f**), can only be found from the real and quadrature reactive power for the added capacitor.
4. The **lettered** solution **steps** are shown in the complete power tabulation (**g**) to assist the reader in following the sequence of the solution. It is imperative that the reader attempt the solution independently, on a separate sheet, to verify the solution and use it as feedback to correct errors along the way.

Now for a somewhat more difficult example, which reveals the advantages of the power tabulation grid in enabling us to systematize our thinking and pose the problem for solution in a straightforward manner.

The utility justified the added expense of a synchronous capacitor in Ex. 17-24 because it reduced the kVA factory load from $3333.\bar{3}$ to 2676.5 kVA. This frees the utility to add additional consumer load (say to 0.8 PF lagging) without having to add any additional generating capacity. How much additional load can be added and what is the final PF of the combination?

EXAMPLE 17-25

The original load on a $3333.\bar{3}$ kVA alternator is a factory whose load, $S_0^* = (2275 + j1410)$ kVA. Draw a power tabulation grid and calculate the

a. Additional kVA load that may be added at a PF of 0.8 lagging, without overloading the alternator
b. Added active power and reactive power of the additional load
c. Final active power and reactive power supplied by the $3333.\bar{3}$ kVA alternator
d. Final power factor of the alternator

Solution

a. The given information is shown in the accompanying power tabulation grid. Assume that x is the additional kVA load. Then the additional real and quadrature powers are $0.8x$ and $j0.6x$, respectively, as shown. Adding each column vertically and using the Pythagorean theorem, we may write $(2275 + 0.8x)^2 + (1410 + 0.6x)^2 = (3333.\bar{3})^2$, and solving this equation yields the quadratic $x^2 + 5352x - 3947163 = 0$. Applying the quadratic formula yields the added kVA load, $x = S_a^* =$ **658.86 kVA**

	P (kW)	Q (kvar)	S^* (kVA)	$\cos\theta$
ORIG	2275	$j1410$	2676.5	0.85 lag
ADDED	$0.8x$	$j0.6x$	x	0.8 lag
FINAL	$(2275 + 0.8x)$	$j(1410 + 0.6x)$	$3333.\bar{3}$	0.841 lag

b. $P_a = 0.8x = 0.8 \times 658.86$ kVA $=$ **527.1 kW** and $Q_a = 0.6 \times 658.86$ kVA $=$ ***j*395.32 kvar**

c. $P_f = P_0 + P_a = 2275 + 527.1 =$ **2802.1 kW** and $Q_f = Q_0 + Q_a = j1410 + j395.32$ kvar $=$ ***j*1805.3 kvar**

d. $F_p = \cos\theta_f = P_f/S_f^* = 2802.1\text{ kW}/3333.\bar{3}\text{ kVA} =$ **0.841 lagging**.

Validity check: $S_f^* = P_f + jQ_f = 2802.1 + j1805.3 = 3333.\bar{3}\angle 32.8^\circ$ and $\cos 32.8^\circ =$ **0.841**

Example 17-25 proves the advantages of PF correction:

1. By improving PF we have reduced the alternator capacity needed and allowed more load to be added without overloading the alternator (in this case $3333.\bar{3}$ kVA).
2. The added load (of approximately 659 kVA) at an average PF of 0.8 lagging does not materially reduce the final PF (in this case 0.841 lagging, which is still high). We know that the solution is correct because of the *validity check* in step (**d**).

Failing this test means that our answers are incorrect. We may also test the answer by using Tellegen's theorem, as shown by Ex. 17-26.

EXAMPLE 17-26

From the solution to Ex. 17-25 we have $\mathbf{S}_0^* = 2676.5\angle 31.79^\circ$ kVA, $\mathbf{S}_a^* = 658.86\angle 36.87^\circ$ kVA, and $-\mathbf{S}_f^* = -(3333.\bar{3}\angle 32.8^\circ)$ kVA. Show that this solution is correct, using Tellegen's theorem. (Hint: use complex addition of apparent powers.)

Solution

$\mathbf{S}_0^* + \mathbf{S}_a^* + (-\mathbf{S}_f^*) = 0 = 2676.5\angle 31.79^\circ + 658.86\angle 36.87^\circ - (3333.\bar{3}\angle 32.8^\circ) = (2275 + j1410) + (527.1 + j395.32) - (2802.1 + j1805.32) = 0$

17-11 GLOSSARY OF TERMS USED

Active (true) power, P_t Power delivered from source to load that is converted from electrical to other energy forms (heat, light, mechanical, chemical, etc.) and that cannot be returned to the source.

Apparent power, S *Scalar* quantity equal to the magnitude of the product of line voltage and line current (in a single-phase circuit).

Complex power, $\mathbf{S}^*$ *Phasor* quantity equal to the phasor product $\mathbf{VI}^*$, which may be expressed in the form $S\angle\theta = P \pm jQ$.

Conjugate match Condition in which the unrestricted load, $\mathbf{Z}_L$, is the conjugate of the internal source impedance, so that maximum power is drawn from an ac voltage or current source.

Correction capacitors High-voltage, high-capacity commercial capacitors, connected across lines of single-phase and multiphase systems, used for power factor correction.

Decibel Unit for designating and measuring the ratio of relative power, voltage, or current gain or loss, equal to one-tenth of a bel.

Effective resistance Total ac resistance covering all losses: eddy current, iron, dielectric and corona (radiation), and transformed power losses as well as conductor ohmic losses. Measured in ohms as the ratio of the power consumed to the absolute value of the current squared.

Half-power points Condition of a resonant system (electrical, mechanical, acoustical, etc.) where the amplitude response is reduced to $1/\sqrt{2}$ of maximum, that is, by 3 dB.

Lagging (PF) load Reactive load whose (phase) current *lags* its supply (phase) voltage.

Lagging vars (kvars) Quadrature (reactive) component of power produced by a lagging inductive load, designated as $+jQ_L$.

Leading (PF) load Reactive load whose (phase) current *leads* its supply (phase) voltage.

Leading vars (kvars) Quadrature (reactive) component of power produced by a leading capacitive load, designated as $-jQ_C$.

Magnitude match Condition in which a restricted load, $\mathbf{R}_L$, having zero reactance, is adjusted to the magnitude of the internal source impedance, $\mathbf{Z}_I$, so that maximum power is drawn by $\mathbf{R}_L$.

Maximum power transfer theorem Given a load $\mathbf{Z}_L$ connected to a practical ac voltage or current source, if the load is *unrestricted*, maximum power is transferred from source to load when the load, $\mathbf{Z}_L$, is the conjugate of the internal source impedance, $\mathbf{Z}_I$.

Millman's theorem Given any number of practical voltage sources and/or current sources *in parallel*, the voltage across all practical sources, converted to current sources, is the phasor sum of the source currents divided by the phasor sum of the source admittances.

Norton current *Short-circuit* current measured at any two terminals of a linear ac (or dc) network containing one or more active dependent (or independent) voltage or current sources. Measurement of Norton current by a zero-impedance ammeter is called the *short-circuit test*.

Norton impedance Internal impedance (or admittance) paralleling a Norton current source. Mathematically, $\mathbf{Z}_n$ is the same as $\mathbf{Z}_{th}$, the Thévenin impedance for a given linear two-terminal network, found by taking the Thévenin voltage, $\mathbf{V}_{th}$, divided by the Norton current, $\mathbf{I}_N$.

Norton's theorem Any combination of linear circuit elements and active ac sources, regardless of connection or complexity, may be replaced by an equivalent two-terminal network consisting of a single ac current source, $\mathbf{I}_N$, paralleled by an equivalent impedance, $\mathbf{Z}_N$, or equivalent admittance, $\mathbf{Y}_N$.

Open-circuit test Determination of the Thévenin voltage, $\mathbf{V}_{th}$, across any two nodes of interest, as measured by an electronic voltmeter (EVM) having infinite internal impedance.

Power factor (PF) Dimensionless ratio of active (true) power, P, to apparent power, S.

Power tabulation grid Tabulation showing (in *rows*) the original, added, and final conditions that occur prior to and after power factor correction. Vertical *columns* usually show active power (P), quadrature power (Q), apparent power (S), and power factor ($\cos\theta$ or F_p). Optional columns may show $\sin\theta$, $\tan\theta$, or θ.

Practical current source A two-terminal independent supply having finite admittance, $\mathbf{Y}_N$, shunting an ideal current source, $\mathbf{I}_N$, whose Thévenin voltage is $\mathbf{I}_N/\mathbf{Y}_N$ and whose terminal voltage varies with load current drawn by some external load, $\mathbf{Z}_L$.

Practical voltage source An independent voltage source having finite internal impedance, $\mathbf{Z}_{th}$, and open-circuit voltage, $\mathbf{V}_{th}$, whose terminal voltage varies with load current drawn by some external load, $\mathbf{Z}_L$.

Quadrature (reactive) power, $\pm jQ$ Power delivered from source to load that is (temporarily) stored in the reactive elements of the load and is returned to the supply.

Reactance-conjugate match For a restricted load, $\mathbf{Z}_L$, having fixed load resistance, $\mathbf{R}_L$, the condition in which the load reactance is the conjugate of the internal source impedance reactance, results in maximum power drawn by the load.

Short-circuit test Determination of the short-circuit (Norton) current, $\mathbf{I}_N$, that flows when an ideal (zero-impedance) ammeter is connected across two terminals.

Source-transformation theorem Given a complex active network containing one or more practical voltage and/or current sources, a given practical voltage (or current) source may be transformed into an equivalent practical current (or voltage) source or vice versa by using the source transformation $\mathbf{V}_{th} = \mathbf{I}_n\mathbf{Z}_n$ or $\mathbf{I}_n = \mathbf{V}_{th}/\mathbf{Z}_{th}$.

Superposition theorem In any multisource network consisting of linear elements and sources at the same or different frequencies, the voltage across or current through a given element of the network is equal to the complex RMS sum of the individual voltages across or currents in that element by each source acting independently.

Synchronous capacitors Overexcited synchronous motors (designed without shaft extensions so that they cannot be coupled to mechanical loads) intended to "float" on a multiphase or single-phase line and draw only leading (capacitive) current from the supply for the purpose of PF correction.

Synchronous motors Constant-speed motors that, when overexcited, are capable of driving mechanical loads and simultaneous drawing leading (capacitive) current from a supply. If not coupled to mechanical load, synchronous motors may be used as synchronous capacitors.

Thévenin current See *Norton current*.

Thévenin impedance Thévenin voltage (open-circuit voltage) divided by Thévenin current (short-circuit current); this is mathematically equal to the Thévenin and/or Norton impedance.

Thévenin's theorem The current that flows in an external impedance, $\mathbf{Z}'$, connected to any two terminals of a complex ac linear network (between which there previously existed a Thévenin voltage, $\mathbf{V}_{th}$, and a Thévenin impedance, $\mathbf{Z}_{th}$) is equal to the Thévenin voltage divided by the complex sum of $\mathbf{Z}'$ and $\mathbf{Z}_{th}$.

Thévenin voltage Open-circuit voltage measured at any two terminals of a complex ac (or dc) linear network containing one or more active dependent or independent voltage (or current) sources. Measurement of the Thévenin voltage by an electronic voltmeter (EVM) is called the *open-circuit test*.

Triangle, power Triangle showing the relation (in quadrature) between active and reactive power, apparent power, and phase angle θ. In the power triangle, leading (capacitive) vars are shown on the negative j axis and lagging (inductive) vars on the positive j axis.

Two-terminal element Device or complex circuit with two (different) points to which electrical connections may be made.

17-12 PROBLEMS

Secs. 17-1 to 17-2

17-1 In the network shown in **Fig. 17-11**, $\mathbf{V}_1 = 10\angle 0^\circ$ V, $\mathbf{V}_2 = 12\angle 30^\circ$ V, $\mathbf{V}_3 = 15\angle 60^\circ$ V, $\mathbf{Z}_1 = 13\angle 67.38^\circ\ \Omega$, $\mathbf{Z}_2 = 25\angle 73.74^\circ\ \Omega$, and $\mathbf{Z}_3 = 41\angle -77.32^\circ\ \Omega$. Calculate in polar form the

a. Equivalent Thévenin voltage, $\mathbf{V}_{TH}$, with due care for source polarity
b. Equivalent Thévenin impedance, $\mathbf{Z}_{TH}$
c. Draw the simplest practical voltage source that replaces the network

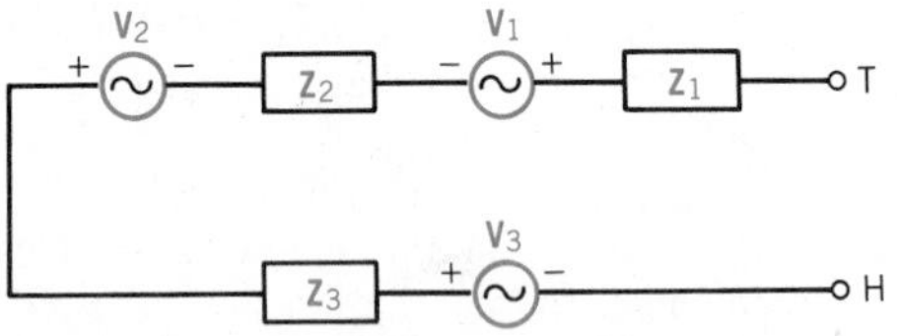

Figure 17-11 Problems 17-1 and 17-2

17-2 Repeat all parts of Prob. 17-1, given $\mathbf{V}_1 = (26 + j15)$ V, $\mathbf{V}_2 = (8 - j6)$ V, $\mathbf{V}_3 = (5 - j12)$ V, $\mathbf{Z}_1 = (3 + j4)\ \Omega$, $\mathbf{Z}_2 = (4 - j5)\ \Omega$, and $\mathbf{Z}_3 = (6 + j8)\ \Omega$.

17-3 Given the data of Ex. 17-5 and the network of Fig. 17-4, calculate

a. $\mathbf{V}_{TH}$ with $\mathbf{Z}_3$ removed, using the voltage divider rule applied to two sources (Eq. 5-8)
b. $\mathbf{Z}_{TH}$ by shorting all voltage sources
c. Replace $\mathbf{Z}_3$ and find current $\mathbf{I}_3$ (compare this value with that found in Ex. 17-5c by superposition and Ex. 17-7b by Millman's theorem)

17-4 In the network shown in **Fig. 17-12**, $\mathbf{V}_1 = 20\angle 0^\circ$ V, $\mathbf{Z}_1 = 8 - j15\ \Omega$, and $\mathbf{Z}_2 = 12 + j5\ \Omega$. With load $\mathbf{Z}_L$ removed, calculate in polar form

a. $\mathbf{V}_{TH}$
b. $\mathbf{I}_{SC}$ (or $\mathbf{I}_N$) when terminals T–H are shorted
c. $\mathbf{Z}_{TH}$, using the relation $\mathbf{V}_{TH}/\mathbf{I}_{SC}$
d. $\mathbf{Z}_{TH}$ from equivalent impedance $\mathbf{Z}_1 \| \mathbf{Z}_2$
e. Draw the single practical voltage source to replace the network seen by $\mathbf{Z}_L$

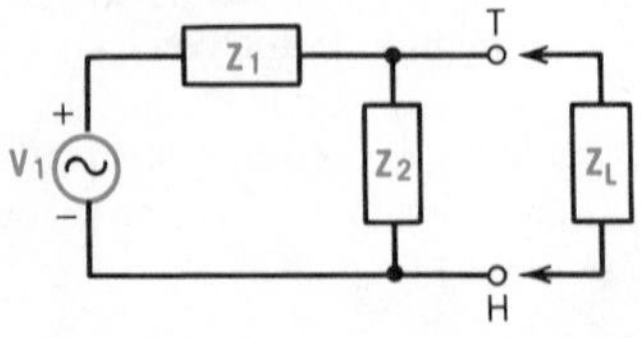

Figure 17-12 Problems 17-4 to 17-7

17-5 Given $\mathbf{Z}_L = 10\angle 45^\circ\ \Omega$ in Prob. 17-4, calculate the

a. Current drawn by $\mathbf{Z}_L$ from the network, $\mathbf{I}_L$
b. Voltage across $\mathbf{Z}_L$, designated as load voltage $\mathbf{V}_l$
c. Power dissipated in $\mathbf{Z}_L$ by two methods

17-6 Repeat all parts of Prob. 17-4 for $\mathbf{V}_1 = 49\angle 0^\circ$ V, $\mathbf{Z}_1 = 34\angle 28.07^\circ\ \Omega$, $\mathbf{Z}_2 = 17\angle 61.93^\circ\ \Omega$, and $\mathbf{Z}_L$ removed.

17-7 Repeat all parts of Prob. 17-5 for $\mathbf{Z}_L = (6 - j8)\ \Omega$ in Prob. 17-6.

17-8 In the network shown in **Fig. 7-13**, $\mathbf{V}_1 = 48\angle 0^\circ$ V, $\mathbf{V}_2 = 12\angle 0^\circ$ V, $\mathbf{Z}_1 = 60\angle 60^\circ$ kΩ, $\mathbf{Z}_2 = 40\angle 40^\circ$ kΩ, and $\mathbf{Z}_L = 16\angle -20^\circ$ kΩ. Calculate in polar form

a. $\mathbf{V}_{TH}$ with $\mathbf{Z}_L$ removed, using the voltage divider rule applied to two sources (Eq. 5-8)
b. $\mathbf{Z}_{TH}$, shorting voltage sources and using $\mathbf{Z}_1 \| \mathbf{Z}_2$
c. Replace $\mathbf{Z}_L$ and find current $\mathbf{I}_L$ and voltage $\mathbf{V}_L$ across all three branches of the network

Sec. 17-3

17-9 Repeat Prob. 17-8 by performing the following steps:

a. Convert all voltage sources into current sources using the source-conversion theorem. (Hint: watch current source polarities and redraw the network appropriately.)
b. Using the current divider rule, find $\mathbf{I}_L$ (current in $\mathbf{Z}_L$)
c. Calculate $\mathbf{V}_L$, the voltage across all branches of the network in Fig. 17-13.

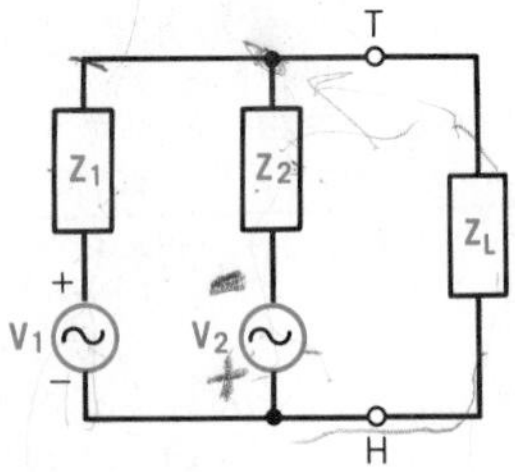

Figure 17-13 Problems 17-8 to 17-11, 17-22, 17-23, 17-27, and 17-28

17-10 In the network shown in Fig. 17-13, $\mathbf{V}_1 = (33.941 - j33.941)$ V, $\mathbf{V}_2 = (60 + j0)$ V, $\mathbf{Z}_1 = \mathbf{Z}_2 = (100 + j0)\ \Omega$, and $\mathbf{Z}_L = (0 - j40)\ \Omega$. Calculate

a. $\mathbf{V}_{TH}$ with $\mathbf{Z}_L$ removed, using any method you prefer
b. $\mathbf{Z}_{TH}$, using any method you prefer
c. Replace $\mathbf{Z}_L$ and find current $\mathbf{I}_L$ and voltage $\mathbf{V}_L$ across all three branches of the network

17-11 Repeat Prob. 17-10 by performing the following steps:

a. Convert all voltage sources into current sources, using the source-conversion theorem (Hint: watch current source polarities and redraw the network appropriately.)
b. Using the CDR appropriately, find $\mathbf{I}_L$ (current in $\mathbf{Z}_L$).
c. Calculate $\mathbf{V}_L$, the voltage across all branches of the network in Fig. 17-13.

17-12 In the network shown in **Fig. 17-14**, $\mathbf{V}_1 = (4 + j0)$ V, $\mathbf{V}_2 = (12 + j0)$ V, $\mathbf{Z}_1 = (0 + j4)\ \Omega$, $\mathbf{Z}_2 = (0 - j2)\ \Omega$, and $\mathbf{Z}_L = (3 + j0)\ \Omega$. Calculate, using any method you prefer,

a. $\mathbf{V}_{TH}$ with $\mathbf{Z}_L$ removed (check your work carefully because this will surprise you)

b. $\mathbf{Z}_{TH}$ with $\mathbf{Z}_L$ removed
c. $\mathbf{I}_L$, current in $\mathbf{Z}_L$, and voltage $\mathbf{V}_L$ with $\mathbf{Z}_L$ restored
d. $\mathbf{P}_L$, active power dissipated in $\mathbf{Z}_L$, by two methods

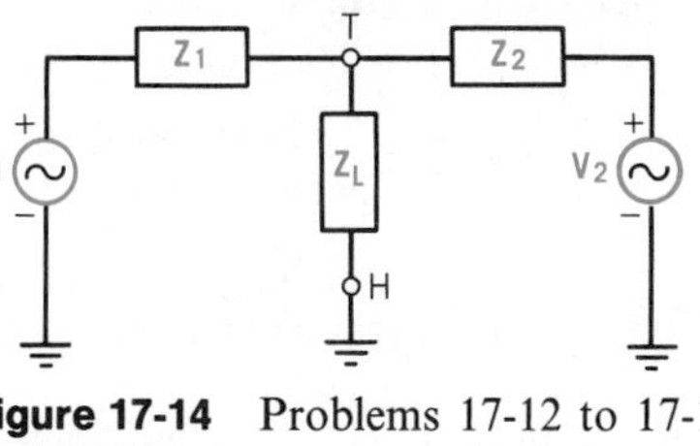

Figure 17-14 Problems 17-12 to 17-17, 17-24 to 17-26, 17-29, and 17-30

17-13 In the network shown in Fig. 17-14, $\mathbf{V}_1 = 50\angle 0^\circ$ V, $\mathbf{V}_2 = 36.06\angle 56.31^\circ$ V, $\mathbf{Z}_1 = 3.606\angle 56.31^\circ\ \Omega$, $\mathbf{Z}_2 = 4.472\angle -63.43^\circ\ \Omega$, and $\mathbf{Z}_L = 10\angle 0^\circ\ \Omega$. Repeat all parts of Prob. 17-12.

17-14 In the network shown in Fig. 17-14, $\mathbf{V}_1 = 100\angle 0^\circ$ V, $\mathbf{V}_2 = 100\angle 90^\circ$ V, $\mathbf{Z}_1 = 10\angle 90^\circ$ kΩ, $\mathbf{Z}_2 = 5\angle 90^\circ$ kΩ, and $\mathbf{Z}_L = 20\angle -90^\circ$ kΩ. Repeat all parts of Prob. 17-12.

17-15 Given the node voltage $\mathbf{V}_N = \mathbf{V}_L = 12\angle 53.13^\circ$ V across all three branches in Prob. 17-12,

a. Calculate all three branch currents
b. Show that the phasor sum of all three branch currents obeys KCL (the sum of the currents entering the node minus the currents leaving the node is zero)

17-16 Given the node voltage $\mathbf{V}_N = \mathbf{V}_L = 16.48\angle -45.9^\circ$ V across all three branches in Prob. 17-13,

a. Calculate all three branch currents
b. Show that the phasor sum of all three branch currents is zero, using KCL

17-17 Given the node voltage $\mathbf{V}_N = \mathbf{V}_L = 89.44\angle 63.435^\circ$ V across all three branches in Prob. 17-14,

a. Calculate all three branch currents
b. Show that the phasor sum of all three branch currents is zero, using KCL

17-18 In the network shown in **Fig. 17-15**, $\mathbf{V}_1 = 10\angle 0^\circ$ V, $\mathbf{V}_2 = 13\angle 0^\circ$ V, $\mathbf{Z}_1 = 3\angle 0^\circ\ \Omega$, $\mathbf{Z}_2 = 5\angle 0^\circ\ \Omega$, $\mathbf{Z}_3 = 4\angle 90^\circ\ \Omega$, $\mathbf{Z}_4 = 12\angle -90^\circ\ \Omega$, and $\mathbf{Z}_5 = 3.835\angle 5^\circ\ \Omega$. Find

a. $\mathbf{V}_{TG}$ and $\mathbf{Z}_{TG}$ with $\mathbf{Z}_5$ removed
b. $\mathbf{V}_{HG}$ and $\mathbf{Z}_{HG}$ with $\mathbf{Z}_5$ removed
c. $\mathbf{I}_{Z_5}$ and $\mathbf{V}_{TH}$ with $\mathbf{Z}_5$ restored. (Hint: draw the equivalent series circuit. Also observe current direction to distinguish between $\mathbf{V}_{HT}$ and $\mathbf{V}_{TH}$.)

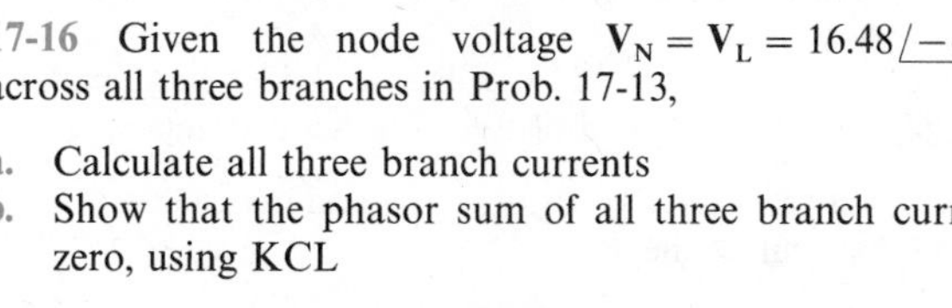

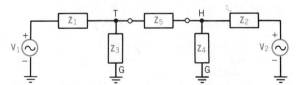

Figure 17-15 Problems 17-18 and 17-19

17-19 Repeat all parts of Prob. 17-18 for $\mathbf{V}_1 = (12 + j35)$ V, $\mathbf{V}_2 = (15 + j8)$ V; $\mathbf{Z}_1 = (3 + j4)\ \Omega$, $\mathbf{Z}_2 = (4 - j3)\ \Omega$, $\mathbf{Z}_3 = (12 - j5)\ \Omega$, $\mathbf{Z}_4 = (5 - j12)\ \Omega$, and $\mathbf{Z}_5 = (1.81 + j6.2)\ \Omega$ in Fig. 17-15.

17-20 An ac capacitance bridge, shown in **Fig. 17-16**, is used to measure unknown capacitance C_x. Whenever the bridge is perfectly balanced, the electronic voltmeter (EVM) reads zero, and the impedance ratios are found similarly to those of an ordinary Wheatstone bridge. But what if the bridge is unbalanced? Can we find the unbalanced current in the EVM and voltage across it? For the unbalanced condition, the values are $\mathbf{V}\angle\theta = 12\angle 0^\circ$ V, $R_s = 3\angle 0^\circ$ kΩ, $R_f = 4\angle 0^\circ$ kΩ, $-jX_{Cs} = -j6$ kΩ, and $-jX_{Cx} = -j5$ kΩ. If the EVM has an impedance $\mathbf{Z} = (11.16 + j3.151)$ kΩ, calculate the

a. Potential at point T with EVM removed
b. Potential at point H with EVM removed
c. Potential difference, V_{TH}, with EVM removed (which terminal is more positive?)
d. Equivalent impedance seen between terminals T–H, with EVM removed
e. Current in EVM with EVM restored
f. True potential difference across EVM when EVM is connected across terminals T–H

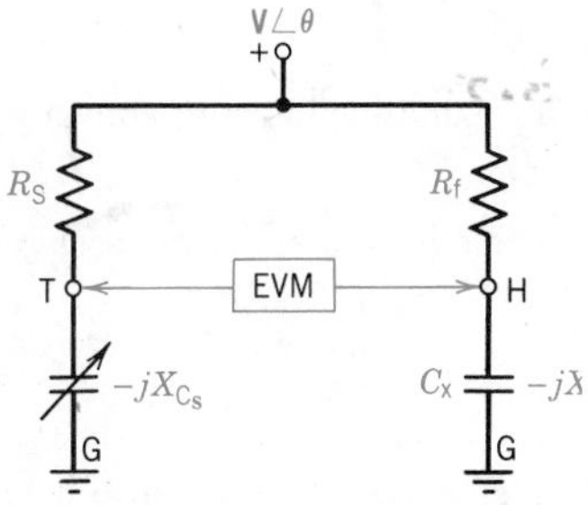

Figure 17-16 Problems 17-20 and 17-21

17-21 Repeat all parts of Prob. 17-20 for $\mathbf{V}\angle\theta = 20\angle 0^\circ$ V, $R_s = 4\angle 0^\circ$ kΩ, $R_f = 6\angle 0^\circ$ kΩ, $-jX_{Cs} = -j3$ kΩ, $-jX_{Cx} = -j4$ kΩ, and an EVM impedance of $10/53.13^\circ$ kΩ.

Secs. 17-4 and 17-5

17-22 Using the superposition theorem, given the data of Prob. 17-8, find the

a. Current, $\mathbf{I}_L$, in $\mathbf{Z}_L$
b. Node voltage across all three branches in parallel

17-23 Using the superposition theorem, given the data of Prob. 17-10, find the

a. Current, $\mathbf{I}_L$, in $\mathbf{Z}_L$
b. Node voltage across all three branches in parallel

17-24 Using the superposition theorem, given the data of Prob. 17-12, find the

a. Current, $\mathbf{I}_L$, in $\mathbf{Z}_L$
b. Node voltage across all three branches in parallel

17-25 Using the superposition theorem, given the data of Prob. 17-13 find the

a. Current, $\mathbf{I}_L$, in $\mathbf{Z}_L$
b. Node voltage across all three branches in parallel

17-26 Given the circuit source polarities shown in Fig. 17-14, $\mathbf{V}_1 = 10\angle 50^\circ$ V at 100 Hz and $\mathbf{V}_2 = 8\angle 20^\circ$ V at 200 Hz. $\mathbf{Z}_1$ is a pure 0.5-H inductor and $\mathbf{Z}_2$ a pure 10-μF capacitor. $\mathbf{Z}_L = 100\angle 0^\circ\ \Omega$. Calculate by superposition the

a. Current $\mathbf{I}'_L$ due to $\mathbf{V}_1$ acting alone
b. Current $\mathbf{I}''_L$ due to $\mathbf{V}_2$ acting alone
c. Actual current in $\mathbf{Z}_L$ as the phasor sum of $\mathbf{I}'_L$ and $\mathbf{I}''_L$

17-27 Given the data of Prob. 17-8, using Millman's theorem, calculate the

a. Node voltage across all three branches in parallel
b. Current, $\mathbf{I}_L$, in $\mathbf{Z}_L$
c. Current delivered to (or received from) the node in practical sources $\mathbf{V}_1$ and $\mathbf{V}_2$
d. Phasor sum of the three branch currents to verify KCL

17-28 Repeat all parts of Prob. 17-27 for the data of Prob. 17-10.
17-29 Repeat all parts of Prob. 17-27 for the data of Prob. 17-12.
17-30 Repeat all parts of Prob. 17-27 for the data of Prob. 17-13.

Sec. 17-6

17-31 Given the data of Prob. 17-4 and the network shown in Fig. 17-12, calculate the

a. Single practical voltage source to replace the network seen by $\mathbf{Z}_L$ (if Probs. 17-4 and 17-5 have not been solved)
b. Value of unrestricted load, $\mathbf{Z}_L$, to draw maximum power from the network seen by $\mathbf{Z}_L$
c. Power dissipated in $\mathbf{Z}_L$ by two methods
d. Compare power dissipated in (**c**) versus power in Prob. 17-5 when $\mathbf{Z}_L = 10\angle 45^\circ\ \Omega$

17-32 Given the data of Prob. 17-6 and the network shown in Fig. 17-12, calculate the

a. Single practical voltage source to replace the network seen by $\mathbf{Z}_L$ (if Probs. 17-6 and 17-7 have not been solved)
b. Value of unrestricted load, $\mathbf{Z}_L$, to draw maximum power from the network seen by $\mathbf{Z}_L$
c. Power dissipated in $\mathbf{Z}_L$, by two methods
d. Compare the power dissipated in (**c**) with the power in Prob. 17-7 when $\mathbf{Z}_L = (6 - j8)\ \Omega$

17-33 Given the data of Prob. 17-10 and the network shown in Fig. 17-13, calculate the

a. Single practical voltage source to replace the network seen by $\mathbf{Z}_L$ (if Prob. 17-10 has not been solved)
b. Value of unrestricted load, $\mathbf{Z}_L$, to draw maximum power from the network seen by $\mathbf{Z}_L$ in (**a**)
c. Power dissipated in $\mathbf{Z}_L$, by two methods
d. Compare the power dissipated in (**c**) with the power in Prob. 17-10 when $\mathbf{Z}_L = (0 - j40)\ \Omega$

17-34 Given the data of Prob. 17-13 and the network shown in Fig. 17-14, calculate the

a. Single practical voltage source to replace the network seen by $\mathbf{Z}_L$ (if Prob. 17-13 has not been solved)
b. Value of unrestricted load, $\mathbf{Z}_L$, to draw maximum power from the network seen by $\mathbf{Z}_L$ in (**a**)
c. Power dissipated in $\mathbf{Z}_L$, by two methods
d. Compare the power dissipated in (**c**) with the power in Prob. 17-13 when $\mathbf{Z}_L = 10 + j0\ \Omega$

17-35 Given a practical voltage source whose open-circuit voltage is $24\angle 0^\circ$ V and whose internal impedance is $(6 + j8)\ \Omega$, calculate the

a. Value of unrestricted load, $\mathbf{Z}_L$, that draws maximum power from the source
b. Maximum power dissipated in $\mathbf{Z}_L$ under any conditions

Sec. 17-7

17-36 Given the source of Prob. 17-35 and a fixed load reactance of $-jX_C = 6\ \Omega$, calculate the

a. Value of $\mathbf{R}_L$ that draws maximum power from the source
b. Maximum power dissipated in $\mathbf{R}_L$ for the above restricted case (of fixed load reactance)
c. dB loss relative to the MPT of Prob. 17-35

17-37 Given the source of Prob. 17-35 and a fixed load resistance of 8 Ω, calculate the

a. Value of jX_L that produces MPT to load R_L and value of $\mathbf{Z}_L$ in rectangular form
b. Power dissipated in $\mathbf{R}_L$ for the above restricted case of fixed load resistance
c. dB loss relative to the MPT of Prob. 17-35, using reactance conjugate match instead of true conjugate match

17-38 Given the source of Prob. 17-35 and a variable resistive load (only), calculate the

a. Value of R_L that draws maximum power from the source, using a magnitude match
b. Maximum power dissipated in R_L for the magnitude match
c. dB loss relative to the MPT of Prob. 17-35, using magnitude match instead of true conjugate match

17-39 Convert the following ratios of gains (or losses) to dB:

a. Voltage gain of 50
b. Current loss ratio of 0.3
c. Power gain of 150
d. Power loss ratio of 0.2

17-40 Convert the following gains (or losses) in dB to ratios:

a. Voltage gain of 34 dB
b. Voltage loss of −10 dB
c. Power gain of 3 dB
d. Power loss of −6 dB

17-41 The input power to a device is 10 kW at a voltage of 1 kV. The output power is 500 W across an output impedance of $20\angle 0^\circ\ \Omega$. Calculate the

a. Absolute power gain or loss in dB
b. Relative voltage gain or loss in dB
c. Input resistance
d. Explain why the voltage loss ratio in (**b**) is not the same as (**a**)

17-42 An amplifier has an input of 12 mV and an output of 25 W across a 4-Ω resistor. Calculate the

a. Output voltage
b. Voltage gain in dB
c. Power gain in dB if the input power is 25 mW

17-43 An amplifier rated at 40 W output is connected to a 10-Ω speaker. The amplifier power gain is 25 dB and its voltage gain is 40 dB. Calculate the

a. Input power needed for full (rated) power output
b. Input voltage needed for full (rated) voltage output

17-44 Complete the following chart by filling in *all* blanks.

Gain or Loss (dB)	Voltage Ratio (V_0/V_{in})	Power Ratio (P_0/P_{in})
		1
	1.414	
−3		
	2	
		0.25
	3.162	
−10		
+20		
	0.1	
		1000
	0.03162	

17-45 Using the table you completed in Prob. 17-44, explain

a. The relation that always exists between the voltage ratio and the power ratio
b. The relation that always exists between the same positive and negative dB values and their corresponding voltage ratios
c. Repeat (**b**) for the corresponding power ratios

17-46 Current, voltage, and power data taken at 60 Hz for a hollow-core inductor are shown below for air, copper and two iron cores. Complete the table by calculating in ohms (for each core material)

a. Effective resistance
b. Coil impedance
c. Coil reactance

Assume all cores have equal diameters.

	Core Material	Air	Solid Copper	Solid Iron	Laminated Iron
	V (volts)	120	120	120	120
	I (amperes)	3	2.96	1.32	1.33
	P (watts)	45	70.3	21.0	19.46
a.	R (Ω)				
b.	Z_L (Ω)				
c.	jX_L (Ω)				

17-47 With respect to the table completed in Prob. 17-46, explain in detail:

a. The increase in effective resistance with a copper core and a solid iron core, respectively
b. The decrease in effective resistance with a laminated iron core compared to a solid iron core
c. Why the power drawn from the supply is higher for a copper core than for air
d. Why the power is lower for solid iron and laminated iron than for air, despite the fact that the effective resistance has increased
e. Why the reactance for a solid copper core is the same as that for air
f. Why the reactance for solid and laminated iron is higher than that for air or copper
g. Why the reactance of laminated iron is slightly lower than that for solid iron

17-48 A laboratory coil draws a current of 1.2 A when 60 V dc is applied across it. The same air-core coil draws 1.9 A when 120 V, 60 Hz is applied and dissipates a power of 183 W. Calculate the

a. Ohmic resistance
b. Effective resistance
c. Impedance
d. Inductive reactance
e. Inductance of the coil

17-49 The coil of Prob. 17-48 is then tested by the voltmeter–ammeter–wattmeter method with three different core materials and the data are tabulated below. Complete the table by calculating for each core material the

a. Effective resistance
b. Impedance
c. Inductive reactance
d. Inductance

Core Material	Coil Voltage at 60 Hz (V)	Coil Current (A)	Power (W)	R_{eff} (Ω)	Z_L (Ω)	jX_L (Ω)	L (H)
Aluminum	120	1.51	159.5				
Copper	120	1.46	155.7				
Iron	120	0.2	4.0				

17-50 With respect to the table completed in Prob. 17-49 and solutions of Prob. 17-48, explain in detail

a. Why the effective resistance of the air coil is higher than its ohmic resistance
b. Why the effective resistance of the aluminum core is higher than the air-core coil
c. Why the effective resistance of the copper core is higher than the aluminum core
d. Why the effective resistance of the iron core is highest despite the smallest power
e. Why the inductive reactance of the iron is considerably higher than that of the other cores
f. Why the iron core produces the highest impedance of all measurements

Sec. 17-9

17-51 Given the four loads shown in **Fig. 17-17**, corresponding voltages and currents are $\mathbf{V}_1 = (0 + j100)$ V, $\mathbf{V}_2 = (100 - j100)$ V, $\mathbf{V}_3 = (42.2 - j7.339)$ V, $\mathbf{V}_4 = (57.8 + j7.339)$ V, $\mathbf{I}_A = (3 + j2)$ A, and $\mathbf{I}_B = (3 - j10)$ A. Calculate the

a. Complex powers $\mathbf{S}_1^*$ through $\mathbf{S}_4^*$, labeling the nature of the load in each
b. Total current drawn from supply, **V**, and complex power of the source, **S***, with due regard for polarity
c. Phasor sum of all five complex powers in the system to show that Tellegen's theorem is verified (if it is not, recheck your work to find your error)

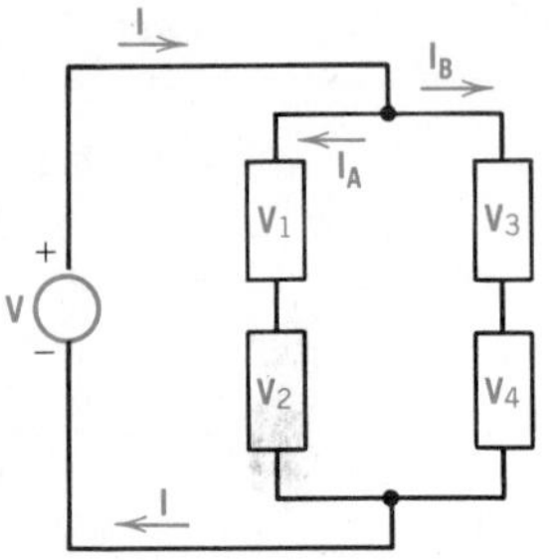

Figure 17-17 Problems 17-51 and 17-52

17-52 Given the four loads shown in Fig. 17-17, $\mathbf{S}_1^* = 500 - j200$ (capacitive), $\mathbf{S}_2^* = 300 + j800$ (inductive), $\mathbf{S}_3^* = 200 + j500$ (inductive), and $\mathbf{S}_4^* = 100 - j100$ (capacitive) volt-amperes. The total voltage $\mathbf{V} = 100\underline{/0^\circ}$ V. Calculate the

a. Conjugate current, $\mathbf{I}_A^*$, and current, $\mathbf{I}_A$, in branch A
b. Conjugate current, $\mathbf{I}_B^*$, and current, $\mathbf{I}_B$, in branch B
c. Total current drawn from the source, **I**
d. Complex power of the source, **S***, with due regard for polarity
e. The phasor sum of all five complex powers in the system, verifying Tellegen's theorem

17-53 Given the four loads shown in Fig. **17-18** and the voltages and currents $\mathbf{V} = (120 + j0)$ V, $\mathbf{I} = (10 + j4)$ A, $\mathbf{I}_B = (8 - j1)$ A, $\mathbf{I}_C = (4 - j4)$ A, and $\mathbf{V}_{CDG} = (15 + j6)$ V, calculate the

a. Voltage across load B, $\mathbf{V}_B$
b. Currents in loads A ($\mathbf{I}_A$) and D ($\mathbf{I}_D$)
c. Complex powers $\mathbf{S}_A^*$ through $\mathbf{S}_D^*$, labeling the nature of the load in each
d. Complex power of the source, **S***, with due regard for polarity
e. Phasor sum of all five complex powers in the system to show that Tellegen's theorem is verified

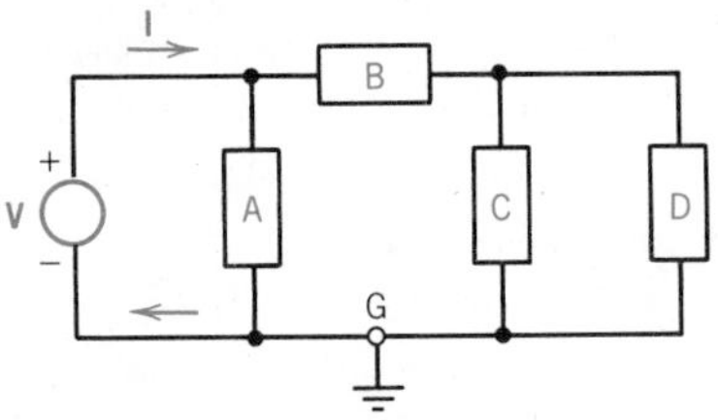

Figure 17-18 Problems 17-53 and 17-54

17-54 Given the four loads shown in Fig. 17-18, the applied voltage is $50\underline{/0^\circ}$ V, $\mathbf{S}_A^* = (500 - j200)$ VA, $\mathbf{S}_B^* = (300 + j800)$ VA, $\mathbf{S}_C^* = (200 + j500)$ VA, and $\mathbf{S}_D^* = (100 - j100)$ VA. Calculate the

a. Total current, **I**, drawn from the supply and the nature of the total load
b. Current, $\mathbf{I}_A$, in load A
c. Current, $\mathbf{I}_B$, in load B
d. Current, $\mathbf{I}_C$, in load C
e. Current, $\mathbf{I}_D$, in load D
f. Current $\mathbf{I}_B$, using $\mathbf{I}_B = \mathbf{I}_C + \mathbf{I}_D$
g. Complex power, **S***, of the source with due regard for polarity
h. Phasor sum of all five complex powers in the system, verifying Tellegen's theorem

Sec. 17-10

17-55 The following loads are paralleled across 208-V single-phase mains: Load A is a 5-kW oven at unity PF. Load B is an 8-kVA motor at a PF of 0.75 lagging. Load C is an overexcited synchronous motor, 10 kW at 0.9 PF leading. Draw a complete power tabulation grid showing these loads and calculate the

a. Total active power drawn from the supply, $\mathbf{P}_T$
b. Total quadrature power drawn from the supply, $\mathbf{Q}_T$
c. Total kVA of the system, $\mathbf{S}_T^*$ (final complex power)
d. Total system power factor, F_p

17-56 A 50 000 kVA three-phase alternator is loaded to rated capacity at 0.7 PF lagging. A synchronous capacitor operating at 0.1 leading PF corrects the system PF to unity. Draw a power tabulation grid and calculate the

a. Original load power, P_o, in kilowatts
b. Original system quadrature power, Q_o
c. Added corrective quadrature power, Q_a, and added active power, P_a
d. kVA rating of the synchronous capacitor, $\mathbf{S}_a^*$, needed to perform correction
e. Final active power, quadrature power, and system kVA, $\mathbf{S}_f^*$ (complex power)
f. Additional load in kilowatts that can be added to the system at unity PF without overloading the alternator

17-57 Calculate the kVA rating of a (lossless) synchronous capacitor needed to raise the PF of a 50-kW load from 0.65 lagging to 0.85 lagging. Draw a power tabulation grid and find in turn (in order to solve the problem)

a. $\mathbf{Q}_0$, original quadrature reactive power
b. $\mathbf{Q}_f$, final quadrature reactive power
c. $\mathbf{Q}_a$, added quadrature correction needed
d. $\mathbf{S}_a^*$, kVA rating of the synchronous capacitor, operating at 0 PF

17-58 The PF of a single-phase load connected to a 120 V, 60 Hz source is raised from 0.6 lagging to 0.8 lagging when a 100-μF lossless capacitor (rated at 200 V dc) is connected across the line. Calculate (letting x = original load power):

a. Reactance of the capacitor, $-jX_C$
b. Leading vars produced by the capacitor, $-jQ_C$
c. Original (and final) active power consumed by the load ($P_0 = P_f$)
d. Original reactive vars of the load (jQ_0)
e. Final reactive vars of the load
f. Final kVA rating of load ($\mathbf{S}_f^*$) or final complex power

17-59 A three-phase synchronous motor added in parallel to the load on a transmission line raises the PF of the line from 0.6 to 0.9 lagging. If the input to the synchronous motor is 200 kVA at 0.8 PF leading, calculate (letting $x = \mathbf{S}_0^*$ in kVA) the

a. Original load, $\mathbf{S}_0^*$, in kVA
b. Final active power consumed, P_f
c. Final quadrature reactive power, jQ_f
d. Final load in kVA, $\mathbf{S}_f^*$ (final complex power)
e. Horsepower of the added motor if its efficiency is 90%

17-60 An alternator supplies a three-phase load consisting of several small induction motors having an average PF of 0.6 lagging. A 500-hp, 90% efficiency synchronous motor having a PF of 0.8 leading is added, raising the system PF to 0.8 lagging. (Hint: let $x = \mathbf{S}_0$ in kVA.) Calculate the

a. Active input power drawn from the lines by the synchronous motor, P_a
b. Correction kvars produced by the synchronous motor, Q_m
c. Original kVA of the induction motors, $\mathbf{S}_0^*$
d. Final active power consumed from the lines, P_f
e. Final quadrature power, jQ_f
f. Final system kVA, $\mathbf{S}_f^*$ (final complex power)

17-61 A 500-kVA three-phase alternator is loaded to capacity with a lagging load of 0.65 PF. A synchronous capacitor of 0.2 PF leading is added to correct the overall system to 0.85 PF lagging. Calculate the

a. kVA rating of the added synchronous capacitor, $\mathbf{S}_a^*$ (let $x = \mathbf{S}_a^*$)
b. Final active power drawn from the mains, P_f
c. Final quadrature reactive power, Q_f
d. Final complex power, $\mathbf{S}_f^*$

17-62 Using the final loading in Prob. 17-61d ($420.5\angle 31.8^\circ$ kVA) on the 500-kVA alternator, calculate the

a. Final system PF with resistive load added
b. Additional *resistive* lighting load that can be added without exceeding the alternator kVA rating
c. Draw the power triangle, showing the original, added, and final system active power, quadrature power, and complex power

17-13 ANSWERS

17-1 a $9.971\angle 44.51^\circ$ V b $21.38\angle -10.78^\circ$ Ω

17-2 a $24.7\angle 21.37^\circ$ V b $14.76\angle 28.3^\circ$ Ω

17-3 a $10.3\overline{5}\angle -8.57^\circ$ V b $2.561\angle 50.195^\circ$ Ω
c $1.273\angle -86.94^\circ$ A

17-4 a $11.63\angle 49.2^\circ$ V b $1.1765\angle 61.9^\circ$ A
c, d $9.88\angle -12.7^\circ$ Ω

17-5 a $667.8\angle 32.87^\circ$ mA b $6.678\angle 77.87^\circ$ V c 3.153 W

17-6 a $16.986\angle 22.72^\circ$ V b $1.441\angle -28.07^\circ$ A
c, d $11.79\angle 50.79^\circ$ Ω

17-7 a $1.258\angle 17.91^\circ$ A b $12.58\angle -35.22^\circ$ V c 9.495 W

17-8 a $12.86\angle -23.22^\circ$ V b $24.35\angle 47.98^\circ$ kΩ
c $0.381\angle -45.2^\circ$ mA, $6.096\angle -65.2^\circ$ V

17-9 b $0.381\angle -45.2^\circ$ mA c $6.096\angle -65.2^\circ$ V

17-10 a $21.395\angle -127.5^\circ$ V b $50\angle 0^\circ$ Ω
c $0.3341\angle -88.84^\circ$ A d $13.4\angle -178.8^\circ$ V

17-11 b $0.334\angle 91.2^\circ$ A c $13.4\angle -178.8^\circ$ V

17-12 a $20\angle 0^\circ$ V b $4\angle -90^\circ$ Ω
c $4\angle 53.13^\circ$ A, $12\angle 53.13^\circ$ V d 48 W

17-13 a $22.89\angle -43.95^\circ$ V b $1.648\angle -45.9^\circ$ Ω
c $1.648\angle -45.9^\circ$ A, $16.48\angle -45.9^\circ$ V d 27.16 W

17-14 a $74.54\angle 63.435^\circ$ V b $3.\overline{3}\angle 90^\circ$ kΩ
c $4.472\angle 153.435^\circ$ mA, $89.44\angle 63.435^\circ$ V d 0

17-15 a $5.3\overline{6}\angle 26.565^\circ$ A; $-8.352\angle -73.3^\circ$ A; $-4\angle 53.13^\circ$ A
b 0

17-16 a $11.18\angle -39.235^\circ$ A; $9.547\angle 141.9^\circ$ A; $-1.648\angle -45.9^\circ$ A
b 0

17-17 a $10\angle -143.13^\circ$ mA; $8.944\angle 63.435^\circ$ mA; $-4.472\angle 153.435^\circ$ mA b 0

17-18 a $8\angle 36.87^\circ$ V; $2.4\angle 36.87^\circ$ Ω
b $12\angle -22.62^\circ$ V; $4.615\angle -22.62^\circ$ Ω
c $1.0513\angle -63.58^\circ$ A; $4.032\angle 121.42^\circ$ V

17-19 a $32\angle 52.27^\circ$ V; $4.324\angle 34.32^\circ$ Ω

b $12.634\angle 19.73^\circ$ V; $3.716\angle -45.21^\circ$ Ω
c $2.24\angle 33.05^\circ$ A, $14.47\angle 106.8^\circ$ V

17-20 a $10.733\angle -26.565^\circ$ V b $9.37\angle -38.66^\circ$ V
c $2.515\angle 24.78^\circ$ V d $5.774\angle -33.07^\circ$ kΩ
e $157.2\angle 24.78^\circ$ μA f $1.823\angle 40.55^\circ$ V

17-21 a $12\angle -53.13^\circ$ V b $11.094\angle -56.3^\circ$ V
c $1.112\angle -19.6^\circ$ V d $5.726\angle -54.98^\circ$ kΩ
e $1.127\angle 39.3^\circ$ μA f $1.127\angle 13.9^\circ$ V

17-22 a $0.381\angle -45.2^\circ$ mA b $6.09\angle -65.2^\circ$ V

17-23 a $0.334\angle 91.2^\circ$ A b $13.4\angle -178.8^\circ$ V

17-24 a $4\angle 53.13^\circ$ A b $12\angle 53.13^\circ$ V

17-25 a $1.645\angle -45.9^\circ$ A b $16.45\angle -45.9^\circ$ V

17-26 a 30.4 mA b 67.71 mA c 74.22 mA

17-27 a $6.093\angle -65.16^\circ$ V b $380.8\angle -45.2^\circ$ μA
c $+0.763\angle -53.06^\circ$ mA; $-0.3895\angle -60.8^\circ$ mA d 0

17-28 a $13.365\angle -178.86^\circ$ V b $0.3341\angle -88.86^\circ$ A
c $580.6\angle -35.\bar{4}^\circ$ mA; $-466.4\angle -0.33^\circ$ mA d 0

17-29 a $12\angle 53.13^\circ$ V b $4\angle 53.13^\circ$ A
c $-2.53\angle -18.4^\circ$ A; $5.3\bar{6}\angle 26.565^\circ$ A d 0

17-30 a $16.476\angle -45.93^\circ$ V b $1.648\angle -45.9^\circ$ A
c $11.18\angle -39.2^\circ$ A; $9.55\angle 141.9^\circ$ A d 0

17-31 a 11.63 V; $9.883\angle -12.74^\circ$ Ω b $9.883\angle 12.74^\circ$ Ω
c 3.508 W

17-32 a $16.99\angle 22.72^\circ$ V; $11.79\angle 50.79^\circ$ Ω
b $11.79\angle -50.79^\circ$ Ω c 9.685 W

17-33 a $21.395\angle -127.5^\circ$ V; $50\angle 0^\circ$ Ω b $50\angle 0^\circ$ Ω
c 2.289 W

17-34 a $22.89\angle -43.95^\circ$ V; $3.911\angle 6.92^\circ$ Ω
b $3.911\angle -6.92^\circ$ Ω c 33.74 W

17-35 a $(6 - j8)$ Ω b 24 W

17-36 a 6.325 Ω b 23.37 W c −0.116 dB

17-37 a $-j8$ Ω; $(8 - j8)$ Ω b 23.51 W c −0.0895 dB

17-38 a 10 Ω b 18 W c −1.25 dB

17-39 a 34 dB b −10.46 dB c 21.76 dB d −7 dB

17-40 a 50.12 b 0.316 c 2 d 0.25

17-41 a −13 dB b −20 dB c 100 Ω

17-42 a 10 V b 58.4 dB c 30 dB

17-43 a 126.5 mW b 200 mV

17-46 a 5 Ω; 8.02 Ω; 12.03 Ω; 11 Ω
b 40 Ω; 40.5 Ω; 90.91 Ω; 90.23 Ω
c $j39.7$ Ω; $j39.74$ Ω; $j90.11$ Ω; $j89.55$ Ω

17-48 a 50 Ω b 50.69 Ω c 63.16 Ω d $j37.7$ Ω
e 100 mH

17-49 a 70 Ω; 73 Ω; 100 Ω b 79.5 Ω; 82.2 Ω; 600 Ω
c $j37.7$ Ω; $j37.7$ Ω; $j591.6$ Ω d 0.1 H; 0.1 H; 1.57 H

17-51 a $(200 + j300)$, $(100 - j500)$, $(200 + j400)$,
$(100 + j600)$ VA b $(6 - j8)$ A, $-(600 + j800)$ VA
c 0

17-52 a $(8 + j6)$, $(8 - j6)$ A b $(3 + j4)$, $(3 - j4)$ A
c $(11 - j10)$ A d $-(1.1 + j1.0)$ kVA e 0

17-53 a $(105 - j6)$ V b $(2 + j5)$ A, $(4 + j3)$ A
c $(240 - j600)$; $(846 + j57)$; $(36 + j84)$;
$(78 - j21)$ VA d $(-1200 + j480)$ VA e 0

17-54 a $(22 - j20)$ A b $(10 + j4)$ A c $(12 - j24)$ A
d $(5.76 - j28.32)$ A e $(6.24 + j4.32)$ A
f $(12 - j24)$ A g $-(1.1 + j1.0)$ kVA h 0

17-55 a 21 kW b $j447$ vars c $21.005\angle 1.22^\circ$ kVA
d 0.9998 lagging

17-56 a 35 MW b $j35.71$ Mvars
c $-j35.71$ Mvars, 3.589 MW d 35 887 kVA
e 38.589 MW, 0, 38.589 MVA
f 11.411 MW

17-57 a $j58.456$ kvars b $j30.99$ kvar c $-j27.47$ kvar
d 27.47 kVA

17-58 a $-j26.525$ Ω b $-j542.9$ var c 930.7 W
d $j1241$ var e $j698$ var f $1163\angle 36.87^\circ$ VA

17-59 a 387.7 kVA b 392.6 kW c $j190.2$ kvar
d $436.25\angle 25.85^\circ$ kVA e 193 hp

17-60 a $414.\bar{4}$ kW b $-j310.8\bar{3}$ kvar c 1775 kVA
d 1480 kW e $j1110$ kvar f $1850\angle 36.87^\circ$ kVA

17-61 a 161.77 kVA b 357.4 kW c $j221.5$ kvar
d $420.5\angle 31.79^\circ$ kVA

17-62 a 0.8965 lagging b 90.9 kW
c $S_i^* = (448.3 + j221.5)$ kVA;
$S_o^* = (357.4 + j221.5)$ kVA

CHAPTER 18

AC Mesh and Nodal Analysis and Dependent Sources

18-1 INTRODUCTION

The previous chapter showed that the network theorems first introduced for dc circuits apply equally to ac circuits with minor modification. Mesh and nodal analysis of dc circuits was introduced in Chapter 8. The so-called *format* (or *inspection*) *method* of writing mesh and nodal equations, used in the dc circuits of Chapter 8, is extended to ac single-source and multisource networks with the same minor modifications: complex impedances are used in place of simple resistances, and all mathematical operations are performed using complex algebra rather than simple algebra.

The chapter begins with mesh and nodal analysis of ac circuits containing only *independent* sources in combination with passive impedances. An *alternative* approach to nodal analysis is also presented to eliminate errors due to source conversions and to expand the reader's horizons.

The concept of *dependence* is introduced, first for dc networks and then ac networks, containing both independent and dependent sources. Less complex circuits such as simple series or parallel ones are analyzed by Kirchhoff's voltage law (KVL) and Kirchhoff's current law (KCL) techniques and the theorems derived from them. More complex circuits are presented that require the use of mesh and nodal analysis.

The chapter concludes with analysis of network elements containing *only* dependent sources in both dc and ac networks. A novel method is introduced enabling solution by conventional series–parallel circuit theory as an alternative verification of the analysis.

18-2 MESH ANALYSIS OF AC NETWORKS CONTAINING ONLY INDEPENDENT SOURCES

Mesh analysis of dc networks was introduced in Sec. 8-4 by the format method, which enabled the writing of mesh equations by inspection. Recall that a mesh was defined as the shortest possible loop in a network. The steps for performing mesh analysis of ac circuits may be summarized as follows:

1. Assign as many clockwise (CW) mesh currents as there are "windows" in the network, one mesh current for each window.

2. Express all impedances in *rectangular* form.
3. Sum the total impedance encountered by each mesh current in each window. Call this *total* impedance the *self-impedance* of the assigned mesh current.
4. The product of the assigned mesh current and its total self-impedance is *always* written as a *positive* volt drop in each equation.
5. Sum the total impedance *mutually* shared with each of the other mesh currents separately. A separate mutual term is written for each mutual mesh current in the impedances that are common to the assigned self-current. The products of all mutual impedances and their respective mutual currents are written as *negative* volt drops in the equation.
6. Sum the voltages in each closed loop. Since these are written on the right-hand (RH) side of the equation, the polarity *last* encountered by the mesh current as it *leaves* the source is the assigned polarity of the voltage source.
7. The number of equations that must be written is always equal to the number of assigned (unknown) mesh currents. Each equation will have the same general form:

$$+\mathbf{I}_s(\sum \mathbf{Z}_s) - \sum(\mathbf{I}_m\mathbf{Z}_m) = \pm\sum \mathbf{V}_s \quad \text{volts (V)} \qquad \text{(18-1)}$$

where $\mathbf{I}_s$ is the particular mesh current assigned for self-impedances, $\mathbf{Z}_s$
$\mathbf{I}_m$ is the mutual mesh current in $\mathbf{Z}_m$, each mutual impedance
$\pm\sum \mathbf{V}_s$ is the sum of the voltages encountered in the assigned mesh

8. The simultaneous set of equations is solved for the unknown currents by using either determinants in complex form (Appendix B-2) or computer programs designed to acccept solution of linear simultaneous equations with complex coefficients.[1]

EXAMPLE 18-1

Given the network shown in **Fig. 18-1**,

a. Write (two) mesh equations by the format method
b. Show the array of impedances and voltages for finding unknown currents
c. Solve for currents $\mathbf{I}_1$ and $\mathbf{I}_2$ with determinants
d. Find node voltage $\mathbf{V}_N$ across all three branches
e. Verify node voltage $\mathbf{V}_N$ by two independent methods, using currents found in (**c**)

Solution

a. For mesh $\mathbf{I}_1$, the voltage drops are $+(19 + j20)\mathbf{I}_1 - (10 + j0)\mathbf{I}_2 = 24\angle 0° - 30\angle 0°$.
For mesh $\mathbf{I}_2$, the voltage drops are $+(10 - j10)\mathbf{I}_2 - (10 + j0)\mathbf{I}_1 = 30\angle 0°$.

b. Array:

$\mathbf{I}_1$	$\mathbf{I}_2$	$\mathbf{V}$
$19 + j20$	$-10 + j0$	$-6 + j0$
$-10 + j0$	$10 - j10$	$30 + j0$

c. $$\text{Determinant } \mathbf{D} = \begin{vmatrix} 19 + j20 & -10 + j0 \\ -10 + j0 & 10 - j10 \end{vmatrix} = (390 + j10) - (100 + j0) = \mathbf{290 + j10}$$

$$\mathbf{I}_1 = \frac{\begin{vmatrix} -6 + j0 & -10 + j0 \\ 30 + j0 & 10 - j10 \end{vmatrix}}{\mathbf{D}} = \frac{(-60 + j60) - (-300 + j0)}{290 + j10}$$
$$= \mathbf{0.8337 + j0.17815 = 0.85255\angle 12.06° \text{ A}}$$

$$\mathbf{I}_2 = \frac{\begin{vmatrix} 19 + j20 & -6 + j0 \\ -10 + j0 & 30 + j0 \end{vmatrix}}{\mathbf{D}} = \frac{(570 + j600) - (60 + j0)}{290 + j10}$$
$$= \mathbf{1.828 + j2.006 = 2.714\angle 47.66° \text{ A}}$$

d. $$\mathbf{V}_N = \mathbf{I}_2(-jX_C) = (2.714\angle 47.66°) \times (10\angle -90°) = \mathbf{27.14\angle -42.34° \text{ V}}$$

e. $$\mathbf{V}_N = \mathbf{V}_1 - \mathbf{I}_1\mathbf{Z}_1 = 24\angle 0° - (0.85255\angle 12.06°)(21.932\angle 65.77°)$$
$$= 24\angle 0° - 18.70\angle 77.83°$$
$$= (24 + j0) - (3.942 + j18.28)$$
$$= \mathbf{20.06 - j18.28 = 27.14\angle -42.34° \text{ V}}$$

[1] Throughout this text examples are solved by using Cramer's rule and the determinant method, since the reader may not have ready access to a personal computer and/or a computer program for solution of simultaneous linear equations in complex form.

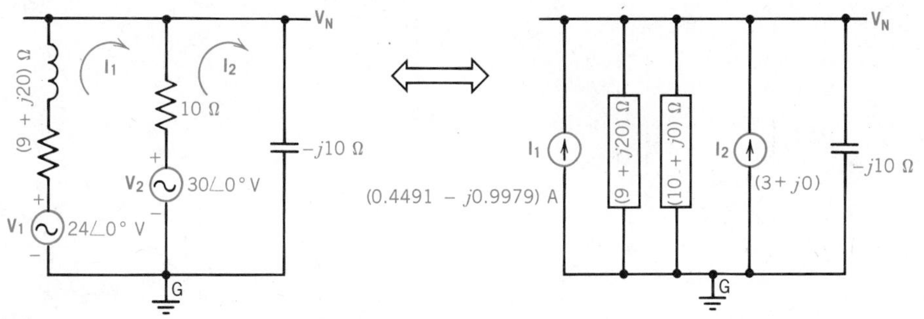

Figure 18-1 Examples 18-1 and 18-5

and $\quad \mathbf{I}_2 = (1.8278 + j2.0059)$ A

and $\quad \mathbf{I}_1 = 0.8337 + j0.17815$ A

from part (**c**). Then

$$\mathbf{I}_2 - \mathbf{I}_1 = 0.9941 + j1.8277 = 2.0806\underline{/61.46^\circ}\ \text{A}.$$

Therefore, in the second branch,

$$\mathbf{V}_\text{N} = \mathbf{V}_2 - (\mathbf{I}_2 - \mathbf{I}_1)\mathbf{Z}_2 = 30\underline{/0^\circ} - (2.0806\underline{/61.46^\circ}) \times 10/0^\circ$$
$$= 30\underline{/0^\circ} - 20.81/61.46^\circ = \mathbf{27.14\underline{/-42.34^\circ}\ V}$$

Let us examine the various parts of Ex. 18-1 to see what insights can be drawn from them. The letters below refer to the parts of the solution.

a. 1. From the two equations written (by inspection) both the self-current impedance volt drops are positive, whereas the mutual volt-drops are both negative. Since the *same* impedance is mutual to both mesh currents, this negative volt drop has the *same* impedance in *both* equations. (See array shown in *solution* part **b**.)
2. The equation for mesh $\mathbf{I}_1$ shows that $\mathbf{V}_2$ is *opposed* to $\mathbf{V}_1$, implying that the net voltage in the $\mathbf{I}_1$ mesh is $\mathbf{V}_1 - \mathbf{V}_2$ or $24\underline{/0^\circ} - 30\underline{/0^\circ}$ V.

b. The array shown exhibits "minor-axis symmetry" (lower left to upper right impedances) for the respective mesh currents. This is an important first check on the accuracy of the equation-writing process, and it always occurs with two or more unknown currents.

c. 1. Using the rules for array expansion, determinant **D** always appears in the **d**enominator and is found by using the impedance coefficients exclusively (without voltages).
2. Mesh currents $\mathbf{I}_1$ and $\mathbf{I}_2$ are found by the usual manner of determinant expansion.
3. $\mathbf{I}_1$ represents the *true* current in branch 1 (Fig. 18-1) and $\mathbf{I}_2$ the *true* current in branch 3 (Fig. 18-1). The current in branch 2 is the difference $\mathbf{I}_2 - \mathbf{I}_1$.

d. The node voltage $\mathbf{V}_\text{N}$ is most easily found as the product $\mathbf{I}_2(-jX_\text{C})$, as shown in step **d**.

e. The value of $\mathbf{V}_\text{N}$ is verified by taking the difference between each source voltage and its series-connected impedance volt drop. This method of validating the accuracy of the mesh currents yields "internal validity" since it uses both KVL and KCL to prove the accuracy of the solution.[2]

In Prob. 17-8, the reader solved the network of Fig. 17-13 by Thévenin equivalent analysis to find node voltage, $\mathbf{V}_\text{L}$, across all three branches. In Prob. 17-22, the reader solved the same network by superposition. In both problems the node voltage solution emerged as $\mathbf{6.09\underline{/-65.2^\circ}\ V}$. Example 18-2 verifies the accuracy of the solution with mesh analysis, as an example of external validation techniques.

EXAMPLE 18-2

For the network shown in **Fig. 18-2** (Probs. 17-8 and 17-22),

a. Write two mesh equations by using the format method for the impedances given (in rectangular form)
b. Show the array of impedances and voltages for finding mesh currents $\mathbf{I}_1$ and $\mathbf{I}_2$
c. Solve for current $\mathbf{I}_2$ (or $\mathbf{I}_\text{L}$) by using determinants
d. Find node voltage, $\mathbf{V}_\text{N}$, across all three branches (or the equivalent terminal voltage of the network)

Solution

First convert all polar impedances to rectangular form: $\mathbf{Z}_1 = 60\underline{/60^\circ}\ \Omega = (30 + j51.96)\ \Omega$, $\mathbf{Z}_2 = 40\underline{/40^\circ}\ \Omega = (30.64 + j25.71)\ \Omega$ and $\mathbf{Z}_\text{L} = 16\underline{/-20^\circ}\ \Omega = (15.035 - j5.472)\ \Omega$

a. For mesh $\mathbf{I}_1$: $+(60.64 + j77.67)\mathbf{I}_1 - (30.64 + j25.71)\mathbf{I}_2 = \mathbf{(60 + j0)}$, and for mesh $\mathbf{I}_2$: $+(45.68 + j20.24)\mathbf{I}_2 - (30.64 + j25.71)\mathbf{I}_1 = \mathbf{(-12 + j0)}$

[2] We could verify the value of $\mathbf{V}_\text{N}$ using "external validity" methods. This involves solving for $\mathbf{V}_\text{N}$ by any one of the following methods: nodal analysis, superposition, Thévenin's theorem, Millman's theorem, and so forth. It is advisable first to try the method of *internal validity* shown in part (**e**) of Ex. 18-1, since it involves the least amount of calculation. If the answers are validated, the reader need go no further since *only* the "correct" current values produce internal validity. See Ex. 18-2.

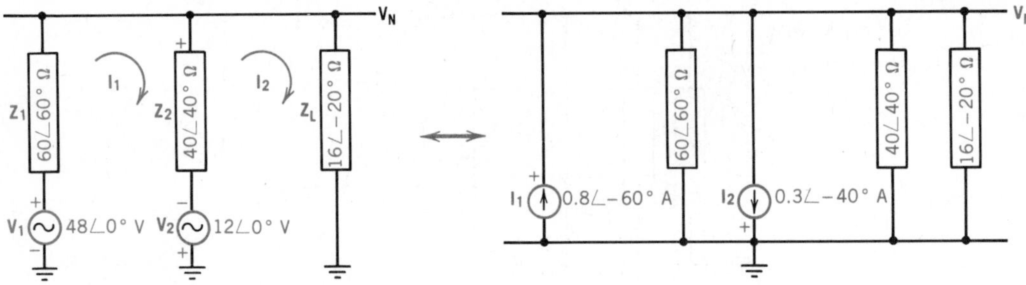

Figure 18-2 Examples 18-2 and 18-6

b.

$\mathbf{I}_1$	$\mathbf{I}_2$	$\mathbf{V}$
$60.64 + j77.67$	$-(30.64 + j25.71)$	$60 + j0$
$-(30.64 + j25.71)$	$45.68 + j20.24$	$-12 + j0$

c. Determinant $\mathbf{D} = \begin{vmatrix} 60.64 + j77.67 & -(30.64 + j25.71) \\ -(30.64 + j25.71) & 45.68 + j20.24 \end{vmatrix}$

$$= 1198 + j4775.3 - (277.8 + j1575.5)$$
$$= \mathbf{920.2 + j3199.8}$$

$$\mathbf{I}_2 = \frac{\begin{vmatrix} 60.64 + j77.67 & 60 + j0 \\ -(30.64 + j25.71) & -12 + j0 \end{vmatrix}}{\mathbf{D}}$$

$$= \frac{-727.68 - j932.04 - (-1838.4 - j1542.6)}{920.2 + j3199.8}$$

$$= \frac{1110.7 + j610.56}{920.2 + j3199.8} = \mathbf{0.2684 - j0.2699}$$

$$= \mathbf{0.3807\angle -45.16^\circ\ A}$$

d. $\mathbf{V}_N = \mathbf{V}_L = \mathbf{I}_2\mathbf{Z}_L = (0.3807\angle -45.16^\circ)(16\angle -20^\circ)$
$= \mathbf{6.091\angle -65.2^\circ\ V}$

Example 18-2 verifies the correct solutions to Probs. 17-8 and 17-22 (see answers in Sec. 17-13), using mesh analysis to verify solutions by Thévenin and superposition techniques. It is an excellent example of verification of solutions by external validity. With regard to the solution of Ex. 18-2, please note that

1. Since we are only interested in node voltage, $\mathbf{V}_N$, it is unnecessary to find $\mathbf{I}_1$.
2. Mesh current $\mathbf{I}_2$ is the "true" current in load $\mathbf{Z}_L$, yielding $\mathbf{V}_N$ directly in part (**d**) of the solution.
3. As in the case of dc mesh and nodal analysis, the reader is cautioned to verify the accuracy of the solution of determinant **D**. An error in the value of **D** affects all results, equally and adversely.
4. The value of $\mathbf{V}_N$ is the terminal voltage of the entire network measured with respect to ground.

In examining Fig. 18-2, the careful (and clever) reader may have observed that the polarity of $\mathbf{V}_2$ is *opposite* to that of $\mathbf{V}_1$. V_1 is attempting to send current *into* node $\mathbf{V}_N$. But V_2 is aided by node $\mathbf{V}_N$ in driving current in a *downward* direction toward ground. What is the true direction of current in $\mathbf{Z}_2$, and can we verify it by a method of internal validity? Example 18-3 answers these (and other) questions.

EXAMPLE 18-3

Using information calculated in Ex. 18-2, calculate the

a. Mesh current $\mathbf{I}_1$
b. "True" current in impedance $\mathbf{Z}_2$
c. Node voltage $\mathbf{V}_N$ from the current in (**b**) as internal validation of (**b**)

Solution

Using the array shown in Ex. 18-2b and the value of **D** in Ex. 18-2c, we may write

$$\text{a.}\quad \mathbf{I}_1 = \frac{\begin{vmatrix} 60 + j0 & -(30.64 + j25.71) \\ -(12 + j0) & (45.68 + j20.24) \end{vmatrix}}{\mathbf{D}}$$

$$= \frac{(2740.8 + j1214.4) - (367.68 + j308.52)}{920.2 + j3199.8}$$

$$= \mathbf{0.4585 - j0.6098 = 0.7629\angle -53.06^\circ\ A}$$

b. $\mathbf{I}_{Z_2} = \mathbf{I}_1 - \mathbf{I}_2 = (0.4585 - j0.6098) - (0.2684 - j0.2699)$
$= 0.1901 - j0.3399 = \mathbf{0.3894\angle -60.78^\circ\ A}\downarrow$

c. $\mathbf{V}_N = \mathbf{I}_{Z_2}\mathbf{Z}_2 - \mathbf{V}_2 = (0.3894\angle -60.78^\circ)(40\angle 40^\circ) - (12 + j0)$
$= (14.56 - j5.526) - (12 + j0)$
$= \mathbf{2.56 - j5.526 = 6.09\angle -65.15^\circ\ V}$

The following insights are to be gleaned from Ex. 18-3:

1. Since $\mathbf{I}_1$ emerges as a larger current than $\mathbf{I}_2$, the "true" current in $\mathbf{Z}_2$ is $\mathbf{I}_1 - \mathbf{I}_2$ in a *downward* direction. (Mesh current $\mathbf{I}_1$ is *downward* through impedance $\mathbf{Z}_2$ and $\mathbf{I}_1$ *predominates*.)
2. To find and verify $\mathbf{V}_N$, we add the volt drops in branch 2. Consequently, $\mathbf{V}_N = +\mathbf{I}_{Z_2}\mathbf{Z}_2 - \mathbf{V}_2$, as shown in (**c**) (i.e., the polarities of the source $\mathbf{V}_2$ and the impedance volt drop across $\mathbf{Z}_2$ are *opposed* to each other.)

3. Since the value of $\mathbf{V}_N$ obtained in (**c**) agrees with Probs. 17-8 and 17-22, we know that our answer is correct (external validation). We also know that the true current in branch 2 obtained in (**b**) is correct since it yields the same value of $\mathbf{V}_N$ as obtained in Ex. 18-2 (internal validation).

Now let us consider a somewhat more difficult network having three "windows" and requiring three mesh currents. As in the previous examples, we select Prob. 17-19, which was solved by an alternative method of analysis, reproduced in Fig. 18-3.

EXAMPLE 18-4

Given the network shown in **Fig. 18-3** and using mesh analysis, calculate the

a. Current in impedance $\mathbf{Z}_5$
b. Potential across $\mathbf{Z}_5$, $\mathbf{V}_{TH}$
c. Compare your answers with those given for Prob. 17-19**c** in Sec. 17-13: **2.24/33.05° A**; **14.47/106.8° V**

Solution

Since $\mathbf{I}_2$ is the "true" current in $\mathbf{Z}_5$ in Fig. 18-3, we need only solve for $\mathbf{I}_2$. But we must write three equations, one for each mesh current:

Mesh $\mathbf{I}_1$: $+(15 - j1)\mathbf{I}_1 - (12 - j5)\mathbf{I}_2 + 0\mathbf{I}_3 = 12 + j35$
Mesh $\mathbf{I}_2$: $+(18.81 - j10.8)\mathbf{I}_2 - (12 - j5)\mathbf{I}_1 - (5 - j12)\mathbf{I}_3 = 0$
Mesh $\mathbf{I}_3$: $+(9 - j15)\mathbf{I}_3 - 0\mathbf{I}_1 - (5 - j12)\mathbf{I}_2 = -(15 + j8)$

Rearranging these equations yields the following array:

$\mathbf{I}_1$	$\mathbf{I}_2$	$\mathbf{I}_3$	$\mathbf{V}$
$15 - j1$	$-12 + j5$	0	$12 + j35$
$-12 + j5$	$18.81 - j10.8$	$-5 + j12$	0
0	$-5 + j12$	$9 - j15$	$-15 - j8$

Checking the array for minor-axis symmetry, we see that our array appears to have met that test.[3] We next solve for determinant **D**, using *only* the coefficients in the three current columns. (The detailed solution of **D** is left as an exercise for the reader.)

$$\mathbf{D} = \begin{vmatrix} 15 - j1 & -12 + j5 & 0 \\ -12 + j5 & 18.81 - j10.8 & -5 + j12 \\ 0 & -5 + j12 & 9 - j15 \end{vmatrix} = \mathbf{2364 - \mathit{j}1151.54}$$

We now solve for the current $\mathbf{I}_2$, replacing the $\mathbf{I}_2$ column with the **V** column, yielding

a. $$\mathbf{I}_2 = \frac{\begin{vmatrix} 15 - j1 & 12 + j35 & 0 \\ -12 + j5 & 0 & -5 + j12 \\ 0 & -15 - j8 & 9 - j15 \end{vmatrix}}{\mathbf{D}}$$

$$= \frac{5846 + j726}{2364 - j1151.54}$$

$$= \mathbf{1.87 + \mathit{j}1.222 = 2.240\angle 33.05^\circ\ A}$$

b. $$\mathbf{V}_{TH} = \mathbf{I}_2\mathbf{Z}_5 = (2.240\angle 33.05^\circ\ \text{A})(6.459\angle 73.73^\circ\ \Omega)$$
$$= \mathbf{14.47\angle 106.8^\circ\ V}$$

c. Answers are identical to those for the solution of Prob. 17-19c by Thévenin's theorem.

With respect to Ex. 18-4, the following should be noted:

1. Had we also solved for currents $\mathbf{I}_1$ and $\mathbf{I}_3$ with the same value of **D** and the array of Ex. 18-4, we could have obtained the complete solution for the network of Fig. 18-3. Using the Thévenin equivalent method of Prob. 17-19 yields only the current $\mathbf{I}_2$. Finding $\mathbf{I}_1$ and $\mathbf{I}_3$ by Thévenin equivalents, although not impossible, would be extremely difficult and laborious.
2. Only by finding the *complete solution* (all voltages and currents) can we validate our answers by internal methods with KVL and KCL. (This is why we used an external validation technique.)
3. Extreme care should be used in expanding the arrays of Ex. 18-4. To avoid errors, the reader is advised to check each operation twice, using a separate sheet to ensure consistency and accuracy.
4. The reader is advised to redo Ex. 18-4 independently, using the answers obtained as feedback to find errors in theory, procedure, and mathematics.

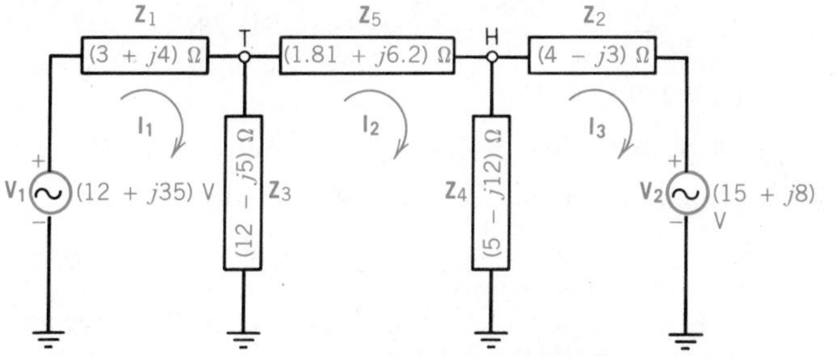

Figure 18-3 Example 18-4

---

[3] As noted previously, the test is performed by drawing a minor-axis diagonal line from the extreme lower left to the upper right (connecting both zero terms). The same impedance appears twice, above the diagonal line and below it. This serves as a check on the accuracy of the equation-writing process.

18-3 NODAL ANALYSIS OF AC NETWORKS CONTAINING ONLY INDEPENDENT SOURCES

Nodal analysis of dc networks was introduced in Sec. 8-6 and enabled the writing of nodal equations by inspection. Since we know that nodal analysis represents the *dual* of mesh analysis in every respect, we may write the following "rules" for the format method of nodal analysis applied to ac circuits:

1. Take the phasor sum of all admittances connected to a given assigned node. Call this total the self-admittance, $\mathbf{Y}_s$, connected to (or dangling from) a given node.
2. The product of the given (assigned) node voltage and its total self-admittance is *always* written as a *positive* current in each nodal equation.
3. Sum the total admittance of all admittances mutually shared with other node voltages assigned to the network. Use a separate shared admittance for each separate node voltage assigned that is mutual to the given node in (1).
4. The products of other assigned node voltages and their respective mutual admittances are always written as *negative* currents on the left side of each node equation.
5. On the right side of each node equation, sum all currents entering and leaving the given (assigned) node of interest (due to *current sources*). Currents *entering* the assigned node are *positive* and currents *leaving* the node are *negative* in accordance with previous KCL convention.[4]
6. The number of nodal equations that must be written is the same as the number of unknown node voltages (apart from the reference or GND node). Each equation will have the same general form:

$$+\mathbf{V}_s(\sum \mathbf{Y}_s) - \sum(\mathbf{V}_m\mathbf{Y}_m) = \pm\sum \mathbf{I} \quad \text{amperes (A)} \qquad (18\text{-}2)$$

7. The simultaneous set of equations is solved for the unknown node voltages, using either determinants in complex form or computer programs designed to accept solution of linear simultaneous equations with complex coefficients.

As in the case of ac mesh analysis, ac nodal analysis by the format method is best illustrated by worked examples. Let us consider the same networks previously solved by mesh analysis.

EXAMPLE 18-5

Given the network shown in Fig. 18-1 having practical current sources,

a. Write one nodal equation for the solution of $\mathbf{V}_N$ by the format method
b. Solve for node voltage $\mathbf{V}_N$, common to all branches
c. Compare the value of $\mathbf{V}_N$ with that obtained in Ex. 18-1e

Solution

a. $$\mathbf{V}_N\left(\frac{1}{9+j20}+\frac{1}{10+j0}+\frac{1}{0-j10}\right) = (0.4491 - j0.9979) + (3 + j0) \qquad (18\text{-}2)$$

b. $\mathbf{V}_N(0.11871 + j0.05842) = 3.449 - j0.9979$

$$\mathbf{V}_N = \frac{3.449 - j0.9979 \text{ A}}{0.11871 + j0.05842 \text{ S}} = \mathbf{20.06 - j18.28}$$

$$= \mathbf{27.14\angle{-42.34^\circ}} \text{ V}$$

c. The answer is identical to that obtained in Ex. 18-1**e**

The following insights can be drawn from Ex. 18-5:

1. Since there is only one unknown node voltage ($\mathbf{V}_N$), only one equation is needed. For the network in Fig. 18-1, nodal analysis provides the most rapid solution involving the least number of mathematical operations.

[4] As in the case of dc nodal analysis, all practical voltage sources may be converted to practical current sources (using the source-conversion theorem) for convenient solution. But see Sec. 18-3.1, which permits nodal analysis to be performed while retaining practical voltage sources *without conversion*!

2. Since there are no mutual (other) nodes, there are no negative current terms on the left side of the equation.
3. In effect, in nodal analysis we are equating currents by using KCL. The currents on the left side of the equation are always those leaving the node. Those on the right side are the phasor sums of the currents provided by sources either entering or leaving the node. The phasor sum of all (current) terms brought to the left side of the equation is zero, per KCL.

EXAMPLE 18-6

Given the network shown in Fig. 18-2 having practical current sources,

a. Write one nodal equation for the solution of $\mathbf{V}_N$ by the format method
b. Solve for node voltage, $\mathbf{V}_N$, common to all branches
c. Compare the value of $\mathbf{V}_N$ with that obtained in Ex. 18-2**d**

Solution

a. $$\mathbf{V}_N\left(\frac{1}{60\angle 60^\circ} + \frac{1}{40\angle 40^\circ} + \frac{1}{16\angle -20^\circ}\right) = 0.8\angle -60^\circ - 0.3\angle -40^\circ \quad (18\text{-}2)$$

b. $\mathbf{V}_N(86.21 - j9.12)\ \text{mS} = (170.2 - j500)\ \text{mA}$;
$\mathbf{V}_N = \mathbf{2.559 - \mathit{j}5.529 = 6.093\angle -65.16^\circ\ V}$

c. The answer is (practically) the same as that obtained in Ex. 18-2**d**.

Note from Ex. 18-6 that

1. The polarity of the voltage and/or current sources determines whether currents are entering or leaving the particular assigned node. This explains the subtraction of current source $\mathbf{I}_2$ from $\mathbf{I}_1$ in solution part **a**. (See Fig. 18-2, right side.)
2. As in the case of Ex. 18-5, *only one* equation is needed. For the network given in Fig. 18-2, therefore, *nodal* analysis provides the most rapid solution with the *least* number of mathematical operations, in comparison to alternative solution methods.

Let us now consider a more difficult nodal analysis example. Recall that the mesh analysis of Fig. 18-3 required the writing of three simultaneous mesh equations. Let us solve the network of Fig. 18-3 by nodal analysis.

EXAMPLE 18-7

Given the network of Fig. 18-3, using nodal analysis

a. Draw the diagram showing all practical current sources, using source conversion
b. Write the (two) nodal equations for unknown node voltages $\mathbf{V}_T$ and $\mathbf{V}_H$
c. Solve for node voltages $\mathbf{V}_T$ and $\mathbf{V}_H$
d. Find voltage $\mathbf{V}_{TH}$
e. Find current in impedance $\mathbf{Z}_5$
f. Compare answers to those obtained in Exs. 18-4**a** and 18-4**b**

Solution

a. The circuit is redrawn in **Fig. 18-4** showing practical current sources. (The values of the current sources are left as an exercise for the reader.)
b. The two nodal equations, written by inspection of Fig. 18-4 by using the nodal format method of writing equations described above, are

$$\mathbf{V}_T\left(\frac{1}{3+j4} + \frac{1}{12-j5} + \frac{1}{1.81+j6.2}\right) - \mathbf{V}_H\left(\frac{1}{1.81+j6.2}\right) = 7.04 + j2.28 \text{ and}$$

$$-\mathbf{V}_T\left(\frac{1}{1.81+j6.2}\right) + \mathbf{V}_H\left(\frac{1}{5-j12} + \frac{1}{4-j3} + \frac{1}{1.81+j6.2}\right) = 1.44 + j3.08$$

c. Simplification of these equations yields the following array:[5]

$\mathbf{V}_T$	$\mathbf{V}_H$	$\mathbf{I}$
$0.2344 - j0.2788$	$-0.04337 + j0.1486$	$7.04 + j2.28$
$-0.04337 + j0.1486$	$0.233 + j0.0424$	$1.44 + j3.08$

From the array, we obtain $\mathbf{D}$, $\mathbf{V}_H$, and $\mathbf{V}_T$:

$$\mathbf{D} = \begin{vmatrix} 0.2344 - j0.2788 & -0.04337 + j0.1486 \\ -0.04337 + j0.1486 & 0.233 + j0.0424 \end{vmatrix} = \mathbf{0.0866 - \mathit{j}0.0421}$$

$$\mathbf{V}_H = \frac{\begin{vmatrix} 0.2344 - j0.2788 & 7.04 + j2.28 \\ -0.04337 + j0.1486 & 1.44 + j3.08 \end{vmatrix}}{\mathbf{D}} = \frac{1.84 - j0.6268}{0.0866 - j0.0421} = \mathbf{20.03 + \mathit{j}2.502}\text{ volts}$$

$$\mathbf{V}_T = \frac{\begin{vmatrix} 7.04 + j2.28 & -0.04337 + j0.1486 \\ 1.44 + j3.08 & 0.233 + j0.0424 \end{vmatrix}}{\mathbf{D}} = \frac{2.0641 + j0.7493}{0.0866 - j0.0421} = \mathbf{15.87 + \mathit{j}16.37}\text{ volts}$$

[5] Reducing the equations to the array shown in Ex. 18-7**c** is left as an exercise for the reader. In the author's experience, most of the errors made by students occur in the steps involving the reduction of equations to an array. The reader may discover a number of errors before arriving at the correct array given above.

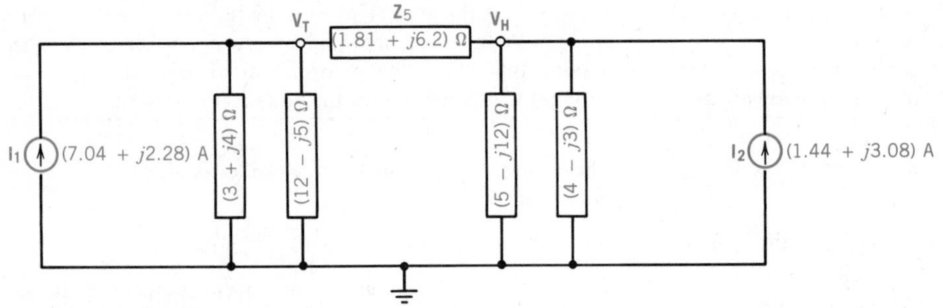

Figure 18-4 Figure 18-3 redrawn for Ex. 18-7

d. Then $\mathbf{V}_{TH} = \mathbf{V}_T - \mathbf{V}_H = (15.87 + j16.37) - (20.03 + j2.502)$
$= \mathbf{-4.16 + j13.87 = 14.48\angle 106.7^\circ\ V}$

e. $\mathbf{I}_{Z_5} = \mathbf{V}_{TH}/\mathbf{Z}_5 = \dfrac{-4.16 + j13.87}{1.81 + j6.2}$
$= \mathbf{1.881 + j1.22 = 2.242\angle 33^\circ\ A}$

f. The answers are essentially the same as and validate the solution of Ex. 18-4 by an alternative method (external validity).

The following insights are to be drawn from Ex. 18-7 (by nodal analysis) versus Ex. 18-4 (by mesh analysis) on the same network:

1. Fewer equations are required for nodal analysis of this network and therefore fewer unknowns occur. This means less mathematical work and less possibility of error. Consequently, in examining a network, the reader should assign nodes *and also* insert mesh currents in windows before deciding *which* of the two methods to use. Obviously, the method involving the *fewest* equations and *unknowns* is the one to be selected.
2. But a *special advantage* of *nodal analysis* (and one that may not be obvious to the reader) is that the node voltages obtained are *true* voltages with respect to ground.[6] Given these voltages, it is possible to find the complete solution of the entire network (all voltages and currents), using KVL and KCL for each branch, by conventional series–parallel circuit techniques.

18-3.1 Alternative Approach to Nodal Analysis

From the foregoing presentation, the reader may have formed the impression that mesh analysis dictates that the sources are practical voltage sources while nodal analysis dictates that the sources are practical current sources.[7] Given network voltage sources, it is actually possible to use them in that form and still write equations to obtain a solution by nodal analysis. This is shown in Ex. 18-8, using the network configurations of Fig. 18-2, redrawn in Fig. 18-5. It will be shown that this alternative approach opens a new door revealing many interesting options that were *not* previously possible.

EXAMPLE 18-8

Using the voltage sources of Fig. 18-2 and their respective polarities,

a. Label and show the directions of current in each branch
b. Using KCL, write the relations between the currents in the network
c. Under each current write the Ohm's law relation for the current in terms of voltage and impedance
d. Solve the equation(s) written for unknown node voltage(s)
e. Substitute node voltage(s) into the current equations to find all branch currents

Solution

a. The circuit of Fig. 18-2 is shown in **Fig. 18-5** with branch currents drawn and labeled, respectively, as $\mathbf{I}_1$, $\mathbf{I}_2$, and $\mathbf{I}_3$
b. $\mathbf{I}_1 = \mathbf{I}_2 + \mathbf{I}_3$

[6] Recall that mesh currents are arbitrarily assigned in mesh analysis in a *clockwise* direction. Consequently, they may not represent the *true* current *direction*. Further, when two mesh currents are shown in opposite directions in the same impedance, *neither* mesh current represents the *true current* in that impedance.

[7] Indeed, many texts give this impression as well. Nothing could be farther from the truth, as will be shown in the discussion that follows.

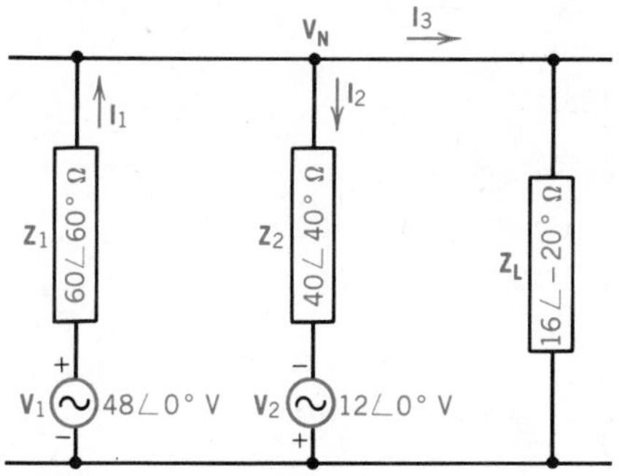

Figure 18-5 Example 18-8

c. $\dfrac{\mathbf{V}_1 - \mathbf{V}_N}{\mathbf{Z}_1} = \dfrac{\mathbf{V}_N + \mathbf{V}_2}{\mathbf{Z}_2} + \dfrac{\mathbf{V}_N}{\mathbf{Z}_L}$ and substituting known values shown in Fig. 18-5 yields:

d. $\dfrac{48\angle 0^\circ - \mathbf{V}_N}{60\angle 60^\circ\ \Omega} = \dfrac{\mathbf{V}_N + 12\angle 0^\circ}{40\angle 40^\circ\ \Omega} + \dfrac{\mathbf{V}_N}{16\angle -20^\circ\ \Omega}.$

Multiplying each term by its denominator's reciprocal yields

$0.8\angle -60^\circ - 0.01\overline{6}\angle -60^\circ\ \mathbf{V}_N = 0.025\angle -40^\circ\ \mathbf{V}_N + 0.3\angle -40^\circ + 0.0625\angle 20^\circ\ \mathbf{V}_N$, and collecting terms gives $\mathbf{V}_N(0.025\angle -40^\circ + 0.0625\angle 20^\circ + 0.01\overline{6}\angle -60^\circ) = 0.8\angle -60^\circ - 0.3\angle -40^\circ$ and $\mathbf{V}_N(0.0862 - j0.009124) = 0.1702 - j0.5$, from which we can find $\mathbf{V}_N$ as

$$\mathbf{V}_N = \frac{0.1702 - j0.5}{0.0862 - j0.009124}$$

$$= \mathbf{2.56 - j5.5295 = 6.093\angle -65.16^\circ\ V}$$

e. Substituting the value of $\mathbf{V}_N$ found in (**d**) into each of the respective current equations of (**d**) yields each branch current and the complete solution of the network:

$$\mathbf{I}_1 = \frac{(48\angle 0^\circ - 6.093\angle -65.16^\circ)\ \text{V}}{60\angle 60^\circ\ \Omega}$$

$$= \frac{(48 + j0) - (2.56 - j5.5295)}{30 + j51.96}$$

$$= \mathbf{0.4585 - j0.6098 = 0.7629\angle -53.1^\circ\ A},$$

$$\mathbf{I}_2 = \frac{6.093\angle -65.16^\circ + 12\angle 0^\circ}{40\angle 40^\circ}$$

$$= \frac{(2.56 - j5.5295) + (12 + j0)}{30.64 + j25.71}$$

$$= \mathbf{0.19 - j0.3399 = 0.3984\angle -60.8^\circ\ A},$$

and $\mathbf{I}_3 = 6.093\angle -65.16^\circ\ \text{V}/16\angle -20^\circ\ \Omega$

$$= \mathbf{0.3808\angle -45.16^\circ\ A} = \mathbf{I}_L$$

The reader should compare the answers of Ex. 18-8 with those obtained in Exs. 18-2, 18-3, and 18-6 both for accuracy and to observe that the method of Ex. 18-8 required *only one* unknown and yielded the complete solution of the network!

Several important insights emerge from Ex. 18-8:

1. It is totally *unnecessary* to convert practical voltage sources to current sources in order to perform nodal analysis. (See Ex. 18-9.)
2. Kirchhoff's current law for currents entering and leaving each node is critical to this alternative method of nodal analysis. These equations must be written first, and they may be used later to yield the complete solution if required.
3. Although the method shown has the advantage of using the practical (and even ideal) voltage sources *directly* without requiring source conversion, it does sacrifice the advantage of writing nodal equations directly by inspection by the format method. This is of little consequence, however, if the voltage sources are ideal or if the voltage at a particular node is given, as shown in Ex. 18-9.

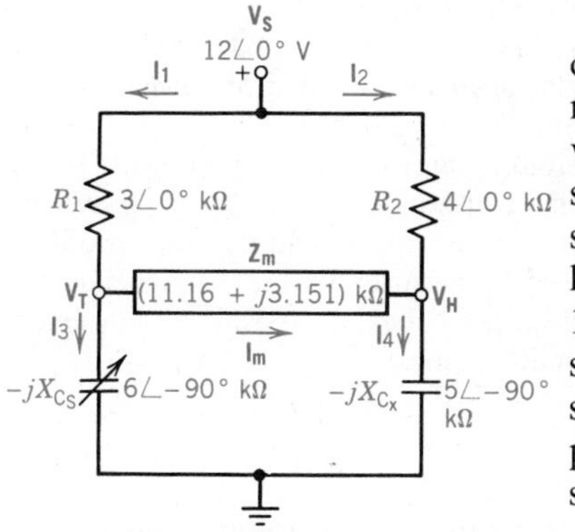

Figure 18-6 Example 18-9

Figure 18-6 shows an unbalanced capacitor bridge used to measure unknown capacitance. Given the parameters shown, we wish to find the current $\mathbf{I}_m$ in the meter whose impedance $\mathbf{Z}_m$ is given. No internal impedance is given for the supply voltage across the bridge. Consequently, we cannot convert this "ideal" voltage source to a practical current source.[8] As shown in Fig. 18-6, the problem could be solved by mesh analysis, but three mesh currents are required, involving three unknowns. (The third mesh must include the source itself, since the circuit of Fig. 18-6 actually has *four* windows[9] but only three are needed to yield a complete solution by *mesh* analysis.) However, if the *alternative* method of *nodal analysis* shown in Ex. 18-8 is used, there are only *two* unknown nodes, $\mathbf{V}_T$ and $\mathbf{V}_H$. As opposed to mesh analysis, we would select the (alternative) nodal method since it simplifies the solution considerably.

[8] It is precisely for this reason that some texts recommend assuming a nominal internal source impedance of, say, 1 Ω, which is negligible compared to all the other impedances. This introduces some error, but it does permit use of "conventional" nodal analysis.

[9] The other two windows are produced by connecting the negative side of the 12-V supply (not shown) to ground.

EXAMPLE 18-9

Given the network shown in Fig. 18-6 and the "assumed" current directions shown,

a. Using KCL, write the relations between currents in the network at each unknown node
b. Under each current, write the Ohm's law relations
c. Solve the equations for the two unknown node voltages, $\mathbf{V}_T$ and $\mathbf{V}_H$
d. Calculate the unbalanced current, $\mathbf{I}_m$, in the meter
e. Compare your answers to parts (**e**) and (**f**) of **Prob. 17-20** (which solves the *same* network by Thévenin equivalent analysis) to validate the accuracy of the solution

Solution

a. $\mathbf{I}_1 = \mathbf{I}_m + \mathbf{I}_3$ (1), $\quad \mathbf{I}_m + \mathbf{I}_2 = \mathbf{I}_4$ (2)

b. $$\frac{12 - \mathbf{V}_T}{3\angle 0^\circ} = \frac{\mathbf{V}_T - \mathbf{V}_H}{11.6\angle 15.77^\circ} + \frac{\mathbf{V}_T}{6\angle -90^\circ} \text{ and}$$

$$\frac{\mathbf{V}_T - \mathbf{V}_H}{11.6\angle 15.77^\circ} + \frac{12 - \mathbf{V}_H}{4\angle 0^\circ} = \frac{\mathbf{V}_H}{5\angle -90^\circ}$$

Note that the kΩ units have been dropped from all denominators. This is permissible since the equations are *unchanged* when multiplied or divided by the same constant. And as in Ex. 18-8, multiplying each term by its denominator's reciprocal yields two equations:

$$4\angle 0^\circ - 0.\bar{3}\angle 0^\circ\, \mathbf{V}_T = 0.0862\angle -15.77^\circ\, \mathbf{V}_T - 0.0862\angle -15.77^\circ\, \mathbf{V}_H + 0.1\bar{6}\angle 90^\circ\, \mathbf{V}_T \qquad (1)$$

$$0.0862\angle -15.77^\circ\, \mathbf{V}_T - 0.0862\angle -15.77^\circ\, \mathbf{V}_H + 3\angle 0^\circ - 0.25\angle 0^\circ\, \mathbf{V}_H = 0.2\angle 90^\circ\, \mathbf{V}_H \qquad (2)$$

c. Collecting common terms and simplifying the equations yields the following array:[10]

$\mathbf{V}_T$	$\mathbf{V}_H$	$\mathbf{I}$
$0.4163 + j0.1432$	$-0.08298 + j0.02344$	$4 + j0$
$-0.008298 + j0.02344$	$0.3330 + j0.1766$	$3 + j0$

$$\mathbf{D} = \begin{vmatrix} 0.4163 + j0.1432 & -0.08298 + j0.02344 \\ -0.008298 + j0.02344 & 0.3330 + j0.1766 \end{vmatrix} = 0.1070 + j0.1251$$

$$\mathbf{V}_T = \frac{\begin{vmatrix} 4 + j0 & -0.08298 + j0.02344 \\ 3 + j0 & 0.3330 + j0.1766 \end{vmatrix}}{\mathbf{D}}$$

$$= \frac{1.5809 + j0.6361}{0.1070 + j0.1251} = \mathbf{9.179 - j4.786}$$

$$= \mathbf{10.35\angle -27.54^\circ\ V}$$

$$\mathbf{V}_H = \frac{\begin{vmatrix} 0.4163 + j0.1432 & 4 + j0 \\ -0.008298 + j0.02344 & 3 + j0 \end{vmatrix}}{\mathbf{D}}$$

$$= \frac{1.5808 + j0.3358}{0.1070 + j0.1251} = \mathbf{7.792 - j5.9717}$$

$$= \mathbf{9.817\angle -37.47^\circ\ V}$$

d. $$\mathbf{V}_{TH} = \mathbf{V}_T - \mathbf{V}_H = (9.179 - j4.786) - (7.792 - j5.972) = \mathbf{1.387 + j1.186 = 1.825\angle 40.53^\circ\ V}$$

$$\mathbf{I}_m = \mathbf{V}_{TH}/\mathbf{Z}_m = 1.825\angle 40.53^\circ \text{ V}/11.6\angle 15.77^\circ \text{ k}\Omega = \mathbf{157.3\angle 24.76^\circ\ \mu A}$$

e. The answer from Prob. 17-20**e**, **f** is $157.2\angle 24.78^\circ$ μA and $1.823\angle 40.55^\circ$ V, showing very slight differences due to rounding errors. This validates the accuracy of the above solution.

Several very important insights emerge from the solution to Ex. 18-9:

1. All branch currents may now be found for the complete solution of the network merely by substituting the values of $\mathbf{V}_T$ and $\mathbf{V}_H$ in the two simple equations of parts (**a**) and (**b**) of the solution.
2. One of the major advantages of both mesh and nodal analysis is that both yield information leading to the *complete* solution of the network. This is not true of the method involving Thévenin analysis, which provides only the current in and voltage across a *single* complex impedance of interest in a given network.
3. As noted earlier, it **is** possible to perform nodal analysis without requiring conversion of voltage to current sources. Nevertheless, as shown in part (**a**) of the solution, we are still *summing currents* by using the KCL relations at each of the two nodes of interest. Thus, Eq. (1) represents KCL at node $\mathbf{V}_T$ and Eq. (2) represents KCL at node $\mathbf{V}_H$. This point will be of major importance in dealing with *dependent* source networks.
4. The technique of multiplying (or dividing) *all* terms in the equations by the same constant to provide more convenient and workable numbers is one that should not be overlooked.
5. The reader is advised to perform Ex. 18-9 independently, using the answers given as feedback to assist in finding errors in procedures and mathematics.

18-4 DEPENDENT SOURCES

Section 2-13 and Fig. 2-6 showed the symbols for dependent voltage and current sources. A diamond is used as the graphical symbol for a dependent source. Figure 2-7 distinguished four types of dependent sources with their associated acronyms:

[10] The simplification is left as an exercise for the reader since this is the step in which most beginners make their errors.

a. Voltage-dependent voltage source (VDVS).
b. Current-dependent voltage source (CDVS).
c. Current-dependent current source (CDCS).
d. Voltage-dependent current source (VDCS).

A dependent source (sometimes called a *controlled* source) is a source that generates either a voltage or a current as a result (or function) of a voltage or current either in a *separate* circuit or elsewhere in the *same* circuit.

Let us first distinguish between independent and dependent sources as *physical* entities. Independent sources actually exist as physical entities (a battery, a dc generator, an ac alternator, an ac oscillator, etc.). But dependent sources are *parts of models* or constructs that are used to represent the electrical properties of such electronic devices as operational amplifiers, transistors, and vacuum tubes. The derivation of such models is beyond the scope of this text since it more properly belongs in an electronics text. But the circuit analysis of dependent sources in combination with or apart from independent sources in networks is a subject that we must consider here. Let us begin with an example involving simple dependence that exists in two separate circuits.

18-4.1 Simple "Separate" Circuit Dependence Configuration

The simplest type of dependence configuration is one where the voltage or current generated by a dependent (controlled) source in "one" circuit is controlled by the voltage or current in "another" circuit. Although the two circuits are separated, they are both part of the *same* system (or device) from which the dependence is modeled. Using the source-conversion theorem, we will also see that it is possible to convert *any dependent* (practical) *voltage* source to a *dependent current* source and vice versa.[11]

EXAMPLE 18-10

In the VDVS of **Fig. 18-7a**, $R_1 = 5\text{ k}\Omega$, $\mathbf{I}_1 = 10\text{ mA}$, $R_0 = 10\text{ k}\Omega$, and the amplification factor $\mu = 20$.

a. Calculate the output voltage $\mathbf{V}_0$ of the VDVS
b. Convert the practical dependent voltage source to a dependent current source, showing dependence amplification factor, **h**, and parallel output impedance, $\mathbf{Z}_p$

Solution

a. $\mathbf{V}_1 = \mathbf{I}_1 R_1 = (10\text{ mA})(5\text{ k}\Omega) = 50\text{ V}$;
$\mathbf{V}_0 = \mu\mathbf{V}_1 = 20 \times \mathbf{V}_1 = 20 \times 50\text{ V} = \mathbf{1000\ V}$
b. $\mathbf{I} = \mu\mathbf{V}_1/\mathbf{R}_0 = 20\mathbf{V}_1/10\text{ k}\Omega = 2 \times 10^{-3}\,\mathbf{V}_1 = \mathbf{hV}_1$,
where $\mathbf{h} = \mathbf{2 \times 10^{-3}}$; $\mathbf{Z}_p = \mathbf{R}_0 = \mathbf{10\angle 0^\circ\ k\Omega}$

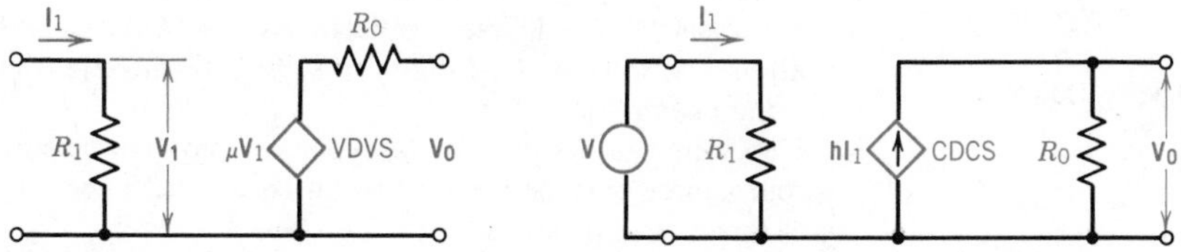

a. Voltage-dependent voltage source (VDVS) b. Current-dependent current source (CDCS) **Figure 18-7** Examples 18-10 and 18-11

EXAMPLE 18-11

In the two-port network of Fig. 18-7b for a CDCS, $\mathbf{V} = 5\text{ V}$, $\mathbf{R}_1 = 10\text{ k}\Omega$, $\mathbf{R}_0 = 5\text{ k}\Omega$, and the current amplification factor $\mathbf{h} = 100$

a. Calculate the output voltage $\mathbf{V}_0$ of the CDCS
b. Convert the practical dependent current source to a dependent voltage source, showing the dependence factor, μ, and series-connected output impedance, $\mathbf{Z}_s$

[11] This is an important and useful technique, which is frequently used in more complex independent and dependent systems. See Sec. 18-8.

Solution

a. $\mathbf{I}_1 = \mathbf{V}/R_1 = 5\text{ V}/10\text{ k}\Omega = 0.5\text{ mA}$;
$\mathbf{V}_0 = (\mathbf{hI}_1)(\mathbf{R}_0) = (100 \times 0.5\text{ mA})(50\text{ k}\Omega) = (50\text{ mA})(5\text{ k}\Omega) = \mathbf{250\ V}$

b. $\mathbf{V}_1 = (\mathbf{hI}_1)(\mathbf{R}_p) = 100\mathbf{I}_1 \times 5\text{ k}\Omega = \mathbf{5 \times 10^5\ I}_1$, where $\mu = 5 \times 10^5$ and $\mathbf{Z}_s = \mathbf{Z}_p = \mathbf{5\angle 0°\ k\Omega}$

18-4.2 Simple Series Configuration Containing both Dependent and Independent Sources

Figure 18-8a shows a voltage-dependent voltage source (VDVS) connected series-*aiding* an (ideal) independent 12-V source supplying two series-connected resistors, R_1 and R_2. By examining this (relatively simple) circuit we will draw several useful insights regarding circuits containing both independent and dependent sources.

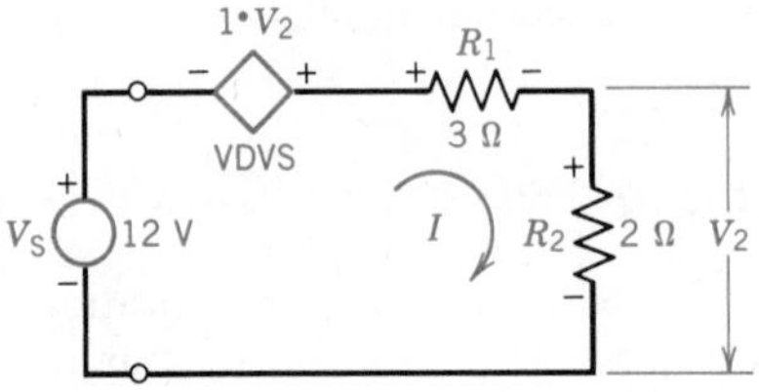

a. Independent and dependent sources in series

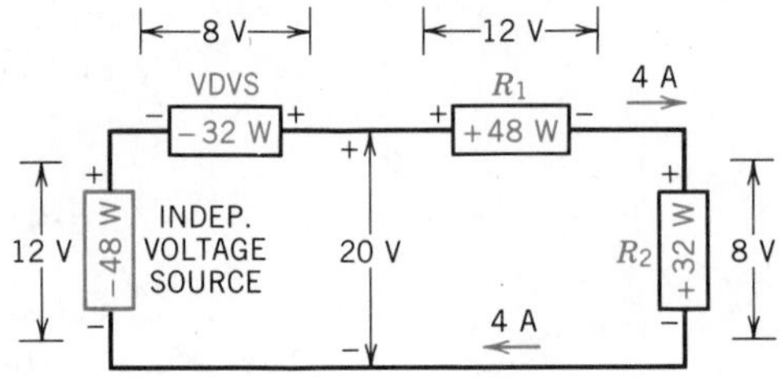

b. Power relations verification using Tellegen's theorem

Figure 18-8 Example 18-12

EXAMPLE 18-12
Given the network shown in Fig. 18-8a, calculate the

a. Clockwise mesh current I in the circuit, using KVL
b. Voltage across each circuit element in Fig. 18-8a
c. Power dissipated in each circuit element and (internal) verification by Tellegen's theorem
d. Total resistance "seen" by the ideal voltage source in the given network

Solution

a. Using KVL for CW mesh current I yields $3I + 2I - 12 - V_2 = 0$; but $V_2 = V_{R_2} = 2I$ and substituting for V_2 yields $5I - 12 - 2I = 0$, from which $I = 12/3 = \mathbf{4\ A}$
b. $V_{R_1} = 3I = 3(4) = \mathbf{12\ V}$, $V_{R_2} = 2I = 2(4) = \mathbf{8\ V}$, VDVS $= V_2 = \mathbf{8\ V}$, and $V_s = \mathbf{12\ V}$, as given (see Fig. 18-8b)
c. $P_1 = V_1 I = (12\text{ V})(4\text{ A}) = \mathbf{+48\ W}$,
$P_2 = V_2 I = (8\text{ V})(4\text{ A}) = \mathbf{+32\ W}$,
$P_s = (-12\text{ V})(4\text{ A}) = \mathbf{-48\ W}$, and
$P_{VDVS} = (-8\text{ V})(4\text{ A}) = \mathbf{-32\ W}$
Verification by Tellegen's theorem: $\sum P = +48 + 32 - 48 - 32 = \mathbf{0}$
d. $R_T = V_s/I = 12\text{ V}/4\text{ A} = \mathbf{3\ \Omega}$

The following important insights may be drawn from Ex. 18-12:

1. KVL in combination with Ohm's law applies as well to networks containing both independent and dependent sources as it does to networks containing only independent sources.
2. The total power dissipated by the resistors in the network is equal to the power supplied by the independent and dependent sources; (i.e., a dependent source supplies power to a network in much the same way as an independent source).[12]
3. Since the sum $R_1 + R_2 = 5\ \Omega$, and the total resistance "seen" by the ideal voltage source is *only* 3 Ω, the dependent voltage source is "behaving" as a resistance of $-2\ \Omega$. This observation leads to two corollary insights, verified in the following examples, that is:
   a. Depending on its polarity, a dependent voltage or current source may be reduced to an equivalent linear resistance or impedance;
   b. Depending on its polarity, the equivalent impedance of a dependent source may have a positive or negative value and always has an equivalent voltage of zero![13]

[12] Indeed, this is fundamental to the property of amplification with dependent sources.

[13] Circuit analysis of those portions of networks containing *dependent* current and/or voltage sources *only* reveals a net voltage of zero. Dependent circuits only contain equivalent positive or negative impedance.

18-5 ANALYSIS OF NETWORKS CONTAINING BOTH DEPENDENT AND INDEPENDENT SOURCES

In this section we analyze series, parallel, and more complex series–parallel networks containing *both* independent and dependent sources. Simpler circuits are solved by using KVL or KCL, or both in combination, as well as the theorems derived from KVL and KCL, namely the superposition, Thévenin, and Norton theorems. More complex networks are analyzed by mesh and/or nodal analysis, depending on whether the sources are voltage or current sources. Circuits having two or more unknown currents or voltages are solved by the determinant method. We will discover that the presence of dependent sources *may* destroy the symmetry existing between the coefficients about the major and minor diagonals. Further, depending on the polarity of the dependent current or voltage sources, the "off-diagonal" coefficients may even be **positive** (rather than negative, as with independent source networks). Consequently, this method of "coefficient inspection" as a check on the equation-writing process should *not be performed.*

As with many of the dc networks analyzed, we will calculate the equivalent voltage and impedance across *two* nodes of interest. The voltage, as we will see, is obtained by conventional methods of analysis. The equivalent impedance (of a network containing *both* dependent and independent sources) is found by applying a short circuit across the two terminals and calculating the short-circuit current that results. The equivalent impedance is then the ratio of the open-circuit voltage to the short-circuit current at the two terminals.[14]

Unlike independent sources, which may be disabled (i.e., voltage sources shorted and current sources opened) for the purpose of finding equivalent impedance, *networks containing dependent sources cannot be disabled.* Consequently, this method of "checking" for the equivalent impedance may not be used. We will discover, however, that the network simplification that results from finding an equivalent voltage and impedance may be used as a check on the accuracy of the analysis.

In each series of examples, we will begin with simple dc circuits and extend the analysis to ac circuits, drawing inferences from each. In this way, as the reader works the examples independently, confidence is gained along with understanding of the solution methods used.

18-5.1 Series Networks Containing both Dependent and Independent Sources

Let us first consider a series circuit similar to that in Ex. 18-12 in which the dependent voltage source *opposes* the independent voltage source.

EXAMPLE 18-13

Given the network shown in **Fig. 18-9a**, calculate the

a. Open-circuit voltage across terminals T–H, V_{TH}
b. Short-circuit current when terminals T–H are shorted
c. Equivalent resistance of the network seen at terminals T–H
d. Draw a simple practical voltage source to replace the entire network at terminals T–H
e. Equivalent resistance of the VDVS

Solution

Using KVL for the CW mesh current I in Fig. 18-9a yields $(12\text{ k})I + (10\text{ k})I + (2\text{ k})I + 3(2\text{ k})I - 60 = 0$ and $(30\text{ k})I = 60$, from which $I = 2$ mA

a. $V_{TH} = V_{oc} = V_1 = I(2\text{ k}) = (2\text{ mA})(2\text{ k}\Omega) = \mathbf{4\ V}$
b. When terminals T–H are shorted, $V_1 = V_{oc} = 0$ and $3 \times V_1 = 0$ across VDVS. The circuit is now shown in Fig. 18-9b, from which $I_{sc} = 60\text{ V}/22\text{ k}\Omega = \mathbf{2.\overline{72}\ mA}$

---

[14] This brings us back to the open-circuit and short-circuit test methods of calculating the impedance of a practical source.

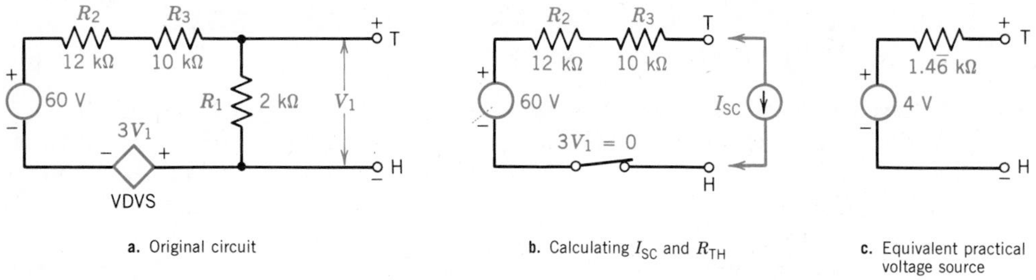

Figure 18-9 Example 18-13

c. $R_{TH} = V_{oc}/I_{sc} = 4\ \text{V}/2.72\ \text{mA} = \mathbf{1.4\bar{6}\ k\Omega}$

d. The equivalent practical voltage source is drawn in Fig. 18-9c.

e. $R_{VDVS} = \dfrac{V_T}{I_T} - R_\Sigma = \dfrac{60\ \text{V}}{2\ \text{mA}} - (12 + 10 + 2)\ \text{k}\Omega = \mathbf{6\ k\Omega}$

The following insights may be drawn from Ex. 18-13:

1. In the previous Ex. 18–12, the VDVS *aided* the independent voltage source and the equivalent resistance of the VDVS was *negative*. In Ex. 18-13, the VDVS *opposed* the independent voltage source and the equivalent resistance is *positive*. This verifies that **polarity** of a dependent source (vis-à-vis an independent source) determines whether the dependent source has *negative* or *positive resistance*.
2. A complex network, regardless of configuration, containing **both** independent and dependent voltage and/or current sources may be reduced to a *single independent practical voltage* (or *current*) *source* across any two nodes of interest in the original complex network.
3. Merely shorting the independent and dependent voltage sources provides an *erroneous* value of $R_{TH} = (2\,\|\,22)\ \text{k}\Omega = 1.8\bar{3}\ \text{k}\Omega$ instead of the *correct* value of $\mathbf{1.4\bar{6}\ k\Omega}$. The method of shorting dependent and independent voltage sources *may not be used*. Instead, we *must use open-circuit* and *short-circuit* test data at the terminals of interest.

EXAMPLE 18-14

Given the network shown in **Fig. 18-10**, calculate

a. Voltage, $\mathbf{V}_{TH}$

b. $\mathbf{Z}_{TH}$ in polar form

c. Dependent current in the time domain and frequency domain

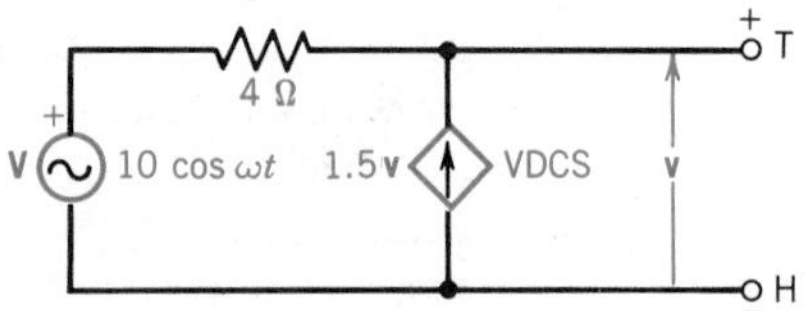

Figure 18-10 Example 18-14

Solution

a. With terminals $\mathbf{V}_{TH}$ open, the CCW current in the 4-Ω resistor is $1.5\mathbf{v}$ since the dependent current source cannot be disabled. Then using KVL for the CCW mesh current, $1.5\mathbf{v}$, yields $+1.5\mathbf{v}(4\ \Omega) + 10\cos\omega t - \mathbf{v} = 0$ and $5\mathbf{v} = -10\cos\omega t$, from which $\mathbf{V}_{TH} = \mathbf{v} = \mathbf{-2\cos\omega t}$ **volts**

b. Shorting $\mathbf{V}_{TH}$ means $\mathbf{v} = 0$ and $1.5\mathbf{v} = 0$. Therefore $\mathbf{I}_{sc} = 10\cos\omega t/4\angle 0^\circ\ \Omega = 2.5\cos\omega t$ and $\mathbf{Z}_{TH} = \mathbf{V}_{oc}/\mathbf{I}_{sc} = -2\cos\omega t/2.5\cos\omega t = \mathbf{-0.8\angle 0^\circ\ \Omega = 0.8\angle 180^\circ\ \Omega}$

c. $\mathbf{I} = 1.5\mathbf{v} = 1.5(-2\cos\omega t) = -3\cos\omega t$

$$= \mathbf{3\sin(\omega t - 90^\circ)\ A} = \frac{3}{\sqrt{2}}\angle -90^\circ\ \text{A}$$

$$= \mathbf{2.121\angle -90^\circ}\ \mathbf{A}$$

The most important insights that emerge from Ex. 18-14 are:

1. "*A dependent current source cannot be denied.*" Its current must appear in the circuit. In Fig. 18-10, the dependent current source sends a CCW current through both the 4-Ω resistor and the ac supply **V**.
2. The above solutions may be verified with frequency-domain values, since $\mathbf{V}_{TH} = \mathbf{IR} + \mathbf{V} = (2.121\angle -90^\circ\ \text{A})(4\angle 0^\circ) + 7.071\angle +90^\circ\ \text{V} = 1.414\angle -90^\circ\ \text{V} = -2\cos\omega t$. This agrees with the solution in (**a**).
3. $\mathbf{Z}_{TH}$ produces a *negative* resistance; but this was anticipated earlier, depending on relative polarities of dependent and independent sources.

EXAMPLE 18-15

Given the network shown in **Fig. 18-11a**, calculate a. $\mathbf{V}_{TH}$ b. $\mathbf{Z}_{TH}$

Solution

It is possible to solve the original network in the form shown;[15] the simplest solution is to convert the independent current source to a voltage source, as shown in Fig. 18-11b. Then using KVL in a CCW mesh yields $-(20\angle -90^\circ)\mathbf{I} + 2\mathbf{V}_R - 25\angle 0^\circ\,\mathbf{I} + 250 = 0$, but $\mathbf{V}_R = 250 - 25\mathbf{I}$ and $2\mathbf{V}_R = 500 - 50\mathbf{I}$. Substituting for $2\mathbf{V}_R$ yields $75\mathbf{I} + 20\angle -90^\circ\,\mathbf{I} = 750$ and $\mathbf{I} = \dfrac{750 + j0}{75 - j20} = 9.336 + j2.490\ \text{A}$

a. Then $\mathbf{V}_C = \mathbf{V}_{TH} = \mathbf{I}(-j\mathbf{X}_C) = 9.662\angle 14.93^\circ \times (20\angle -90^\circ) = \mathbf{193.25\angle -75.07^\circ\ V}$

b. Shorting terminals T–H shorts the capacitor and places $3\mathbf{V}_R = 0$ volts, from which $\mathbf{V}_R = 0$. Then $\mathbf{I}_{sc} = 250\angle 0^\circ\ \text{V}/25\angle 0^\circ\ \Omega = 10\angle 0^\circ\ \text{A}$

[15] This alternative solution is left as an exercise for the reader in Prob. 18-25 at the end of this chapter.

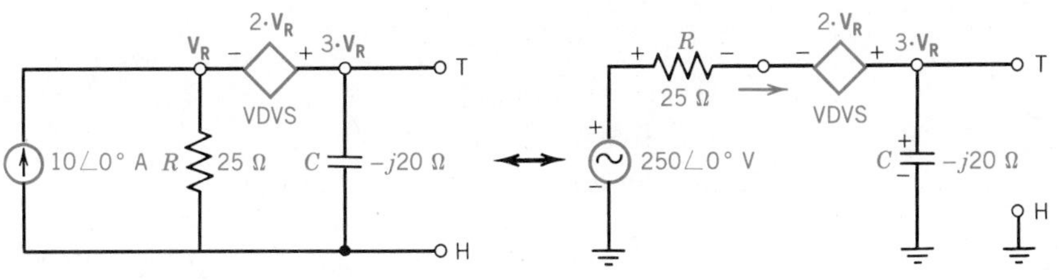

Figure 18-11 Example 18-15 and Prob. 18-25

As an alternative proof, using KVL in a CCW mesh with C shorted yields $2\mathbf{V}_R - 25\mathbf{I} + 250 = 0$, but $2\mathbf{V}_R = 500 - 50\mathbf{I}$ and substituting yields $500 - 50\mathbf{I} - 25\mathbf{I} + 250 = 0$, from which $\mathbf{I} = 750/75 = 10\underline{/0^\circ}$ A. Then

$$\mathbf{Z}_{TH} = \mathbf{V}_{oc}/\mathbf{I}_{sc} = 193.25\underline{/-75.07^\circ}\ \text{V}/10\underline{/0^\circ}\ \text{A}$$
$$= \mathbf{19.325\underline{/-75.07^\circ}\ \Omega}$$

Several insights may be drawn from Ex. 18-15:

1. Converting the current source to a practical independent voltage source produced a simple series circuit having voltage sources only, dictating use of KVL to find the current in both parts (**a**) and (**b**).
2. When C is shorted, the current is $10\underline{/0^\circ}$ A, placing $\mathbf{V}_R$ at ground! Consequently, if $\mathbf{V}_R$ is zero, $2\mathbf{V}_R$ is also zero. No other value of $\mathbf{I}$ can satisfy the volt drops and polarities shown for the VDVS, the source, and the drop across R.
3. *Under no circumstances*, however, should the reader assume (from Fig. 18-11b) that the voltage generated by a VDVS may be ignored. *Only when* that voltage on which it depends is *zero*, as in this case, may we ignore it.

18-5.2 Parallel Networks Containing Dependent and Independent Sources

The previous examples required KVL essentially in solving *series* networks. Let us consider a simple dc *parallel* network and see whether KCL applies.

EXAMPLE 18-16
For the network shown in **Fig. 18-12**, calculate a. V_{TH} b. R_{TH}

Solution

a. Since all sources are current sources, we may use Millman's theorem or nodal analysis, both of which are derived from KCL or $V(\frac{1}{24} + \frac{1}{12} + \frac{1}{8}) = 12 + 2I$, but $2I = 2V/R_2 = 2V/12 = V/6$ amperes and substituting yields $V(0.25 - 0.1\bar{6}) = 12$ and $V = \mathbf{144\ V}$
b. Shorting the output terminals T–H shorts all resistances in parallel. Then $I_2 = 0$ and $2I_2 = 0$. Since the current source must send its current somewhere, $I_{sc} = 12$ A. Then $R_{TH} = V_{oc}/I_{sc} = 144\ \text{V}/12\ \text{A} = \mathbf{12\ \Omega}$

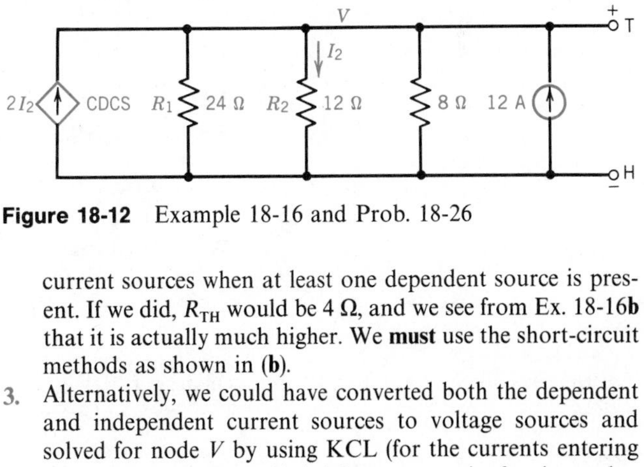

Figure 18-12 Example 18-16 and Prob. 18-26

The following insights are to be drawn from Ex. 18-16:

1. We may use Millman's theorem and/or nodal analysis with independent and dependent sources in much the same way as for networks containing only independent sources.
2. We cannot find the equivalent impedance merely by opening current sources when at least one dependent source is present. If we did, R_{TH} would be 4 Ω, and we see from Ex. 18-16**b** that it is actually much higher. We **must** use the short-circuit methods as shown in (**b**).
3. Alternatively, we could have converted both the dependent and independent current sources to voltage sources and solved for node V by using KCL (for the currents entering and leaving node V). This is left as an exercise for the reader. (See Prob. 18-26 at the end of the chapter.)

EXAMPLE 18-17
For the network shown in **Fig. 18-13** calculate a. V_{TH} b. R_{TH}

Solution
Since the dependent source requires V_2, we evaluate V_2 as $V\frac{8}{12}$ or $2V/3$ volts.

a. Then using either Millman's theorem or nodal analysis, we write $V(\frac{1}{24} + \frac{1}{12}) = 12 - 2(2V/3)$; solving for V yields $V = \mathbf{8.23\ V}$

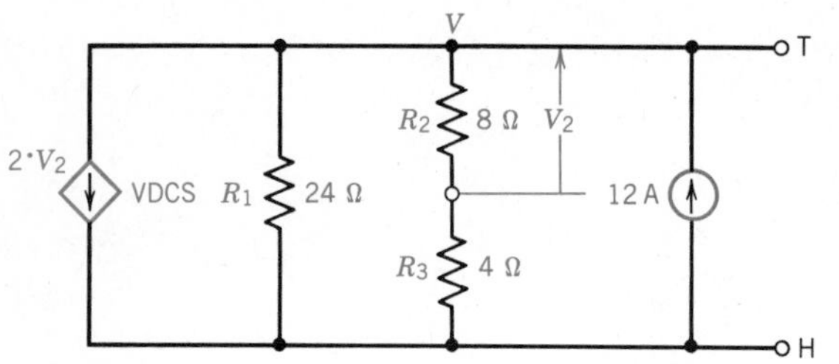

Figure 18-13 Example 18-17

b. Shorting terminals T–H shorts both resistance branches, so $V_2 = 0$ and $2V_2 = 0$. Since the independent current source "cannot be denied," $I_{sc} = 12$ A and $R_{TH} = V_{oc}/I_{sc} = 8.23\text{ V}/12\text{ A} = \mathbf{0.686\ \Omega}$

The insights drawn from Ex. 18-17 are the same as those from Ex. 18-16, with particular emphasis on the fact that KCL is fundamental to the analysis of such parallel networks with both dependent and independent current sources.

Let us now solve an ac problem with an added "wrinkle" of introducing a voltage source along with a current source.

EXAMPLE 18-18

For the network shown in **Fig. 18-14**, calculate a. $\mathbf{V}_{TH}$ b. $\mathbf{Z}_{TH}$

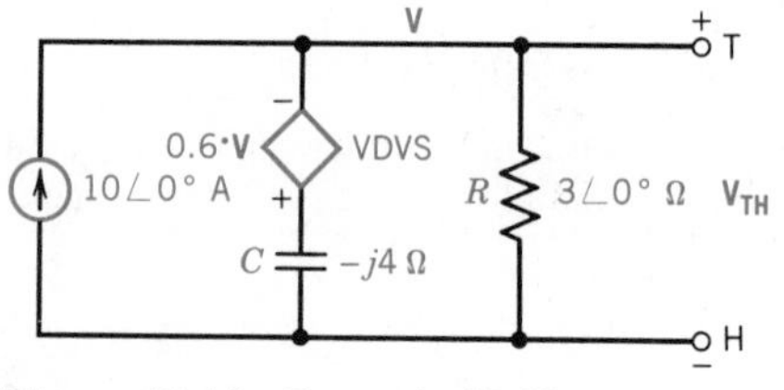

Figure 18-14 Example 18-18

Solution

While it is possible to convert the dependent voltage source to a current source, let us try KCL at node V, using all given and shown values:

a. $$10\underline{/0^\circ}\text{ A} = \frac{\mathbf{V} + 0.6\mathbf{V}}{-j4\ \Omega} + \frac{\mathbf{V}}{3\underline{/0^\circ}\ \Omega} = \mathbf{V}(0.\bar{3} + j0.4),$$

where $\mathbf{V} = \mathbf{V}_{TH}$, from which

$$\mathbf{V} = \frac{(10 + j0)\text{ A}}{(0.\bar{3} + j0.4)\text{ S}} = 12.3 - j14.75 = \mathbf{19.2\underline{/-50.2^\circ}\ V}$$

b. With $\mathbf{V}_{TH}$ shorted, $\mathbf{V} = 0$ and $\mathbf{I}_{sc} = 10/0^\circ$ A, so

$$\mathbf{Z}_{TH} = \mathbf{V}_{oc}/\mathbf{I}_{sc} = 19.2\underline{/-50.2^\circ}/10\underline{/0^\circ} = \mathbf{1.92\underline{/-50.2^\circ}\ \Omega}$$

Example 18-18 shows that we can use KCL to solve a simple three-branch parallel circuit containing a dependent voltage source.

18-5.3 Series–Parallel Networks Containing both Dependent and Independent Voltage Sources

We are now ready to attempt more complex networks. Some of these networks may require either mesh **or** nodal analysis **and** solution of simultaneous equations. Wherever possible, however, we should try to keep the unknown currents and/or unknown voltages to a minimum. This is shown in the examples that follow.

EXAMPLE 18-19

For the network shown in **Fig. 18-15**, calculate: a. $\mathbf{I}_3$ b. $\mathbf{V}_{TH}$ c. $\mathbf{V}_2$ d. $\mathbf{R}_{TH}$

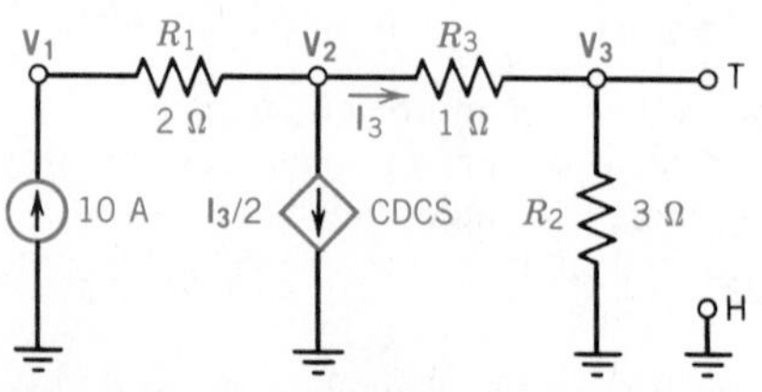

Figure 18-15 Example 18-19

Solution

The circuit has three unknown nodes and two "windows." It would appear that mesh analysis with two unknown mesh currents is indicated. But using nodal analysis *alternatively* (as shown in Sec. 18-3.1), we may write a single equation.

a. $\mathbf{I}_1 = \mathbf{I}_2 + \mathbf{I}_3$, where $\mathbf{I}_1 = 10$ A and $\mathbf{I}_2 = \mathbf{I}_3/2$, and substituting yields $10 = \mathbf{I}_3/2 + \mathbf{I}_3$, from which $\mathbf{I}_3 = \mathbf{6.\bar{6}\ A}$

b. $\mathbf{V}_{TH} = \mathbf{V}_3 = \mathbf{I}_3\mathbf{R}_2 = (6.\bar{6}\text{ A})(3\ \Omega) = \mathbf{20\ V}$

c. $\mathbf{V}_2 = \mathbf{V}_3 + \mathbf{I}_3\mathbf{R}_3 = 20 + (6.\bar{6}\text{ A})(1\ \Omega) = \mathbf{26.\bar{6}\ V}$

d. We now short terminals T–H and note that 10 A enters node $\mathbf{V}_2$, dividing in three parts, that is, one-third in the CDCS and two-thirds in $\mathbf{I}_3$. Therefore, $\mathbf{I}_3 = \frac{2}{3}10$ A $= 6.\bar{6}$ A, as before in (a), and $\mathbf{I}_{sc} = \mathbf{I}_3 = 6.\bar{6}$ A; so $R_{TH} = \mathbf{V}_{oc}/\mathbf{I}_{sc} = 20$ V/$6.\bar{6}$ A $=$ **3 Ω**

The insight to be drawn from Ex. 18-19 is that we should not be tempted to resort to more complex "automatic" methods as a substitute for thinking. By applying KCL at node $\mathbf{V}_2$, the problem parts are easily solved.

EXAMPLE 18-20

For the network shown in **Fig. 18-16**, calculate the time domain

a. Current i
b. v_{TH}
c. $\mathbf{I}_{sc}$
d. $\mathbf{Z}_{TH}$

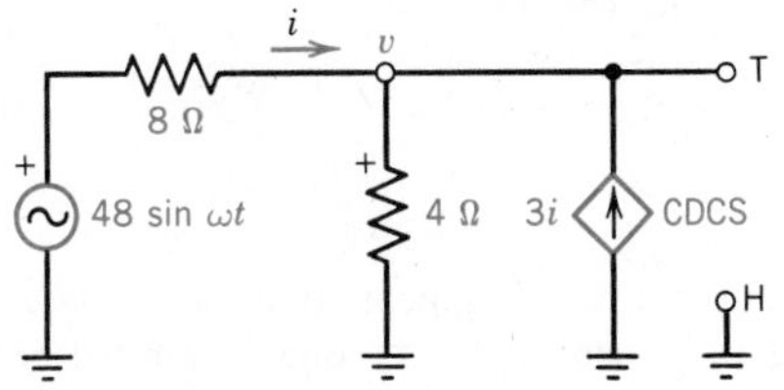

Figure 18-16 Example 18-20

Solution

Using KCL at node v, the current in the 4-Ω resistor is $i + 3i = 4i$. Consequently, using KVL in the LH mesh in a CW direction enables us to write

a. $-48 \sin \omega t + 8i + 4(4i) = 0$, and solving for i yields $i = \mathbf{2 \sin \omega t}$ **amperes**

b. $v_{TH} = v_{4\,\Omega} = v = 4(4i) = 16i = \mathbf{32 \sin \omega t}$ **volts**

c. We short terminals T–H to find $\mathbf{I}_{sc}$ and this shorts the 4-Ω resistor (but not the CDCS). Consideration of Fig. 18-16 shows *two* current sources comprising $\mathbf{I}_{sc}$ (namely i and $3i$), each having *different* values from that in (a). Current i is now $48 \sin \omega t/8\ \Omega = 6 \sin \omega t$, and dependent current $3i = 3(6 \sin \omega t) = 18 \sin \omega t$. Then by superposition, $\mathbf{I}_{sc} = i + 3i = 6 \sin \omega t + 18 \sin \omega t = 24 \sin \omega t$ amperes

d. $\mathbf{Z}_{TH} = \mathbf{V}_{oc}/\mathbf{I}_{sc} = 32 \sin \omega t/24 \sin \omega t = \mathbf{1.\bar{3}\underline{/0°}\ \Omega}$

The following important insights emerge from Ex. 18-20:

1. KCL was used at node v to find the current in the 4-Ω resistor.
2. KVL was used in the LH mesh to find current i in part **(a)**.
3. Superposition was used to find the current $\mathbf{I}_{sc}$ due to two sources.
4. When the output is shorted, any dependent or independent current source across that output must deliver its respective current to the short circuit.

EXAMPLE 18-21

For the network shown in **Fig. 18-17**, calculate a. $\mathbf{V}_{TH}$ b. $\mathbf{I}_{sc}$ c. $\mathbf{Z}_{TH}$

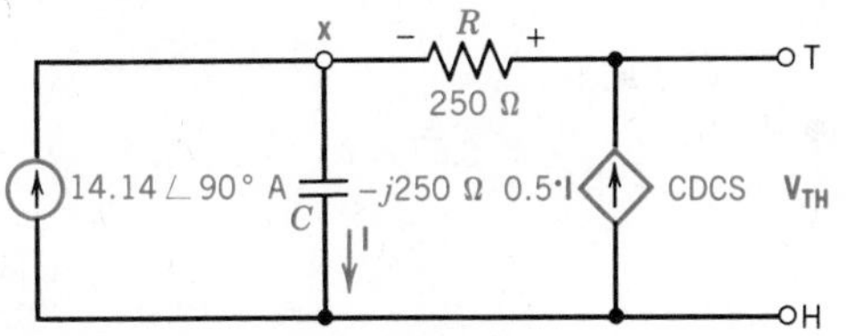

Figure 18-17 Example 18-21

Solution

Using KCL at node **x** yields the current in the capacitor, **I**:

a. $\mathbf{I} = 14.14\underline{/90°} + 0.5\mathbf{I}$, from which $\mathbf{I} = 28.28\underline{/90°}$ A; using KVL in a CCW direction and calling $\mathbf{v} = \mathbf{V}_{TH}$, we write $-\mathbf{v} + 250\underline{/0°}(0.5\mathbf{I}) + 250\underline{/-90°}(\mathbf{I}) = 0$, from which

$$\mathbf{V}_{TH} = (250\underline{/0°})(14.14\underline{/90°}) + 250\underline{/-90°}(28.28\underline{/90°})$$
$$= 7070 + j3535 = \mathbf{7904.5\underline{/26.565°}\ V}$$

b. When we short-circuit terminals T–H, terminal T is connected to ground. The independent current source divides to yield **I** in capacitor C and a current $\mathbf{I}_R$ that is *one* component of the short-circuit current. The CDCS provides the second component of the short-circuit current, and its value is 0.5**I**.

$$\mathbf{I}_C = \mathbf{I} = 14.14\underline{/90°}\,\frac{250\underline{/0°}}{250 - j250} = 10\underline{/135°}\ \text{A}$$

using the current divider rule (CDR)

$$\mathbf{I}_{sc_1} = \mathbf{I}_R = 14.14\underline{/90°}\,\frac{250\underline{/-90°}}{250 - j250} = 10\underline{/45°}\ \text{A}$$
$$= 7.071 + j7.071$$
$$\mathbf{I}_{sc_2} = 0.5\mathbf{I} = 0.5(10\underline{/135°}) = 5\underline{/135°}\ \text{A}$$
$$= -3.5355 + j3.5355$$

Then by superposition, $\mathbf{I}_{sc} = \mathbf{I}_{sc_1} + \mathbf{I}_{sc_2} = 3.5355 + j10.606 = 11.18\underline{/71.56°}$ A

c. $\mathbf{Z}_{TH} = \mathbf{V}_{oc}/\mathbf{I}_{sc} = 7904.5\underline{/26.565°}$ V$/11.18\underline{/71.56°}$ A
$= 707.1\underline{/-45°} = \mathbf{500 - j500\ \Omega}$

Again, in part **(a)** of Ex. 18-21 note that both KCL at node **x** and KVL for the loop including $\mathbf{V}_{TH}$ permitted the solution without having to use mesh or nodal analysis simultaneous equations. Let us now consider some networks requiring such analysis.

18-5.4 Mesh Analysis of DC Networks Containing both Dependent and Independent Sources

As indicated above, we begin with a **dc** circuit requiring mesh analysis. This not only simplifies the arrays but also enables us to draw certain conclusions regarding the techniques used.

EXAMPLE 18-22

For the network shown in **Fig. 18-18a**, calculate the

a. Current furnished by the 18-V source
b. Current furnished by the 9-V source
c. Voltage at node **x**, V_x

Solution

Since all sources are voltage sources, it appears that mesh analysis is indicated and that three simultaneous equations must be written for the three windows. But we can simplify the circuit, as shown in **Fig. 18-18b**, so that *only two* unknowns appear. The mesh equations are

Mesh 1: $50I_1 + 3(10I_1) - 40I_2 = 18$ simplified to $80I_1 - 40I_2 = 18$

Mesh 2: $-40I_1 + 80I_2 + 6(I_1 - I_2)40 = -9$ simplified to $200I_1 - 160I_2 = -9$

The array is

I_1	I_2	V
80	−40	18
200	−160	−9

$$\text{and } \mathbf{D} = \begin{vmatrix} 80 & -40 \\ 200 & -160 \end{vmatrix} = -4800$$

$$\text{a. } I_1 = \frac{\begin{vmatrix} 18 & -40 \\ -9 & -160 \end{vmatrix}}{\mathbf{D}} = \frac{-3240}{-4800} = \mathbf{0.6750\ A}$$

$$\text{b. } I_2 = \frac{\begin{vmatrix} 80 & 18 \\ 200 & -9 \end{vmatrix}}{\mathbf{D}} = \frac{-4320}{-4800} = \mathbf{0.9\ A}$$

$$\text{c. } V_x = (I_1 - I_2)R_x = (0.675 - 0.9)40\ \Omega = -0.2250(40\ \Omega) = \mathbf{-9.0\ V}$$

Example 18-22 reveals the following insights:

1. *Symmetry* about the major and minor diagonals has *disappeared* due to the presence of dependent sources. We may no longer use this as a check on our equation writing!
2. Mesh equations can be written for combinations of independent and dependent voltage sources in the same way as they are written for independent sources only.
3. In determining V_x, the difference between I_1 and I_2 [i.e., $(I_1 - I_2)$] is a *negative* current, which makes $V_x = -9$ V. Since I_2 predominates over I_1 and I_2 is an upward current, V_x emerges as a *negative* voltage with respect to ground.
4. No solution is ever acceptable without verification. In the diagram of Fig. 18-18b, the reader should enter $3V_1 = 20.25$ V; $V_1 = 6.75$ V; $V_x = -9$ V, and $6V_x = -(-54\text{ V}) = 54$ V. These potentials should be used to verify the above currents I_1 and I_2. While this is a method of internal verification, we could also convert the practical independent voltage sources to practical current sources and use nodal analysis, as well. This would be a method of external verification.

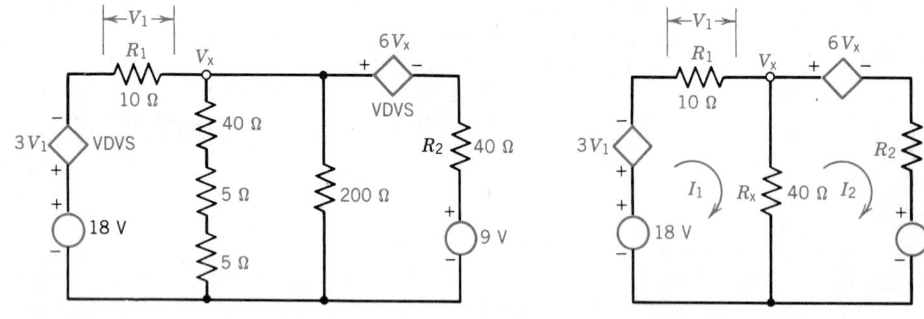

a. Original network b. Simplified network for solution

Figure 18-18 Example 18-22

18-5.5 Nodal Analysis of DC Networks Containing both Dependent and Independent Sources

Now let us consider a problem involving nodal analysis. **Figure 18-19** shows a network containing one practical independent current source and one practical dependent current source. Since there are fewer nodes than windows and the sources are current sources, nodal analysis is indicated.

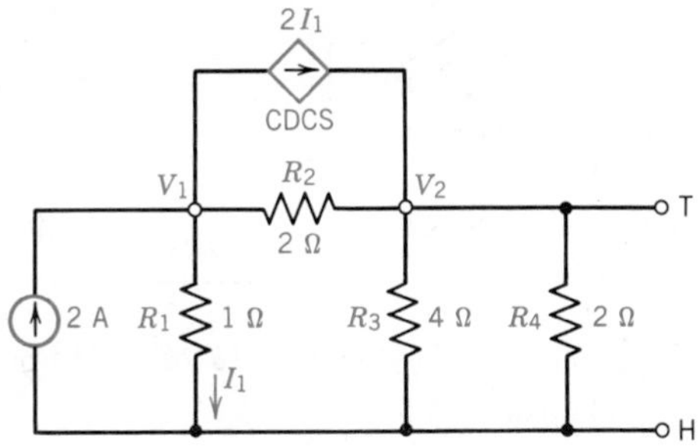

Figure 18-19 Example 18-23

EXAMPLE 18-23

Given the network shown in Fig. 18-19, calculate

a. V_1
b. V_2
c. V_{TH}
d. I_{sc_1} due to the independent current source
e. I_{sc_2} due to the dependent current source
f. R_{TH} of the network

Solution

Since $I_1 = V_1/R_1 = V_1/1 = V_1$, then $2I_1 = 2V_1$ amperes in the nodal equations below.

$V_1\left(\frac{1}{1}+\frac{1}{2}\right) - \frac{V_2}{2} = 2 - 2V_1$, which simplifies to $7V_1 - V_2 = 4$; $\frac{-V_1}{2} + V_2\left(\frac{1}{2}+\frac{1}{4}+\frac{1}{2}\right) = 2I_1 = 2V_1$, which simplifies to $-5V_1 + 2.5V_2 = 0$. The array is

V_1	V_2	I
7	−1	4
−5	2.5	0

and $\mathbf{D} = \begin{vmatrix} 7 & -1 \\ -5 & 2.5 \end{vmatrix} = 12.5$

a. $V_1 = \dfrac{\begin{vmatrix} 4 & -1 \\ 0 & 2.5 \end{vmatrix}}{\mathbf{D}} = 10/12.5 = \mathbf{0.8\ volts}$

b. $V_2 = \dfrac{\begin{vmatrix} 7 & 4 \\ -5 & 0 \end{vmatrix}}{\mathbf{D}} = 20/12.5 = \mathbf{1.6\ volts}$

c. $V_{TH} = V_2 = \mathbf{1.6\ V}$ by inspection

d. Shorting terminals T–H shorts out the 4-Ω and 2-Ω resistors to ground. Using the current divider rule, $I_1 = 2\ A(2/3) = 4/3\ A$ and $I_2 = 2/3\ A$; but I_{sc_1} due to independent current source $I_2 = 2/3\ A$

e. $I_{sc_2} = 2I_1 = 2(4/3) = \mathbf{8/3\ A}$

f. But by superposition, $I_{sc} = I_{sc_1} + I_{sc_2} = 2/3 + 8/3 = \mathbf{10/3\ A}$; then $R_{TH} = V_{oc}/I_{sc} = 1.6\ V/(10/3\ A) = \mathbf{0.48\ \Omega}$

The following insights emerge from Ex. 18-23:

1. In the array in Ex. 18-23, the symmetry about the major and minor diagonals has disappeared due to the presence of dependent sources. Since this first checkpoint is no longer available, we must be doubly sure that our equations are properly written.
2. When terminals T–H are shorted (to find R_{TH}), there are two sources of current, the independent source and the dependent source. Using the CDR in combination with superposition enables us to find each of these component currents, as shown in (**d**), (e), and (**f**).
3. Although current sources are normally open-circuited for the purpose of superposition, they cannot be open-circuited to find the equivalent resistance at terminals T–H whenever dependent sources are present. Recall that dependent sources cannot be disabled.

18-6 MESH ANALYSIS OF AC NETWORKS CONTAINING DEPENDENT AND INDEPENDENT SOURCES

We are now ready and armed to attempt mesh analysis of **ac** networks that cannot be solved by the simpler methods of the previous section. As in the case of mesh analysis of ac networks containing independent sources only (Sec. 18-2), we will use the "format" method of writing equations by inspection. The eight steps of Sec. 18-2 apply to this method of analysis. Ideally, *both* the independent and dependent sources should be voltage sources. If they are not, the source-conversion theorem should be used. Conversion of dependent portions of a network is covered in Sec. 18-8.

EXAMPLE 18-24

For the network shown in **Fig. 18-20**, using mesh analysis calculate

a. $\mathbf{I}_1$
b. $\mathbf{I}_2$
c. $\mathbf{V}_{TH}$
d. $\mathbf{V}_1$
e. Validate the accuracy of your solution by finding the current in the $j10\ \Omega$ impedance by using the voltage obtained in (**c**) and (**d**).

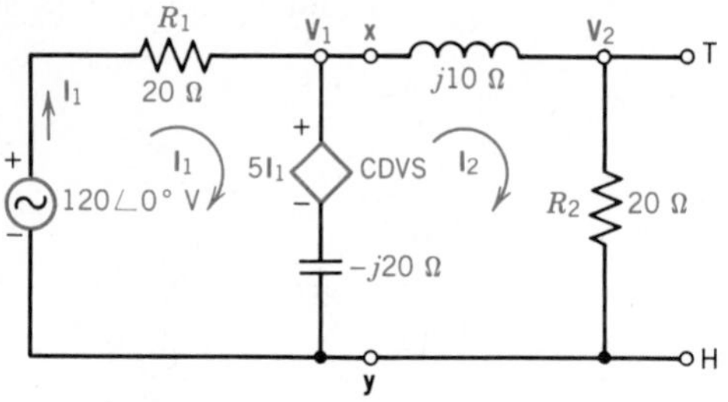

Figure 18-20 Example 18-24, Probs. 18-38 and 18-52

Solution

The only "novelty" in the network of Fig. 18-20 is the dependent voltage source in the second branch, whose voltage depends on the current in R_1. Treating this dependent voltage source as a voltage, we write the equations for each mesh by inspection using the format method.

Mesh 1: $(20 - j20)\mathbf{I}_1 + 5\mathbf{I}_1 - (0 - j20)\mathbf{I}_2 = 120$, which simplifies to $(25 - j20)\mathbf{I}_1 + j20\mathbf{I}_2 = 120$

Mesh 2: $-(0 - j20)\mathbf{I}_1 + (20 - j10)\mathbf{I}_2 - 5\mathbf{I}_1 = 0$, which simplifies to $(-5 + j20)\mathbf{I}_1 + (20 - j10)\mathbf{I}_2 = 0$

Expressing the simplified equations in array form yields

$\mathbf{I}_1$	$\mathbf{I}_2$	$\mathbf{V}$
$25 - j20$	$0 + j20$	$120 + j0$
$-5 + j20$	$20 - j10$	0

and $\mathbf{D} = \begin{vmatrix} 25 - j20 & 0 + j20 \\ -5 + j20 & 20 - j10 \end{vmatrix} = 700 - j550$

a. $\mathbf{I}_1 = \dfrac{\begin{vmatrix} 120 + j0 & 0 + j20 \\ 0 & 20 - j10 \end{vmatrix}}{\mathbf{D}} = \dfrac{2400 - j1200}{700 - j550}$

$= \mathbf{2.953 + \mathit{j}0.6057 = 3.014\angle 11.59^\circ\ A}$

b. $\mathbf{I}_2 = \dfrac{\begin{vmatrix} 25 - j20 & 120 + j0 \\ -5 + j20 & 0 \end{vmatrix}}{\mathbf{D}} = \dfrac{600 - j2400}{700 - j550}$

$= \mathbf{2.196 - \mathit{j}1.7035 = 2.779\angle -37.81^\circ\ A}$

c. $\mathbf{V}_{TH} = \mathbf{V}_{R_2} = \mathbf{I}_2\mathbf{R}_2 = 2.779\angle -37.81^\circ \times 20\angle 0^\circ$

$= \mathbf{55.58\angle -37.8^\circ\ V}$

d. $\mathbf{V}_1 = \mathbf{V} - \mathbf{I}_1\mathbf{R}_1 = 120\angle 0^\circ - (3.014\angle 11.59^\circ)(20\angle 0^\circ)$

$= 120\angle 0^\circ - 60.28\angle 11.59^\circ$

$= 120 + j0 - (59.05 + j12.11)$

$= \mathbf{60.95 - \mathit{j}12.11 = 62.14\angle -11.24^\circ\ V}$

e. $\mathbf{I}_2 = \dfrac{\mathbf{V}_1 - \mathbf{V}_{TH}}{10\angle 90^\circ} = \dfrac{(60.95 - j12.11) - (43.92 - j34.07)}{0 + j10}$

$= \dfrac{17.03 + j21.96}{0 + j10} = \mathbf{2.196 - \mathit{j}1.703\ A}$ (see **b**, above)

The following insights emerge from Ex. 18-24:

1. Mesh analysis of combinations of independent and dependent sources proceeds in the same manner as with independent sources. Essentially, the writing of mesh equations using KVL involves the phasor sum of voltages. Therefore, all sources should be expressed as voltages for use with mesh analysis.
2. Since mesh current $\mathbf{I}_1$ is the "true" current in R_1, the CDVS voltage of $5\mathbf{I}_1$ requires no special treatment. Similarly, since $\mathbf{I}_2$ is the "true" current in R_2, the output voltage $\mathbf{V}_{TH}$ is the product $\mathbf{I}_2 R_2$.
3. Having calculated $\mathbf{V}_1$ and $\mathbf{V}_{TH}$, based on the calculated mesh currents $\mathbf{I}_1$ and $\mathbf{I}_2$, the verification in step (**e**) validates the accuracy of all calculations. Current $\mathbf{I}_2$ in (**e**) is the same as $\mathbf{I}_2$ in (**b**). This method of "internal" validity ensures the accuracy of the solution.

One important precaution is appropriate here. There is a strong temptation to verify the solution to a network like that shown in Fig. 18-20 by **nodal** analysis (i.e., a method of external validity.) The reader might be tempted to convert **both** the practical independent and dependent voltage sources to current sources and repeat the procedure by *nodal* analysis. But that approach is doomed to failure since $\mathbf{I}_1$ is *not* $V_1/20\ \Omega$ and the CDVS depends on $\mathbf{I}_1$. If we cannot convert **both** sources, is there a way that we can validate the solution to Ex. 18-24 by an alternative method?

The answer is that it is possible to validate the solution by an external method. In point of fact, we have already learned it in Sec. 18-5.1. Let us apply Thévenin's theorem to the left of terminals **x**–**y**, a simple series circuit, to find $\mathbf{V}_{xy}$ and $\mathbf{Z}_{xy}$ using previously presented principles.

EXAMPLE 18-25

Given the network of Fig. 18-20, convert the network to the left of terminals **x–y** to a practical ac voltage source by finding

a. $\mathbf{V_{xy}}$
b. $\mathbf{Z_{xy}}$
c. Draw the equivalent network to the left and right of terminals **x–y**

Calculate

d. Current in the 20-Ω resistor, $\mathbf{I_2}$
e. Voltage across terminals T–H, $\mathbf{V_{TH}}$
f. Equivalent impedance, $\mathbf{Z_{TH}}$

Solution

To the left of terminals **x–y** we have a simple series circuit (with the network to the right temporarily removed). The current in that circuit, using KVL, is found by $(20 - j20)\mathbf{I}_1 + 5\mathbf{I}_1 = 120$ and $\mathbf{I}_1 = (120 + j0)$ V$/(25 - j20)$ Ω $= (2.927 + j2.3415)$ A

a. Then $\mathbf{V_{xy}} = \mathbf{V} - \mathbf{I}_1\mathbf{R}_1 = (120 + j0) - (20\underline{/0^\circ})(2.927 + j2.3415) = \mathbf{61.47 - j46.83 = 77.27\underline{/-37.3^\circ}\ V}$

b. We can find $\mathbf{Z_{xy}}$ only by shorting terminals **x–y** and finding the two components of short-circuit current due to the independent and dependent voltage sources. The short-circuit current due to the *independent* voltage source, $\mathbf{I_{sc_1}}$, is $\mathbf{I_{sc_1}} = 120\underline{/0^\circ}$ V$/20\underline{/0^\circ}$ Ω $= 6\underline{/0^\circ}$ A $= \mathbf{I}_1$. Then the dependent voltage source $5\mathbf{I}_1 = 5 \times 6\underline{/0^\circ}$ A $= 30\underline{/0^\circ}$ V and the short-circuit current component due to the dependent voltage source $\mathbf{I_{sc_2}} = 30\underline{/0^\circ}$ V$/20\underline{/-90^\circ}$ Ω $= 1.5\underline{/90^\circ}$ A. So the short-circuit current when terminals **x–y** are shorted is (by superposition) $\mathbf{I_{sc_1}} + \mathbf{I_{sc_2}} = (6 + j1.5)$ A and $\mathbf{Z_{xy}} = \mathbf{V_{oc}}/\mathbf{I_{sc}} = \dfrac{(61.47 - j46.83)\ \text{V}}{(6 + j1.5)\ \text{A}} = \mathbf{7.806 - j9.756\ \Omega}$

c. The network to the left of terminals **x–y** and that restored to the right of **x–y** is shown in **Fig. 18-21**. Observe that Fig. 18-20 has been reduced to a simple series circuit. This has certain great advantages, as we will soon see.

d. $\mathbf{I}_2 = \mathbf{V_{xy}}/\mathbf{Z_T} = \dfrac{(61.47 - 46.83)\ \text{V}}{(27.806 + j0.244)\ \Omega} = \mathbf{(2.196 - j1.703)\ A}$

Note that this answer is exactly the same as that obtained in the solution of Ex. 18-24**e**.

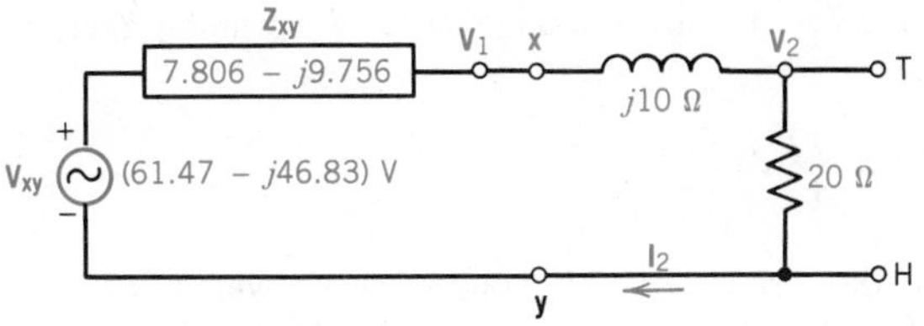

Figure 18-21 Example 18-25

e. $\mathbf{V_{TH}} = \mathbf{V}_{20\,\Omega} = \mathbf{I}_2\mathbf{R}_2 = (2.196 - j1.703)(20\underline{/0^\circ}) = \mathbf{43.92 - j34.06 = 55.58\underline{/-37.8^\circ}\ V}$

f. Had we attempted to find the short-circuit current when terminals T–H are shorted in Fig. 18-20, it would have been difficult because superposition is indicated and the $j20$ Ω impedance occurs in both components. But Fig. 18-21 enables us to find $\mathbf{I_{sc}}$ easily from $\mathbf{I_{sc}} = \mathbf{V_{xy}}/(\mathbf{Z_{xy}} + j10) = \dfrac{61.47 - j46.83}{7.806 + j0.244} = 7.68 - j6.239 = 9.895\underline{/-39.1^\circ}$ A; then equivalent impedance, $\mathbf{Z_{TH}} = \mathbf{V_{oc}}/\mathbf{I_{sc}} = 55.58\underline{/-37.8^\circ}$ V$/9.895\underline{/-39.1^\circ}$ A $= \mathbf{5.617\underline{/1.3^\circ}\ \Omega}$

The following insights are to be drawn from Ex. 18-25:

1. It is possible to verify the accuracy of a mesh analysis solution involving two simultaneous equations by alternative methods other than conventional nodal analysis and/or conversion of independent sources.[16]
2. The solution of Ex. 18-25 is made possible by recognizing that the network to the left of terminals **x–y** in Fig. 18-20 is a simple series circuit whose equivalent output voltage is found by application of KVL.
3. **Superposition** enables us to find $\mathbf{Z_{xy}}$ as the ratio $\mathbf{V_{oc}}/\mathbf{I_{sc}}$ by taking the independent voltage source and dependent voltage **separately** and determining the short-circuit current due to each.
4. Voltages $\mathbf{V}_1$ and $\mathbf{V}_2$ in Fig. 18-20 are easily found as corresponding nodes in Fig. 18-21. The simple series circuit of Fig. 18-21 has simplified the complete solution of Fig. 18-20.

18-7 NODAL ANALYSIS OF AC NETWORKS CONTAINING DEPENDENT AND INDEPENDENT SOURCES

In this section we treat ac networks that may be solved by simultaneous equations, one for each unknown node voltage. As in the case of nodal analysis of ac networks containing independent sources only (Sec. 18-3), we will see the format method of writing equations by inspection. The seven steps of Sec. 18-3 apply to the procedure and method of nodal analysis. Ideally, both the independent and dependent sources should be current sources. *Practical* independent and dependent voltage sources may be converted to current sources. Occasionally, however, the alternative approach to nodal analysis (Sec. 18-3.1) retaining the given voltage sources may be used. As with examples in the previous sections, we will validate our solutions.

[16] See Prob. 18-38 for an alternative approach to nodal analysis by the method given in Sec. 18-3.1

EXAMPLE 18-26

Given the network shown in **Fig. 18-22**, using nodal analysis, calculate

a. $\mathbf{V}_1$ d. $\mathbf{I}_2$
b. $\mathbf{V}_2$ e. $\mathbf{I}_3$
c. $\mathbf{I}_1$ f. $\mathbf{I}_4$
g. Verify the correctness of the solution by showing $\mathbf{I} = \mathbf{I}_1 + \mathbf{I}_2$ and $\mathbf{I}_2 = \mathbf{I}_3 + \mathbf{I}_4$ using the calculated values

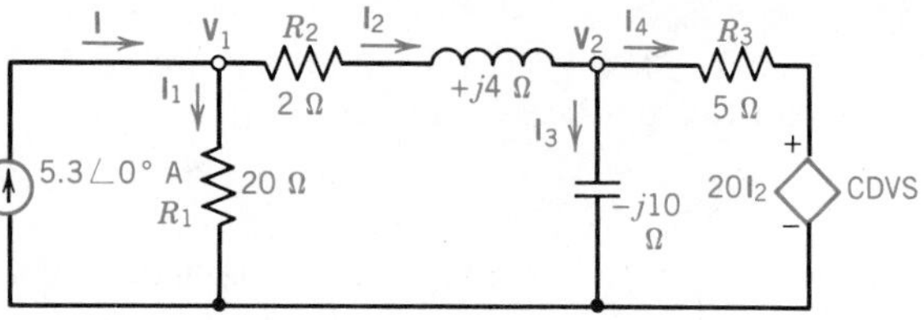

Figure 18-22 Example 18-26 and Prob. 18-54

Solution

Writing nodal equations by inspection (using the format method), we have for currents around node $\mathbf{V}_1$: $\mathbf{V}_1\left(\frac{1}{20}+\frac{1}{2+j4}\right) - \mathbf{V}_2\left(\frac{1}{2+j4}\right) = 5.3 + j0$, which simplifies to $(0.15 - j0.2)\,\mathbf{V}_1 - (0.1 - j0.2)\,\mathbf{V}_2 = 5.3 + j0$. Before writing the second equation we must evaluate the dependent source $20\mathbf{I}_2$ in terms of the node voltages, $\mathbf{V}_1$ and $\mathbf{V}_2$. Since $\mathbf{I}_2 = \frac{\mathbf{V}_1 - \mathbf{V}_2}{2+j4}$ then $20\mathbf{I}_2 = (2 - j4) \times (\mathbf{V}_1 - \mathbf{V}_2)$; currents around node $\mathbf{V}_2$ are $\mathbf{I}_2 = \mathbf{I}_3 + \mathbf{I}_4$ or $\frac{\mathbf{V}_1 - \mathbf{V}_2}{2+j4} = \frac{\mathbf{V}_2}{-j10} + \frac{\mathbf{V}_2 - 20\mathbf{I}_2}{5}$. Substituting for $20\mathbf{I}_2$ and simplifying by collecting common terms yields $-(0.5 - j1)\,\mathbf{V}_1 + (0.7 - j0.9)\,\mathbf{V}_2 = 0$. The array is

$\mathbf{V}_1$	$\mathbf{V}_2$	$\mathbf{I}$
$(0.15 - j0.2)$	$-(0.1 - j0.2) =$	$5.3 + j0$
$-(0.5 - j1.0)$	$(0.7 - j0.9) =$	0

The determinant $\mathbf{D}$ is

$$\mathbf{D} = \begin{vmatrix} 0.15 - j0.2 & -(0.1 - j0.2) \\ -(0.5 - j1) & (0.7 - j0.9) \end{vmatrix} = \mathbf{0.075 - j0.075}$$

a. Then $$\mathbf{V}_1 = \frac{\begin{vmatrix} 5.3 + j0 & -(0.1 - j0.2) \\ 0 & (0.7 - j0.9) \end{vmatrix}}{\mathbf{D}} = \frac{3.71 - j4.77}{0.075 - j0.075}$$
$$= \mathbf{56.5\bar{3} - j7.0\bar{6}} \text{ volts}$$

b. Likewise, $$\mathbf{V}_2 = \frac{\begin{vmatrix} 0.15 - j0.2 & 5.3 + j0 \\ -(0.5 - j1) & 0 \end{vmatrix}}{\mathbf{D}}$$
$$= \frac{2.65 - j5.3}{0.075 - j0.075} = \mathbf{53 - j17.\bar{6}} \text{ volts}$$

c. From Fig. 18-22,
$\mathbf{I}_1 = \mathbf{V}_1/\mathbf{R}_1 = (56.5\bar{3} - j7.0\bar{6})/(20 + j0) = \mathbf{2.82\bar{6} - j0.35\bar{3}\ A}$

d. From Fig. 18-22,
$\mathbf{I}_2 = (\mathbf{V}_1 - \mathbf{V}_2)/\mathbf{Z}_2 = (56.5\bar{3} - j7.0\bar{6}) - (53 - j17.\bar{6})/(2 + j4)$
$= \mathbf{2.47\bar{3} + j0.35\bar{3}\ A}$

e. From Fig. 18-22,
$\mathbf{I}_3 = \mathbf{V}_2/-j10 = (53 - j17.\bar{6})/(0 - j10) = \mathbf{1.7\bar{6} + j5.3\ A}$

f. From Fig. 18-22,
$\mathbf{I}_4 = (\mathbf{V}_2 - 20\mathbf{I}_1)/\mathbf{R}_3 = (3.5\bar{3} - j24.7\bar{3})/(5 + j0)$
$= \mathbf{0.706 - j4.94\bar{6}\ A}$

g. Now we come to the "proof of the pudding," i.e., the internal validation of the computations by checking to see that the calculated currents around each node are verified by KCL:

$$\mathbf{I} = \mathbf{I}_1 + \mathbf{I}_2$$
$$\mathbf{5.3 + j0} = (2.82\bar{6} - j0.35\bar{3}) + (2.47\bar{3} + j0.35\bar{3})$$
$$\mathbf{I}_2 = \mathbf{I}_3 + \mathbf{I}_4$$
$$(\mathbf{2.47\bar{3} + j0.35\bar{3}}) = (1.7\bar{6} + j5.3) + (0.70\bar{6} - j4.94\bar{6})$$

The following insights emerge from Ex. 18-26 containing both independent and dependent sources:

1. As an alternative to dependent voltage source conversion, we retained the dependent voltage source in its original form. But by using the alternative method of nodal analysis (Sec. 18-3.1) we are still able to evaluate current $\mathbf{I}_4$ and the dependent source $20\mathbf{I}_2$ in terms of the unknown node voltages $\mathbf{V}_1$ and $\mathbf{V}_2$.
2. While the first nodal equation was written by the format method, the second was written by the sum-of-the-current method described in Sec. 18-3.1. This was done to simplify the algebraic equation writing.
3. Occasionally, the numbers expressing the coefficients in the array are either very large or very small. The reader has the option of multiplying or dividing all coefficients in any *row* by the same factor. This is a technique that frequently reduces error when entering values into an electronic calculator or a computer.
4. The *internal validation* method shown above (step g of the solution) is an absolute necessity since there are many possibilities for error. If KCL is *not* verified, obviously one or more errors have been made along the way.
5. Alternatively, internal validation is possible using KVL in the center window of Fig. 18-22 (which does not contain a voltage source). The phasor sum of all volt drops, with due regard for current direction, should be zero.
6. It bears repeating that the greatest care must be taken to see that the (two) equations are correctly written, *prior* to the solution steps (**a**) through (**f**). The author recommends repeating these on separate sheets to ensure consistency of results.[17]

[17] As noted previously, equation writing is the step where most students make their errors in solutions. Correct equations are the prime step in the solution process. A minor error in *only one* coefficient invalidates *all* the results.

18-8 ELEMENTS OF DC AND AC NETWORKS CONTAINING DEPENDENT SOURCES ONLY

We now come to *elements* of a complete network containing *only* dependent sources in combination with passive impedances (R, L, and C elements). Such circuit elements containing dependent sources only *do not exist* as *complete* circuits. Any dependent voltage or current source always requires an independent source to produce the requisite voltage or current on which it depends. But it is convenient to analyze such portions or elements *separately* and to draw conclusions about them.

Since such circuit elements contain no active *independent* voltage or current sources, they produce no voltage or current at their terminals (despite the implied dependence). Consequently, circuit elements containing only dependent sources (in combination with R, L, and C) may only be reduced to an equivalent *passive impedance*! How the dependent source affects that impedance is the subject of this section.

A *passive* network may be defined as one containing *no* energy-producing sources. An *active* electric network is defined as one containing one or more sources of energy. Consequently, elements of networks containing *dependent* sources *only* are *passive* networks. They can be reduced to an equivalent impedance *of zero voltage*.

We will also discover that the *source-conversion theorem* may be used with dependent sources in elements of networks containing dependent sources only. This is an important insight since it provides an added technique (which was deliberately avoided until consideration of this topic).

As with the previous circuits analyzed, we begin with simple dc networks and extend the analysis to ac networks. But first let us consider a general method of analysis applicable to all *passive* dc and ac networks.

Consider a bridge network consisting of resistors only, as shown in **Fig. 18-23**. We wish to obtain the equivalent resistance of the entire network. Since no source is shown our procedure will be as follows:

1. Assume either a voltage (or current) source of convenient value applied to the network.
2. Calculate the equivalent current (or voltage) drawn from the source by the network.
3. The equivalent impedance is then the ratio of the applied voltage to the current drawn.

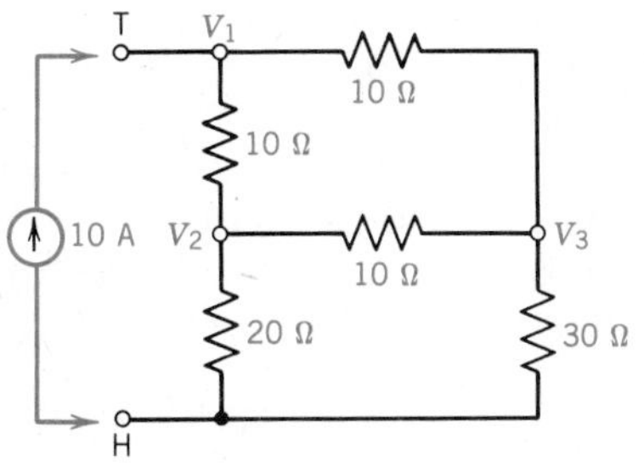

Figure 18-23 Example 18-27A

EXAMPLE 18-27A

Assume a current source of 10 A is applied to terminals T–H of the bridge network of Fig. 18-23. Using nodal analysis, calculate:

a. V_1
b. R_{TH}, the equivalent (passive) resistance of the circuit

Solution

Three nodal equations are required for the three unknown node voltages, namely

$$V_1(\tfrac{1}{10}+\tfrac{1}{10}) - V_2(\tfrac{1}{10}) - V_3(\tfrac{1}{10}) = 10,$$

which reduces to $0.2V_1 - 0.1V_2 - 0.1V_3 = 10$ (1)

$$-V_1(\tfrac{1}{10}) + V_2(\tfrac{1}{10}+\tfrac{1}{10}+\tfrac{1}{20}) - V_3(\tfrac{1}{10}) = 0$$

or $$-0.1V_1 + 0.25V_2 - 0.1V_3 = 0 \quad (2)$$

$$-V_1(\tfrac{1}{10}) - V_2(\tfrac{1}{10}) + V_3(\tfrac{1}{10}+\tfrac{1}{10}+\tfrac{1}{30}) = 0$$

or $$-0.1V_1 - 0.1V_2 + 0.2\bar{3}V_3 = 0 \quad (3)$$

The array of coefficients is

V_1	V_2	V_3	I	
0.2	−0.1	−0.1	10	(1)
−0.1	0.25	−0.1	0	(2)
−0.1	−0.1	$0.2\bar{3}$	0	(3)

Determinant $\mathbf{D} = 2.8\bar{3} \times 10^{-3}$

a. Then $$V_1 = \frac{\begin{vmatrix} 10 & -0.1 & -0.1 \\ 0 & 0.25 & -0.1 \\ 0 & -0.1 & 0.2\bar{3} \end{vmatrix}}{\mathbf{D}} = \frac{0.48\bar{3}}{2.8\bar{3} \times 10^{-3}} = \mathbf{170.6\ V}$$

b. $R_{TH} = V_1/I_T = 170.6\ \text{V}/10\ \text{A} = \mathbf{17.06\ \Omega}$

Example 18-27A reveals the following insights:

1. Any complex passive (or even active) network may be reduced to an equivalent single passive resistance (or impedance) by applying a *known source* and determining the total effect of that source on the network.[18]
2. The opposition of the network is the ratio of the voltage across the entire network divided by the current entering the network.
3. Alternatively, we might have applied a 10-V source across terminals T–H and used *mesh* analysis to find the total current drawn by the network. We would have emerged with an equivalent resistance of 17.06 Ω, as well.

It is not always necessary, however, to use mesh or nodal analysis, particularly with simple series or parallel circuits, as we noticed earlier. The application of KCL around a node or KVL in a closed loop or mesh should be sufficient in many cases.

EXAMPLE 18-27B

Calculate the equivalent resistance of the network elements shown in **Fig. 18-24**.

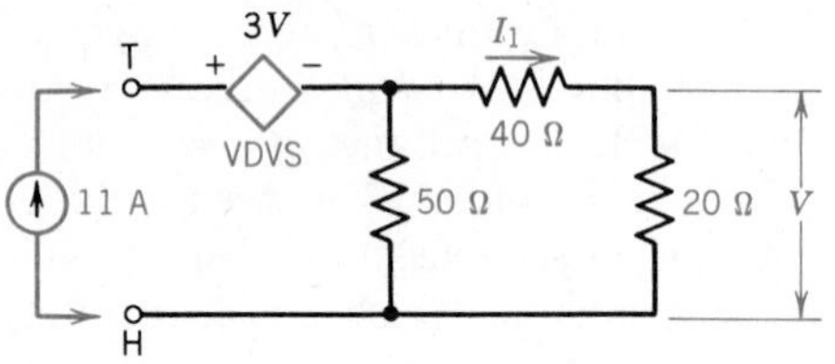

Figure 18-24 Example 18-27B

Solution

Since the dependent voltage source is ideal, no conversion is possible and a current source must be applied to the input terminals T–H. A current of 11 A is selected because it simplifies the math involving the CDR. Using the CDR, $I_1 = 11\text{ A}\dfrac{50\ \Omega}{(110\ \Omega)} = 5\text{ A}$ and $V = 5\text{ A}\,(20\ \Omega) = 100\text{ V}$. Then $3V = 3 \times 100\text{ V} = 300\text{ V}$ which is the potential across the dependent source. Using KVL, the total voltage $V_T = +300\text{ V} + 5\text{ A}(40\ \Omega) + 100\text{ V} = 600\text{ V}$. Now using Ohm's law, $R_{eq} = R_{TH} = V_T/I_T = 600\text{ V}/11\text{ A} =$ **54.54 Ω**

Example 18-27B shows that a dependent source by itself, in combination with passive elements, may be reduced to a single equivalent passive resistance (or impedance) having zero voltage across its terminals.

In the solution to Ex. 18-27B we noted that the dependent voltage source was an ideal source and therefore no source conversion was possible. Let us consider a network element in which source conversion is possible and determine whether source conversion is permissible in passive circuit elements.

EXAMPLE 18-28

Given the passive network elements to the right of terminals T–H in **Fig. 18-25**, calculate the equivalent input resistance R_{TH}

Solution

Since VDCS depends on the input voltage, V, assume a 1-V source at the input. This makes the current source $3 \cdot V$ or 3 A! Then the current in the 50-Ω resistor is $(I - 3)$ A. Now, using KVL in a CW direction through the resistance loop: $30I + 50(I - 3) - 1 = 0$. Solving for I: $I = 151/80 = 1.8875$ A. Then $I_T = I + V/40 = 1.8875\text{ A} + 0.025\text{ A} = 1.9125\text{ A}$, so $R_{eq} = R_{TH} = V_T/I_T = 1\text{ V}/1.9125\text{ A} =$ **0.5229 Ω**

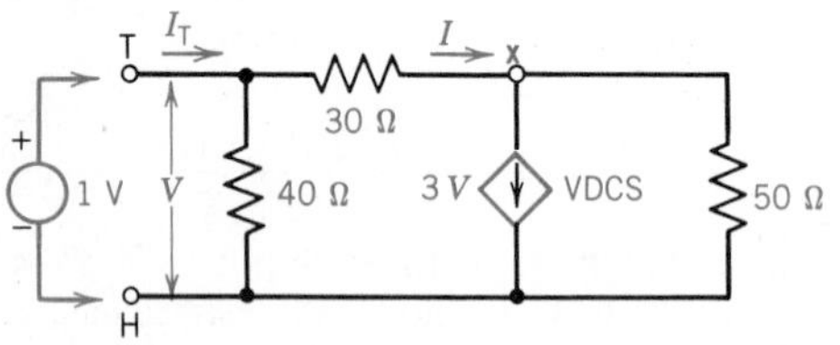

Figure 18-25 Example 18-28 and Prob. 18-55

[18] The author cannot refrain from noting that the method of applying either a test voltage source (or test current source) to a network containing both independent and dependent sources may *also* be used to find the equivalent impedance of the network. But this method dictates "killing" the *independent* sources *only*. Since this often affects the dependent sources, this procedural technique was deliberately avoided in favor of short-circuiting the terminals of interest and finding the short-circuit current (for networks containing both independent and dependent sources). Section 18-8.1 shows an alternative method of finding the impedance of a network containing dependent sources only in combination with impedances.

Using the source-conversion theorem, it is possible to convert the dependent current source into a practical voltage source. Theoretically, this should not affect the total passive resistance of the given network element of Fig. 18-25.

EXAMPLE 18-29

Convert the dependent current source in Fig. 18-25 to a dependent voltage source and calculate the equivalent input resistance, R_{TH}

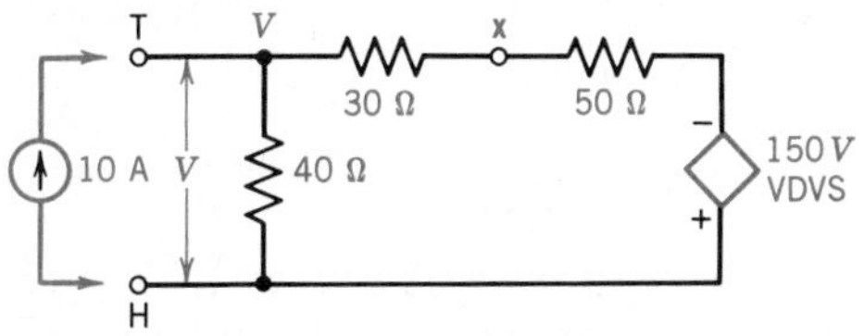

Figure 18-26 Example 18-29

Solution

Using the source-conversion theorem, $V_s = I_sR_s = 3V(50) = 150V$, yielding the practical voltage dependent voltage source shown in **Fig. 18-26**. Since we are now dealing with a dependent voltage source, we select an independent current source of 10 A applied across terminals T–H. Our purpose now is to find V across the 40-Ω resistor (and across the 10-A current source). Using KCL for currents entering and leaving node V, we write $\frac{V}{40} + \frac{(V + 150V)}{80} = 10$, from which $153V = 800$ and $V = 5.229$ V; then $R_{TH} = V_T/I_T = 5.229$ V/10 A = **0.5229 Ω** (the same answer as Ex. 18-28).

Several important insights may be drawn from Exs. 18-28 and 18-29:

1. Given portions of networks containing dependent sources only, the source-conversion theorem may be used to simplify the networks for the purpose of analysis.
2. Dependent sources in combination with passive circuit elements yield an equivalent passive impedance (resistance) whose terminal voltage is zero.
3. Only *practical dependent* current or voltage sources in passive networks may be converted. *Ideal* dependent (and/or independent) voltage or current sources are *incapable* of source conversion.
4. No hard and fast rule applies regarding the nature of the "known" voltage or current source to be applied to the passive network terminals, T–H, as will be shown by the examples below. The circuit may be solved with either assumption.

We are now ready to consider ac networks with *only* dependent sources. As indicated earlier, simpler networks are analyzed by using KCL around a node of interest or KVL around a given closed loop or mesh. More complex networks may be analyzed by nodal or mesh analysis, depending on whether we are dealing with current sources or voltage sources, respectively.

EXAMPLE 18-30

Given the passive network elements shown in **Fig. 18-27**, calculate the equivalent output impedance, $\mathbf{Z}_{TH}$, and equivalent output voltage, $\mathbf{V}_{TH}$

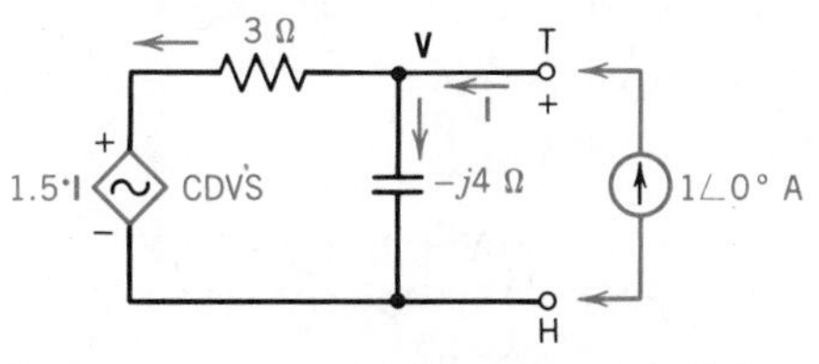

Figure 18-27 Examples 18-30 and 18-35

Solution

Since the CDVS shown depends on **I**, assume a $1\angle 0^\circ$ A source applied to T–H. Since $\mathbf{I} = 1\angle 0^\circ$ A, then $1.5 \cdot \mathbf{I} = 1.5\angle 0^\circ$ volts. We must now find **V**, where $\mathbf{V} = \mathbf{V}_{TH}$. Using KCL for the currents entering and leaving node **V** we write $\frac{\mathbf{V}}{-j4} + \frac{\mathbf{V} - 1.5\angle 0^\circ}{3\angle 0^\circ} = 1\angle 0^\circ$ amperes. Solving this equation for **V** yields $\mathbf{V}_{TH} = \mathbf{V} = \frac{1.5 + j0}{0.\bar{3} + j0.25} = 2.88 - j2.16$ volts $= 3.6\angle -36.87^\circ$ volts. Then the output impedance, $\mathbf{Z}_{TH} = \frac{\mathbf{V}_{TH}}{\mathbf{I}_{TH}} = \frac{3.6\angle -36.87^\circ \text{ V}}{1\angle 0^\circ \text{ A}} = \mathbf{3.6\angle -36.87^\circ\ \Omega}$. The equivalent output voltage, $\mathbf{V}_{TH} = \mathbf{0}$, since there are no independent sources in the passive network.

EXAMPLE 18-31

Repeat Ex. 18-30, using the source-conversion theorem on the CDVS but maintaining the same applied $1\angle 0^\circ$ A current at the input, to verify the solution.

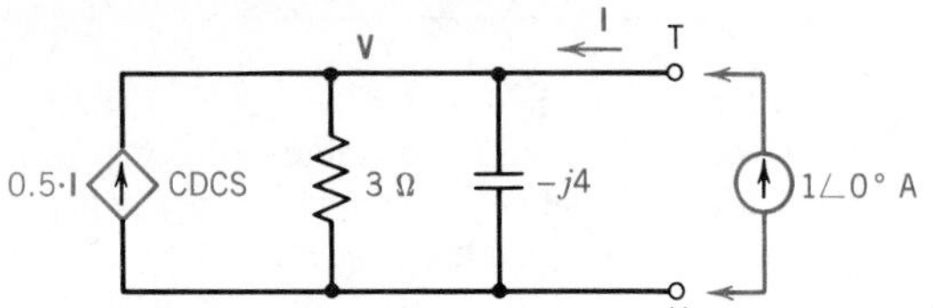

Figure 18-28 Example 18-31

Solution

The CDVS of Fig. 18-27 is converted to a CDCS in **Fig. 18-28**. This produces a simple four-branch parallel circuit whose node voltage is **V**. Using KCL for currents entering and leaving **V** yields $\frac{\mathbf{V}}{3\angle 0^\circ} + \frac{\mathbf{V}}{-j4} = 1\angle 0^\circ + 0.5(1\angle 0^\circ)$, which simplifies to $\mathbf{V}(0.3 + j0.25) = 1.5 + j0$, from which $\mathbf{V} = \frac{1.5 + j0}{0.3 + j0.25} = 3.6\angle -36.87^\circ$ V $= \mathbf{V}_{TH}$. Then $\mathbf{Z}_{TH} = \mathbf{V}_{TH}/\mathbf{I}_{TH} = 3.6\angle -36.87^\circ$ V$/1\angle 0^\circ$ A $=$ $\mathbf{3.6\angle -36.87^\circ}$ **Ω** (same as Ex. 18-30). The equivalent output voltage, $\mathbf{V}_{TH}$ is zero, as noted previously.

Examples 18-30 and 18-31 show, once again, that it is possible to use the source-conversion theorem in passive networks. They also show that we may still maintain the same assumed $1\angle 0^\circ$ A source and solve the network, regardless of the nature of the dependent source. Finally, since no independent sources exist in the network, the output voltage is zero.

EXAMPLE 18-32

For the passive network elements shown in **Fig. 18-29**,

a. Calculate the output impedance $\mathbf{Z}_{TH}$
b. Verify the output impedance by source conversion without changing the assumed source at the input

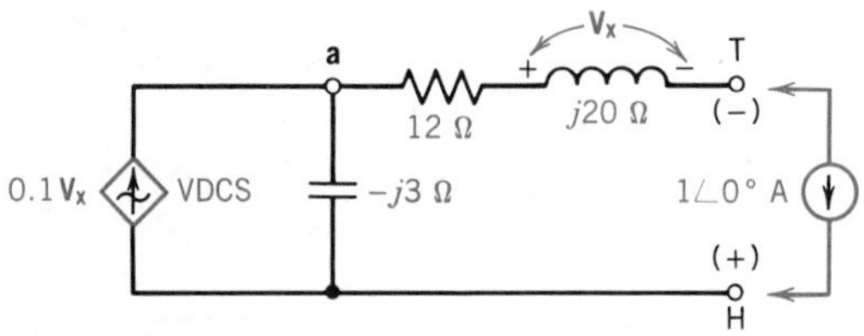

Figure 18-29 Example 18-32

Solution

The VDCS depends on $\mathbf{V}_x$ across the $j20$ Ω impedance. Therefore, if we assume a *downward* current of $1\angle 0^\circ$ A at the input to agree with the polarity given across $\mathbf{V}_x$, we can write the current in the capacitor as $\mathbf{I}_C = 0.1\mathbf{V}_x - 1$ amperes (at node **a**). We also know that $\mathbf{V}_x = (1\angle 0^\circ \text{ A})(20\angle 90^\circ\ \Omega) = 20\angle 90^\circ$ V and therefore $0.1\mathbf{V}_x = 2\angle 90^\circ$ A

a. Using KVL in a CW direction for the RH mesh to find $\mathbf{V}_{TH}$ yields $+1\angle 0^\circ(12 + j20) - (0.1\mathbf{V}_x - 1)(3\angle -90^\circ) - \mathbf{V}_{TH} = 0$, from which $\mathbf{V}_{TH} = 12 + j20 - (6 + j3) = 6 + j17$ volts. Then $\mathbf{Z}_{TH} = \mathbf{V}_{TH}/\mathbf{I}_{TH} = (6 + j17)$ V$/1\angle 0^\circ$ A $=$ $\mathbf{(6 + j17)\ \Omega}$
b. From **a** above, $\mathbf{V}_x = 20\angle 90^\circ$ V and $0.1\mathbf{V}_x = 2\angle 90^\circ$ A. By source conversion, the voltage of the dependent source, $\mathbf{V}_d = (2\angle 90^\circ \text{ A})(3\angle -90^\circ\ \Omega) = 6\angle 0^\circ$ V connected in series with an impedance $-j3$ Ω. This source conversion results in a simple series circuit (not shown) whose impedance $\mathbf{Z}_s = 12 + j20 - j3 = 12 + j17$ Ω carrying a total (assumed) current of $1\angle 0^\circ$ A, resulting in a net voltage of $12 + j17$ V across the impedance (only). Then by KVL for the series mesh in a CW direction we have $-\mathbf{V}_{TH} - 6\angle 0^\circ + (12 + j17) = 0$, from which $\mathbf{V}_{TH} = 6 + j17$ volts. As in **a** above, $\mathbf{Z}_{TH} = \mathbf{V}_{TH}/\mathbf{I}_{TH} = (6 + j17)$ V$/(1 + j0)$ A $=$ $\mathbf{(6 + j17)\ \Omega}$, which agrees with solution in **a**.

EXAMPLE 18-33

For the passive network elements shown in **Fig. 18-30**:

a. Calculate the input impedance, $\mathbf{Z}_{TH}$
b. Verify the answer in **a** by source conversion without changing the assumed source at the input

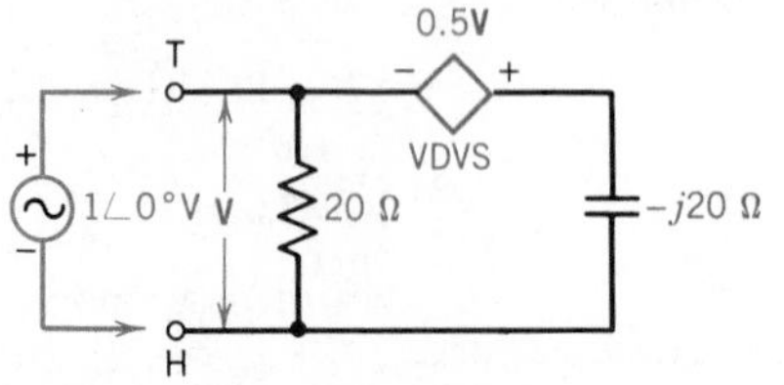

Figure 18-30 Examples 18-33 and 18-36

Solution

a. Since **V** is across the 20-Ω resistor and the VDVS depends on **V**, let us assign $\mathbf{V} = 1\angle 0^\circ$ V at input terminals T–H. Then $0.5\mathbf{V} = 0.5\angle 0^\circ$ V, enabling the solution of currents in each branch and the total current, using KCL at node **V**:

$$\mathbf{I}_{TH} = \frac{1\angle 0^\circ \text{ V}}{20\ \Omega} + \frac{1.5\angle 0^\circ \text{ V}}{-j20\ \Omega} = (0.05 + j0.075) \text{ A}$$

and
$$\mathbf{Z}_{TH} = \mathbf{V}_{TH}/\mathbf{I}_{TH} = (1 + j0) \text{ V}/(0.05 + j0.075) \text{ A} = 6.154 - j9.231 = \mathbf{11.094\angle -56.3^\circ\ \Omega}$$

b. Converting the VDVS to a VDCS by the source-conversion theorem, the current of the dependent source, $\mathbf{I}_{dep} = 0.5$ V$/20\angle -90^\circ$ Ω $= 0.025\angle 90^\circ$ A, paralleled by a $-j20$ Ω impedance. This results in a three-branch parallel circuit (not shown) whose (node) voltage is $1\angle 0^\circ$ V and whose total current $\mathbf{I}_{TH} = 0.05\angle 0^\circ + 0.05\angle 90^\circ + 0.025\angle 90^\circ$ amperes, respectively. The sum $\mathbf{I}_{TH} = (0.05 + j0.075)$ A, from which $\mathbf{Z}_{TH} = \mathbf{V}_{TH}/\mathbf{I}_{TH} = (1 + j0)$ V/(0.05 + j0.075) A = 6.154 − j9.231 = $\mathbf{11.094\angle -56.3^\circ}$ **Ω**, which is identical to the input impedance in **a**.

18-8.1 A Novel Insight into Solutions of Elements of DC and AC Networks Containing Dependent Sources Only Using Conventional Series–Parallel Circuit Theory

Since the previous examples always provided values of impedances and the dependence factor, **h**, the reader may ask, "Is there an alternative method for finding either input or output impedance without having to assume a source at the input?" The answer is that there is such a method.[19] This method enables one to use conventional series–parallel circuit theory in finding the impedance, $\mathbf{Z}_{TH}$, of passive dependent networks.

The equations given for finding either input or output impedance in Table 18-1 are derived in Appendix B-14. The derivations are relatively simple and easy to follow. The results of the derivations appear in the third (last) column of Table 18-1. With respect to the method shown in the last column, please note the following:

1. The method may be used only with *practical* **dependent** voltage and current sources, that is, dependent *voltage* sources having *series* impedance, $\mathbf{Z}_0$, and dependent current sources having parallel impedance, $\mathbf{Z}_0$. (It cannot be used with **ideal** dependent sources or current sources having series impedance or voltage sources shunted by impedance.)
2. The term **h** may be a real number or a complex number.
   - **a.** When **h** is a real number and greater than unity in the expression (1 − **h**), it is always used as a *positive* value of absolute magnitude, that is, (1 − 9) = +8
   - **b.** When **h** is a complex number $\pm j\mathbf{h}$, the expression is treated as a complex term, that is, 1 − **h**, when **h** = $-j5$, now becomes $1 - (-j5) = 1 + j5$
3. $\mathbf{Z}_0$ represents the sum of all impedance in series with a dependent voltage source in a given branch or impedances in parallel with a dependent current source. This also applies to $\mathbf{Z}_2$ in the last four rows of Table 18-1. $\mathbf{Z}_0$ or $\mathbf{Z}_2$ always represents the "lumped" impedance seen by the dependent voltage or current source.
4. The factor containing **h** is always in the *denominator* for *voltage* sources and in the *numerator* for *current* sources. The sign of **h** is always positive for aiding voltage and opposing current dependent sources.
5. Any practical dependent voltage source may be converted to a practical dependent current source and vice versa, using the source-conversion theorem. Frequently, this results in a value of **h** that is complex, but if treated as in (2b) above, it creates no problem.
6. The entire purpose of Table 18-1 is to *reduce* a practical dependent voltage or current source *to an equivalent impedance*. After that, simple series–parallel network theory may be used.

[19] This method of analysis emerged from the author's experience in teaching transistor circuit analysis. It was originally published in I. L. Kosow, *Study guide in alternating current circuits*, Wiley, New York, 1977.

Table 18-1 **Dependency Impedance Method for Finding Input (or Output) Impedance of Passive Dependent Networks**

Network Elements	Relation	Input (or Output) Impedance
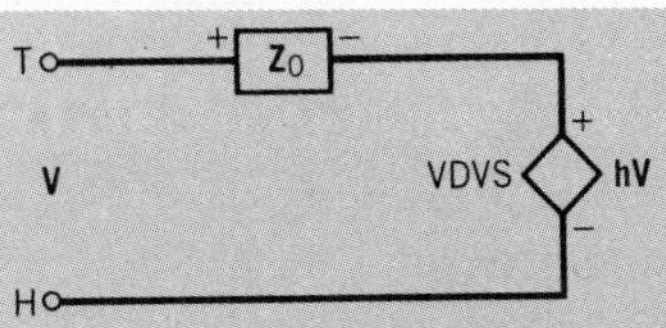	VDVS **opposes** **V** at input. Series opposition increases whenever $\mathbf{h} < 1$ and decreases whenever $\mathbf{h} > 2$.	$\mathbf{Z}_{TH} = \dfrac{\mathbf{Z}_0}{1 - \mathbf{h}}$
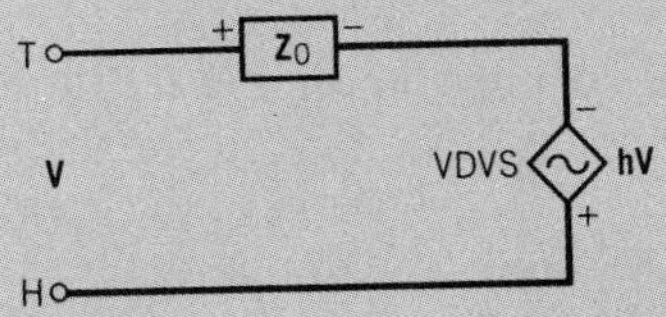	VDVS aids **V** at input. $\mathbf{Z}_{TH}$ is always reduced and is less than $\mathbf{Z}_0$ whether **h** is a fractional or integral factor.	$\mathbf{Z}_{TH} = \dfrac{\mathbf{Z}_0}{1 + \mathbf{h}}$
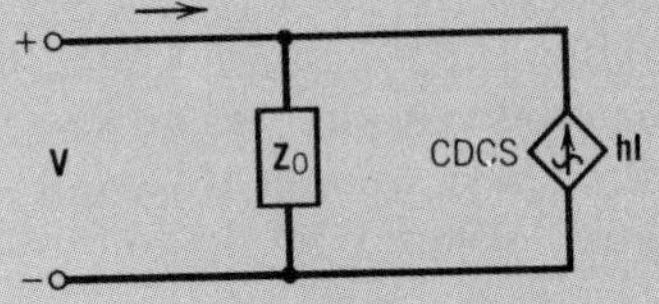	CDCS opposes input **V** and current **I**. Parallel opposition increases $\mathbf{Z}_{TH}$. $\mathbf{Z}_{TH} > \mathbf{Z}_0$ at all times.	$\mathbf{Z}_{TH} = \mathbf{Z}_0(1 + \mathbf{h})$
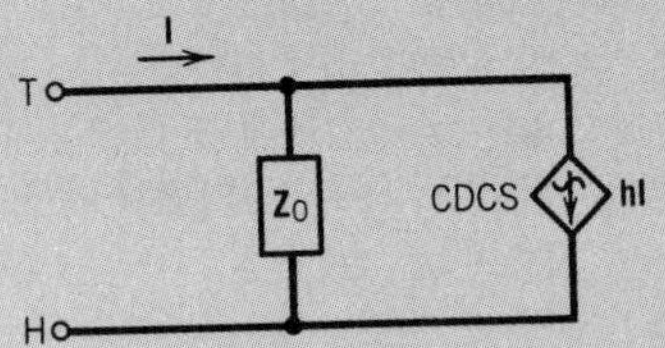	CDCS aids input **V** and current **I**. Whenever $\mathbf{h} < 1$, $\mathbf{Z}_{TH} < \mathbf{Z}_0$. Whenever $\mathbf{h} > 2$, $\mathbf{Z}_{TH} > \mathbf{Z}_0$.	$\mathbf{Z}_{TH} = \mathbf{Z}_0(1 - \mathbf{h})$
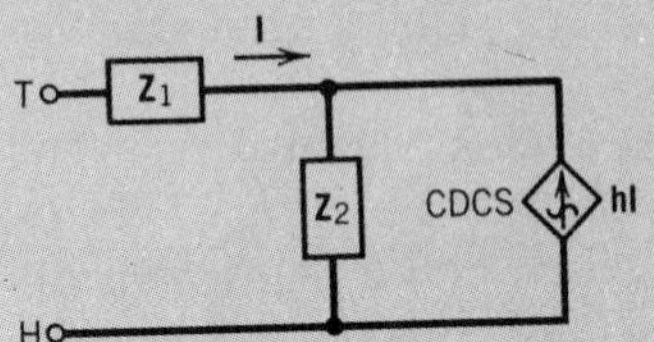	CDCS opposes input **V** and **I**. Parallel opposition increases $\mathbf{Z}_2$, which is in series with $\mathbf{Z}_1$.	$\mathbf{Z}_{TH} = \mathbf{Z}_1 + \mathbf{Z}_2(1 + \mathbf{h})$
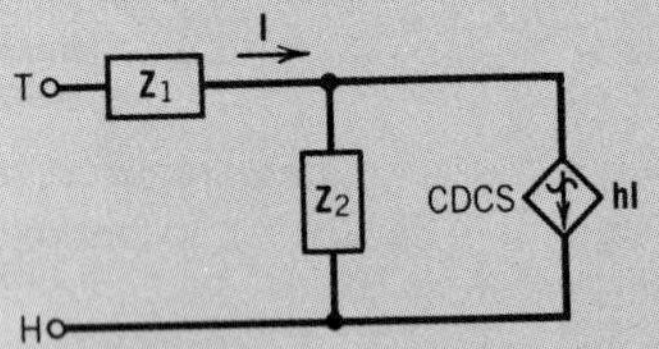	CDCS aids input **V** and current **I**. $\mathbf{Z}_2$ is either increased or decreased depending on whether $\mathbf{h} > 2$ or $\mathbf{h} < 1$.	$\mathbf{Z}_{TH} = \mathbf{Z}_1 + \mathbf{Z}_2(1 - \mathbf{h})$
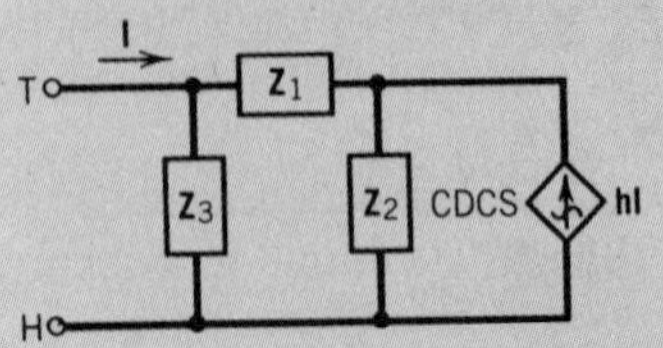	CDCS opposes input **V** and **I** but only $\mathbf{Z}_2$ is increased due to parallel opposition.	$\mathbf{Z}_{TH} = \dfrac{\mathbf{Z}_3}{\mathbf{Z}_1 + \mathbf{Z}_2 + \mathbf{Z}_3}[\mathbf{Z}_1 + \mathbf{Z}_2 + \mathbf{Z}_2(1 + \mathbf{h})]$
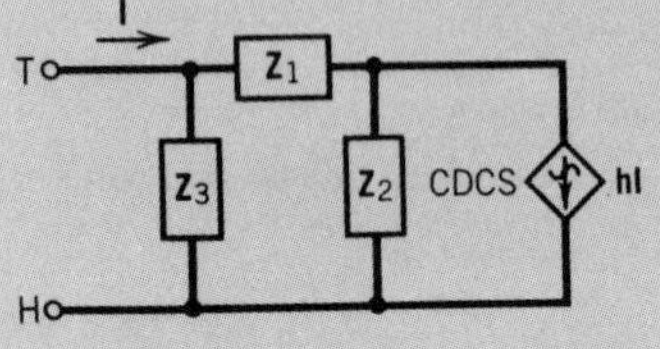	CDCS aids input **V** and **I** but only $\mathbf{Z}_2$ is affected. $\mathbf{Z}_2$ is either decreased or increased depending on whether $\mathbf{h} < 1$ or $\mathbf{h} > 2$.	$\mathbf{Z}_{TH} = \dfrac{\mathbf{Z}_3}{\mathbf{Z}_1 + \mathbf{Z}_2 + \mathbf{Z}_3}[\mathbf{Z}_1 + \mathbf{Z}_2 + \mathbf{Z}_2(1 - \mathbf{h})]$

EXAMPLE 18-34

Calculate the input impedance Z_{TH} for the passive circuit elements shown in Fig. 18-26 using conventional series–parallel circuit theory by

a. Retaining the VDVS
b. Source conversion to a VDCS *and back* to a VDVS

Solution

a. The branch containing the VDVS has a total resistance of 80 Ω. Since the VDVS *aids* the input voltage, $Z_{vdvs} = \frac{80\ \Omega}{(1 + 150)} = 0.5298\ \Omega$. BUT this impedance is in parallel with a 40-Ω resistor. Consequently, $Z_{TH} = (0.5298)\|(40) =$ **0.5229 Ω**. Note that this is exactly the *same* answer as that obtained in Ex. 18-29.
b. This technique is somewhat more complex since it involves *two* source conversions. Converting the VDVS to a VDCS yields a current source of 1.875**V** in parallel with two resistors of 40 Ω and 80 Ω. These, when lumped to a single resistance, produce an equivalent resistance of $26.\bar{6}\ \Omega$ in parallel with a VDCS of 1.875**V**. But this now requires that we reconvert this combination back to a VDVS since **V** is the factor on which the dependency is determined (second row of Table 18-1). The VDVS value is now $I_N R_N = (1.875\mathbf{V})(26.\bar{6}\ \Omega) = 50\mathbf{V}$. Consequently, $Z_{TH} = Z_0/(1 + \mathbf{h}) = 26.\bar{6}\ \Omega/51 =$ **0.5229 Ω**. Note that this is the *same* answer as that obtained in **a** above and also in Ex. 18-29!

EXAMPLE 18-35

Given the network shown in Fig. 18-27, calculate the output impedance Z_{TH} by conventional series–parallel feedback circuit theory

Solution

We must first convert the dependent voltage source to a current source since the factor **h** is a function of current. Therefore $I_{cdcs} = 1.5\mathbf{I}/3\ \Omega = 0.5\mathbf{I}$ paralleled by a 3-Ω resistor which is *also paralleled by* $-j4\ \Omega$. We must now "lump" the two into a single impedance where $Z_0 = 3\ \Omega\| -j4$ or $Z_0 = (3\angle 0^\circ \times 4\angle -90^\circ)/(3 - j4) = 2.4\angle -36.87^\circ\ \Omega$. Since the current source is opposing the voltage source (third row Table 18-1), $Z_{TH} = Z_0(1 + \mathbf{h}) = 2.4\angle -36.87^\circ)(1 + 0.5) =$ **$3.6\angle -36.87^\circ$ Ω**. Note that this is exactly the *same* answer as that obtained in Ex. 18-30!

EXAMPLE 18-36

Given the network shown in Fig. 18-30, calculate the input impedance Z_{TH} by conventional series–parallel feedback circuit theory

Solution

Since the VDVS depends on input voltage **V**, no source conversion is required. The impedance of the branch containing the dependent source is $Z_{vdvs} = Z_s/(1 + \mathbf{h}) = -j20\ \Omega/(1 + 0.5) = -j13.\bar{3}\ \Omega$. Now $Z_{TH} = -j13.\bar{3}\|20\ \Omega = 20\angle 0^\circ \times 13.\bar{3}\angle -90^\circ/(20 - j13.\bar{3}) =$ **$11.09\angle 56.31^\circ$ Ω**. Note that this is exactly the *same* answer as that obtained in Ex. 18-33!

The foregoing examples yield the following important insights with regard to Table 18-1 and its use:

1. We now have an *alternative* method of verification in addition to the method of applying either a known voltage or current to the passive network.
2. When given dependent current sources, these must be dependent on the total circuit current, **I**, and not on currents elsewhere in the network. The same applies to dependent voltage sources shown in the first two rows of Table 18-1. Given other dependences, Z_{TH} must be derived in the way shown in Appendix B-14.
3. The six major *caveats* in the use of Table 18-1 should be reread most carefully before attempting the above examples independently, as well as the problems at the end of this chapter.

18-9 GLOSSARY OF TERMS USED

Active circuit element Source of electrical energy that supplies power to an electric network. In ac circuits such elements may be alternators, transformers, transistors, and so forth.

Bilateral circuit element Element or device that is unaffected by exchanging connections at its terminals. Resistors, inductors, and (most) capacitors are bilateral *passive* devices. Diodes are *unilateral* passive elements.

Controlled source See dependent source.

Dependent source Source whose voltage or current depends on and is proportional to the voltage or current in some other part of a given network. There are four possible types of dependent sources, designated by diamonds in circuit diagrams:

a. Voltage-dependent voltage source (VDVS)

b. Voltage-dependent current source (VDCS)
c. Current-dependent current source (CDCS)
d. Current-dependent voltage source (CDVS)

Ideal current source Two-terminal independent source whose output current remains constant regardless of terminal voltage and load impedance variations. The internal impedance (resistance) of this source is infinity.

Ideal voltage source Two-terminal independent source whose output voltage remains constant regardless of load variations and current drawn by the load. The internal impedance (resistance) of this source is zero.

Independent source Source (either voltage or current and either dc or ac) having one pair of output terminals whose output is independent of the circuit to which it is connected. All independent sources are designated by circles in circuit diagrams.

Linear device (or **element**) Element whose volt-ampere characteristic is constant (*i.e.*, a straight line such that $dv/di = k$) over its entire range of operation.

Loop Closed path in an ac (or dc) network, beginning and ending at the same node.

Mesh analysis Circuit analysis technique in which the mesh (or loop) currents are found through simultaneous KVL equations written for each mesh.

Nodal analysis Circuit analysis technique in which the node voltages are found through simultaneous KCL equations written for the currents entering and leaving each node.

Node Junction where two or more currents merge, or terminal common to two or more branches in a network.

Nonlinear device Device having a volt-ampere characteristic in which a unit change in voltage does not produce a corresponding unit change in current (or vice versa) over its entire range of operation (i.e., $dv/di \neq k$).

Open-circuit voltage, $\mathbf{V}_{oc}$ Voltage measured across the output terminals of an ideal or practical voltage source (either independent or dependent) when no current (or load) is drawn from the source. For an ideal source, $\mathbf{V}_{oc} = \mathbf{V}_L = k$; that is, the voltage across the source is constant regardless of load current drawn.

Passive circuit Circuit whose elements contain no sources.

Practical current source Independent current source *shunted* by some finite internal impedance, $\mathbf{Z}_i$, whose terminal voltage varies with the load current drawn.

Practical voltage source Independent voltage source having finite internal impedance ($\mathbf{Z}_{int} = k$), whose terminal voltage varies with load current drawn from the source. Any practical voltage source may be considered equivalent to an ideal voltage source connected in series with some finite internal impedance, $\mathbf{Z}_i$.

Short-circuit current, $\mathbf{I}_{sc}$ Condition under which the terminals of a source are short-circuited and the short-circuit current that results is limited only by the internal impedance of the source.

Source Active pair of terminals that deliver power to a load.

18-10 PROBLEMS

Sec. 18-2

18-1 Given the network shown in **Fig. 18-31**, $\mathbf{V}_1 = (20 + j0)$ V, $\mathbf{Z}_1 = (8 - j15)\ \Omega$, $\mathbf{Z}_2 = (12 + j5)\ \Omega$, and $\mathbf{Z}_L = 10\angle 45^\circ\ \Omega$. Using mesh analysis, format method, calculate the

a. Mesh current $\mathbf{I}_1$
b. Mesh current $\mathbf{I}_2$
c. $\mathbf{V}_L$ across load $\mathbf{Z}_L$
d. Node voltage across all branches
e. True current in impedance $\mathbf{Z}_2$ by two methods

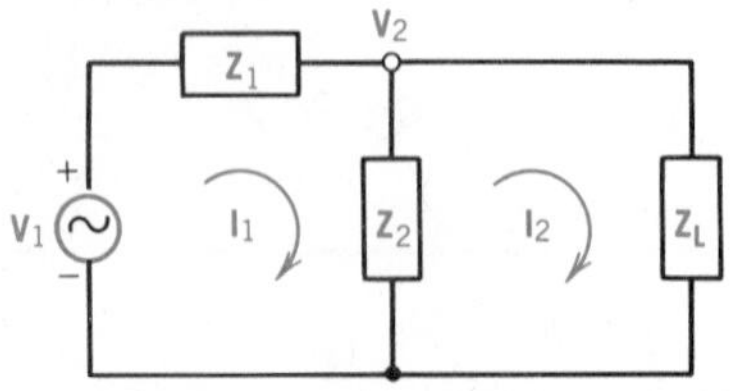

Figure 18-31 Problems 18-1, 18-2, 18-8, and 18-9

18-2 Repeat all parts of Prob. 18-1 given in Fig. 18-31: $\mathbf{V}_1 = 49\angle 0^\circ$ V, $\mathbf{Z}_1 = 34\angle 28.07^\circ\ \Omega$, $\mathbf{Z}_2 = 17\angle 61.93^\circ\ \Omega$, $\mathbf{Z}_L = (6 - j8)\ \Omega$

18-3 Repeat all parts of Prob. 18-1 given in **Fig. 18-32**: $\mathbf{V}_1 = 48\angle -45^\circ$ V, $\mathbf{V}_2 = 60\angle 0^\circ$ V, $\mathbf{Z}_1 = \mathbf{Z}_2 = 100\angle 0^\circ\ \Omega$, and $\mathbf{Z}_L = -j40\ \Omega$

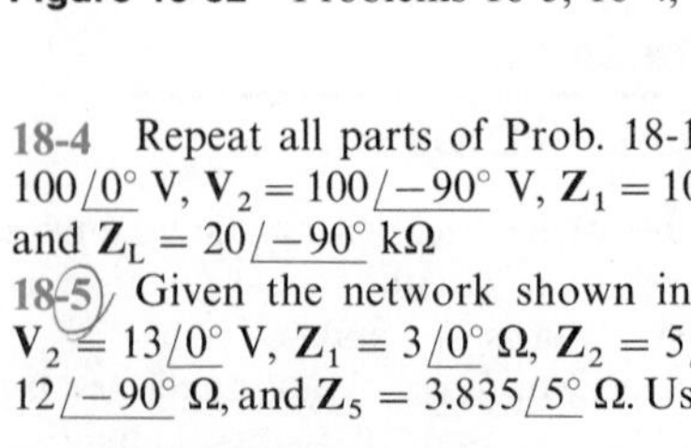

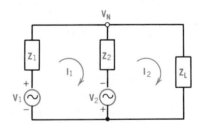

Figure 18-32 Problems 18-3, 18-4, 18-10, and 18-11

18-4 Repeat all parts of Prob. 18-1 given in Fig. 18-32: $\mathbf{V}_1 = 100\angle 0^\circ$ V, $\mathbf{V}_2 = 100\angle -90^\circ$ V, $\mathbf{Z}_1 = 10\angle 90^\circ$ kΩ, $\mathbf{Z}_2 = 5\angle -90^\circ$ kΩ, and $\mathbf{Z}_L = 20\angle -90^\circ$ kΩ

18-5 Given the network shown in **Fig. 18-33**, $\mathbf{V}_1 = 10\angle 0^\circ$ V, $\mathbf{V}_2 = 13\angle 0^\circ$ V, $\mathbf{Z}_1 = 3\angle 0^\circ\ \Omega$, $\mathbf{Z}_2 = 5\angle 0^\circ\ \Omega$, $\mathbf{Z}_3 = 4\angle 90^\circ\ \Omega$, $\mathbf{Z}_4 = 12\angle -90^\circ\ \Omega$, and $\mathbf{Z}_5 = 3.835\angle 5^\circ\ \Omega$. Using mesh analysis, calculate:

a. Mesh current $\mathbf{I}_1$
b. Mesh current $\mathbf{I}_2$

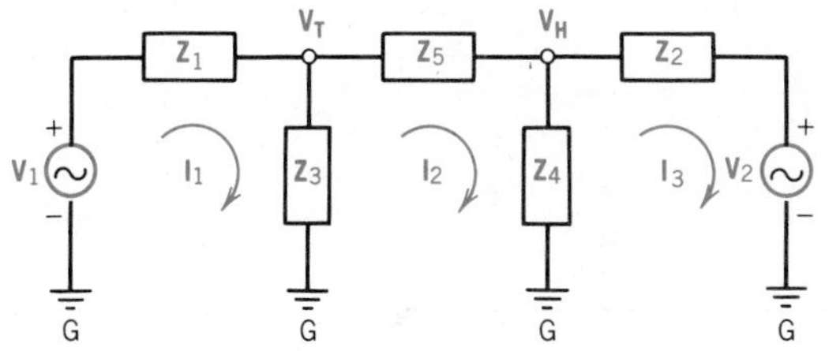

Figure 18-33 Problems 18-5, 18-6, 18-12, and 18-13

c. Mesh current $\mathbf{I}_3$
d. $\mathbf{V}_T$ by two methods
e. $\mathbf{V}_H$ by two methods

18-6 Repeat all parts of Prob. 18-5, given the following values for Fig. 18-33: $\mathbf{V}_1 = (12 + j35)$ V, $\mathbf{V}_2 = (15 + j8)$ V, $\mathbf{Z}_1 = (3 + j4)\ \Omega$, $\mathbf{Z}_2 = (4 - j3)\ \Omega$, $\mathbf{Z}_3 = (12 - j5)\ \Omega$, $\mathbf{Z}_4 = (5 - j12)\ \Omega$, and $\mathbf{Z}_5 = (1.81 + j6.2)\ \Omega$

18-7 Using mesh analysis for the network shown in **Fig. 18-34**,

a. Write four equations in simplest form for finding mesh currents $\mathbf{I}_1$, $\mathbf{I}_2$, $\mathbf{I}_3$, and $\mathbf{I}_4$ *but do not solve array.* Express these equations in a 4 × 4 matrix suitable for computer solution. A computer solution of the above matrix yields the following values for the mesh currents: $\mathbf{I}_1 = (0.5336 + j0.9252)$ A, $\mathbf{I}_2 = (2.0906 + j0.29853)$ A, $\mathbf{I}_3 = (-0.6506 - j0.20031)$ A $\mathbf{I}_4 = (1.5701 + j0.07179)$ A
b. Using your first equation for mesh current $\mathbf{I}_1$ and the above solutions, verify KVL by showing that the algebraic sum of all volt drops in that mesh is zero
c. Repeat (**b**) for mesh $\mathbf{I}_2$
d. Using the given currents, calculate the current in $\mathbf{R}_1$
e. Using any method you choose, calculate the respective voltages at points **y** and **x**
f. Using the voltages found in (**e**), verify the current in $\mathbf{R}_1$ and compare your answer to that found in (**d**) above

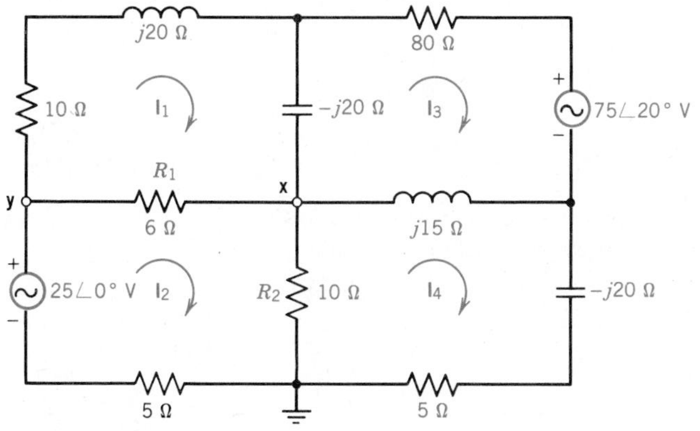

Figure 18-34 Problem 18-7

Sec. 18-3

18-8 Using nodal analysis, format method, given the data of Prob. 18-1 and the network shown in Fig. 18-31, convert the practical independent voltage source to a current source and calculate

a. Node voltage $\mathbf{V}_2$
b. Current $\mathbf{I}_L$ in load $\mathbf{Z}_L$
c. Current in impedance $\mathbf{Z}_2$
d. Compare your answers with those obtained in Prob. 18-1 above and those of Probs. 17-4 and 17-5 of the previous chapter.

18-9 Using nodal analysis and the data of Prob. 18-2, repeat parts **a**, **b**, and **c** of Prob. 18-8
d. Compare your answers with those obtained in Prob. 18-2 using mesh analysis as well as those of Probs. 17-4 and 17-5 of the previous chapter

18-10 Using nodal analysis and the data of Prob. 18-3 for the network shown in Fig. 18-32, convert all practical voltage sources to current sources and calculate the

a. Node voltage $\mathbf{V}_N$
b. Load current $\mathbf{I}_L$ in $\mathbf{Z}_L$
c. Current supplied to the node by practical voltage source $\mathbf{V}_1$ (use Fig. 18-32)
d. Current drawn from the node by practical voltage source $\mathbf{V}_2$ (use Fig. 18-32)
e. Verify your answer by KCL around node $\mathbf{V}_N$
f. Compare above answers with those obtained in Prob. 18-3 using mesh analysis and those of Prob. 17-10 using other methods of analysis

18-11 Using nodal analysis and the given data of Prob. 18-4, repeat parts **a** through **e** of Prob. 18-10
f. Compare your answers with those obtained in Prob. 18-4 using mesh analysis

18-12 Using nodal analysis and the given data of Prob. 18-5 (Fig. 18-33), convert both practical voltage sources to current sources (draw a new circuit diagram) and calculate

a. Node voltage $\mathbf{V}_T$
b. Node voltage $\mathbf{V}_H$
c. Current in $\mathbf{Z}_1$ (use Fig. 18-33)
d. Current in $\mathbf{Z}_2$ (use Fig. 18-33)
e. Current in $\mathbf{Z}_3$
f. Current in $\mathbf{Z}_4$
g. Current in $\mathbf{Z}_5$
h. Compare all currents versus the solutions obtained in Prob. 18-5 using mesh analysis
i. Compare current in $\mathbf{Z}_5$ versus the answer obtained in Prob. 17-18c by Thévenin analysis

18-13 Using nodal analysis and the given data of Prob. 18-6 (Fig. 18-33), repeat parts **a** through **g** of Prob. 18-12
h. Compare all currents versus the solutions obtained in Prob. 18-6 using mesh analysis
i. Compare current in $\mathbf{Z}_5$ versus the answer obtained in Prob. 17-19**c** by Thévenin analysis

Sec. 18-3.1

18-14 Given the network shown in **Fig. 18-35**, $\mathbf{V}\angle\theta = 20\angle 0^\circ$ V, $\mathbf{R}_s = 4\angle 0^\circ$ kΩ, $\mathbf{R}_f = 6\angle 0^\circ$ kΩ, $-j\mathbf{X}_{Cs} = -j3$ kΩ, $-j\mathbf{X}_{Cx} = -j4$ kΩ, and $\mathbf{Z}_{VM} = 10\angle 53.13^\circ$ kΩ. Using the alternative approach to nodal analysis, calculate

a. $\mathbf{V}_T$
b. $\mathbf{V}_H$
c. $\mathbf{V}_{TH}$
d. Current in $\mathbf{Z}_{VM}$
e. Compare the above answers with those found in the solution to Prob. 17-21**e** and **f**
f. Explain why this problem cannot be solved exactly by conventional nodal analysis
g. Explain the major advantage of the nodal analysis solution over the method used in Prob. 17-21

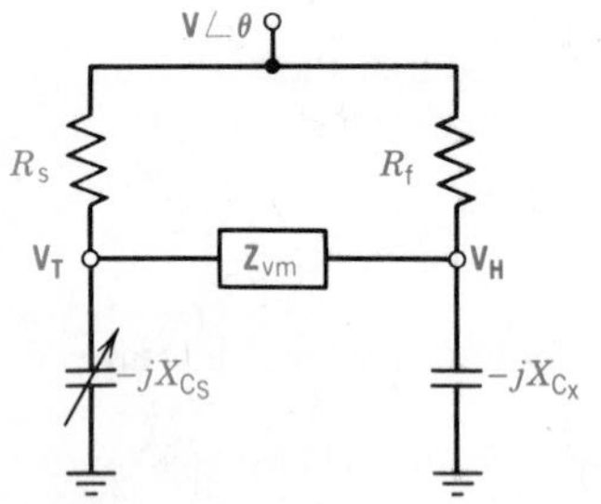

Figure 18-35 Problems 18-14 and 18-15

18-15 a. Verif your solutions to Prob. 18-14 using KCL to show that the currents entering and leaving node $\mathbf{V}_T$ are zero
b. Repeat (**a**) for the currents entering and leaving node $\mathbf{V}_H$
c. Explain why this internal method of validation is *not* possible using the Thévenin analysis technique of Prob. 17-21

Sec. 18-5.1

18-16 Given the network shown in **Fig. 18-36**, calculate

a. **I**
b. $\mathbf{V}_{TH}$

(Hint: convert CDCS to a CDVS and use KVL in series loop to find **I**)

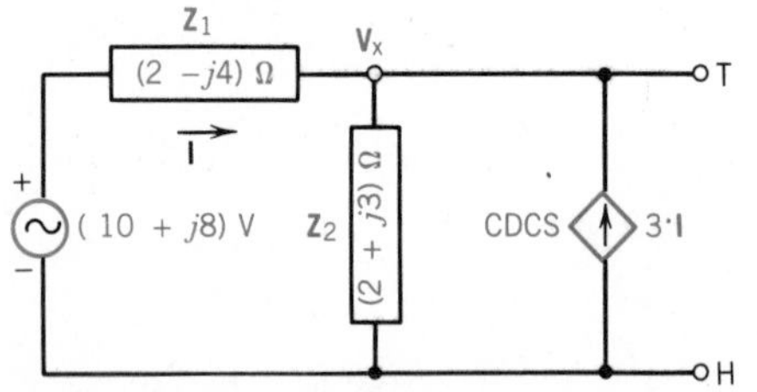

Figure 18-36 Problems 18-16, 18-23, 18-33, and 18-45

18-17 Given the network shown in **Fig. 18-37**, calculate using KVL:

a. **I**
b. $\mathbf{V}_x$

Hint: $\mathbf{V}_x = (8\underline{/90^\circ} - 4\mathbf{I})$

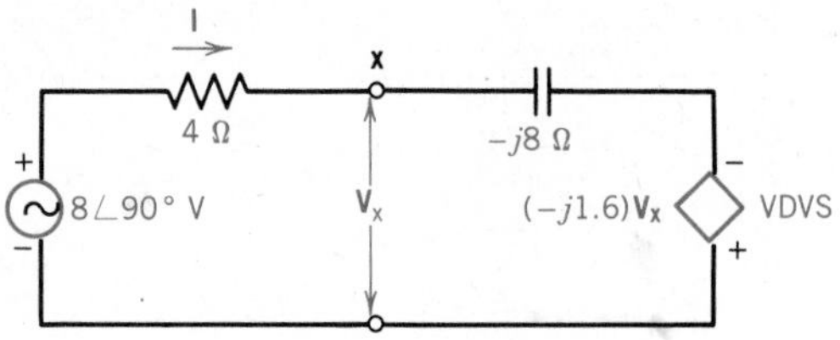

Figure 18-37 Problems 18-17, 18-24, and 18-41

18-18 Given the network shown in **Fig. 18-38**, using source conversion, convert *both* practical current sources to voltage sources and using KVL in a CCW mesh calculate

a. **I**
b. $\mathbf{V}_2$
c. $\mathbf{V}_{TH}$

Hint: $\mathbf{V}_2 = (8\underline{/45^\circ} - 2\mathbf{I})$

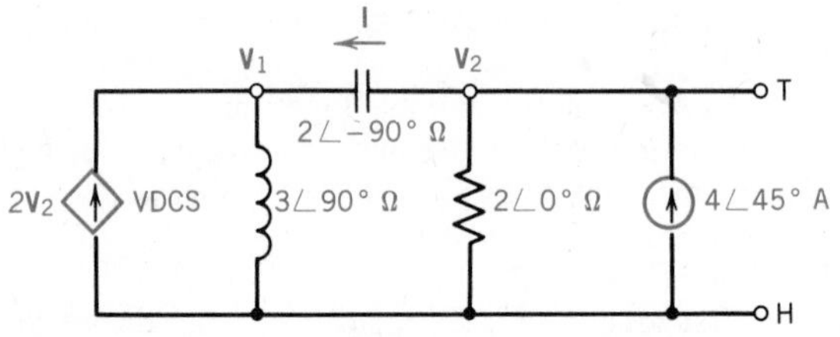

Figure 18-38 Problems 18-18 and 18-30

18-19 Using $\mathbf{V}_{TH}$ as the open circuit voltage, $\mathbf{V}_{oc}$ from Prob. 18-18, calculate

a. Short-circuit current, $\mathbf{I}_{sc}$
b. Output impedance, $\mathbf{Z}_{TH}$

Hint: Use Fig. 18-38 when shorting terminals T–H.

18-20 Given the network shown in **Fig. 18-39**, convert (only) the independent current source to a voltage source, and using KVL calculate

a. **I**
b. **V**
c. $\mathbf{V}_{TH}$
d. Complex power dissipated by each circuit component ($\mathbf{S}^* = \mathbf{VI}^*$)
e. Verify the accuracy of the solution by showing that Tellegen's theorem holds for the network

Hint: $\mathbf{V}_{TH} = 1.5 \times \mathbf{V}$ or $-j5\mathbf{I}$ and $\mathbf{V} = (100\underline{/0^\circ} - 10\mathbf{I})$

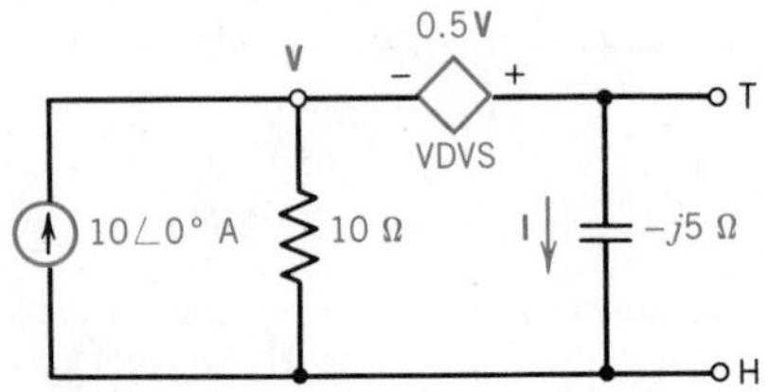

Figure 18-39 Problems 18-20, 18-31, 18-42, and 18-43

18-21 Given the network shown in **Fig. 18-40**, convert CDCS to a practical CDVS and using KVL (in a CW mesh) calculate

a. **I**
b. Node voltage **V**

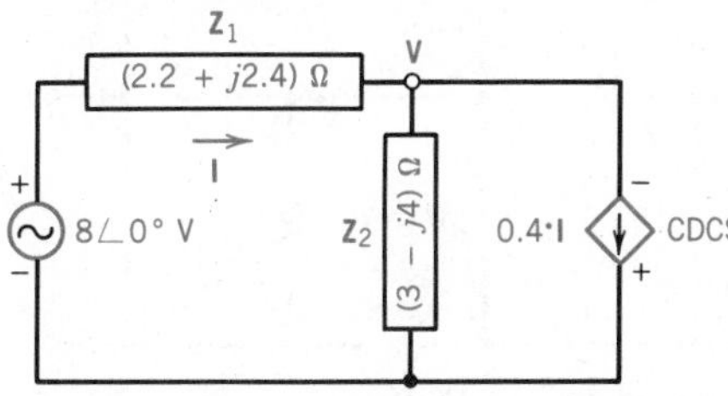

Figure 18-40 Problems 18-21, 18-22, 18-32, and 18-44

Sec. 18-5.2

18-22 For the network shown in Fig. 18-40 (in original form), using parallel circuit relations, calculate

a. **I**
b. Node voltage **V**

Hint: Use KCL at node **V** and KVL for the left-hand mesh

18-23 For the network shown in Fig. 18-36 in its original form, using parallel circuit relations, calculate

a. **I**
b. $\mathbf{V}_x$
c. $\mathbf{V}_{TH}$
d. Verify your solution by comparison with Prob. 18-16

18-24 Given the network shown in Fig. 18-37, convert *both* practical voltage sources to current sources. Using KCL for currents leaving and entering node $\mathbf{V}_x$, calculate

a. $\mathbf{V}_x$
b. Current in 4-Ω resistor shunting independent current source
c. Compare this value of $\mathbf{V}_x$ with the solution for $\mathbf{V}_x$ in Prob. 18-17
d. Explain why the current in the 4-Ω resistor in (**b**) above is not the same as **I** in Fig. 18-37

18-25 Given the original network shown in Fig. 18-11a and without using source-conversion techniques, calculate

a. $\mathbf{V}_R$
b. $\mathbf{V}_{TH}$
c. $\mathbf{Z}_{TH}$
d. Compare your solutions to those given in Ex. 18-15

18-26 Verify the solution of Ex. 18-16 for the network of Fig. 18-12 by converting *all* practical current sources to voltage sources and using KCL at node *V* to calculate

a. $\mathbf{V}_{TH}$
b. $\mathbf{R}_{TH}$
c. Compare your solutions above to those given in Ex. 18-16

Sec. 18-6

18-27 Given the network shown in **Fig. 18-41** and using mesh analysis, calculate

a. $\mathbf{I}_1$
b. $\mathbf{I}_2$
c. $\mathbf{V}_1$
d. $\mathbf{V}_2$
e. $\mathbf{V}_{TH}$

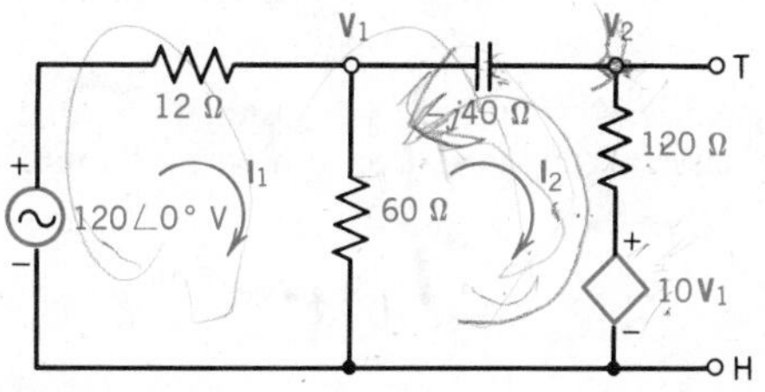

Figure 18-41 Problems 18-27, 18-34, and 18-48

18-28 Given the network shown in **Fig. 18-42** and using mesh analysis, calculate

a. $\mathbf{I}_1$
b. $\mathbf{I}_2$
c. **V**
d. $\mathbf{V}_{TH}$
e. $\mathbf{Z}_{TH}$
f. Draw the simplest practical voltage source to replace the network

Hint: To find $\mathbf{Z}_{TH}$, short terminals T–H and use superposition to find $\mathbf{I}_{sc}$ due to each source

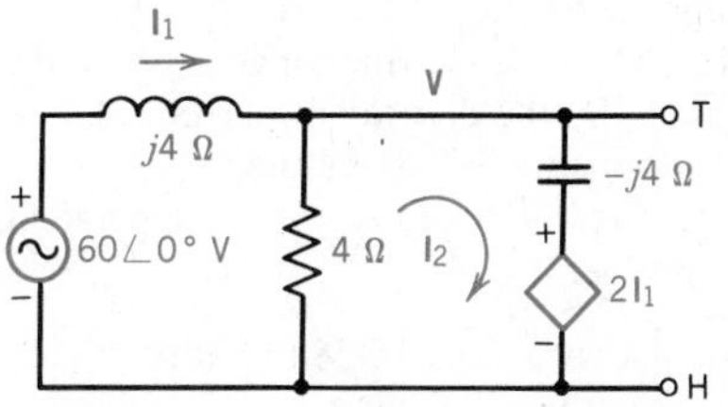

Figure 18-42 Problems 18-28, 18-37, and 18-46

18-29 Given the network shown in **Fig. 18-43**, using mesh analysis, calculate

a. $\mathbf{I}_1$
b. $\mathbf{I}_2$
c. **V**

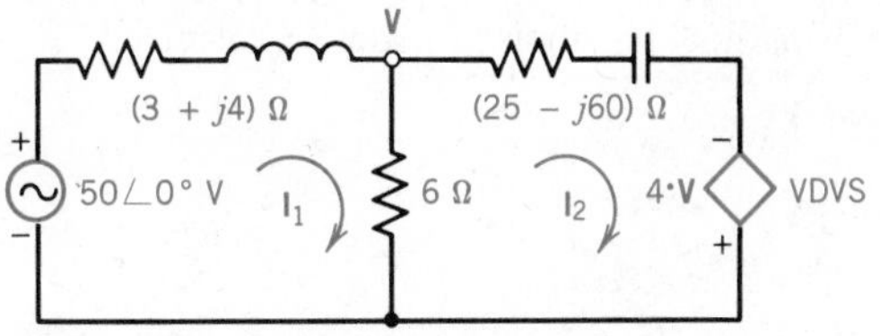

Figure 18-43 Problems 18-29 and 18-47

Sec. 18-7

18-30 Given the network in the form shown in Fig. 18-38, using nodal analysis, calculate

a. $\mathbf{V}_1$
b. $\mathbf{V}_2$
c. $\mathbf{V}_{TH}$
d. $\mathbf{Z}_{TH}$
e. Current in the capacitor, **I**
f. Draw the simplest practical voltage source to replace the network
g. Compare your solutions for **b** and **e** above to those obtained in Prob. 18-18

18-31 Given the network shown in Fig. 18-39, using nodal analysis, convert the dependent voltage source to a current source and calculate

a. Node voltage **V**
b. $\mathbf{V}_{TH}$
c. $\mathbf{Z}_{TH}$

(Hint: $\mathbf{V}_{TH} = \mathbf{V} + 0.5\mathbf{V} = 1.5\mathbf{V}$)

d. Draw the simplest practical voltage source to replace the network
e. Compare your solutions to those obtained in Prob. 18-20

18-32 Given the original network (without source conversion) shown in Fig. 18-40 and using nodal analysis, calculate

a. Node voltage **V**
b. Current **I** supplied by the $8\angle 0^\circ$ V source
c. Current in and voltage across the CDCS
d. Current drawn by the $(3 - j4)$ load
e. Complex powers dissipated by the two loads
f. Complex power generated by the independent voltage source
g. Complex power generated by the dependent current source
h. Verify the above solutions using Tellegen's theorem
i. Explain why it is not permissible to convert the independent voltage source in this problem

18-33 Given the network shown in Fig. 18-36 (without source conversion) and using nodal analysis, calculate

a. $\mathbf{V}_x$
b. **I**, supplied by the independent voltage source
c. Compare the above solutions to those in Probs. 18-16 and 18-23
d. Explain why it is not permissible to convert the independent voltage source in this problem

18-34 Given the network shown in Fig. 18-41 without source conversion but using nodal analysis at node $\mathbf{V}_1$, calculate

a. Node voltage $\mathbf{V}_1$
b. $\mathbf{I}_2$
c. Node voltage $\mathbf{V}_2$
d. $\mathbf{V}_{TH}$
e. $\mathbf{I}_{sc}$
f. $\mathbf{Z}_{TH}$

(Hint: use superposition to find $\mathbf{I}_{sc}$)

g. Draw the simplest practical voltage source to replace the entire network of Fig. 18-41
h. Compare pertinent solution vis-à-vis those in Prob. 18-27

18-35 Using nodal analysis on the network shown in **Fig. 18-44**, calculate

a. $\mathbf{V}_1$
b. $\mathbf{V}_2$
c. $\mathbf{V}_{TH}$

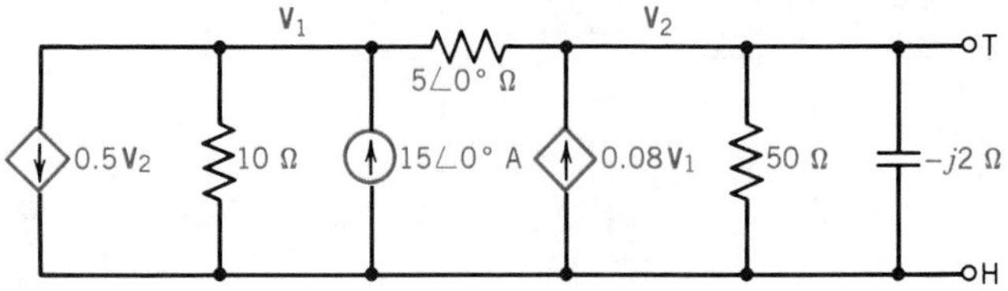

Figure 18-44 Problems 18-35 and 18-36

18-36 Short-circuit terminals T–H in Fig. 18-44 (see Hint, below) and using superposition calculate the

a. Short-circuit current due to independent current source
b. $\mathbf{V}_1$
c. Short-circuit current due to dependent source $0.08\mathbf{V}_1$
d. $\mathbf{Z}_{TH}$ seen at the output terminals of the network
e. Draw the simplest practical voltage source to replace the entire network

Hint: Convert the independent current source to a voltage source and draw the equivalent circuit when terminals T–H are shorted

18-37 Given the network shown in Fig. 18-42, using nodal analysis without source conversion, calculate

a. **V** c. $\mathbf{I}_2$
b. $\mathbf{I}_1$ d. $\mathbf{Z}_{TH}$
e. Draw the simplest equivalent practical voltage source to replace the network
f. Compare the above solutions versus those obtained in Prob. 18-28 using mesh analysis

18-38 In Ex. 18-24 the network of Fig. 18-20 was solved using mesh analysis. Convert only the dependent voltage source to a practical dependent current source and, using nodal analysis, calculate

a. $\mathbf{V}_1$ c. $\mathbf{I}_1$
b. $\mathbf{V}_2$ d. $\mathbf{I}_2$
e. $\mathbf{V}_{TH}$
f. Compare the above solutions with those obtained in Exs. 18-24 and 18-25
g. Explain why the independent voltage source cannot be converted

18-39 One possible equivalent circuit for a transistor amplifier is shown in **Fig. 18-45**.

a. Simplify Fig. 18-45, showing only current sources and only node voltage **V**
b. Calculate **V**, both magnitude and phase (see Hints below)
c. Calculate voltage gain both as a ratio and in dB
d. Explain why $\mathbf{R}_1$ may be deleted for the purposes of circuit analysis

Hints:

1. Find current in the dependent current source, $250I_1$
2. Delete R_1 and convert the practical voltage source to its equivalent current source

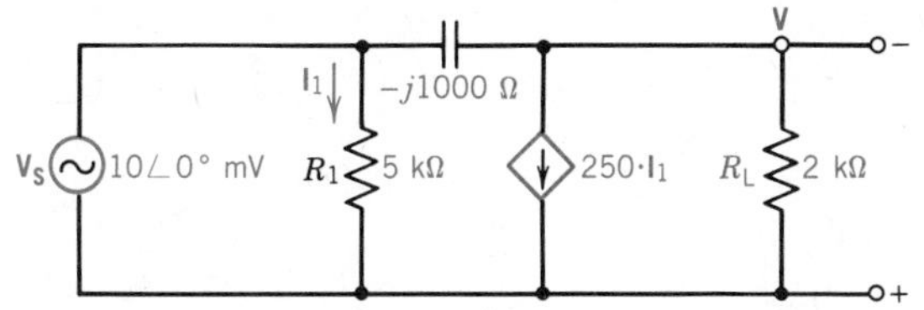

Figure 18-45 Problem 18-39

18-40 An alternative transistor amplifier circuit is shown in **Fig. 18-46**. Calculate

a. $\mathbf{I}_1$
b. $\mathbf{V}_2$ (see Hints, below)
c. Check the value of $\mathbf{I}_1$ by substitution of $\mathbf{V}_2$ in the mesh equation

Hints:

1. There are two unknowns, $\mathbf{I}_1$ and $\mathbf{V}_2$
2. Using nodal analysis, write one equation for the RH portion
3. Using mesh analysis, write one equation for the LH portion
4. Solve the two simultaneous equations for $\mathbf{I}_2$ and $\mathbf{V}_2$

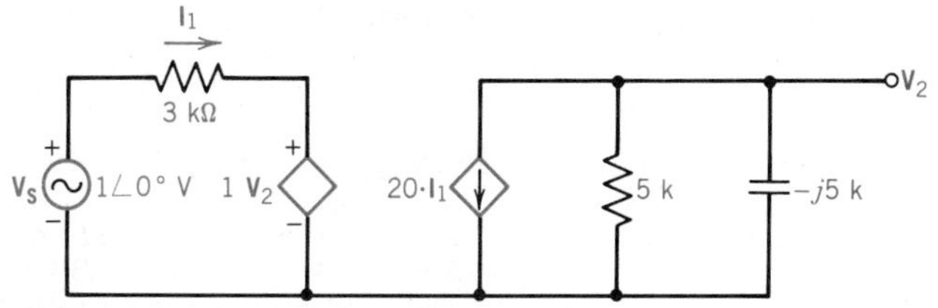

Figure 18-46 Problem 18-40

Sec. 18-8

18-41 The network of Fig. 18-37 was previously analyzed in Probs. 18-17 and 18-24. Using the method shown in Sec. 18-8.1 and conventional ac series–parallel theory, calculate the

a. Equivalent passive impedance "seen" by the practical independent voltage source
b. Current **I** drawn from the independent voltage source
c. Voltage, $\mathbf{V}_x$
d. Compare the above solutions to those of Probs. 18-17 and 18-24 for the same network
e. Comment on the relative ease of this method of analysis

18-42 The network of Fig. 18-39 was analyzed in Prob. 18-20 by mesh analysis and in Prob. 18-31 by nodal analysis. Now, using the method of Sec. 18-8.1, the dependency–impedance method:

a. Convert the practical VDVS to an equivalent dependent passive impedance, $\mathbf{Z}_d$
b. Calculate the total equivalent impedance, $\mathbf{Z}_p$, shunting the independent current source
c. Calculate **V** and $\mathbf{V}_{TH}$
d. Compare the above solution vis-à-vis those of Probs. 18-20 and 18-31 and comment on the ease of solution

18-43 In Fig. 18-39, convert the practical independent current source to a practical voltage source. Using the method of Sec. 18-8.1, the dependency–impedance method:

a. Convert the practical VDVS to an equivalent dependent passive impedance $\mathbf{Z}_d$ and draw the equivalent series circuit
b. Calculate the total series impedance, $\mathbf{Z}_s$, seen by the ideal voltage source
c. Calculate **I**, **V**, and $\mathbf{V}_{TH}$
d. Calculate $\mathbf{Z}_{TH}$ (Hint: short T–H and use KVL in series circuit to find $\mathbf{I}_{sc}$)
e. Draw the practical voltage source that replaces the network of Fig. 18-39 at terminals T–H

18-44 The network of Fig. 18-40 was analyzed in Probs. 18-21, 18-22, and 18-32. Is there a simpler solution involving simple series circuit theory?

a. Convert the practical CDCS to an equivalent dependent passive impedance, $\mathbf{Z}_d$
b. Calculate the total impedance seen by the $8\underline{/0^\circ}$ V source and current **I**
c. Calculate **V** by two alternative methods
d. Compare the above solutions for **I** and **V** to those of Probs. 18-21, 18-22, and 18-32, and comment on the ease of solution by the dependency–impedance method.

18-45 Using the dependency–impedance method for the network given in Fig. 18-36, calculate the

a. Equivalent dependent passive impedance of the CDCS, $\mathbf{Z}_d$
b. Total impedance seen by the ideal independent voltage source, $\mathbf{Z}_T$
c. Current **I**
d. Node voltage $\mathbf{V}_x$
e. $\mathbf{Z}_{TH}$ by superposition
f. Draw the practical voltage source replacing the entire network of Fig. 18-36
g. Compare the simplicity of this solution with methods used in Probs. 18-23 and 18-33 for the network of Fig. 18-36.

18-46 Given the network shown in Fig. 18-42 and using the dependency–impedance method in the following steps, calculate the

a. Equivalent practical dependent current source
b. Total equivalent impedance shunting dependent current source, $\mathbf{Z}_p$
c. Passive dependent impedance of current source $\mathbf{Z}_d$ in series with $j4\ \Omega$
d. Total impedance, $\mathbf{Z}_T$, in series with $60\underline{/0^\circ}$ V ideal source
e. Current $\mathbf{I}_1$
f. Node voltage **V**
g. $\mathbf{Z}_{TH}$ using superposition for currents from the independent and dependent sources
h. Equivalent practical voltage source replacing the entire network

18-47 Given the network shown in Fig. 18-43, using the dependency–impedance method in the following steps, calculate the

a. Equivalent dependent passive impedance of the VDVS, $\mathbf{Z}_d$
b. Parallel impedance in series with the practical independent voltage source, $\mathbf{Z}_p$
c. Total impedance seen by the $50\angle 0°$ V ideal source
d. Current $\mathbf{I}_1$
e. Node voltage $\mathbf{V}$
f. Compare **d** and **e** above with the solution to Prob. 18-29 by mesh analysis on Fig. 18-43

18-48 Verify the solutions to Probs. 18-27 and 18-34 using KVL in a CW mesh for a simple series circuit by converting the network to the left of $\mathbf{V}_1$ in Fig. 18-41 to its Thévenin equivalent voltage and resistance, calculating in turn

a. $\mathbf{I}_2$
b. $\mathbf{V}$
c. $\mathbf{V}_2$
d. Explain why the current in the equivalent series circuit is $\mathbf{I}_2$ and not $\mathbf{I}_1$

18-49 Given the network in the form shown in **Fig. 18-47**, convert the practical VDVS to its equivalent passive impedance in parallel with the 20-Ω resistor and calculate

a. $\mathbf{V}_1$
b. $\mathbf{V}_2$
c. $\mathbf{V}_{TH}$

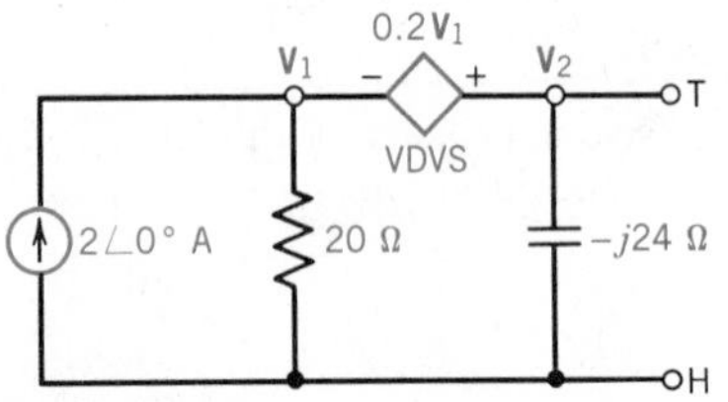

Figure 18-47 Problems 18-49, 18-50, and 18-51

18-50 Verify your solution to Prob. 18-49 using KVL in a simple series mesh by converting the practical independent current source in Fig. 18-47 to a practical voltage source and calculating

a. $\mathbf{I}$ c. $\mathbf{V}_2$
b. $\mathbf{V}_1$ d. $\mathbf{V}_{TH}$

18-51 Verify your solutions to Probs. 18-49 and 18-50 using nodal analysis to find:

a. $\mathbf{V}_1$
b. $\mathbf{V}_2$
c. $\mathbf{V}_{TH}$

Hint: in Fig. 18-47 convert dependent VDVS to a practical current source

18-52 Verify your solutions to Probs. 18-49 and 18-50 by calculating the following complex powers:

a. $\mathbf{S}_T^*$ c. $\mathbf{S}_C^*$
b. $\mathbf{S}_D^*$ d. $\mathbf{S}_R^*$
e. Show that the sum of the complex powers is zero by Tellegen's theorem

18-53 Verify your solutions to Ex. 18-24 and Prob. 18-38 by converting the practical CDVS to a CDCS and then converting it to an equivalent parallel impedance. Using conventional series–parallel circuit theory, calculate

a. Equivalent impedance in series with R_1, the 20-Ω resistor
b. $\mathbf{I}$
c. $\mathbf{V}_1$
d. $\mathbf{V}_2$

(Hint: find $\mathbf{V}_2$ using the VDR)

18-54 Verify the solution to Ex. 18-26 by converting the practical CDVS to a CDCS and then converting it to an equivalent parallel impedance. Using conventional circuit theory, calculate

a. $\mathbf{I}_1$ c. $\mathbf{V}_1$
b. $\mathbf{I}_2$ d. $\mathbf{V}_2$
e. Compare your answers to those given in Ex. 18-26, which used nodal analysis

18-55 Verify the solution to Ex. 18-28 by converting the practical VDCS to a VDVS in Fig. 18-25 and then converting it to an impedance in parallel with the 40-Ω resistor. Using conventional circuit theory, calculate

a. $\mathbf{Z}_{TH}$
b. Compare your solution with that given in Ex. 18-28 for ease of execution and simplicity

18-11 ANSWERS

18-1 a $(0.8537 + j0.7845)$ A b $(0.5609 + j0.3624)$ A
c, d $6.678\angle 77.87°$ V e $0.5137\angle 55.25°$ A

18-2 a $(1.105 - j0.3476)$ A b $(1.1974 + j0.3869)$ A
c, d $12.58\angle -35.22°$ V e $0.7403\angle -97.17°$ A

18-3 a $(0.473 - j0.3367)$ A b $(0.00667 - j0.3341)$ A
c, d $13.36\angle -178.9°$ V e $0.4664\angle -0.32°$ A

18-4 a $(-8 - j16)$ mA b $(-4 + j2)$ mA
c, d $(40 + j80)$ V e $(-4 - j8)$ mA

18-5 a $1.302\angle -69.83°$ A b $1.0513\angle 116.4°$ A
c $0.451\angle 174.3°$ A d $9.4\angle 22.96°$ V e $10.76\angle 1.2°$ V

18-6 a $3.806\angle 48.55°$ A b $2.241\angle 33.05°$ A
c $1.49\angle -10.6°$ A d $22.78\angle 45.9°$ e $20.2\angle 7.2°$ V

18-7 **d** $(1.557 - j0.6267)$ A
e $(14.55 - j1.493)$ V, $(5.205 + j2.267)$ V
f $1.678\angle -21.9^\circ$ A

18-8 **a** $6.678\angle 77.9^\circ$ V **b** $0.6678\angle 32.9^\circ$ A
c $0.5137\angle 55.25^\circ$ A

18-9 **a** $12.58\angle -35.2^\circ$ V **b** $1.258\angle 17.9^\circ$ A
c $0.7403\angle -97.15^\circ$ A

18-10 **a** $13.37\angle -178.9^\circ$ V **b** $0.3341\angle -88.7^\circ$ A
c $0.5807\angle -35.45^\circ$ A **d** $0.4664\angle -0.33^\circ$ A

18-12 **a** $9.401\angle 22.96^\circ$ V **b** $10.76\angle 1.2^\circ$ V **c** $1.302\angle -69.9^\circ$ A
d $0.4503\angle -5.8^\circ$ A **f** $0.897\angle 91.2^\circ$ A

18-14 **a** $(7.212 - j9.331)$ V **b** $(6.12 - j9.605)$ V
c $1.126\angle 14.1^\circ$ V **d** $112.6\angle -39^\circ$ μA

18-16 **a** $1\angle 0^\circ$ A **b** $(8 + j12)$ V

18-17 **a** $(-0.2865 + j0.9685)$ A **b** $4.282\angle 74.5^\circ$ V

18-18 **a** $(3.122 + j3.032)$ A **b, c** $0.7148\angle -145.2^\circ$ V

18-19 **a** $4\angle 45^\circ$ A **b** $0.1787\angle 169.8^\circ$ Ω

18-20 **a** $(9 + j3)$ A **b** $(10 - j30)$ V **c** $(15 - j45)$ V
d $\mathbf{S}_d^* = (0 + j150)$ VA, $\mathbf{S}_T^* = (-100 + j300)$ VA

18-21 **a** $2\angle 0^\circ$ A **b** $6\angle -53.13^\circ$ V

18-22 (see 18-21)

18-23 **a** $1\angle 0^\circ$ A **b, c** $(8 + j12)$ V

18-24 **a** $4.282\angle 74.5^\circ$ V **b** $1.0705\angle 74.5^\circ$ A

18-25 **a** $64.42\angle -75.1^\circ$ V **b** $193.3\angle -75.1^\circ$ V
c $19.325\angle -75.07^\circ$ Ω

18-26 **a** 144 V **b** 12 Ω

18-27 **a** $-(7.\overline{3} + j12)$ A **b** $-(10.8 + j14.4)$ A
c $253\angle 34.7^\circ$ V **d** $835.2\angle -20.2^\circ$ V
e $(784 - j288)$ V

18-28 **a** $(10 - j10)$ A **b** $(5 + j0)$ A **c, d** $(20 - j40)$ V
e $2.\overline{6}\angle 0^\circ$ Ω

18-29 **a** $(6.058 - j1.1913)$ A **b** $(2.028 + j1.1695)$ A
c $30.44\angle -37.4^\circ$ V

18-30 **a** $8.849\angle 138.7^\circ$ V **b, c** $0.715\angle -145.3^\circ$ V
d $0.179\angle 169.7^\circ$ Ω **e** $(3.122 + j3.031)$ A

18-31 **a** $(10 - j30)$ V **b** $(15 - j45)$ V **c** $(1.5 - j4.5)$ Ω

18-32 **a** $(3.6 - j4.8)$ V **b** $2\angle 0^\circ$ A
c $0.8\angle 0^\circ$ A; $(3.6 - j4.8)$ V **d** $1.2\angle 0^\circ$ A
e $\mathbf{S}_1^* = (8.8 + j9.6)$ VA, $\mathbf{S}_2^* = (4.32 - j5.76)$ VA
f $(-16 + j0)$ VA **g** $(2.88 - j3.84)$ VA **h** $0 + j0$

18-33 **a** $(8 + j12)$ V **b** $1\angle 0^\circ$ A

18-34 **a** $(208 + j144)$ V **b** $(10.8 + j14.4)$ A
c, d $(784 - j288)$ V **e** $(8.431 + j0.392)$ A
f $(91.2 - j38.4)$ Ω

18-35 **a** $(36 + j14)$ V **b, c** $(14 - j14)$ V

18-36 **a** $10\angle 0^\circ$ A **b** 50 V **c** $4\angle 0^\circ$ A **d** $(1 - j1)$ Ω

18-37 **a** $(20 - j40)$ V **b** $(10 - j10)$ A **c** $5\angle 0^\circ$ A
d $(2.\overline{6} + j0)$ Ω

18-38 **a** $62.14\angle -11.24^\circ$ V **b** $55.58\angle -37.8^\circ$ V
c $3.014\angle 11.59^\circ$ A **d** $2.78\angle -37.8^\circ$ A
e $55.58\angle -37.8^\circ$ V

18-39 **b** $447.3\angle 115.4^\circ$ mV **c** 44.73; 33.0 dB

18-40 **a** $14.57\angle -133.2^\circ$ μA **b** $1.03\angle 1.77^\circ$ V

18-41 **a** $(3.596 - j2.2472)$ Ω **b** $1.01\angle 106.5^\circ$ A
c $4.282\angle 74.5^\circ$ V

18-42 **a** $-j3.\overline{3}$ Ω **b** $(1 - j3)$ Ω
c $(10 - j30)$ V, $(15 - j45)$ V

18-43 **a** $-j3.\overline{3}$ Ω **b** $(10 - j3.\overline{3})$ Ω
c $(9 + j3)$ A, $(10 - j30)$ V, $(15 - j45)$ V
d $(1.5 - j4.5)$ Ω

18-44 **a** $(1.8 - j2.4)$ Ω **b** $4\angle 0^\circ$ Ω, $2\angle 0^\circ$ A **c** $(3.6 - j4.8)$ V

18-45 **a** $(8 + j12)$ Ω **b** $(10 + j8)$ Ω $1\angle 0^\circ$ A
d $(8 + j12)$ V **e** $(0.878 - j0.902)$ Ω

18-46 **b** $(2 - j2)$ Ω **c** $(3 - j1)$ Ω **d** $(3 + j3)$ Ω
e $(10 - j10)$ A **f** $(20 - j40)$ V **g** $(2.\overline{6} + j0)$ Ω

18-47 **a** $(5 - j12)$ Ω **b** $(4.506 - j1.6302)$ Ω
c $(7.506 + j2.37)$ Ω **d** $(6.057 - j1.913)$ A
e $(24.18 - j18.49)$ V

18-48 **a** $-(10.8 + j14.4)$ A **b** $253\angle 34.7^\circ$ V
c $835.2\angle -20.17^\circ$ V

18-49 **a** $(20 - j20)$ V **b, c** $(24 - j24)$ V

18-50 **a** $(1 + j1)$ A **b** $(20 - j20)$ V **c, d** $(24 - j24)$ V

18-51 **a** $(20 - j20)$ V **b, c** $(24 - j24)$ V

18-52 **a** $(-40 + j40)$ VA **b** $(0 + j8)$ VA **c** $(0 - j48)$ VA
d $(40 + j0)$ VA **e** 0

18-53 **a** $(19 - j8)$ Ω **b** $(2.953 + j0.6057)$ A
c $62.14\angle -11.24^\circ$ V **d** $55.58\angle -37.8^\circ$ V

18-54 **a** $(2.82\overline{6} - j0.35\overline{3})$ A **b** $(2.47\overline{3} + j0.35\overline{3})$ A
c $(56.5\overline{3} - j7.0\overline{6})$ V **d** $(53 - j17.\overline{6})$ V

18-55 **a** 0.5229 Ω

CHAPTER 19

Resonance and Filters

19-1 INTRODUCTION

In the previous ac chapters, the series and parallel networks considered had one fundamental common feature: they could be reduced to a single equivalent impedance that was primarily inductive or capacitive, operating at a lagging or leading power factor. This chapter considers both series and parallel networks operating at a *specific* frequency such that the inductive and capacitive reactances cancel each other. This condition is called "resonance." The frequency at which this condition occurs is called the resonant frequency. At resonance, therefore, the equivalent network impedance is essentially *resistive* in character and the *power factor* is *unity*.

Certain special effects occur as a result of resonance. In the *series-resonant* circuit there *may* be an unusual Q-rise in *voltage* across both the inductor and capacitor that is significantly *higher* than the *supply voltage*. In the *parallel-resonant* circuit there may be an unusual Q-rise in *current* in both the capacitor and inductor that is significantly *higher* than the *total circuit current*. These and other resonant effects have important applications in electronic communications.

One important application of series and parallel resonance occurs in *filters*, that is, transducers or circuits designed to either pass or reject the transfer of energy to a load at certain specific frequencies. Filters are introduced, beginning with simple *low-pass* and *high-pass* filters, to define *passband*, *stopband*, and *cutoff* frequencies. Simple series and parallel resonant bandstop and bandpass filters are considered, including the various ways to predict both *amplitude* and *phase responses*. The effects of filter loading are also discussed, as is bandwidth of these resonant filters.

The chapter concludes with *double-resonant filters*, including their design, prediction of properties and characteristics, and the effect of loading.

19-2 RESONANCE IN AN *RLC* SERIES CIRCUIT

A series *RLC* circuit is shown connected to a variable-frequency supply of constant voltage, **V**, in **Fig. 19-1**. Assume that the frequency is increased and the voltage remains constant. Under these conditions the current continues to rise as the fre-

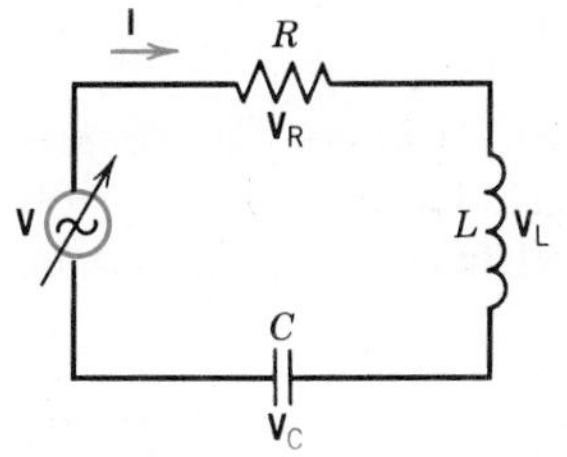

Figure 19-1 *RLC* circuit connected to a variable frequency supply

quency increases until it reaches some maximum value, $\mathbf{I}_0$. Further increases in frequency result in a *decrease* in current (Fig. 19-3). The conditions that occur at resonance (i.e., the frequency, f_0, producing maximum current) and the reasons for it are explained below.

19-2.1 Conditions for Resonance

At the resonant frequency, f_0, maximum current is produced because the inductive reactance is exactly equal and opposite to the capacitive reactance, causing these impedances to cancel each other. In terms of the resonant frequency, f_0, we may say

$$2\pi f_0 L = \frac{1}{2\pi f_0 C} \quad \text{ohms } (\Omega) \qquad (19\text{-}1a)$$

$$\omega_0 L = \frac{1}{\omega_0 C} \quad \text{ohms } (\Omega) \qquad (19\text{-}1b)$$

$$X_{L_0} = X_{C_0} \quad \text{ohms } (\Omega) \qquad (19\text{-}1c)$$

19-2.2 Frequency Relations at Resonance

Solving Eq. (19-1a) for the frequency, f_0, yields the resonant frequency in terms of L and C or

$$f_0 = \frac{1}{2\pi\sqrt{LC}} \quad \text{hertz (Hz)} \qquad (19\text{-}2a)$$

Equation (19-2a) also yields the **radian frequency** at *resonance* or

$$\omega_0 = 2\pi f_0 = \frac{1}{\sqrt{LC}} \quad \text{radians per second (rad/s)} \qquad (19\text{-}2b)$$

If the capacitive reactance (X_C) and the inductive reactance (X_L) are known quantities at some frequency, f, other than the resonant frequency, then by substitution for L and C, the resonant frequency may be found from[1]

$$f_0 = f\sqrt{\frac{X_C}{X_L}} \quad \text{hertz (Hz)} \qquad (19\text{-}2c)$$

[1] Substituting $L = X_L/2\pi f$ and $C = 1/2\pi f \cdot X_C$ in Eq. (19-2a) yields Eq. (19-2c). See Appendix B-15.4 for the derivation of this and Eq. (19-2a).

19-2.3 Resonance at a Constant Frequency

It is also possible to produce resonance at a given (constant) frequency by varying either C or L. For example if C is varied, it is possible to solve Eq. (19-1) for the required capacitance to produce resonance, given a fixed frequency, f_0, and inductance, L, from:

$$C_0 = \frac{1}{(2\pi f_0)^2 L} = \frac{1}{\omega_0^2 L} \quad \text{farads (F)} \qquad \text{(19-3a)}$$

Similarly, the required inductance, L_0, needed to produce resonance is

$$L_0 = \frac{1}{(2\pi f_0)^2 C} = \frac{1}{\omega_0^2 C} \quad \text{henrys (H)} \qquad \text{(19-3b)}$$

19-2.4 Reactance Variations with Changes in Frequency

For the present let us consider how resonance is produced by frequency variation. Since the *RLC* series circuit contains resistance, R, inductive reactance, X_L, and capacitive reactance, X_C, let us consider how these three impedance components vary with frequency. **Figure 19-2** shows this variation, and it should be studied most carefully. The following important points should be noted:

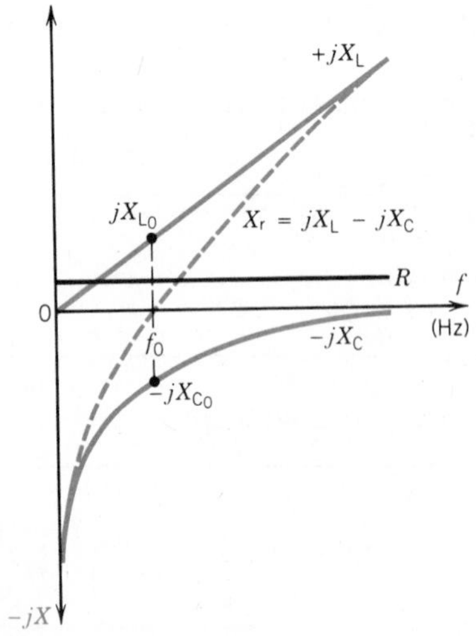

Figure 19-2 Variation of R, $+jX_L$, $-jX_C$, and X_r with frequency in a series *RLC* circuit

1. Resistance, R, is assumed constant over the range at which resonance occurs.[2]
2. Inductive reactance, jX_L, increases *linearly* with frequency, since $jX_L = 2\pi fL = k \cdot f$ for a fixed value of inductance, L.
3. The nonlinear curve for the capacitive reactance, $-jX_C$, is an equilateral parabola, asymptotic to the horizontal frequency axis at high frequencies and asymptotic to the vertical reactance axis at low frequencies. The curve of X_C is parabolic because $X_C = 1/2\pi fC$ and $X_C \times f = 1/2\pi C = k$.
4. The curves bear out conclusions noted in Secs. 13-10 and 14-16, namely
   a. *Inductors* appear to be *short* circuits at *very low* frequencies and *open* circuits at *very high* frequencies.
   b. *Capacitors* appear to be *open* circuits at *very low* frequencies and *short* circuits at *very high* frequencies.
5. The dashed line in Fig. 19-2 represents the resultant (total) reactance, X_r, which is the sum of $+jX_L$ and $-jX_C$ at any frequency. Since X_r represents the *difference* between the two reactances, we can see from Fig. 19-2 that
   a. At frequencies *below* the resonant frequency, f_0, the resultant reactance is *capacitive*, since capacitive reactance exceeds inductive reactance.
   b. At frequencies *above* the resonant frequency, f_0, the resultant reactance is *inductive*, since $X_L > X_C$ and the circuit impedance is $\mathbf{Z} = R + jX_r$.
   c. At the resonant frequency, f_0, the resultant reactance $X_r = 0$. The circuit impedance at f_0 is a *minimum* and purely resistive, $\mathbf{Z}_0 = R$ (see Eq. 19-4b).

[2] Recall that resistance is frequency-sensitive, particularly at high frequencies (Sec. 17-8).

19-2.5 Impedance–Current Variations with Changes in Frequency

Although it is possible to superimpose the circuit impedance variation on Fig. 19-2, it is less confusing to show it separately in **Fig. 19-3**. Since the *RLC* circuit current is at all times a function of total circuit impedance, **Z**, the two are shown in *inverse* relation in Fig. 19-3. The impedance at *any frequency* is

$$\mathbf{Z} = R + j\omega L + \left(\frac{1}{j\omega C}\right) = R + j\left(\omega L - \frac{1}{\omega C}\right) = R \pm jX_r \qquad \text{ohms } (\Omega) \tag{19-4a}$$

And obviously, at the resonant frequency, f_0, the impedance is

$$\mathbf{Z}_0 = R \qquad \text{ohms } (\Omega) \tag{19-4b}$$

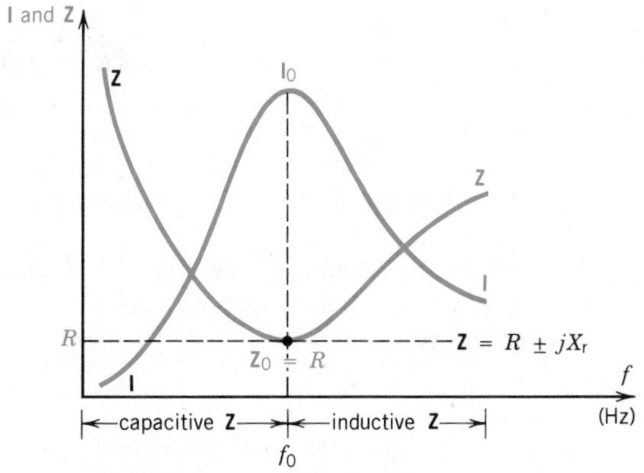

Figure 19-3 Variation of impedance and current with frequency in a series *RLC* circuit

With respect to Fig. 19-3 and the curve of impedance, **Z**, note that

1. Below frequency f_0 the *impedance* is essentially *capacitive*. Above frequency f_0 the *impedance* is essentially *inductive*. (See Fig. 19-2.)
2. As the frequency increases, impedance drops to a minimum at the resonant frequency, where $\mathbf{Z}_0 = R$.
3. The curve of impedance **Z** is *asymmetrical* above and below f_0. The asymmetry is due to the *nonlinearity* of X_C and the linearity of X_L. At low frequencies X_C predominates and **Z** is asymptotic to the impedance axis. At high frequencies X_L predominates and **Z** approaches the straight-line variation of X_L.

With respect to Fig. 19-3 and the curve of the *RLC* circuit current, **I**, note that

1. The current is the reciprocal of the circuit impedance. When the impedance is low the current is high and vice versa, since $\mathbf{I} = \mathbf{V}/\mathbf{Z}$ by Ohm's law.
2. Current at resonance, $\mathbf{I}_0$, is a maximum at f_0, the resonant frequency.
3. The current is *leading* the supply voltage **V** at frequencies **below** f_0 and *lagging* the supply voltage at frequencies **above** f_0.
4. At the resonant frequency, f_0, the current is *in phase* with the supply voltage (since the circuit is *resistive*).

The resonant current, $\mathbf{I}_0$, may be evaluated from

$$\mathbf{I}_0 = \mathbf{I}_{max} = \frac{\mathbf{V}}{R} = \frac{\mathbf{V}}{\mathbf{Z}_0} = \frac{\mathbf{V}_R}{R} \quad \text{amperes (A)} \tag{19-5}$$

At frequencies below and above resonance, the current decreases as shown in Fig. 19-3. The ratio of the current at any other frequency to the maximum current, $\mathbf{I}_0$, is the ratio of the circuit resistance to the circuit impedance or

$$\frac{\mathbf{I}}{\mathbf{I}_0} = \frac{R}{\mathbf{Z}} = \frac{\mathbf{Z}_0}{\mathbf{Z}} = \frac{R}{R \pm jX_r} \quad \text{(dimensionless)} \tag{19-6}$$

Note that Eq. (19-6) is an inverse ratio that enables solution of any off-resonance current in accordance with Ohm's law from

$$\mathbf{I} = \frac{\mathbf{I}_0\mathbf{Z}_0}{\mathbf{Z}} = \frac{\mathbf{V}}{\mathbf{Z}} = \frac{\mathbf{V}}{(R \pm jX_r)} \quad \text{amperes (A)} \tag{19-6a}$$

19-2.6 Phase Angle and Power Factor Variations with Frequency

The variation of circuit phase angle, θ, with frequency is shown in **Fig. 19-4**. Since the current leads the supply voltage at frequencies *below* resonance, the angle θ is *negative*. With respect to Fig. 19-4 note that

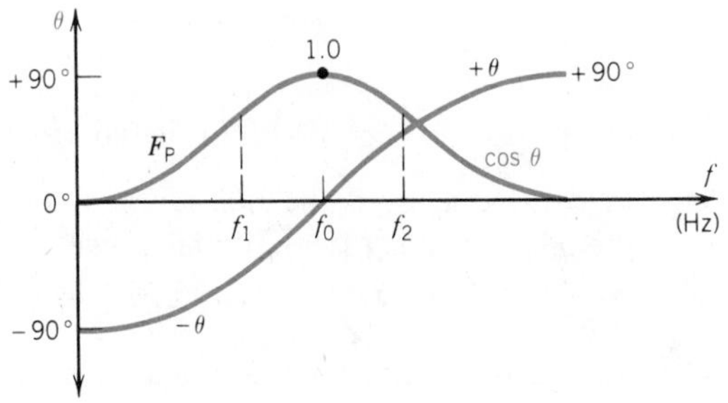

Figure 19-4 Phase angle and power factor variation with frequency using current as a reference for the series *RLC* circuit

1. Since the capacitive reactance is almost infinity at very low frequencies and the inductive reactance is practically zero, the phase angle is $-90°$ since the impedance $\mathbf{Z} = R - jX_C$ and R are negligible compared to the very high capacitive reactance.
2. As the frequency increases, the ratio of net resultant reactance, X_r, to resistance R, decreases. Since phase angle $\theta = \tan^{-1}(X_r/R)$, as X_r decreases (Fig. 19-2), θ decreases. When $X_r = R$, $\theta = -45°$. When $X_r = 0$ at resonance, $\theta = 0°$.
3. At frequencies above f_0, the resultant reactance is inductive and the voltage leads the reference current by some positive phase angle $+\theta$. At extremely high frequencies the resistance is negligible compared to the inductive reactance and the phase angle is $+90°$.

As noted previously, the power factor, F_p, is the cosine of the phase angle, θ. Recall that the cosine of *negative* phase angles is a *positive* value. As shown in Fig. 19-4, the power factor, F_p, varies from a maximum of unity at f_0 to zero at extremely high and extremely low frequencies.

19-2.7 Resonant Q_0 of a Series-Resonant Circuit

The circuit Q was introduced in our discussion of series–parallel equivalence (Sec. 16-6) for the series circuit as the ratio X_L/R or X_C/R. Consequently, at the resonant frequency, f_0, we may define the resonant Q_0 of a series resonant circuit in terms of all of the following:[3]

$$Q_0 = \frac{\omega_0 L}{R} = \frac{1}{R\omega_0 C} = \frac{V_{Lo}}{V} = \frac{V_{Co}}{V} = \frac{1}{R}\sqrt{\frac{L}{C}} = \frac{\sqrt{X_{Co} \cdot X_{Lo}}}{R}$$

$$= \frac{Q_s}{P_0} = \frac{2\pi W_s}{W_d} = \frac{f_0}{B_{hp}} \quad \text{(dimensionless)} \qquad (19\text{-}7)$$

where Q_s is the quadrature reactive power *stored* in either the coil or the capacitor, in vars[4]

W_s is the maximum energy stored in either the coil or the capacitor, in vars

W_d is the energy dissipated in the resistor, R

P_0 is the power dissipated in the resistor R

B_{hp} is the half-power bandwidth of the power, voltage, or current response curves of the resonant circuit

The resonant Q_0 of a series-resonant circuit is essentially a measure of the *selectivity* or *steepness of rise* of the power, voltage, and current response curves, respectively, of the resonant *RLC* series circuit. Numerous examples are given to assist the reader in use of the various equalities of Eq. (19-7). For the present, however, two points are of major significance with respect to Eq. (19-7):

1. Note that R is in the denominator in all the equalities containing R. This means that the higher the resistance, the lower the resonant Q_0, and vice versa.
2. The relation $Q_0 = \frac{1}{R}\sqrt{\frac{L}{C}}$ shows how all three circuit components affect Q_0. The higher the ratio L/C, the *higher* the circuit resonant, Q_0 and the *steeper* the response curve.

Figure 19-5 shows graphically the significance of the equation $Q_0 = \frac{1}{R}\sqrt{\frac{L}{C}}$. The effect of resistance is shown in Fig. 19-5a. A low value of R produces a higher maximum current, I_0, with much *steeper* sides, resulting in a *high* value of Q_0.

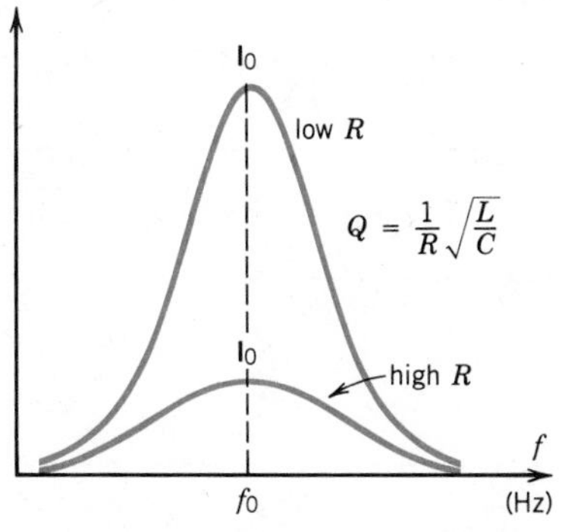

a. Effect of resistance on Q_0 rise

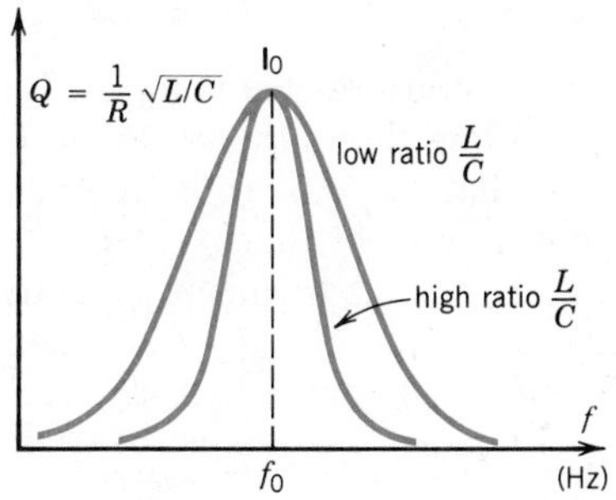

b. Effect of L/C ratio on Q_0 rise. (Resistance constant)

Figure 19-5 Effect of R, L, and C on the response curves of a series-resonant circuit

[3] The equalities shown in Eq. (19-7) are derived in Appendix B-15.

[4] This is one of those instances where the same letter symbol is used to designate two different physical quantities. Q_0 represents the quality or figure of merit of the resonant circuit. Q_s represents the quadrature power stored in reactive elements of the resonant circuit. (The letter Q may also be used to designate capacitive charge in coulombs.)

If the resistance is held constant and the ratio L/C is increased, again Q_0 is increased and the (current) response shows a much steeper rise in Fig. 19-5b.

19-2.8 Resonant Rise of Voltage across *L* and *C* in a Resonant *RLC* Series Circuit at f_0

Perhaps one of the most amazing effects is the "voltage magnification" that may occur in a series-resonant circuit. Whenever the Q_0 is greater than unity, the voltage across L and/or C exceeds the supply voltage! This apparent "violation" of Kirchhoff's voltage law (KVL) is not a violation at all, as we will soon discover. But it does show that this effect is of great significance in electronic circuits and communications in a variety of ways.[5]

At the resonant frequency, f_0, the magnitude of the voltage across the inductor may be derived from

$$\mathbf{V}_{L_0} = \mathbf{I}_0 X_{L_0} = \frac{\mathbf{V}}{\mathbf{Z}_0} X_{L_0} = \frac{\mathbf{V} \cdot jX_{L_0}}{R} = +j\mathbf{V} \cdot Q_0 \qquad \text{volts (V)} \qquad (19\text{-}8a)$$

In the same way, at f_0 the magnitude of the voltage across the capacitor is

$$\mathbf{V}_{C_0} = \mathbf{I}_0 X_{C_0} = \frac{\mathbf{V}}{\mathbf{Z}_0} (X_{C_0}) = \mathbf{V}\left(\frac{-jX_{C_0}}{R}\right) = -j\mathbf{V}Q_0 \qquad \text{volts (V)} \qquad (19\text{-}8b)$$

while the voltage across the resistor at the resonant frequency from Eq. (19-5) is

$$\mathbf{V}_{R_0} = \mathbf{I}_0 R = \frac{\mathbf{V}}{\mathbf{Z}_0} R_0 = \frac{\mathbf{V}}{R_0} \times R_0 = \mathbf{V} \qquad \text{volts (V)} \qquad (19\text{-}8c)$$

Equations (19-8a) and (19-8b) show that the voltages across L and C are "magnified" by a factor of Q_0 times the supply voltage V. If $Q_0 = 50$ (not an unusually high Q for a resonant circuit) and the supply voltage is, say, 100 V, then the voltage across L and C at resonance is as large as 5000 V!

One might assume that $\mathbf{V}_{C_0}$ and $\mathbf{V}_{L_0}$ are the maximum possible voltages. But again we are in for an even greater surprise. **Figure 19-6a** shows the variations of $\mathbf{V}_C$, $\mathbf{V}_L$, and $\mathbf{V}_R$ as a function of frequency. Observe that $\mathbf{V}_{R_0} = \mathbf{V}$ at the resonant frequency f_0. Also note that $\mathbf{V}_{C_0}$ and $\mathbf{V}_{L_0}$ are (indeed) greater than $\mathbf{V}$ at f_0. But these are *not* the maximum values! $\mathbf{V}_C$ is a maximum at some frequency (slightly) lower than f_0 and $\mathbf{V}_L$ is a maximum at a (slightly) higher frequency. The reasons for this are derived and explained in Sec. 19-3.3, but for the present let us merely note that $|\mathbf{V}_{C_0}| = |\mathbf{V}_{L_0}| = \mathbf{V}Q_0$, as shown in Fig. 19-6a and in Eqs. (19-8a) and (19-8b).

A remaining question is whether Kirchhoff's voltage law (KVL) holds for a

[5] The ability of some resonant filters to pass certain frequencies and reject others is but one example of this effect. Similarly, the ability of a receiver to be tuned to a specific frequency while rejecting others is another example.

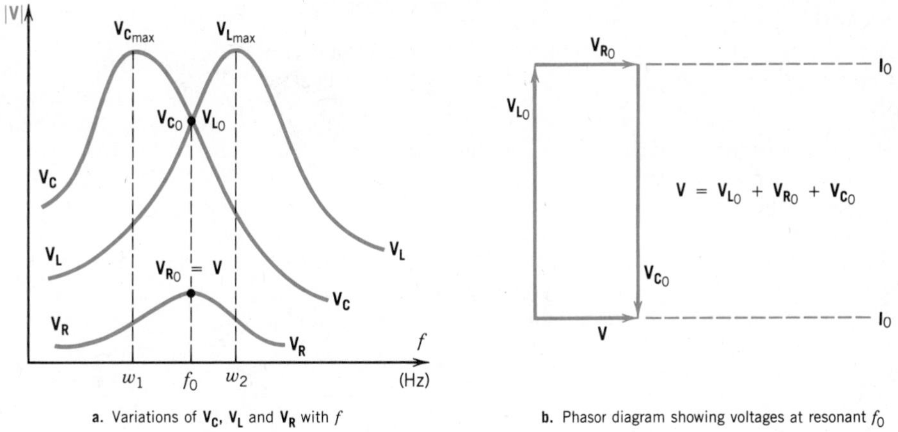

Figure 19-6 Voltage relations in a series *RLC* resonant circuit

series-resonant circuit. The phasor diagram of Fig. 19-6b shows that indeed KVL does hold. Since $\mathbf{V}_{L_0}$ is equal in magnitude *but opposite in phase* to $\mathbf{V}_{C_0}$, the supply voltage $\mathbf{V} = \mathbf{V}_{R_0}$. We may therefore write at the resonant frequency, f_0, from Fig. 19-6b:

$$\mathbf{V} = \mathbf{V}_{R_0} + j\mathbf{V}_{L_0} + (-j\mathbf{V}_{C_0}) = \mathbf{V}_{R_0} \qquad \text{volts (V)} \qquad \textbf{(19-8d)}$$

The following examples illustrate the relations just derived. Readers should attempt the solution independently and use the solution as feedback to verify their understanding of the presentations in Sec. 19-2.

EXAMPLE 19-1

A series *RLC* circuit has $R = 100\ \Omega$, $L = 1$ H, and $C = 1\ \mu$F connected in series to a 10-V variable-frequency supply. Calculate the

a. Frequency at which resonance occurs
b. Ratio L/C
c. Q_0 of the circuit at resonance

Solution

a. $f_0 = 1/(2\pi\sqrt{LC}) = 1/(2\pi\sqrt{10^{-6}}) = \mathbf{159.15\ Hz}$ (19-2a)
b. $L/C = 1\text{ H}/10^{-6}\text{ F} = \mathbf{10^6}$
c. $Q_0 = \sqrt{L/C}/R = \sqrt{10^6}/100 = \mathbf{10}$ (19-7)

EXAMPLE 19-2

Using the given and calculated data of Ex. 19-1, calculate

a. jX_{L_0}, reactance of the inductor at resonance
b. $-jX_{C_0}$, reactance of the capacitor at resonance
c. X_{r_0}, resultant reactance at resonance
d. $\mathbf{Z}_0$, impedance at resonance
e. $\mathbf{I}_0$, maximum current at resonance

Solution

a. $jX_{L_0} = j\omega_0 L = j(2\pi \times 159.15) \times 1 = \boldsymbol{j1000\ \Omega}$
b. $-jX_{C_0} = 1/j\omega C = 1/(2\pi \times 159.15 \times 10^{-6}) = \boldsymbol{-j1000\ \Omega}$
c. $X_{r_0} = (jX_{L_0}) + (-jX_{C_0}) = j1000 - j1000 = \mathbf{0\ \Omega}$
d. $\mathbf{Z}_0 = R + jX_{r_0} = 100 + j0 = \mathbf{100\angle 0^\circ\ \Omega}$ (19-4a)
e. $\mathbf{I}_0 = \mathbf{V}/\mathbf{Z}_0 = 10\angle 0^\circ\text{ V}/100\angle 0^\circ\ \Omega = \mathbf{100\angle 0^\circ\ mA}$

EXAMPLE 19-3

Using the data of the two previous examples, calculate

a. The Q rise in voltage across X_L at resonance, using circuit current
b. Q rise in voltage across X_C at resonance, using circuit current
c. Q rise in voltage across L and C, respectively, using Eqs. (19-8a) and (19-8b)
d. Draw the phasor diagram showing the relation of all four voltages, using $\mathbf{I}_0$ as the reference

Solution

a. $\mathbf{V}_{L_0} = \mathbf{I}_0 jX_{L_0} = 0.1(j1000) = \mathbf{100\angle 90^\circ\ V}$
b. $\mathbf{V}_{C_0} = \mathbf{I}_0(-jX_{C_0}) = 0.1(-j1000) = \mathbf{100\angle -90^\circ\ V}$
c. $\mathbf{V}_{L_0} = j\mathbf{V}Q_0 = j(10)(10) = j100 = \mathbf{100\angle 90^\circ\ V}$ (19-8a)
$\mathbf{V}_{C_0} = -j\mathbf{V}Q_0 = -j(10)10 = -j100$
$= \mathbf{100\angle -90^\circ\ V}$ (19-8b)
d. The phasor diagram is shown in Fig. 19-6b; the reader may add the values obtained for $\mathbf{I}_0$, $\mathbf{V}_{L_0}$, $\mathbf{V}_{C_0}$, $\mathbf{V}_R$, and $\mathbf{V}$

EXAMPLE 19-4

A practical inductor has a resistance of 50 Ω and a Q of 5 at a resonant frequency of 10 kHz from a $5\angle 0°$ V supply. Calculate the

a. Value of series capacitance that produces series resonance with the practical coil
b. Inductance of the coil
c. L/C ratio
d. Q_0, using the L/C ratio
e. Voltage across the capacitor by two methods
f. Voltage across the *practical* inductor (be careful!)

Solution

a. $X_{L_0} = QR = 5 \times 50\ \Omega = j250\ \Omega$ and for resonance $-jX_{C_0} = 250\ \Omega$; $C = 1/2\pi f_0 X_{C_0} = 1/2\pi 10^4 \times 250 = \mathbf{63.66\ nF}$
b. $L = X_L/2\pi f = 250/10^4 \cdot 2\pi = \mathbf{3.98\ mH}$
c. $L/C = 3.98\ \text{mH}/63.66\ \text{nF} = \mathbf{6.25 \times 10^4}$
d. $Q_0 = \sqrt{L/C}/R = \sqrt{6.25 \times 10^4}/50 = \mathbf{5}$ (verifies Eq. 19-7)
e. $\mathbf{V}_{C_0} = -j\mathbf{V}Q_0 = -j(5)(5) = -j25 = \mathbf{25\angle -90°\ V}$ (19-8b)

$$\mathbf{V}_{C_0} = \mathbf{I}_0(-jX_{C_0}) = \frac{\mathbf{V}}{R}(-jX_{C_0}) = \frac{5\angle 0°\ \text{V}}{50\angle 0°\ \Omega}(-j250\ \Omega)$$
$$= \mathbf{25\angle -90°\ V}$$

f. By inspection, $\mathbf{V}_{L_0} = +j25$ V and $\mathbf{V}_R = \mathbf{V} = 5\angle 0°$ V; then $\mathbf{V}_{\text{coil}} = \mathbf{V}_R + \mathbf{V}_{L_0} = 5 + j25 = \mathbf{25.5\angle 78.7°\ V}$

Several interesting conclusions and insights may be drawn from Exs. 19-1 through 19-4:

1. If we know the values of R, L, and C we can compute (and predict) the values of the L/C ratio, the resonant Q_0, and the resonant frequency, f_0 (see Ex. 19-1).
2. Given the applied voltage and R, L, and C, we can compute (and predict) the resonant capacitive and inductive reactances and the maximum (resonant) current, $\mathbf{I}_0$ (Ex. 19-2).
3. Given the applied voltage and R, L, and C, we can determine (predict) resonant voltages ($\mathbf{V}_{C_0}$, $\mathbf{V}_{L_0}$, and $\mathbf{V}_{R_0}$) at the resonant frequencies.
4. Given the Q of a practical coil at a known voltage of given frequency we can predict
   a. The value of the capacitor required for resonance
   b. The L/C ratio
   c. The resonant Q_0 of the coil, which *is the same as the given Q!*[6]
5. Last but hardly least is the realization that the voltage across a *practical coil* in a series resonant circuit, **at the resonant frequency**, f_0, is *always higher* than the voltage across the capacitor. This is due to the added voltage drop across the resistance of the coil. As shown in Ex. 19-4**f**, the voltage across the capacitor is 25 V but the practical coil voltage is 25.5 V! The lower the Q of the practical coil (and the resonant Q_0), the more pronounced is the voltage difference.

EXAMPLE 19-5

The effective resistance of a practical iron core coil is measured at its rated frequency (1 kHz) using the wattmeter–ammeter method shown in **Fig. 19-7a**. The wattmeter reads 10 W and the ammeter reads 100 mA. The coil is then series-connected to a variable capacitor bank across a 1-kHz supply, as shown in Fig. 19-7b. The capacitor bank is varied to produce maximum deflec-

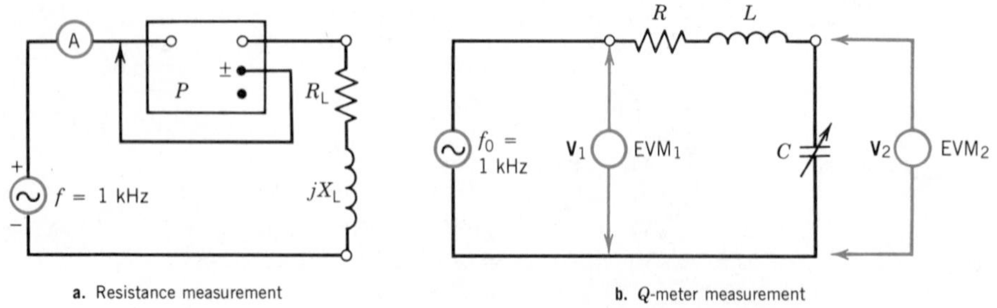

Figure 19-7 Example 19-5; using series resonance to measure Q of a practical coil

[6] The insight of item 4c leads to the design of a commercial Q-meter that uses the principle of series resonance to find the Q of a practical coil. The theory and practice are shown in Ex. 19-5.

tion on the electronic voltmeter, EVM_2. When $EVM_1 = 12$ V, then the maximum voltage across C is $EVM_2 = 60$ V. Calculate the

a. Effective resistance of the coil
b. Q_0 rise in voltage
c. Q of the coil
d. Inductance of the coil, L
e. Capacitance of C in Fig. 19-7b

Solution

a. $R_{ac} = P/I^2 = 10/(0.1)^2 = \mathbf{1000\ \Omega}$
b. $Q_0 = EVM_2/EVM_1 = V_{C_0}/V = 60\text{ V}/12\text{ V} = \mathbf{5}$ (19-7)
c. $Q_L = Q_0 = \mathbf{5}$
d. $L = X_L/\omega = QR/\omega = 5(1000)/2\pi 10^3 = \mathbf{795.8\ mH}$
e. $X_C = QR = 5000\ \Omega$ and $C = 1/2\pi f X_C = 1/2\pi 10^3 \times 5 \times 10^3 = \mathbf{31.83\ nF}$

Notes:

1. When EVM_2 is calibrated directly in Q (rather than volts *rms*) the instrument is called a Q-meter. From step (**b**) above, the higher the voltage across EVM_2, the higher is the Q_0 and the Q_L of the coil that is equal to it.
2. There is an error produced in assuming that the maximum voltage across the capacitor ($\mathbf{V}_{C_{max}}$) is the same as $\mathbf{V}_{C_0}$. This is shown quite clearly in Fig. 19-6a. But this error is compensated in the scale calibration of the Q-meter to ensure a precise Q-measurement.
3. Section 19-3.3 below explains why $\mathbf{V}_{C_{max}} \neq \mathbf{V}_{C_0}$ and, similarly, why $\mathbf{V}_{L_{max}} \neq \mathbf{V}_{L_0}$.

19-3 OFF-RESONANCE RELATIONS IN A SERIES *RLC* CIRCUIT

In Sec. 19-2 our concern was primarily with values of voltage, current, and impedance, all at the resonant frequency, f_0. We are now ready to consider several important relations occurring at frequencies either above or below the resonant frequency but still *in close proximity* to f_0.

As noted earlier, one of widest applications of a tuned or resonant circuit occurs when it is used as a *frequency-selective* or *bandpass* filter. In this application, frequencies above and below f_0 are rejected and only the frequencies at or close to f_0 are amplified as a result of the Q-rise in voltage, as shown in Eq. (19-8a), Eq. (19-8b), and Fig. 19-6a. The term *selectivity* is used as a measure of how well a resonant filter is capable of differentiating between the desired signal (f_0) and unwanted frequencies above and below it. The selectivity, of course, is *directly proportional* to the Q_0. A high-Q resonant circuit or filter is *more selective* than a low-Q circuit.

19-3.1 Half-Power Bandwidth Relations

A measure of the *selectivity* of a *series*-resonant circuit is found in its property to produce a high current (for minimum impedance) over a narrow range of frequencies. This range of frequencies is defined as the bandwidth, B, that is, the range of frequencies from some lower frequency, f_1, below the resonant frequency, f_0, to some higher frequency, f_2, above f_0. These frequencies, f_1 and f_2, are sometimes called corner frequencies or *edge-frequencies* because they represent the extreme lower and upper edges of the bandwidth, B. Quantitatively **ANY** bandwidth, B, as derived in Appendix B-15.3, is

$$B = f_2 - f_1 = \frac{f_0(Q)}{Q_0} = \frac{\sqrt{f_1 f_2}(Q)}{Q_0} \quad \text{hertz (Hz)} \qquad \textbf{(19-9)}$$

where f_2 is some frequency above f_0 and f_1 is some frequency below f_0
Q is the circuit Q (i.e., the tangent of the circuit phase angle) at the off-resonant frequencies f_2 and f_1, respectively, or $\tan\theta$.

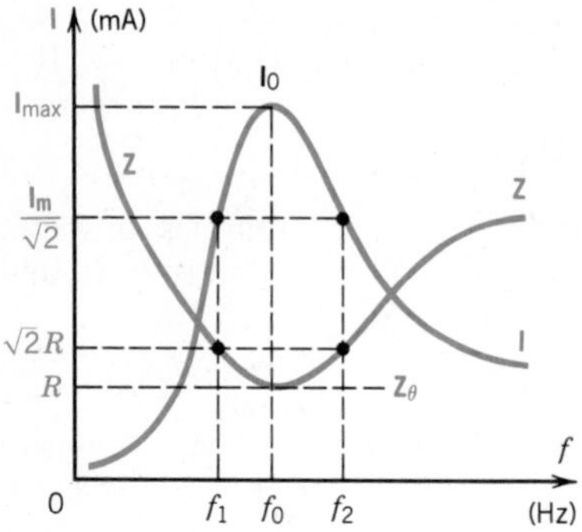

Figure 19-8 Definition of half-power BW

A frequently used selectivity measure is the *half-power* bandwidth or -3 dB bandwidth.[7] The -3 dB bandwidth (BW) is defined as the difference in frequency between the f_1 and f_2 points on the resonance curve, as shown in **Fig. 19-8**, where

1. The current is reduced to $\mathbf{I}_0/\sqrt{2}$ or $0.7071 \cdot \mathbf{I}_0$.
2. The impedance is increased to $\sqrt{2}(R)$ or $\sqrt{2}(\mathbf{Z}_0)$.
3. The power is half the maximum power, P_0, at a level of approximately -3 dB.
4. The circuit phase angle is either leading or lagging by 45°.
5. The tangent of the circuit phase angle (Q as defined in Eq. 19-9) is *unity*.

The half-power or -3 dB bandwidth (BW) is particularly endearing to electrical engineers and technologists because of the various useful equalities noted in Eq. (19-10) below and the five relations just given. As noted in Fig. 19-8, the half-power BW is

$$B_{-3\text{ dB}} = B_{\text{hp}} = f_2 - f_1 = \frac{f_0}{Q_0} = \frac{\sqrt{f_1 f_2}}{Q_0} = \frac{f_0 R}{X_L} = \frac{R}{2\pi L} \quad \text{hertz (Hz)} \quad (19\text{-}10)$$

where f_2 and f_1 are the half-power frequencies defined above and in Fig. 19-8.[8] Of particular interest in Eqs. (19-9) and (19-10) is that

$$f_0 = \sqrt{f_1 f_2} \quad \text{hertz (Hz)} \quad (19\text{-}11)$$

Equation (19-11) implies that f_0 is *not* located midway between f_1 and f_2; that is, it is *not* the *arithmetic* mean of the edge frequencies. Instead, f_0 is actually and precisely the *geometric mean* of the edge frequencies. This is shown most clearly in Fig. 19-8 where the asymmetric impedance and resulting current curve produce a *greater* frequency difference between f_2 and f_0 than between f_0 and f_1. This is also shown in Ex. 19-6**h**.

EXAMPLE 19-6

An *RLC* series circuit connected to a 10-V variable frequency supply has the following values: $R = 10\ \Omega$, $L = 10$ mH, and $C = 1\ \mu$F. Calculate the

a. Resonant frequency, f_0
b. Resonant circuit Q_0
c. Upper edge frequency f_2 if the lower half-power frequency $f_1 = 1514$ Hz
d. Half-power bandwidth, B_{hp}, using the component values given
e. Half-power bandwidth using the resonant frequency and resonant Q_0
f. Half-power bandwidth using the general bandwidth Eq. (19-9) for ANY bandwidth
g. Half-power bandwidth using $f_2 - f_1$
h. Frequency difference $f_2 - f_0$ compared to $f_0 - f_1$
i. Maximum power dissipated at the resonant frequency

Solution

a. $f_0 = 1/2\pi\sqrt{LC} = 1/2\pi\sqrt{10 \cdot 10^{-3} \times 10^{-6}}$ $= \mathbf{1591.55\ Hz}$ (19-2a)
b. $Q_0 = \sqrt{L/C}/R = \sqrt{10\text{ m}/1\ \mu}/10 = \mathbf{10}$ (19-7)
c. $f_2 = f_0^2/f_1 = (1591.55)^2/1514 = \mathbf{1673.1\ Hz}$ (19-11)
d. $B_{\text{hp}} = R/2\pi L = 10/2\pi \times 10^{-2} = \mathbf{159.15\ Hz}$ (19-10)
e. $B_{\text{hp}} = f_0/Q_0 = 1591.55\text{ Hz}/10 = \mathbf{159.15\ Hz}$ (19-10)
f. $B_{\text{hp}} = f_0 Q/Q_0 = 1591.55\ (\tan 45°)/10$ $= \mathbf{159.15\ Hz}$ (19-9)
g. $B_{\text{hp}} = f_2 - f_1 = 1673.1 - 1514 = \mathbf{159.1\ Hz}$ (19-10)
h. $f_2 - f_0 = 1673.1 - 1591.5 = \mathbf{81.6\ Hz}$; $f_0 - f_1 = 1591.5 - 1514 = \mathbf{77.5\ Hz}$
i. $P_0 = \mathbf{V}^2/R = 10^2/10 = \mathbf{10\ W}$

[7] Many texts give the impression that the *only* definition of bandwidth (BW) is the half-power bandwidth. Nothing could be further from the truth. Highly selective bandpass and bandstop filters are frequently designed with narrower BWs. Many wideband amplifiers are designed with wider BWs.

[8] The derivation of Eq. (19-10) appears in Appendix B-15.2.

Example 19-6 verifies the unity of the various equations and relations presented. Part (**h**) shows that f_0 is *not* midway between f_1 and f_2 but is actually closer to f_1 than to f_2, as shown in Fig. 19-8 (and 19-9).

Most communication specialists, however, find it convenient to assume that f_0 *is* midway between f_1 and f_2. Accepting this assumption enables us to evaluate f_1 and f_2 with reasonable accuracy for *any* bandwidth situation, using the simpler approximations

$$f_1 \cong f_0 - \frac{B}{2} \quad \text{hertz (Hz)} \qquad \textbf{(19-12a)}$$

$$f_2 \cong f_0 + \frac{B}{2} \quad \text{hertz (Hz)} \qquad \textbf{(19-12b)}$$

EXAMPLE 19-7

For the value of f_0 and B_{hp} calculated in Ex. 19-6, calculate

a. f_1
b. f_2, using the approximate Eqs. (19-12a, b)
c. Relative errors of each

Solution

a. $f_1 = f_0 - (B_{hp}/2) = 1591.55 - (159.1/2)$
$= \mathbf{1512\ Hz}$ (19-12a)

b. $f_2 = f_0 + (B_{hp}/2) = 1591.55 + (159.1/2)$
$= \mathbf{1671\ Hz}$ (19-12b)

c. $r_e = (1512 - 1514)/1514 = \mathbf{-0.132\%}$ for f_1
$= (1671 - 1673.1)/1673.1 = \mathbf{-0.1255\%}$ for f_2

The relative errors show that **both** *approximate* values are *slightly below* the truer values computed in Ex. 19-6.

Example 19-7 shows that if we accept the assumption of Eqs. (19-12a) and (19-12b) the error produced is less than 0.15% in both cases. We will provide equation 5 (Eqs. 19-16a and 19-16b) that enables the *precise* calculations of the edge frequencies f_1 and f_2 for *any* bandwidth situation. But since the approximate relations are easier to use and provide such a minimal error, we will use Eqs. (19-12a) and (19-12b) throughout this chapter and in the chapters that follow.

19-3.2 Resonant Power, P_0, and Off-Resonant Power, P

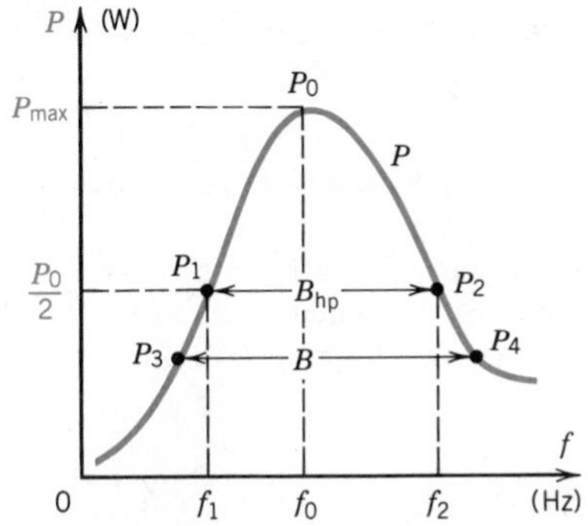

Figure 19-9 Power response as a function of frequency

Figure 19-9 shows the variation of power dissipated in resistor R as a function of frequency, plotted in the range of the resonant frequency, f_0. Maximum power, P_0, occurs at f_0, where the current (from Fig. 19-8) is I_0. At the resonant frequency, f_0, therefore, the maximum power, P_0, is calculated as

$$P_0 = \mathbf{I}_0^2 R = \left(\frac{\mathbf{V}}{\mathbf{Z}_0}\right)^2 R = \left(\frac{\mathbf{V}}{R}\right)^2 R = \frac{\mathbf{V}^2}{R} \quad \text{watts (W)} \qquad \textbf{(19-13)}$$

where $\mathbf{V}$ is the voltage across the resistance R at resonance
$\mathbf{Z}_0$ is R, the circuit impedance at resonance

Equation (19-13) is obvious since at resonance (from Fig. 19-6b) the applied voltage, $\mathbf{V}$, is the voltage across the resistor, R, the only component in the *RLC* series circuit capable of dissipating power.

But now let us consider what happens to the power above and below the resonant frequency, f_0. As derived in Appendix B-15.1, at any other frequency the power P is

$$P = \frac{(V^2/R)}{(1 + Q^2)} = \frac{P_0}{(1 + Q^2)} \quad \text{watts (W)} \qquad (19\text{-}14)$$

where Q is the tangent of the circuit phase angle producing power P at any off-resonance frequencies f_1 and f_2.

Solving this expression for Q yields $Q^2 = (P_0/P) - 1$ or

$$Q = \pm\sqrt{\left(\frac{P_0}{P}\right) - 1} = \frac{X_r}{R} = \tan\theta \qquad (19\text{-}15)$$

Note that $\tan\theta$ in Eq. (19-15), and Q as well, emerge as both a positive and negative value, corresponding to *leading* and *lagging* values of θ, originally shown in Fig. 19-4. Also note that the value of Q obtained in Eq. (19-15) is the same value as the Q given in Eq. (19-9). Consequently, if we know f_0, Q_0, and Q (for a given value of P less than P_0) we can find the bandwidth, B. For example, in Fig. 19-9, given $P_3 = P_4$, we can find Q from Eq. (19-15) and the bandwidth, B, from Eq. (19-9). But we are still unable to evaluate f_1 and f_2 individually (although we can find their difference).

Finding f_1 and f_2 emerges from a reasonably accurate quadratic equation derived from Eq. (19-15):[9]

$$f_1 = \left(\frac{f_0}{2Q_0}\right)\left[\tan(-\theta) + \sqrt{4Q_0^2 + Q^2}\right] \quad \text{hertz (Hz)} \qquad (19\text{-}16a)$$

$$f_2 = \left(\frac{f_0}{2Q_0}\right)\left[\tan(+\theta) + \sqrt{4Q_0^2 + Q^2}\right] \quad \text{hertz (Hz)} \qquad (19\text{-}16b)$$

where Q_0 is the resonant Q at f_0

Q is the tangent of the circuit phase angle at *off-resonant* frequencies f_1, f_2 (note: Q is positive for f_2 and negative for f_1, as shown in Fig. 19-4)

EXAMPLE 19-8

Using the circuit data given and calculated in Ex. 19-6, when the power drops to 2 W on either side of the maximum power at f_0, calculate the

a. Circuit Q
b. Circuit phase angle, θ
c. 2 W bandwidth, B
d. Lower frequency, f_1, using Eq. (19-16a) and the approximate f_1 using Eq. (9-12a)
e. Upper frequency, f_2, using Eq. (19-16b) and the approximate f_2 using Eq. (9-12b)
f. 2 W bandwidth, B, using values calculated in **d** and **e**
g. Verify f_0 using the frequencies computed in **d** and **e**
h. Power in dB (relative to maximum power at f_0) at frequencies f_1 and f_2

[9] The author is indebted to a former colleague, Robert C. Carter, for this insightful presentation of off-resonance conditions, published in R. C. Carter, *Introduction to Electrical Circuit Analysis*, Holt, Rinehart & Winston, 1966, containing Eqs. (19-15) and (19-16), as modified above.

Solution

a. $Q = \sqrt{(P_0/P_1) - 1} = \sqrt{(10/2) - 1} = \mathbf{2}$ (19-15)

b. $\pm\theta = \tan^{-1}(2) = \mathbf{\pm 63.435°}$

c. $B = f_0 Q/Q_0 = 1591.55 \text{ Hz } (2)/10 = \mathbf{318.3 \text{ Hz}}$ (19-9)

d. $f_1 = (f_0/2Q_0)[\tan(-\theta) + \sqrt{4Q_0^2 + Q^2}]$

$$= \frac{1591.55}{20}[-2 + \sqrt{400 + 4}] = \mathbf{1440.35 \text{ Hz}} \quad (19\text{-}16a)$$

$$f_1 = f_0 - B/2 = 1591.55 - (318.3/2) = \mathbf{1432.4 \text{ Hz}} \quad (19\text{-}12a)$$

e. $f_2 = (f_0/2Q_0)[\tan(+\theta) + \sqrt{4Q_0^2 + Q^2}]$

$$= \frac{1591.55}{20}[+2 + \sqrt{400 + 4}] = \mathbf{1758.7 \text{ Hz}} \quad (19\text{-}16b)$$

$$f_2 = f_0 + B/2 = 1591.55 + (318.3/2) = \mathbf{1750.7 \text{ Hz}} \quad (19\text{-}12b)$$

f. $B = f_2 - f_1 = 1758.7 - 1440.35 = \mathbf{318.35 \text{ Hz}}$ (19-10)

g. $f_0 = \sqrt{f_1 f_2} = \sqrt{(1758.7)(1440.35)} = \mathbf{1591.6 \text{ Hz}}$ (19-11)

h. $\mathbf{dB} = 10 \log_{10}(P/P_0) = 10 \log_{10}(2/10) = \mathbf{-7 \text{ dB}}$

Several important insights emerge from Exs. 19-8 and 19-6:

1. The bandwidth in Ex. 19-8 has doubled compared to the half-power BW in Ex. 19-6.
2. The circuit phase angle has increased (from $\pm 45°$ in Ex. 19-6) to $\pm 63.435°$ in Ex. 19-8, which means that the circuit Q has gone from 10 to 2. This reduction in Q accounts for the doubling of the BW in step (**c**) using Eq. (19-9).
3. In steps (**d**) and (**e**) the approximate Eqs. (19-2a and 19-12b) serve as a useful check on accuracy of the more complex and exact calculations of f_1 and f_2 using Eqs. (19-16a and 19-16b).
4. Most important, note that at the -7 dB level the resonant frequency f_0 is still at the geometric mean of frequencies f_1 and f_2. This is shown in step (**g**) of the solution. This relatively simple calculation also serves as a check on the accuracy of the determination of f_1 and f_2. Regardless of BW magnitude, Eq. (19-11) *must always hold*!

19-3.3 Evaluating $\mathbf{V}_{C_{max}}$ and $\mathbf{V}_{L_{max}}$

Figure 19-6a shows that at f_0 there is a Q_0 rise in voltage across C and L such that $\mathbf{V}_{C_0} = Q_0\mathbf{V} = \mathbf{V}_{L_0}$ in magnitude, as shown previously in Eqs. (19-8a) and (19-8b). But the figure also shows that these values, although equal to each other and (usually) larger than the supply voltage, $\mathbf{V}$, are *not the maximum possible* voltages that may occur. As shown in Fig. 19-6a, at some lower frequency below f_0, $\mathbf{V}_{C_{max}}$ occurs at ω_1, and at some higher frequency above f_0, $\mathbf{V}_{L_{max}}$ occurs at ω_2.[10]

Since $\mathbf{V}_{C_{max}}$ and $\mathbf{V}_{L_{max}}$ appear to be equal in magnitude to each other (as are $\mathbf{V}_{C_0}$ and $\mathbf{V}_{L_0}$), as shown in Fig. 19-6a, only one equation is needed to evaluate these maximum values. This equation is derived in Appendix B-15.6 as

$$\mathbf{V}_{C_{max}} = \mathbf{V}_{L_{max}} = \frac{\mathbf{V}Q_0}{\sqrt{(4Q_0^2 - 1)/4Q_0^2}} = \mathbf{V}Q_0\sqrt{\frac{4Q_0^2}{4Q_0^2 - 1}} \quad \text{volts (V)} \quad (19\text{-}17)$$

Both forms of Eq. (19-17) yield the same results, but the last expression is somewhat easier to use and is more revealing. It shows that the factor below the square root sign is *always greater than unity* for values of $Q_0 > 0.5$. A study of Eq. (19-17) reveals that $\mathbf{V}_{C_{max}} > \mathbf{V}_{C_0}$ at extremely low values of Q_0. As Q_0 increases

[10] The radian frequencies ω_1 and ω_2 are used to avoid confusion with f_1 and f_2 in previously derived bandwidth equations.

and approaches 10, for all practical purposes, we may assume that $\mathbf{V}_{C_{max}} = \mathbf{V}_{C0} = \mathbf{V}Q_0$ and $\mathbf{V}_{L_{max}} = \mathbf{V}_{L0} = \mathbf{V}Q_0$.

19-3.4 Frequency ω_1 at Which $\mathbf{V}_{C_{max}}$ Occurs, below ω_0

The radian frequency, ω_1, at which the voltage across the capacitor is maximized is derived in Appendix B-15.5 as

$$\omega_1 = \omega_0 \sqrt{\frac{2Q_0^2 - 1}{2Q_0^2}} = \omega_0 \sqrt{1 - \frac{1}{2Q_0^2}} \quad \text{radians per second (rad/s)} \qquad (19\text{-}18)$$

Again, as in the case of Eq. (19-17), the radian frequency ω_1 in Eq. (19-18) approaches ω_0 at *high* values of Q_0 (where $Q_0 > 10$). The *lower* the value of Q_0, the *greater* the separation between ω_1 and ω_0. Furthermore, ω_0 is multiplied by a factor which is always less than unity in Eq. (19-18). Consequently ω_1 is *always less* than ω_0.

19-3.5 Frequency ω_2 at Which $\mathbf{V}_{L_{max}}$ Occurs, above ω_0

The radian frequency, ω_2, at which the voltage across the inductor is maximized is derived in Appendix B-15.5 as

$$\omega_2 = \frac{\omega_0}{\sqrt{(2Q_0^2 - 1)/2Q_0^2}} = \frac{\omega_0}{\sqrt{1 - \dfrac{1}{2Q_0^2}}} \quad \text{radians per second (rad/s)} \qquad (19\text{-}19)$$

A close study of Eq. (19-19) reveals that

1. ω_0 is divided by a factor that is always less than unity; consequently $\omega_2 > \omega_0$.
2. At high values of Q_0 (above 10), ω_2 approaches ω_0.
3. The lower the value of Q_0, the greater the separation between ω_2 and ω_0.

An examination, therefore, of Eqs. (19-17), (19-18), and (19-19) leads to the following conclusions, summarized in **Fig. 19-10**:

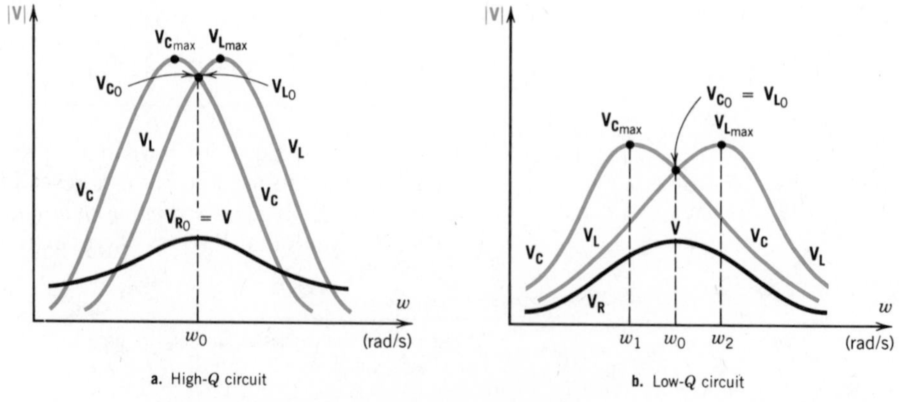

Figure 19-10 Effect of Q_0 in a series *RLC* circuit on ω_1, ω_2, $\mathbf{V}_{C_{max}}$, and $\mathbf{V}_{L_{max}}$

1. In high-Q circuits, the curves of $\mathbf{V}_C$ and $\mathbf{V}_L$ versus frequency cross each other at ω_0, where for all practical purposes $\mathbf{V}_{C_0} = \mathbf{V}_{L_0} \cong \mathbf{V}_{C_{max}} \cong \mathbf{V}_{L_{max}}$ and $\omega_1 \cong \omega_0 \cong \omega_2$, as shown in Fig. 19-10a. Because of the high Q_0, note that $\mathbf{V}_{C_0}$ and $\mathbf{V}_{L_0}$ are much greater in magnitude than $\mathbf{V}_{R_0}$ or $\mathbf{V}$, the supply voltage.
2. As the Q_0 decreases, the curves of $\mathbf{V}_C$ and $\mathbf{V}_L$ are, in effect, "pulled apart" and "flattened" such that there is a greater frequency separation and a smaller Q-rise in voltage at ω_0. For Fig. 19-10b, $\omega_2 > \omega_0 > \omega_1$, $\mathbf{V}_{C_{max}} = \mathbf{V}_{L_{max}}$, but $\mathbf{V}_{C_{max}} > \mathbf{V}_{C_0}$ and $\mathbf{V}_{L_{max}} > \mathbf{V}_{L_0}$.

EXAMPLE 19-9

An *RLC* series circuit consists of $R = 25\ \Omega$, $C = 2\ \mu F$, and $L = 5$ mH, connected across a 10-V ac supply. Calculate

a. Radian frequency, ω_0, at which current $\mathbf{I}_0$ and power P_0 are a maximum
b. L/C ratio and Q_0 at the resonant frequency
c. $\mathbf{V}_{C_0}$ and $\mathbf{V}_{L_0}$ at ω_0
d. $\mathbf{V}_{C_{max}}$ and $\mathbf{V}_{L_{max}}$
e. Radian frequency, ω_1, at which $\mathbf{V}_{C_{max}}$ occurs, and its corresponding frequency f_1 (in Hz)
f. Radian frequency, ω_2, at which $\mathbf{V}_{L_{max}}$ occurs, and its corresponding frequency f_2
g. Radian frequency, ω_0, at which $\mathbf{V}_R$ is a maximum and its corresponding frequency f_1
h. Frequency difference between f_2 (in **f**) and f_1 (in **g**)
i. Half-power bandwidth, B_{hp}, for the current and power response curves

Solution

a. $\omega_0 = 1/\sqrt{LC} = 1/\sqrt{5\text{ mH} \times 2\ \mu\text{F}} = \mathbf{10\,000\ rad/s}$
b. $L/C = 5\text{ mH}/2\ \mu\text{F} = 2500$; $Q_0 = \sqrt{L/C}/R = \sqrt{2500}/25 = \mathbf{2}$
c. $\mathbf{V}_{C_0} = \mathbf{V}_{L_0} = Q_0\mathbf{V} = 2 \times 10\text{ V} = \mathbf{20\ V}$
d. $\mathbf{V}_{C_{max}} = \mathbf{V}_{L_{max}} = \mathbf{V}Q_0\sqrt{\dfrac{4Q_0^2}{4Q_0^2 - 1}} = 20\text{ V}\sqrt{\dfrac{16}{15}}$
$= \mathbf{20.66\ V}$ (19-17)
e. $\omega_1 = \omega_0\sqrt{\dfrac{2Q_0^2 - 1}{2Q_0^2}} = (10\,000\text{ rad/s})\sqrt{\dfrac{7}{8}} = \mathbf{9354\ rad/s}$;
$f_1 = \dfrac{\omega_1}{2\pi} = \mathbf{1489\ Hz}$ (19-18)
f. $\omega_2 = (10\,000\text{ rad/s})\Big/\sqrt{\dfrac{7}{8}} = \mathbf{10\,690.5}$ rad/s;
$f_2 = \dfrac{\omega_2}{2\pi} = \mathbf{1701\ Hz}$ (19-19)
g. From **a** above, $\omega_0 = 10\,000$ rad/s and $f_0 = \omega_0/2\pi =$ **1591.55 Hz**
h. $f_2 - f_1 = (1701 - 1489)\text{ Hz} = \mathbf{212\ Hz}$
i. $B_{hp} = f_0/Q_0 = 1591.55\text{ Hz}/2 = \mathbf{795.8\ Hz}$ (19-10)

Example 19-9 reveals the following important insights:

1. The BW between the maximum values of $\mathbf{V}_{C_{max}}$ and $\mathbf{V}_{L_{max}}$ is much narrower than the half-power bandwidth even when the Q is quite low. (This is why ω_1 and ω_2 are used to avoid confusion with f_1 and f_2.)
2. $\mathbf{V}_{C_{max}}$ is only slightly higher than $\mathbf{V}_{C_0}$ (about 3.3% when the Q_0 is 2, in this example). At smaller values of Q_0, this difference would be more pronounced.
3. At all times $\mathbf{V}_{C_{max}}$ (at ω_1) and $\mathbf{V}_{L_{max}}$ (at ω_2) are equal in magnitude. Similarly, $\mathbf{V}_{C_0} = \mathbf{V}_{L_0}$ at ω_0, as shown in Fig. 19-10b.

Example 19-10 verifies the value of $\mathbf{V}_{C_{max}}$ from Eq. (19-18) using simple series circuit ac theory.

EXAMPLE 19-10

At the off-resonant frequency $\omega_1 = 9354$ rad/s, verify the value of $\mathbf{V}_{C_{max}}$ and B by calculating

a. $-jX_{C_1}$ and $+jX_{L_1}$
b. Resultant X_{r_1} and $\mathbf{Z}_1$
c. Off-resonant current $\mathbf{I}_1$
d. $\mathbf{V}_{C_{max}}$
e. $\mathbf{V}_{L_1}$ at the radian frequency ω_1
f. Bandwidth between maximum values, B

Solution

a. $-jX_{C_1} = 1/\omega_1 C = 1/9354(2\ \mu\text{F}) = \mathbf{-j53.45\ \Omega}$;
$+jX_{L_1} = \omega_1 L = 9354(5\text{ mH}) = \mathbf{+j46.77\ \Omega}$
b. $X_{r_1} = -jX_{C_1} + jX_{L_1} = (-j53.45 + j46.77)\ \Omega = \mathbf{-j6.68\ \Omega}$;
$\mathbf{Z}_1 = R - jX_{r_1} = 25 - j6.68$
$= \mathbf{25.88\angle{-15^\circ}\ \Omega}$ (actually $\angle{-14.96^\circ}$) (19-4a)
c. $\mathbf{I}_1 = \mathbf{V}/\mathbf{Z}_1 = 10\angle 0^\circ\text{ V}/25.88\angle{-15^\circ}\ \Omega$
$= \mathbf{0.3864\angle 15^\circ\ A}$ (19-6a)
d. $\mathbf{V}_{C_{max}} = \mathbf{I}_1(-jX_{C_1}) = (0.3864\angle 15^\circ)(53.45\angle{-90^\circ})$
$= \mathbf{20.66\angle{-75^\circ}\ V}$
e. $\mathbf{V}_{L_1} = \mathbf{I}_1(jX_{L_1}) = 0.3864\angle 15^\circ(46.77\angle 90^\circ) = \mathbf{18.07\angle 105^\circ\ V}$
f. $B = f_0(Q)/Q_0 = 1591.55(\tan 15^\circ)/2 = \mathbf{213\ Hz}$ (19-9)

Example 19-10 verifies Ex. 19-9, *using the value* of ω_1 calculated via Eq. (19-18), and reveals the following:

1. The impedance, $\mathbf{Z}_1$, is *capacitive* with a *negative* phase angle (approximately -15°).
2. The off-resonance impedance, $\mathbf{Z}_1$, is greater than $\mathbf{Z}_0$ at f_0 or $R = 25\angle 0^\circ\ \Omega$.

3. The voltage across the capacitor, $\mathbf{V}_{C_1} = \mathbf{V}_{C_{max}}$, is greater than the voltage across the inductor, $\mathbf{V}_{L_1}$ at radian frequency ω_1, as shown in Fig. 19-10b.
4. The general equation for finding any bandwidth, Eq. (19-9), applies equally to the bandwidth between f_1 and f_2, using the value of $|Q| = \tan 15°$, where Q is the circuit phase angle.
5. The slight difference in B values in Exs. 19-9 and 19-10 is due to the rounding off of the circuit phase angle value to 15°.
6. At frequency f_1 the circuit phase angle is negative since the resultant impedance triangle is negative, due to predominating X_{C_s}. This verifies the negative values of θ in Fig. 19-4 at frequencies below f_0.

19-4 SERIES RLC RESONANCE SUMMARY

We may summarize the foregoing relations briefly, given ideal *RLC* elements in series:

1. At the resonant frequency, $f_0 = 1/2\pi\sqrt{LC}$, the magnitudes X_{L_0} and X_{C_0} are equal and stored energy is oscillating from the electrical field of the capacitor to the magnetic field of the inductor at the resonant frequency. Only R dissipates energy.
2. Current is a maximum, $\mathbf{I}_0$, at the resonant frequency and the circuit impedance is a minimum such that $\mathbf{Z}_0 = R$. Maximum $\mathbf{I}_0 = \mathbf{V}/R = \mathbf{V}/\mathbf{Z}_0$. Maximum power $P_0 = \mathbf{I}_0^2 R$ at resonance.
3. Frequencies below f_0 produce a capacitive impedance, $\mathbf{Z} = R - jX_r$, whereas those above f_0 produce an inductive impedance, $\mathbf{Z} = R + jX_r$. The circuit impedance phase angle is negative below f_0 and positive above f_0.
4. At the resonant frequency f_0, current is in phase with the voltage and the power factor is unity. Below f_0, current is reduced and leads the applied voltage. Above f_0, current is also reduced (due to increased impedance) and lags the applied voltage, **V**.
5. If the circuit Q_0 at resonance exceeds unity, voltage multiplication occurs and the voltage across both the capacitor and the inductor is $\mathbf{V}_{C_0} = \mathbf{V}_{L_0} = Q_0\mathbf{V}$.
6. The half-power bandwidth of the current, resistor voltage, and power response curves is $B_{hp} = f_0/Q_0 = R/2\pi L = \sqrt{f_1 f_2}/Q_0$.
7. At any other power level, the response curve bandwidth is $B = f_0(Q)/Q_0$. See Eq. (19-9).
8. The following conditions exist at the half-power level, $P_0/2 = \mathbf{I}_0^2 R/2$:
   a. At edge frequencies, f_1 and f_2, the power is reduced by -3 dB.
   b. Current is $0.7071\mathbf{I}_0$ and the circuit phase angle is $\pm 45°$.
   c. The tangent of the phase angle, Q, is unity.
   d. The circuit impedance $\mathbf{Z} = \sqrt{2}(R) = \sqrt{2}(\mathbf{Z}_0)\underline{/\pm 45°}\ \Omega$.
9. A sharply rising narrow-bandwidth response curve is produced by a high L/C ratio, a high Q_0, and/or a low R in an *RLC* series circuit.
10. In **high-Q_0** circuits ($Q > 10$), $\mathbf{V}_{C_{max}} \cong \mathbf{V}_{C_0} = \mathbf{V}_{L_0} \cong \mathbf{V}_{L_{max}}$ simultaneously at resonant frequency ω_0.
11. In **low-Q_0** circuits, $\mathbf{V}_{L_{max}}$ occurs at $\omega_2 > \omega_0$ and $\mathbf{V}_{C_{max}}$ occurs at $\omega_1 < \omega_0$. But $\mathbf{V}_{C_0} = \mathbf{V}_{L_0}$ at ω_0, as it does in high-Q_0 circuits.
12. Correction equations are required for low-Q_0 *RLC* series-resonant circuits to yield more precise values of frequencies ω_1 and ω_2, as well as $\mathbf{V}_{C_{max}} = \mathbf{V}_{L_{max}}$.
13. A high Q_0 implies a high selectivity and a correspondingly low (narrow) bandwidth.

19-5 THE THREE-BRANCH IDEAL PARALLEL-RESONANT CIRCUIT

The three-branch parallel-resonant circuit shown in **Fig. 19-11a** is the *dual* of the three-element series *RLC* circuit. As in the case of the series-resonant *RLC* circuit,

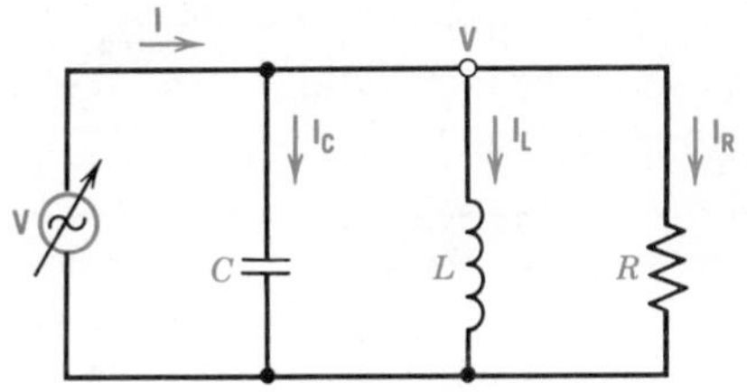

a. Circuit containing ideal elements

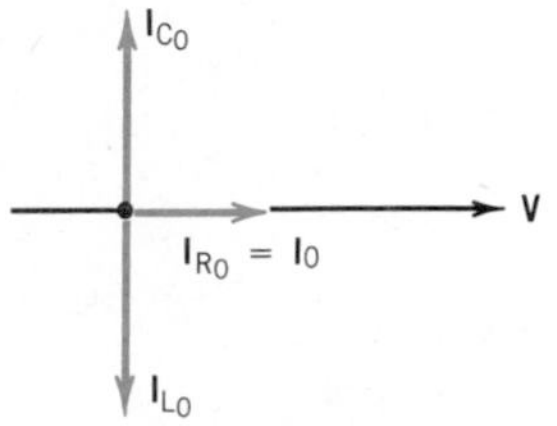

b. Phasor diagram of currents at resonant frequency f_0

Figure 19-11 β-branch parallel-resonant circuit and phasor diagram

we will assume that L is ideal. We begin with the three-branch circuit because (a) it simplifies the analysis with *ideal* components and (b) *all* resonance conditions occur at the *same* resonant frequency, f_0.

19-5.1 Resonant Frequency, f_0, and Radian Frequency, ω_0

As in the case of the series circuit, parallel resonance occurs at frequency

$$f_0 = \frac{1}{2\pi\sqrt{LC}} \quad \text{hertz (Hz)} \qquad \text{(19-20a)}$$

And since resonance is a function of only L and C, the resonant radian frequency is

$$\omega_0 = \frac{1}{\sqrt{LC}} \quad \text{radians per second (rad/s)} \qquad \text{(19-20b)}$$

19-5.2 Current and Impedance Relations

Figure 19-11b shows the current relations at the resonant frequency, f_0. With respect to Fig. 19-11b, note the following current relations:

1. The current in the capacitor, $\mathbf{I}_{C_0}$, leads the reference voltage, **V**, by 90°.
2. The current in the inductor, $\mathbf{I}_{L_0}$, lags the reference voltage, **V**, by 90°.
3. Resonance currents $\mathbf{I}_{C_0}$ and $\mathbf{I}_{L_0}$ are equal and opposite in magnitude with the result that they cancel each other out.
4. Consequently, the total circuit current **I** (in Fig. 19-11a) is $\mathbf{I}_0 = \mathbf{I}_R$. This current is a minimum at f_0 and limited only by the impedance of R.
5. At frequencies above f_0, $\mathbf{I}_C > \mathbf{I}_L$, and below f_0, $\mathbf{I}_L > \mathbf{I}_C$. Consequently, the total circuit current *increases* at frequencies above and below f_0, as shown in **Fig. 19-12**.

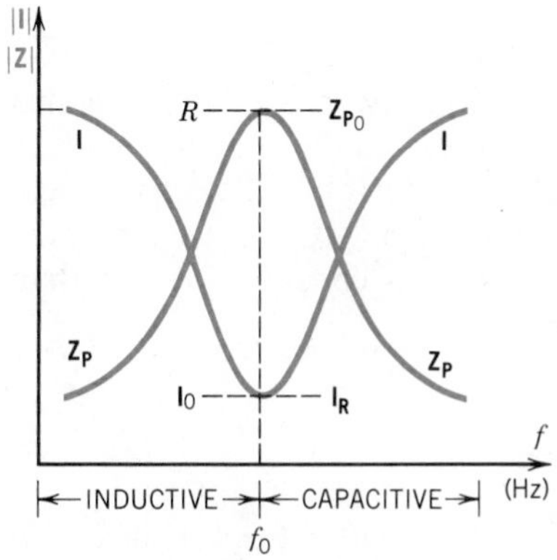

Figure 19-12 Variation of impedance in 3 branch parallel *RLC* circuit with frequency

We may summarize the current relations at resonance by

$$\mathbf{I}_0 = \mathbf{I}_R = \frac{\mathbf{V}}{\mathbf{R}} = \frac{\mathbf{V}}{\mathbf{Z}_{p0}} \quad \text{amperes (A)} \qquad \text{(19-21a)}$$

With respect to impedance relations, since current is a minimum at f_0, then impedance must be a maximum at f_0. We may summarize the equivalent parallel impedance at resonance by

$$\mathbf{Z}_{p0} = \mathbf{Z}_0 = R = \frac{\mathbf{V}}{\mathbf{I}_R} = \frac{\mathbf{V}}{\mathbf{I}_0} \quad \text{ohms } (\Omega) \qquad \textbf{(19-21b)}$$

At frequencies above or below f_0, the capacitor and inductor produce net reactances in parallel with R, thus *reducing* the equivalent parallel impedance. We may summarize the impedance relations as follows:

1. At f_0, the impedance of the "tank" (C in parallel with L) is *infinite* since $\mathbf{I}_C + \mathbf{I}_L = 0$ and only an infinite impedance draws zero current from supply $\mathbf{V}$. Consequently, $\mathbf{Z}_{p0} = (\infty) \| R = R$.
2. At frequencies **below** f_0, $\mathbf{I}_L > \mathbf{I}_C$, and resultant current $\mathbf{I} = \mathbf{I}_R - j\mathbf{I}_X$ is greater in magnitude and *lags* the voltage. This increase in current implies a *reduced* equivalent parallel impedance, $\mathbf{Z}_p$, since there is now a net *inductive* reactance in parallel with R. As shown in Fig. 19-12, this net reactance is *inductive.*
3. At frequencies **above** f_0, $\mathbf{I}_C > \mathbf{I}_L$, and the resultant current $\mathbf{I} = \mathbf{I}_R + j\mathbf{I}_X$ is also greater in magnitude and *leads* voltage $\mathbf{V}$. This increase in current implies a reduced impedance, $\mathbf{Z}_p$, since there is now a net *capacitive* reactance in parallel with R.

19-5.3 Q_0 Rise in Current

The series RLC circuit, at resonant frequency f_0, exhibited a Q-rise in **voltage** across L and C. The parallel RLC three-branch circuit is its *dual.* Consequently, it exhibits a Q-rise in **current** in branches L and C, the so-called "tank." This means that the circulating current $\mathbf{I}_{C0}$ and $\mathbf{I}_{L0}$ in Fig. 19-11b may be *many* times the total current $\mathbf{I}_0$ and many times the resistive current $\mathbf{I}_R$. Mathematically, the current relations for $\mathbf{I}_{C0}$ and $\mathbf{I}_{L0}$ at the resonant frequency f_0 are

$$\mathbf{I}_{C0} = \mathbf{I}_{L0} = Q_0\mathbf{I}_0 = Q_0\mathbf{I}_R = Q_0\left(\frac{\mathbf{V}}{R}\right) \quad \text{amperes (A)} \qquad \text{(19-22)}$$

Equation (19-22) implies that in a high-Q_0 circuit at resonance, large currents are oscillating and circulating between L and C, corresponding to the (lossless) transfer of energy between the magnetic field of the inductor and electrical field of the capacitor.[11] Paradoxically, however, the total current, $\mathbf{I}_0$, is the same as the current in resistor R, as shown by Eq. (19-21a), which is considerably less than $\mathbf{I}_{C0}$ or $\mathbf{I}_{L0}$ in a high-Q_0 circuit.

19-5.4 Circuit Q_0 at Resonant Frequency f_0

Since the three-branch parallel resonant circuit is the *dual* of the three-element series RLC circuit, we may write the resonant Q_0 as

[11] If the resistance, R, in Fig. 19-11a is removed ($R = \infty$) leaving only **ideal** elements L and C, the circulating currents are theoretically infinite and the current from the supply, $\mathbf{I}$, would be zero!

$$Q_0 = \frac{R}{\omega_0 L} = \frac{R}{X_{L_0}} = R\sqrt{\frac{C}{L}} = \frac{R}{X_{C_0}} = \frac{\mathbf{I}_{L_0}}{\mathbf{I}_0} = \frac{\mathbf{I}_{C_0}}{\mathbf{I}_0} = \frac{Q_s}{P_0} = \frac{f_0}{B_{hp}} \qquad (19\text{-}23)$$

where $\mathbf{I}_0$ is $\mathbf{I}_R$, total circuit current
Q_s is the quadrature power stored either in the inductor or the capacitor
P_0 is the power dissipated in the resistor ($\mathbf{I}_0^2 R$) at resonance
B_{hp} is the half-power bandwidth (BW)

Observe that Eq. (19-23) is the *reciprocal* of some of the terms in Eq. (19-7) and with respect to power and bandwidth, is the same as Eq. (19-7). Recall from Eq. (16-25) that the Q of a paralleled resistance and reactance is $Q = R_p/X_p = X_s/R_s$ of an equivalent series circuit, implying that Q is reciprocally related in going from series to parallel circuits. Of particular importance in Eq. (19-23) is that $Q_0 = R\sqrt{C/L}$ for a parallel three-branch circuit (compared to $\sqrt{L/C}/R$ for a three-element series RLC circuit). This relation is important since it enables evaluation of Q_0 from the three-branch parallel circuit elements themselves. Finally, note that since R is now in the *numerator*, the higher the resistance in parallel, the higher the Q!

19-5.5 Effect of Q_0 on Selectivity and Bandwidth

Equation (19-20) showed that the resonant frequency, f_0, depends solely on the component values of L and C. Equation (19-23) showed that the Q-rise in tank current and in impedance depends primarily on the resistance R. **Figure 19-13** shows the effect of a high resistance R in parallel with a given tank ($L \| C$) versus a low resistance R' in parallel with the *same* values of L and C. The high-R response curve shows that the bandwidth (BW) is narrower than the low-R' response. Consequently, the higher the value of R, the higher the Q_0, the smaller the BW, and the greater the selectivity.

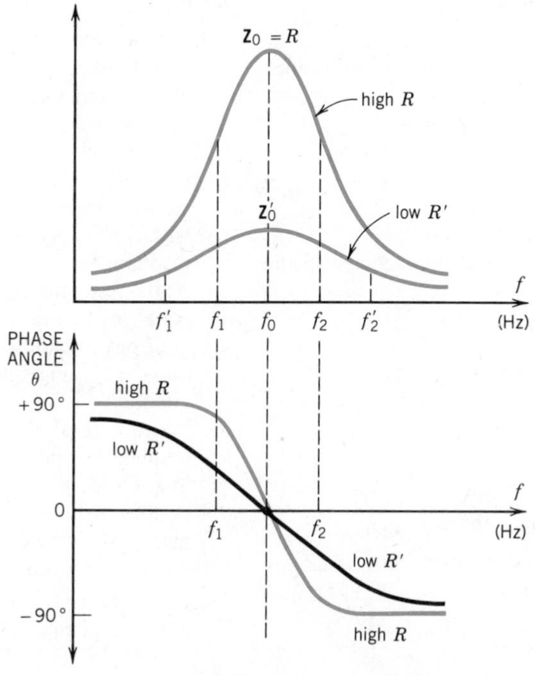

Figure 19-13 Effect of resistance on BW, selectivity, and impedance phase angle

Figure 19-13 also shows the effect of resistance on the impedance and circuit phase angle. From Fig. 19-12 we can (also) see that below f_0 the circuit is inductive, while above f_0 the circuit is capacitive. (An *inductive* impedance triangle has a *positive* phase angle and a *capacitive* impedance triangle has a *negative* phase angle.) Consequently, in Fig. 19-13, θ is always a positive angle between 90° and 0° above f_0. Note that in a *high-resistance* circuit we tend to approach $\pm 90°$ more rapidly as the frequency is varied below or above f_0. A low resistance produces a *more linear* transition, as shown in Fig. 19-13. Consequently, for the same frequency f_1 below f_0, the impedance phase angle is greater for the higher resistance R than the lower resistance R'.

19-5.6 Summary of Conditions in the Ideal Three-Branch Parallel-Resonant Circuit

We may summarize six conditions for the *ideal* three-branch parallel-resonant circuit, *all of which occur at the same* resonant frequency, f_0:

1. Total circuit current, $\mathbf{I}_0$, is a **minimum** and is equal to $\mathbf{I}_R$, the current in resistor R.
2. The resonant frequency is $f_0 = 1/2\pi\sqrt{LC}$.
3. Total circuit impedance is a **maximum** and equal to R at f_0.
4. Minimum total circuit current $\mathbf{I}_0$ is in phase with $\mathbf{V}$ (**unity** PF condition i.e., $F_p = 1$.)
5. Circuit Q_0 is a maximum and equal to $R/\omega_0 L$ and other factors in Eq. (19-23).
6. Circulating tank current in L and C components is a maximum and equal to $Q\mathbf{I}_0$ or $Q\mathbf{I}_R$.

EXAMPLE 19-11

A three-branch ideal parallel *RLC* circuit has a resistance $R =$ 10 kΩ, a 1 H inductance in the second branch, and $C = 1\ \mu$F in the third branch. If the supply voltage is 10 V, constant over a variable frequency range, calculate the

a. Resonant frequency, f_0, and radian frequency, ω_0
b. Inductive and capacitive reactances of each reactive branch
c. Current in each branch $\mathbf{I}_{R_0}$, $\mathbf{I}_{L_0}$, and $\mathbf{I}_{C_0}$
d. Q rise in current in reactance branches (circuit Q_0)
e. Total circuit current, $\mathbf{I}_0$
f. Total circuit impedance, $\mathbf{Z}_0$
g. Circuit half-power bandwidth, B_{hp}
h. Edge frequencies f_1 and f_2 (approximately)
i. Current drawn from the supply at edge frequencies f_1 and f_2

Solution

a. $f_0 = 1/2\pi\sqrt{LC} = 1/2\pi\sqrt{1 \times 10^{-6}} = \mathbf{159.15\ Hz}$,
$\omega_0 = 2\pi f_0 = \mathbf{1000\ rad/s}$ (19-2b)
b. $jX_{L_0} = \omega_0 L = +j1000 \times 1 = \mathbf{1000\angle 90°\ \Omega}$ and $-jX_{C_0} = -j1000 = \mathbf{1000\angle -90°\ \Omega}$
c. $\mathbf{I}_{R_0} = \mathbf{V}/R = 10/0°\ \text{V}/10\ \text{k}\Omega = 1/0°\ \text{mA}$,
$\mathbf{I}_{L_0} = \mathbf{V}/X_{L_0} = 10\angle 0°\ \text{V}/1\angle 90°\ \text{k}\Omega = \mathbf{10\angle -90°\ mA}$,
$\mathbf{I}_{C_0} = \mathbf{V}/X_{C_0} = 10\angle 0°\ \text{V}/1\angle -90°\ \text{k}\Omega = \mathbf{10\angle 90°\ mA}$
d. $Q_0 = \mathbf{I}_{C_0}/\mathbf{I}_{R_0} = 10\ \text{mA}/1\ \text{mA} = \mathbf{10}$ (19-23)
e. $\mathbf{I}_0 = \mathbf{I}_{R_0} = \mathbf{1\angle 0°\ mA}$
f. $\mathbf{Z}_0 = \mathbf{V}/\mathbf{I}_0 = 10\angle 0°\ \text{V}/1\angle 0°\ \text{mA} = \mathbf{10\angle 0°\ k\Omega}$
g. $B_{hp} = f_0/Q_0 = 159.15\ \text{Hz}/10 = \mathbf{15.92\ Hz}$ (19-10)
h. $f_1 \cong f_0 - B_{hp}/2 = 159.2 - (15.92/2) = \mathbf{151.2\ Hz}$ (19-12a)
$f_2 \cong f_0 + B_{hp}/2 = 159.2 + (15.92/2) = \mathbf{167.1\ Hz}$ (19-12b)
i. $\mathbf{I} \cong \sqrt{2}\mathbf{I}_0 = 1.414 \times 1\ \text{mA} = \mathbf{1.414\angle \pm 45°\ mA}$

The following insights are to be drawn from Ex. 19-11:

1. The circulating tank currents, $\mathbf{I}_{C_0}$ and $\mathbf{I}_{L_0}$, are relatively much higher than the current in the resistor or the line current, $\mathbf{I}_0$. Since they are equal in magnitude and opposite in phase, they cancel each other.
2. As a result, the current drawn from the supply, $\mathbf{I}_0$, is the same as the current in the resistor, $\mathbf{I}_R$. This current, $\mathbf{I}_0$, is in phase with the supply voltage, producing unity power factor at the resonant frequency, f_0.
3. At the resonant frequency, impedance, $\mathbf{Z}_0$, is a maximum and equal to the branch resistance.
4. Half-power bandwidth and edge frequencies f_1 and f_2 are all found by using the same relations as for a series-resonant circuit.
5. Because total impedance is a maximum at the resonant frequency, total current $\mathbf{I}_0$ is a minimum. At half-power edge frequencies f_1 and f_2, the total current is $\sqrt{2}\mathbf{I}_0$. At correspondingly higher and lower frequencies, the total current, $\mathbf{I}$, increases as the complex impedance decreases.
6. As shown in Ex. 19-11 step (**i**) at edge frequencies f_1 and f_2 the currents $\mathbf{I}_1$ and $\mathbf{I}_2$, respectively, are $\sqrt{2}\mathbf{I}_0$ in magnitude and out of phase with respect to supply voltage by $\pm 45°$.

19-6 THE TWO-BRANCH PARALLEL-RESONANT CIRCUIT

The three-branch circuit of Sec. 19-5 is "impractical" because the parallel inductor, L, is *never* pure. It was presented as an *ideal* relationship because it enabled the summary that appears in Sec. 19-5.6 and helps us to visualize impedance, current, bandwidth, and phase angle relations in a *parallel-resonant* circuit.

A practical two-branch parallel tank circuit is shown in **Fig. 19-14**. It consists of a practical coil in parallel with a (pure) capacitor. Since most commercial capacitors have a relatively low dissipation factor and resistance, we may *assume* that C is pure. Observe from Fig. 19-14 that the series branch impedance, $R_s + jX_{Ls}$, is easily converted to its parallel three-branch equivalent since both circuits have the same Q, as shown in Eq. (16-25). *This most important insight is fundamental to the two-branch circuit analysis.* Recall from Eqs. (16-23) and (16-24) that

$$R_p = \frac{\mathbf{Z}_s}{\cos\theta} = R_s(1 + Q^2) \qquad \text{and} \qquad X_p = X_s\frac{(1 + Q^2)}{Q^2}$$

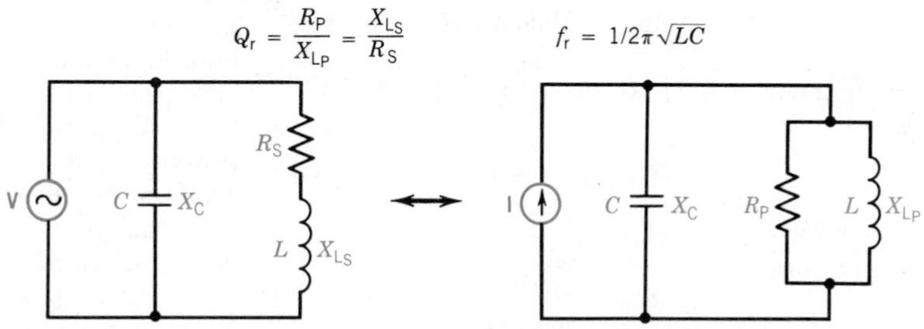

Figure 19-14 A practical two-branch parallel tank circuit and its three-branch parallel equivalent

19-6.1 The High-Q, Two-Branch Parallel Resonant Circuit

For a two-branch parallel resonant tank circuit whose Q exceeds 10, we may treat the two-branch circuit exactly the same as the three-branch circuit. This means that the six conditions summarized in Sec. 19-5.6 for the three-branch circuit apply equally to the two-branch circuit. (Life is very simple, indeed, for solving the **high-Q** two-branch circuit.) Since the *Q is high,* we are entitled to make some *approximations* to simplify our calculations in converting from the practical two-branch to the ideal three-branch circuit:

$$R_{po} = R_s(1 + Q_r^2) \cong R_s(Q_r^2) \qquad \text{ohms } (\Omega) \qquad (19\text{-}24)$$

$$X_{Lp} = X_{Ls}\frac{(1 + Q_r^2)}{Q_r^2} \cong X_{Ls} \qquad \text{ohms } (\Omega) \qquad (19\text{-}25)$$

$$Q_r = \frac{R_{po}}{X_{Lp}} = \frac{X_{Ls}}{R_s} \cong Q_0 \qquad \text{dimensionless} \qquad (19\text{-}26)$$

$$f_r = \frac{1}{2\pi\sqrt{LC}} \cong f_0 \qquad \text{hertz (Hz)} \qquad (19\text{-}27)$$

$$\mathbf{Z}_{po} = \mathbf{Z}_0 = R_{po} \cong Q_r^2 R_s \cong \frac{X_{Ls}^2}{R_s} \cong QX_{Ls} \cong \frac{L}{CR_s} \qquad \text{ohms } (\Omega) \qquad (19\text{-}28)$$

EXAMPLE 19-12

A practical two-branch parallel tank circuit contains an 81-pF capacitor in one branch and a 5-mH inductor whose resistance is 785.4 Ω in the second branch. If the variable frequency supply voltage is $55\angle 0°$ V, calculate the

a. Resonant frequency, f_r, at which parallel resonance occurs
b. Resonant Q of the circuit, Q_r
c. Equivalent resistance, R_p, and impedance, $\mathbf{Z}_p$, of the three-branch equivalent circuit at resonance
d. Reactance of the capacitor at f_r
e. Reactance of the inductor at f_r
f. Impedance of the inductive branch at f_r
g. Current in the capacitive branch at f_r, $\mathbf{I}_{C_0}$
h. Current in the inductive branch at f_r, $\mathbf{I}_{L_0}$
i. Total current drawn from the supply, $\mathbf{I}_0$, as the sum of $\mathbf{I}_{C_0}$ and $\mathbf{I}_{L_0}$
j. Ratio of $\mathbf{I}_{C_0}$ to $\mathbf{I}_0$
k. Total circuit current, $\mathbf{I}_0$, using the equivalent impedance in **c**
l. Verify $\mathbf{Z}_{p_0}$ using the ratio L/CR as shown in Eq. (19-28)

Solution

a. $f_r = 1/2\pi\sqrt{LC} = 1/2\pi\sqrt{5 \times 10^{-3} \times 81 \times 10^{-12}}$
$= \mathbf{250\ kHz}$ (19-27)

b. $Q = X_{Ls}/R_s = 2\pi f_r L/R_s$
$= 2\pi/(250\ \text{k})(5\ \text{m})/785.4 = \mathbf{10}$ (19-26)

c. $R_{p_0} \cong Q^2 R_s = \mathbf{Z}_{p_0} = (785.4) \times (10^2) \cong \mathbf{78.54\ k\Omega}$
(equivalent three-branch circuit Z) (19-24)

d. $-jX_{C_0} = 1/2\pi f_r C = 1/2\pi(250\ \text{k})(81\ \text{p})$
$= \mathbf{-j7859.5\ \Omega}$ (capacitive branch impedance)

e. $+jX_{L_0} = 2\pi f_r L = 2\pi(250\ \text{k})(5\ \text{m})$
$= \mathbf{+j7854\ \Omega}$ (verifying Q of 10)

f. $\mathbf{Z}_{Ls} = R + jX_{Ls} = 785.4 + j7854$
$= \mathbf{7893.2\angle 84.3°\ \Omega}$ (inductive branch impedance)

g. $\mathbf{I}_{C_0} = \mathbf{V}/(-jX_{C_0}) = 55\angle 0°\ \text{V}/-j7859.5 = \mathbf{0 + j6.998\ mA}$

h. $\mathbf{I}_{L_0} = \mathbf{V}/\mathbf{Z}_{Ls} = 55\angle 0°\ \text{V}/7893.2\angle 84.3°\ \Omega$
$= \mathbf{6.968\angle -84.3°\ mA} = \mathbf{(0.692 - j6.934)\ mA}$

i. $\mathbf{I}_0 = \mathbf{I}_{C_0} + \mathbf{I}_{L_0} = (0 + j6.998) + (0.692 - j6.934)$
$= 0.692 + j0.06445 = \mathbf{0.695\angle 5.3°\ mA}$

j. $\mathbf{I}_{C_0}/\mathbf{I}_0 = 6.998/0.695 = 10.07 \cong 10$

k. $\mathbf{I}_0 = \mathbf{V}/\mathbf{Z}_{p_0} = 55\angle 0°\ \text{V}/78.54\angle 0°\ \text{k}\Omega = 0.7\angle 0°\ \text{mA}$

l. $\mathbf{Z}_{p_0} = L/CR = 5\ \text{mH}/81\ \text{pF}(785.4) = 78.594\ \text{k}\Omega$ (19-28)

Example 19-12 shows that the approximations made in Eqs. (19-24) through (19-28) are reasonably *valid* for the *high-Q* tank circuit. With respect to Ex. 19-12 please note that

1. Steps (**a**), (**b**), and (**c**) in the solution enable us to determine the equivalent impedance of the three-branch (and the two-branch practical) circuit as $78.54\angle 0°$ kΩ, using the foregoing approximations.
2. Steps (**d**) through (**l**) verify the assumption using the given two-branch tank circuit.
3. The total current drawn from the supply, $\mathbf{I}_0$, is approximately $0.7\angle 5°$ mA from step (**i**) and $0.7\angle 0°$ mA from the equivalent impedance calculated in steps (**c**) and (**k**).
4. Most important, $\mathbf{Z}_{p_0}$ may be calculated (and verified) directly in step (**l**) from the given components (without the need for Q or f) by using L/CR!
5. Finally, the Q of the circuit is the **same** as the Q of the coil itself!

19-6.2 The Low-Q, Two-Branch Parallel-Resonant Circuit

A high-Q circuit implies, by its very nature, that the practical coil used in the inductive branch has *low* resistance in proportion to its reactance at resonance. The capacitive branch is assumed (for the present) to have no resistance whatever. These assumptions enabled us to equate the high-Q practical two-branch resonant circuit to the three-branch circuit containing ideal elements only in Sec. 19-6.1.

But what if the coil at the resonant frequency, f_r (at which $|jX_L| = |jX_C|$), has a *low* Q? Do the foregoing equations and approximations still hold, or are some modifications required? The answer lies in the realization that the presence of higher resistance in the inductive branch lowers the Q and also the frequencies at which maximum impedance (minimum current) and unity power factor occur, respectively.

As shown in the detailed discussion of Appendix B-16.5, for low-Q circuits we must define three separate and distinct frequencies corresponding to the three resonant conditions:

f_r, the higher frequency at which the magnitude $|jX_L|$ equals $|-jX_C|$

f_m, the middle frequency at which the tank circuit impedance, $\mathbf{Z}_m$, is a maximum and current is a minimum

f_0, the lower frequency at which total tank current is in phase with voltage and unity power factor occurs

As shown in Appendix Table B-16, for the given values of L, C, and R_L we may write

$$f_r = \frac{1}{2\pi\sqrt{LC}} \quad \text{hertz (Hz)} \tag{19-2a}$$

$$Q_r = \frac{2\pi f_r L}{R_L} = \frac{\omega_r L}{R_L} \tag{19-7}$$

$$f_m = f_r\left(\sqrt{\frac{(4Q_r^2 - 1)}{4Q_r^2}}\right) \quad \text{hertz (Hz)} \tag{19-29a}$$

$$f_0 = f_r\left(\sqrt{\frac{(Q_r^2 - 1)}{Q_r^2}}\right) = f_r\left(\sqrt{1 - \frac{CR_s}{L}}\right) \quad \text{hertz (Hz)} \tag{19-29b}$$

and the corresponding values of equivalent circuit impedance may be written, as derived and shown in Appendix Table B-16, as

$$\mathbf{Z}_r = \frac{[(R_L + jX_L)(-jX_C)]}{R_L} \quad \text{ohms } (\Omega) \tag{19-30a}$$

$$\mathbf{Z}_m = \frac{(R_L + jX_{L_m})(-jX_{C_m})}{(R_L + jX_{L_m} - jX_{C_m})} \quad \text{ohms } (\Omega) \tag{19-30b}$$

$$\mathbf{Z}_0 = Q_r^2 R_L = \frac{L}{CR_L} \quad \text{ohms } (\Omega) \tag{19-30c}$$

Similarly, the corresponding values of Q at the corresponding respective frequencies may be written as

$$Q_m = Q_r\left(\frac{f_m}{f_r}\right) \tag{19-31a}$$

$$Q_0 = Q_r\left(\frac{f_0}{f_r}\right) = \sqrt{Q_r^2 - 1} = Q_r\left(\frac{X_{L0}}{X_{L_r}}\right) = \frac{\mathbf{I}_{C0}}{\mathbf{I}_0} = \frac{X_{L0}}{R_L} \tag{19-31b}$$

EXAMPLE 19-13A

A practical inductor has a resistance of 50 Ω and a resonant Q_r of 5 at $f_r = 1$ kHz resonant frequency. It is connected in parallel with a pure capacitor across a $20\underline{/0^\circ}$ V supply. Calculate

a. X_C for resonance at frequency f_r
b. Equivalent impedance of the tank circuit at resonance frequency f_0
c. Frequency, f_0, at which the tank circuit is producing unity PF resonance
d. Circuit Q_0 at frequency f_0
e. Capacitive and inductive reactances at frequency f_0
f. Capacitive branch current at f_0, $\mathbf{I}_{C0}$
g. Inductive branch current at frequency f_0, $\mathbf{I}_{L0}$
h. Total current as the phasor sum of (**f**) and (**g**)
i. Total current using the Q ratio of $\mathbf{I}_{C0}$ to $\mathbf{I}_0$
j. Total current using the value of $\mathbf{Z}_{p0}$ found in (**c**) above

Solution

a. $-jX_{C_r} = |X_{L_r}| = Q_r R_s = 5 \times (50) = \mathbf{-j250\ \Omega}$
(Note $X_{L_r} = +j250\ \Omega$)

b. $\mathbf{Z}_{\mathrm{po}} = Q_r^2 R_s = 5^2(50) = 1250\ \Omega = \mathbf{1.25\underline{/0°}\ k\Omega}$ (19-30c)

c. $f_0 = f_r\sqrt{\dfrac{Q_r^2 - 1}{Q_r^2}} = 1000\ \text{Hz}\sqrt{24/25} = \mathbf{979.8\ Hz}$ (19-29b)

d. $Q_0 = \sqrt{Q_r^2 - 1} = \mathbf{4.9}$ (19-31b)

e. Since the ratio $f_0/f_r = Q_0/Q_r$, use the Q ratio to find reactances, by the ratio method: $X_{C_0} = X_{C_r}(Q_r/Q_0) = -j250\ (5/4.9) = \mathbf{-j255.1\ \Omega}$ (a higher reactance at lower f_0); $X_{L_0} = X_{L_r}(Q_0/Q_r) = +j250\ (4.9/5) = \mathbf{j245\ \Omega}$ (a lower reactance at lower f_0)

f. $\mathbf{I}_{C_0} = \mathbf{V}/-jX_{C_0} = 20\underline{/0°}\ \text{V}/-j255.1\ \Omega = 78.4\underline{/90°}\ \text{mA} = \mathbf{0 + j78.4\ mA}$

g. $\mathbf{Z}_{L_0} = R_s + jX_{L_0} = 50 + j245 = 250\underline{/78.465°}\ \Omega$;
$\mathbf{I}_{L_0} = \mathbf{V}/\mathbf{Z}_{L_0} = 20\underline{/0°}\ \text{V}/250\underline{/78.465°}\ \Omega = 80\underline{/-78.46°}\ \text{mA} = \mathbf{(16 - j78.4)\ mA}$

h. $\mathbf{I}_0 = \mathbf{I}_{C_0} + \mathbf{I}_{L_0} = (0 + j78.4) + (16 - j78.4) = 16 + j0 = \mathbf{16\underline{/0°}\ mA}$

i. $\mathbf{I}_0 = \mathbf{I}_{C_0}/Q_0 = 78.4\ \text{mA}/4.9 = \mathbf{16\ mA}$ (19-31b)

j. $\mathbf{I}_0 = \mathbf{V}/\mathbf{Z}_{\mathrm{po}} = 20\underline{/0°}\ \text{V}/1.25\underline{/0°}\ \text{k}\Omega = \mathbf{16\underline{/0°}\ mA}$

Example 19-13A proves that the correction equations required (because of the degrading of the resonant Q_r by the lower coil Q_0) are valid for the low-Q two-branch parallel-resonant circuit. With respect to Ex. 19-13A, please note that

1. Total circuit current as the sum of $\mathbf{I}_{C_0}$ and $\mathbf{I}_{L_0}$ (step **h**) is in phase with the supply voltage, verifying the unity PF condition.
2. The three methods shown for obtaining the circuit drawn from the supply (steps **h**, **i**, **j**) *all yield identical results.* This verifies the equations given as well as the solution technique.

The question naturally arises of whether the calculated impedance $\mathbf{Z}_{\mathrm{po}}$ at unity PF represents the maximum impedance, since it is evident that the total current is considerably less than the circulating *tank* currents, $\mathbf{I}_{C_0}$ and $\mathbf{I}_{L_0}$. Example 19-13B addresses this question.

EXAMPLE 19-13B

Using the given data of Ex. 19-13A, calculate

a. f_m, frequency at which the tank circuit impedance, $\mathbf{Z}_m$, is a maximum
b. $\mathbf{Z}_m$, equivalent circuit impedance at frequency f_m
c. $\mathbf{I}_{T_m}$, total (minimum) current drawn from supply at frequency f_m
d. Circuit Q_m at the resonant frequency, f_m

Solution

a. $f_m = f_r\sqrt{\dfrac{4Q_r^2 - 1}{4Q_r^2}} = 1\ \text{kHz}\sqrt{\dfrac{4(5)^2 - 1}{4(5)^2}} = \mathbf{995\ Hz}$ (19-29a)

b. $-jX_{C_m} = -jX_{C_r}(f_r/f_m) = -j250(1000/995) = \mathbf{-j251.26\ \Omega}$;
$+jX_{L_m} = +jX_{L_r}(f_m/f_r) = +j250(995/1000) = \mathbf{+j248.75\ \Omega}$;

$$\mathbf{Z}_m = \frac{(R_s + jX_{L_m})(-jX_C)}{R_s + jX_{L_m} - jX_C} = \frac{(50 + j248.75)(0 - j251.26)}{(50 - j2.51)} = \mathbf{1273.42\underline{/-8.49°}\ \Omega} \quad (19\text{-}30b)$$

c. $\mathbf{I}_{T_m} = \mathbf{V}/\mathbf{Z}_m = 20\underline{/0°}/1273.42\underline{/-8.49°} = \mathbf{15.71\underline{/8.49°}\ mA}$

d. $Q_m = Q_r(f_m/f_r) = 5(995/1000) = \mathbf{4.975}$ (19-31a)

Example 19-13B verifies the following for the low-Q parallel-resonant tank circuit:

1. Maximum impedance occurs at a lower frequency (995 Hz) as expressed by Eq. (19-29a) than the frequency given where $|X_L| = |X_C|$ (i.e., 1 kHz) but still higher than f_0 of 979.8 Hz, as expected.
2. At the maximum impedance frequency, the total current is a minimum and slightly leads the voltage by a small angle.
3. The equivalent tank circuit impedance is frequency-sensitive as long as the resonant Q_r exceeds unity. (When Q_r falls below unity, the equivalent impedance no longer peaks at frequency f_m.)
4. At frequency f_m, the circuit Q_m is almost the same as the given Q_r. Even at the lower frequency, f_0, the circuit Q_0 is 4.9 (as shown in Ex. 19-13A, part **d**). This implies that even when Q_r is 5, we may use this value for Q_0 and Q_m with relatively little error.

As a consequence of this last insight, engineers who frequently encounter both high-Q and low-Q two-branch practical tank circuits have attempted to use "quick and easy" shortcut solutions to the low-Q circuit analyzed in Exs. 19-13A and 19-13B. The alternative solution shown in Ex. 19-13C results in a unity power factor current of approximately the same value and impedance but requires much fewer solution steps. The solution retains the two-branch circuit as originally given and assumes that $Q_0 = Q_m = Q_r$ and also that $f_0 = f_m = f_r$. But *it should not be used for extremely low-Q circuits* ranging from approximately $Q_r = 1.1$ to 4. In the event that the calculated Q_r falls between 1.1 and 4, then the correction equations (19-30a) through (19-30c) must be used for impedance computations. Failure to do this results in errors exceeding 5% in computations of equivalent impedance as well as tank and total current.

EXAMPLE 19-13C
(Alternative *approximate* solution to Exs. 19-13A and 19-13B)

a, e. $jX_{L_s} = R_s Q_0 = 50\ \Omega(5) = \mathbf{j250\ \Omega}$;

$$-jX_{C_p} = -jX_{L_s}\left(\frac{Q_0^2+1}{Q_0^2}\right) = -j250\frac{5^2+1}{5^2}$$

$= \mathbf{-j260\ \Omega}$ (compare with **e** above)

b. $\mathbf{Z}_{po} = (Q_0^2 + 1)R_s = (5^2 + 1)50\ \Omega = \mathbf{1300\underline{/0^\circ}\ \Omega}$

f. $\mathbf{I}_{Co} = \mathbf{V}/\mathbf{Z}_{Co} = 20/0^\circ\ \text{V}/(-j260\ \Omega) = 76.9\underline{/+90^\circ}\ \text{mA}$
$= \mathbf{(0 + j76.9)\ mA}$

g. $\mathbf{I}_{Lo} = \mathbf{V}/\mathbf{Z}_{Lo} = 20\underline{/0^\circ}\ \text{V}/(50 + j250) = \mathbf{(15.38 - j76.9)\ mA}$

h, i, j. $\mathbf{I}_0 = \mathbf{I}_{Co} + \mathbf{I}_{Lo} = (0 + j76.9) + (15.38 - j76.9)$
$= \mathbf{15.38\underline{/0^\circ}\ mA}$;

check: $\mathbf{I}_0 = \mathbf{V}/\mathbf{Z}_{po} = 20/0^\circ\ \text{V}/1300/0^\circ\ \Omega = \mathbf{15.38\underline{/0^\circ}\ mA}$

The advantages of this alternative approximate solution are as follows:

1. It is much easier to perform since fewer steps and fewer equations are involved.
2. It yields internally consistent results since the sum of the two branch currents provides an in-phase (unity PF) supply current that is the same as that obtained by the ratio $\mathbf{V}/\mathbf{Z}_{po}$.
3. It retains the original two-branch circuit as part of the solution.
4. The calculated equivalent impedance $\mathbf{Z}_{po}$ is $1300\underline{/0^\circ}\ \Omega$, compared to $1250\underline{/0^\circ}\ \Omega$ (in Ex. 19-13A) and $1273.2\underline{/-8.49^\circ}\ \Omega$ (in Ex. 19-13B), resulting in a small error.
5. Errors below 5% result as long as the Q_r is in excess of 4. (Given a $Q_r < 4$, we can no longer use this approximate method with impunity.)

Because of the relative ease of this method, the end-of-chapter problems will use this approach in the solution of all two-branch **low-Q** circuits, **whenever the Q** falls between 4 and 9.5, approximately.

19-6.3 Summary of Qualifications for Practical Two-Branch Tank Circuits

1. Whenever the circuit Q_r is above 10, for all practical purposes we may assume $Q_r = Q_m = Q_0$ and $f_r = f_m = f_0$. Under these conditions,
   a. Current drawn from the supply, $\mathbf{I}_0$, is a minimum and tank currents $\mathbf{I}_{Lo}$ and $\mathbf{I}_{Co}$ are at a maximum.
   b. Total equivalent circuit impedance $\mathbf{Z}_{po} = Q_r^2 R_s = L/CR_s$ is at a maximum and is resistive. If the tank circuit is fed from a constant-current source, the voltage across the tank circuit is also at a maximum at the resonant frequency, f_r.
   c. Total current, $\mathbf{I}_0$, is in phase with the supply voltage. At frequency f_r, the total current $\mathbf{I}_0 = \mathbf{V}/\mathbf{Z}_{po} = \mathbf{V}/Q_r^2 R_s$.
   d. The capacitive and inductive reactances in each branch are equal.
2. Whenever the circuit Q_r is *below* 10, we must realize that $Q_r \gtreqless Q_m \gtreqless Q_0$ and $f_r \gtreqless f_m \gtreqless f_0$, as shown in **Fig. 19-15**. Under these three separate conditions,[12]
   a. $|X_L|$ is equal to $|X_C|$ only at the resonant frequency, f_r.
   b. Tank circuit impedance, $\mathbf{Z}_m$, is a **maximum** at the (lower) resonant frequency, f_m, and current drawn from the supply is a **minimum**, $\mathbf{I}_m$.
   c. At some (slightly) lower frequency, f_0, the impedance $\mathbf{Z}_0$ is reduced and current $\mathbf{I}_0$ is in phase with the supply voltage, producing the unity PF condition. Impedance $\mathbf{Z}_0$ is resistive and equal to $Q_r^2 R_s$ or L/CR_s.

---

[12] The author is surprised that few textbooks include these three separate distinctions. The only reference which noted them was H. H. Skilling, *Electrical engineering circuits.* 2nd Ed. John Wiley & Sons, 1965, pp. 244 and 399. Skilling verifies his analysis using the transforms of currents in each branch resulting in a minimum current at a frequency slightly higher than f_0 but below f_r.

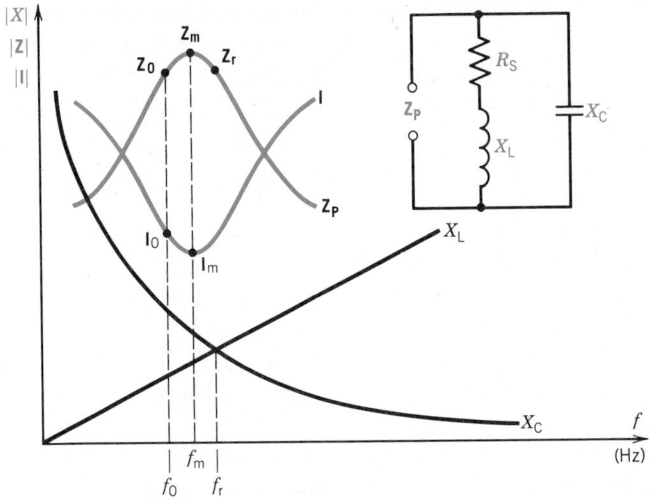

Figure 19-15 Low-Q practical two-branch parallel-resonant circuit

d. The *lower* the Q_r (and Q_m and Q_0), the *greater* the frequency differences between f_0, f_m, and f_r.
e. The lower the Q_r the greater the half-power bandwidth, B_{hp}, as well as the frequency difference between half-power frequencies f_1 and f_2.

3. Bandwidth edge frequencies f_1 and f_2 (for computation of B_{hp}) are determined where the currents are $\sqrt{2}\mathbf{I}_m$ or the tank circuit impedances are $0.7071\mathbf{Z}_m$.
4. The effect of the L/C ratio and/or Q is shown in **Fig. 19-16**. High L/C ratios produce high values of $\mathbf{Z}_{po}$ coupled with sharp response curves of narrow bandwidth and high selectivity, corresponding to a high Q. A low L/C ratio produces a much lower and flatter impedance response, corresponding to a low Q. Since total current $\mathbf{I}$ is the reciprocal of tank circuit impedance, the change in current is much smaller for low L/C ratios.[13]
5. Above f_0, the unity PF frequency, current tends to lead voltage and capacitance predominates. At f_0, the circuit is resistive. Below f_0, the circuit is inductive and current lags voltage. This variation of phase angle with frequency is the same as that shown in Fig. 19-13 for the three-branch parallel resonant circuit and opposite to that shown in Fig. 19-4 for the series RLC circuit. (Recall that phase angle is a function of the impedance triangle.)
6. Since the rise in tank impedance is a function of Q, at extremely low Q values (approaching unity) the impedance and current curves are flattened. Tank circuits with Q values below unity exhibit no rise whatever; that is, $\mathbf{Z}_p$ is constant for all frequencies.
7. For low Q values, between 4 and 10, the approximate method shown in Ex. 19-13C may be used since the errors produced are less than 5%.
8. For *very* low Q values, between 1.1 and 4, the exact method must be used involving Eqs. (19-29), (19-30), and (19-31), if precise values of maximum impedance, minimum current, and resonant frequency, f_m, are desired.

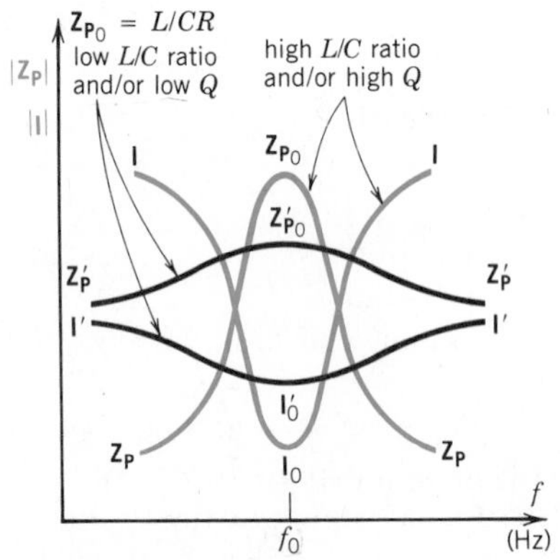

Figure 19-16 Effect of high vs low Q and high vs low L/C ratio on equivalent impedance

19-6.4 The Two-Branch Parallel Circuit with Resistance in Each Branch

The two-branch parallel circuit having resistance in each branch is shown in **Fig. 19-17**. At the outset it should be noted that there are *few* applications, commercially,

[13] Equations (19-28) and (19-30c) show that $\mathbf{Z}_{po}$ is a direct function of the ratio L/C.

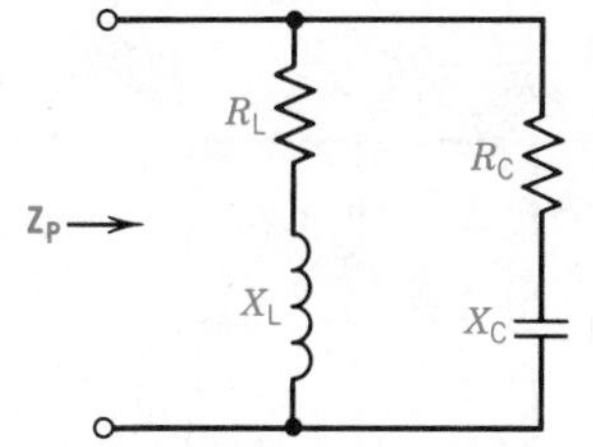

Figure 19-17 The two-branch parallel circuit with resistance in each branch

for this circuit. The presence of resistance in *both* branches reduces the Q and the *selectivity* appreciably. Appendix B-16.1 gives a detailed derivation for finding the resonant frequency, f_0, of this circuit. Recall, by definition, that f_0 is the frequency at which impedance $\mathbf{Z}_{po}$ is a maximum and current is a minimum and in phase with the voltage. From Appendix B-16.1:

$$f_0 = f_r\left(\sqrt{\frac{R_L^2C - L}{R_C^2C - L}}\right) \quad \text{hertz (Hz)} \qquad (19\text{-}32)$$

where f_r is the frequency at which $X_L = X_C$.

As noted in Appendix B-16.1, Eq. (19-32) produces *imaginary* values of f_0 whenever we combine high-resistance capacitors with high-Q coils or low-resistance capacitors with very high resistance coils, that is, coils whose Q values are less than unity. Such combinations fail to maximize impedance and minimize current, for reasons shown in Appendix B-16.1.

On the other hand, the circuit will resonate when we combine low-resistance capacitors with high-Q coils (our practical two-branch circuit covered in Sec. 19-6.3) or high-resistance capacitors with low-Q coils.

Appendix B-16.1 also derives from Eq. (19-32) the relation where X_C and X_L are known at a particular frequency, f, yielding

$$f_0 = f\sqrt{\frac{X_C}{X_L}}\sqrt{\frac{R_L^2 - X_CX_L}{R_C^2 - X_CX_L}} \quad \text{hertz (Hz)} \qquad (19\text{-}33)$$

where f is any frequency producing known values $X_C \neq X_L$

and all other terms have been defined.

Equation (19-33) is important because it shows that when either $R_C = R_L = \sqrt{L/C}$ or $R_C^2 = R_L^2 = X_LX_C$, $f_0 = f$ since the term under the radical is indeterminate. This means that under the foregoing conditions, the circuit is resonant *at all and any* frequencies. Under these conditions, the circuit impedance $\mathbf{Z}_{po} = R_{po} = \sqrt{L/C} = \sqrt{X_C/X_L}$ = a constant value at **all** frequencies.[14] (See Appendix B-16.1.)

Lastly, in Eq. (19-33), if $R_C = R_L$ and the squares $R_C^2 = R_L^2 \neq X_LX_C$, the circuit is resonant at $f_r = 1/2\pi\sqrt{LC}$ in much the same way as a high-Q circuit. At frequency f_r, the circuit impedance is a maximum and the circuit current, $\mathbf{I}_0$, is a minimum.

19-7 LOADING HIGH-Q RESONANT CIRCUITS

Figure 19-18a shows a typical two-branch tank circuit, presumably operating at its resonant frequency f_0 at some relatively high Q, say $Q_0 = Q_r = 20$. The two-branch circuit conceivably might be an oscillator that is connected to an amplifier circuit whose impedance is primarily resistive and shown as R_L. What *is* the effect of *loading* the tank circuit? Does it change the resonant frequency? What happens to bandwidth as a result of loading and the Q_0 of the original tank circuit?

[14] In telephony, an attempt is made to ensure that the lines are loaded in such way that the impedance is constant at all frequencies and current is always in phase with voltage. This ensures maximum power transfer. Recall that maximum power transfer is a unity PF condition.

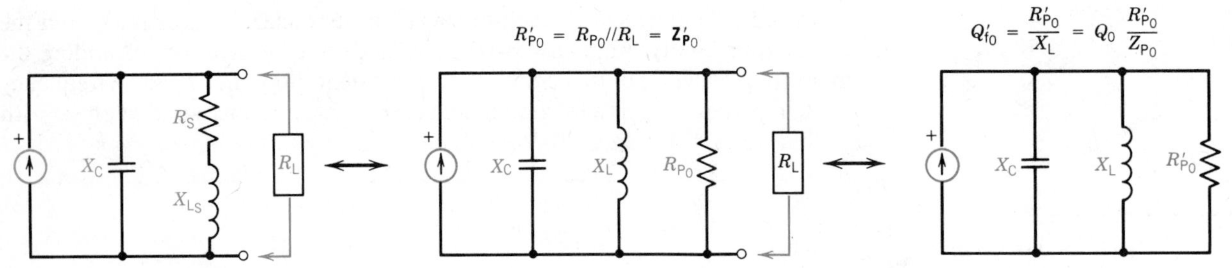

a. Original tank circuit
b. Equivalent 3-branch circuit
c. Loaded tank circuit

Figure 19-18 Steps in calculating final Q of a loaded parallel-resonant two-branch circuit

The solution to these questions is found in converting the two-branch tank circuit into its three-branch equivalent, as shown in Fig. 19-18b. The solution steps are as follows:

1. Convert the two-branch circuit into its three-branch equivalent where $\mathbf{Z}_{po} = R_{po} \cong Q_0^2 R_s$.
2. Take the parallel equivalent of the resistive impedances, $R'_p = (R_{po}) \| R_L$ (Fig. 19-18c).
3. Determine the final Q'_f of the circuit from $Q'_f = R'_p/X_{C_0}$ or $Q'_f = Q_0(R'_p/\mathbf{Z}_{po})$.
4. Determine the resultant half-power bandwidth from $B_{hp} = f_0/Q'_f$.

EXAMPLE 19-14

The practical two-branch tank circuit of Fig. 19-18a has a 16-nF capacitor in one branch and a 10-mH inductor having a resistance of 40 Ω in the other branch. The ideal variable frequency current source delivers a constant current of 1 mA. Calculate the

a. Resonant frequency, f_r
b. Inductive reactance of the coil and the capacitor, respectively
c. Q_r and Q_0, respectively, of the two-branch circuit
d. Equivalent impedance of the two-branch circuit at resonance
e. Voltage across the two-branch circuit at resonance prior to the application of load R_L
f. Half-power bandwidth of the two-branch parallel circuit *prior* to loading

Solution

a. $f_r = 1/2\pi\sqrt{LC} = 1/2\pi\sqrt{(10\text{ m})(16\text{ n})} = \mathbf{12.58\ kHz}$
b. $X_L = 2\pi fL = 2\pi(12.582\text{ k})(10\text{ m}) = \mathbf{+j790.6\ \Omega}$ and $-jX_C = \mathbf{-j790.6\ \Omega}$ since Q_r is high!
c. $Q_r = jX_L/R_L = 790.6\ \Omega/40\ \Omega = \mathbf{19.76}$; $Q_0 \cong 19.76$ since $Q_r > 10$
d. $\mathbf{Z}_{po} = Q_r X_L = 19.76 \times 790.6\ \Omega = \mathbf{15.62\angle 0°\ k\Omega} = R_{po}$
e. $\mathbf{V}_p = \mathbf{I} \cdot \mathbf{Z}_{po} = 1\text{ mA}(15.62\text{ k}\Omega) = \mathbf{15.62\ V}$
f. $B_{hp} = f_0/Q_0 = 12.58\text{ kHz}/19.76 = \mathbf{636.6\ Hz}$

EXAMPLE 19-15

The equivalent impedance of an amplifier connected across the output of the two-branch circuit in Ex. 19-14 is $10\angle 0°$ kΩ. Calculate the

a. Equivalent impedance of the loaded tank circuit, $\mathbf{Z}'_{po}$, at the resonant frequency
b. Final Q of the circuit, Q'_f, as a result of the loading effect
c. Half-power bandwidth of the circuit at the resonant frequency
d. Voltage across the loaded circuit

Solution

a. $R'_{po} = R_{po} \| R_L = (15.62 \| 10)\text{ k}\Omega = \mathbf{5.933\ k\Omega}$
b. $Q'_{f_0} = R'_p/X_L = 5.933\text{ k}/790.6\text{ k} = \mathbf{7.5}$
c. $B_{hp} = f_0/Q'_f = 12.58\text{ kHz}/7.5 = \mathbf{1677\ Hz}$
d. $\mathbf{V}_p = \mathbf{I} \cdot R'_p = 1\text{ mA}(5.933\text{ k}\Omega) = \mathbf{5.933\ V}$

Examples 19-14 and 19-15 lead to the following important insights:

1. The original Q of the circuit was high (19.76), producing a relatively narrow BW (637 Hz) at resonant frequency, *prior to loading*.
2. The effect of loading is to reduce the Q_0 appreciably and, as a result, the BW *increases* in the *same* proportion as the Q reduction (to 1667 Hz).

3. The voltage output of the tank circuit is 15.62 V prior to loading. As a result of loading, the (equivalent impedance and) output voltage is reduced (to 5.933 V).
4. These observations have great significance for either oscillators or tuners using two-branch *LC* circuits. If these resonant circuits are coupled to amplifiers, then the equivalent impedance of these successive amplifier stages must be sufficiently *high* not to degrade appreciably either the output voltage or the bandwidth.

The loading effect is also observed whenever a *practical* constant-current source (such as a transistor) is supplying the two-branch tank circuit. In Fig. 19-18a, we assumed that the constant-current source was ideal. But what if the transistor, which is a *practical* current source, is used, as shown in **Fig. 19-19a**? In this case, the loading of the tank is produced by the transistor collector resistance, R_c, shunting the (ideal) current source. The loading effect on Q_0, the bandwidth, and the voltage across the tank circuit is essentially the same as that described above. Example 19-16 shows this, using a low-Q circuit deliberately.

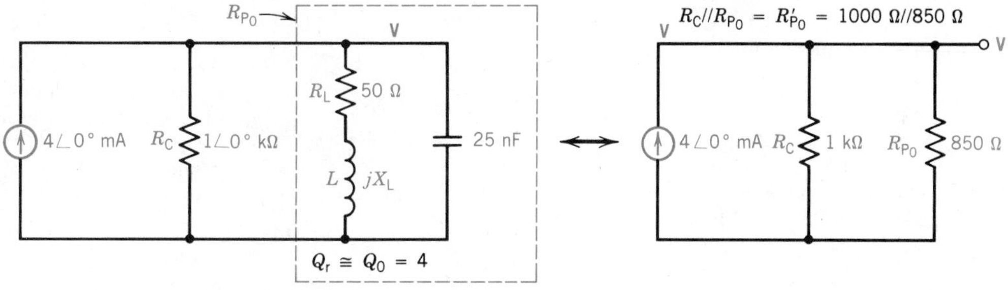

a. Original circuit b. Equivalent of 3-branch circuit

Figure 19-19 Practical current source connected to two-branch practical tank circuit (Ex. 19-16)

EXAMPLE 19-16

Given the *practical* current source connected to the low-Q two-branch practical tank circuit shown in Fig. 19-19, calculate

a. jX_{L_0} at the resonant frequency, f_0
b. $-jX_{C_0}$ at the resonant frequency, f_0
c. Resonant frequency f_0 of the tank circuit
d. Inductance, L, of the coil
e. $\mathbf{Z}_{p0}$ (or R_{p0}) of the two-branch tank circuit only
f. $\mathbf{Z}'_{p0}$ of the equivalent three-branch circuit seen by the ideal source
g. Voltage acrosss all parallel elements in Fig. 19-19, **V**
h. Currents in branches $\mathbf{I}_L$ and $\mathbf{I}_C$ in Fig. 19-19b in rectangular form
i. Resonant Q'_{f0} of the circuit (hint: use R'_{p0}/X_{C0})
j. Half-power bandwidth, B_{hp}
k. Verify all current calculations above using KCL around node **V** in Fig. 19-19a

Solution

a. $+jX_{L_0} = Q_0 R_L = 4 \times 50\ \Omega = \mathbf{+j200\ \Omega}$
b. $-jX_{C_0} = jX_L(Q_0^2 + 1)/Q_0^2 = j200(17/16) = \mathbf{-j212.5\ \Omega}$
c. $f_0 = 1/2\pi X_{C_0} \cdot C = 1/2\pi(212.5)25\ \text{nF} = \mathbf{30\ kHz}$
d. $L = jX_L/2\pi f_0 = j200/2\pi \times 300\ \text{kHz} = \mathbf{1.061\ mH}$
e. $\mathbf{Z}_{p0} = R_L(Q_0^2 + 1) = 50\ \Omega(4^2 + 1) = \mathbf{850\ \Omega}$
f. $\mathbf{Z}'_{p0} = R'_{p0} = R_{p0} \| R_C = (850 \| 1000)\ \Omega = \mathbf{459.5\ \Omega}$
g. $\mathbf{V} = \mathbf{IZ}'_{p0} = 4\underline{/0^\circ}\ \text{mA}(0.4595\underline{/0^\circ}\ \text{k}\Omega) = \mathbf{1.838\underline{/0^\circ}\ V}$
h. $\mathbf{I}_L = \mathbf{V}/\mathbf{Z}_L = 1.838\underline{/0^\circ}\ \text{V}/(50 + j200)\ \Omega$ $= \mathbf{(2.162 - j8.65)\ mA}$
$\mathbf{I}_C = \mathbf{V}/-jX_C = 1.838\underline{/0^\circ}\ \text{V}/(-j212.5)\ \Omega = \mathbf{(0 + j8.65)\ mA}$
i. $Q'_{f0} = R'_{p0}/X_{C_0} = 459.5/212.5 = \mathbf{2.162}$ (note reduction in Q from $Q_0 = 4$)
j. $B_{hp} = f_0/Q'_{f0} = 30\ \text{kHz}/2.161 = \mathbf{13.87\ kHz}$ (note wider BW due to lower Q'_{f0})
k. By KCL, $\mathbf{I}_{L_0} + \mathbf{I}_{C_0} + \mathbf{I}_{R_C} = (2.162 - j8.65) + (0 + j8.65) + (1.838 + j0) = \mathbf{4\underline{/0^\circ}\ mA}$

With respect to Fig. 19-19 and the foregoing solution steps, please note that

1. The resistance, R_C, of the practical current source produces the same loading effect as any load, R_L, connected across the terminals of a two-branch practical tank circuit.

2. The net effect of R_C shunting the two-branch tank circuit is to *reduce* both the equivalent impedance and the (original) Q of the tank circuit and *increase* the bandwidth.
3. Consequently, whenever a transistor current source is "exciting" a tank oscillator or tuning circuit, the transistor output impedance *must be included* in the computation *along with* whatever loads are connected across the practical two-branch circuit.

Example 19-17 shows how a tank circuit may be designed around certain given parameters.

EXAMPLE 19-17

Design a practical two-branch tank circuit to have a maximum impedance of 25 kΩ at resonance, using a coil of resistance 25 Ω and inductance 10 mH, by calculating the following parameters in the order given:

a. Q_r and Q_0
b. X_{L_r} and X_{L_0}
c. Resonant frequency, f_0
d. Required value of C
e. $\mathbf{I}_0$ when tank is connected to a constant supply voltage of $25\angle 0°$ V
f. Circuit bandwidth, B_{hp}, prior to loading
g. Q'_{f0} and its associated BW when a circuit having an input impedance of $25\angle 0°$ kΩ is shunting the tank

Solution

a. $Q_r^2 = R_{po}/R_s$ and $Q_r = \sqrt{R_{po}/R_s} = \sqrt{25\text{ k}/25} = \mathbf{31.62} = Q_0$ since $Q_r > 10$
b. $X_{Lr} = Q_r R_s = 31.62(25) = \mathbf{j790.5} = X_{L0}$ since Q is so high
c. $f_0 = X_L/2\pi L = 790.5/2\pi \cdot 10\text{ m} = \mathbf{12.58\ kHz}$
d. $C = 1/2\pi f_0 X_{C_0} = 1/2\pi \times 12.58\text{ k} \times (790.5) = \mathbf{16\ nF}$
e. $\mathbf{I}_0 = \mathbf{V}/\mathbf{Z}_{po} = 25\angle 0°\text{ V}/25\angle 0°\text{ k}\Omega = 1\angle 0°\text{ mA}$
f. $B_{hp} = f_0/Q_0 = 12.58\text{ kHz}/31.62 = \mathbf{397.85\ Hz}$
g. $\mathbf{Z}'_{po} = (25 \| 25)\text{ k}\Omega = \mathbf{12.5\angle 0°\ k\Omega}$;
$Q'_{f0} = \mathbf{Z}'_{po}/X_{L0} = R'_{po}/X_{L0} = 12\,500/790.5 = \mathbf{15.81}$;
$B_{hp} = f_0/Q'_{f0} = 12.58\text{ kHz}/15.81 = \mathbf{795.7\ Hz}$

Several interesting points emerge from Ex. 19-17:

1. For a given coil, it is possible to design for a given impedance or a given frequency, but *not both simultaneously*! Designing for *both* impedance and frequency involves a new coil selection, since $\mathbf{Z}_{po} = L/RC$, $f_0 = 1/2\pi\sqrt{LC}$, and L/R is already specified. The given design specified impedance. This automatically makes $C = 16$ nF, as shown either by solution step (**d**) or by substitution for C in $\mathbf{Z}_{po} = L/RC$ (left as an exercise for the reader).
2. The shunting impedance in step (**g**) was *equal* to the equivalent tank impedance, R_{po}. This halved the Q and *doubled* the BW, since $\mathbf{Z}'_{po}$ was half of R_p. (It also halves the output voltage of the tank circuit when fed by a constant-current source.)

Consideration of point (2) leads, quite naturally, to the important relation that enables us to find the Q of the *final* (loaded) tank circuit (Q'_{f0}) at the resonant frequency:

$$Q'_{f0} = \frac{(Q_0)(R'_{po})}{R_{po}} \tag{19-34}$$

where Q_0 is the *unloaded* Q of the tank circuit at f_0
R'_{po} is the equivalent tank impedance as a *result* of the *loading* effect
R_{po} is the *unloaded* equivalent impedance of the tank circuit at f_0

Equation (19-34) enables us to predict the loading effect for any load connected across an unloaded (or previously loaded) tank circuit, as shown by Ex. 19-18.

EXAMPLE 19-18

Repeat Ex. 19-15b to find Q'_{f0}, using the given solution data of Ex. 19-14 and Ex. 19-15a

Solution

From Ex. 19-14, $Q_0 = 19.76$ and $R_{po} = 15.62$ kΩ; from Ex. 19-15a, $R'_{po} = 5.933$ kΩ, as a result of a 10-kΩ load across the tank. Then

$$Q'_{f0} = (Q_0)(R'_{po})/R_{po} = 19.76(5.933\text{ k})/15.62\text{ k} = \mathbf{7.5} \quad (19\text{-}34)$$

Note that this is the same answer as obtained in Ex. 19-15**b**. Equation (19-34) shows that the lower the value of R'_{p0} as a result of loading, the lower the final Q'_{f0}. Clearly, decreasing Q'_{f0} increases the bandwidth and decreases the selectivity of the resonant circuit.

This raises the interesting question of whether it is possible to load the tank so heavily that it is no longer resonant (i.e., its Q'_{f0} is *below unity*) and f_0 is an "imaginary" value. This is possible, of course, and it is sometimes done deliberately to "damp out" oscillation or prevent "ringing." The damping technique usually involves adding a series or parallel resistance deliberately to the coils that are to be used. From Eq. (19-29b) we can see that when the ratio CR_s^2/L is **greater than unity**, the circuit is *never* resonant since the expression under the radical is an imaginary value. Consequently, increasing the R_s of a coil *has the same damping effect* as *loading a tank circuit by shunting it* with some parallel resistance, R_p. (This is a *most important insight.*)

19-8 AC WAVE FILTER NETWORKS

In communications, information is transmitted through space or over transmission lines, waveguides, or optical fibers at some carrier frequency that is "modulated" by the information. At the receiving end, it is necessary to *separate* any *unrelated* information that is transmitted simultaneously. A *filter* is defined as a *transducer* for separating waves on the basis of their *frequencies.* Filters are designed to *reject* all *undesired* frequencies or to *pass* (as efficiently as possible) the frequencies containing the *desirable* information to be transmitted.

Filter networks are divided into two major classifications: *passive* networks and *active* filter networks. Passive networks usually consist of series–parallel combinations of R, L, and C. Active networks usually contain sources such as transistors and/or operational amplifiers in combination with R, L and C to obtain the desired filtering effect.

Four very common types of filters are

1. The *low-pass* filter, which allows signals to be transmitted up to a certain maximum or cutoff frequency, f_c. Above f_c, the signals are rejected to a lesser or greater extent depending on the design.
2. The *high-pass* filter, which similarly allows frequencies above the cutoff frequency, f_c, to be transmitted while rejecting frequencies below f_c.
3. The *bandpass* filter, which selects some range of median frequencies to be passed while simultaneously rejecting all frequencies below and above the desired range.
4. The *bandstop* filter, which selects some median range of frequencies to be blocked while simultaneously passing all frequencies below and above the rejected band. Bandstop filters are sometimes called notch filters, wavetraps, or band-elimination, band-suppression, or band-rejection filters.

The subject of active and passive filters would require a separate text. For our purposes we will consider only the *passive* filters as a brief introduction to the subject. Further, the passive filters described below are only the most elementary ones and would hardly serve the more rigorous needs of those used in industry. We will begin with simple low-pass and high-pass filters in this section and treat bandpass and bandstop filters in Sec. 19-9.

19-8.1 Low-Pass *RC* Circuit—Table 19-1a

The low-pass RC circuit is shown in Table 19-1**a**. At all frequencies it is evident that the ratio of output impedance to input impedance is $-jX_C/(R - jX_C)$. But this ratio

Table 19-1 Simple *RC* and *RL* Low-Pass, High-Pass, Bandstop, and Bandpass Filters

Filter Circuit(s)	Configuration(s)	Ratio, $\mathbf{V}_0/\mathbf{V}_{in}$	Cutoff Frequency, f_c	Amplitude-Phase Response
Low pass	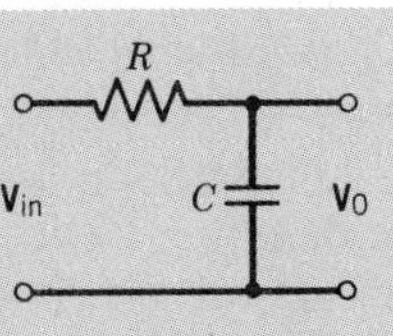 a. *RC* circuit	$\frac{\mathbf{V}_0}{\mathbf{V}_{in}} = \frac{-jX_C}{R - jX_C}$	$f_c = \frac{1}{2\pi CR}$	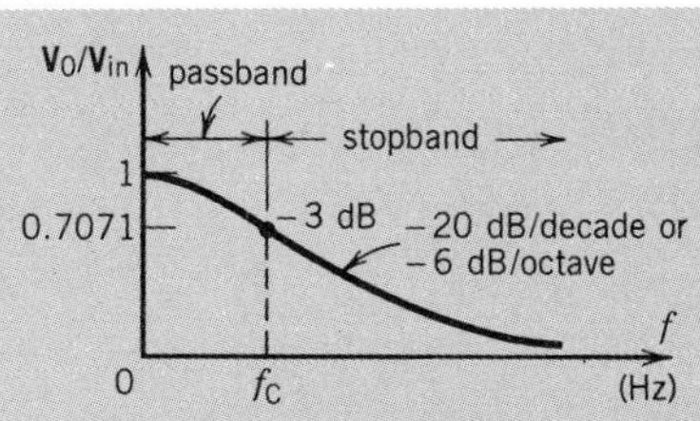
	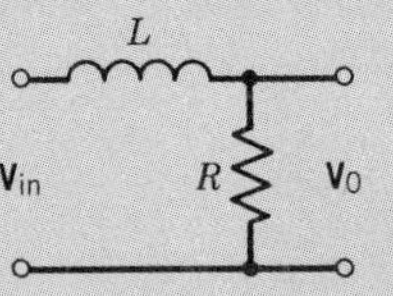 b. *RL* circuit	$\frac{\mathbf{V}_0}{\mathbf{V}_{in}} = \frac{R}{R + jX_L}$	$f_c = \frac{R}{2\pi L}$	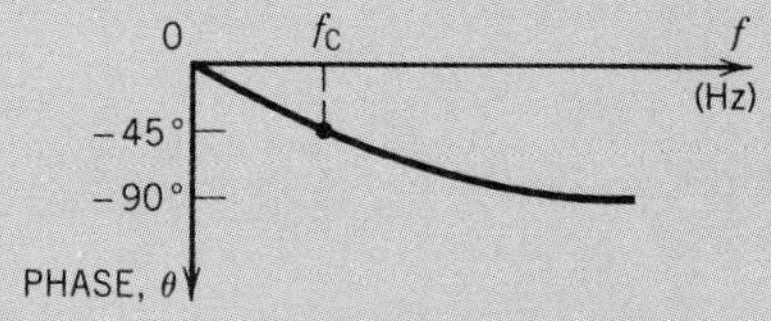
High pass	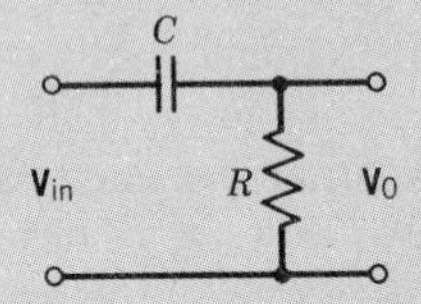 c. *RC* circuit	$\frac{\mathbf{V}_0}{\mathbf{V}_{in}} = \frac{R}{R - jX_C}$	$f_c = \frac{1}{2\pi CR}$	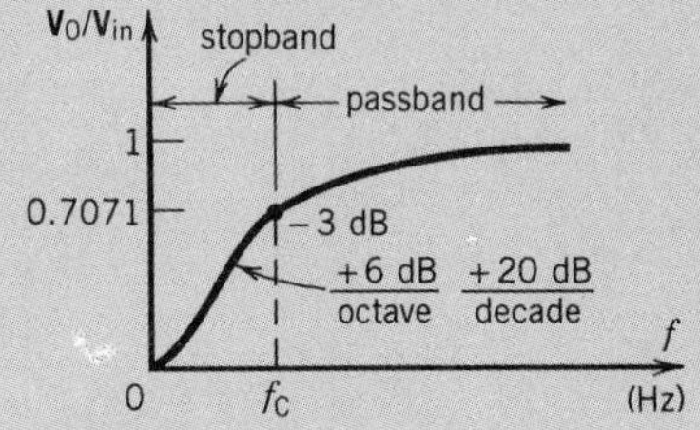
	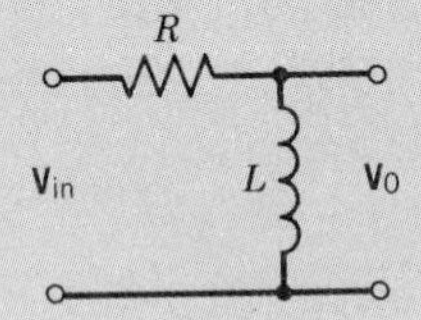 d. *RL* circuit	$\frac{\mathbf{V}_0}{\mathbf{V}_{in}} = \frac{jX_L}{R + jX_L}$	$f_c = \frac{R}{2\pi L}$	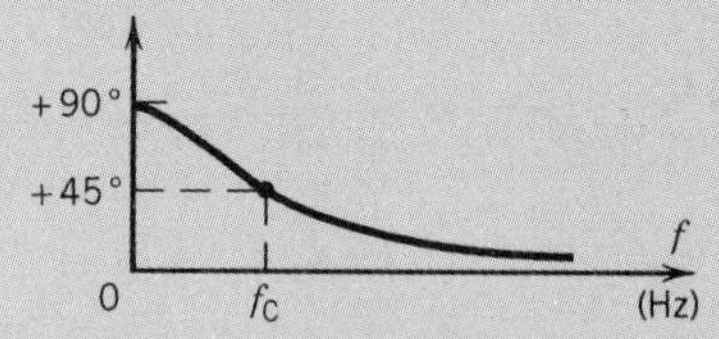
RC bandstop	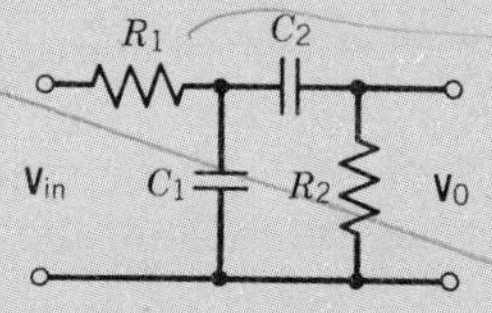 e. *RC* bandstop circuit	$\frac{-jX_{C_1}}{(R_1 - jX_{C_1})}$ from f_{c_1} to f_1 and $\frac{R_2}{(R_2 - jX_{C_2})}$ from f_2 to f_{c_2}	$f_{c_1} = \frac{1}{2\pi C_1 R_1}$ $f_{c_2} = \frac{1}{2\pi C_2 R_2}$	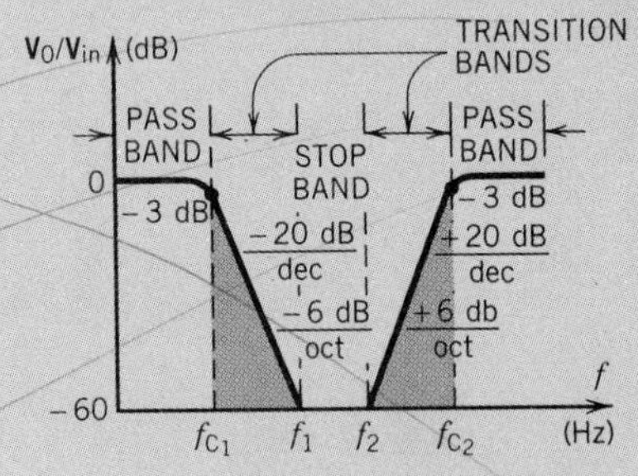
RC bandpass	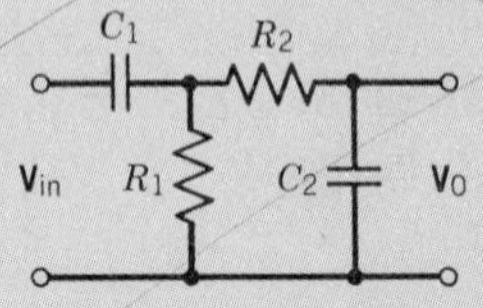 f. *RC* bandpass circuit	$\frac{R_1}{(R_1 - jX_{C_1})}$ from f_1 to f_{c_1} and $\frac{-jX_{C_2}}{(R_2 - jX_{C_2})}$ from f_{c_2} to f_2	$f_{c_1} = \frac{1}{2\pi C_1 R_1}$ $f_{c_2} = \frac{1}{2\pi C_2 R_2}$	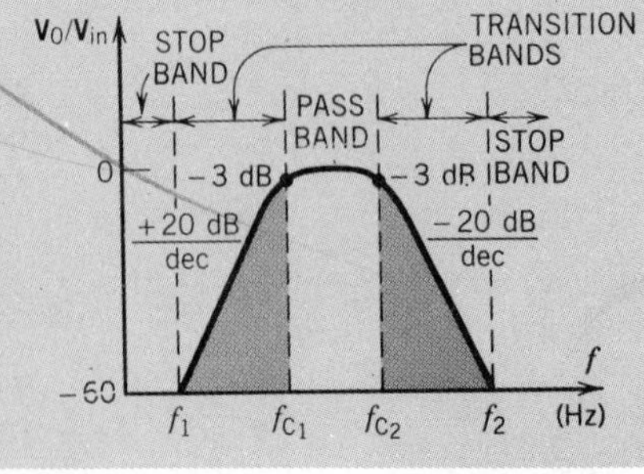

is the same as the ratio $\mathbf{V}_0/\mathbf{V}_{in}$ for the low-pass RC circuit. At low frequencies, because of the high value of X_C, this ratio is practically unity. As the frequency increases, X_C decreases, producing the amplitude response shown in the last column of Table 19-1.

By definition, the cutoff frequency, f_c, occurs whenever

1. $\mathbf{V}_0 = 0.7071\mathbf{V}_{in}$ (i.e. $\mathbf{V}_0$ is -3 dB down from $\mathbf{V}_{in}$)
2. The impedance phase angle $\theta = -45°$ (phase angle between $\mathbf{V}_0$ and $\mathbf{V}_{in}$)
3. $R = -jX_C$ and $V_R = V_C$, in magnitude

But since $X_C = 1/2\pi f_c C = R$, whenever f_c occurs, it follows that

4. $f_c = 1/2\pi CR$, as shown in the first row (column 4) of Table 19-1**a**
5. Conversely, $C = 1/2\pi f_c R$, should we desire to find the required value of C for a given value of R, as shown in Ex. 19-19

EXAMPLE 19-19

A simple low-pass RC filter has a cutoff frequency, f_c, of 1 kHz. Given a constant 10 V, variable frequency ac supply, calculate the

a. Value of C required if R is 10 kΩ
b. Output voltage and dB output when $f = 1$ kHz
c. Output voltage and dB output when $f = 10$ kHz

Solution

a. $C = 1/2\pi f_c R = 1/2\pi 10^3 \times 10 \times 10^3 = \mathbf{15.92\ nF}$

b. $-jX_C = R = 10\angle{-90°}$ kΩ whenever $f = f_c = 1$ kHz and therefore $\mathbf{V}_0 = \mathbf{V}_{in}\dfrac{-jX_C}{R - jX_C} = 10\angle{0°}\ \mathrm{V}\dfrac{-j10\ \mathrm{k}}{(10 - j10)\ \mathrm{k}} =$ $\mathbf{7.071\angle{-45°}\ V}$ (see **1** above); $\mathrm{dB} = 20\log_{10}(\mathbf{V}_0/\mathbf{V}_{in}) =$ $20\log_{10}(7.071/10) = \mathbf{-3.01\ dB}$ (see **1** above)

c. $X_{C_2} = X_{C_1}\dfrac{f_1}{f_2} = -j10\ \mathrm{k}\dfrac{1\ \mathrm{kHz}}{10\ \mathrm{kHz}} = -j1\ \mathrm{k\Omega} = 1\angle{-90°}\ \mathrm{k\Omega}$;

$$\mathbf{V}_0 = V_{in}\frac{-jX_C}{R - jX_C} = 10\angle{0°}\ \mathrm{V}\frac{1\angle{-90°}\ \mathrm{k\Omega}}{(10 - j1)\ \mathrm{k\Omega}} \cong \mathbf{1\angle{-84.3°}\ V};$$

$$\mathrm{dB} = 20\log_{10}(\mathbf{V}_0/\mathbf{V}_{in}) = 20\log_{10}(1/10) \cong \mathbf{-20\ dB}$$

Example 19-19 shows how relatively simple it is to design and/or calculate the amplitude and phase response for an RC low-pass filter. The example shows the following:

a. The verification of the first three points made, namely that whenever $f = f_c$, the phase angle is $-45°$, the amplitude is down by -3 dB, and $R = -jX_C$.
b. Whenever the frequency is increased above f_c, the dB level drops at a -20 dB/decade rate. When the input frequency is 10 kHz, the output is down to -20 dB compared to the zero frequency level.

19-8.2 Low-Pass *RL* Circuit—Table 19-1b

Table 19-1**b** shows that the RL circuit (in the second row) has the same amplitude-phase response as the RC circuit of Ex. 19-19. Let us verify this by using the same value of f_c but different values of R and L.

EXAMPLE 19-20

A simple RL low-pass filter has a cutoff frequency of 1 kHz and a constant 10 V, variable frequency ac supply. Calculate the

a. Required value of L if R is 1 kΩ
b. $\mathbf{V}_0\angle{\theta}$ and dB output when $f = 1$ kHz
c. $\mathbf{V}_0\angle{\theta}$ and dB output when $f = 10$ kHz

Solution

a. $L = R/2\pi f_c = 1\ \mathrm{k\Omega}/2\pi 10^3 = 1/2\pi = \mathbf{159.2\ mH}$ (using equation in second row, fourth column)

b. At $f = f_c = 1$ kHz, $jX_L = R = 1$ kΩ;

$$\mathbf{V}_0 = \mathbf{V}_{in}\frac{R}{(R + jX_L)} = 10\angle{0°}\ \mathrm{V}\frac{1\angle{0°}\ \mathrm{k\Omega}}{(1 + j1)\ \mathrm{k\Omega}} = 7.071\angle{-45°}\ \mathrm{V};$$

$$\mathrm{dB} = 20\log_{10}(7.071/10) \cong \mathbf{-3\ dB}$$

c. At $f = 10$ kHz, $jX_L = \dfrac{10\ \mathrm{kHz}}{1\ \mathrm{kHz}}(1\angle{90°}\ \mathrm{k\Omega}) = \mathbf{10\angle{90°}\ k\Omega}$;

$$\mathbf{V}_0 = \mathbf{V}_{in}\frac{R}{(R + jX_L)} = 10\angle{0°}\ \mathrm{V}\frac{1\angle{0°}\ \mathrm{k\Omega}}{(1 + j10)\ \mathrm{k\Omega}} \cong 1\angle{-84.3°}\ \mathrm{V};$$

$$\mathrm{dB} = 20\log_{10}(\mathbf{V}_0/\mathbf{V}_{in}) = 20\log_{10}(1/10) \cong \mathbf{-20\ dB}$$

Example 19-20 verifies that both the low-pass *RL* and *RC* circuits produce identical amplitude-phase responses, as shown for the first two rows of Table 19-1.

19-8.3 High-Pass *RC* Circuit—Table 19-1c

Table 19-1**c** shows the circuit, the ratio $\mathbf{V}_0/\mathbf{V}_{in}$, the basic cutoff frequency equation, and the amplitude-phase response. Observe that the high-pass *RC* configuration is obtained merely by juxtaposition of *R* and *C* in the low-pass *RC* configuration.

19-8.4 High-Pass *RL* Circuit—Table 19-1d

Table 19-1**d** shows the circuit, the ratio $\mathbf{V}_0/\mathbf{V}_{in}$, the basic cutoff frequency equation, and the amplitude-phase response. The *RL* high-pass circuit is also obtained by "swapping" positions of *R* and *L* in the low-pass circuit.

As in the case of low-pass *RC* and *RL* filters, the amplitude-phase responses of *both* high-pass circuits are *identical.* Verification of high-pass filter networks is left as an exercise for the reader in the end-of-chapter problems.

19-8.5 Cascading *RC* and/or *RL* Filter Networks

As noted in Table 19-1, the low-pass filter configurations produce an idealized "rolloff" of -20 dB per decade or -6 dB per octave. (For each *octave*, the frequency is *doubled*. For each *decade*, the frequency is *multiplied* by a factor of **10**.)

The *low-pass* filter attenuates *high* frequencies. The attenuation is -3 dB at the cutoff frequency, f_c. At twice the cutoff frequency, $2f_c$, therefore, the attenuation is approximately -6 dB. At 10 times the cutoff frequency, the attenuation is -20 dB.

The -20 dB/decade or -6 dB/octave rolloff represents a *linear approximation* of the attenuation shown in amplitude response of the *low-pass* filter of Table 19-1a, within a few dB of the actual or calculated attenuation. The question now arises of what might happen if **two** *low-pass* networks, having the same *R* and *C* values, are *cascaded*; that is, the *output* of one configuration is the *input* of the second (identical) configuration. The answer seems obvious. Such a network would have a rolloff of -40 dB/decade or -12 dB/octave.

In commercial practice, several identical low-pass sections are *cascaded* to produce *composite filters* having sharper rolloffs such as -18 dB/octave or -24 dB/octave. The sharper (or higher) the slope of the rolloff, the smaller is the transition band between the -3 dB frequency and the (higher) frequency at which the output is (say) -60 dB, where the ratio $\mathbf{V}_0/\mathbf{V}_{in}$ is very small (approximately 0.001) and may be considered zero.

All the filter circuits shown in Table 19-1a–d are commonly cascaded in commercial practice to produce either sharper (low-pass) rolloffs or higher *rates of rise*, as in the case of *high-pass* filters.

19-8.6 *RC* Bandstop Filter—Table 19-1e

The question now may be asked, "What happens when a high-pass filter is cascaded to a low-pass filter?" Such a configuration is shown in Table 19-1**e** in its simplest form. Notice that the configuration shown combines the low-pass and high-pass features of the filters of Table 19-1a and 19-1c, as well it should! The *passband* of the *RC* bandstop filter is the portion of the response between 0 and -3 dB. The *stopband* is that portion where the response is below -60 dB. The *shaded* portions represent the *transition* bands of frequencies (between f_{c_1} and f_1 or between f_2 and f_{c_2}) between the stopband and the passbands, respectively. The transition bands may be narrowed by cascading identical elements of each type of filter. Cascading two low-pass filters to two identical high-pass filters produces sharper rolloffs of -40 dB/decade and sharper rises of $+40$ dB/decade, thus *reducing the width of the transition bands.*

For the filter shown in Table 19-1e, f_{c_2} is a higher frequency than f_{c_1}. This means that the *RC* product (or time constant) for the high-pass portion must be *smaller* than for the low-pass portion. When several low-pass elements are cascaded, their elements are identical, but each of these has higher time constants than the (identical) cascaded high-pass elements. Each identical cascaded section uses the equations for f_{c_1} and f_{c_2} given in Table 19-1e, depending on whether we are designing for the low-pass sections or the identical high-pass sections.

19-8.7 *RC* Bandpass Filter—Table 19-1f

By reversing the cascaded sequence of the *RC* bandstop filter, we obtain the *RC* bandpass filter shown in Table 19-1f. This consists of a high-pass filter followed by a low-pass filter, in its simplest form. The amplitude response and controlling equations for this configuration are shown in Table 19-1f. The passband is selected by specifying f_{c_1} and f_{c_2}. Using the equations, appropriate values of R_1 and C_1 as well as R_2 and C_2 are selected. Cascading additional (identical) high-pass elements and a second set of additional (identical) low-pass elements results in sharper rises and rolloffs, narrowing the transition bands.

In commercial practice, several low-pass *RC* elements are cascaded with several high-pass *RC* elements to produce (almost) vertical rolloffs and rises. Unlike *RL* filters, the *RC* filters lend themselves to large-scale integrated circuits (LSICs), where the properties of *R* and *C* are *easily* implemented. Consequently, cascading is rarely done with *RL* filters.

It should be noted that although additional cascaded elements produce extremely sharp rises and rolloffs, the added resistances result in increased power losses. The advent of operational amplifiers in LSI form has produced a class of *active* filters that enable the designer to produce high-pass, low-pass, bandstop, and bandpass characteristics of extremely sharp slope with relatively negligible power loss. The subject of such filters is beyond the scope of this text.

19-9 BANDSTOP AND BANDPASS RESONANT FILTER NETWORKS

Resonant filters find application as either bandstop or bandpass filters, owing to their characteristic *Q*-rise to either current or voltage at the resonant frequency. Recall from our previous study of series and parallel resonant circuits that at their respective resonant frequencies

a. *Series* circuits exhibit *minimum* impedance (*short-circuit*) and *maximum* current.
b. *Parallel* circuits exhibit *maximum* impedance (*open-circuit*) and *minimum* current.

These two relatively simple characteristics enable the configurations shown in Table 19-2 to exhibit their respective bandstop and bandpass curves. Let us briefly examine each, to see how easily the response curves may be predicted from the four configurations shown. We will do this first, followed by a few examples to verify the predictions.

Table 19-2 Simple Series and Parallel Resonant Bandstop and Bandpass Filter Circuits

<table>
<tr><th>Filter Circuit(s) Configuration</th><th>Ratio $\mathbf{V}_0/\mathbf{V}_{in}$</th><th>Circuit Q_0</th><th>Bandwidth, B_{hp}</th><th>Amplitude-Phase Response</th></tr>
<tr><td>BANDSTOP
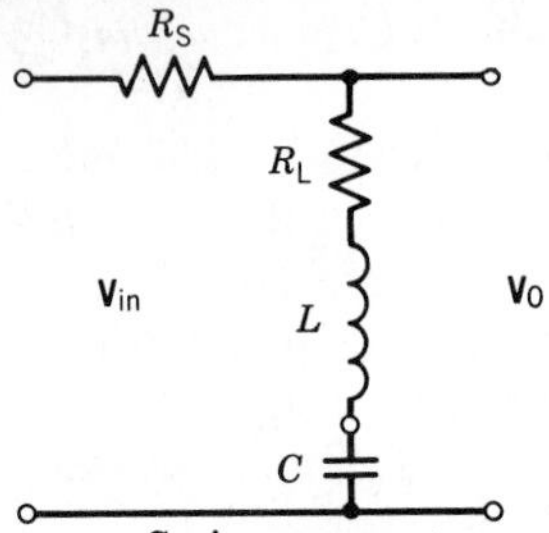

a. Series resonant</td>
<td>At any frequency, f:
$$\frac{\mathbf{V}_0}{\mathbf{V}_{in}} = \frac{R_L + j(X_L - X_C)}{(R_s + R_L) \pm j(X_L - X_C)}$$
At f_0:
$$\frac{\mathbf{V}_0}{\mathbf{V}_{in}} = \frac{R_L}{R_L + R_s}$$
$$R_s \gg R_L$$</td>
<td>$$\frac{\omega_0 L}{(R_s + R_L)}$$</td>
<td>$$\frac{1/2\pi\sqrt{LC}}{Q_0}$$</td>
<td>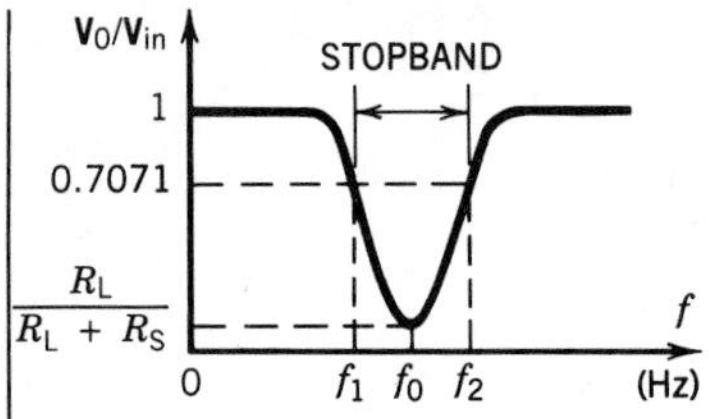
</td></tr>
<tr><td>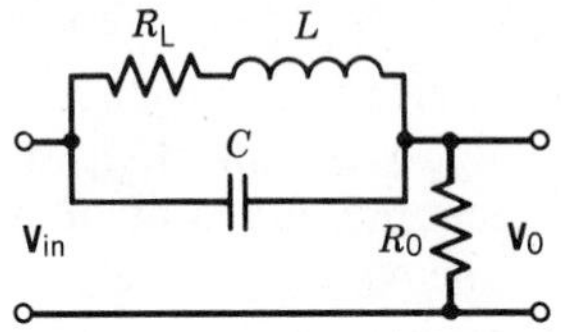

b. Parallel resonant</td>
<td>At any frequency, f:
$$\frac{\mathbf{V}_0}{\mathbf{V}_{in}} = \frac{R_0}{R_0 + \mathbf{Z}_p}$$
where $\mathbf{Z}_p = \dfrac{\mathbf{Z}_L\mathbf{Z}_C}{R_L \pm j(X_L - X_C)}$
At f_0:
$$\mathbf{V}_0/\mathbf{V}_{in} = \frac{R_0}{R_0 + \mathbf{Z}_{p0}}$$
where $\mathbf{Z}_{p0} = Q_0^2 R_L$</td>
<td>$$\frac{\omega_0 L}{R_L}$$</td>
<td>$$\frac{1/2\pi\sqrt{LC}}{Q_0}$$</td>
<td>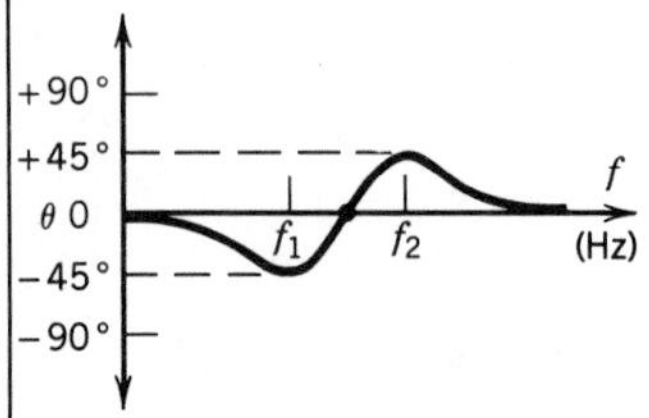
</td></tr>
<tr><td>BANDPASS
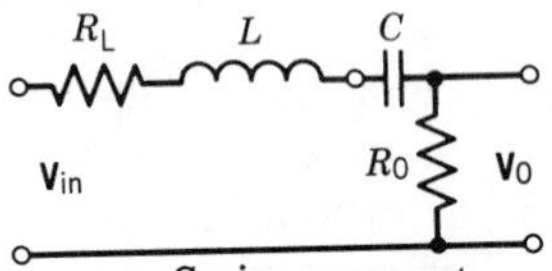

c. Series resonant</td>
<td>At any frequency, f:
$$\frac{\mathbf{V}_0}{\mathbf{V}_{in}} = \frac{R_0}{(R_0 + R_L) \pm j(X_L - X_C)}$$
At frequency f_0:
$$\frac{\mathbf{V}_0}{\mathbf{V}_{in}} = \frac{R_0}{R_L + R_0}$$</td>
<td>$$\frac{\omega_0 L}{R_L + R_0}$$</td>
<td>$$\frac{1/2\pi\sqrt{LC}}{Q_0}$$</td>
<td>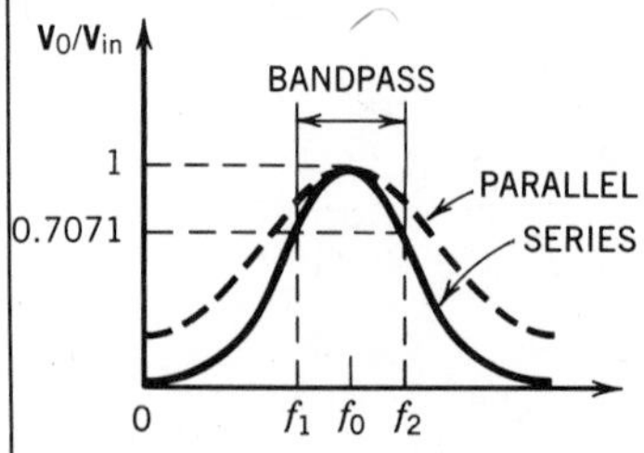
</td></tr>
<tr><td>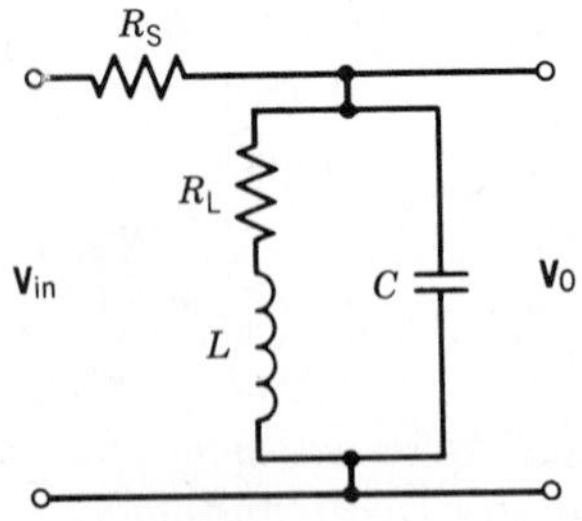

d. Parallel resonant</td>
<td>At any frequency, f:
$$\frac{\mathbf{V}_0}{\mathbf{V}_{in}} = \frac{\mathbf{Z}_p}{\mathbf{Z}_p + R_s}$$
where $\mathbf{Z}_p = \dfrac{\mathbf{Z}_L(-jX_C)}{R_L \pm j(X_L - X_C)}$
At frequency f_0:
$$\mathbf{Z}_{p0} = R_L(Q_0^2 + 1) = R_{p0}$$
where $Q_0 = \dfrac{\omega_0 L}{R_L} = \dfrac{R_{p0}}{X_{C0}}$</td>
<td>Unloaded:
$$\frac{R_{p0}}{X_{C0}}$$
Loaded:
$$\frac{R'_{p0}}{X_{C0}}$$</td>
<td>$$\frac{1/2\pi\sqrt{LC}}{Q_0}$$</td>
<td>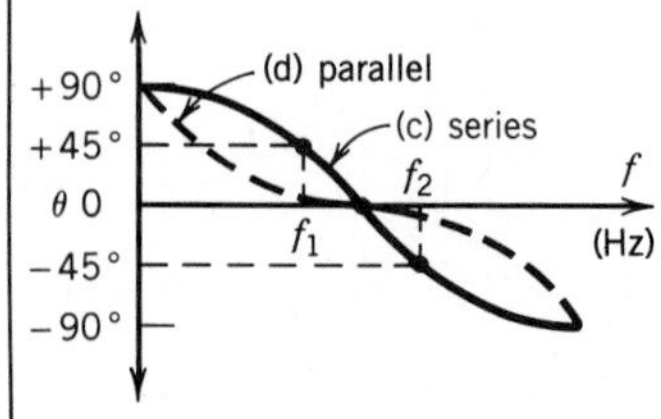
</td></tr>
</table>

19-9.1 Series-Resonant Bandstop Filter—Table 19-2a

The output is shunting the series-resonant circuit. At the resonant frequency, f_0, the output "sees" a *relatively low* resistance, R_L, and the negligible output voltage $\mathbf{V}_0$, as shown in Table 19-2**a** (first row), is $\mathbf{V}_0 = \mathbf{V}_{in}(R_L)/(R_L + R_s)$, where $R_s \gg R_L$. This accounts for the sharp resonant dip in the amplitude response and explains why *bandstop* filters are sometimes called *notch* filters or *wavetraps*. Such filters are used to *reject* a *particular* frequency such as 60-cycle hum produced by transformers or inductors or turntable rumble in recording equipment.

19-9.2 Parallel-Resonant Bandstop Filter—Table 19-2b

In this configuration, the parallel-circuit is in *series* with output resistor R_0. But at resonance, the parallel circuit exhibits almost infinite impedance in comparison to resistance R_0. Consequently, the output voltage across R_0 is relatively small at frequency f_0.

Table 19-2**b** shows the ratio $\mathbf{V}_0/\mathbf{V}_{in}$ for the "unloaded" parallel-resonant bandstop filter. If R_0 is shunted by R_L, then the parallel equivalent, R_0', must be used in all the equations given, where $R_0' = R_0 \| R_L$. Since R_L *reduces* R_0, its effect is to *reduce* the amplitude response at all frequencies. But at the same time, as noted from the relation $Q_0 = (\omega_0 L)^2/R_L R_0$, if R_0 is reduced to R_0', the Q_0 is increased, which narrows the bandwidth and increases the selectivity. This results in a narrower stopband, which is sometimes very desirable.

Of particular interest is that the *same* amplitude-phase response curves apply to both the series-resonant and parallel-resonant *bandstop* filters. Since capacitance predominates at the lower frequencies, the phase angle (of the impedance triangle) is negative below f_0. Above f_0, each resonant circuit behaves as an inductor, producing a positive phase angle. Observe that the cutoff frequency f_1 occurs where $\theta_1 = -45°$ and f_2 where $\theta_2 = +45°$, as in the case of any resonant circuit. Finally, note that the parallel-resonant bandstop filter is the *dual* of the series-resonant bandstop filter, hence the need for the configuration interchange.

19-9.3 Series-Resonant Bandpass Filter—Table 19-2c

As in the case of low-pass and high-pass circuits, merely "swapping" the series-resonant bandstop configuration elements should produce the *bandpass* characteristics. This is indeed the case, as shown in the configuration of Table 19-2**c**.

We may predict the resonant response by noting that at f_0, the series resonant impedance is practically a short circuit and equal to R_L, which is small compared to R_0. Consequently, the output is a maximum at f_0. Since the output is a function of R_0 in the numerator, whenever the filter is loaded by a paralleled R_L the equivalent R_0' is reduced. As in the case of the bandstop filter, this also reduces the amplitude but simultaneously increases the Q, which decreases bandwidth and improves selectivity. This effect narrows the bandpass, which is a desirable characteristic in many instances, particularly where a high degree of selectivity is required.

19-9.4 Parallel-Resonant Bandpass Filter—Table 19-2d

Again, if we transpose the circuit elements of the bandstop parallel-resonant filter (Table 19-2**b**), we obtain the bandpass parallel-resonant filter shown in Table 19-2**d**. Since the output is across the two-branch parallel tank circuit, we know that impedance (and consequently output voltage) is maximized at the resonant frequency, f_0. The amplitude response of this configuration is identical to that of the series configuration whenever the tank is not under load. When loaded, however, the Q'_{f0} is reduced since R'_{po} is reduced (as is the case of the two-branch parallel-resonant tank circuit). Consequently, loading *this* filter tends to reduce the amplitude, reduce the Q, and increase the bandwidth of the bandpass. In some instances, this may be a desirable filter characteristic, where *wider* bandpass is desired.

EXAMPLE 19-21

Given the series-resonant bandstop filter configuration of Table 19-2a; R_s is 2 kΩ, R_L is 10 Ω, L is 320 mH, and C is 198 pF. For the values and equations given in Table 19-2a, calculate the following bandstop parameters:

a. Resonant frequency, f_0
b. Half-power bandwidth, B_{hp}
c. Edge frequencies f_1 and f_2
d. Output voltage $\mathbf{V}_0$ at frequencies f_0, f_1, and f_2
e. Output voltage $\mathbf{V}_0$ at frequencies of 10 kHz and 40 kHz

Solution

a. $f_0 = 1/2\pi\sqrt{LC} = 1/2\pi\sqrt{0.32 \times 198 \times 10^{-12}} = \mathbf{20\ kHz}$

b. $Q_0 = \omega_0 L/(R_s + R_L) = 2\pi(20 \times 10^3)0.32/2010\ \Omega = \mathbf{20}$;
$B_{hp} = f_0/Q_0 = 20\ \text{kHz}/20 = \mathbf{1\ kHz}$

c. $f_1 = f_0 - B/2 = 20\ \text{kHz} - 1\ \text{k}/2 = \mathbf{19.5\ kHz}$;
$f_2 = f_0 + B/2 = 20\ \text{kHz} + 1\ \text{k}/2 = \mathbf{20.5\ kHz}$

d. At f_0 from Table 19-2a, $\mathbf{V}_0 = \mathbf{V}_{in}R_L/(R_L + R_s) = \mathbf{V}_{in} \times 10/2010 = \mathbf{4.975 \times 10^{-3}\ V_{in}}$. At f_1 we calculate X_{L_1}, X_{C_1}, and $j(X_{C_1} - X_{L_1})$ to find $\mathbf{V}_0/\mathbf{V}_{in}$ per Table 19-2a:
$X_{L_1} = 2\pi f_1 L = 2\pi(19.5\ \text{k})(320\ \text{m}) = j39\,207\ \Omega$;
$X_{C_1} = 1/2\pi f_1 C = 1/2\pi(19.5\ \text{k})(198\ \text{p}) = -j41\,221\ \Omega$,
from which $-jX_{01} = -jX_{C_1} + jX_{L_1} = -j41\,221 + j39\,207 = -j2014\ \Omega$.

Then at f_1, $\mathbf{V}_0 = \mathbf{V}_{in}\dfrac{R_L \pm jX_{01}}{(R_L + R_s) \pm jX_{01}} = \mathbf{V}_{in}\dfrac{10 - j2014}{2010 - j2014} = \mathbf{V_{in}(0.708\underline{/-44.7^\circ})}$

Having calculated X_{L_1} and X_{C_1} at a given frequency f_1, we use the ratio method for all other reactances, and therefore $jX_{L_2} = (f_2/f_1)(X_{L_1}) = (20.5/19.5)j39\,207 = j41\,218\ \Omega$, $-jX_{C_2} = (f_1/f_2)X_{C_2} = (19.5/20.5)(-j41\,221) = -j39\,210\ \Omega$, and $jX_{02} = jX_{L_2} - jX_{C_2} = j41\,218 - j39\,210 = j2008\ \Omega$.

Then at f_2, $\mathbf{V}_0 = \mathbf{V}_{in}\dfrac{R_L \pm jX_{02}}{(R_L + R_s) \pm jX_{02}} = \mathbf{V}_{in}\dfrac{10 + j2008}{2010 + j2008} = \mathbf{0.707\underline{/+44.7^\circ}\ V_{in}}$

e. When the input frequency is 10 kHz, using the ratio method:
$jX_L = (10/19.5)(j39\,207) = j20\,106\ \Omega$ and
$jX_C = (19.5/10)(-j41\,221) = -j80\,381\ \Omega$; then
$jX_0\ (\text{at 10 kHz}) = -jX_C + jX_L = -j80\,381 + j20\,106 = -j60\,275\ \Omega$

and $\mathbf{V}_0 = \dfrac{10 - j60\,275}{2010 - j60\,275}\mathbf{V}_{in} = \mathbf{(0.9994\underline{/-1.9^\circ})\ V_{in}}$

Similarly, when the input frequency is 40 kHz, using the ratio method:

$jX_L = (40/19.5)(j39\,207) = j80\,425\ \Omega$;

and $-jX_C = (19.5/40)(-j41\,221) = -j20\,095\ \Omega$;

then jX_0 at 40 kHz $= j80\,425 - j20\,095 = j60\,330\ \Omega$

and $\mathbf{V}_0 = \dfrac{10 + j60\,330}{2010 + j60\,330}\mathbf{V}_{in} = \mathbf{(0.9994\underline{/+1.9^\circ})\ V_{in}}$

Example 19-21 verifies the amplitude-phase response curve for the configuration of Table 19-2**a**. In particular,

1. At the edge frequency f_1 below f_0, the amplitude is $0.707\mathbf{V}_{in}$ and the phase angle is (approximately) -45°, as shown in the phase response for this configuration.
2. Similarly, at edge frequency f_2 above f_0, the amplitude is also $0.707\mathbf{V}_{in}$ but the phase angle is now $+45^\circ$, verifying the phase response shown for this configuration.
3. But when the frequency is reduced considerably from f_0 to 10 kHz, the amplitude $\mathbf{V}_0$ (as shown in **e** of Ex. 19-21) is practically the same as $\mathbf{V}_{in}$ but the phase angle is a *small negative* angle.
4. Similarly, when the frequency is doubled, the amplitude $\mathbf{V}_0 = \mathbf{V}_{in}$ but the phase angle is a *small positive* angle.
5. At the resonant frequency, f_0, $\mathbf{V}_0 \cong (5 \times 10^{-3})\ \mathbf{V}_{in}$, showing a very sharp drop in output voltage.
6. Phase angle is a maximum at the edge frequencies and is approximately 0° at frequencies *significantly* above and below f_0 as well as at f_0, the resonant frequency.

EXAMPLE 19-22

Given the parallel-resonant bandstop filter shown in Table 19-2**b**, R_L is 8 Ω, L is 318.3 μH, C is 7.95 nF, R_0 is 500 Ω, and the input voltage $\mathbf{V}_{in}$ is $10\underline{/0^\circ}$ V. Calculate:

a. f_0
b. Q_r and Q_0
c. B_h
d. f_1 and f_2
e. $\mathbf{Z}_{po}$
f. $\mathbf{V}_0$ at f_0
g. $\mathbf{V}_0$ at f_1
h. $\mathbf{V}_0$ at f_2
i. $\mathbf{V}_0$ at an input frequency of 10 kHz

Solution

a. $f_0 = 1/2\pi\sqrt{LC} = 1/2\pi\sqrt{(318.3\ \mu)(7.95\ \text{n})} = \mathbf{100\ kHz}$

b. $Q_r = \omega_0 L/R_L = 2\pi(100\ \text{k})(318.3\ \mu)/8\ \Omega = j200\ \Omega/8\ \Omega = \mathbf{25} = Q_0$ (high Q)

c. $B_h = f_0/Q_0 = 100\ \text{kHz}/25 = \mathbf{4\ kHz}$
(and $B_h/2 = 4\ \text{kHz}/2 = \mathbf{2\ kHz}$)

d. $f_1 \cong f_0 - B_h/2 = 100\text{ kHz} - 2\text{ kHz} = \mathbf{98\text{ kHz}}$;
$f_2 \cong f_0 + B_h/2 = 100\text{ kHz} + 2\text{ kHz} = \mathbf{102\text{ kHz}}$

e. $\mathbf{Z}_{po} = R_{po} = Q_r^2 R_L = 25^2(8\ \Omega) = \mathbf{5\angle 0^\circ\ k\Omega}$

f. $$\mathbf{V}_0 = \mathbf{V}_{in}\frac{R_0}{(\mathbf{Z}_{po} + R_0)} = 10\angle 0^\circ\text{ V}\,\frac{0.5\text{ k}\Omega}{5.5\text{ k}\Omega} = \mathbf{0.91\angle 0^\circ\ V}$$

g. At the half-power frequency f_1, $\mathbf{Z}_{p_1} = 0.7071(\mathbf{Z}_{po})$, as proved in (**h**) below. Therefore,

$$\mathbf{Z}_{p_1} = 0.7071(5\text{ k}\Omega) = 3535\angle 45^\circ\ \Omega = 2500 + j2500\ \Omega$$

and $$\mathbf{V}_0 = \mathbf{V}_{in}\frac{R_0}{(\mathbf{Z}_{p_1} + R_0)} = 10\angle 0^\circ\text{ V}\,\frac{500 + j0}{500 + (2500 + j2500)}$$

$$= \mathbf{1.28\angle -39.8^\circ\ V}$$

h. At the half-power frequency f_2, $\mathbf{Z}_{p_2}$ may also be found by the reactance method where

$$jX_{L_2} = X_{L0}\frac{f_2}{f_1} = j200\ \Omega\,\frac{102\text{ kHz}}{100\text{ kHz}} = j204\ \Omega$$

and $$jX_{C_2} = X_{C0}\frac{f_1}{f_2} = -j200\ \Omega\,\frac{100\text{ kHz}}{102\text{ kHz}} = -j196\ \Omega$$

Then $$\mathbf{Z}_{p_2} = \mathbf{Z}_L\mathbf{Z}_C/(\mathbf{Z}_L + \mathbf{Z}_C) = \frac{(8 + j204)(-j196)}{8 + j8}$$

$$= 2401 - j2597 = 3536.8\angle -47.2^\circ\ \Omega$$

(Note that this result is approximately the same as **g** above using the half-power relations.)

$$\mathbf{V}_0 = \mathbf{V}_{in}\frac{R_0}{\mathbf{Z}_{p_2} + R_0} = 10\angle 0^\circ\text{ V}\,\frac{500 + j0}{500 + (2401 - j2597)}$$

$$= \mathbf{1.28\angle 41.8^\circ\ V}$$

i. At $f = 10$ kHz,

$$jX_L = j200\ \Omega\,\frac{10\text{ kHz}}{100\text{ kHz}} = j20\ \Omega;$$

$$-jX_C = -j200\,\frac{100\text{ k}}{10\text{ k}} = -j2000\ \Omega.$$

Then $\mathbf{Z}_p = (R + jX_L)\,||\,(-jX_C) = (8 + j20)\,||\,(-j2000)\ \Omega$. But since X_C is so high, we may neglect it and $\mathbf{Z}_p = R + jX_L = 8 + j20\ \Omega$. Then

$$\mathbf{V}_0 = \mathbf{V}_{in}\frac{R_0}{\mathbf{Z}_p + R_0} = 10\angle 0^\circ\text{ V}\,\frac{500 + j0}{508 + j20} = \mathbf{9.83\angle -3.25^\circ\ V}$$

The following important insights may be drawn from Ex. 19-22:

1. The calculated results verify the amplitude-phase response characteristics shown in Table 19-2**b**.
2. The parallel-resonant portion of the circuit does exhibit maximum impedance (5 kΩ) at the resonant frequency. This accounts for the sharp dip in the amplitude response at f_0.
3. At edge frequencies f_1 and f_2, the equivalent parallel circuit impedance is still $0.7071\mathbf{Z}_{po}\angle \pm 45^\circ$, as proved in steps (**g**) and (**h**) of Ex. 19-22. But the amplitude does not quite rise to $0.707\mathbf{V}_{in}$, as *theoretically* shown in the amplitude response of Table 19-2.
4. At frequencies well below and well above the resonant frequency, f_0, as shown by step (**i**), we note that
   a. The amplitude is approaching the off-resonance unity relation for $\mathbf{V}_0/\mathbf{V}_{in}$.
   b. The phase angle is a small negative angle as predicted for the parallel-resonant bandstop filter shown in Table 19-2**b**.

Before leaving this bandstop filter we might ask, "What is the effect of an external load, R_x, connected across R_0?" The answer is almost obvious. R_x only reduces the value of R_0 to some smaller equivalent R_0', for the unloaded relations shown in column 3, row 2 of Table 19-2**b**. Consequently, loading has two effects produced by reducing R_0 to a lower value of R_0':

1. At resonant frequency, f_0, the ratio $\mathbf{V}_0/\mathbf{V}_{in}$ is *reduced* (which is desirable for a bandstop filter).
2. *But* the amplitude of $\mathbf{V}_0$ is reduced at *all* frequencies and the unity ratio is no longer possible. (If R_0' is almost a short circuit, $\mathbf{V}_0$ is practically zero.)

It should be noted from the circuit Q_0 relations shown in the column 4 (row 2) that loading does not degrade the resonant parallel Q_0 at all since neither R_0 nor R_0' appears in these Q relations.

We now turn to two examples illustrating the characteristics of *resonant bandpass* filters.

EXAMPLE 19-23

Given the series-resonant bandpass filter circuit shown in Table 19-2c, R_L is 3.3 Ω, L is 50 mH, C is 202.6 pF, R_0 is 625 Ω, and $\mathbf{V}_{in}$ is $10\angle 0^\circ$ V. Calculate

a. f_0
b. Q_0
c. B_{hp}
d. f_1 and f_2
e. $\mathbf{V}_0$ at f_1

f. $\mathbf{V}_O$ at f_2
g. $\mathbf{V}_O$ at 500 kHz

Solution

a. $f_O = 1/2\pi\sqrt{LC} = 1/2\pi\sqrt{(50\text{ m})(202.6\text{ p})} = \mathbf{50\ kHz}$
b. $Q_O = \omega_O L/(R_T) = \omega_O L/(R_O + R_L)$
$= 2\pi(50\text{ k})(50\text{ m})/(625 + 3.3)$
$= j15\,708/628.3 = \mathbf{25}$
c. $B_{hp} = f_O/Q_O = 50\text{ kHz}/25 = \mathbf{2\ kHz}$ (and $B_{hp}/2 = 1$ kHz)
d. $f_1 = f_O - B_{hp}/2 = 50\text{ kHz} - 1\text{ kHz} = \mathbf{49\ kHz}$;
$f_2 = f_O + B_{hp}/2 = 50\text{ kHz} + 1\text{ kHz} = \mathbf{51\ kHz}$
e. By the ratio method

$$X_{L_1} = (f_1/f_O)X_{L_O} = (49/50)j15\,708 = j15\,394\ \Omega \text{ and}$$

$$X_{C_1} = (f_O/f_1)X_{C_O} = (50/49)(-j15\,708) = -j16\,029\ \Omega$$

Then $X_{C_{O1}} = j(X_{L_1} - X_{C_1}) = j15\,394 - j16\,029 = -j625\ \Omega$

and $$\mathbf{V}_{O1} = \frac{\mathbf{V}_{in}R_O}{(R_O + R_L) + j(X_{L_1} - X_{C_1})}$$

$$= 10\underline{/0^\circ}\text{ V}\,\frac{625}{628.3 - j635} = \mathbf{7\underline{/45.3^\circ}\ V}$$

f. At frequency $f_2 = 51$ kHz:

$$X_{L_2} = (f_2/f_O)X_{L_O} = (51/50)j15\,708 = j16\,022\ \Omega$$

and $X_{C_2} = (f_O/f_2)X_{C_O} = (50/51) - j15\,708 = -j15\,400\ \Omega.$

Then $X_{L_2} - X_{C_2} = j16\,022 - j15\,400 = j622$, so

$$\mathbf{V}_{O2} = \mathbf{V}_{in}\frac{R_O}{(R_O + R_L) + j(X_{L_2} - X_{C_2})}$$

$$= 10\underline{/0^\circ}\text{ V}\,\frac{625}{628.3 + j622} = \mathbf{7.07\underline{/-44.7^\circ}}\text{ V}$$

g. At $f = 500$ kHz, $X_L = (500/50)j15\,708 = j157\,080\ \Omega$ and $X_C = -j1570.8$. Then $X_O = X_L - X_C = j157\,080 - j1570.8 = j155.51\text{ k}\Omega$ and

$$\mathbf{V}_O = \mathbf{V}_{in}\frac{R_O}{(R_O + R_L) + jX_O} = 10\underline{/0^\circ}\,\frac{0.625\text{ k}}{(0.6283 + j155.51)\text{ k}\Omega}$$

$$= \mathbf{0.0042\underline{/-89.77^\circ}}\text{ V}$$

The following insights are to be drawn from Ex. 19-23:

1. At resonant frequency f_O, the output is maximized; $\mathbf{V}_O = (625/628.3)\mathbf{V}_{in} = 0.995\mathbf{V}_{in}$.
2. At edge frequencies f_1 and f_2, the output is approximately $0.707\mathbf{V}_{in}$.
3. At edge frequencies f_1 and f_2, the phase angle is $\pm 45°$. The phase angle is positive for frequencies below f_O and negative for frequencies above f_O.
4. At extremely high frequencies, the output approaches zero and the phase angle approaches $-90°$, as shown in the amplitude phase response of Table 19-2c.

As in the case of the bandstop filters we may ask, "What is the effect of an external load, R_X, connected across R_O?" Again, the answer is fairly evident since R_X is only shunting R_O, producing a lower equivalent resistance, R_O'. Loading consequently has several effects produced by reducing R_O to a lower value of R_O':

1. Since $Q_O = \omega_O L/(R_O + R_L)$ and R_O is reduced to R_O', the Q_O is increased.
2. The increase in Q implies a reduced bandwidth, greater selectivity, and sharper bandpass response curve.
3. The reduction in R_O' produces only a slight reduction in the approximately unity ratio of $\mathbf{V}_O/\mathbf{V}_{in}$ at f_O (unless the external load, R_X, is extremely low).

Let us now turn to the last of the resonant filters shown in Table 19-2**d**.

EXAMPLE 19-24
Given the parallel-resonant bandpass filter shown in Table 19-2d, $\mathbf{V}_{in} = 10\underline{/0^\circ}$ V, $R_s = 2\text{ k}\Omega$, $R_L = 100\ \Omega$, $L = 10$ mH, and $C = 253.3$ pF. Calculate

a. f_O
b. Q_r and Q_O
c. R_{po}
d. $\mathbf{V}_O$ at f_O
e. B_{hp}
f. f_1 and f_2
g. $\mathbf{V}_O$ at f_1
h. $\mathbf{V}_O$ at 50 kHz

Solution

a. $f_O = 1/2\pi\sqrt{LC} = 1/2\pi/(10\text{ m})(253.3\text{ p}) = 100$ kHz
b. $Q_r = 2\pi f_O L/R_L = 2\pi(100\text{ k})(10\text{ m})/100 = j6283/100$
$= \mathbf{62.83} = Q_O$
c. $R_{po} = Q^2 R_L = (62.83)^2 100 = 394.76\text{ k}\Omega$
d. $$\mathbf{V}_O = \mathbf{V}_{in}\frac{R_{po}}{(R_{po} + R_s)} = \frac{394.76\text{ k}\Omega}{396.76\text{ k}\Omega} \times 10\underline{/0^\circ}\text{ V} = \mathbf{9.95\underline{/0^\circ}\ V}$$
e. $B_{hp} = f_O/Q_O = 100\text{ kHz}/62.83 = \mathbf{1.592\ kHz}$ (and $B_{hp}/2 = \mathbf{796\ Hz}$)
f. $f_1 \cong f_O - B_{hp}/2 = 100\text{ k} - 0.796\text{ k} = \mathbf{99.2\ kHz}$ and
$f_2 \cong f_O + B_{hp}/2 = 100\text{ k} + 0.796\text{ k} = \mathbf{100.796\ kHz}$
g. At f_1, $\mathbf{Z}_{p_1} = 0.7071(\mathbf{Z}_{po})\underline{/+45^\circ} = 0.7071 \times (394.76\text{ k})$
$= 279.13/45°\text{ k}\Omega$;

$$\mathbf{V}_{01} = \mathbf{V}_{\text{in}} \frac{\mathbf{Z}_{p_1}}{\mathbf{Z}_{p_1} + R_s} = 10\underline{/0°}\ \text{V} \frac{197.4 + j197.4}{199.4 + j197.4}$$

$$= \mathbf{9.949\underline{/0.3°}\ V}$$

h. At $f = 50$ kHz,

$$jX_L = \frac{50\ \text{kHz}}{100\ \text{kHz}}(j6283) = j3142\ \Omega;$$

$$-jX_C = \frac{100}{50}(-j6283) = -j12\,566\ \Omega;$$

$$X_0 = jX_L - jX_C = j3142 - j12\,566 = -j9424\ \Omega;$$

$$\mathbf{Z}_p = \mathbf{Z}_L\mathbf{Z}_C/(\mathbf{Z}_L + \mathbf{Z}_C) = \frac{(100 + j3142)(0 - j12\,566)}{100 - j9424}$$

$$= 177.8 + j4188 = 4191\underline{/87.6°}\ \Omega;$$

$$\text{and } \mathbf{V}_0 \text{ at } 50\ \text{kHz} = \mathbf{V}_{\text{in}} \frac{\mathbf{Z}_p}{\mathbf{Z}_p + R_s} = 10\underline{/0°}\ \text{V} \frac{177.8 + j4188}{2177.8 + j4188}$$

$$= \mathbf{8.88\underline{/25°}\ V}$$

Example 19-24 yields a number of interesting insights:

1. The circuit parameters are such as to produce an extremely narrow bandwidth due to the high Q. This implies a high degree of selectivity and an extremely sharp-sloped response curve.
2. Because of (1), despite the fact that the impedance has dropped to $0.7071\mathbf{Z}_{po}\underline{/45°}\ \Omega$, the output voltage remains *almost the same* and the phase angle is practically zero!
3. Even when f is half of f_0, the output voltage is still relatively high and the phase angle is only 25°.

We may now raise the question of the effect of loading on the parallel-resonant bandpass filter of Table 19-2d. Let us assume that a (relatively heavy) load of 20 kΩ is connected across the parallel tank circuit output of the filter in Table 19-2d. Example 19-25 shows the effect of such heavy loading.

EXAMPLE 19-25

Assume that a load of 20 kΩ is connected across the tank of the parallel-resonant bandpass filter of Ex. 19-24. Calculate

a. R'_{po} at the resonant frequency, f_0
b. Q'_0
c. B'_{hp}
d. f'_1
e. $\mathbf{V}'_0$ at f_0
f. $\mathbf{V}'_0$ at f'_1

Solution

a. $R'_{po} = R_{po} \| R_X = 394.8\ \text{k}\Omega \| 20\ \text{k}\Omega = \mathbf{19.036\ k\Omega}$
b. $Q'_0 = R'_{po}/jX_{C_0} = 19\,036\ \Omega/6283\ \Omega = \mathbf{3.03}$
c. $B'_{hp} = f_0/Q'_0 = 100\ \text{kHz}/3.03 = \mathbf{33\ kHz}$
d. $f'_1 = f_0 - B_{hp}/2 = 100\ \text{kHz} - 16.5\ \text{kHz} = \mathbf{83.5\ kHz}$
e. $\mathbf{V}'_0 = \dfrac{R'_{po}}{R'_{po} + R_s}\mathbf{V}_{\text{in}} = \dfrac{19\ \text{k}}{21\ \text{k}} 10\underline{/0°}\ \text{V} = \mathbf{9.05\underline{/0°}\ V}$ at f_0
f. At f'_1: $\mathbf{Z}'_{p_1} = 0.7071(\mathbf{Z}'_{po}\underline{/45°}) = 0.7071(19\ \text{k})\underline{/45°}$ $= 13.43\underline{/45°}\ \text{k}\Omega = (9.5 + j9.5)\ \text{k}$

and $\mathbf{V}'_{01} = \mathbf{V}_{\text{in}} \dfrac{\mathbf{Z}'_{p_1}}{\mathbf{Z}'_{p_1} + R_s} = 10\underline{/0°} \dfrac{9.5 + j9.5}{11.5 + j9.5}$

$= \mathbf{9\underline{/5.44°}\ V}$ at 83.5 kHz

Example 19-25 shows how remarkably resistant to loading the circuit of Table 19-2**d** actually is. Despite a reduction in Q_0 from about 63 to 3 and an increase in BW from 1.6 to 33 kHz, the output voltage has only dropped about 1 V at f_0! The same effect occurs at edge frequency f_1, where the voltage drop is approximately 1 V with a small increase in phase shift of about 5°. We may conclude that loading produces only small reductions in output voltage and relatively small increases in phase shift. The only major effect has been an increase in BW, which may be desired in some instances.

19-10 DOUBLE-RESONANT FILTER NETWORKS

Frequently, in communication networks it is desired to selectively pass one band of frequencies and selectively reject a second band. Table 19-3 shows four such double-resonant circuits along with their response curves. The first two configurations show a *higher passband* frequency, f_p, and a *lower stopband* frequency f_s. The last two configurations show a *lower passband* frequency and a *higher stopband* frequency.

Table 19-3 Double-Resonant Configurations and Their Response Curves

Filter Combination	Circuit Configuration	At Lower Frequency	At Higher Frequency	Amplitude Response
a. Single C in parallel with L_1 Two inductors One capacitor	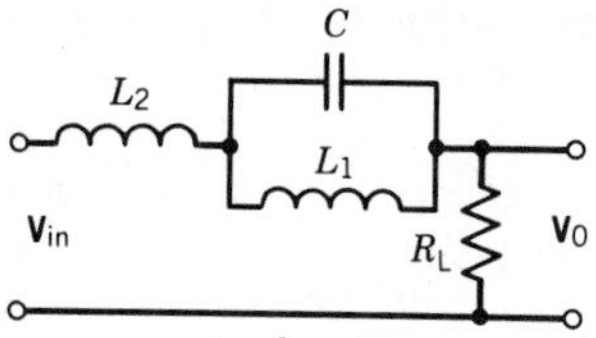 a. $f_p > f_s$	C and L_1 in parallel at resonance, producing high series impedance, low current, and low voltage output	L_2 and C in series resonance, producing low series impedance, high current, and high voltage output	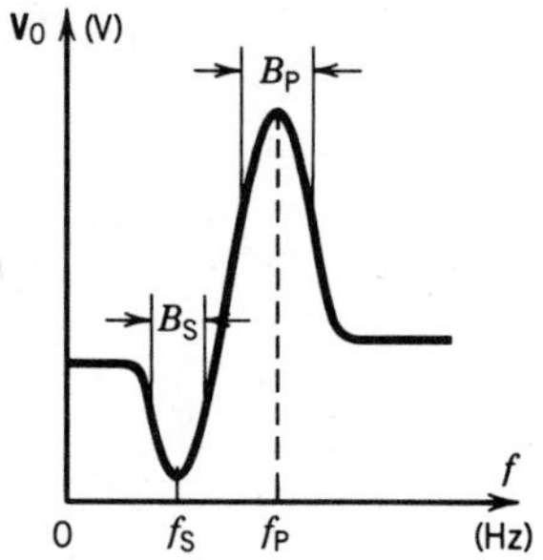
b. Single C in series with L_1 Two inductors One capacitor	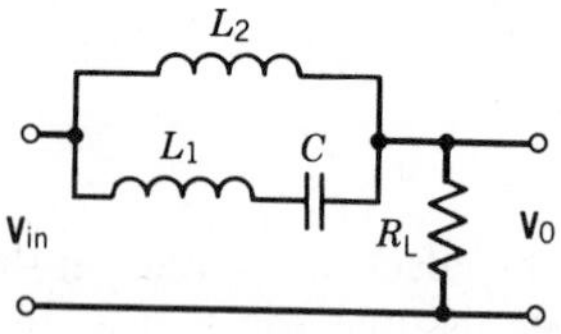 b. $f_p > f_s$	C and L_2 in parallel resonance, producing high series impedance, low current, and low voltage output	C and L_1 in series resonance, producing low series impedance, high current, and high voltage output	
c. Single L in parallel with C_1 Two capacitors One inductor	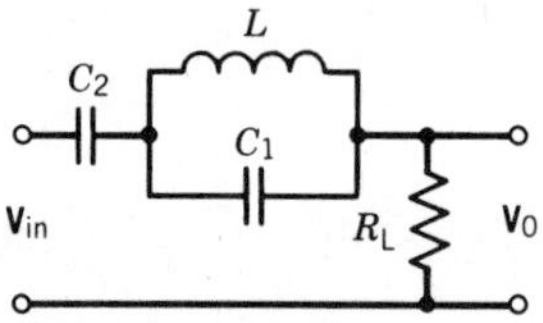 c. $f_s > f_p$	L and C_2 in series resonance, producing low series impedance, high current, and high voltage output	L and C_1 in parallel resonance, producing high series impedance, low current, and low voltage output	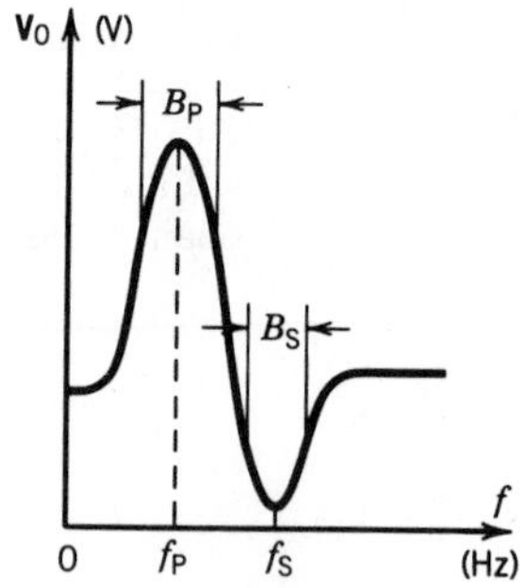
d. Single L in series with C_1 Two capacitors One inductor	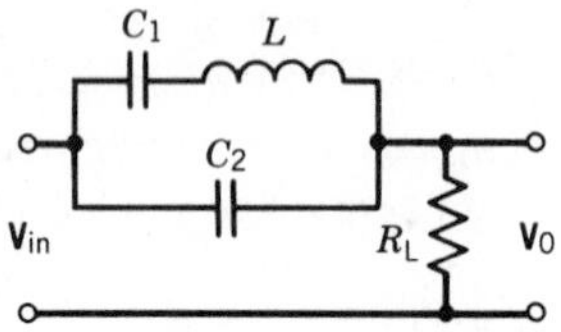 d. $f_s > f_p$	L and C_1 in series resonance, producing low series impedance, high current, and high voltage output	L and C_2 in parallel resonance, producing high series impedance, low current, and low voltage output	

A study of the four configurations in Table 19-3 reveals that *all* double-resonant circuits shown:

1. Contain either one capacitor and two inductors **or** two capacitors and one inductor.
2. Exhibit series resonance at one frequency and parallel resonance at the other frequency.
3. Produce the *stopband* frequency whenever the circuit is *parallel* resonant.
4. Produce the *passband* frequency whenever the circuit is *series* resonant.

A comparison of the configurations and their respective response curves also reveals that

5. Given two inductors and one capacitor, $f_p > f_s$ (i.e., the *passband* frequency is *higher* than the stopband frequency).
6. Given two capacitors and one inductor, $f_s > f_p$ (i.e., the *stopband* frequency is *higher* than the passband frequency).

The fundamental insights in analyzing and predicting the response curves from the given configurations emerge from our earlier study of series and parallel resonance:

A. *Series LC configuration*
   1. Impedance is a minimum at resonance and equivalent to a short circuit.
   2. Below its resonant frequency, C predominates.
   3. Above its resonant frequency, L predominates.

B. *Parallel LC configuration*
   1. Impedance is a maximum at resonance and equivalent to an open circuit.
   2. Below its resonant frequency, L predominates.
   3. Above its resonant frequency, C predominates.

Let us apply these insights to the first configuration shown in Table 19-3**a**:

1. When C and L_1 produce parallel resonance, the high impedance results in frequency f_s, the stopband frequency.
2. With L_2 in series, the resultant reactance of the parallel circuit must be capacitive for series resonance to occur.
3. Since C predominates at a higher frequency, in a parallel circuit, series resonance occurs at the higher frequency, f_p, where the circuit has a low impedance.

Using this analytical technique, it is possible to predict the remaining amplitude-frequency responses from the given configurations.

In the two examples given below we will design two double-resonant filters. The end-of-chapter problems include the design of the remaining two. All designs, however, require *specifying* the two frequencies desired for bandstop and bandpass plus the *single* C or L for the configuration. Given these, the calculations involve finding the remaining component values. The solution steps for the calculation of the network shown in Table 19-3**a**, where C, f_p, and f_s are given, follow:

1. Given C, find L_1 for *parallel* resonance at frequency f_s.
2. Find the resultant parallel reactance, X_{po}, at frequency f_p.
3. Find L_2 for series resonance at frequency f_p.

EXAMPLE 19-26

For the configuration given in Table 19-3**a**, C is 10 nF. Design a double-resonant filter whose stopband frequency is 50 kHz and whose passband frequency is 100 kHz by specifying **a.** L_1 and **b.** L_2

Solution

a. L_1 is (parallel) resonant with C at $f_s = 50$ kHz so $L_1 = 1/(2\pi f_s)^2 C = 1/(2\pi \times 50\text{ k})^2 \times 10\text{ n} = \mathbf{1\ mH}$, and at $f_s = 50$ kHz the reactances are $jX_{L_1} = -jX_C = 2\pi f_s L = 2\pi \times 50\text{ k} \times 1\text{ m} = j314.2\ \Omega$

b. At frequency $f_p = 100$ kHz:

$$jX_{Lp} = jX_{L_1}\frac{f_p}{f_s} = j314.2(100\text{ k}/50\text{ k}) = j628.3\ \Omega \text{ and}$$

$$-jX_{Cp} = -jX_C(f_s/f_p) = -j314.2(50\text{ k}/100\text{ k}) = -j157.1\ \Omega.$$

Then the resultant parallel reactance, X_{po}, at frequency f_p is

$$X_{po} = jX_{Lp}\cdot(-jX_{Cp})/(jX_{Lp} - jX_{Cp}) = \frac{(j628.3)(-j157.1)}{(j628.3 - j157.1)}$$

$$= -j209.5\ \Omega.$$

And for series resonance at frequency f_p, $jX_{L_2} = +j209.5\ \Omega$ and therefore L_2 is $L_2 = jX_{L_2}/2\pi f_p = j209.5/2\pi(100\text{ k}) = \mathbf{33.\bar{3}\ \mu H}$

EXAMPLE 19-27

For the configuration shown in Table 19-3**d**, L is 1 mH. Design a double-resonant filter whose passband frequency is 30 kHz and whose stopband frequency is 100 kHz by specifying **a.** C_1 and **b.** C_2

Solution

a. C_1 is (series) resonant with L at *lower* frequency $f_p = 30$ kHz, so $C_1 = 1/(2\pi f_p)^2 L = 1/(2\pi 30\text{ k})^2\ 1\text{ m} = \mathbf{28.1\ nF}$. Then at the lower frequency, f_p, the respective reactances are $jX_L = 2\pi f_p L = 2\pi(30\text{ k})(1\text{ m}) = j188.5\ \Omega$ and $jX_{C_1} = -j188.5\ \Omega$

b. At the higher frequency, $f_s = 100$ kHz: $jX_{Ls} = jX_L(f_s/f_p) = j188.5(100\text{ k}/30\text{ k}) = j628.3\ \Omega$ and $jX_{C_1s} = jX_{C_1}(f_p/f_s) = -j188.5(30\text{ k}/100\text{ k}) = -j56.55\ \Omega$, so that the resultant reactance in the series branch, $X_{s_0} = j628.3 - j56.55 = j571.75\ \Omega$ Since the resultant reactance is inductive in the series branch, we need capacitive reactance in the parallel branch for parallel resonance and therefore $-jX_{C_2} = -j571.75\ \Omega$. We now calculate C_2 as $C_2 = 1/2\pi f_s X_{C_2} = 1/2\pi(100\text{ k})(571.75) =$ **2.78 nF**

Examples 19-26 and 19-27 show how double-resonant filters are designed. With respect to both of these examples:

1. The solution *always begins* where the *single* component in the configuration is either exclusively in series or in parallel with one of the two remaining opposite components.
2. The relation $LC = 1/\omega_0^2$ is used to find the value of either L or C for series or parallel resonance, depending on the configuration.
3. The next step requires finding the resultant reactance at the remaining frequency.
4. The final step involves finding the remaining component value to produce resonance with the resultant reactance.
5. Designing double-resonant filters, consequently, is a matter of good *bookkeeping* to ensure that *none of the above steps is omitted.*

We have discussed the various ways in which values of C and L are obtained. But no mention was made of R_L. To ensure a reasonable difference between the high and low amplitudes, we should select a value of R_L that is at least one-tenth of the high impedance (produced by the parallel combination) at the parallel-resonant frequency. This technique provides two advantages:

1. It ensures that R_L (so calculated) provides a good *dip* at the *stopband* frequency.
2. In the event of additional loading in parallel with R_L, the value of R_L is sufficiently low so as not to be degraded further.

19-11 GLOSSARY OF TERMS USED

Bandpass filter Frequency band of relatively little attenuation, relative to other regions in the frequency spectrum, designated as stopbands, produced by a given network.

Bandstop filter Frequency band of high attenuation, produced by a given network, relative to other regions designated as passbands.

Bandwidth, *B* *Difference* between any *two* limiting frequencies. As normally applied to the series- or parallel-resonant circuit, the half-power bandwidth, B_{hp}, is the difference between the upper and lower edge frequencies where the power is down -3 dB (reduced by 50%) and the voltage and/or current response is reduced to 0.7071 of its maximum value at resonance.

Circulating (tank) current Current circulating between reactive elements (L and C) in the three-branch and two-branch parallel circuits at resonance due to *resonant current rise.* This current may exceed the supply current by many times.

Cutoff frequency Specific frequency, f_c, identified with the transition between a passband and an adjacent attenuation band of a system, transducer, or filter. The specific attenuation at the cutoff frequency is usually -3 dB.

Filter Transducer for separating waves on the basis of their frequency.

Filter, active Filter containing one or more sources of energy such as transistors and operational amplifiers.

Filter, passive Filter containing only R, L, and C elements.

Half-power frequency Lower or upper limiting frequency or passband frequency of a given bandwidth. Sometimes called the edge frequency.

High-pass filter Filter having a single transmission band extending from some cutoff frequency (not zero) up to infinite frequency.

Ideal parallel-resonant circuit Three-branch circuit containing only pure elements (R, L, and C) in each branch. In practice, this circuit is never realized but it may be represented as equivalent to the practical tank circuit.

***L*/*C* ratio** Another measure of the Q of a series-resonant circuit. A high L/C ratio produces a high-Q series-resonant circuit.

Low-pass filter Filter having a single transmission band extending from zero to some cutoff frequency.

Practical tank circuit Two-branch RLC parallel circuit containing a practical (RL) inductor in one branch and a pure capacitor in the other.

Q_0 Measure of the "quality factor" of a given series- or parallel-resonant circuit. By definition, the ratio of the maximum energy stored times 2π divided by the energy dissipated per cycle at a given frequency. More practical definitions are $Q_0 = X_L/R$, the ratio of inductive reactance to resistance at the resonant frequency of an *RLC* series-resonant circuit, and $Q_0 = R_p/X_L$, the ratio of equivalent parallel resistance to coil reactance of an ideal three-branch parallel-resonant circuit. In both instances, the Q provides an indication of the sharpness of the resonant response curve and its selectivity, varying directly with both.

Resonance In a circuit containing R, L, and C elements, the frequency at which the terminal current and voltage are in phase with each other. In a series *RLC* circuit, at resonance, impedance is a minimum and current is a maximum. In a parallel-resonant circuit, impedance is a maximum and current is minimized. Resonance occurs at a specified resonant frequency for each circuit, depending on component values.

Resonant frequency, f_0 Frequency at which resonance occurs.

Resonant power factor Relation between circuit current and supply voltage at resonance. By definition, a series or a parallel circuit is resonant at unity power factor, when supply voltage and circuit current are in phase.

Resonant current rise In parallel-resonant circuits, the sharp increase in current in either the capacitor or the coil, proportional to the resonant Q_0 of the circuit and approximately $Q_0 I$, where I is the supply current.

Resonant voltage rise In series *RLC* circuits, the sharp increase in voltage across the capacitor or the inductor, proportional to the resonant Q_0 of the circuit and approximately $Q_0 V$, where V is the supply voltage.

Selectivity Measure of the Q of a given resonant series or parallel circuit. Highly selective resonant circuits have high Q values and narrow bandwidth (BW). Low selectivity is associated with low Q and wide BW. A *highly selective* filter has an *abrupt transition* between its passband and stopband regions.

19-12 PROBLEMS

Sec. 19-2 (all).

19-1 For the following values of R, L, and C, calculate the resonant radian frequency, ω_0, the resonant frequency, f_0, inductive reactance, jX_{L_0}, capacitive reactance, $-jX_{C_0}$, circuit impedance, Z_0, and Q_0 at the resonant frequency. Tabulate your results here for ready reference.

	R (Ω)	L (H)	C (μF)	ω_0 (rad/s)	f_0 (Hz)	jX_{L_0} (Ω)	$-jX_{C_0}$ (Ω)	Z_0 (Ω)	Q_0 (−)
a.	5	0.2	5						
b.	20	0.5	0.02						
c.	10	0.002	0.005						

19-2 Given the circuit parameters of Prob. 19-1a and a supply voltage of $10\angle 0^\circ$ V, calculate at the resonant frequency the

a. Total current, $\mathbf{I}_0$
b. Voltage across the resistance, $\mathbf{V}_{R_0}$
c. Voltage across the inductor, $\mathbf{V}_{L_0}$
d. Voltage across the capacitor, $\mathbf{V}_{C_0}$
e. Circuit Q_0 from voltage relations
f. Draw the phasor diagram showing the relation of all four phasor voltages using circuit current as a reference

19-3 Repeat all parts of Prob. 19-2 for the circuit of Prob. 19-1**b**.

19-4 Repeat all parts of Prob. 19-2 for the circuit of Prob. 19-1**c**.

19-5 A coil having an inductance of 100 μH and a resistance of 25 Ω is connected across a $5\angle 0^\circ$ V supply in series with a capacitor to produce resonance at a frequency of 200 kHz. Calculate the

a. Value of the capacitor
b. Resonant series current
c. Voltage across the inductor
d. Voltage across the capacitor
e. Q_0 using voltage relations
f. Q_0 using L/C ratio
g. Q_0 using $\omega_0 L/R$ relations

Secs. 19-3 and 19-4

19-6 Using the tabulation of Prob. 19-1 for each of the three given circuits, calculate and tabulate

a. Half-power bandwidth, B_{hp}, using the relation $B_{hp} = f_0/Q_0$
b. B_{hp}, using the relation $B_{hp} = R/2\pi L$ (see Eq. 19-10)
c. Lower edge frequency, f_1, and upper edge frequency, f_2
d. B_{hp}, using the relation $B_{hp} = f_2 - f_1$

19-7 Given $X_L = j200\ \Omega$, $R = 5\ \Omega$, and a 60-Hz supply voltage $\mathbf{V} = 12\angle 0^\circ$ V applied to an *RLC* series circuit, calculate the

a. Required value of X_C to produce series resonance
b. Required value of C to produce series resonance
c. Maximum current at resonance, $\mathbf{I}_0$
d. Voltage across the inductor at the resonant frequency, f_0, designated $\mathbf{V}_{L_0}$
e. Voltage across the capacitor at f_0, designated $\mathbf{V}_{C_0}$
f. Half-power bandwidth, B_{hp}
g. Maximum power dissipated in the circuit at resonance, P_0

h. Lower edge frequency, f_1, and upper edge-frequency, f_2
i. Power dissipated at frequencies f_1 and f_2
j. Voltage across resistor R at frequencies f_1 and f_2
k. Voltage across inductor at frequencies f_1 and f_2 (be careful)
l. Circuit Q_0 by three methods

19-8 Given the same calculated values of L and C as in Prob. 19-7 but R increased to 200 Ω, calculate the

a. Maximum current at resonance, $\mathbf{I}_0$
b. Circuit Q_0,
c. Voltage across inductor at f_0
d. Maximum possible voltage across inductor $\mathbf{V}_{L_{max}}$
e. Frequency at which $\mathbf{V}_{L_{max}}$ occurs
f. Voltage across capacitor at f_0, $\mathbf{V}_{C_0}$
g. Maximum possible voltage across capacitor, $\mathbf{V}_{C_{max}}$
h. Frequency at which $\mathbf{V}_{C_{max}}$ occurs
i. Half-power bandwidth of current response curve
j. Lower edge frequencies f_1 and f_2
k. Power dissipated at frequencies f_0, f_1, and f_2
l. Circuit current at frequency f_1 (at which $\mathbf{V}_{C_0}$ occurs)
m. Circuit impedance and current at frequency at which $\mathbf{V}_{C_{max}}$ occurs
n. Circuit power at frequency at which $\mathbf{V}_{C_{max}}$ occurs
o. Draw approximate curves of $\mathbf{V}_R$, $\mathbf{V}_C$, and $\mathbf{V}_L$ showing all calculated powers, frequencies, and voltages

19-9 Given a series RLC circuit connected to a $24\angle 0°$ V supply consisting of $R = 10$ Ω, $L = 40$ mH, and $C = 1$ μF, calculate the

a. Resonant frequency f_0
b. Circuit Q_0
c. B_{hp}, half-power BW of current response
d. Approximate maximum voltage across capacitor ($Q_0\mathbf{V}$)
e. Approximate frequency at which maximum voltage occurs
f. Exact frequency at which maximum voltage occurs across capacitor
g. Exact value of maximum voltage across capacitor
h. Relative error between **(e)** and **(f)**
i. Relative error between **(d)** and **(g)**
j. Draw conclusions regarding the need for correction equations in circuits having high Qs versus low Qs

19-10 Design an RLC resonant circuit to operate from a 120 V, 60 Hz supply having a Q_0 of 25 and a total circuit impedance at resonance of 10 Ω by finding the

a. Half-power bandwidth, B_{hp}
b. Edge frequencies f_1 and f_2
c. Required inductance L_0
d. Required capacitance, C_0
e. Maximum current at resonance, $\mathbf{I}_0$
f. Power dissipated at resonance, P_0
g. Power dissipated at frequencies f_1 and f_2
h. Relative power in dB

19-11 An RLC circuit is connected to a 50 kHz, 30 V supply. The circuit is resonant when $X_{C_0} = -j2$ kΩ. If the circuit resistance is 40 Ω, calculate

a. B_{hp}
b. Q_0
c. Half-power frequencies f_1 and f_2
d. Inductance L and capacitance C
e. Resonant rise in voltage $\mathbf{V}_{L_0}$ and $\mathbf{V}_{C_0}$
f. Power at resonance
g. If either the frequency or the reactive components are varied, do the maximum voltages across L and C change from those found in **(e)** above? Explain.

19-12 Design a series resonant circuit to operate from a $24\angle 0°$ V supply having a frequency of 60 kHz. The bandwidth must not exceed 3 kHz and the current drawn from the source must not exceed 100 mA. Specify for the circuit

a. R
b. L
c. C
d. Edge frequencies f_1 and f_2
e. Circuit Q_0

19-13 A circuit consists of a practical inductor connected in series with an ideal capacitor across a 10 V, 1 kHz supply. Measurements at this frequency reveal that $X_C = 2$ kΩ, $X_L = 500$ Ω, and $R_L = 50$ Ω. Calculate the

a. Frequency at which the circuit becomes resonant
b. jX_{L_0} and $-jX_{C_0}$ at the resonant frequency, assuming R_L is constant
c. Q_0 at resonance
d. Power dissipated by the coil at 1 kHz
e. Power dissipated by the coil at f_0
f. dB loss at 1 kHz compared to f_0

19-14 Using a 10-μH coil, design a circuit that is resonant across a 1 V, 5 MHz supply having a Q_0 of 20 by finding

a. B_{hp}
b. R
c. C
d. f_1 and f_2
e. $\mathbf{I}_0$
f. Current at f_1 and f_2
g. P_0
h. dB loss when circuit dissipates 1 mW at a lagging (inductive) PF
i. Current under the conditions in **(h)**
j. Impedance under the conditions in **(h)**
k. Net reactance under the conditions in **(h)**
l. Frequency under the conditions in **(h)**

19-15 A circuit containing a practical inductor (R_L and L), a series resistor, R, and a pure capacitor, C, are connected in series across a constant 120 V, 1 kHz supply. The impedance of the practical inductor is $60\angle 60°$ Ω and the voltage across the practical inductor is $90\angle 0°$ V. The entire circuit dissipates a total power of 100 W. Assuming the given frequency of 1 kHz is *below* the resonant frequency, f_0, calculate the

a. Circuit current $\mathbf{I}$
b. Total circuit resistance, R_t
c. Resistance of the practical inductor R_L
d. Reactance of the practical inductor, X_L
e. Series resistance, R_s
f. Inductance of the practical inductor L
g. Total circuit impedance, $\mathbf{Z}_t$

h. Net circuit reactance, X_0
i. Reactance of the capacitor, X_C
j. Capacitance C
k. Draw the complete series circuit in terms of R, L, and C

19-16 Using given and calculated data from Prob. 19-15, calculate the

a. Resonant frequency, f_0, by two separate methods
b. Resonant Q_0 of the *circuit*
c. Resonant Q rise in voltage across capacitor, $\mathbf{V}_{C_0}$
d. B_{hp} at f_0
e. Maximum possible voltage across inductance (only) if frequency is increased to f_L
f. Frequency, f_L, at which (e) occurs
g. Total circuit impedance at frequency f_L
h. Maximum possible volt drop across practical coil at f_L
i. Maximum possible voltage drop across pure capacitor if frequency is reduced below f_0 to f_C at which $\mathbf{V}_{C_{max}}$ occurs
j. Explain why the maximum volt drop across a practical coil is always higher than the maximum volt drop across a practical capacitor in a series-resonant circuit whose frequency is varied

Sec. 19-5

19-17 A three-branch ideal parallel *RLC* circuit has a resistance R of 20 kΩ in the first branch, L is 20 mH in the second branch, and C is 20 nF in the third branch, all connected across a $5\underline{/0°}$ V, variable frequency supply. Calculate the

a. Frequency at which resonance occurs, f_0
b. Inductive and capacitive reactance of each reactive branch
c. Current in the branches, $\mathbf{I}_{R_0}$, $\mathbf{I}_{L_0}$, and $\mathbf{I}_{C_0}$, respectively, at the resonant frequency f_0
d. Q_0 rise in current in each of the reactive branches
e. Total circuit current at resonance, $\mathbf{I}_0$
f. Total circuit impedance, $\mathbf{Z}_{p_0}$ (or R_{p_0})
g. Half-power bandwidth, B_{hp}
h. Edge frequencies f_1 and f_2
i. Current at edge frequencies f_1 and f_2, respectively, $\mathbf{I}_1$ and $\mathbf{I}_2$
j. Impedance at edge frequencies f_1 and f_2 (include phase angle), by two methods

19-18 A three-branch ideal parallel-resonant circuit has a resistance R of 800 Ω in the first branch, L is 400 mH in the second branch, and C is 10 μF in the third branch, all connected across a $16\underline{/0°}$ V variable frequency supply. Calculate all parts of Prob. 19-17.

Sec. 19-6

19-19 Design a high-Q practical tank (two-branch) circuit having an equivalent impedance (at minimum current) of 25 kΩ and a coil resistance of 25 Ω by calculating the

a. Q of the coil
b. Inductive reactance of the coil at resonance, jX_{L_0}
c. Capacitive reactance of the pure capacitor at resonance, $-jX_{C_0}$
d. Resonant frequency if the coil has a 10-mH inductance
e. Required capacitance, in nF
f. Total circuit current drawn from a $25\underline{/0°}$ V supply (as the sum of the branch currents)
g. Total current drawn from supply using the given equivalent impedance

19-20 Design a high-Q two-branch practical tank circuit to meet the following specifications: minimum current at resonance is 50 mA, bandwidth (B_{hp}) is 1 kHz, Q_0 is 25, and $\mathbf{V}_{C_{max}}$ is 12.5 V. Calculate and specify the

a. Resistance of the coil, R_L
b. $\mathbf{Z}_{p_0}$
c. Resonant frequency, f_0
d. Inductance of the coil, L
e. Shunting tank capacitance C

Hint: Find $\mathbf{I}_{C_0}$ using $\mathbf{I}_{C_0} = Q\mathbf{I}_0$ and then find X_{C_0}

19-21 Given the circuit shown in Fig. 19-20 with the following values: R_s is $1\underline{/0°}$ kΩ, R_L is 2 Ω, L is 150 mH, C is 15 μF, and $\mathbf{V}$ is a $3\underline{/0°}$ V variable frequency supply. Calculate the

a. Resonant frequency, f_0, of the high-Q circuit
b. jX_{L_0} and $-jX_{C_0}$ at f_0
c. Q_0 of the circuit
d. Equivalent tank impedance at resonance, $\mathbf{Z}_{p_0}$
e. Maximum possible output voltage across the tank at f_0, $\mathbf{V}_{0_{max}}$
f. Branch currents $\mathbf{I}_{L_0}$ and $\mathbf{I}_{C_0}$ at the resonant frequency
g. Total tank current drawn from supply at the resonant frequency, $\mathbf{I}_0$, using (f)
h. Total tank current using equivalent tank impedance in series with R_s

19-22 The two-branch, practical, parallel tank circuit shown in the inset of Fig. 19-15 has a resonant frequency, f_r, of 1 kHz, a very low Q_r of 1.2, and a coil resistance, R_s, of 100 Ω. Calculate the

a. Equivalent tank impedance at the frequency f_r (where $|jX_{L_r}| = |-jX_{C_r}|$). Hint: use Eq. (19-30a) to find $\mathbf{Z}_r$.
b. Frequency at which equivalent tank impedance is a maximum, f_m
c. Values of jX_{L_m} and $-jX_{C_m}$ at the frequency calculated in (b)
d. Maximum equivalent tank impedance, $\mathbf{Z}_m$, at frequency f_m. Hint: use Eq. (19-30b).
e. Frequency f_0 at which unity PF occurs
f. Equivalent tank impedance, $\mathbf{Z}_0$, at frequency f_0. Hint: use the product-over-sum method after calculating jX_{L_0} and $-jX_{C_0}$ at frequency f_0.
g. Equivalent tank impedance, $\mathbf{Z}_0$, at frequency f_0 using Eq. (19-30c)
h. Equivalent unity PF tank impedance using the approximate method of Ex. 19-13C. Hint: use $\mathbf{Z}_{p_0} = (Q_0^2 + 1)R_s$.
i. Explain why the approximate method *cannot* be used as a solution to this problem.
j. Half-power bandwidth, B_{hp}, and cutoff frequencies, f_1 and f_2
k. Equivalent tank impedances, $\mathbf{Z}_1$ and $\mathbf{Z}_2$, at frequencies f_1 and f_2, respectively
l. Draw an impedance versus frequency graph similar to Fig. 19-15 showing all five impedances calculated above and their respective frequencies, namely $\mathbf{Z}_1$, $\mathbf{Z}_2$, $\mathbf{Z}_r$, $\mathbf{Z}_m$, and $\mathbf{Z}_0$

19-23 Given the circuit of **Fig. 19-20**, $R_s = 20$ kΩ. The circuit must be resonant at 300 kHz but $\mathbf{V}_0$ cannot be greater than 60% of the input voltage, **V**. If the Q_0 of the coil is 20, design the tank circuit to meet these specifications by finding

a. $\mathbf{Z}_{po}$, the tank impedance at the resonant frequency given
b. jX_{Lo}
c. R_L, resistance of the coil
d. L, inductance
e. C, capacitance of the tank

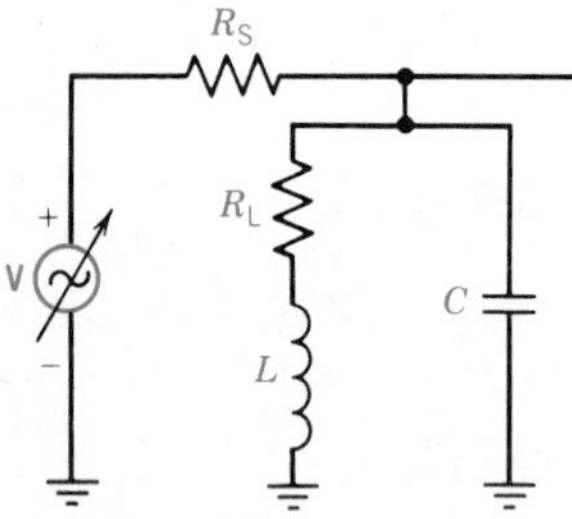

Figure 19-20 Problems 19-21 to 19-24; Problems 19-31 to 19-34

19-24 From the given and computed data of Prob. 19-23, given supply voltage $\mathbf{V} = 5\angle 0°$ V at a frequency of 300 kHz, calculate the

a. Minimum current drawn from supply $\mathbf{I}_0$
b. Half-power BW, B_{hp}
c. Capacitor current, $\mathbf{I}_{Co}$, at f_0 by two independent methods
d. Output voltage at edge frequencies f_1 and f_2
e. Frequencies f_1 and f_2

19-25 In the circuit of **Fig. 19-21**, the current from the ideal current source is $2\angle 0°$ mA. R_L is 250 Ω, L is 1 mH, and C is 1 nF. Calculate the

a. Frequency at which tank impedance and $\mathbf{V}_0$ is a maximum
b. $+jX_{Lo}$ at the above frequency
c. Resonant Q_0 of the circuit
d. $-jX_{Co}$ at resonance
e. $\mathbf{Z}_{po}$, maximum tank impedance
f. Maximum output voltage across the tank at resonance
g. Branch currents $\mathbf{I}_{Lo}$ and $\mathbf{I}_{Co}$ at resonance
h. Verify the source current as the sum of $\mathbf{I}_{Lo}$ and $\mathbf{I}_{Co}$

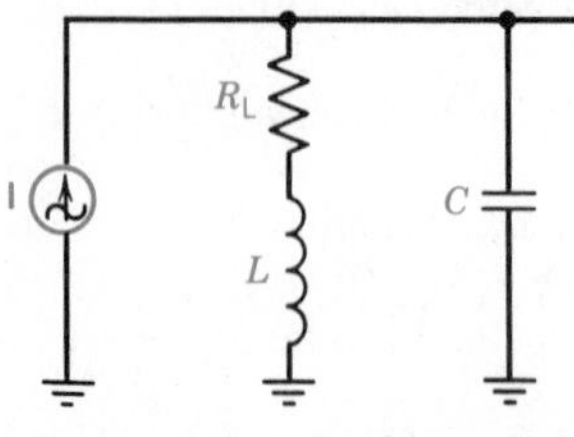

Figure 19-21 Problems 19-25 and 19-26

19-26 For the unloaded low-Q tank circuit of Prob. 19-25, using the given and calculated data, find the

a. Half-power BW, B_{hp}
b. Edge frequencies f_1 and f_2
c. Tank circuit impedance at the corner frequencies
d. Output voltage at the corner frequencies, f_1 and f_2
e. Ratio of the output voltage in (**d**) to the maximum output voltage of the tank at resonance (see **f** in previous problem)

Sec. 19-7

19-27 **Figure 19-22** shows a practical current source connected to a practical two-branch tank circuit. Source values are $R_C = 800$ Ω and $\mathbf{I} = 12\angle 0°$ mA. Tank values are $R_L = 10$ Ω and $jX_L = j50$ Ω. Calculate the

a. Q_0 of the tank
b. $-jX_C$ at the resonant frequency
c. Equivalent impedance of the tank at resonance, $\mathbf{Z}_{po}$
d. Equivalent total impedance, $\mathbf{Z}'_0$ "seen by" the 2-mA "ideal" source
e. Voltage across the equivalent impedance, $\mathbf{V}_0$
f. Branch current $\mathbf{I}_{Lo}$ at resonance
g. Branch current $\mathbf{I}_{Co}$ at resonance
h. Tank current, $\mathbf{I}_0$, as the sum of (**f**) and (**g**)
i. Tank current using the tank impedance in (**c**)
j. Q'_{fo} of the entire circuit as a result of tank loading produced by R_C
k. Half-power BW of the tank circuit prior to loading by R_C if the resonant frequency is 50 kHz
l. Half-power BW of the entire circuit

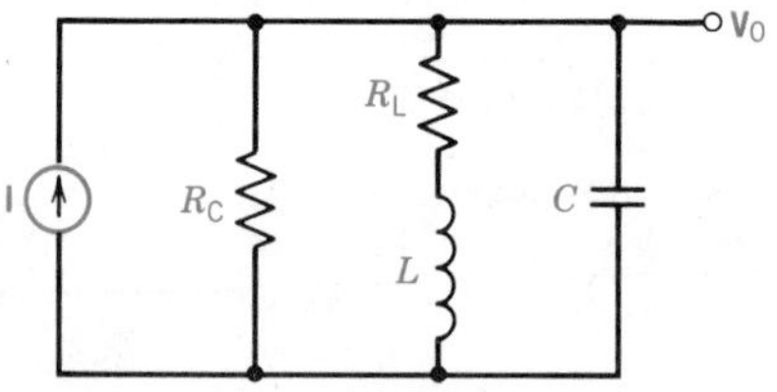

Figure 19-22 Problems 19-27 to 19-29

19-28 In the circuit shown in Fig. 19-22, L is 8 mH, C is 20 pF, R_L is 5 kΩ, R_C is 95.625 kΩ, and **I** is $1\angle 0°$ mA. Calculate the

a. Resonant frequency, f_0
b. jX_{Lo} at f_0
c. Q_0 of the tank
d. $-jX_{Co}$ at f_0
e. Equivalent tank impedance, $\mathbf{Z}_{po}$, at f_0
f. Equivalent total impedance, $\mathbf{Z}'_0$, of the entire circuit
g. Voltage across all branches, $\mathbf{V}_0$
h. Branch current $\mathbf{I}_{Co}$ and $\mathbf{I}_{Lo}$ at resonance
i. Q'_{fo} of the entire circuit
j. Half-power BW of the entire circuit

19-29 In the circuit shown in Fig. 19-22, C is 2.815 nF, L is 10 mH, R_L is 25 Ω, R_C is 32 kΩ, and **I** is $5\angle 0°$ mA. Calculate all parts of Prob. 19-28.

19-30 Repeat all parts of Ex. 19-16 with only the following values changed: R_L is 10 Ω and Q_0 is 20, in Fig. 19-19a.

19-31 Given the circuit shown in Fig. 19-20, R_S is 4.5 kΩ, R_L is 1 kΩ, L is 10 mH, C is 253 pF, and **V** is $10\angle 0°$ V. Calculate

a. f_0
b. Q_0
c. $-jX_{Co}$
d. $\mathbf{Z}_{po}$
e. $\mathbf{V}_0$
f. Half-power BW, B_{hp}, of the unloaded tank circuit

19-32 An external load, $R_X = 39.5$ kΩ, is connected across the tank circuit of Prob. 19-31. Calculate the

a. Equivalent loaded tank impedance at resonance, $\mathbf{Z}'_{po}$
b. Final loaded Q'_{fo} by two methods
c. Output voltage $\mathbf{V}'_0$
d. Half-power BW, B'_{hp}, as a result of load R_X
e. Discuss the effect of external load R_X on each of the parameters calculated above

19-33 In the circuit shown in Fig. 19-20, R_S is 10 kΩ, R_L is 125 Ω, L is 10 mH, C is 2.815 nF, and **V** is $10\angle 0°$ V. Repeat all parts of Prob. 19-31.

19-34 If a load of 20 kΩ is connected across the tank of Prob. 19-33, repeat all parts of Prob. 19-32.

19-35 **Figure 19-23** shows a transistorized (constant) current source supplying $5\angle 0°$ mA to the loaded tank circuit at its resonant frequency, f_0. The collector circuit resistance, R_C, of the transistor is 20 kΩ and the external load resistance, R_X, is 5 kΩ. Tank circuit values are $R_L = 50$ Ω, $L = 2$ mH, and $C = 31.65$ nF. Calculate the

a. Resonant frequency of the tank circuit, f_0
b. Resonant Q_0 of the unloaded tank circuit
c. Equivalent impedance of the unloaded tank circuit, $\mathbf{Z}_{po}$
d. Half-power bandwidth of the unloaded tank circuit, B_{hp}
e. Equivalent impedance of the entire circuit seen by the transistor (include R_C), $\mathbf{Z}'_{po}$
f. Resonant Q'_{fo} of the entire circuit
g. Half-power BW of the loaded tank circuit, B'_{hp}
h. Edge frequencies f_1 and f_2 of the loaded tank circuit
i. Output voltage of the loaded circuit at resonance, $\mathbf{V}_0$
j. Output voltage at edge frequencies f_1 and f_2

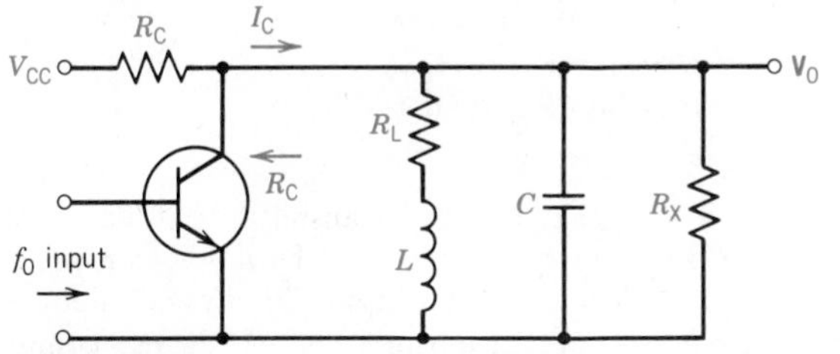

Figure 19-23 Problem 19-35

Sec. 19-8

19-36 Design an RL high-pass filter that will have a cutoff frequency of 4 kHz when connected to a $10\angle 0°$ V variable frequency supply. Calculate the

a. Inductor L, of negligible resistance, given series $R = 3$ kΩ
b. Output voltage, $\mathbf{V}_0$, at frequencies of 0, 1 kHz, 2 kHz, 4 kHz, 8 kHz, and 16 kHz, respectively. Specify both magnitude and phase of the output voltage.

(Hint: Use reactance at f_c to find other reactances quickly at the given frequencies.)

19-37 A high-pass RC filter (Table 19-1c) has a resistance of 2 kΩ and a capacitance of 9.95 nF connected to a $10\angle 0°$ V variable frequency supply. Calculate the

a. Cutoff frequency, f_c
b. Output voltages (both magnitude and phase) at frequencies of 0, 1, 2, 4, 8, 16, and 32 kHz
c. Draw the amplitude-phase response curves for the above filter

19-38 a. Repeat all parts of Prob. 19-37 if a 6-kΩ resistive load is connected across the output of the high-pass RC filter
b. On the same sheet as Prob. 19-37 draw the loaded amplitude-phase response
c. Draw conclusions as to the effect of loading a high-pass RC filter

19-39 Design a simple low-pass RC filter having a 4-kHz cutoff frequency using a $10\angle 0°$ V supply and a 3-kΩ resistor. Specify the

a. Value of C required
b. Output voltage and relative dB output when f is 1 kHz
c. Output voltage and dB output when f is 10 kHz

19-40 A simple RL low-pass filter (Table 19-1b) uses a 5.6-kΩ resistor and 222.8 mH inductor connected to a $10\angle 0°$ V variable frequency supply. Calculate the

a. Cutoff frequency, f_c
b. Output voltages (both magnitude and phase) at frequencies of 0, 1, 2, 4, 8, 16, and 32 kHz
c. Draw the amplitude-phase response curves for the above filter

19-41 a. Repeat all parts of Prob. 19-40 if a 5.6-kΩ resistive load is connected across the output of the low-pass filter
b. Draw the loaded amplitude-phase response on the same sheet as Prob. 19-39
c. Draw conclusions as to the effect of loading a low-pass RL filter

Sec. 19-9

19-42 A series-resonant bandstop filter (Table 19-2a) has the following values: $\mathbf{V}_{in} = 10\angle 0°$ V, $C = 63.3$ pF, $L = 40$ mH, $R_s = 2000$ Ω, and $R_L = 10$ Ω. Calculate the

a. Resonant frequency, f_0
b. Half-power BW, B_{hp}
c. Edge frequencies f_1 and f_2
d. Output voltages at 20 kHz, f_1, f_0, f_2, and 200 kHz (both magnitude and phase)
e. Relative voltage loss in dB at the frequencies given in **(d)** above

19-43 A parallel-resonant bandstop filter (Table 19-2b) has the values $\mathbf{V}_{in} = 10\angle 0°$ V, $R_L = 50$ Ω, $L = 1$ mH, $C = 1$ nF, and $R_0 = 1$ kΩ. Calculate

a. f_0
b. Q
c. B_h
d. Edge frequencies f_1 and f_2
e. $\mathbf{Z}_{po}$
f. $\mathbf{V}_0$ at f_0, f_1, and f_2
g. $\mathbf{V}_0$ at 50 kHz and 500 kHz
h. Relative voltage in dB at all frequencies in **(f)** and **(g)** above

19-44 Given a series-resonant bandpass filter (Table 19-2c) with component values $R_L = 10$ Ω, $L = 200$ mH, $C = 495$ pF, $R_0 = 2$ kΩ, and $\mathbf{V}_{in} = 10\angle 0°$ V. Calculate

a. f_0
b. Q_0
c. B_h
d. f_1 and f_2
e. $\mathbf{V}_0$ at f_0
f. $\mathbf{V}_0$ at f_1
g. $\mathbf{V}_0$ at f_2
h. $\mathbf{V}_0$ at frequencies of 1 kHz and 100 kHz

19-45 Given the parallel-resonant bandpass filter (Table 19-2d) with component values $L = 10$ mH, $C = 25.33$ nF, $R_L = 20\ \Omega$, $R_s = 2\ \text{k}\Omega$, and $\mathbf{V}_{in} = 10\angle 0°$ V. Calculate

a. f_0
b. Q_0
c. R_{po}
d. B_{hp}
e. f_1 and f_2
f. $\mathbf{V}_0$ at f_0
g. $\mathbf{V}_0$ at f_1
h. $\mathbf{V}_0$ at $f = 1$ kHz

19-46 For the *LC* filter shown in **Fig. 19-24**, the component values are $L_1 = 132.6$ mH, $L_2 = 157.9$ mH, $C = 0.133\ \mu$F, and $R_L = 1\ \text{k}\Omega$. The filter is designed to pass one band of frequencies and reject another. Using simple series–parallel ac circuit theory, calculate the

a. Ratio $\mathbf{V}_0/\mathbf{V}_{in}$ at a frequency of $f_1 = 900$ Hz
b. Ratio $\mathbf{V}_0/\mathbf{V}_{in}$ at a frequency of $f_2 = 1200$ Hz
c. dB loss for the ratio found in (**a**)
d. dB loss for the ratio found in (**b**)
e. Frequency f_0 at which the output voltage is zero
f. Based on the above calculations, what class of filter is represented by Fig. 19-24?

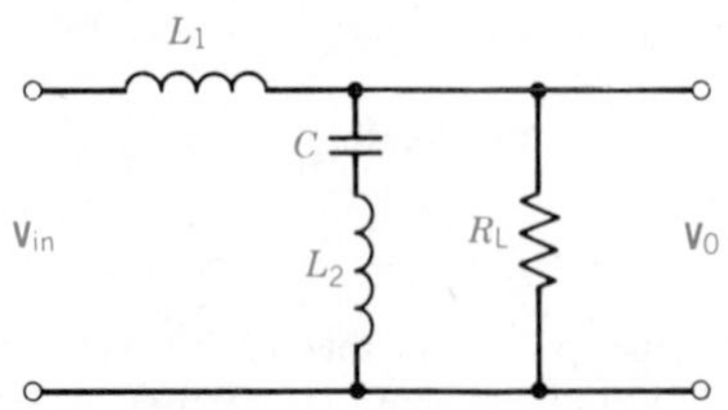

Figure 19-24 Problem 19-46

Sec. 19-10

19-47 In the double-resonant circuit shown in **Fig. 19-25**, the current, $\mathbf{I}_L$, in load R_L is to be a maximum at 1 kHz and a minimum at 900 Hz. If *L* is 50 mH, calculate

a. The value of *C*
b. L_1 or C_1 in the box in the parallel branch

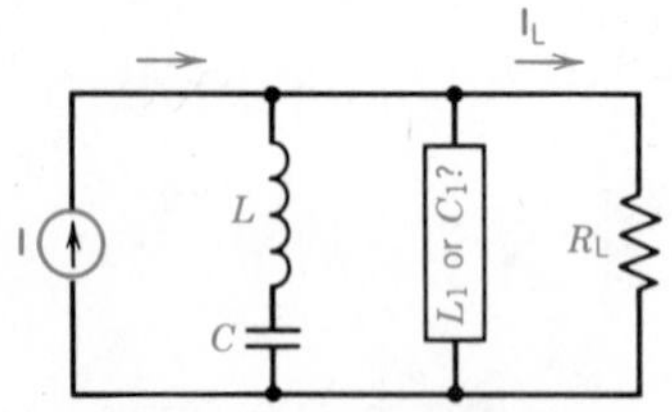

Figure 19-25 Problems 19-47 and 19-48

19-48 Given $C = 500$ pF in Fig. 19-25, $\mathbf{I}_L$ is to be a maximum at 960 kHz and a minimum at 920 kHz. Calculate

a. *L*
b. L_1 or C_1 in the box in the parallel branch

19-49 In the circuit shown in **Fig. 19-26**, R_L is 10 kΩ and *C* is 10 nF. Design a double-resonant circuit whose stopband is 100 kHz and whose passband frequency is 150 kHz by specifying

a. L_1
b. L_2

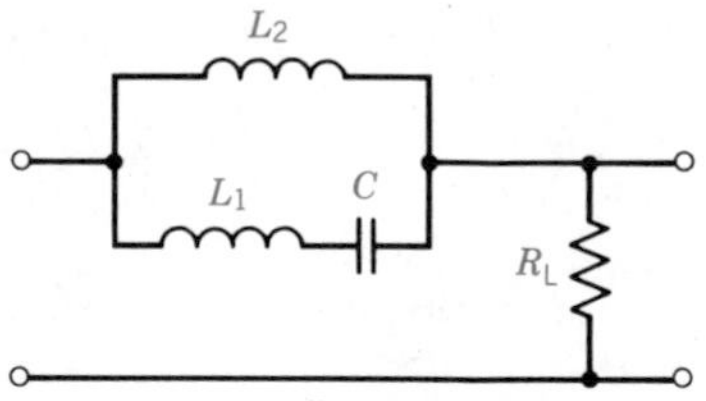

Figure 19-26 Problems 19-49 and 19-50

19-50 Repeat both parts of Prob. 19-49 for a filter whose stopband is 5 kHz and whose passband is 10 kHz, all other values remaining the same.

19-51 Given $C = 5$ nF, design a double-resonant filter whose stopband is 40 kHz and whose passband is 90 kHz. (Hint: see Table 19-3**b**.)

19-52 Given $L = 5\ \mu$H, design a double-resonant filter whose stopband is 200 kHz and whose passband is 100 kHz. (Hint: see **Fig. 19-27** and Table 19-3**c**.)

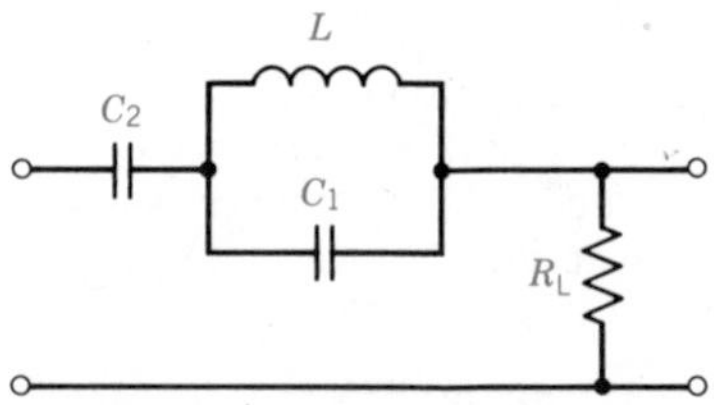

Figure 19-27 Problems 19-52 and 19-53

19-53 Design a filter that passes a transmitter output whose frequency is 500 kHz but rejects the third harmonic content of the fundamental frequency of the output. Specify all values of the filter, which must contain a 1 mH coil. (Hint: use configuration of Table 19-3**c**.)

19-54 Repeat Prob. 19-53 using an alternative design.

19-55 a. Using *RC* filters exclusively, draw the composite filter circuits to implement the block diagrams shown for the four filter circuits in **Fig. 19-28**
b. Identify each filter drawn in (**a**)
c. For each filter draw the amplitude response (dB versus frequency) showing the rate of rise or rolloff in dB per decade and all critical frequency points

19-56 For the three filter circuits shown in **Fig. 19-29**:

a. Identify each type of filter
b. Draw a typical amplitude response (dB versus frequency)

19-57 Repeat both parts of Prob. 19-56 for the three filter circuits shown in **Fig. 19-30**.

19-58 Repeat both parts of Prob. 19-56 for the three filter circuits shown in **Fig. 19-31**.

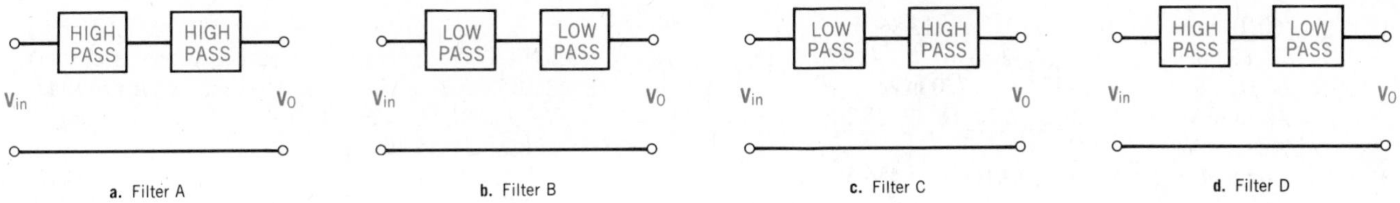

Figure 19-28 Problem 19-55

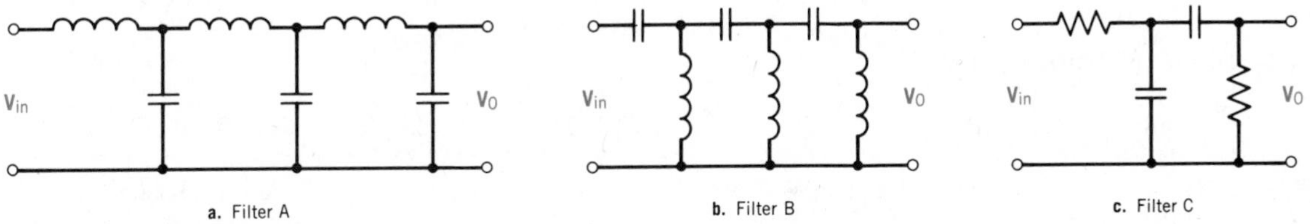

Figure 19-29 Problem 19-56

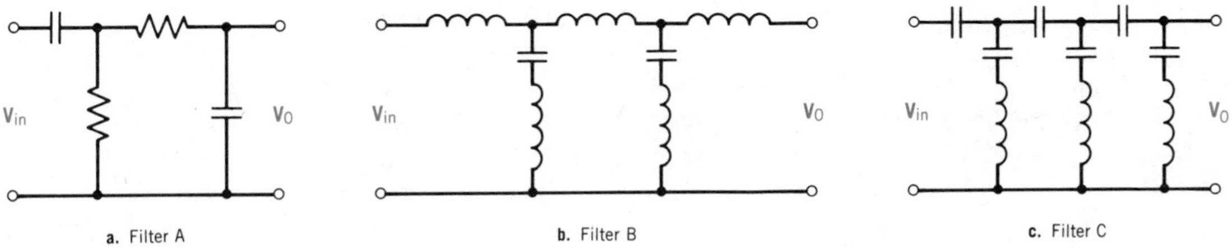

Figure 19-30 Problem 19-57

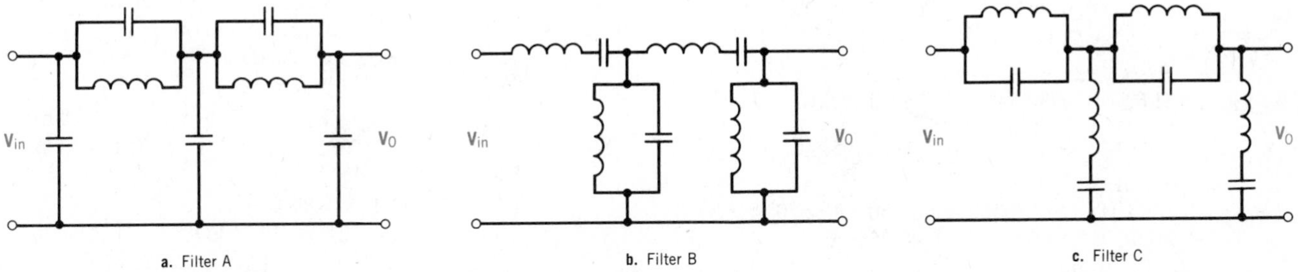

Figure 19-31 Problem 19-58

19-13 ANSWERS

19-1 Q_0 is **a** 40 **b** 250 **c** 63.245
19-2 **a** $2\angle 0^\circ$ A **b** $10\angle 0^\circ$ V **c** $j400$ V **d** $-j400$ V **e** 40
19-3 **a** $0.5\angle 0^\circ$ A **b** $10\angle 0^\circ$ V **c** $j2500$ V **d** $-j2500$ V **e** 250
19-4 **a** $1\angle 0^\circ$ A **b** $10\angle 0^\circ$ V **c** $j632.4$ V **d** $-j632.4$ V **e** 63.24
19-5 **a** 6.33 nF **b** $0.2\angle 0^\circ$ A **c** $j25.13$ V **d** $-j25.13$ V **e, f, g** 5.03
19-6 B_{hp} is **a** 3.98 **b** 6.38 **c** 796
19-7 **a** $-j200\ \Omega$ **b** 13.26 μF **c** $2.4\angle 0^\circ$ A **d** $j480$ V **e** $-j480$ V **f** 1.5 Hz **g** 28.8 W **i** 14.4 W **j** $8.485\angle \pm 45^\circ$ V **k** $335.2\angle 135^\circ$ V; $343.6\angle 45^\circ$ V **l** 40

19-8 a $60\underline{/0^\circ}$ mA b 1 c 12 V d 13.86 V
e 84.85 Hz f 12 V g 13.86 V h 42.43 Hz
i 60 Hz j 30 Hz, 90 Hz k 720 mW
l 42.43 mA m $244.9\underline{/-35.26^\circ}$ Ω, $49\underline{/35.26^\circ}$ mA
n 480.2 mW

19-9 a 795.8 Hz b 20 c 39.8 Hz d 480 V
e 795.8 Hz f 795.3 Hz g 480.15 V
h −0.0629% i 0.0312%

19-10 a 2.4 Hz b 58.8 Hz; 61.2 Hz c 663.1 mH
d 10.61 μF e $12\underline{/0^\circ}$ A f 1.44 kW g 720 W
h −3 dB

19-11 a 1 kHz b 50 d 6.37 mH e 1500 V
f 22.5 W

19-12 a 240 Ω b 12.73 mH c 552.6 pF
d 58.5 kHz, 61.5 kHz e 20

19-13 a 2 kHz b $j1000$; $-j1000$ Ω c 20 d 2.22 mW
e 2 W f −29.5 dB

19-14 a 250 kHz b 15.71 Ω c 101.3 pF
e $63.7\underline{/0^\circ}$ mA f 45 mA g 63.7 mW h −18 dB
i 7.98 mA j 125.3 Ω k $j124.35$ Ω l 1.979 MHz

19-15 a $1.5\underline{/-60^\circ}$ A b $44.\bar{4}$ Ω c 30 Ω d $j51.96$ Ω
e $14.\bar{4}$ Ω f 8.27 mH g 80 Ω h $-j66.52$ Ω
i $-j118.5$ Ω j 1.34 μF

19-16 a 1511 Hz b 1.767 c 212 V d 855.1 Hz
e 221 V f 1649 Hz g $46.54\underline{/17.25^\circ}$ Ω
h 234.1 V i 221 V

19-17 a 7958 Hz b $j1000$ Ω c 0.25, $-j5$, $j5$ mA
d 20 e $0.25\underline{/0^\circ}$ mA f 20 kΩ g 397.9 Hz
h 7759, 8157 Hz i $0.3536\underline{/\pm 45^\circ}$ mA
j $14.14\underline{/\pm 45^\circ}$ kΩ

19-18 a 79.58 Hz b $\pm j200$ Ω c $20\underline{/0^\circ}$ mA, $\pm j80$ mA
d 4 e $20\underline{/0^\circ}$ mA f $800\underline{/0^\circ}$ Ω g 19.9 Hz
h 69.63, 89.53 Hz i $28.28\underline{/\pm 45^\circ}$ mA
j $565.7\underline{/\pm 45^\circ}$ Ω

19-19 a 31.62 b $j790.6$ Ω c $-j790.6$ Ω d 12.582 kHz
e 16 nF f, g $1\underline{/0^\circ}$ mA

19-20 a 0.4 Ω b 250 Ω c 25 kHz d 63.7 μH
e 637 nF

19-21 a 106.1 Hz b $\pm j100$ Ω c 50 d $5\underline{/0^\circ}$ kΩ
e $2.5\underline{/0^\circ}$ V f $0.5 - j25$; $0 + j25$ mA
g, h $0.5\underline{/0^\circ}$ mA

19-22 a $187.45\underline{/-39.8^\circ}$ Ω b 909.06 Hz
c $j109.09$, $-j132$ Ω d $190.41\underline{/-29.6^\circ}$ Ω
e 552.8 Hz, f, g $144\underline{/0^\circ}$ Ω h $244\underline{/0^\circ}$ Ω
j 833 Hz, 492.4 Hz, 1325.8 Hz k $134.6\underline{/\pm 45^\circ}$ Ω

19-23 a $30\underline{/0^\circ}$ kΩ b $j1500$ Ω c 75 Ω d 795.8 μH
e 353.7 pF

19-24 a $100\underline{/0^\circ}$ μA b 15 kHz c $j2$ mA
d $2.12\underline{/\pm 45^\circ}$ V e 292.5; 307.5 kHz

19-25 a 159.155 kHz b $j1$ kΩ c 4 d $-j1062.5$ Ω
e 4250 Ω f $8.5\underline{/0^\circ}$ V g $2 - j8$; $0 + j8$ mA
h $2\underline{/0^\circ}$ mA

19-26 a 39.79 kHz b 139.3; 179.05 kHz
c $3005\underline{/\pm 45^\circ}$ Ω d $6.01\underline{/\pm 45^\circ}$ V e 0.7071

19-27 a 5 b $-j52$ Ω c 260 Ω d 196.2 Ω
e $2.354\underline{/0^\circ}$ V f $9.054 - j45.27$ mA g $j45.27$ mA
h, i 9.054 mA j 3.77 k 10 kHz l 13.26 kHz

19-28 a 397.9 kHz b $j20$ kΩ c 4 d $-j21.25$ kΩ
e 85 kΩ f 45 kΩ g $45\underline{/0^\circ}$ V
h $j2.118$; $0.529 - j2.118$ mA i 2.12 j 187.9 kHz

19-29 a 30 kHz b $j1885$ Ω c 15.08 d $-j1885$ Ω
e 28.55 kΩ f 15.09 kΩ g 75.44 V
h $j40$; $2.64 - j40$ mA i 7.97 j 3.76 kHz

19-30 a $j200$ Ω b $-j200$ Ω c 31.83 kHz d 1 mH
e 4 kΩ f 800 Ω g $3.2\underline{/0^\circ}$ V
h $0.8 - j16$ mA; $+j16$ mA i 4 j 7.96 kHz
k $4\underline{/0^\circ}$ mA

19-31 a 100 kHz b 6.28 c $-j6442$ Ω d 40.5 kΩ
e $9\underline{/0^\circ}$ V f 159.2 kHz

19-32 a 20 kΩ b 3.1 c $8.16\underline{/0^\circ}$ V d 322.6 kHz

19-33 a 30 kHz b 15.08 c $-j1893$ Ω d 28.55 kΩ
e $7.41\underline{/0^\circ}$ V f 2 kHz

19-34 a 11.76 kΩ b 6.212 c 3.703 V d 4.83 kHz

19-35 a 20 kHz b 5.03 c 1315 Ω d 3.98 kHz
e 990 Ω f 3.79 g 5.28 kHz
h 17.36; 22.64 kHz i 4.95 V j $3.5\underline{/\pm 45^\circ}$ V

19-36 a 95.5 mH b at 16 kHz, $\mathbf{V}_0$ is $9.7\underline{/14^\circ}$ V

19-37 a 8 kHz b at 16 kHz, $\mathbf{V}_0$ is $8.94\underline{/26.6^\circ}$ V

19-38 a 10.664 kHz b at 16 kHz, $\mathbf{V}_0$ is $8.32\underline{/33.7^\circ}$ V

19-39 a $41.\bar{6}$ nF b $9.7\underline{/-14^\circ}$ V, −0.265 dB
c $3.71\underline{/-68.2^\circ}$ V, −8.61 dB

19-40 a 4 kHz b at 16 kHz, $\mathbf{V}_0$ is $2.43\underline{/-76^\circ}$ V

19-41 a 2 kHz b at 16 kHz, $\mathbf{V}_0$ is $1.24\underline{/-82.9^\circ}$ V

19-42 a 100 kHz b 8 kHz c 96; 104 kHz
d at both 20 kHz and 200 kHz, $\mathbf{V}_0 = 9.986\underline{/\pm 3^\circ}$ V

19-43 a 159.16 kHz b 20 c 7.96 kHz
d 155.2; 163.1 kHz e 20 kΩ
f $0.476\underline{/0^\circ}$ V; $0.673\underline{/\pm 42.3^\circ}$ V
g $8.95\underline{/-18.1^\circ}$ V; $9.43\underline{/19.5^\circ}$ V h at f_0, −26.4 dB

19-44 a 16 kHz b 10 c 1.6 kHz d 15.2; 16.8 kHz
e 9.95 V f $7.1\underline{/45^\circ}$ V g $7.1\underline{/-45^\circ}$ V
h $62\underline{/89.6^\circ}$ and $163\underline{/-89.1^\circ}$ mV

19-45 a 10 kHz b 31.42 c 19.76 kΩ d 318 Hz
e 9.841; 10.16 kHz f 9.08 V g $9.02\underline{/5.11^\circ}$ V
h $0.33\underline{/70.4^\circ}$ V

19-46 a 0.967 b 0.1597 c −0.29 dB d −15.9 dB
e 1098.3 Hz

19-47 a 625 nF b 2.66 μF

19-48 a 59.9 μH b 5.64 nF

19-49 a 112.6 μH b 140.7 μH

19-50 a 25.33 mH b 76.0 mH

19-51 a 625.4 μH b 2.54 mH

19-52 a 126.7 nF b 380 nF

19-53 101.3 pF; 12.66 pF

19-54 11.26 pF; 90 pF

CHAPTER 20

Transformers

20-1 INTRODUCTION

A transformer is a device for transferring energy from one circuit to another. In almost all kinds of transformers, this energy transfer is accomplished inductively.[1] Basic to the operation of the transformer is the principle of mutual induction. Chapter 12 covered the principle of mutual induction, the dot convention, and mutual inductance between *conductively coupled* series-connected and parallel-connected coils (as illustrations of both Faraday's law of electromagnetic induction and Lenz's law) whenever the current undergoes change, even for a short time interval. Recall that inductance is the property of a circuit or a component to oppose a *change* in current. The treatment of Chapter 12 essentially considered the subject of electromagnetic induction from a dc transient point of view. The transformer is essentially an alternating-current (ac) device, where the *current is constantly changing*. Consequently, mutual induction is occurring constantly to transfer energy *inductively* from one circuit to another.

Because of its ability to step up and step down ac voltages and currents very efficiently, the transformer (more than any other device) is responsible for long-distance transmission of electrical energy throughout the civilized world. Although this is its most extensive application, we will discover that the transformer can also be used as a *circuit-isolation* device (to separate one circuit from another) and as an *impedance-matching* device to ensure maximum power transfer from one circuit to another with relatively little loss of energy.

The discussion begins with the *ideal* transformer, introducing various types of tightly coupled *iron-core* transformers and their applications. The open-circuit and short-circuit tests are introduced to assess transformer efficiency and voltage regulation. Transformer frequency response is considered briefly along with autotransformers.

Loosely coupled transformers are contrasted to tightly coupled transformers, including differences as well as similarities. The various methods of analyzing loosely coupled transformers and circuits are contrasted: mesh analysis, reverse

[1] The only possible exception is the autotransformer, which transfers energy *both inductively and conductively*. See Sec. 20-8.

transfer impedance, Thévenin equivalent, and the reflected impedance method. The chapter concludes with mesh analysis of loosely coupled *multicoil* transformers, using a simplified method of equation writing that facilitates analysis.

20-2 THE IDEAL TRANSFORMER

The theory of transformer operation and its applications are best understood by viewing the transformer first as an ideal device. This simplification enables us to define transformer terms and understand transformer operation. Let us first define the **ideal transformer** as a device that has the following seven properties:[2]

1. Its coefficient of coupling, k, is *unity* (see Eq. 12-12a).
2. Its primary and secondary windings are pure inductors of infinitely large value.
3. Its self and mutual impedances are zero, containing no reactance or resistance.
4. Its leakage flux and leakage inductance are zero.
5. Its power transfer efficiency is 100%; that is, there are no losses due to resistance, hysteresis, or eddy currents.
6. Its transformation turns ratio, α, is equal to the ratio of its primary to secondary terminal voltages and also the ratio of its secondary to primary current.
7. Its core permeability (μ) is infinite.

An *ideal* iron-core transformer is shown in **Fig. 20-1a**, along with certain appropriate electrical symbols, defined below. It consists of two coils *wound in the same direction* on a common magnetic core. The winding connected to the supply, $\mathbf{V}_1$, is called the *primary*. The winding connected to the load, $\mathbf{Z}_L$, is called the *secondary*. Since the ideal transformer has zero primary and secondary impedance, the voltage induced in the primary, $\mathbf{E}_1$, is equal to $\mathbf{V}_1$, the applied voltage. For the same reason, the secondary voltage, $\mathbf{V}_2$, is equal to the secondary induced voltage, $\mathbf{E}_2$. The current $\mathbf{I}_1$ drawn from the supply, $\mathbf{V}_1$, is just sufficient to produce the mutual flux, ϕ_m, and the required magnetomotive force (MMF)—$\mathbf{I}_1 N_1$—to overcome the demagnetizing effect of the secondary MMF—$\mathbf{I}_2 N_2$—as a result of connected load.[3]

Since both windings are wound in the *same* direction, the positive values of instantaneous induced EMF, $\mathbf{E}_1$ and $\mathbf{E}_2$, respectively, are designated by a dot, in

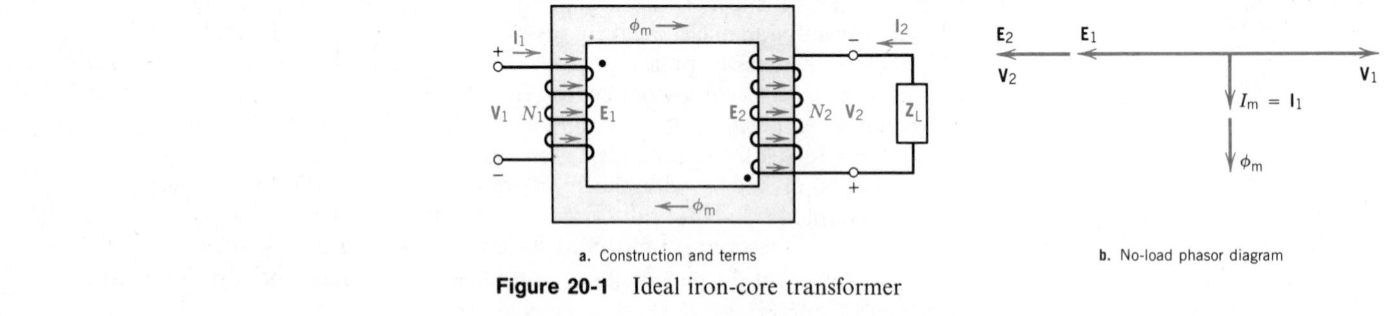

a. Construction and terms

b. No-load phasor diagram

Figure 20-1 Ideal iron-core transformer

[2] It should be noted here that the *practical* (commercial) transformer has *none* of these properties despite the fact that its operation is close to ideal.

[3] Theoretically, since $\mathbf{E}_1 = \mathbf{V}_1$, the current $\mathbf{I}_1$ should be zero. But with $\mathbf{Z}_L$ connected across the secondary, current $\mathbf{I}_2$ is drawn, requiring a corresponding current $\mathbf{I}_1$. Current $\mathbf{I}_1$, therefore, both produces ϕ_m and serves to supply $\mathbf{I}_2$.

accordance with the dot convention (Sec. 12-7). By Lenz's law, induced EMF $\mathbf{E}_1$ opposes $\mathbf{V}_1$. And since $\mathbf{E}_2$ and $\mathbf{E}_1$ are both produced by the same mutual flux, ϕ_m, $\mathbf{E}_2$ is in the same direction as $\mathbf{E}_1$ and also in opposition to $\mathbf{V}_1$. These relations are summarized in the (no-load) phasor diagram shown in Fig. 20-lb. Since $\mathbf{V}_2 = \mathbf{E}_2$, observe that $\mathbf{V}_2$ is displaced 180° from $\mathbf{V}_1$. This accounts for the reversal of polarity shown in Fig. 20-1a between $\mathbf{V}_1$ and $\mathbf{V}_2$.

20-2.1 Ideal Transformer Relations

By definition, the transformer ratio, α, is the ratio of primary turns to secondary turns (i.e., $\alpha = N_1/N_2$). But for the ideal transformer, as noted previously, we may write

$$\alpha = \frac{N_1}{N_2} = \frac{\mathbf{E}_1}{\mathbf{E}_2} = \frac{\mathbf{V}_1}{\mathbf{V}_2} = \frac{\mathbf{I}_2}{\mathbf{I}_1} \quad \text{(a unitless ratio, fractional or integral)} \quad (20\text{-}1)$$

where all terms have been defined for the ideal transformer.

Cross-multiplying terms in Eq. (20-1) yields several interesting equalities:

$$\mathbf{I}_1 N_1 = \mathbf{I}_2 N_2 \quad \text{(ampere-turns)} \quad (20\text{-}1a)$$

$$\mathbf{E}_1\mathbf{I}_1 = \mathbf{E}_2\mathbf{I}_2 = S_2 = S_1 \quad \text{(volt-amperes)} \quad (20\text{-}1b)$$

$$\mathbf{V}_1\mathbf{I}_1 = \mathbf{V}_2\mathbf{I}_2 = S_2 = S_1 \quad \text{(volt-amperes)} \quad (20\text{-}1c)$$

Equation (20-1a) states that the demagnetizing ampere-turns of the secondary are equal and opposite to the magnetizing MMF of the primary of an ideal transformer. The reader may also verify this by using the right-hand finger rule (Fig. 11-4) applied to both the primary turns (producing ϕ_m) and the secondary turns (demagnetizing ϕ_m).

Equation (20-1b) represents the (apparent) power transfer without loss from primary to secondary of the transformer as a result of electromagnetic induction in an ideal transformer.

Equation (20-1c) states that the (apparent) power drawn from the primary supply is equal to the (apparent) power transferred to the secondary load, without any loss whatever, in an *ideal* transformer.

20-2.2 Phasor Diagram—Ideal Transformer, at No Load

Figure 20-1b shows the no-load phasor diagram of the ideal transformer. Examination of this diagram reveals the following important relations:

1. $\mathbf{E}_1$ is equal and opposite in magnitude to $\mathbf{V}_1$.
2. $\mathbf{E}_2$ is in phase with $\mathbf{E}_1$ but opposite (by 180°) to $\mathbf{V}_2$.
3. The magnetizing current, $\mathbf{I}_m$, lags $\mathbf{V}_1$ by 90° and produces ϕ_m, in phase with $\mathbf{I}_m$.
4. $\mathbf{E}_1$ and $\mathbf{E}_2$ lag ϕ_m by 90° and are produced by ϕ_m.
5. $\mathbf{V}_2$ is equal in magnitude to $\mathbf{E}_2$ and opposed to $\mathbf{V}_1$ by 180°.

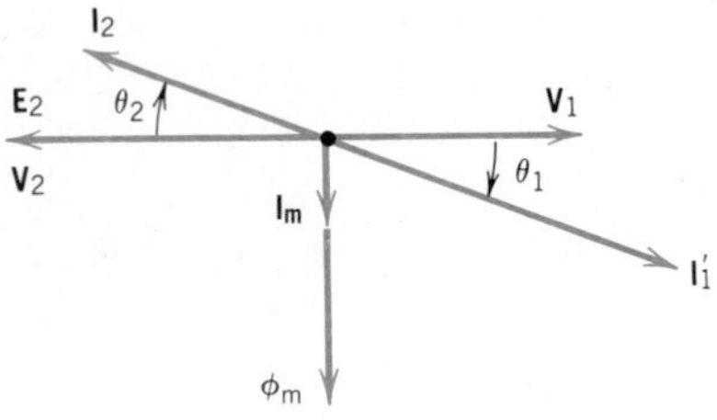

a. Addition of load component of primary current, I'_1

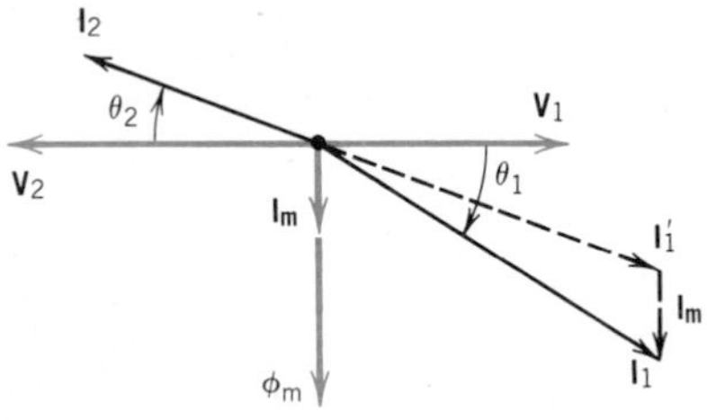

b. Terminal voltage-current-power factor relations of loaded transformer

Figure 20-2 Phasor diagrams, loaded ideal transformer

20-2.3 Phasor Diagrams—Ideal Transformer, Effect of Load Current, $\mathbf{I}_2$

Figure 20-2a shows the phasor diagram that results when load $\mathbf{Z}_L$ is connected across output terminal $\mathbf{V}_2$ in Fig. 20-1a. A current $\mathbf{I}_2$ flows in accordance with Ohm's law, where $\mathbf{I}_2 = \mathbf{V}_2/\mathbf{Z}_L$. Assuming $\mathbf{Z}_L$ is some load containing both resistance and reactance, then current $\mathbf{I}_2$ lags $\mathbf{V}_2$ (and $\mathbf{E}_2$) by some angle θ_2, as shown in Fig. 20-2a. But in accordance with Eq. (20-1a), the demagnetizing effect of $\mathbf{I}_2 N_2$ must be counterbalanced by an equal and opposite magnetizing component of primary current, $\mathbf{I}'_1$, producing $\mathbf{I}'_1 N_1$, as shown in Fig. 20-2a.[4]

Figure 20-2a is most important because it shows the following for the ideal transformer:

1. A *lagging* load applied to the transformer *secondary* produces a *lagging* load on the transformer *primary*.
2. The power product $\mathbf{V}_2\mathbf{I}_2$ must be exactly equal to the product $\mathbf{V}_1\mathbf{I}'_1$.
3. If $\mathbf{V}_2 > \mathbf{V}_1$ (the case of a step-up transformer), then of necessity $\mathbf{I}'_1 > \mathbf{I}_2$, as shown in the scale of phasor diagram 20-2a.
4. Therefore, whenever we step up voltage with a transformer, we also step down current in the secondary, simultaneously.

Figure 20-2b shows the resultant terminal relations whenever the load component of primary current, $\mathbf{I}'_1$, is added to the small component of magnetizing current, $\mathbf{I}_m$. For a lagging load, the primary power factor angle, θ_1, is slightly greater than the secondary power factor angle, θ_2. Similarly, the primary current, $\mathbf{I}_1$, drawn from the supply is slightly greater than the primary load current component, $\mathbf{I}'_1$. But since the magnetizing current of a commercial transformer is relatively small (less than 5% of the full-load current), for the ideal transformer we may assume that θ_1 is equal to θ_2 for either leading or lagging loads.

EXAMPLE 20-1

The high-voltage side of a transformer has 500 turns and the low-voltage side has 100 turns. When connected as a step-down transformer, the load current ($\mathbf{I}_2$) is 12 A. Calculate the

a. Transformation ratio, α
b. Load component of primary current, $\mathbf{I}_1$
c. Transformation ratio if the transformer is used as a step-up transformer

Solution

a. $\alpha = N_1/N_2 = 500\text{ t}/100\text{ t} = \mathbf{5}$ (20-1)
b. $\mathbf{I}_1 = \mathbf{I}_2/\alpha = 12\text{ A}/5 = \mathbf{2.4\ A}$ (20-1)
c. $\alpha = N_1/N_2 = 100\text{ t}/500\text{ t} = \mathbf{0.2}$ (20-1)

Example 20-1 leads to the following insights:

1. Either the high-voltage or the low-voltage side of a transformer may be used as the primary.
2. When the HV side is used as the primary, the transformation ratio, α, is an integral number and the transformer is called a *step-down* transformer.
3. When the LV side is used as the primary, the transformation ratio, α, is a fractional number, as implied in Eq. (20-1), and the transformer is called a *step-up* transformer.

EXAMPLE 20-2

A 2300/115 V, 60 Hz, 4.6 kVA step-down transformer is designed to have an induced EMF of 2.5 V/turn. Assuming an ideal transformer, calculate the

a. Number of high-side and low-side turns
b. Rated primary and secondary currents
c. Step-down and step-up transformation ratios, using **(a)**
d. Step-down and step-up transformation ratios, using **(b)**

[4] The action in a transformer may be likened to an automatic feedback control system. When load $\mathbf{Z}_L$ is applied and current $\mathbf{I}_2$ flows, $\mathbf{I}_2 N_2$ tends to reduce ϕ_m by demagnetization. But as ϕ_m is reduced, it also reduces $\mathbf{E}_1$ in accordance with Faraday's law of electromagnetic (EM) induction (Eq. 12-3). The reduction in $\mathbf{E}_1$ opposing $\mathbf{V}_1$ causes *an increase* in $\mathbf{I}'_1$ just sufficient to restore ϕ_m to its original no load value. Thus, an increase in $\mathbf{I}_2$ must be met by a corresponding increase in $\mathbf{I}'_1$ to maintain ϕ_m constant. The transformer may be viewed as an automatic feedback control system whose purpose is to maintain a constant mutual flux ϕ_m regardless of load variations.

Solution

a. $N_h = V_h/2.5\ V/t = 2300\ V/2.5\ V/t = \mathbf{920\ t} = N_1$ and
$N_l = V_l/2.5\ V/t = 115\ V/2.5\ V/t = \mathbf{46\ t} = N_2$

b. $I_h = I_1 = S_1/V_1 = 4600\ VA/2300\ V = \mathbf{2\ A}$ (20-1c)
$I_l = I_2 = S_2/V_2 = 4600\ VA/115\ V = \mathbf{40\ A}$ (20-1c)

c. $\alpha = N_1/N_2 = 920\ t/46\ t = \mathbf{20}$ (20-1)
$\alpha = N_l/N_h = 46\ t/920\ t = \mathbf{0.05}$ (20-1)

d. $\alpha = I_2/I_1 = 40\ A/2\ A = \mathbf{20}$ (20-1)
$\alpha = I_h/I_l = 2\ A/40\ A = \mathbf{0.05}$ (20-1)

Example 20-2 leads to the following insights:

1. The ratio volts/turn (Faraday's law, Eq. 12-3) is the same for the primary and the secondary, since the same rate of change of the same mutual flux links both primary and secondary.
2. Assuming an ideal transformer, the transformation ratio of any commercial transformer may be obtained from the rated voltages, the rated currents, and/or the turns of the low and high sides of the transformer.
3. The transformation ratio, α, may be either an integral number (HV side as primary) or a fractional number (LV side as primary).

20-2.4 The Ideal Two-Winding Transformer as an Impedance-Matching Device

From Eq. (20-1) we know that the transformation ratios are $\frac{V_1}{V_2} = \frac{N_1}{N_2} = \alpha$, while correspondingly $\frac{I_1}{I_2} = \frac{N_2}{N_1} = 1/\alpha$

If we divide the left-hand voltage ratios by the right-hand current ratios, we obtain the respective impedance ratios, or

$$\frac{V_1/V_2}{I_1/I_2} = \frac{\alpha}{1/\alpha} = \frac{V_1/I_1}{V_2/I_2} = \frac{Z_1}{Z_2} = \alpha^2 \qquad \text{(20-2)}$$

Equation (20-2) establishes the transformer as an impedance-matching device. It states that the impedance on the secondary of a transformer, Z_2, is reflected back to the primary as a primary impedance, which is $Z_2(\alpha^2)$ or

$$Z_1 = \alpha^2 Z_L \quad \text{ohms } (\Omega) \qquad \text{(20-2a)}$$

where Z_L is the matching-transformer secondary load
α is the turns ratio, N_1/N_2

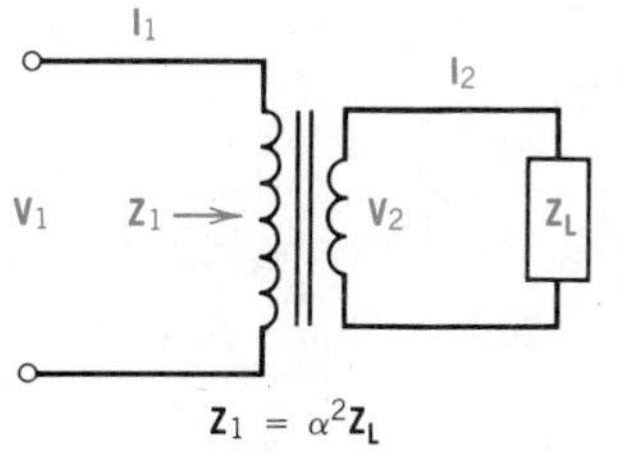

Figure 20-3 Impedance-matching transformer

Figure 20-3 shows an impedance-matching transformer whose load, Z_L, is reflected to the primary as some higher primary impedance, $\alpha^2 Z_L$, using a step-down impedance-matching transformer. Equation (20-2a) enables use of the transformer to maximize the power transfer from the primary to the secondary circuit, as will be shown in Sec. 20-3. But first, some examples to illustrate impedance matching.

EXAMPLE 20-3

An audio output transformer (connected between an audio amplifier and its speaker) has 500 primary turns and 25 secondary turns. If the speaker impedance is 8 Ω, calculate the

a. Impedance reflected to the transformer primary at the output of the audio amplifier
b. Matching transformer primary current if the output of the audio amplifier is 10 V

Solution

a. $\alpha = N_1/N_2 = 500\ t/25\ t = \mathbf{20}$ and
$Z_1 = \alpha^2 Z_L = (20)^2 \times 8\ \Omega = \mathbf{3200\ \Omega}$

b. $I_1 = V_1/Z_1 = 10\ V/3.2\ k\Omega = \mathbf{3.125\ mA}$

20-2.5 The Three-Winding Transformer as an Impedance-Matching Device

Figure 20-4 shows an impedance-matching transformer having two separate secondary windings, N_2 and N_3, each connected to two separate loads, $\mathbf{Z}_2$ and $\mathbf{Z}_3$, respectively. What is the total impedance, $\mathbf{Z}_1$, reflected to the primary by the two separate secondary windings combined? Let us treat each impedance separately.

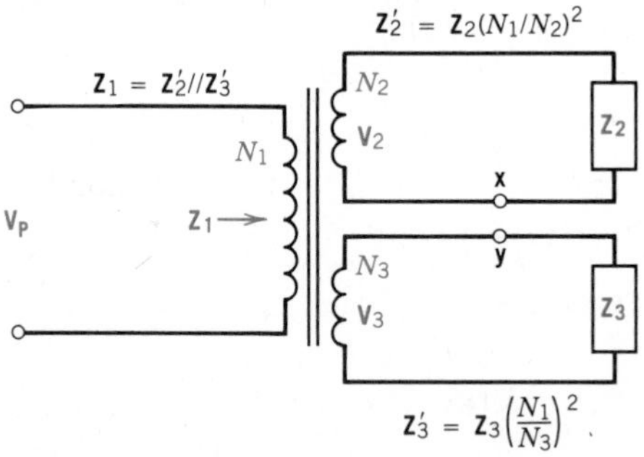

Figure 20-4 Impedance-matching transformer with two separate secondary loads

The impedance reflected by load $\mathbf{Z}_2$ via the ratio N_1/N_2 is $\mathbf{Z}_2' = (N_1/N_2)^2\mathbf{Z}_2$. Similarly, the impedance reflected by load $\mathbf{Z}_3$ via the ratio N_1/N_3 is $\mathbf{Z}_3' = (N_1/N_3)^2\mathbf{Z}_3$. The two reflected impedances, $\mathbf{Z}_2'$ and $\mathbf{Z}_3'$, may be considered as two unequal *parallel* impedances reflected as the primary impedance, where $\mathbf{Z}_1 = \mathbf{Z}_2' \| \mathbf{Z}_3'$ or

$$\mathbf{Z}_1 = \frac{(\mathbf{Z}_2')(\mathbf{Z}_3')}{\mathbf{Z}_2' + \mathbf{Z}_3'} \quad \text{ohms } (\Omega) \qquad \textbf{(20-3)}$$

where $\mathbf{Z}_2' = (N_1/N_2)^2\mathbf{Z}_2$ and $\mathbf{Z}_3' = (N_1/N_3)^2\mathbf{Z}_3$ and voltage $\mathbf{V}_2 = \mathbf{V}_p(N_2/N_1)$ and $\mathbf{V}_3 = \mathbf{V}_p(N_3/N_1)$, respectively.

EXAMPLE 20-4

Given the transformer shown in Fig. 20-4, N_1 is 600 turns, N_2 is 150 turns, and N_3 is 300 turns; $\mathbf{Z}_2$ is a 30-Ω resistive load and $\mathbf{Z}_3$ is a 15-Ω resistive load. The primary voltage applied to the matching transformer is 16 V. Calculate the

a. Impedance $\mathbf{Z}_2'$ reflected to the primary by load $\mathbf{Z}_2$
b. Impedance $\mathbf{Z}_3'$ reflected to the primary by load $\mathbf{Z}_3$
c. Total impedance $\mathbf{Z}_1$ reflected to the primary
d. Total current, $\mathbf{I}_1$, drawn from the supply
e. Total power drawn from the supply at unity PF
f. Voltage $\mathbf{V}_2$ across load $\mathbf{Z}_2$ and power dissipated in load $\mathbf{Z}_2$
g. Voltage $\mathbf{V}_3$ across load $\mathbf{Z}_3$ and power dissipated in load $\mathbf{Z}_3$
h. Total power dissipated in both loads

Solution

a. $\mathbf{Z}_2' = \mathbf{Z}_2(N_1/N_2)^2 = 30\ \Omega(600/150)^2 = \mathbf{480\ \Omega}$ (20-2a)
b. $\mathbf{Z}_3' = \mathbf{Z}_3(N_1/N_3)^2 = 15\ \Omega(600/300)^2 = \mathbf{60\ \Omega}$ (20-2a)
c. $\mathbf{Z}_1 = \mathbf{Z}_2' \| \mathbf{Z}_3' = 480 \| 60 = \mathbf{53.\bar{3}\ \Omega}$ (20-3)
d. $\mathbf{I}_1 = \mathbf{V}_p/\mathbf{Z}_1 = 16\ \text{V}/53.\bar{3}\ \Omega = \mathbf{0.3\ A}$
e. $P_t = \mathbf{V}_p\mathbf{I}_1 \cos\theta = 16\ \text{V} \times 0.3 \times 1 = \mathbf{4.8\ W}$
f. $\mathbf{V}_2 = \mathbf{V}_p(N_2/N_1) = 16\ \text{V}(150/600) = \mathbf{4\ V}$; $P_2 = \mathbf{V}^2/R = (4)^2/30 = \mathbf{0.5\bar{3}\ W}$
g. $\mathbf{V}_3 = \mathbf{V}_p(N_3/N_1) = 16\ \text{V}(300/600) = \mathbf{8\ V}$; $P_3 = \mathbf{V}^2/R = (8)^2/15 = \mathbf{4.2\bar{6}\ W}$
h. $P_t = P_2 + P_3 = 0.5\bar{3} + 4.2\bar{6} = \mathbf{4.8\ W}$ (see **e** above)

Example 20-4 verifies the foregoing equations and also reveals the following important insights:

1. Transformers with separate secondary loads *each* reflect their respective impedances in proportion to the square of their respective turns ratios to the primary.
2. The equivalent primary impedance of the individual reflected impedances is obtained by treating *all* reflected impedances as *impedances in parallel.*
3. Respective secondary voltages across the multiple loads are in direct proportion to their respective turns ratios, as in the case of a simple two-winding transformer.
4. Regardless of the number of loads, the total load power dissipated *must equal the power drawn from the primary*, as-

suming an ideal impedance-matching transformer. This is an important verification of the accuracy of the calculations, as shown in solution steps (**e**) and (**h**).

5. If the transformer of Fig. 20-4 is a tapped transformer (i.e., points **x** and **y** are tied together), the solution is exactly the same as in Ex. 20-4, with loads $\mathbf{Z}_2$ and $\mathbf{Z}_3$ connected as shown in Fig. 20-5.

20-2.6 Tapped Matching Transformers

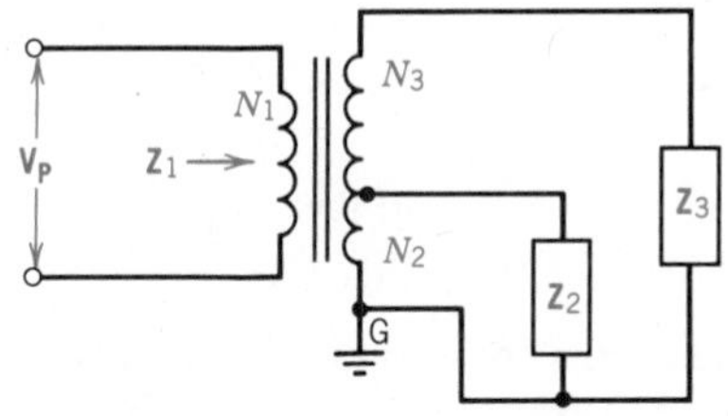

Figure 20-5 Tapped secondary impedance-matching transformer

Frequently, tapped matching transformers are used with their loads connected as shown in **Fig. 20-5**. In this case, load $\mathbf{Z}_3$ "sees" its impedance reflected to the primary as $\mathbf{Z}'_3 = [N_1/(N_2 + N_3)]^2\mathbf{Z}_3$, since $\mathbf{Z}_3$ is connected across the *entire* secondary. The solution of a problem involving Fig. 20-5 is left as an end-of-chapter exercise for the reader, with multiple loads reflected to the primary (Prob. 20-9).

A relatively common type of tapped impedance-matching transformer, used to match the output of a transistorized audio amplifier to a 4-Ω *or* 8-Ω speaker load, is shown in **Fig. 20-6**. The single 4-Ω speaker load *or* 8-Ω load should present the *same* primary impedance, $\mathbf{Z}_p$, to the audio amplifier, in order to match the output impedance of the transistorized audio amplifier, for maximum power transfer (MPT). Example 20-5 reveals some surprising and interesting insights into the configuration of Fig. 20-6.

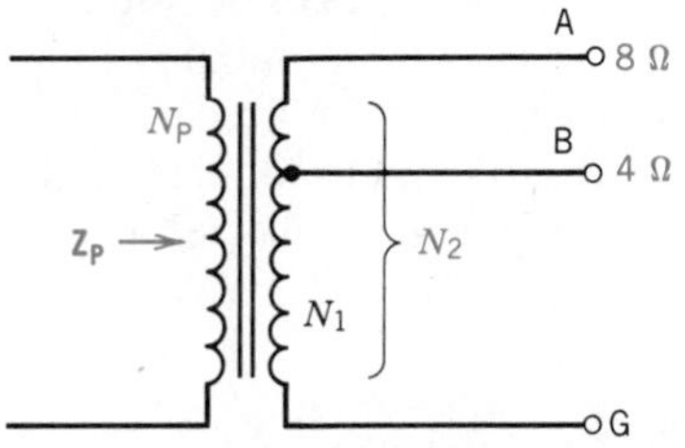

Figure 20-6 Impedance-matching transformer with two taps for a single secondary load (Ex. 20-5)

EXAMPLE 20-5

The output impedance of a (single-channel 100 W) transistorized power amplifier is 3.2 kΩ. A tapped impedance-matching transformer (Fig. 20-6) having 1500 primary turns is used to match the amplifier output to *either* an 8-Ω *or* a 4-Ω speaker. Calculate the

a. *Total* number of secondary turns, N_2, to match an 8-Ω impedance speaker
b. Number of turns, N_1, to match a 4-Ω impedance speaker
c. Impedance that must be connected *between* the 4-Ω and 8-Ω terminals to reflect a primary impedance of 3.2 kΩ

Solution

a. $\alpha = \sqrt{\mathbf{Z}_p/\mathbf{Z}_L} = \sqrt{3200/8} = \mathbf{20}$ (20-2a)
$N_2 = N_p/\alpha = 1500\ \text{t}/20 = \mathbf{75\ t}$ (20-1)

b. $\alpha = \sqrt{\mathbf{Z}_p/\mathbf{Z}_L} = \sqrt{3200/4} = 28.284$
$N_1 = N_p/\alpha = 1500\ \text{t}/28.284 = \mathbf{53\ t}$

c. $N_2 - N_1 = 75\ \text{t} - 53\ \text{t} = 22\ \text{t}$
$\mathbf{Z}_L = \mathbf{Z}_p/\alpha^2 = 3200\ \Omega/(1500/22)^2 = \mathbf{0.69\ \Omega}$

Example 20-5 shows several surprising conclusions:

1. The 4 Ω tap is *not* placed at the center of the secondary winding, N_2, but at some point approximately 0.7 of the total winding with respect to terminal G. (Fig. 20-6).
2. As shown by part (**c**) of the solution, improperly connecting a 4 Ω or 8 Ω speaker between the 8 Ω and 4 Ω taps produces a severe "mismatch". The proper load impedance *between* these connections (for correct impedance match) is 0.69 Ω. Such a mismatch reduces the power (and volume) to the speaker severely.

Example 20-6 shows how Ex. 20-5**c** may be solved without even knowing the primary impedance, $\mathbf{Z}_p$, or the primary turns, N_p, in Fig. 20-6.

EXAMPLE 20-6

John Smith, a bright student and an experimenter, locates a single-channel 100 W transistorized power amplifier whose output terminals, via a matching transformer, are as shown in Fig. 20-6, i.e., 8 Ω, 4 Ω, and G. But he desires to use the amplifier to drive a small 10 W servomotor whose impedance is approximately 0.7 Ω. Show how he calculated the impedance between terminals A and B to determine whether the amplifier can be used to drive the motor.

Solution

$\mathbf{Z}_p = 4\ \Omega(N_p/N_1)^2 = 8\ \Omega(N_p/N_2)^2 = \mathbf{Z}_{AB}(N_p/(N_2 - N_1)^2$

Dividing each of the three numerators by N_p^2 and taking the square root of each term:

$$\frac{\sqrt{\mathbf{Z}_{AB}}}{N_2 - N_1} = \frac{\sqrt{4}}{N_1} = \frac{\sqrt{8}}{N_2}, \text{ yielding } \frac{\sqrt{\mathbf{Z}_{AB}}}{N_2 - N_1} = \frac{\sqrt{8}}{N_2} - \frac{\sqrt{4}}{N_1},$$

yielding $\sqrt{\mathbf{Z}_{AB}} = \sqrt{8} - \sqrt{4} = 0.82843$, from which $\mathbf{Z}_{AB} = (0.82843)^2 = \mathbf{0.69\ \Omega}$

Note:

1. The impedance match is approximately correct so that the amplifier may be used.
2. The answer found algebraically is the same as that in the solution of Ex. 20-5, part **c**.

20-3 MAXIMUM POWER TRANSFER VIA MATCHING TRANSFORMER BETWEEN SOURCE AND LOAD

In the illustrative examples of Sec. 20-2 we learned how to calculate the reflected impedance by using various types of matching transformers (multiple winding, tapped, etc.), as well as their required turns ratio, α. In this section we consider the selection of a matching transformer to maximize the power transfer from a practical source to a load, as well as the related power calculations and power transfer efficiency.

Figure 20-7 shows a *practical* ac source, **V**, whose internal impedance, $\mathbf{Z}_s$, is essentially resistive, R_s. The terminal voltage of the source is $\mathbf{V}_1$. Since the transformer is ideal, the primary induced voltage $\mathbf{E}_1 = \mathbf{V}_1$. The load impedance $\mathbf{Z}_L$ is essentially resistive, R_L, and is coupled to the source via an ideal matching transformer. The load impedance reflected to the primary is $\mathbf{Z}_p$, where $\mathbf{Z}_p = \alpha^2\mathbf{Z}_L = \alpha^2 R_L$.

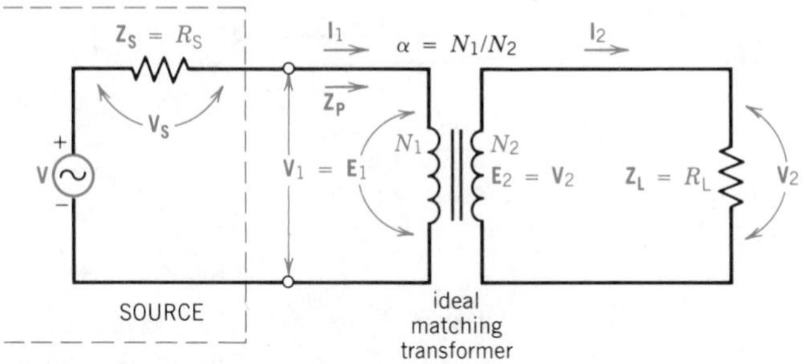

Figure 20-7 Using matching transformer for maximum power transfer from source **V** to load R_L

For MPT in ac circuits (see Sec. 17-6), the following equalities occur:

1. The reflected impedance to the primary must be the conjugate of the source impedance that is, $\mathbf{Z}_p = \mathbf{Z}_s^*$. For the resistive case shown in Fig. 20-7,

$$R_p = R_s = \alpha^2 R_L \quad \textbf{(20-4)}$$

and

$$\alpha = \sqrt{\frac{\mathbf{Z}_p}{\mathbf{Z}_L}} = \sqrt{\frac{R_s}{R_L}} \quad \textbf{(20-4a)}$$

2. Since $\mathbf{Z}_p = \mathbf{Z}_s^*$ and $R_p = R_s$, the terminal voltage $\mathbf{V}_1$ is equal in magnitude to the internal volt drop across the source impedance or half the source voltage **V**:

$$\mathbf{V}_1 = \mathbf{V}_s = \frac{\mathbf{V}}{2} \tag{20-5}$$

3. Then the secondary terminal voltage, $\mathbf{V}_2$, using the transformation ratio of Eq. (20-4a), is

$$\mathbf{V}_2 = \frac{\mathbf{V}_1}{\alpha} = \frac{\mathbf{V}}{2\alpha} \tag{20-6}$$

4. And the secondary current, $\mathbf{I}_2$, and primary current, $\mathbf{I}_1$, are, respectively,

$$\mathbf{I}_2 = \frac{\mathbf{V}_2}{\mathbf{Z}_L} = \frac{\mathbf{V}_1}{\alpha \mathbf{Z}_L} = \frac{\mathbf{V}}{2\alpha \mathbf{Z}_L} = \frac{\mathbf{V}}{2\alpha R_L} \tag{20-7}$$

$$\mathbf{I}_1 = \frac{\mathbf{I}_2}{\alpha} = \frac{\mathbf{V}}{(\mathbf{Z}_s + \mathbf{Z}_p)} = \frac{\mathbf{V}}{2\mathbf{Z}_p} = \frac{\mathbf{V}}{2R_s} \tag{20-8}$$

5. Then the power transferred to the load is

$$P_L = \mathbf{I}_2^2 R_L \tag{20-9}$$

6. The power supplied by the source or the total system power, P_T, is

$$P_T = \mathbf{V}\mathbf{I}_1 \cos\theta_1 = P_L + P_s = \mathbf{I}_2^2 R_L + \mathbf{I}_1^2 R_s \tag{20-10}$$

7. And the power transfer efficiency, η, is

$$\eta = \frac{P_L}{P_T} = \mathbf{50\%} \quad \text{(for the resistive case only)} \tag{20-11}$$

EXAMPLE 20-7

For the circuit shown in Fig. 20-7, the supply voltage of the source is $10\angle 0^\circ$ V, the resistance of the source is 1 kΩ, and the load resistance, R_L is 10 Ω. Calculate the

a. Required transformation ratio of the matching transformer for MPT
b. Terminal voltage of the source at MPT
c. Terminal voltage across the load at MPT
d. Secondary load current $\mathbf{I}_2$ by at least two independent methods
e. Primary load current drawn from the source, $\mathbf{I}_1$, by at least two methods
f. Maximum power dissipated in the load
g. Power dissipated internally within the source
h. Total power supplied by the source by two methods
i. Power transfer efficiency

Solution

a. $\alpha = \sqrt{R_s/R_L} = \sqrt{1000/10} = \mathbf{10}$ (20-4a)
b. $\mathbf{V}_1 = \mathbf{V}/2 = 10\angle 0^\circ\ \text{V}/2 = \mathbf{5\ V}$ (20-5)
c. $\mathbf{V}_2 = \mathbf{V}_1/\alpha = 5\ \text{V}/10 = \mathbf{0.5\ V}$ (20-6)
d. $\mathbf{I}_2 = \mathbf{V}_2/\mathbf{Z}_L = 0.5\ \text{V}/10\ \Omega = \mathbf{50\ mA}$ (20-7)
$\mathbf{I}_2 = \mathbf{V}/2\alpha R_L = 10\ \text{V}/(2 \times 10 \times 10\ \Omega) = \mathbf{50\ mA}$ (20-7)
e. $\mathbf{I}_1 = \mathbf{I}_2/\alpha = 50\ \text{mA}/10 = \mathbf{5\ mA}$ (20-8)
$\mathbf{I}_1 = \mathbf{V}/2R_s = 10\ \text{V}/2 \times 1\ \text{k}\Omega = \mathbf{5\ mA}$ (20-8)
f. $P_L = \mathbf{I}_2^2 R_L = (50\ \text{mA})^2 10\ \Omega = \mathbf{25\ mW}$ (20-9)

g. $P_s = I_1^2 R_s = (5\text{ mA})^2 \times 1\text{ k}\Omega = \mathbf{25\text{ mW}}$ (20-10)
h. $P_T = VI_1 \cos\theta = 10\text{ V}(5\text{ mA})(1) = \mathbf{50\text{ mW}}$ (20-10)
$P_T = P_L + P_s = 25\text{ mW} + 25\text{ mW} = \mathbf{50\text{ mW}}$ (20-10)
i. $\eta = P_L/P_T = 25\text{ mW}/50\text{ mW} = \mathbf{50\%}$ (20-11)

Example 20-7 verifies the validity of Eqs. (20-4) through (20-11). With respect to the above solutions, note that

1. The impedance reflected by the matching transformer to the primary is $\alpha^2 R_L$ or $(10)^2 10\ \Omega$ or $1\text{ k}\Omega$. This is the same as the internal resistance of the practical voltage source, thus ensuring MPT.
2. Since the reflected impedance and the source impedance are equal, the primary terminal voltage of the transformer is half the source voltage, as shown in step (**b**).
3. The matching transformer is a step-down transformer that steps down voltage by a factor of 1/10 and steps up current by a factor of 10.
4. The power transfer efficiency for the resistive case of Ex. 20-7 is 50%. This serves as a check on the accuracy of the calculations.

EXAMPLE 20-8

A transistorized audio power amplifier has a no-load voltage of 20 V and an internal resistance of 18 Ω. It is to be used with an 8-Ω speaker. Calculate the

a. Power delivered to the speaker when connected directly across the amplifier
b. Turns ratio of the matching transformer to maximize speaker power
c. Maximum power delivered to the speaker using the matching transformer of (**b**)

Solution

a. $V_L = \dfrac{8\ \Omega}{(8 + 18)\ \Omega} \times 20\text{ V} = 6.15\text{ V}$ across the 8-Ω speaker;
$P_L = (V_L)^2/R_L = 6.15^2/8 = \mathbf{4.73\text{ W}}$
b. $\alpha = \sqrt{R_s/R_L} = \sqrt{18/8} = \mathbf{1.5} = N_1/N_2$ (20-4a)
c. $V_2 = V/2\alpha = 20\text{ V}/2 \times 1.5 = 6.\bar{6}\text{ V}$ (20-6)
$P_L = (V_2)^2/R_L = 6.\bar{6}^2/8 = \mathbf{5.\bar{5}\text{ W}}$

20-4 ISOLATION TRANSFORMER APPLICATIONS

Because the primary and secondary windings are insulated from each other but are coupled inductively, transformers are often used to *isolate* the secondary from the primary circuit. In most instances the primary circuit is the 120 V, 60 Hz ac power line, which is normally grounded to earth via driven electrodes in the earth or water main pipes electrically connected to one side of the supply.

Figure 20-8a shows three impedances connected in series to a 120 V ac, 60 Hz supply, grounded at point D. An electronic instrument such as an electronic voltmeter (EVM) or cathode ray oscilloscope (CRO) is connected across $\mathbf{Z}_2$. Since these instruments are powered from the same 120 V ac supply, their circuitry and instrument cases are also grounded and a ground terminal is brought out for test purposes, as well. The instrument shown in Fig. 20-8a is shorting impedance $\mathbf{Z}_3$ and, as a result, may damage $\mathbf{Z}_1$ and $\mathbf{Z}_2$ by excessive current, particularly if $\mathbf{Z}_3$ is the current-limiting impedance in the circuit. Furthermore, shorting out one component produces erroneous test measurements as well, since the circuit has been disturbed (changed) in the process. Finally, if a technician making the measurements inadvertently touches both an equipment or conduit ground and points A, B, or C simultaneously, the possibility of electric shock exists, since these points provide a complete circuit to ground.

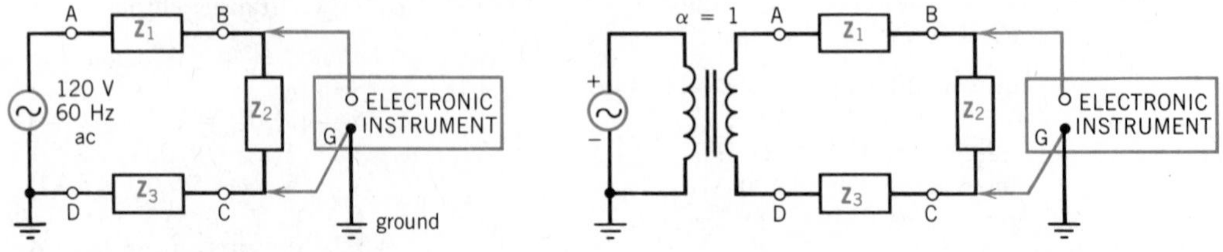

a. Electronic instrument ground shorting out $\mathbf{Z}_3$ and increasing current in $\mathbf{Z}_1$ and $\mathbf{Z}_2$

b. Isolation transformer isolates test circuit from source ground

Figure 20-8 Power line isolation to avoid component damage when using electronic test instruments

All these hazards are eliminated by the 1-to-1 isolation transformer of Fig. 20-8b. The secondary circuit is now isolated from ground. Grounding any terminal, A, B, C, or D, no longer provides a complete path back to the supply. Even touching terminal A, B, or C and an earth ground simultaneously fails to provide a complete path because of the isolation provided by the transformer insulation.

Figure 20-9 shows how an isolation transformer is also used to eliminate dc from waveforms containing both a dc and an ac component. Since the transformer induces voltage only when the current is changing, only the alternating portion of the signal is passed. Similarly, if the ac source contains high-frequency (HF) noise or spikes in its ac waveform, the isolation transformer will pass only the low-frequency (LF) components of the waveform. (Recall that inductors behave as open circuits to HF waveforms.) Such isolation transformers may also contain filter circuits to pass only the desired LF waveforms and ensure that the secondary output is completely noise-free.

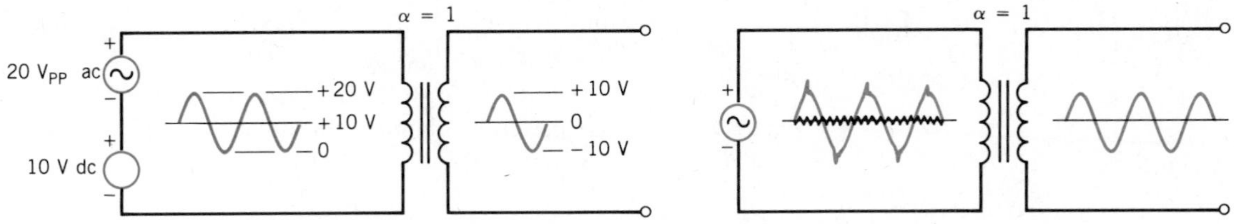

a. Elimination of dc component

b. Elimination of HF noise and spikes in source

Figure 20-9 Elimination of unwanted dc or noise using isolation transformers

Most commercial electrical and electronic equipment is equipped with a three-wire cord so that a fuse or circuit breaker is tripped in the event that an internal connection causes the chassis case to be energized. When the three-wire cord is defeated (by removing the ground prod or using a two-wire adapter) and the plug is accidentally reversed, as shown in **Fig. 20-10a**, the equipment case is energized, providing a shock hazard. If an isolation transformer is used, as shown in Fig. 20-10b, and the plug is accidentally reversed, the chassis case is now isolated from the supply ground, eliminating the shock hazard.

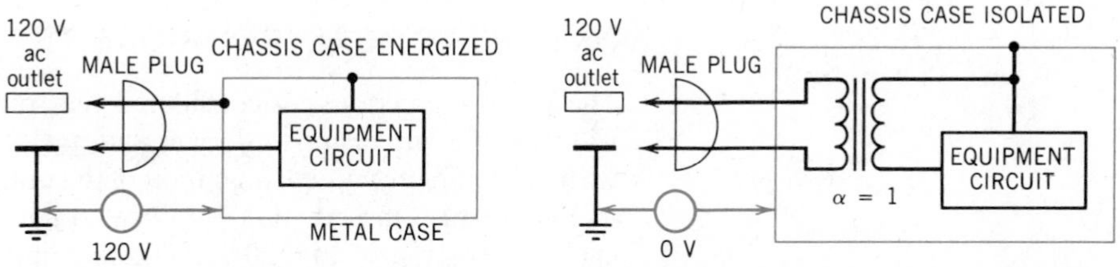

a. Plug reversal to outlet energizes chassis case

b. Isolation transformer maintains chassis case isolation regardless of plug reversal

Figure 20-10 Use of isolation transformer to provide isolation from power line ground

An interesting (and not often realized) application of isolation is found in the three-phase (3ϕ) power distribution system used to supply local single-phase (1ϕ) power to a group of residences in a subdivision. The primary line voltages from the power distribution substation are brought in (at line voltages ranging from about 2 to 10 kV) to a local power distribution 3ϕ transformer. As shown in **Fig. 20-11**, the *secondary* of phase A–C is center-tapped and grounded as a three-wire, 1ϕ, 230/115 V service to one group of residences in a subdivision. Similarly, the secondary of phase B–C also serves a second group of residences with a 230/115 V, 1ϕ, three-wire service. The secondary of phase A–B (not shown) serves a third group of residences. An attempt is made to balance the total 3ϕ load. For example, if there are 30 residences in an entire subdivision, 10 residences are supplied on each phase.

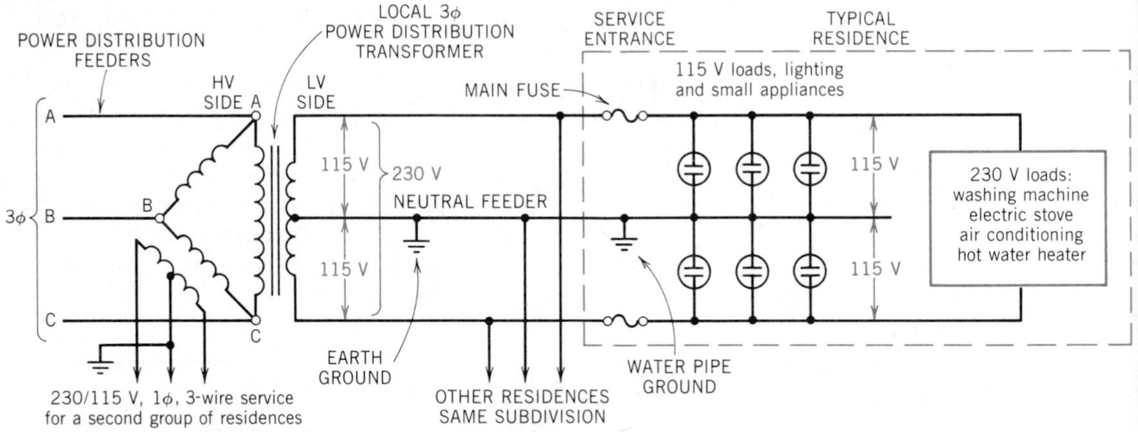

Figure 20-11 Local power distribution transformer for residential service with secondary grounds isolated

Only the secondaries of the three-phase transformer are center-tapped and grounded. If the primaries were center-tapped and grounded, the entire transformer would be shorted out! But the secondaries are *isolated* from each other, which permits each of the three secondaries to be grounded without any possibility of shorting! Note that the neutral feeder is never fused, ensuring that the voltages to ground never exceed 115 V at the secondary. Also observe that the three-wire arrangement permits high-power 1ϕ loads to be fed at 230 V (requiring smaller-capacity wiring due to lower current used for the same power) while small appliances and lighting loads are fed at the safer voltage of 115 V. (See Sec. 21-5 and *ff*. for a more complete discussion of this system.)

20-5 PRACTICAL IRON-CORE TRANSFORMERS

The previous sections all considered the transformer as *ideal*. In this section we will treat the *practical transformer* and its differences from the ideal. We will begin with the complete equivalent circuit of the practical iron-core transformer, as shown in **Fig. 20-12**. Observe that the ideal transformer now occupies only the center enclosed portion of Fig. 20-12. The remaining component values of R, L, and C represent the practical considerations, defined in Table 20-1.

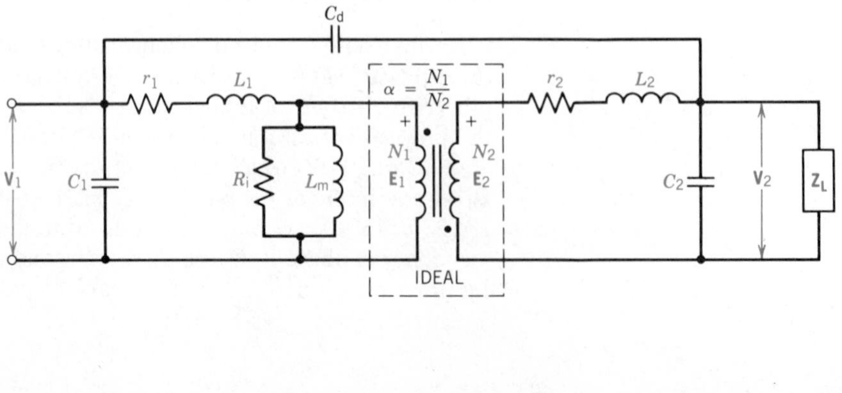

Figure 20-12 Complete equivalent circuit of a practical iron-core transformer

Table 20-1 Definitions of Symbols for the Complete Equivalent Circuit of a Practical Iron-Core Transformer

Symbol	Definition
r_1	Resistance of the primary transformer coil winding
r_2	Resistance of the secondary transformer coil winding
R_i	Equivalent resistance due to *iron* losses in the transformer core as a result of eddy currents and hysteresis of the magnetic laminated core material
L_1	Self-inductance of the primary winding produced by leakage flux that links the primary winding only (see Sec. 12-3)
L_2	Self-inductance of the secondary winding produced by leakage flux that links the secondary winding only (see Sec. 12-3)
L_m	Mutual magnetizing inductance produced by the mutual flux (ϕ_m) linking both the primary and secondary windings (see Sec. 12-6)
C_1	Interwinding capacitance between the insulated turns of the primary coil winding
C_2	Interwinding capacitance between insulated turns of the secondary coil winding
C_d	Distributed capacitance between turns of the primary and turns of the secondary
$\mathbf{E}_1$	Voltage induced in the primary winding by the rate of change of mutual flux (ϕ_m) linking the primary turns; $\mathbf{E}_1$ is always less than $\mathbf{V}_1$ due to the impedance of $r_1 + jX_{L_1}$
$\mathbf{E}_2$	Voltage induced in the secondary winding by the rate of change of mutual flux (ϕ_m) linking the secondary turns
$\mathbf{V}_2$	Secondary terminal voltage of the transformer applied to load $\mathbf{Z}_L$; $\mathbf{V}_2$ is different from $\mathbf{E}_2$ due to the volt drop across secondary coil impedance $r_2 + jX_{L_2}$
$\mathbf{V}_1$	Voltage applied to the transformer at the rated voltage value and frequency

20-5.1 Simplified Equivalent Circuit of the Practical Iron-Core Transformer

From our study of the ideal transformer, we know that resistances, reactances, and impedances are all reflected from the secondary to the primary by α^2 (as shown in Eq. 20-2a). Consequently, $\mathbf{Z}_L$, r_2, and L_2 may all be reflected in Fig. 20-12 to the primary circuit as shown in **Fig. 20-13a**. Note in Fig. 20-13a that all the capacitances have been omitted since at low frequencies (approximately 60 Hz) all capacitors are equivalent to open circuits. (Also note that all inductors have been converted to inductive reactances.) Since the core losses, represented by R_i, account for a relatively small proportion of the total output, for the purpose of *voltage regulation* we may ignore R_i. (It will be included later when we consider transformer efficiency.)

We may now combine resistances and reactances, respectively, as shown in Fig. 20-13b, to find the *equivalent resistance* of the transformer referred to the primary, R_{e_1}, and the *equivalent reactance* of the transformer referred to primary, X_{e_1}:

$$R_{e_1} = r_1 + \alpha^2 r_2 \qquad \text{ohms } (\Omega) \qquad \textbf{(20-12)}$$

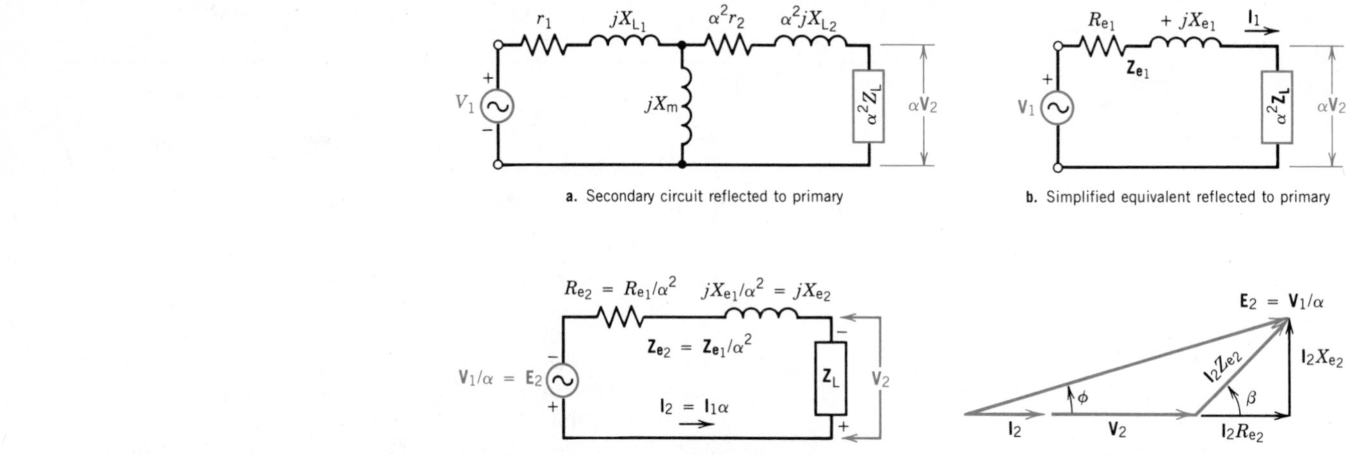

a. Secondary circuit reflected to primary

b. Simplified equivalent reflected to primary

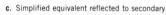

c. Simplified equivalent reflected to secondary

d. Unity PF load phasor diagram reflected to secondary

Figure 20-13 Simplified equivalents reflected to primary and secondary

where r_1 is the primary winding resistance
r_2 is the secondary winding resistance
α is the turns ratio (N_1/N_2)

$$X_{e_1} = X_{L_1} + \alpha^2 X_{L_2} \qquad \text{ohms } (\Omega) \qquad (20\text{-}13)$$

where X_{L_1} is the primary winding reactance
X_{L_2} is the secondary winding reactance

Figure 20-13b also enables us to compute the primary current, $\mathbf{I}_1$, for any value of load, $\mathbf{Z}_L$, from the simple series circuit relation

$$\mathbf{I}_1 = \frac{\mathbf{V}_1}{\mathbf{Z}_{e_1} + \alpha^2 \mathbf{Z}_L} \qquad \text{amperes (A)} \qquad (20\text{-}14)$$

where $\mathbf{Z}_{e_1}$ is $R_{e_1} + jX_{e_1}$ (as defined in Eqs. 20-12 and 20-13)

From Fig. 20-12, the entire primary circuit could have been reflected to the secondary circuit, $\mathbf{V}_2$, using the same method of analysis shown above. $\mathbf{V}_1$ is reflected as $-\mathbf{V}_1/\alpha$ and all impedances (R_{e_1} and jX_{e_1} in Fig. 20-13b) are reflected as R_{e_1}/α^2 and jX_{e_1}/α^2, respectively. The equivalent circuit is shown in Fig. 20-13c. Note the polarity reversal of $\mathbf{E}_2 = \mathbf{V}_1/\alpha$ in this case.[5] With respect to the secondary impedances, we may also write

[5] The polarity reversal stems from the use of the dot convention in Fig. 20-12. Current $\mathbf{I}_1$ is reflected into the secondary as $\alpha\mathbf{I}_1$ and is reversed in direction because the dots in the secondary are reversed with respect to the primary in Fig. 20-12. A review of the dot convention (Sec. 12-7) should clarify this completely.

$$R_{e_2} = r_2 + \frac{r_1}{\alpha^2} = \frac{R_{e_1}}{\alpha^2} \qquad \text{ohms } (\Omega) \qquad \textbf{(20-15a)}$$

$$X_{e_2} = X_{L_2} + \frac{X_{L_1}}{\alpha^2} = \frac{X_{e_1}}{\alpha^2} \qquad \text{ohms } (\Omega) \qquad \textbf{(20-15b)}$$

where all terms have been defined.

Note that for a step-down transformer the secondary impedances are much smaller than the primary impedances.

20-5.2 Voltage Regulation of a Practical Transformer

Figure 20-13d shows the phasor diagram for the unity PF condition where $\mathbf{Z}_L$ is *a purely resistive* load connected across the transformer secondary. $\mathbf{V}_2$ and $\mathbf{I}_2$ are in phase. Volt drops across the equivalent resistance ($\mathbf{I}_2 R_{e_2}$) and reactance ($\mathbf{I}_2 X_{e_2}$) in Fig. 20-13c are shown in Fig. 20-13d. The phasor sum of $\mathbf{V}_2$ and $\mathbf{I}_2\mathbf{Z}_{e_2}$ yields the induced voltage $\mathbf{E}_2$ for the transformer under load (see Fig. 20-12). This simple phasor diagram enables us to predict the voltage regulation of the transformer under rated load conditions, since $\mathbf{E}_2$ is the no-load or open-circuit voltage in Fig. 20-13c. By definition, voltage regulation as described earlier (Sec. 5-6.4) is

$$\text{VR } (\%) = \frac{\mathbf{V}_{NL} - \mathbf{V}_{FL}}{\mathbf{V}_{FL}} \times 100 = \frac{\mathbf{E}_2 - \mathbf{V}_2}{\mathbf{V}_2} \times 100 \qquad \textbf{(20-16)}$$

where all terms have been previously defined.

The problem is slightly more complex when attempting the phasor diagram for the situation where $\mathbf{Z}_L$ is an *inductive–resistive* load ($R_L + jX_L$). In this case the current, $\mathbf{I}_2$, *lags* the voltage, $\mathbf{V}_2$. **Figure 20-14a** shows the phasor diagram for all volt drops that would occur in the circuit of Fig. 20-13c. Note that in this diagram (Fig. 20-14a) $\mathbf{E}_2$ is now much larger than $\mathbf{V}_2$, producing a larger percent voltage regulation. Since $\mathbf{E}_2$ is the open-circuit voltage, this means that as load is applied,

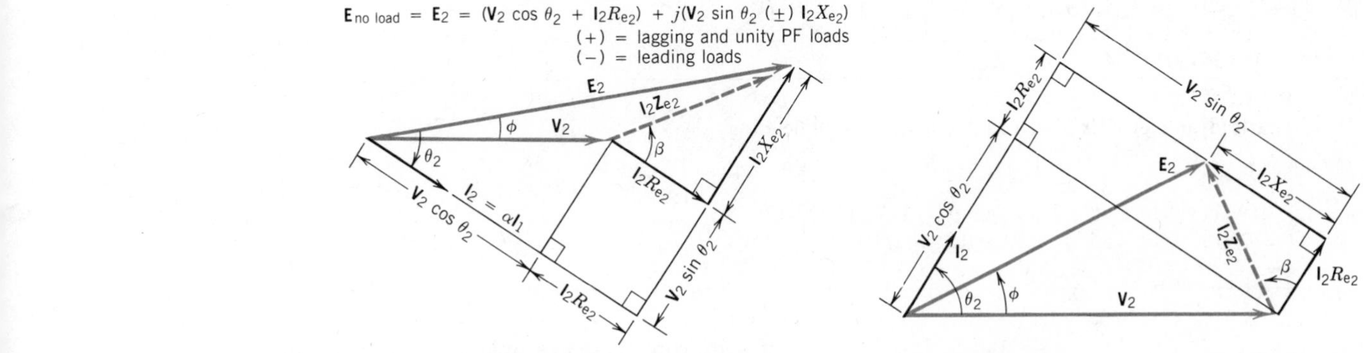

a. Lagging PF load phasor diagram reflected to secondary ($\mathbf{E}_2 > \mathbf{V}_2$)—positive voltage regulation

b. Leading PF load phasor diagram reflected to secondary ($\mathbf{E}_2 < \mathbf{V}_2$)—negative voltage regulation

Figure 20-14 Power transformer secondary voltage regulation. Lagging and leading phasor diagrams the terminal voltage of the transformer *drops* continuously until rated terminal voltage, $\mathbf{V}_2$, is reached at rated load.

The problem is *even slightly more* complex when attempting the phasor diagram for a *leading* load ($R_L - jX_{CL}$) containing capacitive reactance. $\mathbf{I}_2$ leads $\mathbf{V}_2$ by θ_2. $\mathbf{I}_2 R_{e_2}$ is *always in phase* with $\mathbf{I}_2$, as shown in Fig. 20-14b. $\mathbf{I}_2 X_{e_2}$ is always at right angles to $\mathbf{I}_2 R_{e_2}$. The difference between $\mathbf{V}_2$ and $\mathbf{E}_2$ is the impedance drop $\mathbf{I}_2\mathbf{Z}_{e_2}$. Note that in this case of a leading load (Fig. 20-14b) $\mathbf{E}_2$ may be *less* than $\mathbf{V}_2$! This means that as a leading load is applied to the transformer, the terminal voltage tends to *rise*!

Fortunately, only one equation is needed to evaluate the induced voltage, $\mathbf{E}_2$, in order to compute the voltage regulation of any transformer regardless of load PF. It emerges directly from the phasor diagrams of Fig. 20-14:[6]

$$\mathbf{E}_2 = (\mathbf{V}_2 \cos\theta_2 + \mathbf{I}_2 R_{e_2}) + j[\mathbf{V}_2 \sin\theta_2 \,(\pm)\, \mathbf{I}_2 X_{e_2}] \qquad \text{volts (V)} \quad (20\text{-}17)$$

where (+) is used in the j term for *lagging* and *unity* PF loads
(−) is used in the j term for *leading* loads

EXAMPLE 20-9

Short-circuit test measurements made on a 2300/230 V transformer rated at 500 kVA provide the following values of equivalent resistance and reactance, respectively, referred to the low-voltage (secondary) side: $R_{e_2} = 2\ \text{m}\Omega$, $X_{e_2} = 6\ \text{m}\Omega$. Calculate the

a. Rated secondary current, $\mathbf{I}_2$
b. Full load equivalent resistance voltage drop, $\mathbf{I}_2 R_{e_2}$
c. Full load equivalent reactance voltage drop, $\mathbf{I}_2 X_{e_2}$
d. Induced voltage when the transformer is delivering rated current to a unity PF load
e. Voltage regulation at unity PF

Solution

a. $\mathbf{I}_2 = \dfrac{\text{kVA} \times 10^3}{\mathbf{V}_2} = \dfrac{500 \times 10^3\ \text{VA}}{230\ \text{V}} = \mathbf{2.174\ kA}$ (20-1b)

b. $\mathbf{I}_2 R_{e_2} = (2.174\ \text{kA})(2\ \text{m}\Omega) = \mathbf{4.35\ V}$

c. $\mathbf{I}_2 X_{e_2} = (2.174\ \text{kA})(6\ \text{m}\Omega) = \mathbf{13.04\ V}$

d. $\mathbf{E}_2 = (\mathbf{V}_2 \cos\theta_2 + \mathbf{I}_2 R_{e_2}) + j(\mathbf{V}_2 \sin\theta_2 + \mathbf{I}_2 X_{e_2})$
$= (230 \times 1 + 4.35) + j(0 + 13.04)$
$= 234.35 + j13.04 = \mathbf{234.71\ V}$ (20-17)

e. $\text{VR} = \dfrac{\mathbf{E}_2 - \mathbf{V}_2}{\mathbf{V}_2} = \dfrac{234.71 - 230}{230} = \mathbf{2.05\%}$ (20-16)

EXAMPLE 20-10

Repeat parts (**d**) and (**e**) of Ex. 20-9 for rated load at 0.8 PF lagging

Solution

d. $\mathbf{E}_2 = (\mathbf{V}_2 \cos\theta_2 + \mathbf{I}_2 R_{e_2}) + j[\mathbf{V}_2 \sin\theta_2 + \mathbf{I}_2 X_{e_2}]$ (20-17)
$\mathbf{E}_2 = (230 \times 0.8 + 4.35) + j[230 \times 0.6 + 13.04]$
$= 188.35 + j151.04 = \mathbf{241.43\ V}$

e. $\text{VR} = \dfrac{\mathbf{E}_2 - \mathbf{V}_2}{\mathbf{V}_2} = \dfrac{241.43 - 230}{230} = \mathbf{4.97\%}$ (20-16)

EXAMPLE 20-11

Repeat parts (**d**) and (**e**) of Ex. 20-9 for rated load at 0.6 PF leading

Solution

d. $\mathbf{E}_2 = (\mathbf{V}_2 \cos\theta_2 + \mathbf{I}_2 R_{e_2}) + j(\mathbf{V}_2 \sin\theta_2 - \mathbf{I}_2 X_{e_2})$ (20-17)
$= (230 \times 0.6 + 4.35) + j(230 \times 0.8 - 13.04)$
$= 142.35 + j171 = \mathbf{222.5\ V}$

e. $\text{VR} = \dfrac{\mathbf{E}_2 - \mathbf{V}_2}{\mathbf{V}_2} = \dfrac{222.5 - 230}{230} = \mathbf{-3.26\%}$ (20-16)

Insights to be drawn from these examples are the following:

[6] Alternatively, the reader may calculate $\mathbf{E}_2 = \mathbf{V}_2\angle 0^\circ + (\mathbf{I}_2\angle\theta)(\mathbf{Z}_{e_2}\angle\beta)$, using phasor algebra and the diagram of Fig. 20-13d, 20-14a, and/or 20-14b, depending on PF. In this expression, θ is the PF angle and β is $\cos^{-1}(R_{e_2}/\mathbf{Z}_{e_2})$.

1. A unity PF load produces a small (low) percent regulation because $\mathbf{E}_2$ is only slightly larger than $\mathbf{V}_2$, as shown in Ex. 20-9. $\mathbf{E}_2$ leads $\mathbf{V}_2$ by a small angle ϕ, as shown in Fig. 20-13d at unity PF.
2. A 0.8 PF lagging load produces a much higher positive percent regulation because $\mathbf{E}_2$ is somewhat greater than $\mathbf{V}_2$ as shown in Fig. 20-14a, and still leads $\mathbf{V}_2$ by ϕ.
3. A transformer loaded to rated current by a leading load may have a lower percent regulation than unity PF. It may be closer to zero percent regulation or may have negative regulation, depending on the magnitude of the leading PF. In Ex. 20-11 a 0.6 PF leading load produces negative regulation that is greater than regulation at unity PF.
4. The lower or negative regulation of a leading PF load occurs because $\mathbf{E}_2$ may be smaller than $\mathbf{V}_2$, as shown in Fig. 20-14b, and leads $\mathbf{V}_2$ by a larger angle ϕ.

Perhaps the most important insight to be drawn from this discussion is that the voltage regulation of large commercial transformers is *not* obtained by direct loading. In the case of Ex. 20-9, where would one find a 500-kVA load? One would have to "borrow" a small town to find such a load. This is why the values of R_{e_2} and X_{e_2} are always obtained by test procedure in which loading is *simulated*, as in the case of the *short-circuit test*, described below.

20-5.3 Short Circuit Test of a Commercial Transformer

The open-circuit and short-circuit tests to measure the internal resistance of a source were described in Sec. 5-6.3. Example 20-9 provided the values of R_{e_2} and X_{e_2} from "short-circuit test measurements" but did not tell us how they were obtained.

One would have to think twice before daring to short-circuit the secondary of a 2300/230 V 500 kVA transformer with rated voltage applied to the 2300-V side. But if *no* voltage is applied to the 2300-V side and the 230-V side is short-circuited, we can now slowly apply small voltage increments to the HV side until rated current flows in the LV side! And if rated current flows in the LV side, then rated current must also flow in the HV side in accordance with Eq. (20-1). But this means that both windings are carrying rated current and dissipating power due to their primary and secondary winding resistances, $\mathbf{I}_1^2 r_1$ and $\mathbf{I}_2^2 r_2$, respectively. This is the essence of the short-circuit test used to test commercial transformers.

Figure 20-15 shows the typical arrangement of instruments and devices for obtaining short-circuit test data for a commercial transformer. The procedure consists of the following steps:

1. With variable transformer set to zero, short-circuit low-voltage terminals $\mathbf{X}_1 - \mathbf{X}_2$, of the transformer.

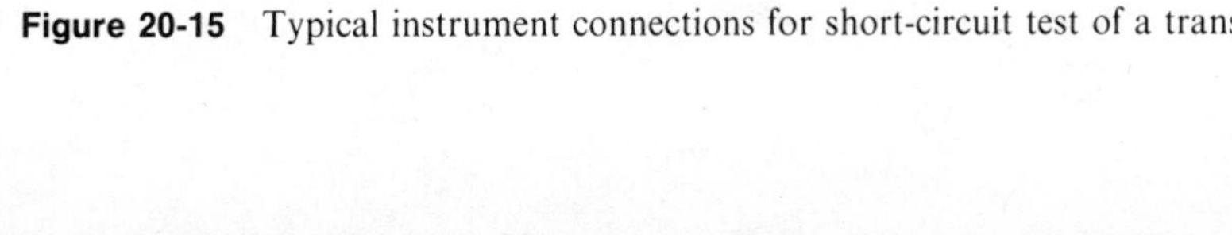

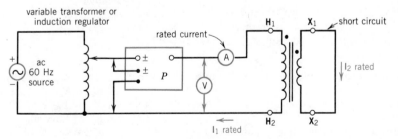

Figure 20-15 Typical instrument connections for short-circuit test of a transformer

2. Increase voltage slowly (and carefully) until rated primary current is recorded on the ammeter. Rated primary current is determined from the transformer kVA rating divided by the voltage of the high-voltage side.
3. From the three instruments on the HV side, record P_{sc}, short-circuit power; $\mathbf{V}_{sc}$, short-circuit voltage; and $\mathbf{I}_{sc}$, short-circuit primary current.

The calculations that are performed on these test data are as follows:

1. Calculate equivalent primary impedance, $\mathbf{Z}_{e_1}$, from the ratio

$$\mathbf{Z}_{e_1} = \frac{\mathbf{V}_{sc}}{\mathbf{I}_{sc}} = \frac{\text{voltmeter reading}}{\text{ammeter reading}} \quad \text{ohms } (\Omega) \qquad \textbf{(20-18)}$$

2. Calculate equivalent primary resistance, $\mathbf{R}_{e_1}$, from the ratio

$$R_{e_1} = \frac{P_{sc}}{(\mathbf{I}_{sc})^2} = \frac{\text{wattmeter reading}}{(\text{ammeter reading})^2} \quad \text{ohms } (\Omega) \qquad \textbf{(20-19)}$$

3. Calculate equivalent primary reactance, X_{e_1}, using either

$$X_{e_1} = \sqrt{(\mathbf{Z}_{e_1})^2 - (R_{e_1})^2} \quad \text{ohms } (\Omega) \qquad \textbf{(20-20)}$$

or

$$\theta = \text{arc cos}\left(\frac{R_{e_1}}{\mathbf{Z}_{e_1}}\right) \quad \text{and} \quad X_{e_1} = \mathbf{Z}_{e_1} \sin\theta \quad \text{ohms } (\Omega) \qquad \textbf{(20-20a)}$$

4. Since we are interested in predicting the voltage regulation at the secondary side, we must reflect R_{e_1} and X_{e_1} to their secondary (low-voltage side) values or

$$R_{e_2} = R_{e_1}/\alpha^2 \quad \text{ohms } (\Omega) \qquad \text{(20-15a)}$$

$$X_{e_2} = X_{e_1}/\alpha^2 \quad \text{ohms } (\Omega) \qquad \text{(20-15b)}$$

EXAMPLE 20-12

A 2300/230 V 20 kVA step-down transformer is connected as shown in Fig. 20-15, with the LV side short-circuited. The test data obtained for the HV side are $P_1 = 250$ W, $\mathbf{I}_1 = 8.7$ A, $\mathbf{V}_1 = 50$ V. Calculate the

a. Equivalent impedance, resistance, and reactance, referred to the HV side
b. Equivalent impedance, resistance, and reactance, referred to the LV side
c. Transformer voltage regulation at unity PF
d. Transformer voltage regulation at 0.7 PF lagging

Solution

a. $\mathbf{Z}_{e_1} = \mathbf{V}_1/\mathbf{I}_1 = 50 \text{ V}/8.7 \text{ A} = \mathbf{5.75\ \Omega}$ (20-18)
$R_{e_1} = P_1/\mathbf{I}_1^2 = 250 \text{ W}/(8.7)^2 = \mathbf{3.3\ \Omega}$ (20-19)
$\theta = \cos^{-1}(R_{e_1}/\mathbf{Z}_{e_1}) = \cos^{-1}(3.3/5.75) = 55°$
$X_{e_1} = \mathbf{Z}_{e_1} \sin\theta = 5.75 \sin 55° = \mathbf{4.71\ \Omega}$ (20-20a)

b. $\mathbf{Z}_{e_2} = \mathbf{Z}_{e_1}/\alpha^2 = 5.75/(10)^2 = \mathbf{57.5\ m\Omega}$
$R_{e_2} = R_{e_1}/\alpha^2 = 3.3/(10)^2 = \mathbf{33\ m\Omega}$ (20-15a)
$X_{e_2} = X_{e_1}/\alpha^2 = 4.71/10^2 = \mathbf{47.1\ m\Omega}$ (20-15b)

c. Rated secondary load current, $\mathbf{I}_2 = S/\mathbf{V}_2 = 20\,000 \text{ VA}/230 \text{ V} = 86.96$ A; $\mathbf{I}_2 R_{e_2} = 86.96 \text{ A}(33 \text{ m}\Omega) = 2.87$ V and $\mathbf{I}_2 X_{e_2} = 86.96 \text{ A}(47.1 \text{ m}\Omega) = 4.1$ V

$$\mathbf{E}_2 = (\mathbf{V}_2 \cos \theta_2 + \mathbf{I}_2 R_{e_2}) + j(\mathbf{V}_2 \sin \theta_2 + \mathbf{I}_2 X_{e_2})$$
$$= (230 \times 1 + 2.87) + j(0 + 4.1) = 232.87 + j4.1$$
$$= 232.9 \text{ V} \quad (20\text{-}17)$$

$$\text{VR at unity PF} = \frac{\mathbf{V}_{NL} - \mathbf{V}_{FL}}{\mathbf{V}_{FL}} = \frac{\mathbf{E}_2 - \mathbf{V}_2}{\mathbf{V}_2}$$
$$= \frac{232.9 - 230}{230} = \mathbf{1.26\%} \quad (20\text{-}16)$$

d. $$\mathbf{E}_2 = (\mathbf{V}_2 \cos \theta_2 + \mathbf{I}_2 R_{e_2}) + j(\mathbf{V}_2 \sin \theta_2 + \mathbf{I}_2 X_{e_2})$$
$$= (230 \times 0.7 + 2.87) + j(230 \times 0.71414 + 4.1)$$
$$= 163.9 + j168.35 = 235 \text{ V} \quad (20\text{-}17)$$

$$\text{VR at 0.7 lagging} = \frac{\mathbf{E}_2 - \mathbf{V}_2}{\mathbf{V}_2} = \frac{235 - 230}{230}$$
$$= \mathbf{2.17\%} \quad (20\text{-}16)$$

Example 20-12 is extremely important because it shows precisely how the short-circuit test data yield the secondary (LV) side resistance and reactance, essential to the calculation of voltage regulation. Only 250 W of power is dissipated in making these measurements, compared to 20 000 W if the regulation is obtained by direct loading! But since the wattmeter records not only the transformer copper losses but also the iron losses, the reader may well ask whether the value of resistance, R_{e_1}, calculated in Ex. 20-12 is too high.[7] This is discussed in Sec. 20-5.4, directly below.

20-5.4 Assumptions Inherent in the Short-Circuit Test

The wattmeter in the short-circuit test of Fig. 20-15 is recording the full-load copper losses of both windings, as well as the iron (or core) losses of the transformer. But since the transformer primary is excited at a voltage considerably below its rated primary voltage, the core losses are reduced significantly. Since the *flux density* of the *transformer varies directly as the applied primary voltage*, we must try to evaluate the core losses at the reduced voltage. In short, we must ask, "What fraction of the wattmeter reading includes the core losses?"

Only two types of iron losses are produced in the core: *eddy current loss* and *hysteresis loss* (Sec. 11-13). Equation (11-15) shows that the eddy current loss is a function of the square of the permissible flux density, or as shown above, the *square* of the *applied voltage.* Equation (11-16) shows that the hysteresis loss also varies *approximately* as the square of the flux density or the *square* of the *applied voltage.* This simple insight enables us to calculate whether the iron or core losses at the reduced voltage of the short-circuit test may be ignored, as shown by Ex. 20-13.

EXAMPLE 20-13

Assuming that the core loss at rated voltage is P_c, evaluate the fraction of P_c measured by the wattmeter in Ex. 20-12

Solution

Since P_c is proportional to the square of the primary voltage, $\mathbf{V}_1$, then under conditions of the short-circuit test,

$$P_{sc} = \left(\frac{\mathbf{V}_{sc}}{\mathbf{V}_1(\text{rated})}\right)^2 \times P_c = \left(\frac{50 \text{ V}}{2300 \text{ V}}\right)^2 \times P_c$$
$$= 4.73 \times 10^{-4} = \mathbf{0.000\,473}\ \boldsymbol{P_c}$$

The conclusions to be drawn from Ex. 20-13 are:

1. The wattmeter reading during the short-circuit test is essentially recording only the copper losses of the windings. The core losses of power transformers are essentially negligible and may be ignored. (The only possible exceptions are either small or high-frequency transformers. Both produce higher core losses at reduced voltages than most commercial power transformers relative to their rated output.)
2. The equivalent resistance and the voltage regulation calculations should not be too far different from those obtained under direct loading of the transformer.

[7] A higher value of R_{e_1} and R_{e_2} yields a higher percent voltage regulation. This means that in actual performance the transformer voltage regulation is better under direct loading than the predictions of the short-circuit test. The designer is assured that the "pessimistic" prediction represents the worst-case situation, which is very useful.

20-6 TRANSFORMER EFFICIENCY

The short-circuit test wattmeter reading yields the copper losses (of both primary and secondary windings) at *rated* load. Since the only other losses that occur in the transformer are the core (or iron) losses, these can be found from the *open-circuit test*.

Figure 20-16 shows the instrument connections for the *open-circuit* test. The test is performed on the low-voltage (LV) side simply because low voltages are more easily available and there is less danger to personnel in making these measurements. In the case of a multiple-winding transformer, the test is performed on the *lowest*-voltage winding available. However, care should be taken to see that the high-voltage terminals are properly insulated from each other and from the personnel making the test. The test procedure consists of the following steps:

1. Bring voltage up from zero until rated voltage for the particular (LV) winding is recorded on the voltmeter.
2. Record the open-circuit power, P, the rated voltage, $\mathbf{V}$, and the magnetizing current, $\mathbf{I}_m$, as measured by wattmeter, voltmeter, and ammeter, respectively.
3. Calculate the core loss from $P_c = P - (\mathbf{I}_m)^2 R_x$, where R_x is the resistance of the low-voltage winding selected.

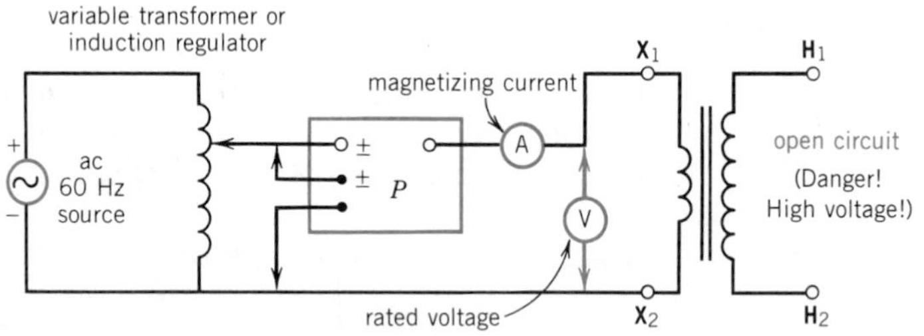

Figure 20-16 Typical instrument connections for open-circuit test of a transformer to determine core losses

Step 3 is performed only in the case of relatively small transformers. Large transformers have relatively low resistance low-voltage windings (since these carry high currents and are wound with heavier wire). Since the winding resistance is low and the magnetizing current very small, the no-load copper loss is usually a small fraction of the reading of wattmeter P and is ignored.

As in the case of the short-circuit test, note that the open-circuit test also consumes a very small fraction of the total rated kVA or VA of the transformer. The core loss usually represents less than 1% of the transformer rating.

Although *voltage regulation* data are obtained from the *short-circuit* test *only*, *both* the open-circuit and short-circuit test data are needed to calculate *transformer efficiency*. The *open-circuit* test yields the *core losses*. The *short-circuit* test yields the *copper losses*. As in the case of voltage regulation, the secondary side (to which load is connected) is used for the calculation of efficiency, η,

$$\eta = \frac{P_0}{P_0 + \text{losses}} = \frac{\mathbf{V}_2\mathbf{I}_2 \cos\theta_2}{\mathbf{V}_2\mathbf{I}_2 \cos\theta_2 + (\underbrace{P_c}_{\text{fixed core losses}} + \underbrace{\mathbf{I}_2^2 R_{e2}}_{\text{variable copper losses}})} \qquad (20\text{-}21)$$

With respect to Eq. (20-21), please note:

1. The numerator represents *useful* output power P_0 transferred from primary to secondary and dissipated in the load.

2. The parenthetical term represents both fixed and variable losses that occur during the power transfer. The *fixed* losses, P_c, are the *core* losses. The *variable* losses, $I_2^2 R_{e_2}$, are the *copper* losses.
3. The copper losses vary with the square of load current, $\mathbf{I}_2$.
4. The denominator or input power is the sum of the output power plus the losses.
5. The core losses, P_c, are fixed as long as $\mathbf{V}_2$ (and $\mathbf{V}_1$) are maintained at rated values.

20-6.1 Maximum Efficiency of a Transformer

Maximum efficiency always occurs at the load point where the fixed losses equal the variable losses.[8] For maximum efficiency in a transformer, copper losses must equal core losses or

$$\mathbf{I}_2^2 \cdot R_{e_2} = P_c \qquad (\text{for } \eta_{max}) \qquad \text{watts (W)} \qquad (20\text{-}22)$$

and the load current at which maximum efficiency occurs is

$$\mathbf{I}_2 = \sqrt{\frac{P_c}{R_{e_2}}} \qquad \text{amperes (A)} \qquad (20\text{-}22a)$$

20-6.2 Efficiency Curves of a Transformer

Equations (20-21) and (20-22a) enable us to predict the shape of the efficiency curves of a transformer under various load and PF conditions.

Equation (20-22a) shows that maximum efficiency is fixed by the ratio P_c/R_{e_2}. Therefore, regardless of the PF of the load, maximum efficiency always occurs at the same load (current) value, $\mathbf{I}_2$. (See **Fig. 20-17**.)

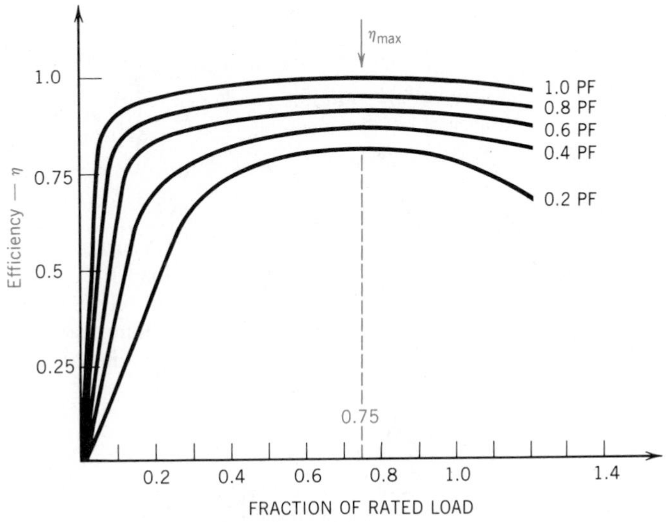

Figure 20-17 Transformer efficiency as affected by load power factor

[8] This statement applies equally to a motor, a generator, a diesel engine, a steam engine, a steam turbine, and so forth and, of course, to a transformer as well.

Equation (20-21) shows that the PF of the load ($\cos\theta_2$) determines the magnitude of the numerator term, $\mathbf{V}_2\mathbf{I}_2\cos\theta_2$. The higher the PF, for the same load current, the higher the efficiency. Alternatively, a reduction in PF, for the same load current, is accompanied by a reduction in efficiency. (See Fig. 20-17.)

Equation (20-21) also shows that at relatively light loads, where P_0 is small in proportion to the losses, the fixed loss (P_c) is high in proportion to the output, P_0. This results in a relatively *low* efficiency (see Fig. 20-17).

Similarly, using Eq. (20-21), under heavy loads where the output is higher than rated load (1.0), the variable copper loss is high; this, combined with the fixed core loss, causes the efficiency to be reduced. (See Fig. 20-17.)

For most transformers, therefore, the efficiency rises from zero (at no load output) to a maximum (at approximately from half to rated load) and then drops off at loads beyond rated load. (See Fig. 20-17.)

20-6.3 Efficiency Calculations

The following examples summarize the use of the short-circuit and open-circuit tests to predict transformer efficiency at various values of load as well as the load at which maximum efficiency occurs for the transformer under test. Note in Ex. 20-14 that the conventional test method requires but a small fraction of the transformer kVA to perform the tests (approximately 1.6% for the short-circuit test and even less for the open-circuit test) in order to predict the transformer efficiency under actual direct loading conditions.

EXAMPLE 20-14

A 2300/208 V, 500 kVA, 60 Hz distribution transformer has just been constructed. It is tested by means of the open-circuit and short-circuit test, prior to being put into service as a step-down transformer. The test data are:

Test	Voltage (V)	Current (A)	Power (W)	Side used
Open circuit	208	85	1800	low-voltage
Short circuit	95	217.4	8200	high-voltage

Calculate from the above test data:

a. Equivalent resistance referred to the low-voltage side, R_{e_2}
b. Resistance of the low-voltage side only, r_2
c. Transformer copper loss of the low-voltage side winding during the open-circuit test
d. Transformer core loss when rated voltage is applied
e. Can the total power measured during the open-circuit test be used as the core loss? Explain.

Solution

a. From the short-circuit test data,

$$R_{e_1} = P/\mathbf{I}_1^2 = 8200\ \text{W}/(217.4)^2 = \mathbf{0.1735\ \Omega}\ \text{and}$$

$$R_{e_2} = R_{e_1}/\alpha^2 = 0.1735\ \Omega/(2300/208)^2 = \mathbf{1.419\ m\Omega}$$

b. $r_2 = R_{e_2}/2 = 1.419\ \text{m}\Omega/2 = \mathbf{0.71\ m\Omega}$[9]
c. $I_m^2 r_2 = 85^2(0.71\ \text{m}\Omega) = \mathbf{5.13\ W}$
d. $P_c = P_{oc} - \mathbf{I}_m^2 r_2 = 1800 - 5.13 = \mathbf{1794.9\ W}$
e. **Yes**. The error is approximately 5/1800 = 0.278%, which is within the error produced by the instruments used in the test. We may assume that the core loss is **1800 W**.

EXAMPLE 20-15

Using the data of Ex. 20-14, calculate

a. Efficiency of the transformer when the secondary is loaded at unity PF to 1/4, 1/2, 3/4, 4/4, and 5/4 rated load. Tabulate all losses, output, and input as a function of each fraction of rated load
b. Repeat (**a**) for the same load conditions but with a load of 0.8 PF lagging
c. Load current at which maximum efficiency occurs regardless of load PF
d. Load fraction at which maximum efficiency occurs
e. Maximum efficiency at unity PF
f. Maximum efficiency at 0.8 PF lagging

[9] It can be shown empirically that each winding resistance is approximately half of the total equivalent resistance referred to that side. Thus, the resistance of the primary winding, r_1, is approximately half of R_{e_1}, and the resistance of r_2 is approximately half of R_{e_2}, as shown above. See Figs. 20-12 and 20-13.

Solution

Preliminary data before tabulating: core loss = fixed loss = **1800 W** from Ex. 20-14; copper loss at rated load = variable loss = **8200 W** from test data of Ex. 20-14; full load output at unity PF = kVA cos θ_2 = 500 kVA × 1 = **500 kW** (from given rating)

a. Tabulation at **unity PF**

LOAD FRACTION OF RATED LOAD	CORE LOSS (kW)	COPPER LOSS (kW)	TOTAL LOSS, P_L (kW)	TOTAL OUTPUT P_0 (kW)	TOTAL INPUT $P_L + P_0$ (kW)	EFFICIENCY P_0/P_{in} (%)
1/4	1.8	0.512	2.312	125.0	127.312	98.18
1/2	1.8	2.050	3.85	250.0	253.85	98.48
3/4	1.8	4.610	6.41	375.0	381.41	98.32
4/4	1.8	8.200	10.00	500.0	510.0	98.04
5/4	1.8	12.800	14.6	625.0	639.6	97.72

b. Tabulation at **0.8 PF lagging**

1/4	1.8	0.512	2.312	100.	102.312	97.74
1/2	1.8	2.05	3.85	200.	203.85	98.11
3/4	1.8	4.61	6.41	300.	306.41	97.91
4/4	1.8	8.2	10.00	400.	410	97.56
5/4	1.8	12.8	14.6	500.	514.6	97.16

c. $\mathbf{I}_2 = \sqrt{P_c/R_{e_2}} = \sqrt{1800/1.417 \times 10^{-3}} = \mathbf{1127.1\ A}$ load at which η_{max} occurs

d. $\mathbf{I}_2 \text{ (rated)} = \dfrac{500\ \text{kVA} \times 1000}{208\ \text{V}} = \mathbf{2403.8\ A}$

Load fraction for $\eta_{max} = 1127.1/2403.8 = \mathbf{0.47}$ (approximately half rated load)

e. $$\eta_{max} = \frac{V_2I_2\cos\theta_2}{V_2I_2\cos\theta_2 + P_c + I_2^2R_{e_2}} = \frac{208 \times 1127.1 \times 1}{208 \times 1127.1 \times 1 + 1800 + (1800)}$$
$= \mathbf{98.49\%}$ at unity PF

f. $$\eta_{max} = \frac{V_2I_2\cos\theta_2}{V_2I_2\cos\theta_2 + P_c + I_2^2R_{e_2}} = \frac{208 \times 1127.1 \times 0.8}{208 \times 1127.1 \times 0.8 + (3600)}$$
$= \mathbf{98.12\%}$ at 0.8 PF lagging

Several useful shortcuts enabled the preparation of the foregoing tables:

1. The copper loss column is easily prepared by using the full-load copper loss of 8.2 kW, which comes directly from the short-circuit test wattmeter reading. Since copper loss varies as the square of the load current, the copper loss at *any* load fraction is (load fraction)^2 × 8.2 kW.
2. The total output column is similarly prepared using the rated output at 4/4 load given in Example 20-14 as 500 kVA. At unity PF, this is 500 kW. Each of the other outputs is obtained from (load fraction) × 500 kW.
3. Since PF does not affect either the copper loss or the core loss, the tabulation of the first four columns of part (**b**) is identical to part (**a**).
4. At 0.8 PF lagging, rated output is now 500 kVA × 0.8 or 400 kW. Using (load fraction) × 400 kW changes each of the values in the P_0 column of part (**b**), which changes the remaining columns, in turn.
5. Observe that *each* efficiency at the *lower* PF is respectively *lower* than the unity PF counterparts at the *same load fraction*, thus verifying Fig. 20-17.
6. The maximum efficiency values obtained in (**e**) and (**f**) are both slightly greater than the half-load efficiency values that appear in the tables, thus verifying the accuracy of the tables and the equations used.

20-7 FREQUENCY RESPONSE OF THE IRON-CORE TRANSFORMER

The clue to the behavior of the iron-core transformer over a range of frequencies is found from an examination of Fig. 20-12, reproduced once again in **Fig. 20-18a**.

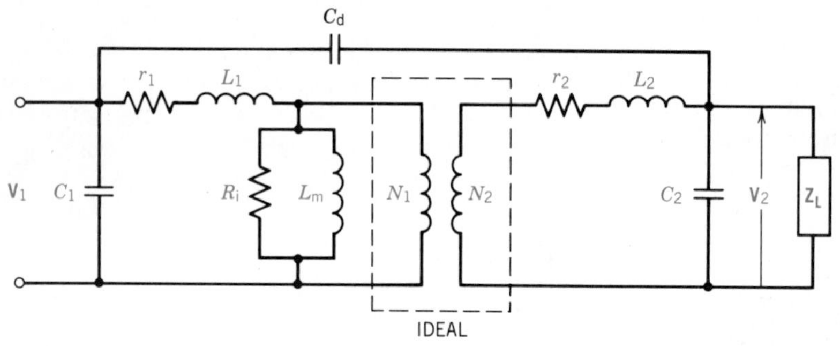

a. Equivalent circuit of iron core transformer

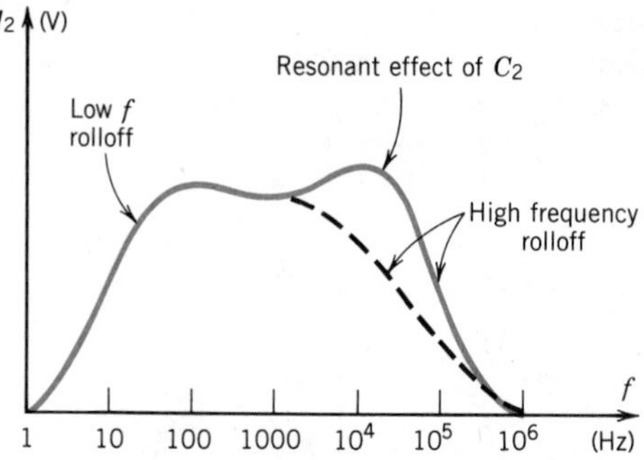

b. Frequency response curve

Figure 20-18 Equivalent circuit and logarithmic frequency response of an iron-core transformer

Most commercial power transformers are designed to operate at a fixed and given frequency. For these, the relations derived for the ideal transformer in the previous sections are sufficiently accurate to predict the secondary output voltage and current at the rated (given) frequency. But what if the iron-core transformer is used as an audio output transformer to transform a wide range of frequencies with equal amplitude to be fed to a speaker from an audio amplifier? Theoretically, the ideal transformer, whose seven characteristics are summarized in Sec. 20-2, should amplify all frequencies (from zero or dc to almost infinity) equally. But the practical iron-core transformer departs considerably from the ideal insofar as frequency response is concerned, as shown in Fig. 20-18b. The response curve exhibits both a low-frequency rolloff and a high-frequency rolloff. Let us see why.

1. At low frequencies, inductances behave as short circuits. The mutual magnetizing inductance, L_m, produced by the mutual flux, ϕ_m, is shunting the input of the ideal transformer. At very low frequencies, say 1 Hz, it is practically a short circuit. As the frequency increases, L_m, a relatively low inductance, increases, accounting for the rise in rolloff from 1 to 100 Hz.
2. At approximately 100 Hz, the reactance of L_m is sufficiently high to limit the magnetizing current $\mathbf{I}_m$ to a relatively small value, causing the response to flatten over the range from 100 Hz to about 5 kHz.
3. As the frequency increases, the voltage drop across the secondary leakage reactance (X_{L_2}) increases, which seriously limits the output and accounts for the high-frequency rolloff shown as a dashed line in Fig. 20-18b.
4. By proper design of the primary and secondary leakage reactances, the interwinding capacitance, C_2, can be made to resonate with the leakage inductance, producing a series-resonant rise across C_2 and across $\mathbf{V}_2$ at frequencies in the neighborhood of 20 kHz. This effect tends to flatten and extend the frequency response curve so that rolloff occurs later at higher frequencies.
5. At extremely high frequencies, C_1 and C_2 both short-circuit the input and output, respectively, accounting for the high-frequency rolloff.

Well-designed audio transformers are relatively expensive since they must be larger and have the following ideal characteristics: extremely low primary and secondary impedances and extremely high mutual magnetizing inductance, L_m. This is achieved by

1. Using heavier conductors to reduce primary and secondary resistance.
2. Using fewer primary and secondary turns to decrease L_1 and L_2.
3. Winding the primary and secondary closely on a common core to reduce leakage and using higher-permeability cores of larger cross-sectional area.

4. Increasing the magnetizing inductance, L_m, by using larger cross-sectional cores, which have relatively short magnetic circuit length and high permeability since $L_m = N^2\mu A/l$. Ideally, the turns should be increased as much as possible.
5. Since L_m is increased by increasing primary and secondary turns, a compromise is effective between many turns in (4) and fewer turns in (2) above.

20-8 AUTOTRANSFORMERS

The autotransformer is one of the most efficient devices known in technology. Typical autotransformer efficiencies range upward from 99% to close to 100%! Furthermore, for the same core size and winding construction, the kVA transfer capability of autotransformers is *much higher* than that of conventional isolation transformers. How can this be?

Figure **20-19a** shows a conventional 1 kVA isolation transformer feeding a load $\mathbf{Z}_L$. If the *undotted* end of the *secondary* winding is connected to the *dotted* end of the *primary* winding (or vice versa), the autotransformer of Fig. 20-19b is obtained. In this mode, *additive polarity* is obtained because (by **KVL**) the instantaneous voltages of the common (1000 V) winding and the low-voltage (100 V) secondary winding (whose voltage is designated as $\mathbf{V}_s$) by phasor addition produce a secondary voltage $\mathbf{V}_2$ of 1100 V. At the same time, the currents (by **KCL**) at the junction of the connection between the two windings produce a primary current of 11 A and a secondary current of 10 A. The kVA transferred from one circuit to another by this transformer, therefore, is $\mathbf{S} = (1000 \text{ V})(11 \text{ A}) = (1100 \text{ V})(10 \text{ A}) = 11$ kVA. Note that this kVA transfer is **11 times** the kVA rating of the same transformer used as an isolation transformer in Fig. 20-19a.

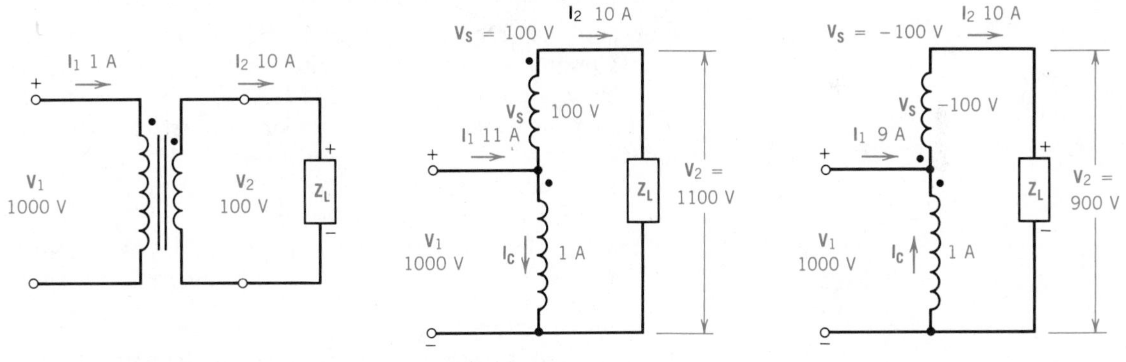

a. 1 kVA isolation transformer

b. Voltages produced by additive polarity. $V_2 = (V_1 + V_S)$; $I_1 = (I_2 + I_C)$ S = 11 kVA

c. Voltages produced by subtractive polarity. $V_2 = (V_1 - V_S)$; $I_2 = (I_1 + I_C)$ S = 9 kVA

Figure 20-19 Autotransformer in additive and subtractive polarity modes

For the *additive-polarity* mode of the step-up autotransformer we may write the following general equations for voltage, current, and apparent power, respectively:

$$\mathbf{V}_2 = \mathbf{V}_1 + \mathbf{V}_s \quad \text{volts (V)} \tag{20-23}$$

$$\mathbf{I}_1 = \mathbf{I}_2 + \mathbf{I}_c \quad \text{amperes (A)} \tag{20-24}$$

$$\mathbf{S} = \mathbf{S}_1 = \mathbf{S}_2 = \mathbf{V}_1\mathbf{I}_1 = \mathbf{V}_2\mathbf{I}_2 \quad \text{volt-amperes (VA) transferred} \tag{20-25}$$

where $\mathbf{I}_c$ is the current in the winding *common* to both primary and secondary
$\mathbf{V}_s$ is the portion of the secondary voltage obtained by *transformation*
$\mathbf{S}$ is the VA transferred by the autotransformer (and not the VA rating of the autotransformer)

With respect to these equations for the *step-up* autotransformer in the *additive-polarity* mode, note that the *secondary voltage* is the phasor sum of the voltages across each winding and the *primary current* is the phasor sum of the currents in each winding.

If the same isolation transformer (Fig. 20-19a) is connected so that *both dotted* ends are connected to the *same* junction, as shown in Fig. 20-19c, we obtain the *subtractive-polarity* connection. In this mode, the transformer is actually behaving as a *step-down* transformer. For the values given in Fig. 20-19c, the kVA transferred by this transformer is (1000 V)(9 A) = (900 V)(10 A) = 9 kVA. Note that this kVA transfer, while less than in the additive-polarity mode, is still **nine times** the kVA rating of the *same* transformer used as an *isolation* transformer.

For the *subtractive-polarity* mode of the autotransformer we may write the following general equations for voltage, current, and apparent power, respectively:

$$\mathbf{V}_2 = (\mathbf{V}_1 - \mathbf{V}_s) \quad \text{or} \quad \mathbf{V}_1 = \mathbf{V}_2 + \mathbf{V}_s \quad \text{volts (V)} \tag{20-26}$$

$$\mathbf{I}_2 = \mathbf{I}_1 + \mathbf{I}_c \quad \text{amperes (A)} \tag{20-27}$$

$$\mathbf{S} = \mathbf{S}_1 = \mathbf{S}_2 = \mathbf{V}_1\mathbf{I}_1 = \mathbf{V}_2\mathbf{I}_2 \quad \text{volt-amperes (VA) } \textbf{transferred} \tag{20-25}$$

With respect to these equations for the autotransformer in the *subtractive-polarity* mode, note that the *primary voltage* is the phasor sum of the voltages across each winding and the *secondary current* is the phasor sum of the currents in each winding. Comparing this to the additive-polarity mode yields the following conclusions, emerging from Eqs. (20-23) to (20-27).

Depending on connection (additive or subtractive), *either* the primary *or* the secondary voltages are added as a *phasor* sum. If we sum the *primary voltages*, then we must sum the *secondary currents*. If we sum the *secondary voltages*, then we must also sum the *primary currents*.[10]

We noted at the outset that, while conventional (isolation) transformers have high efficiencies, *autotransformers have even higher efficiencies*. The reasons for this are as follows:

1. In an isolation transformer, *all* the energy received by the primary must be *transformed* to reach the secondary, and transformation results in losses.
2. But in an autotransformer, only part of the energy is transferred by transformer action. The rest of the energy is transferred *conductively* from primary to secondary. Energy transferred *conductively* produces *no transformer losses*.

For simplicity, let us consider *only* transformers in the *additive*-polarity mode. **Figure 20-20a** shows an autotransformer in the **step-down** mode using the isolation transformer of Fig. 20-19a. For the **step-down** mode we may write the following kVA relations for the transformer of Fig. 20-20a:

---

[10] To some readers this conclusion may seem obvious in order to obtain the *same* kVA transfer of energy from the primary to the secondary side.

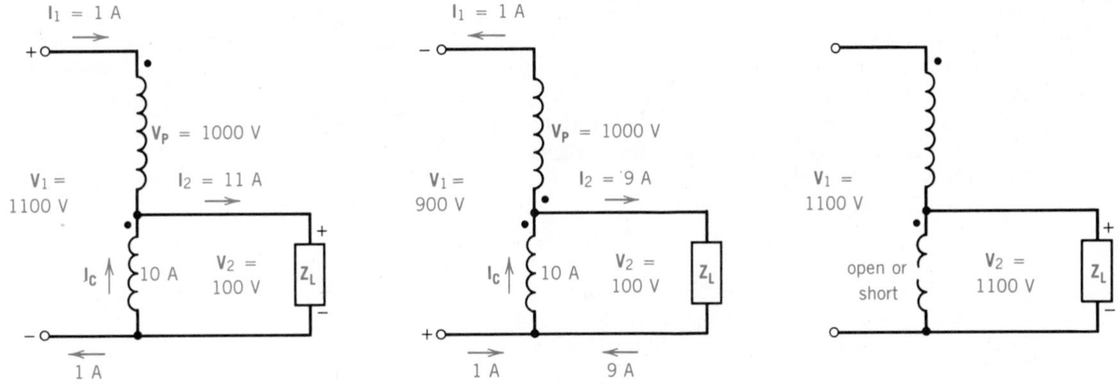

a. Step-down autotransformer, additive polarity mode, $S = 1.1$ kVA

b. Step-down autotransformer subtractive polarity mode; $S = 900$ VA

c. Transformer fault producing both shock hazard and overload on Z_L

Figure 20-20 Autotransformer in additive and subtractive polarity modes and hazards in step-down mode (Ex. 20-16)

Total kVA transfer = **transformed** volt-amperes + **conductive** volt-amperes

$$\mathbf{S}_T = \mathbf{V}_p\mathbf{I}_1 + \mathbf{V}_2\mathbf{I}_1 \tag{20-28}$$

For the **step-up** mode, we may write the following equations for the transformer of Fig. 20-19b:

Total kVA transfer = **transformed** volt-amperes + **conductive** volt-amperes

$$\mathbf{S}_T = \mathbf{V}_s\mathbf{I}_2 + \mathbf{V}_1\mathbf{I}_2 \tag{20-29}$$

EXAMPLE 20-16

For the autotransformer in either the step-down or step-up mode, using Eqs. (20-28) and/or (20-29), respectively, calculate the

a. Total kVA transfer in the *step-down* mode for Fig. 20-20**a**, using given values
b. Total kVA transfer in the *step-up* mode for Fig. 20-19**b**, using given values
c. kVA rating of the autotransformer in (**a**) using Fig. 20-20a
d. kVA rating of the autotransformer in (**b**) using Fig. 20-19b
e. Explain why the kVA transfer in (**b**) is so much higher than in (**a**)

Solution

a. $\mathbf{S}_T = \mathbf{V}_p\mathbf{I}_1 + \mathbf{V}_2\mathbf{I}_1 = (1000\text{ V})(1\text{ A}) + (100\text{ V})(1\text{ A})$
$= 1\text{ kVA transformed} + 0.1\text{ kVA conducted}$
$= \mathbf{1.1\ kVA\ transferred}$ (20-28)

b. $\mathbf{S}_T = \mathbf{V}_s\mathbf{I}_2 + \mathbf{V}_1\mathbf{I}_2$
$= (100\text{ V})(10\text{ A}) + (1000\text{ V})(10\text{ A})$ (20-29)
$= 1\text{ kVA transformed} + 10\text{ kVA conducted}$
$= \mathbf{11\ kVA\ transferred}$

c. $\mathbf{S}_{x\text{-former}} = \mathbf{V}_p\mathbf{I}_1 = \mathbf{V}_2\mathbf{I}_c = (1000\text{ V})(1\text{ A})$
$= (100\text{ V})(10\text{ A}) = \mathbf{1\ kVA}$

d. $\mathbf{S}_{x\text{-former}} = \mathbf{V}_1\mathbf{I}_c = \mathbf{V}_s\mathbf{I}_2 = (1000\text{ V})(1\text{ A})$
$= (100\text{ V})(10\text{ A}) = \mathbf{1\ kVA}$

e. Both transformers have the *same* kVA rating of 1.0 kVA, since the *same* autotransformer is used in both parts. Both transformers *transform* a total of 1 kVA. But the step-down transformer in (**a**) only *conducts* 0.1 kVA while the step-up transformer in (**b**) *conducts* 10 kVA from the primary to the secondary!

Example 20-16 reveals several important insights regarding autotransformers:

1. Autotransformers produce their *highest* kVA transfer at *low* transformation ratios, that is, close to unity, where α is approximately less or greater than one. In Fig. 20-19b $\alpha = 1000\text{ V}/1100\text{ V} = 0.91$ and the total kVA transfer is $\mathbf{S} = 11$ kVA, of which 10 kVA is conducted and 1 kVA is transformed.

2. Autotransformers produce their highest efficiencies when the transformation ratio is close to unity. Under these conditions almost all the energy is conductively transferred and only the portion that is transformed produces losses.
3. If we consider the autotransformer of Fig. 20-19b to consist of only *one* winding, we note that most of the winding (*i.e.*, the common portion) is carrying 1 A while the few remaining turns (the 100-V portion) are carrying 10 A. This reduces the copper losses considerably, increasing the transformer efficiency.
4. Even in the step-down mode of Fig. 20-20b, assuming that only one winding is used, the major portion of that winding carries only 1 A while a small portion carries 10 A. Again, this reduces the copper losses considerably, making for higher autotransformer efficiency.
5. The kVA rating of the autotransformer is not the same as the kVA transferred from one circuit to another.[11] The kVA transferred (at best) may be higher, equal to, or even less than the kVA rating of the transformer, as shown in Ex. 20-16.
6. The kVA *rating* of a standard isolation transformer, reconnected as an autotransformer, *remains the same*! The operation and behavior of any transformer *cannot be changed* merely by reconnecting it in any way. The *losses* are the *same*, the *winding currents* are the *same*, and the *mutual flux* in the core is the *same*.

At this point the reader may ask, "If autotransformers are so superior to conventional isolation transformers, why don't we use autotransformers exclusively?"

One answer is shown in Fig. 20-20c. Should an open occur in the *common* winding, then the secondary voltage, instead of being 100 V, is the *same* as the primary voltage (1100 V)! In other words, *autotransformers provide no isolation between primary and secondary* and, as a result, lack all the advantages of isolation noted in Sec. 20-4. Furthermore, in the event of an open in the common winding (Fig. 20-20c), the extremely high secondary voltage could burn out or seriously damage the equipment connected to the secondary side, not to mention the possibility of being a serious shock hazard to personnel.

Autotransformers, therefore, find their greatest application to situations involving *low voltages*, as in reduced voltage motor starting, for example. Here, the advantages of smaller size and weight, lower cost, and higher efficiency dictate their use with minimum disadvantages or hazards to personnel.

20-9 LOOSELY COUPLED TRANSFORMERS

The conventional power transformers covered in the previous sections were iron-core transformers whose coefficient of coupling was approximately unity ($k \cong 1$). For the remainder of this chapter we will consider either air-core transformers or magnetic-core transformers whose coefficients of coupling are less than unity ($k < 1$). Such transformers are called *loosely coupled* transformers.

Several distinct *differences* exist between tightly coupled (power) transformers and loosely coupled transformers and seriously affect the analysis of the latter:

1. Tightly coupled transformers are sufficiently close to the ideal transformer that the turns ratio ($\alpha = N_1/N_2$) may be used to predict voltages, currents, and reflected impedances. This is not the case for loosely coupled transformers, and other methods must be found involving the use of self- and mutual inductances.
2. Inductive loads on the secondary are reflected as inductive loads on the primary. Capacitive loads on the secondary are reflected as capacitive loads on the primary in conventional *tightly coupled* transformers. Loosely coupled transformers exhibit the reverse effect! *Inductive* loads on the *secondary* are reflected as *capacitive* loads on the *primary*. Similarly, *capacitive* loads on the *secondary* are reflected as *inductive* loads on the *primary* in *loosely coupled* transformers.

[11] This is an error frequently made by students and even in some textbooks.

3. The mutual flux, ϕ_m, in tightly coupled transformers is almost as high as the total flux produced by the primary winding, ϕ_p, where the coefficient of coupling, k, is the ratio $\phi_m/\phi_p \cong 1$; (i.e., the leakage flux is relatively small). But in loosely coupled transformers the mutual flux, ϕ_m, is only a small portion of the total primary flux, ϕ_p, and coefficients of coupling of 0.01 or even 0.001 are not uncommon.
4. In the tightly coupled transformer, the mutual impedance due to mutual inductance could be neglected since the transformer was considered close to ideal (Sec. 20-2). But in the loosely coupled transformer, the only parameter common to *both* the primary and secondary coils is, by definition, the *mutual inductance*, which, however small, *can no longer be ignored*.

Since mutual inductance, M, is an important parameter for the loosely coupled transformer, the relationship between M, and the coefficient of coupling, k, and the self-inductances L_1 and L_2 of two loosely coupled coils, as derived in Sec. 12-6, are repeated here (with a small modification to include ac relations):

$$k = \frac{M}{\sqrt{L_1 L_2}} = \frac{\pm j\omega M}{\sqrt{j\omega L_1 \cdot j\omega L_2}} = \frac{\pm jX_m}{\sqrt{jX_{L_1} \cdot jX_{L_2}}} \qquad \text{(12-12a)}$$

where M is the mutual inductance, in henrys, between two coils, L_1 and L_2

L_1 and L_2 are the respective self-inductances of the primary and secondary coils of a loosely coupled transformer

$\pm j\omega M = \pm jX_m$ is the mutual reactance between the two coils

$+jX_{L_1}$ and $+jX_{L_2}$ are the reactances of the two coils, respectively, at a given frequency

20-9.1 The Dot Convention Applied to Inductively Coupled Coils

In Sec. 12-7 the dot convention was defined. We stated that when the current flows into a dot when *entering* each coil, the polarities are additive and the mutual inductance is *positive*. When current enters the dotted terminal of one coil and leaves at the dotted terminal of a second coil, the mutual inductance between them is *negative*.

The dot convention applied to two inductively coupled coils is shown in **Fig. 20-21**. The mutual inductance between the two coils is positive in Fig. 20-21a because as the current enters *each* coil it encounters a dot. The mutual inductance between the two coils is *negative* in Fig. 20-21b because $\mathbf{I}_1$ encounters a dot *entering* L_1 but $\mathbf{I}_2$ encounters the dot *leaving* L_2. Note that *clockwise reference* currents are used for *both* circuits to enable mesh analysis of both circuits.

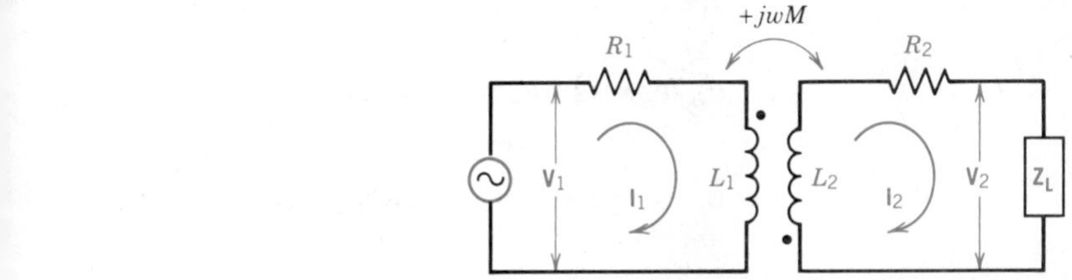

a. Positive mutual inductance: current encounters dot entering each coil

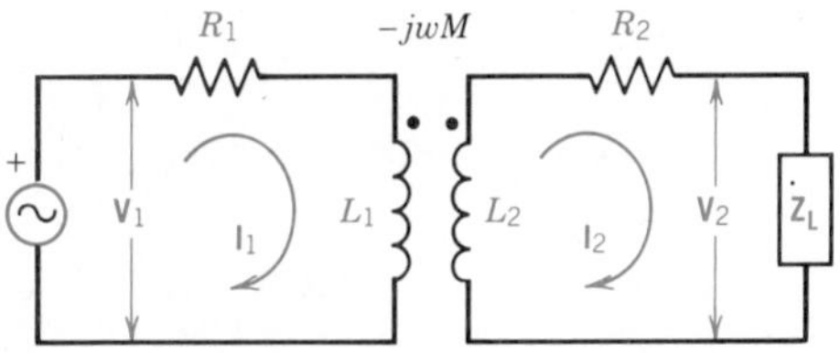

b. Negative mutual inductance-current into dot in one coil only

Figure 20-21 Dot convention to determine whether mutual inductance is positive or negative with respect to reference CW currents

20-9.2 The T-Equivalent of Two Inductively Coupled Coils

Figure 20-22a shows the circuit of Fig. 20-21a reproduced once again, in which the mutual inductance is *negative* by the dot convention. (We will discover later that for *two* inductively coupled coils, it makes little difference whether the mutual inductance is positive or negative.[12] Either assumption yields identical current *magnitudes*.) We noted earlier that the only element common to both the primary and the secondary circuit of Fig. 20-22a is the mutual inductance, whose reactance is $-j\omega M$. The simplest possible network that permits the same component ($-j\omega M$) common to both circuits is the conductively coupled T-network shown in Fig. 20-22b. To make this circuit appear *identical* to Fig. 20-22a, the inductor in series with R_1 is $(L_1 - M)$ and the inductor in series with R_2 is $(L_2 - M)$. The *open-circuit* **input impedance**, $\mathbf{Z}_{11}$, may be written and defined in Fig. 20-22b as

$$\mathbf{Z}_{11} = R_1 + j\omega L_1 - j\omega M + j\omega M$$
$$= R_1 + j\omega L_1 = R_1 + jX_{L_1} = \frac{\mathbf{V}_1}{\mathbf{I}_1} \qquad (\mathbf{I}_2 = 0) \qquad (20\text{-}30)$$

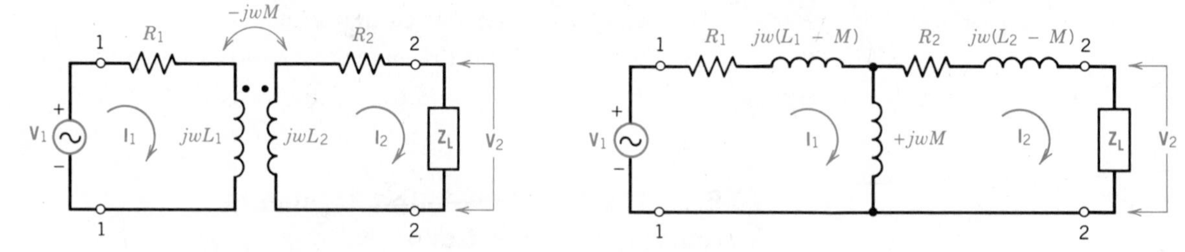

a. Two inductively coupled coils with negative mutual inductance between them

b. Conductively coupled T-equivalent circuit

Figure 20-22 The conductively-coupled T-equivalent of two inductively coupled coils

In the same way, the *open-circuit* **output impedance**, $\mathbf{Z}_{22}$, may be written as

$$\mathbf{Z}_{22} = R_2 + j\omega L_2 - j\omega M + j\omega M$$
$$= R_2 + j\omega L_2 = R_2 + jX_{L_2} = \frac{\mathbf{V}_2}{\mathbf{I}_2} \qquad (\mathbf{I}_1 = 0) \qquad (20\text{-}31)$$

Two other open-circuit parameters may be written, namely

$$\mathbf{Z}_{21} = \frac{\mathbf{V}_2}{\mathbf{I}_1} \qquad (\mathbf{I}_2 = 0), \textbf{ forward transfer impedance} \qquad (20\text{-}32)$$

$$\mathbf{Z}_{12} = \frac{\mathbf{V}_1}{\mathbf{I}_2} \qquad (\mathbf{I}_1 = 0), \textbf{ reverse transfer impedance} \qquad (20\text{-}33)$$

[12] See end-of-chapter Prob. 20-38 where this is proved for *two* inductively coupled coils. It is also proved in Exs. 20-17 through 20-19 by alternative methods of solution.

Our object is to write expressions for the unknown currents, $\mathbf{I}_1$ and $\mathbf{I}_2$. Using mesh analysis for the LH loop, we may write

$$\mathbf{I}_1[R_1 + j\omega(L_1 - M) + j\omega M] - \mathbf{I}_2 j\omega M = \mathbf{V}_1 \quad \text{or} \quad \mathbf{I}_1(R_1 + j\omega L_1) - \mathbf{I}_2 j\omega M = \mathbf{V}_1$$

And for the RH loop, using mesh analysis (format method), we write

$$-\mathbf{I}_1 j\omega M + \mathbf{I}_2[R_2 + j\omega(L_2 - M) + j\omega M + \mathbf{Z}_L] = 0 \quad \text{or}$$
$$-\mathbf{I}_1 j\omega M + \mathbf{I}_2(R_2 + j\omega M + \mathbf{Z}_L) = 0$$

We may now write the equations for each loop in array form as

$\mathbf{I}_1$	$\mathbf{I}_2$	$\mathbf{V}$
$\mathbf{Z}_{11}$	$-j\omega M$	$\mathbf{V}_1$
$-j\omega M$	$(\mathbf{Z}_{22} + \mathbf{Z}_L)$	0

(20-34)

Then, using Cramer's rule for the array, we may write for the primary current, $\mathbf{I}_1$,

$$\mathbf{I}_1 = \frac{\begin{vmatrix} \mathbf{V}_1 & -j\omega M \\ 0 & (\mathbf{Z}_{22} + \mathbf{Z}_L) \end{vmatrix}}{\begin{vmatrix} \mathbf{Z}_{11} & -j\omega M \\ -j\omega M & (\mathbf{Z}_{22} + \mathbf{Z}_L) \end{vmatrix}} = \frac{\mathbf{V}_1(\mathbf{Z}_{22} + \mathbf{Z}_L)}{\mathbf{Z}_{11}(\mathbf{Z}_{22} + \mathbf{Z}_L) + |\omega M|^2} = \frac{\mathbf{V}_1 \mathbf{Z}_2}{(\mathbf{Z}_{11})\mathbf{Z}_2 + |\omega M|^2} \tag{20-35}$$

where $\mathbf{Z}_2$ is the entire secondary circuit impedance.

Similarly, for the array given in Eq. (20-34) we may write for the secondary current, $\mathbf{I}_2$,

$$\mathbf{I}_2 = \frac{\begin{vmatrix} \mathbf{Z}_{11} & \mathbf{V}_1 \\ -j\omega M & 0 \end{vmatrix}}{\begin{vmatrix} \mathbf{Z}_{11} & -j\omega M \\ -j\omega M & (\mathbf{Z}_{22} + \mathbf{Z}_L) \end{vmatrix}} = \frac{j\omega M \cdot \mathbf{V}_1}{\mathbf{Z}_{11}(\mathbf{Z}_{22} + \mathbf{Z}_L) + |\omega M|^2} = \frac{j\omega M \cdot \mathbf{V}_1}{(\mathbf{Z}_{11})\mathbf{Z}_2 + |\omega M|^2} \tag{20-36}$$

where $|\omega M|$ is the **absolute value** of mutual reactance or $|X_m|$.

20-9.3 Input Impedance of a Loosely Coupled Transformer

Equation (20-35) contains within it the expression for the primary input impedance under any conditions of load, $\mathbf{Z}_L$. Transposing terms, we may write for input impedance, $\mathbf{Z}_{in}$,

$$\mathbf{Z}_{in} = \frac{\mathbf{V}_1}{\mathbf{I}_1} = \frac{\mathbf{Z}_{11}(\mathbf{Z}_{22} + \mathbf{Z}_L) + |\omega M|^2}{\mathbf{Z}_{22} + \mathbf{Z}_L} = \mathbf{Z}_{11} + \frac{|\omega M|^2}{\mathbf{Z}_{22} + \mathbf{Z}_L} \tag{20-35a}$$

If we designate $\mathbf{Z}_2 = \mathbf{Z}_{22} + \mathbf{Z}_L$, where $\mathbf{Z}_2$ is the *entire secondary* circuit impedance, we obtain a single complex impedance term that, by conjugate rationalization of the denominator, yields

$$\mathbf{Z}_{in} = \mathbf{Z}_{11} + \frac{|\omega M|^2}{\mathbf{Z}_2} = \mathbf{Z}_{11} + \frac{|\omega M|^2 \times \mathbf{Z}_2^*}{\mathbf{Z}_2 \times \mathbf{Z}_2^*}$$

$$= \mathbf{Z}_{11} + \left|\frac{\omega M}{\mathbf{Z}_2}\right|^2 \times \mathbf{Z}_2^* = \mathbf{Z}_{11} + \alpha^2 \mathbf{Z}_2^* \qquad \textbf{(20-35b)}$$

Observe that the input impedance, $\mathbf{Z}_{in}$, of the transformer in Eq. (20-35b) contains *two* discrete terms: $\mathbf{Z}_{11}$, which is the open-circuit impedance of the primary winding as defined in Eq. (20-30), and the reflected impedance, $\mathbf{Z}_r$, which is

$$\mathbf{Z}_r = \left|\frac{\omega M}{\mathbf{Z}_2}\right|^2 \times \mathbf{Z}_2^* = \alpha^2 \mathbf{Z}_2^* \qquad \textbf{(20-37)}$$

where $\mathbf{Z}_r$ is the secondary-circuit impedance *reflected* to the primary circuit

α is a *reflectance factor*, the ratio of the absolute value $\left|\frac{\omega M}{\mathbf{Z}_2}\right|$

$\mathbf{Z}_2^*$ is the conjugate of the entire secondary-circuit impedance

Equation (20-37) and the general Eq. (20-35b) are *extremely important* relations for the loosely coupled transformer. They provide the following insights:[13]

1. Equation (20-37) shows that the reflected impedance of a loosely coupled transformer may be likened to Eq. (20-2a) of the tightly coupled transformer, thus *unifying the theory* of tightly and loosely coupled transformers.
2. Tightly coupled transformers reflect a secondary to primary impedance of $\alpha^2 \mathbf{Z}_L$, where α is the turns ratio, N_1/N_2. Loosely coupled transformers reflect a secondary to primary impedance of $\alpha^2 \mathbf{Z}_2^*$, where α is the ratio of the absolute value $\left|\frac{\omega M}{\mathbf{Z}_2}\right|$.
3. Since the loosely coupled transformer reflects the *conjugate* of the entire secondary load impedance to the primary by $\alpha^2 \mathbf{Z}_2^*$, *capacitive* loads are reflected *inductively* and inductive loads are *reflected capacitively* (as noted at the beginning of Sec. 20-9).
4. Since $\mathbf{Z}_{11}$ is usually an inductive impedance in a loosely coupled transformer, an inductive or even unity PF load on the secondary reflects a total capacitive impedance that tends to counteract the primary inductive impedance, $\mathbf{Z}_{11}$, producing a total input impedance, $\mathbf{Z}_{in}$, that is closer to unity PF (i.e., an input impedance that is essentially *resistive*).

[13] To the author's knowledge, these first two insights, which unify loosely and tightly coupled transformers, are not published elsewhere.

20-9.4 Reverse Transfer Impedance, $\mathbf{Z}_{12}$, of a Loosely Coupled Transformer

When given $\mathbf{Z}_{11}$, $\mathbf{Z}_{22}$, and $\pm j\omega M$ for a loosely coupled transformer with connected load $\mathbf{Z}_L$, it is possible to use mesh analysis as a general method for finding $\mathbf{I}_1$ and $\mathbf{I}_2$. Alternatively, Eqs. (20-35) and (20-36) may be used, since they were derived by (and from) mesh analysis. Or Eq. (20-35b) may be used to find $\mathbf{I}_1$, given $\mathbf{V}_1$, since by Ohm's law $\mathbf{I}_1 = \mathbf{V}_1/\mathbf{Z}_{\text{in}}$, where $\mathbf{Z}_{\text{in}}$ is expressed in terms of $\mathbf{Z}_{11}$, ωM, and $\mathbf{Z}_2$.

The reader may now ask, "Is there some shortcut expression similar to Eq. (20-35b) that may enable us to find $\mathbf{I}_2$ for any given value of $\mathbf{V}_1$?"

The answer to this question is "Yes, there is. It is found in the *reverse transfer impedance* defined in Eq. (20-33), above."

The reverse transfer impedance is $\mathbf{Z}_{12} = \mathbf{V}_1/\mathbf{I}_2$, and alternatively $\mathbf{I}_2 = \mathbf{V}_1/\mathbf{Z}_{12}$. If we could evaluate $\mathbf{Z}_{12}$, we should be able to find the secondary current, $\mathbf{I}_2$, for any given value of $\mathbf{V}_1$. But, in point of fact, we have evaluated it in Eq. (20-36) already. From Eq. (20-36) transposed, we may write for the **reverse transfer impedance**

$$\mathbf{Z}_{12} = \frac{\mathbf{V}_1}{\mathbf{I}_2} = \frac{(\mathbf{Z}_{11})\mathbf{Z}_2 + |\omega M|^2}{(j\omega M)} \quad \text{ohms } (\Omega) \qquad (20\text{-}36a)$$

where all terms have been defined.

Some important precautions are needed in using **Eq. (20-36a)**:

1. The term $|\omega M|^2$ in the *numerator* is the square of an *absolute number* (not a complex quantity) and is treated as a *real* value to be added to the complex product $(\mathbf{Z}_{11})\mathbf{Z}_2$.
2. The term $(j\omega M)$ in the *denominator* is a **complex** number (i.e., $X_m\angle{+90^\circ}$) and should be treated as such when divided into the numerator term to find $\mathbf{Z}_{12}$.

We are now ready to do a series of examples verifying these relations.

EXAMPLE 20-17

For the original loosely coupled circuit shown in **Fig. 20-23a**,

a. Calculate the mutual reactance, $\pm jX_m$, both magnitude and polarity
b. Draw the T-equivalent, showing all primary and secondary impedances appropriately
c. Using mesh analysis, calculate loop currents $\mathbf{I}_1$ and $\mathbf{I}_2$

Solution

a. $jX_m = k\sqrt{jX_{L_1} jX_{L_2}} = 0.5\sqrt{(j8)(j32)} = \mathbf{-j8\ \Omega}$. Note: since **both** reference loop currents do *not* encounter polarity dots when entering the coil, the mutual reactance $\pm jX_m$ is a **negative** value $-j8\ \Omega$
b. Since $\mathbf{Z}_{11} = 2 + j8\ \Omega$ in Fig. 20-23a, the T-equivalent of Fig. 20-23b must be drawn with $+j8$ as the common, mutual

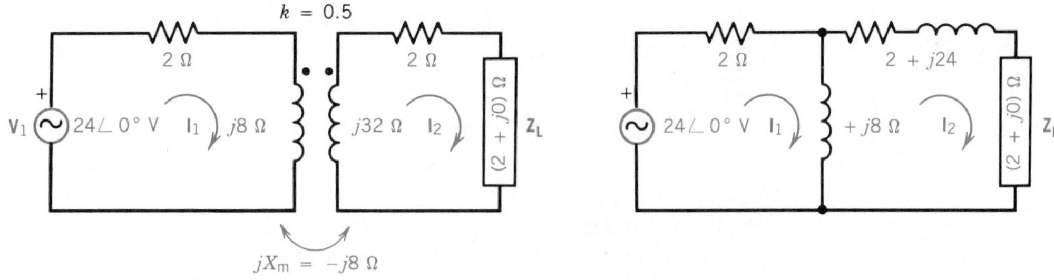

a. Original loosely coupled circuit

b. T-equivalent for mesh analysis yielding same results

Figure 20-23 Examples 20-17 through 20-20

reactance and *no reactance* in series with the 2-Ω resistor in the primary loop, to satisfy this relation. Since $\mathbf{Z}_{22}$ is $(2 + j32)$ Ω, the secondary loop is drawn in Fig. 20-23b with $2 + j24$ Ω as the equivalent secondary reactance. The total impedance, $\mathbf{Z}_2$, of the secondary circuit is $\mathbf{Z}_2 = 4 + j32$ Ω, as verified from Fig. 20-23a and Fig. 20-23b, the T-equivalent circuit.

c. By the format method of mesh analysis for the T-equivalent circuit we may write, by inspection: $\mathbf{I}_1(2 + j8) - \mathbf{I}_2(j8) = 24$ for the LH loop and $-\mathbf{I}_1(j8) + \mathbf{I}_2(4 + j32) = 0$ for the RH loop. These equations yield the following array:

$\mathbf{I}_1$	$\mathbf{I}_2$	$\mathbf{V}$
$2 + j8$	$-j8$	24
$-j8$	$4 + j32$	0

$$\text{determinant } \Delta = \begin{vmatrix} 2 + j8 & -j8 \\ -j8 & 4 + j32 \end{vmatrix} = \mathbf{-184 + j96}$$

$$\mathbf{I}_2 = \frac{\begin{vmatrix} 2 + j8 & 24 \\ -j8 & 0 \end{vmatrix}}{\Delta} = \frac{0 + j192}{-184 + j96} = \mathbf{0.428 - j0.8202}$$

$$= \mathbf{0.9251\underline{/-62.45^\circ}\ A}$$

$$\mathbf{I}_1 = \frac{\begin{vmatrix} 24 & -j8 \\ 0 & 4 + j32 \end{vmatrix}}{\Delta} = \frac{96 + j768}{-184 + j96} = \mathbf{1.302 - j3.495}$$

$$= \mathbf{3.729\underline{/-69.57^\circ}\ A}$$

EXAMPLE 20-18

Verify the **mesh** solution of currents $\mathbf{I}_1$ and $\mathbf{I}_2$, using

a. Eq. (20-35) to find $\mathbf{I}_1$
b. Eq. (20-36) to find $\mathbf{I}_2$

Solution

$$\mathbf{I}_1 = \frac{\mathbf{V}_1\mathbf{Z}_2}{(\mathbf{Z}_{11})\mathbf{Z}_2 + |\omega M|^2} = \frac{24\underline{/0^\circ} \times (4 + j32)}{(2 + j8)(4 + j32) + (8)^2}$$

$$= \frac{96 + j768}{(64 - 248) + j96} = \mathbf{1.302 - j3.495 = 3.729\underline{/-69.57^\circ}\ A}$$

$$\mathbf{I}_2 = \frac{\mathbf{V}_1(j\omega M)}{(\mathbf{Z}_{11})\mathbf{Z}_2 + |\omega M|^2} = \frac{24\underline{/0^\circ}(8\underline{/90^\circ})}{(64 - 248) + j96} = \frac{192\underline{/90^\circ}}{-184 + j96}$$

$$= \mathbf{0.428 - j0.8202 = 0.9251\underline{/-62.45^\circ}\ A}$$

EXAMPLE 20-19

Verify the mesh solution of currents $\mathbf{I}_1$ and $\mathbf{I}_2$, using

a. Input impedance, $\mathbf{Z}_{in} = \mathbf{Z}_{11} + \alpha^2\mathbf{Z}_2^*$, to find $\mathbf{I}_1$
b. Reverse transfer impedance, $\mathbf{Z}_{12} = \dfrac{\mathbf{Z}_{11}\mathbf{Z}_2 + |\omega M|^2}{(j\omega M)}$, to find $\mathbf{I}_2$

Solution

$$\mathbf{Z}_{in} = \mathbf{Z}_{11} + \alpha^2\mathbf{Z}_2^* = (2 + j8) + \left|\frac{8\ \Omega}{32.25}\right|^2 \times (4 - j32)$$

$$= 2.2462 + j6.031 = 6.4354\underline{/69.57^\circ}\ \Omega$$

a. $\mathbf{I}_1 = \mathbf{V}_1/\mathbf{Z}_{in} = 24\underline{/0^\circ}\ \text{V}/6.4354\underline{/69.57^\circ}\ \Omega$

$$= \mathbf{3.729\underline{/-69.57^\circ}\ A}$$

b. $$\mathbf{Z}_{12} = \frac{\mathbf{Z}_{11}\mathbf{Z}_2 + |\omega M|^2}{j\omega M} = \frac{(2 + j8)(4 + j32) + 8^2}{0 + j8}$$

$$= \frac{-184 + j96}{0 + j8} = (12 + j23)\ \Omega$$

$$\mathbf{I}_2 = \mathbf{V}_1/\mathbf{Z}_{12} = \frac{(24 + j0)\ \text{V}}{(12 + j23)\ \Omega} = \mathbf{0.428 - j0.8202}$$

$$= \mathbf{0.9251\underline{/-62.45^\circ}\ A}$$

Examples 20-17 through 20-19 show how it is possible to solve for primary and secondary currents $\mathbf{I}_1$ and $\mathbf{I}_2$ by **three separate** methods, *all* of which yield the *same* answers. The three methods described thus far are:

Example 20-17 The method of mesh analysis requiring solution of complex simultaneous equations to find unknown currents $\mathbf{I}_1$ and $\mathbf{I}_2$.

Example 20-18 Using general equations for $\mathbf{I}_1$ and $\mathbf{I}_2$ derived from mesh analysis techniques.

Example 20-19 Using input impedance $\mathbf{Z}_{in}$ to find $\mathbf{I}_1$ and reverse transfer impedance $\mathbf{Z}_{12}$ to find $\mathbf{I}_2$.

All these methods verify the analyses presented thus far because they yield the same answers *exactly*.

Two other methods are presented below. The first uses a Thévenin analysis technique and requires equations for equivalent voltage and impedance seen by the load, $\mathbf{Z}_L$. The second, and last, uses simple circuit theory. It requires practically no memorization of equations and probably yields a greater insight into the nature of loosely coupled circuits. (The best has been saved for last.)

20-9.5 Thévenin Voltage and Impedance Seen by Load Z_L in a Loosely Coupled Transformer

The generic T-equivalent circuit of Fig. 20-22b is shown once again in **Fig. 20-24a** with minor variation in a form suitable for Thévenin analysis. Our object is to find the equivalent voltage, $\mathbf{V}_{TH}$, and equivalent circuit impedance, $\mathbf{Z}_{TH}$, "seen" by load $\mathbf{Z}_L$, with $\mathbf{Z}_L$ removed. Using the standard voltage-divider technique, we may write the Thévenin voltage by inspection as

$$\mathbf{V}_{TH} = \frac{jX_m \times \mathbf{V}_1}{R_1 + j(X_{L_1} - X_m) + jX_m} = \frac{jX_m \times \mathbf{V}_1}{R_1 + jX_{L_1}} = \frac{jX_m}{\mathbf{Z}_{11}} \times \mathbf{V}_1 \quad \text{volts (V)} \tag{20-38}$$

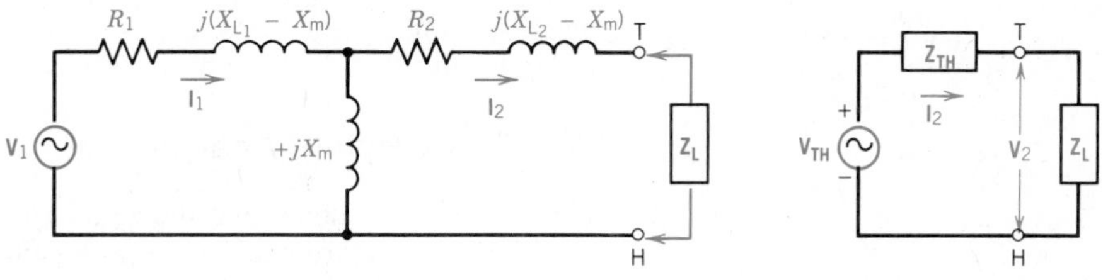

a. T-equivalent for mesh analysis

b. Thévenin equivalent with $\mathbf{Z}_L$ replaced

Figure 20-24 Determining secondary current $\mathbf{I}_2$ and secondary voltage $\mathbf{V}_2$ by Thévenin analysis

The equivalent Thévenin impedance seen by $\mathbf{Z}_L$ requires simple series–parallel circuit theory:

$$\begin{aligned}\mathbf{Z}_{TH} &= R_2 + j(X_{L_2} - X_m) + jX_m \,\|\, [R_1 + j(X_{L_1} - X_m)] \\ &= R_2 + j(X_{L_2} - X_m) + \frac{jX_m[R_1 + j(X_{L_1} - X_m)]}{R_1 + j(X_{L_1} - X_m) + jX_m} \\ &= R_2 + j(X_{L_2} - X_m) + \frac{jX_m[R_1 + j(X_{L_1} - X_m)]}{\mathbf{Z}_{11}} \quad \text{ohms } (\Omega)\end{aligned} \tag{20-39}$$

Let us see if Eqs. (20-38) and (20-39) yield the same answers for the circuit used in Exs. 20-17 through 20-19.

EXAMPLE 20-20

Given the circuit shown in Fig. 20-23a, using the T-equivalent drawn in Fig. 20-23b and Thévenin analysis, calculate the

a. Thévenin voltage seen by load $\mathbf{Z}_L$ using Eq. (20-38)
b. Thévenin impedance seen by load $\mathbf{Z}_L$ using Eq. (20-39)
c. Secondary load current $\mathbf{I}_2$ drawn by load $\mathbf{Z}_L$

Solution

a. $\mathbf{V}_{TH} = \dfrac{jX_m}{\mathbf{Z}_{11}} \times \mathbf{V}_1 = \dfrac{j8}{2 + j8} \times 24\angle 0^\circ$

$= (0.9701\angle 14.04^\circ)(24\angle 0^\circ) = \mathbf{23.283\angle 14.04^\circ}$ **V**

b. $\mathbf{Z}_{TH} = R_2 + j(X_{L_2} - X_m) + \dfrac{jX_m[R_1 + j(X_{L_1} - X_m)]}{\mathbf{Z}_{11}}$

$= (2 + j24) + \dfrac{8\angle 90^\circ(2\angle 0^\circ)}{2 + j8}$

$= (2 + j24) + (1.882\,35 + j0.4706)$

$= \mathbf{(3.882\,35 + \mathit{j}24.4706)\ \Omega}$

c. Then from Fig. 20-24b for the simplified series circuit, we have

$$\mathbf{I}_2 = \frac{\mathbf{V}_{TH}}{\mathbf{Z}_{TH} + \mathbf{Z}_L} = \frac{23.283\angle 14.04^\circ \text{ V}}{(3.882\,35 + j24.4706) + (2 + j0)}$$

$$= \frac{23.283\angle 14.04^\circ}{25.168\angle 76.48^\circ} = \mathbf{0.9251\angle -62.44^\circ\ A}$$

Note that the value of $\mathbf{I}_2$ obtained by Thévenin analysis agrees favorably with that obtained by the three previous methods shown in Exs. 20-17 through 20-19.

One major disadvantage of the Thévenin analysis approach is that it yields *only* the secondary current, $\mathbf{I}_2$. It tells us nothing about the primary circuit voltages and/or currents, not to mention the power transfer conditions of the loosely coupled transformer.

This last method, which we will call the *complete analysis* of the *loosely coupled transformer*, is not only the simplest approach but also the most complete, judging from the information that may be obtained about the loosely coupled circuit. It uses simple conventional ac circuit theory and, therefore, requires practically no memorization of equations. It has the added advantage of yielding the complete power analysis of the loosely coupled transformer, as well.

20-9.6 Complete Analysis of Loosely Coupled Transformer

Up to now we assumed that the voltage source supplying the transformer is ideal. At this point, we are ready to consider that it is a practical source. Nor have we considered, thus far, the power relations that occur in the loosely coupled transformer. The process described below in a series of steps permits the reader to visualize the complete power transfer by using simple circuit theory. We will first develop the relatively simple equations and then illustrate the technique with an example.

Let us assume at the outset that the essential parameters as measured in the laboratory for two inductively loosely coupled coils are L_1, L_2, M, and/or k. The loosely coupled transformer is connected to a load, $\mathbf{Z}_L$, and a source, $\mathbf{V}_g$, whose resistance, R_g, is known. The steps below include some preliminary calculation equations, as well as those Eqs. leading to the complete analysis of the loosely coupled transformer, culminating in the complete power analysis of the loosely coupled transformer.

A. **Preliminary calculations**
1. Find mutual inductance, $M = k\sqrt{L_1 \cdot L_2}$
2. Find primary coil winding reactance, $X_{L_1} = j\omega L_1$
3. Find secondary coil winding reactance, $X_{L_2} = j\omega L_2$
4. Find mutual reactance, $jX_m = j\omega M$
5. Find coefficient of coupling, $k = M/\sqrt{L_2 \cdot L_1} = X_m/\sqrt{X_{L_2} \cdot X_{L_1}}$

B. **Loosely coupled transformer circuit analysis steps (see Fig. 20-25a)**
1. Find total secondary load impedance, $\mathbf{Z}_2 = \mathbf{Z}_{22} + \mathbf{Z}_L = (R_2 + jX_{L_2}) + (R_L \pm jX)$
2. Find reflectance factor, $\alpha = \left|\dfrac{X_m}{\mathbf{Z}_2}\right|$ (a dimensionless ratio)
3. Find reflected impedance, $\mathbf{Z}_r = \alpha^2\mathbf{Z}_2^*$ (secondary impedance reflected to primary)
4. Find total primary input impedance, $\mathbf{Z}_{in} = \mathbf{Z}_{11} + \mathbf{Z}_r = (R_1 + jX_{L_1}) + \alpha^2\mathbf{Z}_2^*$
5. Find total primary current drawn from practical source, $\mathbf{I}_1 = \mathbf{V}_g/(\mathbf{Z}_g + \mathbf{Z}_{in})$

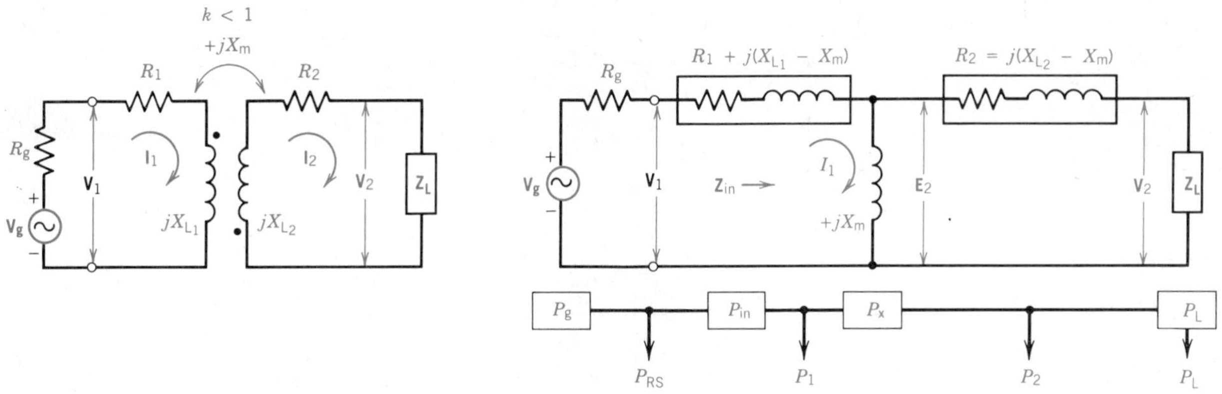

a. Two inductively coupled coils with positive mutual inductance between them

b. T-equivalent and associated power-flow diagram of loosely coupled transformer

Figure 20-25 Complete analysis (voltage, current, and power) of a loosely coupled transformer

6. Find secondary induced voltage, $\mathbf{E}_2 = \mathbf{I}_1(jX_m)$ (see Fig. 20-25b)
7. Find secondary current, $\mathbf{I}_2 = \mathbf{E}_2/\mathbf{Z}_2$ (where $\mathbf{Z}_2$ is obtained in step B-1)
8. Find secondary terminal voltage, $\mathbf{V}_2 = \mathbf{I}_2\mathbf{Z}_L$
9. Find primary transformer terminal voltage, $\mathbf{V}_1 = \mathbf{V}_g - \mathbf{I}_1\mathbf{Z}_g = \mathbf{I}_1(\mathbf{Z}_{in})$ (see step B-4)

C. **Loosely coupled transformer power analysis steps (see Fig. 20-25b)**
1. Find power generated by supply, $P_g = \mathbf{V}_g\mathbf{I}_g = \mathbf{V}_g\mathbf{I}_1$
2. Find power dissipated in resistance of the supply, $P_{Rs} = \mathbf{I}_1^2 R_s$
3. Find input power to the transformer, $P_{in} = P_g - \mathbf{I}_1^2 R_g$
4. Find power dissipated in the transformer primary, $P_1 = \mathbf{I}_1^2 R_1$
5. Find power transformed by the transformer, $P_x = P_{in} - P_1 = \mathbf{E}_2\mathbf{I}_2 \cos\theta_2 = \mathbf{I}_1^2 R_r$
6. Find power dissipated in the transformer secondary, $P_2 = \mathbf{I}_2^2 R_2$
7. Find power dissipated in the load, $P_L = \mathbf{I}_2^2 R_L = P_x \dfrac{R_L}{(R_L + R_2)} = P_x - P_2$

The reader should notice that none of the above equations is complex and that (with a little practice) the complete analysis of voltages, currents, and power is possible without having to memorize or use complicated equations.

EXAMPLE 20-21

Given the loosely coupled air-core transformer shown in **Fig. 20-26a** and all the values given for the practical source and load on the transformer secondary, calculate the

a. Mutual reactance between the two transformer coils, X_m
b. Total impedance of the secondary circuit, $\mathbf{Z}_2$
c. Reflectance factor, α
d. Impedance reflected to the primary, $\mathbf{Z}_r$
e. Input impedance seen at terminals 1–1 of the transformer under the given load, $\mathbf{Z}_{in}$
f. Primary current drawn from the practical source, $\mathbf{I}_1$
g. Voltage induced into the transformer secondary winding, $\mathbf{E}_2$
h. Current in the transformer secondary, $\mathbf{I}_2$, produced by $\mathbf{E}_2$[14]
i. Terminal voltage across the load impedance, $\mathbf{Z}_2$
j. Voltage across the input terminals of the transformer, $\mathbf{V}_1$, by two methods
k. Power generated by the source, P_g
l. Power dissipated within the source, P_{Rg}
m. Power received at the input of the transformer, P_{in}, by two methods
n. Power transformed by the transformer, P_x, by three methods
o. Power dissipated in the secondary of the transformer, P_2

[14] Note the direction of $\mathbf{I}_2$ produced by the polarity of $\mathbf{E}_2$ in Fig. 20-26. If the dot on the secondary coil is reversed, then $\mathbf{I}_2$ must be reversed! The current direction, magnitude, and phase are all verified in end-of-chapter Prob. 20-42 by mesh analysis.

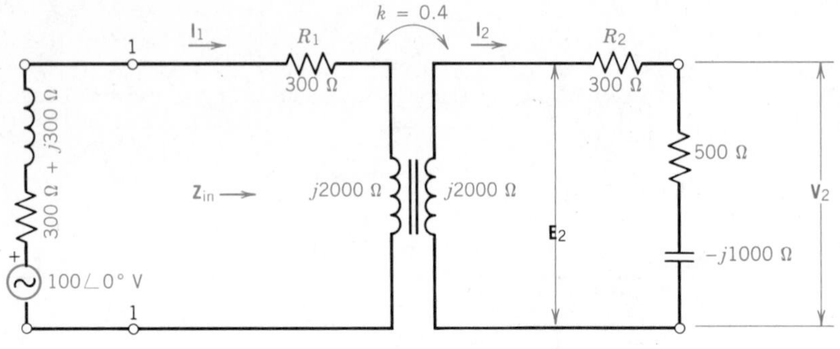

a. Example 20-21

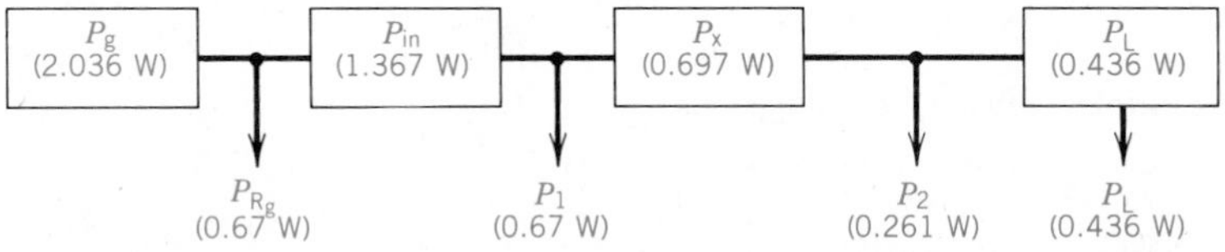

b. Power flow diagram for Ex 20-21

Figure 20-26 Complete phasor domain circuit and power flow diagram for Ex. 20-21

p. Power dissipated in the load by three methods, P_L
q. Efficiency of the power transformation (taken from the input to the transformer)
r. Draw a power flow diagram showing all of the above powers transferred, transformed, and dissipated

Solution

a. $X_m = k\sqrt{X_{L_1} \cdot X_{L_2}} = 0.4\sqrt{(2000)^2} = \mathbf{800\ \Omega}$
b. $\mathbf{Z}_2 = (300 + 2000) + (500 - j1000) = \mathbf{(800 + j1000)\ \Omega}$
c. $\alpha = \left|\dfrac{X_m}{\mathbf{Z}_2}\right| = \dfrac{800}{(800 + j1000)} = \dfrac{800}{1280.6} = \mathbf{0.6247}$
d. $\mathbf{Z}_r = \alpha^2\mathbf{Z}_2^* = (0.6247)^2(800 - j1000) = \mathbf{312.2 - j390.2\ \Omega}$
e. $\mathbf{Z}_{in} = \mathbf{Z}_{11} + \mathbf{Z}_r = (300 + j2000) + (312.2 - j390.2)$
$= \mathbf{(612.2 + j1609.8)\ \Omega}$
f. $\mathbf{I}_1 = \mathbf{V}_g/\mathbf{Z}_T = \dfrac{100\underline{/0°}}{912.2 + j1909.8} = \mathbf{20.36 - j42.64}$
$= \mathbf{47.25\underline{/-64.47°}\ mA}$
g. $\mathbf{E}_2 = \mathbf{I}_1(+jX_m) = (47.25\underline{/-64.47°})(800\underline{/90°})$
$= \mathbf{37.8\underline{/25.53°}\ V}$
h. $\mathbf{I}_2 = \mathbf{E}_2/\mathbf{Z}_2 = (37.8\underline{/25.53°}\ \text{V})\ /\ 1280.6\underline{/51.34°}\ \Omega$
$= \mathbf{29.52\underline{/-25.81°}\ mA}$
i. $\mathbf{V}_2 = \mathbf{I}_2\mathbf{Z}_L = (29.52\underline{/-25.81°}\ \text{mA})(500 - j1000)\ \Omega$
$= \mathbf{33\underline{/-89.24°}\ V}$
j. $\mathbf{V}_1 = \mathbf{I}_1\mathbf{Z}_{in} = (47.25\underline{/-64.47°}\ \text{mA})(612.2 + j1609.8)$
$= \mathbf{81.38\underline{/4.71°}\ V}$ and

$\mathbf{V}_1 = \mathbf{V}_g - \mathbf{I}_1\mathbf{Z}_g = (100 + j0) - (20.36 - j42.64)(300 + j300)$
$= \mathbf{81.1 + j6.684} = \mathbf{81.38\underline{/4.71°}\ V}$
k. $P_g = \mathbf{V}_g\mathbf{I}_1 \cos\theta_1 = 100 \times 47.25\ \text{mA} \times \cos 64.47° = \mathbf{2.036\ W}$
l. $P_{Rg} = \mathbf{I}_1^2R_g = (47.25\ \text{m})^2 \times 300 = \mathbf{0.67\ W}$ (note P_1 also is **0.67 W**)
m. $P_{in} = P_g - P_{Rg} = 2.036 - 0.67 = \mathbf{1.366\ W}$ and
$P_{in} = \mathbf{V}_1\mathbf{I}_1 \cos\theta_d = 81.38 \times 47.25\ \text{m}[\cos(4.71 - (-64.47°))]$
$= \mathbf{1.366\ W}$
n. $P_x = P_{in} - P_1 = 1.366 - 0.67 = \mathbf{0.697\ W}$;
$P_x = \mathbf{I}_1^2R_r = (47.25\ \text{m})^2 312.2 = \mathbf{0.697\ W}$;
$P_x = \mathbf{E}_2\mathbf{I}_2 \cos\theta_2$
$= 37.8 \times 29.52\ \text{m} \times \cos[25.53 - (-25.81°)] = \mathbf{0.697\ W}$
o. $P_2 = \mathbf{I}_2^2R_2 = (29.52\ \text{m})^2 \times 300 = \mathbf{0.261\ W}$
p. $P_L = P_x - P_2 = 0.697\ \text{W} - 0.261\ \text{W} = \mathbf{0.436\ W}$;
$P_L = \mathbf{I}_2^2R_L = (29.52\ \text{m})^2 \times 500 = \mathbf{0.436\ W}$;
$P_L = P_x\dfrac{R_L}{R_L + R_2} = 0.697\ \text{W}\dfrac{500}{(500 + 300)} = \mathbf{0.436\ W}$
q. $\eta = P_o/P_{in} = 0.436\ \text{W}/1.367\ \text{W} = \mathbf{31.9\%}$
r. The power flow diagram is drawn in Fig. 20-26b

Example 20-21 contains many insights into the operation of a loosely coupled two-coil transformer supplied from a practical voltage source. The reader would do well to attempt the complete solution *independently*, using the text solution as feedback to find errors in theory and/or calculation technique. Of particular importance, note that

1. The equations used require little or no memorization since they emerge from standard circuitry applications.
2. The total current drawn from the practical source in **f** must use the total *combined* impedance of the *source* as well as the *input impedance* of the transformer.
3. Similarly, when $\mathbf{I}_2$ is found in **h**, the total secondary circuit impedance, $\mathbf{Z}_2$ must be used in conjunction with $\mathbf{E}_2$, the voltage induced in the secondary winding. This method of find-

ing $\mathbf{I}_2$ is *much simpler* than the transfer impedance or mesh analysis method and requires no memorization whatever.

4. The various alternative methods for finding the various powers dissipated serve as *important checks* on the validity and accuracy of the computations. Should one more of these power equations fail to yield the *same* answer, it means that part (or all) of the previous calculations are in error. Indeed, as we have seen before (in using Tellegen's theorem), *power* computations serve *as an important check* on the accuracy of the voltage and/or current calculations of the complete solution.

20-10 LOOSELY COUPLED MULTICOIL TRANSFORMERS

Mutual inductance between two or more *series*-connected, *conductively coupled* coils was presented in Sec. 12-8. Similarly, mutual inductance between two or more *parallel*-connected, *conductively coupled* coils was presented in Sec. 12-9.

Reflected impedance in tightly coupled *multicoil* transformers, *inductively* coupled, was presented in Sec. 20-2.5. The mutual inductance of such transformers could be neglected since such tightly coupled transformers are considered close to ideal (Sec. 20-2).

In this section, we consider coupling among three or more coils of loosely coupled multicoil transformers. In the previous section, which treated two-coil transformers, if the mutual term between the coils was positive, it was entered into the mesh equation as $(0 + j\omega M)$, and if negative, it was entered as $(0 - j\omega M)$. Using the reflected impedance method, care must be taken to ensure that $\mathbf{I}_2$ flows in the direction indicated by the positive polarity dot on $\mathbf{E}_2$. Consequently, the polarity of the mutual term affects only the phase of $\mathbf{I}_2$. It does not affect the phase or magnitude of $\mathbf{I}_1$ or the magnitude of $\mathbf{I}_2$. Moreover, it does not affect the power dissipated in loop $\mathbf{I}_1$ or $\mathbf{I}_2$. We may summarize, as follows.

Assume a two-coil loosely coupled transformer circuit with positive mutual coupling between the two coils. *If the same circuit parameters* are used but with the *dots reversed*, so as to yield *negative* mutual reactance, all voltage, current, and power **magnitudes** *would remain the same* but the following changes would occur:[15]

1. $\mathbf{I}_2$ would be reversed in phase by 180°.
2. $\mathbf{V}_2$ would be reversed in phase by 180°.

Since the polarity in ac circuits is continuously changing, the difference is relatively unimportant insofar as the two-coil transformer is concerned. Furthermore, the power dissipated in all circuit impedances would not be affected at all.[16]

The foregoing statement also holds for loosely coupled *multicoil* transformers. As long as the load is passive, only the *phase* relations of the secondary currents and voltages are affected. To avoid polarity confusion, however, the safest procedure is mesh analysis. This means *we must apply the dot convention* shown in Fig. 20-21 to determine whether M is positive or negative with respect to the reference clockwise (CW) mesh currents used in the format method of mesh analysis. For three loosely coupled coils we will require three separate mesh equations. For four loosely coupled coils, four equations are needed, and so on. The method of writing the mesh equations is essentially the same, regardless of the number of coils. For this reason, and to reduce the complexity of the arrays, we will deal only with

[15] The proof is left as an end-of-chapter problem for the reader. See Prob. 20-38.

[16] These statements hold only if the secondary circuit contains only *passive* components. If a source is placed in the secondary and the polarity of one coil is reversed, then the currents, voltages, and powers are *no longer the same.* This is proved in Prob. 20-52.

three-coil multicoil transformers. The analysis will be presented algebraically and followed by a worked example.

20-10.1 Writing Mesh Equations for Multicoil Transformers

The writing of mesh equations for inductively coupled coils is the same as that presented in Sec. 20-9.2. Since only the mutual inductance is common to each pair of circuits, it is no longer necessary to draw T-equivalents. Using the format method, equation writing may be done by inspection once we have applied the dot convention to establish the polarity of mutual inductance between any pair of coils.

Figure 20-27a shows a three-coil multiwinding transformer, with the instantaneous polarities shown by dots for each coil. Using CW mesh currents in each circuit, reference mesh current is *entering* the dot in windings L_1 and L_2 and *leaving* the dot in L_3. As shown in Fig. 20-27a, only M_{12} is positive; M_{23} and M_{13} are *both* negative. We may now write the mesh equations by inspection for both self- and mutual inductances for each loop:

$$\text{(Loop 1):}\quad \mathbf{I}_1(R_1 + j\omega L_1) + \mathbf{I}_2(j\omega M_{12}) - \mathbf{I}_3(j\omega M_{13}) = \mathbf{V}_1 \tag{20-40a}$$

$$\text{(Loop 2):}\quad \mathbf{I}_1(j\omega M_{12}) + \mathbf{I}_2(R_2 + j\omega L_2) - \mathbf{I}_3(j\omega M_{23}) + \mathbf{I}_2\mathbf{Z}_2 = 0 \tag{20-40b}$$

$$\text{(Loop 3):}\quad -\mathbf{I}_1(j\omega M_{13}) - \mathbf{I}_2(j\omega M_{23}) + \mathbf{I}_3(R_3 + j\omega L_3) + \mathbf{I}_3\mathbf{Z}_3 = 0 \tag{20-40c}$$

Note that in the format method of equation writing we may generalize:

1. All *self-loop* voltage drops are *positive* for impedances enclosed within the loop.
2. *Positive* mutual inductances between coils yield *positive* voltage drops ($\mathbf{I}X_m$) for the given referenced loop. (Recall that only M_{12} is positive in Fig. 20-27a.)
3. *Negative* mutual inductances between coils yield *negative* voltage drops ($-\mathbf{I}X_m$) for the given referenced loop. (Recall that both M_{23} and M_{13} are negative.)

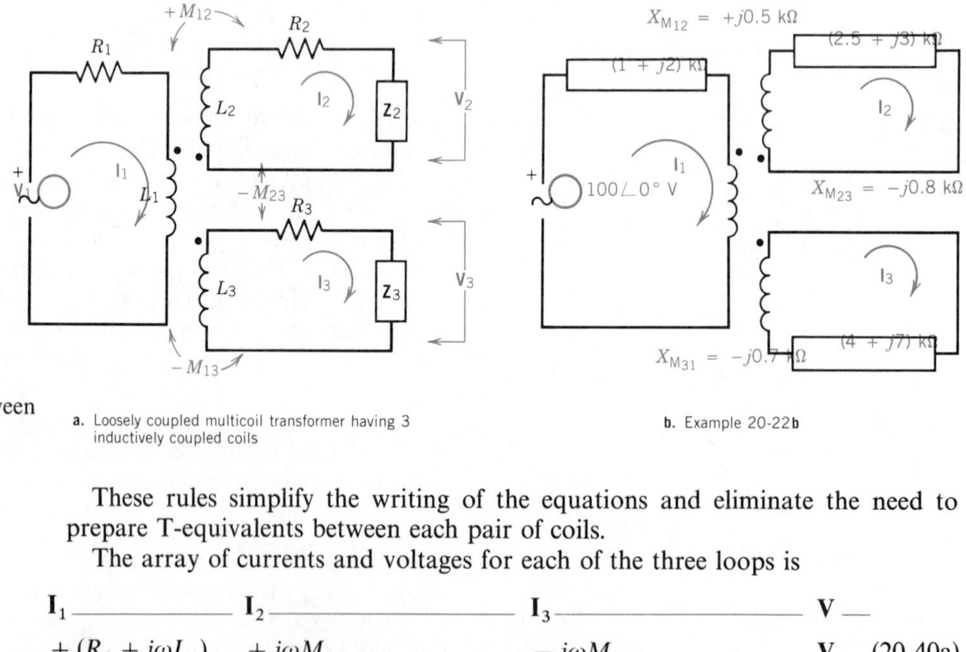

a. Loosely coupled multicoil transformer having 3 inductively coupled coils

b. Example 20-22b

Figure 20-27 Coupling between three inductively coupled coils

These rules simplify the writing of the equations and eliminate the need to prepare T-equivalents between each pair of coils.

The array of currents and voltages for each of the three loops is

$\mathbf{I}_1$	$\mathbf{I}_2$	$\mathbf{I}_3$	$\mathbf{V}$	
$+(R_1 + j\omega L_1)$	$+j\omega M_{12}$	$-j\omega M_{13}$	$\mathbf{V}_1$	(20-40a)
$+j\omega M_{12}$	$+[(R_2 + j\omega L_2) + \mathbf{Z}_2]$	$-j\omega M_{23}$	0	(20-40b)
$-j\omega M_{13}$	$-j\omega M_{23}$	$+[(R_3 + j\omega L_3) + \mathbf{Z}_3]$	0	(20-40c)

In producing this array, before subjecting it to solution either by matrix algebra methods or a computer program, the reader should check all polarities to see that they agree with those given in the three rules generalized above. Observe from the array that *only* $j\omega M_{12}$ *is positive whenever it appears*, whereas $j\omega M_{13}$ and $j\omega M_{23}$ are *both negative whenever they appear*. Further, all *self-loop* voltage drops are *positive* in each loop. (This method of checking the array will avoid a great deal of unnecessary number manipulation.)

One last point concerns voltages $\mathbf{V}_2$ and $\mathbf{V}_3$ shown in Fig. 20-27a. These voltages are actually $\mathbf{I}_2\mathbf{Z}_2$ and $\mathbf{I}_3\mathbf{Z}_3$, respectively. Consequently, once we have evaluated the currents ($\mathbf{I}_2$ and $\mathbf{I}_3$, respectively), these voltages may be computed. Alternatively, if the voltages are given (as known quantities), they would be transposed to the right side of Eqs. (20-40b) and (20-40c) as **negative** values $-\mathbf{I}_2\mathbf{Z}_2$ and $-\mathbf{I}_3\mathbf{Z}_3$, respectively.

20-10.2 Worked Examples

EXAMPLE 20-22

In the circuit shown in Fig. 20-27a, values are $\mathbf{V}_1 = 100\underline{/0^\circ}$ V at $\omega = 100$ kiloradians/second; $L_1 = 20$ mH, $L_2 = 40$ mH, and $L_3 = 50$ mH; $R_1 = 1$ kΩ, $R_2 = 2$ kΩ, and $R_3 = 3$ kΩ; $M_{12} = 5$ mH, $M_{23} = 8$ mH, and $M_{13} = 7$ mH. At the given radian frequency, $\mathbf{Z}_2 = (500 - j1000)$ Ω and $\mathbf{Z}_3 = (1000 + j2000)$ Ω.

a. Calculate the self-reactances of coils L_1, L_2, and L_3 and all three mutual reactances, respectively
b. Redraw Fig. 20-27a showing all the above impedances and reactances to simplify equation writing by mesh analysis, format method
c. Write the three mesh equations and express them in array form for solution by matrix methods. Evaluate the determinant, Δ
d. Show the expressions for currents $\mathbf{I}_1$, $\mathbf{I}_2$, and $\mathbf{I}_3$ in array form, and evaluate $\mathbf{I}_1$, $\mathbf{I}_2$, and $\mathbf{I}_3$, in mA
e. Calculate voltages $\mathbf{V}_2$ and $\mathbf{V}_3$

Solution

a. $X_{L_1} = \omega L_1 = (100\text{ k})(20\text{ m}) = j2$ kΩ; $\mathbf{Z}_{L_1} = \mathbf{(1 + j2)\ k\Omega}$;
$X_{L_2} = \omega L_2 = (100\text{ k})(40\text{ m}) = j4$ kΩ; $\mathbf{Z}_{L_2} = \mathbf{(2 + j4)\ k\Omega}$;
$X_{L_3} = \omega L_3 = (100\text{ k})(50\text{ m}) = j5$ kΩ; $\mathbf{Z}_{L_3} = \mathbf{(3 + j5)\ k\Omega}$;
$X_{12} = \omega M_{12} = (100\text{ k})(5\text{ m}) = \mathbf{+j500\ \Omega}$, $X_{23} = \mathbf{-j800\ \Omega}$, and $X_{31} = \mathbf{-j700\ \Omega}$
b. The circuit is redrawn in Fig. 20-27b, showing the total impedance in each loop as well as the mutual reactances between the three coupled coils. Figure 20-27b permits writing the mesh currents for each loop easily and directly by inspection.
c. Loop 1: $+\mathbf{I}_1(1 + j2) + \mathbf{I}_2(0 + j0.5) + \mathbf{I}_3(0 - j0.7) = 100$
Loop 2: $+\mathbf{I}_1(0 + j0.5) + \mathbf{I}_2(2.5 + j3) + \mathbf{I}_3(0 - j0.8) = 0$
Loop 3: $+\mathbf{I}_1(0 - j0.7) + \mathbf{I}_2(0 - j0.8) + \mathbf{I}_3(4 + j7) = 0$
The three simultaneous equations above are expressed in array form below:

$\mathbf{I}_1$	$\mathbf{I}_2$	$\mathbf{I}_3$	$\mathbf{V}$
$1 + j2$	$0 + j0.5$	$0 - j0.7$	$100 + j0$
$0 + j0.5$	$2.5 + j3$	$0 - j0.8$	0
$0 - j0.7$	$0 - j0.8$	$4 + j7$	0

$$\Delta = \begin{vmatrix} 1 + j2 & 0 + j0.5 & 0 - j0.7 \\ 0 + j0.5 & 2.5 + j3 & 0 - j0.8 \\ 0 - j0.7 & 0 - j0.8 & 4 + j7 \end{vmatrix} = \mathbf{-67.135 + j11.44}$$

d. $$\mathbf{I}_1 = \frac{\begin{vmatrix} 100 + j0 & 0 + j0.5 & 0 - j0.7 \\ 0 & 2.5 + j3 & 0 - j0.8 \\ 0 & 0 - j0.8 & 4 + j7 \end{vmatrix}}{\Delta} = \frac{-1036 + j2950}{-67.135 + j11.44}$$
$$= \mathbf{22.273 - j40.146 = 45.91\underline{/-60.98^\circ}\ mA}$$

$$\mathbf{I}_2 = \frac{\begin{vmatrix} 1 + j2 & 100 + j0 & 0 - j0.7 \\ 0 + j0.5 & 0 & 0 - j0.8 \\ 0 - j0.7 & 0 & 4 + j7 \end{vmatrix}}{\Delta} = \frac{294 - j200}{-67.135 + j11.44}$$
$$= \mathbf{(-4.749 + j2.1698) = 5.221\underline{/155.44^\circ}\ mA}$$

$$\mathbf{I}_3 = \frac{\begin{vmatrix} 1 + j2 & 0 + j0.5 & 100 + j0 \\ 0 + j0.5 & 2.5 + j3 & 0 \\ 0 - j0.7 & 0 - j0.8 & 0 \end{vmatrix}}{\Delta} = \frac{-170 + j175}{-67.135 + j11.44}$$
$$= \mathbf{2.8924 - j2.1138 = 3.5825\underline{/-36.16^\circ}\ mA}$$

e. $\mathbf{V}_2 = \mathbf{I}_2\mathbf{Z}_{L_2} = (-4.749 + j2.1698)\text{ mA} \times (0.5 - j1)\text{ k}\Omega$
$= \mathbf{-0.2047 + j5.8339 = 5.8375\underline{/92^\circ}\ V}$ and
$\mathbf{V}_3 = \mathbf{I}_3\mathbf{Z}_{L_3} = (2.8924 - j2.1138)\text{ mA} \times (1 + j2)\text{ k}\Omega$
$= \mathbf{7.12 + j3.671 = 8.011\underline{/27.275^\circ}\ V}$

Example 20-22 shows a number of techniques that are useful in writing and solving loosely coupled multicoil networks:

1. The format method simplifies the writing completely. Note, that in accordance with the rules given above, **all** voltage drops are written as *positive* values in step (**c**) with the negative mutual terms as $(-j)$ terms and positive mutual terms as $(+j)$ terms. All other self-impedance terms (shown within rectangular blocks in Fig. 20-27b) are represented as positive voltage drops.
2. The summary drawing of Fig. 20-27b consequently permits writing of the three mesh equations in the easiest and most direct way.

3. The dot convention (Sec. 20-9.1) tells us whether the mutual reactances are $(+j)$ or $(-j)$ terms.
4. With the self-impedances and mutual reactances expressed in kilo-ohms in Fig. 20-27b and the voltages in the array of part **(c)** expressed in volts, the currents that emerge are automatically expressed in milliamperes in part **(d)**.

The reader may ask at this point, having obtained values for the mesh currents and load voltages, "How do I know whether my answers are correct?"

Again, the *verification* of the solutions obtained involves *power* calculations. If the total power drawn from the supply is equal to the individual powers dissipated in the three loops, we can be reasonably sure that our procedures and mathematics are both correct.

EXAMPLE 20-23

For the circuit shown in Fig. 20-27b and the currents obtained in Ex. 20-22 for $\mathbf{I}_1$, $\mathbf{I}_2$, and $\mathbf{I}_3$, calculate the

a. Total power drawn from the supply, P_T
b. Power dissipated in loop 1 by $\mathbf{I}_1$, P_1
c. Power dissipated in loop 2 by $\mathbf{I}_2$, P_2
d. Power dissipated in loop 3 by $\mathbf{I}_3$, P_3
e. Total power as the sum $P_1 + P_2 + P_3$ and compare with **a** above

Solution

a. $P_T = \mathbf{V}\mathbf{I}_1 \cos\theta_1 = (100\underline{/0^\circ})(45.91\text{ m})(\cos 60.98^\circ) = \mathbf{2.2272\ W}$
b. $P_1 = \mathbf{I}_1^2 R_1 = (45.91\text{ m})^2 \times 1\text{ k}\Omega = \mathbf{2.1077\ W}$
c. $P_2 = \mathbf{I}_2^2 R_2 = (5.221\text{ m})^2 \times 2.5\text{ k}\Omega = \mathbf{68.15\ mW}$
d. $P_3 = \mathbf{I}_3^2 R_3 = (3.5825\text{ m})^2 \times 4\text{ k}\Omega = \mathbf{51.34\ mW}$
e. $P_t = P_1 + P_2 + P_3 = (2.1077 + 0.06815 + 0.05134)\text{ W}$ $= \mathbf{2.2272\ W}$, which is the same as **a**

The following inferences are to be drawn from Ex. 20-23:

1. Since the individually dissipated powers equal the total power, we know that the solution of Ex. 20-22 is correct. (It is therefore unnecessary to rework Ex. 20-22.)
2. The values of R_1, R_2, and R_3, respectively, used in parts **(b)**, **(c)**, and **(d)** emerge directly from Fig. 20-27b, since these are the resistances of each block. This shows the importance of drawing such a circuit diagram, which not only enables writing of the mesh equations but also simplifies the checking by power calculation.

20-10.3 Commercial Loosely Coupled Transformers

Students occasionally ask "How do you distinguish a loosely coupled from a tightly coupled transformer by appearance only?"

The answer lies (obviously) in the degree of coupling between the transformer coils. Any transformer in which the primary and secondary are separated (i.e., not tightly wound over each other) is loosely coupled. **Figure 20-28a** shows a *varicoupler* transformer in which the air separation between the two coils may be adjusted by a *coupling screw*. Figure 20-28b shows two RF coils having individual variable cores. By varying the inductance of each coil, the coupling may be varied. Figure 20-28c shows the coils wound on a common core. The coupling between the coils is "slug-tuned" by means of a screw that removes or inserts a ferrite core to vary the mutual flux and magnetic coupling between the coils.

The transformers (Figs. 20-28a through 20-28c) are easily identified as loosely coupled. Other transformers with nonmagnetic or high-reluctance magnetic (either powdered iron or ferrite) cores may also result in relatively loose coupling (coefficients of coupling of 0.9 or less). The upper portion of Fig. 20-28d shows the construction stages (from right to left) of a two-winding transformer having a nonmagnetic toroidal (doughnut-shaped) core. The lower portion of Fig. 20-28d shows the construction of a *multicoil* toroidal transformer wound on a *ferrite* core. Coupling between the coils may vary from 0.9 to as low as 0.7 in such transformers, depending on the nature of the ferrite core. Consequently, such transformers (Fig. 20-28d) are still classified as loosely coupled.

By contrast, a tightly coupled transformer designed for printed circuit board (PCB) mounting is shown in Fig. 20-28e. The E-shaped laminations inserted above

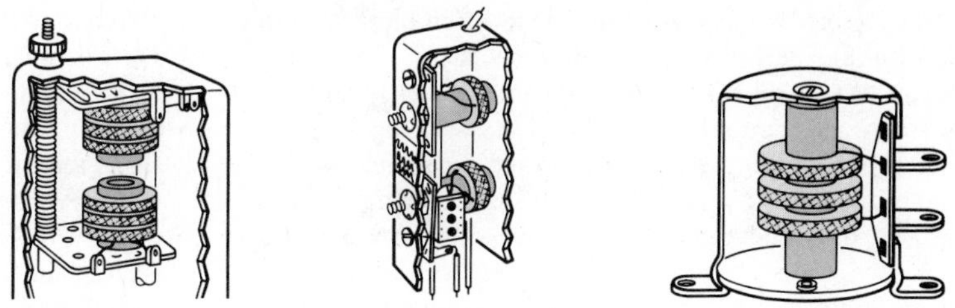

a. Varicoupler transformer b. Individual variable cores c. Common variable core, shielded RF transformer

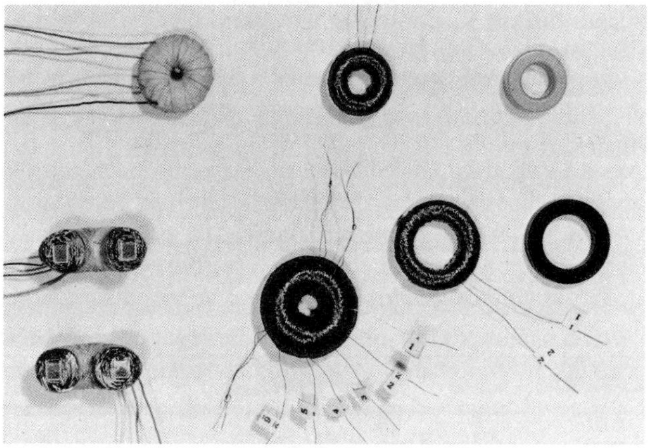

d. Toroidal transformers (Courtesy Microtran Co., Inc.)

e. Tightly coupled, shell-type transformer (Courtesy Microtran Co., Inc.)

Figure 20-28 Distinguishing between loosely coupled and tightly coupled transformers

and below the prewound coils form a shell-type structure that ensures close to unity coupling between the primary and secondary. Tightly coupled transformers are always identifiable by having a higher iron-to-copper ratio than loosely coupled transformers.

20-11 GLOSSARY OF TERMS USED

Autotransformer Multiwinding transformer connected in such a way that it has one continuous winding of which part is common to both the primary and secondary circuits to which the autotransformer is connected.

Coefficient of coupling, *k* Ratio of the mutual inductance, M, to the square root of the product L_1L_2, where L_1 is the total inductance of one mesh and L_2 the total inductance of the second mesh mutually coupled to the first. Also, ratio of mutual flux to total flux in any two coils that are magnetically coupled.

Dot convention Convenient method for specifying winding directions between two or more coils that are magnetically coupled. Currents flowing into dotted terminals produce aiding (adding) magnetic fluxes.

Efficiency Ratio of useful energy or power output to input expressed as a percentage.

Ideal transformer Transformer that neither stores nor dissipates electrical energy, having unity coupling and zero leakage flux and leakage inductance, and whose core permeability is infinite.

Inductor Coil with or without an iron core having the property of inductance.

Leakage flux Flux linking a given coil that is not part of the mutual flux, ϕ_m, common to other coils that are magnetically coupled to the given coil.

Loosely coupled transformer Transformer whose coefficient of coupling is less than unity.

Matching transformer Transformer used as an impedance-matching device to match a high-impedance source to a low-impedance load or vice versa. The matching transformer enables maximum power transfer from source to load with minimum energy loss.

Minimum coupling Location of two or more inductors so that the flux linkage between them is minimized.

Open-circuit test Test to determine the core losses of a conventional iron-core transformer.

Primary Input winding of a transformer, to which a source voltage is applied. Any winding of a transformer may serve as a primary, provided rated voltage and rated frequency are applied.

Reflectance factor, α Ratio of mutual reactance to total secondary impedance in a loosely coupled circuit, determining the magnitude of the impedance reflected to the primary.

Reflected impedance Impedance reflected to the primary due to the load that appears on the secondary.

Secondary One or more output windings of the transformer that are connected to one or more loads, producing a transfer of energy from the primary to the load circuit(s).

Short-circuit test Test to determine the equivalent primary and secondary copper losses of a conventional iron-core transformer, along with its equivalent impedance and reactance, referred to the high-voltage side on which it is performed. Data from the short-circuit test enable calculation of transformer voltage regulation and also full-load copper losses.

Step-down transformer Transformer whose secondary has fewer turns than the primary and whose energy transfer is from high to low voltage.

Step-up transformer Transformer whose secondary has more turns than the primary and whose energy transfer is from low to high voltage.

Transformer Device for transferring energy from one circuit to another.

Transformer ratio, α Ratio of primary to secondary turns in an iron-core transformer, determining the impedance reflected to the primary due to secondary load.

Voltage regulation Change in output (secondary voltage) that occurs when the load (at a specified power factor) is reduced from its rated voltage value to zero, with the primary impressed terminal voltage maintained constant.

20-12 PROBLEMS

Sec. 20-2

20-1 A commercial 60 Hz, 440/110 V, 50 kVA power transformer has 600 turns on its high-voltage (HV) side. Calculate the

a. Number of turns on its low-voltage (LV) side
b. Ratio of transformation, α, when used as a step-down transformer
c. Ratio of transformation, α, when used as a step-up transformer
d. Volts per turn ratio of the HV side
e. Volts per turn ratio of the LV side
f. Rated current of the HV side
g. Rated current of the LV side

20-2 The HV side of a transformer has 750 turns and the LV side has 50 turns. When the HV side is connected to its rated voltage of 120 V, 60 Hz and a rated load of 40 A is connected to the LV side, calculate the

a. Transformation ratio, α
b. Secondary voltage, assuming an ideal transformer
c. Resistance of the load
d. Volts per turn ratio of the secondary and primary, respectively
e. Volt-ampere rating of the transformer

20-3 A commercial 240 V/30 V, 3 kVA, 60 Hz transformer has a ratio of 3 V/turn. Calculate the

a. Number of turns on the HV side and on the LV side
b. Transformation ratio if used as a step-down transformer
c. Transformation ratio if used as a step-up transformer
d. Rated HV-side current and rated LV-side current

20-4 A 10-Ω load draws 20 A from the HV side of a transformer whose transformation ratio $\alpha = 1/8$. Assuming an ideal transformer, calculate the

a. Secondary voltage
b. Primary voltage
c. Primary current
d. Volt-amperes transferred from primary to secondary
e. Transformation ratio when used as a step-down transformer

20-5 A commercial 400 Hz, 220/20 V transformer has 50 turns on its LV side. Calculate the

a. Number of turns on its HV side
b. Transformation ratio, α, when used as a step-up transformer
c. Volts per turn ratio of the HV side and the LV side

20-6 The HV side of a step-down transformer has 800 turns and the LV side has 100 turns. If a rated voltage of 240 V is applied to the HV side and a load impedance of 3 Ω is connected to the LV side, calculate the

a. Secondary voltage and current
b. Primary current
c. Primary input impedance from the ratio of primary voltage and primary current
d. Primary input impedance, using Eq. (20-2a)

20-7 An ac servoamplifier has an output impedance of 250 Ω

and the ac servomotor that it drives has an impedance of 2.5 Ω. Calculate the

a. Transformation ratio, α, of the transformer required to match the two impedances
b. Number of primary transformer turns if the secondary has 10 turns

20-8 In the three-coil matching transformer shown in Fig. 20-4, $N_1 = 1000$ t, $N_2 = 200$ t, $N_3 = 400$ t, $\mathbf{Z}_2 = 10\angle 0°\ \Omega$, $\mathbf{Z}_3 = 16\angle 0°\ \Omega$, and the primary voltage is 30 V RMS. Calculate

a. $\mathbf{Z}_2'$ reflected to the primary
b. $\mathbf{Z}_3'$ reflected to the primary
c. Total equivalent impedance reflected to the primary
d. Primary current drawn from the supply
e. Total power drawn from the supply
f. Voltage $\mathbf{V}_2$ across $\mathbf{Z}_2$ and power dissipated in $\mathbf{Z}_2$
g. Voltage $\mathbf{V}_3$ across $\mathbf{Z}_3$ and power dissipated in $\mathbf{Z}_3$
h. Total power dissipated in loads

20-9 In the *tapped* matching transformer shown in Fig. 20-5, $N_1 = 1000$ t, $N_2 = 200$ t, $N_3 = 300$ t, $\mathbf{Z}_2 = 2\angle 0°\ \Omega$, and $\mathbf{Z}_3 = 5\angle 0°\ \Omega$. The primary voltage is 21 V RMS. Repeat all parts of Prob. 20-8.

20-10 In the three-coil matching transformer shown in Fig. 20-4, $N_1 = 100$ t and $N_2 = N_3 = 50$ t. When a supply voltage of 16 V ac is connected across the primary it draws 16 W from the supply. The load resistance across N_2 dissipates 6.4 W. Calculate the

a. Primary current and input impedance, $\mathbf{Z}_1$
b. Secondary voltages across each load
c. Load resistance across winding N_2
d. Load resistance across winding N_3
e. Verify the input impedance in (**a**) using Eq. (20-3) and the answers to (**c**) and (**d**)

20-11 A tapped impedance-matching transformer found in a stockpile of used equipment has no other information except that its three secondary terminals show 16 Ω, 8 Ω, and G, respectively. Calculate the impedance between the 16-Ω and 8-Ω taps.

Sec. 20-3

20-12 Assume the primary circuit of an audio output transformer has a total equivalent resistance of 100 Ω and is matched to a speaker whose resistance is 4 Ω. Calculate the

a. Transformation ratio that produces maximum power transfer
b. Secondary voltage across the speaker if the primary input voltage is 10 V
c. Maximum power transferred from the supply to the speaker

20-13 Given the circuit shown in Fig. 20-7; the supply voltage of a transistorized voltage source is $48\angle 0°$ V and the source resistance is 2048 Ω. If the load resistance is 8 Ω, calculate the

a. Required transformation ratio of matching transformer for MPT
b. Terminal voltage of the source at maximum power transfer (MPT), $\mathbf{V}_1$
c. Terminal voltage across the load at MPT, $\mathbf{V}_2$
d. Secondary load current $\mathbf{I}_2$ by two independent methods
e. Primary load current drawn from the source, $\mathbf{I}_1$, by two methods
f. Maximum power dissipated by the load, $P_{L_{max}}$
g. Power dissipated internally within the source, P_s
h. Total power supplied by the source by two methods, P_T
i. Power transfer efficiency, η

20-14 Assuming the transistorized source in Prob. 20-13 is connected directly across the 8-Ω load without using a matching transformer, calculate the

a. Voltage across the load, $\mathbf{V}_L$
b. Voltage dropped internally across the source resistance, R_s, of the supply
c. Power dissipated in the load, P_L
d. Power dissipated internally within the source
e. Total power drawn from the source by two methods, P_T
f. Compare P_L in **c** above with P_L in Prob. 20-13**f** and draw conclusions
g. Compare P_T in **e** above with P_T in Prob. 20-13**h** and draw conclusions

20-15 Given the circuit shown in **Fig. 20-29**, calculate the

a. Transformation ratio, α, to ensure MPT from source to complex load impedance $\mathbf{Z}_L$
b. Primary (inductive) reactance reflected to the secondary load circuit
c. Value of $-jX_C$ to ensure MPT from the source
d. Total reactance reflected from secondary to primary at MPT
e. Primary current $\mathbf{I}_1$ and secondary current $\mathbf{I}_2$ at MPT
f. Power dissipated in the load at MPT
g. Power dissipated in $\mathbf{Z}_p$ at MPT
h. Voltage dropped across load impedance, $\mathbf{Z}_L$, at MPT
i. Voltage dropped across primary impedance, $\mathbf{Z}_p$, at MPT

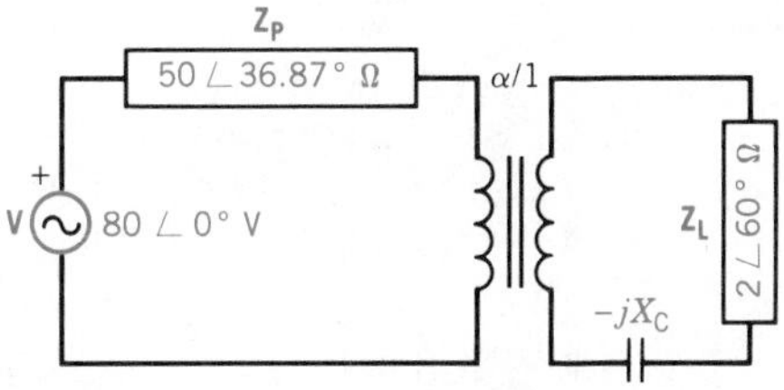

Figure 20-29 Problem 20-15

Sec. 20-5

20-16 A 2300/230 V, 500 kVA, 60 Hz, practical iron-core step-down transformer has the following values: $r_1 = 0.1\ \Omega$, $jX_{L_1} = j0.3\ \Omega$; $r_2 = 1$ mΩ, and $jX_{L_2} = j3$ mΩ. When the transformer is loaded to its rated capacity and used as a step-down transformer, calculate the

a. Rated secondary and primary currents
b. Secondary and primary internal impedances, $\mathbf{Z}_2$ and $\mathbf{Z}_1$ (magnitudes only)
c. Secondary and primary internal voltage drops

d. Secondary and primary induced voltages, assuming that the terminal and induced voltages are in phase
e. Ratio of primary to secondary *induced* voltages, from **d**
f. Ratio of primary to secondary *terminal* voltages from nameplate data

20-17 For the transformer of Prob. 20-16, using the given and calculated data, calculate the

a. Secondary load impedance at rated load, $\mathbf{Z}_L$
b. Primary input impedance at rated load, $\mathbf{Z}_P$, by two methods
c. Compare $\mathbf{Z}_L$ to the secondary internal impedance $\mathbf{Z}_2$ above and explain the difference between them
d. Compare $\mathbf{Z}_P$ to the primary internal impedance $\mathbf{Z}_1$ above and explain the difference

20-18 Using the given data of Prob. 20-16 and the transformation ratio ($\alpha = 10$), calculate the

a. Equivalent resistance of the transformer referred to the HV side
b. Equivalent reactance of the transformer referred to the HV side
c. Equivalent resistance of the transformer referred to the LV side
d. Equivalent reactance of the transformer referred to the LV side
e. Explain why we may assume that the resistance of any coil winding is approximately half of its equivalent resistance (as measured by the short-circuit test)
f. Explain why we may assume that the reactance of any coil winding is approximately half of its equivalent reactance (as calculated from short-circuit test data)

20-19 Measurements made on a 2300/230 V, 50 kVA transformer by the short-circuit test provide the following calculated data of equivalent reactance and resistance referred to the low-voltage side: $X_{e_2} = 40$ mΩ and $R_{e_2} = 20$ mΩ. Calculate the

a. Rated secondary load current, $\mathbf{I}_2$
b. Voltage drops $\mathbf{I}_2 X_{e_2}$ and $\mathbf{I}_2 R_{e_2}$ at full (rated) load
c. Full load secondary induced EMF, $\mathbf{E}_2$, at unity PF
d. Full load secondary induced EMF, $\mathbf{E}_2$, at 0.7 PF lagging
e. Full load secondary induced EMF, $\mathbf{E}_2$, at 0.7 PF leading
f. Voltage regulation at unity PF
g. Voltage regulation at 0.7 PF lagging
h. Voltage regulation at 0.7 PF leading

20-20 Verify the induced voltage $\mathbf{E}_2$ and the voltage regulation at unity PF in Prob. 20-19, using the alternative expression for $\mathbf{E}_2$, i.e., $\mathbf{E}_2 = \mathbf{V}_2\angle 0^\circ + (\mathbf{I}_2\angle\theta)(\mathbf{Z}_{e_2}\angle\beta)$

20-21 A 10 kVA, 60 Hz, 4800/240 V transformer is tested by the open-circuit and short-circuit tests. The test data are

Test	Voltage (V)	Current (A)	Power (W)	Side used
Open-circuit	240	1.5	160	LV
Short-circuit	180	$2.08\overline{3}$	180	HV

Given the above data, calculate the

a. Equivalent resistance and reactance, referred to the HV side
b. Equivalent resistance and reactance, referred to the LV side
c. Voltage regulation of the step-down transformer at unity PF, full load
d. Voltage regulation of the step-down transformer at a PF of 0.8 lagging
e. Voltage regulation of the transformer at a PF of 0.8 leading

Sec. 20-6

20-22 A 20 kVA, 660/120 V transformer has a no-load loss of 250 W and a high-voltage side winding resistance of 0.2 Ω. Assuming that the load losses of the HV and LV sides are equal, calculate the

a. Resistance of the LV side winding
b. Full-load equivalent copper loss referred to the LV side
c. Transformer efficiencies at load values of 25, 50, 75, 100, and 125%, at unity PF load

20-23 Given the open-circuit and short-circuit test data of Prob. 20-21, calculate the

a. Equivalent resistance of the transformer referred to the LV side
b. Full load efficiency at a PF of 0.9 lagging
c. Half-load efficiency at a PF of 0.9 lagging
d. The load fraction and value of secondary load current at which maximum efficiency occurs
e. Maximum efficiency at unity PF
f. Maximum efficiency at a PF of 0.9 lagging

20-24 The efficiency of a 20 kVA, 1200/120 V transformer is a maximum of 98% at exactly 50% of its rated load. Assuming loads at unity PF, calculate the

a. Core loss
b. Efficiency at rated load
c. Efficiency at loads of 75% and 125%

20-25 A 100 kVA, 60 Hz, 12000/240 V transformer is tested by the open-circuit and short-circuit tests and the test data are

Test	Voltage (V)	Current (A)	Power (W)	Side used
Open-circuit	240	8.75	980	low voltage
Short-circuit	600	$8.\overline{3}$	1200	high voltage

For the above transformer in the step-down mode, calculate the

a. Voltage regulation at 0.8 PF lagging
b. Efficiency at 0.8 PF lagging for 1/8, 1/4, 1/2, 3/4, 4/4, and 5/4 rated load
c. Fraction of rated load at which maximum efficiency occurs
d. Maximum efficiency at 0.8 PF lagging

20-26 Repeat Prob. 20-25**b** for efficiencies at a PF of 0.6 leading
20-27 Repeat Prob. 20-25a for the voltage regulation at a PF of 0.6 leading

Sec. 20-8

20-28 A step-up autotransformer is used to supply 3 kV from a 2.4-kV supply line. If the secondary load is 50 A, neglecting losses and magnetizing current, calculate the

a. Current in each part of the transformer
b. Current drawn from the 2.4-kV supply line
c. kVA rating of the autotransformer
d. kVA rating of a comparable conventional two-winding transformer necessary to accomplish the same transformation

20-29 For the transformer in Prob. 20-28, calculate at rated load, **unity PF** the

a. Power transformed from primary to secondary
b. Power transferred conductively from primary to secondary

20-30 A step-down autotransformer is used to supply 100 A at 2 kV from a 2.4-kV supply line. Calculate all parts of Prob. 20-28 for this transformer.

20-31 For the transformer in Prob. 20-30, repeat parts (**a**) and (**b**) of Prob. 20-29.

20-32 For *step-up* autotransformers it can be shown that the power transformed, P_x, is related to the total power entering and leaving the (ideal) autotransformer, P, by the simple relation $P_x = P(1 - \alpha)$, where α is the transformation ratio. Prove algebraically that the power conductively transferred, P_c, is simply $P_c = \alpha P$ (Hint: $P = P_x + P_c$).

20-33 For *step-down* autotransformers it can be shown that the power transformed from primary to secondary, P_x, is related to the total power (entering and leaving the ideal autotransformer) by $P_x = P\left(\frac{\alpha - 1}{\alpha}\right)$. Show algebraically that the power conductively transferred, P_c, in a step-down autotransformer is $P_c = P/\alpha$.

20-34 A step-up 120/180 V autotransformer has a 45-Ω resistive load connected across its output.

a. Draw a circuit diagram showing all voltages and currents
b. Calculate the transformation ratio and the total power transferred
c. Calculate the power conductively transferred and the power inductively transformed (Hint: See Prob. 20-32)

20-35 A step-down 120/80 V autotransformer transfers a total power of 720 W from primary to secondary. Using the relations derived in Prob. 20-33, calculate the

a. Power conductively transferred
b. Power inductively transformed
c. Draw a cricuit diagram showing all voltages and currents

Sec. 20-9

20-36 For the loosely coupled transformer circuit shown in Fig. 20-30

a. Calculate the mutual inductance and reactance. Indicate whether it is aiding (+) or opposing (−) the self-inductances of the two coils.
b. Write the mesh equations for each loop and show the array in its simplest form.
c. Calculate the determinant, Δ.
d. Calculate mesh currents $\mathbf{I}_1$ and $\mathbf{I}_2$, showing both magnitude and phase.

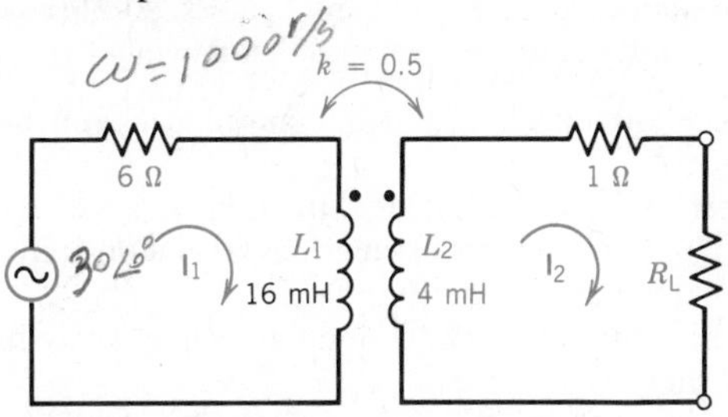

Figure 20-30 Problems 20-36 to 20-41

20-37 Verify the solution of currents in the previous problem by calculating the

a. Total power drawn from the supply, P_T
b. Power dissipated in the primary mesh, P_1
c. Power dissipated in the secondary mesh, P_2
d. Total power dissipated, $P_1 + P_2$, and compare with P_T in (**a**) above

20-38 Repeat *all* parts of Prob. 20-36 with the polarity dot in the secondary of Fig. 20-30 reversed (so that the mutual reactance is *positive*).

e. Compare the calculation of the determinant in part (**c**) and explain why the determinant is the *same* despite the difference in the two arrays.
f. Draw conclusions as to differences in Probs. 20-36 and 20-38 with respect to
   1. magnitudes (only) of currents $\mathbf{I}_1$ and $\mathbf{I}_2$
   2. powers drawn from the supply and dissipated in each loop
   3. *phase* of the primary current $\mathbf{I}_1$
   4. *phase* of the secondary current $\mathbf{I}_2$

20-39 Verify the solutions of Probs. 20-36 and 20-37 (in part) by calculating for the loosely coupled circuit of Fig. 20-30 the

a. Reflected impedance, $\mathbf{Z}_r$, using Eq. (20-37)
b. Input impedance, $\mathbf{Z}_{in}$, using Eq. (20-35a)
c. Primary current, $\mathbf{I}_1$
d. Total circuit power, P_T
e. Power dissipated in the primary, P_1

20-40 Verify the solutions of Probs. 20-36 and 20-37 (in part) by calculating for Fig. 20-30 the

a. Reverse transfer impedance, $\mathbf{Z}_{12}$, using Eq. (20-36a)
b. Secondary current, $\mathbf{I}_2$
c. Power dissipated in the secondary, P_2

20-41 Verify the calculations of $\mathbf{I}_2$ in Probs. 20-36 and 20-40 using the Thévenin equivalent voltage and impedance (seen by load R_L in Fig. 20-30) by calculating

a. $\mathbf{V}_{TH}$ using Eq. (20-38)
b. $\mathbf{Z}_{TH}$ using Eq. (20-39)
c. $\mathbf{I}_2$, secondary current seen by load $\mathbf{Z}_L$

d. Compare the solution of $\mathbf{I}_2$ with that found in Probs. 20-40**b** and 20-36**d**

20-42 Verify the solution of currents $\mathbf{I}_1$ and $\mathbf{I}_2$ in Ex. 20-21 (Fig. 20-26a) using *mesh analysis* by performing the following:

a. Calculate the total equivalent primary impedance seen by the supply
b. Calculate total equivalent secondary circuit impedance, $\mathbf{Z}_2$
c. Redraw Fig. 20-26**a** in terms of calculations (**a**) and (**b**) above for use with mesh analysis (see Fig. 20-31)
d. Write the mesh equations for each loop and show array in its *simplest* form
e. Calculate the determinant, Δ
f. Calculate mesh currents, $\mathbf{I}_1$ and $\mathbf{I}_2$, in polar form
g. Calculate power generated by the source, P_g
h. Calculate total power dissipated in the primary circuit, P_1
i. Calculate total power dissipated in the secondary circuit, P_2
j. Compare these results with those shown in Ex. 20-21 and the power flow diagram of Fig. 20-26**b**

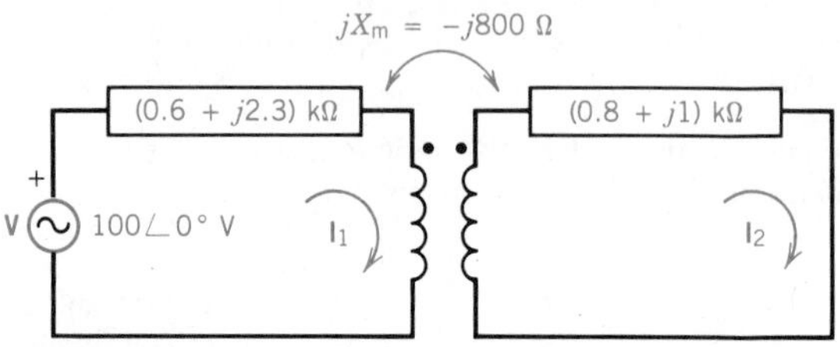

Figure 20-31 Solution to Prob. 20-42**c**

20-43 Given the two-coil loosely coupled transformer shown in Fig. 20-32, $\mathbf{V} = 100\underline{/0^\circ}$ V at $\omega = 10^2$ radians/second, $\mathbf{Z}_1 = (30 + j40)\ \Omega$, $\mathbf{Z}_L = (60 + j80)\ \Omega$, $R_1 = 50\ \Omega$, $L_1 = 0.75$ H, $L_2 = 0.5$ H, $R_2 = 50\ \Omega$, and $k = 0.653$ between coils.

a. Assign polarity dots on Fig. 20-32 based on coil winding directions (Hint: see Fig. 12-11)
b. Using the dot convention for assigned CW mesh currents, determine whether M_{12} is positive or negative
c. Calculate M_{12} and $\pm jX_{m_{12}}$
d. Write mesh equations for each loop and the array in its simplest form
e. Calculate the determinant, Δ
f. Calculate $\mathbf{I}_1$ and $\mathbf{I}_2$
g. Calculate the power drawn from the supply $\mathbf{V}$
h. Calculate the respective powers dissipated in $\mathbf{Z}_1$, coil 1, coil 2, and $\mathbf{Z}_L$

20-44 Using the method shown in Sec. 20-9.6 and Ex. 20-21, given the circuit of Prob. 20-43 and Fig. 20-32, calculate the

a. Reflectance factor, α
b. Reflected impedance, $\mathbf{Z}_r$
c. Total primary input impedance seen by source $\mathbf{V}$
d. Primary current drawn from practical source, $\mathbf{I}_1$
e. Voltage induced into the transformer secondary winding, $\mathbf{E}_2$
f. Current in the transformer secondary, $\mathbf{I}_2$ (Hint: use $-\mathbf{E}_2$ since $\mathbf{I}_2$ opposes $\mathbf{E}_2$)

20-45 Given the loosely coupled transformer shown in Fig. 20-32, $\mathbf{V} = 100\underline{/0^\circ}$ V, $\mathbf{Z}_1 = (30 + j30)\ \Omega$, $R_1 = 20\ \Omega$, $jX_{L_1} = j100\ \Omega$, $R_2 = 50\ \Omega$, $jX_{L_2} = j50\ \Omega$, $jX_m = +j50\ \Omega$, and $\mathbf{Z}_L = (50 - j30)\ \Omega$. Using mesh analysis, calculate

a. $\mathbf{I}_1$
b. $\mathbf{I}_2$
c. Total power drawn from supply
d. Total power dissipated in loop 1
e. Total power dissipated in loop 2

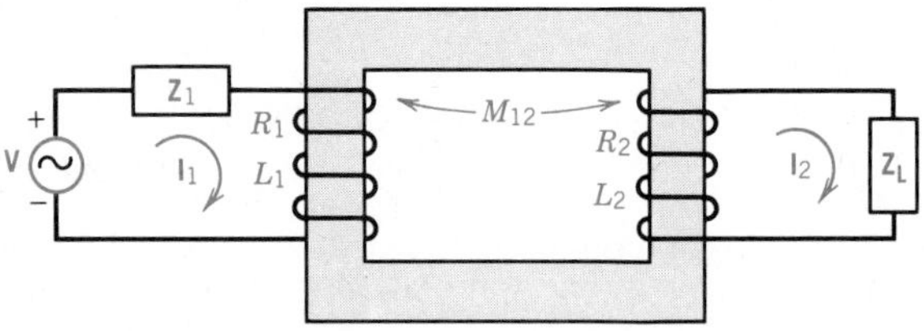

Figure 20-32 Problems 20-43 through 20-46

20-46 Given the values and circuit of Prob. 20-45, calculate all parts of Prob. 20-44.

20-47 Given the loosely coupled autotransformer shown in Fig. 20-33, $\mathbf{Z}_1 = (3 - j9)\ \Omega$, $L_1 = 5$ mH, $L_2 = 8$ mH, $\mathbf{Z}_L = (4 - j5)\ \Omega$, and $\mathbf{V} = 10\underline{/0^\circ}$ V at a frequency $\omega = 10^3$ rad/s. If the coefficient of coupling between coils, $k = 0.6$, calculate

a. M_{12} and $X_{m_{12}}$, indicating whether the reactance is positive or negative
b. Write mesh equations for each loop and express the array in its simplest form
c. Calculate the determinant, Δ. (Hint: if you do not get the correct answer, recheck your equation writing and other calculations.)
d. Calculate mesh currents $\mathbf{I}_1$ and $\mathbf{I}_2$ in polar form
e. Calculate the power drawn from the supply $\mathbf{V}$
f. Calculate the power dissipated in $\mathbf{Z}_1$ and $\mathbf{Z}_L$
g. Calculate the true current in the common winding, L_2
h. Explain why $\mathbf{I}_2$ is greater than $\mathbf{I}_1$ in this circuit of Fig. 20-33

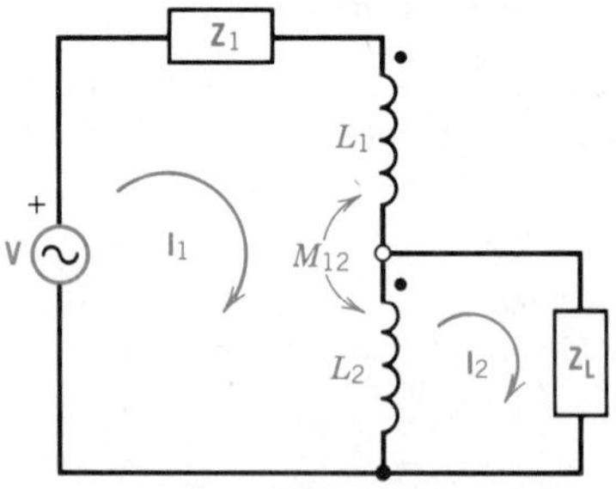

Figure 20-33 Problems 20-47 and 20-48

20-48 Repeat Prob. 20-47 (parts **a** through **f**, *only*) with the dot on L_1 reversed from that shown in Fig. 20-33. (Hint: this produces a loosely coupled *subtractive*-polarity autotransformer.)

20-49 For the two-winding loosely coupled transformer circuit shown in Fig. 20-34, $\mathbf{V} = 10\underline{/0^\circ}$ V at a frequency of 10^3 rad/s,

$\mathbf{Z}_1 = (3 + j4)\ \Omega$, $\mathbf{Z}_c = (1 - j2)\ \Omega$, $\mathbf{Z}_L = (5 - j12)\ \Omega$, $k = 0.6$, $L_1 = L_2 = 4$ mH.

a. Calculate $\mathbf{M}_{12}$ and $\mathbf{X}_{m_{12}}$, indicating whether M_{12} is positive or negative
b. Write the mesh equations for each loop and the array, expressed in simplest form
c. Calculate the determinant, Δ
d. Calculate mesh currents $\mathbf{I}_1$ and $\mathbf{I}_2$
e. Calculate the true current, $\mathbf{I}_c$, in $\mathbf{Z}_c$
f. Calculate the power drawn from the supply, P_t
g. Calculate the power dissipated in $\mathbf{Z}_1$, $\mathbf{Z}_c$, and $\mathbf{Z}_L$

20-50 Repeat all parts of Prob. 20-49 with the dot on L_1 reversed from that shown on Fig. 20-34.

20-51 Given the circuit shown in Fig. 20-35, calculate

a. $\pm jX_m$, indicating whether it is positive or negative
b. $\mathbf{I}_1$ and $\mathbf{I}_2$, using mesh analysis and assumed CW mesh currents
c. Power supplied by source $\mathbf{V}_1$
d. Power supplied by source $\mathbf{V}_2$ (Hint: reverse current direction of $\mathbf{I}_2$ found in **b**)
e. Power dissipated in loop 1
f. Power dissipated in loop 2

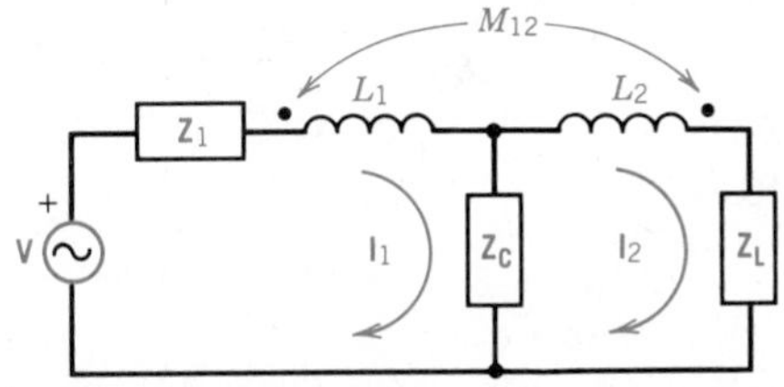

Figure 20-34 Problems 20-49 and 20-50

20-52 1. Repeat all parts of Prob. 20-51 with the dot on the coil in loop 2 reversed.
2. Explain why the currents, component volt drops, and powers are all *different* as a result of the polarity change in Fig. 20-35.

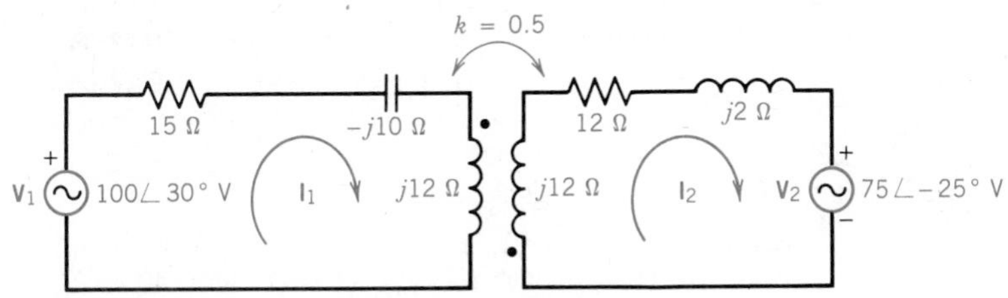

Figure 20-35 Problems 20-51 and 20-52

20-53 In the circuit shown in Fig. 20-36, values are $\mathbf{V}_1 = 25\angle 0^\circ$ V at $\omega = 10^3$ rad/s; $L_1 = 5$ mH, $L_2 = 4$ mH, and $L_3 = 6$ mH; $\mathbf{Z}_1 = (4 - j5)\ \Omega$, $\mathbf{Z}_2 = (5 - j12)\ \Omega$, and $\mathbf{Z}_3 = (3 + j4)\ \Omega$. The coefficients of coupling are $k_{12} = 0.6$, $k_{23} = 0.5$, and $k_{31} = 0.4$. Calculate

a. $X_{m_{12}}$, $X_{m_{23}}$, and $X_{m_{31}}$, indicating whether these are positive or negative
b. Mesh currents $\mathbf{I}_1$, $\mathbf{I}_2$, and $\mathbf{I}_3$, using mesh analysis
c. Voltages $\mathbf{V}_2$ and $\mathbf{V}_3$ across $\mathbf{Z}_2$ and $\mathbf{Z}_3$, respectively
d. Total power supplied by source $\mathbf{V}_1$ to the transformers, P_t
e. Powers dissipated in loops 1, 2, and 3 (P_1, P_2, and P_3, respectively)

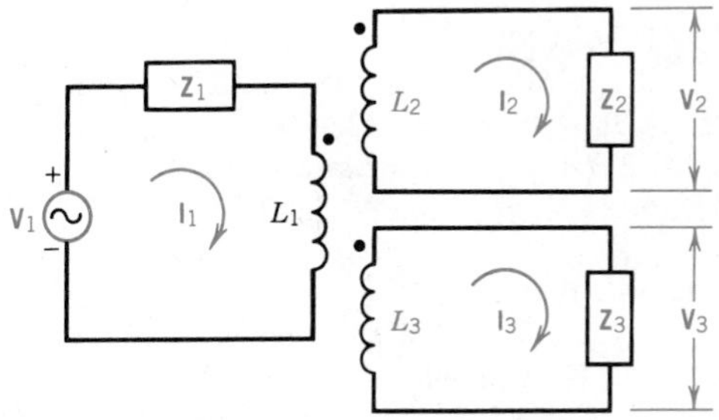

Figure 20-36 Problems 20-53 and 20-54

20-54 Repeat all parts of Prob. 20-53 with the polarity dots on both L_2 and L_3 reversed (such that all mutual inductances are *positive*). After completing all calculations, draw conclusions regarding the effect of positive and/or negative mutual inductances in multicoil transformers on the following:

1. Primary current and power drawn from the supply
2. Secondary current and load voltage magnitudes and powers dissipated
3. Phase of the secondary currents and voltages

20-13 ANSWERS

20-1 a 150 t b 4 c 1/4 d $0.7\overline{3}$ V/t e $0.7\overline{3}$ V/t f $113.\overline{63}$ A g $454.\overline{54}$ A
20-2 a 15/1 b 8 V c 0.2 Ω d 0.16 V/t
20-3 a 80 t, 10 t b 8 c 1/8 d 12.5 A, 100 A
20-4 a 200 V b 25 V c 160 A d 4 kVA e 8/1
20-5 a 550 t b 1/11 c 0.4 V/t

20-6 a 30 V, 10 A b 1.25 A c, d 192 Ω
20-7 a 10 b 100 t
20-8 a 250 Ω b 100 Ω c 71.43 Ω d 0.42 A e 12.6 W f 6 V, 3.6 W g 12 V, 9 W h 12.6 W
20-9 a 50 Ω b 20 Ω c 14.29 Ω d 1.47 Ω e 30.9 W f 4.2 V, 8.82 W g 10.5 V, 22.05 W h 30.9 W
20-10 a 1 A, 16 Ω b 8 V c 10 Ω d $6.\overline{6}$ Ω
20-11 1.37 Ω
20-12 a 5 b 2 V c 1 W
20-13 a 16 b 24 V c 1.5 V d 187.5 mA e 11.72 mA f, g 281.25 mW h 562.5 mW i 50%
20-14 a 186.77 mV b 47.81 V c 4.36 mW d 1.1163 W e 1.1206 W
20-15 a $6.324\overline{5}$ b $j0.75$ Ω c $-j2.482$ Ω d $-j30$ Ω e $6.324\overline{5}$ A f, g 40 W h 12.65 V i 50 V
20-16 a 2.174 kA, 217.4 A b 3.162 mΩ, 0.3162 Ω c 6.874 V, 68.74 V d 236.9 V, 2.231 kV e 9.42 f 10
20-17 a 106 mΩ b 10.6 Ω
20-18 a 0.2 Ω b $j0.6$ Ω c 2 mΩ d $j6$ mΩ
20-19 a 217.4 A b 8.7 V, 4.348 V c 234.5 V d 239.3 V e 227 V f 1.96% g 4.04% h −1.3%
20-21 a 41.47 Ω, $j75.8$ Ω b 104 mΩ; $j189.5$ mΩ c 1.858% d 3.431% e −0.461%
20-22 a $1\overline{66}$ A b 367 W c 94.82; 96.69; 97.05; 97.01; 96.81%
20-23 a 104 mΩ b 96.36% c 95.64% d 0.9413; 39.22 A e 96.7% f 96.36%
20-24 a 102 W b 97.51% c 97.84%; 97.13%
20-25 a 3.916% b 90.92; 94.99; 96.9; 97.32; 97.35; 97.22% c 0.9037 d 97.36%
20-26 b 88.23; 93.43; 95.91; 96.45; 96.49; 96.33%
20-27 −3.086%
20-28 a 12.5, 50 A b 62.5 A c 30 kVA d 150 kVA
20-29 a 30 kW b 120 kW
20-30 a 16.6, $83.\overline{3}$ A b $83.\overline{3}$ A c $3\overline{3}$ kVA d 200 kVA
20-31 a $33.\overline{3}$ kW b $166.\overline{6}$ kW
20-34 b 2/3, 720 W c 480; 240 W
20-35 a 480 W b 240 W
20-36 a $-j4$ Ω c $-30 + j72$ d $1.923\angle{-59.49^\circ}$ A; $1.5385\angle{-22.62^\circ}$ A
20-37 a 29.29 W b 22.19 W; 7.101 W
20-38 a $j4$ Ω c $-30 + j72$ d $1.923\angle{-59.49^\circ}$ A; $1.5385\angle{157.38^\circ}$ A
20-39 a $(1.92 - j2.56)$ Ω b $(7.92 + j13.44)$ Ω c $1.923\angle{-59.49^\circ}$ A d 29.29 W e 22.19 W
20-40 a $(18 + j7.5)$ Ω b $1.5385\angle{-22.62^\circ}$ A c 7.1 W
20-41 a $7.0225\angle{20.556^\circ}$ V b $1.3288 + j3.1233$ Ω c $1.5385\angle{-22.62^\circ}$ A
20-42 a $(0.6 + j2.3)$ kΩ b $(0.8 + j1)$ kΩ e $-1.18 + j2.44$ f $47.25\angle{-64.47^\circ}$ mA, $29.52\angle{-25.81^\circ}$ mA g 2.036 W h 1.3395 W i 0.6971 W
20-43 c 0.4 H, $+j40$ Ω e $-4550 + j23\,050$ f $0.7248\angle{-51.4^\circ}$ A, $0.170\,25\angle{168.8^\circ}$ A g 45.22 W h 15.76; 26.27; 1.45; 1.74 W
20-44 a $0.234\overline{8}$ b $(6.069 - j7.172)$ Ω c $(86.07 + j107.8)$ Ω d $0.7249\angle{-51.4^\circ}$ A e $29\angle{38.6^\circ}$ V f $0.1703\angle{168.8^\circ}$ A
20-45 a $0.6875\angle{-59.4^\circ}$ A b $0.3371\angle{-160.71^\circ}$ A c 35 W d 23.63 W e 11.36 W
20-46 a 0.4903 b $(24.04 - j4.808)$ Ω c $(74.04 + j125.19)$ Ω d $0.6875\angle{-59.4^\circ}$ A e $34.375\angle{30.6^\circ}$ V f $0.3371\angle{-160.7^\circ}$ A
20-47 a 3.795 mH, $-j3.795$ Ω c $139.12 + j25$ d $0.3537\angle{26.68^\circ}$ A, $0.834\,45\angle{79.81^\circ}$ A e 3.16 W f 0.3753 W, 2.785 W g $0.6835\angle{104.3^\circ}$ A
20-48 a $+j3.795$ Ω c $17.68 + j25$ d $1.633\angle{-17.86^\circ}$ A, $1.3733\angle{35.27^\circ}$ A e 15.54 W f 8 W, 7.54 W
20-49 a 2.4 mH, $-j2.4$ Ω c $93.16 + j1.2$ d $1.2517\angle{-59.77^\circ}$ A; $0.1156\angle{21.06^\circ}$ A e $1.2385\angle{-65.06^\circ}$ A f 6.302 W g 4.7 W, 1.534 W, 0.0668 W
20-50 a $+j2.4$ Ω c $112.36 + j0.8$ d $1.033\angle{-64.53^\circ}$ A; $0.4\angle{-82.69^\circ}$ A e $0.665\angle{-53.73^\circ}$ A f 4.442 W g 3.201 W, 0.442 W, 0.8 W
20-51 a $+j6$ Ω b $7.439\angle{39.27^\circ}$ A, $2.1595\angle{134.7^\circ}$ A c 734.18 W d 151.92 W e 830.08 W f 55.92 W
20-52 a $-j6$ Ω b $4.604\angle{46.63^\circ}$ A, $5.509\angle{100.68^\circ}$ A c 441.1 W d 241 W e 317.95 W f 364.19 W
20-53 a $-j2.6833$ Ω, $j2.4495$ Ω, $-j2.1909$ Ω b $5.6365\angle{0.832^\circ}$ A, $1.239\angle{145.8^\circ}$ A, $1.431\angle{10.84^\circ}$ A c $16.11\angle{78.46^\circ}$ V, $7.153\angle{63.97^\circ}$ V d 140.9 W e 127.08 W, 7.676 W, 6.141 W
20-54 a $j2.683$ Ω, $j2.45$ Ω, 2.191 Ω b $5.6365\angle{0.832^\circ}$ A, $1.239\angle{-34.16^\circ}$ A, $1.431\angle{-169.15^\circ}$ A c $16.11\angle{-101.54^\circ}$ V, $7.15\angle{-116^\circ}$ V d 140.9 W e 127.08 W, 7.676 W, 6.141 W

CHAPTER 21

Polyphase Circuits and Systems

21-1 INTRODUCTION

All the ac circuits considered so far have been single-phase (1ϕ) networks. But almost all the commercial energy throughout the world is generated, transmitted, and distributed in the form of polyphase energy, in general, and three-phase (3ϕ) energy, in particular. The reasons for and advantages of the use of polyphase over single-phase systems are given in Sec. 21-2. In general, the advantages emerge because polyphase systems contain two or more sources of the same frequency and amplitude, sharing the same transmission lines but displaced from each in a fixed phase relation. For example, in a 3ϕ, three-wire system, there are three phases displaced 120 electrical degrees apart, transmitting their energy over three lines. Although the power in each phase is pulsating (at twice the system frequency) because of the phase differences, the total power at any instant is never zero but is actually constant!

Since polyphase systems contain two (or more) separate sources, use of *time-domain* representations for them is both cumbersome and meaningless. The *frequency-domain* (phasor) representation is used *exclusively*. But because there may be several multiple sources and two possible time or *phase sequences* (Sec. 21-8), a special type of notation called *double-subscript* notation (Sec. 21-3) is used to designate the respective instantaneous *polarities* of specific *voltages* and/or the instantaneous *directions* of specific currents in the system. This technique enables us not only to add and subtract various voltage and current phasors but also to differentiate between them, when they are affected by phase sequence.

The two basic types of 3ϕ systems, the *delta* and *wye*, are analyzed in terms of their sources and loads under conditions of balanced loads. This is followed by the determination of power in 3ϕ delta or wye balanced systems. The study of unbalanced delta and unbalanced wye systems (both three-wire and four-wire) follows, including the design and construction of commercial phase sequence indicators.

This chapter also covers *measurement* of 3ϕ balanced power by various single wattmeter methods and of both *balanced* and *unbalanced* power with three wattmeters and two wattmeters. A thorough analysis of the two-wattmeter method for measurement of power, power factor, and phase sequence of balanced inductive and capacitive loads and power of unbalanced loads concludes the chapter.

21-2 ADVANTAGES OF POLYPHASE OVER SINGLE-PHASE SYSTEMS

The almost universal use of polyphase power generation is due to the specific advantages of polyphase over single-phase systems:

1. Assuming the same (fixed) voltage between lines, the same kVA transmitted for the same distances, the 3ϕ system, for example, requires approximately 75% of the weight of copper of a comparable 1ϕ system. Less total copper implies use of smaller conductors, less line weight, and fewer towers (poles) to support overhead cables for the same distances of power transmission.
2. Polyphase power is constant, whereas 1ϕ power is pulsating. This produces many advantages for electrical machinery, including (for motors) higher starting torques, quieter operation with less vibration and noise, and improved speed regulation compared to 1ϕ motors of the same rating. For all dynamos (motors and generators) and transformers, the constant power in polyphase equipment results in reduced eddy current and hysteresis losses, tending to increase the efficiency of polyphase over 1ϕ equipment.
3. As a result of (2) above, polyphase equipment (transformers, motors, generators, etc.) is smaller, weighs less, and costs less for the same kVA rating than 1ϕ equipment of the same type.
4. Polyphase motors are generally self-starting, whereas most 1ϕ motors require some type of auxiliary starting device or winding, increasing the cost of the latter for the same horsepower rating. It is precisely for this reason that single-phase motors are limited to the smaller horsepower ratings.
5. Any polyphase system may be transformed (using interphase transformers) into a higher or lower polyphase or single-phase system. Common polyphase transformations from 3ϕ to 6, 12, 18, 24, and even 36ϕ are accomplished with multiwinding transformers. No 1ϕ system can be converted to a true polyphase system with transformers.[1]
6. Because polyphase transformers are so highly efficient, conversion to higher multiple systems (12, 18, 24, 36ϕ, etc.) enables large-scale production of dc (using either large diodes or SCRs) for long-distance power transmission,[2] electrolysis of water, or refining of metals. As the number of secondary phases is increased, the need for smoothing filters is eliminated (since the output contains less ripple) or their size is dramatically reduced compared to 1ϕ rectifiers handling the same average dc power.

For all of these reasons (and others as well), a knowledge of polyphase circuits and systems is fundamental to a knowledge of circuit analysis. Many of the relations and techniques used in solving 1ϕ circuits may be applied equally to polyphase circuits, such as the use of mesh analysis in unbalanced three-wire, 3ϕ systems (Sec. 21-9). In the case of balanced, symmetrical 3ϕ systems, we may use one phase to represent voltages and/or currents of all three phases. The key to understanding polyphase circuits (as noted in the introduction) is that they are composed of separate single-phase systems operating in a predetermined sequence, having a fixed phase relation between them.

---

[1] It is possible to generate polyphase power with a 1ϕ motor to drive a 3ϕ or 6ϕ alternator (called a motor–generator set).

[2] Theoretically, dc transmission has many advantages over ac: lower cost over long distances, lower losses, smaller volt drop, and monopolar (one-line) operation should one side become grounded. Major drawbacks include lack of a highly efficient (and simple) device to change voltages (such as the transformer for ac) and high-capacity, high-speed switching and protective devices (currently available for ac use). High-voltage dc power transmission systems are currently in operation in Sweden, the United States, Great Britain, and France.

21-3 DOUBLE-SUBSCRIPT NOTATION

Double subscripts provide a convenient method for describing a *directed quantity.* In the case of voltages, the notation tells us which voltage is positive at some node compared to another node. Voltage $\mathbf{V}_{12}$ implies that node 1 is *more positive* than node 2 and represents the voltage drop measured across and between nodes 1 and 2. In the case of an ac voltage, this may represent an RMS generated source voltage or an RMS voltage drop between two nodes. Thus, if $\mathbf{V}_{12} = +10$ V instantaneously, then $\mathbf{V}_{21} = -10$ V, that is, $\mathbf{V}_{12} = -\mathbf{V}_{21}$.

When double-subscript notation is applied to currents, it represents the instantaneous *direction of currents* between two nodes. $\mathbf{I}_{12}$ represents the direction of current from reference node 1 to node 2. Thus, $\mathbf{I}_{12}$ is positive whenever current instantaneously flows from node 1 to node 2. If $\mathbf{I}_{12} = 5$ A, then $\mathbf{I}_{21} = -5$ A (i.e., $\mathbf{I}_{12} = -\mathbf{I}_{21}$).

The reader should not be confused when double-subscript notations contain combinations of letters and numbers, such as $\mathbf{V}_{a_1a_2}$ (voltage between nodes $\mathbf{a}_1$ and $\mathbf{a}_2$) or letters containing *primes*, such as $\mathbf{V}_{aa'}$ (voltage between nodes **a** and **a**′).

Figure 21-1 shows two equal-phase voltages displaced by 90°, as in a 2ϕ system. Phase voltage $\mathbf{V}_{ab}$ is the reference voltage of $100\angle 0^\circ$ V in the coil winding whose terminals are **a** and **b**. The voltage in coil **c**–**d** with respect to the reference is $100\angle 90^\circ$ V. Four possible series connections exist, each producing a resultant of the same magnitude but differing in phase with respect to the reference $\mathbf{V}_{ab}$. If we connect nodes **b** and **c**, then by Kirchhoff's voltage law (KVL) the resultant is the phasor sum $\mathbf{V}_{ad}$, as shown by the equality

$$\mathbf{V}_{ad} = \mathbf{V}_{ab} + \mathbf{V}_{cd} = 100\angle 0^\circ + 100\angle 90^\circ = 100 + j100 = \mathbf{141.4\angle 45^\circ\ V} \quad \text{(Fig. 21-1b)}$$

Similarly, if we connect nodes **b** and **d**, then by KVL the resultant is the phasor sum $\mathbf{V}_{ac}$, as shown by the equality

$$\mathbf{V}_{ac} = \mathbf{V}_{ab} + \mathbf{V}_{dc} = 100\angle 0^\circ + 100\angle -90^\circ = 100 - j100 = \mathbf{141.4\angle -45^\circ\ V} \quad \text{(Fig. 21-1c)}$$

Note that in both equalities the resultant voltage of interest dictates the coil connections. If we desired $\mathbf{V}_{da}$ we can obtain it by $\mathbf{V}_{dc} + \mathbf{V}_{ba}$ or alternatively reverse $\mathbf{V}_{ad}$ by 180°. In either case, we still must join nodes **b** and **c**, and the resultant is $141.4\angle -135^\circ$ V.

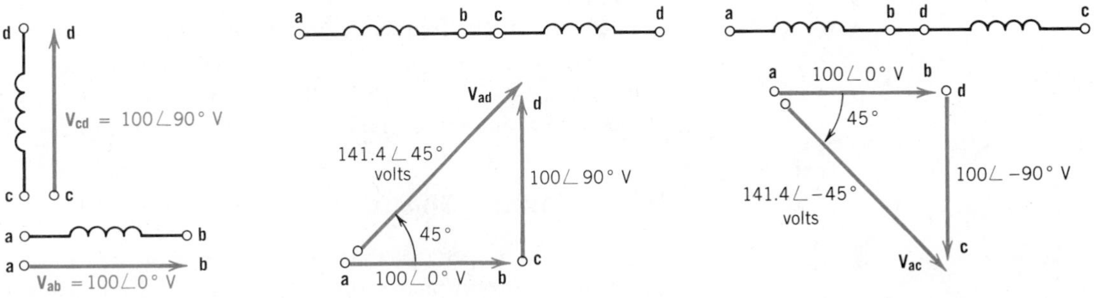

a. Phasors of a 2ϕ system

b. Series connection of terminals b to c

c. Series connection of terminals b to d

Figure 21-1 Use of double-subscript notation showing resultants produced by addition of instantaneous coil voltages

21-4 GENERATION OF THREE-PHASE VOLTAGES

Figure 21-2a shows a cross section of the stator and rotor of a simplified 3ϕ alternator, with the rotor rotating in a counterclockwise (CCW) direction. At the rotor position shown, the voltage-induced coil set A is $\mathbf{E}_{A'A}$. This is a maximum voltage

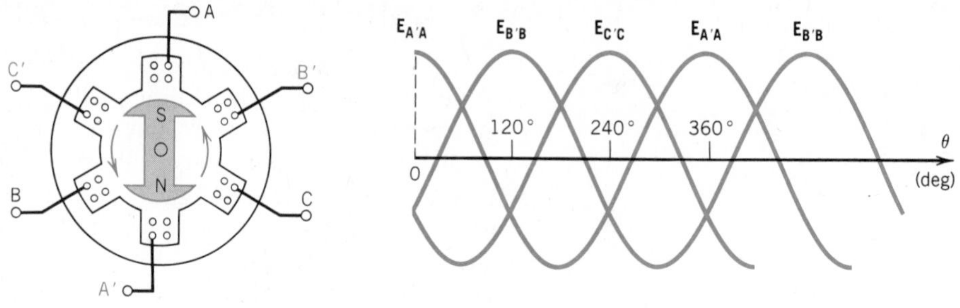

Figure 21-2 Generation of 3ϕ voltages displaced by 120° in a 3ϕ alternator

a. 3ϕ alternator cross section

b. Sequence of induced voltages for 360° of mechanical rotation

since the north and south poles of the rotor are aligned to produce maximum induced voltage in coil set A, as shown in Fig. 21-2b, where the rotation angle, θ, is 0°. As the rotor is rotated CCW by 120°, the north pole moves from coil end A′ to coil end B′, where it induces maximum voltage, $\mathbf{E}_{B'B}$, in coil set B, as shown in Fig. 21-2b. Similarly, the voltage in coil set C is a maximum, $\mathbf{E}_{C'C}$, when the rotor is at its 240° position and the north pole has rotated to coil end C′. The alternator therefore generates three separate (single-phase) stator voltages that are 120° apart in both space and time phase. We may summarize the sinusoidal waveforms of Fig. 21-2b, by noting the following:

1. Each of the waveforms has a *positive* maximum value $\mathbf{E}_{A'A}$, $\mathbf{E}_{B'B}$, and $\mathbf{E}_{C'C}$ that is displaced by 120 *electrical* degrees.
2. The sequence of voltage generation based on the direction of CCW rotation of the rotor is A–B–C. (If the rotor rotation is clockwise, the sequence is A–C–B).
3. When any phase voltage is a positive maximum, the other two phase voltages are negative and each is half its maximum value, as shown in Fig. 21-2b.
4. This also implies that the phasor sum of the three voltages at *any* instant in time is *always zero* in Fig. 21-2b.
5. The three sinusoidal voltages may be represented in the time domain as $\mathbf{E}_{A'A} \sin \omega t$ for phase winding A, $\mathbf{E}_{B'B} \sin(\omega t - 120°)$ for phase winding B, and $\mathbf{E}_{C'C} \sin(\omega t - 240°) = \mathbf{E}_{C'C} \sin(\omega t + 120°)$ for phase winding C. But, as noted above, the time-domain representation is not as useful as the phasor-domain relations.
6. Therefore, in the frequency or phasor domain, the RMS value of phase A is $\mathbf{V}_{A'A}\underline{/0°}$, where $\mathbf{V}_{A'A} = \mathbf{E}_{A'A}/\sqrt{2}$. Similarly, we may write the voltage of phase B as $\mathbf{V}_{B'B}\underline{/-120°}$ and the voltage of phase C as $\mathbf{V}_{C'C}\underline{/+120°}$, where each is the RMS value of its maximum value, as shown in Fig. 21-3a.

21-4.1 Generic Star (or Y) Connection

Figure 21-3a shows the phasor representation of the (RMS) voltages induced in each of the three coil-phase windings of the 3ϕ alternator. Observe that it shows the possibility of bringing out a total of six wires from the alternator of Fig. 21-2a, representing three separate single-phase windings, each displaced by 120° in phase. Although such an arrangement is feasible, it is impractical. We may still obtain the advantages of a 3ϕ six-wire system if we connect coil ends A′, B′, and C′ to a common junction "**n**," as shown in Fig. 21-3b. Whenever the coil ends of any multiphase system are so connected to a common junction, it is called a star system.[3]

[3] In this way, we may have a 6ϕ star, a 12ϕ star, etc. for alternators whose windings are 60° and 30° apart, respectively, on their stators, and so on.

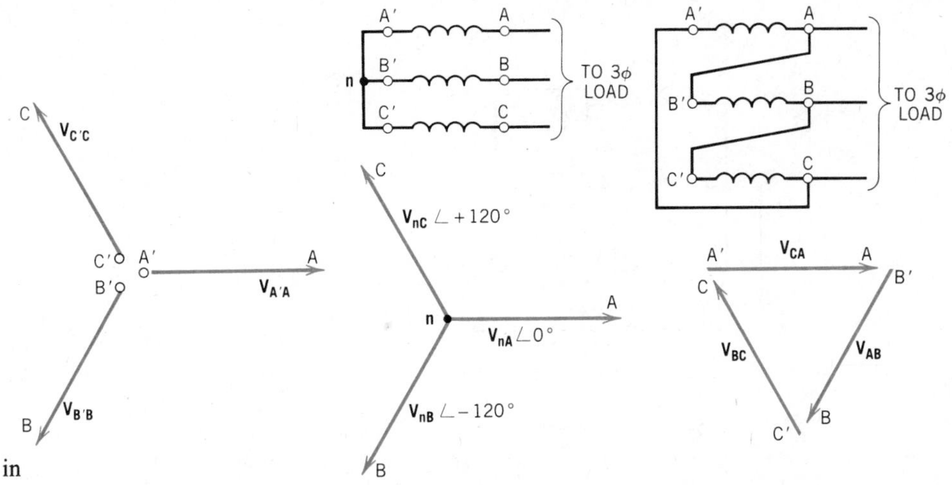

Figure 21-3 Coil phase voltages in generic star and mesh connections producing wye and delta systems

a. Coil voltages as phasors in a 3ϕ alternator

b. Wye (star) connection phase voltages

c. Delta (mesh) connection phase (and line) voltages

Figure 21-3b shows the phasor diagram that results when the three coil ends are connected to a common junction, **n**, producing $\mathbf{V}_{nA}\underline{/0°}$, $\mathbf{V}_{nB}\underline{/-120°}$, and $\mathbf{V}_{nC}\underline{/+120°}$. With respect to the phasor diagram of Fig. 21-3b, please note that

1. The phase voltages are (still) equal to their respective RMS values and are (still) 120° apart.
2. The phasor sum of the three voltages is zero since they are equal in magnitude and 120° apart.
3. The connection is called a wye (Y) connection.

In commercial Y-connected 3ϕ alternators the coil phase ends A′, B′, and C′ are internally connected and only the common junction, **n**, or *neutral*, is brought out along with terminals A, B, and C.

21-4.2 Generic Mesh (or Delta, Δ) Connection

An alternative possible connection of the coil phase windings is the *mesh* connection. In this connection, the three coils are series-connected to form a closed ring or mesh. The delta (Δ) is the simplest form of mesh connection (just as the wye, Y, is the simplest form of the star connection). The coil connections are shown in Fig. 21-3c, where ends A–B′, B–C′, and C–A′ are connected (internally) and terminals A, B, and C are brought out from the alternator stator. As shown in the phasor diagram of Fig. 21-3c, as a result of such internal connections

1. The phase voltages $\mathbf{V}_{CA} = \mathbf{V}_{A'A}$, $\mathbf{V}_{AB} = \mathbf{V}_{B'B}$, and $\mathbf{V}_{BC} = \mathbf{V}_{C'C}$.
2. The phase voltages are (still) equal to their respective RMS values and are (still) 120° apart.
3. The phasor sum of voltages $\mathbf{V}_{CA} + \mathbf{V}_{AB} + \mathbf{V}_{BC}$ is zero since they are equal in magnitude and 120° apart.

21-5 VOLTAGE AND CURRENT RELATIONS OF THE Y-CONNECTED SOURCE WITH BALANCED LOAD

The 3ϕ Y-connected alternator of Fig. 21-3b is shown connected to a balanced Y-connected load in **Fig. 21-4a**. Since the load is "balanced," $\mathbf{Z}_A = \mathbf{Z}_B = \mathbf{Z}_C$. Let us first consider the voltage relations in this 3ϕ, four-wire system, with balanced loads.

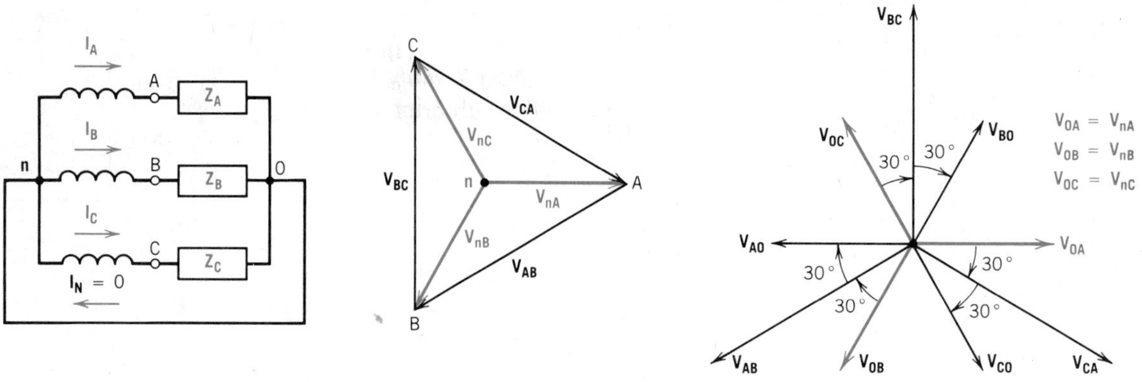

a. 3ϕ, 4-wire, Y-connected system

b. Source phase and line voltages

c. Load phase and line voltages

Figure 21-4 3ϕ four-wire Y-connected balanced load showing phase and line voltages at source and load

21-5.1 Voltage Relations under Conditions of Balanced Load

The phase voltages of the 3ϕ *source*, $\mathbf{V}_{nA}$, $\mathbf{V}_{nB}$, and $\mathbf{V}_{nC}$, as shown in Fig. 21-3b, are shown once again in Fig. 21-4b. Looking back toward the source, we may now use double-subscript notation to find the line voltages $\mathbf{V}_{AB}$, $\mathbf{V}_{BC}$, and $\mathbf{V}_{CA}$, in that sequence.[4] From both Fig. 21-4a and Fig. 21-4b we can see that

$$\mathbf{V}_{AB} = \mathbf{V}_{An} + \mathbf{V}_{nB} \qquad \mathbf{V}_{BC} = \mathbf{V}_{Bn} + \mathbf{V}_{nC} \qquad \mathbf{V}_{CA} = \mathbf{V}_{Cn} + \mathbf{V}_{nA} \tag{21-1}$$

where $\mathbf{V}_{AB}$, $\mathbf{V}_{BC}$, and $\mathbf{V}_{CA}$ are the line voltages.

The reader should note from Fig. 21-4b that merely connecting points A to B, B to C, and C to A (following the phase sequence ABCABC) produces the *same* result as the addition of component phase voltages above. The magnitudes and phase of the respective line voltages are calculated in Example 21-1.

EXAMPLE 21-1

Given phase voltages $\mathbf{V}_{nA} = 100\underline{/0^\circ}$ V, $\mathbf{V}_{nB} = 100\underline{/-120^\circ}$ V, and $\mathbf{V}_{nC} = 100\underline{/120^\circ}$ V, respectively, calculate line voltages a. $\mathbf{V}_{AB}$ b. $\mathbf{V}_{BC}$ c. $\mathbf{V}_{CA}$

Solution

a. $\mathbf{V}_{AB} = \mathbf{V}_{An} + \mathbf{V}_{nB} = 100\underline{/180^\circ} + 100\underline{/-120^\circ}$
$= (-100 + j0) + (-50 - j86.6)$
$= -150 - j86.6 = \mathbf{173.2\underline{/-150^\circ}}$ **V**

b. $\mathbf{V}_{BC} = \mathbf{V}_{Bn} + \mathbf{V}_{nC} = 100\underline{/60^\circ} + 100\underline{/120^\circ}$
$= (50 + j86.6) + (-50 + j86.6)$
$= \mathbf{173.2\underline{/90^\circ}}$ **V**

c. $\mathbf{V}_{CA} = \mathbf{V}_{Cn} + \mathbf{V}_{nA} = 100\underline{/-60^\circ} + 100\underline{/0^\circ}$
$= (50 - j86.6) + (100 + j0)$
$= 150 - j86.6 = \mathbf{173.2\underline{/-30^\circ}}$ **V**

The following insights may be drawn from Ex. 21-1 regarding the source phase and line voltages of Fig. 21-4b for a 3ϕ, Y-connected system:

1. The line voltage across any two alternator terminals is equal to $\sqrt{3}$ times the phase voltage in magnitude.
2. The three line voltages are equally displaced by 120° and equal in magnitude.
3. The line voltages are displaced from their component phase voltages by $\pm 30^\circ$.

---

[4] If the phase sequence is reversed, the order of line voltages is reversed. Since the sequence of voltage generation is ABCABC, the line voltages must follow that sequence as well.

The line voltages found above involved the source phase voltages "looking back" toward the source. Assuming that the load is balanced in Fig. 21-4a, we may determine the line voltages looking toward the load. From the phase voltages of Fig. 21-4b, using double-subscript notation, we may write for each line voltage (in Fig. 21-4c)

$$\mathbf{V}_{AB} = \mathbf{V}_{Ao} + \mathbf{V}_{oB} \qquad \mathbf{V}_{BC} = \mathbf{V}_{Bo} + \mathbf{V}_{oC} \qquad \mathbf{V}_{CA} = \mathbf{V}_{Co} + \mathbf{V}_{oA}$$

From the results of Fig. 21-4c we may conclude that

1. The line voltages across the load are identical to those across the source (since they are the *same*).
2. Source phase voltage $\mathbf{V}_{nA}$ is the same as load phase voltage $\mathbf{V}_{oA}$ because of the neutral connection placing both **n** and **o** at the same potential.
3. The voltage between any line and the neutral is the same as a phase voltage.

EXAMPLE 21-2

The generated line voltage of a 3ϕ Y-connected alternator is 2300 V (between lines). Calculate the

a. Peak value of the line voltage
b. Peak value of the phase voltage
c. Time interval between maximum values of adjacent sinusoidal phase voltages in Fig. 21-2b
d. Phase-to-neutral voltage of the generated phase voltages
e. Line-to-neutral voltage of the system of generated line voltages

Solution

a. $\mathbf{V}_{L_{max}} = \sqrt{2}\,\mathbf{V}_L = \sqrt{2}(2300\text{ V}) = \mathbf{3253\ V}$
b. $\mathbf{V}_{P_{max}} = \mathbf{V}_{L_{max}}/\sqrt{3} = 3253\text{ V}/\sqrt{3} = \mathbf{1878\ V}$
c. $T = 1/f = 1/60\text{ Hz} = \mathbf{16.\bar{6}\ ms}$ for one cycle of alternation or 360°; time between adjacent peaks $= \dfrac{120°}{360°} \times 16.\bar{6}\text{ ms} = \mathbf{5.\bar{5}\ ms}$
d. $\mathbf{V}_{Pn} = \mathbf{V}_L/\sqrt{3} = 2300\text{ V}/\sqrt{3} = \mathbf{1330\ V}$
e. $\mathbf{V}_{Ln} = 2300\text{ V}/\sqrt{3} = 2300\text{ V}/\sqrt{3} = \mathbf{1330\ V}$

21-5.2 Current Relations under Conditions of Balanced Load

Examination of Fig. 21-4a for the currents that flow from the source to the respective loads reveals the following obvious (and not so obvious) relations:

1. The phase currents, $\mathbf{I}_P$, are the *same* as the line currents, $\mathbf{I}_L$, which in turn are the *same* as the phase currents in the load, $\mathbf{I}_{Zp}$, for the simple reason that they are all connected *in series* from terminal **n** to **o**, respectively, in each phase.
2. Each of the three Y-connected loads may be considered as a separate single-phase load across a separate single-phase source whose voltage is $\mathbf{V}_{Pn}$, that is, the phase voltage. Since the load is balanced, any phase is representative of the two remaining phases, but only insofar as current magnitude is concerned.

These relations yield the following for a balanced Y-connected 3ϕ load:

$$\mathbf{I}_L \equiv \mathbf{I}_P \equiv \mathbf{I}_{Zp} = \frac{\mathbf{V}_{Pn}}{\mathbf{Z}_p} \quad \text{amperes (A)} \tag{21-2}$$

where $\mathbf{I}_L$ is the line current, which is the same as $\mathbf{I}_P$
$\mathbf{I}_P$ is the current in each phase of the source and/or the load
$\mathbf{I}_{Zp}$ is the current in the load impedance, per phase
$\mathbf{V}_{Pn}$ is the phase voltage, measured from any line to neutral
$\mathbf{Z}_p$ is the impedance per phase

Since the voltage rise in each phase of the source is equal to the voltage drop across each phase of the load, by Kirchhoff's voltage law, it is conventional to reverse the order of the subscripts in Fig. 21-4a and write $\mathbf{V}_{nA} = \mathbf{V}_{Ao}$. This enables us to use the same phasors for source and load voltages, but with the subscript "o" as the reference node for the loads. This also enables us to write the phase, line, and neutral currents as

$$\mathbf{I}_A = \frac{\mathbf{V}_{nA}}{\mathbf{Z}_A} = \frac{\mathbf{V}_{Ao}}{\mathbf{Z}_A} \qquad \mathbf{I}_B = \frac{\mathbf{V}_{nB}}{\mathbf{Z}_B} = \frac{\mathbf{V}_{Bo}}{\mathbf{Z}_B} \qquad \mathbf{I}_C = \frac{\mathbf{V}_{nC}}{\mathbf{Z}_C} = \frac{\mathbf{V}_{Co}}{\mathbf{Z}_C} \qquad (21\text{-}3)$$

Since by Kirchhoff's current law, the current in the neutral is the phasor sum of the phase and line currents, we may also write

$$\mathbf{I}_N = \mathbf{I}_A + \mathbf{I}_B + \mathbf{I}_C \quad \text{amperes (A)} \qquad (21\text{-}4)$$

In the case of balanced loads, $\mathbf{I}_N = 0$, and this serves as a check on the accuracy of the calculated phase and line currents.

In the event of an *unbalanced* load, $\mathbf{Z}_A \neq \mathbf{Z}_B \neq \mathbf{Z}_C$, the unbalanced current in each line (and phase) is the neutral current, $\mathbf{I}_N$. In effect, as we will see in Sec. 21-10.2, the neutral current serves to maintain the phase voltages across each of the loads equal in magnitude and displaced by 120°. (It also maintains the neutral point, "o," at the geometric center of the three line voltages.)

EXAMPLE 21-3

Given phase voltages of 120 V and balanced phase impedances of $6 + j8\ \Omega$ per phase in Fig. 21-4, assuming the phase sequence ABC as shown, calculate respective line currents in

a. Line A
b. Line B
c. Line C
d. The neutral

Solution

a. $\mathbf{I}_A = \mathbf{V}_{Ao}/\mathbf{Z}_A = 120\angle 0°\ \text{V}/10\angle 53.13°\ \Omega = \mathbf{12\angle -53.13°\ A}$
b. $\mathbf{I}_B = \mathbf{V}_{Bo}/\mathbf{Z}_B = 120\angle -120°\ \text{V}/10\angle 53.13°\ \Omega = \mathbf{12\angle -173.13°\ A}$
c. $\mathbf{I}_C = \mathbf{V}_{Co}/\mathbf{Z}_C = 120\angle 120°\ \text{V}/10\angle 53.13°\ \Omega = \mathbf{12\angle 66.87°\ A}$
d. $\mathbf{I}_N = \mathbf{I}_A + \mathbf{I}_B + \mathbf{I}_C = (7.2 - j9.6) + (-11.914 - j1.4354) + (4.714 + j11.035)\ \text{A} = 0 + j0 = \mathbf{0}$

Example 21-3 shows the following important points regarding *balanced* Y-connected 3ϕ systems:

1. The phase and line currents are the same in magnitude but displaced by 120°.
2. The neutral current is the phasor sum of the line currents. In a balanced load the neutral current is zero.
3. Each phase (and line) current lags (or leads) its respective phase voltage by the same phase angle as the impedance phase angle, θ, where $\theta = \cos^{-1}(R_p/Z_p)$.
4. The power factor angle is also θ, since by definition it is the angle between the phase current and the phase voltage of *each* of the three phases.

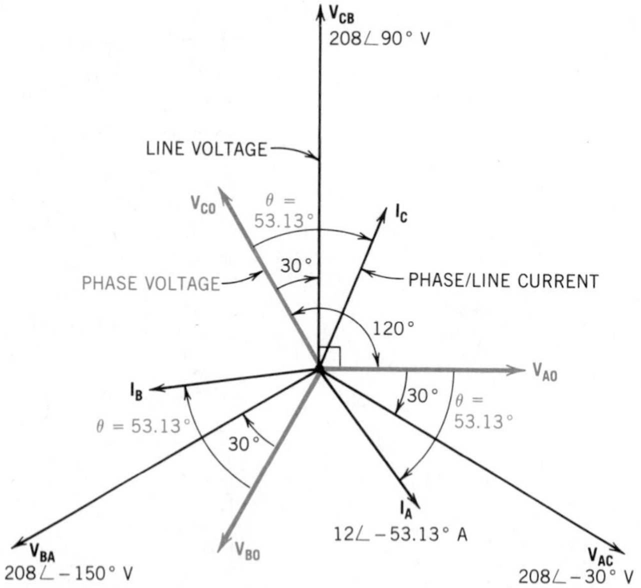

Figure 21-5 Phase and line voltages; phase and line currents in a balanced 3ϕ Y-connected load. ABC phase sequence, lagging load. (Ex. 21-4)

EXAMPLE 21-4

Using $\mathbf{V}_{Ao}$ as the reference voltage phasor, draw all phase and line voltages and phase and line currents for the balanced 3ϕ load of Ex. 21-3

Solution

Shown in **Fig. 21-5**, using phase sequence ABC. With respect to the solution shown in Fig. 21-5 for a balanced 3ϕ, Y-connected system, note that

1. The three phase voltages are displaced by 120°, in proper phase sequence, and equal in magnitude, such that their vector sum is zero.
2. The three line voltages lag (and lead) their component phase voltages by 30°, are equal in magnitude, and are $\sqrt{3}$ times the phase voltage magnitude. Their vector sum is also zero.
3. The three line (and phase) currents (in this case) lag their respective phase voltages by the load impedance phase angle θ (in this case 53.13°). The line currents are equal in magnitude and their phasor sum is zero (i.e., $\mathbf{I}_N = 0$).
4. For purposes of power and power factor, θ, the circuit phase angle, is always the angle between *phase* voltage and *phase* current.
5. Since $\mathbf{I}_N$ is zero, the neutral wire could be removed without affecting any of these relations or changing the phasor diagram shown in Fig. 21-5.

21-6 VOLTAGE AND CURRENT RELATIONS OF THE DELTA-CONNECTED SOURCE WITH BALANCED LOAD

Figure 21-3c showed how the generated coil voltages could be connected to produce a delta-connected source, showing $\mathbf{V}_{CA}$ as the reference voltage and all three line and phase voltages equal in magnitude and displaced by 120°.

Figure 21-6 shows the same delta-connected source feeding a balanced delta-connected load. Load $\mathbf{Z}_{AB}$ means the impedance connected across lines A–B. Since the loads are also connected in a head–tail arrangement (as opposed to a common junction), we may infer that they are also *mesh*-connected in delta. A careful examination of Fig. 21-6 reveals for the balanced Δ-connected system that

1. The phase voltages, $\mathbf{V}_{AB}$, $\mathbf{V}_{BC}$, and $\mathbf{V}_{CA}$, respectively, are the *same* as the voltages between lines A, B, and C, respectively (i.e., $\mathbf{V}_P \equiv \mathbf{V}_L$).
2. The line currents $\mathbf{I}_A$, $\mathbf{I}_B$, and $\mathbf{I}_C$, by Kirchhoff's current law, are the phasor sum of the currents in each phase, respectively, $\mathbf{I}_{AB}$, $\mathbf{I}_{CA}$, and $\mathbf{I}_{BC}$.

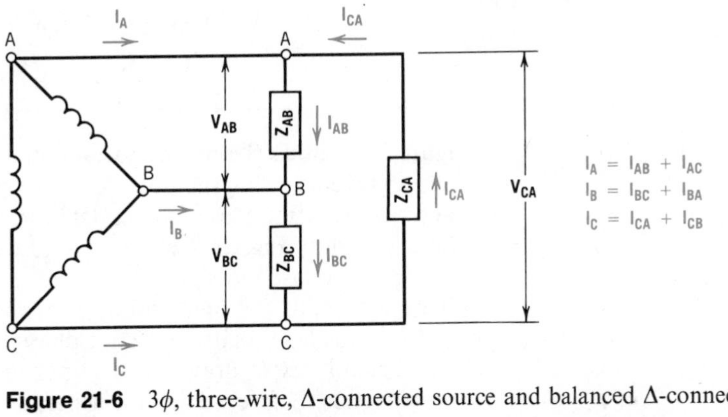

Figure 21-6 3ϕ, three-wire, Δ-connected source and balanced Δ-connected load

Using the phase voltages as our reference, by Ohm's law we know that the phase current $\mathbf{I}_P = \mathbf{V}_P/\mathbf{Z}_P$ (i.e., the phase current is phase voltage divided by the impedance per phase). Using the phase sequence ABC, for the delta-connected system we may write the phase currents for the delta-connected load, shown in **Fig. 21-7a**, as

$$\mathbf{I}_{AB} = \frac{\mathbf{V}_{AB}}{\mathbf{Z}_{AB}} \qquad \mathbf{I}_{BC} = \frac{\mathbf{V}_{BC}}{\mathbf{Z}_{BC}} \qquad \mathbf{I}_{CA} = \frac{\mathbf{V}_{CA}}{\mathbf{Z}_{CA}} \tag{21-5}$$

Finding the line currents (again) requires use of double-subscript notation. For each node, we know that the *line* current entering the node is the *sum* of the *phase* currents *leaving* the node or, as shown in Fig. 21-6, at nodes A, B, and C

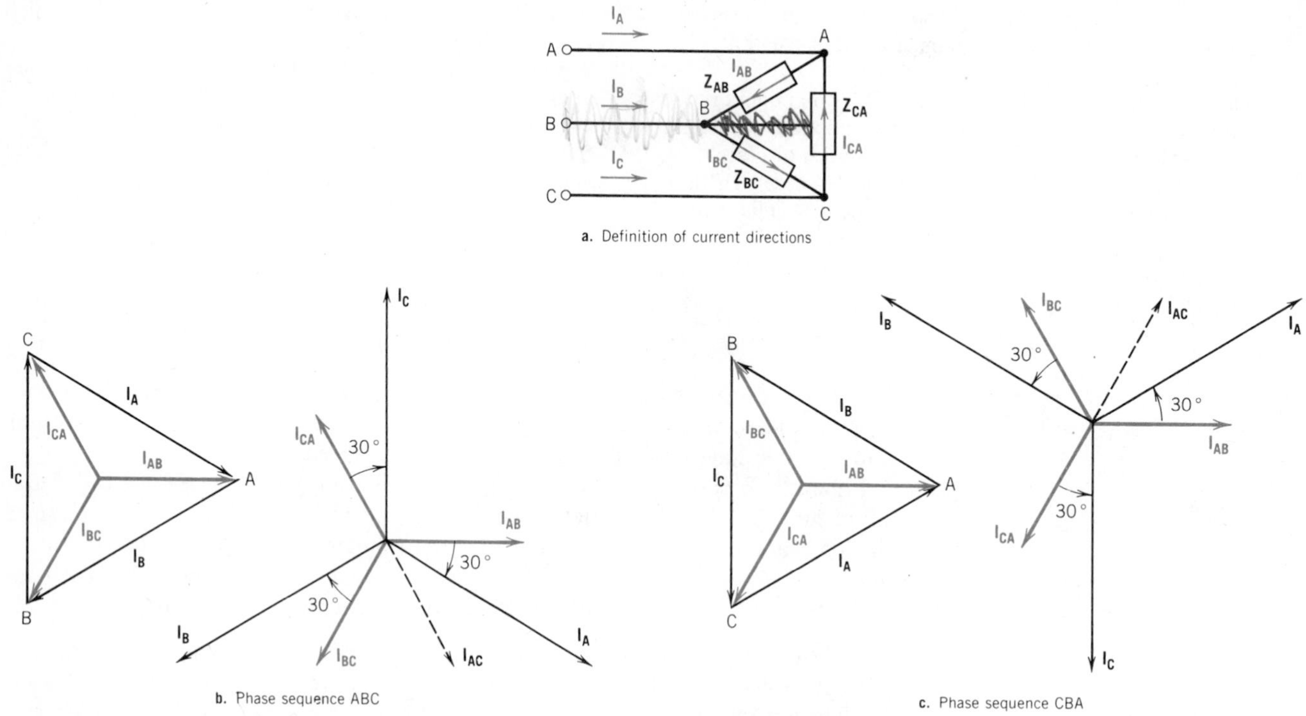

Figure 21-7 Balanced phase and line currents in a Δ-connected load showing effect of phase sequence on phase relations

$$\mathbf{I}_A = \mathbf{I}_{AB} - \mathbf{I}_{CA} = \mathbf{I}_{AB} + \mathbf{I}_{AC} \quad \text{(21-6a)}$$

$$\mathbf{I}_B = \mathbf{I}_{BC} - \mathbf{I}_{AB} = \mathbf{I}_{BC} + \mathbf{I}_{BA} \quad \text{(21-6b)}$$

$$\mathbf{I}_C = \mathbf{I}_{CA} - \mathbf{I}_{BC} = \mathbf{I}_{CA} + \mathbf{I}_{CB} \quad \text{(21-6c)}$$

Figure 21-7 shows the magnitude and phase of the line currents, with respect to the phase currents, using the two possible phase sequences, depending on direction of alternator rotation (described in detail in Sec. 21-8). The magnitudes and phase relations shown in Fig. 21-7 emerge from Eqs. (21-6), using double-subscript notation for either phase sequence.

Figure 21-8 adds the line (and also identical phase) voltages for a load whose phase current is lagging its respective phase voltage by some angle θ. Figures 21-7 and 21-8 enable us to draw many important conclusions regarding *balanced* Δ-connected loads. Before we do, it is necessary to emphasize that the following conclusions hold for **balanced** Δ-connected loads **only**:

1. All line currents are $\sqrt{3}$ times their component phase currents. In Figs. 21-7b and 21-7c the component phase currents $\mathbf{I}_{AB}$ and $\mathbf{I}_{AC}$ produce line current $\mathbf{I}_A$.
2. The line currents lead or lag their component phase currents by 30°, as shown in Figs. 21-7b and 21-7c.
3. The phase angle θ is determined by the voltage across and current in *each phase*, as shown in Fig. 21-8.
4. All phase (and equivalent line) voltages are equal in magnitude and displaced, respectively, by 120°, thus producing a phasor sum of zero.
5. All phase and line currents are equal in magnitude and 120° apart, also producing a phasor sum of zero.

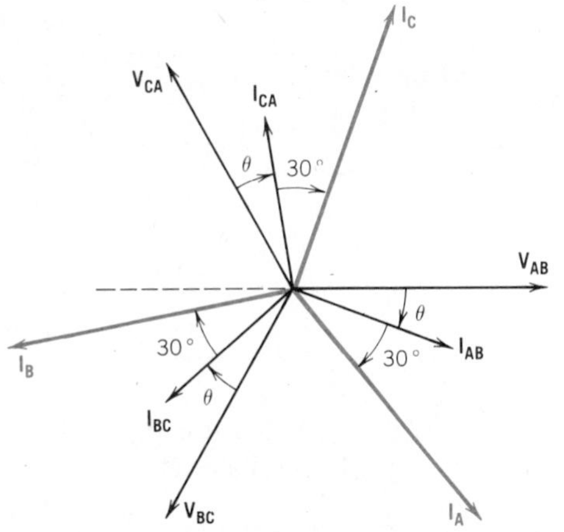

Figure 21-8 Balanced phase and line currents in a Δ-connected load showing their relation to phase (and identical) line voltages for ABC phase sequence

6. For phase sequence ABC (shown in Figs. 21-7b and 21-8), the line currents *lag* their *reference* phase currents by 30°. Thus line current $\mathbf{I}_A$ lags reference phase current $\mathbf{I}_{AB}$ by 30°.
7. For phase sequence CBA, the result opposite to that in (6) occurs. The line currents *lead* their *reference* phase currents by 30°. Note that the same reference phase currents appear in Figs. 21-7b and 21-7c.
8. Phase sequence ABC is sometimes called a *positive* phase sequence since the order of letters increases. (Phase sequence 1–2–3 is the same positive phase sequence as ABC.) Phase sequence CBA (or 3–2–1) is a *negative* phase sequence because the order of letters decreases.
9. Figures 21-7b and 21-7c also show shortcut methods of drawing line currents based on their respective phase sequences. The equilateral triangles show how line currents $\mathbf{I}_A$, $\mathbf{I}_B$, and $\mathbf{I}_C$ may be drawn directly, using Eqs. (21-6) and the respectively known phase sequences.
10. Using the rules established for phase sequence (Sec. 21-8), if the equilateral triangles of Figs. 21-7b and 21-7c are rotated in a CCW direction, they yield the phase sequences ABC and CBA, respectively. This verifies the shortcut method of showing all balanced line currents with respect to their component phase currents.
11. For balanced Δ-connected loads, therefore, once we have found a single reference *phase* current (say $\mathbf{I}_{AB}$) and know the phase sequence, we can find *all* the other phase and line currents! This is shown in Ex. 21-5, which uses a lagging load, as shown in Fig. 21-8.
12. For the condition of loads lagging by some phase angle θ, the *line* currents *lag* their respective line voltages by angle $(30 + \theta)$, as shown in Fig. 21-8. This relation simplifies balanced line current calculations, as shown in Ex. 21-5.
13. For the condition of loads *leading* by some phase angle θ, the line currents either lead their respective line voltages by angle $(\theta - 30°)$ when $\theta > 30°$ or lag their respective line voltages by angle $(30° - \theta)$ when $\theta < 30°$.
14. When the phase angle $\theta = 30°$, the line (and phase) voltages are in phase with the line currents. (Recall that line currents are shifted from phase currents by 30°.)

EXAMPLE 21-5

A delta-connected balanced load of $35 + j120\ \Omega$ is connected across 3ϕ lines whose phase sequence is ABC. If the line voltage of the 3ϕ supply is 125 V and $\mathbf{V}_{AB}$ is used as a reference, calculate the

a. Phase voltages $\mathbf{V}_{AB}$, $\mathbf{V}_{BC}$, and $\mathbf{V}_{CA}$
b. Line currents $\mathbf{I}_A$, $\mathbf{I}_B$, and $\mathbf{I}_C$, without finding individual phase currents
c. Verify the solution in (**b**) by showing that the phasor sum of the three line currents is zero

Solution

a. By definition, $\mathbf{V}_{AB} = 125\angle 0^\circ$ V. Consequently, for phase sequence ABC, $\mathbf{V}_{BC} = 125\angle -120^\circ$ V and $\mathbf{V}_{CA} = 125\angle 120^\circ$ V
b. $\mathbf{Z}_P = 35 + j120 = 125\angle 73.74^\circ$ and $\theta = +73.74^\circ$, so $(30 + \theta) = \mathbf{103.74^\circ}$
Since at all times $\mathbf{I}_P = \mathbf{V}_P/\mathbf{Z}_P = 125\ \text{V}/125\ \Omega = \mathbf{1\ A}$, the three phase currents have magnitudes of 1 A and are each displaced by 120° in phase. Then the line currents $\mathbf{I}_L = \sqrt{3}(\mathbf{I}_P) = \sqrt{3}(1\ \text{A}) = \mathbf{1.732\ A}$, with each line current lagging its respective line voltage by 103.74°, as noted above. We may write the line currents as $\mathbf{I}_A = 1.732\angle(0 - 103.74)^\circ = \mathbf{1.732\angle -103.74^\circ\ A}$, lagging $\mathbf{V}_{AB} = 125\angle 0^\circ$ V; $\mathbf{I}_B = 1.732\angle(-120 - 103.74)^\circ = \mathbf{1.732\angle -223.74^\circ\ A}$, lagging $\mathbf{V}_{BC} = 125\angle -120^\circ$ V; $\mathbf{I}_C = 1.732\angle(120 - 103.74)^\circ = \mathbf{1.732\angle 16.26^\circ\ A}$, lagging $\mathbf{V}_{CA} = 125\angle 120^\circ$ V
c. Since the sum of the balanced line currents must be zero, we may write $\mathbf{I}_A + \mathbf{I}_B + \mathbf{I}_C = (1.732\angle -103.74^\circ) + (1.732\angle -223.74^\circ) + (1.732\angle 16.26^\circ) = (-0.4114 - j1.6825) + (-1.2514 + j1.1975) + (1.6628 + j0.4850) = \mathbf{0 + j0}$

Example 21-5 shows that it is possible to find the line currents drawn by a balanced delta load without having to find the phase currents individually. Although the method of Ex. 21-5 represents a shortcut solution to balanced delta load calculations, the method of Ex. 21-6 is a more fundamental approach since it works for both *balanced* and *unbalanced* loads connected in delta.

EXAMPLE 21-6
Given the data of Ex. 21-5, calculate

a. Phase currents $\mathbf{I}_{AB}$, $\mathbf{I}_{BC}$, and $\mathbf{I}_{CA}$ in polar and rectangular form
b. Line currents $\mathbf{I}_A$, $\mathbf{I}_B$, and $\mathbf{I}_C$

Solution

a. $\mathbf{I}_{AB} = \mathbf{V}_{AB}/\mathbf{Z}_{AB} = 125/0^\circ / 125\angle 73.74^\circ = 1\angle -73.74^\circ = \mathbf{(0.28 - j0.96)\ A}$;
$\mathbf{I}_{BC} = \mathbf{V}_{BC}/\mathbf{Z}_{BC} = 125\angle -120^\circ / 125\angle 73.74^\circ = 1\angle -193.74^\circ = \mathbf{(-0.9714 + j0.2375)\ A}$;
$\mathbf{I}_{CA} = \mathbf{V}_{CA}/\mathbf{Z}_{CA} = 125\angle 120^\circ / 125\angle 73.74^\circ = 1\angle 46.26^\circ = \mathbf{(0.6914 + j0.7225)\ A}$
b. Using the phase sequence ABC, which yields the instantaneous current directions shown in Fig. 21-6, enables us to write the set of Eqs. (21-6), or

$$\mathbf{I}_A = \mathbf{I}_{AB} + \mathbf{I}_{AC} = (0.28 + j0.96) + (-0.6914 - j0.7225) = -0.4114 - j1.6825 = \mathbf{1.732\angle -103.74^\circ\ A} \qquad (21\text{-}6a)$$

$$\mathbf{I}_B = \mathbf{I}_{BC} + \mathbf{I}_{BA} = (-0.9714 + j0.2375) + (-0.28 + j0.96) = -1.2514 + j1.1975 = \mathbf{1.732\angle 136.26^\circ\ A} \qquad (21\text{-}6b)$$

$$\mathbf{I}_C = \mathbf{I}_{CA} + \mathbf{I}_{CB} = (0.6914 + j0.7225) + (0.9714 - j0.2375) = 1.6645 + j0.485 = \mathbf{1.732\angle 16.26^\circ\ A} \qquad (21\text{-}6c)$$

With respect to the solution of Ex. 21-6, note that

1. The line currents obtained are exactly those of the solution of Ex. 21-5b.
2. Since we are solving for the *phase* currents in *each* load *separately*, it does not matter if the phase impedances are balanced (and equal) or different from each other.
3. The line currents are always the phasor sum of the currents entering and leaving nodes A, B, and C in Fig. 21-6. Once we have obtained the respective phase currents, the line currents are obtained from Eqs. (21-6).

21-7 POWER IN 3ϕ DELTA OR WYE BALANCED SYSTEMS

Now that we have established the voltage and current relations in both Y and Δ systems, we may turn to the power relations in these systems. **Figure 21-9** shows five possible combinations of Y–Δ 3ϕ systems. Assuming that the loads are balanced in the five systems of Fig. 21-9, we may write a single set of power equations to satisfy all of them, regardless of connection.

The power dissipated in each phase of a balanced 3ϕ system, or the power per phase, P_p, is

$$P_p = \mathbf{V}_p\mathbf{I}_p \cos\theta = \mathbf{I}_p^2 R_p = \mathbf{I}_p^2(\mathbf{Z}_p \cos\theta) \qquad \text{watts (W)} \qquad (21\text{-}7)$$

where $\mathbf{V}_p$ is the phase voltage and $\mathbf{I}_p$ the phase current
R_p is the resistance of each phase impedance, $\mathbf{Z}_p$
θ is the impedance phase angle or the phase angle between $\mathbf{V}_p$ and $\mathbf{I}_p$

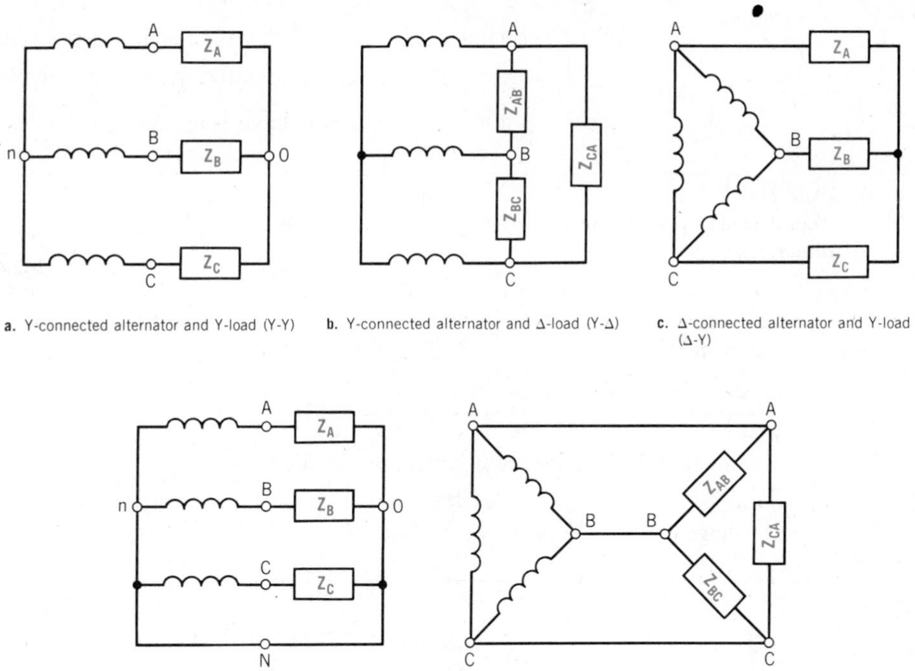

a. Y-connected alternator and Y-load (Y-Y) b. Y-connected alternator and Δ-load (Y-Δ) c. Δ-connected alternator and Y-load (Δ-Y)

Figure 21-9 Five possible 3ϕ system combinations d. Y-Y, 3ϕ, 4-wire system e. Δ-Δ, 3ϕ, 3-wire system

The total power of a 3ϕ balanced system therefore is

$$P_t = 3P_p = 3\mathbf{V}_p\mathbf{I}_p \cos\theta = 3\mathbf{I}_p^2 R_p = \sqrt{3}\mathbf{V}_L\mathbf{I}_L \cos\theta = \mathbf{S}_t^* \cos\theta \qquad \text{watts (W)} \qquad (21\text{-}8)$$

where $\mathbf{S}_t^*$ is the total complex power, as defined in Eq. (21-9) below
$\mathbf{V}_L$ is the line voltage or voltage between lines
$\mathbf{I}_L$ is the line current
$\cos\theta$ is the power factor of the load

Complex power, $\mathbf{S}^*$, was first introduced in Sec. 16-4.7, as a consequence of the power triangle emerging from the impedance triangle. It was also used in connection with Tellegen's theorem in Sec. 17-9. For a 3ϕ balanced load, we may write the total complex power, $\mathbf{S}_t^*$, as

$$\mathbf{S}_t^* = 3\mathbf{I}_p^2\mathbf{Z}_p = 3\mathbf{V}_p\mathbf{I}_p^* = \frac{P_t}{\cos\theta} = P_t \pm j\mathbf{Q}_t = \sqrt{3}\cdot\mathbf{V}_L\mathbf{I}_L \qquad \text{volt-amperes (VA)} \qquad (21\text{-}9)$$

This relation requires the definition of $\mathbf{Q}_t$, the total quadrature or reactive power, which is defined in terms of inductive or capacitive reactive loads as

$$\mathbf{Q}_t = \mathbf{S}_t^* \sin\theta = 3\mathbf{V}_p\mathbf{I}_p^* \sin\theta = +j(3\mathbf{I}_p^2\mathbf{X}_{Lp}) \text{ or } -j(3\mathbf{I}_p^2\mathbf{X}_{Cp}) \qquad \text{(vars)} \qquad (21\text{-}10)$$

where X_{Lp} is the inductive reactance per phase
X_{Cp} is the capacitive reactance per phase

and all other terms have been defined.

EXAMPLE 21-7

A 3ϕ, Y-connected 200 kVA alternator has a voltage of 440 V between lines. Calculate the

a. Rated line current
b. Rated phase voltage
c. Rated phase current

respectively, drawn by *balanced* 3ϕ loads

Solution

a. $I_L = S_t/\sqrt{3}V_L = 200 \times 10^3 \text{ VA}/\sqrt{3} \times 440 \text{ V}$
$= \mathbf{262.43\ A}$ (21-10)
b. $V_p = V_L/\sqrt{3} = 440 \text{ V}/\sqrt{3} = \mathbf{254\ V}$
c. $I_p \equiv I_L = \mathbf{262.43\ A}$

EXAMPLE 21-8

If the alternator in Ex. 21-7 is reconnected in delta, calculate the

a. Rated line current
b. Rated line voltage
c. Rated kVA

Solution

a. $I_L = I_p \times \sqrt{3} = 262.43 \times \sqrt{3} = \mathbf{454.5\ A}$
b. $V_L \equiv V_p = \mathbf{254\ V}$
c. $S = \sqrt{3}V_L I_L = \sqrt{3}(254)(454.5) = \mathbf{200\ kVA}$ (21-10)

Examples 21-7 and 21-8 show several important insights:

1. The kVA rating of an alternator *cannot be changed* by reconnection.
2. The delta-connected alternator provides a higher line current at a lower line voltage.
3. The wye-connected alternator provides a lower line current at a higher line voltage.
4. The phase current and phase voltage of *both* connections are the *same*, since the *same* alternator is used.

So much for the sources shown in Fig. 21-9. Now let us consider the load connections, in turn, via Exs. 21-9 through 21-11.

EXAMPLE 21-9

A 3ϕ, balanced, Y-connected load consists of an impedance per phase of $6 + j10.39\ \Omega$ connected across 208-V, 3ϕ lines. Calculate the

a. Phase voltage
b. Phase current
c. Line current
d. Power per phase
e. Total 3ϕ power by three methods
f. Total quadrature power
g. Total complex power by three methods

Solution

a. $V_p = V_L/\sqrt{3} = 208 \text{ V}/\sqrt{3} = \mathbf{120\ V}$
b. $I_p = V_p/Z_p = 120 \text{ V}/(6 + j10.39) = 120/12\underline{/60^\circ}\ \Omega$
$= \mathbf{10\underline{/-60^\circ}\ A}$
c. $I_L \equiv I_p = \mathbf{10\underline{/-60^\circ}\ A}$ (each phase/line current lags its phase voltage by 60°)
d. $P_p = I_p^2 R_p = (10)^2 6 = \mathbf{600\ W}$
e. $P_t = 3P_p = 3 \times 600 \text{ W} = \mathbf{1800\ W}$ (21-8)
$= \sqrt{3}V_L I_L \cos\theta = \sqrt{3}(208)(10)\cos 60^\circ$
$= \mathbf{1800\ W}$ (21-8)
$= 3V_p I_p \cos\theta = 3(120)(10)\cos 60^\circ = \mathbf{1800\ W}$ (21-8)
f. $Q_t = +j(3I_p^2 X_{Lp}) = +j(3 \times 10^2 \times 10.39)$
$= \mathbf{+j3117\ vars}$ (21-10)
g. $S_t^* = P_t + jQ_t = 1800 + j3117$
$= \mathbf{3600\underline{/60^\circ}\ volt\text{-}amperes}$ (21-9)
$S_t^* = 3I_p^2 Z_p = 3(10^2)(12\underline{/60^\circ}) = \mathbf{3600\underline{/60^\circ}\ VA}$ (21-9)
$S_t^* = 3V_p I_p^* = 3(120)(10\underline{/-60^\circ})^* = \mathbf{3600\underline{/60^\circ}\ VA}$ (21-9)

EXAMPLE 21-10

A 3ϕ, balanced, Δ-connected load is connected across 440-V lines. The impedance per phase is $22\underline{/-36^\circ}\ \Omega$. Calculate all parts of Ex. 21-9.

Solution

a. $V_p \equiv V_L = \mathbf{440\ V}$
b. $I_p = V_p/Z_p = 440\underline{/0^\circ} \text{ V}/22\underline{/-36^\circ}\ \Omega = \mathbf{20\underline{/36^\circ}\ A}$

c. $\mathbf{I}_L = \sqrt{3}\mathbf{I}_p\underline{/\theta - 30^\circ} = \sqrt{3}(20)\underline{/36^\circ - 30^\circ} = \mathbf{34.64\underline{/6^\circ}\ A}$
(each line current leads its respective phase/line voltage by 6°)

d. $P_p = \mathbf{I}_p^2 R_p = (20)^2 17.8\ \Omega = \mathbf{7.119\ kW}$ (21-8)

e. $P_t = 3P_p = 3(7.119\ \text{kW}) = \mathbf{21.358\ kW}$ (21-8)

$P_t = \sqrt{3}\mathbf{V}_L\mathbf{I}_L \cos\theta = \sqrt{3}(440)(34.64)\cos 36^\circ$
$= \mathbf{21.358\ kW}$ (21-8)

$P_t = 3\mathbf{V}_p\mathbf{I}_p \cos\theta = 3(440)(20)\cos 36^\circ$
$= \mathbf{21.358\ kW}$ (21-8)

f. $\mathbf{Q}_t = -j(3\mathbf{I}_p\mathbf{X}_{Cp}) = -j(3 \times 20^2)(12.93)$
$= \mathbf{-j15.52\ kvar}$ (21-10)

g. $\mathbf{S}_t^* = P_t - j\mathbf{Q}_t = 21.36 - j15.52$
$= \mathbf{26.40\underline{/-36^\circ}\ kVA}$ (21-9)

$\mathbf{S}_t^* = 3\mathbf{I}_p^2\mathbf{Z}_p = 3(20)^2(22\underline{/-36^\circ}) = \mathbf{26.4\underline{/-36^\circ}\ kVA}$ (21-9)

$\mathbf{S}_t^* = 3\mathbf{V}_p\mathbf{I}_p^* = 3(440)(20\underline{/36^\circ})^* = \mathbf{26.4\underline{/-36^\circ}\ kVA}$ (21-9)

Examples 21-9 and 21-10 show that for balanced 3ϕ loads

1. It is totally unnecessary to draw phasor diagrams for voltages and currents in order to obtain the total and complex power relations.
2. The total power is most easily found from $P_t = 3\mathbf{I}_p^2 R_p$ with the least amount of error. Furthermore, this relation emerges from the derivation of the power triangle from the impedance triangle. (See Ex. 21-11.)
3. Similarly, the total complex power, $\mathbf{S}_t^*$ is most easily and directly found from $\mathbf{S}_t^* = 3\mathbf{I}_p^2\mathbf{Z}_p$, since it also emerges from the derivation of the power triangle from the impedance triangle. (See Ex. 21-11.)
4. The angles between the line voltages and the line currents are *not* the power factor angles. As shown in Ex. 21-10c, this angle is $(\theta - 30^\circ)$ and not θ itself.
5. The power factor angle is obtained alternatively from the phase impedance angle θ or $\cos^{-1}(R_p/\mathbf{Z}_p)$ in the impedance triangle.
6. The ratio of the total power, P_t, to the total complex power, $\mathbf{S}_t^*$, is the power factor, $\cos\theta$. This emerges from the power triangle and also from Eqs. (21-8) and (21-9).

One last example regarding balanced systems demonstrates the value and importance of the foregoing relations.

EXAMPLE 21-11

Given the combined Y and Δ 3ϕ balanced loads connected to a 220 V 3ϕ source as shown in **Fig. 21-10**, calculate the

a. Phase current of the Y-connected load, $\mathbf{I}_{pY}$
b. Phase current of the Δ-connected load, $\mathbf{I}_{p\Delta}$
c. Total power drawn by the Y-connected load, P_{tY}
d. Total power drawn by the Δ-connected load, $P_{t\Delta}$
e. Total power drawn by the combined load, P_t
f. Total quadrature (reactive) power drawn by the combined load
g. Total complex power, $\mathbf{S}_t^*$, drawn by the combined load
h. Power factor of the combined balanced load by two methods
i. Line current drawn from the 3ϕ source by the combined load by two methods

Solution

a. $\mathbf{I}_{pY} = \dfrac{\mathbf{V}_L/\sqrt{3}}{\mathbf{Z}_p} = \dfrac{220\ \text{V}/\sqrt{3}}{3 - j4} = \mathbf{25.4\ A}$

b. $\mathbf{I}_{p\Delta} = \mathbf{V}_p/\mathbf{Z}_p = 220\ \text{V}/(5 + j12) = \mathbf{16.92\ A}$

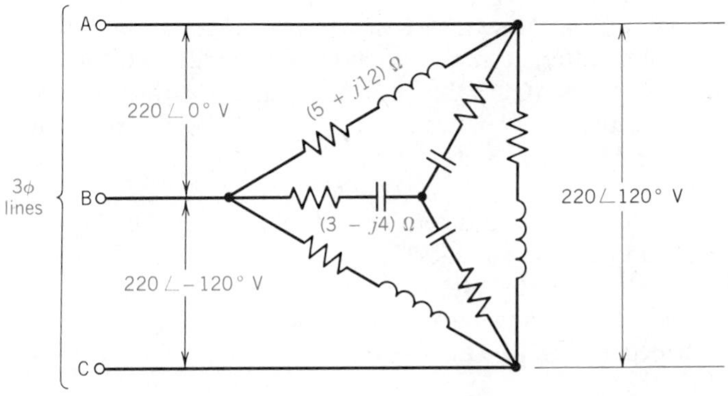

Figure 21-10 Example 21-11. Balanced Y and Δ combinational loads

c. $P_{tY} = 3\mathbf{I}_p^2 R_p = 3(25.4)^2 \times 3 = \mathbf{5806.44\ W}$ (21-8)
d. $P_{t\Delta} = 3\mathbf{I}_p^2 R_p = 3(16.92)^2 \times 5 = \mathbf{4295.86\ W}$ (21-8)
e. $P_t = P_{tY} + P_{t\Delta} = 5806.44 + 4295.86 = \mathbf{10\,102.3\ W}$
f. $\mathbf{Q}_{tY} = 3\mathbf{I}_p^2(-jX_C) = -j3(25.4)^2 4 = \mathbf{-j7741.92\ vars}$ (21-10)
$\mathbf{Q}_{t\Delta} = 3\mathbf{I}_p^2(jX_L) = +j3(16.923)^2 \times 12 = \mathbf{j10\,310\ vars}$ (21-10)
$\mathbf{Q}_t = \mathbf{Q}_{tY} + \mathbf{Q}_{t\Delta} = -j7741.92 + j10\,310 = \mathbf{j2568.05\ vars}$ (21-10)
g. $\mathbf{S}_t^* = P_t + j\mathbf{Q}_t = 10\,102.3 + j2568.05 = \mathbf{10\,423.6\angle 14.26^\circ\ VA}$ (21-9)
h. $\cos\theta = P_t/\mathbf{S}_t^* = 10\,102.3\text{ W}/10\,423.6\text{ VA} = \mathbf{0.9692}$ (21-8)
$\cos\theta = \cos 14.26^\circ = \mathbf{0.9692}$, lagging PF
i. $\mathbf{I}_L = \mathbf{S}_t^*/\mathbf{V}_L \times \sqrt{3} = 10\,423.6\text{ VA}/220 \times \sqrt{3} = \mathbf{27.355\ A}$ (21-9)
$\mathbf{I}_L = P_t/\sqrt{3}\mathbf{V}_L \cos\theta = 10\,102.3/\sqrt{3} \times 220 \times 0.9692 = \mathbf{27.355\ A}$ (21-8)

Example 21-11 yields a number of important insights into 3ϕ power relations:

1. Complex powers (in volt-amperes) may be totaled (added) by adding their true powers (in watts) and quadrature powers (in vars); that is, *complex powers are added in complex form.*
2. The total **complex power** (step **g**) emerges at a *positive* angle, implying that the equivalent total load contains some *inductive* reactive power. Such a load causes the phase current in the *source* to *lag* the generated phase voltage.
3. The balanced line current may be found (step **i**) using either the total true power or the total complex power, Eqs. (21-8) and (21-9), respectively. In both equations use of the factor involving $\sqrt{3}$ appears, as does the line voltage, $\mathbf{V}_L$.
4. The angle of 14.26° obtained in (**g**) is *not* interpreted as the angle by which the line current lags the line voltage. Rather, as noted in (2), it is the angle by which the *phase* current of the combined system lags the *phase* voltage in the *source*.

The reader might wonder why the same relation $P_t = \sqrt{3}\mathbf{V}_L\mathbf{I}_L \cos\theta$ satisfies *both* the wye and delta systems. Since power in either system is $P_t = 3\mathbf{V}_p\mathbf{I}_p \cos\theta$, for the *delta* connection, where $\mathbf{V}_p \equiv \mathbf{V}_L$ and $\mathbf{I}_p = \mathbf{I}_L/\sqrt{3}$, we may write

$$P_t = 3\mathbf{V}_p\mathbf{I}_p \cos\theta = 3\mathbf{V}_L(\mathbf{I}_L/\sqrt{3}) \cos\theta = \sqrt{3}\mathbf{V}_L\mathbf{I}_L \cos\theta \qquad (21\text{-}8)$$

Similarly, for the *wye* connection, where $\mathbf{V}_p = \mathbf{V}_L/\sqrt{3}$ and $\mathbf{I}_p \equiv \mathbf{I}_L$, we may write

$$P_t = 3\mathbf{V}_p\mathbf{I}_p \cos\theta = 3(\mathbf{V}_L/\sqrt{3})\mathbf{I}_L \cos\theta = \sqrt{3}\mathbf{V}_L\mathbf{I}_L \cos\theta \qquad (21\text{-}8)$$

The *universality* of the power relation, Eq. (21-8), enabled the *relatively simple* solution for the total line current, in Ex. 21-11, drawn by combined Y and Δ loads in any combinational system of *balanced* loads.

21-8 PHASE SEQUENCE

Insofar as balanced Y or Δ loads are concerned, the phase sequence of the source and lines has no effect whatever on the magnitude of phase voltages, phase, and line currents of these systems. But in the next sections we will consider *unbalanced* Y and Δ loads. We will discover that phase sequence has a profound effect on the magnitude of the resultant Y and Δ phase currents, as well as on phase voltages of the Y system. For this reason, a discussion of phase sequence is appropriate here.

Phase sequence originates from the sequence in which voltages are induced in each of the three phases of an alternator, as was shown in Fig. 21-2. If the rotor poles are rotating counterclockwise (CCW), as shown in Fig. 21-2a, then the phase sequence is ABCABC. But if the rotor rotates in a clockwise (CW) direction, then the phase sequence becomes ACBACB. Since there are only two possible directions of rotation, there are only two possible phase sequences.

The question arises, "Is it possible to determine the phase sequence from the phasor diagram of phase and/or line voltages?" Figure 21-11a shows the phasor diagram of Fig. 21-3b for a Y-connected alternator whose phase sequence is ABC. The line voltages have been drawn showing $\mathbf{V}_{AB}$, $\mathbf{V}_{BC}$, and $\mathbf{V}_{CA}$, using the shortcut method previously described, in **Fig. 21-11a**. The respective line voltages are separately drawn in Fig. 21-11b.

From Fig. 21-11a we may infer the phase sequence ABC by rotating the equi-

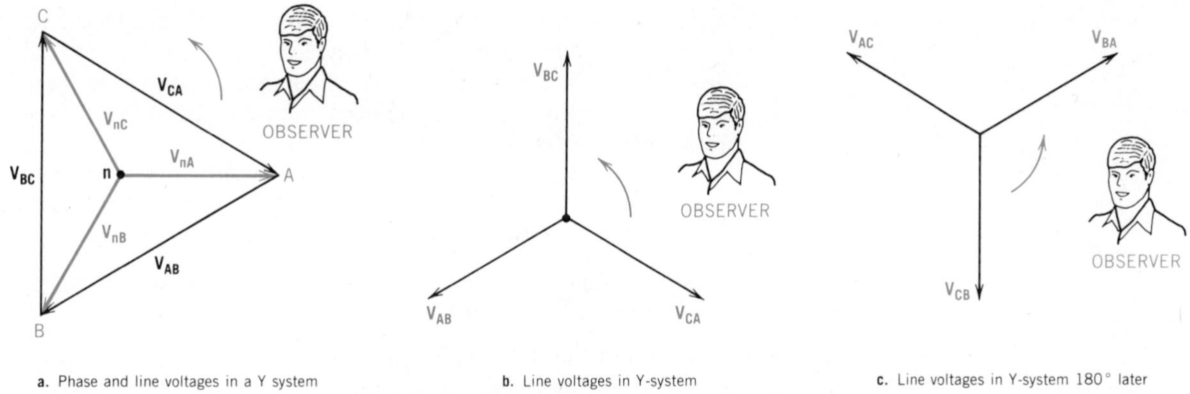

Figure 21-11 Phase sequence ABC as determined from either phase or line voltages

lateral triangle in a CCW direction. The observer notes the sequence ABCABC as a result.

In Fig. 21-11b, if we consider the *second* subscript of each line voltage, the observer must also note the sequence ABCABC. If we consider the first subscript of each line voltage in Fig. 21-11b, the observer must note the sequence CABCABC, which is also the *same*.

Even if we consider the line voltages displaced by 180°, as shown in Fig. 21-11c, taking the *second* subscript yields BCABCA. Taking the *first* subscript of each line voltage yields CABCABC. Consequently, for both phase and line voltages at any instant in time, it is possible to infer the phase sequence from the phasor diagram for a particular 3ϕ system.

The reader may ask at this point, "Given three lines of a 3ϕ system, numbered 1, 2, and 3, how can I determine their phase sequence?"

This question will be answered in Sec. 21-11 after consideration of unbalanced loads, because almost all *phase sequence indicators* are unbalanced loads, *per se*. But first an example to illustrate phase sequence.

EXAMPLE 21-12

One line voltage of a 3ϕ system, $\mathbf{V}_{BC} = 220\underline{/45^\circ}$ V. Determine the other two line voltages when the phase sequence is

a. *ABC*
b. *CBA*

Solution

a. $\mathbf{V}_{CA} = 220\underline{/(45 - 120)^\circ} = \mathbf{220\underline{/-75^\circ}\ V}$ and

$\mathbf{V}_{AB} = 220\underline{/(45 + 120)^\circ} = \mathbf{220\underline{/165^\circ}\ V}$

b. For sequence *CBA*, we require $\mathbf{V}_{CB}$, which is $-\mathbf{V}_{BC} = 220\underline{/(45 + 180)^\circ} = 220\underline{/225^\circ}$ V, so

$\mathbf{V}_{CB} = \mathbf{220\underline{/225^\circ}\ V}$,

$\mathbf{V}_{BA} = 220\underline{/(225 - 120)^\circ} = \mathbf{220\underline{/105^\circ}\ V}$, and

$\mathbf{V}_{AC} = 220\underline{/(225 + 120)^\circ} = 220\underline{/345^\circ} = \mathbf{220\underline{/-15^\circ}\ V}$

21-9 UNBALANCED DELTA-CONNECTED SYSTEMS

We begin our analysis of unbalanced 3ϕ systems with a study of the unbalanced delta-connected system. The unbalanced delta system involves less complexity than the unbalanced wye. This is shown in Table 21-1.

Table 21-1 shows the following summaries regarding delta versus wye systems:

1. There is *only one* kind of *delta* unbalanced system. There are *two* types of *wye unbalanced* systems.

2. In all three systems the **line voltages** *are always balanced*. Since the phase voltage in a delta is the *same* as its line voltage, the phase voltages are also balanced! This reduces the complexity of the unbalanced delta system considerably.
3. The phase currents of an unbalanced delta-connected system are unbalanced and their phasor sum is *not* equal to zero.
4. The line currents of an unbalanced delta-connected system, despite their unbalance, produce *a phasor sum of zero*. This serves as a convenient check on the validity of the solution for unbalanced line currents.

Analysis of unbalanced delta systems involves two basic steps, which we previously encountered in solving balanced delta systems:

1. Find each of the three (unbalanced) phase currents using Eqs. (21-5)
2. Find each of the three (unbalanced) line currents using Eqs. (21-6a), (21-6b), and (21-6c).

Alternatively, as a means of checking on the validity of the foregoing method, mesh analysis may be used. This method requires three mesh currents whose values involve solution of three simultaneous equations with complex coefficients.

Both methods will be shown in the following examples to verify the characteristics of unbalanced delta systems given in Table 21-1, for **unbalanced** systems only.

Table 21-1 Comparison of Unbalanced Delta and Unbalanced Wye Systems

		Wye	
	Delta	3ϕ, Three-Wire System	3ϕ, Four-Wire System
Line voltages	Balanced	Balanced	Balanced
Phase voltages	Balanced	Unbalanced	Balanced
Line currents	Unbalanced but $\sum \mathbf{I}_L = 0$	Unbalanced but $\sum \mathbf{I}_L = 0$	Unbalanced and $\sum \mathbf{I}_L = \mathbf{I}_n$
Phase currents	Unbalanced and $\sum \mathbf{I}_p \neq 0$	Unbalanced but $\sum \mathbf{I}_p \equiv \sum \mathbf{I}_L = 0$	Unbalanced but $\sum \mathbf{I}_p \equiv \sum \mathbf{I}_L = \mathbf{I}_n$

EXAMPLE 21-13

A 3ϕ, three-wire service has a line voltage $\mathbf{V}_{AB} = 200\angle 0^\circ$ V and an ABC phase sequence. Three unbalanced loads are connected in delta, namely $\mathbf{Z}_{AB} = 388\angle -58.9^\circ\ \Omega$, $\mathbf{Z}_{BC} = 468\angle 31.1^\circ\ \Omega$, and $\mathbf{Z}_{CA} = 388\angle -58.9^\circ\ \Omega$. Calculate the

a. Phase currents $\mathbf{I}_{AB}$, $\mathbf{I}_{BC}$, and $\mathbf{I}_{CA}$ in polar and rectangular form using Eqs. (21-5)
b. Line currents $\mathbf{I}_A$, $\mathbf{I}_B$, and $\mathbf{I}_C$
c. Phasor sum of the *phase* currents
d. Phasor sum of the *line* currents
e. Complex power dissipated in each load
f. Total complex power supplied by the service

Solution

a. $\mathbf{I}_{AB} = 200\angle 0^\circ \text{ V}/388\angle -58.9^\circ\ \Omega = \mathbf{0.5155\angle 58.9^\circ\ A}$
$= \mathbf{0.2663 + j0.4414\ A}$
$\mathbf{I}_{BC} = 200\angle -120^\circ \text{ V}/468\angle 31.1^\circ\ \Omega = \mathbf{0.4274\angle -151.1^\circ\ A}$
$= \mathbf{-0.3741 - j0.2065\ A}$
$\mathbf{I}_{CA} = 200\angle 120^\circ \text{ V}/388\angle -58.9^\circ\ \Omega = \mathbf{0.5155\angle 178.9^\circ\ A}$
$= \mathbf{-0.5154 + j0.0099\ A}$

b. Using Eqs. (21-6a) to (21-6c), respectively, for the ABC phase sequence:
$\mathbf{I}_A = \mathbf{I}_{AB} + \mathbf{I}_{AC} = (0.2663 + j0.4414) + (0.5154 - j0.0099)$
$= \mathbf{0.7816 + j0.4315 = 0.893\angle 28.9^\circ\ A}$
$\mathbf{I}_B = \mathbf{I}_{BC} + \mathbf{I}_{BA}$
$= (-0.3741 - j0.2065) + (-0.2663 - j0.4414)$
$= \mathbf{-0.6404 - j0.6479}$
$\mathbf{I}_B = \mathbf{I}_{BC} + \mathbf{I}_{BA} = \mathbf{0.911\angle -134.7^\circ\ A}$
$\mathbf{I}_C = \mathbf{I}_{CA} + \mathbf{I}_{CB} = (-0.5154 + j0.0099) + (0.3741 + j0.2065)$
$= \mathbf{-0.1412 + j0.2164 = 0.258\angle 123.1^\circ\ A}$

c. $\sum \mathbf{I}_p = \mathbf{I}_{AB} + \mathbf{I}_{BC} + \mathbf{I}_{CA}$
$= (0.2663 + j0.4414) + (-0.3741 - j0.2065)$
$+ (-0.5154 + j0.0099)$
$\sum \mathbf{I}_p = \mathbf{I}_{AB} + \mathbf{I}_{BC} + \mathbf{I}_{CA} = \mathbf{-0.623 + j0.245}$
$= \mathbf{0.67\angle 158.6^\circ\ A}$ (not zero! See Table 21-1)

d. $\sum \mathbf{I}_L = \mathbf{I}_A + \mathbf{I}_B + \mathbf{I}_C$
$= (0.7816 + j0.4315) + (-0.6404 + j0.6479)$
$+ (-0.1412 + j0.2164)$
$= 0 + j0$ (sum of line currents is zero! See Table 21-1)

e. $\mathbf{Z}_{AB} = \mathbf{Z}_{CA} = 388\angle -58.9^\circ\ \Omega = \mathbf{200.415 - j332.23\ \Omega}$
$\mathbf{Z}_{BC} = 468\angle 31.1^\circ = \mathbf{400.733 + j241.74\ \Omega}$

$S^*_{AB} = (I_{AB})^2 Z_{AB} = (0.5155)^2(200.415 - j332.23)$
$= \mathbf{53.26 - j88.29\ VA}$
$S^*_{BC} = (I_{BC})^2 Z_{BC} = (0.4274)^2(400.73 + j241.74)$
$= \mathbf{73.20 + j44.16\ VA}$

$S^*_{CA} = (I_{CA})^2 Z_{CA} = (0.5155)^2(200.415 - j332.23)$
$= \mathbf{53.26 - j88.29\ VA}$

f. $S^*_t = S^*_{AB} + S^*_{BC} + S^*_{CA} = \mathbf{179.72 - j132.42}$
$= \mathbf{223.24\underline{/-36.38°}\ VA}$

This solution shows that the total power dissipated is 179.72 W due to the unbalanced load. But owing to the unbalanced line currents, there is no internal method of validation to prove that the answers obtained above are correct.[5] Instead, a method of external validity is used, involving mesh analysis, as shown by Ex. 21-14.

EXAMPLE 21-14

Given the data of Ex. 21-13, using the format method of mesh analysis, as shown in **Fig. 21-12**, calculate

a. Mesh currents $\mathbf{I}_1$, $\mathbf{I}_2$, and $\mathbf{I}_3$
b. Phase currents $\mathbf{I}_{AB}$, $\mathbf{I}_{BC}$, and $\mathbf{I}_{CA}$, using the mesh currents in **(a)**
c. Line currents $\mathbf{I}_A$, $\mathbf{I}_B$, and $\mathbf{I}_C$, using the mesh currents in **(a)**
d. Compare the results of Ex. 21-13 with those found in this example for the solution of unbalanced phase and line currents

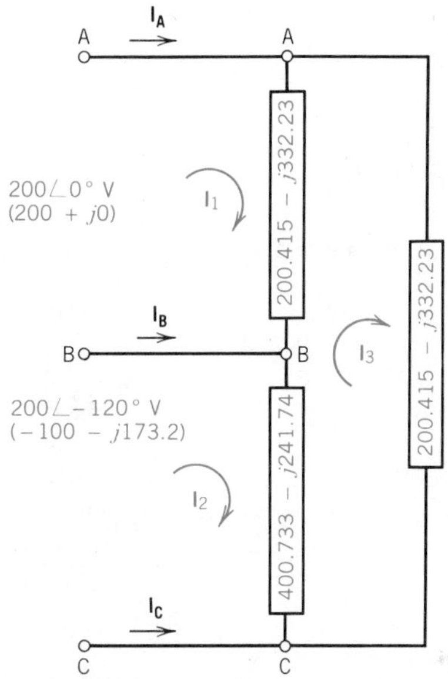

Figure 21-12 Example 21-14 drawn for solution of unbalanced line currents using format method of mesh analysis

Solution

The loop equations written by the format method are

a. Loop $\mathbf{I}_1$: $(200.415 - j332.23)\mathbf{I}_1 + 0\mathbf{I}_2$
$- (200.415 - j332.23)\mathbf{I}_3 = 200 + j0$
Loop $\mathbf{I}_2$: $0\mathbf{I}_1 + (400.733 - j241.74)\mathbf{I}_2$
$-(400.733 - j241.74)\mathbf{I}_3 = -100 - j173.2$
Loop $\mathbf{I}_3$: $-(200.415 - j332.23)\mathbf{I}_1 - (400.733 - j241.74)\mathbf{I}_2$
$+ (200.415 - j332.23)\mathbf{I}_3 = 0$

Solution for currents $\mathbf{I}_1$, $\mathbf{I}_2$, and $\mathbf{I}_3$ by either matrix methods (using determinants and Cramer's rule) or computer program yields $\mathbf{I}_1 = \mathbf{0.7816 + j0.4315}$, $\mathbf{I}_2 = \mathbf{0.1412 - j0.2164}$, and $\mathbf{I}_3 = \mathbf{0.5154 - j0.0099}$

b. For the phase sequence ABC, from Fig. 21-12, we may write the phase current as
$\mathbf{I}_{AB} = \mathbf{I}_1 - \mathbf{I}_3 = (0.7816 + j0.4315) - (0.5154 - j0.0099)$
$= \mathbf{0.2662 + j0.4414\ A}$
$\mathbf{I}_{BC} = \mathbf{I}_2 - \mathbf{I}_3 = (0.1412 - j0.2164) - (0.5154 - j0.0099)$
$= \mathbf{-0.3742 - j0.2065\ A}$, and
$\mathbf{I}_{CA} = -\mathbf{I}_3 = \mathbf{-0.5154 + j0.0099\ A}$

c. Line currents $\mathbf{I}_A$, $\mathbf{I}_B$, $\mathbf{I}_C$, by inspection of Fig. 21-12, are
$\mathbf{I}_A = \mathbf{I}_1 = \mathbf{0.7816 + j0.4315 = 0.893\underline{/28.9°}\ A}$,
$\mathbf{I}_B = \mathbf{I}_2 - \mathbf{I}_1 = (0.1412 - j0.2164) - (0.7816 + j0.4315)$
$= \mathbf{-0.6404 - j0.6479 = 0.911\underline{/-134.7°}\ A}$, and
$\mathbf{I}_C = -\mathbf{I}_2 = \mathbf{-(0.1412 - j0.2164)} = 0.258\underline{/123.1°}\ A$

d. The mesh analysis results are identical to those found in Ex. 21-13 for both phase currents and line currents

With respect to the solution by mesh analysis, please note that

1. Only two line voltages appear in the equations (i.e., $\mathbf{V}_{CA}$ is unnecessary).
2. Loop $\mathbf{I}_3$ does not encounter any sources but instead sees all three impedances.
3. The phase currents in **(b)** are determined by the phase sequence. (If the phase sequence is reversed, these currents are reversed.)
4. In determining $\mathbf{I}_{AB}$, a downward current, we use $\mathbf{I}_1 - \mathbf{I}_3$, since $\mathbf{I}_1$ predominates in a downward direction. Similarly, for $\mathbf{I}_{BC}$ we use $\mathbf{I}_2 - \mathbf{I}_3$ since $\mathbf{I}_2$ predominates.
5. Since *all* line currents are *directed toward the load*, in finding $\mathbf{I}_B$ we use $\mathbf{I}_2 - \mathbf{I}_1$, since $\mathbf{I}_2$ predominates by virtue of its direction toward the load.

[5] The only assurance is that the phasor sum of the line currents is zero in **(d)**. If this sum is *not* zero, we infer that one (or more) calculation steps are in error.

The identical results of Exs. 21-13 and 21-14 show that both the phase currents and the line currents in unbalanced delta systems are no longer displaced from each other by 120° and are unequal in magnitude. Example 21-13 showed that although the phasor sum of the phase currents is not zero, *the phasor sum of the line currents is always zero*. This is true even when we select phase currents at random, as shown by Ex. 21-15.

EXAMPLE 21-15

A 3ϕ, 33 kV, 60 Hz three-wire source supplies the following unbalanced loads across its lines. Load $\mathbf{I}_{AB}$ is 10 kA at unity PF. Load $\mathbf{I}_{BC}$ is 15 kA at a PF of 0.7 lagging. Load $\mathbf{I}_{CA}$ is 12 kA at a PF of 0.9 leading. Using phase sequence ABC, calculate both magnitude and phase of the

a. Current in line A
b. Current in line B
c. Current in line C
d. Phasor sum of the three line currents

Solution

We must first write each of the phase currents in polar form. Since $\mathbf{V}_{AB}$ is assumed as $33\angle 0^\circ$ kV, $\mathbf{I}_{AB} = \mathbf{10\angle 0^\circ}$ **kA** (at unity PF). But $\mathbf{I}_{BC}$ lags $\mathbf{V}_{BC} = 33\angle -120^\circ$ kV by $\theta = \cos^{-1}(0.7) = \mathbf{-45.57^\circ}$, and therefore $\mathbf{I}_{BC} = 15\angle -120^\circ - 45.57^\circ = \mathbf{15\angle -165.57^\circ}$ **kA**. Similarly, $\mathbf{I}_{CA}$ leads $\mathbf{V}_{CA} = 33\angle 120^\circ$ kV by $\theta = \cos^{-1}(0.9) = \mathbf{+25.84^\circ}$, and therefore $\mathbf{I}_{CA} = 12\angle +120^\circ + 25.84^\circ = \mathbf{12\angle +145.84^\circ}$ **kA**

Now that we have each of the unbalanced phase currents, we may use the set of Eqs. (21-6) to find the respective line currents in ABC phase sequence. Writing each of the *phase* currents in rectangular form, we have $\mathbf{I}_{AB} = 10\angle 0^\circ$ kA $= \mathbf{10 + j0}$, $\mathbf{I}_{BC} = 15\angle -165.57^\circ = \mathbf{-14.53 - j3.738}$, and $\mathbf{I}_{CA} = 12\angle 145.84^\circ$ kA $= \mathbf{-9.930 + j6.738}$

a. $\mathbf{I}_A = \mathbf{I}_{AB} + \mathbf{I}_{AC} = (10 + j0) + (9.930 - j6.738)$
$= \mathbf{19.93 - j6.738 = 21.04\angle -18.68^\circ}$ **kA**

b. $\mathbf{I}_B = \mathbf{I}_{BC} + \mathbf{I}_{BA} = (-14.53 - j3.738) + (-10 + j0)$
$= \mathbf{-24.53 - j3.738 = 24.81\angle -171.33^\circ}$ **kA**

c. $\mathbf{I}_C = \mathbf{I}_{CA} + \mathbf{I}_{CB} = (-9.93 + j6.738) + (14.53 + j3.738)$
$= \mathbf{4.6 + j10.476 = 11.44\angle 66.29^\circ}$ **kA**

d. $\sum \mathbf{I}_L = \mathbf{I}_A + \mathbf{I}_B + \mathbf{I}_C = (19.93 - j6.738) + (-24.53 - j3.738)$
$+ (4.6 + j10.476) = \mathbf{0 + j0}$

With respect to this solution, note that

1. The phasor sum of the line currents is zero as shown in **d**.
2. The line currents are unequal in magnitude and not displaced by 120°.
3. If the phase sequence is reversed, Eqs. (21-6) *still* hold and *no new set need be written* for the sequence CBA. (If one was required, we would have included it.)

The effect of phase sequence reversal may be summarized as follows:[6] the phase currents change in phase angle only (magnitudes remain the same); all the line currents change in both magnitude and phase, but the *phase powers remain the same*. Consequently, the total power dissipated is unchanged.

21-10 UNBALANCED WYE-CONNECTED SYSTEMS

Table 21-1 shows that there are two classes of wye-connected systems, namely the 3ϕ, three-wire and the 3ϕ, four-wire. Since the former suffers more as a result of unbalanced loads, we will treat it first. Note that Table 21-1 shows that *only the line voltages are balanced* in a 3ϕ, three-wire system.[7] The phase voltages, line currents, and phase currents are *all* unbalanced.

In contrast, the 3ϕ, four-wire system, as a result of the neutral, manages to maintain its phase voltages balanced, equal, and 120° apart. As we will discover, this modification has several unique advantages over the three-wire system and simplifies the calculations and analysis, as well.

[6] Proof of this is left in end-of-chapter Probs. 21-30 and 21-35.

[7] Because of the relatively high capacity of the power system *lines* supplying most commercial loads, it is virtually impossible for an unbalanced load to alter the magnitude and phase of the incoming *line* voltages.

21-10.1 The 3ϕ, Three-Wire, Unbalanced Y-Connected System

Figure 21-13a shows a very simple unbalanced three-wire system. Unlike the 3ϕ delta unbalanced system of Fig. 21-12, we can see that this system lends itself easily to mesh analysis involving only two unknown currents, $\mathbf{I}_1$ and $\mathbf{I}_2$. Once $\mathbf{I}_1$ and $\mathbf{I}_2$ are calculated, all line and phase currents may be obtained, as well as the phase voltages, $\mathbf{V}_{Ao}$, $\mathbf{V}_{Bo}$, and $\mathbf{V}_{Co}$. Although other methods may be used, because of its relative simplicity, the mesh method is preferable. In fact, by using mesh analysis it is possible to derive a set of general equations for finding $\mathbf{I}_A$, $\mathbf{I}_B$, and $\mathbf{I}_C$, but these equations, containing complex terms, are complicated and overly arduous.

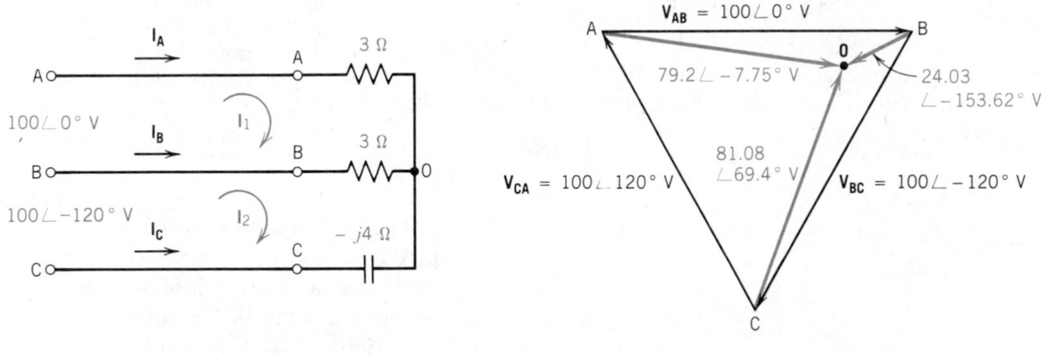

a. 3ϕ, 3-wire unbalanced load

b. Phasor diagram ABC sequence

Figure 21-13 Example 21-16. Unbalanced three-wire circuit and its phasor diagram using ABC phase sequence

A second method involves a Y-to-Δ transformation using the Y–Δ relations described in Sec. 8-7.1. Once we have drawn the equivalent unbalanced delta, the solution follows that described in Sec. 21-9, using Eqs. (21-5) and the set of Eqs. (21-6).

A third and most elegant method is the method of symmetrical components.[8] In this method the unbalanced 3ϕ system is transformed into three sets of balanced 3ϕ phasors of either voltage or current. But again, despite the elegance of this method, it is more involved and more difficult than simple mesh analysis, shown in Ex. 21-16, which is already familiar to us, and involves only two equations.

EXAMPLE 21-16

Given the circuit shown in Fig. 21-13a for a 3ϕ, three-wire, unbalanced Y-connected load, use mesh analysis to calculate

a. Mesh currents, $\mathbf{I}_1$ and $\mathbf{I}_2$, in polar and rectangular form
b. Line currents, $\mathbf{I}_A$, $\mathbf{I}_B$, and $\mathbf{I}_C$, in polar and rectangular form
c. Phase voltages, $\mathbf{V}_{Ao}$, $\mathbf{V}_{Bo}$, and $\mathbf{V}_{Co}$, in polar form
d. Draw the phasor diagram showing all line and phase voltages

Solution

Writing the two mesh equations for $\mathbf{I}_1$ and $\mathbf{I}_2$ produces the following array:

$\mathbf{I}_1$	$\mathbf{I}_2$	$\mathbf{V}$
$6 + j0$	$-3 + j0$	$100 + j0$
$-6 + j0$	$3 - j4$	$-50 - j86.6$

$$\text{Determinant, } \Delta = \begin{vmatrix} 6 + j0 & -3 + j0 \\ -3 + j0 & 3 - j4 \end{vmatrix} = \mathbf{9 - j24}$$

a. $$\mathbf{I}_1 = \frac{\begin{vmatrix} 100 + j0 & -3 + j0 \\ -50 - j86.6 & 3 - j4 \end{vmatrix}}{\Delta} = \frac{150 - j659.8}{9 - j24}$$

$$= \mathbf{26.157 - j3.5589 = 26.4\underline{/-7.75^\circ}\ A}$$

---

[8] Symmetrical components are beyond the scope of this text.

$$\mathbf{I}_2 = \frac{\begin{vmatrix} 6 + j0 & 100 + j0 \\ -3 + j0 & -50 - j86.6 \end{vmatrix}}{\Delta} = \frac{0 - j519.6}{9 - j24}$$

$= \mathbf{18.98 - j7.118 = 20.27\angle -20.56^\circ\ A}$

b. $\mathbf{I}_A = \mathbf{I}_1 = \mathbf{26.4\angle -7.75^\circ\ A = (26.16 - j3.559)\ A}$,
$\mathbf{I}_B = \mathbf{I}_2 - \mathbf{I}_1 = (18.98 - j7.118) - (26.157 - j3.5589)$
$= \mathbf{-7.177 - j3.559 = 8.011\angle -153.62^\circ}$ A, and
$\mathbf{I}_C = -\mathbf{I}_2 = \mathbf{-18.98 + j7.118 = 20.27\angle 159.44^\circ\ A}$

c. $\mathbf{V}_{Ao} = \mathbf{I}_A\mathbf{Z}_A = (26.4\angle -7.75^\circ\ A)(3\angle 0^\circ) = \mathbf{79.2\angle -7.75^\circ\ V}$,
$\mathbf{V}_{Bo} = \mathbf{I}_B\mathbf{Z}_B = (8.011\angle -153.62^\circ\ A)(3\angle 0^\circ\ \Omega)$
$= \mathbf{24.03\angle -153.62^\circ\ V}$, and
$\mathbf{V}_{Co} = \mathbf{I}_C\mathbf{Z}_C = (20.27\angle 159.44^\circ)(4\angle -90^\circ\ \Omega) = \mathbf{81.08\angle 69.44^\circ\ V}$

d. The phasor diagram is drawn in Fig. 21-13b, with the phase voltages inscribed inside the (equilateral) triangle of line voltages.

Example 21-16 shows quite clearly that for unbalanced 3ϕ, three-wire systems,

1. The magnitudes of the line currents (phase currents) are unequal and no longer 120° apart.
2. The magnitudes of the phase voltages are unequal and no longer 120° apart.
3. The phasor sum of any two phase voltages yields the line voltage, respectively, $\mathbf{V}_{Ao} + \mathbf{V}_{oB} = \mathbf{V}_{AB}$, $\mathbf{V}_{Bo} + \mathbf{V}_{oC} = \mathbf{V}_{BC}$, and $\mathbf{V}_{Co} + \mathbf{V}_{oA} = \mathbf{V}_{CA}$.
4. The neutral point "o" has shifted from the geometric center (see Fig. 21-4b) of the equilateral triangle to a point closer to voltage node B. Node B is only 24.03 V above point "o" in voltage, as shown in Fig. 21-13b.
5. At the same time, neutral point "o" has shifted away from nodes A and C. If the loads were balanced, all phase voltages would be $\mathbf{V}_L/\sqrt{3}$ or $100/\sqrt{3}$ or **57.74 V**. Phase voltages $\mathbf{V}_{Ao}$ and $\mathbf{V}_{Bo}$ have increased by 37.2 and 40.4%, respectively.

The last two points above explain why there is a strong need for a fourth neutral wire in an unbalanced 3ϕ Y-connected system. It was noted earlier and in Table 21-1 that the neutral maintains the phase voltages equal in magnitude and 120° apart. In the event that the neutral is accidentally opened, there is a danger that certain loads will experience a *sudden rise* in voltage while others will experience a *sudden drop* in voltage.[9] Both effects are undesirable in industrial, commercial, and residential occupancies. Overvoltage reduces the life of lamps and may damage appliances such as ovens or toasters. Undervoltage causes motors and transformers to draw more current and run hotter, thereby reducing their efficiency and life.

EXAMPLE 21-17

From the data and solutions of Ex. 21-16, calculate the

a. Phasor sum of the phase voltages
b. Phasor sum of the phase (line) currents

Solution

a. $\sum \mathbf{V}_p = \mathbf{V}_{Ao} + \mathbf{V}_{Bo} + \mathbf{V}_{Co}$
$= 79.2\angle -7.75^\circ + 24.03\angle -153.62^\circ + 81.08\angle 69.44^\circ$
$= (78.48 - j10.68) + (-21.53 - j10.68)$
$+ (28.47 + j75.92)$
$= 85.42 + j54.56 = \mathbf{101.36\angle 32.57^\circ\ V}$

b. $\sum \mathbf{I}_L \equiv \sum \mathbf{I}_p = \mathbf{I}_A + \mathbf{I}_B + \mathbf{I}_C$
$= (26.16 - j3.559) + (-7.177 - j3.559)$
$+ (-18.98 + j7.118)$
$= \mathbf{0 + j0}$ (See Table 21-1.)

Example 21-17 clearly shows that for an unbalanced 3ϕ, three-wire system

1. The phasor sum of the phase voltages is no longer zero but may even exceed the line voltage!
2. The phasor sum of the phase and *line* currents *is zero* despite the severe unbalance in their magnitude and phase. This is a partial (internal) verification of the solution accuracy.

It was noted earlier that unbalanced systems are extremely sensitive to a change in phase sequence. Let us repeat Ex. 21-16, using the sequence CBA instead of ABC. We will retain the capacitor in line C and the resistors in lines A and B. But now the line voltages are *reversed*, as shown in **Fig. 21-14a**.

[9] It is precisely for this reason that the neutral is *never* fused in all 3ϕ, four-wire distribution systems.

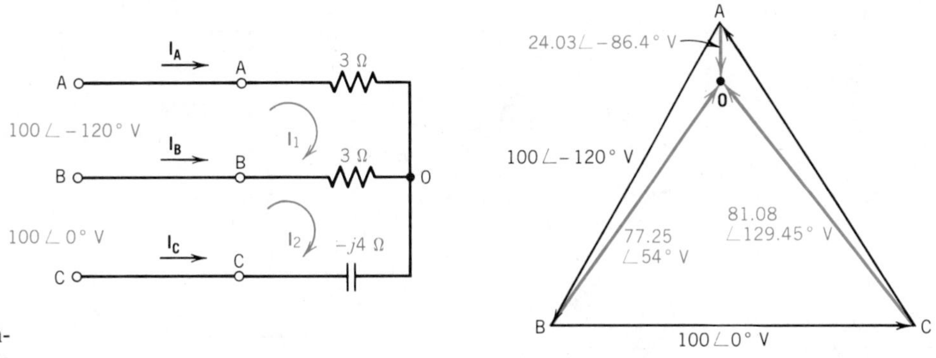

Figure 21-14 Example 21-18. Unbalanced three-wire circuit and its phasor diagram using CBA sequence

a. 3ϕ, 3-wire unbalanced load

b. Phasor diagram CBA sequence

EXAMPLE 21-18

Repeat all parts of Ex. 21-16 for the circuit shown in Fig. 21-14a.

Solution

Writing the two mesh equations for $\mathbf{I}_1$ and $\mathbf{I}_2$ produces the following array:

$\mathbf{I}_1$	$\mathbf{I}_2$	$\mathbf{V}$
$6+j0$	$-3+j0$	$-50-j86.6$
$-3+j0$	$3-j4$	$100+j0$

$$\text{Determinant, } \Delta = \begin{vmatrix} 6+j0 & -3+j0 \\ -3+j0 & 3-j4 \end{vmatrix} = \mathbf{9-j24}$$

a. $$\mathbf{I}_1 = \frac{\begin{vmatrix} -50-j86.6 & -3+j0 \\ 100+j0 & 3-j4 \end{vmatrix}}{\Delta} = \frac{-196.4-j59.8}{9-j24}$$

$$= \mathbf{0.506 - j7.994}$$

$$\mathbf{I}_2 = \frac{\begin{vmatrix} 6+j0 & -50-j86.6 \\ -3+j0 & 100+j0 \end{vmatrix}}{\Delta} = \frac{450-j259.8}{9-j24}$$

$$= \mathbf{15.65 + j12.88}$$

b. $\mathbf{I}_A = \mathbf{I}_1 = \mathbf{0.506 - j7.994 = 8.01\angle{-86.4^\circ}\ A}$,

$\mathbf{I}_B = \mathbf{I}_2 - \mathbf{I}_1 = (15.65 + j12.88) - (0.506 - j7.994)$
$= \mathbf{15.149 + j20.82 = 25.75\angle{53.96^\circ}\ A}$,

$\mathbf{I}_C = -\mathbf{I}_2 = \mathbf{-15.65 - j12.88} = 20.27\angle{-140.55^\circ}$ A

c. $\mathbf{V}_{Ao} = \mathbf{I}_A\mathbf{Z}_A = (8.01\angle{-86.4^\circ}\text{ A})(3\angle{0^\circ}\ \Omega) = \mathbf{24.03\angle{-86.4^\circ}\ V}$,

$\mathbf{V}_{Bo} = \mathbf{I}_B\mathbf{Z}_B = (25.75\angle{53.96^\circ}\text{ A})(3\angle{0^\circ}\ \Omega) = \mathbf{77.25\angle{53.96^\circ}\ V}$,

$\mathbf{V}_{Co} = \mathbf{I}_C\mathbf{Z}_C = (20.27\angle{-140.55^\circ}\text{ A})(4\angle{-90^\circ}\ \Omega)$
$= \mathbf{81.08\angle{+129.45^\circ}\ V}$

d. The phasor diagram is drawn in Fig. 21-14b with the phase voltages inscribed inside the (equilateral) triangle of line voltages in the CBA sequence.

The effect of the change in phase sequence on unbalanced 3ϕ, three-wire systems is observed by comparing the two phasor diagrams of Figs. 21-13b versus 21-14b. The conclusions to be drawn from this comparison are that

1. When the sequence is ABC, the voltage $\mathbf{V}_{Ao}$ (in line A) is *large* and voltage $\mathbf{V}_{Bo}$ (in line B) is *small*, with the capacitor in line C.
2. When the sequence is CBA, the voltage $\mathbf{V}_{Ao}$ (in line A) is *small* and voltage $\mathbf{V}_{Bo}$ (in line B) is *large*, with the capacitor in line C.

These conclusions lead us to the design of a simple *phase-sequence indicator* in which two lamps of identical resistance are used in place of the two resistors. Depending on phase sequence, one lamp burns brightly and one lamp is dim due to the voltage difference. If the sequence is reversed, the brightness of the lamps is reversed. The design and method of predicting phase sequence are described in Sec. 21-11.

21-10.2 The 3ϕ, Four-Wire, Unbalanced Y-Connected System

The 3ϕ, four-wire system is used extensively in industrial and commercial occupancies where both 3ϕ power and 1ϕ power are required. A common type of service

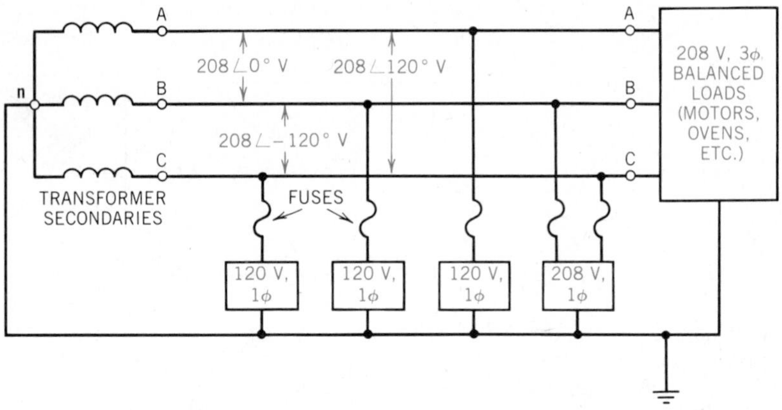

Figure 21-15 3ϕ, four-wire, 208/120 V system

is the 208/120 V, 3ϕ, four-wire system shown in **Fig. 21-15**. This system provides the dual advantages of furnishing both 120 V, 1ϕ loads and 208 V, 1ϕ loads (as shown). In addition, it may supply balanced 3ϕ loads such as larger electric 3ϕ 208 V motors or 3ϕ electric ovens. Typical 120 V loads are small appliances and electric lighting. Typical 208 V, 1ϕ loads are motors from 1 to 25 hp, smaller electric countertop ranges, and ovens. Larger motors would require 208 V, 3ϕ supplies.

In commercial practice, an attempt is made to balance the 120 V, 1ϕ loads on phases A, B, C by the panelboard construction, so that there are limited (but equal) numbers of branch circuits on each phase. But depending on the circuits energized, there is always the possibility of some unbalance.

The presence of the neutral wire ensures that the phase voltages are always equal in magnitude and displaced by 120°, *regardless of load imbalance.* **Figure 21-16** shows how it is possible to draw the phase voltages and determine their magnitude and phase once the phase sequence is known. In Fig. 21-16a, given the sequence ABC for a 3ϕ, 208 V system, the phase voltages $\mathbf{V}_{An}$, $\mathbf{V}_{Bn}$, and $\mathbf{V}_{Cn}$ may be drawn to the geometric neutral inside the equilateral triangle. Similarly, in Fig. 21-16b, given the sequence CBA, the phase voltages may be drawn as a consequence of the line voltages. Note that phase voltage $\mathbf{V}_{An}$ now *leads* reference $\mathbf{V}_{AB}$ in the CBA sequence but *lags* $\mathbf{V}_{AB}$ in the ABC sequence by 30°. (This is a most important and useful insight.)

The solution of an unbalanced 3ϕ, four-wire system generally follows that of the balanced Y-connected source, as described in Sec. 21-5, in that Eqs. (21-2), (21-3), and (21-4) are used. In the case of unbalanced phase (and line) currents, the

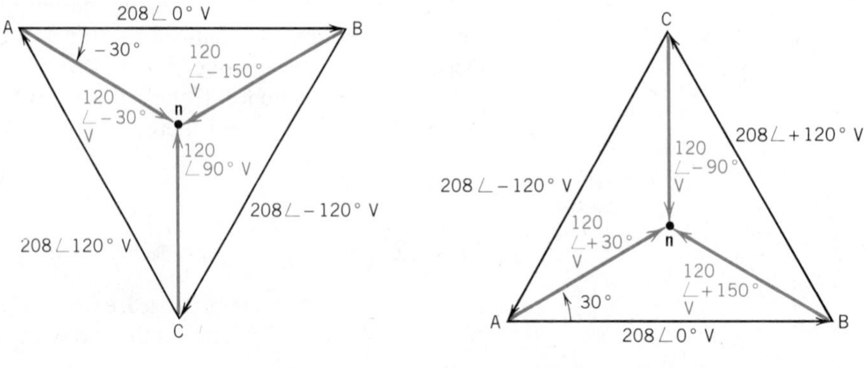

Figure 21-16 Phase and line voltages in a 3ϕ, four-wire system as a result of phase sequence

a. Phase and line voltages; ABC sequence

b. Phase and line voltages; CBA sequence

neutral current is not zero but is the phasor sum of the currents in each line. As shown in Ex. 21-19, the determination of phase voltages (from line voltages) and corresponding phase (and identical line) currents are calculated with little difficulty, even under conditions of severe unbalance. For comparison, we select the unbalanced load of Fig. 21-13a to determine the effect of the neutral on the line currents, as shown in Ex. 21-19.

EXAMPLE 21-19

Given the unbalanced 3ϕ, four-wire load shown in **Fig. 21-17a**, calculate the

a. Relative magnitudes of the phase voltages and show their relation to the line voltages in a phasor diagram
b. Phase (and identical line) currents $\mathbf{I}_A$, $\mathbf{I}_B$, and $\mathbf{I}_C$
c. Neutral current, $\mathbf{I}_N$
d. Draw a phasor diagram showing all line currents and the neutral current
e. Compare line currents and phase voltages found in this example with those found in Ex. 21-16 (Hint: tabulate results for ready comparison)

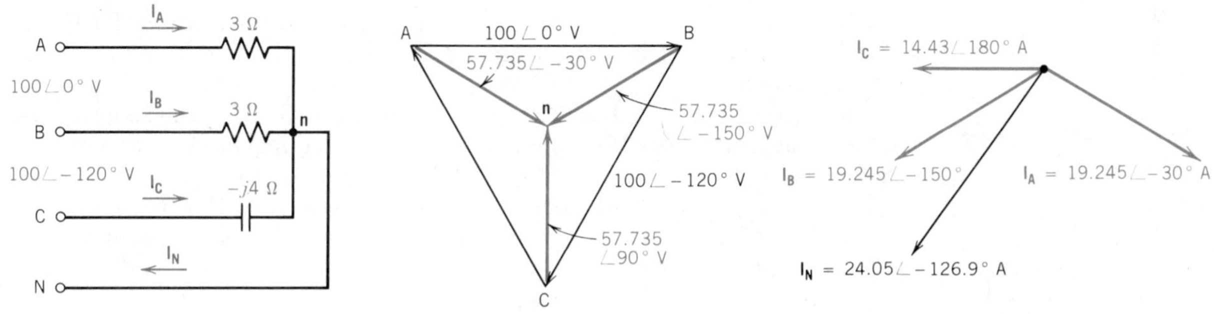

a. 3ϕ, 4-wire unbalanced load b. Line and phase voltages c. Line (phase) and neutral currents

Figure 21-17 Unbalanced 3ϕ, four-wire system, voltages and currents

Solution

a. All phase voltages have magnitude $\mathbf{V}_p = \mathbf{V}_L/\sqrt{3} = 100\ \text{V}/\sqrt{3} = \mathbf{57.735\ V}$, and their relative directions are shown in Fig. 21-17b
b. $\mathbf{I}_A = \mathbf{V}_{An}/\mathbf{Z}_A = 57.735\underline{/-30^\circ}\ \text{V}/3\underline{/0^\circ}\ \Omega = \mathbf{19.245\underline{/-30^\circ}\ A}$
$= 16.\bar{6} - j9.6225\ \text{A},$
$\mathbf{I}_B = \mathbf{V}_{Bn}/\mathbf{Z}_B = 57.735\underline{/-150^\circ}\ \text{V}/3\underline{/0^\circ}\ \Omega$
$= \mathbf{19.245^\circ\underline{/-150^\circ}\ A} = -16.\bar{6} - j9.6225\ \text{A},$
$\mathbf{I}_C = \mathbf{V}_{Cn}/\mathbf{Z}_C = 57.735\underline{/90^\circ}\ \text{V}/4\underline{/-90^\circ}\ \Omega = \mathbf{14.43\underline{/180^\circ}\ A}$
$= -14.43 + j0\ \text{A}$
c. $\mathbf{I}_N = \mathbf{I}_A + \mathbf{I}_B + \mathbf{I}_C$
$= (16.\bar{6} - j9.6225) + (-16.\bar{6} - j9.6225) + (-14.43 + j0)$
$= -14.43 - j19.245 = \mathbf{24.05\underline{/-126.9^\circ}\ A}$
d. See Fig. 21-17c
e.

Same Unbalanced Load	$\mathbf{V}_{Ao}$ (V)	$\mathbf{V}_{Bo}$ (V)	$\mathbf{V}_{Co}$ (V)	$\mathbf{I}_A$ (A)	$\mathbf{I}_B$ (A)	$\mathbf{I}_C$ (A)	$\mathbf{I}_N$ (A)
3ϕ, 3-wire	79.2	24.03	81.08	26.4	8.011	20.27	—
3ϕ, 4-wire	57.735	57.735	57.735	19.25	19.25	14.43	24.05

The tabulation of Ex. 21-19e enables us to draw the following conclusions regarding advantages of 3ϕ, four-wire over 3ϕ, three-wire systems:

1. The phase voltages of the 3ϕ, four-wire system remain *undisturbed* regardless of load unbalance, whereas the three-wire system phase voltages both increase and decrease alarmingly and, possibly, dangerously.
2. The *phase* (and *identical*) *line* currents of the three-phase, four-wire system tend to remain more balanced despite the severely unbalanced load. The three-wire system (lacking a neutral) requires one or more lines to carry the unbalanced currents, resulting in *greater line current imbalance.*

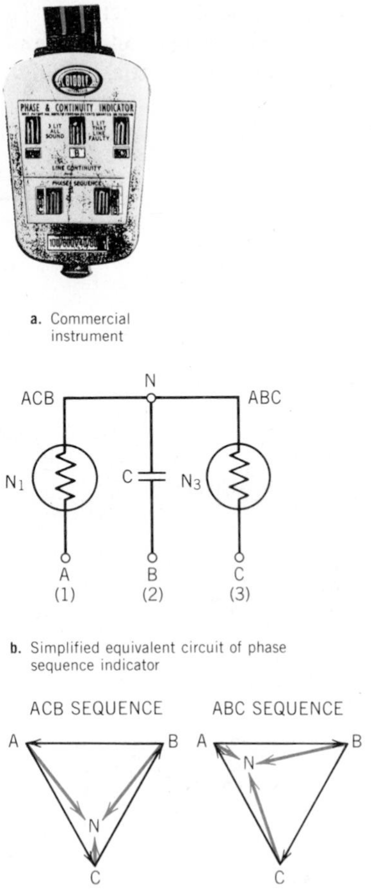

a. Commercial instrument

b. Simplified equivalent circuit of phase sequence indicator

c. Phasor diagrams for each sequence

Figure 21-18 Commercial phase sequence indicator and phasor diagrams resulting from differences in phase sequence (Indicator courtesy Biddle Instruments)

These advantages, coupled with the previously mentioned advantages of supplying *both* lower-voltage 1ϕ loads and higher-voltage 1ϕ and 3ϕ loads, explain why the 3ϕ, four-wire system is used so extensively.

21-11 PHASE SEQUENCE INDICATOR

It was noted at the conclusion to Sec. 21-10.1 that lamps may be used instead of resistors in the circuits shown in Figs. 21-13 and 21-14. If we select the line containing the capacitor, line C, as our first reference, followed by the bright lamp and then the dim lamp (capacitor–bright–dim), we establish the phase sequence automatically.

In Fig. 21-13, voltage $\mathbf{V}_{Bo}$ is small. The sequence therefore is CAB, since $\mathbf{V}_{Ao}$ is large. Rotating the triangle CCW results in CABCAB, verifying the relation.

In Fig. 21-14, voltage $\mathbf{V}_{Ao}$ is small. The sequence therefore is CBA, since $\mathbf{V}_{Bo}$ is large. Rotating the triangle CCW results in CBACBA, verifying the relation.

Figure 21-18a shows a *commercial* phase-sequence indicator. The circuit resistances are designed so as to create a potential above, in one case, and below, in the other case, the ignition potential of the neon lamps. The simplified equivalent circuit, shown in Fig. 21-18b, is essentially the Y-circuit of Figs. 21-13a and 21-14a, whose terminals are marked A, B, and C.[10]

When neon lamp N_1 lights, the sequence of phases is BACBA and so forth; that is, the voltage from phase A to neutral N is much greater than from phase C to N (Fig. 21-18c).

When neon lamp N_3 lights, the sequence of phases is BCABC and so forth; that is, the voltage from phase C to neutral N is much greater than from phase A to N, as shown in the phasor diagrams of Fig. 21-18c.

Table 21-2 summarizes the phase sequence in terms of A, B, C (or 1, 2, 3) for the location of the capacitor terminal in lines B, A, and/or C, respectively. The sequence is always capacitor–bright lamp–dark lamp.

Table 21-2 Phase Sequence Indicator Operation

Capacitor in Line	Bright Lamp Located in Line	(+) Phase Sequence	Bright Lamp Located in Line	(−) Phase Sequence
B	C	BCABC	A	BACBA
A	B	ABCAB	C	ACBAC
C	A	CABCA	B	CBACB

21-12 MEASUREMENT OF 3ϕ POWER DRAWN BY BALANCED LOADS, USING A SINGLE WATTMETER

The basic instrument for 3ϕ power measurement is the wattmeter.[11] This instrument was introduced in connection with single-phase measurement of power in

[10] Frequently, in industry, the terminals of electric equipment to be connected to 3ϕ lines are marked L_1, L_2, and L_3 rather than A, B, C. Some 3ϕ equipment is sensitive to phase sequence. One example is a motor-driven pump, which may be damaged by incorrect motor rotation. Another is a blower, which drives air in the proper (desired) direction.

[11] There are polyphase power instruments such as the polyphase wattmeter, which measures total 3ϕ power. The discussion of this section is confined to the single-phase electrodynamometer that is used to measure polyphase power. See Sec. 21-14.1 for a description and discussion of the *polyphase* wattmeter. Also see Sec. 23-8.5 for a discussion of the electrodynamometer principle and the dynamometer wattmeter.

Chapter 14. The basic theory of electrodynamometer operation is covered in Appendix B-10, which contains the proof that the wattmeter measures true power in ac circuits as $P = \mathbf{VI}\cos\theta$, where **V** is the potential across its potential coil, **I** is the current in its current coils, and θ is the phase angle between the voltage and current.

Figure 21-19a shows the schematic of a four-terminal electrodynamometer wattmeter. Figure 21-19b shows the proper connections for measuring the power drawn by a load from an ac supply. With respect to these figures, please note that

1. The common ($\pm$) terminal is always connected to the same line side, so the instantaneous polarity for both coils is always the same.
2. For the wattmeter (WM) to read upscale, the common ($\pm$) current terminal is connected to the *line* side of the instrument in 3ϕ measurement.
3. If the WM is properly connected in a 3ϕ circuit and the angle θ between the voltage across and current in the WM is greater than either $+90°$ or $-90°$, the WM will read downscale (reversed). This is possible in polyphase measurement (as in the two-wattmeter method), and such a reading is recorded as a *negative* power reading.
4. Whenever a WM reads downscale, in order to obtain the measurement it is necessary to reverse the *current* coil connections.[12]
5. Although not absolutely necessary, it is preferable to connect the common ($\pm$) of the potential coil to the *load* side of the WM. This produces a constant WM error due to current drawn by the *potential* coil. This may be measured by opening switch S, as shown in Fig. 21-19b. (The potential coil error is constant as long as the line voltage is constant.) The (constant) power read by the WM with S open may be subtracted from any WM reading with S closed.

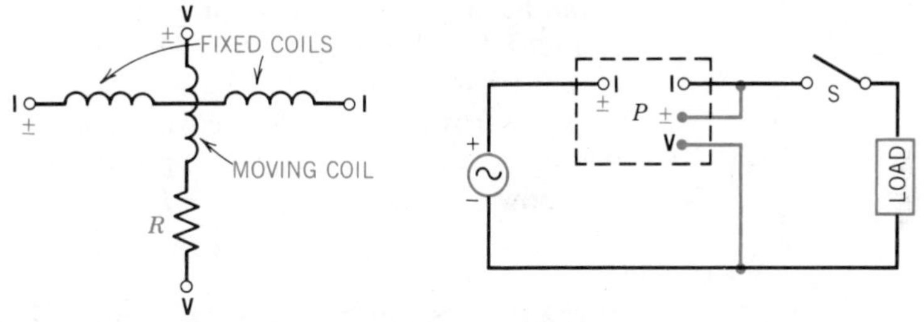

a. Schematic of 4-terminal wattmeter

b. Connection for power drawn by load

Figure 21-19 The electrodynamometer wattmeter construction and connections

21-12.1 Measurement of 3ϕ Power—Balanced Loads Only

A single wattmeter may be used to measure 3ϕ power *provided the load is balanced.* Balanced 3ϕ loads are relatively common and the methods shown in **Fig. 21-20** enable measurement of the total power drawn. Each of these methods requires one auxiliary device in addition to the single dynamometer wattmeter.

---

[12] The potential coils are *never* reversed. Reversing them produces a large potential difference between the current and potential coils. When currents in both these coils are relatively high (due to high load currents at high voltages), this leads to minor electrostatic errors due to the electrical force of attraction between the moving and stationary coils.

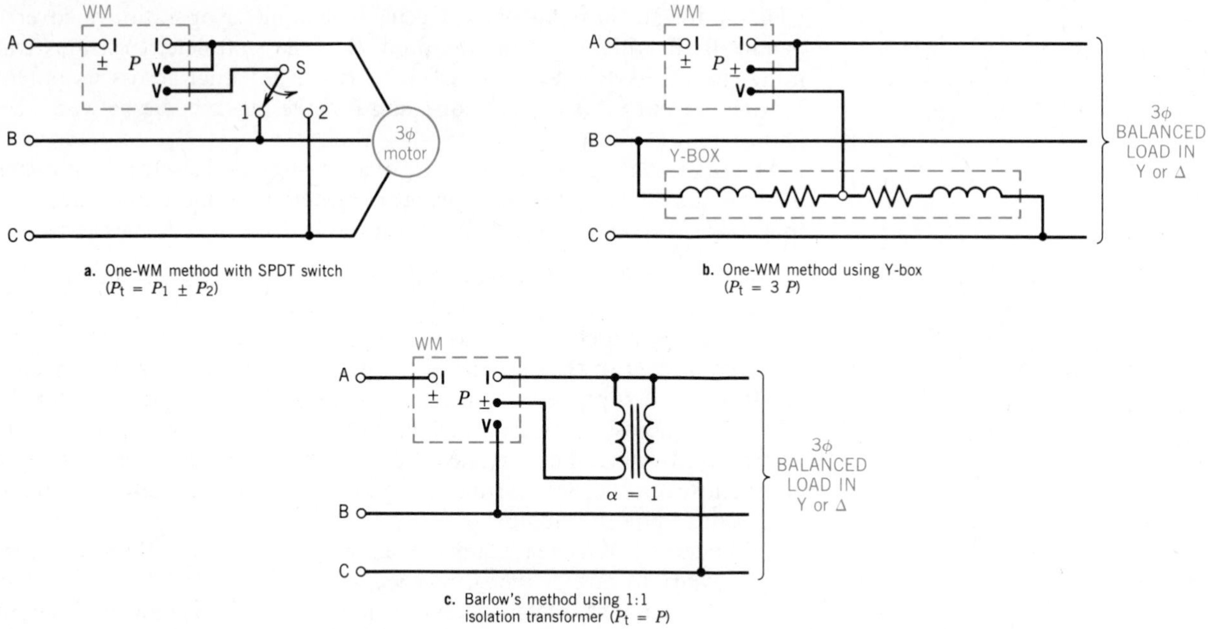

a. One-WM method with SPDT switch ($P_t = P_1 \pm P_2$)

b. One-WM method using Y-box ($P_t = 3\ P$)

c. Barlow's method using 1:1 isolation transformer ($P_t = P$)

Figure 21-20 One-wattmeter methods of measuring balanced 3ϕ power

21-12.2 One-Wattmeter Method with SPDT Switch

Figure 21-20a shows a single-phase WM with current coils in line A and potential coils measuring $\mathbf{V}_{AB}$ in position 1 and $\mathbf{V}_{AC}$ (not $\mathbf{V}_{CA}$) in position 2. The total power of the balanced 3ϕ load is the sum of the wattmeter readings (P_1 and P_2) in positions 1 and 2, respectively. Depending on the power factor of the balanced load, it is possible for one of the WM readings to read reversed. In this case the total power is the *difference* between the two WM readings, since a *reversed* WM reading is recorded as a *negative* power, as noted in (3) above. Using this method, total power is always $P_t = P_1 \pm P_2$.

21-12.3 One-Wattmeter Method Using Y-Box to Measure Balanced 3ϕ Power

A Y-box may be used whose impedances per phase match the moving-coil impedance of the WM exactly. As shown in Fig. 21-20b, the Y-box is connected so that all three impedances represent a Y-connected load across the 3ϕ lines. The moving coil of WM is recording the *phase* voltage of a Y-connected load while its current coils are measuring the phase current drawn by a Y-connected load. The WM reading represents the power per phase of the balanced load, where $P_p = \mathbf{V}_p\mathbf{I}_p \cos\theta$. Consequently, the total 3ϕ power, P_t, is $3P_p$. This means that the total power is three times the power recorded by the WM, or $P_t = 3P$, where P is the WM reading.[13]

[13] In some instrumentation systems, the scale of the WM is *drawn* to display the total power directly (by multiplying its *normal* scale by 3). Such a WM, when permanently used with a Y-box for measuring 3ϕ balanced loads, will read total power directly. (Proof for a Δ-connected load is obtained in Prob. 21-45.)

21-12.4 Barlow's One-Wattmeter Method Using a 1:1 Isolation Transformer

Perhaps the most ingenious method of using a single WM to measure balanced 3ϕ power is Barlow's method, which requires a 1:1 isolation transformer. The transformer secondary reflects $\mathbf{V}_{AC}$ back to the primary, while the transformer is connected across $\mathbf{V}_{AB}$. Consequently, the potential coil of the WM sees the phasor sum of the two line voltages while the current coil sees the current in line A. With this connection, the total 3ϕ power ($\sqrt{3}V_L I_L \cos\theta$) is actually the same as the wattmeter reading, $P_t = P$. Since 1:1 isolation transformers are fairly common, Barlow's method is perhaps the simplest of the three methods shown in Fig. 21-20.

21-13 MEASUREMENT OF 3ϕ POWER—BALANCED OR UNBALANCED LOADS

The methods described in this section enable measurement of 3ϕ power regardless of whether the loads are balanced or unbalanced. To accomplish this, however, *more than one* WM is needed. This raises the question of whether we require three WMs to measure unbalanced 3ϕ power or whether two will suffice. The answer is found in Blondel's theorem.

21-13.1 Blondel's Theorem

Blondel's theorem states that the number of WMs required to measure balanced or unbalanced power, drawn from n lines, in any polyphase system is $n - 1$, *provided* that *one* wire is made *common* to all *potential* coil circuits. Thus, a 3ϕ, three-wire system requires only two WMs and a 3ϕ, four-wire system requires only three WMs, even if the loads are unbalanced. A little reflection explains why Blondel's theorem holds. Consider a 3ϕ, three-wire unbalanced load. If the current coils of each of two WMs are measuring the unbalanced current in two lines, then the current in the third line is also measured, since the phasor sum of the three line currents is zero! This also applies to the line voltages, since the phasor sum of the line voltages is zero. So measuring two of the three line voltages also includes the third.[14]

21-13.2 Measurement of Unbalanced or Balanced Power in a 3ϕ, Four-Wire System

The connections of **Fig. 21-21a** are an illustration of Blondel's theorem. Note that each potential coil of each WM has one wire (the neutral) common to all potential circuits. Since the common (±) terminal of each potential coil is connected to its own line, *each* potential coil measures the *phase* voltage, regardless of the load imbalance. (Recall that the neutral preserves the integrity of the phase voltages, despite imbalance.) Similarly, each current coil of each WM measures phase current. Consequently, each WM is measuring the power dissipated by each phase impedance, regardless of imbalance. The total power is the sum of the three WM readings, or $P_t = P_1 + P_2 + P_3$.

[14] A mathematical proof of Blondel's theorem may be found in F. A. Laws, *Electrical measurements*, 2nd Ed., McGraw-Hill, 1938, p. 337.

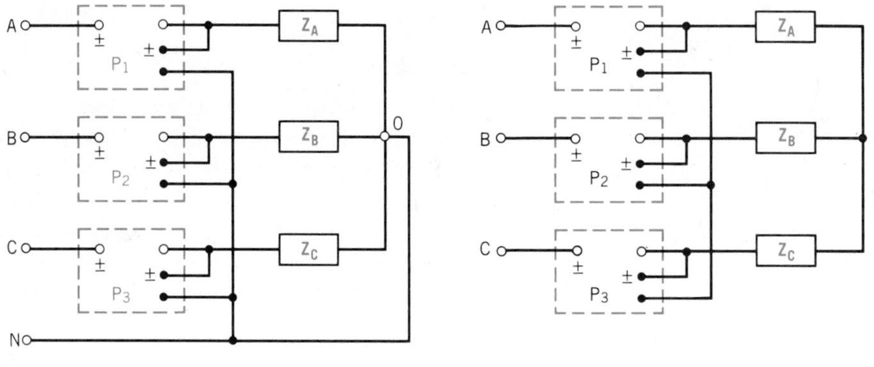

a. 3ϕ, 4-wire power measurement using 3 single-phase WMs

b. 3ϕ, 3-wire power measurement using 3 single-phase WMs

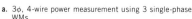

Figure 21-21 Three-wattmeter methods of 3ϕ power measurement ($P_t = P_1 + P_2 + P_3$), balanced and unbalanced loads

21-13.3 Measurement of Unbalanced or Balanced Power in a 3ϕ, Three-Wire System

If the neutral wire is removed, producing a 3ϕ, three-wire system, the same combination of WMs may be used to measure either balanced or unbalanced power. Recall that when the load is unbalanced in a three-wire system, the phase voltages are also unbalanced. But when the potential coils are Y-connected, each potential coil "sees" the phase voltage and phase current and measures $P_p = \mathbf{V}_p\mathbf{I}_p \cos\theta$ for each phase. Consequently, each WM is measuring the true power dissipated by each phase impedance, regardless of imbalance. The total power is the sum of the three WM readings, or $P_t = P_1 + P_2 + P_3$, in the circuit shown in Fig. 21-21b.

21-14 THE TWO-WATTMETER METHOD OF MEASURING 3ϕ POWER, BALANCED OR UNBALANCED

Blondel's theorem, however, tells us that the 3ϕ, three-wire unbalanced (or balanced) power in Fig. 21-21b can be measured with only *two* WMs, since there are three wires. This gives rise to the well-known *two-wattmeter method* used for measuring power to any three-wire 3ϕ load, whether balanced or unbalanced.

The two-wattmeter method does *not* derive its notoriety by virtue of using one less WM. On the contrary, its advantage over all the other methods of measurement previously described is that, in addition to measuring total power dissipated by any 3ϕ load, it *also* yields, by calculation, the power factor of the load, as well as the power factor angle, θ. As we will see, it does this without requiring additional voltmeters and ammeters. Indeed, it is also important because its theory and design give rise to the polyphase wattmeter. Finally, if we know whether the overall load is inductive or capacitive, from the lines in which the wattmeters are located and the respective readings of the wattmeters, we can deduce *the phase sequence*! All this, then, dictates a closer study of the two-wattmeter method.

21-14.1 Relation of the Polyphase Wattmeter to the Two-Wattmeter Circuit

Figure 21-22a shows the typical connections of the two wattmeters for measurement of total 3ϕ power, whether balanced or unbalanced. Observe (*à la* Blondel's theorem) that line B is common to both potential coils and that the WMs are

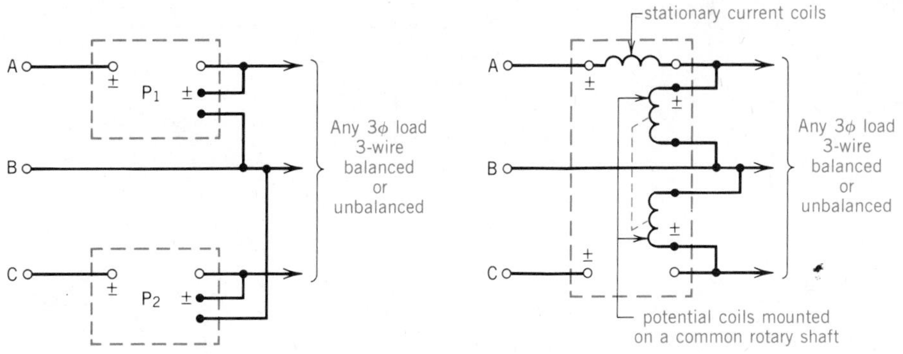

a. Two-wattmeter method of 3ϕ power measurement ($P_t = P_1 \pm P_2$)

b. Polyphase wattmeter ($P_t = P_1 \pm P_2$)

Figure 21-22 Measurement of 3ϕ power (balanced or unbalanced) by the two-wattmeter method and polyphase wattmeter in any 3ϕ three-wire load

measuring line currents $\mathbf{I}_A$ and $\mathbf{I}_C$ and line voltages $\mathbf{V}_{AB}$ and $\mathbf{V}_{CA}$, respectively. Total power is $P_t = P_1 \pm P_2$.

Figure 21-22b shows the similarity in construction of the polyphase wattmeter to the two-wattmeter method. In a single instrument there are two separate sets of current coils, each set connected in a phase line. The potential coils are mounted on a common movable shaft, producing a resultant torque that reflects the product $\mathbf{V}_L\mathbf{I}_L \cos \phi$, where ϕ is the angle between $\mathbf{V}_L$ and $\mathbf{I}_L$ for each wattmeter section. Regardless of whether the load is balanced or unbalanced, the instrumental deflection is proportional to the total power, $\mathbf{P}_t = \mathbf{P}_1 \pm \mathbf{P}_2$.

21-14.2 Phasor Diagrams of Line Currents and Voltages for Both Phase Sequences

To analyze how and why the two-wattmeter method is able to yield so much information about a 3ϕ, three-wire circuit, it is necessary to know the magnitude and direction of the line voltages impressed on and line current in each of the two wattmeters. This means that we must draw phasor diagrams for a typical circuit containing the two instruments.

Figure 21-23a shows two WMs, one in line A and the other in line C, connected to measure power drawn by a hypothetically balanced Y-connected load that is essentially inductive. The Y-connected load is selected because the phase and line currents are the same. As shown in Fig. 21-23a, the reference voltage is $\mathbf{V}_{BC}$, which automatically dictates an ABC phase sequence. Since the load is inductive, we also know that θ is the angle by which *phase* current lags phase voltage. The line and phase voltages are shown in Fig. 21-23a (using the simplistic shortcut method developed earlier in this chapter) for the first phase sequence, ABC.

The first WM, P_1, sees voltage $\mathbf{V}_{AB}$ on its potential terminals and $\mathbf{I}_A$ in its current coils. The second WM, P_2, sees line voltage $\mathbf{V}_{CB}$ on its potential terminals and $\mathbf{I}_C$ in its current coils. Figure 21-23b shows the two line voltages of interest, drawn in *solid* lines. Since $\mathbf{I}_A$ lags $\mathbf{V}_{Ao}$ and $\mathbf{I}_C$ lags $\mathbf{V}_{Co}$, it is also necessary to draw these phase voltages (which also appear in Fig. 21-23a). Once the phase voltages are drawn, it is a simple matter to show $\mathbf{I}_A$ lagging $\mathbf{V}_{Ao}$ and $\mathbf{I}_C$ lagging $\mathbf{V}_{Co}$ by the same angle θ since the loads are balanced. Also, note that phase voltage $\mathbf{V}_{Ao}$ lags $\mathbf{V}_{AB}$ by 30° and $\mathbf{V}_{Co}$ leads $\mathbf{V}_{CB}$ by 30°.

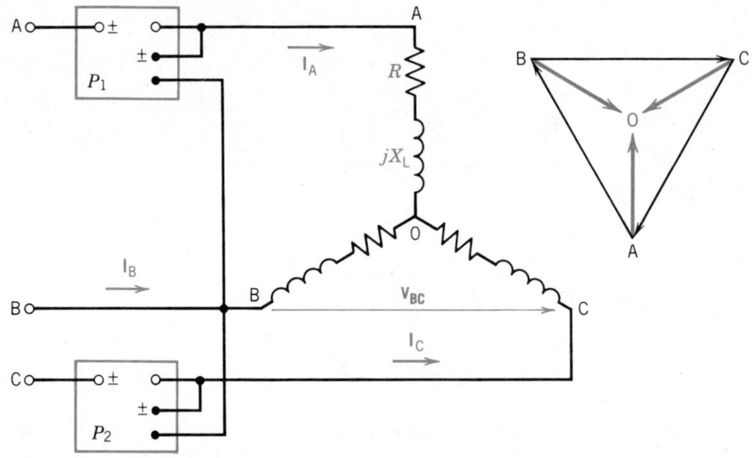

a. Circuit, line voltages and currents using two-wattmeter method. ABC phase sequence using $\mathbf{V}_{BC}$ as reference

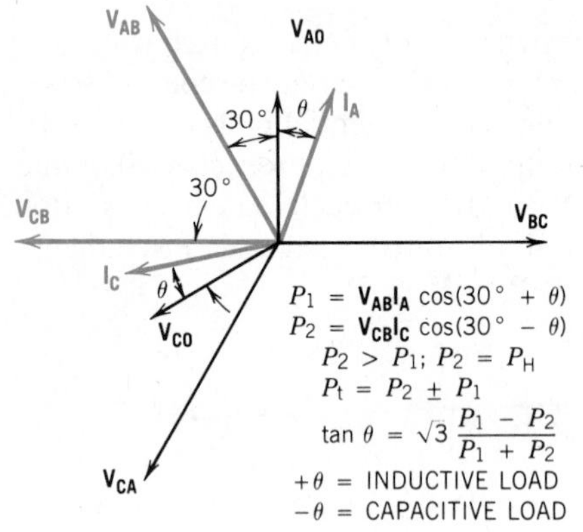

b. Voltages and currents seen by each WM, ABC phase sequence, balanced inductive load

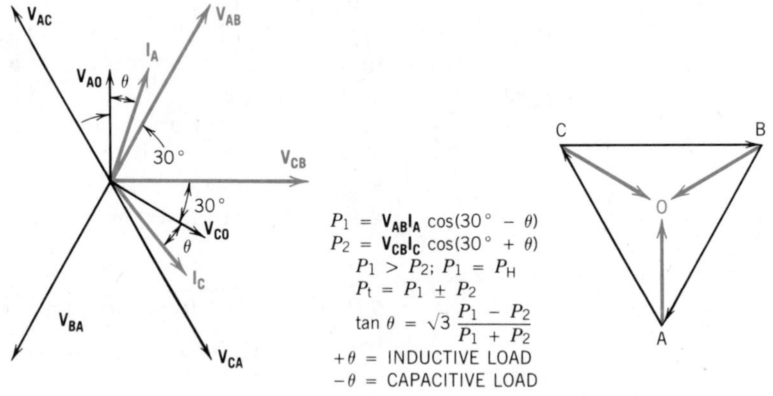

c. Reversed phase sequence CBA showing voltages and currents seen by each WM using same balanced inductive load

Figure 21-23 How wattmeter readings are affected as a result of phase sequence in the two-WM method of power measurement

From Fig. 21-23b, for wattmeter P_1 we can write (for the *lower*-reading WM)

$$P_1 = \mathbf{V}_L\mathbf{I}_L \cos\phi = \mathbf{V}_{AB}\mathbf{I}_A \cos(30+\theta)^\circ = \mathbf{V}_L\mathbf{I}_L \cos(30+\theta)^\circ \quad \text{watts (W)} \qquad (21\text{-}11)$$

Similarly for wattmeter P_2 we can write (for the *higher*-reading WM)

$$P_2 = \mathbf{V}_L\mathbf{I}_L \cos\phi = \mathbf{V}_{CB}\mathbf{I}_C \cos(30-\theta)^\circ = \mathbf{V}_L\mathbf{I}_L \cos(30-\theta)^\circ \quad \text{watts (W)} \qquad (21\text{-}12)$$

And since the total power is the *sum* of the readings of the two WMs,

$$P_t = P_2 \pm P_1 = \mathbf{V}_L\mathbf{I}_L \cos(30 - \theta)^\circ + \mathbf{V}_L\mathbf{I}_L \cos(30 + \theta)^\circ$$
$$= \sqrt{3}\mathbf{V}_L\mathbf{I}_L \cos\theta \qquad (21\text{-}13)$$

EXAMPLE 21-20

Given line voltage $\mathbf{V}_{AB} = 100\underline{/0^\circ}$ V and each phase impedance $\mathbf{Z}_p = 3 + j4\ \Omega$ in Fig. 21-23a, calculate the

a. Magnitude of the phase (and identical line) current
b. Power reading of wattmeter, P_1
c. Power reading of wattmeter, P_2
d. Total power as the sum of the WM readings
e. Total power as $P_t = \sqrt{3}\mathbf{V}_L\mathbf{I}_L \cos\theta$

Solution

a. $\mathbf{I}_p = \mathbf{V}_p/\mathbf{Z}_p = \dfrac{100/\sqrt{3}}{3 + j4} = \dfrac{57.535\text{ V}}{5\underline{/53.13^\circ}\ \Omega} = \mathbf{11.547\underline{/-53.13^\circ}\ A}$

b. $P_1 = \mathbf{V}_L\mathbf{I}_L \cos(30 + \theta)^\circ = 100(11.547)\cos(30 + 53.13)^\circ = \mathbf{138.12\ W}$ (21-11)

c. $P_2 = \mathbf{V}_L\mathbf{I}_L \cos(30 - \theta)^\circ = 100(11.547)\cos(30 - 53.13)^\circ = \mathbf{1061.88\ W}$ (21-12)

d. $P_t = P_1 + P_2 = 138.12 + 1061.88\text{ W} = \mathbf{1200\ W}$ (21-13)

e. $P_t = \sqrt{3}\mathbf{V}_L\mathbf{I}_L \cos\theta = \sqrt{3}(100)(11.547)\cos 53.13^\circ = \mathbf{1200\ W}$ (21-8)

Several important insights can be drawn from Eqs. (21-11) through (21-13) and Ex. 21-20:

1. The angle θ is the phase angle of the impedance and is a *positive* angle (in this example) because the load is *inductive*. Had the load been *capacitive* and balanced, θ would have been used as a *negative* angle.
2. Since the load is inductive, P_1 is the *lower*-reading WM. This results from $\cos(30 + \theta)^\circ$, a term that is the cosine of a *large* angle, thus reducing the power reading.
3. Similarly, since θ is a positive angle, the term $\cos(30 - \theta)$ causes P_2 to be the *higher*-reading WM, since it represents the cosine of a *small* angle.
4. The total power is shown as the **sum** of the power readings, $P_t = P_2 \pm P_1$, in Eq. (21-13). This implies that P_1 may actually be reversed and recorded as a *negative* power reading, depending on the phase angle of load impedance, θ.
5. When $\theta = 60^\circ$, as shown by Eq. (21-11), the power reading of WM P_1 is actually zero! Total power drawn by the load is measured by P_2 only.
6. When $\theta > 60^\circ$, WM P_1 *reverses*, producing a *negative* reading. The total power in this case is $P_t = P_2 - P_1$.

EXAMPLE 21-21

Repeat all parts of Ex. 21-20, given line voltage $\mathbf{V}_{AB} = 208\underline{/0^\circ}$ V and phase impedance $\mathbf{Z}_p = 7 + j24\ \Omega$ in Fig. 21-23a

Solution

a. $\mathbf{I}_p = \mathbf{V}_p/\mathbf{Z}_p = \dfrac{208/\sqrt{3}}{7 + j24} = \dfrac{120.09\text{ V}}{25\underline{/73.74^\circ}\ \Omega} = \mathbf{4.804\underline{/-73.74^\circ}\ A}$

b. $P_1 = \mathbf{V}_L\mathbf{I}_L \cos(30 + \theta) = 208(4.804)\cos(30 + 73.74)^\circ = \mathbf{-237.\bar{3}\ W}$ (21-11)

c. $P_2 = \mathbf{V}_L\mathbf{I}_L \cos(30 - \theta) = 208(4.804)\cos(30 - 73.74)^\circ = \mathbf{721.93\ W}$ (21-12)

d. $P_t = P_2 \pm P_1 = 721.93 + (-237.\bar{3}) = \mathbf{484.6\ W}$ (21-13)

e. $P_t = \sqrt{3}\mathbf{V}_L\mathbf{I}_L \cos\theta = \sqrt{3}(208)(4.804)\cos 73.74^\circ = \mathbf{484.6\ W}$ (21-8)

Example 21-21 reveals that

1. When the impedance phase angle, θ, is greater than 60°, the reading of WM P_1 is actually negative.
2. Total power, P_t, is now the *difference* between the two WM readings.

21-14.3 Power Factor (cos θ) from the Two-Wattmeter Method (Balanced Loads)

There are two situations in which the WM readings are the same. One occurs when the load impedance phase angle $\theta = 0$ (i.e., the load is completely resistive). In this case,

$$P_1 = P_2 = \frac{P_t}{2} \quad \text{when } \theta = 0° \qquad \textbf{(21-14)}$$

The second situation occurs at the opposite extreme, where the load is completely inductive and contains no resistance whatever. In that case, obviously, power is stored and returned to the supply but none is dissipated, that is, $P_t = 0$. For this situation, where $P_1 = -P_2$, we have

$$P_t = P_2 + P_1 = P_2 + (-P_2) = 0 \quad \text{when } \theta = 90° \qquad \textbf{(21-15)}$$

We have also seen, from (5) in the preceding subsection, that when $\theta = 60°$, one WM reads zero. The power factor in this case is cos(60°) or 0.5, either leading or lagging. These three simple facts enable us to develop a curve of power factor (cos θ) versus wattmeter reading ratio, P_L/P_H, shown in **Fig. 21-24**.

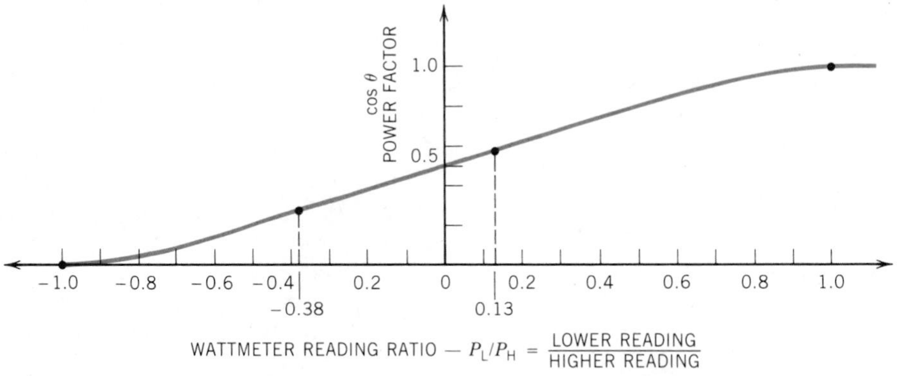

Figure 21-24 Curve for obtaining power factor from ratio of two WM readings

The curve of Fig. 21-24 may be used either with leading or lagging balanced loads. It is obtained from the relations

$$\theta = \tan^{-1} \sqrt{3}\left(\frac{P_H - P_L}{P_H + P_L}\right) = \tan^{-1}\left(\frac{\mathbf{Q}_t}{\mathbf{P}_t}\right) \text{ degrees (°)} \qquad \textbf{(21-16)}$$

where P_H is the power of the *higher*-reading WM

P_L is the power of the *lower*-reading WM (either positive or negative)

Once θ is obtained from Eq. (21-16), the power factor may be computed as

$$F_p = \cos\theta = \frac{(\mathbf{P}_H \pm \mathbf{P}_L)}{\mathbf{S}_t^*} \quad \text{(dimensionless)} \qquad \textbf{(21-17)}$$

where $\mathbf{S}_t^*$ is the total 3ϕ complex power defined in Eq. (21-9).

EXAMPLE 21-22

Calculate the phase angle θ and the power factor, $\cos\theta$, given the wattmeter readings in **a.** Ex. 21-20 **b.** Ex. 21-21

Solution

a. $P_H = P_2 = 1061.88$ W and
$P_L = P_1 = 138.12$ W from Ex. 21-20

$$\theta = \tan^{-1}\sqrt{3}\left(\frac{1061.88 - 138.12}{1200}\right)$$
$$= \mathbf{53.13°} \text{ (see Ex. 21-20)}$$
$$\cos\theta = \cos 53.13° = \mathbf{0.6}$$

b. $P_H = P_2 = 721.93$ W and
$P_L = P_1 = -237.\bar{3}$ W (from Ex. 21-21)

$$\theta = \tan^{-1}\sqrt{3}\left[\frac{721.93 - (-237.3)}{484.6}\right]$$
$$= \mathbf{73.74°} \text{ (see Ex. 21-21)}$$
$$\cos\theta = \cos 73.74° = \mathbf{0.28}$$

EXAMPLE 21-23

Using the higher and lower wattmeter readings in Ex. 21-22, calculate the wattmeter reading ratio and estimate the power factor from the curve of Fig. 21-24

Solution

a. Ratio $= P_L/P_H = 138.12$ W/1061.88 W $= 0.13$
From the curve of Fig. 21-24, a ratio of 0.13 corresponds to 0.6 PF

b. Ratio $= P_L/P_H = -273.\bar{3}$ W/721.93 W $= -0.379$
From the curve of Fig. 21-24, a ratio of -0.379 corresponds to 0.28 PF

21-14.4 Phase Sequence from the Two-Wattmeter Method, Balanced Loads

It is also possible to predict the phase sequence from the locations of the higher- and lower-reading WMs if one knows whether the load is inductive or capacitive. The following rules may be extracted from a careful examination of Figs. 21-23b and 21-23c, shown for inductive loads only, at two sequences:

1. P_H *leads* P_L in phase sequence for *inductively* balanced loads.
2. P_L *leads* P_H in phase sequence for *capacitively* balanced loads.

EXAMPLE 21-24

From the designation of whether P_2 or P_1 is the higher-reading WM in Figs. 21-23b and 21-23c, determine the phase sequence in

a. Fig. 21-23b
b. Fig. 21-23c

Solution

a. In Fig. 21-23b, $P_2 = P_H$ and $P_1 = P_L$. Since the load is inductive, the sequence is P_H followed by P_L or P_2 followed by P_1. Since P_2 is in line C and P_1 is in line A, the sequence is CABC, or ABC phase sequence, as originally specified.

b. In Fig. 21-23c, $P_1 = P_H$ and $P_2 = P_L$. The sequence is P_1 *followed by* P_2 or *A* followed by *C*, that is, *ACBA*, or *CBA* phase sequence, as originally specified.

Example 21-24 proves the rule for *inductively* balanced loads. Intuition tells us that the rule must be *reversed* for *capacitively* balanced loads, as shown in item 2 above.

21-14.5 System Vars and Complex Power from the Two-Wattmeter Method, Balanced Loads

It is also possible to obtain the system $\mathbf{Q}_t$, quadrature reactive power or system vars, from the reading of the individual wattmeters, using

$$\mathbf{Q}_t = \sqrt{3}(P_H - P_L) \quad \text{reactive volt-amperes (vars)} \qquad (21\text{-}18)$$

The astute reader may notice that Eq. (21-18) represents the numerator of Eq. (21-16), while the denominator of Eq. (21-16) represents the total power dissipated. In effect, tan θ is the ratio of the arms of the power triangle or the ratio $\mathbf{Q}_t/\mathbf{P}_t$.

In Eq. (21-18), $\mathbf{Q}_t$ is treated as a positive value for inductively balanced 3ϕ loads and a negative value for capacitively balanced loads.

We may now write complex system power, $\mathbf{S}_t^*$, from the two wattmeter readings as

$$\mathbf{S}_t^* = \mathbf{P}_t \pm j\mathbf{Q}_t = (P_H + P_L) \pm j[\sqrt{3}(P_H - P_L)] \quad \text{volt-amperes (VA)} \qquad (21\text{-}19)$$

where $+j\mathbf{Q}_t$ signifies *inductively* balanced loads
$-j\mathbf{Q}_t$ signifies *capacitively* balanced loads

EXAMPLE 21-25

Calculate the system vars and complex power from the wattmeter readings in

a. Ex. 21-20
b. Ex. 21-21

Solution

a. From Ex. 21-20, $P_H = 1061.88$ W and $P_L = 138.12$ W

$j\mathbf{Q}_t = \sqrt{3}(P_H - P_L) = \sqrt{3}(1061.88 - 138.12)$
$= \mathbf{1600\ vars}$ (21-18)

$\mathbf{S}_t^* = \mathbf{P}_t + j\mathbf{Q}_t = 1200 + j1600$
$= \mathbf{2000\underline{/53.13^\circ}\ VA}$ (21-19)

b. From Ex. 21-21, $P_H = 721.93$ W and $P_L = -237.\bar{3}$ W

$j\mathbf{Q}_t = \sqrt{3}(P_H - P_L) = \sqrt{3}(721.93 + 237.\bar{3})$
$= \mathbf{j1661.5\ vars}$ (21-18)

$\mathbf{S}_t^* = \mathbf{P}_t + j\mathbf{Q}_t = 484.6 + j1661.5$
$= \mathbf{1730.73\underline{/73.74^\circ}\ VA}$ (21-19)

With respect to Ex. 21-25, note that

1. A test of the validity of the calculations is that the phase angle, θ, is *recovered* in the calculation of complex power. This occurs because the complex power triangle is the impedance triangle multiplied by the square of the phase current. The two triangles are *similar* triangles!
2. When the lower-reading WM is negative, the total quadrature reactive power, $\mathbf{Q}_t$, is always higher than the total system power, $\mathbf{P}_t$.

21-14.6 Tests to Determine Whether a Lower-Reading Wattmeter is Positive or Negative

Because of the many interconnections in WMs with both current and potential transformers in the field or the laboratory, to measure high-voltage, high-current systems, it is extremely difficult to trace connections and know whether the WMs are properly connected to read upscale for higher PFs. Consequently, a lower WM may record a negative deflection, but we have no way of knowing whether this lower-reading WM is positive or negative. Two tests are given below to determine the *true* polarity of the *lower*-reading WM. The first is a laboratory method that may be used for smaller loads. The second is a field method suitable for use with larger loads.

A. *Laboratory method*

1. Open the line in which the *current* coil of the *higher*-reading WM is connected.
2. If the lower-reading WM is positive, it should be labeled (originally) as positive, since the WM is now properly recording single-phase power.
3. If the lower-reading WM reads downscale (negative), it should have been labeled (originally) as negative. Note: the original PF of the load might be below 0.5, causing a *negatively* reading WM.

The above test cannot be performed under circumstances where it is either impossible to open a line because continuity of service must be maintained or the nature of the load does not permit single phasing. In this case, the field method below may be performed. This test involves moving the WM potential coils (only) but with *both* WMs *originally reading upscale.*

B. *Field method*

1. Remove the potential lead of the *lower*-reading WM from the common line (which has no WM in series with it).
2. Connect this lead to the line containing the current coil of the *higher*-reading WM.
3. If the lower-reading WM reads *upscale*, record its power as *positive.* (Go to step 5.)
4. If the lower-reading WM reads *downscale*, record its power as *negative.*
5. Replace the potential lead in its original position (to meter the power of the lower-reading WM) in the common line that has no WM in series.

21-14.7 Measurement of Unbalanced 3ϕ, Three-Wire Loads by the Two-Wattmeter Method

It was noted earlier that the two-WM method may be used to measure the total power of unbalanced (as well as balanced) three-wire loads. It must be said at the outset, however, that the *only useful information* obtained from the WM readings is the *total power.* The information regarding phase sequence from the higher versus lower WM *cannot* be used. Similarly, the power factor information of Eqs. (21-16) and (21-17) is meaningless for unbalanced loads, since each load in each phase may be at a different power factor.

Example 21-26 shows that the two-WM method is capable of measuring the total power of an unbalanced 3ϕ, three-wire load, for which we previously calculated the current in each line.

EXAMPLE 21-26

From the solutions of Ex. 21-16 for the unbalanced Y-connected three-wire, 3ϕ system shown in Fig. 21-13, calculate the

a. Power dissipated by the 3-Ω resistor in line A
b. Power dissipated by the 3-Ω resistor in line B
c. Total 3ϕ power dissipated by the unbalanced Y-connected load

If the power is now metered by a WM in line A and line B, respectively, using the two-WM method, calculate the

d. Reading of the WM connected in line A
e. Reading of the WM connected in line B
f. Total power measured by the two WMs

Solution

a. $\mathbf{P_A} = \mathbf{I_A^2 R_A} = (26.4)^2 \times 3 = \mathbf{2090.88\ W}$
b. $\mathbf{P_B} = \mathbf{I_B^2 R_B} = (8.011)^2 \times 3 = \mathbf{192.53\ W}$
c. $\mathbf{P_t} = \mathbf{P_A} + \mathbf{P_B} = 2090.88 + 192.53 = \mathbf{2283.4\ W}$

The voltage across and current in each WM coil from Ex. 21-16 and Fig. 21-13b are

d. $\mathbf{P}_1 = \mathbf{V}_{AC} \times \mathbf{I}_A \cos\phi$
$= (100\underline{/-60^\circ})(26.4\underline{/-7.75^\circ})\cos(-52.25^\circ)$
$= \mathbf{1616.25\ W}$

e. $\mathbf{P}_2 = \mathbf{V}_{BC} \times \mathbf{I}_B \cos\phi$
$= (100/-120^\circ)(8.011\underline{/-153.62^\circ})\cos(-33.62^\circ)$
$= \mathbf{667.1\ W}$

f. $\mathbf{P}_t = \mathbf{P}_1 + \mathbf{P}_2 = 1616.25 + 667.1 = \mathbf{2283.4\ W}$ (21-13)

Several important insights can be drawn from Ex. 21-26:

1. The readings of the WMs, P_1 and P_2, respectively, are *not* the same as the power dissipated in phase A or phase B (i.e., the WMs are *not* reading the power per phase). The reason is fairly obvious because the WMs are metering the line (and not the phase) voltages.
2. But the total power measured by the two-WM method is exactly the same as the sum of the powers dissipated in each phase! This verifies that the two-WM method is capable of measuring *total power* of **unbalanced** loads as well as balanced loads.
3. Note that in step (**d**) of the solution, we used $\mathbf{V}_{AC}$ and not $\mathbf{V}_{CA}$. We must use $\mathbf{V}_{AC}$ because this is the voltage impressed on the WM potential coil located in line A based on the potential coil connection.
4. Since the power factors of phases A and B are unity, while that of C is zero, calculation of power factor from the WM readings is a *meaningless* exercise.
5. Similarly, since the load is *unbalanced*, predicting phase sequence from the WM readings should *not* be attempted.
6. Finally, the two WM readings should *not* be used to obtain total system kvars either, since it also leads to errors. The two WM readings may only be used to verify *total* power of an *unbalanced* Y-connected (or Δ-connected) load.[15]
7. From the foregoing, we may conclude that the greatest *advantages* of the two-WM method of measurement occur when the loads are *balanced*.

21-15 GLOSSARY OF TERMS USED

Balanced polyphase system Load circuit in which *all* phases are alike and symmetrical.

Delta (or Δ) connection 3ϕ circuit that is mesh-connected, having three lines.

Line current Current in the lines between a source and a load.

Line impedance Impedance of the transmission line between source and load.

Line voltage Voltage between lines of a polyphase system. Unless otherwise indicated, *all given voltages are assumed to be line voltages.*

Mesh connection Closed path connection consisting of series-connected branches forming a closed loop. In a 3ϕ mesh, the system is called a *delta* (Δ) connection.

Neutral point Point that has the same potential as the point of junction of a group of *equal* nonreactive resistances whose free ends are connected to appropriate lines of a system.

Phase angle Relation between phasors of phase voltage and phase current in any ac circuit or system, measured in degrees or radians.

Phase current Current in any branch of a polyphase system connected in either a mesh or a star.

Phase impedance Impedance of any branch of a polyphase system, connected in either a mesh or a star.

Phase sequence Order in which successive phase (or line) voltages reach their respective maximum values.

Phase voltage Voltage across any individual branch of a polyphase system connected in either a star or a mesh.

Star connection Polyphase circuit in which each branch of the circuit extends from a respective line terminal and *all* branches meet at a *common* terminal or *neutral* point.

Three-phase system Combination of branches connected in delta (Δ), wye (Y), or both, energized by ac voltages differing in phase by one-third of a cycle or 120°.

Unbalanced polyphase system Polyphase system having branches whose circuits are electrically different compared to a common reference.

Wattmeter (WM) Instrument whose deflection is proportional to the product of voltage across its potential coil and current in its current coil times the cosine of the phase angle between the two. A WM can measure power in dc or ac, single-phase or polyphase circuits.

Wye (or Y) connection Star connection having three branches connected to a common or neutral at one end and whose free ends are each connected to a different line voltage of a 3ϕ system.

[15] It may be of some interest to note that of the three possible WM line locations, one set of locations *always* yields the *correct* reactive power, using $\sqrt{3}(P_H - P_L)$. But this is risky at best, since phase sequence affects the correct location, as well.

21-16 PROBLEMS

Secs. 21-3 and 21-4

21-1 Two coils of a two-phase alternator are designated as $\mathbf{E}_{aA} = 100\underline{/0^\circ}$ V and $\mathbf{E}_{bB} = 100\underline{/90^\circ}$ V. Calculate the magnitudes and phase angles for the following voltages obtained by connecting the two coils in series to yield

a. $\mathbf{E}_{aB}$ c. $\mathbf{E}_{Ab}$
b. $\mathbf{E}_{AB}$ d. $\mathbf{E}_{ab}$

21-2 Three polyphase currents meet at and enter junction x. If $\mathbf{I}_{ax} = 5\underline{/0^\circ}$ A and $\mathbf{I}_{bx} = 5\underline{/-120^\circ}$ A,

a. Calculate current $\mathbf{I}_{cx}$
b. Draw a phasor diagram verifying your solution by showing that your answer in **(a)** satisfies Kirchhoff's current law (KCL)

21-3 The three phase voltages generated in an alternator are $\mathbf{E}_{nA} = 100\underline{/90^\circ}$ V, $\mathbf{E}_{nB} = 100\underline{/-30^\circ}$ V, and $\mathbf{E}_{nC} = 100\underline{/-150^\circ}$ V. Calculate the following line voltages, assuming the alternator is wye-connected:

a. $\mathbf{V}_{AB}$
b. $\mathbf{V}_{BC}$
c. $\mathbf{V}_{CA}$

21-4 The three balanced *phase* currents of a delta-connected alternator are $\mathbf{I}_{CA} = 10\underline{/90^\circ}$ kA, $\mathbf{I}_{AB} = 10\underline{/-30^\circ}$ kA, and $\mathbf{I}_{BC} = 10\underline{/-150^\circ}$ kA. Calculate the line currents leaving the alternator, respectively,

a. $\mathbf{I}_{LA}$
b. $\mathbf{I}_{LB}$
c. $\mathbf{I}_{LC}$

Hint: use KCL for currents leaving and entering each node.

21-5 A 3ϕ, Y-connected alternator with common designated as o has the following phase voltages: $\mathbf{V}_{oA} = 220 + j0$ V, $\mathbf{V}_{oB} = -110 - j190.53$ V, and $\mathbf{V}_{oC} = -110 + j190.53$ V; calculate

a. Line voltage $\mathbf{V}_{AB}$, using double-subscript notation
b. Line voltage $\mathbf{V}_{AB}$, using phasor diagrams for the phase voltages
c. The relation between $\mathbf{V}_{AB}$ and $\mathbf{V}_{oB}$ in terms of magnitude and phase angle

Secs. 21-5 through 21-7

21-6 An 8660 V, 60 Hz balanced 3ϕ, three-wire system has a Y-connected load consisting of 30 + j40 Ω per phase. Calculate the

a. Phase impedance, $\mathbf{Z}_p$
b. Phase voltage, $\mathbf{V}_p = \mathbf{V}_{Ao}\underline{/0^\circ}$
c. Phase and line current, $\mathbf{I}_p$ and $\mathbf{I}_L$
d. Phase angle between $\mathbf{V}_p$ and $\mathbf{I}_p$
e. Draw the complete phasor diagram, showing all three phase voltages and phase currents. Show line voltages using an ABC phase sequence and $\mathbf{V}_{Ao}$ as reference

21-7 Three coils having a **Q** of 1.5 and a resistance of 8 Ω are connected in Y across a 280 V, 3ϕ supply. Calculate all parts of Prob. 21-6.

21-8 Three loads, each having a resistance of 12 Ω and a capacitive series-connected reactance of 5 Ω, are connected in Y across a 440 V, 3ϕ supply. Calculate all parts of Prob. 21-6.

21-9 From the given and calculated data of Prob. 21-6, calculate the

a. Power per phase, P_p
b. Total power by three methods
c. Total quadrature power, $\mathbf{Q}_t$
d. Total complex power, $\mathbf{S}_t^*$ by three methods

21-10 From the given and calculated data of Prob. 21-7, calculate the

a. Power per phase, P_p
b. Total power by three methods
c. Total quadrature power, $\mathbf{Q}_t$
d. Total complex power, $\mathbf{S}_t^*$, by three methods

21-11 From the given and calculated data of Prob. 21-8, calculate the

a. Power per phase, P_p
b. Total power by three methods
c. Total quadrature power, $\mathbf{Q}_t$
d. Total complex power, $\mathbf{S}_t^*$, by three methods

21-12 A balanced 208 V, 3ϕ, three-wire, Y-connected load dissipates a total power of 9 kW at a system PF of 0.8 leading. Calculate the

a. Power per phase, $\mathbf{P}_p$
b. Phase voltage, $\mathbf{V}_p$
c. Phase current, $\mathbf{I}_p$
d. Phase impedance, $\mathbf{Z}_p$
e. Resistance per phase, R_p
f. Series-connected capacitance per phase if the frequency is 60 Hz

21-13 Each phase of a 3ϕ delta-connected load consists of a resistance of 8 Ω in series with a 6-Ω capacitive reactance. All three *line* currents are 30 A. Calculate the

a. Phase impedance, $\mathbf{Z}_p$
b. Phase current, $\mathbf{I}_p$
c. Phase and line voltage, $\mathbf{V}_L$
d. Power factor, F_p, by two different methods
e. Total power by three different methods

21-14 A balanced 3ϕ, Δ-connected load consists of three identical iron-core coils across a 208-V supply. The total power drawn is 1500 W and the line currents are 15 A. Calculate the

a. Resistance of each coil
b. Impedance of each coil
c. Reactance of each coil
d. Load power factor
e. Total complex power, $\mathbf{S}_t^*$

21-15 A 3ϕ, Δ-connected balanced load across a 440-V supply draws a line current of 50 A at a power factor of 0.5 leading. Calculate the

a. Impedance per phase, $\mathbf{Z}_p$
b. Resistance per phase
c. Reactance per phase, $-j\mathbf{X}_p$
d. Total power drawn from the supply, $\mathbf{P}_t$
e. Total complex power, $\mathbf{S}_t^*$, by three methods

21-16 A delta-connected induction motor draws 200 A at 0.85 PF lagging from a 3ϕ 60 Hz, 440 V supply. Calculate the

a. Total system power drawn, $\mathbf{P}_t$
b. Power per phase, $\mathbf{P}_p$
c. Phase current, $\mathbf{I}_p$
d. Impedance per phase, $\mathbf{Z}_p$
e. Resistance per phase, $\mathbf{R}_p$
f. Equivalent series-connected reactance per phase, $j\mathbf{X}_{L_p}$
g. Total complex power, $\mathbf{S}_t^*$, by three methods

21-17 The *line* voltages of a 3ϕ Y-connected alternator are $\mathbf{V}_{AB} = 208\angle 0°$ V, $\mathbf{V}_{BC} = 208\angle -120°$ V, and $\mathbf{V}_{CA} = 208\angle 120°$ V. The alternator is connected to a balanced 3ϕ delta-connected load having an impedance per phase of $20 + j15\ \Omega$. Calculate the

a. Phase currents $\mathbf{I}_{AB}$, $\mathbf{I}_{BC}$, and $\mathbf{I}_{CA}$, respectively, in load
b. Line currents $\mathbf{I}_A$, $\mathbf{I}_B$, and $\mathbf{I}_C$ between alternator and load
c. Phase currents $\mathbf{I}_{nA}$, $\mathbf{I}_{nB}$, and $\mathbf{I}_{nC}$ of Y-connected alternator

Draw a phasor diagram showing all line voltages, line currents, and phase currents. Using the phasor diagram, explain the

d. Relation between phase current and line current magnitudes
e. Phase relation between the line currents with respect to their component phase currents

Sec. 21-8

21-18 Repeat all parts of Prob. 21-17, given the line voltages of a 3ϕ Y-connected alternator as $\mathbf{V}_{AB} = 440\angle -120°$ V, $\mathbf{V}_{BC} = 440\angle 0°$ V, and $\mathbf{V}_{CA} = 440\angle 120°$ V. The 3ϕ balanced delta load has an impedance per phase of $32 - j24\ \Omega$.

f. After doing all parts (**a**) through (**e**), determine the phase sequence of the alternator from phase currents and line voltages, respectively
g. Explain why a reversal in phase sequence does not affect the magnitude and/or phase relation between phase and line currents in a balanced Δ-connected load

21-19 Determine the phase sequence for a 3ϕ Y-connected alternator whose phase voltages are $\mathbf{V}_{An} = 230\angle 90°$ V, $\mathbf{V}_{Bn} = 230\angle 210°$ V, and $\mathbf{V}_{Cn} = 230\angle -30°$ V

21-20 Repeat Prob. 21-19, given only $\mathbf{V}_{An} = 230\angle 15°$ V and $\mathbf{V}_{Cn} = 230\angle 135°$ V

21-21 Given a 3ϕ, Y-connected alternator having an ABC phase sequence. One phase voltage is $\mathbf{V}_{An} = 230\angle 90°$ V. Find

a. The remaining phase voltages
b. Line voltages $\mathbf{V}_{AB}$, $\mathbf{V}_{BC}$, and $\mathbf{V}_{CA}$

21-22 Given line voltage $\mathbf{V}_{AB} = 208\angle -15°$ V and phase sequence ABC in a Y-connected alternator, find the

a. Remaining line voltages
b. Phase voltages $\mathbf{V}_{An}$, $\mathbf{V}_{Bn}$, and $\mathbf{V}_{Cn}$

21-23 Given line voltage $\mathbf{V}_{BC} = 208\angle 15°$ V and phase sequence CBA in a Y-connected alternator, repeat Prob. 21-22.

21-24 The line voltages of a 3ϕ supply are 220 V with phase sequence ABC. A combinational balanced load is connected across the lines. The first is a Δ-connected load having a phase impedance of $60 + j80\ \Omega$. The second is a Y-connected load having a phase impedance of $30 - j45\ \Omega$. Calculate

a. Phase current $\mathbf{I}'_A$ due to balanced Y-connected load only (Hint: draw $\mathbf{V}_{Ao}$ as reference voltage and phasor diagram showing all line voltages in ABC sequence)
b. Phase current $\mathbf{I}_{AB}$ drawn by the balanced Δ-connected load only. (Hint: use $\mathbf{V}_{AB}$ from the diagram drawn in **a.** $\mathbf{V}_{AB} = 220\angle 30°$ V)
c. Line current $\mathbf{I}''_A$ due to balanced Δ-connected load only (Hint: $\mathbf{I}_L$ *lags* $\mathbf{I}_p$ by 30° for the ABC sequence, as shown in Fig. 21-8)
d. Total system line current in lines $\mathbf{I}_A$, $\mathbf{I}_B$, and $\mathbf{I}_C$
e. Total system power drawn from supply, P_t (Hint: assume a Y-connected source whose phase voltage is $\mathbf{V}_{Ao} = 127\angle 0°$ V, as in **a** above)
f. Total system power factor, $\mathbf{F}_p$

21-25 Given the calculated phase currents of Prob. 21-24 and given phase impedances, using the simplified complex power method for the combinational load, calculate the

a. 3ϕ power dissipated in the Y-connected load, P_Y
b. Quadrature 3ϕ reactive power drawn by the Y-connected load, $\mathbf{Q}_Y$
c. 3ϕ power dissipated in the Δ-connected load, $\mathbf{P}_\Delta$
d. Quadrature 3ϕ reactive power drawn by the Δ-connected load, $\mathbf{Q}_\Delta$
e. Total system complex power, $\mathbf{S}_t^*$, in rectangular and polar form
f. Total system line current in lines $\mathbf{I}_A$, $\mathbf{I}_B$, and $\mathbf{I}_C$, using source phase voltage $\mathbf{V}_{Ao} = 127\angle 0°$ V as a reference. (Hint: use $\mathbf{S}_t$ and not $\mathbf{S}_t^*$)
g. Total system power factor, F_p
h. Total system power drawn from the supply

21-26 A 3ϕ, 220 V, 25 hp squirrel cage induction motor has an 85% efficiency and a power factor of 0.75 when connected across the 220 V, 3ϕ supply. Calculate the

a. Magnitude and phase of line currents drawn by the motor, using the ABC phase sequence
b. Complex power drawn from the supply in polar and rectangular form

(Hint: use output horsepower rating and efficiency to find input power in watts)

21-27 Repeat both parts of Prob. 21-26 for a 3ϕ, 440 V, 50 hp motor having an efficiency of 0.9 and a full-load power factor of 0.8, lagging.

21-28 A 220 V, 3ϕ, 20 kW (unity PF) electric oven is connected in combination with the motor of Prob. 21-26 across 220 V, 3ϕ lines. Calculate the

a. Total complex power drawn from the supply
b. Total line current magnitude and phase
c. Overall power factor of the combination

21-29 A 440 V, 3ϕ, 10 kW (unity PF) electric oven is connected in combination with the motor of Prob. 21-27 across 440 V, 3ϕ lines. Calculate all parts of Prob. 21-28.

Sec. 21-9

21-30 Repeat *all* parts of Ex. 21-13, using a CBA phase sequence (Hint: $\mathbf{V}_{BC}$ and $\mathbf{V}_{CA}$ are $200\angle 120°$ V and $200\angle -120°$ V, respectively)
When you complete your calculations, draw conclusions regarding changes in the

1. Magnitude of the phase currents
2. Magnitude and phase of the line currents
3. Phasor sum of the line currents
4. Complex power dissipated in each load
5. Total complex power supplied by the 3ϕ service

21-31 An unbalanced 3ϕ delta-connected load has the following *phase* impedances: $\mathbf{Z}_{AB} = 40\angle 0°\ \Omega$, $\mathbf{Z}_{BC} = 40\angle 0°\ \Omega$, and $\mathbf{Z}_{CA} = 40\angle -90°\ \Omega$. If it is connected to a 2000 V 3ϕ supply having an ABC phase sequence, calculate the

a. Phase currents $\mathbf{I}_{AB}$, $\mathbf{I}_{BC}$, and $\mathbf{I}_{CA}$, using $\mathbf{V}_{AB}$ as a reference voltage
b. Line currents $\mathbf{I}_A$, $\mathbf{I}_B$, and $\mathbf{I}_C$
c. Verify (**b**) by showing that the phasor sum of all line currents is zero

21-32 Repeat all parts of Prob. 21-31, using a CBA phase sequence.

21-33 An industrial building has a balanced 150 kW, 0.7 PF lagging, Δ-connected load, as well as a single-phase, 100 kW, 0.9 PF lagging load connected across a 480 V, 3ϕ, three-wire service. If the single-phase load is connected across lines B and C and the phase sequence is ABC, calculate the

a. Phase current $\mathbf{I}_{AB}$, using $\mathbf{V}_{AB} = 480\angle 0°$ V as a reference
b. Line currents $\mathbf{I}'_{LA}$, $\mathbf{I}'_{LB}$, and $\mathbf{I}'_{LC}$ drawn by the Δ-connected load only
c. Line currents $\mathbf{I}''_{LB}$ and $\mathbf{I}''_{LC}$ drawn by the single-phase load only
d. Line currents $\mathbf{I}_A$, $\mathbf{I}_B$, and $\mathbf{I}_C$
e. Verify (**d**) by showing that the phasor sum of all line currents is zero

21-34 An unbalanced Δ-connected load has the following phase impedances: $\mathbf{Z}_{AB} = 50\angle 40°\ \Omega$, $\mathbf{Z}_{BC} = 20\angle 0°\ \Omega$, and $\mathbf{Z}_{CA} = 40\angle -60°\ \Omega$. If it is connected to a 3ϕ 400 V supply having a CBA phase sequence, calculate the

a. Phase currents $\mathbf{I}_{AB}$, $\mathbf{I}_{BC}$, and $\mathbf{I}_{CA}$, using $\mathbf{V}_{AB}$ as a reference voltage
b. Line currents $\mathbf{I}_A$, $\mathbf{I}_B$, and $\mathbf{I}_C$
c. Complex power dissipated in each phase in rectangular and polar form
d. Total complex power drawn from the supply

21-35 Repeat all parts of Prob. 21-34, using an ABC phase sequence. Then draw conclusions regarding the effect of phase sequence reversal on phase current magnitudes and angles, line current magnitudes and angles, power per phase, and total complex power.

Sec. 21-10

21-36 An unbalanced 3ϕ, three-wire, Y-connected system presents the following phase impedances to a 3ϕ 440 V line having a positive (ABC) phase sequence: $\mathbf{Z}_{Ao} = (60 + j30)\ \Omega$, $\mathbf{Z}_{Bo} = (30 - j90)\ \Omega$, $\mathbf{Z}_{Co} = (80 + j0)\ \Omega$. Using mesh analysis, calculate the

a. Mesh currents $\mathbf{I}_1$ and $\mathbf{I}_2$ in rectangular form
b. Line currents $\mathbf{I}_A$, $\mathbf{I}_B$, and $\mathbf{I}_C$ in rectangular and polar form
c. Phase voltages $\mathbf{V}_{Ao}$, $\mathbf{V}_{Bo}$, and $\mathbf{V}_{Co}$
d. Shift in neutral voltage, $\mathbf{V}_{No}$ (Hint: $\mathbf{V}_{No} = \mathbf{V}_{NA} + \mathbf{V}_{Ao}$)
e. Draw a phasor diagram showing all line and phase voltages and $\mathbf{V}_{No}$

21-37 Repeat all parts of Prob. 21-36, using a CBA phase sequence. When you have completed all parts, draw conclusions regarding the effect of a change in phase sequence on: (1) line (and identical phase) currents, (2) phase voltages, and (3) shift in neutral voltage.

21-38 Using the values of line currents obtained in Prob. 21-36, calculate the

a. Complex power dissipated in each phase impedance, $\mathbf{S}^*_A$, $\mathbf{S}^*_B$, and $\mathbf{S}^*_C$
b. Total 3ϕ complex power drawn by the entire unbalanced 3ϕ, three-wire system

21-39 Repeat both parts of Prob. 21-38 for line currents obtained in Prob. 21-37. When you have completed all parts, draw conclusions regarding the effect of a change in phase sequence on (1) complex power dissipated in each phase impedance and (2) total 3ϕ complex power drawn by the entire unbalanced system.

21-40 If a fourth neutral wire is connected to point "o" in the unbalanced system of Prob. 21-36, calculate the

a. Magnitude and phase of phase voltages $\mathbf{V}_{AN}$, $\mathbf{V}_{BN}$, and $\mathbf{V}_{CN}$
b. Line currents $\mathbf{I}_A$, $\mathbf{I}_B$, and $\mathbf{I}_C$
c. Neutral current, $\mathbf{I}_N$
d. Complex power dissipated in each phase, $\mathbf{S}^*_A$, $\mathbf{S}^*_B$, and $\mathbf{S}^*_C$
e. Total complex power drawn by the entire system, $\mathbf{S}^*_t$
f. Explain why the total complex power drawn is different from that in Probs. 21-38 and 21-39

21-41 A four-wire Y-connected system has $\mathbf{Z}_{Ao} = 100\angle 0°\ \Omega = \mathbf{Z}_{Bo}$, and $\mathbf{Z}_{Co} = 100\angle -90°\ \Omega$, connected to a 440 V, 3ϕ, four-wire source. Using $\mathbf{V}_{AB} = 440\angle 0°$ V as a reference and (positive) phase sequence ABC, calculate the

a. Currents in lines A, B, C, and N
b. Complex power drawn by each load
c. Total complex power drawn by the entire system

21-42 Repeat all parts of Prob. 21-41, using a negative phase sequence. Explain why the total complex power is the same as that in Prob. 21-41.

21-43 Repeat all parts of Prob. 21-41 with the neutral removed. Explain why the total complex power has changed from that in Probs. 21-41 and 21-42.

Sec. 21-11

21-44 Using the currents obtained in Prob. 21-43, calculate phase voltages $\mathbf{V}_{Ao}$ and $\mathbf{V}_{Bo}$. Assuming that the $100/\underline{0^\circ}$ Ω resistors in lines A and B are incandescent lamps, predict the phase sequence in Prob. 21-43. (Hint: see Table 21-2.) Explain why such a circuit could serve as a phase sequence indicator.

Sec. 21-12

21-45 The balanced 3ϕ, three-wire, Δ-connected load of Prob. 21-14 is measured by the one-WM method using a Y-box (Fig. 21-20b).

a. Show mathematically that the power recorded by the WM is $\mathbf{V}_p\mathbf{I}_p \cos\theta$ for a Δ-connected load. (Hint: $\mathbf{V}_p \equiv \mathbf{V}_L$ in a Δ-connected system and the WM is *not* recording $\mathbf{V}_p$!)
b. Assuming an ABC (positive) phase sequence and using $\mathbf{V}_p = \mathbf{V}_{AB} = 108/\underline{0^\circ}$ V as a reference, determine the magnitude and phase of the voltage applied to the voltage coil. (Hint: see Fig. 21-16a.)
c. Given the magnitude of the current in the current coil as 15 A, determine the phase angle of the current.
d. Calculate the power measured by the WM.
e. Calculate total power from (**d**). Compare your answer with that given in the statement of Prob. 21-14.

21-46 The balanced 3ϕ, four-wire Y-connected load of Exs. 21-3 and 21-4 is measured by the one-WM method using an SPDT switch (Fig. 21-20a). Find the

a. Magnitude and phase of current in the WM current coil and voltage across the WM potential coil with the switch in position 1
b. Repeat (**a**) with the switch in position 2
c. Power recorded by the WM in switch position 1
d. Power recorded by the WM in switch position 2
e. Total power from the two WM readings
f. Total power from $\mathbf{P}_t = \sqrt{3}\mathbf{V}_L\mathbf{I}_L \cos\theta$

Sec. 21-14

21-47 If power is measured by the two-WM method with WMs in lines A and C (Fig. 21-23a) for the circuit of Exs. 21-3 and 21-4, find the

a. Reading of the WM located in line A
b. Reading of the WM in line C
c. Total power from the two WM readings, $\mathbf{P}_t$
d. Compare the value of $\mathbf{P}_t$ with that found in Prob. 21-46

21-48 Given the readings of the two WMs in Prob. 21-47, find the

a. Phase sequence (PS), knowing that the load is inductive
b. Power factor angle and power factor
c. Total complex power of the 3ϕ system in rectangular and polar form

21-49 From the given and calculated data for the balanced 3ϕ load of Prob. 21-6, assuming positive phase sequence and WMs in lines A and C, find the

a. Reading of the higher-reading WM, $\mathbf{P}_H$
b. Reading of the lower-reading WM, $\mathbf{P}_L$
c. Total power, $\mathbf{P}_t$, from the WM readings (compare with Prob. 21-9b)
d. Phase sequence from the WM readings
e. Phase angle and power factor from the WM readings
f. Complex power from the WM readings

21-50 Repeat all parts of Prob. 21-49 for the balanced 3ϕ circuit of Prob. 21-8, assuming an ABC (positive) phase sequence.

21-51 Repeat all parts of Prob. 21-49 for the balanced 3ϕ circuit of Prob. 21-12, assuming a CBA (negative) phase sequence.

21-52 For the balanced Δ-connected load of Prob. 21-14, given $\theta = +73.88^\circ$, a positive phase sequence, and regardless of the lines in which the WMs are connected, repeat all parts of Prob. 21-49.

21-53 For the balanced Δ-connected load of Prob. 21-15, given $\theta = -60^\circ$, a positive phase sequence, and regardless of the lines in which WMs are connected, repeat all parts of Prob. 21-49.

21-54 For the Δ-connected induction motor of Prob. 21-16, assuming a positive phase sequence, repeat all parts of Prob. 21-49.

21-55 Power to a 3ϕ, balanced, Y-connected inductive load across 208-V lines is measured by the two-WM method. The WM in line 1 reads 5 kW and that in line 2 reads −3 kW. Calculate the

a. Power factor of the load
b. Complex power drawn from the lines
c. Phase and line currents drawn by the load
d. Load resistance and series reactance per phase
e. Phase sequence of the supply lines

21-56 A small (1 hp) 3ϕ, 220 V delta-connected squirrel-cage induction motor (SCIM) is connected across 220-V lines without mechanical load. The WM in line 1 reads 70 W and that in line 2 reads −20 W. The direction of rotation of the SCIM shaft (viewed from the shaft end) is counterclockwise. Calculate the

a. PF of the motor when unloaded
b. Complex power drawn from the lines
c. Motor line current
d. Phase sequence of the lines
e. Explain how the SCIM may be used as a phase-sequence indicator

21-57 Power to an inductive load connected across 208 V, 3ϕ lines is measured by the two-WM method. The WM in line 1 reads 2400 W and that in line 2 reads 1400 W. The original load is removed and a second (unknown) load is applied without changing WM line connections. The first WM is now −3600 W and the second reads +4800 W. Calculate the

a. Line current and PF of the original load (lagging or leading?)
b. Line current and PF of the second load (lagging or leading?)
c. Total current drawn from the supply with both loads connected and final PF

(Hint: find the complex power for each load)

21-58 Two WMs are properly connected in a balanced 3ϕ circuit to measure total power. One WM reading is one-quarter that of the other. Calculate the load PF when

a. The lower-reading WM is positive
b. The lower-reading WM is negative

21-59 A balanced 208 V, 3ϕ induction motor has a PF of 0.4 lagging when its power is measured by the two-WM method. The power of the *lower*-reading WM in line 2 is -5 kW. The phase sequence is positive. Calculate the

a. Circuit phase angle, θ (positive or negative?)
b. Line current drawn from the supply, $\mathbf{I}_L$
c. Reading of the higher WM in line 1
d. Total complex power drawn from the lines
e. Impedance per phase (if the motor is Δ-connected) in polar and rectangular form

21-60 A highly inductive Δ-connected balanced load having a PF of 0.2 lagging is connected across 220-V 3ϕ lines with power measured by the two-WM method. The lower-reading WM in line 2 reads -8 kW but the higher reading WM (with a maximum deflection of 16 kW) in line 1 is pinned upscale at maximum deflection. Calculate the

a. Reading of the higher WM, $\mathbf{P}_H$
b. Complex power consumed from the supply in polar and rectangular form
c. Impedance per phase of the load in polar and rectangular form

21-61 For the unbalanced Δ-connected load of Ex. 21-13, using WMs connected in lines A and B for measurement of total power (only) by the two-WM method, given a positive phase sequence, find the

a. Reading of the WM in line A
b. Reading of the WM in line B
c. Total (true) power of the unbalanced load from the two WM readings
d. Compare the true power obtained in (**c**) with that obtained in the solution using $\mathbf{I}_p^2\mathbf{R}_p$ for each phase
e. Is it possible to obtain correct total vars from the WM readings? Explain.

21-62 For the unbalanced Δ-connected load of Prob. 21-34, using WMs connected in lines A and B for measurement of total power (only) by the two-WM method, and given phase sequence CBA, repeat all parts of Prob. 21-61.

21-63 Given the unbalanced three-wire Y-connected load of Ex. 21-16, using WMs connected in lines B and C for measurement of total power (only) by the two-WM method, given a positive phase sequence, repeat all parts of Prob. 21-61.

21-64 Given the unbalanced three-wire, Y-connected system of Probs. 21-36 and 21-38, using WMs in lines A and C, repeat all parts of Prob. 21-61, given a positive phase sequence.

21-17 ANSWERS

21-1 a $141.4\angle 45°$ V b $141.4\angle 135°$ V c $141.4\angle -135°$ V d $141.4\angle -45°$ V
21-2 a $5\angle 120°$ A
21-3 a $173.2\angle -60°$ V b $173.2\angle 180°$ V c $173.2\angle 60°$ V
21-4 a $17.32\angle 120°$ kA b $17.32\angle 0°$ kA c $17.32\angle -120°$ kA
21-5 a $381\angle -150°$ V
21-6 a $50\angle 53.13°$ Ω b $5\angle 0°$ kV c $100\angle -53.13°$ A d $-53.13°$
21-7 a $14.42\angle 56.31°$ Ω b $120\angle 0°$ V c $8.32\bar{6}\angle -56.31°$ A d $-56.31°$
21-8 a $13\angle -22.62°$ Ω b $254\angle 0°$ V c $19.54\angle 22.62°$ A d $22.62°$
21-9 a 300 kW b 900 kW c $j1200$ kvars d $1.5\angle 53.13°$ MVA
21-10 a $554.\bar{6}$ W b 1664 W c $j2496$ vars d $3\angle 56.31°$ kVA
21-11 a 4.581 kW b 13.74 kW c $-j5.726$ kvar d $14.89\angle -22.62°$ kVA
21-12 a 3 kW b 120.1 V c 31.23 A d $3.845\angle -36.87°$ Ω e 3.076 Ω f 1150 μF
21-13 a $10\angle -36.87°$ Ω b 17.32 A c 173.2 V d 0.8 e 7.18 kW
21-14 a $6.\bar{6}$ Ω b 24.02 Ω c $j23.075$ Ω d 0.2776 e $5.404\angle 73.885°$ kVA
21-15 a $15.242\angle -60°$ Ω b 7.621 Ω c $-j13.2$ Ω d 19.05 kW e $38.105\angle -60°$ kVA
21-16 a 129.56 kW b 43.186 kW c 115.47 A d 3.8105 Ω e 3.239 Ω f $j2.01$ Ω g $152.4\angle 31.8$ kVA
21-17 a $8.32\angle -36.87°$, $\angle -156.87°$, $\angle 83.13°$ A b $14.41\angle -66.87°$, $\angle 173.13°$, $\angle 53.13°$ A c same as **b** d $\sqrt{3}$ e 30°
21-18 a $11\angle -83.13°$, $\angle 36.87°$, $\angle 156.87°$ A b $19.05\angle -53.13°$, $\angle 66.87°$, $\angle -173.13°$ A c same as **b** d $\sqrt{3}$ e 30° f CBA
21-19 CBA(−)
21-20 ABC(+)
21-21 a $230\angle -30°$, $230\angle -150°$ V b $398\angle 120°$, $398\angle 0°$, $398\angle -120°$ V
21-22 a $208\angle -135°$, $\angle 105°$ V b $120\angle -45°$, $\angle -165°$, $\angle 75°$ V
21-23 a $208\angle 135°$, $\angle -105°$ V b $120\angle -75°$, $\angle 45°$, $\angle 165°$ V
21-24 a $2.348\angle 56.31°$ A b $2.2\angle -23.13°$ A c $3.8105\angle -53.13°$ A d $3.752\angle -17°$, $\angle -137°$, $\angle 103°$ A

e 1367 W f 0.956 lagging

21-25 a 496.2 W b $-j744.3$ var c 871.2 W
d $j1161.6$ var e $1.43\angle 16.97^\circ$ kVA f, g, h same

21-26 a $76.77\angle -41.4^\circ$, $\angle -161.4^\circ$, $\angle 78.6^\circ$ A
b $21.94 + j19.35$ kVA

21-27 a $68\angle -36.9^\circ$, $\angle -156.9^\circ$, $\angle 83.1^\circ$ A
b $41.44 + j31.1$ kVA

21-28 a $46.19\angle 24.77^\circ$ kVA b $121.2\angle -24.77^\circ$ A
c 0.908 lagging

21-29 a $60.11\angle 31.16^\circ$ kVA b $78.87\angle -31.16^\circ$ A
c 0.8557 lagging

21-30 a $0.5155\angle 58.9^\circ$, $\angle 88.9^\circ$, $\angle -61.1^\circ$ A
b $0.8929\angle 88.9^\circ$, $0.2585\angle -176.9^\circ$, $0.911\angle -74.67^\circ$ A
c $0.67\angle 38.56^\circ$ A d 0 e same
f $223.24\angle -36.38^\circ$ VA (same)

21-31 a $50\angle 0^\circ$, $\angle -120^\circ$, $\angle -150^\circ$ A
b $96.6\angle 15^\circ$ A; $86.6\angle -150^\circ$; $25.88\angle 135^\circ$ A c 0

21-32 a $50\angle 0^\circ$, $\angle 120^\circ$, $\angle -30^\circ$ A
b $25.88\angle 75^\circ$, $86.6\angle 150^\circ$, $96.6\angle -45^\circ$ A
c 0

21-33 a $148.81\angle -45.57^\circ$ A
b $257.75\angle -75.6^\circ$, $\angle -195.6^\circ$, $\angle 44.4^\circ$ A
c $231.5\angle -145.8^\circ$, $\angle 34.2^\circ$ A
d $257.75\angle -75.6^\circ$, $444\angle -172.1^\circ$, $487.3\angle 39.57^\circ$ A e 0

21-34 a $8\angle -40^\circ$, $20\angle 120^\circ$, $10\angle -60^\circ$ A
b $3.6945\angle 72.22^\circ$, $27.65\angle 125.7^\circ$, $30\angle -60^\circ$ A
c $3.2\angle 40^\circ$, $8\angle 0^\circ$, $4\angle -60^\circ$ kVA d $12.53\angle -6.45^\circ$ kVA

21-35 a $8\angle -40^\circ$, $20\angle -120^\circ$, $10\angle 180^\circ$ A
b $16.93\angle -17.68^\circ$, $20.21\angle -142.9^\circ$ A, $17.32\angle 90^\circ$ A
c, d same

21-36 a $1.6934 - j1.32\bar{3}$ A, $0.9837 - j4.4056$ A
b $2.149\angle -38^\circ$, $3.163\angle -103^\circ$, $4.514\angle 102.6^\circ$ A
c $144.2\angle -11.435^\circ$, $300\angle -174.6^\circ$, $361.1\angle 102.6^\circ$ V
d $126\angle 128.7^\circ$ V

21-37 a $4.334 + j2.73$; $0.5231 + j1.0906$ A
b $5.122\angle 32.2^\circ$, $4.1485\angle -156.7^\circ$, $1.2096\angle -115.6^\circ$ A
c $343.6\angle 58.8^\circ$, $393.6\angle -228.3^\circ$, $96.77\angle -115.6^\circ$ V
d $172.1\angle 104.1^\circ$ V

21-38 a $309.8\angle 26.56^\circ$, $948.7\angle -71.565^\circ$, $1630.1\angle 0^\circ$ VA
b $2335\angle -19^\circ$ VA

21-39 a $1760\angle 26.565^\circ$, $1632.7\angle -71.565^\circ$, $117.05\angle 0^\circ$ VA
b $2235\angle -19^\circ$ VA

21-40 a $254\angle -30^\circ$, $\angle -150^\circ$, $\angle 90^\circ$ V
b $3.79\angle -56.565^\circ$, $2.68\angle -78.435^\circ$, $3.175\angle 90^\circ$ A
c $3.7\angle -44.84^\circ$ A
d $860.3 + j430$; $215 - j645$; $806.5 + j0$ VA
e $1894\angle -6.51^\circ$ VA

21-41 a $2.54\angle -30^\circ$, $\angle 150^\circ$, $\angle 180^\circ$ A; $3.592\angle -135^\circ$ A
b $645.2\angle 0^\circ$, $\angle 0^\circ$, $\angle -90^\circ$ VA
c $1443\angle -26.57^\circ$ VA

21-42 a $2.54\angle 30^\circ$, $\angle 150^\circ$, $\angle 0^\circ$ A; $3.592\angle 45^\circ$ A b, c same

21-43 a $3.8014\angle -11.565^\circ$ A, $1.019\angle -131.58^\circ$ A,
$3.408\angle 153.43^\circ$ A
b $1445.1\angle 0^\circ$, $\angle 0^\circ$ VA, $1161.45\angle -90^\circ$ VA
c $3114.8\angle -21.89^\circ$ VA

21-44 CAB(+)

21-45 b $120\angle -30^\circ$ V c $15\angle -108.9^\circ$ A d 500 W
e 1.5 kW

21-46 a $12\angle -53.13^\circ$ A, $208\angle 30^\circ$ V
b $12\angle -53.13^\circ$ A, $208\angle -30^\circ$ V
c 298.56 W d 2295.36 W e, f 2593.9 W

21-47 a 298.56 W b 2295.36 W c 2593.9 W d same

21-48 a CAB(+) b 53.13° c 0.6 d $4323.2\angle 53.13^\circ$ VA

21-49 a 796.4 kW b 103.6 kW c 900 kW d (+)
e 53.13° f $1500\angle 53.13^\circ$ kVA

21-50 a 8.526 kW b 5.22 kW c 13.75 kW d (+)
e −22.62°, 0.923 f $14.9\angle -22.62^\circ$ kVA

21-51 a 6.45 kW b 2.55 kW c 9 kW d (−)
e −36.87°, 0.8 f $11.25\angle -36.87^\circ$ kVA

21-52 a 2.25 kW b −0.75 kW c 1.5 kW d (+)
e 73.88° f $5.404\angle 73.88^\circ$ kVA

21-53 a 19.05 kW b 0 c 19.05 kW d (+)
e −60°, 0.5 f $38.105\angle -60^\circ$ kVA

21-54 a 88 kW b 41.6 kW c 129.6 kW d (+)
e 31.79°, 0.85 f $152.42\angle 31.79^\circ$ kVA

21-55 a 0.1429 lagging b $14\angle 81.79^\circ$ kVA
c 38.86 A d 0.4413 Ω, $j3.058$ Ω e 123(+)

21-56 a 0.3054 lagging b $163.7\angle 72.2^\circ$ VA
c 0.43 A d 123(+)

21-57 a 11.59 A, 0.91 lagging b 40.52 A, 0.0822 leading
c 38.19 A, 0.3634 leading

21-58 a 0.6934 b 0.3273

21-59 a +66.42° b 214.9 A c 36 kW
d $77.46\angle 66.42^\circ$ kVA e $1.676\angle 66.42^\circ$ Ω

21-60 a 16.75 kW b $43.75\angle 78.46^\circ$ kVA
c $3.3185\angle 78.46^\circ$ Ω

21-61 a 3.43 W b 176.24 W c 179.7 W

21-62 a 1.44 kW b 11.005 kW c 12.45 kW d same
e no

21-63 a 717.7 W b 1565.4 W c 2283.1 W d same
e no

21-64 a 745.1 W b 1462 W c, d 2207.1 W e no

CHAPTER 22

Nonsinusoidal Waveforms

22-1 INTRODUCTION

Nonsinusoidal waveforms were first introduced in Chapter 14. In Sec. 14-8, we determined the average value of some common nonsinusoidal waveforms. In Sec. 14-10, we determined the RMS value of these nonsinusoidal waveforms. The average and RMS values were summarized in Table 14-1. With the exception of the purely sinusoidal waveform shown in the *first* **row** of Table 14-1, all of the waveforms appearing in Table 14-1 may be classified as *nonsinusoidal* or *complex* waveforms. For simplicity, we may define any waveform that is *not* purely sinusoidal as a *nonsinusoidal* or *complex* waveform.

Our interest in nonsinusoidal waveforms stems from many sources. The full-wave and half-wave rectifier outputs shown in Table 14-1, as well as the SCR outputs of Table 14-2, are examples of common nonsinusoidal waveforms derived from a purely sinusoidal source by nonlinear semiconductors such as diodes and silicon-controlled rectifiers (SCRs). Most testing laboratories use *function* generators, which generate *square*, *triangular*, and *sawtooth* waveforms (or functions) used in testing, for example, the frequency response of amplifiers. It is of some interest to note that the (so-called) sinusoidal waveform produced by each phase of an alternator (shown in Fig. 21-2b) is actually the sum of a series of flat-topped rectangular waveforms, generated in the coil windings, and is *never truly* a *purely* sinusoidal waveform. Devices such as transformers and transistors, assumed to be linear, are in reality nonlinear devices that can distort the "pure" ac sinusoid by acting as "harmonic generators." Distortion, whether due to saturation or clipping, is reduced by a study of its causes and the ways to eliminate or minimize unwanted harmonic generation.

One major concern in Chapter 14 was the evaluation of average and RMS values of *nonsinusoidal* periodic waveforms. In this chapter, we will consider nonsinusoidal waveforms from two opposite yet related viewpoints. We will begin with the *synthesis* of nonsinusoidal waveforms. In so doing, we will discover that *all* periodic complex waveforms are composed of a series of *harmonically related pure sine waves*. The lowest frequency or *fundamental* sine wave determines the fundamental frequency, f, and the period, T, of the complex wave. The remaining sinusoids are *whole-number* multiples of the fundamental frequency, having a fixed magnitude and phase relation to the fundamental. These remaining sinusoids are called *harmonics*. Thus, a sinusoid whose frequency is *twice* its fundamental is called a *second* harmonic or *even* harmonic. A sinusoid whose frequency is *three times* that of the fundamental is called a *third* (or *odd*) harmonic. We will prove that the presence (or absence) of odd and even harmonics, as well as their magnitude and phase with respect to the fundamental sinusoid, produces the waveform which is characteristic of the complex wave. We will discover for example that a square or rectangular waveform is composed only of *odd* harmonics, of decreasing amplitudes, all *in phase* with the *fundamental*. Similarly, a sawtooth waveform is composed of both *odd* (in-phase) harmonics and *even* (out-of-phase) harmonics of decreasing amplitudes in a specific arithmetic progression.

As we develop the "synthesis" of certain waveforms, we will study the various "symmetries" they exhibit. In doing this, we will develop a set of "rules" that enable us to predict the nature of the component harmonic frequencies in any periodic complex wave. This technique, called harmonic wave *synthesis* (in its most perfect form), enables us to determine the relative amplitudes and phase of all harmonics present in comparison to the fundamental. For our purposes, therefore, *synthesis* is a prerequisite to *analysis*. Historically, however, the reverse was the case. The complete *mathematical analysis* of a periodic complex waveform was first published by Baron Jean Baptiste Fourier in 1826, when electricity was in its infancy. Fourier showed that any periodic waveform may be represented as the *sum* of a *trigonometric series* of *harmonically related* sinusoids. This insight is of such pervasive importance to the subject of this chapter that our study of nonsinusoidal waveforms begins with it.

22-2 THE FOURIER SERIES

Fourier showed that *any* periodic function, $f(t)$ may be expressed as[1]

$$f(t) = A_0 + \sum_{n=1}^{\infty} (A_n \sin n\omega_1 t + B_n \cos n\omega_1 t) \qquad (22\text{-}1)$$

where A_0 is the average (dc) value of the waveform

$\omega_1 = 2\pi f_1 = 2\pi/T$ is the fundamental radian frequency (14-5a)

$n\omega_1 t = 2\pi(nf_1 t) = 2\pi nt/T$, where n is the integer sequence 1, 2, 3 etc.

Equation (22-1) may be expanded to the more familiar form:

$$\begin{aligned} f(t) = A_0 &+ A_1 \sin \omega_1 t + A_2 \sin 2\omega_1 t + \cdots + A_n \sin n\omega_1 t + \cdots \\ &+ B_1 \cos \omega_1 t + B_2 \cos 2\omega_1 t + \cdots + B_n \cos n\omega_1 t + \cdots \end{aligned} \qquad (22\text{-}2)$$

Equation (22-2) may now be simplified by combining sine and cosine terms of the same frequency, yielding

[1] Equation (22-1) in the form shown conforms to the conventions established in Sec. 14-6, Fig. 14-10, using the *sine* function as a *reference*. When the *cosine* function is used as a reference, the A_n and B_n magnitudes are *reversed*! Fourier developed his series to study the rate of heat flow in a metal rod. But his insights are currently applicable to problems involving fluid flow, mechanical vibrations, and musical tones, in addition to analysis of periodic electrical waveforms.

$$f(t) = A_0 + C_1 \sin(\omega_1 t + \theta_1) + C_2 \sin(2\omega_1 t + \theta_2) + \cdots + C_n \sin(n\omega_1 t + \theta_n) \qquad (22\text{-}3)$$

where $C = \sqrt{A^2 + B^2}$ and θ is the arc $\tan(B/A)$ at *each harmonic frequency*, respectively, or the phase angle of the harmonic with respect to the fundamental.

Equation (22-2) is sometimes called the "sine–cosine form" of the Fourier series. A number of computer programs use this form to find the Fourier series for such common functions as rectangular pulse waveforms and triangular and sawtooth waveforms.

Equation (22-3) is sometimes called the "amplitude–phase form," since it yields a single amplitude and its phase at a given selective frequency. Most harmonic wave analyzer or spectrum analyzers, used for measuring the magnitudes of the harmonics present, yield the components in the form shown in Eq. (22-3).

EXAMPLE 22-1

Given the following trigonometric Fourier series,

a. Express in amplitude–phase form the equation

$$v(t) = 20 + 24 \cos 1000t - 7 \sin 1000t + 8 \sin 2000t - 15 \cos 2000t \text{ volts}$$

b. Express in sine–cosine form the equation

$$i(t) = 40 - 5 \sin(200t - 20^\circ) + 3 \sin(400t + 40^\circ) + 2 \sin(600t - 36^\circ) \text{ amperes}$$

Solution

a. $C_1 = -7 + j24 = 25\angle 106.3^\circ$ and
$C_2 = 8 - j15 = 17\angle -61.93^\circ$;
$v(t) = 20 + 25 \sin(1000t + 106.3^\circ) + 17 \sin(2000t - 61.93^\circ)$ volts (**Ans.**)

b. $C_1 = -5\angle -20^\circ = -4.698 + j1.71$,
$C_2 = 3\angle 40^\circ = 2.298 + j1.928$, and
$C_3 = 2\angle -36^\circ = 1.618 - j1.176$;
$i(t) = 40 - 4.698 \sin 200t + 1.71 \cos 200t + 2.298 \sin 400t + 1.928 \cos 400t + 1.618 \sin 600t - 1.176 \cos 600t$ amperes (**Ans.**)

The solution to Ex. 22-1a shows that the given signal source may be viewed as containing three separate series-connected sources: a dc source of 20 V, an ac source whose maximum value is 25 V at a frequency of 159.15 Hz, and a second ac source whose maximum value is 17 V at a frequency of 318.3 Hz.

The Fourier trigonometric amplitude–phase form of Eq. (22-3) is a *convergent series* whose frequencies are harmonically related and whose amplitudes are also a function of this relation. In general, we may say that as the *order* of the harmonic (n) *increases*, the amplitude C_n *decreases*. There are three exceptions to this observation:

1. The dc value where $n = 0$ may be smaller than the first harmonic or fundamental, where $n = 1$.
2. If the load to which the waveform is applied contains some combination of L and C, either series or parallel resonance may occur. If *series* resonance occurs, a particular harmonic may be *emphasized*. If *parallel* resonance occurs, a particular harmonic may be *suppressed*. (See Sec. 22-11.)
3. In the case of short-duration **pulses** (see Table 22-1M) the amplitude C_n generally decreases, but at some harmonic (say $n = 10$) the amplitude is increased out of proportion to the others.

The task of determining the Fourier coefficients (A_0, A_n, B_n, and C_n) for any given periodic waveform is mathematically possible but beyond the scope of this work. Instead, we will learn to classify a nonsinusoidal waveform on the basis of its inherent *symmetry*, from which the waveform equations are more easily verified. These insights emerge from a study of the *synthesis* of a number of simplified nonsinusoidal waveforms and the symmetry they exhibit.

22-3 TYPES OF SYMMETRY

A purely *sinusoidal* waveform, for purposes of symmetry analysis, is considered to be *perfectly symmetrical.* Consequently, the addition of odd and/or even harmonics at specific phase angles will affect this symmetry in various ways. The resultant waveform and its symmetry gives us a clue to the *content* of a waveform in terms of the presence and phase angle of the harmonic with respect to the fundamental. We begin with a definition of the three types of symmetry:

1. **x-axis** (or mirror) symmetry (Fig. 22-1b)
2. **y-axis** (or quarter-wave) symmetry (Fig. 22-1c)
3. **z-axis** (or half-wave) symmetry (Fig. 22-1d)

Figure 22-1a permits the *definition* of *each* type of *symmetry*, and the remaining subfigures show *examples* of each. Let us consider each in turn.

22-3.1 X-Axis (or Mirror) Symmetry

The x-axis is defined as the *horizontal* (or *real*) axis in a Cartesian coordinate system. To determine whether a given nonsinusoidal waveform has x-axis symmetry, therefore, we merely displace the *negative-going* half-cycle portion of the waveform to the *left* so that the portion from π to 2π is aligned directly *below* the positive-going portion from 0 to π. As shown in Fig. 22-1b, if the negative half-cycle is the *mirror image* of the positive half-cycle, the waveform possesses x-axis symmetry.

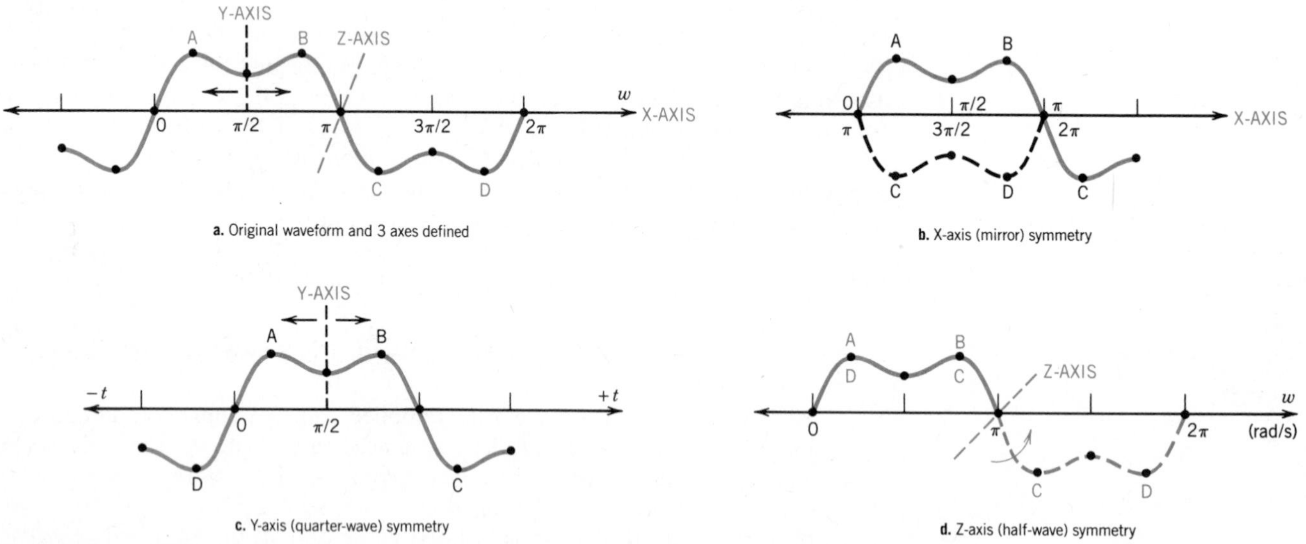

Figure 22-1 Definitions of symmetry

22-3.2 Y-Axis (or Quarter-Wave) Symmetry

The y-axis, defined as the vertical (or imaginary axis), is (usually) located at the *quarter-wave* angular displacement ($\theta = \pi/2$ radians) point. If a vertical line is drawn at this point (or the $3\pi/2$ point), we may test for y-axis symmetry, as shown in Fig. 22-1c. If the waveform has the *same* amplitude *variation* in the *negative* direction (to the *left* of the y-axis) as it does in the *positive* direction (to the *right* of the y-axis), it is said to have y-axis symmetry.

22-3.3 Z-Axis (or Half-Wave) Symmetry

In any coordinate three-dimensional system, the z-axis is an axis perpendicular to *both* the x and y axes. As shown in Fig. 22-1a, the z-axis is drawn perpendicular to the paper plane at the *half-wave* angular displacement ($\theta = \pi$ rads) point. To determine whether a waveform has z-axis symmetry we rotate the negative half-cycle portion (about the z-axis) in a counterclockwise (CCW) direction for a full 180° of rotation. If the negative half-cycle *coincides exactly* with the positive half-cycle, the nonsinusoidal waveform is said to have z-axis symmetry. As shown in Fig. 22-1d, the waveform does have z-axis symmetry since points D and A coincide and points B and C coincide exactly.

We may conclude, therefore, that the waveform of Fig. 22-1a, *like that of a pure sinusoid*, possesses x-axis, y-axis, and z-axis symmetry.

22-4 EFFECT OF ADDITION OF ODD HARMONICS ON WAVEFORM SYMMETRY

We know that a pure sinusoid has all three types of symmetry. What is the effect on the sinusoid if an odd harmonic is added? Synthesizing the resultant waveform should provide the answer. As a prerequisite, the reader should review the first three paragraphs of Sec. 14-7, where phase relations between waveforms are defined.

Figure 22-2a shows the effect of adding a *third* harmonic *in phase* with the fundamental, a pure sinusoid. The resultant complex waveform drawn in Fig. 22-2a is obtained by *graphical addition* of two or more sinusoids at certain specific points, namely

1. When both waveforms cross the zero axis, the resultant is zero ($\omega = 0$, or π or 2π).
2. Whenever one sinusoid is zero, the resultant is the magnitude sum of the remaining sinusoid(s) at that specific angular point ($\omega = \pi/3$, $2\pi/3$, etc.).
3. Whenever one sinusoid is at a maximum, the resultant is that maximum either added or subtracted from the instantaneous value(s) of the remaining sinusoid(s) at the same specific angular point ($\omega = \pi/2$, $3\pi/2$, etc.).
4. Whenever two sinusoids cross each other, the resultant is twice the amplitude at the crossover point.

The use of cross-sectional paper, a pair of dividers, and a millimeter ruler makes the task of graphical addition of sinusoids to produce a nonsinusoidal resultant even easier.

Observe that Fig. 22-2a is exactly the same as Fig. 22-1a. From this, we may conclude that the addition of *odd harmonics* at 0° (or 180°) *with a fundamental sinusoid* does *not* destroy or alter the *symmetry* of the original sinusoid. This holds for all harmonics where $n = 1, 3, 5, 7, 9, \ldots$ (i.e., **odd harmonics only**).

The reader may ask, "But what if the odd harmonic is *not* in phase or 180° out of phase with the fundamental? What if it is at 90° or 120° or some odd angle, say 54°?"

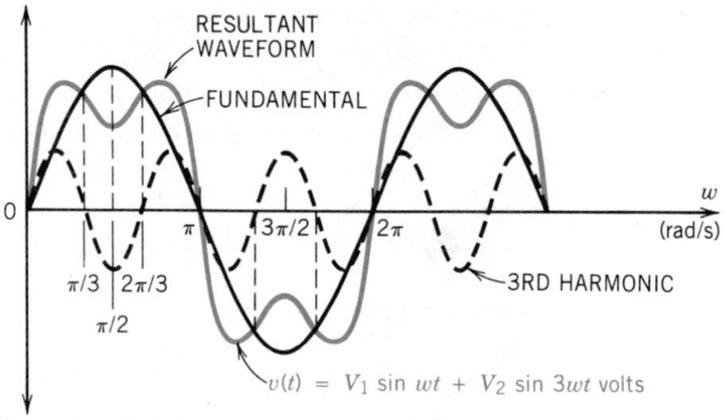

a. Third harmonic in phase with fundamental (x, y, z-axis symmetry)

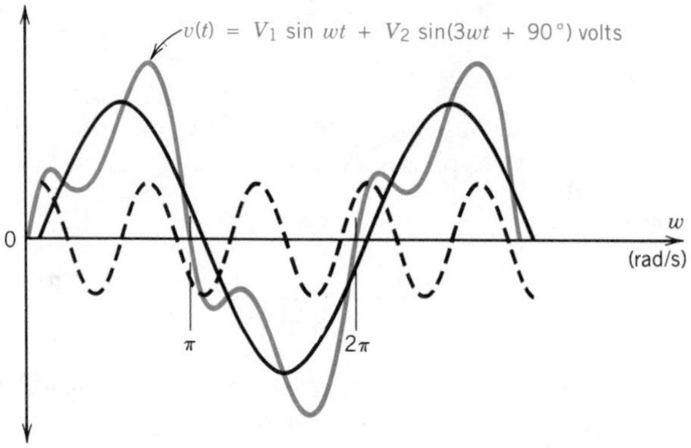

b. Third harmonic leading fundamental by 90° (x-axis symmetry only)

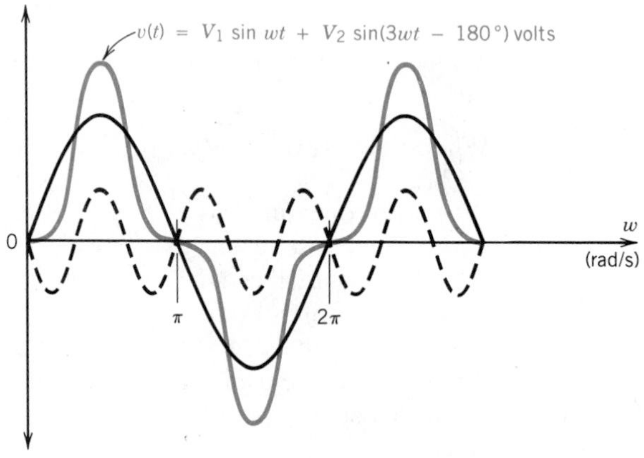

c. Third harmonic 180° out of phase with fundamental (x, y, z-axis symmetry)

Figure 22-2 Effect of odd harmonic phase on symmetry of nonsinusoidal waveform

Figure 22-2b shows the effect of adding a *third* harmonic leading the fundamental by 90°. (Note that the phase of the harmonic is found by using the rules established in Secs. 14-6 and 14-7. The harmonic is at its +90° point when the fundamental is *zero* and *positive-going.*) The resultant nonsinusoidal waveform (whose equation is shown in Fig. 22-2b) has now lost some of its symmetries. It no longer has *y*-axis or *z*-axis symmetry! The only symmetry that has been retained is *x*-axis (mirror) symmetry. Although this is not a conclusive proof, we will state categorically that the addition of odd harmonics to a pure sinusoid at angles other than 0° or 180° results in *mirror* symmetry *only*! Stated another way, if a nonsinusoidal waveform has *only mirror* symmetry, it must contain odd harmonics at angles other than 0° or 180°. It *may also* contain odd harmonics at 0° or 180°, since the presence of these harmonics *does not affect the symmetry* at all, as shown in Fig. 22-2a. Finally, it may *not* contain *even* harmonics.

The proof that odd harmonics at 180° do *not* destroy symmetry at all is shown in Fig. 22-2c. The resultant waveform (whose equation is shown in Fig. 22-2c) has all three types of symmetry (x-axis, y-axis, and z-axis).

EXAMPLE 22-2

For each of the three waveforms shown in **Fig. 22-3**,

a. Identify the various types of symmetries contained in the waveform
b. Identify the harmonics that must be present
c. Identify the harmonics that may be present

Solution

For the **square wave** of **Fig. 22-3a**,

a. The waveform has x-axis, y-axis, and z-axis symmetry
b. It *must* contain *odd* harmonics *only* at 0° or 180° (or both)
c. No other harmonics other than those in (**b**) are permitted

For the **triangular wave** of **Fig. 22-3b**,

a. The waveform has x-axis, y-axis, and z-axis symmetry
b. It *must* contain *odd* harmonics *only* at 0° or 180° (or both)
c. No other harmonics other than those in (**b**) are permitted

For the **half-sawtooth wave** of **Fig. 22-3c**,

a. The waveform has x-axis symmetry *only*!
b. It *must* contain *odd* harmonics *only* at $\pm 90°$
c. It *may* contain *odd* harmonics at 0° or 180° (since these do not destroy symmetry). In fact it does, as shown in Fig. 22-3c.

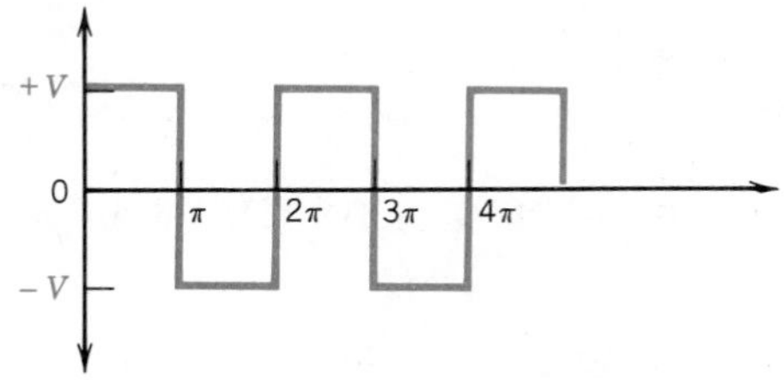

$$v(t) = \frac{4V}{\pi}\left(\sin \omega t + \frac{\sin 3\omega t}{3} + \frac{\sin 5\omega t}{5} + \ldots\right) \text{ volts}$$

a. Square wave and its fundamental equation

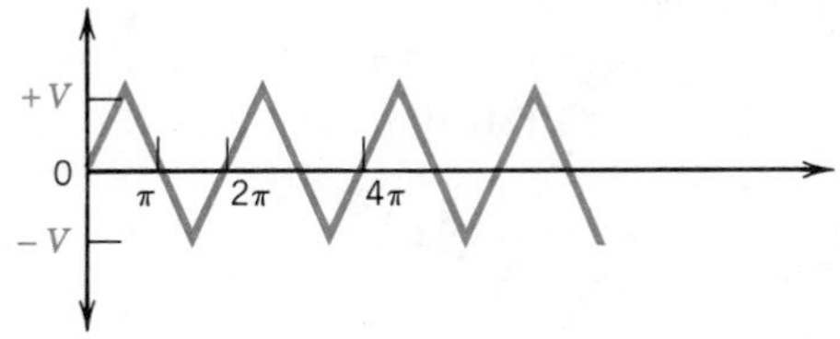

$$v(t) = \frac{8V}{\pi^2}\left(\sin \omega t - \frac{1}{9}\sin 3\omega t + \frac{1}{25}\sin 5\omega t - \frac{1}{49}\sin 7\omega t + \ldots\right)$$

b. Triangular wave and its fundamental equation

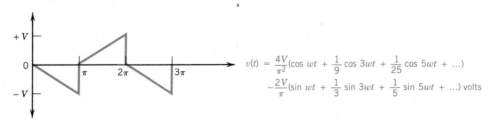

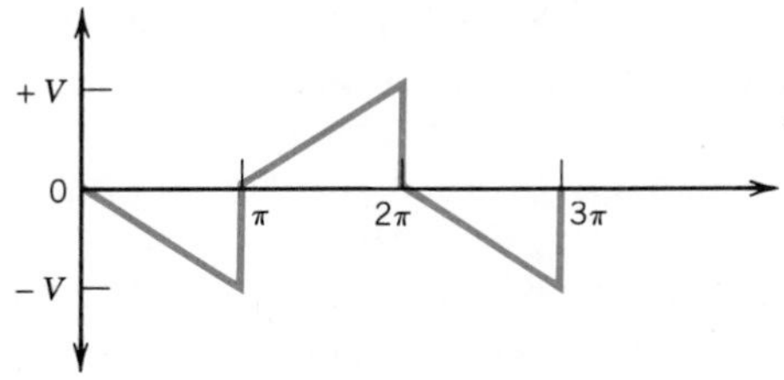

$$v(t) = \frac{4V}{\pi^2}\left(\cos \omega t + \frac{1}{9}\cos 3\omega t + \frac{1}{25}\cos 5\omega t + \ldots\right) - \frac{2V}{\pi}\left(\sin \omega t + \frac{1}{3}\sin 3\omega t + \frac{1}{5}\sin 5\omega t + \ldots\right) \text{ volts}$$

c. Half-sawtooth wave and its fundamental equation

Figure 22-3 Example 22-2

Example 22-2 shows that the conclusions derived from waveform synthesis are verified by the waveform equations emerging from Fourier analysis techniques. The square wave, as shown in Fig 22-3a by its fundamental equation, contains only odd harmonics at 0°. The amplitude of each harmonic varies inversely as the order of the harmonic, *n*.

Similarly, the triangular wave, as shown in Fig 22-3b by its fundamental equation, contains both odd harmonics at 0° and odd harmonics at 180°. The amplitude of each harmonic varies inversely as the *square of the order* of the harmonic.

Finally, the half-sawtooth wave (Fig. 22-3c), as shown by its fundamental equation, contains odd harmonics at 90° and also odd harmonics at 180°. The amplitude of the harmonics at 90° varies inversely as the *square* of the order of the harmonic and the amplitude of those at 180° varies inversely as the order of the harmonic. These equations verify the solutions of Ex. 22-2.

EXAMPLE 22-3

For each of the three waveforms shown in Fig. 22-3, write the magnitude of the harmonic whose order $n = 9$, that is, the ninth harmonic

Solution

a. $v(t)_{(n=9)} = \frac{4V}{\pi}(\sin 9\omega t/9)$

b. $v(t)_{(n=9)} = \frac{8V}{\pi^2}\left(\frac{1}{81}\sin 9\omega t\right)$

c. $v(t)_{(n=9)} = \frac{4V}{\pi^2}\left(\frac{1}{81}\sin 9\omega t\right) - \frac{2V}{\pi}\left(\frac{1}{9}\sin 9\omega t\right)$

22-5 EFFECT OF ADDITION OF EVEN HARMONICS ON WAVEFORM SYMMETRY

We are now ready to investigate the effect of **even-order** harmonics on the symmetry of a pure sinusoid containing *x*-, *y*-, and *z*-axis symmetry. **Figure 22-4** shows a *second* harmonic *in phase* with the fundamental. The resultant waveform is a nonsinusoidal complex wave containing *z*-axis symmetry *only*. The presence of even harmonics (even at 0°) has destroyed *x*-axis and *y*-axis symmetry. Since odd harmonics at 0° or 180° do *not* destroy symmetry, we must conclude that whenever a complex wave exhibits **z-axis** symmetry *only*, it **must** contain **even** harmonics at 0° or 180°. It **may** also contain **odd** harmonics at 0° or 180° since these do not affect waveform symmetry in any way.

Figure 22-4b shows a second harmonic leading its fundamental sinusoid by 90°. The resultant waveform has ***y*-axis** or **quarter-wave** symmetry. The presence of even harmonics at 90° has destroyed *x*-axis and *z*-axis symmetry. From this we may conclude that whenever a complex wave exhibits ***y*-axis** symmetry, it **must** contain **even harmonics at 90°**. It **may** also contain odd harmonics at 0° or 180° since the presence of these does not affect symmetry.

The reader may now ask, "What if a waveform contains no symmetry whatever? Can we predict its harmonic content?"

The answer is found in Fig. 22-4c, which shows an even (second) harmonic leading its fundamental by 45° (i.e., some angle other than 0°, 90°, or 180°). The resultant waveform has neither *x*-axis, *y*-axis, nor *z*-axis symmetry. *All the symmetry of the fundamental has been destroyed.* From this, we may conclude that whenever a waveform displays *no symmetry* whatever, it **must** contain *even har-*

monics at angles *other than* 0°, ±90°, or 180°. It **may** also contain *odd* harmonics at *any* angle, for obvious reasons.

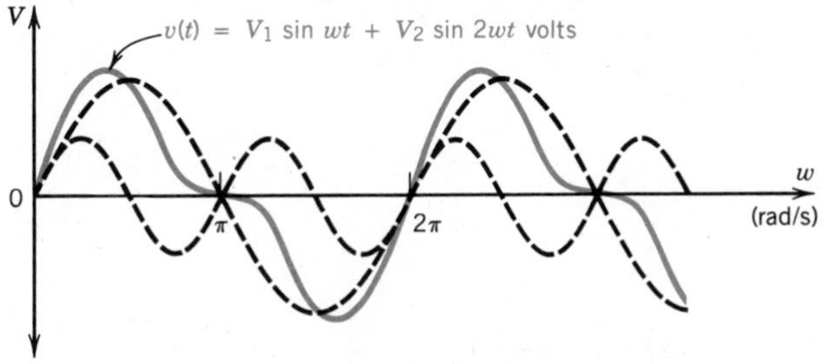

a. Second harmonic in phase with fundamental (z-axis symmetry)

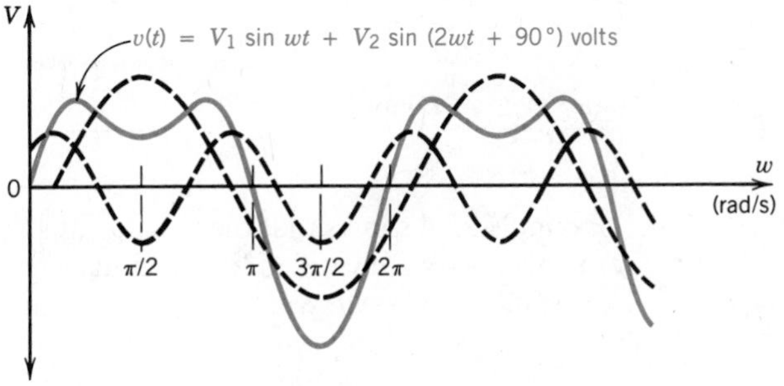

b. Second harmonic leading fundamental by 90° (y-axis symmetry)

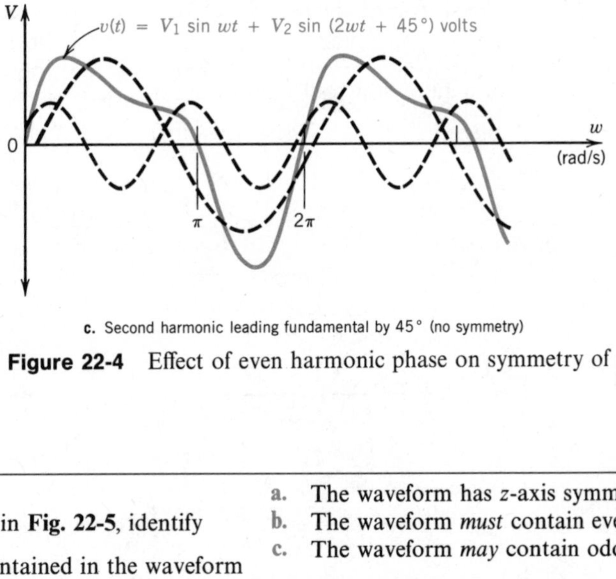

c. Second harmonic leading fundamental by 45° (no symmetry)

Figure 22-4 Effect of even harmonic phase on symmetry of nonsinusoidal waveform

EXAMPLE 22-4

For each of the two waveforms shown in **Fig. 22-5**, identify

a. The various types of symmetries contained in the waveform
b. The harmonics that *must* be present
c. The harmonics that *may* be present

Solution

For the *rising ramp* sawtooth waveform of Fig. 22-5a,

a. The waveform has z-axis symmetry only
b. The waveform *must* contain even harmonics at 0° or 180°
c. The waveform *may* contain odd harmonics at 0° or 180°

For the *descending ramp* sawtooth waveform of Fig. 22-5b,

a. The waveform has z-axis symmetry only
b, c. Same as above

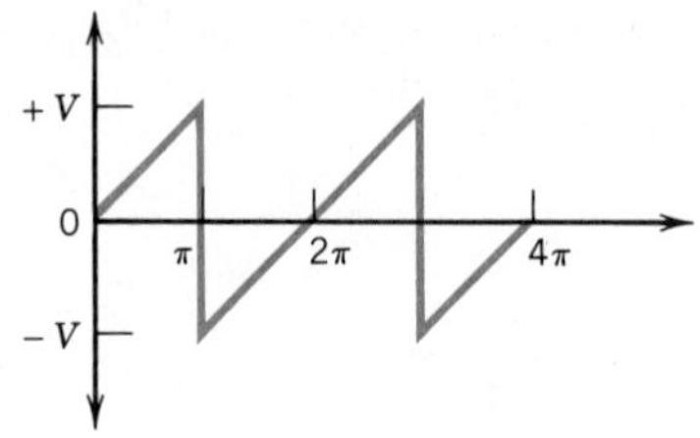

$$v(t) = \frac{2V}{\pi}\left(\sin wt - \frac{1}{2}\sin 2wt + \frac{1}{3}\sin 3wt - \frac{1}{4}\sin 4wt \ldots\right) \text{ volts}$$

a. Sawtooth wave and its fundamental equation (rising ramp)

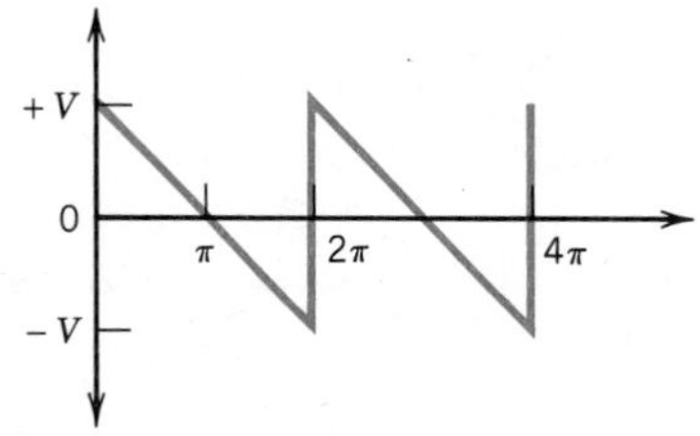

$$v(t) = \frac{2V}{\pi}\left(\sin wt + \frac{1}{2}\sin 2wt + \frac{1}{3}\sin 3wt + \frac{1}{4}\sin 4wt \ldots\right) \text{ volts}$$

b. Sawtooth wave and its fundamental equation (descending ramp)

Figure 22-5 Example 22-4

Example 22-4 shows that the conclusions derived from waveform synthesis using even harmonics are verified by the equations given in Fig. 22-5 for each waveform. The rising ramp sawtooth (Fig. 22-5a) contains *even* harmonics at 180° and *odd* harmonics at 0°. The descending ramp sawtooth (Fig. 22-5b) contains *even* harmonics at 0° and *odd* harmonics at 0°. The amplitudes of both waveforms vary *inversely* as the *order* of their respective harmonics.

Let us now consider two relatively familiar waveforms shown in **Fig. 22-6**. These are shown as they might be displayed on a cathode ray oscilloscope (CRO) using ac coupling (i.e., the dc component is filtered out and the average value of the waveform is zero). Note that now the area above the baseline equals the area below the baseline in both waveforms of Fig. 22-6, meaning that only ac components are left. These waveforms are analyzed for symmetry in Ex. 22-5.

$$v(t) = V_{pp}\left(\frac{\sin wt}{2} - \frac{2\cos 2wt}{1 \times 3 \times \pi} - \frac{2\cos 4wt}{3 \times 5 \times \pi} - \frac{2\cos 6wt}{5 \times 7 \times \pi} - \frac{2\cos 8wt}{7 \times 9 \times \pi} \ldots\right) \text{ volts}$$

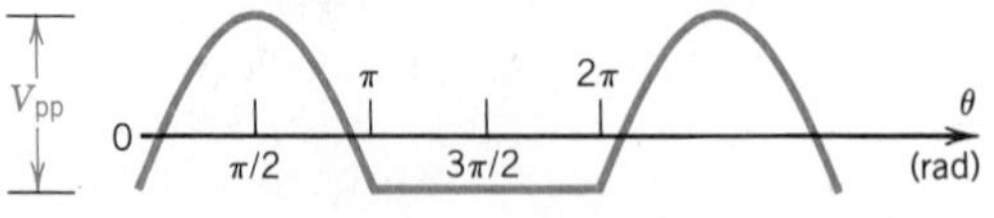

a. HWR, ac component only

$$v(t) = \frac{2V_{pp}}{\pi}\left(-\frac{2\cos 2wt}{1 \times 3} - \frac{2\cos 4wt}{3 \times 5} - \frac{2\cos 6wt}{5 \times 7} \ldots\right) \text{ volts}$$

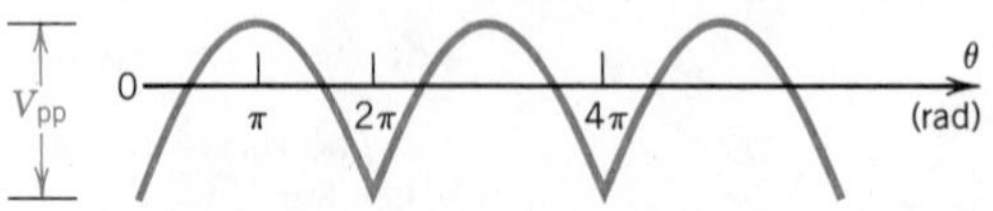

b. FWR, ac component only

Figure 22-6 Example 22-5

EXAMPLE 22-5

For each of the waveforms shown in Fig. 22-6, identify the

a. Various types of symmetries contained in the waveform
b. Harmonics that **must** be present
c. Harmonics that **may** be present

Solution

For the **half-wave**-rectifier (HWR) shown in Fig. 22-6a,

a. The waveform displays y-axis (quarter-wave) symmetry only
b. The waveform **must** contain **even** harmonics at $\pm 90°$
c. The waveform **may** contain **odd** harmonics at 0° or 180°

For the **full-wave**-rectifier (FWR) shown in Fig. 22-6b,

a. The waveform displays y-axis symmetry only
b. The waveform **must** contain **even** harmonics at $\pm 90°$
c. The waveform **may** contain **odd** harmonics at 0° or 180° (or it *may not!*)

Example 22-5 shows that the conclusions derived from waveform synthesis using even harmonics are verified by the equations emerging from Fourier analysis for the waveforms shown in Fig. 22-6. The HWR ac sinusoidal components (Fig. 22-6a) consist of even harmonics at $-90°$ and only one odd harmonic (i.e., the fundamental ac input waveform at 0°). The FWR (Fig. 22-6b) contains even harmonics at $-90°$ *only*. Note that the *fundamental is no longer present* and the first ac component is a *second* harmonic! This explains why the frequency of the FWR (Fig. 22-6b) is *twice* the frequency of the HWR (Fig. 22-6a).

22-6 ANALYSIS OF COMPLEX WAVEFORMS CONTAINING A DC COMPONENT

The Fourier series equations (22-1) through (22-3) all provide for the possibility of a dc component. Up to now, the waveforms we have considered all had an average value of $A_0 = 0$, where A_0 is the amplitude of a zero-frequency term. This was done deliberately because the *analysis for symmetry must be done* with the *dc component removed*. This leads us to a *general procedure* for *analyzing all complex waveforms*:

1. Examine the waveform and determine whether it contains a dc component.[2]
2. If it contains a dc component, determine its value. (See Sec. 14-8 on determining the average value of a symmetrical waveform.)
3. Remove the dc component for purposes of symmetry analysis. If the average value is *positive*, draw the reference (ground) line as a *positive* value *above* the original ground line. If the average value is *negative*, draw the reference ground line as a *negative* value *below* the original reference line.
4. Analyze the waveform to see if it has x-, y-, or z-axis symmetry.
   a. If it has x-, y-, and z-axis symmetry, it must contain only odd harmonics at 0° and/or 180°.
   b. If it has x-axis symmetry *only*, it **must** contain odd harmonics only at $\pm 90°$.
   c. If it has z-axis symmetry *only*, it **must** contain even harmonics at 0° and/or 180°. It may also contain odd harmonics at 0° or 180°.
   d. If it has y-axis symmetry *only*, it **must** contain even harmonics at $\pm 90°$. It may also contain odd harmonics at 0° or 180°.

[2] Using a CRO, this is done by switching alternately from dc to ac coupling and observing the rise or fall of the waveform in a vertical direction. It may also be measured by a dc voltmeter or an electronic voltmeter (EVM) on its dc range. By inspection, it is also found by drawing a line through the waveform such that equal areas occur above and below the line.

e. If it has **no symmetry** whatever, it **must** contain **even** harmonics at (any) angles other than 0, 90°, or 180° and **may** contain odd harmonics at **any** angle.

EXAMPLE 22-6

Given the rectangular pulse waveform shown in **Fig. 22-7a**,

a. Find the average value (or dc component) of the waveform
b. Draw the waveform with the dc component removed
c. Analyze the waveform for symmetry and describe the harmonics contained
d. Write the general equation for the waveform
e. Write the specific equation, given the maximum value of 20 V and a radian frequency of $\omega = 1000$ rad/s, for the first three harmonic terms

Solution

a. For the rectangular wave, $V_{dc} = V_{av} = V/2 = 20\text{ V}/2 = \mathbf{10\ V}$
b. The waveform is drawn in Fig. 22-7b
c. The waveform of Fig. 22-7b contains x-, y-, and z-axis symmetry. It contains only odd harmonics at 0° or 180°. There are *no even harmonics.*
d. The general equation for the waveform of Fig. 22-7a is $v(t) = \frac{V}{2} + \frac{2V}{\pi}(\sin \omega t + \sin 3\omega t/3 + \sin 5\omega t/5 + \cdots + \sin n\omega t/n)$
e. $v(t) = 10 + 12.73(\sin 1000t + \sin 3000t/3 + \sin 5000t/5 \cdots)$ volts

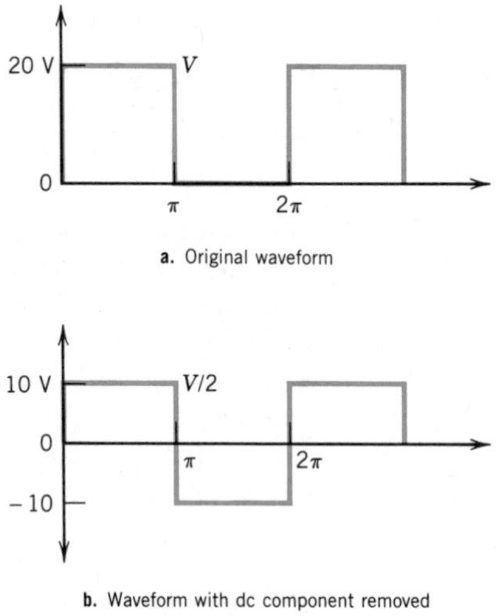

Figure 22-7 Examples 22-6 through 22-8

22-7 RMS VALUE OF A NONSINUSOIDAL WAVEFORM

In Sec. 14-10.3 we observed that the RMS value, A, of a symmetrical waveform containing a dc component of amplitude A_{dc} may be represented by

$$A = \sqrt{A_{dc}^2 + A_{ac}^2} \tag{14-21}$$

where A_{dc} is the amplitude of the dc component

A_{ac} is the equivalent (rectangular) ac amplitude of the ac component

Let us first use this relation to determine the RMS value of the waveform shown in Fig. 22-7.

EXAMPLE 22-7

Using Eq. (14-21), find the RMS value of the waveform shown in Fig. 22-7a

Solution

$$\mathbf{V}_{RMS} = \mathbf{V} = \sqrt{\mathbf{V}_{dc}^2 + \mathbf{V}_{ac}^2} = \sqrt{(10)^2 + (10)^2} \text{ (see Table 14-1)}$$
$$= \mathbf{14.14\ V}$$

The Fourier equation for the waveform provides us with the dc value and the maximum value of each of the harmonics contained within the waveform. If our equation for the square wave of Fig. 22-7a is correct, as written in Ex. 22-6**d** and 22-6**e**, we should obtain approximately the *same* RMS value *using the harmonic component data.* If we treat the harmonics as "separate sources" respectively producing power dissipation in a known resistor, R, we may use the principle of superposition, or

$$\mathbf{V}_{\text{RMS}} = \sqrt{P_t R} = \sqrt{\mathbf{V}_0^2 + \mathbf{V}_1^2 + \mathbf{V}_2^2 + \mathbf{V}_3^2 + \cdots + \mathbf{V}_n^2} \tag{22-4}$$

where P_t is the total power dissipated in resistor R
R is the known resistance to which the complex wave is applied
$\mathbf{V}_0$ is the average value of the dc component of the waveform
$\mathbf{V}_1$, $\mathbf{V}_2$, $\mathbf{V}_3$, and $\mathbf{V}_n$ are the **RMS** values of the fundamental, second, third and nth harmonics, respectively, of the complex wave

Equation (22-4) shows that the simplest way of finding the RMS value of a complex wave (in the absence of, say, a true-reading RMS voltmeter) is to measure the power dissipated by a known resistor to which the waveform is applied.

But in the absence of instrumental methods, since $\mathbf{V}_{\text{RMS}} = \mathbf{V}_m/\sqrt{2}$ and $\mathbf{V}_{\text{RMS}}^2 = (\mathbf{V}_m/\sqrt{2})^2 = \mathbf{V}_m^2/2$, we may write Eq. (22-4) as

$$\begin{aligned}\mathbf{V}_{\text{RMS}} &= \sqrt{(\mathbf{V}_0)^2 + \sum(\text{all RMS values})^2} \\ &= \sqrt{\mathbf{V}_0^2 + \frac{\sum(\text{all maximum values})^2}{2}} \\ &= \sqrt{\mathbf{V}_0^2 + \frac{1}{2}(\mathbf{V}_{m_1}^2 + \mathbf{V}_{m_2}^2 + \mathbf{V}_{m_3}^2 + \cdots + \mathbf{V}_{m_n}^2)}\end{aligned} \tag{22-5}$$

where all terms have been defined.

EXAMPLE 22-8

For the rectangular pulse shown in Fig. 22-7a, using the equation written in the solution of Ex. 22-6e, calculate the

a. RMS value of the waveform, using only the first three harmonic terms
b. Percent relative error (r.e.) between the value obtained in (**a**) and that in Ex. 22-7, using the latter as the true value
c. Explain why the answer in (**a**), of necessity, is slightly lower than the value obtained with Eq. (14-21)

Solution

a. $$\mathbf{V}_{\text{RMS}} = \sqrt{\mathbf{V}_0^2 + \frac{1}{2}(\mathbf{V}_{m_1}^2 + \mathbf{V}_{m_2}^2 + \mathbf{V}_{m_3}^2)}$$
$$= \sqrt{10^2 + \frac{1}{2}\left[12.73^2 + \left(\frac{12.73}{3}\right)^2 + \left(\frac{12.73}{5}\right)^2\right]}$$
$$= \mathbf{13.9\ V} \tag{22-5}$$

b. r.e. = (true value − nominal value)/true value
= (14.14 − 13.90)/14.14 = **1.685%**

c. Example 22-8 proves that the general equation for the waveform of Fig. 22-7a shown in Ex. 22-6d is correct. Had we included *all* the harmonics, the two values would be identical; [i.e., as more harmonics are included, the RMS value using Eq. (22-5) rises!]. But since the amplitudes are decreasing rapidly as the order of the harmonic rises, only the first few harmonics are needed to find the (approximate) RMS value, in most cases, as shown by Exs. 22-9 and 22-10.

EXAMPLE 22-9

Given the triangular pulse waveform shown in **Fig. 22-8a**,

a. Find the average value (or dc component) of the waveform
b. Draw the waveform with the dc component removed
c. Analyze the waveform for symmetry and describe the harmonics contained in it
d. Write the general equation for the waveform shown in Fig. 22-8a
e. Write the specific equation, given a maximum value of 30 V and a radian frequency of $\omega = 100$ rad/s, for the first three harmonic terms.
f. Calculate the RMS value of the waveform using Eq. (14-21)
g. Calculate the RMS value of the waveform using Eq. (22-5)
h. Calculate the relative error between (**f**) and (**g**), using (**f**) as the true value.

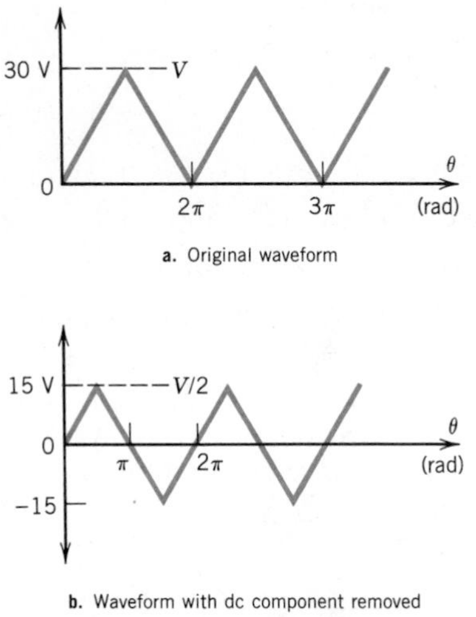

Figure 22-8 Example 22-9

Solution

a. $V_{dc} = V_{av} = V/2 = 30\text{ V}/2 = \mathbf{15\ V}$
b. The waveform is drawn in Fig. 22-8b
c. The waveform of Fig. 22-8b contains x-, y-, and z-axis symmetry. It must contain only odd harmonics at 0° and/or 180°. There are no even harmonics.
d. The general equation for the waveform of Fig. 22-8a is

$$v(t) = V/2 + \frac{4V}{\pi^2}(\sin \omega t - \sin 3\omega t/9 + \sin 5\omega t/25 - \sin 7\omega t/49 + \cdots) \qquad (22\text{-}13)$$

e. The specific equation, given $V = 30$ V and $\omega = 100$ rad/s, is (for the first fours terms)

$$v(t) = 15 + 12.16\left(\sin 100t - \frac{\sin 300t}{9} + \frac{\sin 500t}{25} - \frac{\sin 700t}{49}\right)$$

f. $$\mathbf{V}_{\mathbf{RMS}} = \sqrt{\mathbf{V}_{\mathbf{dc}}^2 + \mathbf{V}_{\mathbf{ac}}^2} = \sqrt{(15)^2 + (15/\sqrt{3})^2} = \mathbf{17.32\ V} \qquad (14\text{-}21)$$

g. $$\mathbf{V}_{\mathbf{RMS}} = \sqrt{\mathbf{V}_0^2 + \frac{1}{2}(\mathbf{V}_{\mathbf{m}_1}^2 + \mathbf{V}_{\mathbf{m}_2}^2 + \mathbf{V}_{\mathbf{m}_3}^2 + \mathbf{V}_{\mathbf{m}_4}^2)}$$
$$= \sqrt{(15)^2 + \frac{1}{2}[12.16^2 + (-1.351)^2 + (0.4864)^2 + (-0.25)^2]}$$
$$= \sqrt{225 + 75} = \mathbf{17.32\ V} \qquad (22\text{-}5)$$

h. r.e. = (17.32 − 17.32)/17.32 = **0%**

With respect to Ex. 22-9, please note the following insights:

1. The general equation for the waveform represents the triangular pulse, as shown in step (**d**) of the solution. This verifies the analysis techniques previously described as well as the equation for the triangular waveform (Fig. 22-8b).
2. Only the fourth-order harmonic is needed to produce *exactly the same answer* by substitution in Eq. (22-5) as that in Eq. (14-21). The reason for this obviously is that the amplitudes fall off so rapidly that those beyond the fourth harmonic are *negligible.*
3. In the solution of part (**f**), recall that a pure triangular waveform has an RMS value of $V_m/\sqrt{3}$. This also verifies the relatively quick solution shown in step (**f**).[3]

EXAMPLE 22-10

Repeat all parts of Ex. 22-9 for the HWR waveform shown in **Fig. 22-9a**.

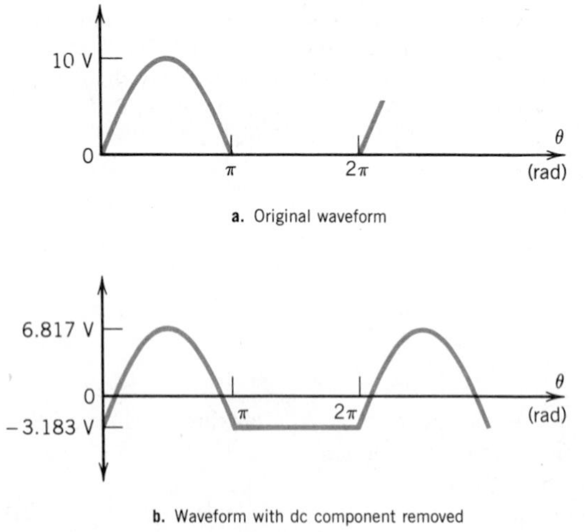

Figure 22-9 Example 22-10

Solution

a. $V_{av} = V_{dc} = V_m/\pi = 10\text{ V}/\pi = \mathbf{3.183\ V}$
b. Fig. 22-9b shows the waveform with this dc component removed.
c. The waveform of Fig. 22-9b contains y-axis (quarter-wave) symmetry only. Therefore it *must* contain even harmonics at ±90° and may contain odd harmonics at 0° and/or 180°.
d. The general equation for the waveform of Fig. 22-9a is $v(t) =$
$$\mathbf{V}\left(\frac{1}{\pi} + \frac{\sin \omega t}{2} - \frac{2\cos 2\omega t}{1 \times 3 \times \pi} - \frac{2\cos 4\omega t}{3 \times 5 \times \pi} - \frac{2\cos 6\omega t}{5 \times 7 \times \pi} - \cdots\right)$$
e. Substituting $\mathbf{V} = 10$ V, the specific equation for the first four terms is

$$v(t) = 3.183 + 5.0 \sin \omega t - 2.122 \cos 2\omega t - 0.4244 \cos 4\omega t - 0.1819 \cos 6\omega t$$

f. $\mathbf{V}_{\mathbf{RMS}} = \mathbf{V}_{\mathbf{m}}/2 = 10\text{ V}/2 = \mathbf{5\ V}$ (from Table 14-1)
g. $$\mathbf{V}_{\mathbf{RMS}} = \sqrt{3.183^2 + \frac{1}{2}(5.0^2 + 2.122^2 + 0.4244^2 + 0.1819^2)} = \mathbf{4.999\ V}$$
h. r.e. = 0% by inspection

[3] An even quicker solution is obtained using $V_m = 30$ V in the original given waveform. The RMS value of the pulse is $\mathbf{V} = V_m/\sqrt{3} = 30\text{ V}/\sqrt{3} = \mathbf{17.32\ V}$, for any triangular waveform using Table 14-1, row 9.

Examples 22-6 through 22-10 verify that the Fourier equations given for the waveforms shown in Figs. 22-7 through 22-9 are correct. These previous examples verified the equations by verifying the RMS value, using the maximum values or harmonic amplitudes contained in the waveform. (As noted earlier, the derivation of these equations is beyond the scope of this text.) It is also possible, however, to obtain and verify these equations experimentally, as noted in Sec. 22-8.

22-8 HARMONIC ANALYSIS AND LINE SPECTRA

Two basic electronic instruments are currently available to measure the specific harmonic frequencies contained within a complex waveform, as well as their appropriate relative amplitudes at their respective frequencies. The first of these is the *harmonic wave analyzer*. The block diagram for the instrument is shown in **Fig. 22-10a**. It consists of a tunable fundamental frequency selector that detects the *lowest* possible frequency contained in the waveform (f_1). Once tuned to this *fundamental* frequency, a selective harmonic filter enables switching to multiples of the fundamental. The RMS amplitude is recorded on an ac voltmeter and the sinusoidal waveform is traced on a recorder, yielding both maximum amplitude and phase of the harmonic. The frequency composition of any nonsinusoidal wave, as expressed by Fourier analysis, is called the *line spectrum* or frequency spectrum. From the recorded data of the harmonic wave analyzer, the line spectrum may be plotted. Figure 22-10b shows the line spectrum for a pulse-type square wave whose maximum value is 100 V.

Harmonic wave analyzers are also used as *distortion-measuring* devices. For example, the 60-Hz signal at 120 V is presumed to be a pure sine wave. The presence of higher-order harmonics, as detected by the wave analyzer, is a measure of the harmonic distortion contained within the supply source sinusoid. Similarly, if when the square wave signal is analyzed, the presence of any even harmonics or odd harmonics that are not in agreement with the general equation given in Table 22-1G (Eq. 22-12) or expressed in Ex. 22-6d represents a measure of the distortion contained within the waveform.

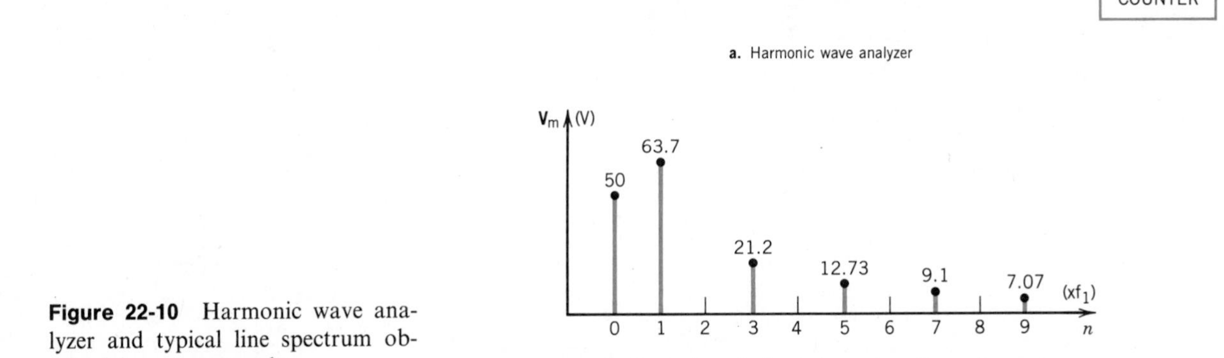

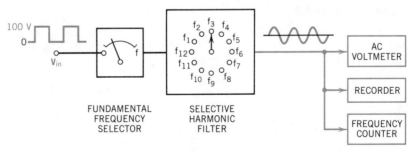

a. Harmonic wave analyzer

b. Line spectrum showing maximum values of harmonics present

Figure 22-10 Harmonic wave analyzer and typical line spectrum obtained for a square pulse

A second instrument, called a *spectrum analyzer*, uses a CRO in combination with what is essentially a narrow-band superheterodyne receiver. The spectrum analyzer produces a display of the line spectrum of a given complex wave on its CRO screen. The block diagram of the spectrum analyzer is shown in **Fig. 22-11**.

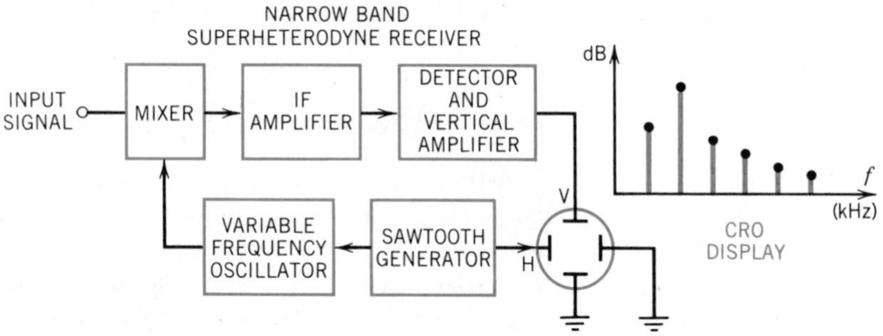

Figure 22-11 Spectrum analyzer

The receiver is tuned by varying the frequency of the voltage-tuned variable frequency oscillator, which also controls the sawtooth generator to sweep the horizontal time base of the CRO deflection plates. As the oscillator is swept through its frequency band by the sawtooth generator, the resultant signal mixes and beats with the incoming input signal to produce an intermediate frequency (IF) in the mixer. Mixer output occurs only when there is a corresponding harmonic in the input signal, matching the sawtooth generator signal. The IF amplifier signals are detected and further amplified and applied to the vertical deflection plates of the CRO. The resultant output in the amplitude–time plane of the CRO represents the amplitude–frequency plot, which is similar to the line spectrum shown in Fig. 22-10b. In effect, the spectrum analyzer provides a *direct display* of the frequency components and their amplitudes on a CRO screen, for any type of complex or nonsinusoidal waveform. A modern digital-readout spectrum analyzer is shown in Fig. 23-30.

EXAMPLE 22-11

Plot the line spectra for the waveforms given in

a. Figure 22-7 (Exs. 22-6 and 22-8)
b. Figure 22-8 (Ex. 22-9)
c. Figure 22-9 (Ex. 22-10)

Solution

a. From Ex. 22-6e, the successive amplitudes are shown plotted in **Fig. 22-12a**
b. From Ex. 22-9g, the successive amplitudes are shown plotted in Fig. 22-12b
c. From Ex. 22-10e, the successive amplitudes are shown plotted in Fig. 22-12c

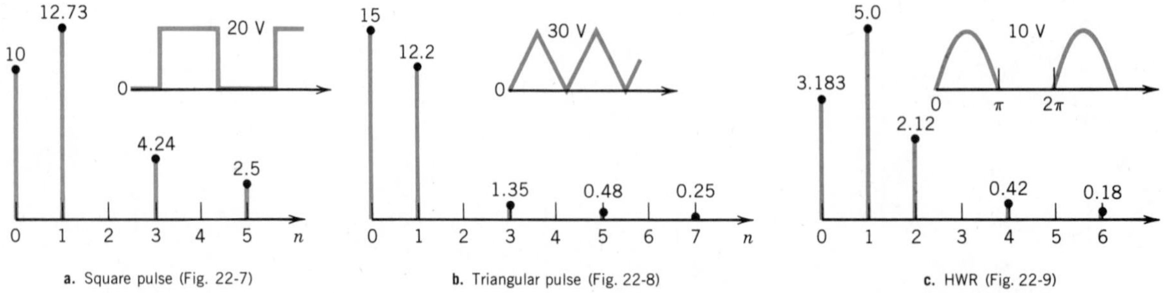

Figure 22-12 Line spectra (for Ex. 22-11) of three waveforms containing dc components

Note that although the line spectra of Fig. 22-12 yield the various dc and ac amplitudes of the odd and even harmonics present, they give us no clue to the *phase* of the harmonics. Our studies of symmetry, however, in combination with the harmonics present, enable us to accept the Fourier series of the next section without the need to derive them by methods of advanced calculus.

22-9 FOURIER EQUATIONS FOR COMMON NONSINUSOIDAL WAVEFORMS

As indicated earlier, deriving the Fourier equations for the various common nonsinusoidal waveforms is a difficult and tedious chore. Our studies of symmetry, combined with data from measurements with harmonic wave analyzers and spectrum analyzers, enables us to infer the various equations. **Table 22-1** shows 14 of some of the most common *nonsinusoidal* waveforms, together with their equations, axial symmetries, and harmonics.

Space does not permit a detailed analysis of each of the waveforms in Table 22-1. Nor is it necessary, particularly when we have verified several of them in previous sections and examples. In each case, however, the reader must remember that if the waveform contains a dc component, it is *first necessary* to *remove that component* before subjecting the waveform to analysis for symmetry. Thus, the common sawtooth pulse (Eq. 22-19, Table 22-1N) displays z-axis symmetry only, whenever the dc component is removed. Recall from Sec. 22-5 that whenever a complex wave displays z-axis symmetry only, it *must* contain even harmonics at 0° and/or 180° and *may* contain odd harmonics at 0° and/or 180°. This may be verified by Eq. (22-19), which shows both odd and even harmonics at 180°.

The same analysis applies to the typical radar pulse (Eq. 22-18, Table 22-1M). After removing the dc component, it is evident that there is *no symmetry whatever*. This implies that the waveform contains **both** *even* and *odd* harmonics at angles other than 0° and/or 180° (i.e., at *any* angles).

Each of the equations in Table 22-1 may be verified by *harmonic synthesis*, using the *graphical techniques* described at the beginning of Secs. 22-4 and 22-5. We noted earlier that since the amplitudes of higher-order harmonics fall off quite rapidly, only the first few terms are needed. This is shown in Ex. 22-12.

Table 22-1 Common Nonsinusoidal Waveforms, Their Equations, and Their Harmonic Content

Waveform	Appearance	Equation for Generation of Waveform	Equation Number	Axial Symmetries Present	Actual Harmonics Present
A. Sine wave	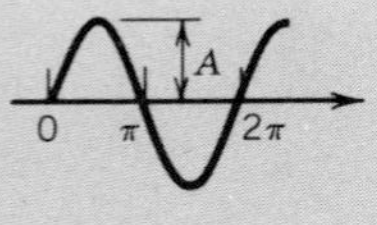	$f(t) = A \sin \omega t$	**22-6**	x, y, z	First at 0°
B. Half-wave rectified sine wave (ideal)	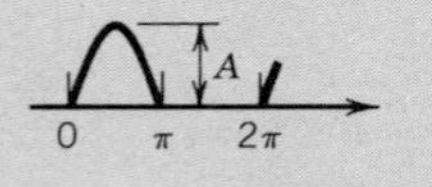	$f(t) = A\left(\frac{1}{\pi} + \frac{1}{2}\sin \omega t - \frac{2}{1 \times 3 \times \pi}\cos 2\omega t - \frac{2}{3 \times 5 \times \pi}\cos 4\omega t - \frac{2}{5 \times 7\pi}\cos 6\omega t - \frac{2}{7 \times 9\pi}\cos 8\omega t - \cdots\right)$	**22-7**	y only	Evens only, lagging by 90° (fundamental at 0°)
C. Full-wave rectified sine wave	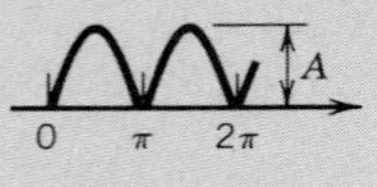	$f(t) = A\frac{2}{\pi}\left(1 - \frac{2}{1 \times 3}\cos 2\omega t - \frac{2}{3 \times 5}\cos 4\omega t - \frac{2}{5 \times 7}\cos 6\omega t - \cdots\right)$	**22-8**	y only	Evens only, lagging by 90° (no fundamental)
D. n-phase rectified sine wave	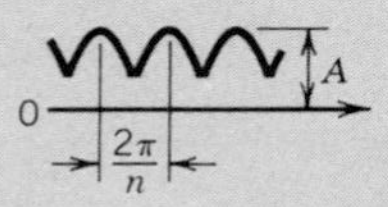	$f(t) = A\frac{n}{\pi}\sin\frac{\pi}{n}\left(1 + \frac{2\cos n\omega t}{n^2 - 1} - \frac{2\cos 2n\omega t}{(2n)^2 - 1} + \frac{2\cos 3n\omega t}{(3n)^2 - 1} - \cdots\right)$	**22-9**	y only	Evens leading and lagging fundamental by 90°
E. Triangular wave	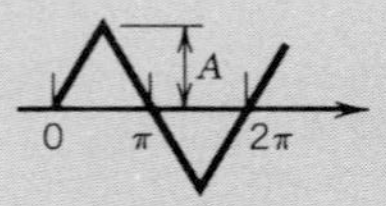	$f(t) = A\frac{8}{\pi^2}\left(\sin \omega t - \frac{1}{3^2}\sin 3\omega t + \frac{1}{5^2}\sin 5\omega t - \frac{1}{7^2}\sin 7\omega t + \cdots\right)$	**22-10**	x, y, z	Odds only at 0° and 180°
F. Rectangular or square wave	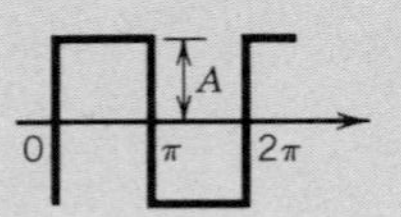	$f(t) = A\frac{4}{\pi}\left(\sin \omega t + \frac{1}{3}\sin 3\omega t + \frac{1}{5}\sin 5\omega t + \frac{1}{7}\sin 7\omega t + \cdots\right)$	**22-11**	x, y, z	Odds only at 0°
G. Rectangular or square wave pulse	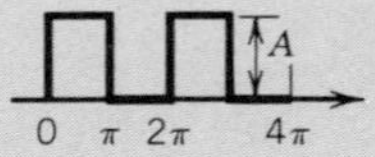	$f(t) = \frac{A}{2} + \frac{2A}{\pi}\left(\sin \omega t + \frac{\sin 3\omega t}{3} + \frac{\sin 5\omega t}{5} + \frac{\sin 7\omega t}{7} + \cdots\right)$	**22-12**	x, y, z	Odds only at 0°

H. Triangular pulse	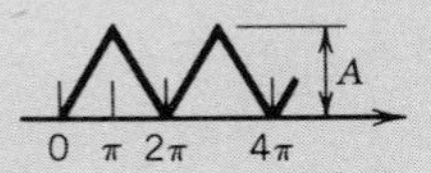	$f(t) = \frac{A}{2} + \frac{4A}{\pi^2}\left(\sin \omega t - \frac{\sin 3\omega t}{9} + \frac{\sin 5\omega t}{25} - \frac{\sin 7\omega t}{49} + \cdots\right)$	**22-13**	x, y, z	Odds only at 0° and 180°
I. Stepped rectangular wave	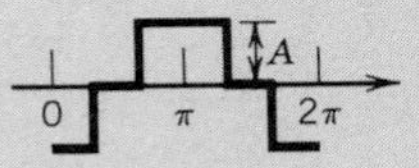	$f(t) = A\frac{2\sqrt{3}}{\pi}\left(\cos \omega t - \frac{1}{5}\cos 5\omega t + \frac{1}{7}\cos 7\omega t - \frac{1}{11}\cos 11\omega t + \cdots\right)$	**22-14**	x only	Odds only at ±90°
J. Sawtooth wave	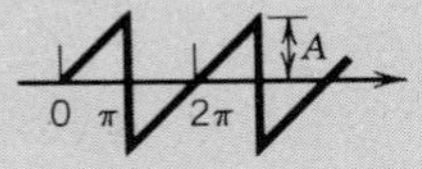	$f(t) = A\frac{2}{\pi}\left(\sin \omega t - \frac{1}{2}\sin 2\omega t + \frac{1}{3}\sin 3\omega t - \frac{1}{4}\sin 4\omega t + \cdots\right)$	**22-15**	z only	Evens only at 180° Odds only at 0°
K. Trapezoidal wave	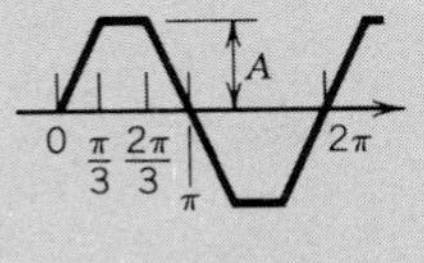	$f(t) = A\frac{6\sqrt{3}}{\pi^2}\left[\left(\sin \omega t - \left(\frac{1}{5^2}\right)\sin 5\omega t + \left(\frac{1}{7^2}\right)\sin 7\omega t - \cdots\right)\right]$	**22-16**	x, y, z	Odds only at 0° and 180°
L. Armature reaction in a three-phase winding waveform	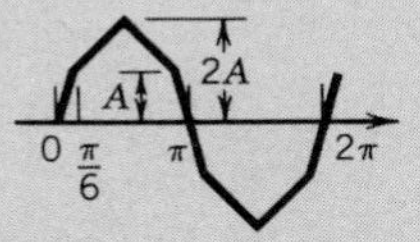	$f(t) = \frac{18}{\pi^2}\left(\sin \omega t + \frac{1}{5^2}\sin 5\omega t - \frac{1}{7^2}\sin 7\omega t + \cdots\right)$	**22-17**	x, y, z	Odds only at 0° and 180°
M. Radar pulse	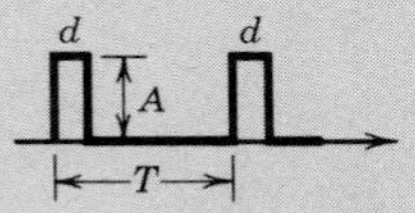	$f(t) = A\frac{d}{T} + 2\frac{A}{\pi}\left[\sin\left(\frac{\pi d}{T}\right)\cos \omega t + \frac{1}{2}\sin\left(2\pi\frac{d}{T}\right)\cos 2\omega t + \frac{1}{3}\sin\left(3\pi\frac{d}{T}\right)\cos 3\omega t + \cdots + \frac{1}{n}\sin\left(n\pi\frac{d}{T}\right)\cos n\omega t\right]$	**22-18**	None	Odds and evens at any angles
N. Sawtooth pulse	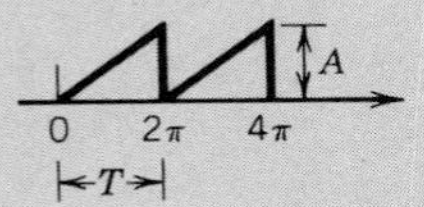	$f(t) = \frac{A}{\pi}\left(\frac{\pi}{2} - \sin \omega_0 t - \frac{\sin 2\omega_0 t}{2} - \frac{\sin 3\omega_0 t}{3} - \frac{\sin 4\omega_0 t}{4} - \cdots\right)$	**22-19**	z only	Evens only at 180° Odds only at 180°

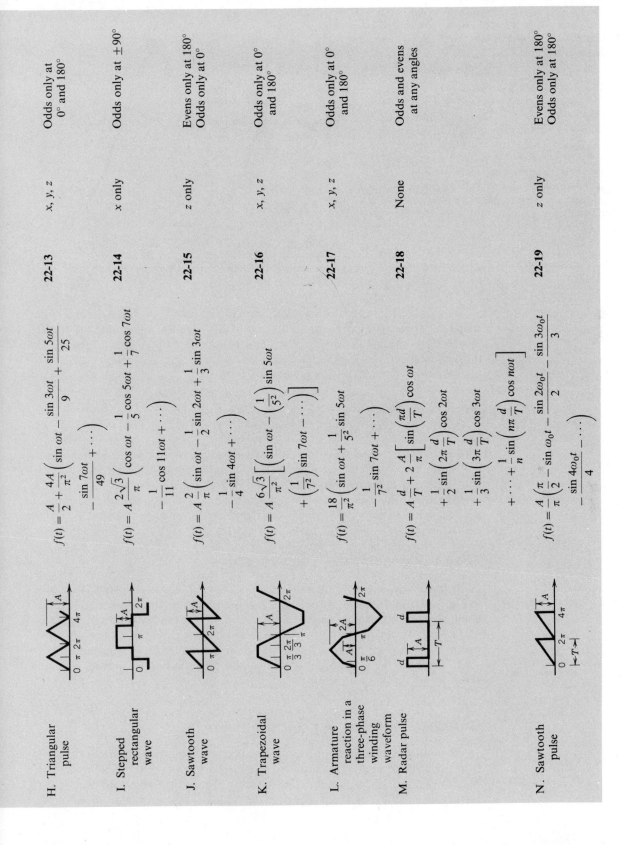

EXAMPLE 22-12

Using only the first three terms of Eq. (22-10) for the triangular waveform (fundamental, third, and fifth harmonics only), show that the equation given in Table 22-1E is valid

Solution

Make $8A/\pi^2 = 10$ so that the fundamental is $10 \sin \omega t$, the third harmonic is $1.111 \sin(3\omega t + 180°)$, and the fifth harmonic is $0.4 \sin 5\omega t$. Solving for A yields $A = 10\pi^2/8 = 12.34$ V, which is the maximum value of the triangular waveform to be produced. **Figure 22-13** shows the graphical addition of the three waveforms: $v(t) = 10 \sin \omega t - 1.111 \sin 3\omega t + 0.4 \sin 5\omega t$, producing a triangular wave. Observe that the positive maxima of the fundamental, third, and fifth harmonics *all coincide*, as do the negative maxima, respectively, producing a maximum positive and negative value of $\pm$**11.5 V**. The remaining harmonics would produce the theoretical maximum of 12.34 V.

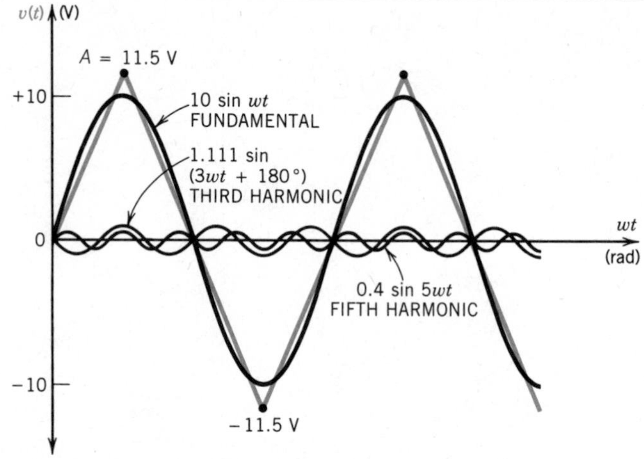

Figure 22-13 Synthesis of triangular waveform using only first three terms of Eq. (22-10). Example 22-12

The graphical synthesis of Ex. 22-12 proves the validity of Eq. (22-10) for the triangular wave. The problems at the end of this chapter will enable the reader to prove the validity of some of the other equations, as well. (See Probs. 22-21 through 22-27.)

22-10 HARMONIC GENERATION AS A FUNCTION OF NONLINEARITY

All harmonics are generated as a result of *nonlinearity* of the electrical components and devices to which a given signal is applied. In some instances this is done deliberately in order to produce a second desirable harmonic waveform from a given nonsinusoidal or sinusoidal waveform. In other instances the harmonics produced are undesirable. We will consider examples of each.

22-10.1 Generation of a Square Wave

Most frequently, square wave generation begins with a pure sinusoid. As shown in **Fig. 22-14a**, a sine wave is applied to a step-up transformer connected to a diode configuration known as a "biased parallel clipper." The diode nonlinear properties are used to advantage to "clip" the positive- and negative-going sinusoids, producing a trapezoidal output. The output is further amplified and clipped, successively, until the final result is a "pure" square wave, having an average value of zero.

22-10.2 Generation of a Positive Pulse Train

In the event that a series of positive pulses is desired, the *nonlinearity* of the *diode* is again used to rectify the output of the square wave generator in Fig. 22-14a. As shown in Fig. 22-14b, a simple half-wave rectifier is used, which, in effect, "clips" the negative half-cycles and passes only the positive half-cycles. The output of Fig.

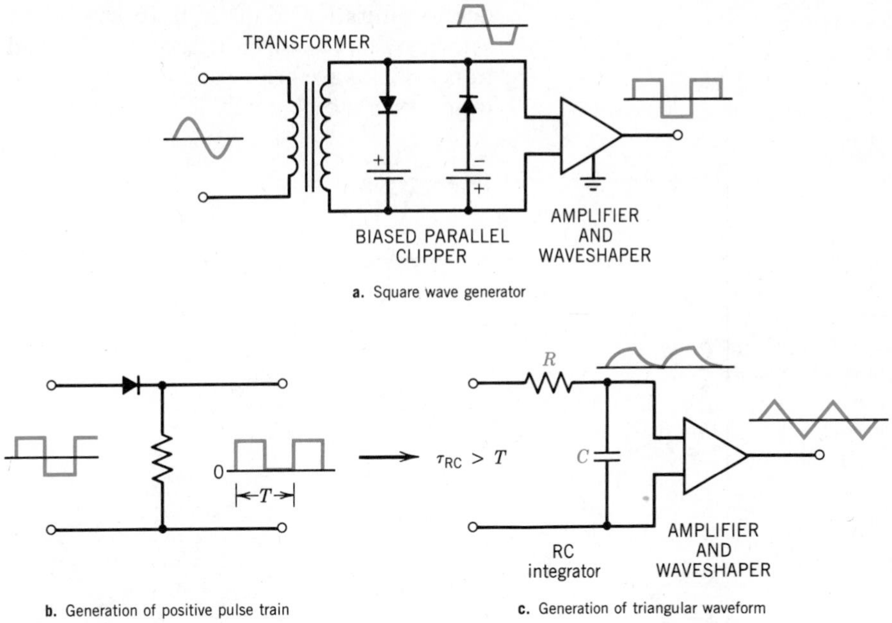

Figure 22-14 Waveform generation of square and triangular waveforms

22-14b, consequently, has a dc component in addition to the various odd harmonics comprising the square wave. (Compare Eqs. 22-11 and 22-12.)

22-10.3 Generation of a Triangular Waveform

The relaxation oscillator (Sec. 10-9, Fig. 10-14) is the customary method of sawtooth generation. **Figure 22-15** shows a solid-state relaxation oscillator with a Shockley diode (instead of a neon gas tube). Alternatively, another solid-state device (the unijunction transistor) may be used. But the basic sawtooth, as shown in Sec. 10-9, is generated by the nonlinearity of the capacitor charging exponentially. As in the case of the square wave generator, the basic sawtooth waveform is further amplified and made more linear to produce the customary positive sawtooth pulse waveform.

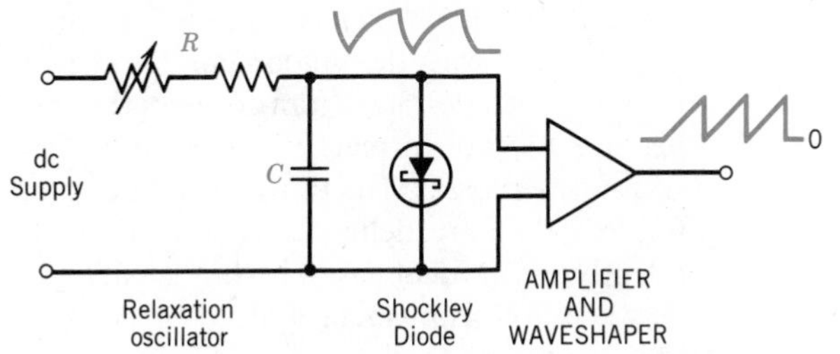

Figure 22-15 Generation of sawtooth waveform

22-10.4 Generation of Nonsinusoidal Magnetizing Current in Commercial Transformers

An interesting application of nonsinusoidal waveform generation occurs in the commercial power transformer. From our study of magnetization (Secs. 11-6, 11-7 and Figs. 11-6, 11-7), we know that the *B–H* curve for a transformer is nonlinear due to saturation of the iron. If the magnetizing current is purely sinusoidal, as shown in **Fig. 22-16a**, the transformer flux and output voltage becomes a flat-topped trapezoidal waveform. Such a waveform is highly undesirable, since we expect transformers to be purely linear elements.

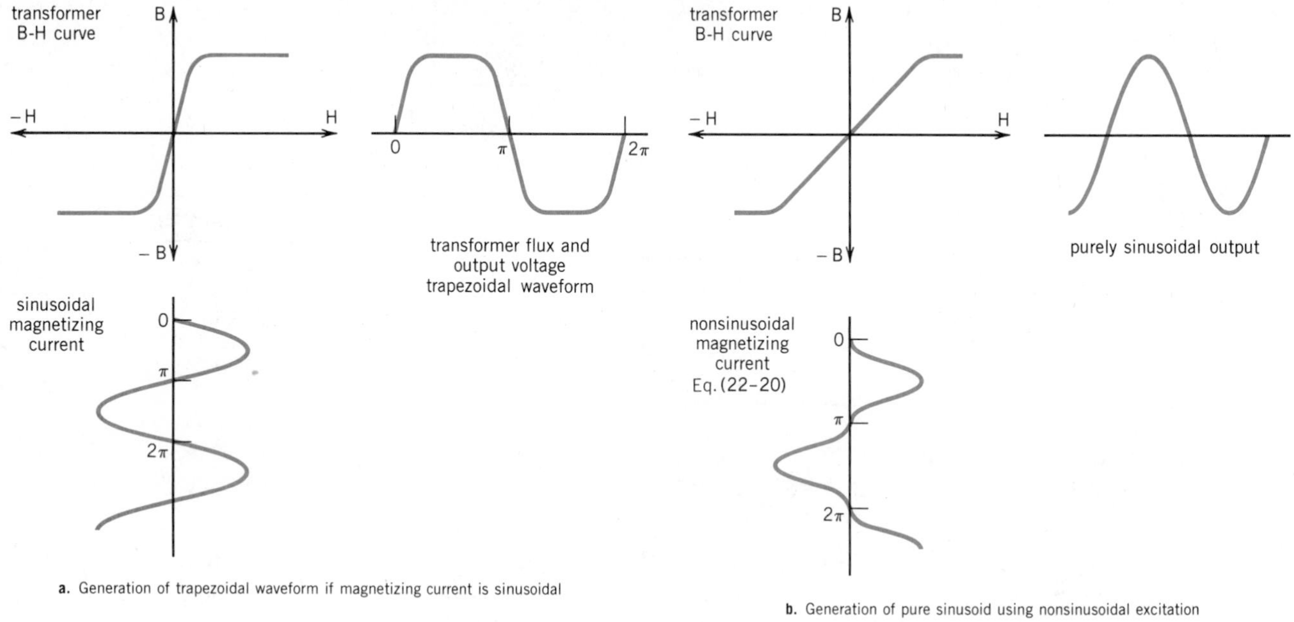

Figure 22-16 Generation of pure sinusoid in commercial transformers subject to magnetic saturation

By changing the magnetic permeability of the iron and reducing hysteresis to a minimum, it is possible to produce the purely sinusoidal output shown in Fig. 22-16b. But this is done at the expense of a nonsinusoidal magnetizing current in the transformer primary. The waveform of the nonsinusoidal current has the following equation at no load:

$$i_0(t) = I_m(\sin \omega t - 0.547 \sin 3\omega t + 0.315 \sin 5\omega t + \cdots) \qquad \textbf{(22-20)}$$

Note that this waveform contains a fairly high third harmonic component 180° out of phase with the fundamental and a fifth harmonic in phase with the fundamental. In single-phase transformers the third harmonic essentially produces overheating of transformer windings and also the load and increased iron (core) transformer losses. The third harmonic is suppressed, however, in polyphase transformers that are delta-connected.[4] The presence of the third harmonic, consequently, is an ideal case of what is called an "engineering trade-off." As a choice between a nonsinusoidal output or the generation of nonsinusoidal exciting current, we accept the latter as the lesser evil. In this way, the power utility company ensures that the waveform delivered to its customers is purely sinusoidal.

[4] The third harmonic can be detected when connecting secondaries of three single-phase transformers in delta. It is customary to measure the terminal voltage *before* closing the delta when the three transformers are mesh-connected. In some transformers, instead of zero, the third harmonic appears as a fairly large voltage. This gives the impression that the transformers are *not* properly phased for mesh connection. Connecting a resistor across the open delta (permitting current flow in the delta winding) immediately suppresses the harmonic, producing zero voltage across it!

22-10.5 Effect of Nonlinear Networks on Complex Waveforms

A circuit containing resistors only has *no effect whatever* on complex waves (assuming that the resistors are linear elements). But the capacitor, as we have seen, is a *nonlinear* element and is also frequency-dependent. (Resistors may be considered independent of frequency, for our purposes.) If a nonsinusoidal waveform is applied to a frequency-dependent network (*RC* or *RL*), as shown in **Fig. 22-17**, the output voltage may *not* be the same as the input. Figure 22-17a shows a square wave applied to a high-pass *RC* filter (Sec. 19-8.3). From previous study, we know that a capacitor acts as a short circuit to high frequencies. Consequently, we would expect *little* or no distortion of the square waveform at *high* frequencies, as shown in Fig. 22-17b, when the input square wave has a *fundamental* frequency of 10 kHz *with higher harmonic multiples.*

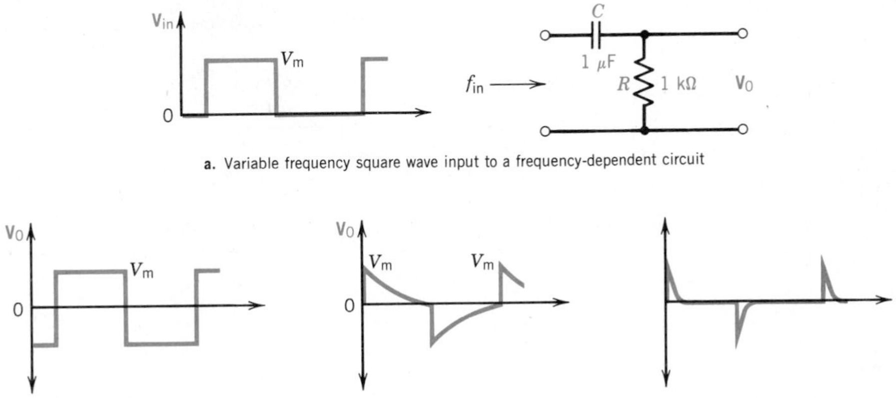

a. Variable frequency square wave input to a frequency-dependent circuit

b. Output when f_{in} = 10 kHz (no distortion)

c. Output when f_{in} = 1 kHz (distortion of input square wave)

d. Output when f_{in} = 100 Hz (distortion)

Figure 22-17 Effect of high-pass *RC* filter in distorting complex wave input at low frequencies

But when the frequency of the input square wave is *reduced* to 1 kHz, as shown in Fig. 22-17c, the output is *distorted* considerably. And when the frequency of the input square wave is reduced to 100 Hz, the output is further distorted, as shown in Fig. 22-17d. In the latter mode, the *RC* circuit is performing as a *differentiator* circuit since its output is differentiating the input square wave.[5]

The reasons for the distortion may be explained as follows:

1. The input square wave contains harmonics of various frequencies that are odd multiples of the fundamental and in phase with the fundamental (Eq. 22-12).
2. Each harmonic frequency produces a different and separate effect on the given capacitance (or inductance), producing *different* capacitive reactances at *various frequencies.*
3. These differences cause changes in amplitude and phase of the original harmonics. The resultant output waveform, therefore, is no longer the same as the original.
4. The degree of distortion of the original input waveform, therefore, depends on the frequency response and phase response characteristics of the network (see Sec. 19-8.3), as well as the network time constant compared to the input period of the waveform.

[5] This occurs whenever the RC circuit time constant is *short* compared to the input square wave period. In this case the input period is 1/100 Hz or 10 ms, whereas the circuit time constant is $\tau = RC =$ (1 kΩ)(1 μF) = 1 ms. The circuit time constant is one-tenth the input period of the waveform.

22-10.6 Square Wave Testing

Since a square wave is rich in *higher-order odd* harmonics, all of which are *in phase* with the fundamental (Eq. 22-11), it serves as an excellent means of testing the linearity of devices designed to be linear, such as transformers and/or amplifiers. If the equipment to be tested is *linear*, therefore, it should pass a square wave *without* the introduction of nonlinearities, that is, *without distortion* of the original square wave.

Figure 22-18a shows the classic circuit used in square wave testing. The square wave generator (SWG) is a variable-frequency source capable of producing a square wave from 10 Hz up to 1 MHz (fundamental frequency).

Figure 22-18b shows ways of checking the high-frequency (HF) response of an amplifier. A small amount of rounding on the *leading edge* of the waveform signifies a *small* loss of HF. On the other hand, if the waveform is very rounded with its corners completely removed, it is an indication of *poor* HF response.

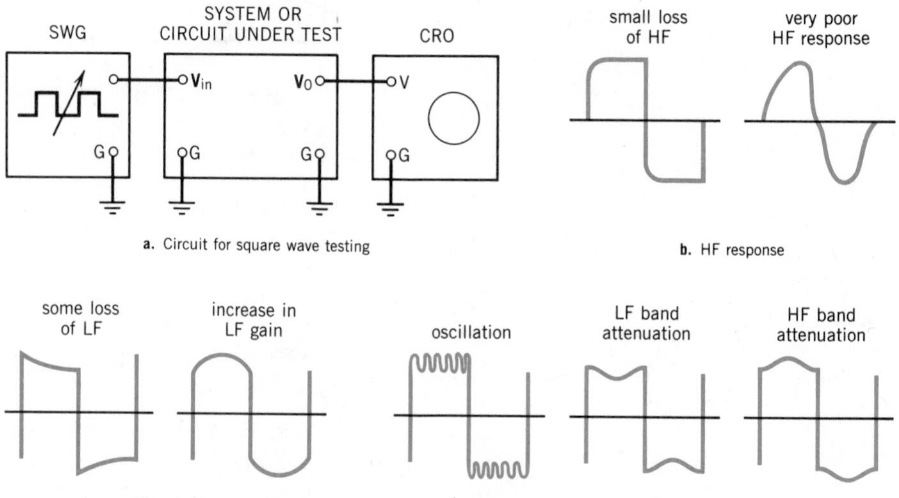

a. Circuit for square wave testing

b. HF response

c. LF response

d. Other distortions

Figure 22-18 Various responses in square wave testing

Figure 22-18c shows the effects of distortion in low-frequency (LF) response. These are similar to those originally given in Figs. 22-17c and 22-17d. In the event that the distortion includes an increase in LF gain, however, the waveform will exhibit straight sides but an increase in rounding in the center of the output waveform.

Other forms of distortion, which may be viewed on a CRO during square wave testing, are shown in Fig. 22-18d. These are oscillation or ringing, LF band attenuation (the reverse of LF gain), and HF band attenuation.

22-10.7 Nonlinearity Produced by a Semiconductor Diode

The commercial semiconductor diode (either germanium or silicon) is *not* an *ideal* diode. The waveforms given in Eqs. (22-8) and (22-9) assume an ideal diode having a barrier voltage of zero, an infinite reverse resistance, and a forward resistance of zero. None of these assumptions is completely valid. Consequently, the output waveform of a half-wave rectifier (HWR) is not exactly that shown in Table 22-1B. The output waveform more closely resembles that shown in **Fig. 22-19**, revealing some small conduction when the diode is reverse-biased due to reverse resistance (R_r). Figure 22-19 shows the classical semiconductor diode characteristic having *low forward* resistance (R_f) and *high reverse* resistance (R_r). The output waveform obtained for a sinusoidal input shows y-axis symmetry only, implying that the waveform must contain even harmonics at $\pm 90°$ but may also contain odd harmonics. Using a wave analyzer, the equation for the practical waveform is shown in Fig. 22-19 for an average value of unity. The reader should compare this equation with that given in Table 22-1B for the ideal HWR.

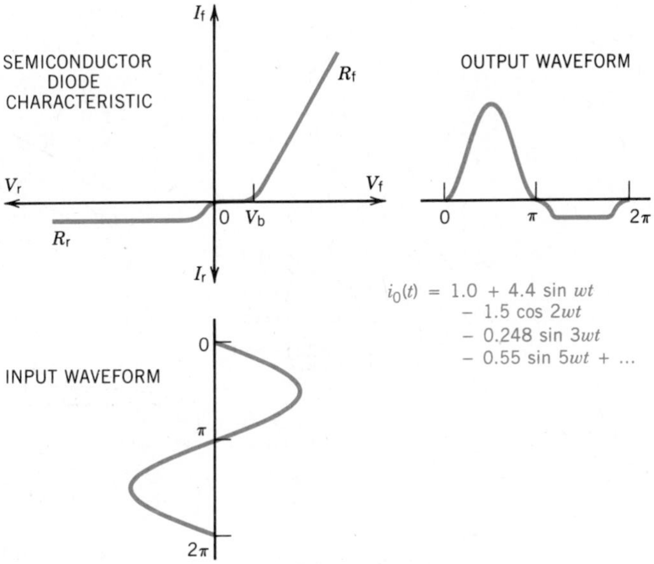

Figure 22-19 Nonlinearity produced by commercial (practical) semiconductor diode

22-10.8 Nonlinearity Produced by Transistors

Figure 22-20 shows the common-emitter configuration collector characteristics of a typical transistor. If the base current variation around the Q-point is excessively high, the collector current is driven into both *saturation and cutoff*. Similarly, the output collector voltage, V_c, would exhibit the same trapezoidal waveform.

But even if the transistor is operated within the "linear" portion of its characteristic, some nonlinearity inevitably results. The reader may ask, "How is this possible, since the load line is perfectly linear?"

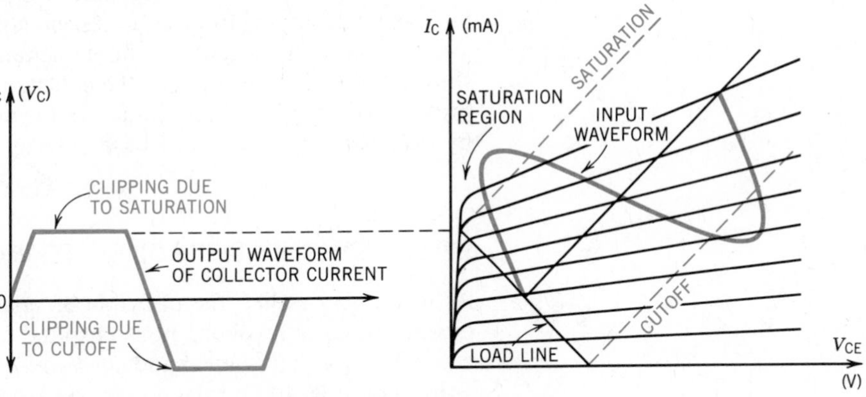

Figure 22-20 Nonlinear (trapezoidal) distortion produced by overdriven transistor

The answer lies in the fact that *equal* increments of *base* current do not produce equal space increments on the collector characteristic. The base current "spreading effect" shown in **Fig. 22-21a** results in *some nonlinearity*, even when the transistor is *not* overdriven. As shown in Fig. 22-21a, at a quiescent base current of 10 μA, for $\beta = 200$, the collector current is 2 mA. But a small base current variation of $\pm 5\ \mu$A produces a maximum collector current of 3.1 mA and a minimum of 1.1 mA. As shown in Fig. 22-21a, the positive variation is 1.1 mA, whereas the negative variation is only 0.9 mA.

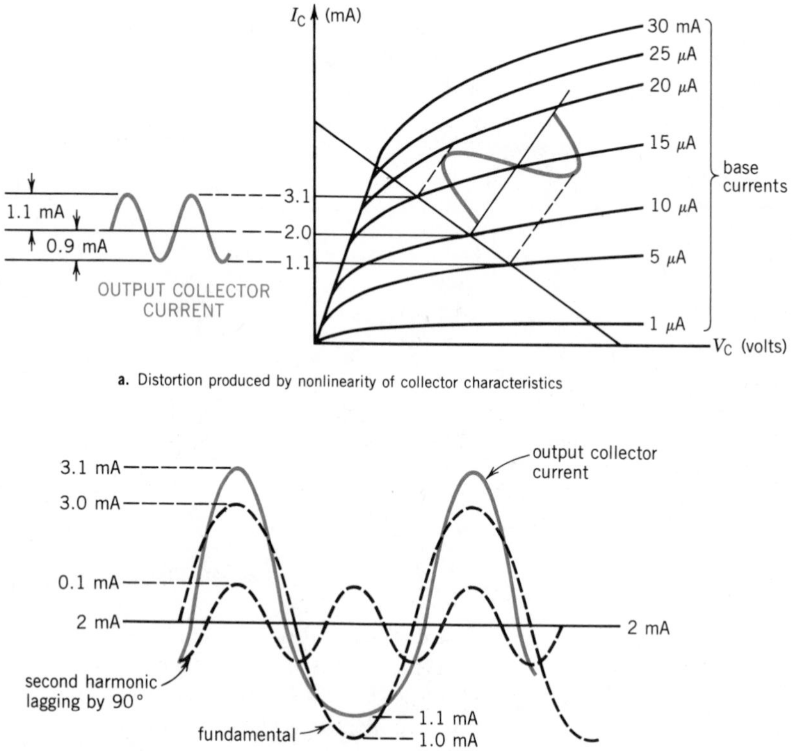

a. Distortion produced by nonlinearity of collector characteristics

b. Detail of output collector current in (a)

Figure 22-21 Waveform distortion due to transistor nonlinearity

Again, a harmonic wave analysis of the output collector current waveform shows that the nonlinear collector current waveform of Fig. 22-21b is composed of a fundamental (varying from the quiescent collector current of 2 mA by ± 1 mA) and a second harmonic lagging the fundamental by 90°. The sum of the two harmonics shown in Fig. 22-21b produce the collector current waveform of Fig. 22-21a. The presence of this second harmonic is a form of distortion that can be reduced only by *reducing* the variation of base current input signal, for any given transistor.

22-11 CIRCUIT RESPONSE TO NONSINUSOIDAL VOLTAGES

As mentioned earlier, the principle of *superposition* is the key to the solution of circuit problems involving nonsinusoidal voltages and/or currents. In the case of a pure resistor that is linear and *not frequency-sensitive*, we need only find the RMS voltage in order to find the power dissipated by the resistor, using $P = V^2/R$. But in the case of a complex impedance containing either an inductor or a capacitor, it is necessary to find the current response at each particular and separate frequency by superposition. The power dissipated by such a circuit is then found from the customary relation $P = (V_m I_m/2) \cos\theta$, derived in Appendix B-10 and expressed in Sec. 14-17 as Eq. (14-39), again using superposition for the powers at each frequency. These points are illustrated in the following examples, along with a number of useful insights.

EXAMPLE 22-13

Given the nonsinusoidal supply voltage $v(t) = 10 \sin 377t + 5 \sin 754t$ volts applied to a complex impedance of $\mathbf{Z} = 3 + j4\ \Omega$ at the fundamental frequency,

a. Sketch the voltage waveform from the equation and determine its symmetry
b. Find the complex current drawn from the supply and express it in Fourier form
c. Sketch the current waveform from the equation and determine its symmetry
d. Calculate the RMS values of current and voltage
e. Calculate the power dissipated by the impedance

Solution

a. The waveform is sketched in **Fig. 22-22a**, showing the z-axis symmetry resulting from an even harmonic in phase with the fundamental.
b. To find the complex current, we must apply superposition, using each frequency contained within the voltage waveform. The use of the following tabulation simplifies matters greatly, since the complex impedance varies as a function of frequency.

ω (rad/s)	R (Ω)	$n\omega L$ (Ω)	$\mathbf{Z}_L$ (Ω)	$\mathbf{V}_m\angle\theta$ (V)	$\mathbf{I}_m\angle\theta$ (A)
377	3	$j4$	$5\angle 53.13°$	$10\angle 0°$	$2\angle -53.13°$
754	3	$j8$	$8.54\angle 69.\bar{4}°$	$5\angle 0°$	$0.585\angle -69.\bar{4}°$ A

From the last column, $i(t) = 2 \sin(377t - 53.13°) + 0.585 \sin(754t - 69.\bar{4}°)$ A

c. The *current* waveform is sketched in Fig. 22-22b. Observe that the resultant current has *no symmetry whatever*, since it contains even harmonics at angles other than 0°, 90°, or 180°. (Actually, the sketch is unnecessary since we could deduce this from the current equation just derived.)

d. $\mathbf{V} = \sqrt{\frac{1}{2}(10^2 + 5^2)} = \mathbf{7.906\ V}$; $\mathbf{I} = \sqrt{\frac{1}{2}(2^2 + 0.585^2)} = \mathbf{1.473\ A}$

e. $$\mathbf{P} = \mathbf{P}_1 + \mathbf{P}_2 = \frac{(10)(2)}{2} \cos 53.13° + \frac{(5)(0.585)}{2} \cos 69.\bar{4}°$$
$$= 6 + 0.5135 = \mathbf{6.5135\ W}$$

Example 22-13 yields the following insights:

1. Although the voltage applied to a complex impedance has z-axis symmetry, the resulting current waveform has *no symmetry whatever*. This comes as no surprise, since the impedance is no longer linear with respect to frequency.
2. To find the complex waveform of current, it was necessary to find the impedance at each frequency, followed by the current at each frequency.
3. Since the resistance is assumed independent of frequency, we might have found the power dissipated in it by two other methods, as shown in Ex. 22-14.

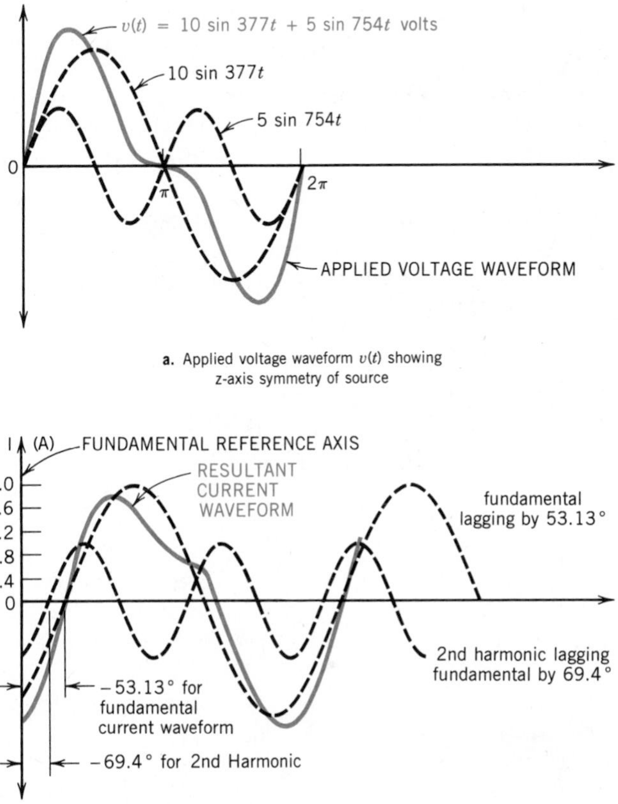

a. Applied voltage waveform $v(t)$ showing z-axis symmetry of source

b. Resultant current waveform $i(t) = 2 \sin(377t - 53.13°) + 0.5852 \sin(754t - 69.4°)$ A

Figure 22-22 Example 22-13 showing effect of nonlinear impedance on current waveform

EXAMPLE 22-14

Verify the power obtained in Ex. 22-13e by using

a. The maximum currents at each frequency
b. The RMS current found for the complex waveform in Ex. 22-13d

Solution

a. $\mathbf{P}_{av} = \mathbf{I}_m^2 R/2 = 3\ \Omega(2^2 + 0.585^2)/2 = \mathbf{6.513\ W}$
b. $\mathbf{P}_{av} = \mathbf{I}^2 R = (1.473)^2 \times 3 = \mathbf{6.51\ W}$

At this point the reader may ask, "If it's so easy to find power using the methods of Ex. 22-14, why bother using the method of Ex. 22-13**e**?" The answer to this question is found in Ex. 22-15.

EXAMPLE 22-15

When a waveform of voltage $v(t) = 10 \sin 377t + 5 \sin 754t$ volts is applied to an unknown impedance, a current waveform $i(t) = 2 \sin(377t - 53.13°) + 0.585 \sin(754t - 69.4°)$ amperes results. Calculate the

a. Total average power dissipated by the impedance
b. Resistance of the complex impedance
c. Inductive reactance of the complex impedance (at the fundamental frequency)
d. Inductance of the circuit

Solution

a. From the Fourier series trigonometric equations given for voltage and current,

$$P_{av} = P_1 + P_2 = \frac{(10)(2)}{2} \cos 53.13° + \frac{(5)(0.585)}{2} \cos 69.\bar{4}°$$

$$= \mathbf{6.513\ W}$$

b. $R = P/\mathbf{I}^2 = 6.513/\frac{1}{2}(2^2 + 0.585^2) = 6.513/2.17\bar{1} = \mathbf{3\ \Omega}$
c. $X_L = R \tan \theta$ and $\theta = 53.13°$, current lagging voltage at $\omega = 377$ rad/s $= 3\ \Omega(\tan 53.13°) = \boldsymbol{j4\ \Omega}$
d. $L = X_L/\omega = 4\ \Omega/377 = \mathbf{10.6\ mH}$

Example 22-15 shows that, given the trigonometric Fourier series for voltage and current, respectively, it is possible to find the total power dissipated, the RMS values of voltage and current, and the nature of the complex impedance to which the voltage was applied! (The observant reader may have noticed that Ex. 22-15 is the *inverse* of Ex. 22-13. This was done deliberately in order to verify the correctness of the solution techniques.)

At this point, the reader may ask, "What if the waveforms of voltage and current contain a dc component? How are these treated?" The answers to these questions are found in the examples of Sec. 22-11.1.

22-11.1 Average Power in Nonsinusoidal Waveforms Containing a DC Component

In effect, the addition of the dc component poses nothing new if we realize that the dc waveforms of applied voltage and resulting current may be treated as components at zero frequency with zero phase shift between them, regardless of the nature of the complex impedance. To simplify matters, let us begin with a purely resistive load.

EXAMPLE 22-16

Given the square wave pulse waveform shown in Table 22-1G with a maximum amplitude of 20 V applied across a 10-Ω resistor,

a. Write the first five terms of the trigonometric Fourier series representing $v(t)$
b. Calculate the average power dissipated by each term in (**a**)
c. Calculate the total average power dissipated in the resistor, using (**b**)
d. Using the theoretical RMS voltage, calculate the true power dissipated
e. Calculate the percentage of actual total power dissipated by the first five terms of the series given in Table 22-1G

Solution

a. $v(t) = \mathbf{V}/2 + (2\mathbf{V}/\pi) \times (\sin \omega t + \sin 3\omega t/3 + \sin 5\omega t/5 + \sin 7\omega t/7)$ (22-12)

$v(t) = 10 + 12.73 \sin \omega t + 4.22 \sin 3\omega t + 2.546 \sin 5\omega t + 1.819 \sin 7\omega t$

b. $P_{dc} = V_{dc}^2/R = 10^2/10 = \mathbf{10\ W}$, $P_1 = V_m^2/2R = 12.73^2/20 = \mathbf{8.106\ W}$, $P_2 = V_m^2/2R = 4.22^2/20 = \mathbf{0.901\ W}$, $P_3 = V_m^2/2R = 2.546^2/20 = \mathbf{0.324\ W}$, $P_4 = V_m^2/2R = 1.819^2/20 = \mathbf{0.165\ W}$
c. $\sum P = P_{dc} + P_1 + P_2 + P_3 + P_4 = \mathbf{19.5\ W}$
d. For a square wave pulse, from Table 14-1, the true RMS voltage is $\mathbf{V} = V_m/\sqrt{2} = 20\ \text{V}/\sqrt{2} = 14.14$ V; $P_t = \mathbf{V}^2/R = (14.14)^2/10 = \mathbf{20\ W}$
e. $\sum P/P_t = 19.5/20 = \mathbf{97.5\%}$ of the total power

Several important insights emerge from Ex. 22-16:

1. Since the load is purely resistive, it is unnecessary to find the currents produced by each harmonic. We are at liberty to use $\mathbf{V}^2/R$ because each voltage component is applied *directly* across the resistance, R.
2. Use of the dc component plus the first four harmonic terms is sufficient to yield an answer within 2.5% of that obtained if *all* the harmonics are used.

EXAMPLE 22-17

Assuming that the fundamental radian frequency of the square pulse in Ex. 22-16 is 100 rad/s and that the pulse is applied across a resistance of 3 Ω in series with a capacitance of 2500 μF,

a. Write the first five terms of the trigonometric Fourier series for $v(t)$
b. Find the complex current drawn from the supply and express it in Fourier form
c. Calculate the RMS value of the current in the complex impedance, using the first four harmonic terms. Predict the symmetry of the current waveform
d. Calculate the power dissipated in the complex impedance, using RMS current in (**c**)
e. Calculate the power dissipated in the complex impedance, using the Fourier components of voltage and current

Solution

a. $v(t) = 10 + 12.73 \sin 100t + 4.22 \sin 300t + 2.546 \sin 500t + 1.819 \sin 700t + \cdots$
b. Apply superposition for each voltage and use a tabulation to find the current:

ω (rad/s)	R (Ω)	$1/n\omega C$ (Ω)	$\mathbf{Z}$ (Ω)	$V_m\angle\theta$ (V)	$I_m\angle\theta$ (A)
0	3	∞	∞	10	0
100	3	$-j4$	$5\angle-53.13°$	$12.73\angle0°$	$2.546\angle53.13°$
300	3	$-j1.\bar{3}$	$3.283\angle-24°$	$4.22\angle0°$	$1.285\angle24°$
500	3	$-j0.8$	$3.105\angle-14.9°$	$2.55\angle0°$	$0.82\angle14.9°$
700	3	$-j0.5714$	$3.054\angle-10.8°$	$1.82\angle0°$	$0.596\angle10.8°$

c. $\mathbf{I} = \sqrt{\tfrac{1}{2}(2.546^2 + 1.285^2 + 0.82^2 + 0.596^2)} = \mathbf{2.14\ A}$
The current waveform must have mirror, x-axis, symmetry only since it contains *only odd* harmonics at angles other than 0° or 180°
d. $P_t = \mathbf{I}^2R = 2.14^2(3) = \mathbf{13.74\ W}$
e. $P_t = 10(0) + \tfrac{1}{2}(12.73)(2.546)\cos 53.13° + \tfrac{1}{2}(4.22)(1.285)\cos 24° + \tfrac{1}{2}(2.55)(0.82)\cos 14.9° + \tfrac{1}{2}(1.82)(0.596)\cos 10.8° = 0 + 9.723 + 2.478 + 1.01 + 0.533 = \mathbf{13.74\ W}$

Several important insights emerge from Ex. 22-17:

1. It is now *necessary* to find the *complex current* in order to find the power dissipated in the complex impedance, since only resistance dissipates power.
2. For the dc component, even though the voltage is 10 V, the current in the circuit is zero because a capacitor has infinite impedance to dc. (The reverse is the case for an inductor, which acts as a short circuit when $\omega = 0$.)
3. The current waveform must have x-axis symmetry since it is composed of odd harmonics only, at various angles.
4. The nonlinearity of the capacitor, however, has distorted the original symmetry of the voltage waveform, which had x-, y-, and z-axis symmetry.
5. It is impossible for the capacitor to introduce second harmonics into the current waveform because there is none in the original voltage waveform. The capacitor is a *passive bilateral* element. (Only active and unilateral passive elements are capable of introducing even harmonics.)
6. The validity of the calculation steps is proved in that solutions (**d**) and (**e**) yield the same total power of 13.74 W.
7. We *did not* use a $\mathbf{V}^2/R$ power calculation (as was done in Ex. 22-16) since the voltage is applied across *both* resistance and capacitive reactance. Although it is possible to compute $\mathbf{V}_R$ at each frequency, we are much safer using I^2R.
8. Had the complex impedance consisted of $R + jX_L$, power would have been dissipated when $\omega = 0$ because there would have been a current component at this frequency! This is an important distinction that must not be overlooked for nonsinusoidal sources containing a dc component.

22-11.2 *RLC* Circuit Response to Nonsinusoidal Voltages

We know from our study of series resonance (Chapter 19) that resonance occurs in a series *RLC* circuit when the circuit is primarily *resistive* and *current* is a *maximum*. One way of producing this effect is to vary the frequency until the current is maximized. Given a series *RLC* circuit to which a nonsinusoidal voltage (rich in harmonics) is applied, resonance or near-resonance effects may occur at some harmonic frequency. This is shown in Ex. 22-18.

EXAMPLE 22-18

Given a series circuit consisting of $R = 5\ \Omega$, $L = 5$ mH, and $C = 50\ \mu$F. The applied voltage waveform is $v(t) = 100 + 150 \sin 1000t + 100 \sin 2000t + 75 \sin 3000t$ volts

a. Calculate the resulting current at each of the four frequency components
b. Calculate the RMS current
c. Calculate the total power dissipated
d. Draw the line spectra for both the applied voltage and the resulting current
e. From (**d**) determine the frequency at which the circuit is resonant
f. Predict the symmetry of the applied voltage waveform
g. Predict the symmetry of the current waveform

Solution

a. Apply superposition for each voltage and use the following tabulation:

ω (rad/s)	R (Ω)	$1/n\omega C$ $-j(\Omega)$	$n\omega L$ $j(\Omega)$	$\mathbf{Z}_t$ (Ω)	$V_m\angle\theta$ (V)	$I_m\angle\theta$ (A)
0	5	∞	0	∞	10	0
1000	5	$-j20$	$j5$	$5 - j15$	$150\angle 0^\circ$	$9.487\angle 71.565^\circ$
2000	5	$-j10$	$j10$	$5 + j0$	$100\angle 0^\circ$	$20.0\angle 0^\circ$
3000	5	$-j6.\bar{6}$	$j15$	$5 + j8.\bar{3}$	$75\angle 0^\circ$	$7.717\angle -59.04^\circ$

b. $\mathbf{I} = \sqrt{\frac{1}{2}(9.487^2 + 20^2 + 7.717^2)} = \mathbf{16.58\ A}$
c. $P = I^2R = (16.58)^2 \times 5 = \mathbf{1373.9\ W}$
d. The line spectra are shown in **Figs. 22-23a** and **22-23b**, respectively.
e. From the tabulation in (**a**) and from Fig. 22-23b, the circuit is resonant at a radian frequency of 2000 rad/s. Note that the current is a maximum at this frequency, as shown in Fig. 22-23b.

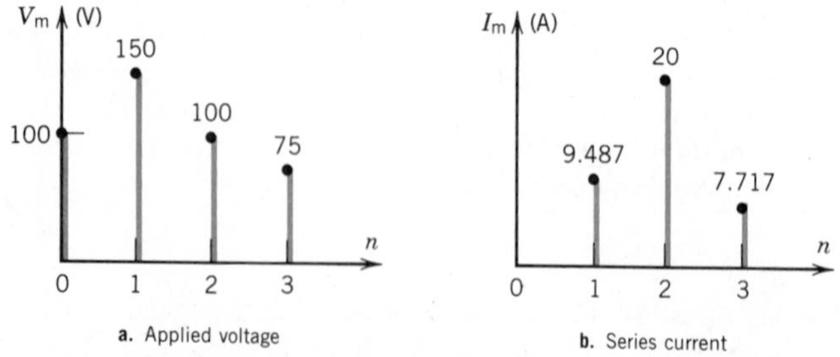

Figure 22-23 Line spectra of applied voltage and resulting current showing effect of series resonance (Example 22-18d)

f. The applied voltage contains odd harmonics at 0° and evens at 0°. The waveform must have **z-axis** symmetry. (See Sec. 22-5.)
g. The resulting current waveform has **no symmetry**. The presence of odd harmonics at angles other than 0° or 180° destroys the original *z*-axis symmetry of the applied voltage waveform.

Example 22-18 yields the following insights:

1. Current is *increased* at the harmonic frequency that produces *series* resonance.
2. Although not proved here, we can assume that current is minimized at the harmonic frequency at which parallel resonance occurs in a two-branch *RLC* parallel circuit. (See Ex. 22-19.)
3. The current is zero at $\omega = 0$, since the capacitor blocks any dc voltage.
4. Dissipated power increases at the resonant frequency (see tabulation in **a**) since current is maximized at that frequency.
5. The nonlinearities of inductance and capacitance with frequency destroy the symmetry and cause the current waveform to have no symmetry whatever.
6. The effects demonstrated and described above become more pronounced whenever the circuit Q increases. If the Q is high and resonance occurs at some high multiple (say $n = 10$), that current is much higher than the currents at $n = 9$ or $n = 11$, assuming the applied voltage contains components at these multiples.
7. Based on (6), therefore, assuming *descending magnitudes in the line spectrum* of applied voltage, any sudden *rise* in the line spectrum of *current* may be attributed to a *series-resonant* effect. Conversely, any sudden *dip* in the line spectrum of *current* may be attributed to a *parallel-resonant* effect. This is shown in Ex. 22-19.

EXAMPLE 22-19

For the series–parallel circuit shown in **Fig. 22-24a**, $jX_L = j6\ \Omega$ and $-jX_C = -j24\ \Omega$ at the fundamental frequency. The applied voltage is $v(t) = 100 + 50 \sin 100t + 25 \sin 200t + 10 \sin 300t$ volts. Calculate the

a. Current waveform, $i(t)$
b. Output voltage waveform, $v_R(t)$
c. Power dissipated in the resistor, R
d. Plot the line spectra of applied voltage and resultant current
e. Predict the symmetry of the input voltage waveform
f. Predict the symmetry of the current and output voltage waveforms

Solution

a. $i(t) = 3.904 \sin(100t - 38.66°) + 0.57 \sin(300t + 55.22°)$ A
b. $v_R(t) = i(R)$
$= 39.04 \sin(100t - 38.66°) + 5.7 \sin(300t + 55.22°)$ V
c. $\mathbf{I}_{\text{RMS}} = \sqrt{\tfrac{1}{2}(3.904^2 + 0.57^2)} = \mathbf{2.79\ A}$
$P = \mathbf{I}^2R = (2.79)^2 10 = \mathbf{77.8\ W}$
d. The line spectra are shown in Figs. 22-24b and 22-24c, respectively.
e. The input voltage waveform has **z-axis** symmetry (evens at 0°, odds at 0°).
f. The current and output voltage have **x-axis** symmetry (odds only at angles other than 0° or 180°).

ω (rad/s)	jX_L (Ω)	$-jX_C$ (Ω)	$\mathbf{Z}_p$ (Ω)	$\mathbf{Z}_t$ (Ω)	$V_m\angle\theta$ (V)	$I_m\angle\theta$ (A)
0	0	∞	0	10	100	10
100	$j6$	$-j24$	$j8$	$10 + j8$	$50\angle 0°$	$3.904\angle -38.66°$
200	$j12$	$-j12$	∞	∞	$25\angle 0°$	0
300	$j18$	$-j8$	$-j14.4$	$10 - j14.4$	$10\angle 0°$	$0.57\angle 55.22°$

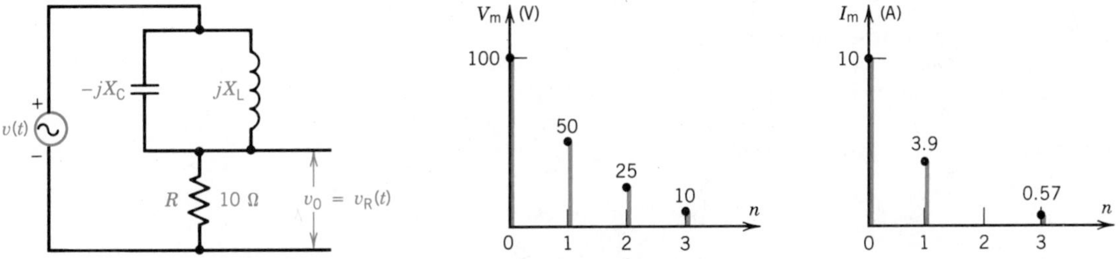

a. Circuit b. Applied voltage c. Total circuit current

Figure 22-24 Circuit and line spectra for Ex. 22-19 showing effect of parallel resonance

Example 22-19 and Figs. 22-24b and 22-24c verify the insight into the effect of parallel resonance noted in (7) above. As shown in Fig. 22-24b, the applied voltage contains a second harmonic of 25 V, maximum amplitude, which is *suppressed* in the line spectrum of current (Fig. 22-24c) due to *parallel* resonance. Also note from the tabulation in Ex. 22-19 that the parallel *LC* circuit containing ideal elements has zero impedance at $\omega = 0$ and infinite impedance at resonance. This is verified by the line spectra of Figs. 22-24b and 22-24c.

22-11.3 Series-Connected and Parallel-Connected Harmonic Sources

From Kirchhoff's voltage law (KVL), we know that source voltages in series may be added to produce a single equivalent voltage source. **Figure 22-25a** shows two sources, $v_1(t)$ and $v_2(t)$, respectively, connected *series-aiding* to a complex impedance $\mathbf{Z}_L$. If the harmonic voltage content of each source is known, it is possible to add the voltages separately at each harmonic frequency to find an equivalent voltage source that may be applied to complex impedance, $\mathbf{Z}_L$, as shown in Ex. 22-20.

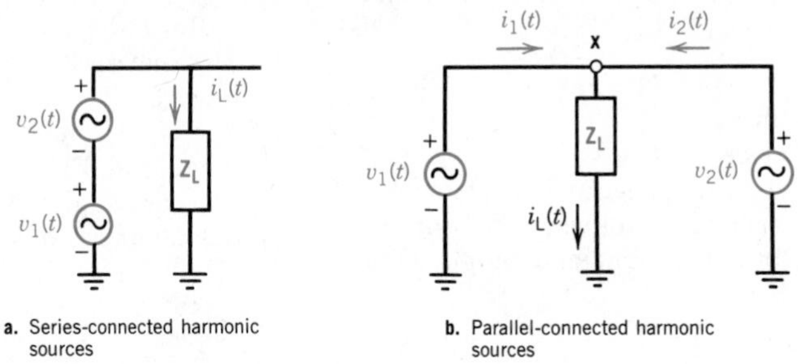

Figure 22-25 Superposition of series-connected and parallel-connected harmonic sources

EXAMPLE 22-20

Complex impedance $\mathbf{Z}_L$ is composed of a resistance of 10 Ω in series with an inductance of 10 mH. The trigonometric Fourier series of the series-connected source voltages are

$v_1(t) = 30 + 30 \sin \omega t + 10 \cos 2\omega t + 3 \sin(3\omega t - 50°)$ volts

$v_2(t) = 10 - 5 \sin \omega t + 5 \sin(2\omega t - 90°) + 4 \sin(3\omega t + 40°)$ volts

Assuming that the fundamental frequency, ω, is 1000 rad/s, calculate the

a. Equivalent harmonic voltage applied to complex load impedance $\mathbf{Z}_L$ expressed in trigonometric Fourier series form
b. Resulting current expressed in trigonometric Fourier series form
c. RMS value of current and power dissipated in the complex impedance, $\mathbf{Z}_L$

Solution

a. By inspection, the sum of the two voltages is[6] $v_1(t) + v_2(t) = 40 + 25 \sin \omega t + 5 \cos 2\omega t + 5 \sin(3\omega t + 3.13°)$ volts

b.

ω (rad/s)	R (Ω)	jX_L (Ω)	$\mathbf{Z}_L$ (Ω)	$V_m\angle\theta$ (V)	$I_m\angle\theta$ (A)
0	10	0	10	40	4
1000	10	$j10$	$10 + j10$	$25\angle 0°$	$1.768\angle -45°$
2000	10	$j20$	$10 + j20$	$5\angle 90°$	$0.2236\angle 26.565°$
3000	10	$j30$	$10 + j30$	$5\angle 3.13°$	$0.1581\angle -68.44°$

From the last column of this tabulation, the harmonic representation of the current waveform is $i(t) = 4 + 1.768 \sin(\omega t - 45°) + 0.2236 \sin(2\omega t + 26.565°) + 0.1581 \sin(3\omega t - 68.44°)$ A

c. $\mathbf{I}_{RMS} = \sqrt{4^2 + \frac{1}{2}(1.768^2 + 0.2236^2 + 0.1581^2)}$
$= \sqrt{4^2 + 1.6} = 4.195$ A and
$P = I^2R = (4.195)^2 \times 10 =$ **176 W**

Example 22-20 is sufficiently straightforward that the solution requires no special comment. But now consider the *same* source voltages connected in *parallel*, as shown in Fig. 22-25b. Should these sources produce the same current in $\mathbf{Z}_L$ as when they were series-connected? If so, what does this prove? Example 22-21 treats each source separately, using superposition in finding the harmonic load current in $\mathbf{Z}_L$.

[6] Proof of this sum is left as an exercise for the reader. See Prob. 22-30.

EXAMPLE 22-21

Given the same source voltages as in Ex. 22-20 connected in parallel as shown in Fig. 22-25b, calculate the

a. Resulting current $i_1(t)$ in Fourier form due to applied voltage $v_1(t)$ only acting across load $\mathbf{Z}_L$
b. Resulting current $i_2(t)$ in Fourier form due to applied voltage $v_2(t)$ only acting across load $\mathbf{Z}_L$
c. Phasor sum of load currents $i_1(t)$ and $i_2(t)$ or true current in $\mathbf{Z}_L$[7]
d. Compare the current obtained in (c) with the current in Ex. 22-20b and draw conclusions about the results

Solution

The following tabulation enables computation of $i_1(t)$ and $i_2(t)$:

ω (rad/s)	$\mathbf{Z}_L$ (Ω)	$v_1(t)$ (V)	$i_1(t)$ (A)
0	10	30	3
1000	$10 + j10$	$30\angle 0°$	$1.5 - j1.5$
2000	$10 + j20$	$10\angle 90°$	$0.4 + j0.2$
3000	$10 + j30$	$3\angle -50°$	$-0.04966 - j0.0808$

ω (rad/s)	$\mathbf{Z}_L$ (Ω)	$v_2(t)$ (V)	$i_2(t)$ (A)
0	10	10	1
1000	$10 + j10$	$-5 + j0$	$-0.25 + j0.25$
2000	$10 + j20$	$0 - j5$	$-0.2 - j0.1$
3000	$10 + j30$	$4\angle 40°$	$0.10\overline{7} - j0.0662$

From the tabulation we may write both currents in Fourier form:

a. $i_1(t) = 3 + 2.121\sin(\omega t - 45°) + 0.4472\sin(2\omega t + 26.565°) + 0.09484\sin(3\omega t - 121.6°)$ A
b. $i_2(t) = 1 + 0.35355\sin(\omega t + 135°) + 0.2236\sin(2\omega t - 153.43°) + 0.1265\sin(3\omega t - 31.56°)$ A
c. $i_L(t) = i_1(t) + i_2(t) = 4 + 1.768\sin(\omega t - 45°) + 0.2236\sin(2\omega t + 26.565°) + 0.1581\sin(3\omega t - 68.44°)$ A
d. The phasor sum of the two currents is identical to that obtained in Ex. 22-20b! The conclusions to be drawn are discussed below.

The following insights are to be drawn from Exs. 22-20 and 22-21:

1. It makes *no difference* whether the ideal harmonic *voltage* sources are connected in *series or parallel.* Both configurations of Fig. 22-25 produce the *same* waveform of nonsinusoidal current in the complex load, $\mathbf{Z}_L$.
2. Consequently, we may treat point x in Fig. 22-25b as a *summing junction.* If two or more *ideal* harmonic sources are connected to the summing junction, they may be added harmonically to produce a *single equivalent* harmonic voltage acting on complex load, $\mathbf{Z}_L$. The total current $i_L(t)$ is then found directly rather than by superposition of separate sources.
3. In summing voltage sources, the *polarity* of the ideal harmonic sources **must** be taken into account. In Fig. 22-25b both sources are positive with respect to summing junction x. This means that their voltages are *added.* But if one of the sources had a negative polarity with respect to x, the resultant equivalent harmonic voltage would be found by subtracting the negative ideal harmonic source from the positive harmonic source.
4. These insights serve to reduce the complexity of calculations, particularly when two or more sources in parallel are summed at the load junction, point x.

22-12 GLOSSARY OF TERMS USED

Amplitude *Maximum* or *peak* value of a waveform.

Aperiodic wave Waveform whose function of voltage is *not* periodic or repeated regularly with time.

Average value of a waveform Average of the absolute value of each instantaneous amplitude, with respect to zero, taken over

[7] Proof of this sum is left as an exercise for the reader. See Prob. 22-31.

the complete period of duration of the waveform, T. In a complex waveform containing many harmonically related frequencies, the zero frequency component is expressed in terms of its average value.

Bandwidth Range of frequencies of a continuous frequency band specified between an upper and lower limiting frequency.

Complex wave See nonsinusoidal wave.

DC component See average value. If a complex wave has an average value, it contains a dc component whose amplitude is that average value.

Even harmonics Sinusoidal components of periodic complex waves having frequencies that are even-number multiples of the fundamental frequency in a trigonometric Fourier series.

Fourier series Representation of a complex periodic function (that fulfills certain mathematical conditions) by a mathematical series in terms of its *average* value, its *fundamental*, and all *harmonically related* sine and cosine terms.

Fundamental frequency Lowest frequency component in the Fourier series and reciprocal of the period of a periodic waveform.

Harmonic Sinusoidal component of a periodic nonsinusoidal waveform, which is an integral multiple of the fundamental frequency component. A harmonic whose frequency is twice the fundamental frequency is called a second harmonic.

Harmonic analysis Technique for determining the frequency components and their relative amplitudes of a periodic nonsinusoidal waveform.

Harmonic analyzer Instrument for measuring the amplitudes of the frequency components present in a periodic complex waveform.

Harmonic content Deviation from the pure sinusoidal waveform expressed in terms of a line spectrum or the order and magnitude of trigonometric Fourier series terms.

Line spectrum Graphical plot showing each of the relative sinusoidal harmonic amplitudes in a nonsinusoidal waveform.

Mirror symmetry Symmetry of a waveform about the x-axis; that is, the (negative) waveform below the x-axis is the mirror image of the (positive) waveform above the x-axis. Waveforms having mirror symmetry contain odd harmonics only. No even harmonics may be present.

Nonperiodic waveform See *aperiodic wave.*

Nonsinusoidal waveform Complex periodic waveform containing three possible types of components: a dc component of zero frequency, a fundamental frequency component, and a potentially infinite number of harmonic multiples of the fundamental frequency. The multiples may be odd or even at any phase with respect to the fundamental.

Odd harmonics Sinusoidal components of the nonsinusoidal waveform having frequencies that are an odd number of integral multiples of the fundamental frequency.

Phase of a harmonic Instantaneous value of the phase angle of a harmonic when the fundamental is at reference zero, that is, the fundamental is at 0° and *positive-going.* A harmonic that is at 0° and negative-going (when the fundamental is at 0°) has a phase of 180°.

RMS value Square root of the average of the *squares* of the instantaneous durations of waveforms taken over an entire period of its fundamental frequency.

Spectrum analyzer Instrument that displays the energy distribution of the frequency components contained in a waveform. The display is usually shown on a CRO screen.

Square wave Periodic waveform having x-, y- and z-axis symmetries as a result of odd harmonics in phase with the fundamental.

Square wave testing Ability of a circuit or component to pass a square wave without distortion. The output waveform is compared to the input square wave on a dual-trace (or dual-beam) CRO. The degree and type of distortion that appears at the output is a measure of the linearity of the device under test.

Symmetry Geometric shape of a waveform with respect to various (x, y, and z) axes.

Waveform generation Production of desired waveforms from other waveforms, either sinusoidal or nonsinusoidal, by suitable combinations of circuits.

***x*-axis symmetry** See mirror symmetry.

***y*-axis symmetry** Symmetry about the y-axis measured at the $\pi/2$ portion of a 2π period, in both a positive and negative direction. Also known as quarter-wave symmetry. A waveform having y-axis symmetry *only* must contain even harmonics at 90°.

***z*-axis symmetry** Symmetry about the z-axis (axis perpendicular to the x- and y-axes). Waveforms symmetrical about the z-axis must contain *even* harmonics at 0° or 180° or both. The symmetry is found by rotating the negative half-cycle 180° about the z-axis to see if it coincides with the positive half-cycle.

22-13 PROBLEMS

Sec. 22-2

22-1 Given the following equations in sine–cosine form, express them in amplitude–phase form:

a. $v(t) = (5 \sin \omega t - 3 \cos \omega t) + (6 \sin 2\omega t - 5 \cos 2\omega t) + (-8 \sin 3\omega t + 10 \cos 3\omega t)$ volts

b. $v(t) = (25 \sin \omega t + 3 \sin 2\omega t + 8 \sin 3\omega t + 10 \sin 4\omega t) + (3 \cos \omega t - 2 \cos 2\omega t - 5 \cos 3\omega t + 4 \cos 4\omega t)$ mV

c. $i(t) = (4 \cos 100t + 3 \cos 500t + 6 \cos 900t) + (3 \sin 100t + 25 \sin 300t + 6 \sin 500t + 4 \sin 700t)$ A

22-2 Given the following equations in amplitude–phase form, express them in sine–cosine form:

a. $v(t) = 25 \sin(\omega t - 53.13°) + 20 \sin(2\omega t - 50°) + 15 \sin(3\omega t + 60°) + 10 \sin(5\omega t - 120°)$ volts

b. $i(t) = 100 \sin(\omega t + 20°) + 60° \sin(3\omega t - 80°)$
$+ 20 \sin(5\omega t + 20°) + 10 \sin(7\omega t - 25°)$
$+ 5 \sin(9\omega t + 45°)$ mA

c. $v(t) = 80 \sin 100t + 70 \sin(200t + 90°) - 60 \sin 300t$
$+ 50 \sin(400t - 90°)$ V

d. $i(t) = 50 \sin 50t - 40 \cos 100t + 30 \sin(150t + 45°)$
$- 20 \sin(200t - 50°)$ mA

22-3 Given the following equation for a nonsinusoidal waveform in sine–cosine form:

$$v(t) = 50 + 30 \sin 377t - 40 \cos 377t + 12 \sin 754t + 5 \cos 754t + 8 \sin 1508t - 6 \cos 1508t + 0.7 \sin 3016t + 2.4 \cos 3016t \text{ volts}$$

a. Determine the frequencies that are present in the signal, in hertz
b. Determine the values of A_0 and C_n at each frequency
c. Express the signal in amplitude–phase form

Secs. 22-3 to 22-5

22-4 Given the equation $v(t) = 10 \sin \omega t - 5 \sin 5\omega t$ volts,
a. Using cross-section paper, sketch the shape of the complex wave by graphical addition. Use a metric ruler and a pair of dividers.
b. Analyze the resultant complex wave for the symmetry(ies) present.
c. Analyze the equation for its harmonic and fundamental components.
d. What does the symmetry of the resultant waveform prove regarding the effect of the given harmonic on the fundamental (x-, y-, and z-axis) symmetry?
e. What is the average value of the resultant waveform?

22-5 Given the equation $v(t) = 10 \sin \omega t + 5 \sin(3\omega t - 90°)$ volts, repeat all parts of Prob. 22-4.
22-6 Given the equation $v(t) = 10 \sin \omega t + 5 \sin 4\omega t$ volts, repeat all parts of Prob. 22-4.
22-7 Given the equation $v(t) = 10 \sin \omega t + 5 \sin(2\omega t - 90°)$ volts, repeat all parts of Prob. 22-4.
22-8 Given the waveform shown in **Fig. 22-26a**,

a. Identify any and all symmetries present
b. Give the order and possible phase(s) of harmonics that *must* be present
c. Give the order and possible phase(s) of harmonics that *may* be present
d. What is the average value of the waveform?

22-9 Repeat all parts of Prob. 22-8 for the waveform shown in Fig. 22-26b.

22-10 Repeat all parts of Prob. 22-8 for the waveform shown in Fig. 22-26c.
22-11 Repeat all parts of Prob. 22-8 for the waveform shown in Fig. 22-26d.
22-12 Repeat all parts of Prob. 22-8 for the waveform shown in Fig. 22-26e.

Secs. 22-6 and 22-7
22-13 Given the equation for the waveform

$$v(t) = 31.83 + 50 \sin 377t - 21.2 \cos 754t - 4.244 \cos 1508t - 1.819 \cos 2262t - 1.0105 \cos 3016t \text{ volts,}$$

find the

a. Average value of the waveform
b. RMS value of the waveform
c. Ratio of the RMS to average value of the waveform (form factor)
d. Symmetry of the waveform based on the nature of the harmonic components in it
e. Fundamental frequency of the waveform
f. Identify the waveform based on the calculations in (**a**), (**b**), and (**c**). (See Table 14-1)

22-14 Repeat all parts of Prob. 22-13, given the equation

$$i(t) = 63.66 - 42.44 \cos 2000t - 8.488 \cos 4000t - 3.638 \cos 6000t - 2.021 \cos 8000t \text{ mA}$$

22-15 Repeat all parts of Prob. 22-13, given the equation

$$v(t) = 100 + 81.06 \sin 314.15t - 9.006 \sin 942.5t + 3.242 \sin 1570.8t - 1.654 \sin 2199t \text{ mV}$$

22-16 Repeat all parts of Prob. 22-13, given the equation

$$i(t) = 50 + 63.66 \sin 1000t + 21.22 \sin 3000t + 12.73 \sin 5000t + 9.095 \sin 7000t \text{ mA}$$

Sec. 22-8
22-17 The line spectrum of the trigonometric Fourier series of a common nonsinusoidal waveform is sometimes called its "harmonic signature." Given the waveform in Prob. 22-13,

a. Plot the line spectrum using the given amplitudes
b. Identify the waveform from part (**f**) of the problem
c. List as many specific characteristics as you can to help you identify the waveform in the future from spectrum analyzer data

22-18 Repeat all parts of Prob. 22-17 for the waveform equation of Prob. 22-14.

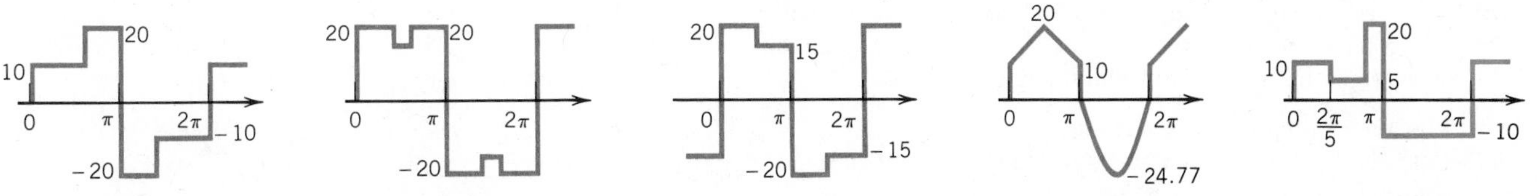

a. Problem 22-8 b. Problem 22-9 c. Problem 22-10 d. Problem 22-11 e. Problem 22-12

Figure 22-26 Problems 22-8 through 22-12

22-19 Repeat all parts of Prob. 22-17 for the waveform equation of Prob. 22-15.

22-20 Repeat all parts of Prob. 22-17 for the waveform equation of Prob. 22-16.

Sec. 22-9

22-21 Given the equation for the sawtooth waveform shown in Table 22-1J (Eq. 22-14), assuming $A = 100$ V,

a. Write the waveform equation so that it contains the first two ac harmonic terms only
b. Use cross-section paper to sketch the resultant by graphical addition
c. Determine the waveform symmetry from the resultant sketched in (**b**)
d. Explain why the symmetry is identical to that of the complete waveform ($n = \infty$) based on only the first two ac terms

22-22 Repeat all parts of Prob. 22-21 for the sawtooth pulse shown in Table 22-1N (Eq. 22-19). Hint: add a dc value of $A_0 = 100$ V to the sketched resultant of Prob. 22-21.

22-23 The triangular waveform shown in Table 22-1E has a maximum value of $A = 100$ V. Calculate

a. Its average value
b. The maximum values of the first four ac components
c. The RMS value of the waveform, using only the values in (**a**) and (**b**)
d. The RMS value of the waveform, using the appropriate equation in Table 14-1

22-24 The sawtooth pulse waveform shown in Table 22-1N has a maximum value of $A = 100$ V. Repeat all parts of Prob. 22-23. (Hint: for part **d** treat the waveform as a triangular wave.)

22-25 The rectangular pulse waveform shown in Table 22-1G has a maximum value of $A = 20$ V. Repeat all parts of Prob. 22-23.

22-26 The ideal HWR of Table 22-1B has a maximum value of 30 V. Repeat all parts of Prob. 22-23.

22-27 The ideal FWR of Table 22-1C has a maximum value of 30 V. Repeat all parts of Prob. 22-24.

Sec. 22-10

22-28 The generated phase voltage of a 3ϕ Y-connected alternator displays a series of odd harmonics that may be considered negligible beyond the fifth. The line voltage of the alternator exhibits the fifth harmonic since the third is canceled as a result of phasor addition. If the alternator line voltage is 208 V and phase voltage is 125 V, calculate the RMS magnitude of the third harmonic generated in the alternator phase winding.

22-29 If perfectly linear, a transistor would produce a pure sinusoidal collector output of 10 mA peak-to-peak above and below its quiescent collector current of 20 mA. Due to base spreading, however, a second harmonic lagging the fundamental by 90° is generated having a maximum value of 1 mA.

a. Plot the resultant output waveform
b. Find the peak maximum and minimum values of the nonsinusoidal collector output current
c. Describe the symmetry of the waveform from your plot and knowledge of its contents
d. Calculate the percent harmonic distortion (PHD) as the ratio of peak-to-peak values of the second harmonic to the fundamental

Sec. 22-11

22-30 Two separate voltage sources are connected in series and applied across a complex impedance: $v_1(t) = 30 + 30 \sin \omega t + 10 \cos 2\omega t + 3 \sin(3\omega t - 50°)$ volts and $v_2(t) = 10 - 5 \sin \omega t + 5 \sin(2\omega t - 90°) + 4 \sin(3\omega t + 40°)$ volts. Calculate the

a. Equivalent harmonic voltage applied across the complex impedance in amplitude–phase form (Hint: express each in sine–cosine form)
b. RMS value of the equivalent voltage of *both* sources

22-31 Two separate current sources deliver current to a complex load impedance:

$$i_1(t) = 3 + 2.121 \sin(\omega t - 45°) + 0.4472 \sin(2\omega t + 26.565°) + 0.09484 \sin(3\omega t - 121.6°) \text{ A and}$$

$$i_2(t) = 1 + 0.35355 \sin(\omega t + 135°) + 0.2236 \sin(2\omega t - 153.43°) + 0.1265 \sin(3\omega t - 31.56°) \text{ A.}$$

Calculate the

a. Equivalent current entering the complex impedance in amplitude–phase form
b. RMS value of the equivalent current

22-32 A 2.5-kΩ resistor is connected across the output of an *ideal* half-wave rectifier circuit whose dc value is 31.83 V. Calculate the

a. Peak value of the rectifier output waveform, V_m
b. RMS value of the output waveform, using only the first four ac terms. See Eq. (22-7), Table 22-1B.
c. RMS value, using the appropriate equation from Table 14-1
d. Average power dissipated in the resistor, using the value from (**c**)

22-33 Given the waveform shown in **Fig. 22-27a**, calculate the

a. Average value
b. RMS value of the ac component (only), using the appropriate equation from Table 14-1
c. RMS value of the nonsinusoidal waveform when applied directly to a 20-Ω linear resistor

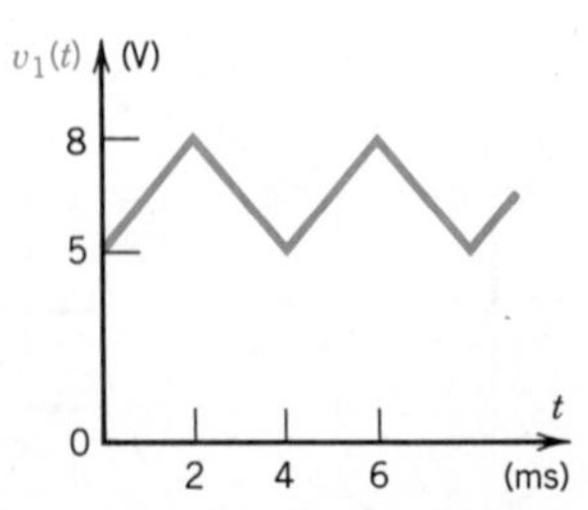

a. Problem 22-33

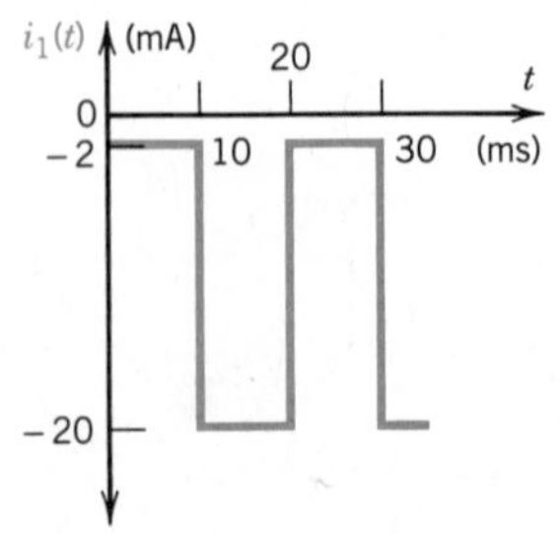

b. Problem 22-34

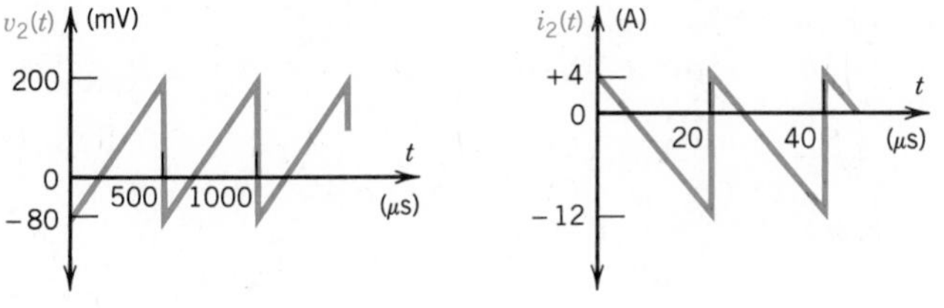

c. Problem 22-35 d. Problem 22-36

Figure 22-27 Problems 22-33 through 22-36

d. Average power dissipated, using the value from (**c**) only
e. Radian frequency of the waveform in radians per second
f. Write the first four terms of the Fourier series for the given waveform in amplitude phase form

22-34 Repeat all parts of Prob. 22-33 for the waveform shown in Fig. 22-27b.
22-35 Repeat all parts of Prob. 22-33 for the waveform shown in Fig. 22-27c.
22-36 Repeat all parts of Prob. 22-33 for the waveform shown in Fig. 22-27d.
22-37 The output of the HWR of Prob. 22-32 having a peak value of 100 V dc is applied to the series combination of an inductor and an output load resistor of 100 Ω. If the inductor has a reactance of $j300$ Ω at the fundamental frequency of 60 Hz, calculate the

a. Equation for the first four terms of voltage produced by the HWR
b. Equation for the nonsinusoidal current in the resistor. (Use a tabulation)
c. Equation for the voltage output across the resistor
d. Compare (**a**) to (**c**). How effective is the given reactor in suppressing the ac terms of the HWR output?

22-38 Repeat parts (**c**) and (**d**) of Prob. 22-37 by adding a 100-μF capacitor in parallel with the 100-Ω resistor with the parallel combination in series with the same inductor. Hint: calculate the equivalent parallel and total impedance at each frequency, total current, and output voltage across the resistor, respectively.
22-39 A pure rectangular square wave has a maximum amplitude of 25π volts. It is connected across a series RLC circuit where $R = 10$ Ω, $jX_L = 10$ Ω, and $-jX_C$ is $-j90$ Ω at the fundamental frequency. Calculate the

a. Equation of applied voltage up to the fourth component, $v(t)$
b. Equation for the resulting current waveform, $i(t)$
c. RMS value of the current
d. Total power dissipated in the circuit
e. Draw the line spectra for both the applied voltage and the resulting current
f. Predict the symmetry of the resulting current waveform

22-40 The output of an ideal full-wave rectifier (FWR) having a maximum amplitude of 50π volts is applied across the series–parallel circuit of Fig. 22-24a. If jX_L is 20 Ω and jX_C is $-j80$ Ω at the fundamental (lowest) frequency, repeat all parts of Prob. 22-39.
22-41 The voltage applied to a complex impedance and the resulting current are, respectively, $v(t) = 200 + 100 \sin(5000t + 120°) + 75 \sin(15\,000t + 150°)$ volts and $i(t) = 3.53 \sin(5000t + 165°) + 3.55 \sin(15\,000t + 168.45°)$ amperes. Determine the

a. Nature of the impedance from the resulting current waveform
b. Average power dissipated in the impedance
c. RMS values of applied voltage and resulting current
d. Resistance of the complex impedance
e. Circuit elements comprising the complex impedance and their values

22-42 In Fig. 22-25a, the complex impedance consists of a 10-Ω resistor in series with a 50-μF capacitor. The voltages applied are, respectively, $v_1(t) = 10 + 12.73 \sin 1000t + 4.24 \sin 3000t + 2.5 \sin 5000t$ volts and $v_2(t) = 15 + 12.27 \sin 1000t - 2.48 \cos 3000t + 6.3 \cos 5000t$ volts. Calculate the

a. Equivalent voltage applied to the complex impedance expressed in Fourier amplitude–phase form and its RMS value.
b. Resulting current in amplitude–phase form
c. RMS value of current and its symmetry
d. Power dissipated in the complex impedance

22-43 In Fig. 22-25b, the complex impedance consists of a 10 Ω resistor in series with a 5-mH inductor. Voltage $v_1(t)$ is the output of an HWR having a maximum amplitude of 50 V. Voltage $v_2(t)$ is the output of an FWR having a maximum amplitude (A) of 50 V also. If the pure sinusoid input to each rectifier has a fundamental radian frequency of $\omega = 1000$ rad/s, using only terms up to 4000 rad/s in each nonsinusoidal rectifier output, calculate all parts of Prob. 22-42.
22-44 Repeat all parts of Prob. 22-43 with the polarity of $v_1(t)$ in Fig. 22-25b reversed from that shown.

22-14 ANSWERS

22-3 a 0, 60, 120, 180, 240 Hz
b 50 V; $50\angle{-53.13°}$ V, $13\angle{22.62°}$ V, $10\angle{-36.87°}$ V, $2.5\angle{73.74°}$ V
22-4 b x, y, z-axis e 0
22-5 b x-axis e 0
22-6 b z-axis e 0
22-7 b y-axis e 0
22-8 a z-axis d 0
22-9 x, y, z-axis d 0
22-10 a x-axis d 0

22-11 a y-axis d 0
22-12 a none d 0
22-13 a 31.83 V b 50 V c 1.57 d y-axis e 60 Hz
22-14 a 63.7 mA b 70.7 mA c 1.11 d y-axis
e 318.3 Hz
22-15 a 100 mV b 115.5 mV c 1.155 d x, y, z
e 50 Hz
22-16 a 50 mA b 69.81 mA c 1.4 d x, y, z
e 159.15 Hz
22-17 b HWR
22-18 b FWR
22-19 b triangular FWR
22-20 b square wave HWR pulse
22-21 c z-axis
22-22 c z-axis
22-23 a 0 b 81.057, 9.006, 3.242, 1.652 V c, d 57.73 V
22-24 a 50 V b 31.83, 15.915, 10.61, 7.96 V c 56.76 V
d 57.74 V
22-25 a 10 V b 12.73, 4.244, 2.55, 1.82 V c 13.96 V
d 14.14 V
22-26 a 9.55 V b 15, 6.37, 1.26, 0.303 V c, d 15 V
22-27 a 19.1 V b 12.7, 2.55, 1.09, 0.606 V c, d 21.21 V
22-28 35 V
22-29 b 11 mA c y-axis d 10%
22-30 b 44.02 V
22-31 b 4.195 A
22-32 a 100 V b 50 V c 50 V d 1 W
22-33 a 6.5 V b 0.866 V c 6.56 V d 2.15 W
e 1571 rad/s
22-34 a −11 mA b 9 mA c 14.21 mA
d 14.21 mA; 4.04 mW e 314.2 rad/s
22-35 a 60 mV b 80.8 mV c 100.7 mV d 0.507 mW
e 12 566 rad/s
22-36 a −4 A b 4.62 A c 6.11 A d 747 W
e 314.2 krad/s
22-37 d 2nd, 4th only suppressed
22-38 d all suppressed
22-39 c 2.556 A d 65.3 W e x-axis only
22-40 c 10.14 A d 1027.5 W f none
22-41 b 251 W c 218.7 V, 3.54 A d 20 Ω e 10 μF
22-42 a 31.2 V c 0.952 A; x-axis d 9.06 W
22-43 a 55.84 V c 5.28 A; none d 278.7 W
22-44 a 25 V c 2.524 A; none d 63.7 W

CHAPTER 23

Analog and Digital Instrumentation

23-1 INTRODUCTION

The previous chapters described and analyzed the effects of combinations of resistance, capacitance, and inductance in dc and ac circuits. This chapter is devoted to devices (instruments) that enable the verification and actual measurement of all the previously described parameters and their circuit effects.

Through a study of the various instruments and their associated devices, we will discover some of the inherent advantages and limitations of specific measurement techniques and instrumental methods.

The chapter begins with the basic permanent-magnet moving-coil (PMMC) meter and its use in ammeters, voltmeters, and ohmmeters (series, shunt, and shunted-series types). Next we consider various methods of measuring resistance with meters and techniques for proper meter placement in high- and low-resistance circuits, culminating in the Wheatstone bridge method of resistance measurement. Other types of meters are considered, such as thermocouple, rectifier, iron vane, and electrodynamometer movements for measurement of voltage, current, and power, including the thermal-watt converter. The volt–ohmist or volt–ohm–milliammeter (VOM) and its advantages and disadvantages are considered.

Modern digital instruments such as the true RMS digital VOM, electronic counters, and waveform function generators are introduced. Various ac impedance bridges are introduced, both earlier and recent digital types. The chapter ends with an introduction to the modern digital-display cathode ray oscilloscope (CRO) and spectrum analyzer.

23-2 THE BASIC PMMC MOVEMENT

All permanent-magnet moving-coil instruments use the principle of motor action first described in Sec. 11-15 and quantified in Eq. (11-23). The basic PMMC movement is shown in **Fig. 23-1**. Its construction is described in Sec. 23-2.4. The *meter current* in the moving coil (Fig. 23-1b) designated as I_m produces a magnetic field, which reacts against the permanent flux density, **B**, of the permanent magnet (Fig. 23-1a). This field interaction produces a twisting mechanical force (torque), which

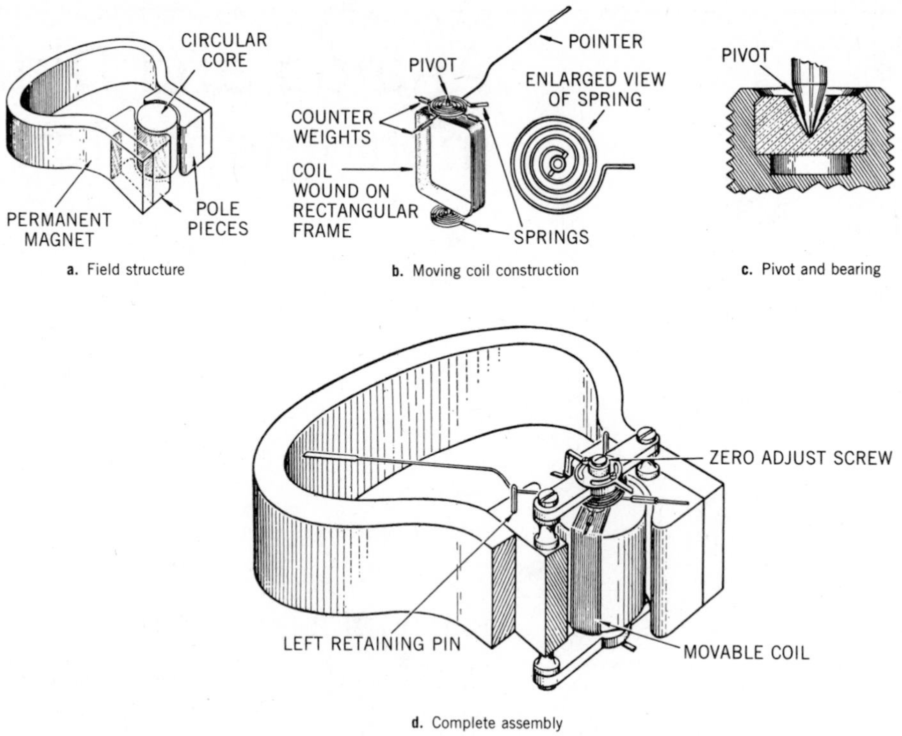

Figure 23-1 Basic PMMC movement using pivoted rectangular coil (U.S. Army TM 11-664, p. 9)

is opposed by the two spiral springs wound in opposite directions (Fig. 23-1b). This technique, known as the D'Arsonval principle, is basic to all PMMC instruments. The force, torque, and deflection of a basic PMMC movement are derived in Sec. 23-2.1.

23-2.1 Factors Affecting PMMC Deflection

The force, **F**, developed by a meter movement carrying current I_m in a field of constant flux density, **B**, is

$$\mathbf{F} = Z\mathbf{B}(I_m l) \sin\theta = 2N\mathbf{B}(I_m l) \sin\theta \qquad \text{newtons (N)} \qquad (23\text{-}1)$$

where Z is the number of conductors in the moving coil of the meter
N is the number of turns of the moving coil ($N = Z/2$)
I_m is the meter current in amperes
l is the average length of the conductors under the magnetic pole
B is the flux density of the permanent magnet in teslas
θ is the angle between **B** and $I_m l$ in degrees

The torque produced by this force is expressed as

$$\mathbf{T} = \mathbf{F} \cdot r = 2N\mathbf{B}(I_m l)r = N\mathbf{B}I_m ld = N\mathbf{B}AI_m$$
$$= N\phi I_m \qquad \text{newton-meters (N·m)} \qquad (23\text{-}2)$$

where d is the diameter of the coil or $2r$, in meters
r is the radius of the coil, in meters
A is the area of the coil, in square meters (m^2)
ϕ is the magnetic flux produced by the permanent magnet, in webers

Equation (23-2) shows that for a given number of turns, since ϕ and N are both constant for a given PMMC meter, the torque produced by the basic PMMC meter (and its pointer deflection) is actually proportional to the value of the current in the meter coil. Most moving-coil instruments are constructed with spiral retarding springs that exert a *countertorque* and maintain the instrument's pointer at zero deflection when no current flows in the meter ($I_m = 0$). Theoretically, if the spring retarding torque is linear, the scale of the PMMC meter should be perfectly linear and a function of meter current, I_m. The instrumental deflection, D, is proportional to and produced by the torque (Eq. 23-2) acting against the retarding springs. But since ϕ is constant in Eq. (23-2), we may also write the deflection, D, as

$$D \propto \mathbf{T} \propto \phi I_m = kI_m \qquad \text{(scale divisions)} \qquad (23\text{-}3)$$

where k is a constant for all the factors in Eq. (23-2) except I_m.

From Eq. (23-3) we may also write the full scale deflection, D_{fs}, of the PMMC meter as

$$D_{fs} = kI_{fs} \qquad \text{(scale divisions)} \qquad (23\text{-}4)$$

where I_{fs} is the meter current for full-scale deflection or the current sensitivity.

Equation (23-4) is most significant since it yields two important insights:

1. It enables evaluation of the meter constant k in terms of the full-scale deflection produced by full-scale current (both of which are readily obvious from the meter face). Recall that k represents all the constant factors in Eq. (23-2).
2. It defines the sensitivity, S, of a given meter, as shown in Eq. (23-5).

23-2.2 Instrumental Sensitivity of a Basic Meter Movement

Sensitivity, S, of an instrument may be defined, generally, as the ratio of instrumental response (in this case, deflection, D) to the cause of the response (meter current, I_m). For a given unit or 100% deflection, k obviously represents the sensitivity, which may be expressed as

$$S = k = \frac{D_{fs}}{I_{fs}} = \frac{1}{I_{fs}} \qquad \text{ohms per volt } (\Omega/\text{V}) \qquad (23\text{-}5)$$

EXAMPLE 23-1

The moving coil of a 60-Ω PMMC movement has 100 turns on a ferromagnetic core, whose diameter is 0.5 in and whose axial length is 1 in. The permanent magnet produces a flux density of 0.5 T in the air gap. When the coil is carrying a full-scale current of 2 mA *perpendicular* to the magnetic field, calculate the

a. Number of conductors (cond)
b. Force per conductor and total force developed by the coil
c. Torque developed by the moving coil
d. Sensitivity of the movement
e. Voltage required to produce full-scale deflection (voltage sensitivity).

Solution

a. $Z = 2N = (2 \text{ cond/t})(100 \text{ t}) = \mathbf{200 \text{ conductors}}$

b. $\mathbf{F} = \mathbf{B}Il \sin\theta = (0.5 \text{ T})(2 \text{ mA/cond})(0.0254 \text{ m/in}) \times (1 \text{ in}) = \mathbf{2.54\ \mu N/cond}$

$\mathbf{F}_t = \mathbf{F} \times Z = (2.54\ \mu\text{N/cond})(200 \text{ cond}) = \mathbf{5.08 \text{ mN}}$

c. $\mathbf{T} = \mathbf{F} \cdot r = F(D/2) = (5.08 \text{ mN})(0.5 \text{ in}/2)(0.0254 \text{ m/in}) = \mathbf{32.36\ \mu N \cdot m}$

d. $S = 1/I_{fs} = 1/2 \text{ mA} = \mathbf{500\ \Omega/V}$

e. $V_{fs} = I_{fs}R_m = (2 \text{ mA})(60\ \Omega) = \mathbf{120 \text{ mV}}$

Example 23-1 shows that the forces and torques developed in analog PMMC meter movements are extremely small. This is why they must be treated with care since they are subject to errors caused by misuse, overheating, friction in bearings, vibration, dampness, and so forth.

23-2.3 Instrumental Accuracy and Error

Accuracy is defined as a measure of how well a meter indicates the true value of the parameter it measures. Accuracy is associated with error, since the smaller the error of a particular measurement, the greater the accuracy of the measurement. Accuracy may be expressed as a percentage. Saying that a PMMC is accurate to 2% implies that any reading taken along any of its scales will not be in error by more than 2% of its full-scale value. In terms of error, we may define accuracy as

$$\text{Percent error} = \frac{\text{error}}{\text{true value}} \times 100 = \frac{\text{true value} - \text{measured value}}{\text{true value}} \times 100 \qquad \text{(23-6)}$$

It is sometimes convenient to find the true value of a measurement when given the percent error as either a positive or a negative value by using

$$\text{True value} = \frac{\text{measured value}}{1 - \%\ \text{error (as a signed decimal value)}} \qquad \text{(23-6a)}$$

EXAMPLE 23-2

A milliammeter has a full-scale deflection current of 1 mA and a meter resistance of 100 Ω. Its scale has 10 major divisions and each major division has 10 minor divisions. If the meter can be read to a precision of $\pm 1/2$ minor division, calculate the

a. Sensitivity of the meter and voltage sensitivity of the meter
b. Total number of minor divisions
c. Milliamperes per major division and microamperes per minor division
d. Reading precision error
e. Lowest possible reading (in microamperes) that can be made accurately
f. Highest possible reading (in milliamperes) that can be made accurately
g. Accuracy due to reading precision error in percent

Solution

a. $S = 1/I_{fs} = 1/1 \text{ mA} = \mathbf{1000\ \Omega/V}$;
$V_{fs} = I_{fs}R_m = 1 \text{ mA} \times 100\ \Omega = \mathbf{100 \text{ mV}}$

b. $N = 10$ maj div(10 min div/maj div) = **100 minor divisions**

c. mA/maj div = 1 mA/10 maj div = **0.1 mA/major division**
μA/min div = 1000 μA/100 min div = **10 μA/min div**

d. (Reading) precision error = $2(\pm 1/2$ min div$) = \mathbf{1}$ minor division

e. Lowest possible reading = (10 μA/min div) × 1 min div = **10 μA**

f. Highest possible reading = 1000 μA − 10 μA = 990 μA = **0.99 mA**

g. % error = error/total divs = 1 min div/100 min div = 0.01 = **1%**

23-2.4 Construction Features of the Basic PMMC Meter

Every analog-indicating instrument must have three properties:

1. An instrumental *deflection* system to produce deflection proportional to some variable parameter (current, voltage, power, etc.).
2. A proportional *control* system to control the deflection to ensure that it is proportional to the variable parameter (spiral springs in the PMMC).
3. A *damping* system to ensure that the pointer does not oscillate and to enable it to reach its final position quickly without overshoot (rectangular aluminum frame in the PMMC).

The *instrumental deflection system* of the PMMC is shown in Figs. 23-1a and 23-1b. The permanent magnetic field structure (Fig. 23-1a) consists of a steel alloy *permanent magnet* (Alnico or Alcomax) with soft-iron *pole pieces*, alloyed to horseshoe magnet by sintering to reduce air gaps. A circular *soft-iron core* reduces the air-gap length and ensures a more uniform flux density in the air gap through which the moving coil rotates.

The *moving coil*, *pointer*, *spiral springs*, and *pivot shafts* complete the instrumental deflection system (Fig. 23-1b). To reduce weight, the coil is wound with fine wire on a rectangular aluminum form. Two oppositely wound spiral springs are connected in series with each coil end and form part of the coil and movement resistance, R_m. In reality, the spiral springs serve as the *proportional control system* by providing a *countertorque* in opposition to the deflection torque of the meter movement quantified in Eq. (23-2). The spring materials used are beryllium–copper, phosphor bronze, or any nonmagnetic spring alloy. Since one spring unwinds while the other winds, the countertorque is directly proportional to the twist angle, which in turn is directly proportional to pointer deflection.

Since the *aluminum frame* (on which the moving coil is wound) is a conductor, *damping* is provided as a result of the *eddy currents* induced in the frame. By Lenz's law, these eddy currents set up a flux to oppose the motion that produced them, thus producing a torque opposing the motion of the coil.

23-2.5 Resistance and Current Ranges of the Basic PMMC Meter

As noted from Eq. (23-5), the smaller the current required for full-scale deflection, the greater the sensitivity of the meter movement and the higher its resistance. High-sensitivity meters are produced by using more turns of finer wire, as noted from Eq. (23-2), to produce the required deflection torque. PMMC meters with the construction of Fig. 23-1 are limited to about 10 μA (sensitivities of 100 kΩ/V) and have resistances of about 5 kΩ for maximum sensitivity.[1] At the low-sensitivity end, PMMC meters are available with full-scale deflection currents as high as 30 mA ($33.\bar{3}$ Ω/V). Higher-current PMMC meters are impractical because the heavier wire increases the coil inertia and weight.

Regardless of construction differences, all PMMC meter movements are essentially *current-responding* devices, as noted from Eq. (23-3). Since we are often required to measure dc currents in excess of 30 mA, we will first consider ways of *extending* the *current* range of basic PMMC meters to produce dc ammeters.

[1] Taut-band meters based on the PMMC principle are available with full-scale deflection currents as low as 1 μA, providing sensitivities up to 1 MΩ/V.

23-3 AMMETERS

The current-divider rule (CDR) is the key to ammeter construction. The only way to increase the current range of any basic movement is to *shunt* (or parallel) the movement with a shunting resistor, R_{sh}. This, in effect, produces a simple parallel circuit with R_{sh} in parallel with the meter resistance, R_m, as shown in **Fig. 23-2a**. The resistance of the shunt, R_{sh}, using KCL and CDR, is

$$R_{sh} = \frac{I_m R_m}{I_{sh}} = \frac{V_{fs}}{I_{sh}} = \frac{I_{fs} R_m}{(I_{sc} - I_{fs})} = \frac{R_m}{(S \times I_{sh})} \quad \text{ohms } (\Omega) \qquad (23\text{-}7)$$

where S is the sensitivity in ohms per volt (Ω/V)
I_{sc} is the *desired scale current* of the ammeter in amperes
R_m is the meter resistance in ohms
I_{sh} is the current in shunt in amperes
I_{fs} is the meter current, I_m, required for full-scale deflection of the meter

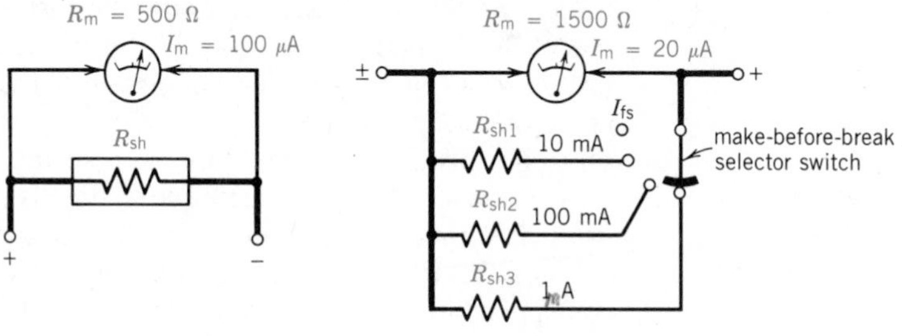

a. Single-range ammeter   b. Multirange ammeter

Figure 23-2 Extension of current range of basic PMMC meter using parallel-connected shunts

---

EXAMPLE 23-3

Given a meter movement whose resistance is 500 Ω and whose full-scale current is 100 μA (Fig. 23-2a), calculate the values of shunt resistance required to convert the meter to read currents of a. 1 mA b. 10 mA c. 100 mA

Solution

a. $I_{sh} = I_{sc} - I_{fs} = 1 \text{ mA} - 0.1 \text{ mA} = 0.9 \text{ mA}$;
$R_{sh} = I_m R_m / I_{sh} = (0.1 \text{ mA})(500\ \Omega)/0.9 \text{ mA} = \mathbf{55.\bar{5}\ \Omega}$

b. $S = 1/I_{fs} = 1/100\ \mu\text{A} = 10\ \text{k}\Omega/\text{V}$;
$I_{sh} = I_{sc} - I_{fs} = 10 \text{ mA} - 0.1 \text{ mA} = 9.9 \text{ mA}$;
$R_{sh} = R_m/(S \times I_{sh}) = 500\ \Omega/(10\ \text{k}\Omega/\text{V})(9.9 \text{ mA}) = \mathbf{5.05\ \Omega}$

c. $V_{fs} = I_m R_m = (100\ \mu\text{A})(500\ \Omega) = 50 \text{ mV}$;
$I_{sh} = I_{sc} - I_m = 100 \text{ mA} - 0.1 \text{ mA} = 99.9 \text{ mA}$;
$R_{sh} = V_{fs}/I_{sh} = 50 \text{ mV}/99.9 \text{ mA} = \mathbf{0.5005\ \Omega}$

Three insights may be drawn from Ex. 23-3 regarding ammeters produced from PMMC meters:

1. As the ammeter current *increases*, the value of shunt resistance *decreases*. Consequently, the *higher* the ammeter *range*, the *lower* the *resistance* of the shunt.
2. Of the various forms of Eq. (23-7) used in the solution of Ex. 23-3, the simplest is $R_{sh} = V_{fs}/I_{sh}$ = (voltage sensitivity)/(current in shunt), shown in part (**c**).
3. The higher the sensitivity, S, the smaller the value of shunt resistance, R_{sh}, as shown in Eq. (23-7).

Figure 23-2b shows the typical construction and circuitry of a *multirange* ammeter, commonly used to extend the range of basic PMMC movements. Observe that the current range selector requires a "make-before-break" switch to prevent *momentary overload* of the basic movement whenever the range is changed. Also note that no shunting is required for the current range marked I_{fs}. In that switch position, the meter current itself (20 μA) provides the minimum current range. Note that the meter switch is positioned at its highest current range (1 A) to prevent meter damage when first connected in a circuit.[2]

Frequently, it is desired to extend the current range of a *commercial* ammeter beyond its *nominal* current range. The ammeter is treated much the same as a basic movement, as shown in Ex. 23-4.

EXAMPLE 23-4

A shunted meter movement has a resistance of 20 mΩ and a full-scale deflection of 1 A. Calculate the value of shunt resistance in parallel with the meter to extend the current range to 10 A.

Solution

$R_{sh} = I_m R_m / I_{sh} = (1\ \text{A})(20\ \text{m}\Omega)/(10 - 1)\ \text{A} = 20\ \text{m}\Omega/9 = \mathbf{2.\bar{2}\ m\Omega}$

23-4 VOLTMETERS

In the same way that the current range of a PMMC movement is extended by means of parallel-connected shunts, the *voltage* range of a PMMC movement may be extended by *series-connected multipliers.* Most basic movements have high sensitivities, which, by definition, require a small current for full-scale deflection. As a result, such movements have a *low* voltage sensitivity, V_{fs}, since by definition $V_{fs} = I_{fs} R_m$. To enable a PMMC meter to measure voltages above V_{fs} (usually not more than a few tenths of a volt) a precision **series resistor** called a *multiplier* is connected as shown in **Fig. 23-3a** to produce a voltmeter. The total resistance of the voltmeter combination is

$$R_t = R_s + R_m = \frac{V_t}{I_m} = S(\rho_V) \qquad \text{ohms}\ (\Omega) \qquad (23\text{-}8)$$

where R_s is the series multiplier resistance (Ω)
R_m is the meter resistance (Ω)
S is the meter sensitivity (Ω/V)
ρ_V is the desired range of the voltmeter (V)
V_t is the total voltage applied to the voltmeter (V)
I_m is the meter current drawn by the voltmeter (A)

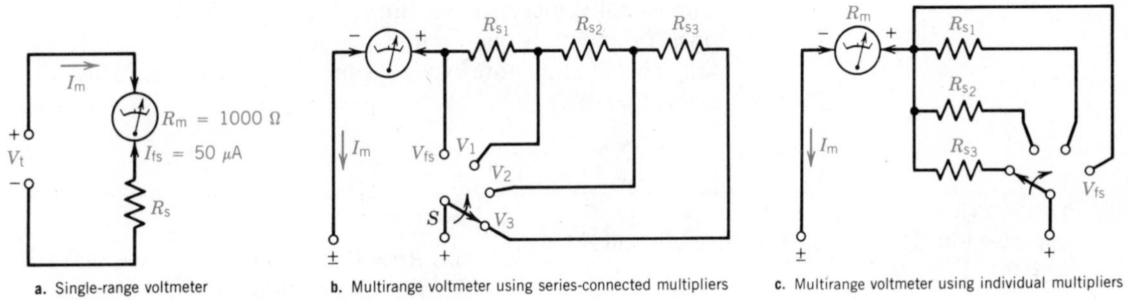

Figure 23-3 Extension of voltage range of basic PMMC meter using series-connected multipliers

---

[2] Calculation of the values of shunts in Fig. 23-2b is left as an end-of-chapter problem for student solution. (See Prob. 23-8.)

Rearranging Eq. (23-8) to solve for the multiplier resistance in series with the PMMC movement yields

$$R_s = R_t - R_m = \left(\frac{V_t}{I_t}\right) - R_m = \frac{(V_t - V_{fs})}{I_{fs}} = S(\rho_V) - R_m \qquad \text{ohms } (\Omega) \quad (23\text{-}9)$$

where all terms have been defined.

EXAMPLE 23-5

Given the basic PMMC movement shown in Fig. 23-3a, calculate the

a. Voltage sensitivity of the PMMC meter, V_{fs}
b. Multiplier resistance, R_s, needed to produce a voltmeter having a full-scale deflection of 150 V dc, by *two independent methods*

Solution

a. $V_{fs} = I_m R_m = I_{fs} R_m = (50\ \mu A)(1\ k\Omega) = \mathbf{50\ mV}$
b. $R_s = (V_t - V_{fs})/I_{fs} = (150 - 0.05)\ V/50\ \mu A = \mathbf{2.999\ M\Omega}$;
$S = 1/I_{fs} = 1/50\ \mu A = 20\ k\Omega/V$;
$R_s = S(\rho_V) - R_m = (20\ k\Omega/V)(150\ V) - 1\ k\Omega$
$= 3\ M\Omega - 1\ k\Omega = \mathbf{2.999\ M\Omega}$

Example 23-5 reveals several important insights:

1. The multiplier resistance, R_s, is many times the meter resistance, R_m, because the voltage range, ρ_V, is many times the voltage sensitivity, V_{fs}.
2. Since a series circuit is involved, the voltage-divider rule (VDR) holds for the relation $(R_s/R_m) = (V_s/V_{fs})$.
3. The sensitivity, S, is a factor in determining multiplier resistance just as it was in determining shunt resistance (see Eq. 23-7). For PMMC voltmeters, the higher the sensitivity, S, the greater the value of multiplier resistance.

Two types of multirange voltmeters are shown in Figs. 23-3b and 23-3c. The first design, shown in Fig. 23-3b, uses *series-connected* multipliers. This design has the advantage that R_{s_2} and R_{s_3} may be commercial precision resistors (see Ex. 23-6). Consequently, it is less expensive to manufacture and repair than the design of Fig. 23-3c. But it does have the disadvantage that if R_{s_1} is open, shorted, or burned out, the voltmeter cannot be used on any of its *higher* voltage scales above V_{fs}.

This disadvantage is overcome in the circuit by using *individual* multipliers (Fig. 23-3c) connected *separately* in series with the PMMC movement. The values of R_{s_2} and R_{s_3} in this design are different from R_{s_2} and R_{s_3} and rarely, if ever, can commercial values be used. But it does have the advantage that a change or loss (open or short) in one multiplier does not affect the other voltage ranges. This design is found in more expensive multirange voltmeters (usually with an accuracy of less than 2%).

EXAMPLE 23-6

The voltage sensitivity of a 1-mA PMMC movement is 100 mV. Calculate the

a. Resistance of the meter, R_m
b. Sensitivity, S
c. Values of R_{s_1}, R_{s_2}, and R_{s_3} in Fig. 23-3b, for ranges of 10 V, 100 V, and 500 V

Solution

a. $R_m = V_{fs}/I_m = 100\ mV/1\ mA = \mathbf{100\ \Omega}$
b. $S = 1/I_{fs} = 1/1\ mA = \mathbf{1000\ \Omega/V}$
c. $R_{s_1} = S \times \rho_V - R_m = (1\ k\Omega/V)(10\ V) - 1\ k\Omega = \mathbf{9\ k\Omega}$;
$R_{t_2} = S \times \rho_V - R_m = (1\ k\Omega/V)(100\ V) - 1\ k\Omega = 99\ k\Omega$;
$R_{s_2} = R_{t_2} - R_{s_1} = 99\ k\Omega - 9\ k\Omega = \mathbf{90\ k\Omega}$;
$R_{t_3} = S \times \rho_V - R_m = (1\ k\Omega/V)(500\ V) - 1\ k\Omega = 499\ k\Omega$;
$R_{s_3} = R_{t_3} - (R_{s_1} + R_{s_2}) = R_{t_3} - R_{t_2} = (499 - 99)\ k\Omega$
$= \mathbf{400\ k\Omega}$

23-5 OHMMETERS

The PMMC movement lends itself, as we have seen, to the construction of ammeters and voltmeters by the selection of appropriately connected shunts and multipliers. Is it also possible to design an ohmmeter using the PMMC movement? We know that *resistance varies inversely as current* by Ohm's law, given a constant-voltage source. Since current serves as a measure of resistance and the PMMC is fundamentally a current-measuring device, it follows that an ammeter could be calibrated in terms of resistance. This is the principle of the *series* ohmmeter.

Three types of ohmmeters are described: the *series* ohmmeter, the *shunted-series* ohmmeter, and the *shunt* ohmmeter. Each has its advantages and disadvantages.

23-5.1 Series Ohmmeter

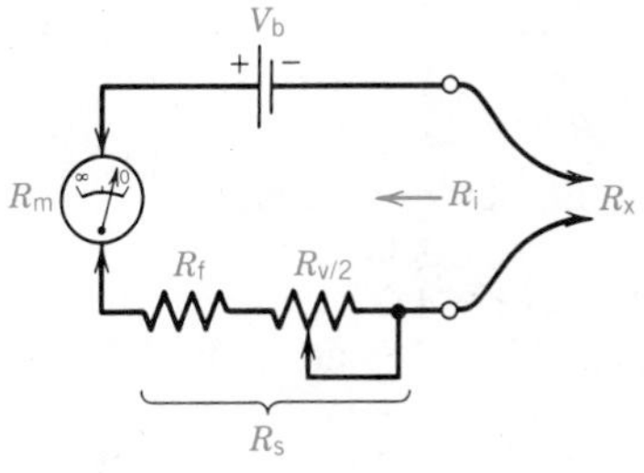

a. Circuit

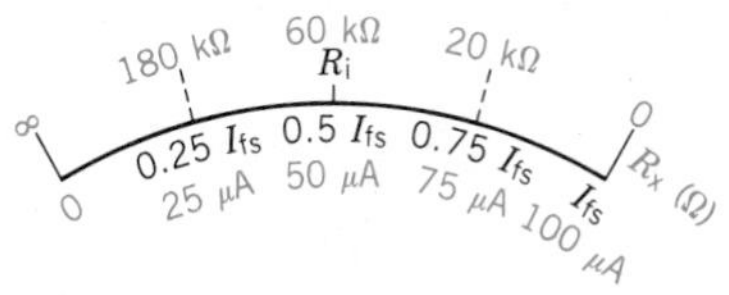

b. Typical scale showing values of Ex. 23-8

Figure 23-4 Series ohmmeter and typical scale

A typical series ohmmeter is shown in **Fig. 23-4a**. Note that the battery supply, V_b, the movement, R_m, and the series connected resistors, R_s, are so connected that when the leads are shorted ($R_x = 0$), the instrument may be "zeroed," that is, adjusted for *full-scale* deflection. Note from the scale shown in Fig. 23-4b that when the ohmmeter is zeroed, full-scale deflection occurs (i.e., $I_m = I_{fs}$) and the external resistance, R_x, is zero. Similarly, if the leads are separated ($R_x = \infty$), no current flows and the scale indicates infinite resistance between the test leads. The scale of Fig. 23-4b therefore is capable, *theoretically*, of reading any value of unknown resistance, R_x, from zero to infinity.

The internal resistance, R_i, of the ohmmeter shown in Fig. 23-4a is

$$R_i = R_m + \left(R_f + \frac{R_v}{2}\right) = R_m + R_s \qquad \text{ohms } (\Omega) \qquad \textbf{(23-10)}$$

where R_f is a fixed series resistance, approximately $0.7R_s$

R_v is a variable resistance used for "zeroing" the ohmmeter to full-scale deflection (fsd). (R_v decreases as V_b ages and decreases)

R_s is the value of series resistance required to zero the ohmmeter regardless of degradation of the battery, V_b

R_m is the resistance of the meter movement

The required battery voltage to operate the ohmmeter is found from

$$V_b = I_{fs}R_i = I_{fs}\left(R_m + R_f + \frac{R_v}{2}\right) \qquad \textbf{(23-11)}$$

where all terms have been defined.

Ohmmeter calculations are simplified by using a deflection factor, d. For any unknown external resistance, R_x, the *deflection factor d* of the ohmmeter is

$$d = \frac{I_m}{I_{fs}} = \frac{R_i}{R_x + R_i} \qquad \text{(unitless ratio)} \qquad \textbf{(23-12)}$$

In terms of instrumental deflection, the ohmmeter scale may be calibrated for *any* ratio of I_m/I_{fs} in terms of external resistance, R_x, where

$$R_x = R_i\left(\frac{1-d}{d}\right) = \frac{V_b}{I_{fs}}\left(\frac{1-d}{d}\right) \quad \text{ohms } (\Omega) \qquad (23\text{-}13)$$

Half-scale deflection in a series ohmmeter always occurs whenever

$$R_x = R_i = R_m + R_s \quad \text{ohms } (\Omega) \qquad (23\text{-}14)$$

EXAMPLE 23-7

A 100-μA, 500-Ω PMMC movement is used to design a series ohmmeter having a center-scale value of 60 kΩ. Calculate the

a. Battery voltage required, V_b
b. Value of R_s
c. Value of R_f, assuming $R_f = 0.7R_s$
d. Value of $R_v/2$ and R_v (assuming $R_v = 0.3R_s$)
e. Draw the ohmmeter circuit showing all the values calculated above

Solution

a. $V_b = I_{fs}R_i = (100\ \mu\text{A})(60\ \text{k}\Omega) = \mathbf{6\ V}$
b. $R_s = R_i - R_m = (60 - 0.5)\ \text{k}\Omega = \mathbf{59.5\ k\Omega}$
c. $R_f = 0.7R_s = 0.7 \times 59.5\ \text{k}\Omega = \mathbf{41.65\ k\Omega}$
d. $R_v/2 = 0.3R_s = 0.3(59.5\ \text{k}\Omega) = 17.85\ \text{k}\Omega$ and $R_v = 2 \times 17.85 \cong \mathbf{35\ k\Omega}$ potentiometer
e. See Fig. 23-4a and enter the calculated values on it

Example 23-7 shows the calculation steps for producing an ohmmeter having a center-scale value of 60 kΩ. Example 23-8 shows how the PMMC scale can be calibrated in steps of external resistance.

EXAMPLE 23-8

Calculate the values of d and R_x for pointer deflections of

a. One-quarter of full-scale deflection
b. Three-quarters of full-scale deflection

Solution

a. $d = I_m/I_{fs} = 0.25I_{fs}/I_{fs} = \mathbf{0.25}$;

$$R_x = R_i\left(\frac{1-d}{d}\right) = 60\ \text{k}\Omega\left(\frac{1-0.25}{0.25}\right) = (60\ \text{k}\Omega) \times 3$$
$$= \mathbf{180\ k\Omega}$$

b. $d = 0.75$ by inspection;

$$R_x = 60\ \text{k}\Omega\left(\frac{1-0.75}{0.75}\right) = 60\ \text{k}\Omega/3 = \mathbf{20\ k\Omega}$$

The values of Ex. 23-8 are shown in Fig. 23-4b. Observe that while the PMMC scale of meter current is *perfectly linear*, the *ohmmeter* scale is *not* and differs in two respects:

1. Values of external resistance increase in a counterclockwise direction.
2. The scale of external resistance becomes more compressed at higher values of resistance than at lower values.

By selecting the value of half-scale deflection as 60 kΩ in Ex. 23-7, we automatically ensured a battery of 6 V for a 100-μA PMMC meter. Had we selected a half-scale deflection of 6 kΩ (in attempting to lower the resistance scale by a factor of 10), we would have required a 0.6-V battery (which is impractical).

The *advantages* of a series ohmmeter are that it lends itself to measurement of high resistances and is relatively simple to design.

The *disadvantages* of the series ohmmeter are as follows:

1. It cannot be used to measure relatively medium or low resistances accurately. (This disadvantage is overcome by the shunted-series and shunt ohmmeters.)
2. The scale is *nonlinear* and crowded to the left but spread out to the right of center scale. This makes interpolation of intermediate values difficult.
3. The scale is *inverse*, with zero resistance at extreme right and infinity at extreme left. This also makes interpolation difficult.
4. When the battery voltage decreases with age, it is necessary to change R_v, which in turn changes R_s and R_i, thus somewhat degrading the half-scale calibration. (This is partly overcome by setting $R_f = 0.7R_s$.)

23-5.2 Shunted-Series Ohmmeter

Most commercial ohmmeters are *shunted-series* ohmmeters, in which a variety of resistance ranges are provided by the selection of various values of shunting resistance and battery voltage. A shunted-series ohmmeter circuit is shown in **Fig. 23-5a** and its scale in Fig. 23-5b. Note that the scale is identical to that of the series ohmmeter in Fig. 23-4b. This shows that the instrument is still fundamentally a series ohmmeter. The major purpose of the added shunt resistance, R_{sh}, is to reduce the meter internal resistance, R_i, thereby producing a lower resistance scale. The internal resistance of the ohmmeter shown in Fig. 23-5a is easily found as

$$R_i = R_s + \left(\frac{R_m R_{sh}}{R_m + R_{sh}}\right) = R_s + R_p \qquad \text{ohms } (\Omega) \qquad (23\text{-}15)$$

where R_{sh} is the resistance shunting the meter resistance, R_m
R_p is the resistance of the parallel combination $R_m \| R_{sh}$

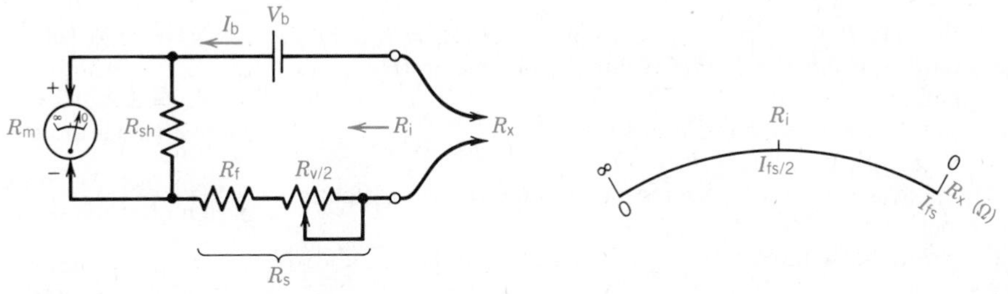

a. Circuit   b. Scale

Figure 23-5 Shunted series ohmmeter and typical scale

The same steps are used in designing the shunted-series ohmmeter (as were previously used in the series ohmmeter) for finding the battery voltage, V_b and deflection factor, *d*. R_p may be selected nominally as any value from $R_m/2$ to a minimum of $R_m/100$.

The battery voltage, V_b, is found from

$$V_b = I_b R_i = I_b(R_s + R_p) \qquad \text{volts (V)} \qquad (23\text{-}16)$$

where I_b is the *total* current supplied by the battery and all other terms have been defined.

The deflection factor, d, is also easily derived from

$$d = \frac{I_m}{I_{fs}} = \frac{R_i}{R_i + R_x} = \frac{R_s + R_p}{R_s + R_p + R_x} \quad \text{(unitless ratio)} \qquad (23\text{-}17)$$

As with the series ohmmeter, half-scale deflection occurs whenever

$$R_x = R_i = R_s + R_p = R_s + (R_m \| R_{sh}) \quad \text{ohms } (\Omega) \qquad (23\text{-}18)$$

The required (*lowered*) series resistance is now determined from

$$R_s = R_i - R_p = R_i - (R_m \| R_{sh}) \quad \text{ohms } (\Omega) \qquad (23\text{-}19)$$

The scale values are calibrated in terms of R_x, using the deflection factor, d, and the internal resistance, R_i, where

$$R_x = R_i\left(\frac{1-d}{d}\right) = (R_s + R_p)\left(\frac{1-d}{d}\right) \quad \text{ohms } (\Omega) \qquad (23\text{-}20)$$

where all terms have been defined.

EXAMPLE 23-9

A 20-μA, 2-kΩ movement is used to design a shunted-series ohmmeter having a center-scale value of 1 kΩ. Calculate the

a. Value of R_{sh} required to produce a value of $R_p = R_m/50$
b. Battery voltage, V_b, required for the ohmmeter. (Hint: first find I_b)
c. Value of R_s needed in Fig. 23-5a
d. Value of R_f, assuming $R_f = 0.7R_s$
e. Value of $R_v/2$ and R_v, the resistance of the variable potentiometer
f. Draw the shunted-series ohmmeter, showing all calculated values

Solution

a. $R_p = R_m/50 = 2000\ \Omega/50 = \mathbf{40\ \Omega}$;
$R_{sh} = (R_pR_m)/(R_m - R_p) = (40 \times 2000)/1960 = \mathbf{40.82\ \Omega}$
b. $I_b = 50(I_{fs}) = 50(20\ \mu\text{A}) = \mathbf{1\ mA}$;
$V_b = I_bR_i = (1\ \text{mA})(1\ \text{k}\Omega) = \mathbf{1\ V}$
c. $R_s = R_i - R_p = 1000\ \Omega - 40\ \Omega = \mathbf{960\ \Omega}$
d. $R_f = 0.7R_s = 0.7(960\ \Omega) = \mathbf{672\ \Omega}$
e. $R_v/2 = 0.3R_s = 0.3(960\ \Omega) = \mathbf{288\ \Omega}$ and
$R_v = 2 \times 288 = 576\ \Omega \cong \mathbf{500\ \Omega}$ potentiometer
f. All values are shown in Fig. 23-6a

Example 23-9 provides the following insights regarding the design of a shunted-series ohmmeter:

1. When we nominally select $R_p = R_m/50$, this automatically yields $I_b = 50(I_{fs})$. Had we selected $R_p = R_m/40$, this would have made $I_b = 40(I_{fs})$, and so forth.
2. When we nominally select the center-scale value as 1 kΩ, this means that $R_i = 1$ kΩ and automatically yields $R_s = R_i - R_p = 1\ \text{k}\Omega - R_p$.

Example 23-10 shows how the scale of a shunted-series ohmmeter may be calibrated.

EXAMPLE 23-10

For each 2-μA step of the basic movement in Ex. 23-9, calculate and tabulate the

a. Deflection factor, d
b. Ratio $(1 - d)/d$
c. External resistance, R_x
d. Show a typical calculation for a meter current of 8 μA

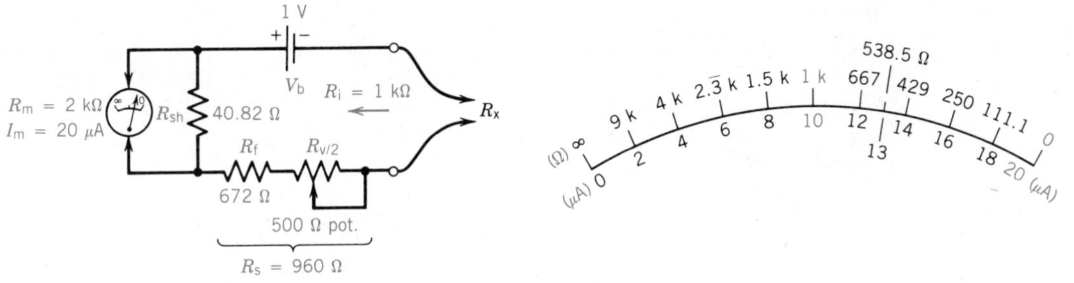

a. Circuit with values b. Scale with values

Figure 23-6 Solutions to Exs. 23-9 and 23-10

e. Draw the complete scale of the shunted-series ohmmeter, showing all values of R_x tabulated

f. Calculate the external resistance when the meter deflects to 13 mA. Verify the value by entering it on the scale drawn in **(e)**

Solution

The tabulation for parts **(a)**, **(b)**, and **(c)** is shown below.

	I_m (μA)	0	2	4	6	8	10	12	14	16	18	20
a.	$d = I_m/I_{fs}$	0	0.1	0.2	0.3	0.4	0.5	0.6	0.7	0.8	0.9	1.0
b.	$(1 - d)/d$	∞	9	4	7/3	3/2	1	2/3	3/7	1/4	1/9	0
c.	R_x (Ω)	∞	9 k	4 k	2.$\bar{3}$ k	1.5 k	1 k	$\overline{666}$	429	250	$\overline{111}$	0

d. $d = I_m/I_{fs} = 8\ \mu A/20\ \mu A = \mathbf{0.4}$ and
$(1 - d)/d = (1 - 0.4)/0.4 = \mathbf{3/2}$ and
$R_x = R_i(1 - d)/d = 1\ k\Omega\ (3/2) = \mathbf{1.5\ k\Omega}$

e. The complete scale is drawn in Fig. 23-6b

f. $d = I_m/I_{fs} = 13\ \mu A/20\ \mu A = \mathbf{0.65}$ and
$(1 - d)/d = (1 - 0.65)/0.65 = \mathbf{0.5385}$;
$R_x = R_i(1 - d)/d = 1\ k\Omega(0.5385) = \mathbf{538.5\ \Omega}$ (see Fig. 23-6b)

These examples reveal the advantages and disadvantages of the shunted-series ohmmeter. The advantages are that the design lends itself to commercial multirange ohmmeters quite nicely and that it permits direct measurement of medium-range resistors. The disadvantages are that it cannot measure extremely high or extremely low values of resistance (see scale in Fig. 23-6b) and that it is slightly more complex to design than simple series ohmmeters and simple shunt ohmmeters.

23-5.3 Shunt Ohmmeter

Any *shunt ohmmeter* is easily and quickly identified by its *direct* scale (i.e., zero ohms on the left and infinite ohms on the right), as shown in **Fig. 23-7b**. In addition, it differs from the two previous ohmmeters in requiring a *momentary-contact switch* (see Fig. 23-7a) to reduce battery wear. The shunt ohmmeter circuit is shown in Fig. 23-7a. Note that the *unknown* resistance, R_x, *shunts* the *meter movement directly*. For this reason, low-sensitivity PMMC movements having low resistances (low R_m) are selected for shunt ohmmeters.[3] The design of the shunt ohmmeter follows that of the two previous series-type instruments.

[3] If high-sensitivity movements are to be used, it is not at all unusual to shunt the movement to produce a lower meter resistance. (See Ex. 23-11.) In this case, the internal resistance of the movement, $R_i = R_m \| R_{sh}$, and the shunt ohmmeter current, $I_m = I_{fs} + I_{sh}$. See Ex. 23-11.

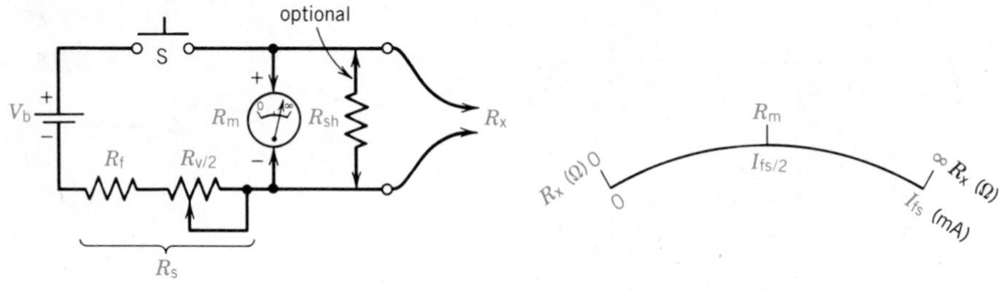

a. Circuit b. Scale

Figure 23-7 Shunt ohmmeter and typical scale

The internal resistance of the instrument, R_i, is essentially and simply[4]

$$R_i = R_m \qquad \text{ohms } (\Omega) \qquad (23\text{-}21)$$

The battery voltage, V_b, is found from the relation (with $R_x = \infty$)

$$V_b = I_{fs}(R_s + R_m) = I_{fs}\left(\frac{R_f + R_v}{2} + R_m\right) \qquad \text{volts (V)} \qquad (23\text{-}22)$$

As for the series and shunted-series ohmmeters, the deflection factor is

$$d = \frac{I_m}{I_{fs}} \qquad (23\text{-}23)$$

For any value of meter current, I_m, or scale deflection, d, the external resistance is

$$R_x = R_m\left(\frac{d}{1-d}\right) = \frac{I_m R_m}{I_x} \qquad \text{ohms } (\Omega) \qquad (23\text{-}24)$$

EXAMPLE 23-11

A high-sensitivity Weston movement having a current sensitivity of 100 μA and a meter of 600 Ω is to be used as a shunt ohmmeter having a midscale value of 10 Ω. Calculate the

a. Value of shunt to be used with the movement
b. Combined full-scale current of both shunt and meter movement
c. Equivalent resistance of the shunted meter and maximum battery current

Solution

a. $R_{sh} = R_p R_m/(R_m - R_p) = (10\ \Omega)(600\ \Omega)/(590\ \Omega)$
$= \mathbf{10.17\ \Omega}$ (5-23a)

b. $I_{sh} = I_m R_m/R_{sh} = (0.1\text{ mA})(600\ \Omega)/10.17\ \Omega = 5.9\text{ mA}$;
$I_m = I_{fs} + I_{sh} = 0.1\text{ mA} + 5.9\text{ mA} = \mathbf{6\ mA}$

c. $R_m = 10\ \Omega$, $I_m = \mathbf{6\ mA}$

---

[4] This equation may not seem obvious at first. But consider that with $R_x = \infty$, R_v in Fig. 23-7a is adjusted for full-scale deflection with momentary switch closed and for zero deflection with the meter shorted by the test leads. Then obviously when $R_x = R_m$, the PMMC meter shows half-scale deflection by current division. Consequently, the internal resistance, R_i, of the instrument is essentially the low meter resistance, R_m.

The insight to be drawn from Ex. 23-11 is that any shunted high-sensitivity meter may be treated as a lower-sensitivity, higher-current meter whose meter resistance is the parallel equivalent $R_p = R_m \| R_{sh}$, as shown in part (**c**). (See Ex. 23-4.)

EXAMPLE 23-12

A model 301 Weston meter movement has a resistance of 2 Ω and a full-scale deflection current of 20 mA. Design a shunt ohmmeter having a midscale resistance value of 2 Ω to operate from a 1.5-V cell. Calculate the

a. Series resistance needed, R_s
b. Value of fixed resistance, R_f
c. Variable resistance $R_v/2$ and R_v
d. Values of resistance for the ohmmeter scale at each 2-mA interval
e. Meter current when measuring an external resistance of 1 Ω
f. External resistance when the meter current is 0.625 mA
g. Draw the shunt ohmmeter showing all values of resistances and potentiometer
h. Draw the complete scale showing all values of current or resistance computed in (**d**), (**e**), and (**f**)

Solution

a. $R_s = V_b/I_{fs} = 1.5\text{ V}/20\text{ mA} = \mathbf{75\ \Omega}$
b. $R_f = 0.7R_s = 0.7 \times 75 = \mathbf{52.5\ \Omega}$
c. $R_v/2 = 0.3R_s = 0.3 \times 75 = \mathbf{22.5\ \Omega}$ and $R_v = 2 \times 22.5 =$ **45 Ω** (use 50-Ω pot)
d.

I_m (mA)	0	2.	4	6	8	10	12	14	16	18	20
$d = I_m/I_{fs}$ (—)	0	0.1	0.2	0.3	0.4	0.5	0.6	0.7	0.8	0.9	1
$d/(1-d)$ (—)	0	1/9	1/4	3/7	2/3	1	1.5	7/3	4	9	∞
$R_x = R_m \dfrac{d}{(1-d)}$ (Ω)	0	$0.\overline{22}$	0.5	0.86	$1.\overline{3}$	2	3	$4.\overline{6}$	8	18	∞

e. $R_p = 1 \| 2 = 2/3\ \Omega$;
$I_t = V_b/(R_s + R_p) = 1.5\text{ V}/(75 + 0.\overline{6})\ \Omega = 19.824\text{ mA}$;

$$I_m = I_t \frac{R_x}{R_x + R_m} = 19.824\text{ mA}\,\frac{1\ \Omega}{3\ \Omega} = \mathbf{6.61\ mA}$$

f. $d = I_m/I_{fs} = 0.625\text{ mA}/20\text{ mA} = 0.03125$

$$R_x = R_m\left(\frac{d}{1-d}\right) = \frac{0.03125}{0.96875} \times 2\ \Omega = \mathbf{0.0645\ \Omega}$$

g. The shunt ohmmeter is shown in **Fig. 23-8a** with all calculated values
h. The complete shunt ohmmeter scale is shown in Fig. 23-8b

The *advantages* of a shunt ohmmeter are that

1. There is no degradation of accuracy or change of the midscale resistance due to battery wear since, as shown in Eq. (23-21), only R_m determines the half-scale deflection.
2. The meter is capable of measuring low resistances directly, without calculation.
3. The scale is a *direct* scale with lower values on the left and higher on the right.

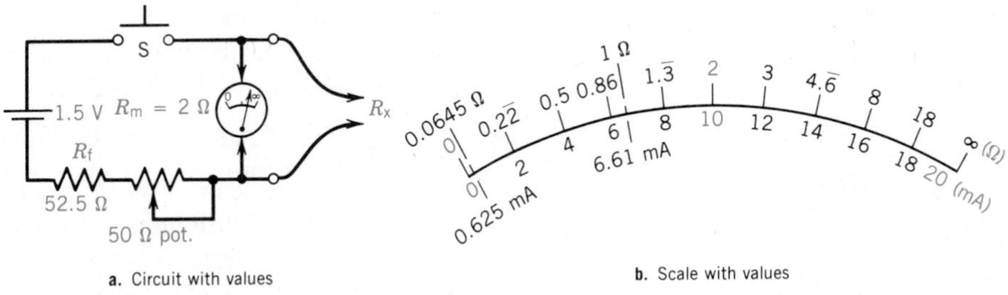

Figure 23-8 Solutions to Exs. 23-11 and 23-12

The *disadvantages* of the shunt ohmmeter are that

1. It cannot be used to measure medium or high resistances.
2. The scale is *nonlinear*; it is crowded (compressed) on the right half and expanded on the lower (left) half.

23-6 USING AMMETERS AND/OR VOLTMETERS TO MEASURE RESISTANCE

There is a limit to the accuracy of measuring resistance with ohmmeters. All series-type ohmmeters, moreover, suffer from degradation due to battery wear. At best, the resistance measured by an ohmmeter is an approximation. All of the following instrumental methods yield more accurate measurements of resistance.

23-6.1 Ammeter Substitution Method

Meter substitution methods require only one meter and a known standard precision resistor, R_s, to measure an unknown resistance, R_x. The ammeter substitution method is shown in **Fig. 23-9a**. The ammeter or milliammeter is an instrument of known accuracy (say 1% or better), and the precision resistor, R_s, is a resistor of known accuracy (say 1% or better). The supply voltage, V_k, is a well-regulated source whose voltage remains constant regardless of loading. Since V_k is constant in Fig. 23-9a, we may write $V_k = I_s R_s = I_x R_x$, and solving for R_x yields

$$R_x = \frac{I_s}{I_x}(R_s) \qquad \text{ohms } (\Omega) \qquad (23\text{-}25)$$

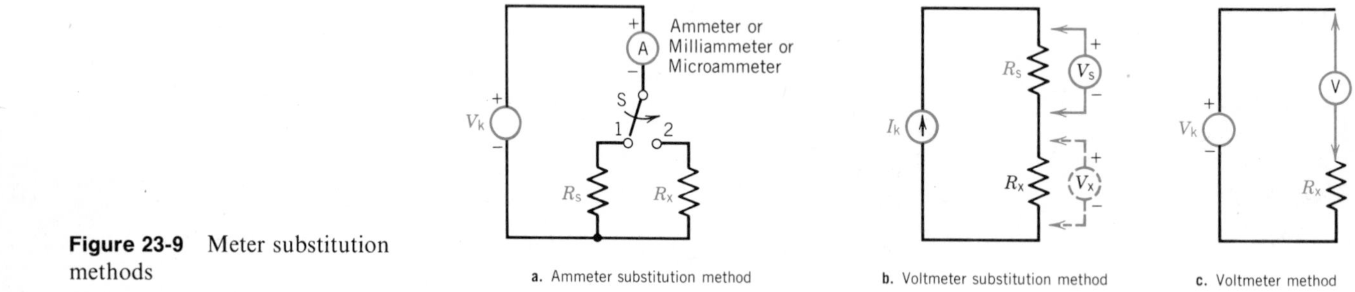

Figure 23-9 Meter substitution methods

The *advantages* of the ammeter substitution method are that

1. It involves the accuracy of only one instrument and one precision resistor. Consequently it is an extremely accurate method of measurement since the *same* instrument is used for measurement of the ratio I_s/I_x in Eq. (23-25).
2. It uses laboratory devices that are readily available, that is, a constant-voltage supply, a precision standard resistor, R_s, and an SPDT switch.

The *disadvantages* of the method are that

1. To use the *same* ammeter or milliammeter, R_s and R_x must be within one order of magnitude. If both are high resistances, a milliammeter is used. If both are low resistances, an ammeter is used. A change in ammeter scale for a second reading degrades the accuracy of the result.

2. To measure a wide range of resistors, several instruments (microammeters, milliammeters, and ammeters) are required along with several precision resistors.
3. Resistance is not obtained directly but by calculation.

23-6.2 Voltmeter Substitution Method

The voltmeter substitution method is analogous to the ammeter substitution method. In this case a constant-current source, I_k, is used, along with a *single* voltmeter that measures voltage across R_s and R_x *on the same scale.* Since the current is the same in each resistor, $I_k = V_s/R_s = V_x/R_x$, and solving for R_x yields

$$R_x = \frac{V_x}{V_s}(R_s) \qquad \text{ohms } (\Omega) \qquad (23\text{-}26)$$

The voltmeter substitution method assumes that the voltmeter used draws negligible current when connected across either R_s or R_x. This works well with *low* values of R_s and R_x. But as these values increase, say to megohms, the voltmeter must be a high-impedance electronic voltmeter (EVM) that draws negligible current on any range. See Fig. 23-9b.

23-6.3 Voltmeter Method

Since most voltmeters of the PMMC type tend to draw current (Sec. 23-6.4), particularly when connected across high resistances, the voltmeter method shown in Fig. 23-9c is a better method to use for unknown high resistance. This method uses the known *voltmeter resistance* as the *standard* resistance. It also requires a constant well-regulated voltage source, V_k. For very high resistances, the voltmeter used should have high sensitivity, as shown in Ex. 23-13.

EXAMPLE 23-13

In Fig. 23-9c, R_x represents a given thickness of rubber insulating tape. The voltmeter is connected first across V_k, which measures 220 V dc. When connected in series with the insulating tape, R_x, the voltmeter measures 5 V on its 250-V scale. The voltmeter sensitivity is 50 000 Ω/V. Calculate the

a. Resistance of the voltmeter
b. Voltage across R_x
c. Resistance R_x

Solution

a. $R_{Vm} = (50\ \text{k}\Omega/\text{V}) \times (250\ \text{V}) = \mathbf{12.5\ M\Omega}$
b. $V_x = V_k - V = 220\ \text{V} - 5\ \text{V} = \mathbf{215\ V}$
c. $R_x = \frac{V_x}{V_s}(R_s) = \frac{215\ \text{V}}{5\ \text{V}}\ 12.5\ \text{M}\Omega = \mathbf{537.5\ M\Omega}$ (23-26)

Example 23-13 shows how ideally suited the voltmeter method (Fig. 23-9c) is for the measurement of unknown high resistances without being subject to the voltmeter loading effect. Since voltmeters are frequently used *across* resistors in circuitry, let us examine the voltmeter loading effect more carefully.

23-6.4 Voltmeter Loading Effect and Voltmeter Sensitivity

The voltmeter loading effect and the property of voltmeter sensitivity are best illustrated by **Fig. 23-10**. In Fig. 23-10a there are three branches in parallel. By

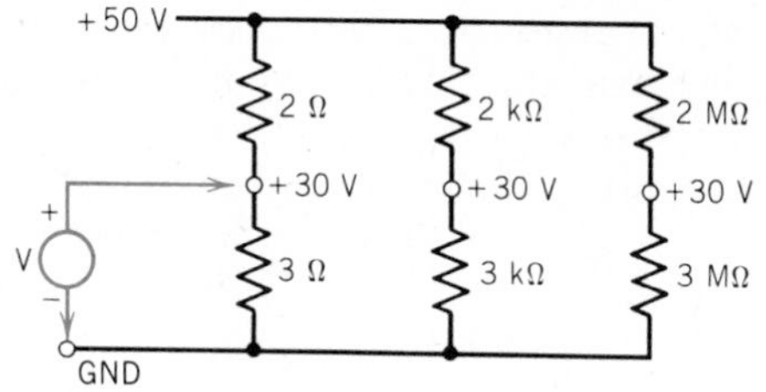

a. Voltages and resistances *before* connecting voltmeter

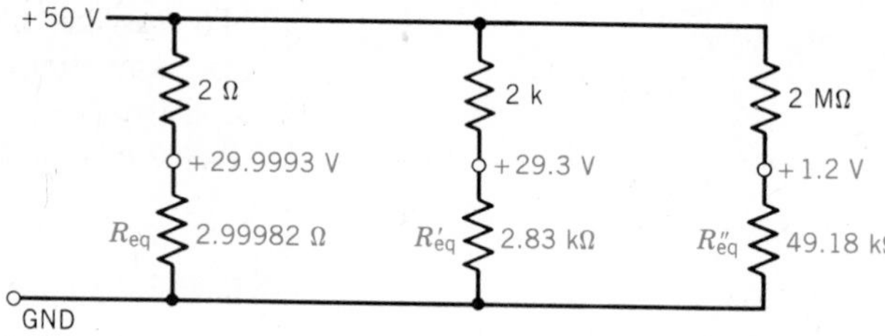

b. Voltages and equivalent resistances *after* connecting voltmeter

Figure 23-10 Voltmeter loading effect due to voltmeter sensitivity

VDR, the voltage between the two series-connected resistors in *each* branch should be +30 V to ground. Let us assume that we use a 1000 Ω/V voltmeter on its 50-V scale to measure and verify the +30 V in each of the three branches. The voltmeter resistance is 50 V × 1 kΩ/V or 50 kΩ. When connected across the 3-Ω resistor, the equivalent resistance is 3 × 50 000/50 003 = 2.999 82 Ω. Using the VDR, the voltmeter will record the true voltage of 29.9993 V or 30 V. Clearly, the original resistance of 3 Ω is unchanged by voltmeter loading.

Now consider the second branch in Fig. 23-10a when a voltmeter of 50 kΩ is connected across the 3-kΩ resistor. The equivalent resistance decreases to 3 × 50/53 or 2.83 kΩ. The voltmeter now reads a true voltage of (2.83/4.83) × 50 V or 29.3 V. Again, although some loading has occurred, it might be considered an acceptable reading, in some circles, since the error is only $2.\bar{3}\%$.

But now consider the third branch. A 3-MΩ resistor now has a 50-kΩ voltmeter connected across it. The equivalent resistance is now 49.18 kΩ! Using the VDR, the voltmeter will record 49.18 kΩ/2049.2 kΩ × 50 V or 1.2 V! In effect, the voltage that previously was 30 V (*before* applying the voltmeter) drops to 1.2 V when the voltmeter is connected. (It also goes back to 30 V whenever the voltmeter is removed.)

This loading effect is caused by using *low*-sensitivity instruments in *high-resistance* circuits. From the foregoing description we may conclude the following:

a. When selecting voltmeters for certain voltage measurements, the *sensitivity of the voltmeter* is an important factor. The *higher* the sensitivity (in Ω/V), the *less* the loading effect.

b. In low-resistance circuits, ordinary (any) voltmeters may be used.

c. In high-resistance circuits, only instruments of high sensitivity or EVMs should be used.

d. When using multirange voltmeters in high-resistance circuits, always use the highest range (even if the deflection is small), since the highest range has the highest resistance and produces the *least* loading effect when connected across high resistances.

e. The type of error created because of instrument sensitivity is called a *loading error*. The *lower* the instrument *sensitivity*, the *greater* the *loading error*.

23-6.5 Ammeter–Voltmeter Method of Measuring Resistance

One of the more popular and common methods of measuring resistance involves the use of Ohm's law, where R_x is (the ratio)

$$R_x = \frac{\text{voltage across resistance } R_x}{\text{current in resistance } R_x} = \frac{\text{voltmeter reading}}{\text{ammeter reading}} = \frac{\text{V}}{\text{A}} \text{ ohms}$$

This method, however, may produce some error because of the factor of the voltmeter resistance (voltmeter sensitivity) previously discussed and also because an error is produced in ignoring the resistance of the ammeter.

Figure 23-11a shows the proper instrumental connections for measuring high resistances. Since the voltmeter V is recording the volt drop across both R_x and the ammeter resistance, R_A, the instrumental connections show that we are actually measuring *both* R_x and R_A. For very precise work, therefore, the precise resistance R_x, a relatively high resistance, is

$$R_x = \frac{V}{A} - R_A \quad \text{ohms } (\Omega) \qquad (23\text{-}27)$$

where all terms have been defined.

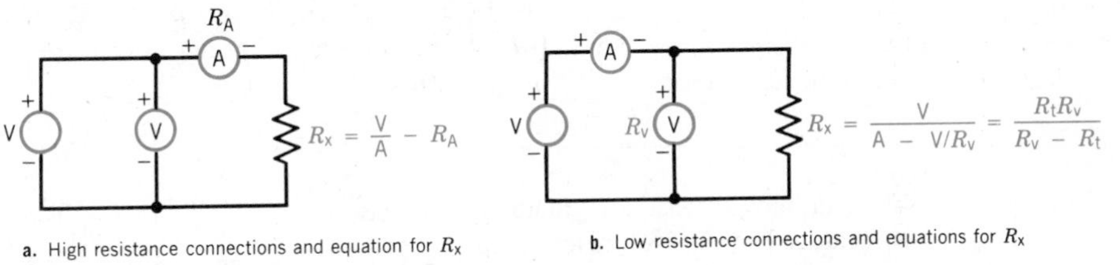

R_A
A
V
V
$R_x = \frac{V}{A} - R_A$
A
V
R_V
V
$R_x = \frac{V}{A - V/R_V} = \frac{R_t R_V}{R_V - R_t}$
a. High resistance connections and equation for R_x
b. Low resistance connections and equations for R_x

Figure 23-11 Proper connections to be used for ammeter–voltmeter method and correction equations

Examination of Eq. (23-27) reveals, however, that since most dc ammeters (and even milliammeters) have fairly low resistances, for most field measurements little error is produced by ignoring the ammeter resistance and merely using the V/A ratio.

Figure 23-11b shows the proper instrumental connections for measuring low resistances. Since the ammeter, A, is recording both the current in R_x and that drawn by the voltmeter, V, the instrumental connections show that we are actually measuring the combined equivalent resistance R_t of the voltmeter, V, in parallel with R_x. For very precise work, therefore, the precise resistance, R_x, a relatively low resistance, is

$$R_x = \frac{V}{A - (V/R_v)} = \frac{R_t \cdot R_v}{(R_v - R_t)} \quad \text{ohms } (\Omega) \qquad (23\text{-}28)$$

where R_t is V/A and R_v is the resistance of the voltmeter.

It should be noted, however, from Fig. 23-11b and Eq. (23-28) that since R_x is a *low* resistance, it draws an appreciable current from the supply. Conversely, R_v, the resistance of the voltmeter, is relatively much higher and draws only a small current from the supply (V/R_v). For *most* field measurements, therefore, little error is produced by *ignoring* the *voltmeter resistance* and merely using the V/A ratio.

23-6.6 Techniques for Determining Proper Voltmeter Placement in Ammeter–Voltmeter Method

Examination of the long-shunt connection (Fig. 23-11a) and the short-shunt connection (Fig. 23-11b) shows that the only difference between the two connections is the placement of the *positive* (+) *voltmeter lead*. For *high*-resistance measurement, the positive voltmeter lead is on the *positive* side of the *ammeter* (long-shunt connection). For *low*-resistance measurement, the positive voltmeter lead is on the *negative* side of the *ammeter* (short-shunt connection).

A question (naturally) arises in situations where R_x is some resistance about which we have no prior knowledge; that is, we do not know whether it is high or low, and we do not know where to connect the positive voltmeter lead.

Figure 23-12 shows the proper technique for determining voltmeter lead connections, by performing the following steps:

1. Place *positive* (+) *voltmeter* lead in position 1 (low-resistance connection), and record readings of V and A.
2. Move *positive* (+) *voltmeter* lead to position 2 and record V and A.
3. If *voltmeter* reading *increases*, restore V to position 1 because R_x is a *low* resistance.
4. If *ammeter* reading *decreases* (after step 2), keep positive voltmeter lead on position 2 because R_x is a *high* resistance.
5. If *both* instruments change, *ignore* the instrument producing the *least change* in deflection.

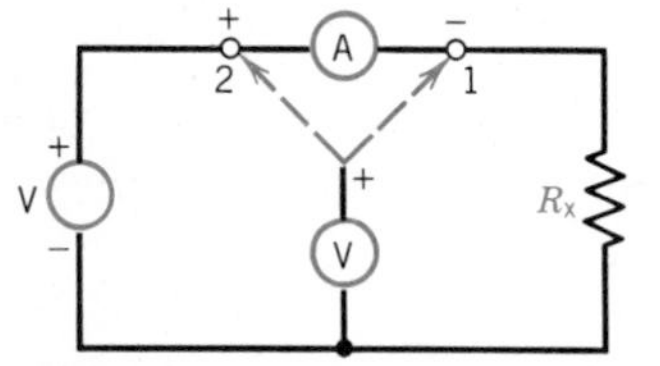

Position 1—low resistance connection
2—high resistance connection

Figure 23-12 Technique to determine proper voltmeter lead placement in using voltmeter-ammeter method

EXAMPLE 23-14

The resistance of an unknown resistor R_x (actually a precision 1.5-Ω resistor) is measured by the voltmeter–ammeter method, using a 1-A ammeter having an internal resistance of 0.5 Ω. The ammeter is connected in series with R_x across a dc supply. A voltmeter (100 Ω/V) connected directly across the supply reads 2.0 V when the ammeter reads exactly 1 A.

a. Explain what is wrong with this method of measurement
b. Calculate R_x by this method of measurement
c. Calculate the percent error produced by **(b)**
d. Explain why the measurement is in error and which connection is causing it
e. The voltmeter on its 3-V scale is now connected directly across R_x. Calculate the equivalent parallel resistance, R_p.
f. Using **(e)** and the VDR, calculate the voltmeter reading
g. Calculate the ammeter reading under the conditions of **(e)**
h. Calculate R_x from the ratio V/A
i. Calculate R_x using the correction Eq. (23-28)

Solution

a. Since R_x is a low resistance, the voltmeter should shunt R_x only
b. $R_x = \text{V/A} = 2\text{ V}/1\text{ A} = \mathbf{2\ \Omega}$
c. PE = (true − experimental) value/true value = (1.5 − 2)/2 = **−33.3̄%** (experimental value is too *high* and should be reduced!)
d. The voltmeter is reading the volt-drop across both ammeter and R_x, instead of the volt-drop across R_x only, as shown in Fig. 23-11b
e. $R_v = 100\ \Omega/\text{V} = \mathbf{300\ \Omega}$;
$R_p = R_x \| R_v = 1.5\ \Omega \| 300\ \Omega = \mathbf{1.4925\ \Omega}$
f. $$V_x = V_p = (R_p/R_t)V_t = \frac{1.4925}{(1.4925 + 0.5)} \times 2\text{ V}$$
$$= \mathbf{1.498\ V} \text{ (voltmeter reading V)}$$
g. $I_t = V/R_t = 2\text{ V}/1.9925\ \Omega = \mathbf{1.0038\ A}$ (ammeter reading A)
h. $R_x = \text{V/A} = 1.498\text{ V}/1.0038\text{ A} = \mathbf{1.492\ \Omega}$
i. $$R_x = \frac{\text{V}}{\text{A} - \text{V}/R_v} = \frac{1.498\text{ V}}{1.0038 - (1.498/300)} = \frac{1.498}{0.9988}$$
$$= \mathbf{1.4998\ \Omega \cong 1.5\ \Omega}$$

The following insights are to be drawn from Ex. 23-14:

1. When the voltmeter is *improperly* connected, a large error is produced (−33.3%).
2. When the voltmeter is properly connected for measuring low resistances, it is *usually unnecessary* to use correction equations, even when the voltmeter sensitivity is low. The ratio of the voltmeter to ammeter reading (V/A) is sufficiently accurate to yield the unknown resistance by Ohm's law (see point 3).
3. The percent error between R_x values in **(h)** and **(i)**, (1.5 − 1.492)/1.5 or 0.5%, is well within the 1 or 2% accuracy of the measuring instruments used.

In comparison to the ohmmeter, resistance measurement by the ammeter–voltmeter method is considerably more accurate. But it still leaves much to be desired, and the ammeter–voltmeter method still suffers from a number of possible errors that are typical of most methods of analog measurement. Some of these are as follows:

1. *Scale errors* The scales of most commercial instruments are printed. But even those made by hand may be marked inaccurately and may not be calibrated.
2. *Parallax errors* Caused by failure to align pointer perpendicular to scale. Mirrored scales help to reduce parallax error.
3. *Accuracy errors* The voltmeter may have 2% accuracy and the ammeter 1% accuracy. At best, then, the limiting accuracy is 2%. But even this is a delusion. The 2% accuracy refers only to the full-scale reading, since 2% of 150 V is ± 3 V. This ± 3 V error may occur even when reading 10 V on the 150-V scale, producing a 30% error!
4. *Loading errors* Noted by correction equations in Fig. 23-11. Even when ammeter and voltmeter are properly connected, some small errors are introduced due to the resistances of the voltmeter and the ammeter, respectively.
5. *Temperature errors* Use of potential sources inevitably results in some I^2R heating of the resistance measured, which changes its value from its resistance at room temperature (20°C).

23-7 WHEATSTONE BRIDGE

Of all the methods of resistance measurement, the Wheatstone bridge is the most accurate and is probably the most widely used for measurement of resistances ranging from 1 Ω to 1 MΩ.[5] The bridge circuit permits comparison of an unknown resistance against standard known values of resistance of extremely high accuracy.

A typical Wheatstone bridge schematic is shown in **Fig. 23-13a**. The upper arms (R_1 and R_2) are *precision* resistors, precisely set in some fixed ratio to each other, called *ratio arms.* The ratio R_2/R_1 is usually set by a single *ratio dial* in the commercial Wheatstone bridge (Fig. 23-13b). The ratio may vary to include 1000/1, 100/1, 10/1, 1/1, 1/10, 1/100, and 1/1000. Resistance R_3 is essentially a variable four-dial decade resistance box yielding values from 1.0 to 9 999 Ω (Fig. 23-13b).

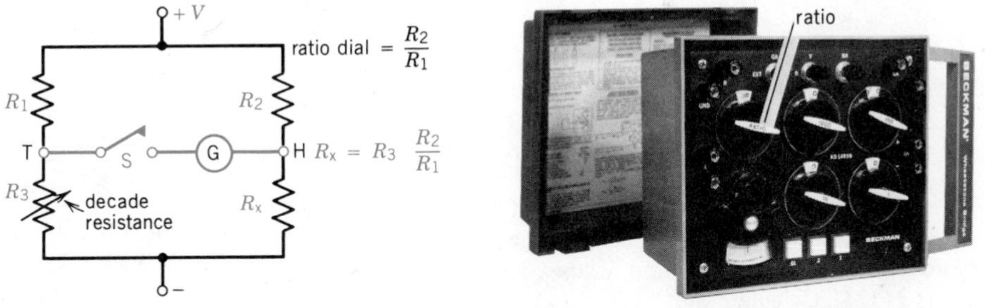

a. Schematic diagram

b. Commercial Wheatstone bridge (Courtesy Beckman Industrial Corp.)

Figure 23-13 Wheatstone bridge

The commercial bridge shown in Fig. 23-13b has a ratio accuracy of $\pm 0.05\%$ and a decade accuracy of $\pm 0.1\%$. The moving-coil galvanometer has a basic sensitivity of 1 μA per division. Terminals are also provided for connecting external battery supplies (BA) and external null detectors for higher sensitivity.

[5] A special modified Wheatstone bridge called a Kelvin bridge is used exclusively for measurement of extremely low resistances, below 1 Ω.

In setting the bridge to a *null* (i.e., the situation where the galvanometer, G, reads zero), the procedure is as follows

1. Determine an approximate ratio of R_2/R_1 to be used. If R_x is small, the ratio should be less than unity. If R_x is large, the ratio dial should be set to greater than unity.
2. The decade resistance, R_3, is varied as momentary contact switch S (Fig. 23-13a) is alternately closed and the zero-center galvanometer deflection is observed.
3. When the potential at point T is the same as that at point H, the bridge is balanced and G reads zero (the null condition).

Using the VDR, we may write $R_1/R_3 = R_2/R_x$, and solving for R_x yields

$$R_x = R_3\left(\frac{R_2}{R_1}\right) \quad \text{ohms } (\Omega) \qquad (23\text{-}29)$$

where R_2/R_1 is the setting of the ratio dial
R_3 is read directly from the decade resistance box

Advantages of this method of resistance measurement by the Wheatstone bridge are that

1. Extremely high accuracy is attained because the measurement is independent of meter accuracy or the accuracy of galvanometer calibration. Accuracies vary from 0.01 to 0.03%!
2. The measurements are made with minimal currents since V is usually very small (about 3–6 V), producing minimal self-heating and/or changes in resistance.

The *disadvantages* of the method include the following:

1. Inability to measure *excessively low* resistances (below 0.1 Ω) or *excessively high* resistances (above 10 MΩ).
2. Limitation of accuracy to the accuracies of the ratio dial (R_2/R_1) and the decades (R_3).

EXAMPLE 23-15

A commercial resistor is coded RED–GREEN–YELLOW–GOLD. It is measured on a commercial Wheatstone bridge whose ratio dial is set at 100/1 and whose decade resistance dial reads 2433 Ω. Calculate the

a. Nominal value of the resistor and its tolerance
b. Resistance as measured by the Wheatstone bridge
c. Percent error between the nominal and the true value
d. Is the resistance within the specified tolerance? Explain.

Solution

a. Nominally 250 kΩ ± 5%
b. $R_x = R_3(R_2/R_1) = 2433(100/1) = 243.3$ kΩ
c. PE = (true − nominal)/true = (243.3 − 250)/243.3 = −2.75%
d. Resistance is within the specified tolerance of ±5%

23-8 OTHER TYPES OF METERS FOR MEASURING VOLTAGE, CURRENT, AND POWER

We have concentrated, so far, on the PMMC meter for measurement of voltage, current, and resistance because of its extensive use. But other types of analog instruments having a pointer and calibrated scale are used, as well, for dc and ac measurement of voltage current and power. These are summarized in Table 23-1.

Table 23-1 Types of Analog Instruments for DC and AC Measurement of Current, Voltage, and Power

Instrument Type[a]	Parameters Measured	Damping Method
PMMC meter	dc current, dc voltage, resistance	Eddy current damping
Thermocouple(s) and PMMC meter	dc/ac current, voltage, and power	Eddy current damping
Rectifier and PMMC meter	ac sinusoidal current and voltage dc current and voltage on second scale	Eddy current damping
Moving-iron-vane meter	dc/ac current and voltage	Air damping
Electrodynamometer	dc/ac current, voltage, and power	Air damping

[a] All of these instruments employ spiral retarding springs for deflection control.

23-8.1 Thermocouple Ammeter and Voltmeter

A *thermocouple* is the junction of two *dissimilar* metals. When the junction is heated, a dc voltage is produced at the free ends. A typical thermocouple ammeter is shown in **Fig. 23-14a**. The dissimilar metals are copper/constantan, iron/constantan, or chromel/alumel, typically. The greater the temperature difference between the hot and cold junctions, the greater the deflection of the PMMC meter. Since the deflec-

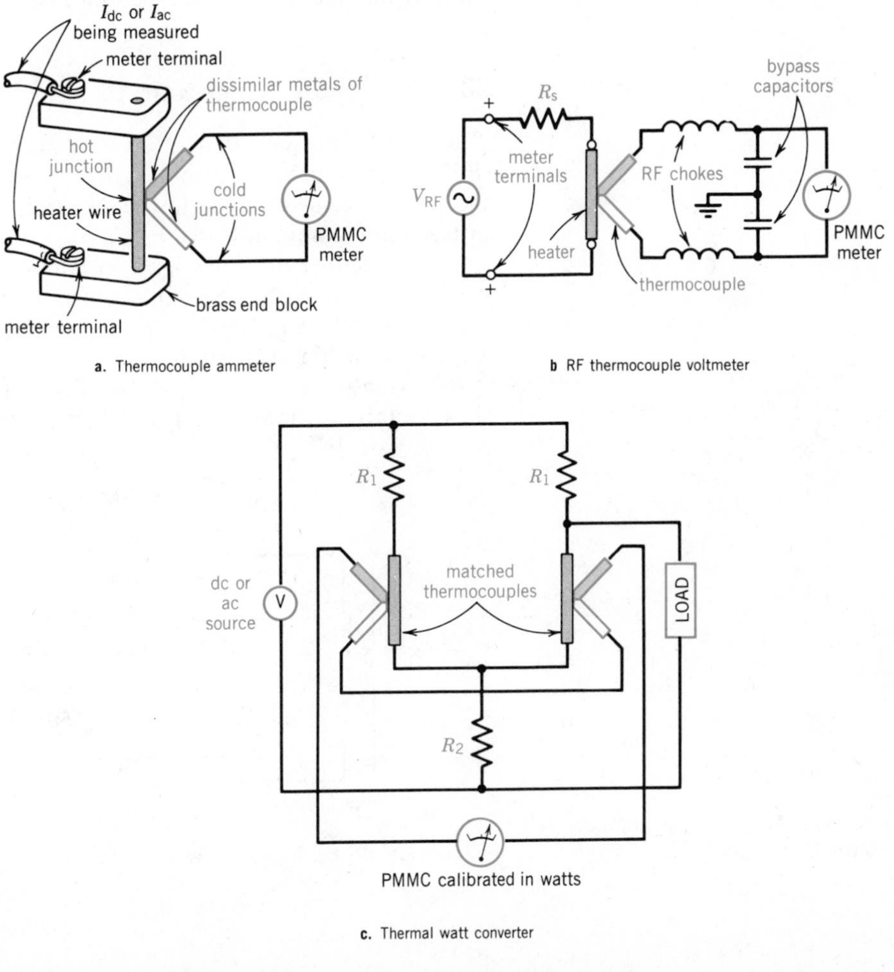

Figure 23-14 Thermocouple instruments

tion is proportional to the heating effect, thermocouple meters are true RMS instruments capable of measuring dc and ac. Since the heat and the generated voltage are proportional to I^2R, the PMMC meter uses a square-law scale, which is more crowded at the high end.

When used as a voltmeter, a series (multiplier) resistance, R_s, is connected in the heater circuit. As shown in Fig. 23-14b, a typical RF voltmeter contains a series choke and parallel (shunting) bypass capacitors to ensure that only dc reaches the PMMC meter by filtering out all stray ac.

The major advantage of thermocouple meters is that they can measure both dc and ac up to frequencies of 50 MHz with errors as low as 1%. A second advantage is that *only one* scale is needed since the instrument is a **true RMS** meter.

The major disadvantage of thermocouple instruments is that they are sluggish in responding to changes in current and, consequently, are subject to burnout if the current rises rapidly above full-scale deflection.

23-8.2 Thermal Watt Converter

Two matched thermocouples are used in a device called a *thermal watt converter*, shown in Fig. 23-14c. The outputs of the thermocouples are connected so that they oppose each other. It can be shown that the output is proportional to $VI \cos \theta$, thus producing a wattmeter with a frequency response from dc up to 15 kHz ac with an accuracy as high as 0.1%. Thermoelement wattmeters (or thermal watt converters) also used for dc and ac instrument calibration because of their inherently high accuracy even when measuring loads at low power factors.

23-8.3 Rectifier Instruments

The PMMC movement may be used in conjunction with a full-wave rectifier (FWR) to measure ac sinusoidal currents and voltages. Because of the linearity of the PMMC scale and the low power consumption of the rectifier, the rectifier instrument is commonly used for ac sinusoidal measurement. **Figure 23-15a** shows a typical rectifier ammeter used in conjunction with a current transformer for measurement of high ac currents. Since the instrument senses average values, the scale is calibrated in RMS, using the form factor ratio $V_{RMS}/V_{AV} = 1.111$, that is, 1.111 times its dc scale calibration.

As current-measuring devices, rectifier ammeters are limited to the lower ranges since shunts are impractical and larger rectifiers are too bulky. Furthermore,

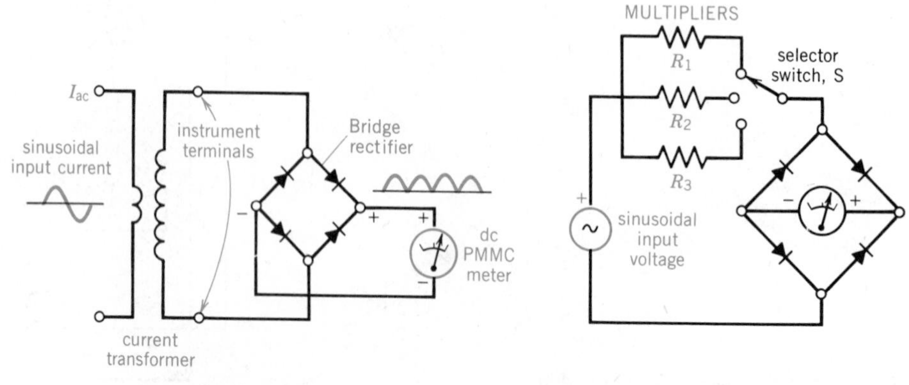

a. Rectifier ammeter

b. Rectifier voltmeter

Figure 23-15 Rectifier ammeter and voltmeter circuits

because of the nonlinearity of the diode characteristic (at low values of forward voltage), the scale is somewhat nonlinear at its lower end.

Rectifier voltmeters (Fig. 23-15b) are more commonly used, particularly in the voltage range from 50 to 500 V, with multipliers similar to those used with dc voltmeters. Rectifier voltmeters are commonly used in VOMs because of their higher sensitivity in comparison to iron vane and/or electrodynamometer movements (described below). Furthermore, the frequency range of the rectifier instrument (although not as high as that of the thermocouple meter) is considerably extended (up to 5 or 10 kHz) compared to these other types, which only respond to a few hundred hertz.

Advantages of rectifier instruments are their use of PMMC movements, which afford (fairly) linear scales and high sensitivities, and their relatively low manufacturing cost. Disadvantages include the *inability* to measure *nonsinusoidal* waveforms, inability to measure *sinusoids* at *high* frequencies (audio and RF ranges), and errors as a result of the temperature sensitivity of the rectifier. Another disadvantage is the need for *two separate* scales when the same meter is used for dc and ac measurement of current and/or voltage.

23-8.4 Iron-Vane Instruments

Deflection torques for iron-vane instruments are obtained either by magnetic attraction or repulsion of a soft-iron vane positioned inside a current-carrying coil. The repulsion type is more commonly used because it provides a much wider radius of deflection (up to 250°). The deflection-type instrument, shown in **Fig. 23-16a**, works equally well for dc, ac, and nonsinusoidal ac waveforms. A fixed stationary soft-iron vane and movable soft-iron vane are both (temporarily) magnetized by the current coil surrounding both vanes, whenever current is in the coil. Since both vanes are magnetized the same way, repulsion occurs between them, producing a torque against the control springs. The deflection torque of this true RMS instrument is proportional to the *square* of the current in the coil, either dc or ac of *any* waveform up to 100 Hz.

Damping in this type of instrument is obtained via an airtight damping chamber (Fig. 23-16b), which prevents oscillation when used on ac, due to nonlinearity of the B–H curve with its attendant hysteresis.

For use as a voltmeter, the coils are wound with many turns of fine wire for the basic meter and multipliers are used for higher voltages. For use as an ammeter, the coils have fewer turns of heavier wire. As ammeters, iron-vane instruments can measure fairly large currents without the need for shunts, but their disadvantages limit the lower current ranges to a minimum of about 15 mA (i.e., 67 Ω/V). Because of this low sensitivity, iron-vane voltmeters cannot be used for either resistance measurement or circuit testing (where their loading effect is prohibitive).

The advantages of iron-vane instruments are their ability to measure dc, ac, and nonsinusoidal waveforms of relatively low frequency, their ruggedness (electrically and mechanically), and the relatively low cost to produce them. Disadvantages are their nonlinear scale, low sensitivity, frequency limitation to below 100 Hz, waveform errors that may be produced by pulse waveforms tending to saturate the iron vanes, and deflection hysteresis (a lower reading with a slowly increasing current and a higher reading with a slowly decreasing current). These result in an accuracy disadvantage that limits this type of instrument to about 5%.

It may be of some interest to note that the iron-vane principle is used in a moving-iron-vane *power factor meter*, in which the phase angle of line current with respect to line voltage is determined by the angular position between the fixed and movable vane.

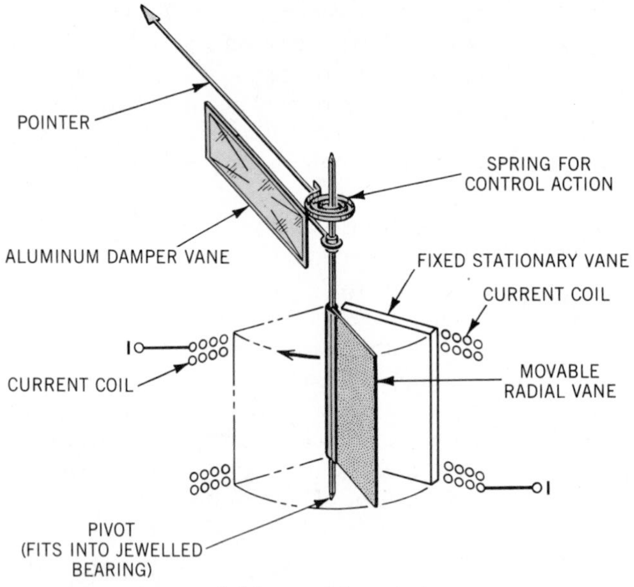

a. Radial-vane repulsion mechanism

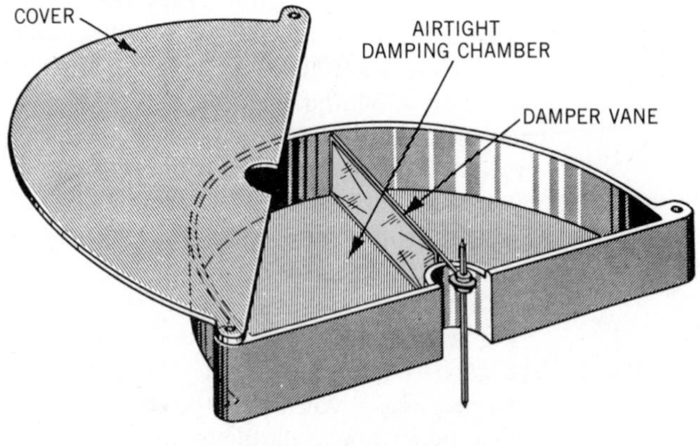

b. Damping mechanism

Figure 23-16 Iron-vane meter mechanisms (U.S. Army TM 11-664)

23-8.5 Electrodynamometer Instruments

Each of the instruments described thus far has some iron incorporated in its deflection structure. Electrodynamometer instruments produce their magnetic fields and deflections with air-core coils. Essentially, an electrodynamometer movement consists of a *pair* of *fixed* air coils and a single movable air coil suspended within the fixed coils. The deflection torque is proportional to the product of current in the stationary coils (I_s) times current in the rotational coil (I_r), since, $T = kI_sI_r$. Consequently, when the movement is used for either ammeters or voltmeters, we find an approximately square-law scale. Conversely, when the movement is used for *wattmeters*, the scale is *linear*. All electrodynamometer instruments are true RMS instruments.

The **electrodynamometer voltmeter** is shown in **Fig. 23-17a**. In this mode, the *stationary* coils are connected in *series*, with the *movable* coil in series with one (or more) multipliers, R_m. Full-scale deflection is obtained with currents as low as

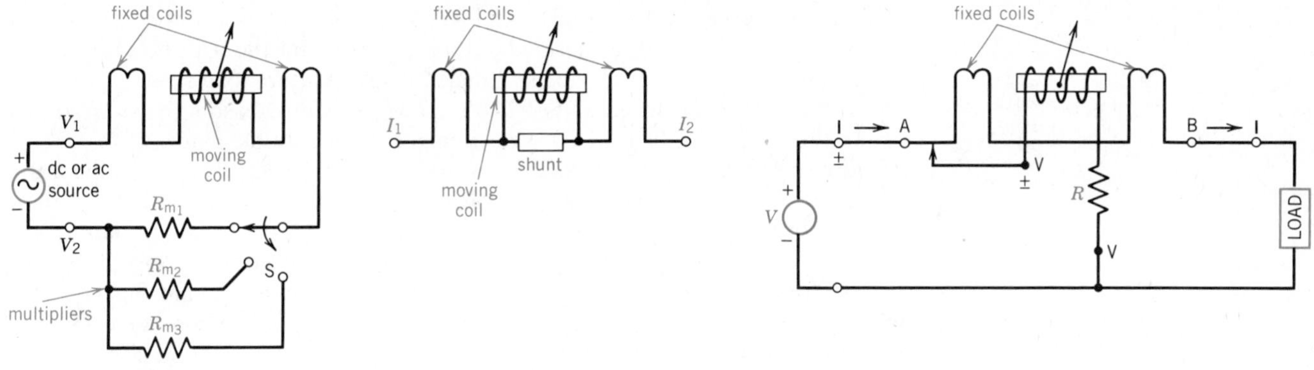

a. Voltmeter connections b. Ammeter connections c. Wattmeter connections (long-shunt connection)

Figure 23-17 Electrodynamometer as a voltmeter, ammeter, and wattmeter showing internal connections

5 mA, producing sensitivities of 200 Ω/V. Voltmeters range from 30 to as high as 750 V, either dc or ac, with true RMS capability. Since all dynamometers act on the principle of interaction between the stationary and movable magnetic fields, the voltmeter measures dc and ac (sinusoidal and nonsinusoidal) waveforms on the *same* scale. Because of their high accuracy and true RMS capability, dynamometer instruments are often used as *standards*, with accuracies as good as 0.1%. It is customary to *calibrate* the instrument scale on dc and then use it to measure ac. Such instruments are called *transfer standards*.

The **electrodynamometer ammeter** is shown in Fig. 23-17b. In this design, the stationary coils are carrying a higher current than the movable coil. Since the stationary coils are fixed, there is no constraint on their weight or bulkiness. The calculation for the shunt resistance is much the same as that for the PMMC meter when used as an ammeter. Current ranges from 1 to 30 A, depending on stationary coil design, with frequency ranges up to 500 Hz. Like the voltmeter, the ammeter is usually manufactured for use as a transfer standard, rather than for field measurement.

The **electrodynamometer wattmeter**, unlike the voltmeter and ammeter, is manufactured primarily as a field instrument for the measurement of power, both dc and ac.[6] Figure 23-17c shows the internal instrumental connections for use of the electrodynamometer as a wattmeter. The instrument has two current terminals and two voltage terminals. The common (±) terminals are usually connected to the *same side* of the *line* and to the same point. This *ensures* that the meter reads *upscale* when measuring dc power.

When using any four-terminal electrodynamometer wattmeter, particularly on ac, care must be taken *not to overload* either the potential or the current coil. At *low* power factors, the deflection is small, but the current coils may be *overloaded.* At low currents and relatively high power factors, the deflection is small and it is possible to overload the potential coil.

With the potential coil connected to point A, as shown in Fig. 23-17c, the wattmeter is not metering the current drawn by the potential coil. This (long shunt) connection is preferable for low-current, high-voltage loads.

If the potential coil is connected to point B (Fig. 23-17c), the wattmeter is including the additional (usually fixed) loss of the potential circuit. This connection

[6] See Appendix B-10 for a proof showing that the average power measured by an electrodynamometer wattmeter is $VI \cos \theta$ when used on an ac single-phase circuit.

is preferred for high-current, lower-voltage loads. Alternatively, when connected (short shunt) to point B, if the load is disconnected and the wattmeter power is recorded, this fixed value may be subtracted from all future load readings, since it represents the wattmeter loss. (See Prob. 23-38.) For this reason, the short-shunt connection is often preferred for measurements with variable loads.

The **electrodynamometer varmeter** has exactly the same construction as the wattmeter (Fig. 23-17c) except that the resistor R in series with the potential coil is replaced by a capacitor. This modification produces a 90° phase shift in the current in the potential coil. Under these circumstances, the deflection is proportional to $VI \sin \theta$ or Q, the quadrature power in vars. It should be noted that the varmeter may be used at only one frequency (usually 60 Hz) because the value of C is computed precisely for a leading current of 90° in the potential coil circuit.

The **electrodynamometer power factor meter**, shown in **Fig. 23-18a**, is different from the previous electrodynamometer instruments in two respects. There are *two* movable coils perpendicular to each other, and these coils are *not* restrained by hairsprings but are free to rotate. The torque produced in each of the two movable coils is a function of the coil impedance (which determines the coil current) and also the mutual inductance between the crossed coils and the stationary field coils. It can be shown that the balanced position of the pointer and its angular displacement, θ, is the phase angle between the applied voltage across the load (energizing the crossed coils) and the current in the load energizing the fixed field coils. The scale may be calibrated in terms of either phase angle θ or $\cos \theta$ or both.

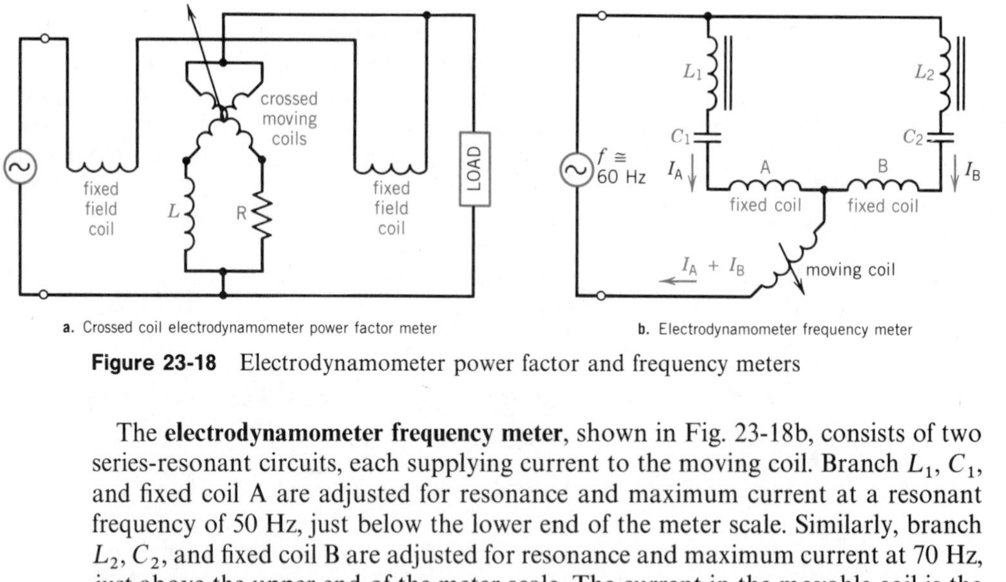

a. Crossed coil electrodynamometer power factor meter

b. Electrodynamometer frequency meter

Figure 23-18 Electrodynamometer power factor and frequency meters

The **electrodynamometer frequency meter**, shown in Fig. 23-18b, consists of two series-resonant circuits, each supplying current to the moving coil. Branch L_1, C_1, and fixed coil A are adjusted for resonance and maximum current at a resonant frequency of 50 Hz, just below the lower end of the meter scale. Similarly, branch L_2, C_2, and fixed coil B are adjusted for resonance and maximum current at 70 Hz, just above the upper end of the meter scale. The current in the movable coil is the sum of the two off-resonance currents supplied by each branch, which in turn determines the torque and deflection of the pointer. Assume the applied frequency is approximately 60 Hz. The circuit in branch A is above resonance and is essentially inductive, whereas the circuit in branch B is below resonance and is essentially capacitive. Current lags voltage in branch A and leads voltage in branch B; consequently, branch A is producing counterclockwise torque and branch B clockwise torque. When the two currents are equal in magnitude, the torques are balanced and the pointer position is approximately 60 Hz. As the frequency drops, current in branch A increases (closer to resonance) while that in branch B decreases and the pointer is deflected downscale.

Electrodynamometer frequency meters are used primarily for checking *low* (power line) *frequencies* of approximately 50, 60, or 400 Hz.

23-9 VOLT–OHM–MILLIAMMETER (VOM) MULTIMETER

The multimeter or VOM (volt-ohmist), despite recent advances in digital instrumentation (Sec. 23-10), remains a most popular instrument for electrical measurements. This is due primarily to its ability to measure dc voltages and currents, ac voltages and currents (see Table 23-2), resistance, and output in dB.

Table 23-2 **Measurement Ranges of the Simpson Model 260, Series 7M**

Measurement	Ranges
Current, dc	0–50 μA; 0–1/10/100/500 mA; 0–10 A (250 mV drop)
Current, ac	Up to 250 A, ac, with optional amp-clamp adapter model 150-2
Voltage, dc	0–250 mV; 0–2.5/10/50/250/500/1000 V at 20 kΩ/V
Voltage, ac	0–2.5/10/50/250/500/1000 V at 5 kΩ/V
Resistance (Ω)	$R \times 1$ (12-Ω center); $R \times 100$ (1.2-kΩ center); $R \times 10\,000$ (120-kΩ center)
dB scales (dB)	-20 to $+10$; -8 to $+22$; $+6$ to $+36$; $+20$ to $+50$ dB
Accuracy	1.5% on 50-μA or 250-mV scale, dc; 2% on all other dc scales; 3.0% on all ac scales

Figure 23-19 shows a typical laboratory quality Simpson model 260, series 7M, with mirrored scale. The instrument contains a 50-μA PMMC meter, which ensures a dc voltage sensitivity of 20 kΩ/V. With a full-wave rectifier for measurement of ac voltage, the sensitivity is reduced to 5 kΩ/V. Table 23-2 summarizes the various measurement and range specifications of the Simpson VOM.

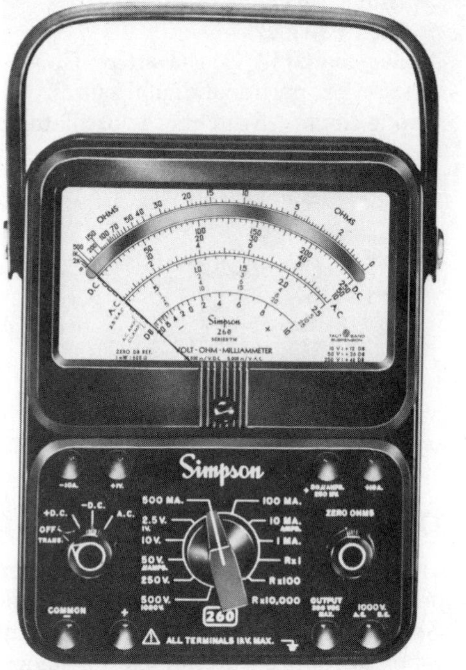

Figure 23-19 Typical mirror-scale laboratory VOM, Simpson model 260, series 7M (Courtesy Simpson Electric Co.)

Despite its extreme popularity and versatility (in containing so many measurement features in a single instrument), the VOM has *two major disadvantages*. The first is that it *only* measures ac of *sinusoidal* waveforms since it incorporates a rectifier for ac measurement (Sec. 23-8.3). The second is that inexperienced users are often confused by the various dc and ac printed scales and take erroneous readings as a result. Both of these disadvantages are overcome in the digital VOM (Sec. 23-11).

23-10 DIGITAL INSTRUMENTS

Perhaps the most significant advance in the last two decades has been the development of *digital* electronic instrumentation. Unlike analog instruments, which require interpolation from pointer position on a graduated scale, digital instruments provide readings *directly* on a display or a screen. Consequently, digital instruments overcome the serious disadvantage of *misreading* one or more analog scales. Furthermore, since digital readouts are observable from a greater distance, digital instruments are far quicker and easier to read than conventional analog instruments. Finally, digital instruments are not subject to parallax errors because no pointer-scale readings are involved.

Almost all digital instruments employ a number of common electronic circuits and devices, shown in the block diagram of **Fig. 23-20** and described below:

1. **Digital display** Typical displays use light-emitting diodes (LEDs), liquid crystal displays (LCDs), and/or electroluminescent displays (ELDs). The digital display in the form of numerals and/or letters represents the instrumental reading of the particular physical quantity measured by the instrument.
2. **Digital output circuit** This is a combination driver/decoder/signal conditioner that operates the digital display to record the decimal equivalent of the quantity measured.
3. **Electronic counter** Counts the pulses proportional to the input signal.
4. **Electronic oscillator** or precise clock Generates a train of digital pulses at a precise frequency.
5. **Analog–digital (A/D) converter** Converts the parameter being measured to a series of proportional digital pulses.
6. **Gating circuit** Switches the oscillator pulses ON and OFF in proportion to the analog signal from the A/D converter.

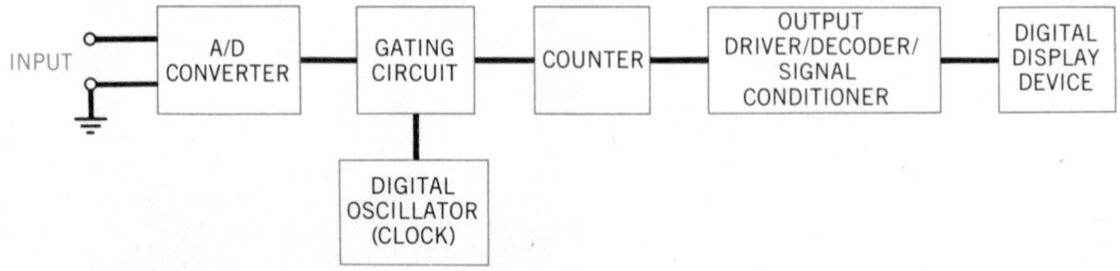

Figure 23-10 Block diagram of typical digital instrument

23-11 TRUE RMS DIGITAL MULTIMETER

The disadvantages of the analog multimeter (Sec. 23-9) are completely overcome in portable (battery-operated) digital multimeters (DMMs). The relatively inexpensive Simpson **true RMS** digital multimeter, model 467, shown in **Fig. 23-21** provides a total of **26** voltage, current, and resistance ranges. It measures dc voltages from

100 μV to 1 kV and ac voltages from 100 μV to 750 V with an input impedance of 10 MΩ on all voltage ranges. It measures ac and dc currents from 100 nA to 2 A, *regardless of waveform.* It reads resistances directly from 0.1 Ω to 20 MΩ. In addition, it incorporates a number of special useful features such as *differential peak hold* (the ability to record extreme positive or negative values of *nonsinusoidal* waveforms), *pulse detection* (which indicates the presence of positive and negative pulses), and a *continuity check* (both visual and audio indication of opens, shorts, and high–low resistance conditions), useful in checking diodes and semiconductors. The basic accuracy of the instrument is $\pm 0.1\%$ on voltage and current readings and $\pm 0.25\%$ on resistance readings.

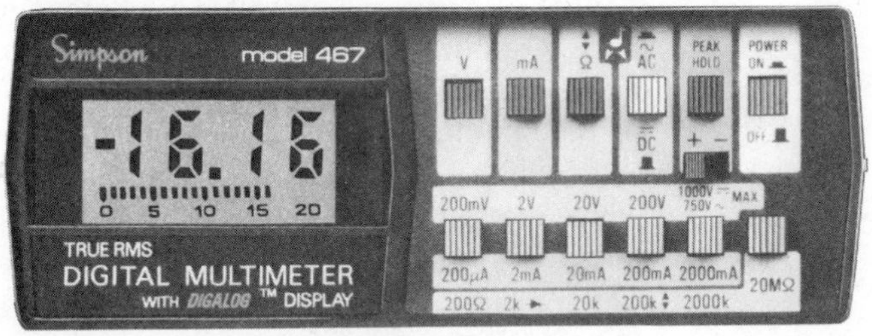

Figure 23-21 True RMS digital multimeter (Courtesy Simpson Electric Co.)

23-12 DIGITAL ELECTRONIC COUNTER

Digital electronic counters are sophisticated frequency meters that are capable of measuring frequencies over an extremely wide range in comparison to the electrodynamometer instrument described in Sec. 23-8.5. Such counters are capable of measuring either periodic or random pulses over a specified period (usually 1 second). A modern digital electronic counter is the Hewlett-Packard model 5386A, shown in **Fig. 23-22**. The counter measures frequency in two ranges. Channel A accepts frequencies from 10 Hz to 100 MHz. Channel B accepts frequencies from 90 MHz to 3 GHz. Frequency is displayed up to eight significant digits. The input impedance of the counter is 1 MΩ shunted by less than 25 pF of capacitance. The counter is fully programmable for use in automatic testing equipment. Gate times of 0.1 and 10 s may be selected in addition to 1 s. Back control terminals also permit remote display of frequency readings, as well as "user-friendly messages" from remote locations.

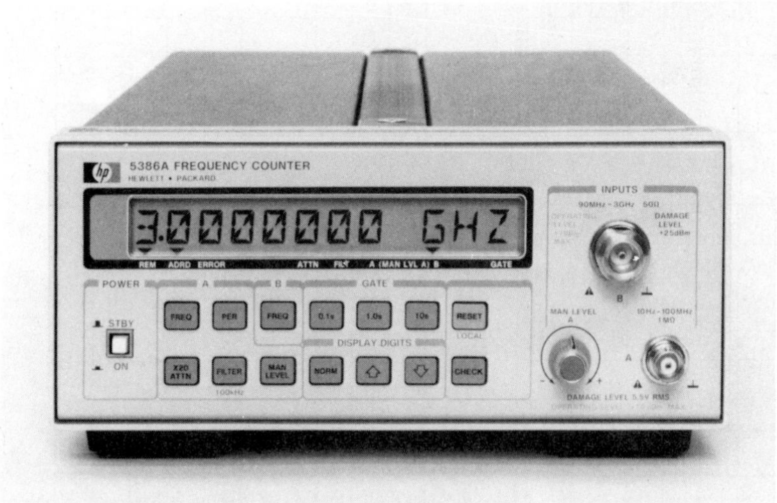

Figure 23-22 Digital electronic counter for precise frequency measurement (Courtesy Hewlett-Packard)

23-13 WAVEFORM FUNCTION GENERATORS

Waveform *function generators* are sophisticated devices that generate sinusoidal, triangular, square, and single or repetitive pulses ranging in frequency from extremely low frequencies up to the megahertz range. A modern digital waveform pulse/function generator is the Hewlett-Packard model 8116A, shown in **Fig. 23-23a**. This instrument can generate frequencies as low as 1 mHz and as high as 50 MHz with amplitudes as high as 32 Vpp (volts peak-to-peak) for all waveform modes (sinusoidal, triangular, square, and pulse). A wide choice of external trigger and modulation modes are incorporated within the instrument, such as amplitude modulation (AM), frequency modulation (FM), pulse width modulation (PWM), and voltage-controlled oscillator (VCO). Figure 23-23b shows some of the more exotic waveforms that can be generated by the model 8116A. Like the HP digital electronic counter, the generator shown in Fig. 23-23a is fully programmable for use in automatic testing equipment.

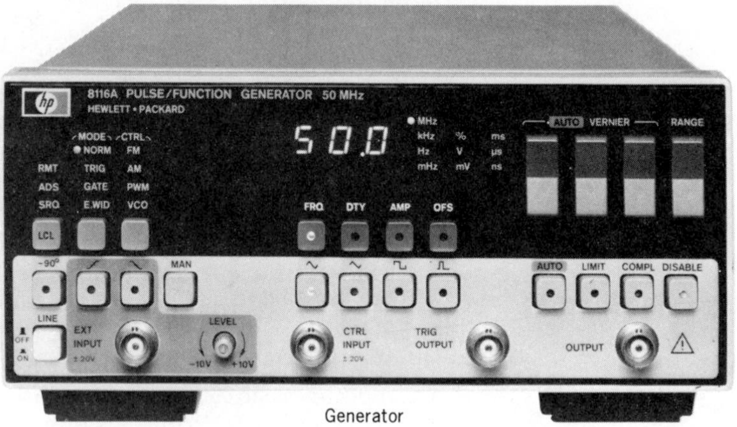

Generator

Figure 23-23 Pulse function generator (Model HP 8116A. Courtesy Hewlett-Packard)

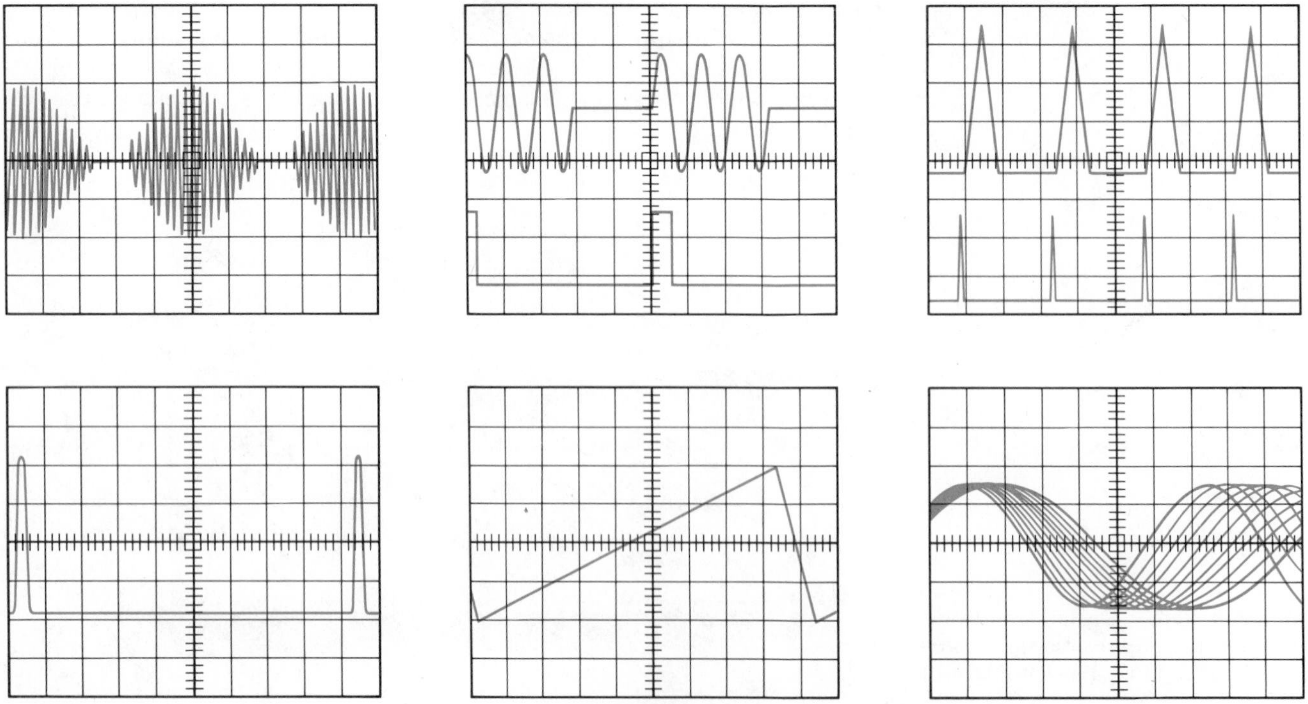

b. Waveform generation capability

23-14 AC BRIDGES

All ac bridges operate on the same ratio-arm principle as the dc Wheatstone bridge (Sec. 23-7) with the following important differences:

1. An ac supply source of constant frequency is used.
2. Impedances replace resistances in at least two of the ratio arms.
3. An **ac** null detector, D, is used instead of a galvanometer to determine when the bridge is balanced. (Usually a high-impedance differential amplifier followed by a center-scale PMMC meter is used as the detector.)
4. An unknown impedance (instead of an unknown resistance) is measured as a series equivalent impedance, $R_x \pm jX_s$.

Table 23-3 shows five of the more commonly used ac bridge circuits for measuring the series equivalent impedance parameters of inductive or capacitive impedances. It should be noted that once these parameters are determined, the equivalent parallel impedance parameters are easily found with the two-element relations covered in Sec. 16-6.

All five circuits shown in Table 23-3 may be reduced to the same general equation when balanced, since the *unknown series* impedance, $\mathbf{Z}_x$, is located in the lower right-hand branch. The general impedance equation is

$$\mathbf{Z}_x = \mathbf{Z}_s\left(\frac{\mathbf{Z}_2}{\mathbf{Z}_1}\right) \quad \text{ohms } (\Omega) \qquad (23\text{-}30)$$

which is essentially the same as Eq. (23-29) for the dc Wheatstone bridge!

Using Eq. (23-30), all of the balance equations for R_x, L_x, and C_x, using the various bridges, may be derived easily.

EXAMPLE 23-16

The bridge circuit of Table 23-3**A** is balanced when $\mathbf{Z}_1 = 5\angle 0^\circ$ kΩ, $\mathbf{Z}_2 = 0.5\angle 0^\circ$ kΩ, $\mathbf{Z}_3 = \mathbf{Z}_s$ is a 10-nF capacitor in series with an 800-Ω resistor, and the frequency of the supply is 1 kHz. *Calculate*

a. $\mathbf{Z}_x$, R_x, and C_x from Eq. (23-30)
b. $\mathbf{Z}_x$, R_x, and C_x from equations given in Table 23-3A
c. Q and D of the unknown series capacitive circuit

Solution

$$\mathbf{Z}_s = R_s - jX_s = R_s - j(1/\omega C)$$
$$= 800 - j(1/2\pi \times 10^3 \times 10 \times 10^{-9})$$
$$= 800 - j15\,915.5\ \Omega$$

a. $\mathbf{Z}_x = \mathbf{Z}_s(\mathbf{Z}_2/\mathbf{Z}_1) = (800 - j15\,915.5)(0.5\angle 0^\circ/5\angle 0^\circ) = \mathbf{80 - j1591.55\ \Omega}$;
$R_x = \mathbf{80\ \Omega}$ and $C_x = 10$ nF$(10/1) = \mathbf{100\ nF}$

b. $\mathbf{Z}_x = \mathbf{Z}_s(R_2/R_1) = (800 - j15\,915.5)(0.5/5) = \mathbf{80 - j1591.55\ \Omega}$;
$C_x = C_s(R_1/R_2) = 10$ nF$(5/0.5) = \mathbf{100\ nF}$;
$R_x = R_s(R_2/R_1) = 800\ \Omega(0.5/5) = \mathbf{80\ \Omega}$

c. $Q_x = jX_s/R_s = 1591.55/80 = \mathbf{19.89}$;
$D_x = 1/Q_x = 1/19.89 = \mathbf{0.005}$

Check using equation from **Table 23-3A**:

$$D_x = \omega C_x R_x = 2\pi 10^3(100\text{ nF})(80) = \mathbf{0.005}$$

Example 23-16 provides the following insights:

1. Both the resistance and the reactance of $\mathbf{Z}_x$ are found to be in the same ratio of R_2/R_1, or 1/10 in this case. (See balance equations in Table 23-3**A**.)
2. The Q of the standard impedance, $\mathbf{Z}_s$, is the same as the Q of the unknown impedance, $\mathbf{Z}_x$. This follows because impedances $\mathbf{Z}_1$ and $\mathbf{Z}_2$ are nonreactive (resistive) components.
3. The dissipation factor, D, which is the reciprocal of Q, may be found independently by using $D = \omega C_x R_x$, as a check on the Q.
4. Each of the remaining bridge circuits may be solved similarly, using either the general Eq. (23-30) or the special balance equations shown in the last column of Table 23-3, which were derived from the general equation for the particular circuit configuration shown.
5. The equivalence of the special balance equations and the general equation (23-30) is proved in end-of-chapter Probs. 23-40 through 23-47. These are solved in a manner similar to Ex. 23-16.

Table 23-3 AC Bridge Circuits

Table Letter	Bridge	Bridge Circuit	Parameters Measured	Balance Equations
A	General form using series impedances	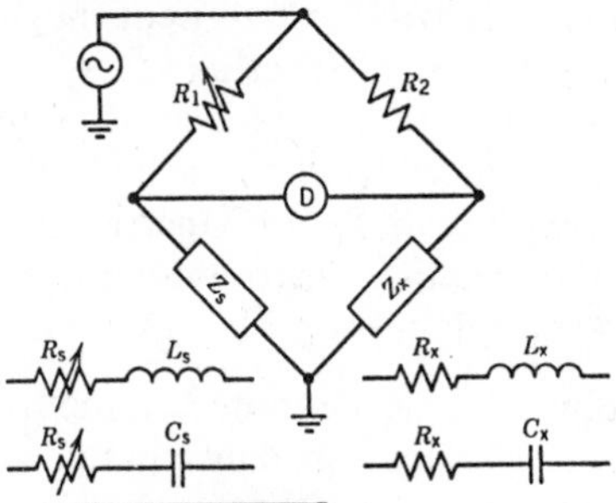	R_x L_x C_x Q D	$Z_x = Z_s(R_2/R_1)$ $C_x = C_s(R_1/R_2)$ $R_x = R_s(R_2/R_1)$ $L_x = L_s(R_2/R_1)$ $Q = \omega L_x/R_x$ $D = \omega C_x R_x$
B	Maxwell bridge	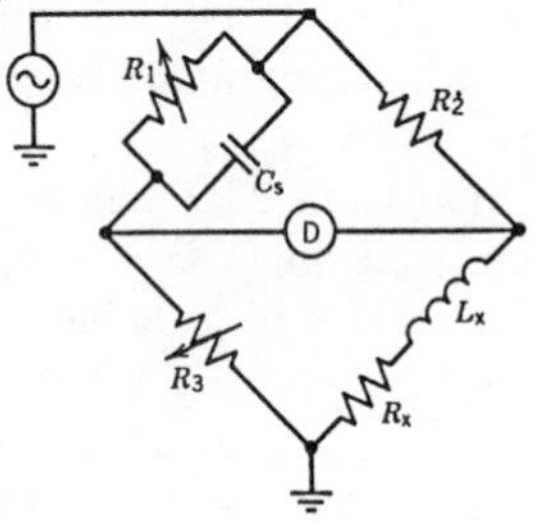	L_x R_x Q	$L_x = R_2 R_3 C_s$ $R_x = R_2 R_3/R_1$ $Q = \omega L_x/R_x$
C	Owen bridge	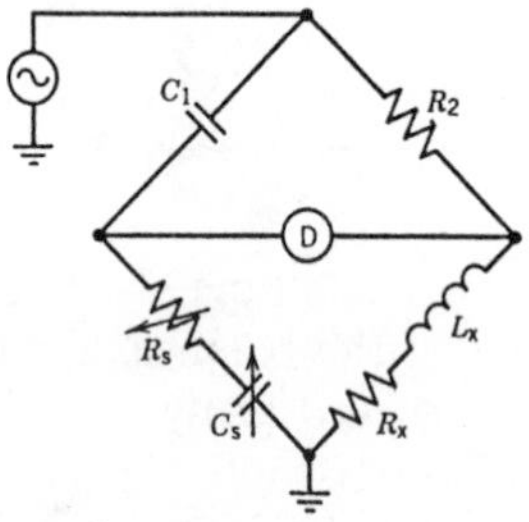	L_x R_x Q	$L_x = R_2 R_s C_1$ $R_x = R_2(C_1/C_s)$ $Q = \omega L_x/R_x$
D	Hay bridge	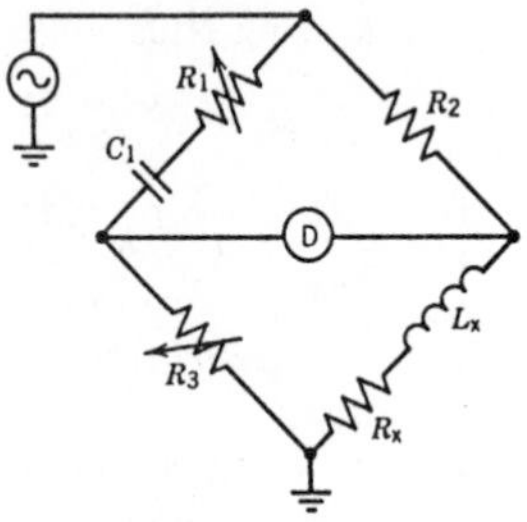	L_x R_x Q	$L_x = \dfrac{R_2 R_3 C_1}{1 + (\omega C_1 R_1)^2}$ $R_x = \dfrac{(\omega C_1)^2 R_1 R_2 R_3}{1 + (\omega C_1 R_1)^2}$ $Q = \omega L_x/R_x$
E	Schering bridge	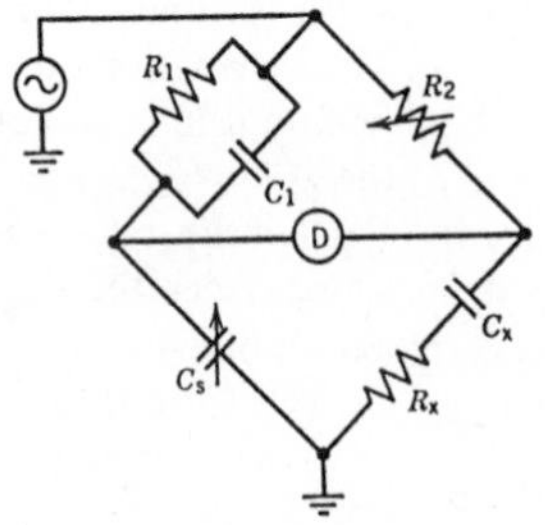	C_x R_x D_x	$C_x = C_s(R_1/R_2)$ $R_x = R_2(C_1/C_s)$ $D_x = \omega C_1 R_1 = \omega C_x R_x$

The bridges of Table 23-3 have been used either singly or in combination in various types of *commercial universal impedance bridges.* Most of these commercial bridges use the Wheatstone bridge for resistance measurements, both dc and ac. The Maxwell bridge has been used for low-Q circuits and the Hay bridge for high-Q measurements (small capacitors or high-Q coils). Instruments of this type required some experience and training for proper use because several and various dial adjustments had to be made in the proper sequence. In several models the scales were reversed for inductance and capacitance, again a source of possible error. Some models even required the use of separate external oscillators.

All this has been changed with the advent of modern *digital impedance bridges.* Most of them feature completely automatic operation. One merely connects the component to the proper terminals and selects the measuring functions desired and one of the test frequencies. The instrument automatically selects the proper measuring range and equivalent circuit mode. The digital display instantly shows the desired parameters.

The HP model 4262A, shown in **Fig. 23-24**, is a typical direct-reading digital LCR meter, capable of a wide range of applications: measuring electrolytic/ceramic capacitors, filter coils, pulse transformers, semiconductor junction capacitance, and internal resistance of dry cells, as well as ordinary resistors, inductors, and capacitors for printed circuits. One $3\frac{1}{2}$ digit display shows LCR and its appropriate units. The other digital display shows either D or Q to four significant digits. Three test signals may be selected: 120 Hz, 1 kHz, and 10 kHz. The capacitance measurement range is from 10 pF to 10 mF. The inductance measurement range is from 10 μH to 1 kH. The resistance measurement range is from 1 mΩ to 10 MΩ. In addition, the instrument will display ESR (equivalent series resistance) and ΔLCR (*deviation* in inductance, capacitance, or resistance, in appropriate units).

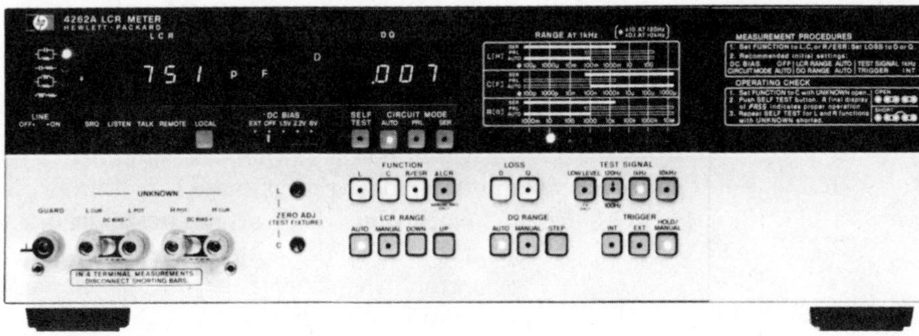

Figure 23-24 Direct-reading digital LCR impedance meter (HP Model 4262A. Courtesy Hewlett-Packard)

In addition to automatic measurements of L, C, and R and the wide range of values that can be measured, the HP 4262A features high accuracy (typically 0.2% of the reading). In Fig. 23-24 it is measuring a capacitance of 751 pF whose dissipation factor, D, is $0.007 \pm 0.2\%$. A deviation measurement is provided in addition to the L, C, R, and ESR functions. By depressing the ΔLCR key, the range is held and the measurement value stored; the display is then reset to zero. The next succeeding measurement is then displayed as the deviation in counts from the stored (standard) value. This is a very useful technique when comparing small trimmer or variable capacitors against a standard.

The instrument also incorporates a SELF-TEST feature, which checks the process amplifier and phase detector/integrator in the analog and digital sections. When operating properly, the word "pass" appears in the LCR window when the SELF-TEST key is depressed.

23-15 CATHODE RAY OSCILLOSCOPE (CRO)

Of all the instruments used in electrical and electronic practice, the cathode ray oscilloscope (CRO) is unquestionably the most versatile. It can measure dc voltage and current as well as ac voltage and current. But more important, it provides a *visual display* of the actual waveforms under consideration. Consequently, it permits measurement of phase shift, rise time, decay time, peak-to-peak voltages and currents, frequency, period, repetition rate, pulse duration, pulse delay time, and a variety of other parameters outside the scope of this brief section.

Figure 23-25 shows a cross-sectional construction of the visual display device, the cathode ray tube (CRT) of a cathode-ray oscilloscope (CRO). A filament and cathode at the base of the CRT generate the electron beam, which is controlled by a grid and accelerated toward the fluorescent screen by a series of accelerating anodes. Two sets of horizontal and vertical deflection plates deflect the beam horizontally and vertically. A phosphor coating on the inside of the CRT face emits light whenever the accelerated electron beam strikes the CRT face, tracing the beam path in proportion to potentials on the horizontal (**H**) and vertical (**V**) deflection plates. Thus the CRT screen, in normal operation, projects a display that shows the voltage applied to the vertical plates and time applied to the horizontal plates via an internal horizontal sweep generator.

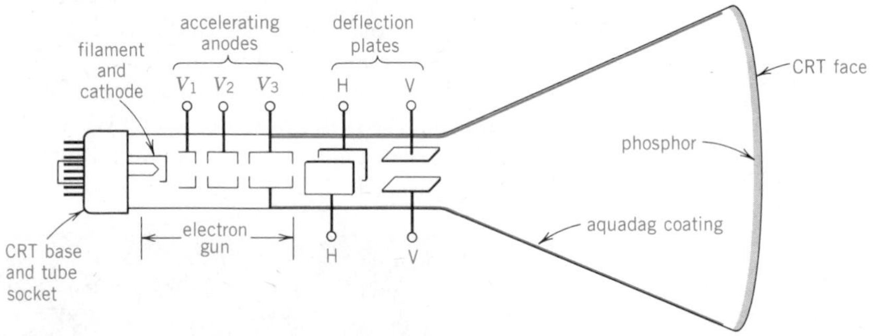

Figure 23-25 Cathode ray tube (CRT) construction

Figure 23-26 shows a block diagram of a relatively simple CRO. The (**V**) (vertical input) signal is applied to the **V**-deflection plates via a multistage vertical amplifier and a delay line. The purpose of the delay line is to delay the rise in vertical signal so that it coincides with the retraced sweep signal of the horizontal time base generator. This enables the start of the vertical signal to coincide with the start of the horizontal time base and also ensures that the waveform is stationary as the moving electron beam repeatedly retraces the same waveform.

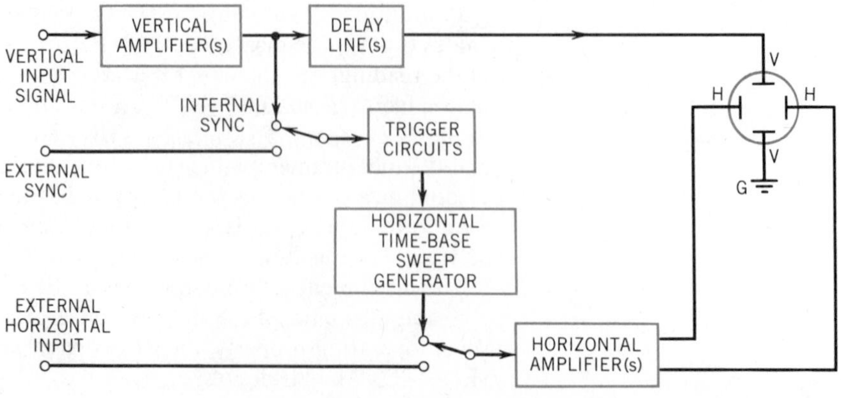

Figure 23-26 Block diagram of CRO circuits

The *internal sync* position also ensures that the time base is triggered at the same frequency as the signal applied to the vertical input. An EXT SYNC position enables selection of the sync signal from some point in the circuit under observation other than the vertical signal to be observed. The EXT HORIZONTAL input enables selection of separate voltages to be applied to the **H**-deflection plates at frequencies and phase other than those applied to the **V**-deflection plates. *Lissajous figures* are obtained in this way, representing the phase differences between two waveforms of the same frequency and/or different frequencies.

Some CROs provide two (or more) simultaneous traces representing different voltages of interest in a given circuit. More expensive CROs accomplish this by a *dual-beam* structure in the electron gun portion of the CRT. But the more commonly used method employs electronic switching to switch the single beam rapidly to two or more vertical input signals, in sequence. The persistence of the CRT phosphor is sufficiently long that it is possible to view as many as four traces simultaneously from four different vertical amplifier sources. Such simultaneous displays enable the CRO operator to measure and compare waveforms of interest with respect to amplitude, phase, frequency, time delay, and so forth.

There are also instances in which the waveform is not periodic but merely a transient burst that lasts for a brief period. When used in the SINGLE SWEEP mode, the CRO makes a single trace of the voltage over a brief time period. The trace may be photographed on Polaroid film using a special camera that fits on the screen face. Alternatively, *storage CROs* are available featuring CRTs that retain the display for several hours after the waveform is first traced on the CRT screen. This storage retention feature is useful not only for recording transients in the single-sweep mode but also for displaying very-low-frequency repetitive waveforms. Such displays are so slow that the start of the repetitive waveform disappears before the end is traced across the screen because of the shorter persistence of conventional CRTs. Storage CRTs permit the operator to examine such signals in detail for long periods of time *without* the need for photography.

In the past few years, the increased use of logic design has produced significant improvements in CROs, including increased accuracy and resolution of image, the availability of more channels, and, most important, the availability of *cursors* in a *digital readout* CRO to display readouts of voltage, phase, time, trigger settings, sweep rate, and so forth in numerical form. This last improvement enables the CRO operator to obtain more precise measurements from CRO waveforms very quickly and accurately without the need for calculation and estimation. Such screen images, when photographed, reveal not only the values measured but also the various CRO channel scale settings, sweep rate, type of coupling, trigger level, and so forth for the waveforms displayed.

At this writing, the 300-MHz Tektronix model 2465 four-channel CRO shown in **Fig. 23-27** is a moderately priced instrument that represents the leading edge of technology in digital readout CRO capability. It features 1% horizontal system and 2% vertical system accuracy on all four channels. On-screen vertical and horizontal cursors enable immediate and accurate measurement of voltage, time, frequency, phase, and ratio measurements. Other measurements include differential voltage, slew rate, rise time, small angle phase differences (using $\times 10$ magnification), delayed sweep, and period of an expanded burst. Other features include the ability to add (or subtract) and/or invert waveforms of CH 2 voltages with respect to CH 1. This feature of invert-and-add enables differential voltage measurements as well as use of the CRO as a null detector.

Lack of space precludes a complete discussion of all the measurements that can be made and/or the various options available, such as a $4\frac{1}{2}$ digit autoranging multimeter, a 150-MHz counter/timer/trigger with 17-bit word recognizer, an IEEE Standard 488 interface bus that provides complete talker/listener control making

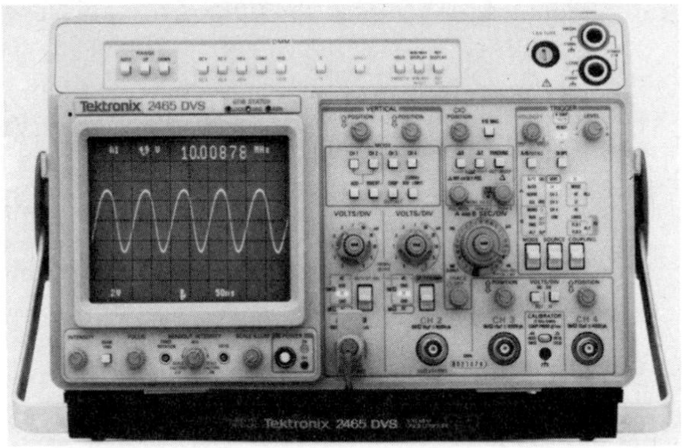

Figure 23-27 Tektronix Model 2465 DVS, four-channel, 300-MHz, digital readout CRO featuring an auto-ranging multimeter, a counter-timer trigger plus word recognizer, GPIB for linking to a controller for system use, and comprehensive TV waveform measurement capability (Courtesy Tektronix, Inc.)

the CRO programmable, and a TV waveform measurement system. One illustration that should be shown, however, is the use of the CRO to measure an unknown impedance at a given frequency. This also shows how current may be measured in addition to voltage and explains the use of the INVERT feature.

Figure 23-28a shows the connections for determining impedance, resistance, and inductance (or capacitance) of an unknown impedance at a given frequency (say 5 kHz). CH 1 of the CRO is measuring the voltage across the unknown impedance (with respect to GND). CH 2 is actually measuring the current in the series circuit of a known resistor connected in series with the unknown impedance. (Recall that current is in phase with voltage across a pure resistor.) Since the GND connection of the CRO is taken at the midpoint of the circuit, the signal generator is used in its push–pull output mode with its own ground connection "floating." Consequently, by inverting the waveform of CH 2 (with respect to GND), we obtain the true phase relation between the current in and voltage across the unknown impedance.

Figure 23-28b shows the CRO display and the digital readouts obtained. The signal generator frequency may be determined from the period, $\Delta t = 200\ \mu s$. The

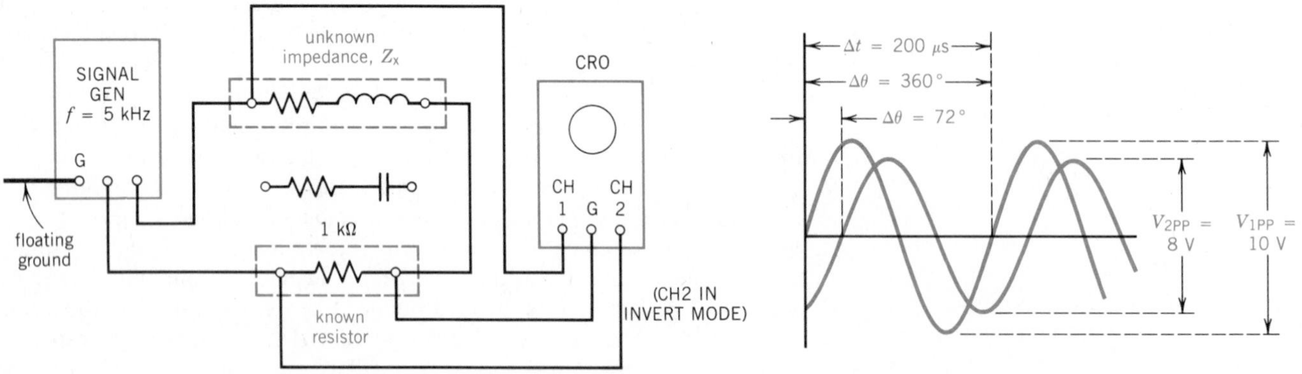

a. Connections to CRO

b. Cursor measurements of ΔV, Δt and $\Delta\theta$ displayed on CRO

Figure 23-28 Using digital readout CRO to measure an unknown impedance phase is found from the cursor position at which each waveform rises positively from its zero value, $\Delta\theta = 72°$. Example 23-17 shows how the observed CRO waveforms yield all the important parameters for the unknown impedance.

EXAMPLE 23-17

From the observed CRO measurements of Fig. 23-28b, calculate all the following parameters *for the unknown impedance* in Fig. 23-28a:

a. Phase angle
b. Voltage and current as RMS values
c. Impedance in rectangular form
d. Real power dissipated in and Q of the impedance
e. Parallel model component values

Solution

a. $\theta = \mathbf{72°}$, current lagging voltage (by observation)

b. $V_x = V_{pp}/2\sqrt{2} = 10\text{ V}/2\sqrt{2} = \mathbf{3.5355\ V}$;
$I_x = 8\text{ V}/2\sqrt{2}(1\text{ k}\Omega) = \mathbf{2.828\ mA}$

c. $Z_x = V_x/I_x = 3.5355\text{ V}/2.828\text{ mA} = 1250\underline{/72°}\ \Omega$
$= \mathbf{(386.3 + j1189)\ \Omega}$

d. $P_{av} = I^2R = (2.828\text{ mA})^2 \times 386.3 = \mathbf{3.09\ mW}$;
$Q = \tan\theta = \tan 72° = \mathbf{3.078}$

e. $R_p = R_s(Q^2 + 1) = 386.3(3.078^2 + 1) = \mathbf{4045\ \Omega}$;
$jX_p = jX_s(Q^2 + 1)/Q^2 = j1189(3.078^2 + 1)/(3.078)^2$
$= \mathbf{j1314\ \Omega}$

The following insights are to be drawn from Ex. 23-17 and Fig. 23-28:

1. By using a known (1-kΩ) resistor in series with the unknown impedance, Z_x, the voltage across that (pure) resistance represents the current waveform.
2. A particular advantage of the use of a known 1-kΩ resistor is that when the voltage across it is measured in volts, the current through it is expressed directly in milliamperes.
3. By inverting CH 2 with respect to GND, we obtain the true phase relation between the current in and voltage across the unknown impedance, Z_x.
4. Consequently, if the signal source becomes a nonsinusoidal waveform (say a square wave), the waveform of CH 2 represents the waveform of current in Z_x, compared to voltage across Z_x (CH 1).
5. Example 23-17 shows how the observed CRO waveform data may be used to calculate both series and model component values of an unknown impedance at the frequency of interest (5 kHz). Although the values are not as precise as the impedance data obtained from ac bridges (Sec. 23-14) and the measurement is not as rapid as the LCR digital bridge (Fig. 23-24), the CRO measurements are sufficiently good for most engineering purposes.

Because the CRO provides a visual display of the particular waveform under consideration, it is also possible to use display data to find the RMS value of a nonsinusoidal waveform containing a dc component. **Figure 23-29a** shows how it is possible to measure the dc voltage of three 1.5-V dry cells connected in series. The CH 1 mode switch is set to GND to position the REF cursor (using the CH 1 vertical position control) at some convenient line at the lower portion of the CRO graticule scale. The CH 1 mode switch is then thrown to its DC (direct coupling) position. The VOLTS/DIV switch is rotated to 2 V/div. A direct reading probe connected across the three 1.5-V cells in series shows the display of Fig. 23-29a. Aligning the ΔV cursor with the dc signal yields the reading of 4.6 V shown in the upper right-hand corner of the CRO display.

Let us assume that we wish to measure the RMS value of the waveform shown in Fig. 23-29b, having *both* an ac and a dc component. The ac component is the output of an FWR measuring 2.4 Vpp. The dc component (which produces an average value of zero for the ac component) is +8 V dc, as shown in Fig. 23-29b. Let us first measure the ac component, followed by the dc component, using the CRO.

The CH 1 mode switch is switched to GND to establish a suitable level and then switched to AC coupling.[7] The VOLTS/DIV switch is rotated to 1 V/div, as shown

[7] In the AC coupled mode, the vertical input is fed to the vertical amplifiers through a coupling capacitor that filters out the dc component of the waveform and passes only the ac component. In the DC (direct-coupled) mode, the capacitor is shorted out and the signal fed directly to the vertical amplifiers.

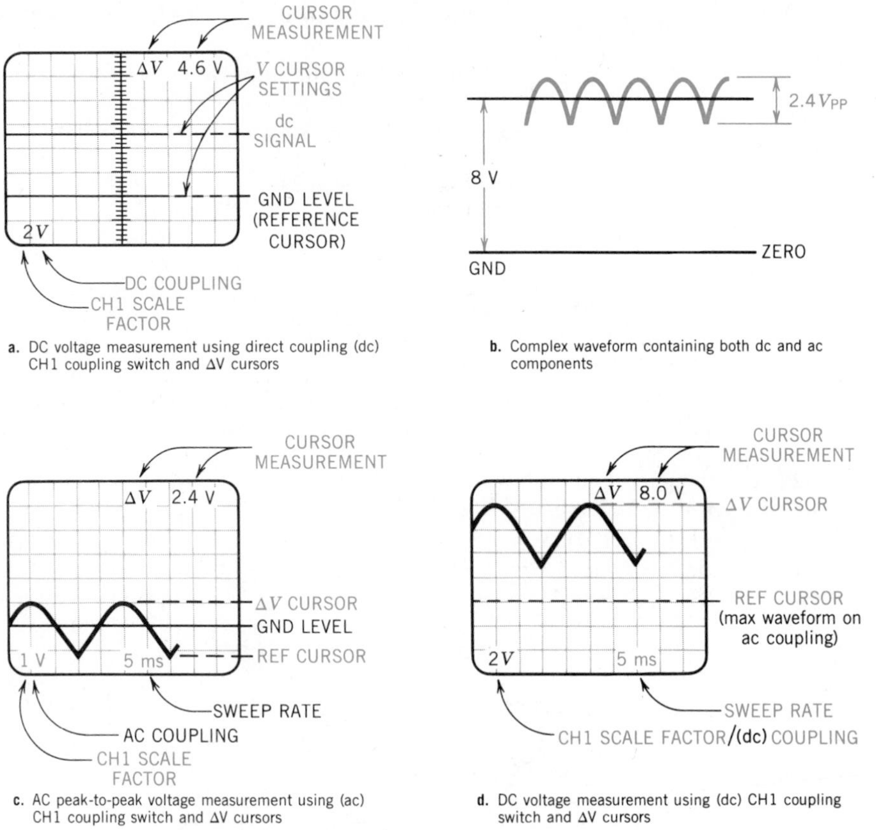

a. DC voltage measurement using direct coupling (dc) CH1 coupling switch and ΔV cursors

b. Complex waveform containing both dc and ac components

c. AC peak-to-peak voltage measurement using (ac) CH1 coupling switch and ΔV cursors

d. DC voltage measurement using (dc) CH1 coupling switch and ΔV cursors

Figure 23-29 Measurement of true RMS voltage of complex waveforms using CRO

in Fig. 23-29c. The ΔV cursors are then positioned to the peak positive and negative excursions of the waveform. The cursor measurement yields $V_{pp} = 2.4$ V ac, as shown in Fig. 23-29c. This peak-to-peak voltage agrees with Fig. 23-29b for the ac component of the waveform.

To measure the dc component, we first position the REF cursor to the maximum positive peak position of the waveform with AC coupling. Next we switch CH 1 mode switch to DC coupling.[8] The CH 1 VOLTS/DIV switch is rotated to 2 V/div and the ΔV cursor set to the positive peak position of the waveform with DC coupling. The cursor measurement yields $V_{dc} = 8$ V, which agrees with Fig. 23-29b. Example 23-18 reviews the calculation for the true RMS value of the waveform.

EXAMPLE 23-18

Calculate the RMS value of the complex waveform measured in Fig. 23-29

Solution

By direct observation, $V_{dc} = 8$ V and $V_{ac} = 2.4\ V_{pp}/\sqrt{2}(2) = 0.8485$ V; $V_{RMS} = \sqrt{V_{dc}^2 + V_{ac}^2} = \sqrt{8^2 + 0.8485^2} = |8 + j0.8485| =$ **8.045 V** (14-21)

---

[8] If the waveform "jumps" *upward*, going from AC to DC coupling, the dc component is *positive*. If it "drops," the dc component is negative. It is customary to set the VOLTS/DIV switch on some higher scale value to prevent the waveform from jumping off the CRO screen.

23-16 SPECTRUM ANALYZER

The last of the modern digital readout instruments discussed here is the spectrum analyzer. The block diagram of the spectrum analyzer was introduced in Sec. 22-8 and shown in Fig. 22-11. We are therefore familiar with the purpose and capability of this instrument.

Figure 23-30 shows a modern Tektronix model 496 spectrum analyzer with seven digital readouts. The instrument performs spectrum analysis in the frequency range from 1 kHz to 1.8 GHz! The dynamic range is 80 dB, and the resolution bandwidth can be varied from 1 MHz to 30 Hz over the entire frequency range. Only three adjustments are made by the operator: the center frequency (97 MHz in Fig. 23-30), the frequency span (2 MHz/div in Fig. 23-30), and the reference level (−20 dBm). The advantages of digital readouts and digital storage are that time-consuming display adjustments are reduced considerably and many of the manual functions are automated through the use of microprocessor controls.

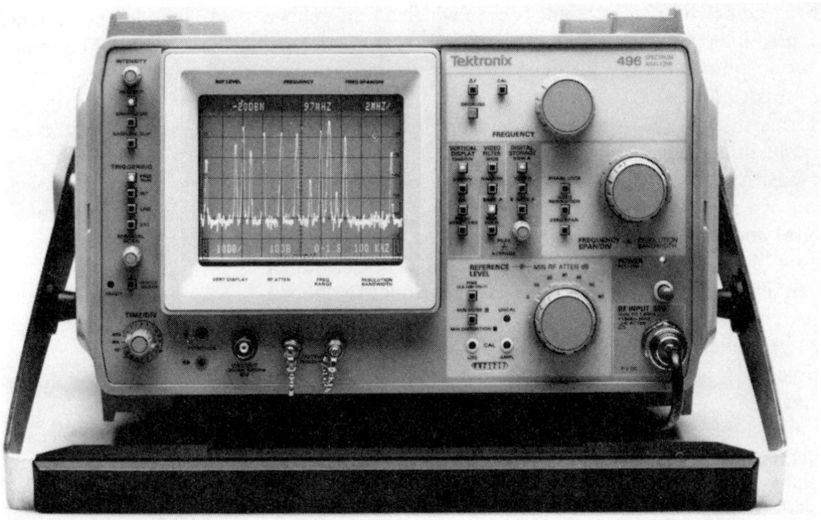

Figure 23-30 Tektronix model 496 spectrum analyzer (Courtesy Tektronix, Inc.)

23-17 GLOSSARY OF TERMS USED

Accuracy Measure of how well an instrument reading or indication approaches the true value of the variable measured by the instrument.

Ammeter Instrument used to measure the magnitude of electric current.

Ammeter substitution method Method of measuring an unknown resistance with only an ammeter, a standard (known) resistance, and a constant-voltage source.

Ammeter-voltmeter method Popular method of measuring resistance by recording the voltage across and current in an unknown resistor and calculating its resistance from Ohm's law.

Analyzer (general) Evaluates and/or measures one or more specific parameters (voltage, current, frequency, logic level, distortion, etc.).

Analyzer, network Evaluates the impedance characteristics of linear networks over a range of frequencies.

Analyzer, spectrum Displays the power or voltage of a time-domain signal as a function of frequency.

Analyzer, transfer function Measures the amplitude and phase response of the output of a circuit relative to the input over a range of frequencies.

Analyzer, waveform Measures the amplitude of a waveform over a specific period, with resolution usually in the low nanosecond range.

Bridge (general) Circuit using a nulling technique for accurate measurement of any parameter (resistance, inductance, capacitance, etc.). Commonly used bridges are the Maxwell bridge, the Wheatstone bridge, and the digital-display LCR bridge.

Bridge, capacitance Measures the capacitance and dissipation factor of a capacitor.

Bridge, impedance Measures the resistance and reactance (or Q) of a component or network, calibrated to read in ohms. For ac bridges, the indicated values may be expressed in polar or rectangular form.

Bridge, inductance Measures the inductance and resistance (or Q) of an inductor.

Bridge, resistance Measures the resistance of a resistor. A Wheatstone bridge and a Kelvin bridge are examples of resistance bridges.

Calibrator Instrument that verifies that the performance of a device or an instrument is within its specified limits of accuracy.

Cathode ray oscilloscope (CRO) Device that displays on a cathode ray tube the instantaneous values of one or more varying electrical quantities as a function of time.

Cathode ray tube (CRT) Electronic tube in which a well-defined and controllable beam of electrons is produced and directed onto a phosphor screen to produce a visible display or effect.

Counter Instrument that performs timing measurements on ac waveforms in addition to determining the number of events that occur in a predetermined time.

Current sensitivity, I_{fs} Current required to produce full-scale deflection of a meter or other type of indicating instrument.

D'Arsonval meter Meter movement consisting of a moving (rotating) coil pivoted in the magnetic field of a permanent magnet. The deflection of the coil is proportional to the meter movement current. The permanent-magnet moving-coil (PMMC) meter is an instrument using the D'Arsonval principle.

Deflection At any part of an instrument scale, the force (for that portion) produced by the electrical quantity to be measured.

Electrometer Highly sensitive instrument for measuring potential difference which draws negligible current from the circuit to which it is applied.

Error Deviation of any measured variable from its actual or true value.

Error, gross Error of human origin, such as misreading of scales, dials, multiplying factors, incorrect adjustments, improper use of instruments, human calculation errors from instrument readings, and improper use of calculators.

Error, instrumental (or systematic) Error in the various instruments or components of an instrumentation system including poor or improper instruments that have been abused or improperly calibrated.

Error, loading Error in a system caused by the process of measurement itself changing the parameter to be measured. A loading error may be included as a gross error.

Error, parallax Error caused by failure of the operator to align the pointer perpendicular to the scale in taking a reading. A parallax error is a gross error.

Error, random Error that is not included in any specific category but is due to unknown causes or variations that are unaccountable.

Error, scale Error due to improperly calibrated or printed scales.

Error, temperature Error introduced by temperature variation in component values of an instrument.

Frequency meter Meter that measures the number of periods or cycles per unit time of a given signal, in hertz.

Galvanometer Instrument designed to measure extremely small currents. An electrometer is a highly sensitive galvanometer.

Generator, ac signal Oscillator producing sine waves whose frequency and amplitude are variable.

Generator, function Instrument that generates sine, square, and ramp (or triangular) signals of varying amplitude and frequency.

Generator, pulse Instrument that supplies pulse waveforms whose period, amplitude, rising and falling edges, and width are usually variable.

Instrument Device for measuring the value of a quantity under observation.

Lissajous figures Figures formed by combining in space two sinusoids at right angles to each other. The form of the figure depends on the relative frequency and time phase of the two sinusoids. The system is widely used for calibrating ac signal generators or comparing integral frequencies of a known signal source against an unknown signal whose frequency is a multiple of the known signal. It is also used for comparing phase relations between two sources of the same frequency.

Meter Any type of measuring device or instrument.

Multimeter Instrument that usually measures dc and ac currents and voltages with a separate scale for measuring resistance in ohms.

Multiplier Resistance connected in series with a basic movement to extend its voltage range when using the movement as a voltmeter.

Nonlinear scale Meter scale that is more compressed (crowded) at one end than the other.

Ohmmeter Direct reading instrument for measuring resistance.

Precision Measure of how well an instrument repeatedly measures the same value when readings are taken independently over a short period. Precision is sometimes indicated by the number of significant digits contained in the individual reading. Precision does not imply accuracy, because an internal defect may result in highly precise readings having large errors. (A seven-place table is more precise than a four-place table, but if a printing error is made in a seven-place value, it is less accurate than the four-place value that contains no error.)

Sensitivity, S Ratio of response to cause, generally. As applied to instruments, the ratio of the instrument indication to a change of input in the measured physical quantity or variable. In moving-coil instruments, sensitivity is the reciprocal of current sensitivity (i.e., $S = 1/I_{fs}$).

Thermal watt converter Instrument having both potential and current input terminals, sometimes containing both current and potential transformers as well as thermal elements. The voltage developed at the output terminals is a measure of the power at the input terminals.

True RMS instruments Instruments providing either a voltage or current indication proportional to the dc heating power of the input voltage or current waveform.

Torque Force tending to produce rotation. As applied to instruments, the turning moment on a moving element of an instrument produced by the physical variable measured, which in turn produces instrumental deflection.

Varmeter Instrument for measuring reactive power, Q, calibrated in vars, kilovars, or megavars.

Volt–ohm–milliammeter (VOM) Single instrument capable of measuring dc and ac voltage and current as well as resistance. (See multimeter.)

Voltmeter Instrument for measuring the magnitude of the potential difference between two points in a dc or ac circuit.

Voltmeter method Special modification of the voltmeter substitution method in which the voltmeter itself serves as a known standard resistor.

Voltmeter-substitution method Method of measuring an unknown resistance using only a voltmeter, standard resistor(s), and a constant-current dc supply.

Wattmeter Instrument for measuring the magnitude of active (average) power in an electric circuit, either dc or ac.

Wheatstone bridge Four-arm bridge used for accurate measurement of resistance, in which all four arms are usually resistive.

23-18 PROBLEMS

Sec. 23-2

23-1 A 1-mA movement consists of 71 turns wound on a core that is 2 cm in diameter and 2 cm long. The permanent magnet sets up an air-gap flux of 0.2 T linking the turns. Assuming the current is perpendicular to the flux, calculate the

a. Number of conductors
b. Force per conductor at full-scale deflection (fsd)
c. Total force developed at fsd
d. Developed torque at fsd
e. Developed torque and half-scale deflection

23-2 A 100-mV movement has a resistance of 20 Ω. At full-scale deflection its retarding springs produce a countertorque of 5×10^{-6} N·m. The coil form placed over the magnetic core has a diameter of 2 cm and an axial length of 1 cm. If the air-gap flux density is 0.05 T, calculate the

a. Current for fsd, I_{fs}
b. Sensitivity
c. Number of turns of wire required for the meter movement
d. Total force developed by the coil
e. Force developed by each conductor of the coil

23-3 Because of a previous excessive overload, the countertorque developed by the retarding springs in Prob. 23-2 is weakened to 4.8 μN·m and, simultaneously, the coil form is misaligned in its pivots so that the conductors make an angle of 70° with respect to the magnetic field of the permanent magnet. Calculate the

a. Deflection when full-scale current (5 mA) flows in the meter, I_{fs}
b. Actual current reading of the pointer on the scale under the conditions of **(a)**
c. Percent error of the instrument at full-scale current, I_{fs}
d. Percent error at a current of 2 mA, assuming the error is linear

23-4 A basic PMMC movement has a full-scale current of 1 mA and a resistance of 100 Ω. Calculate the

a. Sensitivity, S
b. Voltage sensitivity, V_{fs}
c. Current sensitivity, I_{fs}

23-5 The *scale* of a particular voltmeter (VM 1) has 10 major divisions, each of which has 5 minor divisions. If the voltage sensitivity is 100 V and the scale can be read to half a minor division, calculate the

a. Number of minor divisions
b. Volts per minor division
c. Lowest possible reading (above zero) that can be read accurately on the VM
d. Highest possible reading (below fsd) that can be read accurately on the VM
e. Precision error in reading the instrument, in percent

23-6 A second voltmeter (VM 2) has a scale with five major and five minor divisions per major division.

1. Repeat all parts of Prob. 23-5.
2. Compare the precision of VM 2 versus VM 1.

23-7 The VM 1 of Prob. 23-5 and VM 2 of Prob. 23-6 are compared at 50 V and 100 V against a voltmeter standard and the readings are tabulated:

Standard voltmeter (V)	0	50	100
Voltmeter No. 1 (V)	0	46	97
Voltmeter No. 2 (V)	0	52	98

1. Calculate the percent accuracy of
   a. VM 1 at 50 V
   b. VM 1 at 100 V
   c. VM 2 at 50 V
   d. VM 2 at 100 V
2. Compare the accuracy and precision of VM 2 versus VM 1 based on the above calculations and those of Probs. 23-5 and 23-6. Draw conclusions about the distinction between precision and accuracy (see Glossary).

Secs. 23-3 and 23-4

23-8 The 20-μA PMMC movement shown in Fig. 23-2b has a resistance of 1.5 kΩ. Calculate the shunt resistances required to extend the current range to

a. 10 mA
b. 100 mA
c. 1 A

d. Draw a schematic diagram showing all values and required switching for *all four* current scales
e. Explain why "make-before-break" switches are required in most multirange ammeter circuits

23-9 Repeat all parts (**a**) through (**d**) of Prob. 23-8, using a basic PMMC movement having a current sensitivity of 1 mA and a resistance of 50 Ω. Draw conclusions regarding shunt resistance magnitudes versus meter resistances for the same current scales.

23-10 Using the PMMC movement of Prob. 23-8, calculate the *series-connected* multipliers to extend the voltage range of the movement to produce a multirange voltmeter with scales of

a. 50 mA
b. 5 V
c. 150 V
d. 500 V
e. Draw a schematic diagram showing all values and required switching for all four voltage scales (see Fig. 23-3b)

23-11 Repeat all parts of Prob. 23-10, using a 1-mA, 50-Ω, PMMC movement. Draw conclusions regarding multiplier magnitudes versus meter resistances for the same voltage scales.

23-12 Using a 10-μA, 2500-Ω PMMC movement, design a multirange ammeter–voltmeter by calculating all values (R_1 through R_{10}) in **Fig. 23-31**.

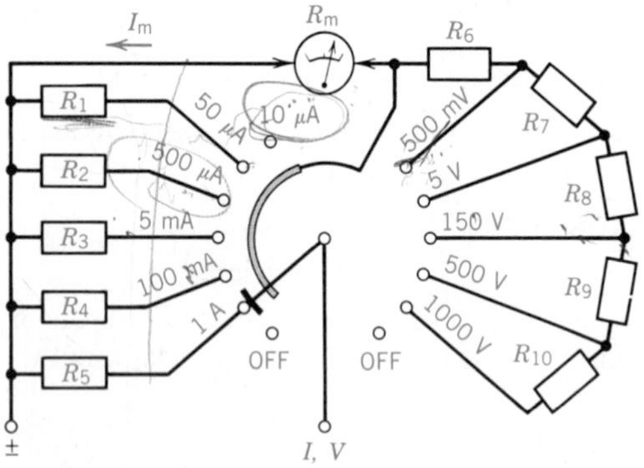

Figure 23-31 Problems 23-12 and 23-13

23-13 Repeat Prob. 23-12, using a 50-μA, 1-kΩ PMMC movement. (Delete the 10-μA terminal.)

23-14 Given a 10-μA, 5-kΩ PMMC movement, design a multirange volt-ammeter using a single switch to produce the following ranges: 10 μA, 1 mA, 10 mA, 100 mA, 1 A, 50 mV, 200 mV, 2 V, 50 V, and 500 V. Specify all shunts and multipliers on a drawn circuit diagram.

23-15 The highest range voltmeter a technician can find in the instrument room is a 0 to 300 V dc instrument with a sensitivity of 1000 Ω/V. Calculate the external series resistance needed to extend the range of the instrument to 1500 V dc.

23-16 A production line run of new 5-mA dc instruments are calibrated with a 5-mA standard. Each new instrument has a 50-μA movement whose resistance is 4000 Ω. One instrument, however, when calibrated against the standard, reads 4.4 mA (indicating an incorrect shunt resistance). Calculate the

a. Proper value of shunt resistance to be used across the PMMC movement
b. Actual value of shunt resistance used in the 4.4-mA instrument
c. Is there any way to *shunt* the 4.4-mA instrument so that it reads correctly? Explain in detail.

23-17 A student purchases a used 150-mA ammeter consisting of a 1-mA, 50-Ω PMMC movement that has been shunted externally. When the student calibrates the meter against a laboratory standard, only 125 mA on the standard produces a deflection of 150 mA on the used ammeter. Calculate the additional shunting resistance that may be used so that the meter will read 150 mA accurately.

Sec. 23-5

23-18 A technician purchases a 50-μA PMMC movement having a resistance of 500 Ω and desires to use it to design a series ohmmeter operating from a 1.5-V penlite (AA) battery. Calculate the

a. Required internal resistance of the ohmmeter, R_i
b. Midscale resistance value of the ohmmeter
c. Required resistance in series with the meter movement, R_s
d. Required fixed resistance R_f
e. Required variable resistance, $R_v/2$ and R_v (potentiometer resistance)
f. Draw a schematic wiring diagram of the meter showing all the values calculated

23-19 The technician desires to calibrate the meter face of the above series ohmmeter, using integral increments of resistance.

a. Calculate current deflections for the following values of *external* resistance: 500 Ω, 1 kΩ, 2 kΩ, 5 kΩ, 10 kΩ, 20 kΩ, 30 kΩ, 50 kΩ, 100 kΩ, 200 kΩ, 500 kΩ, and 1 MΩ.
b. Draw the scale, showing both meter currents and corresponding external resistance values.
c. If the meter has 50 minor divisions and can be read accurately to 1/2 division, calculate the lowest possible and highest possible resistances that can be read. Include these values on the scale drawn in (**b**).

23-20 The (same) technician (of Probs. 23-18 and 23-19) now wishes to use (another) 50-μA, 500-Ω PMMC movement to design a shunted series ohmmeter having a midscale reading of 1000 Ω, also to operate off a 1.5 V AA battery. Calculate the

a. Current drawn from the battery by the ohmmeter when zeroed ($R_x = 0$)
b. R_p and R_{sh} (shunting R_m) based on the ratio of circuit current to meter current
c. Total series resistance, R_s
d. Fixed resistance, R_f, specified to the nearest commercial value
e. Variable resistance $R_v/2$ and R_v (use commercial potentiometer)
f. Draw the schematic of the shunted series ohmmeter showing all the calculated values

23-21 For *each* 5-μA step of the 50-μA movement in Prob. 23-20, calculate and tabulate, respectively, the

a. Deflection factor, d
b. Ratio $(1 - d)/d$
c. R_x

d. Find the external value of R_x whenever the basic movement reads 1 μA (highest measurable value)
e. Find the external value of R_x whenever the basic movement reads 49 μA (lowest measurable value)
f. Draw the complete calibration scale of the shunted series ohmmeter including the values found in (**d**) and (**e**)

23-22 A PMMC 10-mA dc movement has a resistance of 10 Ω. It is desired to use this instrument to design a 10-Ω center-scale shunt ohmmeter operating from a 1.5-V, size C, dry cell. Calculate the

a. Required series resistance, R_s
b. Required fixed resistance, R_f
c. Values of $R_v/2$ and R_v (to nearest commercial values)
d. Draw a circuit diagram of the shunt ohmmeter showing these values.

23-23 Using the shunt ohmmeter designed in Prob. 23-22, calculate the

a. Values of external resistance, R_x for the scale at each 1-mA interval
b. Meter current when measuring an external resistance of 1 Ω
c. Values of external resistance when the meter current is 0.1 mA and 9.9 mA, respectively
d. Draw the complete scale showing both values of R_x and meter current (in milliamperes) for all values calculated in (**a**), (**b**), and (**c**)

Sec. 23-6

23-24 In the ammeter substitution method (Fig. 23-9a), the test instrument is a 10-mA milliammeter having an accuracy of ±2%. The standard resistor R_s is 25 kΩ and the current measured in it is 6.53 mA. When switched to position 2, the current in unknown resistor R_x is 3.65 mA. Calculate the

a. Value of R_x
b. Minimum and maximum possible values of R_x by this method

23-25 Three unknown resistors are measured by the ammeter substitution method. A standard 1-kΩ precision resistor (±1.0%) draws 25.2 mA with a 50-mA ammeter having an accuracy of ±1%. Calculate the resistance ranges of the unknown resistors if

a. R_1 draws 16.4 mA
b. R_2 draws 35.6 mA
c. R_3 draws 21.7 mA

(Hint: calculate the *combined limiting error* due to the precision resistor and ammeter, using the formula 2 × [largest error] + sum of all smaller errors.)

23-26 In the voltmeter substitution method (Fig. 23-9b), the test instrument is an *electrometer* (see Glossary) having an accuracy of ±0.25% and a full-scale deflection of 75 V. A 25-kΩ, 1% precision resistor is connected in series with an unknown resistor, R_x, across a 100-V dc supply of constant voltage. Calculate

a. R_x when the voltage across R_x is 35.85 V as measured by the electrometer
b. Minimum and maximum values of R_x by this method

23-27 A variation of the voltmeter substitution method is the voltmeter method shown in Fig. 23-9c, in which the multiplier of the voltmeter itself serves as the standard resistor. Consider a voltmeter having a sensitivity of 1000 Ω/V connected on its 100-V scale in series with an unknown resistor, R_x, across a 100-V supply. The voltmeter reads 69.1 V and has an accuracy of 1.25%. Calculate

a. R_x
b. Minimum and maximum possible values of R_x, based on the meter accuracy.

23-28 A 1000 Ω/V voltmeter is connected in series with a piece of insulation across a 500-V supply. The voltmeter reads 50 V on its 300-V scale. Calculate the

a. Voltage across the insulation, V_x
b. Insulation resistance, R_x

23-29 A dc voltmeter having an accuracy of ±2% and a sensitivity of 20 kΩ/V is connected in series with a 1.5-MΩ, 1% precision resistor to an unknown dc voltage source. If the voltmeter reads 45 V on its 50-V scale, calculate the

a. Voltage of the dc supply
b. Probable maximum and minimum range of the dc supply voltage

23-30 The resistance of an unknown resistor (actually 50 kΩ) is measured by the voltmeter–ammeter method by placing a 1000 Ω/V voltmeter on its 100-V scale directly across the unknown resistor. A milliammeter in series with the combination draws 3 mA from a 100-V supply. The voltmeter reads 100 V.

a. Explain what is wrong with this method of instrument connection.
b. Calculate R_x by using Ohm's law and the instrument readings.
c. Calculate the percent error (produced by voltmeter loading) in this method of measurement.
d. Explain why this instrument connection is in error and which instrument is reading incorrectly.
e. When the voltmeter is removed from the circuit, the milliammeter reading drops to 2 mA as the voltmeter is reconnected directly across the 100-V supply. Recalculate R_x.
f. Draw conclusions regarding the loading effect on this method of measurement compared to the voltmeter method with the voltmeter in series with unknown resistor R_x.

23-31 An unknown resistance, R_x, is measured by the voltmeter–ammeter method with the voltmeter directly across the supply. The voltmeter reads 6 V and the ammeter reads 3 A. The voltmeter resistance is 1000 Ω/V on its 15-V scale and the ammeter resistance is 0.5 Ω. Calculate the

a. Resistance (apparent) as determined from the meter readings, R'_x
b. True value of R_x, using the given instrumental resistances
c. Percent error between (**a**) and (**b**) produced by improper meter connections
d. Which resistance is of greater significance in producing the error, that of the ammeter or that of the voltmeter? Explain why.

23-32 A 5-mA ammeter having an accuracy of 2% and an internal resistance of 20 Ω is connected in series with an unknown resistance, R_x. A 1% accuracy voltmeter, having a sensitivity of 1000 Ω/V on its 75-V scale, reads 57 V when connected directly across R_x. Determine

a. The true value of R_x if the ammeter reads 4.6 mA on its 5-mA scale
b. The possible range of values of R_x, taking the precision of both instruments into account
c. Why the internal resistance of the ammeter is of no significance in this particular measurement

23-33 A voltmeter having an accuracy of 2% and a sensitivity of 20 kΩ/V is connected on its 1.5-V scale across a supply that measures 1.2 V; a 10-A ammeter connected in series with R_x records 9.5 A when connected across the 1.2-V supply. The ammeter resistance is 10 mΩ and its accuracy is 1.5%. Determine

a. The true value of R_x
b. The possible range of values of R_x, taking the accuracy of *both* instruments into account
c. Why the internal resistance of the voltmeter is of no significance in this particular measurement

23-34 Two views are generally taken to explain the voltmeter loading effect when the voltmeter is connected directly across R_x. One view is that the voltmeter draws current that is recorded by the ammeter. This is shown in the denominator of Eq. (23-28), where $R_x = V/[A - (V/R_v)]$.

The second view is that the voltmeter reduces the equivalent circuit resistance, R_t, thereby drawing more current from the supply, as recorded by the ammeter. This is shown in Eq. (23-28): $R_x = R_t R_v/(R_v - R_t)$, where $R_t = V/A$.

a. Prove algebraically that both forms of Eq. (23-28) are identical by showing that both are equal to $V \cdot R_v/(A \cdot R_v - V)$.
b. Use both forms of Eq. (23-28) to solve Prob. 23-32a to prove the equivalence.
c. Test the equivalence by using $R_x = V \cdot R_v/(A \cdot R_v - V)$ in Prob. 23-32a.

Sec. 23-7

23-35 The Wheatstone bridge shown in Fig. 23-13a is balanced (galvanometer shows a null) when $R_1 = 1$ kΩ, $R_2 = 5$ kΩ, and $R_3 = 1097$ Ω (from decade dials).

a. If a ratio dial is used, what is the ratio setting?
b. If a decade is used for the standard variable resistor, R_3, what are the decade settings?
c. Calculate the value of R_x as read from the bridge settings.
d. If the bridge has an overall accuracy of 0.1%, calculate the upper and lower limits of R_x.

23-36 Assuming the Wheatstone bridge supply voltage, V, is 1.5 V, show that the galvanometer is at a null ($I_G = 0$) by calculating for the given and calculated values in Prob. 23-35:

a. V_T
b. V_H
c. Current in the galvanometer, I_G

(Hint: use Thévenin's theorem)

23-37 In the Wheatstone bridge circuit shown in Fig. 23-13a, $R_1 = 2$ kΩ, $R_3 = 1$ kΩ, $R_2 = 200$ Ω, and $R_x = 101$ Ω. The galvanometer, G, has a resistance of 150 Ω and the supply voltage is 6 V. Calculate (using Thévenin's theorem) the

a. Equivalent resistance of the bridge circuit when switch S (Fig. 23-13a) is open (Hint: use $R_1 \| R_3 + R_2 \| R_x$)
b. Equivalent voltage, V_{TH}, when switch S is open
c. Galvanometer current, I_G, when S is closed
d. Value of R_3 (setting of decade resistances) required to produce a null
e. Limiting error of R_x if each of the known resistors R_1, R_2, and R_3 has a limiting error of ±0.05%

Sec. 23-8

23-38 The electrodynamometer wattmeter of Fig. 23-17c is connected (short shunt) with its potential coil at point B, directly across a load. The total resistance of the wattmeter potential circuit is 2 kΩ. When measuring load power from a 120-V circuit, the wattmeter reads 200 W. Calculate the

a. Wattmeter reading when the load is disconnected
b. True average power drawn by the load and the error if the WM loss is neglected
c. Wattmeter reading if the load power increases to 300 W

23-39 The wattmeter in Prob. 23-38 is now connected exactly as shown in Fig. 23-17c (long shunt). The wattmeter reads 400 W. The current coil resistance is 1.2 Ω and the load current is 3.2 A. Calculate the

a. Supply voltage
b. Power dissipated by the potential coil
c. Power dissipated by the current coil
d. Power dissipated by the load and error if the WM loss is neglected
e. Explain why the potential coil power is not subtracted from the wattmeter reading
f. Explain the advantage of the short-shunt versus the long-shunt connection in measuring power drawn by variable loads

Sec. 23-14

23-40 A Maxwell bridge (Table 23-3**B**) is at a null when $C_s = 0.1$ μF, $R_1 = 15.916$ kΩ, $R_2 = 2$ kΩ, $R_3 = 4$ kΩ, and the supply frequency is 1 kHz. Using the general Eq. (23-30), calculate:

a. The equivalent series impedance of the unknown, Z_x, in polar form
b. Q of the unknown impedance
c. Component values L_x and R_x

23-41 Repeat parts (**c**) and (**b**) of Prob. 23-40, using the balance equations given in the last column of Table 23-3B for the Maxwell bridge.

23-42 An Owen bridge (Table 23-3**C**) is at a null when $C_1 = 0.1$ μF, $C_s = 50$ nF, $R_s = 10$ kΩ, $R_2 = 2$ kΩ, and the supply radian frequency, ω, is 10^3 rad/s. Calculate all parts of Prob. 23-40, using the general impedance Eq. (23-30).

23-43 Repeat parts (**c**) and (**b**) of Prob. 23-42, using the balance equations given in the last column of Table 23-3**C** for the Owen bridge.

23-44 A Hay bridge (Table 23-3**D**) is at a null when R_1 is 2 kΩ, $R_2 = R_3 = 5$ kΩ, C_1 is 2 nF, and the radian frequency ω is 10^5 rad/s. Using the general impedance Eq. (23-30), calculate all parts of Prob. 23-40.

23-45 Repeat parts (**c**) and (**b**) of Prob. 23-44, using the balance equations given in the last column of Table 23-3D for the Hay bridge.

23-46 The Schering bridge (Table 23-3**E**) is at a null when R_1 is 10 kΩ, R_2 is 2 kΩ, C_1 is 0.1 nF, C_s is 1 nF, and the radian frequency ω is 10^6 rad/s. Using the general impedance Eq. (23-30), calculate the

a. Series equivalent impedance, Z_x, in polar and rectangular form
b. Unknown capacitance, C_x, and unknown resistance, R_x
c. Dissipation factor, D_x

23-47 Repeat parts (**b**) and (**c**) of Prob. 23-46, using the balance equations given in the last column of Table 23-3E for the Schering bridge.

Sec. 23-15

23-48 **Figure 23-32a** shows the display of a calibrated CRO whose vertical channel is set to 2 V/div and whose horizontal time base is set to 0.2 μs/div. For the displayed waveform, calculate

a. V_{pp}
b. The period and frequency
c. Average value
d. RMS value
e. Pulse duration of the positive portion

23-49 In Fig. 23-32b, the vertical input of CH 1 is 50 mV/div and of CH 2 is 20 mV/div. The horizontal time base is set to 50 μs/div. Calculate the

a. Period and frequency of both waveforms
b. Peak-to-peak and RMS value of the waveform displayed on CH 1
c. Peak to peak and RMS value of the waveform displayed on CH 2
d. Phase angle by which CH 2 lags CH 1
e. Delay time by which CH 2 lags CH 1

23-50 In Fig. 23-32c, the GND levels of CH 1 and CH 2 originally were set on the same line. When switched to direct coupling (DC), CH 1 displays the waveform shown in Fig. 23-32c, that is, the output of a simple horizontal sweep generator. The vertical CH 1 control is set to 1 V/div and the horizontal time base is set to 0.5 μs/div. Calculate

a. V_{pp}, the peak-to-peak voltage of waveform
b. Period and frequency of the waveform
c. Sweep time (duration of linear rise of ramp), T_s
d. Retrace (or flyback) time, T_r
e. Average value of the waveform (when alternately switched from DC to AC coupling)
f. RMS value of the waveform

23-51 Figure 23-32d shows a negative-going trigger pulse on CH 1 providing a typical triggered sweep output from a UJT (unijunction transistor) which is shown on CH 2. The CRO settings are 2 V/div on both CH 1 and CH 2 with the time base set to 2 μs/div. Find

a. For CH 1: peak pulse value, pulse duration, and pulse repetition rate (PRR)
b. For CH 2: peak output value, minimum (threshold) value, and peak-to-peak voltage
c. For CH 2: sweep interval, T_s (rising ramp only); retrace interval, T_r (falling ramp only); and hold-off interval, T_h (constant portion only)

23-52 The circuit of Fig. 23-28a is used as the CRO input to measure unknown impedance parameters. The waveforms are shown in **Fig. 23-32e**. The CRO time base setting is 200 μs/div. CH 1 is set to 2 V/div and CH 2 to 1 V/div. Calculate the

a. Voltage and current as RMS values
b. Phase angle between parameters in (**a**)
c. Impedance in polar and rectangular form
d. Q of the impedance
e. Parallel model component values

23-53 Repeat all parts of Prob. 23-52, using the same circuit and CRO settings for the waveforms shown in **Fig. 23-32f**.

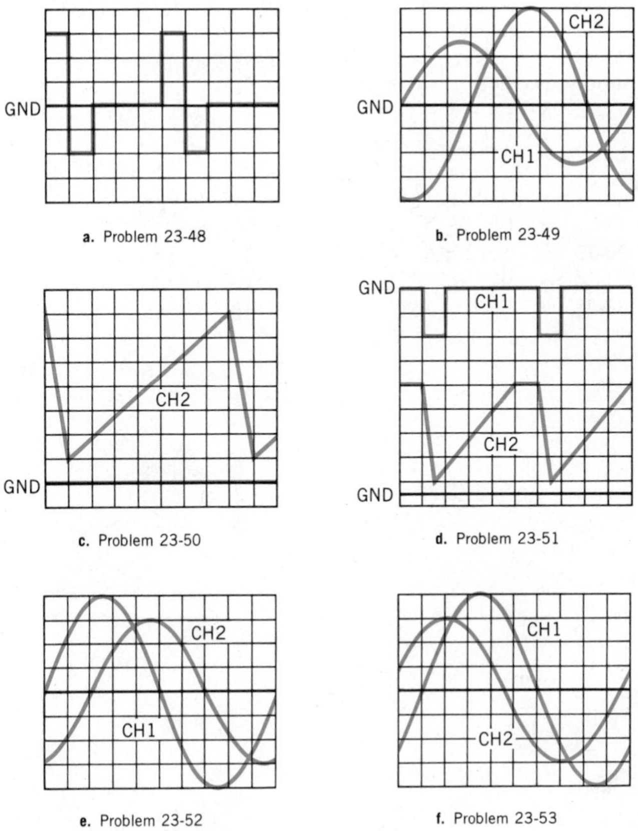

Figure 23-32 Problems 23-48 through 23-53

23-19 ANSWERS

23-1 a 142 b 5 μN c 0.71 mN d 7.1 μN·m
e 3.55 μN·m
23-2 a 5 mA b 200 Ω/V c 100 t d 500 μN
e 2.5 μN/cond
23-3 a 0.9788(fsd) b 4.89 mA c 2.12% d 1.957 mA
e 2.12%
23-4 a 1000 Ω/V b 100 mV c 1 mA
23-5 a 50 minor div b 2 V/m.d. c 1 V d 99 V
e 1%
23-6 a 25 m.d. b 4 V/m.d. c 2 V d 98 V e 2%
23-7 a 8% b 3% c −4% d 2%
23-8 a 3.006 Ω b 0.3 Ω c ~~6 mΩ~~ 30.0006 mΩ
23-9 a $5.\overline{5}$ Ω b 0.5051 Ω c 50.05 mΩ
23-10 a 0 Ω b 99 kΩ c 2.9 MΩ d 7 MΩ
23-11 a 0 Ω b 4.95 kΩ c 145 kΩ d 350 kΩ
23-12 625 Ω; 51.02 Ω; 5.01 Ω; 0.25 Ω; 0.025 Ω; 47.5 kΩ; 450 kΩ; 1 MΩ; 49 MΩ; 50 MΩ
23-13 ∞; 111.1 Ω; 10.1 Ω; 0.50025 Ω; 0.05 Ω; 9 kΩ; 90 kΩ; 2.9 MΩ; 7 MΩ; 10 MΩ
23-14 $\overline{50.5}$ Ω; 5.005 Ω; 0.5 Ω; 50 mΩ; 15 kΩ; 180 kΩ; 4.8 MΩ; 45 MΩ
23-15 1.2 MΩ
23-16 a 40.4 Ω b 35.51 Ω
23-17 2.007 Ω
23-18 a 30 kΩ b 30 kΩ c 29.5 kΩ d 20.65 kΩ
e 8.85 kΩ and 17.7 kΩ
23-19 a, b 500 Ω = 49.2 μA; 1 MΩ = 1.46 μA
c 303 Ω, 2.97 MΩ
23-20 a 1.5 mA b $16.\overline{6}$ Ω, 17.24 Ω c $983.\overline{3}$ Ω
d 680 Ω e 500 Ω pot.
23-21 at I_m = 35 μA: a 0.7 b 3/7 c 429 Ω d 49 kΩ
e 20.4 Ω
23-22 a 140 Ω b 98 Ω c 100 Ω pot.
23-23 b 0.909 mA c 0.1 Ω; 990 Ω
23-24 a 44.726 kΩ b 43.83 kΩ c 45.62 kΩ
23-25 a 1491 to 1538 Ω b 686.7 to 729.1 Ω
c 1126 to 1196 Ω
23-26 a 13.97 kΩ b 13.66 to 14.28 kΩ
23-27 a 44.72 kΩ b 44.16 to 45.28 kΩ
23-28 a 450 V b 2.7 MΩ
23-29 a 112.5 V b 106.9 to 118.1 V
23-30 b $\overline{33}$ kΩ c $\overline{33}$% e 50 kΩ
23-31 a 2 Ω b 1.5 Ω c $-\overline{33}$%
23-32 a 14.84 kΩ b 14.1 to 15.58 kΩ
23-33 a 116.3 mΩ b 109.9 to 122.7 mΩ
23-35 a 5 c 5485 Ω d 5480 to 5490 Ω
23-36 a, b 784.7 mV c 0
23-37 a 733.8 Ω b −13.3 mV c 15.05 μA d 1010 Ω
e 100.85 to 101.15 Ω
23-38 a 7.2 W b 192.8 W, −3.6% c 307.2 W
23-39 a 125 V b 7.81 W c 12.29 W
d 387.7 W, −3.08%
23-40 a $5.\overline{05}\underline{/84.29^\circ}$ kΩ b 10 c 0.8 H; 502.6 Ω
23-41 same answers
23-42 a $4.472\underline{/26.565^\circ}$ kΩ b 0.5 c 2 H; 4 kΩ
23-43 same
23-44 a $4.6425\underline{/68.2^\circ}$ kΩ b 2.5 c 43.1 mH; 1724 Ω
23-45 same
23-46 a $282.8\underline{/-45^\circ}$ Ω, 200 − j200 Ω b 5 nF, 200 Ω
c 1.0
23-47 same
23-48 a 10 V b 1 μs, 1 MHz c 0.4 V d 3.225 V
e 0.2 μs
23-49 a 250 μs, 4 kHz b 250 mV, 88.4 mV
c 160 mV, 56.57 mV d 108° e 150 μs
23-50 a 6 V b 4 μs, 250 kHz c 3.5 μs d 0.5 μs
e 4 V f 4.36 V
23-51 a −4 V, 10 μs, 10^5 PPS b 9 V, 1 V, 8 V_{pp}
c 7 μs, 1 μs, 2 μs
23-52 a 5.657 V, 2.121 mA b 72° c $2.\overline{6}\underline{/72^\circ}$ kΩ
d 3.078 e 89.2 mH, 8629 Ω
23-53 a 5.657 V, 2.121 mA b 54° c $2.\overline{6}\underline{/-54^\circ}$ kΩ
d 1.376 e 9.66 nF, 4536 Ω

APPENDIX

APPENDIX A-1 DEFINITIONS OF THE SI BASE UNITS

Meter Length equal to 1 650 763.73 wavelengths in vacuum of the radiation corresponding to the transition between levels $2p_{10}$ and $5d_5$ of the krypton-86 atom.

Kilogram Unit of mass equal to the mass of the international prototype of the kilogram in the custody of the International Bureau of Weights and Measures, Sevres, Paris, France.

Second Duration of 9 192 631 770 periods of the radiation corresponding to the transition between the two hyperfine levels of the ground state of the cesium-133 atom.

Ampere That constant current which, if maintained in two straight parallel conductors of infinite length, yet negligible cross section, placed 1 meter apart in a vacuum, would produce between these conductors a force of 2×10^{-7} newtons per meter of length.

Kelvin That thermodynamic temperature which is the fraction 1/273.16 of the triple point of water.

Mole Amount of substance of a system that contains as many elementary entities as there are atoms in a 0.012 kg of carbon-12.

Candela Luminous intensity, in the perpendicular direction, of a surface of 1/600 000 square meters of a blackbody at the temperature of freezing platinum under a pressure of 101 325 newtons per square meter.

APPENDIX A-2 DEFINITIONS OF THE SI SUPPLEMENTARY UNITS

Radian Unit of measure of a plane angle with its vertex at the center of a circle and subtended by an arc equal to the radius.

Steradian Unit of measure of a solid angle with its vertex at the center of a spherical surface equal to that of a square with sides equal in length to the radius.

APPENDIX A-3 CONVERSION FACTORS

A. LENGTH (Linear Ratios)

2.54 cm/in	0.1 nm/Å	10^{-3} in/mil	$0.1\bar{6}$ fathom/ft
0.9144 m/yd	1.76×10^3 yd/mi	9.46×10^{12} km/light-year	16.5 ft/rod
1.609 km/mi	6 ft/fathom	3.084×10^{13} km/parsec	10^{-6} m/μm
30.48 cm/ft	3 ft/yd	10^2 cm/m	3.281 ft/m
5.28×10^3 ft/mi	16.5 ft/rod	0.0254 mm/mil	0.001 in/mil

B. AREA (Obtainable by Squaring Linear Ratios, A, Above)

10^4 cm^2/m^2	1.974 10^9 c-mil/m^2	6.944 10^{-3} ft^2/in^2	144 in^2/ft^2
10.76 ft^2/m^2	1.076 10^{-3} ft^2/cm^2	5.454 10^{-9} ft^2/c-mil	5.067×10^{-6} cm^2/c-mil
1.55×10^3 in^2/m^2	10.76 ft^2/m^2	5.067×10^{-10} m^2/c-mil	

C. VOLUME (Obtainable by Cubing Linear Ratios, A, Above)

10^6 cm^3/m^3	10^{-3} m^3/liter
35.31 ft^3/m^3	61.02 in^3/liter
6.102×10^4 in^3/m^3	231 in^3/gal
3.531 10^{-2} ft^3/liter (l)	277.42 in^3/Imp. gal = 10 lb H_2O
10^3 cm^3/liter	1000.028 cm^3/liter = 1 kg H_2O
57.75 in^3/qt	

D. SPEED (Obtainable by Using Above Linear Ratios, A, per Unit Second, Minute, or Hour)

$1.097 \frac{\text{km/h}}{\text{ft/s}}$	$3.6 \frac{\text{km/h}}{\text{m/s}}$	$5.08 \frac{\text{mm/s}}{\text{ft/min}}$	$1.152 \frac{\text{mi/h}}{\text{knot}}$
$0.6818 \frac{\text{mi/h}}{\text{ft/s}}$	$3.281 \frac{\text{ft/s}}{\text{m/s}}$	$0.2778 \frac{\text{m/s}}{\text{km/h}}$	$0.9113 \frac{\text{ft/s}}{\text{km/h}}$
$0.5921 \frac{\text{knot}}{\text{ft/s}}$	$1.467 \frac{\text{ft/s}}{\text{mi/h}}$	$0.5148 \frac{\text{m/s}}{\text{knot}}$	$0.9144 \frac{\text{m/h}}{\text{yd/h}}$
$0.54 \frac{\text{knot}}{\text{km/h}}$	$0.447\,04 \frac{\text{m/s}}{\text{mi/h}}$	$1.853 \frac{\text{km/h}}{\text{knot}}$	$1609.4 \frac{\text{m/h}}{\text{mi/h}}$
$1.944 \frac{\text{knot}}{\text{m/s}}$	$0.01 \frac{\text{m/s}}{\text{cm/s}}$	$0.3048 \frac{\text{m/s}}{\text{ft/s}}$	

E. MASS

2.205 lb/kg*	0.4536 kg/lb*	0.06852 slug/kg	$1 \frac{\text{dyn/cm/s}^2}{\text{g}}$
28.35 g/oz*	1.66×10^{-27} kg/amu	1×10^3 kg/metric ton*	$1 \frac{\text{lb/ft/s}^2}{\text{slug}}$
453.6 g/lb*	907.2 kg/ton*	$1 \frac{\text{N/m/s}^2}{\text{kg}}$	
1.459×10^4 g/slug	10^3 g/kg		
14.594 kg/slug			

F. FORCE AND WEIGHT

10^{-5} N/dyn	3.108×10^{-2} lb/pdl	980.7 dyn/g_f*	$1 \frac{\text{slug}\cdot\text{ft/s}^2}{\text{lb}_f}$
2.248×10^{-6} lb/dyn	9.807×10^{-3} N/g_f*	0.138 26 N/pdl	$1 \frac{\text{kg}\cdot\text{m/s}^2}{\text{N}}$
7.233×10^{-5} pdl/dyn	4.4482 N/lb_f	0.278 N/oz*	
28.35 g/oz	9.807 N/kg_f*	$1 \frac{\text{g}\cdot\text{cm/s}^2}{\text{dyn}}$	
0.2248 lb/N	32.17 pdl/lb	9.807 N/kg_f*	
7.233 pdl/N	2.205 lb/kg_f*		

* On earth only.

G. DENSITY

$515.4\,\frac{kg/m^3}{slug/ft^3}$ $16.02\,\frac{kg/m^3}{lb/ft^3}$ $32.17\,\frac{lb/ft^3*}{slug/ft}$ $62.43\,\frac{lb/ft^3*}{g/cm^3}$

$10\,\frac{kg/m^3}{g/cm^3}$ $2.768\times 10^4\,\frac{kg/m^3*}{lb/in^3}$ $62.43\,\frac{lb/ft^3*}{g/cm^3}$ $1728\,\frac{lb/ft^3}{lb/in^3}$

$$H_2O = 1\,\frac{kg}{liter} = 8.342\,\frac{lb}{gal} = 62.43\,\frac{lb}{ft^3} = 1\ g/cm^3$$

H. ENERGY

10^{-7} J/erg
3.6×10^6 J/kWh
1.602×10^{-19} J/eV
4.186 J/g·cal
1054.8 J/Btu
$1\,\frac{J}{kg\cdot(m/s)^2}$
$1\,\frac{J}{W\cdot s}$

$1\,\frac{J}{N\cdot m}$
$1\,\frac{W\cdot s}{N\cdot m}$
$0.7375\,\frac{ft\cdot lb}{J}$
$3.968\,\frac{Btu}{kg\cdot cal}$
$4.186\times 10^3\,\frac{J}{kg\cdot cal}$

1.49×10^{-10} J/amu
$3413\,\frac{Btu}{kWh}$
$252\,\frac{g\cdot cal}{Btu}$
0.252 kg·cal/Btu
0.2389 g·cal/J
778 ft·lb/Btu

$2.685\times 10^6\,\frac{J}{hp\cdot h}$
$2.93\times 10^{-4}\,\frac{kWh}{Btu}$
$0.7457\,\frac{kWh}{hp\cdot h}$
2.778×10^{-7} kWh/J
0.324 g·cal/ft·lb
1.356 J/ft·lb

I. POWER (Ratios Shown in H Obtain if Each Is Divided by *Same* Unit of Time)

$10^{-7}\,\frac{J/s}{erg/s}$ $4.186\,\frac{W}{g\cdot cal/s}$ $550\,\frac{ft\cdot lb/s}{hp}$ $3.929\times 10^{-4}\,\frac{hp}{Btu/h}$

$745.7\,\frac{W}{hp}$ $1.356\,\frac{W}{ft\cdot lb/s}$ $33\,000\,\frac{ft\cdot lb/min}{hp}$ $5.613\times 10^{-3}\,\frac{hp}{g\cdot cal/s}$

$0.2931\,\frac{W}{Btu/h}$ $0.056\,88\,\frac{Btu/min}{W}$ $1.3405\,\frac{hp}{kW}$ $4.186\,\frac{W}{g\cdot cal/s}$

$1\,\frac{W}{J/s}$ $0.014\,33\,\frac{kg\cdot cal/min}{W}$ $2.260\times 10^{-2}\,\frac{W}{ft\cdot lb/min}$

J. PRESSURE

$1\,\frac{Pa}{N/m}$ 10^2 Pa/mbar $249.1\,\frac{Pa}{in\ (H_2O)}$ $10^6\,\frac{dyn/cm^2}{bar}$

$1.013\times 10^5\,\frac{Pa}{atm}$ $133.322\,\frac{Pa}{mm\ Hg}$ $1333\,\frac{Pa}{cm\ Hg}$ $10^3\,\frac{dyn/cm^2}{m\ bar}$

$6.895\,\frac{kPa}{lb/in^2}$ $0.2491\,\frac{kPa}{in\ (H_2O)}$ $2116\,\frac{lb_f/ft^2}{atm}$ $\frac{760\ mm\ Hg}{1\ atm}$

$47.88\,\frac{Pa}{lb_f/ft^2}$ $2.989\,\frac{kPa}{ft\ (H_2O)}$ $27.85\,\frac{lb_f/ft^2}{cm\ Hg}$

$9.807\,\frac{Pa}{kg_f/m^2}$ $0.1\,\frac{Pa}{dyn/cm^2}$ $10\,\frac{N/m^2}{bar}$

K. LIGHT

10.76 lux/fc $3.426\,\frac{cd/m^2}{fL}$ $1\,\frac{lum/ft^2}{fc}$

* Weight densities, valid on earth only.

L. INERTIA

$10^{-7}\,\dfrac{kg\cdot m^2}{g\cdot cm^2}$ $7.07\times10^4\,\dfrac{g\cdot cm^2}{oz\cdot in\cdot s^2}$ $7.372\times10^{-8}\,\dfrac{slug\cdot ft^2}{g\cdot cm^2}$

$1.829\cdot10^2\,\dfrac{g\cdot cm^2}{oz\cdot in^2}$ $1.829\times10^{-5}\,\dfrac{kg\cdot m^2}{oz\cdot in^2}$ $1.348\times10^{-5}\,\dfrac{slug\cdot ft^2}{g\cdot cm^2}$

$1.357\times10^7\,\dfrac{g\cdot cm^2}{slug\cdot ft^2}$ $70.7\,\dfrac{kg\cdot cm^2}{oz\cdot in\cdot s^2}$ $5.468\times10^{-3}\,\dfrac{oz\cdot in^2}{g\cdot cm^2}$

$1.357\,\dfrac{kg\cdot m^2}{slug\cdot ft^2}$ $2.59\times10^{-3}\,\dfrac{oz\cdot in\cdot s^2}{oz\cdot in^2}$ $1\,\dfrac{dyn\cdot cm\cdot s^2}{g\cdot cm^2}$ $1\,\dfrac{N\cdot m\cdot s^2}{kg\cdot m^2}$

$2.925\times10^{-4}\,\dfrac{kg\cdot m^2}{lb\cdot in^2}$ $7.419\times10^4\,\dfrac{oz\cdot in^2}{lb\cdot ft\cdot s^2}$ $1\,\dfrac{lb\cdot ft\cdot s^2}{slug\cdot ft^2}$ $7.07\times10^4\,\dfrac{dyn\cdot cm\cdot s^2}{oz\cdot in\cdot s^2}$

M. TORQUE

$1.383\times10^4\,\dfrac{g\cdot cm}{lb\cdot ft}$ $7.0612\times10^4\,\dfrac{dyn\cdot cm}{oz\cdot in}$ $1.235\times10^{-5}\,\dfrac{ft\cdot lb}{g\cdot cm}$

$72.01\,\dfrac{g\cdot cm}{oz\cdot in}$ $1.416\times10^{-5}\,\dfrac{oz\cdot in}{dyn\cdot cm}$ $1.389\times10^{-2}\,\dfrac{oz\cdot in}{g\cdot cm}$

$7.0612\times10^{-3}\,\dfrac{N\cdot m}{oz\cdot in}$ $192\,\dfrac{oz\cdot in}{ft\cdot lb}$ $5.208\times10^{-3}\,\dfrac{ft\cdot lb}{oz\cdot in}$

N. TEMPERATURE

$°C = \frac{5}{9}(°F - 32)$ $K = °C + 273.16$ $°R = °F + 459.67$ $°F = \frac{9}{5}\times °C + 32$

O. RESISTIVITY

$6.0153\times10^8\,\dfrac{\Omega\cdot c\text{-}mil/ft}{\Omega\cdot m}$ $6.0153\times10^6\,\dfrac{\Omega\cdot c\text{-}mil/ft}{\Omega\cdot cm}$ $1.662\times10^{-9}\,\dfrac{\Omega\cdot m}{\Omega\cdot c\text{-}mil/ft}$

$10^{-8}\,\dfrac{\Omega\cdot m}{\mu\Omega\cdot cm}$ $0.01\,\dfrac{\Omega\cdot m}{\Omega\cdot cm}$

P. POTENTIAL DIFFERENCE

$1\ V = 1\,\dfrac{J}{C} = 1\,\dfrac{N\cdot m}{C}$ $10^{-8}\,\dfrac{V}{abV}$ $299.8\,\dfrac{V}{esu}$ $1\,\dfrac{Wb/s}{V}$

Q1. MAGNETIC FLUX (ϕ) AND FLUX DENSITY (B)

$10^8\,\dfrac{Mx}{Wb}$ $10^{-4}\,\dfrac{Wb/m^2}{G}$ $10^{-9}\,\dfrac{Wb/m^2}{gamma}$

$10^5\,\dfrac{k\cdot lines}{Wb}$ $1.550\times10^{-2}\,\dfrac{Wb/m^2}{k\cdot line/in^2}$ $2.998\times10^6\,\dfrac{Wb/m^2}{esu}$

$10^{-4}\ T = 1\ G = 1\,\dfrac{Mx}{cm^2}$ $10^{-7}\,\dfrac{Wb/m^2}{mG}$ $\dfrac{10^7}{4\pi}\,\dfrac{G/Oe}{H/m}$ $1\ H = \dfrac{1\ Wb}{A}$

Q2. MAGNETOMOTIVE FORCE, $\mathscr{F}$

$10\,\dfrac{At}{abAt}$ $0.7958\,\dfrac{At}{Gb}$ $1.257\,\dfrac{Gb}{At}$

Q3. MAGNETIC FIELD STRENGTH, H

$10^2 \dfrac{\text{At/m}}{\text{At/cm}}$ $\quad 79.58 \dfrac{\text{At/m}}{\text{Oe}}$ $\quad 1.257 \dfrac{\text{Oe}}{\text{At/cm}}$ $\quad 1.257 \times 10^{-2} \dfrac{\text{Oe}}{\text{At/m}}$

$39.37 \dfrac{\text{At/m}}{\text{At/in}}$ $\quad 2.021 \dfrac{\text{At/in}}{\text{Oe}}$ $\quad 0.4947 \dfrac{\text{Oe}}{\text{At/in}}$ $\quad 0.7958 \dfrac{\text{At/cm}}{\text{Oe}}$

Q4. OTHER MAGNETIC RELATIONS

$\mu_{\text{MKS}} = \mu_{\text{ENG}} (4\pi \times 10^{-7} \times 2.54 \times 10^{-7})$ $\quad 1 \text{ Oe} = 1 \dfrac{\text{Gb}}{\text{cm}} = 1 \dfrac{\text{dyn}}{\text{unit pole}} = 79.6 \dfrac{\text{At}}{\text{m}}$

$\mu_{\text{CGS}} = \mu_{\text{MKS}} \dfrac{10^7}{4\pi}$ $\quad 1 \text{ H} = 1 \dfrac{\text{Wb}}{\text{A}}$ $\quad \dfrac{10^7}{4\pi} \dfrac{\text{G/Oe}}{\text{H/m}}$

R. ELECTRIC CHARGE

$2.778 \times 10^{-3} \dfrac{\text{Ah}}{\text{abC}}$ $\quad 26.81 \dfrac{\text{Ah}}{\text{faraday}}$ $\quad 9.652 \times 10^4 \dfrac{\text{C}}{\text{faraday}}$

2.778×10^{-4} Ah/C $\quad 3600 \dfrac{\text{C}}{\text{Ah}}$

S. ELECTRIC CURRENT

1 abA/emu
10 A/abA
2.998×10^{10} stat A/abA
1 stat A/esu
2.998×10^9 stat A/A
$1 \dfrac{\text{C/s}}{\text{A}}$

T. ANGULAR MEASURE

3600 s/deg
1.745×10^{-2} rad/deg
2.778×10^{-3} rev/deg
0.1592 rev/rad
57.30 deg/rad
6.283 rad/rev
60 min/deg
60 s/min
10^3 mils/rad
3.438 10^3 min/rad
$1.667 \times 10^{-2} \dfrac{\text{deg}}{\text{min}}$
17.45 mils/deg
3.438 min/mil

U. SOLID ANGULAR MEASURE

1 sphere = 4π steradians = 12.57 steradians

V. TIME

365.256 days (solar)/year(siderial)
8.64×10^4 s/day
3.6×10^3 s/h
60 s/min

W. ANGULAR VELOCITY

$9.549 \dfrac{\text{rev/min}}{\text{rad/s}}$ $\quad 0.1047 \dfrac{\text{rad/s}}{\text{rev/min}}$ $\quad 57.3 \dfrac{\text{deg/s}}{\text{rad/s}}$

$0.1592 \dfrac{\text{rev/s}}{\text{rad/s}}$ $\quad 360 \dfrac{\text{deg/s}}{\text{rev/s}}$ $\quad 0.1667 \dfrac{\text{rev/min}}{\text{deg/s}}$

$6 \dfrac{\text{deg/s}}{\text{rev/min}}$ $\quad 6.283 \dfrac{\text{rad/s}}{\text{rev/s}}$ $\quad 60 \dfrac{\text{rev/min}}{\text{rev/s}}$

X. DAMPING

$20.11 \dfrac{\text{oz}\cdot\text{in/rpm}}{\text{ft}\cdot\text{lb/(rad/s)}}$ $\quad 6.75 \times 10^{-3} \dfrac{\text{N}\cdot\text{m/(rad/s)}}{\text{oz}\cdot\text{in/(rev/min)}}$

APPENDIX A-4 TABLES OF RESISTIVITIES AND TEMPERATURE RESISTANCE COEFFICIENT AT 20°C (+ at 20°C Unless Otherwise Specified)

Material	Resistivity Ω·m (ohm-meter)	(ρ) Ω·c-mil/ft	(α_{20}) Temperature Coefficient of Resistance, at 20°C ohms per °C/Ω
Advance (alloy)	48.8 × 10^{-8}	293.5	+0.000 018
Aluminum	2.83 ×	17.02	0.003 91
Antimony	41.7 ×	250.8	0.003 6
Barium	9.8 ×	58.9	0.003 3
Beryl	10.1 ×	60.75	
Bismuth	120.0 ×	721.8	0.004
Brass	7.0 ×	42.1	0.002 1
Calcium	4.59 ×	27.6	0.003 64
Cerium	78 ×	469.1	
Cesium	+0°C 19 ×	114.3	
Chromel	90.0 ×	541.3	0.000 08
Chromium	+0°C 2.6 ×	15.64	
Cobalt	9.7 ×	58.34	0.006 58
Constantan	49 ×	294.7	0.000 008
Copper (std)	1.7241 ×	10.371	0.003 93
German silver	33 ×	198.5	0.000 4
Gold	2.44 ×	14.676	0.003 4
Iron	98.0 ×	589.4	0.006 5
Lead	22 ×	132.3	0.003 9
Lithium	+0°C 8.55 ×	51.4	0.004 7
Magnesium	4.46 ×	26.83	0.004
Manganin	44.0 ×	264.6	0.000 006
Mercury	95.8 ×	576.2	0.000 89
Nichrome	99.6 ×	599.1	0.000 44
Nickel	7.8 ×	46.9	0.005 37
Platinum	+0°C 9.83 ×	59.1	0.003
Silver	1.629 ×	9.805	0.003 8
Steel (rail)	20.0 ×	120.3	0.008
Tin	11.5 ×	69.2	0.004 2
Tungsten	5.52 ×	33.2	0.004 5
Zinc	+0°C 5.75 × 10^{-8}	34.6	0.003 7
Germanium	4.7 × 10^{-1}	2.83 × 10^{8}	Negative
Silicon	2.3 × 10^{3}	1.4 × 10^{12}	Negative

Notes:
1. Temperatures for resistivity at 20°C unless otherwise noted.
2. To find resistivity in Ω·cm, multiply table (value in Ω·m) by 10^2.
3. All temperature coefficients *shown* are positive; that is, resistance increases with increases in temperature in proportion to the coefficient shown per degree celsius.
4. To find the conductivity (σ) in siemens/meter, take the reciprocal of the first column given above.
5. Conversion factors

a. $$\frac{\rho \text{ metric}}{\rho \text{ ENG}} = 1.662 \times 10^{-9} \frac{\Omega\cdot\text{m}}{\Omega\cdot\text{c-mil/ft}}$$

b. $$\frac{\rho \text{ ENG}}{\rho \text{ metric}} = 6.0153 \times 10^{8} \frac{\Omega\cdot\text{c-mil/ft}}{\Omega\cdot\text{m}}$$

APPENDIX A-5 AWG CONDUCTOR SIZES, DIAMETERS, AREAS, AND RESISTANCES OF COMMERCIAL SOLID COPPER CONDUCTORS AT 20°C

	English Units			Equivalent SI Units		
AWG (No.)	Diameter (mils)	Area (c-mils)	Resistance (ohms/1000 ft)	Diameter (mm)	Area (m^2)	Resistance (Ω/km)
0000 (4/0)	460	211 600	0.049	11.68	107.1×10^{-6}	0.160
000 (3/0)	409.64	167 800	0.062	10.40	84.9	0.203
00 (2/0)	364.86	133 080	0.078	9.266	67.43	0.255
0	324.9	105 534	0.098	8.252	53.48	0.316
1	289.3	83 694	0.124	7.348	42.41	0.406
2	257.63	66 373	0.156	6.544	33.63	0.511
3	229.42	52 633	0.197	5.827	26.67	0.645
4	204.31	41 742	0.248	5.189	21.15	0.813
5	181.94	33 102	0.313	4.621	16.77	1.026
6	162.02	26 250	0.395	4.115	13.30	1.29
7	144.28	20 817	0.498	3.665	10.55	1.63
8	128.49	16 510	0.628	3.264	8.367	2.06
9	114.43	13 094	0.792	2.9065	6.635	2.59
10	101.89	10 382	0.999	2.588	5.260	3.27
11	90.74	8234	1.26	2.305	4.173	4.10
12	80.81	6530	1.59	2.0525	3.309	5.20
13	71.961	5178	2	1.828	2.624	6.55
14	64.08	4107	2.52	1.628	2.082	8.26
15	57.07	3257	3.18	1.450	1.651	10.4
16	50.82	2583	4.02	1.291	1.309	13.1
17	45.26	2048	5.06	1.150	1.039×10^{-6}	16.6
18	40.3	1624	6.39	1.024	8.235×10^{-7}	21.0
19	35.9	1288	8.05	0.912	6.533	26.3
20	31.96	1021	10.1	0.812	5.178	33.2
21	28.46	810	12.8	0.723	4.1055	41.9
22	25.35	642.6	16.1	0.644	3.257	52.8
23	22.57	509.4	20.3	0.573	2.579	66.7
24	20.1	404	25.7	0.5105	2.047	83.9
25	17.9	320.4	32.4	0.4547	1.624	106
26	15.94	254.1	41	0.405	1.288	134
27	14.2	201.5	51.4	0.3607	1.022×10^{-7}	168
28	12.64	159.8	64.9	0.321	8.093×10^{-8}	213
29	11.26	126.7	81.4	0.286	6.424	267
30	10.025	100.5	103	0.255	5.107	337
31	8.93	79.7	130	0.227	4.047	425
32	7.95	63.2	164	0.202	3.205	537
33	7.08	50.13	206	0.180	2.545	676
34	6.304	39.74	261	0.160	2.011	855
35	5.614	31.52	329	0.1426	1.597	1071
36	5	25.00	415	0.127	1.267	1360
37	4.453	19.83	523	0.113	1.003×10^{-8}	1715
38	3.965	15.72	655	0.101	8.012×10^{-9}	2147
39	3.531	12.47	832	0.0897	6.319×10^{-9}	2704
40	3.145	9.89	1044	0.0799	5.014×10^{-9}	3422

APPENDIX A-6 MAXIMUM POWER TRANSFER IN DC AND AC CIRCUITS

A-6.1 Derivation of Maximum Power Transfer in DC Circuits

For the practical dc voltage source shown in Fig. 5-10, the power delivered to the load R_L is a variable, p_L, that is

$$p_L = i_L^2 R_L = \frac{E_0^2 R_L}{(r_i + R_L)^2} \qquad \text{since } i = \frac{E_0}{(R_L + r_i)} \text{ from Eq. (5-9b)}$$

To find the value of R_L that can absorb maximum power from the practical source, we maximize p_L by differentiating it with respect to R_L or

$$\frac{dp_L}{dR_L} = \frac{(r_i + R_L)^2 E_0^2 - E_0^2 R_L(2)(r_i + R_L)}{(r_i + R_L)^4} = 0$$

If we equate the derivative to zero as shown, factoring out E_0^2; both E_0^2 and the denominator to the fourth power drop out leaving $2R_L(r_i + R_L) = (r_i + R_L)^2$, and canceling common terms yields

$$R_L = r_i \tag{5-17}$$

As shown in Fig. 5-11, $p_L = 0$ when $R_L = 0$ and $R_L = \infty$. Therefore, p_L is a maximum whenever the load resistance R_L is equal to the internal resistance of the source, r_i.

A-6.2 Derivation of Maximum Power Transfer in AC Circuits

For the practical ac source shown in Fig. 17-6, we may write for the current, **I**,

$$\mathbf{I} = \frac{\mathbf{E}}{(R_i + R_L) + (\pm jX_i + jX_L)}$$

and since we desire an absolute value of **I**, we may write

$$|I| = \frac{|\mathbf{E}|}{\sqrt{(R_i + R_L)^2 + (X_i + X_L)^2}}$$

Since we are interested in obtaining the power transferred to the load, $P_{R_L} = |I|^2 R_L$, then

$$P_{R_L} = \frac{|\mathbf{E}|^2 R_L}{(R_i + R_L)^2 + (X_i + X_L)^2} \tag{A-6.1}$$

Equation (A-6.1) must be differentiated and set equal to zero in order to determine maximum power dissipated in the load resistance, R_L. But since it contains two variables (R_L and X_L), two derivatives must be taken with respect to each of them.

Differentiating Eq. (A-6.1) with respect to R_L yields

$$\frac{dP_{R_L}}{dR_L} = \frac{(R_i + R_L)^2 + (X_i + X_L)^2|\mathbf{E}|^2 - |\mathbf{E}|^2 R_L \times 2(R_i + R_L)}{[(R_i + R_L)^2 + (X_i + X_L)^2]^2} = 0$$

The denominator term drops out by cross-multiplication, and setting the numerator to zero

$$[(R_i + R_L)^2 + (X_i + X_L)^2]|\mathbf{E}|^2 - 2|\mathbf{E}|^2 R_L(R_i + R_L) = 0$$

$$[(R_i + R_L)^2 + (X_i + X_L)^2 - 2R_L(R_i + R_L)]|\mathbf{E}|^2 = 0 \text{ but since } |\mathbf{E}|^2 \text{ is} \neq 0\text{, then}$$

$$(R_i + R_L)^2 + (X_i + X_L)^2 - 2R_L(R_i + R_L) = 0$$

$$R_i + 2R_iR_L + R_L^2 + (X_i + X_L)^2 - 2R_LR_i - 2R_L^2 = 0$$

$$R_i^2 - R_L^2 + (X_i + X_L)^2 = 0$$

$$R_L^2 = R_i^2 + (X_i + X_L)^2$$

$$R_L = \sqrt{R_i^2 + (X_i + X_L)^2}$$

$$= |R_i + j(X_i + X_L)| \qquad \text{(our first expression)} \qquad \text{(A-6.2)}$$

Again, differentiating Eq. (A-6.1) with respect to X_L yields

$$\frac{dP_{R_L}}{dX_L} = \frac{-|\mathbf{E}|^2 R_L^2(X_i + X_L)}{[(R_i + R_L)^2 + (X_i + X_L)^2]^2} = 0$$

and cross-multiplying again yields $-2|\mathbf{E}|^2 R_L(X_i + X_L) = 0$, but since $|\mathbf{E}| \neq 0$ and $R_L \neq 0$, this leaves

$$(X_i \pm X_L) = 0 \qquad \text{and} \qquad j(X_i + X_L) = 0 \qquad \text{(A-6.3)}$$

Returning to Eq. (A-6.2), therefore, and substituting (A-6.3)

$$R_L = |R_i + j(X_i + X_L)| = |R_i| = R_i \qquad \text{(A-6.4)}$$

But since Z_L in Fig. 17-6 is by definition $Z_L = R_L + jX_L$ and $R_L = R_i$ by Eq. (A-6.4) and $jX_L = -jX_i$ by Eq. (A-6.3), then

$$Z_L = R_i - jX_i = Z_i^* \qquad \text{(A-6.5)}$$

Equation (A-6.5) shows that when the load impedance, Z_L, is the conjugate of the internal source impedance, maximum power is transferred from source to load.

APPENDIX A-7 DERIVATION OF RESISTANCE AND CURRENT RELATIONS IN A TWO-BRANCH PARALLEL CIRCUIT

A. Given

$$\frac{1}{R_T} = \frac{1}{R_1} + \frac{1}{R_2} \qquad \text{(5-20)}$$

multiplying all terms by $R_T R_1 R_2$ yields

$$\frac{R_T R_1 R_2}{R_T} = \frac{R_T R_1 R_2}{R_1} + \frac{R_T R_1 R_2}{R_2}$$

$R_1 R_2 = R_T R_2 + R_T R_1$ factoring out R_T and solving for R_T yields

$$R_T = \frac{R_1 R_2}{R_1 + R_2} = R_{eq} \qquad (5\text{-}23)$$

B. Solving Eq. (5-23) above for R_1 and R_2, respectively, yields by simple algebra

$$R_1 = \frac{R_2 R_{eq}}{R_2 - R_{eq}} \quad (5\text{-}23a) \qquad R_2 = \frac{R_1 R_{eq}}{R_1 - R_{eq}} \quad (5\text{-}23b)$$

C. Given $V = V_1 = V_2$ in a two-branch circuit and $I_T = I_1 + I_2$, then $I_2 R_2 = I_1 R_1 = (I_T - I_2)R_1$; multiplying R_1 through the last term at the right yields $I_2 R_2 = I_T R_1 - I_2 R_1$; and solving for I_2 yields

$$I_2 = I_T \frac{R_1}{R_1 + R_2} \qquad (5\text{-}24a)$$

and similarly for I_1,

$$I_1 = I_T \frac{R_2}{R_1 + R_2} \qquad (5\text{-}24b)$$

This relation is known as the current divider rule (CDR).

APPENDIX B-1 DELTA–WYE AND WYE–DELTA TRANSFORMATION

Perhaps the most elegant derivation of relations between the delta (or pi) and wye (or tee) network is obtained from classical admittance parameters of two-port networks.

Figure 1a shows a typical delta network represented as a bilateral two-port. The input ports are between terminals 1–1 and the output ports between terminals 2–2.

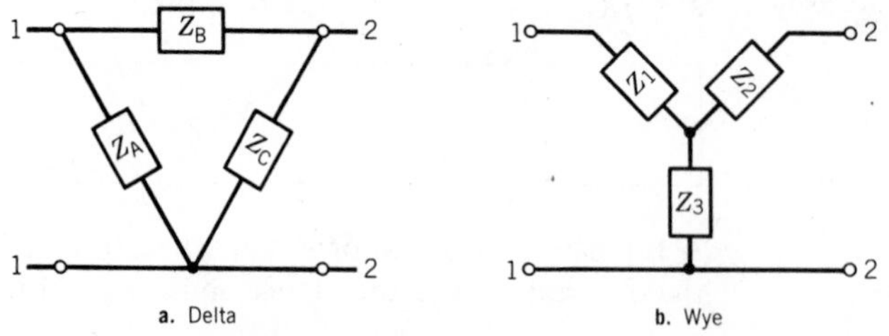

Figure B-1 Delta–wye equivalence

Figure 1b shows the equivalent wye network between the same input and output ports. The network of Fig. B-1b may be replaced by Fig. B-1a (and vice versa) if certain impedance relationships that establish their equivalence are found. These may be found by using short-circuit admittance or y parameters, as follows:

From Fig. B-1a From Fig. B-1b

$$y_{11} = \frac{1}{Z_A} + \frac{1}{Z_B} = \frac{1}{Z_1 + \dfrac{Z_2 Z_3}{Z_2 + Z_3}} \qquad \text{(output terminals shorted)}$$

$$y_{12} = y_{21} = -\frac{1}{Z_B} = \frac{-Z_3}{Z_1Z_2 + Z_2Z_3 + Z_3Z_1} \quad \text{(both input and output terminals shorted)}$$

$$y_{22} = \frac{1}{Z_C} + \frac{1}{Z_B} = \frac{1}{Z_2 + \dfrac{Z_1Z_3}{Z_1 + Z_3}} \quad \text{(input terminals shorted)}$$

Solving these equations for Z_A, Z_B, and Z_C, in terms of their equivalent Y-connected values Z_1, Z_2, and Z_3, respectively, we obtain the wye–delta conversion:

$$Z_A = \frac{Z_1Z_2 + Z_2Z_3 + Z_3Z_1}{Z_2} = \frac{\sum(Z \text{ products})}{Z_2} \tag{B-1.1}$$

$$Z_B = \frac{Z_1Z_2 + Z_2Z_3 + Z_3Z_1}{Z_3} = \frac{\sum(Z \text{ products})}{Z_3} \tag{B-1.2}$$

$$Z_C = \frac{Z_1Z_2 + Z_2Z_3 + Z_3Z_1}{Z_1} = \frac{\sum(Z \text{ products})}{Z_1} \tag{B-1.3}$$

Similarly, if we solve the equations for their inverse relationships, we obtain the delta–wye conversion as

$$Z_1 = \frac{Z_AZ_B}{Z_A + Z_B + Z_C} = \frac{Z_AZ_B}{\sum(Z\text{'s})} \tag{B-1.4}$$

$$Z_2 = \frac{Z_BZ_C}{Z_A + Z_B + Z_C} = \frac{Z_BZ_C}{\sum(Z\text{'s})} \tag{B-1.5}$$

$$Z_3 = \frac{Z_CZ_A}{Z_A + Z_B + Z_C} = \frac{Z_CZ_A}{\sum(Z\text{'s})} \tag{B-1.6}$$

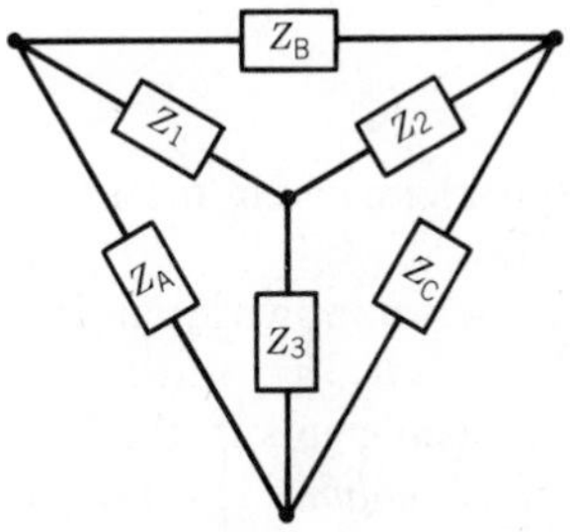

Figure B-2 Superimposition of Δ–Y circuits of Fig. B-1.

The two circuits of Fig. B-1 are superimposed on the same (common) terminals in **Fig. B-2** for the purpose of drawing further inferences from Eqs. (B-1.1) through (B-1.6).

For wye–delta conversion, each respective delta impedance requires the *diagonally opposite* wye impedance of the *denominator* of its respective equation (B-1.1 to B-1.3, respectively). For example, to find Z_A, place Z_2 in the denominator (Eq. B-1.1). Then from Fig. 2, to find Z_B, place Z_3 in the denominator (Eq. B-1.2). And from Fig. B-2, to find Z_C, place Z_1 in the denominator (Eq. B-1.3). For all three equations of the wye–delta conversion, the numerator is merely the *same sum* of the products of three pairs of impedances.

For the delta–wye conversion, the denominator is the same for Eqs. (B-1.4)–(B-1.6). It is the sum of the impedances of the delta loop, $Z_A + Z_B + Z_C$. The numerator of each Y element is the product of adjacent delta impedances connected to the same terminal. Thus, the numerator for finding Z_1 in Eq. (B-1.4) (from Fig. B-2) is the product Z_AZ_B adjacent to Z_1 and connected to the same (common) terminal. The same holds true for finding Z_2 from Eq. (B-1.5) and Z_3 in Eq. (B-1.6).

Although the reader may find Eqs. (B-1.1)–(B-1.6) sufficiently simple in view of Fig. B-2, a further insight is in store for us. We know that the delta (or mesh) is the *dual* of the wye (or star) and vice versa. Since the dual of impedance is admittance, we could use the pattern established in the first three equations (B-1.1 through

B-1.3) to find a second set of equations for the delta–wye conversion to replace Eqs. (B-1.4)–(B-1.6) as

$$Y_1 = (Y_A Y_B + Y_B Y_C + Y_C Y_A)/Y_C = \sum(Y \text{ products})/Y_C \qquad \text{(B-1.4a)}$$

$$Y_2 = \sum(Y \text{ products})/Y_A \qquad \text{(B-1.5a)}$$

$$Y_3 = \sum(Y \text{ products})/Y_B \qquad \text{(B-1.6a)}$$

This insight into duality shows that *only one* equation pattern is required to perform either the delta–wye or wye–delta conversion. As a mnemonic (memory-aiding device), all we need remember is that when trying to find Y-connected elements (given Δ-connected elements), use Y (admittance) values throughout.

The following rules prevail for both Eqs. (B-1.1)–(B-1.3) and (B-1.4a)–(B-1.6a):

1. The numerators are always the sum of the products of the three known pairs.
2. The denominator is always the diagonally opposite known impedance or admittance.
3. To find the DELTA elements (given Y's), use impedances.
4. To find the WYE elements (given delta impedances), convert these to admittances and solve for respective admittances.

APPENDIX B-2 SECOND- AND THIRD-ORDER DETERMINANTS

A. Two Simultaneous Equations and Second-Order Determinants

Simultaneous equations of the form

$$a_1 x + b_1 y = c_1$$

$$a_2 x + b_2 y = c_2$$

frequently occur in electric circuit theory when solving problems using mesh and nodal analysis and/or the combination of KVL and KCL.

In the foregoing set of simultaneous linear equations, the unknown quantities x and y are called variables. The known quantities, a_1, b_1, c_1, a_2, b_2, and c_2, are called constants or coefficients. We may express the set of coefficients in the form of an *array* arranged in terms of their rows and columns. A *column* is a *vertical* set. A *row* is a *horizontal* set. The array, **A**, for the coefficients for two simultaneous equations has three columns and two rows, arranged as $\mathbf{A} = \begin{vmatrix} a_1 & b_1 & c_1 \\ a_2 & b_2 & c_2 \end{vmatrix}$.

The determinant, **D**, is the array of the coefficients of the two *unknown* quantities. Since there are only two unknowns, we call this array a **second-order** determinant: $\mathbf{D} = \begin{vmatrix} a_1 & b_1 \\ a_2 & b_2 \end{vmatrix}$. Note that the determinant of this equation has two columns and two rows, producing a 2 × 2 (two-by-two) array or matrix.

By definition, the value of **D** in terms of its known coefficients (constants) is $\mathbf{D} = a_1 b_2 - a_2 b_1$. In determining the values of x and y, note that **D** ALWAYS appears in the *denominator*.

Rule 1. The determinant, **D**, always appears in the denominator in solving for variables with determinants.

Since x is the variable in the first column of array A, we eliminate this column from the numerator solution of x, leaving columns of c in place of a:

$$x = \frac{\begin{vmatrix} c_1 & b_1 \\ c_2 & b_2 \end{vmatrix}}{\begin{vmatrix} a_1 & b_1 \\ a_2 & b_2 \end{vmatrix}} = \frac{\begin{vmatrix} c_1 & b_1 \\ c_2 & b_2 \end{vmatrix}}{\mathbf{D}} = \frac{b_2c_1 - b_1c_2}{a_1b_2 - a_2b_1} \quad \text{using rules 1 through 5 inclusive}$$

Since y is the variable in the second column of array **A**, we eliminate this column from the numerator solution of y, leaving columns of c in place of b:

$$y = \frac{\begin{vmatrix} a_1 & c_1 \\ a_2 & c_2 \end{vmatrix}}{\begin{vmatrix} a_1 & b_1 \\ a_2 & b_2 \end{vmatrix}} = \frac{\begin{vmatrix} a_1 & c_1 \\ a_2 & c_2 \end{vmatrix}}{\mathbf{D}} = \frac{a_1c_2 - a_2c_1}{a_1b_2 - a_2b_1} \quad \text{using rules 1 through 5, inclusive}$$

Rule 2. In expanding an array to determine its solution we ALWAYS multiply diagonals from LEFT to RIGHT.

Rule 3. Diagonals taken in a DOWNWARD direction are POSITIVE.

Rule 4. Diagonals taken in an UPWARD direction are NEGATIVE.

Rule 5. Only one constant appears from each column and only one constant appears from each row, whenever an array is expanded in its *respective terms.*

EXAMPLE 1

Using second-order determinants, solve for unknown currents I_1 and I_2, given the following pair of simultaneous linear equations:

$$2.6I_1 + 0.6I_2 = 7.2 \qquad \textbf{(1)}$$

$$2.0I_1 - 0.3I_2 = 6.3 \qquad \textbf{(2)}$$

Step 1. Write the complete array:

$$\begin{matrix} & I_1 & I_2 & V \\ \mathbf{A} = & \begin{vmatrix} 2.6 & 0.6 & 7.2 \\ 2.0 & -0.3 & 6.3 \end{vmatrix} \end{matrix}$$

Step 2. Write the determinant, **D**, as the array of coefficients of the variables and evaluate it using rules 2 to 4:

$$\mathbf{D} = \begin{vmatrix} 2.6 & 0.6 \\ 2.0 & -0.3 \end{vmatrix} = (2.6)(-0.3) - (2)(0.6) = -1.98$$

Step 3. Write the value of I_1 and evaluate it using rules 1 to 5:

$$I_1 = \frac{\begin{vmatrix} 7.2 & 0.6 \\ 6.3 & -0.3 \end{vmatrix}}{\mathbf{D}} = \frac{(7.2)(-0.3) - (6.3)(0.6)}{-1.98} = \frac{-5.94}{-1.98} = \mathbf{3.00\ A}$$

Step 4. Write the value of I_2 and evaluate it using rules 1 to 5:

$$I_2 = \frac{\begin{vmatrix} 2.6 & 7.2 \\ 2.0 & 6.3 \end{vmatrix}}{\mathbf{D}} = \frac{2.6 \times 6.3 - 2 \times 7.2}{-1.98} = \frac{3.96}{-1.98} = \mathbf{-2.00\ A}$$

B. Three Simultaneous Equations and Third-Order Determinants

Three simultaneous equations frequently occur in the form

$$\begin{aligned} a_1x + b_1y + c_1z &= k_1 \\ a_2x + b_2y + c_2z &= k_2 \\ a_3x + b_3y + c_3z &= k_3 \end{aligned} \qquad \text{having an array } \mathbf{A} = \begin{vmatrix} a_1 & b_1 & c_1 & k_1 \\ a_2 & b_2 & c_2 & k_2 \\ a_3 & b_3 & c_3 & k_3 \end{vmatrix}$$

with four columns and three rows.

The determinant, **D**, is the array of the coefficients of the three unknown quantities and is called a *third-order* determinant, having a 3 × 3 matrix array:

$$\mathbf{D} = \begin{vmatrix} a_1 & b_1 & c_1 \\ a_2 & b_2 & c_2 \\ a_3 & b_3 & c_3 \end{vmatrix}$$

There are two alternative methods for evaluating the determinant **D** in terms of the known coefficients in the array.

Method 1. Algebraic Expansion of a Third-Order Determinant
A third-order determinant can be expanded into three second-order determinants having the following form:

$$\mathbf{D} = a_1 \begin{vmatrix} b_2 & c_2 \\ b_3 & c_3 \end{vmatrix} - b_1 \begin{vmatrix} a_2 & c_2 \\ a_3 & c_3 \end{vmatrix} + c_1 \begin{vmatrix} a_2 & b_2 \\ a_3 & b_3 \end{vmatrix}$$

$$= a_1b_2c_3 - a_1b_3c_2 - b_1a_2c_3 + b_1a_3c_2 + c_1a_2b_3 - c_1a_3b_2$$

Note:

1. In expanding the three terms, rules 2 through 5 still apply.
2. The product coefficients in the expression are all selected from the first row (a_1, $-b_1$, c_1) but the second, b_1, term is negative.
3. Each of the second-order determinants is chosen from the second and third rows, omitting the column containing the product coefficient.

Method 2. Diagonal Method
The diagonal method should yield the same value of **D** as method 1 using determinant expansion by arrays. It is usually convenient to write the array using rules 2 through 5 as

$$\mathbf{D} = \begin{vmatrix} a_1 & b_1 & c_1 & a_1 & b_1 \\ a_2 & b_2 & c_2 & a_2 & b_2 \\ a_3 & b_3 & c_3 & a_3 & b_3 \end{vmatrix}$$

$$= a_1b_2c_3 + b_1c_2a_3 + c_1a_2b_3 - a_3b_2c_1 - b_3c_2a_1 - c_3a_2b_1$$

Note:

1. Expanding the array to five columns enables us to find all downward (positive) and upward (negative) diagonals in the array.
2. A total of six terms appears in the evaluation of **D** from known coefficients.

We have just evaluated the determinant **D**, which always appears in the denominator by rule 1. We turn to the evaluation of variables x, y, z, which were stated in the given three simultaneous equations.

As before, since x is the variable in the *first* column of array **A**, we eliminate this column from the *numerator* solution for x, leaving columns of k instead:

$$x = \frac{\begin{vmatrix} k_1 & b_1 & c_1 \\ k_2 & b_2 & c_2 \\ k_3 & b_3 & c_3 \end{vmatrix}}{\begin{vmatrix} a_1 & b_1 & c_1 \\ a_2 & b_2 & c_2 \\ a_3 & b_3 & c_3 \end{vmatrix}} = \frac{\begin{vmatrix} k_1 & b_1 & c_1 \\ k_2 & b_2 & c_2 \\ k_3 & b_3 & c_3 \end{vmatrix}}{\mathbf{D}}$$

$$= \frac{k_1b_2c_3 + b_1c_2k_3 + c_1k_2b_3 - k_3b_2c_1 - b_3c_2k_1 - c_3k_2b_1}{\mathbf{D}}$$

Similarly, since y is the variable in the *second* column of array **A**, we eliminate this column from the *numerator* solution for y, leaving columns of k instead:

$$y = \frac{\begin{vmatrix} a_1 & k_1 & c_1 \\ a_2 & k_2 & c_2 \\ a_3 & k_3 & c_3 \end{vmatrix}}{\mathbf{D}} = \frac{a_1k_2c_3 + k_1c_2a_3 + c_1a_2k_3 - a_3k_2c_1 - k_3c_2a_1 - c_3a_2k_1}{\mathbf{D}}$$

Finally, since z is the variable in the *third* column of array **A**, we eliminate this column from the numerator solution for z, leaving columns of k instead:

$$z = \frac{\begin{vmatrix} a_1 & b_1 & k_1 \\ a_2 & b_2 & k_2 \\ a_3 & b_3 & k_3 \end{vmatrix}}{\mathbf{D}} = \frac{a_1b_2k_3 + b_1k_2a_3 + k_1a_2b_3 - a_3b_2k_1 - b_3k_2a_1 - k_3a_2b_1}{\mathbf{D}}$$

It should be noted that method 2 (the diagonal method) for expansion of determinants holds only for second- and third-order determinants.

EXAMPLE 2

Using third-order determinants, solve for unknown currents I_1, I_2, and I_3 given the following set of three simultaneous linear equations:

$$17I_1 - 15I_2 + 0 = 23$$

$$-15I_1 + 19.2I_2 - 1.2I_3 = 0$$

$$0 - 1.2I_2 + 4.2I_3 = -12$$

$$\text{Array, } \mathbf{A} = \begin{array}{c} I_1 \quad I_2 \quad I_3 \quad V \\ \left|\begin{matrix} 17 & -15 & 0 & 23 \\ -15 & 19.2 & -1.2 & 0 \\ 0 & -1.2 & 4.2 & -12 \end{matrix}\right| \end{array}$$

$$\text{Determinant, } \mathbf{D} = \begin{vmatrix} 17 & -15 & 0 \\ -15 & 19.2 & -1.2 \\ 0 & -1.2 & 4.2 \end{vmatrix} = 1370.88 - (969.48)$$

$$= \mathbf{401.4}$$

From array **A**,

$$I_1 = \frac{\begin{vmatrix} 23 & -15 & 0 \\ 0 & 19.2 & -1.2 \\ -12 & -1.2 & 4.2 \end{vmatrix}}{\mathbf{D}} = \frac{1638.72 - (33.12)}{401.4} = \mathbf{4\ A}$$

$$I_2 = \frac{\begin{vmatrix} 17 & 23 & 0 \\ -15 & 0 & -1.2 \\ 0 & -12 & 4.2 \end{vmatrix}}{\mathbf{D}} = \frac{0 - (244.8 - 1449)}{401.4} = \mathbf{3\ A}$$

$$I_3 = \frac{\begin{vmatrix} 15 & -15 & 23 \\ -15 & 19.2 & 0 \\ 0 & -1.2 & -12 \end{vmatrix}}{\mathbf{D}} = \frac{(-3916.8 + 414) - (-2700)}{401.4}$$

$$= \frac{-802.8}{401.4} = \mathbf{-2\ A}$$

C. Four Simultaneous Equations and Fourth-Order Determinants

Given four simultaneous equations of the form

$$\begin{aligned} a_1x + b_1y + c_1z + d_1p &= k_1 \\ a_2x + b_2y + c_2z + d_2p &= k_2 \\ a_3x + b_3y + c_3z + d_3p &= k_3 \\ a_4x + b_4y + c_4z + d_4p &= k_4 \end{aligned} \quad \text{then the array } \mathbf{A} = \begin{vmatrix} a_1 & b_1 & c_1 & d_1 & k_1 \\ a_2 & b_2 & c_2 & d_2 & k_2 \\ a_3 & b_3 & c_3 & d_3 & k_3 \\ a_4 & b_4 & c_4 & d_4 & k_4 \end{vmatrix}$$

$$\text{and the determinant } \mathbf{D} = \begin{vmatrix} a_1 & b_1 & c_1 & d_1 \\ a_2 & b_2 & c_2 & d_2 \\ a_3 & b_3 & c_3 & d_3 \\ a_4 & b_4 & c_4 & d_4 \end{vmatrix}$$

We can no longer use diagonal method 2, but we can use the algebraic expansion method of **four** third-order determinants of the following form, described for method 1 above:

$$\mathbf{D} = a_1\begin{vmatrix} b_2 & c_2 & d_2 \\ b_3 & c_3 & d_3 \\ b_4 & c_4 & d_4 \end{vmatrix} - b_1\begin{vmatrix} a_2 & c_2 & d_2 \\ a_3 & c_3 & d_3 \\ a_4 & c_4 & d_4 \end{vmatrix} + c_1\begin{vmatrix} a_2 & b_2 & d_2 \\ a_3 & b_3 & d_3 \\ a_4 & b_4 & d_4 \end{vmatrix} - d_1\begin{vmatrix} a_2 & b_2 & c_2 \\ a_3 & b_3 & c_3 \\ a_4 & b_4 & c_4 \end{vmatrix}$$

and expanding yields

$$\begin{aligned} \mathbf{D} = {} & a_1b_2c_3d_4 + a_1c_2d_3b_4 + a_1d_2b_3c_4 \\ & - a_1b_4c_3d_2 - a_1c_4d_3b_2 - a_1d_4b_3c_2 && \text{(first term expanded)} \\ & - b_1a_2c_3d_4 - b_1c_2d_3a_4 - b_1d_2a_3c_4 \\ & + b_1a_4c_3d_2 + b_1c_4d_3a_2 + b_1d_4a_3c_2 && \text{(second term expanded)} \\ & + c_1a_2b_3d_4 + c_1b_2d_3a_4 + c_1d_2a_3b_4 \\ & - c_1a_4b_3d_2 - c_1b_4d_3a_2 - c_1d_4a_3b_2 && \text{(third term expanded)} \\ & - d_1a_2b_3c_4 - d_1b_2c_3a_4 - d_1c_2a_3b_4 \\ & + d_1a_4b_3c_2 + d_1b_4c_3a_2 + d_1c_4a_3b_2 && \text{(fourth term expanded)} \end{aligned}$$

With respect to this expansion, and indeed *all* the foregoing expansions, note that

1. Rule 5 obtains in that only one constant appears from each column and only one constant appears from each row in the expansion. This enables a quick check of the accuracy of the expansion, since the *superscript* letters can contain a, b, c, d only once. Similarly, the *subscript* numbers may contain (1, 2, 3, 4) only once.
2. A total of 24 terms are needed merely to find the *denominator* for variables x, y, z, and p in the above four simultaneous equations.

The numerators for evaluating x, y, z, and p are found in the same (classic) manner that was used for third-order determinants, by

1. Eliminating the column in array **A** containing the unknown variable and substituting column k in its place.
2. Using the expansion technique containing 24 terms to evaluate the numerator as a set of four third-order determinants.

It is obvious that the procedure just described is not only tedious but also easily subject to possibility of error, particularly when the coefficients are complex (of the form $a + jb$). Consequently, it is the author's recommendation that whenever simultaneous linear equations containing *more than three* unknowns are to be solved, computer solutions are required using programs written for such solution.

APPENDIX B-3.1 ENERGY STORED IN THE ELECTRIC FIELD OF A CAPACITOR

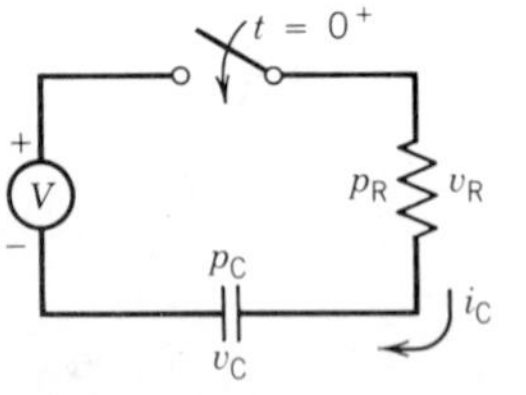

Figure B-3.1 RC circuit

The work done in charging a capacitor may be visualized by some force acting to pull electrons from the positive (neutral) plate of the capacitor and push them onto the negative plate, thereby accomplishing the charge separation typical of a charged capacitor.

If at time, t, a charge $Q'(t)$ has been transferred from one plate to another, the potential difference between the plates, $V(t)$, is

$$V(t) = Q'(t)/C \qquad (9\text{-}1)$$

Now assume that an incremental additional quantity of charge (dq') is transferred. The extra amount of work needed to accomplish this is

$$dW = V\,dq = \frac{Q'(t)}{C} \times dq'$$

If this process is continued until a total charge Q has been transferred, then the total work is

$$W = \int dW = \int_0^q \frac{Q'}{C}\,dq' = \frac{1}{2}\frac{Q^2}{C} \qquad (9\text{-}2)$$

Substituting $Q = CV$ in Eq. (9-2) once and also as a term squared yields

$$W = \frac{Q^2}{2C} = \frac{VQ}{2} = \frac{CV^2}{2} \qquad \text{joules (J)} \qquad (9\text{-}2)$$

where Q is the final charge on the capacitor
V is the final voltage across the capacitor, also shown as V_C

APPENDIX B-3.2 POWER DISSIPATED BY AN *RC* CIRCUIT DURING VOLTAGE BUILDUP AND ENERGY STORAGE IN THE CAPACITOR OF AN *RC* CIRCUIT

Figure B-3.1 shows an *RC* circuit in which the capacitor is initially uncharged, that is, $v_{C_0} = 0$. When the switch is closed at time $t = 0^+$, the voltage v_C across the capacitor *rises* exponentially, as derived in Appendix B-5,

$$v_C = V(1 - \varepsilon^{-t/RC}) = V(1 - \varepsilon^{-t/\tau}) \qquad \text{volts (V)} \qquad (B\text{-}5.5)$$

Simultaneously, the current, i_C, in the capacitor *decays* exponentially in accordance with

$$i_C = \frac{V}{R}(\varepsilon^{-t/RC}) = I_0\varepsilon^{-t/\tau} \qquad \text{amperes (A)} \qquad (10\text{-}3)$$

Since power is the product of voltage and current, the instantaneous power, p_C, delivered to the capacitor by the supply is

$$p_C = v_C \cdot i_C = V(1 - \varepsilon^{-t/\tau}) \times \frac{V}{R}\varepsilon^{-t/\tau} = \frac{V^2}{R}(\varepsilon^{-t/\tau} - \varepsilon^{-2t/\tau}) \qquad (B\text{-}3.1)$$

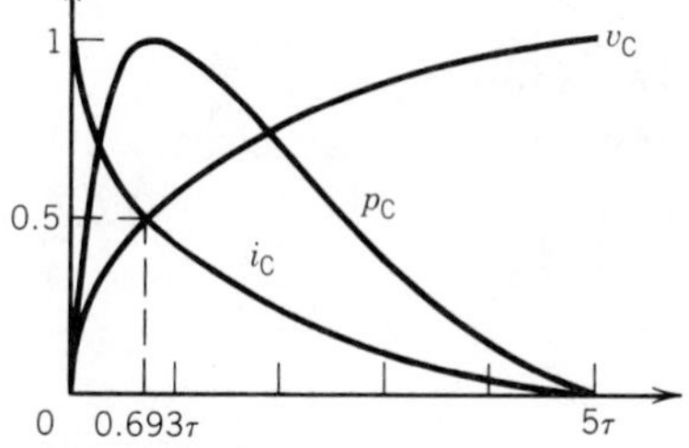

Figure B-3.2 Voltage, current, and power relations

The waveform of Eq. (B-3.1) for p_C is shown in **Fig. B-3.2** along with those of v_C and i_C, on whose product it depends. Observe from Fig. B-3.2 that

1. p_C is a maximum where v_C and i_C cross each other.
2. p_C is a maximum where $t \cong 0.693\tau$.
3. When $t = 0.693\tau$, $v_C = V/2$ and $i_C = I_0/2$.

These relations may be derived and proved by differentiating Eq. (B-3.1) with respect to time and then setting it equal to zero and solving for t, as follows:

$$p_C = \frac{V^2}{R}(\varepsilon^{-t/\tau} - \varepsilon^{-2t/\tau})$$

$$\frac{dp_C}{dt} = \frac{V^2}{R}\left(\frac{-\varepsilon^{-t/\tau}}{\tau} + \frac{2\varepsilon^{-2t/\tau}}{\tau}\right) = \frac{V^2}{R\tau}(2\varepsilon^{-2t/\tau} - \varepsilon^{-t/\tau}) = 0$$

Dividing both sides by $V^2/R\tau$ yields $\varepsilon^{-t/\tau} = 2\varepsilon^{-2t/\tau}$ and taking the natural logarithm of each side yields $-t/\tau = \ln 2 - 2t/\tau$ or $t/\tau = \ln 2$, and so for any RC circuit

$$t = \tau \ln 2 = 0.693\tau \qquad \text{(B-3.2)}$$

When $t = 0.693\tau$, solving for v_C and i_C in their respective exponential equations yields

$$v_C = V(1 - \varepsilon^{-0.693\tau/\tau}) = \frac{V}{2} \quad \text{and} \quad i_C = I_0\varepsilon^{-0.693\tau/\tau} = \frac{I_0}{2} \qquad \text{(B-3.3)}$$

This enables us to write the equation for $p_{C(\max)}$ as

$$p_{C(\max)} = \frac{V}{2} \times \frac{I_0}{2} = \frac{VI_0}{4} = \frac{V^2}{R}(\varepsilon^{-0.693} - \varepsilon^{-1.3863}) = \frac{V^2}{4R} = \frac{I_0^2R}{4} \qquad \text{(B-3.4)}$$

It is most important for the reader to realize that p_C is *not* the power dissipated by R in the RC circuit. It is the power drawn from the supply and delivered *to the capacitor* in order to store energy, W_C, in the capacitor, as derived in Appendix B-3.1. The power dissipated by the resistor in Fig. B-3.1, p_R, during the charging process and as a result of charge transfer to the capacitor is

$$p_R = i_C^2R = (I_0\varepsilon^{-t/\tau})^2R = I_0^2R \cdot \varepsilon^{-2t/\tau} = \left(\frac{V^2}{R}\right)(\varepsilon^{-2t/\tau}) \quad \text{watts (W)} \qquad \text{(B-3.5)}$$

Note that when $t = 0^+$ as the switch is first closed, this power is a maximum and has an instantaneous value of

$$p_{R(\max)} = I_0^2R = \frac{V^2}{R} \qquad \text{(B-3.6)}$$

Note that $p_{R(\max)}$ is four times as large as $p_{C(\max)}$ for any RC circuit.

Appendix B-3.1 derived the total (final) energy stored by a capacitor as $W = CV^2/2$. Since v_C *rises* exponentially from 0 to its final value, V, the energy, w_C, transferred to the capacitor at each instant is

$$\begin{aligned} w_C &= \left(\frac{C}{2}\right)(v_C)^2 = \left(\frac{C}{2}\right)[V(1 - \varepsilon^{-t/\tau})]^2 \\ &= \frac{CV^2}{2}(1 - \varepsilon^{-t/\tau})^2 = \frac{CV^2}{2}(1 - 2\varepsilon^{-t/\tau} + \varepsilon^{-2t/\tau}) \end{aligned} \qquad \text{(B-3.7)}$$

Equation (B-3.7) shows that when $t = 0^+$, $w_C = 0$. When $t = 5\tau$, $w_C = CV^2/2$, which agrees with our derivation (Appendix B-3.1) for the final energy stored in a capacitor that is fully charged.

Figure B-3.3 summarizes and shows the variations of p_C, w_C, and p_R as a function of time when the switch is closed at $t = 0^+$. Note that p_C is a maximum at approximately 0.7τ, whereas p_R is a maximum at $t = 0^+$ s and is four times as great as the maximum value of p_C. Observe that the exponential decay of p_R is twice as fast as that of i_C.

Also observe that w_C, the energy stored in the capacitor, rises exponentially from 0 to its final value of $CV^2/2$.

For comparison, the equations for v_C, i_C, p_C, p_R, p_T, and w_C are all shown in Fig. B-3.3. One item that may not be obvious is that p_R is greater than p_C from 0 to approximately 0.7τ, where $p_R = p_C$. Beyond that time, p_C is always greater than p_R.

$$v_C = V(1 - \varepsilon^{-t/\tau})$$

$$i_C = I_0\, \varepsilon^{-t/\tau}$$

$$p_C = \frac{V^2}{R}(\varepsilon^{-t/\tau} - \varepsilon^{-2t/\tau})$$

$$p_R = \frac{V^2}{R}(\varepsilon^{-2t/\tau})$$

$$w_C = \frac{CV^2}{2}(1 - \varepsilon^{-t/\tau} + \varepsilon^{-2t/\tau})$$

$$p_T = \frac{V^2}{R}\varepsilon^{-t/\tau}$$

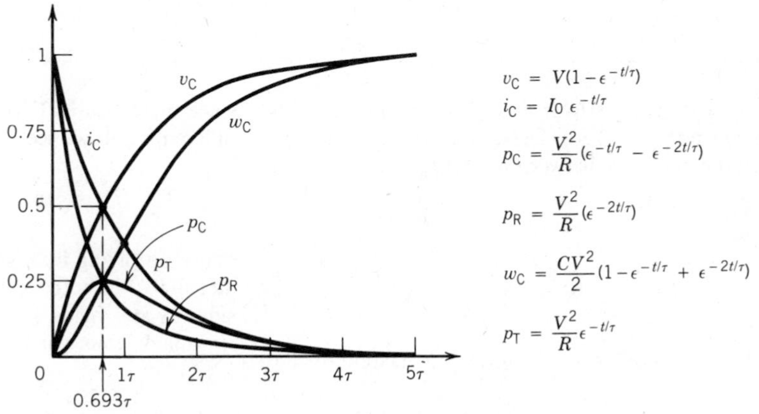

Figure B-3.3   Summary of variations of all parameters with time in charging an *RC* circuit (capacitor initially discharged)

It is obvious that the total power dissipated by the *RC* circuit is at all times $p_T = p_R + p_C$. p_T is a maximum at $t + 0^+$, when it is equal to p_R or V^2/R. Beyond that time, it decreases exponentially to zero in accordance with

$$p_T = p_R + p_C = \frac{V^2}{R}(\varepsilon^{-2t/\tau}) + \frac{V^2}{R}(\varepsilon^{-t/\tau} - \varepsilon^{-2t/\tau}) = \frac{V^2}{R}\varepsilon^{-t/\tau} \quad \text{(B-3.8)}$$

Note that the curve for p_T is the same as that for i_C in Fig. B-3.3.

Now that we have discovered and derived equations for instantaneous values of p_C, p_R, and p_T, we can use them to derive other interesting relationships, as shown by the following four examples.

EXAMPLE 1

In the curve of Fig. B-3.3, derive an expression for the time, t, at which $p_C = 2p_R$

Solution

When $p_C = 2p_R$, substituting Eqs. (B-3.1) and (B-3.5) yields

$$\frac{V^2}{R}(\varepsilon^{-t/\tau} - \varepsilon^{-2t/\tau}) = 2(V^2/R)(\varepsilon^{-2t/\tau}),$$

and canceling V^2/R on both sides yields $\varepsilon^{-t/\tau} = 3\varepsilon^{-2t/\tau}$. Taking the ln of each side produces $-t/\tau = \ln 3 + -2t/\tau$ or $t/\tau = \ln 3$, from which $t = (\textbf{ln 3})\tau = \textbf{1.0986}\tau$

Note that this solution agrees with our observation that above 0.693τ, p_C is always *greater* than p_R.

EXAMPLE 2

In the curve of Fig. B-3.3, derive an expression for the time, t, at which $p_R = 5p_C$

Solution

When $p_R = 5p_C$, then substituting Eqs. (B-3.1) and (B-3.5) yields

$$\frac{V^2}{R}\varepsilon^{-2t/\tau} = 5(\varepsilon^{-t/\tau} - \varepsilon^{-2t/\tau})V^2/R$$

and dividing both sides by V^2/R yields $5\varepsilon^{-t/\tau} = 6\varepsilon^{-2t/\tau}$; dividing both sides by 5, $\varepsilon - t/T = 1.2\varepsilon^{-2t/\tau}$, and then taking ln of each side produces $-t/\tau = \ln 1.2 - 2t/\tau$ or $t/\tau = \ln 1.2$, from which $t = \mathbf{(ln\ 1.2)}\boldsymbol{\tau} = \mathbf{0.1823}\boldsymbol{\tau}$

Note that this solution also agrees with our observation that below 0.693τ, p_C is always *less* than p_R.

EXAMPLE 3

In Fig. B-3.3, $p_C = p_R$ at $t = 0.693\tau$ and also at approximately 5τ, where both p_C and p_R are zero. Derive an expression for the time, t, at which the difference between p_C and p_R is a maximum.

Solution

$$p_C - p_R = \frac{V^2}{R}(\varepsilon^{-t/\tau} - \varepsilon^{-2t/\tau}) - \frac{V^2}{R}\varepsilon^{-2t/\tau} = \frac{V^2}{R}(\varepsilon^{-t/\tau} - 2\varepsilon^{-2t/\tau})$$
$$= \Delta p_{CR}$$

Differentiating the term $d(p_C - p_R)/dt$ and setting it equal to zero yields

$$\frac{dp_{CR}}{dt} = \frac{d(\varepsilon^{-t/\tau} - 2\varepsilon^{-2t/\tau})}{dt} = 0,$$

since the constant V^2/R may be factored out. Carrying out the differentiation we see

$$-\frac{1}{\tau}\varepsilon^{-t/\tau} - 2\left(\frac{-2}{\tau}\right)\varepsilon^{-2t/\tau} = 0 \quad \text{or} \quad -\frac{1}{\tau}\varepsilon^{-t/\tau} + \frac{4}{\tau}\varepsilon^{-2t/\tau} = 0,$$

and multiplying by τ yields $-\varepsilon^{-t/\tau} + 4\varepsilon^{-2t/\tau} = 0$. Now we may add $+\varepsilon^{-t/\tau}$ to both sides so that $4\varepsilon^{-2t/\tau} = \varepsilon^{-t/\tau}$; multiplying both sides by $\varepsilon^{2t/\tau}$ yields $4 = \varepsilon^{t/\tau}$ and taking the ln of both sides yields $t/\tau = \ln 4$, from which

$$t = \tau \ln 4 = 1.3863\tau \qquad \textbf{(B-3.9)}$$

The above relation is shown in Fig. 10-12. Note that the difference $p_C - p_R$ is a maximum when $t = \tau \ln 4 = 1.3863\tau$, as proved by the above derivation.

EXAMPLE 4

Using the value of t obtained in Eq. (B-3.9), express $(p_C - p_R)_{MAX}$ as a ratio of V^2/R

Solution

$p_C = (V^2/R)(\varepsilon^{-t/\tau} - \varepsilon^{-2t/\tau})$ and $p_R = (V^2/R)\varepsilon^{-2t/\tau}$; then at any instant, $p_C - p_R = (V^2/R)(\varepsilon^{-t/\tau} - 2\varepsilon^{-2t/\tau})$. But since the maximum difference occurs when $t = 1.3863\tau$ as derived in Eq. (B-3.9), we obtain by substituting this value for t

$$(p_C - p_R)_{MAX} = \frac{V^2}{R}(\varepsilon^{-1.3863} - 2\varepsilon^{-2.7726}) = \frac{V^2}{R}(0.25 - 0.125)$$
$$= \frac{V^2}{R}(0.125) = \frac{\mathbf{V^2}}{\mathbf{8\,R}} \qquad \textbf{(B-3.10)}$$

APPENDIX B-4 EXPONENTIAL FUNCTIONS USED IN TRANSIENT WAVEFORMS

The exponential function is a function that varies exponentially with time. It has the form

$$y = f(t) = K\varepsilon^{st} \qquad \text{(B-4.1)}$$

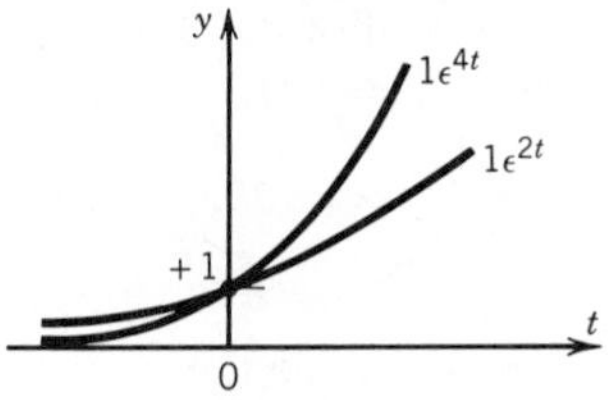

Figure B-4.1 $K = 1$, $s = 2$ and 4

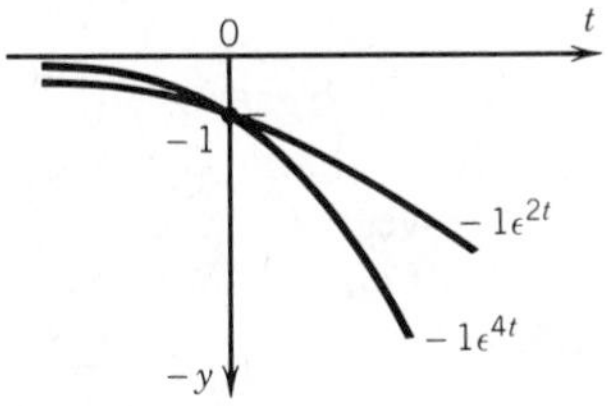

Figure B-4.2 $K = -1$, $s = 2$ and 4

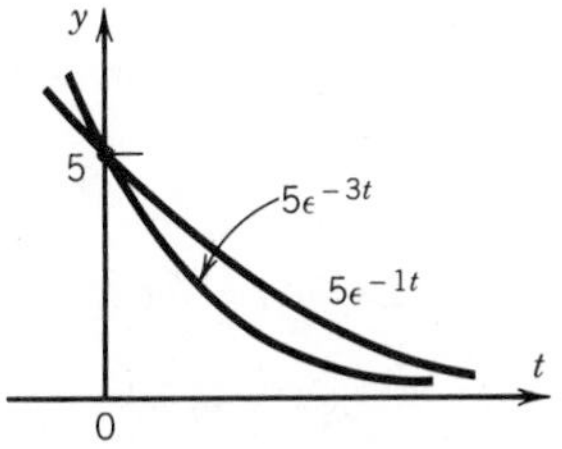

Figure B-4.3 $K = +5$, $s = -1$ and -3

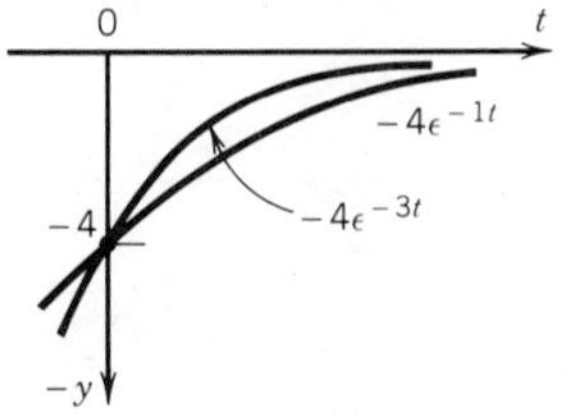

Figure B-4.4 $K = -4$, $s = -1$ and -3

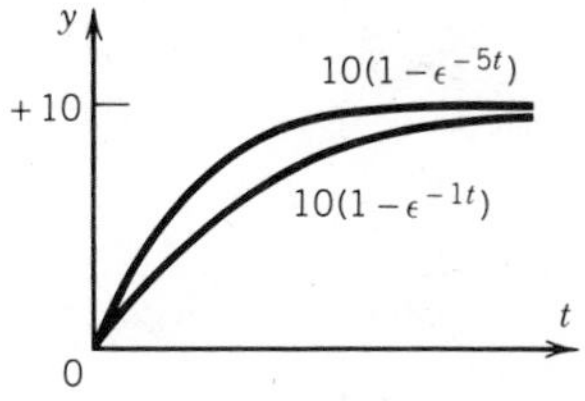

Figure B-4.5 $K = +10$, $s = -1$ and -5

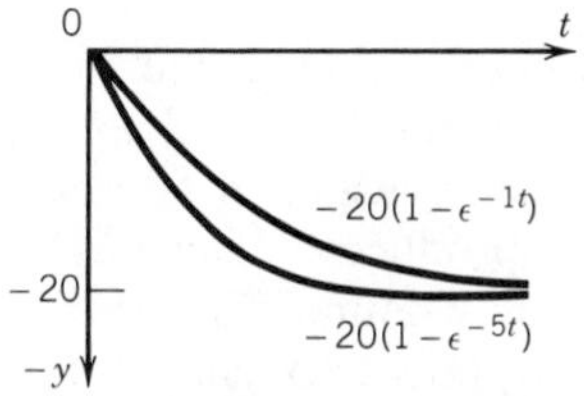

Figure B-4.6 $K = -20$, $s = -1$ and -5

where K is the amplitude of the function (when $t = 0$)
s is the exponential constant of the function
ε is the base of the Napierian or natural logarithmic system,[1] a constant equal to $2.718\,281\,\overline{828}\ldots$

In Eq. (B-4.1), K may be either positive or negative and s may be positive or negative. If s is **positive** we have an *expanding* exponential function, i.e., a function that becomes larger as time increases.

If s is **negative**, we have a *decaying* exponential function, i.e., a function that becomes smaller and decays to zero as time increases.

The following tabulation and the associated figures summarize the graphical plots of y versus time for various positive and negative values of K and s.

K	s	Expansion or decay	Figure
+	+	Expansion to $+\infty$	**B-4.1**
−	+	Expansion to $-\infty$	**B-4.2**
+	−	Decay from + value to 0	**B-4.3**
−	−	Decay from − value to 0	**B-4.4**

Note from Fig. B-4.1 that when $t = 0$, each function crosses the y axis at $K = +1$. Also note that because s is positive, both plots expand to infinity positively. The *greater* the positive value of s, the *higher* the *rate of exponential rise.*

Note from Fig. B-4.2 that because s is still positive, we have two expanding exponential functions. But because K is negative, the expansion is to minus infinity. Again, the *higher* rate of exponential rise occurs when s is a *higher* value.

From Fig. B-4.3, note that when s is negative we have a decaying exponential whose asymptotic value is zero. Again, the *rate* of exponential decay is *greater* for the *higher* value of $-s$.

From Fig. B-4.4, note once again that negative values of both K and s produce decays from negative values of y to zero. Note that when $t = 0$, $y = -4$ because $K = -4$. Lastly, note that the *higher* value of $-s$ produces a *greater* rate of exponential decay.

A second expanding exponential function of importance has the form

$$y = f(t) = \pm K(1 - \varepsilon^{-st}) \qquad \text{(B-4.2)}$$

where K is the amplitude of the function
s is the exponential constant of the function
ε is the base of the Napierian logarithmic system, the constant 2.718 281 8

In Eq. (B-4.2) K may be either positive or negative. When K is positive, the POSITIVE asymptotic value of y, for $t = \infty$, is $+K$ as shown in **Fig. B-4.5**. Conversely, when $t = 0$, the parenthetic expression is zero and $y = 0$, as shown. The variation with s is the same as in Eq. (B-4.1); i.e., the higher the value of s, the higher the rate of exponential rise toward the positive asymptotic value of K.

When K is negative, the NEGATIVE asymptotic value of y, for $t = \infty$, is $y = -K$, as shown in **Fig. B-4.6**. Conversely, when $t = 0$, $y = 0$.

The variation with s is the same. The *greater* the value of s, the *higher* the rate of exponential rise.

[1] To obtain ε as a constant on your electronic calculator insert [1] [ε^x] and read above value.

One final point is worth mentioning: s is always the reciprocal of the time constant τ. In capacitive circuits, $s = 1/RC$. In inductive circuits, $s = 1/(L/R) = R/L$.

APPENDIX B-5 DERIVATION OF EQUATIONS RELATED TO CHARGE AND DISCHARGE OF CAPACITORS IN *RC* CIRCUITS

Given a capacitor, C_1, charged to charge Q. C_2 is completely discharged, as shown in **Fig. B-5.1a**. Assume that we now connect a resistance, R, bridging C_1 to C_2, and complete the circuit. Charge Q must distribute itself between C_1 and C_2 to satisfy the relation $Q = Q_1 + Q_2$, by the law of conservation of charge. We know, of course, that $V_1 = Q_1/C_1$ and $V_2 = Q_2/C_2$ as a result of redistribution. But since current no longer flows after redistribution, $V = V_1 = V_2$ and the volt drop across R is zero.

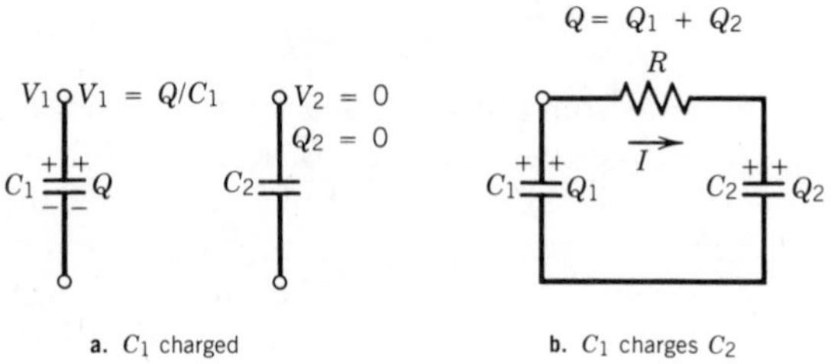

Figure B-5.1 Charge via resistance R

Then simply $V = \dfrac{Q - Q_2}{C_1} = \dfrac{Q_2}{C_2}$ which yields, by solving for Q_2, Q_1, and V, respectively,

$$Q_2 = \frac{C_2}{C_1 + C_2} Q, \qquad Q_1 = \frac{C_1}{C_1 + C_2} Q, \qquad \text{and} \qquad V = \frac{Q}{C_1 + C_2} \tag{B-5.1}$$

The stored energy was originally

$$W_0 = C_1 V_1^2/2 = Q^2/2C_1 = V_1 Q/2 \tag{9-2}$$

and in the final state $W = W_1 + W_2 = Q^2/2(C_1 + C_2)$. This means that there has been an energy loss $W_L = (Q^2/2C_1) - [Q^2/2(C_1 + C_2)]$ or

$$W_L = \frac{Q^2}{2}\left(\frac{1}{C_1} - \frac{1}{C_1 + C_2}\right) = \frac{Q^2}{2C_1}\left(\frac{C_2}{C_1 + C_2}\right) = \frac{C_2}{C_1 + C_2} W_0 \tag{B-5.2}$$

Surprisingly enough, this energy loss in Eq. (B-5.2) is totally independent of resistance, R. (Even if R was zero, the energy loss would be a function strictly of the ratio of C_2 to $C_1 + C_2$.) But the presence of R produces a sustained transient period during which charge distribution occurs, and this produces higher losses.

Let us now begin with charged capacitor, C, and merely connect it across a resistance, R, as shown in Fig. B-5.2. Since I is defined as the rate of charge flow (dQ/dt) and this rate is NEGATIVE because the capacitor is *discharging*, we may write the current during **discharge** as

$$I = -\frac{dQ}{dt} = -C\frac{dv_c}{dt} \qquad \text{(B-5.3)}$$

By Ohm's law, $I = V/R$, and substituting this into Eq. (B-5.3) we get by rearranging

$$\frac{dv_c}{dt} = -\frac{V}{RC} \qquad \text{(B-5.3a)}$$

where V is the voltage across the capacitor initially, or $\frac{dV_c}{V} = -\frac{dt}{RC}$. Integrating the left side with respect to dV_c and the right side with respect to dt for limits of V and t, respectively, yields $\ln(v_c/V) = -t/RC$ and since $\ln x = \varepsilon^x$ we may write $v_c/V = \varepsilon^{-t/RC}$, which when solved for v_c yields

$$v_c = V\varepsilon^{-t/RC} = V\varepsilon^{-t/\tau} \qquad \text{volts (V)} \qquad \text{(B-5.4)}$$

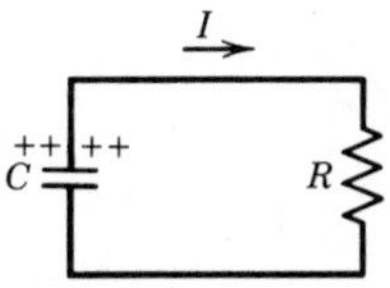

Figure B-5.2 Capacitor discharge via R

where v_c is the capacitor voltage at any instant during transient discharge
V is the supply voltage to which the capacitor was initially charged
RC is the time constant of the circuit in **Fig. B-5.2**
t is any time during the transient in seconds

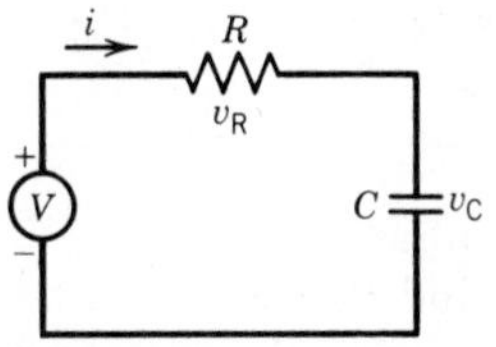

Figure B-5.3 Charging C via R

Now suppose that we charge a capacitor (which has been completely discharged) through a resistor R from a supply V, as shown in **Fig. B-5.3**.

By KVL, at all times during the transient charging period $V = v_R + v_C$ or

$V = iR + v_C = RC\frac{dv_C}{dt} + v_C$ and solving for dv_C/dt yields

$\frac{dv_C}{dt} = -\frac{v_C - V}{RC}$ which by rearrangement of terms yields

$\frac{d(v_C - V)}{dt} = -\frac{(v_C - V)}{RC}$ since V is constant.

This equation has the *same* form as Eq. (B-5.3a) with $(v_C - V)$ as the dependent variable. Hence the solution is (using the same technique) $v_C - V = (v_C - V)\varepsilon^{-t/RC}$, which when solved for v_C yields

$$v_C = V - V\varepsilon^{-t/RC} = V(1 - \varepsilon^{-t/RC}) = V(1 - \varepsilon^{-t/\tau}) \qquad \text{volts (V)} \qquad \text{(B-5.5)}$$

where τ is the product RC in seconds

The charge-sharing problem of Fig. B-5.1b can now be analyzed in light of these transient analysis relations. Since I is a current common to both capacitors C_1 and C_2, we may write

$I = -\dfrac{dQ_1}{dt} = +\dfrac{dQ_2}{dt}$ as C_1 *dis*charges and C_2 charges. But during charge flow $I = (V_1 - V_2)/R$. Eliminating I yields simultaneous equations for V_1 and V_2 as

$$-C_1\frac{dV_1}{dt} = C_2\frac{dV_2}{dt} \quad \textbf{(I)} \qquad \text{and} \qquad -RC_1\frac{dV_1}{dt} = V_1 - V_2 \quad \textbf{(II)}$$

Equation **(I)** can be integrated immediately to yield

$$-C_1V_1 = C_2V_2 + k \quad \textbf{(I)}$$

But since $V_2 = 0$ at $t = 0$ *initially*, the constant $k = -C_1V_{01}$, where V_{01} is the initial value of V_1, across C_1.

Substituting for k and solving for V_2 yields $V_2 = (V_{01} - V_1)C_1/C_2$. If this value of V_2 is substituted into Eq. **(II)**, we obtain

$$-RC_1\frac{dV_1}{dt} = V_1(1 + C_1/C_2) - V_{01}C_1/C_2$$

which, when simplified with least-common denominators (LCDs), yields

$$-\frac{RC_1C_2 \cdot dV_1}{(C_1 + C_2)dt} = V_1 - \frac{V_{01}C_1}{(C_1 + C_2)} \qquad \textbf{(B-5.6)}$$

The above differential equation has a simple exponential solution, which is

$$V_1 = \frac{C_1}{C_1 + C_2}V_{01} + \left(\frac{C_2}{C_1 + C_2}V_{01}\right)(\varepsilon^{-t/RC}) \qquad \textbf{(B-5.7)}$$

where all terms have been defined above except $C = (C_1C_2)/(C_1 + C_2)$.

The first term in Eq. (B-5.7) is the *steady-state* portion and the second is the *transient* portion. When $t = 0$, the exponential portion is unity and the two terms add to $V_1 = V_{01}$, which is as expected, as defined above. When $t = \infty$, the second term of Eq. (B-5.7) is zero and $V_1 = V_{01}(C_1/(C_1 + C_2)$. Note that $\tau = RC$, where C is the equivalent *series* combination of C_1 and C_2, as shown in Fig. B-5.1b. R sees C_1 and C_2 in series by Thévenin's theorem.

Most important, note that when the transient has disappeared and $V_1 = V_{01}(C_1)/(C_1 + C_2)$, this equation is exactly the same as Eq. (9-8) derived in Chapter 9! The equations are identical despite the fact that C_1 and C_2 are in *parallel* in Eq. (9-8).

Frequently Eq. (B-5.5) must be solved for time, t, and we can derive this relation in the following way:

$$\frac{v_C}{V} = 1 - \varepsilon^{-t/\tau} \qquad \text{and} \qquad \varepsilon^{-t/\tau} = 1 - \frac{v_C}{V} = \frac{V - v_C}{V};$$

inverting terms yields $\varepsilon^{t/\tau} = \dfrac{V}{V - v_C}$ and taking the ln of both sides yields $t/\tau = \ln\left(\dfrac{V}{V - v_C}\right)$, which when solved for t yields

$$t = \tau \ln\left(\frac{V}{V - v_C}\right) = RC \ln\left(\frac{V}{V - v_C}\right) \qquad \textbf{(B-5.8)}$$

Equation (B-5.8) may be used to find the time required to charge *any* capacitor from voltage v_C to voltage V, provided v_C is NOT equal to V.

Equation (B-5.5) enables us to find the voltage across a capacitor at any time during its transient rise from zero to a maximum charged steady-state value of V. To find the *rate of change of voltage* across the capacitor at any time t during the transient, we merely differentiate Eq. (B-5.5) and substitute the value of t, or

$$\frac{dv_C}{dt} = \frac{d(V - V\varepsilon^{-t/\tau})}{dt} = 0 - \frac{1}{\tau}(-V\varepsilon^{-t/\tau}) = \frac{V}{\tau}\varepsilon^{-t/\tau} \qquad \textbf{(B-5.9)}$$

Should we desire to find the *rate of change of current* at any time t during the charging (or discharging) period, we may use the same process, or

$$\frac{di_C}{dt} = \frac{d(I_0\varepsilon^{-t/\tau})}{dt} = -\frac{1}{\tau}(I_0\varepsilon^{-t/\tau}) = -\frac{I_0}{\tau}(\varepsilon^{-t/\tau}) \qquad \textbf{(B-5.10)}$$

The insight to be gained from Eqs. (B-5.9) and (B-5.10) is that the rate of change of v_C and i_C at any time, t, during the transient *decreases* from an *initial maximum* rate of change, V_0/τ or I_0/τ, at an exponential decay rate, $\varepsilon^{-t/\tau}$.

APPENDIX B-6.1 DERIVATION OF EQUATION FOR ENERGY STORED IN THE MAGNETIC FIELD OF AN INDUCTOR

At any instant during the time when current is rising in an inductor, the instantaneous power, p, drawn from the supply is

$$p = v_L i \qquad \text{watts or joules/second}$$

During any short time interval, dt, the energy, dw, added to the magnetic field is the product of the power times the time, or

$$dw = v_L i\, dt$$

But $v_L = L\, di/dt$ by definition, and substituting it for v_L yields

$$dw = L\left(\frac{di}{dt}\right) i\, dt = Li\, di \qquad \text{joules}$$

Then the total energy put into the field while i increases from zero to I is

$$W = \int_0^I Li\, di = \left.\frac{Li^2}{2}\right]_0^I = \frac{LI^2}{2} \qquad \text{(12-17)}$$

APPENDIX B-6.2 DERIVATION OF INSTANTANEOUS CURRENT RISE IN AN *RL* CIRCUIT

Given $L\,di/dt = V - iR$ (13-2)

then $\dfrac{di}{V - iR} = \dfrac{dt}{L}$ and multiplying both sides by $-R$ and integrating yields

$\displaystyle\int \frac{-R\,di}{V - iR} = \int \frac{-R\,dt}{L}$ note that the numerator on the left is a differential of the denominator

$\ln(V - iR) = \dfrac{-Rt}{L} + K$ the constant K may be evaluated by inserting values of i and t at $t = 0$. At $t = 0$, $i = 0$ and $\ln V = K$

so $\ln(V - iR) = \dfrac{-Rt}{L} + \ln V$ which yields

$\ln\left(\dfrac{V - iR}{V}\right) = -\dfrac{Rt}{L}$ and taking $\ln^{-1}$ of both sides yields

$\dfrac{(V - iR)}{V} = \varepsilon^{-Rt/L}$ which when solved for i yields

$$i = \frac{V}{R}(1 - \varepsilon^{-Rt/L}) = \frac{V}{R}(1 - \varepsilon^{-t/\tau}) \quad \text{where } \tau = L/R \qquad (13\text{-}3)$$

APPENDIX B-7 AVERAGE AND RMS VALUES OF SINUSOIDAL WAVEFORMS

B-7.1 Average Value of a Half-Cycle of Sine Wave

Given the peak value of the amplitude of either sinusoidal voltage or current as A_m, we wish to find the average value, A_{dc}, of a half-cycle of sine wave, from 0 to π. We may therefore integrate the sine wave to obtain the area under the curve and divide it by its duration, π, or

$$A_{dc} = \frac{1}{\pi}\int_0^{\pi} A_m \sin \omega t\, d(\omega t) = -\frac{1}{\pi} A_m \cos \omega t \Big]_{\omega t = 0}^{\pi}$$
$$= \frac{2}{\pi} A_m = 0.6366 A_m \qquad \text{(B-7.1)}$$

where A_m is the maximum amplitude of either a current or voltage sinusoid.

B-7.2 Average Value of the Output of a Full-Wave Rectifier (FWR)

The rectangular amplitude of the half-cycle average is shown in **Fig. B-7.1a** as derived above, A_{dc}, is $2A_m/\pi$. When full-wave rectification occurs (assuming ideal diodes), the waveform appears as shown in **Fig. B-7.2**. Since each half-cycle has an average value of $2A_m/\pi$, then the average of a full cycle is $2A_m/\pi$ as shown.

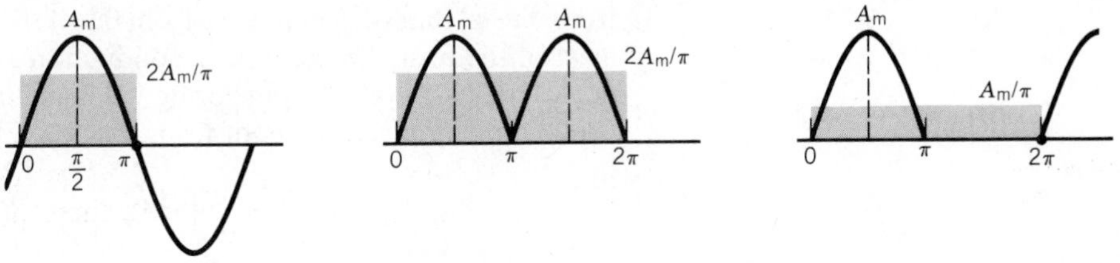

a. Half-cycle average   b. FWR, full-cycle average   c. HWR, full-cycle average

Figure B-7.1 Average values of sinusoids

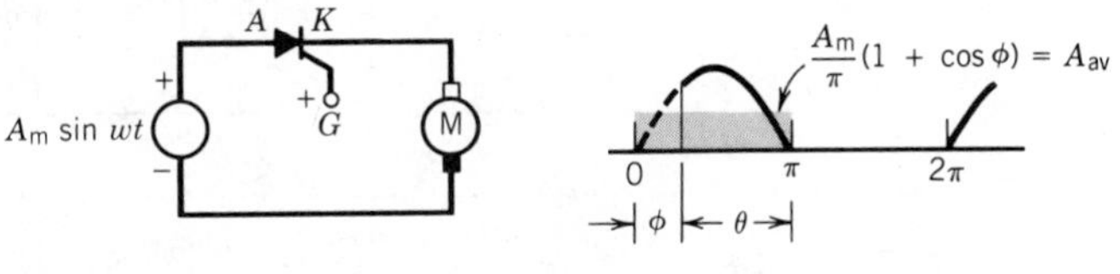

a. HW SCR motor control   b. Waveform showing ϕ and θ

Figure B-7.2 Half-cycle output of half-wave SCR control

B-7.3 Average Value of the Output of a Half-Wave Rectifier (HWR)

A half-wave rectifier conducts for only one half-cycle (i.e., from 0 to π) and blocks the second half-cycle from π to 2π. Since the area under the curve in Fig. B-7.1a is $2A_m$ and this area is now taken over a duration of 2π, then

$$A_{dc} = \frac{2A_m}{2\pi} = \frac{A_m}{\pi} \tag{B-7.2}$$

B-7.4 Average Value of One Quarter-Cycle of Sine Wave, from 0 to $\pi/2$ Radians

In rising from 0 to A_m in a duration of $\pi/2$ radians, from Fig. B-7.1a, the average value, by inspection, **of one quarter-cycle** of sine wave is:

$$A_{dc} = \frac{2A_m}{\pi} \tag{B-7.3}$$

Although Eq. (B-7.3) was derived by inspection, it is proved mathematically in Sec. B-7.5. (See Ex. B-7.1.)

B-7.5 Average Value of the Half-Cycle Output of a Silicon-Controlled Rectifier (SCR) Given Firing Angle, ϕ, and Conduction Angle, θ

A silicon-controlled rectifier (Fig. B-7.2a) conducts only when its anode, A, is positively biased and a positive trigger is applied to its gate, G, at any firing angle from 0 to π; that is, ϕ may vary from 0 to 180°.[1] The conduction angle is the *supplement* of the firing angle, that is, conduction angle $\theta = 180° - \phi = \pi - \phi$. We may derive the half-cycle average or dc value of amplitude as a ratio of maximum amplitude for any given value of firing angle, ϕ, as

$$A_{dc} = A_{hav} = \frac{1}{\pi}\int_{\phi}^{\pi} A_m \sin \omega t \, d(\omega t) = \frac{A_m}{\pi}(-\cos \omega t)\Big]_{\phi}^{\pi}$$

$$A_{dc} = A_{hav} = \frac{A_m}{\pi}(1 + \cos \phi) = \frac{A_m}{\pi}(1 - \cos \theta) \qquad \text{(B-7.4)}$$

EXAMPLE B-7.1

Given $\phi = \theta = 90°$, calculate the average value for one half-cycle of output (duration from 0 to π), using

a. Eq. (B-7.4).
b. Eq. (B-7.3).

Solution

a. $A_{hav} = \frac{A_m}{\pi}(1 + \cos 90°) = \frac{A_m}{\pi}$

b. Since the quarter-cycle average of a sine wave is $2A_m/\pi$, the half-cycle average is (by the area method)

$$A_{hav} = \frac{(2A_m/\pi) \times \pi/2 + (0) \times \pi/2}{\pi} = \frac{A_m}{\pi}$$

EXAMPLE B-7.2

The maximum amplitude of a sine wave is 170 V applied to an SCR, using half-wave control as shown in Fig. B-7.2a. Assuming an ideal SCR, calculate

a. The half-cycle average value if the SCR is fired at an angle of 60°.
b. The average value of dc voltage applied across the motor, M, each cycle (0 to 2π).

Solution

a. $A_{hav} = \frac{A_m}{\pi}(1 + \cos \phi) = \frac{170 \text{ V}}{\pi}(1 + \cos 60°) = 54.11 \times 1.5$

$= \mathbf{81.17\ V}$

b. $A_{dc} = \frac{81.17 \times \pi + 0 \times \pi}{2\pi} = \mathbf{40.58\ V}$

B-7.6 RMS Value of One Cycle of a Pure Sine Wave or Output of the Full-Wave Rectifier (FWR)

By definition, the RMS value is the square root of the sum of the individual instantaneous amplitudes (a_1, a_2, a_3, etc.) taken over an entire cycle, or

$$A_{rms} = \sqrt{\frac{a_1^2 + a_2^2 + a_3^2 + \cdots + a_n^2}{T}} = \sqrt{\frac{1}{T}\int_0^t a^2 \, dt}\text{; squaring both sides,}$$

$$A_{rms}^2 = \frac{1}{T}\int_0^T (A_m \sin \omega t)^2 \, dt = \frac{A_m^2}{2\pi}\int_0^{2\pi} \sin^2 \theta \, d\theta\text{; but } \sin^2 \theta = \frac{1 - \cos 2\theta}{2}\text{, so}$$

$$= \frac{A_m^2}{2\pi}\int_0^{2\pi}\left(\frac{1 - \cos 2\theta}{2}\right) d\theta = \frac{A_m^2}{4\pi}\left[\theta - \frac{\sin 2\theta}{2}\right]_0^{2\pi} \text{ where } \theta = 0, \sin 2\theta = 0 \text{ and when } \theta = 2\pi, \sin 4\pi = 0$$

$$= \frac{A_m^2 \times 2\pi}{4\pi} = \frac{A_m^2}{2} \text{ and taking the square root of each side yields}$$

[1] The SCR is extensively used in a variety of commercial electronic applications: in motor speed control by changing average and RMS values of the waveform applied to the motor, in lamp dimming, heater and oven control, and so forth.

$$A_{\mathrm{rms}} = A_{\mathrm{m}}/\sqrt{2} = 0.7071A_{\mathrm{m}} \qquad \text{(B-7.5)}$$

Although this relation was derived for one full cycle of sinusoid, it also holds for

a. One half-cycle of sinusoid (**Fig. B-7.3b**)
b. One quarter-cycle of sinusoid (Fig. B-7.3b)
c. The output of a full-wave rectifier (FWR) as shown in Fig. B-7.3a

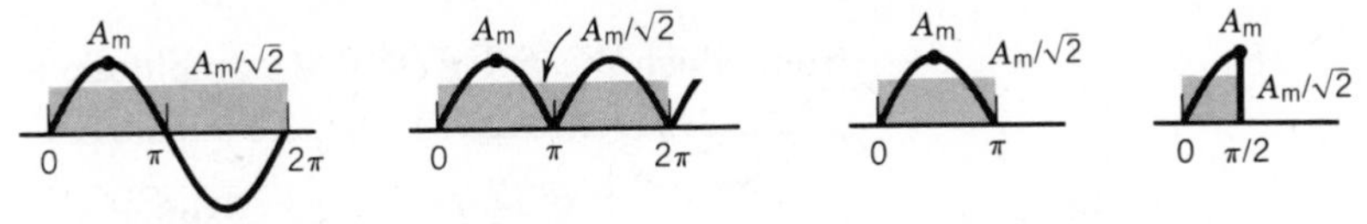

Figure B-7.3 RMS values of sinusoids

In comparing the average value of one half-cycle of sine wave ($0.6366A_{\mathrm{m}}$) to the RMS value of one half-cycle of sine wave ($0.7071A_{\mathrm{m}}$), the reader should note that the RMS value is always greater by the ratio known as the form factor or $F_{\mathrm{f}} = A_{\mathrm{rms}}/A_{\mathrm{dc}} = 0.7071A_{\mathrm{m}}/0.6366A_{\mathrm{m}} = 1.111$. Consequently, for quarter-cycle and half-cycle values, $A_{\mathrm{rms}} = 1.111A_{\mathrm{dc}}$.

B-7.7 RMS Value of One Cycle of a Half-Wave Rectifier (HWR) Output

The derivation of the RMS value of *one cycle* of an HWR output stems from application of the rectangular area method, shown in **Fig. B-7.4**, as follows:

$$A_{\mathrm{rms(HWR)}} = \sqrt{\left[\left(\frac{A_{\mathrm{m}}}{\sqrt{2}}\right)^2 \pi + 0^2\pi\right]\Big/ 2\pi} = \sqrt{\frac{A_{\mathrm{m}}^2}{4}} = \frac{A_{\mathrm{m}}}{2} \qquad \text{(B-7.6)}$$

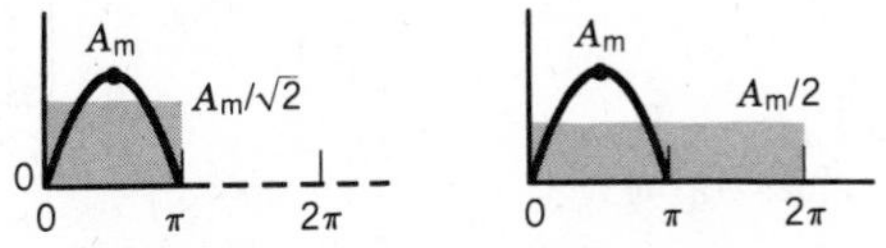

Figure B-7.4 RMS value for one cycle of HWR output

B-7.8 RMS Value of One Cycle of the Half-Cycle Output of a Silicon-Controlled Rectifier, Given Firing Angle ϕ

From the derivation of Eq. (B-7.5) for the RMS value of *one cycle* of *pure sinusoid* we obtain

$$A_{\mathrm{rms}}^2 = \frac{A_{\mathrm{m}}^2}{4\pi}\left[\theta - \frac{\sin 2\theta}{2}\right]_0^{2\pi}, \text{ where } \theta \text{ is the rotation angle}$$

But this general relation must be narrowed so that the upper limit of θ is π and the lower limit of θ is the firing angle ϕ. Therefore, we may rewrite the general equation to show

$$A_{rms}^2 = \frac{A_m^2}{4\pi}\left[\phi - \frac{\sin 2\phi}{2}\right]_\phi^\pi = \frac{A_m^2}{4\pi}\left[(\pi - 0) - \left(\frac{\phi}{180} \times \pi - \frac{\sin 2\phi}{2}\right)\right]$$

$$A_{rms}^2 = \frac{A_m^2}{4\pi}\left[\pi - \frac{\phi}{180^\circ} \times \pi + \sin\frac{2\phi}{2}\right]$$

and multiplying $1/\pi$ through the denominators within the brackets:

$$A_{rms}^2 = \frac{A_m^2}{4}\left(1 - \frac{\phi}{180^\circ} + \frac{\sin 2\phi}{2\pi}\right)$$

from which the *full cycle* RMS amplitude is

$$A_{rms} = \sqrt{\frac{A_m^2}{4}\left(1 - \frac{\phi}{180^\circ} + \frac{\sin 2\phi}{2\pi}\right)} = \frac{A_m}{2}\sqrt{1 - \frac{\phi}{180^\circ} + \frac{\sin 2\phi}{2\pi}} \qquad \text{(B-7.7)}$$

B-7.9 RMS Value of One Half-Cycle of the Half-Cycle Output of an SCR

In the case of SCRs used in the full-wave rectifier configuration, we are interested in the RMS value produced by each SCR. If each SCR gate is *separately* phase-shifted (**Fig. B-7.5b**) then Eq. (B-7.7) is modified by using $A_m/\sqrt{2}$ times the factor under the square root, or

$$A_{rmsH} = \frac{A_m}{\sqrt{2}}\sqrt{1 - \frac{\phi}{180^\circ} + \frac{\sin 2\phi}{2\pi}} \qquad \text{(B-7.8)}$$

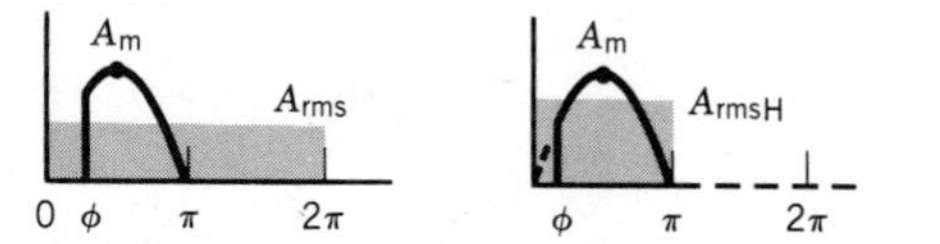

Figure B-7.5 RMS values of HW and FW SCR's

APPENDIX B-8 AVERAGE AND RMS VALUES OF NONSINUSOIDAL WAVEFORMS

B-8.1 Half-Cycle Average Value of a Triangular Waveform

Observe that from 0 to $\pi/2$ radians the curve of $i(t)$ is a straight line of the form $i(t) = mx + b = I_m t/(\pi/2) = 2I_m t/\pi$. Then the average value from 0 to π is twice the area or

$$I_{av} = 2\left(\frac{1}{\pi}\int_0^{\pi/2} \frac{2I_m t}{\pi}\,dt\right) = \frac{4I_m}{\pi^2}\int_0^{\pi/2} t\,dt = \frac{4I_m}{\pi^2}\left[\frac{t^2}{2}\right]_0^{\pi/2}$$

$$= \frac{4I_m}{\pi^2}\left(\frac{\pi^2}{8}\right) = \frac{I_m}{2} \qquad \text{(B-8.1)}$$

B-8.2 Full-Cycle Average Value of the Half-Wave Rectifier Output of a Triangular Waveform

This proof is similar to that of Sec. B-8.1 with the exception that the period is taken over 2π, or

$$I_{\text{av}} = 2\left(\frac{1}{2\pi}\int_0^{\pi/2}\frac{2I_{\text{m}}t}{\pi}\,dt\right) = \frac{2I_{\text{m}}}{\pi^2}\int_0^{\pi/2} t\,dt = \frac{2I_{\text{m}}}{\pi^2}\left[\frac{t^2}{2}\right]_0^{\pi/2}$$

$$= \frac{2I_{\text{m}}}{\pi^2}\left(\frac{1}{8}\times\pi^2\right) = \frac{I_{\text{m}}}{4} \qquad \text{(B-8.2)}$$

Equation (B-8.2) may also be verified by the rectangular area method where

$$I_{\text{av}} = \frac{\dfrac{I_{\text{m}}}{2}\times\pi + 0\times\pi}{2\pi} = I_{\text{m}}/4, \text{ as shown in } \textbf{Fig. B-8.1b}$$

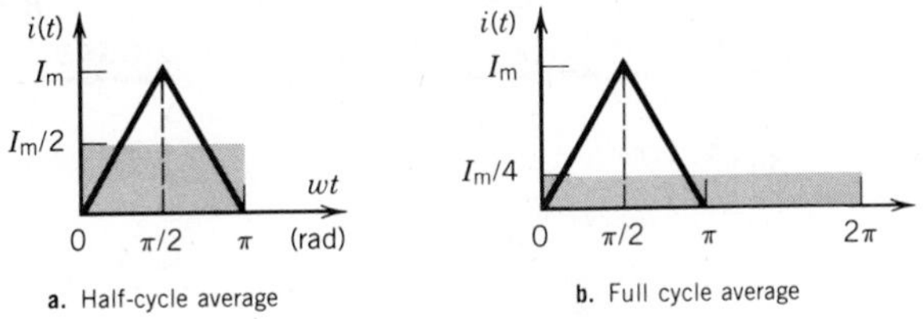

Figure B-8.1 Half-cycle average and full-cycle average of HWR output of a triangular waveform

B-8.3 Half-Cycle RMS Value of a Triangular Waveform

We now have $i(t) = (2I_{\text{m}}/\pi)t$, so using the basic equation for the RMS value and **Fig. B-8.2a** we may write:

$$I^2 = \int_0^{\pi/2}\frac{1}{T}[i(t)]^2\,dt$$

and since we are again seeking twice the area under the curve

$$I^2 = 2\left[\frac{1}{\pi}\int_0^{\pi/2}\left(\frac{2I_{\text{m}}t}{\pi}\right)^2 dt\right] = \frac{2}{\pi}\int_0^{\pi/2}\left(\frac{2I_{\text{m}}t}{\pi}\right)^2 dt = \frac{8I_{\text{m}}^2}{\pi^3}\int_0^{\pi/2} t^2\,dt$$

$$= \frac{8I_{\text{m}}^2}{\pi^3}\left(\frac{t^3}{3}\right)\Bigg]_0^{\pi/2} = \frac{8I_{\text{m}}^2}{\pi^3}\left(\frac{1}{3}\times\frac{\pi^3}{8}\right) = \frac{I_{\text{m}}^2}{3}, \text{from which}$$

$$I = \sqrt{I_{\text{m}}^2/3} = I_{\text{m}}/\sqrt{3} \qquad \text{(B-8.3)}$$

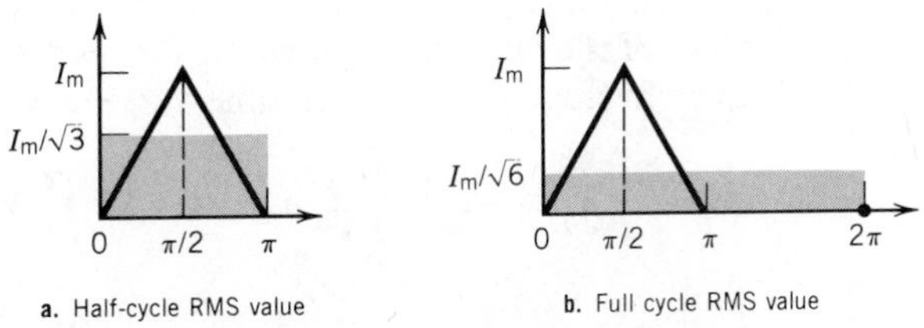

Figure B-8.2 Half- and full-cycle RMS values of HWR output of a triangular waveform

B-8.4 Full-Cycle RMS Value of the HWR Output of a Triangular Waveform

This proof is similar to that of Sec. B-8.3 with the exception that the period is taken over 2π, as shown in Fig. B-8.2b, which yields

$$I^2 = 2 \times \frac{1}{2\pi}\int_0^{\pi/2}\left(\frac{2I_{\mathrm{m}}t}{\pi}\right)^2 dt = \frac{4I_{\mathrm{m}}^2}{\pi^3}\int_0^{\pi/2}(t^2)\,dt$$

$$= \frac{4I_{\mathrm{m}}^2}{\pi^3}\left[\frac{t^3}{3}\right]_0^{\pi/2} = \frac{4I_{\mathrm{m}}^2}{\pi^3}\left[\frac{1}{3}\times\frac{\pi^3}{8}\right] = \frac{I_{\mathrm{m}}^2}{6}, \text{from which}$$

$$I = \sqrt{I_{\mathrm{m}}^2/6} = I_{\mathrm{m}}/\sqrt{6} \qquad \text{(B-8.4)}$$

Table B-8.1 summarizes the average and RMS values for triangular waveforms for a full cycle of input (0 to 2π) under full-wave rectification, half-wave rectification, and quarter-wave rectification conditions.

Table B-8.1 Summary of Full-Wave, Half-Wave, and Quarter-Wave Equations for Average and RMS Values[a]

Waveform	Average Value	RMS Value (per Cycle of Input 0 to 2π)
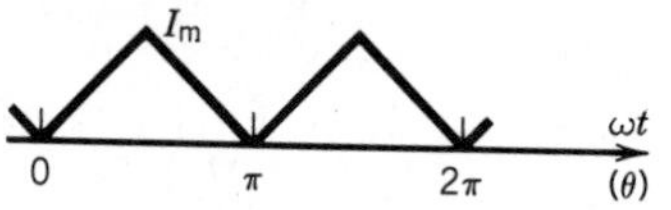	$I_{\mathrm{m}}/2$	$I_{\mathrm{m}}/\sqrt{3}$
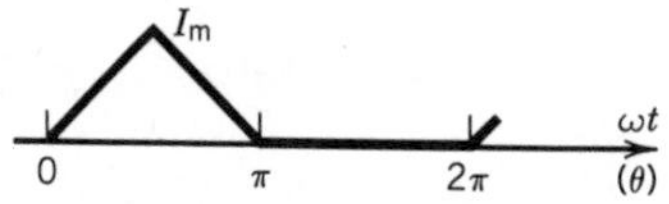	$I_{\mathrm{m}}/4$	$I_{\mathrm{m}}/\sqrt{6}$
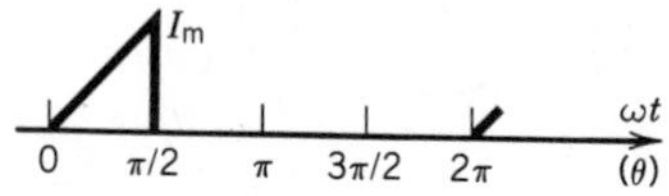	$I_{\mathrm{m}}/8$	$I_{\mathrm{m}}/\sqrt{12}$

[a] Note that in each case, $I_{\mathrm{rms}} > I_{\mathrm{av}}$ at all times.

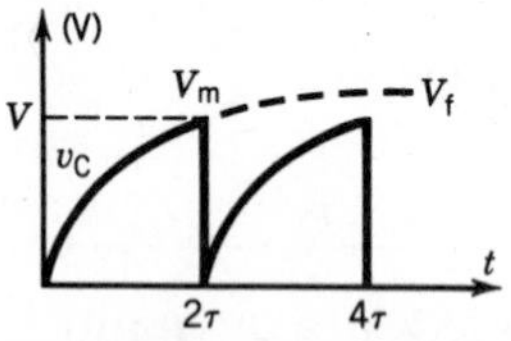

Figure B-8.3 Relaxation oscillator exponential waveform

B-8.5 Average Value of Rising Exponential Waveform

A common rising exponential waveform is that produced by the relaxation oscillator (Fig. 10-14a). The equation for v_C as it attempts to rise to its final value, V_f, is $v_C = V_f(1 - \varepsilon^{-t/\tau})$.

Let us assume that v_C is only permitted to rise to V in 2τ before the cycle is repeated. Since 2τ is the period, we are interested in finding the average and RMS values of the waveform shown in **Fig. B-8.3**.

The average value of any time-varying waveform, by definition, is

$$V_{\mathrm{av}} = \frac{1}{T}\int_0^T v(t)\,dt = \frac{1}{T}\int_0^T v_C(t)\,dt = \frac{1}{T}\int_0^T V_f(1-\varepsilon^{-t/\tau})\,dt = \frac{V_f}{2\tau}\int_0^{2\tau}(1-\varepsilon^{-t/\tau})\,dt$$

$$V_{av} = \frac{V_f}{(2\tau)}\left[t + \tau\varepsilon^{-t/\tau}\right]_0^{(2\tau)} \qquad \textbf{(B-8.5)}$$

where $t = 2\tau$ and $t = 0$, respectively.

Equation (B-8.5) is the generic equation for finding any average value for periods of oscillation from 0 up to 5τ. For example, in Fig. B-8.3, assume v_C is permitted to rise to 3τ before dropping to zero. Then all we need do is substitute (3τ) for the two terms in parentheses in Eq. (B-8.5).

For the case shown in Fig. B-8.3, however, the average value is computed by substituting 2τ for t in the expression enclosed within the *brackets*:

$$V_{av} = \frac{V_f}{2\tau}\left[2\tau + \tau\varepsilon^{-2} - (0 + \tau)\right] = \frac{V_f}{2\tau}\left[2.1353\tau - \tau\right] = \frac{V_f}{2\tau} \times 1.1353\tau = \mathbf{0.5677}\boldsymbol{V_f}$$

Note that the value of V or V_m, the peak value of the waveform, in Fig. B-8.3 is

$$V_m = V_f(1 - \varepsilon^{-t/\tau}) = V_f(1 - \varepsilon^{-2}) = \mathbf{0.8647}\boldsymbol{V_f} \qquad \text{(B-5.5)}$$

B-8.6 RMS Value of Rising Exponential Waveform

To find the RMS value of the waveform shown in Fig. B-8.3, we begin with the definition of any RMS value of a time-varying waveform:

$$V_{rms} = \sqrt{\frac{1}{T}\int_0^T (v_C)^2\,dt} \qquad \text{or} \qquad V_{rms}^2 = \frac{1}{T}\int_0^T (v_C)^2\,dt$$

then
$$V_{rms}^2 = \frac{1}{T}\int_0^T \left[V_f(1 - \varepsilon^{-t/\tau}\right]^2 dt$$

$$= \frac{V_f^2}{T}\int_0^T (1 - \varepsilon^{-t/\tau})^2\,dt \qquad \text{and squaring parenthetical terms}$$

$$= \frac{V_f^2}{2\tau}\int_0^T (1 - 2\varepsilon^{-t/\tau} + \varepsilon^{-2t/\tau})\,dt \qquad \text{and}$$

$$V_{rms}^2 = \frac{V_f^2}{(2\tau)}\left[t + 2\tau\varepsilon^{-t/\tau} - \frac{\tau}{2}\varepsilon^{-2t/\tau}\right]_0^{(2\tau)} \qquad \textbf{(B-8.6)}$$

where $t = (2\tau)$ and $t = 0$.

Equation (B-8.6) is the generic equation for finding any RMS value for periods of oscillation from 0 up to 5τ. As in the case of the average value, assume that v_C is permitted to rise to (3τ) before dropping to zero. Then all we need do is to substitute (3τ) for the *two* terms in parentheses in Eq. (B-8.6) and then take the square root of the result to find the RMS value.

For the case shown in Fig. B-8.3, the RMS value is computed by substituting 2τ for t in the expression enclosed within the *brackets*:

$$V^2 = \frac{V_f^2}{(2\tau)}\left[2\tau + 2\tau\varepsilon^{-2\tau/\tau} - \frac{\tau}{2}\varepsilon^{-4\tau/\tau} - \left(0 + 2\tau - \frac{\tau}{2}\right)\right]$$

$$= \frac{V_f^2}{2\tau}\left[2.2615\tau - (1.5\tau)\right] = \frac{V_f^2}{2\tau} \times 0.7615\tau$$

$$V^2 = 0.380\,76V_f^2 \qquad \text{and} \qquad V = \sqrt{0.380\,76V_f^2} = \mathbf{0.617\,05}\boldsymbol{V_f}$$

Since the RMS value exceeds the average value but is less than the maximum value, the answer is reasonable for the waveform of Fig. B-8.3.

APPENDIX B-9 DERIVATIVE AND INTEGRAL OF THE SINE FUNCTION

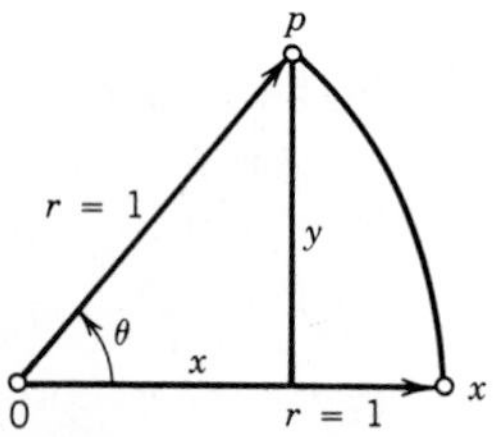

Figure B-9.1 Derivation for derivative of a sine function

In **Fig. B-9.1**, we find the point $p\ (x, y)$ moving along the arc of a circle, describing an angle θ radians at the origin of the circle, O. It is convenient to consider the special case where the radius of the circle is unity. This makes $y = \sin\theta$ at all times.

Then the rate of change of y with respect to time is $\dfrac{dy}{dt} = r\omega\cos\theta$ where $\omega = d\theta/dt$ and $r = 1$, but since $y = \sin\theta$ and $\omega = d\theta/dt$, we may write $\dfrac{d(\sin\theta)}{dt} = \cos\theta\, d\theta/dt$ and multiplying both sides by dt/dx yields

$$\frac{d(\sin\theta)}{dx} = \cos\theta\,\frac{d\theta}{dx} \qquad \text{(B-9.1)}$$

In a like manner we can prove that

$$\frac{d(\cos\theta)}{dx} = -\sin\theta\,\frac{d\theta}{dx} \quad \text{(B-9.2)} \qquad \text{and} \qquad \frac{d(-\sin\theta)}{dx} = -\cos\theta\,\frac{d\theta}{dx} \quad \text{(B-9.3)}$$

and coming around full circle,

$$\frac{d(-\cos\theta)}{dx} = \sin\theta\,\frac{d\theta}{dx} \qquad \text{(B-9.4)}$$

Note that the order of function rotation is the *same* as the axes (real and imaginary) encountered during *counterclockwise* rotation of θ!

Intuition might tell us that *integration* should proceed in a *clockwise* direction of rotation, based on the foregoing differentiation equations, since the process is reversed.

Let us begin with

$$\frac{d(\cos\theta)}{dx} = -\sin\theta\,\frac{d\theta}{dx} \qquad \text{(B-9.2)}$$

or

$$\sin\theta\, d\theta = -d(\cos\theta).$$

Then the integral

$$\int \sin\theta\, d\theta = -\cos\theta + K \qquad \text{(B-9.5)}$$

This is also evident since when we differentiate $d(-\cos\theta)/dx$ in Eq. (B-9.4) we get $\sin\theta(d\theta/dx)$.

In a like manner, then,

$$\int \cos\theta\, d\theta = \sin\theta + K \qquad \textbf{(B-9.6)}$$

APPENDIX B-10 ELECTRODYNAMOMETER WATTMETER MEASUREMENT OF SINGLE-PHASE POWER

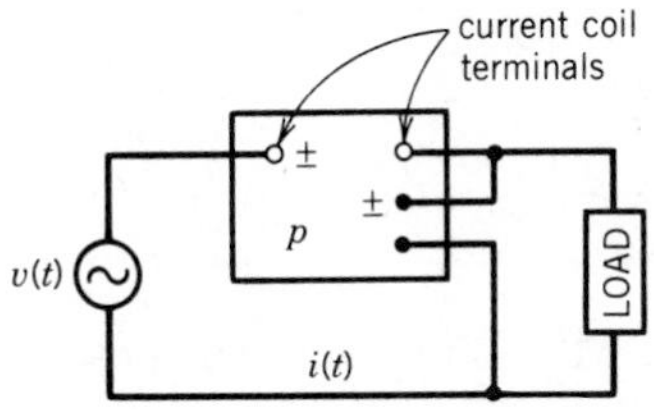

Figure B-10.1 Wattmeter connections

The electrodynamometer movement is basic to the wattmeter. The wattmeter consists of a (pair of) stationary current coil and a movable potential coil suspended in the magnetic field of the stationary current coil. The current coil is connected electrically in series with the load. The potential coil is connected across the load. The common terminals of both current coil and potential coil are connected to the same line side, as shown in **Fig. B-10.1**, with the common current coil always connected to the supply side. Assume that the voltage (applied to the load) as reference is $v(t) = V_m \sin\omega t$. Also assume that the instantaneous current to the load is $i(t) = I_m \sin(\omega t + \theta)$, that is, a leading load. Then the instantaneous power $p(t)$ is $p(t) = v(t) \times i(t) = V_m \sin\omega t \times I_m \sin(\omega t + \theta) = V_m I_m \sin\omega t \times \sin(\omega t + \theta)$

Using the trigonometric identity $\sin A \sin B = \dfrac{\cos(A - B) - \cos(A + B)}{2}$, then

$$\sin\omega t \times \sin(\omega t + \theta) = \frac{\cos(\omega t - \omega t - \theta) - \cos(\omega t + \omega t + \theta)}{2}$$

$$= \frac{\cos(-\theta) - \cos(2\omega t + \theta)}{2}$$

Substituting the identity in the above expression for $p(t)$ yields

$$p(t) = \frac{V_m I_m}{2}[\cos(-\theta) - \cos(2\omega t + \theta)]$$

$$= V \times I[\cos(-\theta) - \cos(2\omega t + \theta)] \text{ and expanding}$$

$$= VI\cos(-\theta) - VI\cos(2\omega t + \theta)$$

$$= VI\cos\theta - VI\cos(2\omega t + \theta) \text{ since } \cos(-\theta) = \cos\theta.$$

But mathematically, the average power may be found as

$$P_{av} = P = \frac{1}{T}\int_0^T P(t)\,dt \text{ so substituting the expression for } p(t) \text{ yields}$$

$$P = \frac{1}{T}\int_0^T [VI\cos\theta - VI\cos(2\omega t + \theta)]\,dt$$

Using the trigonometric identity $\cos(A + B) = \cos A \cos B - \sin A \sin B$, removing constants, and integrating yields

$$P = \frac{VI}{T}\int_0^T (\cos\theta - \cos 2\omega t \cos\theta + \sin 2\omega t \sin\theta)\,dt$$

$$= \frac{VI}{T}\left[t\cos\theta - \frac{\cos\theta \sin 2\omega t}{2\omega} - \frac{\sin\theta\cos 2\omega t}{2\omega}\right]_0^T = \frac{VI}{T}(T\cos\theta - 0 - 0)$$

$$P = VI\cos\theta = \frac{V_m I_m}{2}\cos\theta = S\cos\theta \qquad \textbf{(B-10.1)}$$

where θ is the phase angle between voltage and current
V and I are RMS values of voltage and current

But from Eq. (B-10.1) the power factor, F_p, is the ratio of active power to apparent power:

$$F_p = \cos\theta = \frac{P}{S} = \frac{P}{VI} = \frac{2P}{V_m I_m} \qquad \textbf{(B-10.2)}$$

where S is the apparent power or the product of RMS voltage and current
θ is the phase angle between voltage and current waveforms or phasors

Note: θ may be positive or negative, since $\cos(-\theta) = \cos\theta$

APPENDIX B-11 EULER'S THEOREM AND THE EXPONENTIAL FORM OF A COMPLEX NUMBER

By definition, the exponential function, ε^x, is equal to the limit approached by the infinite series

$$\varepsilon^x = 1 + x + x^2/2! + x^3/3! + \cdots + x^n/n! \qquad \textbf{(B-11.1)}$$

but
$$\cos\theta = 1 - \theta^2/2! + \theta^4/4! - \theta^6/6! \cdots \qquad \textbf{(B-11.2)}$$

and
$$\sin\theta = \theta - \theta^3/3! + \theta^5/5! - \theta^7/7! \cdots \qquad \textbf{(B-11.3)}$$

Note that these sine and cosine series define the trigonometric functions. Note also that the cosine function contains alternately positive and negative *even* powers while the sine function contains alternately positive and negative *odd* powers.

Replacing the exponent x by $j\theta$ in Eq. (B-11.1) yields

$$\varepsilon^{j\theta} = 1 + j\theta + (j\theta)^2/2! + (j\theta)^3/3! + (j\theta)^4/4! + (j\theta)^5/5! \cdots$$

Substituting $(j\theta)^2 = -\theta^2$ and $(j\theta)^3 = -j\theta^3$ while $(j\theta)^4 = \theta^4$ and $(j\theta)^5 = j\theta^5$ etc., then $\varepsilon^{j\theta} = 1 + j\theta - \theta^2/2! - j\theta^3/3! + \theta^4/4! + j\theta^5/5!$ etc. Separating the real from the imaginary terms, we find (to our great surprise):

$$\varepsilon^{j\theta} = (1 - \theta^2/2! + \theta^4/4! + \cdots) + j(\theta - \theta^3/3! + \theta^5/5! + \cdots) \qquad \textbf{(B-11.4)}$$

so that $\varepsilon^{j\theta} = \cos\theta + j\sin\theta$, where θ is the angle in radians.

Equation (B-11.4) is the basis for the exponential form of a complex number, which is derived as shown below.

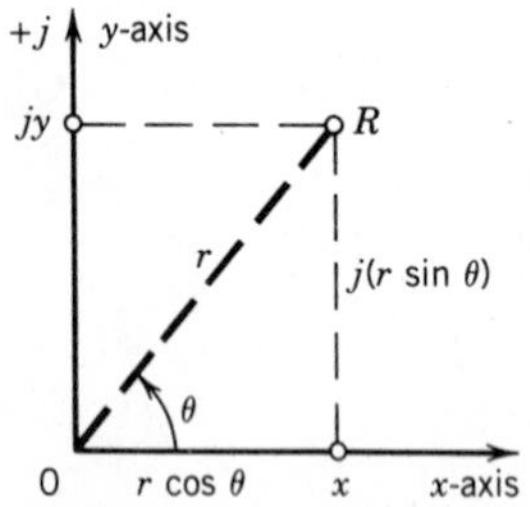

Figure B-11.1 Rectangular and polar forms of $r\angle\theta$

Point R on **Fig. B-11.1** represents a point on the complex plane, which may be written in polar form as $r\angle\theta$ or in rectangular form as $r(\cos\theta + j\sin\theta)$, where θ is the angle in radians. Consequently, using Euler's theorem we may write

$$r\angle\theta = x + jy = r(\cos\theta + j\sin\theta) = r\varepsilon^{j\theta} \qquad \textbf{(B-11.5)}$$

Equation (B-11.5) shows three forms of a complex number: the polar form, the rectangular form, and the exponential form. There is still a fourth form, known as the hyperbolic form, which is derived below. Hyperbolic functions are used extensively in filter networks and in transmission line theory.

From Euler's series for complex exponentials, if $\varepsilon^{j\theta} = \cos\theta + j\sin\theta$, then $\varepsilon^{-j\theta} = \cos\theta - j\sin\theta$. Taking the difference of these two terms yields $\varepsilon^{j\theta} - \varepsilon^{-j\theta} = (\cos\theta + j\sin\theta) - (\cos\theta - j\sin\theta) = j2(\sin\theta)$, from which

$$\sin\theta = \frac{(\varepsilon^{j\theta} - \varepsilon^{-j\theta})}{j^2} \qquad \text{and} \qquad j\sin\theta = \frac{\varepsilon^{j\theta} - \varepsilon^{-j\theta}}{2}$$

But the hyperbolic sine, $\sinh\theta = j\sin\theta$ by definition, and therefore

$$\sinh\theta = \frac{\varepsilon^{j\theta} - \varepsilon^{-j\theta}}{2} \qquad \textbf{(B-11.6)}$$

In the same way, the hyperbolic cosine may be derived by the sum of the terms

$$\varepsilon^{j\theta} + \varepsilon^{-j\theta} = (\cos\theta + j\sin\theta) + (\cos\theta - j\sin\theta) = 2\cos\theta$$

so $\cos\theta = \dfrac{\varepsilon^{j\theta} + \varepsilon^{-j\theta}}{2}$ from which the hyperbolic cosine, cosh, emerges as

$$\cosh\theta = \frac{\varepsilon^{j\theta} + \varepsilon^{-j\theta}}{2} \qquad \textbf{(B-11.7)}$$

The hyperbolic cosine and sine give rise to a fourth form of a complex number

$$r\varepsilon^{j\theta} = r\underline{/\theta} = r(\cos\theta + j\sin\theta) = r\left(\frac{\varepsilon^{j\theta} + \varepsilon^{-j\theta}}{2} + \frac{\varepsilon^{j\theta} - \varepsilon^{-j\theta}}{2}\right)$$
$$= r(\cosh\theta + \sinh\theta) \qquad \textbf{(B-11.8)}$$

The hyperbolic functions also enable a derivation of Euler's transformation by $\cos\theta = \frac{1}{2}(\varepsilon^{j\theta} + \varepsilon^{-j\theta})$ while $j\sin\theta = \frac{1}{2}(\varepsilon^{j\theta} - \varepsilon^{-j\theta})$. Then

$$\cos\theta + j\sin\theta = \tfrac{1}{2}(\varepsilon^{j\theta} + \varepsilon^{-j\theta}) + \tfrac{1}{2}(\varepsilon^{j\theta} - \varepsilon^{-j\theta}) = \tfrac{1}{2}(2\varepsilon^{j\theta}) = \varepsilon^{j\theta} \qquad \text{(B-11.4)}$$

This validates both the hyperbolic functions and Euler's transformation, enabling us to write the fourth form of the complex number given in Eq. (B-11.8).

APPENDIX B-12 SINUSOIDAL RESPONSE OF PURE INDUCTORS AND CAPACITORS

B-12.1 Response of Pure Inductor to Sinusoidal Waveform

For a pure inductor, $v_L = L(di/dt)$ by definition. If we assume the voltage, v_L, across the inductor is a time-varying sinusoid, then

$$v(t) = V_m \sin\omega t = L(di/dt) \text{ and solving for } di \text{ yields}$$

$$di = \frac{V_m}{L}\sin(\omega t)\,dt = \frac{V_m}{\omega L}\sin(\omega t)\,d(\omega t) \text{ and integrating both sides,}$$

$$i(t) = \frac{V_m}{\omega L}\int \sin \omega t \, d(\omega t) = \frac{V_m}{\omega L}(-\cos \omega t) = \frac{V_m}{\omega L}\sin(\omega t - 90°) \quad \text{(B-12.1)}$$

These relations also show that $I_m = V_m/\omega L$, which when transformed to the frequency domain becomes

$$\mathbf{I} = \frac{\mathbf{V}}{jX_L} = \frac{\mathbf{V}}{X_L}\underline{/-90°} \quad \text{(B-12.2)}$$

Equation (B-12.2) shows that sinusoidal current lags applied sinusoidal voltage by 90° for a pure inductor.

B-12.2 Response of Pure Capacitor to Sinusoidal Waveform

For a pure capacitor, by definition, $v_C = q/C$ and the rate of change of voltage across a pure capacitor is

$$\frac{dv_C}{dt} = \frac{1}{C}\frac{dq}{dt}$$

but $i(t) = dq/dt$ and therefore

$$\frac{dv_C}{dt} = \frac{i(t)}{C}$$

Solving for $i(t)$ and substituting $v_C = V_m \sin \omega t$ yields by differentiation

$$i(t) = C\frac{dv_C}{dt} = \frac{C\,d(V_m \sin \omega t)}{dt} = \omega C V_m \cos \omega t$$
$$= \omega C V_m \sin(\omega t + 90°) \quad \text{(B-12.3)}$$

The above relations show also that $I_m = \omega C V_m = V_m/(1/\omega C)$, which, when transformed to the frequency domain, becomes

$$\mathbf{I} = \frac{\mathbf{V}}{-j\mathbf{X}_C} = \frac{\mathbf{V}}{X_C}\underline{/+90°} \quad \text{(B-12.4)}$$

APPENDIX B-13 SERIES–PARALLEL EQUIVALENCE

B-13.1 Basic Equations

Equivalence implies that the series impedance, $\mathbf{Z}_s\underline{/\theta} = R_s + jX_s$, is exactly the same as the impedance of the parallel circuit whose admittance is $\mathbf{Y}_p = G_p - jB_p = (1/R_p) - j(1/X_p)$ shown in **Fig. B-13.1**. Equating the two yields

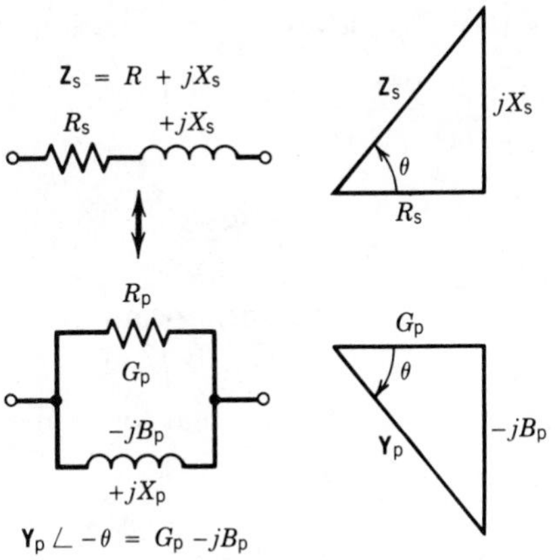

Figure B-13.1 Series–parallel equivalence

$$\mathbf{Y}_p = \frac{1}{\mathbf{Z}_s} = \frac{1}{R_s + jX_s}$$

rationalizing the denominator by conjugate multiplication and expansion of terms gives

$$\mathbf{Y}_p = \frac{R_s - jX_s}{(R_s + jX_s)(R_s - jX_s)} = \frac{R_s}{R_s^2 + X_s^2} - j\frac{X_s}{R_s^2 + X_s^2} = G_p - jB_p$$

Separating real and imaginary terms in the last expression and taking their reciprocals yields

$$R_p = \frac{1}{G_p} = \frac{R_s^2 + X_s^2}{R_s} = \frac{Z_s^2}{R_s} = \frac{Z_s^2}{Z_s \cos\theta} = \frac{Z_s}{\cos\theta} \qquad \text{(B-13.1)}$$

$$X_p = -\frac{1}{jB_p} = \frac{R_s^2 + X_s^2}{X_s} = \frac{Z_s^2}{X_s} = \frac{Z_s^2}{Z \sin\theta} = \frac{Z_s}{\sin\theta} \qquad \text{(B-13.2)}$$

We may also express R_p and X_p in terms of the series circuit Q, defined below:

$$R_p = \frac{1}{G_p} = \frac{R_s^2 + X_s^2}{R_s} \times \frac{R_s}{R_s} = R_s + R_s\frac{X_s^2}{R_s^2} = R_s\left(1 + \frac{X_s^2}{R_s^2}\right) = R_s(1 + Q^2) \qquad \text{(B-13.3)}$$

where $Q = X_s/R_s = \tan\theta$ in the *series* model whose impedance is $Z\angle\theta$. Similarly,

$$X_p = \left(\frac{R_s^2 + X_s^2}{X_s}\right) \times \frac{X_s}{X_s} = X_s\left(\frac{R_s^2}{X_s^2} + \frac{X_s^2}{X_s^2}\right) = X_s\left(\frac{1}{Q^2} + 1\right) = X_s\left(\frac{1 + Q^2}{Q^2}\right) \qquad \text{(B-13.4)}$$

where all terms have been defined.

Equation (B-13.4) leads to the definition of the Q in a parallel circuit, since

$$X_p = X_s\left(\frac{1+Q^2}{Q^2}\right) = \frac{X_s(1+Q^2)\times R_s^2}{X_s^2} \text{ and substituting } R_p = R_s(1+Q^2) \text{ yields}$$

$$= R_s\left(\frac{1+Q^2}{X_s}\right)R_s = R_pR_s/X_s = R_p/Q \text{ from which we obtain}$$

$Q = R_p/X_p$ which is the reciprocal form of the series circuit Q obtained in Eq. (B-13.3) **(B-13.5)**

From the above, it follows that for equivalence to occur

$$Q = \frac{X_s}{R_s} = \frac{R_p}{X_p} = \tan\theta = \frac{B_p}{G_p} \qquad \text{(B-13.6)}$$

in the impedance and admittance triangles of Fig. B-13.1.

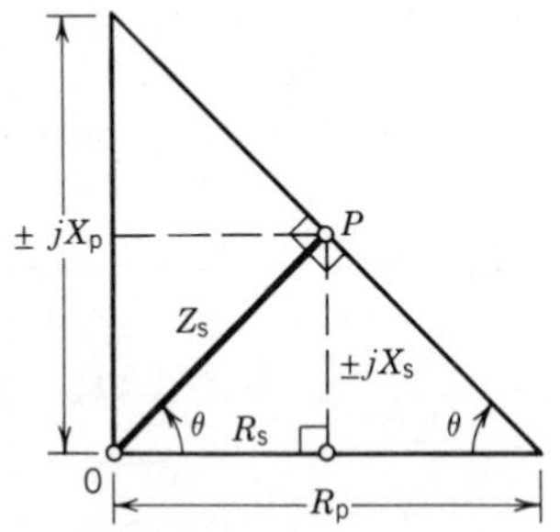

Figure B-13.2 Graphical construction for series–parallel transformation and vice versa

B-13.2 Series–Parallel Equivalence by Graphical Construction

Note from Eq. (B-13.3), since $R_p = R_s(1 + Q^2)$, that R_p is always greater than R_s, even when the Q is small. In the same way, form Eq. (B-13.4), since $X_p = X_s[(1 + Q^2)/Q^2]$, X_p is always greater than X_s. This is shown in the construction of **Fig. B-13.2**, a simple construction technique that enables either series-to-parallel transformation or vice versa, by the following methods.

A. *Series-to-parallel transformation*
   1. On engineering graph paper, lay off series resistance R_s on the horizontal scale and series reactance $\pm jX_s$ on the vertical scale.
   2. Construct the series impedance $Z_s/\theta = R_s + jX_s$ as shown.
   3. Construct a perpendicular at point **P** that is perpendicular to Z_s, extending it to the vertical axis and horizontal axis, respectively, as shown in Fig. B-13.2.
   4. The equivalent parallel resistance, R_p, and equivalent parallel reactance, X_p, represent the horizontal and vertical intersections of the perpendicular line, respectively, and may be read to scale.

B. *Parallel-to-series transformation*

   This transformation is the inverse of the preceding one, using Fig. B-13.2
   1. On engineering graph paper, lay off parallel resistance R_p on the horizontal real axis.
   2. Lay off $\pm jX_p$ on the imaginary vertical axis.
   3. Draw a straight line connecting X_p to R_p as shown in Fig. B-13.2.
   4. Construct a perpendicular from the above line (hypotenuse) to the origin. The length of this line, from O to P, represents the series impedance, Z_s.
   5. Draw a horizontal line from P to the vertical axis. This yields the value of the equivalent series reactance, $\pm jX_s$.
   6. Draw a vertical line from P to the horizontal axis. This yields the value of the equivalent series resistance, R_s.

Note: this construction shows why at all times $R_p > R_s$ and $\pm jX_p > X_s$.

B-13.3 Proof that the Radian Frequency at Which $-jX_s$ is a Maximum is $1/RC$

Since

$$Z_s = \frac{1}{Y_p} = \frac{1}{1/R_p + j\omega C_p} = \frac{R_p}{1 + j\omega R_p C_p}$$

$$= \underbrace{\frac{R_p}{1 + (R_p\omega C_p)^2}}_{\text{(real term)}} - \underbrace{\frac{jR_p^2\omega C_p}{1 + (R_p\omega C_p)^2}}_{\text{(imaginary term)}} = R_s - jX_{C_s} \qquad \text{(B-13.7)}$$

Above expands $Z_s = R_s - jX_{C_s}$ into its equivalent.

From the foregoing expansion, $X_s = R_p^2\omega C_p/[1 + (R_p\omega C_p)^2]$, and to find the frequency at which X_s is a maximum, we must differentiate this expression with respect to ω and set it to zero:

$$\frac{dX_s}{d\omega} = \frac{[R_p^2\omega C_p(2R_p\omega C_p)R_p^2\omega] - [(R_p\omega C_p)^2 + 1] \times [2(R_p\omega C_p) + R_p^2\omega]}{[(R_p\omega C_p)^2 + 1]^2} = 0$$

But for $dX_s/d\omega$ to be zero, the numerator must be zero. This drops out the denominator and sets the numerator equal to zero. Equating the left-hand to the right-hand term yields

$$R_p^2\omega C_p(2R_p\omega C_p)R_p^2\omega = [(R_p\omega C_p) + 1] \times [2(R_p\omega C_p) + R_p^2\omega]$$

$$R_p^2\omega C_p(2\omega R_p^2 C_p) = [(R_p\omega C_p) + 1] \times [R_p^2 C_p]$$

$$2(R_p\omega C_p)^2 = (R_p\omega C_p)^2 + 1$$

$$1(R_p\omega C_p)^2 = 1 \text{ from which}$$

$$\omega^2 = 1/(R_p C_p)^2 \qquad \text{and} \qquad \boldsymbol{\omega = 1/R_p C_p} \text{ (see Fig. 16-13)}$$

B-13.4 Proof that Given Variable ω, When jX_s is a Maximum it is Equal to $R_p/2$

If X_s is a maximum when $\omega = 1/R_p C_p$, since $X_s = R_p^2\omega C_p/[1 + (R_p\omega C_p)^2]$, we may substitute $\omega = 1/R_p C_p$, and X_s becomes

$$X_s = \frac{R_p^2\omega C_p}{1 + (R_p\omega C_p)^2} = \frac{R_p^2 C_p}{R_p C_p\left[\left(\frac{R_p C_p}{R_p C_p}\right) + 1\right]} = \frac{R_p}{1 + 1} = \frac{\mathbf{R_p}}{\mathbf{2}}$$

(see Fig. 16-13)

B-13.5 Proof that as $\omega \to 0$, $R_s \to R_p$, and Similarly as $\omega \to \infty$, $R_s \to 0$

Taking the real term from Eq. (B-13.7) yields $R_s = R_p/[1 + (R_p\omega C_p)^2]$, and as ω approaches 0, $R_s = R_p$ (see Fig. 16-13). Similarly, as ω approaches ∞, $R_s = 0$ (see Fig. 16-13).

B-13.6 Proof that as $\omega \to 0$, $-jX_s \to 0$, and Similarly as $\omega \to \infty$, $-jX_s \to 0$

Taking the imaginary term from Eq. (B-13.7), $jX_s = -j(R_p^2\omega C_p)/[1 + (R_p\omega C_p)^2]$, and as $\omega \to 0$, $-jX_s = 0/1 = 0$ (see Fig. 16-13).

Similarly, as $\omega \to \infty$, $X_s = \infty/\infty^2$, which is *indeterminate*! But using L'Hospital's rule for indeterminate functions yields $d(\omega R_p^2 C_p)/d\omega = R_p^2 C_p$ in the numerator of the foregoing expression and $\dfrac{d[(R_p\omega C_p)^2 + 1]}{d\omega} = 2\omega R_p^2 C_p^2$ in the denominator, so the limit of $R_p^2 C_p/2\omega R_p^2 C_p^2 = k/\infty = 0$. Therefore $-jX_s \to 0$ as $\omega \to \infty$ (see Fig. 16-13).

APPENDIX B-14 VOLTAGE- AND CURRENT-DEPENDENT SOURCES IN PASSIVE AC NETWORKS

The following relations, summarized in Table 18-1, are derived here for finding the input (or output) impedance of passive dependent circuits.

B-14.1 Voltage-Dependent Voltage Sources

B-14.1a *Dependent voltage source opposing independent voltage source*
Figure B-14.1a shows a dependent voltage source opposing an independent voltage source. (For simplicity, both sources are ideal voltage sources.) But a series impedance, Z_0, is connected between the two sources, which, for all practical purposes, may be considered the impedance of the dependent source, $\mathbf{h}\mathbf{V}\angle\theta$. For the polarities shown in Fig. B-14.1a, using KVL for the series mesh, the voltage across $\mathbf{Z}_0$ is $\mathbf{V}\angle\theta - \mathbf{h}\mathbf{V}\angle\theta$ and therefore the circuit current

$$\mathbf{I} = \frac{\mathbf{V}\angle\theta - \mathbf{h}\mathbf{V}\angle\theta}{\mathbf{Z}_0} = \frac{\mathbf{V}\angle\theta(1 - \mathbf{h})}{\mathbf{Z}_0}.$$

Since the application of $\mathbf{V}\angle\theta$ at the input produced current $\mathbf{I}$ as a consequence, then the input impedance is

$$\mathbf{Z}_{\text{in}} = \frac{\mathbf{V}/\theta}{\mathbf{I}} = \frac{\mathbf{V}\angle\theta \cdot \mathbf{Z}_0}{\mathbf{V}/\theta(1 - \mathbf{h})} = \frac{\mathbf{Z}_0}{(1 - \mathbf{h})} \quad \text{ohms } (\Omega) \qquad \text{(B-14.1)}$$

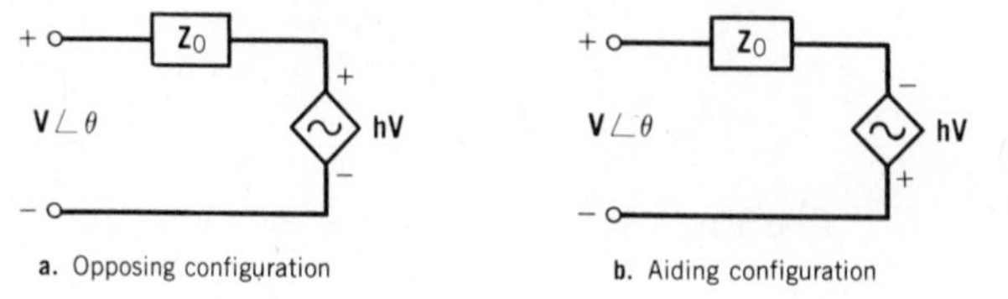

Figure B-14.1 Voltage-dependent voltage sources

B-14.1b *Dependent voltage source aiding independent voltage source*
Figure B-14.1b shows a dependent voltage source aiding an independent voltage source with a total circuit impedance Z_0 connected between them. Using KVL for the instantaneous polarities shown in Fig. B-14.1b, the voltage across $\mathbf{Z}_0$ is $\mathbf{V}\underline{/\theta} + \mathbf{hV}\underline{/\theta}$. The current $\mathbf{I}$ is therefore

$$\mathbf{I} = \frac{\mathbf{V}\underline{/\theta} + \mathbf{hV}\underline{/\theta}}{\mathbf{Z}_0} = \frac{\mathbf{V}\underline{/\theta}(1 + \mathbf{h})}{\mathbf{Z}_0}$$

Then

$$\mathbf{Z}_{\text{in}} = \frac{\mathbf{V}\underline{/\theta}}{\mathbf{I}} = \frac{\mathbf{V}\underline{/\theta} \cdot \mathbf{Z}_0}{\mathbf{V}\underline{/\theta}(1 + \mathbf{h})} = \frac{\mathbf{Z}_0}{(1 + \mathbf{h})} \quad \text{ohms } (\Omega) \qquad \text{(B-14.2)}$$

With respect to these equations, note that

1. *Opposition* of the dependent voltage source results in a *minus* sign in the $1 - \mathbf{h}$ factor.
2. A dependent voltage source *aiding* the supply voltage results in a *plus* sign in the $(1 + \mathbf{h})$ factor.
3. The factor is in the *denominator* in both cases, for dependent *voltage* sources (only).
4. In Eq. (B-14.1), $\mathbf{Z}_{\text{in}}$ may be greater or smaller than $\mathbf{Z}_0$, depending on whether $\mathbf{h}$ is less than unity or greater than 2. In Eq. (B-14.2), $\mathbf{Z}_{\text{in}}$ is always less than $\mathbf{Z}_0$, regardless of the value of $\mathbf{h}$.
5. When $\mathbf{h}$ is greater than unity in Eq. (B-14.1), the factor $(1 - \mathbf{h})$ is always used as its *absolute* magnitude. It may never be negative.

B-14.2 Current Dependent Current Sources—Practical Dependent Source Across Supply

B-14.2a *Current-dependent current source opposing independent current source*
Figure B-14.2a shows a dependent current source *opposing* the input current $\mathbf{I}$. Using KCL for the currents at node $\mathbf{V}\underline{/\theta}$ shows that the current in $\mathbf{Z}_0$ is $\mathbf{I}(\mathbf{h} + 1)$. For the LH mesh, using KVL in a CCW direction yields $-\mathbf{I}(\mathbf{h} + 1)\mathbf{Z}_0 + \mathbf{V}\underline{/\theta} = 0$, from which $\mathbf{V}\underline{/\theta} = \mathbf{I}(\mathbf{h} + 1)\mathbf{Z}_0$. Dividing both sides by $\mathbf{I}$ yields

$$\mathbf{Z}_{\text{in}} = \frac{\mathbf{V}\underline{/\theta}}{\mathbf{I}} = \mathbf{Z}_0(1 + \mathbf{h}) \quad \text{ohms } (\Omega) \qquad \text{(B-14.3)}$$

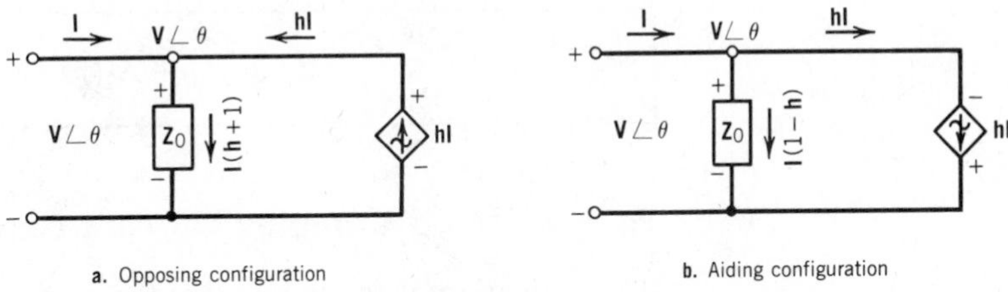

Figure B-14.2 Practical dependent current sources across supply

B-14.2b *Current-dependent current source aiding independent current source*
Figure B-14.2b shows a dependent current source aiding the input current $\mathbf{I}$. Using

KCL for the currents at node $\mathbf{V}\angle\theta$ shows that the current in $\mathbf{Z}_0$ is $\mathbf{I}(1-\mathbf{h})$. For the LH mesh, using KVL in a CCW direction yields $-\mathbf{I}(1-\mathbf{h})\mathbf{Z}_0+\mathbf{V}\angle\theta=0$, from which $\mathbf{V}\angle\theta=\mathbf{I}(1-\mathbf{h})\mathbf{Z}_0$. Dividing both sides by $\mathbf{I}$ yields

$$\mathbf{Z}_{\text{in}}=\frac{\mathbf{V}\angle\theta}{\mathbf{I}}=\mathbf{Z}_0(1-\mathbf{h}) \qquad \text{ohms } (\Omega) \qquad \textbf{(B-14.4)}$$

With respect to Eqs. (B-14.3) and (B-14.4), please note that

1. The feedback factor is always in the numerator.
2. A plus sign is used when the dependent source opposes the current and voltage.
3. A minus sign is used when the dependent source aids the supply current and voltage.
4. $\mathbf{Z}_{\text{in}}$ is always greater than $\mathbf{Z}_0$ when the dependent current source opposes the supply current in Eq. (B-14.3).
5. But whenever $\mathbf{h}<1$, $\mathbf{Z}_{\text{in}}<\mathbf{Z}_0$ and whenever $\mathbf{h}>2$, $\mathbf{Z}_{\text{in}}>\mathbf{Z}_0$ in Eq. (B-14.4).

B-14.3 Current-Dependent Current Sources—Lumped Impedance in Series with Practical Dependent Current Source

B-14.3a *Practical current-dependent source opposing independent current source,* **I**
From Eq. (B-14.3), the impedance seen by $\mathbf{Z}_1$ is $\mathbf{Z}_2(1+\mathbf{h})$ for the opposing configuration shown in **Fig. B-14.3a**. Then the total impedance seen at the input is

$$\mathbf{Z}_{\text{in}}=\mathbf{Z}_1+\mathbf{Z}_2(1+\mathbf{h}) \qquad \text{ohms } (\Omega) \qquad \textbf{(B-14.5)}$$

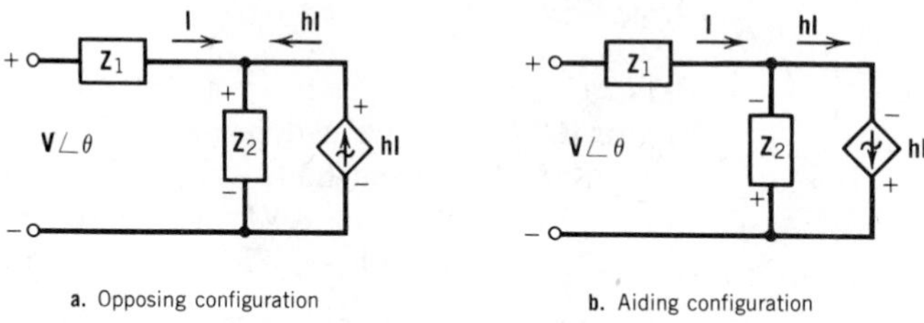

Figure B-14.3 Lumped impedance in series with practical dependent current source

B-14.3b *Practical current-dependent source aiding independent current source,* **I**
By inspection from Fig. B-14.3b, using Eq. (B-14.4), the total input impedance seen at the input is

$$\mathbf{Z}_{\text{in}}=\mathbf{Z}_1+\mathbf{Z}_2(1-\mathbf{h}) \qquad \text{ohms } (\Omega) \qquad \textbf{(B-14.6)}$$

B-14.4 More Complex Current-Dependent Current Sources

The derivation of the input impedance of the two passive circuits shown in **Fig. B-14.4** may be simplified by temporarily eliminating the current source. Under these conditions, $\mathbf{Z}_{\text{in}}=\mathbf{Z}_3\|(\mathbf{Z}_1+\mathbf{Z}_2)$, which yields

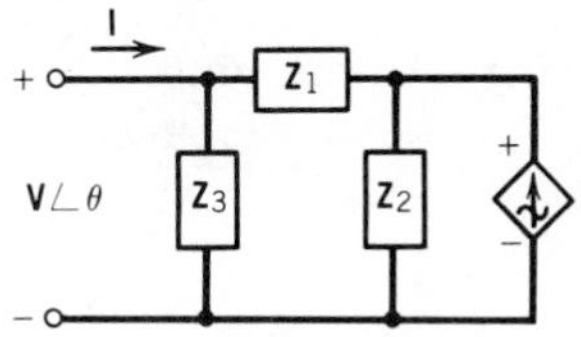

a. Opposing configuration

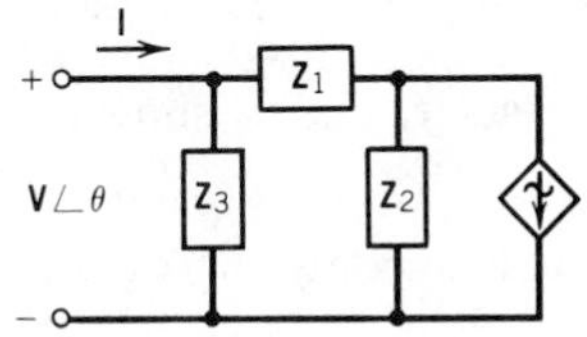

b. Aiding configuration

Figure B-14.4 More complex configuration containing practical dependent current

$$\mathbf{Z}_{\text{in}} = \frac{\mathbf{Z}_3(\mathbf{Z}_1 + \mathbf{Z}_2)}{\mathbf{Z}_1 + \mathbf{Z}_2 + \mathbf{Z}_3},$$

assuming there is no dependent current source.

But we know that if $\mathbf{Z}_2$ is shunted by a dependent current source, the input impedance is *increased* by a factor $\mathbf{Z}_2(1 \pm \mathbf{h})$, depending on whether the dependent source current opposes or aids the supply current.

B-14.4a *Opposing configuration*
We may now modify the relation for the input impedance shown in Fig. B-14.4a by including the effect of the dependent current source in opposition to the total current, **I**, by writing by inspection

$$\mathbf{Z}_{\text{in}} = \frac{\mathbf{Z}_3}{\mathbf{Z}_1 + \mathbf{Z}_2 + \mathbf{Z}_3}[\mathbf{Z}_1 + \mathbf{Z}_2 + \mathbf{Z}_2(1 + \mathbf{h})] \qquad \text{ohms } (\Omega) \qquad \text{(B-14.7)}$$

B-14.4b *Aiding configuration*
In the same way, we may now modify the relation for the input impedance of the configuration shown in Fig. B-14.4b by including the effect of the dependent current source aiding the total current, **I**, and writing by inspection

$$\mathbf{Z}_{\text{in}} = \frac{\mathbf{Z}_3}{\mathbf{Z}_1 + \mathbf{Z}_2 + \mathbf{Z}_3}[\mathbf{Z}_1 + \mathbf{Z}_2 + \mathbf{Z}_2(1 - \mathbf{h})] \qquad \text{ohms } (\Omega) \qquad \text{(B-14.8)}$$

The foregoing equations and their configurations are summarized in Table 18-1.

APPENDIX B-15 SERIES-RESONANT CIRCUIT RELATIONS

B-15.1 Resonant Power, P_0, and Off-Resonant Power, P

In a resonant circuit, the maximum current at the resonant frequency, f_0, is designated as I_0. The maximum power dissipated by an *RLC* series circuit is P_0 at f_0, where

$$P_0 = I_0^2 R = (\mathbf{V}/\mathbf{Z}_0)^2 R = (\mathbf{V}/R)^2 R = \mathbf{V}^2/R \qquad \text{watts (W)} \qquad \text{(19-13)}$$

where **V** is the applied voltage
$\mathbf{Z}_0$ is the circuit impedance at resonance or R at f_0

At any other frequency, either above or below f_0, the power, P, is

$$P = I^2R = (\mathbf{V}/\mathbf{Z})^2 R = \frac{\mathbf{V}^2 R}{\mathbf{Z}^2} = \frac{\mathbf{V}^2 R}{(R + jX_r)^2} = \frac{\mathbf{V}^2 R}{R^2 + (X_L - X_C)^2} = \frac{\mathbf{V}^2 R}{R^2 + X_r^2}$$

$$\frac{\mathbf{V}^2 R}{R^2 + X_r^2 R^2/R^2} = \frac{\mathbf{V}^2 R}{R^2 + R^2 Q^2} = \frac{\mathbf{V}^2 R}{R^2(1 + Q^2)} = \frac{\mathbf{V}^2}{R(1 + Q^2)}$$

$$= P_0/(1 + Q^2) \qquad \text{(19-14)}$$

Equation (19-14) shows that at any frequency other than f_0 (the resonant frequency), the circuit power, P, is reduced by a factor of $(1 + Q^2)$, where Q is the tangent of the circuit phase angle. Note that at f_0, $\theta = 0°$, $Q = 0$, and $P_0 = \mathbf{V}^2/R$.

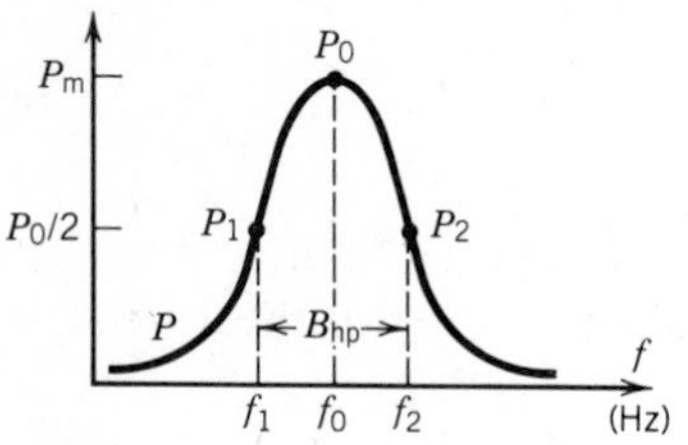

Figure B-15.1 Series RLC resonant power response vs frequency

B-15.2 Definition of Half-Power Bandwidth, B_{hp}, or $B_{-3\text{ dB}}$

By definition, when $P_1 = P_2 = P_0/2 = P_{\max}/2$ the following relations hold:

1. The current is $I_0/\sqrt{2}$ or $0.7071 I_0$.
2. The circuit impedance is $\sqrt{2} \cdot R$ or $\sqrt{2} \cdot Z_0$.
3. The circuit phase angle is $\pm 45°$.
4. Q in Eq. (19-13) is tan 45° or unity.

In terms of dB, the power ratio $P_1/P_0 = P_2/P_0$ is $\text{dB} = 10 \log_{10}(P_1/P_0) = 10 \log_{10}(0.5) = -3.01$ dB

At the half-power or -3 dB bandwidth (BW) frequencies f_1 and f_2 shown in **Fig. B-15.1**, it can be shown that

1. Since the circuit phase angle is 45°, $Q = 1$, the net reactances $X_1 = X_2 = R$, where $X_1 = (X_{C_1} - X_{L_1})$ and $X_2 = (X_{L_2} - X_{C_1})$.
2. $X_{C_2} \cong X_{L_1}$ in magnitude and therefore at frequency f_2 $(X_{L_2} - X_{C_2}) = R = (X_{L_2} - X_{L_1}) = 2\pi L(f_2 - f_1) = 2\pi L(B_{\text{hp}})$, where B_{hp} is the half-power BW
3. From the equality in (2), $B_{\text{hp}} = R/2\pi L$ and dividing both sides by f_0 yields $\dfrac{B_{\text{hp}}}{f_0} = \dfrac{R}{2\pi f_0 L} = \dfrac{R}{X_{L_0}} = \dfrac{1}{Q_0}$ and this yields the fundamental B_{hp} equations:
4. $B_{\text{hp}} = f_0/Q_0 = f_2 - f_1 = R/2\pi L = \sqrt{f_1 f_2}/Q_0$ (19-10)

where f_1 and f_2 are the specific frequencies below and above f_0 producing P_1 and P_2, the half-power points.

B-15.3 Bandwidth at Any Frequency, B

Since the bandwidth (BW) is inversely proportional to Q and f_0, then by the ratio method any bandwidth, B at any Q is $B_{\text{hp}}/B = Q/Q_0 = 1/f_0$, and solving for any bandwidth B,

$$B = f_0 Q/Q_0 = f_2 - f_1 = \sqrt{f_1 f_2}(Q)/Q_0 \qquad (19\text{-}9)$$

where f_2 and f_1, respectively, are any frequencies above and below f_0
Q is the tangent of the circuit phase angle at the off-resonant frequencies f_2 and f_1

B-15.4 Resonant Q_0 Relations

At the resonant frequency, f_0, $X_{L_0} = X_{C_0}$ in magnitude and therefore $\omega_0 L = 1/\omega_0 C$. Solving this equality for ω_0 yields $\omega_0 = 1/\sqrt{LC}$, and since $f_0 = \omega_0/2\pi$ we obtain

$$f_0 = 1/2\pi\sqrt{LC} \qquad (19\text{-}2a)$$

But at any other frequency, f, we have the equalities $L = X_L/2\pi f$ and $C = 1/2\pi f X_C$, and substituting them in Eq. (19-2a) yields

$$f_0 = \frac{1}{2\pi}\sqrt{\frac{X_C}{X_L}(2\pi f)^2} = f\sqrt{\frac{X_C}{X_L}} \qquad (19\text{-}2c)$$

Equation (19-2c) is useful in finding the resonant frequency when the reactances X_L and X_C are known at some other frequency, f, either above or below the resonant frequency, f_0.

But, by definition, the Q_0 at the resonant frequency is $Q_0 = X_{L_0}/R = \omega_0 L/R = X_{C_0}/R = 1/\omega_0 CR$ but since $\omega_0 = 1/\sqrt{LC}$ we may write

$$Q_0 = \omega_0 L/R = L/R\sqrt{LC} = \frac{1}{R}\sqrt{\frac{L}{C}} = \sqrt{X_{C_0} \cdot X_{L_0}}/R$$

$$= f_0/B_{hp} = \frac{\omega_0}{\omega_2 - \omega_1} \tag{19-7}$$

where B_{hp} is the half-power bandwidth
ω_2 and ω_1 are the half-power radian frequencies

In a series-resonant RLC circuit, only R is capable of dissipating energy. The energy stored in L and C, respectively, oscillates at the resonant frequency, f_0, between the magnetic field of the inductor and the electrical (dielectric) field of the capacitor. A broader definition of Q_0, therefore, is the ratio

$$Q_0 = \frac{\text{maximum energy stored per cycle} \times (2\pi)}{\text{energy dissipated in } R \text{ per cycle}} = \frac{2\pi W_s}{W_d} = \frac{Q_s}{P} \tag{19-7}$$

where W_s is the stored energy in joules
W_d is the dissipated energy in joules
Q_s is the reactive power stored in either the coil or the capacitor in vars
P is the power dissipated in the resistor, R, in watts

B-15.5 Evaluating ω_1 and ω_2 at a Given Value of Q_0

To evaluate ω_1, consider the factors affecting $\mathbf{V}_C$ as a function of ω:

$$\mathbf{V}_C = I(-jX_C) = \frac{\mathbf{I}}{\omega C} = \frac{\mathbf{V}}{\omega C \cdot \mathbf{Z}} = \frac{\mathbf{V}}{\omega C\left[R + j\left(\omega L - \dfrac{1}{\omega C}\right)\right]}$$

If this expression is differentiated with respect to ω and $dV_C/d\omega$ is set to zero, solving the expression for ω yields ω_1, the frequency at which V_C is a maximum, or

$$\omega_1 = \omega_0\sqrt{\frac{2Q_0^2 - 1}{2Q_0^2}} = \omega_0\sqrt{1 - \frac{1}{2Q_0^2}} \quad \text{(rad/s)} \tag{19-18}$$

In evaluating ω_2, consider the factors affecting V_L as a function of ω:

$$\mathbf{V}_L = \mathbf{I}(jX_L) = \mathbf{I}(\omega L) = \omega L\frac{\mathbf{V}}{\mathbf{Z}} = \frac{\mathbf{V}\omega L}{[R + j(\omega L - 1/\omega C)]}$$

If this expression is differentiated with respect to ω and $dV_L/d\omega$ is set to zero, solving the expression for ω yields ω_2, the frequency at which $\mathbf{V}_L$ is a maximum, or

$$\omega_2 = \frac{\omega_0}{\sqrt{\dfrac{2Q_0^2 - 1}{2Q_0^2}}} = \frac{\omega_0}{\sqrt{1 - \dfrac{1}{2Q_0^2}}} \quad \text{(rad/s)} \tag{19-19}$$

Observe in both Eqs. (19-18) and (19-19) that when $Q_0 > 10$, the factor below the square root sign is approximately unity and $\omega_0 \cong \omega_1 \cong \omega_2$. But, as shown in Fig. B-15.1, when the Q_0 is low and significantly less than 10, then $\omega_2 > \omega_0 > \omega_1$.

B-15.6 Evaluating $V_{C_{max}}$ and $V_{L_{max}}$ at a Given Value of Q_0

We may evaluate $\mathbf{V}_{C_{max}}$ (and $\mathbf{V}_{L_{max}}$, as well) by using the voltage divider rule (VDR), since at a radian frequency of ω_1 we know that

$$\mathbf{V}_{C_{max}} = \mathbf{I}/\omega_1 C = \mathbf{I}(X_{C_1}) = \frac{\mathbf{V}X_{C_1}}{\mathbf{Z}_1} = \frac{\mathbf{V}X_{C_1}}{R + (-jX_{C_1} + jX_{L_1})}$$

At radian frequency ω_1, $X_{C_1} > X_{C_0}$ by the ratio ω_0/ω_1, and substituting this ratio yields

$$X_{C_1} = X_{C_0}\frac{\omega_0}{\omega_1} = X_{C_0}\frac{1}{\sqrt{1 - \frac{1}{2Q_0^2}}} \quad \text{while} \quad X_{L_1} = X_{L_0}\frac{\omega_1}{\omega_0} = X_{L_0}\sqrt{1 - \frac{1}{2Q_0^2}}$$

Substituting for X_{C_1} and X_{L_1} in the above ($\mathbf{V}_{C_{max}}$) expressions for finding $\mathbf{V}_{C_{max}}$ (or $\mathbf{V}_{L_{max}}$) yields

$$\mathbf{V}_{C_{max}} = \mathbf{V}_{L_{max}} = \frac{\mathbf{V}Q_0}{\sqrt{\frac{4Q_0^2 - 1}{4Q_0^2}}} = VQ_0\sqrt{\frac{4Q_0^2}{4Q_0^2 - 1}} \tag{19-17}$$

where **V** is the applied circuit voltage
Q_0 is the circuit Q at the resonant frequency ω_0
$\mathbf{V}_{C_{max}}$ is the maximum possible voltage across C at ω_1
$\mathbf{V}_{L_{max}}$ is the maximum possible voltage across (ideal) L at ω_2

APPENDIX B-16 PARALLEL-RESONANT CIRCUIT RELATIONS

B-16.1 General Equation for Resonant Frequency of Two-Branch Parallel Circuit with Resistance in Each Branch

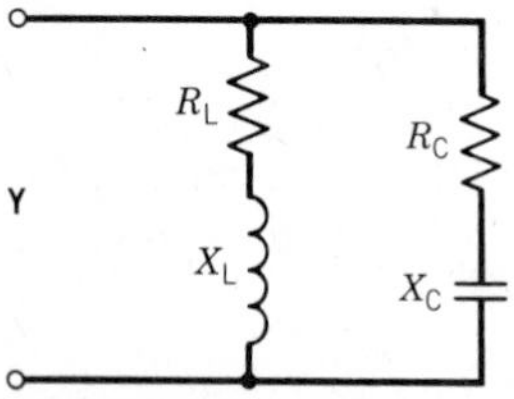

Figure B-16.1 Two-branch parallel resonant circuit

If we can derive a general equation for the resonant frequency, ω_0, for the circuit shown in **Fig. B-16.1**, then we can use this relation to find all other two-branch relations. We begin by writing the admittance relation

$$\mathbf{Y} = \frac{1}{\mathbf{Z}_L} + \frac{1}{\mathbf{Z}_C} = \frac{R_L - jX_L}{R_L^2 + X_L^2} + \frac{R_C + jX_C}{R_C^2 + X_C^2} \quad \text{and collecting } j \text{ terms by expansion}$$

$$\mathbf{Y} = \frac{R_L}{R_L^2 + X_L^2} + \frac{R_C}{R_C^2 + X_C^2} + j\left(\frac{X_C}{R_C^2 + X_C^2} - \frac{X_L}{R_L^2 + X_L^2}\right)$$

But at resonance the admittance **Y** is *pure* conductance, and therefore the j term of susceptances is zero. Setting this j term to zero and then equating each part yields

$$\frac{X_C}{R_C^2 + X_C^2} = \frac{X_L}{R_L^2 + X_L^2} \quad \text{or} \quad \frac{X_L}{X_C} = \frac{R_L^2 + X_L^2}{R_C^2 + X_C^2} \text{ by cross-multiplication.}$$

Extracting ω from the reactances on both sides yields

$$\omega^2 LC = \frac{R_L^2 + \omega^2 L^2}{R_C^2 + (1/\omega^2 C^2)} \quad \text{and clearing the denominator on the right side}$$

$\omega^2 LCR_C^2 + L/C = R_L^2 + \omega^2 L^2$ and collecting terms from both sides

$\omega^2 LCR_C^2 - \omega^2 L^2 = R_L^2 - L/C$ then factoring $\omega^2 LC$ on the left side only

$\omega^2 LC(R_C^2 - L/C) = R_L^2 - L/C$ and dividing both sides by $(R_C^2 - L/C)$ yields

$\omega^2 LC = \dfrac{R_L^2 - L/C}{R_C^2 - L/C}$ which when solved for $\omega = \omega_0$ yields

$\omega_0 = \dfrac{1}{\sqrt{LC}}\sqrt{\dfrac{R_L^2 - L/C}{R_C^2 - L/C}}$ but recall that $\omega_r = 1/\sqrt{LC}$ at which $X_L = X_C$, so therefore

$\omega_0 = \omega_r\sqrt{\dfrac{R_L^2 - L/C}{R_C^2 - L/C}}$ rationalizing denominators and dividing both sides by 2π

$$f_0 = f_r\sqrt{\frac{R_L^2 C - L}{R_C^2 C - L}} \quad \text{hertz (Hz)} \qquad (19\text{-}32)$$

Expressing Eq. (19-32) to show $f_r = f\sqrt{\dfrac{X_C}{X_L}}$ from Eq. (19-2c) where X_C and X_L are known reactances at *any* frequency, f; we express Eq. (19-32) in terms of reactances:

$$f_0 = f\sqrt{\frac{X_C}{X_L}} \times \sqrt{\frac{R_L^2 - X_C X_L}{R_C^2 - X_C X_L}} \quad \text{hertz (Hz)} \qquad (19\text{-}33)$$

A host of interesting possibilities emerge from Eqs. (19-33) and (19-32) namely

1. f_0 is IMAGINARY whenever the expression under the radical is negative or
  a. $R_C^2 > X_C X_L$ *and* $R_L^2 < X_C X_L$
  b. $R_C^2 < X_C X_L$ *and*, simultaneously $R_L^2 > X_C X_L$

The above implies that in order to obtain resonance we must *avoid* high resistance capacitors coupled with low resistance coils or, alternatively, low resistance capacitors coupled with very high resistance coils, *i.e.*, coils whose Qs are less than unity.

2. f_0 is REAL and resonance may occur whenever the expression under the radical is positive or
  a. $R_C^2 < X_C X_L$ *and* $R_L^2 < X_C X_L$
  b. $R_C^2 > X_C X_L$ *and* $R_L^2 > X_C X_L$

The two conditions above raise some interesting possibilities, enabling the following predictions:

2a. Assume $R_C = R_L = 0$, *i.e.*, no resistance in either branch. This yields a 2-branch circuit containing *ideal* elements and $f_0 = f_r = 1/2\pi\sqrt{LC}$, as in the series RLC circuit. As we shall see below, $\mathbf{Z}_{po}$ is infinite at resonance. (Sec. B-16.3)

2b. Assume $R_C^2 = R_L^2 = X_C X_L$ or $R_C = R_L = \sqrt{L/C}$. This combination produces the ratio 0/0 under the radical in Eqs. (19-33) and (19-32), *i.e.*, an *indeterminate* value. Under these circumstances, the circuit is resonant at *all* frequencies and the input current, $\mathbf{I}_0$, is *always in phase* with the input voltage. This makes the circuit impedance, $\mathbf{Z}_{po} = R_{po} = \sqrt{L/C} = \sqrt{X_C X_L}$ *a constant value at all frequencies.*

2c. Assume $R_C^2 = R_L^2 \neq X_C \cdot X_L$. Like 2a above the circuit is also resonant at $f_0 = f_r$ but now the circuit impedance is some maximum value at resonance and the circuit current $\mathbf{I}_0$ is no longer zero but some *minimum* value.

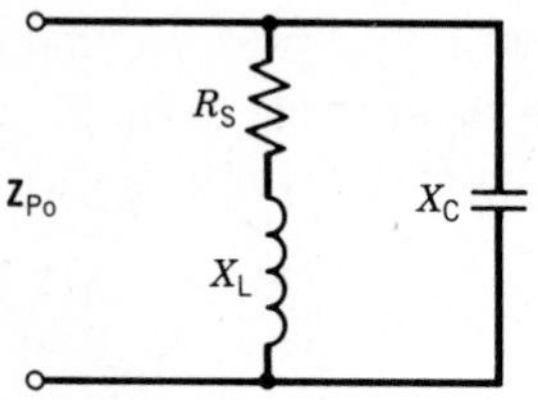

Figure B-16.2 Two-branch practical tank circuit

B-16.2 Two-Branch Practical Resonant with Only Resistance in the Inductive Branch

Commercial coils are never pure and of necessity have some resistance R_s. Most capacitors, however, may be considered to have negligible resistance. This yields the circuit shown in Fig. B-16.2 containing a capacitor whose $R_C = 0$. This circuit is perhaps the most commonly used in both transmitters and receivers in communications, both as an oscillator and a tuner. Setting $R_C = 0$ in the above equation for ω_0 yields

$$\omega_0 = \omega_r \sqrt{\frac{(L/C) - R^2}{L/C}} = \omega_r \sqrt{1 - R^2(C/L)}$$

but $C/L = \omega_0 C/\omega_0 L = 1/\omega_0^2 L^2$. Substituting

$$\omega_0 = \omega_r \sqrt{1 - (R/\omega_0 L)^2} = \omega_r \sqrt{1 - \frac{1}{Q_r^2}} = \omega_r \sqrt{\frac{Q_r^2 - 1}{Q_r^2}}$$

and since $f_0 = f_r \dfrac{Q_0}{Q_r}$

$$f_0 = f_r \sqrt{\frac{Q_r^2 - 1}{Q_r^2}} = f_r \sqrt{1 - \frac{CR_L^2}{L}} = f_r \sqrt{\frac{Q_0^2}{Q_0^2 + 1}} \quad \text{hertz (Hz)} \qquad (19\text{-}29)$$

where f_r is the frequency at which $X_L = X_C$ (only)

f_0 is the frequency at which impedance is a maximum, current is a minimum and the current, I_0, is in phase with the voltage at unity PF

Q_r is the Q of the coil at frequency f_r

Q_0 is the Q of the coil and the circuit at frequency f_0

When the Q_r is low, the frequency difference between f_r and f_0 is relatively significant in that f_0 occurs at a *lower* frequency than f_r. Similarly, when Q_r is low (below 10), the value of Q_0 is relatively *lower* than Q_r. (See B-16.5, below).

But when Q_r is high (10 or more), the frequency difference between f_r and f_0 is almost negligible. The higher the value of Q_r (say 20 or more), the closer the value of f_0 to f_r and the closer is Q_0 to Q_r. For high-Q circuits, therefore we may conclude that $f_0 = f_r = 1/2\pi\sqrt{LC}$ and that at this frequency, $X_L = X_C$, $\mathbf{Z}_{po}$ is a maximum, current is a minimum and in phase with $\mathbf{V}$, the supply voltage.

B-16.3 Impedance Relations of the Two-Branch Parallel Resonant Circuit with Ideal Components

Given the circuit shown in Fig. B-16.2 and using the product-over-the-sum rule for impedance, but assuming $R_s = 0$, i.e., the components in each branch are pure L and C:

$$Z_p = \frac{jX_L(-jX_C)}{j(X_L - X_C)} = \frac{X_L X_C}{j(X_L - X_C)} = \frac{X_L X_C}{0} = \infty$$

when $|X_L| = |X_C|$ at f_r, above.

The equivalent impedance of the 2-branch ideal circuit is infinite and the current $\mathbf{I}_0 = 0$ at the resonant frequency $f_r = f_0 = 1/2\pi\sqrt{LC}$. (Recall that in the ideal series RLC circuit, the resistance is zero and the *current* is infinite). In effect, the 2-branch, high-Q parallel resonant circuit acts as an *open* circuit at its resonant frequency while the low resistance series RLC circuit behaves as a *short* circuit at its resonant frequency.

B-16.4 Impedance Relations of the High-Q Two-Branch Parallel Resonant Circuit

From Sec. B-16.3 above, the equivalent parallel circuit impedance of the **high-Q** 2-branch circuit shown in Fig. B-16.2 is (using the product-over-sum-rule):

$$\mathbf{Z}_p = \frac{(R_s + jX_L)(-jX_C)}{(R_s + jX_L - jX_C)} \text{ and when } X_L = X_C \text{ at } f_r$$

$$\mathbf{Z}_{po} = \frac{X_L X_C - jX_C R}{R_s} = \frac{X_L X_L}{R_s} - jX_C = QX_C - jX_C \quad \text{but when } Q = 10 \text{ or more, } -jX_C \text{ is negligible so}$$

$$\begin{aligned}\mathbf{Z}_{po} &= Q_r X_C = Q_r X_L = Q_r(Q_r R_s) = Q_r^2 R_s \\ &= (Q_0 + 1)^2 R_s \cong Q_0^2 R_s \cong L/CR_s \qquad (19\text{-}31)\end{aligned}$$

where $\mathbf{Z}_{po}$ is a resistive impedance which is a maximum at $f_0 \cong f_r$.

Consequently, when the Q is high ($Q_r > 10$), we may assume that the parallel impedance is at a maximum, is resistive and computed as $\mathbf{Z}_{po} = L/CR_s = Q_r^2 R_s$. *No corrections are necessary* since the three resonant conditions (described below) *all occur at* approximately *the same* frequency.

B-16.5 Impedance Relations of the Low-Q, 2-Branch, Parallel Resonant Circuit

A much more complex situation occurs in the low-Q parallel resonant circuit. Three different resonant conditions occur at three different resonant frequencies. At the highest frequency, $|jX_L| = |jX_C|$. At a somewhat lower frequency, maximum impedance, $\mathbf{Z}_m$, occurs which produces the *minimum current* condition. At the lowest frequency, unity power factor occurs, *i.e.*, current is in phase with the voltage and the resulting impedance is *resistive*. Let us consider each of these, in turn.

A. **At the frequency, f_r, at which $|jX_L| = |jX_C|$**

At this higher frequency, $f_r = 1/2\pi\sqrt{LC}$, and $\omega_r = 1/\sqrt{LC}$. Since both L and R_s are constant, this produces the highest $Q_r = \omega_r L/R_s$. Using the product-over-the-sum rule, the equivalent parallel circuit impedance at this frequency is:

$$\mathbf{Z}_r = \frac{(R_s + jX_L)(-jX_C)}{R_s + jX_L - jX_C} = \frac{(R_s + jX_L)(-jX_C)}{R_s} \text{ where } |jX_L| = |jX_C|$$

B. **At the frequency, f_m, which produces maximum impedance, $\mathbf{Z}_m$, and minimum current**

At this lower frequency, $f_m = f_r\sqrt{\dfrac{4Q_r^2 - 1}{4Q_r^2}}$, the capacitive reactance jX_C is *increased* and the inductive reactance, jX_L is *decreased*. Using the product-over-the-sum rule, the equivalent impedance is:

$$\mathbf{Z}_m = \frac{(R_s + jX_L)(-jX_C)}{R_s + jX_L - jX_C} \text{ where } jX_L = +j\omega_m L \text{ and } -jX_C = -j(1/\omega_m C)$$

C. **At the frequency, f_0, which produces resistive impedance, $\mathbf{Z}_0$ and unity power factor**

At this lowest frequency, $f_0 = f_r\sqrt{\frac{Q_r^2 - 1}{Q_r^2}}$, the capacitive reactance is further increased while the inductive reactance is further reduced until the total circuit current is in phase with the voltage. The circuit impedance, $\mathbf{Z}_0$, is resistive and is found by

$$\mathbf{Z}_0 = Q_r^2 R_s = L/CR_s$$

Table B-16 below summarizes the above equations which enable the solutions for the respective values of parallel circuit impedance, whenever $Q_r < 10$.

Appendix Table B-16 Frequency and Impedance Relations of the Low-Q, Two-Branch, Parallel-Resonant Circuit

RESONANCE CONDITION	FREQUENCY EQUATION	PARALLEL CIRCUIT IMPEDANCE CALCULATION EQUATION	
$\lvert jX_L \rvert = \lvert jX_C \rvert$	$f_r = \frac{1}{2\pi\sqrt{LC}}$	$\mathbf{Z}_r = \frac{(R_s + jX_L)(-jX_C)}{R_s}$	(19-30a)
Maximum $\mathbf{Z}_m$ Minimum current	$f_m = f_r\sqrt{\frac{4Q_r^2 - 1}{4Q_r^2}}$	$\mathbf{Z}_m = \frac{(R_s + jX_L)(-jX_C)}{(R_s + jX_L - jX_C)}$	(19-30b)
Unity power factor	$f_0 = f_r\sqrt{\frac{Q_r^2 - 1}{Q_r^2}}$	$\mathbf{Z}_0 = Q_r^2 R_s = \frac{L}{CR_s}$	(19-30c)

Index

*NOTE: Definitions of terms appear on pages shown in bold type.